THE UNIVERSITY OF ARIZONA SPACE SCIENCE SERIES
RICHARD P. BINZEL, GENERAL EDITOR

The Pluto System After New Horizons
S. A. Stern, J. M. Moore, W. M. Grundy, L. A. Young,
and R. P. Binzel, editors, 2021, 663 pages

Planetary Astrobiology
V. S. Meadows, G. N. Arney, B. E. Schmidt, and
D. J. Des Marais, editors, 2020, 534 pages

Enceladus and the Icy Moons of Saturn
P. M. Schenk, R. N. Clark, C. J. A. Howett, A. J. Verbiscer, and
J. H. Waite, editors, 2018, 475 pages

Asteroids IV
P. Michel, F. E. DeMeo, and W. F. Bottke, editors, 2015, 895 pages

Protostars and Planets VI
Henrik Beuther, Ralf S. Klessen, Cornelis P. Dullemond, and
Thomas Henning, editors, 2014, 914 pages

Comparative Climatology of Terrestrial Planets
Stephen J. Mackwell, Amy A. Simon-Miller, Jerald W. Harder,
and Mark A. Bullock, editors, 2013, 610 pages

Exoplanets
S. Seager, editor, 2010, 526 pages

Europa
Robert T. Pappalardo, William B. McKinnon,
and Krishan K. Khurana, editors, 2009, 727 pages

The Solar System Beyond Neptune
M. Antonietta Barucci, Hermann Boehnhardt, Dale P. Cruikshank,
and Alessandro Morbidelli, editors, 2008, 592 pages

Protostars and Planets V
Bo Reipurth, David Jewitt, and Klaus Keil, editors, 2007, 951 pages

Meteorites and the Early Solar System II
D. S. Lauretta and H. Y. McSween, editors, 2006, 943 pages

Comets II
M. C. Festou, H. U. Keller,
and H. A. Weaver, editors, 2004, 745 pages

Asteroids III
William F. Bottke Jr., Alberto Cellino, Paolo Paolicchi,
and Richard P. Binzel, editors, 2002, 785 pages

Tom Gehrels, General Editor

Origin of the Earth and Moon
R. M. Canup and K. Righter, editors, 2000, 555 pages

Protostars and Planets IV
Vincent Mannings, Alan P. Boss,
and Sara S. Russell, editors, 2000, 1422 pages

Pluto and Charon
S. Alan Stern and David J. Tholen, editors, 1997, 728 pages

Venus II—Geology, Geophysics, Atmosphere,
and Solar Wind Environment
S. W. Bougher, D. M. Hunten,
and R. J. Phillips, editors, 1997, 1376 pages

Cosmic Winds and the Heliosphere
J. R. Jokipii, C. P. Sonett,
and M. S. Giampapa, editors, 1997, 1013 pages

Neptune and Triton
Dale P. Cruikshank, editor, 1995, 1249 pages

Hazards Due to Comets and Asteroids
Tom Gehrels, editor, 1994, 1300 pages

Resources of Near-Earth Space
John S. Lewis, Mildred S. Matthews,
and Mary L. Guerrieri, editors, 1993, 977 pages

Protostars and Planets III
Eugene H. Levy and Jonathan I. Lunine, editors, 1993, 1596 pages

Mars
Hugh H. Kieffer, Bruce M. Jakosky, Conway W. Snyder,
and Mildred S. Matthews, editors, 1992, 1498 pages

Solar Interior and Atmosphere
A. N. Cox, W. C. Livingston,
and M. S. Matthews, editors, 1991, 1416 pages

The Sun in Time
C. P. Sonett, M. S. Giampapa,
and M. S. Matthews, editors, 1991, 990 pages

Uranus
Jay T. Bergstralh, Ellis D. Miner,
and Mildred S. Matthews, editors, 1991, 1076 pages

Asteroids II
Richard P. Binzel, Tom Gehrels,
and Mildred S. Matthews, editors, 1989, 1258 pages

Origin and Evolution of Planetary and Satellite Atmospheres
S. K. Atreya, J. B. Pollack,
and Mildred S. Matthews, editors, 1989, 1269 pages

Mercury
Faith Vilas, Clark R. Chapman,
and Mildred S. Matthews, editors, 1988, 794 pages

Meteorites and the Early Solar System
John F. Kerridge and Mildred S. Matthews, editors, 1988, 1269 pages

The Galaxy and the Solar System
Roman Smoluchowski, John N. Bahcall,
and Mildred S. Matthews, editors, 1986, 483 pages

Satellites
Joseph A. Burns and Mildred S. Matthews, editors, 1986, 1021 pages

Protostars and Planets II
David C. Black and Mildred S. Matthews, editors, 1985, 1293 pages

Planetary Rings
Richard Greenberg and André Brahic, editors, 1984, 784 pages

Saturn
Tom Gehrels and Mildred S. Matthews, editors, 1984, 968 pages

Venus
D. M. Hunten, L. Colin, T. M. Donahue,
and V. I. Moroz, editors, 1983, 1143 pages

Satellites of Jupiter
David Morrison, editor, 1982, 972 pages

Comets
Laurel L. Wilkening, editor, 1982, 766 pages

Asteroids
Tom Gehrels, editor, 1979, 1181 pages

Protostars and Planets
Tom Gehrels, editor, 1978, 756 pages

Planetary Satellites
Joseph A. Burns, editor, 1977, 598 pages

Jupiter
Tom Gehrels, editor, 1976, 1254 pages

Planets, Stars and Nebulae, Studied with Photopolarimetry
Tom Gehrels, editor, 1974, 1133 pages

The Pluto System After New Horizons

The Pluto System After New Horizons

Edited by

S. Alan Stern, Jeffrey M. Moore, William M. Grundy, Leslie A. Young, and Richard P. Binzel

With the assistance of

Renée Dotson

With 74 collaborating authors

THE UNIVERSITY OF ARIZONA PRESS
Tucson

in collaboration with

LUNAR AND PLANETARY INSTITUTE
Houston

About the front cover:

Pluto and Charon as imaged by NASA's New Horizons mission. This enhanced color composite was created from individual New Horizons color images of Pluto and Charon taken on July 14, 2015. The color here is both stretched and was also enhanced through use of wavelengths from 400 to 970 nm, extending beyond the 400- to 700-nm range visible to the human eye. The approximate resolution of these images as presented here is ~10 km/pixel. In contrast, the best New Horizons imaging of each body (although localized, not global) was about 100-fold higher resolution. The Pluto-Charon binary planet can be seen to be a highly dichotomous pair, as had been inferred from earlier datasets, with Pluto's surface being significantly higher in albedo and significantly more variegated than Charon's. For scale, Pluto's diameter is 2376 km; the rotational north poles of both bodies are at the top of each body as shown in this image. The system's four smaller moons are not depicted here.

About the back cover:

Shown here are the less well-resolved, non-encounter (i.e., the so-called "farside") hemispheres of Pluto and Charon in panchromatic, visible light at a pivot wavelength of 606 nm. Even at this resolution, it can be seen that there are dramatic differences in the appearance of Pluto and the distribution of terrain types between its close approach and farside hemispheres. Whereas the image on the front cover is both several times higher in resolution and also is surpassed by still ~100× higher-resolution imagery of each body, these two images represent the best views of the farsides of Pluto and Charon that are likely to be available until an orbiter arrives for more detailed global reconnaissance. The editors of this volume look forward to new exploration revealing these farside hemispheres and their implications for better understanding the origin and evolution of these complex and fascinating worlds, as well as for understanding their kin among the other planets of the Kuiper belt.

The Lunar and Planetary Institute is operated by the Universities Space Research Association under a cooperative agreement with the Science Mission Directorate of the National Aeronautics and Space Administration. Any opinions, findings, and conclusions or recommendations expressed in this volume are those of the author(s) and do not necessarily reflect the views of the National Aeronautics and Space Administration.

The University of Arizona Press
in collaboration with the Lunar and Planetary Institute

∞ This book is printed on acid-free, archival-quality paper.
Manufactured in the United States of America

26 25 24 23 22 21 6 5 4 3 2

Library of Congress Cataloging-in-Publication Data
Names: Stern, Alan, 1957– editor. | Moore, J. (Jeff), editor. |
Grundy, William M., editor. | Young, Leslie A., editor. |
Binzel, Richard P., editor.
Title: The Pluto System After New Horizons / edited by S. Alan Stern, Jeffrey M. Moore, William M. Grundy,
Leslie A. Young, Richard P. Binzel ; with the assistance of Renee Dotson.
Other titles: University of Arizona space science series.
Description: Tucson : University of Arizona Press, 2021. | Series: The University of Arizona Space science series | Includes
bibliographical references and index.
Identifiers: LCCN 2020022869 | ISBN 9780816540945 (hardcover)
Subjects: LCSH: Pluto (Dwarf planet) | Pluto (Dwarf planet)—Exploration.
Classification: LCC QB701 .P586 2021 | DDC 523.49/22—dc23
LC record available at https://lccn.loc.gov/2020022869

This volume and the results herein are dedicated to the entire New Horizons team — all those individuals at every institution who took part in its design, build, launch, and voyage, enabling the first exploration of the Pluto system and the Kuiper belt.

Contents

PART 1: SYSTEM BACKGROUND

PART 2: PLUTO

PART 3: CHARON AND PLUTO'S SMALL SATELLITES

PART 4: ORIGINS, INTERIORS, AND THE BIG PICTURE

List of Contributing Authors

C. J. Ahrens 195
F. Bagenal 379
N. P. Barnes 379
M. A. Barucci 21
T. Bertrand 297, 321, 507
R. A. Beyer 395, 433, 627
R. P. Binzel 3, 9
V. J. Bray 121, 569
M. W. Buie 147, 569
B. J. Buratti 413
R. M. Canup 475
A. Cheng 279
V. F. Chevrier 195
J. C. Cook 433, 457
D. P. Cruikshank 165, 433
C. M. Dalle-Ore 21, 433
P. A. Delamere 379
A. M. Earle 321
H. A. Elliott 379
F. Forget 297, 321
S. Fornasier 21
P. Gao 279
G. R. Gladstone 257
C. R. Glein 507
S. Greenstreet 121
W. M. Grundy 147, 165, 413, 433, 457
D. Hinson 297
J. D. Hofgartner 569
A. D. Howard 55, 105
C. J. A. Howett 147, 413
K.-L. Jessup 279
P. Kollmann 379
K. M. Kratter 475
E. Lellouch 569
I. R. Linscott 165
C. M. Lisse 379
R. M. C. Lopes 55
J. I. Lunine 9
A. Luspay-Kuti 279
K. E. Mandt 279
D. J. McComas 379
W. B. McKinnon 89, 507
R. L. McNutt Jr. 379
J. M. Moore 55, 105
M. J. Neufeld 9
M. Neveu 475
F. Nimmo 89, 395
C. B. Olkin 147, 413, 587
A. H. Parker 545
M. J. Person 257
S. B. Porter 457
S. Protopapa 147, 165, 413, 433
A. R. Rhoden 507
S. J. Robbins 121, 395
P. M. Schenk 55, 121
K. Schindler 3
B. Schmitt 165
M. Showalter 627
B. Sicardy 321
K. N. Singer 55, 121, 395
M. Soluri 603
J. R. Spencer 395, 587
S. A. Stern 9, 587
D. F. Strobel 363
M. E. Summers 257
A. Toigo 297
L. M. Trafton 321
O. M. Umurhan 195
A. J. Verbiscer 147, 413, 457
H. A. Weaver 457, 587, 641
O. L. White 55
D. A. Williams 55
L. A. Young 9, 257, 321
E. J. Zirnstein 379

Scientific Organizing Committee

Richard Binzel (chair and co-editor)
S. Alan Stern (lead editor)
Amanda Bosh
Will Grundy (co-editor)
Carey Lisse
Jeffrey Moore (co-editor)
Silvia Protopapa
Jani Radebaugh
Leslie Young (co-editor)

Acknowledgment of Reviewers

The editors gratefully acknowledge the following individuals, as well as several anonymous reviewers, for their time and effort in reviewing chapters in this volume:

Erik Asphaug
Fran Bagenal
Nadine Barlow
Jason W. Barnes
M. A. Barucci
Chloe B. Beddingfield
Susan Benecchi
Bonnie Buratti
Morgan L. Cable
Richard J. Cartwright
Andrew F. Cheng
Geoffrey Collins
Tom Cravens
Tilmann Denk
Steve Desch
Luke Dones
Martin Duncan
Justin Erwin
Caleb Fassett
Gianrico Filacchione
Scott D. Guzewich
Candice Hansen
Alexander G. Hayes
Bryan J. Holler
Alan Howard
Carly J. A. Howett
Andrew P. Ingersoll
Ross Irwin
Ralf Jaumann
David Jewitt
J J Kavelaars
James T. Keane
Michelle Kirchoff
Tommi T. Koskinen
Vladimir Krasnopolsky
Stamatios Krimigis
Jian-Yang Li
William McKinnon
David Nesvorný
Francis Nimmo
Jani Radebaugh
Darin Ragozzine
Alyssa Rhoden
C T Russell
Paul Schenk
Kevin Schindler
Kelsi N. Singer
Katrin Stephan
Michael E. Summers
Audrey Thirouin
Henry Throop
Steven D. Vance
H. A. Weaver
Don Wilhelms
Sharon A. Wilson

Foreword

This volume is the fruit of the historic New Horizons Pluto flyby of July 2015 — the joyous, frantic, fraught few days in which knowledge of the last of the classically known planets leapt from nearly complete ignorance to in-depth knowledge with more than six gigabytes of data.

Every spacecraft mission embodies the dreams and efforts of large teams of scientists and engineers over years of planning and building, but New Horizons experienced decades of political battles and endured more than the usual share of setbacks, cancellations, new starts, frustrations, and near catastrophes en route to the successful achievement of its goals.

Soon after New Horizons lifted off from Earth in January 2006, astronomers redefined the word "planet" to exclude Pluto. Although this decision had no effect on the spacecraft's trajectory over the course of its nine-year journey, the mission's primary destination shifted in the eyes of many — from the ninth planet to the first world of the outer solar system's vast, unexplored expanse. And yet, when Pluto finally appeared in the high-resolution approach images, first as a dot dancing with its smaller companion, Charon, and then resolving into a variegated heart-encrusted sphere, there was no word more suitable than "planet" to describe what we were seeing.

On July 4, 2015, less than two weeks before the historic encounter with Pluto, the radio signal from New Horizons suddenly disappeared. The team at The Johns Hopkins Applied Physics Laboratory (APL) quickly regained contact, only to face a new glitch: The computer reboot that had caused the loss of signal had also erased all the commands needed to execute the encounter. As the spacecraft barreled on toward Pluto at approximately 50,000 kilometers per hour, dozens of team members raced to restore the erased software. The few remaining days before flyby left them no time to sleep and no margin for error, but they succeeded in setting everything to rights with only hours to spare before the encounter began.

A flyby encounter, unlike an orbiter or lander mission, typically entails long anticipation, followed by a sudden burst of revelation, then off again into the distance. And a first flyby often feels like a preliminary scout for other missions soon to follow. But this encounter, so long in coming, so vitally threatened at its most crucial stage, assumed an air of triumph. This visit to the last of the classically known planets, so unlikely to be repeated within our lifetimes, gained an aura of finality, of completion of a goal sought for a generation.

The encounter itself sparked a joyous congregation of more than 2000 scientists, engineers, space enthusiasts, journalists, and even a few celebrities, all of whom gathered to anticipate, observe, and celebrate the event. Together they watched the little spacecraft, about the size of a baby grand piano, packed with scientific instruments and infused with human ingenuity and curiosity, approach and then pass the farthest world yet reached. The event presented a rare and brief opportunity for the people and activities of the planetary science community to become highly visible. The women and men on the New Horizons teams were applauded, interviewed, and fêted — not just at APL but around the world on social media. The ability to share instantaneous word of their success, along with the first high-resolution pictures as proof, created a new, twenty-first century type of global scientific event.

After the news crews packed up and the crowds dispersed, the New Horizons team caught up (just a little) on sleep and got to work on the data. The camera-ready quick releases and off-the-cuff reactions they had needed to provide at encounter press conferences gave way to deeper analyses. Between then and now, the team and their colleagues in the planetary science community have digested and interpreted the deluge of images and data. This book represents the resultant state of knowledge about the Pluto system in 2020 — worlds newly unveiled, mapped, modeled, and now, to some degree, understood.

Future papers and reviews will be written with new ideas and insights about the Pluto system. Eventually we'll go back. New spacecraft will return to orbit and land there. When that happens, we'll finally see the other side of Pluto in high resolution and monitor the atmosphere with new instruments. Then we'll also see what has changed — how the surface and atmosphere have evolved over the intervening decades. But the chapters in this volume will stand forever as a record of what the planetary science community learned and surmised from the Pluto system on humanity's first foray into the Kuiper belt.

David Grinspoon and Dava Sobel
July 2020

Editor's note: A photographic journal of the journey of New Horizons and its team has been kept by photographer Michael Soluri. Some of his images from this journal are captured in the Epilogue to this volume.

Preface

This is the second Space Science Series volume on the Pluto system. At the time of the publication of the first such volume, *Pluto and Charon,* in 1997, the limitations of our Earth-bound perch for the study of the Pluto system were already well appreciated. It was also clear then that no true picture of Pluto and its then only known moon, Charon, could be achieved without a space mission to reconnoiter them up close. While that previous book was being assembled, a NASA-appointed Science Definition Team was already specifying the goals for such a mission. After many failed directed-mission new start attempts across the 1990s, NASA finally called for principal-investigator-led mission proposals from the scientific community on January 19, 2001, and on November 29 of that same year selected the New Horizons team to carry out the mission.

Over the four years and two months from its selection in November 2001 to its launch in January 2006, New Horizons overcame pernicious funding battles, incredibly challenging nuclear launch approval schedules, the development and NASA certification of a new launch vehicle, and the challenge of creating a Voyager-like mission on a budget five times smaller. New Horizons succeeded at all of this, and successfully entered flight on January 19, 2006 — five years to the day after NASA's call for mission proposals.

The NASA call for proposals that catalyzed New Horizons required intermediate-resolution panchromatic and color imaging, atmospheric ultraviolet and radio science profiling, and infrared composition mapping, with at least 2 gigabits of data storage. New Horizons exceeded those requirements by including much-higher-resolution panchromatic imaging capability, two plasma sensors to enhance atmospheric studies, a significantly higher infrared spectral mapping resolution than NASA required, and 128 gigabits of spacecraft data storage. After a flawless launch in 2006 and flyby of Jupiter in 2007, followed by an essentially textbook cruise across the middle solar system where the giant planets orbit, New Horizons reached closest approach to Pluto and its system of five moons on July 14, 2015.

In many respects, the 1997 *Pluto and Charon* volume played a large role in creating the scientific foundation for sending New Horizons on its journey. A new picture of the Pluto system was emerging as Pluto and Charon became resolved as separate bodies thanks to the use of mutual events, the Hubble Space Telescope (HST), and advancements in groundbased astronomical instrumentation such as adaptive optics. For Pluto, the mutual events that took place a decade before that book had yielded maps showing high-contrast albedo features on the sub-Charon hemisphere. Then in the mid-1990s, HST was able to observe over all longitudes, and Pluto's major albedo units were resolved.

But mysteries continued to arise. As one example, stellar occultations surprisingly showed Pluto's atmospheric pressure was increasing, in spite of Pluto receding from the Sun after having passed perihelion in 1989. Even more puzzling was how long Pluto's atmosphere would hold against collapse on the outbound leg of its orbit. In the midst of these and other tantalizing realizations of how dynamic this distant world might be, the entire Pluto system expanded in its population and complexity. The discovery of Nix and Hydra in 2005 was followed by the discoveries of Styx and Kerberos in 2011 and 2012, each made by New Horizons team members using HST. These discoveries brought about the realization that in going to Pluto, we would not just be exploring a planet, or even a double planet, but a complete system of objects more akin to the giant planets in its satellite compliment than the terrestrial planets.

Many of the questions that inspired and directed the design of New Horizons only deepened as the spacecraft approached the Pluto system in the first half of 2015. Pluto's previously deduced extreme albedo contrasts implied a wide range of terrains, which were seen in increasingly improved detail as New Horizons drew closer and closer. But the resolution necessary for geologic interpretation was impossible until the final few days before the spacecraft's closest approach. And although groundbased visible and infrared spectra had already detected many of the species that were later mapped across Pluto's surface, high-spatial-resolution spectral maps by New Horizons proved crucial to understanding how geology, composition, and surface-atmosphere interactions interplay on Pluto. Pluto's atmosphere was the subject of speculation, conjecture, and only poorly constrained modeling before the flyby; until then, its understanding was noticeably hampered by the lack of knowledge of the nature of the lower atmosphere, the presence of hazes, lower atmospheric thermal structure, and such basic quantities as the surface radius and surface pressure. Charon, which appeared bland from Earth-bound telescopes, began to reveal a complex history of its own as New Horizons approached, as well as fascinating landforms and a completely unique dark polar cap. New Horizons once proved, as past first missions to planets had decades before, that nothing we could do from Earth or space observatories with current technology substitutes for the advantage of close-up scientific reconnaissance.

Data from the mid-2015 Pluto system flyby were spooled back to Earth over the succeeding 16 months. Then in late December 2018 and early January 2019, New Horizons went on, as part of its Decadal Survey-prescribed mission — to explore the Kuiper belt and Kuiper belt objects (KBOs) — to reconnoiter the first KBO observed at close range, Arrokoth.

The scientific discoveries made by New Horizons in these two landmark first flybys of bodies in the Kuiper belt, the third zone of the solar system, are chronicled and reviewed in this book, going to press just five and a half years after the Pluto flyby.

And what did we find? The hundreds of pages of this volume contain that story. In brief, among other things, we found that even a small, cold, isolated planet can be extraordinarily active, with deep implications for other Kuiper belt planets like Eris and Makemake. We also learned that such a world can be as complex as much larger ones, like Earth and Mars. We also found the largest glacier in the solar system, a spectacular convecting N_2-ice sheet known as Sputnik Planitia, as well as both ancient, middle-aged, and young terrains, and multiple lines of evidence pointing to a liquid H_2O ocean in the planet's interior. A wide range of processes are now known to have sculpted Pluto's surface, including glacial flows, a global or quasi-global ridge-trough system, stress fractures, constructional and erosional volatile processes, cryovolcanism, haze fallout, winds, and dunes. The structure of the lower atmosphere was seen to depend on location and time of day, with prominent hazes, and an upper atmosphere that was much colder and escaping far more slowly than expected. Moreover, Pluto's system of satellites also was found to be diverse and fascinating. Among the small satellites, New Horizons found irregular shapes, ammoniated surfaces, complex rotational dynamics, and more. At Charon, New Horizons found evidence for a former liquid water ocean in the interior, unusual landforms, purely ancient surfaces, and no atmosphere whatsoever, although its dark polar cap appears to be sourced by atmospheric transfer from Pluto.

Four years to the day after New Horizons' closest approach, on July 14, 2019, scientists convened at The Johns Hopkins Applied Physics Laboratory's Kossiakoff Center where the spacecraft was built and later operated from, for the "Pluto System After New Horizons" conference. Over the course of four days of scientific talks and wide-ranging discussions, nearly 200 scientists shared results from the analysis and interpretation of New Horizons data, as well as from other observational, modeling, and laboratory studies. Information from diverse subdisciplines of planetary science was brought together to synthesize a much more comprehensive picture of the Pluto system. Most notable was the framing of the Pluto system in the broader context of the outer solar system to a greater extent than had ever been previously possible. Those findings are described in this new volume, *The Pluto System After New Horizons* (*PSANH*).

This book is broadly organized by the objects within the Pluto system, beginning with review chapters on Pluto itself, followed by review chapters on Charon, and then a review of what we know of Pluto's small satellites. But before any of that, the book opens with three chapters on early observations and studies of the Pluto system, as well as mission studies to reconnoiter it, and a chapter on the Pluto system's context within the Kuiper belt. This is followed by the objects within the Pluto system, beginning with review chapters on Pluto itself; then review chapters on its largest moon, Charon; and then a review of what we know of Pluto's four small satellites. In the final section of this volume are review chapters covering the origin, interior, and big picture of Pluto and its place in the Kuiper belt in the post-New Horizons era.

The central section of this volume focuses on Pluto itself, starting with a keystone chapter detailing the geological observations and inferences of the planet, including placing features and processes in a time-stratigraphic setting. This chapter is followed by ones covering geodynamics, the geological evidence for climate-driven evolution of the surface, cratering studies (including chronology), color and photometry, composition, and the rheology of the major ices on Pluto's surface. Next there are chapters covering atmospheric structure, atmospheric photochemistry and haze, atmospheric dynamics, climate cycles, atmospheric escape, and a chapter on the interaction of the solar wind with the environment of the Pluto system. The Charon section follows in an analogous order, starting with the geology of the satellite, followed by color and photometry, and then ending with Charon's atmospheric composition. A chapter covering the four small satellites follows the Charon chapters and concludes this section. As noted above, *PSANH* rounds out with a section of big-picture topics, beginning with a chapter on the system's origin, a geophysical synthesis, which includes the deep interior and its evolution, and a review of light shed on other KBO dwarf planets by our new understanding of the Pluto system. *PSANH* closes with two final chapters, a look at the future of Pluto system observations, including new mission concepts to follow New Horizons, and then finally one covering the results of the New Horizons flyby of the cold classical KBO, named Arrokoth.

No volume on this topic could have occurred without the historic contributions of Lowell Observatory's founder Percival Lowell, Pluto's (and by extension the Kuiper belt's and third zone's) discoverer Clyde Tombaugh, and Charon's discoverers Jim Christy and Robert Harrington.

The editors would like to thank the authors of all the chapters and appendixes in this book for their well-written contributions, and the many chapter referees for their important work to improve the elements of this book. We also thank David Grinspoon and Dava Sobel for their brilliant and heartfelt foreword to this book. We are also thankful to Ms. Cindy Conrad of the Southwest Research Institute for her immense help to the editors throughout the entire development of this book, and want to thank Alissa Earle for the outstanding job she did with the formidable task of indexing this entire volume. And we are especially grateful to Renée Dotson at the Lunar and Planetary Institute in Houston, along with her colleagues Heidi Lavelle, Kevin Portillo, Linda Chappell, and Mercedes Garcia, for their tireless and painstaking efforts in the editing, formatting, and production of this book.

The exploration of Pluto required numerous battles to be fought, and we gratefully acknowledge the historic and pivotal contributions of the following individuals from those battles: the late planetary science giant Michael Belton, who chaired the first planetary decadal survey that recommended a flyby of Pluto as the highest priority; accomplished space physicist Stamatios "Tom" Krimigis, who led The Johns Hopkins Applied Physics Laboratory's space department through the mission's proposal and main development; former U.S. Senator Barbara Mikulski, who spearheaded difficult but ultimately successful funding efforts for the mission; and former NASA Administrator Mike Griffin, who approved the mission's launch despite nuanced launch vehicle concerns. We also thank the many people at NASA and in the planetary science community, as well as the working staff and leadership teams of the Southwest Research Institute and The Johns Hopkins Applied Physics Laboratory, who worked so very long and so very diligently to see New Horizons flown to Pluto and the Kuiper belt, so that the amazing scientific results presented here could be achieved.

Thus, we dedicate this book to the estimated 2500+ men and women of New Horizons, whose work and dedication made these results possible.

S. Alan Stern, Jeffrey M. Moore, William M. Grundy,
Leslie A. Young, and Richard P. Binzel,
July 2020

Part 1:

System Background

Binzel R. P. and Schindler K. (2021) The discoveries of Pluto and Kuiper belt. In *The Pluto System After New Horizons* (S. A. Stern, J. M. Moore, W. M. Grundy, L. A. Young, and R. P. Binzel, eds.), pp. 3–8. Univ. of Arizona, Tucson, DOI: 10.2458/azu_uapress_9780816540945-ch001.

The Discoveries of Pluto and the Kuiper Belt

Richard P. Binzel
Massachusetts Institute of Technology

Kevin Schindler
Lowell Observatory

The discoveries of Pluto and the Kuiper belt draw many parallels to earlier epochs of planetary discovery. In this introductory chapter, we follow common threads in the history of planetary searches, not to retell the story in detail, but to place the discovery of Pluto and the Kuiper belt population in a broad context, creating the imperatives for exploration of the "third zone" of our solar system.

1. PLANETARY DISCOVERY PROLOGUE

If fortune favors the bold, then discovery favors the diligent. While this holds true across broad areas of science, it may apply most aptly to planetary astronomers. The history of telescopic astronomy began unfolding in 1609 with Galileo's examination of the Moon and planets, most notably with the discovery of the moons of Jupiter, now known as the Galilean satellites. Galileo approached his observations with the open-mindedness of a modern scientist, realizing that the application of "new technology" could reveal what was previously unseen and heretofore unknown. The observations themselves were allowed to rank as the ultimate authority, which Galileo dutifully recorded in his notebooks for subsequent interpretation. Deducing that the motions of the "Medician Stars" (Galileo's name for the satellites, honoring his benefactors) were centered around Jupiter, rather than Earth, proved to be a staunch challenge to the then-longstanding concept of a purely geocentric universe. It later fell to Isaac Newton, born within a year of Galileo's death, to deduce the underlying physics of gravity as the central force holding satellites and planets in their orbital paths.

Discovery also requires recognition that what one is seeing is something new or different than expected or predicted. Galileo himself, it seems, actually sighted and recorded Neptune during its conjunction with Jupiter in 1612 (*Kowal and Drake,* 1980), pre-dating its "discovery" by 234 years. To be fair, everything (including the multitude of faint background stars) that Galileo was observing and recording was new. Thus it is not surprising that Neptune slipping past Jupiter could go unrecognized as being *both* new and extraordinary. Such was not the case on March 13, 1781, when a musician turned astronomer, William Herschel, immediately recognized the new and unusual nature of a "large" star in Gemini, at a location that we note is just a few degrees away from where Pluto would be unveiled another century-and-a-half later. Herschel's optics, being good enough to reveal a non-stellar disk, were the key technological advancement enabling the on-the-spot discovery from his backyard garden, which may still be visited today.

Herschel's discovery of Uranus rekindled a curiosity about the regular spacing of the planets traceable back to *Kepler* (1596), who considered there had to be something between Jupiter and Mars: *Inter Jovem et Martem interposui planetem*. Johann Daniel Titius had worked out in 1766 a simple numerical series (communicated by Johann Elert Bode) that matched the spacing of the known planets but with the same incongruity that rankled Kepler: No known planet existed in the Mars-Jupiter gap. For the Titius-Bode "law" this was the series value corresponding to 2.8 AU. The fact that Uranus' orbital distance fit the value "predicted" if one extended the Titius-Bode numerical series beyond Saturn gave credence to the idea that the "missing planet" between Mars and Jupiter must be real. As detailed by *Cunningham* (1988), great confidence for a missing world was held by Baron Franz Xaver von Zach, director of the observatory at Gotha, Germany. Von Zach attempted predictions and conducted observational searches for the missing planet as early as 1785. After a decade of fruitless searches, von Zach organized conferences in 1796 and 1800 with the outcome being a coordinated search by a group of astronomers calling themselves the "Celestial Police." Each was assigned responsibility for searching a one-hour section of the zodiac.

Among those invited to participate in the search was Giuseppe Piazzi, an Italian clergyman serving as director of the Palermo Observatory. However, according to an archive search reported by *Foderà-Serio et al.* (2002), there is no record of the Italian mail ever delivering the invitation. Thus unaware, Piazzi made a true discovery on the night of January 1, 1801, recording the position of a star that had not previously been noted in the star catalog he was compiling. The moment of realization came the following night, when diligent confirmation of the added star revealed that it had moved. While Piazzi first supposed he had discovered a comet, its lack of nebulosity and rate of motion consistent with an orbital distance between Mars and Jupiter led him to surmise he had found *something better than a comet* (letter from G. Piazzi to B. Oriani, 24 January, 1801; *Foderà-Serio et al.,* 2002). The discovery of Ceres, it seemed, solved the problem of the missing planet. While Piazzi announced his findings in publications of the Palermo Observatory, the broadest announcement came fittingly (perhaps ironically) through *von Zach* (1801), as editor of the *Monatliche Correspondenz*.

Still more surprises, and discoveries, followed in rapid succession, with Heinrich Olbers noting on March 28, 1802 a second moving object (Pallas) in the same vicinity of the sky as Ceres. It was *Herschel* (1802) who noted these bodies had star-like (aster-like) appearances, coining the term "asteroids." That the Mars-Jupiter space was a zone comprised by a multitude of bodies became fully apparent within just a few years with the discoveries of Juno (1804) and Vesta (1807).

2. PLANETARY DISCOVERY THROUGH PREDICTION AND SEARCH

It is little recognized that were it not for Carl Friedrich Gauss, the discovery of Ceres might have ended badly for the Celestial Police. The urgency for how to predict where this telescopic planet could be found many months later led Gauss to invent the methodology of least-squares to fit the available data and develop the fundamental methods of orbit solution (*Gauss,* 1809) still in use today. Gauss' method, and prediction (accurate to within 0.5°) for the ephemeris location of Ceres, enabled Olbers to definitively spot it again on what turned out to be its one-year anniversary of discovery, January 1, 1802.

One can perceive that mathematically-minded astronomers became emboldened by the success with Ceres. Contemporaneous with Ceres and the discovery of a brood of asteroids was growing concern that something was askew with the motion of Uranus. Pre-discovery observations of Uranus (unwittingly measured as a fixed star) dating at least to 1690 by the first Astronomer Royal, John Flamsteed, could not be fit with the current measured motion. Restricting the analysis to "post-discovery" measurements from 1781 forward also yielded unsatisfactory orbit solutions, leading to some speculations that a planet beyond Uranus could be exerting a disturbing gravitational force. John Couch Adams was born into this era, where in 1841 (at the young age of 22) he began to consider the problem of the purported irregularities in Uranus' orbit. He reached a conclusion for a predicted location for the unseen planet as early as 1845. As described by the astronomer and historian Patrick Moore (*Tombaugh and Moore,* 1980), a sequence of mishaps and arrogant disregard for the young scientist led to his prediction being ignored by his fellow Englishmen. *Krajnović* (2016) proposes that it was owing to Adams' predictions being imprecise that his compatriots remained unconvinced that an extensive search would bear fruit. Independently and unknowingly in France, Urbain Jean Joseph Le Verrier was simultaneously tackling the problem of the perturbing planet, reaching similar but quite specific solutions for the new planet's ephemeris in the 1845–1846 timeframe. Analogous to the reception Adams received, none of Le Verrier's compatriot astronomers seemed to be convinced that this prediction merited verification — perhaps the possibility of such precision seemed hopelessly optimistic. With his patience exhausted, Le Verrier wrote to Johann Galle at the Berlin Observatory who set to work the very night of the letter's receipt: September 23, 1846. Neptune was discovered within 1° of its predicted position.

Mathematical astronomers did not rejoice for long, as shortly after Neptune was recognized as the eighth member of the planetary family, some scientists analyzing its orbit suggested its apparent mass didn't seem to account for all the irregularities in Uranus' orbit. Ergo, another planet must still be out there somewhere, tugging at Uranus. Le Verrier himself spent varying amounts of time on the problem as did others, including Canadian-American astronomer/mathematician Simon Newcomb of the U.S. Naval Observatory and later Johns Hopkins University and American astronomer David Peck Todd, who spent most of his career at Amherst College. French astronomer Camille Flammarion — later a friend of Percival Lowell's — also played a role, although his reasons for believing a ninth planet existed centered around the apparent relationship between planetary orbits and a seemingly non-random distribution in the heliocentric longitude distribution of comets (*Cruikshank and Sheehan,* 2018). Most of these efforts petered out. Thus, at the opening of the twentieth century, the majority of astronomers doubted the reality of a planet beyond Neptune.

2.1. Percival Lowell and the Search for Pluto

Entering the scene at the turn of the century was Percival Lowell, who in 1894 established his own private observatory initially to prove the existence of intelligent life on Mars. Over the first decade of work at his Flagstaff, Arizona facility, Lowell expanded his research to include characterization of other bodies of the solar system, leading to his development of the concept of planetary evolution that he called "planetology." Not timid to share his ideas and not shy of self-promotion, Lowell summarized his early comparative studies of the planets in a series of public lectures in 1902 at the Massachusetts Institute of Technology (MIT). His engaging style and provocative ideas motivated the publisher Houghton, Mifflin and Company to compile and print the lectures in a diminutive book titled *The Solar System* (*Lowell,* 1903). Of significance toward unraveling the structure of the outer solar system are two passages in which he suggests the possibility of a new outer planet. In the first, Lowell postulates that specific planets to some degree gravitationally control specific meteor showers. Connecting the Andromedids with Jupiter and the Leonids with Uranus, he boldly expounded that the Perseids and Lyrids "go out to meet *the unknown planet* which circles at a distance of about forty-five astronomical units from the Sun." (Italics added.) In typical Lowell pomp, he concludes, "It may seem to you strange to speak thus confidently of what no mortal eye has seen, but the finger of the sign-board of phenomena points so clearly as to justify the definite article. The eye of analysis has already suspected the invisible." It seems interesting that Lowell chose to focus on a distance near 45 AU even though Comet Swift-Tuttle (parent of the Perseids) and Comet C/1861 G1 (Thatcher; parent of the Lyrids) have aphelion distances of 51 and 110 AU, respectively.

While in these initial musings Lowell pointed to these apparent relationships between planets with meteor showers and comets as evidence for a previously undetected planet, he quickly changed focus to evaluating the least-squares of residuals of the orbit of Uranus. This method was reliant on accurate measurements of the position of Uranus as well as sound theoretical estimates. By the time he published his magnum opus on the subject (*Memoirs on a Trans-Neptunian Planet*) in 1915, his estimated distance fell in the 40–50-AU range. Thus Lowell ultimately set 45 AU as the most likely number (*Lowell,* 1915; also see discussion in *Cruikshank and Sheehan,* 2018). Lowell was not completely sold on his own estimates, as indicated by his statement, "Owing to the inexactitude of our data, then, we cannot regard our results with the complacency of completeness we should like" (*Lowell,* 1915).

Later in his book, and in similar fashion as his good friend Flammarion, Lowell likewise connects the orbits of comets to planets, with clusters of comets linked to Jupiter, Saturn, Uranus, and Neptune. Lowell did not stop with the known planets, going on to say he could recognize more comet clusters appearing to clump further from the Sun. Lowell summarizes: "This can hardly be an accident; and if not chance, it means a planet out there as yet unseen by man, but certain sometime to be detected and added to the others. Thus not only are comets a part of our solar system now recognized, but they act as finger-posts to planets not yet known." Lowell's finger pointing based on the orbital distribution of known objects in the outer solar system presages detective work in the current epoch still looking for traces of a massive planet beyond Neptune (*Trujillo and Sheppard,* 2014; *Brown and Batygin,* 2016; *Batygin et al.,* 2019).

In 1905, two years after publication of *The Solar System*, Lowell went from theorizing to searching, quietly beginning what he referred to as "The Invariable Plane Search" (*Schindler and Grundy,* 2018). He tried to mathematically pinpoint the location of the purported planet — with the help of a small team of human "computers." Their objective was to see if adding a perturbing mass could be effective in reducing the residuals of Uranus and Neptune. By this time Lowell had realized the connection of planets to comets and meteor showers was tenuous and thus, in the spirit of Adams and Le Verrier, focused on the apparent unaccounted irregularities in Uranus' orbit, which meant the planet he was looking for was massive enough to tug noticeably on Uranus. Meanwhile, Lowell and his team of observers began an actual search, experimenting with a number of different telescopes, trying to photograph candidate areas of the sky. Lowell's own search effort began in 1905 and faded out in 1909, but the following year he reinvigorated the search upon learning that a former colleague-turned rival, William Pickering at Harvard, was jumping into the planet-search fray. Lowell redoubled his efforts not to be scooped by a competitor by using a suite of new instruments and assistants, doggedly continuing the search until his death in late 1916.

In what Lowell's brother Abbott Lawrence Lowell called Percival's biggest disappointment in life, Percival did not discover a planet (*A. L. Lowell,* 1935). Ironically, however, in a sense Lowell did spot Pluto: One of Percival's assistants, Thomas Gill, unknowingly photographed Pluto twice in 1915 — on March 19 and April 7 (*Schindler and Grundy,* 2018). Thus came an end to this self-taught astronomer's life and career, as well as his personal search for a new world.

2.2. Clyde Tombaugh's Discovery of Pluto

With persistence as a family trait, in 1927 Percival's nephew, Roger Lowell Putnam, assumed the reins as sole trustee of Lowell Observatory and decided to recommence the search for what Percival had long called "Planet X." Observatory director V. M. Slipher formulated a plan for the search, building on the earlier efforts of Lowell but instituting a much more systematic approach. In late 1928, while a new 13-inch astrograph was being constructed at the observatory for use in the search, Slipher received a letter from a 23-year-old Kansas farmer and amateur astronomer, Clyde Tombaugh, who had aspirations as a professional astronomer (*Tombaugh and Moore,* 1980).

Thus began an exchange of correspondence that soon led to Slipher offering Tombaugh a job at the observatory to help with the planet search. Tombaugh arrived at Lowell on January 15, 1929, and several weeks later, on February 11, the primary lens for the 13-inch instrument arrived. Slipher, Tombaugh, and other observatory staff then spent several months installing and testing the astrograph, working through problems such as plate fogging and warping. Commissioning complete, they began the planet search on April 6. Initially, Tombaugh took the plates and the senior astronomers at Lowell then examined them, but they all were also occupied with other duties. Consequently, within a few months the young Tombaugh took over the grueling task of plate examination. In a typical plate of the search area, Tombaugh took one-hour exposures and captured an average of 300,000 star images. During the day, he painstakingly examined pairs of these plates that captured the same portion of the sky but which were taken a few days apart. The challenge of systematically analyzing this sheer number of stellar images was compounded by the appearance of numerous asteroids, whose motion as captured on the plates closely resembled that of the expected planet. For Tombaugh this reinforced the importance of photographing specific areas of the sky only when they were at opposition, when the apparent motion of those interlopers would be maximized and their distance easily measured to ensure they were not a more distant planet (*Hoyt,* 1980; *Tombaugh and Moore,* 1980). The work was tedious and required patience and diligence, traits that Tombaugh had learned growing up on the farm. On February 18, 1930, just 10 months after commencing the search, Tombaugh fulfilled the vision of his fellow self-taught astronomer, Percival Lowell, and discovered emerging "out of the darkness, the planet Pluto." In their book bearing this title, *Tombaugh and Moore* (1980) thoroughly detail the events of the day and following weeks leading to the naming and formal announcement of this newly discovered world.

2.3. James Christy's Discovery of Charon

While Tombaugh's discovery of Pluto is a story of persistence and diligence toward a deliberate search, Jim Christy's discovery of Charon is not unlike that of Herschel or Piazzi, where dedication to the task at hand and open-minded perception proved to be the key ingredients. Under this recipe, what one might think is "accidental" is no accident at all.

Making routine astrometric measurements to track Pluto's orbital motion was one of the duties assigned to James W. Christy, an astronomer at the U.S. Naval Observatory's headquarters in Washington, DC. As part of this task in updating Pluto's ephemeris, Christy had asked astronomers at the Naval Observatory's Flagstaff Station (NOFS) — located just 4 miles west of where Clyde Tombaugh discovered Pluto — to take several images of Pluto using the facility's 61-inch telescope. NOFS astronomer Anthony Hewitt took the requested images on April 13 and May 12, 1978 and then sent them to Christy in Washington, DC. Using a high-precision measuring machine on June 22, Christy began measuring these and other Pluto plates from 1965, 1970, and 1971. Examining these plates under magnification, it seemed many of them were of poor quality because Pluto appeared out-of-round and distinctly asymmetrical, with a bulge in the north-south orientation. Rather than cast the April and May plates aside and ask for new ones, Christy took a broader look at everything the plates had to offer and came to an astounding realization: Background stars on each plate were in fact perfectly round (as they should be) and didn't exhibit the same lopsidedness as the neighboring Pluto images. Elongation of all the plate images — both Pluto and the stars — would have been easily explained as poorly tracked and smeared or out-of-focus images. But with the images revealing that only Pluto itself was appearing elongated, the answer had to be something intrinsic to Pluto itself. That something was presumably the presence of a large satellite, soon to be named Charon (*Smith et al.,* 1978; *Christy and Harrington,* 1980). Any doubts that the photographic anomalies were due to a satellite were removed in 1985 when mutual transits and occultations revealed themselves in time-resolved photometric monitoring of the Pluto system (*Binzel et al.,* 1985). As detailed in the chapter in this volume by Lunine et al., Charon's discovery unlocked the mystery of Pluto's mass and paved the way for the advancing physical studies of Pluto. Fittingly for this volume, with Pluto no longer being a single planetary body, the era of studying the Pluto system was born.

2.4. Revealing the Kuiper Belt: Discovery by David Jewitt and Jane Luu

When Lowell Observatory announced Pluto's discovery on March 13, 1930 — coinciding with the 149th anniversary of Herschel's discovery of Uranus and what would have been Percival Lowell's 75th birthday — Clyde Tombaugh found himself in a very unaccustomed light. Rather than the solitude of diligently photographing the sky or the tedium of examining the resulting plates, as he had regularly done for the past year, Tombaugh found himself in the spotlight speaking with media, answering fan mail, and otherwise taking care of discovery-related requests. Astronomers from around the world wanted details about the new planet's position so they could determine its orbit, while the public just wanted to know what to call this new world, the first planet discovered in the United States. Quite logically, a follow-on question was being asked by scientists and the public alike: Are there other planets like this new one out there, waiting to be discovered? By May 1930, the demands on Tombaugh's time and attention had been resolved and he was finally able to get back to research, allowing him to begin working on the answer to the follow-on question in the most logical way possible: by resuming the search. Thus as a tribute to the finest sense of pure scientific inquiry, Tombaugh continued the systematic search of the skies for the ensuing 13 years. (His only significant time away was to earn his undergraduate degree at the University of Kansas.) During that stretch, Tombaugh and colleagues who occasionally helped him surveyed about 75% of the sky down to a magnitude as faint as 18. All together they captured some 90 million stellar images and recorded about 4000 asteroids, nearly 30,000 extra-galactic nebulae, and 1800 variable stars (*Giclas,* 1997; *Hoyt,* 1980). Adding to the scorecard was one comet, one globular cluster, and five open star clusters.

Nevertheless, as Tombaugh wound down his survey in 1943, Irish-born military officer, electrical engineer, economist, and amateur astronomer Kenneth Edgeworth speculated that comets or other bodies could be expected in the region beyond Neptune. This was the first suggestion [or among the first suggestions; see *Davies et al.* (2008)] of the existence of what later became known as the Kuiper belt. Even though Edgeworth expanded on this idea in 1949, neither of his papers drew much attention (*Edgeworth,* 1943, 1949). Much more well-known Dutch-born astronomer Gerard Kuiper and others subsequently began considering the possibility of a ring of icy bodies beyond Neptune (*Kuiper,* 1951).

So why didn't Tombaugh's painstaking survey find another planet or, for that matter, any other body at or beyond Pluto's distance? The detection limits of the photographic plate/blink comparator technology simply weren't good enough to detect any but the largest and closest members of any transneptunian population. Pluto's large size and high albedo on an inbound trajectory toward perihelion were the combination of factors needed for bringing an outer solar system discovery within the grasp of the technology available in the early twentieth century.

The transformative technical revolution in astronomy (including planetary searches) emerged in the 1980s when the astronomical community broadly adopted the use of digital imaging arrays, specifically charge-coupled devices (CCDs). Rather than a typical photographic plate having a detection efficiency of about 10% for recording an arriving photon, the peak efficiency of a CCD surpasses 90%. The trade-off for search strategy, however, was substantial. CCD detectors were literally postage-stamp-sized or smaller, whereas photographic plates covered substantially greater areas in the focal plane. (For example, the Palomar Schmidt telescope utilized glass plates covering 14 × 14 square inches.)

One of the astronomers who quickly embraced the emerging capability of CCD imaging was British-born David Jewitt. Jewitt made his mark in 1982 by using this cutting-edge detection capability in recovering the nucleus of Comet 1P/Halley on its inbound trajectory toward its 1986 perihelion (*Jewitt and Danielson,* 1984). Around the same time, he began wondering whether the outer solar system was really empty, or just had not been surveyed thoroughly enough to faint enough limits. There was only one way to find out, and that was to look. In some ways this was a bolder step than von Zach (Ceres), Galle (Neptune), or Tombaugh (Pluto) as there was no specific prediction on where to point. Most nearly, it paralleled the dogged determination that Tombaugh had shown 50 years earlier in extending his own survey another 13 years after the accomplishment of its presumed objective. These explorers were rather straightforwardly asking: "What's out there?"

New graduate student Jane Luu arrived at Jewitt's MIT office at an auspicious time (1986) when Jewitt's curiosity and available technology were at a point of convergence. As recounted by *Jewitt* (2010), the pair commenced a two-pronged approach using large-area-format photographic plates (limiting magnitude ~20, two or more magnitudes fainter than Tombaugh) to search broadly and postage-stamp-sized CCDs to search tiny patches more deeply (magnitude limit ~24). The searchers concentrated on the slowest-moving objects, assuring themselves of finding objects in the outermost reaches of our solar system without being distracted by bodies closer in, such as main-belt asteroids. The first of the intermediate objects known as Centaurs, extending beyond the orbit of Saturn, had been detected by Kowal in 1977 (*Kowal et al.,* 1979; *Kowal,* 1989). While new Centaurs would be major discoveries, Jewitt and Luu set their determined gaze at Neptune's distance and beyond, aptly naming their project the Slow Moving Object (SMO) Survey (*Luu and Jewitt,* 1988). Contemporary dynamical analysis (*Duncan et al.,* 1988) as well as work by *Fernández* (1980) pointing to a source region for short-period comets just beyond Neptune gave affirming rationale for optimizing their survey strategy toward this region. After a year or so, the photographic part of the search was abandoned largely due to the physical and mechanical burden of dealing with photographic plates, both in acquiring the images but also in adequately scanning them. In essence, this proved to be the transition from the Tombaugh era and mechanical blink comparators to the modern digital era of comparing images using computer software.

Over the next five years, up to 1992, Jewitt and Luu progressively employed larger-sized CCDs as they became available and migrated to the University of Hawaii enabling more frequent access to larger-aperture telescopes. On the night of August 30, 1992, using the largest CCD they had access to up to that point, they detected a 23rd-magnitude object just beyond the orbit of Neptune (*Jewitt and Luu,* 1993). Computer processing of the succession of four images at the telescope allowed them to realize in real time that they had a potential discovery, with their eyes spotting on the screen an object having the correct rate of motion and expected apparent magnitude. For Jewitt and Luu, their moment of discovery paralleled Herschel, Galle, and Piazzi, who as visual observers naturally made their realizations in real time. Designated 1992 QB_1 (later designated 15760 Albion), the SMO Survey had a long enough track record that reasonable statistics could be projected even from a sample of one; finding 1992 QB_1 meant that thousands of similar objects almost certainly had to be there.

Ignited by the discovery by Jewitt and Luu, additional discoveries beyond Neptune followed and increasing recognition began being cast back to the writings of Edgeworth, Kuiper, and others (*Davies,* 2008). A new "third zone" of our solar system had been discovered, with Pluto being its largest known member. As evidenced by the spawning of this new field in planetary science and its growing to the point of filling its own volume in this book series (*Barucci et al.,* 2008), finding 1992 QB_1 was a game changer in driving the imperative for exploration of the Pluto system. As detailed in the chapter by Lunine et al., exploration of the Kuiper belt and Pluto became the top priority for understanding the newly recognizable structure of the outer solar system. Thus, by the time New Horizons emerged as a candidate mission, its objectives as an explorer of both the Kuiper belt and Pluto were quite clear. Those objectives were realized by the 2015 Pluto system encounter and the 2019 encounter of 2014 MU69 (486958 Arrokoth). In each case, the time span from discovery to *in situ* spacecraft investigation was the shortest in the history of space exploration.

REFERENCES

Barucci M. A., Boehnardt H., Cruikshank D. P., and Morbidelli A., eds. (2008) *The Solar System Beyond Neptune.* Univ. of Arizona, Tucson. 592 pp.

Batygin K., Adams F., Brown M. E., and Becker J. C. (2019) The planet nine hypothesis. *Phys. Rept., 805,* 1–53.

Binzel R. P., Tholen D. J., Tedesco E. F., Buratti B. J., and Nelson R. M. (1985) The detection of eclipses in the Pluto-Charon system. *Science, 228,* 1193–1195.

Brown M. E. and Batygin K. (2016) Observational constraints on the orbit and location of planet nine in the outer solar system. *Astrophys. J. Lett., 824,* L23.

Christy J. W. and Harrington R. S. (1980) The discovery and orbit of Charon. *Icarus, 44*, 38–40.

Cruikshank D. P. and Sheehan W. (2018) *Discovering Pluto: Exploration at the Edge of the Solar System*. Univ. of Arizona, Tucson. 475 pp.

Cunningham C. J. (1988) *Introduction to Asteroids*. Willmann-Bell, Richmond. 208 pp.

Davies J. K., McFarland J., Bailey M. E., and Marsden B. G. (2008) The early development of ideas concerning the transneptunian region. In *The Solar System Beyond Neptune* (M. A. Barucci et al., eds.), pp. 11–23. Univ. of Arizona, Tucson.

Duncan M., Quinn T., and Tremaine S. (1988) The origin of short-period comets. *Astrophys. J. Lett., 328*, L69.

Edgeworth K. (1943) The evolution of our planetary system. *J. Br. Astron. Assoc., 53*, 181–188.

Edgeworth K. (1949) The origin and evolution of the solar system. *Mon. Not. R. Astron. Soc., 109*, 600–609.

Fernandez J. A. (1980) On the existence of a comet belt beyond Neptune. *Mon. Not. R. Astron. Soc., 192*, 481–491.

Foderà-Serio G., Manara A., and Sicoli P. (2002) Giuseppe Piazzi and the discovery of Ceres. In *Asteroids III* (W. F. Bottke Jr. et al., eds.), pp. 17–24. Univ. of Arizona, Tucson.

Gauss C. F. (1809) *Theoria Motus Corporum Coelestium in Sectionibus Conicis Solem Ambientium*. F. Perthes and I. H. Besser, Hamburg. 228 pp.

Giclas H. L. (1997) The 13-inch Pluto Discovery Telescope. Lowell Observatory, Flagstaff. 17 pp.

Herschel W. (1802) Observations on the two lately discovered celestial bodies. *Philos. Trans. R. Soc. London, 2*, 213–232.

Hoyt W. G. (1980) *Planets X and Pluto*. Univ. of Arizona, Tucson. 302 pp.

Jewitt D. (2010) The discovery of the Kuiper belt. *Astron. Beat, 48*, 1–5.

Jewitt D. and Danielson G. E. (1984) Charge-coupled device photometry of comet P/Halley. *Icarus, 60*, 435–444.

Jewitt D. and Luu J. (1993) Discovery of the candidate Kuiper belt object 1992 QB_1. *Nature, 362*, 730–732.

Kepler J. (1596) *Prodromus dissertationum cosmographicarum continens mysterium cosmographicum de admirabili proportione orbium celestium deque causis coelorum numeri, magnitudinis, motuumque periodicorum genuinis et propiis, demonstratum per quinque regularia corpora geometrica. Excudebat Georgius Gruppenbachius, Tubingae*. Translated by Duncan A. M. (1981) *The Secret of the Universe*. Abaris, New York. 267 pp.

Kowal C. T. (1989) A solar system survey. *Icarus, 77*, 118–123.

Kowal C. T. and Drake S. (1980) Galileo's observations of Neptune. *Nature, 287*, 311–313.

Kowal C. T., Liller W., and Marsden B. G. (1979) The discovery and orbit of (2060) Chiron. In *Dynamics of the Solar System* (R. L. Duncombe. ed.), pp. 245–250. IAU Symp. 81, Reidel, Dordrecht.

Krajnovic D. (2016) The contrivance of Neptune. *Astron. Geophys., 57*, 28–34.

Kuiper G. P. (1951) On the origin of the solar system. In *Astrophysics: A Topical Symposium Commemorating the Fiftieth Anniversary of the Yerkes Observatory and a Half Century of Progress in Astrophysics* (J. A. Hynek, ed.), pp. 357–414. McGraw-Hill, New York.

Lowell A. L. (1935) *Biography of Percival Lowell*. Macmillan, New York. 212 pp.

Lowell P. (1903) *The Solar System*. Houghton Mifflin, Boston. 134 pp.

Lowell P. (1915) *Memoir on a Trans-Neptunian Planet*. T. P. Nichols and Son, Lynn. 105 pp.

Luu J. X. and Jewitt D. (1988) A two-part search for slow-moving objects. *Astron. J., 95*, 1256–1262.

Schindler K. and Grundy W. (2018) *Pluto and Lowell Observatory: A History of Discovery at Flagstaff*. History Press, Charleston. 174 pp.

Smith J. C., Christy J. W., and Graham J. A. (1978) 1978 P 1. *IAU Circular 3241*.

Tombaugh C. W. and Moore P. (1980) *Out of the Darkness: The Planet Pluto*. Stackpole, Harrisburg. 221 pp.

Trujillo C. A. and Sheppard S. S. (2014) A Sedna-like body with a perihelion of 80 astronomical units. *Nature, 507*, 471–474.

Von Zach F. X. (1801) Über einen zwischen Mars und Jupiter längst vermuteten, nun wohnscheinlich entdeckten neuen Hauptplaneten unseres Sonnen Systems. *Monatliche Correspondenz zur Beforderung der Erd- und Himmelskunde, 3*, 592–623.

Lunine J. I., Stern S. A., Young L. A., Neufeld M. J., and Binzel R. P. (2021) Early Pluto science, the imperative for exploration, and New Horizons. In *The Pluto System After New Horizons* (S. A. Stern, J. M. Moore, W. M. Grundy, L. A. Young, and R. P. Binzel, eds.), pp. 9–20. Univ. of Arizona, Tucson, DOI: 10.2458/azu_uapress_9780816540945-ch002.

Early Pluto Science, the Imperative for Exploration, and New Horizons

Jonathan I. Lunine
Cornell University

S. Alan Stern and Leslie A. Young
Southwest Research Institute

Michael J. Neufeld
National Air and Space Museum

Richard P. Binzel
Massachusetts Institute of Technology

Like a number of other great discoveries, that of distant Pluto was based upon a misconception, in this case that the orbital motion of Neptune was being influenced by a large, more distant, "Planet X." Nevertheless, Clyde Tombaugh's remarkable discovery launched decades of efforts to push telescopic techniques to the limit, and theoretical speculations as to why the solar system's last planetary outpost should be so small. Those speculations were answered beginning in 1992 with the discovery of the Kuiper belt, of which Pluto is a part. The idea of Pluto as the last planetary frontier galvanized the space science community into pushing for a mission to explore what would turn out to be a body with a wealth of geologic and atmospheric processes and a rich system of satellites. The long odyssey to make the New Horizons mission a reality was the capstone to an era in which the mode of planning planetary exploration was transformed.

1. BEFORE 1930: THE HISTORICAL ASTRONOMICAL CONTEXT OF PLUTO'S DISCOVERY

By the end of the 1920s the United States had become preeminent in the field of observational cosmology, with prominent astronomers such as Milton Humason, Edwin Hubble, Henrietta Leavitt, and Vesto M. Slipher demonstrating the true scale of the cosmos and establishing empirically the relationship between distance and recessional velocities of galaxies. One of those pioneering American astronomers, Slipher, did his work at Lowell Observatory, whose founder, Percival Lowell, began the first — and only systematic — search for a planet beyond Neptune in 1905 (*Reaves,* 1997, and references therein). Lowell's multiyear search, and briefer searches by others, were stimulated by the conclusions of Herschel that the residuals in the orbital elements of Uranus could not be accounted for entirely by Neptune (*Herschel,* 1867), and then by the analyses of *Pickering* (1909) and ultimately of Lowell himself (*Lowell,* 1915). Because such a planet was expected to be massive and hence bright, the much-fainter Pluto escaped detection, although it was later "precovered" in Lowell's archival photographic plates of 1915 (*Lampland,* 1933), and those of Humason in 1919 (*Nicholson and Mayall,* 1931). Tombaugh was the first to identify it, in his inexhaustibly diligent blink comparator analysis of plates taken in 1930 at Lowell Observatory (*Slipher,* 1930).

Although the discovery of Pluto was celebrated at the time as another triumph of American astronomy, it was much fainter than expected and — thanks to the discovery of Charon four decades later (*Christy and Harrington,* 1978) — eventually shown to have a mass much too small to have been responsible for the perturbations of Uranus's orbit that began the hunt decades before. With the analysis by *Standish* (1993) that the apparent residuals in Uranus' orbit were the result of the use of an erroneous mass for Neptune, it became clear that the discovery of the ninth planet was due to hard work by Tombaugh and Lowell's persistence (including bequeathing funds for a new telescope after his passing), but not due to Lowell's theoretical prediction.

One might ask how the history of solar system astronomy might have changed had the search for Pluto failed, or if the search were not mounted in the first place (if, e.g., spurious residuals for Uranus' position had not been obtained). Pluto would very likely have been discovered serendipitously before the 1992 discovery of the first small Kuiper belt object (KBO), labeled 1992 QB_1 (and now called 15760 Albion) (*Jewitt and Luu,* 1993), but plausibly after the first paper predicting a belt of material beyond Neptune (*Edgeworth,* 1943).

In his model for the origin of the solar system, *Edgeworth* (1949) regarded Pluto as an escaped satellite of Neptune, and hence not a constraint on the distribution of material in his planet-forming "annulus." *Kuiper* (1951), on the other hand, explicitly considered Pluto's eccentric orbit in proposing that the solar nebula (and proto-Neptune's orbit) extended out to 50 AU. However, his nebular model is only of historical interest today. Perhaps the best answer to whether planetary astronomy might have changed in the absence of Pluto's discovery until decades later is "not much in the long run," because the need to explain the injection of long period comets into the Oort cloud (*Oort,* 1950) and the planar distribution of short-period comets (*Duncan et al.,* 1987) would each have spurred on searches for objects in a belt beyond Neptune in any case, and would have eventually resulted in Pluto's discovery as well as the discovery of the Kuiper belt and Pluto's cohort of Kuiper belt dwarf planets.

2. 1930–1992: THE NINTH PLANET BEFORE THE KUIPER BELT

2.1. Speculations and Science

The history of Pluto studies prior to the discovery of the next discovered Kuiper belt object, 1992 QB_1, is long and complex (e.g., *Marcialis,* 1997; *Stern and Mitton,* 1998, *DeVorkin,* 2013). Some of the scientific breakthroughs up to the discovery of the Kuiper belt are tabulated in Table 1 and are briefly described below.

Soon after Pluto's discovery (see the chapter by Binzel and Schindler), its orbit was determined to be unusually eccentric and inclined relative to that of the other planets (*Leonard,* 1930). Pluto's dip inward of Neptune led to the idea that the ninth planet might be an escaped moon of Neptune [cf. *Marcialis* (1997) for the complex story of who first proposed this], but some 35 years after the orbit of Pluto was determined, *Cohen and Hubbard* (1965) established by numerical integrations that Pluto and Neptune were locked in a precise 2:3 mean-motion resonance that librates about a center point relative to Neptune. The longitudinal phasing of these two bodies in resonance is such that the two objects can never approach closely and the escaped moon idea is implausible. The libration of the resonance angle prohibits close approaches between Neptune and Pluto; the conjunctions occur near Pluto's aphelion and the strength of the resonance stabilizes Pluto's orbit (cf. *Malhotra and Williams,* 1997). The properties of this resonance would eventually lead to the realization that Pluto was almost certainly formed, and remained in, a belt of primordial bodies beyond the realm of the giant planets.

The physical properties of Pluto, and after its discovery, of Pluto's large satellite Charon, were an active area of research in the decades up to 1990. By mid-century photometric observations determined the rotation period of Pluto (*Walker and Hardie,* 1955), but it was not until the 1970s that Pluto's large obliquity was inferred from improved and extended photometry (*Andersson and Fix,* 1973). With the discovery of Charon (*Christy and Harrington,* 1978), the system mass of just over 10^{25} g was roughly determined. This was crucial, because it established that Pluto was not a massive object. Although its faintness in discovery images had ruled out Pluto as the source of the apparent (and ultimately artefactual) residuals in Uranus' orbital motion, considerable controversy remained for decades over just exactly how small Pluto might be.

Duncombe and Seidelman (1980) tabulated estimates of Pluto's mass, and one can identify three "epochs" between each that the mass declines by an order of magnitude. Prediscovery, planet "X" was on the order of 10 $M_\oplus$; from 1930 to 1955, Pluto is on the order of 1 $M_\oplus$, and from 1968 to 1978, Pluto steadily declines from 0.1 $M_\oplus$ to a final value, given by Charon's discovery, some 50 times smaller. Indeed, a tongue-in-cheek treatment of this history by *Dessler and Russell* (1980) predicted a massless Pluto by 1984, by drawing a best-fit line through all the points. However, it

TABLE 1. Partial list of Pluto/Charon discoveries through 1992.

Year	Discovery	Reference
1930	Pluto discovered; orbit determined	*Bower and Whipple* (1930)
1955	Rotation period of 6.4 days determined	*Walker and Hardie* (1955)
1965	Neptune-Pluto 2:3 orbit resonance found	*Cohen and Hubbard* (1965)
1973	Extreme obliquity of Pluto discovered	*Andersson and Fix* (1973)
1976	Discovery of methane ice on Pluto	*Cruikshank et al.* (1976)
1978	Charon discovered; system mass measured	*Christy and Harrington* (1978)
1985–1990	First maps of Pluto	*Buie and Tholen* (1989)
1986	First reliable radii of Pluto and Charon	*Dunbar and Tedesco* (1986)
1987	Discovery of water ice on Charon	*Marcialis et al.* (1987)
1988	Stellar occultation sees Pluto's atmosphere	*Elliot et al.* (1989)
1991–1993	Pluto-Charon mass ratio	*Young et al.* (1994)
1992	Atmospheric methane	*Young et al.* (1997)
1992	Discovery of N_2 and CO ice on Pluto	*Owen et al.* (1993)

From *Lunine et al.* (1995) and *Stern and Mitton* (1998). Not all references to a given discovery are listed.

is the clustering of the data points, in contrast to a smooth trend, that reveals the underlying cause of the decline: The pre-discovery mass was required to explain the residuals in Uranus' motion, while in the two decades after Pluto's discovery, improved measurements of the motion of Uranus and Neptune reduced the required mass of Pluto assuming it was the perturber. Even more precise measurements in the 1960s and 1970s further reduced Pluto's mass, but were overtaken by the definitive mass given by the orbital periods and separations of Pluto and Charon. That mass eliminated Pluto once and for all as the cause of any residuals in the ice giant orbital motions, which finally were proved erroneous as noted above with the analysis of *Standish* (1993).

As ultimately incorrect as they were, the inflated masses of Pluto up through the mid-1960s combined with imprecise estimates of radii contributed to a science-fictionesque mystique of the ninth planet. Typical values of the density obtained were as large as or much larger than Earth (*Marcialis,* 1997). Was this a new kind of ultradense planet, Mars-sized but made of exotic material? Might it be larger, perhaps Earth-sized after all? Gerard Kuiper instead dismissed the masses as overestimates, and on the basis of B-V colors presciently imagined a smaller body with an ice-covered surface (*Kuiper,* 1950), an idea that was confirmed a quarter century later in a paper led by one of his former students (*Cruikshank et al.,* 1976). However, it would not be until the fortuitous set of "mutual events" [transits of Charon in front of Pluto and occultations of Charon behind Pluto, first detected by *Binzel et al.* (1985)] that highly accurate radii could be obtained (*Buie et al.,* 1992; *Young and Binzel,* 1993) showing Pluto to be 70% the radius of Earth's Moon. Together with the mass given by the orbital separation with Charon and the orbital period, Pluto turned out to have a density approximately close to twice that of water ice — and to be a small rock/ice world at the edge of the solar system.

An interesting perspective regarding the problem of Pluto's bulk properties comes from the history of observations of Neptune's moon, Triton, which orbits the Sun roughly as far as Pluto's perihelion and is just 14% larger in diameter. Discovered in 1846 by William Lassell, Triton is difficult to observe because it is small and close in angle to a larger brighter object, Neptune. Early observations suggested Triton was massive enough to perturb Neptune, and *Alden* (1940) obtained a value of 1.3×10^{26} g: almost twice the mass of Earth's Moon and six times Voyager 2's accurate mass determination in 1989. Visible and infrared observations led to an upper limit for the radius of 2600 km (*Cruikshank et al.,* 1979). Combining the two leads to a density of 1.8 g cm^{-3}, fortuitously close to Triton's actual density and perfectly reasonable for a body with roughly equal amounts of rock and ice. However, subsequent observations shrank Triton until speckle studies produced a radius between 1037 and 1250 km (*Bonneau and Foy,* 1986). Such a value threw the mass determination out because it led to an unphysical density, and it also caused considerable angst among planners for the Voyager 2 encounter with Neptune because the ability to obtain both a UV solar and Earth radio occultation from a possible atmosphere required a larger size for the moon. In the end, as Voyager 2 sped toward Neptune, it became clear that the speckle observations were seeing a bright polar cap and not the darker annulus around it. Triton's numbers from Voyager 2 were a radius of 1353 km, a mass of 2.1×10^{25} kg, and a density of 2.05 g cm^{-3} (*Lodders and Fegley,* 1998).

Although the details are different, the parallels between these Pluto and Triton stories are striking: Both bodies had incorrect bulk parameters thanks to inaccurate observations of Uranus and/or Neptune; both bodies were initially thought to be much bigger than they are, and then began to shrink. However, while Pluto's mass and radius values were settled thanks to the discovery of Charon, Triton's required a spacecraft mission — Voyager — a mission that would presage a remarkable voyage of exploration through the Pluto-Charon system by New Horizons a quarter of a century later.

Crucial to the eventual interest in Pluto as a target of spacecraft exploration was its multiple-component icy surface and the presence of an atmosphere. Cruikshank and colleagues discovered methane by carefully selecting two narrow band filters in the 1–2-μm wavelength region (*Cruikshank et al.,* 1976), concluding it was a surface ice based in part on the absence of a water ice signature. A decade later, water ice was discovered — but on Pluto's companion Charon (*Marcialis et al.,* 1987; *Buie et al.,* 1987). The conclusion that the methane signature was from the surface (*Buie and Fink,* 1987; *Spencer et al.,* 1990) would be confirmed years later (*Stern et al.,* 1993). A stellar occultation by Pluto was observed in 1988, definitively establishing the presence of an atmosphere (*Elliot et al.,* 1989; *Hubbard et al.,* 1988). Analysis of airborne and groundbased occultation data combined with consideration of the energy balance in the tenuous atmosphere led to the conclusion that this atmosphere could not be mostly methane, but rather dominated by a heavier molecule (*Yelle and Lunine,* 1989), as suggested some years earlier (*Trafton,* 1981). A few years later, nitrogen and carbon monoxide ices were detected in the 2.1–2.4-μm part of the near infrared spectrum of Pluto, with N_2 dominating over CO (*Owen et al.,* 1993). Thus, because N_2 is also the most volatile of the three detected ices, Pluto's atmosphere — like that of Earth, Titan, and Triton — turns out to be mostly molecular nitrogen. The main differences of Pluto from Triton known by 1992 were the absence of both CO_2 and detectable H_2O ice from Pluto's surface (*Brown et al.,* 1995; *Cruikshank et al.,* 1997) and different insolation patterns thanks to Pluto's large obliquity and orbital eccentricity. Indeed, as Pluto retreats from the Sun, its atmosphere may begin to thin or even collapse (not settled, even today!), with uncertain timing thanks to the inertial effects of surface ices (*Stern and Trafton,* 1984; see also the chapter in this volume by Young et al.). Because chemical and physical processes in an ultracold atmosphere are of keen scientific interest, this in turn would provide another imperative for exploring Pluto before its orbit took it too far from the Sun.

Mapping of what would turn out to be a compositionally and spatially complex surface began with the 1985 mutual events (*Buie et al.,* 1992; *Young and Binzel,* 1993), which showed stark albedo contrasts later confirmed by Hubble imaging (*Stern et al.,* 1997). These contrasts, combined with the knowledge that Pluto has an atmosphere, made it clear that Pluto could be a very dynamic world for volatile transport (*Hansen and Paige,* 1996). By the early 1990s Pluto had become known as an intriguing ice-rich world much like its slightly larger cousin Triton — with volatile ices and a thin atmosphere, bound not to a giant planet but rather to a moon within an order of magnitude the same mass. Like Earth and Venus, Uranus and Neptune, or Ganymede and Callisto, Triton appeared to have a near twin in terms of bulk properties and surface composition, but in a very different dynamical configuration. The scientific impetus for spacecraft exploration of Pluto became strong. The discovery of the Kuiper belt dwarf planet cohort a few years later made that strong impetus become compelling (*National Research Council,* 2003).

2.2. Missed Opportunities to Explore Pluto by Spacecraft

Just four years after the first successful planetary flyby by Mariner 2 at Venus, Gary Flandro of the California Institute of Technology's Jet Propulsion Laboratory (JPL) showed that an upcoming alignment of all the giant planets would allow a spacecraft launched in the 1975–1980 timeframe to use Jupiter's gravitational field to explore them all (*Flandro,* 1966). Among the sample missions he calculated were a 1978 launch to fly past Jupiter, Saturn, Uranus, and Neptune, and a 1977 launch to visit Jupiter and Pluto. JPL quickly proposed a "grand tour" robotic mission to exploit Flandro's trajectories. In the years after Apollo, steeply declining space program budgets doomed this and other ambitious robotic missions, but did allow for a scaled back "Mariner Jupiter-Saturn" twin-spacecraft mission to be launched in 1977 as Voyagers 1 and 2 (*Schurmeier,* 1974). (Earlier, NASA had launched a scientifically less ambitious pair of Ames Research Center probes, Pioneers 10 and 11, to Jupiter in 1972 and 1973.) Of the four spacecraft, two went beyond their original targets: After its 1974 Jupiter flyby, Pioneer 11 also encountered Saturn in 1979; after its 1981 Saturn flyby, Voyager 2 completed the giant planet part of the grand tour by also visiting Uranus in 1986 and Neptune in 1989.

However, in all of this, Pluto was missing. The Pioneer remote sensing payload was not built to operate at Pluto's distance from the Sun, although in the end the two spacecraft did send back space plasma physics data from distances even beyond Pluto's orbit. Pioneer 10's trajectory was designed conservatively to avoid excessive radiation during the Jupiter flyby, preventing a Pluto flyby, and Pioneer 11's was designed to get it to Saturn, which also precluded going onward to Pluto. So neither trajectory allowed a trip to the vicinity of Pluto. Once Voyager 1 was successful, Voyager 2's trajectory was reshaped to allow it to reach Uranus and then Neptune from Saturn, again precluding a trip to Pluto. Voyager 1's trajectory through the Saturn system could have been redirected to allow a Pluto encounter, completing the original grand tour goals. However, one of the most important moons of Saturn, Titan, was known since the 1940s to have an atmosphere with methane as a minor or major component, and a 1978 conference on the Saturn system made the case for Titan as a scientifically important target in its own right (*Hunten and Morrison,* 1978). Furthermore, the haziness of the atmosphere made it impossible to determine Titan's true size, but the extent of the haze layers suggested it might well be the largest moon in the solar system.

Thus, the decision was made to direct Voyager 1 toward an extremely close flyby of Titan, allowing a radio occultation measurement of its atmosphere and physical size, and potentially detailed views of the surface from the TV cameras. While the high optical thickness of the atmosphere precluded such views, the other measurements were successful — revealing a body just slightly smaller than Ganymede (which earned the title of largest moon), with a dense and mostly nitrogen atmosphere that at Titan's surface is four times denser than sea level air on Earth. The greenhouse-warmed surface temperature of 94 K and presence in the atmosphere of methane suggested methane seas, or even a global ocean of liquid ethane and methane (cf. *Coustenis and Taylor,* 2008). The scientific interest generated by Titan's large size and nitrogen atmosphere, with the possibility of surface seas, helped to propel NASA and the European Space Agency to jointly agree to a Saturn orbiter with Titan probe, a mission that would come to be called Cassini-Huygens when it was authorized in 1989. The most ambitious planetary mission to date, Cassini-Huygens discovered over its 13 years of flying within the Saturn system a methane hydrologic cycle on Titan's variegated surface, with lakes, seas, and methane rivers, and convincingly established the presence of a liquid water ocean beneath Titan's crust (*Hayes et al.,* 2018).

While the scientific payoff of the decision to send Voyager 1 to Titan was inarguably stupendous, it prevented Voyager 1 from being sent to Pluto, and ultimately delayed a mission to Pluto by roughly a quarter of a century. It also necessitated a heroic effort on the part of a dedicated group of planetary scientists to make such a mission — New Horizons ultimately happen. It is quite possible, as we show below, that the effort might have failed. However, improvement in instrument technology between the 1970s and the late 1990s (instrument technologies are frozen a decade before launch) allowed a much richer dataset at Pluto than Voyager could have provided, as (in just one example of many) one can see by comparing the Voyager 2 vidicon images of Triton with the New Horizons solid-state detector images of Pluto and Charon.

3. 1992–2006: KUIPER BELT, PLUTONIAN RECLASSIFICATION BY ASTRONOMERS, AND NEW HORIZONS

3.1. Discovery of the Kuiper Belt, and Implications for Pluto

With the demise of the escaped satellite model for Pluto's origin, the ninth planet became an enigmatic outpost in a solar system that had otherwise seemed so well organized into an inner realm of rocky planets and a much vaster outer solar system of giant planets and their extensive satellite systems. It was perhaps surprising, then, that more attention was not paid to the possibility that Pluto might not be unique. *Edgeworth*'s (1949) and *Kuiper*'s (1950) papers were considered highly speculative, while *Oort* (1950) and *Whipple* (1951, 1964) focused more on the question of the source region of comets. However, spurred by dynamical considerations, searches for a transneptunian belt of material became more frequent and more sensitive through the 1980s, culminating in the discovery of a 100-km-sized body, 1992 QB_1 (*Jewitt and Luu,* 1993). From there, the pace of discoveries picked up, with tens, then hundreds, then even more bodies found in the Kuiper belt [see the review by *Jewitt* (1999) and the chapter in this volume by Barucci et al.]. Further into the period 2002–2005, a number of bodies with diameters between 1000 km and 2300 km were discovered (*Brown et al.,* 2005). The last of these, Eris, is practically the same size as Pluto and, based on the orbit of its moon Dysnomia, is about 25% more massive than Pluto (and 29% less massive than Triton).

Because of both its retrograde orbit and Neptune's lack of a regular satellite system, even before the discovery of other large objects in the Kuiper belt, Triton was shown to be a captured body formerly in heliocentric orbit (e.g., *Goldreich et al.,* 1989). While Pluto is in a 2:3 orbital resonance with Neptune, Eris is part of the "scattered disk" of Kuiper belt objects without any dynamical relationship to Neptune. Triton, Pluto, and Eris all have surface ices more volatile than water ice (*Tegler et al.,* 2010, 2012), at least the first two likely have atmospheres continuously around their orbit, and all are the same size and (within a factor of 1.6) the same mass. Remarkably, these seem to be versions of the same type of body sitting in dynamically distinct environments. Numerous workers made arguments for declaring a third solar system class of "ice dwarf" planets [beginning with *Stern* (1991)] like the terrestrial and giant planets, some of which (e.g., Triton) have been lost from solar orbit due to interactions with Neptune. However, rather than accommodate this, the International Astronomical Union (IAU) voted in August 2006 to move Pluto from the category of planet to that of dwarf planet, which it shares in the view of the IAU with the asteroid Ceres, with Eris, and with several other large Kuiper belt dwarf planets. The ill wisdom of such a designation, and its somewhat awkward definition (dwarf planets are round but not massive enough to fully clear planetesimals from their vicinity), has been and will continue to be debated (*DeVorkin,* 2013). The reassignment, however, was irrelevant to the success of the New Horizons mission, which had launched to its target eight months before. In the end, Pluto by itself and with its system of satellites would turn out to be every bit as interesting as any larger planet, and the discovery of the Kuiper belt provided the cosmogonic context that made its exploration of fundamental importance.

3.2. The 1990s: Attempts to Explore Pluto and the Kuiper Belt

Much of what is described in this section comes from the accounts by *Stern* (2008) and *Neufeld* (2014a, 2016). The story of the first mission to Pluto really begins at the end of the grand tour mission — with Voyager 2's flyby of Neptune and Triton in 1989. One of the authors of this review, S.A.S., then completing his doctorate at the University of Colorado Boulder, had already become an advocate for a Pluto mission, proposing it on behalf of an informal scientific interest group (aka the "Pluto Underground") in a May 4 meeting that year to the NASA Solar System Exploration Director at the time, Geoffrey Briggs.

The context then for planetary exploration was grim. Voyager 2 was a legacy of the program of the 1960s, when large missions such as Viking were thought to be the only mode of business. These missions, costing hundreds of millions in then-year dollars (which today well exceeds $1 billion), have come to be known as Flagship missions (although they were not called so at the time). However, the declining space budgets of the 1970s, along with the delays to the Galileo mission caused by the 1986 loss of the space shuttle Challenger, had put the planetary program in crisis. At the time of the August 1989 Voyager 2 flyby of Neptune, NASA had launched only one planetary mission in the preceding 11 years — Magellan to Venus that same May. In that year, however, Briggs initiated studies within the Solar System Exploration Division on smaller, cheaper missions, akin to NASA's Astrophysics Explorers. Stamatios "Tom" Krimigis of the Applied Physics Laboratory (APL) of the Johns Hopkins University was a strong advocate for smaller spacecraft, and APL had developed a rendezvous mission to an asteroid under the short-lived Planetary Observer program. In late 1989 Briggs proposed the Discovery program for small planetary missions. His successor, Wesley Huntress, ultimately succeeded in getting the Discovery program funded in 1993 (*Neufeld,* 2014b).

Under the then-nascent 1990 Discovery program, Briggs asked Robert Farquhar of Goddard Spaceflight Center, an expert in celestial mechanics, to conduct a study of an inexpensive mission to Pluto; Stern was appointed study scientist. Called Pluto350 for the proposed spacecraft dry mass in kilograms, the intent was to fly past Pluto and Charon with a minimal payload. Meanwhile, since the early 1980s JPL had been advancing a concept for a new class of outer solar system Flagship missions called Mariner Mark II (*Neugebauer,* 1983). These would be highly capable spacecraft, larger and heavier than the Voyagers and capable of executing a number of different missions in the outer solar system, including the potential exploration of Pluto-Charon. Ultimately, only one of these would be built and flown — the Cassini Saturn orbiter — by which point the concept of a series of spacecraft had evaporated. But in application to Pluto, the Mariner Mark II would require a very heavy and expensive launch vehicle, the Air Force's Titan IV Centaur, which NASA would have to pay for. In this incarnation, which NASA favored in 1991, a Pluto mission would have a major impact on the agency's solar system exploration budget.

Another issue also reared its head at the time. If the nation were to send a highly capable and complex spacecraft all the way across the solar system, why not send it to a giant planet system? Triton — as large and chemically complex as Pluto — was only briefly explored by Voyager 2. And detailed exploration of Neptune could be part of the package. But this would leave Pluto out in the cold — again. The Pluto Underground lobbied for Pluto based not only on the novelty of the science — Pluto was to be the last planet to be explored — but also because of the progressive loss of surface illumination over the southern hemisphere and the possible dramatic loss of the atmosphere as Pluto retreated from the Sun [the latter turned out not to happen, probably due to nitrogen-ice-covered Sputnik Planitia (*Meza et al.,* 2019)]. In other words, time was believed to be of the essence.

The idea of sending a Mariner Mark II to the outermost solar system was short-lived, however, as the first two planned missions of that line, Cassini and Comet Rendezvous/Asteroid Flyby (CRAF), proved to be too expensive in the budget environment of the early 1990s. NASA soon cancelled CRAF and ordered Cassini descoped by removing a scan platform and making other cuts.

Pluto advocates shifted their attention back to a Pluto350-like mission, but this was displaced by an even more radical concept that JPL engineers Robert Staehle and Stacy Weinstein called Pluto Fast Flyby (PFF). Conceived to have a spacecraft wet mass (with propellant) of only 160 kg, the design required an extremely limited scientific payload, but piqued the interest of new NASA Administrator Dan Goldin. By mid-1992, this JPL concept included two spacecraft for redundancy and to provide complete coverage of both hemispheres of Pluto and Charon. Although the mission cost was less than half the billion dollars proposed for a single Mariner Mark II, the added launch costs for two spacecraft were too formidable to afford.

Several events in 1993 conspired to doom a new start for PFF. First, Mars Observer was lost to a propellant line explosion as it neared Mars in August. Then that same month Goldin told the PFF team that he could not afford to budget for two Titan IV Centaur launch vehicles, which were required because the lack of Jupiter gravity-assist opportunities in the later 1990s necessitated a direct launch to Pluto. That fall, Congress provided money for two Discovery missions, APL's Near Earth Asteroid Rendezvous (NEAR) mission and JPL's Mars Pathfinder, but did so over the wishes of Goldin. Caught between a fatally expensive dual heavy-lift launch for PFF, the need for a replacement Mars orbiter for Observer, and Goldin's very negative reaction to the extra Discovery mission, Huntress — by then the Associate Administrator for Space Science — terminated the idea of bringing forward PFF as a new start.

Efforts by the Pluto Underground then shifted to the possibility of an international launch provider, to remove the substantial costs of a U.S. vehicle. One of us (S.A.S.) took advantage of a major new initiative in 1993 between the U.S. and Russia to cooperate in space, and met with the director of the Russian Academy of Sciences' Space Research Institute (IKI), Albert Galeev, to gauge interest in providing the launch vehicles. As a scientist, Galeev was interested, if IKI could have a significant scientific role in the mission. This led to a proposed Russian-built atmospheric probe for Pluto. The advocacy group the Planetary Society delivered its own version of a Pluto collaboration with Russia to Huntress in early 1994, which helped to force an initially reluctant NASA Headquarters to pursue the possibility.

But momentum was diffused yet again when Huntress went to Moscow with a much broader palette of cooperation: Mars exploration, a close-approach solar mission (Solar Probe), and the PFF. The Russian scientists seemed more interested in Solar Probe, but by drawing on Farquhar's earlier proposal to use similar spacecraft and Jupiter flyby trajectories, the two missions could be packaged as a "Fire and Ice" program. But after multiple scientific and technical meetings and conferences in Russia, Germany, and the U.S., it was clear that whether these missions would go ahead depended ultimately on the Russian Space Agency's willingness to provide Proton launch vehicles for free. By 1996 it was evident they were not. This and the upper stage failure of a Russian Proton rocket in late 1996, leading to the complete loss of the IKI-led Mars 96 mission, made Russian collaboration in a Pluto launch no longer viable.

Meanwhile, PFF went through multiple design changes and iterations as NASA Administrator Goldin kept moving the goalposts on cost and mass as he sought to make the mission a poster child for his "better, faster, cheaper" (BFC) approach. The failure of two BFC Mars missions in 1999 would ultimately hobble Goldin's campaign, but that was years in the future. The scientific community, JPL, and Huntress at NASA Headquarters all thought they were ready to end trade studies for PFF in 1994, but Goldin insisted on another two years of technology work. At the end of the year, PFF became Pluto Express, the novelty of which was to design the spacecraft architecture around the science. This "sciencecraft" concept was small enough to allow launch on a single large booster, but the low overall spacecraft weight was extremely demanding technologically.

Although a well-attended scientific workshop took place in July 1993 in Flagstaff, leading to the previous Pluto system book in this Space Science Series (*Stern and Tholen,* 1997), NASA's Pluto mission did not then have the universal support of the science community. The National Academy of Science's Committee on Lunar and Planetary Exploration (COMPLEX) gave short shrift to Pluto in a 1994 report, and without an overarching once-a-decade strategy like the Academy's Decadal Survey in Astronomy and Astrophysics, the planetary community had a difficult time arriving at a consensus set of priorities. Nevertheless, after community urging, NASA convened a Pluto Express Science Definition Team (SDT) in 1995 under the chairmanship of one of us (J.I.L.) in an effort to pull together the science case for the mission, both in the context of what was known of Pluto and the increasing pace of discoveries of KBOs and Pluto's relationship to them.

By that point, the following was known about the Kuiper belt: (1) The count of transneptunian bodies exceeding 100 km in diameter that had been directly observed from groundbased telescopes had reached 28 (*Jewitt and Luu,* 1995). (2) Like Pluto-Charon, the orbits of many KBOs cluster near the 2:3 mean-motion resonance with Neptune at a = 39 AU, with these orbits stabilized by the resonance. Other objects were found in other mean-motion resonant relationships with Neptune, such as the 3:4 resonance. (3) The idea that the Kuiper belt is likely a remnant of the much more extensive (and long gone) protoplanetary disk of gas and dust from which the solid objects of the solar system formed was strengthened by then-new dynamical simulations (*Duncan et al.*, 1995). (4) The inferred spatial density of KBOs was known to be sufficiently high to make it highly likely that Pluto Express could be redirected to pass by at least one other KBO after flying through the Pluto-Charon system.

The SDT report (*Lunine et al.*, 1995) cited the uniqueness of several aspects of Pluto, including its atmospheric energy balance, a possible comet-like interaction with the solar wind, and binary planet nature of Pluto and Charon. The data at hand then also hinted that overall the physical and chemical processes on Pluto are complex and hence demand close up exploration. The SDT report concluded that the opportunity to visit one or more KBO's beyond Pluto, in context with the exploration of the Pluto/Charon system itself, would be of keen scientific interest and exciting to the public. In sum, the presentations to the SDT and consequently the ensuing SDT report made a compelling case that a mission to Pluto and beyond could be done at low cost and yet have extremely high scientific and public interest value.

As compelling as the Pluto and Kuiper belt science were, events once again conspired to overtake the mission. In 1995, *Mayor and Queloz* (1995) detected the first extrasolar planet, and the following year NASA announced evidence of former biologic activity in a meteorite from Mars (*McKay et al.,* 1996). Although the latter discovery was soon rejected by most in the scientific community, the two events stimulated a presidential statement and repackaging of NASA funds into the newly named Origins program. Over the next two years, the magnetometer onboard the Galileo spacecraft provided compelling evidence for a salty, liquid-water ocean beneath the ice crust of Jupiter's moon Europa (*Kivelson et al.,* 2000). The close juxtaposition of these events boosted interest in the search for life, and as a result suddenly Europa was competing with Pluto for a new mission opportunity.

Capitalizing on public excitement and Congressional interest, Goldin and Huntress got a new start in 1998 for an Outer Planets/Solar Probe program, wrapping together a Europa orbiter with the now renamed Pluto-Kuiper Express (PKE), plus a Solar Probe. The project office was established at JPL (*Neufeld,* 2014a). Like Fire and Ice, a common spacecraft was planned, ostensibly because all three missions had to go to Jupiter, but the other requirements for each were quite different. A harsh radiation environment was the huge challenge for Europa Orbiter, which had to remain deep within the jovian magnetosphere, while PKE and Solar Probe simply did fast flybys through the Jupiter system. Nonetheless, all three missions were saddled with the radiation-hardening requirements of Europa Orbiter, and JPL tied the missions to an effort to develop radiation-hard electronics in a program called X-2000. Furthermore, it was stipulated that Europa Orbiter would go first in 2003, followed by PKE in 2004 and then Solar Probe in 2007, to the significant dismay of both Pluto supporters and the space physics community.

The requirements imposed by the environment around Europa, plus difficulties reaching the radiation-hardening goals set by the X-2000 program, led to significant mass and hence cost growth in all three spacecraft. Complicating matters further was the departure from NASA Headquarters of Associate Administrator Huntress, who had been so intimately involved in and supportive of the discussions surrounding international cooperation on a Pluto mission. In his place as Associate Administrator stepped Edward Weiler, an astronomer who had been chief scientist of Hubble Space Telescope before ascending the ranks at NASA Headquarters. Much of Weiler's career had been spent at NASA Goddard Spaceflight Center, in contrast to Huntress, who had been at JPL prior to moving to NASA headquarters in Washington DC. Furthermore, Weiler's first years as Associate Administrator saw the inflight loss of two JPL Mars probes, victims of Goldin's "better, faster, cheaper" philosophy, which Weiler had not been party to. Thus, he had every reason to be suspicious of JPL, and was also wary of programs that posed a danger of exceeding the available funding.

As the combined costs of Europa Orbiter and PKE exceeded $1.4 billion and continued to climb, Weiler's relationship with JPL became adversarial as he pressured the Laboratory to reign in costs. A positive development was the formation of Science Definition Teams and Announcements of Opportunity for Europa Orbiter and PKE instruments. Regardless, the concept of a common spacecraft was unraveling through 1999 owing to the cost increases it resulted in, as well as delays in the availability of radioisotopic power and uncertainties in launch vehicle costs. Moreover, NASA's space science program was now under pressure from the Office of Management and Budget (OMB), which saw the program as out of control. With the growth of astrobiology research at NASA, Congress and the OMB saw the science that might be done at Europa to assess its ocean's suitability for life as the higher priority. PKE's planetary formation focus, plus its "last planet yet to be explored" public appeal, were seen as less important. On September 12, 2000, Weiler, with OMB backing, told JPL to stop work on PKE — canceling the mission.

PKE's cancellation fractured the outer planets community. Anticipating cancellation, the Planetary Society began a campaign in July to highlight Pluto as a key planetary target. Nine days after cancellation, the Division for Planetary Sciences of the American Astronomical Society (DPS), the professional organization of planetary scientists, issued a press release highlighting the imperative of reaching Pluto before the collapse of the atmosphere. NASA's Solar System Exploration Subcommittee (SSES), then chaired by the University of Arizona's Michael Drake, met in Pasadena at the end of October 2000, with the Pluto cancellation as a centerpiece topic on its agenda. Two of the authors of this chapter (S.A.S. and J.I.L.) gave presentations on the value of Pluto science, based on the findings of the 1995 Science Definition Team report and subsequent research developments. The importance of the Kuiper belt for understanding planet formation was emphasized, specifically that the presence of large numbers of icy bodies created in the outer regions of new systems, subsequently disturbed by the migration of the orbits of the major planets, presented a new picture of planet formation that begged for investigation by spacecraft. JPL's presentation to the committee made it clear that, by stipulating a common spacecraft bus, the design of PKE had been compromised and its costs driven up in order to satisfy the radiation parts selection and heavy shielding requirements of Europa Orbiter.

The net result of the meeting was a letter of strong SSES support to NASA for a Pluto mission and skepticism about the Europa Orbiter. As the costs of the latter soared due to the technological challenge of Jupiter's harsh radiation environment, Weiler cancelled it in 2003, in part because Congress kept appropriating funds for Pluto. That in turn set off a 12-year-long odyssey through a variety of mission concepts until Congress stipulated a new start for Europa Clipper. (Although this Outer Planets/Solar Probe program never came to fruition, New Horizons did launch to Pluto in 2006. Parker Solar Probe followed in 2018; and Europa Clipper is planned for launch by the mid 2020s.) The rancor of the Pluto-Europa debate would be reflected in conflicts between astrobiology and planetary science goals in the outer solar system, some up to the present. But at the close of the millennium, as Cassini-Huygens was sailing toward Saturn, the immediate effect was that the U.S. planetary program had no follow-on mission anywhere in the outer solar system.

3.3. After the Millennium: Origin and Development of New Horizons

In the wake of DPS, Planetary Society, public, and SSES protests against the cancellation of PKE, Weiler approached space scientist Tom Krimigis, then head of the APL Space Department, about the feasibility of a competitive, relatively low-cost Pluto program. With the success of APL's low-cost NEAR Discovery-class spacecraft, then orbiting the asteroid Eros, Krimigis was in a position to do a quick study showing how such a spacecraft could be based on NEAR and another APL Discovery spacecraft then in development, the ill-fated COmet Nucleus TOUR (CONTOUR). Following optimism resulting from the APL study Krimigis led, and with Goldin's assent, Weiler had his Solar System Exploration Division Director Colleen Hartman quickly put together an Announcement of Opportunity (AO) for a competitively bid Pluto mission, similar to but on a larger scale than the PI-led Discovery missions.

NASA released the AO on the last day of the Clinton administration, January 19, 2001. Several teams formed to respond with proposals, but the Bush administration's OMB canceled the Pluto mission the next month — forcing NASA to suspend the AO. The issue was not only budgetary; OMB still favored Europa Orbiter. APL was, however, in Maryland, and Krimigis called upon the powerful NASA Appropriations Committee member Barbara Mikulski to intervene. The resulting letter from her office instructed NASA to restart the AO process. Ultimately, five proposals were received in early April 2001. NASA ranked and downselected to two proposals in June 2001: the APL New Horizons proposal, with one of the authors as PI (S.A.S.), and a JPL proposal called POSSE led by Larry Esposito of the University of Colorado. Each submitted second-step proposals in September 2001, and on November 29, NASA selected New Horizons for development.

Details of the differences between the proposals are left to the other reviews cited in this chapter, but the innovative payload, low-cost approach, and the PI's community leadership in pushing for a Pluto mission for over a decade were surely positive factors. However, despite its win, the odds remained against New Horizons — the budget profile provided by NASA and the overall budget cap were extremely challenging, the availability of nuclear fuel for the radioisotopic power sources was extremely limited thanks to the cessation of its production by the Federal government, the timetable to make the necessary Jupiter gravity assist flyby was very tight, and the lower-cost booster that would launch the mission was only just becoming available. Within the new administration, the mission was politically radioactive; having had the mission rammed down its throat by a powerful senator, the OMB eventually threatened publicly to withdraw support for the planetary program in general, and then left the mission out of the 2003 President's budget.

While Senator Mikulski worked to get the project funded again through the Congress, the prospect of year-to-year political heroics to sustain it cast a deep shadow over its viability. Two events pushed the mission back from its abyss to reality. First, the inaugural planetary Decadal Survey came out in 2003 with its community consensus prioritization, ranking a Pluto mission at the very top of the medium-class (0.5–1 billion dollars) mission list, specifically emphasizing the value of the mission as a Kuiper belt explorer as the Pluto Express SDT had done eight years before. This was especially pivotal to the resuscitation of the mission, because it gave Mikulski a powerful scientific rationale for her efforts to get the mission funded. And second, Hartmann had crafted within her Solar System Exploration Division and successfully sold to upper management at NASA a proposed new line of PI-led missions, with a cap twice that of Discovery, called New Frontiers. Unlike Discovery's wide-open landscape for proposing any mission within its cost cap, New Frontiers PIs would be given a limited list of high-priority targets; this program and the Decadal's Pluto Kuiper belt recommendation made it natural to combine Pluto and Kuiper belt on the list and designate New Horizons as the first New Frontiers mission. The only remaining political hurdle was to convince Weiler and Hartmann that the APL team could bring the mission to launch readiness by the beginning of 2006, when the next Jupiter gravity assist was possible. This having been accomplished — although not without many interesting wrinkles and numerous developmental challenges (*Neufeld,* 2014a; see also *Stern and Grinspoon,* 2018) — the mission was finally put on solid political and programmatic ground. That enabled a very challenging but ultimately successful spacecraft development, leading to a successful launch on January 19, 2006, just five years to the day after the proposal AO was released.

4. 2006–PRESENT: THE SUCCESS OF NEW HORIZONS AND THE FUTURE

4.1. New Horizons in Flight

After its launch, New Horizons made a 9.5-year crossing of the solar system, which included a gravity assist, flight test, and scientific flyby of Jupiter in February and March 2007. It then explored the Pluto system in early and mid-2015, reaching closest approach to the planet and all of its satellites on July 14, 2015; data transmission from that flyby was completed in October 2016. Meanwhile, in late 2015, New Horizons targeted its first KBO flyby, to explore a small (~36-km-long) cold classical KBO designated 2014 MU_{69} Arrokoth (formerly Ultima Thule). That flyby culminated in a closest approach on January 1, 2019; data from that flyby are still being transmitted to Earth as of this writing and are not expected to be complete until 2021.

The New Horizons spacecraft carries a payload that significantly broadened the minimum mission scientific objectives over those specified by the Pluto Kuiper Express SDT (*Lunine et al.,* 1995). This, along with spacecraft pointing and other capabilities that well matched the payload's highest-resolution sensors, and the inclusion of two 64-GB solid-state flight recorders, allowed the mission to far exceed the baseline scientific objectives of both the SDT and the NASA AO (01-OSS-01) that called for flyby mission proposals to Pluto and the Kuiper belt.

The enhanced payload and spacecraft capabilities of New Horizons, combined with a very highly optimized flyby observing plan at the Pluto system, allowed New Horizons to make numerous groundbreaking discoveries about Pluto and its satellites, as this Space Science Series volume details at length. Among the most significant findings from this exploration are the following: Pluto is far more complex than similar-sized icy satellites explored by other missions and it remains intensely geologically active 4.5 **b.y.** after its formation. The planet displays evidence for all of the following: geologic activity even to the recent past or today; extensive tectonics and true polar wander, and strong evidence for cryovolcanism; a global internal liquid water ocean; a water ice crust; several styles of putative volcanism; ancient terrains that date back close to the planet's formation epoch; a haze-filled atmosphere with a rich hydrocarbon and nitrile minor species composition accompanying the major gas, molecular nitrogen; strong mega-seasonal cycles and epochs of much higher atmospheric pressure than the current epoch, owing to obliquity cycles; and a lack of newly detected satellites or rings.

Pluto's giant satellite Charon, the other object in the binary planet pair, displays evidence for a former liquid water ocean in its interior, early epoch tectonics, an age as great as Pluto's, a lower bulk density than Pluto, several kinds of unique geological expressions not yet seen elsewhere, and a lack of detectable atmosphere at a level orders of magnitude below Earth-based limits. Pluto's four small satellites, which orbit the binary pair, were found to each be irregular in shape, to be complex in their rotational dynamics, and to have surface ages (where observed) as old as Pluto and Charon's oldest terrains, as well as surface compositions that include both water ice and ammonia or ammoniated species. Additionally, a major advance in understanding of the small-diameter size-frequency distribution of KBOs was made using crater statistics on both Pluto and Charon.

The flyby exploration of Arrokoth also yielded numerous discoveries (see the chapter in this volume by Stern et al.). In brief, these include discovering that the object is a contact binary with unexpectedly flattened lobes that formed near one another, very likely in a local pebble collapse cloud, then became an orbiting pair and then gently merged into the contact binary configuration; it also displays discrete geological units and significant albedo heterogeneity, but only small color variegation and spectroscopic evidence for methanol on its surface. The paucity of detected craters on Arrokoth implies a relative dearth of KBOs <1 km in diameter and collisionally benign ancient and present-day Kuiper belt environments. No satellites, orbiting rings/dust structures, or evidence of atmosphere were found to accompany this fascinating object.

4.2. The Contextual Successes of New Horizons

Results from the New Horizons mission are detailed in the chapters in this volume on geology (White et al. and Spencer et al.), color (Olkin et al. and Howett et al.), surface composition (Cruikshank et al. and Protopapa et al.), atmospheric structure (Summers et al.), and plasma environment (Bagenal et al.), and in the chapter on its small satellites (Porter et al.).

As a first mission to both a previously unexplored planet and two new types of body (the ice dwarf planets and KBOs of the outer solar system), New Horizons very much falls in context with the earliest Mariners, Pioneers, and Voyagers, which each undertook similarly groundbreaking first reconnaissance of closer planets. However, in large measure because the New Horizons spacecraft and its instrument payload were based on advanced technologies not available to those much earlier missions, the scientific return from the first flybys New Horizons conducted of the Pluto system and Arrokoth generated much larger datasets and accomplished the collection of dataset resolutions and types (e.g., surface composition mapping) that predecessor first flyby missions could not. For example, if one compares the New Horizons Pluto flyby datasets to the exploration of Mars, it can fairly be said that New Horizons took the Pluto system to a state of knowledge crudely equivalent to Mars after about a generation of spacecraft exploration from the 1960s to the 1990s.

New Horizons demonstrated that the reconnaissance exploration of planets and smaller bodies in the Kuiper belt need not incur the multi-billion-dollar cost (adjusted to today's dollars) of Voyagers 1 and 2 despite yielding similarly spectacular results. New Horizons also demonstrated the viability of the New Frontiers class of PI-led missions, roughly twice the cost of the Discovery missions with a commensurately higher science return. This in turn not only opened the door to further exploration of the Kuiper belt and its planets with such missions, but also a variety of other targets throughout the solar system.

The long battle for a Pluto mission accelerated a change around the year 2000 in how planetary missions were advocated and selected (*Neufeld,* 2014a). Previously, NASA Headquarters, usually the Associate Administrator for Space Science in cooperation with the Director of Solar System Exploration (the division's name at the time), routinely assigned missions to JPL. But Goldin's "faster, better, cheaper" approach and the rise of competed missions in the Discovery Program opened up the planetary mission selection process to a larger number of actors. Notably, it made APL and later Goddard Spaceflight Center into viable competitors with JPL for planetary missions, which in turn encouraged political intervention by the Maryland congressional delegation, especially Mikulski. Around the same time, the planetary science community made its advocacy voice much more effective through the DPS and the National Academy of Science's Decadal Survey process. The fight over New Horizons sped the 2002 to 2003 decadal process toward prioritizing Pluto and hence the resurrection of the mission. It also impelled the creation of the New Frontiers line, expanding competitive missions. The decadal process now reigns more or less supreme in prioritization, and competition remains central outside of Flagship missions. The fight over a mission to Pluto may therefore have been one of the most consequential episodes in the history of planetary science policy in the last 30 years.

4.3. Where Do We Go from Here in the Post-New Horizons Era?

In a broad sense, the era of the initial reconnaissance of all the planets and the major types of solar system objects known at the dawn of the Space Age came to a close with New Horizons. That said, the exploration of the solar system in general and Pluto and the Kuiper belt in particular all remain very far from complete.

Going forward, much more detailed exploration of the bodies of our solar system lies ahead. Such exploration is already yielding a wide variety of orbital, surface, and even subsurface robotic exploration of increasing diversity and capability. Similarly, we are now also seeing the beginnings of sample return from many more bodies than simply Earth's Moon, and plans for new human exploration of the Moon and the first human explorations of Mars are now underway.

As to Pluto and the Kuiper belt, there is little debate that further exploration is necessary (see the chapter in this volume by Buie et al.). This is likely to come in at least two forms in the next few decades, including the flyby reconnaissance of more (current and former if one includes Centaur missions) KBOs and a wider variety of dwarf planets, and to more indepth studies of the Pluto system with an orbiter bringing time domain, complete mapping, and new kinds of investigations New Horizons could not (or did not know to) bring.

Acknowledgments. We thank Fran Bagenal and Kevin Schindler for careful reviews. S.A.S., L.A.Y., and R.P.B. thank NASA for funding support via the New Horizons project.

REFERENCES

Alden H. L. (1940) The mass of the satellite of Neptune. *Astron. J., 49,* 71–72.

Andersson L. E. and Fix J. D. (1973) Pluto: New photometry and a determination of the axis of rotation. *Icarus, 20,* 279–283.

Binzel R. P., Tholen D. J., Tedesco E. F., Buratti B. J., and Nelson R. M. (1985) The detection of eclipses in the Pluto-Charon system. *Science, 228,* 1193–1195.

Bonneau D. and Foy R. (1986) First direct measurements of the diameters of the large satellites of Uranus and Neptune. *Astron. Astrophys., 161,* L12–L13.

Bower E. C. and Whipple F. (1930) The orbit of Pluto. *Publ. Astron. Soc. Pac., 42,* 236–240.

Brown M., Trujillo C. A., and Rabinowitz D. L. (2005) Discovery of a planetary-sized object in the scattered Kuiper belt. *Astrophys. J. Lett., 635,* L97–L100.

Brown R. H., Cruikshank D. P., Veverka J., Helfenstein P., and

Eluszkiewicz J. (1995) Surface composition and photometric properties of Triton. In *Neptune and Triton* (D. P. Cruikshank, ed.), pp. 991–1030. Univ. of Arizona, Tucson.
Buie M. W. and Fink U. (1987) Methane absorption variations in the spectrum of Pluto. *Icarus, 70,* 483–498.
Buie M. W. and Tholen D. J. (1989) The surface albedo distribution of Pluto. *Icarus, 79,* 23–37.
Buie M. W., Cruikshank D. P., Lebofsky L. A., and Tedesco E. F. (1987) Water frost on Charon. *Nature, 329,* 522–523.
Buie M. W., Tholen D. J., and Horne K. (1992) Albedo maps of Pluto and Charon: Initial mutual event results. *Icarus, 97,* 211–227.
Christy J. W. and Harrington R. S. (1978) The satellite of Pluto. *Astrophys. J., 93,* 1005–1008.
Cohen C. J. and Hubbard E. C. (1965) Libration of the close approaches of Pluto to Neptune. *Astron. J., 70,* 10–13.
Coustenis A. and Taylor F. W. (2008) *Titan: Exploring an Earthlike World.* World Scientific, Singapore. 412 pp.
Cruikshank D. P., Plicher C. B., and Morrison D. (1976) Pluto: Evidence for methane frost. *Science, 194,* 835–837.
Cruikshank D. P., Stockton A., Dyck H. M., Becklin E. E., and Macy W. Jr. (1979) The diameter and reflectance of Triton. *Icarus, 40,* 104–114.
Cruikshank D. P., Roush T. L., Moore J. M., Sykes M. V., Owen T. C., Bartholomew M. J., Brown R. H., and Tryka K. A. (1997) The surfaces of Pluto and Charon. In *Pluto and Charon* (S. A. Stern and D. J. Tholen, eds.), pp. 221–267. Univ. of Arizona, Tucson.
Dessler A. J. and Russell C. T. (1980) From the ridiculous to the sublime: The pending disappearance of Pluto. *Eos Trans. AGU, 61,* 690.
DeVorkin D. H. (2013) Pluto: The problem planet and its scientists. In *Exploring the Solar System: The History and Science of Planetary Exploration* (R. D. Launius, ed.), pp. 323–362. Palgrave Macmillan, New York.
Dunbar R. S. and Tedesco E. F. (1986) Modeling Pluto-Charon mutual eclipse events. I. First order models. *Astron. J., 92,* 1201–1209.
Duncan M., Quinn T., and Tremaine S. (1987) The formation and extent of the solar system comet cloud. *Astron. J., 94,* 1330–1338.
Duncan M. J., Levison H. F., and Budd S. M. (1995) The dynamical structure of the Kuiper belt. *Astron. J., 110,* 3073–3081.
Duncombe R. L. and Seidelmann P. K. (1980) A history of the determination of Pluto's mass. *Icarus, 44,* 12–18.
Edgeworth K. E. (1943) The evolution of our planetary system. *J. British Astron. Assoc., 53,* 181–188.
Edgeworth K. E. (1949) The origin and evolution of the solar system. *Mon. Not. R. Astron. Soc., 109,* 600–609.
Elliot J. L., Dunham E. W., Bosh A. S., Slivan S. M., Young L. A., Wasserman L. H., and Millis R. L. (1989) Pluto's atmosphere. *Icarus, 77,* 148–170.
Flandro G. (1966) Fast reconnaissance missions to the outer solar system utilizing energy derived from the gravitational field of Jupiter. *Acta Astronaut., 12,* 329–337.
Goldreich P., Murray N., Longaretti P. Y., and Banfield P. (1989) Neptune's story. *Science, 245,* 500–504.
Hansen C. J. and Paige D. A. (1996) Seasonal nitrogen cycles on Pluto. *Icarus, 120,* 247–265.
Hayes A. G., Lorenz R. D., and Lunine J. I. (2018) A post-Cassini view of Titan's methane-based hydrologic cycle. *Nature Geosci., 11,* 306–313.
Herschel J. F. W. (1867) *Outlines of Astronomy, 10th edition.* Blanchard and Lea, Philadelphia.
Hubbard W. B., Hunten D. M., Dieters S. W., Hill K. M., and Watson R. D. (1988) Occultation evidence for an atmosphere on Pluto. *Nature, 336,* 452–454.
Hunten D. M. and Morrison D., eds. (1978) *The Saturn System.* NASA Conf. Publ. 2068, Washington, DC. 447 pp.
Jewitt D. J. (1999) Kuiper belt objects. *Annu. Rev. Earth Planet. Sci., 27,* 287–312.
Jewitt D. J. and Luu J. (1993) Discovery of the candidate Kuiper belt object 1992 QB1. *Nature, 362,* 730–732.
Jewitt D. J. and Luu J. (1995) The solar system beyond Neptune. *Astron. J., 109,* 1867–1876.
Kivelson M., Khurana K. K., Russell C. T., Volwerk M., Walker R. J., and Zimmer C. (2000) Galileo magnetometer measurements: A stronger case for a subsurface ocean at Europa. *Science, 289,* 1340–1343.
Kuiper G. P. (1950) The diameter of Pluto. *Publ. Astron. Soc. Pac., 62,* 133–137.
Kuiper G. P. (1951) On the origin of the solar system. *Proc. Natl. Acad. Sci. U.S.A., 37,* 1–14.
Lampland C. O. (1933) Lowell's photographic observations of Pluto in 1915, 1929 and 1930. *Publ. Am. Astron. Soc., 7,* 7–8.
Leonard F. C. (1930) The new planet Pluto. *Astron Soc. Pac. Leaflet, 30,* 121–124.
Lodders K. and Fegley B. Jr. (1998) *The Planetary Scientist's Companion.* Oxford, New York. 400 pp.
Lowell P. (1915) Memoir on a trans-neptunian planet. *Mem. Lowell Obs., 1,* 1.
Lunine J. I., Cruikshank D., Galeev A. A., Jennings D., Jewitt D., Linkin S., McNutt R., Neubauer F., Soderblom L., Stern S. A., Terrile R., Tholen D., Tyler L., and Yelle R. V. (1995) *Pluto Express: Report of the Science Definition Team.* NASA, Washington, DC.
Malhotra R. and Williams J. G. (1997) Pluto's heliocentric orbit. In *Pluto and Charon* (S. A. Stern and D. J. Tholen, eds.), pp. 127–157. Univ. of Arizona, Tucson.
Marcialis R. L. (1997) The first 50 years of Pluto-Charon research. In *Pluto and Charon* (S. A. Stern and D. J. Tholen, eds.), pp. 27–83. Univ. of Arizona, Tucson.
Marcialis R. L., Rieke G. H., and Lebofsky L. A. (1987) The surface composition of Charon: Tentative identification of water ice. *Science, 237,* 1349–1351.
Mayor M. and Queloz D. (1995) A Jupiter-mass companion to a solar-type star. *Nature, 378,* 355–359.
McKay D., Gibson E. K. Jr., Thomas Keprta K. L., Vali H., Romanek C. S., Clemett S. J., Chillier D. F., Maechling C. R., and Zare R. N. (1996) Search for past life on Mars: Possible relic biogenic activity in martian meteorite ALH84001. *Science, 273,* 924–930.
Meza E. and 150 colleagues (2019) Pluto's lower atmosphere and pressure evolution from ground-based stellar occultations, 1988–2019. *Astron. Astrophys., 625,* A42.
National Research Council (2003) *New Frontiers in the Solar System: An Integrated Exploration Strategy.* National Academies Press, Washington, DC (e-Book), DOI: 10.17226/10432.
Neufeld M. J. (2014a) First mission to Pluto: Policy, politics, science, and technology in the origins of New Horizons, 1989–2003. *Hist. Stud. Nat. Sci., 44,* 234–276.
Neufeld M. J. (2014b) Transforming solar system exploration: The origins of the Discovery program, 1989–1993. *Space Policy, 30,* 5–12.
Neufeld M. J. (2016) The difficult birth of NASA's Pluto mission. *Phys. Today, 69,* 40–47.
Neugebauer M. (1983) Mariner Mark II and the exploration of the solar system. *Science, 219,* 443–449.
Nicholson S. B. and Mayall N. U. (1931) Positions, orbit and mass of Pluto. *Astrophys. J., 73,* 1–12.
Oort J. H. (1950) The structure of the cloud of comets surrounding the solar system and a hypothesis concerning its origin. *Bull. Astron. Inst. Netherlands, 11,* 91–110.
Owen T. C., Roush T. L., Cruikshank D. P., Elliot J. L., Young L. A., de Bergh C., Schmitt B., Brown R. H., and Bartholomew M. J. (1993) Surface ices and atmospheric composition of Pluto. *Science, 261,* 745–748.
Pickering W. H. (1909) A search for a planet beyond Neptune. *Ann. Harvard College Obs., 61,* 113–162.
Reaves G. (1997) The predictions and discoveries of Pluto and Charon. In *Pluto and Charon* (S. A. Stern and D. J. Tholen, eds.), pp. 3–25. Univ. of Arizona, Tucson.
Schurmeier H. M. (1974) The Mariner Jupiter/Saturn 1977 mission. *Space Congress Proc., 4,* 28–41.
Slipher V. M. (1930) *Lowell Obs. Observation Circ.,* May 1.
Spencer J. R., Buie M. W., and Bjoraker G. L. (1990) Solid methane on Triton and Pluto: 3–4 μm spectrophotometry. *Icarus, 88,* 491–496.
Standish E. M. Jr. (1993) Planet X: No dynamical evidence in the optical observations. *Astron. J., 105,* 2000–2006.
Stern S. A. (1991) On the number of planets in the outer solar system: Evidence of a substantial population of 1000-km bodies. *Icarus, 90,* 271–281.
Stern S. A. (2008) The New Horizons Pluto Kuiper belt mission: An overview with historical context. *Space Sci. Rev., 140,* 3–21.

Stern S. A. and Grinspoon D. (2018) *Chasing New Horizons: Inside the Epic First Mission to Pluto.* Picador, London. 320 pp.
Stern S. A. and Mitton J. (1998) *Pluto and Charon: Ice Worlds on the Ragged Edge of the Solar System.* Wiley, New York. 216 pp.
Stern S. A. and Tholen D. J., eds. (1997) *Pluto and Charon.* Univ. of Arizona, Tucson. 728 pp.
Stern S. A. and Trafton L. (1984) Constraints on bulk composition, seasonal variation, and global dynamics of Pluto's atmosphere. *Icarus, 57,* 231–240.
Stern S. A., Weintraub D. A., and Festou M. C. (1993) Evidence for a low surface temperature on Pluto from millimeter-wave thermal emission measurements. *Science, 261,* 1713–1716.
Stern S. A., Buie M. W., and Trafton L. M. (1997) HST high-resolution images and maps of Pluto. *Astron. J., 113,* 827–843.
Tegler S. C., Cornelison D. M., Grundy W. M., Romanishin W., Abernathy M. R., Bovyn M. J., Burt J. A., Evans D. E., Maleszewski C. K., Thompson Z., and Vilas F. (2010) Methane and nitrogen abundances on Pluto and Eris. *Astrophys. J., 725,* 1296–1305.
Tegler S. C., Grundy W. M., Olkin C. B., Young L. A., Romanishin W., Cornelison D. M., and Khodadadkouchaki R. (2012) Ice mineralogy across and into the surfaces of Pluto, Triton, and Eris. *Astrophys. J., 751,* 76.
Trafton L. M. (1981) Pluto's atmospheric bulk near perihelion. *Adv. Space Res., 1,* 93–97.
Walker M. F. and Hardie R. H. (1955) A photometric determination of the rotational period of Pluto. *Publ. Astron. Soc. Pac., 67,* 224–231.
Whipple F. L. (1951) A comet model. I. The acceleration of comet Encke. *Astrophys. J., 111,* 375–394.
Whipple F. L. (1964) Evidence for a comet belt beyond Neptune. *Proc. Natl. Acad. Sci. U.S.A., 51,* 711–718.
Yelle R. V. and Lunine J. I. (1989) Evidence for a molecule heavier than methane in Pluto's atmosphere. *Nature, 339,* 288–290.
Young E. F. and Binzel R. P. (1993) Comparative mapping of Pluto's sub-Charon hemisphere: Three least squares models based on mutual event lightcurves. *Icarus, 102,* 134–139.
Young L. A., Olkin C. B., Elliot J. L., Tholen D. J., and Buie M. W. (1994) The Charon-Pluto mass ratio from MKO astrometry. *Icarus, 108,* 186–199.
Young L. A., Elliot J. L., Tokunaga A., de Bergh C., and Owen T. (1997) Detection of gaseous methane on Pluto. *Icarus, 127,* 258–262.

Barucci M. A., Dalle Ore C., and Fornasier S. (2021) The transneptunian objects as the context for Pluto. In *The Pluto System After New Horizons* (S. A. Stern, J. M. Moore, W. M. Grundy, L. A. Young, and R. P. Binzel, eds.), pp. 21–52. Univ. of Arizona, Tucson, DOI: 10.2458/azu_uapress_9780816540945-ch003.

The Transneptunian Objects as the Context for Pluto: An Astronomical Perspective

M. A. Barucci
LESIA, Observatoire de Paris, Université PSL, CNRS, Université de Paris, Sorbonne Université

C. M. Dalle Ore
NASA Ames Research Center and SETI Institute

S. Fornasier
LESIA, Observatoire de Paris, Université PSL, CNRS, Université de Paris, Sorbonne Université, Institut Universitaire de France

---

A detailed study of transneptunian objects (TNOs) is essential in understanding the processes that occurred in our solar nebula at large heliocentric distance as well as in other planetary systems. Although TNOs are challenging objects, often at the limit of observability by groundbased telescopes, they show high heterogeneity in albedos, colors, compositions, densities, and satellite system architectures. An overview of the state of the art of the physical properties and surface composition of these faint and distant objects as acquired from Earth-based astronomical assets is presented. In particular, we will discuss how the TNOs provide context for Pluto and, conversely, how the Pluto observations from New Horizons help to better understand the TNO population, particularly the larger objects. Overall the results from New Horizons provide feedback for testing many groundbased techniques even though ground results often lack precision. The lessons learned will help to better characterize our continuing investigations of other bodies in the outer regions of the solar system.

1. INTRODUCTION

The region of the solar system beyond Neptune is populated by a vast quantity of objects called transneptunians or Kuiper belt objects (TNOs or KBOs; in this paper we refer to them as TNOs). (134340) Pluto belongs to this population. These objects formed in the external part of our planetary system where planetary nebular density was too low to allow objects to grow into larger planets. In this region beyond the planets, accretion proceeded at least up to the size of about 2700 km in diameter, limited by local dynamical properties. Neptune's moon, Triton, believed to be an escaped TNO captured by Neptune (*McKinnon et al.,* 1995, *Agnor and Hamilton,* 2006), is the largest of these objects.

After the discovery of Pluto in 1930 by Clyde Tombaugh, followed only in 1992 by that of another object of the same population, (15670) Albion (previously unnamed and called 1992 QB1; *Jewitt and Luu,* 1992), the finding of TNOs (with distinct orbital and physical properties) increased rapidly. The discovery of these bodies (at present more than 3000 have been detected and listed in the Minor Planet Center) revolutionized our understanding of the processes and evolution of the protoplanetary nebula in our solar system and, by extension, of other exoplanetary systems. Small bodies provide the richest information for studying a planetary system. In particular, TNOs give information on the accretion and evolutionary processes that sculpted the present outer solar system. Thanks to the continuous observation of the transneptunian region, the science of the solar system beyond Neptune evolved rapidly and a completely new view of our system has been revealed, allowing the development of new models of formation and evolution of our and other planetary systems. Thus, the entire population of TNOs must be the foundational starting point for understanding Pluto and its context in our solar system.

Based on their dynamical properties (*Elliot et al.,* 2005; *Gladman et al.,* 2008), the population has been divided into several groups: classical, resonant, scattering (or scattered) disk, and detached objects. *Gladman et al.* (2008) also designated objects with orbits a >2000 AU as part of the inner Oort cloud. The objects with semimajor axes and perihelia between the orbits of Jupiter and Neptune are part of another population called Centaurs expected to be escaped TNOs or collisional remnants of TNOs (*Levison and Duncan,* 1997; *Pan and Sari,* 2005).

The *classical* population includes bodies with semimajor axis between 40 and 48 AU and e <0.2. It has been divided into two subpopulations, cold and hot classical objects, based on the inclination. Cold classical objects have low i, and low e, while hot classical objects have high inclination (>5°).

Gomez (2003) showed by numerical simulation that the hot population could find its origin in the migration of Neptune, which scattered planetesimals originally formed inside 30 AU. As such, the current classical population could be the superposition of these objects with the local cold objects, believed to have formed *in situ*, beyond 30 AU.

Resonant objects are locked in mean-motion resonances with Neptune and their configuration provides them dynamical stability. Although many resonances exist, the 3:2 mean-motion resonance with Neptune (a = 39.4 AU) is the most densely populated, and objects trapped in it are called plutinos as they include Pluto. The origin of these resonant objects is also closely related to Neptune's migration: Indeed, while Neptune was migrating outward, objects that followed this migration were captured in a resonance (*Malhotra,* 2019).

Scattering/scattered disk objects have very eccentric orbits and perihelion distances near that of Neptune, while *detached objects* are those in very eccentric orbits with large perihelion distances but away from Neptune's influence (*Gladman et al.,* 2008). They have long lifetimes. (136199) Eris is part of the scattered population, while an example of a detached object is (90377) Sedna, whose orbit has a perihelion distance at 76 AU and semimajor axis a = 506.8 AU.

The availability of large groundbased telescopes (8–10 m) and telescopes out of Earth's atmosphere (Hubble, Spitzer, and Herschel) has allowed the start of the investigation of the physical properties and composition of these distant objects (*Jewitt and Luu,* 2001; *Brown,* 2000), obtaining good accuracy for the brightest ones.

A few TNOs are relatively large (>1400 km). In fact, in 2005 three large icy bodies were discovered — (136108) Haumea, (136472) Makemake, and (136199) Eris — that are comparable to (134340) Pluto in size. After the discovery of Eris, with an initial size estimation larger than Pluto, the International Astronomical Union (IAU) revised the definition of a planet, introducing the new category of "dwarf planets" as "objects large enough to be in hydrostatic equilibrium but not large enough to have cleaned their orbit of other minor bodies." These three TNOs and Pluto are included in this category and many other are candidates. The asteroid (1) Ceres, thought (although controversially) to have originated as part of the TNO population (*McKinnon,* 2012; *De Sanctis et al.,* 2015), is also a member of the dwarf planets' class.

In general, because of their size and distance, understanding the detailed composition of TNOs is difficult. The majority of objects are small and far away and particularly difficult to study even with the largest available ground telescopes. Knowledge of TNOs' physical properties, in particular, their composition, yields information on the thermal and chemical processes in the outer protoplanetary disk, sets constraints on the conditions of the early solar system, and informs us about their dynamical evolutionary history.

In this chapter we will give a state-of-the-art overview of the physical properties and composition of these faint and distant objects. This is complementary to the Parker chapter in this volume, which will focus on the dynamical TNO properties and the connection to the Pluto system. We will describe what has been confirmed about Pluto from the ground knowledge and, in particular, give more details on the larger objects closer to Pluto in their characteristics. The results obtained on TNOs will be coupled with those obtained by the New Horizons mission, to outline what has been learned from the TNO population at large and where we have to go from here.

2. TRANSNEPTUNIAN OBJECT PHYSICAL PROPERTIES

2.1. Observational Techniques: Occultations and Radiometry

The most precise technique to determine the size and shape of a TNO is by means of stellar occultation, i.e., when a TNO passes in front of a star. This technique has only been applied to determine the size and shape of TNOs in the past 10 years. In fact, TNOs have angular diameters on the order of ~10 milliarcseconds, and therefore high accuracy is required in both the TNO's ephemeris and the stellar catalogs (at the milliarcsecond level). So far ~60 occultation events by ~25 TNOs and Centaurs have been recorded (*Ortiz et al.,* 2019, and references therein), yielding a very precise size estimation for about 15 TNOs and Centaurs. This technique has also allowed scientists to discover the first ring systems not circling a giant planet: a double ring around the Centaur Chariklo (*Braga-Ribas et al.,* 2014) and a single ring around the dwarf planet Haumea (*Ortiz et al.,* 2017); ring systems are also inferred to be around other bodies (*Ortiz et al.,* 2015).

Radiometry is the most powerful technique for investigating sizes and albedos of the transneptunian population at a statistically significant level. This technique combines thermal infrared and visible observations. The absolute H magnitude alone, being proportional to the area and albedo of a body, gives only a crude estimate of the size assuming a given geometric albedo range. To determine the individual diameter and albedo values of a TNO, observations in the visible range should be coupled with thermal infrared observations. For the TNOs, the peak of the thermal emission ranges between ~40 and 100 μm. Observations in the thermal infrared are done mostly from space, because of the telluric atmosphere's low transmission in most of the near-, mid-, and far-infrared range, and because of the faint thermal flux of the transneptunian population, related to their large heliocentric distances and to their low surface temperature. Groundbased observations covering the N and Q bands at 7–14 μm and 17–25 μm, respectively, and the submillimeter and millimeter range permitted the determination of the thermal properties of few Centaurs and TNOs, including the Pluto-Charon system (*Stern et al.,* 1993, and references therein). The very first infrared space telescope observations of Centaurs and TNOs were performed with the Infrared Astronomical Satellite (IRAS) and the Infrared

Space Observatory (ISO). These included the thermal characterization of the Pluto-Charon system (*Aumann and Walker,* 1987; *Tedesco et al.,* 1987; *Sykes et al.,* 1987; *Sykes,* 1999; *Lellouch et al.,* 2000a,b).

The next big step forward in the determination of the transneptunian population size, albedo, and thermal properties was achieved using the Spitzer Space Telescope, the Wide-field Infrared Survey Explorer (WISE), and the Herschel infrared space telescope. With Spitzer the albedo and size of 47 objects were derived (*Stansberry et al.,* 2008, 2012; *Brucker et al.,* 2009). Another 52 Centaurs and scattered disk objects (SDOs) located at heliocentric distances closer than about 20 AU were investigated with WISE. Finally, 170 TNOs and Centaurs were studied with Herschel (*Müller et al.,* 2010, 2019) (see references in Table 1), with partial overlap with the Spitzer and WISE TNOs sample but covering longer wavelengths.

2.1.1. Thermal properties. To interpret the thermal data, different models can be applied. The most popular ones are the standard thermal model (STM) (cf. *Lebofsky et al.,* 1986; *Lebofsky and Spencer,* 1989) and the near-Earth asteroid thermal model (NEATM) (*Harris,* 1998). The former assumes a smooth, spherical asteroid, not rotating, with zero thermal inertia, observed at zero phase angle and with a fixed beaming factor η, and a grey emissivity constant for all wavelengths. The latter is a hybrid version of the STM, which also assumes a spherical shape of the body and constant emissivity, but in which η is a free parameter that empirically represents the combined effects of thermal inertia, spin state, and surface roughness. High η values at large heliocentric distances unambiguously indicate high thermal inertia and the lack of a dusty regolith, while $\eta < 1$ indicates a strong "beaming" effect due to surface roughness.

The NEATM is the model most often used to interpret TNO thermal data. In the case of limited wavelength coverage, it is run with a fixed beaming factor. Alternatively, the thermophysical model (TPM) (*Lagerros,* 1996, 1997, 1998; *Müller and Lagerros,* 1998) can ingest a non-spherical shape and a wavelength-dependent emissivity, and can provide temperature distribution, thermal inertia, and constraints on the spin axis, but requires some physical properties information as input.

Lellouch et al. (2013) investigated the thermal properties of 85 TNOs and Centaurs from Spitzer and Herschel observations (reporting a relatively high variability of the beaming factor). With the use of a statistical approach on the bodies spin rate and surface roughness, they estimated the thermal inertia from the η value. They found a mean thermal inertia value of 2.5 ± 0.5 J m^{-2} s$^{-0.5}$ K^{-1}, i.e., 2–3 orders of magnitude lower than expected for compact ices (*Ferrari and Lucas,* 2016), and lower than that measured for the water-rich saturnian satellites (5–20 J m^{-2} s$^{-0.5}$ K^{-1}) (*Howett et al.,* 2011, 2012) or for Pluto and Charon (16–26 and 9–14 J m^{-2} s$^{-0.5}$ K^{-1} respectively) (*Lellouch et al.,* 2011, 2016), whose surface texture might, however, be affected by the condensation/sublimation cycles of volatiles. Low TNO thermal inertia indicates surfaces with a very high porosity (*Lellouch et al.,* 2013). Low thermal inertia values were found also for 52 Centaurs and SDOs from WISE observations (*Bauer et al.,* 2013), and, for the TNOs, derived also from thermophysical models (*Müller et al.,* 2020). *Lellouch et al.* (2013) also found an anticorrelation with the heliocentric distances, with the thermal inertia decreasing by more than a factor of 2 from the inner (8–25 AU) to the outer (41–53 AU) regions of the solar system. They suggest that the thermal inertia is inversely proportional to the heliocentric distance, that in these bodies the heat transfer is affected by radiative conductivity within pores and increases with depth in the subsurface.

Overlapping Spitzer, WISE, and Herschel observations in the 20–500-μm range are available for a tenth of the objects, permitting the first study of the spectral energy distribution and submillimeter emissivity (*Fornasier et al.,* 2013; *Lellouch et al.,* 2017). When calculating the local temperatures and monochromatic fluxes with NEATM models, the emissivity (ε) is usually assumed constant for all wavelengths. The far-infrared observations clearly indicate a significant decrease in the spectral emissivity longward of ~250 μm and especially at 500 μm for most of the bodies investigated (*Fornasier et al.,* 2013). This behavior is also confirmed at the millimeter wavelength as observed with the Atacama Large Millimeter/submillimeter Array (ALMA) (*Lellouch et al.,* 2016; *Brown and Butler,* 2017). *Fornasier et al.* (2013) interpreted the low emissivity at submillimeter and millimeter wavelengths as attributed to surfaces having absorption coefficients 10–20× larger than those of pure water ice (*Matzler,* 1998), probably due to impurities within the ices. The Pluto-Charon system also shows a decrease of the emissivity with wavelength, interpreted as a combination of a high dielectric constant (3–5) and the transparency of the surface material, which has a typical penetration depth of ~1 cm at 500 μm (*Lellouch et al.,* 2016).

2.1.2. Size and albedo. Size and albedo values, derived from Herschel, Spitzer, and WISE observations, and a few from occultations, are available for a total of 170 TNOs and Centaurs and are reported in Table 1. Eight additional bodies have lower and upper limit estimations for the diameter and albedo values, respectively. The most striking observation is the huge variation of the albedo value across the TNO population, which includes both extremely dark surfaces with geometric albedo of 2–3% and high reflective bodies. The latter objects are the volatile-rich dwarf planets and Haumea family members, with a geometric albedo greater than 50% and as high as 96% for Eris. However, excluding the peculiar volatile-rich bodies and those with upper/lower limit estimation only, the albedo ranges between 2% and 33%, with a mean value of $9.9 \pm 0.5\%$ on a sample of 160 objects. Thus globally the TNO population is relatively dark. For newly discovered objects, their diameter can be estimated with an error lower than a factor of 1.6 from their absolute magnitude assuming an albedo value of 10%. Figure 1 reports the geometric albedo vs. diameter for the different dynamical classes of the transneptunian population (Table 1), defined according to the *Gladman et al.* (2008) dynamical classification. The size of known TNOs ranges

TABLE 1. Physical properties of the TNOs and Centaurs have been reported for all objects with diameter and albedo determined by space telescopes.

Object	Dyn.	D (km)	errD+ (km)	errD– (km)	Pv	err_ pv+	Err_ pv–	Period (hr)	Amplitude (Δm_v)	Taxa	References
(2060) Chiron	Cen	215.6	9.9	–9.9	0.167	0.037	–0.03	5.918 ± 0.0001	0.088 ± 0.003	BB	*Fornasier et al.* (2013); *Bus et al.* (1989)
(5145) Pholus	Cen	99	15	–14	0.155	0.076	–0.049	9.98	0.6	RR	*Duffard et al.* (2014); *Tegler et al.* (2005)
(7066) Nessus	Cen	57	17	–14	0.086	0.075	–0.034			RR	*Duffard et al.* (2014)
(8405) Asbolus	Cen	85	8	–9	0.056	0.019	–0.015	8.9351 ± 0.003	0.55	BR	*Duffard et al.* (2014); *Kern et al.* (2000)
(10199) Chariklo	Cen	238	10	–10	0.042	0.005	–0.005	7.004 ± 0.036	0.11	BR	*Fornasier et al.* (2013, 2014)
(10199) Chariklo*	Cen	246	6	–6	0.042	0.001	–0.001				*Leiva et al.* (2017)
(10370) Hylonome	Cen	74	16	–16	0.051	0.03	–0.017			BR	*Duffard et al.* (2014)
(31824) Elatus	Cen	49.8	10.4	–9.8	0.049	0.028	–0.016	13.41 ± 0.04	0.102 ± 0.005	RR	*Duffard et al.* (2014); *Bauer et al.* (2002)
(32532) Thereus	Cen	62	3	–3	0.083	0.016	–0.013	8.3091 ± 0.001	0.16 ± 0.02	BR	*Duffard et al.* (2014); *Ortiz et al.* (2003)
(52872) Okyrhoe	Cen	35	3	–3	0.056	0.012	–0.01	6.08	0.07 ± 0.01	BR	*Duffard et al.* (2014); *Thirouin et al.* (2010)
(52975) Cyllarus	Cen	56	21	–18	0.139	0.157	–0.064			RR	*Duffard et al.* (2014)
(54598) Bienor	Cen	198	6	–7	0.043	0.016	–0.012	9.14 ± 0.04	0.75 ± 0.09	BR	*Duffard et al.* (2014); *Ortiz et al.* (2003)
(55576) Amycus	Cen	104	8	–8	0.083	0.016	–0.015	9.76	0.16 ± 0.01	RR	*Duffard et al.* (2014); *Thirouin et al.* (2010)
(60558) Echeclus	Cen	64.6	1.6	–1.6	0.052	0.007	–0.007	26.802	0.24 ± 0.06	BR	*Duffard et al.* (2014); *Rousselot et al.* (2005)
(63252) 2001 BL41	Cen	34.6	6.6	–6.1	0.043	0.028	–0.014			BR	*Duffard et al.* (2014)
(83982) Crantor	Cen	59	11	–12	0.121	0.064	–0.038	6.97 ± 0.04	0.14 ± 0.04	RR	*Duffard et al.* (2014); *Ortiz et al.* (2003)
(95626) 2002 GZ32	Cen	237	8	–8	0.037	0.004	–0.004	5.8 ± 0.03	0.15 ± 0.03	BR	*Duffard et al.* (2014); *Dotto et al.* (2008)
(119315) 2001 SQ73	Cen	90	23	–20	0.048	0.03	–0.018			BR	*Duffard et al.* (2014)
(119976) 2002 VR130	Cen	24.4	5.4	–4.6	0.093	0.066	–0.036				*Duffard et al.* (2014)
(120061) 2003 CO1	Cen	94	5	–5	0.049	0.005	–0.006	4.51	0.06 ± 0.01	BR	*Duffard et al.* (2014); *Thirouin et al.* (2010)
(136204) 2003 WL7	Cen	105	6	–7	0.053	0.01	–0.01	16.48	0.05 ± 0.01	BB	*Duffard et al.* (2014); *Thirouin et al.* (2010)
(145486) 2005 UJ438	Cen	16	1	–2	0.256	0.097	–0.076	8.32	0.11 ± 0.01	RR-IR	*Duffard et al.* (2014); *Thirouin et al.* (2010)
(148975) 2001 XA255	Cen	37.7	10.5	–10.5	0.041	0.014	–0.014				*Bauer et al.* (2013)
(248835) 2006 SX368	Cen	76	2	–2	0.052	0.007	–0.006			BR	*Duffard et al.* (2014)
(250112) 2002 KY14	Cen	47	3	–4	0.057	0.011	–0.007	7.12	0.11 ± 0.01	RR	*Duffard et al.* (2014); *Thirouin et al.* (2010)
(281371) 2008 FC76	Cen	68	6	–7	0.067	0.017	–0.011			RR	*Duffard et al.* (2014)
(309139) 2006 XQ51	Cen	39.1	15.7	–15.7	0.139	0.058	–0.058				*Bauer et al.* (2013)
(310071) 2010 KR59	Cen	110.1	30.8	–30.8	0.121	0.037	–0.037				*Bauer et al.* (2013)
(309737) 2008 SJ236	Cen	17.7	1.5	–1.5	0.074	0.021	–0.021				*Bauer et al.* (2013)
(328884) 2010 LJ109	Cen	44.2	3.8	–3.8	0.083	0.021	–0.021				*Bauer et al.* (2013)

TABLE 1. (continued).

Object	Dyn.	D (km)	errD+ (km)	errD– (km)	Pv	err_pv+	Err_ pv–	Period (hr)	Amplitude (Δm_v)	Taxa	References
(330759) 2008 SO218	Cen	11.8	0.4	–0.4	0.097	0.017	–0.017				*Bauer et al.* (2013)
(332685) 2009 HH36	Cen	33	2.8	–2.8	0.078	0.018	–0.018				*Bauer et al.* (2013)
(342842) 2008 YB3	Cen	67.1	1	–1	0.062	0.012	–0.012				*Bauer et al.* (2013)
(346889) 2009 QV38	Cen	23.2	9.5	–9.5	0.062	0.049	–0.049				*Bauer et al.* (2013)
(447178) 2005 RO43	Cen	194	10	–10	0.056	0.036	–0.021				*Duffard et al.* (2014)
(900391) 29P/SW 1	Cen	46	13	–13	0.033	0.015	–0.015				*Bauer et al.* (2013)
(901056) 167P/CINEOS	Cen	66.2	22.9	–22.9	0.053	0.019	–0.019				*Bauer et al.* (2013)
(903956) C/2011 KP36	Cen	55.1	19.4	–19.4	0.101	0.062	–0.062				*Bauer et al.* (2013)
2000 GM137	Cen	8.6	1.5	–1.5	0.043	0.026	–0.016				*Duffard et al.* (2014)
2004 QQ26	Cen	79	19	–19	0.044	0.039	–0.014				*Duffard et al.* (2014)
2013 AZ60	Cen	62.3	5.3	–5.3	0.029	0.006	–0.006	9.39	0.013 ± 0.008		*Pal et al.* (2015a)
2008 JS14	Cen	14.5	1.8	–1.8	0.044	0.019	–0.019				*Bauer et al.* (2013)
2010 CR140	Cen	7.5	1.4	–1.4	0.02	0.01	–0.01				*Bauer et al.* (2013)
2010 HU20	Cen	10.5	1.1	–1.1	0.101	0.024	–0.024				*Bauer et al.* (2013)
2010 LG61	Cen	0.9	0.2	–0.2	0.089	0.056	–0.056				*Bauer et al.* (2013)
2010 OR1	Cen	3.3	0.6	–0.6	0.055	0.013	–0.013				*Bauer et al.* (2013)
2010 OM101	Cen	3.1	0.2	–0.2	0.029	0.005	–0.005				*Bauer et al.* (2013)
2010 PO58	Cen	8.9	0.6	–0.6	0.035	0.007	–0.007				*Bauer et al.* (2013)
2007 VH305	Cen	23.8	8	–8	0.07	0.036	–0.036				*Bauer et al.* (2013)
2008 HY21	Cen	24	1.5	–1.5	0.044	0.01	–0.01				*Bauer et al.* (2013)
2010 BL4	Cen	15.7	3.2	–3.2	0.114	0.052	–0.052				*Bauer et al.* (2013)
2010 ES65	Cen	26.9	7.9	–7.9	0.049	0.024	–0.024				*Bauer et al.* (2013)
2010 FH92	Cen	28	0.6	–0.6	0.047	0.007	–0.007				*Bauer et al.* (2013)
2010 RM64	Cen	21	2	–2	0.159	0.048	–0.048				*Bauer et al.* (2013)
2010 TH	Cen	69.9	24.2	–24.2	0.078	0.033	–0.033				*Bauer et al.* (2013)
2011 MM4	Cen	63.7	6.2	–6.2	0.083	0.024	–0.024				*Bauer et al.* (2013)
2000 CN105	CC	247	63	–40	0.151	0.07	–0.059			RR	*Lacerda et al.* (2014a)
2001 QS322	CC	186	99	–24	0.095	0.531	–0.06				*Vilenius et al.* (2014)
2001 RZ143	CC	140	39	–33	0.191	0.066	–0.045				*Vilenius et al.* (2012)
2001 XR254	CC	221	41	–71	0.136	0.168	–0.044				*Vilenius et al.* (2014)

TABLE 1. (continued).

Object	Dyn.	D (km)	errD+ (km)	errD– (km)	Pv	err_pv+	Err_ pv–	Period (hr)	Amplitude (Δm_v)	Taxa	References
2003 BF91	CC							9.1	1.09 ± 0.25		*Trilling and Bernstein* (2006)
2003 BH91	CC							2.8	0.42		*Trilling and Bernstein* (2006)
2003 BG91	CC							4.2	0.18 ± 0.075		*Trilling and Bernstein* (2006)
2003 FM127	CC							6.22 ± 0.02	0.46 ± 0.04		*Kern* (2006)
2003 QY90A	CC							3.4 ± 1.1	0.34 ± 0.06		*Kern and Elliot* (2006)
2003 QY90B	CC							7.1 ± 2.9	0.9 ± 0.18		*Kern and Elliot* (2006)
2003 QR91	CC	280	27	–30	0.054	0.035	–0.028				*Vilenius et al.* (2014)
2003 WU188	CC	<220			>0.15						*Vilenius et al.* (2014)
(486958) Arrokoth[†]	CC	30			0.165	0.01	–0.01				*Stern et al.* (2019)
(66652) Borasisi	CC	163	32	–66	0.236	0.438	–0.077	6.4 ± 1	0.08 ± 0.02	IR-RR	*Vilenius et al.* (2014); *Kern* (2006)
(79360) Sila-Nunam	CC	343	42	–42	0.09	0.027	–0.017	150.149	0.12 ± 0.01	RR	*Vilenius et al.* (2012); *Rabinowitz et al.* (2014)
(88611) Teharonhiawako	CC	220	41	–44	0.145	0.086	–0.045				*Vilenius et al.* (2014)
(119951) 2002 KX14	CC	455	27	–27	0.097	0.014	–0.013				*Vilenius et al.* (2012)
(119951) 2002 KX14[*]	CC	365	30	–21	0.15	0.04	–0.03				*Alvarez-Candal et al.* (2014)
(120181) 2003 UR292	CC	136	16	–26	0.105	0.081	–0.033				*Vilenius et al.* (2014)
(135182) 2001 QT322	CC	159	30	–47	0.085	0.424	–0.052				*Vilenius et al.* (2014)
(275809) 2001 QY297	CC	229	22	–108	0.152	0.439	–0.035			BR	*Vilenius et al.* (2014)
(385266) 2001 QB298	CC	196	71	–53	0.167	0.162	–0.082				*Mommert* (2013)
(385437) 2003 GH55	CC	178	21	–56	0.15	0.182	–0.031			RR	*Vilenius et al.* (2014)
(469438) 2002 GV31	CC	<180			>0.019			29.2	0.35 ± 0.06		*Vilenius et al.* (2014); *Pàl et al.* (2015b)
(469514) 2003 QA91	CC	260	30	–36	0.13	0.119	–0.075				*Vilenius et al.* (2014)
(469705) 2005 EF298	CC	174	27	–32	0.16	0.13	–0.07	9.65	0.31 ± 0.04		*Vilenius et al.* (2012); *Benecchi and Sheppard* (2013)
(508869) 2002 VT130	CC	324	57	–68	0.097	0.098	–0.049				*Mommert* (2013)
1996 TS66	HC	159	44	–46	0.179	0.173	–0.07			RR	*Vilenius et al.* (2014)
2001 KA77	HC	310	170	–60	0.099	0.052	–0.056	>6	>0.14	RR	*Vilenius et al.* (2012); *Kern* (2006)
2001 QC298	HC	303	27	–30	0.061	0.027	–0.017	12	0.4		*Vilenius et al.* (2014); *Snodgrass et al.* (2010)
2001 QD298	HC	233	27	–63	0.067	0.062	–0.014				*Vilenius et al.* (2012)
2002 GH32	HC	<180			>0.13			3.98	0.36 ± 0.02	IR	*Vilenius et al.* (2014); *Thirouin et al.* (2016)
2003 SQ317	HC							7.21 ± 0.001	0.85 ± 0.05		*Lacerda et al.* (2014b)
2010 FX86	HC							15.8	0.26 ± 0.04		*Benecchi and Sheppard* (2013)

TABLE 1. (continued).

Object	Dyn.	D (km)	errD+ (km)	errD– (km)	Pv	err_pv+	Err_ pv–	Period (hr)	Amplitude (Δm_v)	Taxa	References
2010 VK201	HC							7.59	0.3 ± 0.02		*Benecchi and Sheppard* (2013)
(19308) 1996 TO66	HC	<330			>0.20			7.92 ± 0.04	0.26 ± 0.03	BB	*Vilenius et al.* (2018); *Sheppard and Jewitt* (2003)
(19521) Chaos	HC	600	140	–130	0.05	0.03	–0.016			IR	*Vilenius et al.* (2012)
(20000) Varuna	HC	668	154	–86	0.127	0.04	–0.042	6.3436 ± 0.0001	0.41 ± 0.09	IR	*Lellouch et al.* (2013); *Ortiz et al.* (2003)
(24835) 1995 SM55	HC	<280			>0.360			8.08	0.05 ± 0.02	BB	*Vilenius et al.* (2018); *Thirouin et al.* (2016)
(35671) 1998 SN165	HC	393	39	–38	0.06	0.019	–0.013	8.84	0.16 ± 0.01	BB	*Vilenius et al.* (2012); *Lacerda and Luu* (2006)
(50000) Quaoar	HC	1073.6	37.9	–37.9	0.127	0.01	–0.009	17.6788 ± 0.0004	0.13 ± 0.03	RR	*Fornasier et al.* (2013); *Ortiz et al.* (2003)
(50000) Quaoar*	HC	1110	5	5	0.109	0.007	–0.007			RR	*Braga-Ribas et al.* (2013)
(55565) 2002 AW197	HC	768	39	–38	0.112	0.012	–0.011	8.86 ± 0.01	0.08± 0.02	IR	*Vilenius et al.* (2014); *Ortiz et al.* (2006)
(55636) 2002 TX300	HC	323	95	–37	0.76	0.15	–0.45	16.24 ± 0.08	0.08 ± 0.02	BB	*Vilenius et al.* (2018); *Sheppard and Jewitt* (2003)
(55636) 2002 TX300*	HC	286	10	–10	0.88	0.15	–0.06				*Elliot et al.* (2010)
(55637) 2002 UX25	HC	165	36	–42	0.049	0.038	–0.017	6.55	0.09 ± 0.03	IR	*Lellouch et al.* (2013); *Thirouin* (2013)
(78799) 2002 XW93	HC	565	71	–73	0.038	0.043	–0.025				*Vilenius et al.* (2012)
(79983) 1999 DF9	HC							6.65 ± 0.4	0.02	RR	*Lacerda and Luu* (2006)
(86177) 1999 RY215	HC	263	29	–37	0.039	0.012	–0.007			BR	*Vilenius et al.* (2012)
(90568) 2004 GV9	HC	680	34	–34	0.077	0.008	–0.008	5.86 ± 0.03	0.16 ± 0.03	BR-IR	*Vilenius et al.* (2012); *Dotto et al.* (2008)
(120178) 2003 OP32	HC	274	47	–25	0.54	0.11	–0.15	9.71	0.18 ± 0.01	BB	*Vilenius et al.* (2018); *Benecchi and Sheppard* (2013)
(120347) Salacia	HC	901	45	–45	0.044	0.004	–0.004	6.61	0.04 ± 0.02	BB	*Fornasier et al.* (2013); *Thirouin et al.* (2014)
(136108) Haumea	HC	1240	69	–58	0.804	0.062	–0.095	3.9155 ± 0.0001	0.29 ± 0.02	BB	*Fornasier et al.* (2013); *Lacerda et al.* (2008)
(136108) Haumea*	HC	1595	11	–11	0.51	0.02	0.02			BB	*Ortiz et al.* (2017)
(136472) Makemake	HC	1440	9	–9	0.77	0.02	–0.02	7.771 ± 0.003	0.029 ± 0.002	BB	*Lellouch et al.* (2013); *Heinze and de Lahunta,* (2009)
(136472) Makemake*	HC	1440	9	–9	0.77	0.03	–0.03				*Ortiz et al.* (2012)
(138537) 2000 OK67	HC	164	33	–45	0.169	0.159	–0.052			RR	*Vilenius et al.* (2014)
(145452) 2005 RN43	HC	679	55	–73	0.107	0.029	–0.018	13.89	0.06 ± 0.01	IR-RR	*Vilenius et al.* (2012); *Benecchi and Sheppard* (2013)
(145453) 2005 RR43	HC	300	43	–34	0.44	0.12	–0.1	7.87	0.06 ± 0.01	BB	*Vilenius et al.* (2018); *Thirouin et al.* (2010)

TABLE 1. (continued).

Object	Dyn.	D (km)	errD+ (km)	errD– (km)	Pv	err_pv+	Err_pv–	Period (hr)	Amplitude (Δm_v)	Taxa	References
(148780) Altjira	HC	331	51	–187	0.043	0.183	–0.009			RR-IR	*Vilenius et al.* (2014)
(150642) 2001 CZ31	HC							4.71	0.21 ± 0.02	BB/BR	*Lacerda and Luu* (2006)
(174567) Varda	HC	792	91	–84	0.102	0.024	–0.02	5.91	0.02 ± 0.01	IR	*Vilenius et al.* (2014); *Thirouin et al.* (2014)
(182934) 2002 GJ32	HC	416	81	–78	0.035	0.019	–0.011			RR	*Vilenius et al.* (2014)
(202421) 2005 UQ513	HC	498	63	–75	0.202	0.084	–0.049	7.03	0.05 ± 0.02		*Vilenius et al.* (2012); *Thirouin et al.* (2012)
(230965) 2004 XA192	HC	339	120	–95	0.26	0.34	–0.15	7.88	0.07 ± 0.02		*Vilenius et al.* (2012); *Thirouin et al.* (2012)
(307251) 2002 KW14	HC	161	35	–40	0.31	0.281	–0.094	13.25	0.25 ± 0.03		*Vilenius et al.* (2012); *Benecchi and Sheppard* (2013)
(307261) 2002 MS4	HC	934	47	–47	0.051	0.036	–0.022	7.33	0.05 ± 0.01		*Vilenius et al.* (2012); *Thirouin* (2013)
(307616) 2003 QW90	HC	401	63	–48	0.084	0.026	–0.022			RR	*Lacerda et al.* (2014a)
(308193) 2005 CB79	HC							6.76	0.13 ± 0.02		*Thirouin et al.* (2010)
(416400) 2003 UZ117	HC	222	57	–42	0.29	0.16	–0.11	11.29	0.09 ± 0.01	BB	*Vilenius et al.* (2018); *Thirouin et al.* (2016)
(444030) 2004 NT33	HC	423	87	–80	0.125	0.069	–0.039	7.87	0.04 ± 0.01	BB-BR	*Vilenius et al.* (2014); *Thirouin et al.* (2012)
(469306) 1999 CD158	HC	<310			>0.130			6.88	0.49 ± 0.03	BR-IR	*Vilenius et al.* (2018); *Thirouin et al.* (2016)
(469615) 2004 PT107	HC	400	45	–51	0.032	0.011	–0.007	20	0.05		*Vilenius et al.* (2014); Snodgrass et al. (2010)
(134340) Pluto†	Plu	2376	2	2	0.52	0.14	–0.03	153.2935	0.3	BB	*Stern et al.* (2018); *Tholen and Buie,* 1997
Charon†	Plu	1212	1	1	0.41	0.02	–0.02	153.2935			*Stern et al.* (2018); *Tholen and Buie,* 1997
2001 KD77	Plu	232.3	40.5	–39.4	0.089	0.044	–0.027			IR-RR	*Mommert et al.* (2012)
2002 XV93	Plu	549.2	21.7	–23	0.04	0.02	–0.015				*Mommert et al.* (2012)
2003 UT292	Plu	185.6	17.9	–18	0.067	0.068	–0.034				*Mommert et al.* (2012)
(15789) 1993 SC	Plu							7.7	0.04	RR	*Tegler et al.* (1997)
(15810) Arawn	Plu							5.47 ± 0.33	0.58		*Porter et al.* (2016)
(15820) 1994 TB	Plu	85	36	–28	0.172	0.258	–0.097	6	0.26	RR	*Lellouch et al.* (2013); *Romanishin and Tegler,* 1999
(15875) 1996 TP66	Plu	154	28.8	–33.7	0.074	0.063	–0.031	1.96	<0.04	RR	*Mommert et al.* (2012); *Collander-Brown et al.* 1999
(28978) Ixion	Plu	617	19	–20	0.141	0.011	–0.011	12.4 ± 0.3		IR	*Lellouch et al.* (2013); *Galiazzo et al.* (2016)
(32929) 1995 QY9	Plu							7.3 ± 0.1	0.6 ± 0.04	BR	*Sheppard and Jewitt* (2002)
(33340) 1998 VG44	Plu	248	43	–41	0.063	0.026	–0.017			IR	*Lellouch et al.* (2013)
(38628) Huya	Plu	458	9.2	–9.2	0.083	0.004	–0.004	5.21	0.02 ± 0.01	IR	*Fornasier et al.* (2013); *Thirouin et al.* (2014)
(47171) 1999 TC36	Plu	393.1	25.2	–26.8	0.079	0.013	–0.011	6.21 ± 0.02	0.06	RR	*Mommert et al.* (2012); *Ortiz et al.* (2003)
(47932) 2000 GN171	Plu	147.1	20.7	–17.8	0.215	0.093	–0.07	8.329 ± 0.005	0.61 ± 0.03	IR	*Mommert et al.* (2012); *Sheppard and Jewitt* (2002)

TABLE 1. (continued).

Object	Dyn.	D (km)	errD+ (km)	errD– (km)	Pv	err_pv+	Err_pv–	Period (hr)	Amplitude (Δm_v)	Taxa	References
(55638) 2002 VE95	Plu	249.8	13.5	–13.1	0.149	0.019	–0.016	9.97	0.05 ± 0.01	RR	*Mommert et al.* (2012); *Thirouin et al.* (2010)
(84719) 2002 VR128	Plu	448.5	42.1	–43.2	0.052	0.027	–0.018				*Mommert et al.* (2012)
(84922) 2003 VS2	Plu	523	35.1	–34.4	0.147	0.063	–0.043	7.4208	0.224 ± 0.013		*Mommert et al.* (2012); Thirouin (2013)
(90482) Orcus	Plu	958.4	22.9	–22.9	0.231	0.018	–0.011	10.08 ± 0.01	0.04 ± 0.02	BB	*Fornasier et al.* (2013); *Ortiz et al.* (2006)
(120216) 2004 EW95	Plu	291.1	20.3	–25.9	0.044	0.021	–0.015			BB	*Mommert et al.* (2012)
(120348) 2004 TY364	Plu	512	37	–40	0.107	0.02	–0.015	11.7 ± 0.01	0.22 ± 0.02	BR-IR	*Lellouch et al.* (2013); Sheppard 2007
(133067) 2003 FB128	Plu	186	27	–29	0.074	0.035	–0.021				*Lacerda et al.* (2014a)
(139775) 2001 QG298	Plu							13.7744 ± 0.0004	1.14 ± 0.04	BR	*Sheppard and Jewitt* (2004)
(144897) 2004 UX10	Plu	398.1	32.6	–39.3	0.141	0.044	–0.031	7.58 ± 0.05	0.14 ± 0.04	BR-IR	*Mommert et al.* (2012); *Perna et al.* 2009
(175113) 2004 PF115	Plu	468.2	38.6	–49.1	0.123	0.043	–0.033				*Mommert et al.* (2012)
(208996) 2003 AZ84	Plu	727	62	–67	0.107	0.023	–0.016	6.72 ± 0.05	0.14 ± 0.03	BB	*Mommert et al.* (2012); *Sheppard and Jewitt* (2003)
(208996) 2003 AZ84*	Plu	772	12	–12	0.097	0.009	–0.009			BB	*Dias-Oliveira et al.* (2017)
(307463) 2002 VU130	Plu	252.9	33.6	–31.3	0.179	0.202	–0.103				*Mommert et al.* (2012)
(341520) 2007 TY430	Plu							9.28	0.24 ± 0.0		*Thirouin et al.* (2014)
(450265) 2003 WU172	Plu	312			0.039						*Lacerda et al.* (2014a)
(455502) 2003 UZ413	Plu	670	84	–82	0.07	0.022	–0.015	4.13 ± 0.05	0.13 ± 0.13	BB	*Lacerda et al.* (2014a); *Perna et al.* 2009
(469372) 2001 QF298	Plu	408.2	40.2	–44.9	0.071	0.02	–0.014			BB	*Mommert et al.* (2012)
(469708) 2005 GE187	Plu							11.99	0.29 ± 0.02		*Thirouin et al.* (2016)
(469987) 2006 HJ23	Plu	216.4	29.7	–34.2	0.281	0.259	–0.152				*Mommert et al.* (2012)
2003 HA57	Plu							6.44	0.31 ± 0.03		*Thirouin et al.* (2016)
2007 JF43	Plu							9.52	0.22 ± 0.02		*Benecchi and Sheppard* (2013)
2010 EL139	Plu							6.32	0.15 ± 0.03		*Benecchi and Sheppard* (2013)
2014 JL80	Plu							34.87	0.55		*Thirouin and Sheppard* (2018)
2014 JO80	Plu							6.32	0.6 ± 0.05		*Thirouin and Sheppard* (2018)
2014 JQ80	Plu							12.16	0.76 ± 0.04		*Thirouin and Sheppard* (2018)
2014 KC102	Plu							9	0.2		*Thirouin and Sheppard* (2018)
2015 BA519	Plu							8	0.16		*Thirouin and Sheppard* (2018)
(126154) 2001 YH140	Res	349	81	–81	0.08	0.05	–0.05	8.45 ± 0.05	0.34 ± 0.06	IR	*Müller et al.* (2010); *Ortiz et al.* (2006)
(119066) 2001 KJ76	Res							3.38 ± 0.39	0.34 ± 0.06		*Kern* (2006)

TABLE 1. (continued).

Object	Dyn.	D (km)	errD+ (km)	errD– (km)	Pv	err_pv+	Err_ pv–	Period (hr)	Amplitude (Δm_v)	Taxa	References
2002 GP32	Res	201	25	–29	0.091	0.061	–0.024				*Lacerda et al.* (2014a)
(26308) 1998 SM165	Res	291	22	–26	0.083	0.018	–0.013	7.1 ± 0.01	0.45 ± 0.03	RR	*Lellouch et al.* (2013); *Sheppard and Jewitt* (2002)
(26375) 1999 DE9	Res	311	29	–32	0.163	0.041	–0.026			IR	*Lellouch et al.* (2013)
(42301) 2001 UR163	Res	352	85	–53	0.209	0.082	–0.074			RR	*Lacerda et al.* (2014a)
(82075) 2000 YW134	Res	<500			>0.08					IR	*Müller et al.* (2010)
(84522) 2002 TC302	Res	165	36	–42	0.049	0.038	–0.017	5.41	0.04 ± 0.01	RR	*Fornasier et al.* (2013); *Thirouin et al.* (2012)
(119979) 2002 WC19	Res	348	45	–45	0.167	0.052	–0.037			RR	*Lellouch et al.* (2013)
(143707) 2003 UY117	Res	247	30	–29	0.126	0.039	–0.028			IR	*Lacerda et al.* (2014a)
(182294) 2001 KU76	Res							5.27 ± 0.02	0.28 ± 0.05		*Kern* (2006)
(225088) 2007 OR10	Res	1230	50	–50	0.14	0.01	–0.01	44.81	0.0444 ± 0.085		*Kiss et al.* 2019; *Pàl et al.* (2016)
(278361) 2007 JJ43	Res							12.097	0.1 ± 0.005	IR	*Pàl et al.* 2015b
(308379) 2005 RS43	Res	271	45	–40	0.193	0.071	–0.053			BR	*Lacerda et al.* (2014a)
(312645) 2010 EP65	Res							14.97	0.17 ± 0.03		*Benecchi and Sheppard* (2013)
(469505) 2003 FE128	Res	157	60	–7	0.167	0.085	–0.072	5.85 ± 0.15	0.5 ± 0.14		*Lacerda et al.* (2014a); *Kern* (2006)
(471143) 2010 EK139	Res	470	35	–10	0.250	0.02	–0.05	7.07	0.12 ± 0.02		*Pàl et al.* (2012); *Benecchi and Sheppard* (2013)
2012 DR30	SDO	188	9.4	–9.4	0.076	0.031	–0.025			BR-IR	*Kiss et al.* (2013)
(15874) 1996 TL66	SDO	339	20	–20	0.11	0.021	–0.015	12	0.07 ± 0.02	BB	*Santos-Sanz et al.* (2012); *Thirouin et al.* (2010)
(26181) 1996 GQ21	SDO	349	43	–49	0.127	0.043	–0.026			RR	*Mommert* (2013)
(29981) 1999 TD10	SDO	103.7	13.6	–13.5	0.044	0.014	–0.01	15.42 ± 0.02	0.65 ± 0.05	BR	*Stansberry et al.* 2008; *Ortiz et al.* (2003)
(33128) 1998 BU48	SDO							12.6 ± 0.1	0.68 ± 0.04	RR	*Sheppard and Jewitt* (2002)
(42355) Typhon	SDO	185	7	–7	0.044	0.003	–0.003	9.67	0.07 ± 0.01	BR	*Santos-Sanz et al.* (2012); *Thirouin et al.* (2010)
(44594) 1999 OX3	SDO	135	13	–12	0.081	0.018	–0.015			RR	*Lellouch et al.* (2013)
(48639) 1995 TL8	SDO	244	82	–63	0.231	0.189	–0.102			RR	*Lellouch et al.* (2013)
(65489) Ceto	SDO	281	11	–11	0.056	0.006	–0.006	4.43 ± 0.03	0.13 ± 0.02	BR-IR	*Santos-Sanz et al.* (2012); *Dotto et al.* 2008
(73480) 2002 PN34	SDO	112	7	–7	0.049	0.006	–0.006	10.22 ± 0.06	0.18 ± 0.04	BR	*Santos-Sanz et al.* (2012); *Ortiz et al.* (2003)
(82158) 2001 FP185	SDO	332	31	–24	0.046	0.007	–0.007			IR	*Santos-Sanz et al.* (2012)
(87555) 2000 QB243	SDO							9.01 ± 0.04	0.2 ± 0.02		*Kern* (2006)
(127546) 2002 XU93	SDO	164	9	–9	0.038	0.004	–0.004			BB-BR	*Santos-Sanz et al.* (2012)
(309239) 2007 RW10	SDO	247	30	–30	0.083	0.068	–0.039				*Santos-Sanz et al.* (2012)

TABLE 1. (continued).

Object	Dyn.	D (km)	errD+ (km)	errD– (km)	Pv	err_pv+	Err_ pv–	Period (hr)	Amplitude (Δm_v)	Taxa	References
(145451) 2005 RM43	SDO							6.71	0.05 ± 0.01	BB	*Thirouin et al.* (2010)
(336756) 2010 NV1	SDO	44.2	8	–8	0.057	0.03	–0.03				*Bauer et al.* (2013)
(445473) 2010 VZ98	SDO							9.72	<0.18		*Benecchi and Sheppard* (2013)
(471137) 2010 ET65	SDO							7.88	0.13 ± 0.02		*Benecchi and Sheppard* (2013)
2001 KG77	SDO							4.8 ± 2.2	0.8 ± 0.26		*Kern* (2006)
2005 VJ119	SDO	28.5	6.9	–6.9	0.126	0.06	–0.06				*Bauer et al.* (2013)
2010 BK118	SDO	46.4	1.8	–1.8	0.068	0.013	–0.013				*Bauer et al.* (2013)
2010 GW64	SDO	6.4	0.4	–0.4	0.047	0.012	–0.012				*Bauer et al.* (2013)
2010 GW147	SDO	15.9	0.7	–0.7	0.037	0.006	–0.006				*Bauer et al.* (2013)
2010 JH124	SDO	7	0.7	–0.7	0.052	0.024	–0.024				*Bauer et al.* (2013)
C/(2010) KW7	SDO	4.9	0.2	–0.2	0.047	0.011	–0.011				*Bauer et al.* (2013)
2010 WG9	SDO	112.7	61.9	–61.9	0.074	0.08	–0.08	263.78 ± 0.12	0.14		*Bauer et al.* (2013); *Benecchi and Sheppard* (2013)
2010 PU75	SDO							12.39	0.27 ± 0.03		*Benecchi and Sheppard* (2013)
2008 OG19	SDO							8.727	0.437 ± 0.011	IR-RR	*Fernández-Valenzuela et al.* (2016)
(40314) 1999 KR16	Det.	232	34	–36	0.105	0.049	–0.027	5.8	0.12 ± 0.06	RR	*Vilenius et al.* (2018); *Thirouin et al.* (2012)
(90377) Sedna	Det.	995	80	–80	0.32	0.06	–0.06	10.273 ± 0.003	0.02	RR	*Pàl et al.* (2012); *Gaudi et al.* (2005)
(120132) 2003 FY128	Det.	460	21	–21	0.079	0.01	–0.01	8.54	0.12 ± 0.02	IR	*Santos-Sanz et al.* (2012); *Thirouin et al.* (2010)
(136199) Eris	Det.	2326	12	–12	0.96	0.04	–0.04	25.92	0.1	BB	*Lellouch et al.* (2013); *Roe et al.* 2008
(136199) Eris*	Det.	2326	12	–12	0.96	0.09	–0.04				*Sicardy et al.* 2011
(145480) 2005 TB190	Det.	464	62	–62	0.148	0.051	–0.036	12.68	0.12 ± 0.01	IR	*Santos-Sanz et al.* (2012); *Thirouin et al.* (2012)
(229762) 2007 UK126	Det.	599	77	–77	0.167	0.058	–0.038	11.05	0.03 ± 0.01		*Santos-Sanz et al.* (2012); *Thirouin et al.* (2014)
(229762) 2007 UK126*	Det.	638	28	–14	0.159	0.007	–0.013				*Benedetti-Rossi et al.* (2016)
(303775) 2005 QU182	Det.	416	73	–73	0.328	0.16	–0.109	9.22	0.12 ± 0.02	IR-RR	*Santos-Sanz et al.* (2012); *Benecchi and Sheppard* (2013)
(470316) 2007 OC10	Det.	309	37	–37	0.127	0.04	–0.028			IR	*Santos-Sanz et al.* (2012)

* Objects with size derived from occultations.
† From the New Horizon mission.

Taxonomy is derived from *Barucci et al.* (2005a) and *Belskaya et al.* (2015).
Dyn. indicates dynamical class: Cen = Centaurs, CC = cold classicals, HC = hot classicals, Plu = plutinos, Res = resonants other than plutinos, SDO = scattered, Det = detached.

from a few tenths of a kilometer to ~2380 km for the dwarf planet Pluto. The size distribution is affected by discovery and selection biases in observations, discriminating against the smallest TNOs, which are the most challenging to detect and to precisely characterize.

2.1.3. Analysis of the dynamical populations. Size and albedo values have been determined for 55 Centaurs (*Bauer et al.,* 2013; *Duffard et al.,* 2014; *Tegler et al.,* 2016; *Romanishin and Tegler,* 2018). Their sizes range from ~1 km to ~240 km. Chariklo is the largest Centaur known thus far. Most of the Centaurs are smaller than ~120 km in diameter. They typically have dark surfaces with a mean albedo of 7.4 ± 0.6%. *Duffard et al.* (2014) and *Bauer et al.* (2013) found no correlation between the size and albedo of the investigated Centaurs, and no correlation of these two physical parameters with their orbital elements.

The sizes of 44 classicals were obtained by *Vilenius et al.* (2012, 2014) from Herschel and Spitzer thermal data. Their diameters range from 130 to 930 km (Table 1), excluding the dwarf planet Makemake (D = 1430 km). Thermal measurements (Fig. 1) confirm a distinct behavior between cold and hot classicals in both their albedo values and size distributions (*Grundy et al.,* 2005; *Brucker et al.,* 2009; *Vilenius et al.,* 2014). The cold classicals have an average albedo value of 13.2 ± 1.1% (on a sample of 17 objects) and are smaller than 400 km. The hot classicals have a wider size distribution, with diameters ranging from 159 to 934 km (excluding Makemake), and a lower mean albedo (10.2 ± 1.4%, on a sample of 26 objects), excluding the ice-rich Haumea family members. The cumulative size distribution is also clearly different: The cold classicals have a size distribution with a steeper slope than the hot classicals (*Bernstein et al.,* 2004; *Petit et al.,* 2008; *Fraser et al.,* 2010, 2014; *Adams et al.,* 2014; *Vilenius et al.,* 2014). (486958) Arrokoth (previously called 2014 MU_{69}), target of the New Horizons flyby, is classified as a cold classical based on its orbit characteristics. Its shape is elongated, as it composed of two lobes with sizes of about 20 and 15 km each pushing the size lower limit of the cold classicals to a much smaller value. Its V-band geometric albedo is 0.165 ± 0.01 (see the chapter by Stern et al. in this volume).

Thirty-eight resonant TNOs were investigated in the thermal wavelengths (Fig. 1): 25 plutinos populating the 3:2 resonance with Neptune (excluding the Pluto-Charon system), 11 objects populating the resonances with Neptune outside the classical TNO belt, and 1 object, 2001 YH_{140}, populating the 3:5 resonance with Neptune, which sits inside the classical population. The size of the plutinos, with the exception of Pluto (2376 km) (*Stern et al.,* 2018), ranges from 85 km to 960 km for the largest body investigated (Orcus). The mean albedo value of the plutinos is 11.4 ± 1.3% (excluding Pluto, which has the highest albedo). From an analysis of 18 plutinos, *Mommert et al.* (2012) found no correlation between different physical parameters, except for an anti-correlation between eccentricity and diameter. However, in their sample, the fact that highly eccentric objects tend to be smaller is interpreted as a discovery bias, because objects on more eccentric orbits come closer to the Sun, therefore improving their detectability. Objects in resonance with Neptune outside the classical region are typically bigger and brighter than the plutinos, also likely due to discovery and selection biases. In this sample the size of the objects range from 157 to 1230 km for the biggest body, 2007 OR_{10} (*Kiss et al.,* 2019), and their mean albedo is 15.3 ± 2.1%.

Sizes and albedos were also obtained for 12 SDOs and 8 detached TNOs, including the dwarf planet Eris (Table 1). SDOs are smaller than 400 km (excluding 2007 OR_{10}, which is also classified as an outer resonant), and have an average albedo of 7.5 ± 1.0%. Detached objects in this sample are all larger than 250 km, which is expected based on discovery and selection biases. Their mean albedo (19.5 ± 4.7%), with the exception of Eris, is also higher than that of SDOs. The different size and albedo values for detached and SDO objects has been previously noted in the literature (*Santos-Sanz et al.,* 2012; *Stansberry et al.,* 2008). A positive correlation between the geometric albedo and the perihelion distance was reported by *Stansberry et al.* (2008) on a sample of 50 SDOs and detached objects observed with Spitzer. They explain this correlation by arguing that objects closer to the Sun may more easily lose volatiles because of sublimation, and experience a higher degree of space weathering, which in term may produce surface darkening. On a smaller sample (15 objects), *Santos-Sanz et al.* (2012) also found a strong correlation between albedo and size for SDOs/detached objects, i.e., the larger objects have higher albedo. In fact, larger bodies might more easily retain the volatiles released by intrinsic activity or collisions. In particular, they found no objects ≤300 km with an albedo higher than 10%.

2.2. Density

The bulk density is the most fundamental physical parameter containing information on the internal composition and structure of an object, and can provide insight on the origin and evolution of TNOs for planet-formation models. The density of TNOs can be derived when mass and size estimations are available. Mass can be determined quite accurately in the case of binary or multiple systems from satellite orbital properties and perturbations. Density limits may also be inferred from visible lightcurves, assuming the hydrostatic equilibrium for the body (*Sheppard et al.,* 2008). In Fig. 2 we review densities measured for 27 binary systems.

Most of the TNOs smaller than 400 km have densities less than 1 g cm^{-3}, as already noticed in the literature (*Stansberry et al.,* 2006). A few have densities close to 1 g cm^{-3}, while Borasisi and 2002 WC_{19} have much higher values with much less accuracy. These TNOs smaller than 400 km have a median density of 0.79 g cm^{-3}, slightly higher than that of the 67P comet nucleus (0.538 g cm^{-3}) (*Preusker et al.,* 2017). A density lower than that of water ice (0.945 g cm^{-3}) for TNOs smaller than 400 km implies a small rock-to-ice ratio and substantial porosity. Altjira and Typhon have the lowest density values (~0.3 g cm^{-3}). Such low bulk density requires substantial porosity (40–70%), even for material densities typical of methane and water ice (0.5–1 g cm^{-3}). *Vilenius et al.* (2014) suggested that these extremely low-density TNOs might be rubble piles of icy components with important macroporosity.

Conversely, all the TNOs larger than ~750 km in diameter have density greater than 1 g cm^{-3}, indicating a larger rock-to-ice ratio and/or less porosity. For the TNOs larger than 400 km, there is a strong correlation between density and increasing effective diameter. Mid-sized objects in the 400–800-km-diameter range (2002 UX_{25}, 2003 AZ_{84}, and 2007 UK_{126}) have density values in the 0.7–1.2 range. *Grundy et al.* (2019) proposed that these objects mark the transition between small, porous objects and the larger bodies, which have reduced macroporosity.

The fact that the density increases with object size is partially expected because of gravitational self-compaction, which reduces the macroporosity (*Brown,* 2012). However, this scenario alone cannot explain the huge variation in density observed for TNOs. In fact, as shown in Fig. 2, the density varies by a factor of ~5 between small and large bodies. If we assume a similar composition in terms of rock-to-ice ratio across the Kuiper belt, such huge density variation would require very high porosity values for the small bodies, larger than 80%. An alternative scenario envisions a larger rock-to-ice ratio for the dwarf planets. All these objects have high densities, which are reflective of different formation locations/times/processes than the smaller bodies. In this scenario, dwarf planets would have been rapidly formed by gravitational collapse in overdense regions of the protoplanetary disk (*Johansen et al.,* 2007). Giant impacts have also been proposed to explain the high density of dwarf planets and 1000-km-sized TNOs (*Brown and Schaller,* 2007; *Brown et al.,* 2007a; *Fraser et al.,* 2013). In fact, *Barr and Schwamb* (2016) noticed that among the dwarf planets, those with larger densities have small moons. They suggested that the much denser dwarf planets formed from differentiated bodies, which experienced high-velocity collisions. These high energetic impacts removed large amounts of water ice from the differentiated mantle, resulting in a dense primaries and small icy satellites.

2.3. Rotational Periods and Lightcurve Amplitude

Table 1 lists the rotational period, lightcurve amplitude, and associated references of TNOs and Centaurs. We refer

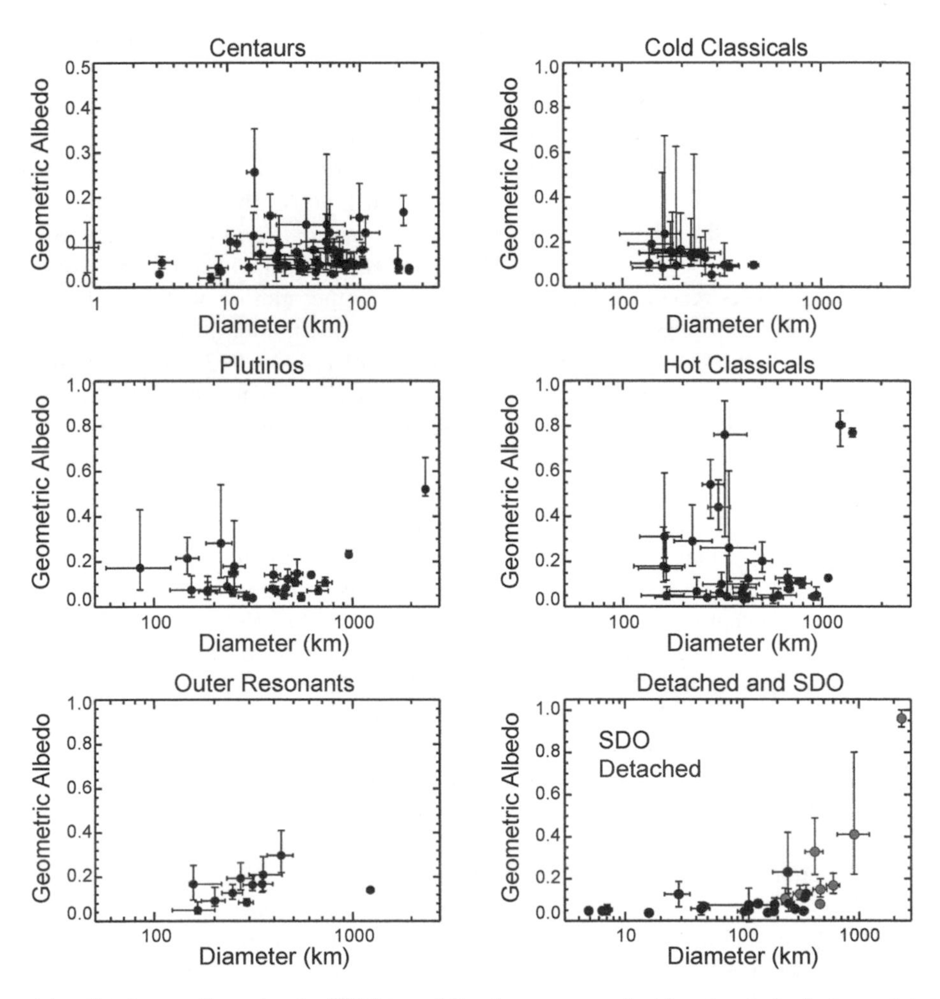

Fig. 1. Geometric albedo vs. diameter for TNOs and Centaurs separated by dynamical class. A total of 170 objects are included (see Table 1 for the individual values). Diameter is given in logarithmic scale.

the reader to *Sheppard et al.* (2008) for a review on the associated methodology. In case of single- and double-peak solutions, we indicate in Table 1 the double-peak spin rate. Although more than 100 objects have light curves reported in the literature, only about 50 have a unique solution for the rotational period.

Duffard et al. (2009) derived a mean rotational period of 7.35 hours for the TNOs and Centaurs from the analysis of 91 light curves. In a more recent analysis, *Thirouin et al.* (2016) analyzed the rotational properties of the Haumea family members, and compared them to those of the overall TNO population. They reported a mean rotational period of 7.65 ± 0.54 hours on a sample of 90 TNOs, excluding the Haumea family members, thus consistent with *Duffard et al.* (2009) findings. For objects associated with the Haumea family, they found a rotational period of 6.27 ± 1.19 hours for the confirmed members of the family (11 objects), and of 6.44 ± 1.16 hours for the other 12 family member candidates. *Thirouin et al.* (2016) thus suggest that Haumea family members may possibly rotate faster than other TNOs; however, the limited sample and the large rotational period uncertainties of Haumea family members prevent a definite conclusion.

Thirouin et al. (2016) also found that smaller members of the Haumea family rotate slower than the larger fragments. It is worth noting that Haumea, with a rotational period of 3.9155 ± 0.0001 (*Lacerda et al.,* 2008), rotates faster than any other known body larger than 100 km in diameter. This high spin rate is proposed to be the result of the impact that created its satellites, rings, and collisional family. On the other extreme, Pluto has a rotational period of 153.2935 hours (*Stern et al.,* 2015), longer than the average values of TNOs and Centaurs, because of the tidal spindown to reach synchronicity with Charon.

The amplitude of the TNO lightcurves reaches values as high as 1.1 mag for 2003 BF_{91} and 2001 QG_{298} (*Trilling and Bernstein,* 2006; *Sheppard and Jewitt,* 2004). Amplitude variations are attributed to varying projected cross-sections for elongated/non-spherical objects, albedo heterogeneities, or variations in topography across the surface of an object. It has been suggested that amplitudes <0.15 mag may be attributed to albedo variations of a body across its surface, while amplitudes >0.15 mag are more likely due to elongated objects (*Sheppard et al.,* 2008; *Thirouin et al.,* 2010). However, geometry can play a significant role in what we see from Earth, resulting in ambiguous interpretations in particular for low-amplitude lightcurves (*Lacerda and Luu,* 2006; *Benecchi et al.,* 2019). In their survey of 65 TNOs, *Alexandersen et al.* (2019) observed objects smaller than approximately 160 km (assuming a mean albedo of 10%) and found that fainter, and thus likely smaller, bodies have larger lightcurve amplitudes. They also found large brightness variations, which are important to take into account when determining their color properties.

Pluto has a lightcurve amplitude <0.3 mag (*Tholen and Buie,* 1997), and direct imaging from New Horizons provides the unambiguous attribution of these variations to its surface albedo. Very large amplitudes, combined with slow rotational periods, are often attributed to contact binaries with similar-sized components, such as 2001 QG_{298} (*Sheppard and Jewitt,* 2004). From the analysis of 12 plutino lightcurves, *Thirouin and Sheppard* (2018) found large-amplitude light curves for three objects, which they interpret as likely contact binary systems. From their study they estimate the fraction of contact binaries in the plutino population to be as large as 40–50%. A large fraction of contact binaries also exists in the cold population, and it was suggested that nearly all initial cold population TNOs were binaries (*Fraser et al.,* 2017). The discovery that Arrokoth is a contact binary (*Stern et al.,* 2019) strengthens this conclusion.

3. COMPOSITION

Transneptunian objects are challenging observation targets because they are distant, faint, and show brightness variations with their rotation (*Alexandersen et al.,* 2019). The majority of TNOs are too faint to be studied by spectroscopy, and therefore photometry was the natural first step for investigating a global view of their surface properties. In fact, accurate magnitudes and colors for a large number of objects provide information and context for more indepth techniques like spectroscopy and spacecraft flybys.

3.1. Color/Reflectance Slope Classifications

Multicolor band photometry (colors) can provide the first characterization of objects but only limited constraints on the surface composition. However, they can be used to classify objects into groups. The first to analyze color properties, *Tegler and Romanishin* (1998), analyzed the BVR colors of 16 TNOs and Centaurs and found two groups: one very red and the other almost neutral with respect to the Sun. The bimodal distribution in colors was later confirmed by *Peixinho et al.* (2003, 2012) and *Tegler et al.* (2008) for the Centaur population. With increasing numbers of objects, the bimodality has become less obvious; nevertheless, some clustering is found when considering different dynamical populations (see *Lacerda et al.,* 2014a).

Large photometric surveys (in the visible and near-infrared) of TNOs continue to provide surface colors, yielding constraints on the surface properties of a wide range of objects. About 300 objects have been observed as part of photometric surveys, revealing a broad diversity in colors. Many TNOs have surface colors significantly redder than any other solar system bodies. Statistical analyses have been employed in looking for possible correlations across colors, orbital parameters (e.g., *Doressoundiram et al.,* 2008, and references therein), and sizes, and comparing them with other similar objects like Trojans and comet nuclei (*Hainaut et al.,* 2012). Highly inclined classical objects have diverse colors ranging from gray to red and have large sizes, while the cold classical objects (low-i and low-e) are generally very red and small and thought to be primordial.

The first attempt at producing a taxonomy was made by *Barucci et al.* (2001), applying the same techniques used

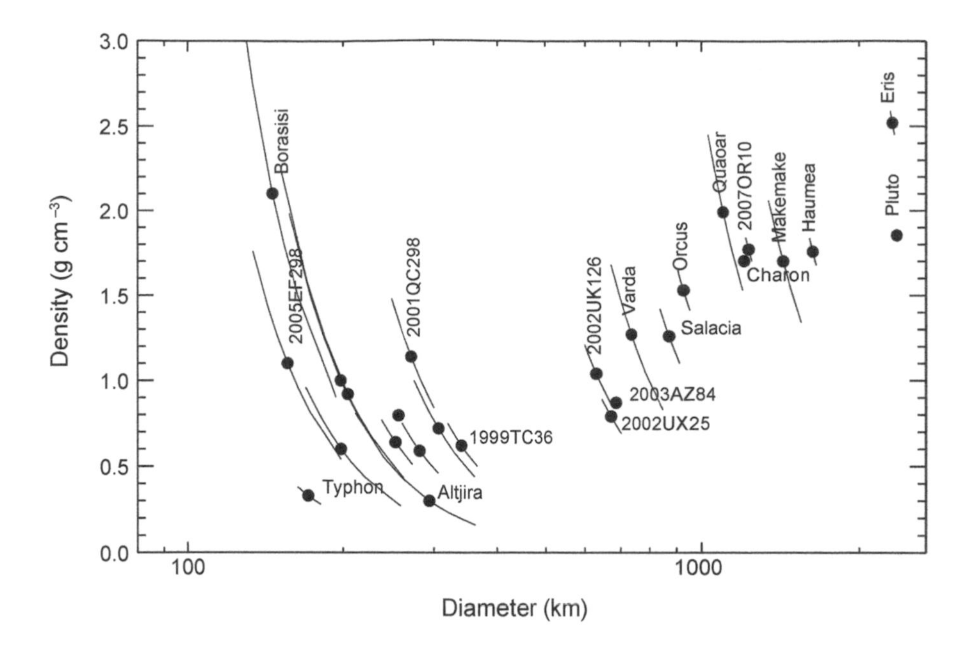

Fig. 2. Bulk density vs. the effective diameter for 25 TNOs and the Centaur Chariklo. 2002 WC19 has been excluded due to large uncertainties (3.5 ± 1.7 g cm^{-3}) (*Kovalenko et al.*, 2017, and references therein; *Brown and Butler,* 2017; *Ortiz et al.*, 2017; *Stern et al.*, 2018; *Dias-Oliveira et al.*, 2017; *Kiss et al.*, 2019; *Grundy et al.*, 2019; *Leiva et al.*, 2017; *Sicardy et al.*, 2011).

to classify asteroids on a sample of 22 TNOs and Centaurs characterized by four colors (B-V, V-R, V-I, and V-J). The analysis yielded four homogeneous classes with different surface properties. With the increase of available data, *Barucci et al.* (2005a) analyzed a sample of 51 objects, confirming the previous results at a high confidence level and underlining the importance of the near-infrared colors in the class separation. The four classes are identified by a two-letter designation (RR, RI, BR, BB), which indicates possible differences in the surface nature of these objects. The class RR (for red) comprise objects with the reddest colors of the whole population; the BB group (for blue) contains objects with neutral colors, constant with wavelength; while the two intermediate groups RI and BR include objects with intermediate color behaviors (Fig. 3).

With the increase of photometric data also at longer wavelengths, *Fulchignoni et al.* (2008) enlarged the number of objects to be classified and extended the colors to the K band (B-V, V-R, V-I, V-H, V-K), confirming the previously defined four groups. The result, which was highly statistically significant, provided a strong indication that colors highlight differences in the surface nature and are connected to their physicochemical evolution. *Fulchignoni et al.* (2008) further extended the taxonomy to 135 objects even though only three colors (B-V, V-R, V-I) were available for part of the dataset. They used two different statistical methods and studied the distribution with respect to the dynamical classes and orbital elements: RR object inclinations are low, BB objects have high inclinations, and Centaurs show a clear bimodality (BR or RR taxonomy).

An attempt to include albedo in classification scheme was made by *Dalle Ore et al.* (2013). They applied statistical clustering to a sample of 43 objects using the same colors BVRIJ and albedo data available in the literature and obtained 10 classes, of which only 6 classes were significantly populated. Unfortunately, more than half the albedos used for this classification were considerably revised by the new Herschel survey data. *Belskaya et al.* (2015), following the increase in number and quality of colors in the visible and near-infrared, again reanalyzed the available sample, extending the classification to 258 objects. They confirmed the previous classification into the BB, BR, IR, and RR taxonomic groups with a better definition and separation of the classes. One of the objectives of the analysis by Belskaya et al. was also to include albedo, obtained by Herschel, in their analysis. The analysis was applied on a sample of 49 objects for which both BVRIJ colors and albedos were available. The albedo did not show a noticeable impact on the classification other than the separation of a subgroup of high-albedo objects within the BB group associated with the largest TNOs. Furthermore, it was found that objects in the BR group had lower values of albedo compared to other groups. Overall, the low impact of the albedo values on the later taxonomy of TNOs is due to the comparatively small statistics and particularly large uncertainties in albedos.

Groundbased observations of Pluto had it originally classified as a BB object in the TNO taxonomy. In Fig. 3 one of the *Grundy et al.* (2013) spectra is plotted along with its corresponding photometric bands overlain onto the taxonomic classes, confirming its original classification. However, when one of the reddest New Horizons spectra is used, its classification changes to a hybrid case, whereas in the visible it corresponds to RR and at longer wavelengths changes to IR/BR and bluer. This is due to the strong influence of methane ice, which, with its near-infrared bands, introduces a blue slope in the spectrum.

Lacerda et al. (2014a), using Herschel and Spitzer albedos (Table 1), analyzed the visible spectral slope (measured from optical spectra or derived from broadband BR colors) and albedo for 109 objects. They found two large groups of objects corresponding to the dark-neutral (mean geometric albedo ~5%, mean visible spectral slope 10% (100 nm^{-1}) and the bright-red objects (mean geometric albedo ~15%, mean spectral slope 35% (100 nm^{-1}).

New color surveys, in particular the Colours of the Outer Solar System Origins Survey (Col-OSSOS) (*Schwamb et al.,* 2019), focus on discovering objects by performing observations on outer solar system objects brighter than magnitude 23.6. *Marsset et al.* (2019), analyzing a set of 229 colors in the visible wavelength range (combined by Col-OSSOS and other surveys) of dynamically excited orbits and Centaurs, found that the overall color distribution is bimodal with two distinct classes, gray and red.

We note that past attempts at classifying TNOs based on colors and albedo do not contradict each other. When considering only visible colors two large groups of objects are well separated (corresponding to BR and RR classes as in the Barucci et al. taxonomy). More classes are identified when near-infrared colors are included in the analysis.

3.2. Correlations

Lacerda et al. (2014a) also investigated the distribution in the albedo-slope space of the different dynamical classes. They found that cold classicals, detached, and outer resonant objects all belong to the bright, red group, while the other dynamical classes (Centaurs, plutinos, hot classicals, and SDOs) have bodies in both groups, thus showing a variability both in terms of albedo and slope/colors. This supports the scenario that cold classical, detached, and outer resonant objects probably originated far from the Sun (*Batygin et al.,* 2011; *Gladmann et al.,* 2002; *Malhotra,* 1995). Their surfaces are therefore representative of outer solar system planetesimals: They are redder in color because they formed far from the Sun, beyond the stability line of methanol and/or hydrogen sulfide, which are strong reddening agents (*Brown et al.,* 2011; *Fraser and Brown,* 2012; *Wong and Brown,* 2016). Conversely, variability in terms of surface colors and albedo is observed among dynamically excited TNOs such as hot classicals, plutinos, Centaurs, and SDOs, which likely formed across a wider range of heliocentric distances, from 20 AU to about 48 AU (*Petit et al.,* 2011), and which also in some cases (e.g., Centaurs) now are back closer to the Sun and potentially evolved.

Particularly striking are the differences between the hot and cold classicals populations. They are distinct in term of their colors/spectral slopes, the cold objects being generally dominated by red colors. Cold classicals have also a steeper slope in the size distribution than the hot population (*Petit et al.,* 2011). The number of binary systems is three times larger among the cold classical population than in the hot population (30% vs. 10%, respectively) (*Noll et al.,* 2008; *Grundy et al.,* 2011). All these differences support the idea that these two populations formed in different regions. The hot classical objects formed closer to the Sun then were later injected into the TNO population through interactions with the giant planets during the planetary migration, while the cold classicals presumably formed *in situ.*

Photometric results looking at binary TNOs observed by Hubble (*Benecchi et al.,* 2009) demonstrate a correlation between primary and satellite colors. The colors of binaries as a group are indistinguishable from that of a large population of apparently single TNOs. They concluded that the colors of TNO population could be primordial and characteristic of the compositional diversity present in the solar nebula.

Other than the unique color properties of the cold classicals and Centaurs, local conditions appear to have no primary influence on the TNOs' surface colors. No other correlations have been found with the dynamical properties (*Barucci et al.,* 2011). With Hubble observations in the near-infrared region, *Fraser and Brown* (2012) found a bimodality among small TNOs with perihelia <35 AU. On the basis of the first Col-OSSOS data, *Pike et al.* (2017) discriminate three different surface groups: red cold classical, dynamical excited red TNOs, and dynamical excited neutral TNOs. *Fraser et al.* (2017) identified some blue-colored objects in the cold classicals, all of them binaries. They suggested that blue binaries are contaminants that were pushed out during Neptune's migration on to cold classical orbits. *Marsset et al.* (2019) showed that the red objects have a lower orbital inclination distribution than their gray counterparts. This was interpreted in favor of the hypothesis of different formation locations for these objects, and a compositional gradient in the original protodisk of planetesimals rather than different evolutionary pathways.

Many controversies still exist on the nature of TNOs. Variations in the nature of the surface of TNOs might also be due to different exposure to the various alteration processes, and the objects might be at different stages of evolution. In fact, the taxonomic groups can be interpreted in terms of evolution of the TNO population and/or in terms of different original composition. Another hypothesis on color variation has been suggested by *Brown et al.* (2011) and consists of early evaporation and irradiation of volatiles followed by dynamical mixing. The objects formed in the early solar system in the region at 15–30 AU, at the time outside Neptune's orbit, were later exposed to sunlight, which drove off volatiles. After being irradiated the objects were then dispersed and mixed.

The color interpretation is still a matter of debate, but to constrain the composition and the evolution of these objects only by color variation remains very limiting.

3.3. Spectral Properties

Even though spectra provide information only about the first thin layer of the surface, observations at wavelengths

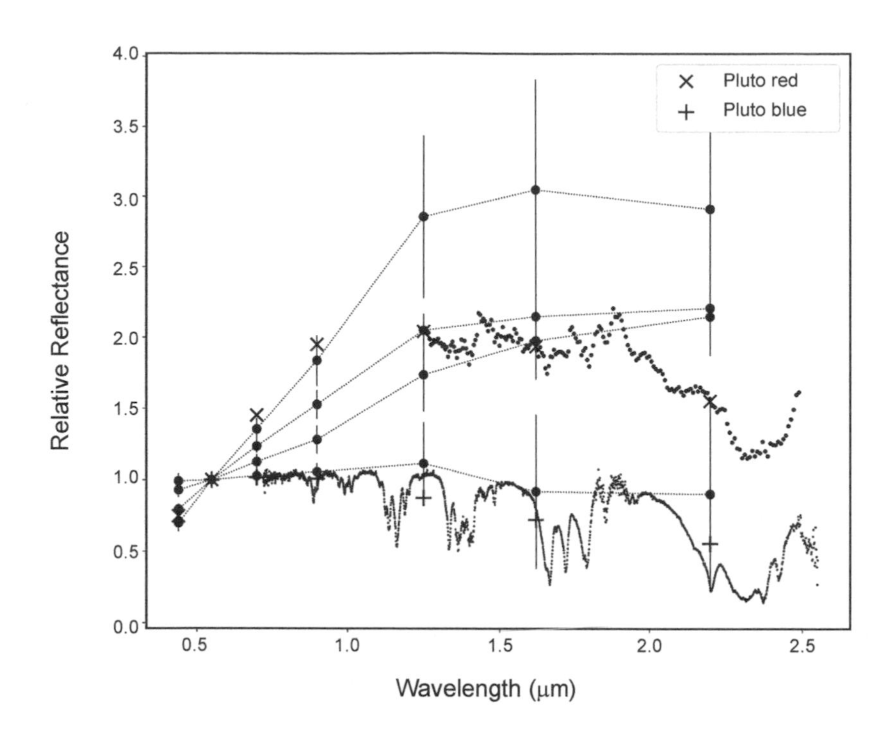

Fig. 3. Average reflectance of the four taxonomical classes (BB, BR, IR, and RR from bottom to the top) with the relative error bars from *Perna et al.* (2010). Two Pluto spectra (normalized at 0.55 μm) are compared to the TNO taxonomy. The bottom spectrum is a groundbased observation of Pluto from *Grundy et al.* (2013), which, according to the original classification of Pluto, is part of the BB class. The top spectrum is an average of New Horizons' darkest areas mostly from the dark equatorial regions (informally called Cthulhu).

between 0.4 and 2.5 μm provide the best technique available at the present time from the ground to characterize the main minerals, ices, and organics present on their surfaces. To obtain a good interpretation, the spectra have to include diagnostic features. Generally, TNOs are too faint for groundbased spectroscopic observations even with the world's largest telescopes such as Keck, Subaru, Gemini, and and the Very Large Telescope (VLT), and only the brightest objects have been observed thus far.

Several objects have been studied with space telescopes. A few authors (*Noll et al.,* 2000; *Kern et al.,* 2000) observed TNOs and Centaurs, with Hubble obtaining near-infrared spectra or broadband photometry while a few other objects have been observed in the far-infrared with the Spitzer telescope (*Barucci et al.,* 2008).

The mid-infrared in the spectral range 2.5–40 μm is also important in studying the fundamental vibrations and associated rotational-vibrational structures of most molecules. These include H_2O and CH_3OH ices, which show features in this wavelength range that are also useful in constraining, in addition to composition, grain size. However, TNOs are too faint for this type of observation and only a few Trojan asteroids (*Emery et al.,* 2006a) and bright Centaurs (*Barucci et al.,* 2008) were observed by Spitzer, showing emissivity peaks at around 10 and 20 μm, interpreted as the presence of silicates with small grain sizes. The broadband reflectivity has been measured for about 30 objects at 3.55 and 4.5 μm and for a few of them at 5.8 μm as well (*Emery et al.,* 2006b). These data were used as a proxy to extend near-infrared spectra obtained from the ground to provide additional constraints in discriminating the presence of different ices on the surface (*Emery et al.,* 2007; *Dalle Ore and Morea et al.,* 2015; *Barucci et al.,* 2015).

To investigate the composition of a planetary body, spectral data at sufficiently high S/N and knowledge of its albedo are necessary to properly constrain radiative transfer modeling. The presence of absorption features allows detection of different compounds on the surface, determination of their pure or diluted state, and the temperature of ices. The knowledge of albedo at a given wavelength will also be necessary to constrain the amount of ices, organics, carbons, and silicates, which have different reflectivity. For the modeling (for details, see *Barucci et al.,* 2008), knowledge of optical constant is needed to generate synthetic spectra making use of different types of mixtures like intimate, geographical, and molecular — such as the Maxwell-Garnett approach (*Maxwell Garnett,* 1904; *Hage and Greenberg,* 1990) — based on diffusive medium theory.

Good spectra in the visible and near-infrared are available for a limited number of objects and only about 50 of them (TNOs and Centaurs) show clear or possible absorption features. Others show almost featureless spectra, which make any attempt at deriving surface composition particularly challenging.

The visible part of the spectrum is generally featureless but can show a gradient in reflectivity with different degrees of variation with wavelength. As discussed above, it can be flat neutral to very red, making some of these objects among the reddest in the solar system. Red spectral slopes can be associated with the presence of organic or other irradiated compounds. In this spectral region a few faint bands associated with the presence of aqueously altered minerals such as phyllosilicates have been detected on a few objects along with CH_4. The aqueously altered silicates have been reported by several authors (*de Bergh et al.,* 2004; *Alvarez-Cantal et al.,* 2008) with bands at 0.60, 0.70–0.74, and 0.82–0.83 μm but were never confirmed when reobserved. The difficulty in detecting shallow features is often connected with the quality and noise of the spectra. The presence of aqueously altered silicates might seem surprising, but the possibility of their presence is particularly interesting as hydrous materials seem to be present in debris disks, interplanetary dust particles collected in Earth's stratosphere (see, e.g., *Mackinnon and Rietmeijer,* 1987). *Rietmeijer and Nuth* (2001) have suggested that serpentine could in fact form in the solar nebula by the hydrogenation of very small Fe-rich silicate grains condensing from the gas phase.

The study of wavelength shifts of the absorption methane band has been used to detect the presence of methane diluted in nitrogen and studied by several authors. *Licandro et al.* (2006b) studied the shift of the strong methane band around 0.89 μm for Eris. *Tegler et al.* (2007) observed the shift of the CH_4 bands in the visible regions for Makemake and interpreted it as the possible presence of N_2 ice, CO, or Ar. The shift of the bands on Makemake was later confirmed by *Lorenzi et al.* (2015) and *Perna et al.* (2017). Laboratory experiments (*Quirico and Schmitt,* 1997) confirmed that a blueward spectral shift indicates that methane is in solid solution with nitrogen.

Near-infrared observations in the range 1–2.4 μm are the most diagnostic from groundbased observations for detecting the presence of ices and volatiles. Due to their brightness Pluto and Triton (possibly a captured TNO) (*Brown et al.,* 1995; *Cruikshank et al.,* 1997) were the first objects to show clear absorption features due to CH_4, N_2, and CO (Fig. 4). The first TNO spectra were exceedingly noisy, but as soon as large objects were discovered, the quality increased, particularly when the 8–10-m-class telescopes are used. The presence of CH_4 was easily identified on other large TNOs (*Brown et al.,* 2005; *Barucci et al.,* 2005b; *Licandro et al.,* 2006a).

About 80 objects have been observed in total, even if some of these objects remain at the limit of observability by the available ground instrumentations. More than half of the observed objects show signatures of ices on their surfaces. H_2O ice is present at 1.5, 165, and 2.0 μm depending on its phase (amorphous or crystalline); CH_4 at 1.7 and 2.3 μm; NH_3 at 2.0 and 2.22 μm; N_2 at 2.21 μm; and CH_3OH at 2.27 μm. Other minor species have been reported, but their presence requires confirmation. The presence of N_2 and CO is very difficult to detect due to the weaker absorption bands of these species. The only object (after the detection on Pluto and Triton) showing the signature of N_2 in their spectra is Sedna. The relative abundance of volatile content varies clearly on the surface of these objects and changes the albedo values. One explanation for this is volatile loss and retention as a function of size and temperature/location (*Schaller and Brown,* 2007).

Following their spectral characterization, the TNO population has been grouped as a function of the content of their surface components as follows.

3.3.1. Water-ice-rich objects. Water ice is thought to be an important and widespread component of TNOs and indeed many of them show spectra with moderate to deep absorption bands at 1.5 and 2.2 μm. The majority of objects belonging to the BB taxonomical class seem to have H_2O-ice-rich surfaces. Most spectra are of poor quality, but those with a sufficiently good S/N also show the band at 1.65 μm, characteristic of crystalline water ice. Knowledge of the H_2O ice phase gives important constraints on the formation and evolution of these objects as the presence of ice in crystalline form implies that it has been heated to temperatures above 100 K. This heating could have originated from impacts, generated in the deep interiors, or during the solar nebula formation phase. Crystallization can occur on a mantle of water ice and amorphous ice can still survive after collisions (*Steckloff and Sarid,* 2019).

Barucci et al. (2011) and *Brown et al.* (2012) analyzed a large collection of high-quality spectra obtained at VLT and Keck, respectively. They identified the presence of water ice with the measurement of the depth of the 2-μm absorption feature. The results show that objects with abundant water-ice content (depth of the band >20%) have a smaller absolute magnitude, which corresponds to larger-sized objects. While water ice is present in all dynamical classes of objects, no objects (except in the Haumea family) smaller than about 500 km in diameter have been found with strong water-ice absorption bands.

(136108) Haumea is a unique dwarf planet boasting a system of satellites, rings, and family of objects all with almost pure water-ice surfaces. Although, following the model by *Schaller and Brown* (2007), its large size would allow the presence of abundant volatiles, its surface is instead dominated by almost pure water ice. Its moons and its dynamical family objects follow the same behavior. Unlike other members of its family and the satellites, Haumea is large enough to be differentiated, i.e., to have an icy layer surrounding a rocky body. The mutual orbital perturbation of the two satellites allows the estimate of its density of 1.9 g cm^{-3} (*Ortiz et al.,* 2017). *Brown et al.* (2007a) suggested that Haumea and its family are the product of a giant impact onto a larger proto-Haumea. The fact that the surface of these bodies remains still uncontaminated by dust, dark impactors, or irradiated hydrocarbons is still a puzzle. In addition, all spectra show the band of crystalline water ice, which is also surprising as irradiation should transform crystalline water ice into amorphous ice on a short timescale, according to laboratory experiments (*Dartois et al.,* 2015). It is clear that the physics of crystallization on Haumea still needs to be thoroughly understood. Cryovolcanism is a possibility but, as far as the smaller objects are concerned, its presence is unlikely.

With the New Horizons flyby of Pluto a detection of water ice on Pluto was confirmed (*Cook et al.,* 2019b) in spite of the heavy contamination of CH_4 ice. On Pluto water ice is considered to be the bedrock surface on which volatile ices deposit and recondense after diurnal and seasonal migration. Water ice mixed with ammonia and a red colorant was also found in a few craters and troughs in the region informally known as Cthulhu and is associated with cryovolcanism on Pluto (*Cruikshank et al.,* 2019; *Dalle Ore et al.,* 2019).

3.3.2. Methane-rich objects. Methane has strong absorption bands in the near-infrared that make it an easy detection. Similar to Pluto and Triton, the largest TNOs, i.e., Eris, Makemake, and Sedna, have spectra dominated by methane (Fig. 4). Some objects show methane dissolved in nitrogen. The presence of CH_4 on these large objects is possible because their large mass, and therefore high gravity allows retention of such a volatile. However, smaller quantities of CH_4 can also be present on smaller objects, unlike N_2 and CO, which have only weak features and are harder to detect. In all these large objects the presence of N_2 has also been inferred or recognized in the small wavelength shifts of visible and/or near-infrared spectral bands. Quaoar has a low surface abundance of CH_4 on a water ice substrate and only the strongest features of CH_4 are detected (*Dalle Ore et al.,* 2009).

3.3.3. Methanol-rich objects. The presence of a methanol signature is considered an indication of a chemically primitive surface and this molecule is an abundant component of active comets and of the interstellar medium. The Centaur Pholus, one of the reddest objects in the solar system, was the first object for which the absorption band at 2.27 μm typical of methanol was detected (*Cruikshank et al.,* 1998). The plutino (55638) 2002 VE_{95} also shows a typical signature of CH_3OH and/or a photolytic product of methanol (*Barucci et al.,* 2012). Many other objects seem to have spectra with a reduced albedo beyond 2.2 μm, suggesting the possible presence of CH_3OH (or similar molecule), but the spectral quality is too low for confirmation. Pholus, 2002 VE_{95}, and Sedna also show very red spectral slopes in the visible, which is in agreement with the picture of irradiated hydrocarbons. *Dalle Ore and Morea et al.* (2015) analyzed the composition of ultrared objects (following the taxonomic class RR) using available colors up to 4.5 μm (obtained with Spitzer) and inferred that methanol/hydrocarbon ices can be present in the surface of ultrared objects. Small TNOs are too faint for spectroscopic work from Earth, so methanol has been discovered so far only on tiny Arrokoth, the 2019 flyby target of the New Horizons mission.

3.3.4. Ammonia-trace objects. Ammonia has a band at 2.22 μm and was first detected on Charon (*Brown and Calvin,* 2000; *Merlin et al.,* 2010), later on Orcus (*Barucci et al.,* 2008), and recently on Pluto from New Horizons observations (*Dalle Ore et al.,* 2019). In all cases the presence of ammonia has been suggested as a flow of ammonia-rich interior liquid water onto the surface. The detection of ammonia is very important for the geological study of TNOs, particularly in investigating the possibility of cryovolcanism.

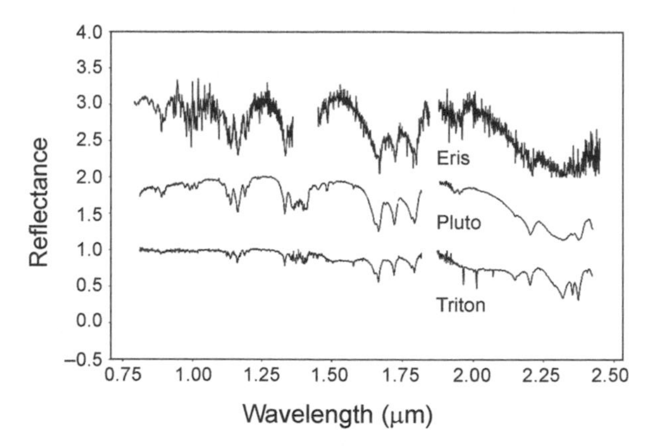

Fig. 4. A selection of groundbased observations. Eris as observed by *Merlin et al.* (2009) compared to Pluto (*Grundy et al.,* 2013), taken at longitude 180° and latitude 44°, and Triton as observed by *Grundy et al.* (2010) at longitude 60° and latitude –47°.

3.3.5. Featureless-spectrum objects. Many objects display featureless spectra in the visible and near-infrared, with a wide range of colors. Some spectra are featureless as a result of low S/N; however, many could be genuinely featureless if their surface is rich in organic material and/or carbon. Irradiation processes are thought to be responsible for the coloration of objects as a result of dehydrogenation of hydrocarbon or carbon-containing ices such as methanol. In the case of red spectra and moderately high albedo, the spectral behavior is often explained by the presence of tholins or kerogen as shown on the basis of laboratory experiments.

3.3.6. Irradiation processes. All surfaces in the solar system are modified by different kinds of irradiation, or space weathering, such as solar wind, ultraviolet photons, and cosmic rays. The effect of radiation changes as a function of heliocentric distance and the reactions are also different as a function of the chemical composition of the surfaces. *Hudson et al.* (2008) estimated the irradiation at different distances from the Sun with different energies. This process forms a crust, which hides the real composition of the objects. Complex materials like tholins (produced by spark discharge in low-pressure gas) or kerogen are often used as analogs to interpret spectra. Tholins have been associated with extended irradiation of hydrocarbons and nitrogen-containing components. Irradiation changes the visible slope of the spectra, affects the albedo, and can remove or modify spectral bands (*Brunetto et al.,* 2006). Irradiation of CH_4 (*Bennett et al.,* 2006) can yield C_2H_6 (ethane), C_2H_2 (acetylene), and C_2H_4 (ethylene). Laboratory experiments also show amorphization of the crystalline water ice from ion irradiation, especially at very low temperatures (*Mastrapa and Brown,* 2006).

3.3.7. Silicates and aqueous alteration materials. Silicates have not been clearly detected on TNO surfaces although they could be present in addition to ices and their irradiated products. Silicates like olivine and pyroxenes have often been included in models for these objects (*Barucci et al.*, 2011) even if their specific features are not detected in the wavelength range 0.4–2.4 μm. However, a positive detection of silicate emission was been identified at 10 and 20 μm in the mid-infrared of the centaur Asbolus (*Barucci et al.*, 2008).

Aqueous alteration is a low-temperature chemical alteration of materials that takes place in the presence of liquid water. How aqueous alteration could have occurred at such low temperatures far from the Sun is not well understood, but it cannot be excluded that hydrated minerals could have been formed directly in the early solar nebula. Aqueously altered materials seem to be present on some TNOs and some of their bands have been detected on visible spectra at 0.6, 0.70–0.74, and 0.82–0.83 μm (*de Bergh et al.*, 2004; *Alvarez-Cantal et al.*, 2008; *Fornasier et al.*, 2009).

3.3.8. Resurfacing processes. Collisions have played an important role in the evolution of TNOs and different regimes of collisions (superficial or deep impact) can change the resurfacing process. The consequence of collisions is not only the alteration of the surface properties but also the modification of the internal structure of the targets. Activity induced by intrinsic processes, such as a tenuous atmosphere or outgassing from the interior, is also important and is presumed to play a role in the resurfacing of TNOs. Finally, the internal release (like cryovolcanism) should also be taken into account (*McKinnon et al.*, 2008) as a possible mechanism that can resurface and mask surface composition.

4. THE LARGEST OBJECTS

The largest known objects have been studied the most extensively, as they are in general brighter than the rest of the TNO population. Three of them (Eris, Pluto, and Makemake) fall into the official category of dwarf planets.

(136472) Makemake has one known satellite, provisionally designated S/2015 (136472) 1. It is the largest in the classical dynamical population, the brightest TNO after Pluto, and consequently easily studied. The methane absorption bands in the spectra surface of Makemake are deeper and broader than those of the other TNOs. Its spectral signature was modeled with methane in large grains (about 1 cm), with ethane (a product of UV photolysis of methane) and an irradiated tholin-like material (*Brown et al.*, 2007b). In 2015 Brown et al. obtained high signal-to-noise (S/N) spectra in the range 1.4–2.5 μm and confirmed — on 80% of the surface — clear evidence of methane and ethane, and suggested the presence of ethylene, acetylene, and high-mass alkanes, all expected products of the continued irradiation of methane. It is impossible to precisely identify the quantity of these hydrocarbon products from spectra, even though spectral models imply a few percent contribution to the surface composition from these irradiation products. A possible explication of the abundance of methane can be linked to Makemake's high eccentric orbit, which allows recycling and refreshing of the surface on orbital timescales (309 years, more than Pluto's 248 years and Haumea's 283 years). Methane destroyed by irradiation becomes a dark residue while fresh methane migrates seasonally, as has been closely observed on Pluto.

(136199) Eris is the most massive object ($1.67 \pm 0.02 \times 10^{22}$ kg) (*Brown and Schaller*, 2007) of the TNO population with a mass ~1.27× that of Pluto. It belongs to the scattered dynamical objects with a highly eccentric orbit (perihelion at 37.9, aphelion at 97.6). Its orbital period is 558 years and it is the second largest dwarf planet. Its satellite Dysnomia is 500× fainter than Eris. The visible and near-infrared spectra are dominated by absorptions due to methane. The spectra are very similar to those of Pluto (Fig. 4), even though in the visible its spectrum is less red than that of Pluto. The weak N_2 band at 2.15 μm is not detected. However, the presence of N_2 is inferred as a consequence of the spectral band shifts of CH_4, as *Merlin et al.* (2009) suggested and as had been proposed by *Licandro et al.* (2006b) and *Brown et al.* (2005). The fact that the band of N_2 is not detectable is probably due to the fact that at Eris' temperatures N_2 ice is in the α-phase and not in the β-phase as observed on Triton or Pluto (the transition between the α- and β-phases occurs near 35.6 K). The quantity of nitrogen on Eris' surface could also be less dominant than on Triton's and Pluto's surfaces. Spectral models of Eris' surface suggest the presence of large and small particles of methane ice and a possible stratification of methane ice diluted in N_2, with a layer of diluted methane ice between two layers of pure ice. The presence of other volatiles is also a possibility. The high albedo and the lack of compositional rotational variation could imply a surface dominated by a seasonal cycling atmosphere. Eris just passed aphelion and, as noted by *Brown et al.* (2005), CH_4 at ~30 K is practically non-volatile. Therefore, at this time, no CH_4 would be expected to be exchanged between Eris' N_2 atmosphere and the surface; CH_4 is inferred to be segregated from N_2 into layers. As Eris moves toward perihelion, which will be reached around the year 2257, its atmosphere may become more Pluto-like. The temperature excursion between aphelion and perihelion for Eris is more extreme than that of Pluto, providing a good test for atmospheric studies. Monitoring of the behaviors of the atmosphere of the two dwarf planets will therefore be fundamental in understanding the processes that govern atmospheric phenomena at extremely low temperatures, such as "freeze-out," N_2 phase transition, volatile mixing and migration, and ice chemistry in general.

(50000) Quaoar is dynamically a classical TNO with a small satellite, Waywot. The multi-chord stellar occultation by *Braga-Ribas et al.* (2013) measured an effective diameter of 1111 ± 5 km and a density of 1.99 ± 0.46 g cm^{-3}. *Fraser et al.* (2013) detected a higher value of 2.7–5.0 g cm^{-3} from measurements of the Quaoar-satellite orbit. Quaoar has been widely observed and spectral modeling by *Dalle Ore et al.* (2009) portrays a surface containing crystalline water ice, CH_4, and possibly C_2H_6. Further model interpretation of the highest-quality spectra available, including Spitzer data (*Barucci et al.*, 2015), confirmed the presence of CH_4 and C_2H_6, as previously reported, as well as measurements of crystalline H_2O ice in abundance over the amorphous H_2O ice.

An indication of the possible presence of $NH_3 \cdot H_2O$ is also reported with the new evidence of N_2 from a shift in the CH_4 bands. The albedo at the two Spitzer bands suggests there may also be CO diluted in N_2, along with CO_2 in one part of the surface. The composition detected on the surface of Quaoar may suggest the presence of an efficient renewal mechanism. Moreover, the presence of the fragile $NH_3 \cdot H_2O$ ice on its surface could imply that Quaoar's surface might be relatively young. Quaoar shows more similarities than differences when compared with Pluto. In particular, the fact that CO_2 is only present on Quaoar at particular longitudes (as Triton) may provide clues about Pluto where this ice is still undetected.

(90377) Sedna is one of the reddest objects among the most distant known and dynamically belongs to the Detached category. It has a perihelion at 76 AU and aphelion at 956 with an orbital period of 11,704 years. Its spectra show a heterogenic surface composition. From the first set of observations, a similarity with Triton was suggested (*Barucci et al.,* 2005b) in particular based on the presence of CH_4 and N_2 on its surface. A temporary N_2 atmosphere was also inferred. More recently, further observations and modeling by *Barucci et al.* (2010), including Spitzer photometric data, support large heterogeneity. Spectral modeling attributed to the surface organic materials (Triton and Titan tholin), serpentine, H_2O, CH_4, N_2, and C_2H_6 in varying trace amounts making this dwarf planet very different from Pluto. As such, it is an excellent candidate for future observations and monitoring of atmosphere and surface. In fact, based on Pluto's experience, the detection of H_2O ice and minerals is challenging, as volatile migration tends to cover the stable surface constituents.

5. (134340) PLUTO

A large amount of data exists for Pluto and it has given us an opportunity to study this object as an individual entity, by analyzing its different terrains and their variations over time, and as a member of the outer solar system, by comparing it to other TNOs.

Furthermore, Pluto has given us a chance to evaluate the quality and limitations of the observations obtained from the different observing platforms and the ability to tune groundbased instruments by identifying the best wavelengths to target for particular kinds of observations. Pluto has given us full appreciation of the value of photometric data and highlighted the best way to make use of them. In the following sections we will outline the main points in Pluto's history with a view to its geological, physical, and chemical evolution framed in the context of TNOs.

5.1. Density Determination

Null et al. (1993) were the first to measure the densities of Pluto and Charon independently. The measurements were done using the Hubble widefield CCD camera for seven independent observations made over 3.2 days in order to capture Pluto's barycentric motion and deduce masses and densities for Pluto and Charon. They obtained for the first time the Charon/Pluto mass ratio as well as the individual masses MP = $(13.10 \pm 0.24) \times 10^{24}$ g and MC = $(1.10 \pm 0.18) \times 10^{24}$ g for Pluto and Charon respectively. Adopting radius measurements from *Tholen and Buie* (1990), they obtained a density $\rho P = 2.13 \pm 0.04$ g cm^{-3} and $\rho C = 1.30 \pm 0.23$ g cm^{-3}. Measurements from stellar occultations (*Elliot and Young,* 1992) yielded very different results but always found Pluto's density to be significantly higher than Charon's.

The assessment was repeated by *Foust et al.* (1997), who made use of the offset of the center of light of the blended image of Charon and Pluto with respect to the ephemeris of the system. They modeled the offsets and found a mean mass ratio of 0.117 ± 0.006, coming very close to the new value of 0.124 ± 0.008 by *Null and Owen* (1996). From the mass ratio they deduced masses of $\sim 1.32 \pm 0.01 \times 10^{25}$ g for Pluto and $1.5 \pm 0.1 \times 10^{24}$ g for Charon. From these they found a density of 1.94 ± 0.12 g cm^{-3} for Pluto and 1.5 ± 0.2 g cm^{-3} for Charon, which were in general agreement with one of the estimates of *Elliot and Young* (1992) based on stellar occultations and adopting a "haze model."

Olkin et al. (2003) made use of the fine guidance sensors on the Hubble Space Telescope (HST), an astrometric instrument, to obtain a more precise ratio. The observations, spanning 4.4 days, yielded a ratio of 0.122 ± 0.008, corresponding to a Pluto density of 1.8–2.1 g cm^{-3} depending on the adopted radius value.

Finally, New Horizons *in situ* measurement of Pluto and Charon's radii (1188 ± 4 km and 660 ± 3 km respectively) provided the most precise measurement of the density of the two bodies: 1.860 ± 13 g m^{-3} for Pluto and 1.702 ± 21 g m^{-3} for Charon (*Stern et al.,* 2018). When looking at Fig. 2, Pluto is one of the TNOs with the highest densities and largest radii, supporting the correlation between density and size. The precise measurement of Pluto's density is essential in determining its bulk composition, which according to calculations by *McKinnon et al.* (2017) amounts to a rock-to-ice ratio of 2:1. This is in agreement with the general behavior of larger TNOs having higher densities and lower macroporosity as discussed in section 2.2.

5.2. Geology and Albedo Maps

Stern et al. (1997) were the first to describe features on the surface of Pluto, making use of HST images of Pluto at 410 and 278 nm. They reported bright polar regions, a darker equatorial area interrupted by four bright subunits and with the darkest part corresponding roughly with the sub-Charon point and interrupted by the "cleanest ice" at longitude approximately 180°, i.e., the anti-Charon point. They further announced the brightness of both poles, with the northern one being brighter and larger. With the exception of the southern terrains, which were in darkness, and the farside, which was imaged only at "albedo" scales, these surface features were unveiled in their entirety by the 2015 New Horizons flyby.

Buie et al. (2010a) announced the detection of temporal changes on Pluto's surface. This was supported by observations made with the high-resolution camera on HST between June 2002 and June 2003. The observations consisted of 12 distinct exposures at sub-Earth longitudes over a range of solar phase angles not to exceed 2°. The resulting light curve showed a decrease in amplitude with respect to that obtained in the late 1980s. Analogous measurements of Charon do not show the same behavior, where Charon's light curve is practically the same as that recorded in the early 1990s. Variations in two of the passbands also implied substantial reddening of Pluto's surface-reflected light in the recent years that cannot be explained by changes in viewing geometry and therefore support the idea that Pluto's surface is evolving over time. The evolution is thought to be linked to volatile transport processes. While this concept could not be directly confirmed by New Horizons observations, a snapshot at a certain point in time, the surface distribution of ices and their relative volatility with respect to their geographical location amply confirmed this first interpretation. *Bertrand et al.* (2018) were able to model the New Horizons observations with the aid of a three-dimensional global climate model (GCM). The model incorporated atmospheric dynamics and transport, turbulence, radiative transfer, and molecular conduction, as well as phase changes for N_2, CH_2, and CO. With the assumption of a topographic depression at Sputnik Planitia the model reproduces, at least in a qualitative way, the distribution of ice, the temperature profile in the atmosphere, and the surface pressure variations measured with occultations between 1988 and 2015.

Buie et al. (2010b) made use of the same dataset as in *Buie et al.* (2010a) along with the reprocessed Hubble Faint Object Camera images taken in 1994 (*Stern et al.,* 1997), which covered UV, blue, and visible wavelengths to produce two color-albedo maps of Pluto and Charon. The Pluto maps confirmed the wide color and albedo variations with longitude first reported by *Stern et al.* (1997). They reported a neutrally colored higher-albedo northern hemisphere and darker regions with a more remarkable color diversity implying compositionally more varied terrains.

Figure 5 shows a direct comparison between the map obtained by *Buie et al.* (2010b) and that from New Horizons data. The latter is a mosaic (adapted from Fig. 1 of *Schenk et al.,* 2018) consisting of both Long Range Reconnaissance Imager (LORRI) and Multispectral Visible Imaging Camera (MVIC) panchromatic imaging ranging in pixel scale from 234 m/pixel to 35 km/pixel, but projected at 300 m/pixel. It is understood that regions extending downward from the east-west boundary between the illuminated and unilluminated areas are brightened by light scattered by atmospheric haze. They are processed to have brightness properties similar to areas otherwise illuminated by sunlight. Enhanced colors in the map are the product of MVIC filters blue, red, and CH_4, chosen to provide the greatest color contrast.

When comparing the two maps the obvious correspondence between the icy and bright Sputnik Planitia in the New Horizons mosaic and the bright patch in the *Buie et al.* (2010b) map is an impressive demonstration of the high quality of the Hubble data, but also a sobering reminder that only a spacecraft can at this time deliver the level of detail revealed by the New Horizons data in the overall encounter hemisphere. It is also intriguing to point out that the second very bright patch in the *Buie et al.* (2010b) map located at longitude 270° and latitude about –45° does not have a corresponding feature in the New Horizons mosaic, as it falls beyond the encounter hemisphere horizon. At this time no interpretation has been given to the feature and it might take another mission to Pluto to unveil its true nature.

5.3. Composition

In the ~25 years preceding New Horizons, advances in telescope technology and investigations through the Hubble Space Telescope allowed a series of discoveries about the Pluto system that were confirmed during the flyby. After the initial discoveries of CH_4 frost on the surface of Pluto by *Cruikshank et al.* (1976), and of CO and N_2 by *Owen et al.* (1993), the first spectroscopic measurements of Pluto's tenuous atmosphere were given by *Young et al.* (1992). On the eve of the New Horizons flyby, a few authors gave new constraints (e.g., *Grundy et al.,* 2013; *Merlin,* 2015) on the volatile-ices distribution on the surface of the dwarf planet.

The presence and distribution of these very volatile ices was confirmed at once when the first Lisa Hardaway Infrared Mapping Spectrometer (LEISA) and MVIC CH_4 band imagery were returned by the spacecraft. LEISA is a short-wavelength, infrared, spectral imager (*Reuter et al.,* 2008) covering the wavelength region between 1.25 and 2.5 μm with a spectral resolving power ($\lambda/\Delta\lambda$) of 240. While the spatial resolution of the first images to reach the ground was not as impressive as those to follow in the next few months, it was sufficient to allow detection and distribution mappings of all three ices (*Stern et al.,* 2015). Furthermore, the geology of the surface clearly evident in the LORRI images hinted at the presence of a H_2O bedrock supporting the mountains in the vicinity of Sputnik Planitia (*Stern et al.,* 2015).

H_2O was spectroscopically detected in the LEISA scans by *Cook et al.* (2019b). Applying a technique based on the comparison of a template spectrum with the observed pixels, Cook was also able to determine the likelihood of the presence of minor species such as CH_3OH, C_2H_6, and minor hydrocarbon species.

In their analysis of the LEISA closest-approach scans, *Grundy et al.* (2016) and *Protopapa et al.* (2017) highlighted the widely varying spectral signature of the various terrains that characterize Pluto's surface at the time of the 2015 flyby. One of the unexpected results of the encounter was indeed the extreme diversity of Pluto's surface, past *Buie's* (2010b) somewhat anticipated predictions shown in the Fig. 5a. However, the extreme spectral variations on Pluto's surface could only be revealed by the spectral imaging instruments onboard New Horizons. The right panel of Fig. 6 (adapted from *Protopapa et al.,* 2017, their Fig. 2) shows spectra sampled across a few of the different terrains, as marked on the left panel. The spectra range from fairly smooth, sloped, and almost devoid of any features (spectrum c on the right panel) to strongly affected by hydrocarbon bands (spectra d and e).

While the hydrocarbon signature had been clearly seen in the groundbased data, the other spectra lacking strong features were completely new. As shown in the left panel of Fig. 6, these featureless spectra belong to the dark equatorial region informally named Chtulhu and illustrate the signature of a dark refractory material that is thought to originate from irradiation of ices native to Pluto's surface (spectrum also shown in Fig. 3). There are three possible sources of the reddish material: deposition of the atmospheric haze over geological timescales (*Grundy et al.,* 2018), irradiation of surface ices from cosmic-rays secondary product irradiation, and cryovolcanic activity (*Cruikshank et al.,* 2019). It has been suggested that more than one material might be present at Cthulhu at this time.

A similarity in the spectral shape of the average reddish-dark material to that coloring Arrokoth has been noted by *Cook et al.* (2019b). This small, pristine TNO was observed *in situ* for the first time during the third New Horizons flyby on January 1, 2019. The similarity in color and spectral shape of the colored materials hints at the common processes that might have acted on both Pluto and Arrokoth and many other TNOs showing the same coloration.

In the years between 2001 and 2012, *Grundy et al.* (2013) monitored Pluto's spectral signature in the 0.8- to 2.4-μm range, making use of the Infrared Telescope Facility (IRTF) SpeX instrument with the goal of mapping and monitoring the longitudinal distribution of the volatile ices on the surface. They described a concentration of CO and N_2 on Pluto's anti-Charon hemisphere preferentially at equatorial latitudes, while CH_4 appeared to gather mostly near the northern latitudes. They reported that, of the volatile ices, CH_4 was the one yielding the strongest evidence for changes related to secular evolution. The bottom left panel of Fig. 6 shows a selection of spectra from *Grundy et al.* (2013). The spectra shown were selected as sampling areas closest to the regions chosen by *Protopapa et al.* (2017). Those areas are marked in colors on the LORRI base map in the top left panel of Fig. 6, corresponding to the colors of the spectra in the lower panel. As an example, spectrum c in *Protopapa et al.* (2017) corresponds to the green trace in Fig. 6 identified by coordinates (121, 38).

Two issues clearly arise when examining the left and right panels of Fig. 6. The first is the degree of spectral variation seen in the two panels. On the right side of Fig. 6, the spectra showing the largest difference are c and e, roughly corresponding to locations (110, 49) and (180, 44). These spectra have very little in common, as the strong hydrocarbon bands that characterize spectrum e are almost completely absent in spectrum c. The groundbased spectra that geographically correspond to c and e, namely (110, 49) and (180, 44) respectively, shown in the left panel of Fig. 6 do not show the same kind of behavior. In particular, (110, 49) shows the same strong CH_4 bands as (180, 44), not as strikingly different from the other spectra as the c spectrum. Aside from temporal variations that might be affecting the strength of the lines to some degree, this is probably mostly due to the lack of spatial resolution of the groundbased observations. In fact, *Grundy et al.* (2013) could not resolve Pluto and therefore the observations, timed to be centered at specific longitudes, covered the entire sub-Earth hemisphere. The dotted segments overlain on the map and corresponding to the colored dots approximate the range in latitudes covered by each observation and give an idea of the vast area sampled by each observation.

As a result, the terrains contributing to each spectrum are more varied than those chosen by *Protopapa et al.* (2017) and include large areas rich in hydrocarbon ices. The reason why the spectra on the left side of Fig. 6 are not very distinct is due to the fact that in a spatial mixture a spectrum with strong bands — such as those from the regions in the northern areas rich in CH_4 — will prevail over one with weak or no bands. The large spectral variation is also reported in Fig. 3, showing the average spectrum of New Horizons' reddest area from Cthulhu compared to the spectra obtained by groundbased observations.

Triton has often been touted to be Pluto's sibling (*Cruikshank,* 2005) because of size, albedo, density, similar atmosphere, and the spectral similarity between the two objects, displaying some of the same volatile ices. Figure 4 shows one of the Triton spectra taken by *Grundy et al.* (2010) on one of 53 nights from 2000 to 2009 at NASA's IRTF. The spectrum shown was taken at longitude (60, –47). As in the case of Pluto, Triton also shows spectral variations typical of an average of the signature of many different terrains. The comparison of Pluto's to one of Triton's spectra from *Grundy et al.* (2010) (Fig. 4) shows at least three obvious differences. Pluto's spectrum has markedly stronger CH_4 bands but weaker N_2 bands (as shown by the band at 2.15 μm). Furthermore, Pluto's continuum shows indications of red coloration at shorter wavelengths, up to ~1.2 μm, which changes into a blue slope at longer than 1.4 μm but no apparent H_2O signature. Triton's coloration is instead not as apparent and consists only of a weak blue slope at wavelengths longer than 1.9 μm. On the other hand, small depressions in the Triton continuum at ~1.5 μm and 2.0 μm are indicative of the possible presence of H_2O. Along with the weaker CH_4 bands the somewhat stronger H_2O signature in Triton's spectra, compared to those in the groundbased spectra for Pluto (Fig. 4), suggests the possibility that Triton might have more H_2O relative to CH_4 ice on its surface than does Pluto. As another comparison, a spectrum of Eris (*Merlin et al.,* 2009) was also included in Fig. 4. Eris appears similar to the two objects, in particular Pluto, even if at lower S/N. As in the case of Pluto, N_2 appears weaker on Eris than on Triton. In both Pluto and Eris the CO_2 bands are not as apparent as they are in the Triton spectrum and on Eris there is even less evidence for presence of H_2O ice. All in all, Pluto seems to be the intermediate step between Triton and Eris as far as composition is concerned.

Among its differently colored terrains, Triton shows a number of features that point to the presence of processed hydrocarbons in the form of refractory materials. Furthermore, the presence of streaks near the southern pole has been interpreted

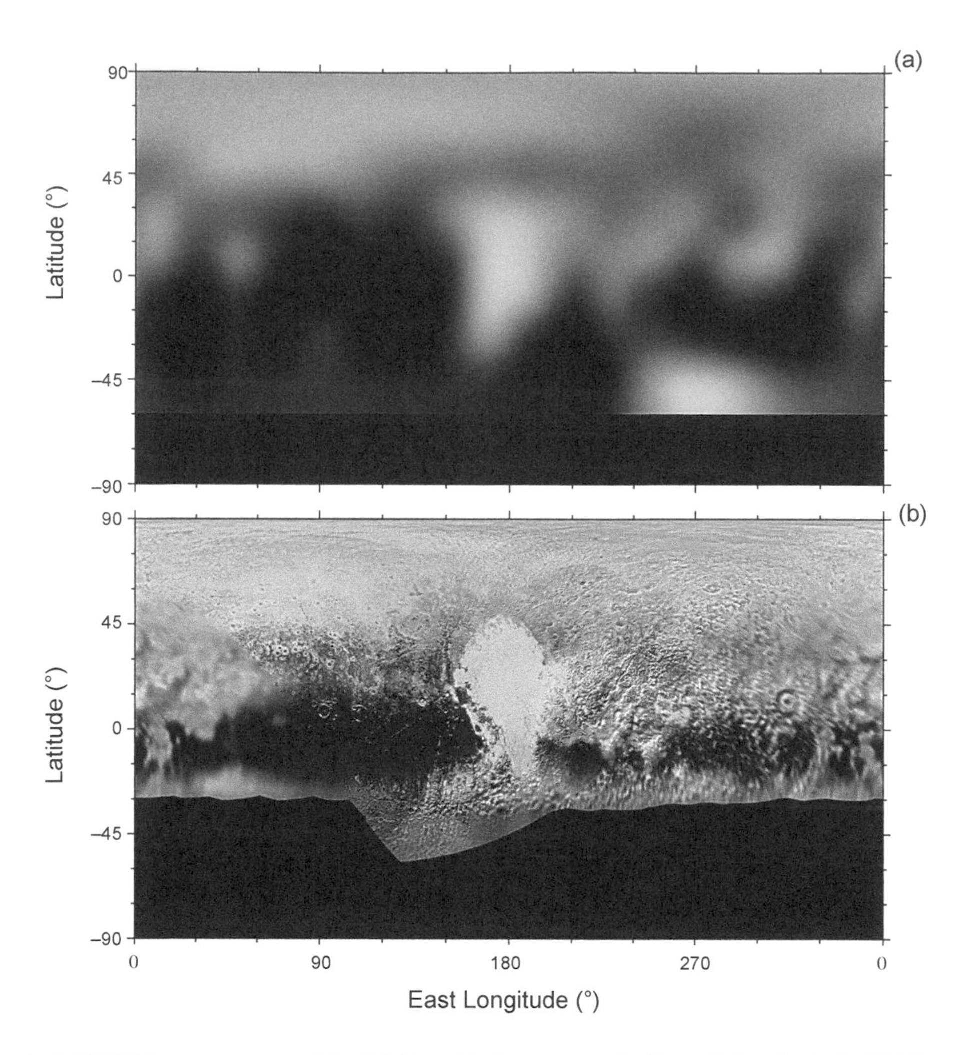

Fig. 5. **(a)** *Buie et al.* (2010b) map compared to **(b)** New Horizons mosaic (from *Schenk et al.*, 2018).

as an indication of material being ejected from a subcrustal source. This and other indications of cryovolcanism on the surface are other features in common with Pluto. However, Voyager had no composition mapping spectrometer capable of studying Triton, so no compositional maps at high resolution yet exist to compare to Pluto.

5.4. Atmospheric Measurements

After years of attempted measurements, *Hubbard et al.* (1988) and *Elliot and Young* (1992) were the first to detect a tenuous CH_4 atmosphere around Pluto. Their measurements were obtained from an occultation by Pluto of a nearby star. Since then many additional measurements were made to determine the stability of the atmosphere over time. While the expectation was for a decrease in pressure due to Pluto moving away from the Sun (e.g., *Hansen et al.*, 2015; *Young,* 2013; *Olkin et al.*, 2015), it was shown through several occultation studies that this was an oversimplification (*Buie,* 2002). Indeed, independently, *Elliot et al.* (2003) found a slight expansion with respect to the measurements done in 1988. *Sicardy et al.* (2006) also reported no substantial decrease in atmospheric pressure and vertical structure with respect to measurements obtained from a previous occultation in 2002. In 2015 the New Horizons flyby was indeed able to detect a tenuous extended atmosphere, mostly composed of CH_4 and N_2, and traces of their irradiation products, along with a haze layer. Furthermore, *Bertrand et al.* (2018), based on their GCM and flyby observations of an accumulation of N_2 and CO ice in Sputnik Planitia, predicted a seasonal atmospheric pressure. They also forecasted that the seasonal frost will disappear in the following decades. For a more detailed and complete description of Pluto's atmospheric work we refer the reader to the chapters in this volume by Young et al., Summers et al., and Mandt et al.

6. MODELS OF SURFACE COMPOSITION

6.1. Analytical Tools

Data analysis before New Horizons was mostly centered on scattering radiative transfer modeling, often adopting the *Hapke* (1993) approach or the similar approach by

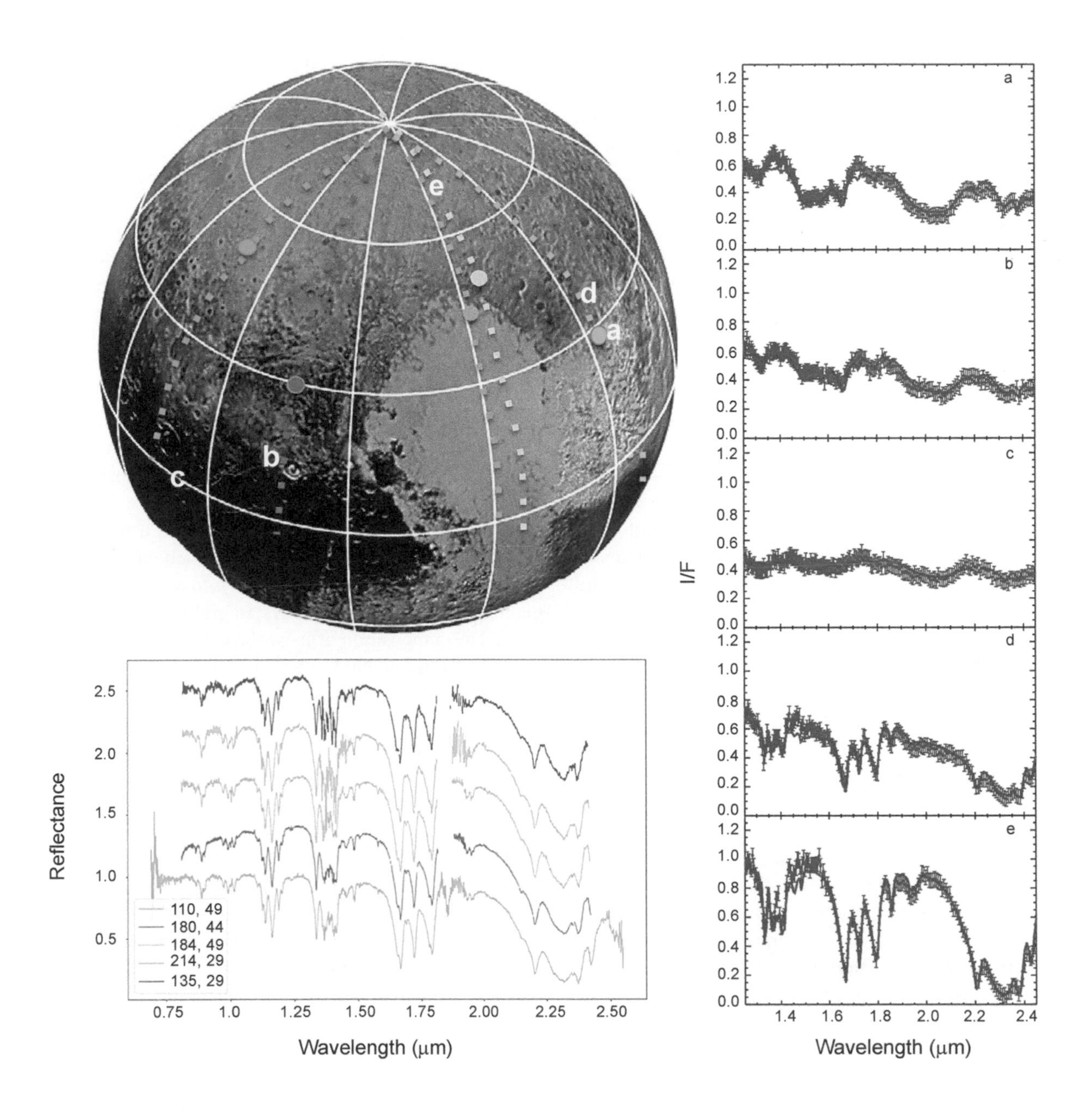

Fig. 6. Spectral signature of different terrains on Pluto's encounter hemisphere during the 2015 New Horizons flyby. (Adapted from *Protopapa et al.*, 2017). The left panel shows the geographical location of the sampled spectra shown on the right panel in black letters. The colored dots represent the location of the closest groundbased observations (*Grundy et al.*, 2013). On the left a selection of ground observations of Pluto at different longitudes are shown, as observed by *Grundy et al.* (2013).

Shkuratov et al. (1999). The Hapke approximation consists of calculating the bidirectional reflectance of a medium composed of closely packed particles of a single component, taking into account the effects of surface roughness, opposition effect, and anisotropic scattering. Similarly, the *Shkuratov et al.* (1999) method consists of calculating the bidirectional reflectance of a system of parallel plates, also taking into account the effects of surface roughness but not illumination angles.

When computing the reflectance of mixtures of different types of particles, the choice of combination is most often either areal or intimate. An areal (also called geographical) mixture consists of materials spatially segregated as it would be in fields of different crops when observed from above. On the other hand, in an intimate mixture the surface consists of different types of particles mixed homogeneously together in close proximity as in a salt and pepper mix. A third kind of mixing has also been employed in the study of TNOs, often called the "molecular" mix or Maxwell Garnett model (*Maxwell Garnett,* 1904; *Hage and Greenberg,* 1990). It adopts the effective medium approximation in the handling of the optical constants before the radiative transfer calculations are performed. This approximation represents the case of contamination from particles of diameter much smaller than the wavelength, enclosed in a matrix of uniform composition. It was adopted in a study of very red TNOs by *Grundy* (2009).

With the advent of the high-spatial-resolution datasets obtained with LEISA and MVIC, the radiative transfer models used to interpret the spectra are the same even though other tools have become necessary in order to deal with the overwhelmingly large number of pixels. They include clustering tools, the pixel-by-pixel Hapke modeling code, and cross-correlation applications, adopted by *Dalle Ore et al.* (2018), *Protopapa et al.* (2017), and *Cook et al.* (2019a) respectively to analyze the wealth of data. These tools make use of mathematical and statistical approaches that reduce dimensionality without compromising information by employing modern computer operating systems or networks of computers that allow for fast computation and large storage of data.

6.2. Laboratory

High-spatial-resolution data have allowed the exploration of a variety of terrains in conditions that are unique. In order to be able to interpret what we see in detail at close range in environments that are new and groundbreaking, more laboratory work has become necessary. Optical constants are a fundamental input into the modeling of the observations and therefore to the interpretation of the physical, geological, and chemical evolution active on Pluto. Limitations in the variety of available indices of refraction for ices as well as refractory materials remain a problem. While ice mixtures have already been tackled to some degree, refractory material related studies are still lagging behind. Depending on the source, the materials and processes involved in the formation differ and therefore require a different set of laboratory experiments. Some of the molecules to be considered as candidates to be included to properly interpret the dark material include hydrocarbons C_2H_2, C_2H_4, and C_2H_6, all known to be present in Pluto's atmosphere (*Gladstone et al.,* 2016), and molecules produced by hydrolysis of tholins, molecules of simple polyaromatic and aliphatic hydrocarbons (PAHs) as in the dataset of *Izawa et al.* (2014) with red coloration. Optical constants obtained at relevant, very low temperatures for all of the above and probably others will be needed to fully model the complex spectral signature of the dark material and ultimately understand its composition.

7. CONCLUSIONS

In the last 20 years of observations, TNOs revealed a great diversity of surface properties. Albedo value shows the huge variation across the TNO population going from 2–3% to more than 50% for the volatile-rich objects, and up to 96% for Eris. The sizes measured for known TNOs range from a few tenths of a kilometer to Pluto-sized (over 1000 km). A large variation has also been estimated for the density, which increases with object size. Broad surface color diversity has been observed thanks to the numerous photometric surveys, with many TNOs having colors significantly redder than any other solar system bodies. A few TNOs have spectral properties similar to Pluto, such as Eris and Makemake, and some others could have undergone a similar geological evolution. The surface composition varies a lot across the TNO population, with many TNOs showing absorption features related to different ices, while others have featureless spectra with composition richer in organics and carbon. For the latter, irradiation processes played a fundamental role in their surface evolution. We also know that, beyond space weathering, many other processes can affect the TNO's surface, such as impacts, atmospheric loss activities, and cryovolcanism.

The New Horizons data have revealed a much greater variety of terrains and detail on Pluto than would have ever been detected without the flyby. Pluto's surface is rich in geological features from craters, to grabens, to "bladed" terrains. The composition of its surface shows mixtures of ices such as CH_4 diluted in N_2, which is present in abundance in Sputnik Planitia, a basin of freshly deposited ice renewed constantly by the volatile ices migrating from the northern terrains. The ice in Spuntnik Planitia is stable and yet shows evidence of convection. At the edges of the basin there are mountains of H_2O ice towering over the basin and topped by CH_4 ice. This amazing landscape could not have been imagined before the New Horizons flyby. Yet Sputnik Planitia is not the only peculiarly unique place on Pluto. Virgil Fossae is another interesting location; it is a brightly-red-colored graben in the middle of a dark terrain displaying evidence of cryovolcanism. Other areas show evidence of stratification of dark and icy materials interleaved. Pluto has offered us more unexpected features, e.g., a constantly mutating distribution of ices tied to its seasons and diurnal variations, triggering a continuous flow of volatile ices, or, in the atmosphere, a layer of haze contributing over time to the production of the dark material covering a part of the equator.

The diversity of composition and surface morphology revealed on Pluto holds promise of similar findings on all the dwarf and candidate dwarf planets and in particular on those that are more similar to Pluto, like Eris and Makemake, which show high methane content. The detected complexity on Pluto will help to motivate more detailed studies of surface variations for other objects by the groundbased larger telescopes. The high eccentric orbit of Makemake could allow fresh methane to migrate seasonally as observed on Pluto and might explain the methane abundance. The high albedo and the lack of compositional rotational variation on Eris could imply a surface dominated by a seasonal cycling atmosphere. The extreme temperature variation for Eris between aphelion and perihelion can provide important input for atmospheric studies. Quaoar also shows some similarities with Pluto. The fact that CO_2 is present on Quaoar as on Triton at particular longitudes may provide clues to detect it on Pluto. Surface heterogeneity could be also present in objects with a composition mainly including stable ices such as H_2O, for which variations can still be possible even if the changes are not connected to the variations of an atmosphere.

However, while it would be ideal to be able to visit and explore every dwarf planet reachable from Earth and return to Pluto with an orbiter, monetary and temporal mission costs limit our choices. The majority of these objects can only be observed from the ground. The Pluto flyby has demonstrated that observational techniques used up to now from the ground are excellent, even though they have some limitations on precision and surface details. Some remotely observable features should continue to be investigated for a statistically significant number of objects. The techniques/tools of interest include:

- Long-term photometry to study spatial as well as temporal variations (large-amplitude light curves, varied low-resolution albedo maps as in the case of Pluto). This requires detailed photometric observations over a long period of time, which also permits constraining the targets' shape.
- Occultations aimed at more precise measurements of size as well as detection of an atmosphere, ring, and satellites. This technique can be applied to many new objects as soon as the high-precision stellar catalogs produced by Gaia will be available (*Brown,* 2019). The occultation will give more precise information on the shape. The detection of an atmosphere by occultation can also assess surface compositional groups when no spectroscopy or poor spectroscopy is available.
- Spectroscopy by new-generation instruments with new-class telescopes (30–40 m), particularly the James Webb Space Telescope (JWST), will revolutionize our understanding about the surface composition and the atmosphere.
- Laboratory experiments are still needed to improve input parameters for modeling the surface composition/properties. A more complete database of optical constants is necessary to obtain detailed information on the surface components, in particular for refractory materials. New laboratory experiments with more sophisticated instrumentation could also be useful to better understand the various physical processes (as irradiation, deposit of materials, etc.) to better constrain surface properties.

Some of the most fundamental properties of TNOs are still quite poorly constrained on these distant objects, including diameter and density. Observations at long thermal wavelengths, in the millimeter and submillimeter, are well suited for determining these quantities, in particular for those objects with companions. ALMA, an astronomical interferometer of 66 radio telescopes in the Atacama Desert of northern Chile, will be able to determine albedos and diameters for many Centaurs and TNOs.

The future new generation of telescopes of extremely large size coupled with a new generation of instruments will produce a wealth of new discoveries on TNOs. Some of these are the U.S. Giant Magellan Telescope (GMT), the Thirty Meter Telescope (TMT) to be built in Chile and on Mauna Kea (Hawai'i) respectively, and the Extremely Large Telescope (ELT), the world's largest optical/near-infrared telescope, currently under construction by the European Southern Observatory agency and located on top of the Cerro Armazones in the Atacama Desert of northern Chile.

Significant progress will be possible, in particular with the JWST under the operation and support by the NASA-ESA Space Telescope Science Institute (STScI). Its launch is expected in 2021. It will provide groundbreaking science particularly for studies of TNOs. The telescope will be equipped with four science instruments for imaging (NIRCam, NIRISS, and MIRI) and spectroscopy (the previous three plus NIRSpec), covering wavelengths from 0.7 to 28 μm (*Parker et al.,* 2016).

The New Horizons flyby unveiled an unexpectedly diversified Pluto. With the extended mission the spacecraft arrived at 50 AU, giving us groundbreaking information about Arrokoth. The spacecraft has enough fuel to operate until the 2030s, but it is unlikely to have another TNO flyby.

We know to expect many degrees of diversity among TNOs and dwarf planets. Many mysteries still remain unsolved, but a lot will be learned with the new-generation telescopes and future space missions that will bring many exciting discoveries in the future.

REFERENCES

Adams E. R., Gulbis A. A. S., Elliot J. L., et al. (2014) De-biased populations of Kuiper belt objects from the Deep Ecliptic Survey. *Astron. J., 148(55).*

Agnor C. B. and Hamilton D. P. (2006) Neptune's capture of its moon Triton in a binary-planet gravitational encounter. *Nature, 441,* 192–194.

Alexandersen M., Benecchi S. D., Chen Y.-T., et al. (2019) OSSOS. XII. Variability of 65 transneptunian objects using the Hyper Suprime-Cam. *Astrophys. J., 244,* 19–36.

Alvarez-Candal A., Fornasier S., Barucci M. A., et al. (2008) Visible spectroscopy of the new ESO large program on transneptunian objects and Centaurs. *Astron. Astrophys., 487,* 741.

Alvarez-Candal A., Ortiz J. L., Morales N., et al. (2014) Stellar occultation by (119951) 2002 KX14 on April 26, 2012. *Astron. Astrophys., 571(A48).*

Aumann H. H. and Walker R. G. (1987) IRAS observations of the Pluto-Charon system. *Astron. J., 94,* 1088–1091.

Barr A. C. and Schwamb M. E. (2016) Interpreting the densities of the Kuiper belt's dwarf planets. *Mon. Not. R. Astron. Soc., 460,* 1542–1548.

Barucci M. A., Fulchignoni M., Birlan M., et al. (2001) Analysis of transneptunian and Centaur colours: Continuous trend or grouping? *Astron. Astrophys., 371,* 1150.

Barucci M. A., Belskaya I., Fulchignoni M., and Birlan M. (2005a) Taxonomy of Centaurs and transneptunians objects. *Astron. J., 130,* 1291.

Barucci M. A., Cruikshank D. P., Dotto E., et al. (2005b) Is Sedna another Triton? *Astron. Astrophys., 439,* L1–L4.

Barucci M. A., Brown M. E., Emery J. P., and Merlin F. (2008) Composition and surface properties of transneptunian objects and Centaurs. In *The Solar System Beyond Neptune* (M. A. Barucci et al., eds.), pp. 507–524. Univ. of Arizona, Tucson.

Barucci M. A., Dalle Ore C. M., Alvarez-Cantal A., et al. (2010) Sedna: Investigation of surface compositional variation. *Astron. J., 140,* 2095–2100.

Barucci M. A., Alvarez-Cantal A., Merlin F., et al. (2011) New insights on ices in Centaurs and transneptunians populations. *Icarus, 214,*

297–307.
Barucci M. A., Merlin F., Perna D., et al. (2012) The extra red plutino (55638) 2002 VE95. *Astron. Astrophys., 539,* 152.
Barucci M. A., Dalle Ore C. M., Perna D., et al. (2015) (50000) Quaoar: Surface composition variability. *Astron. Astrophys., 584,* A107.
Batygin K., Brown M. E., and Fraser W. C. (2011) Retention of a primordial cold classical Kuiper belt in an instability-driven model of solar system formation. *Astrophys. J., 738,* 13–20.
Bauer J. M., Meech K. J., Fernández Y. R., et al. (2002) Observations of the Centaur 1999 UG5: Evidence of a unique outer solar system surface. *Publ. Astron. Soc. Pac., 114(802),* 1309–1321.
Bauer J. M., Grav T., Blauvelt E., et al. (2013) Centaurs and scattered disk objects in the thermal infrared analysis of WISE/NEOWISE Observations. *Astrophys. J., 773,* 22.
Belskaya I. N., Barucci M. A., and Fulchignoni M. (2015) Updated taxonomy of transneptunian objects and centaurs: Influence of albedo. *Icarus, 250,* 482–491.
Benecchi S. D. and Sheppard S. (2013) Light curves of 32 large transneptunian objects. *Astron. J., 145,* 124.
Benecchi S. D., Noll K. S., Grundy W. M., et al. (2009) The correlated colors of transneptunians binary. *Icarus, 200,* 292–303.
Benecchi S. D., Porter S. B., Buie M. W., et al. (2019) The HST lightcurve of (486958) 2014 MU69. *Icarus, 334,* 11.
Benedetti-Rossi G., Sicardy B., Buie M. W., et al. (2016) Results from the 2014 November 15th multi-chord stellar occultation by the TNO (229762) 2007 UK126. *Astron. J., 152(6),* 156.
Bennett C. J., Jamieson C. S., Osamura Y., and Kaiser R. I. (2006) Laboratory studies on the irradiation of methane in interstellar, cometary, and solar system ices. *Astrophys. J., 653,* 792–811.
Bernstein G. M., Trilling D. E., Allen R. L., Brown M. E., Holman M., and Malhotra R. (2004) The size distribution of transneptunian bodies. *Astron. J., 128,* 1364–1390.
Bertrand T., Forget F., Umurhan O. M., et al. (2018) The nitrogen cycles on Pluto over seasonal and astronomical timescales. *Icarus, 309,* 227.
Braga-Ribas F., Sicardy B., Ortiz J. L., et al. (2013) The size, shape, albedo, density, and atmospheric limit of transneptunian object (50000) Quaoar from multi-chord stellar occultations. *Astrophys. J., 773,* 26.
Braga-Ribas F., Sicardy B., Ortiz J. L., et al. (2014) A ring system detected around the Centaur (10199) Chariklo. *Nature, 508(7494),* 72–75.
Brown A. G. A. (2019) The future of Gaia universe. Presented at the 53rd ESLAB Symposium: The Gaia Universe, ESTEC/ESA, Noordwijk, The Netherlands, 8–12 April 2019.
Brown M. E. (2000) Near-infrared spectroscopy of Centaurs and irregular satellites. *Astron. J., 119,* 977–983.
Brown M. E. (2012) The composition of the Kuiper belt objects. *Annu. Rev. Earth Planet. Sci., 40,* 467–494.
Brown M. E. and Butler B. J. (2017) The density of mid-sized Kuiper belt objects from ALMA thermal observations. *Astron. J., 154(19).*
Brown M. E. and Calvin W. M. (2000) Evidence for crystalline water and ammonia ices on Pluto's satellite Charon. *Science, 287,* 107–109.
Brown M. E. and Schaller E. L. (2007) The mass of dwarf planent Eris. *Science, 316,* 1585.
Brown M. E., Trujillo C. A., and Rabinowitz D. L. (2005) Discovery of a planetary-sized object in the scattered Kuiper belt. *Astrophys. J. Lett., 635,* L97–L100.
Brown M. E., Barkume K. M., Ragozzine D., et al. (2007a) A collisional family of icy objects in the Kuiper belt. *Nature, 446,* 294–296.
Brown M. E., Barkume K. M., Blake G. A., et al. (2007b) Methane and ethane on the bright Kuiper belt object 2005 FY9. *Astron. J., 133,* 284–289.
Brown M. E., Schaller E. L., and Fraser W. C. (2011) A hypothesis for the color diversity of the Kuiper belt. *Astrophys. J. Lett., 739,* L60–L64.
Brown M. E., Schaller E. L., and Fraser W. C. (2012) Water ice in the Kuiper belt. *Astron. J., 143,* 146.
Brown M. E., Schaller E. L., and Blake G. A. (2015) Irradiation products on dwarf planet Makemake. *Astron. J., 149,* 105.
Brown R. H., Cruikshank D. P., Veverka J., et al. (1995) Surface composition and photometric properties of Triton. In *Neptune and Triton* (D. P. Cruikshank, ed.), pp. 991–1030. Univ. of Arizona, Tucson.
Brucker M. J., Grundy W. M., Stansberry J. A., et al. (2009) High albedos of low inclination classical Kuiper belt objects. *Icarus, 201,* 284–294.
Brunetto R., Barucci M. A., Dotto E., and Strazzulla G. (2006) Ion irradiation of frozen methanol, methane, and benzene. Linking to the colors of Centaurs and trans-neptunian objects. *Astrophys. J., 644,* 646–650.
Buie M. W. (2002) Seasonal atmospheric and surface changes on Pluto. American Astronomical Society, 201st AAS Meeting, Abstract #61.04. *Bull. Am. Astron. Soc., 34,* 1212.
Buie M. W., Grundy W. M., Young E. F., et al. (2010a) Pluto and Charon with the Hubble Space Telescope. I. Pluto and Charon with the Hubble Space Telescope: Monitoring global change and improved surface properties from light curves. *Astron. J., 139,* 1117–1127.
Buie M. W., Grundy W. M., Young E. F., et al. (2010b) Pluto and Charon with the Hubble Space Telescope. II. Resolving changes on Pluto's surface and a map for Charon. *Astron. J., 139,* 1128–1143.
Bus S. J., Bowell E., Harris A. W., et al. (1989) 2060 Chiron — CCD and electronographic photometry. *Icarus, 77,* 223–238.
Collander-Brown S. J., Fitzsimmons A., Fletcher E., et al. (1999) Light curves of the trans-neptunian objects 1996 TP66 and 1994 VK8. *Mon. Not. R. Astron. Soc., 308(2),* 588–592.
Cook J. C. and 23 colleagues (2019a) Comparison of near infrared spectra between Pluto-system objects and 486958 2014 MU69: Analysis of New Horizons spectral images. *Lunar and Planetary Science L,* Abstract #2818. Lunar and Planetary Institute, Houston.
Cook J. C. and 27 colleagues (2019b) The distribution of H_2O, CH_3OH, and hydrocarbon-ices on Pluto: Analysis of New Horizons spectral images. *Icarus, 331,* 148–169.
Cruikshank D. P. (2005) Triton, Pluto, Centaurs, and trans-neptunian bodies. *Space Sci. Rev., 116,* 421–439.
Cruikshank D. P., Pilcher C. B., and Morrison D. (1976) Pluto: Evidence for methane frost. *Science, 194,* 835–837.
Cruikshank D. P., Roush T. L., Moore J. M., et al. (1997) The surfaces of Pluto and Charon. In *Pluto and Charon* (S. A. Stern and D. J. Tholen, eds.), pp. 221–267. Univ. of Arizona, Tucson
Cruikshank D. P., Roush T. L., and Bartholomew M. J. (1998) The composition of Centaur 5145 Pholus. *Icarus, 135,* 389–407.
Cruikshank D. P. and 28 colleagues (2019) Recent cryovolcanism in Virgil Fossae on Pluto. *Icarus, 330,* 155–168.
Dalle Ore C. M., Barucci M. A., Emery J. P., et al. (2009) Composition of KBP (50000) Quaoar. *Astron. Astrophys., 501,* 349–357.
Dalle Ore C. M., Morea L. V., Roush T. L., et al. (2013) A compositional interpretation of trans-neptunian objects taxonomies. *Icarus, 222,* 307–322.
Dalle Ore C. M., Barucci M. A., Emery J. P., et al. (2015) The composition of "ultra-red" TNOs and Centaurs. *Icarus, 252,* 311.
Dalle Ore C. M., Protopapa S., Cook J. C., et al. (2018) Ices on Charon: Distribution of H_2O and NH_3 from New Horizons LEISA observations. *Icarus, 300,* 21–32.
Dalle Ore C. M. and 17 colleagues (2019) Detection of ammonia on Pluto's surface in a region of geologically recent tectonism. *Sci. Adv., 5,* eaav5731.
Dartois E., Augé B., Boduch P., et al. (2015) Heavy ion irradiation of crystalline water ice. *Astron. Astrophys., 576,* A125.
De Bergh C., Boehnhardt H., Barucci M. A., et al. (2004) Aqueous altered silicates in the surface of two plutinos? *Astron. Astrophys., 416,* 791.
De Sanctis M. C., Ammannito E., Raponi A., et al. (2015) Ammoniated phyllosilicates with a likely outer solar system origin on (1) Ceres. *Nature, 528,* 241–244.
Dias-Oliveira A., Sicardy B., Ortiz J. L., Braga-Ribas F., et al. (2017) Study of the plutino object (208996) 2003 AZ84 from stellar occultations: Size, shape, and topographic features. *Astron. J., 154(22).*
Doressoundiram A., Boehnhardt H., Tegler S. C., and Trujillo C. (2008) Color properties and trends of the transneptuian objects. In *The Solar System Beyond Neptune* (M. A. Barucci et al., eds.), pp. 91–104. Univ. of Arizona, Tucson.
Dotto E., Perna D., Barucci M. A., et al. (2008) Rotational properties of Centaurs and trans-neptunian objects — Lightcurves and densities. *Astron. Astrophys., 490,* 829–833.
Duffard R., Ortitz J. L., Thirouin A., et al. (2009) Transneptunian objects and Centaurs from lightcurve. *Astron. Astrophys., 505,* 1283–1295.
Duffard R., Pinilla-Alonso N., Santos-Sanz P., et al. (2014) TNOs are Cool: A survey of the transneptunian region: A Herschel-PACS

view of 16 Centaurs. *Astron. Astrophys., 564(A92).*

Elliot J. L. and Young L. A. (1992) Analysis of stellar occultation data for planetary atmospheres. I. Model fitting application to Pluto. *Astron. J., 103,* 991.

Elliot J. L., Ates A., Babcock B. A., et al. (2003) The recent expansion of Pluto's atmosphere. *Nature, 424,* 165–168.

Elliot J. and 10 colleagues (2005) The deep ecliptic survey: A search for Kuiper belt objects and Centaurs. II. Dynamical classification, the Kuiper belt plane, and the core population. *Astron. J., 129,* 1117–1162.

Elliot J. L., Person M. J., Zuluaga C. A., et al. (2010) Size and albedo of Kuiper belt object 55636 from a stellar occultation. *Nature, 465(7300),* 897–900.

Emery J. P., Cruikshank D. P., and Van Cleve J. (2006a) Thermal emission spectroscopy (5.2–38 micron) of three Trojan asteroids with the Spitzer Space Telescope. *Icarus, 182,* 496–512.

Emery J. P., Dalle Ore C., Cruikshank D. P., et al. (2006b) Reflectances of Kuiper belt objects at lambda > 2.5 microns. Abstract #P13C-0183 presented at 2006 Fall Meeting, AGU, San Francisco, California, 11–15 December.

Emery J. P., Dalle Ore C., Cruikshank D. P., et al. (2007) Ices on (90377) Sedna: Confirmation and compositional constraints. *Astron. Astrophys., 466,* 395.

Fernandez-Valenzuela E., Ortiz J. L., Duffard R., Santos-Sanz P., and Morales N. (2016) 2008 OG19: A highly elongated trans-neptunian object. *Mon. Not. R. Astron. Soc., 456,* 2354–2360.

Ferrari C. and Lucas A. (2016) Low thermal inertias of icy planetary surfaces. Evidence for amorphous ice? *Astron. Astrophys., 588,* A133.

Fornasier S., Barucci M. A., de Bergh C., et al. (2009) Visible spectroscopy of the new ESO large programme on transneptunian objects and Centaurs. *Astron. Astrophys., 508,* 457–465.

Fornasier S., Lellouch E., Müller T., et al. (2013) TNOs are Cool program: Combined observations of 9 Centaurs and TNOs with PACS and SPIRE instruments onboard the Herschel Space Telescope. *Astron. Astrophys., 555,* A15.

Fornasier S., Lazzaro D., Alvarez-Candal A., et al. (2014) The Centaur 10199 Chariklo: Investigation into rotational period, absolute magnitude, and cometary activity. *Astron. Astrophys., 568,* L11.

Foust J. A., Elliot J. L., Olkin C. B., et al. (1997) Determination of the Charon/Pluto mass ratio from center-of-light astrometry. *Icarus, 126(2),* 362–372.

Fraser W. C. and Brown M. E. (2012) The Hubble Wide Field Camera 3 test of surfaces in the outer solar system: The compositional classes of the Kuiper belt. *Astron. J., 749(33).*

Fraser W. C., Brown M. E., and Schwamb M. E. (2010) The luminosity function of the hot and cold Kuiper belt populations. *Icarus, 210,* 944.

Fraser W. C., Batygin K., Brown M. E., and Bouchez A. (2013) The mass, orbit, and tidal evolution of the Quaoar-Weywot system. *Icarus, 222,* 357.

Fraser W. C., Brown M. E., Morbidelli A., et al. (2014) The absolute magnitude distribution of Kuiper belt objects. *Astron. J., 782,* 100.

Fraser W. C., Bannister M. T., Pike R. E., et al. (2017) All planetesimals born near the Kuiper belt formed as binaries. *Nature Astron., 1,* 0088.

Fulchignoni M., Belskaya I., Barucci M. A., et al. (2008) Transneptunian objects taxonomy. In *The Solar System Beyond Neptune* (M.A. Barucci et al., eds.), pp. 181–192. Univ. of Arizona, Tucson.

Galiazzo M., de la Fuente Marcos C., de la Fuente Marcos R., et al. (2016) Photometry of Centaurs and trans-neptunian objects: 2060 Chiron (1977 UB), 10199 Chariklo (1997 CU26), 38628 Huya (2000 EB173), 28978 Ixion (2001 KX76), and 90482 Orcus (2004 DW). *Astrophys. Space Sci., 361(7),* 212.

Gaudi B. S., Stanek K. Z., Hartman J. D., Holman M. J., and McLeod B. A. (2005) On the rotation period of (90377) Sedna. *Astron. J., 629(1),* L49–L52.

Gladman B., Holman M., Grav T., Kavelaars J., Nicholson P., Aksnes K., et al. (2002) Evidence for an extended scattered disk. *Icarus, 157,* 269–279.

Gladman B., Marsden B. G., and Van Laerhoven C. (2008) Nomenclature in the outer solar system. In *The Solar System Beyond Neptune* (M. A. Barucci et al., eds.), pp. 43–58. Univ. of Arizona, Tucson.

Gladstone G. R., Stern S. A., Ennico K., Olkin C. B., Weaver H. A., et al. (2016) The atmosphere of Pluto as observed by New Horizons. *Science, 351(6279),* 1284–1293.

Gomez R. S. (2003) The origin of the Kuiper belt high-inclination population. *Icarus, 161,* 404–418.

Grundy W. M. (2009) Is the missing ultra-red material colorless ice? *Icarus, 199,* 560–563.

Grundy W. M., Noll K. S., and Stephens D. C. (2005) Diverse albedos of small trans-neptunian objects. *Icarus, 176,* 184–191.

Grundy W. M., Young L. A., Stansberry J. A., Buie M. W., Olkin C. B., and Young E. F. (2010) Near-infrared spectral monitoring of Triton with IRTF/SpeX II: Spatial distribution and evolution of ices. *Icarus, 205,* 594–604.

Grundy W. M. et al. (2011) Five new and three improved mutual orbits of transneptunian binaries. *Icarus, 213,* 678–692.

Grundy W. M., Olkin C. B., Young L. A., Buie M. W., and Young E. F. (2013) Near-infrared spectral monitoring of Pluto's ices: Spatial distribution and secular evolution. *Icarus, 223,* 710–721.

Grundy W. M. and 33 colleagues (2016) Surface compositions across Pluto and Charon. *Science, 351,* aad9189.

Grundy W. M., Bertrand T., Binzel R. P., et al. (2018) Pluto's haze as a surface material. *Icarus, 314,* 232.

Grundy W. M., Noll K. S., Buie M. W., Benecchi S. D., et al. (2019) The mutual orbit, mass, and density of transneptunian binary G!kún'hòmdímà (229762 2007 UK126). *Icarus, 334,* 30–38.

Hage J. I. and Greenberg J. M. (1990) A model for the optical properties of porous grains. *Astrophys. J., 361,* 251.

Hainaut O. R., Boehnhardt H., and Protopapa S. (2012) Colours of minor bodies in the outer solar system. II. A statistical analysis revisited. *Astron. Astrophys., 546,* 115.

Hansen C. J., Paige D. A., and Young L. A. (2015) Pluto's climate modeled with new observational constraints. *Icarus, 246,* 183–191.

Hapke B. (1993) *Theory of Reflectance and Emittance Spectroscopy.* Topics in Remote Sensing, Vol. 3, Cambridge Univ., Cambridge.

Harris A. W. (1998) A thermal model for near-Earth asteroids. *Icarus, 131,* 291.

Heinze A. N. and de Lahunta D. (2009) The rotation period and light-curve amplitude of Kuiper belt dwarf planet 136472 Makemake (2005 FY9). *Astron. J., 138,* 428–438.

Howett C. J. A., Spencer J. R., Schenk P., et al. (2011) A high-amplitude thermal inertia anomaly of probable magnetospheric origin on Saturn's moon Mimas. *Icarus, 216,* 221.

Howett C. J. A., Spencer J. R., Hurford T., Verbiscer A, and Segura M. (2012) PacMan returns: An electron-generated thermal anomaly on Tethys. *Icarus, 221,* 1084.

Hubbard W. B., Hunten D. M., Dieters S. W., Hill K. M., and Watson R. D. (1988) Occultation evidence for an atmosphere on Pluto. *Nature, 336,* 452–454.

Hudson R. L., Palumbo M. E., Strazulla G., et al. (2008) Laboratory studies of the chemistry of transneptunian object surface materials. In *The Solar System Beyond Neptune* (M. A. Barucci et al., eds.), pp. 507–524. Univ. of Arizona, Tucson.

Izawa M. R. M., Applin D. M., Norman L., and Cloutis E. A. (2014) Reflectance spectroscopy (350–2500 nm) of solid-state polycyclic aromatic hydrocarbons (PAHs). *Icarus, 237,* 159–181.

Jewitt D. C. and Luu J. (1992) 1992 QB1. *IAU Circular 5611.*

Jewitt D. C. and Luu J. (2001) Color and spectra of Kuiper belt objects. *Astron. J., 122,* 2099–2114.

Johansen A., Oishi J. S., Mac Low M. M., et al. (2007) Rapid planetesimal formation in turbulent circumstellar disks. *Nature, 448,* 1022–1025.

Kern S. D. (2006) A study of binary Kuiper belt objects. Ph.D. thesis, Massachusetts Institute of Technology. Proquest Section 0753, Part 0606, Publication Number AAT 0809060. Source: DAI-B 67/06.

Kern S. D. and Elliot J. L. (2006) Discovery and characteristics of the Kuiper belt binary 2003 QY90. *Icarus, 183,* 179–185.

Kern S. D., McCarthy D. W., Buie M. W., et al. (2000) Compositional variation on the surface of Centaur 8405 Asbolus. *Astrophys. J. Lett., 542,* L155–L159.

Kiss C., Szabò G., Horner J., et al. (2013) A portrait of the extreme solar system object 2012 DR30. *Astron. Astrophys., 555,* A3.

Kiss C., Marton G., Parker A. H., et al. (2019) The mass and density of the dwarf planet (225088) 2007 OR10. *Icarus, 334,* 3.

Kovalenko I. D., Doressoundiram A., Lellouch E., et al. (2017) TNOs are Cool: A survey of the trans-neptunian region. XIII. Statistical analysis of multiple trans-neptunian objects observed with Herschel Space Observatory. *Astron. Astrophys., 608,* A19.

Lacerda P. and Luu J. (2006) Analysis of the rotational properties of

Kuiper belt objects. *Astron. J., 131,* 2314–2326.
Lacerda P., Jewitt D., and Peixinho N. (2008) High-precision photometry of extreme KBO 2003 EL61. *Astron. J., 135(5),* 1749–1756.
Lacerda P., Fornasier S., Lellouch E., et al. (2014a) The albedo-color diversity of transneptunian objects. *Astrophys. J. Lett., 793,* L2.
Lacerda P., McNeill A., and Peixinho N. (2014b) The unusual Kuiper belt object 2003 SQ317. *Mon. Not. R. Astron. Soc., 437,* 3824–3831.
Lagerros J. S. V. (1996) Thermal physics of asteroids. I. Effects of shape, heat conduction, and beaming. *Astron. Astrophys., 310,* 1011–1020.
Lagerros J. S. V. (1997) Thermal physics of asteroids. III. Irregular shapes and albedo variegations. *Astron. Astrophys., 325,* 1226–1236.
Lagerros J. S. V. (1998) Thermal physics of asteroids. IV. Thermal infrared beaming. *Astron. Astrophys., 332,* 1123–1132.
Lebofsky L. A. and Spencer J. R. (1989) Radiometry and thermal modeling of asteroids. In *Asteroids II* (R. P. Binzel et al., eds.), pp. 128–147. Univ. of Arizona, Tucson.
Lebofsky L. A., Sykes M. V., Tedesco E. F., et al. (1986) A refined "standard" thermal model for asteroids based on observations of 1 Ceres and 2 Pallas. *Icarus, 68,* 239.
Leiva R., Sicardy B., Camargo J. I. B., et al. (2017) Size and shape of Chariklo from multi-epoch stellar occultations. *Astron. J., 154,* 159.
Lellouch E., Laureijs R., Schmitt B., et al. (2000a) Pluto's non-isothermal surface. *Icarus, 147,* 220–250.
Lellouch E., Paubert G., Moreno R., Schmitt B. (2000b) NOTE: Search for variations in Pluto's millimeter-wave emission. *Icarus, 147,* 580–584.
Lellouch E., Stansberry J., Emery E., et al. (2011) Thermal properties of Pluto's and Charon's surfaces from Spitzer observations. *Icarus, 214,* 701.
Lellouch E., Santos-Sanz P., Lacerda P., Mommert M., et al. (2013) TNOs are Cool: A survey of the trans-neptunian region. IX. Thermal properties of Kuiper belt objects and Centaurs from combined Herschel and Spitzer observations. *Astron. Astrophys., 557,* A60.
Lellouch E., Santos-Sanz P., Fornasier S., et al. (2016) The long-wavelength thermal emission of the Pluto-Charon system from Herschel observations. Evidence for emissivity effect. *Astron. Astrophys., 588,* A2.
Lellouch E., Moreno R., Müller T., et al. (2017) The thermal emission of Centaurs and trans-neptunian objects at millimeter wavelengths from ALMA observations. *Astron. Astrophys., 608,* A45.
Levison H. F. and Duncan M. J. (1997) From the Kuiper belt to Jupiter-family comets: The spatial distribution of ecliptic comets. *Icarus, 127,* 13–32.
Licandro J., Pinilla-Alonso N., Pedani M., et al. (2006a) Methane ice rich surface of large TNO 2005 TY9. *Astron. Astrophys., 445,* L35–L38.
Licandro J., Grundy W. M., Pinilla-Alonso N., et al. (2006b) Visible spectroscopy of 2003 UB313: Evidence for N_2 ice on the surface of the largest TNO? *Astron. Astrophys., 480(1),* L5–L8.
Lorenzi V., Pinilla-Alonso N. and Licandro J. (2015) Rotationally resolved spectroscopy of dwarf planet (136472) Makemake. *Astron. Astrophys., 577,* A86.
Mackinnon I. D. R. and Rietmeijer F. J. M. (1987) Mineralogy of condritic interplanetary dust particles. *Rev. Geophys., 25,* 1527.
Malhotra R. (1995) The origin of Pluto's orbit: Implications for the solar system beyond Neptune. *Astron. J., 110,* 420–429.
Malhotra R. (2019) Resonant Kuiper belt objects. *Geosci. Lett., 6,* 12.
Marsset M., Fraser W. C., Pike R. E., et al. (2019) Col-OSSOS: Color and inclination are correlated throughout the Kuiper belt. *Astron. J., 157,* 94–111.
Mastrapa R. M. E. and Brown R. H. (2006) Ion irradiation of crystalline H_2O-ice: Effect on the 1.65-μm band. *Icarus, 183,* 207–214.
Matzler C. (1998) Microwave properties of ice and snow. In *Solar System Ices* (B. Schmitt et al., eds.), pp. 241–257. Astrophysics and Space Science Library, Vol. 227, Springer, Dordrecht.
Maxwell Garnett J. C. (1904) Colours in metal glasses and in metallic films. *Philos. Trans. R. Soc., A, 203,* 385.
McKinnon W. B. (2012) Where did Ceres accrete? In *Asteroids, Comets, Meteors 2012,* Abstract #6475. LPI Contribution 1667, Lunar and Planetary Institute, Houston.
McKinnon W. B., Lunine J. I., and Banfield D. (1995) Origin and evolution of Triton. In *Neptune and Triton* (D. P. Cruikshank, ed.), pp. 807–877. Univ. of Arizona, Tucson.
McKinnon W. B., Prialnik D., Stern S. A., and Coradini A. (2008) Structure and evolution of Kuiper belt objects and dwarf planets. In *The Solar System Beyond Neptune* (M. A. Barucci et al., eds.), pp. 213–241. Univ. of Arizona, Tucson.
McKinnon W. B., Stern S. A., Weaver H. A., et al. (2017) Origin of the Pluto-Charon system: Constraints from the New Horizons flyby. *Icarus, 287,* 2–11.
Merlin F. (2015) New constraints on the surface of Pluto. *Astron. Astrophys., 582,* A39.
Merlin F., Dumas C., Barucci M. A., et al. (2006) Spectroscopic analysis of the bright TNOs 2003 UB313 and 2003 EL61. *Bull. Am. Astron. Soc., 38,* 556.
Merlin F., Alvarez-Candal A., Delsanti. A., et al. (2009) Stratification of methane ice on Eris' surface. *Astron. J., 137,* 315–328.
Merlin F., Barucci M. A., de Bergh C., et al. (2010) Chemical and physical properties of the variegated Pluto and Charon surfaces. *Icarus, 210,* 930–943.
Mommert M. (2013) Remnant planetesimals and their collisional fragments: Physical characterization from thermal infrared observations. Ph.D. thesis, Freie Universitat Berlin, DOI: 10.17169/refubium-6484.
Mommert M., Harris A. W., Kiss C., et al. (2012) TNOs are Cool: A survey of the trans-neptunian region. V. Physical characterization of 18 plutinos using Herschel-PACS observations. *Astron. Astrophys., 541,* A93.
Müller T. G. and Lagerros J. S. V. (1998) Asteroids as far-infrared photometric standards for ISOPHOT. *Astron. Astrophys., 338,* 340–352.
Müller T. G., Lellouch E., Stansberry J., et al. (2010) TNOs are Cool: A survey of the trans-neptunian region. I. Results from the Herschel science demonstration phase (SDP). *Astron. Astrophys., 518,* L146.
Müller T., Kiss C., Alí-Lagoa V., et al. (2019) Haumea's thermal emission revisited in the light of the occultation results. *Icarus, 334,* 39–51.
Müller T., Lellouch E., and Fornasier S. (2020) Trans-neptunian objects and Centaurs at thermal wavelengths. In *The Transneptunian Solar System* (D. Prialnik et al., eds.), pp. 153–182. Elsevier, Netherlands.
Noll K. S., Luu J., and Gimore D. (2000) Spectrophotometry of four Kuiper belt objects with NICMOS. *Astron. J., 119,* 970–976.
Noll K. S., Grundy W. M., Chiang E. I., et al. (2008) Binaries in the Kuiper belt. In *The Solar System Beyond Neptune* (M. A. Barucci et al., eds.) pp. 345–363. Univ. of Arizona, Tucson.
Null G. W. and Owen T. C. (1996) Charon/Pluto mass ratio obtained with HST CCD observations in 1991 and 1993. *Astron. J., 111,* 1368–1381.
Null G. W., Owen T. C., and Synnott S. P. (1993) Masses and densities of Pluto and Charon. *Astron. J., 105(6),* 2319–2335.
Olkin C. B., Wasserman L. H., and Franz O. G. (2003) The mass ratio of Charon to Pluto from Hubble Space Telescope astrometry with the fine guidance sensors. *Icarus, 164(1),* 254–259.
Olkin C. B. and 23 colleagues (2015) Evidence that Pluto's atmosphere does not collapse from occultations including the 2013 May 04 event. *Icarus, 246,* 220–225.
Ortiz J. L., Gutiérrez P. J., Casanova V., and Sota A. (2003) A study of short term rotational variability in TNOs and Centaurs from Sierra Nevada Observatory. *Astron. Astrophys., 407,* 1149–1155.
Ortiz J. L, Gutiérrez P. J., Santos-Sanz P., Casanova V., and Sota A. (2006) Short-term rotational variability of eight KBOs from Sierra Nevada Observatory. *Astron. Astrophys., 447,* 1131–1144.
Ortiz J. L., Sicardy B., Braga-Ribas F., et al. (2012) Albedo and atmospheric constraints of dwarf planet Makemake from a stellar occultation. *Nature, 49,* 566.
Ortiz J. L., Duffard R., Pinilla-Alonso N., et al. (2015) Possible ring material around Centaur (2060) Chiron. *Astron. Astrophys., 576,* A18.
Ortiz J. L., Santos-Sanz P., Sicardy B., et al. (2017) The size, shape, density, and ring of the dwarf planet Haumea from a stellar occultation. *Nature, 550,* 219–223.
Ortiz J. L., Sicardy B., Camargo J. I. B., et al. (2019) Stellar occultations by transneptunians objects: From predictions to observations and prospects for the future. In *The Transneptunian Solar System* (D. Prialnik et al., eds.), pp. 413–438. Elsevier, Netherlands.
Owen T. C., Roush T. L., Cruikshank D. P., et al. (1993) Surface ices and atmospheric composition of Pluto. *Science, 261,* 745–48.
Pàl A., Kiss C., Müller T. G., et al. (2012) TNOs are Cool: A survey of the trans-neptunian region. VII. Size and surface characteristics of (90377) Sedna and 2010 EK139. *Astron. Astrophys., 541,* L6.
Pàl A., Kiss C., Horner J., et al. (2015a) Physical properties of the

extreme Centaur and super-comet candidate 2013 AZ60. *Astron. Astrophys., 583,* A93.

Pàl A., Szabó R., Szabó G. M., et al. (2015b) Pushing the limits: K2 observations of the trans-neptunian objects 2002 GV31 and (278361) 2007 JJ43. *Astron. J., 804,* 45.

Pàl A., Kiss C., Müller T., et al. (2016) Large size and slow rotation of the trans-neptunian object (225088) 2007 OR10. Discovered from Herschel and K2 observations. *Astron. J., 151,* 117.

Pan M. and Sari R. (2005) Shaping the Kuiper belt size distribution by shattering large but strengthless bodies. *Icarus, 173,* 342–348.

Parker A., Pinilla-Alonso N., Santos-Sanz P., et al. (2016) Physical characteristics of TNOs with the James Webb Space Telescope. *Publ. Astron. Soc. Pac., 128,* 018010.

Peixinho N., Doressoundiram A., Delsanti A., et al., (2003) Reopening the TNOs color controversy: Centaurs bimodality and TNOs unimodality. *Astron. Astrophys., 410,* L29–L32.

Peixinho N., Delsanti A., Guilbert-Lepoutre A., et al. (2012) The bimodal colors of Centaurs and small Kuiper belt objects. *Astron. Astrophys., 546,* A86.

Perna D., Dotto E., Barucci M. A., et al. (2009) Rotations and densities of trans-neptunian objects. *Astron. Astrophys., 508,* 451–455.

Perna D., Barucci M. A., Fornasier S., et al. (2010) Colors and taxonomy of Centaurs and trans-neptunian objects. *Astron. Astrophys., 510,* 53.

Perna D., Hromakina T., Merlin F., et al. (2017) The very homogeneous surface of the dwarf planet Makemake. *Mon. Not. R. Astron. Soc., 3,* 3594.

Petit J.-M., Kavelaars J. J., Gladman B., and Loredo T. (2008) Size distribution of multikilometer transneptunian objects. In *The Solar System Beyond Neptune* (M. A. Barucci et al., eds.), pp. 71–87. Univ. of Arizona, Tucson.

Petit J.-M., Kavelaars J. J., Gladman B. J., et al. (2011) The Canada-France Ecliptic Plane Survey : Full data release: The orbital structure of the Kuiper belt. *Astron. J., 142,* 131.

Pike R. E., Fraser W. C., Schwamb M. E., et al. (2017) Col-OSSOS: Z-band photometry reveals three distinct TNO surface type. *Astron. J., 154,* 101–109.

Porter S. B., Spencer J. R., Benecchi S., et al. (2016) The first high-phase observations of a KBO: New Horizons Imaging of (15810) 1994 JR1 from the Kuiper belt. *Astrophysical. J. Lett., 828,* L15.

Preusker F., Scholten F., Matz K.-D., et al. (2017) The global meter-level shape model of comet 67P/Churyumov-Gerasimenko. *Astron. Astrophys., 607,* L1.

Protopapa S., Grundy W. M., Reuter D. C., et al. (2017) Pluto's global surface composition through pixel-by-pixel Hapke modeling of New Horizons Ralph/LEISA data. *Icarus, 287,* 218–228.

Quirico E. and Schmitt B. (1997) A spectroscopic study of CO diluted in N_2 ice: Applications for Triton and Pluto. *Icarus, 128,* 181.

Rabinowitz D. L., Benecchi S. D., Grundy W. M., et al. (2014) The rotational light curve of (79360) Sila-Nunam, an eclipsing binary in the Kuiper belt. *Icarus, 236,* 72–82.

Reuter D. C., Stern S. A., Scherrer J., et al. (2008) Ralph: A visible/infrared imager for the New Horizons Pluto/ Kuiper belt mission. *Space Sci. Rev., 140,* 129–154.

Rietmeijer F. J. M. and Nuth J. A. (2001) Serpentine by hydrogenation of Fe-rich ferromagnesiosilica PCs in aggregate IDPs. In *Lunar and Planetary Science XXXII,* Abstract #1219. Lunar and Planetary Institute, Houston.

Roe H. G., Pike R. E., and Brown M. E. (2008) Tentative detection of the rotation of Eris. *Icarus, 198(2),* 459–464.

Romanishin W. and Tegler S. C. (1999) Rotation rates of Kuiper belt objects from their light curves. *Nature, 398(6723),* 129–132.

Romanishin W. and Tegler S. C. (2018) Albedos of Centaurs, jovian Trojans, and Hildas. *Astron. J., 156,* 19.

Rousselot P., Petit J.-M., Poulet F., et al. (2005) Photometric study of Centaur (60558) 2000 EC 98 and trans-neptunian object (55637) 2002 UX 25 at different phase angles. *Icarus, 176,* 478–491.

Santos-Sanz P., Lellouch E., Fornasier S., et al. (2012) TNOs are Cool: A survey of the transneptunian region. IV. Size/albedo characterization of 15 scattered disk and detached objects observed with Herschel Space Observatory-PACS. *Astron. Astrophys., 541,* A92.

Schaller E. L. and Brown M. E. (2007) Volatile loss and retention on Kuiper belt objects. *Astrophys. J., 659,* 61.

Schenk P. M. and 19 colleagues (2018) Basins, fractures, and volcanoes: Global cartography and topography of Pluto from New Horizons. *Icarus, 314,* 400–433.

Schwamb M. E., Fraser W. C., Bannister M. T., et al. (2019) Co-I-OSSOS: The Colors of the Outer Solar System Origins Survey. *Astrophys. J., 243,* 12.

Sheppard S. (2007) Light curves of dwarf plutonian planets and other large Kuiper belt objects: Their rotations, phase functions, and absolute magnitudes. *Astron. J., 134,* 787–798.

Sheppard S. and Jewitt D. (2002) Time-resolved photometry of Kuiper belt objects: Rotations, shapes, and phase functions. *Astron. J., 124,* 1757–1775.

Sheppard S. and Jewitt D. (2003) Hawaii Kuiper Belt Variability Project: An update. *Earth Moon Planets, 92,* 207–219.

Sheppard S. and Jewitt D. (2004) Extreme Kuiper belt object 2001 QG298 and the fraction of contact binaries. *Astron. J., 127,* 3023–3033.

Sheppard S., Lacerda P, and Ortiz J. L. (2008) Photometric lightcurves of transneptunians objects and Centaurs: Rotations, shapes, and densities. In *The Solar System Beyond Neptune* (M. A. Barucci et al., eds.), pp. 129–142. Univ. of Arizona, Tucson.

Shkuratov Y., Starukhina L., Hoffmann H., and Arnold G. (1999) A model of spectral albedo of particulate surfaces: Implications for optical properties of the Moon. *Icarus, 137,* 235–246.

Sicardy B. and 44 colleagues (2006) Charon's size and an upper limit on its atmosphere from a stellar occultation. *Nature, 439,* 52–54.

Sicardy B., Ortiz J. L., Assafin M., et al. (2011) A Pluto-like radius and a high albedo for the dwarf planet Eris from an occultation. *Nature, 478,* 493–496.

Snodgrass C., Carry B., Dumas C. et al. (2010) Characterisation of candidate members of (136108) Haumea's family. *Astron. Astrophys., 511,* A72.

Stansberry J. A., Grundy W. M., Margot J. L., et al. (2006) The albedo, size, and density of binary Kuiper belt object (47171) 1999 TC36. *Astrophys. J., 643,* 556–566.

Stansberry J., Grundy W., Brown M., et al. (2008) Physical properties of Kuiper belt and Centaur objects: Constraints from the Spitzer Space Telescope. In *The Solar System Beyond Neptune* (M. A. Barucci et al., eds.), pp. 161–179. Univ. of Arizona, Tucson.

Stansberry J. A., Grundy W. M., Mueller M., Benecchi S. D., Rieke G. H., Noll K. S., et al. (2012) Physical properties of trans-neptunian binaries (120347) Salacia-Actaea and (42355) Typhon-Echidna. *Icarus, 219,* 676–688.

Steckloff J. and Sarid G. (2019) Amorphous water ice within icy planetesimals survives collision evolution in the early solar system. *EPSC Abstracts, 13,* EPSC-DPS 2019-568-2.

Stern S. A., Weintraub D. A., and Festou M. C. (1993) Evidence for a low surface temperature on Pluto from millimeter-wave thermal emission measurements. *Science, 261,* 1713–1716.

Stern S. A., Buie M. W., Trafton L. M. (1997) HST high-resolution images and maps of Pluto. *Astron. J., 113,* 827–843.

Stern S. A., Bagenal F., Ennico K., Gladstone G. R., Grundy W. M., et al. (2015) The Pluto system: Initial results from its exploration by New Horizons. *Science, 350,* 292.

Stern S. A., Grundy W. M., McKinnon W. B., Weaver H. A., and Young L. A. (2018) The Pluto system after New Horizons. *Astron. Astrophys., 56,* 357–392.

Stern S. A., Weaver H. A., Spencer J. R., Olkin C. B., et al. (2019) Initial results from the New Horizons exploration of 2014 MU69, a small Kuiper belt object. *Science, 364(6441),* aaw9771.

Sykes M. V. (1999) IRAS survey-mode observations of Pluto-Charon. *Icarus, 142,* 155–159.

Sykes M. V., Cutri R. M., Lebofsky L. A., and Binzel R. (1987) IRAS serendipitous survey observations of Pluto and Charon. *Science, 237,* 1336–1340.

Tedesco E. F., Veeder G. J., Dunbar R. S., and Lebofsky L. A. (1987) IRAS constraints on the sizes of Pluto and Charon. *Nature, 327,* 127–129.

Tegler S. C. and Romanishin W. (1998) Two distinct population of Kuiper belt objects. *Nature, 392,* 49–51.

Tegler S. C., Romanishin W., Stone A., Tryka K., Fink U., and Fevig R. (1997) Photometry of the trans-neptunian object 1993 SC. *Astron. J., 114,* 1230–1233.

Tegler S. C., Romanishin W., Consolmagno G. J., Rall J., Worhatch R., Nelson M., and Weidenschilling S. (2005) The period of rotation, shape, density, and homogeneous surface color of the Centaur 5145 Pholus. *Icarus, 175(2),* 390–396.

Tegler S. C., Grundy W. M., Romanishin W., et al. (2007) Optical spectroscopy of the large Kuiper belt objects 136472 (2005 FY9) and 136108 (2003 EL61). *Astron. J., 133,* 526.

Tegler S. C., Bauer J. M., Romanishin W., and Peixinho N. (2008) Colors of Centaurs. In *The Solar System Beyond Neptune* (M. A. Barucci et al., eds.), pp. 105–114. Univ. of Arizona, Tucson.

Tegler S. C., Romanishin W., and Consolmagno G. (2016) Two color populations of Kuiper belt and Centaur objects and the smaller orbital inclinations of red Centaur objects. *Astron. J., 152,* 210.

Thirouin A. (2013) Study of trans-neptunian objects using photometric techniques and numerical simulations. Ph.D. thesis, Granada University, Spain, available online at *http://hdl.handle.net/10481/30832.*

Thirouin A. and Sheppard S. S. (2018) The plutino population: An abundance of contact binaries. *Astron. J., 155,* 248.

Thirouin A., Ortiz J. L., Duffard R., et al. (2010) Short-term variability of a sample of 29 trans-neptunian objects and Centaurs. *Astron. Astrophys., 522,* A93.

Thirouin A., Ortitz J. L., Campo Bagatin A., et al. (2012) Short-term variability of 10 trans-neptunian objects. *Mon. Not. R. Astron. Soc., 424,* 3156–3177.

Thirouin A., Noll K. S., Ortiz J. L., et al. (2014) Rotational properties of the binary and non-binary populations in the trans-neptunian belt. *Astron. Astrophys., 569,* A3.

Thirouin A., Sheppard S. S., Noll K. S., et al. (2016) Rotational properties of the Haumea family members and candidates: Short-term variability. *Astron. J., 151,* 148.

Tholen D. J. and Buie M. W. (1990) Further analysis of Pluto-Charon mutual event observations. *Bull. Am. Astron. Soc., 22,* 1129–1129.

Tholen D. and Buie M. W. (1997) Bulk properties of Pluto and Charon. In *Pluto and Charon* (S. A. Stern and D. J. Tholen, eds.), pp. 193–220. Univ. of Arizona, Tucson.

Trilling D. E. and Bernstein G. M. (2006) Light curves of 20–100 km Kuiper belt objects using the Hubble Space Telescope. *Astron. J., 131,* 1149–1162.

Vilenius E., Kiss C., Mommert M., Müller T., Santos-Sanz P., Pàl A., et al. (2012) TNOs are Cool: A survey of the trans-neptunian region. VI. Herschel/PACS observations and thermal modelling of 19 classical Kuiper belt objects. *Astron. Astrophys., 541,* A94.

Vilenius E., Kiss C., Müller T., Mommert M., Santos-Sanz P., Pàl A., et al. (2014) TNOs are Cool: A survey of the trans-neptunian region. X. Analysis of classical Kuiper belt objects from Herschel and Spitzer observations. *Astron. Astrophys., 564,* A35.

Vilenius E., Stansberry J., Müller T., Mueller M., Kiss C., Santos-Sanz P., et al. (2018) TNOs are Cool: A survey of the trans-neptunian region. XIV. Size/albedo characterization of the Haumea family observed with Herschel and Spitzer. *Astron. Astrophys., 618,* A136.

Wong I. and Brown M. E. (2016) A hypothesis for the color bimodality of Jupiter Trojans. *Astron. J., 152,* 90.

Young E. F., Binzel R. P., and Stern S. A. (1992) A frost model for Pluto's surface: Implications for Pluto's atmosphere. In *Lunar and Planetary Science XXIII,* Abstract #1563. Lunar and Planetary Institute, Houston.

Young L. A. (2013) Pluto's seasons: New predictions for New Horizons. *Astrophys. J. Lett., 766,* L22.

Part 2:

Pluto

White O. L., Moore J. M., Howard A. D., Schenk P. M., Singer K. N., Williams D. A., and Lopes R. M. C. (2021) The geology of Pluto. In *The Pluto System After New Horizons* (S. A. Stern, J. M. Moore, W. M. Grundy, L. A. Young, and R. P. Binzel, eds.), pp. 55–87. Univ. of Arizona, Tucson, DOI: 10.2458/azu_uapress_9780816540945-ch004.

The Geology of Pluto

O. L. White
SETI Institute, NASA Ames Research Center

J. M. Moore
NASA Ames Research Center

A. D. Howard
Planetary Science Institute

P. M. Schenk
Lunar and Planetary Institute

K. N. Singer
Southwest Research Institute

D. A. Williams
Arizona State University

R. M. C. Lopes
Jet Propulsion Laboratory, California Institute of Technology

---

Pluto's complex surface geology records nearly its entire history, from the heavy bombardment era more than 4 b.y. ago to the present day. The highly partitioned nature of Pluto's geology can be attributed to combinations of exogenic and endogenic processes acting to varying degrees upon the landscape; the primary influence appears to be atmosphere-surface volatile transport, which is strongly controlled by Pluto's eccentric seasons and climate zones. The hemisphere seen at close range during the New Horizons flyby shows ongoing surface geological activity centered on a vast basin containing a thick deposit of predominantly nitrogen ice that is undergoing thermal convection. Surrounding terrains show active glacial flow of volatile ices and apparent transport and rotation of large, buoyant water ice crustal blocks. A sequence of methane ice deposits displaying a complex bladed texture stretches around much of Pluto's equatorial region. Tall, enigmatic mounds with large central depressions are conceivably cryovolcanic. The most ancient and cratered terrains, seen in the northern and western regions of the nearside hemisphere, are extensionally faulted, partly eroded by glaciation and sublimation, and mantled by volatile ices to varying degrees at high latitudes and by a dark blanket of atmospheric haze particles in equatorial regions.

1. INTRODUCTION

Prior to 2015, our understanding of Pluto was limited only to its orbital characteristics, bulk geophysical properties (i.e., size and density), and albedo and surface composition (*Moore et al.,* 2015). Modeling of Pluto's interior and thermal history in the previous decades (*McKinnon et al.,* 1997; *Hussmann et al.,* 2006; *Robuchon and Nimmo,* 2011) had shown that Pluto might have experienced significant geologic activity, and various studies theorized as to how Pluto had been affected by processes such as impact cratering, sedimentary processes (including volatile migration), aeolian processes, tectonism, cryovolcanism, and viscous relaxation of topography (e.g., *Kamata and Nimmo,* 2014; *Barr and Collins,* 2015; *Bierhaus and Dones,* 2015; *Bray and Schenk,* 2015; *Greenstreet et al.,* 2015; *Moore et al.,* 2015; *Stern et al.,* 2015a; *Neveu et al.,* 2015). Detailed examination of Pluto's geologic record, however, remained the only realistic means of reconciling theory and observation.

The flyby of NASA's New Horizons spacecraft (*Stern,* 2008) past Pluto at a closest approach distance of 12,500 km above its surface on July 14, 2015 yielded robust datasets that permitted detailed geological analysis. New Horizons imaged ~78% of Pluto's surface, with the nearside (i.e., closest approach) hemisphere being the ~50% of the surface that was imaged between 76 and 850 m/pixel (Fig. 1). The remaining >25% of the imaged surface is the farside, imaged at pixel scales between 2.2 and 32 km/pixel, typically allowing only surface features on a scale of tens of kilometers to be discerned for this hemisphere. In addition to the portion of Pluto that was directly illuminated during flyby, the southern edge of the nearside hemisphere incorporates a swath of haze-lit imaging in which surface features beyond the terminator are revealed by dim light scattered by Pluto's atmospheric haze. This haze-lit region is a valuable bonus to the area of Pluto mapped at high resolution, but the low signal-to-noise ratio hinders visual assessment of surface relief. A crucial data product for interpreting Pluto's geology is a digital elevation model (DEM) of the nearside hemisphere, created using stereophotogrammetric analysis of New Horizons images (Fig. 1), and that covers >42% of Pluto's surface (*Schenk et al.*, 2018).

These datasets revealed an unexpectedly diverse range of terrains, implying a complex geological history (*Stern et al.*, 2015b; *Moore et al.*, 2016). Pluto's geological provinces are often highly distinct, can exhibit disparate crater spatial densities (*Moore et al.*, 2016; *Robbins et al.*, 2017; *Singer et al.*, 2019), and display evidence for having been affected by both endogenic and exogenic energy sources (including internal heating and insolation/climatic effects). Most surprisingly, large-scale surface renewal in response to internal heating is ongoing through the present day, as demonstrated compellingly by the sprawling, convecting nitrogen ice plains of Sputnik Planitia (*McKinnon et al.*, 2016; *Trowbridge et al.*, 2016; *Vilella and Deschamps*, 2017; *Buhler and Ingersoll*, 2018; *Wei et al.*, 2018). This landform, which dominates Pluto's nearside, has likely been the single most powerful influence on Pluto's geological and atmospheric evolution for much of its history. It has been the subject of investigation focusing on its implications for the physical state of Pluto's interior, as well as Pluto's tectonism and polar orientation (*Johnson et al.*, 2016; *Keane et al.*, 2016; *Nimmo et al.*, 2016; *Hamilton et al.*, 2016; *Conrad et al.*, 2019; *Kamata et al.*, 2019; *Kimura and Kamata*, 2020). At both global and local scales, mobilization and transport of volatile ices and atmospheric haze particles across Pluto on geological timescales appear to have played a prominent role in determining the appearance and distribution of Pluto's highly varied landscapes (*Bertrand and Forget*, 2016; *Binzel et al.*, 2017; *Stern et al.*, 2017; *Earle et al.*, 2017, 2018a,b; *Forget*

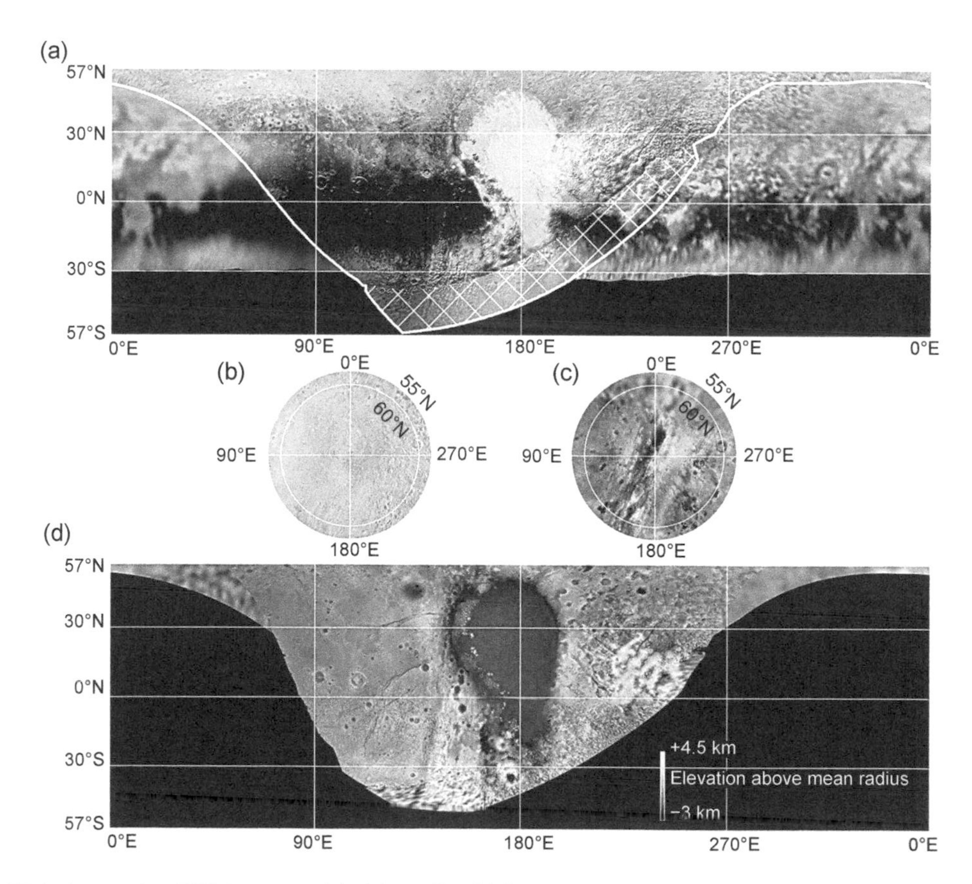

Fig. 1. (a),(b) Global mosaic of Pluto assembled from the highest-resolution New Horizons imaging available at all locations, shown in simple cylindrical projection between 57°N and 57°S, and in polar projection north of 55°N. Black areas were not imaged. The nearside hemisphere is above the white line and the farside is below it. The cross-hatched zone indicates the region of haze-lit imaging. **(c),(d)** Stereo DEM of Pluto's nearside hemisphere, projected in the same manner as the mosaic. Black areas were not imaged or do not have resolvable stereogrammetric data. Mean radius of Pluto is 1188.3 km.

et al., 2017; *Grundy et al.,* 2018; *Bertrand et al.,* 2018, 2019, 2020a,b; *Johnson et al.,* 2021; *Lewis et al.,* 2021). Some terrain types on Pluto appear to be unique within the solar system, and frequently reveal the importance of nitrogen ice glaciation and surface-atmosphere interactions throughout Pluto's history (*Moore et al.,* 2016, 2017, 2018; *Moores et al.,* 2017; *Howard et al.,* 2017a,b; *Umurhan et al.,* 2017; *White et al.,* 2017, 2019; *Schenk et al.,* 2018; *Telfer et al.,* 2018; *Cruikshank et al.,* 2019a,b, 2021). Perhaps the most intriguing of Pluto's landforms are the putative cryovolcanic edifices of Wright and Piccard Montes (*Moore et al.,* 2016; *Singer et al.,* 2016; *Martin and Binzel,* 2021).

To date, the International Astronomical Union (IAU) has assigned official names to 40 surface features on Pluto. The nomenclature section at the end of this volume (Appendix A by Beyer and Showman) lists the names and coordinates of these features and maps their locations across Pluto, as well as those of other features that as yet only have informal names that have not been sanctioned by the IAU. The section also defines the landform descriptor terms.

As the introduction to the Pluto section of this volume, this chapter will provide the reader with a broad overview of Pluto's geological units, and relate Pluto's global geological history as determined from stratigraphic relations and crater counts. Subsequent chapters focus on more specific topics associated with Pluto's geology, surface properties and composition, and atmosphere.

2. MATERIAL UNITS

New Horizons imaging and data products including the global DEM (*Schenk et al.,* 2018) and compositional mapping (*Grundy et al.,* 2016; *Protopapa et al.,* 2017; *Schmitt et al.,* 2017; *Dalle Ore et al.,* 2019; *Cook et al.,* 2019) allow geological mapping of Pluto. The fundamental application of this is to determine the sequence of geological events represented by the mapped units and the derivation of the geological history expressed on the surface. Mapped units are considered to be composed of three-dimensional materials (i.e., they have thickness that extends below the surface), which can be placed in a stratigraphic position relative to other units and structures on the basis of crosscutting, superposition, and embayment relationships (*Wilhelms,* 1990). Consequently, geological mapping is a tool for understanding the evolution of a planetary surface.

This section describes six geological groups that have been identified for Pluto's nearside hemisphere (Fig. 2), with each group consisting of units that represent a distinct episode of geological activity on Pluto's surface (section 4 separately discusses the geology of Pluto's less-well-resolved farside). The groups are described in chronological order from youngest to oldest, with descriptions and interpretations of their constituent terrains. Pluto's nearside features a depression ~1300 km × ~1000 km wide that hosts the nitrogen-ice plains of Sputnik Planitia. This depression is interpreted to be an impact basin, the elliptical planform of which implies an oblique impact either from the south-southeast or north-northwest (*Elbeshausen et al.,* 2013; *Stern et al.,* 2015b, 2021; *Moore et al.,* 2016; *McKinnon et al.,* 2016; *Schenk et al.,* 2018). The original basin rim is among the oldest geological features identified on Pluto, on account of its enormous size and the fact that the basin rim is crosscut by a graben that is itself highly eroded (*Schenk et al.,* 2018). In its present form, however, it is merely a topographic feature and not a group in its own right. It has been variously and extensively modified around its perimeter since its formation, with none of the original rim material remaining in a pristine state. The southern rim is absent, with the nitrogen ice plains of Sputnik Planitia that fill the basin extending to and covering low-elevation terrain to the south.

2.1. Sputnik Group

The Sputnik Group (Fig. 3) represents the youngest surface on Pluto and includes units that are experiencing resurfacing into the present day. The rate of resurfacing, however, does not produce surface changes on a scale resolvable by New Horizons imaging of the nearside (hundreds of meters to >1 km per pixel) to be discerned across the multi-hour timescale of the encounter (*Hofgartner et al.,* 2018). It consists of deposits of nitrogen and carbon monoxide ice in solid solution (*Stern et al.,* 2015b; *McKinnon et al.,* 2016),

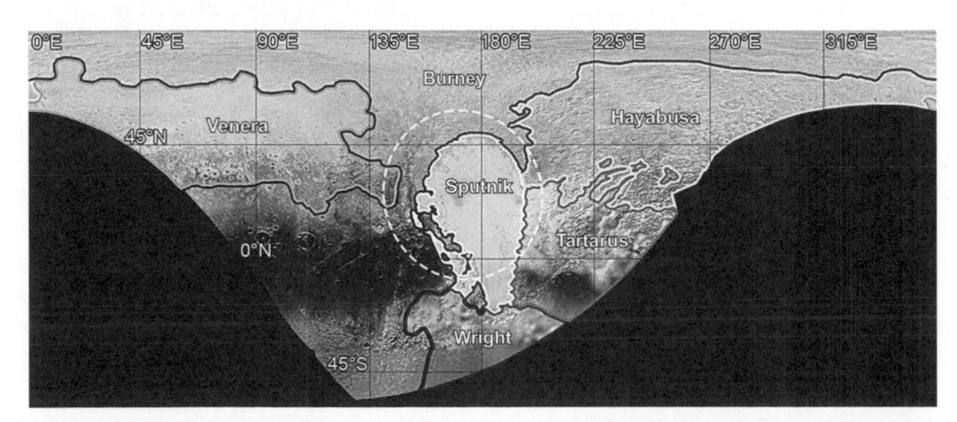

Fig. 2. Mapping of the approximate boundaries of the six geological groups identified in the nearside hemisphere (i.e., the portion of Pluto covered by imaging between 76 and 850 m/pixel). The white dashed line indicates the approximate original rim of the interpreted Sputnik impact basin.

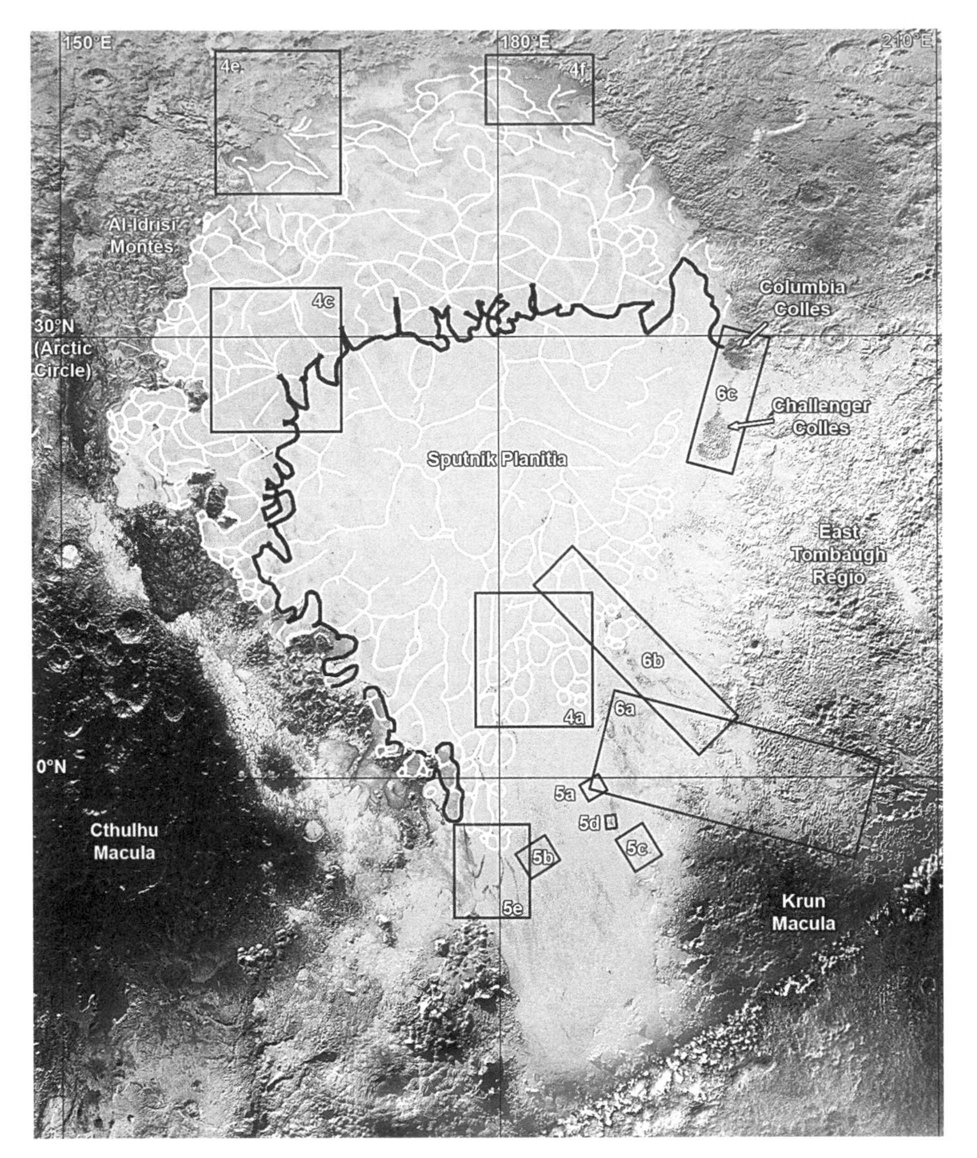

Fig. 3. Sputnik Planitia and East Tombaugh Regio. White lines mark troughs defining cell boundaries in the cellular plains. The black line marks the boundary between the bright plains to the south and the dark plains to the north and west. Numbered boxes refer to elements in Figs. 4, 5, and 6.

in addition to a surficial dusting and possibly a mixed-in element of trace amounts of methane ice. The deposits partially fill the giant Sputnik impact basin (*Moore et al.,* 2016; *Schenk et al.,* 2018) and depressed terrain to its south, forming Sputnik Planitia, which represents the western portion of the bright, heart-shaped feature of Tombaugh Regio that dominates Pluto's nearside.

The surface of Sputnik Planitia is 2.5 to 3.5 km lower in elevation relative to the rim of the basin, and ~2 km lower than Pluto's mean radius. On a scale of hundreds of kilometers, the plains are very flat, but dip down to form a trough at their northern and northeastern margin that is ~1 km below the level of the central plains. The overwhelming concentration of Pluto's nitrogen and carbon monoxide ice (heretofore referred to simply as "nitrogen ice") within the Sputnik basin (e.g., *Grundy et al.,* 2016; *Lewis et al.,* 2021) is a consequence of its very low elevation (the basin's depth is estimated to be as much as ~10 km) (*McKinnon et al.,* 2016), and location in the low- to mid-latitudes, which receive the lowest insolation on Pluto (*Bertrand and Forget,* 2016; *Hamilton et al.,* 2016; *Earle et al.,* 2017), averaged over a ~2.8-Ma obliquity cycle (*Dobrovolskis and Harris,* 1983; *Dobrovolskis et al.,* 1997). These factors combine to make the Sputnik basin a very powerful "cold trap" and a natural setting for the accumulation of volatiles. The basin has been interpreted to be one of the oldest geologic features on Pluto's nearside ($\gtrsim$4 G.y.) (*Moore et al.,* 2016; *Schenk et al.,* 2018), and modeling of volatile behavior in response to topography has shown that infilling of the basin with the majority of surface nitrogen ice would be complete by tens of millions of years after its formation (*Bertrand et al.,* 2018).

Sputnik Planitia has therefore existed on Pluto's surface for much of its history, but the sensitivity of its nitrogen ice to mobilization by climate change and atmospheric circulation, as well as by Pluto's internal heat flux, has caused what has probably been continuous resurfacing of the plains since its formation. No impact craters have been identified in Sputnik Planitia or any other occurrences of ponded nitrogen ice in contiguous mapping coverage of 315 m/pixel or in any higher-resolution strips (*Moore et al.,* 2016; *Robbins et al.,* 2017; *Singer et al.,* 2019), giving Sputnik Planitia a maximum crater retention age of only 30 to 50 m.y. (see the chapter by Singer et al. in this volume). It should be noted that assessment of the timescales of various geological processes affecting Sputnik Planitia is largely based on the present understanding of nitrogen-ice rheology, which to date has not been extensively investigated, and of which there are contradictory assessments (*Eluszkiewicz and Stevenson,* 1990; *Yamashita et al.,* 2010). The Sputnik Group (Fig. 3) can be broadly divided into cellular and non-cellular plains, with the former being restricted to central and northern Sputnik Planitia, and the latter occurring in southern Sputnik Planitia, its eastern and western margins, and all the uplands settings.

2.1.1. Cellular plains. The central and northern portions of Sputnik Planitia display a reticulate pattern of troughs a few kilometers wide that divide the plains into ovoid cells (white lines in Fig. 3) (*Moore et al.,* 2016). One hundred ninety-nine cells have been identified, with a mean cell diameter of 31 km and minimum and maximum cell diameters of 8.5 km and 110 km respectively (*White et al.,* 2017). The trough network is often fragmentary and incomplete, especially within a ~500-km-wide-region in the center of the cellular plains where cell size tends to increase.

The troughs generally separate the cells from each other, except in some instances where wider expanses of non-cellular plains separate the cells ("1" labels in Fig. 4a). A few isolated cells surrounded by non-cellular plains and mountains are seen at the southeastern and southwestern boundaries of the cellular plains (e.g., "2" labels in Fig. 4a). The cellular plains display a pitted texture, with individual pits reaching a few hundred meters across. The troughs commonly display a low, medial ridge ("3" labels in Fig. 4b). The pits and troughs are particularly well-defined in the southeastern bright cellular plains, primarily due to the high-solar-incidence angles encountered here (~70°). Preliminary photoclinometric, or shape-from-shading, DEMs generated for this area indicate that the cells are elevated in their central portions by 100 to 150 m, and that the troughs and pits display depths of some tens of meters (*Beyer et al.,* 2019). The surface texture at the cell centers is typically smooth in 76 m/pixel imaging, becoming more densely pitted toward the edges, although instances are encountered where smooth-textured zones extend all the way to the cell boundaries ("8" labels in Fig. 4d) and even cut across troughs (label "4" in Fig. 4b). The boundary between the smooth and pitted textures is generally sharply defined and not gradational. Isolated troughs are sometimes seen within cell interiors (e.g., the "X"-shaped junction of four troughs indicated by label "5" in Fig. 4b).

The cellular morphology of these plains is interpreted to be the consequence of active solid-state convection from thermal input from within or below the glacier (*Stern et al.,* 2015b; *Moore et al.,* 2016; *McKinnon et al.,* 2016; *Trowbridge et al.,* 2016). The medial ridges result from the compressive forces that prevail at the cell boundaries, where laterally migrating nitrogen ice is converging and being pushed up at the centers of the troughs. Instances of smooth or pitted regions crosscutting troughs dividing cells, as well as isolated troughs within cells, may indicate instability in the convective system, with cell boundaries decaying as small cells merge to form larger ones. The nitrogen ice of the cellular plains is filling the Sputnik impact basin, where the thickness attained by the nitrogen ice is interpreted to exceed ~1 km, which is the minimum thickness necessary to support convection for estimated present-day heat flow conditions on Pluto (*McKinnon et al.,* 2016). The increase in cell size and decrease in trough connectivity toward the center of the plains is consistent with a thickened nitrogen-ice layer there, where the depth of the basin is expected to reach a maximum (*McKinnon et al.,* 2016). The pitted texture is caused by sublimation of nitrogen ice from the surface of the plains.

Independent estimates of the respective rates of convective flow and nitrogen-ice sublimation (*McKinnon et al.,* 2016; *Moore et al.,* 2017) indicate that the nitrogen-ice layer probably convects not in an isoviscous regime, but in a sluggish lid regime, whereby the surface is in motion but moves at a much slower rate than the deeper, warmer subsurface. Numerical modeling of sluggish lid convection indicates average surface horizontal velocities of a few centimeters to a few tens of centimeters each terrestrial year for the nitrogen ice within the cells, implying a surface renewal timescale of ~420,000 to ~890,000 years (*McKinnon et al.,* 2016; *Trowbridge et al.,* 2016; *Buhler and Ingersoll,* 2018). There are competing hypotheses regarding the energy source of the convection, including radiogenic heat production in Pluto's rock fraction (which presently yields a conductive heat flow of ~3 mW m^{-2}) (*McKinnon et al.,* 2016; *Trowbridge et al.,* 2016) and a volumetrically heated system within Sputnik Planitia itself (*Vilella and Deschamps,* 2017). See the chapter by McKinnon et al. in this volume for further discussion.

Whereas a cellular pattern exists in the northern, dark, trough-bounding plains, there are indications that convective activity in this marginal region of Sputnik Planitia is less vigorous than in the south. Coarsely pitted, non-cellular plains are distributed among the cells and, as is the case for the non-cellular plains in southern Sputnik Planitia, the nitrogen deposits here may be thin enough such that convective overturn cannot be supported. At some locations proximal to the edge of Sputnik Planitia, the cellular configuration gives way to lobate flow patterns as revealed by albedo contrasts in the plains ("11" labels in Fig. 4e), and in some cases flow is seen to occur between nunataks (exposed peaks surrounded by glacial ice) of uplands material that are protruding above the plains ("14" labels in Fig. 4f) (*Stern et al.,* 2015b; *Umurhan et al.,* 2017). The nitrogen ice is mobile here, but is experiencing lateral flow toward the edges of Sputnik Planitia rather than convective overturn (*Umurhan et al.,* 2017). Such behavior is consistent with a regional

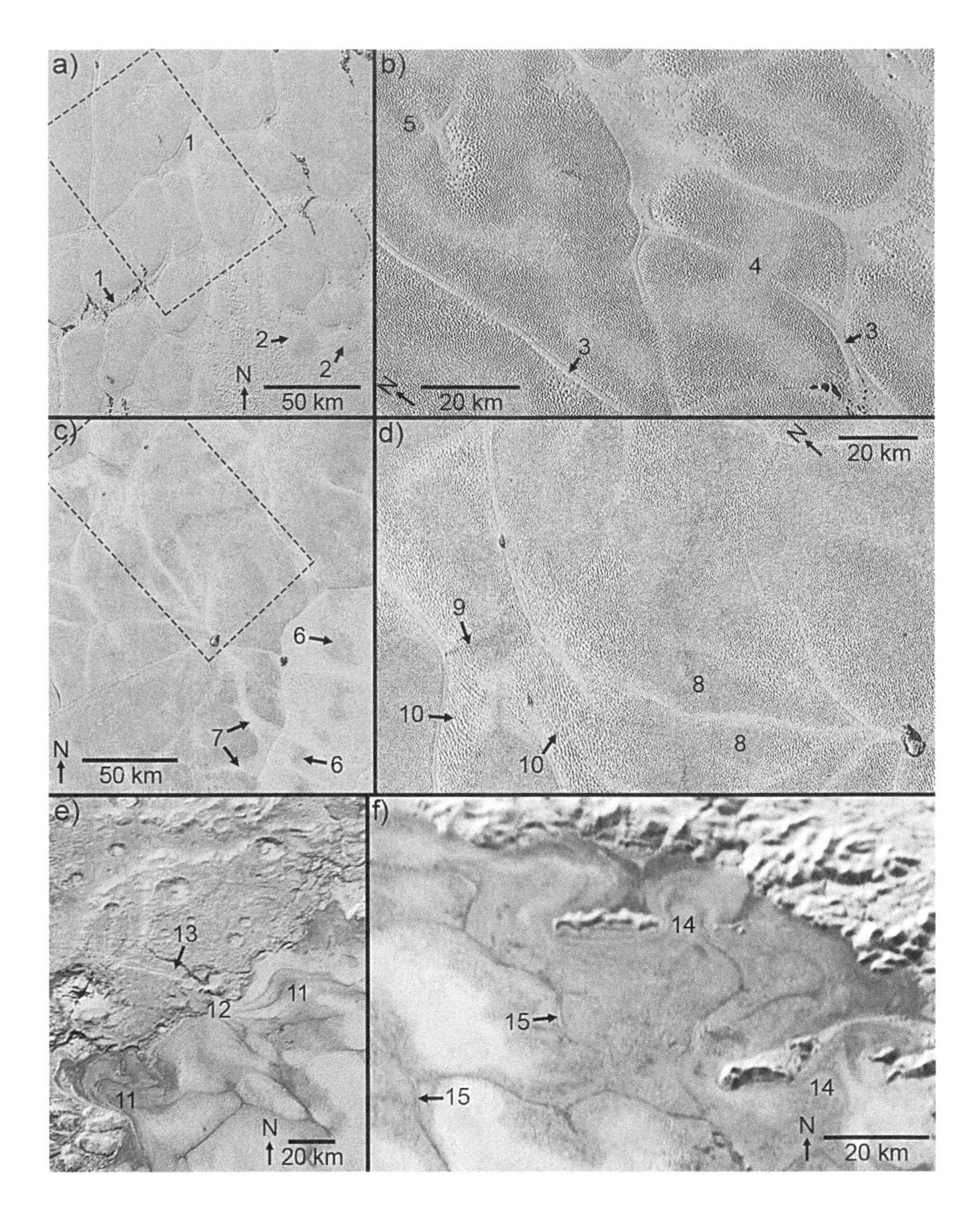

Fig. 4. Cellular plains in Sputnik Planitia. See Fig. 3 for context. **(a)** Bright cellular plains. Illumination is from the top left. Dashed box indicates boundary of area shown in **(b)**. 1: Interstitial expanses of noncellular plains. 2: Isolated cells at the southeast margin of the cellular plains. **(b)** Detail of bright cellular plains. Illumination is from the left. 3: Low, medial ridges within troughs separating the cells. 4: Smooth-textured zone cutting across a boundary between cells. 5: "X"-shaped junction of four troughs. **(c)** Contact between the dark (top left) and bright (bottom right) cellular plains. Illumination is from the top left. Dashed box indicates boundary of area shown in **(d)**. 6: Enclaves of dark plains surrounded by bright plains. 7: Filaments of bright plains extending into the dark plains. **(d)** Detail of dark cellular plains. Illumination is from the left. 8: Smooth-textured zones extending to the cell boundaries. 9: Dark streak emanating from a cell boundary. 10: Linear texture in pitted plains. **(e),(f)** Dark, trough-bounding plains in the northwest and northeast of Sputnik Planitia respectively. Illumination is from the top left. 11: Lobate flow patterns. 12: Convergence of flow bands toward the edge of Sputnik Planitia. 13: Uplands graben intersecting Sputnik Planitia. 14: Lobate flow patterns between nunataks. 15: Bright central lineations within dark cell boundaries.

downslope gradient approaching the northern boundary of Sputnik Planitia (*Howard et al.*, 2017a), the cause of which may be related to the current high rate of nitrogen-ice sublimation from these northern plains (*Bertrand et al.*, 2018). A strong convergence of banding (label "12" in Fig. 4e) toward a graben in the surrounding uplands that intersects Sputnik Planitia (label "13" in Fig. 4e) suggests that appreciable localized flow may have occurred toward the edge of Sputnik Planitia here, implying localized mass loss of nitrogen ice, perhaps to the subsurface through fracturing of the crust (*Howard et al.*, 2017a).

*2.1.2. **Non-cellular plains.*** These plains occur within the southern regions and periphery of Sputnik Planitia, and in the pitted uplands of East Tombaugh Regio. They are also seen on the floors of a few low-latitude impact craters to the west of Sputnik Planitia and in depressions within the bladed terrain deposits (*Moore et al.,* 2018). As with the cellular plains, these plains exhibit pitted textures, but the pits can reach much larger diameters (typically >1 km) and also show a broader range of morphologies. The transition from cellular to non-cellular plains is interpreted to reflect thinning of the volatile ice layer, and perhaps also declining heat flow (*McKinnon et al.,* 2016). The non-cellular plains are not convecting, but they have nevertheless very recently resurfaced. Whereas the maximum crater retention age of Sputnik Planitia is 10 m.y. (*Greenstreet et al.,* 2015), *Wei et al.* (2018) estimate that viscous relaxation of the nitrogen ice can eliminate crater morphology on a timescale as small as years to 10^4 years, depending on the ice viscosity.

The larger sizes attained by pits in the non-cellular plains compared to those in the cellular plains can be attributed at least in part to the <1-m.y. timescale predicted for resurfacing by convective overturn, which limits the sizes to which pits in the cellular plains can grow before they are eliminated by resurfacing (*Moore et al.,* 2017). But growth of pits by sublimation erosion in both the cellular and non-cellular plains must compete with other processes that reduce pit relief, including mantling by newly-deposited volatiles and infilling of the pits through glacial flow of the nitrogen ice, and the relative rates of these competing processes play an important role in controlling pit size and morphology (*Moore et al.,* 2017). The non-cellular plains in Fig. 5a demonstrate this explicitly. Plains in the north and west of the image display densely spaced pits that reach a few hundred meters across (label "1"), indicating that sublimation is the dominating process. But to the southeast, the texture transitions to smooth, sparsely pitted plains, where pits tend to be shallow and rounded (label "2"). Here, viscous relaxation of the ice may be occurring at a rate that rivals that of pit excavation via sublimation, resulting in decay and consequential subdued relief of the pits (*Moore et al.,* 2017).

Extending across both these types of plains are dense fields of large pits typically reaching ~1 km in diameter. These pits can be single or merge to form doublets, as well as chains that align parallel to each other (Fig. 5b), or sometimes combine to form curved, wavelike swarms (Fig. 5c). These patterns probably develop in response to non-homogenous, anisotropic substrate properties, such as compositional or textural (grain size or grain orientation) layering, fractures, or crevasses (*Moore et al.,* 2017). Some of these pit swarms are associated with inferred active glacial flows, and are elongated along flow margins, implying that the pits nucleated from aligned structures in the surficial ices. Instigation of shallow crevasse formation via shearing of flowing nitrogen ice could be exploited by sublimation to widen these small fractures into large, elongated and aligned pits. *Moore et al.* (2017) and *Wei et al.* (2018) argue that enhanced viscosity of the ice layer is necessary to support the relief of these large pits, which can attain depths of 100–200 m (*White et al.,* 2017; *Beyer et al.,* 2019). *Moore et al.* (2017) suggested that such enhancement might be due to an elevated concentration of a "stiffer" volatile ice such as methane. A higher viscosity would likely also be necessary to permit shearing of the ice for plausible strain rates, as otherwise the ice would merely deform via diffusive creep.

Wherever the floors of these pits are resolved, they always display a low albedo, and a rubbly texture can be discerned on the floor of the largest pit of all (label "3" in Fig. 5d). *White et al.* (2017) interpreted this to be dark material covering the pit floors, specifically haze particles entrained within the ice underneath a surficial layer of bright, newly deposited volatiles (section 2.1.4). As sublimation of the nitrogen ice progresses and pits deepen, the entrained

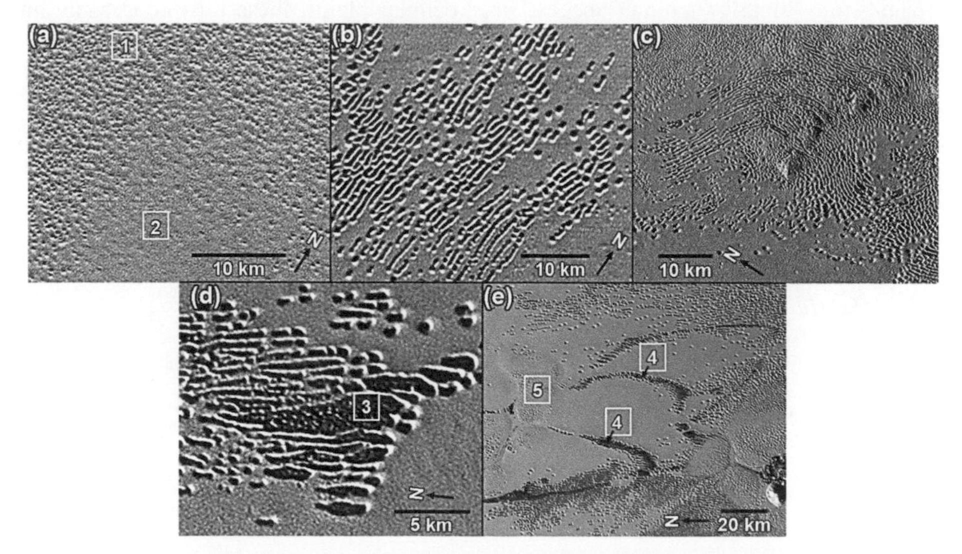

Fig. 5. Non-cellular plains in Sputnik Planitia. See Fig. 3 for context. **(a)** Transition from pitted plains (1) to smooth, sparsely pitted plains (2). Illumination is from the top. **(b)** Field of large, elongate, and aligned pits. Illumination is from the top. **(c)** Field of elongate and aligned pits forming a wavelike pattern. Illumination is from the left. **(d)** Field of especially wide, dark-floored pits. Illumination is from the bottom-left. 3: The largest pit, measuring 11.5 km long by 2.5 km wide. **(e)** Chains of joined pits with dark material covering their floors that form polygonal outlines (4), located south of the southernmost cells of the bright cellular plains (5). Illumination is from the bottom left.

haze particles are liberated and build up on the floors of the pits as a lag deposit. Eventual accumulation of a threshold thickness of haze particles on their floors would act to suppress further sublimation (e.g., *Moore et al.,* 1999; *White et al.,* 2016). In addition, some of the haze particles may be sourced directly from the atmosphere itself: Climate modeling indicates that accumulation of such particles in these southern regions of Sputnik Planitia has been especially high over the last few decades due to strong katabatic winds and condensation flows sweeping in from the surrounding dark uplands (*Bertrand et al.,* 2020a). For dark-floored pits directly adjacent to Sputnik Planitia's boundaries with dark uplands such as Cthulhu and Krun Maculae, where ice thicknesses are expected to be thinner than a few hundred meters, sublimation of the nitrogen ice may have penetrated all the way through the ice layer, revealing the dark substrate (*Moore et al.,* 2017; *White et al.,* 2017). At the southern extent of the bright cellular plains (label "5" in Fig. 5e), wide chains of large, joined, dark-floored pits trace the outlines of vaguely triangular polygons, with sparsely pitted and very flat plains existing within their interiors ("4" labels in Fig. 5e). *White et al.* (2017) interpreted these to be formerly active convection cells that were moved southward and their planforms distorted by lateral flow of the ice in this region of Sputnik Planitia. This would have shifted the cells away from the convective zone, causing their relief to relax and flatten.

2.1.3. Valley glaciers and rafting blocks. The most compelling evidence for ongoing mobilization of nitrogen ice of the non-cellular plains is seen at the boundary between Sputnik Planitia and the bright, pitted uplands of East Tombaugh Regio. When seen at high-solar-incidence angles (>70°), dark bands reaching a few kilometers across are often seen where the non-cellular plains occur in valleys leading from the uplands into Sputnik Planitia (label "1" in Fig. 6a). The bands are aligned parallel to the valley walls, with smaller bands sometimes converging to form larger ones. The bright, pitted uplands are interpreted to be where nitrogen sublimated from Sputnik Planitia has redeposited across the landscape immediately to the east, coating the rugged terrain with a continuous veneer of nitrogen ice, which flows downslope to pond in depressions throughout the uplands ("2" labels in Fig. 6a) (*Moore et al.,* 2016; *Howard et al.,* 2017a; *White et al.,* 2017; *Bertrand et al.,* 2018). At the western extreme of East Tombaugh Regio, the surface elevation of these ponded nitrogen-ice deposits is 1.5–2 km above that of Sputnik Planitia. The dark marginal bands in valleys connecting these deposits with Sputnik Planitia have been interpreted as fragments of the pitted uplands that have eroded and accumulated on glacial nitrogen ice flowing downslope through the valleys into Sputnik Planitia, in a manner similar to lateral moraines on terrestrial valley glaciers (*Moore et al.,* 2016; *Howard et al.,* 2017a; *White et al.,* 2017). After debouching into Sputnik Planitia, the dark bands radiate outward, indicating that the glacial flow is spreading out into the plains in a manner somewhat akin to piedmont glaciers on Earth (approximate boundaries indicated by "3" labels in Fig. 6a).

The moraine trails that form these dark flow bands consist of debris that is too small for individual objects to be resolved by the 315 m/pixel New Horizons imaging that covers these glaciers. But in addition to the moraine trails, chains of kilometer-sized hills that align along the paths of some valley glaciers ("4" labels in Fig. 6b) are also seen, indicating that glacial erosion of uplands material (likely methane ice) has caused large blocks to break off and then "float" in the denser nitrogen ice as they are rafted into the interior of Sputnik Planitia by the glaciers (*Moore et al.,* 2016; *Howard et al.,* 2017a; *White et al.,* 2017). Within the cellular plains, these hills congregate into densely packed

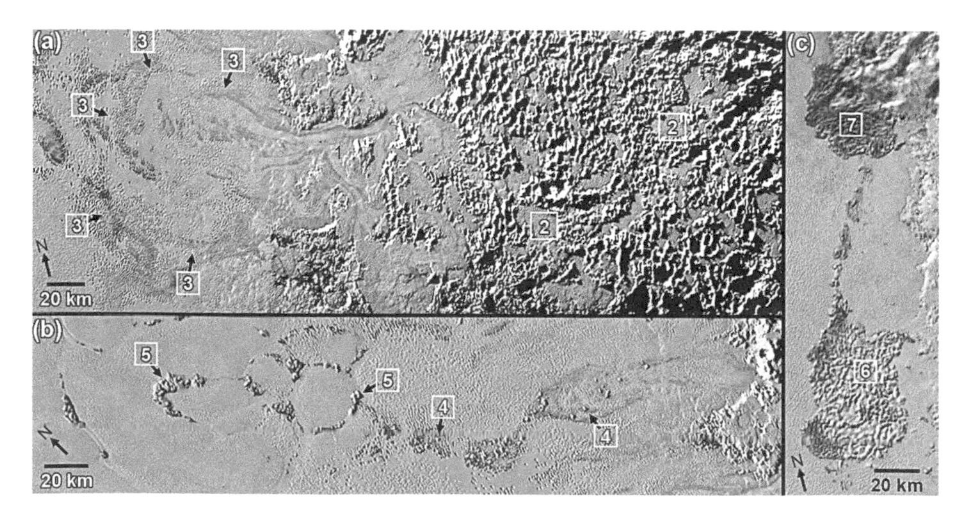

Fig. 6. Glacial landforms at the eastern margin of Sputnik Planitia. See Fig. 3 for context. **(a)** Glacial flow of nitrogen ice from the bright, pitted uplands of East Tombaugh Regio into Sputnik Planitia. Illumination is from the top left. 1: Dark, converging flow bands in valleys that connect upland flats to Sputnik Planitia. 2: Nitrogen ice ponded in depressions in East Tombaugh Regio. 3: Approximate boundaries of a glacier that has debouched into Sputnik Planitia. **(b)** A chain of hills (4) that coincides with the path of a glaciers entering Sputnik Planitia. Hills in the cellular plains cluster at cell boundaries (5). Illumination is from the left. **(c)** Challenger (6) and Columbia (7) Colles. Illumination is from the top left.

clusters at the cell boundaries that reach several to a few tens of kilometers across ("5" labels in Fig. 6b) as a consequence of the convective motion causing the buoyant blocks to be pushed to the edges of the cells. The largest conglomerations of hills, Challenger Colles and Columbia Colles (labels "6" and "7" respectively in Fig. 6c), cluster adjacent to the edge of Sputnik Planitia, and form roughly rectangular masses measuring several tens of kilometers across. These great accumulations of eroded blocks are likely grounded in the shallow nitrogen-ice deposits here (*White et al.,* 2017).

2.1.4. Surface albedo across Tombaugh Regio. The albedo of Tombaugh Regio is not uniform across its expanse. Much of it is uniformly bright (Bond albedo of 0.9–1), including Sputnik Planitia's bright cellular plains, its southern, eastern, and western non-cellular plains, and the pitted uplands and interstitial plains of East Tombaugh Regio (*Buratti et al.,* 2017), but the northern and northwest cellular plains of Sputnik Planitia are darker (Bond albedo of ~0.8). The contact between the bright and dark plains is seen in Fig. 4c. Spectral mapping of Tombaugh Regio indicates compositional differences between the bright and dark regions. The bright regions exhibit stronger nitrogen and carbon monoxide absorption relative to the darker ones (*Grundy et al.,* 2016; *Protopapa et al.,* 2017; *Schmitt et al.,* 2017). The darker regions in turn show a higher absorption corresponding to the raw spectral indicator of "dark red material" (*Schmitt et al.,* 2017), interpreted to be haze particles that have settled out of the atmosphere, and are sometimes referred to as tholins (*Stern et al.,* 2015b; *Grundy et al.,* 2016, 2018). Ongoing convection of the nitrogen ice would lead to continuous entrainment of surface haze particles within the ice throughout Pluto's history, causing their concentration within the ice to increase over time (*White et al.,* 2017; *Grundy et al.,* 2018).

The northern boundary of the bright portion of Tombaugh Regio is ~30°N, which is the present Arctic Circle at Pluto's current obliquity of 120° (Fig. 3). Terrain north of 30°N currently experiences continuous illumination over a diurnal period at least once during an orbit, whereas terrain between 30°N and 30°S experiences a consistent diurnal cycle. A consequence of this is that, averaged across an entire Pluto year, Pluto's Arctic experiences a higher insolation than terrain in the diurnal band (*Hamilton et al.,* 2016; *Earle et al.,* 2017). Therefore, regions of Tombaugh Regio to the north and south of 30°N respectively represent zones of net sublimation and deposition of nitrogen ice (*Bertrand and Forget,* 2016; *Forget et al.,* 2017; *White et al.,* 2017; *Lewis et al.,* 2021). The haze particle-enriched nitrogen ice of Sputnik Planitia is therefore only exposed at the surface in the northern zone of net sublimation, forming the dark plains. The southern portion of Sputnik Planitia is where a thin veneer of bright nitrogen ice has deposited onto the surface, which is thick enough (estimated at several meters) to obscure the lower albedo of the darker nitrogen ice beneath it, but not thick enough to mask or soften the topographic relief of the troughs, ridges, and pits. Proximal to the main contact between the two units (black line in Fig. 3), the dark cellular plains form enclaves reaching tens of kilometers across that are embayed by the bright cellular plains ("6" labels in Fig. 4c), and that become smaller and less numerous with increasing distance to the southeast. The main contact itself shows "filaments" of the bright cellular plains extending into the dark cellular plains (label "7" in Fig. 4c). The configuration of this contact is indicative of the bright veneer becoming gradually thinner as the zone of net deposition transitions to the zone of net sublimation.

The band of dark cellular plains extending south of 30°N down the western margin of Sputnik Planitia (Fig. 3) indicates that factors besides latitude also affect nitrogen-ice sublimation and deposition. This region hosts dark streaks (e.g., label "9" in Fig. 4d) that are often seen in the lee of isolated mountain blocks, and that are interpreted to be wind streaks (*Telfer et al.,* 2018; *Bertrand et al.,* 2020a). In addition, the pitted appearance of the plains transitions to a distinct texture of regularly spaced (~0.4–1 km), linear ridges ("10" labels in Fig. 4d) close to the boundary between Sputnik Planitia and al-Idrisi Montes (*Telfer et al.,* 2018). These are oriented orthogonally to nearby wind streaks, and are hypothesized to be dunes composed of windblown, sand-sized particles of methane ice (*Telfer et al.,* 2018). The dark plains and these surface features are interpreted to occur here due to localized, topographically controlled atmospheric effects (see the chapter by Moore and Howard in this volume).

The dark trough-bounding plains at the northern extent of Sputnik Planitia have a lower albedo than the dark cellular plains, and the lowest albedo of any category of plains in Tombaugh Regio (Figs. 4c,d). These dark plains congregate around cell boundaries, which here appear as dark lineations reaching ~1 km across. The trough/ridge structure is rarely apparent, although in a few rare instances, thin, bright central lineations are resolved within the dark cell boundaries ("15" labels in Fig. 4f). These plains may be where an especially high concentration of haze particles is entrained within the ice, and the lineations are probably partly obscured medial ridges due to the trough being clogged by haze particles that have accumulated within it (*White et al.,* 2017). The enhanced sublimation of nitrogen ice currently taking place within these plains (*Bertrand et al.,* 2018) would have the effect of increasing the concentration of the refractory haze particles within the ice. This concentrating effect may be amplified by the fact that there are no mountain ranges separating these plains from the uplands north of Sputnik Planitia, meaning that the nitrogen ice is thinner here than elsewhere in the cellular plains. Convective motion of the nitrogen ice may play a role in causing the haze particles to accumulate at the edges of cells (*White et al.,* 2017).

2.2. Wright Group

Located to the south and southwest of Sputnik Planitia, the Wright Group (Fig. 7) is distinguished by several circular, rimless depressions and only one possible 5.5-km-diameter impact crater, and is dominated by the twin, broad,

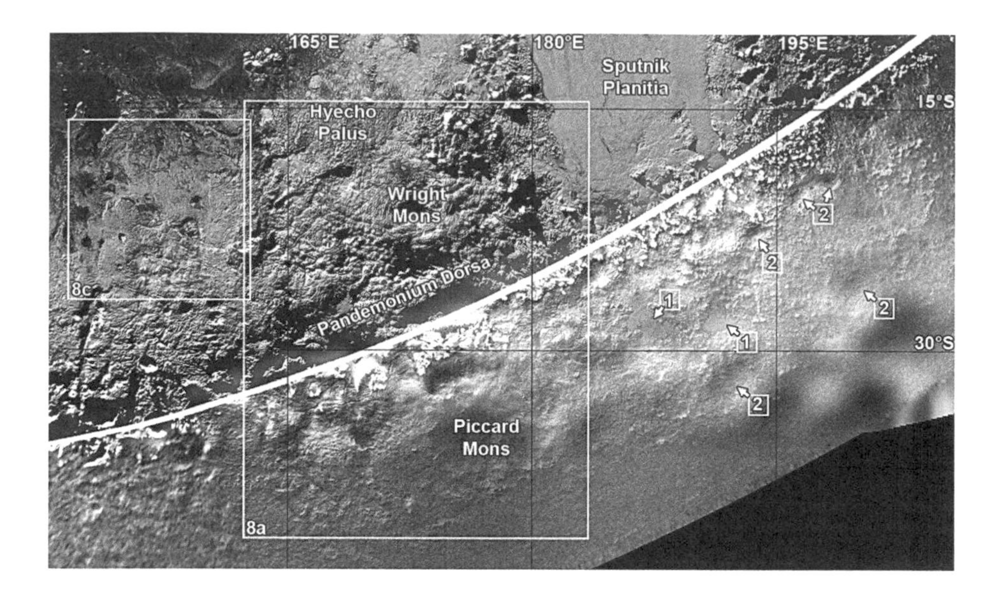

Fig. 7. The Wright Group. White line marks the northern boundary of the haze-lit zone. Numbered boxes refer to elements in Fig. 8. 1: Deposits of nitrogen ice ponded in depressions. 2: Rimless, dark-floored depressions (cavi).

quasicircular edifices of Wright and Piccard Montes (*Moore et al.,* 2016; *Singer et al.,* 2016; *Schenk et al.,* 2018). During the New Horizons flyby, Wright Mons and its immediate surroundings were illuminated at a high-solar-incidence angle (>80°), accentuating their topography, but the majority of the visible portion of the Group, including Piccard Mons, was contained within the haze-lit zone. To a degree, the very different illumination conditions can complicate consistent definition and interpretation of geomorphology across the Group, but it can be established that the terrain south of Sputnik Planitia, including the edifices themselves and their surroundings, are rough-textured on a scale of several kilometers. The portion of the Wright Group that exists to the southwest of Sputnik Planitia consists of a large plateau that is generally flatter and smoother than the southern terrain at this scale.

Wright and Piccard Montes (Fig. 8a) can be referred to as annular massifs, as they consist of very large mounds with enormous central depressions (Fig. 8b) that can reach as deep or deeper than the edifices are tall (*Schenk et al.,* 2018), the combination of which forms a quasi-annular shape. Wright Mons measures 155 km across, rises 3.5–4.7 km above its base, and has a central depression 45 km across and 3.5–4.5 km deep with respect to the uneven annular crest of the massif. Piccard Mons (its topography more poorly resolved than Wright's) measures 200 × 280 km across, rises 5.0 km above its base, and has a central depression 100–120 km across and 11.3 km deep with respect to the annular crest. Separating the two edifices is a pair of broad, curvilinear ridges (Pandemonium Dorsa) that display relief of ~4 km. The flanks of Wright Mons from its rim to its base are relatively shallow (~5°) and consist of hummocky terrain of semi-regular hills ~8–10 km in diameter, which themselves display a rubbly texture at a scale of ~1 km. A single, 5.5-km-diameter impact crater is observed on Wright Mons' summit region. The hummocky morphology is clearly discernible as far as the terminator (where it is exhibited by Pandemonium Dorsa). The annular summit region displays a lineated fabric that is impressed upon the hummocky texture, and that appears to be concentric about the central depression. The walls of the depression itself reach slopes of 30°, with the southern sunlit walls showing a blocky texture. The morphology of the haze-lit Piccard Mons is more difficult to discern. Its flanks display a rough texture at a scale of a few kilometers, but they do not seem to have a hummocky texture comparable to that seen at Wright.

In addition to the central depressions of Wright and Piccard Montes, their flanks and the surrounding terrain feature several other depressions, termed "cavi," that frequently display circular planforms, do not exhibit raised rims or ejecta (distinguishing them from impact craters), and are shallower than the depressions of Wright and Piccard ("1" labels in Figs. 8a,b). Portions of the walls and floors of many of these depressions, including those of Wright and Piccard, appear darker than their surroundings, which is likely due to an overlay of dark material. An exception is one cavus that has pitted nitrogen-ice deposits covering its floor (label "2" in Figs. 8a,b), and similar deposits are seen ponding in depressions to the north and east of Wright and Piccard Montes, some of which connect to Sputnik Planitia ("1" labels in Fig. 7 and "3" labels in Figs. 8a,b). The plain of Hyecho Palus to the north of Wright displays an etched appearance, being broadly flat but with a rough, blocky texture at a scale of ~1 km (label "4" in Figs. 8a,b). Hyecho Palus is embayed by the southwestern nitrogen-ice plains of Sputnik Planitia, above which it is elevated by 0.5–1 km. The boundary between Hyecho Palus and the hummocky terrain of the northern flank of Wright Mons is quite gradational, meaning that it may represent a degraded form of the Wright Mons material (*Singer et al.,* 2016) that has been overridden and glacially scoured by the nitrogen ice, leaving behind ponded deposits after its recession. Some locations

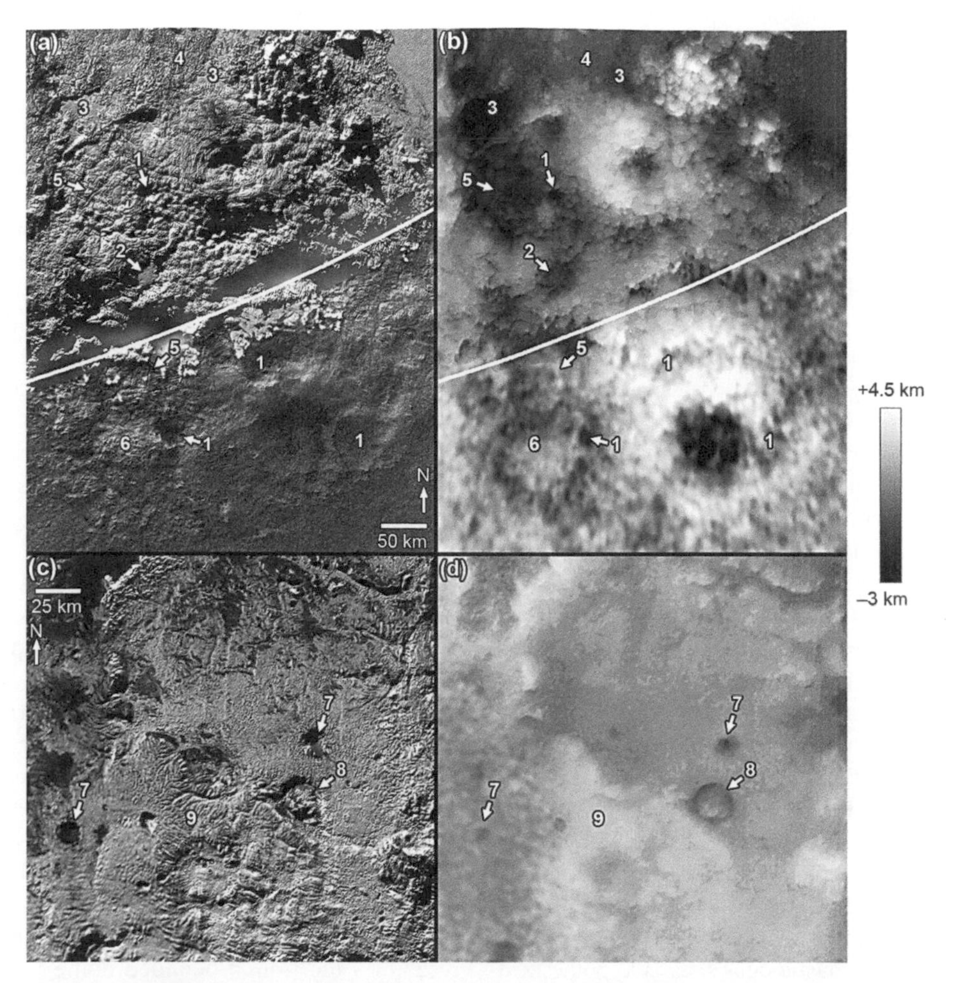

Fig. 8. Terrains within the Wright Group. See Fig. 7 for context. Images are shown at left, corresponding DEMs at right. **(a),(b)** Wright and Piccard Montes and surrounding terrain. White line marks the northern boundary of the haze-lit zone. Illumination is from the top left. 1: Rimless, dark-floored depressions (cavi). 2: Ponded nitrogen ice deposit on the floor of a cavus. 3: Nitrogen ice deposits ponded in depressions. 4: The rough, blocky plain of Hyecho Palus. 5: Localized fracture networks. 6: Rough-textured dome surrounded by a trough. **(c),(d)** Western plateau region. Illumination is from the top left. 7: Rimless, dark-floored depressions (cavi). 8: Cavus with a domical feature on its floor. 9: Platform dissected by branching valley networks.

in the terrain surrounding Wright and Piccard Montes show reticulate networks of fractures that typically reach a few kilometers across, and that appear to indicate localized tectonic disruption ("5" labels in Figs. 8a,b).

The height, overall structure, and surface texture of Wright and Piccard Montes is not suggestive of erosion, and their origin is more likely to be constructional. With no obvious source for the material in surrounding areas, and the presence of the large central depressions, it appears that subsurface material was mobilized onto the surface of Pluto to build these features. As such, Wright and Piccard Montes have been tentatively interpreted to be cryovolcanic in origin (*Moore et al.,* 2016; *Singer et al.,* 2016; *Schenk et al.,* 2018), even though the bulk morphologies of Wright and Piccard Montes are considerably different than tall shield volcanos of comparable size on terrestrial planets (the summit calderas of which do not occupy such a large fraction of the edifice). The near-total absence of impact craters across the Wright Group suggests that such cryovolcanism occurred relatively recently: Ages derived for Wright Mons range from ~1–4 G.y. (see the chapter by Singer et al. in this volume).

The bulk composition of Wright and Piccard Montes is an important factor in assessing the feasibility of the cryovolcanism hypothesis, but it is not precisely known. Based on the volume of material involved (~4×10^4 km^3 for the main structure of Wright Mons alone) and its apparent high strength (sufficient to maintain slopes up to 30°); it is probably water-ice-based. Overpressurization of a subsurface ocean through partial freezing is likely the only mechanism that can feasibly generate such a large volume of water-ice-based cryomagma at depth and cause it to ascend to the surface, especially if an antifreeze is present such as ammonia (e.g., *Fagents,* 2003; *Manga and Wang,* 2007; *Neveu et al.,* 2015; *Nimmo et al.,* 2016; *Cruikshank et al.,* 2019a,b; *Lesage et al.,* 2020; *Martin and Binzel,* 2021; see also the chapter by Nimmo and McKinnon in this volume).

In addition to the source of the material, a cryovolcanic hypothesis must also explain how it is emplaced, which likely depends in large part on the concentration of dissolved volatiles within the cryomagma. Radiation to space and rapid exsolution of dissolved gases will chill the surface of an effusing cryolava to form a solid crust with a liquid interior, while the base of the flow will freeze from contacting the substrate at ~40 K (*Cruikshank et al.,* 2019a). *Neveu et al.* (2015) demonstrated through geochemical modeling that such volatile exsolution within an ascending cryomagma may even be sufficient to drive upward crack propagation and eject the fluid explosively onto the surface. Central to the issue of emplacement is the role of the central depressions in the development of these edifices, which is presently not well understood. They may represent vents that erupted material to form annular massifs around their peripheries, and therefore grew deeper as the massifs surrounding them grew higher. As far as can be resolved (down to hundreds of meters scale), the flanks of these edifices are not made of multiple thin flows, as terrestrial shield volcanos are. The hummocky morphology of Wright Mons is more suggestive of emplacement as a viscous flow rather than a pyroclastic fall sourced from the depressions. Material forming the hummocks may be emplaced as a solid-state flow or as a partial melt, but it is not readily apparent as to whether each hummock represents an individual flow front (perhaps effused somewhat in the style of a pillow lava), and if they do, what the flow directionality is and whether they are originating from the depressions. As such, the material forming the bulk of these edifices cannot be directly traced back to the depressions as their source. The dark material that appears to cover portions of the walls and floors of the depressions may be some sort of erupted, cryoclastic debris, as is hypothesized for dark material that mantles topography around a segment of Virgil Fossae (*Cruikshank et al.,* 2019a,b; *Dalle Ore et al.,* 2019) (see section 2.6.3) and a smaller trough in Viking Terra (*Cruikshank et al.,* 2021) (see section 2.5).

Given their scale, it is likely that surface collapse has contributed at least in part to the formation of the depressions, but is probably not the sole explanation for them. If formed entirely due to collapse of the summits of conical edifices, this would represent removal of >10% of the edifices' volumes, a vast fraction, and with the possible exception of the concentric fabric at Wright Mons, the interiors of these depressions do not appear to have prominent terraced faults as commonly exhibited by collapsed summit calderas on the terrestrial planets. The smaller cavi in surrounding terrains have likely formed by collapse, however, as is particularly evident of the ~55-km-diameter cavus that superposes the eastern rim of the central depression of Piccard Mons. Just to the west of Piccard Mons there is a rough-textured dome ~80 km in diameter and 3–4 km high that is surrounded by a trough, which is itself superposed by a 30 km cavus on its northeast rim (label "6" in Figs. 8a,b). The dome itself does not show a central depression, and whether it represents an evolutionary precursor to the giant edifices of Wright and Piccard Montes, or is an unrelated feature, is unknown.

The haze-lit terrain extending to the east of Piccard Mons is rugged and knobby at a scale of several kilometers, appearing distinct from the hummocky terrain at Wright Mons, but also exhibits some dark-floored cavi ("2" labels in Fig. 7) and a few <10-km-diameter impact craters. The presence of cavi like those in the vicinity of Wright and Piccard Montes, and the very low crater count, suggest that this terrain is part of the Wright Group. An intermediate-albedo region within the poorly resolved farside, displaying a mottled texture on a scale of tens of kilometers, bounds the Wright Group to its east (section 4). The knobby terrain likely continues eastwards into the farside to form at least the westernmost portion of this zone (*Stern et al.,* 2021). If this terrain is also cryovolcanic in origin, then this raises the issue of how it was emplaced relative to Wright and Piccard Montes, and whether its rubbly, degraded appearance indicates that it is a heavily modified variant of the material forming those edifices, or instead has a separate genesis.

To the west of Wright Mons is an intermediate-albedo, generally flat plateau that is bordered on its northern, western, and southern margins by rugged, cratered uplands that prevail to the southwest of Sputnik Planitia, and on its eastern margin by a ~4-km-high scarp that separates it from the hummocky terrain of Wright Mons and ponded nitrogen-ice deposits of Sputnik Planitia (Fig. 8c). In terms of regional topographic setting, this area appears to belong to the Burney Group, but it can be considered as belonging to the Wright Group as it also displays several cavi and a low impact crater count. Like those farther east, these cavi form rimless, quasi-circular depressions that display dark floors and walls ("7" labels in Figs. 8c,d), and one also includes a domical feature on its floor (label "8" in Figs. 8c,d) that resembles the trough-bounded dome to the west of Piccard Mons. Smooth plains that display subkilometer-sized boulders strewn across their surface occupy much of the plateau. A platform that is raised 1–2 km above the plains in the southern portion of the plateau has been dissected by dense, branching networks of ~1-km-wide valleys that converge into deeply incised troughs 2–4-km-wide (label "9" in Figs. 8c,d), and are interpreted to have been glacially scoured by former nitrogen-ice coverage (*Howard et al.,* 2017a).

A unique combination of factors may account for the unusual smoothness and low crater count of this plateau. The eastern-bounding scarp terminates at the southeast margin of the plateau, where it makes a much more gradual contact with the hummocky terrain of Wright Mons and almost grades into it in terms of albedo, topography, and surface texture. In addition to the cavi shared by the two terrains, this would seem to indicate that the plateau is genetically related to the lower-elevation cryovolcanic province to the east, but where the cryovolcanism has instead modified an uplands setting. The relative flatness of the plateau compared to the knobby terrain east of Piccard Mons (which is otherwise similar in that it displays scattered cavi but no large edifices) may be a consequence of glacial scouring if the nitrogen ice that carved the dendritic valleys flowed eastward to spread out across the plains en route to Sputnik Planitia, into which most of the eroded debris would ultimately be deposited. Formation of the cavi themselves, which do not appear to have been substantially eroded or infilled, would have postdated this glacial scouring.

Precisely how cryovolcanic resurfacing would proceed and relate to the cavi distributed across the varied terrains of the Wright Group remains enigmatic. But perhaps the major obstacle to resolving the many questions regarding the origin of the Wright Group is how extensive it is. It obviously continues farther into the poorly resolved farside and into the shadowed southern regions, but without knowing its total extent, it is impossible to even roughly estimate the volume of its constituent materials and to comprehensively categorize their morphological variation, which are essential steps for evaluating whether a cryovolcanic hypothesis for their formation is feasible.

2.3. Tartarus Group

The Tartarus Group consists of a variety of rough-textured uplands units located to the east of Sputnik Planitia (Fig. 9). The most characteristic of these is the bladed terrain deposits of Tartarus Dorsa, which are separated from Sputnik Planitia by the bright, pitted uplands of East Tombaugh Regio. Both of these terrains are bordered to the south by the dark, heavily eroded terrain of Krun Macula, and to the north by undulating plains of Hayabusa Terra. Very few possible impact craters exist within this Group (*Robbins et al.,* 2017; *Singer et al.,* 2019). Tartarus Dorsa and East Tombaugh Regio are nearly crater-free, although a notable characteristic of East Tombaugh Regio and Krun Macula is the prevalence of dense clusters of deep pits, which complicates the recognition of impact craters within these terrains. A crater retention age of 200–300 m.y. has been derived for the bladed terrain, and depending on the crater diameter under consideration (10–40 km) the bright, pitted uplands show a surface age between ~200 m.y. and ~2 G.y. (*Moore et al.,* 2018; see also the chapter by Singer et al. in this volume).

2.3.1. Bladed terrain deposits. Tartarus Dorsa consist of broad, elongate swells that are typically ~400 km long and ~100 km wide, and that tend to be separated from one another by steep-walled troughs that include Sun Wukong Fossa, Sleipnir Fossa, and others that form the southern components of Mwindo Fossae (Fig. 9) (*Moore et al.,* 2018). These swells display relief of 2–5 km above the rough plains to the north, and form some of the highest-standing terrain in the close approach hemisphere, reaching 4.5 km above mean radius. The surface of these swells has a unique texture consisting of dense fields of roughly evenly spaced, often subparallel sets of steep ridges that are characterized by sharp crests and divides, which is referred to as "bladed terrain" ("1" labels in Fig. 10a) (*Moores et al.,* 2017; *Moore et al.,* 2018). The blades display flank slopes of ~20°, are typically spaced 3–7 km crest-to-crest, and exhibit relief of ~300 m from crest to base, although a few along the margins of the blade fields show local relief up to ~1000 m (*Moore et al.,* 2018). This texture is most prominent at elevations higher than 2 km above mean radius. The blades usually become smaller and lower relief as they approach lower-lying areas adjacent to the steep-walled troughs, and at the boundaries of blade fields ("2" labels in Fig. 10a). Blades

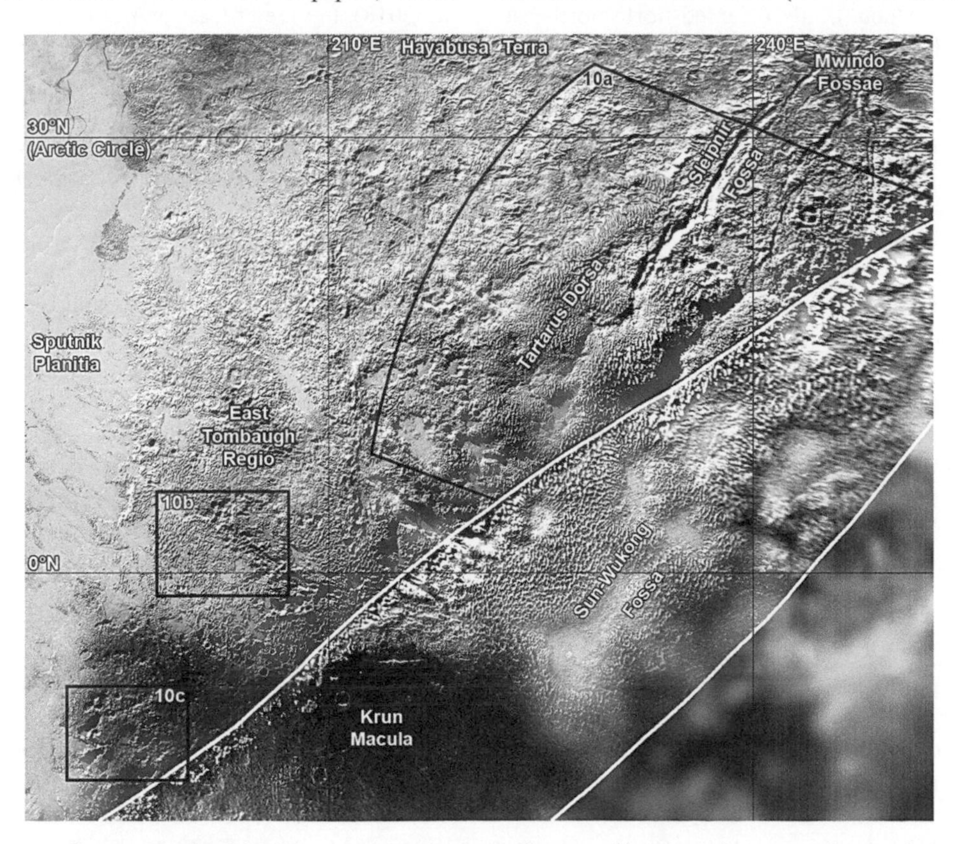

Fig. 9. The Tartarus Group. Thick white lines enclose the haze-lit zone. Numbered boxes refer to elements in Fig. 10.

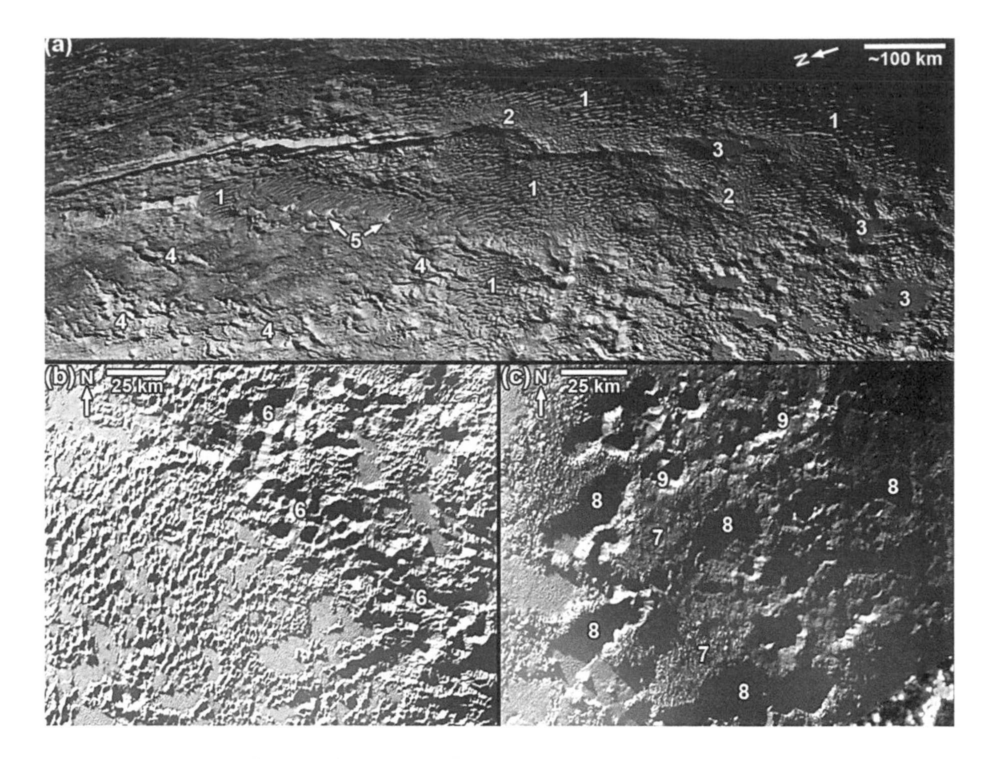

Fig. 10. Terrains within the Tartarus Group. See Fig. 9 for context. **(a)** Highly oblique view of bladed and arcuate terrain. Illumination is from the lower left. 1: Steep, sharp-crested ridges characteristic of bladed terrain. 2: Lower-relief bladed terrain. 3: Nitrogen ice deposits ponded in depressions. 4: Scarps of arcuate terrain. 5: Partly exhumed, arcuate scarps overlain by bladed terrain deposits. **(b)** Southern region of the bright, pitted uplands. Illumination is from the top left. 6: Complexes of deep pits organized into a northwest-southeast-aligned linear chain. **(c)** Dark, eroded uplands. Illumination is from the top left. 7: Etched, rough upland plateau. 8: Complexes of deep pits. 9: Interconnected pits forming an east-northeast-/west-southwest-aligned trough.

north of ~10°N are commonly oriented north-northeast-south-southwest, but those nearer the equator show a more reticulate pattern similar to that of the bright, pitted uplands, with a wider range of blade orientations. Nitrogen ice has ponded on the floors of some depressions between blade complexes, perhaps infilling structural troughs ("3" labels in Fig. 10a).

Spectral data from New Horizons (*Grundy et al.,* 2016; *Protopapa et al.,* 2017; *Schmitt et al.,* 2017) indicate that the bladed terrain deposits are composed primarily of methane ice. The texture of the bladed terrain is characteristic of decrescence, whereby uniform removal of material from a landscape makes outward-facing projections increasingly pointed and angular while initial inward-facing indentations become rounded. Methane ice at Pluto surface conditions can evidently support the several-hundred-meter relief and steep slopes of the blades, consistent with the rheological conclusions of *Eluszkiewicz and Stevenson* (1990). The occurrence of the bladed terrain deposits at high elevation, when considered with their distinct decrescence texture, indicates that they record a complex history of climate change, with mass precipitation of methane ice followed by a transition to climatic conditions favoring sublimation (*Moore et al.,* 2018; see the chapter by Moore and Howard in this volume). The blades are considered to be partially analogous to aligned penitentes on terrestrial, low-latitude, high-elevation ice fields (*Moores et al.,* 2017; *Moore et al.,* 2018), but are at least two orders of magnitude larger, and the processes that control the amplitude and spacing of the blades are not yet fully understood. The troughs that separate the swells are of tectonic origin (section 3) and postdate the emplacement of the bladed terrain deposits.

2.3.2. Arcuate terrain. This generally light-toned, high-relief terrain, located within undulating plains to the north and northwest of the bladed terrain deposits, shows pronounced arcuate scarps, with gentle, convex, north-facing slopes and steeper, concave, south-facing slopes ("4" labels in Fig. 10a) (*Moore et al.,* 2018). The scarp crests are dominantly oriented east-west and display a methane-rich spectral signature. The bladed terrain deposits appear to be superimposed on the arcuate terrain (label "5" in Fig. 10a), indicating that emplacement of the bladed terrain deposits postdates the formation of the arcuate terrain. The methane-rich spectral signatures exhibited by both terrain types, and the observation that in some cases the arcuate terrain forms curvilinear outcrops extending from the bladed terrain deposits that mimic the elongate planforms of the swells, suggest that a genetic relationship between these terrains may exist. The arcuate ridges may be remnants of a formerly greater extent of bladed terrain deposits, but which formed under different climatic conditions. Alternatively, they may be marginal deposits left behind after retreat of more extensive bladed terrain deposits (*Moore et al.,* 2018).

2.3.3. Bright, pitted uplands. This terrain forms the majority of the eastern lobe of the high-albedo Tombaugh Regio, and consists of a >500-km-wide rugged upland plateau that ranges in elevation between –0.5 and 2.5 km above mean radius (*Moore et al.,* 2016, 2017, 2018; *Howard et al.,* 2017a; *White et al.,* 2017). Its surface texture transitions from rugged hills and scarps in its northern region to dense pitting in the south, where individual pits reach 5–10 km across and form a network of sharp-crested, reticulate ridges (Fig. 10b) (*Moore et al.,* 2018). The high albedo of the region is due to a continuous veneer of volatile ice (primarily nitrogen) blanketing the terrain, which has ponded in depressions to form expanses of pitted plains that are exhibiting glacial flow, but the hills and reticulate ridges show a dominant methane spectral signature (*Schmitt et al.,* 2017; *Moore et al.,* 2018). Along the eastern boundary of these uplands, the reticulate ridges become integrated to form distinct northeast-southwest-trending ridge-and-trough terrain (*White et al.,* 2017), and they appear to transition to the bladed texture of the bladed terrain deposits to the east, a transition that is accompanied by a decrease in albedo in low-phase imaging. This can be explained by the initial methane ice deposition east of Sputnik Planitia occurring at the highest elevations to form the present bladed terrain deposits, with subsequent deposition expanding to cover lower-elevation areas, forming the reticulate portion of the bright, pitted uplands that are low enough to be coated by depositing nitrogen ice. These terrains therefore appear to manifest a surficial composition sequence from dominance by nitrogen closest to Sputnik Planitia (including return-flow nitrogen glaciation) to increasing dominance of methane ice to the east culminating in the bladed terrain deposits (*Moore et al.,* 2018), a sequence that corresponds to an altitudinal control on ice stability (see the chapter by Moore and Howard in this volume).

The bright, pitted uplands also host several complexes of deep pits, where individual pits can reach >10 km across and >1.5 km deep, and are bounded by sharp marginal scarps that can rise more than 1 km above surrounding terrain ("6" labels in Fig. 10b) (*Howard et al.,* 2017b). The pits are organized into northwest-southeast-aligned linear chains along structural trends that can reach >200 km long, and that likely represent grabens. These pit clusters are distinguished from the smaller-scale reticulate landforms of the bright, pitted uplands themselves, and their large depths and frequent association with tectonic lineations indicates that they are formed via collapse of surface materials into subsurface voids opened up by such tectonism, rather than sublimation (*Howard et al.,* 2017b).

2.3.4. Dark, eroded uplands. Krun Macula consists of a rough upland plateau that displays an etched appearance, with a faint north-south-aligned fabric ("7" labels in Fig. 10c). Interconnected complexes of pits, troughs, and basins up to 3 km deep and 20 km wide (*Howard et al.,* 2017b) invariably interrupt this texture (e.g., "8" labels in Fig. 10c). The scale of these depressions implies formation via surface collapse. In places, conglomerations of pits form large troughs with quasi-parallel walls (the two ends of one example marked by "9" labels in Fig. 10c), suggesting that, as in the bright, pitted uplands, some of the surface collapse has occurred in response to tectonic fracturing. The better-preserved portions of Krun Macula can rise up to 4 km in elevation, similar to the bladed terrain deposits, but more typically are at 1.5–2.5 km (*Stern et al.,* 2021), comparable to the higher portions of the bright, pitted uplands that are themselves interpreted to be modified bladed terrain deposits (section 2.3.3). There are no obvious constructional landforms and only a handful ($\gtrsim$10) of depressions that could be interpreted as impact craters (*Robbins et al.,* 2017; *Singer et al.,* 2019). Like Cthulhu Macula, Krun Macula displays a continuous blanket of dark haze particles that have settled onto it from the atmosphere, and that is thin compared to topographic relief (*Grundy et al.,* 2018). Krun Macula shows that a macula with a relatively young crater age can occur at elevations within the permanent diurnal zone in uplands directly to the east of Sputnik Planitia that would otherwise seem to be a natural location for deposition of methane to form the bladed terrain deposits. Possible explanations for this are that Krun Macula represents a region where bladed terrain deposits have accumulated a blanket of dark haze particles on their surface, or receded to reveal a dark substrate (*Stern et al.,* 2021). These hypotheses are discussed in section 4 in the context of the farside maculae.

2.4. Hayabusa Group

Mostly located north of 37°N, the southernmost limit of the permanent Arctic zone (*Binzel et al.,* 2017), the Hayabusa Group is situated in the northeast of the nearside hemisphere. The defining morphological characteristic of this Group (Fig. 11) is rounded terrain that is smooth at a scale of a few kilometers, which is best expressed in Pioneer Terra (Figs. 12a,b). The distinct appearance of this terrain is most apparent when comparing its topography to that of the adjacent uplands that are finely dissected by dendritic valleys to the northwest ("1" labels in Figs. 12a,b). *Howard et al.* (2017b) described this region as smooth-textured uplands, which display broad and rounded topography and rise abruptly up to 3 km above irregular, flat-floored depressions and scarp-bounded lowlands that reach tens of kilometers wide ("2" labels in Figs. 12a,b). The scale of the depressions suggests that surface collapse played a role in their formation. There is a strong geographic association between the depressions and the smooth uplands, and their formation may be intimately related (*Howard et al.,* 2017b). The smooth, rounded morphology of the uplands is diagnostic of accrescence, whereby addition of material has occurred uniformly over a regional surface, causing projections to become rounded and inward facing, and valleys to become sharply indented. Several compositional endmembers have been identified in this region, including water ice forming the crustal bedrock, which is mantled by other spectral units including nitrogen- and methane-rich ices (*Grundy et al.,* 2016; *Protopapa et al.,* 2017; *Schmitt et al.,* 2017;

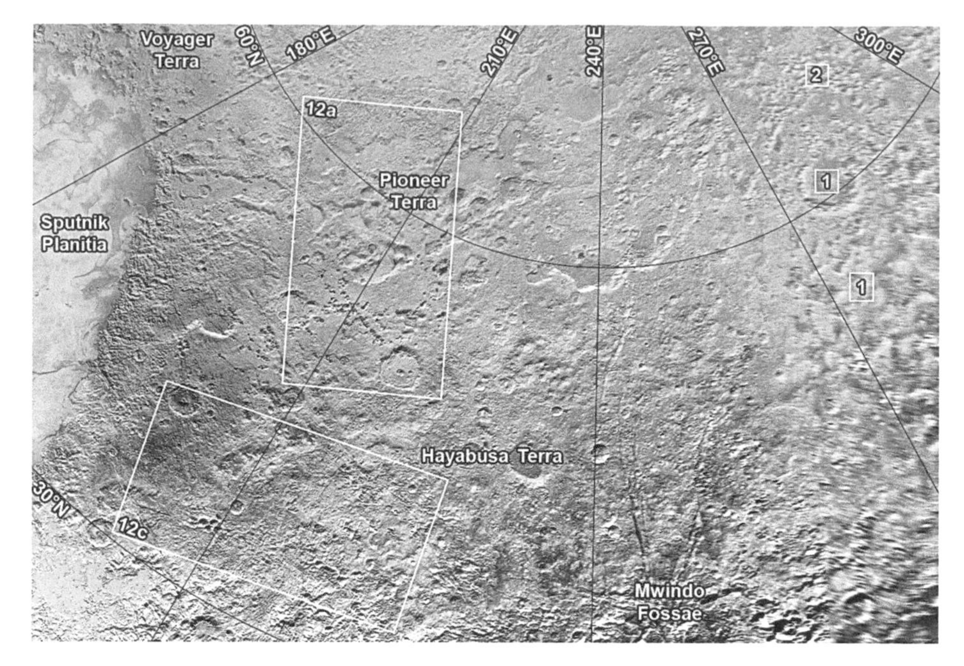

Fig. 11. The Hayabusa Group. Numbered boxes refer to elements in Fig. 12. 1: Irregular, lobate, flat-topped plateaus separated by lower-albedo depressions. 2: Pitted terrain.

Lewis et al., 2021). Gaseous emissions from the subsurface emanating from the depressions may have led to gradual emplacement of a mantle, with a methane-rich component depositing on the surrounding terrain, forming the broadly rounded divides that characterize the smooth uplands and depressions (*Howard et al.,* 2017b).

Terrains to the south and east of Pioneer Terra display morphological variants of the smooth-textured uplands and depressions. To the south, the eroded smooth uplands of Hayabusa Terra resemble in broad topographic pattern the smooth uplands and depressions, with broad, rolling topography at a scale of tens of kilometers, but are distinguished by a rough, pitted texture at a scale of hundreds of meters to kilometers (Figs. 12c,d). *Howard et al.* (2017b) tentatively identified this as an older landscape that was covered by the same methane-rich mantling deposit as the smooth uplands terrain, but that has been more extensively modified by sublimation erosion. The generally lower albedo of the eroded smooth uplands relative to the smooth uplands indicates that a darker substrate is being exposed across the area (e.g., label "5" in Figs. 12c,d). Both the bladed terrain deposits and the arcuate terrain of the Tartarus Group superpose the eroded smooth uplands at their southern boundary, and given that the bladed terrain deposits show evidence for having undergone recession, the southernmost portions of the eroded smooth uplands would therefore have been overlain by the deposits and subsequently exhumed as the deposits receded.

East of 260°E, the terrain takes on a mottled appearance, with the lighter-toned mantling deposit forming irregular, lobate, flat-topped plateaus that are separated by lower-albedo depressions ("1" labels in Fig. 11). The lobate morphology of the plateaus is an accrescence texture, implying that this terrain is where mantling of the surface has occurred to an incomplete degree, leaving gaps where the darker substrate remains exposed (*Stern et al.,* 2021). But in other localities, the terrain appears pitted at a scale of several kilometers (label "2" in Fig. 11), indicating that decrescence may also have affected parts of it. This partly mantled terrain appears to display characteristics of both the smooth uplands and depressions of Pioneer Terra (a thick, light-toned mantle with smooth, rounded textures) and the eroded smooth uplands of Hayabusa Terra (darker-toned areas and occasional rough, pitted relief), and so may record a history of alternating deposition and erosion of the mantling material (*Stern et al.,* 2021).

In addition to the large, flat-floored, irregular depressions of Pioneer Terra, terrain to the north and northeast of Sputnik Planitia also displays clusters of smaller, sharp-edged, relatively circular pits that are 5–20 km in diameter and often display conical profiles (*Howard et al.,* 2017b). The pits tend to adjoin each other and can form dense complexes tens of kilometers across (label "3" in Figs. 12a,b), whereas in other locations they are organized into linear chains aligned parallel to structural trends (unlike the larger depressions) (label "4" in Figs. 12a,b). These pit clusters resemble those in the Tartarus Group, and a broad region containing such pits can be identified stretching from Krun Macula to Voyager Terra north of Sputnik Planitia. Few pits are identified east of 230°E, although this may partly be due to decreasing image resolution. Collapse of surface materials into subsurface voids appears to be the best explanation for most of these pits, as supported by the association between pits and tectonic lineations, as well as the observation that most of the pit complexes are clearly indented below the surrounding terrain (*Howard et al.,* 2017b). The origin of other pit complexes, which display markedly elevated rims that sometimes rise more than 1 km above the surrounding terrain, is more ambiguous. These complexes may represent

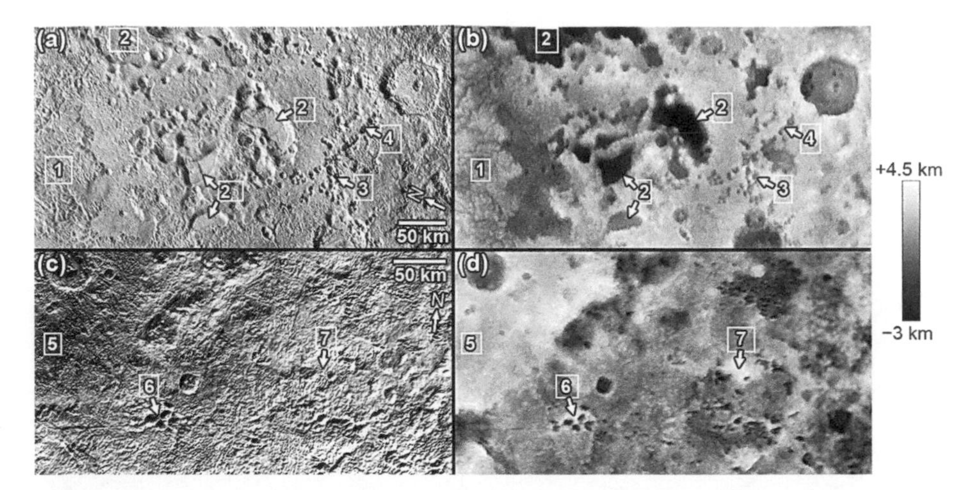

Fig. 12. Terrains within the Hayabusa Group. See Fig. 11 for context. Images are shown at left, corresponding DEMs at right. **(a),(b)** Smooth uplands of Pioneer Terra. Illumination is from the left. 1: Dendritic valley networks. 2: Irregular, deep, flat-floored depressions. 3: Complex of sharp-edged, circular pits. 4: Pits aligned in a linear chain. **(c),(d)** Eroded smooth uplands of Hayabusa Terra. Illumination is from the top left. 5: Dark substrate revealed by erosion of intermediate albedo surface material. 6: Complex of pits with rims elevated >1 km above the surrounding terrain. 7: Isolated mound reaching >1.5 km above the surrounding terrain that has been partially affected by pit collapse.

remnants of >1-km-thick bladed terrain deposits/arcuate terrain left isolated after the recession of the bladed terrain deposits, with the remnants subsequently experiencing collapse to varying degrees: extensively in the case of the complex at label "6" in Figs. 12c,d, relatively minimally in the case of that at label "7".

Like the Venera and Burney Groups (sections 2.5 and 2.6), the Hayabusa Group is heavily cratered. Modification of crater morphology is a common phenomenon across Pluto (see the chapter by Singer et al. in this volume for more details), but of these Groups, mantling has substantially modified a higher proportion of craters within the Hayabusa Group, where relatively few craters display sharp raised rims compared to the others. As such, whereas the formation of the Venera and Burney Groups likely date to ≳4 G.y., the mantling episode that defines the Hayabusa Group is interpreted to have occurred predominantly after substantial resurfacing of the other Groups had ceased, and possibly as late as Pluto's middle age (*Howard et al.,* 2017b).

2.5. Venera Group

The mid-latitudes of the northwest of Pluto's nearside hemisphere are characterized by a high-albedo region (*Buratti et al.,* 2017) extending in a band from ~40°N–~70°N and ~0°E–~135°E (Fig. 13). On a scale of hundreds of kilometers, the terrain is mostly flat, featuring undulating plains that display total relief of about 1 km. This bright region is bounded to the south by a band of darker, mottled terrain that separates it from Cthulhu Macula, with the several-hundred-kilometers-long crenulated scarp of Piri Rupes forming a major topographic boundary. This region also contains the major tectonic systems of Djanggawul Fossae and Inanna and Dumuzi Fossae. The entire area is highly cratered (*Singer et al.,* 2019), with craters reaching a few tens of kilometers in diameter and showing bright, sharp rims and dark floors. Most conspicuous among these are the bright craters in Vega Terra (label "1" in Fig. 13).

This region appears less obviously mantled than the smooth, rounded accrescence morphology exhibited by the well-preserved portions of the Hayabusa Group. Yet it has distinctive features that do seem to indicate the presence of a mantling deposit, albeit one that is thinner and was emplaced earlier than that in Hayabusa. The landforms that are most diagnostic of a mantling deposit are those that are interpreted to be erosional. In the vicinity of Inanna and Dumuzi Fossae (Fig. 14a), the mottled texture is resolved to be finely dissected terrain with massifs of bright, rounded hills separated by short, stubby, valleys and more expansive, dark, enclosed plains (e.g., "1" labels in Fig. 14a). The angular, truncated and poorly developed morphology of the valleys is diagnostic of decrescence, implying that the bright material has undergone sublimation erosion. This bright material is exposed in the sunlit, north-facing walls of Inanna and Dumuzi Fossae, where it is seen to overlie a darker layer ("2" labels in Fig. 14a). The low albedo of some of the plains and valleys separating the bright massifs are interpreted to be where the bright layer has eroded away, exposing this darker substrate. The very dark valley floors, plains, and craters, however, are blanketed by accumulated haze particles (*Grundy et al.,* 2018). Based on profiles taken across the Fossae in a DEM with a lateral resolution of 315 m/pixel (*Schenk et al.,* 2018), the bright layer is estimated to be ~1 km thick where exposed in the walls of Inanna and Dumuzi Fossae.

These observations suggest that the entire band of mottled terrain that separates the bright terrain of Venera Terra from Cthulhu Macula is a marginal zone where the bright mantling material has been partly removed by sublimation erosion. The ~500 km × ~260 km depression of Piri Planitia, bounded along its southern and western edges by the ~600 m–~1 km-high Piri Rupes (Fig. 14b), represents

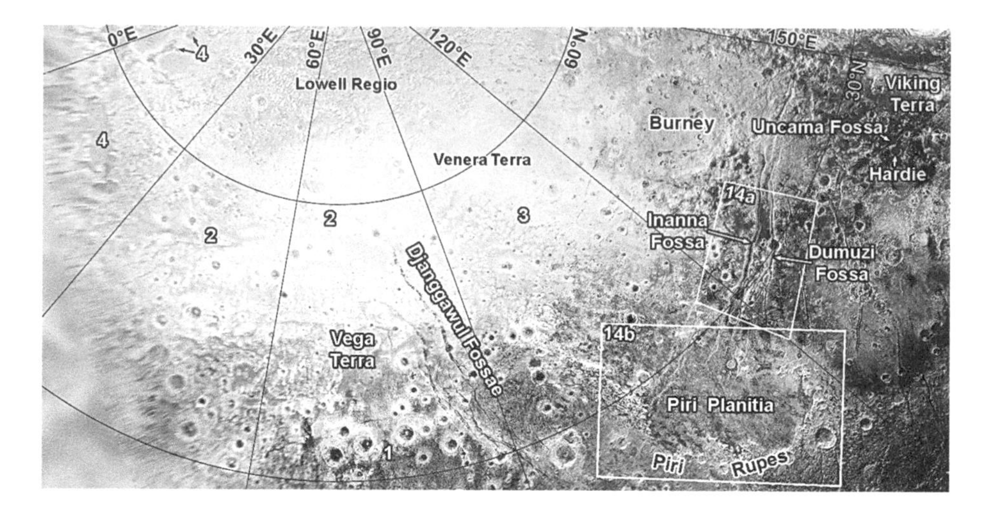

Fig. 13. The Venera Group. Numbered boxes refer to elements in Fig. 14. 1: Bright-rimmed, dark-floored craters. 2: Shallow, linear and curvilinear troughs. 3: Polygonal pattern formed by a reticulate network of troughs. 4: Irregular, ameboid, dark-floored depressions.

a distinctive subsection of the mottled terrain where such erosion has caused large-scale scarp retreat. Piri Planitia's southern portion (label "3" in Fig. 14b) is distinguished by the absence of massifs of hills, consisting instead of mottled plains. The scarps of Piri Rupes display sharply pointed projections into Piri Planitia and are broken up into isolated mesas in some places ("4" labels in Fig. 14b). The scarps stand higher than the plateau surfaces surrounding Piri Planitia, implying that the scarps may incorporate segments of older, high-standing topography (*Moore et al.,* 2017). A portion of the western scarp exhibits prominent pitting and dark-floored, truncated valleys within bright material along its crests (label "5" in Fig. 14b), similar to the bright, finely dissected terrain around Inanna and Dumuzi Fossae.

The higher albedo of Venera Terra compared to the mottled terrain suggests that the coverage of the mantling deposit is more continuous here, although erosion has still affected it. Localized networks of shallow, linear, and curvilinear troughs (e.g., "2" labels in Fig. 13) are likely tectonic in origin. In Venera Terra, an especially large, reticulate network of troughs 3–4 km wide separates the bright plains into polygons that reach a few tens of kilometers across (label "3" in Fig. 13). *Moore et al.* (2016) speculated that this network might have originated as tectonically disrupted blocks, the margins of which were subsequently widened by sublimation erosion. Irregular, ameboid, dark-floored depressions existing at the western margin of this bright mantle ("4" labels in Fig. 13) may indicate the onset of large-scale deterioration of the bright mantle that culminates in the coarser-textured mottled terrain of eastern Hayabusa Terra ("1" labels in Fig. 11) (*Stern et al.,* 2021).

The strong spectral signature of water ice is seen within low-albedo, low-elevation portions of the mottled terrain separating Venera Terra and Cthulhu Macula (*Grundy et al.,* 2016; *Protopapa et al.,* 2017; *Schmitt et al.,* 2017; *Cook et al.,* 2019), and excepting the haze particle coverage, the darker material seen across the Venera Group likely represents exposed water ice crust that underlies the mantling material. Determining the identity of the material forming the mantle of the Venera Group is complicated by the fact that this region of Pluto is where it is most difficult to distinguish what albedo and compositional features are representative

Fig. 14. Terrains at the southern edge of the Venera Group. See Fig. 13 for context. **(a)** Inanna and Dumuzi Fossae. Box indicates the boundaries of the 234 m/pixel imaging in the inset at top right. Illumination is from the top. 1: Dark plains separating massifs of bright, rounded hills. 2: Bright layer overlying a darker layer on sunlit walls of Inanna and Dumuzi Fossae. **(b)** Piri Planitia and Piri Rupes. Illumination is from the top left. 3: Mottled plains of Piri Planitia. 4: Promontories and mesas of the Piri Rupes scarp. 5: Pitting and dark-floored, truncated valleys within bright material along western Piri Rupes.

of the underlying geology vs. later veneers of volatile ices that cover the surface. The bright material of the mottled terrain (Fig. 14a) displays the spectral signature of methane (*Grundy et al.,* 2016; *Protopapa et al.,* 2017; *Schmitt et al.,* 2017). This is interpreted to be the predominant material forming the Venera Group mantle, but it is likely mixed with a stronger, refractory material (i.e., water ice) that allows it to support high-relief topography, including the scarps of Piri Rupes and the walls of the Fossae (*Moore et al.,* 2017).

Not all terrain across this region displaying a methane signature necessarily forms part of the Venera Group mantle, however, specifically those in elevated, topographically prominent locations. This includes the rims of the bright craters in Vega Terra, where the strong methane signature is more likely caused by seasonal deposition of methane ice, rather than them being isolated, remnant outcrops of the eroded Venera Group mantle. Similarly, whereas sublimation of methane ice has likely been instrumental in the erosion that has exhumed Piri Planitia, the methane signature seen at the high-standing scarp crests of Piri Rupes may also represent such seasonal deposition. The high-albedo Venera Terra does not display a pronounced methane signature, but rather that of nitrogen, forming part of a mid-latitude belt of elevated surface nitrogen abundance that also characterizes the Hayabusa Group (*Grundy et al.,* 2016; *Protopapa et al.,* 2017; *Schmitt et al.,* 2017; *Lewis et al.,* 2021). Since neither the Venera nor Hayabusa Groups show any geomorphic evidence for substantial deposits of ponded nitrogen ice, like those forming the Sputnik Group, this signature is most likely caused by a thin veneer of nitrogen ice covering the mid-latitudes.

The Venera and Burney Groups are heavily cratered to a similar degree (*Singer et al.,* 2019). The mantle of the Venera Group is younger than the water ice crustal material of the Burney Group, however, as it embays the polar uplands at its northern boundary, but the abundance of large, well-preserved impact craters suggests that it is still ancient. The deposition event that formed it probably occurred in the immediate aftermath of the Sputnik basin impact event (~4 G.y.), perhaps concurrently with the nitrogen-ice glaciation that shaped the uplands to the north of Sputnik Planitia. Piri Planitia displays an abnormally low crater count (*Moore et al.,* 2017), which likely indicates that the sublimation erosion event that took place here removed a sufficient amount of the overlying mantle to obliterate the impact craters that had formed within it.

Evidence for recent cryovolcanic activity may exist in Viking Terra, at the southeastern edge of the Venera Group (Fig. 13). Here, *Cruikshank et al.* (2021) identify morphological evidence of infilling of the ~9-km-wide Uncama Fossa and the adjacent 28-km-diameter Hardie impact crater with ammoniated, dark material, and suggest that the crater and fossa trough may have been flooded by a cryolava debouched along fault lines in the trough and in the floor of the impact crater. The now-frozen cryolava consisted of liquid water infused with the dark material (presumed to be tholins) and one or more ammoniated compounds. *Cruikshank et al.* (2021) interpreted similarly ammoniated dark material surrounding a nearby trough to represent cryoclastic material erupted during a fountaining event. This activity would have taken place ~1 G.y. due to the instability of ammonia compounds when exposed to Pluto's space environment.

2.6. Burney Group

The Sputnik, Wright, Tartarus, Hayabusa, and Venera Groups represent regional-scale deposits that have been emplaced during or since the era of high impact rate on Pluto (i.e., ~4 G.y.). The Burney Group (named after the ~290-km-diameter Burney multi-ring impact basin, the largest post-Sputnik impact feature yet identified on Pluto) covers the remainder of Pluto's nearside hemisphere, and even though it contains a great variety of terrains distributed across a vast latitude range (Fig. 15), no part of it is interpreted to have experienced a regional-scale depositional event since that time. This means that its surface geology mostly dates from the era immediately following the formation of the Sputnik impact basin and primarily consists of exposed water ice crust, with varying degrees of shallow mantling by volatile ices and haze particles. Its terrains are among the most cratered on Pluto's nearside, and variations in crater spatial density appear to derive from differential erosion of them. Much of the Burney Group is therefore a record of Pluto's surface environment during the earliest observable portion of its history, and of the landform evolving processes that operated then.

2.6.1. Glacially modified terrains. Voyager Terra, located to the north of Sputnik Planitia, is a zone of uplands featuring a variety of intergrading erosional morphologies (*Moore et al.,* 2016; *Howard et al.,* 2017a,b). To the northeast of Sputnik Planitia, dendritic valley networks are seen to dissect plateaus that are bounded by scarps of ~1 km relief ("1" labels in Figs. 16a,b). The networks are typically at low elevations of <1 km above mean radius. These valleys are organized much as terrestrial drainage networks, reach 2–5 km in width, and (where resolved by the DEM) are some hundreds of meters deep. Most of the valleys terminate in broad depressions tens of kilometers across ("2" labels in Figs. 16a,b) but lack obvious depositional landforms extending beyond the terminus (*Howard et al.,* 2017a). Impact craters up to a few tens of kilometers across interrupt the networks ("3" labels in Figs. 16a,b). To the south is an area of rugged, deeply dissected terrain that slopes steeply toward Sputnik Planitia, which it borders (Fig. 16c). Valley relief is typically 0.5–1 km. The major valleys are organized into crude dendritic patterns, but some terminate in blind depressions and others have irregular gradients. Nitrogen ice of Sputnik Planitia intrudes into some valleys ("4" labels in Fig. 16c).

To the northwest and west of Sputnik Planitia, there is a transition to a region displaying massifs reaching >3 km high that are separated by broad, sometimes sinuous valleys and interconnected basins. This region displays distinctive landscapes termed fluted and washboard terrains (Fig. 16d)

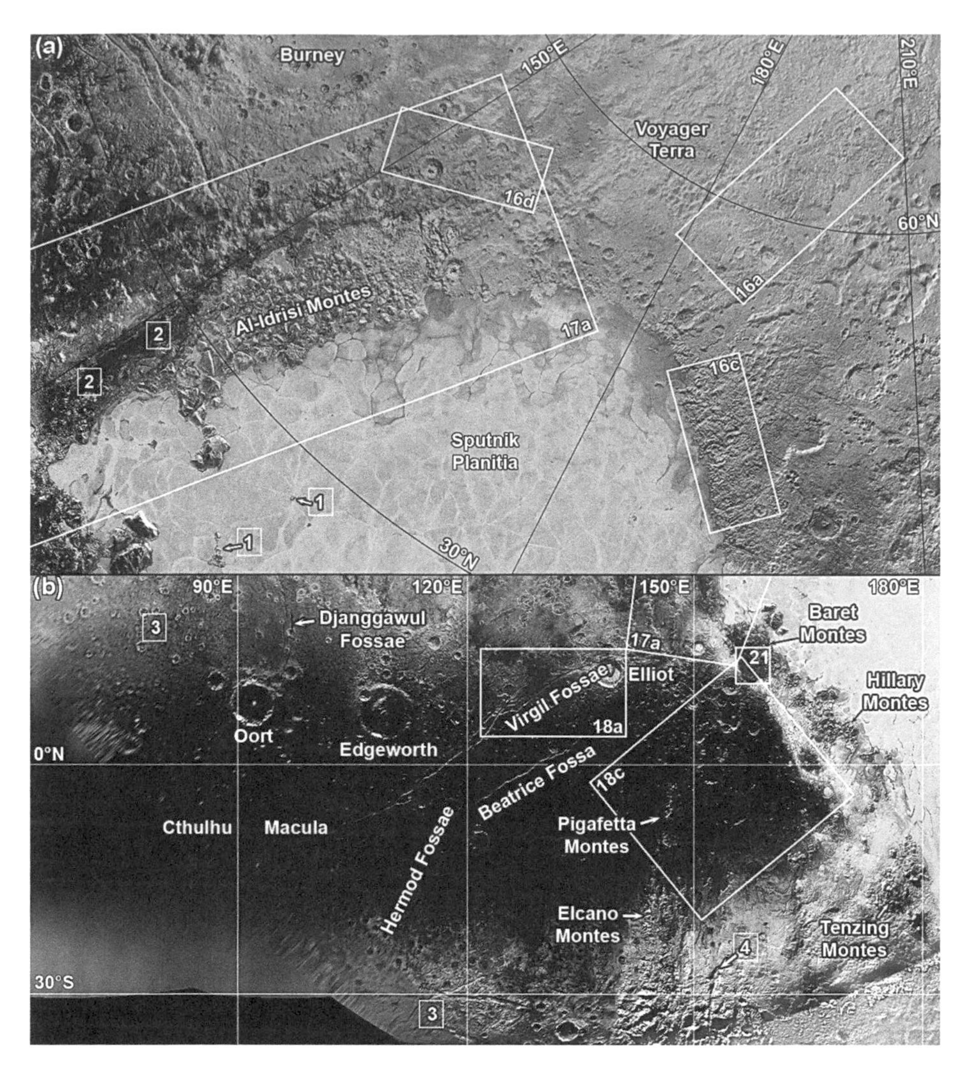

Fig. 15. The non-polar portions of the Burney Group. Numbered boxes refer to elements in Figs. 16–18 and Fig. 21. **(a)** Terrains of the Burney Group to the north and northwest of Sputnik Planitia. 1: Isolated mountains within Sputnik Planitia. 2: Low-albedo, north-south-aligned, subparallel ridges separating al-Idrisi Montes from the uplands. **(b)** Terrains of the Burney Group to the west and southwest of Sputnik Planitia. 3: Intermediate-albedo plains bordering Cthulhu Macula to its north and south. 4: Irregular and ameboid depression.

(*Moore et al.,* 2016; *Howard et al.,* 2017a; *White et al.,* 2019). These consist of parallel to subparallel ridge and trough sets that occur on the flanks of spurs and massifs ("5" labels in Fig. 16d) as well as level topographic settings of the valley and basin floors proximal to Sputnik Planitia ("6" labels in Fig. 16d). Ridges are spaced 1–3 km crest-to-crest, and individual ridges can exceed 20 km in length. A distinct type of fluted terrain is the "fluted craters," the walls of which display a pattern of radially aligned ridges and troughs ("7" labels in Fig. 16d). This morphology extends beyond the lowland valleys and basins to a plateau to the northwest, but washboard and fluted terrains are mostly located <1 km above mean radius. A defining characteristic of these ridges is their common east-northeast/west-southwest orientation, regardless of latitude or location relative to Sputnik Planitia.

The origin of these various landscapes has been interpreted to be paleo-glacial nitrogen-ice flows (*Moore et al.,* 2016; *Howard et al.,* 2017a; *White et al.,* 2019). Such glaciation is likely to have been confined to an era soon after the formation of the Sputnik basin, partly because these landscapes display a high crater count, but also due to consideration of how surface nitrogen ice would redistribute after the basin's formation. Modeling of volatile behavior in response to topography has shown that infilling of the Sputnik basin with all available surface nitrogen ice could be complete by tens of millions of years after its formation (*Bertrand et al.,* 2018). During this period, substantial nitrogen-ice glaciation could have been supported within high-latitude, relatively low-elevation uplands adjacent to Sputnik Planitia, but would be unstable both due to nitrogen migrating to the floor of the newly formed Sputnik Planitia basin (*Bertrand and Forget,* 2016), as well as the high latitudes experiencing high insolation and annual mean surface temperatures on account of Pluto's high obliquity (*Earle et al.,* 2017). Glacial recession from these regions therefore took place

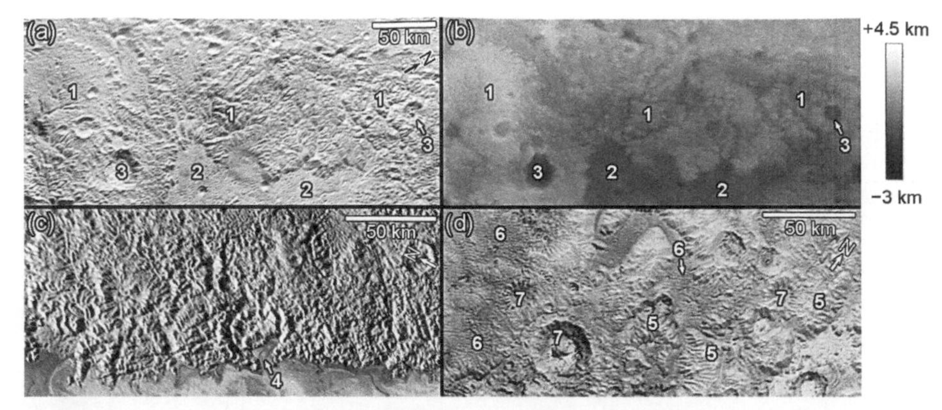

Fig. 16. Glacially modified terrains in uplands north of Sputnik Planitia. See Fig. 15a for context. **(a,b)** Dissected terrain of Voyager Terra. Image is shown at left, corresponding DEM at right. Illumination is from the top. 1: Dendritic valley networks. 2: Broad, flat-floored depressions in which the valley networks terminate. 3: Impact craters superposing the valley networks. **(c)** Rugged, deeply dissected terrain. Illumination is from the left. 4: Glacially-flowing nitrogen ice of Sputnik Planitia intruding into a valley formed by an eroded impact crater. **(d)** Washboard and fluted terrain. Illumination is from the top-right. 5: Fluted flanks of spurs and massifs. 6: Washboard texture on the floors of valleys and basins. 7: Impact craters with fluted walls.

as most of Pluto's nitrogen ice migrated to the very deep, low-latitude Sputnik Planitia basin.

Nitrogen ice glaciation would differ in important respects from terrestrial water ice glaciation, particularly in the absence of freeze-thaw weathering of adjacent topography and the higher density of nitrogen ice relative to probable substrates (*Howard et al.,* 2017a), and these landscapes accordingly lack many distinctive features of terrestrial glaciation. Whether nitrogen ice accumulating on undissected uplands would sculpt valley networks is uncertain; if the glaciers were frozen throughout, they would be cold-based and therefore inefficient in eroding their beds (*Head and Marchant,* 2003; *Gjermundsen et al.,* 2015). But sufficiently thick accumulations of nitrogen ice (~1–2 km for reasonable geothermal gradients) (*McKinnon et al.,* 2016) could result in basal melting (*Howard et al.,* 2017a), which would lubricate the base of the glacier and substantially increase its capacity for erosion. This effect would be enhanced if the substrate is porous, fine-grained, less dense than nitrogen ice, and poorly indurated.

Howard et al. (2017a) interpreted the troughs of fluted terrain to be incised into the substrate by glacial erosion, like the dendritic valleys and dissected terrain to the east. Alternatively, by noting their coincidence with the tectonically deformed northwestern rim of Sputnik Planitia, and the similarity of their texture to fields of sublimation pits in southern Sputnik Planitia, *White et al.* (2019) suggested that both washboard and fluted ridges are tectonically-liberated crustal debris that were buoyant in pitted glacial nitrogen ice that formerly covered the area, and that were deposited after the nitrogen ice receded via sublimation. The cause of the common orientation of the ridges, however, remains unknown. More generally, there is a notable discrepancy between the volume of material excavated to form the dissected landscapes and the lack of identifiable corresponding deposits: Moraines and outwash plains analogous to those seen at terrestrial glaciated landscapes have not been identified (*Howard et al.,* 2017a). Overall, the spatial variation in morphology of the glacial features north of Sputnik Planitia is likely to be a response to local topographic setting, substrate properties, latitudinal variations in insolation, and variation in depths and durations of ice accumulation (*Howard et al.,* 2017a).

2.6.2. Sputnik mountain ranges. A discontinuous chain of mountains extends for hundreds of kilometers along the western boundary of Sputnik Planitia (*Moore et al.,* 2016; *White et al.,* 2017; *Schenk et al.,* 2018). From north to south, these are al-Idrisi Montes (Figs. 15a and 17), Baret Montes, Hillary Montes, and Tenzing Montes (Fig. 15b). Individual mountains are typically between 5 and 13 km in diameter and reach elevations of 0.5–1.5 km above the surrounding plains (*Skjetne et al.,* 2021), although a few represent some of the locally highest and steepest relief on Pluto, reaching tens of kilometers across and >5 km in height (*Schenk et al.,* 2018), and with slopes up to 40°–50°. Planform aspect ratios of the mountains are mostly between 1:1 and 2:1 (*Skjetne et al.,* 2021). The mountains are mostly close-packed in discrete ranges, where they display apparently random orientations, and tend to be segregated according to size, with smaller mountains clustering separately from larger ones ("1" and "2" labels respectively in Figs. 17a,b). The nitrogen-ice plains of Sputnik Planitia embay the ranges, and often separate individual mountains, especially in the southern ranges. The mountains have a blocky appearance and tend to display flat or sloping faces with a texture comparable to that seen in the uplands to the west (inset in Fig. 17a). Spectral data has confirmed that the mountains are composed of water ice (*Grundy et al.,* 2016; *Protopapa et al.,* 2017; *Schmitt et al.,* 2017; *Cook et al.,* 2019), which is relatively strong and inert at plutonian conditions and can support their great relief and steep slopes (*Stern et al.,* 2015b; *Moore et al.,* 2016).

The spectral signature of water ice is also seen in the uplands to the west of these mountain ranges, and is

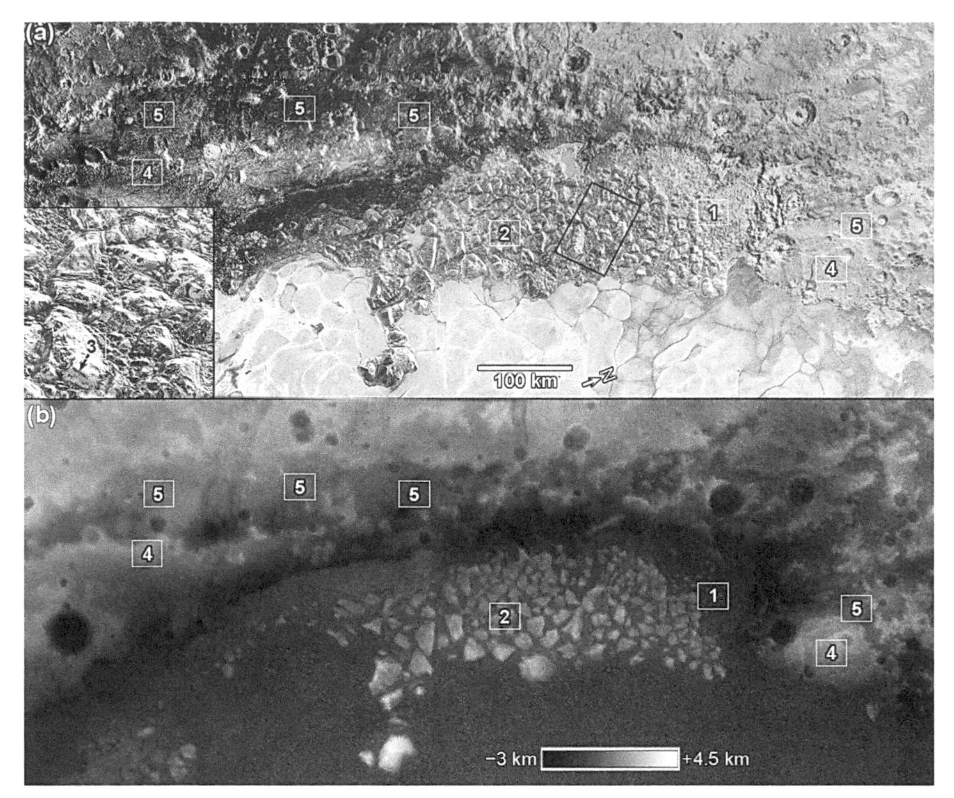

Fig. 17. The mountain range of al-Idrisi Montes and adjacent uplands. See Fig. 15 for context. Image is shown in **(a)**, corresponding DEM in **(b)**. Illumination is from the right. Box in **(a)** indicates the boundaries of the 76 m/pixel imaging in the inset at bottom left. 1,2: Agglomerations of smaller and larger mountains, respectively. 3: Dark lineation crossing the face of a tilted block. 4: Tilted crustal blocks separated from the uplands by broad troughs (5).

hypothesized to be the main crust-forming material on Pluto (e.g., *McKinnon et al.,* 1997; *Hussmann et al.,* 2006). The mountains are interpreted to be fragments of pre-existing water ice crust that have been detached from the surrounding uplands, and subsequently rotated and transported, by the action of denser, flowing nitrogen ice that has intruded and exploited weaknesses in the crust (*Moore et al.,* 2016; *White et al.,* 2017). Dark lineations that cross the faces of some of the tilted blocks (label "3" in Fig. 17a inset) may represent layering within the original crustal material that is now exposed. Fragmentation of crustal material would be especially efficient if the nitrogen ice were to infiltrate tectonized crust: The uplands immediately to the west of al-Idrisi Montes are where fractures of a major north-south-trending tectonic system intercept Sputnik Planitia. Being located around the rim of the Sputnik impact basin, this crust may also have been weakened through brecciation. *Howard et al.* (2017a) hypothesized that glacial nitrogen ice that formerly covered the uplands dislodged blocks of fractured crust and translated them by traction forces as it flowed toward Sputnik Planitia, thereby forming al-Idrisi Montes. For reasonable slopes and friction coefficients, the largest blocks can be transported from their source over very short (10^2–10^5-yr) timescales, and the influence of viscous drag would have a size-filtering effect, with the largest blocks migrating farthest from the uplands, which is consistent with observations (*O'Hara and Dombard,* 2021). Within the uplands to the north and west of al-Idrisi Montes there are tilted crustal blocks ("4" labels in Figs. 17a,b) that are separated from the main uplands by troughs of the trough-ridge system that reach tens of kilometers across and >1 km deep ("5" labels in Figs. 17a,b). These blocks slope sharply toward Sputnik Planitia and display high scarps on their anti-Sputnik sides. The northern block is crossed by a graben toward which nitrogen ice in Sputnik Planitia is evidently flowing. These tilted blocks may represent partial crustal detachments, but did not ultimately fragment to form ranges of mountains (*Howard et al.,* 2017a).

Compared to al-Idrisi Montes, the more southerly Baret, Hillary, and Tenzing Montes are farther removed from the uplands and are not obviously proximal to tectonic lineations. These mountains may have originated due to the action of nitrogen ice in Sputnik Planitia itself on crustal material, rather than uplands glaciation. Consideration of isostasy within Sputnik Planitia means that it is highly probable that the mountain ranges have essentially remained in place (to within some tens of kilometers) since the crust from which they are formed was fragmented. Due to the minimal density contrast between nitrogen and water ice (0.942–1.0 g cm^{-3} and 0.935 g cm^{-3} respectively), and considering that many of the mountains display elevations above the nitrogen-ice plains of a few kilometers, the roots of most mountains are likely grounded on the base of the shallow (<1-km) nitrogen-ice deposits at the edge of Sputnik Planitia (*Moore et al.,* 2016; *White et al.,* 2017; *Skjetne et al.,* 2021). Mobilization of most mountains by the nitrogen ice is therefore limited to undermining, shifting, and rotating, but not regional-scale transportation. An exception is the handful of smaller mountains located within the interior of Sputnik Planitia that are separated from the main ranges (e.g. "1" labels in Fig. 15a). Given that Sputnik Planitia is interpreted to be filling a basin, the nitrogen ice would be deeper here, and the roots of these mountains are small enough to allow them to float within it and to be transported by its convective motion (*Moore et al.,* 2016; *White et al.,* 2017).

These mountains are interpreted to be composed of ancient uplands crust that dates from immediately after the Sputnik basin impact, but exactly when they were dislodged from the crust to form the ranges is an unresolved issue, and not all the mountain ranges may necessarily have formed in a single episode. The hypothesis that uplands glaciation was instrumental in the formation of at least some of the mountains (al-Idrisi Montes) would imply an ancient origin, as glaciation on such a scale would most likely have been limited to the tens of millions of years following the formation of the Sputnik impact basin (*Bertrand et al.,* 2018; *White et al.,* 2019). This scenario may be contradicted by the apparently young age of the mountains, as no unambiguous craters have been identified in contiguous 315 m/pixel imaging or in any higher-resolution strips (*Robbins et al.,* 2017; *Singer et al.,* 2019). The steep slopes of the mountains, however, may be conducive to mass wasting, which would act to obliterate any craters (*Howard et al.,* 2017a), making the mountains appear more youthful than they actually are: The absence of craters suggest a surface age of 200–300 m.y. (see the chapter by Singer et al. in this volume).

2.6.3. Dark, cratered plains and rugged uplands. The equatorial region to the west of Sputnik Planitia is mostly occupied by Cthulhu Macula (Fig. 15b), the largest of the discontinuous chain of dark maculae that circle Pluto's equator (section 4). It consists of rugged, mountainous terrain proximal to Sputnik Planitia and smooth plains in its western portion, which extend for >1000 km into the farside, where the plains are superposed by the bladed terrain deposits. Cthulhu Macula is among the most cratered terrains on Pluto's nearside (*Robbins et al.,* 2017; *Singer et al.,* 2019) and displays little evidence for modification by exogenic geological processes. The cratered plains of the far western nearside hemisphere may even predate the Sputnik impact (*Schenk et al.,* 2018). As is the case for all maculae on Pluto, the low albedo is a consequence of the accumulation of a blanket of haze particles on the surface. Its landforms appear sharp and well defined, suggesting that the blanket is a superficial coating that does not mask or degrade underlying topographic relief, rather than a distinct exposed geological stratum (*Grundy et al.,* 2018). There is no obvious morphological or topographic contrast between Cthulhu Macula and adjacent terrains to the north and south (*Schenk et al.,* 2018), which present an intermediate albedo ("3" labels in Fig. 15b). Pluto's maculae are mostly confined to the permanent diurnal zone, which experiences the mildest seasons and overall lowest insolation on Pluto (*Binzel et al.,* 2017). The diurnal Sun is able to keep the low-albedo terrain warm enough to not become an attractive cold trap for new volatiles (*Bertrand and Forget,* 2016; *Earle et al.,* 2017), explaining why it is mostly free of the seasonally mobile volatile ices that mantle much of Pluto's surface, and instead records long-term accumulation of haze particles. *Grundy et al.* (2018) determined that unperturbed haze particle accumulation would coat the surface to a thickness of ~14 m over the age of the solar system, although fluctuating atmospheric pressure in response to the obliquity cycle may cause haze production to be periodically interrupted (*Johnson et al.,* 2021).

Exhibiting less than a kilometer of topographic relief over hundreds of kilometers (*Schenk et al.,* 2018), the western smooth plains are the flattest terrain outside of Sputnik Planitia yet seen on Pluto (west of ~140°E in Fig. 15b). Even where they are not covered by the haze particle blanket, they appear relatively featureless, and the absence of landforms diagnostic of accrescence or decrescence processes, as well as the lack of valley networks suggestive of glacial erosion, confirms that these plains have not been visibly modified by volatile mobilization since they formed ~4 G.y. The plains feature two of the largest impact craters seen on Pluto's nearside: 120-km-diameter Oort and 145-km-diameter Edgeworth. Besides impact cratering, the plains have been affected by regional-scale tectonism in the form of Djanggawul Fossae and the northeast-southwest-trending Virgil and Beatrice Fossae, which cross Hermod Fossae that extend from the north and belong to the same tectonic system as Inanna and Dumuzi Fossae ("1" label in Figs. 18a,b). Virgil Fossae crosscuts the 85-km-diameter impact crater Elliot, which displays ponded nitrogen-ice deposits on its floor ("2" label in Figs. 18a,b). A possible instance of recent geological activity within these plains is the mantled terrain in and around a segment of Virgil Fossae (white arrows in Figs. 18a,b). This localized zone of mantling diverges from the haze particle blanket that defines Cthulhu Macula in that it displays a lighter color, and the topographic relief of small-scale features appears muted, including a subparallel chain of subrounded, rimless depressions several kilometers across (*Cruikshank et al.,* 2019b). Ammoniated water ice deposits have been identified here, and the mantling material may be cryoclastic materials recently (<1 G.y.) erupted by fountaining events from this segment of Virgil Fossae (*Cruikshank et al.,* 2019a,b; *Dalle Ore et al.,* 2019).

East of ~140°E, the plains transition to a much more mountainous and rugged environment (Figs. 18c,d). This terrain forms the broad, raised southwestern rim of the Sputnik impact basin and it is crosscut by the ancient north-south-trending ridge-trough system (*Schenk et al.,* 2018). It includes large craters that are the deepest yet seen on Pluto (>4 km) ("3" labels in Figs. 18c,d). The eastern margin of Cthulhu Macula abuts the nitrogen-ice plains of Sputnik Planitia, which superpose and embay the uplands. The ridge-trough system encompasses a variety of landforms here, including degraded scarps, graben, and troughs; dissected, linear, mountainous ridges; and four elongate, flat-floored depressions that reach 4 km deep (*Schenk et al.,* 2018).

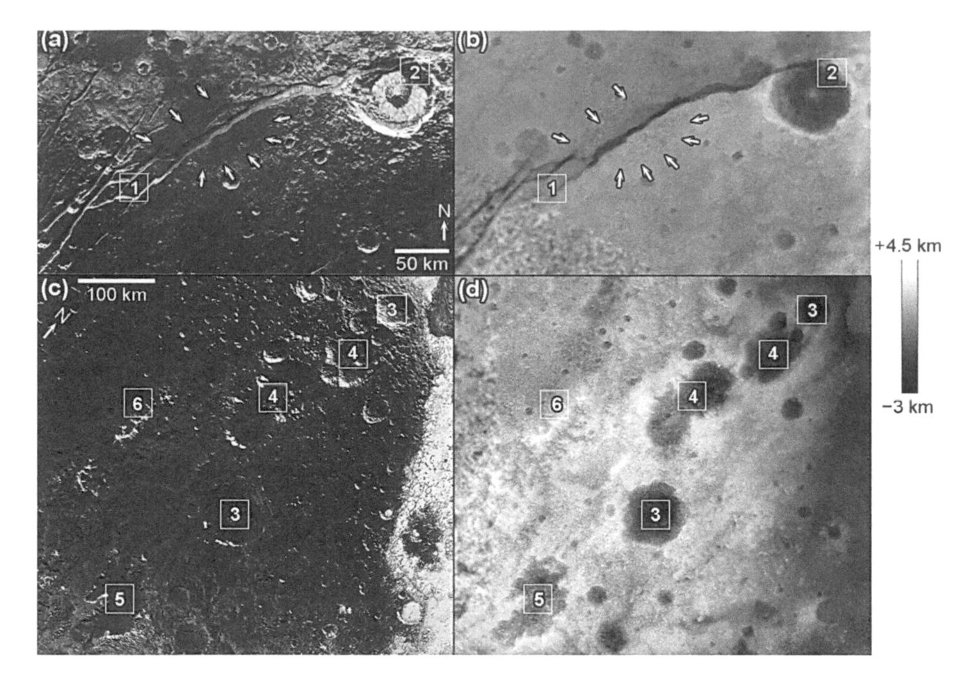

Fig. 18. Terrains within Cthulhu Macula. See Fig. 15b for context. Images are shown at left, corresponding DEMs at right. **(a),(b)** Smooth, cratered plains in the vicinity of Virgil Fossae. Illumination is from the top. 1: Fractures of Hermod and Virgil Fossae crosscutting each other. 2: Virgil Fossae crosscutting Elliot crater. **(c),(d)** Cratered and tectonized terrain in eastern Cthulhu Macula. Illumination is from the top right. 3: Large and deep craters. 4: Large, deep, flat-floored depressions with ovoid planforms. 5: Irregular, ameboid, flat-floored depression. 6: Pigafetta Montes with bright summit regions.

The two northern depressions, the larger and deeper of the four, are ovoid in planform with fluted walls ("4" labels in Figs. 18c,d). The two southern depressions are more irregular and ameboid and display lobate reentrants (label "4" in Fig. 15b and label "5" in Figs. 18c,d). Shallow, degraded graben, scarps, and troughs link all four depressions. Given the large scale of these depressions and their elongation in the same direction as the ridge-trough system, explosive cryovolcanism and surface collapse facilitated by tectonism is a potential origin (*Schenk et al.,* 2018; *Ahrens and Chevrier,* 2021).

The mountainous ridges of the ridge-trough system include Pigafetta Montes (Fig. 15b and "6" label in Figs. 18c,d), which are located entirely within Cthulhu Macula, and Elcano Montes (Fig. 15b), which extend beyond Cthulhu Macula's southern boundary. The largest of the Elcano Montes massifs rise up to 4.5 km above local surfaces and are among the highest-standing individual features relative to Pluto mean radius. The summit regions of both ridges (i.e., elevations higher than ~2 km above mean radius), as well as the north-facing slopes of crater walls in the vicinity of Pigafetta Montes, are capped by bright methane ice. The low overall insolation received in this low-latitude region (*Earle et al.,* 2017), in combination with the high elevations of these peaks, makes them suited for net methane deposition despite the low albedo of the haze blanket (which has the effect of raising the daytime surface temperature), and such deposits are likely to be perennially stable (*Bertrand et al.,* 2020b). Those occupying lower-elevation crater walls of a particular aspect are likely ephemeral, however, disappearing and reforming on the opposite walls as the seasons change.

Pigafetta Montes display fluted dissection on their slopes, and are surrounded by 5–6-km-wide, flat-floored valleys in a crude dendritic pattern. Elcano Montes are ringed by low-relief plains, into which are incised several valley systems that drain westward. *Howard et al.* (2017a) described these mountainous ridges as "alpine terrain" and interpreted them as having been sculpted by the flow of glacial nitrogen ice, noting that the landscapes here are those that are most similar to terrestrial glaciation. Like the glacial features north of Sputnik Planitia, any such glaciation must have been ancient given that these ridges and valleys display several large impact craters.

2.6.4. Polar uplands. Pluto's north polar region (named Lowell Regio) features a 2–3-km-high, ~600-km-wide dome centered on the north pole (*Schenk et al.,* 2018) (Fig. 19). It has no discrete margin and tends to grade into the surrounding terrain. It consists of heavily dissected uplands featuring a rough-hewn series of ridges, plateaus, and craters that are bisected by fractures of the north-south-trending ridge-trough system (*Howard et al.,* 2017b). The tectonism is expressed as subparallel, very eroded troughs and graben, the most prominent of which is a ~150-km-wide, flat-floored graben that reaches ~6 km deep. In the global color map, these uplands display a distinct golden hue (*Olkin et al.,* 2017) that corresponds to high methane absorption (*Protopapa et al.,* 2017; *Schmitt et al.,* 2017; *Earle et al.,* 2018b). The valley floors of the ridge-trough system are lighter toned than the dissected uplands that surround them, and feature

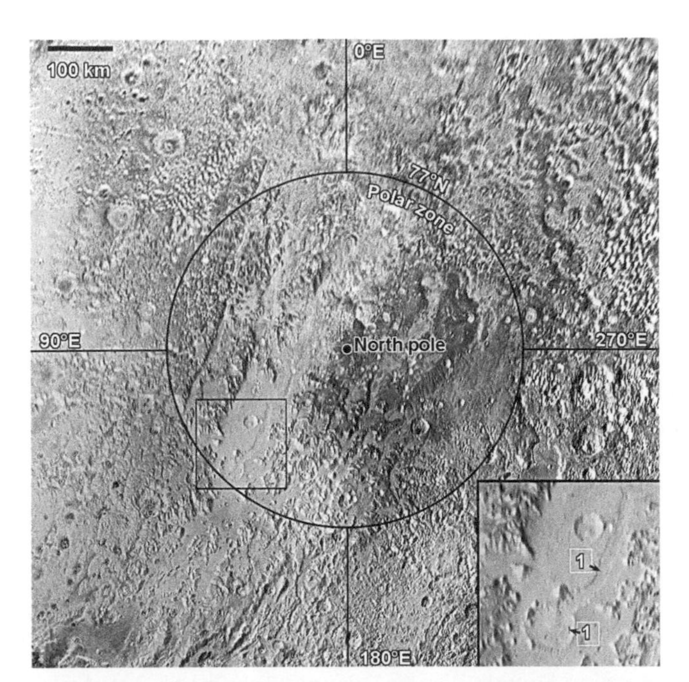

Fig. 19. Pluto's north polar region, seen in polar stereographic projection centered over the north pole. Box indicates the area covered by the inset at bottom right. Illumination is from the left. 1: Impact crater and valley with eroded and muted topographic relief.

some landforms with eroded and muted topographic relief ("1" labels in Fig. 19 inset). The polar region is slightly less cratered than elsewhere in the Burney Group (*Singer et al.,* 2019). The mere presence of several large craters, and its deformation by the stratigraphically low ridge-trough system, does, however, indicate that the polar uplands are ancient and their very rough and fretted appearance shows that they have primarily been modified by erosion rather than deposition (*Howard et al.,* 2017b).

Howard et al. (2017b) hypothesized that the polar uplands are a dissected mantle composed of friable, methane ice-based sediment up to 3 km thick. The uplands, however, are mostly contained within the permanent polar zone north of 77°N (*Binzel et al.,* 2017), where insolation on Pluto is highest and arctic seasons are most consistently experienced (*Hamilton et al.,* 2016; *Earle et al.,* 2017), and therefore where seasonal, insolation-driven volatile mobilization is most intense. Numerical climate modeling of the methane cycles on Pluto (*Bertrand et al.,* 2019) indicates that this zone will on average experience net sublimation of methane ice over hundreds of millions of years, thereby precluding the accumulation of kilometers-thick methane ice deposits at any time in Pluto's history. Additionally, it is uncertain as to whether the yield strength of methane ice (*Eluszkiewicz and Stevenson,* 1990; *Yamashita et al.,* 2010) could allow it to support the high topographic relief exhibited by the uplands. *Howard et al.* (2017b) offered an alternative hypothesis whereby the material of the polar uplands was emplaced as a locally thick accumulation of water ice-rich ejecta from the oblique Sputnik impact: If the impact was incident from the south-southeast, then a high proportion of ejecta would have been deposited downrange, i.e., to the north-northwest of the basin (e.g., *Gault and Wedekind,* 1978; *Herrick and Hessen,* 2006; *Shuvalov,* 2011). If water ice does form the bedrock of this region, it must be covered by at least several centimeters or even meters of methane ice (*Bertrand et al.,* 2019), as the spectral signature of water ice is absent. Such methane-rich deposits may have accumulated over intermediate timescales possibly associated with Pluto's 2.8-m.y. obliquity cycle (*Grundy et al.,* 2018).

If the polar uplands are composed of refractory water ice and never accumulated a thick mantle of volatile ices, then their eroded state and relative paucity of impact craters may be a consequence of weakening of the bedrock by countless cycles of deposition and sublimation of volatile ices, primarily methane and nitrogen. The erosional power would be enhanced if depositing volatiles penetrated structural weaknesses within the water ice bedrock, causing it to disaggregate and lose mechanical strength upon their sublimation (*Moore et al.,* 1996, 1999). Such erosion would make the uplands more disposed to mass wasting, debris from which may be covering the floors of the lowlands: The infilled appearance of some landforms on the valley floors (Fig. 19 inset) is diagnostic of vertical sedimentation, whereby depressions are preferentially filled with a deposit, while ridges remain exposed. This process would cause minimal degradation of the uplands, but if it has operated continuously across Pluto's history, then their present-day appearance represents the cumulative effect of billions of years of gradual erosion. An alternative, more intense mechanism to erode the polar uplands is via flow of glacial nitrogen ice (*Howard et al.,* 2017b): Their dissected, sometimes fluted, appearance resembles that of glacially dissected terrain on the northeast rim of Sputnik Planitia (Fig. 16c). As with methane ice, however, the high mean insolation experienced at the pole makes it an unnatural location for accumulation of thick deposits of nitrogen ice (*Bertrand et al.,* 2018).

3. TECTONISM

Tectonic fracturing is a key indicator of the level of endogenic activity within a planetary body, and Pluto's surface displays extensive tectonic deformation across the nearside hemisphere. Mapping studies (*Keane et al.,* 2016; *McGovern et al.,* 2019) have revealed that Pluto possesses a global, non-random system of extensional faults (Fig. 20), indicating global expansion due to partial freezing of a subsurface ocean as the overarching driver of tectonism (*Hammond et al.,* 2016; *Nimmo et al.,* 2016). The tectonism appears as fault scarps and graben that can reach >10 km wide and hundreds of kilometers long; shorter, narrower troughs; and chains of collapsed pits. The variety of configurations and preservation states of the various fault systems suggest multiple deformation episodes and prolonged tectonic activity (*Moore et al.,* 2016), with localized effects governing the nature and timing of how the expansion is expressed tectonically. Compressional tectonism is not evident, with the possible

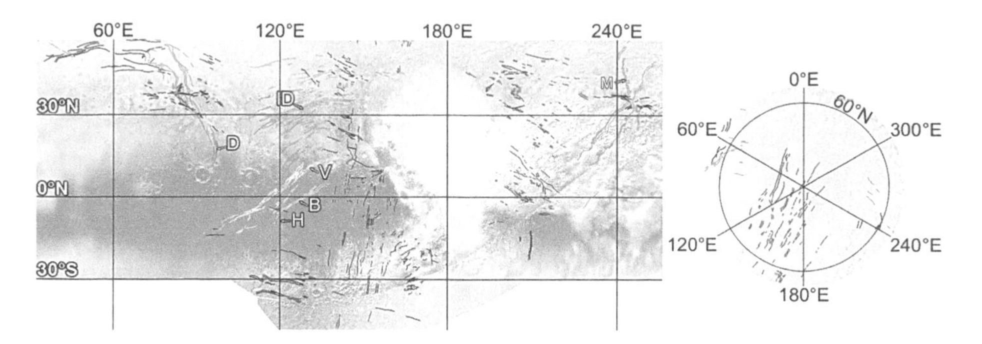

Fig. 20. Mapped extensional tectonic lineations across Pluto's nearside hemisphere (*McGovern et al.,* 2019). Colored lineations represent tectonic systems described in this section. Labels indicate individual fossae: ID = Inanna and Dumuzi, V = Virgil, B = Beatrice, H = Hermod, D = Djanggawul, M = Mwindo.

exception of localized instances expressed as low-albedo, subparallel ridges separated by several hundred meters to a few kilometers, which occur between Baret (Fig. 21) and al-Idrisi Montes ("2" labels in Fig. 15a) and the uplands of Viking Terra (*White et al.,* 2017; *Ahrens and Chevrier,* 2019). The ridges tend to align parallel to the edge of Sputnik Planitia and have been interpreted to form due to the buoyant, mobile water ice mountains being pushed toward the uplands by convective motions in Sputnik Planitia, with a weaker, darker, overlying layer folding in response to the compressive forces (*Ahrens and Chevrier,* 2019).

The terrain to the west of Sputnik Planitia (between 100°E and 150°E) displays the most prominent and well-preserved tectonism on Pluto's nearside. These fractures

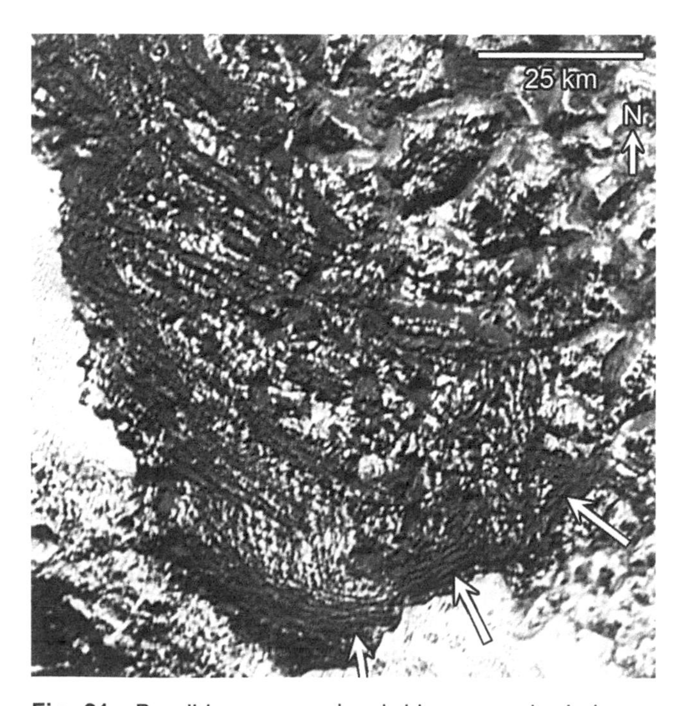

Fig. 21. Possible compressional ridges seen in dark material covering southwestern Baret Montes. The ridges are predominantly oriented northwest-southeast, although those proximal to the southern edge of the mountain range (white arrows) tend to align east-northeast/west-southwest, roughly parallel to the border of the range. See Fig. 15b for context. Illumination is from the top left.

are subparallel, segmented graben and troughs that group together in several commonly aligned belts reaching hundreds of kilometers long. The largest graben display throws of >2 km. In the northern portion of the system that contains Inanna and Dumuzi Fossae (Fig. 14a), fractures are quasi-radial to Sputnik Planitia (orange lines in Fig. 20). As the system extends southward, it curves to adopt a north-south orientation where it crosses into the southern hemisphere (forming Hermod Fossae). Between 3°N and 5°S, these fractures intersect with the northeast-southwest-aligned fractures of Virgil and Beatrice Fossae (yellow lines in Fig. 20), which along with a system located at 30°S,120°E are approximately radial to Sputnik Planitia along their entire length (>1000 km for Virgil Fossae). Fractures belonging to Virgil and Hermod Fossae appear to crosscut each other (label "1" in Fig. 18a), suggesting that, despite their different orientations, these two sets of fractures likely formed concurrently. The southern scarp of Virgil Fossae is ~1 km higher than its northern scarp, indicating that it may be a half-graben (*Schenk et al.,* 2018; *Conrad et al.,* 2019). The sharply defined scarp crests of fractures within this western group, and the observation that they crosscut impact craters (most notably Virgil Fossae cutting Elliot; Fig. 18a), is an indication of their relative youth. Zones of continuous mantling of the surface by dark material that surround a segment of Virgil Fossae as well as a trough to the northeast may indicate recent (<1 G.y.) cryovolcanic activity (*Cruikshank et al.,* 2019a,b, 2021; *Dalle Ore et al.,* 2019). Lithospheric stresses related to the formation of the Sputnik basin and its infilling with nitrogen ice may enhance subsurface fluid transport to the surface through radially oriented fractures such as Virgil Fossae (*Cruikshank et al.,* 2019a,b; *McGovern and White,* 2019).

Terrain to the east of Sputnik Planitia (between 185°E and 225°E) features scarps, troughs, and pit chains (Figs. 10c and 12a, as well as red lines in Fig. 20). The alignment of the pit chains parallel to structural trends in the vicinity suggests that they are found in a location where surface collapse has occurred as tectonism disturbs an overlying mantle (*Howard et al.,* 2017b). These fractures are predominantly oriented northwest-southeast, with those in the south being quasi-radial to Sputnik Planitia, whereas those in the north are closer to tangential. Great circles extrapolated from the fractures align fairly well with the northern fractures west of Sputnik Planitia, raising the possibility that these geographically separate sets may represent a single tectonic system, the central portion of which is covered by Sputnik Planitia. The eastern fractures appear more eroded than those west of Sputnik Planitia, and do not crosscut impact craters so commonly, but their degraded appearance is at least partly due to the major mantling and erosional episodes that have affected this region, which the fractures west of Sputnik Planitia have not been subjected to.

In the far west and far east of the nearside hemisphere lie Djanggawul (label "2" in Fig. 13) and Mwindo (Figs. 9 and 11) Fossae respectively, tectonic systems that are oriented azimuthally about a pole located very close to the geographic center of Sputnik Planitia (green lines in Fig. 20). The configuration of these systems implies an origin tied to Sputnik Planitia. The modeling of *Keane et al.* (2016) found that reorientation of Pluto in response to infilling of the Sputnik basin with nitrogen ice would generate stresses that are approximately consistent with the Sputnik-azimuthal orientations of these systems, and that the azimuthal fractures might be crosscut by the quasi-radial ones proximal to Sputnik Planitia. These two sets of fractures are sufficiently removed from each other such that they do not intersect, meaning that crosscutting relationships cannot be established, although the azimuthal fractures do appear less well preserved than those nearer Sputnik Planitia. Mwindo Fossae are unusual in that they converge to a nexus, implying that a localized stress field caused them to diverge from the Sputnik-azimuthal orientation (*McGovern et al.,* 2019). The geology at the nexus is unremarkable relative to its surroundings, providing no clue as to what endogenic process may have contributed to this convergence. What does distinguish this terrain is that bladed terrain deposits are hypothesized to have previously covered it (*Moore et al.,* 2018). The southern fractures of Mwindo Fossae cut through existing deposits for hundreds of kilometers. Loading imparted by the deposits, and possibly the lithospheric response to their recession, may form part of the explanation for this unique feature. *Cruikshank et al.* (2019a) highlighted the exposure of dark material along fractures of Mwindo Fossae, as they did for a segment of Virgil Fossae, and noted that on Earth and Mars, radiating graben are often associated with magmatic dike swarms at depth and possible subsurface fluid expulsion.

One of the most dramatic physiographic features on Pluto is a complex, eroded band of graben, troughs, ridges, plateaus, tilted blocks, and elongate depressions that extends at least 3200 km from the north pole to the limit of New Horizons coverage at ~45°S, and that is ~300–400 km wide (*Schenk et al.,* 2018) (Fig. 22 and purple lines in Fig. 20). Crossing the equator at ~150°E, this north-northeast/south-southwest-trending "ridge-trough system" traverses many terrain types and manifests differently along its length: degraded graben in Lowell Regio ("1" labels in Fig. 22); graben and tilted blocks ("2" labels) where the system intersects Sputnik Planitia and extends beneath the nitrogen-ice sheet; and sporadic, fragmentary scarps, graben, and troughs aligned with degraded, mountainous ridges (label "3" in Fig. 22) as well as elongate, flat-floored, possibly cryovolcanic depressions ("4" labels) in Cthulhu Macula. The structure may well extend further into the poorly resolved farside (section 4) and into the shadowed southern regions (*Schenk et al.,* 2018; *Stern et al.,* 2021). The system certainly represents the earliest evidence of tectonism yet seen on Pluto due to its highly eroded state, and because its elements are invariably crosscut by other tectonic lineations. It may even predate the Sputnik basin-forming impact, but since it crosscuts the broad raised rim of the Sputnik basin and terrain leading down to Sputnik Planitia itself, some deformation did still occur after the basin formed (*Schenk et al.,* 2018). Equatorial crustal thickening has been hypothesized to be the cause of such an immense tectonic feature aligned along a great circle (*McGovern et al.,* 2019), although this would require the system to be aligned along a "paleo-equator" prior to reorientation of Pluto.

4. THE FARSIDE OF PLUTO

The great contrast in pixel scale between New Horizons imaging of Pluto's nearside (76–850 m/pixel) and farside (2.2–32 km/pixel) means that the delineation and interpretation of farside terrains are based largely on albedo variations as seen in low-phase approach imaging. In addition, contacts between separate terrains that are easily defined in high-resolution nearside imaging that abuts the farside can be extrapolated to a certain extent into the farside. While only sporadically distributed across the farside, limb profiles allow assessment of topographic relief at a scale of hundreds of meters. This section summarizes farside research to date, and is primarily sourced from the detailed examination contained in *Stern et al.* (2021).

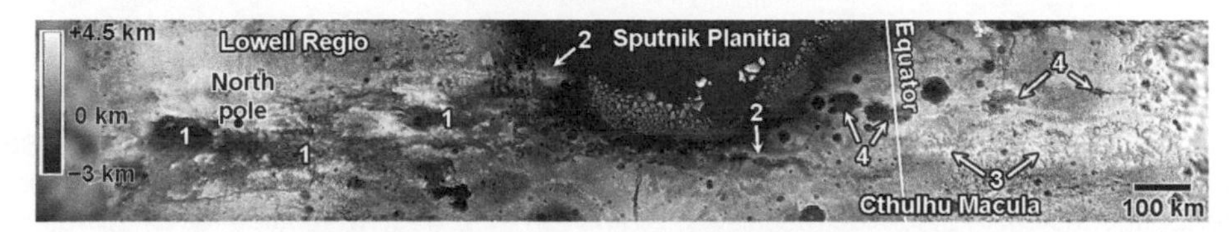

Fig. 22. Reprojection of the stereo DEM in Fig. 1 such that the north-northeast/south-southwest-oriented ridge-trough system tracks horizontally across the map (*Schenk et al.,* 2018). The map extends across ~160° of latitude and is centered at 151.5°E at the equator. 1: Degraded graben. 2: Tilted blocks. 3: Mountainous ridges. 4: Elongate, flat-floored depressions.

Moore et al. (2018) identified farside expanses of material with high methane absorption values that correlate to a generally intermediate-albedo unit located within 30° of Pluto's equator. Limb profiles extracted from imaging of better than 1 km/pixel that pass over this unit show that it displays relief of 2–4 km above adjacent terrain (*Schenk et al.,* 2018; *Stern et al.,* 2021). The composition, elevation, and latitudinal distribution of this material are therefore comparable to those of the bladed terrain deposits of Tartarus Dorsa that border the eastern edge of the nearside hemisphere (section 2.3.1), indicating that Tartarus Dorsa form the western extreme of a vast, low-latitude belt of bladed terrain deposits extending across >220° of longitude (outlined in black in Fig. 23). The easternmost occurrence of these deposits extends just inside the nearside hemisphere and is seen to superpose the dark plains of Cthulhu Macula (label "1" in Fig. 23) (*Stern et al.,* 2021).

Extending east from the eroded uplands of Krun Macula (section 2.3.4) into the farside is a discontinuous chain of dark maculae ("2" labels in Fig. 23), which at 70°E link up with the smooth, cratered plains of Cthulhu Macula at the western boundary of the nearside hemisphere (section 2.6.3), a geological environment that is very different to Krun Macula. There must be at least one morphological transition within the farside between these two macula types, and *Stern et al.* (2021) identified a transition between ~335°E and ~10°E based on the differing appearance of the maculae in approach imaging of 15–20 km/pixel to the west (dark/intermediate-albedo mottled texture) and east (dark and homogeneous) of this longitude. The close spatial association of Krun Macula and the western farside maculae to the bladed terrain deposits (whereby an intricate and angular contact separates the two terrain types, and both are seen to embay occurrences of the other) suggests that a genetic relationship between the two may exist, which would likely involve recession of the bladed terrain deposits by sublimation erosion in response to secular climate change, as is hypothesized to have occurred for Tartarus Dorsa (*Moore et al.,* 2018) (section 2.3). A hypothesized scenario involves recession causing the configuration and surface elevation of the bladed terrain deposits to change over time, leading to suppression of mobilization of surface methane ice for large expanses of relatively low-elevation bladed terrain deposits (*Stern et al.,* 2021). Mobilization would have been interrupted for a period of time that was long enough (millions of years) (*Grundy et al.,* 2018) to allow accumulation of a continuous layer of haze particles that would inhibit any further large-scale volatile mobilization. Alternatively, the maculae may represent regions where the deposits have been entirely stripped away by the recession, thereby exposing the water ice crust onto which they were emplaced, in which case the low albedo of the exhumed terrain represents the original blanket of haze particles that the crust accumulated prior to emplacement of the bladed terrain deposits. In either scenario, the haze blanket would prevent further large-scale exogenic modification of the maculae, leaving only endogenic processes to subsequently affect them, e.g., the extensive surface collapse seen at Krun Macula.

The intermediate-albedo terrain that prevails to the north of ~15°N on the farside is interpreted to be an extension of the plains that are variably mantled by predominantly methane ice-rich material in the same latitude range on the nearside, including Hayabusa, Vega, and Venera Terrae and Lowell Regio (*Stern et al.,* 2021) (sections 2.4 and 2.5). The mottled appearance of Hayabusa Terra where it borders the eastern edge of the nearside is seen to extend into the farside as far as 310°E to 320°E (label "3" in Fig. 23), east of which there is a transition to a more homogeneous texture characteristic of Venera Terra. Intermediate albedo terrain to the south of ~20°S was only ever viewed at very high emission angles (>80°), which along with the coarse pixel scale of the imaging, confounds determination of its nature. It probably represents a methane ice-rich mantle like that

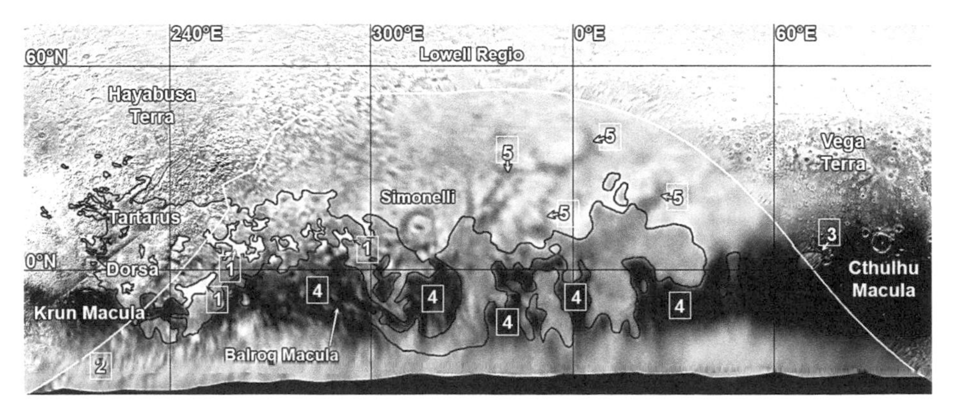

Fig. 23. Mosaic of Pluto's farside (2.2–32 km/pixel) with high-resolution nearside hemisphere imaging (850 m/pixel and better) north of the white line. The boundaries of mapped bladed terrain deposits are outlined in black (*Stern et al.,* 2020); these boundaries are approximate within the farside. 1: Bladed terrain deposits at the western margin of the nearside hemisphere. 2: Discontinuous chain of farside maculae. 3: Transition from western mottled terrain to eastern homogeneous terrain. 4: Intermediate-albedo terrain of the eastern Wright Group. 5: Bright regions interpreted as ponded nitrogen ice deposits. 6: Bright floor of Simonelli crater, interpreted to be ponded nitrogen ice deposits. 7: Dark lineations interpreted as valleys.

seen at similar latitudes in the northern hemisphere, although the possibly cryovolcanic, intermediate-albedo terrain of the Wright Group (label "4" in Fig. 23) likely extends into its far western region to some degree (section 2.2).

A succession of irregular, high-albedo regions that reach some tens of kilometers across is seen extending into the farside from the eastern edge of the nearside hemisphere ("5" labels in Fig. 23). These are interpreted to be ponded nitrogen-ice deposits like those seen occurring interstitially to Tartarus Dorsa (*Stern et al.,* 2021) (section 2.3.1). The most easterly deposit that can be confidently identified is at 294°E, although the bright annular floor (label "6" in Fig. 23) of the Simonelli central peak impact crater at 314°E may also indicate the presence of such deposits. Together, these deposits form a low-latitude band of uplands nitrogen-ice deposition, the areal coverage of which diminishes from the continuous blanket of nitrogen ice that covers East Tombaugh Regio to the sporadic, angular deposits in the western farside. The pixel scale of imaging to the east of Simonelli is too coarse (11–32 km/pixel) to confidently determine whether high-albedo areas here are ponded nitrogen-ice deposits or something else, such as seasonal methane-ice deposits. *Stern et al.* (2021) hypothesize that the mottled texture of Balrog Macula (imaged at ~7–10 km/pixel) may represent seasonal accumulations of methane ice on a scale of several kilometers to tens of kilometers across that have deposited onto the dark haze blanket (like that seen at Pigafetta Montes on the nearside, section 2.6.3), although the texture may alternatively be caused by outcrops of bladed terrain that are rising above the surrounding haze particle blanket.

Tectonism has not been conclusively identified on Pluto's farside, but a network of low-albedo lineations observed between 330°E and 30°E in the low- to mid-latitudes ("7" labels in Fig. 23), which are determined to be low elevation in limb topography, may represent fractures (*Schenk et al.,* 2018; *Stern et al.,* 2021). If they are, then some may represent an extension of the ridge-trough system seen on the nearside (section 3). Alternatively, the lineations may collectively form a >1000-km-diameter zone of crustal disruption that is located antipodal to the location of the initial impact that carved out the Sputnik basin (*Stern et al.,* 2021), similar to zones of chaotic terrain that occur antipodal to large impact basins elsewhere in the solar system, notably Mercury (e.g., *Schultz and Gault,* 1975).

5. GEOLOGICAL HISTORY OF PLUTO

Among the earliest geological episodes recorded on Pluto's surface is the formation of the ~1300-km-wide by ~1000-km-long depression containing Sputnik Planitia, interpreted to be an impact basin that likely dates to >4 G.y. (*Moore et al.,* 2016; *Schenk et al.,* 2018). The newly formed basin overwhelmingly dominated Pluto's nearside topography and represented a powerful cold trap for nitrogen ice (*Hamilton et al.,* 2016). Infilling of the basin with most of Pluto's nitrogen ice would be achieved by tens of millions of years after its appearance (*Bertrand et al.,* 2018). The period of infilling was accompanied by the recession of glacial nitrogen ice from uplands to the north and west of Sputnik Planitia. These areas were erosionally sculpted into a variety of dissected terrains and dendritic valley networks by the flowing nitrogen ice (*Howard et al.,* 2017a), which may also have been responsible for the emplacement of deposits forming the washboard and fluted terrain (*White et al.,* 2019). The timing of the dislodgement of blocks of water ice crust to form the mountain ranges that line Sputnik Planitia's western edge is not known, but may have been enabled by intrusion of the denser glacial nitrogen ice covering the uplands into fractured and brecciated crust at this time (*Howard et al.,* 2017a). The massive north-south ridge-trough system also appeared around this time as Pluto expanded due to partial freezing of its subsurface ocean (*Keane et al.,* 2016; *Nimmo et al.,* 2016). Initiation of the tectonism that formed it may even predate the Sputnik basin-forming impact (*Schenk et al.,* 2018).

This era (≳4 G.y.) was also characterized by a high impact rate (*Singer et al.,* 2019) and the formation of most of Pluto's craters. Due to blanketing by haze particles, the heavily cratered landscapes of Cthulhu Macula are those that appear to have been least modified since the end of this cratering episode. Haze particles within Pluto's atmosphere have settled out across its entire surface throughout its history (*Grundy et al.,* 2018), but only form continuous, dark blankets in the equatorial diurnal zone, where weak seasonal mobilization of volatiles means that there are no obstacles to the accumulation of haze particles (*Binzel et al.,* 2017; *Earle et al.,* 2017), thereby resulting in the discontinuous chain of maculae of the equatorial uplands. Pluto's north polar region is interpreted to be formed of ancient water ice crust, like Cthulhu Macula, but has instead experienced highly active seasonal, insolation-driven volatile mobilization throughout Pluto's history, which prevents thick deposits of volatile ices from ever accumulating there (*Bertrand et al.,* 2019).

The immediate post-Sputnik impact era also witnessed emplacement of the bright, methane-rich mantle of the Venera Group that covers Venera and Vega Terrae to the northwest of Sputnik Planitia. The northern portion of the mantle has remained relatively uneroded (north of ~43°N), but has experienced extensive sublimation erosion in its southern portion. The high count of relatively well-preserved craters that are superimposed upon the mantle indicates that both its emplacement and erosion took place while the impact rate was still high. An exception is the sparsely cratered Piri Planitia, where sublimation erosion may have taken place after the impact rate waned and/or the erosion was of sufficient magnitude to completely erase the topographic expressions of existing impact craters.

A thicker mantling deposit that postdates that in Venera and Vega Terrae has covered Pioneer and Hayabusa Terrae to the northeast of Sputnik Planitia. Fewer impact craters are seen here than to the northwest (*Robbins et al.,* 2017; *Singer et al.,* 2019), and the morphologies of many of them have been smoothened and rounded by the mantling deposit, as have landforms within this region in general. The paucity of well-preserved impact craters suggests that the mantling material of the Hayabusa Group was likely emplaced as the impact rate was waning (i.e., <4 G.y.). Pioneer Terra features large and deep pits that may have been the source of gaseous emissions, with a methane-rich component depositing onto the surrounding landscape (*Howard et al.,* 2017b). This mantle also shows evidence of sublimation erosion in its southern portion in Hayabusa Terra. The preservation of old (~4 G.y.), cratered terrains at high latitudes attest that relatively stable cyclical processes giving neither a substantial net volatile deposition nor net volatile erosion have likely operated throughout Pluto's history, consistent with the extreme seasonal insolation and temperature variation encountered here as a consequence of Pluto's high obliquity (*Binzel et al.,* 2017).

The methane ice-rich bladed terrain deposits of the Tartarus Group represent a massive, high-elevation counterpart to the low-elevation, predominantly nitrogen-ice deposit of Sputnik Planitia. Stretching around the equator from the uplands east of Sputnik Planitia to the cratered plains of far-western Cthulhu Macula, these deposits superpose all terrains that they encounter and, at least within Tartarus Dorsa (the best-imaged portion), are nearly free of impact craters. The bladed terrain deposits originated during an episode of massive precipitation of atmospheric methane at high elevation within Pluto's low-latitude diurnal zone, where averaged insolation is at a minimum (*Binzel et al.,* 2017; *Moore et al.,* 2018). Their high stratigraphic position and lack of craters suggests that these deposits were emplaced after the impact rate had waned. Subsequent excursions in Pluto's climate partially eroded the deposits via sublimation into their current bladed texture, and may also have erased impact craters. Tartarus Dorsa have a surface age no older than 200–300 m.y. (*Moore et al.,* 2018; see also the chapter by Singer et al. in this volume). In contrast to the mantled terrains of the Arctic, the bladed terrain deposits are considered to primarily record secular, long-term changes in Pluto's climate, with only modest resurfacing occurring in response to the weak volatile mobilization that takes place at low latitudes. Portions of the bladed terrain deposits seem to have experienced histories that diverge from that of the deposits at large: Some of the bright, pitted uplands of East Tombaugh Regio, possibly as old as a few billion years (see the chapter by Singer et al. in this volume), seem to be bladed terrain deposits that are currently experiencing mantling by a continuous veneer of volatile ices. The exact nature of Krun Macula and the western farside maculae is uncertain, but they may be relatively low-elevation bladed terrain deposits where mobilization of the surface methane ice was inactive for a sufficiently long period to allow accumulation of a continuous blanket of dark haze particles onto the surface, or water ice substrate with a haze particle covering that was exhumed after recession of the bladed terrain deposits (*Stern et al.,* 2021).

Tectonism continued to affect Pluto's surface following the decay of the impact rate. Djanggawul and Mwindo Fossae are oriented azimuthally to Sputnik Planitia, and perhaps formed due to reorientation of Pluto associated with the infilling of the Sputnik basin (*Keane et al.,* 2016). Mwindo Fossae crosscut Tartarus Dorsa, indicating that the tectonic episode that formed them postdated emplacement of the deposits. The youngest, most sharply-defined fractures are those belonging to the systems west of Sputnik Planitia that include Inanna, Dumuzi, Virgil, Hermod, and Beatrice Fossae, which crosscut all impact craters that they encounter. Fractures within East Tombaugh Regio and Hayabusa Terra to the east of Sputnik Planitia may have been formed by the same tectonic episode that produced those to the west, although they appear more eroded. These eastern fractures include pit chains, implying that they formed during or after emplacement of the mantles of the Hayabusa and Tartarus Groups, with the fracturing causing surface collapse of the mantling material.

The landforms of the Wright Group, on the southern boundary of Sputnik Planitia, represent the most recent, regional-scale emplacement of material onto Pluto's surface. The twin edifices of Wright and Piccard Montes, with their extremely large central depressions, have been tentatively interpreted to be cryovolcanic (*Moore et al.,* 2016; *Singer et al.,* 2016; *Schenk et al.,* 2018). Both these and the surrounding landscape show very few, small impact craters, and the Wright Group as a whole probably dates from Pluto's middle age (~1–3 G.y.). Rimless depressions that may represent vents are seen across this region, and if the whole Group is a cryovolcanic province, then it may have formed due to massive eruption of cryomagmas that ascended from Pluto's interior due to ongoing freezing and overpressurization of Pluto's subsurface ocean (*Nimmo et al.,* 2016; *Hammond et al.,* 2016; *Martin and Binzel,* 2021). More recent (~1 G.y.) cryovolcanic activity may also be recorded in accumulations of dark material that fill Virgil and Uncama Fossae and mantle terrain surrounding them. These may represent ammoniated water ice deposits erupted as flows and fountains of cryoclastic materials from fissures along these troughs (*Cruikshank et al.,* 2019a,b, 2021; *Dalle Ore et al.,* 2019).

The most obvious manifestation of present-day geological activity on Pluto is the ongoing resurfacing of Sputnik Planitia and East Tombaugh Regio via convection, glacial flow, and sublimation/redeposition of nitrogen ice, which has likely continued unabated since Sputnik Planitia's formation ~4 G.y. ago (*Moore et al.,* 2016, 2017; *McKinnon et al.,* 2016; *Howard et al.,* 2017a; *White et al.,* 2017). No impact craters have been identified in Sputnik Planitia, where solid-state convection and viscous relaxation of the nitrogen ice causes surface renewal on a timescale of less than a million years (*McKinnon et al.,* 2016; *Trowbridge et al.,* 2016; *Buhler and Ingersoll,* 2018; *Wei et al.,* 2018). East Tombaugh Regio is presently experiencing deposition of primarily nitrogen ice fed by sublimation from Sputnik Planitia (*Howard et al.,* 2017a; *Moore et al.,* 2017; *White et al.,* 2017). The nitrogen ice subsequently reenters Sputnik Planitia via glacial flow. Mobilization of nitrogen and methane ice is also ongoing across Pluto's uplands (most significantly in the mid-latitudes and north polar region), where these species form thin, transient layers that sublimate and redeposit in response to insolation and temperature changes tied to seasonal and obliquity cycles (*Forget et al.,* 2017; *Bertrand et al.,* 2018, 2019).

6. SUMMARY

The New Horizons flyby of Pluto clearly revealed it to have a richly diverse geology, the complex nature of which is the result of multiple environmental factors (both exogenic and endogenic) combining to govern the distribution and behavior of different surface compositional suites to strongly varying degrees across even small lateral distances. Consideration of Pluto's location within the solar system and its physical attributes are key to understanding its active geological history: It orbits at a distance where volatile ices like nitrogen, methane, and carbon monoxide are stable across its surface, but which are easily mobilized by climatic cycles associated with Pluto's elliptical orbit, the seasons across its 248-Earth-year orbital period, and its high obliquity (103°–127°), which varies on a 2.8-m.y. cycle. Pluto's highly disparate climate zones and considerable overall topographic relief (~11 km) are instrumental in determining the boundaries of where certain volatile species are stable on the surface across Pluto's history. In addition, Pluto is large enough to have retained a subsurface ocean, the ongoing freezing of which has likely powered extensive tectonism and possibly also cryovolcanism. The heat flow emanating from Pluto's interior, while totaling only a few milliwatts per square meter, is still sufficient to cause convection within the giant nitrogen-carbon monoxide ice deposit of Sputnik Planitia.

Data returned by New Horizons has inspired a new era of Pluto research, but a number of scientific questions remain unresolved that may only be thoroughly addressed by returning to Pluto (see the chapter by Buie et al. in this volume). These include identifying what environmental factors unrelated to long-term climate, such as topography, elevation, thermal inertia, and prevailing winds, are most influential in governing volatile mobilization across Pluto. An important aspect of this is recognizing which compositional and morphological signatures of a landscape are due to the existence of a distinct, exposed geological stratum, and which that are due to "overprinting" by a seasonal veneer of deposited volatile ices.

Also important is distinguishing the relative effects of secular vs. cyclical climatic variations on Pluto's geology (see the chapter by Moore and Howard in this volume). Features of particular interest in this respect are Tombaugh Regio and the bladed terrain deposits, which respectively form massive concentrations of nitrogen and methane ice. These volatile repositories were established early on in Pluto's history, and the extent to which they have responded to Pluto's dramatic, obliquity-governed climatic cycles is uncertain. Fluctuations in surface temperature-pressure conditions stemming from such cycles may have caused the nitrogen-ice coverage of Tombaugh Regio to expand and contract, and volatile mobilization has possibly ceased for some parts of the bladed terrain deposits and remained active for others.

The effect of endogenic processes on Pluto's geology has not yet been well quantified. For a number of terrains, there is uncertainty as to the extent to which endogenic vs. exogenic processes have contributed to their appearance, such as the mantled terrains of the Hayabusa Group. Wright and Piccard Montes and their surrounding terrains seem to be more explicitly endogenic in origin, and may represent a sprawling cryovolcanic province. But cryovolcanism is a nascent field of research within planetary science, and further investigation of how cryomagmas may be generated, ascend, and erupt at plutonian conditions is warranted.

REFERENCES

Ahrens C. J. and Chevrier V. F. (2019) Compressional ridges on Baret Montes, Pluto as observed by New Horizons. *Geophys. Res. Lett., 46(24),* 14328–14335.

Ahrens C. J. and Chevrier V. F. (2021) Investigation of the morphology and interpretation of Hekla Cavus, Pluto. *Icarus, 356,* 114108.

Barr A. C. and Collins G. C. (2015) Tectonic activity on Pluto after the Charon-forming impact. *Icarus, 246,* 146-155.

Bertrand T. and Forget F. (2016) Observed glacier and volatile distribution on Pluto from atmosphere-topography processes. *Nature, 540,* 86–89.

Bertrand T., Forget F., Umurhan O. M., et al. (2018) The nitrogen cycles on Pluto over seasonal and astronomical timescales. *Icarus, 309,* 277–296.

Bertrand T., Forget F., Umurhan O. M., et al. (2019) The CH_4 cycles on Pluto over seasonal and astronomical timescales. *Icarus, 329,* 148–165.

Bertrand T., Forget F., White O. L., and Schmitt B. (2020a) Pluto's beating heart regulates the atmospheric circulation: Results from high-resolution and multiyear numerical climate simulations. *J. Geophys. Res.–Planets, 125,* e2019JE006120.

Bertrand T., Forget F., Schmitt B., White O. L., and Grundy W. M. (2020b) Equatorial mountain chains on Pluto are covered by methane frosts resulting from a unique atmospheric process. *Nature Commun., 11,* 5056.

Beyer R. A., Schenk P. M., Moore J. M., et al. (2019) High-resolution pixel-scale topography of Pluto and Charon. In *Pluto System After New Horizons,* Abstract #7042. LPI Contribution No. 2133, Lunar and Planetary Institute, Houston.

Bierhaus E. B. and Dones L. (2015) Craters and ejecta on Pluto and Charon: Anticipated results from the New Horizons flyby. *Icarus, 246,* 165–182.

Binzel R. P., Earle A. M., Buie M. W., et al. (2017) Climate zones on Pluto and Charon. *Icarus, 287,* 30–36.

Bray V. J. and Schenk P. M. (2015) Pristine impact crater morphology on Pluto — Expectations for New Horizons. *Icarus, 246,* 156–164.

Buhler P. B. and Ingersoll A. P. (2018) Sublimation pit distribution indicates convection cell surface velocities of ~10 cm per year in Sputnik Planitia, Pluto. *Icarus, 300,* 327–340.

Buratti B. J., Hofgartner J. D., Hicks M. D., et al. (2017) Global albedos of Pluto and Charon from LORRI New Horizons observations. *Icarus, 287,* 207–217.

Conrad J. W., Nimmo F., Schenk P. M., et al. (2019) An upper bound on Pluto's heat flux from a lack of flexural response of its normal faults. *Icarus, 328,* 210–217.

Cook J. C., Dalle Ore C. M., Protopapa S., et al. (2019) The distribution of H_2O, CH_3OH, and hydrocarbon-ices on Pluto: Analysis of New Horizons spectral images. *Icarus, 331,* 148–169.

Cruikshank D. P., Materese C. K., Pendleton Y. J., et al. (2019a) Prebiotic chemistry of Pluto. *Astrobiology, 19(7),* 831–848.

Cruikshank D. P., Umurhan O. M., Beyer R. A., et al. (2019b) Recent cryovolcanism in Virgil Fossae on Pluto. *Icarus, 330,* 155–168.

Cruikshank D. P., Dalle Ore C. M., Scipioni F., et al. (2021) Cryovolcanic flooding in Viking Terra on Pluto. *Icarus, 356,* 113786.

Dalle Ore C. M., Cruikshank D. P., Protopapa S., et al. (2019) Detection of ammonia on Pluto's surface in a region of geologically recent tectonism. *Sci. Adv., 5,* eaav5731.

Dobrovolskis A. R. and Harris A. W. (1983) The obliquity of Pluto. *Icarus, 55,* 231–235.

Dobrovolskis A. R., Peale S. J., and Harris A. W. (1997) Dynamics of the Pluto-Charon binary. In *Pluto and Charon* (S. A. Stern and D. J. Tholen, eds.), pp. 159–167. Univ. of Arizona, Tucson.

Earle A. M., Binzel R. P., Young L. A., Stern S. A., Ennico K., Grundy W., Olkin C. B., Weaver H. A., and the New Horizons Geology and Geophysics Imaging Team (2017) Long-term surface temperature modeling of Pluto. *Icarus, 287,* 37–46.

Earle A. M., Binzel R. P., Young L. A., Stern S. A., Ennico K., Grundy W., Olkin C. B., Weaver H. A., and the New Horizons Surface Composition Theme Team (2018a) Albedo matters: Understanding runaway albedo variations on Pluto. *Icarus, 303,* 1–9.

Earle A. M., Grundy W., Howett C. J. A., et al. (2018b) Methane distribution on Pluto as mapped by the New Horizons Ralph/MVIC instrument. *Icarus, 314,* 195–209.

Elbeshausen D., Wünnemann K., and Collins G. S. (2013). The transition from circular to elliptical impact craters. *J. Geophys. Res., 118,* 2295–2309.

Eluszkiwicz J. and Stevenson D. J. (1990) Rheology of solid methane and nitrogen: Applications to Triton. *Geophys. Res. Lett., 17,* 1753–1756.

Fagents S. A. (2003) Considerations for effusive cryovolcanism on Europa: The post-Galileo perspective. *J. Geophys. Res., 108,* E12.

Forget F., Bertrand T., Vangvichith M., Leconte J., Millour E., and Lellouch E. (2017) A post-New Horizons global climate model of Pluto including the N_2, CH_4, and CO cycles. *Icarus, 287,* 54–71.

Gault D. E. and Wedekind J. A. (1978) Experimental studies of oblique impact. *Proc. Lunar Planet. Sci. Conf. 9th,* pp. 3843–3875.

Gjermundsen E. F., Briner J. P., Akçar N., et al. (2015) Minimal erosion of Arctic alpine topography during late Quaternary glaciation. *Nature Geosci., 8,* 789–793.

Greenstreet S., Gladman B., and McKinnon W. B. (2015) Impact and cratering rates onto Pluto. *Icarus, 258,* 267–288.

Grundy W. M., Binzel R. P., Buratti B. J., et al. (2016) Surface compositions across Pluto and Charon. *Science, 351(6279),* aad9189.

Grundy W. M., Bertrand T., Binzel R. P., et al. (2018) Pluto's haze as a surface material. *Icarus, 314,* 232–245.

Hamilton D. P., Stern S. A., Moore J. M., Young L. A., and the New Horizons Geology, Geophysics and Imaging Theme Team (2016) The rapid formation of Sputnik Planitia early in Pluto's history. *Nature, 540,* 97–99.

Hammond N. P., Barr A. C., and Parmentier E. M. (2016) Recent tectonic activity on Pluto driven by phase changes in the ice shell. *Geophys. Res. Lett., 43,* 6775–6782.

Head J. W. and Marchant D. R. (2003) Cold-based mountain glaciers on Mars: Western Arsia Mons. *Geology, 31,* 641–644.

Herrick R. R. and Hessen K. K. (2006) The planforms of low-angle impact craters in the northern hemisphere of Mars. *Meteoritics & Planet. Sci., 41,* 1483–1495.

Hofgartner J. D., Buratti B. J., Devins S. L., et al. (2018) A search for temporal changes on Pluto and Charon. *Icarus, 302,* 273–284.

Howard A. D., Moore J. M., Umurhan O. M., et al. (2017a) Present and past glaciation on Pluto. *Icarus, 287,* 287–300.

Howard A. D., Moore J. M., White O. L., et al. (2017b) Pluto: Pits and mantles on uplands north and east of Sputnik Planitia. *Icarus, 293,* 218–230.

Hussmann H., Sohl F., and Spohn T. (2006) Subsurface oceans and deep interiors of medium-sized outer planet satellites and large trans-neptunian objects. *Icarus, 185,* 258–273.

Johnson B. C., Bowling T. J., Trowbridge A. J., and Freed A. M. (2016) Formation of the Sputnik Planum basin and the thickness of Pluto's subsurface ocean. *Geophys. Res. Lett., 43,* 10068–10077.

Johnson P. E., Young L. A., Protopapa S., et al. (2021) Modeling Pluto's minimum pressure: Implications for haze production. *Icarus, 356,* 114070.

Kamata S. and Nimmo F. (2014) Impact basin relaxation as a probe for the thermal history of Pluto. *J. Geophys. Res.–Planets, 119,* 2272–2289.

Kamata S., Nimmo F., Sekine Y., Kuramoto K., Noguchi N., Kimura J., and Tani A. (2019) Pluto's ocean is capped and insulated by gas hydrates. *Nature Geosci., 12,* 407–410.

Keane J. T., Matsuyama I., Kamata S., and Steckloff J. K. (2016) Reorientation and faulting of Pluto due to volatile loading within Sputnik Planitia. *Nature, 540,* 90–93.

Kimura J. and Kamata S. (2020) Stability of the subsurface ocean of Pluto. *Planet. Space Sci., 181,* 104828.

Lesage E., Massol H., and Schmidt F. (2020) Cryomagma ascent on Europa. *Icarus, 335,* 113369.

Lewis B. L., Stansberry J. A., Holler B. J., et al. (2021) Distribution and energy balance of Pluto's nitrogen ice, as seen by New Horizons in 2015. *Icarus, 356,* 113633.

Manga M. and Wang C.-Y. (2007) Pressurized oceans and the eruption of liquid water on Europa and Enceladus. *Geophys. Res. Lett., 34,* L07202.

Martin C. R., and Binzel R. P. (2021) Ammonia-water freezing as a mechanism for recent cryovolcanism on Pluto. *Icarus, 356,* 113763.

McGovern P. J. and White O. L. (2019) Stress-enhanced ascent of cryomagmas through Pluto's ice shell from nitrogen ice loading of a Sputnik Planitia basin. In *Lunar and Planetary Science L,* Abstract #2994. Lunar and Planetary Institute, Houston.

McGovern P. J., White O. L., and Schenk P. M. (2019) Tectonism across Pluto: Mapping and interpretations. In *Pluto System After New Horizons,* Abstract #7063. LPI Contribution No. 2133, Lunar and Planetary Institute, Houston.

McKinnon W. B., Simonelli D. P., and Schubert G. (1997) Composition, internal structure, and thermal evolution of Pluto and Charon. In *Pluto and Charon* (S. A. Stern and D. J. Tholen, eds.), pp. 295–343. Univ. of Arizona, Tucson.

McKinnon W. B., Nimmo F., Wong T., et al. (2016) Convection in a volatile nitrogen-ice-rich layer drives Pluto's geological vigour. *Nature, 534,* 82–85.

Moore J. M., Mellon M. T., and Zent A. P. (1996) Mass wasting and ground collapse in terrains of volatile-rich deposits as a solar system-wide geological process: The pre-Galileo view. *Icarus, 122,* 63–78.

Moore J. M., Asphaug E., Morrison D., et al. (1999) Mass movement and landform degradation on the icy Galilean satellites: Results of the Galileo nominal mission. *Icarus, 140,* 294–312.

Moore J. M., Howard A. D., Schenk P. M., et al. (2015) Geology before Pluto: Pre-encounter considerations. *Icarus, 246,* 65–81.

Moore J. M., McKinnon W. B., Spencer J. R., et al. (2016) The geology of Pluto and Charon through the eyes of New Horizons. *Science, 351,* 1284–1293.

Moore J. M., Howard A. D., Umurhan O. M., et al. (2017) Sublimation as a landform-shaping process on Pluto. *Icarus, 287,* 320–333.

Moore J. M., Howard A. D., Umurhan O. M., et al. (2018) Bladed terrain on Pluto: Possible origins and evolution. *Icarus, 300,* 129–144.

Moores J. E., Smith C. L., Toigo A. D., and Guzewich S. D. (2017) Penitentes as the origin of the bladed terrain of Tartarus Dorsa on Pluto. *Nature, 541,* 188–190.

Neveu M., Desch S. J., Shock E. L., and Glein C. R. (2015) Prerequisites for explosive cryovolcanism on dwarf planet-class Kuiper belt objects. *Icarus, 246,* 48–64.

Nimmo F., Hamilton D. P., McKinnon W. B., et al. (2016) Reorientation of Sputnik Planitia implies a subsurface ocean on Pluto. *Nature, 540,* 94–96.

O'Hara S. and Dombard A. J. (2021) Downhill sledding at 40 AU: Mobilizing Pluto's chaotic mountain blocks. *Icarus, 356,* 113829.

Olkin C. B., Spencer J. R., Grundy W. M., et al. (2017) The global color of Pluto from New Horizons. *Astron. J., 154,* 258.

Protopapa S., Grundy W. M., Reuter D. C., et al. (2017) Pluto's global surface composition through pixel-by-pixel Hapke modeling of New Horizons Ralph/LEISA data. *Icarus, 287,* 218–228.

Robbins S. J., Singer K. N., Bray V. J., et al. (2017) Craters of the Pluto-Charon system. *Icarus, 287,* 187–206.

Robuchon G. and Nimmo F. (2011) Thermal evolution of Pluto and implications for surface tectonics and a subsurface ocean. *Icarus, 216,* 426–439.

Schenk P. M., Beyer R. A., McKinnon W. B., et al. (2018) Basins, fractures, and volcanoes: Global cartography and topography of Pluto from New Horizons. *Icarus, 314,* 400–433.

Schmitt B., Philippe S., Grundy W. M., et al. (2017) Physical state and distribution of materials at the surface of Pluto from New Horizons LEISA imaging spectrometer. *Icarus, 287,* 229–260.

Schultz P. H. and Gault D. E. (1975) Seismic effects from major basin formations on the Moon and Mercury. *Moon, 12,* 159–177.

Shuvalov V. (2011) Ejecta deposition after oblique impacts: An influence of impact scale. *Meteoritics & Planet. Sci., 46,* 1713–1718.

Singer K. N., White O. L., Schenk P. M., et al. (2016) Pluto's putative cryovolcanic constructs. In *Lunar and Planetary Science XLVII,* Abstract #2276. Lunar and Planetary Institute, Houston.
Singer K. N., McKinnon W. B., Gladman B., et al. (2019) Impact craters on Pluto and Charon indicate a deficit of small Kuiper belt objects. *Science, 363,* 955–959.
Skjetne H. L., Singer K. N., Hynek B. M., et al. (2021) Morphological comparison of blocks in chaos terrains on Pluto, Europa, and Mars. *Icarus, 356,* 113866.
Stern S. A. (2008) The New Horizons Pluto Kuiper Belt mission: Overview with historical context. *Space Sci. Rev., 140,* 3–21.
Stern S. A., Porter S., and Zangari A. (2015a) On the roles of escape erosion and the viscous relaxation of craters on Pluto. *Icarus, 250,* 287–293.
Stern S. A., Bagenal F., Ennico K., et al. (2015b) The Pluto system: Initial results from its exploration by New Horizons. *Science, 350,* aad1815.
Stern S. A., Binzel R. P., Earle A. M., et al. (2017) Past epochs of significantly higher pressure atmospheres on Pluto. *Icarus, 287,* 47–53.
Stern S. A., White O. L., McGovern P. J., et al. (2021) Pluto's far side. *Icarus, 356,* 113805.
Telfer M. W., Parteli E. J. R., Radebaugh J., et al. (2018) Dunes on Pluto. *Science, 360,* 992–997.
Trowbridge A. J., Melosh H. J., Steckloff J. K., and Freed A. M. (2016) Vigorous convection as the explanation for Pluto's polygonal terrain. *Nature, 534,* 79–81.
Umurhan O. M., Howard A. D., Moore J. M., et al. (2017) Modeling glacial flow on and onto Pluto's Sputnik Planitia. *Icarus, 287,* 301–319.
Vilella K. and Deschamps F. (2017) Thermal convection as a possible mechanism for the origin of polygonal structures on Pluto's surface. *J. Geophys. Res.–Planets, 122,* 1056–1076.
Wei Q., Hu Y., Liu Y., Lin D. N. C., Yang J., Showman A. P. (2018) Young surface of Pluto's Sputnik Planitia caused by viscous relaxation. *Astrophys. J. Lett., 856,* L14.
White O. L., Umurhan O. M., Moore J. M., et al. (2016) Modeling of ice pinnacle formation on Callisto. *J. Geophys. Res.–Planets, 121,* 21–45.
White O. L., Moore J. M., McKinnon W. B., et al. (2017) Geological mapping of Sputnik Planitia on Pluto. *Icarus, 287,* 261–286.
White O. L., Moore J. M., Howard A. D., et al. (2019) Washboard and fluted terrains on Pluto as evidence for ancient glaciation. *Nature Astron., 3,* 62–68.
Wilhelms D. E. (1990) Geologic mapping. In *Planetary Mapping* (R. Greeley and R. M. Batson, eds.), pp. 208–260. Cambridge Univ., Cambridge.
Yamashita Y., Kato M., and Arakawa M. (2010) Experimental study on the rheological properties of polycrystalline solid nitrogen and methane: Implications for tectonic processes on Triton. *Icarus, 207,* 972–977.

Nimmo F. and McKinnon W. B. (2021) Geodynamics of Pluto. In *The Pluto System After New Horizons* (S. A. Stern, J. M. Moore, W. M. Grundy, L. A. Young, and R. P. Binzel, eds.), pp. 89–103. Univ. of Arizona, Tucson, DOI: 10.2458/azu_uapress_9780816540945-ch005.

Geodynamics of Pluto

Francis Nimmo
University of California Santa Cruz

William B. McKinnon
Washington University in St Louis

In this article we summarize our understanding of Pluto's internal structure and evolution following the New Horizons mission. Pluto's density implies it is roughly 70% rock and 30% ice by mass, although it may plausibly also contain a sizeable fraction of carbon compounds. The lack of compressional structures indicates that it is fully differentiated, although this differentiation most likely cannot have been complete prior to the Charon-forming impact. Heat flux estimates are conflicting: One relaxed crater suggests high heat fluxes, while the absence of observed flexure suggests low heat fluxes. Pluto's energy budget is dominated by slow radioactive decay, and is sufficient to form and maintain a present-day subsurface ocean roughly 100 km thick beneath an ice shell. Four lines of circumstantial evidence point toward such an ocean; although none is definitive, taken together we think it probable that an ocean existed, and may still be present. These are (1) surface extension and (2) possible cryovolcanism (both likely due to a thickening ice shell), (3) an absence of a detectable fossil bulge, and (4) reorientation driven by the nitrogen-filled Sputnik Planitia basin. If an ocean exists, the ice shell above must be cold and rigid. These conditions could be maintained if the ocean is ammonia-rich, or if there is a layer of clathrates at the base of the ice shell. At least the second half of Pluto's history consisted of slow cooling and thickening of an ice shell and continued reactions between the warm silicate core and the overlying ocean.

1. INTRODUCTION

The New Horizons mission promoted Pluto from a pixelated blob — the province of astronomers — to a complex and active world. The aim of this chapter is to review what we now think about Pluto's geodynamics: its structure, its history, and what drives its activity. Many of the conclusions we draw will be uncertain or provisional, and in some cases we have scarcely improved on our pre-New Horizons knowledge. Nonetheless, a significant amount has been learned, in particular the strong (although circumstantial) evidence for a subsurface ocean. Such an argument could certainly not have been made before 2015.

The rest of this chapter is organized as follows. We will begin with the observational constraints (section 2) and then discuss inferences arising from these observations. The topics covered, in decreasing order of certainty, are Pluto's bulk structure (section 3.1), whether it possesses a subsurface ocean (section 3.2), the properties of the ice shell (section 3.3), and Pluto's long-term evolution (section 3.4). With these results in hand, we will then discuss Pluto and Charon in the context of the menagerie of icy worlds elsewhere in the solar system (section 4). A brief discussion of possible future studies (and future spacecraft exploration) follows (section 5). Finally, we will summarize our findings.

Several other chapters in this book are relevant to the material presented here. In particular, Pluto's geology (see the chapter by White et al. in this volume), impact history (see the chapter by Singer et al.), and surface composition (see the chapter by Cruikshank et al.) are all covered more extensively than in section 2 below, while section 3.4 has overlaps with the chapter by McKinnon et al.

2. OBSERVATIONAL CONSTRAINTS

New Horizons provided surface images, spectroscopic mapping data, and topography, all of which provide constraints on Pluto's geophysical behavior. Although separating observations from interpretation is never entirely possible, this section attempts to summarize the former without delving too much into the latter.

2.1. Shape and Density

The density and shape of Pluto and Charon give zeroth-order constraints on their interior structures (section 3.1). Prior to New Horizons, Pluto's relatively thick atmosphere and hazes inhibited confident determination of Pluto's radius in stellar occultation and mutual event campaigns (e.g., *Lellouch et al.,* 2009; *Tholen,* 2014). Pluto and Charon's masses were already well-known thanks to the motions of the small satellites, which allowed accurate determination of the barycenter (*Brozović et al.,* 2015).

Approach images taken by the New Horizons Long Range Reconnaissance Imager (LORRI) combined with occultation measurements from several instruments yielded a mean radius of 1188.3 ± 1.6 km, and revealed no sign of oblateness (<0.6%; section 3.1.4) (*Nimmo et al.,* 2017). Combined with its known mass, Pluto's inferred density was 1854 ± 11 kg m^{-3}; for Charon the results were 606.0 ± 1.0 km and 1701 ± 33 kg m^{-3} respectively (these are actual, as opposed to "uncompressed," densities). The masses (and thus the densities) may change slightly following incorporation of the encounter observations (*Jacobson et al.,* 2015), but it is unlikely that they will change enough to alter any of the discussion presented below.

2.2. Surface Geology

Descriptions of the surface geology and composition of Pluto may be found in *Moore et al.* (2016); more recent summaries are in the chapters in this volume by White et al. and Cruikshank et al. From the point of view of Pluto's bulk structure and evolution, the following observations are the most pertinent. Pluto's rugged surface is judged to be composed primarily of water ice (*Moore et al.,* 2016). More volatile nitrogen and methane deposits are found in some areas, notably the Sputnik Planitia basin, which is dominated by N_2 ice (*Grundy et al.,* 2016; *Protopapa et al.,* 2017; *Schmitt et al.,* 2017). Ammoniated species are rare on Pluto's surface, although some traces are found around Virgil Fossae (*Cruikshank et al.,* 2019); ammonia signatures are much more common on Charon, especially associated with certain impact craters (see the chapter by Protopapa et al. in this volume). Much of Pluto is heavily cratered; although the impact flux is imperfectly known, the bulk of the surface has inferred ages of several gigayears (see the chapter by Singer et al. in this volume). The principal exception is the nitrogen plains of Sputnik Planitia, which show no craters at all and on this basis are likely <100 m.y. old (*Moore et al.,* 2016; see also the chapter by Singer et al.). Sputnik Planitia itself probably represents an impact basin (*Johnson et al.,* 2016), which, given its size (D ≈ 1000 km), is likely a very ancient feature.

Pluto shows a striking array of extensional tectonic features, some of which are stratigraphically young (crosscutting impact craters) and have sharp edges suggesting only limited degradation. A degraded north-south trending "ridge-trough system" might be evidence of more ancient extension (see the chapter by White et al.). There is very little evidence for compression, although ~100-km-scale Tartarus Dorsae ridges east of Sputnik Planitia and the north-south methane-capped mountains of eastern Cthulhu could conceivably have a compressional origin (*McGovern and White,* 2019).

Two enigmatic features south of Sputnik Planitia, Wright and Piccard Montes, have elevated topography of several kilometers, central depressions, and a lumpy surface texture (see the chapter by White et al.). These features do not show any unusual spectroscopic characteristics, but based on height and observed slopes (up to 30°), their bulk composition must be dominated by water ice. Their origin is uncertain, but it has been argued that they may be "cryovolcanic" constructs (*Stern et al.,* 2015; *Moore et al.,* 2016). Similarly, the stratigraphically young fractures at Virgil Fossae are at the center of an ammonia signature, and it has been argued that this signature is the result of a recent cryovolcanic eruption (*Cruikshank et al.,* 2019), although discrete flow boundaries are not evident in panchromatic images. While present-day cryovolcanism at Enceladus evidently produces diffuse surface deposits (*Scipioni et al.,* 2017), we remind the reader that initial identification of volcanic or cryovolcanic features using images alone has frequently been overturned when different or better data became available (*Moore and Pappalardo,* 2011). As a result, these interpretations should be treated with caution.

2.3. Heat Flux

Geological activity is ultimately driven by heat from the interior. Accordingly, obtaining estimates of present-day (or ancient) heat flux is very important for understanding the geodynamics of a planet. On Pluto there are several ways of doing so.

2.3.1. Convecting nitrogen. One of the most unexpected discoveries of New Horizons was the existence of ~30-km-wide cellular patterns in the nitrogen plains of Sputnik Planitia. These were immediately identified as evidence for thermally-driven convection (*Stern et al.,* 2015; *McKinnon et al.,* 2016; *Trowbridge et al.,* 2016) and thus provided one potential avenue to determine the local heat flux.

Experimental measurements of solid nitrogen viscosity (*Yamashita et al.,* 2010) show that it is mildly temperature- and stress-dependent; at stresses of 0.1 MPa the viscosities are 2.5×10^9 and 0.6×10^9 Pa s, respectively, at temperatures of 45 K and 56 K. Such viscosities are very low compared to, for example, water ice, which has a viscosity of 10^{13-14} Pa s near its melting temperature (*Goldsby and Kohlstedt,* 2001). Although there are many uncertainties in its rheology (see the chapter by Umurhan et al. in this volume), and new lab work is necessary, it is clear that nitrogen is very mobile at Pluto's surface conditions.

Convection involving either small viscosity variations, or extreme variations in which the near-surface lid is immobile, results in cell widths roughly double the convecting layer thickness. However, if the viscosity contrasts are intermediate, such that the motion of the lid is non-negligible compared to the interior velocities, the cell aspect ratio can be much larger. This arises because larger aspect ratios minimize viscous dissipation in the system (*Lenardic et al.,* 2006). Nitrogen convection on Pluto very likely falls in this intermediate regime and so the layer depth is probably significantly less than 30 km (*McKinnon et al.,* 2016).

We can gain insight about the dynamics of convection in nitrogen ice by considering scaling arguments. The surface heat flux, velocity, and dynamic topography can be written (*Solomatov,* 1995)

$$F \sim \frac{k\Delta T}{\delta_0} \quad , \quad u \sim \frac{\kappa d}{\delta_0^2} \quad , \quad h \sim \alpha \Delta T \delta_0$$

where d is the layer thickness, k and κ are the thermal conductivity and diffusivity, respectively, ΔT is the temperature contrast across the layer, α is the thermal expansivity, and δ_0 is the thickness of the sluggish lid. The quantity δ_0 in turn depends on the vigor of convection as measured by the so-called Rayleigh number. The thickness of the lid depends on the heat flux; this factor also controls the rise speed of the convective flow, and the stress and topography it imposes on the lid.

Assuming that the surface temperature is 38 K (e.g., *Gladstone et al.,* 2016), and that the nitrogen is entirely solid (with a melting point of 63 K; see the chapter by Umurhan et al.), $\Delta T < 25$ K, but neither F nor d are known a priori. Guessing that $d \approx 5$ km (based on *McKinnon et al.,* 2016) and taking the heat flux $F \approx 3$ mW m^{-2} (see section 3.3.2), we obtain $\delta_0 \approx 1.7$ km, $u \approx 0.7$ cm yr^{-1}, and $h \approx 80$ m. Numerical simulations produce very similar results (*McKinnon et al.,* 2016), with the exception of somewhat higher velocities (a few centimeters per year).

High-resolution images provide additional, albeit incomplete, constraints on the character of convection in Sputnik Planitia. Photoclinometry-derived topography suggests that the cell centers are elevated by roughly 100–150 m (*Schenk et al.,* 2018a), consistent with the numerical simulations. Small, kilometer-scale pits are observed on many of the convection cells in Sputnik Planitia (see the chapter by White et al.). These pits increase in size with distance from the cell center. Assuming that these are sublimation pits, this trend has been used to infer surface velocities of 10 cm yr^{-1}, or cell ages of about 0.5 m.y. (*Buhler and Ingersoll,* 2018). This age is consistent with the total absence of impact craters on Sputnik Planitia at New Horizons resolution. The inferred surface velocity hints at either a larger layer depth or a higher heat flux than assumed above, but uncertainties in nitrogen rheology preclude a definitive statement.

2.3.2. Elastic thickness. Because the heat flux determines the temperature gradient, it also controls the mechanical properties of outer layers of planetary bodies. Geological materials typically lose their elastic strength at roughly 50% of the melting temperature, depending on the strain rate. Hence, a higher heat flux implies a thinner elastic layer (*Watts,* 2001). If the thickness of this elastic layer can be estimated, the heat flux can therefore be constrained (e.g., *McNutt,* 1984).

The large extensional graben west of Sputnik Planitia impose upward stresses on the surrounding lithosphere. In principle, the lithosphere should bend upward in response, generating rift-flank uplift as observed at some rifts on Earth (*Brown and Phillips,* 1999). Since no such uplift is seen, the lithosphere must be thick enough to avoid detectable deformation. In practice, this places a lower bound on the lithospheric elastic thickness of about 10 km (*Conrad et al.,* 2019).

Loading of an elastic plate (e.g., by deposition of nitrogen ice) results in extensional fractures that can be either radial or concentric depending on the elastic thickness and body radius compared with the load radius (*Janes and Melosh,* 1990; *Freed et al.,* 2001). Sputnik Planitia may be one area where this kind of analysis could be used to place further constraints on Pluto's elastic thickness (*Keane et al.,* 2016; *Mills and Montesi,* 2018), although as with many large basins it may prove difficult to distinguish any flexural topography from topography due to the ejecta blanket.

2.3.3. Viscous relaxation. The viscosity of water-rich ice crusts is strongly temperature-dependent, so that high heat fluxes result in relatively low viscosities at shallow depths. Topographic features like craters produce stress gradients in the subsurface; if the viscosity is low enough, lateral flow will occur, resulting in viscous relaxation (shallowing) of these craters (e.g., *Kamata and Nimmo,* 2014). Viscously-relaxed craters are observed on bodies like Enceladus and Ganymede and have been used to estimate local heat fluxes (e.g., *Dombard and McKinnon,* 2006; *Bland et al.,* 2012). Areas of thinned crust are likewise expected to thicken due to lateral flow at a rate depending on the heat flux (*Stevenson,* 2000).

To date, only one anomalously shallow crater (Edgeworth, D = 145 km) has been identified on Pluto (*McKinnon et al.,* 2018). If it is shallow because of viscous relaxation, numerical models for relaxation of similar-sized craters (e.g., Alcander on Dione) (*White et al.,* 2017) suggest that the heat flux must have been in excess of roughly 50 mW m^{-2}. Although Edgeworth appears to possess the bowed-upward shape typical of relaxed craters, it is puzzling that nearby Oort (D = 120 km) shows no such signs of relaxation. It may simply be that Oort formed later, when heat fluxes had declined.

3. INFERENCES

3.1. Bulk Structure

A pre-New Horizons view of possible Pluto structures is given in *McKinnon et al.* (1997). This work presented a baseline Pluto model rather similar to the structure we infer below, but also considered possible alternatives, including carbon-rich, low-density, and undifferentiated versions. The latter two can now be ruled out.

3.1.1. Rock mass fraction. If Pluto is assumed to consist mostly of silicates and water/ice, then its bulk density of 1854 ± 11 kg m^{-3} indicates that it is roughly two-thirds rock and one-third water/ice by mass. Taking the H_2O density to be 0.95 g cc^{-1}, a rock density of 3.5 g cc^{-1} would imply a rock mass fraction of 67% and an H_2O layer 340 km thick. Similarly, a "rock" density of 2.5 g cc^{-1} (with the lower density implying a substantial non-silicate component) would yield 79% and 196 km, respectively. As illustrated by these end-member examples, precise estimates are difficult because of several uncertain factors.

1. "Rock" density. The jovian moon Io (density 3.5 g cc^{-1}) is often assumed to represent the density of "rock" (actually silicates + iron) in the outer solar system. But at least for Enceladus, gravity measurements suggest that the "rock" interior has a density more like 2.5 g cc^{-1}, probably as a result of some combination of porosity and/or hydrothermal alteration (e.g., *Hemingway et al.,* 2018). The extent of hydrothermal alteration and/or porosity (presumably water-filled) in Pluto's "rock" is unknown, but is expected on theoretical grounds (*Vance et al.,* 2007; *Neveu et al.,* 2017); analogous hydrothermal alteration has been suggested for Titan (*Castillo-Rogez and Lunine,* 2010).

2. Ice porosity. As argued below, the ice shell of Pluto may also have retained some porosity. If so, then the actual rock mass fraction would be higher than the estimates above. Much of the density difference between Pluto and Charon can be explained by Charon retaining more ice-shell porosity (as a result of its lower heat flux and pressure) (*Bierson et al.,* 2018).

3. Carbon (and other) compounds. As discussed by *Simonelli and Reynolds* (1989) and *McKinnon et al.* (1997), Pluto's interior might contain a non-negligible quantity of carbon compounds. Cometary elemental abundances certainly suggest the existence of substantial amounts of carbon, far more than the amounts found in the most carbon-abundant chondritic meteorites (~5 wt.%). As such, a major carbonaceous and/or graphitic component may exist within Pluto. Although there is no direct evidence for such a component, its presence would have potentially profound implications for Pluto's structure and evolution (see the chapter by McKinnon et al.).

3.1.2. Energy sources. To a large extent the evolution of a body is determined by the energy sources available. In particular, Pluto differs from many icy objects in that it is unlikely to have experienced significant tidal heating, if any, since it reached its synchronous state with Charon (*Barr and Collins,* 2015). Given Pluto's density and assuming that its silicate portion has chondritic radiogenic abundances (*Robuchon and Nimmo,* 2011), the likely energy sources can be determined (Table 1). The accretion estimate is a lower bound, obtained by assuming zero velocity at infinity; the actual value could be somewhat higher. We also included an estimate for the mean energy delivered by a Charon-forming impact (*Canup,* 2011), although in reality the the temperature change due to such an impact will be highly spatially variable.

The main take-away from Table 1 is that Pluto's evolution is most likely dominated by radioactive decay. Other energy sources (accretion, despinning, even a Charon-forming impact) are smaller in comparison, although not negligible.

3.1.3. Differentiation state. Both Pluto and Charon are clearly ice-dominated at the surface (*Grundy et al.,* 2016). There is no evidence supporting the suggestion that both bodies might possess an undifferentiated "carapace" consisting of a mixture of rock and ice (*Desch et al.,* 2009). Prior to New Horizons it was unclear whether Pluto and Charon were fully differentiated, into a deep "rock" core and a shallower ice shell. Triton, a comparably-sized body, is certainly differentiated, but that is because it experienced extreme heating during its capture into Neptune's orbit (e.g., *Goldreich et al.,* 1989; *McKinnon et al.,* 1995). The moon Callisto, larger than Pluto but with a comparable density (1.83 g cc^{-1}) is sometimes referred to as only

TABLE 1. List of energy sources and sinks for Pluto, modified from *Robuchon and Nimmo* (2011).

Energy	Value (J)	ΔT_{eff} (K)	Commentary
Radioactive	1.3×10^{28}	780	Released between 30 m.y. and 4.5 G.y. after solar system formation
Accretion	5.7×10^{27}	340	$3GM^2/5R$
Thermal	3.3×10^{27}	210	To warm up Pluto from 40 K to 250 K
Charon-forming impact	1.5×10^{27}	89	Collision between bodies with radii 1106 and 836 km and densities 1.8 g cc^{-1} (*Canup,* 2011); evaluated using $GM_1M_2/(R_1 + R_2)$
Despinning	9.2×10^{26}	55	To slow down Pluto from 3 h to the present-day period
Latent heat	9.5×10^{26}	55	To melt the top 200 km of water ice
Differentiation	8.4×10^{26}	49	To differentiate into a two-layer structure (*Hussmann et al.,* 2010)

partially-differentiated (*Anderson et al.,* 2001). However, we do not know this for sure; it is based on the assumption that Callisto is hydrostatic (see below), which has not been demonstrated and may not be correct (*McKinnon,* 1997; *Schubert et al.,* 2004).

The total gravitational energy released during Pluto's accretion is larger than the amount required to heat it up to 250 K (Table 1 above). Radioactive decay releases more than twice as much total energy, but the rate of heat release is much slower. Because differentiation requires melting, all that theory can tell us is that differentiation of Pluto is plausible but not assured (*McKinnon et al.,* 1997).

However, in addition there are three pertinent observational constraints. The first is that, if Pluto were undifferentiated then as it cooled over time, deeply-buried ice would convert from ice I to ice II (*McKinnon et al.,* 1997). The accompanying reduction in volume would lead to compressional tectonics on the surface — which are not observed. Thus, a completely undifferentiated body can be ruled out (and the same goes for Charon).

Second, Charon and Pluto have roughly the same rock mass fraction (see above). Assuming that Charon formed via an impact, if Pluto had been totally differentiated at the time of impact then Charon would be mostly composed of ice (since the impact mainly excavates near-surface material). Because this is not the case, then assuming an impactor composition similar to Pluto's, Pluto must have been at most partially differentiated when the impact happened (*Canup,* 2011). Subsequent differentiation could have happened either as a consequence of the impact itself, or due to later heating via radioactive decay.

Last, there is at least circumstantial evidence that Pluto has a subsurface ocean (see below). If so, the high temperatures required suggest complete differentiation.

Overall, it seems almost certain that Pluto consists of separate ice and "rock" layers. Whether a separate, central iron core is also present is unknown. There is no evidence for a core-generated magnetic field (*McComas et al.,* 2016). A dynamo can operate if the heat flux extracted from the core exceeds the adiabatic value; scaling the results of *Nimmo and Stevenson* (2000) to the lower gravity of Pluto, the minimum heat flux required is thus roughly 1–4 mW m^{-2}. The expected deep mantle heat flux at present is roughly 3 mW m^{-2} (see Fig. 2 below), so a dynamo is possible if a liquid iron core exists. More likely, given the relatively modest amount of heat generated during accretion and subsequently (Table 1), is that a dynamo is absent because separation of iron and silicates never occurred, or was sufficiently limited that whatever small core formed is now solid or nearly so.

3.1.4. Shape. In the absence of gravity measurements, the shape of a planetary body can sometimes be used to determine its moment of inertia and thus constrain its internal structure (e.g., *Dermott and Thomas,* 1988). The main requirement is hydrostatic equilibrium, i.e., that the body has no long-term elastic strength and adopts the shape a fluid body would have.

The shape of a hydrostatic body, in terms of its ellipsoidal axes a > b > c, is determined by tidal and rotational parameters and its internal structure as described by the fluid Love number h_2. Under the assumption of hydrostatic equilibrium, this fluid Love number can be directly related to the moment of inertia via the so-called Radau-Darwin relation (*Munk and MacDonald,* 1960).

It can be shown that

$$a = R_p h_2 \left[\frac{1}{6} q_{rot} + q_{tid} \right]$$

$$b = R_p h_2 \left[\frac{1}{6} q_{rot} - \frac{1}{2} q_{tid} \right]$$

$$c = R_p h_2 \left[-\frac{1}{3} q_{rot} - \frac{1}{2} q_{tid} \right]$$

where $q_{rot} = R_p^3 \omega^2 / G m_p$ and $q_{tid} = R_p^3 m_s / a^3 m_p$ denote the rotational and tidal potentials. Here R_p and m_p are the body radius and mass, Ω_p is its angular rotation frequency, m_s is the mass of the tide-raising body and a is its distance. For a synchronous satellite $q_{tid} = q_{rot}$.

In the absence of tidal effects, a hydrostatic planet takes the shape of an oblate spheroid with oblateness $\frac{(a-c)}{R} = \frac{1}{2} q_{rot} h_2$ and $\frac{(a-c)}{(b-c)} = 1$. For a hydrostatic synchronous satellite with $m_s \ll m_p$ the shape is that of a triaxial ellipsoid for which $\frac{(a-c)}{R} = 2 q_{rot} h_2$ and $\frac{(a-c)}{(b-c)} = 4$. In its present orbital and spin state, a hydrostatic Pluto should be only modestly triaxial with a predicted $\frac{(a-c)}{(b-c)} = 1.32$ and an oblateness of 539 m (0.05%) for $h_2 = 2.5$. The equivalent gravitational flattening, referred to as J_2, would be about 2×10^{-4}, or roughly 12 mGal at the surface. The actual distortions will be smaller because h_2 is reduced by any degree of central mass concentration. For example, the high and low core density Pluto structures discussed in section 3.1.1 above would yield normalized moments of inertia of 0.323 and 0.345, respectively, and fluid h_2 values of 1.88 and 2.04.

The upper bound on Pluto's oblateness of 0.6% is thus consistent with the expected present-day oblateness. At earlier times, when Charon was closer and Pluto was spinning faster, the oblateness would have been larger. Charon's present distance to Pluto is 16.5 Pluto radii (R_P); any bulge locked in prior to Charon reaching 14 R_P should have been detectable by this method (*Nimmo et al.,* 2017). Either Pluto's interior was too soft to retain such a fossil bulge, or any initial bulge was later removed (e.g., by an ocean or failure of the lithosphere; see below).

3.1.5. Summary. Pluto has a "rock" mass fraction of ≈70% and is probably fully differentiated at the present day. Because its predicted present-day oblateness (flattening) is too small to detect, no direct estimates of its moment of inertia are available. Significant uncertainties in the silicate porosity, extent of hydrothermal alteration and abundance of organic compounds leave open a fairly wide range of acceptable structural models. Figure 1 shows three possible models, including one in which an ocean is not present at all.

3.2. Is There an Ocean?

Whether or not Pluto has a present-day subsurface ocean is a key question. As discussed below, direct evidence of an ocean is lacking, but there are several circumstantial lines of evidence that point toward such a feature. In section 5 we discuss how confirmation of such an ocean's existence could be obtained.

3.2.1. ***Theory.*** Table 1 shows that energy produced by radioactive decay comfortably exceeds that required to heat and melt Pluto's entire ice inventory. The real issue is whether or not this heat can be removed rapidly enough to avoid melting. *Robuchon and Nimmo* (2011) investigated this question in some detail. They assumed a differentiated Pluto with an ammonia-free ice shell very similar to that shown in Fig. 2 and modeled its thermal evolution for different ice basal viscosities. If this viscosity were low enough ($<2 \times 10^{16}$ Pa s), the ice shell underwent convection and removed heat rapidly enough to remain below the melting point, so that no ocean ever developed. Conversely, with higher ice viscosities the ice shell remained conductive throughout and experienced melting due to heating from below, resulting in an ocean surviving to the present day.

Other groups assumed convection was not operating and investigated whether an ocean could survive. For Pluto, a present-day ocean is a common outcome of models, especially if ammonia (which acts as an antifreeze) is present (*Hussmann et al.,* 2006; *Desch et al.,* 2009; *Hammond et al.,* 2016). Although *Hammond et al.* (2016) argued that a sufficiently conductive core could result in the ocean completely refreezing, this appears to be the result of a minor error in their code (see *Bierson et al.,* 2018).

Theory thus permits the existence of a subsurface ocean. But since the ice viscosity depends on the (unknown) grain size, the temperature at which melting begins, and the potential role of silicate fines in the ice (*Desch and Neveu,* 2017), an ocean is not *required* (in the sense of being inevitable). Observations are thus required to resolve the issue.

3.2.2. ***Surface stresses.*** An important observation is the predominance of extensional tectonic features on Pluto (section 2.2). Such extension is commonly observed on icy satellites (*Collins et al.,* 2010) and was mentioned as a possibility for Pluto (*Cruikshank et al.,* 1997; *Moore et al.,* 2015) prior to the New Horizons flyby. Particularly for bodies in which higher-pressure ice phases are not involved, a simple explanation of this extension is the partial refreezing of a subsurface ocean. Since ice takes up more volume than water, freezing of an ocean requires the icy surface to move outward; because it is doing so on a spherical body, this outward motion automatically produces extension. Thus, the existence on Pluto of relatively young extensional tectonics is consistent with a refreezing ocean.

Furthermore, *Hammond et al.* (2016) argued that complete freezing of an ocean would likely result in formation of ice II and young contractional features (see above); because such features are not observed, they concluded that a subsurface ocean is present now. We note, however, that for a sufficiently large (low-density) core, pressures would not be high enough for ice II stability and so this argument is not conclusive.

In the absence of an ocean, the dominant effect on surface stresses is cooling of the interior, which will lead to present-day compression. Similarly, progressive serpentinization of the silicate core will tend to increase the core density (*Beyer et al.,* 2017; *Bierson et al.,* 2018) and lead to overall compression. The absence of recent compression strongly favors the existence of an ocean.

This conclusion is relatively insensitive to different thermal evolution scenarios. If Pluto starts cold but differentiated, formation of an ocean leads to stresses that are initially compressional. They then switch to extensional as the ocean begins to freeze. If an ocean never develops, the stresses are comparable in magnitude but the sign is opposite: initial extensional followed by later compression, as the interior cools (*Robuchon and Nimmo,* 2011). If Pluto begins life warm, the stresses are monotonically extensional (if an ocean develops) and monotonically

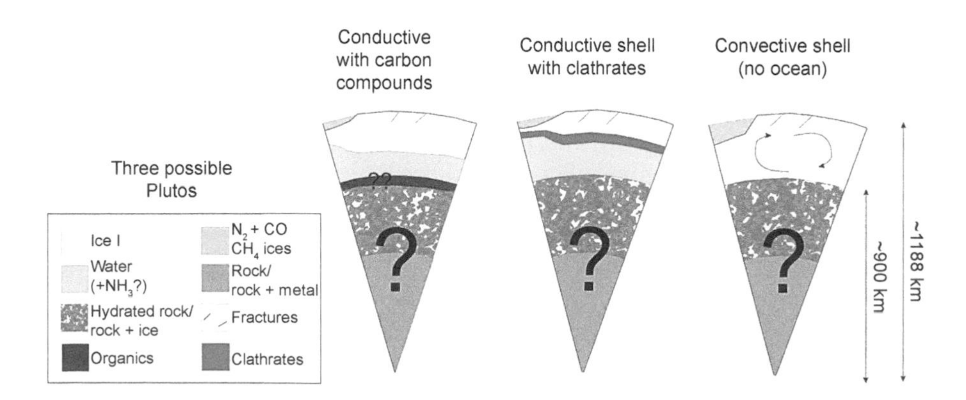

Fig. 1. Schematic view of three possible Pluto interiors (section 3.1). Interfaces are to scale, except the depth of the surface Sputnik Planitia basin is exaggerated for clarity. Question marks denote inferences (e.g., the presence of carbon compounds) that are uncertain (see text). In the absence of an ocean, the Sputnik Planitia basin would have to be deeper to explain its location (see text).

compressional (if not). We remind the reader that recent tectonic features are certainly extensional (section 2.2), while older features are more ambiguous, and discuss the issue of Pluto's thermal history further below (section 3.4). The apparently young nature of some extensional faults suggests ongoing ocean freezing.

3.2.3. Fossil bulge. Prior to New Horizons, the possibility existed that Pluto and/or Charon might preserve a "fossil bulge," a frozen relic of a distorted shape established earlier in their spin-orbit evolution similar to that seen at Iapetus (*Castillo-Rogez et al.,* 2007) or the Moon (*Garrick-Bethell et al.,* 2014; *Keane and Matsuyama,* 2014). However, no such bulges were detected (section 2.1).

Maintaining a bulge imparted by an earlier, faster spin rate requires a body to be able to withstand the stresses associated with the bulge. This then depends on the thickness and strength of Pluto's icy lithosphere, and the rigidity of its core if Pluto is differentiated (*Robuchon and Nimmo,* 2011). That no fossil oblateness can be detected for Pluto implies that its icy lithosphere must have been thin and/or weak enough during or after spindown to relax (cf. *Singer and McKinnon,* 2011). Moreover, any oblate core must also have been soft enough to relax. Modeling by *Robuchon and Nimmo* (2011) showed that the development of an ocean always led to the loss of a fossil bulge. The presence of an ocean would also have accelerated Pluto's tidal and spin evolution (*Barr and Collins,* 2015); the reduced time available for cooling would have limited the ability of Pluto to freeze in a fossil shape during spindown. Thus the absence of an observed fossil bulge is consistent with the presence of an ocean, although it does not require it.

3.2.4. Cryovolcanism? Cryovolcanic features are typically assumed to require the presence of liquid water or an ice-water slurry (*Kargel,* 1995), in which case a subsurface ocean would probably be required. However, liquid water is denser than the surrounding water ice crust, thus presenting a barrier to cryovolcanism. One result of partial ocean freezing is to pressurize the water beneath, which can in principle overcome the density barrier (*Manga and Wang,* 2007; *Martin and Binzel,* 2021). Thus, if the enigmatic features seen south of Sputnik Planitia are indeed cryovolcanos (section 2.2), ocean freezing would be one way to explain their presence. Furthermore, the combination of stresses arising from ocean freezing, true polar wander, and loading by the nitrogen ice of Sputnik Planitia shows that substantial tectonic stresses are generated within an annulus around Sputnik Planitia that incorporates Wright and Piccard Montes (*Cruikshank et al.,* 2019; *McGovern and White,* 2019), potentially explaining their location. The detection of ammonia at Virgil Fossae (section 2.2) could be explained by the eruption of NH_3-rich material, perhaps driven by pressurization arising from partial ocean freezing.

In short, a present-day subsurface ocean provides a simple (although not unique) explanation for the proposed cryovolcanic features. However, since we do not know the origin of these features, this argument for a subsurface ocean remains provisional.

3.2.5. Sputnik Planitia. A final line of evidence arises from the location of the predominantly nitrogen-filled Sputnik Planitia. The center of this basin (roughly 175°E, 18°N) is located close to the anti-Charon point at 0° latitude, 180° longitude. Although this could simply be a coincidence (probability of roughly 5%), an alternative is that Sputnik Planitia caused Pluto to reorient such that it moved closer to the anti-Charon point. This reorientation would have generated stresses that are approximately consistent with the orientation of tectonic features around Sputnik Planitia (*Keane et al.,* 2016). However, in order for such reorientation to occur, Sputnik Planitia would have to represent a mass excess, or positive gravity anomaly, despite being a negative topographic feature. *Nimmo et al.* (2016) argued that the most likely way of making a positive gravity anomaly at Sputnik Planitia was to invoke a subsurface ocean.

The argument proceeds as follows. Sputnik Planitia is lower than the surrounding regions by at least 3 km, which would result in a negative gravity anomaly of about 110 mGal (assuming a surface density of 0.9 g cc^{-1}). Nitrogen ice is denser than water ice by about 0.1 g cc^{-1}, so this negative gravity anomaly could be just overcome if the nitrogen layer were about 27 km thick. As explained in section 2.3.1, such a layer would imply a convective aspect ratio significantly less than that expected based on the (admittedly poorly-known) nitrogen ice rheology.

Since the density contrast between water and ice is also about 0.1 g cc^{-1}, an ice shell thinned by 27 km beneath the Sputnik Planitia basin can also provide the required gravity anomaly if an ocean is present. Numerical models (*Johnson et al.,* 2016) show that shell thinning by 20–60 km is expected for a Sputnik Planitia-forming impact. In this picture the Sputnik Planitia basin is isostatically compensated immediately after the impact (i.e., the gravity anomaly is ~0). The lithosphere then cools and is subsequently loaded with nitrogen, in a manner analogous to the mascon basins on the Moon. The final gravity anomaly then depends on the nitrogen thickness d and the lithospheric elastic thickness. Although neither is known, a nitrogen layer 5 km thick would yield a surface gravity anomaly of about +200 mGal if the lithosphere were infinitely rigid.

Additional evidence that Sputnik Planitia is a positive gravity load comes from the radial orientation of multiple normal fault valleys surrounding Sputnik Planitia (*Keane et al.,* 2016). Such a fault pattern is expected for a positive load on a spherical shell or lithosphere when the horizontal scale of the load is sufficiently large (*Janes and Melosh,* 1990).

Keane et al. (2016) point out that an impact redistributes mass from the center to the periphery, in the form of an ejecta blanket and crustal thickening. Depending on the details of this redistribution, if no mass is lost it can result in a positive gravity anomaly at spherical harmonic degree-2 (which is what matters for reorientation), without requiring any additional effects. But mass can be lost (either vaporized or ejected from Pluto). So whether or not an ocean is really required depends on how mass was redistributed during the impact, and the non-negligible mass contribution from the impactor itself. This question can be addressed using a combination of existing topographic data (*Schenk et al.,* 2018b) and models (*Johnson et al.,* 2016). At present, it seems almost certain that Sputnik Planitia caused reorientation, and likely (but not certain) that an ocean was required.

3.2.6. Summary. Thermal evolution models permit a present-day ocean to exist, but do not require it. Four observations — extensional tectonics, cryovolcanism, absence of a fossil bulge, and the location of Sputnik Planitia — can all be explained if Pluto has a subsurface ocean. None of these observations *requires* such an ocean to be present, and each can be explained by alternative processes. Nonetheless, taken together it seems probable, although not certain, that Pluto does indeed possess a subsurface ocean at present. A corollary is that the ice shell above is conductive rather than convecting.

3.3. What Is the Ice Shell Structure?

3.3.1. Temperature structure. Considering the range of possible interior models outlined above, it is possible to infer some aspects of the temperature structure of Pluto. Figure 2a shows one possible present-day Pluto temperature structure, taken from the model shown in Fig. 1 of *Bierson et al.* (2018). In this model, the core is heated by radiogenic decay but is probably not convecting, although advection of heat via fluid circulation might be important (see below). The ice shell is about 200 km thick and conductive, with an ocean about 150 km thick beneath. The conductive core is close to the melting point ($\approx$1400 K for peridotite) at its center.

Figure 2b shows the ice shell temperature structure in more detail. The curvature of the profile arises because ice has a thermal conductivity that varies as 1/T; the inflection point at 60 km depth (100 K) signals the transition from shallow, low-conductivity porous ice to deeper, pore-free ice. The main uncertainties in these models are the conductivity values assumed: That for ice depends on the porosity (discussed in more detail below), while that for the core is uncertain (*Hammond et al.,* 2016) and may be effectively increased by hydrothermal circulation.

The profile shown in Fig. 2b has two important implications. The first is that the ice shell of Pluto is strong. Ice typically behaves elastically at low temperatures and viscously at high temperatures. Calculations presented in *Conrad et al.* (2019) suggest that the base of the elastic layer on Pluto should occur at a temperature of 118 ± 15 K. Comparison with Fig. 2b shows that the elastic layer will be roughly 100 km thick. This estimate is uncertain because of factors such as porosity (which will weaken the ice), yielding, and what strain rate to use. Nonetheless, this result indicates that the present-day ice shell should have little difficulty in supporting topographic loads.

The second, paradoxically, is that the base of the ice shell is too warm. First, ice at 270 K typically has a viscosity of 10^{13-14} Pa s, less than the minimum required to permit ice shell convection (section 3.2.1) and implying that the lower portion of this shell could be convective (*McKinnon,* 2006). An ice shell this warm will flow rapidly so as to erase any lateral variations in shell thickness. The timescale for such lateral flow depends on the wavelength over which flow occurs, the viscosity at the base of the shell η_b, and the effective channel thickness over which flow occurs (*Stevenson,* 2000). Because it was argued above that Sputnik Planitia represents an area of significantly thinner ice, we need to be able to maintain these shell thickness variations. This in turn requires a high ice viscosity at the base of the shell (where flow will occur).

One possible way of maintaining a high-viscosity shell is to require a very cold ocean. Because η_b depends on the temperature at the ice-ocean interface, a sufficiently cold ocean will shut down lateral flow. For the same heat flux, a thinner ice shell will also result. However, *Nimmo et al.* (2016) found that that the ocean had to be 200 K or colder to avoid lateral flow (this would also prevent convection). Such

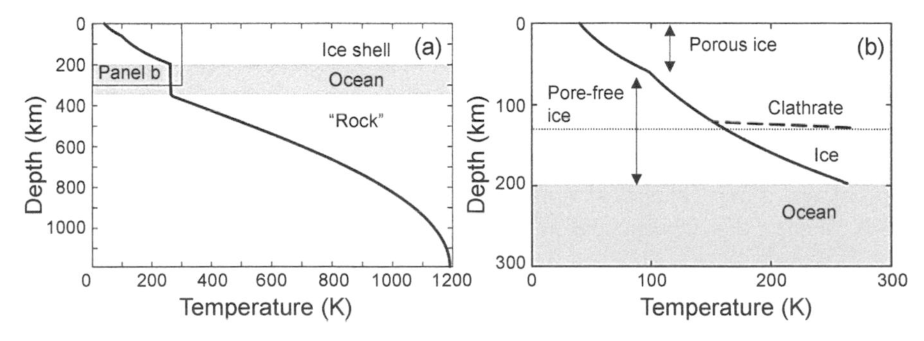

Fig. 2. Possible present-day Pluto temperature structure, taken from the model shown in Fig. 1 of *Bierson et al.* (2018). **(a)** Overall view of temperature profile. **(b)** Inset showing ice-shell temperature profile. The curvature arises because conductivity is temperature-dependent, and the slope break denotes the transition from porous to pore-free ice. The dashed line indicates the profile if a thin basal layer of clathrates is present (section 3.3.1), resulting in a thinner ice shell overall (dotted line).

a temperature requires an ocean with an NH_3 concentration of around 25 wt.% (*Leliwa-Kopystyński et al.,* 2002). Even for an initial NH_3 bulk composition of 5 wt.% (*McKinnon et al.,* 2008), such a high value is difficult to achieve and would require almost complete ocean freezing (at least 80%). In addition, such an amount of NH_3 would so reduce the ocean density that its ability to produce a positive gravity anomaly would be eliminated.

A perhaps more attractive alternative is to invoke a layer of clathrates (*Kamata et al.,* 2019). Because clathrates have a thermal conductivity almost an order of magnitude lower than that of regular water ice, even a thin clathrate layer can greatly change the ice shell temperature structure. Figure 2b shows that the result is a large temperature drop across the clathrate layer, resulting in a shell that is thinner (dotted line) and an ice portion that has a much lower basal temperature (in this case 150 K as opposed to 270 K for the clathrate-free case). Lateral flow in such an ice shell is not significant over the age of the solar system.

An additional implication of the clathrate hypothesis is that it may naturally explain the relative paucity of CO in Pluto's atmosphere compared to N_2 (*Glein and Waite,* 2018). Because CO clathrates form more readily than N_2 clathrates when multiple guest molecules are involved (*Mousis et al.,* 2012), the low relative abundance of atmospheric CO may be a natural outcome of the clathration process (see the chapter in this volume by McKinnon et al.). Clathrates have also been invoked as a way of removing noble gases from Pluto's atmosphere (*Mousis et al.,* 2013).

3.3.2. Heat flux and elastic thickness. As noted above, the temperature structure of the shell depends on the heat flux and conductivity structure. For ice, which has a thermal conductivity that varies as roughly k_0/T with k_0 ($\approx$567 W m^{-1}) a constant (*Klinger,* 1980), we may write

$$F = k_0 \frac{\ln(T_b/T_s)}{d}$$

where d is the total shell thickness, T_b and T_s are the surface and basal temperatures, and here we are neglecting sphericity.

The mean effective thermal conductivity of the shell will be reduced in the presence of either a near-surface porous layer or a clathrate layer at the base. Porosity may gradually decrease due to annealing (*Bierson et al.,* 2018). Independent of this effect, clathrates are likely to grow with time (*Kamata et al.,* 2019). The consequences can be pronounced: A clathrate layer 10 km thick with a conductivity one-tenth of that of ice would reduce shell thickness from 200 km to 138 km (cf. Fig. 2b).

Once the temperature structure of the ice shell is known, its mechanical properties can be predicted. As argued above (section 3.3.1), by taking the base of the elastic layer to be defined by a particular temperature, the elastic thickness T_e of the ice shell can be deduced from the heat flux, or vice versa.

Figure 3 plots the expected relationship between heat flux (log scale) and T_e. The shaded region shows the expected surface heat flux arising from radioactive decay, with the lower limit being the expected present-day value of $\approx$3 mW m^{-2}. For a Pluto with a porous, low-conductivity surface layer (dashed lines) the expected present-day elastic thickness is then roughly 100 km, which agrees with Fig. 2a, while it could have been as low as 50 km at earlier times.

Figure 3 also shows the lower bound on T_e of 10 km inferred by *Conrad et al.* (2019) based on the absence of rift-flank uplift observed (section 2.3.2). This is not a strong constraint, but is at least consistent with the expected (radiogenic) heat flux. Also shown on Fig. 3 is the heat flux in excess of 50 mW m^{-2} indicated by the apparently relaxed 145-km-diameter crater Edgeworth (section 2.3.3). This is not remotely consistent with the expected radiogenic heat flux, and only marginally (at best) with the absence of rift flank uplift.

3.4. How Has Pluto Evolved Over Time?

Icy satellites can experience complicated and nonmonotonic thermal histories because of tidal heating arising from proximity to a much more massive primary (e.g., *Hussmann and Spohn,* 2004). Pluto does not suffer from this complication. Nonetheless, our relative ignorance of its present-day structure and (even more so) its initial conditions make consideration of Pluto's thermal evolution fraught with uncertainty.

Pluto probably experienced a fairly dramatic early history: a Charon-forming impact, followed by completion of

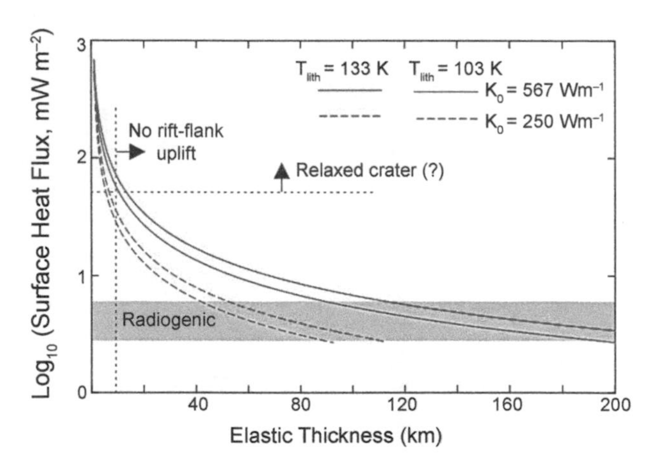

Fig. 3. Relationship between heat flux and elastic thickness (section 3.3.2). The temperature defining the base of the ice shell is denoted T_{lith}. Conductivity is calculated by k_0/T where k_0 is a constant chosen to represent either pore-free or porous ice. The shaded region denotes the expected heat flux range over Pluto's history from radiogenic elements. The two dashed lines denote observational constraints derived from the absence of rift-flank uplift (section 2.3.2) and the presence of one relaxed crater (section 2.3.3).

differentiation, despinning, and tidal bulge collapse (*Barr and Collins,* 2015) as Charon moved outward. However, we see no fossil bulge and no tectonic features obviously associated with despinning: in short, no evidence of this early history. This is most likely for three reasons. First, Pluto has a "short memory": The conductive timescales of the core and shell are sufficiently short that present-day conditions are very insensitive to the initial conditions (*Nimmo and Spencer,* 2015). Second, the energy budget of Pluto is dominated by radioactive decay, rather than these early processes (Table 1), so their influence would in any case have been limited. Third, Pluto's active surface geology (see the chapter in this volume by White et al.) has likely served to erase or obscure its earliest tectonic signatures.

Nonetheless, we can make two deductions. One relies on ^{26}Al: If Pluto had formed early enough (within roughly the first 5 m.y.), heating from ^{26}Al decay would have ensured complete differentiation. But since Charon's composition requires an incompletely-differentiated proto-Pluto (section 3.1.3), the Charon-forming impact would have to have occurred in a very narrow time window. More likely, Pluto formed after ^{26}Al was extinct, as *Bierson and Nimmo* (2019) concluded for Kuiper belt objects in general. The other deduction has to do with the absence of compressional features (section 2.2). A Pluto that began frozen and then formed an ocean would have experienced early compression, and then later extension (*Robuchon and Nimmo,* 2011; *Hammond et al.,* 2016). Although these early features might have been overprinted, and in any case are often harder to identify than extension, the absence of compression hints that Pluto might have started hot and then gradually cooled down (*Bierson et al.,* 2020). On the other hand, Table 1 makes it clear that a hot start is not assured, and depends in particular on the details of Pluto's accretion.

Figure 4 shows an example "cold-start" Pluto evolution (albeit one in which full differentiation is assumed) from *Hammond et al.* (2016). Radioactive decay warms the core; heat transferred out of the core causes the ice to melt, resulting in initial compression. The ocean reaches a maximum thickness at about 2 G.y. and then begins to refreeze, transitioning to extensional tectonics at about 1 Ga. In this model the present day heat flux is about 3 mW m^{-2} and the present-day ocean thickness about 50 km. If an insulating clathrate layer grew, this cooling and thickening will have been slower than in the clathrate-free models, delaying the onset of surface activity and resulting in a thicker ocean at the present day.

Assuming an ocean exists, the high temperatures predicted for Pluto's core (Fig. 2a) suggest that reactions would take place between the rock, water, and any organics present, as long as there is some permeability present. By analogy with other ocean worlds like Enceladus or Europa, one would expect a salty ocean to develop (*Zolotov,* 2007). Whether such salts could be detectable in any cryovolcanic deposits is an open question. Outgassing of gaseous species initially trapped in the interior is also likely, unless they were sequestered by clathration (*Mousis et al.,* 2013; *Kamata et al.,* 2019) or converted to other, more stable forms (*Glein and Waite,* 2018).

Another consequence of water-rock interactions is that the core might lose its heat more effectively than simple conductive models (Fig. 2a) assume. The importance of such advection depends on the unknown permeability of the core, but the effect on Pluto's overall evolution may be significant (*Gabasova et al.,* 2018) and is certainly worth

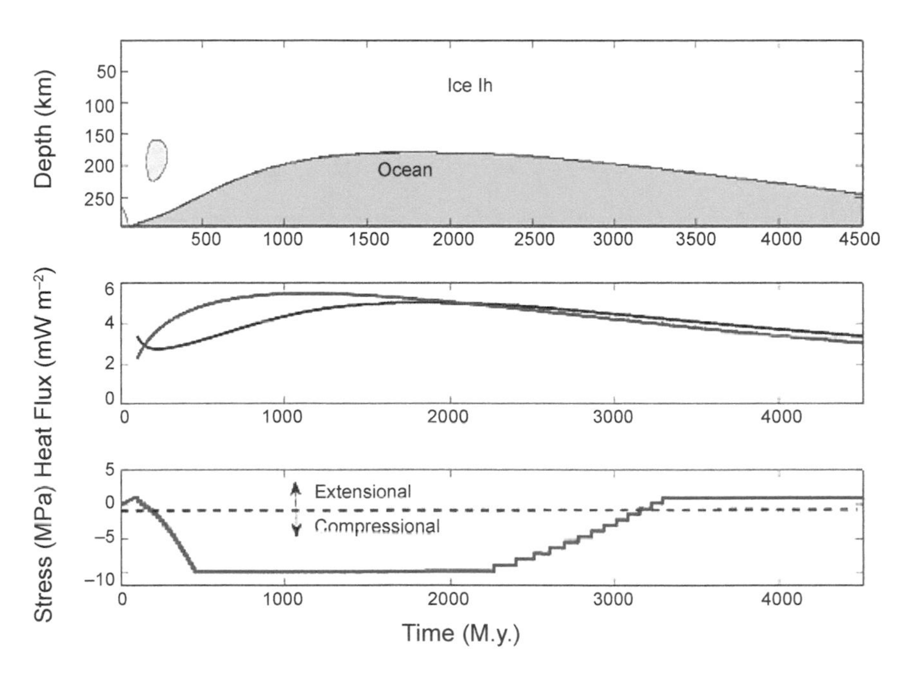

Fig. 4. "Cold-start" Pluto thermal evolution model, from *Hammond et al.* (2016). **(a)** Ice shell thickness evolution, with green blobs indicating where ice II is stable. **(b)** Heat fluxes out of the core (red line) and at the surface (black line). **(c)** Predicted surface stresses.

further study. Furthermore, if the core can dehydrate, then higher-temperature processes may be possible (graphitization of organics, smelting of sulfides, etc.).

4. PLUTO AND CHARON IN CONTEXT

An important question is how Pluto and Charon compare with the icy moons orbiting the giant planets of the outer solar system. Perhaps the most important difference is that Pluto and Charon are not subjected to large tidal stresses and heating, which can be a major driver for geological activity among the moons. Triton is thought to be a captured dwarf planet from the Kuiper belt, but it was exposed to massive tidal heating during the capture process (*Goldreich et al.,* 1989).

Figure 5 is a summary plot of the inferred structures of a variety of icy worlds, including Pluto and Charon [for a review on how these structures are determined, see *Nimmo and Pappalardo* (2016)]. A rock:ice ratio of 2:1 by mass is roughly correct for many of the moons, as well as Pluto and Charon. Exceptions include Tethys, which appears to be almost pure ice and may be a relic of a disruptive impact (*Asphaug and Reufer,* 2013), and Europa, which may have been volatile-depleted by tidal heating or forming in a warm part of the protosatellite disk (*Canup and Ward,* 2009). Triton is only slightly denser than Pluto, suggesting that volatile loss driven by tidal heating was limited at least in this case. Although we have only very limited information on the structures of other Kuiper belt objects, their inferred densities suggest a generally similar rock:ice ratio to that of Pluto (*Bierson and Nimmo,* 2019).

As reviewed in *Schubert et al.* (2004) and *Nimmo and Pappalardo* (2016), the only icy world known to be fully differentiated is Ganymede, with an active dynamo that indicates an iron core. Europa and Enceladus have ice shells overlying oceans, with a deeper interior that is presumed to be mostly rock; it is not known if an iron core is present. At both Titan and Callisto, it is possible that the rock and ice have not fully separated. However, this conclusion is quite uncertain: We don't know if Callisto is hydrostatic (section 3.1.3), while Titan is close to, but not exactly, hydrostatic (*Durante et al.,* 2019). Furthermore, a measured moment of inertia is non-unique in terms of possible internal structures. Triton is presumed to be differentiated due to ancient tidal heating, but there is no direct evidence for this. As reviewed in *Nimmo and Pappalardo* (2016), the three Galilean satellites plus Titan are all thought to have subsurface oceans maintained by a combination of tidal and radiogenic heating. Enceladus' ocean can only be maintained by tidal heating because the radiogenic contribution is negligible. Triton, being almost identical in size and density to Pluto, could in principle maintain an ocean, but there is no direct evidence for one. Smaller moons like Dione might also possess present-day oceans (*Zannoni et al.,* 2019).

There is a general expectation that more volatile materials are likely to condense and survive at greater distances from the Sun. The jovian satellites show no obvious signatures

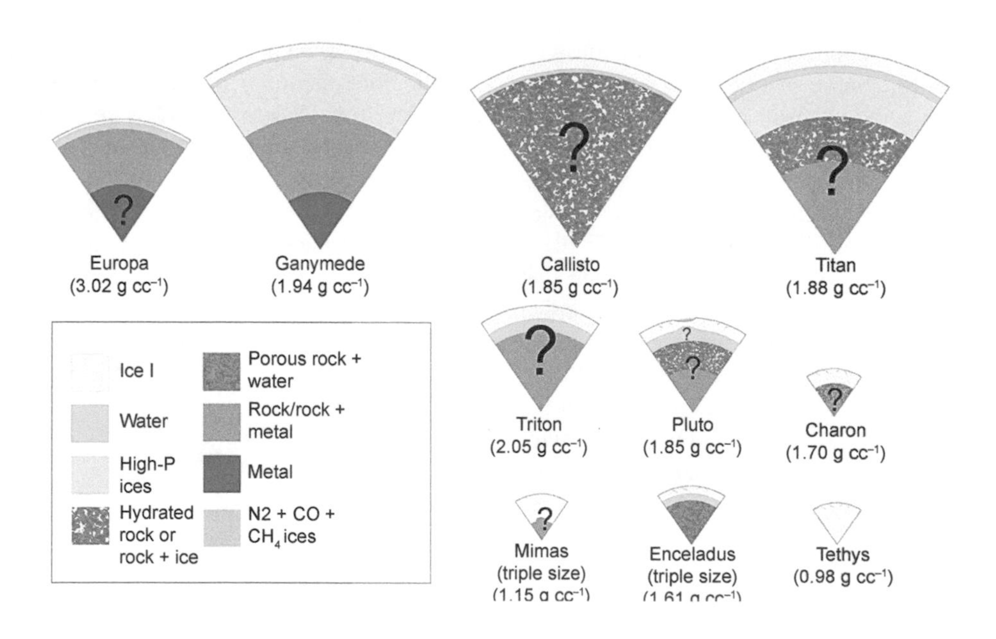

Fig. 5. Pluto compared with other icy worlds. All these bodies are to the same relative scale, except for Mimas and Enceladus. Question marks denote uncertainties in structures depicted (see text).

of NH_3, while measurements of Titan's atmosphere (*Niemann et al.,* 2005) and the plumes of Enceladus (*Waite et al.,* 2009) indicate it is present in saturnian satellites. The fact that Pluto possesses large surface reservoirs of volatile materials (N_2, CH_4, CO) not stable at the inner moons is in accordance with this expectation. Clathrates (section 3.3.1) have been proposed as a driver of Titan's evolution (*Tobie et al.,* 2006). Except at Enceladus, however, direct evidence of the chemistry of subsurface oceans is presently almost non-existent.

5. GOING FORWARD

5.1. Near-Term Advances With Existing Data

Although a fair amount has already been gleaned from the existing observations, there are several areas where more work could be done.

Getting better constraints on heat flux and how it has changed with time is an obvious topic. Apparently relaxed craters (section 2.3.3) represent one option, and so does flexure at Sputnik Planitia (section 2.3.2) or elsewhere. Perhaps other aspects of crater morphometry or topography can also be used to deduce surface heat fluxes and/or mechanical properties.

The interaction between atmospheric, geodynamic, and rotational processes driven by Sputnik Planitia over a variety of timescales has been the subject of some study (*Keane et al.,* 2016; *Bertrand et al.,* 2018), but more could be done. For instance, the fact that Sputnik Planitia is so flat, despite removal and addition of nitrogen in different locations (*Bertrand et al.,* 2018), might be used to deduce a lower bound on the thickness of the material there.

This of course raises another topic: that of nitrogen (and methane) rheology. Some of our current uncertainty arises from ignorance of how these materials deform; more measurements are critical (see the chapter in this volume by Umurhan et al.).

Finally, the interaction between chemistry and geophysics at Pluto is likely to be a fruitful field (*Neveu et al.,* 2017; *Glein and Waite,* 2018). For instance, the suggestion that clathration is responsible for the lack of atmospheric CO (*Kamata et al.,* 2019) emphasizes the point that atmospheric evolution is tied to processes in the interior (*Stern,* 1989). Similarly, the role of water-rock interactions on Pluto's overall thermal evolution (section 3.4), and how they might influence the putative ocean (salts?) and atmosphere (outgassing), have so far barely been explored.

5.2. Future Missions

Looking further ahead, a Pluto orbiter would be desirable, and is achievable with current propulsion systems (*Stern et al.,* 2020). However, as *Howett et al.* (2020) document, Pluto is a challenging geophysical target: Standard ways of probing for an ocean, such as measuring the tidal response (*Moore and Schubert,* 2000) or looking for an induction signature (*Zimmer et al.,* 2000), will not work. On the other hand, the idea of Sputnik Planitia being a positive gravity anomaly (section 3.2.5) is eminently testable, while measuring the static gravity or topography should yield a moment of inertia (section 3.1.4). Probing the N_2 plains with sounding radar could determine their thickness, while measuring atmospheric ^{40}Ar (a decay product of ^{40}K) would place an important constraint on time-integrated outgassing from the silicate interior.

6. SUMMARY

Despite the wealth of data returned by New Horizons, studies of Pluto's interior are still in their infancy, and we can say very little with certainty. Pluto is $\approx$70% "rock" by mass with the remainder some mix of ice and carbon compounds, and Charon does not need to have a different bulk composition. We can rule out an undifferentiated Pluto based on the apparent absence of compressional features. Four lines of circumstantial evidence point toward a present-day ocean roughly 50–150 km thick: extension, possible cryovolcanism, no detectable fossil bulge, and Sputnik Planitia. However, none of them is definitive. If an ocean exists (which we think is likely but not certain), the ice shell above must be cold and rigid. Although this could be due to an ammonia-rich ocean, a perhaps more likely possibility is a layer of clathrates at its base (or distributed throughout). The ocean may have been present since Pluto formed, or it may only have appeared hundreds of millions of years later as radioactive heat built up. The ocean reached a maximum thickness midway through Pluto's evolution; it subsequently slowly thinned and the ice shell thickened — driving extension and perhaps cryovolcanism — while reactions between the warm silicate core and the overlying ocean continued.

Our understanding of geodynamics on Pluto is still primitive. Nonetheless, Pluto has certainly shown itself to be far more interesting than expected, expanding the range of likely ocean worlds out to 40 AU. The Kuiper belt is likely a menagerie of similarly fascinating worlds, and we eagerly await the next opportunity to take a close-up look.

Acknowledgments. We thank A. Hayes and J. Keane for thoughtful and thorough reviews. Parts of this review were supported by NASA's New Horizons mission and NASA grant 80NSSC18K0549.

REFERENCES

Anderson J. D., Jacobson R. A., McElrath T. P., Moore W. B., Schubert G., and Thomas P. C. (2001) Shape, mean radius, gravity field, and interior structure of Callisto. *Icarus, 153,* 157–161, DOI: 10.1006/icar.2001.6664.

Asphaug E. and Reufer A. (2013) Late origin of the Saturn system. *Icarus, 223(1),* 544–565, DOI: 10.1016/j.icarus.2012.12.009.

Barr A. C. and Collins G. C. (2015) Tectonic activity on Pluto after the Charon-forming impact. *Icarus, 246,* 146–155, DOI: 10.1016/j.icarus.2014.03.042.

Bertrand T., Forget F., Umurhan O. M., Grundy W. M., Schmitt B., Protopapa S., Zangari A. M., et al. (2018) The nitrogen cycles on Pluto over seasonal and astronomical timescales. *Icarus, 309,* 277–

296, DOI: 10.1016/j.icarus.2018.03.012.
Beyer R. A., Nimmo F., McKinnon W. B., Moore J. M., Binzel R. P., Conrad J. W., Cheng A., et al. (2017) Charon tectonics. *Icarus, 287*, 161–174, DOI: 10.1016/j.icarus.2016.12.018.
Bierson C. J. and Nimmo F. (2019) Using the density of Kuiper belt objects to constrain their composition and formation history. *Icarus, 326*, 10–17.
Bierson C. J., Nimmo F., and McKinnon W. B. (2018) Implications of the observed Pluto-Charon density contrast. *Icarus, 309*, 207–219, DOI: 10.1016/j.icarus.2018.03.007.
Bierson C. J., Nimmo F., and Stern S. A. (2020) Evidence for a hot start and early ocean formation on Pluto. *Nature Geosci., 13*, 468–472.
Bland M. T., Singer K. N., McKinnon W. B., and Schenk P. M. (2012) Enceladus' extreme heat flux as revealed by its relaxed craters. *Geophys. Res. Lett., 39*, L17204, DOI: 10.1029/2012GL052736.
Brown C. D. and Phillips R. J. (1999) Flexural rift flank uplift at the Rio Grande rift, New Mexico. *Tectonics, 18*, 1275–1291, DOI: 10.1029/1999TC900031.
Brozović M., Showalter M. R., Jacobson R. A., and Buie M. W. (2015) The orbits and masses of satellites of Pluto. *Icarus, 246*, 317–329, DOI: 10.1016/j.icarus.2014.03.015.
Buhler P. B. and Ingersoll A. P. (2018) Sublimation pit distribution indicates convection cell surface velocities of ~10 cm per year in Sputnik Planitia, Pluto. *Icarus, 300*, 327–340, DOI: 10.1016/j.icarus.2017.09.018.
Canup R. M. (2011) On a giant impact origin of Charon, Nix, and Hydra. *Astron. J., 141*, 35, DOI: 10.1088/0004-6256/141/2/35.
Canup R. M. and Ward W. R. (2009) Origin of Europa and the Galilean satellites. In *Europa* (R. T. Pappalardo, et al., eds.), pp. 59–83. Univ. of Arizona, Tucson.
Castillo-Rogez J. C. and Lunine J. I. (2010) Evolution of Titan's rocky core constrained by Cassini observations. *Geophys. Res. Lett., 37(20)*, L20205, DOI: 10.1029/2010GL044398.
Castillo-Rogez J. C., Matson D. L., Sotin C., Johnson T. V., Lunine J. I., and Thomas P. C. (2007) Iapetus' geophysics: Rotation rate, shape, and equatorial ridge. *Icarus, 190*, 179–202, DOI: 10.1016/j.icarus.2007.02.018.
Collins G. C., McKinnon W. B., Moore J. M., Nimmo F., Pappalardo R. T., Prockter L. M., and Schenk P. M. (2010) Tectonics of the outer planet satellites. In *Planetary Tectonics* (T. R. Watters and R. A. Schultz, eds.), pp. 264–350. Cambridge Univ., Cambridge.
Conrad J. W., Nimmo F., Schenk P. M., McKinnon W. B., Moore J. M., Beddingfield C. B., Beyer R. A., et al. (2019) An upper bound on Pluto's heat flux from a lack of flexural response of its normal faults. *Icarus, 328*, 210–217, DOI: 10.1016/j.icarus.2019.03.028.
Cruikshank D. P., Roush T. L., Moore J. M., Sykes M., Owen T. C., Bartholomew M. J., Brown R. H., and Tryka K. A. (1997) The surfaces of Pluto and Charon. In *Pluto and Charon* (S. A. Stern and D. J. Tholen, eds.), pp. 221–267. Univ. of Arizona, Tucson.
Cruikshank D. P., Umurhan O. M., Beyer R. A., et al. (2019) Recent cryovolcanism in Virgil Fossae on Pluto. *Icarus, 330*, 155–168, DOI: 10.1016/j.icarus.2019.04.023.
Dermott S. F. and Thomas P. C. (1988) The shape and internal structure of Mimas. *Icarus, 73*, 25–65, DOI: 10.1016/0019-1035(88)90084-X.
Desch S. J. and Neveu M. (2017) Differentiation and cryovolcanism on Charon: A view before and after New Horizons. *Icarus, 287*, 175–186, DOI: 10.1016/j.icarus.2016.11.037.
Desch S. J., Cook J. C., Doggett T. C., and Porter S. B. (2009) Thermal evolution of Kuiper belt objects, with implications for cryovolcanism. *Icarus, 202*, 694–714, DOI: 10.1016/j.icarus.2009.03.009.
Dombard A. J. and McKinnon W. B. (2006) Elastoviscoplastic relaxation of impact crater topography with application to Ganymede and Callisto. *J. Geophys. Res., 111*, E01001, DOI: 10.1029/2005JE002445.
Durante D., Hemingway D. J., Racioppa P., Iess L., and Stevenson D. J. (2019) Titan's gravity field and interior structure after Cassini. *Icarus, 326*, 123–132, DOI: 10.1016/j.icarus.2019.03.003.
Freed A. M., Melosh H. J., and Solomon S. C. (2001) Tectonics of mascon loading: Resolution of the strike-slip faulting paradox. *J. Geophys. Res., 106*, 20603–20620, DOI: 10.1029/2000JE001347.
Gabasova L. R., Tobie G., and Choblet G. (2018) Compaction-driven evolution of Pluto's rocky core: Implications for water-rock interactions. In *Lunar and Planetary Science XLIX*, Abstract #2512. LPI Contribution No. 2083, Lunar and Planetary Institute, Houston.
Garrick-Bethell I., Perera V., Nimmo F., and Zuber M. T. (2014) The tidal-rotational shape of the Moon and evidence for polar wander. *Nature, 512(7513)*, 181–184, DOI: 10.1038/nature13639.
Gladstone G. R., Stern S. A, Ennico K., Olkin C. B., et al. (2016) The atmosphere of Pluto as observed by New Horizons. *Science, 351(6279)*, aad8866, DOI: 10.1126/science.aad8866.
Glein C. R. and Waite J. H. (2018) Primordial N_2 provides a cosmochemical explanation for the existence of Sputnik Planitia, Pluto. *Icarus, 313*, 79–92, DOI: 10.1016/j.icarus.2018.05.007.
Goldreich P., Murray N., Longaretti P. Y., and Banfield D. (1989) Neptune's story. *Science, 245(4917)*, 500–504, DOI: 10.1126/science.245.4917.500.
Goldsby D. L. and Kohlstedt D. L. (2001) Superplastic deformation of ice: Experimental observations. *J. Geophys. Res., 106*, 11017–11030, DOI: 10.1029/2000JB900336.
Grundy W. M., Binzel R. P., Buratti B. J., Cook J. C., Cruikshank D. P., Dalle Ore C. M., Earle A. M., et al. (2016) Surface compositions across Pluto and Charon. *Science, 351*, aad9189, DOI: 10.1126/science.aad9189.
Hammond N. P., Barr A. C., and Parmentier E. M. (2016) Recent tectonic activity on Pluto driven by phase changes in the ice shell. *Geophys. Res. Lett., 43*, 6775–6782, DOI: 10.1002/2016GL069220.
Hemingway D., Iess L., Tajeddine R., and Tobie G. (2018) The interior of Enceladus. In *Enceladus and the Icy Moons of Saturn* (P. M. Schenk et al., eds.), pp. 57–77. Univ. of Arizona, Tucson.
Howett C. J. A., Robbins S., Elliot H., and Ernest C. M. (2020) Combined Pluto orbiter and Kuiper belt exploration mission. In *Lunar and Planetary Science LI*, Abstract #1342. LPI Contribution No. 2326, Lunar and Planetary Institute, Houston.
Hussmann H. and Spohn T. (2004) Thermal-orbital evolution of Io and Europa. *Icarus, 171*, 391–410, DOI: 10.1016/j.icarus.2004.05.020.
Hussmann H., Sohl F., and Spohn T. (2006) Subsurface oceans and deep interiors of medium-sized outer planet satellites and large trans-neptunian objects. *Icarus, 185*, 258–273, DOI: 10.1016/j.icarus.2006.06.005.
Hussmann H., Choblet G., Lainey V., Matson D. L., Sotin C., Tobie G., and van Hoolst T. (2010) Implications of rotation, orbital states, energy sources, and heat transport for internal processes in icy satellites. *Space Sci. Rev., 153*, 317–348, DOI: 10.1007/s11214-010-9636-0.
Jacobson R. A., Brozović M., Buie M., Porter S., Showalter M., Spencer J., Stern S. A., et al. (2015) The orbits and masses of Pluto's satellites after New Horizons. *AAS/Division for Planetary Sciences Meeting Abstracts, 47*, 102.08.
Janes D. M. and Melosh H. J. (1990) Tectonics of planetary loading: A general model and results. *J. Geophys. Res., 95*, 21345–21355, DOI: 10.1029/JB095iB13p21345.
Johnson B. C., Bowling T. J., Trowbridge A. J., and Freed A. M. (2016) Formation of the Sputnik Planum basin and the thickness of Pluto's subsurface ocean. *Geophys. Res. Lett., 43*, 10068–10077, DOI: 10.1002/2016GL070694.
Kamata S. and Nimmo F. (2014) Impact basin relaxation as a probe for the thermal history of Pluto. *J. Geophys. Res., 119*, 2272–2289, DOI: 10.1002/2014JE004679.
Kamata S., Nimmo F., Sekine Y., Kuramoto K., Noguchi N., Kimura J., and Tani A. (2019) Pluto's ocean is capped and insulated by gas hydrates. *Nature Geosci., 12*, 407–410, DOI: 10.1038/s41561-019-0369-8.
Kargel J. S. (1995) Cryovolcanism on the icy satellites. *Earth Moon Planets, 67*, 101–113.
Keane J. T. and Matsuyama I. (2014) Evidence for lunar true polar wander and a past low-eccentricity, synchronous lunar orbit. *Geophys. Res. Lett., 41(19)*, 6610–6619, DOI: 10.1002/2014GL061195.
Keane J. T., Matsuyama I., Kamata S., and Steckloff J. K. (2016) Reorientation and faulting of Pluto due to volatile loading within Sputnik Planitia. *Nature, 540*, 90–93, DOI: 10.1038/nature20120.
Klinger J. (1980) Influence of a phase transition of ice on the heat and mass balance of comets. *Science, 209*, 271–272, DOI: 10.1126/science.209.4453.
Leliwa-Kopystyński J., Maruyama M., and Nakajima T. (2002) The water-ammonia phase diagram up to 300 MPa: Application to icy satellites. *Icarus, 159*, 518–528, DOI: 10.1006/icar.2002.6932.
Lellouch E., Sicardy B., de Bergh C., Käufl H.-U., Kassi S., and Campargue A. (2009) Pluto's lower atmosphere structure and methane abundance from high-resolution spectroscopy and stellar occultations. *Astron. Astrophys., 495(3)*, L17–L21, DOI:

10.1051/0004-6361/200911633.

Lenardic A., Richards M. A., and Busse F. H. (2006) Depth-dependent rheology and the horizontal length scale of mantle convection. *J. Geophys. Res., 111*, B07404, DOI: 10.1029/2005JB003639.

Manga M. and Wang C.-Y. (2007) Pressurized oceans and the eruption of liquid water on Europa and Enceladus. *Geophys. Res. Lett., 34*, L07202, DOI: 10.1029/2007GL029297.

Martin C. B. and Binzel R. P. (2021) Ammonia-water freezing in Pluto's subsurface ocean as a mechanism for recent cryovolcanism. *Icarus, 356,* 113763, DOI: 10.1016/j.icarus.2020.113763.

McComas D. J., Elliott H. A., Weidner S., Valek P., Zirnstein E. J., Bagenal F., Delamere P. A., et al. (2016) Pluto's interaction with the solar wind. *J. Geophys. Res., 121*, 4232–4246, DOI: 10.1002/2016JA022599.

McGovern P. J. and White O. L. (2019) Stress-enhanced ascent of cryomagmas through Pluto's ice shell from nitrogen ice loading of a Sputnik Planitia basin. In *Lunar and Planetary Science L*, Abstract #2994. LPI Contribution No. 2132, Lunar and Planetary Institute, Houston.

McKinnon W. B. (1997) Mystery of Callisto: Is it undifferentiated? *Icarus, 130(2)*, 540–543, DOI: 10.1006/icar.1997.5826.

McKinnon W. B. (2006) On convection in ice I shells of outer solar system bodies, with detailed application to Callisto. *Icarus, 183(2)*, 435–450, DOI: 10.1016/j.icarus.2006.03.004.

McKinnon W. B., Lunine J. I., and Banfield D. (1995) Origin and evolution of Triton. In *Neptune and Triton* (D. P. Cruikshank et al., eds.), pp. 807–877. Univ. of Arizona, Tucson.

McKinnon W. B., Simonelli D. P., and Schubert G. (1997) Composition, internal structure, and thermal evolution of Pluto and Charon. In *Pluto and Charon* (S. A. Stern and D. J. Tholen, eds.), pp. 295–343. Univ. of Arizona, Tucson.

McKinnon W. B., Prialnik D., Stern S. A., and Coradini A. (2008) Structure and evolution of Kuiper belt objects and dwarf planets. In *The Solar System Beyond Saturn* (M. A. Barucci et al., eds.), pp. 213–241. Univ. of Arizona, Tucson.

McKinnon W. B., Nimmo F., Wong T., et al. (2016) Convection in a volatile nitrogen-ice-rich layer drives Pluto's geological vigour. *Nature, 534*, 82–85, DOI: 10.1038/nature18289.

McKinnon W. B., Schenk P. M., Bland M. T., Singer K. N., White O. L., Moore J. M., Spencer J. R., et al. (2018) Pluto's heat flow: A mystery wrapped in an ocean inside an ice shell. In *Lunar and Planetary Science XLIX*, Abstract #2715. LPI Contribution No. 2083, Lunar and Planetary Institute, Houston.

McNutt M. K. (1984) Lithospheric flexure and thermal anomalies. *J. Geophys. Res., 89*, 11180–11194, DOI: 10.1029/JB089iB13p11180.

Mills A. C. and Montesi L. (2018) Determining the elastic thickness of Sputnik Planitia on Pluto and its surrounding using topography and inverse theory. Abstract P311-3832 presented at 2018 Fall Meeting, AGU, Washington, DC, 10–14 December.

Moore J. M. and Pappalardo R. T. (2011) Titan: An exogenic world? *Icarus, 212*, 790–806, DOI: 10.1016/j.icarus.2011.01.019.

Moore J. M., Howard A. D., Schenk P. M., et al. (2015) Geology before Pluto: Pre-encounter considerations. *Icarus, 246*, 65–81, DOI: 10.1016/j.icarus.2014.04.028.

Moore J. M., McKinnon W. B., Spencer J. R., et al. (2016) The geology of Pluto and Charon through the eyes of New Horizons. *Science, 351*, 1284–1293, DOI: 10.1126/science.aad7055.

Moore W. B. and Schubert G. (2000) The tidal response of Europa. *Icarus, 147*, 317–319, DOI: 10.1006/icar.2000.6460.

Mousis O., Guilbert-Lepoutre A., Lunine J. I., Cochran A. L., Waite J. H., Petit J.-M., and Rousselot P. (2012) The dual origin of the nitrogen deficiency in comets: Selective volatile trapping in the nebula and postaccretion radiogenic heating. *Astrophys. J., 757*, 146, DOI: 10.1088/0004-637X/757/2/146.

Mousis O., Lunine J. I., Mandt K. E., Schindhelm E., Weaver H. A., Stern S. A., Waite J. H., Gladstone R., and Moudens A. (2013) On the possible noble gas deficiency of Pluto's atmosphere. *Icarus, 225*, 856–861, DOI: 10.1016/j.icarus.2013.03.008.

Munk W. H. and MacDonald G. J. F. (1960) *The Rotation of the Earth: A Geophysical Discussion*. Cambridge Univ., Cambridge. 323 pp.

Neveu M., Desch S. J., and Castillo-Rogez J. C. (2017) Aqueous geochemistry in icy world interiors: Equilibrium fluid, rock, and gas compositions, and fate of antifreezes and radionuclides. *Geochim. Cosmochim. Acta, 212*, 324–371, DOI: 10.1016/j.gca.2017.06.023.

Niemann H. B., Atreya S. K., Bauer S. J., Carignan G. R., Demick J. E., Frost R. L., Gautier D., et al. (2005) The abundances of constituents of Titan's atmosphere from the GCMS instrument on the Huygens probe. *Nature, 438(7069)*, 779–784, DOI: 10.1038/nature04122.

Nimmo F. and Pappalardo R. T. (2016) Ocean worlds in the outer solar system. *J. Geophys. Res., 121*, 1378–1399, DOI: 10.1002/2016JE005081.

Nimmo F. and Spencer J. R. (2015) Powering Triton's recent geological activity by obliquity tides: Implications for Pluto geology. *Icarus, 246*, 2–10, DOI: 10.1016/j.icarus.2014.01.044.

Nimmo F. and Stevenson D. J. (2000) Influence of early plate tectonics on the thermal evolution and magnetic field of Mars. *J. Geophys. Res., 105(E5)*, 11969–11980, DOI: 10.1029/1999JE001216.

Nimmo F., Hamilton D. P., McKinnon W. B., et al. (2016) Reorientation of Sputnik Planitia implies a subsurface ocean on Pluto. *Nature, 540*, 94–96, DOI: 10.1038/nature20148.

Nimmo F., Umurhan O., Lisse C. M., Bierson C. J., Lauer T. R., Buie M. W., Throop H. B., et al. (2017) Mean radius and shape of Pluto and Charon from New Horizons images. *Icarus, 287*, 12–29, DOI: 10.1016/j.icarus.2016.06.027.

Protopapa S., Grundy W. M., Reuter D. C., Hamilton D. P., Dalle Ore C. M., Cook J. C., Cruikshank D. P., et al. (2017) Pluto's global surface composition through pixel-by-pixel Hapke modeling of New Horizons Ralph/LEISA data. *Icarus, 287*, 218–228, DOI: 10.1016/j.icarus.2016.11.028.

Robuchon G. and Nimmo F. (2011) Thermal evolution of Pluto and implications for surface tectonics and a subsurface ocean. *Icarus, 216*, 426–439, DOI: 10.1016/j.icarus.2011.08.015.

Schenk P., Beyer R., Moore J., McKinnon W., Spencer J., Stern S., Olkin C., Ennico K., and Weaver H. (2018a) High-resolution topography of Pluto and Charon: Getting down to details. In *Lunar and Planetary Science XLIX*, Abstract #2300. LPI Contribution No. 2083, Lunar and Planetary Institute, Houston.

Schenk P. M., Beyer R. A., McKinnon W. B., et al. (2018b) Basins, fractures and volcanoes: Global cartography and topography of Pluto from New Horizons. *Icarus, 314*, 400–433, DOI: 10.1016/j.icarus.2018.06.008.

Schmitt B., Philippe S., Grundy W. M., Reuter D. C., Côte R., Quirico E., Protopapa S., et al. (2017) Physical state and distribution of materials at the surface of Pluto from New Horizons LEISA imaging spectrometer. *Icarus, 287*, 229–260, DOI: 10.1016/j.icarus.2016.12.025.

Schubert G., Anderson J. D., Spohn T., and McKinnon W. B. (2004) Interior composition, structure and dynamics of the Galilean satellites. In *Jupiter: The Planet, Satellites and Magnetosphere* (F. Bagenal et al., eds.), pp. 281–306. Cambridge Univ., Cambridge.

Scipioni F., Schenk P., Tosi F., D'Aversa E., Clark R., Combe J.-P., and Dalle Ore C. M. (2017) Deciphering sub-micron ice particles on Enceladus surface. *Icarus, 290*, 183–200, DOI: 10.1016/j.icarus.2017.02.012.

Simonelli D. P. and Reynolds R. T. (1989) The interiors of Pluto and Charon: Structure, composition, and implications. *Geophys. Res. Lett., 16(11)*, 1209–1212, DOI: 10.1029/GL016i011p01209.

Singer K. N. and McKinnon W. B. (2011) Tectonics on Iapetus: Despinning, respinning, or something completely different? *Icarus, 216*, 198–211, DOI: 10.1016/j.icarus.2011.08.023.

Solomatov V. S. (1995) Scaling of temperature- and stress-dependent viscosity convection. *Phys. Fluids, 7*, 266–274, DOI: 10.1063/1.868624.

Stern S. A. (1989) Pluto: Comments on crustal composition, evidence for global differentiation. *Icarus, 81*, 14–23, DOI: 10.1016/0019-1035(89)90121-8.

Stern S. A., Bagenal F., Ennico K., et al. (2015) The Pluto system: Initial results from its exploration by New Horizons. *Science, 350*, aad1815, DOI: 10.1126/science.aad1815.

Stern S. A., Tapley M. B., Finley T. J., and Scherrer J. R. (2020) Pluto orbiter–Kuiper belt explorer: Mission design for the gold standard. *J. Spacecr. Rockets, 57(5)*, 956–963, DOI: 10.2514/1.A34658.

Stevenson D. J. (2000) Limits on the variation of thickness of Europa's ice shell. In *Lunar and Planetary Science XXXI*, Abstract #1506. LPI Contribution No. 1000, Lunar and Planetary Institute, Houston.

Tholen D. J. (2014) The size of Pluto. *AAS/Division for Planetary Sciences Meeting Abstracts, 46*, 404.01.

Tobie G., Lunine J. I., and Sotin C. (2006) Episodic outgassing as the origin of atmospheric methane on Titan. *Nature, 440(7080)*, 61–64, DOI: 10.1038/nature04497.

Trowbridge A. J., Melosh H. J., Steckloff J. K., and Freed A. M. (2016) Vigorous convection as the explanation for Pluto's polygonal terrain. *Nature, 534*, 79–81, DOI: 10.1038/nature18016.

Vance S., Harnmeijer J., Kimura J., Hussmann H., DeMartin B., and Brown J. M. (2007) Hydrothermal systems in small ocean planets. *Astrobiology, 7*, 987–1005, DOI: 10.1089/ast.2007.0075.

Waite J. H., Lewis W. S., Magee B. A., Lunine J. I., McKinnon W. B., Glein C. R., Mousis O., et al. (2009) Liquid water on Enceladus from observations of ammonia and ^{40}Ar in the plume. *Nature, 460*, 487-490, DOI: 10.1038/nature08153.

Watts A. B. (2001) *Isostasy and Flexure of the Lithosphere*. Cambridge Univ., Cambridge. 478 pp.

White O. L., Schenk P. M., Bellagamba A. W., Grimm A. M., Dombard A. J., and Bray V. J. (2017) Impact crater relaxation on Dione and Tethys and relation to past heat flow. *Icarus, 288*, 37–52, DOI: 10.1016/j.icarus.2017.01.025.

Yamashita Y., Kato M., and Arakawa M. (2010) Experimental study on the rheological properties of polycrystalline solid nitrogen and methane: Implications for tectonic processes on Triton. *Icarus, 207*, 972–977, DOI: 10.1016/j.icarus.2009.11.032.

Zannoni M., Hemingway D., Gomez Casajus L., and Tortora P. (2020) The gravity field and interior structure of Dione. *Icarus, 345*, 113713, DOI: 10.1016/j.icarus.2020.113713.

Zimmer C., Khurana K. K., and Kivelson M. G. (2000) Subsurface oceans on Europa and Callisto: Constraints from Galileo magnetometer observations. *Icarus, 147*, 329–347, DOI: 10.1006/icar.2000.6456.

Zolotov M. Y. (2007) An oceanic composition on early and today's Enceladus. *Geophys. Res. Lett., 34*, L23203, DOI: 10.1029/2007GL031234.

Moore J. M. and Howard A. D. (2021) The landscapes of Pluto as witness to climate evolution. In *The Pluto System After New Horizons* (S. A. Stern, J. M. Moore, W. M. Grundy, L. A. Young, and R. P. Binzel, eds.), pp. 105–120. Univ. of Arizona, Tucson, DOI: 10.2458/azu_uapress_9780816540945-ch006.

The Landscapes of Pluto as Witness to Climate Evolution

Jeffrey M. Moore
NASA Ames Research Center

Alan D. Howard
Planetary Science Institute

The New Horizons mission revealed a dynamic planet whose complex landscape is dominated by volatile redistribution controlled by climatic cycles on seasonal and Milankovitch-like scales, superimposed upon probably non-cyclic long-term (multi-gigayear-scale) trends in the surface-atmosphere volatile inventory. Many of these volatile-related landforms have familiar analogs on Earth and Mars despite different ices and a much colder plutonian environment. Landforms include glaciers, valleys possibly at least in part formed by fluid flow, and formation of condensation deposits that then develop ridges, pinnacles, and pits through sublimation controlled by variations in solar illumination. The details of the formative processes controlling the scale and morphology of the landforms are poorly constrained, but mechanistic modeling may help to clarify them. A more complete understanding will await future missions giving a global temporal picture of landforms and processes and further evidence for their variability.

1. INTRODUCTION

Pluto is one of a half dozen worlds in the solar system known to have evolving surface-atmosphere-mediated climates. Climate and climate evolution are of intense interest because of their implications for planetary habitability, surface activity, and fate. Worlds with atmospheres and solid surfaces invariably have landscapes that manifest the effects of climate evolution (e.g., *Mackwell et al.,* 2013). Pluto is certainly no exception. The two most important abundant volatiles on Pluto's surface and atmosphere (methane and nitrogen) are close to their phase change temperatures, so these transformations drive its geological processes, much as H_2O does on Earth. Water is also abundant on Pluto but is so far below its melting point there that it is considered a refractory solid, like silicate rock in Earth's crust. In this chapter we will characterize Pluto's diverse landforms and terrains, and what they tell us about the past and present climate over the last 4 G.y.

We have organized our review of landforms, and the landscapes in which they are situated, from the youngest to the oldest features. We necessarily rely on crater retention and superposition to determine relative ages as reported in the chapters by White et al. and Singer et al. in this volume. Likewise, chapters by Forget et al., Umurhan et al., and Young et al. review the physical, rheological, and chemical mechanisms and processes of atmospheric and climate evolution. As the White et al. chapter provides detailed observations and descriptions of landforms and landscapes, we will only provide observation and description summaries, specifically as they illuminate formation mechanisms and implications for climate evolution.

We conclude with a discussion of the several timescale cycles of climate evolution that we infer for the history of Pluto, and a few tests for future investigations.

2. OBSERVATIONS

2.1. Young Landforms

This discussion focuses on landforms that are inferred to largely reflect climate variations ranging over timescales from one Pluto year (248 Earth years) to the last few obliquity-precession cycles (~3 m.y. each) (*Earle and Binzel*, 2015; *Earle et al.*, 2017).

The youngest, indeed currently active, unambiguously recognized landforms on Pluto are the flow and sublimation features found in and around Sputnik Planitia and East Tombaugh Regio, along with the convection cells on Sputnik Planitia. The convection cells on Sputnik Planitia are young constructional landforms, but they are of endogenic origin and discussed elsewhere in this volume (see the chapters by White et al. and Umurhan et al.).

2.1.1. Pits within Sputnik Planitia. Sputnik Planitia is an enormous (800 × 1450 km) deposit of primarily nitrogen mixed with small amounts of CO and CH_4 ices that fills an impact basin, and it represents the main known surface reservoir of nitrogen ice on Pluto (*Stern et al.*, 2015; *Moore et al.*, 2016).(Some place names used here, such as Sputnik Planitia, are official. Others are informal, such as Cthulhu Macula. See also Appendix A of this volume for the location of named features.) Consisting of a giant expanse of plains, the nearly level surface of Sputnik Planitia is a reflection of the low viscosity of its constituent ices (*Eluszkiewicz and Stevenson*, 1990; *Yamashita et al.*, 2010), which flow readily at obliquity-cycle timescales and causes Sputnik Planitia to be most likely an equipotential surface (*Schenk et al.,* 2018). The plains display two textural elements: (1) polygonal cells tens of kilometers across with well-defined edges, and (2) abundant pits 300–500 m in width, up to several kilometers in length, generally superimposed on the cells (Fig. 4 in the chapter by White et al.). The geometry and size of the cells are consistent with convective overturn of Sputnik Planitia nitrogen ice to a depth of hundreds of meters driven by the geothermal heat flux (*McKinnon et al.*, 2016; *Trowbridge et al.*, 2016; see the chapter by White et al.). Pit morphology ranges from space-filling pitting to isolated, shallow depressions. A few localities on Sputnik Planitia appear to be free of pits and other regions, particularly in the north, are imaged at a resolution at which pit recognition is problematic. The volatility of N_2 ice is consistent with pits forming through enhanced sublimation in depressions due to reflected solar illumination (*Moore et al.*, 2017), analogous to penitentes on terrestrial high-altitude, tropical glaciers (*Amstutz*, 1958; *Bergeron et al.*, 2006; *Claudine et al.*, 2015). The scale and morphology of the pits may evolve through time into an equilibrium shape (*Buhler and Ingersoll*, 2017). Alternatively, the scale of the pits may be determined by a balance between sublimation deepening and N_2 ice flow infilling. The small size and depth of the pits on Sputnik Planitia are consistent with formation in N_2 ice, which, in solid solution with CO ice, is the most volatile of the major ices on Pluto's surface. *Moore et al.* (2017) concluded that fields of pits in Sputnik Planitia, both large and small, are landscapes in which the balance between sublimation and N_2 ice flow drives their formation and evolution. Local depressions receive reflected light energy from surrounding slopes, enhancing sublimation rates on depression floors, leading to a runaway evolution of a deeply pitted surface. This feedback is analogous to terrestrial penitentes, in which the surface is incredibly rough at the scale of a few meters with knife-edged divides and rounded pits. In contrast, however, the pits on Pluto have lower slope gradients and are considerably larger than their terrestrial analogs. That difference suggests that the scale and relief of the pitting are set by the competition between depression-focused sublimation tending to deepen pits, balanced by infilling by diffusive N_2 ice flow. The heuristic model of pit formation within *Moore et al.* (2017) closely mimics the form, spacing, and arrangement of a variety of Sputnik Planitia's pits. In their model, the temporal evolution of pitted surfaces is such that considerable time passes with little happening, followed eventually by very rapid development of relief and rapid sublimation (Fig. 1).

Sublimation pits are typically 300–500 m across and on the order of 50 m deep, based upon preliminary photoclinometric techniques (*Beyer et al.,* 2019). The maximum sublimation net loss over the 2.8 m.y. obliquity cycle is about 1 km (*Bertrand et al.*, 2018). If pit development occurred linearly in time during the cycle, then a 50-m-deep pit could form on the order of 140 k.y. (or about 480 Pluto years). But this timescale is probably a minimum, because simulations (*Moore et al.*, 2017) suggest a substantial lag time for pits to deepen appreciably from an initially low-relief surface. Divides between pits also presumably sublimate as well, further increasing the timescale for pit development. Thus pit development and erasure likely relate to obliquity timescales of a few million years, and individual pits might persist through more than one obliquity cycle. This formation timescale is order-of-magnitude consistent with the

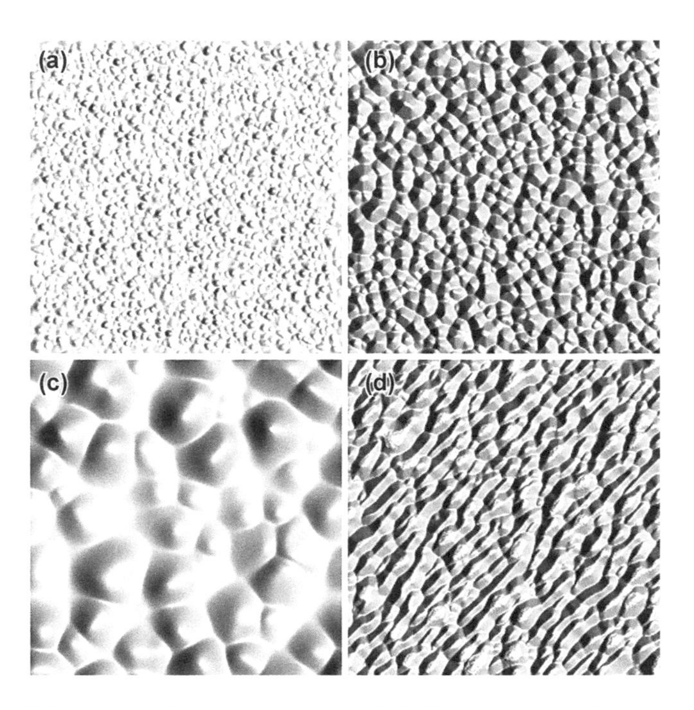

Fig. 1. Landform evolution models illustrating the formation of sublimation pits in soft nitrogen ice. Sublimation pits develop as a competition between focusing of light energy at the floor of depressions, which tends to deepen pits, and infilling by flow of nitrogen ice, which tends to flatten pits and destroy small pits. **(a)** Early stage of pit formation. **(b)** Late stage of pit development in which pits reach an equilibrium size with the characteristic scale determined by the balance between solar illumination intensity and ice diffusivity. **(c)** Increasing N_2 diffusivity relative to illumination by a factor of 10 increases the equilibrium pit size. **(d)** Elongated pit chains are created when slopes facing to upper left sublimate 3× more rapidly than comparable slopes facing in other directions. All model landscapes are shaded by illumination from the left to show relief. Adapted from *Moore et al.* (2017).

500-k.y. to 1-m.y. overturn times for the large convection cells in Sputnik Planitia (*McKinnon et al.*, 2016; *Trowbridge et al.*, 2016), although the rate of surface overturn could be much longer if the situation is that of sluggish-lid convection (*McKinnon et al.*, 2016).

The sublimation pitting on large portions of Sputnik Planitia has a high albedo and thus appears to have developed into ices that are relatively free of refractory residues with lower albedo (e.g., dust, dark red material). The floors of some deep pits on Sputnik Planitia, however, are distinctly dark and relatively flat (Fig. 5d in the chapter by White et al.). Some of these dark-floored pits are seen in the marginal areas of Sputnik Planitia, where the nitrogen ice is expected to be thinner than in the center. One possibility is that deep sublimation is accompanied by accumulation of dark refractory material, perhaps swept in by wind. Alternatively, sublimation of the nitrogen ice here has exposed a distinct, dark-albedo layer with reduced susceptibility to sublimation. In particular, extensive fields of dark-bottomed pits are seen in western Sputnik Planitia directly adjacent to the dark highlands of Cthulhu Macula, and the floors may be exposing such highlands terrain, which appear to be covered by relatively shallow ice here.

However, this is likely not the case for dark-bottomed pits seen within the cellular plains closer to the center of Sputnik Planitia, where the nitrogen ice may reach kilometers thick, based on considerations of solid-state convection here (*McKinnon et al.*, 2016; *Moore et al.*, 2016), too thick for sublimation to deepen pits to the base of the nitrogen ice. Dark-bottomed pits here invariably concentrate in the troughs that mark the cell boundaries (e.g., Fig. 4a in White et al. chapter). This distribution raises the possibility that the dark material here originates from atmospheric fallout and becomes trapped through saltation within these troughs, or alternatively, that the sublimation forming the pits may be liberating refractory dark red material entrained within the nitrogen ice that builds up on the pit floors (*White et al.*, 2017) (see section 2.1.2). In either case, the low albedo may initially cause preferential heating of the ice underneath it, and therefore may increase sublimation and pit growth rates, at least until the dark floor material is thick enough to provide an insulating lag, or else further pit growth is suppressed by inward flow of soft N_2 ice.

The climatic implication of the dark-floored pits may be that there are long-term (obliquity-scale or longer) alternations between net sublimation on Sputnik Planitia (with surface accumulation of non-volatile components such as dark red material hypothesized to be haze deposits) and episodes of net accumulation of N_2 and CO, implying that the current conditions are undergoing net sublimation (see Fig. 6c of *Bertrand et al.*, 2018). Most of the dark-floored pits on Sputnik Planitia are between 10°N and 10°S, which under orbital conditions during the last few hundred Pluto years is a latitudinal zone of net sublimation (*Bertrand et al.*, 2018).

2.1.2. Latitudinal variation across Sputnik Planitia. Sputnik Planitia displays notable variations in albedo, topography, and pit morphology across its great latitudinal extent, which ranges between 50°N and 23°S. These variations are in part the result of Sputnik Planitia extending across very different climate zones, from the permanent Arctic in its northern reaches to the permanent diurnal zone that covers much of its southern half (*Binzel et al.*, 2017) (Fig. 2). The different average insolation experienced within these zones affects the average surface thermal balance within them (*Earle et al.*, 2017), and accordingly the magnitude of sublimation and condensation of volatile ices across the surface of the plains. *White et al.* (2017) identified plains with three distinct albedo characteristics: consistently bright plains covering much of Sputnik Planitia south of 30°N; darker, patchier plains north of 30°N; and the darkest plains at the northern boundary of Sputnik Planitia, where they concentrate around the troughs defining cell boundaries. Mapping of the boundary between the bright and darker plains (Fig. 3 in the chapter by White et al.) indicates that the bright plains occur where a thin veneer of fresh nitrogen ice has condensed onto the plains, superposing darker ice that forms the bulk of Sputnik Planitia (*White et al.*, 2017). The low albedo of the darker ice is most likely due to entrainment of dark haze particles settling from Pluto's atmosphere over the course of Sputnik Planitia's history, and which are also responsible for the very low albedo of the maculae in Pluto's equatorial regions (see section 3.1). This interpretation of the bright and dark plains is supported by their correlation with present climate zones: 37°N corresponds to the present Arctic Circle. The bright plains are therefore contained within the present diurnal zone south of this latitude, whereas the dark plains are mostly contained within the Arctic zone to the north of that latitude (with the exception of an expanse of dark plains that extends down Sputnik Planitia's western margin; see section 2.1.3). Climate modeling indicates that the low-insolation, diurnal portion of Sputnik Planitia should currently, during northern spring/summer, be a zone of net condensation of nitrogen ice, forming the bright surface veneer, and that the high-insolation, Arctic portion should be a zone of net sublimation of nitrogen ice, meaning that no veneer is deposited here and the darker ice intrinsic to Sputnik Planitia is left exposed (*Bertrand et al.*, 2018). Given that the boundary between the Arctic and diurnal zones oscillates in accordance with Pluto's ~2.8 m.y. obliquity cycles, the boundary between the dark and bright plains in northern Sputnik Planitia is also expected to oscillate between 13°N and 37°N. The present Antarctic Circle is at 30°S, meaning that all of southern Sputnik Planitia is covered by this bright veneer. But the Antarctic Circle will reach 23°S, the southernmost extent of Sputnik Planitia, in <1.5 m.y., after which it will travel further toward the equator, with the high insolation associated with the expanded Antarctic zone possibly removing the surface veneer from southern Sputnik Planitia and exposing darker ice beneath (see Figs. 3 and 4 in the White et al. chapter).

In the northern latitudes of Sputnik Planitia, pitting is dense to the point of universal where imaged at sufficient resolution. North of 35°, however, the best image resolution

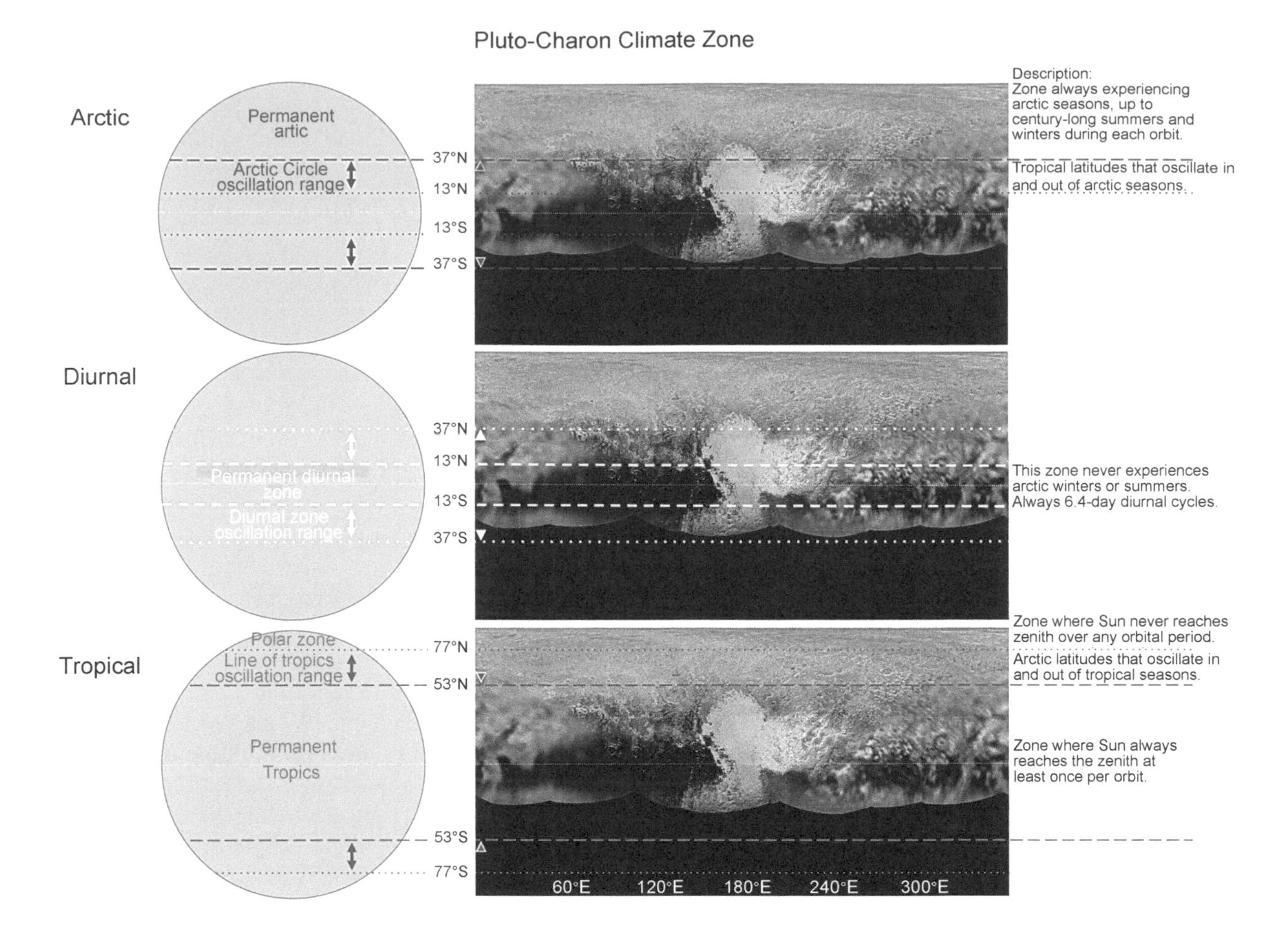

Fig. 2. Pluto "climate zones" as defined by astronomical cycles. For clarity these zones are depicted in three panels for each of the predominant seasonal effects. Pluto's New Horizons base map is shown for reference. The vertical double arrows show the range of oscillation for each boundary created by the 2.8-Ma period of Pluto's obliquity cycle, having a mean obliquity of 115° and an amplitude of ±12°. The base of each triangle (superimposed on the maps) indicates the current location of each zone boundary, while pointing in the direction of the boundary's current migration. Explanatory labels appear at right; where written to describe northern latitudes, note that they apply identically to their southern hemisphere analogs. From *Binzel et al.* (2017).

(~380 m/pixel) is approximately equal to the spatial scale of pitting, but high-frequency filtering of imaging suggests that the dense pitting extends to the northern extent of Sputnik Planitia at ~50°. Between 35°N and 17°S, high-resolution imaging (80–125 m/pixel) reveals the scale, morphology, and density of pitting as a function of latitude. Dense pitting extends southward to about 15°N. Between 15°N and 10°N, shallow, sparse pitting begins to appear near convective cell boundaries. At about 6°N, large expanses of sparse shallow pitting become prominent. In this same equatorial belt, broad pits with dark flows occur along cell margins, as do long, nearly parallel "bacilli" pits that also commonly have dark floors (see Fig. 5d in the White et al. chapter). These features are often associated with the outer margins of N_2 glacial flows debouching onto Sputnik Planitia from East Tombaugh Regio. This pattern persists until the southern limit of Sputnik Planitia at about 23°S. Over the last 500 k.y. the equatorial latitudes have received >50 m of net deposition of N_2-rich ices, and as much as 200 m over the last 2.8-m.y. obliquity cycle (*Bertrand et al.*, 2018). This accumulation suggests that sublimation pitting may be inhibited or degraded by high rates of ice deposition, whereas pit development is able to persist or initiate in more northerly latitudes, where up to 1 km of N_2-rich ices sublimate over the obliquity cycle (*Bertrand et al.*, 2018).

The contrast in net sublimation and condensation between the equatorial and high-latitude regions of Sputnik Planitia has also influenced the large-scale topography of the plains. Along the northern margin of Sputnik Planitia, the outer 10–20 km of the nitrogen ice sheet are depressed a few hundred meters relative to the interior of the ice sheet (*Schenk et al.*, 2018). These depressed plains roughly coincide with the occurrences of dark, trough-bounding plains, which display evidence of lateral flow toward the northern edge of Sputnik Planitia (Figs. 4e and 4f in the chapter by White et al.), and the two may be related to the especially high sublimation rate of nitrogen ice within this northern region, which experiences the highest insolation of any part of Sputnik Planitia (*Bertrand et al.*, 2018). Such sublimation has been modeled to have removed a thickness of ~1 km of nitrogen from the northern edge of Sputnik Planitia during the last 2 m.y. (*Bertrand et al.*, 2018). This loss may account for the depressed plains and northwards lateral flow of the ice, as the low-viscosity nitrogen ice flows to fill the void left by the sublimated ice, as well as the especially dark plains, as the enhanced sublimation of nitrogen ice causes the concentration of entrained haze particles to increase to levels higher than anywhere else in Sputnik Planitia (*White et al.*, 2017). The lower latitude of the southernmost edge of Sputnik Planitia relative to the northernmost suggests that less sublimation has taken place than in the north across a similar timescale, but the ice sheet here does still reach >1 km below the interior plains (*Schenk et al.*, 2018), indicating southward flow (*Howard et al.*, 2017a). Corresponding evidence of southward flow within Sputnik Planitia's southern region includes chains of pits marking the outlines of inactive cells that have been moved away from the zone of active convection (*White et al.*, 2017) (see also Fig. 5e in the chapter by White et al.) and the common north-south alignment of pits within southern Sputnik Planitia, which may be caused by anisotropic stress conditions related to southward flow of the ice (*Moore et al.*, 2017). These various observations indicate that nitrogen ice flows from the center of Sputnik Planitia toward its northern and southern edges, in accord with modeling by *Bertrand et al.* (2018).

2.1.3. Streaks and dune-like forms. South of the Arctic Circle (30°N), Sputnik Planitia mostly presents a high albedo, which has been interpreted to indicate recent condensation of a veneer of bright nitrogen-dominated ice onto the plains (*White et al.*, 2017; *Bertrand et al.*, 2018), with the dark cellular plains to the north of this latitude indicating where prevailing sublimation has exposed dark red material entrained in nitrogen ice (*White et al.*, 2017; *Bertrand et al.*, 2018; see the chapter by White et al.). A band of dark cellular plains extending south of 30°N down the western margin of Sputnik Planitia (Fig. 3 in the chapter by White et al.), however, indicates that factors besides latitude also affect nitrogen sublimation and condensation. This region of Sputnik Planitia is proximal to the large mountain ranges of Al-Idrisi and Baret Montes, which can display topographic relief reaching several kilometers. High-resolution climate modeling of Sputnik Planitia with this regional topography has confirmed that windier conditions occur here, and it can explain the extension of the dark cellular plains south of 30°N, because the strong winds at the surface prevent the establishment of a stable, consistent veneer of bright nitrogen ice (*Bertrand et al.*, 2020). Other surface features attest to the prevalence of stronger winds relative to areas that are further to the interior of Sputnik Planitia. These features include dark streaks (Fig. 3) that are often seen in the lee of isolated mountain blocks, and which are interpreted to be wind streaks (*Stern et al.*, 2015; *Telfer et al.*, 2018; *Bertrand et al.*, 2020). In addition, the pitted appearance of the plains transitions to a distinct texture of aligned, regularly spaced (~0.4–1 km), linear ridges (Fig. 3) close to the boundary between Sputnik Planitia and Al-Idrisi Montes (*Telfer et al.*, 2018). These ridges are oriented orthogonally to nearby wind streaks and are comparable in morphology to terrestrial transverse dune fields. *Telfer et al.* (2018) hypothesize that the ridges are dunes composed of sand-sized particles of methane ice that have been lofted from the surface of Sputnik Planitia by sublimation of N_2, transported, and then deposited. Modeled wind speeds under present conditions appear to be unable to directly entrain sand-sized particles, although they could be transported by saltation once lofted by some other mechanism such as upward nitrogen gas flux from sublimation. These features, if indeed they are eolian dunes, could also be relics of an epoch of higher surface pressure at other times during the quasi-periodic Milankovitch climate cycle, when entrainment would be more likely (*Stern et al.*, 2017).

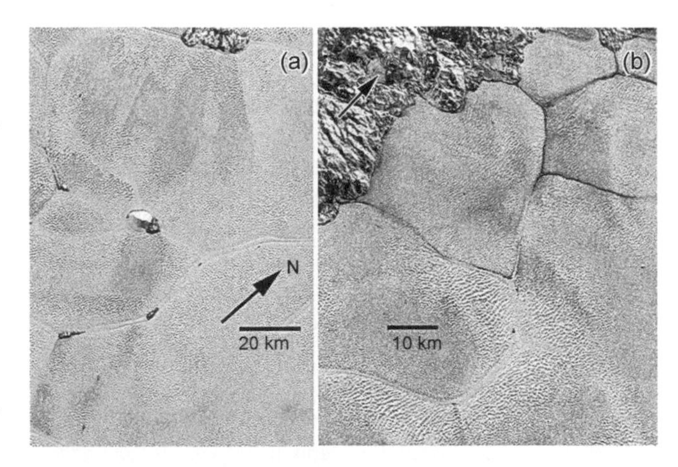

Fig. 3. Eolian features on Pluto. **(a)** Wind streaks generally trending south-southeastward behind topographic obstacles in western Sputnik Planitia (excerpted from the 117 m/pixel P_MPAN_1 LORRI observation, centered on ~19°N, 166°E). **(b)** Fields of low, similarly sized ridges exhibiting a highly regular spatial patterning, which *Telfer et al.* (2018) attribute to aeolian dune formation located in western Sputnik Planitia (excerpted from the 76 m/pixel P_MVIC_LORRI_CA LORRI observation, centered on ~34°N, 160°E).

2.1.4. Methane ice coatings on equatorial mountain peaks and crater rims. The deposits of bright methane frost seen on the peaks of Pigafetta Montes and the north-facing slopes of impact craters in their vicinity indicate that the low albedo of the accumulations of dark red deposits (hypothesized to be atmospherically deposited) does not entirely preclude condensation of methane (*Moore et al.*, 2016; *Grundy et al.*, 2018). These haze particles settle out over the entirety of Pluto's surface, at a rate that is fairly rapid compared to geological timescales (but not volatile transport timescales): A 35-cm coating can be accumulated in ~100 m.y. (*Grundy et al.*, 2018). They can only accumulate to form continuous haze deposits, however, on surfaces that are not subject to regular, ongoing volatile mobilization.

This process limits their accumulation to regions within the diurnal zone, with its low, and minimally varying, insolation and low surface temperatures (*Hamilton et al.*, 2016; *Binzel et al.*, 2017; *Earle et al.*, 2017). These regions are not already occupied by the bladed terrain deposits or the nitrogen-rich ice of Sputnik Planitia. The peaks of Pigafetta Montes extend 2 km or more above surrounding terrain (Fig. 4). The warmer atmospheric temperatures at higher altitude inhibit nitrogen condensation but permit methane deposition on high-elevation surfaces such as the bladed terrain and Pigafetta Montes. Upon establishment of a thin methane frost, the albedo of the surface is raised, reducing the surface temperature and encouraging further condensation of methane onto the deposit slopes (*Bertrand and Forget*, 2019). The occurrence of the methane deposits on lower-elevation, north-facing crater rims may instead be an insolation effect. All surfaces within the permanent diurnal zone receive comparatively little insolation relative to higher latitudes across a plutonian year, but during Pluto's northern fall and winter (which extended, most recently, from 1865 to 1989), these north-facing slopes would have received less insolation than the south-facing slopes, with temperatures being low enough to cold-trap methane frost. Given that the northern hemisphere is now entering summer, these deposits are expected to sublimate away over the coming decades, with fresh methane deposits instead appearing on south-facing slopes (*Bertrand and Forget*, 2019).

2.1.5. N_2 ice deposition and glaciation in East Tombaugh Regio. Like the bright plains of Sputnik Planitia, the bright, pitted uplands of East Tombaugh Regio (Fig. 5; also Figs. 9 and 10b in the chapter by White et al.) extend roughly to the present Arctic Circle at 30°N. They appear to represent a continuation of the zone of recent condensation of a veneer of bright nitrogen-rich ice across the landscape, in this case an expanse of relatively low-elevation bladed and reticulate terrain deposits formed of methane ice (and probably also arcuate terrain; see section 2.3.2 in the chapter by White et al.), found on the hilly northern region of the bright, pitted uplands. The climate simulations of *Bertrand et al.* (2019) show that relatively bright methane deposits can create cold traps for nitrogen ice outside of Sputnik Planitia, but the closer they are to the equator, the more long-lived are the nitrogen ice reservoirs. The bladed and reticulate methane deposits located along Sputnik Planitia's eastern border are therefore a natural location to host a long-lived reservoir of condensed nitrogen ice that had sublimated from Sputnik Planitia. This condensed, low-viscosity nitrogen ice has extensively modified and eroded these deposits through glacial flow: Active N_2-rich ice glaciation along the eastern boundary of Sputnik Planitia is indicated by flow lines converging to 2–5-km-wide troughs and spreading downstream onto Sputnik Planitia, implying that nitrogen ice is being funneled through the troughs from upland plateaus onto Sputnik Planitia ~1 km below (*Howard et al.*, 2017a) (see Figs. 6a,b in the chapter by White et al.). The nitrogen ice that had sublimated from Sputnik Planitia and condensed onto East Tombaugh Regio is therefore glacially flowing back into Sputnik Planitia, forming a "nitrogen cycle" between Sputnik Planitia and East Tombaugh Regio (*Moore et al.*, 2016; *Howard et al.*, 2017a). The higher-elevation bladed terrain deposits farther east have formed at an elevation where only methane ice is stable (*Moore et al.*, 2018; *Bertrand et al.*, 2019), and so they remain comparatively untouched by nitrogen ice precipitation, with the exception of lower-lying areas adjacent to the troughs and at the boundaries of blade fields.

Moore et al. (2018) suggest that the west-to-east sequence of landform elements from the lowlands of Sputnik Planitia, through the bright pitted uplands, to the bladed terrain are genetically related and may be driven by N_2-CO sublimated from Sputnik Planitia and condensed (and further modified) on the uplands to the east. The bright pitted uplands and the bladed terrain methane ice deposits occupy the same latitude belt and manifest a surficial compositional sequence from dominance by N_2 ice closest to Sputnik Planitia (including return-flow N_2 glaciation) to increasing dominance of CH_4 ice to the east, culminating in the bladed terrain CH_4 ice deposits. This compositional sequence corresponds to an altitudinal control on ice stability, with only CH_4 being stable at high relative elevations. However, modeling by *Bertrand et al.* (2019) suggests that CH_4 sublimating off the bladed terrain deposits is instead transported westward by a retrograde zonal flow, and due to downward winds triggered by the eastern high-relief boundary of Sputnik Planitia basin, initially resulting in deposits of bright CH_4 frost on East Tombaugh Regio. Thus brightened and hence chilled, East Tombaugh Regio subsequently cold-traps N_2

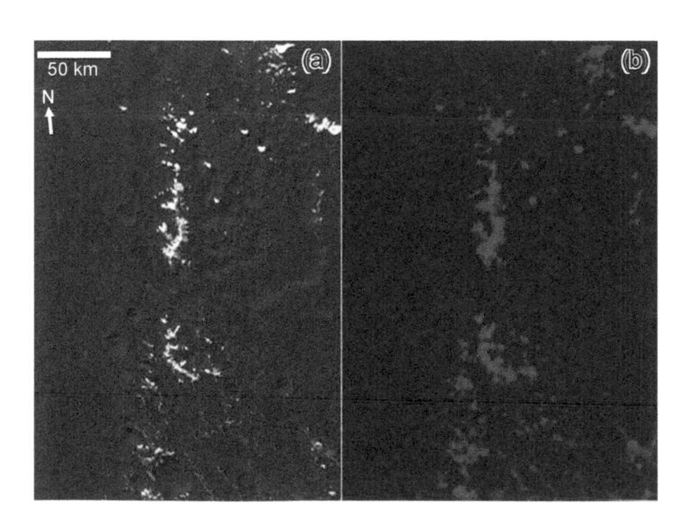

Fig. 4. **(a)** An oblique view of Pigafetta Montes (formerly informally named Enrique Montes) and craters with bright thin (probably seasonal) deposits of CH_4 ice on north-facing crater rims within Cthulhu Macula, seen in an enhanced Multispectral Visible Imaging Camera (MVIC) color image with a pixel scale of 680 m/pixel. Image is centered at 147.0°E, 7.0°S. **(b)** MVIC CH_4 spectral index map of the same scene, with purple indicating CH_4 absorption. The bright-capped summits of the Montes and crater rim deposits in **(a)** correlate to strong CH_4 absorption in **(b)**. Adapted from *Moore et al.* (2018).

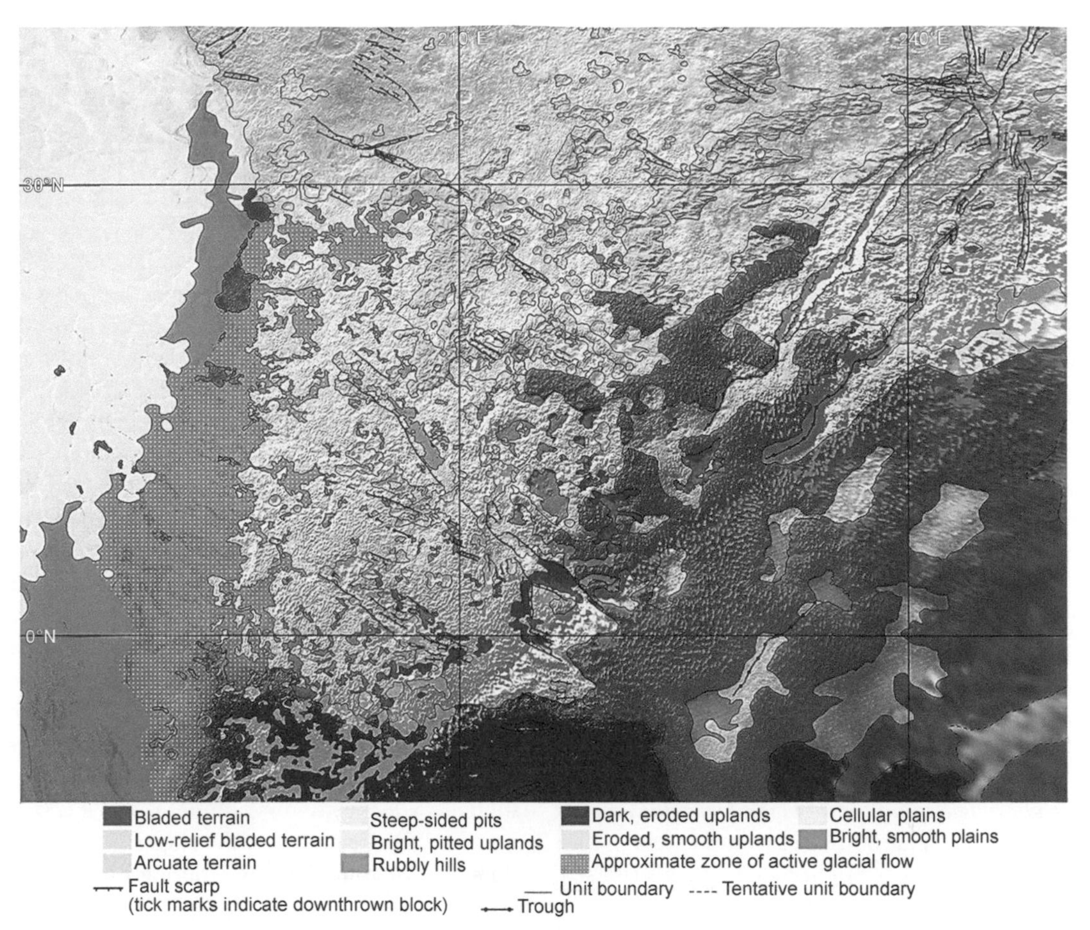

Fig. 5. Climate-related deposits and features in eastern Tombaugh Regio. Geological map of the encounter hemisphere region that includes eastern Sputnik Planitia, East Tombaugh Regio, southern Hayabusa Terra, and the bladed terrain deposits (BTD) of Tartarus Dorsa. Map is overlain on the global mosaic (Fig. 1a in the chapter in this volume by White et al.) and is centered at 14.7°N, 215.7°E. Tectonic lineations mapped in *McGovern et al.* (2019) are indicated. Bladed terrain: Roughly evenly spaced, often subparallel sets of sharp, steep ridges situated on high ground of the BTD, which display a strong methane spectral signature. The ridges tend to be oriented north-south. Impact craters are virtually absent. Low-relief bladed terrain: Closely associated with the bladed terrain, but displays a lower-amplitude bladed texture and sometimes a non-bladed, rubbly texture at the several-decameter scale of the images. Shows lobate boundaries at the northern edge of the bladed terrain deposits. Lower in elevation than the bladed terrain and tends to appear within depressions surrounded by the bladed terrain. Arcuate terrain: High-relief terrain showing pronounced arcuate scarps, with gentle, convex, north-facing slopes and steeper, concave, south-facing slopes. The scarp crests are dominantly oriented east-west. *Dark, eroded uplands:* Low-albedo, rugged terrain displaying a rubbly texture and pervasive surface collapse, forming pit complexes tens of kilometers across. Few obvious impact craters are observed. Eroded, smooth uplands: Broadly contoured uplands over distances of tens of kilometers, interrupted by pits, lowlands, and ancient crater basins. Displays a rough texture at decameter to kilometer scale. Steep-sided pits: Pits typically reaching >10 km across that form deep (up to ~3 km), conical depressions. They tend to cluster to form complexes, with the pits being separated by sharp ridges. These complexes are typically aligned along structural trends, including fault scarps, and can form northwest-southeast-trending trouvghs, especially within East Tombaugh Regio and southern Hayabusa Terra. Bright, pitted uplands: High-albedo, rugged terrain featuring individual pits that reach 5–10 km across, and that form a network of sharp-crested, reticulate ridges. The pits can locally intersect to form distinct northeast-southwest-trending ridge-and-trough terrain. Impact craters are virtually absent. Rubbly hills: Hills scattered across eastern Sputnik Planitia that reach a few kilometers across and tend to collect into densely packed conglomerations reaching tens of kilometers across. Cellular plains: High-albedo plains within central and northern Sputnik Planitia that display a finely pitted texture and are divided into ovoid cells by a network of troughs. Impact craters are absent. *Bright, smooth plains:* High-albedo, generally smooth but sometimes slightly hummocky plains that occur at the eastern margin of Sputnik Planitia and on the floors of basins within the bright pitted uplands and the bladed and low-relief bladed terrain. Displays a lightly pitted texture, like the cellular plains, with pits reaching <1 km in diameter. Dark flow lines are seen in the plains where they occur in valleys linking East Tombaugh Regio with Sputnik Planitia, and this material is interpreted to be nitrogen ice undergoing glacial flow (indicated by the white stipple pattern). Impact craters are absent. Adapted from *Moore et al.* (2018) and *McGovern et al.* (2019). The *bright, smooth plains* and *cellular plains* have a strong nitrogen spectral signature, and all other units have a dominant methane signature.

from the atmosphere, which accumulates there finally in adequately thick deposits to begin flowing glacially into Sputnik Planitia. In this scenario, some initial frosting of East Tombaugh Regio by CH_4 presumably during some intermediate age in Pluto's history initializes the subsequent long-term sustained and current precipitation of N_2 there. One possible counter example to this hypothesis is the observation of N_2 glaciers currently occupying valleys along the west rim of a plateau at ~20°S latitude (Fig. 6). There is a high albedo band presumably of N_2-rich frost starting at the southern terminus of Sputnik Planitia and extending westward, crossing the location of the glacially occupied valleys (*Howard et al.*, 2017a). This arrangement seems consistent with simple advection and reprecipitation of N_2 on the plateau's edge, feeding the glaciers in the valleys. All these relationships remain to be investigated in greater detail with additional volatile transport modeling.

2.2. Intermediate-Age Landforms

In this section we discuss landforms that have likely persisted through multiple obliquity cycles but are young enough to have few impact craters (see the chapter by Singer et al. in this volume).

2.2.1. Bright pitted uplands and bladed terrains on East Tombaugh Regio. The plateau-like eastward extent of East Tombaugh Regio, mapped as bright pitted uplands (Fig. 5), features reticulate kilometer-wide ridges of CH_4-dominated ice interspersed with flat depressions several kilometers in diameter floored by N_2 ice (*Howard et al.*, 2017b; *Moore et al.*, 2017, 2018) (Fig. 5). This landscape apparently has been created by spatially patterned condensation and sublimation of N_2 and CH_4 ices, but the process interactions controlling the landscape form and scale are presently uncertain. One possibility is that the ridges forming bright pitted uplands, which are mostly composed of reticulate ridges on East Tombaugh Regio, and the arcuate terrain, seen mostly northward of the present boundaries of bladed terrain, are remnants of an early and larger extent of massive CH_4 deposits that have undergone extensive erosion and retreat. If this is so, then this retreat and degradation of formerly more extensive (and reaching lower elevations) massive CH_4 deposits might be the result of a general warming of Pluto, perhaps in response to the increasing output of the Sun over the age of the solar system, and resulting in secular loss, such as from the top of the atmosphere.

Bladed terrain features prominent, largely north-south-oriented ridges spaced a few kilometers apart with relative relief of several hundred meters and a dominant CH_4 spectral signature (Fig. 5; Fig. 10a in the chapter by White et al.) (*Moore et al.*, 2018). Its occurrence is limited to high elevations and low latitudes (*Moore et al.*, 2018). The north-south orientation of the ridges is likely related to the dominant illumination angle during Pluto's obliquity cycle

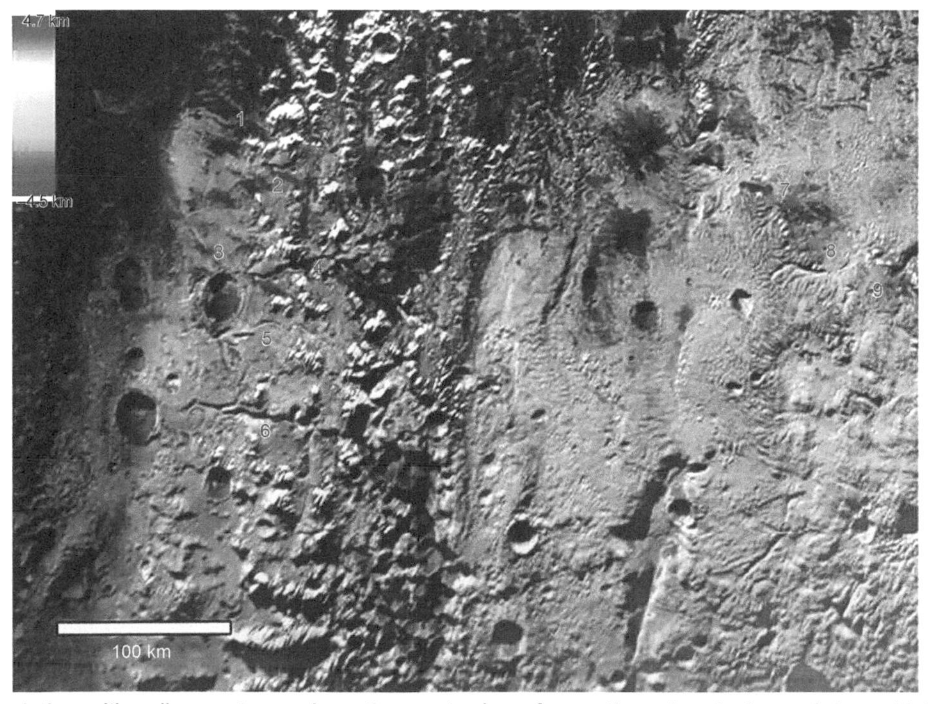

Fig. 6. Alpine glaciation with valley systems along the west edge of a southern hemisphere plateau. Light-toned dendritic valleys 1–3 lead downslope from alpine peaks to light-toned deposits. These valleys appear not to be deeply incised and may contain N_2 ice deposits. The N_2 deposits in the valleys and the light-toned deposits at their termini are probably active N_2 glaciers. Deeply dissected valleys 4–6 similarly lead from alpine mountains and plains toward the west, terminating in the cratered plains. Valleys 7–9 exhibit plateau-style dissection. MVIC 500 m/pixel image PEMV_01_P_MPAN1 centered at 148°E, 23°S. North is up. Topographic information used throughout this chapter is taken from *Schenk et al.* (2018). From *Howard et al.* (2017a).

(*Moores et al.*, 2017) but the relative roles of sublimation, condensation, elevation, and subsurface structure in determining blade size and the history of blade formation are potentially complex (*Moore et al.*, 2018). Additional description of these landscapes is contained in section 2.3.1 of the chapter by White et al.

Analysis by *Moore et al.* (2018) and *Bertrand et al.* (2019) indicates that methane in Pluto's atmosphere will currently preferentially precipitate at low latitudes, where net annual solar energy input is lowest. Whereas both atmospheric nitrogen and methane will precipitate at low elevations, the much higher abundance of nitrogen in the atmosphere means that precipitation of nitrogen ice dominates at low elevations, leaving methane ice to precipitate at higher elevations where atmospheric concentration of CH_4 is high (see the chapter by Forget et al. in this volume). Following a probably early time of massive emplacement of these Bladed terrain CH_4 ice deposits, however, excursions in Pluto's climate appear to have resulted in a transition from conditions favoring precipitation to those favoring sublimation of methane ice, resulting in the partial erosion of the deposits to form the observed bladed surface texture, or this terrain may possibly be caused by cyclical alternation between erosion and deposition. Methane-cycle simulations by *Bertrand et al.* (2019) indicate that Milankovitch-cycle net loss from bladed terrain is less than a meter, confirming its persistence across many such cycles (i.e., scales of 10^8 to 10^9 years).

2.2.1.1. Climate implications. The distribution and morphology of landforms on Sputnik Planitia resulting from glacial flow could potentially result either from steady-state or declining glacial flow rates. In Fig. 5, the dark blue unit with a stipple overprint is the transition zone from East Tombaugh Regio to western Sputnik Planitia in which active glacial flow is evident, based on convergent flow lines on the East Tombaugh Regio uplands and divergent flow on Sputnik Planitia, the latter indicated by flow lines marked by elongated pitting as well as accumulations of methane or water ice blocks rafted onto Sputnik Planitia from East Tombaugh Regio (pink unit in Fig. 5). Some accumulations of the rafted blocks extend up to 260 km onto Sputnik Planitia, well beyond the glacial flow lines evidenced by elongated pitting. A steady-state hypothesis (*Howard et al.*, 2017a; *White et al.*, 2017) suggests that glacial flow continuously occurs toward the center of Sputnik Planitia but convection cell overturn on older N_2 ice reorganizes the flow into convective patterns with the rafted blocks accumulating along convective cell margins (light green region in Fig. 5). An alternative hypothesis is that the extent of glacial flow onto Sputnik Planitia has diminished through time, with the more extensive earlier flows (light green in Fig. 5) being modified by cellular convection (*Howard et al.*, 2017a). By this scenario, the uplands east of Sputnik Planitia were once covered with more extensive N_2-rich ice deposits, which have gradually been depleted, partly via sublimation (and then possible ultimate loss to space), and partly by glacial flow back into Sputnik Planitia. The large collection of blocks (Challenger Colles) at the mouth of the current trough would likely have required extended flow from the uplands, which does not appear to be happening at present. The low viscosity of N_2 ice suggests that any large reservoir of such ice on east Tombaugh Regio would be depleted by glacial flow over timescales much shorter than Pluto's age. A cycle of N_2 ice accumulation and strong glacial flow followed by waning activity might be favored by the strong astronomical forcing of Pluto's solar illumination, with the longest cyclical forcing corresponding to the 2.8-m.y. obliquity cycle and possibly the 3.7-m.y. longitude of perihelion cycle (*Earle and Binzel*, 2015).

2.2.1.2. Widespread distribution of bladed terrain. Bladed terrain deposits appear to be widespread in the low latitudes of the poorly resolved sub-Charon hemisphere, based on circumstantial evidence from albedo mapping, spectral observations, and topography within limb profiles (*Moore et al.*, 2018; *Stern et al.*, 2020). Every place in the low latitudes where high regional elevations have been observed show evidence of bladed terrain deposits. Thus bladed terrain, along with other deposits of volatiles in Tombaugh Regio proper (including Sputnik Planitia), represents an active response of the landscape to current and past climates, and it is very likely a major terrain type on Pluto (see Fig. 23 in the chapter by White et al.).

2.2.2. Alcyonia Lacus. This unusual feature is an irregular, 30-km-long by 10-km-wide expanse of nitrogen ice plains that is surrounded by mountains and rubbly plains of Al-Idrisi Montes, on the northwestern rim of Sputnik Planitia (Fig. 7). The lacus lies in a linear depression surrounded by terrain rising locally up to 2.5 km above the lacus. It displays faint, light-toned, concentric banding; it has a smooth boundary with the surrounding terrain; and pitting is absent from its surface. These characteristics set it apart from other small expanses of non-cellular plains that occur within Al-Idrisi Montes and at the northern and western margins of Sputnik Planitia, where they display angular boundaries and pitted surfaces, implying a distinct origin for Alcyonia Lacus. The absence of pits indicates that it is currently being resurfaced by some means other than sublimation, and/or that its rheology is unable to support pit topography. The relatively smooth boundaries, whereby the surrounding mountains and hills do not intrude into its interior, indicate that it may not have been a passive feature, but that it may have actively modified its boundaries. Accordingly, *Stern et al.* (2017) hypothesize that this feature may once have been a "pond" of liquid nitrogen, the stability of which was maintained during a past era of especially high atmospheric pressure (120–150 mbar), and that has since frozen. High obliquity induces volatile migration that might cause atmospheric pressure excursions several orders of magnitude greater than the present value, perhaps allowing liquid N_2 or CO on the surface (*Stern et al.*, 2017).

Alternatively, Alcyonia Lacus may be an active glacial feature in which flow of the nitrogen ice is sufficient to close pits that form through sublimation, as well as to define a smooth margin for the lacus by shepherding hilly and

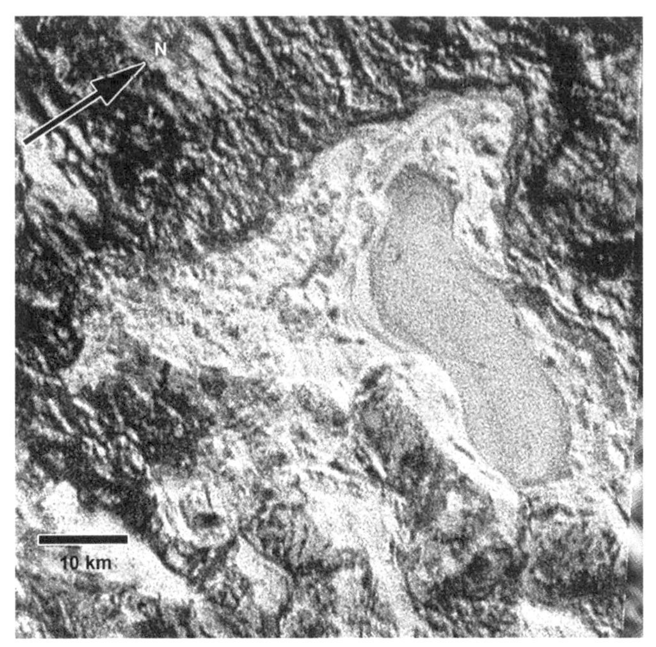

Fig. 7. Alcyonia Lacus. An isolated, ponded, lake-like feature just north of Sputnik Planitia, perhaps a relic of an era when the atmospheric surface pressure was sufficient to permit liquid N_2 at the surface. An alternative explanation is provided in the text. 36°N, 151°E, ~130 m/pix P_MPAN_1 LORRI observation. North is up.

mountainous material to its edges. The dark lineations may be manifestations of flow within the lacus. The lacus may be situated away from the main body causing any blocky material within the zone of mobilization to drift to its edges (*White et al.*, 2017).

3. OLDER UPLAND LANDFORMS

In this section we discuss landforms that likely formed over many obliquity cycles or are remnant from early in Pluto's history.

3.1. Maculae

The bladed terrain deposits form one of the two main landforms that characterize uplands within the permanent diurnal zone (between ±13° latitude). The second is the discontinuous chain of expanses of dark terrain (informally termed "maculae"), composed of haze deposits, that encircle Pluto's equator and have the lowest albedo of any terrain type seen to date on Pluto (see Fig. 23 in the White et al. chapter). Given that the methane ice of the bladed terrain deposits preferentially condenses at high elevation (*Moore et al.*, 2018), the crustal water ice at lower elevation within the diurnal uplands is left exposed, and because volatile mobilization on Pluto is at a minimum here, it accumulates a continuous haze deposit. Once a continuous layer is established, the low albedo of the surface and its consequential high surface temperature (*Earle et al.*, 2017) mean that any subsequent volatile mobilization is further inhibited, but not entirely prevented, as evidenced by seasonal methane frost observed on the haze deposit in the vicinity of Pigafetta Montes (see section 2.2.1). In contrast to Pluto's high latitudes, where continuous volatile mobilization powered by extreme seasonal variations in insolation and surface temperature means that no single material tends to accumulate and dominate in any particular location, the milder conditions of the equatorial latitudes allow preservation of whatever material is initially "seeded" at the surface (whether it is the nitrogen ice of Sputnik Planitia, the methane ice of the bladed terrain deposits, or the haze deposit of the maculae). The equatorial zone is therefore capable of maintaining stark, longitudinal albedo and volatile abundance variations over longer timescales (*Binzel et al.*, 2017; *Earle et al.*, 2018).

Morphological and crater density variation among the maculae, however, indicates that the terrains covered by the haze deposit have been formed over a range of ages. The maculae that are well-resolved within the close approach hemisphere include Cthulhu Macula (the largest of the maculae) in the west and Krun Macula in the east. Easternmost Cthulhu Macula displays a region with a high count of impact craters (*Singer et al.*, 2019), and accordingly it has been interpreted to be ancient crustal material where the haze has simply accumulated since Pluto's early history (*Grundy et al.*, 2018). Krun Macula appears much more degraded and displays many fewer depressions that can be interpreted as impact craters; it therefore appears younger and significantly more modified than Cthulhu Macula. *Stern et al.* (2020) identified a probable transition between these different types of maculae on Pluto's sub-Charon side at ~15°E, and they noted that the maculae to the west of this point display a close spatial association to the farside bladed terrain deposits. These observations suggest that Krun Macula and the western farside maculae may have evolved as a consequence of recession and/or evolution of the bladed terrain deposits in response to presumably long-term secular climate change, which is also thought to be responsible for the development of the bladed texture via sublimation of the methane ice (see section 2.2.1). *Stern et al.* (2020) offer a hypothesis based on this idea: that Krun and the western maculae represent relatively low-elevation bladed terrain deposits where, at some point in their history, mobilization of the methane ice in response to mild seasonal- and obliquity-driven climate change was interrupted for long enough to allow a layer of haze particles to accumulate and reach a sufficient thickness to prevent further volatile mobilization, as had already occurred for Cthulhu Macula.

3.2. Smooth Uplands Mantling Deposit and Depressions

The permanent Arctic region of Pluto (>37°N) has experienced extreme insolation and seasonal cycles across its history [assuming that obliquity cycles have remained relatively stable (*Sussman and Wisdom*, 1988, 1992)], and prior to encounter, the Arctic was expected to show geological evidence for mostly seasonal volatile exchange in response to such cycles (e.g., *Moore et al.*, 2015). These expectations have been largely fulfilled by New Horizons observations, as much of these latitudes exhibit abundant craters that indicate relatively modest modification of the landscape since the cessation of heavy bombardment, which is consistent with minimal long-term volatile deposition and erosion resulting from the seasonal cycles that prevail here (*Binzel et al.*, 2017). An exception is the region to the northeast of Sputnik Planitia (encompassing Pioneer and Hayabusa Terrae), which displays a strong CH_4 spectral signature and textures suggestive of a thick deposited mantle and its possible subsequent erosional modification (*Howard et al.*, 2017b). The smooth, mantled landscape alternates with impact craters and large, flat-floored depressions, the latter up to 3 km deep and tens of kilometers across (Figs. 12a,b in the chapter by White et al.). The decrease in thickness of the CH_4-rich mantle away from the depression edges suggests that the depressions may have sourced the mantle, through either low-power explosive emplacement or through long-term accumulation of deposits from gaseous emanations and volatile condensation (*Howard et al.*, 2017b; *Cruikshank et al.*, 2019; see the chapter by White et al.).

The floors of these depressions, which can reach almost 3 km below the global mean radius, show a N_2 spectral signature and, where observed in higher-resolution imaging, extensive relief at the scale of 100–300 m and a wavelength of 2–5 km (*Howard et al.*, 2017b). These observations suggest that nitrogen ice has condensed from the atmosphere onto these low-elevation areas (*Bertrand and Forget*, 2016). The existence of this subtle relief and small impact craters on the depression floors, however, indicate that a thin, possibly seasonal accumulation of nitrogen ices discontinuously mantles a rigid substrate (*Howard et al.*, 2017b).

Within Hayabusa Terra, to the southeast of the large depressions, the terrain appears rougher and more eroded (Figs. 12c,d in the chapter by White et al.). The mantle here has largely been removed by sublimation erosion (*Howard et al.*, 2017b), with a darker substrate (likely crustal water ice) being exposed where the lighter-toned mantle is absent. Farther to the east ("1" labels in Fig. 11 in the chapter by White et al.), the terrain displays a coarse, mottled texture, with lobate, high-standing occurrences of the mantling material separated by darker substrate. This region may therefore record a history of alternating deposition and sublimation erosion of the mantling material (*Stern et al.*, 2020).

3.3. Arctic Glacial Terrains

The uplands to the north and northeast of Sputnik Planitia have been dissected by valleys with diverse morphologies (*Moore et al.*, 2016; *Howard et al.*, 2017a) (see Figs. 6 and 8, as well as Fig. 16 in the chapter by White et al.). Some valleys may be traced for 150–200 km, often in dendritic patterns (*Howard et al.*, 2017a). A similar belt of glacial features occurs south of –20° to the southwest of Sputnik Planitia (*Howard et al.*, 2017a) (Fig. 6). An origin of these valleys through runoff from liquid N_2 precipitation is disfavored by the low temperature and low atmospheric pressure under modern seasonal and obliquity climate cycling, although arguments have been advanced that short-duration warming epochs during the multi-million-annum, long-term orbital and obliquity cycles (Fig. 9) may have led to significantly higher pressures and episodic liquid N_2 precipitation (*Stern et al.*, 2017). These valley features have alternatively been attributed to sculpting by past N_2 glaciers earlier in Pluto's history with a larger N_2 inventory (*Howard et al.*, 2017a). There is also the possibility that liquid N_2 flows resulting from basal glacier melting may have sculpted the valleys, interspersed with direct scour of the surface by flowing ice, analogous to terrestrial valley glaciation (see chapters by White et al. and Umurhan et al. for further details regarding glacial erosion on Pluto).

In addition to the valleys, portions of the northwest rim of Sputnik Planitia display aligned ridges (termed washboard and fluted ridges) [Fig. 8 (*White et al.*, 2019); also see Fig. 16d in the chapter by White et al.] that are superimposed on underlying topography. Covering an area of at

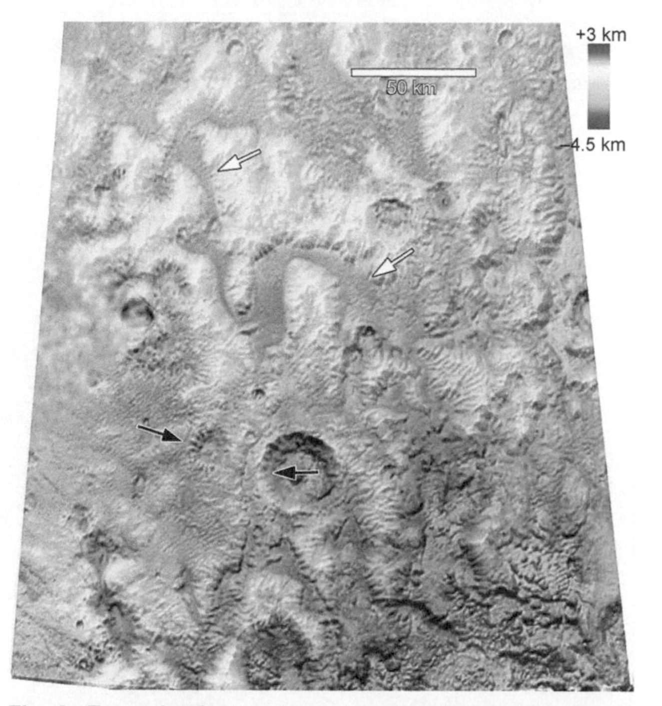

Fig. 8. Example of large sinuous canyons carved by N_2 glaciers into the highlands north of Sputnik Planitia, exhibiting washboard or fluted texture morphology. Black arrows point to interior crater walls with fluted ridge-and-trough texturing. The ridges composing the elements of fluted texture on crater walls are typically spaced 3 km apart. Along canyon walls fluting extends from summits of ~1.5–2-km-high ridges downslope to end abruptly at the floor of depressions. White arrows point to washboard texture oriented perpendicular to the downslope direction of the canyons. The flatter upland areas also have an overprint of washboard texture oriented northeast-southwest. Topographic information used throughout this chapter is taken from *Schenk et al.* (2018). Image centered at 153°E, 50°N; MVIC 340 m/pixel PEMV_01_P_MVIC_LORRI_CA. Lambertian projection. North is up.

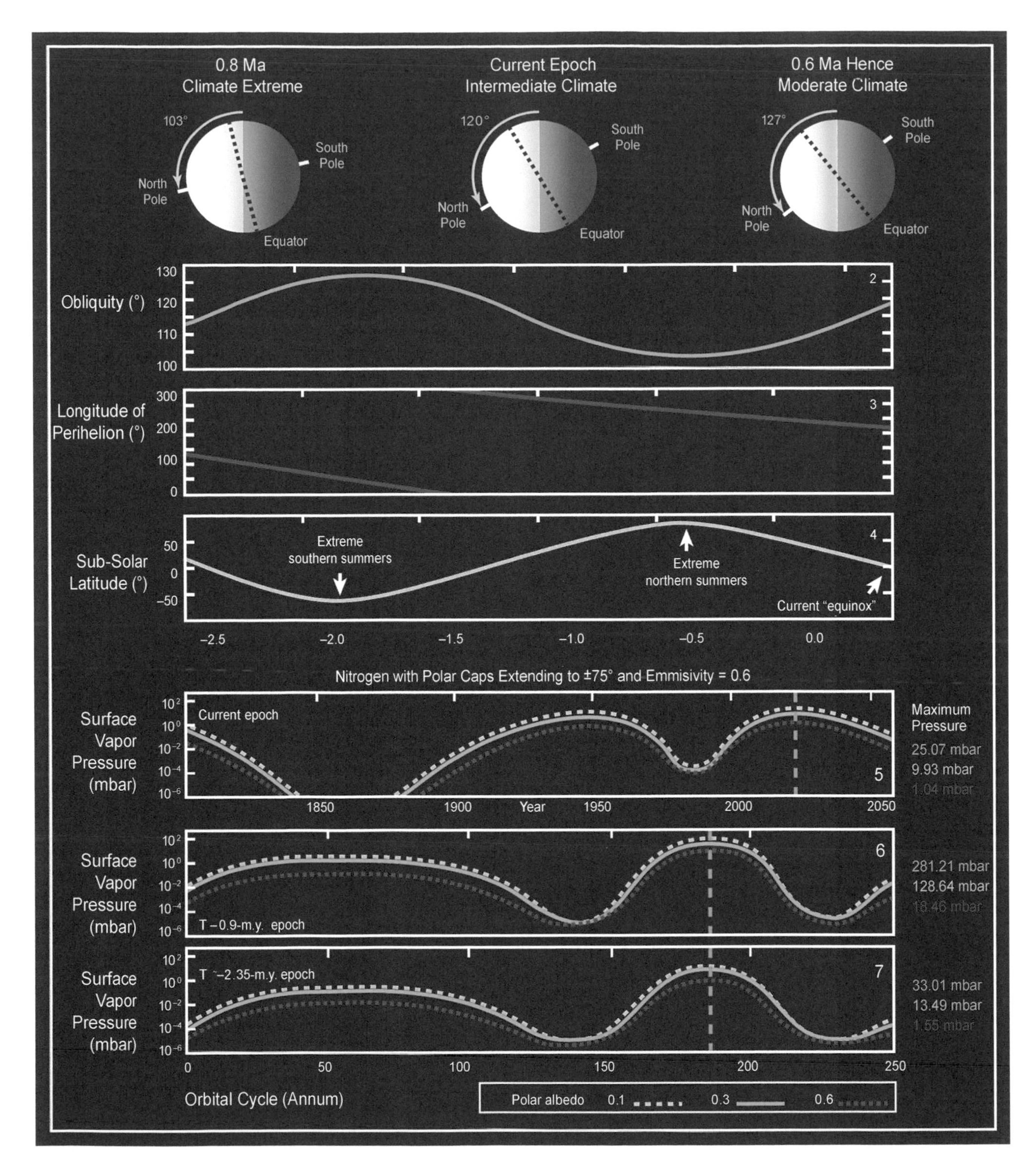

Fig. 9. Preliminary simple models of atmospheric molecular nitrogen pressure over a single Pluto orbit at three different epochs used to explore the possibility of substantial atmospheres in the past. Note, for instance, these models use a static (i.e., non-time-dependent) polar cap extent, despite growing and decreasing atmospheric pressures. *Top row:* Schematic representations of the obliquity states of Pluto at the minimum, median, and maximum values. The middle globe shows Pluto near its current state of 119°. *Rows 2–4:* The effects of obliquity, longitude of perihelion, and subsolar latitude with time over a "Milankovitch-like" cycle. Model Results — *Row 5:* A Pluto year in the current epoch where perihelion occurs at "equinox." *Row 6:* The "extreme northern summer" epoch occurring 0.9 m.y. ago, where the subsolar point is at high northern latitudes at perihelion. *Row 7:* The "extreme southern summer" epoch occurring 2.35 m.y. ago, where the subsolar latitude is furthest south at perihelion. These model polar caps extend to ±75 °latitude and have an emissivity of 0.6. From *Stern et al.* (2017).

least 70,000 km^2, the ridges are dominantly aligned about 72°E of north and spaced 1–3 km crest to crest (*White et al.*, 2019). The origin of these features is uncertain, but based on their occurrence in generally low-elevation, low-relief settings adjacent to Sputnik Planitia that coincide with the large north-south-trending ridge-trough system (see section 3 in the chapter by White et al.), and through comparison with fields of sublimation pits seen in southern Sputnik Planitia, *White et al.* (2019) concluded that they represent crustal debris that was buoyant in pitted glacial N_2 ice that formerly covered this area, and that was deposited after the N_2 ice receded via sublimation.

The nitrogen ice deposits of Sputnik Planitia are interpreted to fill (or occupy) an ancient impact basin that formed ~4 Ga, with the highlands surrounding it on three sides as the rim complex (*Stern et al.*, 2015; *McKinnon et al.*, 2016; *Moore et al.*, 2016; *Schenk et al.*, 2018). A well-defined rim has not been identified on the southern side. This enormous basin, the depth of which may be as much as ~10 km (*McKinnon et al.*, 2016), was a powerful cold trap for volatiles across Pluto (*Hamilton et al.*, 2016; *Bertrand and Forget*, 2016), especially nitrogen, the precipitation of which dominates over that of methane at low elevations (*Moore et al.*, 2018). Prior to the impact, surface nitrogen ice would presumably have been distributed quite broadly across low-lying areas of Pluto, but the majority of it would have migrated to and infilled the basin over tens of millions of years following its formation (*Bertrand et al.*, 2018). The timing of the original (initial) accumulation of volatiles (N_2, CO, CH_4) in the surface inventory is uncertain. Much of this accumulation presumably occurred very early during and after the final stages of accretion of Pluto and Charon, but some release may have continued to later in Pluto's history (e.g., *Cruikshank et al.*, 2019). If volatile accumulation on the surface occurred primarily before the Sputnik-forming impact, the glacial features surrounding the basin could have formed during a relatively short interval during which nitrogen migrated toward the basin (*White et al.*, 2019). One possibility for formation of the crustal water ice blocks that form the mountain ranges of Hillary, Tenzing, Baret, and Al-Idrisi Montes is that glacial nitrogen ice deposits in the uplands to the west and northwest of Sputnik Planitia were thick enough such that the glacial flow into Sputnik Planitia contributed to their displacement and transport (*Howard et al.*, 2017a). Although superposition relations between the glacial terrains and their surroundings are difficult to assess with only New Horizons flyby data, the observation that a number of impact craters reaching a few tens of kilometers across are superimposed upon the glacial landforms supports the hypothesis that nitrogen ice glaciation in the uplands around Sputnik Planitia began relatively soon after the basin's formation early in Pluto's history. Crater surface age estimates of the washboard and fluted terrain by *White et al.* (2019) indicate that the ridges date to ~4 Ga. *Howard et al.* (2017a), however, suggest that glacial scouring northeast of Sputnik Planitia modified a substantial CH_4 mantle, indicating an extended period of glaciation. Glaciation continues to the present along the eastern edge of Sputnik Planitia, and glacial ice may still infill glacially carved valleys southwest of Sputnik Planitia (Fig. 6).

Recondensation of nitrogen ice ultimately derived from the sublimation of Sputnik Planitia is hypothesized to occur onto the low-insolation diurnal zone in the uplands to its east (East Tombaugh Regio), which hosts active, ongoing nitrogen ice glaciation (see section 2.1.5). Such glacial activity has likely not returned to the higher latitudes of the permanent Arctic (>37°N) since the era of the recession of early glaciation. This absence is primarily because the higher averaged insolation experienced here precludes large-scale precipitation of nitrogen ice (*Bertrand et al.*, 2018), although the confinement of uplands nitrogen ice glaciation to a very localized area on Pluto in the present era may also be a consequence of the inventory of nitrogen ice declining over geologic time, probably by loss to space (*Gladstone et al.*, 2016). In the north polar region of Pluto, the landscape appears highly eroded (Fig. 19 in the chapter by White et al.), with the walls of valleys and craters displaying a distinct fluted appearance that is comparable to fluted terrains farther south, interpreted to be glacial in origin (*Howard et al.*, 2017a). The low elevation of the valley floors here (reaching more than 3 km below mean radius) would have facilitated precipitation of large-scale nitrogen ice deposits here. This effect, however, would have been offset by the very high average insolation and consequential high temperatures experienced at the pole (*Hamilton et al.*, 2016; *Earle et al.*, 2017), as well as the extreme seasonal climatic cycles (*Binzel et al.*, 2017), which would inhibit the establishment of thick deposits of volatile ices. Any accumulation of nitrogen ice here would have been transient and would likely not have reached sufficient thickness for long enough to carve valleys via glacial flow. The polar region displays a strong methane spectral signature (*Protopapa et al.*, 2017; *Schmitt et al.*, 2017; *Earle et al.*, 2018), but likewise, long-term climatic conditions here preclude the establishment of a thick deposit (*Bertrand et al.*, 2019). The observed methane signature more likely indicates a layer that is centimeters to meters thick and subject to climatic changes on seasonal and obliquity timescales (*Grundy et al.*, 2018). The absence of thick mantling deposits in the polar region throughout its history suggests that its high topographic relief is a reflection of the rigid, inert water ice crust and bedrock that underlies seasonal veneers of volatile ices (*Howard et al.*, 2017a). The eroded appearance of this terrain may be a consequence of the constant volatile mobilization in response to seasonal and obliquity cycles that characterize this region of Pluto.

3.4. Degradation of Methane Mantles

Several landform types dominated by methane spectral signatures suggest net erosion occurring during later Pluto history. These landforms include the pitted mantle on the Hayabusa Terra highlands (*Howard et al.*, 2017b), the retreating scarps of the Piri Rupes terrain (*Moore et al.*, 2017), and the troughs of the fretted terrain (see the chapter by White et al.). Erosion of the north-facing scarps of Piri Rupes (Fig. 14b in the White et al. chapter) would be expected from Pluto's climate zonation, but the more general degradation of the other mantles, together with the apparent erosion of the bladed terrain (*Moore et al.*, 2018), suggests a net loss of methane, perhaps to space or to surface reservoirs. This erosion may be driven by the slowly increasing luminosity of the Sun, or perhaps by the long-term development of locally dark surfaces that promote sublimation. The current knowledge of the geologic record prevents definitive elucidation of these relatively early events, with solutions to be gleaned by future exploration.

4. DISCUSSION AND CONCLUDING THOUGHTS

4.1. Four Timescales of Pluto's Climate

Like all other worlds in the solar system with surface-atmosphere mediated climates, the climate of Pluto has operated over several different, sometimes cyclic, timescales. In this section we summarize and organize our preceding discourse into four timescale categories of increasing seniority.

4.1.1. Annual. Pluto's annual climate cycle is extreme over its 248-Earth-year duration due to 0.25 orbital eccentricity, which makes solar illumination vary by a factor of 2.77. Over annual time periods, it is likely that only minor cycles of sublimation of volatiles would occur — insufficient to modify topographic features including the sublimation pits. Thin CH_4 deposits on crater rims at low latitudes may come and go on seasonal timescales (*Bertrand and Forget*, 2019). *Bertrand et al.* (2018) calculate that the yearly maximum net sublimation and deposition of N_2 at Sputnik Planitia is measured in tens of millimeters. This annual cycle will express itself with regional changes in surface color and brightness with the seasons.

4.1.2. Milankovitch cycles. Pluto's eccentric orbit coupled with its high obliquity produce a complex variation in the seasonality and latitudinal distribution of illumination with a 2.8-m.y. periodicity (Figs. 2 and 9) (*Dobrovilskis*, 1989; *Dobrovilskis and Harris*, 1989; *Earle and Binzel*, 2015; *Binzel et al.*, 2017; *Earle et al.*, 2017). This periodicity greatly influences the latitudinal redistribution of N_2 and CH_4 (*Bertrand et al.*, 2018, 2019), with net sublimation or deposition of N_2-rich ices reaching hundreds of meters, depending upon latitude (*Bertrand et al.*, 2018). The beat frequency of the combined 2.8-m.y. obliquity period and the 3.7-m.y. longitude of perihelion precession period results in cycles of extreme seasonal solar insolation that likely cause substantial changes in regional surface temperature and atmospheric pressure, which may even permit liquid N_2 (*Earle and Binzel*, 2015; *Stern et al.*, 2017), among other phenomena (e.g., *Moore et al.*, 2018; *Telfer et al.*, 2018). Formation and eradication of pitting of Sputnik Planitia may occur over this timescale. Glacial activity from East Tombaugh Regio onto Sputnik Planitia may also wax and wane over this timescale.

4.1.3. Long-term evolution. Few constraints exist concerning the early climate and volatile inventory of Pluto, although volatile redistribution is likely to have occurred over timescales significantly longer than that of Milankovitch-like cycles (*Bertrand et al.*, 2018, 2019). Observations from the New Horizons encounter suggest low rates of N_2 loss to space (*Gladstone et al.*, 2016), although evidence for extensive upland glaciation implies a formerly much larger abundance of N_2, which may have undergone appreciable long-term depletion of its inventory (*Howard et al.*, 2017a; *White et al.*, 2017). Volatile exchange rates for CH_4 are sufficiently low that formation and evolution of bladed terrain has occurred over timescales considerably longer than that of obliquity cycles (*Bertrand et al.*, 2019). Bladed terrain may thus contain a potentially interpretable record of the long-term evolution of Pluto's climate. Evidence for possible formerly greater extents of massive CH_4 deposits (the reticulate ridges of bright pitted uplands in East Tombaugh Regio and arcuate terrain) may imply a slow secular loss of CH_4, perhaps forced by increasing solar luminosity. Similarly, the formation and subsequent modification of the extensive mantles of CH_4 and the formation of the dichotomous landscape of N_2-rich depressions and the CH_4-rich uplands probably occurred over this long timescale.

Interpretation of long-term (gigayear-scale) variation in landform evolution due to volatile activity (mantles, glaciers, etc.) should be tempered by uncertainties about the orbital characteristics of Pluto over long timescales (*Sussman and Wisdom*, 1988, 1992). Most planetary obliquity stability models are only considered valid out to a few 10^8 annum, at most.

4.1.4. Climate shortly after heavy bombardment. Superposed craters on glacial terrains indicate that N_2 glaciation began early in Pluto's history. Some of the inferred paleoglacial features, including washboard terrain (*Howard et al.*, 2017a; *White et al.*, 2019) and the extensive mantles (*Howard et al.*, 2017b), may have formed in the first billion years after the formation of Pluto. These early-period climatically driven deposits may have been affected by postulated outward orbital migration (e.g., *Levison et al.*, 2008) and the evolving faint young Sun.

4.2. Unresolved Issues and Opportunities for Future Exploration

The New Horizons mission viewed only about 30% of Pluto's surface at resolutions greater than 500 m/pixel and only narrow strips at resolutions better than 200 m/pixel. As a result, many issues about climate and landform evolution remain unresolved. In this section, we review these uncertainties and possibilities for future spacecraft missions to clarify these uncertainties, particularly if they can provide high-resolution (e.g., meter-scale) global data.

- Most of the dynamic landforms on Pluto are products of volatile exchanges and phase changes, such as the glaciers, surficial mantles, Sputnik Planitia pits, and bladed terrain. Many uncertainties remain about the volatile inventories of Pluto, specifically their size and location, as well as their evolution over seasonal, Milankovitch-cycle, and longer timescales. Continued analysis of New Horizons data, modeling of atmospheric and surface processes, and remote observations can refine our understanding, but many questions will inevitably remain unanswered pending future spacecraft missions.
- The history and formational processes of bladed terrain and the original CH_4 deposits in which they occur remain uncertain. This terrain is a major reservoir of Pluto's methane, but its total thickness, areal extent, and CH_4 composition are only known within order-of-magnitude limits. Are the blades primarily erosional, like ice penitentes on terrestrial glaciers, or are they primarily constructional, like the ice pinnacles on Callisto (*White et al.*, 2016)? What limits their relief — material strength, volatile processes? What processes determine the spacing and orientation of the blades — solar illumination direction, or atmospheric process such as prevailing winds? Did the bladed terrain deposits form early in Pluto's history, or are they more recent?
- Nitrogen ice glaciers are active at present on Pluto, and there are ancient landforms (dissected massifs, dendritic valley networks, washboard terrain) that are inferred to have been sculpted by glaciation early in Pluto's history. At present there is insufficient evidence to provide a definitive glacial timeline. Was there glacial activity during Pluto's 3+-G.y. middle history? Likewise, there remain uncertainties about the early glacial activity. Were the valleys seen widespread across the near-encounter hemisphere sculpted in part by subglacial melting and fluid nitrogen flows? Although the characteristics of the valley systems are unlike terrestrial drainage networks, there remains the possibility that global temperatures may have been episodically high enough to create a fluid nitrogen precipitation cycle. No depositional landforms such as outwash plains, moraines, or eskers have been identified in association with the inferred early glaciation. On Earth and Mars, such glacial features have low relief, are readily modified by non-glacial processes, and are often resolvable only in high-resolution imaging. For instance, observations of outwash deposits at the bases of glacial valleys (which on Earth and Mars require meter-scale resolution) would demonstrate that liquid N_2 played a role in glaciation and valley formation.
- The existence, morphology, and extent of eolian dunes will serve as strong indicators of episodes of substantial variations in atmospheric surface pressure and wind velocities. Much of the dark patterning seen in the region around Inanna and Dumuzi Fossae at ~300 m/pixel (see Fig. 14a in the White et al. chapter) strongly resembles patterning seen in the equatorial regions of Mars at similar resolutions that, on Mars, is consistently revealed to be eolian deposits forming dunes, ripples, and drifts in high-resolution (meter-scale) images. Also, the sources of eolian deposits could be evaluated. For instance, are they all derived from atmospheric haze particles, or alternatively from putative cryopyroclastic deposits such as those proposed by *Cruikshank et al.* (2019)?
- The spatial patterning of surficial volatiles (primarily nitrogen and methane) as inferred from multi-spectral observations at the time of the New Horizons mission was complex. The spatial distribution results from the influence of latitude, elevation, seasonal insolation, Milankovitch cycles, and longer-term volatile exchanges. Although theoretical modeling, including global circulation model simulations, continues to provide constraints on responsible processes, a return visit 20–30 years after the 2015 encounter would help to separate seasonal effects from longer-term controls. Although most landforms with appreciable relief are likely to have evolved over timescales longer than seasonal, wind streaks and possible dune features on Sputnik Planitia, and possibly elsewhere, may have responded to seasonal changes. A search for changes in pit morphology on Sputnik Planitia may confirm or challenge the conclusion that these features evolve over Milankovitch-cycle timescales. Likewise, the hypothesis that bright CH_4 coatings of north-facing crater rims in Cthulhu Macula are seasonal can be verified or refuted.
- Mars and Titan, like Pluto, have a 4-G.y. geologic and climatic history writ in their landforms, surficial materials, and subsurface rocks and sediments. Later processes tend to obscure the record of early events. The origin of enigmatic landforms such as the possible nitrogen paleolake Alcyonia Lacus cannot be uniquely determined from existing New Horizons observations. Many of the clues to past and present processes and environments can only be resolved by global-scale, high-resolution observations, up to or exceeding meter-scale imaging, such as acquired by the Context Camera (CTX) and High Resolution Imaging Experiment (HiRISE) camera orbiting Mars onboard the Mars Reconnaissance Orbiter. Pluto presents a particular challenge to fully resolve seasonal effects, given the 248-Earth-year orbit. It is premature to specify instrumentation for a future mission, but we can list the types of datasets that would help to clarify the mysteries of Pluto. These include high-resolution imaging, global high-resolution topography (meter-scale, both vertically and horizontally), spectral and thermal surface and atmospheric observations, and subsurface sounding of volatile reservoirs.

REFERENCES

Amstutz G. C. (1958) On the formation of snow penitentes. *J. Glaciol.*, *3*, 304–311.

Bergeron V., Berger C., and Betterton M. (2006) Controlled irradiative formation of penitentes. *Phys. Rev. Lett.*, *96(9)*, DOI: 10.1103/PhysRevLett.96.098502.

Bertrand T. and Forget F. (2016) Observed glacier and volatile distribution on Pluto from atmosphere-topography processes. *Nature*,

540, 86–89, DOI: 10.1038/nature19337.

Bertrand T. and Forget F. (2019) How seasonal methane snow forms on Pluto on mountaintops, crater rims, and slopes. Abstract P42C-06 presented at 2019 Fall Meeting, AGU, San Francisco, California, 9–13 December.

Bertrand T., Forget F., Umurhan O. M., et al. (2018) The nitrogen cycles on Pluto over seasonal and astronomical time scales. *Icarus, 308*, 277–296, DOI: 210.1016/j.icarus.2018.1003.1012.

Bertrand T., Forget F., Umurhan O. M., et al. (2019) The CH_4 cycles on Pluto over seasonal and astronomical timescales. *Icarus, 329*, 148–165, DOI: 110.1016/j.icarus.2019.1002.1007.

Bertrand T., Forget F., White O., et al. (2020) Pluto's beating heart regulates the atmospheric circulation: Results from high-resolution and multiyear numerical climate simulations. *J. Geophys. Res., 125*, e2019JE006120.

Beyer R. A., Schenk P. M., Moore J. M., et al. (2019) High-resolution pixel-scale topography of Pluto and Charon. *Pluto System After New Horizons*, Abstract #7042. LPI Contribution No. 2133, Lunar and Planetary Institute, Houston.

Binzel R. P., Earle A. M., Young L. A., et al. (2017) Climate zones on Pluto and Charon. *Icarus, 287*, 30–36, DOI: 10.1016/j.icarus.2016.1007.1023.

Buhler P. B. and Ingersoll A. P. (2017) Sublimation pit distribution indicates convection cell surface velocity of ~10 centimeters per year in Sputnik Planitia, Pluto. *Icarus, 300*, 327–340, DOI: 310.1016/j.icarus.2017.1009.1018.

Claudine P., Jarry H., Vignoles G., et al. (2015) Physical processes causing the formation of penitentes. *Phys. Rev. E, 92*, 033015, DOI: 033010.031103/PhysRevE.033092.033015.

Cruikshank D. P., Umurhan O. M., Beyer R. A., et al. (2019) Recent cryovolcanism in Virgin Fossae on Pluto. *Icarus, 330*, 155–160, DOI: 110.1026/j.icarus.2019.1004.1023.

Dobrovilskis A. R. (1989) Dynamics of Pluto and Charon. *Geophys. Res. Lett., 16*, 1217–1220, DOI: 1210.1016/GL1016i1011p01217.

Dobrovilskis A. R. and Harris A. W. (1989) The obliquity of Pluto. *Icarus, 55*, 231–235, DOI: 210.1016/0019-1035(1083)90077-90075.

Earle A. M. and Binzel R. P. (2015) Pluto's insolation history: Latitudinal variation and effects on atmospheric pressure. *Icarus, 250*, 405–412.

Earle A. M., Binzel R. P., Young L. A., et al. (2017) Long-term surface temperature modeling of Pluto. *Icarus, 287*, 37–46, DOI: 10.1016/j.icarus.2016.1009.1036.

Earle A. M., Binzel R. P., Young L. A., et al. (2018) Albedo matters: Understanding runaway albedo variations on Pluto. *Icarus, 303*, 1–9, DOI: 10.1016/j.icarus.2017.1012.1015.1012.

Eluszkiewicz J. and Stevenson D. J. (1990) Rheology of solid methane and nitrogen: Applications to Triton. *Geophys. Res. Lett., 17(10)*, 1753–1756.

Gladstone G. R., Stern S. A., Ennico K., et al. (2016) The atmosphere of Pluto as observed by New Horizons. *Science, 351(6279)*, 1280, DOI: 1210.1127/science.aad8866.

Grundy W. M., Bertrand T., Binzel R. P., et al. (2018) Pluto's haze as a surface material. *Icarus, 314*, 232–245, DOI: 210.1016/j.icarus.2018.1005.1019.

Hamilton D. P., Stern S. A., Moore J. M., et al. (2016) The rapid formation of Sputnik Planitia early in Pluto's history. *Nature, 540*, 97–99, DOI: 10.1038/nature20586.

Howard A. D., Moore J. M., Umurhan O. M., et al. (2017a) Present and past glaciation on Pluto. *Icarus, 287*, 287–300, DOI: 210.1016/j.icarus.2016.1007.1006.

Howard A. D., Moore J. M., White O. L., et al. (2017b) Pluto: Pits and mantles on uplands north and east of Sputnik Planitia. *Icarus, 293*, 218–230, DOI: 210.1016/j.icarus.2017.1002.1027.

Levison H. F., Morbidelli A., Van Loerhoven C., et al. (2008) Origin of the structure of the Kuiper belt during a dynamical instability in the orbits of Uranus and Neptune. *Icarus, 196*, 258–273, DOI: 210.1016/j.icarus.2007.1011.1035.

Mackwell S. J., Simon-Miller A. A., Harder J. W., and Bullock M. A., eds. (2013) *Comparative Climatology of Terrestrial Planets*. Univ. of Arizona, Tucson. 712 pp.

McGovern P. J., White O. L., and Schenk P. M. (2019) Tectonism across Pluto: Mapping and interpretations. *Pluto System After New Horizons*, Abstract #7063. LPI Contribution No. 2133, Lunar and Planetary Institute, Houston.

McKinnon W. B., Nimmo F., Wong T., et al. (2016) Convection in a volatile nitrogen-ice-rich layer drives Pluto's geological vigour. *Nature, 534*, 82–85, DOI: 10.1038/nature.18289.

Moore J. M., Howard A. D., Schenk P. M., et al. (2015) Geology before Pluto: Pre-encounter considerations. *Icarus, 246*, 65–81, DOI: 10.1016/j.icarus.20144.04028.

Moore J. M., Howard A. D., Umurhan O. M., et al. (2017) Sublimation as a landform-shaping process on Pluto. *Icarus, 287*, 320–333, DOI: 310.1016/j.icarus.2016.1008.1025.

Moore J. M., Howard A. D., Umurhan O. M., et al. (2018) Bladed terrain on Pluto: Possible origins and evolution. *Icarus, 300*, 129–144, DOI: 110.1016/j.icarus.2017.1008.1003.

Moore J. M., McKinnon W. B., Spencer J. R., et al. (2016) The geology of Pluto and Charon through the eyes of New Horizons. *Science, 351(6279)*, 1284–1293.

Moores J. E., Smith C. L., Toigo A. D., et al. (2017) Penitentes as the origin of the bladed terrain of Tartarus Dorsa on Pluto. *Nature, 541*, 188–190, DOI: 110.1038/nature20779.

Protopapa S., Grundy W. M., Reuter D. C., et al. (2017) Pluto's global surface composition through pixel-by-pixel Hapke modeling of New Horizons Ralph/LEISA data. *Icarus, 248*, 218–228, DOI: 210.1016/j.icarus.2016.1011.1028.

Schenk P. M., Beyer R. A., McKinnon W. B., et al. (2018) Basins, fractures and volcanoes: Global cartography and topography of Pluto from New Horizons. *Icarus, 314*, 400–433, DOI: 410.1016/j.icarus.2018.1006.1008.

Schmitt B., Philippe S., Grundy W. M., et al. (2017) Physical state and distribution of materials at the surface of Pluto from New Horizons LEISA imaging spectrometer. *Icarus, 287*, 229–260, DOI: 210.1016/j.icarus.2016.1012.1025.

Singer K. N. et al. (2019) Impact craters on Pluto and Charon indicate a deficit of small Kuiper belt objects. *Science, 363*, 955–959, DOI: 10.1126/science.aap8628.

Stern S. A., Bagenal F., Ennico K., et al. (2015) The Pluto system: Initial results from its exploration by New Horizons. *Science, 350*, 292, 291–298.

Stern S. A., Binzel R. P., Earle A. M., et al. (2017) Past epochs of significantly higher pressure atmospheres on Pluto. *Icarus, 287*, 47–53, DOI: 10.1016/j.icarus.2016.1011.1022.

Stern S. A., White O. L., McGovern P. J., et al. (2020) Pluto's far side. *Icarus, 356(113805)*, DOI:10.1016/j.icarus.2020.113805.

Sussman G. J. and Wisdom J. (1988) Numerical evidence that the motion of Pluto is chaotic. *Science, 241*, 433–437, DOI: 410.1126/science.1242.4684.1433.

Sussman G. J. and Wisdom J. (1992) Chaotic evolution of the solar system. *Science, 257*, 56–62.

Telfer M. W., Parteli E. J. R., Forget F., et al. (2018) Dunes on Pluto. *Science, 360*, 992–997.

Trowbridge A. J., Melosh H. J., Steckloff J. K., et al. (2016) Vigourous convection as the explanation for Pluto's polygonal terrain. *Nature, 534*, 79–81, DOI: 10.1038/nature18016.

White O. L., Umurhan O. M., Moore J. M., et al. (2016) Modeling of ice pinnacle formation on Callisto. *J. Geophys. Res., 121*, 21–45, DOI: 10.1002/2015JE004846.

White O. L., Moore J. M., McKinnon W. B., et al. (2017) Geological mapping of Sputnik Planitia on Pluto. *Icarus, 287*, 261–286.

White O. L., Moore J. M., Howard A. D., et al. (2019) Washboard and fluted terrains on Pluto as evidence for ancient glaciation. *Nature Astron., 3*, 62–68, DOI: 10.1038/s41550-018-0592-z.

Yamashita Y., Kato M., and Arakawa M. (2010) Experimental study on the rheological properties of polycrystalline solid nitrogen and methane: Implications for tectonic processes on Triton. *Icarus, 207*, 972–977, DOI: 910.1016/j.icarus.2009.1011.1032.

Singer K. N., Greenstreet S., Schenk P. M., Robbins S. J., and Bray V. J. (2021) Impact craters on Pluto and Charon and terrain age estimates. In *The Pluto System After New Horizons* (S. A. Stern, J. M. Moore, W. M. Grundy, L. A. Young, and R. P. Binzel, eds.), pp. 121–145. Univ. of Arizona, Tucson, DOI: 10.2458/azu_uapress_9780816540945-ch007.

Impact Craters on Pluto and Charon and Terrain Age Estimates

K. N. Singer
Southwest Research Institute

S. Greenstreet
University of Washington

P. M. Schenk
Lunar and Planetary Institute

S. J. Robbins
Southwest Research Institute

V. J. Bray
University of Arizona

Pluto's terrains display a diversity of crater retention ages ranging from areas with no identifiable craters to heavily cratered terrains. This variation in crater densities is consistent with geologic activity occurring throughout Pluto's history as well as a variety of resurfacing styles, including both exogenic and endogenic processes. Using estimates of impact flux and cratering rates over time, Pluto's heavily cratered terrains appear to be relatively ancient, 4 Ga or older. Charon's smooth plains, informally named Vulcan Planitia, did experience early resurfacing, but there is a relatively high spatial density of craters on Vulcan Planitia and almost all overprint the other types of volcanic or tectonic features. Both Vulcan Planitia and the northern terrains on Charon are also estimated to be ancient, 4 Ga or older. The craters on Pluto and Charon also show a distinct break in their size-frequency distributions (SFDs), where craters smaller than ~ 10–15 km in diameter have a shallower SFD power-law slope than those larger than this break diameter. This SFD shape on Pluto and Charon is different than what is observed on the Earth's Moon, and gives the Kuiper belt impactor SFD a different shape than that of the asteroid belt.

1. INTRODUCTION

One of the goals of the New Horizons mission was to use impact craters observed on Pluto and Charon to both understand geologic surface processes and learn about the size-distribution of the greater population of Kuiper belt objects (KBOs) (*Stern,* 2008; *Young et al.,* 2008). It was not known if there would be many impact craters on Pluto, or if craters would be resurfaced and effectively erased by atmospheric and surface processes. Charon, however, was expected to have craters. Craters were discovered on both worlds and provided a wealth of knowledge about the surface ages and geologic processes operating on Pluto and Charon. Additionally, the impact craters on Pluto and Charon yielded insights into the population of impactors in the outer solar system and the origin and evolution of planetesimals.

Craters are useful geologic tools for investigating a number of topics. Because the initial shape of a crater of a given size is fairly well known for many surface materials, deviations from that initial form can reveal information either about the crater formation conditions (e.g., formation in a thin shell) or the geologic processes that later modified a crater. Both the depth and morphology of modified impacts can be compared to the same properties for relatively fresh or unmodified impacts to produce a qualitative and/or quantitative estimate of the processes occurring over time. Craters also excavate the surface, allowing a view of the near-surface interior. Both the walls and ejecta deposits of craters can reveal subsurface layering and materials that may not be detectable on the surface. Ejected material, being a thinner deposit, erodes more quickly, and thus can serve as a time marker sensitive to relatively recent epochs. Larger craters form more frequently in early solar system history, and their large size often allows them to persist longer under erosive processes, thus they can be a witness to the more distant past and longer timescales.

Because craters are thought to have formed with a relatively steady rate over the last ~ 4 G.y. (there are exceptions to this), the variation in the spatial density of craters between different geologic units speaks to their relative age differences. Younger terrains have fewer craters, and older terrains generally have higher crater densities and more large craters. Crater densities combined with models of the impact flux and cratering rates over time can give an estimate of the age of the surface in units of time, usually expressed in millions (m.y.) or billions (G.y.) of years.

Through scaling laws, the craters can also be related to the impactors that made them. The size distribution of the impacting population can be derived. This requires taking into account any resurfacing that may have occurred once the craters formed, and accounting for any secondary crater or circumplanetary populations from debris in the system, either created by fragments ejected from primary impacts or from breaking up of small moons. Having multiple terrains — and in the case of the Pluto-system, multiple bodies — to compare among can help discriminate between a primary impacting population and other populations/geological effects.

Here we describe the above type of investigations performed using New Horizons data of Pluto and Charon. Note some feature names used in this chapter are formal and others are informal. (For reference, see Appendix A in this volume.) The image sequences or scan names used here refer to the Request ID for each observation, which is a unique identifier that can be found in the headers of the data as archived on the official repository for New Horizons data, the Planetary Data System (PDS) Small Bodies Node (*https://pds-smallbodies.astro.umd.edu/data_sb/missions/newhorizons/index.shtml*). The images come from two instruments on New Horizons: the LOng Range Reconnaissance Imager (LORRI) and the Multispectral Visible Imaging Camera (MVIC). Details about the relevant datasets can be found in Table 1. New Horizons flew through the Pluto-system and thus observed one side of Pluto better than the other. We define the encounter hemisphere of both Pluto and Charon as the surface area of each body observed at higher resolution (pixel scales of ~ 76–865 m px^{-1}) during the closest approach of the spacecraft during its flyby (Fig. 3). The non-encounter hemisphere consists of portions of Pluto and Charon that were seen at lower resolution (pixel scales

TABLE 1. Image datasets and mosaics (modified from *Singer et al.,* 2019).

Request ID	Plot Legend Short Title	Instrument Mode	~Pixel Scale (m px^{-1})	Mosaic Size or Scan	Exposure or Scan Rate
Pluto					
PELR_P_LORRI		1 × 1	850 ± 30	4 × 5	150 ms
PELR_P_LEISA_HIRES*		1 × 1	234 ± 13	1 × 12	50 ms
PELR_P_MPAN_1*		1 × 1	117 ± 2	1 × 27	10 ms
PELR_P_MVIC_LORRI_CA*	LCA	1 × 1	76	1 × 35	10 ms
PEMV_P_MPAN1	MPAN	Pan TDI 1	480 ± 5	Scan	1600 μrad s^{-1}
PEMV_P_MVIC_LORRI_CA	MCA	Pan TDI 2	315 ± 8	Scan	1000 μrad s^{-1}
Charon					
PELR_C_LORRI	C_LORRI	1 × 1	865 ± 10	2 × 4	150 ms
PELR_C_LEISA_HIRES*	C_LEIII	1 × 1	410 ± 5	1 × 7	60 ms
PELR_C_MVIC_LORRI_CA*	LORRI_CA	1 × 1	154	1 × 9	10 ms
PEMV_C_MVIC_LORRI_CA	MVIC_CA	Pan TDI 1	622	Scan	1000 μrad s^{-1}

* Higher-resolution strips taken as sequential ride-along LORRI frames during scanning of another instrument.

of ~ 2.2–32 km px^{-1}) during New Horizons' approach to the system. One Pluto or Charon day is 6.4 Earth days. As they rotated, New Horizons was able to see different portions of the bodies on approach, hence the wide range of resolutions for the non-encounter hemisphere.

2. CRATER MORPHOLOGIES

The final morphology of a pristine impact crater is a result of the collapse of a geometrically simple, bowl-shaped "transient crater" that forms shortly after impact from a combination of excavation and compression of the surface (e.g., *Melosh,* 1989). For simple craters this involves minor rim debris sliding, but for complex craters, floor uplift and rim failure are involved. This collapse is controlled by a complex interplay of crustal material strength, target body gravity, and impactor energy. Although ice in many outer solar system planetary settings (at temperatures of tens of Kelvins to 100 K) is significantly stronger than ice on the surface of the Earth, which is much closer to its melting point of 273 K (e.g., *Durham and Stern,* 2001), it is still weaker than rock in terms of tensile and compressive strength (also see the chapter by Umurhan et al. in this volume). On account of this basic strength difference, the amount of collapse of the transient cavity during crater modification is greater and occurs at smaller diameters (D) for craters forming in ice than on a rocky body of similar gravity (e.g., *Schenk et al.,* 2004).

Craters on Pluto and Charon display many similarities to craters on other icy worlds, including a transition from smaller, bowl-shaped craters referred to as simple craters, to larger, flatter (e.g., more pie-pan shaped) craters, some with central peaks (Fig. 1; also see figures in section 3). There is one crater on Pluto with a deep central depression (an eroded 85-km crater at 5.7°S, 155.3°E; Figs. 1e–1h) that appears similar to central pit craters seen on icy satellites such as Ganymede and also on Mars and Ceres (e.g., *Schenk,* 1993; *Alzate and Barlow,* 2011; *Bray et al.,* 2012; *Conrad et al.,* 2019). The expected transition diameter from central peak to central pit craters on Pluto, assuming g^{-1} scaling from Ganymede and Callisto, is ~ 55 km. Thus, the lack of prevalence for central pit morphologies at larger diameters is somewhat surprising. However, there are only a handful of craters larger than 55 km in diameter and smaller than the very largest basins, so this limits the possible examples.

The largest impact feature seen on Pluto is the very large (D ~ 800–1000 km) Sputnik basin, which is partly filled by the nitrogen-rich ice sheet of Sputnik Planitia; the second largest is Burney basin (~240 km in diameter). Burney basin lies just north of Sputnik Planitia and exhibits multiple ring structures, the only structure confirmed to do so (*Moore et al.,* 2016). Both Sputnik and Burney are extensively eroded and modified. Burney basin preserves a depth of 2–3 km (*Schenk et al.,* 2018a). The surface of Sputnik Planitia is 2.5–3.5 km below the eroded edge of the basin, but the initial, unfilled depth of Sputnik basin may have been as deep as 10 km (*McKinnon et al.,* 2016). The circular Simonelli feature (Fig. 1d) on the non-encounter hemisphere of Pluto (which was only seen at lower resolution) is similar in size to Burney and appears to show a concentric, ring-like structure, and potentially also a large central peak. Some of Simonelli's appearance is due to topography, and some may be due to deposits of bright ice in topographic lows emphasizing the concentric nature (*Stern et al.,* 2020).

After a crater forms, geologic processes can reshape or erode the crater over time and most craters on Pluto show signs of at least some modification. Geologic processes acting on Pluto are described in detail in many sources (e.g., *Moore et al.* 2016; see also the chapters by White et al. and by Moore and Howard in this volume) and those processes specific to crater modification are discussed in the main text and supplement of *Singer et al.* (2019). These processes act in some areas to erode craters while in other areas craters can be either mantled or infilled (e.g., a few craters have nitrogen-rich ice deposits on their floors similar to the plains of Sputnik Planitia). Erosion of ejecta deposits appears to occur quickly on Pluto as very few can be easily identified. The processes affecting individual regions are discussed in section 3.

Craters on Charon also generally exhibit the progression from bowl-shaped to complex morphologies with increasing size expected for an icy body (Fig. 2; see also figures in section 3). The craters on Charon can serve as a good reference for what a more pristine crater would look like on Pluto, although the gravity is lower on Charon (g ~ 0.3 m s^{-2}) than on Pluto (g ~ 0.6 m s^{-2}), and this difference must be taken into account when comparing crater morphologies between the two bodies. On Charon many of the craters larger than D ~ 10 km have extensively ridged crater floors, similar to those seen on icy saturnian satellites (*White et al.,* 2013, 2017; *Schenk et al.,* 2018b,c) and on ice-rich Ceres (*Schenk et al.,* 2019), and are diagnostic of floor uplift and deformation. Depths of relatively unmodified complex craters on Pluto and Charon have been measured (*Schenk et al.,* 2018a,2018c). These depths indicate simple-to-complex transition diameters (from an inflection in a plot crater depth vs. diameter) of ~4.3 km and ~5.3 km on Pluto and Charon, respectively. The morphological transition to central peaks occurs at D ~ 12.5 km on Charon. These transition diameters and the range of depths on the two bodies are consistent with gravity scaling of crater dimensions in hypervelocity craters on ice-rich targets (*Schenk et al.,* 2018c).

Ejecta deposits around Charon craters take a variety of forms. Many craters with a distinct albedo pattern that is a combination of dark inner ejecta and bright outer ejecta/rays are found in Oz Terra where the overhead lighting is well suited for observing albedo variation (*Robbins et al.,* 2019). Some craters on Vulcan Planitia (Fig. 2) also display thicker ejecta deposits with distinct margins similar to those found on other icy bodies (*Robbins et al.,* 2018), including Ganymede and Dione (*Horner and Greeley,* 1982; *Schenk et al.,* 2004, 2018b; *Boyce et al.,* 2010) as well as Mars (*Mouginis-Mark,* 1979; *Costard,* 1989; *Barlow and Bradley,* 1990; *Barlow et al.,* 2000; *Robbins et al.,* 2018). These thicker ejecta deposits are identified in areas (such as on Vulcan Planitia) where the lighting is oblique and enables the topography of more subtle features to be seen. Thus, the fact that they are observed more on Vulcan Planitia may be due to lighting geometry effects. Secondary cratering (or the lack thereof) is discussed in section 5.1 below.

3. REGIONAL CRATER SIZE-FREQUENCY DISTRIBUTIONS

Impact crater size-frequency distributions (SFDs) are useful for understanding both the geologic histories of Pluto and Charon and the impactor populations that formed the craters (*Moore et al.*, 2016; *Robbins et al.*, 2017; *Singer et al.*, 2019). The plots below are displayed in an R-plot format (*Crater Analysis Techniques Working Group*, 1979), where the "R" stands for relative. The differential number of craters for a given diameter (D) bin is proportional to a power law with an exponent of q ($dN/dD \propto D^q$), where q is often referred to as the log-log slope. The R-plot SFD divides this differential SFD by D^{-3} such that a differential distribution with q = −3 is a horizontal line, and q = −4 and −2 form lines that slope downward and upward with increasing D, respectively (see guide in Figs. 4d and 8b).

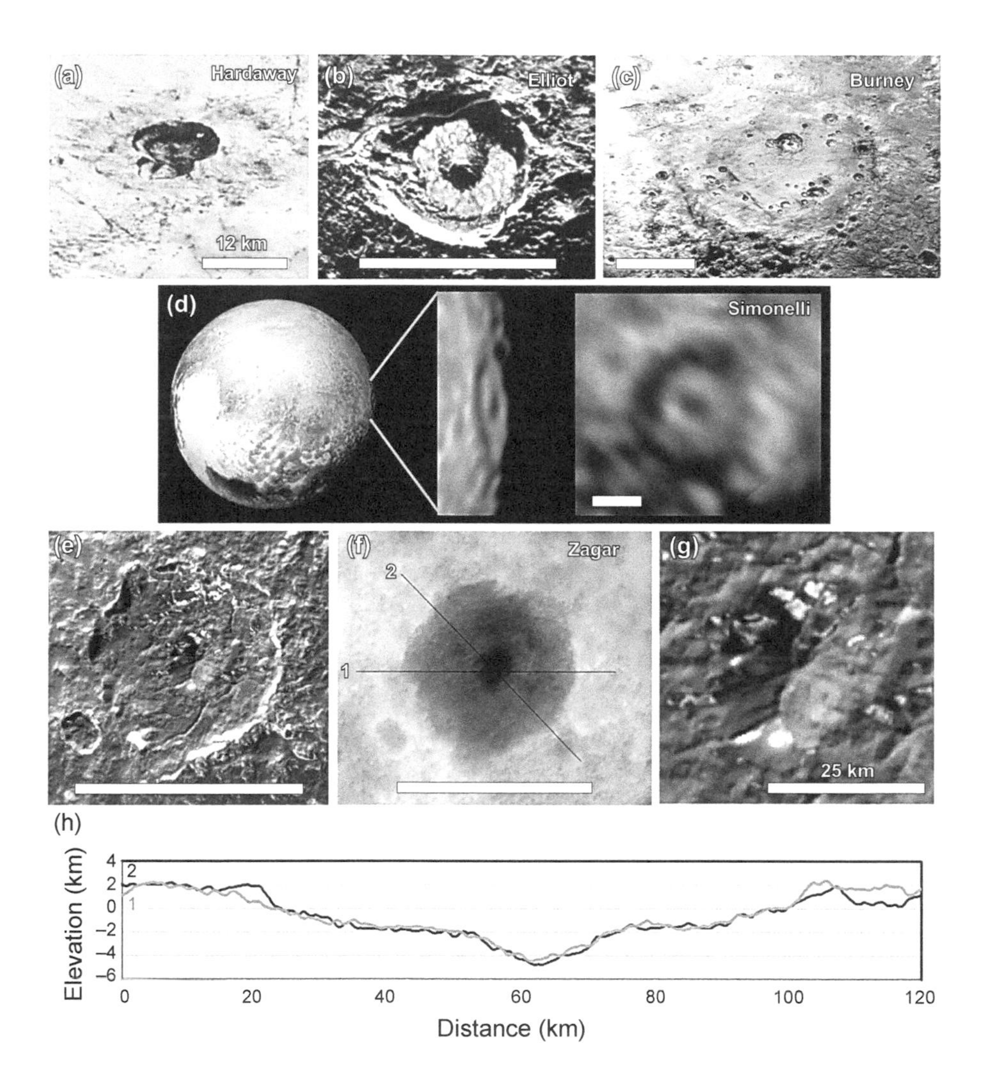

Fig. 1. Examples of Pluto crater morphologies, including **(a)** portion of the highest resolution image strip showing a double crater with distinct layering in the walls and hints of dark material in an ejecta deposit (this is one of only a few examples where ejecta deposits can be easily detected on Pluto); **(b)** central peak with wall terraces and nitrogen-rich ice deposits in the floor; **(c)** basin with multiple ring structures; **(d)** three views of a possible multi-ring structure with a large central peak on the non-encounter hemisphere of Pluto; **(e)** crater with deep central pit; **(f)** topography of central pit crater and surroundings, where white is high and black is low, and the elevations range from approximately +3.8 km to −5.3 km (a linear stretch is applied); **(g)** close-up of central pit structure; and **(h)** topographic profiles as shown in **(f)** (the profiles start at the numbered side). All scale bars are 100 km except where noted. Image sources: **(a)** from PELR_P_MVIC_LORRI_CA (76 m px^{-1}); **(b)**, **(c)**, **(e)**, **(g)** from PEMV_P_MVIC_LORRI_CA (315 m px^{-1}); **(d)** from PELR_PC_MULTI_MAP_B_12_L1AH (16.7 km px^{-1}), where the rightmost panel is a simple cylindrical reprojection. Note that in this figure and subsequent figures, different stretches of the pixel values have been applied to different panels, therefore absolute brightness cannot be compared across the frames.

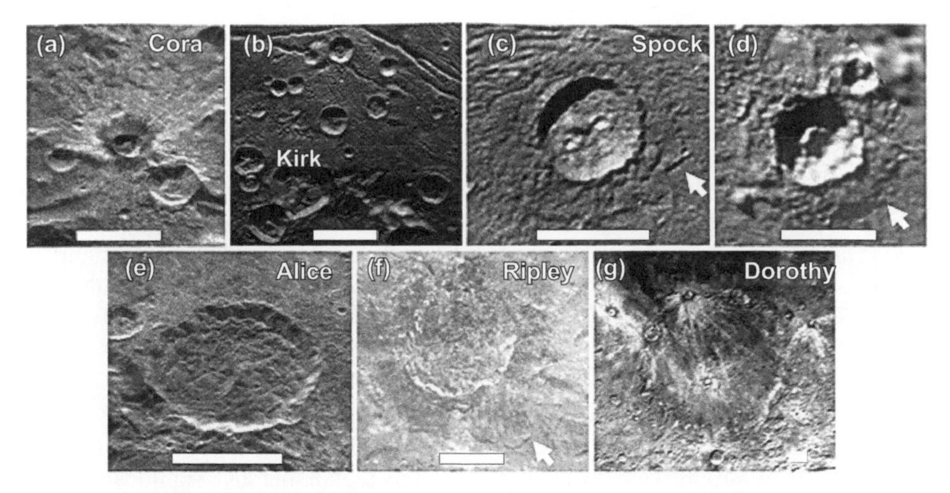

Fig. 2. Examples of Charon crater morphologies, including **(a)** distinctive ejecta albedo patterns, **(b)** smaller bowl-shaped simple craters and craters with mass wasted material on their floors, **(c)–(d)** complex craters with central peaks and thick ejecta deposits, **(e)–(f)** complex craters with heavily terraced floors, and **(g)** the largest identifiable crater on Charon's encounter hemisphere (~250 km in diameter). All scale bars are 30 km in length. White arrows indicate the margins of thicker ejecta deposits and in the case of **(f)**, a landslide-like ejecta deposit. Image sources: **(a)**, **(b)**, and **(e)** from PELR_C_MVIC_LORRI_CA (160 m px^{-1}); **(c)** and **(d)** from PEMV_C_MVIC_LORRI_CA (622 m px^{-1}); and **(f)** and **(g)** from PELR_C_LORRI (865 m px^{-1}).

Because a slope of q = −3 is commonly seen, the R-plot helps visually distinguish changes from this slope as a function of diameter and between different crater populations. For each distribution, we normalize the number of craters per bin by the mapped area to give the density of craters per size bin. In this section, we also follow the convention of making all of the plots square (e.g., 1 order of magnitude on the x-axis is the same physical length as 1 order of magnitude on the y-axis) so that the slopes can be directly visually compared between different plots. The data and areas for the distributions are described in *Singer et al.* (2019) and the full dataset is also archived at the following DOI: *10.6084/m9.figshare.11904786*.

3.1. Pluto Terrains

Here we describe the broad physiographic provinces on Pluto (Fig. 3) and present their crater SFDs for terrain age analysis. We briefly comment on the general geologic context of the regions, especially as related to degradation

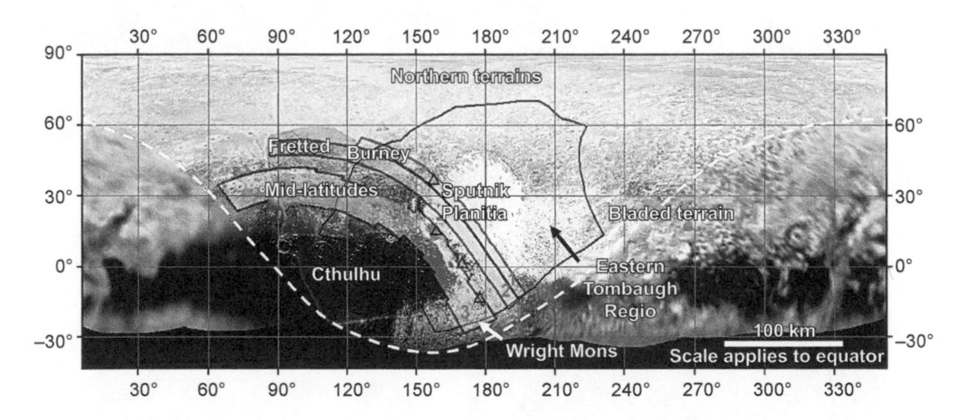

Fig. 3. Simple cylindrical projection of the Pluto basemap mosaic available on NASA's Planetary Data System (PDS) website (*https://pds-smallbodies.astro.umd.edu/holdings/nh-p_psa-lorri_mvic-5-geophys-v1.0/dataset.shtml*), with the shaded areas demarcating the broad regions described in the text. See sections 3 and 5 for details of each region. Colored areas indicate extent of mappable PEMV_P_MPAN1 (480 m px^{-1}) and solid black outline encompasses the more limited area from PEMV_P_MVIC_LORRI_CA (315 m px^{-1}). Small black triangles indicate the chaotic mountain block regions. The high-resolution strips are also shown and labeled 1, 2, and 3 in blue text for PELR_P_MVIC_LORRI_CA (76 m px^{-1}), PELR_P_MPAN_1 (117 m px^{-1}), and PELR_P_LEISA_HiRES (234 m px^{-1}), respectively. White dashed curve separates the encounter hemisphere from the non-encounter hemisphere. The black regions at the bottom of the map were regions in an extended period of darkness (Pluto night) during the time of the New Horizons flyby. Modified from *Singer et al.* (2019).

of craters, or superposition relationships of other geologic processes compared with cratering. More details on the geologic context for different regions on Pluto can be found in the chapter by White et al. in this volume.

3.1.1. Cthulhu Macula. Pluto's dark equatorial band has as complicated and diverse a geologic history as any other location on Pluto. Although Cthulhu Macula is unified in its dark albedo, likely a tholin coating of variable thickness (*Cruikshank et al.,* 2015; *Stern et al.,* 2015; *Grundy et al.,* 2016, 2018; *Protopapa et al.,* 2017; *Schmitt et al.,* 2017; *Cook et al.,* 2019), terrains vary from heavily cratered, apparently ancient regions, to smooth, more lightly cratered plains (Fig. 4). Here we use the entire broad physiographic province of Cthulhu (Fig. 3) on the crater plots (Fig. 4d), meaning that this represents an average crater spatial density for the entire region. The two largest craters after Burney basin (D ≈ 240 km), Edgeworth (D ≈ 140 km) and Oort (D ≈ 110 km), are located in Cthulhu. The existence of many large craters in Cthulhu with varying preservation states indicates a relatively ancient surface overall. In addition to higher resolution, MVIC instrument scans of Cthulhu have better signal to noise than the LORRI instrument coverage, allowing a more detailed look at this dark terrain.

3.1.2. Western mid-latitudes. The mid-latitude region described here lies north of Cthulhu and represents a zone of transitional albedo between the dark Cthulhu and the brighter northern terrains (Fig. 5). This region contains both relatively heavily cratered areas and large eroded plains (i.e., Piri and Bird Planitiae) that show few craters (*Moore et al.,* 2016, 2017). The latitude of 40° was selected as the northern boundary, but the albedo transition is gradual. Distinctive "bright-halo" craters with methane deposits (*Grundy et al.,* 2016) exist both here and in the fretted terrain to the north (see next section).

The mid-latitudes exhibit the same paucity of small craters (D < 10 km) as seen on other terrains on Pluto (and Charon as described at length in section 3.3), but are also deficient in the largest craters compared to Cthulhu. The surface area of Cthulhu seen in the MVIC close approach scan (MCA_Cthulhu in Fig. 4d) is similar to that of the mid-latitude region seen in the same scan (MCA_Mid-lat in Fig. 5b) and the Cthulhu region has four craters larger than D = 50 km, while the mid-latitude region has none.

It could be by chance that not many large craters occur in this region, but also significant resurfacing, possibly via an early mantling episode, may have erased some large craters in this region. The lack of large craters suggest early (rather than later) resurfacing, because large craters should occur more frequently in Pluto's early history. Eroded scarps (notably Piri Rupes) and bright-halo craters suggest sublimation and redeposition of frosts have played a large role in modifying this region over various timescales, possibly related to Pluto's different seasonal zones (*Binzel et al.,* 2017; *Earle et al.,* 2017). The exact configuration of large smooth regions adjacent to heavily cratered regions is not easily explained, however. Overall this region appears middle-aged to old, but large areas have been, and possibly continue to be, resurfaced.

3.1.3. Fretted terrain, Burney Basin, and other northern terrains. North of the 40th parallel Pluto has yet another diverse set of terrains (Fig. 6). Starting in the west,

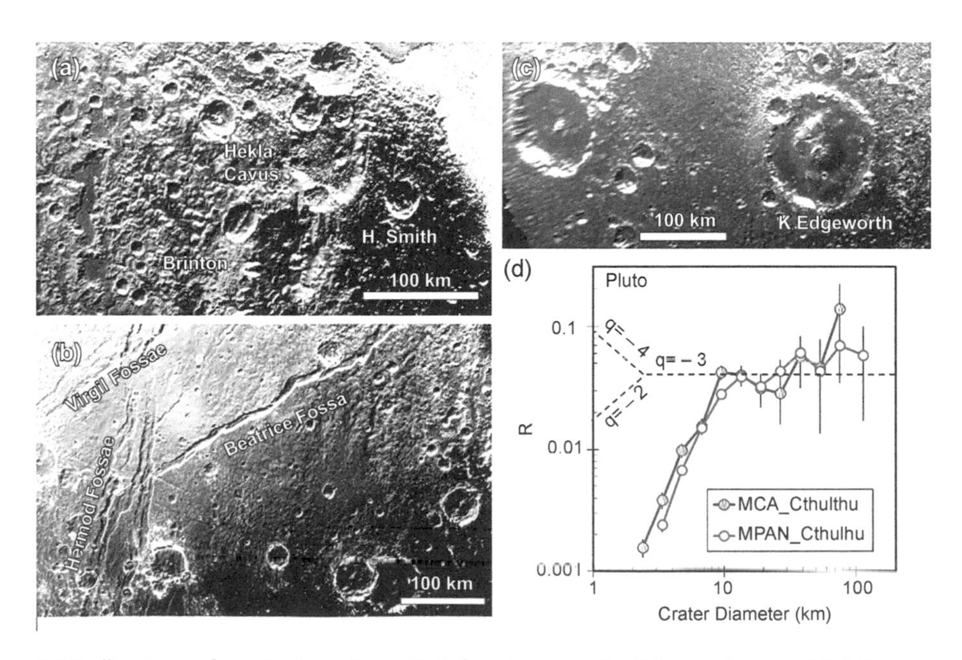

Fig. 4. Surfaces of Cthulhu range from **(a)** heavily cratered and rugged to **(b)** smoother and lightly cratered. **(c)** Two large craters on the encounter hemisphere. **(d)** R-plot of the Cthulhu region in two different datasets [see Fig. 3 for dataset extents and *Singer et al.* (2019) for details]. Also shown in **(d)** are dashed lines representing a differential slope of –4, –3, and –2 for reference. Image sources: **(a)** from PEMV_P_MVIC_LORRI_CA (315 m px^{-1}), noted as MCA in the legend of this plot and subsequent plots; **(b)** and **(c)** from PEMV_P_MPAN1 (480 m px^{-1}), noted as MPAN in plot legends.

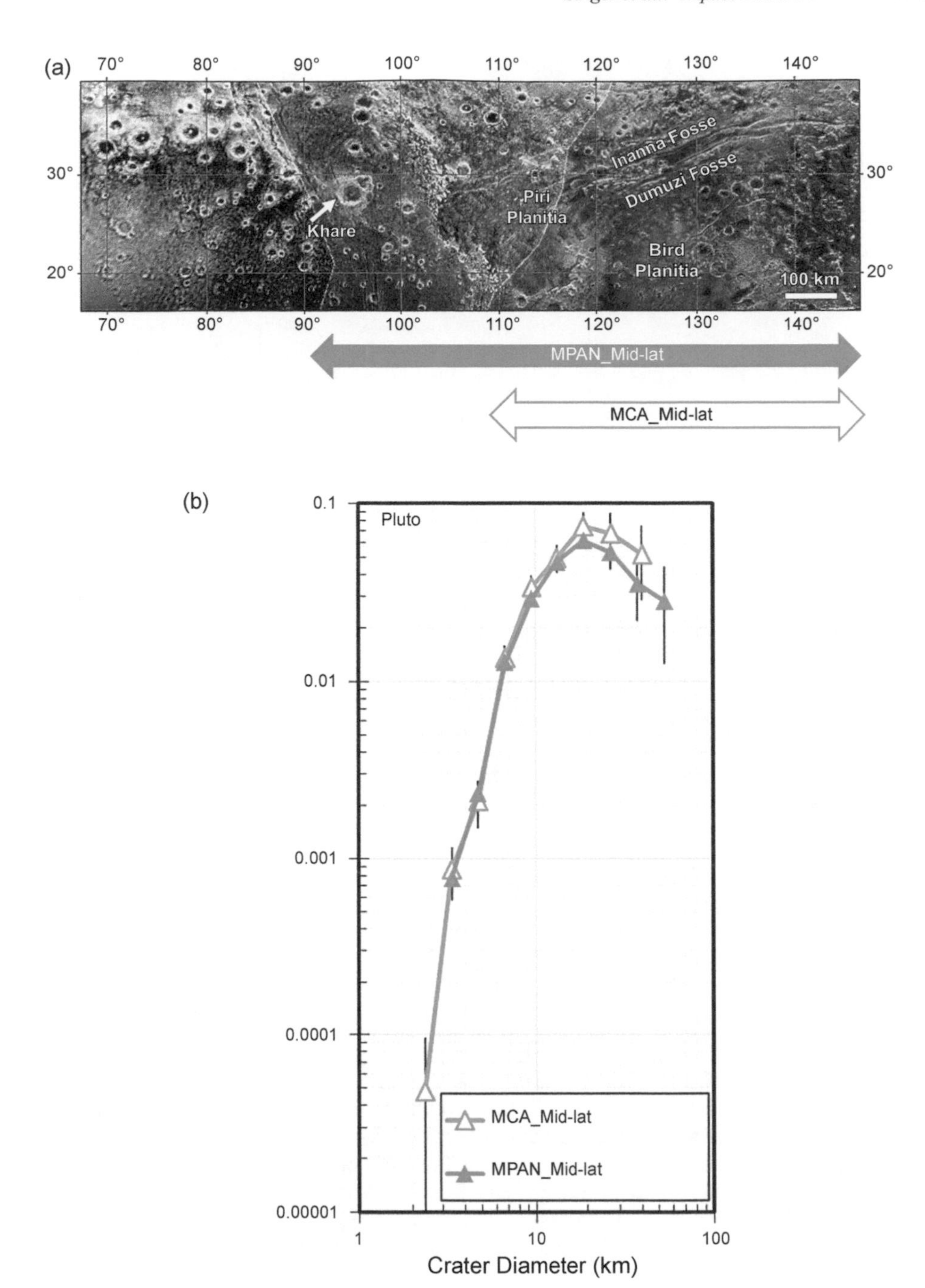

Fig. 5. (a) Simple cylindrical zoom of the mid-latitude region west of Sputnik Planitia showing varied crater densities and distinctive "bright halo craters." Background map from the 850 m px^{-1} (PELR_P_LORRI) mosaic, with black outlines indicating the terrain covered by the higher-resolution datasets whose crater measurements are shown in **(b)**. **(b)** SFDs from this region from the two different resolution datasets that cover different subsets of the area: PEMV_P_MVIC_LORRI_CA (315 m px^{-1}) and PEMV_P_MPAN1 (480 m px^{-1}).

a set of unique conspicuous eroded valleys termed "fretted terrain" (*Howard et al.,* 2017; see also the chapter by Moore and Howard in this volume) cover an approximately 250 × 450-km area. Although there are very few, if any, cases where the valleys can be seen to directly cut through or breach crater rims, in some areas the valleys appear to be diverted around crater rims. This pattern indicates the crust itself is old and the valleys are a later feature, consistent with some of the craters' degraded appearance.

Burney basin lies to the east of the fretted terrain and represents a unique surface. The highest resolution LORRI strip (pixel scale of 76 m px^{-1}) passes over Burney. The area inside the basin rim is generally smoother than much of the surrounding terrain. There are many possible reasons for this. The interior of the basin may erode differently than its surroundings due to the basin formation altering the material properties in this location. Alternately, or in addition, the basin may collect more atmospheric deposits

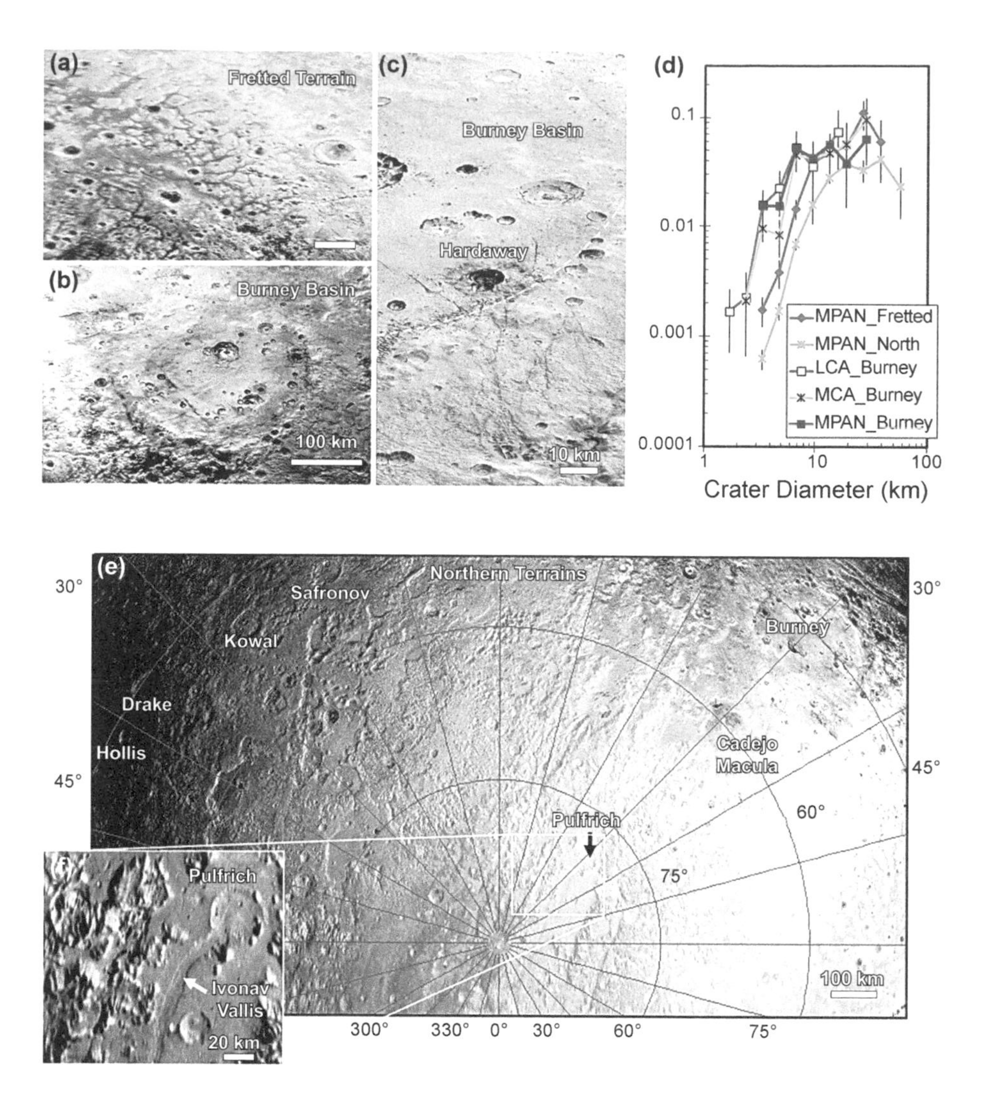

Fig. 6. **(a)** Fretted terrain and **(b)** Burney basin shown at 480 m px^{-1} (PEMV_P_MPAN1). **(c)** A portion of the highest-resolution strip at a pixel scale of 76 m px^{-1} (PELR_P_MVIC_LORRI_CA) with part of Burney basin (dashed outline indicates approximate rim location) and superimposed younger craters, some with dark/bright layers visible in the walls, and one with distinct dark ejecta (Hardaway crater). **(d)** SFDs for the two northern regions (fretted terrain and the north polar area), and a comparison of the craters superimposed over Burney basin mapped at two different pixel scales. **(e)** North polar stereographic view of the varied northern terrain at 480 m px^{-1} (PEMV_P_MPAN1), most of which is relatively lightly cratered and heavily mantled or resurfaced. **(f)** Several infilled craters exist in a large valley near Pluto's north pole.

because of its bowl-like geometry. A more thorough future investigation involving modeling and topographic data may shed light on this topic. Burney basin is also a site of so-called washboard terrain, which has been suggested to be an ancient texture, fitting with the likely old age of the basin (*White et al.,* 2019).

Nitrogen ice (N_2), methane ice (CH_4), and carbon monoxide (CO) sublimate and redeposit in Pluto's seasonal and mega-seasonal cycles (e.g., *Stansberry,* 1994; *Spencer et al.,* 1997; *Young,* 2013; *Bertrand et al.,* 2018, 2019; see also the chapter by Young et al. in this volume), and radiolytic processing of CH_4 either in the atmosphere or on the surface creates heavier, darker, and generally redder complex molecules over time, which have been referred to as haze particles or tholins (e.g., *Grundy et al.,* 2018). To this point, several alternating bright and dark layers are seen in a few craters in the highest-resolution strip (Figs. 1a and 6c), hinting at previous epochs of deposition. These alternating dark/bright layers do not look like the product of mass wasting alone: The layers are seen at approximately the same elevation in several craters, as well as in fault and mountain block walls seen elsewhere on Pluto (*Moore et al.,* 2016). This image strip also reveals dark ejecta around the freshest crater (Hardaway crater in Figs. 1a and 6c) and possible dark ejecta blocks around a few others. The older, more degraded craters show hints of dark material, but it appears that bright volatile deposition has occurred over the entire region, covering any darker material ejected by the older craters. Therefore, the general albedo of craters in this region gives an indication of relative age for a given impact. With a refined cratering rate model, this information might put constraints on the rate of volatile deposition in this region. It should be noted, however, that different regions on Pluto do not follow this same pattern (e.g., in Cthulhu where the surface is primarily dark).

3.1.4. Lightly cratered terrains. Pluto has several regions with few to no identifiable craters. Eastern Tombaugh Regio (ETR) lies to the east of Sputnik Planitia and is dominated by rough, likely sublimation-driven, pitted terrain and perched, ponded glacial N_2 (*Moore et al.,* 2016, 2017; see also the chapter by Moore and Howard in this volume). One large potential crater in the center of ETR (D = 30 km, labeled "A" in Fig. 7a) and several more along the edges are being eroded by the pit-forming process. On the few- to 10-km scale many pits form quasi-circular features. Most of these pits do not appear to be impact craters due to their lack of ejecta or distinct rims and the fact that they share septa (dividing walls) with the surrounding pits. For the few circular features that appear to stand on their own, it is harder to determine if they are impact craters. Several deeper pit-like features (termed cavi) lie near Sputnik Planitia (labeled "B" and "C" in Fig. 7a). They are ~16 km and ~10 km in diameter, respectively, and ~3.5 km and ~2.3 km in average depth based on stereo topography (*Schenk et al.,* 2018a), giving them depth-to-diameter ratios of ~0.22. This is somewhat deep for craters of this size for icy surfaces, which are generally closer to a depth-to-diameter ratio of 0.1, but not implausibly so. The larger pit is not radially symmetrical in the height of the "rim" compared to the base. These features share the same craggy appearance as much of the rest of ETR, but they stand out because they are not filled with nitrogen ice, whereas much of the surrounding terrain is. Although the floors of these features are at or below the level of much of the surrounding smoother, nitrogen-ice-filled plains nearby, they do not appear to be filled with nitrogen ice themselves because they do not have flat floors. It is possible the surrounding craggy terrain is simply tall enough to keep N_2-rich ice from flowing into these depressions, whereas in other areas the N_2-rich ice is able to flow into the lows between the craggy peaks. Alternatively, the N_2-rich ice was removed from these pits by some mechanism (drainage or venting of pressurized gas are some possible mechanisms). These larger pits could be modified craters, or perhaps they formed through a completely different mechanism such as erosion, collapse, or venting from the subsurface. If they are modified craters, no ejecta or distinct rims are left for these features. Given the morphology of these deep pits is dissimilar in many respects from an impact crater, and is similar in many respects to the craggy terrain found

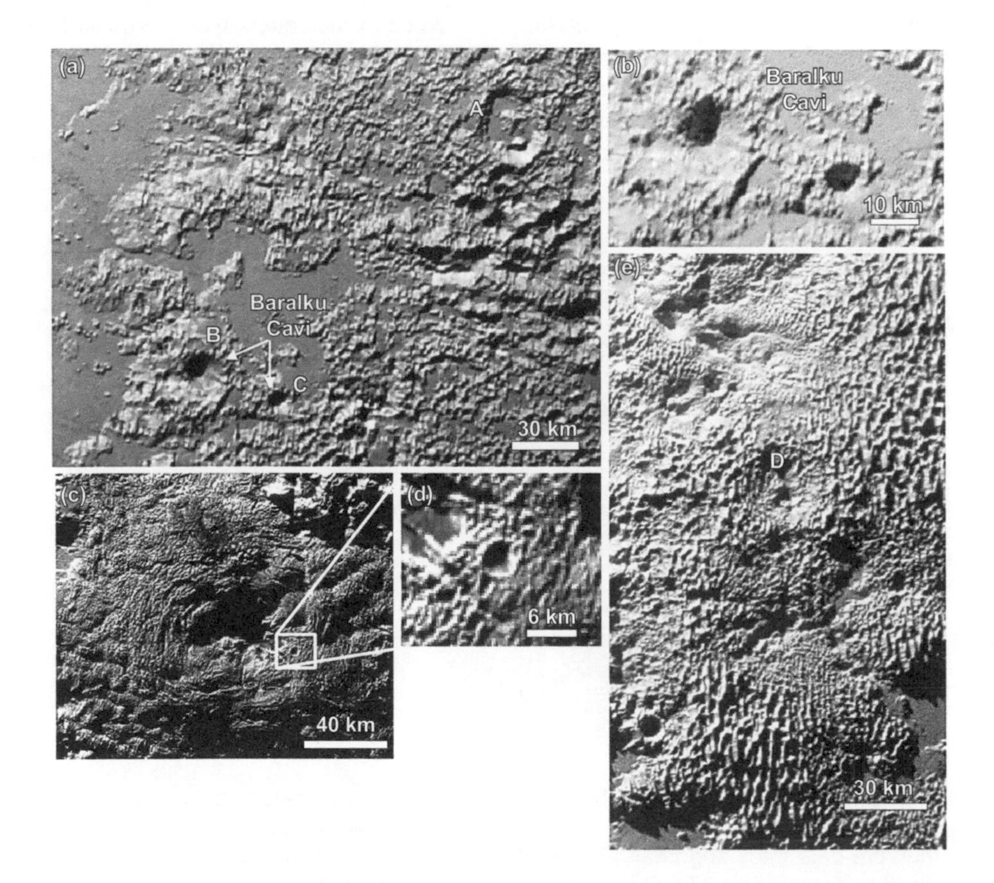

Fig. 7. Lower-density cratered terrains include **(a),(b)** eastern Tombaugh Regio; **(c),(d)** Wright Mons; and **(e)** bladed terrain. A few of the larger ambiguous features are labeled with uppercase letters, including several pit-like structures informally named the Baralku Cavi described at length in section 3.1.4; **(b)** displays a different image stretch of Baralku Cavi to show the nature of the structures inside the shadowed regions. Images are all from PEMV_P_MVIC_LORRI_CA (315 m px^{-1}).

elsewhere in ETR, we have not included them in the SFD displayed in Fig. 13b.

There are no distinct, unambiguous impact craters anywhere on Sputnik Planitia. The main body of bladed terrain (located on Tartarus Dorsa) is also devoid of obvious impacts, although one ambiguous ~32-km-diameter subcircular feature (labeled "D" in Fig. 7e) and several smaller ones exist closer to the edge of this terrain.

Few, if any, possible craters superpose the main mound of Wright Mons (*Singer et al.*, 2016). One possible 5.5-km-diameter crater sits in the ridged terrain near the rim of the Wright Mons central depression. This one possible crater on the main mound has no obvious ejecta and is only quasi-circular with a non-continuous rim. If it was formed as an impact crater, it would be degraded and/or deformed from its original form. The wrinkly texture around the central depression is made of ridged structures that often look arcuate on small scales, and the ridges are emphasized in the roughly east-west direction by the oblique lighting (Wright Mons lies near the terminator). Thus, it is not clear if this feature on Wright Mons is an impact crater, and it could alternatively be a collapse pit, ridge structure, or cavus similar to others around the Mons. There are a few small, more-circular possible craters in nearby terrains, but no other crater-like features are visible on the mound itself. If this one 5.5-km-diamater feature is a crater, it is not fresh. In addition to the characteristics described above, it has an accumulation of dark material in its interior. It is possible, although unlikely, to have one 5.5-km-diameter crater form on Wright Mons, and for it to have time to degrade to its present state, while no other visible craters form. We present two cases for Wright Mons crater SFDs in Fig. 13b: One case where an upper limit is set by the lack of craters, and another case with this one 5.5-km-diameter feature as the only crater on Wright Mons.

3.2. Charon Terrains

3.2.1. Vulcan Planitia. Vulcan Planitia is a large plain that dominates the southern portion of Charon's encounter hemisphere (Fig. 8). In addition to craters, enigmatic "mountains in moats," narrow ridges and troughs, and wider ropy structures decorate the smooth plains of Charon's Vulcan Planitia (Fig. 8a). In general, where craters occur, they seem to overprint all other features, and only a few examples of craters that may have been cut by tectonic activity exist (Figs. 8d,e). Alternatively, it is possible the craters formed over a tectonic feature, which affected their final forms. This may apply to the craters shown with an arrow in Fig. 8e where the craters are more comparable in size to the fracture they interact with. Note that these two crater "halves" are unlikely to be part of a pit chain (formed through drainage of regolith into a fracture, rather than by impact) because they are offset from each other, are slightly different sizes, and have distinct rims. Given there are very few examples of this type of feature, it is clear that most craters formed after tectonic episodes.

None of the craters within Vulcan Planitia are partially filled, obviously embayed, or breached with smooth later flows (e.g., in the lunar mare style). Just north of Vulcan Planitia on the eastern side (north of the large tectonic scarps defining the border) there are some partially filled

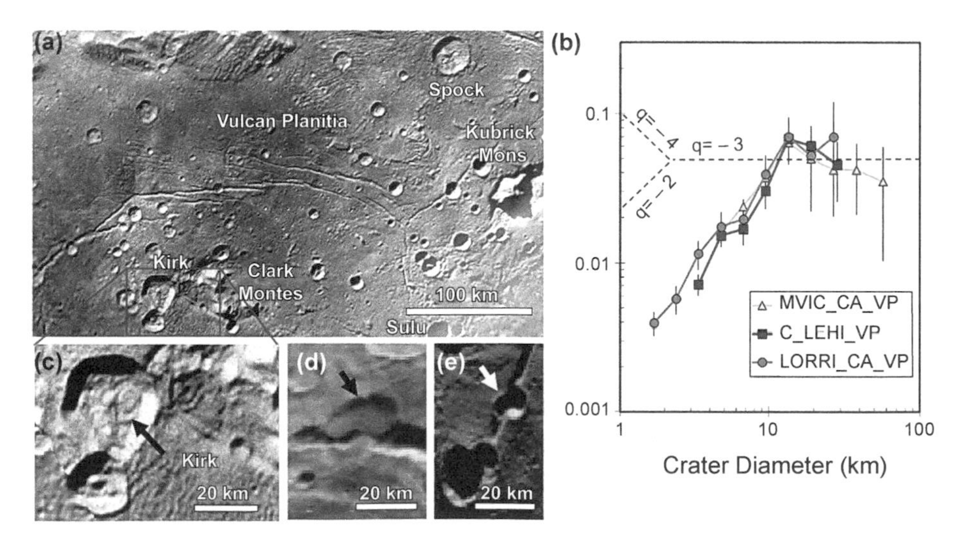

Fig. 8. **(a)** Charon's Vulcan Planitia, **(b)** SFDs for three different-resolution datasets that cross Vulcan Planitia (also shown are dashed lines representing a differential slope of −4, −3, and −2 for reference), **(c)** close up of hummocky-floored craters, **(d)** one crater that may pre-date fracturing near 13°N/328°E, and **(e)** one or two craters that may have been affected by faulting on Vulcan Planitia (centered at 27°S/328°E) (but see text for discussion). Image sources: **(a)** from a mosaic of PELR_C_LORRI (865 m px^{-1}) and PELR_C_LEISA_HIRES (234 m px^{-1}; noted as C_LEHI in the legend); **(c)–(e)** from PEMV_C_MVIC_LORRI_CA (622 m px^{-1}; noted as MVIC_CA in the legend). In the plot legend, LORRI_CA refers to crater data collected from the PELR_C_MVIC_LORRI_CA image set. Figure modified from *Singer et al.* (2019).

craters; it is unclear if these flows were contemporaneous with the emplacement of Vulcan Planitia (and not so thick that the craters remain) or if they occurred somewhat later. Some large craters within Vulcan Planitia do have hummocky floors, but they resemble landslide material seen in some lunar crater floors (see Fig. S4 in *Singer et al.*, 2019). Some more lightly cratered areas exist across the plain. These areas are conspicuous to the eye, but further statistical analysis is needed to determine if these low-density areas could be stochastic or if they may indicate some later resurfacing after the majority of Vulcan Planitia was emplaced.

Crater identification on Vulcan Planitia is relatively straightforward (*Robbins et al.*, 2017; *Singer et al.*, 2019) given the favorable lighting geometry (oblique Sun) and the mostly smooth surface (Fig. 8a). Small craters do exist, but are not abundant, and D $\lesssim$ 10-km craters are deficient compared to a constant logarithmic slope distribution extrapolated from larger craters, similar to what is seen on Pluto. The highest-resolution strip on Charon (pixel scale of 155 m px^{-1}) yields similar results to the lower-resolution datasets (*Robbins et al.*, 2017; *Singer et al.*, 2019).

3.2.2. Oz Terra. The terrain north of Charon's smooth plains, collectively termed Oz Terra (Fig. 9; here we include the dark Mordor Macula north polar region in our SFDs), is a complex terrain with numerous scarps, valleys, large depressions, and a number of craters (*Beyer et al.*, 2017; *Robbins et al.*, 2019). The nearly overhead (low phase angle) lighting and lower signal-to-noise of the LORRI images of Charon's generally dark terrains make crater identification difficult in Oz Terra. Thus, we restricted the R-values as seen in Fig. 14 to D > 50 km. Large structures were also examined with stereo topography. The large arcuate scarp in Mordor Macula (McCaffrey Dorsum indicated with small white arrows in Fig. 9) at first glance appears to enclose a basin, but topography reveals that the area interior to the scarp does not have a bowl-shaped, negative topographic expression (as would be expected for a basin; Figs. 9c,d), indicating the scarp may be instead a tectonic feature (*Beyer et al.*, 2017). A few smaller craters (D < 30 km) can be seen in Oz Terra, including distinctive craters that have dark inner-ejecta near the crater rim, and bright rays farther from the crater (some examples can be seen in Fig. 2). This ejecta pattern is consistent with layering in the near-subsurface of Charon, which has been excavated by these craters (e.g., *Robbins et al.*, 2019). The material closer to the surface is ejected farther forming the bright rays, and deeper material (in these cases darker material) is ejected at lower speeds, landing closer to the crater. Several of these craters are associated with a stronger ammonia and water

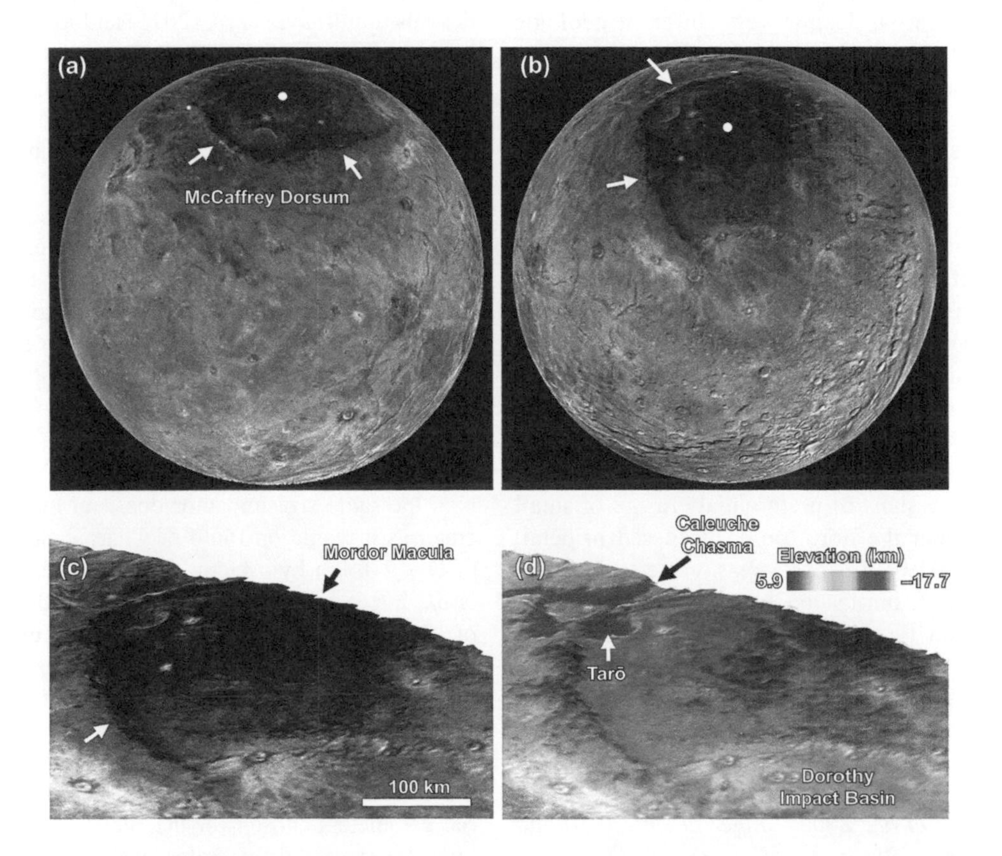

Fig. 9. **(a)** Western portion and **(b)** eastern portion of Oz Terra on Charon. White dot indicates Charon's northern or positive pole location. **(c)**,**(d)** Perspective views of Mordor Macula with heights projected by stereo topography (*Beyer et al.*, 2017; *Schenk et al.*, 2018c). Small white arrow in **(a)**–**(c)** indicates arcuate ridge discussed in section 3.2.2. Images are all from PELR_C_LORRI (865 m px^{-1}). The crater SFD for Oz Terra can be seen in Fig. 14.

ice signatures in the New Horizons spectral data, implying this material is excavated from below (*Grundy et al.,* 2016; *Dalle Ore et al.,* 2018; see the chapter by Protopapa et al. in this volume).

3.3. Summary of Pluto and Charon Crater Size-Frequency Distributions and Relation to Impactor Populations

As can be seen above (and in Figs. 13a and 14), the crater SFDs on Pluto and Charon all show a break or "elbow" in the distributions, where larger craters have an average differential slope more similar to a power law exponent of q = −3 (*Robbins et al.,* 2017; *Singer et al.,* 2019), and craters smaller than ~10–15 km in diameter have a slope that is more shallow (closer to 0) than q = −2. A full table of slopes for each region is given in Table S2 of *Singer et al.* (2019). For example, on Charon, the average slope for D < 10 km is −1.7 ± 0.2. The Pluto regions show a larger variation in the slope below ~10 km diameter — consistent with the wider range of geologic processes occurring on Pluto. We believe this slope is representative of the impactor population, and not solely a product of differential erosion of smaller craters. This conclusion comes from several lines of evidence:

1. The crater SFDs on both Pluto and Charon show the same break location despite very different geologic histories between the two bodies (e.g., Charon does not have volatile ices on its surface or an atmosphere like Pluto does).
2. The same slope is seen for the eastern and western halves of Vulcan Planitia if it is split down the middle vertically (*Singer et al.*, 2019, Fig. S6).
3. The craters on Vulcan Planitia do not show any obvious signs of preferential erasure of small craters. Typically, a process that preferentially erases smaller craters would also leave some partially affected craters (e.g., intermediate-sized craters that are partially filled, embayed, or mantled), and those are not seen on Vulcan Planitia.
4. Although there are more geologic processes occurring on Pluto that could modify craters, there are only a few areas that show signs of preferential erasure of small craters [such as near the north pole; discussed in detail in *Singer et al.* (2019)]. In most areas, craters are either completely erased or, similar to Charon, the more heavily cratered regions on Pluto do not all show intermediate-sized craters that are partially erased.
5. A similar shallow SFD slope for the equivalent size range of craters (taking into account scaling for different impactor velocities and surface gravities) is seen on some outer planet satellite surfaces such as on Europa and Ganymede (e.g., *Zahnle et al.,* 2003; *Singer et al.,* 2019) and the uranian satellites (*Kirchoff and Dones,* 2018).

Singer et al. (2019) examined the break location for Charon's Vulcan Planitia, because this is the surface that shows the least signs of crater modification between Pluto and Charon. They found the break to be at a crater diameter of ~13 km, which equates to an impactor size of d ~ 1–2 km depending on which scaling parameters are used (see supplement of *Singer et al.,* 2019).

A distinct break in slope around these sizes (d ~ 1–2 km) is not seen for the asteroid belt, which in contrast retains a steeper differential slope (close to −3 or even steeper) at smaller sizes (see discussion in *Singer et al.*, 2019). This means the Kuiper belt has a large deficit of smaller objects compared to what would be expected if the SFD had a steeper slope (*Singer et al.,* 2019). In the smallest size bin measurable by the New Horizons crater data for Charon (the bin centered at 1.7 km in diameter in Fig. 8 for the LORRI_CA_VP distribution), there are only 30 craters found. There would be approximately an order of magnitude more craters expected (~300 total) if there were no "elbow" in the SFD.

New Horizons encountered the cold classical KBO (486958) Arrokoth at 43 AU on January 1, 2019 (*Stern et al.,* 2019; see also the chapter by Stern et al. in this volume). Arrokoth is a contact binary ~35 km across. By all indications the surface of Arrokoth is ancient, but it is not heavily cratered (*Singer et al.,* 2020a; *Spencer et al.,* 2020). This finding supports a lack of small objects in the Kuiper belt.

Nix was the best-imaged small satellite during New Horizons' 2015 flyby (*Weaver et al.,* 2016; see also the chapter by Porter et al. in this volume). The Nix crater population is described in *Weaver et al.* (2016) and *Robbins et al.* (2017). Nix's surface has a relatively high spatial density of craters, equivalent to or higher than most heavily cratered terrains on Pluto or Charon, implying it is an old surface. Both the spatial density and number of craters are more uncertain on Nix than on Pluto or Charon because of the uncertainty in estimating the surface area on which craters were emplaced and the ambiguity of which circular, crater-like features are truly impact craters. Fitting a power law to the data is also uncertain based on what data should or should not be included and an appropriate minimum diameter to which the data should be fit. Based on an examination of the most reliable features, fitting craters with diameters 2–6 km (N = 15 features) from *Robbins et al.* (2017) yields a shallow differential slope of q = −1.19 ± 0.82. Because of the lower gravity on Nix, the same size impactor does not make the same size crater as it would on Pluto or Charon. Scaling Nix craters of D = 2–6 km by a factor of 2.1 (as described in *Weaver et al.,* 2016) would yield an equivalent diameter of craters of ~1–3 km on Pluto or Charon. While this uncertainty is large, the shallow slope is consistent with the shallow slope for D < 10–15-km craters on Pluto and Charon.

The equivalent SFD slope in the impactor population is, based on scaling theory, slightly shallower than that of the craters and mildly dependent on the choice of material parameters chosen for the scaling from crater sizes to impactor sizes. All the craters we can observe on Pluto or Charon should be in the gravity regime, and no matter which endmember materials are chosen (hard, non-porous ice or a more porous material like regolith/sand), only a small slope change is introduced. The full description of the scaling law derivation is in the supplement of *Singer et al.* (2019) for Pluto and Charon. For the final power law scaling form of $d = a*D^b$ where d is the impactor diameter, D is the crater diameter, and a is a scaling factor, b comes out to be either 1.151 or 1.088 for a non-porous or porous target surface respectively. The change in slope follows $q_{impactors} = q_{craters}/b$. For Charon's small crater slope of -1.7 ± 0.2, the equivalent differential impactor or KBO slope is approximately –1.5 (for b = 1.151) or –1.6 (for b = 1.088). This difference in slope between the impactor and crater distributions due to scaling is within the error bars of the slope itself but is mentioned here for completeness. The equivalent H-magnitude slope for a luminosity distribution of observed KBOs is described below, where we use the –1.5 value for the impactor slope for objects smaller than the "elbow" in subsequent analysis in this chapter.

This same conversion between crater and impactor SFD slopes applies for the larger craters, which display an average slope close to –3. However, there is a range of slopes for larger craters that also varies over the range of diameters available, as can be seen in the SFD plots, and the error bars are large in some cases (*Singer et al.,* 2019, their Table S2). Thus, we continue to use the slopes as in *Greenstreet et al.* (2015) for modeling the cratering rates of larger craters (described more below), which is consistent with the crater data [see also discussion in the supplement of *Singer et al.* (2019)].

4. IMPACTOR POPULATIONS AND RATES

4.1. Current Structure of the Kuiper Belt

The Kuiper belt is a reservoir of small bodies located beyond the orbit of Neptune, extending from 30 AU to ~1000 AU with the majority of the classical and resonant subpopulations falling between roughly 30 AU and 85 AU. Pluto is the largest known KBO. There are currently ~3300 known KBOs with diameters $d \gtrsim 100$ m, but much of the population remains undiscovered due to their large distance from Earth. Thus, debiased surveys (those accounting for their observational biases) are currently the only available method for determining the intrinsic population of KBOs.

As more KBOs have been discovered, it has become clear that the Kuiper belt is divided into several dynamical subpopulations as described in *Gladman et al.* (2008), including objects in a mean-motion resonance (MMR) with Neptune, scattering objects that are actively experiencing deviations in semimajor axes due to current dynamical interactions with Neptune, and the classical or detached objects, which are neither resonant nor currently scattering off Neptune and are divided into several subcomponents.

Pluto's orbit (semimajor axis $a \approx 39.5$ AU, eccentricity $e \approx 0.24$, and inclination $i \approx 17°$) puts it in the 3:2 MMR with Neptune (*Cohen and Hubbard,* 1965). Although Pluto's orbit (perihelion $q \approx 30.0$ AU and aphelion $Q \approx 49.0$ AU) crosses that of Neptune ($a \approx 30.1$ AU), the resonance protects Pluto from planetary close encounters with Neptune, allowing its orbit to remain stable for billion-year timescales.

4.2. Giant Planet Migration

The complex structure in the Kuiper belt, including many objects in MMR with Neptune, the scattering and detached subpopulations on high-eccentricy (e), high-inclination (i) orbits, and the excited (moderate-e and moderate-i) hot component of the main classical belt, has numerous cosmogonic implications for solar system formation. The migration of the giant planets through a massive (10–100-$M_\oplus$) disk of planetesimals located from the giant planet region to the outer edge of the primordial Kuiper belt (*Fernandez and Ip,* 1984; *Hahn and Malhotra,* 2005), which is only halted when Neptune reaches the outer edge of the disk (*Gomes et al.,* 2005), is the leading explanation for much of the current structure of the Kuiper belt.

The Nice model is a currently heavily explored model of giant planet migration in the early solar system that aims to reproduce the current orbital architecture of the giant planet system (*Tsiganis et al.,* 2005), the capture of the Jupiter (*Morbidelli et al.,* 2005) and Neptune (*Tsiganis et al.,* 2005) Trojan populations, and the precipitation of the late heavy bombardment of the terrestrial planets (*Gomes et al.,* 2005). The largest difficulty with the Nice model, however, is getting the massive planetesimal disk to remain for several hundred million years without either accreting into planets or collisionally grinding itself down into dust before the planets can disperse it into the current structure seen today. The 500-m.y. delay in instability in the Nice model is thus now giving way (*Mann,* 2018) to a shorter phase with most of the rearrangement occurring in $\lesssim 1\%$ the age of the solar system (see *Nesvorný,* 2018, for a recent review).

Once the migration of the giant planets ended, the subpopulations of the Kuiper belt have since naturally dynamically depleted at differing rates due to their differing orbital parameters over the past ≈ 4 G.y. During this time period, it is assumed the orbital distribution of each population has not changed. A summary of the estimated subpopulation decay rates from the literature can be found in *Greenstreet et al.* (2015). Due to the lack of knowledge about the detailed structure of the region beyond Neptune during the giant planet migration process, impact and cratering rates for the Pluto system can only solidly rely on the orbital structure of the Kuiper belt currently known, which is believed to have been unchanged for the past ≈4 G.y.

4.3. Size Distributions of Kuiper Belt Subpopulations

There are large uncertainties in the Kuiper belt size distribution for objects with g-band absolute magnitude $H_g > 9.16$, where the g-band has an effective wavelength of 463.9 nm and a width of 128.0 nm. Absolute magnitude (H) is defined to be the visual magnitude an observer would record if an object were placed both one astronomical unit

(AU) away from the observer and 1 AU from the Sun at zero phase angle where the asteroid is fully illuminated. $H_g >$ 9.16 corresponds to a diameter $d < 100$ km for a g-band albedo p = 5% using the equation $d \simeq 100 \text{ km } \sqrt{(0.05/p)} * 10^{0.2(9.16-H_g)}$. The differential number of objects N as a function of H-magnitude is defined by $dN \propto 10^{(aH)}$, where a is the logarithmic "slope" (hereafter simply referred to as the slope) and maps to the differential distribution in object diameter d, $dN \propto d^{(q)}$, by $-q = 5a + 1$.

The Kuiper belt size distribution for $H_g \gtrsim 8$–9 has been absolutely calibrated by the Canada France Ecliptic Plane Survey (CFEPS) (*Petit et al.,* 2011; *Gladman et al.,* 2012) and is well represented by a single logarithmic "slope" a for all subpopulations with the exception of the hot and cold components of the main classical belt, which appear to have different values of a (*Bernstein et al.,* 2004, 2006; *Petit et al.,* 2011; *Adams et al.,* 2014; *Fraser et al.,* 2014). Extending to KBOs smaller than $H_g = 9$, it is clear a single power law does not fit the observations and a break in the differential size distribution at this H_g-magnitude is needed, which is explored in great detail in *Jewitt et al.* (1998), *Gladman et al.* (2001), *Bernstein et al.* (2004), *Fraser and Kavelaars* (2008), *Fuentes and Holman* (2008), *Shankman et al.* (2013), *Adams et al.* (2014), and *Fraser et al.* (2014). (See also the chapter by Parker in this volume for additional information on observation surveys of KBOs.)

Due to the uncertainty in the Kuiper belt size distribution for $H_g \lesssim 9$, one must assume a size distribution when computing impact and cratering rates onto Pluto and Charon, which provides the largest source of uncertainty for the resulting rates. In the past, groups have modeled a variety of assumed impactor size distribution slopes and compared the resulting crater SFDs, each producing slightly different predictions. The uncertainties in the impactor size distribution are manifested in calculated terrain ages that use the various predictions for absolute calibration. Terrain age estimates for Pluto and Charon will be discussed in section 5.

4.4. Computing Collision Probability

Soon after the Kuiper belt was discovered, impact rates onto Pluto and Charon began emerging in the literature. *Weissman and Stern* (1994), *Durda and Stern* (2000), and *Zahnle et al.* (2003) were some of the first to produce estimates. Their methods consisted of particle-in-a-box calculations or approximations of the KBO number density for those that intersect Pluto's orbit at an average impact speed. *De Elia et al.* (2010) computed the impact flux of plutinos (KBOs other than Pluto located in the 3:2 MMR with Neptune) onto Pluto assuming typical impact speeds of 1.9 km s^{-1} and used this to calculate cratering rates onto Pluto from the plutinos alone.

Dell'Oro et al. (2013) computed impact probabilities of the individual KBO subpopulations from the CFEPS L7 model (*Petit et al.,* 2011; *Gladman et al.,* 2012) through collisional evolution but did not extend their analysis to the cratering rate onto Pluto. The extension of the *Dell'Oro et al.* (2013) cratering rate of plutinos onto Pluto assuming a mean impact velocity was performed by *Bierhaus and Dones* (2015), who also included the production of secondary and sesquinary craters onto the surfaces of Pluto and Charon into their analysis (see section 4.6 for more on secondary and sesquinary craters). Around the same time, *Greenstreet et al.* (2015, 2016) computed the impact and cratering rates onto Pluto and Charon using a similar combination of KBO subpopulations to those used in *Bierhaus and Dones* (2015) but with a different assumed KBO size distribution for objects smaller than telescopic surveys have observed ($d \lesssim$ 100 km), taking into consideration the unique dynamics of Pluto's orbit in the Kuiper belt. *Greenstreet et al.* (2015) computed the impact probability onto Pluto and Charon by modifying a version of the Öpik collision probability code that implements the method described in *Wetherill* (1967) and is based on *Dones et al.* (1999).

Most of the impact and cratering rates onto Pluto and Charon in the literature assume an average impact speed. *Zahnle et al.* (2003) and *Dell'Oro et al.* (2013) quote typical impact speeds from KBOs onto Pluto (or the plutinos) to be approximately 2 km s^{-1}. *Bierhaus and Dones* (2015) adopt this average impact speed for their analysis. *Greenstreet et al.* (2015) modified their analysis to produce a spectrum of impact speeds for each Kuiper belt subpopulation. Compared to other cratered bodies in the outer solar system that have been studied to date, Pluto uniquely sits within the Kuiper belt. Due to the detailed orbital architecture of the various subpopulations within the belt, each subpopulation has a different impact probability and thus impact speed onto Pluto. This is emphasized by the complex dynamics Pluto's orbit experiences over time (see *Greenstreet et al.,* 2015, for a more detailed discussion). Figure 10 shows the impact velocity spectrum onto Pluto from *Greenstreet et al.* (2015). They find a mean impact speed of 2 km s^{-1} for the combined KBO population, but show that the various subpopulations produce a wide spread in impact speeds from Pluto's escape speed at 1.2 km s^{-1} out to a tail at 5 km s^{-1}. Before turning their impact speeds into averages, *Dell'Oro et al.* (2013) computed impact speed distributions for the collisional evolution of the KBO subpopulations. Their impact speed distribution for the plutinos onto each other independently reproduces the same main trends found by *Greenstreet et al.* (2015) for the plutinos impacting Pluto.

4.5. Impact Rates Onto Pluto and Charon

Table 2 lists the computed impact rates for Pluto and Charon described below. *Weissman and Stern* (1994) provided estimates of $d > 2.4$-km comets impacting Pluto and Charon at present rates. For KBOs with $d > 2$ km, *Durda and Stern* (2000) estimated impacts occur on Pluto and Charon on shorter timescales. In addition, they computed the impact rate for KBOs with $d > 100$ km on Pluto.

Zahnle et al. (2003) computed impact rates onto Pluto and Charon for $d > 1.5$-km comets. For $d > 100$-km impactors, *Zahnle et al.* (2003) provided an impact rate scaled

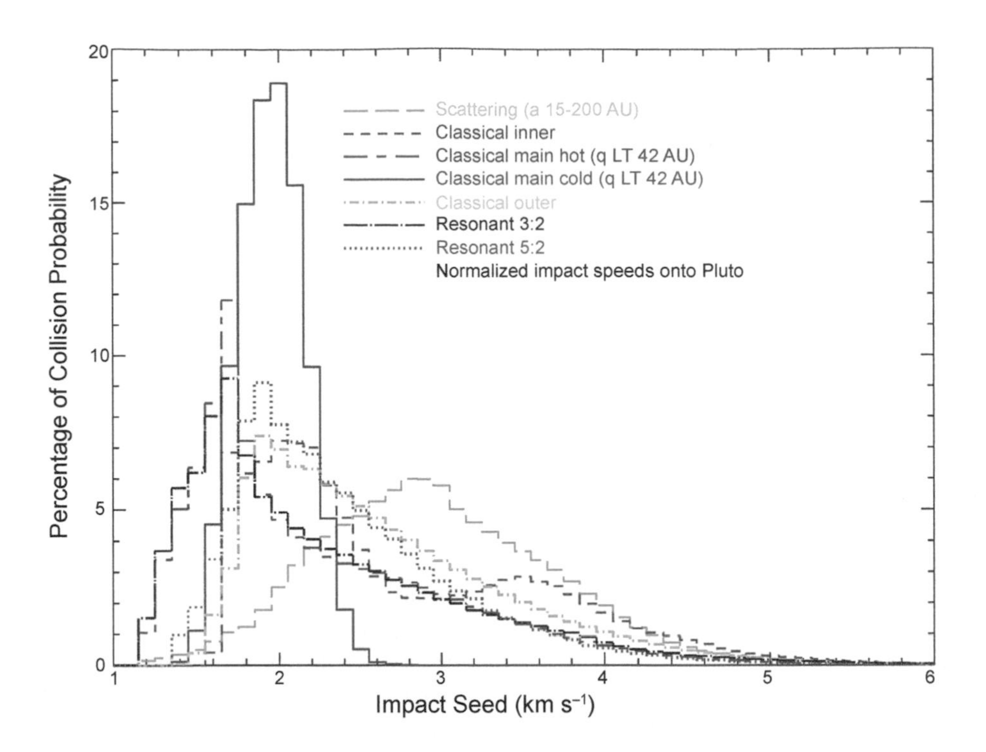

Fig. 10. Impact speed distributions onto Pluto with data from *Greenstreet et al.* (2015). Escape speed from Pluto is 1.2 km s^{-1}. Each subpopulation's distribution is separately normalized.

from the calculations of *Nesvorný et al.* (2000) for plutino impacts onto Pluto and an Öpik-style collision probability estimate from W. Bottke (personal communication). *Zahnle et al.* (2003) also reported an impact rate on Charon that is 16% that on Pluto for d > 100-km impactors. *De Elía et al.* (2010) calculated impact rates for Pluto and Charon for d > 1-km plutinos.

Bierhaus and Dones (2015) estimated impact rates for Pluto and Charon using the estimates from *Dell'Oro et al.* (2013) for plutinos with d > 1 km colliding with each other. *Greenstreet et al.* (2015) provided an impact rate for d > 100-km impactors onto Pluto from a similar combination of Kuiper belt subpopulations to those used in *Bierhaus and Dones* (2015). In addition, *Greenstreet et al.* (2015)

TABLE 2. Pluto and Charon impact rates for a range of impactor diameters from the literature (see text for more details on each study).

Reference	Impactor Diameter Range (km)	Pluto Impact Rate ($\times 10^{-8}$ yr^{-1})	Charon Impact Rate ($\times 10^{-8}$ yr^{-1})
Weissman and Stern (1994)	>2.4	53	10
Durda and Stern (2000)	>2	260	31
Zahnle et al. (2003)	>1.5	100–250	16–42
de Elia et al. (2010)	>1	36–160	10–44
Bierhaus and Dones (2015)	>1	46	10
Durda and Stern (2000)	>100	0.012	—
Zahnle et al. (2003)	>100	0.0023	0.00037
Greenstreet et al. (2015)	>100	0.0048	0.00092

reported an impact rate 19% that on Pluto for d > 100-km impactors on Charon.

Both *Bierhaus and Dones* (2015) and *Greenstreet et al.* (2015) broke down their impact rates by Kuiper belt subpopulation and noted the importance of each to the total impact rate on Pluto. Both papers found that the cold classical KBOs dominate the Pluto impact flux, but by vastly differing amounts [≈84% (*Bierhaus and Dones,* 2015) vs. ≈37% (*Greenstreet et al.*, 2015) for d ≥ 10 km] given their different assumed KBO size distributions.

Although these published estimates span a time period of more than two decades, they are surprisingly in agreement to within a factor of 3 for both Pluto and Charon. This means terrain ages calculated from these estimates agree to within a factor of 3–4. Although this disagreement is relatively small, this could mean a range of 1–4 G.y. in the age estimate of a given surface area on either body, for example. Thus, the assumptions and considerations that went into the predictions should be considered when used for interpretation of the observational data.

4.6. Cratering Rates Onto Pluto and Charon

To convert impact rates to cratering rates, a crater scaling law is needed. The properties of the target and impactor both play a role in the size of crater produced for a given size impactor (e.g., *Holsapple,* 1993). Crater scaling laws are developed from a combination of empirical data and physics principles (*Housen and Holsapple,* 2011). The empirical data includes both laboratory work and measurements of observed craters and ejecta, both from natural impacts and manmade explosion cratering. The crater scaling law used in *Greenstreet et al.* (2015) from *Zahnle et al.* (2003) was developed for solid, non-porous geological materials (relevant to the icy surfaces of Pluto and Charon). As an example, the scaling from *Zahnle et al.* (2003) produces roughly 3% larger simple crater diameters than the scaling law from *Housen and Holsapple* (2011), which was used in *Bierhaus and Dones* (2015). However, as discussed in section 3.3, the difference in slope between the impactor and crater distributions due to different scalings is small and within the error bars of the slope itself.

As mentioned above, *Bierhaus and Dones* (2015) included the production of secondary and sesquinary craters in their computation of the cratering rates onto Pluto and Charon. Secondary craters are produced by debris fragments ejected during the formation of the primary crater that impact outside the primary crater at high enough velocity to form their own crater. Sesquinary craters are made when ejected material is traveling fast enough to escape the source body and subsequently reimpact that body or another object at a later time. *Bierhaus and Dones* (2015) concluded that sesquinaries are not expected to be an important component of the overall Pluto or Charon crater SFDs and secondaries should be visible as a steeper branch of the crater SFD at diameters less than a few kilometer on Pluto and are likely not to be visible in Charon's observed crater SFD.

To produce a predicted crater SFD, one must integrate over the impactor size distribution to convert impactor diameter to crater diameter. The resulting crater SFD then depends on the assumed KBO size distribution. If an average impact speed is used, the crater scaling laws give a single crater diameter for each impactor diameter. In the case of an impact velocity spectrum, the impactor-to-crater diameter conversion is no longer a one-to-one relationship. A given size impactor produces craters with a variety of diameters when traveling at a range of impact speeds. To account for this, one must integrate down the impact speed distribution as well as the impactor size distribution in the conversion of impact rates to cratering rates onto Pluto and Charon (*Greenstreet et al.,* 2015).

Bierhaus and Dones (2015) suggested two possible crater SFD slopes: (1) slope of $q \sim -2$ ($a = 0.2$) from the young terrains of the Galilean and saturnian satellites, which indicates a shallow distribution; (2) slope of $q \sim -3$ ($a = 0.4$) from observational data of KBOs extended from $d \approx 100$ km down to smaller sizes. *Greenstreet et al.* (2015) used several KBO size distribution models to illustrate the uncertainty in the crater SFD slope. Their models included a power law with a "knee" at $H_g = 9.0$ that has a sudden slope change from $a_{bright} = 0.8$ ($q = -5$) to $a_{faint} = 0.4$ ($q = -3$), a power law with a sudden drop in the differential number of objects (i.e., a "divot") by a factor of 6 with the same a_{bright} and a_{faint} as the "knee" distribution, and the "wavy" size distributions from *Minton et al.* (2012) and *Schlichting et al.* (2013). The resulting crater SFDs for Pluto and Charon are shown in Fig. 11 in the form of a relative crater frequency R-plot, which is normalized to a differential D^{-3} distribution (as described in section 3). The predicted crater SFD from *Zahnle et al.* (2003) based on young surfaces on Europa and Ganymede is also shown in Fig. 11. Note that the two slopes in the *Zahnle et al.* (2003) broken power law are the same as the two separate slopes used in *Bierhaus and Dones* (2015). It should be also noted that because the escape speeds (Pluto: $v_{esc} = 1.2$ km s^{-1}, Charon: $v_{esc} = 0.675$ km s^{-1}) and gravitational accelerations (Pluto: $g = 64$ cm s^{-2}, Charon: $g = 26$ cm s^{-2}) are different for Pluto and Charon and both are factors in the crater-scaling laws, a given impactor will produce a slightly larger crater on Charon than on Pluto, shifting the crater SFDs. A discussion of the uncertainties in the crater SFDs can be found at the end of section 4.7.

4.7. Predictions and Implications for Crater Size-Frequency Distributions in the Pluto System

In anticipation of the New Horizons flyby of Pluto and Charon in 2015, *Bierhaus and Dones* (2015) and *Greenstreet et al.* (2015, 2016) made predictions about the cratered surfaces of both bodies. They reported that the expected visible range of craters on both Pluto and Charon is D ~ 1–100 km given the uncertainty in the impactor size distribution. *Bierhaus and Dones* (2015) pointed out that if the crater SFDs are like those of the Galilean and saturnian satellites, young unsaturated terrains should show

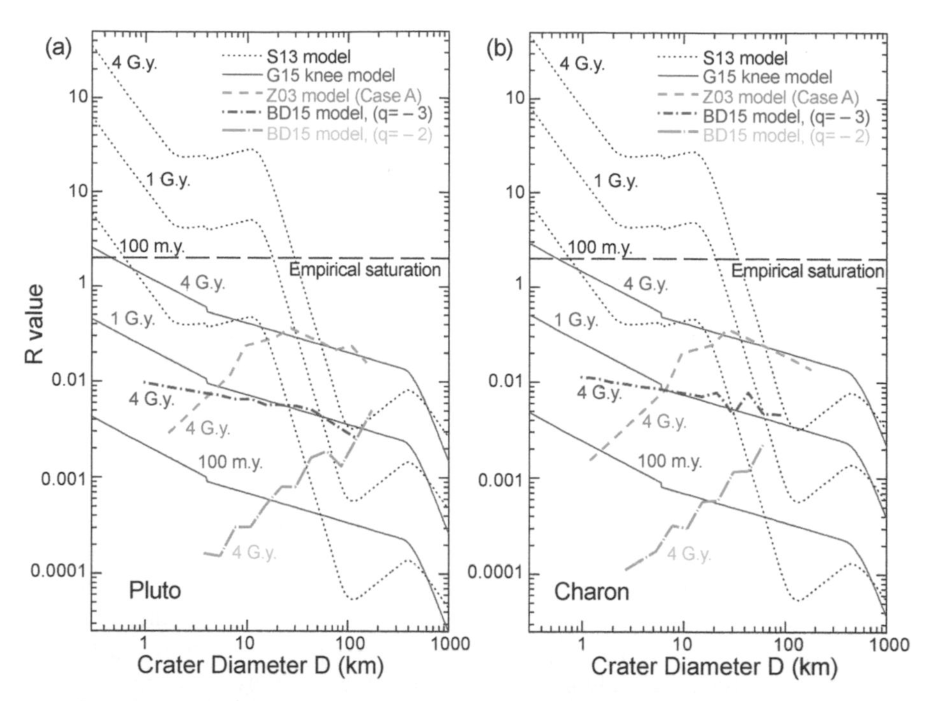

Fig. 11. Relative crater frequency R-plot including KBO SFD predictions from *Zahnle et al.* (2003) (*Z03*), *Schlichting et al.* (2013) (*S13*), *Bierhaus and Dones* (2015) (*BD15*), and *Greenstreet et al.* (2015, 2016) (*G15*) for **(a)** Pluto and **(b)** Charon. Predicted crater SFDs from the *G15* and *S13* models show three different age surfaces. The empirical crater saturation density calculated in *Greenstreet et al.* (2015) from *Melosh* (1989) is shown for reference, where crater densities typically do not increase due to the erasure of previous craters from newly forming ones.

a shallow differential slope of q ~ –2 (a ~ 0.2). This would imply there is no process present that erodes small KBOs as they migrate inward from the region beyond Neptune to the giant planet region. Likewise, it would also imply that small KBOs are deficient relative to an extrapolation of larger KBOs visible to telescopic surveys down to smaller sizes. This last point would have important implications for constraints on the evolution and possible formation mechanism of the KBO subpopulations. However, if the q ~ –3 (a ~ 0.4) slope from large (d ~ 100 km) KBOs extends down to small sizes (d < 1 km), then there will be a steep (q ≳ –3) crater SFD observed across all crater sizes (*Bierhaus and Dones*, 2015; *Greenstreet et al.*, 2015). This would mean some process could exist that destroys small KBOs as they move to smaller heliocentric distances near Jupiter and Saturn, making them underabundant at these distances (*Bierhaus and Dones,* 2015).

Both *Bierhaus and Dones* (2015) and *Greenstreet et al.* (2015) argue that even in 4 G.y. of bombardment, Pluto and Charon are not expected to be saturated for D > a-few-kilometer craters, making it possible to at least relatively (or model-dependently) date their surfaces. *Bierhaus and Dones* (2015) added that if few secondary craters are visible, surfaces may not be saturated at any crater diameters observable by New Horizons. Only one of the impactor size distributions would predict saturation of Pluto's surface: the *Schlichting et al.* (2013) size distribution as explored in *Greenstreet et al.* (2015), which if correct would mean Pluto's surface would saturate for D ≲ 15 km in only 1 G.y. of bombardment. However, as shown in the observed crater SFDs for Pluto and Charon provided in section 3, a deficit of craters is found for this size range (D ≲ 15 km), ruling out the *Schlichting et al.* (2013) size distribution model for impactors of this size.

There are various sources of uncertainty in the predicted cratering rates: (1) the choice of the various impactor size distributions assumed (which translate into the slopes of the predicted crater SFDs), (2) the impactor population estimates used for normalizing the size distributions (manifested in the vertical scaling of the crater SFDs), and (3) the different crater scaling laws, which shift the crater SFDs to slightly different crater diameters. The compilation of these leads to a roughly factor of 2 uncertainty in predictions of the crater SFDs on Pluto and Charon.

The largest uncertainty in the cratering rates and thus terrain ages computed for the Pluto system is the extrapolation of the KBO size distribution to diameters d ≲ 100 km [corresponding to a crater diameter of D ≲ 400 km on Pluto (*Bierhaus and Dones*, 2015; *Greenstreet et al.*, 2015)]. New Horizons observations probe impactor diameters below the current observational limits through the craters observed on both Pluto and Charon, providing the first observational opportunity to determine relative slopes and population estimates of the d ≲ 100-km KBO impactor population. As discussed in section 3.3, the information from New Horizons revealed an additional break in the impactor/crater SFD at sizes smaller than can be seen with Earth- or spacebased telescopes. We have added this new information into the terrain age predictions discussed in the next section.

5. TERRAIN AGES

Here we combine the measured crater SFDs for Pluto and Charon (section 3) and predictions for the crater spatial density of a given age terrain (section 4) to produce estimates of the crater retention age for different terrains. The crater retention age refers to how old the observed surface (and near-subsurface) is with respect to formation and removal of craters. A surface with no identifiable craters (down to the limits of our image resolution) means that all craters have been removed by resurfacing processes that may have been ongoing for a long period of time, or may have been removed in one or more discrete events in the relatively recent past. These surface age estimates use the starting place that all craters on Pluto and Charon were formed by single heliocentric impactors and not from secondary or sesquinary debris, or from binary primary impactors. As discussed below in section 5.1 (and in *Singer et al.,* 2019), we see no strong signs of secondary craters or circumbinary debris.

Binary objects are observed in the Kuiper belt by telescopic surveys, but observational constraints limit our knowledge of their occurrence rate (see discussion in the chapter by Parker in this volume). Statistical analysis of whether a binary impactor population can be seen in the crater populations on Pluto and Charon is also discussed in the chapter by Parker. A large binary fraction in the impactor population would affect the resulting crater densities on Pluto and Charon and thus the estimated terrain ages. If, for an extreme case, 50% of the impactor population is a near-equal-sized binary, this binary fraction could shift observed crater densities to higher values by as much as 50% more than crater densities produced by only single impactors. Not all binaries would necessarily produce distinct craters, because that would depend on the orientation of the binary as it impacted and the separation between the two objects, but the example here gives the maximum effect. This would result in the corresponding terrain ages shifting to 50% shorter than that for an impactor population consisting of only single impactors due to the decreased amount of time required to reach the same spatial density of craters. In simple terms, the equation for including the influence of binaries is (single impactor terrain age estimates)/(1 + binary fraction) = (binary impactor terrain age estimates). For example, a terrain age estimate of 4 G.y. for a population of only single impactors would correspond to a terrain age estimate of 4/1.5 = 2.7 G.y. for a population consisting of 50% binaries. This simple equation can easily be scaled for any binary fraction within the impacting population.

In the more realistic case that all binaries do not consist of objects of the same size, the above scaling is an upper limit of the affect of binary impactors on terrain age estimates. If binary pairs are not the same size, the craters each binary produce will not be the same size, decreasing the rate at which craters of a given size are accumulated on the surface of either Pluto or Charon. This results in a smaller shift in crater densities compared to an impactor population with only single impactors (i.e., by <50% for a 50% binary impactor fraction) and thus a smaller shift in the corresponding terrain age estimates. The size distribution of the binaries would be required to determine the quantitative shift in the terrain age estimates for an impactor population consisting of any fraction of binaries compared to that of a single-impactor-only impacting population.

We emphasize that all ages reported here are rough, order-of-magnitude estimates, and the uncertainties in these results depend on several factors. There is uncertainty in the predicted cratering rate (described above in section 4.6) of roughly a factor of 2 (*Greenstreet et al.*, 2015). The impactor flux and cratering rate models will be continually refined as more observations of the Kuiper belt are made. Here we do not pick a specific crater size to pin our estimates to (as is done for some other bodies in the solar system) because that is not practical with our datasets. Thus, the estimates below are either given as upper or lower limits, or as rough best answers given the limitations of the data and models.

Determining the age of surfaces with few or no craters is an additional challenge, but has been discussed for other bodies (e.g., *Michael et al.,* 2016). It is clear that the large areas of Pluto's encounter hemisphere that are devoid of craters are younger than the cratered regions, and some have additional age constraints based on modeling of their ongoing activity, such as Sputnik Planitia (*McKinnon et al.,* 2016). Here we discuss age estimates for the younger terrains based on the *Greenstreet et al.* (2015, 2016) "knee" model modified with an additional break in slope for craters D < 13 km (we will call this slope break the "elbow") (see Figs. 13 and 14 and *Singer et al.*, 2019). We use the average slope of craters on Charon's Vulcan Planitia, q = –1.7, as a representative slope for this additional piece of the SFD power law to modify the result of *Greenstreet et al.* (2015, 2016). We use a modified version of the *Greenstreet et al.* (2015, 2016) "knee" model as our reference model here, because it is the most recent cratering rate prediction based on telescopic observations of the Kuiper belt subpopulations, and because this extrapolation of the impactor size distribution most closely matches the constraints found by New Horizons (e.g., the actual crater distributions found). The *Zahnle et al.* (2003) base prediction is fairly similar to that of *Greenstreet et al.* (2015, 2016), as can be seen in Fig. 11, and thus would produce similar results. All these models, and the terrain ages predicted by them, are subject to future revisions.

Each case is described below, but for surfaces with no craters, we give an *upper limit* constraint on the crater retention age based on the smallest crater that could be seen with the available image resolution and the area of that unit. This age constraint is an upper limit because there may, or may not be, smaller craters on these terrains that we cannot see because our image resolution is too coarse. If, in reality, no smaller craters exist on these surfaces, the derived age constraint would be even younger than what we report here. This method is conceptually similar to that in *Michael et al.* (2016) for terrains with no craters, but we do not have a full chronology function and so our approach is somewhat simpler. Several of the terrains described below have similar upper age constraints because the image resolution is similar over these terrains (and they are also somewhat similar in areal extent). This does not necessarily mean that they were all resurfaced at the same time, but rather that they were resurfaced sometime between the present and the upper age constraint. In a few cases superposition relationships or other information can be used for additional terrain timing relationships (also described in section 3 and in the chapter by White et al. in this volume).

5.1. A Note about Secondary Craters and Crater Ejecta

Pluto has a striking lack of obvious secondary craters, or even noticeable crater ejecta (except for a handful of cases; see Fig. 1), at all the available image resolutions. Obvious secondary craters are defined here as craters in clusters or chains or craters/clusters with radial indicators such as v-shaped ejecta or elongation/asymmetry that point back to a primary crater (some examples are given in Fig. 12). Not all secondary craters form with these morphological features, and some of these features can be modified or lost over time, but they are general indicators of secondary craters. Secondary craters also tend to have steep SFD slopes (as steep as $q = -6$), specific spatial patterns (e.g., radial distribution around a primary crater, a chain or cluster that points to a larger primary crater), and a specific size relationship to the primary crater: The very largest secondaries are generally not more than ~5–8% the size of the primary (e.g., *Melosh,* 1989; *Singer et al.,* 2013). There may be secondary craters below our resolution limits or non-obvious secondary craters. Both obvious and non-obvious secondary craters would tend to steepen the SFD at small crater sizes, as is seen on Europa (*Bierhaus et al.,* 2009), but we observe no steepening of the SFD. Although pits exist on Pluto at many scales, and pit chains occur over likely fractures, none of these pits appear to be secondary craters. On Charon, ejecta deposits are more visible, indicating a slower process of erosion than on Pluto. However, there are no features resembling traditional secondary craters on the smooth plains of Vulcan Planitia. There is only one possible feature seen on Oz Terra that could represent a crater chain: a narrow catena-like chain located at ~30°N/10°E (Fig. 12d). Lighting geometry, image resolution, and terrain characteristics (such as large albedo variations or local geology) can also affect how all features,

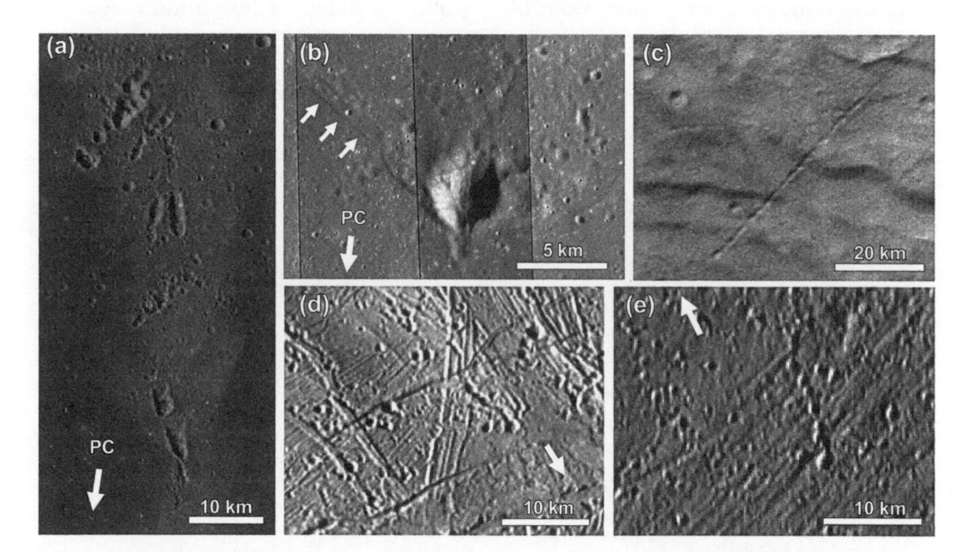

Fig. 12. Example secondary craters from other bodies in the solar system. Very few features on Pluto or Charon appear to be secondary craters, but they may exist below the resolution limit of our images. Note that in addition to being clusters/chains, these features are found either radial to or nearby a primary crater and are the expected sizes for secondary craters around their respective primary. The arrows noted with "PC" indicates the direction to the primary crater. **(a)** Classic examples of secondary crater clustering and morphology around the lunar crater Copernicus (93 km in diameter) (more details in *Singer et al.,* 2020a) as seen in the Lunar Reconnaissance Orbiter Camera (LROC) Wide Angle Camera mosaic at 100 m px^{-1}. **(b)** Closeup of a relatively unmodified lunar secondary crater (also thought to be from Copernicus) where v-shaped ejecta can be seen (three arrows in a row give an example). Additionally, the rim of the crater in the downrange direction is less well formed, which is a common feature of secondary craters that are found relatively close to their primary. Image from LROC Narrow Angle Camera images shown at ~1.3 m px^{-1} (modified from *Singer et al.,* 2020b). **(c)** Charon crater chain on Oz Terra centered at 10.2°E, 30.8°N. Image from PEMV_C_MVIC_LORRI_CA, 622 m px^{-1}. **(d)** Secondary craters around the Tyre impact basin on Jupiter's moon Europa shown at ~210 m px^{-1} (*Bierhaus et al.,* 2009; *Singer et al.,* 2013). **(e)** Secondary craters around the Achelous crater on Ganymede shown at ~180 m px^{-1} (*Singer et al.,* 2013).

including secondary craters, appear. For example, some image conditions can make the smaller, more subtle feature such as v-shaped ejecta more difficult to see. However, the resolution and lighting geometry varies greatly across both Pluto and Charon, and at the sizes of the mapped craters, few to no secondary-crater-like features are seen. Because of the reasons listed here, we believe secondary craters are not a large contribution to the crater data presented here.

Here we also consider the size of secondary craters that might be expected to be seen with the New Horizons data if they were present. Conservatively, a secondary crater and distinctive secondary morphology (such as v-shaped ejecta) as small as ~0.8 km (~10 pixels) across could be seen in the highest-resolution, closest-approach strip, which would be produced by an ~16-km-diameter crater nearby [if the largest secondary craters are ~5% of the parent crater size (*Melosh,* 1989; *Singer et al.,* 2013, 2020b)] or a larger, more distant crater. There are several craters with D $\gtrsim$ 16 km in the neighborhood of the highest-resolution strip, but very few fresh craters that have the best chance of exhibiting visible ejecta (not yet hidden by mantling and erosion). Although Pluto and Charon were predicted to have low secondary crater densities given the low primary impact speeds and the bodies' relatively low escape velocities (*Bierhaus and Dones,* 2015) (see also section 4.6), it is still somewhat surprising to find no hint of them.

The 10 largest primary craters on Pluto's encounter hemisphere (excepting the Sputnik basin) range from 60 to 240 km in diameter. The largest secondary craters from these could be D ~ 3–12 km, which would be resolvable over much of Pluto. These largest craters all appear to be relatively old and degraded on Pluto, with no clear signs of ejecta or radial scouring left. There are no clustered craters (or craters with secondary morphology) of the appropriate secondary crater sizes (D ~ 3–12 km) around these 10 largest primary craters, or anywhere else, on the encounter hemisphere. On Charon, the largest primary craters are in a similar range to Pluto, with D ~ 55–230 km, and some of their secondaries should also be visible if they exist on the encounter hemisphere. For Vulcan Planitia, where the lighting is optimal, the five largest primary craters are D ~ 30–65 km and could produce secondary craters as large as D ~ 1.5–3 km (resolvable in the mid- to high-resolution datasets). Although we do see more proximal ejecta deposits on Charon, the secondary craters on Pluto and Charon appear to be erased (for older craters), not visible at our current resolution (as most young/fresh craters are smaller), and/or are not produced in identifiable numbers to begin with. This issue is worthy of future study from a cratering mechanics standpoint.

5.2. Terrain Ages

5.2.1. Summary of terrain ages. Pluto displays a wide variety of terrain ages, including several types of young surfaces, a few middle-aged regions, and several

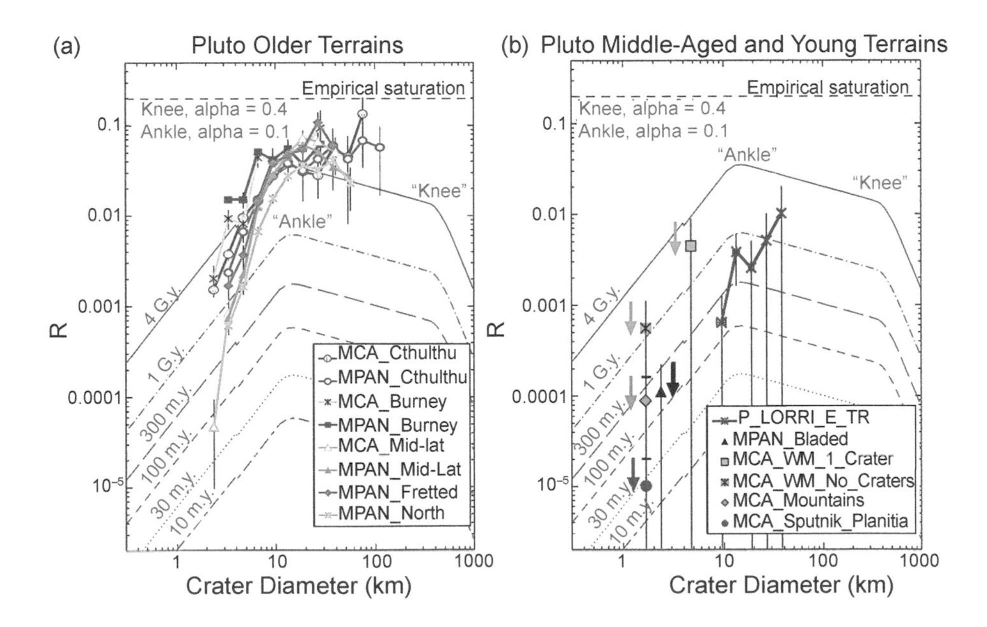

Fig. 13. A comparison of crater densities with predictions modified from *Greenstreet et al.* (2015) for **(a)** Pluto's older terrains and **(b)** Pluto's younger and middle-aged terrains. Modified prediction curves include the additional "elbow" break in slope matching Charon's crater distribution (see text for more information). In **(b)** the single points with arrows next to them represent upper limit ages for Pluto's younger terrains, which display few if any craters. The top of the arrow is placed at the top of the one-sigma error bar for each point. For the legends, each dataset is identified by both the image/scan craters were mapped on and the geologic unit (see Fig. 3): MCA = MVIC closest approach scan, MPAN = MVIC PANframe scan, P_LORRI = Pluto LORRI encounter hemisphere mosaic, E_TR = Eastern Tombaugh Regio, and WM = Wright Mons. Additional details about each dataset and the outlines for each region are given in *Singer et al.* (2019; see main text and supplement). The empirical saturation line is explained in Fig. 11.

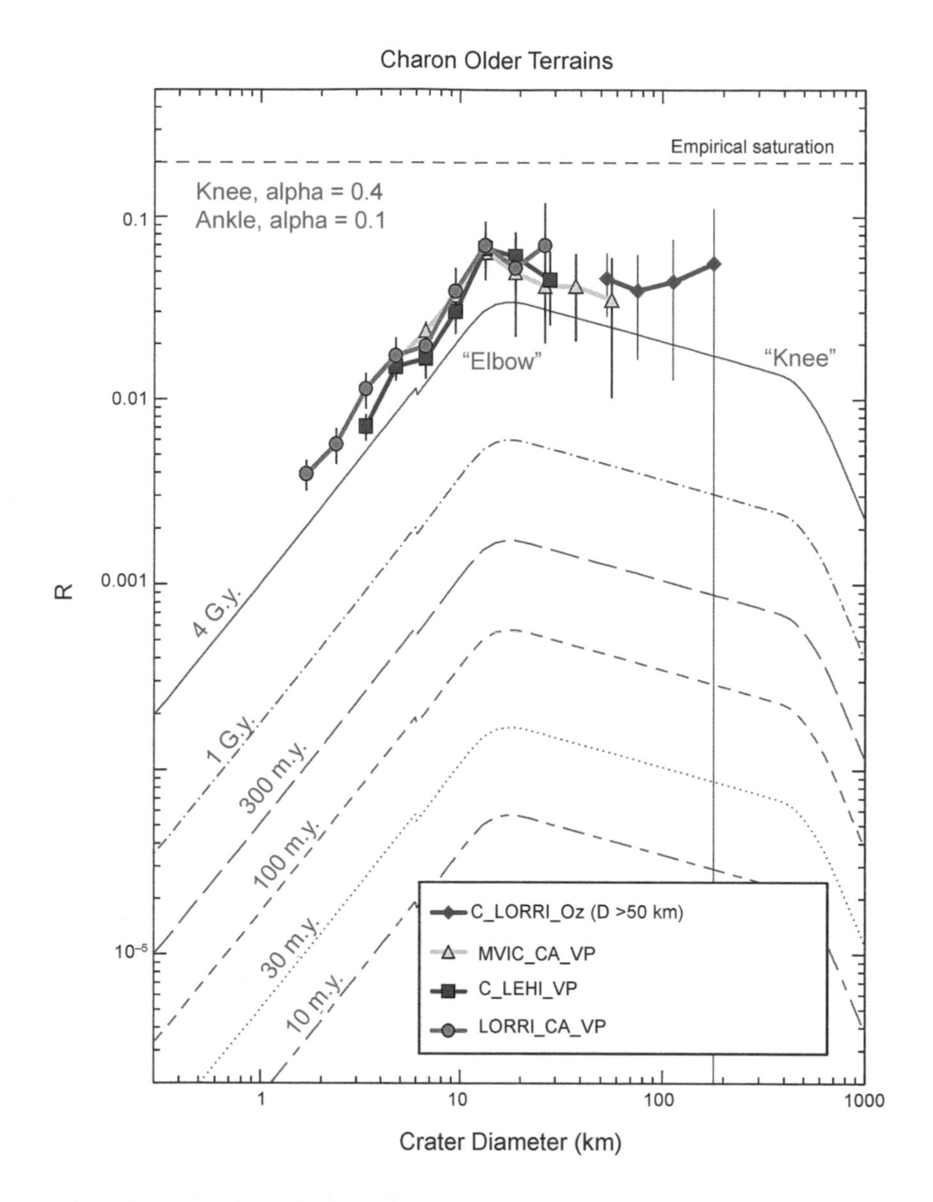

Fig. 14. A comparison of crater densities with predictions modified from *Greenstreet et al.* (2015, 2016) for Charon's terrains. C_LORRI = Charon LORRI encounter hemisphere mosaic, MVIC_CA = MVIC closest approach scan, C_LEHI = Charon LORRI mosaic associated with the LEISA instrument high-resolution scan, LORRI_CA = LORRI closest-approach high-resolution mosaic, Oz = Oz Terra, and VP = Vulcan Planitia. Additional details about each dataset and the outlines for each region are given in sections 1 and 3 and in *Singer et al.* (2019; see main text and supplement).

heavily-cratered, ancient surfaces (Fig. 13). Pluto's terrains with high crater densities appear to be quite old, as their crater densities are above the 4-G.y. prediction for the *Greenstreet et al.* (2015) "knee" model (see additional discussion below).

All regions on Charon appear relatively old (~4 G.y. or older; Fig. 14). Although Vulcan Planitia was resurfaced, creating the relatively smooth plain (*Stern et al.,* 2015; *Moore et al.,* 2016; *Beyer et al.,* 2019; *Robbins et al.,* 2019), the plain still has a relatively large number of craters overall, indicating this resurfacing occurred early in Charon's history. Crater densities for Oz Terra and Vulcan Planitia are similar for the small range of crater diameters where they overlap, thus small differences in age cannot be distinguished with the current crater data. We include some information about Charon here for comparison to Pluto, and additional discussion about Charon's terrain ages can be found in the chapter by Spencer et al. in this volume.

5.2.2. ***Pluto older terrains.*** Many of Pluto's terrains have relatively high crater densities (Fig. 13a) and are consistent with a surface age estimate of 4 G.y. or older. These older terrains show some spread in their crater densities, consistent with Pluto's complex geologic history. All the terrains discussed here and the Burney basin are presumed to post-date the Sputnik-basin-forming impact event, which likely created a large resurfaced area on the encounter hemisphere of Pluto from both the topographic feature of the basin itself and its ejecta. There are a few other stratigraphic relationships we can infer. As described above in section 3.1.3, dark/bright layers are seen in the walls of craters that superpose Burney, implying these alternating layers formed prior to the superposing craters. Because the superposing crater spatial density inside Burney is high, this suggests both Burney and these layers are relatively ancient (again 4 G.y. or older). Extensive descriptions of these regions and their geologic context can be found in section 3 and also in the chapter by White et al. in this volume.

5.2.3. Pluto young and middle-aged terrains. Eastern Tombaugh Regio (the eastern side of Pluto's bright heart-shaped region; Figs. 7a,b) is a middle aged terrain (*Moore et al.*, 2016) with a few large but eroded craters, giving a possible upper age limit of a few billion years (Fig. 13b). Eastern Tombaugh Regio is likely resurfaced by an *ongoing* combination of volatile sublimation and deposition (*Moore et al.,* 2017), which appears more active in the middle portions of the terrain, thus it is difficult to tie this crater retention age to a single geologic event.

Sputnik Planitia has no obvious craters, in either the ~315 m px^{-1} MVIC mosaic covering the entirety of the feature or any of the higher-resolution strips crossing it (down to a pixel scale of ~80 m px^{-1}). We can estimate an approximate, upper-limit crater retention age for Sputnik Planitia by calculating the R-value for the hypothetical scenario where one crater exists somewhere on the planitia just below the resolution limit of the images. This implies a surface model age of ~30–50 m.y. for the revised power law, including the break at D ~ 13 km to a shallow slope, if one crater ~1.6 km across is hidden below the resolution limit (~5 pixels in the 315 m px^{-1} PEMV_P_MVIC_LORRI_CA scan). The polygonal features found there are likely created by sluggish lid convection (*Stern et al.*, 2015; *Moore et al.*, 2016), which could resurface on timescales of roughly 500,000 years, limiting the age of the surface.

Other young areas with few if any craters are the chaotic mountain blocks in Sputnik Plantia, the putative cryovolcanic construct Wright Mons to the southwest of Sputnik Planitia, and Pluto's "bladed terrain" to the east of Tombaugh Regio. The mountain blocks and Wright Mons share the same D = 1.6-km crater size constraint as Sputnik Planitia but cover a smaller area, thus we derive an older maximum age estimates for these terrains.

The chaotic mountain blocks (*Moore et al.,* 2016; *Skjetne et al.,* 2020) have no obvious craters at a pixel scale of 315 m px^{-1} (the pixel scale available for all mountainous regions combined). Although they do not represent a continuous surface, there are some flat block tops that could host pre- or post-disruption craters, and craters that are large enough could still form and be visible on the uneven terrain. In the highest-resolution images (pixel scale of 77 m px^{-1}) that cover one region of chaotic blocks, there are some small (D < 1 km) circular or subcircular features that may be small impacts and one or two larger features that may be craters. With all of the disruption, however, it is difficult to distinguish between a remnant impact crater and a later collapse feature (e.g., the alcove of a mass-wasting event can have a curved upper scarp). Additionally, the morphology of any post-disruption craters could be affected by formation on the many slopes or uneven terrains that make up the chaotic blocks, making them more difficult to identify. In this work we use the absence of clear craters at 315 m px^{-1} for an upper limit, because large post-disruption craters could in principle still be visible. This gives an upper limit age of ~200–300 m.y. for the mountain blocks (top of the error bar); however, it is also not entirely clear what event is being dated. The block material generally resembles the nearby terrain just outside of Sputnik Planitia (that is not broken up). Thus, this is likely an older surface/crust that was emplaced as part of the Sputnik basin ejecta deposits, with a partial resurfacing from the disruption into chaotic blocks.

An upper limit for crater spatial density of the bladed terrain, from the lack of any distinct craters 2.5 km across or larger (~5 pixels in the 480 m px^{-1} PEMV_P_ MPAN1 scan), is plotted in Fig. 13b. This yields upper limit ages of ~200–300 m.y. old for the bladed terrain. (See additional discussion in the chapter by White et al. in this volume.)

With no craters on the main topographic mound of Wright Mons, this yields a 1–2-G.y. upper age limit for the surface (if using the top of the error bar). If the one possible 5.5-km-diameter feature is indeed a crater on Wright Mons (see discussion in section 3.1.4), it would yield an age closer to 3–4 G.y. These relatively high upper limits on Wright Mons' age do not mean that the feature itself is that old or that the process that built it necessarily only operated early in Pluto's history. First, these are extreme upper limits based on the image resolutions available and the top of the error bar. Second, because the number of craters is divided by the area in order to calculate crater densities, the area of measurement also matters. For two regions, both with no craters, the larger area will give a younger age. This can be seen in Fig. 13b where the points for the three terrains of Sputnik Planitia, the chaotic mountain blocks, and Wright Mons stack on top of each other. The image resolution is the same; each region just covers a different area. Another way to understand this is that it is statistically less likely for no craters to form over time in a larger area than a smaller one. Wright Mons covers a smaller area than some other regions measured at the same image resolution, thus we cannot constrain the age as well with crater measurements. Work on other geologic indicators in the future may give a better estimate of a minimum age, or at least a more realistic "most likely" age, for Wright Mons.

6. SUMMARY AND FUTURE WORK

The New Horizons flyby of Pluto and Charon provided the first detailed surface images of KBOs that are still residing in the Kuiper belt today. This provides unique information about the population of smaller KBOs that impacted Pluto and Charon to make craters. We find a range of older, middle-aged, and younger terrains on Pluto. The oldest terrains on Pluto have modeled crater retention ages of ~4 G.y. or older. For the youngest terrains on Pluto — those with no identifiable craters — we can only put a rough upper limit on the model ages. The very youngest terrains such as Sputnik Planitia are likely continually resurfaced into the present. On Charon, we find both Oz Terra and Vulcan Planitia have fairly high crater densities that imply relatively old crater retention ages, ~ 4 G.y. or older. The craters in the Pluto-system also showed that the Kuiper belt has a different SFD shape than the asteroid belt, with significantly fewer small (D < 10–5-km) objects.

For Pluto, one overarching topic for future research is continued use of crater densities and morphologies along with geologic mapping and stratigraphic relationships to better understand Pluto's geologic history. We have identified many of the major processes operating on Pluto, and they encompass a wide range of process types, from tectonism to volcanism to sublimation/deposition features. Can we quantify the extent of erosion, mantling, or infilling of Pluto's craters? Future classification, measurements, and modeling of geologic processes will all bear on this topic. Additionally, modeling of crater mechanics necessary to reproduce the depth-to-diameter trends and other morphological aspects of craters on Pluto and Charon (including the lack of obvious secondary craters) may also produce insights into surface and subsurface material properties on Pluto. The information regarding small KBOs from New Horizons can also be used with updated cratering rate models to improve estimates of terrain age. For Charon, some areas of lower crater spatial density exist across the plain. These areas are conspicuous to the eye, but further statistical analysis is needed to determine if these low-density areas could be stochastic or if they may indicate some later resurfacing after the majority of Vulcan Planitia was emplaced.

Continued comparison of the Pluto and Charon data to the crater populations of the moons of Jupiter, Saturn, Uranus, and Neptune can provide information on those impactor populations and help constrain which components of the satellite crater populations came from heliocentric vs. planetocentric impactors. And finally, comparing what we have learned about small KBOs directly to other small-body populations, such as the asteroid belt, Centaurs (objects in the giant planet region), or comet populations can provide information about how these bodies form initially and evolve over the history of the solar system.

REFERENCES

Adams E. R. et al. (2014) De-biased populations of Kuiper belt objects from the deep ecliptic survey. *Astron. J., 148,* 55–71.

Alzate N. and Barlow N. G. (2011) Central pit craters on Ganymede. *Icarus, 211,* 1274–1283, DOI: 10.1016/j.icarus.2010.10.015.

Barlow N. G. and Bradley T. L. (1990) Martian impact craters: Correlations of ejecta and interior morphologies with diameter, latitude, and terrain. *Icarus, 87,* 156–179, DOI: 10.1016/0019-1035(90)90026-6.

Barlow N. G. et al. (2000) Standardizing the nomenclature of martian impact crater ejecta morphologies. *J. Geophys. Res., 105,* 26733–26738, DOI: 10.1029/2000JE001258.

Bernstein G. M., Trilling D. E., Allen R. L., Brown M. E., Holman M., and Malhotra R. (2004) The size distribution of trans-neptunian bodies. *Astron. J., 128,* 1364–1390.

Bernstein G. M., Trilling D. E., Allen R. L., Brown M. E., Holman M., and Malhotra R. (2006) Erratum: "The size distribution of trans-neptunian bodies." *Astron. J., 131,* 2364–2364, DOI: 10.1086/503194.

Bertrand T. et al. (2018) The nitrogen cycles on Pluto over seasonal and astronomical timescales. *Icarus, 309,* 277–296, DOI: 10.1016/j.icarus.2018.03.012.

Bertrand T. et al. (2019) The CH_4 cycles on Pluto over seasonal and astronomical timescales. *Icarus, 329,* 148–165, DOI: 10.1016/j.icarus.2019.02.007.

Beyer R. A. et al. (2017) Charon tectonics. *Icarus, 287,* 161–174, DOI: 10.1016/j.icarus.2016.12.018.

Beyer R. A. et al. (2019) The nature and origin of Charon's smooth plains. *Icarus, 323,* 16–32, DOI: 10.1016/j.icarus.2018.12.036.

Bierhaus E. B., Zahnle K., and Chapman C. R. (2009) Europa's crater distributions and surface ages. In *Europa* (R. T. Pappalardo et al., eds.), pp. 161–180. Univ. of Arizona, Tucson.

Bierhaus E. B. and Dones L. (2015) Craters and ejecta on Pluto and Charon: Anticipated results from the New Horizons flyby. *Icarus, 246,* 165–182, DOI: 10.1016/j.icarus.2014.05.044.

Binzel R. P. et al. (2017) Climate zones on Pluto and Charon. *Icarus, 287,* 30–36, DOI: 10.1016/j.icarus.2016.07.023.

Boyce J., Barlow N., Mouginis-Mark P., and Stewart S. (2010) Rampart craters on Ganymede: Their implications for fluidized ejecta emplacement. *Meteoritics & Planet. Sci., 45,* 638–661, DOI: 10.1111/j.1945-5100.2010.01044.x.

Bray V. J., Schenk P. M., Melosh H. J., Morgan J. V., and Collins G. S. (2012) Ganymede crater dimensions: Implications for central peak and central pit formation and development. *Icarus, 217,* 115–129, DOI: 10.1016/j.icarus.2011.10.004.

Cohen C. J. and Hubbard E. C. (1965) Libration of the close approaches of Pluto to Neptune. *Astron. J., 70,* 10–13, DOI: 10.1086/109674.

Conrad J. W. et al. (2019) An upper bound on Pluto's heat flux from a lack of flexural response of its normal faults. *Icarus, 328,* 210–217, DOI: 10.1016/j.icarus.2019.03.028.

Cook J. C. et al. (2019) The distribution of H_2O, CH_3OH, and hydrocarbon-ices on Pluto: Analysis of New Horizons spectral images. *Icarus, 331,* 148–169, DOI: 10.1016/j.icarus.2018.09.012.

Costard F. M. (1989) The spatial distribution of volatiles in the martian hydrolithosphere. *Earth Moon Planets, 45,* 265–290, DOI: 10.1007/BF00057747.

Crater Analysis Techniques Working Group (1979) Standard techniques for presentation and analysis of crater size-frequency data. *Icarus, 37,* 467–474, DOI: 10.1016/0019-1035(79)90009-5.

Cruikshank D. P. et al. (2015) The surface compositions of Pluto and Charon. *Icarus, 246,* 82–92, DOI: 10.1016/j.icarus.2014.05.023.

Dalle Ore C. M. et al. (2018) Ices on Charon: Distribution of H_2O and NH_3 from New Horizons LEISA observations. *Icarus, 300,* 21–32, DOI: 10.1016/j.icarus.2017.08.026.

de Elia G. C., Di Sisto R. P., and Brunini A. (2010) Impactor flux and cratering on the Pluto-Charon system. *Astron. Astrophys., 521,* A23.

Dell'Oro A., Campo Bagatin A., Benavidez P. G., and Alemañ R. A. (2013) Statistics of encounters in the trans-neptunian region. *Astron. Astrophys., 558,* A95.

Dones L., Gladman B., Melosh H. J., Tonks W. B., Levison H. F., and Duncan M. (1999) Dynamical lifetimes and final fates of small bodies: Orbit integrations vs Opik calculations. *Icarus, 142,* 509–524.

Durda D. D. and Stern S. A. (2000) Collision rates in the present-day Kuiper belt and Centaur regions: Applications to surface activation and modification on comets, Kuiper belt objects, Centaurs, and Pluto-Charon. *Icarus, 145,* 220–229.

Durham W. B. and Stern L. A. (2001) Rheological properties of water ice-applications to satellites of the outer planets. *Annu. Rev. Earth Planet. Sci., 29,* 295–330, DOI: 10.1146/annurev.earth.29.1.295.

Earle A. M. et al. (2017) Long-term surface temperature modeling of

Pluto. *Icarus, 287,* 37–46, DOI: 10.1016/j.icarus.2016.09.036.
Fernandez J. A. and Ip W.-H. (1984) Some dynamical aspects of the accretion of Uranus and Neptune: The exchange of orbital angular momentum with planetesimals. *Icarus, 58,* 109–120, DOI: 10.1016/0019-1035(84)90101-5.
Fraser W. C. and Kavelaars J. J. (2008) A derivation of the luminosity function of the Kuiper belt from a broken power-law size distribution. *Icarus, 198,* 452–458.
Fraser W. C., Brown M. E., Morbidelli A., Parker A., and Batygin K. (2014) The absolute magnitude distribution of Kuiper belt objects. *Astrophys. J., 782,* 100–113.
Fuentes C. I. and Holman M. J. (2008) A SUBARU archival search for faint trans-neptunian objects. *Astron. J., 136,* 83–97.
Gladman B., Kavelaars J. J., Petit J. M., Morbidelli A., Holman M. J., and Loredo T. (2001) The structure of the Kuiper belt: Size distribution and radial extent. *Astron. J., 122,* 1051–1066.
Gladman B., Marsden B. G., and Vanlaerhoven C. (2008) Nomenclature in the outer solar system. In *The Solar System Beyond Neptune* (M. A. Barucci et al., eds.), pp. 43–57. Univ. of Arizona, Tucson.
Gladman B. et al. (2012) The resonant trans-neptunian populations. *Astron. J., 144,* 23–47.
Gomes R., Levison H. F., Tsiganis K., and Morbidelli A. (2005) Origin of the cataclysmic late heavy bombardment period of the terrestrial planets. *Nature, 435,* 466–469.
Greenstreet S., Gladman B., and McKinnon W. B. (2015) Impact and cratering rates onto Pluto. *Icarus, 258,* 267–288, DOI: 10.1016/j.icarus.2015.05.026.
Greenstreet S., Gladman B., and McKinnon W. B. (2016) Corrigendum to "Impact and cratering rates onto Pluto" [Icarus 258 (2015) 267–288]. *Icarus, 274,* 366–367, DOI: 10.1016/j.icarus.2016.03.003.
Grundy W. M. et al. (2016) Surface compositions across Pluto and Charon. *Science, 351,* aad9189, DOI: 10.1126/science.aad9189.
Grundy W. M. et al. (2018) Pluto's haze as a surface material. *Icarus, 314,* 232–245, DOI: 10.1016/j.icarus.2018.05.019.
Hahn J. and Malhotra R. (2005) Neptune's migration into a stirred-up Kuiper belt: A detailed comparison of simulations to observations. *Astron. J., 130,* 2392–2414.
Holsapple K. A. (1993) The scaling of impact processes in planetary sciences. *Annu. Rev. Earth Planet. Sci., 21,* 333–373.
Horner V. M. and Greeley R. (1982) Pedestal craters on Ganymede. *Icarus, 51,* 549–562, DOI: 10.1016/0019-1035(82)90145-2.
Housen K. R. and Holsapple K. A. (2011) Ejecta from impact craters. *Icarus, 211,* 856–875, DOI: 10.1016/j.icarus.2010.09.017.
Howard A. D. et al. (2017) Present and past glaciation on Pluto. *Icarus, 287,* 287–300, DOI: 10.1016/j.icarus.2016.07.006.
Jewitt D., Luu J., and Trujillo C. (1998) Large Kuiper belt objects: The Mauna Kea 8K CCD survey. *Astron. J., 115,* 2125–2135.
Kirchoff M. and Dones L. (2018) Impact crater distributions of the uranian satellites: New constraints for outer solar system bombardment. *42nd COSPAR Scientific Assembly,* B5.4-7-18.
Mann A. (2018) Bashing holes in the tale of Earth's troubled youth. *Nature, 553,* 393–395, DOI: 10.1038/d41586-018-01074-6.
McKinnon W. B. et al. (2016) Convection in a volatile nitrogen-ice-rich layer drives Pluto's geological vigour. *Nature, 534,* 82–85, DOI: 10.1038/nature18289.
Melosh H. J. (1989) *Impact Cratering: A Geologic Perspective.* Oxford Univ., New York. 245 pp.
Michael G. G., Kneissl T., and Neesemann A. (2016) Planetary surface dating from crater size-frequency distribution measurements: Poisson timing analysis. *Icarus, 277,* 279–285, DOI: 10.1016/j.icarus.2016.05.019.
Minton D. A., Richardson J. E., Thomas P., Kirchoff M., and Schwamb M. E. (2012) Combining saturnian craters and Kuiper belt observations to build an outer solar system impactor size-frequency distribution. *Asteroids, Comets, Meteors 2012,* Abstract #6348. LPI Contribution No. 1667, Lunar and Planetary Institute, Houston.
Moore J. M. et al. (2016) The geology of Pluto and Charon through the eyes of New Horizons. *Science, 351,* 1284–1293, DOI: 10.1126/science.aad7055.
Moore J. M. et al. (2017) Sublimation as a landform-shaping process on Pluto. *Icarus, 287,* 320–333, DOI: 10.1016/j.icarus.2016.08.025.
Morbidelli A., Levison H. F., Bottke W. F., Dones L., and Nesvorný D. (2005) Chaotic capture of Jupiter's trojan asteroids in the early solar system. *Nature, 435,* 462–465.
Mouginis-Mark P. (1979) Martian fluidized crater morphology: Variations with crater size, latitude, altitude, and target material. *J. Geophys. Res., 84,* 8011–8022, DOI: 10.1029/JB084iB14p08011.
Nesvorný D. (2018) Dynamical evolution of the early solar system. *Annu. Rev. Astron. Astrophys., 56,* 137–174, DOI: 10.1146/annurev-astro-081817-052028.
Nesvorný D., Roig F., and Ferraz-Mello S. (2000) Close approaches of trans-neptunian objects to Pluto have left observable signatures on their orbital distribution. *Astron. J., 119,* 953–969.
Petit J. M. et al. (2011) The Canada-France ecliptic plane survey–full data release: The orbital structure of the Kuiper belt. *Astron. J., 142,* 131–155.
Protopapa S. et al. (2017) Pluto's global surface composition through pixel-by-pixel Hapke modeling of New Horizons Ralph/LEISA data. *Icarus, 287,* 218–228, DOI: 10.1016/j.icarus.2016.11.028.
Robbins S. J. et al. (2017) Craters of the Pluto-Charon system. *Icarus, 287,* 187–206, DOI: 10.1016/j.icarus.2016.09.027.
Robbins S. J. et al. (2018) Investigation of Charon's craters with abrupt terminus ejecta, comparisons with other icy bodies, and formation implications. *J. Geophys. Res., 123,* 20–36, DOI: 10.1002/2017je005287.
Robbins S. J. et al. (2019) Geologic landforms and chronostratigraphic history of Charon as revealed by a hemispheric geologic map. *J. Geophys. Res., 124,* 155–174, DOI: 10.1029/2018JE005684.
Schenk P. M. (1993) Central pit and dome craters: Exposing the interiors of Ganymede and Callisto. *J. Geophys. Res., 98,* 7475–7498, DOI: 10.1029/93JE00176.
Schenk P. M., Chapman C. R., Zahnle K., and Moore J. M. (2004) Ages and interiors: The cratering record of the Galilean satellites. In *Jupiter: The Planet, Satellites and Magnetosphere* (F. Bagenal et al., eds.), pp. 427–456. Cambridge Univ., Cambridge.
Schenk P. M. et al. (2018a) Basins, fractures and volcanoes: Global cartography and topography of Pluto from New Horizons. *Icarus, 314,* 400–433, DOI: 10.1016/j.icarus.2018.06.008.
Schenk P. M., White O. L., Byrne P. K., and Moore J. M. (2018b) Saturn's other icy moons: Geologically complex worlds in thier own right. In *Enceladus and the Icy Moons of Saturn* (P. M. Schenk et al., eds.), pp. 237–265. Univ. of Arizona, Tucson.
Schenk P. M. et al. (2018c) Breaking up is hard to do: Global cartography and topography of Pluto's mid-sized icy moon Charon from New Horizons. *Icarus, 315,* 124–145, DOI: 10.1016/j.icarus.2018.06.010.
Schenk P. et al. (2019) The central pit and dome at Cerealia Facula bright deposit and floor deposits in Occator crater, Ceres: Morphology, comparisons and formation. *Icarus, 320,* 159–187, DOI: 10.1016/j.icarus.2018.08.010.
Schlichting H. E., Fuentes C. I., and Trilling D. E. (2013) Initial planetesimal sizes and the size distribution of small Kuiper belt objects. *Astron. J., 146,* 36.
Schmitt B. et al. (2017) Physical state and distribution of materials at the surface of Pluto from New Horizons LEISA imaging spectrometer. *Icarus, 287,* 229–260, DOI: 10.1016/j.icarus.2016.12.025.
Shankman C., Gladman B. J., Kaib N., Kavelaars J. J., and Petit J. M. (2013) A possible divot in the size distribution of the Kuiper belt's scattering objects. *Astrophys J. Lett., 764,* L2–L5.
Singer K. N., McKinnon W. B., and Nowicki L. T. (2013) Secondary craters from large impacts on Europa and Ganymede: Ejecta size-velocity distributions on icy worlds, and the scaling of ejected blocks. *Icarus, 226,* 865–884, DOI: 10.1016/j.icarus.2013.06.034.
Singer K. N. et al. (2016) Pluto's putative cryovolcanic constructs. *Lunar and Planetary Science XLVII,* Abstract #2276. Lunar and Planetary Institute, Houston.
Singer K. N. et al. (2019) Impact craters on Pluto and Charon indicate a deficit of small Kuiper belt objects. *Science, 363,* 955–959, DOI: 10.1126/science.aap8628.
Singer K. N. et al. (2020a) Impact craters on 2014 MU69: Implications for Kuiper belt object size-frequency distributions and planetesimal formation. *AAS Meeting Abstracts, 235,* Abstract# 419.06.
Singer K. N., Jolliff B. L., and McKinnon W. B. (2020b) Lunar secondary craters and scaling to ejected blocks reveals scale-dependent fragmentation trend. *J. Geophys. Res., 125,* e2019JE006313, DOI: 10.1029/2019JE006313.

Skjetne H. L. et al. (2020) Morphological comparison of blocks in chaos terrains on Pluto, Europa, and Mars. *Icarus,* 113866, DOI: 10.1016/j.icarus.2020.113866.

Spencer J. R., Stansberry J. A., Trafton L. M., Young E. F., Binzel R. P., and Croft S. K. (1997) Volatile transport, seasonal cycles, and atmospheric dynamics on Pluto. In *Pluto and Charon* (S. A. Stern and D. J. Tholen, eds.), pp. 435–474. Univ. of Arizona, Tucson.

Spencer J. R. et al. (2020) The geology and geophysics of Kuiper belt object (486958) Arrokoth. *Science, 367,* aay3999, DOI: 10.1126/science.aay3999.

Stansberry J. A. (1994) Surface-atmosphere coupling on Triton and Pluto. Ph.D. thesis, Univ. of Arizona, Tucson. 162 pp.

Stern S. A. (2008) The New Horizons Pluto Kuiper belt mission: An overview with historical context. *Space Sci. Rev., 140,* 3–21.

Stern S. A. et al. (2015) The Pluto system: Initial results from its exploration by New Horizons. *Science, 350,* aad1815, DOI: 10.1126/science.aad1815.

Stern S. A. et al. (2019) Initial results from the New Horizons exploration of 2014 MU69, a small Kuiper belt object. *Science, 364,* eaaw9771, DOI: 10.1126/science.aaw9771.

Stern S. A. et al. (2020) Pluto's far side. *Icarus,* 113805, DOI: 10.1016/j.icarus.2020.113805.

Tsiganis K., Gomes R., Morbidelli A., and Levison H. F. (2005) Origin of the orbital architecture of the giant planets of the solar system. *Nature, 435,* 459–461.

Weaver H. A. et al. (2016) The small satellites of Pluto as observed by New Horizons. *Science, 351,* aae0030, DOI: 10.1126/science.aae0030.

Weissman P. R. and Stern S. A. (1994) The impactor flux in the Pluto-Charon system. *Icarus, 111,* 378–386.

Wetherill G. W. (1967) Collisions in the asteroid belt. *J. Geophys. Res., 72,* 2429–2444.

White O. L., Schenk P. M., and Dombard A. J. (2013) Impact basin relaxation on Rhea and Iapetus and relation to past heat flow. *Icarus, 223,* 699–709, DOI: 10.1016/j.icarus.2013.01.013.

White O. L., Schenk P. M., Bellagamba A. W., Grimm A. M., Dombard A. J., and Bray V. J. (2017) Impact crater relaxation on Dione and Tethys and relation to past heat flow. *Icarus, 288,* 37–52, DOI: 10.1016/j.icarus.2017.01.025.

White O. L. et al. (2019) Washboard and fluted terrains on Pluto as evidence for ancient glaciation. *Nature Astron., 3,* 62–68, DOI: 10.1038/s41550-018-0592-z.

Young L. A. (2013) Pluto's seasons: New predictions for New Horizons. *Astrophys. J. Lett., 766,* L22, DOI: 10.1088/2041-8205/766/2/L22.

Young L. A. et al. (2008) New Horizons: Anticipated scientific investigations at the Pluto system. *Space Sci. Rev., 140,* 93–127.

Zahnle K., Schenk P., Levison H., and Dones L. (2003) Cratering rates in the outer solar system. *Icarus, 163,* 263–289.

Olkin C. B., Howett C. J. A., Protopapa S., Grundy W. M., Verbiscer A. J., and Buie M. W. (2021) Colors and photometric properties of Pluto. In *The Pluto System After New Horizons* (S. A. Stern, J. M. Moore, W. M. Grundy, L. A. Young, and R. P. Binzel, eds.), pp. 147–163. Univ. of Arizona, Tucson, DOI: 10.2458/azu_uapress_9780816540945-ch008.

Colors and Photometric Properties of Pluto

C. B. Olkin, C. J. A. Howett, S. Protopapa
Southwest Research Institute

W. M. Grundy
Lowell Observatory

A. J. Verbiscer
University of Virginia

M. W. Buie
Southwest Research Institute

With the flyby of New Horizons through the Pluto system in July 2015, we have greatly advanced our understanding of the color and photometric properties of Pluto. A combination of Earth-based and New Horizons observations is critical to understand the photometric properties of Pluto. From low-phase-angle Earth-based observations, Pluto's global opposition surge can be modeled. Adding higher-phase-angle observations from the New Horizons flyby allows us to quantify the photometric properties across the diversity of terrains on Pluto. There are insolation-driven patterns of color variegation across Pluto with dark red terrain dominating the equatorial region where there is recently the most intense insolation. Diverse tools to assess color differences including color ratios, principal component analysis, and spectral slopes all depict a surface with a diversity of terrains. High-resolution color imaging can inform interpretation of observations on the lower-resolution non-encounter hemisphere.

1. INTRODUCTION

1.1. Earth-Based Observations of Pluto for Photometry and Color Maps

Less than 100 years ago, Pluto was discovered by Clyde Tombaugh using photographic plates and the 13-inch Lowell refractor at Lowell Observatory in Arizona, USA (*Tombaugh*, 1960). This discovery was made by meticulously comparing images taken nights apart to search for a moving object using a Zeiss blink microscope. As technology improved from photographic plates to charge-coupled device (CCD) cameras and orbiting telescopes like the Hubble Space Telescope (HST) came on line, we learned more about Pluto. Early photometric light curves allowed us to understand Pluto's color as a function of rotational phase (*Walker and Hardie*, 1955) but, unknown at the time, the light curves were actually the combined light of Pluto and its large moon Charon.

Following the discovery of Charon in 1978 (*Christy and Harrington,* 1978), it became apparent that Pluto's heliocentric motion and the orientation of Charon's orbit plane with respect to Pluto would present a season of mutual events between the two bodies starting in 1984 and ending in 1990 (*Andersson et al.,* 1978). Charon would undergo a series of transits and occultations of Pluto as viewed from Earth. These mutual events enabled the first spatially-resolved, photometric measurements of Pluto by means of a subtractive technique that could isolate the individual contributions from each component of the binary system (e.g., *Binzel*, 1988; *Buie et al.*, 1992; *Reinsch et al.,* 1994). However, this technique had its own limitations: Owing to the synchronous nature of the system, only one hemisphere of each body could be derived. *Binzel* (1988) and *Reinsch et al.* (1994) measured the B–V color index of Pluto's anti-Charon-facing hemisphere (180°E longitude) in statistical agreement, as 0.867 ± 0.008 and 0.871 ± 0.014, respectively, and significantly redder than the Pluto-facing hemisphere of Charon, whose B–V color is 0.700 ± 0.010. The B–V color index is the difference of the B magnitude and the V magnitude of an object, where B and V are two filters in the Johnson photometric system where the B filter has a mean wavelength of 442 nm and the V (or visual) filter has a mean wavelength of 540 nm. The difference in magnitudes is equivalent to a ratio of intensities. For reference, the B–V color index of the Sun is 0.65, so a neutral surface reflecting sunlight would have a B–V color index of 0.65. The individual color measurements of Pluto and Charon enabled by the series of mutual events showed that Pluto is red and has a distinctly different color than Charon (which is closer to neutral).

The availability of the HST in the 1990s enabled the measurement of Pluto's disk-integrated color across the full illuminated surface by obtaining light curves of Pluto separate from Charon. Free from atmospheric distortion, HST easily resolved the binary system. Using HST observations acquired between 1992 and 1993, *Buie et al.* (1997a) found that Pluto's V-band light curve has total amplitude 0.33 mag with minimum and maximum light at longitudes 100°E and 220°E, respectively. *Buie et al.* (1997a) also measured Pluto's V-band phase coefficient, $\beta = 0.0294 \pm 0.0011$ mag degree^{-1} at solar phase angles between g = 0.6° and g = 2°. The phase coefficient measures the decrease in reflectance of an object as the phase angle increases and provides a metric by which the scattering properties of surface particles on Pluto can be compared to those of other solar system objects.

The state of the knowledge of Pluto's surface prior to 1997 is described in more detail in *Buie et al.* (1997b). Thus, the remainder of this chapter will focus on knowledge gained since 1997.

Buie et al. reported a full disk color for Pluto alone of (B–V) = 0.868 ± 0.002 from 1992 to 1993 averaged from just two central longitudes where (B–V) = 0.873 ± 0.002 at 123°E longitude and (B–V) = 0.862 ± 0.002 at 289°E longitude. *Buie et al.* (2010a) reported a rotational average of (B–V) = 0.954 ± 0.001 from 2002–2003 with a variation in color of 0.02 with rotation where the subsolar latitude had increased to 29°N. During these color measurements there appears to have been a steady evolution toward redder surface coloration on Pluto. This is consistent with results from *Buratti et al.* (2015, 2003), who used light curve data from the Pluto system obtained between 2008 and 2012 and by comparing to previous data concluded that Pluto's surface may have become more red from 1999 to 2012.

Hubble Space Telescope observations using two filters (435 nm and 555 nm) in 2002–2003 were used to synthesize a true color map of Pluto shown in Fig. 1 (*Buie et al.*, 2010b). From the two-color photometry, *Buie et al.* (2010b) modeled two color units on the surface, a gray unit and a tholin unit, to produce the color map.

1.2. New Horizons Observations

New Horizons' Ralph instrument is actually two instruments: The Linear Etalon Imaging Spectral Array (LEISA) is an infrared imaging spectrometer, and the Multispectral Visible Imaging Camera (MVIC) is an optical panchromatic and color imager (cf. *Reuter et al.*, 2008). In this section we focus upon the latter instrument, MVIC, which provides color images using four color filters (see Table 1) and CCDs that operate in time delay and integration mode (TDI). Results of the composition of the surface of Pluto from LEISA are in the chapter in this volume by Cruikshank et al. Each CCD has 5024 × 32 pixels, with a single pixel being 19.77 mrad × 19.77 mrad, so the field of view (FOV) of the TDI array is 5.7 × 0.037 degrees. TDI is a way of building up a large-format image as the FOV is quickly scanned across a scene. It works by syncing the transfer rate between 32 TDI pixels to the spacecraft's scan rate, thus the same scene passes through each of the 32 pixels, accumulating signal before it is read out and effectively increasing the integration time. The calibration of MVIC is detailed in *Howett et al.* (2017).

The four filters span the visible spectrum and into the near-infrared spectrum. The pivot wavelength λ_{pivot} (*Tokunaga and Vacca*, 2005) is an estimate of the effective wavelength of the system

$$\lambda_{pivot} = \sqrt{\frac{\int \lambda \chi(\lambda) S(\lambda) d\lambda}{\int (\chi(\lambda) S(\lambda)/\lambda) d\lambda}}$$

The pivot wavelength was calculated using the throughput of the system, $\chi(\lambda)$ (*Howett et al.*, 2017) and the spectrum of the Sun, $S(\lambda)$, at given wavelength (λ).

The color observations made by MVIC of Pluto can be separated into unresolved approach observations, resolved close-approach observations, and departure observations. The most useful observations for investigating surface color are the approach observations, as they imaged the sunlit surface of Pluto, whereas the departure observations are most useful for understanding Pluto's atmosphere, as detailed in the chapter by Young et al. in this volume.

The unresolved observations (i.e., where Pluto fills less than an MVIC pixel) begin on April 9, 2015, and run through July 2, 2015, after which Pluto is resolved by MVIC (i.e., Pluto is bigger than a pixel). The resolved observations of Pluto begin on July 9, 2015, and continue until the flyby, July 14, 2015 (see Fig. 2). The reason for the gap between unresolved and resolved observations was a spacecraft safe-mode event, which occurred on July 4, 2015, after which it took several days for the nominal observing sequence to be resumed (*Johns Hopkins University Applied Physics Laboratory*, 2015). A detailed analysis of the resolved color observations is provided in section 3.

2. PLUTO PHOTOMETRIC PROPERTIES

Pluto's photometric properties are derived from the quantitative measurement of reflected radiation from its surface. Both Earth-based and spacecraft measurements inform Pluto's global photometric properties, but only spatially resolved observations, such as those obtained by New Horizons (and to a limited extent, those obtained during mutual events and HST), can elucidate the scattering properties of its distinct terrains. Due to the size of the Earth's orbit, Pluto can never be observed from Earth at solar phase angles greater than 2°. Only spacecraft observations acquired from heliocentric distances greater than 1 au can achieve higher solar phase angles. In fact, one of New Horizons' mission goals was to obtain imaging of Pluto and Charon at moderate and high solar phase angles (*Young et al.*, 2008).

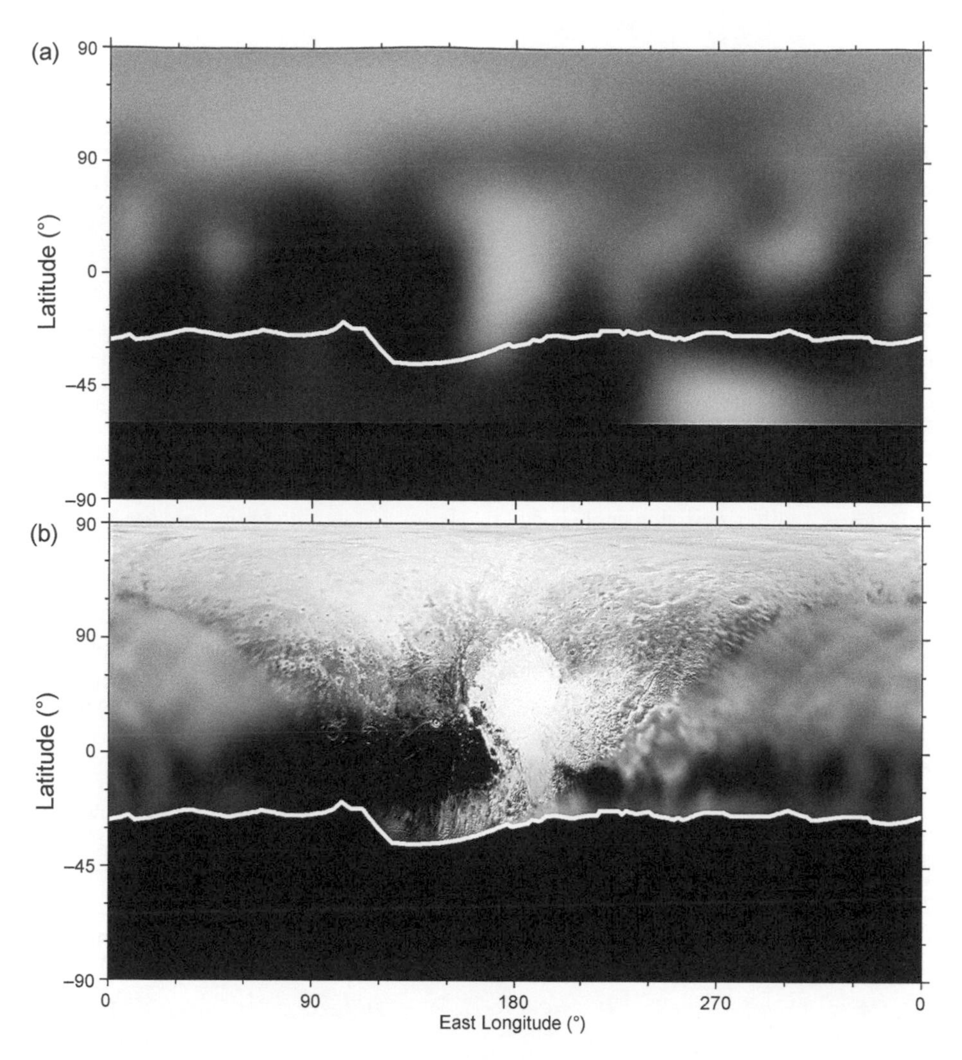

Fig. 1. (a) A true color map of Pluto from Hubble Space Telescope observation in 2002–2003. The area below 60°S latitude was not visible. The white line indicates the region not visible during the New Horizons flyby. **(b)** An enhanced color map of Pluto from New Horizons observations in 2015. This enhanced color image is produced from the three broadband filters of MVIC (MVIC B in the blue channel, MVIC R in the green channel, and MVIC NIR in the red channel). Comparing these two maps, despite the widely different spatial resolutions, there are features in common. The bright region near 180° longitude, that we now know as Spunik Planitia, is evident in both maps. Additionally, the dark equatorial band is present in both maps and even the larger extent of the Cthulhu relative to the Macula to the east of Sputnik Planitia is apparent in **(a)**. The difference in color at the high northern latitudes is also evident. The named features here and throughout the chapter can be found on the maps in Appendix A in this volume by Beyer and Showalter.

2.1. Hapke Radiative Transfer Model to Derive Photometric Properties

The scattering theory of *Hapke* (1981, 1993, 2012) is the most widely used for modeling reflected radiation from a particulate planetary surface with the goal of retrieving information about its photometric and physical properties. We describe below the Hapke radiative transfer model, which has been used to conduct the quantitative analysis of the reflectance measurements of Pluto. For more detailed descriptions of the Hapke model as applied to outer solar system surfaces, the reader is referred to summaries by *Verbiscer et al.* (2013) and *Li et al.* (2015).

The term reflectance refers to the fraction of incident radiation scattered into many directions by a material. There are several kinds of reflectance according to the degree of collimation of the source and the detector (or observer, in case for example of groundbased observations). In this chapter we will consider bidirectional reflectance (r), the total fraction of radiation scattered into a single direction by a surface illuminated from above by a highly collimated source. The bidirectional reflectance is a physically idealized concept. In reality, the solid angles for both collimated source and detector are finite, and what we can measure is actually biconical reflectance. However, in most cases of remote sensing, both the angular sizes of source and detector are very small as seen from the object.

TABLE 1. Details of the MVIC color arrays.

Array Name	Array Description	Wavelength Range (nm)	Pivot Wavelength (nm)	Array Size (pixels)
Blue	Blue TDI	400–550	492	5024 × 32
Red	Red TDI	540–700	624	5024 × 32
NIR	Near-infrared TDI	780–975	861	5024 × 32
CH_4	Methane-band TDI	860–910	883	5024 × 32

The bidirectional reflectance is therefore a good approximation, and an important simplification in theoretical analysis. The bidirectional reflectance of a rough surface r is a function of incidence angle relative to zenith, i, emission angle relative to zenith, e, and solar phase angle, g. The solar phase angle is the angle between the directions to the source and detector as seen from the surface. The approximate expression for the bidirectional reflectance of anisotropic scatterers is

$$r(i, e, g) = \frac{w}{4\pi}\frac{\mu_{0e}}{\mu_{0e} + \mu_e}$$
$$\{[1 + B_S(B_{S0}, h_S, g)]\, p(\xi, g) + H(\mu_{0e}, w)\, H(\mu_e, w) - 1\}$$
$$[1 + B_C(B_{C0}, h_c, g)]\, S(i, e, g, \theta).$$

The single-scattering albedo, w, is the ratio of the scattering to extinction coefficient of the medium. Therefore, w ranges between 0 and 1 and w = 0 implies that the particles do not scatter but instead absorb all incident radiation. The *Chandrasekhar* (1960) H-functions describe multiple scattering by regolith grains and are functions of the single-scattering albedo and the cosines of the effective incident and emergent angles, given by μ_{0e} and μ_e respectively. *Hapke* (2002) adopted an approximation to the H-functions for anisotropic multiple scattering; however, approximations to the H-functions deviate most significantly from their exact values for high single-scattering albedos. Single-scattering albedos on Pluto vary considerably from the dark Cthulhu (w ~ 0.2) to Sputnik Planitia (w ~ 1.0) (*Protopapa et al.*, 2020); therefore, approximations to the H-functions are sufficient to model the scattering properties of the darker terrains on Pluto, such as Cthulhu and Krun Macula, but exact H-functions would be best to model the brightest terrains, such as Sputnik Planitia and Lowell Regio.

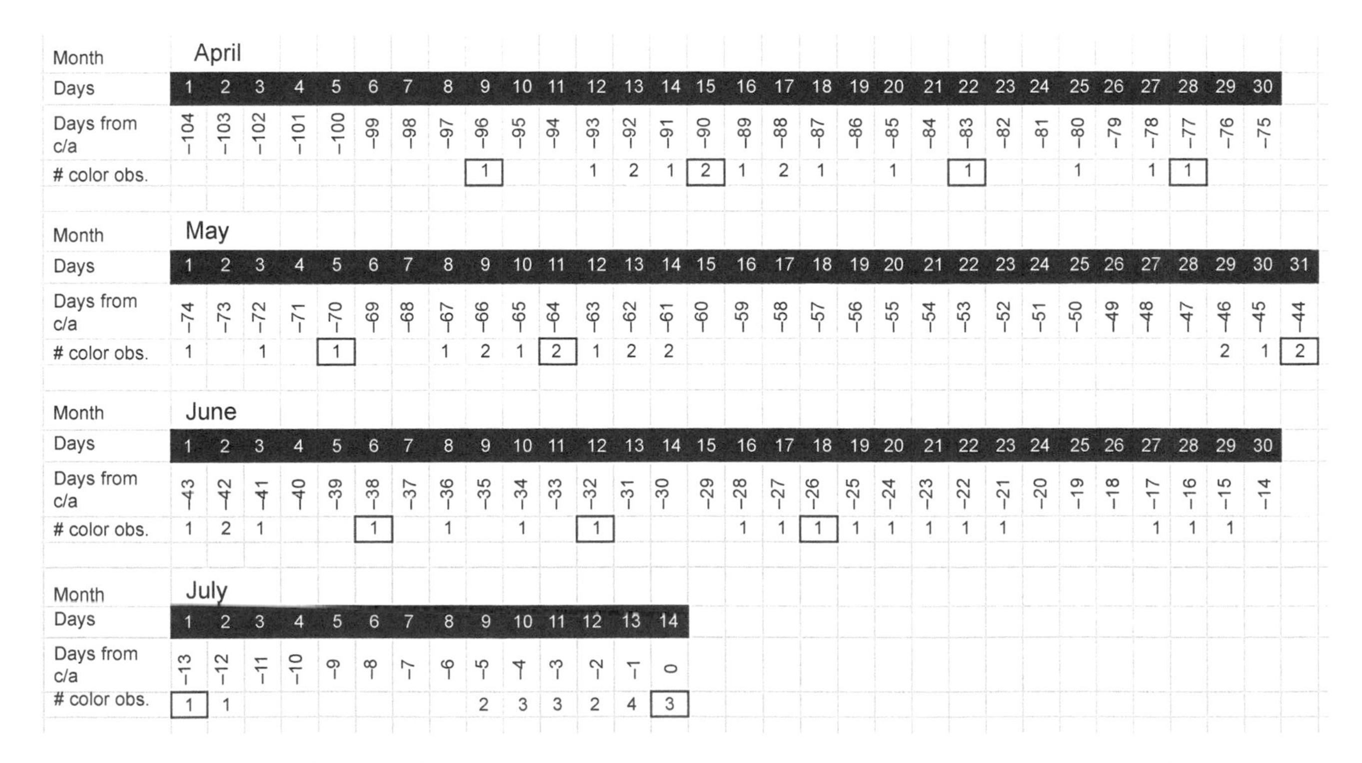

April

Days	Days from c/a	# color obs.
1	−104	
2	−103	
3	−102	
4	−101	
5	−100	
6	−99	
7	−98	
8	−97	
9	−96	[1]
10	−95	
11	−94	
12	−93	1
13	−92	2
14	−91	1
15	−90	[2]
16	−89	1
17	−88	2
18	−87	1
19	−86	
20	−85	1
21	−84	
22	−83	[1]
23	−82	
24	−81	
25	−80	1
26	−79	
27	−78	1
28	−77	[1]
29	−76	
30	−75	

May

Days	Days from c/a	# color obs.
1	−74	1
2	−73	
3	−72	1
4	−71	
5	−70	[1]
6	−69	
7	−68	
8	−67	1
9	−66	2
10	−65	1
11	−64	[2]
12	−63	1
13	−62	2
14	−61	2
15	−60	
16	−59	
17	−58	
18	−57	
19	−56	
20	−55	
21	−54	
22	−53	
23	−52	
24	−51	
25	−50	
26	−49	
27	−48	
28	−47	
29	−46	2
30	−45	1
31	−44	[2]

June

Days	Days from c/a	# color obs.
1	−43	1
2	−42	2
3	−41	1
4	−40	
5	−39	
6	−38	[1]
7	−37	
8	−36	1
9	−35	
10	−34	1
11	−33	
12	−32	[1]
13	−31	
14	−30	
15	−29	
16	−28	1
17	−27	1
18	−26	[1]
19	−25	1
20	−24	1
21	−23	1
22	−22	1
23	−21	1
24	−20	
25	−19	
26	−18	
27	−17	1
28	−16	1
29	−15	1
30	−14	

July

Days	Days from c/a	# color obs.
1	−13	[1]
2	−12	1
3	−11	
4	−10	
5	−9	
6	−8	
7	−7	
8	−6	
9	−5	2
10	−4	3
11	−3	3
12	−2	2
13	−1	4
14	0	[3]

Fig. 2. Schedule of MVIC color observations of Pluto. All observations were made with all four color filters. The days when New Horizons was imaging the closest approach hemisphere are denoted by squares around the number of color observations taken that day. The number of days from closest approach (c/a) is also given.

The shadowing function S describes the large-scale roughness of a particulate surface and is a function of the viewing geometry (i, e, g), and the mean topographic slope angle θ. For the derivation of large-scale roughness the reader is referred to Chapter 12 of *Hapke* (2012). It is important to note that in Hapke's derivation of macroscopic roughness, photons that enter shadows do not emerge. While this limitation does not affect the measurement of roughness on Pluto's darker terrains, it does restrict the extent to which roughness can be determined on the brightest terrains such as Sputnik Planitia and Lowell Regio.

Particulate surfaces often exhibit a nonlinear increase in reflectance as the solar phase angle decreases. This nonlinear trend is the opposition surge or opposition effect. Two mechanisms produce the opposition effect: interparticle shadow-hiding and a constructive interference phenomenon known as coherent backscatter. The contributions of both mechanisms are included in the bidirectional reflectance equation.

The shadow hiding opposition effect (SHOE) is produced by surface particles casting shadows at solar phase angles between 0° and 20°. At opposition, or zero phase, surface particles hide their own shadows, and the reflectance increases.

The SHOE backscatter function is given by

$$B_S(g) = \frac{B_{S0}}{(1 + 1/h_S)\tan(g/2)}$$

where B_{S0} and h_S are the amplitude and angular width of the opposition effect (also known as the compaction factor), respectively. The SHOE amplitude B_{S0} measures regolith grain transparency and ranges from nearly zero for transparent grains to unity for opaque grains. In general, the angular width of the SHOE depends on the size distribution of regolith particles as well as how their packing varies with depth. Assuming, however, that regolith grains have uniform size and packing with depth, then the SHOE angular width h_S is a function of regolith porosity alone.

According to *Helfenstein and Shepard* (2011), $h_S = -0.302\phi^{1/3} \ln(1-1.209\phi^{2/3})$ where ϕ is the packing factor and the porosity $p = (1-\phi)$. Readers are referred to Chapter 9 of *Hapke* (2012) for the derivation of the SHOE.

The coherent backscatter opposition effect (CBOE) is most pronounced at the smallest phase angles, between 0° and 2°. CBOE occurs when incident and emergent rays constructively interfere with one another, enhancing the reflectance by up to a factor of 2 (*Hapke,* 2002). Like the SHOE, CBOE is described by two parameters: the amplitude B_{C0} and an angular width h_C. The angular width of the CBOE h_C is related to the transport mean free optical path length of a photon Λ_T: $h_C = \lambda/(4\pi\Lambda_T)$ where λ is the wavelength of light. The amplitude of the CBOE B_{C0} is related to the intensity of multiple scattering and should be highest for high-albedo surfaces (or surface particles).

The single particle phase function p(ξ, g) describes the direction into which incident radiation is preferentially scattered by surface particles and is usually represented by a one-term *Henyey-Greenstein* (1941) phase function

$$p(\xi, g) = \frac{1 - \xi^2}{(1 + 2\xi \cos g + \xi^2)^{3/2}}$$

The asymmetry factor ξ is the average value of the cosine of the scattering angle $\Theta = \pi - g$. Figure 3 shows the behavior of the one-term Henyey-Greenstein phase function for several values of the asymmetry parameter ξ. For a summary of other formulations of the single particle phase function (e.g., variants of the two-term Henyey-Greenstein function), the reader is referred to *Verbiscer et al.* (2018, Appendix A). Regolith particles usually have both forward- and back-scattering lobes that are better fit with a two-term Henyey-Greenstein function if there are data available at both large and small phase angles.

These so-called Hapke parameters ω, ξ, θ, B_{S0}, h_s, B_{C0}, and h_c characterize the scattering properties of a particulate planetary surface. It is possible to solve the bidirectional reflectance equation uniquely for the bidirectional reflectance when reflectance measurements covering a wide range of viewing geometries (i, e, g) are available. Note that all Hapke parameters are wavelength dependent except the mean roughness slope angle, θ, which describes surface topography at spatial scales below the resolution limit of the detector.

2.2. Global Photometric Properties

HST's Advanced Camera for Surveys High-Resolution Channel (ACS-HRC) observed the Pluto-Charon system

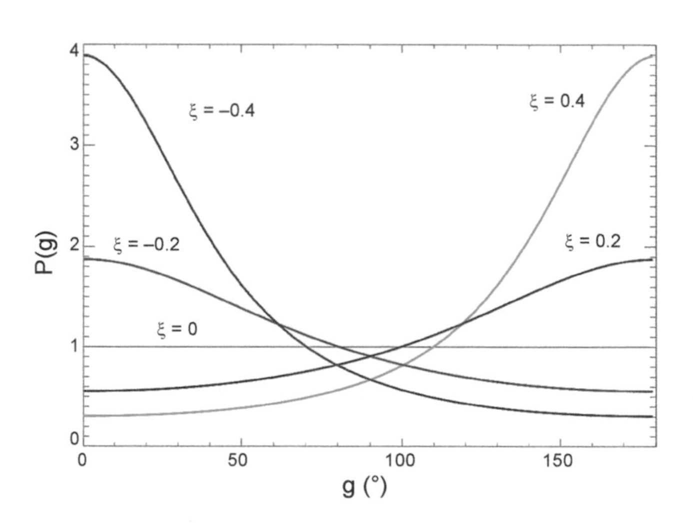

Fig. 3. One-term *Henyey-Greenstein* (1941) phase function for several values of the asymmetry parameter ξ. Surface particles scatter isotropically if ξ = 0 and p(ξ, g) = 1. If ξ > 0, (ξ, g) increases monotonically between 0 and π, and decreases monotonically if ξ < 0.

from 2002 to 2003 at 12 different longitudes at solar phase angles ranging between g = 0.34° and 1.74°. *Buie et al.* (2010a) found from the analysis of these observations that the amplitude of Pluto's light curve had decreased since the mutual event season during the 1980s. In addition, the V-band reflectance had increased while the B-band reflectance decreased, indicating that Pluto's color had reddened significantly since the mid-1980s, interpreted as evidence of volatile transport on its surface (*Buie et al.,* 2010a). Pluto's B–V color was now 0.9540 ± 0.0010, higher than the previously measured 0.867 ± 0.008. Pluto's V-band phase coefficient increased by 2σ since the 1992–1993 measurements to β = 0.0355 ± 0.045 mag degree^{-1}.

Buie et al. (2010b) applied an early version of the Hapke photometric model (*Hapke,* 1993) to the HST's ACS-HRC whole-disk observations of Pluto, deriving the single-scattering albedo, macroscopic roughness, compaction state of the optically active portion of the regolith, and single-particle phase function in the B and V bands. These physical surface quantities, derived from the application of a photometric model based on radiative transfer, offer insights into geophysical processes, both past and present, on Pluto's surface.

Owing to the eccentricity of Pluto's orbit, the minimum solar phase angle at which it can be observed from Earth is not once per half orbital period, as it is for other planets further from the Sun than Earth. Pluto can only be observed at the minimum phase angle when it crosses the line of nodes at 87- and 161-year intervals (Fig. 4). Since shortly after the discovery of Charon in 1978, the minimum phase angle at which Pluto can be observed at each opposition decreased, and after Pluto crossed the line of nodes in 2018, the minimum phase angle increases at each opposition. The minimum phase angle at which Pluto can be observed from Earth is equivalent to the angular size of the solar radius viewed from Pluto, g = 0.0079°. The HST observations in 1992–1993 (*Buie et al.,* 1997a) were acquired at solar phase angles g > 0.6° and at g > 0.34° in 2002–2003.

The HST observed the Pluto system during 40 orbits throughout 2015 (HST Program 13667, M. Buie, PI), coincident with the New Horizons flyby, at phase angles g > 0.061°. *Verbiscer et al.* (2019) produced independent solar and rotational phase curves (Fig. 6) using all V-band data from this HST program. The total amplitude of Pluto's V-band light curve in 2015 was 0.15 mag, 0.18 mag less than the total amplitude measured in 1992–1993 when the subobserver point was closer to Pluto's equator. Pluto's solar phase curve in 2015 has a phase coefficient β = 0.013 mag degree^{-1} between phase angles g > 0.5° and g < 1.75°, shallower than the β = 0.0355 mag degree^{-1} measured in 2002–2003 and the β = 0.0294 measured in 1992–1993, but within the uncertainties of both previous measurements. *Verbiscer et al.* (2019) applied the current Hapke photometric model (*Hapke,* 2012) to these observations to derive a set of disk-integrated Hapke parameters for Pluto (see Table 2). Care must be taken, however, in the interpretation of these parameters because they were derived using a limited range of phase angles that did not exceed g = 1.75°. Observations at higher phase angles are required to constrain both macroscopic roughness and the asymmetry parameter; thus for this fit to global, low-phase-angle data, the roughness parameter is held fixed at 20°, a typical value for outer solar system bodies.

The surface of Pluto is moderately backscattering with an asymmetry parameter ξ = –0.36, similar to that of many other icy surfaces in the outer solar system. Charon has a much stronger opposition surge in the V-band than Pluto (see

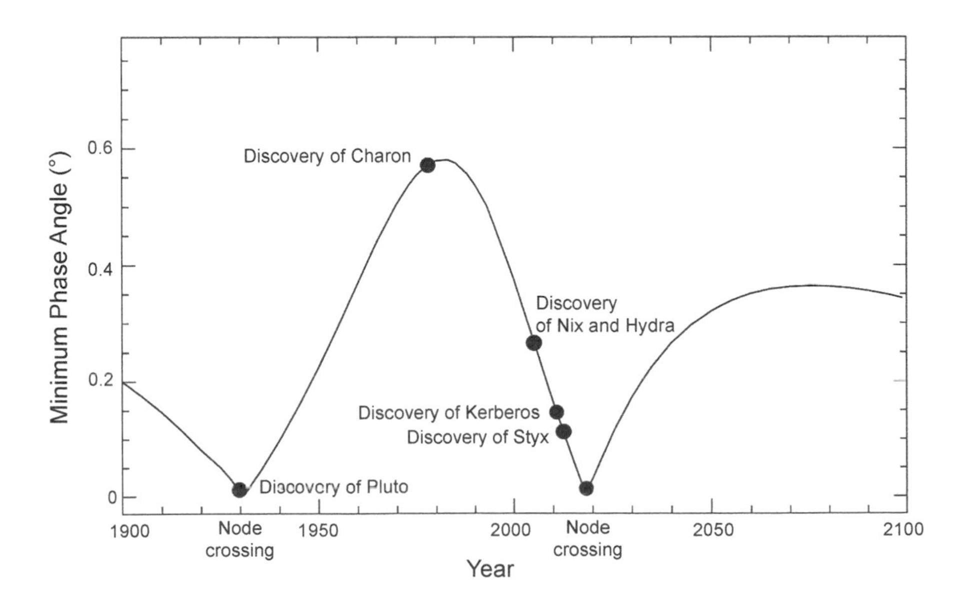

Fig. 4. Minimum solar phase angle at which Pluto and Charon can be observed at opposition from Earth. The minimum phase angle at opposition steadily decreased since 1980, reaching the minimum on July 12, 2018. The next node crossing for Pluto and Charon does not occur until 2179 (not shown).

TABLE 2. Global photometric parameters of Pluto.

Photometric Parameter	Value
ω single-scattering albedo	0.528
θ macroscopic roughness	20° (fixed)
ξ asymmetry parameter	–0.36
B_{S0} amplitude, SHOE	0.307
h_S angular width, SHOE	0.206 radians
B_{C0} amplitude, CBOE	0.074
h_C angular width, CBOE	0.0017 radians

Fig. 5). The shadow hiding opposition surge amplitude is a measure of grain transparency, with perfectly opaque particles having B_{S0} = 1.0 and transparent particles having B_{S0} = 0. Therefore, on average, particles on Pluto's surface are more transparent than on Charon's surface. The angular width of the shadow hiding opposition surge is related to the surface porosity (the ratio of the volume of empty space to the total volume). From h_s = 0.206, the average surface porosity on Pluto is 58%, compared to 95% on Charon. Globally, Pluto has a more compact surface than Charon, which has a more porous surface of predominantly water ice particles that appear opaque in the sense that the material is so strongly scattering that light is unable to pass through it. Pluto's surface is also largely comprised of transparent grains, similar to those found on Triton. Ongoing or more recent geological activity on Pluto may account for its more compact surface, on average, than that of Charon, which has a more ancient surface that has been subject to micrometeoroid bombardment and impact gardening, resulting in higher surface porosity.

2.3. Hapke Modeling of MVIC Color

The spatially resolved New Horizons Ralph/MVIC images in the visible wavelength range 400–910 nm from the Pluto flyby (*Stern et al.,* 2015) enabled derivation of disk-resolved photometric properties of Pluto (*Protopapa et al.,* 2020). Decoupling the intrinsic surface albedo variability from effects related to the observing geometry permits the quantitative investigation of the heterogeneity of Pluto's surface by removing the effects of viewing geometry (i, e, g). To derive these photometric properties, the same surface regions need to be observed at different viewing geometries. Observations of Pluto's encounter hemisphere recorded on July 13–14, 2015, have been used for this analysis because they are the highest resolution and the phase angle changes fastest near closest approach.

Four distinct terrains or regions of interest (ROIs) across Pluto's encounter hemisphere have been analyzed: (a) the terrains of Cthulhu and Krun Macula (red in Fig. 6), (b) the yellow hue on Pluto's north pole (yellow in Fig. 6), (c) the icy terrains including but not limited to Sputnik Planitia (magenta in Fig. 6), and (d) the less red terrains to the north of Cthulhu and south of 35°N (green in Fig. 6). The color analysis of Pluto by *Olkin et al.* (2017) prompted the choice of these ROIs, which span almost entirely the full color-color diagram (Fig. 6c). Therefore, analysis of these four ROIs provide a sense of the range of Pluto's photometric properties in the visible wavelength range. These four different ROIs cover a variety of geologic terrains (see the chapter by White et al. for details about geologic mapping). ROIa consists of dark ancient terrain covered in tholin. ROIb corresponds to pixels that appear yellow in the enhanced color image of Pluto (Fig. 1). ROIc covers much of the encounter hemisphere of Pluto and corresponds to the presence of volatile ices. ROId covers two different terrains (Baret Montes and part of Viking Terra) from a geologic perspective, but that have the same color properties. This unit is located near Pluto's north pole on the higher-altitude terrain.

The Hapke radiative transfer model described in section 2.1 is applied to conduct a multi-wavelength photometric analysis of Pluto, meaning that each MVIC color filter is treated separately. This is done for each ROI. Because of the limited phase angle coverage (the MVIC Pluto data cover a phase angle range between 10° and 40°), the coherent backscattering opposition effect is ignored and the parameters θ, B_{S0}, h_s are held constant and equal to

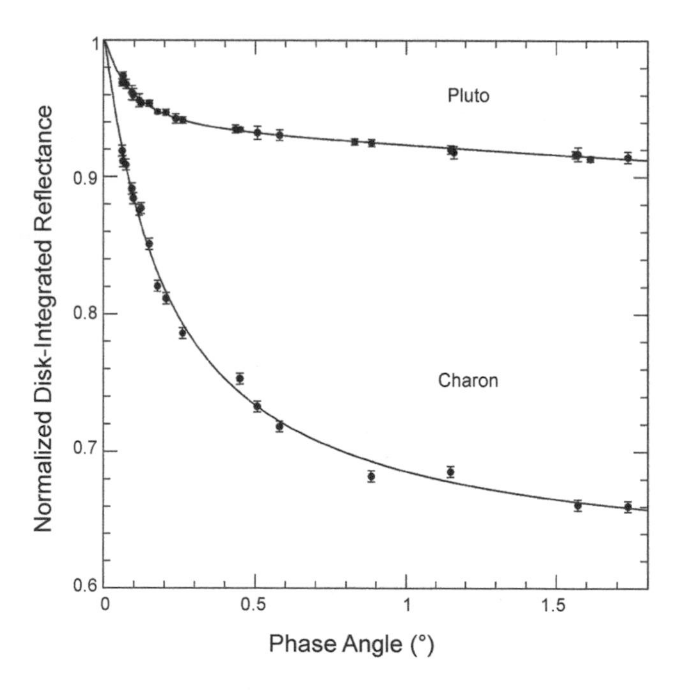

Fig. 5. Normalized disk-integrated solar phase curves of Pluto and Charon from HST Program 13667 (M. Buie, PI) acquired in 2015 in the F555W filter (V-band) at phase angles between g = 0.06° and g = 1.75°. Pluto's opposition surge is much narrower and of lower amplitude than Charon's. Pluto's phase curve is shallower than Charon's and thus has a smaller phase coefficient. Error bars represent 1σ uncertainties. From *Verbiscer et al.* (2019).

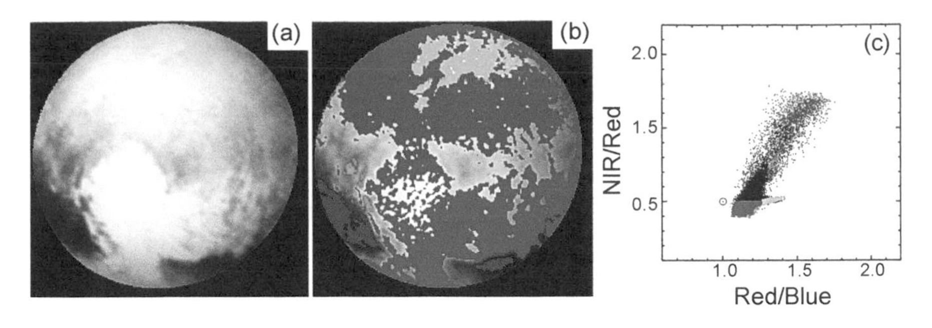

Fig. 6. **(a)** Enhanced color image of Pluto with MVIC's blue, red, and NIR filter images displayed in blue, green, and red color channels, respectively. The image scale is 21 km/pixel. **(b)** Red, yellow, magenta, and green points represent ROIs a, b, c, and d, respectively. These regions are overlaid on the color image shown in **(a)**. **(c)** Color-color plot for the color image in **(a)** (black points). The pixels corresponding to the four ROIs are shown with the same color convention adopted in **(b)**. The areas of Cthulhu and Krun Macula (ROI a, red points) and the volatile-rich terrains (ROI c, magenta points), correspond to the end members on the top right side and bottom left side of the steep-sloped mixing line of the color-color diagram, respectively, while the regions of Baret Montes and Viking Terra (ROI d, green points) fall among the intermediate pixels along this mixing line. The yellow hue on Pluto's north pole (ROI b, yellow points) lays across the less-steep color mixing line of the color-color diagram. Adapted from *Protopapa et al.* (2020).

20°, 0.307°, and 0.206° respectively. These values have been obtained from disk-integrated analysis (*Verbiscer et al.,* 2019) (section 2.2). The free parameters in the model are single-scattering albedo w and the Henyey-Greenstein asymmetry parameter ξ. It should be noted that there is a correlation between the macroscopic roughness and the asymmetry parameter. *Protopapa et al.* (2020) investigated the joint error distribution for the single-scattering albedo, the asymmetry parameter, and the macroscopic roughness and found that the solutions are not sensitive to the macroscopic roughness because of the limited phase angle coverage. A one-term Henyey-Greenstein function (*Henyey and Greenstein,* 1941) is used. Figure 7 shows their values and the correspondent 1σ errors for each ROI.

Figure 7a shows the wide diversity of single-scattering albedo across Pluto's surface. The darkest and reddest terrains are those of Cthulhu and Krun Macula (ROI a, red points in Figs. 6 and 7). These terrains display in fact a very steep single-scattering albedo as a function of wavelength. Additionally, w is very low, on the order of ~0.2 in the blue filter. In contrast, regions such as Sputnik Planitia (ROI c) or the polar cap terrains characterized by a yellow coloration in enhanced color images (ROI b) appear neutral in color and extremely bright with single-particle-scattering albedos of ~0.98 or higher. The single-scattering albedo is a function exclusively of the optical constants n and k, the real and imaginary part of the refractive index respectively of the materials out of which the terrain is made of, their relative abundance, and particle size. Therefore, the observed differences across Pluto's terrains can be interpreted only in terms of composition.

Pluto's surface is dominated by ices of methane, nitrogen, carbon monoxide, water ice and refractory materials known as tholins (see the chapter by Cruikshank et al. in this volume for a description of the surface composition of Pluto and a definition of tholins). While ices are generally neutral in the visible wavelength range, tholin materials absorb strongly in the visible wavelength range (e.g., *Brassé et al.*, 2015).

Therefore, variations in albedo and spectral slope between different terrains across the surface of Pluto are attributed to changes in abundance and grain size of Pluto's coloring agent. *Protopapa et al.* (2017, 2020) show, through radiative transfer modeling of Pluto's reflectance measurements, that high concentration of dark compounds is found in Cthulhu while Lowell Regio is highly depleted in tholins.

Pluto's terrains are backscattering, with ξ constant, within the error, throughout the wavelength range and equals to an average of -0.21 ± 0.07. This value for ξ is relatively close to the global value of -0.36 found by *Verbiscer et al.* (2019) for Pluto and -0.23 at 0.56 µm by *Hillier et al.* (1994) for Triton from Voyager. Given the similarities between Pluto and Triton composition, this is not surprising. However, it is important to note that in the range between 10° and 40° (phase angle range of our MVIC data), the backscattering lobe dominates (see Fig. 3). Therefore, it is not unexpected to find ξ to be negative from this analysis. Also, it is important to note that the Triton and Pluto data spanned different phase angles and as a result the Pluto analysis had to assume a constant value for the macroscopic roughness.

2.4. Bond Albedo

The Bond albedo contains information about the energy balance of and volatile transport on any planetary surface. It is the ratio of the power of the radiation scattered back into space to the power of the total incident radiation on the surface. The Bond albedo A_B is the product of the geometric albedo p and the phase integral q: $A_B = pq$ where the phase integral $q(\lambda) = \int_0^\pi \Phi(g, \lambda) \sin g \, dg$ and $\Phi(g, \lambda)$ is the normalized disk-integrated phase curve.

Integrating the Bond albedo over the entire electromagnetic spectrum yields the bolometric Bond albedo. *Buratti et al.* (2017) constructed a preliminary map of Bond albedos on Pluto from New Horizons LORRI images from their map of normal reflectances on Pluto (see Fig. 8). They approximated

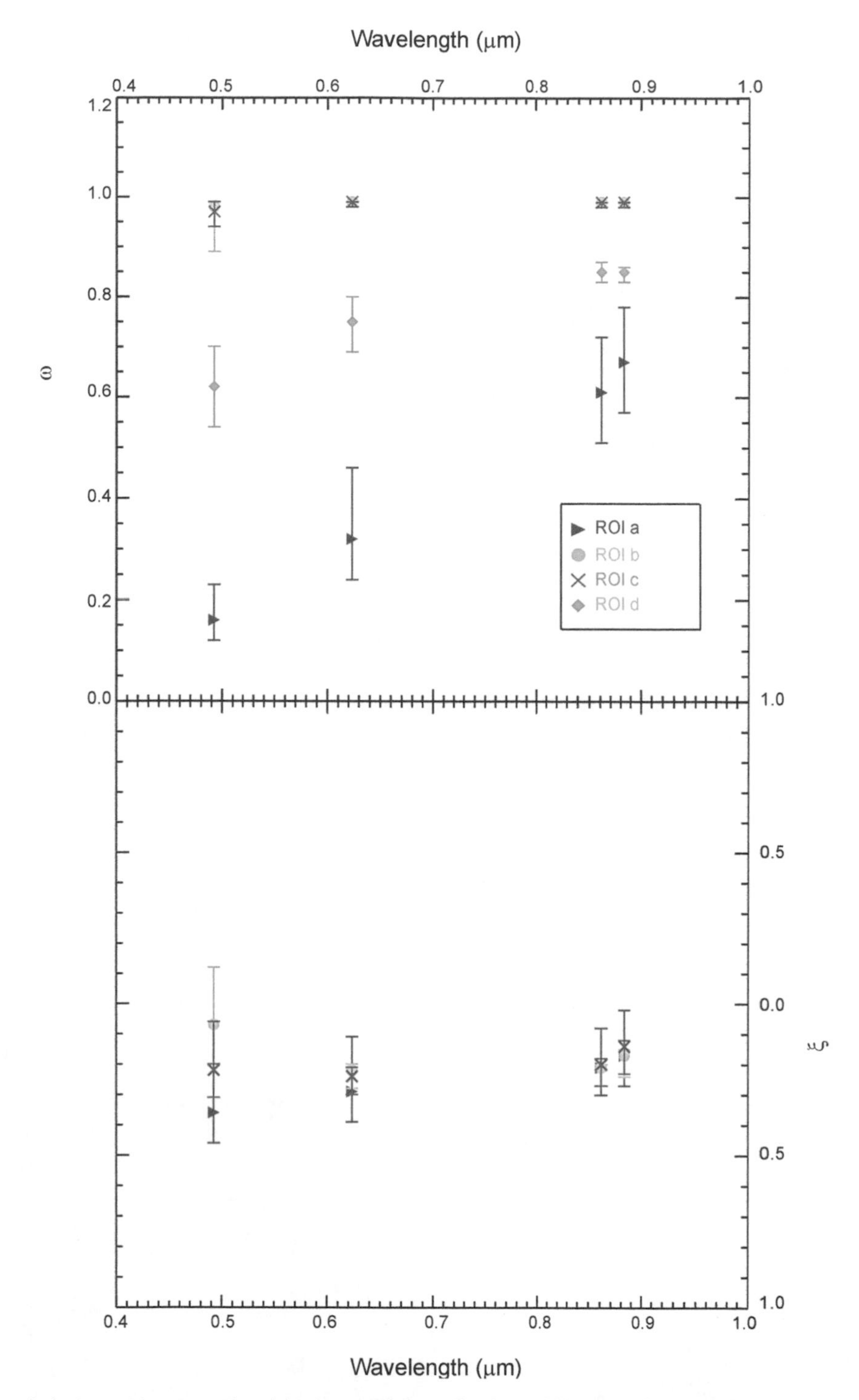

Fig. 7. Disk-resolved photometric properties (single-scattering albedo and Henyey-Greenstein asymmetry parameter) of four distinct terrains across Pluto's encounter hemisphere shown at each filter's pivot wavelength. From *Protopapa et al.* (2020).

the Bond albedo at each point on Pluto's surface by multiplying an approximated geometric albedo at each point by Triton's phase integral at 0.55 μm, an approximation for Pluto's phase integral from LORRI, which has a pivot wavelength 0.607 μm. Although the geometric albedo is a disk-integrated quantity and the normal reflectance is a disk-resolved measurement, for low-albedo surfaces such as Cthulhu and regions east of Sputnik Planitia in Pluto's southern hemisphere, the normal reflectance equals the geometric albedo. For brighter areas, the geometric albedo can be estimated using $p = 2\frac{(1-A)}{3} + A\frac{f(0°)}{2}$ where f(0°) is the surface phase function at g = 0° and A is the fraction of singly-scattered radiation (*Buratti and Veverka,* 1983). Bond albedos on Pluto span the entire range from <0.1 on Cthulhu to nearly 1.0 on Sputnik Planitia. Regions such as Voyager, Venera, and Hayabusa Terrae have intermediate Bond albedos A_B ~0.5, while Lowell Regio has a higher Bond albedos A_B ~0.8; therefore, surface temperatures on Lowell Regio and Sputnik Planitia are lower

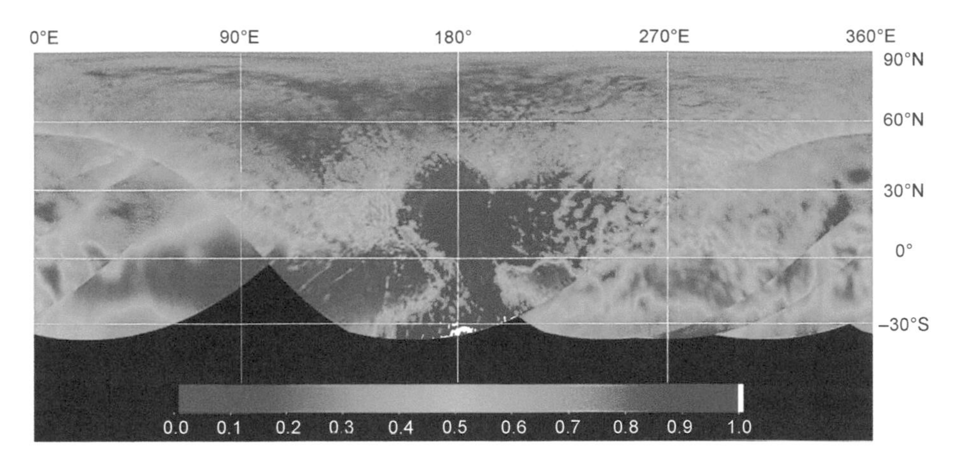

Fig. 8. Preliminary map of Bond albedos on Pluto from New Horizons LORRI images (*Buratti et al.,* 2017).

than those on Voyager, Venera, and Hayabusa Terrae. Owing to its low Bond albedo, Cthulhu likely has the highest surface temperatures on Pluto.

3. COLOR OF PLUTO FROM NEW HORIZONS

3.1. Significant Color Trends Across Pluto

Using the highest-resolution color images of Pluto, a near-global color map has been constructed and is displayed in Fig. 1b. The southern extent of Pluto was not imaged by New Horizons. It was in arctic winter at Pluto's south pole at the time of the flyby (the subsolar latitude was 51.5°) and the subspacecraft point is generally ~40°N for all the observations except at closest approach, where it was further south (26°N), and therefore the color map does not extend beyond about 30°S.

The highest-resolution color imaging occurred ~40 min from closest approach (at 11:10:52 UT on July 14, 2015) with an image scale of 660 m/pixel and a subspacecraft latitude of 168° and subspacecraft longitude of 26°. The non-encounter hemisphere was observed days before on July 10, 2015, with a subspacecraft latitude of 0° and an image scale of 90 km per pixel.

There are some large-scale color trends immediately apparent in Fig. 1b. The darkest and most red material is preferentially located near the equator. The high northern latitudes are extensively covered by a yellow terrain type. The latitudinal variations of albedo and the long-term

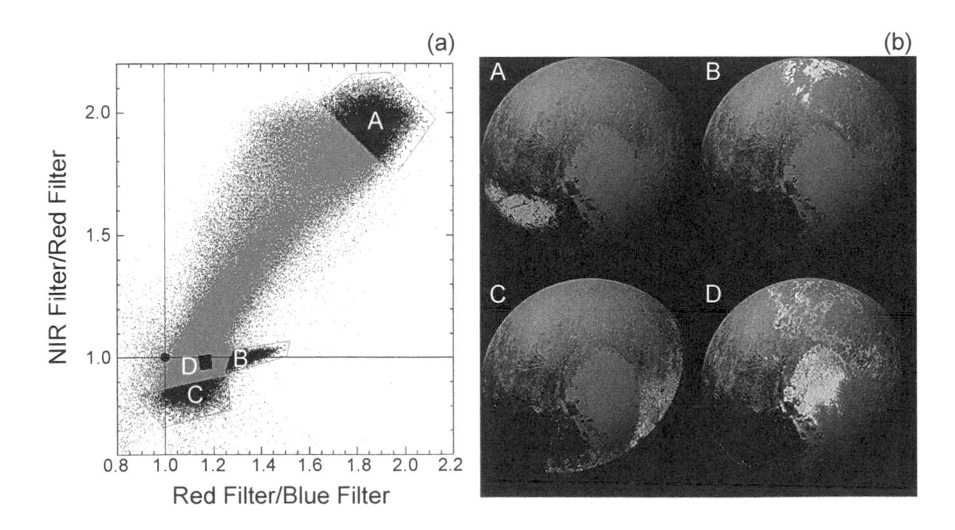

Fig. 9. **(a)** The color-color diagram of the highest-resolution color observation of Pluto after binning by a factor of 3. The two colors are constructed from the three broadband filters: specifically, the y-axis is the ratio of the I/F from the NIR filter to the I/F of the red filter, and the x-axis is the ratio of the I/F of the red filter to the I/F of the blue filter data. The intersection at (1,1) is the neutral (or solar) color. Four regions, labeled A, B, C, and D, are indicated in the color-color diagram. **(b)** The points in the Pluto image that correspond to the regions indicated in **(a)**. Region A is in Cthulhu, region B is the yellow terrain of Lowell Regio, region C is East Tombaugh Regio and south of Cthulhu, and region D is the northern extent of Sputnik Planitia and the margins of the yellow terrain of Lowell Regio.

seasonal pattern of insolation is discussed in the chapter led by Young. These color units and others will be investigated below through the use of color-color plots, principal component analysis (PCA), and color slope.

3.2. Color-Color Diagrams

A color-color diagram of the highest-resolution Pluto color data is shown in Fig. 9. Solar color (neutral color terrain) would fall on the point (1,1) in the color-color diagram. The color-color diagram shows two color mixing lines. Both originate in the lower left of the figure, which corresponds to relatively neutral terrain, associated with volatile ices. The predominant color mixing line extends to point A in the upper right, which corresponds to the dark red material in the center of Cthulu (located at Pluto's equator). The points between the neutral icy end member and the dark red end member extend in latitude away from the equator.

Binzel et al. (2017) describe the climate zones and identifies the correlation of albedo with latitude. Figure 10 shows that the correlation with latitude extends to color as well as albedo. There is likely a causal relationship between the insolation and latitudinal pattern of albedo/color. For the last 2.8 m.y., the region between 13°S and 13°N has received sunlight each Pluto day (*Binzel et al.,* 2017), while the more extreme latitudes have undergone prolonged (~100 yr) arctic seasons with no insolation. Seasonal models have predicted volatile transport over a Pluto year or longer with the poles exchanging volatile ices (see the chapter by Young et al. in this volume for more about Pluto's seasonal cycles). The low albedo of the equatorial region, once established, helps to reinforce the sequestration of volatiles outside of the equatorial band because the low-albedo terrain will absorb more insolation and preferentially warm the equatorial terrain compared to the higher-albedo terrain. This pattern of regular insolation and the positive feedback of a low-albedo surface results in the predominant mixing line in the color-color diagram that spans from relatively neutral volatile ices to dark red processed organic material (similar to tholins produced in labs on Earth).

The minor mixing ratio is identified in box B of Fig. 9 and is visually seen as a yellow terrain in Lowell Terra at northern latitudes on Pluto. The most extreme units of this color mixing line cluster near Pluto's north pole. This terrain is associated with northern latitude at moderate to high elevations and the presence of CH_4 ice (*Olkin et al.,* 2017).

The colors identified in Fig. 9 as D map to both northern Sputnik Planitia and the less-yellow terrain in the northern latitudes. One attribute that both of these terrains have in common is that they are sublimating (*Protopapa et al.,* 2017).

Region C in Fig. 9 shows the stark delineation of the bluest terrain on Pluto. This color unit covers the eastern half of Tombaugh Regio. This color covers the eastern edge of Tombaugh Regio and south of Cthulu to the west of Sputnik Planitia. This terrain has a spectral slope of $-1.4 \pm 0.2\%$ per 100 nm (normalized at 550 nm following

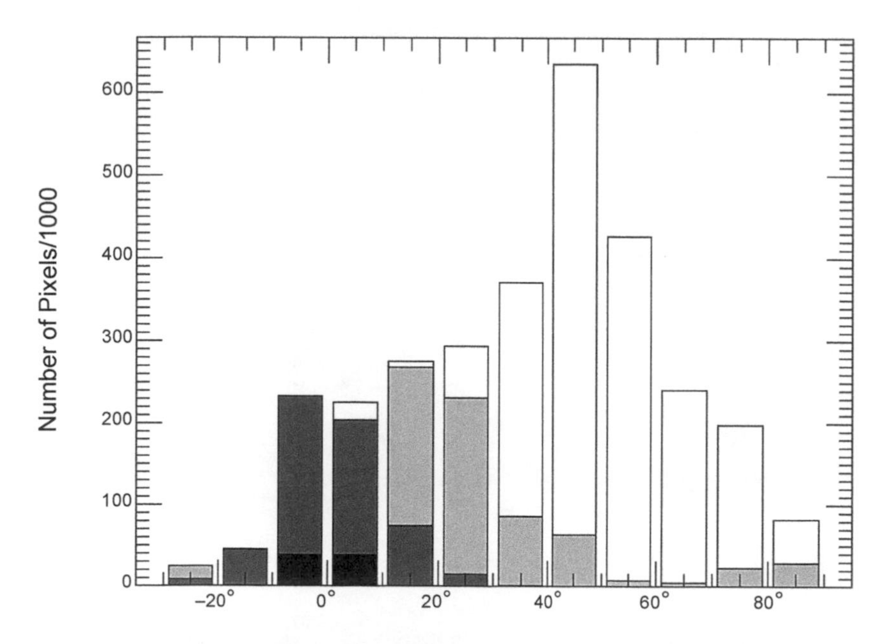

Fig. 10. Pluto's surface color as a function of latitude on Pluto. The surface color was divided into four bins based on the ratio of the I/F in the MVIC NIR filter to the I/F in the MVIC red filter (NIR/red). Black corresponds to NIR/red values greater than 2 (the most red terrain), dark gray is between 2 and 1.5, light gray is between 1.5 and 1.0, and white is less than 1.0 (the least-red terrain). There is a strong correlation between surface color and latitude with the most-red terrain located near the equator.

Doressoundriam et al., 2002). For comparison, the dark red terrain of Cthulhu has a spectral slope of 50.1 ± 0.1% per 100 nm. The spectral slope is the slope of the best fit line through the I/F of the three broadband filters. This terrain (region C) is also associated with the presence of methane ice (*Protopapa et al.,* 2017).

3.3. Principal Component Analysis

An image with n color planes can be thought of as an n-dimensional dataset. For n up to 3, it is straightforward to display each color plane in a separate color channel of a three-primary-color display. These primary color channels are usually red-green-blue or cyan-magenta-yellow for light emitting or reflecting displays respectively. The human eye is well-suited to evaluate the information content in such a display. But when n is greater than 3, as in the case of MVIC, other methods are helpful for exploring the information contained in the image dataset. A widely used method for this purpose is PCA and has been applied to Pluto (*Grundy et al.,* 2016). This linear algebra technique reprojects the n-dimensional dataset onto a new n-dimensional space in which all axes are mutually orthogonal. The first principal component (PC1) corresponds to the axis of maximum variance in the original data. The second principal component (PC2) is the axis of maximum variance orthogonal to the first. Each successive PC captures the maximum remaining variance orthogonal to the prior ones. The direction of the unit vector in wavelength space corresponding to each PC is the eigenvector, and the projection of the dataset along that unit vector is the PC image.

Another way to think of the process is to picture each pixel of the image as one point in an n-dimensional cloud containing all the pixels. The direction in which this cloud is most dispersed is the first principal component. Collapsing the cloud along that axis leaves an n–1-dimensional cloud, upon which the same operation can be performed, and so on until no dimensions remain.

In typical multi-wavelength imaging applications, PC1 corresponds to variations in illumination across the target scene. These geometric factors tend to affect all wavelengths very similarly, resulting in an eigenvector that is flat across all wavelengths and an image that looks like an average over all wavelengths, with lower noise than the individual color planes since they all contribute to it, just as in an average. Subsequent PCs provide information about color variations across the scene. In the case of Pluto (Fig. 11), PC2 captures color slope, as can be seen from the eigen-

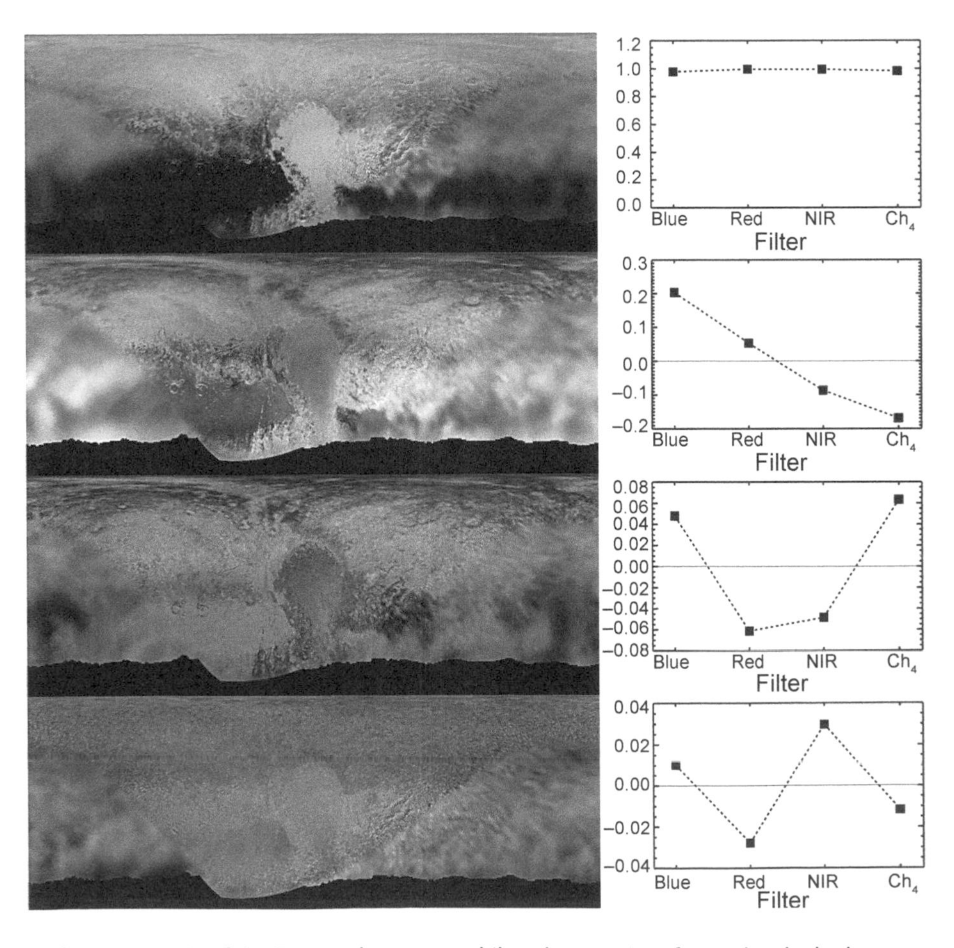

Fig. 11. The principal components of the Pluto color map and the eigenvectors for each principal component. The principal components (left) and eigenvectors (right) are ordered from 1 to 4 from the top to the bottom of the figure.

vector. The least-red areas of Pluto's surface show up as bright in the PC2 image while the reddest areas show up as dark. Virgil Fossae exhibits one of the reddest colorations by this measure while the bladed terrain, as exemplified by Tartarus Dorsa, is the least red. PC3 corresponds to differences between (red, NIR) and (blue, CH_4) filters, where red and NIR behave similarly to one another and blue and CH_4 behave similarly to one another. The PC3 image is dark where there is stronger absorption in blue and CH_4 than there is in red and NIR, such as in the bladed terrain, and to a lesser extent in Sputnik Planitia and the yellow terrain near the north pole. PC4 corresponds to differences between (red, CH_4) and (blue, NIR). The amount of variance captured by each PC is a useful measure of different types of variability in the data. For the Pluto global MVIC mosaic, PC1 accounts for 97.64% of the variance, with PC2, PC3, and PC4 accounting for 2.00%, 0.31%, and 0.05%, respectively.

Since PC1 corresponds to overall brightness differences, and PCs 2–4 correspond to color differences, a natural way to display the four PCs is to use PCs 2–4 to determine a hue (by assigning PC2, PC3, and PC4 to red, green, and blue channels, respectively) and then use PC1 to set the brightness for each pixel in the scene. (This is done in Fig. 12, resulting in a garish image that highlights how colors vary across Pluto.) The same trends discussed above are seen here including the yellow terrain in the enhanced color image appears purple here due to the combination of PCs 2 and 4.

4. SELECTED REGIONS

In this section, we will focus on selected regions of Pluto's surface with notable color units. For reference, a map with surface names can be found in Appendix A by Beyer and Showalter in this volume.

4.1. Virgil Fossae

There is a distinct red color on Pluto associated with Virgil Fossae, a trough system that cuts through the prominent central-peaked crater, Elliot Crater (Fig. 13). The distinct red color is associated most strongly with Virgil Fossae, but it can also be seen in local craters and Beatrice Fossae. The spectral slope of this color unit is 28.7 ± 0.02% per 100 nm (normalized at 550 nm), which is significantly less red than Cthulhu. Compositionally, this is also a distinctive area on Pluto. This localized area near the Virgil Fossae has water ice and ammonia on the surface (*Dalle Ore et al.,* 2019). The cause of the distinctive coloration is not known. The unique coloration could be a result of a distinct texture, mechanical or scattering configuration, and/or the combination of tholin, water ice, and ammonia or ammoniated compounds, but other regions on Pluto such as Viking Terra also show evidence of tholin, water ice, and ammonia without the same distinctive red color (*Cruikshank et al.,* 2019b). The difference could be the form of the ammoniated compound or the relative fraction of tholins, water ice, and ammoniated compounds. *Cruikshank et al.* (2019a,b) suggest subsurface sources and cryovolcanic processes for the red material in both Viking Terra and Virgil Fossae.

4.2. Al Idrisi Montes

In some cases, a combination of color and geology can reveal historical processes that occurred on Pluto. The Al Idrisi Montes could be produced from the breakup of larger blocks (*White et al.*, 2017). Figure 14 shows a small area of Al Idrisi Montes with two blocks that likely used to be joined along the edges indicated by the arrows. The edges of these blocks have the same shape and they also both exhibit a red layer below the surface that is visible

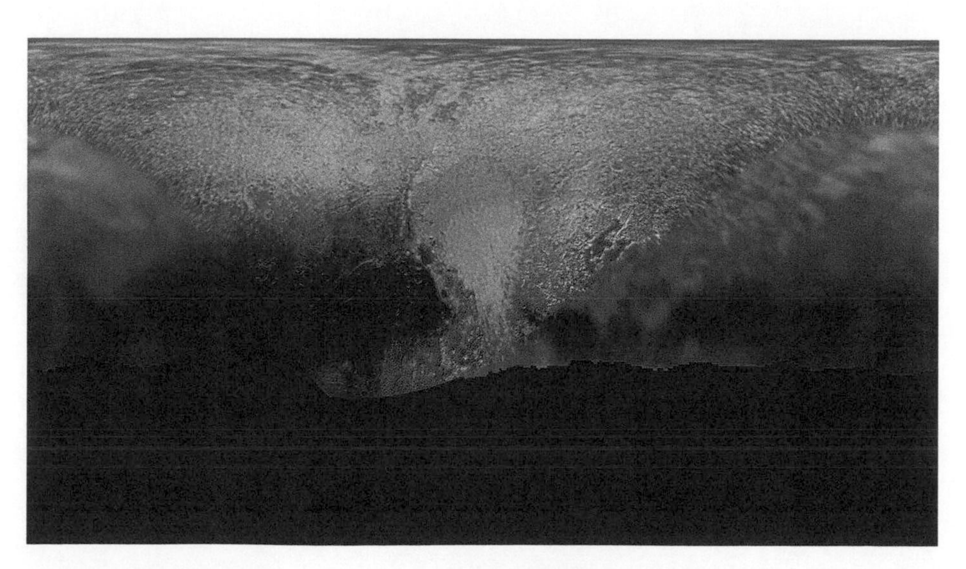

Fig. 12. The principal components rendered in RGB (PC2-4) and brightness (PC1).

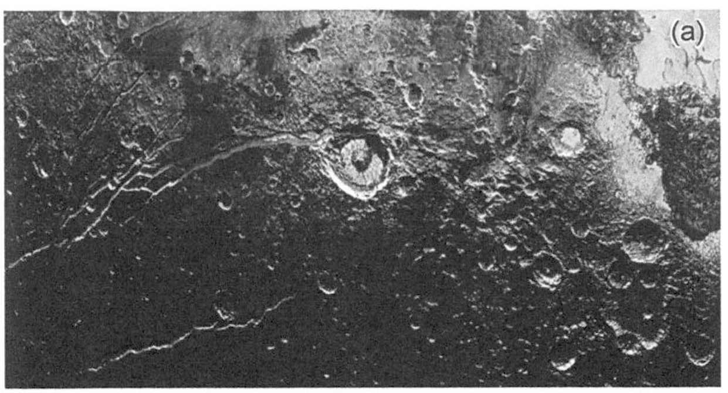

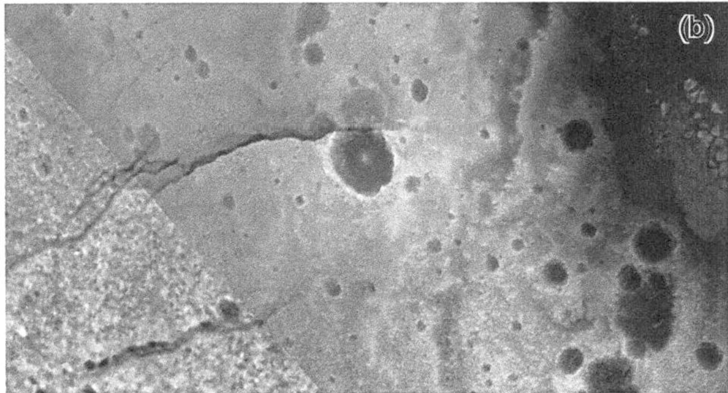

Fig. 13. (a) The enhanced color image of the Virgil Fossae region. A distinct red color can be seen in the trough even as it intersects Elliot crater, indicating that the red coloration was implanted after the impact crater formed. For scale, Elliot crater has a diameter of ~90 km. **(b)** The digital elevation map (*Schenk et al.,* 2018) of the same region. At the deepest parts of Virgil Fossae where the trough is more than 2.5 km deep, the distinctive red coloration is the most significant.

in the vertical face of the blocks (near the arrows). The southern extent of Al Idrisi Montes shows more space between the blocks, as evidenced by Sputnik Planitia-like material between the blocks and more sharp edges on the blocks. It could be that the blocks in the north of Al Idrisi Montes have interacted with each other more than the blocks in the south, resulting in a more rounded surface on the blocks in the north.

The red material on Pluto is typically associated with a tholin. The source of red layer in the Al Idrisi Montes, if it is tholins, could be deposition from the atmospheric haze (*Grundy et al.,* 2018), production on the surface, or upwelling from a subsurface reservoir as postulated for Virgil Fossae (see *Dalle Ore et al.,* 2019; *Cruikshank et al.,* 2019b). The layer extends across the entire face of both blocks with a layer of neutral color layer between it and the surface. One possible hypothesis for the formation of the layer is that the tholin layer formed on the surface and was later covered by volatile ices during the large-scale seasonal transport of volatile ices on Pluto. It is unlikely that the layer formed from a subsurface reservoir as seen at Virgil Fossae because there is no indication of red material below the very stratified layer.

4.3. Lowell Regio

This region located near Pluto's north pole is dominated by color units ranging from neutral to yellow (Fig. 15) and exemplifies the terrain along the minor mixing line in the color-color plots. This region is rich in CH_4 ice and depleted in N_2 ice (*Schmitt et al.,* 2017). The yellow terrain is located at higher elevations. Tholins have been made in the lab with different colors ranging from yellow to red to brown (see the chapter in this volume by Cruikshank et al.). It could be that the yellow coloration of Lowell Regio is due to formation of tholins under different circumstances than the widespread dark red tholins found elsewhere on Pluto. But what those circumstances could be is not known at this time; it could be a result of

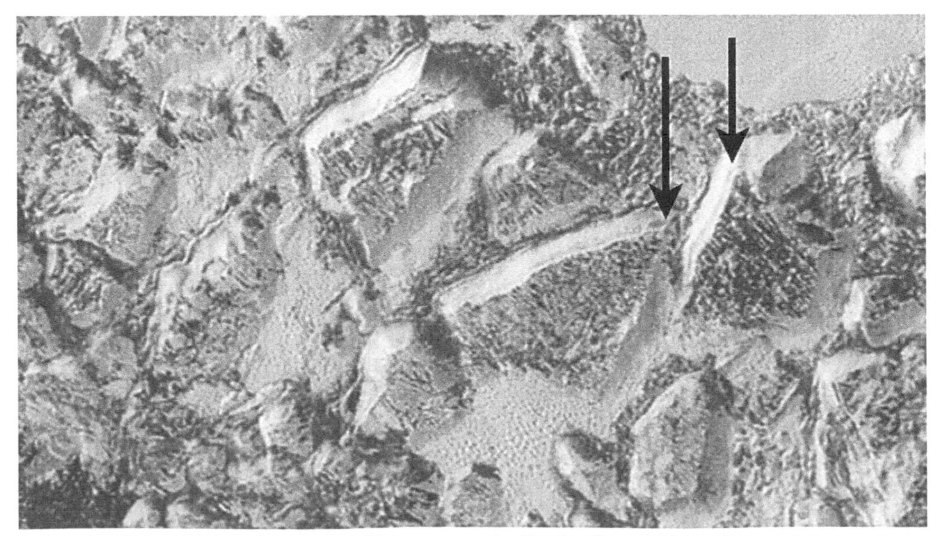

Fig. 14. A part of the southern Al Idrisi Montes. The arrows indicate the similar shape of two adjacent blocks and the red layer visible in both vertical faces.

Fig. 15. Pluto's north pole. There is a clear distribution of yellow and neutral terrain units.

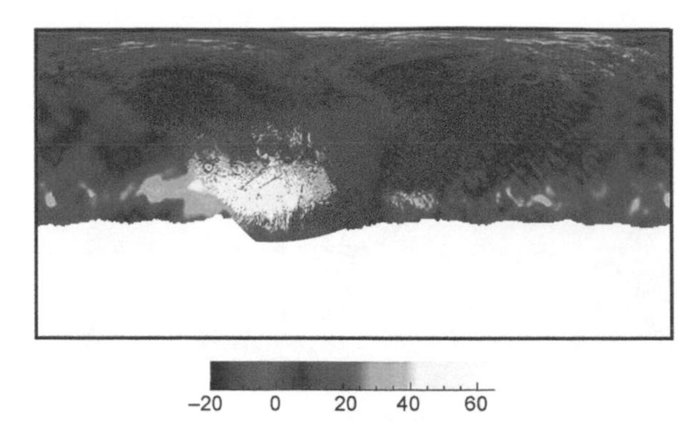

Fig. 17. Spectral slope map of Pluto using the MVIC B and R filters from New Horizons.

the pigment originating in the haze layers (*Grundy et al.,* 2018). It is worth noting that the terrain at the margin of the yellow terrain of Lowell Regio (Fig. 9, region D) is the same color as the northern part of Sputnik Planitia. These regions occupy the same small region of the color-color plot (Fig. 9a) and have a lighter yellow coloration.

It is interesting to consider the similarities and differences between some terrains. The yellow terrain of Lowell Regio is methane-rich and high-altitude. These two traits are also shared with the bladed terrain of Tartarus Dorsa (Fig. 16), which has a very different coloration than the yellow terrain of Lowell Regio. This tells us that the methane abundance and altitude alone do not determine the visible color. Other minor compositional constituents can have impact the color of a terrain.

Fig. 16. Pluto's bladed terrain. This high-elevation, CH_4-rich terrain located near the equator does not exhibit the yellow hue common in the high-elevation, CH_4-rich polar region.

5. COMPARISON WITH TRITON

Before New Horizons reached Pluto, the best analog for Pluto was thought to be Triton. Both objects have a similar size, tenuous atmospheres, and the same volatile ices (N_2, CH_4, and CO) that drive activity and support their atmospheres by sublimating and condensing on seasonal timescales. Additionally, Triton is likely a captured Kuiper belt object (*Agnor and Hamilton,* 2006). Following the New Horizons flyby of Pluto, it is possible to compare the resolved color imaging of Pluto and Triton; however, the Voyager and New Horizons imagers did not have identical filter passbands. To facilitate comparison, filters that match passbands as closely as possible were chosen for our analysis. For Pluto, the MVIC B (effective wavelength 492 nm) and R (effective wavelength 624 nm) are used to produce a spectral slope map (Fig. 17) that gives the spectral slope per 100 nm and normalized at 550 nm. For Triton, the Voyager blue (effective wavelength 475 nm) and orange (effective wavelength 589 nm) are used to construct the spectral slope map in Fig. 18.

Pluto shows a wider range of colors than Triton with the maximum extending to 75% (minimum is –23%) while the maximum slope for Triton reaches only 28% (minimum is –5%). The larger slope is indicative of the redder terrain on Pluto like Cthulhu and equatorial maculae. A similar terrain type to the maculae on Pluto doesn't exist on Triton.

6. CONCLUSIONS

New Horizons has revealed Pluto in detail not achievable from Earth. Just as these data answered questions about the nature of the Pluto system, they have spawned new questions. We knew before the New Horizons encounter that there was great variety in the colors on the surface of Pluto. Now we can see the correlation of these colors with geology and composition.

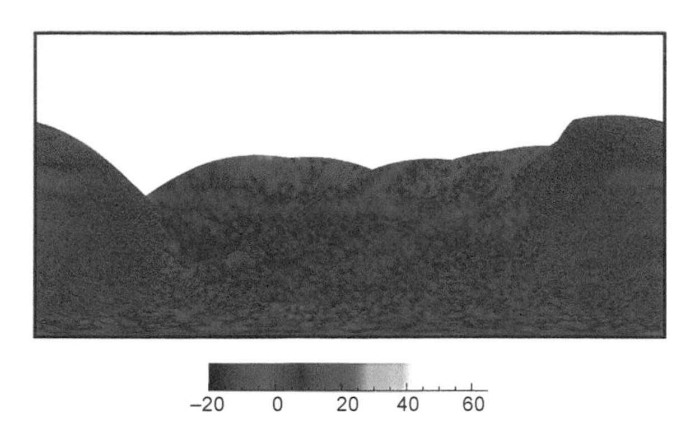

Fig. 18. Spectral slope map of Triton using the blue and orange filters from Voyager. The spectral slope map is centered on 180° longitude. The southern hemisphere of Triton was explored with Voyager 2.

Pluto's color falls predominantly on two different color mixing lines. The end member for both mixing lines is the same — the relatively neutral color of volatile ices. Most terrain on Pluto falls along a color mixing line that extends from the neutral volatile ices to the dark red terrain, which is consistent with tholins in color. The latitudinal pattern of the distribution of color is consistent with solar insolation patterns. Specifically, near-equatorial regions have received more insolation recently than the polar regions and the nonvolatile tholins remain in the equatorial region while the volatile ices have migrated to regions with less insolation.

Additionally, the low albedo of the equatorial maculae increases the amount of sunlight absorbed by the surface, increasing its temperature, which would also drive the volatile migration.

The minor color mixing line is confined to the northern terrain of Pluto and extends from the north of Sputnik Planitia to the north pole. This color unit could be another type of tholin but why it differs from the dark red terrain is not known yet.

Future work involves further lab studies of volatile and non-volatile materials on Pluto, continued groundbased observations of Pluto, and continued exploration of the Kuiper belt by spacecraft missions. Laboratory studies of tholins and volatile ices will help us understand the color agents on Pluto's surface, including the dark red of Cthulhu, the distinctive red of Virgil Fossae, and the yellow terrain of Lowell Regio. Continued groundbased observations are essential for monitoring seasonal change on Pluto, which is best understood by monitoring the atmospheric pressure (via occultations) and surface changes in color and composition. Further exploration of the Kuiper belt by spacecraft missions will advance our understanding of these objects that are the remnants of solar system formation.

REFERENCES

Agnor C. B. and Hamilton D. P. (2006) Neptune's capture of its moon Triton in a binary-planet gravitational encounter. *Nature, 441,* 192–194.

Andersson L. E., Kunkle T., and Kasle D. (1978) IAU Circ. 3286.

Binzel R. P. (1988) Hemispherical color differences on Pluto and Charon. *Science, 4869,* 1070–1072.

Binzel R. P., et al. (2017) Climate zones on Pluto and Charon. *Icarus, 287,* 30–36.

Brassé C., Muñoz O., Coll P., and Raulin F. (2015) Optical constants of Titan aerosols and their tholins analogs: Experimental results and modeling observational data. *Planet. Space Sci., 109,* 159–174.

Buie M. W., Tholen D. J., and Horne K. (1992) Albedo maps of Pluto and Charon: Initial mutual event results. *Icarus, 97,* 211–227.

Buie M. W., Tholen D. J., and Wasserman L. H. (1997a) Separate lightcurves of Pluto and Charon. *Icarus, 125,* 233–244.

Buie M. W., Young E. F., and Binzel R. P. (1997b) Surface appearance of Pluto and Charon. In *Pluto and Charon* (S. A. Stern and D. J. Tholen, eds.), pp. 269–294. Univ. of Arizona, Tucson.

Buie M. W. et al. (2010a) Pluto and Charon with the Hubble Space Telescope. II. Resolving changes on Pluto's surface and a map for Charon. *Astron. J., 139,* 1128–1143.

Buie M. W. et al. (2010b) Pluto and Charon with the Hubble Space Telescope. I. Monitoring global change and improved surface properties from light curves. *Astron. J., 139,* 1117–1127.

Buratti B. J. and Veverka J. (1983) Voyager photometry of Europa. *Icarus, 55,* 93–110.

Buratti B. J. et al. (2003) Photometry of Pluto in the last decade and before: Evidence for volatile transport. *Icarus, 162,* 171–182.

Buratti B.J., et al. (2015) Photometry of Pluto 2008–2014: Evidence of ongoing seasonal volatile transport and activity. *Astrophys. J. Lett., 804,* L6–L12.

Buratti B. J., Hofgartner J. D., Hicks M. D., et al. (2017) Global albedos of Pluto and Charon from LORRI New Horizons images. *Icarus, 287,* 207–217.

Chandrasekhar S. (1960) *Radiative Transfer.* Dover, New York. 416 pp.

Christy J. W. and Harrington R. S. (1978) The satellite of Pluto. *Astron. J., 83,* 1005–1008.

Cruikshank D. et al. (2019a) Prebiotic chemistry of Pluto. *Astrobiology, 19,* 831–848.

Cruikshank D. et al. (2019b) Recent cryovolcanism in Virgil Fossae on Pluto. *Icarus, 330,* 155–168.

Dalle Ore C. M. et al. (2019) Detection of ammonia on Pluto's surface in a region of geologically recent tectonism. *Science Adv., 5,* eaav5731.

Doressoundriam A., et al. (2002) The color distribution in the Edgeworth-Kuiper belt. *Astron. J., 124,* 2279–2296.

Grundy W. M., et al. (2016) Surface compositions across Pluto and Charon. *Science, 351,* aad9189.

Grundy W. M., et al. (2018) Pluto's haze as a surface material. *Icarus, 314,* 232–245.

Hapke B. (1981) Bidirectional reflectance spectroscopy. 1. Theory. *J. Geophys. Res., 86,* 3039–3054.

Hapke B. (1993) *Theory of Reflectance and Emittance Spectroscopy.* Cambridge Univ., Cambridge. 455 pp.

Hapke B. (2002) Bidirectional reflectance spectroscopy 5. The coherent backscatter opposition effect and anisotropic scattering. *Icarus, 157,* 523–534.

Hapke B. (2012) *Theory of Reflectance and Emittance Spectroscopy, 2nd edition.* Cambridge Univ., Cambridge. 513 pp.

Helfenstein P. and Shepard M. K. (2011) Testing the Hapke photometric model: Improved inversion and the porosity correction. *Icarus, 215,* 83–100.

Henyey L. G. and Greenstein J. L. (1941) Diffuse radiation in the galaxy. *Astrophys. J., 93,* 70–83.

Hillier J., Veverka J., Helfenstein P., and Lee P. (1994) Photometric diversity of terrains on Triton. *Icarus, 109,* 296–312.

Howett C. J. A. and 37 colleagues (2017) Inflight radiometric calibration of New Horizons' Multispectral Visible Imaging Camera (MVIC).

Icarus, 287, 140–151.

Johns Hopkins University Applied Physics Laboratory (2015) New Horizons team responds to spacecraft anomaly. Press release issued July 4, 2015, available online at *http://pluto.jhuapl.edu/News-Center/News-Article.php?page=20150704*.

Li J. Y., Helfenstein P., Buratti B. J., Takir D., and Clark B. E. (2015) Asteroid photometry. In *Asteroids IV* (P. Michel et al., eds.), pp. 129–150. Univ. of Arizona, Tucson.

Olkin C. B. and 25 colleagues (2017) The global color of Pluto from New Horizons. *Astron. J., 154*, 258.

Protopapa S. and 22 colleagues (2017) Pluto's global surface composition through pixel-by-pixel Hapke modeling of New Horizons Ralph/LEISA data. *Icarus, 287,* 218–228.

Protopapa S., et al. (2020) Disk-resolved photometric properties of Pluto and the coloring materials across its surface. *Astron. J., 159*, 74.

Reinsch K., Burwitz V., and Festou M. C. (1994) Albedo maps of Pluto and improved physical parameters of the Pluto-Charon system. *Icarus, 108*, 209–218.

Reuter D. C., et al. (2008) Ralph: A visible/infrared imager for the New Horizons Pluto/Kuiper belt mission. *Space Sci. Rev., 140*, 129–154.

Schenk P. M. et al. (2018) Basins, fractures and volcanoes: Global cartography and topography of Pluto from New Horizons. *Icarus, 314*, 400–433.

Schmitt B. et al. (2017) Physical state and distribution of materials at the surface of Pluto from New Horizons LEISA imaging spectrometer. *Icarus, 287*, 229–260.

Stern S. A. et al. (2015) The Pluto system: Initial results from its exploration by New Horizons. *Science, 350,* aad1815.

Tokunaga A. T. and Vacca W. D. (2005) The Mauna Kea Observatory near-infrared filter set. III. Isophotal wavelengths and absolute calibration. *Publ. Astron. Soc. Pac., 117*, 421–426.

Tombaugh C. W. (1960) Reminiscences of the discovery of Pluto. *Sky & Telescope, 19*, 264–270.

Verbiscer A. J., Helfenstein P., and Buratti B. J. (2013) Photometric properties of solar system ices. In *The Science of Solar System Ices* (M. S. Gudipati and J. Castillo-Rogez, eds.), pp. 47–72. Astrophysics and Space Science Library Vol. 356, Springer, New York.

Verbiscer A. J., Helfenstein P., Buratti B. J., and Royer E. (2018) Surface properties of Saturn's icy moons from optical remote sensing. In *Enceladus and the Icy Moons of Saturn* (P. Schenk et al., eds.), pp. 323–341. Univ. of Arizona, Tucson.

Verbiscer A. J., Showalter M. R., Buie M. W., and Helfenstein P. (2019) The Pluto system at true opposition. In *Pluto System After New Horizons*, Abstract #7050. Lunar and Planetary Institute, Houston.

Walker M. F. and Hardie R. (1955) A photometric determination of the rotational period of Pluto. *Publ. Astron. Soc. Pac., 67*, 224–231.

White O. L. et al. (2017) Geologic mapping of Sputnik Planitia on Pluto. *Icarus, 287*, 261–286.

Young L. A. et al. (2008) New Horizons: Anticipated scientific investigations at the Pluto system. *Space Sci. Rev., 140*, 93–127.

Cruikshank D. P., Grundy W. M., Protopapa S., Schmitt B., and Linscott I. R. (2021) Surface composition of Pluto. In *The Pluto System After New Horizons* (S. A. Stern, J. M. Moore, W. M. Grundy, L. A. Young, and R. P. Binzel, eds.), pp. 165–193. Univ. of Arizona, Tucson, DOI: 10.2458/azu_uapress_9780816540945-ch009.

Surface Composition of Pluto

Dale P. Cruikshank
NASA Ames Research Center

William M. Grundy
Lowell Observatory

Silvia Protopapa
Southwest Research Institute

Bernard Schmitt
Université Grenoble Alpes, Centre National de la Recherche Scientifique (CNRS)

Ivan R. Linscott
Stanford University

Pluto's bulk density (1854 ± 6 kg m^{-3}) indicates a large interior component of rocky material, but the surface is covered with a suite of volatile (primarily N_2, CH_4, CO) and involatile (H_2O) ices that form the planet's crust and all its topographic structures. Those structures include impact craters, at least one large ice-filled basin, known active glaciers, mountain ranges, and a global pattern of stress-induced faulting. The ices are colored with complex organic materials from exogenic (atmospheric aerosols, surface irradiation) and endogenic (processed ices and cryovolcanic) sources. Using data from the New Horizons spacecraft, this chapter explores the physics and chemistry of Pluto's ices and organics and their interactions. Models, maps, and statistical analysis of the spectral bands define details of the mixing, evolution, and distribution of the surface components through long- and short-term cyclic variations over time. Observations at a radio wavelength probes the ices below the visible surface.

1. INTRODUCTION

Pluto was discovered at Lowell Observatory in 1930 (*Tombaugh and Moore,* 1980), and while its orbit, rotation period (*Walker and Hardie,* 1955), and a few other characteristics could be determined, as a point-source body of stellar magnitude ~15, it did not readily yield information on its size, mass, or composition. Its disk was unresolved even in very large telescopes (but see *Kuiper,* 1950), and it was too faint for diagnostic spectroscopy. Consequently, details of its composition, size, and mass were not discovered until the late 1970s and 1980s. However, in the absence of direct information about the nature of Pluto, *Kuiper* (1950) reasoned that because the planet is small and cold

> "Such a body must have some atmosphere, though most of its original atmosphere will have frozen out owing to the low equilibrium temperature at Pluto, 40–50 K. Both the atmosphere and the condensation products will prevent the albedo from being extremely low: Nearly all snows are white, the crystals being small (H_2O, CO_2, CH_4, etc.). On the other hand, the albedo need not be that of freshly fallen snow, 0.7–0.8, because of several effects, including grit deposited by comets and meteors, which will darken snow over the ages. However, the rocky surface of Pluto would be expected to be invisible, which may explain why its color index is only slightly different from the Sun, quite contrary to the results for Mercury, Mars, and the Moon. For Pluto . . . (it) is consistent with a gritty snow surface."

As Earth-based observational data became available, but especially with the flyby of the New Horizons spacecraft in 2015, Kuiper's thinking 65 years earlier is correctly seen as prescient.

Prior to the development of a near-infrared (1–3 μm) spectrometer with sufficient sensitivity to probe the spectrum of Pluto where strong diagnostic bands of the ices Kuiper had noted, plus CO, N_2, and others, a multi-filter photometric technique was applied by *Cruikshank et al.* (1976). Guided in part by the theoretical work of *Lewis* (1972) on condensation of volatiles in the solar nebula and outer solar system, filters were designed to distinguish H_2O, NH_3, or CH_4 in the solid state by means of their strong absorption bands in the region 1.0–2.5 μm. Observations in March 1976 with the Kitt Peak National Observatory 4-m Mayall telescope gave a strong indication of the presence of solid CH_4 on Pluto's surface; the identification of CH_4 was confirmed by spectra four years later (*Cruikshank and Silvaggio,* 1980; *Soifer et al.,* 1980). Gaseous CH_4 in Pluto's atmosphere was identified spectroscopically by *Young* (1994) and *Young et al.* (1997).The early photometry of Pluto (*Walker and Hardie,* 1955) demonstrated that the albedo of Pluto's surface is not uniform, and the epoch of mutual transits and occultations of the planet and the satellite (1985–1990) afforded an opportunity to map the albedo variations across the surface (*Buie et al.,* 1992; *Young and Binzel,* 1993).

Following the detection of condensed N_2 on Triton (*Cruikshank et al.,* 1984, 1988, 1993) investigators were alerted to the possibility of solid N_2 on Pluto, and an absorption band at 2.15 μm, initially identified as the 0–2 overtone in N_2 ice, was discovered spectroscopically by *Owen et al.* (1993). This band is now known to be a dimol band (see below) and is designated 2(0–1).

The 2(0–1) band in N_2 absorbs light relatively weakly, having a Lambert absorption coefficient of only about 0.015 cm^{-1} at the band center (*Grundy et al.,* 1993). For such a weak band to be visible in the *Owen et al.* (1993) spectrum of Pluto, prior to escaping from the surface the light reflected from Pluto's surface must on average have passed through a substantial thickness of N_2 ice, on the order of several decimeters, as in the case of Triton. This pathlength is considerably greater than the millimeter pathlengths in pure CH_4 ice that would explain the numerous CH_4 absorptions in Pluto's near-infrared spectrum. The discovery of the necessarily long pathlengths in N_2 ice, which is also directly relevant to the appearance of N_2 on Triton (*Cruikshank et al.,* 1984), changed a prevalent perception of Pluto's surface in which CH_4 is the most abundant ice to one in which CH_4 and CO (see next section) are minority contaminants in a surface dominated by N_2 ice, as for Triton.

The spectral shape of the N_2 bands can be used as a surface thermometer (*Grundy et al.,* 1993). Below 35.6 K, the cubic α-phase of N_2 is stable and displays very narrow absorptions, the main ones peaking at 2.148 μm and in the range between 4.15 and 4.30 μm, requiring high spectral resolution and signal precision for detection. The hexagonal β-phase, stable above 35.6 K, has broad bands around 2.15, 4.18, and 4.29 μm with temperature-sensitive widths (*Grundy et al.,* 1993; *Quirico et al.,* 1996). The analysis of the profile of the 2.15-μm band of nitrogen led *Tryka et al.* (1993, 1994) to propose a value of 40 ± 2 K for the temperature of nitrogen ice on Pluto, closely similar to temperatures derived from millimeter-wave detections by *Altenhoff et al.* (1988) and subsequent long-wavelength measurements.

Continued improvements in near-infrared spectrometer performance eventually enabled a detailed investigation of the spectrum of Pluto (*Douté et al.,* 1999) and modeling of the surface properties. Douté et al. used a modeling algorithm that considered compact and stratified media, finding that three distinct geographical units on the planet were required to explain all the structures found in the spectrum. The unit covering ~70% of the surface was seen to be a fine-grained layer of pure CH_4 lying upon a compact polycrystalline substratum of N_2-CH_4-CO in a molecular mixture. A second unit covering ~20% of the surface was regarded as either a single thick layer of pure, large-grained CH_4 or a unit with large-grained CH_4 forming a substratum to a surficial layer of a mixture of N_2-CH_4-CO. In that scenario, the remainder of the surface is covered by a third unit that is highly reflective and spectrally neutral, as expected for nearly pure N_2. *Grundy et al.* (2013, 2014) conducted a synoptic study of Pluto's spectrum over the full wavelength range 0.8–2.45 μm, resolving patterns of the distribution of condensed CH_4, N_2, and CO on the planet on a hemispheric scale. An updated grand-average spectrum of Pluto is given in Fig. 1, which shows the spectrum of Pluto from 0.8 to 2.45 μm with the individual absorption bands or band complexes identified.

The spectral signature of ethane (C_2H_6) was found in Pluto's spectrum in studies by several investigators, although it is difficult to recognize the contribution of this molecule to the spectrum because it overlaps with CH_4 (*Nakamura et al.,* 2000; *Sasaki et al.,* 2005; *Cruikshank et al.,* 2006; *DeMeo et al.,* 2010; *Holler et al.,* 2014).

Two studies extended the wavelength range of Pluto spectroscopy beyond 2.5 μm, but still within the range in which reflected sunlight dominates the radiation from the planet. *Olkin et al.* (2007) recorded the spectrum out to ~4.2 μm, which is the limit of Earth's atmospheric transparency in the standard photometric L band (2.9–4.1 μm). The components of the best-fitting radiative transfer model for Pluto's anti-Charon hemisphere had three-terrain units consisting of 21% coverage by Titan tholin (*Khare et al.,* 1984), 37% coverage by CH_4 at 60 K, and 42% coverage by CH_4 diluted in N_2 at 41 K with very large grains (~95 mm). *Olkin et al.* (2007) were unable to distinguish C_2H_6 or other hydrocarbons in this extended spectral region, presumably due to the strong CH_4 spectral bands.

Protopapa et al. (2008) also obtained a spectrum in the L band, as well as data of lesser quality in the M band (4.45–5 μm). These spectroscopic measurements were obtained with adaptive optics techniques and refer to Pluto only, without unresolved contamination by light from Charon. Structure in the M band is consistent with the CO fundamental at 4.67 μm; the first overtone, 2ν, is clearly present at 2.35 μm. In addition, a clear absorption at 4.76 μm appears to be consistent with the presence of HCN ice (*Masterson and Khanna,* 1990). This band has been found in the spectrum of Triton (*Burgdorf et al.,* 2010) and identified as solid HCN. Other nitriles also absorb in this region; confirmation of HCN ice requires additional data, but we note that HCN in Pluto's atmosphere has been reported by *Lellouch et al.* (2017). The best fit to the data of *Protopapa et al.* (2008) is a model with a geographic mixture of pure CH_4 ice, CH_4, and CO diluted in N_2, acrylonitrile CH_2CHCN, and the Titan tholin of *Khare et al.* (1984).

The results of the *Olkin et al.* (2007) and *Protopapa et al.* (2008) investigations are similar to one another in that satisfactory models of the spectra employ the same molecules

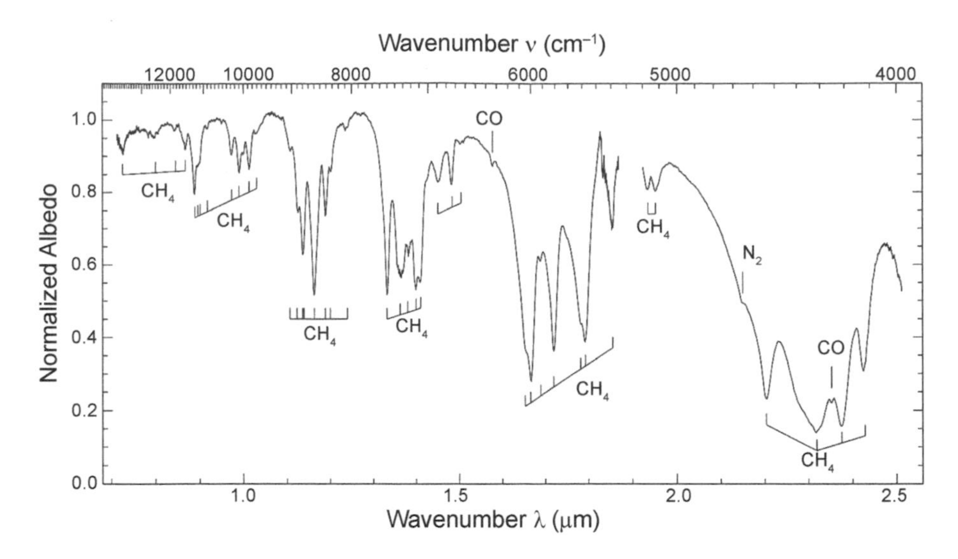

Fig. 1. Average of many spectra of Pluto, including negligible light from Charon, from 2001 to 2013; based on the composite spectrum by *Grundy et al.* (2013) with additional data from subsequent years. Data from Holler et al. (in preparation).

and solid solutions detected at shorter wavelengths. There is evidence for one or more nitriles in the data of Protopapa et al., but confirmation is needed. Both investigations include a non-ice component in the form of a tholin for which the optical and chemical properties determined by *Khare et al.* (1984) are representative.

In their 11-year monitoring study of Pluto's spectrum, *Grundy et al.* (2013, 2014) found that the strengths of the absorption bands of N_2, CH_4, and CO are modulated by the planet's 6.39-day rotation period in ways that constrain the zonal (longitudinal) distributions of these three components of the surface. While the bands of CO and N_2 are strongest on the anti-Charon hemisphere, the strongest CH_4 absorption is offset by ~90° from the longitude of the strongest CO and N_2 bands. In the Grundy et al. study, longer-term trends showed the CH_4 bands increasing in strength as the amplitude of their diurnal variation diminished, suggesting that the northern latitudes coming into view as the sub-Earth latitude moves north have higher methane abundance. Over the same time interval, the bands of CO and N_2 were declining, suggesting that these ices were preferentially distributed at more southern or equatorial latitudes. *Grundy et al.* (2014) suggested that scattering by the formation of small crystals of CH_4 in the surface ices might mask the bands of CO and N_2 as an alternative explanation for their observed decline with the advancing seasons on the planet.

2. THE ICES OF PLUTO AND THEIR DISTRIBUTION

2.1. Compositional Mapping of Pluto with New Horizons

The New Horizons spacecraft was launched on January 19, 2006, and flew past Pluto and Charon on July 14, 2015, at a distance of ~12,500 km from the Pluto's surface and ~17,900 km from Charon's. The initial results of the multi-faceted investigation of Pluto and its satellites are reported in *Stern et al.* (2015), while the early analysis of the surface composition and distribution of the components are found in *Grundy et al.* (2016), *Schmitt et al.* (2017), and *Protopapa et al.* (2017), and that of the small satellites in *Weaver et al.* (2016).

The Ralph instrument package on New Horizons consists of the Multispectral Visible Imaging Camera (MVIC) and the Linear Etalon Imaging Spectral Array (LEISA) (*Reuter et al.,* 2008), which share a common telescope. Together they provided the data for spectral mapping in the wavelength region 0.4–2.5 μm. With LEISA, the full hemisphere of Pluto visible at the encounter was spectrally mapped with spatial sampling ~6.2–7 km per pixel, and as sharp as ~2.8 km per pixel over a portion of the planet during the closest approach phase. Images in the four filters of MVIC achieved spatial sampling up to 0.7 km per pixel for color and CH_4 abundance.

MVIC has four separate 5024 × 32-pixel charge-coupled device (CCD) arrays, each with a different interference filter covering the detector. Both Pluto and Charon were imaged with MVIC in three broadband color filters (400–550 nm, 540–700 nm, 780–975 nm) and a narrow- band filter (860–910 nm) centered on the 890-nm absorption band of CH_4). MVIC is operated in time-delay integration (TDI) mode in which the image is read out along the short axis of each chip at a rate corresponding to the scan rate, so the effective exposure time is 32× the time needed to scan across a single pixel. In LEISA, a pair of filters is affixed to the 256 × 256-pixel HgCdTe array, but the LEISA filters are wedged interference filters, so that each row of pixels is sensitive to a different wavelength. One of the LEISA filters covers wavelengths from 1.25 to 2.5 μm at a resolution (λ/Δλ) 240 while the other covers 2.1 to 2.25 μm at a resolution of 560. The whole LEISA array is repeatedly read out and saved at a frame rate corresponding to the time needed to shift the field of view by one LEISA pixel.

To collect a multispectral dataset with Ralph, the field of view is scanned across the target scene by slewing the spacecraft. During such a scan, each point in the scene is eventually imaged at each wavelength, but since the spacecraft is moving relative to the target, the geometry changes over the course of the scan. The result is a point cloud of individual footprints, each with a distinct wavelength and geometry. These datasets are archived and publicly accessible through the small bodies node of the Planetary Data System (PDS). With suitable software, such as the Integrated Software for Imagers and Spectrometers (ISIS3, available from the U.S. Geological Survey), they can be re-projected to create multispectral maps.

2.2. Nitrogen, Methane, Carbon Monoxide, and the Nitrogen:Methane Solid Solution System

The three compounds N_2, CH_4, and CO each play an especially important role in shaping Pluto's surface owing to their abundance and to their volatility at Pluto surface temperatures. At room temperature, these small molecules exist in a gaseous state, but at Pluto's low temperatures, they condense to form van der Waals bonded crystalline solids, referred to as ices. The weakness of the bonding between molecules in these ices and their low masses allows molecules to escape readily from the solid to the gas phase, even at Pluto's very low surface temperatures in the 35 to 60 K range, depending on the surface albedo. The gas pressure at which condensation balances sublimation is known as the vapor pressure and is shown as a function of temperature in Fig. 2 (*Fray and Schmitt,* 2009). Small changes in temperature translate to large changes in vapor pressure. Nitrogen has the highest vapor pressure of the three ices, explaining why it dominates Pluto's lower atmosphere. Methane has the lowest vapor pressure of the three, while carbon monoxide has a vapor pressure intermediate between the other two. This volatility and the weak intermolecular bonds enable these ices to evolve over time, sublimating and recondensing elsewhere (e.g., *Spencer et al.*, 1997), flowing as glaciers, and undergoing processes such as grain growth and sintering (e.g., *Zent et al.,* 1989; *Eluszkiewicz and Stevenson,* 1990; *Eluszkiewicz et al.,* 1998; see also the chapter by Umurhan et al. in this volume).

The three volatile ices can be remotely detected and mapped by near-infrared reflectance spectroscopy thanks to vibrational transitions that absorb light at characteristic wavelengths for each molecule. The propensity of a material to absorb light as a function of wavelength (λ) is expressed as an absorption coefficient, which comes in two forms. The Lambert absorption coefficient has units of inverse length and is often written as $\alpha(\lambda)$. It indicates the probability of absorption as a function of distance traversed through the material, as expressed in the Beer-Lambert absorption law, which says that the intensity of a beam of light is attenuated as $I_x = I_0 e^{-x\alpha(\lambda)}$, where I_0 is the initial intensity and I_x is the intensity after traversing distance x through the material. Another form of absorption coefficient is the dimensionless imaginary part of the refractive index, often written as $k(\lambda)$. This form is related to $\alpha(\lambda)$ through the equation $\alpha(\lambda) = 4\pi k(\lambda)/\lambda$ [or equivalently, $k(\lambda) = \alpha(\lambda)\lambda/4\pi$].

In order for a photon to excite a molecular vibration, the molecule must have a dipole moment. As a homonuclear molecule, N_2 has no permanent dipole moment. But when a nitrogen molecule collides with another molecule, a temporary dipole moment is induced that enables "collision-induced" absorption of photons to excite the molecule's stretching vibrational mode (*Shapiro and Gush,* 1966). The same thing occurs in solid or liquid N_2, where random thermal motions cause transient distortion of molecules that are in close proximity to one another, enabling them to absorb light.

Molecular vibrations are quantized, with the lowest excited state referred to as the fundamental. The fundamental stretching mode of N_2 is excited by 4.292-μm photons. It is sometimes referred to as the 0–1 transition, to indicate that those photons are exciting the molecule from the ground state to the first vibrational excited state. Higher-energy photons can excite additional quanta, causing absorption bands at shorter wavelengths (higher energies), known as overtone bands. In N_2, the first overtone or 0–2 transition is at 2.166 μm, approximately half the wavelength of the fundamental band. Molecules containing more atoms have additional vibrational modes corresponding to the various ways the molecule can bend or stretch. CH_4 has four such modes, although the symmetry of the molecule allows only two of them (at 3.322 and 7.700 μm in ice) to be excited by photons interacting with an isolated molecule that is not collisionally distorted. A single photon can simultaneously excite more than one vibrational mode, producing combination bands. CH_4 has numerous overtone and combination bands, decreasing in strength of absorption toward shorter wavelengths (e.g., *Quirico and Schmitt,* 1997a; *Grundy et al.,* 2002). The CO molecule has a permanent dipole moment, with its stretching fundamental band at 4.68 μm. The 0–2 overtone of CO is at 2.352 μm and the 0–3 overtone is at 1.578 μm in CO ice (*Legay and Legay-Sommaire,* 1982; *Legay-Sommaire and Legay,* 1982; *Quirico and Schmitt,* 1997b).

N_2 and CO ices exhibit another type of absorption in which a single photon simultaneously excites the fundamental vibration in two adjacent molecules. This is referred to in the literature as a two-phonon, two-vibron, or dimol absorption. In N_2 ice, the dimol band at 2.148 μm is considerably stronger than the 0–2 overtone at 2.166 μm (*Grundy et al.,* 1993; *Quirico et al.,* 1996), while in CO ice the dimol band at 2.337 μm is weaker than the 0–2 overtone at 2.352 μm (*Legay and Legay-Sommaire,* 1982, *Quirico et al.,* 1996). If a CO molecule is in close proximity to an N_2 molecule, it is also possible for a 2.24-μm photon to excite the 0–1 fundamental vibrations simultaneously in both molecules (*Tegler et al.,* 2019).

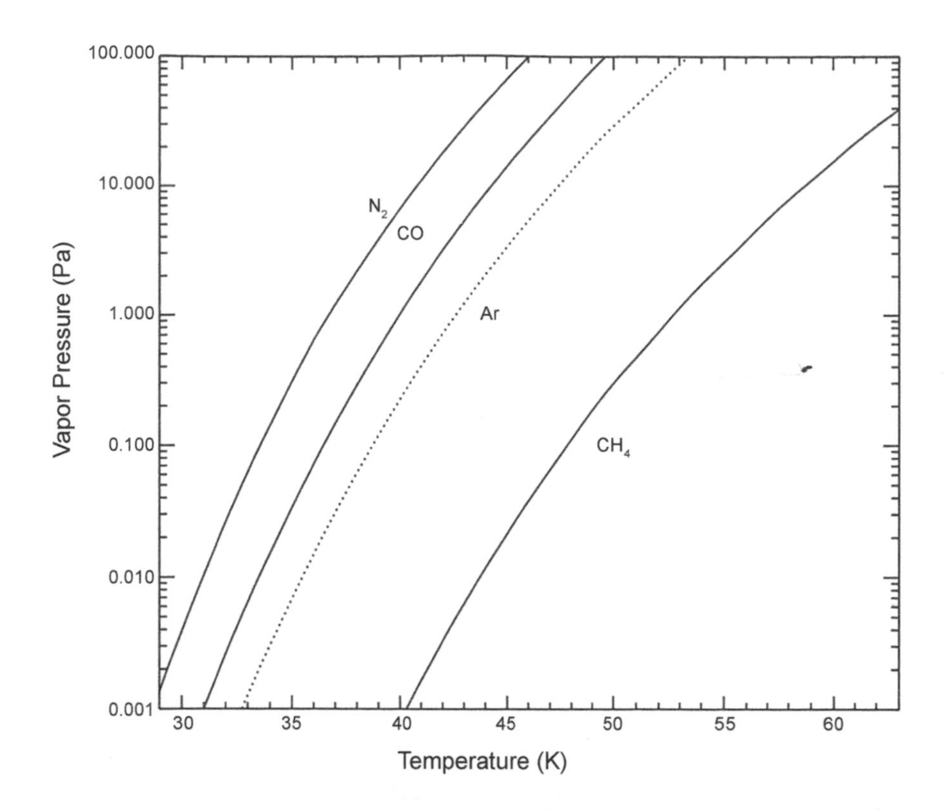

Fig. 2. Vapor pressures of N_2, CO, Ar, and CH_4 as functions of temperature. The atmospheric pressure at Pluto's surface at the New Horizons encounter was ~1 Pa in 2015 (*Hinson et al.,* 2017; *Young et al.,* 2018). Argon has not been detected on Pluto; its vapor pressure is intermediate between CO and CH_4. From *Fray and Schmitt* (2009).

Substitution of rare isotopes shifts the wavelengths of vibrational absorptions by changing the mass of the molecule (e.g., *Brown and Cruikshank,* 1997). The most abundant rare isotopes in N_2, CO, and CH_4 are ^{13}C in CO and CH_4, ^{18}O in CO, D in CH_4, and ^{15}N in N_2. The shifted absorptions could potentially be used for remote sensing isotopic ratios in these ices (e.g., *Grundy et al.,* 2011), but this possibility has not been exploited much to date, since high signal precision and high spectral resolution are required. We note, however, that *Clark et al.* (2019) have determined the D/H and $^{13}C/^{12}C$ ratios in the ices of Phoebe and Iapetus from Cassini VIMS data of high signal precision.

It is notable that argon is another volatile species that could potentially exist on Pluto's surface. As shown in Fig. 2, its vapor pressure is between those of CO and CH_4. As a single atom, argon has no vibrational features, nor does another potential component, Xe, and consequently neither argon ice nor xenon can be readily identified using infrared reflectance spectroscopy.

The crystal structures of volatile ices depend on their temperature. Above 35.6 K, nitrogen forms a hexagonal crystal structure β-N_2, while below that temperature it forms a denser, cubic crystal structure α-N_2 (*Scott,* 1976). CO molecules are very similar in size and shape to N_2 molecules, and CO ice too has warmer hexagonal β and colder cubic α phases, with the transition temperature between them being 61 K (e.g., *Vetter et al.,* 2007). CH_4 ice also has two phases known as CH_4 I and CH_4 II. Both CH_4 phases have a cubic crystal structure, but in warmer CH_4 I, the molecules are orientationally disordered (*Fabre et al.,* 1979). The transition between CH_4 phases I and II occurs at 20.4 K, well below surface temperatures on Pluto. The vibrational absorptions of N_2, CO, and CH_4 molecules all depend to some extent on the crystal structure and temperature because the crystal lattice enables phonon interaction and because the crystal structure affects the proximity and orientation of neighboring molecules, which are the source of induced dipole moments. This dependence means that some information about temperature and crystal structure can be derived from remote spectroscopic observations.

So far, we have described pure materials, but in nature, mixtures are common. With their similar molecular sizes and shapes, N_2 and CO are fully miscible in one another. As one type of molecule is exchanged for the other in a mixture, phase transition temperatures shift with their relative concentration, as shown in the binary phase diagram for condensed CO and N_2 (Fig. 3b). This phase diagram is relatively simple, being dominated by the stability fields of three phases: liquid, β ice, and α ice. Between these phases are narrow regions where two condensed phases are simultaneously stable. The binary phase diagram for N_2 and CH_4 shows a eutectic system with the lowest temperature liquid forming at 62 K, at a composition of about 75% CH_4 and 25% N_2 (Fig. 3a). Solubility of CH_4 in N_2 ice phases is limited, as is solubility of N_2 in CH_4 ice phases, especially at lower temperatures. When CH_4 is

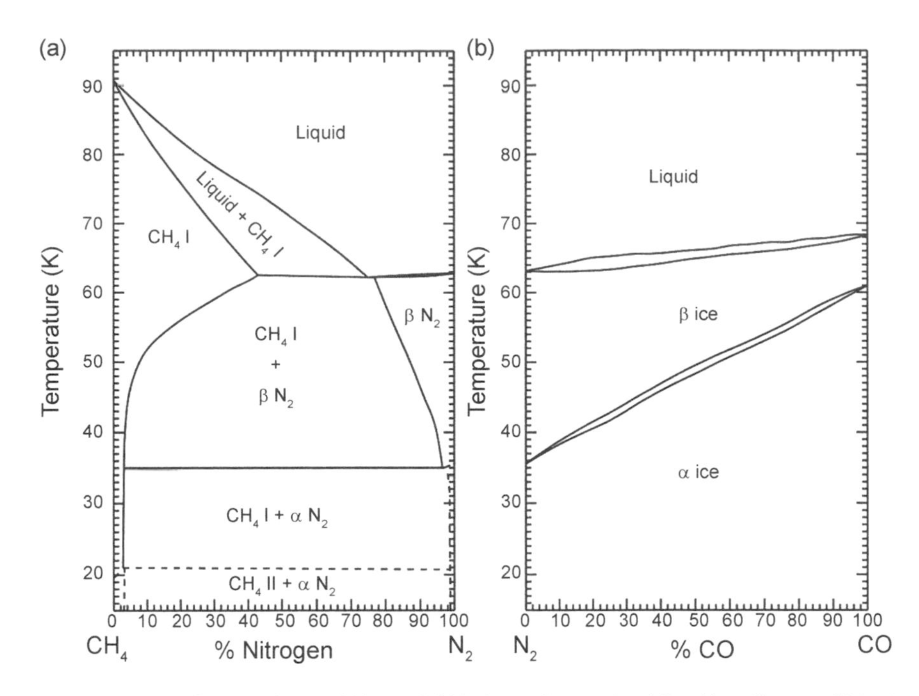

Fig. 3. (a) Binary phase diagram for condensed N_2 and CH_4 from the work of *Prokhvatilov and Yantsevich* (1983). Implications for Triton and Pluto were discussed by *Lunine and Stevenson* (1985), *Protopapa et al.* (2015), and *Trafton* (2015). **(b)** Binary phase diagram for condensed N_2 and CO, based on studies by *Vetter et al.* (2007) and *Tegler et al.* (2019). In the narrow regions between the two sets of lines two condensed phases are simultaneously stable.

diluted in N_2 ice, its vibrational absorption bands shift to slightly shorter wavelengths (*Quirico and Schmitt,* 1997a). Similar shifts are seen in spectra of Pluto, indicating that at least some of Pluto's CH_4 ice is diluted in N_2 ice (*Douté et al.,* 1999; *Tegler et al.,* 2010, 2012). *Protopapa et al.* (2015) showed that smaller shifts are found for solid solutions of N_2 diluted in CH_4. For CO, the band shifts in N_2 ice are very small, on the order of 0.3 nm (*Quirico and Schmitt,* 1997b), but the increase in the width of the band upon dilution allows the identification of diluted CO with very high spectral resolution observations (*Merlin et al.,* 2018). Vapor pressures above mixtures of ices are not necessarily simply related to the vapor pressures of the pure species (*Trafton,* 2015; *Tan and Kargel,* 2018) and a phase diagram for the N_2 + CO + CH_4 ternary system has not yet been published, so more laboratory work is needed.

As noted, when seen at high spectral resolution, Pluto's spectrum displays CH_4 bands shifted toward shorter wavelengths compared to the central wavelengths of pure CH_4 measured in the laboratory. This led to the conclusion that CH_4 is dissolved at low concentrations in a matrix of solid N_2 (*Quirico and Schmitt,* 1997a). The 1.69-μm band in Pluto's spectrum has been instead attributed to pure CH_4 only because this feature is present in pure CH_4 ice but has never been observed in any sample of CH_4 diluted at low concentrations in N_2 (neither in α- nor β-N_2 phases) at any temperature. These arguments, as well as the lack of optical constants for CH_4 ice with a small fraction of N_2, justified the modeling approach to Pluto spectra in recent years (*e.g., Douté et al.,* 1999; *Olkin et al.,* 2007; *Protopapa et al.,* 2008), employing pure CH_4 as a proxy of CH_4 with diluted N_2 and CH_4 diluted at low concentrations in N_2. The N_2-CH_4 binary phase diagram at thermodynamic equilibrium (*Prokhvatilov and Yantsevich,* 1983) (Fig. 3a) dictates that if both CH_4 and N_2 are present on Pluto's surface within a certain range of abundance (CH_4/N_2 between 5% and 97%), two distinct phases must co-exist, i.e., one phase when methane ice is saturated with nitrogen (CH_4:N_2) and the other when nitrogen ice is saturated with methane (N_2:CH_4). This has been suggested by *Trafton* (2015) and *Cruikshank* et al. (2015).

But thermodynamical equilibrium does not always apply in a natural setting. The coexistence of various processes having different timescales and different spatial scales makes thermodynamical equilibrium uncommon; this is especially the case on a locally actively sublimating surface in contact with compact polycrystals with diffusion time constants much higher than other processes. Specifically, the N_2-CH_4 phase diagram (Fig. 3) shows that the solubility limits of CH_4 and N_2 in each other are temperature dependent and equal at 40 K [Pluto's plausible surface temperature as inferred by *Tryka et al.* (1994)] to about 5% (N_2:CH_4 with 5% CH_4) and 3% (CH_4:N_2 with 3% N_2). New optical constants for solid solutions of methane diluted in nitrogen (N_2:CH_4) and nitrogen diluted in methane (CH_4:N_2) are presented by *Protopapa et al.* (2015) at temperatures between 40 K and 90 K. Those data reveal that the 1.69-μm band of Pluto's spectrum cannot be considered definitive evidence for the presence of pure CH_4 on Pluto's surface, as this feature is observed not only in samples of pure CH_4 (at temperatures below 60 K), but also in CH_4:N_2 samples.

2.3. Nitrogen, Methane, and Carbon Monoxide on Pluto

The spatially resolved spectral images of Pluto obtained with LEISA have enabled a far more detailed investigation of Pluto's surface ices in various combinations than could be established from Earth-based spectroscopy. A first series of maps of the band depths of CH_4 (over the 1.3–1.4-μm band complex), N_2 (at 2.15 μm), and CO (at 1.57 μm) was derived by *Grundy et al.* (2016) using the pair of LEISA spectral images recorded just prior Pluto's encounter (6–7 km spatial resolution). Although still affected by some calibration issues and intrinsic noise, these maps already showed the major spatial units of volatile reservoirs and their strong correlation with geological features and latitude. In particular, Sputnik Planitia hosts all three volatile molecules, while N_2 and CH_4 dominate the middle north latitudes and CH_4 ice covers the north polar regions. See Appendix A by Beyer and Showalter in this volume for maps and nomenclature for Pluto and Charon.

Further indepth studies of the distribution and state of these volatile molecules and their correlation with one another were conducted by *Schmitt et al.* (2017) on the initial same dataset by first performing a principal component analysis (PCA) that allowed a significant reduction of the noise level and instrumental artifacts of the spectral data. Although the PCA axes are generally difficult to attribute unambiguously to a single physical parameter (Fig. 4), their analysis enabled pinpointing a number of spatial units with specific spectral behaviors and correlating them using the first 10 principal axes. This preliminary analysis also led to the refinement of the band depth definitions for CH_4, N_2, and CO and to the definition of new spectral indicators for the identification and qualitative mapping of H_2O ice and the non-ice red material.

The integrated band depths of CH_4 have been derived over the four methane absorption band complexes (around 1.30–1.43, 1.58–1.83, 1.90–2.00, and 2.09–2.48 μm) within the LEISA spectral range and having different average band depths. The 1.7-μm band complex, with the highest signal precision and unsaturated bands, has been chosen to produce the "nominal" CH_4 map (Fig. 5a) (*Schmitt et al.*, 2017).

The resulting maps show that CH_4 is present in various amounts over most of the anti-Charon side of Pluto, except in the central part of Cthulhu Regio and Krun Macula, which are covered by the dark-red material, and in a number of smaller places such as Kiladze Crater or Baret Montes, which are dominated by water ice (see section 3 below). The CH_4 bands are particularly strong over the north polar area, over most of Tombaugh Regio, and extend to the east in Tartarus Dorsa where ridges of CH_4 ices have been identified (*Moore et al.*, 2018).

Different bands probing different average depths within the surface provide somewhat different distribution maps of band intensity. In particular, Sputnik Planitia displays the greatest band depth for the strongest CH_4 bands (2.09–2.48 μm) but more shallow depth for the weakest bands

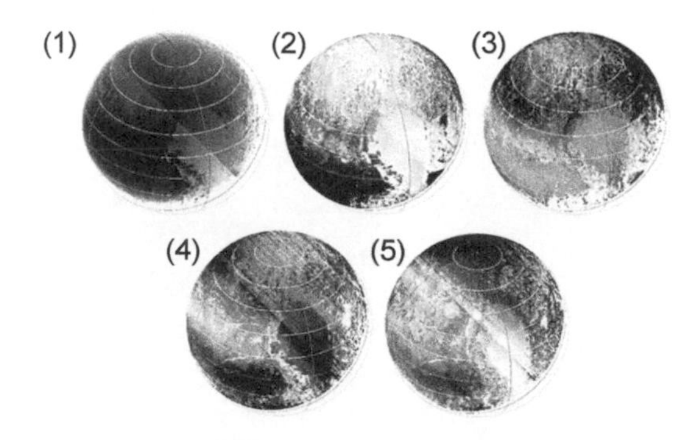

Fig. 4. The five first axes of the PCA of two LEISA spectro-image cubes. Axis 1 mostly reflects continuum reflectance variations, axis 2 shows CH_4 absorption, axis 3 is similar to the band depth map of the weak CH_4 bands (around 1.95 μm) and appears to be correlated with CH_4 grain size, axis 4 is a combination of H_2O ice and CH_4-rich ice, and axis 5 mostly reflects nitrogen ice distribution (plus some H_2O ice contribution). Diagonal linear streaks are artifacts in the processing of the spectral images. From *Schmitt et al.* (2017).

(1.9–2.0 μm). This difference may result from a vertical stratification of CH_4 within the first tens of centimeters of the ice layer and the current diurnal/annual/climatic cycles of sublimation and condensation.

The integrated depth of the N_2 band is more difficult to map because this weak band spans only a few spectral channels and is situated on the wing of the strong, nearly saturated, 2.2-μm CH_4 band. However, with the use of a local PCA, a large fraction of the noise and instrumental effects have been removed and a much higher S/N map than the initial one has been derived (Fig. 6b). It should be noted that in the figure displaying the N_2 ice band depth the values between 0.005 and about 0.020 (dark blue) are not easily distinguishable from the N_2 free area (black), although they cover a relatively wider area around the nitrogen ice spots. The reader can refer to the original data posted in PDS to locate precisely the extent of nitrogen ice.

It is important to note that the nitrogen band depth is highly sensitive to various parameters that include grain size and methane concentration in the N_2 crystal structure. Indeed, *Schmitt et al.* (2017) have shown using radiative transfer simulations that show that above about 1% CH_4 diluted in N_2 ice, the measured N_2 band depth starts to drop to zero and can even give negative values due to the curvature of the underlying 2.2-μm CH_4 band (i.e., the continuum below the N_2 band cannot be assumed to be a straight line). As a result, the area covered by nitrogen ice may be more extended than that which is mapped by the presence of the 2.15-μm band.

To overcome this limitation, another way to map the presence of nitrogen ice has been devised using the spectral shift of the methane bands occurring upon dilution in nitrogen ice (*Quirico and Schmitt,* 1997a) compared to the position of pure CH_4 ice (*Grundy et al.*, 2002) or to the CH_4:N_2

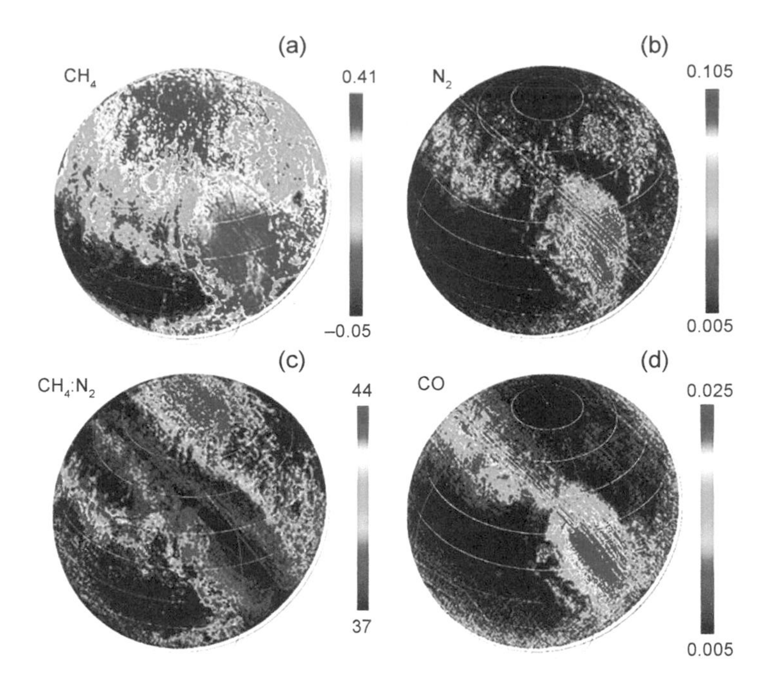

Fig. 5. **(a)** Band depth map of CH_4 (over the 1.58–1.83-μm band complex), **(b)** N_2 (at 2.15 μm), and **(d)** CO (at 1.57 μm), as well as (c) the "phase index" of CH_4:N_2 ice (values around 37 depict N_2:CH_4-dominated ice, and around 42 for CH_4:N_2 ice). For N_2 and CO the detection limits have both been set to a band depth of 0.005, but the color scale of these figures does not clearly allow the distinction between 0.005 and about 0.020 (dark blue) from the nitrogen free area (black). From *Schmitt et al.* (2017).

ice (*Protopapa et al.,* 2015). A "CH_4 band position," or "phase index," has been defined using the seven methane bands with the sharpest peaks to best allow the separation of the N_2:CH_4 ice and CH_4:N_2 ice phases (*Schmitt et al.,* 2017). The N_2:CH_4 ice phase should have an index of about 36.5 and mostly positive N_2 band depths, and the N_2:CH_4 ice phase has a value around 42 and a negative N_2 band depth. The spectral smile slightly affects this index by spreading its values by ±2 units. The map of this index is shown in Fig. 5c, and demonstrates that N_2, when diluted in CH_4, is more widespread than indicated in Fig. 6b, which shows the distribution of the 2.15-μm band. The 2.15-μm band represents pure N_2 or a solid solution of N_2:CH_4 in which the N_2 molecule is most abundant.

From this map it clearly appears that the CH_4-rich ice zones are mostly located in the north polar region (above ~60°N), northeast of Sputnik Planitia, and in the east and southwest parts of Tombaugh Regio, as well as along the northern (~15°–35°N) border of Cthulhu. Conversely, N_2:CH_4 ice is ubiquitous over the whole Sputnik Planitia and the northwest mid-latitudes (between 30° and 60°–70°), but also dominate the temperate northeast regions (Hayabusa Terra), although dotted with a number of CH_4-rich spots. A number of isolated N_2-rich ice spots also exist in many localized places, such as East Tombaugh Regio.

Figures 6a,b are correlation plots of integrated band depths vs. the phase index defined above, and they show the density of points mapped in Fig. 5.

The correlation of this "phase index" with the nitrogen and methane band depth maps is very instructive and shows that there is a continuous transition between areas covered with ices purely or mostly condensed in the N_2:CH_4 phase and those dominated by CH_4-rich ice (Fig. 5c). This transition most probably reflects a progressive physical transition between these phases during their fractional sublimation or condensation (see section 2.4). It should be noted that along this transition the N_2 band depth first strongly decreases (and even reaches negative values) for indexes (37–38), which are only slightly larger than for pure N_2:CH_4 phase (36.5). This is because of the effect previously described: Above about 1% CH_4 (i.e., well below the saturation threshold of CH_4 in N_2 ice, which is 5% at 40 K), the N_2 band disappears in the wing of the strong 2.2-μm methane band.

The map in Fig. 5c also shows that this spectral transition is reflected in a spatial transition between these phases. One major transition with phase gradation is seen at global scale from the mid-latitudes dominated by N_2:CH_4 ice, toward the polar regions to the north and toward the northern fringes of Cthulhu Regio to the south, both dominated by CH_4:N_2 ice.

The correlation plot in Fig. 5c is, however, perturbed by the other materials, mostly H_2O ice and the red material, which generally have very low phase indexes ($<\sim34$) but can have higher values due to noise. Fortunately, these materials have both negative N_2 band depth and negative integrated CH_4 band depths allowing us to isolate them. By combining the three parameters, the N_2 and CH_4 band depth

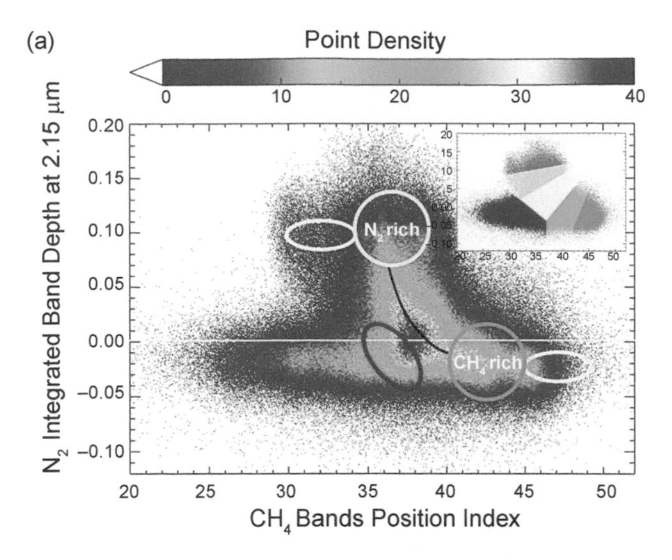

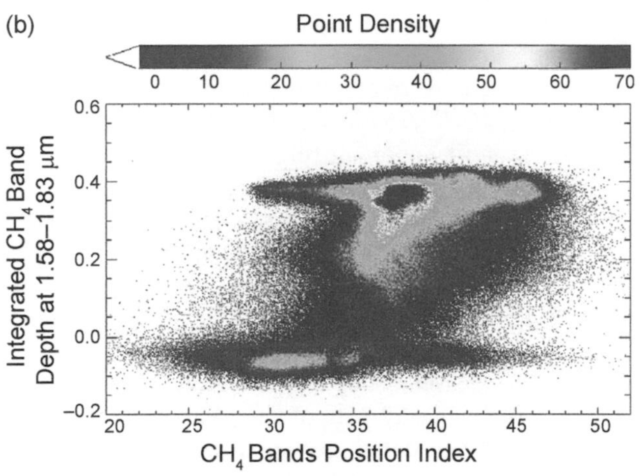

Fig. 6. Correlation plots of N_2 and CH_4 ice integrated band depths vs. the "phase index." **(a)** N_2 ice band depth at 2.15 µm. **(b)** CH_4 band depth around 1.7 µm. The detection thresholds of N_2 and CH_4 are given by the yellow horizontal lines. N_2:CH_4 ice phase is around 36.5 (yellow circle). CH_4:N_2 ice phase is around a value of 42 (green circle). N_2:CH_4 ice with a large concentration of CH_4 (negative N_2 band depth) is located in the red ellipse. The progressive transition from N_2:CH_4 ice to CH_4:N_2 ice occurs along the curved black line in **(a)**. From *Schmitt et al.* (2017).

maps and the phase index map, a first "presence map" of nitrogen ice, has been produced for the anti-Charon hemisphere (*Schmitt et al.,* 2018). In particular this presence map shows that the N_2:CH_4 phase is almost ubiquitous over the northwest mid-latitudes, much more like the phase index map (Fig. 5c) than the spottier distribution depicted by the N_2 band depth map (Fig. 5b). In view of their critical importance for the determination of the global radiative equilibrium and temperature of the volatiles on Pluto, other presence maps (e.g., *Lewis et al.,* 2021; *Johnson et al.,* 2021) have been devised using various combinations of these band depth maps and the quantitative abundance maps obtained by *Protopapa et al.* (2017).

The map of CO (Fig. 5d) is especially difficult to construct because at the resolution of the LEISA spectra the band of the first overtone at 2.35 µm is blended between the two almost saturated CH_4 bands at 2.32 and 2.37 µm, and the very narrow and weak band of the second overtone at about 1.58 µm is expressed mostly over a single spectral channel on the wing of the 1.66-µm CH_4 band. Also, this is a slightly noisier spectral region, despite the local PCA applied. In addition, the spectral smile that affect the LEISA spectral images may shift the spectral scale by up to one to two spectral elements on the side of the images, making a single band depth criterion based on only 2–3 spectral elements very sensitive to this artifact. The best compromise between these effects provided a CO map (Fig. 5d) that confirms the presence of CO but is rather qualitative in terms of intensity of the band (*Schmitt et al.,* 2017). The CO map displays a distribution similar to that of the N_2 band depth. Its correlation plot with the phase index is also similar, displaying an "L"-shaped transition from N_2:CH_4 to the CH_4:N_2 ice phase. A closer comparison of the two maps shows that in the mid-latitude northwest region, while the CO abundance also decreases toward both the north and the south, its maximum appears to be shifted to higher latitude compared to N_2. The correlation of the CO band depth with the N_2 band depth also displays two distinct behaviors, one strongly correlated with N_2 and another that points to some exposures of nitrogen ice devoid of CO (Fig. 7a). The correlation plots with CH_4 (Fig. 7b) and with the phase index shows that CH_4-rich ice has a very small amount of CO remaining trapped, and the CO/CH_4 band depth ratio also decreases when the ice evolves from ternary N_2:CO:CH_4 ice to the CH_4:CO:N_2 phase.

A pixel-by-pixel Hapke radiative transfer model (*Hapke,* 2012), considering global thermodynamic equilibrium, has been applied by *Protopapa et al.* (2017) to two resolved scans of Pluto's encounter hemisphere acquired with LEISA. An areal mixture of H_2O ice, N_2:CH_4 ice, CH_4:N_2 ice, and Titan tholin (*Khare et al.,* 1984) was employed, a choice that relies on the spectral evidence collected over the course of several years and the condition of global thermodynamic equilibrium. The free parameters in this analysis are the fractional area (F_i) and effective grain diameter (D_i) of each surface terrain in the mixture. Therefore, this model yields compositional maps defining the spatial distribution of the abundance and textural properties of the materials present on the near-encounter hemisphere of Pluto. This approach, in contrast to the analysis of spectral parameters and PCA, provides a quantitative assessment of the surface composition of Pluto so that grain size effects can be distinguished from the relative abundances of the components. Because a pixel-by-pixel modeling analysis is highly computationally expensive, the LEISA data, acquired at the native resolution of 6–7 km/pixel, have been degraded to ~12 km/pixel (for a total of ~4×10^4 pixels to model). Figure 8 shows the maps for the volatiles' methane and nitrogen ices resulting from this analysis.

We note that this approach has some limitations: (1) The visible wavelength range, which is disregarded by *Protopapa et al.* (2017), contains important compositional information

(a)

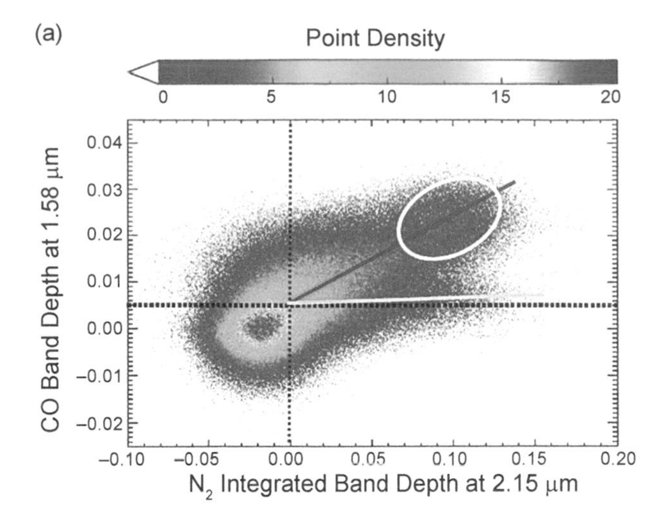

(b)

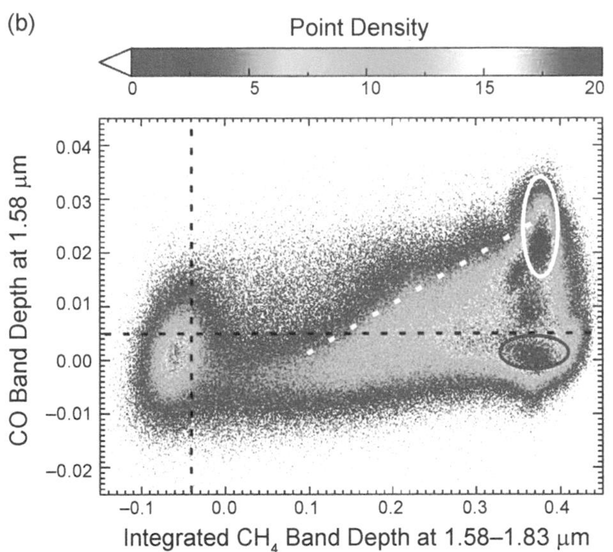

Fig. 7. Correlation plot between the 1.58-µm CO band depth and **(a)** the depth of the 2.15-µm N_2 band and **(b)** the 1.7-µm CH_4 band depth. CO is absent or at a very low level (undetectable) in CH_4-rich ice (blue ellipse, N polar regions) but also in some N_2:CH_4 ice (yellow line). The white ellipse mostly corresponds to Sputnik Planitia, with both CO and CH_4. The dotted horizontal lines represent the detection threshold of CO (0.005) and the vertical lines are the detection thresholds of N_2 and CH_4, respectively. The blue ellipse corresponds to the north polar region: CH_4-rich ice without detectable CO. From *Schmitt et al.* (2017).

since low-albedo organic compounds display in this region the most diagnostic spectral signatures (e.g., low albedo and red slope; see section 5). (2) The estimates of the concentration and particle size of each surface compound strongly rely on the choice of Pluto's photometric properties (Hapke parameters such as the cosine asymmetry factor, compaction parameter, amplitude of the opposition effect, and mean roughness slope). *Protopapa et al.* (2017) applied an early version of the Hapke parameters with respect to the latest values derived by *Verbiscer et al.* (2019) and *Protopapa et al.* (2020) (see the chapter by Olkin et al. in this volume). (3) A correction factor in the LEISA radiometric calibration was discovered subsequent to the 2017 analysis (*Protopapa et al.*, 2020). (4) A slight global shift in the LEISA spectral calibration was recently discovered (B. Schmitt, personal communication). Implementing these limitations in a deeper analysis would possibly lead to different percentages and grain sizes of the surface components with respect to those presented in the 2017 paper and shown in their Fig. 2. Despite these limitations, the quantitative comparative study of Pluto's principal surface units remains of value.

Protopapa et al. (2017) initially adopted a constant solubility of CH_4 in N_2 equal to its limit, 5%, as dictated by the binary phase diagram (Fig. 3) for a temperature of 40 K and for most of the range of possible nitrogen/methane relative abundances (between ~5% and 97%; see section 2.2). While this assumption satisfactorily reproduced the spectral behavior of several regions across Pluto's surface (e.g., Lowell Regio; see Fig. 8e,), it failed to reproduce the strong 2.15-µm N_2 band in regions like Sputnik Planitia. The latter is a deep reservoir of convecting ices (*McKinnon et al.,* 2016a,b; *Trowbridge et al.*, 2016), with nitrogen playing the most important role (regions within Sputnik Planitia are labeled "i" and "j" in Fig. 8b). When relaxing this hypothesis and letting the concentration of CH_4 in N_2 be a free parameter only constrained by the saturation value (i.e., $<5\%$), amounts close to 0.3% are found to reproduce the 2.15-µm N_2 band (Fig. 8e). Such low dilution content of CH_4 in N_2 was also adopted in the modeling analysis of groundbased measurements of Pluto. For example, *Douté et al.* (1999) obtained 0.36% of CH_4 diluted in N_2 in their best fit model considering geographical mixture. Because Sputnik Planitia displays the presence of both CH_4-rich (~50%; see Fig. 8a) and N_2-rich components (~50%; see Fig. 8c), *Protopapa et al.* (2017) put forth the idea that these findings could be suggestive of the presence of a cold trap in Sputnik Planitia, in agreement with that reported by *Hamilton* (2016). However, as noted by *Protopapa et al.* (2017), such a low temperature is difficult to reconcile with the state of N_2 on Pluto's surface. In fact, the phase diagram dictates that a saturation value of 0.36% of CH_4 diluted in N_2 implies temperatures much lower than 40 K. This would suggest that N_2 is in the α-phase, which occurs below 35.6 K, and is inconsistent with spectroscopic evidence collected so far. Such a low temperature may also be inconsistent with local and global energy balance at the surface and with atmospheric equilibrium. But colder ice at some depth below the subsurface could be a possibility, a scenario that still needs to be investigated. The temperature interpretation is therefore left to a further effort that will include the analysis of data in the higher-resolution 2.1–2.25-µm LEISA segment, as well as a more accurate CH_4-N_2 phase diagram that includes the contribution of CO ice. The consequences for Pluto's global thermal balance and volatile transports also need to be studied.

Protopapa et al. (2017) suggested layering within Sputnik Planitia as an alternative idea to justify the modeling results. The authors further point out that the physical

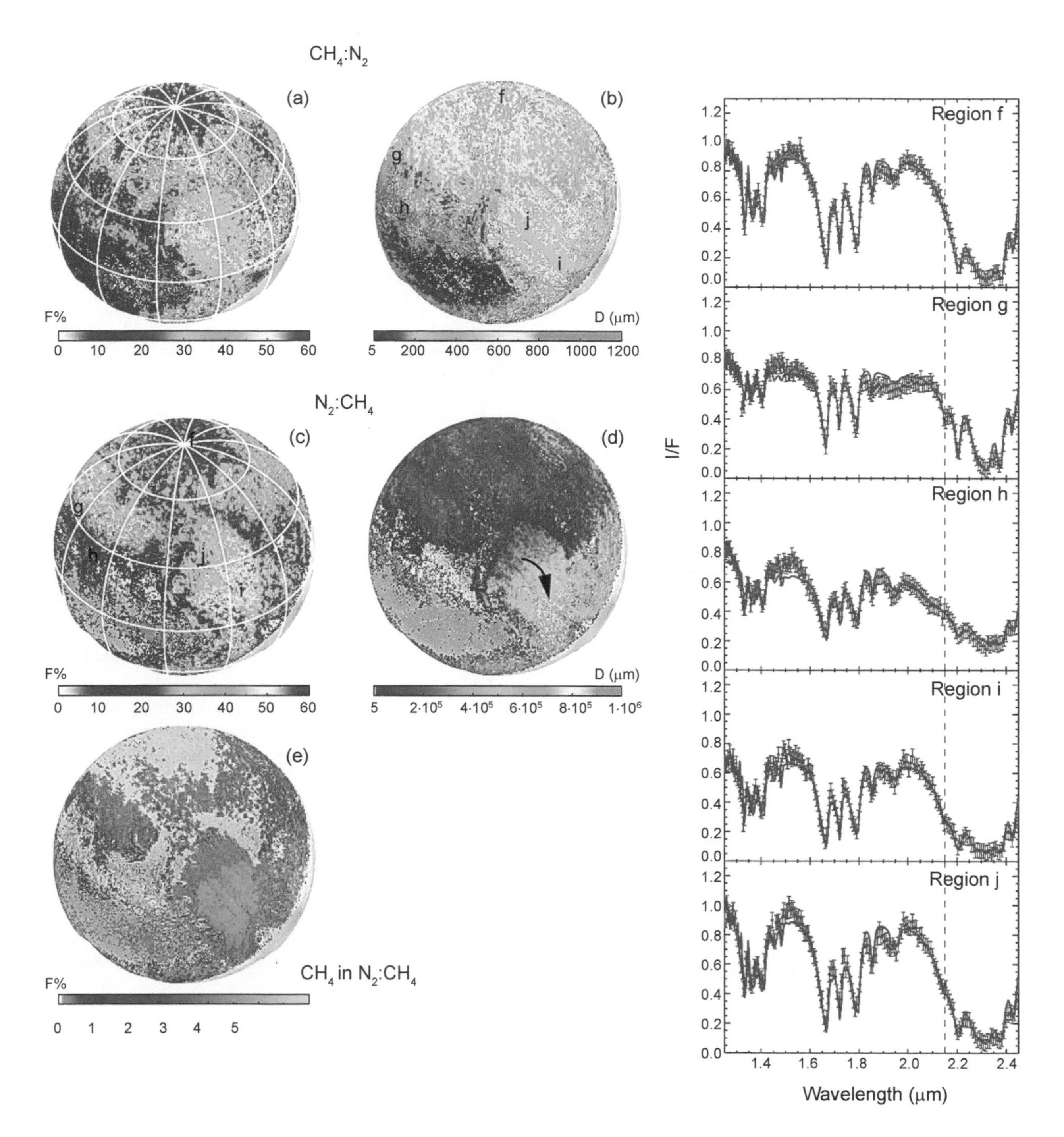

Fig. 8. Compositional maps of Pluto obtained by applying a pixel-by-pixel Hapke radiative transfer model to the LEISA scans. CH_4-rich component displaying **(a)** its relative abundance and **(b)** its relative pathlength. **(c)** and **(d)** show the relative abundance and pathlength of the N_2-rich component, respectively, while **(e)** shows the dilution fraction of CH_4 in N_2. The composition maps are superposed on the reprojected Long Range Reconnaissance Imager (LORRI) base map. Spectra extracted in regions of interest and labeled in **(b)** and **(c)** are shown as blue dots on the righthand side. The spectra are compared with their corresponding best fit modeling (solid red line). Adapted from *Protopapa et al.* (2017).

properties of N_2 across Sputnik Planitia suggest a possible sublimation transport of this volatile from the northwest to the center of the basin (see section 2 for details and arrow in Fig. 8d). This could lead to a dynamical differentiation between N_2-rich and CH_4-rich components (*Stansberry et al.,* 1996), which would possibly produce layering within Sputnik Planitia. This idea is supported by global circulation models (*Bertrand et al.,* 2018). A possible stratification of ice phases was initially suggested by *Douté et al.* (1999) from the analysis and modeling of Pluto's hemispheric spectra obtained from groundbased observations. The continuous spatial transitions from methane-poor nitrogen ice to methane rich ice evidenced at global and local scales by *Schmitt et al.* (2017) in a number of places point to the macroscopic and microphysical processes that can lead to such a vertical stratification (see next section) and maintain the "whole optical layer" (top meter) far from global thermodynamic equilibrium.

The weak absorption coefficient of the N_2 ice band requires especially large pathlengths for the absorption band to appear so prominent in Pluto's spectrum. The large equivalent grain sizes obtained by *Protopapa et al.* (2017) for N_2 ice suggest that at least the top meter of the surface is most probably in the form of a compact polycrystal with very few or no bubbles or voids that serve as scattering centers, analogous to the surface of Triton (*Brown et al.*, 1995). So if a CH_4 phase is present at the top of the surface, and not intimately mixed in the bulk of the surface, then on a timescale on the order of Pluto's seasons it can be thermodynamically disconnected from the compact nitrogen ice below, as no rapid exchange through the gas phase by sublimation/condensation is possible. Additionally, solid-state diffusion of methane is probably very low, although currently unknown.

2.4. Differentiation Processes of Ices

The correlations between the different band depths and phase index of the volatile molecules on Pluto shows that there are progressive transitions occurring between the different ice phases and that these transitions occur at global and local scales. These factors point to various processes that may affect the distribution, composition, and phases of the volatiles with latitude, longitude, and time. One of the main processes that can lead to differentiation of the ices is sublimation driven by the heat absorbed from insolation.

Considering the vapor pressures of N_2, CO and CH_4 and the phase diagrams of N_2:CH_4 and N_2:CO (but in the absence of knowledge of the binary CO:CH_4 and ternary N_2:CO:CH_4 phase diagrams) (Figs. 2 and 3) we can describe what happens to an initially condensed N_2:CO:CH_4 phase unsaturated with CH_4 (CO has no saturation limit) upon progressive sublimation (Fig. 9).

The first sublimation step primarily involves the loss of most N_2 and part of CO, thus increasing the CO/N_2 ratio (on which there is no limit) and the CH_4/N_2 ratio up to a saturation level of 5% (at 40 K) where the CH_4:CO:N_2 phase, saturated in N_2 (3%) and CO (unknown saturation), should appear. We have then a mixture of two crystalline phases, one dominated by N_2 and the other by CH_4. Then, upon further sublimation of N_2 and CO, the amount of the CH_4:CO:N_2 phase increases to the detriment of the N_2:CO:CH_4 phase, which at some point should disappear. This disappearance may be only very local, as the sublimation process mostly affects the top surface of the layer and may be much faster than the other processes that can tend to equilibrate the composition of the whole layer, such as molecular diffusion in the gas or in the solid. Consequently, the layer can be at thermodynamic equilibrium at the microscale, but not necessarily at larger scales, either vertical or horizontal. For example, a thin layer of CH_4:CO:N_2 ice may grow over the initial compact centimeter-sized polycrystalline N_2:CO:CH_4 ice deposit. The formation of such a layer upon sublimation of N_2:CH_4 ice has been demonstrated in the laboratory by *Stansberry et al.* (1996) and the occurrence of such a CH_4-rich layer on Pluto has been inferred from the analysis of groundbased, high-spectral-resolution observations (*Douté et al.*, 1999) and suggested by *Protopapa et al.* (2017) (see section above).

As sublimation of N_2 and CO continue (and also CH_4, but at a much lower rate), the N_2:CO:CH_4 ice phase may eventually disappear from the whole layer (or from some optically and thermally thick layer), leaving only the CH_4:CO:N_2 phase visible and in thermodynamic and thermal contact with the atmosphere. This phase will then heat up to adjust both the temperature of the surface and thus the sublimation rate of the main molecule, CH_4, to equilibrate the absorbed solar radiation flux with thermal emission and latent heat

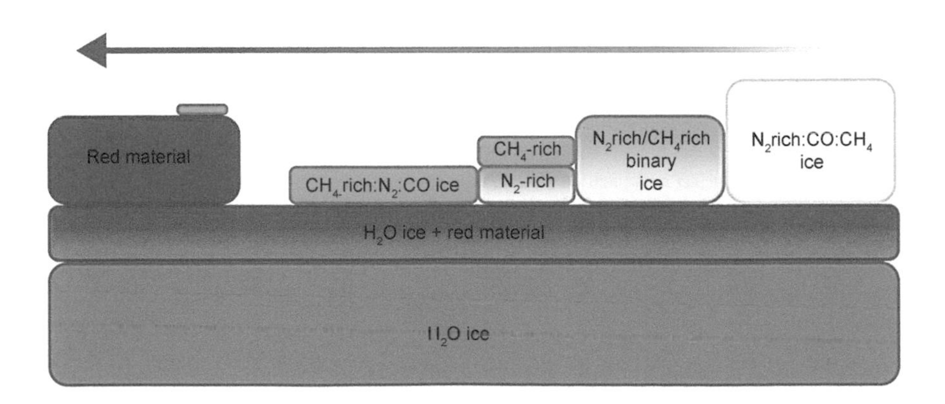

Fig. 9. Schematic representation of the various materials and their possible mixing states at the surface of Pluto upon progressive sublimation of N_2:CO:CH_4 ice. The first sublimation step mostly removes N_2 and part of the CO, thus increasing CH_4/N_2 ratio up to saturation, where the CH_4:CO:N_2 phase appears, and progressively forming a thin layer at the surface. Upon further sublimation of N_2 and CO, the N_2:CO:CH_4 phase disappears. The final CH_4:CO:N_2 deposit may still contain a small amount of CO and N_2.

fluxes. The next steps in the evolution of its composition and phase will depend on the N_2:CO:CH_4 phase diagram, which is currently unknown. However, it is highly probable that the amount of CO dissolved in the structure of the CH_4 crystal is, as N_2, also low, and therefore the CO/CH_4 ratio should also strongly decrease compared to its initial ratio in the N_2:CO:CH_4 phase. The CH_4-rich ice top layer and the final lag deposit may still contain some small amount of CO and N_2, but they should be undetectable with LEISA. The question of how the CH_4:CO:N_2 phase appearing upon sublimation segregates from the initial N_2:CO:CH_4 ice deposit is critical, as it constrains the physical and thermal evolution of the upper layer and its exchange of matter and heat with the atmosphere.

The occurrence of a vertical stratification of methane over wide areas of the surface of Pluto was suggested by *Douté et al.* (1999) from radiative transfer modeling of high-spectral-resolution telescopic observations. However, the slightly smaller band shifts in the strong CH_4 bands relative to the weak ones that allowed this inference cannot be used in the case of the LEISA data both because of its lower spectral resolution and its less accurate spectral calibration. A study is underway to find other criteria, using the relative CH_4 band depth intensities to pinpoint the presence of such stratification and possibly map their local occurrence and progressive formation (*Schmitt et al.,* 2019). This may help to constrain the transport of volatiles resulting from the current diurnal/annual sublimation/condensation cycles.

2.5. Circulation of Volatile Ices

Protopapa et al. (2017) identify in the composition maps shown in Fig. 8 the presence of large-scale latitudinal variations of methane and nitrogen, which have been confirmed at a global scale by *Gabasova et al.* (2021, e.g., their Fig. 9). Specifically, three latitudinal bands are identified: the first extending from the pole to 55°N is enriched in methane (spectrum labeled "f" in Fig. 8), the second extends south to 35°N and is dominated by nitrogen (identified with "g" in Fig. 8), and the third reaches 20°N and is mainly composed of methane (region "h"). The spectra extracted in regions "f" and "g" (Fig. 8, right column) clearly differ for the 2.15-μm absorption band due to N_2, displayed by the spectrum "g" but not shown by its counterpart "f." The strength of this feature marks the presence of the N_2-rich component: the deeper the band, the higher the N_2 abundance and/or the larger its pathlength. Another spectroscopic marker for N_2 is the absolute level of the continuum, being higher where N_2 occurs. These lines of evidence suggest a decreasing relative amount of the CH_4-to-N_2-rich components with decreasing latitude from the pole to 35°N (Figs. 8a,c). The area of interested labeled "h" in Fig. 8 stands out instead for particularly large pathlengths of CH_4 (~1000 μm, see Fig. 8b). In LEISA observations the diagnostic is the breadth of the 2.3-μm CH_4 band (see spectrum marked "h" in Fig. 8).

Protopapa et al. (2017) identified a correlation with the expectations of vigorous spring sublimation after a long polar winter. The boundaries between these latitudinal large-scale regions are somewhat ragged and not perfectly aligned with latitude, however, suggesting that other effects, such as surface composition and topography, are important as well. To demonstrate the role of insolation in shaping the transitions found in the maps of volatile cover, the average energy flux at a given latitude of Pluto's hemisphere over the time periods 1995–2005 and 2005–2015 is computed (Fig. 10). Pluto's north pole has been sunlit since 1987. Regions northward of 75° and northward of 55° have received more solar illumination than any other latitude over the past 20 and 10 years, respectively (Fig. 12; sunlit fraction equals one). This continuous illumination at Pluto's north pole seems to have sublimated N_2 into the atmosphere, with the most likely redeposition occurring at points southward (Fig. 13). As noted above, N_2, CO, and CH_4 ices are all identified on Pluto's surface, and are volatile at Pluto's surface temperatures, with N_2 being the most volatile of the three.

Protopapa et al. (2017) argue that the southward-moving polar sublimation front is not the only volatile transport process occurring on the surface of Pluto. A slowly moving N_2 sublimation front expanding northward from Cthulhu Regio is also suggested. From 1975 to 1995, regions south of 35° have been exposed to much stronger heating than regions to the north, leading to sublimation of N_2 ice and its redeposition in regions northward (Fig. 11).

On the hemisphere best imaged in the New Horizons encounter, the latitudinal pattern described above is interrupted by Sputnik Planitia. The large abundance of N_2:CH_4 (Fig. 8c), its large pathlength (Fig. 10d), and the small amount of CH_4 diluted in N_2 (Fig. 8c and section 2.2) demonstrate the major role of N_2 ice in this region of Pluto's surface. Both an albedo and a compositional dichotomy are observed within the basin. *Protopapa et al.* (2017) identify two regions within Sputnik Planitia (see labels "i" and "j" in Fig. 8c), with the northwest part being depleted in N_2 with respect to the center. A shallower N_2 absorption band, and therefore a lower abundance of N_2:CH_4, is observed in spectra extracted from the northwest region. The central part of Sputnik Planitia ("i" in Fig. 8b) presents smaller concentrations of CH_4 in N_2 with respect to the northwest region, and larger optical pathlengths in the N_2-rich component. These physical properties have been interpreted by *Protopapa et al.* (2017) as suggestive of a possible transport of N_2 from the northwest to the south of Sputnik Planitia (see arrows in Fig. 8d and Fig. 11). Using a volatile transport model, *Bertrand et al.* (2018) show that the northwest region of Sputnik Planitia is indeed losing ice both on both annual and climatic timescales, while net condensation should occur below 25°N (see *Bertrand et al.,* 2018) (Fig. 6). Sublimation of N_2 in the northwest part of Sputnik Planitia leaves behind CH_4, which is less volatile, justifying the larger amount of CH_4 incorporated in the N_2-rich component (Fig. 10e) as well as the smaller grains (Fig. 10d). Condensation of N_2

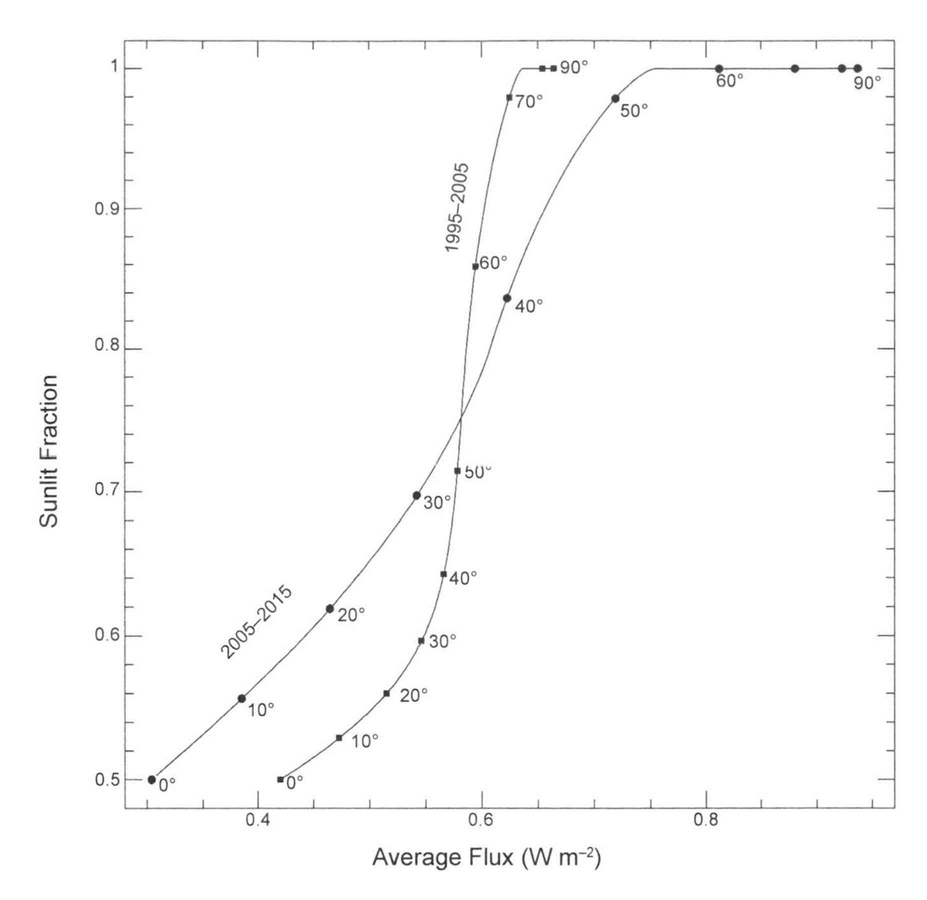

Fig. 10. The average energy flux to a given latitude of Pluto's hemisphere over the decades 1995–2005 and 2005–2015. Adapted from *Protopapa et al.* (2017).

would instead explain the larger amount of the N_2-rich component in the center of the basin (Fig. 10c). *White et al.* (2017) identify an albedo variation within Sputnik Planitia, identifying regions "j" and "l" as dark and bright cellular plains, respectively. *White et al.* (2017) consider it likely that the lower albedo of the dark cellular plains arises from an enhanced concentration of tholin resulting from long-term deposition from an atmospheric source and entrainment in the ice, together with prolonged convection in the ice sheet and sublimation of N_2 from its surface. The lower albedo of the dark cellular plains would enhance N_2 sublimation by raising the surface temperature, even if very slightly.

The non-encounter Charon-facing hemisphere of Pluto has also been mapped, but at a much lower resolution, using the image recorded during the three days (half a rotation period of Pluto, ~3.2 days) on approach to the planet and prior to the closest point of the flyby. This was accomplished first from the MVIC data (*Earle et al.,* 2018) (Fig. 12), and then using the LEISA spectral images (*Gabasova et al.,* 2021) (Fig. 13). The MVIC 0.89-µm CH_4 band map has spatial resolution ranging between 0.65 and ~43 km/pixel, while the map of *Gabasova et al.* (2021) using the CH_4 band complex at 1.7 µm from LEISA data was mapped with resolution ranging from 2.7 to ~400 km/pixel (Fig. 13). Both maps point to the presence of an almost continuous belt of CH_4 ice around the equator, stretching between about 220° and 60°E longitude from Tartarus Dorsa to the western tip of Cthulhu Regio.

3. WATER ICE, ITS DISTRIBUTION, AND ITS ROLE AS THE BASEMENT ROCK

3.1. Detection and Distribution of Water Ice

In view of Pluto's bulk density (1.854 g cm^{-3}), water ice should account for about 55–60% of the volume of Pluto and should form a mantle about 300 km thick if Pluto's interior is completely differentiated. But at Pluto's surface temperatures H_2O ice behaves more as a rock with extremely low volatility and very high viscosity (see the chapter by McKinnon et al. in this volume). It should then be considered the basement rock in which most of the geological features are expressed (faults, craters, cryovolcanos, etc.), except those built or filled by volatiles, such as the nitrogen ice plain of Sputnik Planitia, the methane ice ridges of Tartarus Dorsa, and regions in and around fossae west of Sputnik Planitia.

Water ice was not detected on Pluto's surface from groundbased observations and was not clearly seen until the New Horizons encounter. Its broad spectral absorption

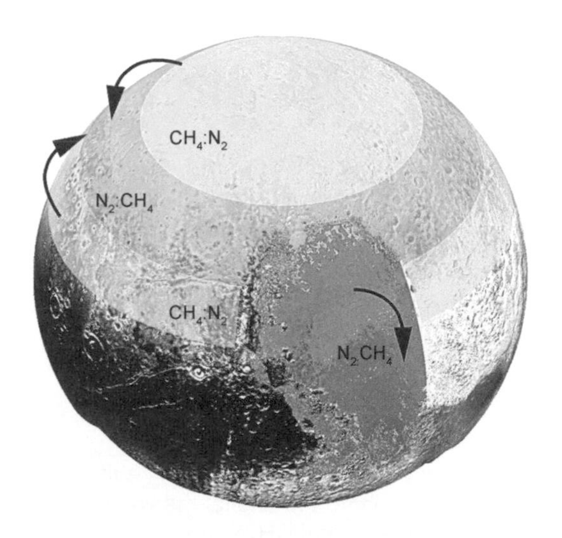

Fig. 11. Schematic view of the large-scale variations of the major ices identified across the surface of Pluto. The arrows indicate the direction of the nitrogen sublimation transport discussed in the text. Adapted from *Protopapa et al.* (2017).

bands around 1.3, 1.5, 1.65, and 2 μm are shallow and overlain by numerous strong CH_4 bands, making them indiscernible as spectral signatures, contrary to the case of Triton. Consequently, prior to New Horizons, exposed water ice was expected to be absent or extremely sparse (*Grundy et al.,* 2013). This expectation was borne out by the LEISA data showing the widespread condensation of N_2:CH_4 and CH_4:N_2 ices on the surface (see Figs. 5 and 10), as well as the large and thick deposits of organic material. These materials mostly hide the bedrock both at visible and near-infrared wavelengths. Only a few spots at specific geological features, such as in fossae walls and surrounding regions, some of which may be geologically active, display clear and prominent water ice signatures. At other places its fractional area or mixing ratio is smaller and the signature is weaker but present, indicating a widespread but patchy distribution of H_2O ice.

Several studies have addressed the issue of the spatial distribution and abundance of water ice on Pluto using the LEISA data, each using a different technique (*Grundy et al.,* 2016; *Schmitt et al.,* 2017; *Protopapa et al.,* 2017; *Cook et al.,* 2019). Because the classical technique of band depth determination from local continuum cannot work properly for H_2O ice on Pluto, due to the almost ubiquitous interfering CH_4 bands absorbing both in the continuum and in the bands of the pure water ice spectrum, *Schmitt et al.* (2017) devised a specific spectral indicator to minimize interference with CH_4 ice bands as well as with the organic material absorption. At the time of Schmitt's 2017 paper the photometric calibration of the spectra was in question, so this relative spectral indicator (a ratio of several spectral pixels around 2.02–2.09 μm to spectral pixels around 1.365–1.41 μm) was also intended to limit calibration biases. From this spectral indicator they obtained a qualitative map of occurrence of water ice (Fig. 14) that sharply defines a number of spatial units, with varying water ice abundance and/or grain size, strongly correlated with geological features.

Grundy et al. (2016) and *Cook et al.* (2019) produced a H_2O map by correlating every observed spectrum with a crystalline H_2O-ice model of a Charon-like spectrum. Although it is more sensitive to photometry biases, this technique provides a qualitative distribution map very similar to the previous one (Fig. 14b).

The pixel-by-pixel Hapke radiative transfer model (at 12 km/pixel) provided the first quantitative maps of abundance and grain size of water ice on Pluto (*Protopapa et al.,* 2017). With a detection threshold around 10% fractional abundance the resulting distribution map (Fig. 14c) correlates well with the previous maps except at a few places (in particular the middle and upper right stripes) where spectrophotometric calibration uncertainties may have introduced biases. The abundance and grain sizes are correlated with geological

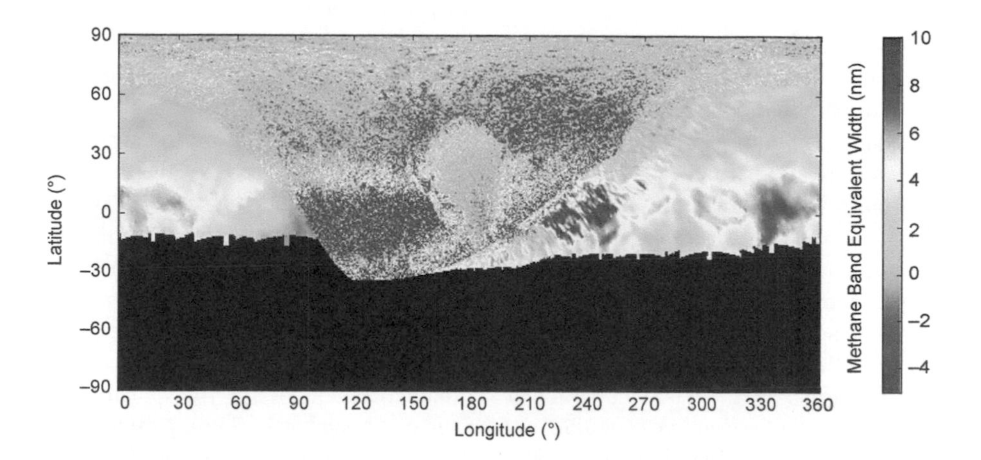

Fig. 12. Global map in cylindrical projection of the equivalent width of the 0.89-μm CH_4 band derived from MVIC images of the near- and far-encounter hemispheres of Pluto in the filter that isolates that absorption band. In the near-encounter hemisphere, the maximum spatial resolution achieved is 0.66 km/pixel. From *Earle et al.* (2018).

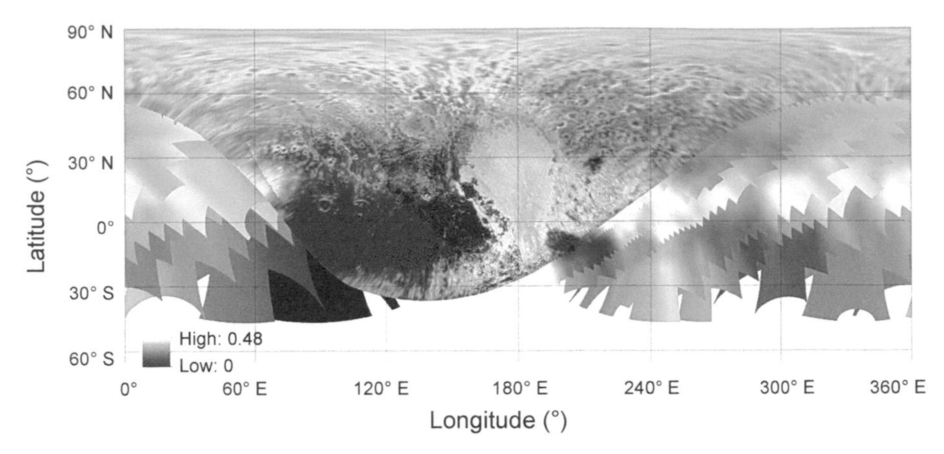

Fig. 13. Global map in cylindrical projection of the integrated 1.7-µm CH_4 absorption band with spatially registered LEISA data from the near- and far-encounter hemispheres. In the near-encounter hemisphere the spatial resolution is between 2.7 and 7 km/pixel and decreases to 400 km/pixel to the east. This map demonstrates that CH_4 abundance is widespread across most of Pluto, with enhanced absorption at the north pole (Lowell Regio) and at the longitude of Sputnik Planitia (~180°E), and a significant depletion over Cthulhu (~70°–160°E). From *Gabasova et al.* (2021).

features and are typically 20–40% with grain sizes in the 100–500-µm range where the ice is the most abundant, and less than 20 µm when mixed with organic material all over Cthulhu Regio (Figs. 14c,d).

Several interesting features emerge from these maps. First, there is a clear dichotomy in terrains with exposures of H_2O ice between the western and eastern sides of Sputnik Planitia, with an extended area containing water ice over most of the western area (especially along fossae and in the mountain ranges along the west side of Sputnik Planitia: Hillary Montes, Baret Montes, and other unnamed topography) but only a faint and spotty distribution and limited between about 25° and 45°N on the eastern side. A major exception is the region in the vicinity of Kiladze Crater that stands out with a particularly high abundance of H_2O ice, close to 60% (*Protopapa et al,* 2017). The spectra of this area clearly display the 1.65-µm absorption band, which is particularly strong at low temperature and is characteristic of the crystalline phase of the H_2O ice (*Grundy and Schmitt,* 1998).

Second, water ice is mostly not found above 45°N latitude (especially on the west side) in Sputnik Planitia as well as over most of Tombaugh Regio. This lack of bedrock outcrop is mostly explained by thick layers of volatiles deposited over these regions (CH_4:N_2 in the polar area and in Tartarus Dorsa, and N_2:CH_4 over the whole Tombaugh Regio).

Over Cthulhu the water ice spectral signature weakens when going from the perimeter toward the center, straddling the equator. This behavior is correlated with the increasing amount of organic material that covers this region (Fig. 17), probably reducing the fractional area of visible bedrock at the subpixel scale. A few spots in the center of Cthulhu and at its northern fringe are also completely devoid of water ice. These spots in fact correspond to mountain ranges (e.g., Pigafetta Montes) and crater rims (e.g., Oort, K. Edgeworth) fully covered by CH_4:N_2 ice (Fig. 5c).

Another striking feature of water ice distribution is its strong signature along several fossae, in particular Virgil Fossae, Inanna Fossae, and Dumuzi Fossae. Water ice, mixed with some organic material, clearly dominates the composition of the bottom and possibly north-facing walls of these deep faults while no CH_4 ice is detected there (Fig. 15).

3.2. Ammonia in Water Ice

Ammonia is an important molecule in planetary interiors and surfaces. In a fluid mix of H_2O and NH_3, in sufficient concentrations (~30% NH_3) the freezing temperature of the liquid is depressed by as much as 100°C. In icy bodies, the freezing point depression allows for the maintenance of a liquid interior for some significant time interval as the body loses the heat from its collisional formation and decay of included radioactive isotopes. The presence of NH_3 in H_2O also increases the viscosity of the fluid (*Kargel,* 1992), a factor that must be included in calculations of the physics of cryovolcanism. In terms of the chemistry of a fluid in the interior of a planetary body, ammonia is important in the formation of organic molecules and contributing to the gas pressure that can result in the ejection of cryolavas in cryovolcanic eruptions on the surface (*Neveu et al.,* 2015, 2017). A full discussion of these two effects is beyond the scope of this chapter, but see *Cruikshank et al.* (2019b,c). In the present context, we focus on the detection of ammonia or an ammoniated compound on Pluto's surface, and the dimension of time that it inserts into the discussion of the formation of some regions on the planet.

Some, but not all, exposures of H_2O ice on Pluto are found to exhibit the spectral signature of NH_3 or an ammoniated compound at ~2.2 µm, particularly when the interfering effects of CH_4 absorption bands are minimal (*Dalle Ore et al.,* 2019). In all cases of the ammonia occurrence investigated so far, the H_2O ice layer is colored; the distinctive red color in the trough and surroundings of Virgil Fossae is the

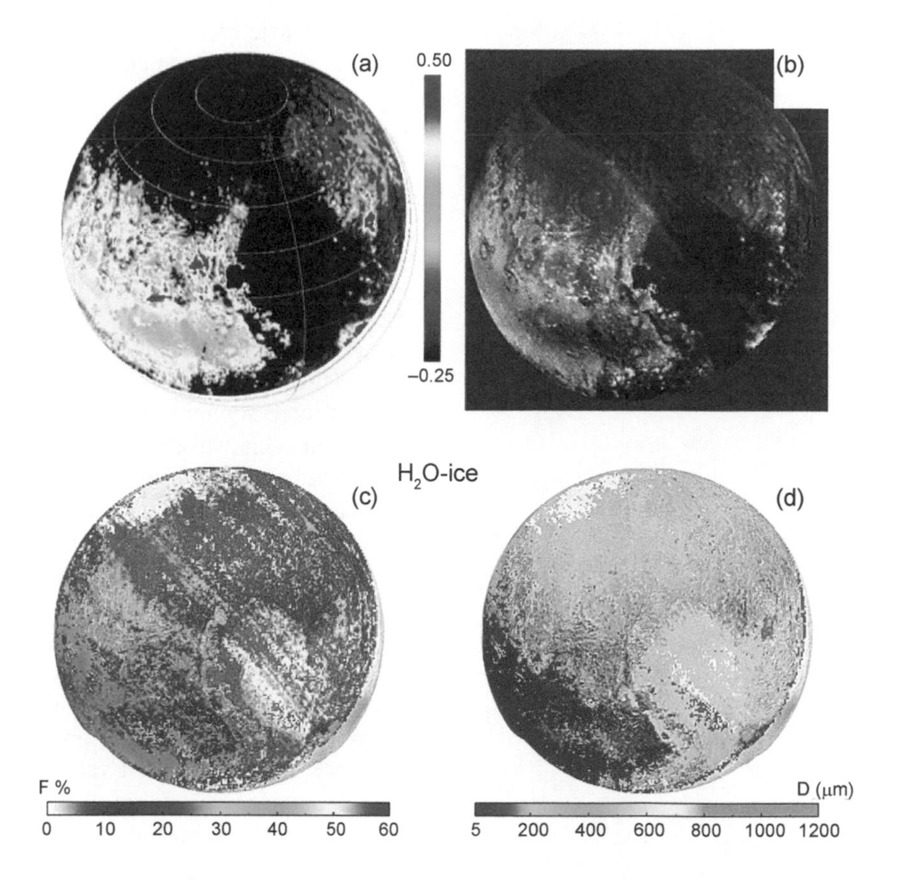

Fig. 14. Four maps of H_2O ice derived from different techniques. **(a)** Spectral indicator optimizing the detection of H_2O ice among the spectral absorptions of CH_4 and the organic material. Detection threshold is –0.23 (*Schmitt et al.,* 2017). **(b)** Correlation map between observed spectra and a crystalline H_2O-ice model spectrum (*Cook et al.,* 2019). **(c)** Fractional area coverage with H_2O ice. **(d)** Corresponding grain size derived from a pixel-by-pixel Hapke radiative transfer model applied to the LEISA scans (Protopapa et al., 2017).

clearest example. In this region and elsewhere, for example in exposures in Viking Terra, the coloration is attributed to a tholin-like complex organic component embedded in the H_2O ice (*Cruikshank et al.,* 2019b,c).

An absorption band at 2.2 μm is also found on Charon and Nix, but as in the case of the band on Pluto, the identification of the exact chemical nature of the ammonia component is ambiguous (see the chapter by Protopapa et al. in this volume). The ambiguity arises in part because only one absorption band is detected, and that band is broad and lies on the long-wavelength wing of a prominent H_2O ice band.

Because NH_3 in H_2O significantly lowers the freezing point of the mixture, the liquid can be maintained in a cooling body for substantially longer than H_2O without the NH_3. Eventually, however, a small planetary body will cool to the freezing point of the mixture, and as it freezes, crystals of ammonia hydrate are formed (*Uras and Devlin,* 2000). Ammonia hydrates can take at least three forms: $NH_3 \cdot 2H_2O$, $NH_3 \cdot H_2O$, or $2NH_3 \cdot H_2O$. Hydrates also form as an ice composed of H_2O and NH_3 is warmed (*Moore et al.,* 2007). Complex refractive indexes for the hydrates in the spectral region of interest are not currently available for inclusion in computational models, and consequently the hydrates cannot be distinguished from one another.

Another possibility for the identification of the band seen on Pluto is one or more ammoniated salts, and while reflectance spectra for several of these salts are available, the band near 2.2 μm is broad and indistinct, precluding a clear identification. *Cook et al.* (2018) found that NH_4Cl is a good spectral match to the 2.2-μm absorption band in the spectrum of Pluto's satellite Nix. In studies of the Ceres spectra from the Dawn mission, *DeSanctis et al.* (2015, 2016) included ammoniated salts NH_4Cl and $(NH_4)_2CO_3$ in their models.

Ammonia (NH_3) and the ammonium ion (NH_4^+) are destroyed by ultraviolet (UV) radiation and by charged particles in the solar wind and cosmic rays (e.g., *Cruikshank et al.,* 2019b), although the destruction rates vary with the circumstances. In a pure mixture of NH_3 in H_2O, dissociated NH_3 molecules can readily recombine, while in a multicomponent mixture NH_2^- will combine with other molecular fragments, resulting in a net loss of ammonia. The hydrates and ammoniated salts are expected to be more resilient to destruction because the incoming energy can be distributed among other bonds in the molecular structure.

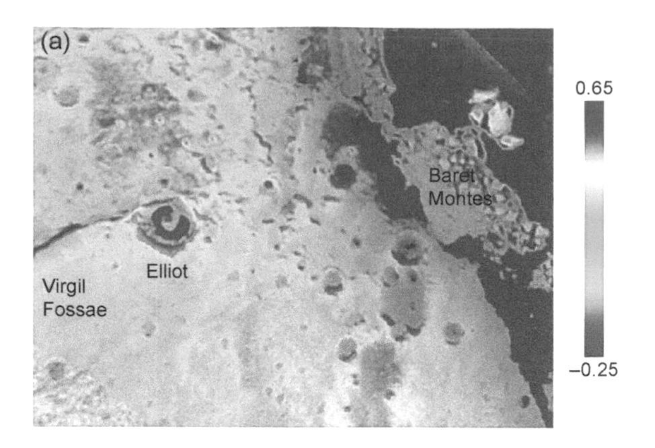

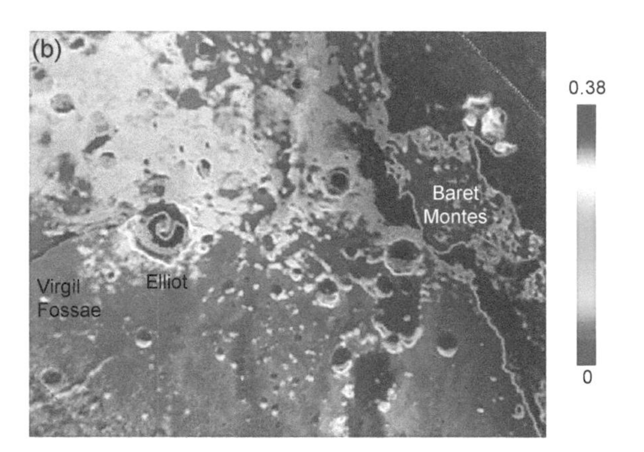

Fig. 15. High-resolution maps (at 2.8 km/pixel) of part of Viking Terra (Elliot crater, Virgil Fossae, and Baret Montes area). **(a)** H_2O spectral index and **(b)** 1.70-μm CH_4 band depth overlaid on Pluto's digital elevation model (*Schenk et al.,* 2018) in cylindrical projection (derived from maps available on the Planetary Data System) (*Schmitt et al.,* 2017).

See *Cruikshank et al.* (2019b) and the Protopapa et al. chapter in this volume for a more detailed discussion of ammonia in the context of Pluto's chemistry and cryovolcanism.

4. SURFACE COMPONENTS PRECIPITATED FROM THE ATMOSPHERE

The principal molecular ices on Pluto's surface are present in their gaseous state in the planet's atmosphere. Diurnal, seasonal, and long timescale exchanges of these components between atmosphere and surface, as well as the deposition of other molecular solids precipitated from the atmosphere where they are produced by photochemistry, are fundamental to Pluto's chemical dynamics. This section is focused on the nature of atmospheric photoproducts and their precipitation to the surface, while section 5 expands on the nature of the refractory components of atmospheric origin and those produced directly by chemical reactions in the surface ices.

After the flyby, New Horizons looked back toward the dark side of Pluto and recorded a spectacular blue glow encircling the planet. This was sunlight, forward scattered by haze particles in Pluto's atmosphere. The haze was detected by four instruments on the spacecraft. It was seen in absorption by the Alice UV spectrometer at altitudes up to 350 km above the surface (*Young et al.,* 2018), and the LEISA infrared spectroimager detected the haze via forward scattered near-infrared light (*Grundy et al.,* 2018). The LORRI panchromatic CCD imager and MVIC visual color imager did the same at visible wavelengths, and their much higher spatial resolution revealed the complex spatial structure of the haze, with numerous layers extending long distances around the limb of the planet (e.g., *Stern et al.,* 2015; *Gladstone et al.,* 2016; *Cheng et al.,* 2017). The haze scatters more sunlight at shorter wavelengths, causing its blue color. This wavelength dependence implies Rayleigh scattering by particles smaller than the wavelength, but the angular dependence of the haze scattering, being strongly forward scattering, requires larger particles. To reconcile these seemingly conflicting constraints, micrometer-sized aggregates of much smaller constituent particles have been proposed (*Gladstone et al.,* 2016; *Cheng et al.,* 2017).

Pluto's haze is the product of chemistry enabled by energetic radiation impinging on the upper atmosphere. Pluto's lower atmosphere is predominantly composed of N_2, since it has the highest vapor pressure of Pluto's three volatile surface ices, and small amounts of CO and CH_4 are also present. But at high altitudes, the lighter CH_4 molecule becomes increasingly abundant, and above 1400 km, CH_4 surpasses N_2 to become the primary atmospheric constituent and also the primary feedstock for haze-forming chemistry (*Young et al.,* 2018; see also the chapter by Summers et al. in this volume).

Solar UV light is thought to be the primary driver of Pluto's haze production, with Lyman-α emission (121.6 nm, 10.2 eV) being an especially important source. A CH_4 molecule will readily interact with a Ly-α photon, to be fragmented into radicals and ions that can go on to chemically react with other atmospheric constituents. Unlike CH_4, N_2 and CO molecules cannot be photodissociated by Ly-α. More-energetic, shorter-wavelength UV photons can do so, although such photons are less abundant than Ly-α, and their flux is more variable over the solar cycle. Solar wind is another potential instigator of atmospheric chemistry. Typical energies of solar wind particles (protons and electrons) are in the kiloelectron volt range, a hundred times more energetic than a Ly-α photon, and a minority have much higher energies. It is unclear how important the solar wind is as a driver of Pluto's atmospheric chemistry, since much of it is diverted around Pluto by the planet's conductive ionosphere (*Cravens and Strobel,* 2015; *Bagenal et al.,* 2016). But thanks to the higher energies of solar wind particles, they may still play a role if some fraction can reach the atmosphere. Even more energetic protons and heavier atomic nuclei arrive in the form of cosmic rays, with the megaelectron-volt through gigaelectron-volt energy range being most relevant. Destroying a small molecule consumes a minuscule fraction of such a particle's energy, so it goes on to interact with many more molecules, as do fragments from the destroyed molecules, themselves accelerated to high energies. The resulting cascade of progressively lower-energy particles can ultimately process a very large number of molecules in the atmosphere and on the surface.

Radicals and ions in the atmosphere can encounter and react with other atmospheric constituents to build larger, more complex molecules. Pluto's upper atmosphere is thin with large mean-free-paths, so encounters take time. Electrostatic attraction between oppositely charged ions favors their interaction, so chemistry between negative and positive ions is especially important. Similar processes occur in the atmosphere of Titan and have been studied extensively in that context (e.g., *Vuitton et al.,* 2009; *Lavvas et al.,* 2013; *Hörst et al.,* 2018). Progressively larger molecules form through consecutive accretion of small ions and radicals, eventually reaching thousands of daltons in mass with sizes on the order of tens of nanometers. Such macromolecular monomers are thought to be the small particles from which are assembled the larger micrometer-sized aggregates required to account for the forward-scattering nature of the haze.

Micrometer-sized particles cannot stay aloft in Pluto's thin atmosphere indefinitely. They gradually settle to the surface, with a settling timescale on the order of Earth days to months, depending on the size and density of the particles (*Cheng et al.,* 2017; *Gao et al.,* 2017; *Zhang et al.,* 2017). The rapid settling implies rapid replenishment and a potentially important source of chemically processed material at Pluto's surface. Estimates of production rates at the time of the 2015 New Horizons flyby work out to over 10 m of accumulation over the age of the solar system, enough to significantly influence surface composition (*Cheng et al.,* 2017; *Wong et al.,* 2017; *Grundy et al.,* 2018).

The nature of the bonds between the small constituent molecules and fragments that form the haze particles is unclear but is critically important to understanding their chemical contribution to Pluto's surface environment (*Grundy et al.,* 2018). Tholins produced in terrestrial laboratory environments are often described as the refractory residues that remain after irradiated material has been raised to room temperature, driving off more weakly bound constituents. Such a material should be inert under Pluto surface conditions. But at the low temperature of Pluto's upper atmosphere (~70 K) (*Hinson et al.,* 2017; *Young et al.,* 2018), much more weakly bonded components may be important, and could perhaps even dominate the composition of the monomers. *Wong et al.* (2017) simulated the production of small molecules through photochemistry in Pluto's atmosphere, finding that abundant light hydrocarbons and nitriles are produced, including C_2H_2, C_2H_4, C_2H_6, CH_3C_2H, HCN, C_6H_6, and C_4H_2, some of them having been observed by New Horizons (*Gladstone et al.,* 2016, *Young et al.,* 2018). It is not known to what extent these small organic molecules accrete to the growing monomers through chemical reactions vs. through adsorption mechanisms that preserve their molecular integrity. *Lavvas et al.* (2013) estimated that as little as 7% to 10% of the content of Titan's haze could be considered as tholin. *Luspay-Kuti et al.* (2017) observed that the propensity of small molecules to adhere to the haze particles was a function of altitude, with greater "stickiness" at higher altitudes. Such stickiness might be related to the presence of complicated charge distributions and/or dangling bonds on the surfaces of the monomers resulting from the haphazard accretion of ions and fragments at high altitude, while by the time a particle reaches lower altitude, so many small molecules have been adsorbed that they dominate its surface. Beyond that point, accreting more molecules would be more like condensation than adsorption, and could thus account for the reduced stickiness at lower altitudes.

Through much of a haze particle's descent through Pluto's upper atmosphere, the temperature remains nearly constant. But in the lower atmosphere, the temperature changes significantly with altitude. Around 20–40 km above the surface, Pluto's atmospheric temperature reaches a maximum of approximately 110 K (*Hinson et al.,* 2017). This temperature is well above the melting points of several of the simple hydrocarbons predicted to be incorporated into the haze (*Wong et al.,* 2017). To the extent that such molecules retain their independent existence, haze particles that include a significant payload of such simple hydrocarbons should partially melt, or at least anneal, and would likely evolve from complex fractal shapes to more compacted shapes, due to surface tension. They may also adhere to one another, and some sublimation loss of small molecules could also occur. Below the altitude of this temperature maximum, Pluto's atmosphere cools rapidly toward surface temperatures around 40 K. As their temperature drops on descent to the surface, any component of haze particles that had previously melted would refreeze, and additional new condensation could cause the haze particles to grow. Photometric evidence for changes in the scattering behavior of the haze at low altitudes may be related to this thermal history (*Cheng et al.,* 2017).

The issues of quantity and timescale for haze production are difficult to assess with much certainty. Energy to drive atmospheric production of complex organics has been available throughout solar system history in the form of UV light, solar wind, and cosmic rays. But they are all expected to vary in intensity with the solar cycle and as the Sun ages and passes through distinct galactic environments. Occasionally, far more radiation becomes available as a result of astrophysical events like nearby supernovae or the compression of the heliosphere as the solar system passes through dense hydrogen clouds. The existence of suitable chemical feedstock is even less certain over time. Consider Triton, with a similar inventory of surface volatile ices to Pluto but comparatively little atmospheric haze, at least at the time of the 1989 Voyager encounter. The difference is attributed to the fact that Triton has much less atmospheric CH_4 and is photolytically destroyed at lower altitude, rather than enriching the high altitudes where abundant ions exist to facilitate haze-building chemistry (*Strobel and Zhu,* 2017). The example of Triton raises the question of whether similarly unproductive circumstances could have prevailed during some of Pluto's history. Or conversely, atmospheric production of complex organic material could have been more intense at certain times in the past when even more favorable atmospheric configurations prevailed.

Grundy et al. (2018) attempted unsuccessfully to model the visible to near-IR reflectance of three distinct areas on Pluto using volatile ices plus a common set of tholin optical constants, presumed to originate from the haze. They argued that their inability to match the spectra of the different regions using a common tholin pointed to further environment-dependent chemical evolution of the haze particles, subsequent to their arrival at the surface. More recently, *Protopapa et al.* (2020) were able to match MVIC + LEISA spectra of distinct regions over the wavelength range 0.4–2.5 μm, using a common Titan-like tholin (*Khare et al.,* 1984), after accounting for the very different scattering properties of the regions. This result suggests that the strongly pigmented tholin material delivered by haze is relatively inert at Pluto's surface. However, to the extent that the haze also delivers a payload of more weakly bound small molecules along with the tholin component, those smaller molecules may behave very differently in Pluto's diverse surface environments and might play important geological roles in some settings, for instance, by alloying with and modifying the rheological properties of the volatile ices that dominate Pluto's active surface geology.

In the lowest regions of the atmosphere where the temperature drops below the condensation temperature for the hydrocarbons and nitriles produced by photochemistry at higher altitudes (listed above), these molecules may condense on Pluto's surface as ices. *Krasnopolsky and Cruikshank* (1999) calculated the deposition rates for several ices, finding that the rates are reflective of the molecular abundances of the species at the bottom of the atmosphere. The deposition rate was found to be highest for C_2H_2 at 65 g cm^{-2} G.y.$^{-1}$ and C_4H_2 (representing the heavier hydrocarbons in the model) at 58 g cm^{-2} G.y.$^{-1}$. The rates for HCN and HC_3N were 14 and 23 g cm^{-2} G.y.$^{-1}$, respectively. *Wong et al.* (2017) refined the calculations of deposition rates using information from New Horizons on the atmospheric temperature structure and improved molecular abundances, but found the same basic condensation sequence, with C_2H_2 leading at 179 and HCN at 35 g cm^{-2} G.y.$^{-1}$. Wong et al. found that the precipitation rates of the C_2 hydrocarbons were higher than in the earlier work, and as a consequence of more rapid removal of the C_2 hydrocarbons, their model yields a lower flux of higher-order hydrocarbons than the model of *Krasnopolsky and Cruikshank* (1999).

Of the several molecules modeled by *Krasnopolsky and Cruikshank* (1999) and *Wong et al.* (2017), the only one actually detected by near-infrared spectroscopy on Pluto surface is C_2H_6, although as we have noted above, HCN may have been marginally detected in spectra presented by *Protopapa et al.* (2008); HCN has been identified in Triton's spectrum (*Burgdorf et al.,* 2010). However, the abundance profiles of gaseous C_2H_2, C_2H_4, and C_2H_6 in the atmosphere of Pluto clearly demonstrate that these molecules condense on the haze between 400 and 200 km and precipitate to the surface (*Gladstone et al.,* 2016; *Young et al.,* 2018). At least two factors are at play in limiting our capability to detect more of the several molecules predicted to be condensing on Pluto's surface. First, the spectral signatures of these molecules in the wavelength region currently available for observations from Earth and space, as well as with the New Horizons spacecraft, are relatively weak, and they are greatly confused by the dominance of the spectral bands of CH_4 over most of the available wavelengths. In addition, modest spectral resolution and the relative weakness of the overtone and combination bands further limit the detection capability. Second, the dynamic processes of sublimation, condensation, and mass movement of some components of Pluto's atmosphere on annual, seasonal, and Milancović timescales (see the chapter by Forget et al. in this volume) essentially ensure that minor molecular components precipitating from the atmosphere are unlikely to accumulate in large (and detectable) amounts in most regions of the surface.

5. NON-ICE COMPONENTS OF PLUTO'S SURFACE

Non-ice components of Pluto's surface are seen in natural color in images derived from the four color filters in MVIC (Olkin et al., this volume). Pluto's surface is covered with a range of brown to yellow- and red-brown hues that largely conform to geological and geomorphological structures. When the image data are enhanced through digital processing, a wide range of colors emerges, ranging from near-white through yellow, orange, brown, and red (*Stern et al.,* 2015; *Olkin et al.,* 2017; see also the chapter by Olkin et al. in this volume). These colors, whether enhanced or not, are not the intrinsic colors of the ices detected spectroscopically on the surface, but instead represent one or more non-ice components that are pervasive across the entire planet's surface.

The surfaces of many other solar system bodies are colored to varying degrees, and a similar orange-brown color is seen in the atmosphere of Titan. Similarly, as discussed above (section 4), chemical processes in the atmosphere of Pluto produce complex organic tholins. On Titan, the coloration arises from aerosol particles consisting of complex molecules resulting from the photolysis of methane and nitrogen (e.g., *Cable et al.,* 2012). In early laboratory experiments synthesizing the color of Titan by exposing a gaseous mixture of CH_4 and N_2 to electrical discharges, *Khare et al.* (1981) reported the production of a refractory solid having a brown-red color. Finding similar results with the energetic processing by electrical discharge and UV photons of the same and other gas mixtures, but always those containing a source of carbon (usually CH_4), *Sagan and Khare* (1979) coined the term "tholin" for the material produced in such experiments (see also *Sagan et al.,* 1984). The term "Titan tholin" was applied to the material made from the Titan mixture of gas: 10% CH_4 and 90% N_2. In order to facilitate the use of Titan tholin in radiative transfer models of Titan's atmosphere (and other colored bodies), *Khare et al.* (1984) measured and reported the complex refractive indices over a wide range of wavelength. Subsequently, "Titan tholin" has been used to model not only Titan's atmosphere, but the colored solid surfaces of many solar system bodies, resulting in a voluminous literature on the subject (see review by *Cable et al.,* 2012). For more on this subject, see also *Brassé et al.* (2015) and *Cruikshank et al.* (2005, 2019c).

A simplified and generic definition is that tholins are disordered polymer-like materials made of repeating chains of linked subunits and complex combinations of functional groups that contain carbon. Nitrogen and other elements can be incorporated to varying degrees in the aromatic and aliphatic subunits. With this broad definition, the term tholin can be applied to such materials synthesized in the laboratory or occurring as a result of natural chemical reactions in a planetary environment. This definition is also a suitable working description of tholins produced by the UV- and charged-particle irradiation of ices (e.g., *Materese et al.,* 2014, 2015) described below.

Pluto's atmospheric aerosol particles were described in section 4, and here we note that these particles of atmospheric origin appear to be closely analogous to the *Khare et al.* (1981) Titan tholin. Some of the studies of their material, particularly the complex refractive indices (*Khare et al.,* 1984), are useful in describing the material on Pluto, as shown by *Protopapa et al.* (2017, 2020). Specifically, *Protopapa et al.* (2020) use a tholin material with optical constants very similar to that of Titan tholin by *Khare et al.* (1984) to reproduce the spectral properties of Pluto's regions having diverse coloration, such as Lowell Regio and Cthulhu Regio. Variations of Pluto's coloration at large scale are attributed by *Protopapa et al.* (2020) to changes in abundance and grain size of the Titan-like tholin material.

As demonstrated in laboratory experiments with an ice composed of CH_4, N_2, and CO, direct irradiation of surface ices by UV photons and charged particles from the Sun also produce reactions resulting in the formation of colored, refractory tholins (e.g., *Materese et al.,* 2014, 2015). The color of the tholin produced by radiolysis is a good but not perfect match to the visible colors found in some regions on Pluto (*Spencer et al.,* 2016). Analysis of the Pluto ice tholins show the presence of carboxylic acids, ketones, alcohols, aldehydes, amines, and nitriles in a solid having small aromatic rings connected with short aliphatic bridges. In the electron irradiation experiments, the nitrogen incorporation was significant, with the atomic ratios showing N/C ~0.9 and O/C ~0.2. The UV experiments showed a lower N concentration (N/C ~0.4) because UV is less efficient than electrons in breaking the N≡N bond in N_2. In terms of the molecular structure, the Pluto ice tholin is closely comparable to that of Titan tholin (*Lebreton et al.,* 2009) and an organic component of interstellar dust (*Kwok and Zhang,* 2011).

The formation of complex organics in Pluto's surface ices by UV exposure depends on the transparency of the atmosphere, which is variable over annual and long-term timescales (*Bertrand et al.,* 2019). At times when atmospheric CH_4 is abundant, it effectively blocks solar Ly-α photons from reaching the surface. However, UV photons at other wavelengths penetrate the atmosphere and have sufficient energy to photolize methane ice on the surface (*Steffl et al.,* 2020). Several laboratory experiments in the UV photolysis of CH_4 in the gas and solid phases have demonstrated the rapid photolysis of the molecule and its recombination into more complex configurations, resulting in the formation of a colored refractory solid tholin (e.g., *Thompson et al.,* 1987; *Khare et al.,* 1984; *Imanaka et al.,* 2004, 2012; *Materese et al.,* 2014, 2015). Figure 16 shows the reflectance spectrum and color image of the *Materese et al.* (2015) Pluto ice tholin made by electron irradiation of a co-deposited mixture of $N_2 + CH_4 + CO$ (100:1:1).

As noted in section 3.2, a third source of complex organic compounds has been proposed to explain an exposure of colored H_2O ice that also carries the spectral signature of one or more ammoniated materials. In Virgil Fossae and surroundings, *Cruikshank et al.* (2019b,c) studied the surface exposure of a unique colored material and associated it with an effusion of a cryolava, proposing that it emerged along faults in Pluto's crust that define the graben structure, probably including both surface flows and a fountaining episode (*Cruikshank et al.,* 2019a). The unusual color of the deposit may indicate a composition different from that of tholins in atmospheric aerosols and surface ices processed by factors in Pluto's space environment. Alternatively, it may indicate a difference in age of the emplaced material, a scenario in which prolonged exposure of recently debouched and uniquely colored material on the surface might alter its color through interaction with UV light, cosmic rays, and/or charged particles from the solar wind.

Tholins exhibit strong absorption in the UV and violet spectral regions, with decreasing absorption toward longer wavelengths, giving the material a yellow, red, or brown color. The absorbance is usually minimum at ~1 μm. The upward slope in reflectance from short to long wavelengths varies from one tholin to another, depending on the starting materials and conditions of formation. Figure 16a from *Imanaka et al.* (2004) demonstrates the change in absorbance, hence color, for gas-phase Titan tholin made from the same mix of CH_4 and N_2 at different pressures. The variation in color reflects different proportions of N_2 in the resulting solid (N_2 is more efficiently incorporated at low pressure), the proportions and sizes of aromatic rings, the abundance of saturated C-H bonds, and other factors.

The distribution of tholins and H_2O from the pixel-by-pixel modeling of *Protopapa et al.* (2017) is shown in Fig. 17a, and the wide range of albedo, with spectra corresponding to six different regions, is shown in Figs. 17b,c.

The tholin components of the surface span a wide range in albedo, with a minimum of <0.1 (*Buratti et al.,* 2017; see also the chapter by Olkin et al. in this volume). Together with the ice-covered regions of the surface, the planet-wide range of albedo is a factor of 10, which is greater than any other planetary body except Iapetus. In a given season and time of day, the surface temperature is regulated by the local albedo, thus affecting the ability to retain deposits of the volatile ices, particularly N_2 but also CO and CH_4. The

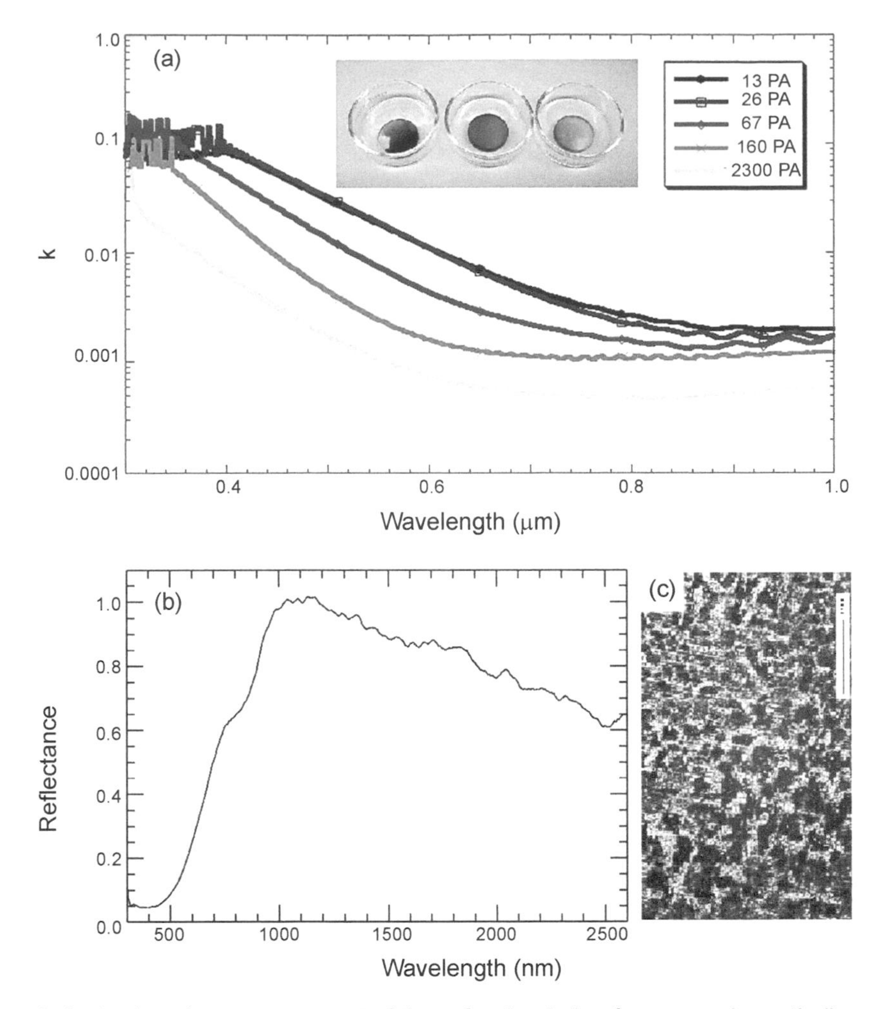

Fig. 16. **(a)** Variations in k, the imaginary component of the refractive index for a gas-phase tholin produced with a cold-cathode discharge in a gaseous mixture of 90% N_2 and 10% CH_4 at five different pressures to explore tholin formation in Titan's atmosphere at different altitudes (*Imanaka et al.,* 2004). The color of the tholin tracks the change in pressure in the reaction chamber. **(b)** Reflectance spectrum of an ice-phase tholin made by 1.2 keV electron radiolysis of a codeposited ice mixture of N_2, CH_4, and CO (proportions 100:1:1) (spectrum courtesy of T. Hiroi, Brown University). **(c)** The tholin shown in color *(Materese et al.,* 2015; *Cruikshank et al.,* 2016).

surface temperature in Cthulhu can reach 72 K at certain seasons (*Grundy et al.,* 2018).

6. TEMPERATURE SOUNDING OF THE UPPER FEW METERS OF THE SURFACE

6.1. Measurements with the Radio Science Experiment

With the objective of better understanding the distribution of temperatures and providing insight into the nature of Pluto's ices, New Horizons carried the Radio Science Experiment (REX) to conduct a limited radiometric survey of Pluto. REX used the spacecraft's high-gain antenna (HGA) to study the radiometric brightness of Pluto and Charon at 4.17 cm wavelength (7.18 GHz, X-band), with a temperature precision of 0.1 K and spatial resolution ~0.12 of Pluto's diameter when the measurements were made shortly after the spacecraft's closest approach. The flux density at 4.17 cm received from Pluto is gray-body thermal emission from a depth of less than ~1 m in ice-free regions, to as much as ~500 m for methane ice (*Bird et al.,* 2019). A description of the design and implementation for REX is given in *Tyler et al.* (2008).

The radio brightness was measured in a scan when the spacecraft was ~14,400 km from the center of Pluto, outbound at 14 km s^{-1}. The "footprint" of the HGA on Pluto's surface, defined by the 3-dB boundary of the HGA beam pattern, illuminated a circle mid-scan of 12% of Pluto's diameter, or equivalently 288 km in diameter. The scans were performed by a uniform rotation of the spacecraft along first a diametric chord, and then crossing the chord over Pluto's winter pole.

The scans are illustrated in Fig. 18. Pluto's terminator was near the middle of the diametric chord, causing much

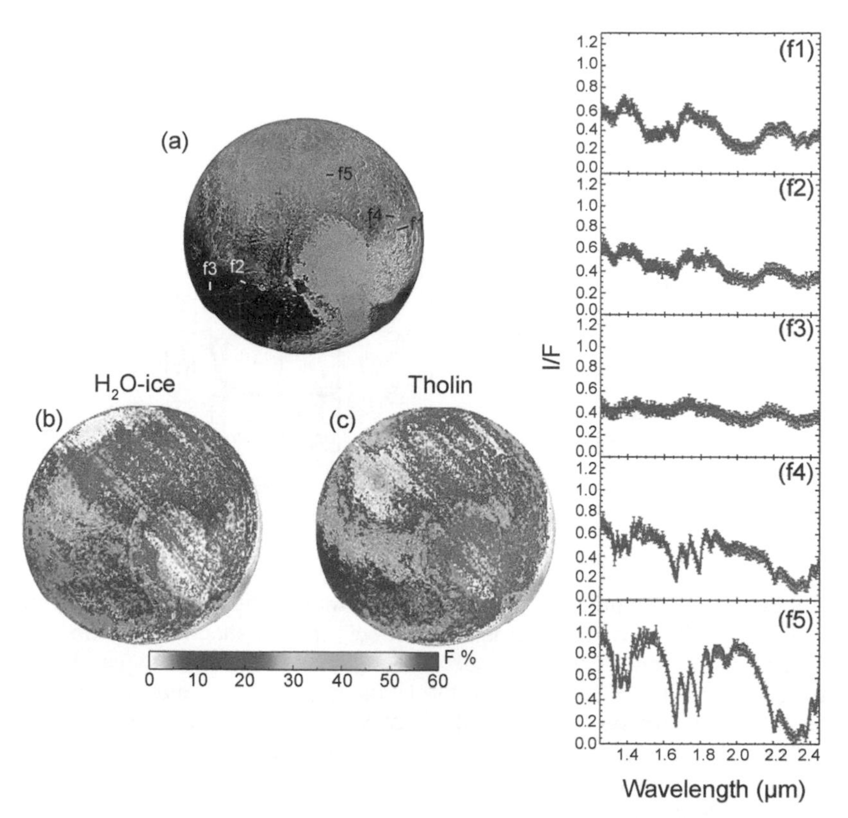

Fig. 17. (a) Base map of Pluto, showing the wide range of albedo [normal albedo range 0.08–1.0 (*Buratti et al.,* 2017)] across the surface of the hemisphere observed at the New Horizons encounter. The map identifies five regions for which the extracted spectra are shown in the adjacent panels. The abundances (F%) of **(b)** H_2O ice and **(c)** tholin are projected on the LORRI base map. These values are derived from pixel-by-pixel modeling of the LEISA data and are adapted from *Protopapa et al.* (2017). In the spectra panel, f1 shows the characteristic signatures of H_2O ice at the location of Kiladze Crater, while f2 comes from a region in Cthulhu Regio and presents, in comparison with f1, shallower H_2O absorption bands and a neutral spectral slope consistent with smaller H_2O-ice particle diameters (~10 μm). The signature absorption band of N_2 at 2.15 μm, although intrinsically weak, is strongest in regions f4 and f5. These and other regions show variable strength of the CH_4 bands. The wavelength positions of the CH_4 and N_2 bands are illustrated in Fig. 1 [Pluto spectral data are from *Protopapa et al.* (2017)].

of the chord to be in darkness: some in diurnal night, and the remainder within seasonal darkness. A special condition existed near the sunlit end as the diametric chord scan traversed the nitrogen ice glacier in Sputnik Planitia. Somewhat near the glacier's midpoint, the angles of incidence and reflectance between the spacecraft and the sub-Earth direction became equal. This condition identifies the specular point on Pluto's surface. Here, an X-band transmission from Earth reflected and scattered toward the spacecraft, performing a bistatic radar experiment (see section 6.3 below). The winter pole chord crossed the terminator as well, with a small portion in diurnal night and the larger portion within the seasonal darkness of the winter pole.

Additional radio brightness measurements were made during the Pluto encounter, consisting of (1) nightside scans of both Pluto and Charon, between ingress and egress of the radio occultation experiment (*Hinson et al.,* 2017), although not covering the same region as the diametric chord or the winter pole chord, and (2) full-disk measurements of Pluto and Charon, before and after closest approach, when the sizes of Pluto and Charon were smaller than the HGA beam. The nightside scan and the full-disk radiometry observations and their interpretations are in *Bird et al.* (2019).

The radio brightness temperature, T_b, is related to the physical temperature of Pluto, T_P, through the emissivity, ε_r, $T_b = \varepsilon_r T_p$. Radio emissions can arise from much greater depths than reflected light at optical wavelengths, possibly meters to kilometers, depending on the transparency of the material. The radio wavelengths therefore afford the advantage of probing to greater depth than the optical wavelengths, but the calculation of kinetic temperatures at depth is complicated by the fact that that the emissivity of the component materials is largely unknown for radio wavelengths at the expected temperatures of Pluto. The emissivity is also known to be dependent on the emission angle and the components of liner polarization parallel and perpendicular to the plane containing the surface normal and the direction to the spacecraft (*Heilels and Drake,* 1963). The radio measurements were made over a wide range of emission angles, or phase, adding extra

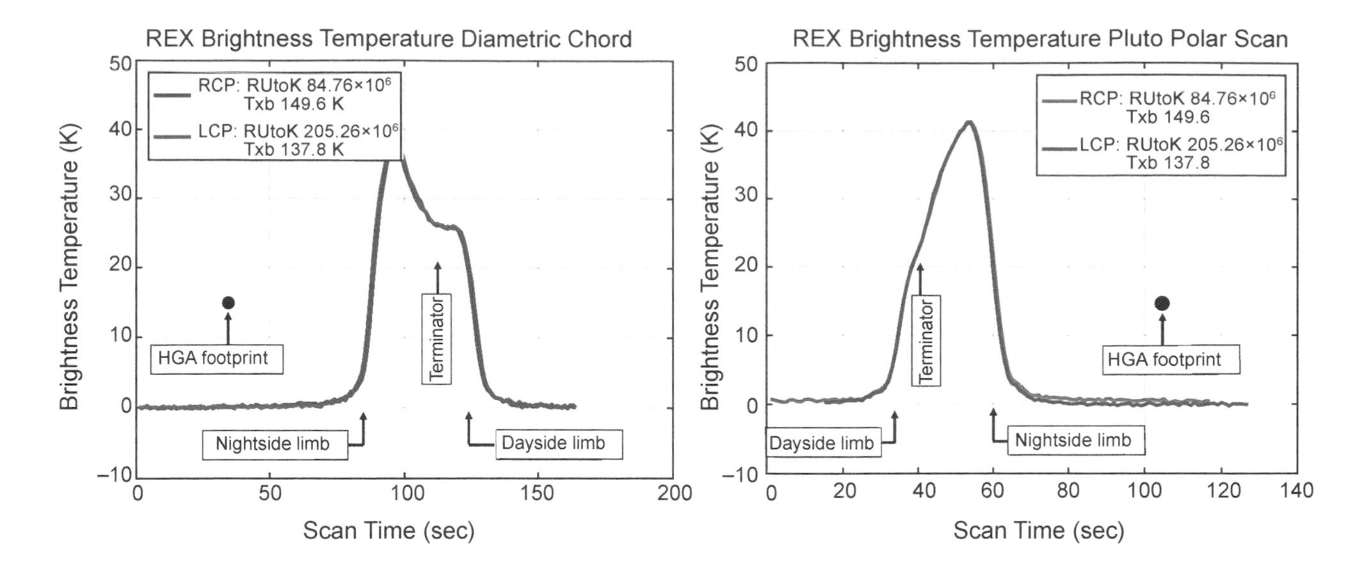

Fig. 18. Radiometric brightness temperatures for the diametric and winter pole scans. The temperature profiles are in blue for righthand circular polarization (RCP) and red for lefthand circular polarization (LCP). The polarization measurements are plotted on top of each other and are nearly identical. The brightness temperatures use conversion constants in *Bird et al.* (2019). Pluto fills the beam of the HGA interior to the limbs, but the filling is a function of the scan exterior to the limbs. The X-band receiver noise increases the standard deviation (STD) of the temperature measurements in the scan, but the increase is less than the width of the lines in the plots. Arrows show the locations of the dayside and nightside limbs, as well as the day-to-night terminator. The temperature scales are the same, but the timescales are not, due to the differing duration of the two scans.

constraints to physical models of the emissivity and other properties of the surface materials.

The REX radio brightness temperatures in Fig. 18a demonstrate that the warmest brightness temperatures are found in the darkest places. To illustrate this situation, Fig. 17a shows the REX scans of the visible imaged portions of Pluto overlaid on a rectangular projection of Pluto's whole surface. The warmest temperatures, i.e., between 40 K and 50 K, are below –40° latitude, low enough in latitude to be within Pluto's winter night. In addition to the polar terrains illuminated by scattered light from atmospheric haze, the REX radiometry experiment provides direct observations of a part of the surface in Pluto's winter night.

6.2. Nature of the Surface from Pluto's Thermal Emission

Thermal electromagnetic radiation from surfaces like Pluto would nominally be expected to be unpolarized. This is canonically the case at normal incidence, i.e., zero phase. However, at oblique incidence, say, for angles greater than 45°, the thermal radiative process involves propagation of electromagnetic waves within the subsurface toward the surface, and then transmission through the boundary surface interface. Differences result from polarization-dependent scattering and transmission through interfaces characterized by Fresnel propagation. The polarization differences in Fig. 19b, although on the order of a few percent, are much larger than the measurement errors (~0.15%). (Note, for example, the 3σ symbol in the lower righthand side of the figure.) Differences of ~1% are then ~10× the error. Such high measurement precision provides information about the nature of the material responsible for the thermal emission. For example, the nightsides of both the diametric and polar scans exhibit both signs of the polarization difference. Enigmatically, the radiometer temperature is warm in the polar scan near the winter pole with an excess of LCP, while the warmest temperature in the diametric scan is closer to the equator but has an excess of RCP. In pursuit of these peculiarities, further analysis will incorporate phase angular dependence with X0-band emission and scattering to model temperature and polarization measurements.

6.2.1. Sputnik Planitia. Although Sputnik Planitia, with its reservoir of nitrogen ice, ought to have a temperature very close to the sublimation temperature of 37 K, the measured brightness temperature in Sputnik Planitia is 27 ± 1 K, dropping some 3 K in the interior where the large-scale convection indicates a greater thickness of the ice. The REX occultation found Pluto's atmosphere's temperature at the ingress point to be 38.9 ± 2.1 K (*Hinson et al.,* 2017), consistent with the surface temperature of the nitrogen ice. The radio brightness temperature is equivalent to the physical temperature only if the atmosphere's emissivity is unity; in Sputnik Planitia the discrepancy with the sublimation temperature of nitrogen ice is resolved if the emissivity is 0.73.

The REX occultation found Pluto's atmosphere's temperature at the ingress point to be 38.9 ± 2.1 K (*Hinson et al.,* 2017), consistent with the surface temperature of the nitrogen ice. The radio brightness temperature is equivalent

(a)

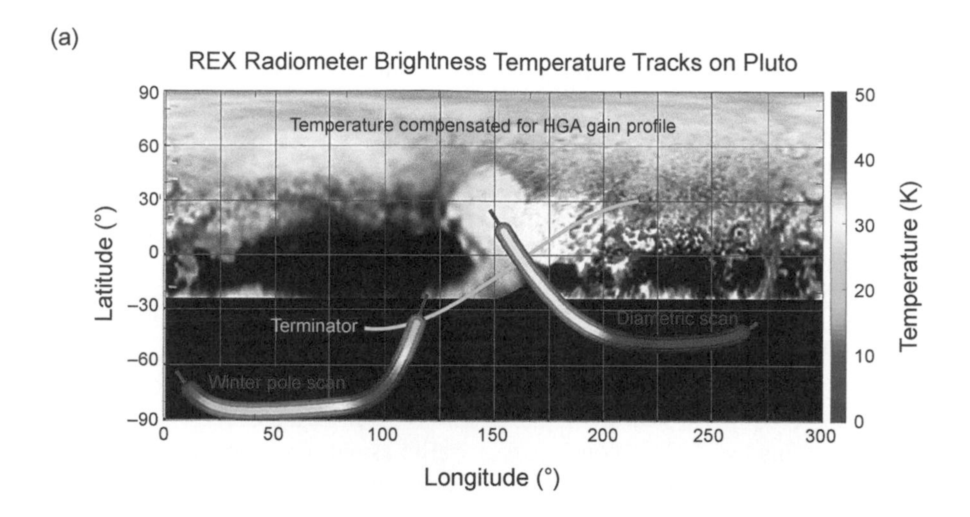

(b)

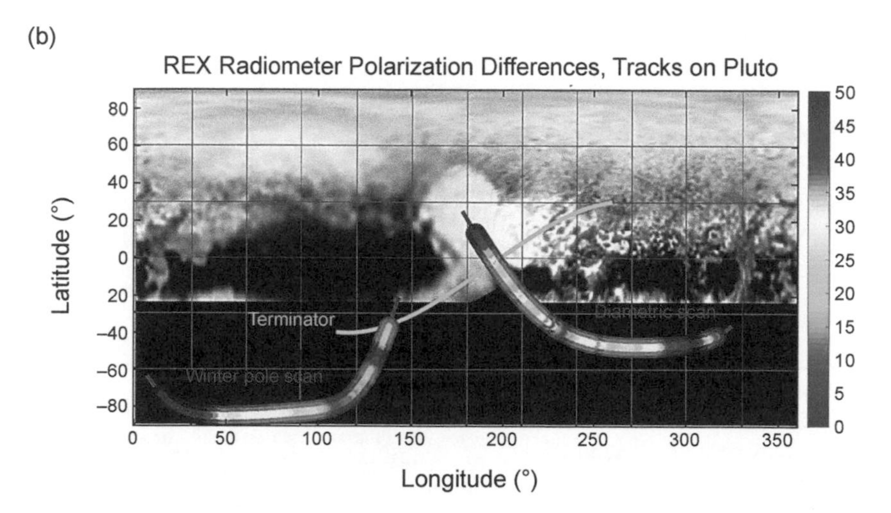

Fig. 19. (a) REX radio brightness temperatures. Both the diametric and polar scans are located on the longitude vs. latitude map of Pluto's surface. The brown stubs at the ends of the scans are where the HGA gain is very low and side-lobe irregularities result in unreliable temperatures. **(b)** Difference in brightness temperature between the RCP and LCP. The diametric scan is on the right, and the winter pole scan is on the left. The brightness temperature profiles of Fig. 18a are included for reference as the red dotted line (scale also in red). The approximate locations of the dayside and nightside limbs are included, as well as the day-to-night terminator. The two filled dots represent the 3-dB diameters of the HGA footprint on Pluto. The precision of the radiometric measurement is 0.07 K, indicated by the vertical brackets in the plots for the differences only. A test is included to assess if the polarization differences result from small time offsets between RCP and LCP. Each color is the polarization difference, overlaid on the plot, for time offsets, stepped by 0.01 s, from 0.0 s to 0.2 s. The variability at the limbs is likely due to scan-time offsets. However, the polarization differences in the interior of the scans are not so affected.

to the physical temperature only if the emissivity is unity; in Sputnik Planitia the discrepancy with the sublimation temperature of nitrogen ice is resolved if the emissivity is 0.73.

Although the microwave emissivity of nitrogen ice at its sublimation temperature and the atmospheric pressure of Pluto's surface are largely unknown, an emissivity value of 0.73 is perhaps consistent with an irregular icy surface of low microwave absorption. Although the nitrogen ice in Sputnik Planitia is not pure, with methane and trace amounts of water ice and CO, the mixture still could be transparent to microwaves. If relatively transparent, the skin depth (or pathlength) would be many times the 4.17-cm wavelength. If the pathlength is indeed that large, say tens or perhaps hundreds of wavelengths, then the thermal radiation originates from the glacier's interior, likely with the temperature of the deeper material. As Pluto's interior temperature would be expected to increase with depth, the emissivity would correspondingly be even smaller.

*6.2.2. **Pluto's winter pole.*** The radiometric brightness profile of the winter pole in Fig. 16a reveals an additional temperature dichotomy. Most of the winter pole profile lies within Pluto's unilluminated hemisphere and consequently out of sight of the optical and infrared imagers and compositional mapping. The portion of the winter pole scan on the dayside is consistent with the lower temperatures associated with the nitrogen-ice-rich regions found in the diametric scan. The small portion of the scan still in the visible but entering into Pluto's nightside covers an optically darker region where tholins are concentrated (*Protopapa et al.,* 2017). The highest brightness temperature in the winter pole scan is 40 K, found centered on the winter pole itself.

The seemingly enigmatic condition where Pluto's nightside is much warmer than its dayside, and where the polar region now in permanent shadow is the warmest of all, was anticipated by *Leyrat et al.* (2016, 2017), using a thermally forced model and large skin depths to model microwave emission. In contrast with the diametric scan, the winter pole scan's brightness temperature does decrease markedly toward the ends of the scan where the emission angle becomes 90°. This behavior is consistent with the general expectation for thermal emission (*Heiles and Drake,* 1963), indicative of an emissivity and material character substantially different from the emissivity and material properties within the diametric scan.

7. SUMMARY AND IMPLICATIONS FOR PLUTO'S ORIGIN AND EVOLUTION

Pluto's surface is seen to consist of a complex and changing mosaic of ices ranging in volatility over ~40 orders of magnitude at the planet's surface temperature, from the refractory H_2O to the hypervolatile N_2 (*Fray and Schmitt,* 2009). In this chapter we have focused on the composition of Pluto's surface and the identification and dynamics of the volatile components detected by spectroscopic techniques. However, with a mean density of 1.854 ± 0.006 g cm^{-3} (*McKinnon et al.,* 2017), the bulk composition of Pluto is dominated by a rocky component, giving an overall rock-ice mass distribution of rock/rock + ice = 0.655 ± 0.005. The high abundance of rock appears to be consistent with the view that Pluto and other dwarf planets as well as smaller Kuiper belt objects now orbiting beyond ~40 AU originated somewhat closer to the Sun, and were later perturbed outward by the outward migration of Neptune early in solar system history.

Pluto may or may not be fully differentiated, but in any event the crust supporting the topography and hosting the veneer of volatiles is composed of frozen H_2O. We have seen that H_2O ice is exposed in some regions of Pluto's surface either as original bedrock from which the volatiles have been stripped away or as a superficial deposit from one or more geologically recent cryovolcanic events. Apart from impact craters, basin-forming impacts, and crustal fracturing, all of Pluto's surface structures result from or have been modified by the inventory of volatiles. The sublimation-condensation cycles of the atmosphere, seasonal and secular redistribution of ices, and the deposition of solid aerosol particles formed in the atmosphere all have served to create Pluto's highly varied landscape and paint it with a range of colors. Chemical changes in surface ices induced by factors in the space environment have modified the texture, color, and composition of some of the structures, particularly the oldest ones. In some regions of the surface, cryovolcanic fluids, possibly carrying complex organic compounds from some unknown depth in the crust, appear to have filled low-lying topography and have built a few tall structural edifices.

The effects of the physical properties, evolution, and mobility of Pluto's volatiles are seen in every topographical feature and structural province identified and discussed in the chapters in this volume by Moore et al. and White et al., including layered blocks of H_2O ice floating in solid N_2, retreating scarps, and the layered walls of impact craters and fossae. Of special note is the nitrogen ice sheet in Sputnik Planitia, the inflow of N_2 from the surrounding highlands to the east, and the convective cells in motion across much of the basin.

The astonishing variety of the structures and compositional provinces of surface of Pluto revealed by New Horizons seems quite extraordinary but are likely a portent of future discoveries among the highly varied populations of dwarf planets and smaller bodies in the vast region of the solar system beyond Neptune.

REFERENCES

Altenhoff W. J. et al. (1988) First radio astronomical estimate of the temperature of Pluto. *Astron. Astrophys., 190,* L15–L17.

Bagenal F., Horányi M., McComas D. J., et al. (2016) Pluto's interaction with its space environment: Solar wind, energetic particles, and dust. *Science, 351,* 1282.

Bertrand T., Forget F., Grundy W., et al. (2018) The nitrogen cycles on Pluto over seasonal and astronomical timescale. *Icarus, 309,* 277–296.

Bertrand T., Forget F., Umurhan O. M., et al. (2019) The CH_4 cycles on Pluto over seasonal and astronomical timescales. *Icarus, 329,* 148–165.

Bird M. K., Linscott I. R., Tyler G. L., et al. (2019) Radio thermal emission from Pluto and Charon during the New Horizons encounter. *Icarus, 322,* 192–209.

Brassé C., Muñoz O., Coll P., and Raulin F. (2015) Optical constants of Titan aerosols and their tholins analogs: Experimental results and modeling/observational data. *Planet. Space Sci., 109,* 159–174.

Brown R. H. and Cruikshank D. P. (1997) Determination of the composition and state of icy surfaces in the outer solar system. *Annu. Rev. Earth Planet. Sci., 25,* 243–277.

Brown R. H., Cruikshank D. P., Veverka J., Helfenstein P., and Eluszkiewicz J. (1995) Surface composition and photometric properties of Triton. In *Neptune and Triton* (D. P. Cruikshank, ed.), pp. 991–1030. Univ. of Arizona, Tucson.

Buie M. W., Tholen D. J., and Horne K. (1992) Albedo maps of Pluto and Charon: Initial mutual event results. *Icarus, 97,* 211–227.

Buratti B. J., Hofgartner J. D., Hicks M. D., et al. (2017) Global albedos of Pluto and Charon from LORRI New Horizons observations. *Icarus, 287,* 207–217.

Burgdorf M. J., Cruikshank D. P., Dalle Ore C., et al. (2010) A tentative identification of HCN ice on Triton. *Astrophys. J. Lett., 718,* L53–L57.

Cable M. L., Hörst S. M., Hodyss R., Beauchamp P. M., Smith M. A., and Willis P. A. (2012) Titan tholins: Simulating Titan organic chemistry in the Cassini-Huygens era. *Chem. Rev., 112,* 1882–1909.

Cheng A. F., Summers M. E., Gladstone G. R., et al. (2017) Haze in Pluto's atmosphere. *Icarus, 290,* 112–133.

Clark R. N., Brown R. H., Cruikshank D. P., and Swayze G. A. (2019) Isotopic ratios of Saturn's rings and satellites: Implications for the origin of water and Phoebe. *Icarus, 321,* 791–802.

Cook J. C., Dalle Ore C. M., Protopapa S., et al. (2018) Composition of Pluto's small satellites: Analysis of New Horizons spectral images. *Icarus, 315,* 30–45.

Cook J. C., Dalle Ore C. M., Protopapa S., et al. (2019) The distribution of H_2O, CH_3OH, and hydrocarbon-ices on Pluto: Analysis of New Horizons spectral images. *Icarus, 331,* 148–169.

Cravens T. E. and Strobel D. F. (2015) Pluto's solar wind interaction:

Collisional effects. *Icarus, 246,* 303–309.

Cruikshank D. P. and Silvaggio P. M. (1980) The surface and atmosphere of Pluto. *Icarus, 41,* 96–102.

Cruikshank D. P., Pilcher C. B., and Morrison D. (1976) Pluto: Evidence for methane frost. *Science, 194,* 835–837.

Cruikshank D. P., Brown R. H., and Clark R. N. (1984) Nitrogen on Triton. *Icarus, 58,* 293–305.

Cruikshank D. P., Brown R. H., Tokunaga A. T., Smith R. G., and Piscitelli J. R. (1988) Volatiles on Triton: The infrared spectral evidence, 2.0–2.5 micrometers. *Icarus, 74,* 413–423.

Cruikshank D. P., Roush T. L., Owen T. C., et al. (1993) Ices on the surface of Triton. *Science, 261,* 742–745.

Cruikshank D. P., Imanaka H., and Dalle Ore C. M. (2005) Tholins as coloring agents on outer solar system bodies. *Adv. Space Res., 36,* 178–183.

Cruikshank D. P., Mason R. E., Dalle Ore C. M., et al. (2006) Ethane on Pluto and Triton. *Bull. Am. Astron. Soc., 38,* 518.

Cruikshank D. P., Grundy W. M., DeMeo F. E., et al. (2015) The surface compositions of Pluto and Charon. *Icarus, 246,* 82–92.

Cruikshank D. P., Clemett S. J., Grundy W. M., et al. (2016) Pluto and Charon: The non-ice surface component. *Lunar Planetary Science XLVII,* Abstract #1700. Lunar and Planetary Institute, Houston.

Cruikshank D. P., Moroz L. V., and Clark R. N. (2019a) Visible and infrared spectroscopy of ices, volatiles, and organics. In *Remote Compositional Analysis: Techniques for Understanding Spectroscopy, Mineralogy, and Geochemistry of Planetary Surfaces* (J. L. Bishop et al., eds.), pp. 102–119. Cambridge Univ., Cambridge.

Cruikshank D. P., Umurhan O. M., Beyer R. A., et al. (2019b) Recent cryovolcanism in Virgil Fossae on Pluto. *Icarus, 330,* 155–168.

Cruikshank D. P., Materese C. K., Pendleton Y. J., et al. (2019c) Prebiotic chemistry of Pluto. *Astrobiology, 17,* 7.

Dalle Ore C. M., Cruikshank D. P., Protopapa S., et al. (2019) Detection of ammonia on Pluto's surface in a region of geologically recent tectonism. *Sci. Adv., 5,* eaav5731.

DeMeo F. E., Dumas C., de Bergh C., et al. (2010) A search for ethane on Pluto and Triton. *Icarus, 208,* 412–424.

De Sanctis M. C., Ammannito E., Raponi A., et al. (2015) Ammoniated phyllosiliates with a likely outer solar system origin on Ceres. *Nature, 528,* 241–244.

De Sanctis M. C., Raponi A., Ammannito E., et al. (2016) Bright carbonate deposits as evidence of aqueous alteration on Ceres. *Nature, 536,* 54–58.

Douté S., Schmitt B., Quirico E., et al. (1999) Evidence for methane segregation at the surface of Pluto. *Icarus, 142,* 421–444.

Earle A. M., Grundy W., Howett C. J. A., et al. (2018) Methane distribution on Pluto as mapped by the New Horizons Ralph/MVIC instrument. *Icarus, 314,* 195–209.

Eluszkiewicz J. and Stevenson D. J. (1990) Rheology of solid methane and nitrogen: Applications to Triton. *Geophys. Res. Lett., 17,* 1753–1756.

Eluszkiewicz J., Leliwa-Kopystynski J., and Kossacki K. J. (1998) Metamorphism of solar system ices. In *Solar System Ices* (B. Schmitt et al., eds.), pp. 119–138. Kluwer, Boston.

Fabre D., Thiéry M. M., Vu H., and Kobashi K. (1979) Raman spectra of solid CH_4 under pressure. I. Phase transition between phases II and III. *J. Chem. Phys., 71,* 3081–3088.

Fray N. and Schmitt B. (2009) Sublimation of ices of astrophysical interest: A bibliographic review. *Planet. Space Sci., 57,* 2053–2080.

Gabasova L. R., Schmitt B., Grundy W. M., et al. (2021) Global compositional cartography of Pluto from intensity-based registration of LEISA data. *Icarus,* in press, DOI: 10.1016/j.icarus.2020.113833.

Gao P., Fan S., Wong M. L., et al. (2017) Constraints on the microphysics of Pluto's photochemical haze from New Horizons observations. *Icarus, 287,* 116–123.

Gladstone G. R., Stern S. A., Ennico K., et al. (2016) The atmosphere of Pluto as observed by New Horizons. *Science, 351,* aad8866.

Grundy W. M. and Schmitt B. (1998) The temperature-dependent near-infrared absorption spectrum of hexagonal H_2O ice. *J. Geophys. Res.–Planets, 103,* 25809–25822.

Grundy W. M., Schmitt B., and Quirico E. (1993) The temperature dependent spectra of α and β nitrogen ice with application to Triton. *Icarus, 105,* 254–258.

Grundy W. M., Schmitt B., and Quirico E. (2002) The temperature-dependent spectrum of methane ice I between 0.7 and 5 μm and opportunities for near-infrared remote thermometry. *Icarus, 155,* 486–496.

Grundy W. M., Morrison S. J., Bovyn M. J., Tegler S. C., and Cornelison D. M. (2011) Remote sensing D/H ratios in methane ice: Temperature-dependent absorption coefficients of CH_3D in methane ice and in nitrogen ice. *Icarus, 212,* 941–949.

Grundy W. M., Olkin C. B., Young L. A., et al. (2013) Near-infrared spectral monitoring of Pluto's ices: Spatial distribution and secular evolution. *Icarus, 223,* 710–721.

Grundy W. M., Olkin C. B., Young L. A., and Holler B. J. (2014) Near-infrared spectral monitoring of Pluto's ices II: Recent decline of CO and N_2 ice absorptions. *Icarus, 235,* 220–224.

Grundy W. M., Cruikshank D. P., Gladstone G. R., et al. (2016) Formation of Charon's red polar caps. *Nature, 539,* 65–68.

Grundy W., Bertrand T., Binzel R. P., et al. (2018) Pluto's haze as a geological material. *Icarus, 314,* 232–245.

Hamilton D. P., Stern S. A., Moore J. M., and Young L. A. (2016) The rapid formation of Sputnik Planitia early in Pluto's history. *Nature, 540,* 97–99.

Hapke B. (2012) *Theory of Reflectance and Emittance Spectroscopy, 2nd edition.* Cambridge Univ., Cambridge. 513 pp.

Heiles C. E. and Drake F. C. (1963) The polarization and intensity of thermal radiation from a planetary surface. *Icarus, 2,* 281–292.

Hinson D. P., Linscott I. R, Young L. A., et al. (2017) Radio occultation measurements of Pluto's neutral atmosphere with New Horizons. *Icarus, 290,* 96–211.

Holler B. J., Young L. A., Grundy W. M., Olkin C. B., and Cook J. C. (2014). Evidence for longitudinal variability of ethane ice on the surface of Pluto. *Icarus, 243,* 104–110.

Hörst S. M., Yoon Y. H., Ugelow M. S., et al. (2018) Laboratory investigations of Titan haze formation: *In situ* measurement of gas and particle composition. *Icarus, 301,* 136–151.

Imanaka H., Khare B. N., Elsila J. E., et al. (2004) Laboratory experiments of Titan tholin formed in cold plasma at various pressures: Implications for nitrogen-containing polycyclic aromatic compounds in Titan haze. *Icarus, 168,* 344–366.

Imanaka H., Cruikshank D. P., Khare B. N., and McKay C. P. (2012) Optical constants of laboratory synthesized complex organic materials: Part 1. Titan tholins at mid-infrared wavelengths (2.5–25 μm). *Icarus, 218,* 247–261.

Johnson P. E., Young L. A., Protopapa S., et al. (2021) Modeling Pluto's minimum pressure: Implications for haze production. *Icarus, 356,* 114070.

Kargel J. S. (1992) Ammonia-water volcanism on icy satellites: Phase relations at 1 atmosphere. *Icarus, 100,* 556–574.

Khare B. N., Sagan C., Zumberge J. E., Sklarew D. S., and Nagy B. (1981) Organic solids produced by electrical discharge in reducing atmospheres: Tholin molecular analysis. *Icarus, 48,* 290–297.

Khare B. N., Sagan C., Arakawa E. T., et al. (1984) Optical constants of organic tholins produced in a simulated Titanian atmosphere: From soft X-ray to microwave frequencies. *Icarus, 60,* 127–137.

Krasnopolsky V. A. and Cruikshank D. P. (1999) Photochemistry of Pluto's atmosphere and ionosphere near perihelion. *J. Geophys. Res., 104,* 21979–21996.

Kuiper G. P. (1950) The diameter of Pluto. *Publ. Astron. Soc. Pac., 62,* 133–137.

Kwok S. and Zhang Y. (2011) Mixed aromatic-aliphatic organic nanoparticles as carriers of unidentified infrared emission features. *Nature, 479,* 80–83, DOI: 10.1038/nature10542.

Lavvas P., Yelle R. V., Koskinen T., et al. (2013) Aerosol growth in Titan's ionosphere. *Proc. Natl. Acad. Sci. U.S.A., 110,* 2729–2734.

Lebreton J.-P., Coustenis A., Lunine J., Raulin F., Owen T., and Strobel D. (2009) Results from the Huygens probe on Titan. *Astronomy Astrophys. Rev., 17,* 149–179.

Legay F. and Legay-Sommaire N. (1982) Vibrational absorption spectrum of solid CO in the first harmonic region. Two-phonon transition. *Chem. Phys., 65,* 49–57.

Legay-Sommaire N. and Legay F. (1982) Analysis of the infrared emission and absorption spectra from isotopic CO molecules in solid α-CO. *Chem. Phys., 66,* 315–325.

Lellouch E., Gurwell M., Butler B., et al. (2017). Detection of CO and HCN in Pluto's atmosphere with ALMA. *Icarus, 286,* 289–307.

Lewis B. L., Stansberry J. A., Holler B. J., et al. (2021) Distribution and energy balance of Pluto's nitrogen ice, as seen by New Horizons in 2015. *Icarus, 356,* 113633.

Lewis J. S. (1972) Low-temperature condensation from the solar nebula. *Icarus, 16,* 241–252.

Leyrat C., Lorenz R. D., and Le Gall A. (2016) Probing Pluto's underworld: Ice temperatures from microwave radiometry decoupled from surface conditions. *Icarus, 268,* 50–55.
Leyrat C., Le Gall A., Lorenz R., and Boomi S. (2017) Predicted antenna temperatures measured by REX/New Horizons during the Pluto's flyby probing the sub-surface in microwave. *AAS/Division for Planetary Sciences Meeting Abstracts, 49,* #215.03.
Lunine J. I. and Stevenson D. J. (1985) Physical state of volatiles on the surface of Triton. *Nature, 317,* 238–240.
Luspay-Kuti A., Mandt K., Jessup K. L., et al. (2017) Photochemistry on Pluto — I. Hydrocarbons and aerosols. *Mon. Not. R. Astron. Soc., 472,* 104–117.
Masterson C. M. and Khanna R. K. (1990) Absorption intensities and complex refractive indices of crystalline HCN, HC_3N, and C_4N_2 in the infrared region. *Icarus, 83,* 83–92.
Materese C. K., Cruikshank D. P., Sandford S. A., et al. (2014) Ice chemistry on outer solar system bodies: Carboxylic acids, nitriles and urea detected in refractory residues produced from the UV-photolysis of N_2:CH_4:CO containing ices. *Astrophys. J., 788,* 111.
Materese C. K., Cruikshank D. P., Sandford S. A., et al. (2015) Ice chemistry on outer solar system bodies: Electron radiolysis of N_2-, CH_4-, and CO-containing ices. *Astrophys. J., 812,* 150.
McKinnon W. B., Nimmo F., Wong T., et al. (2016a) Thermal convection in solid nitrogen, and the depth and surface age of cellular terrain within Sputnik Planum, Pluto. *Lunar and Planetary Science XLVII,* Abstract #2921. Lunar and Planetary Institute, Houston.
McKinnon W.B., Nimmo F., Wong T., et al. (2016b) Convection in a volatile nitrogen-ice-rich layer drives Pluto's geological vigour. *Nature, 534,* 82–85.
McKinnon W. B., Stern S. A., Weaver H. A., et al. (2017) Origin of the Pluto-Charon system: Constraints from the New Horizons flyby. *Icarus, 287,* 2–11.
Merlin F., Lellouch E., Quirico E., and Schmitt B. (2018) Triton's surface ices: Distribution, temperature, and mixing state from VLT/SINFONI observations. *Icarus, 314,* 274–293.
Moore J. M., Howard A. D., Umurhan O. M., et al. (2018) Bladed terrain on Pluto: Possible origins and evolution. *Icarus, 300,* 129–144.
Moore M. H., Ferrante R. F., Hudson R. L., and Stone J. N. (2007) Ammonia-water ice laboratory studies relevant to outer solar system surfaces. *Icarus, 190,* 260–273.
Nakamura R., Shinji S., Masateru I., et al. (2000) Subaru infrared spectroscopy of the Pluto-Charon system. *Publ. Astron. Soc. Japan, 52,* 551–556.
Neveu M., Desch S. J, Shock E. L., and Glein C. R. (2015) Prerequisites for explosive cryovolcanism on dwarf planet-class Kuiper belt objects. *Icarus, 246,* 48–64.
Neveu M., Desch S. J., and Castillo-Rogez J. C. (2017) Aqueous geochemistry in icy world interiors: Equilibrium fluid, rock, and gas compositions, and fate of antifreezes and radionuclides. *Geochem. Cosmochim. Acta, 212,* 324–371.
Olkin C. B., Young E. F., Young L. A., et al. (2007) Pluto's spectrum from 1.0–4.2 μm: Implications for surface properties. *Astron. J., 133,* 420–431.
Olkin C. B., Spencer J. R., Grundy W. M., et al. (2017) The global color of Pluto from New Horizons. *Astron. J., 154,* 258.
Owen T. C., Roush T. L., Cruikshank D. P., et al. (1993) Surface ices and atmospheric composition of Pluto. *Science, 261,* 745–748.
Prokhvatilov A. I. and Yantsevich L. D. (1983) X-ray investigation of the equilibrium phase diagram of CH_4-N_2 solid mixtures. *Low Temp. Phys., 9,* 94–98.
Protopapa S., Boehnhardt H., Herbst T. M., et al. (2008) Surface characterization of Pluto and Charon by L and M band spectra. *Astron. Astrophys., 490,* 365–375.
Protopapa S., Grundy W. M., Tegler S. C., and Bergonio J. M. (2015) Absorption coefficients of the methane-nitrogen binary ice system: Implications for Pluto. *Icarus, 253,* 179–188.
Protopapa S., Grundy W. M., Reuter D. C., et al. (2017) Pluto's global surface composition through pixel-by-pixel Hapke modeling of New Horizons Ralph/LEISA data. *Icarus, 287,* 218–228.
Protopapa S., Olkin C., Grundy W. M., et al. (2020) Disk-resolved photometric properties of Pluto and the coloring materials across its surface. *Astron.J., 159,* 74.
Quirico E. and Schmitt B. (1997a) Near-infrared spectroscopy of simple hydrocarbons and carbon oxides diluted in solid N_2 and as pure ices: Implications for Triton and Pluto. *Icarus, 127,* 354–378.
Quirico E. and Schmitt B. (1997b) A spectroscopic study of CO diluted in N_2 ice: Applications for Triton and Pluto. *Icarus, 128,* 181–188.
Quirico E., Schmitt B., Bini R., and Salvi P. R. (1996) Spectroscopy of some ices of astrophysical interest: SO_2, N_2 and N_2:CH_4 mixtures. *Planet. Space Sci., 44,* 973–986.
Reuter D. C., Stern S. A., Scherrer J., et al. (2008) Ralph: A visible/infrared imager for the New Horizons Pluto/Kuiper belt mission. *Space Sci. Rev., 140,* 129–154.
Sagan C. and Khare B. N. (1979) Tholins: Organic chemistry of interstellar grains and gas. *Nature, 277,* 102–107.
Sagan C., Khare B. N., and Lewis J. S. (1984) Organic matter in the Saturn system. In *Saturn* (T. Gehrels and M. S. Matthews, eds.), pp. 788–807. Univ. of Arizona, Tucson.
Sasaki T., Kanno A., Ishiguro M., et al. (2005) Search for nonmethane hydrocarbons on Pluto. *Astrophys. J. Lett., 618,* L57–L60.
Schenk P. M., Beyer R. A., McKinnon W. B., et al. (2018) Basins, fractures and volcanoes: Global cartography and topography of Pluto from New Horizons. *Icarus, 314,* 400–433.
Schmitt B., Philippe S., Grundy W. M., et al. (2017) Physical state and distribution of materials at the surface of Pluto from New Horizons LEISA imaging spectrometer. *Icarus, 287,* 229–260.
Schmitt B., Gabasova L., Sylvain P., et al. (2018) Evidence of local CH_4 stratification on Pluto from LEISA data and a complete N_2 ice map. *AAS/Division for Planetary Sciences Meeting Abstracts, 50,* #506.02.
Schmitt B., Gabasova L., Bertrand T., et al. (2019) Methane stratification on Pluto inferred from New Horizons LEISA data. *Pluto System After New Horizons,* Abstract #7004. LPI Contribution No. 2133, Lunar and Planetary Institute, Houston.
Scott T. A. (1976) Solid and liquid nitrogen. *Phys. Rept., 27,* 89–157.
Shapiro M. M. and Gush H. P. (1966) The collision-induced fundamental and first overtone bands of oxygen and nitrogen. *Can. J. Phys. 44,* 949–963.
Soifer B. T., Neugebauer G., and Matthews K. (1980) The 1.5–2.5 microns spectrum of Pluto. *Astron J., 85,* 166–167.
Spencer J. R., Stansberry J. A., Trafton L. M., et al. (1997) Volatile transport, seasonal cycles, and atmospheric dynamics on Pluto. In *Pluto and Charon* (S. A. Stern and D. J. Tholen, eds.), pp. 435–473. Univ. of Arizona, Tucson.
Spencer J. R., Stern S. A., Olkin C., et al. (2016) The colors of Pluto: Clues to its geological evolution and surface/atmospheric interactions. Abstract P54A-01 presented at 2016 Fall Meeting, AGU, San Francisco, California, 12–16 December.
Stansberry J. A., Spencer J. R., Schmitt B., et al. (1996) A model for the overabundance of methane in the atmospheres of Pluto and Triton. *Planet. Space Sci., 44,* 1051–1063.
Steffl A. J., Young L. A., Strobel D. F., et al. (2020) Pluto's ultraviolet spectrum, surface reflectance, and airglow emissions. *Astron. J., 159,* 224.
Stern S. A., Bagenal F., Ennico K., et al. (2015) The Pluto system: Initial results of its exploration by New Horizons. *Science, 350(6258),* aad1815.
Strobel D. F. and Zhu X. (2017) Comparative planetary nitrogen atmospheres: Density and thermal structures of Pluto and Triton. *Icarus, 291,* 55–64.
Tan S.P. and Kargel J. S. (2018) Solid-phase equilibria on Pluto's surface. *Mon. Not. R. Astron. Soc., 474,* 4254–4263.
Tegler S. C., Cornelison D. M., Grundy W. M., et al. (2010) Methane and nitrogen abundances on Pluto and Eris. *Astrophys. J., 725,* 1296–1305.
Tegler S. C., Grundy W. M., Olkin C. B., et al. (2012) Ice mineralogy across and into the surfaces of Pluto, Triton, and Eris. *Astrophys. J., 751(1),* 76.
Tegler S. C., Stufflebeam T. D., Grundy W. M., et al. (2019) A new two-molecule combination band as a diagnostic of carbon monoxide diluted in nitrogen ice on Triton. *Astron. J., 158(1),* 17.
Thompson W. R., Murray B. G. J. P. T., Khare B. N., and Sagan C. (1987) Coloration and darkening of methane clathrate and other ices by charged particle irradiation: Applications to the outer solar system. *J. Geophys. Res., 92,* 14933–14947.
Tombaugh C. W. and Moore P. (1980) *Out of the Darkness: The Planet Pluto.* Stackpole Books, Lanham. 222 pp.
Trowbridge A. J., Melosh H. J., Steckloff J. K., and Freed A. M. (2016) Vigorous convection as the explanation for Pluto's polygonal terrain. *Nature, 534,* 79–81.
Trafton L. A. (2015) On the state of methane and nitrogen ice on Pluto

and Triton: Implications of the binary phase diagram. *Icarus, 246,* 197–205.
Tryka K. A., Brown R. H., Anicich V., Cruikshank D. P., and Owen T. C. (1993) Spectroscopic determination of the phase composition and temperature of nitrogen on Triton. *Science, 261,* 751–754.
Tryka K. A., Brown R. H., Cruikshank D. P., et al. (1994) Temperature of nitrogen ice on Pluto and its implications for flux measurements. *Icarus, 112*, 513–527.
Tyler G. L., Linscott I. R., Bird M. K., et al. (2008) The New Horizons Radio Science Experiment (REX). *Space Sci. Rev., 140,* 217–259.
Uras N. and Devlin J. P. (2000) Rate study of ice particle conversion to ammonia hemihydrate: Hydrate crust nucleation and NH_3 diffusion. *J. Phys. Chem., 104,* 5770–5777.
Verbiscer A. J., Showalter M. R., Buie M. W., and Helfenstein P. (2019) The Pluto System at true opposition. Pluto System After New Horizons, Abstract #7050. LPI Contribution No. 2133, Lunar and Planetary Institute, Houston.
Vetter M., Jodl H. J., and Brodyanski A. (2007) From optical spectra to phase diagrams — The binary mixture N_2-CO. *Low Temp. Phys., 33,* 1052–1060.
Vuitton V., Lavvas P., Yelle R. V., et al. (2009) Negative ion chemistry in Titan's upper atmosphere. *Planet. Space Sci., 57,* 1558–1571.
Walker M. F. and Hardie R. H. (1955) A photometric determination of the rotational period of Pluto. *Publ. Astron. Soc. Pac., 67,* 224–231.
Weaver H. A., Buie M. W., Buratti B. J., et al. (2016) The small satellites of Pluto as observed by New Horizons. *Science, 351(6279),* aae0030.
White O. L., Moore J. M., McKinnon W. B., et al. (2017) Geological mapping of Sputnik Planitia on Pluto. *Icarus, 287,* 261–286.
Wong M. L., Fan S., Gao P. et al. (2017) The photochemistry of Pluto's atmosphere as illuminated by New Horizons. *Icarus, 287,* 110–115.
Young E. F. and Binzel R. P. (1993) Comparative mapping of Pluto's sub-Charon hemisphere: Three least squares models based on mutual event lightcurves. *Icarus, 102,* 134–149.
Young L. A. (1994) Bulk properties and atmospheric structure of Pluto and Charon. Ph.D. thesis, Massachusetts Institute of Technology, available online at *https://core.ac.uk/download/pdf/4394723.pdf.*
Young L. A., Elliot J. L., Tokunaga A., et al. (1997) Detection of gaseous methane on Pluto. *Icarus, 127,* 258–262.
Young L. A., Kammer J. A., Steffl A. J., et al. (2018) Structure and composition of Pluto's atmosphere from the New Horizons solar ultraviolet occultation. *Icarus, 300,* 174–199.
Zent A. P., McKay C. P., Pollack J. B., and Cruikshank D. P. (1989) Grain metamorphism in polar nitrogen ice on Triton. *Geophys. Res. Lett., 16,* 965–968.
Zhang X, Strobel D. F., and Imanaka H. (2017) Haze heats Pluto's atmosphere yet explains its cold temperature. *Nature, 551,* 352–355.

Umurhan O. M., Ahrens C. J., and Chevrier V. F. (2021) Rheological and thermophysical properties and some processes involving common volatile materials found on Pluto's surface. In *The Pluto System After New Horizons* (S. A. Stern, J. M. Moore, W. M. Grundy, L. A. Young, and R. P. Binzel, eds.), pp. 195–255. Univ. of Arizona, Tucson, DOI: 10.2458/azu_uapress_9780816540945-ch010.

Rheological and Thermophysical Properties and Some Processes Involving Common Volatile Materials Found on Pluto's Surface

O. M. Umurhan
SETI Institute at NASA Ames Research Center

C. J. Ahrens and V. F. Chevrier
Arkansas Center for Space and Planetary Sciences, University of Arkansas

Pluto's surface is composed of several different icy volatile materials constituting a diversity of unique peculiar landforms as revealed by New Horizons. This chapter reviews the laboratory measured thermophysical and rheological properties of these non-H_2O ices — which include $N_2$and CH_4, among others — in the temperature and pressure ranges of relevance to the volatile ice harboring surface environment of Pluto and other icy bodies of the outer solar system. This review also offers a primer to some of the physical processes likely shaping the appearance of the observed geological structures and, furthermore, relates the process inputs to the currently known and unknown values of these thermophysical and rheological quantities. We highlight avenues for future laboratory work. Thus, this chapter serves as an encyclopedia for volatile ice modelers and experimentalists alike.

1. INTRODUCTION

Planetary bodies of the outer solar system, including satellites of the gas giants, dwarf planets, comets and Kuiper belt objects (KBOs), harbor surface ices made of volatile molecules like N_2, CO, CH_4, C_2H_6, clathrates, and various hydrocarbon complexes. Imaging of these surfaces, brought to us by spacecraft like New Horizons, Cassini, Rosetta, and others, reveal complex and often-mysterious geological structures composed of these materials and shaped by processes involving all three states of matter. This chapter focuses on the surface volatile-ice-rich geological structures observed on Pluto by New Horizons with an eye toward their realistic modeling. The structures of interest include Sputnik Planitia, the bladed terrain, and the observed ongoing N_2 glacial flow on the eastern slopes of Tombaugh Regio as well as on landscapes exhibiting evidence of past glaciation (*Howard et al.,* 2017a; *Moore et al.,* 2018; see also the chapter by Moore and Howard in this volume).

A major step toward explaining the nature of these bodies' surface geologies and establishing their correct evolutionary timescales requires accurate knowledge of the thermophysical and rheological properties of their composite materials. Rheological information of these volatile ices, like strength and dynamic deformation, currently rests on laboratory derived empirical relationships, which are, in turn, essential for accurate physical modeling of these surfaces using state of the art landform evolution modeling. Laboratory research offers unique components to help answer the interdisciplinary nature of planetary sciences, some of which are listed here:

Planetary Science Questions	Laboratory Measurements/ Properties
What is the composition and structure of the subsurface material?	Viscoelastic, density, dielectric
What is the composition and structure of the surface material?	Strain rates, density, thermodynamic phase transitions
What is the interior structure?	Conductivity, specific heat, viscoelastic properties
What are the power processes driving geological evolution?	Diffusivity, conductivity, phase transitions
What are the surface-atmosphere interactions?	Phase stability, saturation pressure
What are the radiative processes and surface evolution over short timescales?	Heat capacity, saturation pressures, phase stability, latent heat

While the search for answers to the above questions is certainly continuous, only some of the necessary laboratory measurements have been accomplished, which we discuss later in this chapter. With a particular emphasis placed on Pluto — but certainly not restricted to it only — the purpose of this review is twofold. First, to provide a fundamental primer on some of the geological processes probably at play in creating the observed morphologies and chemical signatures of cold ($T < 100$ K) icy surfaces, including identifying the input physical quantities that go into such physical models. Second, to present a current up-to-date compilation of these temperature- dependent input rheological and thermophysical quantities, such as density, heat capacity, coefficients of thermal expansion, conductivity, latent energies of fusion and sublimation, chemical diffusion coefficients, vapor-pressure equilibrium curves, elastic and Young's moduli, grain-size-dependent viscosity and elastic-to-ductile-to-brittle transition lines. In most cases, this information is relatively more complete for these molecules in the pure state while relatively little is currently known about these properties (especially the rheological ones) when these volatiles exist as mixtures and alloys, which we know to be the case on these cold outer solar system bodies. Acquiring this data for alloys and mixtures is an essential frontier for planetary science as an increasing number of NASA and European Space Agency (ESA) missions head into the outer solar system in the next 2–3 decades.

2. RELEVANT PHYSICAL PROCESSES

2.1. A Survey of Observed Landforms of Interest

We share here a few illustrative example surface features observed on planetary bodies in the outer solar system with a particular emphasis on Pluto. In speaking about surface geologic constructs, it is common to refer to the deformation (deviatoric) stresses experienced by a structure (mountain, pits, valleys, etc.), gauged by estimating both the typical depth of the structure H and its surface elevation slope tan θ. The applied basal-stress is then estimated by $\tau_b = \rho g H \tan\theta$ where the ice density is ρ and the surface gravity is g (more formally discussed in section 2.2).

2.1.1. Glacial flow of volatile ices. Along the eastern shoreline of Pluto's Sputnik Planitia (surface $g \sim 0.642$ m s^{-2}, during New Horizons flyby: surface T = 37 K, surface P = 1 Pa) lie several examples of recent glacial flow from the highlands of Eastern Tombaugh Regio (Fig. 1). The flats of Sputnik Planitia, some 2 km below Pluto's mean radius, exhibit flow-lobe features indicating the farthest extent of glacial debris deposition. The darkened debris can be traced into the eastern highlands, indicating that the flows originated much higher up and traveled down through several networked channel valleys. These channel valleys extend 50–100 km in length, indicating a typical down-gradient angle of about $\theta \sim 3°–4°$, and are anywhere from 50 to 400 m deep. Spectroscopic measurements indicate that Sputnik Planitia contains primarily N_2 ice ($\rho \approx 0.95$ g cm^{-3}), with a small fraction (5%) of CH_4 ice ($\rho \approx 0.4$ g cm^{-3}) and trace amounts (<1%) of CO ice ($\rho \approx 0.95$ g cm^{-3}), while the highlands of Eastern Tombaugh Regio predominantly contain CH_4 ice with a small smattering of dissolved N_2 ice at similar fractions (*Schmitt et al.,* 2017; *Grundy et al.,* 2016). We note that observations of NH_3 appear to be restricted to various landforms found on Cthulu Macula, including regions in and around Virgil Fossae and others (*Cruikshank et al.,* 2019).

Of the many important questions, perhaps most critical for understanding Pluto's geologic evolution is that regarding the timescales on which this flow occurred (see the

Fig. 1. A section of Pluto displaying evidence of glacial flow. The map shows relative elevation. Inset shows undoctored figure. Yellow arrows indicate location furthest extent of glacial flow. White arrows indicate possible rafting of H_2O ice blocks. From *Umurhan et al.* (2017) and *Howard et al.* (2017b).

chapter by Moore and Howard in this volume). In order to answer this question, one needs information about how readily a layer of volatile ice — potentially N_2, or mixtures of binary N_2:CH_4, or ternary N_2:CH_4:CO — creeps down such chutes. Essential information required to determine the rate of creep includes knowledge of the material's rheology and an estimate of the basal stresses. The corresponding basal stresses experienced on one of these typical channels filled with some kind of N_2-CH_4 mixture (or alloy) is in the range of $\rho g \tan\theta \approx 1$–10 kPa (0.01–0.1 atm). However, laboratory data pertaining to the ice rheology N_2 or CH_4 are restricted to these materials in their pure state only and do not extend down to such small stresses (*Eluszkiewicz and Stevenson,* 1990; *Yamashita et al.,* 2010; more on these studies below). The behavior of NH_3-H_2O slurries, which are relevant to cryovolcanic processes occurring at temperatures ~180 K, are a helpful guide in conceiving how such processes may play out for N_2/CH_4 mixtures. For associated laboratory work and modeling, see extensive work by *Croft et al.* (1988) and *Kargel et al.* (1991). For more on the subject of NH_3-H_2O slurries and its role in related phenomena like cryovolcanism, see the chapter by Singer et al. in this volume.

2.1.2. Sputnik Planitia (Pluto). The western lobe of the heart-shaped region dubbed Tombaugh Regio, Sputnik Planitia, is a large high-albedo basin (approximately 8.75×10^5 km^2) set about 2 km beneath Pluto's mean radius, and is roughly located between 150°E and 200°E and 20°S and 45°N (*Stern et al.,* 2015; *Moore et al.,* 2016; *White et al.,* 2017). It consists mostly of N_2 ice with trace elements of CO and CH_4 according to their permissible stable mixing ratios (*Grundy et al.,* 2016; *Schmitt et al.,* 2017; *Protopapa et al.,* 2017). The region contains possibly-floating angular blocks of water ice dozens of kilometers across, plains with 1–2-km-scale pits with depths from 50 to 200 m, and most importantly, regions of bright and dark cellular plains separated by troughs roughly 50–100 m-deep (see the chapter by White et al. in this volume). The cell patterns have typical horizontal scales ranging from 20 to 35 km and are reminiscent of buoyant convection (*Moore et al.,* 2016; *McKinnon et al.,* 2016). Owing to its low elevation and overall large expanse (~10% of Pluto's surface landmass), the N_2 it contains controls the temperature and surface pressure of Pluto's atmosphere, causing it to be in vapor-pressure equilibrium (*Young,* 2017; *Young et al.,* 2018; *Bertrand et al.,* 2018). Why is there a textural/shading variation across the cellular plains, and might this have something to do with liquids flowing over its surface (*Stern et al.,* 2017)?

2.1.3. Bladed terrain (Pluto). East of Pluto's Tombaugh Regio — in regions generally located in a latitudinal band between ±25° — sit deep structures (500 m) composed of methane, with relatively steep grades $\theta \sim 10°$–$20°$ (Fig. 2) that exhibit a predominantly northerly orientation. How this bladed terrain came into being remains a puzzle, but current thinking suggests that these structures look like the product of sublimation-shaped processing, much like "penitentes" and "suncups" seen on H_2O ices in very cold and dry climes like the Atacama Desert (*Moore et al.,* 2016, 2017; *Moores et al.,* 2017). Adding to the mystery is the question of why they appear only on parts of Pluto that are at least 1–1.5 km above its mean radius. The answer to this question probably lies in a combination of atmospheric structure and near-surface flow coupled with sensitivities in the vapor-pressure-equilibrium curves of both N_2 and CH_4 (*Bertrand et al.,* 2019). With basal deviatoric stresses on the order of 20–50 kPa (0.2–0.5 atm), how long can such a structure

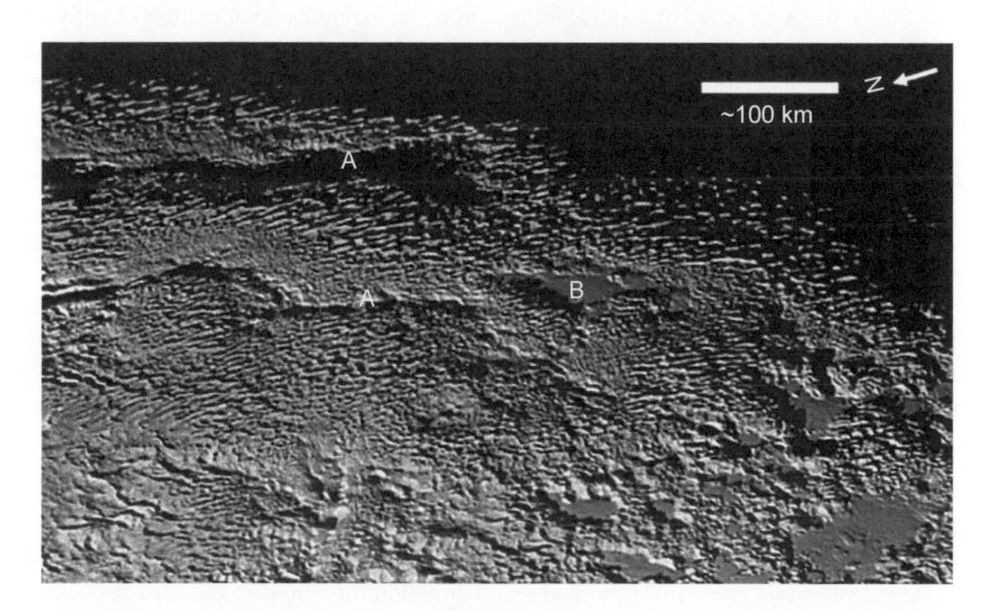

Fig. 2. Bladed terrain on Pluto. Image taken with New Horizons. The patterned CH_4 covered regions are slightly red and showcase elongated ridging often times abutting deep fracture/faults (labeled "A"). Smooth patches of N_2 ice are found in various low-lying regions (labeled "B").

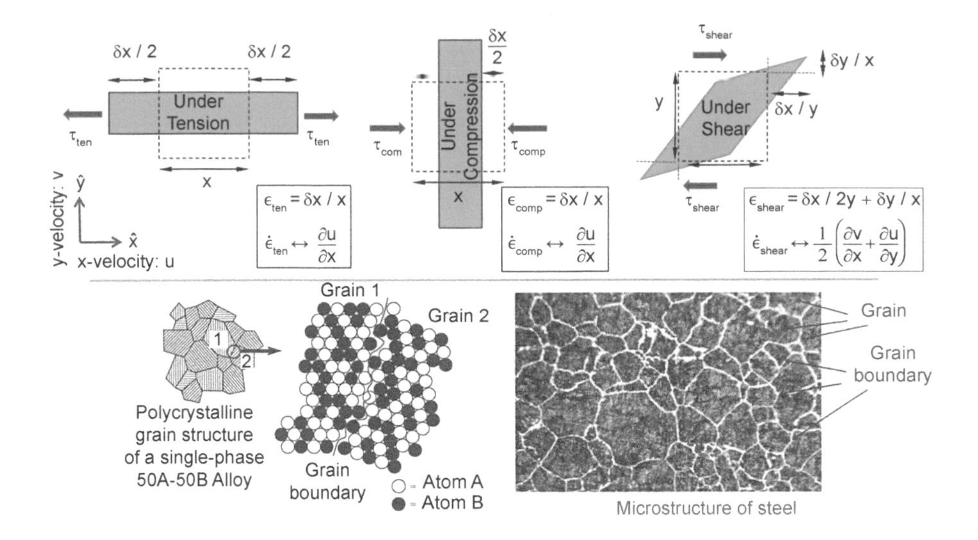

Fig. 3. Diagrams on the top row illustrate each of the three kinds of stresses possible on a block of ice: tensile, compressive, and shear. The corresponding strain ϵ (defined as the differential change in the length of the body divided by the length) is labeled according to the type of stress applied, τ. In general, each type of applied stress will give rise to differing strain deformations — although in modeling considerations they are equated. Strain rates $\dot{\epsilon}$ translate to velocity gradients. Bottom panels illustrate the nature of polycrystalline ices. Each grain will have a preferred symmetry axis, but these will vary in orientation from grain to grain. The lines within indicate basal slip plane orientations, showing the directions along which dislocations preferentially glide. Grain motion can often occur along grain boundaries. The bottom right shows the microstructure of steel, an alloyed polycrystalline material. Used under image license from Creative Commons(Wikipedia, CC BY 2.0), *https://creativecommons.org/licences/by/2.0/legalcode*.

survive? The answer to this requires assessing the strain rate of CH_4 under such stresses, either as a pure ice or as an alloy with N_2, as discussed further in section 2.4.3.

2.2. Rheology Fundamentals for Polycrystalline Ices

From a modeling standpoint, as pertaining to planetary conditions, ice blocks are either composed of a single large crystal lattice, an amorphous ice, or as a polycrystalline ice (i.e., a collection of grains that individually are mini-crystal lattice blocks with given crystal lattice orientation; see Fig. 3). In the context of planetary ices, where the scales of interest are large, it is generally assumed that ice complexes are polycrystalline and we assume this to be characteristic of volatile ices of the outer solar system throughout this review, keeping in mind that future observations may invalidate this assumption. Generally speaking, a block of polycrystalline ice subject to an amount of applied stress (whether it be compressive, tensile, or shear) will respond with (1) ductile (or plastic) flow or will experience brittle fracture failure when the applied stress exceeds the material's yield stress, or (2) elastic motions when the applied stress is below the yield stress. The latter instance means that the material will deform from its equilibrium configuration according to the applied stresses, but it will return to this equilibrium when the applied stress is removed. This physical picture of an elastic medium may be envisioned as a complex of spring-connected masses — with restoring force proportional to the local curvature of the deformed configuration — collectively responding to some applied stress (*Proulx,* 2011). A fundamental key feature of the elastic regime is that in the absence of cracks and other forms of dissipation, all motions are reversible.

Ductile flow results in the deformation (i.e., strain, ε) at a rate (i.e., strain rate, ε) that depends upon several factors, including the temperature (T) of the ice, the typical or effective size of the grains (d) comprising the block, and perhaps other factors. Under brittle failure, which tends to occur if the applied stress or driven strain rates are high, the specimen will rapidly develop cracks and fissures that spread through the block section and result in structural disintegration (*Petrenko and Whitworth,* 1999). We will focus mainly on the ductile flow regime of solids, but knowledge of the brittle limits of a polycrystalline ice is also important and is usually determined from experimental techniques (see further explanation below).

2.2.1. Strain, strain rate, and stress: Some fundamentals. Strain is a relative measure of how much a body of a certain size (say, L) stretches or deforms (δλ). In its basic abstract form strain is defined as $\varepsilon \equiv \delta\lambda/L$, while the strain rate is defined as its instantaneous time derivative, i.e., $\dot{\epsilon} \equiv d\varepsilon/dt$. The strain and strain rate of a body are further distinguished based on its deformation in a given direction measured relative to the body's size in each of its three dimensions. This means that strain and strain rate are deformational tensor quantities, and should each be written in the more generalized form $\varepsilon_{ij}, \dot{\varepsilon}_{ij}$, where the subscripts i, j = 1,2,3 designates the unit normal vector directions $\hat{x}$, $\hat{y}$, $\hat{z}$ When the indices are identical the response is compressional, e.g., $\varepsilon_{11} \leftrightarrow \delta x/x$ measures the relative strain of the sample of length x that deforms an amount δx. If the strain is purely in one direction like the previous example, then when its sign is positive it is referred to as being in tension, while a negative sign indicates compression. On physical and geometric grounds, both ε_{ij} and $\dot{\varepsilon}_{ij}$ must be symmetric tensors (see more information below). Thus, when the indices are different, the resulting strain and strain rate has shear-like characteristics and is correspondingly defined as $\varepsilon_{ij} \equiv (1/2)\ (\delta x_i/x_j + \delta x_j/x_i)$, where deformations ($\delta x_i$) and sizes/scales ($x_j$) are measured in the normal direction indicated by the subscripted index. [In more general terms, disturbances are thought of as being a sum of local deformation and local rotation. In this case, the total strain would normally be written as a distortion tensor, $\varepsilon_{ij} \equiv \epsilon_{ij} + (1/2)\quad (\delta x_i/x_j – \delta x_j/x_i)$, where ϵ_{ij} (the deformation tensor) captures strain while the remainder quantifies local rotation (e.g., *Landau and Lifschitz,* 1987; *Regev et al.,* 2016). The latter terms are not considered relevant for surface geologic materials. Internal (unresolved) rotational distortions are appropriate for biological materials and polymers that exhibit such interior degrees of freedom. Additionally, the deformation tensor should contain an expression representing volume contraction, but geologic materials are by and large treated as incompressible.] For example, for i = 1, j = 2, the strain is written as $\epsilon_{12} \equiv (1/2)\quad (\delta x/y + \delta y/x)$, which ensures that $\epsilon_{12} = \epsilon_{21}$. When the externally applied stresses force the material into the ductile regime and when *approaching the continuum limit*, the strain rate is expressed in terms of derivatives of fluid velocities. Thus, for example

$$\frac{d\epsilon_{11}}{dt} \leftrightarrow \frac{du}{dx},\ \frac{d\epsilon_{22}}{dt} \leftrightarrow \frac{dv}{dy},\ \frac{d\epsilon_{12}}{dt} \leftrightarrow \frac{1}{2}\left(\frac{du}{dy} + \frac{dv}{dx}\right)$$

and so on, where u and v denote the fluid velocity in the $\hat{\mathbf{x}}$ and $\hat{\mathbf{y}}$ directions, respectively.

In geophysical settings and primarily owing to the relative stiffness of the underlying materials — whether they be particles held together by loose cohesion of blocks of polycrystalline ices — a "material element" is always in exact total force balance with no dynamical inertia, i.e., the (mass) × (acceleration) terms are absent from the equations of motion. The total forces acting upon the material element will be the sum of two types of forces: (1) all long-range body forces (e.g., gravity) denoted with $\mathbf{F}_b$, and (2) short-range intermolecular "internal" forces that hold the solid or liquid together, such as lattice potentials in a crystal (*Turcotte and Schubert,* 2002, *Regev et al.,* 2016). [Like the Dude's "rug" that "holds the room together" (*The Big Lebowski*, Working Title Films, Universal Pictures, 1998).] Stress is a measure of the internal forces and is thought of in terms of a force per unit area (a pressure). To see why, consider the following: A material element will have bounding surface and if the element is small enough the bounding surfaces can be thought of as a collection of intersecting planes forming the shape of, for example, a tetrahedron. Choosing one of these flat surfaces with some outwardly pointing unit normal (say $\hat{\mathbf{n}}$ **)**, one speaks of the internal forces as having a vector with components along the surface normal as well as in the plane of the surface (i.e., perpendicular to the normal). In this sense, the forces are known as normal and sheared with respect to the surface with unit normal $\hat{\mathbf{n}}$. Typically, the forces are expressed as a force per unit volume and written as $f = f_i\hat{x}_i$; i = 1,2,3 where the unit normal vectors cover the three mutually orthogonal spatial directions. This force may be written as the divergence of a rank 2 tensor, i.e., $\sigma \leftrightarrow \sigma_{ij}$, where $\mathbf{f_i} = \Sigma_j(\delta\partial_{ij}/\partial x_j)$. [This relation is a consequence of the Cauchy integral theorem, i.e., any scalar may be written as the divergence of a vector quantity. The expression $\nabla \cdot \sigma$ is shorthand for $\Sigma_j(\partial\sigma_{ij}/\partial x_j)$.] It is customary to write the total stresses as a sum of a deviatoric stress tensor, τ, minus a term expressing the isotropic effect of a hydrostatic pressure, P, i.e., $\sigma_{ij} \to -P\delta_{ij} + \tau_{ij}$. The deviatoric term captures all the forces that hold the material element together and is the main cause of body distortion, while pressure — having its familiar fluid dynamical meaning — acts in the opposite sense, causing changes in volume. On the assumption that there are no net internal torques anywhere within the material, it follows that the deviatoric stress must be a symmetric tensor like the strain and strain rate. [This simplification is based on the assumption of microscale isotropy, which is not valid for complex materials like polymers in which orientation matter. The assumption of isotropy that leads to the symmetry of the stress tensor is sometimes known as Cauchy's second law (see also *Landau and Lifschitz,* 1986). On the assumption of material isotropy and total force balance we have Cauchy's first law of motion, which states that

$$0 = -\nabla P + \nabla \cdot \vec{\tau} + \mathbf{F}_b \qquad (1)$$

In the case when the material behaves as a plastic ductile flow, the relationship between stress and strain rate depends on the constituent materials (e.g., type and size of grains, temperature, etc.). In practice, these relationships are assessed empirically through laboratory experiments as a general microscopic theory establishing such correspondences from first principles does not currently exist. These experiments involve subjecting an ice sample — with scales larger than the scales of, say, the comprising grains — to an externally applied force, often referred to as an applied stress, while simultaneously monitoring the resulting strain and strain rate of the sample (see Fig. 3). (It is unfortunate that "applied stress" and "stress" do not refer to the same phenomenon. The latter refers to the internal forces holding the material element together, while the former indicates the internal forces acting in the opposite sense. This nomenclature often leads to confusion.) For example, such experiments may involve subjecting an ice sample to an applied compressive stress, i.e., an external force per unit area directed in a direction along one of the sample's surfaces (depicted in Fig. 4 as τ_{com}). Similarly, an applied shear stress might be exerted upon the sample's surface but in a direction tangential to its surface normal (depicted in Fig. 4 as τ_{shear}). In both experimental procedures, the resulting strain and strain rate of the deforming sample are recorded as a function of time. This empirical "constitutive" relationship is often cast in the generic form of a Glen law, in which

$$\dot{\varepsilon}_{ij} = A(T, d_g, \cdots)\tau^{n-1}\tau_{ij} \quad (2)$$

with

$$\begin{aligned} \tau^2 &\equiv \frac{1}{6}\left[(\tau_{11} - \tau_{22})^2 + (\tau_{22} - \tau_{33})^2 + (\tau_{33} - \tau_{11})^2\right] \\ &+ \tau_{12}^2 + \tau_{23}^2 + \tau_{31}^2 \\ \dot{\epsilon}^2 &\equiv \frac{1}{6}\left[(\dot{\epsilon}_{11} - \dot{\epsilon}_{22})^2 + (\dot{\epsilon}_{22} - \dot{\epsilon}_{33})^2 + (\dot{\epsilon}_{33} - \dot{\epsilon}_{11})^2\right] \\ &+ \dot{\epsilon}_{12}^2 + \dot{\epsilon}_{23}^2 + \dot{\epsilon}_{31}^2 \end{aligned} \quad (3)$$

where τ and $\dot{\epsilon}$ are scalar quantities denoting the second invariant of the deviatoric stress and strain rate tensors (respectively). [A physical quantity that expresses itself as a tensor has quantities that are invariant under any sensible linear coordinate transformation. In linear algebra, a normal 3×3 matrix **S** has three preserved quantities: a first invariant, which is its trace, $J_1 = \mathrm{tr}(\mathbf{S})$; a second invariant, which is the trace of its square, $J_2 = \mathrm{tr}(\mathbf{S}^2)$; and the third invariant, which is its determinant, $J_3 = \det(\mathbf{S})$. In the case of the deviatoric stress tensor, J_1 is always 0 because the body deformations represented by the deviatoric stress is presumed to be volume preserving. Thus, J_2 is generally non-zero and can be thought of as quantifying the "square magnitude" of the stress tensor. Empirically motivated nonlinear stress-strain rate relationships for plastic flows represent this nonlinearity with a coefficient of proportionality, which is some power of τ that is the square root J_2, like in the Glen Law.]

The function A depends upon the temperature of the sample as well as the other factors including a characteristic grain-sized d_g. Often in the literature these second invariant quantities are written with subscripts, i.e., $\tau_2 \leftrightarrow \tau$, $\dot{\epsilon}_2 \leftrightarrow \dot{\epsilon}$. We often refer to these scalars in a more abstract sense to denote stress in generalized terms.

Viscosity may be defined for a ductile material as the scalar function measuring the proportionality between the strain rate and stress tensors. There are several ways to define this relationship, but in practical geophysical applications it is usually written in terms of the scalar relationship between the two respective second invariants (*Solomatov,* 1995): If the stress is a function of the strain rate, i.e., if $\tau = \tau(\dot{\epsilon})$, then the viscosity is simply

$$\eta \equiv \frac{\tau}{\dot{\epsilon}} \quad (4)$$

[On the assumption that this relationship is monotonic, one can also equivalently speak of the more intuitively obvious relationship, $\dot{\epsilon} = \dot{\epsilon}(\tau)$ (e.g., *Solomatov,* 1995).]

When a relationship between the two stress tensors is known or assumed, e.g., equation (2), the stress term in Cauchy's equation is written in the familiar form

$$0 = \frac{\partial P}{\partial x_i} + F_i + \sum_{j=1,2,3} \frac{\partial}{\partial x_i}\left[\eta(d_g, \dot{\epsilon}, T, \cdots)\dot{\epsilon}_{ij}\right] \quad (5)$$

*2.2.2. **Constitutive relations.*** In the ductile regime, a Newtonian rheology refers to a solid-state material whose strain rate linearly depends upon the applied stress, i.e., n = 1 in the Glen law (equation (2)), and the viscosity is independent of the stress/strain rate. Some ices like H_2O and N_2 can also exhibit strain rates that nonlinearly depend upon the applied stress, and such rheology is termed non-Newtonian $n \neq 1$. In general, the exponent in the behavior will be a function of the applied stress for any given material. Furthermore, based on previous laboratory analysis and various theoretical considerations, A(T) takes on an "Arrhenius form." Speaking now about the strain rate-stress relationship in generic terms and dropping the tensor notation on ϵ, $\dot{\epsilon}$, and τ, an ice block's stress-strain rate relationship may be characterized by a more specific instance of the empirical Glen law (*Durham et al.,* 2010)

$$\dot{\epsilon} = A d^{\,p} \tau^{n} \exp\left[\, Q_{act} / RT \right] \quad (6)$$

where the exponents p, n, and the coefficient A are experimentally determined quantities. Q_{act} is the ice-block sample's activation energy per mole and it indicates the material's Arrhenius temperature dependence reflecting the interior grain-collective's susceptibility to deformation. [The activation energy is actually the sum of a free energy activation energy (E_{act}) plus activation volume V_{act}, which takes into account the pressure dependences in the underlying processes. Formally, then $Q_{act} = E_{act} + PV_{act}$, but in practice we work with Q_{act} only as it is the quantity that is experimentally measured. Additionally, for near-surface processes the work term likely has little influence. However, this distinction is important for processes deep within planetary interiors.] Often in the literature an *activation temperature* is used, and it is defined as $T_{act} \equiv Q_{act}/R$, where the gas constant $R = 8.314$ J (mol^{-1} K). In the n = 1 case the material is understood to be isoviscous when $Q_{act} = 0$. In the n = 1 case and for a given restricted temperature range ΔT centered on T, the material is effectively isoviscous provided $(\Delta T/T)\ (Q_{act} / RT) << 1$.

As alluded to above, the problem of characterization is subtler than it might appear because the material's response will depend upon which mechanism of deformation is active under the given conditions, e.g., memory effects, among many other factors. On the long timescales believed to be relevant for outer solar system ices, the deformation mechanism is probably most central. In general, a polycrystalline specimen comprising grains of a particular size will deform by "creep mechanisms" that depend upon the applied stress as well as the temperature. As such, the values of the experimentally determined exponents and activation temperatures will vary as a function of τ and T. Referencing Fig. 3, it is often the case that the response of an ice block will be different whether a sample is in compression vs. tension. Similarly, the strain rate may be different from the compression/tension case vs. the sheared case. A terrestrial example is the formation of crevasses on the surfaces of rapidly expanding glaciers, wherein the ice on the glacier's surface begins to fracture under the intense tensile strain rates to which it is subjected. In this case the tensile stresses exceed the range permissible for ductile flow.

As written, the empirical relationship in equation (6) expresses the stress-strain rate dependence in a general fashion, but it should be kept in mind that it is desirable to directly examine the response of these volatile ices under these differing types of applied stresses. Laboratory experiments done on volatile ices appropriate to planetary surface conditions have been conducted exclusively under compression — generally extending these results as being appropriate for the two other stress conditions (*Durham et al.,* 2010).

2.2.3. Creep mechanisms: Types, forms, and microphysics. The types of ice creep mechanisms, which are varied and often interdependent, can be broadly broken down as grain-size-sensitive (GSS) diffusion flow and grain-size-independent (GSI) plastic flow. GSS regimes are generally active when the applied stresses are low. They involve several different grain-deforming mechanisms including grain boundary sliding, dislocation slip, and others, with stress exponents in the linear or near-linear (superplastic) regime. For deeper coverage, see *Petrenko and Whitworth* (1999) and the reviews of *Durham et al.* (2010) and *Kohlstedt and Hansen* (2015). We also recommend *Barr and Pappalardo* (2005) and *Barr and McKinnon* (2007) for an applied review of H_2O ice rheology and its dependence on grain size on Europa.

Broadly speaking, the microphysics of grain deformation and metamorphism involves the transport and reorientation of vacancies and defects within a crystal. From a fundamental thermodynamics point of view and given enough time, a collection of multiple grains will eventually turn into a single large perfectly ordered crystal, this being the lowest possible quantum energy state (*Eluszkiewicz,* 1991). In the case of both crystalline and amorphous solids, it is accepted that the interatomic cohesive forces are generally quite strong in H_2O ice crystals owing to the strength and complexity of hydrogen bonds, and that the presence of dislocations or surface heterogeneities is the main cause of the relative weakness of most such specimens (*Macmillan,* 1972). In volatile ice crystals, especially those with predominantly van der Walls intermolecular bonding or non-polar covalent bonding (CH_4), the relative importance of the role that dislocations and surface heterogeneities play in weakening a crystal sample is still poorly constrained. Appealing by analogy to H_2O polycrystalline grains, even a single defect can drastically reduce the strength of a specimen, which explains why very high strength is normally a structure-dependent non-reproducible property. The motion and transport of imperfections (defects) within a crystalline grain expresses this long-term drive of the physical system. Defects also play an essential role in allowing for plastic deformation of rocks and ices and they come in three general types: (1) point defects, which are vacancies or impurities within a crystal matrix that move about by means of diffusion; (2) one-dimensional line defects, often referred to as dislocations; and (3) two-dimensional planar defects involving grain-grain interfaces and interphase boundaries (*Kohlstedt and Hansen,* 2015). The planar boundaries between grains that ultimately get removed through this principle of energy minimization effectively results in larger grains consuming their smaller neighbors owing to the preferred lower-energy state associated with larger grains [known as the "Ostwald ripening and abnormal grain growth" (*Ostwalt,* 1897; *Hillert,* 1955; *Ratke and Voorhees,* 2002); see also the discussion in section 2.2.5].

A given grain in a polycrystalline ice may have many dislocations that individually propagate along their slip planes (see Fig. 4 for further details including types and geometries of dislocations). As a general principle, an applied stress placed upon — or the mere presence of dislocations within — a grain will elastically deform the grain and possibly also create more dislocations. The dislocations within each grain then effectively interact with one another through the stress field that arises from this distortion. The propagation of dislocations within the grain complex operates to relieve the aggregate stress within, which contributes to (equivalently) reducing the energy in the system or subsystem (*Weertman and Weertman,* 1992; *Kohlstedt and Hansen,*

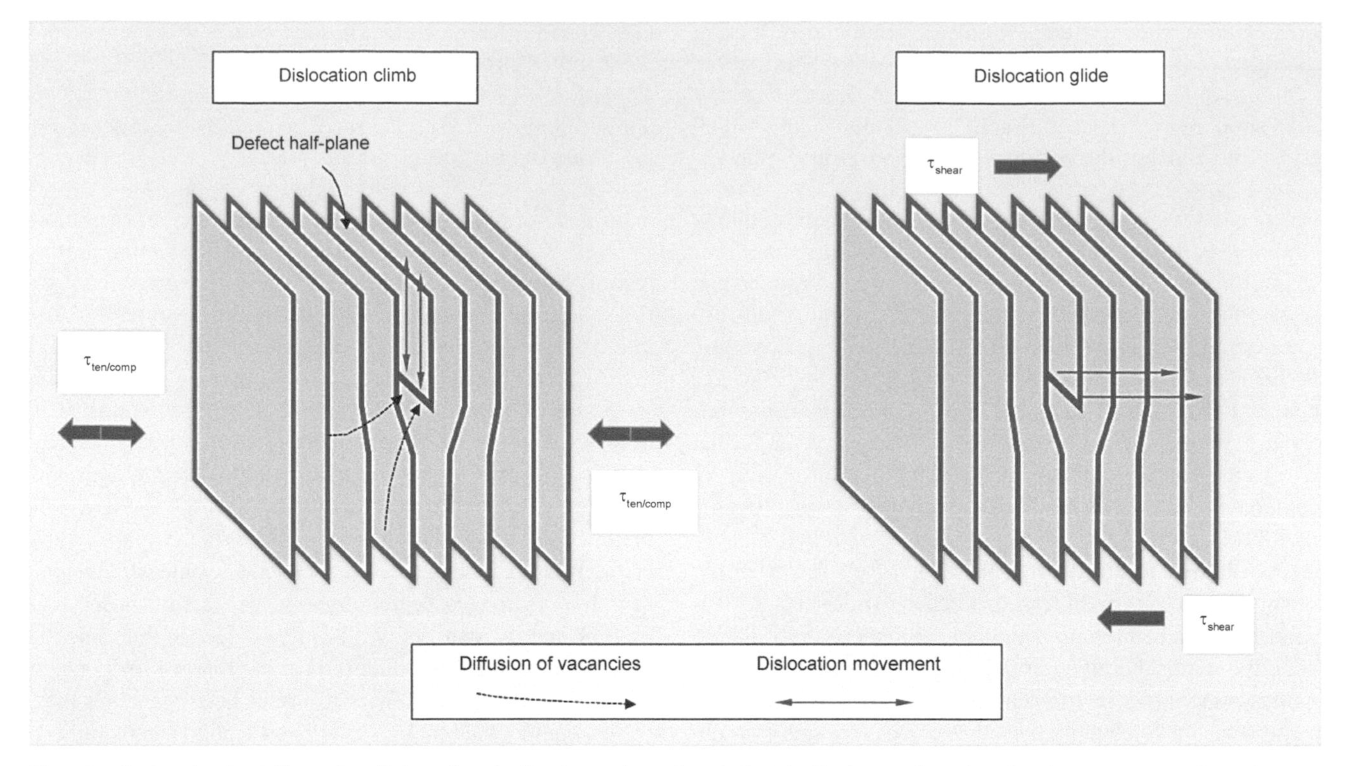

Fig. 4. A classic depiction of a dislocation is the insertion of a defect half plane of molecules between an otherwise ordered configuration. Two types of dislocations are shown. *Left:* Climb dislocations propagate based on the availability of and the manner in which vacancies are drawn toward dislocation line. As such, it is temperature sensitive. The net result will be motion of the dislocation in the defect half plane. Under compression, the defect will move upward until it reaches grain boundary while under tension it moves downward. *Right:* When the grain is under shear and the deformation is strong enough, a neighboring full lattice plane (say, to the right of the defect plane) will reconfigure itself onto the defect because it is energetically preferable, resulting in a net translation of the dislocation toward the right. This continues until the dislocation (and half plane) reaches the edge of the grain. This process is not temperature dependent as it does not rely on the diffusion of vacancies. See also *Petrenko and Whitworth* (1999).

2015). The concomitant response of defect-laden grains to applied stress, elastic distortions, and their natural tendency to minimize the aggregate energy in the system will result in one or more of the defect modes (vacancies, lines, and grain boundaries) dynamically adjusting to achieve this purpose.

The mechanism of deformation will depend upon the amplitude of the applied stress, the typical grain size, and the temperature of the sample. We speak of them here in terms of increasing values of this stress. For relatively low applied stress situations (i.e., conditions relevant for the near-surface environments of the icy bodies of the outer solar system), the ductile deformation of polycrystalline ices occurs under the process generally referred to as grain-boundary sliding (GBS). An externally applied force will produce stress concentrations where the angular tips and corners of grains meet the boundaries of others. If the deformation translates into sliding without induced grain-scale microfracturing or localized cavitation, then there must be some kind of mass transfer away from the stress concentration to permit translation.

The processes accommodating this mass transfer effect are broadly referred to either as diffusional-creep — which comes in two concomitantly active modes — or dislocation slide involving lattice dislocations (see review of *Kohlstedt and Hansen,* 2010). Diffusional creep involves the diffusion of vacancies (holes) from distant parts within a grain crystal and toward the concentrated stress location. This diffusion, which is temperature activated (i.e., Arrhenius in form), proceeds both through the grain's interior and over its bounding surface. This vacancy migration occurs due to stress variations that emerge due to elastic distortions of the crystal's molecular matrix induced by an applied (external) stress. The two diffusion pathways are known respectively as Nabarro-Herring creep and Coble creep (*Nabarro,* 1948; *Herring,* 1950; *Coble,* 1963).

A functional relationship quantifying the rate of diffusion can be motivated from first principles but must be supplemented with some laboratory measurements. In the generalized abstract sense, then the Newtonian stress-strain rate/viscosity relationship is expressed as

$$\begin{aligned} \frac{1}{\eta} = \frac{\dot{\epsilon}}{\tau} &= \frac{14}{RT}\left(\frac{D_v}{d_g^2} + \pi\frac{\delta D_b}{d_g^3}\right), \; D_{\{v,b\}} \\ &= D_{0,\{v,b\}} \exp\left(\frac{Q_{\{v,b\}}}{RT}\right) \end{aligned} \tag{7}$$

where δ measures the diffusion width between neighboring grains, usually ~10^{-9} m (*Durham et al.,* 2010; *Kohlstedt and Hansen,* 2015). While the volume (lattice) and body diffusion coefficients (D_v and D_b respectively) and their attendant total activation energies (Q_v and Q_b respectively) could in principle be estimated from physical arguments, in practice they are determined from direct laboratory measurements utilizing a variety of techniques, including nuclear magnetic resonant imaging. Which process dominates as a function of temperature depends upon the particular measured values of $D_{0\{v,b\}}$ and $Q_{\{v,b\}}$. In general, Coble creep preferentially operates at lower temperatures compared to Nabarro-Herring creep. It should be kept in mind that the diffusion of vacancies toward the stress point promotes the motion of one grain along the boundary of the other. In this sense, then, vacancy diffusion accommodates GBS and this highlights how many of the processes involved work in tandem with one another (*Raj and Ashby,* 1971).

As the applied stress increases, one approaches the GSI regime, in which deformations occur by several process involving the movement of dislocations. These processes are generally broken down into three types: (1) dislocation-accommodated GBS, (2) dislocation creep, and (3) low-T plasticity (*Kohlstedt and Hansen,* 2015). In this regime the creep processes are nonlinear with respect to the applied stress (i.e., $n > 1$) and this is generally referred to as the plastic or power-law regime. Dislocations within grains are both created in response to an applied stress or nucleated (*Frank and Read,* 1950), and they play a role in maintaining the ongoing deformation of the grain. Multiple converging dislocations can impede their mutual advance, annihilate one another upon encounter, become incorporated onto the boundary of a grain, or continue to propagate into the interior of another grain once having emerged onto the boundary of the parent primary. Nothing is known about this regime for the volatile ices of interest beyond the initial exploratory laboratory work of *Yamashita et al.* (2010) and *Bol'shutkin et al.* (1968), and short of dedicated laboratory effort any inferences about plastic behavior must be drawn from water ice or other materials [a good resource for more information is *Barr and McKinnon* (2007) and references therein].

Similar to how diffusive vacancy migration aids GBS, dislocations similarly provide the mechanism of motion for grains (*Raj and Ashby,* 1971). This motion is achieved either through glide (low-temperature limit) or climb (high-temperature limit). Glide and climb both occur simultaneously, but climb is more efficient at higher temperatures because dislocation climb requires sufficient vacancy diffusion, which is enhanced at higher temperatures. This mode of motion is observed to be responsible for very large non-fractured strains of materials in tension (see, e.g., *Ridley,* 1995). The constitutive equation in this regime is "superplastic," i.e., it is nonlinear with stress and grain size exponents roughly around n = 2 and m = 2 (respectively), viz. equation (6). This relation is supported both by theoretical argument (*Kaibyshev,* 2002) and by experimental results involving various silicate rocky materials (e.g., *Hiraga et al.,* 2010) and H_2O ice (*Goldsby and Kohlstedt,* 2001). However, if subgrains exist within grains the exponents shift toward n = 3 and m = 1, the idea being that subgrains consume dislocations, robbing the larger grain of its means to deform.

The superplastic regime may have relevance to CH_4 ices on Pluto where organic tholins, thought to be manufactured in the upper atmosphere and subsequently rained out onto the surface, could form into grains that steadily grow an outer CH_4 ice shell. If such compound nested grains make it to the surface, they could become a compact surface polycrystalline tholin-CH_4 ice after enough grain-grain sintering removes intergrain porosity (*Eluszkiewicz,* 1991). For a relatively recent and fairly complete discussion of sintering processes, see *Molaro et al.* (2019). The exponent trends observed in the compaction experiments of both CH_4 and N_2 ice (*Yamashita et al.,* 2010) seem to suggest that they are in this so-called superplastic regime. Follow-up verification studies are critically needed.

At larger grain sizes and larger temperatures, the process shifts to dislocation creep. This process involves a complicated steady-state interplay of applied stress-induced creation of dislocations that subsequently propagate across the grain and on toward its boundary through climb and glide with occasional dislocation annihilations occurring along the way (*Nabarro,* 1967; *Weertman,* 1999). This regime is characterized by exponents in the range $3 < n < 5$ and shows relatively little sensitivity on grain size, i.e., $m \approx 5$. The mechanism tends to be operative at sufficiently large grain sizes and higher temperatures.

The low-T plastic regime occurs at sufficiently high stresses where the primary motion is dislocation glide as the low temperatures provide insufficient vacancies to promote climb. In this case, glide is facilitated by the creation of kinks, which allow a pathway for dislocations to propagate against the stiffness of the lattice itself (*Frost and Ashby,* 1982; *Kohlstedt and Hansen,* 2015). Kink creation is temperature activated, which, in practice, means that the strain rate will have both an algebraic and an exponential dependence on the applied stress. For example, for rocky materials made of olivine (Mg_2SiO_4, also known as dunite), the strain rate dependence has the form

$$\dot{\epsilon} \sim \tau^2 \exp\left\{-\frac{T_{act,k}}{T} \cdot \left[1 - \left(\tau/\tau_p\right)^r\right]^s\right\} \qquad (8)$$

where $T_{act,k}$ is the activation temperature for kink creation, r and s are model-dependent parameters, and τ_p is a characteristic stress associated with interatom potentials within the lattice. The main trend here is that this behavior is grain size independent. Furthermore, the strain rate increase with applied stress will rapidly lead to fracture. Therefore, the low-T plasticity regime operates under a very narrow range of applied stresses bounded above by the fracture stress of the material itself. Whether or not this behavior is expected for volatile ices of interest remains to be seen.

Deformation maps are a common way to depict which mechanism should be operative as a function of applied stress, temperature, and grain size. In Fig. 5, we depict the deformation map of olivine taken from the review of *Kohlstedt and Hansen* (2015). Except for solid CH_4 (discussed in section 3), no other deformation maps exist for the ices of interest.

Thus, in general, the stress-dependent quality of ice deformation (i.e., which mechanisms are active) can result in extreme variations of the resulting strain rate. Characterizing this feature of volatile ices, their mixtures, and their alloys is of singular importance toward reconstructing the histories of icy bodies of the outer solar system. By way of example, Fig. 6 qualitatively depicts the complex rheology of solid H_2O (ice I). As can be seen from the figure, the strain rate response as a function of applied stress does not fall on a single line in the (logarithmic) plot but instead shows significant changes across the range of applied stresses. For example, it would be foolhardy to perform experiments in the easily accessible high-stress laboratory range and then extrapolate these results to predict the strain rates when the stresses are 2–3 orders of magnitude less since the behavior of the materials can drastically change across this range of stresses. To reliably assess the rheology of volatile ices requires mapping out and identifying all the above-described response regions — from the dislocation creep regime down to the diffusion regime.

Developing accurate empirically derived constitutive relationships for these solids is central to assessing the longevity of high-standing structures, such as Pluto's bladed terrain and the steep side-walled shallow pitting observed on Sputnik Planitia under various types of applied stress. In the case of H_2O ice in some of its various ice phases and clathrates (especially under conditions appropriate for both terrestrial and outer solar system conditions), these constitutive relations are fairly well-defined, as illustrated in Fig. 6. No such information exists for volatile ices under the conditions appropriate for the satellites and solid surfaces of the outer solar system. Figure 7 also identifies stress regions expected for the icy satellites of the Galilean and saturnian systems. By extension, the conditions for the outer solar system objects will be in similar ranges of 1–100 kPa. The corresponding predicted strain rates are on the order of thousands to billions of Earth years. Thus, the predicted evolutionary timescales are intimately tied to the temporal response of a given material under appropriate applied stress. Accurate measurements of their responses will provide reliable estimates for the ages of the variety of ice structures seen on the surfaces of these outer solar system bodies. At present, these figures are not reliably known for volatile ices existing under the pressure-temperature conditions of the outer solar system.

Much in the way steel alloys and cast iron derives its strength from various phases of iron (ferrite, austenite, martensite, etc.) containing stabilizing agents like C, Ni (ranging from 1% to 10% by molar weight in these impurities), we emphasize the central importance of assessing the strengths of analogous kinds of "volatile ice

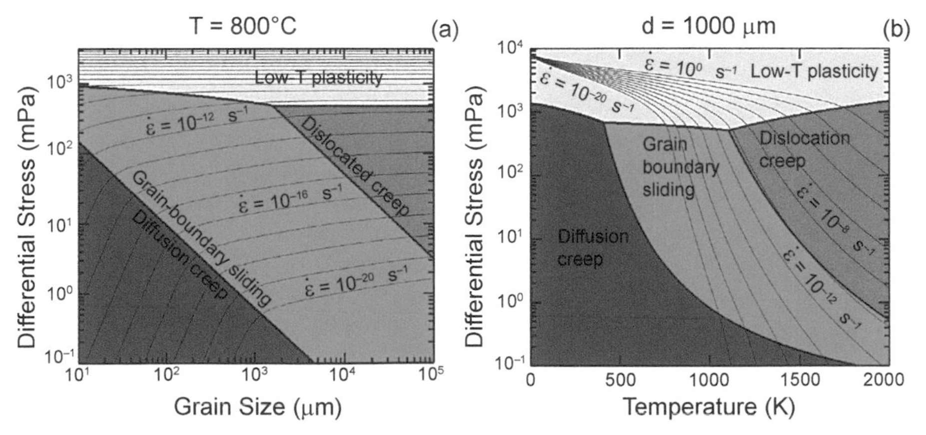

Fig. 5. An example of a deformation map. The deformation map of olivine (Mg_2O_4) is shown depicting the various processes at play adopted from *Kohlstedt and Hansen* (2015). Diffusion creep generally occurs at lower temperatures and stresses and is fundamentally Newtonian. Grain boundary sliding (GBS) comes in two flavors: diffusion accommodated and dislocation accommodated, where the latter is nonlinear. The remainder mechanisms are nonlinear "power-law" processes. Dislocation creep occurs generally at higher temperatures. Low-T plasticity occurs at the highest differential stresses. In general, strain rate is a function of temperature, grain size, and stress. The figures also depict strain-rate isolines. **(a)** Deformation map at constant temperature. Low-T plasticity and dislocation creep are nearly insensitive to grain size. **(b)** Deformation map at constant grain size d. In this example, for the given grain size (1 mm) the strain rates of diffusion creep are so long that the material effectively does not move at low temperatures and at differential stresses under tens of gigapascals. It is important to develop a comprehensive array of deformation maps for the wide variety volatile ices of the outer solar system.

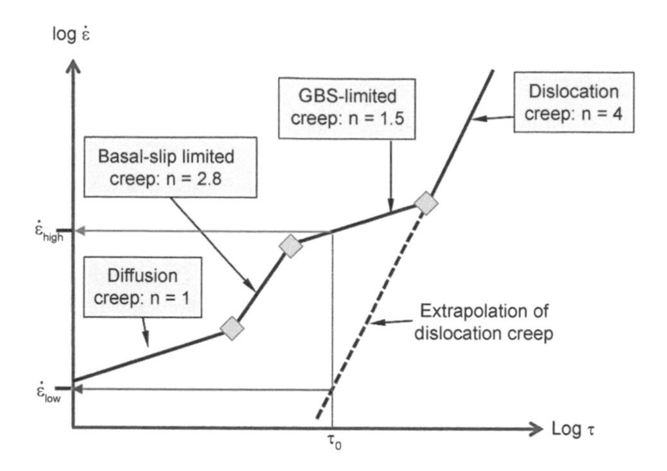

Fig. 6. Schematic of the stress-strain rate of H_2O ice (ice I) across a range of applied stresses. Various flow/creep regimes of H_2O are shown. Lines of extrapolation lines of each of the four creep regions are also shown. There is danger in extrapolating too far in either direction of sampled stress regimes. For example, if data were available only from the dislocation regime and one wants to know the strain rate for a given stress τ_0, one would severely underestimate the strain rate based on the extrapolated rheology based on its measured rheology. Peach-colored diamonds designate stress values at which creep mechanisms change type. To reliably assess the rheology of volatile ices requires mapping out and identifying all the regions connecting the plastic dislocation creep (n = 4) regime down to the diffusion (n = 1) regime. Adapted from *Goldsby and Kolstadt* (2001).

alloys" as geologic formations on places like Pluto have been spectroscopically determined to be CH_4-dominated composites with a smattering of N_2. As such, the nature and age of high-standing structures like the bladed terrain on Pluto depends very much on assessing strength of CH_4 both in its pure and adulterated state, the latter state including the possible role of strengthening and/or stabilizing additives like H_2O. Additionally, the near-surface behavior of structures like pits on Sputnik Planitia may involve a complex alloy-like behavior between tholins and N_2:CO:CH_4 ice mixtures (*Yu et al.,* 2018; see the chapter by Moore and Howard in this volume).

2.2.4. Brittle fracture of polycrystalline ice. The preceding discussion has emphasized the fact that various processes maintain ductile flow, including diffusion, grain-boundary sliding, and dislocation glide and creep. All these mechanisms are affected by the applied stress and are rate limited by temperature. By contrast, pushing toward higher strain rates leads to the creation of cracks, which may or may not propagate through the material. When the cracks do spread they lead to fracture, also known as brittle-fracture failure. In our usage, a crack is understood to be a rip in the lattice fabric that does not induce structural failure of the sample, while a fracture indicates either a single crack or collection of cracks working in concert that precipitates significant structural weakening or outright failure. The critical criterion for brittle fracture failure is a function of the kind of material (i.e., the type of bonds holding the crystal together), temperature, the strain rate, and the type of stress applied (e.g., tension, compression, or shear). In a polycrystalline ice, cracks will appear if there are insufficient grains with basal slip planes (i.e., glide planes) oriented with the applied stress (for example, see the schematic in Fig. 3). The transition from ductile behavior to fracturing corresponds to the inability of dislocations to move across the grain to relieve the imposed external stresses. When dislocations are mobile, they accumulate along grain boundaries, and their collective stresses induce a crack. When cracks appear (i.e., nucleate) along the grain boundaries, they generally have lengths that are a small multiple (3–5×) of the typical grain

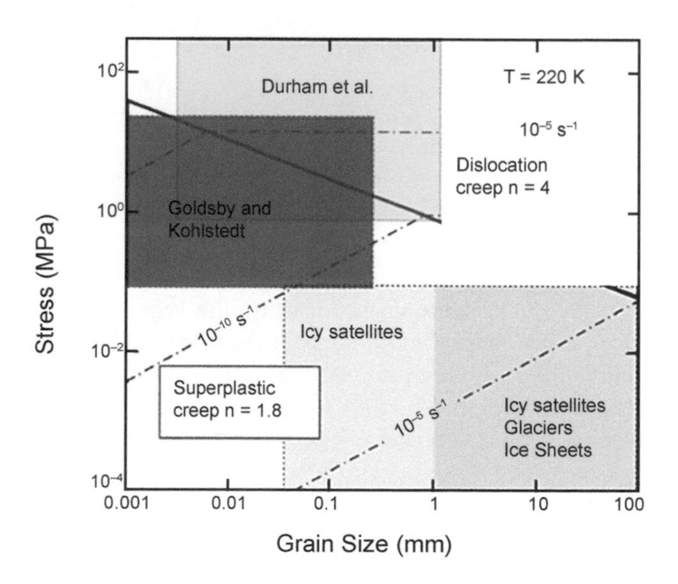

Fig. 7. Regimes of H_2O ice I examined by various laboratory experiments (authors of the studies are shown on the figure). The regimes appropriate to the icy satellites are indicated. Values of the strain rates in H_2O ice are extrapolated here. Similar parameter maps are needed for volatile ices N_2, CO and CH_4, and their alloys, under conditions appropriate for the Kuiper belt and TNOs. Image from *Durham et al.* (2010).

size. The presence of cracks alone does not mean the sample will fracture. Often, cracks can be present, but if they do not grow or propagate, ductile flow may continue. The problem occurs when cracks nucleate at a rate faster than they heal by the propagation of dislocations, or if cracks begin to propagate or spread. Following the discussion found in *Petrenko and Whitworth* (1999), we expand this discussion with some further physical details.

The response of a polycrystalline ice sample is a function of the total strain of the sample as well as the strain rate it experiences. Large strains can be achieved by a sample if the strain rate is low (i.e., general plastic flow). However, if the strain rate is too rapid, then strain may continue but the process will terminate when fracture occurs. The various types of responses are shown in Fig. 6. In practical laboratory experiments testing the response of ices and metals, the strain vs. stress is monitored [e.g., see the experiments on CH_4 and Ar of *Leont'eva et al.* (1970)]. If the sample survives straining beyond its yield stress and if the strain grows at a more or less constant rate and fixed stress, then a stress-strain rate data point characterizing ductile flow becomes available (Fig. 6). The further response of the material depends on the type of applied stress. In either case, however, there will be a critical condition for brittle fracture failure that we generically refer to as τ_f (d_g, $\dot{\epsilon}$, T, $\cdots$). It is important to keep in mind that the grain size dependence can be very important and that the mechanism driving forth fracture will be different depending on the strain rate experienced by the sample. Whether the cause of crack nucleation is due to dislocation (creep) pileup from within individual grains or from grain boundary sliding is not obvious. Work involving H_2O ice (*Schulson,* 1990; *Schulson and Duval,* 2009; *Tanabe,* 2013) suggests cracks nucleate by grain boundary sliding, the likely reason being that dislocations do not propagate fast enough to account for the rate of nucleation.

For samples in tension subjected to relatively high strain rates, cracks can nucleate in the form of rings with axes aligned parallel to the applied stress and the nucleation rate is a function of $\dot{\epsilon}$. The diameter of the rings may be a factor of 1–10 d_g. Fracture ensues because the large number of small cracks appearing throughout the interior overwhelms the sample, i.e., they proliferate and accumulate faster than they can anneal. In the lower strain rate case, cracks nucleate at a lower rate, but failure will occur due to crack propagation (i.e., a given crack grows).

In the study of H_2O ice, *Schulson* (1987, 2006)showed that τ_f ($\dot{\epsilon}$T) ~ τ_{f0} ($\dot{\epsilon}$) + K_f ($\dot{\epsilon}$, T)/$\sqrt{d_g}$, where K_f is known as the fracture toughness, a quantity that is uncovered in controlled laboratory experiments and depends upon the type of underlying mechanism at play. τ_{f0} is zero-point offset. Each of these two quantities is inherently a function of the strain rate.

2.2.5. Grain growth and dynamic recrystalization. The foregoing discussion demonstrates that polycrystalline grain size and matrix sample strength is an important input parameter in determining how ice structures and samples respond to stress and strain, yet is also often the most poorly constrained quantity. We present a short discussion of sintering and the process of dynamic recrystallization, with relevant references to their appropriate modeling and application to planetary ices.

Sintering is the evolutionary process by which a frosty collection of polycrystalline grains metamorphoses; moreover, it "plays a key role in the microstructural evolution of planetary ices and that understanding its nature is critical to understanding the landscape evolution of icy worlds" (*Molaro et al.,* 2019). Sintering results in the increased cohesion between neighboring grains through the growth of a "neck" at grain-grain contact points. The most modern and complete mathematical modeling framework for sintering is laid out in *Swinkels and Ashby* (1981, hereafter *SA81*), which is based on the careful observations of grain growth in very porous aluminum oxide powders (*Greskovitch and Lay,* 1972) and has been used in numerous terrestrial metallurgical applications. The *SA81* approach has been applied to a variety of planetary settings — both surface and subsurface environments alike — including modeling the volatile ices on Triton (*Eleuskiewicz,* 1991) and H_2O ice on Europa (*Molaro et al.,* 2019) among many other locales across the solar system [see further comprehensive review and the most up-to-date model discussion in *Molaro et al.* (2019)]. The *SA81* picture posits that sintering undergoes four stages: Stage 0, the initial emplacement and adhesion of grains; Stage 1, neck growth at grain-grain contact points; Stage 2, a transition phase once the neck radius is about 0.5–0.8 d_g, which leads to Stage 3, the densification stage, in which the remaining pore space in the sintering-transfigured grain matrix steadily reduces. The sintering process might be considered largely complete once the neck radius reaches 90% of d_g (*Molaro et al.,* 2019). There are six concurrently operative mass transfer processes that are sensitive to several measures of a grain's surface and boundary curvatures. These effects are in turn grouped into two categories. There are three non-densifying mechanisms involving (exposed) surface mass sources: (1) surface diffusion, (2) lattice/volume diffusion, and (3) vapor transport, together with three densifying mechanisms involving non-exposed mass sources: (4) grain boundary diffusion from grain boundaries, (5) lattice diffusion from grain boundaries, and (6) lattice diffusion from dislocation sources. All the above processes depend on the various diffusion coefficients and activation energies discussed in section 2.2.3: Processes (2) and (5)–(6) require knowledge of volume diffusion D_v and activation temperature Q_v, process (3) requires information on vapor pressures, while processes (1) and (4) require knowledge of body/surface diffusion coefficents $D_{b,s}$ (respectively) and their attendant activation temperatures $Q_{b,s}$. These quantities require laboratory measurement or heavy numerical computations to determine, especially for grains composed of mixtures and alloys of the volatiles of interest.

The physical picture of grain growth for planetary ices follows the behavior uncovered by *Greskovitch and Lay* (1972) (see also their laboratory images), which is conceptually summarized in Fig. 1 of *Eluszkiewicz* (1991). Moreover, as noted by *Kirk* (1990) and *Eluszkiewicz* (1991), the neck growth phase (Stage 1), which leaves grain sizes largely unaltered, is the rate-limiting step of grain evolution, while densification, porosity reduction (d_g nearly constant, Stages 2–3) and subsequent grain-coarsening (d_g growth) all happen on relatively fast timescales. For example, in the theoretical analysis by *Eluszkiewicz* (1991), assessing the sintering times (t_s) for pure N_2 or pure CH_4 ice showed that $t_s \sim d_g^4$ (see Fig. 4 of that study). The latter phase grain-coarsening process involves a low-porosity (<8%) matrix of grains, where the grain boundary of a large grain migrates at the expense of a smaller neighbor eventually swallowing it up. The grain migration is sensitive to grain boundary curvature and boundary diffusion $D_b(Q_b)$ (see further explanation below). In the limit of low porosity, this growth is satisfactorily modeled according to the theory of *Hillert* (1965) in which the grain size as a function of time follows the form $d_g^m - d_{g0}^m = (t - t_s) K_g$, where d_{g0} is the initial grain size (i.e., when sintering is complete) and the constant K_g represents the grain boundary mobility given by the expression $K_g \equiv (16/81) \cdot \gamma D_b \Omega^{2/3}/kT$, in which Ω is the molecular volume and γ is surface tension (not discussed here). [Number-size analyses of polycrystalline samples generally show that the particles follow a log- normal distribution centered on a characteristic grain size d_g. The *Hillert* (1965) model addresses the secular growth of this characteristic grain size. Furthermore, at any one instant the model posits that particles with $d < d_g$ will shrink while those with $d > d_g$ will grow, keeping in mind that all throughout this process the target grain size d_g slowly increases as well.] The index m depends on the type of material under consideration but is typically 2. The implied grain sizes inferred from Pluto's surface imaging using New Horizons Ralph/LEISA data indicates while CH_4-dominant (i.e., CH_4:N_2-rich) surface volatile ices have grain sizes around 0.1–1 mm, N_2 ice dominant surface volatile ices (i.e., N_2:CH_4-rich) appear to be significantly larger, ~1 m (*Protopapa et al.,* 2017; see also the chapter by Protopapa et al. in this volume). A detailed theoretical model for this curious trend in the N_2 rich volatile ice mix remains to be elucidated.

Dynamic recrystallization (DRX) aims to address the question of what (if any) equilibrium grain size characterizes a given ice sample [following early formulations by *Derby and Ashby* (1987) and *Derby* (1991)]. As the foregoing discussion shows, with grains continuously growing at the expense of their neighbors, an otherwise dynamically inactive ice layer ought to transform into a collection of very large and substantially transparent crystals, e.g., like the nearly 1-m-sized transparent volatile ice blocks predicted for Triton by *Kirk* (1990) and *Eluszkiewicz* (1991) and others. However, ice samples continuously subjected to unsteady dynamic stresses can nucleate "sub"grains inside already established larger grains. Thus, the inexorable march toward ever larger structures is tempered by the creation of small-scale grains, which leads to an equilibrium grain size characterizing a statistically steady state. Which subgrain nucleation process holds depends upon how close a sample is to its melt temperature and the magnitude and quality of the strainfield (*Drury and Urai,* 1990). *Shimizu* (1998) constructed a theoretical model to predict the equilibrium grain size where the nucleation mechanism is governed by the high-temperature (i.e., ~0.3–0.5 < T/T_m < 1) low-$\dot{\epsilon}$ subgrain-rotation (SGR) mechanism, which involves a time-dependent strainfield that leads to bending along a symmetry axis followed by a rotation of the bending axis, which promotes the emergence of subgrain walls that eventually leads to the creation of a whole new grain [e.g., for an instructive discussion on SGR see *Bestmann and Prior* (2003), paying particular attention to their Fig. 14]. [We also note that there are other nucleation mechanisms appropriate to other stress temperature conditions (e.g., see discussion in *Drury and Urai,* 1990), but these are not discussed here largely because the volatile ices of interest are generally in their "high"-temperature regimes, i.e., sufficiently close to T_m.] In the Shimizu model, the rate at which SGR leads to new grains is a function of the speed at which dislocations migrate toward the wall, which in turn is a product of the volume diffusion rate (D_v) of dislocations and the applied stress field τ. The rate at which densified grains grow is controlled by the rate at which grain boundaries propagate, which in turn is proportional to the boundary diffusion rate (D_b) of dislocations. As noted previously, the rate of these processes is strongly sensitive to temperature as it controls the speed at which dislocations propagate in the crystal matrix. Putting these together results in the expression for the mean grain size

$$\frac{d_g}{b} = A'\left(\frac{\tau}{\mu}\right)^{-p} \exp\left[-\frac{Q_b - Q_v}{RT}\right]$$

where b is the Burger vector's length, μ is the shear modulus, the coefficient A' and index p are nondimensional quantities depending on the material of interest and generally assessed through laboratory testing, where the p has been observed to take on values ranging from 0.85 to 2 and A' can run from 10 to 100. *Rozel et al.* (2011) showed how the above quoted equilibrium grain-size relationship is a robust solution of properly treating grain-size evolution in a self-consistent manner.

Barr and McKinnon (2007) have applied DRX toward making predictions about the typical size of water grains on Europa and the subsequent effect it has on the heat transport through its icy shell by dictating material viscosity, which has direct consequences for the existence or absence of interior oceans. From a practical standpoint they further argue that in certain circumstances the above formula may be applied in a more simplified form by setting the exponential term to 1 and assuming a slightly modified value for A'. Following *Barr and McKinnon* (2007), in addition to its application to terrestrial mantle convection (*Rozel,* 2012) and Europa (e.g., *Han and*

Showman, 2010), the role of DRX has been discussed and/or applied to numerous icy planetary bodies of the outer solar system including several recent studies pertaining to water ice grains in Ganymede (*Bland et al.,* 2017), in Enceledus (*Behounkova et al.,* 2013; *Rozel et al.,* 2014), on Titan (*Litwin et al.,* 2012; *Sotin et al.,* 2010), in Triton and Pluto (*Hammond et al.,* 2016, 2018), and for N_2 ice grains in Pluto's Sputnik Planitia (*McKinnon et al.,* 2016).

2.3. Sublimation-Deposition Modeling Fundamentals

Sublimation-deposition physics is controlled by the flow of mass and energy toward the surface upon which the action takes place. We describe a relatively simplified physical setting to showcase the processes operating following the example of *Claudine et al.* (2015) (see Fig. 8): We imagine an initially flat landscape with its surface at $z = 0$ that subsequently distorts into a spatiotemporally evolving shape with function, $z = Z_s(x,t)$, once deposition-sublimation proceeds. The ground below is taken as semi-infinite, unmoving, and composed of a volatile ice with solid density ρ_s, thermal conductivity K, and specific heat C_p. The temperature of the ice layer is a function of the vertical and horizontal coordinates as well as time, i.e., T(x, z, t). The temperature at the surface is denoted T_0. The evolution of the temperature in the interior — i.e., all points below the surface Z_s — is then governed by

$$\rho_s C_p \frac{\partial T}{\partial t} = \nabla K \cdot \nabla T + \psi + Q_{int} \quad (9)$$

$Q_{int} = Q_{int}(\rho_v,\phi,\cdots)$ represents an additional source of energy/gain loss within the interior, which may be a function of the interior porosity (ϕ), the vapor content (ρ_v), and other physical characteristics describing the interior structure of the ice (e.g., its permeability; see further information below). The function $\psi(x,z,t)$ denotes the amount of irradiant energy (e.g., visual light) absorbed within the interior. The idea is that light enters the surface and penetrates the interior via scattering and eventually gets absorbed on some characteristic absorption length scale called Λ. A crude way to characterize this penetration physics is using a steady-state fluence model (*Claudine et al.,* 2015)

$$\nabla \cdot \Lambda^2 \nabla \psi - \psi = 0 \quad (10)$$

valid everywhere within the subsurface below $z = Z_s$. A fluence model describes the way radiation field ψ (quantified in terms of a "fluence," which has the units radiative power flux, e.g., W m^{-2}) propagates through a scattering medium: The first term on the LHS of the equation represents photon diffusion via scattering while the second term represents the absorption of radiation. $\Lambda^2 \equiv \ell D/c_m$ where D is the scattering diffusion coefficient, c_m is the speed of light through the medium, and ℓ is the absorption mean free path. The two equations require boundary conditions that contain all the information describing what energy comes into and out of the surface Z_s.

Defining $\mathbf{\hat{n}_s}$ as the unit normal of the deforming surface Z_s, the boundary condition on the fluence model is

$$\mathbf{\hat{n}_s} \cdot \nabla\psi = (1 - A_V)\mathbf{\hat{n}_s} \cdot \mathbf{F_v} \quad (11)$$

where $\mathbf{F_v}$ denotes the ambient incident visual radiation flux and A_V is the material's surface visual albedo. For the temperature evolution equation (9), we have

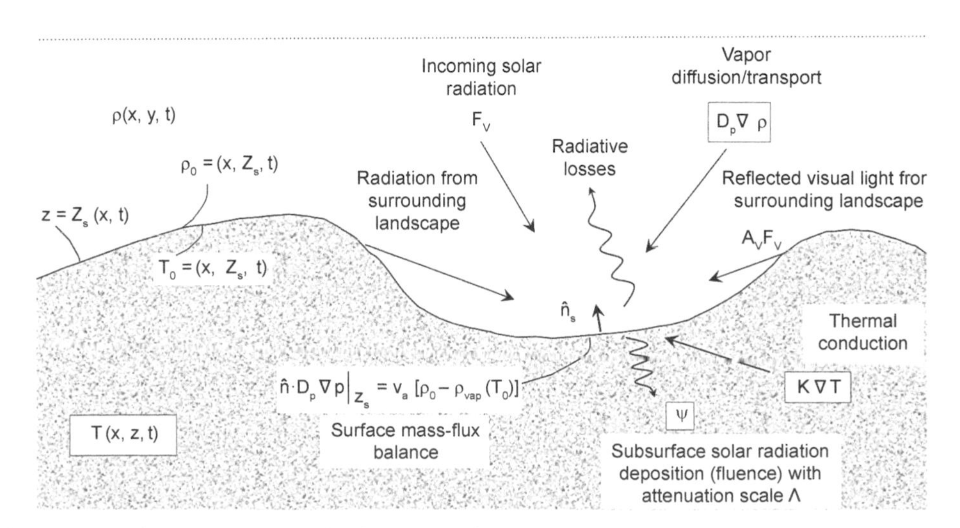

Fig. 8. A simplified schematic representing physics governing the sublimation deposition processes. Adapted and expanded from *Claudine et al.* (2015).

$$\hat{\mathbf{n}}_s \cdot K\nabla T = -\sigma T_0^4 + (1 - A_{IR})\hat{\mathbf{n}}_s \cdot \mathbf{F_{IR}} + L_s q \quad (12)$$

In order, the terms on the right in equation (12) are (1) radiative losses; (2) the absorbed incident non-visual radiation perhaps in the form of thermal re-radiation from other parts of the landscape, F_{IR}, with associated albedo A_{IR}; and (3) thermal gains/losses from the deposition. The final expression is written in terms of a heat of sublimation (L_s, in units of energy per mass) multiplied by a mass flux q. The mass flux at the surface accounts for the vapor pressure of the solid, $P_{vap}(T_0)$, which constantly transforms molecules of mass m_s from the solid into gas and sends them into the atmosphere above with typical thermal (Boltzmann) speeds $v_a = \sqrt{8kT_0/\pi m_s}$ against the rate at which atmospheric gaseous molecules strike the surface

$$q = v_n\left[\rho_0 - \rho_{vap}(T_0)\right],\ \rho_{vap} = P_{vap}/c_s^2;\ c_s^2 = kT/m_s \quad (13)$$

in which ρ_0 is the atmospheric gas density of the volatile species at the evolving surface Z_s, whose boundary always grows in a direction normal to itself, but a simplified reasonable approximation is to express its development simply as

$$\rho_s \frac{\partial Z_s}{\partial t} = v_a\left[\rho_0 - \rho_{vap}(T_0)\right] \quad (14)$$

We assume that the icy surface is unmoving, but this assumption may be relaxed by adding a term to the righthand side of equation (14) representing the flow of ice, like a glacier law (e.g., see section 2.4.2).

While q remains non-zero, the atmosphere must supply molecules to (or remove them from) the surface. If the volatiles are a minor constituent of the atmosphere, then this supply is advected across the near surface by the atmospheric flow or diffuses toward/from the surface. In the classic approach, with no flow replenishment, the latter is assumed (*Claudine et al.,* 2015). Thus, the equation for the concentration of the volatile gas and its diffusion is written on the assumption that the density of the ambient gas is nearly constant. In this case the concentration evolution can be written as an evolution equation for the gas density

$$\frac{\partial \rho}{\partial t} + u \cdot \nabla\rho = \nabla \cdot D_\rho \nabla\rho \quad (15)$$

where u is the bulk motion of the ambient gas and D_ρ is the diffusion coefficient (units of $m^2\ s^{-1}$) — the former is usually ignored while the latter depends upon the turbulent state of the atmospheric flow. In either fluid dynamical state, an estimated value for the diffusion coefficient comes from either experimental measurement or empirical extrapolation from known relationships. The final step toward a complete description involves linking the sublimation/deposition flux into and out of the surface Z_s, equation (14), to the diffusive gradient of the volatile species by imposing at the surface the condition

$$\hat{\mathbf{n}}_s \cdot D_\rho \nabla\rho_{z=Z_s} = q = v_a\left[\rho_0 - \rho_{vap}(T_0)\right] \quad (16)$$

noting that the gradient of the density is evaluated at the surface. The total evolution involves the simultaneous solution of equations (9)–(16), subject to far-field boundary conditions.

The visual light attenuation depth, Λ, also depends on direct measurements or detailed radiative transfer calculations of the penetration of light into the ice in question — whether it has the consistency of loosely packed porous snow [e.g., like the snow measurements by *Warren et al.* (2006)] or hard-packed crystalline or polycrystalline ice. When the materials are mixtures of multiple subliming species, equations monitoring the evolution of each of the constituent species must also be considered.

The focus of this discussion is not to develop solutions to these fundamental model equations but rather to highlight the importance of having reliable laboratory measurements of all the input quantities to produce reliable predictions of the processes at play. In other words, it is important to know precise values of temperature-pressure-dependent thermophysical quantities like ice densities, specific heats, diffusion rates, vapor pressures, thermal conductivities, and heats of phase changes from solid-liquid-gas. Most important is to understand what these thermophysical quantities are for mixtures, especially the characterization of vapor pressures.

2.4. Short Survey of Relevant Surface Phenomena

2.4.1. Solid-state convection. A viscous layer of depth H heated from below in a gravitational field (*g*) will undergo buoyant motions owing to the temperature dependence of a material's density. In classic isoviscous circumstances, the criterion for the onset of convection from the purely static conducting state depends upon the Rayleigh number, a non-dimensional measure of the destabilizing buoyant motions working against the stabilizing forces of viscous and thermal diffusion, which is defined as

$$Ra \equiv \frac{\rho^2 C_p g \alpha \Delta T H^3}{K\eta} \quad (17)$$

where α is the thermodynamically defined coefficient of thermal expansion at constant pressure, in units of K^{-1}, which is temperature and composition dependent, and is routinely experimentally determined. At onset, the temperature difference across the layer (ΔT) will depend upon incoming geothermal flux F_{geo} from beneath the layer, the layer's depth, and the material's conductivity according to $\Delta T = HF_{geo}/K$. It is generally assumed that at onset the material's rheology is Newtonian (*Turcotte and Schubert,* 2002) but not necessarily isoviscous. In the isoviscous case, the criterion for onset is for $Ra > Ra_c^{(0)}$, where the critical value $Ra_c^{(0)}$ depends upon both the thermal and stress boundary conditions employed for a model's linear analysis, but generally hovers around the value of 1000 for stress conditions resembling those for ice layers like Sputnik Planitia (see also *Chandrasekhar,* 1961). Theoretical studies of onset in Newtonian fluids when viscosity is temperature dependent show significant differences in the critical criterion (*Stengel et al.,* 1982; *Richter et al.,* 1983; *White,* 1988). When the temperature range across the layer corresponds to a significant difference in viscosity, the criterion for onset is considered in terms of the layer's Ra based on its basal viscosity, and is well approximated with

$$Ra \equiv \frac{\rho^2 C_p g \alpha \Delta T H^3}{K\eta_b} > Ra_c \quad (18)$$

in which the critical Ra (Ra_c) is Ra evaluated at the critical value of the temperature difference and layer depth, i.e., $Ra_c \equiv Ra(H = H_c)$. Ra_c must satisfy

$$\frac{\rho^2 C_p g \alpha \Delta T H_c^3}{K\eta_b} = Ra_c^{(0)} + 20.9 \ln\left(\frac{\eta_t}{\eta_b}\right) \quad (19)$$

where η_b and η_t respectively denote the viscosity at the bottom and top of the layer (*Turcotte and Schubert,* 2002; for a recent analysis see *Umurhan et al.,* 2016). As such, it is essential to ascertain an accurate measure of the activation energies and associated diffusion constants appearing in equation (7).

Several studies have investigated the fate of solid-state convection of N_2 ice as applied to Pluto's Sputnik Planitia. At the time of this writing many facets regarding the correct physical characterization and fundamental nature of the underlying convection remain either unsettled or unknown (*McKinnon et al.,* 2016; *Trowbridge et al.,* 2016; *Vilella and Deschamps,* 2017). Of primary concern is the unknown explanation for the apparently low cell aspect ratio, which is intimately tied to questions concerning the correct characterization of the interior ice's constituency and associated rheology (*McKinnon et al.,* 2016). An estimate might be made of the temperature difference across a layer of depth H undergoing solid-state convection by appealing to the defined Nusselt number (Nu) relationship: When convection is vigorous, the primary pathway for heat transport is through rising plumes. The efficiency of this heat transport, in terms of a horizontal average, is a relationship between Ra and Nu, the latter of which is defined as the mean thermal heat flux (F) coming up through the layer divided by the enhanced heat flux due to convection motions, i.e., $Nu \equiv FH/K\Delta T$. In the purely conductive state Nu = 1. The relationship between Nu and Ra is generally determined from extensive numerical simulations (e.g., *Ahlers et al.,* 2009) and in the infinite Pr limit, where Pr is the Prandtl number defined as the ratio between the kinematic viscosity and the coefficient of thermal diffusion; this relationship is expressed by

$$Nu = \left(\frac{Ra}{Ra_c}\right)^{\beta} \quad (20)$$

in which the exponent, β, is also a weak function of Ra. The actual value of β depends on how large Ra is, but it generally lies between 2/7 for intermediate values $10^5 < Ra < 10^{10}$ (*Johnston and Doering,* 2009; *Sotin and Labrosse,* 1999) and asymptoting toward 1/3 for values $Ra >> 10^{10}$ based on experimental evidence and variational theoretical arguments (*Ahlers et al.,* 2009). [It should be noted that in regular convection, as this exponent approaches 1/3 convection becomes extremely vigorous and the thermal boundary layers adjust their thicknesses so as to be marginally stable to convection themselves (*Otero et al.,* 2004; *Doering and Constantin,* 1998; Howard, 1964). In practice, this means that the temperature difference across the layer is whatever it is at marginal onset.] Once a reliable estimate for the exponent is known, an actual temperature difference across a convecting layer may be estimated using the Ra-Nu relationship in equation (20) together with the Ra approximation of equation (18). Based on the analysis of numerical simulations of *Niemala et al.* (2000), *Trowbridge et al.* (2016) assumed $\beta \approx 0.313$ to estimate the temperature differences across the depth of Sputnik Planitia, assuming that the thermal heat flux through the layer is given by the geothermal heat flux estimated by *Robuchon and Nimmo* (2011), i.e., $F = F_{geo}$. However, it is important to emphasize that that a reliable estimate hinges on accurate information of the actual rheology of the underlying ice. Future refinements in this information will revise the current estimates of temperature differences across layers, surface temperature gradients, and associated rheological, structural, and transport properties and the like.

2.4.2. Glacial flow. A polycrystalline ice layer with a non-flat surface elevation, i.e., $Z_s(x,y,t)$, sitting atop bedrock surfaces Z_b, which may or may not itself be sloped, will flow like a glacier on some timescale. There are a variety of modeling methods and approaches to treat this kind of flow, but the simplest and sufficiently revealing tactic is via the vertically integrated approach (*Hindmarsh,* 2004; *Benn and Evans,* 2010). If the aspect ratio of the glacier dynamics is low, then using the shallow sheet approximation (SSA) is justified (*Hindmarsh,* 2004). The aspect ratio is defined by

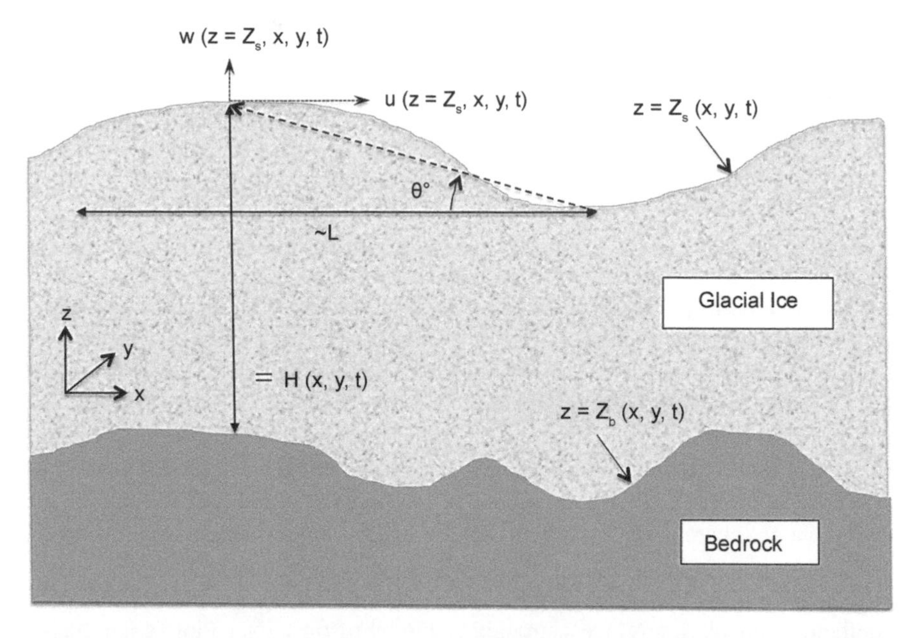

Fig. 9. Glacier flow modeling schematic. The dashed arrows indicate the local vertical and horizontal velocity of the glacial flow evaluated at the interface position $z = Z_S$. Adapted from *Umurhan et al.* (2017).

the vertical thickness (H) divided by the horizontal scale (L) of the ice layer (see Fig. 9), i.e., $\gamma \equiv H/L$ where H is the ice thickness and is related to the other variables according to $Z_s(x, t) = H(x, t) + Z_b(x, t)$. For this illustration we consider ice flowing in the horizontal x and vertical z directions only, keeping in mind that the results henceforth developed generalize into arbitrary horizontal directions (see further explanation below). Thus, when $\gamma \ll 1$, then in the SSA, to lowest-order terms in powers of γ, the stress-strain rate relationship provides a differential relationship between the horizontal velocity and the applied tangential stress arising from horizontal gradients (due to variations in Z_s) of ice overburden pressure. Integration of the lowest-order tangential stress equation and imposing the condition that it be zero at the surface results in the following relationship for the horizontal velocity

$$\eta \frac{\partial \mathbf{u}}{\partial z} = -\rho g(Z_s - z)\tan\theta; \ \tan\theta = \nabla Z_s \quad (21)$$

On the assumption that the glacier is dry and the ice is incompressible, this equation is integrated upward to develop a solution $\mathbf{u}(x,z,t)$ subject to the boundary condition that $u(z = Z_b) = 0$. Based on this solution for u and the incompressibility relationship, a solution to the vertical velocity is readily derived, i.e.,

$$\frac{\partial u}{\partial x} + \frac{\partial w}{\partial z} = 0, \rightarrow w(z) = -\int_{Z_b}^{z} \frac{\partial u(z', x, t)}{\partial x} dz' \quad (22)$$

As written, the vertical velocity is zero at the bedrock interface $z = Z_b$.

In the terrestrial setting, a glacier is either wet or dry depending on whether the base of the glacier supports liquid H_2O or some kind of slushy wet H_2O ice state. In such circumstances, the presence of a low-viscosity material at the base changes the flow characteristics of the glacier and plays an important role in eroding the underlying bedrock below (see below). What a wet N_2 ice glacier (i.e., a wet slushy N_2 base) would look like and what its erosive characteristics would be is currently unknown and not intuitively obvious because the relative densities between the liquid and solid state of N_2 is opposite to that of H_2O. Further exploratory theoretical research is required to address this matter.

We review the procedure to obtain an evolution equation for the ice thickness H, which requires following the evolution of the surface interface position $z = Z_s(x, t)$. On the physically motivated basis that the surface interface is a material boundary and that the flow beneath it is incompressible with zero porosity, the Lagrangian time derivative of the material boundary gives the relationship kinematic $dZ_s/dt = w(z = Z_s)$ or

$$\frac{\partial Z_s}{\partial t} + u(z = Z_s, x, t)\frac{\partial Z_s}{\partial x} = w(z = Z_s, x, t) \quad (23)$$

meaning that the material boundary is advected by the glacier's internal velocity field evaluated at the boundary's position (e.g., *Chandrasekhar,* 1961; *Whitham,* 1999). This sequence of calculations is a mathematically rigorous approach to derive a glacier flow evolution equation to first-order accuracy, and formally generalizes the vertically integrated approach of previous derivations (*Balmforth and Craster,*

1999, 2000). Expressing the kinematic boundary condition in terms of H and Z_b, assuming Z_b is time independent, and utilizing the generic Glen law form for the constitutive stress-strain rate relationships (equation (2)), yields

$$\frac{\partial H}{\partial t} = \nabla \cdot \Xi \nabla Z_s - \mathbf{u}_s \cdot \nabla Z_s \quad (24)$$

where the surface horizontal velocity field is now generalized into both horizontal directions and written as the vector $u_s = u(x,y,t)\hat{x} + v(x,y,t)\hat{y}$. The first term on the righthand side of the above equation, which arises from the righthand side of equation (23), gives the evolution equation a diffusive character with a nonlinear diffusion coefficient (Ξ) that is, in general, a function of several functions, including H, Z_b, and other features that are characteristic of the local thermophysical environment.

For example, *Umurhan et al.* (2017) developed a functional form for this coefficient considering N_2 ice's strongly insulating characteristics and using the constitutive relationship reported by the experiments of *Yamashita et al.* (2010). In this study, the authors assumed that the ice layer is purely conductive and that its temperature profile adjusts to the impingent geothermal flux from the interior, estimated by *Robuchon and Nimmo* (2011) to be $F_{geo} \approx 0.004$ W m^{-2}. Applying the above procedure yields the Arrhenius-Glen form for N_2 ice on Pluto

$$\Xi = \Xi_{dry};\ \underbrace{g_Q \exp\left[\frac{H/H_a}{1 + H/H_{\Delta T}}\right] \Xi_{glen}}_{\text{Arrhenius-Glen form}}, \qquad \Xi_{glen} \equiv \frac{A_s(\rho g H)^n H^2}{n+2}(\tan\theta)^{n-1} \quad (25)$$

The coefficient A_s is the same prefactor coefficient A appearing in the generic Glen Law constitutive relationship in equation (6). [Note: *Umurhan et al.* (2017) erroneously expressed the tanθ term as sinθ. This approximation to the correct term is only appropriate because θ denotes surface gradients on the ice that are generally low ($\theta < 20°$). It follows that $\tan\theta \approx \sin\theta$ and the errors in the simulation results displayed in the previous study are relatively inconsequential, not affecting the qualitative trends reported therein.] T_s is a reference temperature usually taken to be the surface temperature of the glacial ice. The value of A_s, the exponent n, and the effective activation energies are based on the fitted values based on the one-rheology experiment results for N_2 carried out by *Yamashita et al.* (2010) (see in particular equation (36)). The Arrhenius prefactor contains information about the activation energy and underlying thermal gradient, wherein

$$H_{\Delta T} \equiv T_s / T_{0z},\ H_a \equiv T_s^2 / (T_{0z} T_{act})$$

where a specific value for the mean temperature gradient inside the flowing ice is given in terms of the typical nonporous conductivity of solid N_2, i.e., $T_{0z} = F_{geo}H/K(N_2)$. The non-dimensional number, g_Q, is an approximate scaling factor that ranges between 0.5 and 1. When updated experimental values or new constitutive relationships are developed for N_2 ices and their alloys, the input parameters for the above Arrhenius Glen law form will have to be updated.

The second term on the righthand side of equation (24), which physically captures the advection of the surface pattern by the horizontal velocity field, is usually dropped (as it was in *Umurhan et al.,* 2017) on account of the assumption that it is dwarfed in magnitude by the diffusive effect. This may not always be justified, especially if there are strong variations of viscosity with depth as noted by, e.g., *Balmforth and Craster* (1999). Future theoretical reexamination of this matter is warranted.

Figure 10 shows an application of this flow law for a model of flow over Pluto's landscape considering realistic N_2 deposition rates over the course of one Pluto Milankovitch cycle (e.g., *Bertrand et al.,* 2018). We note that further refinements of this flow law should take into account features like unsteady porosity of the ice layer and thermal expansion, as is done when modeling terrestrial glaciers and ice sheets (*Cuffey and Paterson,* 2010).

2.4.3. Ice penitentes and suncups. Through atmospheric exchange, icy surfaces in relatively dry environs can evolve into textured forms called suncups and penitentes. Suncups and penitentes develop under conditions of relative high surface temperatures and low atmospheric humidity over troughs, together with relatively low temperatures and high humidity over peaks and crests (*Amstutz et al.,* 1958; *Lliboutry,* 1954). For example, troughs deepen when the vapor pressure of the ice is high and relative vapor content in the atmosphere above is low, permitting a net loss over the localized bit of surface (a similar logic applies for what happens over peaks). In the terrestrial context, suncups are morphologically distinguished from penitentes owing to the former's relatively shallow degree of concavity and general roundedness while the latter is characterized by sharply pointed divides and rounded troughs (see Fig. 11). The emergence of and the relative spacing between structures also depends upon several input thermophysical effects and quantities, outlined in section 2.3, but perhaps most influential is the role of reflected and reradiated solar radiation. Theoretical modeling of penitentes and suncups considers their development under two general circumstances: (1) during periods of relatively rapid sublimation and deposition (*Betterton,* 2006; *Claudine et al.,* 2015) or (2) as a quasi-steady growth process primarily driven by differential insolation effects amplified by the surrounding landscape (*Moore et al.,* 2017). The physical regimes might be divided according to whether feature growth timescales are short or long compared to the net rate of deposition/sublimation of volatiles to/from the surface (see also discussion in *Saarloos,* 1987).

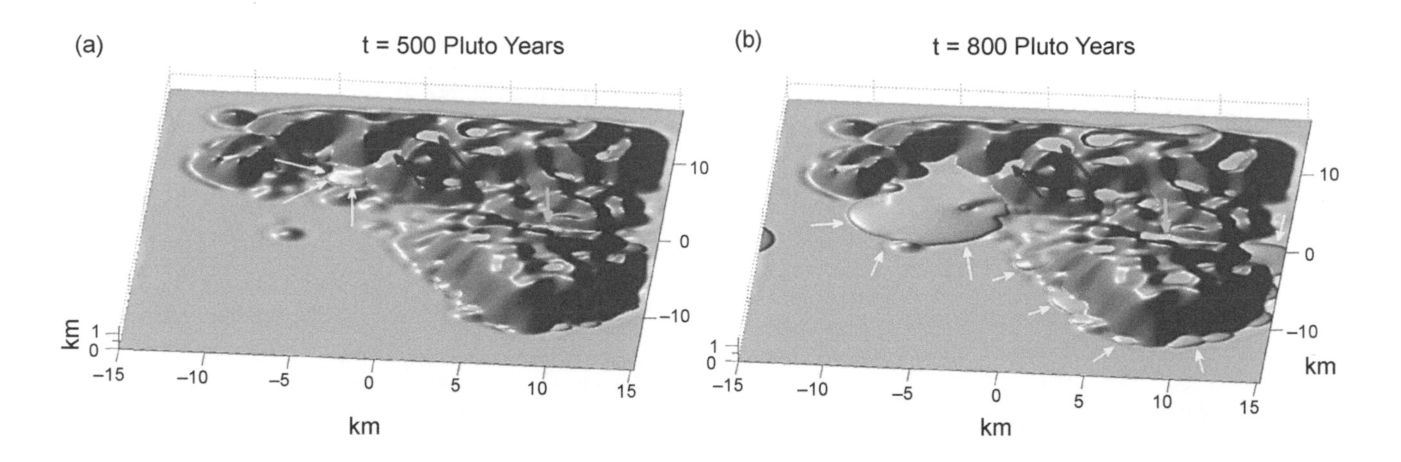

Fig. 10. A model of accumulation and flow of N_2 glacial ice over a model surface of Pluto's Eastern Tombaugh Regio using the simulation platform MARSSIM (*Howard,* 1994). The initially bare H_2O experiences steady N_2 deposition with no bedrock erosion but with N_2 deposition above 200 m at a rate m yr^{-1} following the climatic cycle modeling of *Bertrand et al.* (2018). N_2 ice is treated as flowing as a dry glacier according to the Arrhenius-Glen law formulation with a steady emergent geothermal flux of 4 mW m^{-2}. At early stages, N_2 ice from the surrounding topography steadily accumulates into depressions, which temporarily act as cold traps, e.g., like those indicated by red/aquamarine arrows. By a little under 500 Pluto years, the lower catchment [red arrow in **(a)**] containing an ever-deepening pool of N_2 ice brims over and begins the process of flooding the lower flats whose extent is indicated by the long yellow arrows in **(a)**. **(b)** After a few hundred orbits the glacial flow has expanded out onto the plain by several kilometers. The advancing glacial front experiences a boost: The rate of ice supplied by the lower catchment increases when the upper catchment [red arrow in **(b)**] starts to spill over. The steep hills facing the lower plains on the lower right of each frame eventually leaks N_2 as well (short yellow arrows). The aquamarine arrow indicates another elongated depression collecting N_2 ice from the surroundings.

The high-standing structures of the bladed terrain have been explained in terms the physics controlling the emergence of penitentes. Two studies have considered this to some extent: (1) *Moores et al.* (2017) applied the theory of *Claudine et al.* (2015) to explain the observed relative spacing of bladed terrain structures (dozens of kilometers). The theory of *Claudine et al.* (2015) examines the response of the surface while it undergoes net volatile deposition from the atmosphere using the model setting described in section 2.3. *Moores et al.* (2017) showed that three competing factors — thermal diffusion, enhanced self-illumination, and molecular diffusivity — conspire to induce a Mullins-Sekerka type of instability that amplifies crenellations on moving interfacial fronts (*Mullins and Sekerka,* 1964; *Saarloos,* 1987). [The Mullins-Sekerka instability involves the dynamics of an advancing front, e.g., involving a phase change effect like going from gas into solid. The speed of the propagating front normal to itself is often a monotonically increasing function of the gradient of some diffusive quantity (like temperature) found in the medium into which the front is propagating. For example, if the front defines the separation between a solid and gas, a forward protuberance from the solid side of the front induces distortions in the isolines of the diffusive quantity on the gassy side of the front that causes the tip of the protuberance to advance a little faster than other parts: If the diffusive quantity is temperature, then their temperature isolines crowd around the tip of the proterberance thereby inducing enhanced heat loss from the solid, cooling it, and consequently promoting gas to solidify on it. The effect induces exponential runaway pattern development (fractality). The process is generally controlled by some restoring effect, like surface tension, whose stabilizing action gets increasingly effective at increasingly finer scales. This effect has been used to model the formation of snowflakes and explain their fractal qualities. See cited references for more information.] *Moores et al.* (2017) can explain the observed relative separation of the structures in terms of an unknown length scale parameter indicating the height above the atmosphere on which the humidity is fixed, most likely related to the level of atmospheric turbulence of the CH_4 laden N_2 air flowing over the topography. However, *Moore et al.* (2017) examined the emergence of crenellated ice structures on a model with a non-advancing surface (cf. *Moores et al.,* 2017) with no differential thermal heat transfer and uniform albedo, but that the sublimation is proportional to the total local illumination, which depends upon the Sun's location, the shape of local topography, and the amount of reflected and reradiated light. The authors conclude that even in this minimal model one can generate high-standing structures like penitentes as well as low-aspect-ratio ones with features like the pits of Sputnik Planitia resembling terrestrial suncups.

A central mystery that remains is the question of how high-standing structures like the bladed terrain can

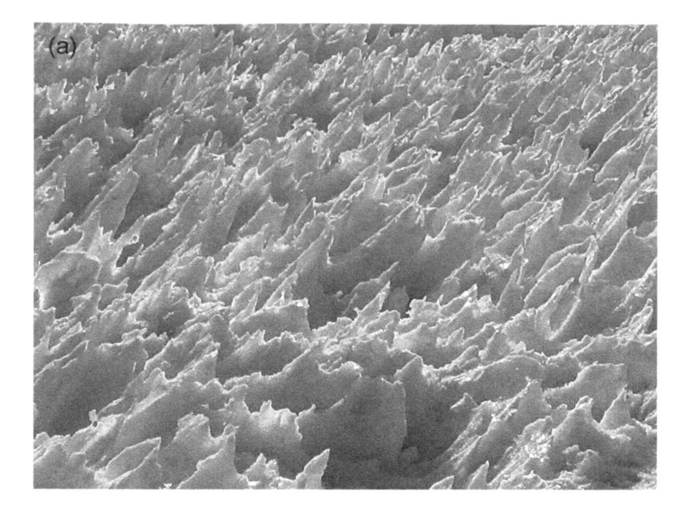

Fig. 11. **(a)** Penitentes growing on south-facing slopes of Mount Rainier. Structures exhibit 0.5-m vertical scales and their orientation and growth are southerly in the direction of the noon Sun Credit: Mark Sanderson, September 2006 (Wikipedia, CC BY 2.0) License, *https://creativecommons.org/licences/by/2.0/legalcode*. **(b)** Suncup field in Yosemite Valley taken during melt season (June 2011). Horizontal scales are 50 cm. Concavities are far wider than deep. Credit: Photo by Ben Ward and Sam Ward, from *On the Pacific Crest Trail* (January 2015), *http://www.theperipherymag.com/essay-pacific-crest-trail-part-2*.

remain so without flowing, which is really a question of the unknown rheology of the ice comprising them. If the bladed terrains are composed of CH_4 ice, then if the ice grains are larger than 1 cm, and based on the estimated rheologies of polycrystalline CH_4 ice grains, the aggregate ice structures are strong enough to support the towering structures for several tens of millions of years (*Moore et al.,* 2017). A similar open question pertains to the unknown rheology of N_2 ice pits on Sputnik Planitia: Are the pits in steady state or are they growing, and is the shape of the pit governed by the tendency for the ice to fill the pit back in while concomitantly subliming? This too centers on the unknown characterization of the rheology of N_2 ice or N_2-CO-CH_4 ice alloy. Furthermore, is it possible that the tholins act as a hardening agent for N_2 ice [e.g., consider the strength characteristics of laboratory generated tholins (*Yu et al.,* 2018)]? A comprehensive mathematical model that takes into account the flow of the underlying ice for a given rheologic constitutive relationship would involve adding glacial ice mass-flux terms like the righthand side of equation (24) to the righthand side of the equation set modeling the near surface physics, i.e., equations (9)–(16).

2.4.4. Solid-state greenhouse. The Voyager 2 flyby of Triton showed evidence of geyser-like plume activity — both prior to and ongoing during the encounter — that is largely associated with Triton's high-albedo southern polar terrain (SSG) (*Smith et al.,* 1989; *Soderblom et al.,* 1990; *Hansen et al.,* 1990). The physical driver of the plumes remains a mystery. The several proposed mechanisms, both endogenic and exogenic, are reviewed in *Kirk et al.* (1995). One process of possible relevance to Pluto's surface ices is the solid-state greenhouse effect (*Kirk et al.,* 1990), in which solid N_2 or CH_4 experiences a solid-to-gas phase transition at some shallow depth beneath the surface followed by pressurized escape. While New Horizons did not observe plumes or evidence of plume-like activity on Pluto, given that Triton's surface composition is probably like Pluto's, a process like SSG should be feasible on Pluto as well.

There are broadly three guises of the SSG that may be applicable (*Kirk et al.,* 1995). The first of these is the classical SSG, in which the translucent ice is homogenous with a single characteristic absorptivity, e.g., pure N_2 ice. The second is the "super" SSG (sSSG), in which an overlying nearly transparent layer of N_2 ice sits atop a highly absorbing layer of dirty ice, possibly containing organic material. With respect to these bookend scenarios, the presence of such transparent N_2 ice on Triton is circumstantially supported by the presence of N_2 ice blocks with scattering/absorption length scales of nearly 1 m (or larger) on the surface of Pluto's Sputnik Planitia (*Protopapa et al.,* 2017), as well as evidence wherein a several-meters-thick layer of transparent N_2 ice would explain the 2.15-mm absorption of Triton's spectrum (*Cruikshank et al.,* 1984; *Cruikshank and Apt,* 1984). Indeed, the possibility that near-surface layers of N_2 ice on Pluto may contain substantial amounts of low-albedo organics is also suggested by the darkened materials found in the shallow pits all over the southern reaches of Sputnik Planitia (*Moore et al.,* 2017; *Buhler and Ingersoll,* 2017). In either case, the temperature of the ice below even a few meters of translucent ice can result in the effective increase of N_2 vapor pressure. In the sSSG, an increase of 4–6 K underneath as little as 2–5 m of ice can result in vapor pressures exceeding the overburden pressure of the overlying ice layer (*Smith et al.,* 1989; *Matson and Brown,* 1989; *Kirk et al.,* 1990). The energy budget is also central: There must be a concomitant rise in the temperature of the ice and sufficiently high transport of insolation-sourced energy to drive the solid-vapor transition.

Once the conditions for subsurface vapor over-pressurization are met, the other half of the problem involves properly ascribing the means by which the gas makes its way to the surface. In the context of Triton's plumes — in which the plumes must emerge from an underground escape chute that is relatively small (<1–2 km in diameter) — this remains an unsolved matter. Currently there are at least three possible pathways, two involving direct vertical transport of gas, while the third requires some amount of lateral motion (*Kirk et al.,* 1995): The first involves the so-called leaky greenhouse, in which a sufficiently porous surface permits the overpressurized N_2 gas to escape directly to the surface. However, given the power constraints based on observations of Triton's plumes, the required emitting area is much larger than the plume source's diameter upper limit.

The champagne cork mode — inline with the image of a rapid depressurization — involves the catastrophic failure of the overlying ice's structural integrity together with sublimation-driven rapid outgassing emerging from a restricted cross-section. Numerical estimates show that while this is a plausible pathway, the sheer volume of N_2 ice required to sublime to supply the minimum amount of material is thought to be too large (i.e., requiring ice blocks of tens of meters). The third pathway, called the heat-pipe model, was proposed to circumvent the shortcomings of the first two methods wherein a subsurface vertically thin, cylindrically symmetric, and sufficiently porous "collector area," itself containing a smaller emitter disk region, produces highly pressurized vapor that laterally transports toward the centralized emitter region as a converging flow (*Kirk et al.,* 1990). This scenario addresses the small geyser region and the sublimating ice volume issue of the straightforward solar-geyser scenarios.

In the champagne cork model, undersurface complex porosity promotes lateral mass-energy transport over vertical gas and thermal diffusion. Lateral flow follows pathways excavated by sublimed ice. The model also posits a stark difference in the ice's effective thermal conductivity along the lateral conduits as compared to the effectively impermeable and insulating ice above. However, the means by which this state of affairs is reached have yet to be demonstrated. One possibility is that strong vertically confined variations in the vapor pressure promote flow and thermal transport asymmetries.

Kirk et al. (1990) propose to take the energy properties of these variations into account by positing that the gas obeys a Darcy's law, in which

$$\frac{\eta u}{k_m} = -\nabla P_{vap} - \rho_{vap} g \hat{z} \qquad (26)$$

wher η is the gas viscosity, k_m is the material permeability (in units of meters squared), and P_{vap} and ρ_{vap} are the vapor pressure and densities (respectively). The vapor pressure/densities and viscosities are strongly temperature dependent. Moreover, permeability is generally a function of porosity, as is known from the study of terrestrial ices (e.g., *Costa,* 2006; *Adolph and Albert,* 2014). For recent theoretical work on this see *Nishiyama and Yokohama* (2017). We note that porosity itself evolves over time due to compaction and sublimation (*Bierson and Nimmo,* 2019). What might be the functional form for the permeability of N_2 or CH_4 ices, or their mixtures, is currently not known. Solving the above for u, equation (26) may be expressed as a vapor mass flux $q_m \equiv \rho_{vap}u$. Invoking the sublimation processes described in section 2.3, if L denotes the enthalpy of vaporization, then the amount of heat harnessed to and from the ice (per second) in the subsurface sublimation process is. From mass continuity

$$\frac{\partial \rho_{vap}}{\partial t} + \nabla \cdot \rho_{vap}\mathbf{u} = 0 \qquad (27)$$

the righthand side of equation (9) must include the expression

$$Q_{int} = \phi L V \cdot (k_m/\eta)\left(\rho_{vap}\nabla P_{vap} + \rho_{vap}^2 g\,\hat{z}\right) \qquad (28)$$

To completely specify the evolution of the system requires taking into account the evolution of the porosity (see *Bierson and Nimmo,* 2019), which changes on account of the viscous flow of the ice as well as the interior sublimation into pore spaces that causes the porosity to increase. A useful setting showing how the various processes unfold together is the water-driven slotted pore model explaining the maintenance of tidally sourced plume activity on Enceladus (*Kite and Rubin,* 2016).

2.4.5. Erosion. Erosion occurs when the stresses on a surface exceed the strength of the bedrock material. On Pluto, the fundamental bedrock is assumed to be primarily a H_2O ice base with a surficial cover of CH_4 ice (*Grundy et al.,* 2016; *Protopapa et al.,* 2017; *Moore et al.,* 2017). While the strength of H_2O ice is roughly known for the temperatures appropriate to Pluto, the strength of the materials doing the erosion (whether they be CH_4 or some organic tholin-CH_4 mixture) is not known and remains to be determined from laboratory experiments under the appropriate conditions. Some relevant erosional processes (borrowed in part from the literature on terrestrial glaciation) are briefly summarized below.

2.4.5.1. Abrasion. A basally wet moving glacier will drive abrasion through the scraping of surfaces by debris in the basal ice (*Sugden and John,* 1976; *Hallet,* 1979; *Iverson,* 1991; *Clarke,* 2005). In Pluto's context, clasts are assumed to be H_2O. The degree to which H_2O ice debris abrades the bedrock depends upon debris concentration (*Hallet,* 1981), the hardness of the ices involved (e.g., *Fish and Zaretsky,* 1997; *Boulton,* 1974), the speeds by which the debris moves over the bed — slightly less than the speed of the glacier itself, depending heavily on the nature of the near-bed glacial ice — and the contact forces that debris fragments exert downward against the substrate (e.g., *Hallet,* 1979, 1981). Additionally, circumstances may arise in which fine debris polishes the bedrock by removing small-scale bed protuberances (*Benn and Evans,* 2010). Terrestrial glaciation models of abrasion have posited the erosion rates to depend on ~u_s^2, where u_s is the basal sliding speed (e.g., *Macgregor et al.,* 2009). Sliding speed and other variables (e.g., the amount of abrasive debris available) depend upon the density and dynamical state of the wet layer, i.e., whether the basal ice layer is wet and whether the liquid remains bound at depth or instead rises due to buoyancy effects.

2.4.5.2. Phase-change-induced hydrofracture. The strength of porous rock infiltrated with liquid H_2O depends on effective stress, which is the difference between the total stress from grain-grain friction and pore-water pressure. Hydrofracture occurs when the water pressure increases, thereby causing the rock strength to decrease past the applied stress (*Kamb and Engelhardt,* 1987; *van der Veen,* 1996). An analogous effect can occur if porous or cracked H_2O ice bedrock becomes infiltrated by solid N_2. If the overlying N_2 ice layer thickens, then owing to the low thermal conductivity of the solid-state N_2, solid-liquid phase transition could occur for N_2 ice within the bedrock. This transition will increase the pressure inside pores and cracks, and this in turn should promote a hydrofracture, with maximum destruction at the N_2 ice-bedrock interface (*Umurhan et al.,* 2016). A similar process should occur for replacing N_2 with CH_4 as well as an N_2/CH_4 mixture.

2.4.5.3. Quarrying. Quarrying involves subjecting fracture-weakened bedrock to enhanced basal stress by the overlying ice (e.g., *Clarke,* 2016). In terrestrial glaciation, quarrying is especially effective for glacier ice moving over step cavities containing liquid H_2O (*Iverson,* 1991, 2012; *Ugelvig et al.,* 2016): If the pressure in the liquid state is less than the overburden pressure of the ice, then the part of the bedrock directly supporting glacier ice must compensate by providing additional pressure support and will consequently experience elevated stress. This stress can cause fracture-laden bedrock to approach its brittle failure point. This process is likely most relevant for Pluto in the immediate aftermath of its early bombardment, when the H_2O ice crust contained fresh surface fractures. The applicability of this process and its corresponding erosion rates depends upon the stability of similar pockets of liquid N_2 (i.e., longevity before refreezing) and upon bedrock drainage and crack size distributions (e.g., *Keane et al.,* 2016, and references therein). Estimates for the latter effect might be guided by terrestrial glaciation studies (e.g., *Jaeger and Cook,* 1979) and examined for Pluto's H_2O (*Nimmo et al.,* 2016; *Hamilton et al.,* 2016). Assuming a stable wet base, an erosion-by-quarrying model for Pluto's N_2 glaciations could be based on the formalism of *Iverson* (2012). Terrestrial glaciation theory indicates erosion rates by quarrying should be ~u_s^2 (e.g., *Macgregor et al.,* 2009; *Ugelvig et al.,* 2016).

3. THERMOPHYSICAL AND RHEOLOGICAL DATA

3.1. Presentation of Review Collection

Here we wish to provide an easy, accessible collection of thermophysical and rheological calculations from several experimental techniques and low temperature conditions <100 K. We have surveyed experimental work for 14 different thermophysical properties for ices that could be present in the outer solar system and astrophysical environments. Of course, there is an impressive amount of dedicated literature on these ices; here we will give experimental values only for <100 K temperature conditions.

3.1.1. Crystal structure types. The processes and scales of interest covered in this review pertains primarily to polycrystalline ices. The crystal structure within the grain shapes many features of the grain itself, including how deformation may occur across the grain as well as serving as a source of latent heat when the crystal changes from one phase to another. We briefly review these fundamentals. In a basic definition, crystal structure of a crystalline material is the ordered arrangement of the molecular constituents. These ordered structures occur in a mineralogical nature to form symmetric patterns that repeat along a three-dimensional space in matter (*Hook and Hall,* 1991). Two main components create the blueprint for a crystal structure: the basis (chemistry of the building blocks) and the lattice (the translational symmetry, or arrangement in two-dimensional space). The two-dimensional Bravais lattice shape (square, hexagonal, oblique, etc.) is then repeated to a three-dimensional Bravais lattice shape (cubic, tetragonal, monoclinic, hexagonal, etc.). A note on the cubic crystal structure is that it depends on the innermost molecule and symmetry of the constituents involved that the cubic phase could be deemed simple, face-centered, or body-centered. Some crystal structures could also be considered disordered, caused by the partial (or complete) loss of the three-dimensional periodicity of a physical property, typically the center of mass (*Prager et al.,* 1988). For ices in the outer solar system, cubic and hexagonal are the typical crystalline structures of ice.

3.1.2. Ice crystallinity. Water ice has two arrangements in the low-temperature, low-pressure regime, namely Ih [hexagonal symmetry in a tetrahedral arrangement (*Brumberg et al.,* 2017)] and Ic [metastable cubic variant at <140 K, where oxygen atoms are arranged in a diamond structure (*Efimov et al.,* 2011)]. Below conditions ≤77 K, water ice can exist in the amorphous phase (although it is not the thermodynamically favored state), where there is no distinct, organized crystalline atomic structure (*Efimov et al.,* 2011; *Yokochi et al.,* 2012). In thermal equilibrium, CH_4 II ≥ 21 K the molecules are partially ordered while the center of mass structure is unchanged when transitioning to CH_4 I, a face-centered cubic (orientationally disordered) phase (*Prager,* 1988; *Khanna and Ngoh,* 1990). CO_2 is an amorphous phase solid <30 K, then transitions to a cubic phase (*de Bergh et al.,* 2008). The α-CO phase is cubic, then transitions to β-CO at ~61.6 K to the hexagonal crystalline structure (*Quirico and Schmitt,* 1997b). NH_3 is mostly amorphous <80 K, then a metastable cubic phase at ~120 K (*de Bergh et al.,* 2008). C_2H_6 (ethane) <25 K is in the amorphous phase, 25–60 K is metastable, and there are two currently observed crystalline phases at 88.9 K (*de Bergh et al.,* 2008). From experimental research, several properties of the ice can be observed by various techniques to describe crystallographic directions of the ice: optical constants (and refractive properties), adsorption, surface tension, plastic deformation, and dislocation. However, more work on the crystalline changes that take place at the phase transitions of these ices (including pure, complex mixtures and clathrate hydrate formations) must be continued to be comprehensive.

3.2. Comments on Specific Parameters

3.2.1. Triple points. In a thermophysical definition, the triple point of a substance is the junction where the respective phases (solid, liquid, vapor) can coexist in thermodynamic equilibrium. Table 1 summarizes the values of the triple points for several ice species. This information can be found in the National Institute of Standards and Technology (NIST) Chemistry WebBook (*http://webbook.nist.gov/chemistry*), the Air Liquide database, and the CRC *Handbook of Chemistry and Physics* (cited in *Lide,* 2006; *Fray and Schmitt,* 2009). Figure 12 displays the values from Table 1.

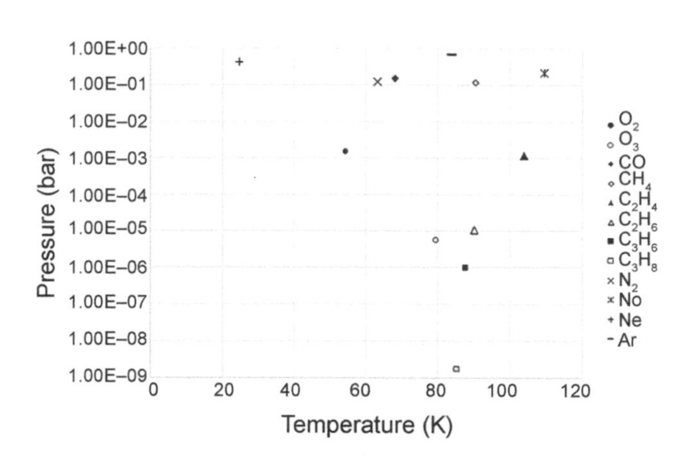

Fig. 12. Triple points of selected ice species from Table 1. Values listed in *Fray and Schmitt* (2009).

TABLE 1. Specific ice species with triple points at <100 K temperatures.

Species	Temperature (K)	T ± (K)	Pressure (bar)	P ± (bar)
O_2	54.33	0.006	0.00149	0.03
O_3	79.6	0.3	0.0000056	0.6
CO	68.1	0.1	0.1537	0.0003
CH_4	90.6854	0.0003	0.11696	0.00002
C_2H_4	104	0.1	0.0012	0.0001
C_2H_6	90.34	0.01	0.000011	0.1
C_3H_6	87.8	0.8	0.000001	–
C_3H_8	85.47	0.05	1.685E-09	0.001
N_2	63.14	0.06	0.1255	0.0005
NO	109.5	0.1	0.219	0.001
Ne	24.56	0.05	0.433	0.003
Ar	83.8058	–	0.68891	0.00002
2H	18.63	–	0.1712	–
CO_2	216.58	0.03	5.185	0.005
NH_3	195.41	0.01	0.0609	0.0003

TABLE 2. Density laboratory measurements <100 K temperatures.

O_2	T (K)	Density (g cm^{-3})	Reference	CO_2	T (K)	Density (g cm^{-3})	Reference	N_2	T (K)	Density (g cm^{-3})	Reference
	4.2	1.542	*Roder* (1978)		10	1	*Satorre et al.* (2008)		8	1.032	*Scott* (1976)
	18.75	1.542	*Roder* (1978)		12	1	*Satorre et al.* (2008)		10	0.93	*Satorre et al.* (2008)
	20	1.542	*Roder* (1978)		14	1	*Satorre et al.* (2008)		10	1.031	*Scott* (1976)
	22	1.54	*Roder* (1978)		18	1.05	*Satorre et al.* (2008)		12	1.031	*Scott* (1976)
	23.88	1.537	*Roder* (1978)		30	1.17	*Satorre et al.* (2008)		13	0.95	*Satorre et al.* (2008)
	23.88	1.527	*Roder* (1978)		35	1.19	*Satorre et al.* (2008)		14	1.03	*Scott* (1976)
	24	1.527	*Roder* (1978)		40	1.25	*Satorre et al.* (2008)		16	0.93	*Satorre et al.* (2008)
	26	1.522	*Roder* (1978)		45	1.35	*Satorre et al.* (2008)		16	1.029	*Scott* (1976)
	28	1.518	*Roder* (1978)		50	1.45	*Satorre et al.* (2008)		18	1.028	*Scott* (1976)
	30	1.512	*Roder* (1978)		55	1.49	*Satorre et al.* (2008)		19	0.92	*Satorre et al.* (2008)
	32	1.507	*Roder* (1978)		65	1.49	*Satorre et al.* (2008)		20	1.027	*Scott* (1976)
	34	1.5	*Roder* (1978)		70	1.48	*Satorre et al.* (2008)		22	0.97	*Satorre et al.* (2008)
	36	1.494	*Roder* (1978)		75	1.56	*Satorre et al.* (2008)		22	1.025	*Scott* (1976)
	38	1.487	*Roder* (1978)		77	1.54	*Satorre et al.* (2008)		24	1.023	*Scott* (1976)
	40	1.479	*Roder* (1978)		80	1.51	*Satorre et al.* (2008)		26	1.021	*Scott* (1976)
	42	1.471	*Roder* (1978)		85	1.46	*Satorre et al.* (2008)		28	1.019	*Scott* (1976)
	43.801	1.463	*Roder* (1978)						30	1.016	*Scott* (1976)
	43.801	1.388	*Roder* (1978)						32	1.013	*Scott* (1976)
	44	1.388	*Roder* (1978)						34	1.009	*Scott* (1976)
	46	1.38	*Roder* (1978)						35	1.007	*Scott* (1976)
	48	1.373	*Roder* (1978)						37	0.995	*Scott* (1976)
	50	1.366	*Roder* (1978)						38	0.993	*Scott* (1976)
	52	1.359	*Roder* (1978)						40	0.989	*Scott* (1976)
	54	1.352	*Roder* (1978)						42	0.986	*Scott* (1976)
	54.361	1.351	*Roder* (1978)						44	0.984	*Scott* (1976)
									48	0.974	*Scott* (1976)
									52	0.966	*Scott* (1976)
									56	0.958	*Scott* (1976)
									60	0.949	*Scott* (1976)

TABLE 2. (continued).

CH_4	T (K)	Density (g cm^{-3})	Reference	NH_3	T (K)	Density (g cm^{-3})	Reference	C_2H_4	T (K)	Density (g cm^{-3})	Reference
	10	0.48	*Satorre et al.* (2008)		15	0.67	*Satorre et al.* (2017)		13	0.47	*Satorre et al.* (2017)
	15	0.45	*Satorre et al.* (2008)		20	0.73	*Satorre et al.* (2017)		16	0.51	*Satorre et al.* (2017)
	17	0.46	*Satorre et al.* (2008)		40	0.82	*Satorre et al.* (2017)		18	0.53	*Satorre et al.* (2017)
	22	0.48	*Satorre et al.* (2008)		60	0.9	*Satorre et al.* (2017)		20	0.56	*Satorre et al.* (2017)
	25	0.46	*Satorre et al.* (2008)		90	0.91	*Satorre et al.* (2017)		22	0.57	*Satorre et al.* (2017)
	27	0.48	*Satorre et al.* (2008)		100	0.85	*Satorre et al.* (2017)		26	0.58	*Satorre et al.* (2017)
	30	0.46	*Satorre et al.* (2008)						28	0.59	*Satorre et al.* (2017)
	35	0.49	*Satorre et al.* (2008)						30	0.58	*Satorre et al.* (2017)
									33	0.63	*Satorre et al.* (2017)
									35	0.65	*Satorre et al.* (2017)
									40	0.64	*Satorre et al.* (2017)
									45	0.66	*Satorre et al.* (2017)
									50	0.63	*Satorre et al.* (2017)
									55	0.64	*Satorre et al.* (2017)
									60	0.63	*Satorre et al.* (2017)
									65	0.64	*Satorre et al.* (2017)

For Pluto, the relatively low triple point of N_2 (63.14 K) and CO (68.1 K) facilitates melting, which can lead to formation (and erosion) of glacial structures at Sputnik Planitia, along with the initiation of a volatile cycle that results from pitting and deposition of frosts (*Buratti et al.,* 2017; *Howard et al.,* 2017b; *Moore et al.,* 2017; *Glein and Waite,* 2018).

3.2.2. Density. Crystallization is a process that can occur in different ways and can lead to differences in density profiles of various ices. Such differences can arise from the influences of phases and impurities in the ice matrix. Table 2 shows a collection of 100 experimental determinations of density, and they are plotted together in Fig. 13. O_2 has a slight decrease at ~23.88 K and a relatively larger decrease in density at ~43.8 K, which are the α-β and β-γ solid O_2 phase transitions, respectively (*Roder,* 1978; *Fray and Schmitt,* 2009). Mixed ices could present a difference in densities of the ice matrix (*Satorre et al.,* 2017), as tabulated in Table 3.

The density measurements from *Satorre et al.* (2008) show an interesting density change for carbon dioxide as compared to methane and nitrogen. The CO_2 density increases by 50% from 10 K to 80 K, with a more compact structure at the ≥50 K formation, implying a crystallization process or modification. There are also slight differences in N_2 experimental quantities <25 K (*Scott,* 1976; *Satorre et al.,* 2008).

In the case of Pluto, the density measurements are imperative for modeling several physical processes, such as basal stress, sublimation-deposition models (as previously discussed in section 2.3), solid-state convection where buoyant motions are dependent on density of the material (section 2.4.1), and abrasion (section 2.4.5). The current knowledge for the density measurements of Pluto-relevant ices is lacking the carbon monoxide constituent and water ice

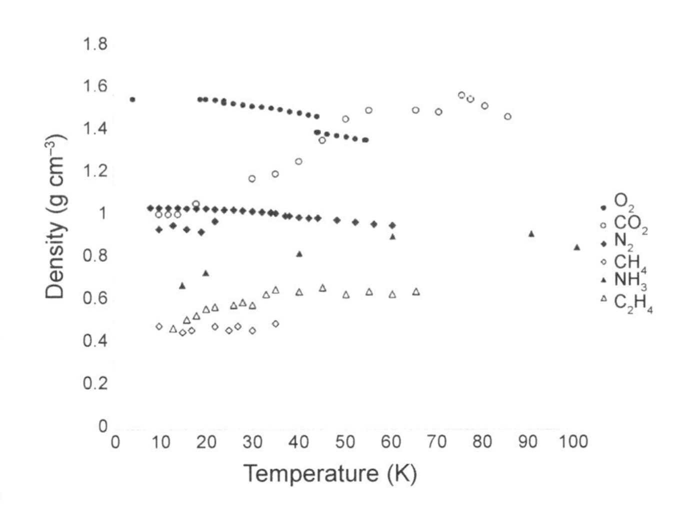

Fig. 13. Visual collection of the density measurements from Table 2. O_2 from *Roder* (1978); CO_2 and CH_4 from *Satorre et al.* (2008); N_2 from *Scott* (1976) and *Satorre et al.* (2008); NH_3 and C_2H_4 from *Satorre et al.* (2017). Note that the size of the symbols shown are larger than the errors presented in the respective references.

<100 K conditions, as well as different (relevant) mixtures of ices (e.g., CH_4-N_2-CO or H_2O-based mixtures).

3.2.3. Sublimation pressures. We have compiled a total of 140 experimental measurements for sublimation pressures for temperatures <100 K of 11 different ice species (see Table 4). Solid oxygen (O_2) has three distinct crystalline phases, with temperatures of the phase transitions of 23.781 K (α-β transition) and 43.772 K (β-γ transition) (*Mullins et al.,* 1963). Although the β and γ crystalline phases are mainly

TABLE 3. Ice mixture densities respective to molar ratios from *Satorre et al.* (2018).

CO_2-N_2	T (K)	mol% CO_2	Density (g cm^{-3})
	14	10	0.94
	14	20	0.87
	14	30	0.87
	14	50	0.89
	14	70	0.92
	14	90	0.95
CH_4-C_2H_6	T (K)	mol% C_2H_6	Density (g cm^{-3})
	30	0	0.44
	30	10	0.47
	30	30	0.51
	30	40	0.53
	30	50	0.55
	30	60	0.56
	30	70	0.58
	30	90	0.57
	30	100	0.54

TABLE 4. Table of experimental measurements of sublimation pressure as a function of 2138 temperature < 100 K.

Species	Temperature (K)	Pressure (Bar)	Reference	Species	Temperature (K)	Pressure (bar)	Reference	Species	Temperature (K)	Pressure (bar)	Reference
O_2	37	0.000000158	*Aoyama* (1934)	CO	55	0.00316	*Lide* (2006)	CH_4	45	0.00000126	*Tickner and Lossing* (1951)
	39.5	0.000001	*Aoyama* (1934)		57	0.01	*Lide* (2006)		49	0.00000199	*Tickner and Lossing* (1951)
	41.5	0.0000063	*Aoyama* (1934)		59	0.0158	*Shinoda* (1969)		53	0.000005012	*Tickner and Lossing* (1951)
	42	0.00000398	*Aoyama* (1934)		60	0.0251	Lide (2006)		55	0.0000158	*Tickner and Lossing* (1951)
	44	0.0000125	*Aoyama* (1934)		63	0.0398	*Shinoda* (1969)		58	0.0000794	*Armstrong et al.* (1955)
	45	0.000025	*Aoyama* (1934)		65	0.063	Lide (2006)		60	0.000107	*Armstrong et al.* (1955)
	47	0.000063	*Aoyama* (1934)		67	0.1047	*Shinoda* (1969)		65	0.000631	*Armstrong et al.* (1955)
	49.5	0.000158	*Aoyama* (1934)		69	0.12589	*Lide* (2006)		70	0.00199	*Lide* (2006)
	51	0.000398	*Aoyama* (1934)		70	0.158	*Lide* (2006)		75	0.00794	*Lide* (2006)
	54	0.001	*Aoyama* (1934)		75	0.2512	*Lide* (2006)		80	0.01584	*Lide* (2006)
	55	0.00158	*Lide* (2006)		80	0.6309	*Lide* (2006)		85	0.05011	*Lide* (2006)
	60	0.00631	*Lide* (2006)		81	1	*Lide* (2006)		90	0.1	*Lide* (2006)
	61	0.01	*Lide* (2006)		85	1.5848	*Lide* (2006)		95	0.1584	*Lide* (2006)
	65	0.016	*Lide* (2006)		90	2.512	*Lide* (2006)		100	0.2512	*Lide* (2006)
	70	0.063	*Lide* (2006)		95	3.981	*Lide* (2006)				
	71	0.1	*Lide* (2006)		100	6.3096	*Lide* (2006)				
	75	0.126	*Lide* (2006)								

TABLE 4. (continued).

Species	Temperature (K)	Pressure (Bar)	Reference	Species	Temperature (K)	Pressure (bar)	Reference	Species	Temperature (K)	Pressure (bar)	Reference
	80	0.316	*Lide* (2006)								
	85	0.631	*Lide* (2006)								
	90	1	*Lide* (2006)								
	95	1.58	*Lide* (2006)								
	100	2.512	*Lide* (2006)								

Species	Temperature (K)	Pressure (bar)	Reference	Species	Temperature (K)	Pressure (bar)	Reference	Species	Temperature (K)	Pressure (bar)	Reference
O_3	65	0.00000001	*Hanson and Mauersberger* (1986)	CO_2	70	1E-13	*Bryson et al.* (1974)	C_2H_2	98	0.000001122	*Tickner and Lossing* (1951)
	66	0.000000025	*Hanson and Mauersberger* (1986)		73	3.16E-13	*Bryson et al.* (1974)				
	67	3.98E-08	*Hanson and Mauersberger* (1986)		75	1.58E-12	*Bryson et al.* (1974)				
	68	6.31E-08	*Hanson and Mauersberger* (1986)		77	7.94E-12	*Bryson et al.* (1974)				
	69	0.000000079	*Hanson and Mauersberger* (1986)		79	2.51E-11	*Bryson et al.* (1974)				
	70	0.0000001	*Hanson and Mauersberger* (1986)		80	3.981E-11	*Bryson et al.* (1974)				
	72	0.000000158	*Hanson and Mauersberger* (1986)		83	1.58E-10	*Bryson et al.* (1974)				
	74	0.000000398	*Hanson and Mauersberger* (1986)		85	5.011E-10	*Bryson et al.* (1974)				

TABLE 4. (continued).

Species	Temperature (K)	Pressure (bar)	Reference	Species	Temperature (K)	Pressure (bar)	Reference	Species	Temperature (K)	Pressure (bar)	Reference
	75	0.000001	Hanson and Mauersberger (1986)		87	0.000000001	Bryson et al. (1974)				
	78	0.00000199	*Hanson and Mauersberger* (1986)		89	3.16E-09	*Bryson et al.* (1974)				
	79	0.00000398	*Hanson and Mauersberger* (1986)		90	6.31E-09	*Bryson et al.* (1974)				
	85	0.00001	*Lide* (2006)		93	1.26E-08	*Bryson et al.* (1974)				
	91	0.0001	*Lide* (2006)		95	5.012E-08	*Bryson et al.* (1974)				
	100	0.001	*Lide* (2006)		97	0.0000001	*Bryson et al.* (1974)				
					99	0.000000126	*Bryson et al.* (1974)				
					100	1.995E-07	*Bryson et al.* (1974)				

Species	Temperature (K)	Pressure (bar)	Reference	Species	Temperature (K)	Pressure (bar)	Reference	Species	Temperature (K)	Pressure (bar)	Reference
N_2	21	1E-13	*Borovik et al.* (1960)	C_2H_4	75	2.511E-07	*Menaucourt* (1982)	C_2H_6	85	0.000001	*Tickner and Lossing* (1951)
	22	1.58E-13	*Borovik et al.* (1960)		80	0.000001	*Menaucourt* (1982)		87	0.000001585	*Tickner and Lossing* (1951)
	23	6.31E-13	*Borovik et al.* (1960)		83	0.000002511	*Menaucourt* (1982)		89	0.00000631	*Tickner and Lossing* (1951)
	24	1E-12	*Borovik et al.* (1960)		85	0.000007943	*Menaucourt* (1982)		90	0.00001	*Lide* (2006)
	25	1.58E-12	*Borovik et al.* (1960)		87	0.00001122	*Menaucourt* (1982)		100	0.0001	*Lide* (2006)
	26	1E-11	*Borovik et al.* (1960)		89	0.0000199	*Menaucourt* (1982)				
	27	6.31E-11	*Borovik et al.* (1960)		91	0.0000631	*Menaucourt* (1982)				

TABLE 4. (continued).

Species	Temperature (K)	Pressure (bar)	Reference	Species	Temperature (K)	Pressure (bar)	Reference	Species	Temperature (K)	Pressure (bar)	Reference
	28	2.512E-10	*Borovik et al.* (1960)		95	0.000126	*Menaucourt* (1982)				
	35	0.00000631	*Frels et al.* (1974)		97	0.0002512	*Menaucourt* (1982)				
	38	0.00001	*Lide* (2006)		100	0.0005012	*Menaucourt* (1982)				
	41	0.0001	*Lide* (2006)								
	45	0.001	*Lide* (2006)								
	50	0.00398	*Lide* (2006)								
	55	0.01258	*Lide* (2006)								
	60	0.1	*Lide* (2006)								
	65	0.1258	*Lide* (2006)								
	70	0.5012	*Lide* (2006)								
	75	0.6309	*Lide* (2006)								
	80	1	*Lide* (2006)								
	85	1.585	*Lide* (2006)								
	90	3.981	*Lide* (2006)								
	95	3.981	*Lide* (2006)								
	100	10	*Lide* (2006)								

Species	Temperature (K)	Pressure (bar)	Reference	Species	Temperature (K)	Pressure (bar)	Reference
NO	63	5.012E-07	*Ernst* (1961)	N_2O	65	1E-12	*Bryson et al.* (1974)

TABLE 4. (continued).

Species	Temperature (K)	Pressure (bar)	Reference	Species	Temperature (K)	Pressure (bar)	Reference
	65	0.000001	*Ernst* (1961)		70	2.512E-12	*Bryson et al.* (1974)
	70	0.00000631	*Ernst* (1961)		73	1E-11	*Bryson et al.* (1974)
	72	0.00001	*Ernst* (1961)		75	3.98E-11	*Bryson et al.* (1974)
	76	0.0000631	*Ernst* (1961)		77	1.585E-10	*Bryson et al.* (1974)
	78	0.000126	*Ernst* (1961)		79	6.31E-10	*Bryson et al.* (1974)
	84	0.000631	*Ernst* (1961)		80	0.000000001	*Bryson et al.* (1974)
	85	0.001	*Ernst* (1961)				
	91	0.00794	*Ernst* (1961)				
	95	0.0126	*Ernst* (1961)				
	97	0.0158	*Johnston and Giauque* (1929)				
	100	0.0398	*Lide* (2006)				

listed, there are currently no experimental measurements performed for T < 23.781 for the α phase. Solid nitrogen (N_2) has two crystalline phases, α and β, with the solid-solid transition at 35.61 ± 0.5 K (*Giauque and Clayton,* 1933; *Prokhvatilov and Yantsevich,* 1983; *Grundy et al.,* 1993). In Fig. 14, there is a gap of experiments between 28 and 35 K. CO also has two crystalline phases, with the temperature of the phase transition at 61.55 K ± 0.5 K (α-β) (*Clayton and Giauque,* 1932; *Angwin and Wassermann,* 1966). Methane also goes through a phase transition (α-β) at ~21 K, particularly when mixed with nitrogen (*Prokhvatilov and Yantsevich,* 1983; *Tegler et al.,* 2010), but no sublimation-pressure experimental data have yet been recorded. Only one experimental point regarding the sublimation pressure was recorded for acetylene C_2H_2 for T < 100 K (*Tickner and Lossing,* 1951). Ethylene (C_2H_4) and ethane (C_2H_6) behave similarly in the temperature range (83–100 K) and pressures (10^{-6}–10^{-4} bar), although more work needs to be done at T < 65 K.

Nitrogen easily sublimates at Pluto's surface owing to its relatively high vapor pressure, which can lead to pitting and frost deposition at Sputnik Planitia, and largely contributing to the existence of an atmosphere (*Gladstone et al.,* 2016; *Buhler and Ingersoll,* 2017; *Glein and Waite,* 2018). However, as noted from the experimental collection, that data is lacking at 28–35 K for relevant plutonian ices, where CO could lead to better understanding of ices at Sputnik Planitia. Figure 14 only displays pure ice samples, while N_2-CO or CH_4-dominant mixtures are also lacking, which could influence the sublimation behavior at the surface of dwarf planets and KBOs.

3.2.4. Specific heat capacity. In general, the value of specific heat c depends on the ice matrix, the change of state involved, and the particular state of the system at the time of transferring heat, thus also dependent on temperature. Specific heats of solids, in the case of all ices, are only an approximation of temperature and pressure (*Journaux et al.,* 2019). Specific heat concerns the storage of thermal energy within the molecules of the ice substrate.

We have compiled a total of 392 experimental measurements for specific heat capacity ranging from 2 K to 100.74 K temperatures in Table 5. We note increases in the heat capacity of nitrogen (N_2) and methane (CH_4) at the triple points [63.14 K and 90.68 K, respectively (*Fray and Schmitt,* 2009)]. As shown in Fig. 15, there is also an increase in

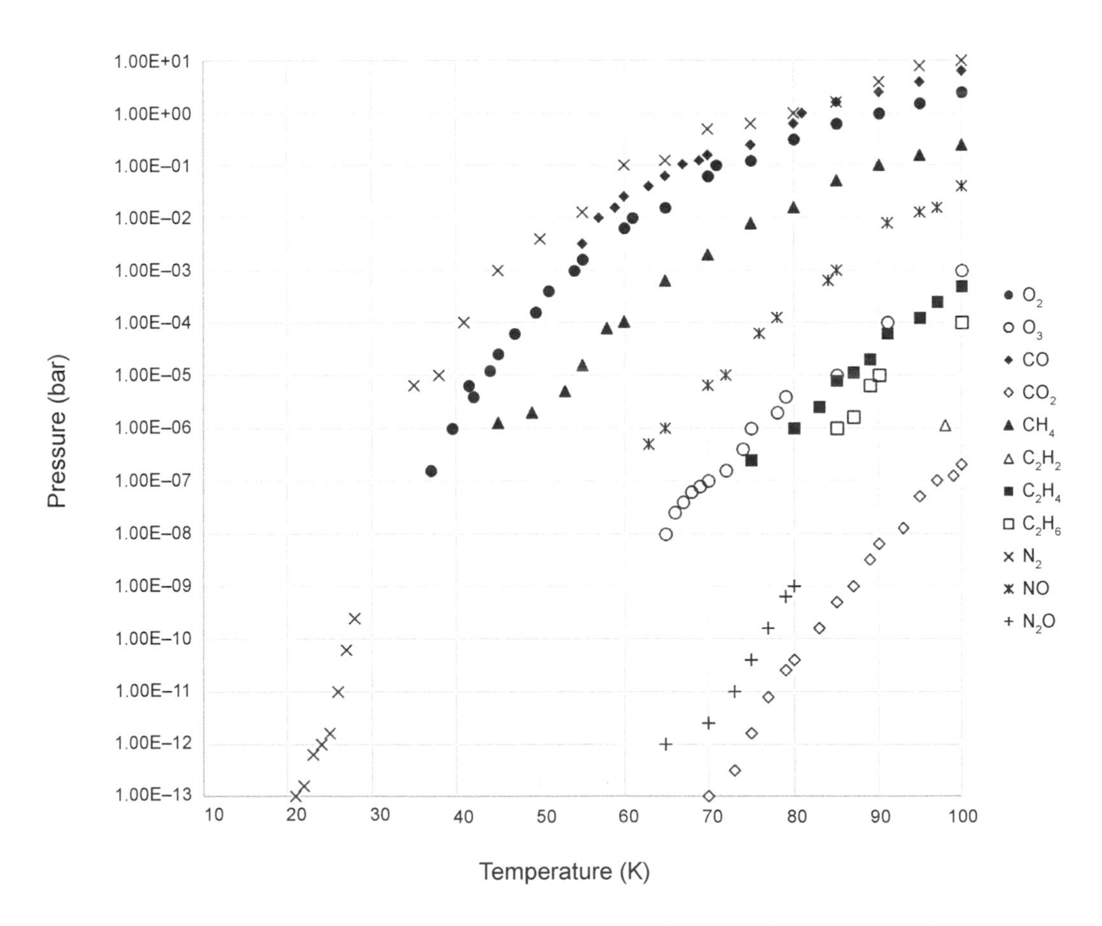

Fig. 14. Sublimation pressures of various ice compounds as a function of temperature. O_2 measurements from *Aoyama and Kanda* (1934) and *Lide* (2006); CO from *Shinoda* (1969) and *Lide* (2006); CO_2 and N_2O from *Bryson et al.* (1974); O_3 from *Hanson and Mauersberger* (1986) and *Lide* (2006); CH_4 from *Tickner and Lossing* (1951), *Armstrong et al.* (1955), and *Lide* (2006); C_2H_2 from *Tickner and Lossing* (1951); N_2 from *Borovik et al.* (1960), *Frels et al.* (1974), and *Lide* (2006); C_2H_4 from *Menaucourt* (1982); C_2H_6 from *Tickner and Lossing* (1951) and *Lide* (2006); and NO from *Johnston and Giauque* (1929), *Ernest* (1961), and *Lide* (2006).

TABLE 5. Table of experimental measurements of specific heat capacity [specific heat in units of (J mol^{-1} K^{-1})].

T	Ice Ih	References	T	N_2	References	T	CH_4	References	T	CO_2	Reference
2.144	0.002	*Flubacher* (1960)	8	2.34	*Scott* (1976)	5.493	2.26	*Colwell et al.* (1963)	90.05	0.97	*Maass and Barnes* (1926)
2.415	0.003	*Flubacher* (1960)	10	4.73	*Scott* (1976)	5.651	2.47	*Colwell et al.* (1963)	T	CO	References
2.548	0.003	*Flubacher* (1960)	12	7.75	*Scott* (1976)	5.881	2.39	*Colwell et al.* (1963)	8	1.26	*Krupskii et al.* (1973)
2.586	0.003	*Flubacher* (1960)	14	10.76	*Scott* (1976)	6.282	2.49	*Colwell et al.* (1963)	10	2.61	*Krupskii et al.* (1973)
2.686	0.003	*Flubacher* (1960)	15.82	13.08	*Giauque and Clayton* (1933)	7.014	3.34	*Colwell et al.* (1963)	12	4.33	*Krupskii et al.* (1973)
2.798	0.004	*Flubacher* (1960)	16	13.90	*Scott* (1976)	8.061	3.70	*Colwell et al.* (1963)	14	6.33	*Krupskii et al.* (1973)
2.895	0.005	*Flubacher* (1960)	17.66	16.08	*Giauque and Clayton* (1933)	9.409	4.50	*Colwell et al.* (1963)	14.36	6.85	*Clayton and Giauque* (1932)
3.089	0.005	*Flubacher* (1960)	18	16.62	*Scott* (1976)	9.741	4.78	*Colwell et al.* (1963)	16	8.81	*Krupskii et al.* (1973)
3.114	0.005	*Flubacher* (1960)	19.51	19.16	*Giauque and Clayton* (1933)	10.801	5.30	*Colwell et al.* (1963)	16.94	10.29	*Clayton and Giauque* (1932)
3.182	0.006	*Flubacher* (1960)	20	19.85	*Scott* (1976)	11.818	6.40	*Colwell et al.* (1963)	18	11.36	*Krupskii et al.* (1973)
3.397	0.008	*Flubacher* (1960)	21.94	22.81	*Giauque and Clayton* (1933)	11.904	6.26	*Colwell et al.* (1963)	19.37	13.68	*Clayton and Giauque* (1932)
3.582	0.008	*Flubacher* (1960)	22	23.03	*Scott* (1976)	12.065	6.52	*Colwell et al.* (1963)	20	13.92	*Krupskii et al.* (1973)
3.651	0.009	*Flubacher* (1960)	24	25.79	*Scott* (1976)	12.816	7.40	*Colwell et al.* (1963)	21.93	16.65	*Clayton and Giauque* (1932)
3.669	0.009	*Flubacher* (1960)	24.49	26.51	*Giauque and Clayton* (1933)	13.549	8.40	*Colwell et al.* (1963)	22	16.20	*Krupskii et al.* (1973)
3.725	0.010	*Flubacher* (1960)	24.85	26.71	*Giauque and Clayton* (1933)	14.046	9.06	*Colwell et al.* (1963)	24	18.21	*Krupskii et al.* (1973)

TABLE 5. (continued).

T	Ice Ih	References	T	N_2	References	T	CH_4	References	T	CO_2	Reference
3.914	0.012	*Flubacher* (1960)	26	28.39	*Scott* (1976)	15.137	10.74	*Colwell et al.* (1963)	24.31	19.15	*Clayton and Giauque* (1932)
4.044	0.012	*Flubacher* (1960)	27.14	30.02	*Giauque and Clayton* (1933)	15.918	12.16	*Colwell et al.* (1963)	26	20.10	*Krupskii et al.* (1973)
4.145	0.013	*Flubacher* (1960)	28	30.90	*Scott* (1976)	16.741	14.07	*Colwell et al.* (1963)	26.64	21.41	*Clayton and Giauque* (1932)
4.151	0.013	*Flubacher* (1960)	28.32	31.57	*Giauque and Clayton* (1933)	17.822	17.30	*Colwell et al.* (1963)	28	21.96	*Krupskii et al.* (1973)
4.65	0.020	*Flubacher* (1960)	29.89	34.07	*Giauque and Clayton* (1933)	18.274	18.53	*Colwell et al.* (1963)	29.01	23.79	*Clayton and Giauque* (1932)
4.741	0.021	*Flubacher* (1960)	30	33.91	*Scott* (1976)	19.315	25.56	*Colwell et al.* (1963)	30	23.61	*Krupskii et al.* (1973)
4.759	0.020	*Flubacher* (1960)	31.29	36.19	*Giauque and Clayton* (1933)	19.321	25.97	*Colwell et al.* (1963)	31.56	26.26	*Clayton and Giauque* (1932)
5.182	0.027	*Flubacher* (1960)	32	37.43	*Scott* (1976)	19.829	32.94	*Colwell et al.* (1963)	32	25.23	*Krupskii et al.* (1973)
5.272	0.029	*Flubacher* (1960)	32.84	39.34	*Giauque and Clayton* (1933)	20.003	37.24	*Colwell et al.* (1963)	34	26.74	*Krupskii et al.* (1973)
5.737	0.042	*Flubacher* (1960)	34	41.87	*Scott* (1976)	20.587	40.81	*Colwell et al.* (1963)	36	28.15	*Krupskii et al.* (1973)
5.738	0.038	*Flubacher* (1960)	34.42	43.04	*Giauque and Clayton* (1933)	20.697	26.40	*Colwell et al.* (1963)	38	29.45	*Krupskii et al.* (1973)
5.799	0.039	*Flubacher* (1960)	34.68	43.92	*Giauque and Clayton* (1933)	20.711	20.26	*Colwell et al.* (1963)	39.85	33.96	*Clayton and Giauque* (1932)
6.258	0.053	*Flubacher* (1960)	35	45.01	*Scott* (1976)	21.049	19.37	*Colwell et al.* (1963)	40	30.65	*Krupskii et al.* (1973)
6.724	0.068	*Flubacher* (1960)	35.05	45.38	*Giauque and Clayton* (1933)	22.248	19.89	*Colwell et al.* (1963)	42	31.82	*Krupskii et al.* (1973)

TABLE 5. (continued).

T	Ice Ih	References	T	N_2	References	T	CH_4	References	T	CO_2	Reference
6.735	0.069	*Flubacher* (1960)	35.33	44.67	*Giauque and Clayton* (1933)	23.536	20.51	*Colwell et al.* (1963)	44	32.88	*Krupskii et al.* (1973)
7.246	0.089	*Flubacher* (1960)	39.13	37.46	*Giauque and Clayton* (1933)	24.222	20.89	*Colwell et al.* (1963)	44.21	37.91	*Clayton and Giauque* (1932)
7.799	0.114	*Flubacher* (1960)	40	37.76	*Scott* (1976)	25.804	21.76	*Colwell et al.* (1963)	44.71	38.05	*Clayton and Giauque* (1932)
7.907	0.120	*Flubacher* (1960)	42	38.43	*Scott* (1976)	28.2	23.19	*Colwell et al.* (1963)	46	33.81	*Krupskii et al.* (1973)
8.786	0.173	*Flubacher* (1960)	43.27	39.04	*Giauque and Clayton* (1933)	30.789	24.66	*Colwell et al.* (1963)	47.9	41.40	*Clayton and Giauque* (1932)
8.898	0.180	*Flubacher* (1960)	44	39.15	*Scott* (1976)	33.47	26.11	*Colwell et al.* (1963)	48	34.64	*Krupskii et al.* (1973)
9.871	0.261	*Flubacher* (1960)	48	40.61	*Scott* (1976)	36.269	27.50	*Colwell et al.* (1963)	48.34	41.60	*Clayton and Giauque* (1932)
9.955	0.268	*Flubacher* (1960)	48.07	40.83	*Giauque and Clayton* (1933)	39.25	28.86	*Colwell et al.* (1963)	50	35.48	*Krupskii et al.* (1973)
11.098	0.383	*Flubacher* (1960)	51.88	42.24	*Giauque and Clayton* (1933)	50.572	32.65	*Colwell et al.* (1963)	52	36.21	*Krupskii et al.* (1973)
11.109	0.383	*Flubacher* (1960)	52	42.20	*Scott* (1976)	52.638	33.47	*Colwell et al.* (1963)	52.34	46.10	*Clayton and Giauque* (1932)
11.462	0.426	*Flubacher* (1960)	53.55	42.96	*Giauque and Clayton* (1933)	54.768	34.17	*Colwell et al.* (1963)	54	36.94	*Krupskii et al.* (1973)
12.228	0.522	*Flubacher* (1960)	55.88	43.71	*Giauque and Clayton* (1933)	56.813	34.78	*Colwell et al.* (1963)	55.07	49.11	*Clayton and Giauque* (1932)
12.431	0.556	*Flubacher* (1960)	56	43.71	*Scott* (1976)	58.784	35.33	*Colwell et al.* (1963)	56	37.46	*Krupskii et al.* (1973)

TABLE 5 (continued).

T	Ice Ih	References	T	N_2	References	T	CH_4	References	T	CO_2	Reference
12.549	0.570	*Flubacher* (1960)	57.99	44.59	*Giauque and Clayton* (1933)	60.69	35.84	*Colwell et al.* (1963)	56.82	53.21	*Clayton and Giauque* (1932)
13.525	0.716	*Flubacher* (1960)	60	45.55	*Scott* (1976)	85.099	42.37	*Colwell et al.* (1963)	58	38.12	*Krupskii et al.* (1973)
13.539	0.723	*Flubacher* (1960)	61.4	46.43	*Giauque and Clayton* (1933)	88.44	43.58	*Colwell et al.* (1963)	59	38.48	*Krupskii et al.* (1973)
14.509	0.887	*Flubacher* (1960)	61.41	46.35	*Giauque and Clayton* (1933)	90.37	55.73	*Colwell et al.* (1963)	59.04	56.98	*Clayton and Giauque* (1932)
14.625	0.903	*Flubacher* (1960)	65.02	55.81	*Giauque and Clayton* (1933)	93.391	52.96	*Colwell et al.* (1963)	60	38.73	*Krupskii et al.* (1973)
15.522	1.072	*Flubacher* (1960)	68.41	56.31	*Giauque and Clayton* (1933)				61.55	50.33	*Clayton and Giauque* (1932)
15.609	1.084	*Flubacher* (1960)	70.28	56.31	*Giauque and Clayton* (1933)				66.02	50.91	*Clayton and Giauque* (1932)
16.502	1.270	*Flubacher* (1960)	72.69	56.77	*Giauque and Clayton* (1933)				68.09	51.50	*Clayton and Giauque* (1932)
16.567	1.273	*Flubacher* (1960)	74.57	56.90	*Giauque and Clayton* (1933)				70.02	60.37	*Clayton and Giauque* (1932)
17.437	1.448	*Flubacher* (1960)	76.58	57.28	*Giauque and Clayton* (1933)				72.17	60.42	*Clayton and Giauque* (1932)
17.829	1.535	*Flubacher* (1960)	77.74	57.11	*Giauque and Clayton* (1933)				75.47	60.62	*Clayton and Giauque* (1932)
18.957	1.769	*Flubacher* (1960)							75.8	60.25	*Clayton and Giauque* (1932)
19.742	1.941	*Flubacher* (1960)							78.78	60.33	*Clayton and Giauque* (1932)

TABLE 5. (continued).

T	Ice Ih	References	T	N_2	References	T	CH_4	References	T	CO_2	Reference
21.06	2.239	*Flubacher* (1960)							79.06	60.62	*Clayton and Giauque* (1932)
23.108	2.700	*Flubacher* (1960)							80.61	60.71	*Clayton and Giauque* (1932)
25.089	3.141	*Flubacher* (1960)							83.39	60.29	*Clayton and Giauque* (1932)
27.034	3.563	*Flubacher* (1960)							84.66	60.50	*Clayton and Giauque* (1932)
T	CD_4	References	T	CH_4*	References	T	NH_3	References			
2.402	0.028	*Colwell et al.* (1963)	2.3	0.018	*Colwell et al.* (1963)	15.04	0.737	*Overstreet and Giauque* (1937)			
2.503	0.028	*Colwell et al.* (1963)	2.326	0.021	*Colwell et al.* (1963)	15.71	0.837	*Overstreet and Giauque* (1937)			
2.925	0.032	*Colwell et al.* (1963)	2.558	0.030	*Colwell et al.* (1963)	17.26	0.992	*Overstreet and Giauque* (1937)			
3.006	0.035	*Colwell et al.* (1963)	2.698	0.385	*Colwell et al.* (1963)	19.75	1.499	*Overstreet and Giauque* (1937)			
3.453	0.046	*Colwell et al.* (1963)	2.939	0.055	*Colwell et al.* (1963)	22.74	2.215	*Overstreet and Giauque* (1937)			
3.495	0.047	*Colwell et al.* (1963)	3.056	0.064	*Colwell et al.* (1963)	26.08	3.140	*Overstreet and Giauque* (1937)			
3.98	0.065	*Colwell et al.* (1963)	3.37	0.094	*Colwell et al.* (1963)	29.74	4.241	*Overstreet and Giauque* (1937)			
4.033	0.069	*Colwell et al.* (1963)	3.464	0.105	*Colwell et al.* (1963)	33.46	5.497	*Overstreet and Giauque* (1937)			
4.583	0.108	*Colwell et al.* (1963)	3.989	0.184	*Colwell et al.* (1963)	37.78	6.975	*Overstreet and Giauque* (1937)			

TABLE 5. (continued).

T	CD_4	References	T	CH_4^*	References	T	NH_3	References
4.613	0.113	*Colwell et al.* (1963)	4.274	0.234	*Colwell et al.* (1963)	42.32	8.503	*Overstreet and Giauque* (1937)
5.488	0.228	*Colwell et al.* (1963)	4.529	0.282	*Colwell et al.* (1963)	46.79	10.011	*Overstreet and Giauque* (1937)
6.421	0.457	*Colwell et al.* (1963)	4.662	0.310	*Colwell et al.* (1963)	51.23	11.547	*Overstreet and Giauque* (1937)
7.442	0.885	*Colwell et al.* (1963)	5.079	0.419	*Colwell et al.* (1963)	56.11	13.285	*Overstreet and Giauque* (1937)
8.525	1.489	*Colwell et al.* (1963)	5.346	0.444	*Colwell et al.* (1963)	61.24	14.934	*Overstreet and Giauque* (1937)
9.473	2.193	*Colwell et al.* (1963)	5.662	0.548	*Colwell et al.* (1963)	65.47	16.383	*Overstreet and Giauque* (1937)
10.372	2.994	*Colwell et al.* (1963)	5.834	0.641	*Colwell et al.* (1963)	69.69	17.656	*Overstreet and Giauque* (1937)
11.241	3.890	*Colwell et al.* (1963)	6.057	0.590	*Colwell et al.* (1963)	74.63	19.104	*Overstreet and Giauque* (1937)
11.835	4.593	*Colwell et al.* (1963)	6.127	0.695	*Colwell et al.* (1963)	79.56	20.624	*Overstreet and Giauque* (1937)
12.38	5.234	*Colwell et al.* (1963)	6.664	0.942	*Colwell et al.* (1963)	84.05	21.914	*Overstreet and Giauque* (1937)
11.835	4.593	*Colwell et al.* (1963)	6.127	0.695	*Colwell et al.* (1963)	79.56	20.624	*Overstreet and Giauque* (1937)
12.38	5.234	*Colwell et al.* (1963)	6.664	0.942	*Colwell et al.* (1963)	84.05	21.914	*Overstreet and Giauque* (1937)
13.986	7.503	*Colwell et al.* (1963)	7.536	1.223	*Colwell et al.* (1963)	88.7	23.174	*Overstreet and Giauque* (1937)
15.894	10.697	*Colwell et al.* (1963)	8.461	1.725	*Colwell et al.* (1963)	93.44	24.413	*Overstreet and Giauque* (1937)
17.769	14.541	*Colwell et al.* (1963)	9.739	2.567	*Colwell et al.* (1963)	98.32	25.841	*Overstreet and Giauque* (1937)
17.769	14.541	*Colwell et al.* (1963)	9.739	2.567	*Colwell et al.* (1963)	98.32	25.841	*Overstreet and Giauque* (1937)
19.756	19.665	*Colwell et al.* (1963)	T	NO	References	T	C_3H_8O	References
21.357	26.599	*Colwell et al.* (1963)	15.57	3.915	*Johnston and Giauque* (1929)	19.49	8.106	*Kelley* (1929)
22.379	45.259	*Colwell et al.* (1963)	17.75	5.455	*Johnston and Giauque* (1929)	23.82	11.510	*Kelley* (1929)
22.412	45.887	*Colwell et al.* (1963)	19.51	6.657	*Johnston and Giauque* (1929)	26.88	14.290	*Kelley* (1929)
22.671	25.824	*Colwell et al.* (1963)	22.27	8.223	*Johnston and Giauque* (1929)	29.99	16.919	*Kelley* (1929)
22.847	26.168	*Colwell et al.* (1963)	25.36	10.002	*Johnston and Giauque* (1929)	34.03	20.059	*Kelley* (1929)
23.57	28.751	*Colwell et al.* (1963)	28.59	11.752	*Johnston and Giauque* (1929)	38.68	23.496	*Kelley* (1929)
24.005	30.744	*Colwell et al.* (1963)	32.08	13.708	*Johnston and Giauque* (1929)	43.02	26.548	*Kelley* (1929)
24.516	33.808	*Colwell et al.* (1963)	35.5	15.453	*Johnston and Giauque* (1929)	46.9	29.006	*Kelley* (1929)
25.503	42.580	*Colwell et al.* (1963)	39.2	16.998	*Johnston and Giauque* (1929)	51.04	31.501	*Kelley* (1929)
25.876	48.316	*Colwell et al.* (1963)	43.51	18.761	*Johnston and Giauque* (1929)	55.41	33.871	*Kelley* (1929)
26.36	60.625	*Colwell et al.* (1963)	48.16	20.494	*Johnston and Giauque* (1929)	68.51	40.980	*Kelley* (1929)
27.425	34.742	*Colwell et al.* (1963)	52.64	23.207	*Johnston and Giauque* (1929)	72.44	42.747	*Kelley* (1929)
27.85	33.453	*Colwell et al.* (1963)	57.25	23.806	*Johnston and Giauque* (1929)	77.64	45.385	*Kelley* (1929)
28.621	32.644	*Colwell et al.* (1963)	62.32	25.188	*Johnston and Giauque* (1929)	82.27	47.227	*Kelley* (1929)

TABLE 5. (continued).

T	CD_4	References	T	CH_4*	References	T	NH_3	References
28.815	32.464	*Colwell et al.* (1963)	67.9	26.762	*Johnston and Giauque* (1929)	90.44	50.200	*Kelley* (1929)
29.756	32.351	*Colwell et al.* (1963)	73.24	28.303	*Johnston and Giauque* (1929)	99.62	53.214	*Kelley* (1929)
31.59	32.477	*Colwell et al.* (1963)	78.84	29.969	*Johnston and Giauque* (1929)	T	C_3H_6O	References
35.362	33.226	*Colwell et al.* (1963)	83.83	31.468	*Johnston and Giauque* (1929)	17.77	7.951	*Kelley* (1929)
38.906	34.252	*Colwell et al.* (1963)	88.96	32.636	*Johnston and Giauque* (1929)	20.99	11.141	*Kelley* (1929)
51.897	37.304	*Colwell et al.* (1963)	94.04	34.039	*Johnston and Giauque* (1929)	23.98	13.775	*Kelley* (1929)
53.708	37.882	*Colwell et al.* (1963)	98.51	35.157	*Johnston and Giauque* (1929)	27.05	17.216	*Kelley* (1929)
55.596	38.133	*Colwell et al.* (1963)	T	CH_3OH	References	30.68	20.733	*Kelley* (1929)
56.819	38.167	*Colwell et al.* (1963)	5	0.063	*Carlson and Westrum* (1971)	34.24	24.547	*Kelley* (1929)
57.559	38.736	*Colwell et al.* (1963)	10	0.703	*Carlson and Westrum* (1971)	37.55	27.687	*Kelley* (1929)
58.603	38.862	*Colwell et al.* (1963)	15	2.445	*Carlson and Westrum* (1971)	41.07	31.104	*Kelley* (1929)
59.454	39.113	*Colwell et al.* (1963)	20	5.104	*Carlson and Westrum* (1971)	47.38	36.036	*Kelley* (1929)
60.463	39.331	*Colwell et al.* (1963)	25	8.302	*Carlson and Westrum* (1971)	54.9	41.772	*Kelley* (1929)
62.268	39.641	*Colwell et al.* (1963)	30	11.623	*Carlson and Westrum* (1971)	61.34	46.557	*Kelley* (1929)
64.022	40.005	*Colwell et al.* (1963)	35	14.930	*Carlson and Westrum* (1971)	67.03	49.865	*Kelley* (1929)
60.463	39.331	*Colwell et al.* (1963)	25	8.302	*Carlson and Westrum* (1971)	54.9	41.772	*Kelley* (1929)
62.268	39.641	*Colwell et al.* (1963)	30	11.623	*Carlson and Westrum* (1971)	61.34	46.557	*Kelley* (1929)
64.022	40.005	*Colwell et al.* (1963)	35	14.930	*Carlson and Westrum* (1971)	67.03	49.865	*Kelley* (1929)
65.731	40.377	*Colwell et al.* (1963)	40	18.129	*Carlson and Westrum* (1971)	70.47	51.665	*Kelley* (1929)
60.463	39.331	*Colwell et al.* (1963)	25	8.302	*Carlson and Westrum* (1971)	54.9	41.772	*Kelley* (1929)
62.268	39.641	*Colwell et al.* (1963)	30	11.623	*Carlson and Westrum* (1971)	61.34	46.557	*Kelley* (1929)
64.022	40.005	*Colwell et al.* (1963)	35	14.930	*Carlson and Westrum* (1971)	67.03	49.865	*Kelley* (1929)
65.731	40.377	*Colwell et al.* (1963)	40	18.129	*Carlson and Westrum* (1971)	70.47	51.665	*Kelley* (1929)
72.096	41.659	*Colwell et al.* (1963)	45	21.143	*Carlson and Westrum* (1971)	74.31	54.010	*Kelley* (1929)
74.383	41.441	*Colwell et al.* (1963)	50	23.944	*Carlson and Westrum* (1971)	78.41	56.103	*Kelley* (1929)
76.202	42.119	*Colwell et al.* (1963)	60	28.910	*Carlson and Westrum* (1971)	82.18	57.820	*Kelley* (1929)
77.863	42.496	*Colwell et al.* (1963)	70	33.164	*Carlson and Westrum* (1971)	86.15	59.285	*Kelley* (1929)
83.717	43.752	*Colwell et al.* (1963)	80	36.949	*Carlson and Westrum* (1971)	93.44	62.718	*Kelley* (1929)
85.517	44.171	*Colwell et al.* (1963)	90	40.478	*Carlson and Westrum* (1971)	100.74	65.984	*Kelley* (1929)
87.274	44.882	*Colwell et al.* (1963)	100	43.836	*Carlson and Westrum* (1971)			

*Non-equilibrium solid CH_4.

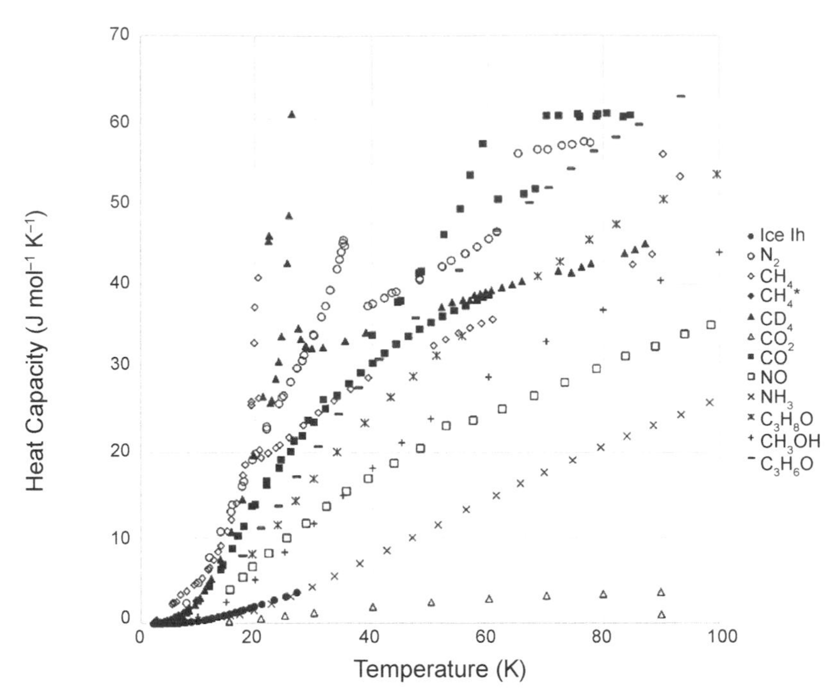

Fig. 15. Heat capacity measurements. Ice Ih data points from *Flubacher* (1960); N_2 from *Giauque and Clayton* (1933) and *Scott* (1976); CH_4, CD_4, and CH_4* (non-equilibrium solid-state CH_4) from *Colwell et al.* (1963); CO_2 from *Maass and Barnes* (1926); CO from *Clayton and Giauque* (1932) and *Krupskii et al.* (1973); NH_3 from *Overstreet ad Giauque* (1937); NO from *Johnston and Giauque* (1929); CH_3OH from *Carlson and Westrum* (1971); and C_3H_8O and C_3H_6O from *Kelley* (1929).

heat capacity of N_2 and CH_4, where these increases are also observed at α-β phase changes [$\Delta c \approx 7$ J mol^{-1} K^{-1} ~35 K and $\Delta c \approx 14$ J mol^{-1} K^{-1} ~21 K, respectively (*Prokhvatilov and Yantsevich,* 1983)]. Only one observation was made available below 100 K for carbon dioxide (CO_2) (*Maass and Barnes,* 1926). *Colwell et al.* (1963) inspected solid CH_4, solid CH_4 in non-equilibrium, and deuterated-methane (CD_4). Non-equilibrium CH_4 solids have a lower heat capacity than equilibrated solid CH_4. Two separate experiments measured the heat capacity of carbon monoxide (CO) (*Clayton and Giauque,* 1932; *Krupskii et al.,* 1973). These two experiments fit well but have discrepancies ($\Delta c \approx 7$ J mol^{-1} K^{-1} > 40 K; Fig. 15), probably due to the various crystallization orientations that CO may manifest above this temperature (*Quirico and Schmitt,* 1997b). Further experimental studies are required to determine the heat capacities of more exotic ices, including alcohols and hydrocarbons <100 K, in the exploration of Pluto and outer solar system ices.

In the case of Pluto, the determination of specific heat in icy mixtures (beyond the pure ices tabulated here) would help expand current models in determining Pluto's phase changes in the ice shell (*Nimmo et al.,* 2016), climate models for condensation cycles [namely CH_4 and N_2 (*Forget et al.,* 2017)], and hydrocarbon relations to the icy substrate prevalent in the southwest close-encounter hemisphere of Pluto [C_2-dominant hydrocarbons relevant to Pluto listed in *Grundy et al.* (2018)].

3.2.5. Elastic moduli. The bulk elastic properties of an ice determine how much the ice will compress given an external pressure. The ratio of the pressure change to the fractional volume compression is the bulk modulus. The ratio of the stress to the strain, regarding compressional or elongational forces, defines the Young's modulus. Both variables give us an idea about the planar forces that can be supported and the pressure-temperature conditions that an icy substrate can endure (refer to section 2.2).

Young's modulus, E; Poisson's ratio, ν; and shear modulus, G, can be found using the following relationships (from, e.g., *Yasui et al.,* 2017; *Vance et al.,* 2018a; and references therein)

$$E = \rho V_T^2 \frac{3V_L^2 - 4V_T^2}{V_L^2 - V_T^2} \quad (29)$$

$$\nu = \frac{1}{2} \frac{V_L^2 - 2V_T^2}{V_L^2 - V_T^2} \quad (30)$$

$$G = \rho V_T^2 \quad (31)$$

where ρ is the density of the sample and V_L and V_T are the longitudinal and transverse sound speeds respectively.

A collection of results from 14 experimental measurements are given in Table 6. Included in these results are measurements from *Huo et al.* (2011), who documented two different methane clathrate structures, sI and sII, from experimental sources. The ammonia-water ice (NH_3:H_2O) mixtures from *Lorenz and Shandera* (2001) are 16 mol% NH_3, compared to pure water ice Ih.

TABLE 6. Young's (E) and bulk moduli measurements at relevant temperatures.

Species	Modulus Type	Temperature (K)	Modulus (GPa)	Reference
Ice Ih	E	<273	0.009–0.0112	*Durham and Stern* (2001)
	E	90	0.1–0.5	*Lorenz and Shandera* (2001)
	E	100	0.2–0.9	*Lorenz and Shandera* (2001)
CH_4	Bulk	15	3.08	*Yamashita et al.* (2010)
	Bulk	40	2.12	*Yamashita et al.* (2010)
N_2	Bulk	20	2.16	*Yamashita et al.* (2010)
	Bulk	44	1.47	*Yamashita et al.* (2010)
CO_2	E	80	13.12	*Musiolik et al.* (2016)
NH_3:H_2O	E	90	0.85	*Lorenz and Shandera* (2001)
	E	100	0.1	*Lorenz and Shandera* (2001)
CH_4 clathrate sI	E	<273	8.4	*Huo et al.* (2011)
	Bulk	<273	8.76	*Huo et al.* (2011)
CH_4 clathrate sII	E	<273	8.2	*Huo et al.* (2011)
	Bulk	<273	8.48	*Huo et al.* (2011)

For Pluto models, the elastic and shear moduli have been used to investigate Pluto's ice shell (*Nimmo et al.,* 2016), subsurface clathrates (*Durham et al.,* 2003; *Kamata et al.,* 2019), and flexural deformation of Pluto's water ice lithosphere (*Conrad et al.,* 2019). The collected measurements here are only representative of pure ices (except the H_2O-NH_3), where more complex mixtures and alcohols (e.g., methanol, ethylene) are lacking for the range of ices in the outer solar system and the Kuiper belt, or the understanding of subsurface cryovolcanic materials. Knowledge of mixtures or alcohols can lead to insights in the crystalline structure (e.g., grain deformation) or macroscale structural processes (e.g., tectonism, ice shell evolution) under low temperatures and pressures.

3.2.6. Ductility in tension and compression. *Leont'eva et al.* (1970) present the results of tension experiments measuring the ductility of CH_4 and Ar. Their experiments were designed after the Bol'shutkin apparatus (*Bol'shutkin,* 1975). This tension experiment is somewhat similar to the experimental apparatus of *Yamashita et al.* (2010). The axial length of the ice samples was 5 cm with a diameter of 10 cm. The grains within the ice samples were about

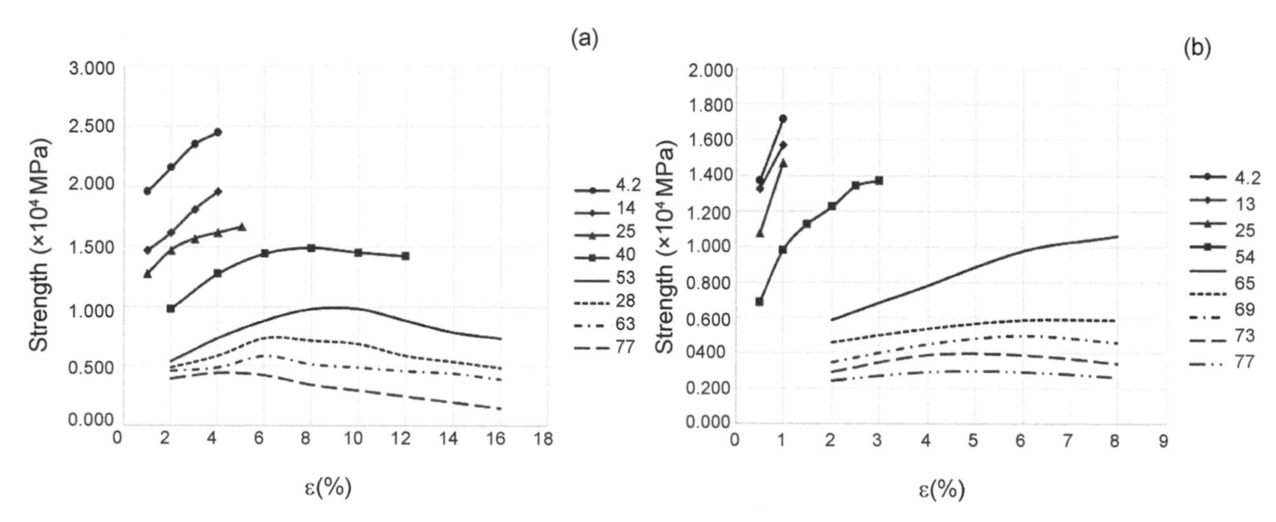

Fig. 16. (a) Solid methane stress-strain curves. **(b)** Argon stress-strain curves. Both sets of strain curves from *Leont'eva et al.* (1970).

TABLE 7. Stress-strain measurements of CH_4 and Ar (from *Leonteva et al.*, 1970).

CH_4	Temperature (K)	ε (%)	Strength (×10^4 MPa)	Ar	Temperature (K)	ε (%)	Strength (× 10^4 MPa)
	4.2	1	1.961		4.2	0.5	1.373
	4.2	2	2.157		4.2	1	1.716
	4.2	3	2.354		13	0.5	1.324
	4.2	4	2.452		13	1	1.569
	14	1	1.471		25	0.5	1.079
	14	2	1.618		25	1	1.471
	14	3	1.814		44	0.5	0.686
	14	4	1.961		44	1	0.981
	25	1	1.275		44	1.5	1.128
	25	2	1.471		44	2	1.226
	25	3	1.569		44	2.5	1.344
	25	4	1.618		44	3	1.373
	25	5	1.667		54	2	0.588
	40	2	0.981		54	4	0.785
	40	4	1.275		54	6	0.981
	40	6	1.442		54	8	1.065
	40	8	1.491		65	2	0.461
	40	10	1.451		65	4	0.539
	40	12	1.422		65	6	0.588
	53	2	0.539		65	8	0.588
	53	4	0.735		69	2	0.350
	53	6	0.883		69	4	0.452
	53	8	0.981		69	6	0.500
	53	10	0.981		69	8	0.461
	53	12	0.883		73	2	0.294
	53	14	0.785		73	4	0.392
	53	16	0.735		73	6	0.392
	58	2	0.490		73	8	0.343
	58	4	0.588		77	2	0.245
	58	6	0.735		77	4	0.294
	58	8	0.716		77	6	0.294
	58	10	0.686		77	8	0.265
	58	12	0.588				
	58	14	0.539				
	58	16	0.490				
	63	2	0.461				
	63	4	0.490				
	63	6	0.588				
	63	8	0.520				
	63	10	0.490				
	63	12	0.461				
	63	14	0.441				
	63	16	0.392				
	77	2	0.392				

TABLE 7. (continued).

CH_4	Temperature (K)	ε (%)	Strength (×10⁴ MPa)	Ar	Temperature (K)	ε (%)	Strength (× 10⁴ MPa)
	77	4	0.441				
	77	6	0.422				
	77	8	0.343				
	77	10	0.294				
	77	12	0.245				
	77	14	0.196				
	77	16	0.147				

0.2–0.3 mm in size. In each experiment the long axis of the sample was deformed at a rate of 1 mm min^{-1}, i.e., $\dot{\epsilon} = 3.33 \times 10^{-4}$ s^{-1}. The strain of the samples was monitored over time and several other properties were either directly measured (yield stress, fracture stress; summarized in Table 7, displayed in Fig. 16) or in the case like activation energies, inferred from theoretical predictions based on observed timescales of stress relaxation. Two clear regimes were observed, one for low temperatures ($T < 0.5\ T_m$ for CH_4 and $T < 0.75\ T_m$ for Ar, where T_m is melting temperature) where the yield stress and the activation energies are vastly different depending on this temperature divide. The observed plastic ductile behavior in the low-temperature range for CH_4 (5 K< T < 25 K) was conjectured to be the result of low-T plasticity (dislocation glide mitigated by Peierls barriers) as the specimens deformed steadily at values just below the fracture stress. In the higher-temperature range for CH_4 (50 K< T < 77 K) the authors suggest that dislocation creep was responsible for the observed behavior. Note that the stresses quoted in the *Leont'eva et al.* (1970) study are in units of p mm^{-2}, where p is the force imparted by m = 1 g in Earth's surface gravity field, i.e., p = mg. *Yamashita et al.* (2010) examined the compressive response of CH_4 ice. The grain sizes in their experiments were not measured. The compression experiments were run in the strain rate range $10^{-4}\ s^{-1} < \dot{\epsilon} < 0.02\ s^{-1}$. They collected several stress-strain rate ductile flow data points at three temperatures: T = 45 K, 64 K, and 77 K. They also measured brittle fracture failure at several lower temperatures. Given that there is some overlap between the two experiments (especially where the strain rates are coincident), we estimate the activation energies from fitting the function form

$$\dot{\epsilon} = A_d(CH_4)\exp(-Q_{act}/RT)\tau^n \quad (32)$$

to the T = 64 K and T = 77 K data acquired by *Yamashita et al.* (2010). The coefficient A_d is given the subscript d owing to the unknown value of the grain size. We restrict our attention to these temperature values because the inferred activation energies reported by *Leont'eva et al.* (1970) in this same temperature range is constant, which suggests that the operative deformation mechanism is the same too. Indeed, a least-squared fit indicates that

$$\begin{aligned} A_d(CH_4) &\approx 5.5\times10^4\ MPa^{-n}s^{-1}, \\ n \approx 1.72,\ Q_{act} &\approx 8.51\ kJ\cdot K^{-1}\ mol^{-1} \end{aligned} \quad (33)$$

It lends confidence that the value of the Q_{act} is about the same as the value estimated by *Leont'eva et al.* (1970), about 9.2 kJ K^{-1} mol^{-1}. These values translate to effective activation temperatures T_a = 1020 ± 4 K and 1100 ± 6 K respectively. The similarity between the estimated activation energies *could* mean that the operating deformation mechanism is the same in both experiments and either the mechanism is insensitive to grain size or the grain sizes of the two experiments are the same. However, it should be kept in mind that the two experiments are fundamentally different, in that one was in tension while the other was in compression and thus any strong conclusions should be tempered until more dedicated experiments are done. We also note that the fitted numbers calculated here for CH_4 are at variance with those presented by *Wei et al.* (2018). *Yamashita et al.* (2010) estimate that the brittle fracture strength at $\dot{\epsilon} = 3.33 \times 10^{-4}\ s^{-1}$ is about 4 MPa for a sample at 20–30 K while it is nearly half that in the measurements made by *Leont'eva et al.* (1970). This is perhaps a testament to the different processes contributing to fracture depending on whether the specimen is subjected to tension or compression (see earlier discussion).

We provide a similar fit to the N_2 ductile flow data (T = 45 K, 54 K) of Yamashita using the same above formulation and therefore we find that

$$\begin{aligned} A_d(N_2) &\approx 31.8\ MPa^{-n}s^{-1},\ n \approx 2.0, \\ Q_{act} &\approx 3.25\ kJ\cdot K^{-1}\ mol^{-1} \end{aligned} \quad (34)$$

which corresponds to an activation temperature T = 390 K. These numbers should be considered revisions to the numbers quoted in *Umurhan et al.* (2017).

TABLE 8. Table of experimental measurements of thermal conductivity (k).

Ice Ih	T (K)	k (W mK^{-1})	References
	5	100	*Ahmad and Phillips* (1987)
	10	120	*Slack* (1980)
	10	100	*Ahmad and Phillips* (1987)
	15	60	*Slack* (1980)
	20	38	*Slack* (1980)
	25	40	*Ahmad and Phillips* (1987)
	40	16.1	*Slack* (1980)
	50	20	*Ahmad and Phillips* (1987)
	75	15	*Ahmad and Phillips* (1987)
	80	8.1	*Slack* (1980)
	100	6.5	*Slack* (1980)
	100	9	*Ahmad and Phillips* (1987)
	100	6	*Lorenz and Shandera* (2001)

N_2	T (K)	k (W mK^{-1})	References
	5	10	*Scott* (1976)
	15	1	*Scott* (1976)
	14.5	0.7*	*Krivchikov et al.* (2005)
	15.5	0.6*	*Krivchikov et al.* (2005)
	15.5	0.15	*Krivchikov et al.* (2005)
	16.5	0.17	*Krivchikov et al.* (2005)
	17.5	0.15	*Krivchikov et al.* (2005)
	20	0.392	*Scott* (1976)
	25	0.33	*Scott* (1976)
	30	0.294	*Scott* (1976)
	35	0.27	*Scott* (1976)
	36	0.213	*Scott* (1976)
	40	0.215	*Scott* (1976)
	45	0.213	*Scott* (1976)
	50	0.211	*Scott* (1976)
	55	0.208	*Scott* (1976)

CH_4 Clathrate	T (K)	k (W mK^{-1})	References
	2	0.006	*Krivchikov et al.* (2005)
	3	0.009	*Krivchikov et al.* (2005)
	4	0.015	*Krivchikov et al.* (2005)
	5	0.2	*Krivchikov et al.* (2005)
	6	0.2	*Krivchikov et al.* (2005)
	7	0.23	*Krivchikov et al.* (2005)
	8	0.25	*Krivchikov et al.* (2005)
	9	0.27	*Krivchikov et al.* (2005)
	10	0.3	*Krivchikov et al.* (2005)
	20	0.3	*Krivchikov et al.* (2005)
	30	0.4	*Krivchikov et al.* (2005)
	40	0.5	*Krivchikov et al.* (2005)
	50	0.5	*Krivchikov et al.* (2005)

	T (K)	k (W mK^{-1})	References
	5	2	*Ahmad and Phillips* (1987)
	10	6	*Ahmad and Phillips* (1987)
	25	5.5	*Ahmad and Phillips* (1987)
	50	5	*Ahmad and Phillips* (1987)
	75	4.7	*Ahmad and Phillips* (1987)
	100	4	*Ahmad and Phillips* (1987)

NH_3	T (K)	k (W mK^{-1})	References
	97	3.1	*Lorenz and Shandera* (2001)

TABLE 8. (continued).

CH_4 Clathrate	T (K)	k (W mK^{-1})	References	T (K)	k (W mK^{-1})	References
	60	0.5	*Krivchikov et al.* (2005)			
	70	0.5	*Krivchikov et al.* (2005)			
	80	0.4	*Krivchikov et al.* (2005)			
	90	0.37	*Krivchikov et al.* (2005)			
	100	0.4	*Krivchikov et al.* (2005)			

3.2.7. Creep. There are only two published semi-empirical studies examining the diffusion creep limit for pure N_2 and CH_4 ices (*Kirk,* 1990; *Eluszkiewicz and Stevenson,* 1990). The inferred volumetric quantities $D_{0,v}$ (volume diffusivity) and $Q_{0,v}$ (activation energy) from NMR observations of spin-lattice relaxation times of N_2 ice samples containing Ar impurities that hop lattice sites in an essentially random diffusive manner (*Esteve and Sullivan,* 1981). Relaxation time measurements of CH_4 mixtures with its deuterated molecule (CD_4) as the impurity are in *Gerritsma et al.* (1971; see also *Chezeau and Strange,* 1979). Although the boundary diffusion amplitudes and their activation energies were not directly measured from these experiments, they have been estimated by *Eluszkiewicz and Stevenson* (1991) by relating them to their volume counterparts $\delta D_{0,b} = 10^{-9} D_{0,v}$ and $Q_{0,b} \approx (2/3)Q_{0,v}$ following the physical arguments of *Ashby and Verall* (1978). Nothing is currently known about the diffusive behaviors of these volatile ices as mixtures and alloys.

For Pluto, the determination of creep mechanisms (as explained in sections 2.1.1 and 2.2.3) are lacking for CO and particular care has been taken in current models to have CO equivalent to N_2 rheological properties (*Umurhan et al.,* 2017). However, these experiments are still needed.

3.2.8. Thermal conductivity. The rate of heat transfer through an ice depends on the temperature gradient and the thermal conductivity of the material. There are several factors that also influence the thermal conductivity value, such as grain size and porosity, due to heat transfer by lattice vibrations (*Ross and Kargel,* 1998). There is no net motion of the icy medium as the energy propagates through

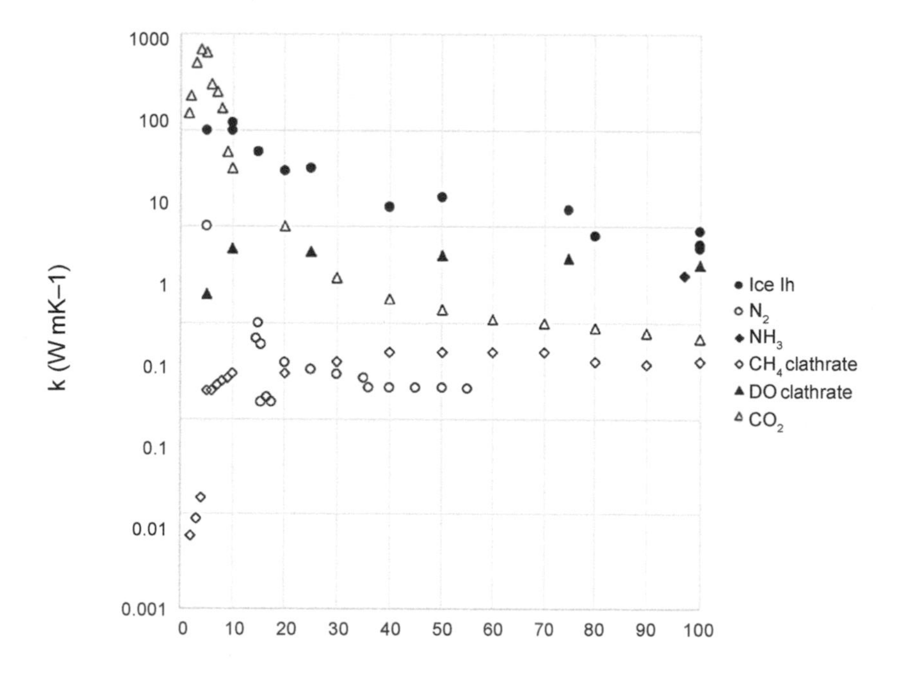

Fig. 17. Experimental measurements of thermal conductivity (k). Ice Ih measurements from *Slack* (1980), *Ahmad and Phillips* (1987), and *Lorenz and Shandera* (2001); N_2 from *Scott* (1976) and *Krivchikov et al.* (2005); CH_4 clathrate from *Krivchikov et al.* (2005); DO clathrate from *Ahmad and Phillips* (1987); and NH_3 from *Lorenz and Shandera* (2001).

TABLE 9. Table of experimental measurements of thermal expansion (α).

Ice Ih	T (K)	a (×10^{-6}) K^{-1}	Reference	CH_4	T (K)	a (×10^{-3}) K^{-1}	Reference	N_2	T (K)	a (×10^{-4}) K^{-1}	Reference
	80	7.7	*Corriccini and Gniewek* (1961)		10	0.1	*Heberlein and Adams* (1970)		10	2.5	*Scott* (1976)
	90	10.2	*Corriccini and Gniewek* (1961)		15	0.2	*Heberlein and Adams* (1970)		15	5	*Scott* (1976)
	100	12.7	*Corriccini and Gniewek* (1961)		18	0.5	*Heberlein and Adams* (1970)		18	7	*Scott* (1976)
CO_2	T (K)	a (×10^{-4}) K^{-1}	Reference		19	0.7	*Heberlein and Adams* (1970)		19	10	*Scott* (1976)
	20	0.32	*Corriccini and Gniewek* (1961)		20	3	*Heberlein and Adams* (1970)		30	13.5	*Scott* (1976)
	30	0.49	*Corriccini and Gniewek* (1961)		21	0.3	*Heberlein and Adams* (1970)		35	25	*Scott* (1976)
	40	0.67	*Corriccini and Gniewek* (1961)		22	0.3	*Heberlein and Adams* (1970)		40	18	*Scott* (1976)
	50	0.93	*Corriccini and Gniewek* (1961)		23	0.3	*Heberlein and Adams* (1970)		45	18.5	*Scott* (1976)
	60	0.98	*Corriccini and Gniewek* (1961)		25	0.3	*Heberlein and Adams* (1970)		50	20	*Scott* (1976)
	70	1.16	*Corriccini and Gniewek* (1961)						55	22	*Scott* (1976)
	80	1.32	*Corriccini and Gniewek* (1961)						60	25	*Scott* (1976)
	90	1.47	*Corriccini and Gniewek* (1961)						63.14	0.44	*Grilly and Mills* (1957)
	100	1.6	*Corriccini and Gniewek* (1961)						64.84	0.403	*Grilly and Mills* (1957)
									67.4	0.341	*Grilly and Mills* (1957)

TABLE 9. (continued).

T (K)	a (×10⁻⁶) K⁻¹	Reference	T (K)	a (×10⁻³) K⁻¹	Reference	N_2	T (K)	a (×10⁻⁴) K⁻¹	Referencev
							70.46	0.312	*Grilly and Mills* (1957)
							73.36	0.288	*Grilly and Mills* (1957)
							77.24	0.237	*Grilly and Mills* (1957)
							83.57	0.202	*Grilly and Mills* (1957)
							89.54	0.168	*Grilly and Mills* (1957)

it. As for the exploration of Pluto, thermal conductivity is essential for investigating sublimation deposition processes (see section 2.3) and erosional mechanisms (section 2.4.5). The recent study of a possible clathrate layer on Pluto (and other icy bodies, such as Titan) should also be further investigated based on laboratory experiments. For example, the experimental work by *Vu et al.* (2020) in determining the formation of clathrate hydrates from liquid ethane and water ice relevant to Titan provides an insight to the complexity of clathrate formation that should be extrapolated to Pluto's lower temperature and pressure conditions.

Table 8 has 66 collected experimental measurements for thermal conductivity (k; in units of W mK^{-1}) of various materials ranging from 1.5-K to 100-K temperatures, and they are plotted in Fig. 17. We have also included the DO clathrate, a 1.3-dioxolane clathrate type, from *Ahmad and Phillips* (1987). The DO clathrate uniformly has a higher thermal conductivity than solid CH_4 by a factor of more than 10. This difference could indicate how porosity and the cage-like structure of a clathrate can influence thermal conductivity, although more work is recommended to understand clathrate structures. Only one laboratory measurement on NH_3 has been recorded with a 16% molar ratio of NH_3:H_2O ice sample (*Lorenz and Shandera,* 2001), being ~48% lower in thermal conductivity than pure water ice. *Sumarokov et al.* (2003) notes that CO_2 reaches high thermal conductivity values ~700 W mK^{-1}, unusual for simple molecular crystal structures.

3.2.9. Thermal expansivity. Knowledge of the thermal expansion coefficients of astrophysically-relevant ices is needed to solve a variety of planetary problems. These include estimating thermal ice pressures, thermal cracking and weakening of ice sheets, and examining the effects of differential thermal expansion between varying ice components and glacial structures, like those observed on Pluto. For Pluto, thermal expansivity is used namely for solid-state

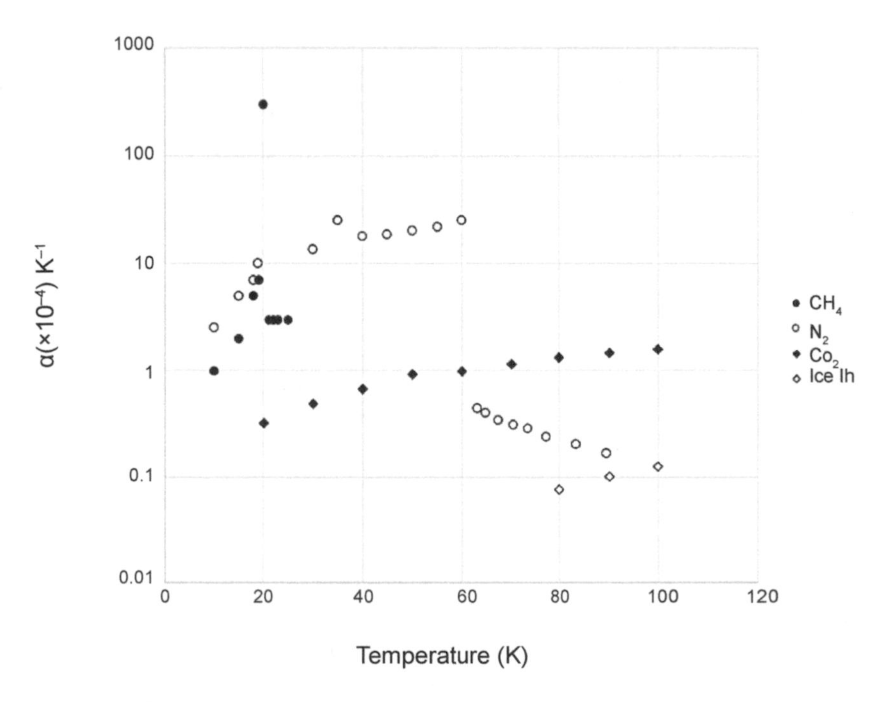

Fig. 18. Thermal expansion laboratory measurements. Data points on ice Ih and CO_2 from *Corriccini and Gniewek* (1961); CH_4 from Heberlein and Adams (1970); and N_2 from *Grilly and Mills* (1957) and *Scott* (1976).

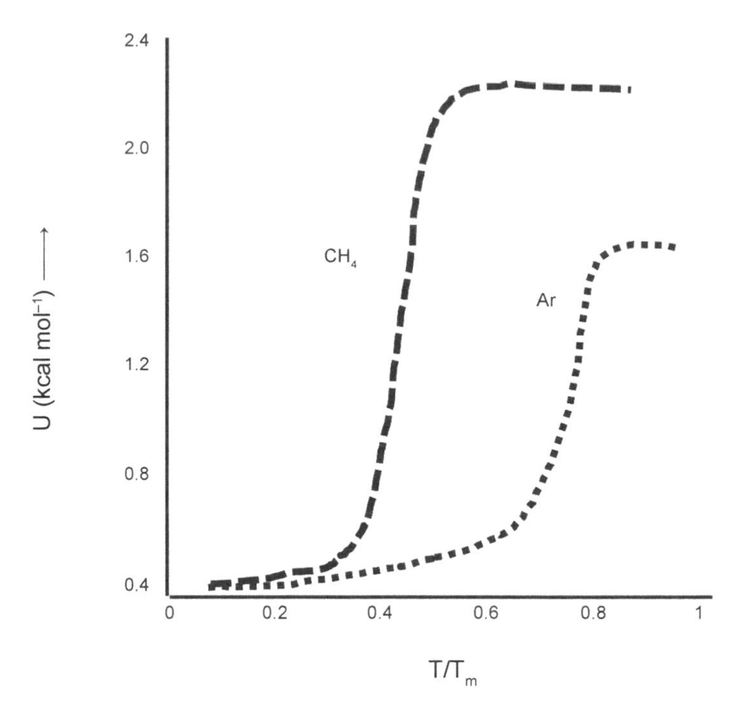

Fig. 19. Plastic deformation of CH_4 and Ar in temperature dependence of activation energy (U). Adapted from *Leont'eva et al.* (1970).

convection processes (discussed in section 2.4.1) and used in recent models to investigate Sputnik Planitia (*Nimmo et al.,* 2016) and the thermal evolution of Pluto's surface-ocean interactions (*Robuchon and Nimmo,* 2011).

We have assembled 50 experimental measurements for thermal expansivity of ices ranging from 8.5-K to 100-K temperatures (Table 9) and plotted them in Fig. 18. The data show an increase of ~85% at the α-β phase change of N_2 at ~35 K (*Scott,* 1976) and a sharp decrease in thermal expansivity of ~98% at the nitrogen triple point (63.14 K) (*Grilly and Mills,* 1957). At ~20 K, CH_4 thermal expansivity increases rapidly by nearly a factor of 10, then decreases ~75% at the α-β transition 21 K (*Heberlein and Adams,* 1970) (Fig. 18), although more experimental work needs to be done in the β-CH_4 regime (>21 K).

More work is also needed for more complex mixtures such as N_2-CH_4 as well as water-ice mixtures. These would provide an idea of how the crystal structure in mixtures/non-pure mineralogy can differ in thermal expansivity and thermal stability. For example, the different phases of N_2 are potentially found in different regions (including subsurfaces) on Pluto (discussed in section 2.4.5) (see *Stansberry and Yelle,* 1999; *Glein and Waite,* 2018). This alone can suggest variations in the thermal expansivity across Pluto, even on long geologic timescales, but N_2 mixed with CH_4 or CO can expand

TABLE 10. Temperature-dependent parameters for solid nitrogen and methane, where n and A are discussed in section 2.4.2 (values from *Yamashita et al.,* 2010).

	T(K)	n	$\log_{10}A$ (Mpa^{-n} s^{-1})
N_2	45	2.10 ± 0.08	–2.3
	56	2.27 ± 0.06	–1.5
CH_4	45	1.89 ± 0.08	–4
	64	1.68 ± 0.07	–2.3
	77	1.75 ± 0.05	–1.2

TABLE 11. Table of experimental measurements for sublimation energy at temperatures <100 K [measurements found in *Luna et al.* (2014) and references within].

CH_4	T (K)	E (kJ mol^{-1})	CO_2	T (K)	E (kJ mol^{-1})	CO	T (K)	E (kJ mol^{-1})
	35.5-36.5	8.5		91.5-92.5	29.3		33.5-34.5	6.3
	54-90	9.2		80-90	22.37		54-61	7.6
	79-89	10					40-50	7.98
	53-91	9.7					28.5-29.5	7.11
							28.5-29.5	7.13
							28-29	6.94
N_2	T (K)	E (kJ mol^{-1})	NH_3	T (K)	E (kJ mol^{-1})			
	24.5-25.5	4.3		100-105	25.57			
	26-27	6.65		100-108	23.15			

our understanding even further of N_2-dominant regions on Pluto (e.g., Sputnik Planitia).

3.2.10. Sublimation, activation, and desorption energies. Activation energy is the energy that an ice must possess to go through a specific physical or chemical reaction. From *Leont'eva et al.* (1970), the temperature-dependent activation energy for CH_4 and Ar shows that a decrease in temperature causes a sharp decrease of the activation energy of deformation both in CH_4, and Ar (Fig. 19). The activation energy at $T > 0.3\ T_m$ (melting temperature) in CH_4 and $0.65\ T_m$ in Ar did not exceed 0.10–0.05 of the activation energy of self-diffusion, which is typical for the Peierls mechanism, or the thermally-dependent dislocation movement of crystals. Note that the low-temperature character of plastic deformation in CH_4 and Ar is observed at the same temperatures for which $\varepsilon_t/\varepsilon = 1$.

The activation energies pertinent to the ductile flow experiments of N_2 and CH_4 reported in *Leont'eva et al.* (1970) and *Yamashita et al.* (2010) are discussed in section 3.2.6.

The work of *Yamashita et al.* (2010) has yielded stress exponents (n and A as described in section 2.4.2) for pure N_2 and CH_4 ices under very strong stressing (in the vicinity of 0.1–10 MPa; Table 10). But what is not certain is what type of stress regime the ices were exhibiting under those reported laboratory conditions and what the typical grain size was that was contained in their sample. Nonetheless, under those conditions the ice response was observed to be

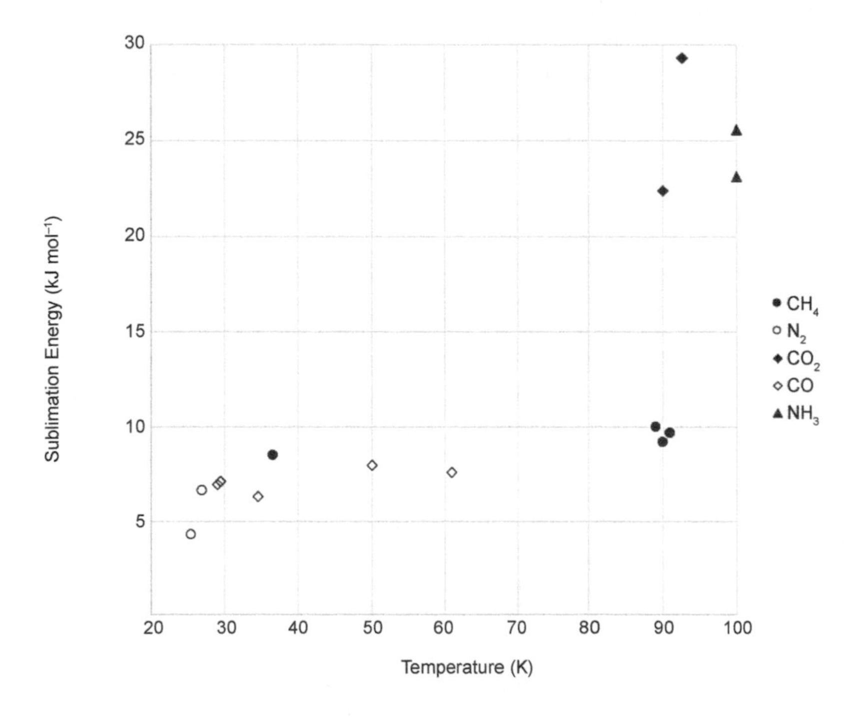

Fig. 20. Visual plot of the sublimation energy of ices. Data points from *Luna et al.* (2014).

TABLE 12. Carbon dioxide crystalline types with different measured desorption energies (*Galvez et al.*, 2007).

Temperature	CO_2 Type	E_d (kJ mol^{-1})
90	amorphous	20.8
90	crystalline	19.9
92	amorphous	20.7
93	crystalline	20
95	crystalline	19.8

relatively rapid with strain rates around 10^{-4}–10^{-6} s^{-1}, i.e., requiring experiments to be run from hours to a week or more. Conditions on outer solar system bodies require assessing the strain rate response at far lower stresses, in the range of 1–100 kPa (0.01–1 atm). So far researchers have made predictions regarding the responses of each ice by linearly extrapolating the above response down by several orders of magnitude. This may be risky to do so as one may wildly over- or underestimate the temporal response based on this incomplete information. No follow-up laboratory experiments examining this regime for these ices and different mixtures for either the high-stress or low-stress regimes are currently available.

Sublimation energy is the heat required to change an ice substance from solid phase to gas phase. This is the more favorable energy to measure for ice in a laboratory setting. *Luna et al.* (2014) collected 16 relevant experimental measurements (Table 11 and Fig. 20). CO_2 and NH_3 have higher sublimation energies than CH_4, N_2, and CO by a factor of ≥2. More measurements should be taken at the phase transitions and triple points of CH_4 and N_2, which would allow our understanding of sublimation energies at plutonian conditions, especially at the more extreme temperatures either seasonally on the surface or warmer putative ocean conditions, to be complete.

Another activation-type energy to measure with relevant ices is the desorption energy, or the energy required to release a substance from a surface. This energy was measured by *Gálvez et al.* (2007) for CO_2 ices, amorphous and crystalline, at <100 K temperatures (Table 12).

3.2.11. ***Seismoacoustic properties.*** Seismology in the realm of icy worlds has been growing recently, especially to understand the deposition of materials on ocean-laden worlds and surface-ocean interactions (*Vance et al.*, 2018b). Fluid motions, fracturing, and other physical processes need defined seismoacoustic variables. Such properties can give us insight to the flexural responses across Pluto and the possibility of a subsurface ocean (*Hammond et al.*, 2016; *Kamata et al.*, 2019). Recent models using ice Ih (and other higher-pressure forms of water ice) sound velocities in fluid-solid interactions of icy bodies, namely Europa, are found in *Panning et al.* (2017), *Brown* (2018), and *Journaux et al.* (2019).

Sound velocities, more specifically the longitudinal, V_L, and transverse, V_T, wave velocities, can be expressed by a relationship between density and elastic moduli as expressed by *Terasaki and Nishida* (2018)

$$V_L = \sqrt{\frac{K_S + \frac{4\mu}{3}}{\rho}}, \quad V_T = \sqrt{\frac{\mu}{\rho}} \qquad (35)$$

where K_S and μ are the adiabatic bulk and shear moduli, respectively. Thus, elastic moduli can be determined from sound speeds (V_L and V_T) and density.

We have compiled 44 (0.1 MPa) experimental measurements for sound speeds ranging from 10 K to 100 K (Table 13 and Fig. 21). The data show a slight decrease in sound speed ~6% in the N_2 samples, more prominent in the longitudinal speeds from *Scott* (1976) and *Yamashita et al.* (2010). This decrease is consistent with the phase change from α-β N_2 at ~35 K (*Prokhvatilov and Yantsevich,* 1983). However, sound velocities of pure samples (e.g., CO) and non-pure samples (e.g., CH_4-N_2 mixtures, water-ice-dominant mixtures, or alcohol/liquid-mixtures) are currently lacking. Pluto models may benefit from such experimental data to investigate surface-ocean interactions and the proposed clathrate layers.

3.2.12. ***Viscosity of cryoliquids.*** Cryovolcanism, including effusive and explosive eruptive deposits, is apparent in the outer solar system, including at Titan (*Lopes et al.*, 2007), Triton (*Kargel and Strom,* 1990), and Pluto (*Kargel and Strom*, 1990; *Schenk,* 1990; *Kargel,* 1998). *Kargel* (1998) reported several viscosities (Table 14), noting that the ammonia-water (NH_3-H_2O) peritectic liquid near its freezing point has a similar viscosity to alkali basalt (at terrestrial conditions), but at much lower temperatures (<200 K vs. ≥1000 K, respectively).

We have compiled 25 experimental measurements for viscosity, ranging from 70 K to 281 K (Table 14 and Fig. 22). Pure water ice Ih holds the highest brittleness compared to nitrogen and the cryolava materials. The H_2O-NH_3-CH_3OH mixture shows the highest viscosity compared to the other ice mixtures given by *Kargel et al.* (1991). The

TABLE 13. Collected experimental measurements of acoustic speeds, both longitudinal (V_L) and transverse (V_T).

Ice Ih	Temp (K)	Pressure (GPa)	v_L (m s^{-1})	v_T (m s^{-1})	Reference
	77	0		2100	*Gromnitskaya et al.* (2001)
	77	0.2		2050	*Gromnitskaya et al.* (2001)
	77	0.4		2000	*Gromnitskaya et al.* (2001)
	77	0.6		1950	*Gromnitskaya et al.* (2001)
	77	0.8		1900	*Gromnitskaya et al.* (2001)
	77	1		1850	*Gromnitskaya et al.* (2001)
	100	0.5	3730		*Gromnitskaya et al.* (2001)

CH_4	Temp (K)	v_L (m s^{-1})	v_T (m s^{-1})	Reference	N_2	Temp (K)	v_L (m s^{-1})	v_T (m s^{-1})	Reference
	10	3400	-	*Yamashita et al.* (2010)		10	1800	900	*Yamashita et al.* (2010)
	10	2540	1650	*Leont'eva et al.* (2014)		15	1780		*Yamashita et al.* (2010)
	13	2520	-	*Leont'eva et al.* (2014)		16	1775	884	*Scott* (1976)
	15	3350	2000	*Yamashita et al.* (2010)		18	1774	884	*Scott* (1976)
	15	2500	-	*Leont'eva et al.* (2014)		20	1771	883	*Scott* (1976)
	16	2540	1630	*Leont'eva et al.* (2014)		20	1750	900	*Yamashita et al.* (2010)
	17	2480	1600	*Leont'eva et al.* (2014)		22	1767	882	*Scott* (1976)
	18	2450	1570	*Leont'eva et al.* (2014)		24	1761	881	*Scott* (1976)
	20	3100	1900	*Yamashita et al.* (2010)		25		890	*Yamashita et al.* (2010)
	20	2400	1540	*Leont'eva et al.* (2014)		26	1752	880	*Scott* (1976)
	21	2420	1550	*Leont'eva et al.* (2014)		28	1741	878	*Scott* (1976)
	25	3080	1900	*Yamashita et al.* (2010)		30	1729	875	*Scott* (1976)
	30	3000	1900	*Yamashita et al.* (2010)		30	1700	890	*Yamashita et al.* (2010)
	35	2950	-	*Yamashita et al.* (2010)		32	1711	871	*Scott* (1976)
	40	2900	1850	*Yamashita et al.* (2010)		34	1690	865	*Scott* (1976)
	45	2850	1840	*Yamashita et al.* (2010)		35	1670	840	*Yamashita et al.* (2010)
	50	2800	1830	*Yamashita et al.* (2010)		37	1561	839	*Scott* (1976)

TABLE 13. (continued)

CH_4	Temp (K)	v_L (m s^{-1})	v_T (m s^{-1})	Reference	N_2	Temp (K)	v_L (m s^{-1})	v_T (m s^{-1})	Reference
	55	2700	1750	*Yamashita et al.* (2010)		40	1550	837	*Scott* (1976)
	60	2650	1700	*Yamashita et al.* (2010)		40		800	*Yamashita et al.* (2010)
	60	2300	1450	*Leont'eva et al.* (2014)		42	1542	836	*Scott* (1976)
	63	2260	1420	*Leont'eva et al.* (2014)		44	1532	834	*Scott* (1976)
	65	2600	1600	*Yamashita et al.* (2010)		45	1530	790	*Yamashita et al.* (2010)
	65	2250	1400	*Leont'eva et al.* (2014)		48	1511	832	*Scott* (1976)
	70	2510	1550	*Yamashita et al.* (2010)		50	1500	790	*Yamashita et al.* (2010)
	70	2210	1350	*Leont'eva et al.* (2014)		52	1487	829	*Scott* (1976)
	75	2500	1530	*Yamashita et al.* (2010)		55	1490	790	*Yamashita et al.* (2010)
	75	2240	1330	*Leont'eva et al.* (2014)		56	1460	826	*Scott* (1976)
	80	2210	1330	*Leont'eva et al.* (2014)		60	1420	780	*Yamashita et al.* (2010)

N_2 data taken at 1 MPa show a more consistent trend compared to the various pressure data. N_2 at 0.5-MPa pressure has a drastic decrease in viscosity at temperatures >90 K. The N_2 data at 0.1-MPa pressures show a slight increase of viscosity at conditions >80 K. More tabulation of the N_2 viscosity >70-K conditions are found in *Lemmon and Jacobson* (2004), but these are mainly calculated extrapolations to low-temperature conditions.

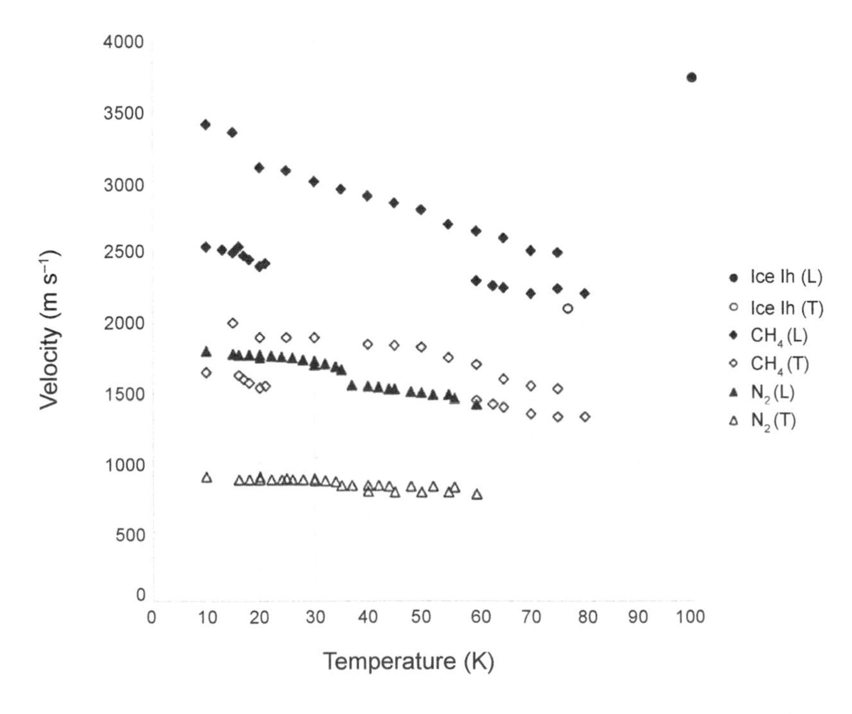

Fig. 21. Sound velocity (longitudinal L and transverse T) laboratory measurements. Ice Ih from *Gromnitskaya et al.* (2001); CH_4 from *Yamashita et al.* (2010) and *Leont'eva et al.* (2014); and N_2 from *Scott* (1976) and *Yamashita et al.* (2010).

TABLE 14. Table of experimental measurements of viscosity.

Ice Species	Temperature (K)	Viscosity (μ poise^{-1})	Reference
N_2*	75	1658	*Hanley et al.* (1974)
N_2*	80	54.6	*Hanley et al.* (1974)
N_2*	85	58	*Hanley et al.* (1974)
N_2*	90	61.3	*Hanley et al.* (1974)
N_2*	95	64.7	*Hanley et al.* (1974)
N_2*	100	68	*Hanley et al.* (1974)
N_2†	75	1667	*Hanley et al.* (1974)
N_2†	80	1384	*Hanley et al.* (1974)
N_2†	85	1165	*Hanley et al.* (1974)
N_2†	90	990.5	*Hanley et al.* (1974)
N_2†	95	67	*Hanley et al.* (1974)
N_2†	100	70.2	*Hanley et al.* (1974)
N_2‡	80	1395	*Hanley et al.* (1974)
N_2‡	85	1176	*Hanley et al.* (1974)
N2‡	90	1000	*Hanley et al.* (1974)
N_2‡	95	855.6	*Hanley et al.* (1974)
N2‡	100	731.3	*Hanley et al.* (1974)
Ice Ih	281.8	0.02	*Kargel* (1998)
NH_3-H_2O	176	63.10	*Kargel* (1998)
H_2O-NH_3-CH_3OH	158	63095.7	*Kargel* (1998)
H_2O-$CaCl_2$-$MgCl_2$	208	1.00	*Kargel* (1998)
H_2O-$CaCl_2$	223	0.40	*Kargel* (1998)
H_2O-$MgCl_2$	240	0.20	*Kargel* (1998)
H_2O-$MgSO_4$	251	0.125	*Kargel* (1998)
H_2O-NaCl	251	0.1	*Kargel* (1998)

*N_2 is data points taken at 0.1-MPa pressure.
†N_2 taken at 0.5-MPa pressure.
‡N_2 taken at 1 MPa pressure.

A note on other possible liquids that are relevant to icy bodies, mostly those observed on Titan, are the methane and ethane compositions depending on the concentration of dissolved nitrogen (*Hanley et al.,* 2019). From *Cordier et al.* (2017), previous models predicted the existence of two liquid layers in equilibrium with the vapor phase under certain temperature and pressure conditions. *Hanley et al.* (2019) confirmed the presence of the two-liquid phase, but at cooler temperatures (82 K rather than the predicted 85 K) and higher pressures (0.183 MPa rather than the predicted 0.17 MPa) than those on the surface of Titan. With this discovery, experimental cryoliquids and complex mixtures could also be further explored to expand future plutonian cryovolcanic models.

3.2.13. Shear and breakaway. Although O_2 ice is not a major astrophysical ice on outer solar system bodies, O_2 ice is still relevant to the (irradiative) formation of other species such as ozone, water, and carbon oxides (*Dodson-Robinson et al.,* 2009; *Muntean et al.,* 2015). The thermophysical processes of O_2 ices can thus influence the crystalline stability of more dominant ices or sputtering effects [e.g., O_2 ice sputters more efficiently than N_2 ice by a factor of ~2 (*Ellegaard et al.,* 1994)]. This sputtering effect is predicted toward the development of thin atmospheres such as those found around Pluto and icy moons of Jupiter and Saturn (*Johnson,* 1989; *Muntean et al.,* 2015). On a solid surface, this instability influences the shear breakage, or simply ice ejection from built-up pressures. For example, amorphous CO_2 ice (~50 K) shatters (a shearing of the ice) and collapses (ice ejection) to form non-circular depressions by the buildup of dynamic percolation of gas releases (*Laufer et al.,* 2005). Such experimental work could help us understand the non-circular patterns found on Pluto and other dwarf planets, although this work should be repeated to investigate other relevant ices.

The most collected laboratory measurements of a solid (O_2) at low temperatures <100 K were by *Bates and*

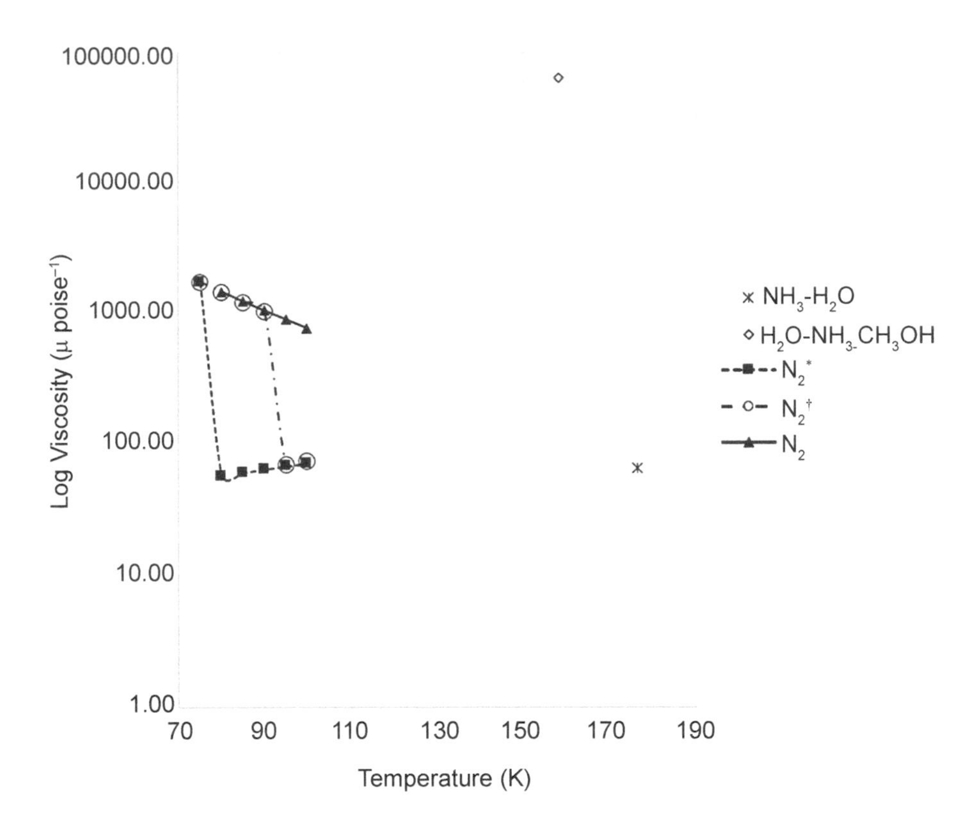

Fig. 22. Viscosities of cryoliquids (<200 K) from *Kargel* (1998) and N_2 from *Hanley et al.* (1974). N_2^*: data points taken at 0.1-MPa pressure; $N_2^\dagger$: data points taken at 0.5-MPa pressure; $N_2^\ddagger$: data points taken at 1-MPa pressures.

Altshuler (1995) with 11 experimental points (Table 15 and Fig. 23). The shear stress, or the slippage deformation along a plane, is compared to the breakaway stress, a type of dislocation mechanism. The shear stress increases ~18 K, then decreases past the α-β transition (23.781 K), although more experimental work should be done at the phase transi-

TABLE 15. Solid O_2 with respective phases and specific stresses measured from laboratory experiments from *Bates and Altshuler* (1995).

Temperature (K)	Phase	Shear Stress (MPa)	Breakaway Stress (MPa)
4	α	3.49	—
4	α	—	0.57
4	α	—	0.52
18	α	4.46	—
30	β	2.7	—
30	β	—	0.63
40	β	—	0.58
40	β	1.4	—
45	γ	0.31	—
45	γ	—	0.18
45	γ	—	0.19

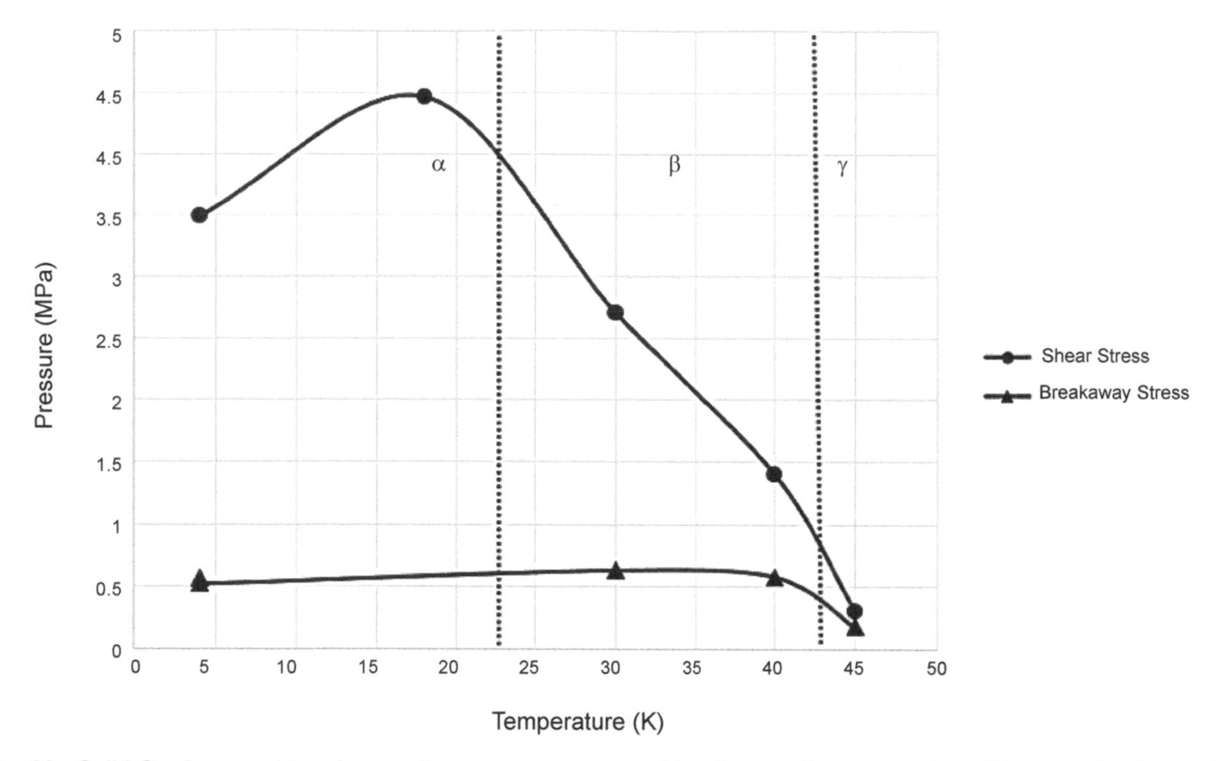

Fig. 23. Solid O_2 shear and breakaway stresses over a range of low temperatures spanning all three major O_2 crystalline phases. Data points from *Bates and Altshuler* (1995).

tions of O_2, particularly the α-β transition (*Mullins et al.,* 1963; *Bates and Altshuler,* 1995).

3.2.14. Shock properties of water ice. The response of ice to shock compression regulates the outcome of collisions between planetary bodies. The thermodynamic work from an impact event may be derived from a material's shock Hugoniot, which describes the point of all possible thermodynamic states a material can exist in behind a shock. The Hugoniot refers to the dynamic strength and compression states as represented by the ice material state variables such as stress tensor, specific volume, mass, density, particle velocity, and shock speed (*Stewart and Ahrens,* 2005). For Pluto, these experimental studies are relevant for the modeling of the impact-basin-forming Sputnik Planitia (*Keane et al.,* 2016) and the creation of a Charon-sized satellite and frequent, stochastic impacts for surficial reddening (*Sekine et al.,* 2017).

Derived from the shock wave measurements of *Stewart and Ahrens* (2005) centered at initial temperatures (T_0) of 100 K, the water-ice Hugoniot is composed from the elastic shocks in ice Ih, along with other water ice phases at higher pressures. The dynamic strength of ice Ih is strongly dependent on initial temperature, and the Hugoniot elastic limit varies from 0.05 to 0.62 GPa as a function of temperature and peak shock stress. *Stewart and Ahrens* (2005) presented new bulk sound speed measurements from shock pressures between 0.4 and 1.2 GPa and reported revised values for

TABLE 16. Titan-relevant organic tholin-like material with measured elastic modulus, hardness, and fracture toughness (values from *Yu et al.,* 2018).

Organic Species	Chemical Formula	E (GPa)	Hardness (GPa)	K_c (MPa–m$^{1/2}$)
Naphthalene	$C_{10}H_8$	7.5–9.5	0.07–0.1	—
Biphenyl	$C_{12}H_{10}$	1.9–3.0	0.03–0.08	—
Phenanthrene	$C_{14}H_{10}$	5.0–7.0	0.1–0.2	0.09–0.18
Coronene	$C_{24}H_{12}$	1.5–3.1	0.3–0.8	—
Adenine	$C_5H_5N_5$	3.7–5.0	0.1–0.2	—
Melamine	$C_3H_6N_6$	6.0–12.0	0.3–0.7	—
Tholin	CxHyNz	10.4	0.53–0.83	0.036

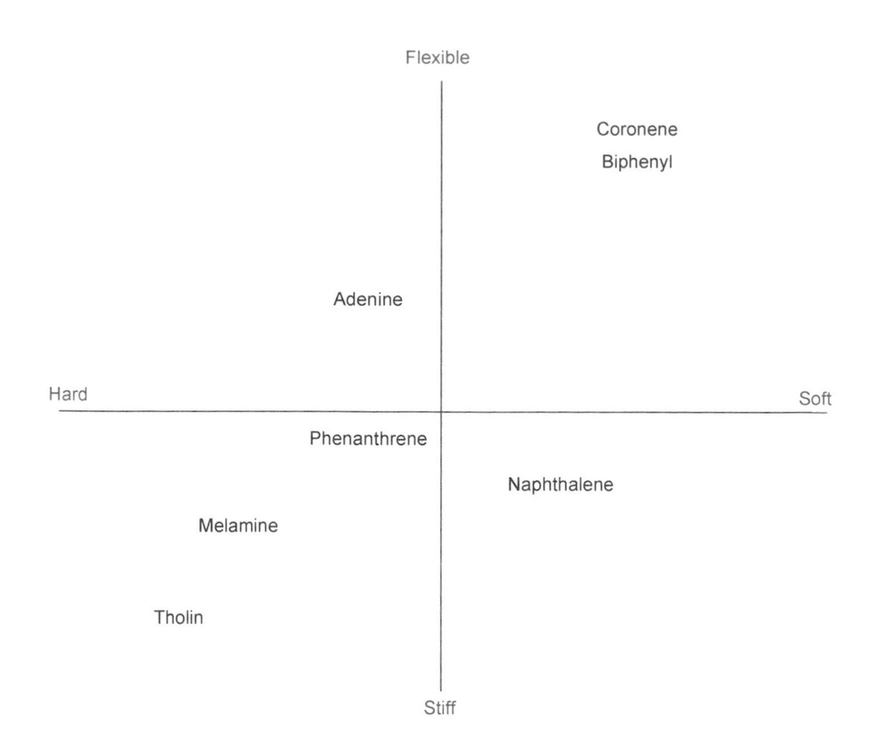

Fig. 24. Stiffness-hardness levels of different organic and tholin materials relevant to Titan, as addressed in *Yu et al.* (2018).

the shock pressures required to induce incipient melting (1.6 ± 0.3 GPa) and complete melting (4.1 ± 0.3 GPa) upon isentropic release from the shock state (for T_0 = 100 K).

3.2.15. Tholin and organic hardness. Pluto's large regions of rusty red presents an interesting mix of chemistry and physics at work. For chemistry, this haze material as a surface component may not be purely cosmetic but may influence the accumulation of ices over short geologic timescales (*Grundy et al.,* 2018). This tholin-like material provides an insight into the possible thermophysical processes involved, such as the stability of the tholins mixed with the water ice lithosphere, and the influence of the hardness of such material. However, this work at relevant Pluto conditions are still being developed and current experiments are instead using Titan-like materials as analogs (*Ahrens et al.,* 2018). *Yu et al.* (2018) used a nanoindentation technique to study the mechanical properties of Titan-relevant organics and tholin-like materials. Table 16 shows the Titan-relevant candidates measured in their study, along with respective elastic modulus (E), hardness, and fracture toughness (K_c). As displayed in Fig. 24, it is interesting to note the range of hardness and stiffness of these materials, mainly how tholins have the most hardness and stiffness of the materials, implying brittleness mechanics that still need to be explored.

4. FUTURE LABORATORY PROSPECTS

The need for continuous use and improvement of laboratory facilities in the analysis of the outer solar system and astrophysical environments is critical to understanding material properties. There is a need to fill in the gaps in planetary material and ice databases, especially various phases of ices and mixed compositions. Additionally, stability, vapor pressures, and other variables of such binary or complex ice mixtures are lacking. A complexity to laboratory prospects includes mineralogical and chemical roles defined by texture; grain shape and size and growth; crystallographic orientations; lattice-preferred orientations; distribution of phases, especially grain boundaries and polarity; porosity; diffusion; etc.; these can all have major effects on flows, localization of constituents, cyclicity or seasonality, and transport properties of plutonian materials (*Ahrens et al.,* 2018). Determining the fundamental rheology, heat transport, and other dynamical processes are key to understanding Pluto's and other icy bodies' evolution and ice behavior over time. We hope that this chapter gives experimentalists a resource to not only improve upon laboratory work, but to expand the lack of experiments accomplished for other ice compounds (pure and complex mixtures) and tholin organics.

5. SUMMARY

In this chapter, we have shown the wide variety of thermophysical and rheological properties of plutonian and outer solar system relevant ices through experimental techniques. Several key questions remain unanswered and can only be obtained by the advancement of laboratory research. While most of the ices documented in this chapter are pure ice samples, we hope that this collection can provide resources for a number of icy planetary bodies and geologic processes and further motivate laboratory research to extend to complex ice mixtures.

The diversity in the mineralogy, chemistry, and geology — to name just a few disciplines — in the research of Pluto and other dwarf planets and KBOs presents an opportunity to sample the range of qualities in the laboratory where previous and present missions, like NASA New Horizons, or models are incapable. That complexity also gives us a wealth of information to expand our current knowledge on the Kuiper belt and the formation of such icy bodies. Each of these *in situ* analyses in a low-temperature simulated environment raise new questions, which also lead to next-generation Pluto science.

REFERENCES

Adolph A. C. and Albert M. R. (2014) Gas diffusivity and permeability through the firn column at Summit, Greenland: Measurements and comparison to microstructural properties. *Cryosphere, 8*, 319–328.

Ahlers G., Grossmann S., and Lohse D. (2009) Heat transfer and large scale dynamics in turbulent Rayleigh-Bénard convection. *Rev. Mod. Phys., 81*, 503–537.

Ahmad N. and Phillips W. (1987) Thermal conductivity of ice and ice clathrate. *Solid State Commun., 63(2)*, 167–171, DOI: 10.1016/0038-1098(87)91189-6.

Ahrens C., Grundy W., Mandt K., et al. (2018) Recent advancements and motivations of simulated Pluto experiments. *Space Sci. Rev., 214(8)*, 130, DOI: 10.1007/s11214-018-0558-6.

Amstutz G. C. (1958) On the formation of snow penitentes. *J. Glaciol., 3*, 304–311.

Angwin M. J. and Wasserman J. (1966) Nitrogen-carbon monoxide phase diagram. *J. Chem. Phys., 44(1)*, 417–418, DOI: 10.1063/1.1726486.

Aoyama S. and Kanda E. (1934) Vapor pressures of solid O_2 and N_2. *J. Chem. Soc. Japan, 55*, 23–29.

Armstrong G., Brickwedde F., and Scott R. (1955) Vapor pressures of methanes. *J. Res. Natl. Bur. Stand. 55(1)*, 39–52.

Ashby M. F. and Verall R. A. (1978) Micromechanisms of flow and fracture, and their relevance to the rheology of the upper mantle. *Philos. Trans. R. Soc. London A, 288*, 59–95, DOI: 10.1098/rsta.1978.0006.

Balmforth N. J. and Craster R. V. (1999) A consistent thin-layer theory for Bingham fluids. *J. Non-Newtonian Fluid Mech., 84*, 65–81.

Balmforth N. J. and Craster R. V. (2000) Dynamics of cooling domes of viscoplastic fluid. *J. Fluid Mech., 422*, 225–248.

Barr A. C. and McKinnon W. B. (2007) Convection in ice I shells and mantles with self-consistent grain size. *J. Geophys. Res., 112*, E02012.

Barr A. C. and Pappalardo R. T. (2005) Onset of convection in the icy Galilean satellites: Influence of rheology. *J. Geophys. Res., 110*, E12005, DOI: 10.1029/2004JE002371.

Bates S. and Altshuler T. (1995) Shear strength testing of solid oxygen. *Cryogenics, 35(9)*, 559–566, DOI: 10.1016/0011-2275(95)91254-I.

Běhounková M., Tobie G., Choblet G., and Čadek O. (2013) Impact of tidal heating on the onset of convection in Enceladus' ice shell. *Icarus, 226*, 898–904, DOI: 10.1016/j.icarus.2013.06.033.

Benn D. I. and Evans D. J. A. (2010) *Glaciers and Glaciation, 2nd edition*. Routledge, London. 816 pp.

Bertrand T., Forget F., Umurhan O. M., et al. (2018) The nitrogen cycles on Pluto over seasonal and astronomical time scales. *Icarus, 308*, 277–296, DOI: 10.1016/j.icarus.2018.03.012.

Bertrand T., Forget F., Umurhan O. M., et al. (2019) The CH_4 cycles on Pluto over seasonal and astronomical timescales. *Icarus, 329*, 148–165, DOI: 10.1016/j.icarus.2019.02.007.

Bestmann M. and Prior D. J. (2003) Intragranular dynamic recrystallization in naturally deformed calcite marble: Diffusion accommodated grain boundary sliding as a result of subgrain rotation recrystallization. *J. Struct. Geol., 25*, 1597–1613, DOI: 10.1016/S0191-8141(03)00006-3.

Betterton M. D. (2001) Theory of structure formation in snow-fields motivated by penitentes, suncups, and dirt cones. *Phys. Rev. E, 63*, 056129.

Bierson C. J. and Nimmo F. (2019) Using the density of Kuiper belt objects to constrain their composition and formation history. *Icarus, 326*, 10–17.

Bland M. T., Singer K. N., McKinnon W. B., and Schenk P. M. (2017) Viscous relaxation of Ganymede's impact craters: Constraints on heat flux. *Icarus, 296*, 275–288.

Bol'shutkin D. N., Borisova L. I., and Leonteva A. V. (1968) Creep of crystalline methane. *Sov. Phys. Solid State, Engl. Transl., 10*, 1248–1249.

Bol'shutkin D. N., Grushko V. I., Karpenko I. V., et al. (1975) Method for investigating the mechanical properties of gases solidified under compression. *Mater. Sci., 9*, 566–567, DOI: 10.1007/BF00715531.

Borovik E., Grishin S., and Grishina E. (1960) The vapor pressure of nitrogen and hydrogen at low pressures. *Tech. Phys., 5(5)*, 506–511.

Boulton G. S. (1974) Processes and patterns of glacial erosion. In *Glacial Geomorphology* (D. R. Coates, ed.), pp. 41–87. Allen and Unwin, London.

Brown J. (2018) Local basis function representations of thermodynamic surfaces: Water at high pressure and temperature as an example. *Fluid Phase Equilib., 463*, 18–31.

Brown R. H. and Matson D. L. (1987) Thermal effects of insolation propagation into the regoliths of airless bodies. *Icarus, 72*, 84–94.

Brumberg A., Hammonds K., Baker I., et al. (2017) Single-crystal I_h ice surfaces unveil connection between macroscopic and molecular structure. *Proc. Natl. Acad. Sci. U.S.A., 114(21)*, 5349–5354.

Bryson C., Cazarra V., and Levenson L. (1974) Sublimation rates and vapor pressures of water, carbon dioxide, nitrous oxide, and xenon. *J. Chem. Eng. Data, 19(2)*, 107–110, DOI: 10.1021/je60061a021.

Buhler P. B. and Ingersoll A. P. (2017) Sublimation pit distribution indicates convection cell surface velocity of ~10 centimeters per year in Sputnik Planitia, Pluto. *Icarus, 300*, 327–340, DOI: 10.1016/j.icarus.2017.09.018.

Buratti B. et al. (2017) Global albedos of Pluto and Charon from LORRI New Horizons observations. *Icarus, 287*, 207–217.

Carlson H. and Westrum J. (1971) Methanol: Heat capacity, enthalpies of transition and melting, and thermodynamic properties from 5–300°K. *J. Chem. Phys., 54(4)*, 1464–1471, DOI: 10.1063/1.1675039.

Chandrasekhar S. (1961) *Hydrodynamic and Hydromagnetic Stability*. Clarendon, Oxford. 652 pp.

Chezeau J. M. and Strange J. H. (1979) Diffusion in molecular crystals. *Phys. Rept., 53*, 1–92.

Clarke G. K. C. (2005) Subglacial processes. *Annu. Rev. Earth Planet. Sci., 33*, 247–276, DOI: 10.1146/annurev.earth.33.092203.122621.

Claudine P., Jarry H., Vignoles G., et al. (2015) Physical processes causing the formation of penitentes. *Phys. Rev. E, 92*, 033015, DOI: 10.1103/PhysRevE.92.033015.

Clayton J. and Giauque W. (1932) The heat capacity and entropy of carbon monoxide. Heat of vaporization. Vapor pressures of solid and liquid. Free energy to 5000°K from spectroscopic data. *J. Am. Chem. Soc., 54(7)*, 2610–2626, DOI: 10.1021/ja01346a004.

Coble R. L. (1963) A model for boundary diffusion controlled creep in polycrystalline materials. *J. Appl. Phys., 34*, 1679–1682.

Colwell J., Gill E., and Morrison J. (1963) Thermodynamic properties of CH_4 and CD_4: Interpretation of the properties of the solids. *J. Chem. Phys., 39*, 635–653, DOI: 10.1063/1.1734303.

Conrad J. W., Nimmo F., Schenk P. M., et al. (2019) An upper bound on Pluto's heat flux from a lack of flexural response of its normal faults. *Icarus, 328*, 210–217.

Cordier D., Garcia-Sanchez F., Justo-Garcia D., and Liger-Belair G. (2017) Bubble streams in Titan's seas as a product of liquid N_2+CH_4+C_2H_6 cryogenic mixture. *Nature Astron., 1*, 0102, DOI: 10.1038/s41550-017-0102.

Corriccini R. and Gniewek J. (1961) *Thermal Expansion of Technical Solids at Low Temperatures: A Compilation from the Literature*. National Bureau of Standards Monograph 29, Washington, DC. 22 pp.

Costa A. (2006) Permeability-porosity relationship: A reexamination of the Kozeny-Carman equation based on a fractal pore-space geometry assumption. *Geophys. Res. Lett., 33*, L02318.

Croft S. K., Lunine J. I., and Kargel J. (1988) Equation of state of ammonia-water liquid: Derivation and planetological applications. *Icarus, 73*, 279–293.

Cruikshank D. P., Brown R. H., and Clark R. N. (1984) Nitrogen on Triton. *Icarus, 58*, 293–305.

Cruikshank D. P. and Apt J. (1984) Methane on Triton: Physical state and distribution. *Icarus, 58*, 306–311.

Cruikshank D. P., Umurhan O. M., Beyer R. A., et al. (2019) Recent cryovolcanism in Virgil Fossae on Pluto. *Icarus, 330*, 155–168, DOI: 10.1016/j.icarus.2019.04.023.

Cuffey K. M. and Paterson W. S. B. (2010) *The Physics of Glaciers, 4th edition*. Butterworth-Heinemann, Burlington. 704 pp.

de Bergh C. et al. (2008) Laboratory data on ices, refractory carbonaceous materials, and minerals relevant to transneptunian objects and Centaurs. In *The Solar System Beyond Neptune* (M. A. Barucci et al., eds.), pp. 483–506. Univ. of Arizona, Tucson.

de Bresser J. H. P., Peach C. J., Reijs J. P. J., and Spiers C. J. (1998) On dynamic recrystallization during solid state flow: Effects of stress and temperature. *Geophys. Res. Lett., 25*, 3457–3460, DOI: 10.1029/98GL02690.

Derby B. (1991) The dependence of grain size on stress during dynamic recrystallisation. *Acta Metall. Mater., 39*, 955–962.

Derby B. and Ashby M. F. (1987) On dynamic recrystallization. *Scr. Metall., 21*, 879–884.

Dodson-Robinson S., Willacy K., Bodenheimer P., et al. (2009) Ice lines, planetesimal composition and solid surface density in the solar nebula. *Icarus, 200(2)*, 672–693.

Doering C. R. and Constantin P. (1998) Bounds for heat transport in a porous layer. *J. Fluid Mech., 376*, 263–296.

Drury M. R. and Urai J. L. (1990) Deformation-related recrystallization processes. *Tectonophys., 172*, 235–253, DOI: 10.1016/0040-1951(90)90033-5.

Durham W. B. and Stern L. A. (2001) Rheological properties of water ice: Applications to satellites of the outer planets. *Annu. Rev. Earth Planet. Sci., 29*, 295–330, DOI: 10.1146/annurev.earth.29.1.295.

Durham W. B., Kirby S. H., Stern L. A., and Zhang W. (2003) The strength and rheology of methane clathrate hydrate. *J. Geophys. Res., 108*, 2182.

Durham W. B., Prieto-Ballesteros O., Goldsby D. L., and Kargel J. S. (2010) Rheological and thermal properties of icy materials. *Space Sci. Rev., 153*, 273–298, DOI: 10.1007/s11214-009-9619-1.

Effimov V., Izotov A., Levchenko A., Mezhov-Deglin L., and Khasanov S. (2011) Structural transitions in ice samples at low temperatures and pressures. *JETP Lett., 94*, 621–625.

Ellegaard O., Schou J., Stenum B., et al. (1994) Sputtering of solid nitrogen and oxygen by keV hydrogen ions. *Surf. Sci., 302(3)*, 371–384.

Eluszkiewicz J. (1991) On the microphysical state of the surface of Triton. *J. Geophys. Res., 96*, 19217–19229.

Eluszkiewicz J. and Stevenson D. J. (1990) Rheology of solid methane and nitrogen: Applications to Triton. *Geophys. Res. Lett., 17(10)*, 1753–1756.

Ernest H. (1961) Purification and vapor pressure of nitric oxide. *J. Chem. Phys., 35(4)*, 1531–1532, DOI: 10.1063/1.1732104.

Esteve D. and Sullivan N. S. (1981) N. M. R. study of self-diffusion in solid N_2. *Solid State Commun., 39*, 969–971, DOI: 10.1016/0038-1098(81)90067-3.

Fish A. M. and Zarestsy Y. K. (1997) *Ice Strength as a Function of Hydrostatic Pressure and Temperature*. CRREL Rept. 97, U.S. Army Corps of Engineers, Cold Regions Research and Engineering Laboratory, Hanover, New Hampshire.

Flubacher P., Leadbetter A. J., and Morrison J. A. (1960) Heat capacity of ice at low temperatures. *J. Chem. Phys., 33(6)*, 1751–1755, DOI: 10.1063/1.1731497.

Forget F., Bertrand T., Vangvichith M., et al. (2017) A post-new horizons global climate model of Pluto including the N_2, CH_4 and CO cycles. *Icarus, 287*, 54–71.

Frank F. C. and Read W. T. (1950) Multiplication processes for slow moving dislocations. *Phys. Rev., 70*, 722 723.

Fray N. and Schmitt B. (2009) Sublimation of ices of astrophysical interest: A bibliographic review. *Planet. Space Sci., 57*, 2053–2080, DOI: 10.1016/j.pss.2009.09.011.

Frels W., Smith D. R., and Ashworth T. (1974) Vapour pressure of nitrogen below the triple point. *Cryogenics, 14(1)*, 3–7, DOI: 10.1016/0011-2275(74)90035-6.

Frost H. J. and Ashby M. F. (1982) *Deformation-Mechanism Maps: The Plasticity and Creep of Metals and Ceramics*. Pergamon, Oxford. 166 pp.

Gálvez O. et al. (2007) A study of the interaction of CO_2 with water ice. *Astron. Astrophys., 472*, 691–698.

Gerritsma C. J., Oosting P. H., and Trappeniers N. J. (1971) Proton-spin-lattice relaxation and self-diffusion in methanes: II. Experimental results for proton-spin-lattice relaxation times. *Physica, 51*, 381–394, DOI: 10.1016/0031-8914(71)90048-6.

Giauque W. F. and Clayton J. O. (1933) The heat capacity and entropy of nitrogen. Heat of vaporization. Vapor pressures of solid and liquid. The reaction ½ N_2 + ½ O_2 = NO from spectroscopic data. *J. Am. Chem. Soc., 55(12)*, 4875–4889, DOI: 10.1021/ja01339a024.

Gladstone G. R., Stern S. A., Ennico K., et al. (2016) The atmosphere of Pluto as observed by New Horizons. *Science, 351(6279)*, aad8866.

Glein C. R. and Waite J. H. (2018) Primordial N_2 provides a cosmochemical explanation for the existence of Sputnik Planitia, Pluto. *Icarus, 313*, 79–92.

Goldsby D. L. and Kohlstedt D. L. (2001) Superplastic deformation of ice: Experimental observations. *J. Geophys. Res., 106(B6)*, 11017–11030, DOI: 10.1029/2000JB900336.

Greskovich C. and Lay K. W. (1972) Grain growth in very porous Al_2O_3 compacts. *J. Am. Ceram. Soc., 55*, 142–146, DOI: 10.1111/j.1151-2916.1972.tb11238.x.

Grilly E. R. and Mills R. L. (1957) Volume change on melting of N_2 up to 3500 kg/cm^2. *Phys. Rev., 105(4)*, 1140–1145, DOI: 10.1103/PhysRev.105.1140.

Gromnitskaya E. L., Stal'gorova O. V., Brazhkin V. V., and Lyapin A. G. (2001) Ultrasonic study of the nonequilibrium pressure-temperature diagram of H_2O ice. *Phys. Rev. B, 64*, 094205, DOI: 10.1103/PhysRevB.64.094205.

Grundy W. M., Schmitt B., and Quirico E. (1993) The temperature-dependent spectra of α and β nitrogen ice with application to Triton. *Icarus, 105(1)*, 254–258, DOI: 10.1006/icar.1993.1122.

Grundy W. M. et al. (2016) Surface compositions across Pluto and Charon. *Science, 351*, aad9189, DOI: 10.1126/science.aad9189.

Grundy W. M., Bertrand T., Binzel R. P., et al. (2018) Pluto's haze as a surface material. *Icarus, 314*, 232–245.

Hallet B. (1979) A theoretical model of glacial abrasion. *J. Glaciol., 23*, 39–50.

Hallet B. (1981) Glacial abrasion and sliding: Their dependence on the debris concentration in basal ice. *Ann. Glaciol., 2*, 23–28.

Hamilton D. P. et al. (2016) The rapid formation of Sputnik Planitia early in Pluto's history. *Nature, 540*, 97–99, DOI: 10.1038/nature20586.

Hammond N. P., Barr A. C., and Parmentier E. M. (2016) Recent tectonic activity on Pluto driven by phase changes in the ice shell. *Geophys. Res. Lett., 43*, 6775–6782, DOI: 10.1002/2016GL069220.

Hammond N. P., Parmentier E. M., and Barr A. C. (2018) Compaction and melt transport in ammonia-rich ice shells: Implications for the evolution of Triton. *J. Geophys. Res., 123*, 3105–3118, DOI: 10.1029/2018JE005781.

Han L. and Showman A. P. (2010) Coupled convection and tidal dissipation in Europa's ice shell. *Icarus, 207*, 834–844.

Hanley H. J. M., McCarty R. D., and Haynes W. M. (1974) The viscosity and thermal conductivity coefficients for dense gaseous and liquid argon, krypton, xenon, nitrogen, and oxygen. *J. Phys. Chem. Ref. Data, 3(4)*, 979–1018, DOI: 10.1063/1.3253152.

Hanley J., Groven J. J., Grundy W. M., et al. (2019) Characterization of possible two liquid layers in Titan seas. *Lunar and Planetary Science L*, Abstract #1712. Lunar and Planetary Institute, Houston.

Hansen C. J., McEwen A. S., Ingersoll A. P., and Terrile R. J. (1990) Surface and airborne evidence for plumes and winds on Triton. *Science, 250*, 421–424.

Hanson D. and Mauersberger K. (1986) The vapor pressures of solid and liquid ozone. *J. Chem. Phys., 85(8)*, 4669–4672, DOI: 10.1063/1.451740.

Heberlein D. C. and Adams E. D. (1970) Thermal expansion of solid methane. *J. Low Temp. Phys., 3(2)*, 115–121, DOI: 10.1007/BF00628321.

Herring C. (1950) Diffusional viscosity of a polycrystalline solid. *J. Appl. Phys., 21*, 437–444.

Hillert M. (1965) On the theory of normal and abnormal grain growth. *Acta Metall., 13*, 227–238.

Hindmarsh R. C. A. (2004) A numerical comparison of approximations to the Stokes equations used in ice sheet and glacier modeling. *J. Geophys. Res., 109*, F01012, DOI: 10.1029/2003JF000065.

Hiraga T., Miyazaka T., Tasaka M., and Yoshida H. (2010) Mantle superplasticity and its self-made demise. *Nature, 468*, 1091–1094.

Hook J. R. and Hall H. E. (1991) *Solid State Physics, 2nd edition*.

Wiley, Chichester. 496 pp.
Howard A. D. (1994) A detachment-limited model of drainage basin evolution. *Water Resour. Res., 30,* 2261–2285.
Howard A. D., Moore J. M., Umurhan O. M., et al. (2017a) Present and past glaciation on Pluto. *Icarus, 287,* 287–300, DOI: 10.1016/j.icarus.2016.07.006.
Howard A. D., Moore J. M., White O. L., et al. (2017b) Pluto: Pits and mantles on uplands north and east of Sputnik Planitia. *Icarus, 293,* 218–230, DOI: 10.1016/j.icarus.2017.02.027.
Howard L. N. (1966) Convection at high Rayleigh numbers. In *Applied Mechanics* (H. Görtler, ed.), pp. 1109–1115. Springer, Berlin.
Huo H., Liu Y., Zheng Z., Zhao J., Jin C., and Lv T. (2011) Mechanical and thermal properties of methane clathrate hydrates as an alternative energy resource. *J. Renewable Sustainable Energy, 3,* 063110, DOI: 10.1063/1.3670410.
Iverson N. R. (1991) Potential effects of subglacial water-pressure fluctuations on quarrying. *J. Glaciol., 37(125),* 27–36.
Iverson N. R. (2012) A theory of glacial quarrying for landscape evolution models. *Geology, 40(8),* 679–682.
Jaeger J. and Cook N. (1979) *Fundamentals of Rock Mechanics.* Chapman and Hall, London. 593 pp.
Johnson R. (1989) Effect of irradiation on the surface of Pluto. *Geophys. Res. Lett., 16(11),* 1233–1236.
Johnston H. L. and Giauque W. F. (1929) The heat capacity of nitric oxide from 14°K to the boiling point and the heat of vaporization. Vapor pressures of solid and liquid phases. The entropy from spectroscopic data. *J. Am. Chem. Soc., 51(11),* 3194–3214, DOI: 10.1021/ja01386a004.
Johnston H. and Doering C. R. (2009) A comparison of turbulent thermal convection between conditions of constant temperature and constant flux. *Phys. Rev. Lett., 102,* 064501, DOI: 10.1103/PhysRevLett.102064501.
Journaux B., Brown J., Pakhomova A., et al. (2019) Holistic approach for studying planetary hydrospheres: Gibbs representation of ices thermodynamics, elasticity, and the water phase diagram to 2,300 MPa. *J. Geophys. Res., 125(1),* e2019JE006176.
Kaibyshev O. A. (2002) Fundamental aspects of superplastic deformation. *Mater. Sci. Eng., A, 324,* 96–102.
Kamata S., Nimmo F., Sekine Y., et al. (2019) Pluto's ocean is capped and insulated by gas hydrates. *Nature Geosci., 12,* 407–410.
Kamb B. and Engelhardt H. (1987) Waves of accelerated motion in a glacier approaching a surge: The mini-surges of Variegated Glacier, Alaska, U.S.A. *J. Glaciol., 33,* 27–46.
Kargel J. S. (1998) Physical chemistry of ices in the outer solar system. In *Solar System Ices* (B. Schmitt et al., eds.), pp. 3–32. Astrophysics and Space Science Library Vol. 227, Springer, Dordrecht.
Kargel J. and Strom R. (1990) Cryovolcanism on Triton. *Lunar and Planetary Science XXI,* pp. 599-600. Lunar and Planetary Institute, Houston.
Kargel J. S., Croft S. K., Lunine J. I., and Lewis J. S. (1991) Rheological properties of ammonia-water liquids and crystal-liquid slurries: Planetological applications. *Icarus, 89(1),* 93–112.
Keane J. T., Matsuyama I., Kamata S., and Steckloff J. K. (2016) Reorientation and faulting of Pluto due to volatile loading within Sputnik Planitia. *Nature, 540,* 90–93, DOI: 10.1038/nature20120.
Kelley K. K. (1929) The heat capacity of methyl alcohol from 16°K to 298°K and the corresponding entropy and free energy. *J. Am. Chem. Soc., 51(1),* 180–187, DOI: 10.1021/ja01376a022.
Khanna R. K. and Ngoh M. (1990) Crystal field effects on the infrared spectra of phones I and II of crystalline CH_4. *Spectrochim. Acta, Part A, 46(7),* 1057–1063.
Kirk R. L. (1990) Diffusion kinetics of solid methane and nitrogen: Implications for Triton. *Lunar and Planetary Science XXI,* pp. 631–632. Lunar and Planetary Institute, Houston.
Kirk R. L., Brown R. H., and Soderblom L. A. (1990) Subsurface energy storage and transport for solar-powered geysers on Triton. *Science, 250,* 424–429.
Kirk R. L., Soderblom L. A., Brown R. H., Kieffer S. W., and Kargel J. S. (1995) Triton's plumes: Discovery, characteristics, and models. In *Neptune and Triton* (D. P. Cruikshank, ed.), pp. 949–989. Univ. of Arizona, Tucson.
Kite E. S. and Rubin A. M. (2016) Sustained eruptions on Enceladus explained by turbulent dissipation in tiger stripes. *Proc. Natl. Acad. Sci. U.S.A., 113,* 3972–3975.
Kohlstedt D. L., and Hansen L. N. (2015) Constitutive equations, rheological behavior, and viscosity of rocks. In *Treatise on Geophysics, 2nd Edition, Vol. 2* (G. Schubert, ed.), pp. 441–472. Elsevier, Oxford.
Krivchikov A. I., Gorodilov B. Y., Korolyuk O. A., Manzhelii V. G., Conrad H., and Press W. (2005) Thermal conductivity of methane-hydrate. *J. Low Temp. Phys., 139(5–6),* 693–702, DOI: 10.1007/s10909-005-5481-z.
Krupskii I. N., Prokhvatilov A. I., Erenburg A. I., and Yantsevich L. D. (1973) Structure and thermal expansion of alpha CO. *Phys. Status Sol. A, 19,* 519–527.
Landau L. D. and Lifshitz E. M. (1986) *Elasticity: Course of Theoretical Physics, Volume 7, 2nd edition.* Pergamon, Oxford. 165 pp.
Landau L. D. and Lifshitz E. M. (1987) *Fluid Mechanics: Course of Theoretical Physics, Volume 6, 2nd edition.* Butterworth-Heinemann, Amsterdam. 551 pp.
Laufer D., Pat-El I., and Bar-Nun A. (2005) Experimental simulation of the formation of non-circular active depressions on Comet Wild-2 and of ice grain ejection from cometary surfaces. *Icarus, 178(1),* 248–252.
Lemmon E. W. and Jacobsen R. T. (2004) Viscosity and thermal conductivity equations for nitrogen, oxygen, argon, and air. *Intl. J. Thermophys., 25(1),* 21–69, DOI: 10.1023/B:IJOT.0000022327.04529.f3.
Leont'eva A. V., Stroilov Y. S., Lakin E. E., and Bolshutkin D. N. (1970) Zero point oscillation energy effect on plastic deformation in solidified gases. *Phys. Status Sol., 42,* 543–549.
Leont'eva A. V., Prokhorov A. Y., Zakharov A. Y., and Erenburg A. I. (2014) Abnormal behavior of crystalline methane in temperature interval 60–70 K: From experiment to theory. *ArXiV e-prints,* arXiv:1404.5145.
Lide D. R., ed. (2006) *CRC Handbook of Chemistry and Physics, 87th edition.* CRC, Boca Raton. 2388 pp.
Litwin K. L., Zygielbaum B. R., Polito P. J., Sklar L. S., and Collins G. C. (2012) Influence of temperature, composition, and grain size on the tensile failure of water ice: Implications for erosion on Titan. *J. Geophys. Res., 117,* E08013, DOI: 10.1029/2012JE004101.
Lliboitry L. (1954) The origin of penitents. *J. Glaciol., 2,* 331–338.
Lopes R. M. C., Mitchell K. L., Stofan E. R., et al. (2007) Cryovolcanic features on Titan's surface as revealed by the Cassini Titan Radar Mapper. *Icarus, 186,* 395–412, DOI: 10.1016/j.icarus.2006.09.006.
Lorenz R. D. and Shandera S. E. (2001) Physical properties of ammonia-rich ice: Application to Titan. *Geophys. Res. Lett., 28(2),* 215–218, DOI: 10.1029/2000GL012199.
Luna R., Satorre M. A., Santonja C., and Domingo M. (2014) New experimental sublimation energy measurements for some relevant astrophysical ices. *Astron. Astrophys., 566,* A27, DOI: 10.1051/0004-6361/201323249.
Maass O. and Barnes W. H. (1926) Some thermal constants of solid and liquid carbon dioxide. *Proc. R. Soc. London Ser. A, 111,* 224–244, DOI: 10.1098/rspa.1926.0065.
MacGregor K. R., Anderson R. S., and Waddington E. D. (2009) Numerical modeling of glacial erosion and headwall processes in alpine valleys. *Geomorphology, 103,* 189–204.
MacMillan N. H. (1972) The theoretical strength of solids. *J. Mater. Sci., 7,* 239–254.
Matson D. L. and Brown R. H. (1989) Solid-state greenhouse and their implications for icy satellites. *Icarus, 77,* 67–81.
McKinnon W. B., Nimmo F., Wong T., et al. (2016) Convection in a volatile nitrogen-ice-rich layer drives Pluto's geological vigour. *Nature, 534,* 82–85, DOI: 10.1038/nature18289.
Menaucourt J. (1982) Pression de vapeur saturante de l'éthylène entre 77 K et 119 K. *J. Chim. Phys., 79(6),* 531–535, DOI: 10.1051/jcp/1982790531.
Molaro J. L., Choukroun M., Phillips C. B., Phelps E. S., Hodyss R., Mitchell K. L., et al. (2019) The microstructural evolution of water ice in the solar system through sintering. *J. Geophys. Res., 124,* 243–277, DOI: 10.1029/2018JE005773.
Moore J. M., McKinnon W. B., Spencer J. R., et al. (2016) The geology of Pluto and Charon through the eyes of New Horizons. *Science, 351(6279),* 1284–1293.
Moore J. M., Howard A. D., Umurhan O. M., et al. (2017) Sublimation as a landform-shaping process on Pluto. *Icarus, 287,* 320–333, DOI: 10.1016/j.icarus.2016.08.025.
Moore J. M., Howard A. D., Umurhan O. M., et al. (2018) Bladed terrain on Pluto: Possible origins and evolution. *Icarus, 300,* 129–144, DOI: 10.1016/j.icarus.2017.08.031.

Moores J. E., Smith C. L., Toigo A. D., et al. (2017) Penitentes as the origin of the bladed terrain of Tartarus Dorsa on Pluto. *Nature, 541,* 188–190, DOI: 10.1038/nature20779.

Mullins J. C., Ziegler W. T., and Kirk B. S. (1963) The thermodynamics properties of oxygen from 20° to 100°K. In *Advances in Cryogenic Engineering, Vol. 8* (K. D. Timmerhaus, ed.), pp. 123–134. Springer, Boston.

Mullins W. W. and Sekerka R. F. (1988) Stability of a planar interface during solidification of a dilute binary alloy. In *Dynamics of Curved Fronts* (Pierre Pelcé, ed.), pp. 345–352. Academic, San Diego.

Muntean E. A., Lacerda P., Field T. A., et al. (2015) Sputtering of oxygen ice by low energy ions. *Surf. Sci., 641,* 204–209.

Musiolik G., Teiser J., Jankowski T., and Wurm G. (2016) Collisions of CO_2 ice grains in planet formation. *Astrophys. J., 818(1),* 16, DOI: 10.3847/0004-637X/818/1/16.

Nabarro F. R. N. (1948) Deformation of crystals by the motion of single ions. In *Report of a Conference on Strength of Solids Held at the H. H. Wills Physical Laboratory, University of Bristol, on 7–9 July 1947,* pp. 75–90. Physical Society, London.

Nabarro F. R. N. (1967) Steady-state diffusional creep. *Philos. Mag., 16,* 231–237.

Niemela J. J., Skrbek L., Sreenivasan K. R., and Donnelly R. J. (2000) Turbulent convection at very high Rayleigh numbers. *Nature, 404,* 837–840.

Nimmo F. et al. (2016) Reorientation of Sputnik Planitia implies a subsurface ocean on Pluto. *Nature, 540,* 94–96.

Nishiyama N. and Yokoyama T. (2017) Permeability of porous media: Role of the critical pore size. *J. Geophys. Res., 122,* 6955–6971.

Ostwald W. (1897) Studien über die bildung und umwandlung fester körper. [Studies on the formation and transformation of solid bodies.] *Z. Phys. Chem., 22,* 289–330.

Otero J., Dontcheva L. A., Johnston H., et al. (2004) High-Rayleigh-number convection in a fluid-saturated porous layer. *J. Fluid Mech., 500,* 263–281.

Overstreet R. and Giauque W. F. (1937) Ammonia. The heat capacity and vapor pressure of solid and liquid. Heat of vaporization. The entropy values from thermal and spectroscopic data. *J. Am. Chem. Soc., 59(2),* 254–259, DOI: 10.1021/ja01281a008.

Panning M. P., Stahler S. C., Huang H., et al. (2017) Expected seismicity and the seismic noise environment of Europa. *J. Geophys. Res., 123(1),* 163–179.

Petrenko V. F. and Whitworth R. W. (1999) *Physics of Ice.* Oxford Univ., New York. 373 pp.

Prager M. (1988) Rotational tunneling in a disordered system: Nonequilibrium methane CH_4 II. *Can. J. Chem., 66(4),* 570–574.

Prokhvatilov A. I. and Jancevic L. D. (1983) X-ray investigation of the equilibrium phase diagram of CH_4-N_2 solid mixtures. *Sov. J. Low Temp. Phys., 9(2),* 94–98.

Protopapa S. et al. (2017) Pluto's global surface composition through pixel-by-pixel Hapke modeling of New Horizons Ralph/LEISA data. *Icarus, 287,* 218–228, DOI: 10.1016/j. icarus.2016.11.028.

Proulx T., ed. (2011) *Time Dependent Constitutive Behavior and Fracture/Failure Processes, Vol. 3: Proceedings of the 2010 Annual Conference on Experimental and Applied Mechanics.* Springer, New York. 406 pp.

Quirico E. and Schmitt B. (1997a) Near-infrared spectroscopy of simple hydrocarbons and carbon oxides diluted in solid N_2 and as pure ices: Implications for Triton and Pluto. *Icarus, 127,* 354–378.

Quirico E. and Schmitt B. (1997b) A spectroscopic study of CO diluted in N_2 ice: Applications for Triton and Pluto. *Icarus, 128,* 181–188.

Raj R. and Ashby M. F. (1971) On grain-boundary sliding and diffusional creep. *Metall. Trans. A, 2,* 1113–1127.

Ratke L. and Voorhees P. W. (2002) *Growth and Coarsening: Ostwald Ripening in Material Processing.* Springer, New York. 298 pp.

Regev O., Umurhan O. M., and Yecko P. A. (2016) *Modern Fluid Dynamics for Physics and Astrophysics.* Graduate Texts in Physics Series, Springer, New York. 680 pp.

Richter F. M., Nataf H.-C., and Daly S. F. (1983) Heat transfer and horizontally averaged temperature of convection with large viscosity variations. *J. Fluid Mech., 129,* 173–192.

Ridley N. (1995) C. E. Pearson and his observations of superplasticity. In *Superplasticity: 60 Years after Pearson* (N. Ridley, ed.), pp. 1–5. Institute of Materials, London.

Robuchon G. and Nimmo F. (2011) Thermal evolution of Pluto and implications for surface tectonics and a subsurface ocean. *Icarus, 216,* 426–439, DOI: 10.1016/j.icarus. 2011.08.015.

Roder H. M. (1978) The molar volume (density) of solid oxygen in equilibrium with vapor. *J. Phys. Chem. Ref. Data, 7(3),* 949–957, DOI: 10.1063/1.555582.

Ross R. G. and Kargel J. S. (1998) Thermal conductivity of solar system ices, with special reference to martian polar caps. In *Solar System Ices* (B. Schmitt et al., eds.), pp. 33–62. Astrophysics and Space Science Library Vol 227, Springer, Dordrecht.

Rozel A. (2012) Impact of grain size on the convection of terrestrial planets. *Geochem., Geophys., Geosyst., 13,* Q10020, DOI: 10.1029/2012GC004282.

Rozel A., Ricard Y., and Bercovici D. (2011) A thermodynamically self-consistent damage equation for grain size evolution during dynamic recrystallization. *Geophys. J. Intl., 184,* 719–728, DOI: 10.1111/j.1365-246X.2010.04875.x.

Rozel A., Besserer J., Golabek G. J., Kaplan M., and Tackley P. J. (2014) Self-consistent generation of single-plume state for Enceladus using non-Newtonian rheology. *J. Geophys. Res., 119,* 416–439, DOI: 10.1002/2013JE004473.

Satorre M. Á., Domingo M., Millán C., Luna R., Vilaplana R., and Santonja C. (2008) Density of CH_4, N_2, and CO_2 ices at different temperatures of deposition. *Planet. Space Sci., 56,* 1748–1752, DOI: 10.1016/j.pss.2008.07.015.

Satorre M. Á, Luna R., Milláan C., Domingo M., and Santonja C. (2017) Density of ices of astrophysical interest. In *Laboratory Astrophysics* (G. M. Munoz Caro and R. Escribano, eds.), pp. 51–69. Springer, Cham, Switzerland.

Schmitt B. et al. 2017. Physical state and distribution of materials at the surface of Pluto from New Horizons LEISA imaging spectrometer. *Icarus, 287,* 229–260, DOI: 10.1016/j.icarus.2016.12.025.

Schulson E. M. (1987) The fracture of ice Ih. *J. Phys., Colloq. 48, C1,* 207–220.

Schulson E. M. (1990) The brittle compressive fracture of ice. *Acta Metall. Mater., 30,* 1963–1976.

Schulson E. M. (2006) The fracture of water ice Ih: A short overview. *Meteoritics & Planet. Sci., 41,* 1497–1508.

Schulson E. M. and Duval P. (2009) *Creep and Fracture of Ice.* Cambridge Univ., Cambridge. 401 pp.

Scott T. A. (1976) Solid and liquid nitrogen. *Phys. Rept., 27(3),* 89–157.

Sekine Y., Genda H., Kamata S., and Funatsu T. (2017) The Charon-forming giant impact as a source of Pluto's dark equatorial regions. *Nature Astron., 1(2),* 1–6.

Shinado T. (1969) Vapor pressure of carbon monoxide in condensed phases. *Bull. Chem. Soc. Japan., 42,* 1815–1820, DOI: 10.1246/bcsj.42.2815.

Shimizu I. (1998) Stress and temperature dependence of recrystallized grain size: A subgrain misorientation model. *Geophys. Res. Lett., 25,* 4237, DOI: 10.1029/1998GL900136.

Slack G. A. (1980) Thermal conductivity of ice. *Phys. Rev. B., 22(6),* 3065–3071, DOI: 10.1103/PhysRevB.22.3065.

Smith B. A. and 64 colleagues (1989) Voyager 2 at Neptune: Imaging science results. *Science, 246,* 1422–1449.

Soderblom L. A. and 8 colleagues (1990) Triton's geyser-like plumes: Discovery and basic characterization. *Science, 250,* 410–415.

Solomatov V. S. (1995) Scaling of temperature- and stress-dependent viscosity convection. *Phys. Fluids, 7,* 266–274, DOI: 10.1063/1.868624.

Sotin C. and Labrosse S. (1999) Three-dimensional thermal convection in an iso-viscous, infinite Prandtl number fluid heated from within and from below: Applications to the transfer of heat through planetary mantles. *Phys. Earth Planet. Inter., 112,* 171–190.

Sotin C., Mitri G., Rappaport N., Schubert G., and Stevenson D. (2010) Titan's interior structure. In *Titan from Cassini-Huygens* (R. Brown et al., eds.), pp. 61–74. Springer, Heidelberg.

Stansberry J. A. and Yelle R. V. (1999) Emissivity and the fate of Pluto's atmosphere. *Icarus, 141(2),* 299–306.

Stengel K. C., Oliver D. C., and Booker J. R. (1982) Onset of convection in a variable viscosity fluid. *J. Fluid Mech., 120,* 411 431.

Stern S. A., Bagenal F., Ennico K., et al. (2015) The Pluto system: Initial results from its exploration by New Horizons. *Science, 350,* aad1815.

Stern S. A., Binzel R. P., Earle A. M., et al. (2017) Past epochs of significantly higher pressure atmospheres on Pluto. *Icarus, 287,* 47–53, DOI: 10.1016/j.icarus.2016.11.022.

Stewart S. and Ahrens T. (2005) Shock properties of H_2O ice. *J Geophys. Res.–Planets, 110,* E3.

Sugden D. E. and John B. S. (1976) *Glaciers and Landscape: A Geomorphological Approach.* Edward Arnold, London. 376 pp.

Sumarokov V. V., Stachowiak P., and Jezowski A. (2003) Low-temperature thermal conductivity of solid carbon dioxide. *Low Temp. Phys., 29(5),* 449–450, DOI: 10.1063/1.1542510.

Swinkels F. B. and Ashby M. F. (1981) A second report on sintering diagrams. *Solid State Commun., 29,* 259–281.

Tanabe Y. (2013) Fracture toughness for brittle fracture of elastic and plastic materials. *Mater. Trans., 54(3),* 314–318.

Tegler S. C., Cornelison D. M., Grundy W. M., Romanishin W., Abernathy M. R., Bovyn M. J., Burt J. A., Evans D. E., Maleszewski C. K., Thompson Z., and Vilas F. (2010) Methane and nitrogen abundances on Pluto and Eris. *Astrophys. J., 725(1),* 1296–1305, DOI: 10.1088/0004-637X/725/1/1296.

Terasaki H. and Nishida K. (2018) Chapter 9 — Density and elasticity measurements for liquid materials. In *Magmas under Pressure, Advances in High-Pressure Experiments on Structure and Properties of Melts* (Y. Kono and C. Sanloup, eds.), pp. 237-260. Elsevier, Amsterdam.

Tickner A. W. and Lossing F. P. (1951) The measurement of low vapor pressures by means of a mass spectrometer. *J. Phys. Chem., 55(5),* 733–740, DOI: 10.1021/j150488a013.

Trowbridge A. J., Melosh H. J., Steckloff J. K., et al. (2016) Vigourous convection as the explanation for Pluto's polygonal terrain. *Nature, 534,* 79–81, DOI: 10.1038/nature18016.

Turcotte D. L. and Schubert G. (2002) *Geodynamics, 2nd edition.* Cambridge Univ., New York. 456 pp.

Ugelvig S. V., Egholm D. L., and Iverson N. R. (2016) Glacial landscape evolution by subglacial quarrying: A multiscale computational approach. *J. Geophys. Res., 121,* 2042–2068, DOI: 10.1002/2016JF003960.

Umurhan O. M. et al. (2016) An expanded analysis of nitrogen ice convection in Sputnik Planum. *AAS/Division for Planetary Sciences Meeting Abstracts, 48,* 213.06.

Umurhan O. M., Howard A. D., Moore J. M., et al. (2017) Modeling glacial flow on and onto Pluto's Sputnik Planitia. *Icarus, 287,* 301–309.

Van Der Veen C. J. (1996) Tidewater calving. *J. Glaciol., 42,* 375–385.

Van Saarloos W. (1998) Three basic issues concerning interface dynamics in nonequilibrium pattern formation. *Phys. Rep., 301,* 9–43.

Vance S. D., Panning M. P., Stähler S., et al. (2018a) Geophysical investigations of habitability in ice-covered ocean worlds. *J. Geophys. Res., 123(1),* 180–205.

Vance S. D., Kedar S., Panning M. P., et al. (2018b) Vital signs: Seismology of icy ocean worlds. *Astrobiology, 18(1),* 37–53.

Vilella K. and Deschamps F. (2017) Thermal convection as a possible mechanism for the origin of polygonal structures on Pluto's surface. *J. Geophys. Res., 122,* 1056–1076, DOI: 10.1002/2016JE005215.

Vu T. H., Choukroun M., Sotin C., et al. (2020) Rapid formation of clathrate hydrate from liquid ethane and water ice on Titan. *Geophys. Res. Lett., 47(4),* e2019GL086265.

Warren S. G., Brandt R. E., and Grenfell T. C. (2006) Visible and near-ultraviolet absorption spectrum of ice from transmission of solar radiation into snow. *Appl. Opt., 45,* 5320–5334.

Weertman J (1999) Microstructural mechanisms of creep. In *Mechanics and Materials: Fundamentals and Linkages* (M. A. Meyers et al., eds.), pp. 451–488. Wiley, New York.

Weertman J. and Weertman J. R. (1992) *Elementary Dislocation Theory.* Oxford Univ., Oxford. 228 pp.

Wei Q. et al. (2018) Young surface of Pluto's Sputnik Planitia caused by viscous relaxation. *Astrophys. J. Lett., 856,* L14.

White D. B. (1988) The planforms and onset of convection with a temperature-dependent viscosity. *J. Fluid. Mech., 191,* 247–286.

White O. L. et al. (2017) Geological mapping of Sputnik Planitia on Pluto. *Icarus, 287,* 261–286, DOI: 10.1016/j.icarus.2017.01.011.

Whitham G. B. (1999) *Linear and Nonlinear Waves.* Wiley, London. 660 pp.

Yamashita Y., Kato M., and Arakawa M. (2010) Experimental study on the rheological properties of polycrystalline solid nitrogen and methane: Implications for tectonic processes on Triton. *Icarus, 207,* 972–977, DOI: 10.1016/j.icarus.2009.11.032.

Yasui M., Schulson E. M., and Renshaw C. E. (2017) Experimental studies on mechanical properties and ductile-to-brittle transition of ice-silica mixtures: Young's modulus, compressive strength, and fracture toughness. *J. Geophys. Res., 122,* 6014–6030, DOI: 10.1002/2017JB014029.

Yokochi R., Marboeuf U., Quirico E., and Schmitt B. (2012) Pressure dependent trace gas trapping in amorphous water ice at 77 K: Implications for determining conditions of comet formation. *Icarus, 218(2),* 760–770.

Young L. A. (2017) Volatile transport on inhomogeneous surfaces: II. Numerical calculations (VT3D). *Icarus, 284,* 443–476.

Young L. A. and 25 colleagues (2018) Structure and composition of Pluto's atmosphere from the New Horizons solar ultraviolet occultation. *Icarus, 300,* 174–199.

Yu X., Hörst S. M., He C., McGuiggan P., and Crawford B. (2018) Where does Titan sand come from: Insight from mechanical properties of Titan sand candidates. *J. Geophys. Res., 123,* 2310–2321.

Summers M. E., Young L. A., Gladstone G. R., and Person M. J. (2021) Composition and structure of Pluto's atmosphere. In *The Pluto System After New Horizons* (S. A. Stern, J. M. Moore, W. M. Grundy, L. A. Young, and R. P. Binzel, eds.), pp. 257–278. Univ. of Arizona, Tucson, DOI: 10.2458/azu_uapress_9780816540945-ch011.

Composition and Structure of Pluto's Atmosphere

Michael E. Summers
George Mason University

Leslie A. Young, G. Randall Gladstone
Southwest Research Institute

Michael J. Person
Massachusetts Institute of Technology

Observations of Pluto's atmosphere made by the New Horizons spacecraft during its flyby of Pluto on July 14, 2015, have revolutionized our understanding of Pluto's atmospheric composition and structure. Building upon an extensive series of Earth-based telescopic observations, and supplemented by theory and models, it is now possible to develop a reasonably coherent picture of the vertical structure of Pluto's atmosphere. The New Horizons mission answered many of the longstanding questions about Pluto's atmosphere. The Radio Science Experiment (REX) revealed that the atmosphere exhibits a narrow, and presumably localized, temperature boundary layer over Sputnik Planitia, but there was no evidence of a troposphere as had been proposed. Additionally, Pluto's lower atmosphere has both a strong positive vertical temperature gradient as well as a complex haze layer, both of which would have had an impact on the interpretation of earlier stellar occultation measurements. All pre-encounter photochemical models of Pluto's middle atmosphere correctly predicted that there would be a variety of hydrocarbons, and the Alice ultraviolet (UV) spectrograph found the major ones: acetylene (C_2H_2), ethylene (C_2H_4), ethane (C_2H_6), and methylacetylene (CH_3C_2H). However, pre-encounter models of the thermal structure suggested a high-temperature extended atmosphere with rapid hydrodynamic outflow as the escape process. Instead, the Alice solar occultation measurement determined that the upper atmosphere was colder and more compact, implying slower thermal Jeans escape. Overall, New Horizons found that many aspects of Pluto's atmosphere are more complex, and the interactions more subtle, than we once believed. These results have also revealed some challenging new questions. For example, what is the nature of the surface/atmosphere interaction — in particular that of the temperature boundary layer over Sputnik Planitia, and how does that interaction vary across Pluto's terrain and with time of day? What causes the highly structured vertical profiles of the C_2H_x hydrocarbons in the middle atmosphere, and how are they connected to the haze particles via heterogeneous processes? Are the haze layers triggered by nucleation processes in the ionosphere, and does the sedimentation and growth of haze particles mirror those in Titan's detached haze layer? What is the radiative cooling agent for the upper atmosphere? How do atmospheric waves, e.g., buoyancy and Rossby waves, modulate the condensation of photochemically produced hydrocarbons? Atmospheric structure and composition play key roles in all these questions. We close by discussing some important future directions for research.

1. INTRODUCTION

Speculation about the composition and structure of a possible atmosphere on Pluto date back nearly half a century (e.g., *Hart,* 1974; *Golitsyn,* 1975; *Stern and Trafton,* 1984). After the discovery of Charon (*Christy and Harrington,* 1980) and detection of gaseous methane (*Cruikshank and Silvaggio,* 1980; *Fink et al.,* 1980), theoretical models suggested that Pluto's atmosphere was more extended than other known planetary atmospheres (*Stern and Trafton,* 1984, 2008). A stellar occultation in 1988 (*Elliot et al.,* 1989; *Hubbard et al.,* 1988) confirmed the existence of Pluto's atmosphere suggested from vapor pressure arguments and surface volatile ice detection [an ambiguous single-chord occultation was previously observed in 1985 (*Brosch,* 1995)]. Further analysis of the 1988 occultation showed that the dominant atmospheric gas could not be methane (CH_4), and provided the first clear evidence of vertical structure in the atmosphere, with models suggesting that the atmospheric temperature increased rapidly above a cold surface (37–40 K) to over 100 K within the first scale height (*Yelle and Lunine,* 1989). Molecular nitrogen (N_2) ice was detected on Pluto by *Owen et al.* (1993). Vapor pressure equilibrium at a surface temperature in the 36–40-K range indicated an N_2 surface pressure in the range ~3–60 μbar.

This knowledge, along with the detection of CH_4 as a minor constituent (*Young et al.,* 1997), set the stage for the first one-dimensional photochemical/vertical diffusive transport models (*Summers et al.,* 1997; *Lara et al.,* 1997; *Krasnopolsky and Cruikshank,* 1999). These models showed that solar UV radiation would drive methane destruction with concomitant production of hydrocarbons such as C_2H_2, C_2H_4, C_2H_6, and nitriles such as HCN and H_2CN, building up these species as important chemical components of the atmosphere. These higher hydrocarbons and nitriles were expected to deposit as frosts on the surface, although they have yet to be detected there.

Thus, prior to the New Horizons flyby of Pluto in July 2015, there were the beginnings of a coherent picture of the overall vertical temperature profile and the composition of Pluto's atmosphere. However, there were still several puzzles. First, there was controversy over whether the near-surface "kink" in the stellar occultation observations actually indicated a large vertical temperature gradient in the lower atmosphere, or whether it was caused by a haze layer or deep troposphere (*Elliott et al.,* 1989; *Yelle and Lunine,* 1989). Second, the high upper atmosphere temperature suggested that the atmosphere was highly extended — with an exobase at a distance of several times Pluto's radius (*Strobel et al.,* 1996). Such an extended atmosphere would imply very large escape rates, perhaps via rapid hydrodynamic outflow — so rapid that the escape rate would be limited by available solar UV energy (*Hunten and Watson,* 1982), and that could lead to the loss of a large portion of Pluto's volatile inventory over the history of the solar system. Third, the N_2 atmospheres of Triton and Pluto are very different yet are of comparable size; are the differences related to their disparate atmospheric methane abundances, or something else? And fourth, stellar occultation observations indicated that the total atmospheric pressure continued to increase *after* Pluto's perihelion in 1988, which suggested a delayed atmospheric collapse as Pluto moved further from the Sun (*Elliott et al.,* 2007; *Young,* 2013; *Hansen et al.,* 2015; *Olkin et al.,* 2015). Thus, by the time of the New Horizons flyby in 2015 it was clear that progress on understanding these mysteries would require comprehensive observations of Pluto's atmosphere.

A central goal of the New Horizons mission was to characterize the structure and composition of Pluto's atmosphere (*Spencer and Stern,* 2003; *L. A. Young et al.,* 2008) using flyby observations provided by the spacecraft's Radio Experiment (REX) instrument (*Hinson et al.,* 2017, 2018; *Tyler et al.,* 2008) used for radio occultation measurements of the pressure and temperature structure of the atmosphere; the Alice instrument that provided extreme-ultraviolet (EUV) and far-ultraviolet (FUV) solar and stellar occultation measurements, as well as airglow and reflected sunlight observations (*Stern et al.,* 2008; *Young et al.,* 2018); and searches for hazes (and possibly clouds or plumes) from the Long-Range Reconnaissance Imager (LORRI) and the Multispectral Visible Imaging Camera (MVIC).

Many of the atmospheric observations were carried out during the few hours around the closest approach of New Horizons to Pluto at 11:48 UT on July 14, 2015 (*Stern et al.,* 2015). Pressure and temperature profiles of the lower atmosphere at ingress and egress locations were derived from the REX data during the radio occultation (*Gladstone et al.,* 2016; *Hinson et al.,* 2017), and information on the vertical variation of N_2, CH_4, C_2H_x photochemical products/haze and temperature structure of the atmosphere were derived from the Alice data (*Stern et al.,* 2015; *Gladstone et al.,* 2016; *L. A. Young et al.,* 2008). Images of the distributions and properties of the global hazes were obtained from LORRI and MVIC images (*Gladstone et al.,* 2016; *Cheng et al.,* 2017).

The New Horizons observations of Pluto's atmosphere confirmed some aspects of our understanding of its structure and composition and also produced some surprises. The lower atmosphere (below ~25 km altitude; see Fig. 1) was found to be dominated by a strong positive temperature gradient, as expected from Earth-based stellar observations. Yet the REX observations showed the existence of a thin temperature boundary layer over Sputnik Planitia — at the ingress occultation point, and not the deep tropopause suggested by early modelers (e.g., *Stansberry et al.,* 1994) who were attempting to reconcile the inconsistent radii estimates from the mutual events and stellar occultations.

As expected, the middle atmosphere (from ~25 km to ~425 km) was found to be a region of complex photochemical activity driven by solar UV radiation that produced the copious C_2H_x hydrocarbons detected by New Horizons and the HCN detected contemporaneously from Earth (*Lellouch et al.,* 2017). However, the diffuse haze with embedded thin bright layers was a surprise, as was the inference that the middle atmosphere is largely stagnant. Although an atmospheric haze produced by condensation of photochemical products/haze had been expected (*Summers et al.,* 1997; *Krasnopolsky and Cruikshank,* 1999), the large vertical extent and the numerous bright embedded thin layers that extended horizontally for many hundreds of kilometers raised questions about how atmospheric waves might influence these distributions of photochemical products. For instance, atmospheric gravity-buoyance waves could modulate C_2H_x condensation onto haze particles and focus the haze layers (*Cheng et al.,* 2017). Additionally, while the middle atmosphere was found to have a lower overall eddy mixing rate than expected, it might be more significant near the surface (*Krasnopolsky,* 2020).

Finally, the Alice solar occultation measurements showed that the upper atmosphere (above ~425 km) was found to have very low N_2 densities at high altitudes, implying lower

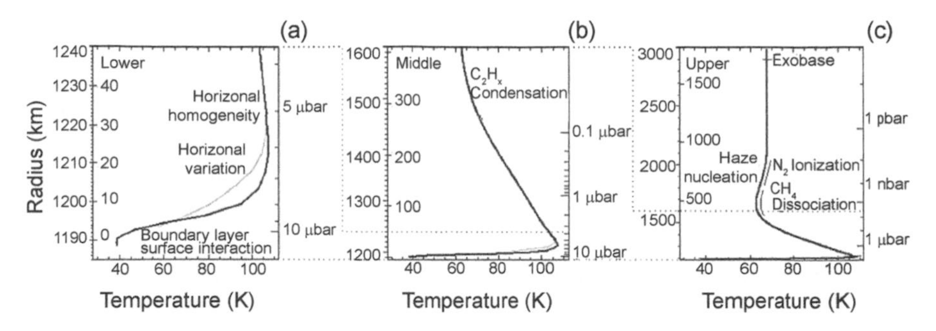

Fig. 1. Pluto's atmospheric regions as defined by characteristics of the temperature profile. **(a)** The lower atmosphere is characterized by a positive temperature gradient from the surface to a peak of ~107 K at ~25 km altitude. Over Sputnik Planitia there is a narrow (<5 km) and presumably localized temperature boundary layer, as discussed in the text. **(b)** The middle atmosphere extends from ~25 km to a temperature minimum of ~60 K at ~425 km. **(c)** In the upper atmosphere, above ~425 km, the temperature increases slightly until it reaches a nearly isothermal region at ~67 K above ~900 km. This combined temperature structure was produced by joint analysis of the Alice solar occultation observations and the REX radio occultation (REX ingress profile in black, REX egress in gray) as described by *Young et al.* (2018). The approximate altitudes are referenced to a surface radius of 1190 km. The dominant thermodynamic processes in each region are indicated. Adapted from *Young et al.* (2018).

temperatures (by ~30 K) than expected. This has major implications for atmospheric escape and thus the integrated loss of volatiles from Pluto over the course of its history (*Gladstone et al.,* 2016).

The purpose of this review is to survey our understanding of the structure and composition of Pluto's atmosphere based upon analyses carried out since the New Horizons flyby. Understanding the composition and structure of Pluto's atmosphere overlaps the topics of atmospheric chemistry, dynamics, atmospheric escape, and surface interactions, which are themes of other chapters in this volume. Here we focus on describing the observed structure and composition of Pluto's atmosphere and how the radiative heating/cooling balance leads to that observed structure. For more detailed discussion of the composition and structure of Pluto's atmosphere as determined by New Horizons observations the reader is referred to the recent reviews by *Stern et al.* (2015), *Young et al.* (2018), and *Gladstone and Young* (2019). Other chapters in this volume by Mandt et al., Forget et al., Strobel et al., and Young et al. cover closely related topics.

The characteristics of the vertical temperature structure allow for a ready division into layers that are somewhat analogous to those in other planetary atmospheres. The lower atmosphere consists of a region of positive temperature gradient from the surface to ~25 km altitude, although the exact altitude is slightly model dependent — this region is termed the lower atmosphere and is superficially analogous to Earth's stratosphere. The middle atmosphere is characterized by a negative temperature gradient from the top of the lower atmosphere up to ~425 km altitude, analogous to Earth's mesosphere. Above ~425 km, the atmospheric temperature increases slightly until it reaches a nearly isothermal region (~900 km) that is termed the upper atmosphere and is analogous to Earth's thermosphere. These analogies are defined by the observed structure — the physical and chemical processes in each layer of Pluto's atmosphere are much different from those in Earth's atmosphere.

We will begin with a discussion of the composition of the lower atmosphere as revealed by New Horizons, as supplemented by contributions from Earth-based near-infrared (IR) and microwave observations. We will then discuss the vertical temperature and pressure structures, beginning with the lower atmosphere, followed by a discussion of the middle atmosphere where most photochemistry and haze particle condensation occurs. Finally, we will discuss the upper atmosphere where ionization peaks and haze particle nucleation and coagulation occur. At the end we will summarize some important questions and avenues of research needed to further our understanding of Pluto's atmospheric composition and structure.

2. LOWER ATMOSPHERE

2.1. Lower Atmosphere — Composition

The exchange of mass, momentum, and energy between Pluto's atmosphere and surface occurs in the lower atmosphere, so understanding its composition and temperature structure with variability both vertically and horizontally and across Pluto's surface is central to the discussion of almost any type of surface/atmosphere interaction. This discussion includes important topics such as volatile transport, haze particle growth and sedimentation in the lowest layer, and temporal variations of the atmosphere driven by seasonal (and perhaps longer-term) changes in surface ices and their temperatures. Pluto's lower atmosphere is unique in the solar system, so comparisons with other planets are not necessarily helpful in understanding the specific processes at work on Pluto.

The first unambiguous detection of Pluto's atmospheric methane was made in 1992 (*Young et al.,* 1997) using the CSHELL echelle spectrograph on NASA's Infrared Telescope Facility (IRTF) telescope on Mauna Kea. The resulting spectrum in the 1661.8–1666.9-nm wavelength range revealed lines of the $2\nu_3$ band of methane, indicating gaseous methane with an atmospheric mixing ratio in the range 0.03–0.45%, assuming an N_2 surface pressure of 58 µbar at 40 K temperature. *Lellouch et al.* (2009) found a methane abundance of 0.5 ± 1%, and later *Lellouch et al.* (2011) found a value of 0.6 +0.6%, –0.3%.

This amount of gaseous CH_4 is several hundred times larger than that predicted for vapor pressure equilibrium above an ideal mixture of N_2-CH_4-CO ices on Pluto's surface as determined by Raoult's law. This overabundance of CH_4 in Pluto's atmosphere has been a persistent puzzle and has motivated detailed studies of the phase diagram of N_2:CH_4:CO ices (see the chapter by Young et al. in this volume).

For pure ices, the N_2 vapor pressure is ~10^4× larger than that of CH_4 at 37 K (*Fray and Schmidt,* 2009). *Prokhvatilov and Yantsevich* (1983) determined that the N_2-CH_4 binary phase diagram shows that N_2 and CH_4 ices cannot mix with arbitrary ratios at equilibrium but have maximum ratios of 5% CH_4 in a N_2-ice rich surface or 3% N_2 in a CH_4-ice rich surface, respectively.

Recent theoretical work (*Tan and Kargel,* 2018; *Trafton,* 2015) showed that for a broad range of CH_4 mole fractions and plausible Pluto temperatures (e.g., mole fractions between ~5% and ~97% at 37 K), N_2 and CH_4 will exist in equilibrium as two species (N_2 and CH_4) in three states (N_2:CH_4 gas, N_2-rich ice saturated with CH_4, and CH_4-rich ice saturated with N_2), the equilibrium partial pressures of each state being a function of temperature. At equilibrium, the N_2-rich ice state will be saturated with CH_4, the CH_4-rich ice state will be saturated with N_2, and the temperature-dependent gaseous mole fraction of CH_4 will range from 0.0036% to 0.01% from 36 to 40 K (*Tan and Kargel,* 2018). The *Young et al.* (1997) determination of the methane mixing ratio of 0.03–0.45% is thus significantly higher than that predicted for global vapor pressure equilibrium with uniform mixtures of surface ices.

The New Horizon's Alice experiment carried out solar UV occultation measurements as the spacecraft flew behind Pluto during both entry and exit, and these occultations provided detailed information on atmospheric composition from the surface to near 1500 km altitude (*Gladstone et al.,* 2016). Methane absorption was measured from ~80 km altitude (below which methane was opaque) to near ~1200 km altitude (where methane absorption was too small to produce good retrievals). In order to use the Alice observations to quantify the near-surface methane abundance, extrapolation to the surface requires an atmospheric model that includes upward methane transport above the surface by either eddy or molecular diffusion (or both) as well as photochemical destruction at higher altitudes (*Young et al.,* 2018).

Several studies have used one-dimensional photochemical models to simulate the methane vertical profile (*Luspay-Kuti et al.,* 2017; *Mandt et al.,* 2017; *Young et al.,* 2018; *Strobel and Zhu,* 2017; *Krasnopolsky,* 2020). These models include vertical diffusive transport and generally obtain good fits to the observed methane profile above ~80 km altitude for assumed eddy diffusion coefficients [denoted by K_{zz}, which is the z component (vertical) of eddy mixing due to vertical gradients in the mixing ratio] between 550 and 4000 cm^2 s^{-1}, and assuming a near-surface methane mixing ratio of 0.28% and 0.35%, respectively, the lower value obtained when methane is allowed to flow vertically in the model. *Wong et al.* (2017) obtained a slightly larger methane mixing ratio of 0.4% with an assumed comparable eddy diffusion coefficient of K_{zz} ~ 4000 cm^2 s^{-1}.

The recent photochemical modeling results mentioned above suggest that eddy mixing in Pluto's atmosphere is extremely slow compared to molecular diffusion. The timescale of eddy mixing $\tau_e \sim H^2/K_{zz}$ where H is the background scale height = kT/mg, and T is temperature, g is the local gravitational acceleration, and m is the molecular mass. The molecular diffusion timescale is $\tau_m \sim H^2/D_{CH_4\text{-}N_2}$, where $D_{CH_4\text{-}N_2}$ is the molecular diffusion coefficient for CH_4 in a background N_2 atmosphere. The homopause is the location where $\tau_e \sim \tau_m$, which is at or very near (below ~12 km) the surface. However, it is not known how K_{zz} depends upon density (or pressure) in such a low-density, nearly inviscid atmosphere like Pluto's.

A different result is given in the recent model by *Krasnopolsky* (2020), who found that the eddy K_{zz} ~ 3×10^4 cm^2 s^{-1}, i.e., an order of magnitude larger than the above studies, with an assumed near-surface methane mixing ratio of 0.45%, larger than those referenced above. If the eddy mixing rate is indeed that large, then the location of the homopause is near ~110 km altitude, and thus extends well into the middle atmosphere.

Thus, while determinations of diffusion coefficients are currently inconsistent, models of the Alice methane vertical distribution support the upper end of the initial *Young et al.* (1997) methane mixing ratio determination, as well as the conclusion that there is an overabundance of methane in Pluto's atmosphere — under the assumption of globally uniform distributions of CH_4 and N_2 ices.

Trafton et al. (1997) proposed that the overabundance of atmospheric methane could be explained as a process of "detailed balancing," in which the atmosphere was mostly N_2 with a mixing ratio of ~0.3% CH_4 that is in vapor pressure equilibrium with a surface that has a very thin layer (too thin to be detected in the IR) of mostly CH_4 ice (98.2%) and that has a small amount (1.8%) of N_2 ice (assuming the vapor pressure is in accord with Raoult's law). However, to obtain 0.3% CH_4 the *Trafton et al.* (1997) model required a N_2 surface pressure of ~0.2 µbar — much smaller than the 11.7-µbar pressure observed by REX. *Stansberry et al.* (1996) showed that the overabundance of CH_4 could also be produced if Pluto's surface was mostly N_2 ice, with CH_4 ice confined to small essentially pure "patches" that were at warmer temperatures (>5 K) than the surrounding surface. The higher temperature of the CH_4 ice patches would lead to a large sublimated upward mass flux of CH_4 that would then mix throughout the atmosphere by turbulent diffusion, leading to the atmospheric overabundance of methane as observed by *Young et al.* (1997, 2018), which would be consistent with the extrapolations of the Alice CH_4 observations to the surface.

The New Horizons LEISA surface spectral mapping observations (*Protopapa et al.,* 2017; *Schmitt et al.,* 2017) found areas [potentially like the patches of the *Stansberry et al.* (1996) patch model] of nearly pure CH_4 ice. The latitude band 15°–30°N (exclusive of Sputnik Planitia), the northern polar area north of ~75°N, and the bladed terrain of eastern Tombaugh Regio are mostly CH_4. As noted by *Young et al.* (2018), this CH_4 is found in areas of higher elevations (*Howard et al.,* 2017; *Moore et al.,* 2018), such as on the top of the mountain ridges in the (informally named) Cthulhu Regio. These CH_4 patches could be time dependent due to seasonal transport, such that the resulting influx into the atmosphere — and thus the overabundance of CH_4 — could be variable on seasonal and longer timescales.

Three-dimensional time-dependent atmosphere global climate models (GCMs) have been used to study volatile transport in Pluto's atmosphere (*Zalucha and Michaels,* 2013; *Bertrand and Forget,* 2016, 2017), and have simulated the effects of patches of warm CH_4 ice and turbulent boundary layer transport. Depending upon the assumed location, size, and temperature of the warm CH_4 patches, and how the southern polar N_2 ice cap is characterized, an atmospheric CH_4 mixing ratio of 0.5–1% will result. These models also predict factor of 2 variations in the methane mixing ratio between hemispheres. When applied to the time period of the New Horizons flyby in 2015, the GCMs produce only slight variation in CH_4 mixing ratio with altitude in the lower atmosphere above the location of the ingress and egress occultations. This provides a useful constraint that confirms the extrapolations of the methane mixing ratio of *Young et al.* (2018).

All existing photochemical models of Pluto's atmosphere (e.g., *Summers et al.,* 1997; *Lara et al.,* 1997; *Krasnopolsky and Cruikshank,* 1999; *Gao et al.,* 2017; *Wong et al.,* 2017; *Luspay-Kuti et al.,* 2017; *Mandt et al.,* 2017; *Krasnopolsky,* 2020) showed robust production of C_2 hydrocarbons, C_4H_2, C_3H_4, and nitriles such as HCN and H_3CN as a consequence of effective methane photolysis in the middle atmosphere. These models indicate that many photochemically produced species would be supersaturated and thus are expected to condense onto the haze particles as they settle through the middle and lower atmosphere to the surface.

Alice observations of lower- and middle-atmosphere CH_4 and C_2 hydrocarbons show near homogeneity between in their profiles at ingress and egress (see discussion below). However, there are less useful constraints on the abundances of CH_4 and C_2H_6 in the lowest part of the atmosphere (below ~80 km) because of insufficient signal to noise in the UV absorption. Knowledge of the abundance profiles of these important photochemical species in this low-altitude region of atmospheric condensation where haze particle growth and sedimentation occur is thus currently provided only by the output of photochemical-vertical transport models.

2.2. Lower Atmosphere — Pressure and Temperature Structure

Prior to the New Horizons flyby observations of Pluto, models of Pluto's lower atmosphere were largely constructed upon the assumptions that the total pressure was buffered by the N_2 saturation vapor pressure and that the abundances of the minor species CH_4 and CO were controlled by equilibrium vapor pressures above a mix of N_2, CH_4, and CO ices (*Hansen et al.,* 2015; *Spencer et al.,* 1997; *Toigo et al.,* 2015; *Young,* 2013). The low-spatial-resolution albedo and composition maps that were available (*Lellouch et al.,* 2011; *Hansen et al.,* 2015; *Young,* 2012; *Young,* 2013) helped guide the construction of surface ice distribution models. Speculations on the existence and nature of an Earth-analog convective tropopause (*Stansberry et al.,* 1994) were severely hampered by lack of accurate knowledge of Pluto's radius (pre-encounter estimates of Pluto's radius ranged from 1150–1210 km) and by the ambiguity about whether the kink in the stellar occultation observations near the surface was a consequence of a steep positive thermal gradient or due to absorption by a thin haze layer. The New Horizons observations have enabled decisive progress in each of these areas.

The REX radio occultation (*Stern et al.,* 2015; *Hinson et al.,* 2017) was implemented as a radio uplink from Earth using 4.2-cm-wavelength signals transmitted by antennas of the NASA Deep Space Network (DSN) and received by the REX instrument (*Tyler et al.,* 2008) as it went behind Pluto. The ingress occultation occurred at sunset near the center of the anti-Charon hemisphere over the southeast corner of Sputnik Planitia, and the egress occultation at sunrise near the center of the Charon-facing hemisphere. Because the occultations were near diametric behind Pluto, and because optical navigation (OpNav) imaging allowed a precise reconstruction of the position of New Horizons relative to Pluto, very precise determinations of both the ingress and egress radii separately were obtained. The mean radio occultation radius, 1189.9 ± 0.2 km, is consistent with that obtained from imaging analysis 1188.3 ± 1.6 km (*Nimmo et al.,* 2017).

Analysis of the REX occultation measurements revealed a strong positive temperature gradient (below ~30 km altitude) above both ingress and egress locations (Fig. 2). The average lower-atmosphere temperature gradient was found to be somewhat larger at ingress (6.4 ± 0.9 K km^{-1}) than at egress (3.4 ± 0.9 K km^{-1}). These results are in general agreement with the results retrieved from Earth-based stellar occultation measurements (3–10 K km^{-1}) using retrieval models that assume that the "kink" in the lower atmosphere stellar occultation is due to a temperature inversion (*Dias-Oliveira et al.,* 2015; *Elliot et al.,* 2003, 2007; *Hubbard et al.,* 1990; *Sicardy et al.,* 2003, 2016; *E. F. Young et al.,* 2008).

This large positive temperature gradient in the lower atmosphere is consistent with the picture of solar heating in the lower and middle atmosphere balanced by thermal conduction downward to the surface where the near-surface cooling is accomplished by either CH_4 (*Yelle and Lunine,*

(a)

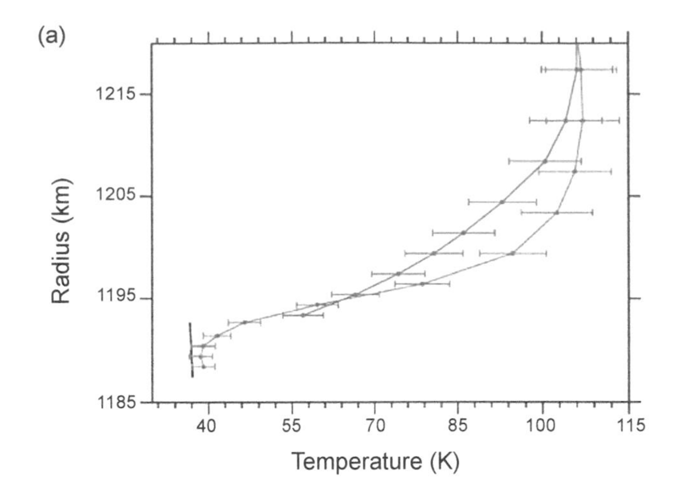

(b)

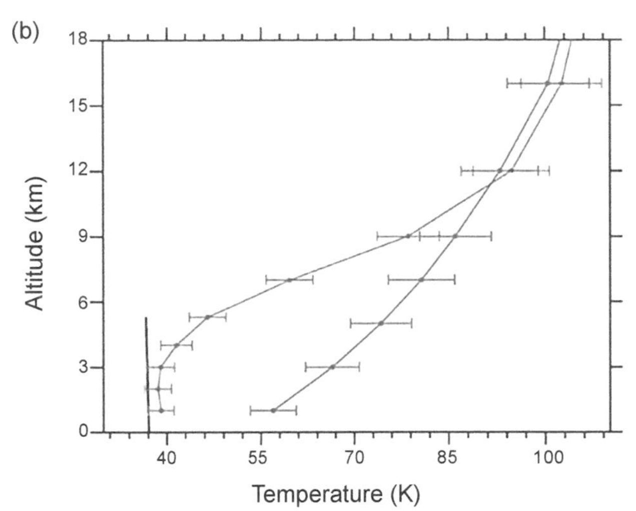

Fig. 2. Temperature profiles in Pluto's lower atmosphere from the REX uplink radio occultations for ingress near Sputnik Planitia (yellow) and egress (blue), shown for **(a)** radial distance and **(b)** altitude. The accuracy in temperature is indicated by 1-σ error bars and near the surface is ~3 K. The ingress and egress profiles above 25 km are nearly identical and consistent with groundbased stellar occultation measurements. The ingress profile shows a cold boundary layer where the temperature is nearly constant, 38.9 ± 2.1 K (close to the saturation temperature of N_2). This boundary layer is missing in the egress profile, where the surface air temperature is 51.6 ± 3.8 K. The mean values of the pressure and radius were determined to be 11.5 ± 0.7 μbar at 1189.9 ± 0.2 km. The solid black line shows the N_2 dew point as a function of altitude. From *Hinson et al.* (2017).

1989), CH_4 and CO together (*Strobel et al.,* 1996; *Zhu et al.,* 2014), HCN, or perhaps aerosols (*Zhang et al.,* 2017). At altitudes above ~30 km, both ingress and egress temperature measurements converge to a common value between 105 and 110 K at 50 km altitude. These large thermal gradients lead to large values of the Brunt-Väisälä (buoyancy) frequency — a measure of atmospheric stability. Thus, the large thermal gradient acts as a "lid," suppressing vertical mixing between the lower and upper atmosphere.

However, there is a difference in the lowest altitude probed by the REX temperature inversion which at ingress ends approximately 4 km lower than that at egress (*Hinson et al.,* 2017). This distinctive difference in the two temperature profiles is illustrated in Fig. 2, and has been reasonably assumed to be a consequence of a narrow boundary layer at the ingress site over Sputnik Planitia that is composed of N_2 that was sublimated earlier during the day of the encounter due to solar heating (*Hinson et al.,* 2017). The egress occultation profile shows no suggestion of a boundary layer. It should be noted that the vertical spatial resolution of the REX observations may not be sufficient to resolve a temperature boundary layer that is on the order of ~1 km or smaller, since the Fresnel scale at egress was 1.55 km (*Hinson et al.,* 2017).

The REX determination of surface pressure and near-surface temperature are consistent with vapor-pressure equilibrium over N_2 ice at ingress (12.8 ± 0.7 μbar and 38.9 ± 2.1 K), but the near-surface temperature at egress is warmer, consistent with the lower-albedo N_2 ice-free terrain (perhaps covered by tholins — carbon-enriched amorphous hydrocarbons) and the surface pressure is also lower than that at ingress (10.2 ± 0.7 μbar and 51.6 ± 3.8 K).

Precise knowledge of the surface temperature and pressure, as well as the vertical temperature gradient within the boundary layer, are critical for understanding the vertical flow of energy and volatiles. The REX observations of the temperature profile over Sputnik Planitia are important guides for atmospheric thermal models, but the vertical resolution is too low to accurately resolve the vertical temperature gradient within a boundary layer that is only 1–4 km in vertical extent, and thus it is difficult to determine whether the boundary layer is governed by the dry N_2 adiabat [$\Gamma_d(N_2) = -g/C_p = -0.063$ K km^{-1}; where g is the local gravitational acceleration and C_p is the heat capacity of N_2], or the saturated lapse rate [$\Gamma_{sat}(N_2) = -gT/L = -0.0996$ K km^{-1}, where L is the latent heat of sublimation] or perhaps something else entirely. This leads to an important weakness in our understanding of the energy, momentum, and mass flux due to surface-atmosphere interactions over Sputnik Planitia. This problem is compounded by the likelihood that the boundary layer evolves over diurnal and seasonal timescales.

Global climate models of the diurnal variation of the atmosphere over Sputnik Planitia suggests that the basin acts as a "leaky N_2 piston" (*Forget et al.,* 2017) for the atmosphere. During the day, solar-heating-driven sublimation produces an "excess" N_2 mass that, if constrained locally, would be equivalent to about 10% of the local mass of the atmosphere (*Young et al.,* 2018). This allows outflow of N_2 from Sputnik Planitia during the day, with wind speeds of a few meters per second. This also implies that the observed boundary layer might never actually be in equilibrium with local atmospheric and surface properties, and conversely the non-local nature of the control of the boundary layer inhibits equalization of temperatures at a given pressure level.

The *Forget et al.* (2017) GCM has successfully simulated some characteristics of the atmosphere over Sputnik Planitia, such as the steep vertical temperature gradient, but its low vertical resolution prevents accurate simulation of boundary layer structure. Models with much higher vertical resolution, with resolutions at least 100 m (or perhaps 10 m or even smaller), would be needed to more fully explore the formation and time evolution of the boundary layer over Sputnik Planitia, as well as the potential for a much smaller boundary layer elsewhere that could not be seen in the REX egress data.

3. MIDDLE ATMOSPHERE

In Pluto's middle atmosphere (see Fig. 1) solar UV radiation dissociates methane (wavelengths <~140 nm), which initiates the photochemical production of higher hydrocarbons and nitriles that can condense either in Pluto's atmosphere, producing haze particles, or on its surface. Photochemical production of these condensable species is accompanied by release of atomic and molecular hydrogen that diffuses upward and (along with other species) thermally escapes from the top of the atmosphere. Pluto's small size — and thus low gravitational binding energy to atoms and molecules — leads to efficient escape from the top of the atmosphere. The composition and structure of Pluto's middle atmosphere is directly and strongly connected to the nature of both Pluto's surface and upper atmosphere via a range of processes including chemistry, dynamics, thermodynamics, and phase changes.

Water vapor has been proposed as the cooling agent for Pluto's atmosphere (*Strobel and Zhu,* 2017), although it would require supersaturation of H_2O by several orders of magnitude. More recently, photochemical haze particles have been proposed as the primary cooling agent in the upper atmosphere (*Zhang et al.,* 2017). Figure 3 shows the timescales for energy transfer by a variety of processes. The timescale for particle cooling is much shorter than that for gas cooling or conduction (*Zhang et al.,* 2017). Gas and particles are expected to be in thermal equilibrium below ~700 km, and above that altitude heat transfer is by conduction (not escape), producing a near-isothermal upper atmosphere at ~70 K (see Fig. 4).

In pre-encounter thermal models, the predicted peak temperature was ~130 K at ~100 km altitude and was a result of large inferred vertical temperature gradient in the lower atmosphere. Rapid gas to haze energy transfer and efficient haze particle cooling brings this peak temperature down to ~107 K at an altitude of ~25 km. Cooling by HCN and C_2H_2 cannot produce the low upper atmospheric temperature determined by Alice observations (*Gladstone et al.,* 2016); however, haze particle cooling and heating have been shown to be plausible processes to explain the low temperatures (*Zhang et al.,* 2017).

The atmospheric temperature profile is a critical input to photochemical models both because many gas phase chemical reaction rates are temperature dependent, and because most of the photochemically produced species are supersaturated — implying rapid loss of these species from the gas phase due to condensation on haze particles.

Prior to the New Horizons flyby of Pluto, knowledge of the chemical structure of Pluto's middle atmosphere was mostly based upon theoretical photochemical models (*Summers et al.,* 1997; *Lara et al.,* 1997; *Krasnopolsky and Cruikshank,* 1999), and also generally premised on the idea that Pluto's atmosphere, which was composed of

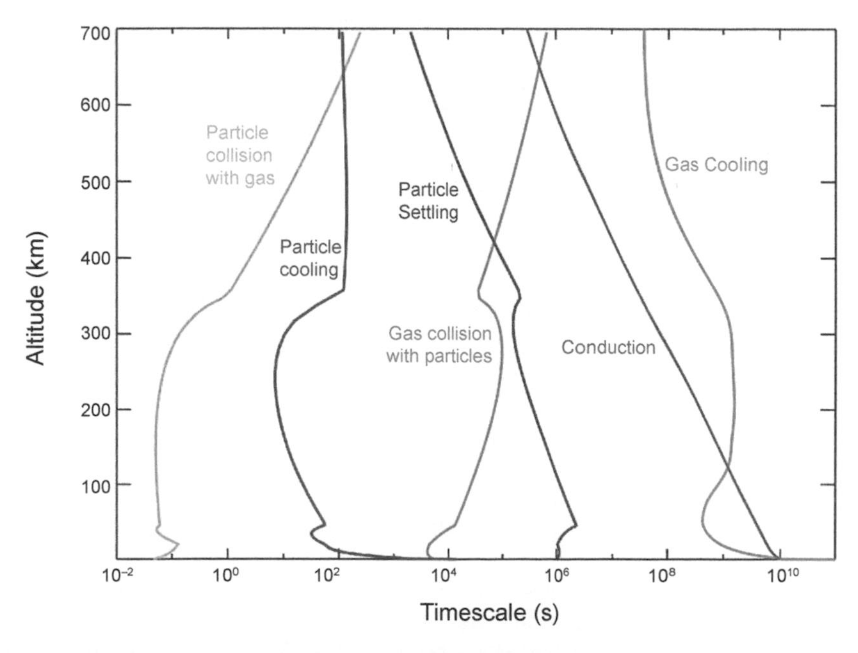

Fig. 3. Important timescales for processes in Pluto's atmosphere. From *Zhang et al.* (2017).

primarily N_2 with traces of CH_4, would be similar to the N_2 atmospheres of Triton and Titan. Although there were some Earth-based column measurements of CH_4 and limits on CO prior to the New Horizons flyby, direct observations from which the chemical and thermal vertical structure of the middle atmosphere could be inferred were limited.

The New Horizons flyby of Pluto included the Alice instrument — a UV spectrograph with a bandpass of 52–187-nm spectral range (*Stern et al.,* 2008). This bandpass includes wavelengths at which solar radiation dissociates methane, as well wavelengths sensitive to absorption by the predicted C_2H_x hydrocarbons. In addition, the Alice spectral range is bounded by N_2 absorption continuum on the short end and significant extinction by haze on the long end. Ultraviolet occultations have proven invaluable for measuring the structure and composition of the other two N_2-rich atmospheres in the outer solar system, Titan (*Smith et al.,* 1982; *Herbert et al.,* 1987; *Vervack et al.,* 2004; *Koskinen et al.,* 2011; *Kammer et al.,* 2013; *Capalbo et al.,* 2015) and Triton (*Broadfoot et al.,* 1989; *Herbert and Sandel,* 1991; *Stevens et al.,* 1992; *Krasnopolsky et al.,* 1992). Thus, the Alice instrument was the central platform for New Horizons observations of the composition and structure of the atmosphere. During the solar occultation, Alice measured how the absorption by molecular species and extinction by haze particles varied with altitude as the Sun passed behind the atmosphere (Fig. 5) (*Gladstone et al.,* 2016). From those measurements the vertical density profiles of the absorbing species and haze particles were determined (*Young et al.,* 2018).

The absorption of molecular species in Pluto's atmosphere during the solar occultation is shown in Fig. 6 for (1) the pre-encounter model and (2) the New Horizons Alice data. Figure 6a shows the line-of-sight (LOS) transmission (Tr) as a function of wavelength and altitude. From the LOS transmission the line-of-sight optical depth was determined, and from the optical depth and the absorption cross sections of the molecules the LOS column abundances were determined (see Fig. 7). The Alice solar occultation showed absorption from N_2, CH_4, C_2H_2, C_2H_4, and C_2H_6 as well as haze (*Gladstone et al.,* 2016; *Young et al.,* 2018).

In Fig 7b the trace gases show a range of behaviors with altitude. Methane shows a near-exponential decrease with geopotential height (equivalent constant gravity height) and thus appears to be in diffusive equilibrium. Strictly speaking, this means that any methane chemical loss is slow enough that vertical transport from the surface can replenish it without changing its scale height.

Of the measured photochemically produced hydrocarbons (i.e., C_2H_2, C_2H_4, C_2H_6), none of their density profiles show a simple exponential decrease of density with altitude as would be expected for diffusive equilibrium. Their more complex vertical profiles reflect highly altitude dependent production and loss processes (see the chapter by Mandt et al. in this volume for discussion of photochemistry and haze interactions). The most abundant C_2 hydrocarbon predicted by pre-encounter photochemical models, C_2H_6, has an abundance that shows the same vertical slope as that of CH_4 and N_2 and approximately

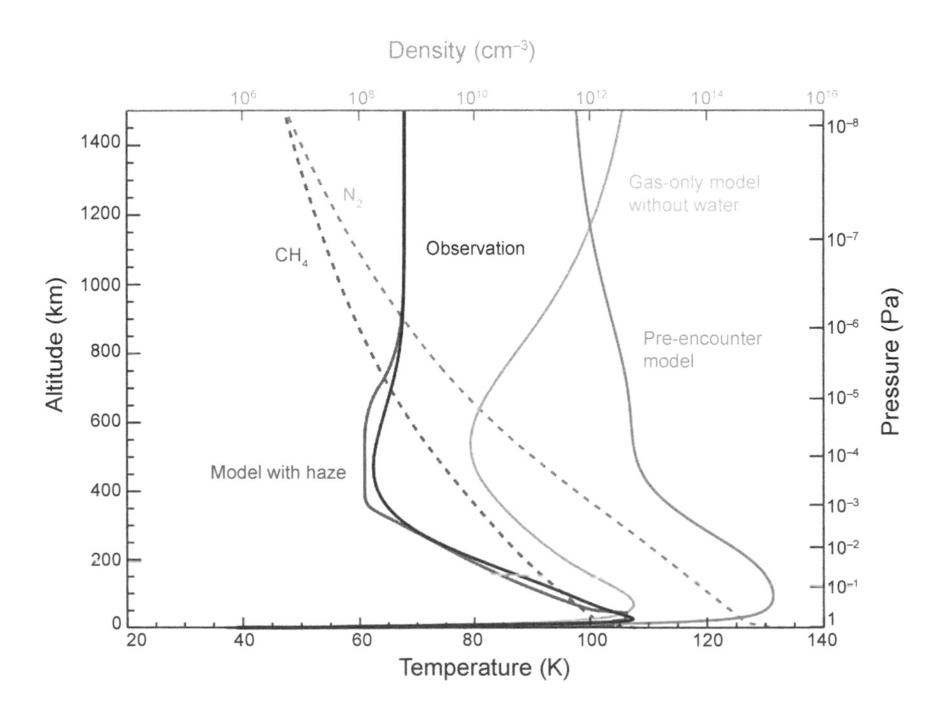

Fig. 4. Temperature and CH_4 and N_2 density profiles. The solid black line shows the temperature profile determined by analysis of New Horizons observations. Models of the pre-encounter temperature (blue) with gas-phase cooling only, including HCN and C_2H_2 (yellow), and with haze-mediated heating and cooling (red). From *Zhang et al.* (2017).

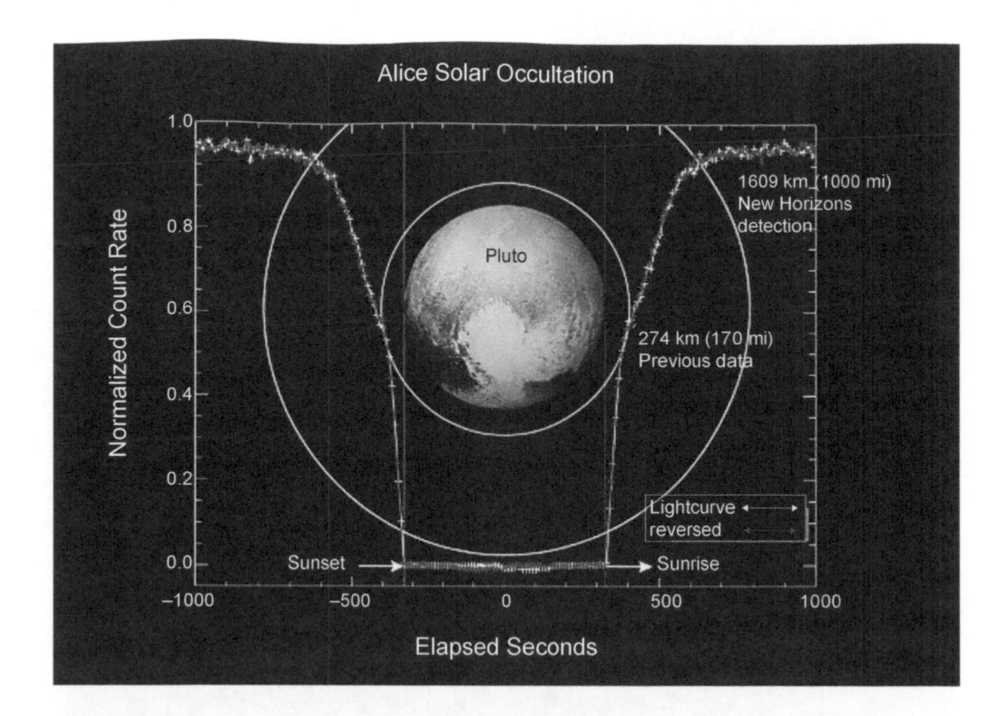

Fig. 5. The Alice solar occultation of Pluto is illustrated by showing the total Alice count rate on the vertical axis as the New Horizons spacecraft passed into the shadow of the planet. The decrease in total count rate on the ingress occultation began ~800 s before closest approach, which corresponds to an altitude ~1200 km above the surface. The absorption of solar radiation at the highest altitudes was mostly by N_2. As the occultation probed lower altitudes, absorption was also by methane and photochemically produced hydrocarbons. In the lowest ~100 km, absorption was due to a combination of atmospheric nitrogen, methane, higher-order hydrocarbons, and haze. Upon emergence from Pluto's shadow (the egress occultation) the count rate profile showed almost exactly the reverse of the entry profile — indicating that Pluto's atmosphere is globally symmetric. Courtesy of NASA/New Horizons Project.

tracks the mean atmospheric scale height below ~150 km altitude, although the uncertainty of the retrievals below ~80 km is large. The C_2H_6 density profile has a hint of a maxima just above ~200 km — although that could be a region of slower loss — and at higher altitudes shows a more rapid decrease with altitude at a slower rate than that of CH_4, as would be expected because of its higher mass and thus lower scale height. The other hydrocarbons C_2H_4 and C_2H_2 show analogous behaviors to C_2H_6.

Of special interest is the density of C_2H_4, which shows a rapid decrease with altitude from above the surface to ~150 km altitude then increases to a clear maximum near ~300 km, above which its abundance and its rate of decrease mirrors that of C_2H_6. Acetylene has somewhat similar profile but with a minimum density near 150 km and a maximum near 250 km altitude (Fig. 8). The chemical models of *Wong et al.* (2017) show clear minima and maxima for C_2H_4 and C_2H_2, where they assume dominant chemical loss condensation onto haze particles and also find the local maximum in density for C_2H_4 above that of C_2H_2 — although both maxima are above the observations. These compare somewhat favorably with models for hydrocarbon production near 300–400 km and haze condensation near 200 km, especially for C_2H_2 and C_2H_4 (*Wong et al.,* 2017).

3.1. Hydrogen Cyanide and Carbon Monoxide

Atacama Large Millimeter Array (ALMA) observations taken near the time of the New Horizons flyby retrieved HCN and CO profiles (*Lellouch et al.,* 2017). Both CH_4 and CO had been previously detected in high-resolution near-IR spectra, which probed the lowest scale height of the atmosphere (*Lellouch et al.,* 2011, 2015). *Lellouch et al.* (2017) detected the CO (3-2) and HCN (4-3) rotational transitions using the ALMA interferometer and confirming CO emission and reporting the first detection of HCN. These emission lines probe the abundances of CO and HCN from the surface to ~450 km and ~900 km altitude, respectively. The CO line measurement provides temperature constraints on Pluto's dayside atmosphere from ~50 to 400 km altitude and is consistent with an upper atmospheric temperature of ~70 ± 2 K near 300 km altitude, in agreement with the New Horizons Alice solar occultation data (*Gladstone et al.,* 2016). The ALMA CO measurement implies a CO mixing ratio of 515 ± 40 ppm.

The HCN rotational line shape is consistent with high supersaturation by perhaps 7–8 orders of magnitude, with an observed column abundance of $1.6 \pm 0.4 \times 10^{14}$ cm^{-2} and inferred mixing ratios $>1.5 \times 10^{-5}$ above 450 km, and $\sim 4 \times 10^{-5}$ near 800 km. If HCN were at its saturation

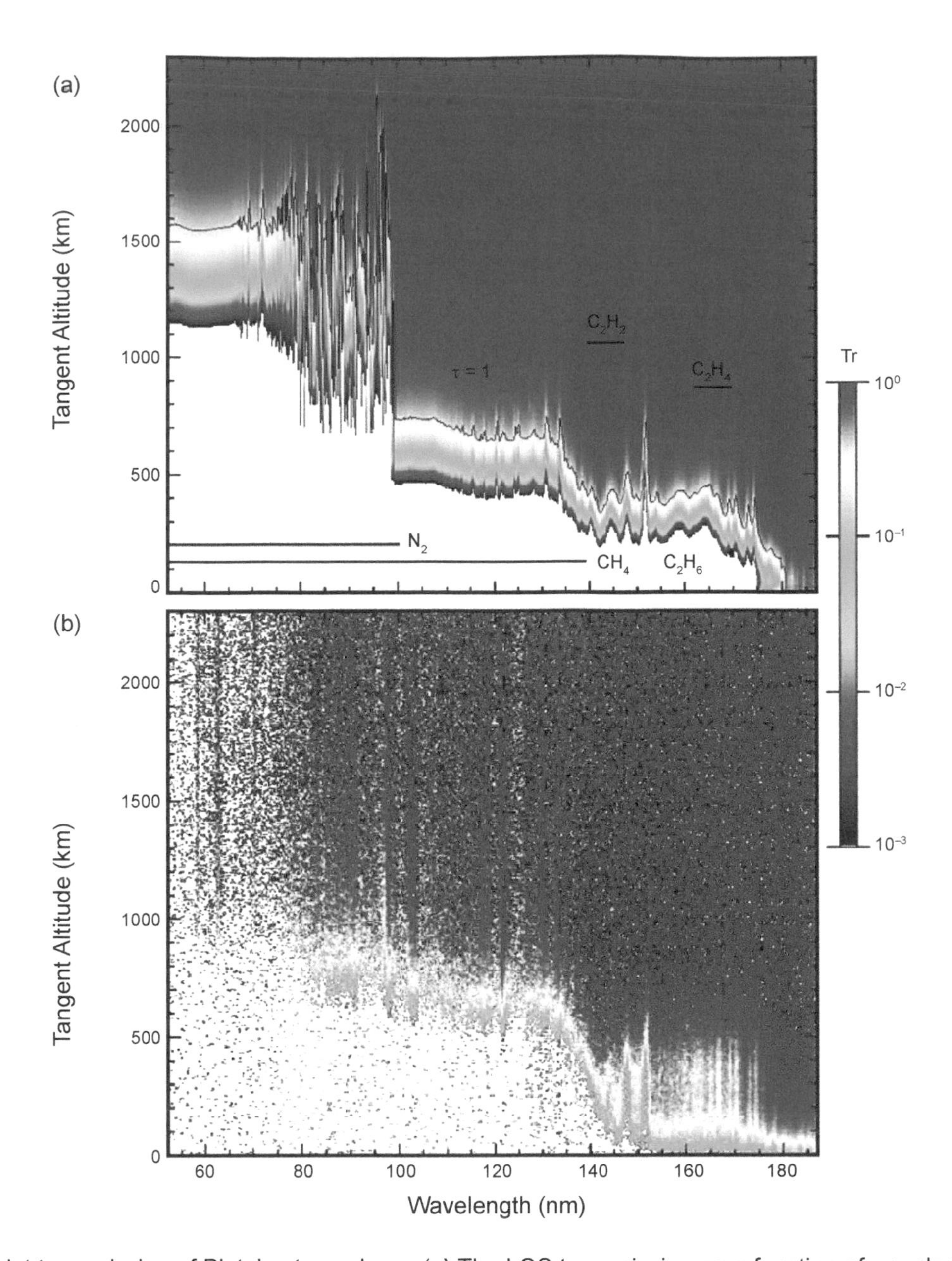

Fig. 6. Ultraviolet transmission of Pluto's atmosphere. **(a)** The LOS transmission as a function of wavelength and tangent altitude for a pre-encounter model of Pluto's atmosphere, with the optical depth τ = 1 line (black) indicated. The regions where N_2, CH_4, C_2H_2, C_2H_4, and C_2H_6 contribute significantly to the opacity are indicated. N_2 absorbs in discrete bands for wavelengths 80–100 nm, with bands and an underlying continuum absorbing at wavelengths between 65 and 80 nm, and an ionization continuum at wavelengths <65 nm. CH_4 otherwise dominates the opacity at wavelengths <140 nm. C_2H_6 has a similar cross section to CH_4, but absorbs out to longer wavelengths near 145 nm, where it also contributes to the opacity. C_2H_2 has strong absorption bands at 144, 148, and 152 nm. C_2H_4 is expected to account for much of the opacity at 155–175 nm. C_4H_2 accounts for much of the opacity at wavelengths 155–165 nm. **(b)** The LOS transmission of Pluto's atmosphere determined from the Alice solar occultation data. The Alice data are normalized (at each ultraviolet wavelength) to unocculted and unabsorbed levels at high altitude. In comparison with the model transmission, N_2 opacity begins at much lower altitudes (~500 km lower), while CH_4 opacity begins about 100 km higher than in the model. The Alice data do not have adequate spectral resolution to capture the structure of many of the intense lines. Adapted from Fig. 6 of *Gladstone et al.* (2016).

vapor pressure, this would require a much higher temperature (>92 K) than that inferred from Alice observations. *Lellouch et al.* (2017) proposed that such supersaturations might be due to lack of condensation nuclei, and thus extremely slow condensation loss from the gas phase (*Lellouch et al.,* 2011, 2015).

3.2. Constraints on Atmospheric Structure from Wave Observations

The presence of distinct haze layering in the lower and middle atmosphere haze could be due to the effect of gravity (or internal buoyancy) waves, perhaps generated by flow

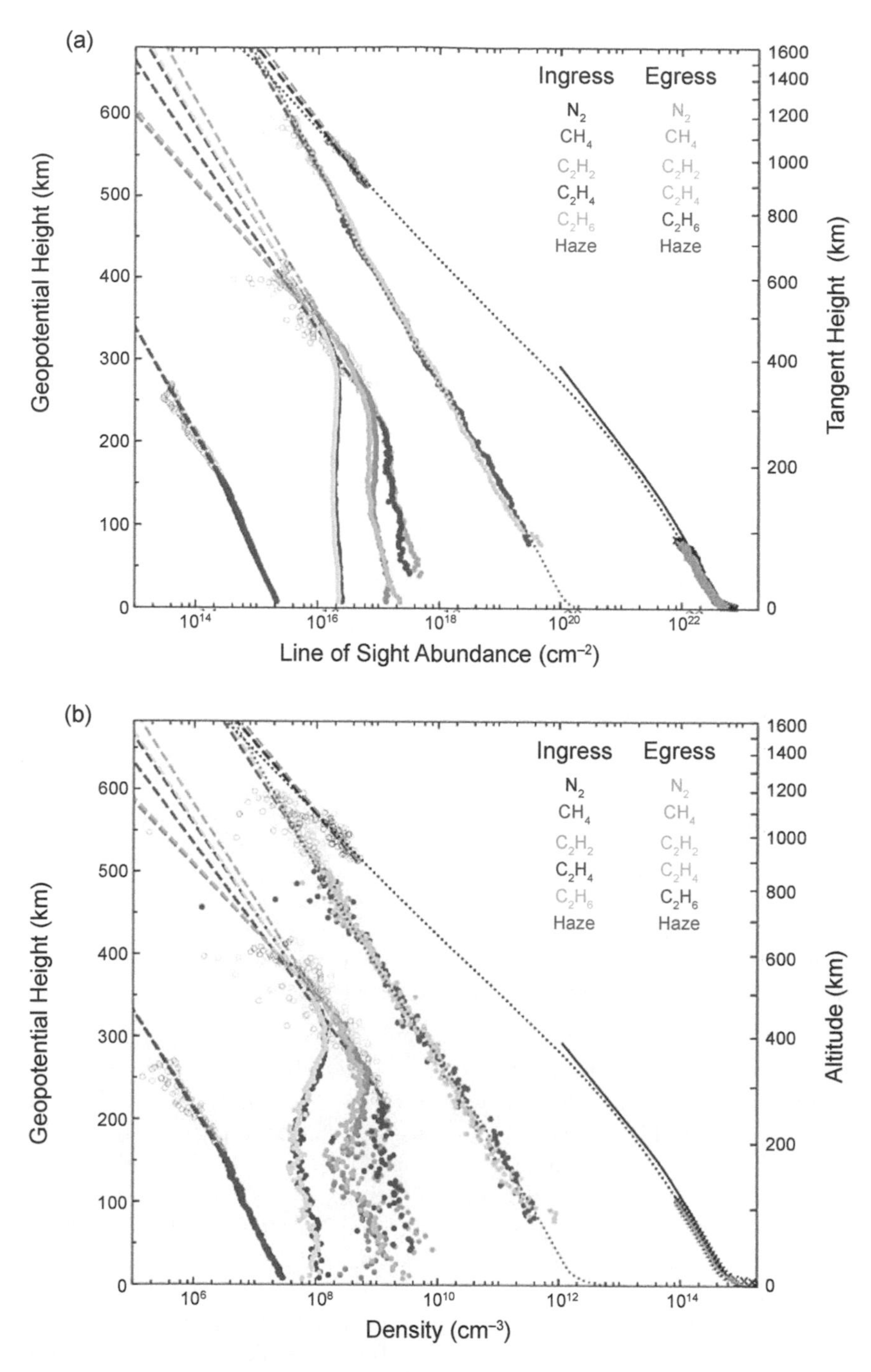

Fig. 7. Observed components in Pluto's atmosphere from the Alice solar occultation. **(a)** LOS abundance (molecules cm^{-2}); **(b)** retrieved local density (molecules cm^{-3}). Circles represent the points used to derive a functional approximation to the line-of-sight abundance, and the dashed lines represents the fitted function vs. geopotential height. The darker colors denote the ingress occultation and the lighter colors the egress occultation. The haze values are plotted as the optical depth times 10^{15} cm^{-2} in top panel, and as the extinction coefficient multiplied by 10^{15} cm^{-2} in the bottom panel. Dashed lines show model CH_4 and N_2 (dotted lines) and measurements of N_2 from the REX radio occultation below 110 km altitude as black crosses for ingress and gray crosses for egress. The 2015 groundbased stellar occultation is shown as the solid black line. The vertical scale is geopotential height. Adapted from Fig. 17 in *Young et al.* (2018).

over surface topography (*Gladstone et al.,* 2016) or generated by the diurnal sublimation cycle (*Toigo et al.,* 2010; *French et al.,* 2015) concentrated over Sputnik Planitia.

Waves can be inferred from the layering observed the New Horizons images (*Cheng et al.,* 2017) from very near the surface (within ~5–8 km) to above 200 km altitude. The observed haze is global [Alice-inferred haze extinction profiles are very similar at sunset and sunrise (*Young et al.,* 2018)] with a thin background component that has a scale height of 40–50 km (*Cheng et al.,* 2017). The numerous

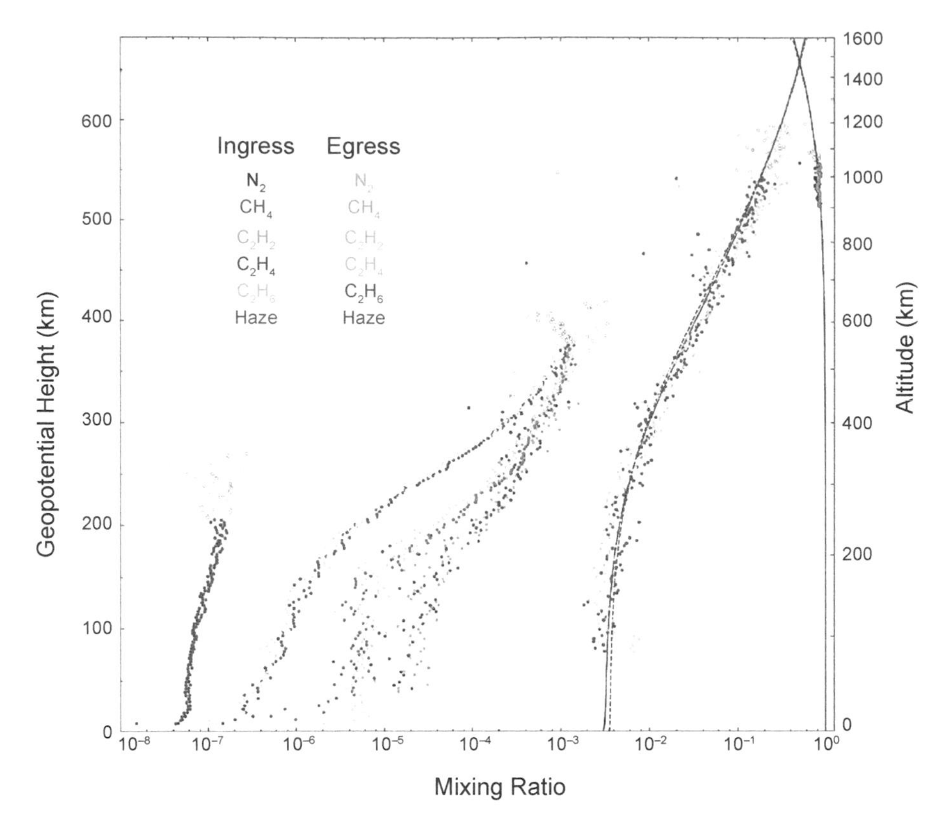

Fig. 8. Plot of retrieved mixing ratio (n/n_{total}) vs. geopotential height on the left axis and altitude on the right axis. For haze values, that plotted is $\varepsilon * 10^{15}$ cm^2/n_{total}, where ε is the extinction coefficient. Solid lines are based on the nominal model interpolated between Alice and REX observation for N_2 (black), and for that obtained by applying a one-dimensional transport model to the Alice observations of CH_4 (red) (details given in *Young et al.*, 2018). Dots are observed densities divided the modeled total density: N_2 (black for ingress; gray for egress), CH_4 (red for ingress; pink for egress), C_2H_6 (brown for ingress; mustard for egress), C_2H_2 (dark green for ingress; light green for egress), C_2H_4 (dark blue for ingress; light blue for egress), and haze (dark purple for ingress; light purple for egress). Adapted from Fig. 21 of *Young et al.* (2018).

(as many as 20) embedded thin (~1–3 km thickness) layers are observed between the surface and 200 km altitude, separated on average by ~10 km (see Fig. 9). These layers are too narrow to be detected in Alice observations. Time evolution of the haze was not observed, although the observation window for observing haze changes was short. *Cheng et al.* (2017) studied pairs of high-resolution images taken 2–5.5 hr apart that showed no clear differences in haze brightness, suggesting that the haze appears to be constant in time over timescales of a few hours or less.

Layers in the vertical density and temperature profiles have also been seen from groundbased stellar occultations, both in the lower atmosphere below 40 km altitude (e.g., *Pasachoff et al.*, 2005; *E. F. Young et al.*, 2008) and in the middle atmosphere at 150–300 km altitude (*Person et al.*, 2008; *Hubbard et al.*, 2009). The waves in the lower atmosphere from stellar occultations have vertical wavelengths near 3–6 km (*L. A. Young et al.*, 2008) and projected aspect ratios near 1:30 (*Pasachoff et al.*, 2005). In the middle/upper atmosphere, the layers have vertical wavelengths near 5–40 km (*McCarthy et al.*, 2008), and are also mainly horizontal and remained coherent across ~1200 km between ingress and egress (Fig. 10) (*McCarthy et al.*, 2008; *Person et al.*, 2008).

Unlike direct imaging, the occultation wave analysis (e.g., *Person et al.*, 2008) requires subtracting the background atmosphere to reveal the vertical waves (see Fig. 11). The vertical wavelengths seen are ~25–35 km, generally consistent with a Rossby wave model between 140 and 222 km altitude. Future grazing occultation observations of this nature could be very helpful for elucidating the global structure and for long-term monitoring of the mean state of the upper atmosphere.

The waves near 150–300 km altitude were interpreted as a combination of internal buoyancy waves (*McCarthy et al.*, 2008), which would be more prevalent in the lower atmosphere but are difficult to make coherent over large distances, and Rossby waves (*Person et al.*, 2008), which are strongly coherent over the relevant distance scales but are subject to dissipation (*Hubbard et al.*, 2009) and do not easily propagate vertically (*Toigo et al.*, 2010). *Hubbard et al.*'s (2009) power spectrum decomposition of the likely wave components suggests the majority of the power in the waves is consistent with buoyancy waves for higher wave numbers but is more consistent with Rossby waves at longer wavelengths. The key point for global atmospheric structure is that the observed waves are nearly symmetric about the central time of the observed occultation (and distance from

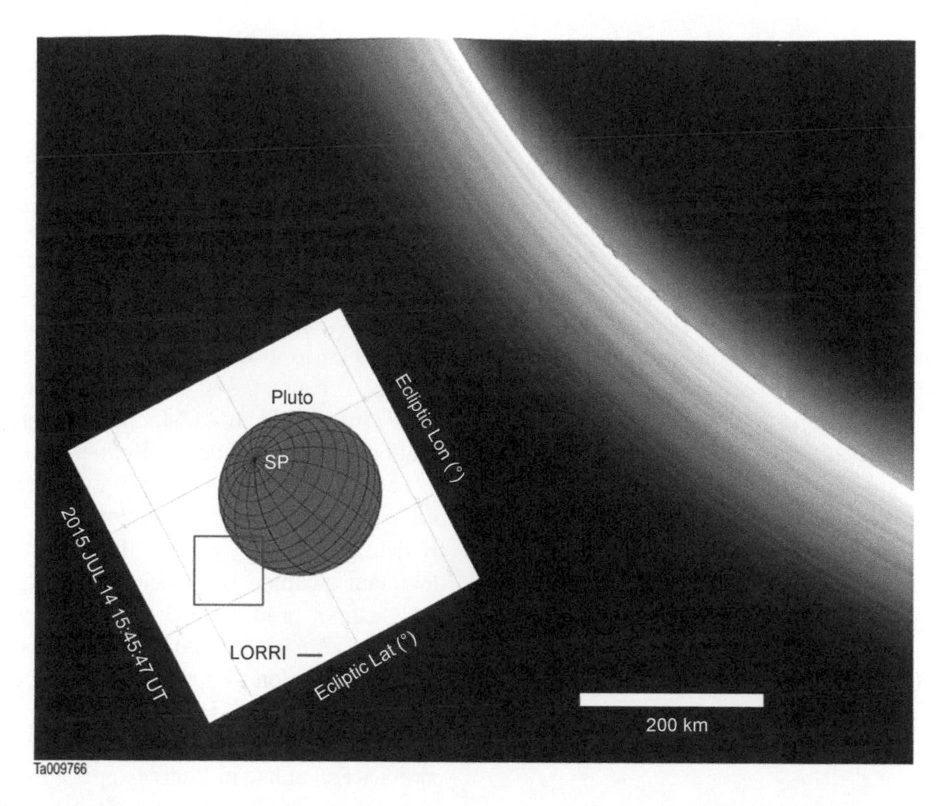

Fig. 9. Image of haze in Pluto's atmosphere obtained from a two-image stack made of LORRI observations obtained on July 14, 2015. The spatial resolution is ~0.95 km pixel^{-1}, showing many bright and very thin haze layers embedded in the background haze, which extends to at least ~300 km (*Cheng et al.,* 2017). Insert shows the location of the haze image superimposed on the scale of Pluto's disk, with the South Pole (SP) location noted. Adapted from *Cheng et al.* (2017).

the closest tangent point), implying a remarkable degree of uniformity in the upper atmosphere (*Person et al.,* 2008; *Hubbard et al.,* 2009), an idea consistent with the large-scale haze structures seen from New Horizons imaging (*Cheng et al.,* 2017).

If gravity waves are the cause for the layering in haze brightness, especially at the lower altitudes, perhaps through focusing of the particle density in the vertical direction, there remains the question of the nature of the mechanism generating the waves, i.e., whether the gravity wave are produced by the diurnal cycle of sublimation/condensation of the N_2 ice (*Forget et al.,* 2017; *French et al.,* 2015; *Toigo et al.,* 2010) or by orographic forcing from winds of ~1–10 m s^{-1} blowing over mountainous topography of ~1.5 km height, width, and consistent horizontal spacing of ~120 km as in the *Cheng et al.* (2017) model, or perhaps, by some other still unknown mechanism.

3.3. Middle Atmosphere Summary

The New Horizons Alice UV spectrograph observations of Pluto's middle atmosphere, along with the Earth-based observations, provided detailed information on the chemical components, temperature, and haze extinction profile of this region of the atmosphere. The REX radio science observations provided the first temperature and pressure profiles that connect the middle atmosphere to the lower atmosphere, and that to the atmosphere near the surface. These observations have shown that (1) the upper atmosphere is colder than expected, (2) the global haze has an important impact on the chemical structure, (3) the middle atmosphere is globally symmetric, and (4) waves have important influence on haze structure and the background atmosphere.

The temperature profiles observed by the Alice solar occultations show that the atmosphere appears spherically symmetric in the middle and upper atmosphere (*Kammer et al.,* 2020). However, the atmospheric haze appears to show a north/south asymmetry that may be a consequence of a meridional variation in photochemistry and/or haze production.

The vertical profiles of the observed C_2H_x species show depletions that may indicate rapid condensation onto the haze particles and/or heterogeneous reactions on the haze particle surfaces. In the middle atmosphere the temperature decreases with altitude. The gas phase and heterogeneous processes that control the loss of the C_2H_x constituents from the gas phase are thought to include condensation and/or adsorption onto haze particles, but the precise pathways between creation of condensation of nuclei in Pluto's ionosphere and their ultimate growth and sedimentation onto the surface as chemically complex conglomerates is largely unknown. However, recent models have begun to clarify the intermediate steps (see the chapter by Mandt et al. in this volume). Pluto's atmosphere is colder than both Titan's and Triton's, which would facilitate more rapid condensation onto condensation nuclei and thus increase loss rates from the photochemical region of the atmosphere. Photochemically produced, highly condensable species such as diacetylene and benzene, and perhaps methyl acetylene (*Steffl et al.*, 2020), could coat small haze particles, forming surfaces on which all the C_2H_x hydrocarbons can directly and rapidly condense at low atmospheric temperatures (*Lavvas et al.*, 2008, 2009, 2013).

The evidence for the local removal of C_2H_x species is seen in plots of the mixing ratios (see *Young et al.*, 2018). Condensation of the C_2H_x hydrocarbons onto haze particles as they sediment downward between 400 and 200 km altitude changes both their size and their sphericity via coagulation. Haze modeling by *Gao et al.* (2017) suggested that the effective haze radii are larger at lower altitudes and would thus be more efficient scavengers in that region.

The coherent haze layering structures imaged in the middle atmosphere could be explained by vertical waves, although the genesis of these waves is still poorly understood.

4. UPPER ATMOSPHERE

4.1. Temperature

The upper atmosphere is thought to be dominated by just two species, N_2 and CH_4. As noted above, the New Horizons Alice measurements (*Gladstone et al.*, 2016; *Young et al.*, 2018) revealed much lower N_2 LOS absorption in Pluto's upper atmosphere than expected based on the pre-encounter model of the energetics and atmospheric escape.

Young et al. (2018) inferred temperatures for the upper atmosphere of 76 ± 16 K at ingress and 79 ± 17 K at egress. The Alice solar occultation measured N_2 from 900 to 1100 km altitude, and the REX radio occultation measured N_2 from 0 to 100 km altitude. A modeled temperature minimum near 470 km altitude is required in order to decrease the N_2 quickly enough to satisfy the observed N_2 line-of-sight abundances in the lower atmosphere (from REX observations) and the atmosphere above 900 km altitude (Alice observations). This temperature minimum near 470 km is also compatible with the broad temperature minimum spanning 300–500 km altitude that *Lellouch et al.* (2017) inferred from ALMA observations of CO and HCN rotational line emissions in Pluto's atmosphere.

The New Horizons Solar Wind Around Pluto (SWAP) observations corroborated the low atmospheric escape rates by detecting mass 16 rather than mass 28 particles as the dominant escaping ion species. Cooling by CO, HCN, and C_2H_2 cannot produce such a low upper atmospheric temperature (*Zhang et al.*, 2017). Water has not been detected, so its contribution to upper atmosphere cooling is yet to be determined, although H_2O must be present at some (perhaps very small) concentration due to infalling interplanetary dust particles that ablate in Pluto's upper atmosphere (*Poppe and Horanyi*, 2018; *Krasnopolsky*, 2020).

4.2. Airglow

The Alice instrument measured several airglow features during the New Horizons flyby of Pluto. The results discussed here are reported by *Steffl et al.* (2020) and were

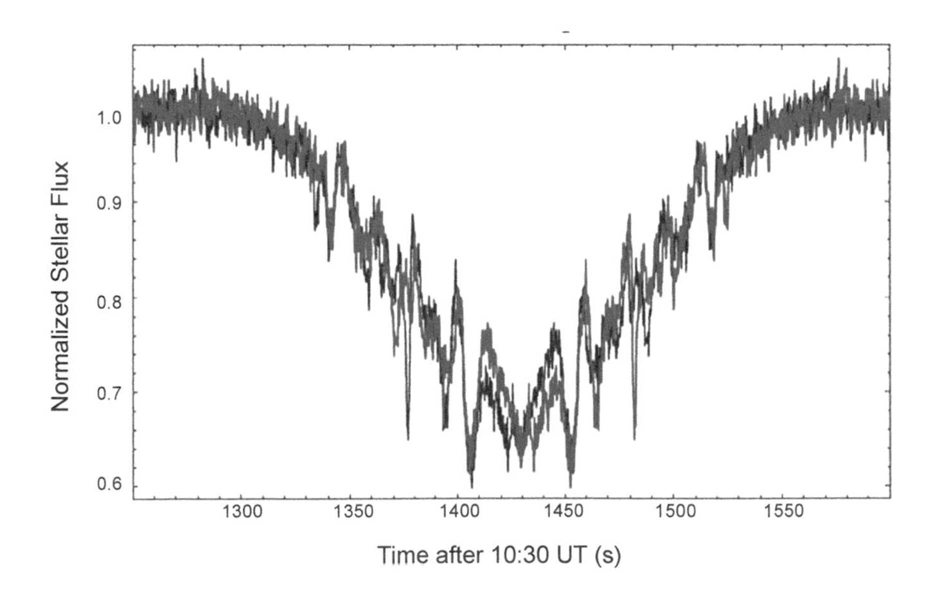

Fig. 10. Light curve retrieved during the March 18, 2007, stellar occultation by Pluto. The black line is a plot of the full-resolution (4 Hz) light curve obtained at the 6.5-m MMT (*Person et al.*, 2008). The red line is the same light curve reversed in time and overlaid on the original. Note the extremely close correspondence between the individual features on the opposite ends of the light curve. For example, the oscillations seen at 1330 s and 1520 s appear almost identical even though, with an occultation velocity of 6.8 km s^{-1}, they occurred over 1200 km apart in Pluto's atmosphere. The most striking differences occur in the center of the occultation, where the atmosphere is probed most deeply, indicating that the higher-level structure is more coherent than at the lower altitudes. From *Person et al.* (2008).

obtained when the disk of Pluto completely filled the instrument field of view. The brightest airglow feature measured was H I Lyman-α emission at 121.6 nm as predicted by pre-encounter models (*Summers et al.,* 1997), but the observed emission was 29.3 ± 1.9 R (where 1 Rayleigh is a unit of column emission rate equivalent to 10^6 photons cm^{-1} s^{-1}), not much smaller than the predicted emission of 37 R. At that wavelength the atmospheric transmission is near zero, mostly due to methane absorption, so the feature is most likely produced in Pluto's upper atmosphere.

H I Lyman-β at 102.6 nm was also detected with an intensity of 0.20 ± 0.04 R. These column integrated emission rates for Lyman-α and Lyman-β can be translated to atomic hydrogen column abundances of 4.2×10^{10} and 1.1×10^{11} cm^{-2}, respectively. Atomic hydrogen in Pluto's upper atmosphere is a consequence of photochemistry that acts to destroy CH_4 and convert its hydrogen to both H and H_2 that escape from the top of the atmosphere.

N II at 108.5 nm was also observed, with an intensity of 1.0±0.26 R compared to a pre-encounter prediction of 0.4 R (*Summers et al.,* 1997). This line is produced by dissociative photoionization of N_2 by solar EUV and X-rays. As of this writing, this N II airglow feature represents the only detection of ions in Pluto's atmosphere. In addition, at least two emission lines from N I are present in the airglow. The brightest is at 149.3 nm at 1.28 ± 0.33 R, and another at 113.4 nm with intensity of 0.46 ± 0.16 R.

Argon is important for constraining cosmic chemistry models of the outer solar system. A search was made for resonant scattering of solar radiation by argon, but emission was not detected at 104.8 nm (a 3-σ upper limit of 0.26 R, compared with a pre-encounter prediction of 0.3 R). The pre-encounter prediction had been made assuming an atmospheric abundance of 6% corresponding to the upper limit for its abundance in Titan's atmosphere at the time. Argon is important for constraining cosmic chemistry models of the outer solar system.

By modeling Alice observations in the 140–185-nm spectral region, *Steffl et al.* (2020) detected methylacetylene (propyne) in absorption, with an inferred atmospheric column density of ~5×10^{15} cm^{-2}, corresponding to a column-integrated mixing ratio of 1.6×10^{-6}. This represents the first detection of a C_3 hydrocarbon in Pluto's atmosphere and is an important new constraint for photochemical models. *Steffl et al.* (2020) found that Pluto's atmosphere has an optical depth <1 in the 140–185-nm region, implying that UV photons reaching the surface may play a role in darkening of photochemical deposits.

4.3. Ionosphere

In Pluto's atmosphere solar EUV and cosmic X-rays ionize N_2 with peak ionization at ~700 km altitude at a rate comparable to that of Triton, so it is reasonable to expect that Pluto should have an ionosphere. Pre-encounter models of Pluto's ionosphere (*Summers et al.,* 1997; *Krasnopolsky and Cruikshank,* 1999) assumed that the ionospheric models developed to explain the Triton ionosphere observed by Voyager 2 would also be generally applicable to Pluto. These models predicted peak electron densities of ~800–1100 cm^{-3} at altitudes ~600–1200 km, depending on the assumed methane abundance and temperature.

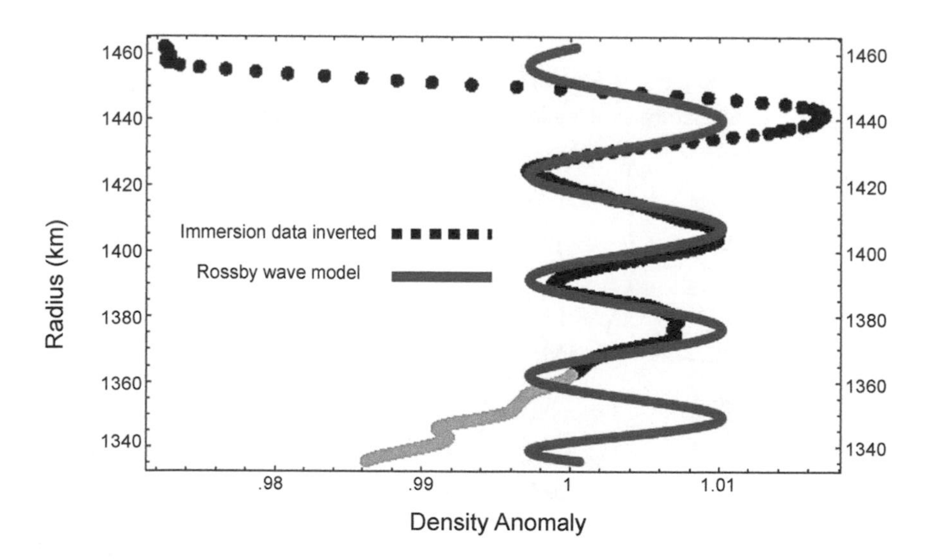

Fig. 11. Plot of the density anomaly obtained by inverting the curve displayed in Fig. 7. The data show the normalized number density excursions of Pluto's atmosphere from a smooth exponential in the 1340–1460-km radius range. This excursion value is the result of the inverted number density profile divided by the best-fitting exponential profile. The red line shows an empirical model of a vertically propagating wave, with a wavelength of approximately 35 km at 1460 km radius. Note that the wavelength decreases slightly with decreasing altitude reaching 25 km at 1340 km radius. The deviations in the lower atmosphere may indicate a breakdown in the coherent wave structures at lower altitudes (region emphasized in gray). After *Person et al.* (2008).

The REX sunset (ingress) radio occultation occurred at a solar zenith angle of 90.2° and the sunrise (egress) at 89.9°. Although an ionosphere was not detected, the REX radio occultation provided the most rigorous upper limit yet available for the integrated column density of Pluto's ionosphere. The egress occultation was slightly more sensitive, with a reported 1-σ electron density column upper limit of 2.3×10^{11} cm^{-2}. In the *Hinson et al.* (2018) model this corresponded to a peak electron density of ~1000 cm^{-3}.

This non-detection of ionospheric electrons is likely due to the high abundance of CH_4 in Pluto's upper atmosphere (*Summers et al.,* 1997; *Krasnopolsky and Cruikshank,* 1999; *Krasnopolsky,* 2020). The rapid charge change reaction of N_2^+ and N^+ with CH_4 produces an ionosphere dominated by molecular ions. Rapid recombination of molecular ions leads to the low expected electron densities comparable to, or smaller than, the REX detection threshold.

The dominant ion in the pre-encounter ionosphere models varied, although *Summers et al.* (1997) produced a range a range of types of ionospheres depending upon the assumed methane abundance and found that the more relevant high-methane case had $C_2H_5^+$ as dominant ion.

The more recent *Krasnopolsky* (2020) model accounts for the New Horizons results and the progress in the problem based on the Cassini-Huygens studies of Titan. The predicted ionospheric electron density peak was ~800 cm^{-3} at 750 km, with $HCNH^+$, $C_2H_5^+$, and $C_3H_5^+$ as the most abundant ions near the peak, and with $C_9H_{11}^+$ the dominant ion below 600 km. The more recent *Krasnopolsky* (2020) model has a much larger number of neutral and ion-neutral chemical reactions and predicts $C_3H_5^+$ and $C_5H_5^+$ as the dominant ion. Observational determination of the ionospheric makeup may be one of the most challenging aspects of future Pluto atmosphere studies, at least until an orbiter is sent that can make *in situ* measurements.

5. FINAL REMARKS AND FUTURE PROGRESS

Comparisons with Titan, and especially Triton, helped guide the early chemical and thermal balance models of Pluto's atmosphere and ionosphere, atmospheric interactions with its surface, and escape processes. However, the uniqueness of each atmosphere has produced widely different pressure and temperature profiles (Fig. 12). In each case, the surface acts as a source of methane for a primarily N_2 atmosphere, where photochemistry produces higher hydrocarbons and nitriles that condense onto haze particles and ultimately sediment onto their surfaces. Each surface also shows evidence of dark material, thought to be tholin deposits, which is consistent with this picture.

Yet the manner in which these overall processes are individually expressed has been seen to critically depend upon both the surface temperature and the methane concentrations in the atmosphere. The relatively higher temperature of Titan leads to much higher atmospheric CH_4 and therefore (despite its higher gravity) a much more extended stratosphere wherein photochemistry produces a much more diverse set of hydrocarbons and nitriles (*Krasnopolsky,* 2014) in the atmosphere. The longer sedimentation time leads to growth of haze particles by condensation to larger sizes, polymerization, and subsequently the formation of different haze particle composition than are found on Pluto (*Lavvas et al.,* 2013; *Krasnopolsky,* 2014; *Stern et al.,* 2015; *Person et*

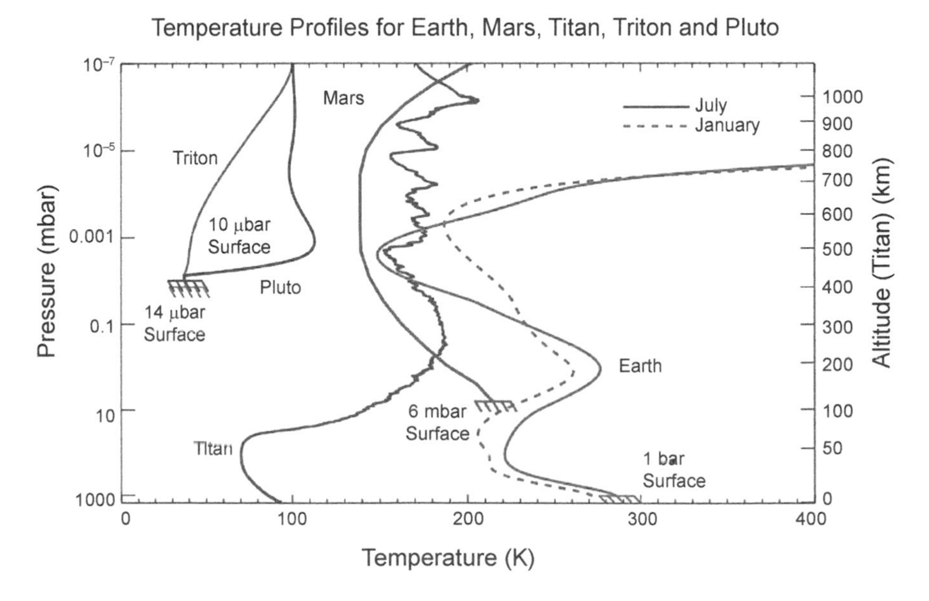

Fig. 12. Plot shows a comparison of the temperature profiles of Pluto, Triton, Titan, and the Earth, all of which have predominantly N_2 atmospheres. Titan's atmospheric temperature is from the Cassini-Huygens probe. The blue and black curves are models of Pluto and Triton temperature profiles, respectively. The profiles for the Earth and Mars are based upon climatological averages. Adapted from Fig. 22 of *Cheng et al.* (2017).

al., 2020). The much higher methane abundance in Titan's atmosphere also leads, in part, to larger atmospheric net heating and generates the higher atmospheric temperature. Titan's higher ionospheric density than Pluto is in part because of higher solar input, although magnetospheric charged particle ionization from Saturn's magnetosphere may also play a role.

The lower temperature and lower abundance of CH_4 on Triton leads to a substantially lower pressure N_2 atmosphere than on Titan (*Elliott et al.,* 1998), with only trace amounts of CH_4 and consequently smaller overall photochemical production of organic compounds that condense and produce a barely detectable haze layer (*Smith et al.,* 1989), and a smaller net deposition of organics and/or tholin-type compounds onto its surface. The lower amount of CH_4 also leads to smaller net atmospheric heating in Triton's atmosphere than on Pluto and thus lower atmospheric temperatures. Triton's higher ionospheric density may be due to charged particle precipitation and ionization from Neptune's magnetosphere (*Lyons et al.,* 1992; *Krasnopolsky and Cruikshank,* 1995; *Summers et al.,* 1997; *Mandt and Luspay-Kuti,* 2019).

Pluto's ionosphere, although not detected by the New Horizons REX experiment, is still thought to be sufficiently dense to cause the production of charged monomers that form the seeds of the haze particles. The larger haze abundances on Titan speak to an extended altitude region over which nucleation begins in the ionosphere at high altitudes, sedimentation of charged macromolecules into the regions where they grow by accumulation of photochemically produced hydrocarbons, and ultimately complex fractal haze particle geometries. In this sense Pluto is a lower-methane version of Titan's atmosphere.

These worlds also have distinctive mysteries in their middle and upper atmospheres. What is the source of ionization for Triton's ionosphere and how does the complex magnetospheric interaction modulate that ionosphere? On Titan, what is the degree of gas phase and solid phase processing of the condensed hydrocarbons that result in the extended haze? For Pluto, the high middle and upper atmospheric temperatures indicates an unknown heating agent (although there are several good candidates — see previous discussion). The seasonal and longer-term evolution of the chemical and temperature structure of each body are also expected to follow greatly divergent pathways because of these key differences in surface temperature and atmospheric CH_4 abundance, as well as varying orbital, seasonal, and obliquity variation timescales and amplitudes.

The status of our understanding of the composition and structure of Pluto's atmosphere was more easily articulated before the New Horizons flyby than afterward, largely because of our ignorance of the complexity of the atmosphere. Although our knowledge of the atmospheric composition has greatly improved as a consequence of the extensive observational datasets obtained, there are key weaknesses in certain areas and a need for additional observations in order to aid further progress. Below we list some of these key needs, focusing the remaining science questions, and then we identify some of the laboratory, modeling, and observational developments that could help make progress over the coming decades.

5.1. Lower Atmosphere

The near-surface mixing ratio of CH_4 is key for developing models of the chemical processes acting in the atmosphere. When applied to the time period of the New Horizons flyby in 2015, the GCMs produce only slight variation in CH_4 mixing ratio with altitude in the lower atmosphere at the location of the ingress and egress occultations. This result is consistent with the conclusions of *Young et al.* (2018), who used a one-dimensional vertical flux model to extrapolate the CH_4 abundance from 80 km down to the surface.

Improvements in our understanding of its abundance may prove difficult, yet would be highly useful. In addition, the CH_4 abundance may vary locally and diurnally due to local temperature variations and horizontal dynamical transport.

The characteristics of Pluto's surface boundary layer are important for understanding CH_4 variability. Even though a temperature boundary layer was observed by REX at the one location over the southern portion of Sputnik Planitia, there must be some boundary layer effect on mass, energy, and momentum everywhere — even though this boundary layer might be very narrow, perhaps only meters thick. There are no obvious Earth-based observations, at least in the near future, that can give us direct observations of the boundary layer, and we will need to improve the ability of GCMs to resolve fine-scale vertical structure and to follow potential time dependence. Such improvements require much-higher-resolution models to resolve the structure and time evolution of the boundary layer.

Long-term monitoring of the surface pressure and haze as Pluto moves along its orbit further from the Sun is also of key significance in examining how the surface pressure and methane abundance varies with time.

5.2. Middle Atmosphere

The *Young et al.* (2018) retrieval of hydrocarbon profiles relied upon C_2H_2, C_2H_4, and C_2H_6 cross-sections that were measured at laboratory temperatures of 150 K, 140 K, and 150 K, respectively, which is much warmer than the 60–70 K temperatures present throughout much of the Pluto's middle and upper atmosphere. Although the estimated errors from using the higher-temperature cross-sections is only ~10–20%, these hydrocarbon profiles are central to understanding atmospheric chemistry, and therefore lower-temperature cross-sections would greatly improve the constraints on these retrievals.

The pre-encounter photochemical models (*Summers et al.,* 1997; *Lara et al.,* 1997; *Krasnopolsky and Cruikshank,* 1999) were based upon analogy with Titan and Triton (*Young et al.,* 1994; *Strobel et al.,* 1990) and predicted other photochemically produced species that were not detected by New Horizons, such as C_4H_2 and H_2CN. Further searches from Earth-based observations might be warranted. Also, on the modeling front, it would be useful to quantify the constraints of their non-detection on atmospheric composition. *Mandt et al.* (2019) and *Krasnopolsky* (2020) produced models with updated and much-expanded kinetics bases and are steps in the right direction.

The early models also assumed an influx of H_2O from the top of Pluto's atmosphere via delivery by interplanetary dust, albeit at with fluxes about 2 orders of magnitude smaller than that at Triton and Titan (*Horanyi et al.,* 2016; *Poppe and Horanyi,* 2018). This influx should accompany other minor species such as CO_2, CO, and H_2O_2 (*Krasnopolsky,* 2020). Water vapor may also be important for atmospheric thermal balance (*Strobel and Zhu,* 2017). A detection or upper limit on these species might help set limits on the cosmic influx of interplanetary dust.

5.3. Upper Atmosphere

The upper atmosphere is thought to be dominated by just two species, N_2 and CH_4, although it contains minor photochemically produced species, such as methylacetylene (*Steffl et al.,* 2020). This is the region where ionization is thought to produce the nuclei for haze particle nucleation, so understanding the structure and composition of the upper atmosphere is foundational for understanding the haze and its evolution. Furthermore, there remain unanswered questions about the identity of the cooling agents that cause the lower temperature observed there.

Methane absorption is generally too weak to detect in the Alice observations above ~1200 km (*Young et al.,* 2018). Can that be improved? Similarly, can our analysis of the N_2 absorption, and thus inferred density profile, in the upper atmosphere be improved? *Young et al.* (2018) suggest that the electronic bands of N_2 between 80 and 100 nm, particularly at 97–98 nm where the solar spectrum has some strong lines, could be used to increase the accuracy of inferred N_2 densities in the upper atmosphere and potentially extend the region of sensitivity to lower altitudes.

Is Pluto's atmospheric radiative equilibrium controlled almost entirely by haze particles instead of gas molecules as suggested by *Zhang et al.* (2017)? The clear prediction that Pluto would be brighter at infrared wavelengths if its radiative energy balance is controlled by haze can be put to the test with IR observations from future telescopes, which will be able to distinguish water rotational line cooling from haze particle cooling (*Zhang et al.,* 2017).

5.4. Seasonal and Longer-Term Changes in the Atmospheric Composition and Temperature

The New Horizons results set the stage for numerous computational/modeling studies of the diurnal, seasonal, and longer-term studies of Pluto's atmosphere. There are many questions that can be addressed with such comprehensive models. How does the atmospheric composition, especially CH_4, change over seasonal and longer timescales? The CH_4 column abundance influences the ability of solar UV light to reach the surface and to drive chemical processing of CH_4 (as well as other chemicals) in surface ices. The CH_4 column abundance also impacts the degree of photochemical processing within the atmosphere itself and, indirectly, the sedimentation rate of tholin-like particles to the surface. The degree of photochemical processing of atmospheric CH_4 potentially influences the long-term integrated loss of CH_4 from Pluto's surface.

Some of the remaining uncertainties in knowledge of the abundance profiles of CO, HCN, and CH_4 might be removed through collaborative analysis of the Alice solar occultation data with near-contemporaneous groundbased IR observations (*Lellouch et al.,* 2015), groundbased stellar occultations (e.g., *Sicardy et al.,* 2016; *Person et al.,* 2020), and groundbased radio observations (*Lellouch et al.,* 2017).

The extensive observational results that characterize the chemical composition and temperature structure from the New Horizon flyby have set the stage for many more comprehensive studies that also overlap other aspects of Pluto's atmosphere, such as haze and tholin production and sedimentation onto the surface, atmospheric thermodynamics and upper atmospheric escape, surface/atmosphere boundary layer interactions, and seasonal and long-term evolution of Pluto's atmosphere. Comprehensive modeling efforts are underway in several of these areas and model improvements in spatial and temporal resolution will likely reveal further insights into the complexity of Pluto's atmosphere.

5.5. Future Observations and Observational Platforms

Continued space-based and Earth-based stellar occultation measurements are essential for understanding the seasonal variation of Pluto's surface pressure and haze burden, as well as the variation of near-surface atmospheric temperature. These are absolutely essential inputs for climate models of the atmospheric system. Stellar occultations over the next couple of decades will reveal the details of the surface pressure decrease, as well as other atmospheric changes that occur, as Pluto continues receding from the Sun.

The unexpectedly complex atmosphere observed by New Horizons begs further exploration by orbiters and landers. The Alice solar occultations probed just two points above Pluto's surface at two local times. A dedicated Pluto orbiter with an *in situ* ion/neutral mass spectrometer would help elucidate the full three-dimensional structure of Pluto's upper atmosphere, the haze nucleation mechanism(s), the escape mechanism(s), the nature of interplanetary dust impact and ablation, and atmospheric wave breaking in the upper atmosphere. An orbiter could also provide the high-resolution measurements of the near-surface regions where volatile transport occurs and where energy and momentum are exchanged with the surface. Such high-resolution mapping would provide the surface boundary constraints required for further development of high-resolution GCMs.

A Pluto lander, equipped with surface sampling and GCMS analysis capabilities, would reveal the chemical nature of the extensive plains of dark material that covers Cthulhu Regio. The chemical makeup of this dark material is completely unknown. Understanding the nature of this dark material may be key to understanding its ultimate source, which is presumably nucleation and growth to monomers in the upper atmosphere/ionosphere. However, the degree of chemical processing of photochemical products by solar UV as it grows into haze particles and sediments onto Pluto's surface as dark material is largely unknown. The dark material on Pluto's surface may also be key to understanding the astrobiological potential of Pluto.

As large space-based telescopes are brought online, further avenues of detailed observational and theoretical studies of Pluto's atmospheric structure and composition will become possible. With the launch of the James Webb Space Telescope, now scheduled for mid-2021, near-IR observations of Pluto's atmosphere will allow thermal emission monitoring of Pluto's upper atmosphere. This will provide the necessary measurements needed to model of the effects of lower atmosphere changes in, e.g., pressure, temperature, and buoyancy wave generation on upper atmosphere temperature structure. The IR observations of Pluto's atmosphere will finally allow a test of the haze heating/cooling proposal of *Zhang et al.* (2017). With even larger telescopes, such as the 30-m telescope and the Large UV/Optical/IR Surveyor (LUVOIR), the spatial resolution of Earth-based images of Pluto are expected to approach those obtained by New Horizons during its historic flyby.

REFERENCES

Bagenal F., Horanyi M., McComas D. J., McNutt R. L., Elliott H., et al. (2016) Pluto's interaction with its space environment: Solar wind, energetic particles, and dust. *Science, 351*, aad9045.

Bell J. M., Waite J. H., Westlake J. H., Bougher S. W., Ridley A. J., et al. (2014) Developing a self-consistent description of Titan's upper atmosphere without hydrodynamic escape. *J. Geophys. Res., 119*, 4957–4972.

Bertrand T. and Forget F. (2016) Observed glacier and volatile distribution on Pluto from atmosphere-topography processes. *Nature, 540*, 86–89.

Bertrand T. and Forget F. (2017) 3D modeling of organic haze in Pluto's atmosphere. *Icarus, 287*, 72–86.

Bertrand T., Forget F., Umurhan O. M., Grundy W. M., Schmitt B., et al. (2018) The nitrogen cycles on Pluto over seasonal and astronomical timescales. *Icarus, 309*, 277–296.

Bosh A. S., Person M. J., Levine S. E., Zuluaga C. A., Zangari A. M., et al. (2015) The state of Pluto's atmosphere in 2012–2013. *Icarus, 246*, 237–246.

Broadfoot A. L. and 21 colleagues (1989) Ultraviolet spectrometer observations of Neptune and Triton. *Science, 246*, 1459–1466.

Brosch N. (1995) The 1985 stellar occultation by Pluto. *Mon. Not. R. Astron. Soc., 276*, 571–578.

Brown G. N. Jr. and Ziegler W. T. (1980) Vapor pressure and heats of vaporization and sublimation of liquids and solids of interest in cryogenics below 1-atm pressure. *Adv. Cryog. Eng., 25*, 662–670.

Cable M. L., Vu T. H., Hodyss R., Choukroun M., Malaska M. J., et al. (2014) Experimental determination of the kinetics of formation of the benzene-ethane co-crystal and implications for Titan. *Geophys. Res. Lett., 41*, 5396–5401.

Capalbo F. J., Bénilan Y., Yelle R. V., and Koskinen T. T. (2015) Titan's upper atmosphere from Cassini/UVIS solar occultations. *Astrophys. J., 814*, 86.

Catling D. C. and Zahnle K. J. (2009) The planetary air leak. *Sci. Am., 300*, 36–43.

Cheng A. F., Weaver H. A., Conard S. J., Morgan M. F., Barnouin-Jha O., et al. (2008) Long-Range Reconnaissance Imager on New Horizons. *Space Sci. Rev., 140*, 189–215.

Cheng A. F., Summers M. E., Gladstone G. R., Strobel D. F., Young L., et al. (2017) Haze in Pluto's atmosphere. *Icarus, 290*, 112–133.

Christy J. W. and Harrington R. S. (1978) The satellite of Pluto. *Astron. J., 83*, 1005–1008.

Christy J. W. and Harrington R. S. (1980) The discovery and orbit of Charon. *Icarus, 44*, 38–40.

Cruikshank D. P. and Silvagio P. M. (1980) The surface and atmosphere of Pluto. *Icarus, 41*, 96–102.

Cruikshank D. P., et al. (2019) Prebiotic chemistry on Pluto. *Astrobiology, 19*, 831–848.

Desai R. T., Coates A. J., Wellbrock A., Vuitton V., Crary F. J., et al. (2017) Carbon chain anions and the growth of complex organic molecules in Titan's ionosphere. *Astrophys. J. Lett., 844*, L18.

Dias-Oliveira A., Sicardy B., Lellouch E., Vieira-Martins R., et al. (2015) Pluto's atmosphere from stellar occultations in 2012 and 2013. *Astrophys. J., 811*, 53.

Elliot J. L., Dunham E. W., Bosh A. S., Slivan S. M., Young L. A., Wasserman L. H., and Millis R. L. (1989) Pluto's atmosphere. *Icarus, 77*, 148–170.

Elliot J. L., et al. (1998) Global warming on Triton. *Nature, 393*, 765–767.

Elliot J. L., Ates A., Babcock B. A., Bosh A. S., Buie M. W., et al. (2003) The recent expansion of Pluto's atmosphere. *Nature, 424*, 165–168.

Elliot J. L., Person M. J., Gulbis A. A. S., Souza S. P., Adams E. R., et al. (2007) Changes in Pluto's atmosphere: 1988–2006. *Astron. J., 134*, 1–13.

Erwin J., Tucker O. J., and Johnson R. E. (2013) Hybrid fluid/kinetic modeling of Pluto's escaping atmosphere. *Icarus, 226*, 375–384.

Fink U., Smith B. A., Johnson J. R., Reitsema H. J., Benner D. C., and Westphal J. A. (1980) Detection of a CH_4 atmosphere on Pluto. *Icarus, 44*, 62–71.

Forget F., Bertrand T., Vangvichith M., Leconte J., Millour E., and Lellouch E. (2017) A post-New Horizons global climate model of Pluto including the N_2, CH_4 and CO cycles. *Icarus, 287*, 54–71.

Fray N. and Schmitt B. (2009) Sublimation of ices of astrophysical interest: A bibliographic review. *Planet. Space Sci., 57*, 2053–2080.

French R. G., Toigo A. D., Gierasch P. J., Hansen C. J., Young L. A., et al. (2015) Seasonal variations in Pluto's atmospheric tides. *Icarus, 246*, 247–267.

Gao P., Fan S., Wong M. L., Liang M. C., Shia R. L., et al. (2017) Constraints on the microphysics of Pluto's photochemical haze from New Horizons observations. *Icarus, 287*, 116–123.

Gladstone G. R. and Young L. A. (2019) New Horizons observations of the atmosphere of Pluto. *Annu. Rev. Earth Planet. Sci., 47*, 119–140.

Gladstone G. R., Pryor W. R., and Stern S. A. (2015) Lyα@Pluto. *Icarus, 246*, 279–284.

Gladstone G. R., Stern S. A., Ennico K., Olkin C. B., Weaver H. A., et al. (2016) The atmosphere of Pluto as observed by New Horizons. *Science, 351*, aad8866.

Golitsyn G. S. (1975) A possible atmosphere on Pluto. *Sov. Astron. Lett., 1*, 19–20.

Grundy W. M., Olkin C. B., Young L. A., Buie M. W., and Young E. F. (2013) Near-infrared spectral monitoring of Pluto's ices: Spatial distribution and secular evolution. *Icarus, 223*, 710–721.

Grundy W. M., Cruikshank D. P., Gladstone G. R., Howett C. J. A., and Lauer T. R. (2016) The formation of Charon's red poles from seasonally cold-trapped volatiles. *Nature, 539*, 65–68.

Grundy W. M., Bertrand T., Binzel R. P., Buie M. W., Buratti B. J., et al. (2018) Pluto's haze as a surface material. *Icarus, 314*, 232–245.

Guo Y. and Farquhar R. W. (2007) New Horizons mission design. *Space Sci. Rev., 240*, 49–74.

Hansen C. J. and Paige D. A. (1996) Seasonal nitrogen cycles on Pluto. *Icarus, 120*, 247–265.

Hansen C. J., Paige D. A., and Young L. A. (2015) Pluto's climate modeled with new observational constraints. *Icarus, 246*, 183–191.

Hart M. H. (1974) A possible atmosphere for Pluto. *Icarus, 21*, 242–247.

Herbert F. and Sandel B. R. (1991) CH_4 and haze in Triton's lower

atmosphere. *J. Geophys. Res., 96*, 19241–19252.
Herbert F., Sandel B. R., Yelle R. V., Holberg J. B., Broadfoot A. L., Shemansky D. E., Atreya S. K., and Romani P. N. (1987) The upper atmosphere of Uranus: EUV occultations observed by Voyager 2. *J. Geophys. Res., 92*, 15093–15109.
Hinson D. P., Simpson R. A., Twicken J. D., Tyler G. L., and Flasar F. M. (1999) Initial results from radio occultation measurements with Mars Global Surveyor. *J. Geophys. Res., 104*, 26997–27012.
Hinson D. P., Linscott I. R., Young L. A., Tyler G. L., Stern S. A., et al. (2017) Radio occultation measurements of Pluto's neutral atmosphere with New Horizons. *Icarus, 290*, 96–111.
Hinson D. P., Linscott I. R., Strobel D. F., Tyler G. L., Bird M. K., et al. (2018) An upper limit on Pluto's ionosphere from radio occultation measurements with New Horizons. *Icarus, 307*, 17–24.
Hofgartner J. D., Buratti B. J., Devins S. L., Beyer R. A., Schenk P., et al. (2018) A search for temporal changes on Pluto and Charon. *Icarus, 302*, 273–284.
Horanyi M., Poppe A., and Sternovsky Z. (2016) Dust ablation in Pluto's atmosphere. *EGU General Assembly Conference Abstracts, 18*, Abstract #3652. Copernicus, Göttingen.
Hörst S. M. (2017) Titan's atmosphere and climate. *J. Geophys. Res., 122*, 432–482.
Hörst S. M. and Tolbert M. A. (2013) *In situ* measurements of the size and density of Titan aerosol analogs. *Astrophys. J. Lett., 770*, L10.
Howard A. D., Moore J. M., Umurhan O. M., White O. L., Anderson R. S., et al. (2017) Present and past glaciation on Pluto. *Icarus, 287*, 287–300.
Hubbard W. B., Yelle R. V., and Lunine J. I. (1990) Non-isothermal Pluto atmosphere models. *Icarus, 84*, 1–11.
Hubbard W. B., Hunten D. M., Dieters S. W., Hill K. M., and Watson R. D. (1988) Occultation evidence for an atmosphere on Pluto. *Nature, 336*, 452–454.
Hubbard W. B., McCarthy D. W., Kulesa C. A., Benecchi S. D., Person M. J., et al. (2009) Buoyancy waves in Pluto's high atmosphere: Implications for stellar occultations. *Icarus, 204*, 284–289.
Hunten D. M. (1973) The escape of light gases from planetary atmospheres. *J. Atmos. Sci., 30*, 1481–1494.
Hunten D. M. and Watson A. J. (1982) Stability of Pluto's atmosphere. *Icarus, 51*, 665–667.
Ingersoll A. P., Summers M. E., and Schlipf S. G. (1985) Supersonic meteorology of Io: Sublimation-driven flow of SO_2. *Icarus, 64*, 375–390.
Johnson R. E., Tucker O. J., Michael M., Sittler E. C., Smith H. T., Young D. T., and Waite J. H. (2009) Mass loss processes in Titan's upper atmosphere. In *Titan from Cassini-Huygens* (R. H. Brown et al., eds.), pp. 373–391. Springer, New York.
Kammer J. A., Shemansky D. E., Zhang X., and Yung Y. L. (2013) Composition of Titan's upper atmosphere from Cassini UVIS EUV stellar occultations. *Planet. Space Sci., 88*, 86–92.
Kammer J. A., et al. (2019) New Horizons upper limits on O_2 in Pluto's present-day atmosphere. *Astron. J., 154(2)*, 55–58.
Kammer J. A., Gladstone G. R., Young L. A., Steffl A. J., Parker J. W., Greathouse T. K., Retherford K. D., et al. (2020) New Horizons observations of an ultraviolet stellar occultation and appulse by Pluto's atmosphere. *Astron. J., 159,* 1.
Khare B. N., Sagan C., Arakawa E. T., Suits F., Callcott T. A., and Williams M. W. (1984) Optical constants of organic tholins produced in a simulated titanian atmosphere: From soft X-ray to microwave frequencies. *Icarus, 60*, 127–137.
Koskinen T. T., Yelle R. V., Snowden D. S., Lavvas P., Sandel B. R., Capalbo F. J., Benilan Y., and West R. A. (2011) The mesosphere and lower thermosphere of Titan revealed by Cassini/UVIS stellar occultations. *Icarus, 216*, 507–534.
Krasnopolsky V. A. (1999) Hydrodynamic flow of N_2 from Pluto. *J. Geophys. Res., 104*, 5955–5962.
Krasnopolsky V. A. (2014) Chemical composition of Titan's atmosphere and ionosphere: Observations and the photochemical model. *Icarus, 236*, 83–91.
Krasnopolsky V. A. (2018) Some problems in interpretation of the New Horizons observations of Pluto's atmosphere. *Icarus, 301*, 152–154.
Krasnopolsky V. A. (2020) A photochemical model of Pluto's atmosphere and ionosphere. *Icarus, 335*, 113374.
Krasnopolsky V. A. and Cruikshank D. P. (1995) Photochemistry of Triton's upper atmosphere and ionosphere. *J. Geophys. Res., 100 (E10)*, 21271–21286.
Krasnopolsky V. A. and Cruikshank D. P. (1999) Photochemistry of Pluto's atmosphere and ionosphere near perihelion. *J. Geophys. Res., 104*, 21979–21996.
Krasnopolsky V. A., Sandel B. R., and Herbert F. (1992) Properties of haze in the atmosphere of Triton. *J. Geophys. Res., 97*, 11695–11700.
Lara L. M., Ip W. H., and Rodrigo R. (1997) Photochemical models of Pluto's atmosphere. *Icarus, 130*, 16–35.
Larson E. J. L., Toon O. B., and Friedson A. J. (2014) Simulating Titan's aerosols in a three-dimensional general circulation model. *Icarus, 243*, 400–419.
Lavvas P. P., Coustenis A., and Vardavas I. M. (2008) Coupling photochemistry with haze formation in Titan's atmosphere, Part I: Model description. *Planet. Space Sci., 56*, 27–66.
Lavvas P., Yelle R. V., and Vuitton V. (2009) The detached haze layer in Titan's mesosphere. *Icarus, 201*, 626–633.
Lavvas P., Yelle R. V., Koskinen T., Bazin A., Vuitton V., et al. (2013) Aerosol growth in Titan's ionosphere. *Proc. Natl. Acad. Sci. U.S.A., 110*, 2729–2734.
Lellouch E., de Bergh C., Sicardy B., Käufl H. U., and Smette A. (2011) High resolution spectroscopy of Pluto's atmosphere: Detection of the 2.3 μm CH_4 bands and evidence for carbon monoxide. *Astron. Astrophys., 530*, L4.
Lellouch E., Gurwell M., Butler B., Fouchet T., and Lavvas P. (2017) Detection of CO and HCN in Pluto's atmosphere with ALMA. *Icarus, 286*, 289–307.
Lellouch E., et al. (2009) Pluto's lower atmosphere structure and methane abundance from high-resolution spectroscopy and stellar occultations. *Astron. Astrophys., 495*, L17.
Lellouch E., de Bergh C., Sicardy B., Forget F., Vangvichith M., and Käufl H. U. (2015) Exploring the spatial, temporal, and vertical distribution of methane in Pluto's atmosphere. *Icarus, 246*, 268–278.
Lindal G. F., et al. (1983) The atmosphere of Titan: An analysis of the Voyager 1 radio occultation measurements. *Icarus, 53*, 348–363.
Lindzen R. S. and Hong S. S. (1974) Effects of mean winds and horizontal temperature gradients on solar and lunar semidiurnal tides in the atmosphere. *J. Atmos. Sci., 31*, 1421–1446.
Lodders K. and Fegley B. (1998) *The Planetary Scientist's Companion.* Oxford, New York. 400 pp.
Luspay-Kuti A., Mandt K., Jessup K. L., Kammer J., Hue V., et al. (2017) Photochemistry on Pluto — I. Hydrocarbons and aerosols. *Mon. Not. R. Astron. Soc.*, 472, 104–117.
Lyons J. R., Yung Y. L., and Allen M. (1992) Solar control of the upper atmosphere of Triton. *Science, 256*, 204–206.
Mandt K. E. and Luspay-Kuti A. (2019) Comparative planetology of the ion chemistry at Pluto, Titan, and Triton. *Pluto System After New Horizons*, Abstract #7047. LPI Contribution No. 2133, Lunar and Planetary Institute, Houston.
Mandt K., Luspay-Kuti A., Hamel M., Jessup K. L., Hue V., et al. (2017) Photochemistry on Pluto: Part II HCN and nitrogen isotope fractionation. *Mon. Not. R. Astron. Soc., 472*, 118–128.
Mandt K. E., *et al.* (2019) Ion densities and composition of Titan's upper atmosphere derived from the Cassini Ion Neutral Mass Spectrometer: Analysis methods and comparison of measured ion densities to photochemical model simulations. *J. Geophys. Res., 117(E10)*, E10006.
Marouf E. A., Tyler G. L., and Rosen P. A. (1986) Profiling Saturn's rings by radio occultation. *Icarus, 68*, 120–166.
McCarthy D. W., Hubbard W. B., Kulesa C. A., Benecchi S. D., Person M. J., Elliot J. L., and Gulbis A. A. S. (2008) Long-wavelength density fluctuations resolved in Pluto's high atmosphere. *Astron. J., 136*, 1519–1522.
McComas D. J., Elliott H. A., Weidner S., Valek P., Zirnstein E. J., et al. (2016) Pluto's interaction with the solar wind. *J. Geophys. Res., 121*, 4232–4246.
McNutt R. L. (1989) Models of Pluto's upper atmosphere. *Geophys. Res. Lett., 16*, 1125–1228.
Moore J. M., Howard A. D., Umurhan O. M., White O. L., Schenk P. M., et al. (2018) Bladed terrain on Pluto: Possible origins and evolution. *Icarus, 300*, 129–144.
Moores J. E., Smith C. L., Toigo A. D., and Guzewich S. D. (2017) Penitentes as the origin of the bladed terrain of Tartarus Dorsa on Pluto. *Nature, 541*, 188–190.
Nimmo F., Umurhan O., Lisse C. M., Bierson C. J., Carver J., et al. (2017) Mean radius and shape of Pluto and Charon from New Horizons images. *Icarus, 28*, 12–29.
Olkin C. B., Young L. A., French R. G., Young E. F., Buie M. W., et

al. (2014) Pluto's atmospheric structure from the July 2007 stellar occultation. *Icarus, 239*, 15–22.
Olkin C. B., Young L. A., Borncamp D., Pickles A., Sicardy B., et al. (2015) Evidence that Pluto's atmosphere does not collapse from occultations including the 2013 May 04 event. *Icarus, 246*, 220–225.
Owen T. C., Roush T. L., Cruikshank D. P., Elliot J. L., Young L. A., de Bergh C., Schmitt B., Geballe T. R., Brown R. H., and Bartholomew M. J. (1993) Surface ices and atmospheric composition of Pluto. *Science, 261*, 745.
Pasachoff J. M., Souza S. P., Babcock B. A., Ticehurst D. R., Elliot J. L., et al. (2005) The structure of Pluto's atmosphere from the 2002 August 21 stellar occultation. *Astron. J., 129*, 1718–1723.
Person M. J., Elliot J. L., Gulbis A. A. S., Zaluaga C. A., Babcock B. A., et al. (2008) Waves in Pluto's upper atmosphere. *Astrophys. J., 136*, 1510–1518.
Person M. J., Bosh A. S., Zaluaga C. A., Sickafoose A. A., Levine S. E., et al. (2020) Haze in Pluto's atmosphere: Results from SOFIA and ground-based observations of the 2015 June 29 Pluto occultation. *Icarus*, in press.
Poppe A. R. and Horanyi M. (2018) Interplanetary dust delivery of water to the atmospheres of Pluto and Triton. *Astron. Astrophys., 617*, L5.
Prokhvatilov A. I. and Yantsevich L. D. (1983) X-ray investigation of the equilibrium phase diagram of CH_4-N_2 solid mixtures. *Sov. J. Low Temp. Phys., 9*, 94–98.
Protopapa S. and 22 colleagues (2017) Pluto's global surface composition through pixel-by-pixel Hapke modeling of New Horizons Ralph/LEISA data. *Icarus, 287*, 218–228.
Rannou P. and West R. (2018) Supersaturation on Pluto and elsewhere. *Icarus, 312*, 36–44.
Reuter D. C., et al. (2008) Ralph: A visible/infrared imager for the New Horizons Pluto/Kuiper belt mission. *Space Sci. Rev., 140*, 129–154.
Schenk P., Beyer R. A., McKinnon W. B., Moore J. M., Spencer J. R., et al. (2018) Basins, fractures and volcanoes: Global cartography and topography of Pluto from New Horizons. *Icarus, 314*, 400–433.
Schmitt B. and 28 colleagues (2017) Physical state and distribution of materials at the surface of Pluto from New Horizons LEISA imaging spectrometer. *Icarus, 287*, 229–260.
Sicardy B., Widemann T., Lellouch E., Veillet C., Cuillandre J. C., et al. (2003) Large changes in Pluto's atmosphere as revealed by recent stellar occultations. *Nature, 424*, 168–170.
Sicardy B., Talbot J., Meza E., Camargo J. I. B., Desmars J., et al. (2016) Pluto's atmosphere from the 2015 June 29 ground-based stellar occultation at the time of the New Horizons flyby. *Astrophys. J. Lett., 819*, L38.
Smith G. R., Strobel D. F., Broadfoot A. L., Sandel B. R., Shemansky D. E., and Holberg J. B. (1982) Titan's upper atmosphere: Composition and temperature from the EUV solar occultation results. *J. Geophys. Res., 87*, 1351–1359.
Smith B. A., et al. (1989) Voyager 2 at Neptune: Imaging science results. *Science, 246(4936)*, 1422–1449.
Spencer J. R., Stansberry J. A., Trafton L. M., Young E. F., Binzel R. P., and Croft S. K. (1997) Volatile transport, seasonal cycles, and atmospheric dynamics on Pluto. In *Pluto and Charon* (S. A. Stern and D. J. Tholen, eds.), pp. 435–473. Univ. of Arizona, Tucson.
Stansberry J. A., Yelle R. V., Lunine J. I., and McEwen A. S. (1992) Triton's surface-atmosphere energy balance. *Icarus, 99*, 242–260.
Stansberry J. A., Lunine J. I., Hubbard W. B., Yelle R. V., and Hunten D. M. (1994) Mirages and the nature of Pluto's atmosphere. *Icarus, 111*, 503–513.
Stansberry J. A., Spencer J. R., Schmitt B., Benchkoura A. I., Yelle R. V., and Lunine J. I. (1996) A model for the overabundance of methane in the atmospheres of Pluto and Triton. *Planet. Space Sci., 44*, 1051–1063.
Steffl A. J., Young L. A., Strobel D. F., Kammer J. A., Evans J. S., Stevens M. H., Schindhelm, Rebecca N., et al. (2020) Pluto's ultraviolet spectrum, surface reflectance, and airglow emissions. *Astron. J., 159(6)*, 274.
Stern S. A. and Trafton L. M. (1984) Constraints on bulk composition, seasonal variation, and global dynamics of Pluto's atmosphere. *Icarus, 57*, 231–240.
Stern S. A. and Trafton L. M. (2008) On the atmospheres of objects in the Kuiper Belt. In *The Solar System Beyond Neptune* (M. A. Barucci et al., eds.), pp. 365–380. Univ. of Arizona Press, Tucson.
Stern S. A., Slater D. C., Scherrer J., Stone J., Dirks G., et al. (2008) Alice: The ultraviolet imaging spectrograph aboard the New Horizons Pluto-Kuiper Belt mission. *Space Sci. Rev., 140*, 155–187.
Stern S. A., Bagenal F., Ennico K., Gladstone G. R., Grundy W. M., et al. (2015) The Pluto system: Initial results from its exploration by New Horizons. *Science, 350*, aad1815.
Stern S. A., Kammer J. A., Gladstone G. R., Steffl A. J., Cheng A. F., et al. (2017a) New Horizons constraints on Charon's present-day atmosphere. *Icarus, 287*, 124–130.
Stern S. A., Binzel R. P., Earle A. M., Singer K. N., Young L. A., et al. (2017b) Past epochs of significantly higher-pressure atmospheres on Pluto. *Icarus, 287*, 47–53.
Stern S. A., Kammer J. A., Barth E. L., Singer K. N., Lauer T., et al. (2017c) Evidence for possible clouds in Pluto's present-day atmosphere. *Astron. J., 154*, 43.
Stevens M. H., Strobel D. F., Summers M. E., and Yelle R. V. (1992) On the thermal structure of Triton's thermosphere. *Geophys. Res. Lett., 19*, 669–672.
Strobel D. F. (2009) Titan's hydrodynamically escaping atmosphere: Escape rates and the structure of the exobase region. *Icarus, 202*, 632–641.
Strobel D. F. and Zhu X. (2017) Comparative planetary nitrogen atmospheres: Density and thermal structures of Pluto and Triton. *Icarus, 291*, 55–64.
Strobel D. F., Cheng A. F., Summers M. E., and Strickland D. J. (1990) Magnetospheric interaction with Triton's ionosphere. *Geophys. Res. Lett., 17*, 1661–1664.
Strobel D. F., Zhu X., Summers M. E., and Stevens M. H. (1996) On the vertical thermal structure of Pluto's atmosphere. *Icarus, 120*, 266–289.
Summers M. E., Strobel D. F., and Gladstone G. R. (1997) Chemical models of Pluto's atmosphere. In *Pluto and Charon* (S. A. Stern and D. J. Tholen, eds.), pp. 391–434. Univ. of Arizona, Tucson.
Tan S. P. and Kargel J. S. (2018) Solid-phase equilibria on Pluto's surface. *Mon. Not. R. Astron. Soc., 474*, 4254–4263.
Telfer M. W., Parteli E. J. R., Radebaugh J., Beyer R. A., Bertrand T., et al. (2018) Dunes on Pluto. *Science, 360*, 992–997.
Tian F. and Toon O. B. (2005) Hydrodynamic escape of nitrogen from Pluto. *J. Geophys. Res., 32*, L18201.
Toigo A. D., Gierasch P. J., Sicardy B., and Lellouch E. (2010) Thermal tides on Pluto. *Icarus, 208*, 402–411.
Toigo A. D., French R. G., Gierasch P. J., Guzewich S. D., Zhu X., and Richardson M. I. (2015) General circulation models of the dynamics of Pluto's volatile transport on the eve of the New Horizons encounter. *Icarus, 254*, 306–323.
Tomasko M. G. and West R. A. (2009) Aerosols in Titan's atmosphere. In *Titan from Cassini Huygens* (R. H. Brown et al., eds.), pp. 297–321. Springer, New York.
Tomasko M. G., et al. (2008) A model of Titan's aerosols based on measurements made inside the atmosphere. *Planet. Space Sci., 56*, 669–707.
Trafton L. M. (1980) Does Pluto have a substantial atmosphere? *Icarus, 44*, 53–61.
Trafton L. M. (2015) On the state of methane and nitrogen ice on Pluto and Triton: Implications of the binary phase diagram. *Icarus, 246*, 197–205.
Trafton L. M. and Stern S. A. (1983) On the global distribution of Pluto's atmosphere. *Astrophys. J., 267*, 872–881.
Trafton L. M., Matson D. L., and Stansberry J. A. (1998) Surface/atmosphere interaction and volatile transport (Triton, Pluto, and Io). In *Solar System Ices* (B. Schmitt et al., eds.), pp. 773–812. Kluwer, Dordrecht.
Trafton L. M., Hunten D. M., Zahnle K. J., and McNutt R. L. (1997) Escape processes at Pluto and Charon. In *Pluto and Charon* (S. A. Stern and D. J. Tholen, eds.) pp. 347–390. Univ. of Arizona, Tucson.
Tucker O. J., Johnson R. E., and Young L. A. (2015) Gas transfer in the Pluto-Charon system: A Charon atmosphere. *Icarus, 246*, 291–297.
Tucker O. J., Erwin J. T., Deighan J. I., Volkov A. N., and Johnson R. E. (2012) Thermally driven escape from Pluto's atmosphere: A combined fluid/kinetic model. *Icarus, 217*, 408–415.
Tyler G. L., Linscott I. R., Bird M. K., Hinson D. P., Strobel D. F., et al. (2008) The New Horizons radio science experiment (REX). *Space Sci. Rev., 140*, 217–259.
Vervack R. J., Sandel B. R., and Strobel D. F. (2004) New perspectives on Titan's upper atmosphere from a reanalysis of the Voyager 1 UVS solar occultation. *Icarus, 170*, 91–112.
Volkov A. N., Johnson R. E., Tucker O. J., and Erwin J. T. (2011) Thermally driven atmospheric escape: Transition from hydrodynamic

to Jeans escape. *Astrophys. J. Lett., 729*, L24.
Vuitton V., Tran B. N., Persans P., and Ferris J. P. (2009) Determination of the complex refractive indices of Titan haze analogs using photothermal deflection spectroscopy. *Icarus, 203*, 663–671.
Watson A. J., Donahue T. M., and Walker J. C. G. (1981) The dynamics of a rapidly escaping atmosphere: Applications to the evolution of Earth and Venus. *Icarus, 48*, 150–166.
Westlake J. H., Waite J. H., Carrasco N., Richard M., and Cravens T. (2014) The role of ion-molecule reactions in the growth of heavy ions in Titan's ionosphere. *J. Geophys. Res., 119*, 5951–5963.
Willacy K., Allen M., and Yung Y. L. (2016) A new astrobiological model of the atmosphere of Titan. *Astrophys. J., 829*, 79.
Wong M. L., Fan S., Gao P., Liang M. C., Shia R. L., et al. (2017) The photochemistry of Pluto's atmosphere as illuminated by New Horizons. *Icarus, 287*, 110–115.
Yelle R. V. and Lunine J. I. (1989) Evidence for a molecule heavier than methane in the atmosphere of Pluto. *Nature, 339*, 288–290.
Yelle R. V. and Elliot J. L. (1997) Atmospheric structure and composition: Pluto and Charon. In *Pluto and Charon* (S. A. Stern and D. J. Tholen, eds.), pp. 347–390. Univ. of Arizona, Tucson.
Yelle R. V., Cui J., and Müller-Wodarg I. C. F. (2008) Methane escape from Titan's atmosphere. *J. Geophys. Res., 113*, E10003.
Yelle R. V., Lunine J. I., Pollack J. B., and Brown R. H. (1995) Lower atmospheric structure and surface-atmosphere interactions on Triton. In *Neptune and Triton* (D. P. Cruikshank, ed.), pp. 1031–1105. Univ. of Arizona, Tucson.
Young E. F., French R. G., Young L. A., Ruhland C. R., Buie M. W., et al. (2008) Vertical structure in Pluto's atmosphere from the 2006 June 12 stellar occultation. *Astron. J., 136*, 1757–1769.
Young L. A. (1994) Bulk properties and atmospheric structure of Pluto and Charon. Ph.D. thesis, Massachusetts Institute of Technology, Cambridge. 124 pp.
Young L. A. (2012) Volatile transport on inhomogeneous surfaces: I — Analytic expressions, with application to Pluto's day. *Icarus, 221*, 80–88.
Young L. A. (2013) Pluto's seasons: New predictions for New Horizons. *Astrophys. J. Lett., 766*, L22.
Young L. A., Elliot J. L., Tokunaga A., de Bergh C., and Owen T. (1997) Detection of gaseous methane on Pluto. *Icarus, 127*, 258–262.
Young L. A. and 27 colleagues (2008) New Horizons: Anticipated scientific investigations at the Pluto system. *Space Sci. Rev., 140*, 93–127.
Young L. A., Kammer J. A., Steffl A. J., Gladstone G. R., Summers M. E., et al. (2018) Structure and composition of Pluto's atmosphere from the New Horizons solar ultraviolet occultation. *Icarus, 300*, 174–199.
Yung Y. L. and Lyons J. R. (1990) Triton: Topside ionosphere and nitrogen escape. *Geophys. Res. Lett., 17*, 1717–1720.
Yung Y. L. and DeMore W. B. (1999) *Photochemistry of Planetary Atmospheres*. Oxford, New York. 480 pp.
Zalucha A. M. and Michaels T. I. (2013) A 3D general circulation model for Pluto and Triton with fixed volatile abundance and simplified surface forcing. *Icarus, 223*, 819–831.
Zalucha A. M., Gulbis A. A. S., Zhu X., Strobel D. F., and Elliot J. L. (2011) An analysis of Pluto occultation light curves using an atmospheric radiative-conductive model. *Icarus, 211*, 804–818.
Zhang X., Strobel D. F., and Imanaka H. (2017) Haze heats Pluto's atmosphere yet explains its cold temperature. *Nature, 551*, 352–355.
Zhu X., Strobel D. F., and Erwin J. T. (2014) The density and thermal structure of Pluto's atmosphere and associated escape processes and rates. *Icarus, 228*, 301–314.

Mandt K. E., Luspay-Kuti A., Cheng A., Jessup K.-L., and Gao P. (2021) Photochemistry and haze formation. In *The Pluto System After New Horizons* (S. A. Stern, J. M. Moore, W. M. Grundy, L. A. Young, and R. P. Binzel, eds.), pp. 279–296. Univ. of Arizona, Tucson, DOI: 10.2458/azu_uapress_9780816540945-ch012.

Photochemistry and Haze Formation

K. E. Mandt, A. Luspay-Kuti, A. Cheng
The Johns Hopkins University Applied Physics Laboratory

K.-L. Jessup
Southwest Research Institute

P. Gao
University of California Berkeley

One of the many exciting revelations of the New Horizons flyby of Pluto was the observation of global haze layers at altitudes as high as 200 km in the visible wavelengths. This haze is produced in the upper atmosphere through photochemical processes, similar to the processes in Titan's atmosphere. As the haze particles grow in size and descend to the lower atmosphere, they coagulate and interact with the gases in the atmosphere through condensation and sticking processes that serve as temporary and permanent loss processes. New Horizons observations confirm studies of Titan haze analogs suggesting that photochemically produced haze particles harden as they grow in size. We outline in this chapter what is known about the photochemical processes that lead to haze production and outline feedback processes resulting from the presence of haze in the atmosphere, connect this to the evolution of Pluto's atmosphere, and discuss open questions that need to be addressed in future work.

1. INTRODUCTION

The New Horizons flyby of Pluto revealed global haze layers extending up to altitudes as high as 200 km in visible images as discussed in *Cheng et al.* (2017). Although early observations of Pluto's atmosphere from groundbased and Earth-orbiting telescopes indicated that haze may be present, the extent of the haze layers and the role of haze feedback in Pluto's atmosphere was completely unexpected. We outline here what was known about the photochemistry and potential for haze formation in Pluto's atmosphere before the New Horizons flyby of Pluto, how the observations from New Horizons have impacted our understanding of Pluto's atmospheric photochemistry and haze formation, feedback occurring due to the haze, and how these revelations impact our understanding of the evolution of Pluto's atmosphere.

1.1. Pre-New Horizons Picture of Pluto's Atmosphere

1.1.1. Detection of Pluto's atmosphere. Prior to the New Horizons flyby of Pluto, understanding of Pluto's atmosphere was limited to information gleaned from groundbased and Earth-orbiting telescopes. The first sign that Pluto may have an atmosphere was the detection of methane frost on the surface (*Cruikshank et al.,* 1976). This detection was made by making photometric observations of Pluto's reflectance in the infrared at 1.55 and 1.73 μm, which are the diagnostic wavelengths for methane and water ice absorption. By comparing these observations with laboratory measurements of the reflectance of methane ice in the wavelength range between 1 and 4 μm the authors determined that methane frost must be present on the surface. This detection meant that Pluto should have an atmosphere because methane ices could remain on the surface long term only if an atmosphere was present in vapor phase equilibrium with the surface (e.g., *Trafton,* 1980; *Fink et al.,* 1980; *Cruikshank and Silvaggio,* 1980). The first definitive detection of an atmosphere on Pluto was made with observations of an occultation of Pluto by a relatively bright (magnitude +12.8) star (*Brosch,* 1995). However, the first published detection of Pluto's atmosphere was during an occultation of Pluto by another relatively bright star that was observed by multiple telescopes in the southern hemisphere and the Kuiper Airborne Observatory (KAO). A stellar occultation occurs when Pluto passes between a star and the Earth. Measuring the reduction in brightness of the star as Pluto passes in front of it produces a "light curve" that provides information about the atmosphere and the solid body radius of Pluto. We illustrate in Fig. 1 the spatial coverage of Pluto's atmosphere for the telescopes observing the 1988 occultation (adapted from *Millis et al.,* 1993) and give an example light curve from this occultation (from *Elliot et al.,* 1989).

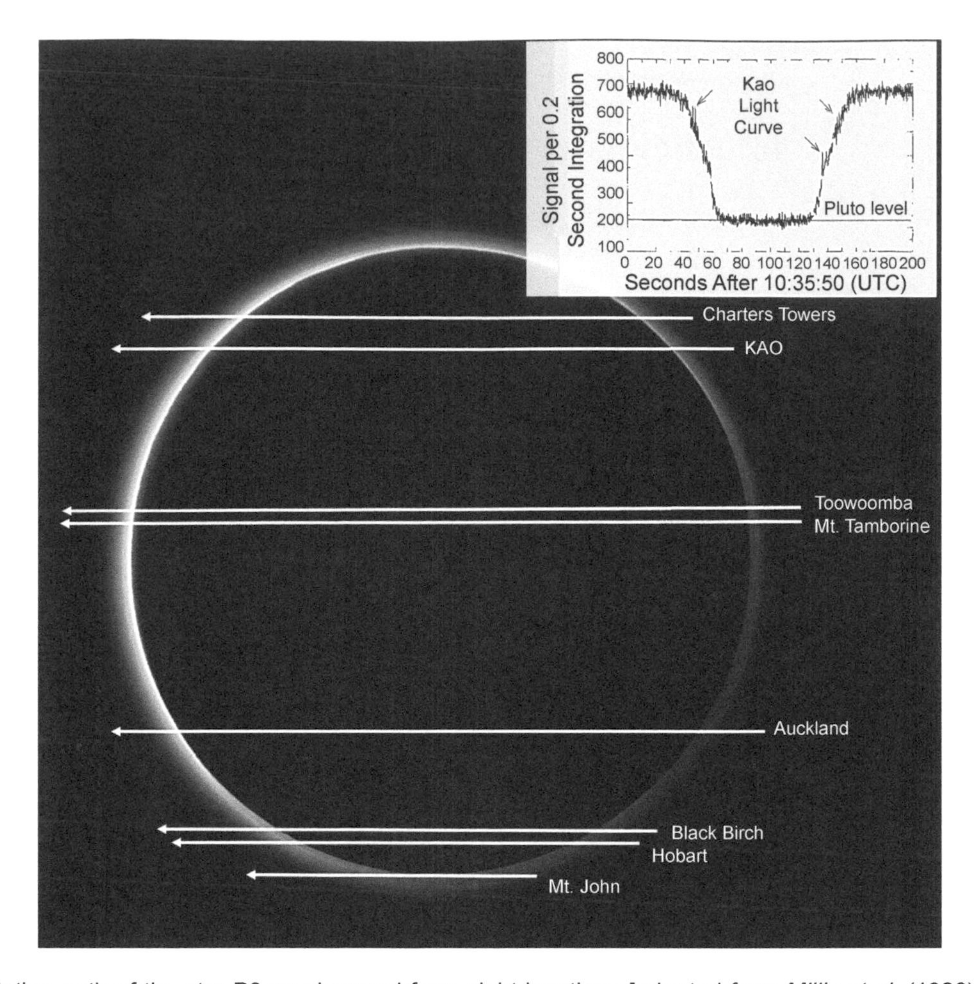

Fig. 1. Occultation path of the star P8 as observed from eight locations [adapted from *Millis et al.* (1993) to show paths over a New Horizons image of Pluto's global haze]. The light curve from Kuiper Airborne Observatory (KAO) in the inset shows time along the horizontal vs. signal level along the vertical axis (adapted from *Elliot et al.,* 1989). The gradual decline at the beginning of the occultation indicates detection of an isothermal atmosphere. Around the halfway point between the beginning of the occultation and complete occultation of the stellar signal at the "Pluto level" (solid line) there is an inflection point where the slope of the light curve increases. Arrows point to spikes that could indicate temperature variations in the atmosphere.

1.1.2. Characteristics of Pluto's atmosphere. The light curve provides a scale height, H, of the neutral atmosphere

$$H = \frac{kT}{m_a g}$$

where T is the atmospheric temperature, m_a is the mean mass of the molecules in the atmosphere, k is the Boltzmann constant, and g is the force of gravity. The slope of the light curve is used to derive the scale height, obtaining information about atmospheric temperature and/or composition. Initial analysis of the occultation observation proposed that Pluto had an extended methane atmosphere as thick as Pluto's radius (*Hubbard et al.,* 1988). However, observations from KAO could be interpreted as an extensive atmosphere made up of either methane or nitrogen, depending on the assumed temperature (*Elliot et al.,* 1989). *Yelle and Lunine* (1989) studied the energy balance of the atmosphere by simulating how energy input into and radiation from the atmosphere affects the temperature and concluded that a molecule heavier than methane must be present. The authors proposed that this molecule was either carbon monoxide (CO), molecular nitrogen (N_2), or argon (Ar), with a preference for CO because it was predicted to be the most dominant carbon-bearing molecule in the protosolar nebula (PSN). This was further demonstrated by thermal observations made between 1991 and 1993, which showed that the surface temperature of Pluto was too low to support a methane-dominated atmosphere, meaning that the atmosphere must be predominantly N_2 or CO (*Stern et al.,* 1993; *Jewitt,* 1994). Later detection of CO and N_2 ice on the surface showed that N_2 was the most abundant ice and that Pluto's atmosphere was likely to be predominantly N_2 with trace amounts of methane (CH_4) and CO (*Owen et al.,* 1993).

The light curve observations from KAO also showed a clear inflection point that could be interpreted as either a reversal in temperature variation as a function of altitude, also known as a temperature inversion, or a layer of haze near Pluto's surface, or some combination of both. This is illustrated in the inset to Fig. 1, where the slope of the light curve changes between the initial decline of the stellar signal and complete loss of the signal from the star behind Pluto (solid line in figure inset). Further analysis of the compiled observations found that the higher-quality observations confirmed either the presence of haze or a steep thermal gradient caused by a temperature inversion near the surface (*Millis et al.,* 1993). The authors suggested that haze provided a better explanation of the total quenching of light observed by KAO, but without two-color measurements a haze could not be proven. The next successful multichord stellar occultation of Pluto's atmosphere was in 2002 by the star P131.1. This occultation and subsequent ones extended observations to multiple wavelengths in the visible and infrared range, and showed a significant increase in surface pressure and further indications of a possible near-surface haze layer (*Elliot et al.,* 2003).

The potential for haze to form in Pluto's atmosphere was not surprising. Early studies of the photochemistry in Jupiter's (*Gladstone,* 1982; *Strobel,* 1983) and Titan's (*Yung et al.,* 1984) atmospheres discussed pathways for the production of haze initiated by photolysis of CH_4 that could logically be applied to Pluto (e.g., *Stansberry et al.,* 1989). The Voyager 2 flyby of Neptune detected haze below 30 km altitude in Triton's atmosphere (*Smith et al.,* 1989) and sparked comparative planetology studies between Triton and Pluto that led to improved models for photochemistry, the production of haze, and potential haze feedback in Pluto's atmosphere (e.g., *Krasnopolsky and Cruikshank,* 1995, 1999; *Summers et al.,* 1997).

Just as the results of the Voyager 2 flyby of Triton fed into an improved understanding of Pluto's atmosphere, the groundbreaking *in situ* and remote observations of Titan's atmosphere by the Cassini-Huygens mission (e.g., *Coates et al.,* 2007; *Crary et al.,* 2009; *Wahlund et al.,* 2009; *Brown et al.,* 2009) inspired valuable laboratory studies (e.g., *Dimitrov and Bar-Nun,* 2003; *Imanaka et al.,* 2004; *Trainer et al.,* 2013) and improved modeling for radiative transfer (e.g., *Yelle,* 1991; *Bampasidis et al.,* 2012) and photochemical processes (e.g., *Wilson and Atreya,* 2004; *Vuitton et al.,* 2006; *Lavvas et al.,* 2008; *De La Haye et al.,* 2008; *Krasnopolsky,* 2009; *Dobrijevic et al.,* 2016). These efforts resulted in significant advancements in understanding the photochemical processes relevant to Pluto's atmosphere leading into the New Horizons flyby.

The thermal structure of Pluto's atmosphere based on the occultation observations indicated the temperature was predicted to be high enough in the upper atmosphere to allow either hydrodynamic (e.g., *Stern and Trafton,* 1984; *Trafton,* 1990; *Krasnopolsky,* 1999; *Tian and Toon,* 2005; *Strobel,* 2008) or enhanced thermal escape of the high energy range of the velocity distribution function, known as Jeans escape (*Tucker et al.,* 2012), of the main constituent, N_2. The thermal structure plays an important role in photochemical processes and in the long-term evolution of the atmosphere depending on whether escape or photochemistry have a major impact on the production and loss of volatiles in Pluto's atmosphere. A review of the composition and structure of Pluto's atmosphere is provided in the chapter in this volume by Summers et al., while Pluto's atmospheric escape is reviewed in the chapter by Strobel.

1.1.3. Seasonal variation of Pluto's atmosphere. Pluto has a highly elliptical orbit with an orbital period of 248 years. The current (i.e., vs. long-term average) perihelion and aphelion distances are 29.66 and 49.31 astronomical units (AU) (distance of Earth from the Sun) respectively. This drastic difference in maximum and minimum distance from the Sun means that the flux of solar photons reaching Pluto is almost 3× greater at perihelion than at aphelion. This difference will result in variation of the surface pressure over a Pluto year, with debate as to whether or not the atmosphere would collapse at aphelion (*Stern and Trafton,* 1984; *Hanson and Paige,* 1996; *Young,* 2013; *Olkin et al.,* 2015). A review of the volatile and climate cycles on Pluto is provided in the chapter in this volume by Young et al., while the dynamics of Pluto's lower atmosphere are reviewed in Forget et al.

Pluto's atmosphere was first detected shortly before perihelion when Pluto was located at a distance of 29.76 AU from the Sun. Later occultations provided observations of the atmosphere from 2002 leading up to the New Horizons flyby on July 14, 2015, and covering a Pluto-Sun distance from 30.54 to 32.69 AU. Although the structure of the upper atmosphere did not appear to change significantly (*Elliot et al.,* 2007), the surface pressure increased until 2008 before beginning to decrease (see *Young,* 2013, and references therein), even though Pluto was moving away from the Sun during this entire time period.

Changes in surface pressure over Pluto's long year will influence the photochemical production of haze in several ways. First, a reduction in surface pressure decreases the amount of N_2 and CH_4 available for haze production. Reduced surface pressure may also indicate lower temperatures. This influences neutral reactions, which have a rate that depends on temperature, and increases condensation near the surface. Finally, the increasing distance from the Sun will decrease the flux of ultraviolet (UV) and extreme ultraviolet (EUV) photons available to initiate chemistry by photodissociation and photoionization of N_2 and CH_4, processes that processes that support Pluto's haze formation (see section 2). Although the 1988 occultations showed possible evidence of a haze layer near the surface, and occultations in 2003 showed indications of an extinction layer near the surface potentially attributable to aerosol condensation droplets (*Rannou and Durry,* 2009), other observations appeared to be a better fit with temperature inversion rather than haze (e.g., *Elliot et al.,* 2007; *Young et al.,* 2008). *Elliot et al.* (2007) proposed that haze may have formed around the 1988 occultations and later cleared. Leading up to the New Horizons flyby, one of the main questions related to photochemistry in Pluto's atmosphere was whether or not Pluto would have a near-surface haze layer similar to that observed at Triton (*Smith et al.,* 1989).

1.2. New Horizons Revelations

The New Horizons flyby provided several surprising discoveries about Pluto's atmosphere. The first was that the temperature in the upper atmosphere was significantly cooler than predicted, meaning that the atmosphere was being cooled by an unexpected process and was not escaping hydrodynamically (*Gladstone et al.,* 2016, hereafter *G2016*; *Young et al.,* 2018, hereafter *Y2018*). This discovery has major implications for the loss of volatiles and long-term evolution of the atmosphere, because escape had been predicted to be the dominant influence on how the atmosphere evolves (e.g., *Lunine et al.,* 1989; *Mandt et al.,* 2016). Additionally, photochemical reactions between neutrals are temperature-dependent, so a colder temperature profile influences photochemical production of complex organics that would lead to haze in Pluto's atmosphere.

The second surprise was the densities of hydrocarbons containing two carbon atoms (C_2H_x), or acetylene (C_2H_2), ethylene (C_2H_4), and ethane (C_2H_6), as a function of altitude. As shown in Fig. 2a, the densities of these species show an unexpected drop in density, or a density inversion, below 300–400 km. This density profile indicates that a loss process is at work removing these molecules from the atmosphere below 400 km. We explore this observation in detail in sections 2 and 4.

The final surprise from the New Horizons flyby was the extensive global layers of haze in the atmosphere (*Stern et al.,* 2015), illustrated in the full disk view of Pluto in Fig. 1 and in detail in Fig. 2b. We discuss these observations in further detail in section 3. Although the New Horizons observations confirmed the existence of haze in Pluto's atmosphere, the analysis of haze extinction profiles and the atmospheric temperature profile indicated that the inflection points recorded in the occultation light curves described above result from the strong temperature inversion in the lower atmosphere and not haze extinction (*Cheng et al.,* 2017).

1.3. Exploring the New Horizons Revelations

We discuss the implications of the New Horizons observations for understanding photochemistry and haze production in Pluto's atmosphere. Section 2 outlines the first steps in photochemistry, followed by section 3, which describes the haze observations and discusses what these observations indicate about molecular growth in Pluto's atmosphere. The presence of haze leads to feedback processes that are discussed in section 4. Then, in section 5 we discuss the implications of these new revelations for our understanding of how Pluto's atmosphere has evolved over time. We finish the chapter with a summary of what was learned, and a discussion of what work remains to be done through further analysis of the New Horizons datasets, future groundbased observations, and any potential future mission to the Pluto system.

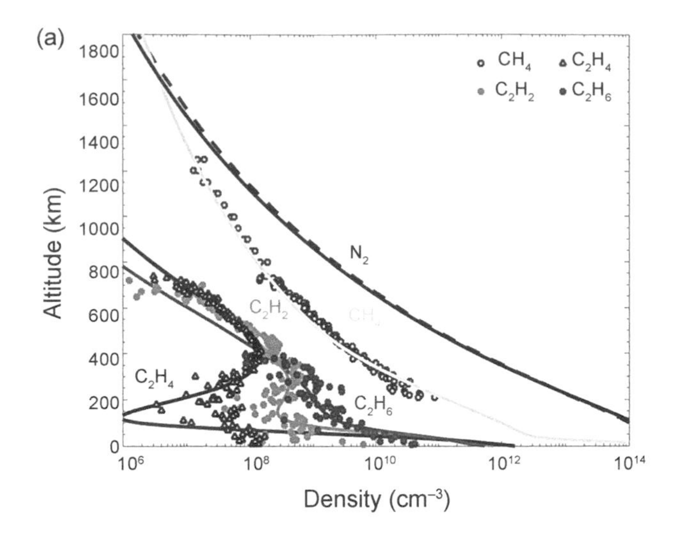

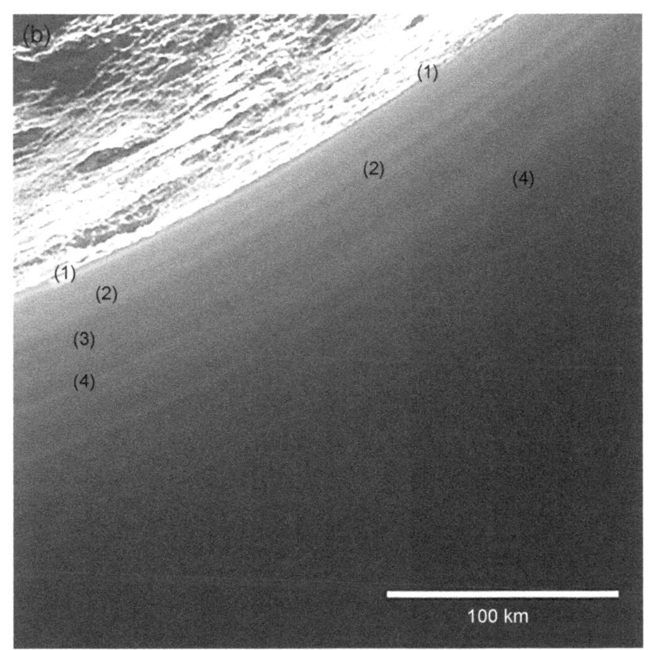

Fig. 2. (a) Densities of the main atmospheric species, molecular nitrogen (N_2) and methane (CH_4), and the C_2H_x species, acetylene (C_2H_2), ethylene (C_2H_4), and ethane (C_2H_6), as a function of altitude showing a density inversion below 400 km. Model results are shown with the lines and suggest that haze particle hardening with aging is both necessary and sufficient to explain the inversion in the vertical profiles of C_2 hydrocarbons in Pluto's atmosphere [adapted from *Luspay-Kuti et al.* (2017) to use data from *Young et al.* (2018)]. **(b)** Several haze layers evaluated in *Cheng et al.* (2017).

2. PHOTOCHEMISTRY: FIRST STEPS

One-dimensional photochemical models provide a theoretical basis for observed expected atmospheric composition as a function of altitude by tracking the rate of production and loss for a given species through a system of chemical reactions. Model results vary relative to specified atmospheric conditions — including the vertical temperature profile, the incident solar flux at the top of the atmosphere, and the presumed upper and lower boundary conditions. Depending on the model application, the assumed atmospheric state may be defined relative to a specific latitude and local time, or relative to global average values. To retrieve the anticipated atmospheric composition profiles, photochemical models solve the one-dimensional continuity equation simultaneously for all specified atmospheric constituent species (*Banks and Kockarts,* 1973)

$$\frac{\partial n_i}{\partial t} + \frac{\partial \Phi_i}{\partial z} = P_i - L_i$$

where n_i, Φ_i, P_i, L_i, are the species number density (in cm^3), vertical diffusive flux (in $cm^2\ s^{-1}$), and the chemical production and loss rates ($cm^3\ s^{-1}$), respectively, for each species i. Through this equation, the one-dimensional models trace the relative roles of the thermal structure, transport, and active chemical reactions between neutral and ion species created and/or destroyed by photolytic and kinetic processes on the overall vertical structure and composition of the atmosphere.

2.1. Pluto-Specific Photochemical Modeling

Prior to the New Horizons flyby of Pluto, the composition of Pluto's atmosphere was inferred from groundbased observations of occultations as described above in section 1, and by applying photochemical models that had been validated for Triton using Voyager 2 observations (e.g., *Krasnopolsky and Cruikshank,* 1995, 1999; *Summers et al.,* 1997). The New Horizons flyby provided the first-ever altitude profiles of major constituents, N_2 and CH_4, as well as the minor constituent C_2 hydrocarbons (*G2016*).

Around the time of the New Horizons flyby, observations with the Atacama Large Millimeter Array (ALMA) provided an abundance measurement for CO, an altitude profile of hydrogen cyanide (HCN), and an upper limit for the detection of $HC^{15}N$ in Pluto's atmosphere (*Lellouch et al.,* 2017, hereafter *L2017*). The HCN observations are illustrated in Fig. 3. These observations provided new constraints for photochemical modeling of Pluto's atmosphere, inspiring three key modeling efforts reported to date in the literature. These models are (1) the Caltech/JPL KINETICS chemistry-transport model first developed by *Allen et al.* (1981) and updated for Pluto calculations by *Wong et al.* (2017) (hereafter *W2017*), which focuses solely on chemical processes occurring among the neutral gas species; (2) the coupled ion-neutral photochemical (INP) model developed initially for Titan by *De La Haye et al.* (2008) to evaluate the coupled ion and neutral chemistry and adapted to Pluto by *Luspay-Kuti et al.* (2017) (hereafter *LK2017*) and *Mandt et al.* (2017) (hereafter *M2017*); and (3) a coupled ion-neutral photochemical model developed by *Krasnopolsky* (2020) (hereafter *KR2020*) as an update to *Krasnopolsky and Cruikshank* (1999), which was developed prior to the New Horizons era and draws on knowledge gained from modeling Triton (*Krasnopolsky and Cruikshank,* 1995) and Titan (*Krasnopolsky,* 2009).

The comparative planetology leveraging modeling experience with Triton's and Titan's atmospheres has proven to be of high value, as understanding of each atmosphere

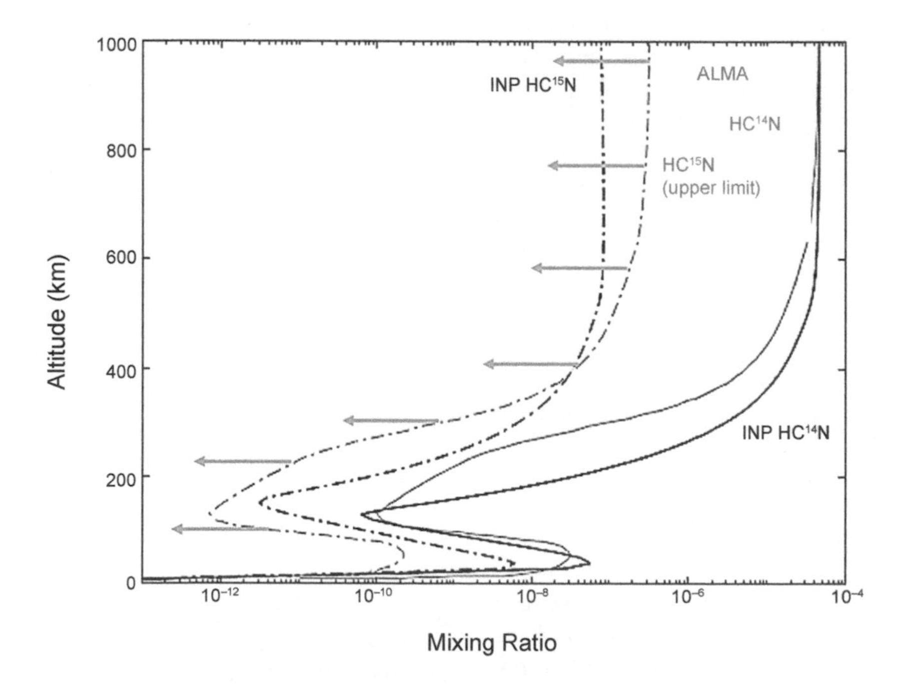

Fig. 3. Observed and modeled altitude profiles of hydrogen cyanide (HCN), comparing the measured $HC^{14}N$ profile and estimate $HC^{15}N$ upper limit (indicated with gray vertical arrows) from the Atacama Large Millimeter Array (ALMA) observations with the coupled ion-neutral photochemical model (INP) simulations of these species. Adapted from *Mandt et al.* (2017).

is advanced by observations made by multiple missions. A model for Titan or Triton is converted to Pluto's conditions by adjusting the surface gravity, the vertical temperature profile, the heliocentric distance (which defines the incident solar flux at the top of the atmosphere), the anticipated incident energetic particles at the top of the atmosphere, the assumed surface pressure (which along with the surface T determines the sublimated N_2 and CH_4 density at the surface), and the assumed planet radius. Thin atmospheres can be simulated using a "plane-parallel" model that assumes all layers of the atmosphere are flat because cross sections show different levels of the atmosphere that appear flat. However, in the case of an atmosphere that extends far from the surface of the planet, the layers of the atmosphere curve along with the surface requiring spherical corrections. Pluto, Titan, and Triton all have extended atmospheres, so in all cases the model calculations are completed assuming a spherical atmosphere.

The vertical diffusive flux, Φ_i in equation (2), plays an important role in determining minor species densities as a function of altitude. It is calculated based on understanding of vertical eddy diffusion, which describes the turbulent mixing of the atmosphere, and molecular diffusion, which is specific to each species in the atmosphere. This flux varies as a function of temperature, thermal diffusion factor, and atmospheric scale height. Eddy diffusion dominates at lower altitudes where the total neutral density is greatest and maintains constant abundance with altitude for each species unless the constituent is significantly influenced by production and loss processes represented in the right side of equation (2). When molecular diffusion dominates, the species in the atmosphere will separate according to mass, allowing lighter species to become more abundant than heavier species with increasing distance from the surface. The altitude above which molecular diffusion becomes more dominant than eddy diffusion is called the homopause, and this is different for each species. This different homopause for each species occurs because each species has a unique value for molecular diffusion resulting in a different crossover altitude, or homopause, for each species (see example in Fig. 7 for H_2 and CH_4). Constraints for molecular diffusion are based on laboratory measurements made at temperatures much greater than those in Pluto's atmosphere and extrapolated to Pluto conditions with an uncertainty of less than 7% (*Plessis et al.,* 2015). Eddy diffusion is constrained by fitting an equation for the vertical flux, Φ_i, to the altitude profile measured for a long-lived constituent.

In each model, boundary conditions are defined relative to the observed or expected atmospheric conditions. Each of the published models use the CH_4 mixing ratio derived from New Horizons (*G2016*; *Y2018*) as a fixed lower boundary condition. Although *LK2017* does not consider the incident anticipated H_2O flux at the top of the atmosphere like the other two modeling efforts, the chemistry due to H_2O is not expected to have a significant impact on the hydrocarbon or nitrile profiles because H_2O primarily interacts with CO and its dissociative products (*Rodrigo and Lara,* 2002). *LK2017* begins with the surface N_2 gas density equivalent to the value expected for N_2 gas in vapor pressure equilibrium with the N_2 ice. The N_2 value near the surface is found in this model to increase with time in response to chemically driven N_2 production in the lower atmosphere. The other two modeling efforts assume the N_2 density at the surface is fixed. The extension of the N_2 vertical profile above the surface is a combination of the values observed by New Horizons at high altitude and the values calculated based on the temperature profile derived from the New Horizons stellar and radio occultation datasets (*G2016*; *Y2018*).

In each of the published models, solar EUV and UV radiation, interplanetary Lyman-α emission, and galactic cosmic rays initiate photochemical processes starting with photodissociation and photoionization. Although N_2 is the dominant gas in Pluto's atmosphere, CH_4 photodissociation is critical for the formation of more complex molecules that eventually lead to haze within the atmosphere. Each model also explores the role of gas species condensation as a function of altitude relative to New Horizons-era temperature profiles and laboratory measurements of the saturation vapor pressure curves, although there are some differences in the anticipated sensitivity of individual species to condensation and vertical mixing as a function of altitude as discussed in section 4.

Figures 4 and 5 illustrate how the photodissociation of CH_4 and N_2 initiates a chain of reactions leading to the production of hydrocarbons and nitriles [based on photochemical studies of Titan's atmosphere using the precursor model for *LK2017* (*De La Haye et al.,* 2008)]. These initial schemes of reactions are followed by further photodissociation and ionization processes, as well as neutral reactions

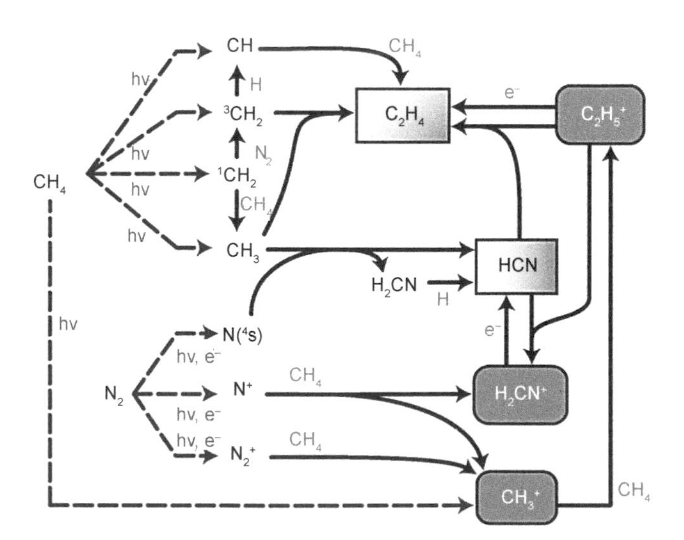

Fig. 4. Chain of reactions initiated by photodissociation of CH_4 and N_2 to produce the most basic nitrile, HCN, and one of the more basic hydrocarbons, C_2H_4. Ions that are produced in this scheme are shown in the dark gray squares and neutrals produced in the scheme are shown in the light gray squares. Adapted from chemical scheme I from *De La Haye et al.* (2008).

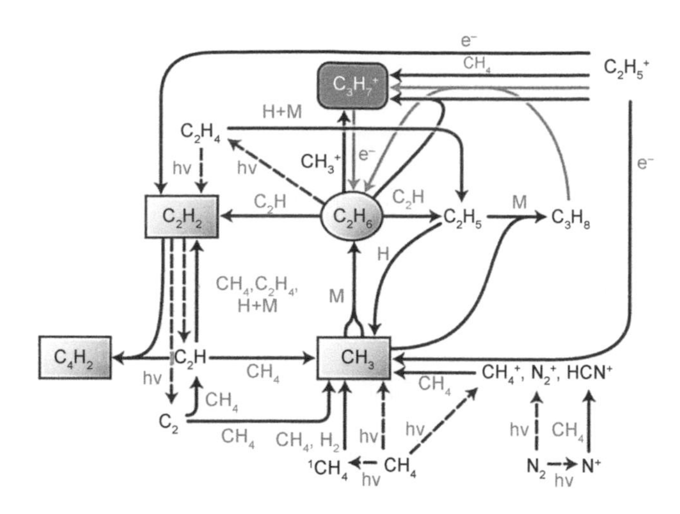

Fig. 5. Chain of reactions initiated by photodissociation of CH_4 and N_2 to produce more complex hydrocarbons. Ions that are produced in this scheme are shown in the dark gray squares and neutrals produced in the scheme are shown in the light gray squares. Adapted from chemical scheme II from *De La Haye et al.* (2008).

and ion-neutral reactions eventually forming very large negatively charged macromolecules. It is important to note that the neutral reaction rates are temperature-dependent, so the chemistry at Titan and Pluto will have similar reactions but the rate of the reactions will differ as a result of the different temperatures in each atmosphere. The presence of CO and CO_2 in Titan's atmosphere has been shown to influence the composition of aerosols produced in laboratories simulating Titan-like conditions by allowing oxygen to be incorporated into the aerosols (*Trainer et al.,* 2004; *Hörst and Tolbert,* 2014). The presence of CO in Pluto's atmosphere should also allow for incorporation of oxygen into the haze as illustrated in Fig. 6 (adapted from *Trainer et al.,* 2004).

Although there are important differences between the three modeling studies described above, there is some agreement between them on what processes are involved in the production and loss of the species observed by New Horizons and ALMA in Pluto's atmosphere. We outline these processes in Table 1. The primary source of N_2 and CH_4 is the sublimation of ices on the surface. Methane is also produced by the photodissociation of larger hydrocarbons, and by bimolecular neutral reactions that transfer a hydrogen atom from one molecule to CH_3, an example of this type of reaction would be

$$CH_3 + N_2H \rightarrow CH_4 + N_2$$

The primary loss processes for both species are photodissociation and photoionization, escape from the top of the atmosphere. *LK2017* and *KR2020* both include ion chemistry and find that reactions between ions and molecules play a role in the production and loss of CH_4. Hydrogen cyanide is primarily produced through electron recombination of $HCNH^+$ and through the transfer of a proton from $HCNH^+$ to a neutral with a higher proton affinity. An example of this type of reaction is

$$C_4H_2 + HCNH^+ \rightarrow HCN + C_4H_3^+$$

The primary loss process for HCN is condensation onto aerosols and the surface as well as sticking to aerosols. The interaction with aerosols is discussed in greater detail in section 4. Finally, the C_2 hydrocarbons are produced primarily by photodissociation of larger hydrocarbons in the case of C_2H_2 and C_2H_4, where C_2H_6 is produced by three-body, or trimolecular, reactions. The main reaction producing C_2H_6 is

$$CH_3 + CH_3 + M \rightarrow C_2H_6 + M$$

where M is a molecule that facilitates the reaction but is not influenced by it. These types of reactions are only effective in the denser part of the atmosphere near the surface.

The primary loss process for the C_2 hydrocarbons is where the three modeling studies differ the most. *W2017* suggests that these species are lost by condensation onto aerosols while *LK2017* found that they could not condense and had to be lost by sticking to aerosols as described in section 4. Finally, *KR2020* suggests that these molecules flow to the surface where they condense as ices on the surface, forcing an inversion in the altitude profile. More

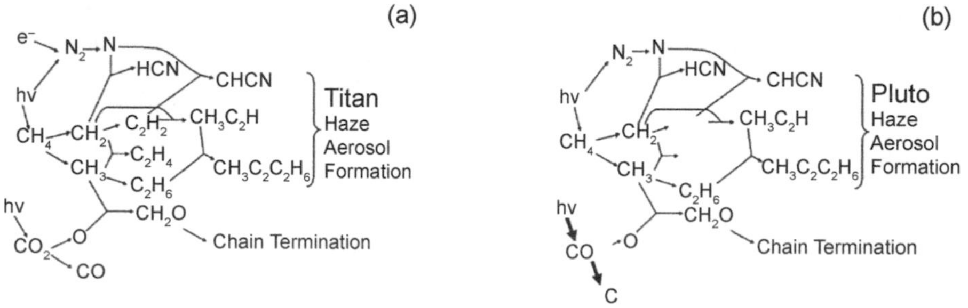

Fig. 6. Illustration of the incorporation of oxygen into the haze in **(a)** Titan's and **(b)** Pluto's atmosphere. Differences are in bold. Adapted from *Trainer et al.* (2004).

TABLE 1. Main production and loss processes for species observed by New Horizons and ALMA in Pluto's atmosphere.

	Production	Loss
CH_4	Sublimation from surface	Escape
	H-transfer reactions	Photodissociation
	Ion-neutral reactions	Ion reactions
	Photodissociation of larger hydrocarbons	
N_2	Sublimation from surface	Escape
HCN	Ion-neutral reactions	Sublimation to surface
	Electron recombination of $HCNH^+$	Condensation onto aerosols
		Adsorption onto aerosols
C_2H_2	Photodissociation of larger hydrocarbons	Sublimation to surface
		Adsorption onto aerosols
		Ion-neutral reactions
C_2H_4	Photodissociation of larger hydrocarbons	Sublimation to surface
		Adsorption onto aerosols
		Ion-neutral reactions
		Photodissociation
C_2H_6	Trimolecular neutral reactions	Sublimation to surface
		Adsorption onto aerosols
		Ion-neutral reactions
		Photodissociation

work is needed to better understand loss processes for C_2 hydrocarbons in Pluto's atmosphere.

2.2. Vertical Dynamics Represented by the Eddy Diffusion Coefficient

Methane has a relatively long chemical lifetime in Pluto's atmosphere and is sensitive to transport processes. This makes the altitude profile observed by New Horizons the most reasonable constraint available for inferring vertical eddy diffusion, or K_{zz}, in Pluto's atmosphere. Unfortunately, as illustrated in Fig. 7, the published literature presents K_{zz} values as low as 10^3 (*W2017*; *Y2018*) and 10^4 (*KR2020*), and as high as 10^6 cm^2 s^{-1} (*G2016*; *L2017*; *LK2017*). This broad range of inferred diffusion rates is difficult to evaluate in detail due to limited information provided in the published literature. It is possible that the differences are linked to how the boundary conditions assumed in each model are applied to the diffusion calculation.

W2017 reports the lowest values using the KINETICS 1-D PCM, which calculates transport independent of any ion or neutral chemistry or any consideration of CH_4 escape rates. KINETICS uses the classical formulation of the diffusion equation (*Banks and Kockarts,* 1973) like *G2016*, but does not provide sufficient explanation as to why their result is so much lower than the initial estimate reported in *G2016*. This makes it difficult to determine why the result changed and is so different from the values derived by *L2017* and *LK2017*, which use a slightly modified one-dimensional diffusion equation. *Y2018* also report a very low value using the *Strobel et al.* (2009) one-dimensional transport model, which includes escape and requires an upward flux of methane from the surface to balance escape at the top of the atmosphere. On the other hand, *KR2020* derives a value that is an order of magnitude larger by fitting not only to CH_4, but also to the C_2 hydrocarbons assuming a downward flux of the hydrocarbons due to condensation at the lowest altitudes. Finally, the calculations completed by *LK2017* apply a fixed lower boundary mixing ratio for CH_4 relative to N_2 and allow thermal escape at the top of the atmosphere.

There are slight differences in the mixing ratio for methane at the surface in each of the models. Although there is some sensitivity in the inferred K_{zz} values to the

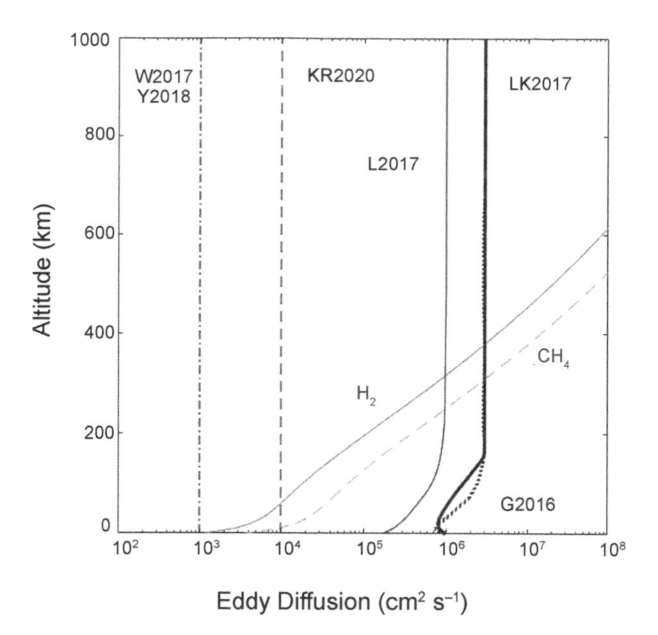

Fig. 7. Reported eddy diffusion coefficients in the published literature for the post-New Horizons analysis of Pluto's atmosphere compared with molecular diffusion for H_2 and CH_4. *W2017* and *Y2018* have the lowest reported eddy diffusion, while *G2016*, *L2017*, and *LK2017* have the highest. *KR2020* has an intermediate value.

conditions in the surface boundary layer, it is not sufficient to cause variation in the inferred K_{zz} value by ~2 to 3 orders of magnitude. There is also a small difference in the formulation for the diffusive flux equation in *W2017* compared to *LK2017*. The formulation used in *W2017* is the classical equation (*Banks and Kockarts,* 1973), while INP uses a formulation modified for a two-component atmosphere (*De La Haye,* 2005; *De La Haye et al.,* 2007). The primary difference between these formulations is that the thermal diffusion coefficient is multiplied by a factor of $[(n-n_i)/n]$, where n is the total number density and n_i is the number density of species i. In the case where the CH_4 and N_2 atmospheric densities are comparable, the thermal diffusion coefficient would be multiplied by a factor of ~0.5. However, this is only relevant above 800 *km* in Pluto's atmosphere. At all lower altitudes, the CH_4 and N_2 line-of-sight abundance profiles differ by 2 orders of magnitude, so the $(n-n_i)/n$ term is ~1 at these altitudes and there is no real difference in the classical and modified diffusive flux calculations. Therefore, this difference in the diffusive flux equation cannot explain the difference in results between *W2017* and *LK2017*. Aside from the formulation of the diffusion equation the main differences between the models is in the handling of the methane lower boundary, and whether CH_4 loss to photochemical processes (photoloysis, neutral, and ion-neutral reactions) is included in determining the eddy diffusion profile. These two factors are likely the main source of difference in the results in these models.

2.3. Including Gas-Aerosol interaction Proxies Within the Pluto Photochemical Models

A complete description of the processes impacting Pluto's atmospheric structure would require that the continuity equation is defined as a function of losses and gains occurring as a result of solar driven chemistry, kinetic chemistry, polymerization (leading to aggregate macromolecule formation and sedimentation), condensation (i.e., gas species phase changes), gas-involatile interactions (which may proceed by two different processes: either condensation onto condensation nuclei, or bonding via coagulation to and coalescence into a haze particle), diffusion, escape, sublimation, and condensate precipitation. Interactions between gas and hazes are discussed in greater detail in section 4.

3. MOLECULAR GROWTH LEADING TO AEROSOL AND HAZE FORMATION

Haze formation in Pluto's atmosphere is the natural endpoint of the photodissociation of methane and nitrogen and the production of more complex species. Although the molecules detected by New Horizons are limited to N_2, CH_4, and C_2 hydrocarbons, with HCN and CO detections from ALMA (section 2), photochemical models of Pluto's atmosphere predict a panoply of higher-order hydrocarbons and nitriles (e.g., *Krasnopolsky and Cruikshank,* 1999; *W2017*) that could act as the precursors to the global haze.

3.1. New Horizons Observations of Pluto's Global Haze

Pluto's haze was observed by several instruments onboard New Horizons, spanning more than an order of magnitude in wavelength and multiple scattering angles around the planet. *Cheng et al.* (2017) summarized the findings from the Long Range Reconnaissance Imager (LORRI) instrument, which observed at wavelengths between 350 and 850 nm and scattering angles from 20° (nearly backscattering) to 169° (nearly forwardscattering). From full-disk images, these workers found that the haze extends completely around the limb of Pluto with greater haze brightness in Pluto's northern hemisphere than its equatorial latitudes or southern hemisphere. The scale height of the haze brightness was ~50 km and decreased as the altitude increased. The most striking feature of the haze is the ~20 concentric layers (Fig. 2b) that merge, separate, and appear or disappear when traced around the limb; their average thickness is on the order of kilometers, but spatially variable (*G2016*).

In an effort to explain the greater haze brightness in the northern hemisphere rather than the subsolar point during the occultation observations, *Bertrand and Forget* (2017) coupled a three-dimensional global climate model (GCM) for Pluto with a parameterization for haze production. This GCM tracks the sublimation, transport, and condensation of nitrogen and methane on the surface and was updated to evaluate the production and transport of haze up to altitudes of 600 km. The authors were able to demonstrate that the methane loss rates would be highest at northern latitudes around 250 km in altitude leading to the greatest haze production at this location rather than at the subsolar point. This occurs because the flux of photons at Lyman-α wavelengths was greatest at these latitudes during the New Horizons flyby.

The LORRI haze observations cannot be explained by a single haze particle shape (*Cheng et al.,* 2017). Near the surface, the haze is strongly forwardscattering while also exhibiting a significant backscattering lobe, similar to comet dust (e.g., *Ney and Merrill,* 1976; *Kolokolova et al.,* 2004), which is made up of spherical particles that contain ices. However, at 45 km above the surface, the backscattering lobe is missing while the forwardscattering remains intense (*Cheng et al.,* 2017), signaling a transition in shape from spherical to fractal aggregates — fluffy conglomerations of smaller particles with significant void space (e.g., *West and Smith,* 1991; *Lavvas et al.,* 2010). The existence of aggregates is also supported by the Multispectral Visible Imaging Camera (MVIC), which observed the haze in blue (400–550 nm) and red (540–700 nm) wavelengths. *G2016* found that the haze was brighter in the blue channel, suggesting scattering by small (~10 nm) particles, which contrasts with the intense forwardscattering typical of larger (>100 nm) particles. This seeming discrepancy can be resolved by invoking aggregates, since they are made of small particles but their bulk radius can be much larger (*G2016*).

Y2018 summarized the findings of the Alice ultraviolet spectrograph, which observed solar occultations of Pluto's atmosphere from 52 to 187 nm. At these shorter wavelengths, the haze is much more opaque, and as such was detectable up to an altitude of 350 km. The scale height of the haze in the UV is also larger, at ~70 km. Importantly, like the forward- and backscattering LORRI observations, the Alice and LORRI data also cannot be explained by a single haze particle shape: Spherical particles that fit the LORRI data at all phase angles underestimate the UV extinction seen by Alice, while aggregate particles that fit the UV extinction seen by Alice and the forwardscattering data from LORRI underestimate the backscattering observations (*Cheng et al.,* 2017). This suggests that the shape — and thus the formation process — of near-surface haze particles are more complex than assumptions of only spheres or only aggregates and that the reality is likely a combination of both.

3.2. Lessons from Titan: Haze Precursors and the Formation of Monomers

We show in Fig. 8 a comparison of the temperature vs. pressure in the atmosphere of Pluto compared to that of Titan, Saturn's largest moon, and Triton, Neptune's largest moon. Pluto's atmospheric composition is similar to both atmospheres, as all three are composed primarily of N_2, with CH_4 as the second most abundant constituent. The Cassini-Huygens mission's exploration of Titan revealed a complex web of neutral and ion reactions beginning above an altitude of 1000 km. There, the photodissociation and ionization of N_2 and CH_4 due to EUV and FUV photons and energetic particles from Saturn's magnetosphere create an ionosphere, which is also fed by metallic ions from the ablation of meteorites. The detection of aerosols at altitudes as high as the ionosphere (*Liang et al.,* 2007) and large ions in the upper atmosphere (e.g., *Waite et al.,* 2007; *Crary et al.,* 2009; *Coates et al.,* 2009) suggests that haze formation

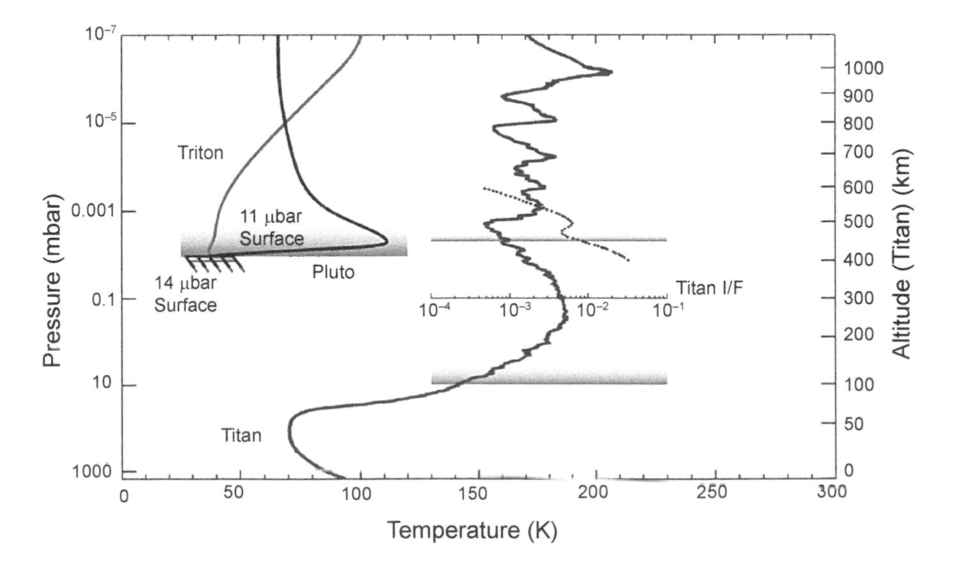

Fig. 8. Comparison of the atmospheric temperature and pressure profiles for Titan based on Cassini Huygens probe observations, Triton from Voyager 2, and Pluto from New Horizons. Also shown are the relevant altitudes for these pressures in Titan's atmosphere and the Titan I/F for the altitude range ~400–600 km. The shaded regions indicate the locations of the haze layers in Titan's and Pluto's atmospheres. Because Triton's haze was only observed up to ~30 km, which is very close to the surface, the Triton haze layer is not shown. Adapted from Fig. 22 of *Cheng et al.* (2017).

begins in the ionosphere, although the production pathway is not fully understood.

One proposed production pathway consists of ion and neutral chemical reactions rapidly producing haze precursor molecules such as benzene (C_6H_6) and heavier polycyclic aromatic hydrocarbons (PAHs) (*Wilson and Atreya,* 2004, *Trainer et al.,* 2013, *Yoon et al.,* 2014), followed by collision and sticking of the PAHs that then create small spherical particles with radii of ~0.1–0.5 nm (*Lavvas et al.,* 2011). These particles become negatively charged due to the absorption of free electrons in the ionosphere. The negative charge attracts positive ions, leading to rapid growth of particles to radii >1 nm (*Lavvas et al.,* 2013). Finally, neutral heterogeneous reactions on the surfaces of these particles result in the formation of spherical particles with radii of several nanometers (*Lavvas et al.,* 2011), which can then collide and stick to form the observed haze particles. A review of Titan's atmospheric chemistry and haze formation processes can be found in *Hörst* (2017).

There also exist, however, several key differences between Pluto and Titan that call into question the applicability of haze formation processes inferred in Titan's atmosphere to Pluto. The most striking difference is the energetics of haze formation. New Horizons failed to detect an ionosphere at Pluto and constrained the ion density to be <1000 e^- cm^{-3} (*Hinson et al.,* 2018), 5× less than that of Titan (*Gladstone and Young,* 2019). This may not be surprising, since Pluto receives less than a tenth of the solar UV flux received by Titan and there is no giant planet magnetosphere to source energetic particles. On the other hand, Lyman-α radiation scattered from H atoms in the local interstellar medium could be a significant driver of photochemistry (*W2017*), and deposition of ions stemming from the ablation of water ice and silicate interplanetary dust particles could feed a low-density ionosphere (*Poppe and Horányi,* 2018). These differences in upper atmosphere energetics could lead to variations in haze formation pathways and composition. In addition, the CO mixing ratio of Pluto's atmosphere [~0.05% (*Gladstone and Young,* 2019)] is ~10× larger than that of Titan. To explore this, laboratory investigations (e.g., *Hörst and Tolbert,* 2014; *Fleury et al.,* 2014, *He et al.,* 2017) have shown that changing CO mixing ratios in a N_2-CH_4 gas mixture exposed to high-energy sources could lead to dramatically different haze production rates, particle sizes, and composition.

3.3. Aggregation and Transport of Aerosols

The particle size and number density of aerosols in planetary atmospheres are controlled by the balance between microphysical and transport processes. Microphysical processes refer to those that directly impact the size of particles, including growth by uptake of volatile vapors (condensation) and the adsorption of gases (sticking), or by colliding and sticking with other aerosol particles (coagulation and coalescence), as well as loss due to evaporation or fragmentation. Transport processes include sedimentation of aerosols under the action of gravity, as well as the diffusion of particles due to Brownian motion (Brownian diffusion) and large-scale atmospheric dynamics (e.g., eddy diffusion and advection).

The low temperatures and densities of Pluto's atmosphere means that only a subset of microphysical and transport processes need to be considered. In particular, sedimentation of aerosols due to gravity dominates over diffusion and dynamics for most of Pluto's atmosphere (*Gao et al.,* 2017, *Bertrand and Forget,* 2017). As a result, small spherical particles growing through ion and neutral chemistry in the putative Pluto ionosphere can only grow to a certain size before their sedimentation timescale becomes shorter than their growth timescale, and they begin their journey toward the surface. This size should be on the order of a few to 10 nm, as inferred from MVIC observations (section 3.1), similar to analogous particles in Titan's atmosphere (section 3.2). Furthermore, because the mean free path of aerosols is large compared to their size — their Knudsen numbers are large — the sedimentation velocity becomes inversely proportional to the atmospheric density and the square root of the temperature, resulting in decreasing fall speeds toward Pluto's surface (*Gao et al.,* 2017).

Nascent haze particles can collide with each other during descent and stick together with weak intermolecular adhesion forces in the process of coagulation (*Dimitrov and Bar-Nun,* 1999). For sufficiently rigid particles, sticking may occur at only one point between them, forming fractal aggregates with the constituent particles termed "monomers." The rate of coagulation is proportional to the square of the particle number density and the coagulation kernel, which is a rate coefficient used to calculate coagulation probabilities that takes into account the thermal velocity (Brownian coagulation) and relative fall velocities (gravitational collection) of the colliding aerosol particles. At large Knudsen numbers, the Brownian coagulation kernel becomes proportional to the square root of temperature (*Lavvas et al.,* 2010), suggesting increasing coagulation rates with decreasing altitude between a few hundred to tens of kilometers above Pluto's surface, where the temperature also increases with decreasing altitude. In contrast, laboratory studies conducted in preparation for the Cassini-Huygens mission observations at Titan found that aerosols produced in the laboratory hardened as they grew in size and "aged," making them less sticky (*Dimitrov and Bar-Nun,* 2002). This effect would lead to a decrease in coagulation rates with altitude.

Gao et al. (2017) computed the haze particle number density and size distribution in Pluto's atmosphere using an aerosol physics model and compared their results to haze extinction observations from the Alice ultraviolet spectrograph (section 3.1). They assumed that the haze production rate was equal to the methane photolysis rate computed by the photochemical model of *W2017*. *Gao et al.* (2017) found that, above 200 km, haze particles remained small, either as individual monomers or small aggregates, due to rapid sedimentation and low coagulation rates (Fig. 9). Below 200 km, decreasing fall speeds and increasing coagulation kernels result in the growth of aggregates through coagulation, ultimately forming aggregates with radii ~100–200 nm near the surface, consistent with LORRI data (*G2016*; *Cheng et al.,* 2017). *Gao et al.* (2017) also simulated an alternative case where all haze particles were spherical, but found that they underestimated the particle size and ultraviolet extinction.

The ultraviolet extinction predicted by the aggregate model of *Gao et al.* (2017) increases with decreasing altitude faster than that observed by the New Horizons Alice UV spectrograph, as illustrated in Fig. 10. Curiously, this can be explained if the monomers within the fractal aggregates grew with decreasing altitude. While 5–10-nm monomers can reproduce the data above 150 km, 25-nm monomers are needed below 100 km. Monomers can grow through coalescence: If there is enough excess energy available after coagulation, then the intermolecular adhesion forces can be replaced by stronger, cohesive intermolecular bonds (*Dimitrov and Bar-Nun,* 1999). At this stage, the monomers are partially fused together while still keeping their chemical structures. If there is still considerable energy left in the system then coalescence will continue and the particles will fuse together, forming spheres. Another way for monomers to grow is through condensation and/or adsorption of atmospheric gases onto the aggregate, which will be elaborated on in section 4. The growth and possible coalescence of monomers with time and decreasing altitude may provide an explanation for the discrepancy between the multi-wavelength, multi-phase angle observations of Alice and LORRI on New Horizons (section 3.1).

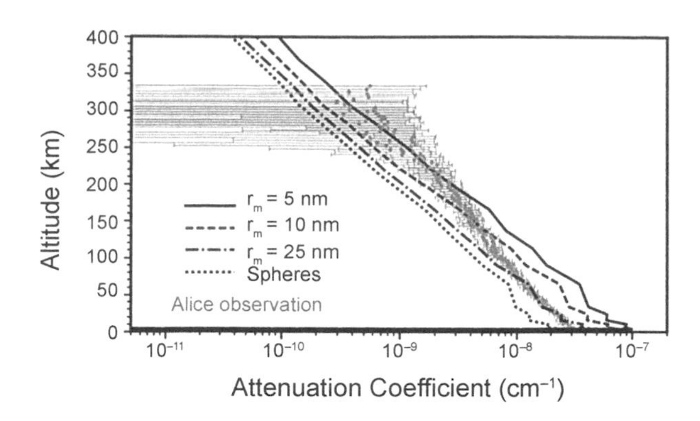

Fig. 10. The New Horizons Alice observed (points) and model attenuation coefficient (curves) of Pluto's haze. Models with monomer radii of 5 nm (solid), 10 nm (dashed), and 25 nm (dash-dot) are presented along with the attenuation coefficient of spherical haze particles. All haze models have the same production rate. Updated from *Gao et al.* (2017).

4. HAZE FEEDBACK PROCESSES IN PLUTO'S ATMOSPHERE

As described in the previous sections, there are many similarities in photochemistry and haze formation in Pluto's and Titan's atmospheres that allow application of lessons learned from Titan modeling to Pluto. Of particular interest is how the formation and growth of haze feeds back into atmospheric processes. Observations made during the New Horizons flyby provide important constraints on these feedback processes. Studies of Titan's atmosphere found that the haze particles play an important role in shaping the vertical abundance profiles of hydrocarbon and nitrile species when the molecules condense onto and irreversibly stick to the haze particles (*Willacy et al.,* 2016). This influence of haze in Titan's atmosphere raises the question of whether haze particles play a similar role in Pluto's atmosphere.

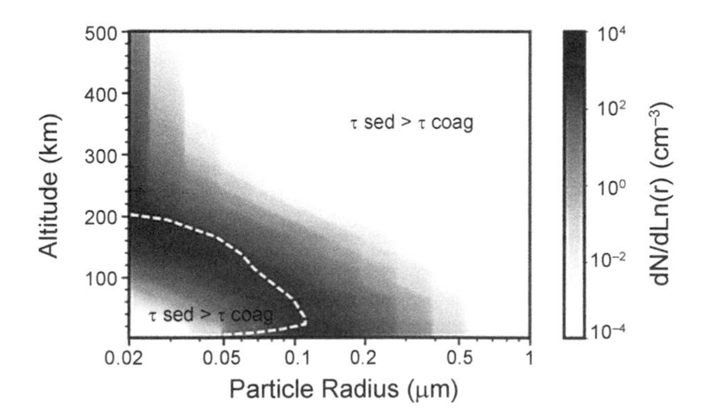

Fig. 9. Particle number density as a function of altitude and particle radius, assuming a fractal aggregate shape for the particles. The peak of the particle size distribution can be traced by balancing the sedimentation timescale with the coagulation timescale (dashed line), indicating that the particles can grow until sedimentation transports them to lower altitudes. Updated using the model from *Gao et al.* (2017) described in section 3.3.

An early indication for possible haze feedback in Pluto's atmosphere was the shape of the vertical C_2 hydrocarbon profiles retrieved from New Horizons. Instead of continuously increasing toward the surface, as would be expected if photochemistry were the only process affecting their profiles, the densities of C_2H_2, C_2H_4 and C_2H_6 showed an inversion between 100 and 500 km [illustrated in Figs. 2 and 11 (*G2016*; *W2017*; *LK2017*; *Y2018*)]. The observed inversion in the altitude range where haze particles were also observed suggests that haze-molecule interactions dominate hydrocarbon loss in this altitude range. Furthermore, the altitude profile of HCN determined with the ALMA observations (*L2017*) also shows haze-particle interaction because HCN is supersaturated, or has partial pressures that are orders of magnitude larger than the saturation vapor pressure, above 150 km and drops to subsaturated, or partial pressures below the saturation vapor pressure, around ~50 km.

4.1. Condensation onto Haze Particles

Whether a species in Pluto's atmosphere will condense onto haze particles or not depends on the collision rate between the molecule and the particles, the availability of condensation nuclei, and the molecule's partial and saturated vapor densities. Condensation requires that the density of an atmospheric constituent be greater than its saturation vapor density. The saturation vapor density of a species is derived from its saturation vapor pressure. The saturation vapor pressures of the species observed in Pluto's atmosphere have been calculated at their vapor-solid equilibrium based on limited laboratory data. However, the temperature range over which these laboratory measurements were performed does not extend down to the lowest temperatures observed in Pluto's atmosphere. Thus, extrapolation to low temperatures is often needed based on expressions determined from laboratory experiments at higher temperatures.

Figure 12 shows the calculated saturation vapor pressures and densities for species observed in Pluto's atmosphere using equilibrium expressions available in the literature. When comparing the saturation vapor densities to the atmospheric densities derived from New Horizons measurements, it is apparent that for the most part, the hydrocarbon densities are significantly lower than their saturation vapor densities through most of the atmosphere. The HCN altitude profile was observed by ALMA and shows that HCN is supersaturated above 150 km (*L2017*). *Rannou and West* (2018) evaluated condensation processes in Pluto's atmosphere using a model based on the classical laws of cloud nucleation, or the formation of clouds by the phase transition from supersaturated air into droplets. Nucleation can occur either by heterogenous nucleation, which is nucleation onto a surface, or through homogenous nucleation, which occurs away from a surface. Heterogenous nucleation is more common than homogenous nucleation, and in the case of Pluto's atmosphere would take place on the surfaces of photochemically produced haze particles. *Rannou and West* (2018) found that homogenous nucleation is unlikely in Pluto's atmosphere, requiring the presence of aerosols as nucleation sites for heterogenous nucleation, thus allowing HCN to reach high supersaturation above the layers of haze.

Based on this, HCN is able to condense throughout the atmosphere of Pluto, but the C_2 hydrocarbons cannot condense except for within a thin, near-surface layer. The one exception is C_2H_2, which may condense between 300 and 500 km if the saturation vapor pressure expression from *Moses et al.* (1992) is used (*LK2017*). This implies that condensation onto haze particles does not affect the C_2H_4 and C_2H_6 vertical density profiles. Furthermore, the photochemical model of *LK2017* suggests that condensation onto haze particles is negligible even for C_2H_2, which may become saturated over a narrow altitude range. Thus, *LK2017* conclude that condensation onto the haze particles is not responsible for the observed density inversion in the hydrocarbon density profiles.

These results are in contradiction to the model results of *W2017*, who conclude that condensation is the major loss process for the hydrocarbons, and that in order to reproduce the inversion of C_2H_4, its saturation vapor pressure must be much lower than what would result from extrapolation of laboratory data to the low temperatures of Pluto's atmosphere. *W2017* suggest that the saturation vapor pressure of C_2H_4 must be the same as that of C_2H_2 at the low temperatures of Pluto, although they provide no physical explanation as to why this would be the case. The currently available published laboratory data, and empirical and thermodynamic relations show no indication for such behavior of the C_2H_4 vapor pressure.

The atmospheric profile of HCN appears to be strongly influenced by condensation onto the haze particles (*M2017*; *W2017*; *L2017*). Based on Fig. 11, any HCN densities greater than 10^8 cm^{-3} are subject to condensation, given sufficient availability of condensation nuclei. The efficiency with which HCN molecules condense onto haze particles is roughly a factor of 2 smaller determined by *M2017*, who used the photochemical model described in *LK2017*, than that found by *W2017*. Although it is important to note that while the treatment of condensation is stated to be the same based on the approach of *Willacy et al.* (2016) in both photochemical models, the equation given by *W2017* does not include a term for the saturation vapor pressure, and more closely resembles the equation for incorporation into aerosols (see below in section 4.2). Because of this inconsistency, a direct comparison of the results of *W2017* to the results of *LK2017* and *M2017* is challenging.

4.2. Incorporation into Haze Particles

Atmospheric molecules may irreversibly stick to and become incorporated into haze particles, leading to their permanent loss (*Willacy et al.,* 2016). As such, this haze feedback process has the potential to influence the altitude profiles of species in Pluto's atmosphere. This incorporation of molecules into haze particles is independent of the saturation vapor density. The loss rate due to incorporation into haze particles depends on the mean surface area of haze particles per volume as a function of altitude, the sticking efficiency of the molecules, the thermal velocity of the given molecule, and the atmospheric density of the molecule at a given altitude (*Willacy et al.,* 2016).

The formation and distribution of haze particles in Pluto's atmosphere were recently modeled based on New Horizons measurements by *Gao et al.* (2017) and show that the haze particles' surface area increases with decreasing altitude (Fig. 12a). Using this altitude profile of the haze particles, recent photochemical models quantified the loss of the detected molecules in Pluto's atmosphere due to their irreversible sticking to haze particles. As discussed in detail section 3.3, haze particles are sticky when they first form, and become less sticky as they age due to hardening. The sticking efficiency of a given molecule factors in both the ability of a given molecule to stick to the haze particles and the stickiness of the haze particles themselves according to their hardening/aging. It is important to note when comparing Pluto and Titan that Pluto's atmosphere receives lower UV radiation and experiences significantly lower temperatures than Titan's atmosphere. This leads to lower collision rates between the molecules and the haze particles and lower-energy deposition rates compared to Titan. Thus, we expect the hardening and aging of haze particles to be slower at Pluto, which could potentially

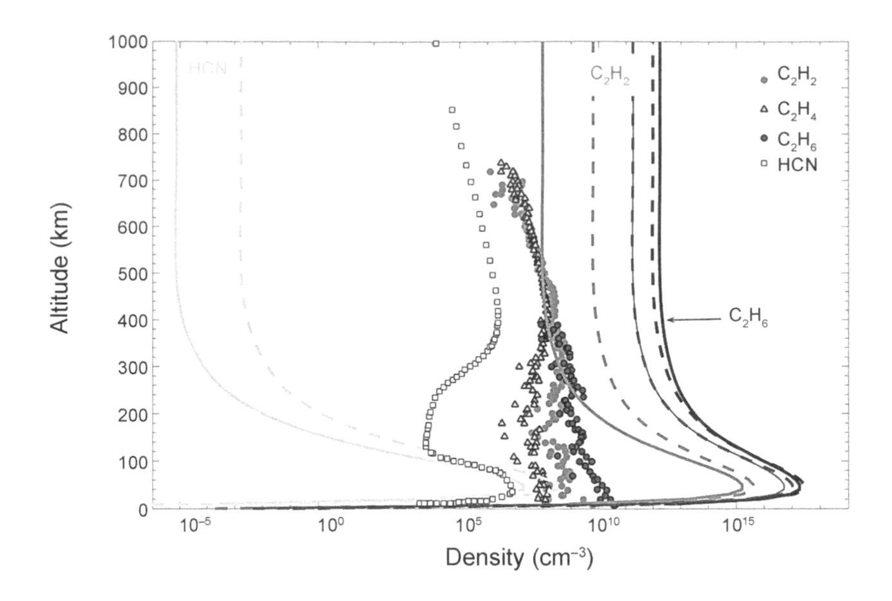

Fig. 11. Comparison of the saturation vapor densities to the New Horizons C_2H_2 (medium gray), C_2H_4 (dark gray), and C_2H_6 (black) densities and the ALMA HCN (light gray) densities. The solid lines are saturation vapor densities as listed in Table 2 of *Willacy et al.* (2016), while the dashed lines are saturation vapor densities from *Fray and Schmitt* (2009). This comparison demonstrates that HCN is supersaturated above 150 km and subsaturated around 50 km, while the C_2 hydrocarbons are subsaturated in almost all altitude ranges. Modified from *LK2017* and *M2017*.

increase the rate of hydrocarbon and nitrile incorporation into Pluto's haze (*LK2017*).

The sticking efficiencies of the C_2 hydrocarbons and HCN were empirically determined by fitting the densities observed by New Horizons and ALMA with modeled densities. The sticking efficiencies show that the stickiness of haze particles in Pluto's atmosphere is inversely proportional to their mean surface area (Fig. 12b). Thus, the ability of molecules to stick to them also decreases with increasing particle size (*LK2017*; *M2017*). This means that the larger that the aerosol surfaces available to collide with and stick to are, the fewer hydrocarbons and HCN are actually able to stick to them. The decreasing stickiness of haze particles with increasing haze particle size (which increases with decreasing altitude) in Pluto's atmosphere is demonstrated in Fig. 12. With the sticking efficiencies and haze stickiness empirically derived for Pluto's hydrocarbons, the observed inversion in their vertical density profiles by New Horizons was successfully reproduced by INP [Fig. 2a (*LK2017*)]. The fact that the condition of molecules sticking to the haze is necessary to reproduce the density inversion in Pluto's atmosphere indicates that haze particles do indeed harden, or become less sticky as they age, similar to what was predicted by *Dimitrov and Bar-Nun* (2002) for Titan's aerosols.

4.3. Atmospheric Thermal Balance

In addition to the inversion of the vertical density profiles of the C_2 hydrocarbons, another issue raised by New Horizons measurements was the significantly lower atmospheric temperature than predicted by pre-New Horizons theory (*G2016*; *Zhu et al.*, 2014), as shown by the comparison of temperature profiles in Fig. 13. The colder-than-expected temperatures suggest some kind of a cooling mechanism in Pluto's atmosphere. While typically atmospheric gases determine the radiative energy balance of planetary atmospheres, they were found to be insufficient to cause the observed low temperatures in Pluto's atmosphere (*Strobel and Zhu,* 2017). However, hazes are another source of heating and cooling in planetary atmospheres. *Zhang et al.* (2017) assessed the radiative effects of hazes in Pluto's atmosphere by a multi-scattering radiative transfer model and found that the heating and cooling rates caused by Pluto's hazes are 1 to 2 orders of magnitude larger than the gas cooling rates. Thus, Pluto's atmosphere is unique among planetary atmospheres in the sense that the total radiative energy budget is driven by haze particles instead of gas molecules. In other words, Pluto's atmospheric gas temperature appears to be controlled by the haze particles instead of the gases themselves below an altitude of about 700 km. Above 700 km, heat conduction drives the temperature, according to *Zhang et al.* (2017).

A caveat for this thermal balance model is that it must rely on estimated optical properties of the haze, called optical constants. Currently no observational constraints on the composition of these haze particles exist. The lack of detailed knowledge in the haze composition may lead to large uncertainties in the modeled heating and cooling rates. If hazes are indeed the drivers of atmospheric cooling on Pluto, then they are expected to radiate large amounts of heat into space. In that case, Pluto would appear several orders of magnitude brighter in the mid-infrared wavelength range than previously thought. The James Webb Space Telescope (JWST), expected to be launched in 2021, should be able to detect this predicted brightening (*Zhang et al.,* 2017).

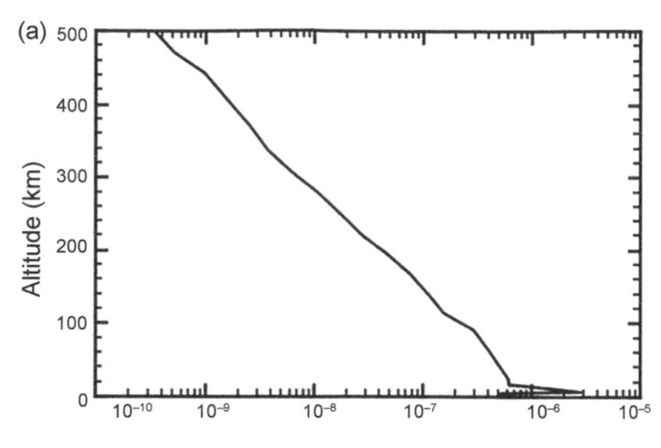

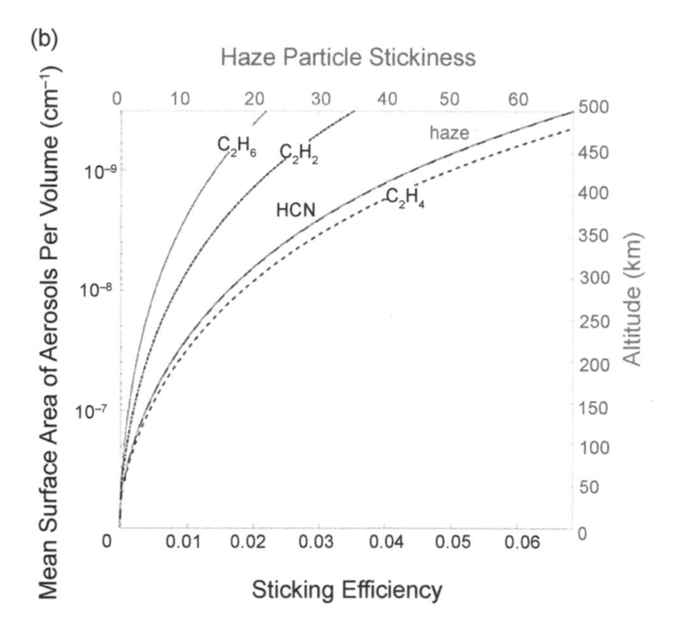

Fig. 12. **(a)** Total surface area of aerosols per volume of atmosphere from *Gao et al.* (2017). **(b)** Sticking efficiency (dimensionless) of the C_2 hydrocarbons and HCN as a function of haze particle surface area from *Gao et al.* (2017) and altitude (black lines), and the ability of haze particles to physically trap molecules through sticking, described as stickiness (gray dashed line). Created with data obtained from *LK2017* and *M2017*.

Because JWST could directly detect the radiative effects of haze, future observations would help to differentiate between the effects of haze and gas molecules such as water vapor. This would be a giant leap toward settling the issue of Pluto's atmospheric temperature.

5. VARIABILITY IN PHOTOCHEMISTRY AND LONG-TERM EVOLUTION

Pluto's atmosphere could provide important clues to the origin of the planet's volatiles, but only if the evolution of Pluto's atmosphere is well understood (*Mandt et al.,* 2016; *M2017*). More specifically, its building blocks are likely to have formed in the outer solar system, but it is not clear if the temperature and pressure conditions allowed the building blocks to trap enough nitrogen in the form of N_2 to produce Pluto's atmosphere, or if the nitrogen was originally trapped as NH_3 or organics and later converted to N_2. Although the ratio of N_2 to NH_3 is believed to have been ~10 in the PSN (*Lewis and Prinn,* 1980), N_2 requires much colder temperatures to be trapped in either amorphous (*Bar-Nun et al.,* 1985, 1988) or crystalline (*Mousis et al.,* 2012, 2014) water ice. If Pluto's building blocks formed at temperatures below ~40 K, they would have accreted N_2 ice in greater abundance than NH_3 ice, but ices formed at higher temperatures would have been deficient in N_2. Because NH_3 has not been detected in the atmosphere and no upper limit has been published, we do not yet have a ratio of N_2 to NH_3 for Pluto's atmosphere.

Comets are a reasonable analog for the building blocks of Pluto because their building blocks may have formed in similar conditions. The Rosetta spacecraft provided the first detailed measurements of N_2 in a coma of a comet (*Rubin et al.,* 2015, 2018, 2019), providing a ratio of N_2 to NH_3 of 0.13 at Comet 67P/Churyumov-Gerasimenko (67P/C-G) (*Rubin et al.,* 2019). This ratio and the lack of detection of N_2 in other comets (e.g., *Cochran,* 2002) suggests that they are deficient in N_2 relative to NH_3, either as a result of their formation temperature (*Iro et al.,* 2003) or because they did not retain N_2 (1) beyond their first pass through the solar system (*Owen et al.,* 1993) or (2) due to internal radiogenic heating at early epochs after formation (*Mousis et al.,* 2012). The detection of N_2 in 67P/C-G is important because 67P/C-G is a short-period comet thought to originate in the Kuiper belt. Comparison of its composition with Pluto suggests that Pluto could have formed in a similar temperature range and may have retained some N_2 from the PSN. However, the greater abundance of NH_3 relative to N_2 indicates that Pluto could have obtained a larger abundance of nitrogen in its primordial ices in the form of NH_3.

Although organics have been proposed as a significant source (~50%) of Titan's N_2 (*Miller et al.,* 2019), a reasonable process for efficiently breaking the carbon-nitrogen bond in organics that is necessary for formation of N_2 has not been identified. Therefore, the most likely source of Pluto's nitrogen was either N_2 or NH_3 that was trapped in ices in the PSN. *Glein and Waite* (2018) showed through simple mass-balance calculations that the initial N_2/H_2O ratio reported by Rosetta for 67P/C-G (*Rubin et al.,* 2015) could explain the current inventory of N_2 on the surface and in the atmosphere of Pluto. However, the bulk abundance of N_2 relative to CO in the coma of 67P/C-G is a factor of ~5 lower than the protosolar value (*Rubin et al.,* 2019), and the authors note that they were not able to explain the fact that Pluto's atmosphere and surface should have much more CO than has been detected when comparing with the N_2/CO ratio observed at 67P/C-G.

Because it is important to determine the source of Pluto's nitrogen, a valuable constraint for the origin of nitrogen

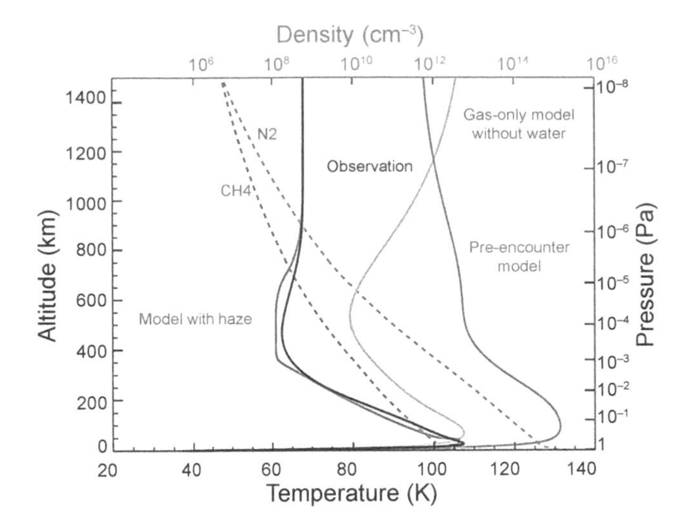

Fig. 13. Observed temperature and density profiles in Pluto's atmosphere (black and dashed lines), and modeled temperature profiles showing higher theoretical temperatures before the New Horizons flyby, model temperatures with atmospheric gas molecules as the only coolant, and model temperatures with haze feedback. From *Zhang et al.* (2017).

would be the current nitrogen isotope ratio in N_2 in Pluto's atmosphere. Although this measurement cannot be taken as primordial, it can be used to constrain the origin of nitrogen if the evolution of the atmosphere is understood [e.g., *Mandt et al.* (2014) for Titan]. Once the evolution of the atmosphere is constrained, the primordial ratio derived from modeling the evolution of the atmosphere can be compared with other primordial measurements in the Sun, comets, meteorites, and giant planet atmospheres.

Mandt et al. (2016) and *M2017* evaluated the evolution of the nitrogen isotope ratio, $^{14}N/^{15}N$, in Pluto's atmosphere based on both pre-New Horizons information and updated based on results of the New Horizons flyby. The nitrogen isotope ratio when Pluto formed, i.e., the primordial ratio, can indicate whether the majority of Pluto's nitrogen came from N_2, which would have a ratio of ~450, or NH_3, which would have a ratio of ~130 (see *M2017* and references therein). The pre-New Horizons study found that escape of particles from the top of the atmosphere would have a dominant effect on how the isotope ratio would change over time because high escape rates in a thin atmosphere would remove more of the light isotope compared to the heavy isotope, resulting in a current ratio that is significantly lower than the primordial value. However, when applying the discovery by New Horizons that the escape rate was much lower than predicted (*G2016*) with photochemical models attempting to reproduce the ALMA upper limit for the $HC^{15}N$ abundance (*L2017*), *M2017* found that photochemistry and haze feedback play a more important role in the evolution of the nitrogen isotopes over time. This is an important consideration for any studies that attempt to determine the origin of nitrogen at Pluto using future observations of the nitrogen isotopes.

6. SUMMARY AND CONCLUSIONS

Prior to the New Horizons flyby of Pluto, understanding of Pluto's atmosphere was limited to information gleaned from groundbased and Earth-orbiting telescopes. This information was obtained primarily by occultations of Pluto by relatively bright stars measuring the reduction in brightness of the star as Pluto passed in front of it. Based on this information, the atmosphere was known to be primarily composed of N_2 with trace amounts of CH_4 and CO, and was presumed to have hydrocarbons and nitriles based on understanding of atmospheric chemistry derived from observations of Triton's and Titan's atmospheres. Some observations suggested the possibility of haze in the lower atmosphere, but could also have been explained by a temperature gradient near the surface. Additionally, Pluto's upper atmospheric temperature was predicted to be high enough to allow for a large escape rate of N_2 at the top of the atmosphere.

The New Horizons flyby revolutionized our understanding of Pluto's atmosphere by showing evidence of haze layers and strong interactions between the atmospheric gas and haze particles. The observations showed that these interactions remove large amounts of HCN through condensation and C_2 hydrocarbons and HCN through molecules sticking to aerosols as they form and descend through the atmosphere. Physical characteristics of the aerosols can be derived from comparing models with the observations. These comparisons show that the aerosol shapes are not strictly spherical or aggregate and that their ability to adsorb molecules, or stickiness, decreases as their size increases and they age. These results are groundbreaking and will have implications for future work studying the haze in Titan's and Triton's atmospheres.

Another important outcome of the New Horizons flyby was the discovery that the temperature in the upper atmosphere was significantly lower than predicted, and that a haze feedback of cooling by radiation to space may be the cause. This has implications for the ongoing chemistry in, the escape of particles from, and the long-term evolution of Pluto's atmosphere. Reduced escape rates allow Pluto to retain volatiles on the surface and in the atmosphere long-term.

Several open questions remain to be explored by future ground- and spacebased observations as well as any future spacecraft mission to Pluto. Observations with JWST could confirm that the cooling in the atmosphere is caused by thermal radiation of aerosols. Long-term monitoring of the atmosphere using ground- and spacebased telescopes can search for evidence of haze, monitor surface pressure and methane abundance, and track temperature changes as the distance of Pluto from the Sun increases. Intentional collaboration between modeling groups is needed to resolve the differences in eddy diffusion coefficients illustrated in Fig. 7. Comparative studies of the effects of condensation and sticking to aerosols in the atmospheres of Titan, Pluto, and Triton are also needed and of high value for evaluating haze in planetary atmosphere. This will have implications for exoplanets as well. Finally, any future mission should be capable of measuring the nitrogen isotope ratio in N_2 to provide the needed constraints for understanding the origin on Pluto's nitrogen.

REFERENCES

Allen M., Yung Y. L., and Waters J. W. (1981) Vertical transport and photochemistry in the terrestrial mesosphere and lower thermosphere (50–120 km). *J. Geophys. Res.–Space Phys., 86(A5),* 3617–3627.

Bampasidis G., Coustenis A., Achterberg R. K., et al. (2012) Thermal and chemical structure variations in Titan's stratosphere during the Cassini mission. *Astrophys. J., 760(2),* 144.

Banks P. M. and Kockarts G. (1973) *Aeronomy, Part A.* Academic, New York. 430 pp.

Bar-Nun A., Herman G., Laufer D., and Rappaport M. L. (1985) Trapping and release of gases by water ice and implications for icy bodies. *Icarus, 63(3),* 317–332.

Bar-Nun A., Kleinfeld I., and Kochavi E. (1988) Trapping of gas mixtures by amorphous water ice. *Phys. Rev. B, 38(11),* 7749.

Bertrand T. and Forget F. (2017) 3D modeling of organic haze in Pluto's atmosphere. *Icarus, 287,* 72–86.

Brosch N. (1995) The 1985 stellar occultation by Pluto. *Mon. Not. R. Astron. Soc., 276(2),* 571–578.

Brown R., Lebreton J. P., and Waite H., eds. (2009) *Titan from Cassini-Huygens.* Springer, Netherlands. 535 pp.

Cheng A. F., Summers M. E., Gladstone G. R., et al. (2017) Haze in Pluto's atmosphere. *Icarus, 290,* 112–133.

Coates A. J., Crary F. J., Lewis G. R., Young D. T., Waite J. H., and Sittler E. C. (2007) Discovery of heavy negative ions in Titan's ionosphere. *Geophys. Res. Lett., 34(22).*

Coates A. J., Wellbrock A., Lewis G. R., Jones G. H., Young D. T., Crary F. J., and Waite J. H. Jr. (2009) Heavy negative ions in Titan's ionosphere: Altitude and latitude dependence. *Planet. Space Sci., 57(14–15),* 1866–1871.

Cochran A. L. (2002) A search for N_2^+ in spectra of comet C/2002 C1 (Ikeya-Zhang). *Astrophys. J. Lett., 576(2),* L165.

Crary F. J., Magee B. A., Mandt K., Waite J. H. Jr., Westlake J., and Young D. T. (2009) Heavy ions, temperatures, and winds in Titan's ionosphere: Combined Cassini CAPS and INMS observations. *Planet. Space Sci., 57(14–15),* 1847–1856.

Cruikshank D. P. and Silvaggio P. M. (1980) The surface and atmosphere of Pluto. *Icarus, 41(1),* 96–102.

Cruikshank, D. P., Pilcher C. B., and Morrison D. (1976) Pluto: Evidence for methane frost. *Science, 194(4267),* 835–837.

De La Haye V. (2005) Coronal formation and heating efficiencies in Titan's upper atmosphere: Construction of a coupled ion, neutral and thermal structure model to interpret the first INMS Cassini data. Ph.D. thesis, University of Michigan, available online at *http://hdl.handle.net/2027.42/125340.*

De La Haye V. et al. (2007) Cassini Ion and Neutral Mass Spectrometer data in Titan's upper atmosphere and exosphere: Observation of a suprathermal corona. *J. Geophys. Res.–Space Phys., 112(A7).*

De La Haye V., Waite J. H. Jr., Cravens T. E., Robertson I. P., and Lebonnois S. (2008) Coupled ion and neutral rotating model of Titan's upper atmosphere. *Icarus, 197(1),* 110–136.

Dimitrov V. and Bar-Nun A. (1999) A model of energy-dependent agglomeration of hydrocarbon aerosol particles and implication to Titan's aerosols. *J. Aerosol Sci., 30*(1), 35–49.

Dimitrov V. and Bar-Nun A. (2002) Aging of Titan's aerosols. *Icarus, 156(2),* 530–538.

Dimitrov V. and Bar-Nun A. (2003) Hardening of Titan's aerosols by their charging. *Icarus, 166(2),* 440–443.

Dobrijevic M., Loison J. C., Hickson K. M., and Gronoff G. (2016) 1D-coupled photochemical model of neutrals, cations and anions in the atmosphere of Titan. *Icarus, 268,* 313–339.

Elliot J. L., Dunham E. W., Bosh A. S., Slivan S. M., Young L. A., Wasserman L. H., and Millis R. L. (1989) Pluto's atmosphere. *Icarus, 77(1),* 148–170.

Elliot J. L., et al. (2003) The recent expansion of Pluto's atmosphere. *Nature, 424(6945),* 165.

Elliot J. L., Person M. J., Gulbis A. A. S., et al. (2007) Changes in Pluto's atmosphere: 1988–2006. *Astron. J., 134(1),* 1.

Fink U., Smith B. A., Benner D. C., Johnson J. R., Reitsema H. J., and Westphal J. A. (1980) Detection of a CH_4 atmosphere on Pluto. *Icarus, 44(1),* 62–71.

Fleury B. et al. (2014) Influence of CO on Titan atmospheric reactivity. *Icarus, 238,* 221–229.

Fray N. and Schmitt B. (2009). Sublimation of ices of astrophysical interest: A bibliographic review. *Planet. Space Sci., 57(14–15),* 2053–2080.

Gao P., Fan S., Wong M. L., et al. (2017) Constraints on the microphysics of Pluto's photochemical haze from New Horizons observations. *Icarus, 287,* 116–123.

Gladstone G. R. (1982) Radiative transfer with partial frequency redistribution in inhomogeneous atmospheres: Application to the jovian aurora. *J. Quant. Spectrosc. Radiat. Transfer, 27(5),* 545–556.

Gladstone G. R. and Young L. A. (2019) New Horizons observations of the atmosphere of Pluto. *Annu. Rev. Earth Planet. Sci., 47,* 119–140.

Gladstone G. R., Stern S. A., Ennico K., et al. (2016) The atmosphere of Pluto as observed by New Horizons. *Science, 351(6279),* aad8866.

Glein C. R. and Waite J. H. Jr. (2018) Primordial N_2 provides a cosmochemical explanation for the existence of Sputnik Planitia, Pluto. *Icarus, 313,* 79–92.

Hansen C. J. and Paige D. A. (1996) Seasonal nitrogen cycles on Pluto. *Icarus, 120(2),* 247–265.

He C., Hörst S. M., Riemer S., Sebree J. A., Pauley N., and Vuitton V. (2017) Carbon monoxide affecting planetary atmospheric chemistry. *Astrophys. J. Lett., 841(2),* L31.

Hinson D. P., Linscott I. R., Strobel D. F., Tyler G. L., Bird M. K., Pätzold M., and the New Horizons Science Team (2018) An upper limit on Pluto's ionosphere from radio occultation measurements with New Horizons. *Icarus, 307,* 17–24.

Hörst S. M. (2017) Titan's atmosphere and climate. *J. Geophys. Res.–Planets, 122(3),* 432–482.

Hörst S. M. and Tolbert M. A. (2014) The effect of carbon monoxide on planetary haze formation. *Astrophys. J., 781(1),* 53.

Hubbard W. B., Hunten D. M., Dieters S. W., Hill K. M., and Watson R. D. (1988) Occultation evidence for an atmosphere on Pluto. *Nature, 336*(6198), 452.

Imanaka H., Khare B. N., Elsila J. E., et al. (2004) Laboratory experiments of Titan tholin formed in cold plasma at various pressures: Implications for nitrogen-containing polycyclic aromatic compounds in Titan haze. *Icarus, 168(2),* 344–366.

Iro N., Gautier D., Hersant F., Bockelée-Morvan D., and Lunine J. I. (2003) An interpretation of the nitrogen deficiency in comets. *Icarus, 161(2),* 511–532.

Jewitt D. C. (1994) Heat from Pluto. *Astron. J., 107,* 372–378.

Kolokolova L., Hanner M. S., Levasseur-Regourd A. C., and Gustafson B. Å. (2004) Physical properties of cometary dust from light scattering and thermal emission. In *Comets II* (M. C. Festou et al., eds.), pp. 577–604. Univ. of Arizona, Tucson.

Krasnopolsky V. A. (1999) Hydrodynamic flow of N_2 from Pluto. *J. Geophys. Res.–Planets, 104(E3),* 5955–5962.

Krasnopolsky V. A. (2009) A photochemical model of Titan's atmosphere and ionosphere. *Icarus, 201(1),* 226–256.

Krasnopolsky V. A. (2020) A photochemical model of Pluto's atmosphere and ionosphere. *Icarus, 335,* 113374.

Krasnopolsky V. A. and Cruikshank D. P. (1995) Photochemistry of Triton's atmosphere and ionosphere. *J. Geophys. Res.–Planets, 100(E10),* 21271–21286.

Krasnopolsky V. A. and Cruikshank D. P. (1999) Photochemistry of Pluto's atmosphere and ionosphere near perihelion. *J. Geophys. Res.–Planets, 104(E9),* 21979–21996.

Lavvas P. P., Coustenis A., and Vardavas I. M. (2008) Coupling photochemistry with haze formation in Titan's atmosphere, Part I: Model description. *Planet. Space Sci., 56(1),* 27–66.

Lavvas P., Yelle R. V., and Griffith C. A. (2010) Titan's vertical aerosol structure at the Huygens landing site: Constraints on particle size, density, charge, and refractive index. *Icarus, 210(2),* 832–842.

Lavvas P., Sander M., Kraft M., and Imanaka H. (2011) Surface chemistry and particle shape: Processes for the evolution of aerosols in Titan's atmosphere. *Astrophys. J., 728(2),* 80.

Lavvas P. et al. (2013) Aerosol growth in Titan's ionosphere. *Proc. Natl. Acad. Sci. U.S.A., 110(8),* 2729–2734.

Lellouch E., Gurwell M., Butler B., et al. (2017) Detection of CO and HCN in Pluto's atmosphere with ALMA. *Icarus, 286,* 289–307.

Lewis J. S. and Prinn R. G. (1980) Kinetic inhibition of CO and N_2 reduction in the solar nebula. *Astrophys. J., 238,* 357–364.

Liang M. C., Yung Y. L., and Shemansky D. E. (2007) Photolytically

generated aerosols in the mesosphere and thermosphere of Titan. *Astrophys. J. Lett., 661(2),* L199.
Lunine J. I., Atreya S. K., and Pollack J. B. (1989) Present state and chemical evolution of the atmospheres of Titan, Triton, and Pluto. In *Origin and Evolution of Planetary and Satellite Atmospheres* (S. K. Atreya et al., eds.), pp. 605–665. Univ. of Arizona, Tucson.
Luspay-Kut A., Mandt K., Jessup K. L., Kammer J., Hue V., Hamel M., and Filwett R. (2017) Photochemistry on Pluto — I. Hydrocarbons and aerosols. *Mon. Not. R. Astron. Soc., 472*(1), 104–117.
Mandt K. E., Mousis O., Lunine J., and Gautier D. (2014) Protosolar ammonia as the unique source of Titan's nitrogen. *Astrophys. J. Lett., 788(2),* L24.
Mandt K. E., Mousis O., and Luspay-Kuti A. (2016) Isotopic constraints on the source of Pluto' s nitrogen and the history of atmospheric escape. *Planet. Space Sci., 130,* 104–109.
Mandt K., Luspay-Kuti A., Hamel M., Jessup K. L., Hue V., Kammer J., and Filwett R. (2017) Photochemistry on Pluto: Part II HCN and nitrogen isotope fractionation. *Mon. Not. R. Astron. Soc., 472(1),* 118–128.
Miller K. E., Glein C. R., and Waite J. H. Jr. (2019) Contributions from accreted organics to Titan's atmosphere: New insights from cometary and chondritic data. *Astrophys. J., 871(1),* 59.
Millis R. L., Wasserman L. H., Franz O. G., et al. (1993) Pluto's radius and atmosphere: Results from the entire 9 June 1988 occultation data set. *Icarus, 105(2),* 282–297.
Moses J. I., Allen M., and Yung Y. L. (1992) Hydrocarbon nucleation and aerosol formation in Neptune's atmosphere. *Icarus, 99(2),* 318–346.
Mousis O., Lunine J. I., Madhusudhan N., and Johnson T. V. (2012) Nebular water depletion as the cause of Jupiter's low oxygen abundance. *Astrophys. J. Lett., 751(1),* L7.
Mousis O. et al. (2014) Scientific rationale for Saturn's *in situ* exploration. *Planet. Space Sci., 104,* 29–47.
Ney E. P. and Merrill K. M. (1976) Comet West and the scattering function of cometary dust. *Science, 194(4269),* 1051–1053.
Olkin C. B., Young L. A., Borncamp D., et al. (2015) Evidence that Pluto's atmosphere does not collapse from occultations including the 2013 May 04 event. *Icarus, 246,* 220–225.
Owen T. C., Roush T. L., Cruikshank D. P., et al. (1993) Surface ices and the atmospheric composition of Pluto. *Science, 261(5122),* 745–748.
Plessis S., McDougall D., Mandt K., Greathouse T., and Luspay-Kuti A. (2015) Uncertainty for calculating transport on Titan: A probabilistic description of bimolecular diffusion parameters. *Planet. Space Sci., 117,* 377–384.
Poppe A. R. and Horányi M. (2018) Interplanetary dust delivery of water to the atmospheres of Pluto and Triton. *Astron. Astrophys., 617,* L5.
Rannou P. and Durry G. (2009) Extinction layer detected by the 2003 star occultation on Pluto. *J. Geophys. Res.–Planets, 114(E11).*
Rannou P. and West R. (2018) Supersaturation on Pluto and elsewhere. *Icarus, 312,* 36–44.
Rodrigo R. and Lara L. M. (2002) Photochemistry of planetary atmospheres. In *The Evolving Sun and Its Influence on Planetary Environments* (B. Montesinos et al., eds.), p. 133. ASP Conf. Ser. 269, Astronomical Society of the Pacific, San Francisco.
Rubin M. et al. (2015) Molecular nitrogen in comet 67P/Churyumov-Gerasimenko indicates a low formation temperature. *Science, 348(6231),* 232–235.
Rubin M. et al. (2018) Krypton isotopes and noble gas abundances in the coma of comet 67P/Churyumov-Gerasimenko. *Sci. Adv., 4(7),* eaar6297.
Rubin M. et al. (2019) Elemental and molecular abundances in comet 67P/Churyumov-Gerasimenko. *Mon. Not. R. Astron. Soc., 489(1),* 594–607.
Smith B. A. et al. (1989) Voyager 2 at Neptune: Imaging science results. *Science, 246(4936),* 1422–1449.
Stansberry J. A., Lunine J. I., and Tomasko M. G. (1989) Upper limits on possible photochemical hazes on Pluto. *Geophys. Res. Lett., 16(11),* 1221–1224.
Stern S. A. and Trafton L. (1984) Constraints on bulk composition, seasonal variation, and global dynamics of Pluto's atmosphere. *Icarus, 57(2),* 231–240.
Stern S. A., Weintraub D. A., and Festou M. C. (1993) Evidence for a low surface temperature on Pluto from millimeter-wave thermal emission measurements. *Science, 261(5129),* 1713–1716.
Stern S. A., Bagenal F., Ennico K., et al. (2015) The Pluto system: Initial results from its exploration by New Horizons. *Science, 350(6258),* aad1815.
Strobel D. F. (1983) Photochemistry of the reducing atmospheres of Jupiter, Saturn, and Titan. *Intl. Rev. Phys. Chem., 3(2),* 145–176.
Strobel D. F. (2008) N_2 escape rates from Pluto's atmosphere. *Icarus, 193(2),* 612–619.
Strobel D. F. (2009) Titan's hydrodynamically escaping atmosphere: Escape rates and the structure of the exobase region. *Icarus, 202*(2), 632-641.
Strobel D. F. and Zhu X. (2017) Comparative planetary nitrogen atmospheres: Density and thermal structures of Pluto and Triton. *Icarus, 291,* 55–64.
Summers M. E., Strobel D. F., and Gladstone G. R. (1997) Chemical models of Pluto's atmosphere. In *Pluto and Charon* (S. A. Stern and D. J. Tholen, eds.), pp. 391–434. Univ. of Arizona, Tucson.
Tian F. and Toon O. B. (2005) Hydrodynamic escape of nitrogen from Pluto. *Geophys. Res. Lett., 32(18).*
Trafton L. (1980) Does Pluto have a substantial atmosphere? *Icarus, 44(1),* 53–61.
Trafton L. (1990) A two-component volatile atmosphere for Pluto. I — The bulk hydrodynamic escape regime. *Astrophys. J., 359,* 512–523.
Trainer M. G., Pavlov A. A., Jimenez J. L., McKay C. P., Worsnop D. R., Toon O. B., and Tolbert M. A. (2004) Chemical composition of Titan's haze: Are PAHs present? *Geophys. Res. Lett., 31(17).*
Trainer M. G., Sebree J. A., Yoon Y. H., and Tolbert M. A. (2013) The influence of benzene as a trace reactant in Titan aerosol analogs. *Astrophys. J. Lett., 766(1),* L4.
Tucker O. J., Erwin J. T., Deighan J. I., Volkov A. N., and Johnson R. E. (2012) Thermally driven escape from Pluto's atmosphere: A combined fluid/kinetic model. *Icarus, 217(1),* 408–415.
Vuitton V., Yelle R. V., and Anicich V. G. (2006) The nitrogen chemistry of Titan's upper atmosphere revealed. *Astrophys. J. Lett., 647(2),* L175.
Wahlund J. E., Galand M., Müller-Wodarg I., et al. (2009) On the amount of heavy molecular ions in Titan's ionosphere. *Planet. Space Sci., 57(14–15),* 1857–1865.
Waite J. H., Young D. T., Cravens T. E., Coates A. J., Crary F. J., Magee B., and Westlake J. (2007) The process of tholin formation in Titan's upper atmosphere. *Science, 316(5826),* 870–875.
West R. A. and Smith P. H. (1991) Evidence for aggregate particles in the atmospheres of Titan and Jupiter. *Icarus, 90(2),* 330–333.
Willacy K., Allen M., and Yung Y. (2016) A new astrobiological model of the atmosphere of Titan. *Astrophys. J., 829(2),* 79.
Wilson E. H. and Atreya S. K. (2004) Current state of modeling the photochemistry of Titan's mutually dependent atmosphere and ionosphere. *J. Geophys. Res.–Planets, 109(E6).*
Wong M. L., Fan S., Gao P., et al. (2017) The photochemistry of Pluto's atmosphere as illuminated by New Horizons. *Icarus, 287,* 110–115.
Yelle R. V. (1991) Non-LTE models of Titan's upper atmosphere. *Astrophys. J., 383,* 380–400.
Yelle R. V. and Lunine J. I. (1989) Evidence for a molecule heavier than methane in the atmosphere of Pluto. *Nature, 339(6222),* 288.
Yoon Y. H., Hörst S. M., Hicks R. K., Li R., de Gouw J. A., and Tolbert M. A. (2014) The role of benzene photolysis in Titan haze formation. *Icarus, 233,* 233–241.
Young E. F., French R. G., Young L. A., et al. (2008) Vertical structure in Pluto's atmosphere from the 2006 June 12 stellar occultation. *Astron. J., 136(5),* 1757.
Young L. A. (2013) Pluto's seasons: New predictions for New Horizons. *Astrophys. J. Lett., 766(2),* L22.
Young L. A., Kammer J. A., Steffl A. J., et al. (2018) Structure and composition of Pluto's atmosphere from the New Horizons solar ultraviolet occultation. *Icarus, 300,* 174–199.
Yung Y. L., Allen M., and Pinto J. P. (1984) Photochemistry of the atmosphere of Titan: Comparison between model and observations. *Astrophys. J. Suppl., 55(3),* 465–506.
Zhang X., Strobel D. F., and Imanaka H. (2017) Haze heats Pluto's atmosphere yet explains its cold temperature. *Nature, 551(7680),* 352.
Zhu X., Strobel D. F., and Erwin J. T. (2014) The density and thermal structure of Pluto's atmosphere and associated escape processes and rates. *Icarus, 228,* 301–314.

Forget F., Bertrand T., Hinson D., and Toigo A. (2021) Dynamics of Pluto's atmosphere. In *The Pluto System After New Horizons* (S. A. Stern, J. M. Moore, W. M. Grundy, L. A. Young, and R. P. Binzel, eds.), pp. 297–319. Univ. of Arizona, Tucson, DOI: 10.2458/azu_uapress_9780816540945-ch013.

Dynamics of Pluto's Atmosphere

François Forget
Laboratoire de Météorologie Dynamique, Centre National de la Recherche Scientifique

Tanguy Bertrand
NASA Ames Research Center

David Hinson
SETI Institute

Anthony Toigo
The Johns Hopkins University Applied Physics Laboratory

Pluto provides an exceptional natural laboratory to study atmospheric dynamics. It is an intermediate case between fast rotators like Earth and Mars and slow rotators like Venus and Titan. It is also uniquely characterized by a near-surface stratosphere, long radiative timescales and strong N_2 condensation flows. Yet, not many observations are available. Numerical climate models are therefore used to explore the atmospheric circulation. Near the surface, local downslope katabatic winds are superimposed on the flows from the subliming ice to the condensing deposits, with a Coriolis turning of the wind. Different general circulation regimes are possible, including superrotation. However, simulations performed with surface conditions most consistent with New Horizons observations suggest that the most likely regime is a global "retrorotation" forced by the conservation of angular momentum in the condensation-sublimation flows. Tidal and gravity waves are also present, creating fluctuations in the observed density profiles and the spectacular layering in the haze.

1. INTRODUCTION

The circulation of Pluto's atmosphere is of great interest for at least two reasons. First, it plays a key role in the various processes that have sculpted the surface and its volatile reservoirs, as discussed in several other chapters of this book. The nitrogen, methane, and carbon monoxide cycles, the evolution of their spectacular surface reservoirs, the distribution of organic material on the surface, and the possibility of dunes, wind streaks, and volcanic deposits all directly depend on the dynamics of Pluto's atmosphere at the diurnal, seasonal, and astronomical timescales. Second, Pluto is one of only 10 solar system bodies with an appreciable atmosphere, and only 6 of these are terrestrial (Earth, Venus, Mars, Titan, Triton, Pluto). With its unique characteristics, Pluto provides an invaluable natural laboratory to test theories, concepts, and tools that have been developed to understand the dynamics of planetary atmospheres in the past few decades.

Pluto's rotation period (6.39 Earth days) is intermediate between the fast rotators like Earth and Mars (where baroclinic waves and jet streams play a key role) and the slow rotators like Venus and Titan (characterized by the superrotation of the atmosphere). However, Pluto does not fit easily in such a classification. It is far from the typical "terrestrial atmosphere" circulation regimes that are explored in theoretical studies in which one varies the rotation rate, size, or even obliquity of a transparent, Earth-like atmosphere (*Williams,* 1988; *Kaspi and Showman,* 2015; *Wang et al.,* 2018). Pluto's atmosphere is distinctive in its long radiative timescale (see Table 1), its unique near surface stratosphere-like thermal structure, and the key role played by the condensation-sublimation flows of nitrogen. Pluto's atmosphere is rarefied, 6000× less massive than Earth's atmosphere per unit area, yielding a surface pressure is only 1 Pa. However, as the atmosphere remains in the continuum regime, the primitive equations of meteorology remain valid (down to pressure levels several orders of magnitude lower than 1 Pa). Nevertheless, unlike in a 1-bar atmosphere, processes like molecular conduction and viscosity are relevant throughout Pluto's atmosphere. Similarly, while Pluto's surface gravity is 16× less than on Earth, this has no effect on the circulation except for a vertical rescaling (*Thomson and Vallis,* 2019).

In spite of the wealth of new data provided by New Horizons and by complementary state-of-the-art telescopic observations, many key observations required to fully characterize the atmospheric dynamics (winds, temperature gradients) are not yet available. Pluto's atmospheric dynamic studies therefore heavily rely on theoretical and numerical

TABLE 1. Properties of Pluto and its atmosphere.

Symbol	Value	Description	Reference
r_p	1188.3 ± 1.6 km	Global mean radius	*Nimmo et al.* (2017)
GM	869.6 ± 1.8 km^3 s^{-2}	Gravitational parameter	*Brozović et al.* (2015)
g_o	0.616 m s^{-2}	Surface gravity	
P	6.3872 days	Rotation period	*Tholen et al.* (2008)
Ω	1.1386 × 10^{-5} s^{-1}	Rotation rate	
β_o	1.9 × 10^{-11} m^{-1} s^{-1}	Meridional gradient of Coriolis parameter at equator, $2\Omega/r_p$	
R	296.8 J kg^{-1} K^{-1}	Gas constant (N_2)	
c_p	1039 J kg^{-1} K^{-1}	Specific heat at constant pressure	
g_o/c_p	0.59 K km^{-1}	Adiabatic lapse rate at surface	
τ_g	~10 Earth years	Radiative timescale (gas alone)	*Zhang et al.* (2017)
τ_h	~1 Earth year	Radiative timescale (including haze)	*Zhang et al.* (2017)
	Properties of the Lower Atmosphere in Sputnik Planitia		
p_s	1.28 ± 0.07 Pa	Surface pressure	*Hinson et al.* (2017)
T_{sat}	37.1 K	Saturation temperature of N_2 ice	*Fray and Schmitt* (2009)
H	18 km	Scale height near surface ($=RT_{sat}/g_0$: lowest value on Pluto)	
D	3 km	Depth of basin	*McKinnon et al.* (2016)
f_c	1.14 × 10^{-5} s^{-1}	Coriolis parameter at 30°N	
R_d	4000 km	Rossby deformation radius, $(g_oD)^{1/2}/f_c$	
	Properties of the Atmosphere at 100 km Altitude		
T	100 K	Temperature	*Dias-Oliveira et al.* (2015)
dT/dz	–0.2 K km^{-1}	Vertical gradient of T	*Dias-Oliveira et al.* (2015)
H	57 km	Scale height at 100 km	
N	1.3 × 10^{-3} s^{-1}	Buoyancy frequency	

All uncertainties are 1σ.

model studies, notably using global climate models. Yet what we have observed on Pluto is of unique interest and often enigmatic. We must decode these rich datasets to make the most of Pluto as a new natural laboratory for testing our understanding of planetary atmospheres.

In section 2, we review the different sources of information that are available to understand Pluto's atmospheric dynamics, including theories, observations, and numerical models. In section 3 we present theoretical models of the near-surface winds, and we discuss different scenarios for the general circulation at the time of the New Horizons encounter in section 4. In section 5 we discuss gravity waves and tides, which were detected in some observations. Section 6 comments upon variations of the atmospheric circulation over seasonal and astronomical timescales and section 7 discusses the influence of the atmospheric circulation on Pluto's surface ices. We conclude with a summary and a look toward future research and observations.

2. HOW CAN WE STUDY THE DYNAMICS OF PLUTO'S ATMOSPHERE?

2.1. Initial Investigation of Pluto's Atmosphere and Its Dynamics

Pluto's discovery led to speculation about the existence of an atmosphere (*Kuiper,* 1950), but it is only in 1976 that a volatile — methane ice — was first detected on Pluto's surface (*Cruikshank et al.,* 1976). The presence of an appreciable atmosphere was proven by a 1988 stellar occultation (*Hubbard et al.,* 1988; *Elliot et al.,* 1989) (see section 2.2.1 below), while CO and N_2 ices were detected from UKIRT observations by *Owen et al.* (1993). As N_2 is the most volatile of the three ices, this implied that Pluto's atmosphere is dominated by N_2, and that CH_4 and CO are only minor constituents. Gaseous CH_4 was directly observed in its 1.66-μm band by *Young et al.* (1997) and CO near 2.25 μm by *Lellouch et al.* (2011).

Within that context, limited research on Pluto's atmospheric dynamics was conducted before 2010. A first realistic study assuming an N_2-dominated atmosphere (among other possibilities) was presented by *Stern and Trafton* (1984). Similar to the discussion in section 2.1, they tried to estimate whether Pluto has an "axially symmetric circulation" regime like Titan or a "wave circulation regime" in which heat transfer from the sunlit latitude to the colder regions would be achieved by baroclinic waves, like Earth. Their non-dimensional similarity parameter analysis showed that the regime depended on certain assumptions, although in the case of an N_2-dominated atmosphere they concluded that the wave circulation regime was likely. It now seems that they underestimated the "Rossby radius" (see section 2.2.3). In fact, Pluto's atmospheric dynamics regime is subtle in this regard, because it can be considered intermediate between "fast rotators," like Earth and Mars, and "slow rotators," like Venus and Titan. This condition was also noted by *Del Genio and Zhou* (1996) in their reference study on superrotation. They wrote that Pluto "is of considerable interest" because its atmosphere is in "the transition between the baroclinic and quasi-barotropic regimes, somewhere between Earth and Titan." Based on estimations of the Rossby number, they could not conclude whether Pluto's atmosphere was likely to superrotate like Titan and Venus. As discussed in section 4, this still remains a possibility. Additionally, the atmospheric circulation on Pluto is strongly influenced by the N_2 condensation and sublimation flows (see section 2.2.2). This point was actually the focus of the 1997 review chapter on volatile transport, seasonal cycles, and atmospheric dynamics by *Spencer et al.* (1997).

2.2. What Drives the Circulation?

The general circulation of any planetary atmosphere is fundamentally driven by horizontal gradients of pressure (as measured at a given altitude or, more precisely, geopotential). In the absence of major composition variations, such gradients are primarily induced by temperature gradients resulting from differential solar heating (the atmosphere expands where it is warmer) and, in the case of Pluto, by condensation-sublimation processes at the surface. The resulting circulation is in turn modified by the rotation of the planet (centrifugal and Coriolis forces; see Table 1), by the convergence and divergence of mass, and by the different dynamical waves induced by local or regional perturbations, which can, under some conditions, influence the zonal mean circulation through "wave-mean flow" interaction.

2.2.1. Atmospheric temperature gradients. Both surface heating and direct absorption of sunlight can generate temperature gradients and dynamically force the atmosphere. Compared to Earth and Mars, the thermal flux between the surface and the atmosphere is relatively small on Pluto, because the atmosphere is largely transparent in the thermal infrared, and because the sensible heat flux is limited by the static stability of the atmosphere adjacent to the ground. That leaves direct absorption of solar radiation by methane [in the near-infrared (*Strobel et al.,* 1996)] and by haze particles [at all wavelengths (*Zhang et al.,* 2017)] as the primary source of atmospheric heating on Pluto.

However, horizontal temperature gradients are expected to remain very small in Pluto's atmosphere because of its slow rotation, its small size, and the long radiative timescale of the atmosphere.

On rapidly rotating planets, large meridional gradients in the zonal-mean temperature field can persist in a state of geostrophic balance in which the Coriolis force approximately balances the pressure gradients in the horizontal momentum equation. With its rotation period of 6.39 Earth days, Pluto is a relatively slow rotator on which the Coriolis force is weaker than on Earth or Mars, but not as small as on Titan and especially Venus. However, theoretical considerations as well as laboratory and numerical experiments show that the relevant parameter to estimate the role of rotation in the building of meridional temperature gradients is the product of the rotation rate Ω and the planetary radius r_p [the key figure used for instance in the Thermal Rossby number — see, e.g., *Read* (2011) — defined by $R\Delta\theta_h/(\Omega r_p)^2$ with R the specific gas constant and $\Delta\theta_h$ a proxy of the radiative-convective equilibrium horizontal temperature contrast]. The parameter Ωr_p is 34× smaller for Pluto than for Earth. Numerical simulations of the atmosphere of an Earth-like planet with a rotation rate decreased by that factor (e.g., *Williams,* 1988; *Kaspi and Showman,* 2015; *Wang et al.,* 2018) show that in such conditions the effect of rotation is too weak to maintain geostrophic balance. Any horizontal gradients in the zonal-mean temperature field are efficiently removed by atmospheric motion and meridional temperature gradients are consequently very small, except near the poles.

In addition, compared to Earth or Mars, the radiative timescale τ_R of the atmosphere is long: (1–10 Earth years depending on the effect of the haze; see Table 1). This is because the atmosphere is a very weak emitter in the thermal infrared (*Strobel et al.,* 1996; *Forget et al.,* 2017; *Zhang et al.,* 2017). How efficiently the heat is spatially redistributed can be estimated by comparing this radiative timescale with a dynamical timescale (see, e.g., *Selsis et al.,* 2011) defined by $\tau_D = r_p/V$ where r_p is the planetary radius and V is the average of the wind speed. Even assuming a conservative $V = 0.1$ m s^{-1} (see section 4) and $\tau_R = 1$ Earth year, this still yields $\tau_R/\tau_D > 2.5$, suggesting that solar heating is efficiently redistributed by the circulation.

For these reasons, on Pluto one can estimate that the temperature gradients remain very small except in the first few kilometers above the surface where differences in surface temperatures and nitrogen fluxes can locally influence the atmospheric temperatures.

2.2.2. Condensation-sublimation processes. When volatile species condense or sublime at the surface, they cause a local deficit or excess of mass that creates a pressure gradient (condensation within the atmosphere has the same effect but is negligible on Pluto, as discussed below). Among the three major species that can condense on Pluto (N_2, CH_4, CO), only nitrogen (N_2) has a significant effect because it is by far the dominant constituent of the atmosphere at all times. Once again, the induced pressure gradients are very rapidly compensated by the atmospheric circulation and remain very small as demonstrated by numerical simulation of the process (*Toigo et al.,* 2015; *Forget et al.,* 2017).

Condensation-sublimation flows are negligible in all solar system atmospheres except on Mars and perhaps Triton (whose atmosphere is poorly understood) and Io (whose atmosphere is orders of magnitude thinner than Pluto's). The martian atmosphere is composed of 95% CO_2, which seasonally condenses in the polar regions down to about 50° latitude in fall, winter, and spring in each hemisphere. About 30% of the atmosphere is converted seasonally into polar ice, resulting in large seasonal pressure variations and a significant condensation-sublimation flow. But this flow remains much smaller than the thermally induced circulation: Mars global circulation models can ignore this flow entirely and still capture the basic aspects of the atmospheric circulation (e.g., *Haberle et al.,* 1993, 1997).

The situation on Pluto is much different: The condensation-sublimation flux is the major driver of the atmospheric circulation (*Toigo et al.,* 2015; *Forget et al.,* 2017; *Bertrand et al.,* 2020a). This is not due to the fact that the atmospheric forcing from sublimation and condensation is stronger on Pluto than on Mars. Indeed, the ratio of the condensation-sublimation flux (in kilograms per second of CO_2 and N_2, respectively) to the total mass of the atmosphere (in kilograms) is about the same on both planets, near 2×10^{-8} s^{-1} [estimation based on the global climate simulations of Mars' CO_2 cycle from *Forget et al.* (1998), and of Pluto's N_2 cycle in 2015 from *Bertrand et al.* (2020a)]. However, as discussed above, the thermally induced circulation is weaker on Pluto than on Mars. The dominance of the condensation flow and the uncertainties in the distribution and seasonal evolution of surface N_2 ice reservoirs means that the condensation flow also presents the greatest uncertainty in modeling and predicting atmospheric dynamical behavior. This uncertainty results mainly from the lack of observations in the southern polar night, the unknown properties of the observed N_2 ice deposits outside Sputnik Planitia (are they seasonal or perennial?), and the very high sensitivity of the modeled nitrogen cycle to the assumed ground and ice properties (see the chapters by Young et al. and Umurhan et al. in this volume). As detailed below, this prevents us from determining the general circulation with greater confidence.

2.2.3. Atmospheric waves. As on any planet, Pluto's atmosphere is a medium through which any local, regional, or planetary-scale perturbation will propagate. Atmospheric waves can influence the circulation of Pluto's atmosphere by transporting energy and momentum from the region where they are excited to the region where they dissipate. The character of these waves is constrained by the size and rotation rate of Pluto. Rotation effects become important when the scale of the motion exceeds the Rossby radius, which can be calculated in several ways. The value listed in Table 1 (4000 km) is a lower limit based on an assumed vertical scale D of only 3 km (much less than the scale height H, which exceeds 18 km). Even with this conservative estimate the Rossby radius is more than half the circumference of Pluto. Hence, wave motion on Pluto is not strongly influenced by rotation.

As discussed later in this chapter, observations and numerical simulations suggest that several types of waves are present in Pluto's atmosphere. The excitation comes from several sources: wind flowing over topography, the diurnal cycle of sublimation and condensation of N_2, and (possibly) barotropic instability. Owing to the large Rossby radius, this forcing is likely to excite inertia-gravity waves (including tides), whereas Rossby waves, which play a central role in atmospheric dynamics on Earth and Mars, are unlikely to be present on Pluto.

2.2.4. The gravitational influence of Charon. One can wonder if the tidal forces resulting from the presence of Charon could influence the circulation of the atmosphere of Pluto, in spite of the synchronous rotation of the system. A similar problem has been addressed in the case of Titan, which is locked in synchronous rotation about Saturn. *Tokano and Neubauer* (2002) showed that gravitational tides should have a large impact on the dynamic meteorology down to the surface. However, the effect results from the eccentricity of Titan's orbit around Saturn, which is significant (0.0292), whereas the mutual orbits of Pluto and Charon around their barycenter are almost circular. Therefore, this effect is negligible on Pluto (*Vangvichith,* 2013).

2.3. Clues from Observations

Direct observations of atmospheric dynamics remain very limited for Pluto. Global wind measurements would be ideal to determine the global circulation, but unfortunately they are very difficult to obtain, even for more easily observed worlds like Mars. Instead, the most must be made of the available datasets, which provide indirect clues on wind processes and atmospheric waves as well as constraints for numerical simulations of Pluto's atmospheric dynamics.

2.3.1. Stellar occultations. As mentioned above, Pluto's atmosphere was discovered through stellar occultation measurements. In fact, the first direct detection of atmospheric gases was obtained through an occultation in 1985 and reported by *Brosch et al.* (1995). However, a subsequent occultation event in 1988 yielded the first published results (*Hubbard et al.,* 1988; *Elliot et al.,* 1989, 2003; *Millis et al.,* 1993). Another stellar occultation event was observed in 2002 (*Sicardy et al.,* 2003; *Pasachoff et al.,* 2005), and then occultation events became more frequent as Pluto fortuitously passed between Earth and the galactic plane, providing a denser background of stars to act as occultation light sources (*Young et al.,* 2008). Subsequent stellar occultation events were observed in 2006 (*Elliot et al.,* 2007; *Young et al.,* 2008), 2007 (*McCarthy et al.,* 2008; *Person et al.,* 2008; *Olkin et al,* 2014), 2008 (*Gulbis et al.,* 2011), 2011 (*Person et al.,* 2013; *Gulbis et al.,* 2015), 2012–2013 (*Dias-Oliveira et al.,* 2015; *Olkin et al.,* 2015; *Bosh et al.,* 2015), 2014 (*Pasachoff et al.,* 2016), 2015 (*Sicardy et al.,* 2016; *Pasachoff et al.,* 2017), and 2016 (*Meza et al.,* 2019).

Groundbased stellar occultation measurements provide information on how the light fluxes from a star decrease as a function of radial distance from the center of the planet as the planet moves in front of the star. The decrease in flux is primarily a consequence of refraction, which depends on the atmospheric composition and the vertical profile of number density. Both model fitting and numerical inversion have been used to analyze and interpret stellar occultation light curves. The results range from simple estimates for the atmospheric scale height (hence measuring the ratio of temperature to mean molecular weight) to profiles of refractivity, number density, pressure, and temperature. Inversion of the light curve yields a refractivity profile, number density is obtained by assuming a composition of pure N_2, pressure is obtained by assuming hydrostatic balance, and temperature is obtained from the ideal gas law. Data recorded by large telescopes under optimum viewing conditions are sensitive to atmospheric structure at altitudes as high as about 400 km and as low as about 5 km above the surface (*Meza et al.,* 2019).

Profiles of temperature retrieved from stellar occultations (see Fig. 1 in the chapter by Summers et al. in this volume) show that Pluto's atmosphere is layered, with a middle atmosphere (between 30 and ~400 km altitude) that has a base temperature of around 100–110 K with a weak lapse of temperature with height above 30 km. Below that level, the atmosphere cools at lower altitudes to about 40–50 K near the surface, which is close to the condensation temperature of N_2 at the surface pressure. Occultation observations further indicate the presence of oscillations in the vertical temperature profile, and *Sicardy et al.* (2003), *Pasachoff et al.* (2005), *McCarthy et al.* (2008), *Person et al.* (2008), and *Hubbard et al.* (2009) have proposed that the observed oscillations could be due to atmospheric waves. The fact that the different occultation observations of these oscillations sampled different locations on the planet implies that the waves are global in scale, at least at the terminator local times of day. This is further discussed in sections 5.2 and 5.3.

The lowest several kilometers of Pluto's atmosphere are not currently accessible to Earth-based stellar occultation measurements, but the surface pressure can be calculated through downward extrapolation. For example, *Lellouch et al.* (2009) considered a range of models for the lower atmosphere and constrained the surface pressure to be 0.7–2.4 Pa. In addition, stellar occultation measurements have revealed a remarkable threefold increase in the total mass of the atmosphere between 1988 and 2016 (e.g., *Meza et al.,* 2019). Some of these results were used to constrain thermal models of Pluto's volatile surface-atmosphere exchange system (e.g., *Young,* 2013; *Hansen et al.,* 2015).

Finally, stellar occultations can be used to search for any deformation of the atmosphere caused by centrifugal forces acting on the zonal winds. This can be done by comparing occultation profiles at various latitudes — or by modeling the "central flash" that occurs when the observer is close to the center of Pluto's occultation shadow. This type of centrifugal deformation was considered by *Elliot et al.* (1997) as a possible interpretation for a large atmospheric asymmetry observed in a stellar occultation at Triton, but their conclusions have not been confirmed. No significant departure from spherical symmetry has been observed on Pluto, as reflected by the remarkable degree of consistency among light curves observed at many different locations (e.g., *Dias-Oliveira et al.,* 2015).

2.3.2. Radio science from New Horizons. On July 14, 2015, New Horizons performed an uplink radio occultation of Pluto's neutral atmosphere using the Radio Science instrument (REX) (*Tyler et al.,* 2008; *Hinson et al.,* 2017). The entry occultation (193.5°E, 17.0°S) sounded the atmosphere at sunset [16:31 Local Time (LT)] in southeast Sputnik Planitia, an enormous topographic basin that contains kilometer-deep deposits of N_2 ice (*Grundy et al.,* 2016; *McKinnon et al.,* 2016; *Moore et al.,* 2016). The exit occultation (15.7°E, 15.1°N) occurred on the opposite side of Pluto at sunrise (4:42 LT) near the center of the Charon-facing hemisphere. This section is a brief summary of REX results that pertain to atmospheric dynamics; see *Hinson et al.* (2017) for a more detailed discussion.

The local radius of Pluto as determined from the REX data is 1187.4 ± 3.6 km at entry and 1192.4 ± 3.6 km at exit. The dominant error source is the uncertainty in spacecraft position, which can be removed through averaging; the mean radius is 1189.9 ± 0.2 km, in agreement with the global average 1188.3 ± 1.6 km derived by *Nimmo et al.* (2017) from images taken by New Horizons.

The atmospheric profiles retrieved from the REX data extend downward to within 1 km of the surface (see Tables 5 and 6 from *Hinson et al.*, 2017), and yield essentially direct measurements of surface pressure on Pluto. There is a significant difference between the results at entry (1.28 ± 0.07 Pa) and exit (1.02 ± 0.07 Pa) owing to the 5-km difference in radius at the two locations. The best pressure reference is the average of the results at entry and exit: 1.15 ± 0.07 Pa at 1189.9 ± 0.2 km.

At altitudes above 25 km, the REX profiles are consistent with Earth-based stellar occultation measurements (e.g., *Meza et al.*, 2019). At lower altitudes the REX profiles reveal horizontal variations in atmospheric structure that had not been measured previously, as shown in Fig. 1. The entry profile contains a cold, dense layer in the lowest 3 km above the ground; its mean temperature (38.9 ± 2.1 K) is close to the condensation temperature of N_2. The temperature gradient at the base of the entry (sunset) profile is –0.5 ± 0.7 K km^{-1}, as compared with a dry adiabat of –0.6 K km^{-1}. The cold layer is capped by a strong inversion with a peak gradient of +9.5 ± 1.2 K km^{-1}.

Unlike the entry (sunset) profile, the exit (sunrise) profile has a much weaker inversion (+4.7 ± 0.9 K km^{-1}) that extends to the surface with no sign of a cold boundary layer. The temperature 1 km above the ground at exit (57.0 ± 3.7 K) is more than 10 K warmer than at entry, which suggests that the local surface there is devoid of N_2 ice.

The difference between the two REX profiles, particularly the presence of a cold boundary layer at occultation entry in Sputnik Planitia, is a valuable constraint on the atmospheric dynamics. *Forget et al.* (2017) were able to reproduce the main features of the profiles in their global climate model (see section 2.3) and suggested that the cold layer forms in response to the combined effects of N_2 daytime sublimation and topographic confinement. Moreover, subsequent simulations by *Bertrand* (2017) confirmed that atmospheric winds play an important role in the formation of the cold boundary layer by transporting freshly sublimed N_2 from northern Sputnik Planitia (much closer to the subsolar latitude) to the measurement location in southeast Sputnik Planitia, where local condensation is occurring.

2.3.3. Alice ultraviolet observations from New Horizons. The Alice UV spectrometer on New Horizons obtained two atmospheric profiles (at locations near the REX profiles described above) by observing how the absorption by molecular species and extinction by haze particles vary with altitude as the Sun passes behind the atmosphere as seen from the spacecraft (*Gladstone et al.*, 2016; *Young et al.*, 2018). A few hours later, Alice observed occultations by two bright stars, providing data over different areas of Pluto from those probed by the solar occultation (*Kammer et al.*, 2020).

The observations were able to measure the vertical density profiles of CH_4, C_2H_2, C_2H_4, and C_2H_6 and also served to constrain haze properties. Pressure and temperature could be derived from the density of N_2 retrieved from the solar occultations but only in the 900–1100-km altitude range. The Alice observations are discussed in more detail elsewhere in this book (see the chapter by Summers et al. in this volume). With regard to atmospheric dynamics, Alice provided several interesting results. First, the measurements of temperature and density in the upper atmosphere complement the stellar occultations and REX lower atmosphere datasets. Second, the vertical distribution of methane and other species suggests that both eddy mixing and mixing by the atmospheric circulation are limited, resulting in a homopause below 12 km, above which molecular diffusion seems to dominate turbulent mixing (*Young et al.*, 2018). The corresponding pressure level is about 0.7 Pa, whereas the homopause is usually around 10^{-4} Pa on terrestrial planets like Earth, Mars, and Venus. This is probably a consequence of the very high static stability of the atmosphere, notably below 20 km where Pluto's atmosphere can be considered as a near-surface stratosphere with air temperatures strongly

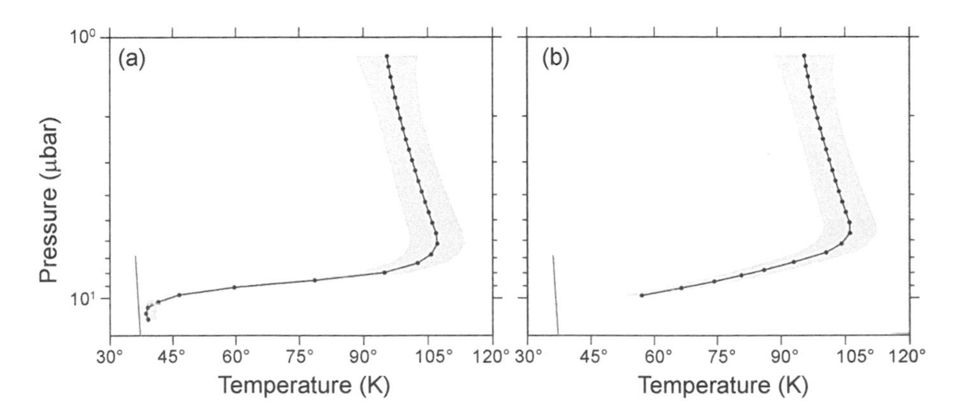

Fig. 1. Temperature profiles of the lower atmosphere at **(a)** entry/sunset and **(b)** exit/sunrise (black), as derived from REX radio occultation measurements. Gray shading denotes the standard deviation of temperature. The base of each profile is 1 km above the local surface. The blue line on the left is the condensation temperature of N_2 (*Fray and Schmitt*, 2009). The cold boundary layer in the entry profile arises from atmospheric dynamics within Sputnik Planitia (see text). From *Hinson et al.* (2017).

increasing with altitude. Finally, the comparison of profiles from different locations indicates that the composition of the upper atmosphere is horizontally homogeneous, at least below 40° latitude (*Kammer et al.,* 2020).

2.3.4. Atmospheric and surface imaging by New Horizons. Images captured by New Horizons' LORRI and RALPH instruments also provided clues of dynamical activity in Pluto's atmosphere. They revealed an extensive circumplanetary atmospheric haze with tens of embedded layers (*Gladstone et al.,* 2016). This reinforced the idea that orographic or tidal atmospheric waves are at play on Pluto, as suggested before the flyby by Earth-based stellar occultation measurements (*Person et al.,* 2008; *McCarthy et al.,* 2008; *Hubbard et al.,* 2009; *Toigo et al.,* 2010; *Gladstone et al.,* 2016). These atmospheric waves are discussed in more detail in section 5.

Several images taken by LORRI at high phase angle (near the terminator) showed small, bright, low-altitude features resembling discrete clouds or ground fogs (*Stern et al.,* 2017). However, it was not possible to confirm the presence of clouds at altitude with the available observations, and these features may also be confused with haze or bright patches of surface ice.

New Horizons observations of Pluto's surface also revealed geological features whose formation could be related to the atmosphere circulation. For instance, New Horizons observed dark wind streaks (Fig. 2a) located on the western side of Sputnik Planitia between 15°N and 25°N (*Stern et al.,* 2015) and oriented northwest-southeast (153 ± 10°). The streaks form an elongated albedo contrast with the surrounding N_2-rich ice plains, slightly darkening the ice. Similar streaks have been observed on Mars (*Thomas et al.,* 1981; *Greeley et al.,* 1993; *Geissler,* 2005). On Pluto the streaks appear to stem from isolated water ice blocks. Modification of wind flow by these topographic obstacles is interpreted to be the cause of the surface albedo contrast (*Stern et al.,* 2015). Two separate wind streaks with different orientations sometimes stem from a single block, which could reflect recent circulation changes. In addition, the observation of regularly spaced, linear morphological features on the western side of Sputnik Planitia (Fig. 2b) has been interpreted as icy dunes (*Stern et al.,* 2015), which could have formed in the recent geological past from the deposition of sand-sized particles of CH_4-rich ice (*Telfer et al.,* 2018).

Finally, longitudinal asymmetries in ice distribution, composition, texture, or color observed on Pluto's surface could reflect the effect of regional dynamics causing these differences. For instance, Cthulhu Regio, the volatile-free region west of Sputnik Planitia, is thought to be covered by a dark mantle of organic materials (tholins), whereas the regions east of Sputnik Planitia are covered by N_2-rich and CH_4-rich ices (see the chapter by Cruikshank et al. in this volume), including the bladed terrain deposits, which extend between 210°E and 40°E (*Moore et al.,* 2018). The preferential condensation of methane frost on mountain tops may also result from atmospheric dynamical processes (*Bertrand et al.,* 2020b). Another example is also found

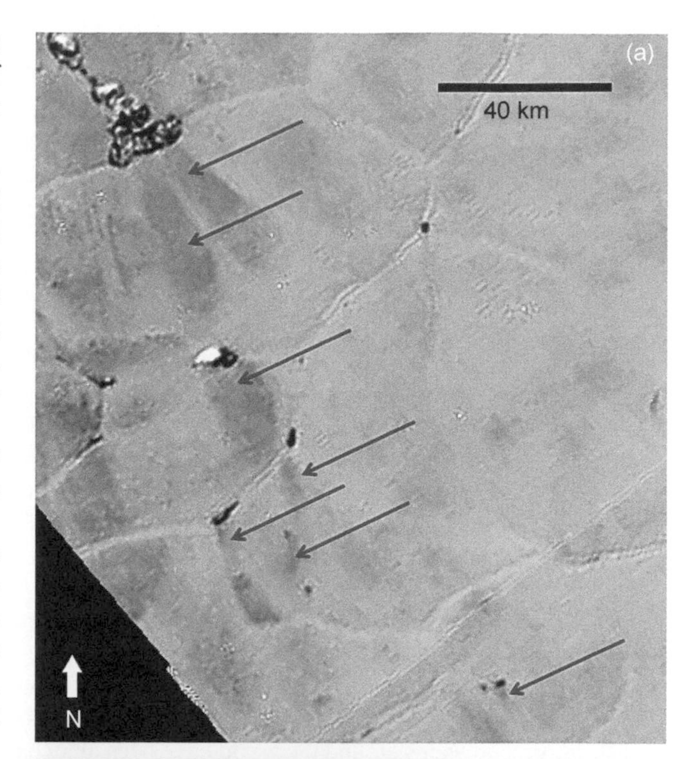

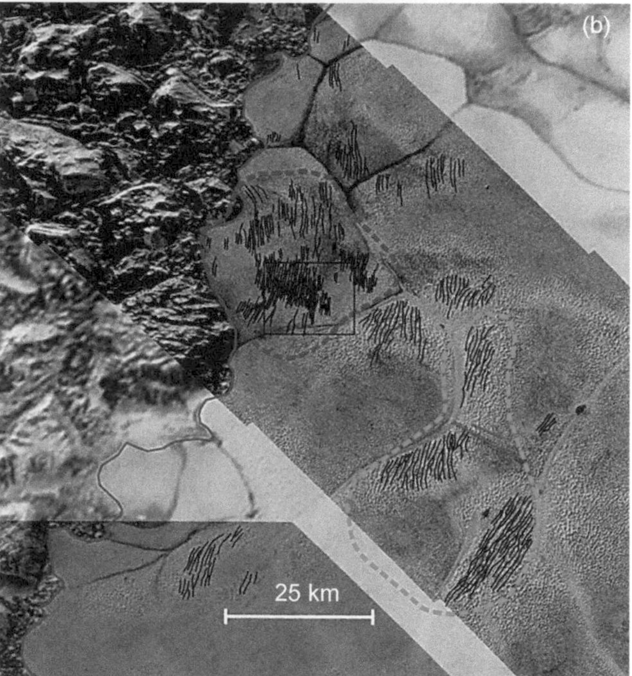

Fig. 2. (a) Possible wind streaks on western Sputnik Planitia observed by the LORRI camera on New Horizons. They are associated with dark spots or hills (*Stern et al.,* 2015). **(b)** LORRI observations of dunes (black lines) at the margins of western Sputnik Planitia thought to be composed of sand-sized (~200 to ~300 μm) particles of methane (*Telfer et al.,* 2018). Prominent wind streaks are marked with orange lines. The dashed lines illustrate dune fields with different orientations.

in Sputnik Planitia, where the relatively dark plains in its north and northwestern regions contrast with the brighter plains in its center. The difference of albedo between the dark and bright plains is ~0.05 (*Buratti et al.,* 2017). The darker color correlates with a weaker spectral signature of N_2 and CH_4, which has been attributed to a decrease in the size of ice grains rich in both CH_4 and N_2, coexisting with a larger amount of CH_4-rich ice grains (*Schmitt et al.,* 2017; *Protopapa et al.,* 2017).

2.4. Atmospheric Models

2.4.1. Three-dimensional "global climate models." In the absence of direct observations of the dynamics of Pluto's atmosphere, a large part of our knowledge of the general and local atmospheric circulation has been derived from three-dimensional models of the atmosphere, in particular general circulation models (GCMs) (also referred to as global climate models). A GCM computes the temporal evolution of the variables that control the meteorology and the climate of the planet at different points on a three-dimensional grid spanning the entire atmosphere. On Earth, GCMs have been applied to weather forecasting and climate change projections. Because these models are built from fundamental physical equations rather than empirical relationships, several teams around the world have been able to successfully adapt them to the other planets (Mars, Venus) and satellites (Titan, Triton) that have a solid surface and a substantial atmosphere. This expertise could be applied to Pluto. Pluto's atmosphere is thick enough for such models: The fluid dynamic equations solved in GCMs remain valid as long as collisions dominate, i.e., up to the exobase. On Mars for instance, GCMs extend up to a pressure level around 10^{-8} Pa (*Bougher et al.,* 1990; *Gonzalez-Galindo et al.,* 2009).

The list of processes that must be taken into account in a Pluto GCM includes (1) the large-scale dynamical motions that can be calculated at the model resolution using a "dynamical core" solving the "primitive equation of meteorology" (a simplified version of the Navier-Stokes equation); (2) the mixing, transport, and drag resulting from subgrid-scale dynamical processes such as turbulence, convection, or gravity waves; (3) radiative transfer through atmospheric gases, hazes, and clouds at solar and thermal wavelengths; (4) the molecular diffusion processes enabled by the long mean-free paths of air molecules that are significant at low pressure (thermal conduction, molecular viscosity, diffusion of molecules); (5) surface temperatures and thermal conduction in the subsurface; (6) the condensation and sublimation of volatiles on the surface and in the atmosphere and the resulting microphysical and dynamical effects; and (7) photochemistry and the formation of hazes.

A first three-dimensional simplified GCM (without phase changes) was presented by *Zalucha and Michaels* (2013) for Pluto and Triton. It was built around the dynamical core of the Massachusetts Institute of Technology (MIT) GCM and included a realistic treatment of radiative transfer. A more realistic model was developed by *Toigo et al.* (2015) a few months before the New Horizons encounter. This "PlutoWRF" GCM was the Pluto-specific implementation of the planetWRF model (*Richardson et al.,* 2007), which itself is an expansion and extension of the National Center for Atmospheric Research (NCAR) Weather Research and Forecasting (WRF) model (*Skamarock et al.,* 2008). This model included a "robust treatment of nitrogen volatile transport" but of course could not take into account the constraints later provided by New Horizons, in particular the distribution of ices and the topography on the anti-Charon hemisphere. After New Horizons, the Laboratoire de Météorologie Dynamique (LMD) team presented a Pluto version of their LMD GCM (*Forget et al.,* 2017; *Bertrand et al.,* 2020a; following earlier studies by *Vangvichith,* 2013) already adapted to other atmospheres in the solar system (e.g., *Forget et al.,* 1999; *Lebonnois et al.,* 2010, 2012). The LMD GCM includes most physical processes needed to calculate the atmospheric circulation (including N_2 phase changes) as well as the CH_4 and CO cycles. It was used to perform three-dimensional modeling of organic haze in Pluto's atmosphere (*Bertrand and Forget,* 2017).

2.4.2. Challenges to modeling Pluto's atmospheric dynamics: Long timescales and model initialization. Even if one could design a perfect model of the processes at work in the Pluto environment, simulating Pluto would remain challenging. This is because unlike on Mars, Earth, or even Venus, the timescales involved in the evolution of the climate system at Pluto are so long that it is difficult to reach a realistic model state insensitive to the initial state, even after running the model for months of computer time.

A first major such problem is related to the distribution of surface ices and subsurface temperatures, which plays a key role in Pluto's climate and atmospheric dynamics. Even if one could prescribe an initial state based on the surface ice reservoirs mapped by New Horizons, this would be a false solution because what is important is not only the locations of the ice reservoirs, but also the fact that they are somewhat in equilibrium with the atmosphere (with a slow evolution resulting from the seasonal and astronomical variations). Initializing the model with prescribed subsurface temperatures and surface ice deposits unrelated to a natural physical evolution of the surface would likely result in a very strong overestimation of the condensation or sublimation fluxes from reservoirs that would most likely be out of equilibrium at any timescales. As detailed in the chapter by Young et al. in this volume, the surface reservoir of N_2, CH_4, and CO ices are the outcome of thousands of years of evolution (*Hansen and Paige,* 1996; *Young,* 2013; *Toigo et al.,* 2015; *Bertrand and Forget,* 2016; *Bertrand et al.,* 2018, 2019). Running the GCM for such a long duration is not yet feasible.

To deal with this issue, the strategy followed by *Vangvichith* (2013), *Toigo et al.* (2015), *Forget et al.* (2017), and *Bertrand et al.* (2020a) has been to run a reduced version of the GCM in which the three-dimensional atmospheric transport and dynamics are replaced by a simple global mixing of the volatile gases. Such models work well on Pluto because the surface energy balance is not significantly sensitive to the atmospheric sensible heat flux and to the radiative transfer through the air. Without computationally expensive calculations to explicitly simulate the atmosphere, much faster numerical simulations can be performed on the same horizontal grid, the same subsurface model, and the same surface/atmosphere volatiles exchange parameterizations as in the full GCM. *Toigo et al.* (2015) used the outcome of a 20-Pluto-year simulation; *Forget et al.* (2017) used a 168-Pluto-year simulation (more time was required to include the effect of topography on the volatile cycles) and *Bertrand et al.* (2020a) used a 125,000-Pluto-year simulation [to account for astronomical variations of the orbit and obliquity as in *Bertrand et al.* (2018, 2019)]. The initial states used for the three-dimensional GCM were selected among many possible results on the basis of their agreement with the observed pressure evolution and the distribution of surface ices in 2015.

Once the surface ices and subsurface temperatures have been initialized with these reduced GCMs, another problem is the long radiative timescale of Pluto's atmosphere (several Earth years; see Table 1) and the time required to reach established CH_4 and CO atmospheric cycles in equilibrium with the surface reservoir. For instance, *Forget et al.* (2017) show that it takes about 20 Earth years for two simulations initiated with two temperature profiles chosen among plausible possibilities (e.g., differing by 30 K) to evolve into states that differ by less than 2 K (hazes that may shorten the radiative timescale were not taken into account). There again, it is important to study the atmospheric circulation once the atmosphere has reached a multi-Earth-year steady-state. Otherwise the modeled results could represent the transient atmospheric circulation required to evolve toward this steady state, and not the reality. To deal with this issue, *Zalucha and Michaels* (2013) and *Toigo et al.* (2015) initialized their models with temperature profiles derived from one-dimensional radiative-conductive-convective calculations. These one-dimensional models were primarily used to explore the effects of varying input parameters in an efficient manner, allowing the most significant or interesting choices to be further explored with the full three-dimensional model. Ultimately, most three-dimensional GCMs simulations begin with an initial state as realistic as possible (based on observations or one-dimensional models) and then run for a sufficient spin-up period: 15 years in *Zalucha and Michaels* (2013) and about 30 years in *Toigo et al.* (2015), *Forget et al.* (2017), and *Bertrand et al.* (2020a). This last study also included a 3-Pluto-year (750-Earth-year) low-resolution simulation, which is discussed in section 7.

3. NEAR-SURFACE CIRCULATION

In the absence of topography and surface ice deposits, theoretical simulations have shown that the modeled near-surface winds are very weak (*Zalucha and Michaels,* 2013; *Toigo et al.,* 2015; *Forget et al.,* 2017) with much less than 1 m s^{-1} at 20 m above the surface. This notably results from the large static stability of Pluto's near surface "stratosphere," which decouples the lowest atmospheric levels from the general circulation above. In reality, models that include more detailed surface properties show that significant surface winds will be created by the effect of slopes and/or the presence of N_2 ice deposits, as detailed below.

3.1. Slope Winds

Forget et al. (2017) performed the first GCM simulations considering surface topography on Pluto. This study found that the near-surface circulation is dominated by the presence of katabatic (downslope) winds. These winds result from the fact that the surface is much colder than the atmosphere. The air close to the slopes is cooled and tends to flow down because it is denser than the air away from the slope at the same level. Figure 3 illustrates the formation of such winds on two (hypothetical) 4-km-high, 800-km-wide mountains. The wind at 20 m above the surface reaches 4 m s^{-1}. Strong katabatic winds exist on Earth in specific locations (notably on the elevated ice sheets of Antarctica and Greenland). They also play a key role on Mars during nighttime and reverse during daytime when the Sun warms the surface (anabatic upslope wind). On Pluto, because the atmosphere is always warmer than the surface, and because of its long radiative timescale, the diurnal variations of surface temperature have a limited effect on the katabatic winds, which only increase by 20% during the night compared to the day (*Forget et al.,* 2017).

3.2. Winds Induced by Sublimation-Condensation Flow of Nitrogen and Surface Temperature Contrasts

The second key process that controls the near-surface circulation is the condensation and sublimation of nitrogen, as discussed above in section 2.1.1. Several factors are important here.

First, these flows must have a considerable effect on local dynamics because they involve relatively large volumes of air. For instance, *Forget et al.* (2017) estimated that at 45°N in 2015 (just south of the subsolar-point), about 230 g m^{-2} of ice sublimates every Pluto day in 2015. This corresponds to a 2.5-km-deep layer of N_2 gas per day, or 15% of the atmospheric column. Locally, it not only changes the surface temperature and pressure, but it also modifies the dynamical structure of the boundary layer by "pumping" the air when condensation occurs on the surface, and by injecting a large amount of cold, pure N_2 gas (with no horizontal velocity) when N_2 sublimes. *Forget et al.* (2017) proposed an algorithm to take this effect into account in atmospheric models. Second, the condensation and sublimation of N_2 induces a surface flow from the sublimating region to the condensing region. Typically, near-surface winds follow a sublimation flow from the subliming sunlit N_2 ice to the condensing regions (near or in the polar night), with a Coriolis turning of the wind as the air travels from one hemisphere to the other. For instance, *Toigo et al.* (2015) predicted interhemisphere diurnal-mean near surface winds of the order of 1–2 m s^{-1} assuming a large N_2 ice polar cap in each hemisphere in 2015

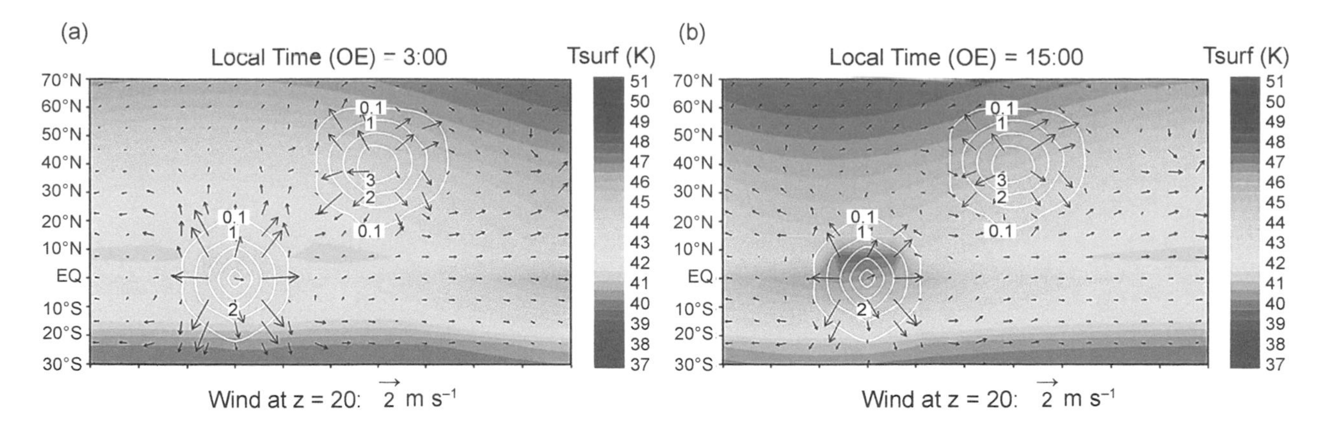

Fig. 3. Maps of LMD GCM modeled surface temperature and winds at 20 m above the surface in July 2015 on a region of Pluto where two artificial 4-km-high mountains have been added to illustrate the formation of downslope winds on Pluto. The topography is shown by white contours. The local time at longitude 0°E is 3:00 h (nighttime) and 15:00 h (daytime). From *Forget et al.* (2017).

(Fig. 4), a distribution that was disproven by New Horizons observations. *Forget et al.* (2017) found similar velocities across the planet in their "alternative scenario" with N_2 ice currently subliming at mid-latitude and in Sputnik Planitia, and condensing in a hidden polar cap in the southern polar night (Fig. 5). Both models also showed that locally, near the boundary between subliming nitrogen deposits (where surface temperature are maintained at the nitrogen frost point) and sunlit bare ground (where surface temperatures can be 10–15 K warmer), the condensation flow is superimposed and often reinforced by a thermal breeze inducing a near-surface flow component from the N_2-ice-covered surface toward the bare ground with the usual Coriolis turning of the wind. The subliming and thermal components usually combine and are both related to nitrogen deposits. This is why we describe them in the same section here.

3.3. The Case of Sputnik Planitia

Sputnik Planitia is a key region on Pluto in many respects and provides a particularly interesting case because it is in a deep basin surrounded by slopes and filled with subliming and condensing N_2 ice. In fact, it is thought to control the general circulation on Pluto, as detailed in section 4.2.

Bertrand et al. (2020a) explored the near-surface circulation in the 3-km-deep Sputnik Planitia basin by performing high-resolution simulations (96 × 72 grid points, about 50 km at the equator) with the LMD GCM. In 2015 (northern spring), these simulations predict a near-surface anticlockwise atmospheric current flowing over Sputnik Planitia from its northeast to its southwest side, triggered by the N_2 diurnal sublimation-condensation cycle (see Fig. 6b). Most of the sublimation occurs in the northern part of the ice sheet (in particular above 30°N where constant insolation across a diurnal cycle results in large N_2 sublimation rates), whereas most of the condensation occurs in the southern part, close to the winter polar night. This leads to a net north-to-south sublimation flow of cold air that is deflected westward by the Coriolis effect as it moves toward the equator, like trade winds on Earth and on Mars. Then, as it reaches the high relief western boundary of the basin (defined by mountain ranges that reach elevations up to 5 km above the plains), the flow is deflected and follows the boundary down to the southern latitudes of the basin, similar to terrestrial or martian western boundary currents (*Anderson,* 1976; *Joshi et al.,* 1994, 1995). The wind directions and the western boundary current predicted by the GCM are consistent with the wind directions derived from the wind streaks mentioned in section 2.2.4 (Fig. 6a). This model also predicts a small eastern boundary current that forms in the late afternoon as the flow reaches the southeastern edge of Sputnik Planitia and is guided back toward northern latitudes.

4. THE GLOBAL CIRCULATION DURING THE NEW HORIZONS ENCOUNTER

4.1. Three Possible General Circulation Regimes

In the absence of any direct observational data, the circulation regime on Pluto can only be studied with the help of models, and major uncertainties remain. As mentioned above, *Forget et al.* (2017) used a post-New Horizons version of the LMD Pluto GCM and performed a comprehensive characterization of the dynamics within Pluto's atmosphere. They highlighted the sensitivity of the general circulation to the atmospheric transport of N_2 and therefore to the locations of the sources and sinks of N_2 on the surface. They obtained three possible different dynamical circulation regimes for 2015, depending on the initial location of the N_2 ice deposits:

1. If N_2 ice was placed on the poles or at high latitudes in both hemispheres, there was a "retrorotation" as in *Toigo et al.*'s (2015) prediction for the New Horizons encounter in 2015 shown in Fig. 4a. In such a regime, the mean me-

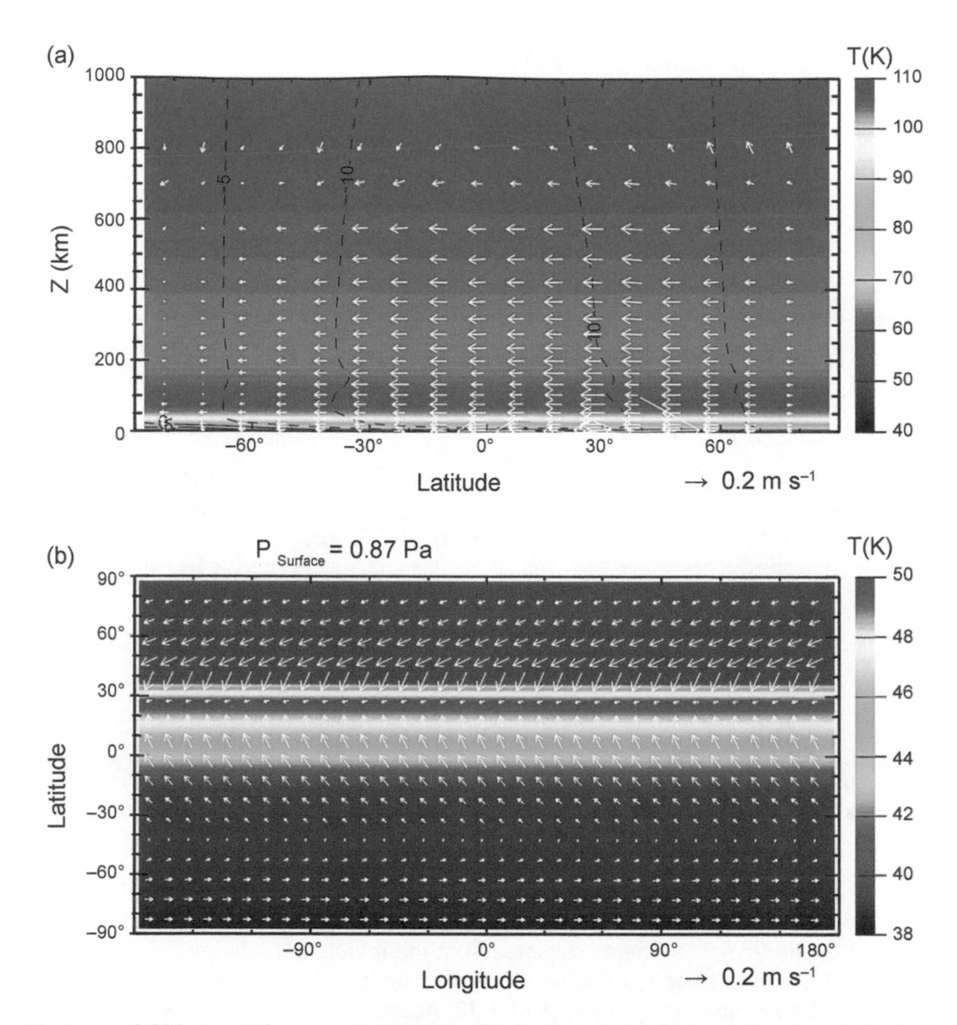

Fig. 4. Pre-New Horizons GCM simulation results showing the "retro-rotation" circulation (westward winds at all latitudes) in 2015 on a Pluto-like planet assuming that it is covered by large polar caps of nitrogen ice in both hemispheres, with only bare ground in the 10°S–35°N latitudinal range. **(a)** Diurnal- and zonal-average temperature and wind profiles. Temperature (K) is shown as the background colors, black contours are magnitude of the zonal wind (positive eastward values are solid lines, negative westward values are dashed lines), and the arrows are the meridional (north-south) and vertical winds (scaled by a factor of 10) expressed together as a vector. **(b)** Map of near-surface diurnal-average air temperature and near-surface winds. Temperature (K) is shown as background colors, and arrows are near-surface horizontal winds. Adapted from *Toigo et al.* (2015).

ridional circulation corresponds to an open north-to-south circulation cell as net amounts of N_2 are transferred from the subliming northern cap to the condensing southern cap (there is no return flow). The zonal circulation is then characterized by a global *retrorotation* with westward (retrograde, in the direction opposite to the planet rotation) winds at most latitude. Such winds result from the conservation of angular momentum of the air particles as they flow from the sunlit pole to the polar night above the equator, where they are farther from the rotation axis than where they started from. Above the equator the zonal winds reach at most 15 m s^{-1}, which corresponds to the velocity of air particles with zero angular momentum (i.e., starting at rest from high latitudes) 100 km above the equator.

2. If N_2 ice was placed in Sputnik Planitia and at mid-northern-latitudes (as observed by New Horizons) and at the south pole, then the model predicted a zonal wind field characterized by an intense prograde jet-stream poleward of 40°S and a prograde *superrotation* at most other latitudes (Fig. 7b). The high-latitude jet is a classical feature of terrestrial atmospheres and is triggered by the conservation of angular momentum as air parcels move from the N_2 deposits at equatorial and mid-latitude to the south polar region (which is closer to the axis of rotation). Superrotation was more surprising. It is also observed on Venus and Titan. *Forget et al.* (2017) showed that the process at work in these simulations was probably the same as the mechanism usually proposed for these bodies: Atmospheric waves generated by barotropic instabilities in the high-latitude jet redistribute angular momentum equatorward. A study of the variations of the jet by *Forget et al.* (2017) showed that it is subject to instabilities that create a wavenumber 1 wave that propagates eastward with a 0.5–0.8-Pluto-day period (Fig. 7d). At 60°N, such waves are clearly visible

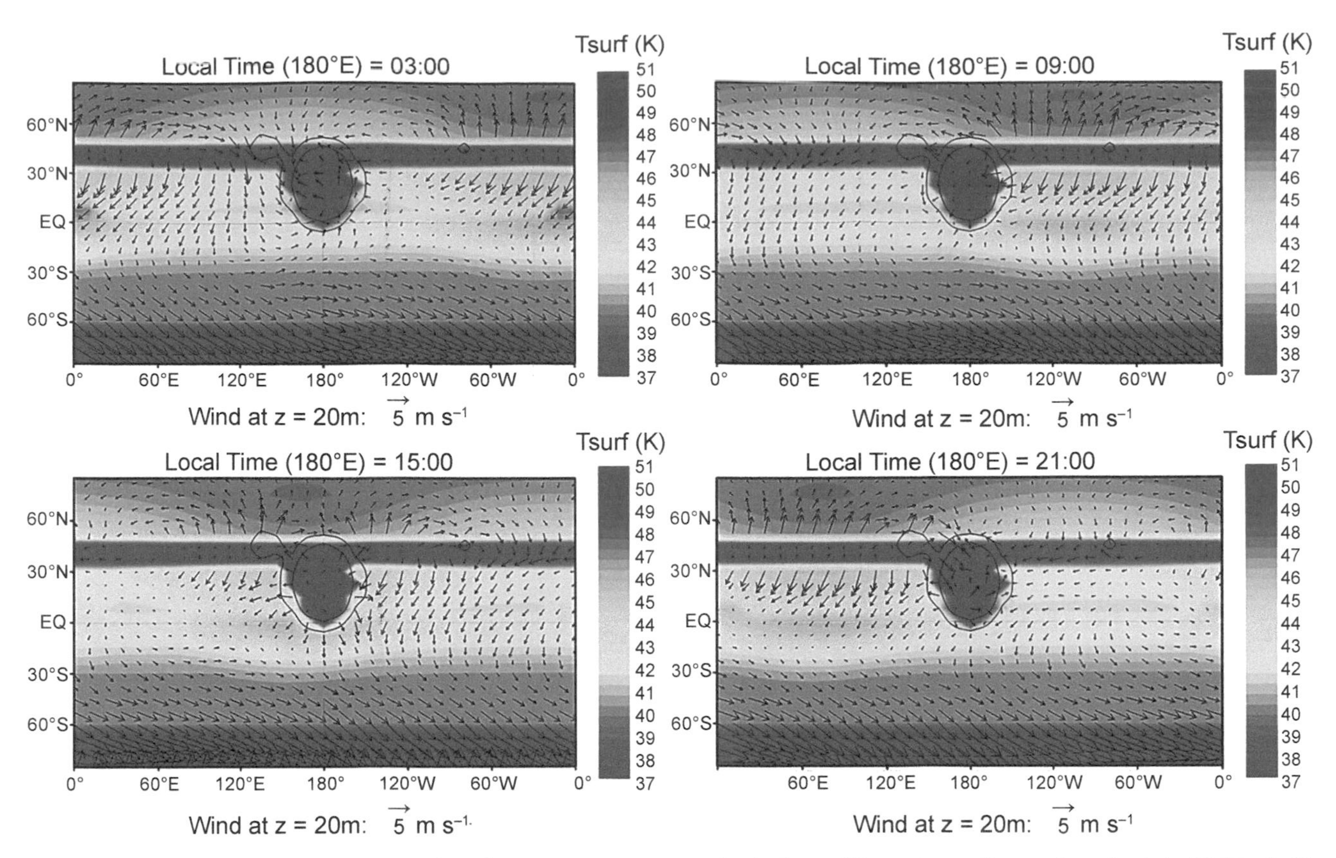

Fig. 5. LMD GCM simulated maps of surface temperatures and winds at 20 m above the surface in July 2015 at different local times. N_2 ice deposits are at the N_2 frost point near 37 K and appear dark blue. In this particular simulation N_2 ice present in Sputnik Planitia (simplified here) sublimes from the mid-latitude deposits and moderately condenses in the south polar region. The local time LT in the middle of the map (longitude 180°) is 3:00, 9:00, 15:00, and 21:00, with LT (hours) = [longitude (°)–subsolar point longitude (°)]/15 + 12. Adapted from *Forget et al.* (2017).

at an altitude of 140 km in the temperature and meridional wind fields (Fig 7d). In Fig. 7c, the extension of this wave is mapped by plotting the meridional wind variability as a function of latitude and height showing that it propagates to the equator. *Toigo et al.* (2015) also found superrotation during the spinup of their simulation (i.e., before 2015), but it is likely that the atmospheric circulation was not yet equilibrated.

3. If N_2 ice was placed in Sputnik Planitia only, the zonal winds obtained were weak (Fig. 7a) and induced by the gradient of radiative heating by the Sun between the fall and spring hemispheres inducing a weak (a few meters per second) retrograde jet at the equator and in the sunlit hemisphere and a weak jet stream at 100 km at the edge of the polar night. Similar results were obtained by *Toigo et al.* (2015) when they assumed that nitrogen ice was only present in the southern tropics (see their Fig. 20).

4.2. The Most Likely 2015 Circulation Regime: Retrorotation Induced by Sputnik Planitia

Forget et al. (2017) concluded that retrorotation regime #1 was unlikely in 2015 because it required a strong cross-equatorial sublimation flow that they could not create in the model while remaining consistent with the N_2 ice distribution mapped by New Horizons and the observed increase of surface pressure in recent years [e.g., *Meza et al.* (2019); this increase indicates that N_2 condensation at the south pole could not be very intense and therefore that the coverage of the south hemisphere by N_2 ice could be limited]. They were wrong. Indeed, in these early LMD GCM simulations designed just after the New Horizons encounter and at relatively low resolution, the Sputnik Planitia basin was represented as a simple circular crater located north of the equator. In reality, Sputnik Planitia extends southward down to 25°S, and taking that into account dramatically changes the global circulation. *Bertrand et al.* (2020a) performed higher-resolution LMD GCM simulations (~50 km at the equator), including the New Horizons topography data (*Schenk et al.,* 2018) and realistic surface ice distributions obtained from analyses of LEISA's near-infrared spectra of Pluto's surface (*Schmitt et al.,* 2017; *Protopapa et al.,* 2017). In these new simulations, a strong, global retrorotation was obtained. What happens is that the sublimation-condensation flow inside Sputnik Planitia is sufficient to drive the entire atmosphere into a global retrorotation regime (Fig. 8) because of the conservation of angular momentum as the N_2 molecules

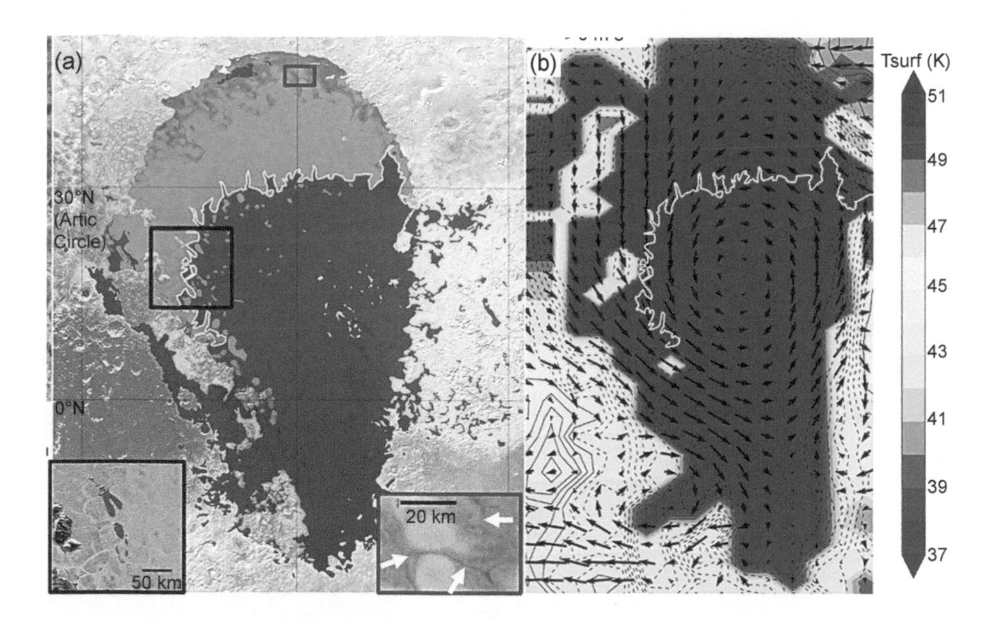

Fig. 6. (a) Simplified version of the geological map of *White et al.* (2017) depicting bright N_2 ice plains (red), dark N_2 ice plains (blue), mountains and hills lining the western rim of Sputnik Planitia (green), and bright pitted uplands of east Tombaugh Regio (cyan). Yellow line maps the continuous boundary between the bright and dark plains, as well as the northern boundary of the bright pitted uplands. Black box indicates the location of features in Sputnik Planitia interpreted as wind streaks, as mapped in purple in the inset. Blue inset box and white arrows indicate the location of dark troughs, possibly filled with dark materials (*White et al.,* 2017). **(b)** Map of diurnal mean surface temperatures and horizontal winds in Sputnik Planitia obtained with the LMD GCM simulations of *Bertrand et al.* (2020a) for July 2015 at 1000 m above the surface. The yellow line replicates the bright/dark boundary in **(a)**. Figure adapted from *Bertrand et al.* (2020a); inset adapted from *Stern et al.* (2015) and *Telfer et al.* (2018).

are transported from one hemisphere to the other and move away from the rotation axis as they cross the equator (Fig. 8). The zonal wind reaches ~8–13 m s^{-1} at altitudes 20–250 km. All their simulations showed that the meridional circulation in the upper atmosphere remains weak at all longitudes, with winds lower than 1 m s^{-1}, and that the zonally averaged meridional circulation is dominated by a southward flow, which is strengthened by the potential presence of mid-latitude N_2 deposits outside Sputnik Planitia and weak thermal gradients between both hemispheres. In their model, most of the cross-equatorial transport of air occurs in Sputnik Planitia, which seems to be an efficient channel to transport freshly sublimed air, gaseous CH_4, and other atmospheric constituents from the northern to the southern hemisphere in 2015.

5. GRAVITY WAVES AND TIDES

5.1. Introduction and Theoretical Predictions

As with other atmospheres, local or regional perturbation in Pluto's atmosphere creates waves. In section 4 we mentioned the possible role of barotropic waves induced within a possible southern jet stream circulation in 2015 (Figs. 7b–d). In the absence of further observations, such waves remain hypothetical and model-dependent. Nevertheless, observations (stellar occultations and images of haze layering) suggest the presence of several types of waves in the atmosphere of Pluto: small-scale gravity waves and, on larger scales, inertia-gravity waves and/or atmospheric tides.

5.1.1. Gravity waves. Gravity or buoyancy waves typically result from a displacement induced by topographic variations or by a non-topographic process like N_2 condensation and sublimation). These waves can vertically propagate. On Pluto, because the lower atmosphere is strongly stratified, the waves are easily excited. To conserve momentum in a medium in which density exponentially decreases with altitude, such waves can achieve very large amplitudes and break. On Pluto, however, damping due to kinematic viscosity and thermal diffusivity should be extremely effective in suppressing vertical propagation of waves with vertical wavelengths of a few kilometers or less (*Hubbard et al.,* 2009; *French et al.,* 2015). Breaking (and even non-breaking) gravity waves can act to decelerate the large-scale flow toward their phase-speed (zero for topography-induced gravity waves) and to produce turbulence and mixing at altitudes near their breaking level. In some models, inertia-gravity waves have been invoked to explain the New Horizons observations of layered haze, as detailed below in section 5.3.

5.1.2. Atmospheric tides. Atmospheric tides are planetary-scale waves excited by the diurnal cycle of the insolation. In fact, one can view them as planetary-scale gravity waves with periods that are harmonics of the solar day. Such waves are a major component of the global circulation

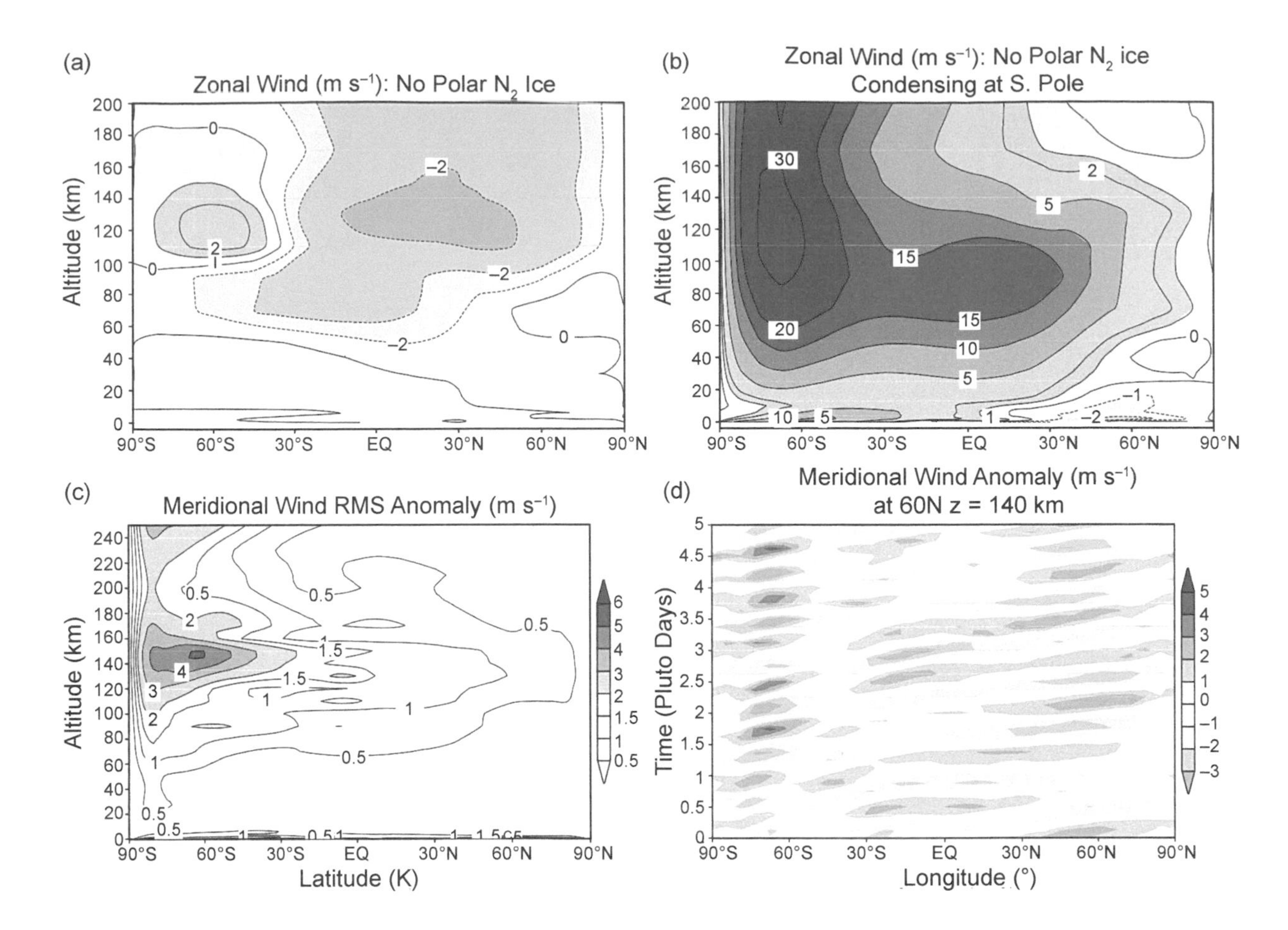

Fig. 7. **(a)** Zonal-mean zonal wind in the *Forget et al.* (2017) simulation with N_2 ice only in a small Sputnik Planitia-like basin in the northern hemisphere. **(b)** Same, but for the simulation with south pole N_2 condensation forcing a light condensation flow (as presented in Fig. 5) and superrotation of the atmosphere. **(c)** Same simulation as **(b)**; this image shows the vertical-latitudinal structure of the waves thought to force the superrotation in **(b)** by plotting the zonal average of the root-mean-square standard deviation of the local meridional wind from the zonal-averaged meridional wind. **(d)** Same simulation as **(b)** and **(c)**: Hövmoller diagram of the meridional wind anomaly (difference between the local and the zonal mean wind, in m s^{-1}) at 0°N, revealing the details of the propagating waves thought to force the superrotation. Adapted from *Forget et al.* (2017).

on Mars and have been studied in the upper atmosphere of Earth. On these planets, the tides are mostly forced by the heating and cooling of the surface and the atmosphere and are therefore described as "thermal tides." On Pluto, given the long-radiative timescale of the atmosphere and the limited heat transfer from nitrogen-ice-free surfaces to the atmosphere above, it seems that the primary source of the diurnal forcing will be the condensation and sublimation of the N_2 (notably in Sputnik Planitia), which induces large diurnal variations in all fields as explained above.

On Mars and on Earth, the behavior of the tides has been very well predicted and understood on the basis of the classical tidal theory that was developed 50 years ago (e.g., *Chapman and Lindzen,* 1970), in which the horizontal and vertical structure of the tides is described using Hough functions. This theory has been applied to understand the basic characteristics of the tides on Pluto (*Toigo et al.,* 2010; *French et al.,* 2015) and to understand the density fluctuations in stellar occultations (see section 5.2 below). As for the gravity waves, damping due to kinematic viscosity and thermal diffusivity should be extremely effective in suppressing vertical propagation of waves with vertical wavelengths of a few kilometers or less (*Hubbard et al.,* 2009; *French et al.,* 2015). The dominant surviving tidal modes on present day Pluto have characteristic vertical wavelengths of 10–13 km.

The behavior of the tides can also be simulated by GCMs that compute the diurnal cycle of the N_2 condensation and sublimation (*Forget et al.,* 2017; *Bertrand et al.,* 2020a). Figure 9 shows the complex diurnal oscillations in temperature that appear at the equator in the reference simulations of *Bertrand et al.* (2020a). Their amplitudes are limited to 0.2 K, but the same kind of structure affects the different components of the wind and density at a given altitude. Tidal amplitudes are 4× weaker if N_2 condensation-sublimation processes are switched off. A detailed analysis of these simulations in light of tidal theory has not been performed yet, but at first glance the GCM simulations appear to be

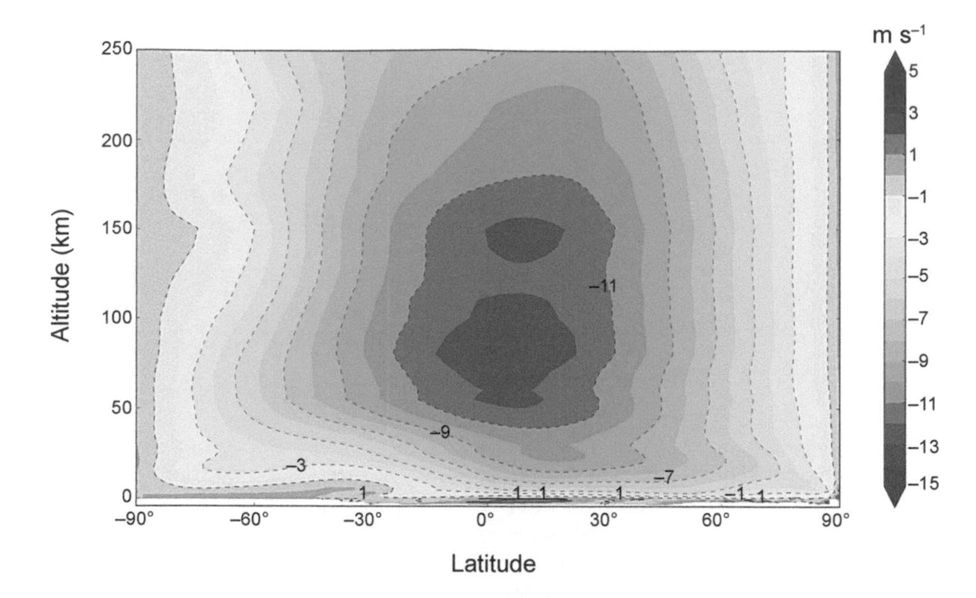

Fig. 8. Zonally averaged zonal winds (in m s^{-1}) obtained in the state-of-the-art GCM simulations of *Bertrand et al.* (2020a). This is the best available estimate of the circulation regime in 2015, which exhibits global retrorotation. This particular case assumes that nitrogen ice is only present in Sputnik Planitia, but essentially the same results are obtained with other realistic nitrogen ice reservoirs, showing that Sputnik Planitia itself actually drives the zonal mean circulation.

consistent with what is expected from upward-propagating atmospheric tides (with downward phase velocity but upward group velocity): Below 80 km, the strongest mode is diurnal with zonal wavenumber 1 and a vertical wavelength around 20 km. Above 150 km, the semi-diurnal wavenumber 2 tide starts to dominate with a much longer vertical wavelength.

5.2. Waves Revealed by Density Fluctuations in Stellar Occultations

As groundbased stellar occultation measurements became more frequent around the turn of the century, distinct oscillations in the observed curves of light flux that were above the background noise of the data became apparent. As the mean light flux profile can be mapped into a vertical density profile and, with application of the hydrostatic equilibrium equation, into a vertical thermal profile, these oscillations in light flux can thus be mapped into oscillations around mean density and thermal profiles.

Elliot et al. (1989, 2003), *Yelle and Elliott* (1997) (reporting on data from *Millis et al.,* 1993), *Sicardy et al.* (2003), *Pasachoff et al.* (2005), *Gulbis et al.* (2015), and *Pasachoff et al.* (2017) all reported the presence of "spikes" in their light curves (Fig. 10). Some form of vertically propagating wave was typically cited as the source, although most of these observations suffered from low signal-to-noise ratios and sparsely observed spikes, hindering more detailed investigation of their overall dynamical structure. Despite these limitations, *Yelle and Elliott* (1997) noted that the general absence of spikes, as well as their small magnitude when present, implied a stable atmosphere. *Sicardy et al.* (2003) found fluctuations in their thermal profile on the order of 0.5–1%, and *Pasachoff et al.* (2005) noted that while disturbances appeared only in the lower half of the atmosphere, their distribution in altitude is not globally uniform.

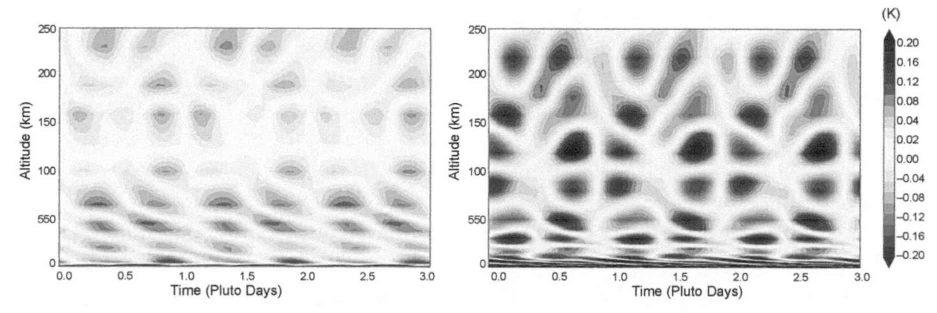

Fig. 9. Temperature anomaly (difference between instantaneous value and diurnal average) showing the signature of atmospheric tides at 0°E–0°N in GCM simulations obtained for July 2015 with **(a)** N_2 ice in Sputnik Planitia only or **(b)** N_2 ice in the mid-latitudes in addition to Sputnik Planitia. From *Bertrand et al.* (2020a).

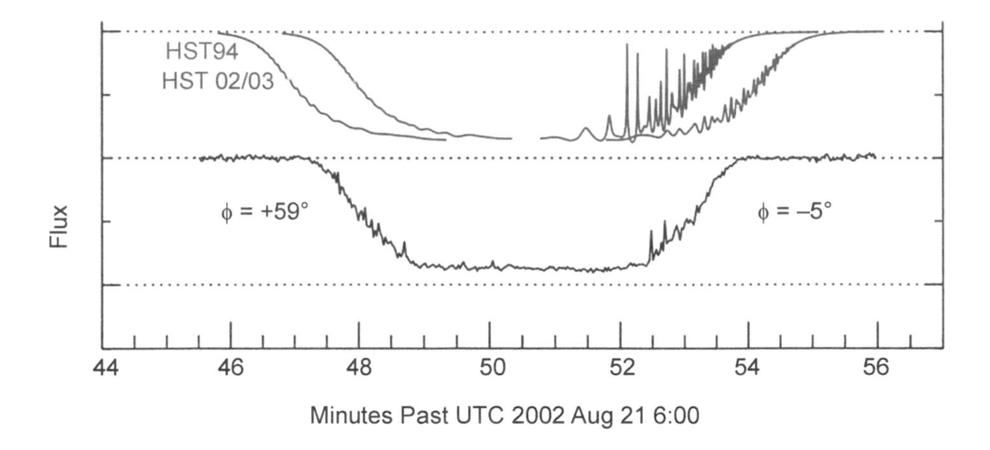

Fig. 10. Observations and model light curves for the August 21, 2002, occultation by Pluto showing "spikes" in the light curves that are attributed to atmospheric waves. The CFHT observations (*Sicardy et al.*, 2003) are shown at the bottom, and synthetic light curves calculated from tidal models (*French et al.*, 2015) are shown at the top. These are based on the atmosphere predicted assuming two N_2 frost maps derived from HST in 1994 and 2002–2003 (the calculations using HST94 frost map are shown in blue and are displaced horizontally). Ingress occurred at high northern latitude (φ = 59°) and egress was nearly equatorial (φ = –5°). The spikes in the actual occultation data are more prominent during egress than ingress. The model light curves show this dichotomy even more strongly, which suggests that this can be attributed to the difference in latitudes sampled during the event. Adapted from *French et al.* (2015).

McCarthy et al. (2008) and *Person et al.* (2008) reported on near-infrared and visible observations of a 2007 stellar occultation event, respectively. They found fluctuations in their derived temperature profiles with wavelengths of 8–20 km and 25–35 km respectively over an approximate range of 150–300 km altitude above Pluto's surface, consistent with the presence of an atmospheric wave. The similarities in the spikes in the ingress and egress light curves, which are separated by 1200 km along the limb of Pluto, implies that the horizontal wavelength is large. The papers discussed inertia-gravity waves and Rossby waves as the cause for the oscillations. Additionally, *McCarthy et al.* (2008) noted that the waves are close to breaking at the altitudes where they were observed. *Hubbard et al.* (2009) investigated the gravity wave hypothesis further, with the same observational dataset, and found the power spectrum of observed oscillations to be consistent with the gravity-wave dispersion relation, but could not rule out a small contribution from Rossby waves.

Young et al. (2008) also observed spikes in their light curves during a 2006 stellar occultation event. They derived temperature fluctuations of typically less than 0.5 K in the lowest 100 km of the atmosphere to be associated with vertical wavelengths of about 2.7–11 km. They also described the observed oscillations as being suggestive of gravity waves, with increasing vertical wavelengths with altitude.

While vertically propagating gravity waves generated by air flow over topographic obstacles (e.g., mountains) are common on Earth and thus also a potential source of gravity wave generation on Pluto as well, *Toigo et al.* (2010) and *French et al.* (2015) investigated solar-induced diurnal cycles of sublimation and condensation of N_2 ice-covered regions as an alternate method of gravity wave generation on Pluto. As discussed above, while the forcing period of such oscillations would be limited to that of the diurnal cycle of heating, the spatially inhomogeneous distribution of ice sheets could generate a wide variety of waves. Both studies found good agreement with previously described observations of the vertical oscillation wavelengths, amplitudes, and occultation ingress and egress asymmetries of fluctuations in the occultation profiles (Fig. 10).

5.3. Waves Revealed by the Layered Structure of Atmospheric Hazes

High-phase-angle images from New Horizons revealed an atmosphere suffused with optically thin haze whose distinctive structure implies the presence of atmospheric waves (*Gladstone et al.*, 2016; *Cheng et al.*, 2017). The haze is organized into about 20 concentric layers, with a typical separation of 10 km and a horizontal extent of hundreds of kilometers, as shown in Fig. 11. Structure that appears in individual profiles of I/F (the ratio of the observed haze brightness by the incident solar flux) also suggests the presence of waves with somewhat larger vertical wavelengths (tens of kilometers).

The structure of the haze layers is reminiscent of planetary-scale waves observed at similar altitudes and latitudes in the grazing stellar occultation of March 18, 2007 (*Person et al.*, 2008; *McCarthy et al.*, 2008; *Hubbard et al.*, 2009). Spectral analysis of temperature profiles retrieved from the stellar occultation data indicates the presence of a superposition of waves with vertical wavelengths of 10–40 km. The horizontal scale of the waves must be large, on the order of 1000 km, to account for the high degree of symmetry between the entry and exit segments of the stellar occulta-

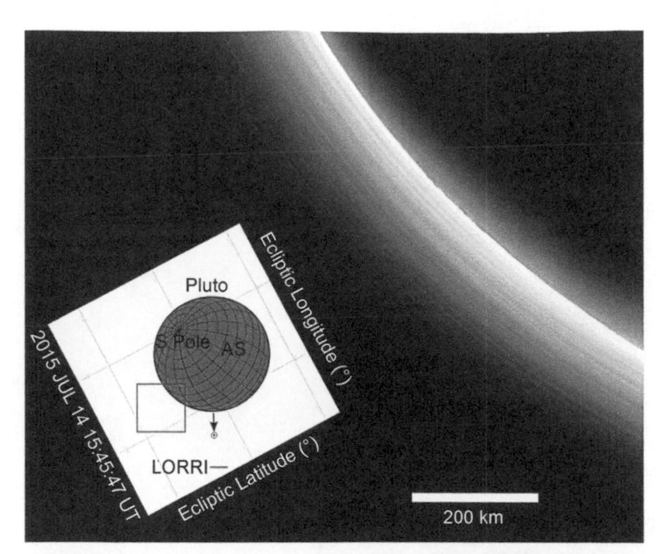

Fig. 11. Composite LORRI image at 0.95 km pixel^{-1} resolution, showing layered haze extending from the surface to an altitude of about 200 km (*Gladstone et al.,* 2016) (Fig. 5). The phase angle is 169°. The inset shows the geometry of the observation, where SP is Pluto's south pole and AS is antipodal to the subsolar point. This segment of Pluto's limb is at roughly 40°E.

tion light curve. These wave properties are consistent with the observed separation of the haze layers and their alignment with Pluto's limb.

The New Horizons images provide additional clues about the identity of the waves. The haze layers are more distinct near the equator than at mid-latitudes (*Cheng et al.,* 2017) (Figs. 8 and 12), which suggests the presence of a planetary-scale wave that is confined to low latitudes. Moreover, there is no evidence for vertical motion of the haze layers in images acquired up to 5.4 h apart at the same location. This implies either that the waves are stationary with respect to the surface or that the period is much longer than 5 h (e.g., atmospheric tides, as discussed in sections 5.1 and 5.2). It is not yet clear which explanation is correct.

Gladstone et al. (2016) and *Cheng et al.* (2017) constructed a detailed model for stationary inertia-gravity waves that accounts for many features of the haze layers. Pluto's surface provides strong orographic forcing, particularly at short horizontal wavelengths, which causes the waves to saturate at low altitudes (below 10 km). This produces density modulation that is sufficient to generate haze layers over an extended altitude range, consistent with the observations. The model requires a background horizontal wind speed of about 2 m s^{-1} to ensure that the vertical wavelength matches the observed separation of the haze layers, about 10 km. (The vertical wavelength is proportional to the wind speed.) However, the near-surface winds in GCM simulations are highly variable, both diurnally (Fig. 5) and spatially (Figs. 3 and 6). This could radically change the type of wave excited by orographic forcing.

As discussed in section 5.1.2, atmospheric tides on Pluto are excited primarily by the diurnal cycle of N_2 sublimation and condensation. Both GCM simulations and models based on classical tidal theory predict vertical wavelengths of about 10–20 km, plausibly consistent with the separation of the haze layers. But the tidal amplitude in the most recent GCM simulations is only about 0.2 K (Fig. 9), which may not be sufficient to generate haze layers as distinct as those in Fig. 11. Hence, the source of the haze layers remains uncertain.

6. VARIATIONS OF THE ATMOSPHERIC CIRCULATION OVER SEASONAL TIMESCALES

Bertrand et al. (2020a) performed multi-Pluto-year simulations of Pluto's atmosphere at relatively low resolution (~150 km at the equator). By assuming permanent mid-latitudinal bands of N_2 deposits outside Sputnik Planitia, they show that the retrorotation regime of the atmosphere is maintained during most of Pluto's year, with a maximum westward wind of ~10–12 m s^{-1} centered above Sputnik Planitia during northern spring and summer (Fig. 12). This is because in their simulations, there is always enough cross-equatorial transport of gaseous N_2 in Sputnik Planitia (and outside), from north to south during northern spring and summer or south to north during the opposite season, to trigger westward winds. Around Ls = 270–300° (southern summer), the zonal winds at 20 km are still directed westward but are significantly weaker. This has been interpreted as due to the larger extent of the ice sheet in the northern hemisphere compared to the southern (Sputnik Planitia is not symmetrical about the equator). Because of this asymmetry, the sources of N_2 are weaker than the sinks of N_2 during Ls = 270–300°, and significant meridional transport during this period occurs in the northern part of the ice sheet (the cross-equatorial transport of gaseous N_2 is much weaker during this season, hence the weaker winds). Interestingly, their model predicted that the retrorotation regime is currently at its highest intensity, because the subsolar point is at ~50°N and there is preferential sublimation of N_2 from the mid-latitudinal deposits and from the northern part of Sputnik Planitia. Another maximum is obtained around solar longitude Ls = 218°, when the subsolar point is above the latitude ~33°S and the southern N_2 deposits are preferentially sublimating. These results are highly model-sensitive and strongly depend on the assumed ice distribution.

7. THE INFLUENCE OF CIRCULATION ON THE SURFACE ICES

Pluto displays a surprisingly heterogeneous surface with dramatic albedo, color, and composition contrasts. On the one hand, some of these contrasts are organized in latitudes and are therefore likely to be controlled by the insolation-driven changes in volatile sublimation-condensation rates (see the chapter by Young et al. in this volume). On the other hand, other features are asymmetric in longitude. This

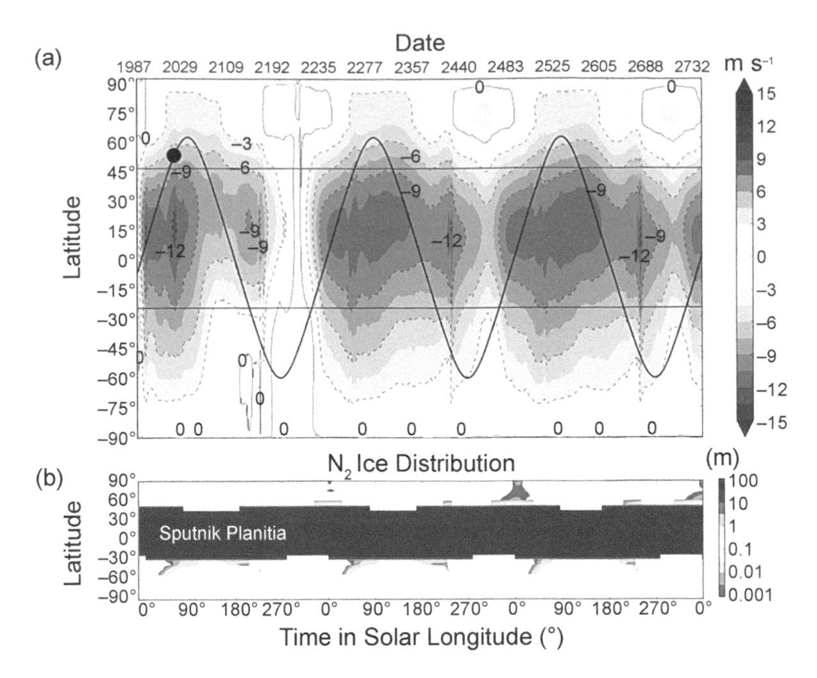

Fig. 12. Annual evolution of Pluto's general atmospheric circulation obtained with the 11.25° × 7.5° LMD GCM simulations over three consecutive Pluto years reported by *Bertrand et al.* (2020a). **(a)** Zonal mean zonal winds at 20 km above local surface. The black solid line indicates the latitude of the subsolar point and its position in 2015 is shown by the black circle. The red horizontal solid lines indicate the bounding latitudes of Sputnik Planitia. The general circulation is dominated by a retrorotation (westward winds) during most of the year. **(b)** Zonal mean N_2 ice distribution (Sputnik Planitia is a permanent equatorial kilometer-thick N_2 ice sheet). From *Bertrand et al.* (2020a).

could be due to interactions with atmosphere dynamical processes and could therefore provide clues about the circulation of Pluto's atmosphere. Examples of such features, which we discuss next, include the contrast between the dark northwestern plains and the bright central plains of Sputnik Planitia (see section 7.1), the presence of equatorial CH_4-rich bladed terrain deposits and volatile-free tholins-covered regions (Cthulhu) east and west of Sputnik Planitia respectively (section 7.2), the bright East Tombaugh Regio (see section 7.3), and CH_4 frost on top of the highest mountains in the equatorial regions (see section 7.4).

7.1. The Dark and Bright Plains of Sputnik Planitia

White et al. (2017) hypothesized that the contrast between the dark and bright plains of Sputnik Planitia in 2015 (Fig. 6a) is due to net N_2 sublimation north of 30°N, revealing and concentrating darker and older material-infused ice that forms the bulk of the N_2-rich ice filling the Sputnik Planitia basin, while net condensation south of 30°N is depositing a thin veneer of fresh, bright N_2 ice onto the southern plains. This has been supported by volatile transport simulations, showing that the latitudes north of 15°N were sublimation-dominated areas in 2015, whereas latitudes south of 15°N were condensation-dominated areas (*Bertrand et al.,* 2018). However, the band of dark plains extending south of 30°N down the western margin of Sputnik Planitia indicates that factors besides latitude-dependent insolation are also influential in defining the albedo contrast in the plains, and the observation of dark wind streaks (Fig. 6a) there suggests that near-surface winds may play a role (*Stern et al.,* 2015; *Telfer et al.,* 2018).

The high-resolution simulations performed by *Bertrand et al.* (2020a) with the LMD Pluto GCM are consistent with the hypothesis that aeolian activity is uncovering dark plains in this western region of Sputnik Planitia, as these simulations predict windier conditions roughly above the dark plains (induced by the sublimation flow and the western boundary current; see section 3 and Fig. 6b). *Bertrand et al.* (2020a) investigated different mechanisms that could contribute to an increase of sublimation or erosion of N_2 ice at these dark plains. These are detailed below.

*7.1.1. **Effect of sensible heat flux**.* One mechanism is the heating of the surface by the nighttime downward sensible heat flux, which is enhanced along the northern and western boundary of Sputnik Planitia in the LMD model simulations, because winds are stronger there and because ~3 K warmer air is injected at night from the surrounding dark material-covered slopes and terrains. In fact, there seems to be a pattern on New Horizons images whereby the darkest plains in the midwestern part of Sputnik Planitia tend to be proximal to tall mountains, which is consistent with a larger sensible heat flux induced by downslope transport of adiabatically warmed air from the top of these mountains to the plains by stronger katabatic winds. *Bertrand et al.* (2020a) found that the western terrains of the ice sheet could have lost about 10% more N_2 ice than the central terrains from this process. This corresponds to about 3 mm in 2015 and 45 mm over the last 15 Earth years.

7.1.2. ***Effect of accumulation of surface haze particles.*** A second mechanism is the effect of accumulation of haze particles in Sputnik Planitia, which could contaminate the N_2-rich ice and lower its surface albedo, impacting the sublimation-condensation rates. *Bertrand et al.* (2020a) used the haze parameterization described in *Bertrand and Forget* (2017) to estimate, at a global scale, how the settling of haze particles onto the surface is impacted by the N_2 condensation and sublimation flows. They found that haze particles tend to be repelled from the N_2 ice plains by N_2 sublimation flows and drawn toward the surface by N_2 condensation flows. However, the katabatic winds predicted on the depression's slopes (where N_2 ice tends to be distributed) balance the daytime sublimation flows and strengthen the nighttime condensation flows, thus favoring the settling of haze particles in low-altitude N_2 ice reservoirs. *Bertrand et al.* (2020a) estimated that the accumulation of 10-nm haze particles can be up to ~10× larger in depressions containing N_2 ice than elsewhere during the period 1984–2015. In particular, their GCM simulations predicted a larger haze accumulation in the western regions of Sputnik Planitia than in the eastern regions during this period, with large amounts of haze particles deposited onto the N_2 ice plains surrounding the Al-Idrisi, Zheng He, Baret, and Hillary Montes (due to strong katabatic winds on the slopes of these reliefs). In general, the agreement with the western dark plains observed by New Horizons is not excellent, because the observed boundaries between bright and dark plains are more abrupt than what the simulation produced.

Alternatively, or in addition to that, large amounts of dark materials could be mobilized from the northern edge of Sputnik Planitia. In this region, very dark troughs forming the edge of some cellular features are observed and could be a source of dark materials blown away by N_2 ice sublimation. *Bertrand et al.* (2020a) tested this hypothesis by adding in the model a source of dark materials roughly at the location of the very dark plains. The source was set proportional to the sublimation rate of N_2 (injection only during daytime). Results showed that most of the dark materials thus injected accumulate in the western and southern sides of Sputnik, in a good agreement with the dark and bright plains observed by New Horizons, suggesting that this process could trigger the albedo contrast such as observed.

7.1.3. ***Other effects.*** Other possible mechanisms disrupting N_2 ice include enhanced mechanical erosion or transport of CH_4-rich ice grains by the winds over the ice sheet. However, the surface stress on Pluto is on the order of $\mu N\ m^{-2}$. This is 100–1000× weaker than on Mars, making mechanical erosion unlikely on Pluto, at least over the timescales involved here to explain the albedo contrast in Sputnik Planitia. As for the transport of CH_4-rich ice grains, the size of these grains in Sputnik Planitia seem to be on the order of a millimeter (see Fig. 4b in *Protopapa et al.*, 2017), which may be too large to be transported far by Pluto's winds and saltation processes (*Telfer et al.*, 2018). In addition, the observed dunes, thought to be composed of CH_4-rich ice particles, seem to correspond to brighter areas on the ice sheet's surface and therefore this process cannot explain the contrast between bright N_2-rich or dark CH_4-rich surface.

7.1.4. ***Surface albedo feedback.*** Runaway forcing (positive feedback) by albedo and composition have also been suggested as efficient mechanisms to maintain albedo contrasts in the equatorial regions (*White et al.*, 2017; *Moore et al.*, 2018; *Earle et al.*, 2018; *Bertrand et al.*, 2020a). *Bertrand et al.* (2020a) showed that the slightly lower albedo of N_2 ice and the subsequent increased sunlight absorption in the western regions of Sputnik Planitia is sufficient to invert the net surface energy balance (from negative to positive) and lead to a net diurnal sublimation in these regions, whereas the bright central and eastern regions (located at same latitudes) remain dominated by net condensation. Initially, a variation in downward nighttime sensible heat flux and/or haze deposition (and eventually erosion and ice grains transport, but less likely) could have triggered a slight contrast in albedo and composition between the northwestern and central plains of Sputnik Planitia, then boosted and maintained by runaway mechanisms.

7.2. Methane-Rich Bladed Terrain Deposits

A striking longitudinal asymmetry observed on Pluto by New Horizons is the presence of the CH_4-rich bladed terrain deposits east of Sputnik Planitia (*Moore et al.*, 2018), while the uplands of Cthulhu Macula to the southwest of the ice sheet are mostly volatile-free and covered by a thick mantle of dark red material, probably several meters thick. It is possible that CH_4-rich ice accumulated at the location of the bladed terrains in Pluto's past because of the initially higher topography of these regions (see the chapter by Moore et al. in this issue). However, could the atmospheric circulation also play a role in the processes leading to this asymmetry? For instance, during periods of equatorial accumulation of CH_4-rich ice, the retrorotation regime of the atmosphere and the injection of cold N_2-rich air from Sputnik Planitia could transport and push gaseous CH_4 westward (*Bertrand et al.*, 2019), thus favoring the accumulation of CH_4 ice at the westernmost longitudes (i.e., east of Sputnik Planitia). This remains to be explored with climate models. A very small difference in accumulation between east (Cthulhu) and west (Tartarus Dorsa) longitudes could have been sufficient at first to trigger this asymmetry, because CH_4-rich ice accumulation in the west and haze accumulation darkening the surface of Cthulhu in the east would induce very efficient positive amplifying feedbacks strengthening these resurfacing processes (*Earle et al.*, 2018).

The ridges (or "blades") of the bladed terrain deposits display a dominant north-south orientation (*Moores et al.*, 2017; *Moore et al.*, 2018), which may also originate in part from atmosphere dynamical processes, although it may also be a north-south-aligned sublimation texture due to its equatorial location, as suggested by *Moore et al.* (2018). *Moores et al.* (2017) showed that these ridges may form like penitentes and that their orientation would be controlled by the wind direction varying through time. Further investigations are needed to fully understand the processes leading to these peculiar geological formations.

7.3. East Tombaugh Regio

Another region of interest is the eastern part of Tombaugh Regio (the right lobe of the heart). This surface is relatively bright and covered by N_2-rich and CH_4-rich frosts. The formation of permanent and seasonal volatile ice deposits in this region is not yet understood. We could imagine that the gaseous CH_4 injected in the atmosphere from the sublimation of the CH_4-rich bladed terrain deposits could be transported westward by the retrograde winds predicted by the LMD Pluto GCM, and could quickly recondense in Tombaugh Regio, forming bright ice deposits there. Albedo feedbacks could then be sufficient to trigger more CH_4 and N_2 condensation in this region (*Bertrand et al,* 2020a; *Earle et al.,* 2018). However, the condensation of CH_4 west of the bladed terrain is not verified everywhere on Pluto. For instance, bladed terrain deposits are observed east of the Krun Macula region (southeast of Sputnik Planitia), but this region remains dark and is not covered by bright CH_4 frosts.

7.4. Methane-Rich Frost on Mountain Tops

In the equatorial regions, CH_4 condensation seems to be favored at high altitude. Whereas most of Cthulhu appears depleted in volatile ice (*Schmitt et al.,* 2017), the LORRI and MVIC instruments revealed the presence of small amounts of CH_4-rich frost on top of the highest peaks (*Earle et al.,* 2018; *Protopapa et al.,* 2017). For instance, the isolated high-altitude mountain chains, known as Pigafetta Montes, are capped with bright CH_4-rich frosts ($A \approx 0.65$; *Buratti et al.,* 2017) above ~1.5 km altitude with a striking resemblance to terrestrial alpine landscapes.

Bertrand et al. (2020b) found that high-resolution simulations performed with the LMD GCM could predict high-altitude CH_4 ice accumulation where the frost-capped mountains are observed, and that the process at work is related to atmospheric dynamics. This process is completely different from what is observed on Earth, where atmospheric temperatures decrease with altitude (mostly because of adiabatic cooling), inducing condensation at higher altitude. Conversely, on Pluto, atmospheric temperatures strongly increase with altitude in the first kilometers above the surface. The ground temperature itself is not influenced by the altitude of the surface because the atmosphere is very thin. During the night, the surface is therefore a cold trap where methane will condense if its mixing ratio is high enough. The GCM predicts that the near-surface atmospheric CH_4 mixing ratio strongly varies with altitude in the equatorial regions, with an enrichment in gaseous CH_4 above ~4 km altitude and a depletion in the lowest levels of the atmosphere. This explains why the highest mountains, which peak into the CH_4-enriched air, get capped by frost. Why is Pluto's equatorial atmosphere depleted in gaseous CH_4 near the surface and enriched at higher altitudes? The model shows that ascending motions in the western regions of Sputnik Planitia are responsible for this enrichment by transporting gaseous CH_4 upward. *Bertrand et al.* (2020b) suggested that two different mechanisms are involved. First, as mentioned above, as N_2 sublimes in the north of Sputnik Planitia and condenses in the south, it creates a circulation cell that transports methane southward and toward higher altitudes above the equatorial regions. Second, the interaction between the zonal flow and the high western boundary of Sputnik Planitia's basin produces vertical upward motions that also contribute to the vertical transport of gaseous CH_4.

8. SUMMARY AND CONCLUSIONS

Pluto's atmosphere provides an exceptional natural laboratory to test understanding of atmospheric dynamics, as it differs from other terrestrial atmospheres in the solar system owing to a unique near-surface stratosphere, with a very long radiative timescale and a general circulation that is probably controlled by N_2 condensation-sublimation flows. However, in spite of the New Horizons data revolution, not much observational data on the circulation of the atmosphere is available. Three-dimensional numerical global climate models have therefore been used to provide a basic knowledge of the atmospheric circulation. Near the surface, local katabatic winds flowing down most slopes on Pluto are superimposed on a regional and local sublimation component flowing from the subliming sunlit nitrogen ice to the condensing deposits, with a Coriolis turning of the wind as the air travels from one hemisphere to the other. At higher altitude, different circulation regimes are possible, from a prograde superrotation to a retrograde global rotation. However, the simulations performed with surface conditions most consistent with New Horizons observations suggest that the most likely regime is a global "retrorotation" forced by the conservation of angular momentum of air particles flowing above the equator in the condensation-sublimation flow. In this regard, the circulation across the Sputnik Planitia basin seems sufficient to force the entire zonal circulation independently of what is assumed regarding N_2 condensation in the south polar region. Pluto's atmosphere likely contains tides, gravity waves, and barotropic waves induced by jet instabilities. This is strongly suggested by the density fluctuations observed in stellar occultations and by the layered structure of atmospheric hazes. However, much remains to be done to understand these observations and their implications for Pluto's atmospheric dynamics. In particular, new very-high-resolution models may be needed to reproduce these enigmatic observations.

Clearly, new observations are needed to confirm our view of the general circulation of Pluto's atmosphere. In the near future, telescopic observations may reveal spatial variations in the temperature field (*Gurwell et al.,* 2019).

It might even be possible to retrieve the magnitude and the direction of the zonal wind by measuring the Doppler shift induced by the zonal circulation, e.g., by monitoring one of the numerous atmospheric CO lines at submillimeter wavelengths with radio telescopes. However, to fully measure the time-evolving atmospheric circulation and thermal structure, a Pluto orbiter would be required. For example, a New Horizons-like payload (imaging, NIR spectroscopy, radio science) could be complemented by thermal infrared mapping of the surface and thermal sounding of the atmosphere by using CH_4 or CO lines. In fact, sounding could ideally be performed by a submillimeter heterodyne instrument resolving CO line emissions, which could also make Doppler wind measurements at the limb and provide a temporal and spatial monitoring of the circulation. This could also be done by wind light detection and ranging (LiDAR) using the reflection of a laser signal on the aerosols to measure their velocity, and by extension, the ambient winds.

Another source of information might come from the exploration of the largest satellite of Neptune, Triton, which resembles Pluto in many aspects as both bodies share similar sizes, densities, atmospheric composition, pressures and temperatures, and types of surface ices (N_2, CO, CH_4, H_2O). If Triton has a permanent northern polar cap of N_2 ice (in addition to the southern cap observed by Voyager 2), then the mean meridional circulation should be dominated by an open south-to-north (in current season) or north-to-south (in the opposite season) circulation cell leading to retrorotation, as suggested for Pluto. However, Triton's atmosphere differs from Pluto's by its thermal structure. An 8-km-deep cold troposphere has been revealed by radio occultations and confirmed by the observations of geysers that are blown away once they reach this altitude (*Yelle et al.,* 1995). This structure is probably related to Triton's atmosphere containing less CH_4 (by a factor of 10) and less haze than Pluto's atmosphere. If hazes can decrease the radiative timescale from 10 to 1 Earth on Pluto, as hypothesized by *Zhang et al.* (2017), this could lead to fundamental differences in the atmosphere dynamics of both objects.

Until future missions to Pluto and Triton becomes possible or likely, the analysis of the New Horizons dataset will continue for many years. In particular, it is likely that an improved understanding of the formation of the various surface deposits will provide key information on Pluto's atmospheric dynamics.

Acknowledgments. We are grateful to the New Horizon's team and in particular to the Atmospheres Science Theme Team (especially R. Gladstone, M. Summers, and L. Young) for their early interest in our work. We thank A. Ingersoll, A. Stern, M. Summers, L. Young, and an anonymous reviewer for their reviews and advice on this chapter. F. F. and T. B. benefited from support from CNES. T. B. was also supported for this research by an appointment to the National Aeronautics and Space Administration (NASA) Post-doctoral Program at the Ames Research Center administered by the Universities Space Research Association (USRA) through a contract with NASA. D. H. was supported by the New Horizons Project and A. T. acknowledges the previous support of NASA research program grants to enable his contributions.

REFERENCES

Anderson D. L. T. (1976) The low-level jet as a western boundary current. *Monthly Weather Review, 104,* 907.

Bertrand T. (2017) Préparation et analyses des observations de l'atmosphère et des glaces de Pluton par la mission NASA New Horizons à l'aide de modèles numériques de climat. Ph.D. thesis, Université Pierre et M. Curie, France, available online at *https://tel.archives-ouvertes.fr/tel-01746075v2.*

Bertrand T. and Forget F. (2016) Observed glacier and volatile distribution on Pluto from atmosphere-topography processes. *Nature,* 987.

Bertrand T. and Forget F. (2017) 3D modeling of organic haze in Pluto's atmosphere. *Icarus,* 287, 72–86, DOI: 10.1016/j.icarus.2017.01.016.

Bertrand T., Forget, F., Umurhan O. M., Grundy W. M., Schmitt B., Protopapa S., Zangari A. M., et al. (2018) The nitrogen cycles on Pluto over seasonal and astronomical timescales. *Icarus,* 309, 277–296.

Bertrand T., Forget F., Umurhan O. M., Moore J. M., Young L. A., Protopapa S., Grundy W. M., Schmitt B., Dhingra R. D., and Binzel R. P. (2019) The CH_4 cycles on Pluto over seasonal and astronomical timescales. *Icarus, 329,* 148.

Bertrand T., Forget F., White O., Schmitt B., Stern S. A., Weaver H. A., Young L. A., Ennico K., and Olkin C. B (2020a) Pluto's beating heart regulates the atmospheric circulation: Results from high-resolution and multiyear numerical climate simulations. *J. Geophys. Res.–Planets, 125,* e06120.

Bertrand T., Forget F., Schmitt B., White O. L., and Grundy W. M. (2020b) Equatorial mountains on Pluto are covered by methane frosts resulting from a unique atmospheric process. *Nature Commun., 11,* 5056.

Bosh A. S., Person M. J., Levine S. E., Zuluaga C. A., Zangari A. M., Gulbis A. A. S., Schaefer G. H., et al. (2015) The state of Pluto's atmosphere in 2012–2013. *Icarus, 246,* 237–246.

Bougher S. W., Roble R. G., Ridley E. C., and Dickinson R. E. (1990) The Mars thermosphere: 2. General circulation with coupled dynamics and composition. *J. Geophys. Res., 95,* 14811–14827.

Brosch N. (1995) The 1985 stellar occultation by Pluto. *Mon. Not. R. Astron. Soc., 276(2),* 571–578.

Brozović M., Showalter M. R., Jacobson R. A., and Buie M. W. (2015) The orbits and masses of satellites of Pluto. *Icarus, 246,* 317–329.

Buratti B. J. and 16 colleagues (2017) Global albedos of Pluto and Charon from LORRI New Horizons observations. *Icarus 287,* 207.

Chapman S. and Lindzen R. S. (1970) *Atmospheric Tides — Thermal and Gravitational.* Gordon and Breach, London. 200 pp.

Cheng A. F. and 13 colleagues (2017) Haze in Pluto's atmosphere. *Icarus, 290,* 112–133.

Cruikshank D. P., Pilcher C. B., and Morrison D. (1976) Pluto: Evidence for methane frost. *Science, 194,* 835–837.

Del Genio A. D. and Zhou W. (1996) Simulations of superrotation on slowly rotating planets: Sensitivity to rotation and initial condition. *Icarus, 120,* 332.

Dias-Oliveira A., Sicardy B., Lellouch E., et al. (2015) Pluto's atmosphere from stellar occultations in 2012 and 2013. *Astrophys. J., 811(1),* 53.

Earle A. M. and 8 colleagues (2018) Albedo matters: Understanding runaway albedo variations on Pluto. *Icarus 303,* 1.

Elliot J. L., Dunham E. W., Bosh A. S., Slivan S. M., Young L. A., Wasserman L. H., and Millis R. L. (1989) Pluto's atmosphere. *Icarus, 77(1),* 148–170.

Elliot J. L., Stansberry J. A., Olkin C. B., Agner M. A., and Davies M. E. (1997) Triton's distorted atmosphere. *Science,* 278, 436–439.

Elliot J. L., Ates A., Babcock B. A., Bosh A. S., Buie M. W., Clancy K. B., Dunham E. W., et al. (2003) The recent expansion of Pluto's atmosphere. *Nature, 424(6945),* 165–168.

Elliot J. L., Person M. J., Gulbis A. A. S., Souza S. P., Adams E. R., Babcock B. A., Gangestad J. W., et al. (2007) Changes in Pluto's atmosphere: 1988–2006. *Astron. J., 134(1),* 1–13.

Forget F., Hourdin F., and Talagrand O. (1998) CO_2 Snow fall on Mars: Simulation with a general circulation model. *Icarus, 131,* 302–316.

Forget F., Hourdin F., Fournier R., Hourdin C., Talagrand O., Collins M., Lewis S. R., Read P. L., and Huot J.-P. (1999) Improved general circulation models of the martian atmosphere from the surface to above 80 km. *J. Geophys. Res., 104 (24),* 155–176.
Forget F., Bertrand T., Vangvichith M., Leconte J., Millour E., and Lellouch E. (2017) A post-New Horizons global climate model of Pluto including the N_2, CH_4, and CO cycles. *Icarus, 287,* 54–71.
Fray N. and Schmitt B. (2009) Sublimation of ices of astrophysical interest: A bibliographic review. *Planet. Space Sci., 57,* 2053–2080.
French R. G., Toigo A. D., Gierasch P. J., Hansen C. J., Young L. A., Sicardy B., Dias-Oliveira A., and Guzewich S. D. (2015) Seasonal variations in Pluto's atmospheric tides. *Icarus, 246,* 47–267.
Geissler P. E. (2005) Three decades of martian surface changes. *J. Geophys. Res.–Planets, 110,* E02001.
Gladstone G. R. and 32 colleagues (2016) The atmosphere of Pluto as observed by New Horizons. *Science, 351,* aad8866.
Gonzalez-Galindo F., Forget F., Lopez-Valverde M. A., Angelats I Coll M., and Millour E. (2009) A ground-to-exosphere martian general circulation model: 1. Seasonal, diurnal, and solar cycle variation of thermospheric temperatures. *J. Geophys. Res., 114,* E04001.
Greeley R., Skypeck A., and Pollack J. B. (1993) Martian aeolian features and deposits: Comparisons with general circulation model results. *J. Geophys. Res., 98,* 3183.
Grundy W. M. and 33 colleagues (2016) Surface compositions across Pluto and Charon. *Science, 351,* aad9189.
Gulbis A. A. S., Bus S. J., Elliot J. L., Rayner J. T., Stahlberger W. E., Rojas F. E., Adams E. R., et al. (2011) First results from the MIT Optical Rapid Imaging System (MORIS) on the IRTF: A stellar occultation by Pluto and a transit by exoplanet XO-2b. *Publ. Astron. Soc. Pac., 123(902),* 461–469.
Gulbis A. A. S., Emery J. P., Person M. J., Bosh A. S., Zuluaga C. A., Pasachoff J. M., and Babcock B. A. (2015) Observations of a successive stellar occultation by Charon and graze by Pluto in 2011: Multiwavelength SpeX and MORIS data from the IRTF. *Icarus, 246,* 226–236.
Gurwell M. A. and 6 colleagues (2019) The atmospheres of Pluto and Triton: Investigations with ALMA. In *Pluto System After New Horizons,* Abstract #7060. LPI Contribution No. 2133, Lunar and Planetary Institute, Houston.
Haberle R. M., Pollack J. B., Barnes J. R., et al. (1993) Mars atmospheric dynamics as simulated by the NASA Ames General Circulation Model: 1. The zonal-mean circulation. *J. Geophys. Res., 98(E2),* 3093, DOI: 10.1029/ 92JE02946.
Haberle R. M., Houben H. C., Barnes J. R., and Young R. E. (1997) A simplified three-dimensional model for martian climate studies. *J. Geophys. Res., 102(E4),* 9051, DOI: 10.1029/ 97JE00383.
Hansen C. J. and Paige D. A. (1996) Seasonal nitrogen cycles on Pluto. *Icarus, 120,* 247.
Hansen C. J., Paige D. A., and Young L. A. (2015) Pluto's climate modeled with new observational constraints. *Icarus,* 246, 183–191.
Hinson D. P. and 15 colleagues (2017) Radio occultation measurements of Pluto's neutral atmosphere with New Horizons. *Icarus, 290,* 96–111.
Hubbard W. B., Hunten D. M., Dieters S. W., Hill K. M., and Watson R. D. (1988) Occultation evidence for an atmosphere on Pluto. *Nature, 336(6198),* 452–454.
Hubbard W. B., McCarthy D. W., Kulesa C. A., Benecchi S. D., Person M. J., Elliot J. L., and Gulbis A. A. S. (2009) Buoyancy waves in Pluto's high atmosphere: Implications for stellar occultations. *Icarus, 204,* 284–289.
Joshi M. M., Lewis S. R., Read P. L., and Catling D. C. (1994) Western boundary currents in the atmosphere of Mars. *Nature, 367,* 548.
Joshi M. M., Lewis S. R., Read P. L., and Catling D. C. (1995) Western boundary currents in the Martian atmosphere: Numerical simulations and observational evidence. *J. Geophys. Res., 100,* 5485.
Kammer J. A. and 15 colleagues (2020) New Horizons Observations of an ultraviolet stellar occultation and appulse by Pluto's atmosphere. *Astron. J. 159,* 26.
Kaspi Y. and Showman A. P. (2015) Atmospheric dynamics of terrestrial exoplanets over a wide range of orbital and atmospheric parameters. *Astrophys. J., 804(60).*
Kuiper G. P. (1950) The diameter of Pluto. *Publ. Astron. Soc. Pac., 62,* 133-137.
Lebonnois S., Hourdin F., Eymet V., Crespin A., Fournier R., and Forget F. (2010) Super-rotation of Venus' atmosphere analyzed with a full general circulation model. *J. Geophys. Res.–Planets, 115,* 6006.
Lebonnois S., Burgalat J., Rannou P., and Charnay B. (2012) Titan global climate model: A new 3-dimensional version of the IPSL titan GCM. *Icarus, 218,* 707–722.
Lellouch E. et al. (2009) Pluto's lower atmosphere structure and methane abundance from high-resolution spectroscopy and stellar occultations. *Astron. Astrophys., 495,* L17–L21.
Lellouch E., de Bergh C., Sicardy B., Käufl H. U., and Smette A. (2011) High resolution spectroscopy of Pluto's atmosphere: Detection of the 2.3 μm CH_4 bands and evidence for carbon monoxide. *Astron. Astrophys., 530,* L4, DOI:10.1051/0004-6361/201116954.
McCarthy D. W., Hubbard W. B., Kulesa C. A., Benecchi S. D., Person M. J., Elliot J. L., and Gulbis A. A. S. (2008) Long-wavelength density fluctuations resolved in Pluto's high atmosphere. *Astron. J., 136(4),* 1519–1522.
McKinnon W. B. and 14 colleagues (2016) Convection in a volatile nitrogen-ice-rich layer drives Pluto's geological vigour. *Nature, 534,* 82–85.
Meza E., Sicardy B., Assafin M., et al. (2019) Lower atmosphere and pressure evolution on Pluto from ground-based stellar occultations, 1988–2016. *Astron. Astrophys., 625,* A42.
Millis R. L. , Wasserman L. H., Franz O. G., Nye R. A., Elliot J. L., Dunham E. W., Bosh A. S., et al. (1993) Pluto's radius and atmosphere: Results from the entire 9 June 1988 occultation data set. *Icarus, 105(2),* 282–297.
Moore J. M. and 40 colleagues (2016) The geology of Pluto and Charon through the eyes of New Horizons. *Science, 351,* 1284–1293.
Moore J. M. and 25 colleagues (2018) Bladed terrain on Pluto: Possible origins and evolution. *Icarus, 300,* 129.
Moores J. E., Smith C. L., Toigo A. D., and Guzewich S. D. (2017) Penitentes as the origin of the bladed terrain of Tartarus Dorsa on Pluto. *Nature,* 541, 188.
Nimmo F. and 16 colleagues (2017) Mean radius and shape of Pluto and Charon from New Horizons images. *Icarus, 287,* 12–29.
Olkin C. B., Young L. A., French R. G., Young E. F., Buie M. W., Howell R. R., Regester J., Ruhland C. R., Natusch T., and Ramm D. J. (2014) Pluto's atmospheric structure from the July 2007 stellar occultation, *Icarus,* 239, 15–22.
Olkin C. B., Young L. A., Borncamp D., Pickles A., Sicardy B., Assafin M., Bianco F. B., et al. (2015) Evidence that Pluto's atmosphere does not collapse from occultations including the 2013 May 04 event. *Icarus,* 246, 220–225.
Owen T. C., Roush T. L., Cruikshank D. P., et al. (1993) Surface ices and atmospheric composition of Pluto. *Science, 261,* 745–748.
Pasachoff J. M., Souza S. P., Babcock B. A., Ticehurst D. R., Elliot J. L., Person M. J., Clancy K. B., Roberts L. C., Hall D. T., and Tholen D. J. (2005) The structure of Pluto's atmosphere from the 2002 August 21 stellar occultation. *Astron. J., 129(3),* 1718–1723.
Pasachoff J. M., Person M. J., Bosh A. S., Sickafoose A. A., Zuluaga C., Kosiarek M. R., Levine S. E., et al. (2016) Trio of stellar occultations by Pluto one year prior to New Horizons' arrival. *Astron. J., 151(4),* 97.
Pasachoff J. M., Babcock B. A., Durst R. F., Seeger C. H., Levine S. E., Bosh A. S., Person M. J., et al. (2017) Pluto occultation on 2015 June 29 UTC with central flash and atmospheric spikes just before the New Horizons flyby. *Icarus,* 296, 305–314.
Person M. J., Elliot J. L., Gulbis A. A. S., Zuluaga C. A., Babcock B. A., Mckay A. J., Pasachoff J. M., et al. (2008) Waves in Pluto's upper atmosphere. *Astron. J., 136(4),* 1510–1518.
Person M. J., Dunham E. W., Bosh A. S., et al. (2013) The 2011 June 23 stellar occultation by Pluto: Airborne and ground observations. *Astron. J., 146(4),* 83.
Protopapa S. and 22 colleagues (2017) Pluto's global surface composition through pixel-by-pixel Hapke modeling of New Horizons Ralph/LEISA data. *Icarus, 287,* 218.
Read P. L. (2011) Dynamics and circulation regimes of terrestrial planets. *Planet. Space Sci., 59,* 900–914.
Richardson M. I., Toigo A. D., amd Newman C. E. (2007) PlanetWRF: A general purpose, local to global numerical model for planetary atmospheric and climate dynamics. *J. Geophys. Res., 112,* E09001.
Schenk P. M. and 19 colleagues (2018) Basins, fractures, and volcanoes: Global cartography and topography of Pluto from New Horizons. *Icarus, 314,* 400.
Schmitt B. and 28 colleagues (2017) Physical state and distribution of materials at the surface of Pluto from New Horizons LEISA imaging spectrometer. *Icarus, 287,* 229.
Selsis F., Wordsworth R. D., and Forget F. (2011) Thermal phase curves

of nontransiting terrestrial exoplanets. I. Characterizing atmospheres. *Astron. Astrophys., 532,* A1.
Sicardy B., Widemann T., Lellouch E., et al. (2003) Large changes in Pluto's atmosphere as revealed by recent stellar occultations. *Nature, 424(6945),* 168–170.
Sicardy B., Talbot J., Meza E., et al. (2016) Pluto's atmosphere from the 2015 June 29 ground-based stellar occultation at the time of the New Horizons flyby. *Astrophys. J. Lett., 819(2),* L38.
Skamarock W. C. et al. (2008) *A Description of the Advanced Research WRF Version 3.* Tech. Rept. No. NCAR/TN-475+STR, National Center for Atmospheric Research, Boulder, DOI: 10.5065/D68S4MVH.
Spencer J. R., Stansberry J. A., Trafton L. M., Young E. F., Binzel R. P., and Croft S. K. (1997) Volatile transport, seasonal cycles, and atmospheric dynamics on Pluto. In *Pluto and Charon* (S. A. Stern and D. J. Tholen, eds.), pp. 435–473. Univ. of Arizona, Tucson.
Stern S. A. and Trafton L. (1984) Constraints on bulk composition, seasonal variation, and global dynamics of Pluto's atmosphere. *Icarus, 57,* 231.
Stern S. A. and 150 colleagues (2015) The Pluto system: Initial results from its exploration by new horizons. *Science, 350,* 1815.
Stern S. A. and 12 colleagues (2017) Evidence for possible clouds in Pluto's present-day atmosphere. *Astron. J., 154,* 43.
Strobel D. F., Zhu X., Summers M. E., and Stevens M. H. (1996) On the vertical thermal structure of Pluto's atmosphere. *Icarus, 120,* 266–289.
Telfer M. W. and 19 colleagues (2018) Dunes on Pluto. *Science, 360,* 992.
Tholen D. J., Buie M. W., Grundy W. M., and Elliott G. T. (2008) Masses of Nix and Hydra. *Astron. J., 135,* 777–784.
Thomas P., Veverka J., Lee S., and Bloom A. (1981) Classification of wind streaks on Mars. *Icarus, 45,* 124.
Thomson S. I. and Vallis G. K. (2019) The effects of gravity on the climate and circulation of a terrestrial planet. *Q. J. R. Meteor. Soc., 145,* 2627.
Toigo A. D., Gierasch P. J., Sicardy B., and Lellouch E. (2010) Thermal tides on Pluto. *Icarus, 208(1),* 402–411.
Toigo A. D., French R. G., Gierasch P. J., Guzewich S. D., Zhu X., Richardson M. I. (2015) General circulation models of the dynamics of Pluto's volatile transport on the eve of the new horizons encounter. *Icarus, 254,* 306–323.
Tokano T. and Neubauer F. M. (2002) Tidal winds on Titan caused by Saturn. *Icarus, 158(2),* 499–515.
Tyler G. L., Linscott I. R., Bird M. K., Hinson D. P., Strobel D. F., Pätzold M., Summers M. E., and Sivaramakrishnan K. (2008) The New Horizons Radio Science Experiment (REX). *Space Sci. Rev., 140,* 217–259.
Vangvichith M. (2013) Modélisation des Atmosphères et des Glaces de Pluton et Triton. Ph.D. thesis, Ecole Polytechnique France, available online at *http://www.sudoc.fr/17695547X.*
Wang Y., Read P. L., Tabataba-Vakili F., Young R. M. B. (2018) Comparative terrestrial atmospheric circulation regimes in simplified global circulation models. Part I: From cyclostrophic super-rotation to geostrophic turbulence. *Q. J. R. Meteor. Soc., 144,* 2537.
Williams G. P. (1988) The dynamical range of global circulations I. *Climate Dynam., 2,* 205.
White O. L. and 24 colleagues (2017) Geological mapping of Sputnik Planitia on Pluto. *Icarus, 287,* 261.
Yelle R. V. and Elliot J. L. (1997) Atmospheric structure and composition: Pluto and Charon. In *Pluto and Charon* (S. A. Stern and D. J. Tholen, eds.), pp. 347–390. Univ. of Arizona, Tucson.
Yelle R. V., Lunine J. L., Pollack J. B., and Brown R. H. (1995) lower atmospheric structure and surface-atmosphere interaction on Triton. In *Neptune and Triton* (D. P. Cruikshank, ed.), pp. 1047–1122. Univ. of Arizona, Tucson.
Young E. F., French R. G., Young L. A., Ruhland C. R., Buie M. W., Olkin C. B., Regester J., et al. (2008) Vertical structure in Pluto's atmosphere from the 2006 June 12 stellar occultation. *Astron. J., 136(5),* 1757–1769.
Young L. A. (2013) Pluto's seasons: New predictions for New Horizons. *Astrophys. J. Lett., 766(2),* L22.
Young L. A., Elliot J. L., Tokunaga A., de Bergh C., and Owen T. (1997) Detection of gaseous methane on Pluto. *Icarus, 127,* 258, DOI: 10.1006/icar.1997.5709.
Young L. A. et al. (2018) Structure and composition of Pluto's atmosphere from the New Horizons solar ultraviolet occultation. *Icarus, 300,* 174–199.
Zalucha A. M. and Michaels T. I. (2013) A 3D general circulation model for Pluto and Triton with fixed volatile abundance and simplified surface forcing. *Icarus,* 223, 819–831.
Zhang X., Strobel D. F., and Imanaka H. (2017) Haze heats Pluto's atmosphere yet explains its cold temperature. *Nature, 551,* 352–355, DOI: 10.1038/nature24465.

Young L. A., Bertrand T., Trafton L. M., Forget F., Earle A. M., and Sicardy B. (2021) Pluto's volatile and climate cycles on short and long timescales. In *The Pluto System After New Horizons* (S. A. Stern, J. M. Moore, W. M. Grundy, L. A. Young, and R. P. Binzel, eds.), pp. 321–361. Univ. of Arizona, Tucson, DOI: 10.2458/azu_uapress_9780816540945-ch014.

Pluto's Volatile and Climate Cycles on Short and Long Timescales

Leslie A. Young
Southwest Research Institute

Tanguy Bertrand
NASA Ames Research Center

Laurence M. Trafton
McDonald Observatory, University of Texas at Austin

François Forget
Laboratoire de Météorologie Dynamique (LMD/IPSL), Sorbonne Université

Alissa M. Earle
Massachusetts Institute of Technology

Bruno Sicardy
LESIA, Observatoire de Paris, Université PSL, CNRS, Sorbonne Université, Univ. Paris Diderot, Sorbonne Paris Cité

The volatiles on Pluto's surface — N_2, CH_4, and CO — are present in its atmosphere as well. The movement of volatiles affects Pluto's surface and atmosphere on multiple timescales. On diurnal timescales, N_2 is transported from areas of high to low insolation, and the latent heat of sublimation or condensation maintains a nearly isobaric atmosphere. On seasonal (orbital) timescales, Pluto's atmosphere changes its pressure by orders of magnitude, but most models predict that it is unlikely to collapse even at aphelion due to the equatorial N_2 source in Sputnik Planitia and the high thermal inertia of the subsurface. On seasonal timescales, meters of N_2 ice are transported across Pluto's surface, but it is not yet clear from models how much of this transport is between areas that maintain N_2 over an entire year (such as Sputnik Planitia) and to what extent deposition creates new volatile-covered areas (of either N_2-rich or CH_4-rich ice) or sublimation reveals underlying terrain. Pluto's orbit and obliquity variations on timescales of ~3 million years (m.y.) (a Milankovitch cycle) induce considerable climate changes along with local accumulation or erosion of meter- to kilometer-thick layers of volatile ice. In a non-cyclical process, volatiles filled the large depression that is now Sputnik Planitia.

1. INTRODUCTION

Much of Pluto's surface and atmosphere can be understood by considering its volatile and climate cycles (e.g., *Spencer et al.,* 1997). Pluto has three volatile ices on its surface and in its atmosphere: N_2, CH_4, and CO. N_2 is the main constituent in Pluto's atmosphere, and is present on large areas of its surface. Unlike Earth, Venus, or Titan, Pluto's main atmospheric constituent is able to condense on its surface. This leads to important interactions between the atmosphere and surface via relaxation to solid-gas equilibrium. There are similarities with other atmospheres whose main constituent is also solid on the surface: Mars, Triton, Io (e.g., *Trafton et al.,* 1988) to some extent, and probably other volatile-bearing dwarf planets in the Kuiper belt at some point in their orbits (e.g., *Young et al.,* 2020). Despite their differences in temperature, surface pressure, composition, and miscibility of the main and secondary constituents, these vapor-pressure-equilibrium atmospheres share three important characteristics. First, because pressure is very sensitive to temperature, and temperature is sensitive to insolation, albedo, and volatile distribution, we find that vapor-pressure-equilibrium atmospheres vary widely over time. For example, the martian surface pressure varies by ~30% over its year (e.g., *Hourdin et al.,* 1993), while those of Pluto and Triton vary by orders of magnitude (e.g., *Trafton et al.,* 1988). Second, because of latent heat exchanges, sublimation and condensation can make up for energy imbalances between insolation and thermal emission, so there is transport of volatiles keeping a globally near-uniform surface pressure if the atmosphere is thick enough (*Leighton and Murray,* 1966; *Trafton and Stern,* 1983). Third, because the ices are subliming and condensing, they are inherently out of thermodynamic equilibrium. The atmospheres supported by vapor pressure are sensitive to departures of the volatile system from thermal equilibrium, which is otherwise prescribed by the equilibrium phase diagram.

Insolation differences across Pluto's surface are the major driver for the sublimation and condensation of volatiles (see section 3.1), their transport across Pluto, and the net exchange of volatiles between the surface and atmosphere. The volatile cycle evolves over different timescales. Diurnal sublimation and condensation occur over Pluto's 6.4-day rotation. Seasonal transport and large pressure changes occur over Pluto's 248-year orbit, made more pronounced by Pluto's large eccentricity of 0.25 and obliquity of 119.6°. The character of its seasonal behavior changes over its ~3-m.y. Milankovitch-like cycles in the orbit and pole orientation (*Dobrovolskis et al.,* 1997). Finally, Pluto experiences changes over the 4.5 G.y. of solar system history.

These continual cycles of sublimation and condensation of volatile ices induce geological processes such as resurfacing, glacial flow, bladed terrain, and pit formation, and contribute indirectly to dunes of ice and perhaps to the erosion of the water ice bedrock (e.g., on Al-Idrisi, section 5.2). Because of the volatile cycles, parts of Pluto's surface are among the youngest in the solar system (see the chapter by Moore and Howard in this volume). The volatile cycles are also key drivers of Pluto's atmosphere dynamics, as the condensation and sublimation of N_2-induced winds that strongly impact the general circulation (see the chapter by Forget et al. in this volume).

Several properties of Pluto's surface and subsurface affect and complicate the volatile cycles (section 3). Unlike H_2O and CO_2 ice on Mars, Pluto's volatile ices (N_2, CH_4, CO) mix together to form solid solutions. There is no pure phase for these species on Pluto, owing to their attraction at the molecular level. In thermodynamic equilibrium, the state of Pluto's volatiles thus depends on a ternary phase diagram. Changes in the composition of these mixtures impact the volatile cycles (see section 3.1). In addition, the ices are contaminated by the accumulation of dark materials from sedimentation of haze particles or from processing of the surface by UV irradiation or energetic particles. Nitrogen is likely to transition between its current β crystalline phase to the colder α crystalline phase on seasonal timescales, affecting its emissivity and other properties (*Stansberry and Yelle,* 1999). New Horizons also revealed that the topography is playing a significant role in the distribution of volatile ice (see section 3), as N_2-rich ice is mostly found in the depressions, whereas CH_4-rich ice has been detected at higher altitude, particularly in the equatorial regions. Finally, climate-modeling studies have highlighted how sensitive the sublimation-condensation rates of volatile ices are to the surface albedo, emissivity, thermal inertia, and solid phase of the ice. These properties of the volatile ices lead to complex interactions between the N_2 and CH_4 cycles involving positive or negative feedbacks between changes of surface properties and sublimation-condensation rates.

A clearer insight into the processes controlling Pluto's volatile cycles can help us understand the complex present-day Pluto and the Pluto of the past, as well as make testable predictions for Pluto's near future. Models have been used to simulate Pluto's volatile cycles at different timescales with different levels of simplification.

We review the observations relevant to volatile and climate cycles in section 2, and cover the basic physics of volatile cycles in section 3. In section 4, we review various models, and then present the volatile cycles over diurnal, seasonal, and longer scales in section 5. We explore open questions and future work in section 6.

2. OBSERVATIONS

2.1. Earth-Based Observations

Observations made by groundbased telescopes or space-based observatories such as the Hubble Space Telescope (HST) are important for volatile cycles even in the post-flyby era. They provide the setting for the history of work on Pluto's volatile cycles; some observations are only possible from Earth-based telescopes; and telescopic observations are the only means of assessing long-term changes on Pluto.

2.1.1. Albedo and color maps. While New Horizons cameras resolved geological features over much of Pluto's surface, they could not image the southern latitudes that were in darkness during the New Horizons flyby. In addition, because New Horizons was a flyby, it only observed a snapshot in time of Pluto. For changes over decades and for southern coverage, both important for volatile cycles, we turn to maps from Earth-based telescopes or HST, even though their resolutions to date are at best 250–500 km per resolution element, or a few resolution elements across Pluto's 2380-km diameter. *Buie et al.* (1997) reviewed mapping efforts prior to 1996. *Buie et al.* (2010b) provided important mapping improvements.

The first maps were constructed from the changes in the mean brightness and amplitude of rotational light curves since 1954, or from the series of Pluto and Charon transits and eclipses — collectively termed mutual events — that occurred within a few years of Pluto's 1987 equinox. [Mutual events also gave the first good constraints on Pluto's radius, e.g., 1178 ± 23 km when limb darkening is included (*Young and Binzel,* 1994; see also *Tholen and Buie*, 1997).] Subsequently, Pluto maps were constructed from direct imaging with HST in 1994 and 2002–2003. The most complete analyses from rotational light curves were by *Drish et al.* (1995), from mutual events were by *E. F. Young et al.* (1999, 2001) (Fig. 1a), and from HST maps was by *Buie et al.* (2010b) (Fig. 1b). Pluto's disk was also marginally resolved by modern speckle imaging (*Howell et al.,* 2012), but no maps have been constructed yet from speckle data.

Pluto's changing rotational light curve gives clear evidence for a bright southern pole prior to the most recent equinox (*Andersson and Fix,* 1973; *Buie and Tholin,* 1989; *Drish et al.,* 1995; *Buie et al.,* 1992). Between 1954 and 1986, when the sub-Earth latitude changed from 55°S to 3°S, Pluto's rotational light curve decreased in brightness and generally increased in amplitude, as would occur if a bright and uniform southern pole were seen more directly in 1954 than in 1986, and a variegated darker equator were seen more directly in 1986 than 1954. [The International Astronomical Union (IAU), the New Horizons mission, and many Pluto researchers all now use the rotational north pole (see *Zangari,* 2015).] The conclusion of a bright southern pole depends very little on underlying model assumptions, particularly Pluto's radius. As discussed in *Drish et al.* (1995), all maps that use the 1954 rotational light curve include a bright southern pole. For example, the anti-Charon hemisphere (latitude 180°) of *Buie et al.* (1992) is based on rotational light curves, and presents a bright southern pole for that hemisphere. Mutual events also showed bright southern terrains on the sub-Charon hemisphere centered at 0° longitude (*E. F. Young et al.,* 1999, 2001; *Buie et al.,* 1992). Some of the bright southern terrains presented in the mutual event maps may be affected by the then-uncertain radii and limb-darkening parameters of Pluto and Charon. For example, the mutual event map of *Young et al.* (1999) shows brighter albedos near the limb for ~one-fifth of the projected radius (Fig. 1a), raising concerns about the accuracy of the derived albedos of southernmost latitudes. By the time of the HST observations in 1992 (sub-Earth latitude 12.7°N), the southern polar region was more foreshortened and therefore difficult to image; HST-derived maps for 1992 indicated that the terrains at the southern terminator, 45°S to 78°S, were bright at the sub-Charon hemisphere (0° longitude) but dark at the anti-Charon hemisphere. The HST observations of 2002–2003 (sub-Earth latitude near 30°N) showed even less of the southern terrains and similar longitudinal variation at the southern terminator.

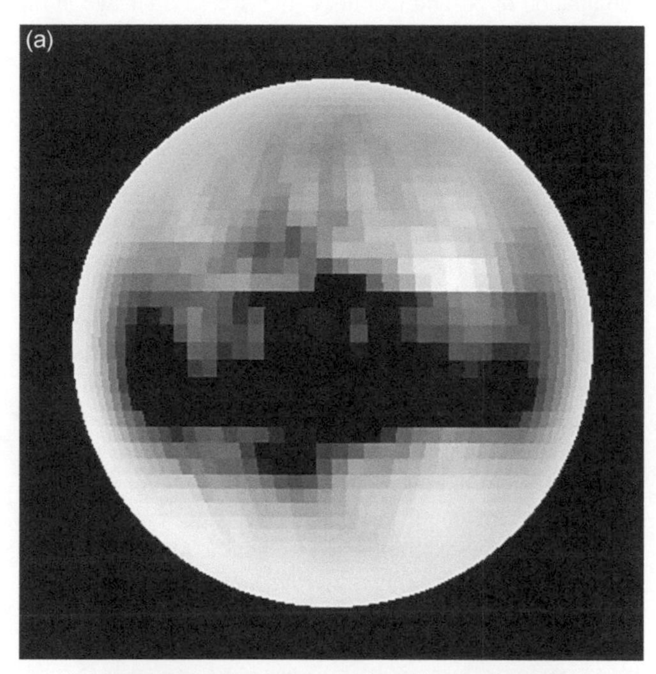

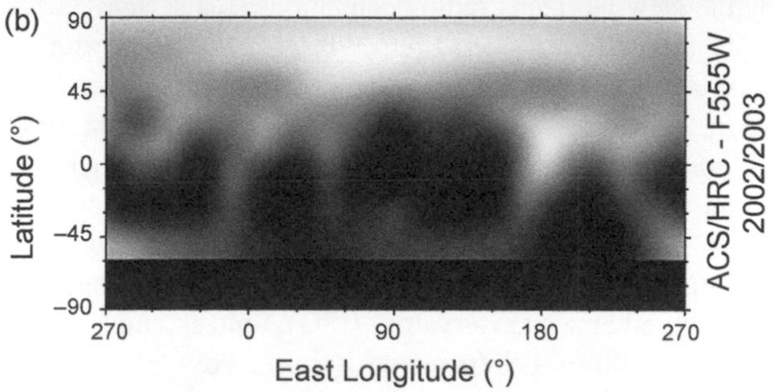

Fig. 1. Pre-encounter maps of Pluto. **(a)** Map of the sub-Charon hemisphere centered on 0° longitude, 0° latitude ca. 1987 (*Young et al.,* 1999). **(b)** Maps of all longitudes (centered on 90°E) and latitudes north of 60°S, ca. 2002–2003 (*Buie et al.,* 2010).

The HST maps and rotational light curves suggest changes in Pluto's surface albedo pattern [not just changes due to viewing geometry, as discussed in *Buie et al.* (2010a,b)]. Pluto darkened by 5% between 1933–1954 in the B filter (445 nm); since the sub-Earth latitude was near 53°S for both epochs, this implies real darkening of the southern terrains, not a mere geometric effect (*Schaefer et al.,* 2008). *Drish et al.* (1995) were unable to make a single static map consistent with rotational light curves from 1954 to 1986; in particular, snapshot maps using data from 1980 to 1986 did not show a bright and extended southern cap, but this snapshot included only a 10° change in subobserver latitude. *Buratti et al.* (2003) compared rotational light curves from 1954 to 1999 with Faint Object Camera (FOC) HST maps (*Stern et al.,* 1997), and concluded that lightcurve amplitudes were consistent with a static model; they also noted that the observed 1999 light curve was darker than models predict, but they took this as only weak evidence for volatile transport. *Buie et al.* (2010b) saw evidence for a reddening trend from 2000 through 2002/2003, and darkening in B between 2002 and 2003, although a static model was consistent with the data in V (551 nm) over the year of data collection. *Buratti et al.* (2015) show a decrease in the lightcurve amplitude from primarily 2013 observations that were much smaller than predicted from the FOC HST maps. New or newly remeasured photometry now exists from 1930 to 1951 (*Buie and Folkner,* 2015); 1961 to 1966 (*Eglitis et al.,* 2018); and 1990 to 1993, 1999, and 2013 (*Buratti et al.,* 2003, 2015). All these datasets can be usefully reassessed now that radii, orbits, photometric properties (e.g., limb-darkening), and a snapshot of a high-resolution map are better known from the New Horizons encounter.

2.1.2. Surface composition and temperature. *Cruikshank et al.* (2015) and the chapter in this volume by Cruickshank et al. review the spectroscopy and composition of Pluto's surface. As with albedo maps, groundbased and HST observations are useful for historical context, for spectral resolution and spectral ranges not reached by New Horizons, and for time variability. Spectra with $\lambda/\Delta\lambda \sim 350$, similar to that of the Linear Etalon Imaging Spectral Array (LEISA) on New Horizons, gave the first evidence for N_2 as a major species, with CO and CH_4 as minor species (*Owen et al.,* 1993). Many of the observations of Pluto's ices have been taken in the near-infrared (IR) (e.g., *Grundy et al.,* 2013; *Cruikshank et al.,* 2015; Cruikshank et al., this volume), but CH_4 is also detectable at visible wavelengths, which gives a longer timebase of observations [e.g., beginning in 1980 (*Grundy and Fink,* 1996)]. Observations in the mid-IR give constraints on the fraction of pure N_2 ice, the nature of the tholins on the surface, and the mixing state of CH_4 (*Olkin et al.,* 2007). Pre-encounter observations were often of the combined spectra of Pluto and Charon. There are no published spectra that resolve the disk of Pluto [although the potential exists, as the OH-Suppressing Infrared Integral Field Spectrograph (OSIRIS) instrument on Keck has resolution comparable to the visible HST maps (*Holler et al.,* 2017)]. Rather, many programs obtain longitudinal information by observing how the spectrum changes as Pluto rotates (*Grundy et al.,* 2013). In Pluto's near-IR spectrum, CH_4 dominates, N_2 is detectable at 2.15 μm, and CO bands are seen at 1.58 and 2.35 μm. Water was not detected pre-encounter (*Grundy and Buie,* 2002). Ethane (C_2H_6) and tholins were also detected spectroscopically prior to New Horizons (*DeMeo et al.,* 2010; *Olkin et al.,* 2007). Non-volatiles, such as H_2O, C_2H_6, and tholins, affect the volatile cycles indirectly; while these components are not available for volatile sublimation, the non-volatile surface albedos and temperatures control which areas are available for volatile condensation. Carbon-dioxide-rich ice has not been detected on Pluto, or elsewhere in the Kuiper belt. A near-IR spectrum of Pluto is presented in the chapter in this volume by Cruikshank et al.

Methane ice on Pluto is seen to be in two states: diluted in N_2-rich ice or as CH_4-rich ice. These can be distinguished in three ways. First, the bands of CH_4 diluted in N_2 are shifted to shorter wavelengths by ~4–6 nm in H and K bands compared to pure CH_4 or CH_4-rich ice (*Douté et al.,* 1999; *Protopapa et al.,* 2015). Second, there is a feature at 1.69 μm that is only present in pure CH_4 ice or CH_4-rich ice (*Douté et al.,* 1999, *Protopapa et al.,* 2015). Third, the spectrum of dilute CH_4 has a shallower slope from 2.9 to 3.2 μm than pure CH_4, and presumably shallower than CH_4-rich ice as well (*Olkin et al.,* 2007).

Modeling of spectra has typically found roughly equal areas of fine-grained CH_4-rich ice and larger-grained N_2-rich ice. The amount of CH_4 diluted in the large-grained N_2-rich ice has been modeled at 0.36% to 0.5% (*Douté et al.,* 1999; *Olkin et al.,* 2007; *Protopapa et al.,* 2008; *Merlin et al.,* 2010). Laboratory work shows N_2 becomes saturated in CH_4:N_2 ice at ~3% at Pluto temperatures (*Prokhvatilov and Yantsevich,* 1983; see phase diagram in the chapter by Cruikshank et al.). At these small amounts, the effect of the diluted N_2 is not detectable in the spectrum of the fine-grained CH_4-rich ice. Although many early papers (e.g., *Douté et al.,* 1999) refer to areas of "pure CH_4 ice," spectroscopically these ices can be either pure CH_4 or CH_4-rich. Thermodynamic arguments suggest that all the areas described as "pure CH_4" are actually CH_4-rich (see section 3.1.1).

Pluto's Earth-based disk-integrated spectrum shows a clear variation with subsolar longitude, including a maximum in the N_2 and CO absorption signature at the longitudes now identified with Sputnik Planitia (e.g., *Grundy and Buie,* 2001; *Grundy et al.,* 2013). Compared to rotational variation, decade-long changes are subtle. To measure how Pluto's surface changes with time requires monitoring with a consistent instrument and observing procedure to minimize systematic errors as well as the careful removal of the rotational signature. Two such programs are detailed in *Grundy and Fink* (1996) and *Grundy et al.* (2013, 2014). Together, these show a long-term rise in the CH_4 absorption features and a recent decline in the strength of the N_2 and CO absorption features. *Grundy et al.* (2014) argue that these changes are not explained by changes in viewing geometry alone, using pre-encounter albedo and composition maps. Future work is needed to determine if the observed decrease is due to the loss of N_2 and CO ices from the visible hemisphere, preferential sublimation of N_2 and CO in N_2-rich ice, or a change in the surface texture that decreases the absorption band, or if the changes are consistent with the New Horizons composition map and changing viewing geometry only.

The temperature of the N_2-rich ice is critical for the study of volatile cycles, because the atmospheric surface pressure can be inferred from vapor-pressure equilibrium and the temperature of the N_2 ice. *Tryka et al.* (1994) report 40 ± 2 K from the temperature-dependent N_2 band, implying pressures of 1.9–15.9 Pa (19–159 μbar). Certainly, N_2 is in its β crystalline phase. For pure N_2, this implies a temperature above the 35.62 K phase transition (0.42 Pa). The temperature of the N_2 phase transition is lowered slightly when diluted with trace CH_4 (*Prokhvatilov and Yantsevich,* 1983) and raised slightly when diluted with trace CO (*Tegler et al.,* 2019). A sideband of N_2 at 2.162 μm appears at temperatures colder than 41 K and is apparent in Pluto's spectrum. Thus, even though a correlation between grain size and temperature complicates temperature retrieval (*Grundy,* 1995), near-IR spectra imply that the temperature of the surface N_2 ice is between 35.62 and 41 K. This is consistent with the surface pressure observed by the New Horizons radio occultations (see section 2.2).

In thermal emission, Pluto's brightness temperature is seen to be colder at longer wavelengths (*Stern et al.,* 1993; *Jewitt,* 1994; *Sykes et al.,* 1987; *Aumann and Walker,* 1987; *Lellouch et al.,* 2000, 2011, 2016; B. J. Butler, personal communication), which is explained by a combination of Pluto's variegated terrain and a wavelength-dependent spectral emissivity (*Lellouch et al.,* 2011, 2016). Data from the Infrared Space Observatory (ISO), Spitzer, and Herschel measured Pluto's thermal emission across its emission peak, near 100 μm. These data show rotational variation, with colder brightness temperatures at 180°E to 240°E, the longitudes associated with bright albedos and N_2 absorption (*Lellouch et al.,* 2000, 2011, 2016). These measurements imply N_2 temperatures near 37 K, and low thermal inertia values within the diurnal skin depth (i.e., the uppermost few centimeters) for the CH_4-rich and volatile-free terrains of ~20–30 J m^{-2} $s^{-0.5}$ K^{-1} [hereafter referred to as thermal inertial units (TIU), as introduced by *Putzig* (2006)]. The low thermal inertia implies a porous structure for the top few centimeters, which are affected by the diurnal temperature cycle. As with albedo and IR spectra, decade-long changes in thermal emission are much more subtle than rotational changes. Observations were taken prior to equinox and perihelion at ~100 μm (*Sykes et al.,* 1987; *Aumann and Walker,* 1987) and at millimeter/submillimeter (*Stern et al.,* 1993; *Jewitt,* 1994), but these were not rotationally resolved. Models based on pre-encounter composition maps were able to match ISO and Spitzer data with static composition maps and thermophysical parameters of bolometric Bond albedo, emissivity, and thermal inertia that were static from 1997 to 2004 (*Lellouch et al.,* 2011b). Hints of change between 2004 and 2007 (*Lellouch et al.,* 2011b) were not supported by later data from 2012 (*Lellouch et al.,* 2016). No further observations of Pluto near its emission peak have been made since 2012, although observations at 0.8 mm to 0.9 cm have been taken from the Atacama Large Millimeter Array (ALMA), Submillimeter Array (SMA), and Very Large Array (VLA) (B. J. Butler, personal communication).

2.1.3. Atmospheric structure, composition, and variation. Stellar occultations observed from Earth have proven to be a powerful technique to probe the changes in Pluto's atmosphere over the span of decades, occasionally at high spatial resolution. Stellar occultations are more sensitive than near-IR spectra for detecting changes in N_2-ice temperature (under the assumption of vapor-pressure equilibrium). More information can be found in the chapter in this volume by Summers et al. Stellar occultations work as follows: The differential refraction of stellar rays causes a drop of flux, which can be used to retrieve the vertical refractivity profile of the atmosphere via an inversion technique for observations with sufficiently high cadence and signal-to-noise ratio. Assuming a composition (a predominately N_2 atmosphere in the case of Pluto), plus the ideal gas law and hydrostatic equilibrium, local profiles of density, pressure, and temperature can be derived (see, e.g., *Vapillon et al.,* 1973; *Elliot and Olkin,* 1996; *Elliot et al.,* 2003a), often at a vertical resolution of a few kilometers or better. For lower-quality light curves, typically a temperature and pressure at a reference altitude can be derived through forward modeling.

Due to refraction effects, the deepest atmospheric layers accessible to groundbased occultations of Pluto do not reach the surface but probe the atmosphere down to typically 10–20 km altitude (pressure ~1 Pa). For high-quality light curves, levels up to 380 km altitude (~1 mPa) can be probed.

Pluto's atmosphere was first glimpsed during a stellar occultation in 1985 (*Brosch,* 1995) and fully recognized as such after a high-quality occultation was observed by multiple sites in 1988 (*Hubbard et al.,* 1988; *Elliot et al.,* 1989). Beginning in 2002, Pluto started to move in front of the galactic center, increasing the rate of successful events to a few per year.

As of 2016, more than 20 stellar occultations by Pluto have been documented (*Young,* 2013; *Bosh et al.,* 2015; *Meza et al.,* 2019; *Desmars et al.,* 2019). Some occultations had sufficient quality to provide a good assessment of the global pressure changes over time. For instance, an occultation observed in 2002 showed that Pluto's atmosphere underwent an increase in pressure by a factor of ~2 between 1988 and 2002 (*Elliot et al.,* 2003b; *Sicardy et al.,* 2003). This increase was confirmed as more occultations were observed (*Elliot et al.,* 2007; *Olkin et al.,* 2015; *Sicardy et al.,* 2016). As Pluto had continuously receded from the Sun since its perihelion in 1989, more-complex effects were clearly at work besides heliocentric distance (see section 4.2).

Studies of the changes in Pluto's climate were confounded by a fundamental uncertainty in the nature of Pluto's lower atmosphere and by differences in analysis technique (see *Bosh et al.,* 2015, for a review of these complications). The first complication is that Pluto's radius was not directly measured until the New Horizons flyby of 2015 (as noted, stellar occultations do not probe to Pluto's surface). Occultation work prior to the flyby perforce presented results as a function of radius (distance to Pluto center) rather than altitude (distance to Pluto surface). A second complication is the unknown cause of the low fluxes reached by the occultation light curves, corresponding to the lower tens of kilometers in Pluto's atmosphere, affecting fluxes below half-light in 1988 or roughly quarter-light in 2002 and beyond (i.e., the portions of the light curve where the stellar flux is less than half or a quarter of its unocculted value). While the upper portions of the light curves were well described by a clear atmosphere with at most a small thermal gradient, the lower fluxes were modeled as caused by haze absorption ["haze/no-steep-gradient" (*Elliot et al.,* 1989)], a steep thermal gradient below ~1215 km radius ["clear/steep-gradient" (*Hubbard et al.,* 1990; *Elliot et al.,* 2003a; *Dias-Oliveira et al.,* 2015)], or a hybrid.

One solution to the haze/thermal-gradient puzzle is to focus on only the upper portions of the light curve, sidestepping the haze/gradient question. Even this is problematic, as described by *Bosh et al.* (2015). By one measurement — the radius in Pluto's shadow at which the stellar flux drops to half its unocculted value — the large increase between 1988 and 2002 had effectively leveled off after 2002 (*Bosh et al.,* 2015). By another measurement — a two- or three-parameter model of pressure and temperature and optional (small) thermal gradient at a reference radius (typically 1275 km), fit to only the upper portions of the light curve — the pressure has increased steadily at 3.5%–7.5% per year since 1988 (*Bosh et al.,* 2015; *Olkin et al.,* 2015; B.J. Butler, personal communication).

Other approaches look for changes using the entire light curve. For very high-quality data, the light curve can be inverted to derive temperature, pressure, and density vs. radius, under the assumptions of a clear N_2 atmosphere and hydrostatic equilibrium (e.g., *Elliot et al.,* 2003a; *Sicardy et al.,* 2003; *Young et al.,* 2008). For data of high or moderate quality, *Elliot and Young* (1992) developed a six-parameter model that included the three upper-atmosphere parameters describing (1) temperature, (2) a small temperature gradient, and (3) a reference pressure, plus three parameters describing a haze with (4) an onset radius, (5) an onset absorption coefficient, and (6) a haze scale height. This model (hereafter *EY92*) was applied to many observations from 1988 and beyond (*Elliot and Young,* 1992; *Young et al.*, 2008; *Gulbis et al.,* 2015). While the *EY92* model matched the shape of many light curves, it failed to reproduce the central flash sometimes observed near the center of Pluto's shadow, and did not allow for the ~60 K drop in temperature from the upper atmosphere to a surface near 40 K (see below). Other researchers (*Hubbard et al.,* 1990; *Zalucha et al.,* 2011a,b) constructed thermal profiles from physically motivated radiative-conductive models, dependent on the surface radius (again, this was unknown before the New Horizons flyby) and CH_4 or CO mixing ratios. A difficulty with applying physically motivated models is that the energetics of Pluto's atmosphere is not well understood (see the chapter in this volume by Summers et al.).

A promising approach is the technique of fitting with a temperature template, which has several advantages: It uses the entire light curve, it can describe even the central portions of the occultation light curve (unlike the *EY92* model), and it does not rely on an accurate understanding of Pluto's atmospheric energy balance. *Dias-Oliveira et al.* (2015) constructed a temperature template from a high-quality occultation light curve, which was able to be fit to various occultation events assuming no significant haze absorption (Fig. 2). The template includes a steep thermal gradient below ~1215 km, a maximum near 1215 km, and a mild negative gradient of dT/dr = –0.2 K km^{-1} between 1215 and 1390 km. Above 1400 km, the temperature is poorly constrained, and the template adopts an isothermal upper atmosphere at 81 ± 6 K. The shape of the temperature template is shown in Fig. 13 of *Dias-Oliveira et al.* (2015). This 10-parameter temperature-template model (9 temperature parameters plus a reference pressure) gives a good agreement between the pressures derived from Earth in 2015 (*Sicardy et al.,* 2016; *Meza et al.,* 2019) and from the REX experiment onboard New Horizons (*Hinson et al.,* 2017), which validates the applicability of the temperature template. In the most comprehensive occultation survey made to date, *Meza et al.* (2019) used the temperature template to trace back Pluto's atmospheric changes between 1988 and 2016, revealing an increase in pressure by a factor of about 3 during that period.

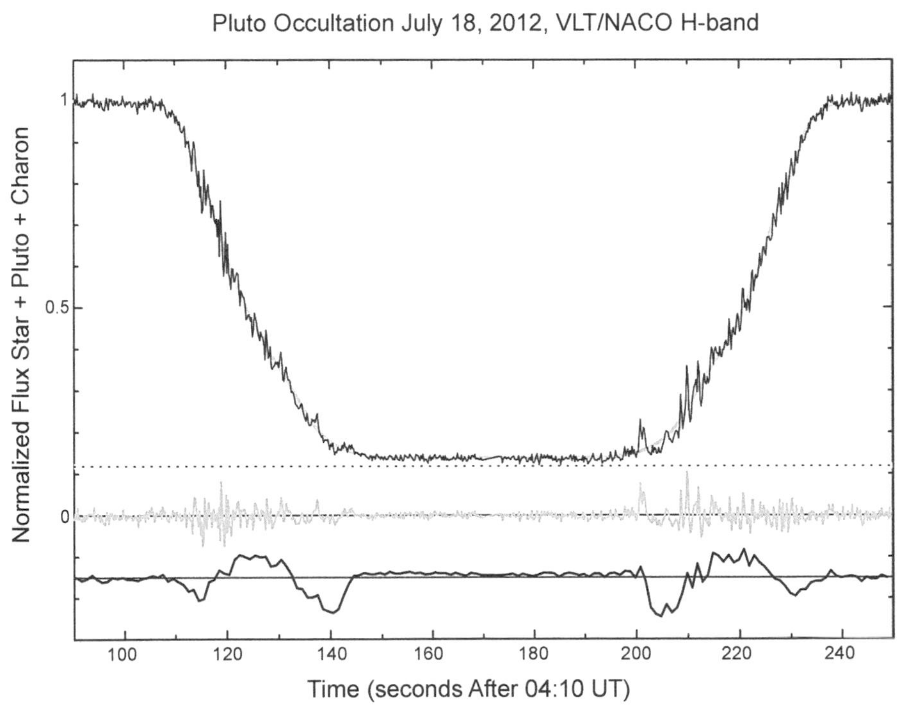

Fig. 2. Pluto occultation from Paranal on July 18, 2012, adapted from *Dias-Oliveira et al.* (2015) (converted to grayscale). The gray curve in the upper panel is from the template thermal profile. The data plotted spans about 3 minutes (see the UT time plotted along the horizontal axis) and includes flux from Pluto, Charon, and the star. From top to bottom, the horizontal lines indicate (1) the measured zero stellar flux (dotted), (2) the zero flux, together with the residuals (data minus model), and (3) the residuals (shifted by –0.15 for better viewing) obtained by forcing an isothermal mesosphere at 95.5 K above the stratopause, near 1215 km. The large discrepancy then observed shows that a mesosphere with a mild negative temperature gradient is necessary to explain the data. This result is independent of the particular isothermal mesosphere considered (here at 95.5 K), as the same kind of discrepancies appear with isothermal mesospheres at other temperatures.

Stellar occultations have proven the most sensitive way of measuring seasonal change in Pluto's atmosphere. This in turn allowed modelers to constrain certain physical parameters that enter into models of seasonal volatile transport (*Bertrand and Forget,* 2016; *Johnson et al.,* 2021; *Bertrand et al.,* 2018) (see section 4.3). These models predict that Pluto's atmosphere is near its maximum pressure and will start to decline in the upcoming decades. Thus, it is important to continue observing stellar occultations from Earth or Earth orbit to confirm that prediction. Unfortunately, as Pluto moves in front of more depleted stellar fields, such occultations have already become rarer.

Groundbased observations confirmed CH_4 as a minor species from IR spectroscopy at high spectral resolution (*Young et al.,* 1997; *Lellouch et al.,* 2011a,b, 2015) and CO as a minor species from IR and radio spectroscopy (*L. A. Young et al.,* 2001; *Bockelée-Morvan et al.,* 2001; *Lellouch et al.,* 2010; *Greaves et al.,* 2011; *Lellouch et al.,* 2017), as well as the detection of HCN (*Lellouch et al.,* 2016). Because the total surface pressure was not known pre-encounter, the derived mixing ratio of the minor species was also uncertain. For N_2 pressures of 6.5–24 μbar, the best-fit mixing ratios are 0.06% to 0.53% for CH_4 (with a preferred value of 0.44%), and 0.04% to 0.07% for CO (*Lellouch et al.,* 2011a, 2015). The precision has not yet been good enough to detect changes with time. This may change, with new telescopes or instrumentation, and also now that we can use New Horizons measurements of the temperature structure to improve models of CH_4 line absorption.

2.2. New Horizons Observations

The New Horizons spacecraft made many observations relevant to volatile transport. Other chapters in this volume cover this in detail: White et al. (geology), Moore and Howard (the geologic evidence for sublimation processes), Olkin et al. (photometric properties and color), Cruikshank et al. (volatile distributions), and Summers et al. (atmospheric structure). New Horizons observations revealed latitudinal variation in albedo and composition, and albedo variations larger than any in the solar system apart from Iapetus, as well as dramatic topography, glaciers, and a range of surface ages. The presence of the deep N_2 plains of Sputnik Planitia was completely unexpected, and dominates much of the post-encounter thinking about volatile transport and volatile cycles.

New Horizons measurements yielded surface radii and pressures, ending years of ambiguous interpretation of near-surface conditions from groundbased stellar occultations. The mean radius from imaging is 1188.3 ± 1.6 km (*Nimmo et al.,* 2017). The observed topography on Pluto varies about this mean by roughly ±5 km (*Schenk et al.,* 2018). Specifically, the radii probed with precision by the REX radio occultation were 1187.4 ± 3.6 km at ingress and 1192.4 ± 3.6 km at egress (*Hinson et al.,* 2017). The mean pressure at 1189.9 km was 1.15 ± 0.07 Pa (*Hinson et al.,* 2017), and the variation between ingress and egress is as expected given their different altitudes. The pressure of 1.15 ± 0.07 Pa is consistent with N_2 ice in equilibrium at 36.9 ± 0.1 K (*Fray and Schmitt,* 2009). The CH_4 number density was measured by solar occultation (*Young et al.,* 2018). Models were needed to extrapolate CH_4 number density to the surface from 80 km, the lowest altitude at which CH_4 was well measured, resulting in a CH_4 surface mixing ratio of 0.28–0.35%.

Sputnik Planitia is a feature spanning roughly 1200 km in longitude and 2000 km in latitude, centered at roughly 15°N and 180°E (Fig. 3), depressed several kilometers below the surrounding terrain. It covers 5% of Pluto's surface (*White et al.,* 2017) and is likely to be 3–10 km deep (*McKinnon et al.,* 2016; *Trowbridge et al.,* 2016), which makes it the major reservoir of N_2 on Pluto (*Glein and Waite,* 2018), with a mass equivalent to a global layer ~60–500 m deep (6×10^4 to 5×10^5 kg m^{-2}). Both N_2-rich and CH_4-rich ice phases are present in Sputnik Planitia (*Protopapa et al.,* 2017), with traces of CO (*Grundy et al.,* 2016; *Schmitt et al.,* 2017). Consistent with groundbased spectra, the CH_4 is seen at only 0.3%–0.5% in the N_2-rich ice, which is much lower than its saturation limit (see section 3.1). There is a gradient of properties within Sputnik Planitia, with the northwest area having a smaller grain size or path-length and relatively less N_2 (*Protopapa et al.,* 2017).

Outside of Sputnik Planitia, Pluto's surface shows strong latitudinal patterns in albedo (*Schenk et al.,* 2018), color

Fig. 3. Orthographic view of the encounter hemisphere of Pluto, centered on 160°E, 45°N, with 30° lines of latitude and longitude. Hue and saturation indicate composition and brightness is set by the basemap (*Schenk et al.,* 2018). The red channel is the H_2O spectral index of *Schmitt et al.* (2017) scaled from –0.00 to 0.25. The blue channel is from the N_2 spectral index of *Schmitt et al.* (2017) scaled from –0.005 to 0.105. The green channel is the average of the 1.7-μm CH_4 spectral index of *Schmitt et al.* (2017) scaled from –0.05 to 0.41 and the 0.89-μm CH_4 spectral index of *Earle et al.* (2018b) scaled from –3 to 4.

(*Olkin et al.*, 2017; *Earle et al.*, 2018b) and composition (*Protopapa et al.*, 2017; *Schmitt et al.*, 2017; *Earle et al.*, 2018; *Gabasova et al.*, 2021). Near the equator, Pluto's surface shows high-albedo contrast, with Cthulhu and other dark, reddish areas girdling the equator to roughly 10°N. These dark areas are mostly volatile free, most likely covered by tholins (see the chapter by Cruikshank et al.), but New Horizons detected CH_4-rich ice at some high-altitude ridgetops and northward slopes within these dark areas. At low to mid northern latitudes (10°N–35°N), LEISA detected abundant CH_4-rich ice in the same pixels as H_2O ice. The CH_4-rich bladed terrain is also at these latitudes. In the mid-latitudes (35°N–55° N), N_2-rich ice is common. In both the midlatitudes and the bladed terrain, N_2 favors low altitudes (*Lewis et al.*, 2021). In the N_2-rich ice, trace amounts of CO are present, and the CH_4 is diluted at well below its saturation value, typically 0.1%–0.5% (*Protopapa et al.*, 2017). At high latitudes (55°N–90°N), there is increasingly more CH_4-rich ice with increasing northern latitude. The presence or amount of N_2 in the CH_4-rich ice is not measurable. The composition of terrains south of the dark equatorial areas is difficult to measure in the New Horizons data.

For all volatile ices, inside or outside of Sputnik Planitia, the derived grain sizes for the CH_4-ich ice is relatively small, 0.2–1 mm, while the derived grain sizes for the N_2-ich ice is quite large, 0.1–1 m (*Protopapa et al.*, 2017). Groundbased observations also found small CH_4 grains and large N_2 grains (*Merlin et al.*, 2010; *Grundy and Buie*, 2001; *Douté et al.*, 1999). The large N_2 grain sizes reported in these works are more usefully thought of as distances between scattering centers than the size of an individual grain (*Grundy and Buie*, 2001).

Ethane could not be detected by LEISA because the spectral resolution is too low to observe its weak signature at 2.405 μm. Although CO_2 is present on Triton, this nonvolatile ice is not detected on Pluto.

3. CHARACTERISTICS OF THE VOLATILE CYCLES ON PLUTO

3.1. Generalities

Several bodies in the solar system have atmospheres supported hydrostatically by the vapor pressure of ices on the surface, including Mars, Triton, and Pluto. On Pluto, global hydrostatic equilibrium at a nearly uniform temperature of the N_2-rich ice is enforced primarily by the high volatility of N_2. Since the ice temperature drives the vapor pressure that determines the bulk of the atmosphere, Pluto's diurnal, seasonal, and longer-term volatile cycles are sensitive to the geographically changing insolation.

3.1.1. Pure ices and ice mixtures. The equilibrium vapor pressure of the ice phase of a pure species is determined uniquely by its temperature according to the Clausius-Clapeyron equation. This pressure is often useful in scenarios that depart from equilibrium as well, such as the situation on Pluto of subliming or condensing ice, as long as the temperature is stable. At the low temperatures of Triton and Pluto, the volatile species are N_2, CO, and CH_4, in descending order of volatility (Fig. 4). The contrast between the volatile pressures of these three ices becomes more extreme with decreasing temperature. At 40 K, the N_2 equilibrium pressure is 6.55 Pa and the CO and CH_4 pressures are smaller by factors of 0.16×10^{-4} and 1.2×10^{-4}, respectively [using the compilation of *Fray and Schmitt* (2009)]. At 36.9 K, in equilibrium with the 1.15 Pa from the New Horizons radio occultation, the CO and CH_4 pressures are smaller than that of N_2 by factors of 0.12×10^{-5} and 6.0×10^{-5}, respectively.

These vapor pressures are extremely sensitive functions of temperature. For example, at 37 K, an increase of only 0.16 K (0.4%) increases the N_2 pressures by 10%. This is because of the large latent heat of sublimation, L, which depends weakly on temperature, and which in this temperature range is ~2.5×10^5 J kg^{-1}, 2.8×10^5 J kg^{-1}, and 6.1×10^5 J kg^{-1} for N_2, CO, and CH_4 respectively. The equilibrium pressure increases with temperature according to the Clausius-Clapeyron equation $dp_{eq}/dT_{eq} = (p_{eq}/T_{eq})(L/RT_{eq})$, where R is the specific gas constant. For N_2, the unitless ratio L/RT_{eq} is much larger than 1, and ranges from 31 to 18 for T_{eq} between 30 and 45 K; for CO and CH_4, the ratios are similarly large (32–21 for CO, and 39–26 for CH_4).

Hydrostatically supported atmospheres demonstrate the different behavior of immiscible volatile ices (such as H_2O and CO_2 on Mars) and miscible ones (such as N_2 and CH_4 on Pluto or Triton). When ice includes immiscible species, the total vapor pressure is the sum of the vapor pressures of the pure constituents. This is because their self-attraction at the molecular level is stronger than their mutual attraction. They therefore do not mix at the molecular level and so the species do not mutually perturb each other's energy levels. Their spectral lines thus remain unshifted. Their vapor pressure depends on a single thermodynamic variable, temperature, and it does not depend on the relative proportions of the species present.

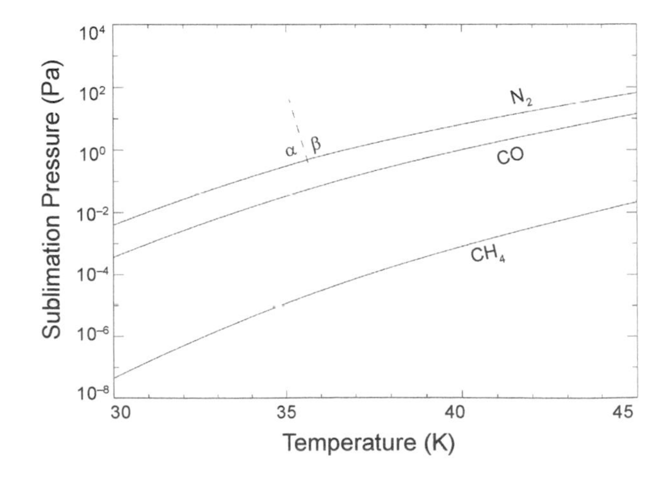

Fig 4. Equilibrium vapor pressure vs. ice temperature over pure ices. From *Fray and Schmitt* (2009).

Miscible species (e.g., N_2 and CH_4), for which their mutual attraction at the molecular level is greater than their self-attraction, exhibit the opposite — and a more complicated — behavior. In the solid phase, these species are found to be mixed at the molecular level, forming a solid solution (*Prokhvatilov and Yantsevich,* 1983; *Trafton,* 2015). This results in the perturbation of each other's molecular energy levels, with spectral lines of the participating species being shifted relative to the spectra of the pure solid species. In general, the total pressure will not be the sum of the partial pressures of the pure species, but will depend on the composition of the system through the equation of state (EOS). The thermodynamic state of the system determines the phases and the total vapor pressure, and in thermodynamic equilibrium, the state depends on the equilibrium phase diagram for the solid solution. In general, an EOS will be needed to calculate the equilibrium pressure and phase compositions. With three volatiles, a ternary phase diagram is needed to describe the phases of Pluto's solid solution (*Tan and Kargel,* 2018). However, isobaric calculations of the ternary $N_2/CO/CH_4$ diagram in Table 2 of *Tan and Kargel* (2018) show that the equilibrium composition of CO is insensitive to temperature changes. Moreover, CO and N_2 are similar in properties (e.g., molecular weight and sublimation pressure) and the phase behavior of their ices. This is the rationale for neglecting CO in Pluto's phase equilibrium and adopting the N_2-CH_4 binary phase diagram for EOS calculations (*Trafton et al.,* 2019).

The binary phase diagram (*Prokhvatilov and Yantsevich,* 1983; see also the chapter by Cruikshank et al. in this volume) shows that the solubility of the solid solution (CH_4 in N_2-rich ice, or N_2 in CH_4-rich ice) is only a few percent at Pluto's low ice temperatures. If the solution were saturated, in equilibrium two separate solid phases would form: a N_2-rich phase and a CH_4-rich phase, each at the saturation concentration. No pure ice of either species would exist in thermodynamic equilibrium. The partial pressures at fixed temperature in equilibrium do not depend on the relative proportions of these phases, which is determined by the overall composition of the system. In this saturated case there are two species (N_2 and CH_4) and three phases (N_2-rich ice saturated with CH_4, CH_4-rich ice saturated with N_2, and vapor), and Gibb's phase rule says there should be only 1 degree of thermodynamic freedom. Temperature thus suffices to determine the equilibrium state of the system including the vapor pressure.

As described by *Trafton* (2015, his section 5.2), in thermodynamic equilibrium the three-phase saturated solution might be expected to have some similarities to an immiscible solution, namely the dependence on a single thermodynamic variable. But the component partial pressures fall short of those for the pure components as calculated by *Tan and Kargel* (2018) (and tabulated in *Young et al.,* 2019); In the saturated three-phase system at 35.6–40.0 K, the ratio of an ice's partial pressure to its pure vapor pressure is 98–95% for N_2 and 74–79% for CH_4. If the solid solution were unsaturated in thermodynamic equilibrium, then only a single solid phase would be present: a "homogeneous" phase that is either N_2-rich or CH_4-rich, analogous to the homogeneous gas phase. In this case (two species, two phases), Gibb's phase rule says there should be 2 degrees of thermodynamic freedom, and the vapor pressure would depend on two thermodynamic variables: e.g., temperature and ice mole fraction, x_{CH_4} in N_2-rich ice or x_{N_2} in CH_4-rich ice.

A common approximation in lieu of the EOS has been to use Raoult's law (*Owen et al.,* 1993; *Trafton,* 1990; *Forget et al.,* 2017; *Bertrand et al.,* 2019), which states that the partial pressure of the dominant component of a solution is the vapor pressure of the pure component times its mole fraction in the solution. While Raoult's law also applies to solid solutions, it does not strictly apply to the minor component of the solution, especially when the phase changes with concentration, as for Pluto's volatiles. For minor components of a solution, Henry's law is more appropriate, which does not reference the pure species vapor pressure. Take, for example, N_2-rich ice at 38 K with CH_4 undersaturated at x_{CH_4} = 0.3% [e.g., at the sublimation point, or at V-VS$_1$ boundary in the terminology of *Tan and Kargel* (2018)], for which the application of Raoult's law as an approximation would predict a partial pressure of 0.5 μPa, while the EOS of *Tan and Kargel* (2018) predicts a partial pressure of 13 μPa. At the CH_4 saturation value of x_{CH_4} = 3.9%, the partial pressure via Raoult's law would be 6.6 μPa, vs. 130 μPa for the EOS. Similarly, for CH_4-rich ice at 38 K with N_2 undersaturated [similarly, at the sublimation point, or at V-VS$_2$ boundary in the terminology of *Tan and Kargel* (2018)], a partial pressure for N_2 of 0.045 Pa would be reached at $x_{N_2} \approx 2.5\%$ applying Raoult's law but only $x_{N_2} \approx 0.08\%$ using the EOS. However, as shown in section 3.4.1, the historical misapplication of Raoult's law to approximate the vapor pressure of the minor component of the solution has little practical impact on calculations of mass fluxes.

3.1.2. Insolation at various timescales. The intensity and distribution of the incident solar flux depends on the instantaneous heliocentric distance and subsolar latitude (*Levine et al.,* 1977). Since these parameters are well known, insolation on Pluto can be easily calculated before making any assumptions about thermal properties, surface composition, and reflectivity, etc., thus making incident insolation one of the basic inputs for the various models described below. At the time of the New Horizons encounter, Pluto was at 32.9 AU from the Sun, for a normal incident solar flux of 1263 mW m^{-2}. Pluto was experiencing northern spring (Fig. 5d), with its subsolar latitude at 52°N. During this time all latitudes south of 38°S were in constant winter shadow, while all areas north of 38°N received constant sunlight, with the maximum diurnal average insolation (995 mW m^{-2}) occurring at the north pole (Fig. 6a,b).

Pluto obliquity is currently 119.51° and the solar longitude at perihelion is ~3.8°, which means that equinoxes occur near perihelion and aphelion (Fig. 5d). Therefore both hemispheres receive a similar distribution of insolation over Pluto's year (*Earle et al.,* 2017) (Fig. 6c,d). Each pole receives the same maximum insolation, and comparable latitudes spend the same amount of time in continuous darkness. Because Pluto

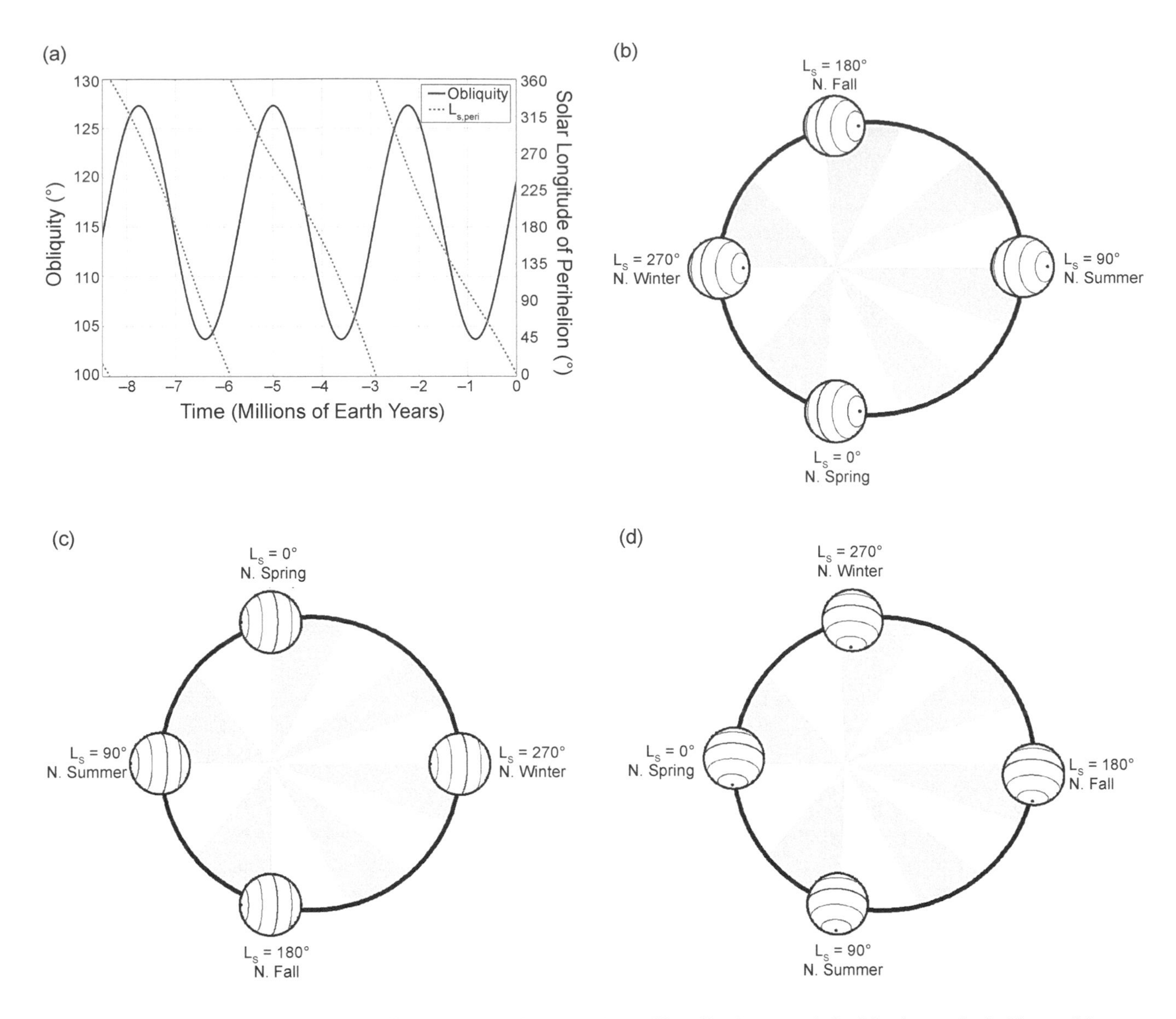

Fig. 5. (a) The astronomical cycles of Pluto during the past 8.5 million Earth years (~3 obliquity cycles). The solid curve indicates obliquity (left axis), and the dashed curve indicates the longitude of perihelion ($L_{s,peri}$, right axis). Over 2.8 m.y., Pluto's obliquity ranges between a high value (i.e., closer to 90°) of 104°, most recently occurring at 0.9 Ma, and a low value (i.e., closer to 180°) of 127°, occurring at 2.35 Ma. $L_{s,peri}$ cycles over 360° over ~2.8 m.y. as well. The slight curvature of $L_{s,peri}$ is due to 24° of modulation by the 3.96-m.y. libration of the argument of perihelion. **(b)–(d)** Pluto's orbit and polar orientation as seen from above the plane of the orbit. The thick ellipse represents Pluto's orbit. Pluto orbits counterclockwise. The gray wedges show 12 equally spaced divisions in time (roughly 20.7 yr per division). Pluto's obliquity is greater than 90°, so the southern pole is indicated, as well as latitudes –60°, –30°, the equator (bolder), and 30°. Pluto's location is indicated for four solar longitudes (L_S): L_S = 0° (northern spring), L_S = 90° (northern summer), L_S = 180° (northern fall), and L_S = 270° (northern winter). **(b)** At 2.34 Ma, Pluto experienced southern summer at perihelion. **(c)** At 0.86 Ma, Pluto experienced northern summer at perihelion. **(d)** In its current orbit, Pluto experiences northern spring equinox near perihelion.

has a high orbital eccentricity (approximately 0.25), its heliocentric distance varies from 29.6 to 49.9 AU, which leads to the normal incident solar flux varying by a factor of ~3 over the course of its orbit (see review by *Spencer et al.,* 1997).

When averaged over a full orbit, insolation is symmetric across the equator (*Nadeau and McGehee,* 2017; *Nadeau and Jaschke,* 2019; *Hamilton et al.,* 2016) (Fig. 6e,f). Due to Pluto's high obliquity, the average insolation over one Pluto year in its current orbit is highest at the poles and lowest at 25°N and 25°S.

Also shown (Fig. 6) are the climate zones on Pluto as defined by *Binzel et al.* (2017). In Pluto's current orbit, the subsolar latitude reaches ±61° at the solstices. Between these extremes are the tropics, where the Sun is overhead at some time in the year, equivalent to Earth's tropics of Cancer and Capricorn. The complement, poleward of

±61°, is termed the polar climate zone. Latitudes poleward of ±29° experience days where the Sun never sets in the summer or never rises in the winter, otherwise known as the arctic (and the complement, between ±29°, is termed the diurnal climate zone). Unlike on Earth, the tropics and artic latitudes overlap, for a tropical arctic climate zone. The climate zones appear associated with the gross geology of Pluto (Fig. 7). In particular, the diurnal zone is a zone where albedo feedback is most favored, leading to strong albedo contrasts (*Earle et al.,* 2018a).

The obliquity, solar longitude of perihelion, and eccentricity vary over timescales of several million years (equation (1)). This is known as a Milankovitch mechanism, which also occurs on other planetary bodies such as Earth, Mars, and Titan. Recent years have emphasized the importance of these cycles, but have also introduced some elements of confusion. We present here the equations controlling the insolation at Pluto, using the equations of *Dobrovolskis et al.* (1997). *Earle* (2018) showed that estimates of the important dynamical parameters (the principle moments of inertia, the semi-major axis of the Pluto-Charon system, and the radii and masses of Pluto and Charon) have changed by less than 6% since the work of *Dobrovolskis et al.* (1997). Until such time as the numerical integrations are repeated, we also use the numerical results from *Dobrovolskis et al.* (1997). We define an intermediate angle (ψ), which varies with a period of 360°/(91° m.y.$^{-1}$) = 3.96 m.y., and a phase angle (ϕ), defined as the angle, projected onto Pluto's orbital plane, between the ascending node on the invariable plane and Pluto's spin vector, which varies with a period of 360°/(130.2° m.y.$^{-1}$) = 2.76 m.y. From these we can calculate the eccentricity (e), obliquity (θ), and solar longitude of perihelion [$L_{s,peri}$, related to the parameters in *Dobrovolskis et al.* (1997) via $L_{s,peri} = -\chi$, where χ is the angle, measured in Pluto's orbital plane, from Pluto's longitude of perihelion to its vernal equinox]. The subsolar latitude (δ_S) varies over Pluto's orbit depending on the true anomaly (ν). These can be expressed as a function of time (t) in millions of years C.E. (i.e., the flyby on July 14, 2015, was at t = 0.0020155):

$$\psi = 72.8° + 91.0°t$$

$$\phi = 19.5° + 130.2°t$$

$$e = 0.244° + 0.022° \cos\psi + 0.005° \cos 3\psi$$

$$\theta = 115.5° + 11.8° \sin\phi \quad (1)$$

$$L_{s,peri} = -\phi + 24.0° \sin\psi$$

$$\sin\delta_S = \sin\theta \sin(\nu + L_{s,peri})$$

Pluto's obliquity varies by about 23° between 104° and 127° over a period of 2.8 m.y. (Fig. 5a), thus affecting the latitudes of the climate zones and the annually averaged insolation. Over 2.8-m.y. timescales, the latitude of minimum insolation moves from ±30° to the equator, and the difference between average insolation at the equator vs. the poles becomes more or less dramatic. The eccentricity oscillates between 0.222 and 0.266 with a 3.95-m.y. period.

The longitude of the ascending node (Ω) regresses over 360° with a period of 3.7 m.y., given by $\Omega = 111.428° - 97.209°t - 1.5° \sin\psi$, while the argument of perihelion (ω_p) librates with a period of 3.96 m.y. The longitude of perihelion, being the sum of Ω and ω_p, also regresses with a period of 3.7 m.y. However, these angles are of minor importance for the insolation at Pluto. Numerical integration shows Pluto's spin pole advancing with a precession period near 11.5 m.y. (*Dobrovolskis et al.,* 1997). This combines with the change in the longitude of the ascending node to result in a period for ϕ of 2.76 m.y., which modifies the solar longitude of perihelion ($L_{s,peri}$) (Fig. 5a). In Pluto's current orbit, the northern spring equinox was nearly coincident with perihelion ($L_{s,peri} \approx 0$) (Fig. 5d). When a solstice is at perihelion [$L_{s,peri}$ = 90° for northern summer solstice at perihelion (Fig. 5c) or $L_{s,peri}$ = 270° for southern (Fig. 5b)], one hemisphere experiences a short, intense summer and long winter, while the other hemisphere experiences a short winter and long, less intense summer (Fig. 8) (see also *Bertrand et al.,* 2019, their Fig. 5). Because of the prominence of the N_2 deposits in Sputnik Planitia near 15°N, the solar forcing of the N_2-rich ices is most extreme when northern solstice is at perihelion (Fig. 8a) and nearly uniform when northern solstice is at aphelion (Fig. 8b). Perihelion northern summer solstice ($L_{s,peri}$ ~90°) happens at times of high obliquity, and perihelion southern summer solstice ($L_{s,peri}$ ~270°) happens at times of low obliquity, with some small variation due to the 24.0° libration of ω_p (Fig. 5).

3.1.3. Thermal inertia. Surface temperatures depend on more than just the average insolation. The substrate is able to store the heat accumulated during the day or summer and release it during night or winter, an ability quantified by its thermal inertia. Thermal inertia dampens and retards the response of the surface temperature to insolation and raises the mean temperatures (*Spencer et al.,* 1989). The effects are very different on diurnal and seasonal timescales. Using the definition of *Spencer et al.* (1989), the subsurface thermal wave has a characteristic skin depth $Z = \Gamma / (\rho c \sqrt{\omega})$, where Γ is the thermal inertia, ρ is density, c is the specific heat, and ω is frequency (2π/period). The importance of the thermal inertia is parameterized by $\Theta = \Gamma\sqrt{\omega}/(\varepsilon\sigma T_{SS}^3)$, where ε is emissivity, σ is the Stephan-Boltzmann constant, and T_{SS} is the temperature in instantaneous equilibrium with the normal insolation (*Spencer et al.,* 1989). For small values of Θ ($\Theta \ll 1$), the temperature responds quickly to changes in insolation, and little heat is conducted into or out of the subsurface. For large values of Θ ($\Theta \gg 4$), subsurface conduction is comparable to the peak-to-peak variation in the solar heating, and the amplitude of the temperature variation is small.

While Γ, ρ, and c are properties of the material (including porosity), Z and Θ also depend on the forcing timescale.

From the change in observed thermal emission as Pluto rotates, *Lellouch et al.* (2011b, 2016) reported a low diurnal thermal inertia of $\Gamma = 16–26$ J $s^{-1/2}$ m^{-2} K^{-1} (TIU) for the CH_4-rich and non-volatile terrain types. The thermal inertia for the N_2-rich terrain types could not be measured from this data, since the temperature of the N_2-rich terrain is nearly isothermal over a Pluto day, regardless of the thermal inertia (e.g., *Young,* 2012). The thermal inertia derived for the CH_4-rich and non-volatile terrains was much smaller than the thermal inertia for solid ices near 40 K (*Spencer and Moore,* 1992), whether H_2O (2200 TIU), N_2 (530 TIU), or CH_4 (630 TIU). The diurnal skin depths are ~1 cm (for a near-surface thermal inertial of 20 TIU), and so the reported thermal inertia relates to the thermophysical properties of the topmost layers. The small thermal inertia implies a substrate with high porosity at the top-most layers of Pluto's surface (*Carson et al.,* 2005). The seasonal thermal wave penetrates deeper into the substrate, below this high-porosity crust. Several models predict much larger thermal inertias at the depths probed by the seasonal wave. For example, *Bertrand and Forget* (2016) favor 800 TIU, for which the seasonal skin depth is 40 m.

For the diurnal thermal wave near perihelion, the thermal parameter Θ is about 4.5–6 for the CH_4-rich and non-

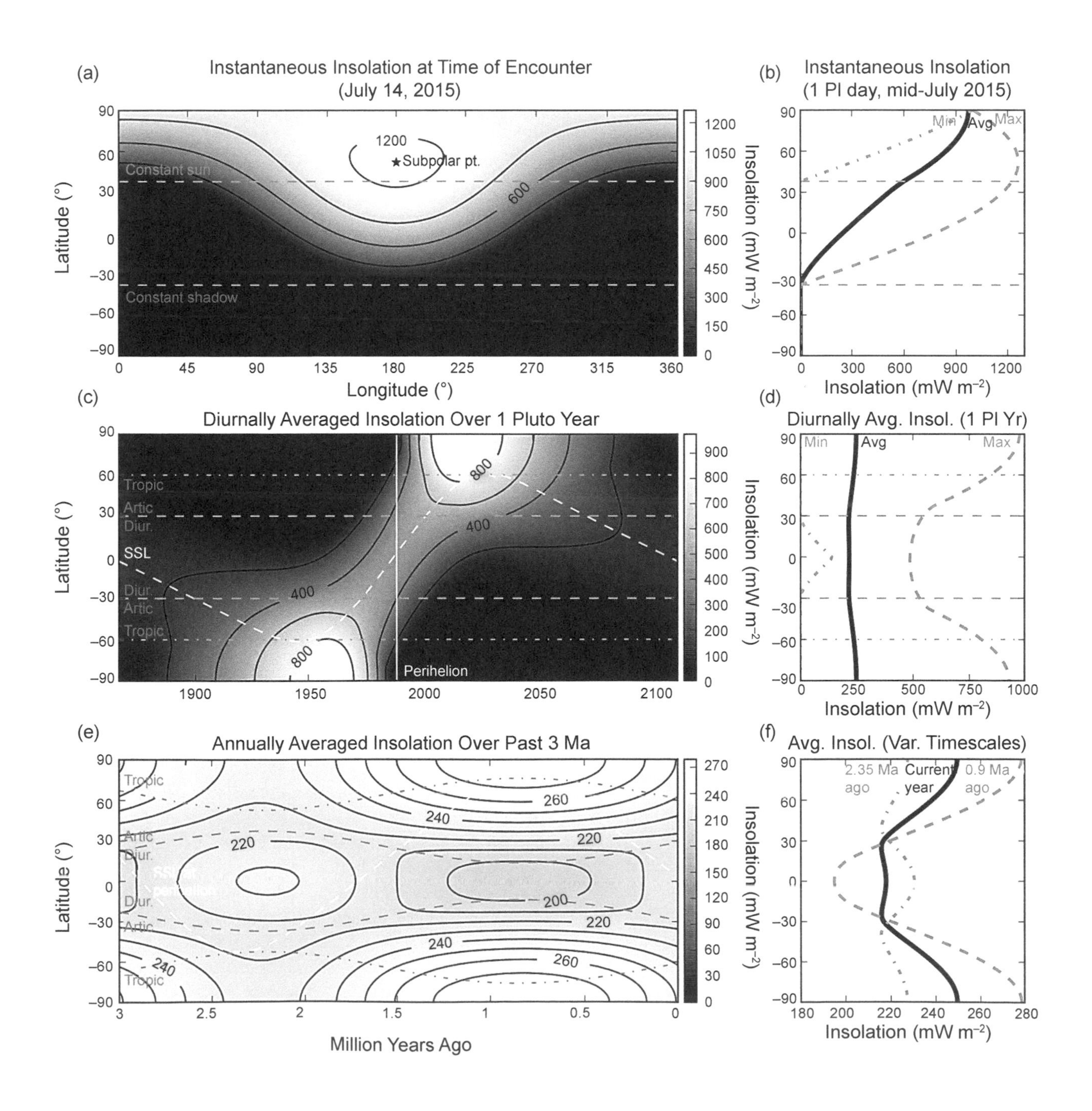

volatile terrain types (*Lellouch et al.,* 2016). Surprisingly, for seasonal thermal wave, the thermal parameter is likely to be similar. Pluto's year is 1.4×10^4 times longer than its day, so $\sqrt{\omega}$ is 118 times smaller, which should decrease Θ. Countering this is the fact that the thermal inertia is larger at the depths probed by the seasonal wave than at the depths probed by the diurnal wave, and that the relevant temperature is not the subsolar temperature, T_{SS} (as in *Spencer,* 1989), but rather the colder global N_2 temperature (as in *Young,* 2012). Near the times of maximum insolation, the thermal conduction is comparable to the energy lost by thermal emission for thermal parameters greater than about unity. When there is no insolation (e.g., during polar winter), the heat stored in the substrate can have a significant effect.

3.1.4. Atmospheric effects on surface energy balance. Pluto's atmosphere is so transparent and rarefied that any atmospheric processes (clouds, hazes, convection, radiative transfer within the atmosphere) have no significant impact on the radiative thermal and conductive balance at the surface (*Stansberry et al.,* 1994; *Forget et al.,* 2017). Thermal conduction to the surface given a near-surface gradient of ~5 K km^{-1} (*Hinson et al.,* 2017) is only ~0.02 mW m^{-2}, and the heat transfer through turbulent diffusion is only a few milliwatts per square meter (*Stansberry et al.,* 1992), so the ambient atmospheric temperature hardly affects the local surface temperature. In addition, most of Pluto's thermal emission is longward of 52 μm, which is not absorbed by the atmosphere. Nevertheless, the atmosphere affects the surface energy balance through changes to the albedo (as newly deposited ice is likely to be bright), or to the thermal inertia (as the sublimation and condensation can affect porosity). Most significantly, the atmosphere affects surface temperatures through latent heat of sublimation, particularly of N_2-rich ice, and is the medium for N_2 moving from subliming spots to condensing locations (see the chapter in this volume by Forget et al.).

3.2. Nitrogen Cycle

It is useful to begin with the simple picture of Pluto's N_2 atmosphere in global balance with its surface N_2 ice (e.g., *Spencer et al.,* 1997). For a sufficiently high surface pressure, the pressure at a near-surface reference altitude will be nearly the same everywhere, and the surface pressures and temperatures of the N_2 ice will be close to vapor-pressure equilibrium. For a surface without topography, the N_2 ice temperatures will be close to isothermal across the surface, at a value that is determined by the radiative balance between the absorbed insolation and thermal radiation from all the N_2 ice. Sublimation occurs from areas of high insolation, with condensation onto areas of low insolation, and the sublimation winds transport mass and latent heat of sublimation from areas of net sublimation to areas of net condensation. Because N_2 is so much more volatile than CO or CH_4, the small amount of CH_4 or CO diluted in the N_2-rich ice does not change this basic picture.

Putting these considerations into equation form, the local energy balance can be written

$$c_s \frac{dT}{dt} = \mu(1 - A)\frac{S_{1AU}}{\Delta^2} - \varepsilon\sigma T^4 - k\frac{dT}{dz} - L\dot{m} \quad (2)$$

where c_s is the surface heat capacity [in J m^{-2} K^{-1} (see *Forget et al.,* 2017; *Hansen and Paige,* 1996)], t is time, μ is the cosine of the incidence angle, A is the bolometric Bond

---

Fig. 6. *(facing page)* Incident solar flux (insolation) on Pluto at various timescales. Solid curves are contours, labeled in milliwatts per square meter (mW m^{-2}). **(a)** Insolation over one Pluto rotation at the time of New Horizons flyby, with longitudes relative to midnight. The star indicates the subsolar point at a relative longitude of 180° (e.g., noon) and a subsolar latitude of 52°N. The thin dashed lines indicate the latitudes of constant sunlight (northward of 38°N) or constant shadow (southward of 38° S) at flyby. **(b)** The thick gray dashed curve shows the maximum instantaneous insolation during a Pluto day, and the thick gray dash-dot curve shows the minimum. The thick black solid curve shows the diurnal average. The thin dashed lines indicate the latitudes of constant sunlight and constant shadow, as in **(a)**. **(c)** Diurnally averaged insolation over the current Pluto year, as a function of Earth year CE. The white dashed curve indicates the subsolar latitude (SSL). The thin gray dot-dashed lines indicate the tropics at ±61°, between which the Sun is overhead at some time in Pluto's year. The thin gray dashed lines indicate the artic/diurnal division at ±29°, between which the Sun rises and sets every day of Pluto's year. Latitudes north of 29°N or south of 29°S experience arctic night sometime during Pluto's year. The thin solid vertical line marks the time of perihelion. At perihelion, Pluto's subsolar latitude was close to 0°. **(d)** The thick gray dashed curve shows the maximum diurnally averaged insolation during a Pluto year, and the thick gray dash-dot curve shows the minimum. The thick black solid curve shows the annual average. The thin gray dot-dashed lines indicate the tropics, and the thin gray dashed lines indicate the artic/diurnal division, as in the middle left plot. **(e)** Annually averaged insolation as a function of time before the present. The white dashed curve indicates the subsolar latitude (SSL) at perihelion. The thin gray dot-dashed curves indicate the latitudes of the tropics, and the thin gray dashed curves indicate the latitudes of the artic/diurnal division. The latitudes of the tropics and the artic/diurnal division both vary over Milankovitch cycles. **(f)** The thick gray dashed curve shows the average insolation from 0.9 Ma to the present day, and the dash-dotted curve shows the average insolation from 2.35 Ma to the present day. The thick black solid curve shows the annual average for the current Pluto year, identical to the thick black solid curve in **(d)**.

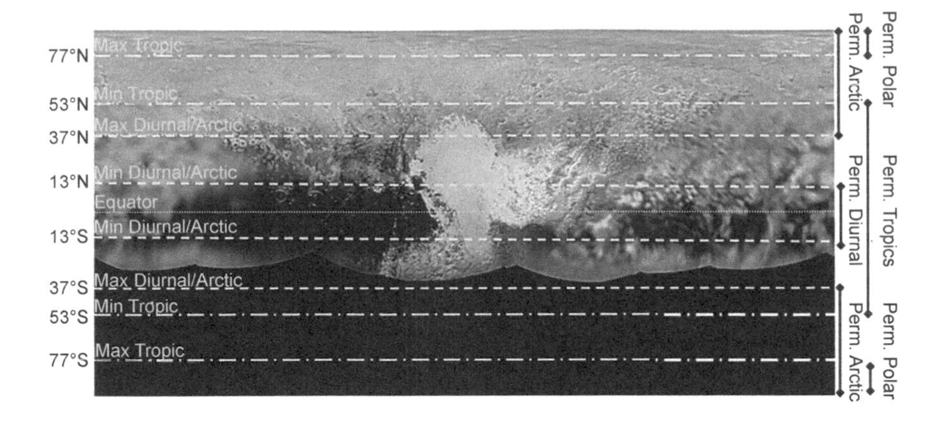

Fig. 7. Climate zones on Pluto and their geographic expression. Over the last Milankovitch cycle, the division between the arctic and diurnal climate zones (dashed lines) oscillates between ±13° and ±37°, and the division between the tropic and polar climate zones (dash-dot lines) oscillates between ±53° and ±77°. The area between ±13° is a permanent diurnal zone, which experiences sunrise and sunset every Pluto day. The area between ±53° is the zone of permanent tropics, where the Sun always reaches the zenith at least once per orbit. Poleward of ±37° is the permanent arctic zone, which always experiences arctic seasons, up to century-long summers and winters during each orbit. Poleward of ±77° is the permanent polar zone, where the Sun never reaches zenith over any orbital period. Adapted from *Binzel et al.* (2017) (see also the larger figure in the chapter by Moore and Howard in this volume).

albedo, S_{1AU} is the solar constant (i.e., the normal solar flux at 1 AU, 1361 W m^{-2}), Δ is the heliocentric distance in AU, ε is the emmissivity, σ is the Stefan-Boltzmann constant, k is the conductivity, z is the substrate depth (0 at the surface, negative below), L is the latent heat of sublimation, and $\dot{m}$ is the net is the net sublimation rate of N_2 ice (kg m^{-2} s^{-1}). Locally, the mass fluxes for N_2 are energy limited, and the lefthand side is small compared with the other terms, so that $L\dot{m}$ balances insolation, thermal emission, and conduction to the subsurface. Globally, what goes up mainly comes down; on seasonal timescales, and sometimes on diurnal timescales, the global average of $L\dot{m}$ can be neglected (e.g., *Young,* 2012).

Consider an ideal Pluto. If Pluto's atmosphere were always dense enough to ensure efficient global transport of N_2, if the surface were entirely covered by N_2, if the thermal inertia were negligible, if the global average of $L\dot{m}$ can be neglected, and if there were no topography, then the global average of equation (2) would simplify to $\varepsilon\sigma T_{eq}^4 \approx \bar{\mu}(1-A) S_{1AU}/\Delta^2$, where $\bar{\mu} = 1/4$ for spatially uniform ice. To first order, the N_2 frost temperature depends on $(1-A)/\varepsilon$. However, the real Pluto deviates from this ideal because the atmosphere may or may not occasionally be non-global, Pluto's surface is heterogeneous, thermal inertia is likely to be significant on diurnal and seasonal timescales, and New Horizons detected significant topography on Pluto (e.g., *Trafton,* 1990, 2015; *Trafton et al.,* 1998; *Grundy and Buie,* 2002; *Hansen and Paige,* 1996; *Stern et al.,* 2015). We deal with each of these items in turn.

3.2.1. Departure from global equilibrium (or "Will Pluto's atmosphere collapse?"). For the New Horizons flyby conditions (subsolar latitude = 51.6°, p = 1.15 Pa, T_{eq} = 36.9 K), Pluto's N_2 ice temperatures should be nearly isothermal across the globe, and the N_2 ice should be nearly in equilibrium with the atmosphere, despite local sublimation or condensation and the variation of insolation with time. Near-equilibrium is a consequence of the high volatility and latent heat of sublimation of N_2 at Pluto's temperatures, as can be demonstrated several ways. First, the number of molecules leaving the N_2-rich ice nearly equals the number of molecules impinging on and sticking to it. That is, the net sublimation or condensation rate is much smaller than the one-way upward or downward flux. Roughly estimating the mass flux by $L\dot{m} \approx \varepsilon\sigma T^4$ gives $\dot{m} \approx 4.2 \times 10^{-7}$ kg m^{-2} s^{-1}. By kinetic theory (the Knudsen-Langmuir equation), and assuming unit sticking coefficient, the one-way upward mass flux is $\dot{m}_{up} = p_{eq}/\sqrt{(2\pi RT)}$ [where, as in equation (3), R is the specific gas constant]. At flyby conditions, this gives $\dot{m}_{up} = 4.4 \times 10^{-3}$ kg m^{-2} s^{-1}, about 10,000 times larger than the net flux. Second, the timescale for N_2 to relax toward an equilibrium pressure is very short. By hydrostatic equilibrium and kinetic theory, $dp/dt = g\dot{m} = -(p-p_{eq})g/\sqrt{(2\pi RT)}$, where g is the surface gravity. This equation is one of relaxation to an equilibrium, $p = p_{eq} + (p(t_0)-p_{eq})\exp(-(t-t_0)/\tau))$, with a timescale $\tau = \sqrt{(2\pi RT)}/g$. For Pluto's flyby conditions, this timescale is 7 minutes, about 1300 times shorter than Pluto's 6.4-day rotation. Third, the isothermality of the surface requires that the sublimation wind, v_{sw}, be much smaller than the sound speed, $v_s = \sqrt{(\gamma RT)}$, where γ is the adiabatic constant (*Trafton and Stern,* 1983; *Spencer et al.,* 1997). The sublimation wind is the wind needed to transport atmosphere from areas of greater to lesser insolation. For an ideal surface covered uniformly with N_2 ice, if the subsolar latitude is at the pole, then the sublimation wind can be calculated by imagining a girdle at the equator with area

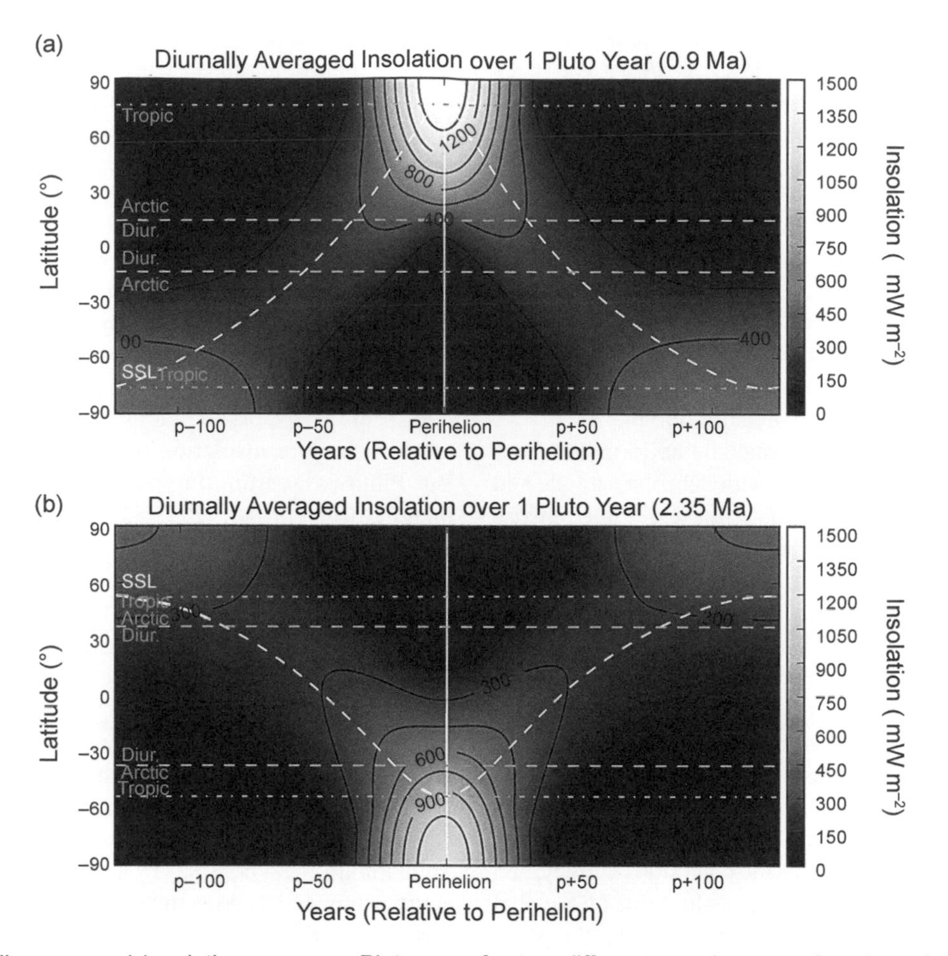

Fig. 8. Diurnally averaged insolation over one Pluto year for two different epochs, as a function of Earth years since perihelion. Curves and lines are as in Fig. 6c. **(a)** Northern summer solstice at perihelion ($L_{s,peri}$ = 90°) at a time of high obliquity. **(b)** Southern summer solstice at perihelion ($L_{s,peri}$ = 270°) at a time of low obliquity. The white dashed curve indicates the subsolar latitude (SSL). The thin gray dot-dashed lines indicate the tropics, between which the Sun is overhead at some time in Pluto's year. The thin gray dashed lines indicate the arctic/diurnal division, between which the Sun rises and sets every day of Pluto's year. The thin solid vertical line marks the time of perihelion.

equal to the circumference times the scale height (*Trafton and Stern,* 1983; *Spencer et al.,* 1997). Multiplying this area by the number density and sublimation wind speed gives a mass flux, which must equal the net condensation in the winter hemisphere. Different subsolar latitudes and different choices of where to place the girdle affects the calculated sublimation wind. The maximum sublimation wind is smaller for equator-to-pole flow than for pole-to-pole flow. This effect can be expressed by $v_{sw} = 4\xi\varepsilon\sigma T^4 r_0 g/(Lp_{eq})$, where r_0 is Pluto's surface radius and ξ is a unitless constant that varies between 0.044 and 0.30 depending on illumination (*Young et al.,* 2020; *Stern and Trafton,* 2008), giving $v_{sw} \approx 0.20$ m s^{-1} at the flyby, in rough agreement with the meridional speeds in *Forget et al.* (2017). At T = 36.9 K, the sound speed, v_s, is 123 m s^{-1}, or ~550× larger than the sublimation speed.

This picture of an isothermal, isobaric surface breaks down with lower temperatures and pressures. *Spencer et al.* (1997) and *Trafton and Stern* (1983) described the breakdown of the global atmosphere at temperatures below ~30.5 K, for p_{eq} = 6.4 mPa for pole-to-pole flow (e.g., from the subsolar to the antisolar point). At that temperature, the atmosphere is still in local equilibrium with the ice: The timescale for equilibrating is still only 7 minutes, and the imbalance between the upward and downward mass flux is only 1 part in 150. However, N_2 still must flow from latitudes of greater insolation to latitudes of lesser insolation. A thinner atmosphere requires faster sublimation winds to transport the same mass. For a surface covered completely with volatiles, at 30.5 K, the sound speed is only ~5× larger than the sublimation wind, which is too small to maintain an isothermal surface (*Stern and Trafton,* 1984; *Trafton,* 1984). At these low pressures, the conditions are more like Io's atmosphere, where the atmosphere is collisional, supersonic winds flow, and the principal volatile is transported across the globe, but the rate of transport is not enough to regulate the surface temperature. For such cases, simplified hydrodynamics equations can be applied, as has been done for Io (*Ingersoll et al.,* 1985), Eris (*Hofgartner et al.,* 2019), and previously for Pluto and Triton under the assumption of a thick CH_4 atmosphere (*Trafton and Stern,* 1983; *Trafton,* 1984). As we discuss in section 4, for many assumptions of albedo, emissivity, and thermal inertia, Pluto's atmosphere stays global throughout its entire orbit.

3.2.2. Surface heterogeneity and the nitrogen cycle. It was well established that Pluto's surface was varied in albedo and composition before the New Horizons flyby and confirmed spectacularly by the LORRI and Ralph instruments of New Horizons. An inhomogeneous N_2 ice distribution lowers the pressure at which an atmosphere transitions from global to local, because a smaller net sublimation or condensation flux needs to move across the planet.

Because Pluto's N_2 ice is not spatially uniform, the mean insolation onto the N_2 ice varies with subsolar latitude and longitude, altering $\bar{\mu}$. On diurnal timescales, the energy balance is controlled by the diurnally averaged insolation onto the N_2 ice, not the instantaneous insolation, because the latent heat required to increase the column mass of the atmosphere in response to the changing surface temperature dampens how quickly temperatures can change over a rotation. *Young* (2012, 2017) defined a unitless parameter, analogous to the Spencer thermal parameter, that dampens the change in surface temperature because of the "atmospheric breathing," which can be written as $\Theta_A = (\omega/fg)(d\ \ln p_{eq}/d\ \ln T) L p_{eq}/(\varepsilon\sigma T^4)$, where f is the fraction of the surface covered by volatiles. For the conditions at flyby, and using $f \approx 1/2$, $\Theta_A \approx 2300$. Even near the limit of a global atmosphere, at 6.4 mPa, $\Theta_A \sim 40$. For large Θ_A, the peak-to-peak temperature is smaller by a factor of $4/\Theta_A$ compared with what it would be if the thermal emission were in instantaneous equilibrium with the average insolation. In Pluto's current configuration, the N_2 ice temperatures warm up and cool off by only millikelvins as Sputnik Planitia — the bulk of Pluto's known N_2 — approaches local noon or midnight.

On seasonal timescales, the mean insolation on the N_2 ice is greatly influenced by the large reservoir of N_2 in Sputnik Planitia, which is primarily in the northern hemisphere, centered on 15°N. In Pluto's current orbit, this implies the greatest insolation happens partway between perihelion and summer solstice, near the time of maximum insolation at 15°N (e.g., *Johnson et al.,* 2021) (see Fig. 6c).

When perihelion coincided with northern summer solstice, 0.9 Ma, each Pluto orbit experienced a period of short and intense insolation onto Sputnik Planitia. In contrast, when perihelion coincided with southern summer solstice, 2.3 Ma, the competing effects of heliocentric distance and subsolar latitude nearly balance, leading to mild changes in temperatures over a Pluto orbit.

And how do volatile cycles lead to Pluto's surface heterogeneity? The areas not covered by N_2 are seen by New Horizons to be CH_4-rich (section 3.4), or covered by the non-volatiles: H_2O and a dark material, presumably tholin (section 3.5). Areas covered with N_2-rich ice can lose N_2 through preferential sublimation, first becoming CH_4-rich ice and then revealing a substrate of H_2O or other material. Sublimation and processing by UV or energetic particles could perhaps form a surface lag deposit of non-volatile organics. Conversely, areas will become new N_2 condensation sites if they become colder than the equilibrium N_2 temperature for the local surface pressure (see section 3.5). Because N_2 ice has high albedo and the non-volatiles are generally low albedo, heterogeneity, once established, maintains itself in a positive feedback (*Young,* 1993; *Earle et al.,* 2018a).

3.2.3. Impact of topography on the nitrogen cycle. If Pluto's surface were entirely flat (i.e., had no topographic relief), then those areas that were covered by N_2 ice would be nearly isothermal across the globe, due to the transport of latent heat of sublimation from areas with more insolation to areas with less insolation. But Pluto is far from flat, as revealed by New Horizons imaging (*Stern et al.,* 2015; *Moore et al.,* 2016; *Schenk et al.,* 2018), with altitudes as much as 5 km above and below the mean level (see section 2.2). The altitude compared to the mean level affects the stability of N_2 ice. Hydrostatic equilibrium requires that the surface pressure is higher at lower altitudes, and so therefore is the equilibrium ice temperature. Since the N_2 ice and gas reach equilibrium quickly, the actual N_2 ice temperature will also be warmer at low altitudes and cooler at high altitudes. The warmer, low-altitude N_2 ice radiates more heat to space than does the cooler, high-altitude N_2 ice. To maintain local conservation of energy, vapor-pressure equilibrium, and hydrostatic equilibrium, the difference in thermal radiation needs to be balanced by latent heat of sublimation through the increased deposition of volatile ice at low altitude and the increased sublimation at high altitude. Consequently, N_2 ice tends to accumulate and be more stable at low altitude.

This atmospheric-topographic process explains why N_2 ice preferentially accumulates in the depressions on Pluto. The atmospheric-topographic process was described by *Trafton et al.* (1998) and *Stansberry et al.* (2014), and was used by *Trafton and Stansberry* (2015), *Bertrand and Forget* (2016), *Bertand et al.* (2018, 2019), and *Young et al.* (2017). *Trafton and Stansberry* (2015) quantified the effect of topography by considering idealized mountains and valleys, based on preliminary topographic relief results from the New Horizons flyby.

Table 1 updates the results of *Trafton and Stansberry* (2015) using improved parameters from New Horizons, illustrating the above processes and the extension of volatile transport to the vertical dimension for day-night, seasonal, or astronomical cycles. Table 1 shows the ice temperature and mass transport flux for topographic relief of ±4 km at the subsolar point and dark side (night or winter) at perihelion, during the New Horizons flyby, and at 42 AU. In the model, Pluto's atmosphere is supported by the vapor pressure of N_2 ice in global radiative equilibrium at a reference level taken to be near the mean of the topographic radial variations (approximating that of the mean level of volatile ice) with parameters determined from the New Horizons flyby and adjusted for other solar

TABLE 1. Topographic temperature differences and transport of N_2 ice on Pluto.

	Perihelion: $T_{ref} = 38$ K		31.9 AU: $T_{ref} = 36.9$ K		42 AU: $T_{ref} = 32.2$ K	
Dark side	T_i (K)	$\dot{m}$ (kg m^{-2} s^{-1})	T_i (K)	$\dot{m}$ (kg m^{-2} s^{-1})	T_i (K)	$\dot{m}$ (kg m^{-2} s^{-1})
Mountain (+4 km)	37.6203	-0.45×10^{-6}	36.5361	-0.40×10^{-6}	31.8647	-0.23×10^{-6}
Valley (–4 km)	38.3874	-0.48×10^{-6}	37.2791	-0.43×10^{-6}	32.4563	-0.25×10^{-6}
Subsolar	T_i (K)	$\dot{m}$ (kg m^{-2} s^{-1})	T_i (K)	$\dot{m}$ (kg m^{-2} s^{-1})	T_i (K)	$\dot{m}$ (kg m^{-2} s^{-1})
Mountain (+4 km)	37.6205	0.44×10^{-6}	36.5365	0.40×10^{-6}	31.8700	0.23×10^{-6}
Valley (–4 km)	38.3875	0.40×10^{-6}	37.2793	0.36×10^{-6}	32.4595	0.21×10^{-6}

distances as needed. The topographic variations are measured with respect to a N_2 atmospheric reference level at $p_{ref} = 1.15$ Pa (*Hinson et al.,* 2017), which, in the absence of topographic relief, would correspond to an isothermal ice surface in global radiative balance and vapor phase equilibrium near 36.9 K. The atmospheric pressure at the topographic ice surface is evaluated by $p_s = p_{ref} \exp(-h/H_{ref})$. H_{ref} is the effective pressure scale height spanning the approximately isothermal topographic relief and h is the displacement of the surface from the reference layer, positive upward. $H_{ref} = 18$ km during the New Horizons flyby (*Hinson et al.,* 2017). The temperature at the reference level T_{ref} is determined from $p_{ref} = p_{eq}(T_{ref})$, but for other solar distances, T_{ref} and the scale height are varied according to the insolation flux [weighted by *Tryka et al.* (1993, 1994) near perihelion]. The temperature just above the surface, T_s, is determined by $p_s = p_{eq}(T_s)$ since the atmospheric column is supported hydrostatically by the ice equilibrium vapor pressure. T_i is the local ice temperature. Unit emissivity is assumed, and an ice albedo A = 0.85 (*Burrati et al.,* 2017). The equation that balances the difference between the one-way thermal fluxes from ice evaporation and sticking atmospheric gas against the difference between the local absorbed insolation and ice radiation fluxes is given, for a unit sticking coefficient, by

$$L\dot{m} = L\left(\frac{p_{eq}(T_i)}{\sqrt{2\pi RT_i}} - \frac{p_{eq}(T_s)}{\sqrt{2\pi RT_s}}\right) = \mu(1 - A)\frac{S_{1AU}}{\Delta^2} - \varepsilon\sigma T_i^4 \quad (3)$$

The cosine of the incidence angle, μ, is 1 for the subsolar point and 0 for the dark side (night or arctic winter).

Table column headings are the ice temperature (T_i) and the local vertical mass flux ($\dot{m}$), positive for sublimation. Table 1 quantifies the preferential accumulation of N_2 ice in the depressions on Pluto, as predicted by *Trafton et al.* (1988), a result that has since been observed. Note that the dark side deposition flux is greater in the valleys than in the mountains, while the subsolar sublimation flux is greater in the mountains than in the valleys. This reflects the net volatile transport to the dark hemisphere as well as to local depressions. The sublimation winds arise to maintain global hydrostatic equilibrium. Finally, note that the volatile transport between hemispheres is balanced mountain to mountain, but is unbalanced valley to valley in favor of deposition into the dark side valleys. Net deposition in the topographic depressions is more efficient than net deposition at higher altitudes at maintaining the darkside temperature (i.e., warming the surface). Conversely, net sublimation from the mountains is more efficient than from the depressions for maintaining the dayside temperature (i.e., cooling the surface).

At perihelion conditions, the equilibrium temperature in the valley is 0.77 K warmer than on the mountain, and the nighttime condensation rate is higher in the valley by 3.7×10^{-8} kg m^{-2} s^{-1}, equivalent to a rate of 0.117 cm yr^{-1} or 30 cm per Pluto year, or about 3.6 cm (Pluto yr)$^{-1}$ km^{-1}.

Young et al. (2017) found a similar result, that 1.5-km-deep depressions at 57°N are 0.13 K warmer than the surrounding uplands, with a net condensation rate of 5 cm per Pluto year, also giving 3 cm (Pluto yr)$^{-1}$ km^{-1}. Over one Pluto year, N_2 accumulation at depth can affect albedo and thus energy balance. It may also be detectable in near-IR spectra through wavelength shifts in the CH_4 absorption. It is likely too small to affect the detectability of N_2 by its weak 2.15-μm feature.

The enhanced condensation rate fills up depressions at a rate proportional to the depth of the depression, or $\dot{h} = h/t_{infill}$. The scale factor can be found from simple scale analysis (*Trafton et al.,* 1998). In a depression of depth h, the pressure will be higher because of hydrostatic equilibrium, by dp/dz = –p/H, where H = RT/g is the pressure scale height, and g is the gravitational acceleration, 0.62 m s^{-1}. Since the depressions are shallow, typically a few kilometers deep, the increase in the pressure will depend only weakly on the details of the atmospheric temperature structure. The equilibrium temperature will also be higher, according to the Clausius-Clapeyron equation, $dp_{eq}/dT_{eq} = (p/T)(L/RT)$. Together, the equilibrium temperature follows the wet adiabat, with lapse rate $-dT_{eq}/dz = \Gamma_w = T_{eq}\, g/L$. For Pluto at 37 K, $\Gamma_w \approx 0.09$ K km^{-1}. The surface at the bottom of the depression, being warmer by $\Delta T = \Gamma_w$ h, emits more thermal radiation than the surface at the top of the depression by $\Delta F = 4\varepsilon\sigma T^3 \Gamma_w$ h, or, for T = 37 K and ε = 1, an additional energy flux of ΔF = 1.05 mW m^{-2} per 1 km of depth. If the absorbed insolation is the same at the top and bottom of a depression, and ignoring thermal conduction, then the additional thermal emission drives a larger condensation rate (*Trafton et al.,* 1998; *Trafton and Stansberry,* 2015). We divide by the latent heat, L, to get mass flux (mass per area per time), and again by density ρ (~1 g cm^{-3}) to get the rate at which a depression fills up: $\dot{h} = \Delta F/(\rho L)$. Altogether, the infilling time constant is

$$t_{infill} = \frac{\rho L^2}{g 4\varepsilon\sigma T^4} \quad (4)$$

The infilling can be expressed as a decaying exponential, $h(t) = h(0) \exp(-t/t_{infill})$, demonstrating that t_{infill} is a characteristic timescale for infilling that is independent of the depression depth. For temperatures of 30–45 K, t_{infill} is 19–3 m.y. That is, scale analysis predicts that it took ~7 m.y. to fill up Sputnik Planitia from a depth of 5.5 km to its current depth of 2 km. The infilling of the deep Sputnik Planitia basin could also be explained by this process (independent of the Milankovitch cycles), with N_2 filling the basin in simulations in typically in 10 m.y. (*Bertrand and Forget,* 2016; *Bertrand et al.,* 2018).

Comparing the infilling timescale to the rate of glacial flow (Fig. 9), we see that infilling of depressions by atmospheric transport is more important than glacial flow on regional to global scales (hundreds to thousands of kilometers), both because of timescale arguments and because the atmospheric transport can fill in depressions even when high topography separates the depression sinks from N_2-ice sources.

3.3. Carbon Monoxide Cycle

Carbon monoxide is almost as volatile as N_2, but it is found only in trace amounts in the atmosphere as a consequence of its low mixing ratio in the N_2-rich ices and the ternary phase diagram (*Tan and Kargel,* 2018). No CO-rich ices have been detected on Pluto from either New Horizons or groundbased observations, and the strong CO bands are seen co-located with the N_2-rich ice (*Schmitt et al.,* 2017). This is also seen in how CO and N_2 vary with subobserver longitude in groundbased spectra, for both Pluto (*Grundy et al.,* 2013) and Triton (*Holler et al.,* 2016). On Triton (*Tegler et al.,* 2019), a rarely seen N_2-CO combination band at 2.239 μm was detected, indicating simultaneous excitation of nearby N_2 and CO in ice. The existence of the band indicates that CO and N_2 molecules are intimately mixed in the ice on Triton, rather than existing as separate regions of pure CO and pure N_2 deposits.

3.4. Methane Cycle

Great progress has been made on the equilibrium behavior of N_2:CO:CH_4 mixtures (*Trafton,* 2015; *Tan and Kargel,* 2018). However, CH_4 is observed to be in disequilibrium on Pluto, and the CH_4 cycle is still not well understood. The cycles of N_2 and CH_4 are entwined, and the drivers of

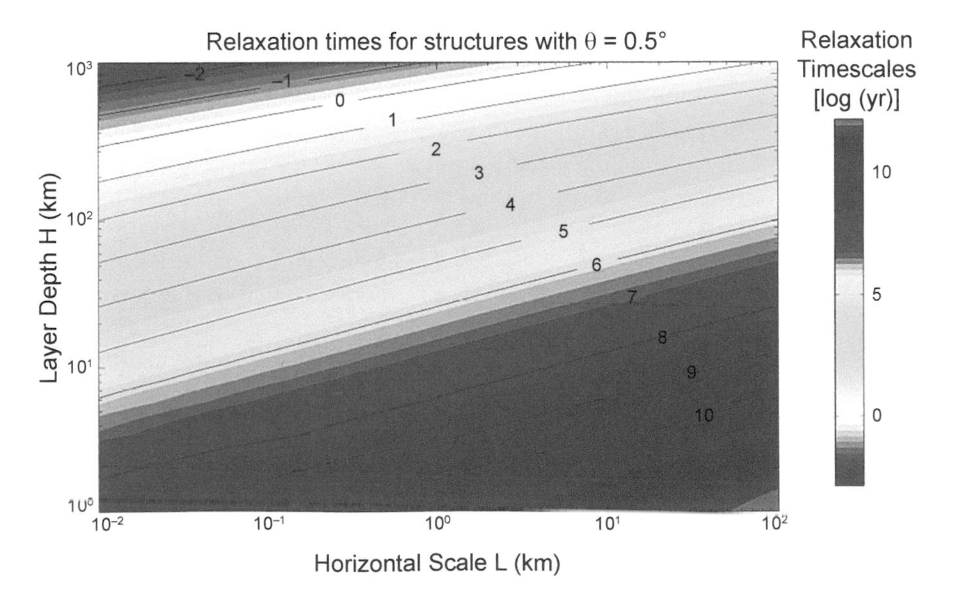

Fig. 9. Relaxation timescale for glacial flow over shallow slopes. The timescale for filling in depressions by atmospheric transport at current Pluto temperatures is about 10^7 yr. Thus, atmospheric transport is quicker (more efficient) than glacial flows for shallow deposits or deposits separated by large distances. Moreover, atmospheric transport operates even when the source and sink areas are discontinuous, as is more likely for shallow deposits and large separations. From *Umurhan et al.* (2017).

disequilibrium of CH_4 occurs on multiple timescales (*Young et al.,* 2019; *Trafton et al.,* 2019). Methane can condense at warmer temperatures than N_2. Since fresh ice of either species is probably bright, areas of high-albedo CH_4 may then be able to cool down enough to allow N_2 condensation. This process could explain the presence of N_2-rich deposits observed by New Horizons outside the deepest depressions, such as Eastern Tombaugh Regio, if the deposited N_2 were also high-albedo (*Bertrand et al.,* 2019). Also, the CH_4 cycle affects the atmospheric chemistry. In Pluto's current atmosphere, the first step in haze production is the absorption of Lyman-α by CH_4. If the atmosphere becomes optically thin to Lyman-α, because of lower CH_4 mixing ratio or smaller surface pressure, this will fundamentally change haze formation. Since haze can alter surface properties (albedo, emissivity), this can have a positive or negative feedback on volatile cycles. Processing of the surface by energetic particles or UV photons can also affect surface properties (see reviews by *Salama,* 1998; *Strazulla,* 1998). Thus, the CH_4 cycle is important for understanding how Pluto works.

Models of the CH_4 volatile cycle should strive to explain the following observed phenomena: (1) The atmospheric mixing ratio of CH_4 is ~0.3% (*Young et al.,* 1997, 2018), 10–100× higher than expected by three-phase equilibrium (where the three phases are N_2-rich ice saturated with CH_4, CH_4-rich ice saturated with N_2, and vapor). (2) The dilution of CH_4 in the N_2-rich terrains is only ~0.3–0.5% (*Protopapa et al.,* 2017; *Douté et al.,* 1999) rather than the ~4% expected for three-phase equilibrium. (The dilution of N_2 within CH_4 is not measurable from near-IR spectra.) (3) Methane-rich terrains dominate the surface northward of 55°N, with significant CH_4-rich terrain at low northern latitudes (20°N–35°N) (*Protopapa et al.,* 2017; *Schmitt et al.,* 2017). (4) The fraction of CH_4-rich ice is slightly larger in northern (vs. southern) Sputnik Planitia, where the dilution of CH_4 in N_2-rich ice is also larger (*Protopapa et al.,* 2017; *Schmitt et al.,* 2017). (5) Methane is seen near the equator in eastern Tombaugh Regio and in the bladed terrain. (6) Methane is seen at high altitude, particularly striking on Pigafetta Montes (previously termed Enrique Montes).

3.4.1. Enhanced methane atmospheric mixing ratio. The enigma of Pluto's high atmospheric CH_4 mixing ratio has been recognized since the early models of atmospheric energy balance, and since gaseous CH_4 was first discovered on Pluto (*Lellouch,* 1994; *Stansberry et al.,* 1996b; *Young et al.,* 1997; *Spencer et al.,* 1997; *Trafton et al.,* 1998). In the context of observations by New Horizons and new N_2-CO-CH_4 thermodynamics, the observed atmospheric CH_4 mixing ratio [0.28%–0.35% (*Young et al.,* 2018)] is much higher than the CH_4 mixing ratio for three-phase equilibrium [0.003%–0.013% (*Tan and Kargel,* 2018; *Young et al.,* 2019)]. As reviewed in *Trafton et al.* (1998), two models were proposed that alternatively explain the observations, termed "detailed balance" (*Trafton,* 1990) and "methane patch" (*Stansberry et al.,* 1996b).

The term "detailed balance" refers to a near-balance between the competing processes of sublimation and condensation on the molecular scale, all occurring in a state near thermal equilibrium, where every kinetic ice-gas transition at the surface is balanced by its inverse transition. Owing to the coupling of hydrostatic and phase equilibrium in this model, a thin layer of CH_4-rich ice is maintained dynamically on the surface of the volatile ice. In this model, CH_4- and N_2-rich ice in the layer are separately in phase equilibrium, where their equilibrium vapor pressures balance the atmospheric partial pressures at the surface. During net sublimation or deposition, this solid layer "rides" on the ice surface, with its molecular membership continuously changing. The thickness of the layer is limited by the depth to which it interacts with the atmosphere. A surface discontinuity in the ice composition thus results from satisfying hydrostatic and phase equilibrium simultaneously. This detailed balance layer may be too thin to be detected spectroscopically.

N_2 may diffuse through this layer during net sublimation. However, its vapor pressure will be reduced by the ratio of the mole fraction of N_2 at the surface vs. below the layer. With fewer N_2 molecules on the surface, the N_2 vapor pressure must be lower. Since N_2 is the dominant species in Pluto's atmosphere, the "detailed balance" model predicts that the total atmospheric pressure could be smaller than that over pure N_2 ice by one or two orders of magnitude. Take for example CH_4-rich ice at 38 K, for which the CH_4 equilibrium pressure is 0.17 mPa (*Fray and Schmitt,* 2009). For a 0.3% gaseous CH_4 mixing ratio and CH_4-ice at 38 K, the N_2 pressure would be suppressed by a factor of 49, reduced from a pure vapor pressure of 2.2 Pa to a partial pressure of only 0.045 Pa (S. P. Tan, personal communication). But the measurements of Pluto's actual surface pressure and radius from the REX instrument (*Hinson et al.,* 2017) is consistent with a surface in equilibrium with N_2-rich or pure N_2 ice. How much could the N_2 pressure be reduced and still be in equilibrium? As discussed in section 2.1.2, the N_2 ice temperature is less than 41 K (p_{eq} = 10.9 Pa) from the shape of the 2.15-μm feature, and the 3σ lower limit on the surface pressure was measured to be 0.9 Pa, for a reduction of pressure by at most a factor of 12.

Calculations of the three-phase equilibrium state of Pluto's N_2 and CH_4 ice vs. temperature by Tan (*Young et al.,* 2019), based on the EOS of *Tan and Kargel* (2018), revealed an equilibrium atmospheric CH_4 ratio several orders of magnitude lower than the ~0.3% observed (*Young et al.,* 2018), implying that Pluto's ice is not in a state of three-phase thermal equilibrium, where both saturated phases of N_2 and CH_4 exist with the atmospheric gas phase. Up to this this time, the EOS for these ices was unknown, so vapor pressures of the solid solution were approximated by Raoult's law, which was, in some instances, an oversimplification given the easily saturated solid phases. Using the EOS of *Tan and Kargel* (2018) and the detailed balancing model, *Trafton et al.* (2019) extended this calculation to the two-phase equilibrium case, covering the N_2-rich ice with an unsaturated CH_4-rich layer. They confirmed the disequilibrium for CH_4: The observed partial pressure of Pluto's CH_4 remained far in excess of the two-phase equilibrium CH_4 vapor pressure in this model. The CH_4-rich layer would have to have a temperature of 42.5 K to explain the observed mixing ratio. This disequilibrium is likely a result of the much lower volatility of CH_4 than N_2. Moreover, as a minor constituent, CH_4 gas may not be fully in hydrostatic equilibrium.

The CH_4 patch model (*Stansberry et al.*, 1996b) considers the impact on the atmospheric CH_4 mixing ratio of pure or CH_4-rich areas, perhaps 15 K warmer than the N_2 ice temperature. As a minor species, the CH_4 mass flux is not given by energy balance (equation (2)) or by the Knudsen-Langmuir expression, but by turbulent diffusion of CH_4 through N_2

$$\dot{m} = -\rho U_* C_m (\chi_z - \chi_0) \quad (5)$$

where ρ is the atmospheric density at the surface, χ_z is the atmospheric molar mixing ratio of CH_4 at a level near the surface, z (described below), χ_0 is the atmospheric molar mixing ratio in equilibrium with the surface (i.e., the equilibrium partial pressure of CH_4 at the local surface temperature and composition divided by the local surface pressure, $\chi_0 = p_{eq,CH_4}(T)/p_s$), U_* is the friction velocity, and C_m is the mass transfer coefficient. *Forget et al.* (2017) uses a similar formulation, $\dot{m} = -\rho U_1 C_d (q_1 - q_0)$, where U_1 is the horizontal wind, C_d is the drag coefficient computed for the first atmospheric layer, and q_1 and q_0 are the mass mixing ratios at the first atmospheric layer and at the surface, respectively, defined by $q = (\mu_{CH_4}/\mu_{N_2})\chi$, where μ is the molecular weight of CH_4 or N_2. *Stansberry et al.* (1996b) and *Forget et al.* (2017) took different approaches to the choice of z. *Stansberry et al.* (1999b) integrated the empirical expressions used by Monin-Obukhov similarity theory from the surface to z to find C_m. The upper boundary, z, was chosen to be large enough that the computed C_m was insensitive to the choice of z. They used z = 2 km, which was much larger than the depth of the sublayer dominated by eddy diffusion (~14 cm). *Forget et al.* (2017) explicitly computed the exchange very near the surface. In their model, z is the altitude of the first layer of the three-dimensional GCM, or 7 m.

For typical values of $U_* C_m$, turbulent diffusion is about 2000–10,000× slower than Knudsen-Langmuir sublimation (as used by *Moore et al.*, 2018). The observed atmospheric mixing ratio at the time of the New Horizons flyby was $\chi_z \sim 0.3\%$, and models typically predict $\chi_z \sim 0.001\%–1\%$ (*Bertrand et al.*, 2019). While modelers (*Stansberry et al.*, 1996b; *Forget et al.*, 2017; *Bertrand et al.*, 2019) have not yet applied the new EOS results, this will have little effect on the calculated mass flux from equation (5). First consider the current atmosphere above N_2-rich ice. Even with CH_4 saturated in N_2-rich ice, the EOS gives atmospheric mixing ratios of $\chi_0 \sim 50$ ppm; smaller mixing ratios or the (inappropriate) use of Raoult's law will only decrease this value. Since $\chi_z \approx 0.3\% \gg \chi_0$, the deposition rate onto N_2-rich ice calculated in equation (5) depends negligibly on the value of χ_0. For sublimation of CH_4 from warm CH_4-rich patches, χ_0 is calculated from the ratio of CH_4 partial pressure at some elevated temperature to the local surface pressure (e.g., the REX measurement of 1.15 Pa). Sublimation would occur for pure CH_4 at 42.1 K or hotter, but it is thermodynamically unlikely that pure CH_4 exists on Pluto. If the CH_4-rich ice were saturated with N_2, the calculated critical temperature for CH_4 sublimation would be 42.4 K, only 0.3 K warmer than if the CH_4-rich ice were pure. Considering the uncertainties in albedo and emissivity, this is a small difference. At just a few degrees warmer, $\chi_0 \gg \chi_z$ for all pure CH_4 or CH_4-rich terrains.

Observations appear to support some version of a warm CH_4 patch model: New Horizons directly detected distinct areas of CH_4-rich and N_2-rich ice (*Schmitt et al.*, 2017; *Protopapa et al.*, 2017; *Moore et al.*, 2018), and models of globally isothermal N_2-rich ice and locally warmer CH_4 ice are consistent with ISO, Spitzer, and Heschel thermal measurements (*Lellouch et al.*, 2000, 2011b, 2016). Because Pluto's thermal light curve is likely to be heavily influenced by the thermal contrast between the non-volatile and volatile terrains, it will be useful to model existing or new thermal measurements in light of New Horizons observations of albedo and surface composition, and to model separate thermal inertias for non-volatile and CH_4-rich areas. In this context, there is no mystery in the observed atmospheric methane mixing ratio, because there are many CH_4-rich deposits that are too warm to allow much N_2 in them. Rather the enigma is why all CH_4 does not get permanently cold-trapped in cold N_2-rich deposits.

Stansberry et al. (1996b) showed that sublimation from warm CH_4-rich patches was, in general, much more efficient than the condensation onto N_2-rich areas, both because the mass transfer coefficient is larger for sublimation than condensation, but also because χ_0 increases exponentially with temperature. Equating the global integral of CH_4 subliming from patches to the global integral of condensation, *Stansberry et al.* (1996b) showed cases for which a patch area covering only a few percent of Pluto's surface sufficed to explain atmospheric CH_4 mixing ratios between 0.2% and 1%, given N_2 ice at 38–40 K with warm CH_4 patches at 53–54 K. Colder N_2 ice temperatures and lower atmospheric mixing ratios favored smaller areas covered by the patches.

However, Pluto's surface has CH_4 patches on regional scales, as deduced from Earth-based observations and confirmed by New Horizons (*Schmitt et al.*, 2017; *Protopapa et al.*, 2017). *Trafton* (2015) argued that continent-sized CH_4 patches are too big to be explained by the CH_4 patch model. This conclusion would also apply to Pluto's CH_4-rich polar and high latitudes, for which the CH_4-rich ice would need to be at ~42.5 K (*Trafton et al.*, 2019) to explain the global CH_4 mixing ratio of ~0.3% (*Young et al.*, 2018).

Thus, some version of the CH_4 patch model remains a viable explanation of Pluto's CH_4 enhancement, as it is supported by the observed CH_4-rich ices (*Protopapa et al.*, 2017) and works well in some post-flyby models (*Forget et al.*, 2017; *Bertrand et al.*, 2019). However, small temperature deviations from thermal equilibrium are still capable of explaining Pluto's atmospheric CH_4 enrichment and surface pressure, in terms of warmer and cooler regional temperature deviations that give rise to the distinct CH_4-rich and N_2-rich ice deposits, respectively (*Trafton et al.,* 2019). Thus, the detailed balance model may also play a role but with separate thermal equilibria for the N_2-rich and CH_4-rich ice, respectively, having different temperatures and possibly confined to specific regions or latitudes. But, the detailed balance model cannot explain the observed CH_4 enhancement in terms of global thermal equilibrium covering both N_2-rich and CH_4-rich ice deposits simultaneously.

3.4.2. Undersaturated methane in nitrogen-rich ices. Because all three phases exist at Pluto (CH_4-rich ice, N_2-rich ice, and vapor), in equilibrium the CH_4-rich ice would be saturated with a few percent N_2, and the N_2-rich ice would be saturated with a few percent CH_4. This is not what was seen in Hapke modeling of the New Horizons observations (*Protopapa et al.,* 2017), in which there is only ~0.3–0.5% CH_4 in N_2-rich ice. The disequilibrium is aided by the longer timescale for equalizing CH_4 between vapor and ice. The equilibrium timescale is over a Pluto day in length because atmospheric CH_4 has to diffuse through N_2 (*Young et al.,* 2019). While the long timescale can help maintain the disequilibrium, the initial cause of the disequilibrium is not yet clear. It is intriguing that the CH_4 mixing ratio in the atmosphere is the same as in the N_2-rich ice. This might have been the case had the atmosphere been escaping while being replaced by the volatile reservoir, or subsequently condensing from a large atmospheric mass in the past. What is unclear is why this has not evolved toward equilibrium, and the answer may be a matter of timescales (*Young et al.,* 2019).

3.4.3. Formation of methane-rich ice. Outside of Sputnik Planitia, there is a striking latitudinal dependence of the CH_4-rich ice. Methane-rich ice dominates from 20°N to 35°N and 55°N to 90°N, with N_2 dominating the intervening latitudes (*Protopapa et al.,* 2017; *Schmitt et al.,* 2017); the interpretation is that Pluto is showing two sublimation fronts, with N_2 subliming from low latitudes near equinox and more recently from high latitudes (Fig. 6c).

Methane-rich ice is also seen at high altitude in various locations (Fig. 2): on isolated mountaintops within Cthulhu, such as Pigafetta Montes; on crater rims and the edges of scarps west of northern Sputnik Planitia; on the uplands in the N_2-rich latitudes east of northern Sputnik Planitia [the "smooth uplands" and "eroded smooth uplands" of *Howard et al.* (2017b)]; and on the low-latitude bladed terrains (*Moore et al.,* 2018).

From the equilibrium phase diagram (*Tan and Kargel,* 2018; *Trafton et al.,* 2019), whether CH_4-rich ice or N_2-rich ice dominates locally depends on the direction that the thermodynamic state of the ice deviates from the globally set thermal equilibrium. When the state locally deviates isothermally to lower pressures relative to the global equilibrium hydrostatic pressure, or deviates isobarically to higher temperatures, the CH_4-rich phase of ice forms as the local N_2-rich ice sublimes. Conversely, opposite local deviations from global equilibrium drive the formation of the N_2-rich ice phase, as the local atmosphere N_2 condenses. Thus, CH_4-rich ice may form where N_2-rich ice preferentially sublimes, such as at subsolar latitudes, where the insolation is greater, or for ice at high elevations, where the equilibrium temperature is lower. Pressure deviations could result from regional circulation patterns while temperature deviations could result from the arrival of winds warmed over broad non-icy areas that drive the regional formation of the CH_4-rich phase — or from the arrival of cooler dense air, such as occurs at Sputnik Planitia, driving in the formation of the N_2-rich phase there.

Another mechanism may be at play for equatorial CH_4-rich deposits at altitude (*Bertrand et al.,* 2020b; see also the chapter in this volume by Forget et al.), involving an enhancement of atmospheric CH_4 (and therefore higher condensation rates of CH_4) at higher altitudes. The New Horizons UV occultations cannot constrain this, since atmospheric CH_4 was too opaque to be measured below 80 km.

3.5. Water and Tholin-Rich Areas

The nonvolatile areas of Pluto are warmer by 10–20 K than the N_2-rich ice (and can be also warmer than the CH_4-rich ice) because they tend to be covered by materials with a bolometric Bond albedo near 0.1 (*Buratti et al.,* 2017), which is much lower than that of N_2-rich areas. As nonvolatiles, they are not cooled by the latent heat of sublimation. A positive feedback can be established, in which dark areas can get warm and continue to stay free of volatiles, while bright areas stay cold, and become volatile condensation sites.

The temperature for non-volatile-covered areas are dictated by the insolation (including the effect of local slopes and shadowing), the albedo, the thermal inertia, and the emissivity. All areas with very low albedos are volatile free, but not all volatile-free areas have very low albedos. If the thermal inertia and the mean insolation are both high enough then the non-volatiles will remain clear over an entire Pluto orbit. Otherwise, if a nonvolatile area becomes colder than the mean N_2 ice temperature, then the nonvolatile areas will become condensation sites (so there should be no location colder than the global N_2 ice temperature adjusted for altitude). Because fresh N_2 or CH_4 ice is likely to be bright, new fresh frost may allow areas to remain cold even at summer seasons. In this way, the thermal inertia may be important for the creation of annual nitrogen deposits and the transition of annual deposits into perennial deposits.

4. A REVIEW OF MODELS

The processes controlling the volatile and climate cycles, described in section 3, have been used to model the behavior of volatiles on Pluto over multiple timescales since the first volatiles were detected in Pluto spectra. In this section, we concentrate on those models with simulations of the cyclical or long-term evolution of Pluto's volatiles. This section is roughly chronological, since there is a general pattern of new observations spurring new models of Pluto's volatile cycles, leaving the field with new paradigms (Table 2).

4.1. Spurred by Methane: The First Volatile Transport Models

Volatile transport models for Pluto were inspired by the detection of CH_4 frost on Pluto by *Cruikshank et al.* (1976), the large amplitude of Pluto's rotational light curve, and decade-long changes in Pluto's light curve with changes in subsolar latitude. After the discovery of CH_4 ice on Pluto by *Cruikshank et al.* (1976), it was assumed that there would be a CH_4 atmosphere based on the volatility of CH_4. Subsequent observations confirmed the detection of CH_4 absorption in spectra with increasingly better spectral resolution and data quality. *Fink et al.* (1980) reported an atmospheric CH_4 abundance of 27 ± 7 m-Am (19.3 ± 5.0 kg m^{-2}; 12.0 ± 3.1 Pa). [Here and throughout this chapter we convert column densities in meter-Amagats (m-Am) or kilograms per square meters to pressures using the current best value of surface gravity, g = 0.62 m s^{-2}.] *Cruikshank and Silvaggio* (1980) estimated a vertical column of ~7 m-Am CH_4 (5.0 kg m^{-2}; 3.1 Pa) for an atmosphere if all the observed absorption were gaseous, but concluded that it was impossible to derive a gas abundance from the data. The first generation of models studied the implications of either a massive CH_4 atmosphere, or a tenuous CH_4 atmosphere supported by the CH_4 ices, with or without another gas species. Although the CH_4 spectral signature turned out to be almost entirely due to the surface ice, not gas (*Spencer et al.,* 1990; *Young et al.,* 1997), a series of early models mainly by Trafton, Stern, and coauthors established some of the critical concepts for volatile transport on Pluto, and were motivators for the 1988 occultation observation that finally definitively detected the atmosphere.

The observed CH_4 spectral signature raised the challenge of volatile retention (see the chapter in this volume by Strobel). By 1980, rough values of Pluto's mass and radius were available. Using a steady-state model of Pluto's atmosphere supported by the vapor pressure of CH_4 ice and a bolometric Bond albedo between 0.19 and 0.52 to constrain the ice temperature, *Trafton* (1980) concluded that Pluto's mass was too small to hold an isothermal CH_4 atmosphere in vapor equilibrium with an ice warmer than 45 K. The excess CH_4 would overflow the gravitational potential

TABLE 2. Models for volatile cycles and trends.

Motivating Observations	Models of Volatile Evolution*	New Paradigms of Volatile Evolution
CH_4 spectra	TS: *Trafton* (1980, 1981, 1984, 1989); *Trafton and Stern* (1983); *Stern and Trafton* (1984); *Trafton et al.* (1988); *Stern et al.* (1988)	Escape and replenishment of the atmosphere; vapor-pressure and hydrostatic equilibrium; mass and energy transport; conditions for a global atmosphere; atmospheric "laundering"
Stellar occultation (1988); surface N_2, CO, CH_4 (CH_4-rich and diluted in N_2-rich); surface heterogeneity; gaseous CH_4	TS: *Trafton* (1989, 1990); HP: *Hansen and Paige* (1996) (HP96); Other: *Young* (1993); *Stansberry and Yelle* (1999)	Albedo feedback; two-volatile (N_2/CH_4) interaction; N_2 emissivity and phase change; thermal inertia; loss of summer N_2 and the importance of subsolar latitude
Gaseous CH_4, CO; surface N_2, CO, CH_4 (variation with time); thermal emission; stellar occultations (2002 and beyond).	TPM: *Lellouch et al.* (2000, 2011b, 2016); HP: *Elliot et al.* (2007); *Bosh et al.* (2015); *Hansen et al.* (2015); VT3D: *Young* (2012, 2013); *Olkin et al.* (2015); LMD: *Vangvichith* (2013); TF/PlutoW-RF: *Toigo et al.* (2010, 2015); *French et al.* (2015)	Low thermal inertia (diurnal skin depth); high thermal inertia (seasonal skin depth); CH_4 variation (surface and gas); atmospheric "breathing"
New Horizons flyby, including composition maps on geologic spatial scales, surface radius, atmospheric structure, composition, and surface pressure	PlutoWRF: *Moores et al.* (2017); VT3D: *Young et al.* (2015); *Lewis et al.* (2021); *Johnson et al.* (2021); EB: *Stern et al.* (2017), *Binzel et al.* (2017); *Earle et al.* (2017, 2018a); LMD: *Forget et al.* (2017); *Bertrand and Forget* (2016); *Bertrand et al.* [2018 (B18), 2019 (B19)]; *Meza et al.* (2019)	Sputnik Planitia as equatorial reservoir; role of topography; geologic expression of volatile cycles; other paths for volatile transport; albedo feedback in the equatorial regions; Milankovitch cycles; question of southern N_2; puzzle of widespread CH_4; cold CH_4 ice enabling N_2 condensation

*Sorted by model category: TS: Papers by Trafton, Stern, and colleagues. HP: Papers by Hansen, Paige, and colleagues. TPM: Thermophysical models by Lellouch and colleagues. VT3D: Three-dimensional volatile transport model. LMD: Laboratoire de Météorologie Dynamique. TF: Papers by Toigo and French. EB: Papers by Earle and Binzel. PlutoWRF: National Center for Atmospheric Research (NCAR) Weather Research and Forecasting GCM. Other papers also mention volatile evolution. An emphasis is given in this table to papers with predictions of volatile behavior over various cyclical timescales.

well ("blowoff") and escape from Pluto. Three solutions contributed to the resolution to this dilemma. First was the deduction that a heavier gas (such as Ar, N_2, CO, or O_2) must be present in greater abundance to reduce the scale height enough to gravitationally bind the atmosphere, and to provide a diffusive barrier great enough to slow the escape rate (*Trafton,* 1980, 1981) — and, in fact, N_2 dominates over CH_4 in Pluto's atmosphere. The second was that Pluto's surface gravity is closer to 0.51 m s^{-2} than 0.30 m s^{-2} (*Fink et al.,* 1980) — in agreement with the measured value of 0.62 m s^{-2}. The third was that the blowoff predicted by the isothermal atmospheric escape model of *Trafton* (1980) would not occur in such a rapidly expanding atmosphere owing to adiabatic cooling at altitude (*Hunten and Watson,* 1982). Once *Hunten and Watson* (1982) pointed this out, the relative retention of Pluto's atmosphere was no longer in doubt; the cold upper atmosphere, as measured by New Horizons, confirms that Pluto is not undergoing blowoff.

Seminal models describing the structure of the volatile atmospheres of Pluto and Triton (Table 2) assumed both hydrostatic equilibrium and thermodynamic equilibrium for the state of the volatile ice-atmosphere system. Some of these began by taking the *Fink et al.* (1980) measurement of a thick CH_4 atmosphere (e.g., *Trafton and Stern,* 1983), while others considered a range of volatile gases (e.g., *Stern and Trafton,* 1984). These papers established several key concepts.

First, papers presented arguments for the presence of other volatiles (N_2, Ar, CO, or O_2) in addition to CH_4 (*Trafton,* 1981; *Stern and Trafton,* 1984). They discussed the retention and loss of various volatiles. *Trafton et al.* (1988) calculated that ~0.05 cm of CH_4 would be lost per orbit to escape or chemical destruction (assuming a CH_4 atmosphere with a surface temperature of 43.2 K), and concluded that replenishment is a key element of any volatile cycle.

Second, these papers derived the expressions for how surface pressures and temperatures vary across the surface, including variation with time of day or latitude (in these papers, termed the longitudinal and latitudinal tides). The variation for the latitudinal tide was derived for arbitrary subsolar latitude in *Trafton* (1984). These papers found, for example, that 7 cm-Am or more of CH_4 (0.05 kg m^{-2}; 0.03 Pa) would maintain global hydrostatic and vapor-pressure equilibrium. This established the central paradigm for Pluto's surface and atmosphere, in which winds transport both mass and energy (through latent heat of sublimation) across the globe, maintaining a surface that is nearly isobaric and isothermal.

Third, these papers predicted that vapor-equilibrium pressures would vary by about a factor of 100 between perihelion and aphelion (a little less variation for CH_4, a little more for N_2). They posed a central question that still drives our studies of the volatile cycles on Pluto: When is global uniformity a good approximation, and when is it not? *Stern and Trafton* (1984) calculated that even CH_4, the least volatile of the ices they considered, should allow for global uniformity of pressure and volatile ice temperature throughout Pluto's orbit, if the surface were evenly covered by volatile ices. If the surface were unevenly covered by ices, then freeze-out could occur if only a thin veneer of temporary seasonal volatile ices existed at each pole, and sublimed from the summer hemisphere.

Fourth, the aptly titled article "Why is Pluto Bright" (*Stern et al.,* 1988) described how the seasonal cycle of volatile sublimation and deposition can drastically alter the surface appearance. Comets and other bodies are darkened as energetic particles and UV photons process ices. Methane in particular is converted to larger, darker, less-volatile hydrocarbons. *Stern et al.* (1988) calculated ~1 cm of clean CH_4 would be deposited each Pluto orbit, freshening or "laundering" the surface, with a possible seasonal phase lag of the refreshening. Thus, they showed that the movement of volatiles can fundamentally alter Pluto's surface.

4.2. Atmosphere Confirmed: The First Constrained Volatile Transport Models

An atmosphere was definitively discovered by occultation in 1988 (*Hubbard et al.,* 1988, *Elliot et al.,* 1989) (section 2.1.3), yielding refractivity and scale height vs. distance from Pluto's center (not vs. altitude, since the surface radius was still unknown), with the caveat of the unknown influence of hazes on the interpretation of the stellar occultation. After Pluto's N_2 ice was discovered (*Owen et al.,* 1993), refractivity and scale height could be interpreted as pressure and temperature. Nitrogen had also been detected as the dominant species in Pluto's sister world, Triton, in 1989. The invaluable 1988 occultation, combined with other breakthrough observations in the 1990s, allowed modeling of Pluto's volatile cycles pinned to concrete observations. Note, however, that the true surface pressure and surface radius were not known until the New Horizons flyby, and the 1988 occultation gave only a single snapshot of Pluto's atmosphere.

Trafton (1990) quantitatively calculated the impact of a two-component atmosphere on Pluto, N_2, and CH_4. The thermodynamics of that work were discussed in section 3.4.1. Here, we note the predictions for the time evolution of the volatiles, namely that (1) the escaping atmosphere must be resupplied from the frozen reservoir, with the prediction that the atmospheric bulk composition should resemble that of the reservoir, and (2) for some bulk compositions, the partial pressure of N_2 will not be much different between perihelion and aphelion. *Young* (1993) modeled an N_2 surface and atmosphere given Pluto mutual-event albedo maps. This work derived the "girdling latitude" approach described in section 3.2.1 to calculate sublimation wind speeds as a function of latitude. He found that pressures would drop, in one reference case, to a few millipascals by 2070, at which time the sublimation winds would have to be supersonic in order to maintain an isobaric surface, likely an untenable situation. His transport model demonstrated the power of albedo feedback, in which bright areas can stay volatile-covered (and bright), even when they are close to the subsolar latitude. The albedo feedback allows for strong albedo contrasts of bright areas covered with N_2 ice neighboring dark areas of substrate at the same latitude, as seen in mutual event and HST maps. *Stansberry and Yelle* (1999) described the effect of N_2 ice emissivity, ε, on the N_2-ice temperature. In general, a lower emissivity requires a higher temperature, T. Because the low-temperature phase of N_2 ice (the α-phase) has a lower emissivity than the β-phase (*Stansberry et al.,* 1996a; *Lellouch et al.,* 2000), *Stansberry and Yelle* (1999) argued that Pluto's atmosphere would remain at 35.6 K, the temperature of the α-β phase transition, from a few decades after perihelion through to aphelion. The latent heat difference between the α and β phases also stabilizes the N_2 frost (*Duxbury and Brown,* 1993).

The most comprehensive and concrete model of this period was by *Hansen and Paige* (1996, hereafter *HP96*), which built on a similar model applied to Triton (*Hansen and Paige,* 1992), in turn based on a martian model. For 17 years, *HP96* was the only model of Pluto's seasonal behavior that included thermal inertia and mobile N_2 ice. Over 50 cases were run, with thermal inertia ranging from 42 to 2093 TIU, substrate albedo of 0.2–0.8, N_2-ice albedo of 0.2–0.8, frost emissivity (constant with crystalline phase) of 0.2–1.0, and average global N_2 inventory of 50–200 kg m^{-2}. This inventory is now known to be very small compared to the N_2 residing in Sputnik Planitia (e.g., *Glein and Waite,* 2018). An important simplification was that only N_2 was included. Seven simulations were plotted in detail (with an eighth showing an example of solstice at perihelion), showing temperatures at selected latitudes, surface pressures, albedo, and frost distribution. Key concepts from the *HP96* models were the importance of thermal inertia, especially for the timing of the formation of winter polar caps, and that the effect of subsolar latitude can be more important than the change in heliocentric distance (including delaying the atmospheric collapse until after perihelion).

4.3. Signs of Change: Modeling the Observed Volatile Cycles

As observations of Pluto's surface spectra improved in quality and frequency and laboratory data also improved, it became clear that at least three surface components existed and covered large areas on Pluto: N_2-rich ice, with dilute amounts of CH_4 and CO; CH_4-rich ice (indistinguishable in spectra from pure CH_4); and a third component, presumably tholin (e.g., *Grundy and Fink,* 1996; *Grundy and Buie,* 2002; *Grundy et al.,* 2002, 2014). Rough composition maps were made by combining HST maps with the changes of spectra vs. subsolar longitude as Pluto rotated (*Grundy and Fink,* 1996; *Lellouch et al.,* 2000).

Lellouch et al. (2000, 2011b, 2016) modeled surface temperatures using these rough composition maps and compared results to observed thermal emission. These thermal models assumed that areas free of volatiles varied their temperatures with latitude and time of day according to thermophysical models that balance insolation, thermal emission, and subsurface conduction. Temperatures of CH_4-rich ices (or pure CH_4 ices if any) were also varied in the thermal models according to thermophysical models. As shown by *Stansberry et al.* (1996a), sublimation rates from CH_4-rich ice at 54 K were large enough that the energy expended for the solid-gaseous phase change (the latent heat of sublimation of CH_4) slowed down further heating, so the CH_4-rich ices were capped at 54 K. The N_2-rich ice was assumed to be isothermal with location and time of day, as required by N_2's large latent head of sublimation at Pluto temperatures (*Trafton and Stern,* 1983; *Young,* 2012). *Lellouch et al.* (2000, 2011b, 2016) found a surprisingly small thermal inertia for the diurnal variation, of only ~25 TIU, nearly 100× smaller than the thermal inertia of pure water ice, the likely bedrock. The derived thermal properties applied to the diurnal skin depth (~3 cm). The tholins in these models reached temperatures as high at 63 K, while the CH_4-rich areas reached their maximum of 54 K, 19 K above the 35 K N_2 temperature (*Lellouch et al.,* 2000).

In 2002, two stellar occultations by Pluto were observed (*Elliot et al.,* 2003b; *Sicardy et al.,* 2003), revealing that Pluto's atmospheric pressure had roughly doubled in the 14 years since the previously observed occultation. To many, this was extraordinary (*Hubbard,* 2003), although, as mentioned by both *Elliot et al.* (2003b) and *Sicardy et al.* (2003), this post-perihelion rise was predicted in many *HP96* simulations.

As more stellar occultation observations were observed and analyzed, researchers compared trends with the predictions of *HP96* (Fig. 10) (*Pasachoff et al.,* 2005; *Elliot et al.,* 2007; *Young et al.,* 2008; *Zalucha et al.,* 2011a,b; *Bosh et al.,* 2015; *Gulbis et al.,* 2015; *Dias-Oliveira et al.,* 2015). Among the runs favored by these authors (*Elliot et al.,* 2007; *Young et al.,* 2008) were Figs. 6 and 7 in *HP96*, which had small thermal inertia (42 TIU) and moderate N_2 inventory (50–100 kg m^{-2}). The comparisons were hampered by the uncertainty in Pluto's surface pressure and temperature. By 2006, the trends in pressure could be used to eliminate some of the *HP96* runs. The persistent questions — When will Pluto's atmosphere begin its bulk decline? What will the pressure be for the New Horizons flyby? — could not be answered with the small number of cases presented in detail in *HP96* and the many unconstrained parameters of the model.

Young (2013) developed a new model whose physics was very similar to that of *HP96* (mobile N_2 ices, subsurface conduction, globally isobaric and isothermal N_2 ice), but with much faster computation time (VT3D) (*Young,* 2017). This allowed *Young* (2013) to cover a grid of thermal inertia from 1 to 3162 TIU, N_2-ice albedo of 0.2–0.8, frost emissivity of 0.55–0.8, and average global N_2 inventory of 20–640 kg m^{-2}. Of the 672 cases run, 52 were consistent with the 1988, 2002, and 2006 occultations, and from these, only 19 were also consistent with thermal, visible, and near-IR data. The 52 cases fell into three broad classes. One of these [exchange with early collapse (EEC)] had a very small N_2 inventory, in which

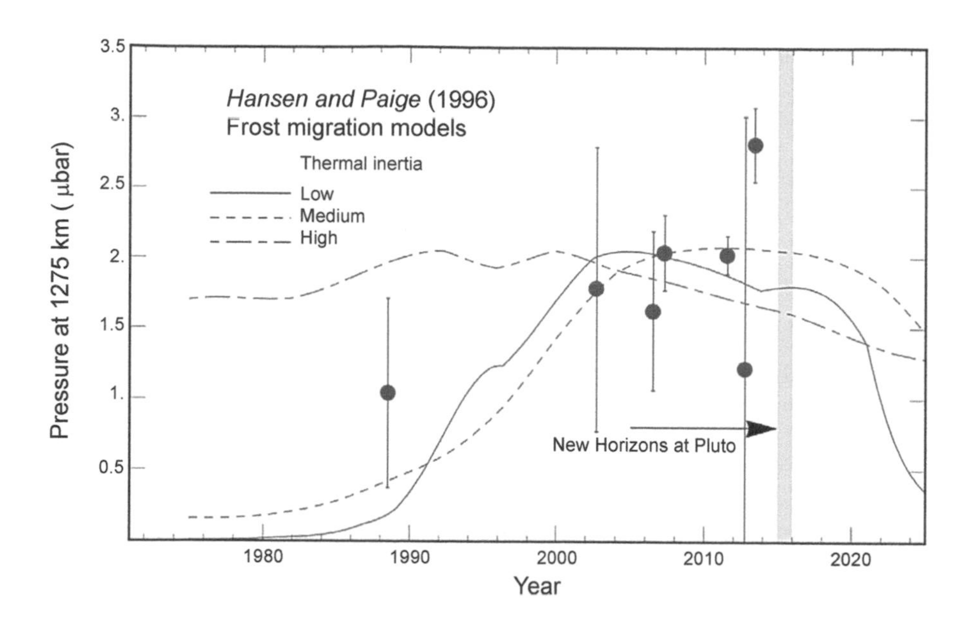

Fig. 10. Measurements of the extent of Pluto's atmosphere, as determined via several stellar occultations between 1988 and 2013 from *Bosh et al.* (2015). The frost migration model predictions of *Hansen and Paige* (1996) are also included, adjusting the pressure to a radius of 1275 km, to provide context for the differing measurements. The arrival date of the New Horizons spacecraft at Pluto is indicated by the gray vertical bar. The pressure at 1275 km radius from the REX radio occultation (*Hinson et al.*, 2017) was 1.948 ± 0.309 μbar (0.1948 ± 0.0309 Pa). Figure adapted from *Bosh et al.* (2015).

the spatially confined northern polar cap was modeled to sublime shortly after equinox. In EEC, the peak occurred before the 2002 occultations, and the predicted pressure plummeted by 2015. The second class [exchange with pressure plateau (EPP)] shared many characteristics with Fig. 7 from *HP96* [one of the favored runs of *Elliot et al.* (2007) and *Bosh et al.* (2015)]: small thermal inertia (10 TIU) and small N_2 inventory (8–16 kg m^{-3}), an exchange of N_2 between northern and southern polar caps, two strong pressure maxima per Pluto orbit, minimum surface pressures below 1 mPa, and a peak pressure near 2002 followed by a very slow decrease. The final class [permanent northern volatiles (PNV)] had no equivalent in *HP96*, and was characterized by high thermal inertia (typically ≥1000 TIU); large N_2 inventory (typically >100 kg m^{-2}); the presence of N_2 in the northern hemisphere at every point in Pluto's orbit; one pressure peak per Pluto orbit; minimum surface pressures near 1 Pa, well above the pressure for which the atmosphere becomes local; and pressures that increase from 1988 through 2015. On the basis of a 2013 occultation that showed that the pressure was still increasing (*Olkin et al.*, 2015), the occultation record eliminated the EEC class and favored the PNV class, which predicted pressures still rising at flyby, with surface pressures near 1.5 to 10 Pa.

Hansen et al. (2015) used the identical model as *Hansen and Paige* (1996), updated with new observational constraints. As discussed by *Hansen et al.* (2015), the physics is similar to *Young* (2013) and *Olkin et al.* (2015), but the engine was completely independent, and the set and weighting of observational constraints differed. While *Young* (2013) emphasized the occultation record, and the increase of occultation pressures from 1988 to 2013, *Hansen et al.* (2015) emphasized the albedo record (section 2.1.1), especially that (1) HST did not detect bright zonal bands, disallowing models where N_2 ice persisted at a given latitude over a Pluto orbit; (2) a southern polar cap was detected before and during equinox; (3) the northern cap was larger than the southern cap at equinox; and (4) the infrared spectral evidence pointed toward a shrinking northern polar cap from 2001 to 2012. They found it very difficult to find combinations of parameters that had pressures rising past 2013 and a south polar cap in 1988. Their favored model (run 22) had a thermal inertia of 42 TIU and an N_2 inventory of 50 kg m^{-2}, both higher than Fig. 7 of *HP96*, but much lower than the PNV runs of *Young* (2013). Similar in character to the *HP96* models and the *Young* (2013) EPP models, *Hansen et al.* (2015) runs 22 and 48 had two pressure peaks per Pluto orbit, a complete exchange between summer and winter poles, and pressure minima near the local/global cutoff. *Hansen et al.* (2015) predicted, for the time of the flyby, a large summer polar cap extending to low- to midnorthern latitudes that would be in the process of subliming, and pressures between 0.14 and 3.2 Pa that were falling slowly.

Both the *Hansen et al.* (2015) and *Young* (2013) papers had trouble reconciling the occultation record with the evidence for a bright southern cap prior to and at equinox. *Young* (2013) did not try to reproduce that cap, and found high-thermal-inertia solutions with pressures rising past 2015 and northern N_2 ice deposits that persisted through a Pluto year. *Hansen et al.* (2015) required the southern cap, leading to lower thermal inertia solutions with pressures dropping slowly in 2015, pressures at flyby near 2.4 Pa, and a large but mobile northern cap. Neither model anticipated the large reservoir of N_2 in Sputnik Planitia. Both models highlight one of the persistent questions: the amount and role of volatiles in Pluto's southern latitudes.

Several factors between the 2002 occultations and the 2015 flyby inspired new studies of Pluto's atmospheric dynamics. The anticipated arrival of spacecraft sparked interest. Several occultations showed the signature of wave-like temperature fluctuations. The seasonal models of *Young* (2013) and *Hansen et al.* (2015) could provide GCMs with initial conditions of volatile distributions and surface or subsurface temperatures. And some simplified dynamical models could shed light on seasonal effects. Much of this is treated in the chapter in this volume by Forget et al., and so we focus here on where dynamics touch on volatile cycles. In brief, *Toigo et al.* (2010) and *French et al.* (2015) calculated the thermal tides on Pluto, quantitatively predicting vertical winds at the surface due to the diurnal cycle of sublimation and condensation. Two GCMs were developed pre-encounter [PlutoWRF (*Toigo et al.,* 2015); LMD (*Vangvichith,* 2013)], which also ran their own predictions for seasonal volatile cycles. *Vangvichith* (2013) showed the interaction between the N_2 and CH_4 cycles. *Toigo et al.* (2015) showed double-peaked pressures over a Pluto year for lower thermal inertia and single-peaked for higher, and suggested that the mobile ices in the PNV class of *Young* (2013) would become static if the model were run for more Pluto years. As with *Young* (2013) and *Hansen et al.* (2015), the PlutoWRF and LMD models pre-encounter could not anticipate the N_2 reservoir of Sputnik Planitia (although some of their simulations used the joint spectra/HST maps that include an equatorial N_2 source near 180° longitude). The themes of southern N_2, perennial vs. seasonally mobile ices, diurnal tides, minimum pressure, and entwined N_2-CH_4 cycles all proved to be as important after the New Horizons flyby as they were before.

4.4. Pluto Volatiles After New Horizons

In Table 3, we list those post-encounter models that calculated sublimation/condensation rates or changes in volatile distributions or surface pressures. They are organized roughly by research group, to more clearly show the hierarchy of models. We present these in this section in roughly chronological order, emphasizing their assumptions and their broad conclusions, particularly how they alter our understanding of the processes most affecting Pluto's interacting surface and atmosphere. A consolidated summary of the model results sorted by the different timescales is given in section 5.

4.4.1. First models of Sputnik Planitia. The discovery of Sputnik Planitia changed everything. An enormous reservoir of N_2-rich ice in the diurnal zone affects the volatile and climate cycles on all timescales, from non-cyclic, to Milankovitch cycles, Pluto's seasonal cycle, or Pluto's day.

In pre-encounter models (*Young,* 2013; *Hansen et al.,* 2015), surface pressures plummeted when the last of the N_2 ice left the summer hemisphere, because the globally averaged insolation onto the N_2 ice dropped to nearly zero. But on the real Pluto, the N_2-rich Sputnik Planitia straddles the equator, and so there is some insolation onto N_2-rich ices every Pluto day. Early post-flyby models that imposed an N_2-ice reservoir at Sputnik Planitia (*Young et al.,* 2015) showed that adding this reservoir delayed the sublimation from the summer hemisphere and raised the minimum pressure for a sample case with low seasonal thermal inertia [*Young* (2013); EPP14, with 3 TIU], but had little effect on the volatile distribution and pressures for a case with high thermal inertia [*Young* (2013); PNV23, with 1000 TIU].

After the New Horizons flyby, researchers soon recognized that Sputnik Planitia was a natural place for N_2 to accumulate, both because of its low elevation (*Stern et al.,* 2015; *Moore et al.,* 2016) (cf. section 3.2.3) and its lower insolation, averaged over a Pluto year (*Hamilton et al.,* 2016) (cf. section 3.1.2).

Two important early post-flyby models (*Forget et al.,* 2017; *Bertrand and Forget,* 2016, hereafter *BF16*) included an idealized Sputnik Planitia as a circular depression, 3.8 km deep and 1000 km wide. These papers arose from Laboratoire de Météorologie Dynamique (LMD), which developed two related volatile transport models: a full global climate model (GCM) (*Forget et al.,* 2017; see also the chapter by Forget et al. in this volume); and a reduced version that used a simple global mixing function for N_2, CH_4, and CO in place of three-dimensional atmospheric transport and dynamics (*BF16*). These two models were complementary. The reduced model ran much more quickly and was therefore suitable for simulations on the timescale of Pluto's 248-year orbit to tens of thousands of years. The reduced model was used to initialize the surface ice distribution and subsurface temperatures for the GCM. The GCM calculated the transport between surface and atmosphere, and within the atmosphere from one location to another, on the timescale of Pluto's day to several Earth decades. Tests done with the GCM determined the timescales for atmospheric transport of CH_4 (10^7 s, i.e., about 4 Earth months) and N_2 (1 s, instantaneous mixing) used in the reduced model.

The main result from *Forget et al.* (2017) for the purpose of this chapter was that the cold atmospheric layer seen in the lowest kilometers above Sputnik Planitia was due in part to the daytime sublimation of N_2 over Sputnik. The winds and atmospheric temperature results from the GCM are covered in the chapter by Forget et al.

The volatiles on *BF16* were initialized with 50 kg m^{-2} of each of N_2 and CH_4 distributed uniformly over Pluto. The seasonal model of *BF16* captured many of the important points of Pluto's surface volatiles: a migration of N_2 into the Sputnik Planitia basin after several thousands of years, the role of Sputnik Planitia as the dominant N_2 reservoir, the presence of CH_4-rich ice at higher latitudes, and volatile-free areas near the equator. The model predicted seasonal variation in the CH_4 distribution, with the northern CH_4

TABLE 3. Post-encounter models of volatile cycles and trends.

Model	Diurnal	Seasonal	Milankovitch	Topography	Subsurface Conduction	Mobile N_2	CH_4	Resolution (km)	Tuned to occultations
Moores et al. (2017)	Yes	Yes	Yes	No	Yes	Yes	No	100	Yes
Earle et al. (2017)	No	Yes	Yes	No	No	No	Yes§	20	No
Stern et al. (2017)	No	Yes	Yes	No	No	No	Yes§	150	No
Earle et al. (2018a)	No	Yes	Yes	No	No	Yes	Yes§	10	No
Young et al. (2015)	No	Yes	No	No	Yes	Yes	No	208	Yes
Lewis et al. (2021)	No	No*	No	Yes	No	No	No	1	Yes
Johnson et al. (2021)	No	Yes	Yes	No	Yes	No	No	1	No
Bertrand and Forget (2016)	Yes	Yes	No	Yes	Yes	Yes‡	Yes	190	Yes
Forget et al. (2017)	Yes	No†	No	Yes	Yes	Yes‡	Yes	190	Yes
Bertrand et al. (2018)	Yes	Yes	Yes	Yes	Yes	Yes	No	95 and 190	No
Bertrand et al. (2019)	Yes	Yes	Yes	Yes	Yes	Yes	Yes	190	
Meza et al. (2019)	Yes	Yes	No	Yes	Yes	Yes‡	Yes¶	95	Yes
Bertrand et al. (2020a)	Yes	Yes	No	Yes	Yes	Yes	Yes	63 and 190	Yes

*Snapshot of mass fluxes at flyby.
†Calculated on timescale of several Earth decades.
‡N_2 allowed to be mobile, but evolves to a static configuration.
§N_2 and CH_4 modeled independently.
¶CH_4 is implemented, but does not interact with the N_2 cycle in practice.

cap disappearing by 2030. The modeled CH_4 variation and perhaps the N_2 confinement were a consequence of the model's small volatile inventory.

BF16 were also able to find combinations of thermal inertia and N_2-ice albedo that reproduced the surface pressures seen by New Horizons and the occultation record. These models favored high thermal inertias for roughly the upper 40 m of the subsurface; the reference model was 800 TIU, and the *BF16* model ruled out thermal inertias lower than 500 TIU.

4.4.2. Geological considerations. Pluto's geology revealed evidence of ongoing, recent, and past volatile cycles (see the chapter in this volume by Moore and Howard). These include cryovolcanism (Wright and Picard Mons); the pits within Sputnik Planitia; the albedo variations within Sputnik Planitia; possible wind streaks and dunes in western Sputnik Planitia; high-altitude methane deposits; N_2 glaciers flowing into eastern Sputnik Planitia from the neighboring highlands; kilometer-wide ridges of CH_4-rich ice in the bladed terrain; the smooth Alcyonia Lacus; albedo contrast in the equatorial regions (maculae vs. Tombaugh Regio); evidence for thick, eroded CH_4-rich mantles in Pioneer and Hayabusa Terrae northeast of Sputnik Planitia; the eroded scarps of Peri Planitia; and erosional valleys north, northeast, and southwest of Sputnik Planitia. The rich geology at Pluto proved that, once again, our imaginations were less wide-ranging than reality. Pre-encounter, modelers considered volatile transport involving sublimation/condensation and atmospheric winds. Because of New Horizons, we now know that Pluto's volatile transport processes also include large-scale convection within Sputnik Planitia, ancient and present-day glacial flow, liquification of N_2 at the base of ancient or current deep N_2 ice, and possible ancient cryovolcanism.

Some of the geologic features observed by New Horizons motivated the possibility that Pluto's pressure may have been much higher in the past (*Stern et al.,* 2017; *Telfer et al.,* 2018). The importance of the Milankovitch cycles was reinforced by Pluto's latitudinal variation in albedo (*Binzel et al.,* 2017), which raised the possibility that obliquity and the solar longitude at perihelion, $L_{s,peri}$, could be important for Pluto's volatile cycles. This was investigated by *Earle et al.* (2017), who defined polar caps extending to ±45°, ±60°, or ±75°, and by *Stern et al.* (2017), who included a rectangular Sputnik Planitia 45° wide from 45°N to 30°S. Given emissivity of 0.6 or 0.9, and albedo from 0.1 to 0.6, these models calculated pressures vs. time for three epochs: the current epoch; the extreme northern summer at 0.9 Ma when perihelion coincided with northern summer solstice; and the extreme southern summer at 2.35 Ma when perihelion coincided with southern summer solstice. They found that the Milankovitch cycles can have a significant effect, which has been borne out by subsequent modelers (e.g., *Bertrand et al.,* 2018, 2019).

The relationship between geology and the Milankovitch cycles (*Binzel et al.,* 2017) was explored quantitatively by *Earle et al.* (2018a), who calculated the stability of N_2 and CH_4 ices. Nitrogen and CH_4 were modeled independently, treating each in turn as the dominant volatile. For the N_2 cycle, they found two very different styles of variation, depending on albedo. For dark N_2 ice ($A < 0.3$), the equator lost its N_2, and the seasonal cycle was an exchange of N_2 between northern and southern polar caps. For bright N_2 ice ($A > 0.6$), there was still some exchange of the polar caps, since the deposition rate onto a latitude in Artic night is independent of albedo. However, the main result for bright N_2 ice was that the N_2 was trapped near the latitudes where the annual insolation was least. Thus, for high obliquity (103°, 0.9Ma), when the insolation minimum was at the equator, the N_2-ice distribution was sharply peaked around the equator, mainly confined to ±30°. For low obliquity (126°, 2.35 Ma), when the insolation minimum was at ±30°, the N_2-ice distribution was very broad, with maxima near ±45°. Comparing the low- and high-albedo cases, *Earle et al.* (2018a) concluded that, qualitatively, N_2 ice was subject to runaway albedo at low- to midlatitudes.

The bladed terrain may be a sublimation effect, in analogy with terrestrial penitentes (*Moores et al.,* 2017; *Moore et al.,* 2018; see also the chapter by Moore and Howard in this volume). This motivated the climate calculations in *Moores et al.* (2017) using the PlutoWRF GCM engine described in *Toigo et al.* (2015). *Moores et al.* (2017) incorporated some new model simulations based on early Pluto results, including some latitudinal information but no topography. Their models predicted not just pressure vs. time at different epochs of Pluto's Milankovitch cycles, but also parameters related to the growth of penitentes. The *Moores et al.* (2017) result that the pressure is a maximum near equinox may be related to early Pluto results that showed the 2015 surface pressure and the latitudinal distribution of surface ices (*Stern et al.,* 2015; *Moore et al.,* 2016; *Gladstone et al.,* 2016), but did not include topography, such as a bowl-shaped depression for Sputnik Planitia (A. D. Toigo, personal communication). However, the *Moores et al.* (2017) models also show the importance of the Milankovitch cycles for Pluto climate studies and the geologic expression of those cycles.

4.4.3. Nitrogen cycles on seasonal and Milankovitch timescales. Higher-resolution New Horizons data products followed, in 2017 and 2018, including the near-surface pressures vs. altitude, and high-resolution geologic, topographic, color, and composition surface maps (see section 2.2).

Using these products, *Bertrand et al.* (2018, hereafter *B18*) modeled the N_2 cycle on Pluto with a model based on *BF16*, including thermal conduction into the substrate. Additions to the *BF16* model included the Milankovitch cycles, topographic information, glacial flow, a larger N_2 inventory, and higher spatial resolution for some runs focused on Sputnik Planitia. Simplifications include averaging insolation over diurnal cycle [which should have minimal impact (*Young,* 2012)] and consideration of only a small set of seasonal thermal inertias (generally 800 TIU, with a few cases at 400 or 1200 TIU) and albedo (generally 0.7, with a few cases at 0.4). *B18* included a much larger N_2 inventory than *BF16*, which takes much longer to equilibrate. Thus, *B18* considered two related problems: a study of the sublimation patterns within Sputnik Planitia, run at higher spatial resolution, with an initial distribution of N_2 confined to the Sputnik Planitia basin; and a study of the stability of N_2 outside Sputnik Planitia, with a variety of initial N_2 distributions. The experiments of *B18* on N_2 stability included N_2 inventories with global averages of 200 m, 500 m, or 1000 m (2×10^5, 5×10^5, or 10^6 kg m^{-2}), consistent with the expected volume of N_2 ice within Sputnik Planitia (*Glein and Waite,* 2018), with various initial distributions of N_2 that were global, confined to low latitudes, or confined to high latitudes.

Lewis et al. (2021) calculated energy balance and mass fluxes using high-resolution maps of bolometric albedo (*Buratti et al.,* 2017) and N_2 presence (B. Schmitt, personal communication), based in part on the CH_4 spectral shifts, and extrapolating ice coverage to unmapped areas. By matching the observed 2015 surface pressure, they derived N_2 emissivity of 0.47–0.78, with the exact value dependent on assumptions of the unmapped hemisphere (the non-encounter longitudes and the southern latitudes). From this, they produced a map of the current sublimation and condensation rates. Like *B18*, they found a trend of north-to-south transport within Sputnik Planitia. However, the maps of *Lewis et al.* (2021) also included N_2 observed outside Sputnik Planitia. They found high sublimation rates of 3 g cm^{-2} yr^{-1} [0.52 mm (Pluto day)$^{-1}$] for N_2 at midlatitudes.

Johnson et al. (2021) used similar maps as *Lewis et al.* (2021), and calculated pressure vs. time over a Pluto year given four assumptions of a static N_2 ice distribution. This work made a systematic search over albedo (A), emissivity (ε), and thermal inertia (Γ), and identified those triplets of A,ε, and Γ that satisfied the New Horizons pressure constraints (surface pressure 1.15 Pa) and 1988 pressure constraints ($3.14 > p_{2015}/p_{1988} > 1.82$). The motivation for this work was to study the minimum pressures reached over Milankovitch timescales to see if haze production would be disrupted. As described in section 5.2 (Milankovitch cycles) and section 5.3 (seasonal cycles), these models found that the presence of Sputnik Planitia tended to maintain a minimum pressure that was high enough to allow haze production.

While the LMD models in *Meza et al.* (2019) formally include CH_4 ice, we treat them in this section. The three simulations with rising pressures 1988–2015 in *Meza et al.* (2019) had perennial N_2 in Sputnik Planitia and in the depressions in the northern hemisphere, with low-albedo CH_4 ice ($A_{CH_4} = 0.5$). Because the low-albedo CH_4 remained warm, there was very little seasonal deposition of N_2 ice. In the southern hemisphere, the CH_4 deposits were very thin and the albedo remained low. Thus, the N_2 ice distribution in *Meza et al.* (2019) was essentially static, and directly comparable to the runs of *Johnson et al.* (2021).

4.4.4. Nitrogen-methane cycles on seasonal and Milankovitch timescales. *Bertrand et al.* (2019, hereafter *B19*) addressed Pluto's climate with both the N_2 and CH_4 volatile cycles. Surface units are treated as either volatile-free, N_2-rich with 0.5% CH_4, or pure CH_4. In reality, the CH_4 mixing ratio within N_2-rich ice was observed to vary with location on Pluto (*Protopapa et al.,* 2017), which is not tracked by this model. As described in section 3.4.1, this simplification should have little effect on the calculated N_2 or CH_4 mass fluxes over N_2-rich areas or the CH_4 mass fluxes over CH_4-rich areas. However, N_2 fluxes over CH_4-rich areas is more problematic. For condensation, the model simply assumes that a cold CH_4-rich area can become a N_2 condensation site when the temperature is low enough, at which point the area is modeled as N_2-rich. N_2 sublimation from CH_4-rich areas is not modeled. *B19* explored the complex ways in which modeled changes in surface albedos, emissivities, and thermal inertias affected volatile transport, surface distributions, and atmospheric pressure and composition. *B19* presented a variety of runs to show the range of modeled outcomes; in *Meza et al.* (2019), runs were fine-tuned (e.g., in N_2 albedo) to match the occultation record.

This work demonstrates the challenge of how to explain the observed widespread presence of CH_4 outside of Sputnik Planitia, as models such as *B19* predict that CH_4 would accumulate in Sputnik Planitia on the timescale of several million years if CH_4 were allowed to condense into Sputnik Planitia. The mechanism for preventing all the CH_4 from accumulating into Sputnik Planitia is unknown, but *Bertrand et al.* (2019) speculate that the mixing state of the upper layers of the volatiles may be responsible (section 3.4). It may also be related to the cycle of net N_2 into or out of Sputnik Planitia over Milankovitch cycles (*Bertrand et al.,* 2018, 2019).

One initial state in *B19* began with N_2 ice filling Sputnik Planitia basin, and with a 4-m-thick layer of CH_4 ice covering the entire globe. This was particularly applicable to long-term trends, including over a timespan of several Milankovitch cycles. An alternate initial state also began with N_2 in Sputnik Planitia but included an infinite reservoir of CH_4 roughly at the latitudes of the bladed terrains (excluding Cthulhu). This initial state represented the bladed terrains as a perennial CH_4 source. These runs allowed different CH_4 albedos for equatorial CH_4 vs. CH_4 at midlatitudes or the poles, and are particularly applicable to the current Pluto season or Milankovitch cycle. The results of these model runs are described in section 5.2 (Milankovitch cycles) and section 5.3 (seasonal cycles).

5. VOLATILE TRANSPORT AT SHORT AND LONG TIMESCALES

In this section, we summarize the results of the preceding models on several timescales (Table 4). We avoid the term "secular," which has been applied both to evolutionary changes over 4 G.y. and observed changes on decadal timescales. Sputnik Planitia dominates much of the movement of volatiles at all timescales, as described in section 4.4, as the surface pressure is heavily influenced by the distribution of N_2. The N_2 cycle in turn interacts with the CH_4 cycle.

The various timesscales are intertwined as well. We choose to present the timescales in the order listed in Table 4, because each timescale provides the context for the next-shorter timescale. A non-cyclical process filled in Sputnik Planitia, the presence of which affects all of the subsequent cycles.

Where possible, we emphasize observational constraints. Complementary perspectives on volatile processes can be found in other chapters: The chapter by Forget et al. covers many issues relating to the diurnal cycles, and the chapters by White et al. and Moore and Howard cover aspects of the volatile activity on astronomical timescales.

5.1. Non-Cyclical Timescales

A singular event in Pluto's volatile history is the infilling of Sputnik Planitia. If Sputnik formed as an impact basin (*Stern et al.,* 2015; *Moore et al.,* 2016) in its current location, then the combined effects of enhanced condensation in a topographic low and insolation minimum near the equator

TABLE 4. Timescales for volatile transport.

Timescale	Cycle Duration	Examples
Non-cyclical (aka evolutionary)	10 m.y. to 4.5 G.y. duration, not cyclic	Infilling of Sputnik Planitia; atmospheric escape; formation of bladed terrain
Milankovitch cycles (aka orbit/pole cycles, mega-seasons)	2.8-m.y. obliquity cycle; 3.7-m.y. eccentricity and libration of argument of perihelion	Variation of pressure extreme; Sputnik Planitia shorelines and erosion; northern and southern boundaries of bladed terrain; glacial flow
Seasonal (aka orbital, annual)	248 Earth years	North-south flow within Spunik Planitia; pressure changes; transport of seasonal volatile ice; albedo/color
Diurnal (aka rotational, daily)	6.4 Earth days	Global atmosphere; cold troposphere above Sputnik Planitia

would cause Sputnik Planitia to fill in about 10 m.y. (Fig. 11) (*Bertrand and Forget,* 2016; *Bertrand et al.,* 2018), or a rate of ~1 km m.y.$^{-1}$ (10^6 kg m^{-2} m.y.$^{-1}$). Some modelers find that the insolation effect might be enough, without an impact basin (*Hamilton et al.,* 2016). Some model runs (*Bertrand et al.,* 2018) found that N_2 could be temporarily stable in low latitudes outside of Sputnik Planitia for tens of millions of years, especially for large N_2 inventories, before finally condensing into Sputnik Planitia. *Earle et al.* (2018a) found that both dark, N_2-free areas (such as Cthulhu), and bright, N_2-rich areas (such as Tombaugh Regio) were stable near the equator on million-year timescales. Considering N_2 alone, *B18* found N_2 was never stable at high latitudes, but CH_4 complicates this picture (*B19*).

Methane is much less mobile than N_2. CH_4 also preferentially condenses near the latitudes of long-term average minimum insolation, but it does it slowly (*Bertrand et al.,* 2019), at a rate near 1 m m.y.$^{-1}$ (1×10^3 kg m^{-2} m.y.$^{-1}$). The model of *Bertrand and Forget* (2016) had a millimeter-thick veneer of CH_4 ice, and simulated a pole-to-pole seasonal exchange. Initializing Pluto with N_2 in Sputnik Planitia and a globally uniform distribution of 4 m of CH_4 with a uniform CH_4 albedo (*Bertrand et al.,* 2019, their section 3), the CH_4 is modeled to slowly migrate to the equator, mainly by loss of CH_4 from the poles (Fig. 12). If the modeled CH_4 ice is more mobile (darker CH_4 albedo, or lower substrate thermal inertia), then the equatorial deposits cover a wider latitude range.

The modeled CH_4 deposits tended to occur slightly north of the equator, at the latitudes of the bladed terrain on the encounter hemisphere. At these slow rates, it would have had to take millions of years to build up the bladed terrains to the observed depths of hundreds of meters crest to base (*Moore et al.,* 2018), and longer if the mass of the bladed terrain deposits were kilometers deep. Even though CH_4 is less mobile than N_2, why CH_4 is not also mainly sequestered within Sputnik Planitia remains a puzzle. Simulations over tens of millions of years (*Bertrand et al.,* 2019) revealed two potential reservoirs of CH_4: either at low latitudes similar to the bladed terrain, or at midlatitudes, similar to the CH_4-rich mantled terrains. However, only one reservoir or the other is stable at any given time. For current epoch, the models of

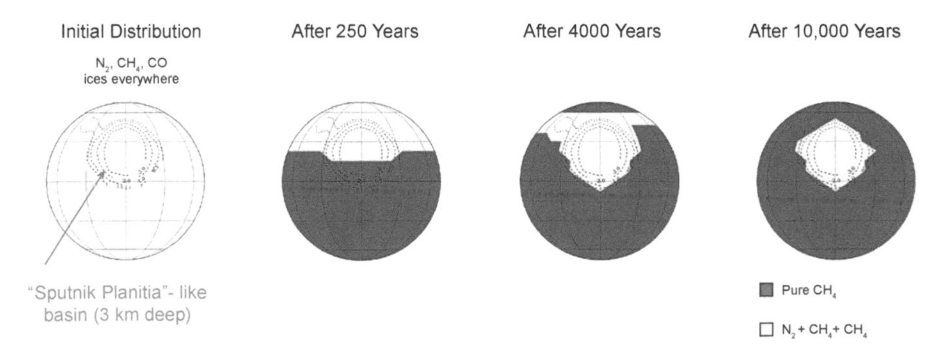

Fig. 11. Surface maps from the reference simulation of *Bertrand and Forget* (2016) with thermal inertia = 800 TIU. The simulation began with all ices uniformly distributed.

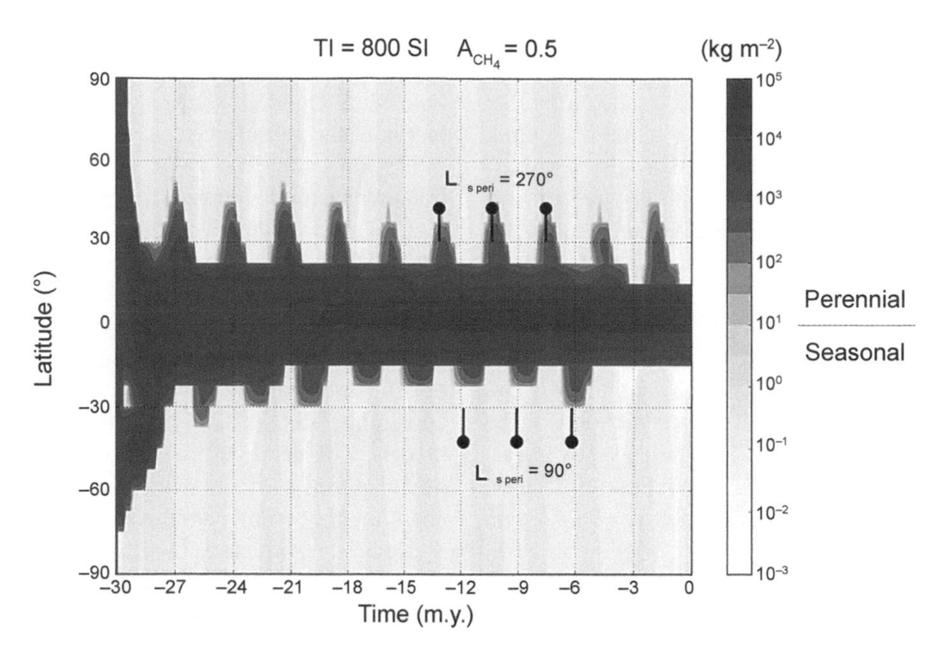

Fig 12. Example run for an initially global CH_4 with A_{CH_4} = 0.5, showing the zonal annual mean of CH_4 outside of Sputnik Planitia. Figure from *Bertrand et al.* (2019).

Bertrand et al. (2019) suggested that the low-latitude CH_4 is subliming, carving out the blades of the bladed terrain and feeding the midlatitude CH_4. At rates of 1 m m.y.$^{-1}$, if there is a cyclic exchange between low- and midlatitude CH_4, it must occur over hundreds of millions of years. The mechanism for such a cycle is unclear.

Prior to the New Horizons encounter, the escape of volatiles over the age of the solar system was expected to be a major component in shaping the landscape of Pluto. However, the escape rate of N_2 [~5×10^{22} N_2 s^{-1} (*Young et al.,* 2018)] was found from New Horizons measurements to be ~3000× lower than expected pre-encounter (*Gladstone et al.,* 2016). Furthermore, the major escaping species was found to be CH_4 [~6×10^{25} CH_4 s^{-1} (*Young et al.,* 2018)], which was also escaping too slowly to sculpt the landscape (see the chapter in this volume by Strobel). Less atmospheric N_2 or CH_4 is lost due to escape than is lost by incorporation into precipitating hydrocarbons and nitriles (*Krasnopolsky,* 2020).

5.2. Milankovitch Cycles

Since the flyby, there has been a new appreciation for the impact of the Milankovitch cycles, particularly the 2.8-m.y. obliquity cycle, and the variation of the solar longitude at perihelion (varying by 360° over 2.8 m.y., with a 24° oscillation superimposed that varies with the 3.95-m.y. eccentricity cycle). The location of the equatorial CH_4 deposits vary with the Milankovitch cycle too. For example, some models of *Bertrand et al.* (2019) show a steady accumulation of bladed terrains, with the boundary of the terrains varying such that it favors the perihelion winter hemisphere (e.g., equator and low southern latitudes when $L_{s,peri}$ = 90°; Fig. 12).

Even without any variation in the surface distribution of volatile ices, the changing insolation would affect atmospheric pressure and winds (*Moores et al.,* 2017; *Earle et al.,* 2017; *Meza et al.,* 2019; *Johnson et al.,* 2021). In particular, the annual pressure minima and maxima vary with obliquity.

In models based on the observed N_2-ice distribution, *Johnson et al.* (2021) found the minimum seasonal pressures varied between 50 and 300 mPa, with lower minimum pressures (<6 mPa) possible for some N_2 albedos and distributions. *B18* found minima of ~10 mPa for nominal N_2 albedo A_{N_2} = 0.7 and higher minimum pressures of ~1 Pa for darker A_{N_2} = 0.4. This implies that, for the great majority of parameter space, Pluto's atmosphere remains global throughout its entire year over multiple obliquity cycles, and sublimation winds are able to keep N_2 ice temperatures nearly isothermal (cf. section 3.2.1).

Earle et al. (2017) and *Stern et al.* (2017) concluded that maximum pressures could be on the order of kilopascals, much higher than the current pressure of 1.1 Pa, perhaps tied to lines of geologic evidence for periods of high surface pressure. However, their results required a mechanism for confining N_2 to Pluto's poles, and no such confining mechanism is identified for Pluto's current Milankovitch cycle (i.e., the last 2.8 m.y.). In contrast, *B18* found maximum seasonal pressures of ~10 Pa for A_{N_2} = 0.7, and only 50 Pa even for dark N_2 with A_{N_2} = 0.4. *B18* argued that periods of high surface pressure (*Stern et al.,* 2017) were not recent. Alternatives include higher pressure very early in Pluto's evolution, or ancient flows at the bottom of deep glaciers (*B18*).

Regardless of how Sputnik Planitia was established, it is now a deep reservoir of N_2 straddling Pluto's equator, experiencing insolation changes. The difference in insolation at the northern and southern extremes of Sputnik Planitia vs. its center causes differential sublimation or condensation; even with glacial flow, the model of *B18* predicts variation of the N_2 topography of up to 300 m over Milankovitch cycles (Fig. 13) (*B18*), but it could be less if a kilometer-deep bedrock is assumed near the edges of the ice sheet. Averaged over 1.3 to 0.1 Ma, *B18* predicted that the center of Sputnik Planitia accumulated N_2, so that that the ice sheet currently covers a minimum area, with depressions at the northern and southern boundaries. The modeled rise and fall of the edges of Sputnik Planitia, especially the northern boundary, may have contributed to erosion of the water-ice border, particularly near the enigmatic Al-Idrisi region. Averaged over a 2.8-m.y. obliquity cycle, *B18* found that Sputnik Planitia lost 1 km from its northern edge by sublimation and 0.15 km from its southern edge, with net transport to the center. These changes in ice thickness are balanced by glacial flow. By some models, the deep pits within Sputnik Planitia probably formed on timescales of 100,000 yr (*Buhler and Ingersoll,* 2018) — longer than Pluto's orbit, but short on Milankovitch timescales (but see the chapter by Moore and Howard for suggested formation timescales comparable to or longer than the Milankovich cycle). A snapshot of the annually averaged insolation suggests a correlation with pit formation (a sublimation process; see the chapter by Moore and Howard), so the latitudes of pit formation may well vary on Milankovitch timescales.

Outside of Sputnik Planitia, N_2 can be stable over a season at the equator or at midlatitudes, but the stability varies with obliquity (*B19*, their Fig. 4).

5.3. Seasonal Cycles

Pluto's present obliquity (119.6°) and orbital eccentricity results in pronounced seasonal variations in illumination. These drive significant changes in sublimation and condensation rates of the volatile ices over the course of a Pluto year, and thus affect the atmospheric pressure and composition and the volatile surface ices. Volatile transport models highlight the sensitivity of the volatile cycles to several unconstrained parameters, such as the initial ice distribution, bolometric Bond albedo, emissivity, and seasonal thermal inertia of the subsurface (*Hansen and Paige,* 1996; *Young,* 2013; *Stansberry and Yelle,* 1999). An important observational constraint for these free parameters came from the large number of stellar occultations acquired between 1988 and 2016, showing a threefold increase in surface pressure during this period (see section 2.1.3) (see also *Olkin et al.,* 2015; *Bosh et al.,* 2015; *Meza et al.,* 2019). In July 2015, New Horizons observations directly measured the surface pressure and volatile distribution, which brought stronger constraints for the free parameters in volatile transport models. The deep permanent reservoirs of volatile ice in Sputnik Planitia fundamentally changed our understanding about the motion of volatiles on Pluto.

New Horizons flew by Pluto during northern spring. Models predicted N_2 transport within Sputnik Planitia from the north (15°N–50°N) to the south (30°S–15°N) in 2015, and more generally a north-to-south transport between equinoxes in 1988 and 2109 (Fig. 14) (*B18*, their Figs. 7 and 8), and south-to-north transport during the opposite seasons (the preceding 124 years). *Protopapa et al.* (2017) found that the abundance and grain size of the N_2-rich ice increased from the northwest to southeast within Sputnik Planitia, and *Schmitt et al.* (2017) found a similar trend with the N_2 band depth at 2.15 μm. *B18* derived that over the last 30 years, a few centimeters of N_2-rich ice has moved north-to-south, which is comparable to the derived grain size of CH_4 in northwest Sputnik Planitia (*Protopapa et al.,* 2017), and may indicate the formation of CH_4-rich ice during sublimation (*Stansberry et al.,* 1996b; *Tan and Kargel,* 2018). It is less clear whether the deposition of a few centimeters of N_2 will affect the IR spectra of the N_2-rich ice in central and southern Sputnik Planitia, where the ice exhibits decimeter-to-meter-scale

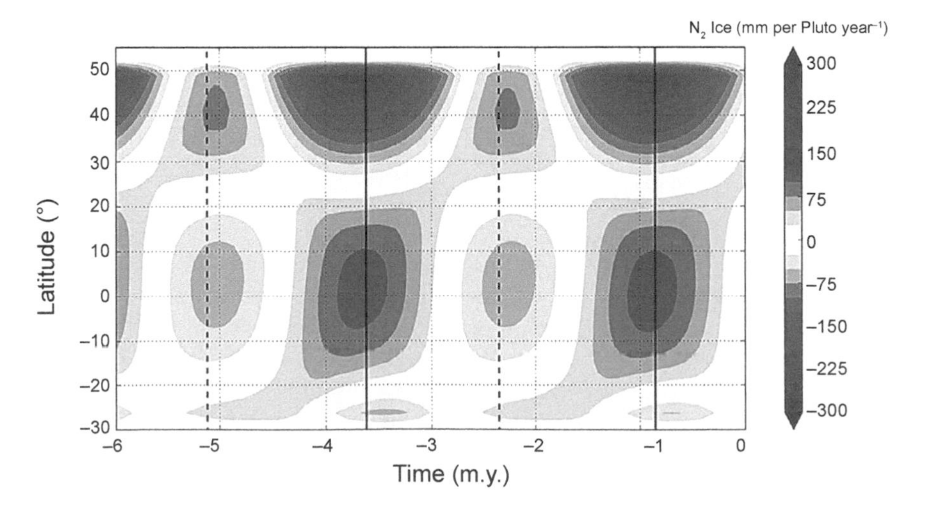

Fig. 13. Evolution of the annual mean condensation-sublimation rate of N_2 ice in Sputnik Planitia with time (millimeters per Pluto year; red is condensation), assuming that the ice sheet remains flat (same as Fig. 8). The vertical solid and dashed lines correspond to the periods of high (104°) and low (127°) obliquity, respectively. Figure from *Bertrand et al.* (2018).

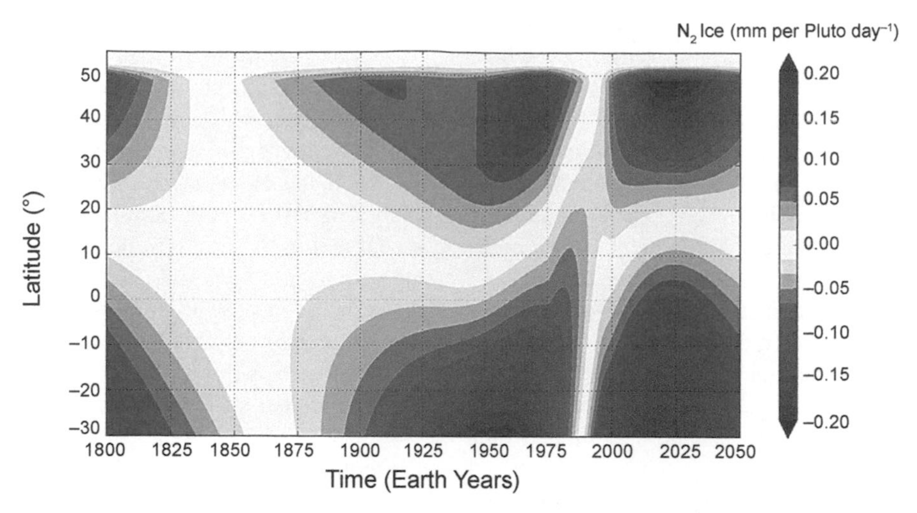

Fig. 14. Evolution of the diurnal mean condensation-sublimation rate within Sputnik Planitia (millimeters per Pluto day; red is condensation), in current orbital conditions, from 1800 to 2050 assuming that the ice sheet remains flat. Figure from *Bertrand et al.* (2018).

"grain sizes." The deposited ice may affect the albedo, emissivity, or diurnal thermal inertia.

The very simple calculation of how a uniform surface covered with N_2-ice reacts to the heliocentric distance varying from 29.7 to 49.9 AU, and ignoring thermal inertia, indicates that the N_2 equilibrium temperature would change by over 8 K, and the equilibrium pressure would change by 3 orders of magnitude. Some pre-encounter models predicted pressures that could drop precipitously and dramatically if the ices in the summer hemisphere were allowed to sublime completely, leaving only ices on the winter hemisphere (*Trafton et al.,* 1998, *Hansen and Paige,* 1996; *Young,* 2013; *Hansen et al.,* 2015). The equator receives less sunlight when averaged over a Pluto year, which led *Trafton et al.* (1988) to predict that the perennial N_2 ice (that persists over a Pluto year) would reside within 32° of the equator. *HP96* simulated this with the high-thermal inertia run that they plotted (their Fig. 3, 2093 TIU). In the other runs plotted in *HP96*, the equator is rarely the coldest latitude, and N_2 ice jumps over the equator to the winter polar cap, leaving the equator bare (again, Sputnik Planitia was unknown at this time). Very few runs in *HP96* predicted perennial N_2 ice (termed "permanent zonal band" in *HP96*). Rather, the *HP96* models generally showed two polar caps exchanging

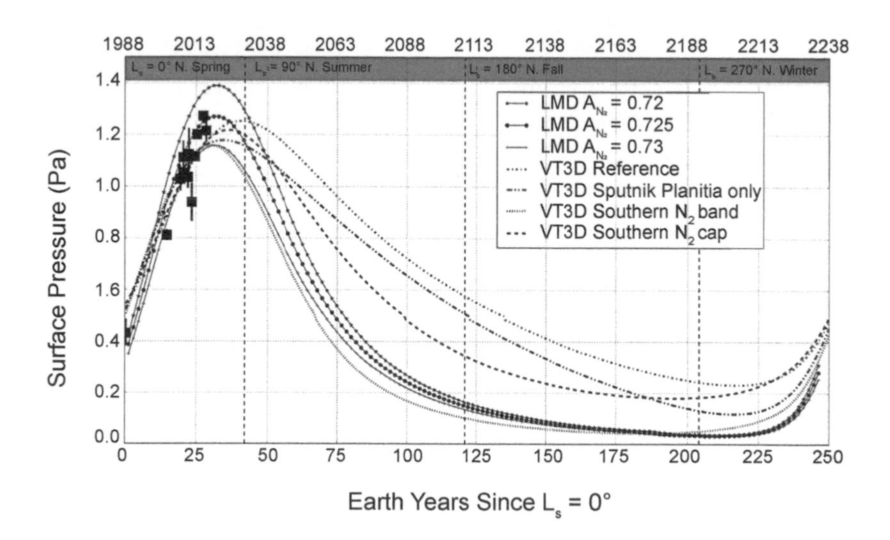

Fig. 15. Dashed lines give results for four models of pressure over a Pluto season for four assumptions of a static N_2-ice distribution from *Johnson et al.* (2021): Sputnik only, the reference N_2-presence map of *Lewis et al.* (2021), the N_2-presence map plus a southern N_2 polar cap, or the presence map plus a southern N_2 mid-latitude band. Solid lines give data for three models of pressure over a Pluto season from *Meza et al.* (2019). The vertical dashed lines mark the timing of solstices and equinoxes.

volatiles, with flow from the summer to the winter cap. The thermal inertia determined the timing of the new winter pole. Most *HP96* models showed a steady increase in pressure after each equinox, as the old winter pole became the new summer pole and received more direct sunlight. Then, when the last of the N_2 from the summer pole sublimed away, sometime between perihelion and northern summer solstice, the pressure plummeted. The typical *HP96* model runs therefore showed two pressure peaks per Pluto year.

Post-encounter, we see that the N_2 never leaves either summer hemisphere completely, as Sputnik Planitia ensures that there is always N_2 absorbing sunlight every day (*Young et al.*, 2015; *Bertrand and Forget*, 2016; *Meza et al.*, 2019; *Johnson et al.*, 2021). This decreases the magnitude of the allowable pressure variation.

Models that are tuned to observed pressure and have N_2 confined to Sputnik Planitia (Fig. 15) (*BF16*; *Johnson et al.*, 2021; *Meza et al.*, 2019) have typical solutions with $\Gamma \sim 700$–800 TIU, $A \sim 0.7$, and $\varepsilon \sim 0.65$–0.9 [or, more generally, $(1-A)/\varepsilon \sim 0.4$]. In Pluto's current orbit, these models find that (1) there is only one minimum and maximum every Pluto year, while pre-encounter models with completely mobile ices often had one maximum at each of the southern or northern summers; (2) the pressure maximum occurs near the time of maximum insolation on Sputnik Planitia, sometime between perihelion and northern solstice (i.e., near the time of the New Horizons flyby and the writing of this chapter), with a maximum pressure near 1.1 to 1.3 Pa (with a good match achieved by tuning the albedo and emissivity); and (3) the pressure minimum occurs near northern winter solstice, but remains usually large enough to maintain a global atmosphere. Differences between the *Meza et al.* (2019) and *Johnson et al.* (2021) predictions likely arose from the differences in the observational constraints. To some extent, the spatial resolution may play a role, since many of the N_2-rich midlatitude depressions were observed to be tens of kilometers across.

Surface pressures were modeled to peak within a decade of flyby (*B16*; *Meza et al.*, 2019; *Johnson et al.*, 2021). It is intriguing to think that we can test this observationally by measuring the time of the peak pressure from groundbased occultations, looking for a zero or negative slope in the change of pressure vs. time. Care must be taken in this search. Measuring a slope is more difficult than measuring a value. This is exacerbated because different methods for measuring the pressure have yielded different results (section 2.1.3).

Outside of Sputnik Planitia, copious N_2-rich ice is observed at midlatitudes, and in isolated spots at equatorial or high northern latitudes. Including these more-northerly deposits in models (*Meza et al.*, 2019; *Johnson et al.*, 2021) delays the timing of the maximum pressure by 5–15 years vs. a Sputnik-only N_2 reservoir and raises the highest pressure by ~30% (a consequence of the delayed date of the pressure peak, and the constraint that the 2015 pressure is near 1.1 Pa). Typical thermal inertias are similar to the Sputnik-only case, with very slightly larger albedos, and still with $(1-A)/\varepsilon \sim 0.4$. *Johnson et al.* (2021) assumed the N_2-rich ices controlling the pressure did not change in their locations, equivalent to stating that the pressure is controlled by the perennial N_2 deposits that remain over the course of a Pluto year. This is also the case favored by *Meza et al.* (2019). If the N_2-free areas (including CH_4 ice) are relatively dark ($A \sim 0.5$ for both polar and equatorial), then N_2 is confined in the northern hemisphere to Sputnik Planitia and permanent midlatitude N_2 deposits in depressions, with almost no condensation in the southern hemisphere. In this case, the *Johnson et al.* (2021) reference case and *Meza et al.* (2019) describe similar physical situations. The differences in the predicted surface pressure variation between these two models may be due to differences in the details of the distribution of the perennial N_2 deposits, as *Johnson et al.* (2021) constructed a map based on infrared spectra and *Meza et al.* (2019) predicted where perennial N_2 would form self-consistently based on simulations and topographic models.

We have no compositional information south of 40°S. There are three approaches to the question of the distribution and evolution of southern N_2. One is observational, from the historical albedo record. As described in section 2.1, there is evidence for a bright southern pole in 1988. But since CH_4-rich ice and N_2-rich ice have similar albedos in Pluto's northern hemisphere, deriving composition from only albedo is difficult. For this reason, the historical albedo record has not yet been used to constrain any post-flyby volatile transport models. A second approach is to impose N_2 below 40°S, calculate its effects on pressures or haze production, and compare results to observations (*Forget et al.*, 2017; *Bertrand and Forget*, 2017; *Meza et al.*, 2019; *Johnson et al.*, 2021, *Lewis et al.*, 2021). If too much N_2 is in the southern hemisphere, then pressures would not be increasing since 1988 as observed (*Meza et al.*, 2019; *Johnson et al.*, 2021), but solutions can be found with a southern cap (N_2 from 90°S to 60°S) or southern band (N_2 from 35°S to 55°S) that contains 20% N_2-rich ice (*Johnson et al.*, 2021). The southern-band solutions have substantially lower thermal inertias than other solutions (*Johnson et al.*, 2021), consistent with the conclusions of *Lewis et al.* (2021) that solutions that ignore thermal inertia require southern N_2-rich ice. The third approach is to derive the conditions under which the southern latitudes should be frost-covered or clear — i.e., to see if models predict the formation of southern N_2-rich ice.

Most pre-flyby models predicted seasonal southern N_2 (*Hansen and Paige*, 1996; *Young*, 2013; *Hansen et al.*, 2015), but included no topographical data, as none was available. Post-flyby models with only the N_2 cycle have N_2-rich ice only in the depression that is Sputnik Planitia (*Bertrand et al.*, 2018), but the addition of CH_4 to models allows CH_4 condensation sites to then become N_2 condensation sites (*Bertrand et al.*, 2019), unless the CH_4 is dark. The stability of N_2 outside of Sputnik Planitia or CH_4 at the equator or mid to high latitudes depends critically on the albedo (Fig. 16). The effect of albedo is particularly important in the diurnal zone (*Earle et al.*, 2018a). *Earle*

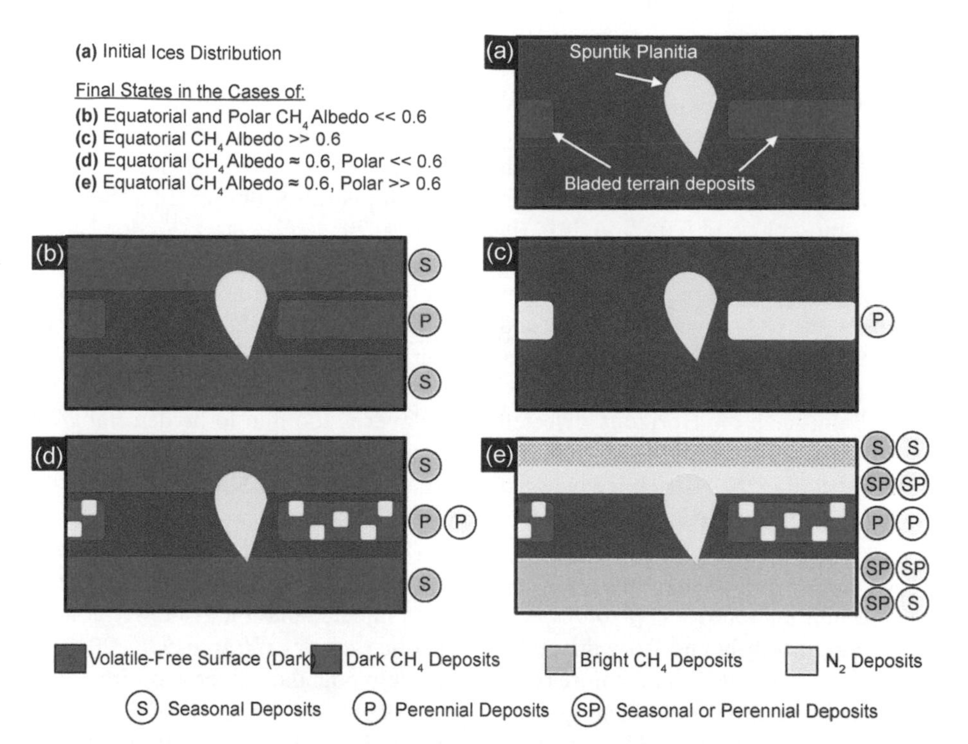

Fig 16. Summary of the simulation results for runs with that were initialized with N_2 in Sputnik Planitia, and an infinite reservoir of CH_4 at the latitudes and longitudes of the bladed terrain (*Bertrand et al.,* 2019, their section 4).

et al. (2018a) treated the darkening of ices with time due to haze deposition and the subsequent brightening when new ice is laid down, and found that runaway albedo may get triggered over a single Pluto season.

For the class of runs with uniformly dark CH_4 ($A_{CH_4} \ll 0.6$, Fig. 16b), *B19* found that the CH_4 remained too warm for N_2 condensation. At mid to high latitudes, dark CH_4 was not stable against sublimation, and formed only thin seasonal deposits. This class appeared to match the occultation record. Whether or not this class is viable depends critically on whether the observational evidence is for perennial or seasonal N_2 or CH_4 deposits at mid to high latitudes.

For the class of runs with bright equatorial CH_4 ($A_{CH_4,eq} \gg 0.6$, Fig. 16c), *B19* found that N_2 was able to condense on the bladed terrains and to remain very stable in these equatorial regions. Nitrogen thus covered CH_4 and shut down the CH_4 cycle. No volatile deposits — either seasonal or perennial — formed outside of the equator in these models. This did not match the volatile distribution seen at Pluto.

For the class of runs with moderately bright equatorial CH_4 ($A_{CH_4,eq} \approx 0.6$, Fig. 16d,e), *B19* found that N_2 was able to condense onto equatorial CH_4, but only in depressions, leaving high-altitude areas CH_4-rich, as observed. If the run also had dark midlatitude/polar CH_4 ($A_{CH_4,polar} \ll 0.6$, Fig. 16d), then the CH_4 can form seasonal deposits at mid to high latitudes. *B19* found that if the run had bright midlatitude/polar CH_4 (Fig. 16e), the N_2 was able to condense onto it, trap more CH_4, and result in more perennial CH_4 ice deposits at midlatitudes (the CH_4 deposits remained seasonal at the poles in all cases). The moderately bright equatorial, bright polar CH_4 class produced some of the observed trends in surface volatiles: N_2 in depressions, seasonal N_2 and CH_4 at midlatitudes, and seasonal N_2 at the poles.

Returning to pressure over a season, *Bertrand et al.* (2019) found that southern N_2 tended to lead to an earlier peak in pressure, near 2010, but *Johnson et al.* (2021) were able to find solutions that matched the 1988–2015 occultation record, with either a southern N_2 cap or a southern midlatitude band with N_2 areal fraction of 20%. Even with some southern N_2-rich ice, most post-flyby models predict that Pluto's atmosphere will maintain a global atmosphere over Pluto's year with current obliquity and orbital conditions (*Johnson et al.,* 2021; *Bertrand and Forget,* 2016; *Meza et al.,* 2019). This was also indicated by some of the pre-New Horizons models (*Trafton,* 1990; *Stansberry and Yelle,* 1999; *Olkin et al.,* 2015).

Higher minimum pressure is partly due to the effect of the high seasonal thermal inertia of the substrate, which enables the poles to store the heat accumulated during summer and to release it during winter. The poles thus remain too warm for N_2 condensation. It is also partly due to the near-equatorial location of Sputnik Planitia, ensuring insolation on N_2-ices even during northern winter. The pressure maximum and minimum can be more extreme during times when northern solstice coincides with perihelion (*Johnson et al.,* 2021; *Bertrand et al.,* 2018). In the LMD and VT3D models, the minimum surface pressure generally stays above 6 mPa, and so is likely always global, although it is possible to have lower pressures in some Pluto orbital configurations for some model inputs (*Johnson et al.,* 2021; *Bertrand et al.,* 2018).

In those post-flyby models that include CO and CH_4 (*Bertrand and Forget,* 2016; *Bertrand et al.,* 2019), the CO tends to follow N_2 and condense where N_2-rich deposits are already present, in agreement with observations. *Bertrand and Forget* (2016) showed that 0.3% CO (*Merlin,* 2015) in N_2-rich ice filling Sputnik Planitia leads to an atmospheric mean gas volume mixing ratio close to 0.04% in 2015, in good agreement with telescopic measurements (*Lellouch et al.,* 2011a). They also showed that without the presence of permanent N_2-rich deposits, the CO gas mixing ratio would reach 20–30% in 2010, which is unrealistic and reinforces the picture of a permanent cold trap in Sputnik Planitia.

Outside of Sputnik Planitia, New Horizons detected clear latitudinal trends for the volatiles on Pluto. Methane-rich ice was seen northward of 55°N, N_2-rich ice between 35°N and 55°N, and CH_4-rich ice between 20°N and 35°N (*Schmitt et al.,* 2017; *Protopapa et al.,* 2017; *Earle et al.,* 2018b). The N_2-rich band may be the seasonally varying remnant of two opposing sublimation fronts (*Protopapa et al.,* 2017): a slowly moving north-bound front driven by low-latitude heating from 1975 to 1995, and a more rapid south-bound front driven by high-latitude heating from 1995 to 2015, particularly from 2005 to 2015. The effect of solar insolation on the CH_4 distribution is supported by simulations (*Bertrand and Forget,* 2016; *Bertrand et al.,* 2019) that reproduce, to first order, the spatial distribution of N_2 and CH_4 ices observed in 2015 outside Sputnik Planitia. In these simulations, when the midlatitude/polar CH_4-rich ice is bright ($A > 0.65$), it gets cold enough in autumn and winter that N_2 condenses on it and forms seasonal or perennial N_2-rich deposits at midlatitudes, as observed. Seasonal or perennial CH_4 deposits may even form on dark, low-latitude terrains during winter, consistent with the low-latitude band of CH_4-rich ice (*Protopapa et al.,* 2017; *Earle et al.,* 2018b).

Whether the two sublimation fronts will meet (*Protopapa et al.,* 2017) is related to the core question of whether the midlatitude N_2 ice is seasonal (varying greatly — perhaps completely subliming — every Pluto orbit) or perennial (slowly growing or shrinking over the timescale of many Pluto orbits).

Bertrand and Forget (2016) predicted a range of atmospheric CH_4 mixing ratio from 0.1% to 5% over the current Pluto year. A larger ratio could be expected in northern winter, when most post-flyby models predict that the atmosphere will be thinner by orders of magnitudes and would thus provide less dilution of methane.

5.4. Diurnal

Much of the diurnal volatile cycle involves the winds transporting heat and mass across the surface, and the vertical winds (also sometimes called "atmospheric breathing") from local sublimation and condensation, as described in the chapter on atmospheric dynamics by Forget et al. There are few direct observational constraints on the diurnal cycle.

Sputnik Planitia is mainly within the diurnal zone, and so experiences sunrise and sunset every single Pluto day for the last tens of millions of years or longer. The higher noontime insolation is balanced by sublimation, with diurnal variation of only ~millikelvins for N_2-rich ice temperature, or ~millipascals for surface pressure (*Young,* 2012; *Forget et al.,* 2017). At the time of flyby, superimposed on the net north-south transport within Sputnik Planitia, tens of grams per square meter of fresh N_2 condensed and sublimed each day (*Forget et al.,* 2017, their Fig. 8), equivalent to a depth of several tens of micrometers of N_2-rich ice. It is likely that this thin diurnal ice layer is too thin to be detectable spectroscopically in the near-IR. The N_2-rich ice outside of Sputnik Planitia, like the ice within Sputnik Planitia, is also likely to stay nearly isothermal.

The daily sublimation of tens of grams per square meter of N_2 corresponds to about 2 km of cold N_2 gas being injected into Pluto's lower atmosphere over Sputnik Planitia each day (*Hinson et al.,* 2017; *Forget et al.,* 2017). The cold boundary layer was directly detected in the REX ingress profile (*Hinson et al.,* 2017; *Forget et al.,* 2017). The daily sublimation and condensation cycles induce vertical winds that vary in magnitude on a daily cycle. These may force waves that are detectable as layers seen in haze images (*Forget et al.,* 2017) or the density and temperature variations seen in groundbased occultations (*Toigo et al.,* 2010; *French et al.,* 2015).

Areas free of N_2 over a day will be warmer than the N_2-rich ice. The temperatures of non-volatile terrains are not buffered by the latent heat of sublimation, and tend to be warmer due to their lower albedo. The temperatures of CH_4-rich ices are only somewhat buffered by latent heat. *Forget et al.* (2017) modeled CH_4 temperatures that range from just slightly warmer than the N_2-rich ice to 52 K in isolated polar areas. Thermal models based on the pre-encounter composition maps and constrained by ISO, Spitzer, and Herschel thermal observations favor low thermal inertia for the diurnal skin depth, and predict CH_4 temperatures that also reach a localized maximum of 54 K (*Lellouch et al.,* 2000, 2011, 2016).

If a CH_4-rich or non-volatile area becomes cold enough, it would become an N_2 condensation site. Would volatiles condense overnight and be visible at the dawn limb? Such a search will probably wait for an orbiter, since the equatorial dawn limb was only imaged by New Horizons after the solar occultation, at high phase angle and low spatial resolution.

Methane-rich areas that do not become N_2 condensation sites will be warmer than the N_2-rich ice, and will be CH_4 condensation or sublimation sites depending on whether they are cooler or warmer than the critical temperature (equation (5)), which is 42.1 K for pure CH_4 ice (slightly warmer for CH_4-rich ice saturated with N_2). The CH_4 mass flux will vary with location and time of day. The diurnal variation at a given location is only 3–6 K (*Forget et al.,* 2017). Considering a 3 K temperature range around 42.1 K, and using $U_* C_m \sim 0.22$ cm s^{-1} [U_* is the friction velocity, and C_m is the mass transfer coefficient (*Forget et al.,* 2017)], the mass rates are about 0.1 μm of condensation per Pluto day at 40.6 K, and 0.3 μm of sublimation per Pluto day at 43.6 K. Thus, the daily cycle of CH_4 sublimation and condensation is small. It is also asymmetric, with sublimation being more efficient than condensation, which will contribute to the net mass flux of CH_4 into the atmosphere averaged over many Pluto days. This asymmetry might impact Pluto's surface and atmosphere on longer timescales (seasonal or longer), but this effect has not yet been studied.

6. CONCLUSIONS AND FUTURE WORK

Pluto complements Triton, Io, and Mars as examples of vapor-pressure-equilibrium atmospheres. In the outer solar system, Pluto is unique, both for its intrinsic properties (the surface complexity, the presence of Sputnik Planitia, the prevalence of both CH_4 and N_2) and for the rich collection of observations (the New Horizons surface maps of geology, topography, and critically, composition; the UV and radio occultations and a nearly simultaneous groundbased stellar occultation; the long time-base of occultations and other Earth-based observations). Pluto has already allowed us to study the effects of albedo, emissivity, thermal inertia, topography, and Milankovitch-like orbital variation on the volatile and climate cycles. It is poised to be a laboratory in which to study the physics of ice mixtures, the N_2 α-β phase transition, albedo feedbacks, and the role of ice contamination by haze particles.

While great strides have been already made, important open questions remain.

1. *How do the CH_4-N_2 thermodynamics impact the volatile transport on Pluto and the vertical structure of Pluto's ices?* New Horizons found that while both N_2-rich and CH_4-rich phases of ice are present, the N_2-rich ice is not saturated with CH_4. This is surprising because the N_2-rich ice is expected to saturate before any CH_4-rich ice forms. The lack of saturation is even more surprising given the excess of CH_4 in Pluto's atmosphere. Part of the solution may lie in the formation of CH_4-rich ices during sublimation, as calculated by *Tan and Kargel* (2018). Part of the solution may also be the complex interplay between the creation of a CH_4-rich layer at the surface-atmosphere interface (*Schmitt et al.,* 2017; *Trafton et al.,* 1998, 2019), the annealing rate of newly deposited N_2-rich ice (*Eluszkiewicz,* 1991; *Young et al.,* 2019), and the sublimation of N_2 some centimeters below the surface and redeposition near the surface from solar gardening (*Grundy and Stansberry,* 2000).

2. *Where are the seasonal and perennial ices? Which ices have built up over millions of years?* Prior to New Horizons, many models imagined a small inventory of N_2 ice moving across Pluto, alternately covering up and revealing substrate. One of the many ways in which Sputnik Planitia has shaken our concepts of "how Pluto works" is the new picture of a large reservoir of N_2 and the transport of atmospheric N_2 between perennial deposits of N_2 ice that last over many Pluto orbits. The CH_4 ice also appears to have some deposits that come and go over a Pluto orbit, but also has large deposits, such as the bladed terrain, that must have taken millions of years to build. When did the bladed terrain form, and is it current disappearing? Why is the bladed terrain located east of Sputnik Planitia, whereas the regions west of Sputnik Planitia are depleted in volatile ices? How and when did the extensive CH_4 mantles north and east of Sputnik Planitia form? Does the process of CH_4 segregation feed the atmosphere with more gaseous CH_4 that will form CH_4 rich deposits elsewhere, incidentally preventing CH_4 from entirely migrating into Sputnik Planitia? These questions are also tied to estimates of the current total volatile inventory derived from composition maps, since seasonal deposits will be shallower than perennial ones. The depth of current or past deposits is also connected to the question of glacial flows, since N_2 is more ductile (or perhaps even liquid) under the pressure of deep ice deposits.

3. *What is the composition, history, and evolution of the southern hemisphere?* Historical photometry argues for a bright southern polar cap, pre-equinox. If the bright southern cap was N_2-rich, then it should have remained an N_2-rich condensation site during the ensuing darkness. However, many models made after New Horizons suggest that the unseen southern hemisphere should be free of N_2 in order to match the occultation record. It is not clear how to reconcile these. The south pole will not be seen again from Earth until 2109.

4. *What are the processes that change albedo, composition, or other thermophysical properties of the volatile ices? What drives the strong contrasts in color on Pluto's surface?* Albedo sets up a positive feedback, particularly in the diurnal climate zone, but, with the exception of Sputnik Planitia, it is unclear how this is triggered. Albedo could be a function of age if ices are processed by UV and energetic particles, or if ices are contaminated by sedimenting haze particles — all of which are also affected by the atmospheric conditions, so there could be complex feedbacks. Thin (seasonal) deposits may allow the albedo of the underlying material to control the energy balance. The N_2 grain size depends on the deposition rate, and the effect of "solar gardening" can increase porosity (and thermal inertia) at depth and decrease it at the uppermost layer.

One approach to addressing these questions is to improve future models to account for additional processes: the alteration of albedo and other surface properties, the interaction of CH_4 and N_2, and the CH_4 cycle over astronomical timescales. With higher fidelity comes increased complexity, as many elements of these processes remain unknown or not well constrained. In order to further understand the processes at play on Pluto (and on many other KBOs), observations, modeling, and experiments will need to be combined. Experiments on ice mixtures are needed, to test the predictions of *Tan and Kargel* (2018) and to study the behavior of N_2, CO, and CH_4 in disequilibrium. We only have atmospheric data over a tenth of Pluto's year, during which time we observed a tripling of the surface pressure. Groundbased observations of pressure (stellar occultations), surface temperature (thermal radiometry), atmospheric and surface composition (spectroscopy), and surface albedo (visible imagery) are needed at least every few Earth years in order to further constrain the models. Our current puzzlement would serve well in the design of any future mission to Pluto. While our understanding of Pluto's volatile cycles has considerably increased thanks to the New Horizons observations, the flyby of Pluto in 2015 only provided a snapshot in time of Pluto's dynamic and evolving surface and atmosphere, only 40% of the total surface was imaged, and the atmosphere was probed at only a few distinct locations. An orbiter mission is needed to continue to explore and understand Pluto, and in particular how its surface and atmosphere evolve with time and are shaped by complex volatile transport processes. As the breakthrough observations of the 1980s and 1990s allowed us to take the first steps toward understanding Pluto's volatiles, and New Horizons gave Pluto science one giant leap, the combination of new modeling, lab work, observations, and a possible return to Pluto will lift us to new heights.

Acknowledgments. We are grateful to C. Hansen, A. Howard, and A. Stern for their reviews of this chapter. L.Y. was supported in part by NASA ROSES SSW grant NNX15AH35G and by NASA's New Horizons mission to the Pluto system. T.B. was supported for this research by an appointment to the National Aeronautics and Space Administration (NASA) Post-Doctoral Program at the Ames Research Center administered by the Universities Space Research Association (USRA) through a contract with NASA. L.M.T. was supported by NASA Grant 80-NSSC18K0860. F.F. and T.B. benefited from support from CNES. A.M.E. was supported in part by the NASA New Horizons mission to Pluto and The Kuiper Belt under SwRI Subcontract 299433Q. The work by B.S. leading to these results has received funding from the European Research Council under the European Community's H2020 2014-2020 ERC Grant Agreement no. 669416 "Lucky Star."

REFERENCES

Andersson L. E. and Fix J. D. (1973) Pluto: New photometry and a determination of the axis of rotation. *Icarus, 20*, 279–283.

Aumann H. H. and Walker R. G. (1987) IRAS observations of the Pluto-Charon system. *Astron J., 94*, 1088–1091.

Bertrand T. and Forget F. (2016) Observed glacier and volatile distribution on Pluto from atmosphere-topography processes. *Nature, 540*, 86–89.

Bertrand T. and Forget F. (2017) 3D modeling of organic haze in Pluto's atmosphere. *Icarus, 287*, 72–86.

Bertrand T. and 14 colleagues (2018) The nitrogen cycles on Pluto over seasonal and astronomical timescales. *Icarus, 309*, 277–296.

Bertrand T. and 16 colleagues (2019) The CH_4 cycles on Pluto over seasonal and astronomical timescales. *Icarus, 329*, 148–165.

Bertrand T. et al. (2020a) Pluto's beating heart regulates the atmospheric circulation: Results from high resolution and multi-year numerical climate simulations. *J. Geophys. Res., 125*, e2019JE006120.

Bertrand T., Forget F., Schmitt B., White O., and Grundy W. (2020b) Equatorial mountains on Pluto are covered by methane frosts resulting from a unique atmospheric process. *Nature Commun., 11*, DOI: 10.1038/s41467-020-18845-3.

Binzel R. P. and 13 colleagues (2017) Climate zones on Pluto and Charon. *Icarus, 287*, 30–36.

Bockelée-Morvan D., Lellouch E., Biver N., Paubert G., Bauer J., Colom P., and Lis D. C. (2001) Search for CO gas in Pluto, Centaurs and Kuiper belt objects at radio wavelengths. *Astron. Astrophys., 377*, 343–353.

Bosh A. S. and 25 colleagues (2015) The state of Pluto's atmosphere in 2012–2013. *Icarus, 246*, 237–246.

Brosch N. (1995) The 1985 stellar occultation by Pluto. *Mon. Not. R. Astron. Soc., 276*, 571–578.

Buhler P. B. and Ingersoll A. P. (2018) Sublimation pit distribution indicates convection cell surface velocities of ~10 cm per year in Sputnik Planitia, Pluto. *Icarus, 300*, 327–340.

Buie M. W. and Folkner W. M. (2015) Astrometry of Pluto from 1930–1951 observations: The Lampland plate collection. *Astron. J., 149*, 22.

Buie M. W. and Tholen D. J. (1989) The surface albedo distribution of Pluto. *Icarus, 79*, 23–37.

Buie M. W., Young E. F., and Binzel R. P. (1997) Surface appearance of Pluto and Charon. In *Pluto and Charon* (S. A. Stern and D. Tholen, eds.), pp. 269–294. Univ. of Arizona, Tucson.

Buie M. W., Grundy W. M., Young E. F., Young L. A., and Stern S. A. (2010a) Pluto and Charon with the Hubble Space Telescope. I. Monitoring global change and improved surface properties from light curves. *Astron. J., 139*, 1117–1127.

Buie M. W., Grundy W. M., Young E. F., Young L. A., and Stern S. A. (2010b) Pluto and Charon with the Hubble Space Telescope. II. Resolving changes on Pluto's surface and a map for Charon. *Astron. J., 139*, 1128–1143.

Buratti B. J., Hillier J. K., Heinze A., Hicks M. D., Tryka K. A., Mosher J. A., Ward J., Garske M., Young J., and Atienza-Rosel J. (2003) Photometry of Pluto in the last decade and before: Evidence for volatile transport? *Icarus, 162*, 171–182.

Buratti B. J., Hicks M. D., Dalba P. A., Chu D., O'Neill A., Hillier J. K., Masiero J., Banholzer S., and Rhoades H. (2015) Photometry of Pluto 2008–2014: Evidence of ongoing seasonal volatile transport and activity. *Astrophys. J. Lett., 804*, L6.

Buratti B. J. and 16 colleagues (2017) Global albedos of Pluto and Charon from LORRI New Horizons observations. *Icarus, 287*, 207–217.

Carson J. K., Lovatt S. J., Tanner D. J., and Cleland A. C. (2005) Thermal conductivity bounds for isotropic, porous materials. *Intl. J. Heat Mass Transfer, 48*, 2150–2158.

Cruikshank D. P. and Silvaggio P. M. (1980) The surface and atmosphere of Pluto. *Icarus, 41*, 96–102.

Cruikshank D. P., Pilcher C. B., and Morrison D. (1976) Pluto: Evidence for methane frost. *Science, 194*, 835–837.

Cruikshank D. P. and 12 colleagues (2015) The surface compositions of Pluto and Charon. *Icarus, 246*, 82–92.

DeMeo F. E., Dumas C., de Bergh C., Protopapa S., Cruikshank D. P., Geballe T. R., Alvarez-Candal A., Merlin F., and Barucci M. A. (2010) A search for ethane on Pluto and Triton. *Icarus, 208*, 412–424.

Desmars J. and 15 colleagues (2019) Pluto's ephemeris from ground-based stellar occultation (1988–2016). *Astron. Astrophys., 625*, A43.

Dias-Oliveira A. and 45 colleagues (2015) Pluto's atmosphere from stellar occultations in 2012 and 2013. *Astron. J., 811*, 53.

Dobrovolskis A. R., Peale S. J., and Harris A. W. (1997) Dynamics of the Pluto-Charon binary. In *Pluto and Charon* (S. A. Stern and D. Tholen, eds.), pp. 159–190. Univ. of Arizona, Tucson.

Douté S., Schmitt B., Quirico E., Owen T. C., Cruikshank D. P., de Bergh C., Geballe T. R., and Roush T. L. (1999) Evidence for methane segregation at the surface of Pluto. *Icarus, 142*, 421–444.

Drish W. F., Harmon R., Marcialis R. L., and Wild W. J. (1995) Images of Pluto generated by matrix light curve inversion. *Icarus, 113*, 360–386.

Duxbury N. S. and Brown R. H. (1993) The phase composition of Triton's polar caps. *Science, 261*, 748–751.

Earle A. M. (2018) Spectral mapping and long-term seasonal evolution of Pluto. Ph.D. thesis, Massachusetts Institute of Technology, available online at *https://dspace.mit.edu/handle/1721.1/117915*.

Earle A. M., Binzel R. P., Young L. A., Stern S. A., Ennico K., Grundy W., Olkin C. B., Weaver H. A., and the New Horizons Geology and Geophysics Imaging Team (2017) Long-term surface temperature modeling of Pluto. *Icarus, 287*, 37–46.

Earle A. M. and 7 colleagues (2018a) Albedo matters: Understanding runaway albedo variations on Pluto. *Icarus, 303*, 1–9.

Earle A. M. and 19 colleagues (2018b) Methane distribution on Pluto as mapped by the New Horizons Ralph/MVIC instrument. *Icarus, 314*,

195–209.
Eglitis I., Eglite M., Kazantseva L. V., Shatokhina S. V., Protsyuk Y. I., Kovylianskaya O. E., and Andruk V. M. (2018) Astrometric and photometric processing of Pluto digitized photographic observations 1961–1996. In *Astroplate 2016* (P. Skala, ed.), pp. 5–8. Czech Technical Univ., Prague.
Elliot J. L. and Olkin C. B. (1996) Probing planetary atmospheres with stellar occultations. *Annu. Rev. Earth Planet. Sci., 24*, 89–124.
Elliot J. L. and Young L. A. (1992) Analysis of stellar occultation data for planetary atmospheres. I. Model fitting, with application to Pluto. *Astron. J., 103*, 991–1015.
Elliot J. L., Dunham E. W., Bosh A. S., Slivan S. M., Young L. A., Wasserman L. H., and Millis R. L. (1989) Pluto's atmosphere. *Icarus, 77*, 148–170.
Elliot J. L., Person M. J., and Qu S. (2003a) Analysis of stellar occultation data. II. Inversion, with application to Pluto and Triton. *Astron. J., 126*, 1041–1079.
Elliot J. L. and 28 colleagues (2003b) The recent expansion of Pluto's atmosphere. *Nature, 424*, 165–168.
Elliot J. L. and 19 colleagues (2007) Changes in Pluto's atmosphere: 1988–2006. *Astron. J.*, 134, 1–13.
Eluszkiewicz J. (1991) On the microphysical state of the surface of Triton. *J. Geophys. Res., 96*, 19217–19229.
Fink U., Smith B. A., Johnson J. R., Reitsema H. H., Benner D. C., and Westphal J. A. (1980) Detection of a CH_4 atmosphere on Pluto. *Icarus, 44*, 62–71.
Forget F., Bertrand T., Vangvichith M., Leconte J., Millour E., and Lellouch E. (2017) A post-New Horizons global climate model of Pluto including the N_2, CH_4 and CO cycles. *Icarus, 287*, 54–71.
Fray N. and Schmitt B. (2009) Sublimation of ices of astrophysical interest: A bibliographic review. *Planet. Space Sci., 57*, 2053–2080.
French R. G., Toigo A. D., Gierasch P. J., Hansen C. J., Young L. A., Sicardy B., Dias-Oliveira A., and Guzewich S. D. (2015) Seasonal variations in Pluto's atmospheric tides. *Icarus, 246*, 247–267.
Gabasova L., et al. (2021) Global compositional cartography of Pluto from intensity-based registration of LEISA data. *Icarus, 356,* DOI: 10.1016/j.icarus.2020.113833.
Gladstone G. R. et al. (2016) The atmosphere of Pluto as observed by New Horizons. *Science, 351*, aad8866.
Glein C. R. and Waite J. H. (2018) Primordial N_2 provides a cosmochemical explanation for the existence of Sputnik Planitia, Pluto. *Icarus, 313*, 79–92.
Greaves J. S., Helling C., and Friberg P. (2011) Discovery of carbon monoxide in the upper atmosphere of Pluto. *Mon. Not. R. Astron. Soc., 414,* L36–L40.
Grundy W. M. (1995) Methane and nitrogen ices on Pluto and Triton: A combined laboratory and telescope investigation. Ph.D. thesis, Univ. of Arizona, Tucson. 185 pp.
Grundy W. M. and Buie M. W. (2001) Distribution and evolution of CH_4, N_2, and CO Ices on Pluto's surface: 1995 to 1998. *Icarus, 153,* 248–263.
Grundy W. M. and Buie M. W. (2002) Spatial and compositional constraints on non-ice components and H_2O on Pluto's surface. *Icarus, 157*, 128–138.
Grundy W. M. and Fink U. (1996) Synoptic CCD spectrophotometry of Pluto over the past 15 years. *Icarus, 124*, 329–343.
Grundy W. M. and Stansberry J. A. (2000) Solar gardening and the seasonal evolution of nitrogen ice on Triton and Pluto. *Icarus, 148,* 340–346.
Grundy W. M., Buie M. W., and Spencer J. R. (2002) Spectroscopy of Pluto and Triton at 3–4 microns: Possible evidence for wide distribution of nonvolatile solids. *Astron. J., 124,* 2273–2278.
Grundy W. M., Olkin C. B., Young L. A., Buie M. W., and Young E. F. (2013) Near-infrared spectral monitoring of Pluto's ices: Spatial distribution and secular evolution. *Icarus, 223*, 710–721.
Grundy W. M., Olkin C. B., Young L. A., and Holler B. J. (2014) Near-infrared spectral monitoring of Pluto's ices II: Recent decline of CO and N_2 ice absorptions. *Icarus, 235*, 220–224.
Grundy W. M. et al. (2016) Surface compositions across Pluto and Charon. *Science, 351*, aad9189.
Gulbis A. A. S., Emery J. P., Person M. J., Bosh A. S., Zuluaga C. A., Pasachoff J. M., and Babcock B. A. (2015) Observations of a successive stellar occultation by Charon and graze by Pluto in 2011: Multiwavelength SpeX and MORIS data from the IRTF. *Icarus, 246*, 226–236.
Hamilton D. P. and 19 colleagues (2016) The rapid formation of Sputnik Planitia early in Pluto's history. *Nature, 540*, 97–99.
Hansen C. J. and Paige D. A. (1992) A thermal model for the seasonal nitrogen cycle on Triton. *Icarus, 99,* 273–288.
Hansen C. J. and Paige D. A. (1996) Seasonal nitrogen cycles on Pluto. *Icarus, 120*, 247–265.
Hansen C. J., Paige D. A., and Young L. A. (2015) Pluto's climate modeled with new observational constraints. *Icarus, 246*, 183–191.
Hinson D. P. and 15 colleagues (2017) Radio occultation measurements of Pluto's neutral atmosphere with New Horizons. *Icarus, 290*, 96–111.
Hofgartner J. D, Buratti B. J., Hayne P. O., and Young L. A. (2019) Ongoing resurfacing of KBO Eris by volatile transport in local, collisional, sublimation atmosphere regime. *Icarus, 334*, 52–61.
Holler B. J., Young L. A., Grundy W. M., and Olkin C. B. (2016) On the surface composition of Triton's southern latitudes. *Icarus, 267,* 255–266.
Holler B. J., Young L. A., Buie M. W., Grundy W. M., Lyke J. E., Young E. F., and Roe H. G. (2017) Measuring temperature and ammonia hydrate ice on Charon in 2015 from Keck/OSIRIS spectra. *Icarus, 284,* 394–406.
Hourdin F., Le Van P., Forget F., and Talagrand O. (1993) Meteorological variability and the annual surface pressure cycle on Mars. *J. Atmos. Sci., 50*, 3625–3640.
Howell S. B., Horch E. P., Everett M. E., and Ciardi D. R. (2012) Speckle camera imaging of the planet Pluto. *Publ. Astron. Soc. Pac., 124*, 1124–1131.
Hubbard W. (2003) Planetary science: Pluto's atmospheric surprise. *Nature, 424*, 137–138.
Hubbard W. B., Hunten D. M., Dieters S. W., Hill K. M., and Watson R. D. (1988) Occultation evidence for an atmosphere on Pluto. *Nature, 336*, 452–454.
Hubbard W. B., Yelle R. V., and Lunine J. I. (1990) Nonisothermal Pluto atmosphere models. *Icarus, 84*, 1–11.
Hunten D. M. and Watson A. J. (1982) Stability of Pluto's atmosphere. *Icarus, 51*, 665–667.
Ingersoll A. P., Summers M. E., and Schlipf S. G. (1985) Supersonic meteorology of Io: Sublimation-driven flow of SO_2. *Icarus, 64*, 375–390.
Jewitt D. C. (1994) Heat from Pluto. *Astron. J., 107,* 372–378.
Johnson P. E., Young L. A., Protopapa S., Schmitt B., Gabasova L. R., Lewis B. L., Stansberry J. A., Mandt K. E., and White O. L. (2021) Modeling Pluto's minimum pressure: Implications for haze production. *Icarus, 356,* DOI: 10.1016/j.icarus.2020.114070.
Krasnopolsky V. A. (2020) A photochemical model of Pluto's atmosphere and ionosphere. *Icarus, 335*, 113374.
Leighton R. B. and Murray B. C. (1966) Behavior of carbon dioxide and other volatiles on Mars. *Science, 153*, 136–144.
Lellouch E. (1994) The thermal structure of Pluto's atmosphere: Clear vs. hazy models. *Icarus, 108*, 255–264.
Lellouch E., Laureijs R., Schmitt B., Quirico E., de Bergh C., Crovisier J., and Coustenis A. (2000) Pluto's non-isothermal surface. *Icarus, 147*, 220–250.
Lellouch E., de Bergh C., Sicardy B., Ferron S., and Käufl H.-U. (2010) Detection of CO in Triton's atmosphere and the nature of surface-atmosphere interactions. *Astron. Astrophys., 512*, L8.
Lellouch E., de Bergh C., Sicardy B., Käufl H.-U., and Smette A. (2011a) High resolution spectroscopy of Pluto's atmosphere: Detection of the 2.3 μm CH_4 bands and evidence for carbon monoxide. *Astron. Astrophys., 530*, L4.
Lellouch E., Stansberry J., Emery J., Grundy W., and Cruikshank D. P. (2011b) Thermal properties of Pluto's and Charon's surfaces from Spitzer observations. *Icarus, 214*, 701–716.
Lellouch E., de Bergh C., Sicardy B., Forget F., Vangvichith M., and Käufl H.-U. (2015) Exploring the spatial, temporal, and vertical distribution of methane in Pluto's atmosphere. *Icarus, 246*, 268–278.
Lellouch E. and 11 colleagues (2016) The long-wavelength thermal emission of the Pluto-Charon system from Herschel observations: Evidence for emissivity effects. *Astron. Astrophys., 588*, A2.
Lellouch E. and 18 colleagues (2017) Detection of CO and HCN in Pluto's atmosphere with ALMA. *Icarus, 286*, 289–307.
Levine J. S., Kraemer D. R., and Kuhn W. R. (1977) Solar radiation incident on Mars and the outer planets: Latitudinal, seasonal, and atmospheric effects. *Icarus, 31,* 136–145.
Lewis B. L., Stansberry J. A., Holler B. J., Grundy W. M., Schmitt B., Protopapa S., Lisse C., Stern S. A., Young L., Weaver H. A., Olkin C., Ennico K., and the New Horizons Science Team

(2021) Distribution and energy balance of Pluto's nitrogen ice, as seen by New Horizons in 2015. *Icarus, 356,* DOI: 10.1016/j.icarus.2020.113633.

McKinnon W. B. and 61 colleagues (2016) Convection in a volatile nitrogen-ice-rich layer drives Pluto's geological vigour. *Nature, 534,* 82–85.

Merlin F. (2015) New constraints on the surface of Pluto. *Astron. Astrophys., 582,* A39.

Merlin F., Barucci M. A., de Bergh C., DeMeo F. E., Alvarez-Candal A., Dumas C., and Cruikshank D. P. (2010) Chemical and physical properties of the variegated Pluto and Charon surfaces. *Icarus, 210,* 930–943.

Meza E. and 169 colleagues (2019) Lower atmosphere and pressure evolution on Pluto from ground-based stellar occultations, 1988–2016. *Astron. Astrophys., 625,* A42.

Moore J. M. and 153 colleagues (2016) The geology of Pluto and Charon through the eyes of New Horizons. *Science, 351,* 1284–1993.

Moore J. M. and 25 colleagues (2018) Bladed terrain on Pluto: Possible origins and evolution. *Icarus, 300,* 129–144.

Moores J. E., Smith C. L., Toigo A. D., and Guzewich S. D. (2017) Penitentes as the origin of the bladed terrain of Tartarus Dorsa on Pluto. *Nature, 541,* 188–190.

Nadeau A. and Jaschke E. (2019) Stable asymmetric ice belts in an energy balance model of Pluto. *Icarus, 331,* 15–25.

Nadeau A. and McGehee R. (2017) A simple formula for a planet's mean annual insolation by latitude. *Icarus, 291,* 46–50.

Nimmo F. and 16 colleagues (2017) Mean radius and shape of Pluto and Charon from New Horizons images. *Icarus, 287,* 12–29.

Olkin C. B., Young E. F., Young L. A., Grundy W., Schmitt B., Tokunaga A., Owen T., Roush T., and Terada H. (2007) Pluto's spectrum from 1.0 to 4.2 μm: Implications for surface properties. *Astron. J., 133,* 420–431.

Olkin C. B. and 23 colleagues (2015) Evidence that Pluto's atmosphere does not collapse from occultations including the 2013 May 04 event. *Icarus, 246,* 220–225.

Olkin C. B. and 25 colleagues (2017) The global color of Pluto from New Horizons. *Astron. J., 154,* 258.

Owen T. C., Roush T. L., Cruikshank D. P., Elliot J. L., Young L. A., de Bergh C., Schmitt B., Geballe T. R., Brown R. H., and Bartholomew M. J. (1993) Surface ices and atmospheric composition of Pluto. *Science, 261,* 745–748.

Pasachoff J. M., Souza S. P., Babcock B. A., Ticehurst D. R., Elliot J. L., Person M. J., Clancy K. B., Roberts L. C., Hall D. T., and Tholen D. J. (2005) The structure of Pluto's atmosphere from the 2002 August 21 stellar occultation. *Astron. J., 129,* 1718–1723.

Prokhvatilov A. L. and Yantsevich L. D. (1983) X-ray investigations of the equilibrium phase diagram of CH_4-N_2 solid mixtures. *Soviet J. Low Temp. Phys., 9,* 94–98.

Protopapa S., Boehnhardt H., Herbst T. M., Cruikshank D. P., Grundy W. M., Merlin F., and Olkin C. B. (2008) Surface characterization of Pluto and Charon by L and M band spectra. *Astron. Astrophys., 490,* 365–375.

Protopapa S., Grundy W. M., Tegler S. C., and Bergonio J. M. (2015) Absorption coefficients of the methane-nitrogen binary ice system: Implications for Pluto. *Icarus, 253,* 179–188.

Protopapa S. and 22 colleagues (2017) Pluto's global surface composition through pixel-by-pixel Hapke modeling of New Horizons Ralph/LEISA data. *Icarus, 287,* 218–228.

Putzig N. E. (2006) Thermal inertia and surface heterogeneity on Mars. Ph.D. thesis, Univ. of Colorado, Boulder, available online at *https://www.proquest.com/docview/305354067.*

Salama F. (1998) UV photochemistry of ices: The role of photons in the processing of ices. In *Solar System Ices* (B. Schmitt et al., eds.), pp. 259–280. Kluwer, Dordrecht.

Schaefer B. E., Buie M. W., and Smith L. T. (2008) Pluto's light curve in 1933–1934. *Icarus, 197,* 590–598.

Schenk P. M. and 19 colleagues (2018) Basins, fractures and volcanoes: Global cartography and topography of Pluto from New Horizons. *Icarus, 314,* 400–433.

Schmitt B. and 28 colleagues (2017) Physical state and distribution of materials at the surface of Pluto from New Horizons LEISA imaging spectrometer. *Icarus, 287,* 229–260.

Sicardy B. and 40 colleagues (2003) Large changes in Pluto's atmosphere as revealed by recent stellar occultations. *Nature, 424,* 168–170.

Sicardy B. et al. (2016) Pluto's atmosphere from the 2015 June 29 ground-based stellar occultation at the time of the New Horizons flyby. *Astrophys. J. Lett., 819,* L38.

Spencer J. R. and Moore J. M. (1992) The influence of thermal inertia on temperatures and frost stability on Triton. *Icarus, 99,* 261–272.

Spencer J. R., Lebofsky L. A., and Sykes M. V. (1989) Systematic biases in radiometric diameter determinations. *Icarus, 78,* 337–354.

Spencer J. R., Buie M. W., and Bjoraker G.L. (1990) Solid methane on Triton and Pluto: 3- to 4-μm spectrophotometry. *Icarus, 88,* 491–496.

Spencer J. R., Stansberry J. A., Trafton L. M., Young E., Binzel R. P., and Croft S. K. (1997) Volatile transport, seasonal cycles, and atmospheric dynamics on Pluto. In *Pluto and Charon* (S. A. Stern and D. Tholen, eds.), pp. 435–473. Univ. of Arizona, Tucson.

Stansberry J. A. and Yelle R. V. (1999) Emissivity and the fate of Pluto's atmosphere. *Icarus, 141,* 299–306.

Stansberry J. A., Yelle R. V., Lunine J. I., and McEwen A. S. (1992) Triton's surface-atmosphere energy balance. *Icarus, 99,* 242–260.

Stansberry J. A., Lunine J. I., Hubbard W. B., Yelle R. V., and Hunten D. M. (1994) Mirages and the nature of Pluto's atmosphere. *Icarus, 111,* 503–513.

Stansberry J. A., Pisano D. J., and Yelle R. V. (1996a) The emissivity of volatile ices on Triton and Pluto. *Planet. Space Sci., 44,* 945–955.

Stansberry J. A., Spencer J. R., Schmitt B., Benchkora A., Yelle R. V., and Lunine J. I. (1996b) A model for the overabundance of methane in the atmospheres of Pluto and Triton. *Planet. Space Sci., 44,* 1051–1063.

Stansberry J. A., Grundy W., and Young L. A. (2014) The influence of topography on volatile transport. *AAS/Division for Planetary Sciences Meeting Abstracts, 46,* 401.08.

Stern S. A. and Trafton L. M. (1984) Constraints on bulk composition, seasonal variation, and global dynamics of Pluto's atmosphere. *Icarus, 57,* 231–240.

Stern S. A. and Trafton L. M. (2008) On the atmospheres of objects in the Kuiper belt. In *The Solar System Beyond Neptune* (M. A. Barucci et al., eds.), pp. 365–380. Univ. of Arizona, Tucson.

Stern S. A., Trafton L. M., and Gladstone G. R. (1988) Why is Pluto bright? Implications of the albedo and lightcurve behavior of Pluto. *Icarus, 75,* 485–498.

Stern, S. A., Buie M. W., and Trafton L. M. (1997) HST high-resolution images and maps of Pluto. *Astron. J., 113,* 827.

Stern S. A., Weintraub D. A., and Festou M. C. (1993) Evidence for a low surface temperature on Pluto from millimeter-wave thermal emission measurements. *Science, 261,* 1713–1716.

Stern S. A. et al. (2015) The Pluto system: Initial results from its exploration by New Horizons. *Science, 350,* aad1815.

Stern S. A. and 13 colleagues (2017) Past epochs of significantly higher pressure atmospheres on Pluto. *Icarus, 287,* 47–53.

Strazulla G. (1998) Chemistry of ice induced by bombardment with energetic charged particles. In *Solar System Ices* (B. Schmitt et al., eds.), pp. 281–301. Kluwer, Dordrecht.

Sykes M. V., Cutri R. M., Lebofsky L. A., and Binzel R. P. (1987) IRAS serendipitous survey observations of Pluto and Charon. *Science, 237,* 1336–1340.

Tan S. P. and Kargel J. S. (2018) Solid-phase equilibria on Pluto's surface. *Mon. Not. R. Astron. Soc., 474,* 4254–4263.

Tegler S. C. and 13 colleagues (2019) A new two-molecule combination band as a diagnostic of carbon monoxide diluted in nitrogen ice on Triton. *Astron. J., 158,* 17.

Telfer M. W. and 19 colleagues (2018) Dunes on Pluto. *Science, 360,* 992–997.

Tholen D. J. and Buie M. W. (1997) Bulk properties of Pluto and Charon. In *Pluto and Charon* (S. A. Stern and D. Tholen, eds.), pp. 193–220. Univ. of Arizona, Tucson.

Toigo A. D., Gierasch P. J., Sicardy B., and Lellouch E. (2010) Thermal tides on Pluto. *Icarus, 208,* 402–411.

Toigo A. D., French R. G., Gierasch P. J., Guzewich S. D., Zhu X., and Richardson M. I. (2015) General circulation models of the dynamics of Pluto's volatile transport on the eve of the New Horizons encounter. *Icarus, 254,* 306–323.

Trafton L. M. (1980) Does Pluto have a substantial atmosphere? *Icarus, 44,* 53–61.

Trafton L. M. (1981) Pluto's atmospheric bulk near perihelion. *Adv. Space Res., 1,* 93–97.

Trafton L. M. (1984) Large seasonal variations in Triton's atmosphere. *Icarus, 58,* 312–324.

Trafton L. M. (1989) Pluto's atmosphere near perihelion. *Geophys. Res.*

Lett., 16, 1213–1216.
Trafton L. M. (1990) A two-component volatile atmosphere for Pluto I. The bulk hydrodynamic escape regime. *Astrophys. J., 359,* 512–523.
Trafton L. M. (2015) On the state of methane and nitrogen ice on Pluto and Triton: Implications of the binary phase diagram. *Icarus, 246,* 197–205.
Trafton L. M. and Stansberry J. A. (2015) On the departure from isothermality of Pluto's volatile ice due to local isolation and topography. *AAS/Division for Planetary Sciences Meeting Abstracts, 47,* 210.02.
Trafton L. M. and Stern S. A. (1983) On the global distribution of Pluto's atmosphere. *Astrophys. J., 267,* 872–881.
Trafton L. M., Stern S. A., and Gladstone G. R. (1988) The Pluto-Charon system: The escape of Charon's primordial atmosphere. *Icarus, 74,* 108–120.
Trafton L. M., Matson D. L., and Stansberry J. A. (1998) Surface/atmosphere interaction and volatile transport (Triton, Pluto, and Io). In *Solar System Ices* (B. Schmitt et al., eds.), pp. 773–812. Kluwer, Dordrecht.
Trafton L. M., Tan S., and Stansberry J. H. (2019) On the equilibrium state of Pluto's surface ice. In *Pluto System After New Horizons,* Abstract #7070. LPI Contribution No. 2133, Lunar and Planetary Institute, Houston.
Trowbridge A. J., Melosh H. J., Steckloff J. K., and Freed A. M. (2016) Vigorous convection as the explanation for Pluto's polygonal terrain. *Nature, 534,* 79–81.
Tryka K. A., Brown R. H., Anicich V., Cruikshank D. P., and Owen T. C. (1993) Spectroscopic determination of the phase composition and temperature of nitrogen ice on Triton. *Science, 261,* 751–754.
Tryka K. A., Brown R. H., Cruikshank D. P., Owen T. C., Geballe T. R., and de Bergh C. (1994) Temperature of nitrogen ice on Pluto and its implications for flux measurements. *Icarus, 112,* 513–527.
Umurhan O. M. and 14 colleagues (2017) Modeling glacial flow on and onto Pluto's Sputnik Planitia. *Icarus, 287,* 301–319.
Vangvichith M. (2013) Modélisation des atmosphères et des glaces de Pluton et Triton. Ph.D. thesis, Université Pierre et Marie Curie, Paris, available online at *https://www.theses.fr/2013EPXX0021.*
Vapillon L., Combes M., and Lecacheux J. (1973) The beta Scorpii occultation by Jupiter. II. The temperature and density profiles of the Jupiter upper atmosphere. *Astron. Astrophys., 29,* 135.
White O. L. and 24 colleagues (2017) Geological mapping of Sputnik Planitia on Pluto. *Icarus, 287,* 261–286.
Young E. F. (1993) An albedo map and frost model of Pluto. Ph.D. thesis, Massachusetts Institute of Technology, Cambridge. 125 pp.
Young E. F. and Binzel R. P. (1994) A new determination of radii and limb parameters for Pluto and Charon from mutual event lightcurves. *Icarus, 108,* 219–224.
Young E. F., Galdamez K., Buie M. W., Binzel R. P., and Tholen D. J. (1999) Mapping the variegated surface of Pluto. *Astron. J., 117,* 1063–1076.
Young E. F., Binzel R. P., and Crane K. (2001) A two-color map of Pluto's sub-Charon hemisphere. *Astron. J., 121,* 552–561.
Young E. F. and 13 colleagues (2008) Vertical structure in Pluto's atmosphere from the 2006 June 12 stellar occultation. *Astron. J., 136,* 1757–1769.
Young L. A. (2012) Volatile transport on inhomogeneous surfaces: I — Analytic expressions, with application to Pluto's day. *Icarus, 221,* 80–88.
Young L. A. (2013) Pluto's seasons: New predictions for New Horizons. *Astrophys. J. Lett., 766,* L22.
Young L. A. (2017) Volatile transport on inhomogeneous surfaces: II. Numerical calculations (VT3D). *Icarus, 284,* 443–476.
Young L. A., Elliot J. L., Tokunaga A., de Bergh C., and Owen T. (1997) Detection of gaseous methane on Pluto. *Icarus, 127,* 258–262.
Young L. A., Cook J. C., Yelle R. V., and Young E. F. (2001) Upper limits on gaseous CO at Pluto and Triton from high-resolution near-IR spectroscopy. *Icarus, 153,* 148–156.
Young L. and 15 colleagues (2015) Volatile transport implications from the New Horizons flyby of Pluto. *AAS/Division for Planetary Sciences Meeting Abstracts, 47,* 101.04.
Young L. A. and 25 colleagues (2018) Structure and composition of Pluto's atmosphere from the New Horizons solar ultraviolet occultation. *Icarus, 300,* 174–199.
Young L. A., Tan S. P., Trafton L. M., Stansberry J. A., Grundy W. B., Protopapa S., Schmitt B., Umurhan O. M., and Bertrand T. (2019) On the disequilibrium of Pluto's volatiles. In *Pluto System After New Horizons,* Abstract #7039. LPI Contribution No. 2133, Lunar and Planetary Institute, Houston.
Young L. A., Braga-Ribas F., and Johnson R. E. (2020) Volatile evolution and atmospheres of trans-neptunian objects. In *The Trans-Neptunian Solar System* (D. Prialnik et al., eds.), pp. 127–151. Elsevier, Amsterdam.
Zalucha A. M., Gulbis A. A. S., Zhu X., Strobel D. F., and Elliot J. L. (2011a) An analysis of Pluto occultation light curves using an atmospheric radiative-conductive model. *Icarus, 211,* 804–818.
Zalucha A. M., Zhu X., Gulbis A. A. S., Strobel D. F., and Elliot J. L. (2011b) An investigation of Pluto's troposphere using stellar occultation light curves and an atmospheric radiative-conductive-convective model. *Icarus, 214,* 685–700.
Zangari A. (2015) A meta-analysis of coordinate systems and bibliography of their use on Pluto from Charon's discovery to the present day. *Icarus, 246,* 93–145.

Strobel D. F. (2021) Atmospheric escape. In *The Pluto System After New Horizons* (S. A. Stern, J. M. Moore, W. M. Grundy, L. A. Young, and R. P. Binzel, eds.), pp. 363–377. Univ. of Arizona, Tucson, DOI: 10.2458/azu_uapress_9780816540945-ch015.

Atmospheric Escape

Darrell F. Strobel
The Johns Hopkins University

Of the historic solar system planets with surface pressures exceeding 1 μbar, Pluto's atmosphere was the least gravitationally bound, and originally thought to be escaping hydrodynamically or even unstable against blowoff. But when constrained by available solar power, estimated escape rates were in the range of (1–10) × 10^{27} molecules s^{-1}. Our actual knowledge of escape rates is limited to the period of the New Horizons flyby in July 2015, when Pluto was post-perihelion at ~32.9 AU from the Sun. The New Horizons data yielded a cold (~65–68 K), compact upper atmosphere (exobase radius ~2900 km) with escape rates of N_2 = (3–8) × 10^{22} s^{-1} and CH_4 = (4–8) × 10^{25} s^{-1} at very subsonic velocities, rendering the atmosphere essentially hydrostatic. For Charon the New Horizons upper limits on probable N_2 and CH_4 atmospheres yield solar energy limited escape rates of ~1 × 10^{25} molecule s^{-1}.

1. INTRODUCTION

Pluto's atmosphere has the distinction of being the least gravitationally bound atmosphere in the solar system at the 1-μbar level. It was widely believed to be escaping hydrodynamically (*Trafton,* 1980; *Trafton et al.*, 1997; *Krasnopolsky,*1999; *Strobel*, 2008b). [In this chapter, "hydrodynamic escape" means that the escape process can be described as a fluid characterized by an organized bulk outflow velocity. For slow hydrodynamic escape, the fluid bulk velocity at the top of the atmosphere (formally known as the exobase, rigorously defined later, where the atmosphere transitions from collisional to collisionless) is subsonic and characterized by gravitational potential energy that is much larger than the thermal energy that greatly exceeds the bulk flow kinetic energy. For fast (or full) hydrodynamic escape the bulk velocity equals or exceeds the speed of sound at the exobase.] The essential physics governing Pluto's hydrodynamic escape was thought to be found in the classic paper by *Parker* (1958) on coronal expansion into the solar wind. *Trafton* (1980) argued that Pluto's atmosphere was unstable against blowoff, i.e., fast hydrodynamic escape. *Hunten and Watson* (1982) noted, however, from the thesis research of Watson (*Watson et al.*, 1981) that there is an energy limit on atmospheric escape, and in the case of Pluto this is based on the absorption rate of solar extreme ultraviolet (EUV) and ultraviolet (UV) radiation. This energy-limited escape rate rendered Pluto's atmosphere stable against blowoff and they estimated an escape rate ~1 × 10^{28} CH_4 s^{-1}, as at the time only CH_4 ice had been detected on the surface (*Cruikshank et al.,* 1976) and CH_4 gas was assumed to be the major atmospheric constituent.

Parker (1964) derived special solutions for stellar coronas where their flows could become supersonic if they possessed large finite ratios of high gas density to thermal heat conductivity at the sonic point and even larger ratios at their lower boundary. He found a limiting case of supersonic expansion for asymptotically large density at the lower boundary that had a finite energy flux at infinity if the temperature decreased to zero there. Assuming this ratio is large implies in the total energy equation that the radial velocity is small, and that velocity squared terms may be neglected. In the limit that the ratio goes to infinity, the radial expansion velocity goes to zero, whereas for a finite ratio at infinity, the thermal heat conduction flux remains finite. In *Parker's* (1964) concluding remarks he suggested that for planetary coronas the flow would transition to supersonic "so far out as to be of no consequence." This solution is referred to as slow hydrodynamic escape in this chapter.

McNutt (1989) applied the *Parker* (1964) high-density limit to Pluto's hydrodynamic escape with an analytic treatment based on the ordering of non-dimensional terms in the total energy conservation equation. *McNutt* (1989) emphasized that Pluto's gravitational well is deep and significantly larger than its atmospheric enthalpy and substantially larger than the outflow kinetic energy below the exobase. For an admixture of CH_4-CO gas, he estimated an escape rate of ~8 × 10^{27} s^{-1}. Likewise, *Hubbard et al.* (1990) obtained a comparable CH_4 escape rate of ~1 × 10^{28} s^{-1}. *McNutt* (1989) omitted the enthalpy term in the total energy equation, c_pT, which represents the atmosphere's internal energy that can contribute to escape. *Krasnopolsky* (1999) included it rigorously above the EUV heating level and approximately below the EUV heating level in an atmosphere composed of N_2-CH_4 rather than CH_4-CO. *Krasnopolsky* (1999) underscored the importance of CH_4 absorption of solar Lyman-radiation in driving hydrodynamic escape, which he found to be in the slow outflow regime, and obtained escape rates of 1.55 × 10^{27}, 2.6 × 10^{27}, and 5.75 × 10^{27} N_2 s^{-1} for solar minimum, medium, and maximum conditions, respectively, with the CH_4 mixing ratio equal to 0.028. The corresponding CH_4 escape rates were in the range of $(1.1–1.2) \times 10^{27}$ s^{-1}. In the papers discussed so far, Pluto's EUV and UV heat inputs were represented by delta-functions rather than the more realistic Chapman functions in Pluto's atmosphere.

In the previous Pluto system volume in this series, the chapter by *Trafton et al.* (1997) gave an extensive discussion and review on escape processes from Pluto's atmosphere and Charon's potentially very tenuous atmosphere. They favored strongly the interpretation that Pluto's atmosphere was escaping hydrodynamically and that escape was energy limited in accordance with *Hunten and Watson* (1982). In contrast to prior studies, *Tian and Toon* (2005) relaxed many of the assumptions previous authors had made by solving the full equations for conservation of mass, momentum, and total energy for a one-component gas in one-dimensional radial hydrodynamic escape with distributed heating. They obtained a N_2 escape rate of $\sim 1.5 \times 10^{28}$ s^{-1} for solar minimum conditions at 30 AU, an order of magnitude larger than what *Krasnopolsky* (1999) calculated. Both studies had common lower boundary conditions and heating efficiencies for N_2 EUV absorption, but *Tian and Toon* (2005) omitted CH_4 heating. Their exobases were located at 10–13 r_p (r_p = 1400 km assumed, thus intersecting Charon's Hill sphere), where their flows were still very subsonic, Ma $\ll$ 1, and density profiles were almost hydrostatic. The escaping N_2 only went supersonic for radial distances greater than 30 r_p, although their fluid equations were no longer valid beyond the exobase. This predicted, inflated atmosphere of Pluto would hence present a large cross-sectional area to absorb solar EUV photons to heat the atmosphere, especially as the N_2 number density at the exobase was $\sim 10^5$ cm^{-3} at r ~ 16,000 km and the heating efficiency was assumed constant out to the exobase.

Common to all these previous studies was the adoption of a lower boundary condition for temperature (= 97 K) without any physics in the models that would yield this boundary temperature. The studies of *Krasnopolsky* (1999) and *Tian and Toon* (2005) adopted this temperature inferred from the stellar occultation measurements of *Elliot and Young* (1992) and *Millis et al.* (1993).

The question of whether Pluto's atmosphere was undergoing slow hydrodynamic, organized outflow driven below by thermospheric heating in the sense of *Parker's* (1964) high-density regime was addressed by *Tucker and Johnson* (2009). Although this paper treats the same hypothesized outflow from Titan's atmosphere, their analysis is equally applicable to Pluto. They adopted Direct Simulation Monte Carlo (DSMC) to describe atmospheric outflow, as this method makes no assumptions of how processes (e.g., convection, heat conduction, etc.) drive the outflow. They adopted heating rates supplied by *Strobel* (2008a) to fairly compare his predictions of a slow hydrodynamic escape model with their DSMC calculations. Their principal conclusion was that upward heat conduction at and from below the exobase cannot significantly enhance thermal escape beyond what one calculates by Jeans escape rate. Thus, slow hydrodynamic escape as defined by *McNutt* (1989), *Krasnopolsky* (1999), and *Strobel* (2008a,b) does not accelerate the thermal escape rate.

On the theoretical front the two papers by *Volkov et al.* (2011a,b) have clarified the general problem of thermal escape from planetary atmospheres and the transition, as the gravitational binding energy relative to thermal energy increases, from organized supersonic outflow to random thermal evaporation of individual atoms/molecules at the exobase known as Jeans escape. This transition is narrow and very abrupt. They also found that escape rates were enhanced over traditional Jeans escape rates by approximately a factor of 2 in thermal escape regimes for moderately gravitationally bound atmospheres such as Pluto's.

Tucker et al. (2012) followed up their study specifically for Pluto with a fluid model for its atmosphere above a radial distance of 1450 km (altitude = 260 km) and a DSMC model for the exobase region and exosphere. With atmospheric heating rates supplied by *Strobel* (2008b), *Tucker et al.* (2012) calculated for solar minimum conditions an N_2 escape rate of 1.2×10^{27} N_2 s^{-1} in comparison with 1.8×10^{27} N_2 s^{-1} (*Strobel,* 2008b), because both were constrained by the same energy limit on atmospheric escape. But the predicted structures of the upper atmosphere and exosphere had substantially different number density and temperature profiles, resulting in the locations of the exobase at r ~ 6200 km with T = 87K (*Tucker et al.,* 2012) and ~3700 km with T = 65 K (*Strobel*, 2008b). *Tucker et al.* (2012) estimated that ~1% of the escaping N_2 molecules would impact Charon.

Erwin et al. (2013) developed a hybrid fluid DSMC model, like *Tucker et al.* (2012), but added time dependence to the fluid equations to integrate solutions to steady state rather than solving steady-state equations. Their goal was to investigate whether Pluto's atmosphere was capable, with increasing solar heating, to make the transition from Jeans escape rates via slow hydrodynamic escape to full hydrodynamic escape. For plausible solar heating rates, their Pluto solutions never yielded escape rates in the transonic regime, where the Mach number approached 1. Both *Erwin et al.* (2013) and *Tucker et al.* (2012) confirmed that fluid equations are valid only up to Knudsen number (Kn, definition below) = 0.2 and to get up to the exobase (Kn = 0.7–1) requires alternate approaches in section 3. Both also confirmed that the effective escape rate for Pluto's atmospheric structure is about twice the Jeans escape rate, by comparison of escape rates calculated with a combination fluid-DSMC model and fluid-Jeans model.

In *Zhu et al.* (2014), the *Strobel et al.* (1996) one-dimensional radiative-conductive model for Pluto's stratospheric density and thermal structure was converted to a fluid model with a non-hydrostatic radial momentum equation associated with atmospheric escape of N_2. The energy equation was augmented with adiabatic cooling due to hydrodynamic expansion and solar far-ultraviolet (FUV) and EUV heating in the upper atmosphere as well as the previous improved treatment of energy transfer within the CH_4 molecule among vibrational levels (*Zalucha et al.,* 2011a,b). For the upper boundary conditions at the exobase a parameterization of enhanced Jeans escape rates was developed by offline kinetic DSMC model simulations (*Erwin et al.,* 2013). The key fundamental assumption was that the departures from thermal equilibrium at the molecular level near and above the exobase could be fully represented by this upper boundary condition. For the New Horizons Pluto flyby in July 2015, *Zhu et al.* (2014) predicted escape rates $\sim 3.5 \times 10^{27}$ N_2 s^{-1}, $\leq 1 \times 10^{26}$ CH_4 s^{-1}, exobase at 8 $r_0 \sim 9600$ km. Zhu et al. also investigated escape as a hydrodynamic process and found that Pluto is locked into an elevated Jeans escape regime and never makes a sharp transition to full hydrodynamic escape even with increases in solar heating by factors of 3. For their Pluto model atmosphere, the hydrodynamic approach underestimates the escape rate by about 13% but asymptotically approaches the enhanced Jeans escape rate under elevated energy input and/or when the exobase coincides with the transonic level, with a common escape rate $\geq 10^{28}$ N_2 s^{-1}.

During the New Horizons Pluto flyby period in July 2015, *Koskinen et al.* (2015) submitted a paper where they adapted their multispecies escape model for extrasolar planets to predict escape rates $\sim(1.8–2.3) \times 10^{27}$ N_2 s^{-1}, $\sim(1–2) \times 10^{26}$ CH_4 s^{-1}. Their total mass loss rate was energy (power) limited and analogous to expectations for close-in extrasolar giant planets. Like *Zhu et al.* (2014), their model yielded an expanded atmosphere with an exobase located at $\sim(5–6)$ $r_0 \sim$ (6000–7200) km. In summary, most models for the upper atmosphere, prior to the New Horizons flyby, were constrained by energy-limited solar EUV-FUV heating rates. Consequently, these models had low-Mach-number (Ma $\ll$ 1) outflow radial velocities and yielded N_2 escape rates in the range of $(1–10) \times 10^{27}$ s^{-1} $\sim$ $(3–30) \times 10^{28}$ amu s^{-1} and were more extended than inferred from New Horizons data discussed in section 2.

The purpose of this chapter is to revisit the question of escape rates from Pluto's upper atmosphere now that we have the scientific results from the New Horizons mission. Several topics discussed in this chapter were also reviewed in a recent book chapter on Titan atmospheric escape (*Strobel and Cui,* 2014) and are repeated here for background material.

Escape takes place mostly from a level in the upper atmosphere known as the exobase, where the transition from a collisional thermosphere to a collisionless exosphere occurs. Below the exobase the atmosphere can be treated as a fluid, because the mean free path, the distance a molecule or atom travels before making a collision, is shorter than the smallest macroscopic length scale, which is the pressure scale height, H, that characterizes the exponential decay of pressure with altitude. Above the exobase is a quasi-collisionless region known as the exosphere where the mean free path exceeds the atmospheric scale height. In the former region individual atoms and molecules undergo frequent collisions in macroscopic timescales, and in a volume H^3, there are enough atoms and molecules that, for all macroscopic purposes, the atmosphere can be viewed as a continuum fluid. In the latter region, collisions are sufficiently infrequent that atoms and molecules execute dynamical trajectories that are influenced mostly by Pluto's gravitational field.

The exobase is located where the mean free path is approximately equal to the atmospheric scale height. Its classical definition is where the escape probability is e^{-1} for an atom or molecule traveling upward in excess of the escape velocity without suffering a collision, and given by

$$\text{Probability} = \int_{\text{exobase}}^{\infty} \sigma n(r) dr = \exp[-\zeta(r_{\text{exobase}})^{-1}] = e^{-1} \quad (1)$$

where $\zeta(r) = [\sigma\ n(r)\ H(r)]^{-1}$ and σ is the collision cross section with a typical value of $\sigma \sim (3–4) \times 10^{-15}$ cm^2 for molecules. For hard-sphere elastic collisions, the mean free path, l(r), is given by $1/[\sqrt{2}\ \sigma\ n(r)]$. The above definition of the exobase implies that $l(r) = H(r)/\sqrt{2}$ at the exobase, whereas if the probability of escape were 50%, the two length scales would be equal. Atoms and molecules are not hard spheres with isotropic scattering independent of energy. When the particle's potential is steeply varying and characterized by forward scattering, the mean free path becomes $l(r) = 1/\ [0.5\ \sigma\ n(r)]$ and σ is the momentum transfer cross section, also known as the diffusion cross section (cf. *Johnson et al.*, 2008). In terms of column density above the exobase, $n(r_{exobase}) \times H(r_{exobase})$, it is now equal to $2\sqrt{2}/\sigma$, rather than just $1/\sigma$, as a consequence of a highly forward scattering cross section. Finally, it follows that the location of the exobase is constituent specific in a multi-species atmosphere, and, to be more precise, should be regarded as a transition region rather than a distinct level.

In molecular gas dynamics another relevant non-dimensional number is the local Knudsen number, $Kn(r) = l(r)/H(r)$. If the exobase is defined in terms of a probability of e^{-1}, then $Kn(r) = 1/\sqrt{2}$ at the exobase, whereas if the probability were 50%, then Kn = 1. In terms of the Knudsen number the Navier-Stokes equations describing the atmosphere as a continuum fluid break down when $Kn \geq (0.1–0.2)$ (*Bird,* 1994).

In a gravitationally bound atmosphere/exosphere, the key non-dimensional parameter governing escape is the Jeans λ parameter

$$\lambda(r) = \frac{v_{esc}^{\ 2}}{U(r)^2} = \frac{G\,M\,m}{r\,k\,T(r)} = \frac{r}{H(r)} = \frac{\text{gravitational PE}}{\text{random thermal KE}} \quad (2)$$

where G is Newton's gravitational constant, M is the mass of Pluto (or Charon), m is the mass of the atom/molecule, k is Boltzmann's constant, T(r) is the atmospheric temperature, U(r) is the most probable velocity $= \sqrt{2kT/m}$, v_{esc} is the escape velocity $= \sqrt{2GM/r}$, H(r) is the scale height (= kT/mg), and g is the gravitational acceleration (= GM/r^2). For the main constituents at Pluto's exobase ($r_{exo} \sim 2900$ km), the values of the Jeans λ parameter at the following temperature

T = 68 K are N_2 ~ 15; CH_4 ~ 8.5, indicating that they are moderately bound gravitationally [Fig. 1 and cf. Table 5 and section 8.4 in *Young et al.* (2018)].

2. REVIEW OF RELEVANT DATA

Because globally averaged escape rates cannot be directly measured, they must be calculated. The relevant data needed are the location of the exobase, the number density profiles of the escaping species, and the major constituent along with the temperature profile within a couple of scale heights of the exobase. The only specific data that we have for Pluto's upper atmosphere were acquired during the New Horizons Pluto flyby in July 2015. In comparison to expectations and predictions by various papers reviewed in the Introduction, Pluto's upper atmosphere was surprisingly cold and compact (*Stern et al.*, 2015; *Gladstone et al.*, 2016).

The two key experiments of the Pluto flyby were the radio occultation (REX), which yielded the N_2 density, pressure, and temperature profiles from the surface to ~110 km (*Hinson et al.,* 2017), and the Alice UV upper atmospheric solar occultation (*Young et al.*, 2018). These two sets of data allowed the construction of a model atmosphere consisting of the number density profiles of N_2 and CH_4 and the temperature profiles from the surface (radius r_p = 1190 km, actual average radius: 1189.9 ± 0.2 km) to r = 2300 km based on data (Fig. 1). For the model atmosphere to reach the exobase (r = 2900 km for Kn = 0.7) it was necessary to extrapolate the data about 600 km based on the assumption that upper atmosphere was isothermal (*Strobel and Zhu,* 2017), spherically symmetric (both REX and Alice ingress and egress data were the same within their error bars above r = 1220 km), and N_2 and CH_4 were essentially hydrostatic, in gravitational diffusive equilibrium.

The isothermal temperature at the exobase was constrained by the Alice data and theoretical considerations to be in the range of 65–68 K with a nominal model temperature of 67.8 K and a constraint on the upward CH_4 flux by a model temperature of 65 K (*Young et al.*, 2018). A temperature as low as 61 K can be ruled out as the corresponding N_2 scale height is too small to fit the Alice N_2 data above r = 2100 km. Likewise, a temperature of 71 K would present a challenge to connect the Alice data with REX data in the first 110 km and likewise ruled out.

With the New Horizons density and temperature structure illustrated in Fig. 1, *Young et al.* (2018) calculated escape rates of N_2 = (3–8) × 10^{22} s^{-1} and CH_4 = (4–8) × 10^{25} s^{-1}. The former was about a factor of 10^5 lower than pre-encounter predictions due to the rapid decrease in temperature from its peak value ~107 K at 1215 km to a broad secondary temperature minimum ~63 K between 1550 and 1800 km, which rendered the atmosphere more compact than expected, as well as a lower exobase with a cold temperature. For CH_4 the New Horizons inferred escape rates were only a factor of 2 lower than predicted as its escape rate was constrained by a principle known as the Hunten limiting rate (*Hunten,* 1973), discussed in section 4.1.

Above the exobase, the New Horizon spacecraft acquired only plasma data with the Pluto Energetic Particle Spectrometer Science Investigation (PEPPSI) (*Kollmann*

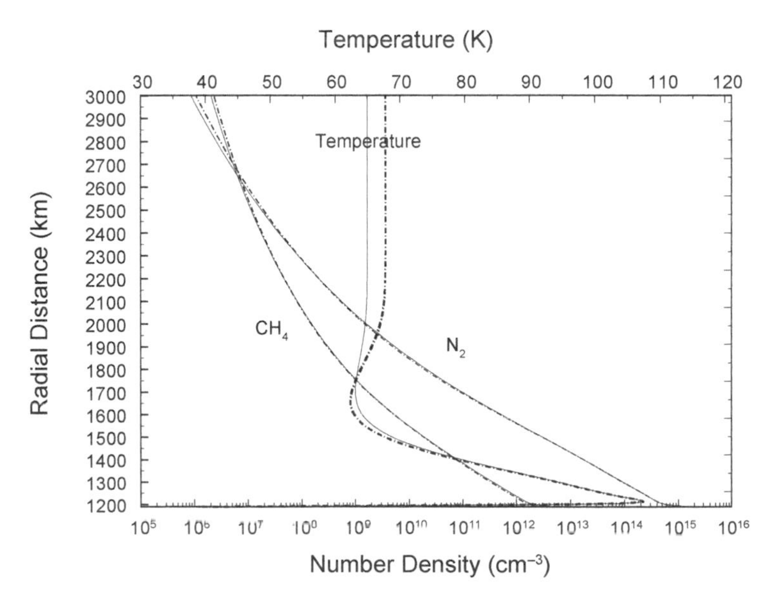

Fig. 1. Nominal (dash-dot line) and flux-constrained (solid line) Pluto atmosphere analytic temperature profiles consistent with N_2 and CH_4 line-of-sight abundances from the New Horizons Alice solar occultation and the REX radio occultation experiments. The N_2 density profiles were calculated assuming hydrostatic balance and the CH_4 density profiles were calculated for nominal profile (dash-dot lines) with an analytic expression for mixing ratio and for the flux constrained profile as a steady-state solution to the equations of continuity with eddy diffusion and the two species' molecular diffusion equations (see *Young et al.*, 2018, for details).

et al., 2019) and the Solar Wind Around Pluto (SWAP) (*McComas et al.*, 2016) instruments and particle data from the Venetia Burney Student Dust Counter (SDC). From the SWAP data, *McComas et al.* (2016) derived a CH_4^+ ion loss rate ~5 × 10^{23} s^{-1}, integrated along the New Horizons trajectory through Pluto's heavy ion tail, and approximately 1% of the calculated CH_4 escape rate from Alice data (*Young et al.*, 2018) (cf. section 4).

3. REVIEW OF ATMOSPHERIC ESCAPE

It has been known for a long time that atmospheric escape proceeds in an organized, hydrodynamic outflow for small values of λ (e.g., for comets) and by Jeans escape for large values of λ (e.g., giant and terrestrial planets), where individual atoms/molecules escape in a random fashion based on their leakage from the high-energy Maxwellian tail of their velocity distribution if their outward directed velocity exceeds the escape velocity in a quasi-collisionless exosphere (cf. *Hunten*, 1982). Until very recently, there were no definitive calculations concerning where the transition from hydrodynamic to Jeans escape occurs in λ space.

A rigorous solution to the full Boltzmann equation

$$\frac{\partial f_s}{\partial t} + \overleftarrow{v_s} \cdot \nabla f_s + \ddot{g} \cdot \nabla_{v_s} f = \left(\frac{\delta f_s}{\delta t} \right) \quad (3)$$

could answer the question but would be exceedingly difficult, as the lefthand side is a differential equation and the righthand side is an integral equation in both space and velocity variables. In practice one needs to solve it in a statistical manner. Here $f_s = f_s(\ddot{r}, \overleftarrow{v_s}, t)$ is the distribution function of species *s*, $\overleftarrow{v_s}$ is the vector velocity, and $\left(\frac{\delta f_s}{\delta t}\right)$ is the Boltzmann collision integral. Numerical approaches are described next.

3.1. Direct Simulation Monte Carlo Kinetic Model

The preferred statistical approach is known as the Direct Simulation Monte Carlo (DSMC) particle model, which solves the full non-linear Boltzmann equation including collisions and the physics of collisions by calculation of probabilities with enough test particles to get acceptable statistics (e.g., *Bird*, 1994). The key assumption is that an atmosphere/exosphere can be represented by a finite number of cells and in each cell, the large number of real atoms/molecules can be represented by many, but fewer, simulation particles. Thus, the atmosphere/exosphere can be divided into a number of cells, each of which contain enough simulation particles at each instance of time sufficient to represent properties of the much larger number of real particles. Each simulation particle has an individual internal energy, position, and velocity and is integrated forward in time in sufficiently small intervals that are shorter than the local collision time in each cell under the action of gravity and binary collision physics. After each time step the simulation particle has an updated internal energy, position, and velocity. With enough simulation particles, the spatial and time evolution of the distribution function, $f_s(\ddot{r}, \overleftarrow{v_s}, t)$, should closely approximate the rigorous solution to the non-linear Boltzmann equation. The creation and removal of exospheric atoms/molecules in elliptical orbits is fully accounted for in a DSMC calculation and hence this population of particles is completely characterized.

The two recent papers by *Volkov et al.* (2011a,b), the former with more emphasis on fundamental fluid physics and the latter on applications to planetary atmospheres, contain the most accurate, thorough solutions to the planetary escape problem. To accomplish this, about 1 billion test atoms/molecules were needed. The key result from their research is that transition from organized flow to random flow is abrupt. For atoms, which can be treated as elastic hard spheres (HS), the transition occurs in the range 2.1 < λ < 2.8; while for diatomic molecules, represented by the Larsen-Borgnakke variable hard sphere model (VHS-LB) (cf. *Bird*, 1994), the transition occurs at slightly larger values of λ, 2.4 < λ < 3.6, due to the two rotational degrees of freedom that can contribute to the expansion. Note that these results are restricted to a single constituent atmosphere with no internal heating and cooling processes, e.g., solar and plasma heating below and above the exobase The advantage of DSMC calculations is that they yield the same results as fluid calculations in the collision limit (Kn ≪ 1) as well as the same results as collisionless solutions to the Boltzmann equation, when Kn ≫ 1, in addition to the transition region characteristic of the exobase region (Kn ~ 1). The DSMC models do have an upper density limit, though, due to local collision times in each cell decreasing inversely as density. Higher densities demand ever-increasing computational resources. Thus, in practice one needs to couple a fluid model for the higher-density upper atmosphere with a DSMC model for the lower densities in the exobase region, where local thermodynamic equilibrium breaks down, as in *Tucker et al.* (2012) and *Erwin et al.* (2013), or offline as *Zhu et al.* (2014).

3.2. The Collisionless Boltzmann Equation Model: Thermal Jeans Escape in a Collisionless Exosphere

An analytic method is to treat the exosphere as collisionless immediately above the exobase, even though at the exobase the probability of escape is e^{-1}. Then the righthand side of equation (3) is thus 0 and this equation can be written as the Lagrangian derivative of the distribution function, $Df_s / Dt = 0$, and yields Liouville's theorem that $f_s(\ddot{r}, \overleftarrow{v_s}, t)$ is a constant along particle trajectories (*Herring and Kyle*, 1961; *Chamberlain*, 1963). An alternate approach is to solve the collisionless Boltzmann equation $Df_s/Dt = 0$ exactly, where the solution is a discontinuous, multivalued function of the constants of motion: conservation of energy and angular momentum (*Aamodt and Case,* 1962). With either approach, the distribution function is a truncated Maxwellian velocity distribution at the exobase, because incoming particles with velocities exceeding the escape velocity are assumed absent due to some unspecified sink far from a planet/satellite. Their absence prevents the eventual build-up of a full Maxwellian distribution above the escape level, and hence an extension of the barometric, hydrostatic law beyond the base of the exosphere. The truncated Maxwellian velocity distribution ensures that the integrated radial neutral column density in the exosphere is bounded by ~1.45 × $n(r_{exo})$ $H(r_{exo})$ (*Herring and Kyle,* 1961). Not included in these collisionless solutions are particles in elliptic orbits that never (cannot) intersect the exobase as there is no mechanism to create and remove these particles. The density for a spherically symmetric exosphere may be written with subscript "**0**" denoting evaluation at exobase $\mathbf{r_0}$

$$n_{exo}(r) = n_0\left[\exp\left[-\lambda_0\left(1-\frac{r_0}{r}\right)\right]\left(1-\frac{1}{2}\mathrm{erf\,c}\left(\sqrt{\lambda_0\frac{r_0}{r}}\right)\right)\right]$$

$$-n_0\left[\sqrt{1-\left(\frac{r_0}{r}\right)^2}\exp\left(-\frac{\lambda_0}{1+\frac{r_0}{r}}\right)\left(1-\frac{1}{2}\mathrm{erf\,c}\left(\sqrt{\frac{\lambda_0\frac{r_0}{r}}{1+\frac{r_0}{r}}}\right)\right)\right] \quad (4)$$

$$-n_0\left[\sqrt{\frac{\lambda_0\frac{r_0}{r}}{\pi}}\left[\left(1-\sqrt{1-\frac{r_0}{r}}\right)\exp(-\lambda_0)\right]\right]$$

for particles in ballistic and outgoing hyperbolic orbits only (*Herring and Kyle,* 1961). The subscript "0" denotes values at the lower boundary and n is the number density which is a function of λ and/or also r; no specific numerical value is implied. Higher order velocity moments of a distribution function solution to the collisionless Boltzmann equation have substantial anisotropies in the pressure and temperature components with the "horizontal components" asymptotically approaching 0 as $r \to \infty$, whereas the radial components asymptotically approach constant positive values (*Lemaire,* 1966; *Volkov et al.*, 2011a).

For particles in escaping hyperbolic orbits, $\mu > 0$, where μ is cosine of the angle between the particle velocity and the radial outward direction, the classic Jeans escape flux follows directly from the first velocity moment of the collisionless distribution function solution, f_{exo}, and is in terms of quantities evaluated at the exobase

$$\Phi(r\to\infty) = \iiint v\mu f_{exo} d^3v = n_{exo}(r_0)\frac{nv_{th}}{4}(1+\lambda_0)e^{-\lambda_0} \quad (5)$$

where the integration is over the velocity volume $d^3v = 2\pi v^2\, dv d\mu$, $v_{th} = \sqrt{8kT/\pi m}$ is the thermal speed, and the transmission of escaping particles from the exobase is assumed to be unity, in spite of the definition of the exobase as the level where the transmission is e^{-1}.

Note that here f_{exo} is a truncated Maxwellian velocity distribution that includes only particles with ballistic and escaping hyperbolic trajectories and excludes those with incoming hyperbolic trajectories and those in elliptical orbits. The usual derivation of the classic Jeans escape flux, the righthand side of equation (5), adopts a full Maxwellian velocity distribution with no bulk radial velocity and evaluates the escape flux at the exobase and not infinity. Jeans escape takes place only when there exists an exobase located at Kn ~ 1, where $\lambda > 2.8$ for atoms and $\lambda > 3.5$ for diatomic molecules.

3.3. Fluid/Hydrodynamic Models

In the fluid approach, one takes various velocity moments of solutions to the Boltzmann equation (equation (3)) to obtain the equations for conservation of mass, momentum, and energy, and with higher moments, equations for the pressure tensor and heat flow vector (cf. *Chapman and Cowling,* 1970; *Schunk and Nagy,* 2009). Closure of these equations requires assumptions about the velocity distribution function. Generally, it is assumed that solutions $f_s(\vec{r}, \vec{v_s}, t)$ are small departures from the zeroth term, a Maxwellian distribution, which implies that collision frequencies are much larger than macroscopic frequencies or, equivalently, the interval between particle collisions is much shorter than any macroscopic timescale. In terms of length scales, this assumption is stated as Kn < 0.2 (*Bird,* 1994). For the Navier-Stokes fluid equations, this implies a collision-dominated fluid where the stress tensor and heat flow vector terms are much smaller than the density, velocity, and temperature terms (*Schunk and Nagy,* 2009).

Strobel (2008b) invoked the high-density, slow-outflow regime of *Parker* (1964) and *Krasnopolsky* (1999) as the most plausible mechanism to explain a Pluto N_2 escape at a rate greater than 10^{26} s^{-1} powered by net solar UV heating due to mostly CH_4 absorption. But *Strobel* (2008b) and *Krasnopolsky* (1999) did not provide a mechanism to accelerate molecules to velocities required to escape Pluto's gravitational pull, before collisions in the exobase/lower exosphere region became too infrequent. In fact, their solutions at the exobase had Mach numbers ≪ 1, were essentially hydrostatic, and had bulk velocities that were insufficient to escape Pluto's gravitational potential well, thus providing confirmation that the high-density, slow-outflow regime of *Parker* (1964) was not applicable to planetary escape.

The key problem for fast hydrodynamic escape is that atoms/molecules must be accelerated to supersonic velocities at a critical point before the atmosphere becomes essentially a collisionless fluid (Kn ~ 0.2). For hydrodynamic escape with a bulk outflow velocity, v, driven by heating of the atmosphere, the relevant steady-state equations in non-dimensional variables are (cf. *Watson et al.*, 1981; *Johnson et al.*, 2008)

$$r^2\, n\, v = F\,, \qquad \text{mass} \quad (6)$$

$$\frac{d\psi}{d\lambda} = 2\,\frac{\dfrac{d\tau}{d\lambda} + 2\dfrac{\tau}{\lambda} - 1}{\dfrac{\tau}{\psi} - 1}\,, \qquad \text{momentum} \quad (7)$$

$$\frac{d}{d\lambda}\left(\lambda - \tilde{c}_p\tau - \frac{1}{2}\psi - \frac{\tau^s}{\zeta}\frac{d\tau}{d\lambda}\right) = \frac{(r_0\lambda_0)^3}{\lambda^4}\,\frac{Q(n\,(\lambda,\mu_i))}{F\,kT_0}\,, \qquad \text{energy} \quad (8)$$

where Q is the net heating/cooling, a function of n, the mixing ratios of radiatively active constituents, μ_i,

$$\lambda = \frac{GMm}{rkT_0},\ \tau = \frac{T}{T_0},\ \psi = \frac{mv^2}{kT_0} = \left(\frac{v}{c}\right)^2,\ \zeta = \frac{Fk}{\kappa_0 r_0 \lambda_0},\ \tilde{c}_p = \frac{c_p}{k/m} \quad (9)$$

where $c = (kT_0/m)^{1/2}$ is the isothermal speed of sound, c_p, the specific heat at constant p, $\kappa = \kappa_0\,(T/T_0)^s$ is the thermal conductivity, and F the outward flux. The subscript "0" denotes values at the lower boundary.

The equation of motion, equation (7), has a singularity at $\tau = \psi$, where the kinetic energy is equal to the thermal energy. To remove the singularity and go supersonic in steady-state, the numerator of this equation must vanish: $d\tau/d\lambda + 2\,\tau/\lambda - 1 = 0$, which holds if the flow were adiabatic ($Q = 0$, $\kappa = 0$, $\zeta = \infty$) at $\tau = \psi$. This implies that thermal heat conduction must vanish at the critical point. But heat conduction was hypothesized to deliver the energy to escape in slow hydrodynamic escape (e.g., *McNutt,* 1989; *Krasnopolsky,* 1999; *Strobel,* 2008a,b). According to *Volkov et al.* (2011a,b), heat conduction, which depends on collisions, becomes extremely inefficient in transporting heat as the exobase is approached. Based on theory (*Zhu et al.*, 2014), molecules are still expanding at low Mach numbers, $\psi \ll 1$, near Pluto's exobase. Thus, a critical point cannot be reached in the collisional atmosphere and slow hydrodynamic escape cannot be the mechanism to account for large N_2 and CH_4 outflow rates in excess of the enhanced Jeans escape rates.

The total energy equation for a spherically symmetric, ideal gas atmosphere when only gravity is an external force can be written as

$$\frac{1}{r^2}\frac{\partial}{\partial r}\left[r^2\left\{\rho\, v\left(\frac{1}{2}v^2 + c_p T + \Phi_g\right) - \kappa\frac{\partial T}{\partial r}\right\}\right] = Q \quad (10)$$

where gravity is written as a gradient of a potential field, $\nabla\Phi_g = GM\,\vec{r}\,/\,r^3$, and $h = c_pT$ is the enthalpy. In a conservative, dissipation-less atmosphere, ($Q = 0$, $\kappa = 0$), the conservation of total energy reduces with mass conservation ($r^2\,\rho\,v = r_0^2\rho_0\,v_0$ = constant) to Bernoulli's equation for potential flow of a compressible gas

$$\frac{1}{2}v^2 + c_pT + \Phi_g \approx c_pT_0 + \Phi_{g0} \quad (11)$$

where at very low altitudes, the first term on the lefthand side is much smaller than the other terms and the conserved total energy is the righthand side of equation (11) evaluated at a suitably low altitude. If $v(r = \infty) > 0$, the atmosphere is escaping. The threshold conditions for atmospheric blowoff are $v(r = \infty) = 0$ at $r = \infty$ and $h_0 = -\Phi_{g0}$, if all the enthalpy of the atmosphere were converted into directed radial outflow kinetic energy that is subsequently converted into gravitational potential energy. This requires that the temperature decreases to $T = 0$ from T_0, which is only possible for adiabatic, isentropic supersonic expansion into a vacuum. The condition $h_0 = -\Phi_{g0}$ can be rewritten as

$$\frac{GM\,/\,r_0}{c_pT_0} = \frac{R\lambda_0}{c_p} = 1 \quad (12)$$

where R = k/m, the gas constant, and it is assumed that as the atmosphere expands into a vacuum its internal energy is accessible and convertible into translational energy with escape velocity $v_{esc} = (c_p/R)^{0.5}\,U = (c_p/R)^{0.5}\,(2kT/m)^{0.5}$. For atoms, $v_{esc} = 1.6\,U$, $\lambda_0 = 5/2$, and for diatomic molecules, $c_p = 7/2\,R$, $v_{esc} = 1.8\,U$, $\lambda_0 = 7/2$. Integration of equation (10), assuming mass conservation, yields

$$\left(r_0^2\rho_0 v_0\right)\left[\left(\frac{1}{2}v^2 + c_pT + \Phi_g\right)\right]_{r_0}^{r_{Kn=0.2}} - \left[r^2\kappa\frac{\partial T}{\partial r}\right]_{r_0}^{r_{Kn=0.2}} = \int_{r_0}^{r_{Kn=0.2}} Q r^2 dr \quad (13)$$

where $r_{Kn\,=\,0.2}$ is the radial distance above which the fluid equations are no longer valid, and, in most atmospheres, the radiative heating and cooling is negligible. Thus, the mass escape rate is proportional to the integrated net heating of the atmosphere by solar radiation, mostly UV absorbed in the upper atmosphere and converted with some efficiency into heat. Equation (13) illustrates mathematically what quantities are needed to estimate the energy-limited escape rate. While thermal heat conduction only appears as boundary conditions in equation (13), it can redistribute radiative heating and cooling within the atmosphere in a manner that may affect the net heating (righthand side of equation (13)) available to drive escape.

The calculation of the actual heating efficiency is not a trivial exercise (e.g., *Stevens et al.*, 1992; *Krasnopolsky*, 1999; *De La Haye et al.*, 2008), and their consensus heating efficiencies are approximately 0.5 and 0.25 for CH_4 and N_2, respectively. In addition, any radiative cooling process can lower the available power to drive escape. This in fact was the surprising result inferred from the New Horizon solar occultation data that Pluto has a cold, compact upper atmosphere (*Stern et al.*, 2015; *Gladstone et al.*, 2016; *Young et al.,* 2018).

3.4. Non-Thermal Escape Processes

The discussion so far has focused on thermal processes where the atoms and molecules are close to thermal equilibrium. The situation at Pluto is more complex than this because although Pluto is not imbedded in a magnetosphere, it is potentially exposed to the solar wind. However, from the New Horizons SWAP data, we find that there is a bow shock and boundary region called the "Plutopause" that directs solar wind plasma (protons and alpha particles) around Pluto (*McComas et al.*, 2016). Only the heavy ions, produced from escaping N_2 and CH_4 molecules by charge exchange with protons and photoionization, can penetrate through these barriers to form an extended heavy ion tail downstream (*Barnes et al.*, 2019). Thus, in the atmosphere, N_2 and CH_4 molecules are mostly exposed to energetic photoelectrons, which can dissociate molecules and can ionize atoms and molecules, but this occurs too deep in the atmosphere below the exobase to sputter the atmosphere away. Energetic electrons can dissociate N_2 and CH_4 to produce translationally hot N atoms, and dissociative recombination of N_2^+ ions also yield hot N atoms, such as

$$e^* + (N_2, CH_4) \rightarrow e + (N^* + N^*, CH_3 + H^*)$$
$$N_2^+ + e \rightarrow N^* + N^*$$

where "*" denotes translationally hot. The hot atoms can continue these interactions until their energy is sufficiently degraded and they become thermalized. Generally, the production of hot recoils increases with decreasing altitude, while the probability of escape decreases. These two factors lead to the maximum generation of escaping atoms ~1 atmospheric scale height (~150 km) below the exobase, whereas the peak production of energetic neutrals is more than 1000 km below the exobase.

The number of pickup heavy ions, produced by solar wind protons that have charge exchanged with escaping CH_4 molecules, that can penetrate the atmosphere is unknown. But *Cravens and Strobel* (2015) have predicted that these heavy pickup ions could further undergo charge exchange with neutrals in Pluto's upper atmosphere/exosphere and yield a slower cold ion and generate a translationally hot neutral. The latter can potentially escape depending on its path, but generally will be forward scattered into the atmosphere. The new cold ions will thermalize with atmospheric neutrals.

4. ESCAPE RATE CALCULATIONS

4.1. Pluto

Both *Erwin et al.* (2013) and *Tucker et al.* (2012) confirmed that the fluid equations are valid only up to Kn = 0.2; to get up to the exobase (Kn = 0.7–1) requires alternate approaches. For the reported escape rates in *Young et al.* (2018), we adopted the method of *Zhu et al.* (2014), based on parameterizations from offline kinetic DSMC model simulations that form an upper boundary condition at the exobase. The fluid equations are solved up to the Kn = 0.7 exobase, where an enhanced Jeans escape rate is applied for the boundary conditions on CH_4 and N_2

$$F_{\text{enhanced J}} = \Gamma(\lambda) F_J = \left[\Gamma(\lambda)\frac{nv_{th}}{4}(1+\lambda)e^{-\lambda}\right]_{r=r_{exo}} \quad (14a)$$

$$\Gamma(\lambda) = \exp\left(4.086446916 - 0.870018743 + \lambda\right) + \exp\left(0.63554463 - 0.019922313 + \lambda\right) \quad (14b)$$

The quantity $\Gamma(\lambda)$ is a numerical curve fit to this function plotted in Fig. 1 of *Zhu et al.* (2014). One should not construe the accuracy of enhancement factor by the number of significant figures in the expression, as it is based on model runs with the hybrid model of *Erwin et al.* (2013) for five values of λ in a pure N_2 atmosphere and is most accurate between λ = 4.5–11.

The Jeans escape rate is enhanced here over the normal Jeans escape rate when λ ~ 3–15 at the exobase because molecules no longer possess an equilibrium Maxwellian distribution due in part to a nonnegligible bulk velocity (cf. Fig. 2 of *Volkov et al.*, 2011a). Consequently, the molecules are characterized by an outward drifting Maxwellian velocity distribution function that yields an enhanced escape rate (their Fig. 3). Also, enhancement of the Jeans escape rate occurs, principally in some cases, due to collisions above the exobase (*Tucker et al.*, 2015). In Fig. 2 both $\lambda(T)$ and $\Gamma(\lambda(T))$ are illustrated as a function of temperature for CH_4 at its exobase radius of 2920 km. For the "nominal" Pluto atmosphere with T = 67.8 K (cf. Fig. 1) at the exobase, we have r_{exo} = 2780, 2920 km; λ = 15, 8.5; $\Gamma(\lambda)$ = 1.4, 1.6; $n_{rexo} = 3.4 \times 10^6, 2.9 \times 10^6$ cm^{-3}; $4\pi\, r_{exo}^2\, F_{\text{enhanced J}}$ =

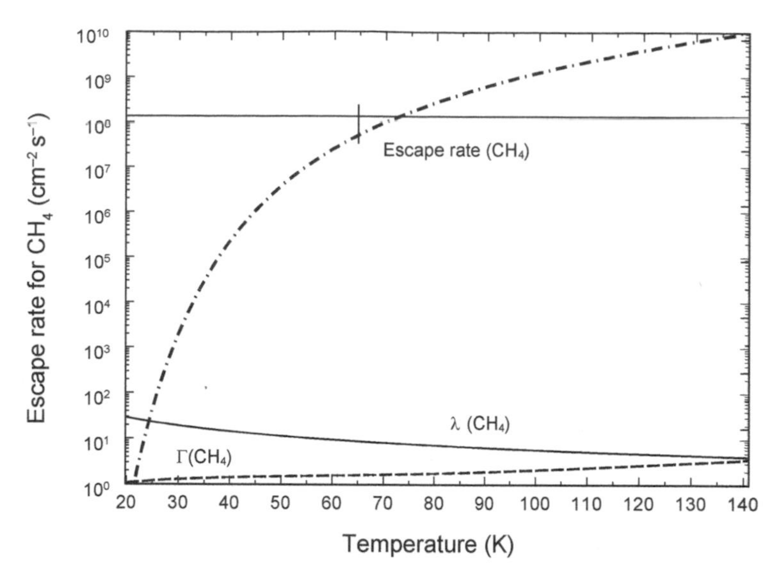

Fig. 2. Pluto escape calculations for CH_4, at an exobase radius = 2920 km, of its Jeans λ parameter (equation (2)), its enhanced escape parameter Γ (equation (14b)), and its escape rate/flux (cm^{-2} s^{-1}, dash-dot line) at 2920 km as a function of temperature for a nominal exobase density = 3×10^6 cm^{-3}. The solid horizontal line at 1.35×10^8 cm^{-2} s^{-1} is the Hunten limiting flux constraining the upward flow of CH_4 due to escape and photochemical loss for CH_4 volume mixing ratio = 0.003. The short vertical line at T = 65 K intersects the NH escape rate at 5.9×10^7 cm^{-2} s^{-1}. The surface area of the exobase is 1.1×10^{18} cm^2 and the NH actual escape rate ~5.5×10^{25} s^{-1} in comparison to the limiting rate of 1.4×10^{26} s^{-1}.

8×10^{22}, 7.6×10^{25} s^{-1} respectively for N_2 and CH_4. The "flux constrained" atmosphere assumes that Pluto's atmosphere, as it travels along in its highly elliptical orbit around the Sun, adjusts its density profiles and upward fluxes driven by photodissociation, photoionization, and escape to satisfy the steady-state balance equations of continuity and momentum at each instance in time. For this atmosphere with T = 65 K (cf. Fig. 1) at the exobase, we have r_{exo} = 2740, 2860 km; λ = 16, 9; $\Gamma(\lambda)$ = 1.4, 1.6; $n_{exo}(r_{exo}) = 3.8 \times 10^6$, 3.1×10^6 cm^{-3}; $4\pi\, r_{exo}^2\, F_{enhanced\,J} = 3.4 \times 10^{22}$, 4.6×10^{25} s^{-1} respectively for N_2 and CH_4. Note that all entries have been rounded off to at most two significant figures, except exobase locations, to communicate the level of accuracy given the time constants driving Pluto's atmosphere. In Fig. 2 one can infer the sensitivity of the CH_4 escape rate to temperature when the exobase radius and surface area is held fixed.

Because CH_4 is a minor constituent throughout most of the atmosphere, except near the exobase, there is the additional constraint of the maximum rate that it can diffuse through the major constituent N_2 to reach the exobase. This maximum rate is known as the Hunten limiting (flux) rate (*Hunten,* 1973) and for CH_4 it is given by

$$(CH_4) = 4\pi r^2 \mu(CH_4)\, n(N_2)\, D_{CH_4\text{-}N_2}\left(\frac{1}{H_{N_2}} - \frac{1}{H_{CH_4}}\right) \quad (15)$$

where the CH_4 mixing ratio, $\mu(CH_4)$, is constant when it is diffusing upward at the maximum rate. The molecular diffusion coefficient, $D_{CH_4\text{-}N_2}$, has $n(N_2)$ in its denominator to cancel the N_2 density dependence in the equation and $D_{CH_4\text{-}N_2}\,(1/H_{N_2} - 1/H_{CH_4}) \propto T^{-0.25}\, r^{-2}$ for a weak temperature dependence, and the r^{-2} from the scale height terms cancels the explicit r^2 in the equation. The limiting choke point in the atmosphere generally occurs in the vicinity of the homopause. This limit is shown in Fig. 2 for $\mu(CH_4) = 0.003$ with a very weak dependence on temperature when the exobase radius and surface area are held fixed and may inhibit excessive escape of light minor species.

For the range of $\lambda > 4.5$ (Jeans and enhanced Jeans escape regimes), the exact location of the exobase is not critical as atmospheric escape is at low Mach number outflow and the atmosphere is essentially hydrostatic, isothermal, and the density variation in the vicinity of the exobase is $n(r) = n_0 \exp(\lambda - \lambda_0)$. Thus, the escape rate becomes

$$4\pi r_{exo}^2\, F_{enhanced\,J} \simeq 4\pi r_{exo}^2\, n_0 e^{\lambda_{r_{exo}} - \lambda_0}\left[\Gamma(\lambda)\frac{v_{th}}{4}(1+\lambda)e^{-\lambda}\right]_{r = r_{exo}} \quad (16)$$

with the exponential dependence on λ removed and only a very weak dependence on λ, as $\lambda(r_{exo}) \propto 1/r_{exo}$, in the vicinity of the exobase from the quantity $\Gamma(\lambda)(1+\lambda)$, which is bounded by 14 and 22 over the λ range of 4–15 (Fig. 2).

For the New Horizons epoch, the inferred escape rates are N_2 = (3–8) $\times 10^{22}$ s^{-1} and CH_4 = (4–8) $\times 10^{25}$ s^{-1}. The range reflects the sensitivity to and uncertainty of the extrapolated number densities and temperature at the exobase from New Horizons solar occultation data acquired some 600 km below.

The New Horizons results are in sharp contrast with the predictions for the July 2015 Pluto flyby of escape rate ~3.5×10^{27} N_2 s^{-1}, exobase at 8 r_P ~ 9600 km, with

Jeans $\lambda \sim 5$ by *Zhu et al.* (2014). Zhu et al. applied the Hunten limiting flux principle (*Hunten,* 1973) with a CH_4 mixing ratio of 0.0025 (close to nominal New Horizons value of 0.003) and estimated a maximum escape rate of $\sim 1 \times 10^{26}$ CH_4 s^{-1}, consistent with the derived rate given above. At the actual exobase of 2920 km ($r_P \sim 2.45$) the atmospheric temperature is ~(65–68) K, not the predicted ~95 K. Although *Zhu et al.* (2014) included CO rotational line cooling with the correct CO mixing ratio ~500 ppmv (*Lellouch et al.*, 2017), their predicted upper atmosphere was still too hot despite significant adiabatic cooling associated with their large escape rate. The addition of supersaturated HCN cooling constrained by ALMA observations (*Lellouch et al.*, 2017) was still not enough to bring the temperature down to the solar occultation inferred values, but yielded temperatures ~105 K at 2920 km and N_2 escape rates $>2 \times 10^{27}$ s^{-1}at the exobase with calculated location at ~5500 km in the model.

Strobel and Zhu (2017) proposed one possible solution to cooling Pluto's upper atmosphere, namely a modest influx of H_2O molecules $\sim 3.4 \times 10^{19}$ s^{-1}, a volatile component of the dust grains intercepted by Pluto and detected by New Horizons' Venetia Burney Student Dust Counter compared with the actual flux of 1.5×10^{21} H_2O s^{-1} or 3.8 kg day^{-1} (*Poppe and Horanyi*, 2018). The problem with H_2O molecules cooling Pluto's atmosphere is H_2O must be highly supersaturated by some 15–18 orders of magnitude based on ambient temperatures, although during ablation their temperature would be elevated. Regarding this, from ALMA observations we know that HCN can be supersaturated by 7 orders of magnitude in the upper atmosphere (*Lellouch et al.*, 2017). *Rannou and West* (2018) state that "Because H_2O vapor pressure is lower than HCN vapor pressure, we expect that its nucleation is even more difficult than for HCN and that high supersaturation may also be possible." They argue that supersaturation by 15 orders of magnitude is possible, but not by a factor of almost 10^{18}. Even with ALMA located in the very dry Atacama Desert in Chile, it would be exceedingly difficult and more probably virtually impossible to detect any water vapor in Pluto's atmosphere through the overlying water vapor in Earth's atmosphere above ALMA.

As an alternative explanation for the cold temperatures, *Zhang et al.* (2017) reasoned that Pluto's haze particles can explain the cold 65–68 K upper atmosphere, if the complex indices of refraction are the same measured values of *Khare et al.* (1984) for Titan laboratory analog particles (a.k.a. tholins). The critical requirement is that the haze particles have very low, continuous single-scattering albedos throughout the thermal infrared. Hydrocarbon ices do not meet this requirement. If the haze particles do possess continuous low single-scattering IR albedos, then Pluto would appear increasingly brighter than its surface blackbody spectrum starting at wavelengths shorter than 25 μm and reaching several orders of magnitude by 15 μm. Going forward, the James Webb Space Telescope's Mid-Infrared Instrument imager should be able to detect this infrared excess on Pluto and distinguish between the effects of haze vs. cooling by water rotational lines.

For a single-component atmosphere, Fig. 3 for CH_4 and Fig. 4 for N_2 illustrate Jeans escape rates in molecules per second from Pluto respectively as a function of the exobase location, temperature in units of degree K divided by their mass in amu. For example, for CH_4, T/amu = 5 represents a temperature of 80 K. The exobase density is estimated from the definition of the Jeans λ parameter at the exobase, $\lambda_{exo} = r_{exo}/H(r_{exo}) = (GMm/kT_{exo}\, r_{exo})$ (cf. equation (2)), in combination with the definition of the location of the exobase, $l(r) = 1/[\sqrt{2}\, \sigma\, n(r)] = H(r)/\sqrt{2}$, to obtain $n(r_{exo}) = 1/\sigma\, H(r_{exo})$. One notes the general trend that to generate larger escape rates one needs higher temperatures and/or expanded atmospheres with very large exobase radii. *Zhu et al.* (2014) predicted a N_2 escape rate of 3.5×10^{27} s^{-1}, for $\lambda = 5.3$, T/amu = 2.1, exobase radius = 9440 km. From Fig. 4 one infers approximately the correct escape rate. But one must be cautious to use these figures for inferring escape rates from the New Horizon data where the CH_4 and N_2 densities are comparable in the vicinity of the exobase and the above expression for the exobase density, $n(r_{exo})$, will overestimate the individual density of each escaping constituent and the location of the exobase. In this situation overestimates by up to a factor of 4 are possible and the reader is encouraged to do an actual calculation for a specific atmosphere.

These contour plots can be adapted to other molecules. For example, H_2O at 18 amu, the T/amu = 5 on the x axis in the CH_4 Fig. 3 corresponds to T = 90 K, etc. For heavier molecules one can use Fig. 4 for N_2 and scale appropriately. But it is important to confirm that the Jeans parameter λ is greater than 4.5 at the exobase to obtain a valid escape rate. And for minor constituents there is the additional constraint on the maximum rate that any gas can diffuse through the major constituent to reach the exobase, i.e., the Hunten limiting rate (*Hunten,* 1973), given for CH_4 by equation (15) and applicable to other lighter than N_2 species by obvious substitution of their mole fractions, diffusion coefficients through N_2, and their scale heights.

As an aside, Pluto's atmosphere provides an important lesson for studying escape from exoplanetary atmospheres where it is common to use the energy-limited escape rate principle (*Hunten and Watson,* 1982) to calculate the escape rate (e.g., *Lammer et al.,* 2009). Generally, the composition of the atmosphere is limited to the major species and the power is assumed to be supplied by the major species absorbing stellar EUV, FUV, and maybe soft X-rays. Estimates must be made for heating efficiencies. On Pluto, the minor constituent CH_4 is a more important heat supplier than N_2 because it absorbs solar radiation over a broader portion of the UV spectrum and the fact that its heating efficiency is 50% in comparison to 25% for N_2. In exoplanetary atmospheres few minor constituents are known with any certainty and less so their density distributions.

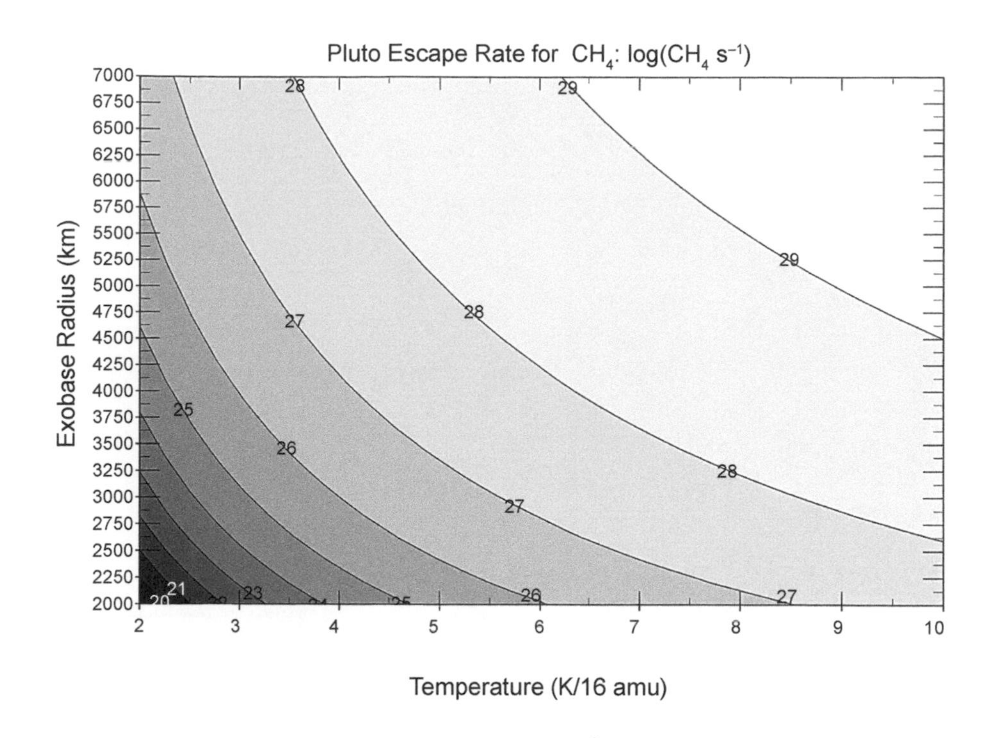

Fig. 3. Pluto escape rates for CH_4 (log CH_4 s^{-1}) as a function of exobase radius (km) and temperature (K amu^{-1}; e.g., for CH_4, T/amu = 5 ⇒ 80 K). This contour plot can be adapted to other molecules. For example, H_2O at 18 amu, T/amu = 5 on the x axis corresponds to T = 90 K, etc.

Even after the New Horizons Pluto flyby, we still do not know with certainty what renders its upper atmosphere colder than expected and reduces its overall escape rate from the predicted N_2 rate of 3.5×10^{27} s^{-1} and estimated CH_4 rate of 1×10^{26} s^{-1} for a total of 10^{29} amu s^{-1} to 5×10^{22} N_2 s^{-1} and 6×10^{25} CH_4 s^{-1} for a total of 10^{27} amu s^{-1}.

4.2. Charon

New Horizons performed a solar occultation experiment and a search for airglow emissions during its Charon flyby in July 2015 (*Stern et al.,* 2017). This solar occultation observation provided stringent upper limits on the two most plausible atmospheric constituents N_2 and CH_4 of 4.2 and 0.3 picobars and vertical column densities of 3.1×10^{15} and 4.0×10^{14} cm^{-2}, respectively. If these species were in vapor pressure equilibrium with their pure surface ices, their respective ice temperatures would be ~23.1 and 29.3 K. REX radiometry observations of Charon at 4.2 cm wavelength yielded dissimilar brightness temperature between dayside and nightside. The mean brightness temperature value on the dayside was ~47 K, whereas the observed nightside brightness temperature mean was ~41 K (*Bird et al.,* 2019). The emissivity for Charon's surface is unknown, but is clearly <1, so the actual surface temperatures would be higher. Thus, from these data it is unlikely that either N_2 or CH_4 has significant ice patches on the surface.

The *Stern et al.* (2017) analysis assumed an atmospheric temperature of 60 K for each species for which an upper limit was inferred. The author constructed a series of isothermal N_2-CH_4 atmospheres in the range 30–60 K and first assumed each constituent was in gravitational diffusive equilibrium. The locations of the respective exobases and the associated enhanced Jeans escape rates were calculated. For 60 K, the CH_4 escape rate exceeded the Hunten limiting rate (equation (15)) through the N_2 background atmosphere (cf. Fig. 5) and it became clear that CH_4 would be well mixed with a N_2 scale height. An even more stringent constraint on their escape rates is the energy (power) limit (*Hunten and Watson,* 1982). These upper limit atmospheres are optically thin to solar EUV and FUV radiation and, when coupled to the heating efficiencies of 25% for N_2 and 50% for CH_4 for an illustrative realistic 50 K atmosphere, the energy-limited escape rates are 1×10^{25} N_2 and 1.4×10^{25} CH_4 s^{-1}, as illustrated in Fig. 5. Now, at these energy-limited escape rates, the initial time constant to remove the N_2 would be 0.5 Earth year and to remove CH_4 only 0.1 Earth year, both much shorter than a Pluto year. The key point is that any atmosphere/exosphere on Charon would be very short-lived.

4.3. Escaping Pluto Nitrogen and Methane Molecules Captured as a Source for a Charon Atmosphere

Tucker et al. (2012) estimated that ~1% of Pluto's escaping N_2 molecules would impact Charon. *Tucker et al.* (2015) investigated in more detail how Charon's gravity modifies Pluto's escape rate and the global structure of its exosphere. While Charon perturbs the trajectories of

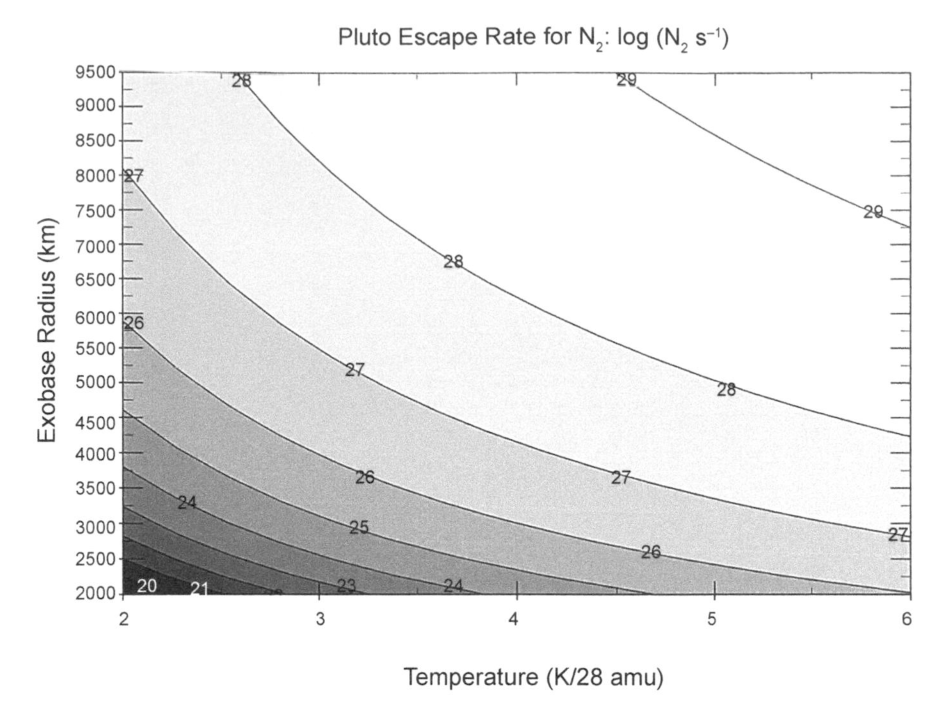

Fig. 4. Pluto escape rates for N_2 (log N_2 s^{-1}) as a function of exobase radius (km) and temperature (K amu^{-1}; e.g., for N_2, T/amu = 3 ⇒ 84 K). This contour plot can be adapted to other molecules as described in Fig. 3 and text.

molecules ejected from the exobase, the overall Pluto escape rate is only lowered by ~ 5% from the calculated escape rate in the absence of Charon. Because *Tucker et al.* (2015) was published prior to the New Horizons Pluto flyby, here we scale their deposition rate of N_2 molecules on Charon's surface of ~5.7×10^{25} N_2 s^{-1} by their calculated escape rate of 2.3×10^{27} s^{-1} and multiply by the New Horizons inferred N_2 escape rate ~5×10^{22} s^{-1} to obtain an adjusted N_2 deposition rate ~1.2×10^{21} s^{-1} on Charon's surface. Their globally averaged column density of ~1×10^{13} N_2 cm^{-2} and line of sight (LOS) column densities up to ~3×10^{14} N_2 cm^{-2} would likewise have to be scaled down to ~2.8×10^{8} N_2 cm^{-2} and LOS column densities up to ~6.5×10^{9} N_2 cm^{-2}, clearly beyond the realm of detection by the New Horizons solar occultation experiment described in the previous section of this review. *Tucker et al.* (2015) did not perform calculations for Pluto's escaping CH_4.

While *Tucker et al.* (2012, 2015) were published prior to the New Horizon Pluto flyby, the comprehensive three-dimensional DSMC simulations by *Hoey et al.* (2017) were made post flyby and used the preliminary New Horizons Alice data analysis reported in *Gladstone et al.* (2016) rather than the full occultation data analysis (*Young et al.*, 2018) from datasets that took much longer to download from the spacecraft. The key atmospheric difference between *Young et al.* (2018) and *Gladstone et al.* (2016) is that the former LOS N_2 column densities were ~80% of the latter.

Hoey et al. (2017) calculated a Pluto system escape rate of 7×10^{25} CH_4 s^{-1}, about 40% larger than *Gladstone et al.* (2016). The *Hoey et al.* (2017) result provides by modeling strong confirmation of the *Young et al.* (2018) nominal escape rate of 7.6×10^{25} CH_4 s^{-1}. The latter was based on the procedure implemented by *Zhu et al.* (2014) to estimate what a rigorous DSMC calculation would yield.

Hoey et al. (2017) calculated that ~3% of the escaping CH_4 would be deposited on the surface of Charon, which is an ~3× larger percentage than *Tucker et al.* (2012) obtained for N_2. These calculations were performed for the binary system of Pluto and Charon with their individual gravitational potentials. *Hoey et al.* (2017) included not only the L1 Lagrange point but the other four Lagrange points as well, and found a preferential deposition of Pluto's escaping CH_4 onto Charon's leading hemisphere, peaking at 315°E (45°W) along the equator. They further note that Charon could support a thin exosphere with a maximum column density ~1.5×10^{13} cm^{-2} in simulations with New Horizon's exobase boundary conditions, if Charon's surface acted as a diffuse reflector. This predicted column density is more than an order of magnitude lower than the New Horizons Alice upper limits on plausible atmospheric constituents (*Stern et al.*, 2017) and as such is consistent with the New Horizons non-detection reviewed above. The *Hoey et al.* (2017) simulations also suggest that the binary system rotation around a barycenter outside of Pluto's atmosphere may potentially leave a detectable, spiral corkscrew tail of escaping and/or weakly bound CH_4 molecules subject to the nature of the solar wind interaction with the extended exosphere of the system.

As evidence that some CH_4 escaping Pluto is deposited on Charon, *Grundy et al.* (2016b) cite its unique dark red northern polar cap, which was its cold winter pole, with temperatures below 25 K from about 1860 to 1989 (their

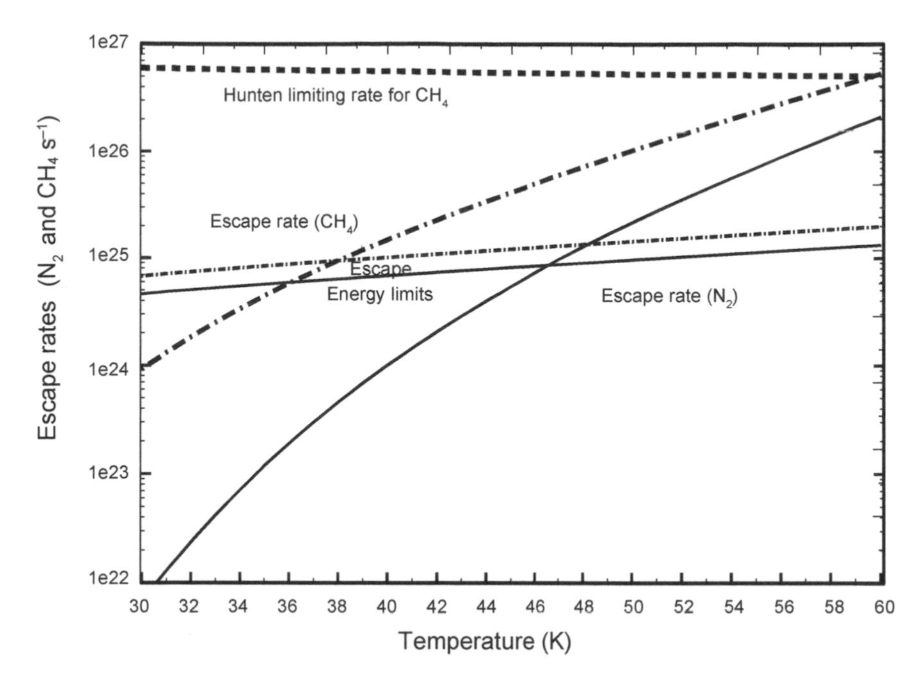

Fig. 5. Charon escape rates for N_2 (solid lines) and CH_4 (dash-dot lines) as a function of isothermal atmospheric temperatures for respective surface number densities of 4×10^8 and 4×10^7 cm^{-3}, where CH_4 outflow is sufficiently large to approach the Hunten limiting rate and render it well mixed with N_2 and attain N_2's scale height. The vertical column densities are close to the inferred upper limits from *Stern et al.* (2017). The Hunten limiting rate for CH_4 is plotted for an assumed mixing ratio of 0.1. This rate scales linearly with mixing ratio. The solar energy (power) provides more controlling limits on escape and restricts their rates to ~10^{25} s^{-1}. The calculated rates include the variation of exobase radii as a function of temperature due to the expanding atmosphere.

Fig. 3). There, one would expect captured CH_4 gas would be cold trapped over its long polar winter (~1 century). Grundy et al. show that this CH_4 ice could be energetically processed by cosmic rays, interplanetary Lyman-α, solar/stellar UV radiation, etc., to generate complex organic molecules evolving into solid deposits like laboratory-produced tholins that are also dark, reddish material. This subject is treated in more depth in the chapter by Cruikshank et al. in this volume.

5. CONCLUDING REMARKS

In conclusion, our most constraining knowledge of escape rates from Pluto's atmosphere is limited to the epoch of the New Horizons flyby in July 2015, when Pluto was post-perihelion at ~32.9 AU from the Sun. Perihelion occurred on September 5, 1989, when Pluto was at 29.66 AU and aphelion will occur in February 2114 when Pluto is at 49.3 AU. For the New Horizons epoch, the inferred escape rates were N_2 = (3–8) × 10^{22} s^{-1} and CH_4 = (4–8) × 10^{25} s^{-1}, for a total of ~10^{27} amu s^{-1}.

Before discussing escape on geological timescales, it is instructive to review various timescales applicable to the current atmosphere. Pluto's atmospheric radiative time constant away from the surface is ~30 Earth years (*Zhang et al.,* 2017) and at the CH_4 homopause, r ~ 1200 km, the diffusion time constant is ~$H^2/D_{CH_4\text{-}N_2}$ ~ (20 km)^2/4000 cm^2 s^{-1} ~ 10^9 s or ~30 yr. If CH_4 ice patches on the surface supporting the CH_4 in the atmosphere were to suddenly cool, atmospheric CH_4 would collapse by diffusion through N_2 down to the surface at an initial rate of 7 × 10^{32} CH_4 s^{-1}, with time constant ~3000 s divided by the fractional surface coverage of CH_4 ice. The only time constant shorter in the atmosphere is the hydrostatic adjustment time constant, which is the speed of sound (0.18 km s^{-1}) divided by the gravitational acceleration (60 cm s^{-2}) or ~300 s. Thus, the survivability of the atmosphere is controlled by the surface ices and their temperatures.

Loss of the atmosphere occurs due to irreversible escape, photodissociation, and photoionization. For CH_4 the irreversible photochemical loss probability is ≥93% of the total rate, whereas for N_2 it is about 50% of the total rate (*Krasnopolsky,* 2020). To derive atmospheric residence time constants with respect to these rates we take a typical CH_4 escape rate ~5 × 10^{25} s^{-1} and a net dissociation rate of ~8 × 10^{25} s^{-1}, for a current total removal rate of ~10^{26} s^{-1}. Given the current total number of CH_4 molecules in the atmosphere ~2 × 10^{36}, the time constant to deplete an atmosphere of CH_4 is ~720 Earth years or ~3 Pluto years. For N_2, if we take a typical N_2 escape rate ~5 × 10^{22} s^{-1} and a net photoionization rate ~5 × 10^{24} s^{-1}, for a current total removal rate ~5 × 10^{24} s^{-1} with a current total number of N_2 molecules in the atmosphere ~7 × 10^{38}, the time constant to deplete an atmosphere of N_2 is ~4.4 million Earth years or ~18,000 Pluto years, which is comparable to Pluto's Milankovitch-like seasonal cycles.

With the large reservoir of N_2 ice in Sputnik Planitia, retention of a N_2 and buffered atmosphere should not be currently a problem. *Glein and Waite* (2018) estimate (0.4–3) × 10^{20} moles = (2.4–18) × 10^{43} N_2 molecules on the surface of Pluto as ice and most is in Sputnik Planitia. At the current total removal rate with an assumed 10 × 10^{43} N_2 molecules, it would take at least 600 b.y. to remove all the N_2 and at the largest predicted escape rates of ~1 × 10^{28} s^{-1} it would take 0.3 b.y.

From New Horizons there is abundant evidence for methane ice widely distributed over Pluto's low latitudes of ±30° (*Moore et al.,* 2018). In addition, Pluto's northern polar region, including Lowell Regio, where thick deposits rich in methane ice were found (*Grundy et al.,* 2016a), has accumulated methane ice probably over timescales associated with Pluto's 3-m.y. Milankovitch-like seasonal cycles. From *Moore et al.* (2018) and as suggested by editor Leslie Young, if the bladed terrain deposits are predominately CH_4 ice, then Pluto currently has at least a few times 10^{17} kg of CH_4 ice on Pluto's surface. This estimate is based on a volume of 400 × 600 × 2 = 480,000 km^3. With the density of CH_4 ice = 500 kg m^{-3}, this yields approximately 10^{43} CH_4 molecules (= 2.4 × 10^{17} kg). At the current CH_4 removal rate of ~10^{26} s^{-1}, this yields a depletion time of 3 b.y. Because CH_4 is a minor constituent in Pluto's atmosphere by virtue of CH_4's ice having a lower-equilibrium vapor pressure than N_2's ice, CH_4 escape rates are constrained by the Hunten limiting rate and proportional to the CH_4 volume mixing ratio (cf. equation (15) and Fig. 2). If the CH_4 mole fraction at the surface were sufficiently low ≤0.01% then CH_4 would be photodissociated near the surface and its resupply to the lower atmosphere would be constrained by the Hunten limiting rate of upward diffusive transport. As a result, CH_4 would never reach the exobase to escape (*Strobel and Zhu,* 2017). To reduce the depletion time substantially the CH_4 atmospheric mixing ratio would need to be elevated beyond the assumed 0.003 in this estimate.

Predicting atmospheric escape rates occurring in the past and future require knowing what mystery cooling agent(s) keeps Pluto's upper atmosphere cold. The consequence of the unknown cooling agent(s) is a suppression of the N_2 escape rate by a factor of ~40,000 (New Horizons N_2 actual escape rate ~5 × 10^{22} s^{-1}, compared to predicted rates >2 × 10^{27} s^{-1}). One hypothesized cooling agent, HCN, is known from ALMA observations (*Lellouch et al.,* 2017) to be supersaturated by ~7 orders of magnitude at some altitudes, yet still is inadequate to cool the upper atmosphere. Other suggestions are (1) a modest influx of H_2O molecules (*Strobel and Zhu,* 2017), which in the atmosphere would have to be supersaturated by ~18 orders of magnitude; and (2) Pluto's haze particles (*Zhang et al.*, 2017) if they have very low, continuous single-scattering albedos throughout the thermal infrared. In an evolving climate that removes or enhances the abundance of the dominate cooling agent there could be a dramatic effect on escape rates, but not necessarily the irreversible photochemical loss rates as most of the driving solar UV and interplanetary Lyman-α radiation is currently absorbed above r = 1500 km. For example, to bring the densities there down to the surface would require a reduction in the N_2 density by a factor of 1000 and in the CH_4 density by a factor of 300.

Acknowledgments. This research was supported in part by the New Horizons Mission through SWRI Contract No. 277043Q. D.F.S. acknowledges beneficial discussions with the New Horizons Atmospheric Theme Team in many telecoms and thanks the three referees for their very helpful comments.

REFERENCES

Aamodt R. E. and Case K. M. (1962) Density in a simple model of the exosphere. *Phys. Fluids, 5(1962),* 1019–1021.

Barnes N., Delamere P., Strobel D., et al. (2019) Constraining the IMF at Pluto using New Horizons SWAP data and hybrid simulations. *J. Geophys. Res.–Space Phys., 124,* DOI: 10.1029/2018JA026083.

Bird G. A. (1994) *Molecular Gas Dynamics and the Direct Simulation of Gas Flows.* Oxford Univ., Oxford. 458 pp.

Bird M. K., Linscott I. R., Tyler G. L., et al. (2019) Radio thermal emission from Pluto and Charon during the New Horizons encounter. *Icarus, 322,* 192–209.

Chamberlain J. W. (1963) Planetary coronae and atmospheric evaporation. *Planet. Space Sci., 11,* 901–960.

Chapman S. and Cowling T. G. (1970) *The Mathematical Theory of Non-Uniform Gases, 3rd edition.* Cambridge Univ., Cambridge. 423 pp.

Cravens T. E. and Strobel D. F. (2015) Pluto's solar wind interaction: Collisional effects. *Icarus, 246,* 303–309.

Cruikshank D. P., Pilcher C. B., and Morrison D. (1976) Pluto: Evidence for methane frost. *Science, 194,* 835–837.

De La Haye V., Waite J. H. Jr., Cravens T. E., et al. (2008) Heating Titan's upper atmosphere. *J. Geophys. Res., 113,* A11314.

Elliot J. L. and Young L. A. (1992) Analysis of stellar occultation data for planetary atmospheres. 1. Model fitting, with application to Pluto. *Astron. J., 103,* 991–1015.

Erwin J., Tucker O. J., and Johnson R. E. (2013) Hybrid fluid/kinetic modeling of Pluto's escaping atmosphere. *Icarus, 226,* 375–384.

Gladstone G. R. and 159 colleagues (2016) The atmosphere of Pluto as observed by New Horizons. *Science, 351,* aad8866.

Glein C. R. and Waite J. H. Jr. (2018) Primordial N_2 provides a cosmochemical explanation for the existence of Sputnik Planitia, Pluto. *Icarus, 313,* 79–92.

Grundy W., Binzel R. P., Buratti B. J., et al. (2016a) Surface compositions across Pluto and Charon. *Science, 351,* aad9189.

Grundy W., Cruikshank D. P., Gladstone G. R., et al. (2016b) Formation of Charon's red poles from seasonally cold-trapped volatiles. *Nature, 539,* 65–68.

Herring J. and Kyle L. (1961) Density in a planetary exosphere. *J. Geophys. Res., 66,* 1980–1982.

Hinson D. P., Linscott I. R., Young L. A., et al. (2017) Radio occultation measurements of Pluto's neutral atmosphere with New Horizons. *Icarus, 290,* 96–111.

Hoey W. A., Yeoh S. K., Trafton L. M., et al. (2017) Rarefied gas dynamic simulation of transfer and escape in the Pluto-Charon system. *Icarus, 287,* 87–102.

Hubbard W. B., Yelle R. V., and Lunine J. I. (1990) Nonisothermal Pluto atmosphere models. *Icarus, 84,* 1–11.

Hunten D. M. (1973) The escape of light gases from planetary atmospheres. *J. Atmos. Sci., 30,* 1481–1494.

Hunten D. M. (1982) Thermal and nonthermal escape mechanisms for terrestrial bodies. *Planet. Space Sci., 30,* 773–783.

Hunten D. M. and Watson A. J. (1982) Stability of Pluto's atmosphere. *Icarus, 51,* 665–667.

Johnson R. E., Combi M. R., Fox J. L., et al. (2008) Exospheres and atmospheric escape. *Space Sci. Rev., 139,* 355–397.

Khare B. N., Sagan C., Arakawa E. T., et al. (1984) Optical constants of organic tholins produced in a simulated titanian atmosphere: From soft X-ray to microwave frequencies. *Icarus, 60,* 127–137.

Kollmann P., Hill M. E., McNutt R. L. Jr., et al. (2019) Suprathermal ions in the outer heliosphere. *Astrophys. J., 876,* 46.

Koskinen T. T., Erwin J. T., and Yelle R. V. (2015) On the escape of

CH_4 from Pluto's atmosphere. *Geophys. Res. Lett., 42,* 7200–7205.
Krasnopolsky V. A. (1999) Hydrodynamic flow of N_2 from Pluto. *J. Geophys. Res., 104,* 5955–5962.
Krasnopolsky V. A. (2020) A photochemical model of Pluto's atmosphere and ionosphere. *Icarus, 335,* 113374.
Lammer H., Odert P, Leitzinger M., et al. (2009) Determining the mass loss limit for close-in exoplanets: What can we learn from transit observations? *Astron. Astrophys., 506,* 399–410.
Lellouch E., Gurwell M., Butler B., et al. (2017) Detection of CO and HCN in Pluto's atmosphere with ALMA. *Icarus, 286,* 289–307.
Lemaire J. (1966) Evaporative and hydrodynamical atmospheric models. *Ann. Astrophys., 29,* 197–203.
McComas D. J., Elliott H. A., Weidner S., et al. (2016) Pluto's interaction with the solar wind. *J. Geophys. Res.–Space Phys., 121(5),* 4232–4246.
McNutt R. L. (1989) Models of Pluto's upper atmosphere. *Geophys. Res. Lett., 16,* 1225–1228.
Millis R. L., Wassermann L. H., Franz O. G., et al. (1993) Pluto's radius and atmosphere. Results from the entire 9 June 1988 occultation data set. *Icarus, 105,* 282–297.
Moore J. M., Howard A. D., Umurhan O. M., et al. (2018) Bladed terrain on Pluto: Possible origins and evolution. *Icarus, 300,* 129–144.
Parker E. N. (1958) Dynamics of the interplanetary gas and magnetic fields. *Astrophys. J., 128,* 664–676.
Parker E. N. (1964) Dynamical properties of stellar coronas and stellar winds. II. Integration of the heat-flow equation. *Astrophys. J., 139,* 93–122.
Poppe A. R and Horányi M. (2018) Interplanetary dust delivery of water to the atmospheres of Pluto and Triton? *Astron. Astrophys., 617,* L5.
Rannou P. and West R. (2018) Supersaturation on Pluto and elsewhere. *Icarus, 312,* 36–44.
Schunk R. W. and Nagy A. F. (2009) *Ionospheres — Physics, Plasma Physics, and Chemistry.* Cambridge Univ., Cambridge. 628 pp.
Stern S. A. and 150 colleagues (2015) The Pluto system: Initial results from its exploration by New Horizons. *Science, 350,* aad1815.
Stern S. A., Kammer J. A., Gladstone G. R., et al. (2017) New Horizons constraints on Charon's present-day atmosphere. *Icarus, 287,* 124–130.
Stevens M. H., Strobel D. F., Summers M. E., and Yelle R. V. (1992) On the thermal structure of Triton's thermosphere. *Geophys. Res. Lett., 19,* 669–672.
Strobel D. F. (2008a) Titan's hydrodynamically escaping atmosphere. *Icarus, 193,* 588–594.
Strobel D. F. (2008b) N_2 Escape rates from Pluto's atmosphere. *Icarus, 193,* 612–619.
Strobel D. F. and Cui J. (2014) Titan's upper atmosphere/exosphere, escape processes and rates. In *Titan: Interior, Surface, Atmosphere and Space Environment* (I. Mueller-Wodarg et al., eds.), pp. 355–375. Cambridge Univ., Cambridge.
Strobel D. F. and Zhu X. (2017) Comparative planetary nitrogen atmospheres: Density and thermal structures of Pluto and Triton. *Icarus, 291,* 55–64.
Strobel D. F., Zhu X., Summers M. E., and Stevens M. H. (1996) On the vertical thermal structure of Pluto's atmosphere. *Icarus, 120,* 266–289.
Tian F. and Toon O. B. (2005) Hydrodynamic escape of nitrogen from Pluto. *Geophys. Res. Lett., 32,* L18201.
Trafton L. M. (1980) Does Pluto have a substantial atmosphere? *Icarus, 44,* 53–61.
Trafton L. M., Hunten D. M., Zahnle K. J., and McNutt R. L. Jr. (1997) Escape processes at Pluto and Charon. In *Pluto and Charon* (S. A. Stern and D. J. Tholen, eds.), pp. 475–522. Univ. of Arizona, Tucson.
Tucker O. J. and Johnson R. E. (2009) Thermally driven atmospheric escape: Monte Carlo simulations for Titan's atmosphere. *Planet. Space Sci., 57,* 1889–1894.
Tucker O. J., Erwin J. T., Deighan J. I., et al. (2012) Thermally driven escape from Pluto's atmosphere: A combined fluid/kinetic model. *Icarus, 217,* 408–415.
Tucker O. J., Johnson R. E., and Young L. A. (2015) Gas transfer in the Pluto-Charon system: A Charon atmosphere. *Icarus, 246,* 291–297.
Volkov A. N., Johnson R. E., Tucker O. J., and Erwin J. T. (2011a) Thermally-driven atmospheric escape: Transition from hydrodynamic to Jeans escape. *Astrophys. J. Lett., 729,* L24.
Volkov A. N., Tucker O. J., Erwin J. T., and Johnson R. E. (2011b) Kinetic simulations of thermal escape from a single component atmosphere. *Phys. Fluids, 23,* 066601.
Watson A. J., Donahue T. M., and Walker J. C. G. (1981) The dynamics of a rapidly escaping atmosphere: Applications to the evolution of Earth and Venus. *Icarus, 48(1981),* 150–166.
Young L. A., Kammer J. A., Steffl A. J., et al. (2018) Structure and composition of Pluto's atmosphere from the New Horizons solar ultraviolet occultation. *Icarus, 300,* 174–199.
Zalucha A. M., Gulbis A. A. S., Zhu X., et al. (2011a) An analysis of Pluto occultation light curves using an atmospheric radiative-conductive model. *Icarus, 211,* 804–818.
Zalucha A. M., Zhu X., Gulbis A. A. S., et al. (2011b) An analysis of Pluto's troposphere using stellar occultation light curves and an atmospheric radiative-conductive-convective model. *Icarus, 214,* 685–700.
Zhang X., Strobel D. F., and Imanaka H. (2017) Haze heats Pluto's atmosphere yet explains its cold temperature. *Nature, 551,* 352–355.
Zhu X., Strobel D. F., and Erwin J. T. (2014) The density and thermal structure of Pluto's atmosphere and associated escape processes and rates. *Icarus, 228,* 301–314.

Bagenal F., McComas D. J., Elliott H. A., Zirnstein E. J., McNutt R. L. Jr., Lisse C. M., Kollman P., Delamere P. A., and Barnes N. P. (2021) Solar wind interaction with the Pluto system. In *The Pluto System After New Horizons* (S. A. Stern, J. M. Moore, W. M. Grundy, L. A. Young, and R. P. Binzel, eds.), pp. 379–392. Univ. of Arizona, Tucson, DOI: 10.2458/azu_uapress_9780816540945-ch016.

Solar Wind Interaction with the Pluto System

F. Bagenal
University of Colorado Boulder

D. J. McComas
Princeton University

H. A. Elliott
Southwest Research Institute

E. J. Zirnstein
Princeton University

R. L. McNutt Jr., C. M. Lisse, P. Kollmann
The Johns Hopkins University Applied Physics Laboratory

P. A. Delamere and N. P. Barnes
University of Alaska

The scientific objectives of NASA's New Horizons mission include quantifying the rate at which atmospheric gases are escaping Pluto and describing its interaction with the surrounding space environment. The two New Horizons instruments, Solar Wind Around Pluto (SWAP) and Pluto Energetic Particle Spectrometer Science Investigation (PEPSSI), measured particle fluxes as the spacecraft approached Pluto on the flank of the interaction, passed behind the planet, and traversed the tail of the interaction. The SWAP instrument revealed an interaction region much smaller than predictions with upstream bow shock and boundary (Plutopause) at distances of ~4.5 and ~3.5 R_P (Pluto radii), respectively. The surprisingly small size of the interaction is consistent with a reduced atmospheric escape rate, as well as a particularly high solar wind flux. SWAP detected a flux of heavy ions (most likely CH_4^+) streaming down the tail that extends >100 R_P downstream. Observations from the PEPSSI instrument suggest that ions are accelerated and/or deflected around Pluto. In the wake of the interaction region, PEPSSI observed suprathermal particle fluxes equal to about one-tenth of the flux in the interplanetary medium and increasing with distance downstream. Models consistent with these measurements indicate a weak interplanetary magnetic field and a highly kinetic interaction.

1. INTRODUCTION

Prior to the New Horizons flyby, speculations about the nature of the solar wind interaction with Pluto ranged from a vast comet-style interaction with an effusive escaping atmosphere to a compact Mars- or Venus-style interaction with a tightly-confined atmosphere (*Bagenal and McNutt,* 1989; *Sauer et al.,* 1997; *Bagenal et al.,* 1997; *Ip,* 2000). Over the years, as the effects of the weak interplanetary field of the outer heliosphere were considered, expectations grew for an even more complicated interaction, with large (comparable to the scale of the Pluto system) ion gyroradii of the pickup ions and significant effects on kinetic scales, where ions and sometimes electrons need to be considered as particles rather than acting as a fluid (*Kecskemety and Cravens,* 1993; *Shevchenko et al.,* 1997; *Delamere and Bagenal,* 2004; *Harnett et al.,* 2005; *Delamere,* 2009).

Figure 1 shows examples of the solar wind interactions with Pluto anticipated before the Pluto flyby, evolving from fluid to kinetic. *Bagenal et al.* (2015) reviewed the literature and made pre-flyby predictions of the types of interactions Pluto could have with the solar wind. It was clear that the nature and scale of the interaction of the solar wind with Pluto's atmosphere depends on the kinetic pressure of

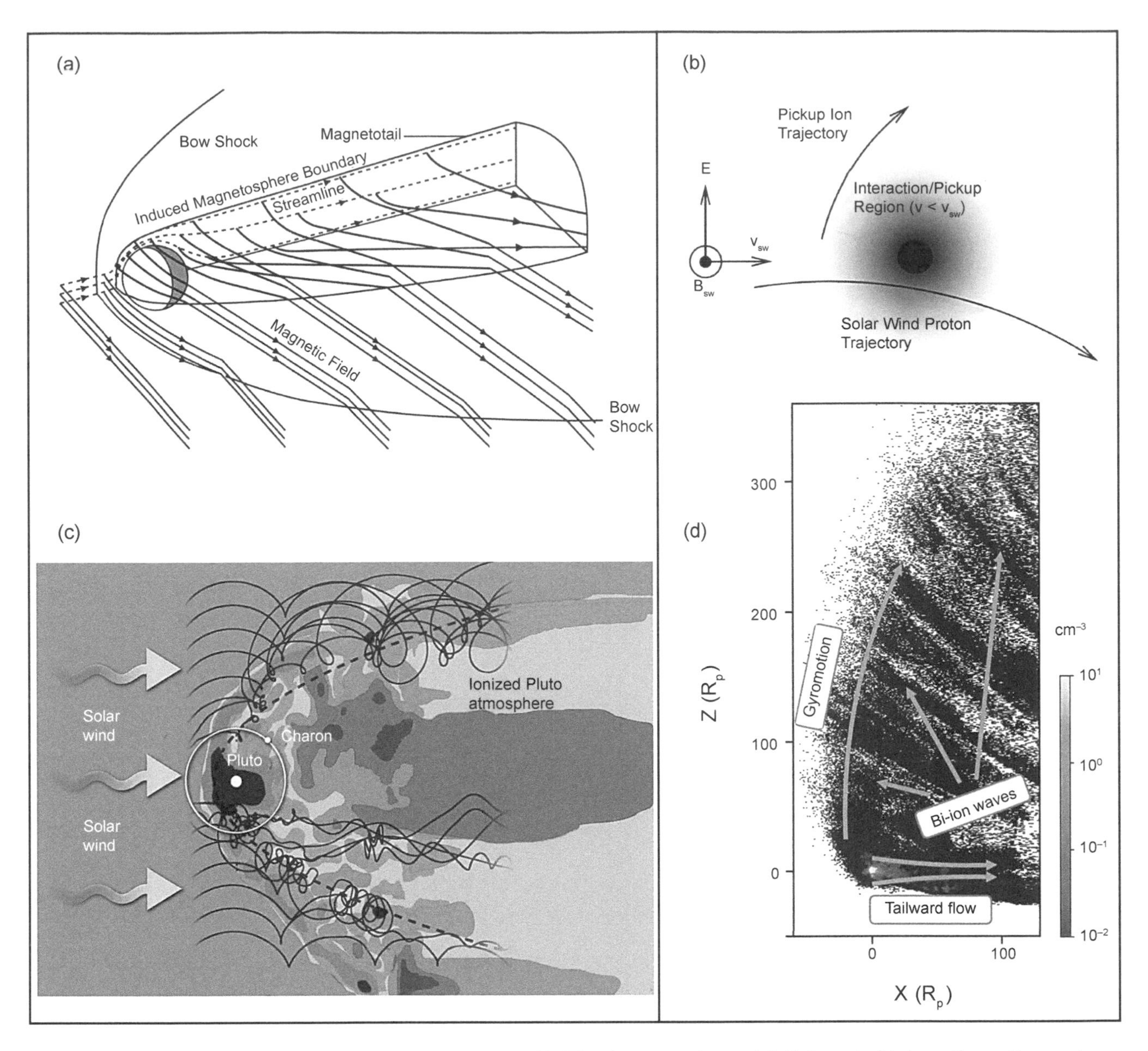

Fig. 1. Overview of physics of solar wind interaction with Pluto's atmosphere. **(a)** Solar wind interaction with an atmosphere illustrated by early studies of the Venus interaction (from *Saunders and Russell,* 1986). **(b)** Recognition of the weak interplanetary magnetic field meant that the electrodynamics produce an asymmetric interaction (from *Delamere and Bagenal,* 2004). **(c)** Numerical simulations of the interaction suggested a variable, kinetic interaction (based on *Delamere,* 2009). **(d)** Recent models matching New Horizons data show small interaction region with large, asymmetric, variable, tenuous flows of plutogenic methane ions upstream (from *Barnes et al.,* 2019).

the tenuous and supersonic solar wind, which varies by a factor of ~5 in the outer heliosphere. Before the New Horizons flyby, the estimates of the atmospheric escape rate ranged from as low as 1.5×10^{25} molecules s^{-1} to as high as 2×10^{28} molecules s^{-1} (*Krasnopolsky,* 1999; *Tian and Toon,* 2005). Combining these wide-ranging predictions of atmospheric escape rates with Voyager and New Horizons observations of extensive variability of the solar wind at 33 AU produced estimates of the scale of the interaction region that spanned all the way from 7 to 1000 R_P, where the Pluto radius is taken for this chapter to be R_P = 1190 km [the actual average radius is 1188.3 ± 0.8 km (*Stern et al.* 2015; *Nimmo et al.,* 2017)].

In this chapter we first summarize the evidence for atmospheric escape from Pluto (for full details, see the chapter in this volume by Strobel). Next, we review the solar wind conditions observed by New Horizons on approach to Pluto and compare them with expectations based on Voyager measurements. In section 2 we provide an overview of the observations made at Pluto by the Solar Wind Around Pluto (SWAP) instrument (*McComas et al.,* 2008b) and the Pluto Energetic Particle Spectrometer Science Investigation

(PEPSSI) instrument (*McNutt et al.,* 2008). Puzzling X-ray measurements are then summarized. In section 3 the particle measurements are compared with theoretical models before we wrap up with discussions and conclusions. Finally, we consider valuable measurements that could be made in the future.

1.1. Atmospheric Escape and Ionosphere

Pluto's atmosphere was first detected in 1988 during a stellar occultation (*Elliot et al.,* 1989) and was later determined to be primarily composed of N_2 with minor abundances of CH_4 and CO, with surface pressures of ~17 μbar (*Young et al.,* 2001; *Lellouch et al.,* 2015, 2017). Pluto's low gravity (surface gravity is 0.063 g and escape speed is 1.2 km s^{-1}) implies that a large flux of atmospheric neutrals can escape. Most pre-New Horizons models for the upper atmosphere were constrained by solar extreme ultraviolet (EUV) to far ultraviolet (FUV) heating rates and yielded N_2 escape rates in the range of (1–10) × 10^{27} molecules s^{-1} ~ (3–30) × 10^{28} amu s^{-1}. The most recent atmospheric model pre-New Horizons of *Zhu et al.* (2014) indicates a dense and expanded atmosphere with an escape rate of ~3.5 × 10^{27} N_2 molecules s^{-1} and an exobase at 8 R_P ~ 9600 km. These are the conditions that were anticipated on arrival at Pluto.

While atmospheric escape is hard to measure directly, measurements of the upper atmospheric composition and structure provide key constraints on atmospheric escape rates. The New Horizons trajectory was designed to provide solar and Earth occultations of Pluto's atmosphere by the ultraviolet spectrometer (Alice) and Radio Science Experiment (REX) instruments, respectively (*Stern et al.,* 2008; *Young et al.,* 2008). The radio occultation by REX yielded the N_2 density, pressure, and temperature profiles from the surface to ~110 km (*Hinson et al.,* 2017).

The Alice UV solar occultation provided densities of N_2 from 900 to 1100 km altitude, of CH_4 for 80–1200 km, and the temperature profiles out to ~1600 km (*Young et al.,* 2018). In comparison to expectations, Pluto's upper atmosphere was surprisingly cold and compact (*Stern et al.,* 2015; *Gladstone et al.,* 2016).

The two sets of occultation data allowed *Strobel and Zhu* (2017) to construct a model atmosphere from the surface to R = 2300 km. For the model atmosphere to reach the exobase, it was necessary to assume that upper atmosphere was isothermal, comprised N_2 and CH_4, and was in hydrostatic equilibrium in order to extrapolate the data about 600 km to R ~ 2900 km. Since the REX and Alice data on both ingress and egress were the same within their error bars above R = 1220 km, it is reasonable to model the atmosphere as spherically symmetric. For the New Horizons epoch, the inferred escape rates were N_2 = (3–8) × 10^{22} s^{-1} and CH_4 = (4–8) × 10^{25} s^{-1}, for a total of ~10^{27} amu s^{-1} (*Strobel and Zhu,* 2017). This limited atmospheric escape drastically reduces the neutral material upstream of Pluto available for ionization and mass-loading the solar wind. A detailed review of the physical processes of atmospheric escape and current constraints based on New Horizons data are provided in the chapter in this volume by Strobel.

The REX radio occultation of Pluto also set an upper limit on the density of the ionosphere (*Hinson et al.,* 2017, 2018). *Hinson et al.* (2018) developed a model of Pluto's ionosphere that is consistent with both the solar occultation measurements made by the New Horizons Alice UV instrument (*Young et al.,* 2018) and the REX pressure and temperature profiles at altitudes below about 100 km (*Hinson et al.,* 2017). The derived model ionospheric subsolar profile peaks with a density of ~3000 cm^{-3} at an altitude of 600 km with lower values (~800 cm^{-3} at 800 km) at the terminator. These numbers are consistent with the upper limit of ~1000 cm^{-3} for the terminator obtained by the REX measurements (*Hinson et al.,* 2018). These ionospheric densities are substantially lower than those measured by Voyager 2 at Triton (*Tyler et al.,* 1989), which *Hinson et al.* (2018) attribute to a higher abundance of methane in Pluto's atmosphere at ionospheric altitudes, producing molecular ions that undergo rapid dissociative recombination, resulting in low electron densities. Further discussion of Pluto's atmosphere and ionosphere are found in the chapter in this volume by Summers et al.

1.2. Solar Wind at 33 AU

On its eccentric orbit around the Sun, Pluto experiences a wide range of conditions in the interplanetary medium between ~30 and 50 AU. The Sun and the outflowing solar wind vary on a wide range of timescales. The ~11-year solar cycle is associated with reversals of the polarity of the Sun's magnetic field. The solar cycle has a major impact on the coronal structure, which in turn drives the three-dimensional solar wind that fills and inflates the heliosphere. In July 2015 New Horizons flew past Pluto as the Sun was in the descending phase of the solar cycle. To predict conditions at the time of the flyby, *Bagenal et al.* (2015) took Voyager 2 data from 1988 to 1992 when it traversed from 25 to 39 AU at ecliptic latitudes of +4° to –8° (see their Table 1). The maximum of solar cycle 22 was in ~1990, around the middle of the sample period. The New Horizons flyby of Pluto occurred as the Sun was in the declining phase of a remarkably low sunspot cycle. Superimposed on the solar cycle is a longer-term trend of weakening solar wind (*McComas et al.,* 2008a, 2013). To estimate conditions at the time of the Pluto flyby, *Bagenal et al.* (2015) scaled down the solar wind parameters (see their Table 2) by appropriate factors from *McComas et al.* (2013). These Voyager 2 values, both original and scaled down, are shown in Table 1 for comparison with the data obtained by the New Horizons SWAP instrument.

In this region, the solar wind flow is nearly radial at speeds between ~350 and ~550 km s^{-1}. Densities and dynamic pressures roughly follow a 1/distance^2 profile. Note that interstellar neutral atoms penetrate the solar system, and when they are ionized they are picked up and carried along

TABLE 1. Solar wind conditions at 33 AU.

Plasma Property	Formula (units)	10th Percentile V2 V2 scaled	Median V2 V2 scaled	90th Percentile V2 V2 scaled	Observed by SWAP
Solar wind speed	V_{SW} (km s^{-1})	380 340	430 382	480 430	403
Proton density	n (cm^{-3})	0.0020 0.0015	0.0058 0.0042	0.014 0.01	0.025
Proton flux	nV (km s^{-1} cm^{-3})	0.84 0.55	2.4 1.55	7.0 4.6	10
Proton temperature	T (eV)	0.26 0.16	0.57 0.34	1.5 0.86	0.66 (6.0*)
Proton thermal pressure (IPUIs*)	P = nkT (fPa)	0.12 0.053	0.53 0.24	2.1 0.92	2.5 (28*)
Proton ram pressure	$P = rV^2$ (pPa)	0.55 0.32	1.7 1.00	4.0 2.3	6.0
Sound speed (IPUIs*)	$V_s = (\gamma kT/m_i)^{1/2}$	6.3 4.9	9.4 7.3	15 12	10 (28*)
Sonic Mach # (IPUIs*)	$M_s = V_{SW}/V_s$	60 69	46 52	32 36	40 (14*)
Magnetic field strength	B (nT)	0.08 0.05	0.15 0.10	0.28 0.19	0.3+ 0.1–
Pickup CH_4^+ R_{gyro}	R_{gyro} (R_p)	670	400 450	240	190+ 570–
Alfven speed	$V_A = B/(\mu_o\rho)^{1/2}$ (km s^{-1})	22	45 36	96	41+ 14–
Alfven Mach no.	$M_A = V_{SW}/V_A$	4.6	9.5 11	20	9.8+ 29–
Magnetosonic Mach no.	$M_{MS} = V_{SW}/(V_A^2 + V_s^2)^{1/2}$	17	9 10.4	5	9.5+ 23–

*Includes thermal pressure of IPUIs from *McComas et al.* (2010, 2017) and *Randol et al.* (2012, 2013).

Voyager 2 plasma data obtained from 1988 to 1992 between 25 and 39 AU (upper values) plus parameters scaled for long-term trends (lower values) (*Bagenal et al.*, 2015). Right column has quantities observed by the New Horizons SWAP instrument from *Bagenal et al.* (2016). γ is the adiabatic index, m_i is ion mass, μ_o is the permeability, and ρ is mass density. "+" shows the upper value of the interplanetary magnetic field (0.3 nT) and "–" gives the lower value of the interplanetary magnetic field (0.1 nT).

by the solar wind. These interstellar pickup ions (IPUIs) gain a thermal energy comparable to the local solar wind bulk speed and are therefore ~kiloelectron-volt energies [see the review of IPUIs by *McComas et al.* (2017) and references therein]. These IPUIs have a noticeable effect, decreasing the speed of the solar wind by several percent (*Elliott et al.*, 2019) and increasing the average temperature, counteracting adiabatic cooling on expansion and resulting in a relatively flat temperature profile (*Richardson et al.*, 2008).

In Table 1, we compare the interplanetary plasma conditions based on Voyager 2 data for 33 AU with the observations made by the New Horizons SWAP instrument on approach to Pluto (from *Bagenal et al.*, 2016). In the absence of a direct measurement of the local magnetic field, we examine two cases: (1) an interplanetary magnetic field (IMF) of ~0.3 nT, at the upper end of the range observed by Voyager 2, and (2) an IMF of ~0.1 nT that is consistent with the scaled-down value.

We note that New Horizons did not carry a magnetometer because it was not required to fulfill the primary New Horizons Pluto-Charon and KBO measurement objectives; it was one of several instruments considered but not included in the payload (*Stern et al.,* 2008). Not only was Pluto not judged to have any intrinsic magnetic field (*Trafton et al.,* 1997) but also, given the knowledge of the IMF from Voyager 2 at Pluto's distance from the Sun, a good magnetic field measurement would require an extended boom to remove interference from the magnetic field generated by the spacecraft. In addition to cost, the implementation of a boom would have been problematic given the need to maneuver the spacecraft during the Pluto system flyby to accommodate the high-priority, and body-fixed, remote sensing instruments. With the expectation (proven correct) that the only magnetic field would be that induced by either Pluto's outflowing atmosphere or an ionosphere (*Bagenal et al.,* 1997), the indirect deduction of the presence of such a field was considered adequate (*Weaver et al.,* 2008), especially given the higher-priority science goals and resource and cost constraints placed on the mission by NASA.

At the time of the flyby, the solar wind conditions near Pluto derived from SWAP measurements (*Elliott et al.,* 2016; *McComas et al.,* 2016) were nearly constant with a solar wind speed of ~403 km s^{-1}, a proton density of ~0.025 cm^{-3}, a proton temperature of ~7700 K (0.7 eV), a proton dynamic (or ram) pressure of ~6 pPa, and a core solar wind proton thermal pressure of ~2.5 fPa (Table 1). From the properties of just the thermal solar wind, we calculate a sonic Mach number of ~40. The sonic Mach number is substantially reduced to 14 if we include the IPUIs, which provide thermal pressure that is roughly an order of magnitude greater than the thermal population at these heliocentric distances (*McComas et al.,* 2010, 2017; *Randol et al.,* 2012, 2013). Particularly important for the observed solar wind interaction with the Pluto system is the fact that the measured solar wind density and associated pressures were significantly higher than usually found at this distance from the Sun, probably due to a relatively strong traveling interplanetary shock that passed over the spacecraft 5 days before the flyby.

2. NEW HORIZONS PARTICLE OBSERVATIONS AT PLUTO

As New Horizons flew past the Pluto system on July 14, 2015, the SWAP and PEPSSI instruments measured changes in the particle fluxes associated with a complicated interaction between the solar wind and gases emanating from Pluto (*Bagenal et al.,* 2016; *McComas et al.,* 2016; *Zirnstein et al.,* 2016; *Kollmann et al.,* 2019b). The Alice UV instrument on New Horizons included a mode that could detect high-energy (>1 MeV) electrons (*Stern et al.,* 2008). No obvious signal was detected from any high-energy penetrating electrons other than during the Jupiter encounter (*Steffl et al.,* 2012).

Figure 2 shows the geometry of the flyby with the spacecraft approaching on the flank of the interaction with a closest approach to Pluto at a distance of ~10 R_P, when crossing the terminator. The spacecraft then passed behind the planet, crossing the Pluto-Sun line at ~50 R_P downstream. Along the trajectory the spacecraft made multiple turns so that the different remote-sensing instruments could observe the Pluto system. These changes in attitude complicate the analysis of particle data. Figure 2 shows a summary of the main observations.

2.1. Solar Wind Around Pluto (SWAP) Observations of Interaction Structure

This section summarizes the observations made by the SWAP instrument (*McComas et al.,* 2008b) at Pluto (*McComas et al.,* 2016). The solar wind conditions were nearly constant around the flyby, which was advantageous for interpreting the solar wind interaction with Pluto, although the high dynamic pressure probably made the region smaller than usual during the flyby. Figure 2b shows a color spectrogram of SWAP coincidence counts as a function of energy per charge (E/q) and time. On the left and right sides of the plot, away from the Pluto interaction, the red and yellow bands located slightly under 1 and 2 keV/q are solar wind protons and alpha particles, respectively, which show the ambient solar wind conditions (similar upstream and downstream) listed in Table 1. Figure 2c shows light ions (blue) and heavy ions (red) throughout the interaction region. A heavy ion burst (HI, shown in red) was seen ahead of the shock when the spacecraft turned so that SWAP was viewing in the right direction to see newly ionized material, plutogenic heavy ions, beginning to be picked up.

McComas et al. (2016) used SWAP ion composition data to establish the location of ~10% slowing inbound as the location of the bow shock upstream of the interaction. Extrapolating from the spacecraft location at the crossing (~8 R_P downstream and ~9 R_P transverse) using an Earth-based conic approximation for the shape of the bow shock (*Sibeck et al.,* 1991), they derived a distance of just 4.5 R_P upstream of Pluto at the nose.

The SWAP data indicates that New Horizons started entering Pluto's heavy ion tail at ~8.4 R_P transverse and ~13.1 R_P tailward from Pluto. *McComas et al.* (2016) call this boundary the Plutopause and applied the *Sibeck et al.* (1991) conic shape of Earth's magnetopause to find a rough standoff distance at the nose. Using this approach, the start of significant heavy ions (Fig. 2) corresponds to an upstream distance at the nose of ~3.4 R_P, while the exit of the Plutopause boundary and entrance into the heavy ion tail scales to a standoff distance ahead of Pluto of only ~2.5 R_P [smaller than the *Bagenal et al.* (2016) initial estimate]. The thickness of the Plutopause boundary, where both light and heavy ions are seen to slow simultaneously and rapidly, scales to only ~0.9 R_P thick, spanning 3.4–2.5 R_P in distance

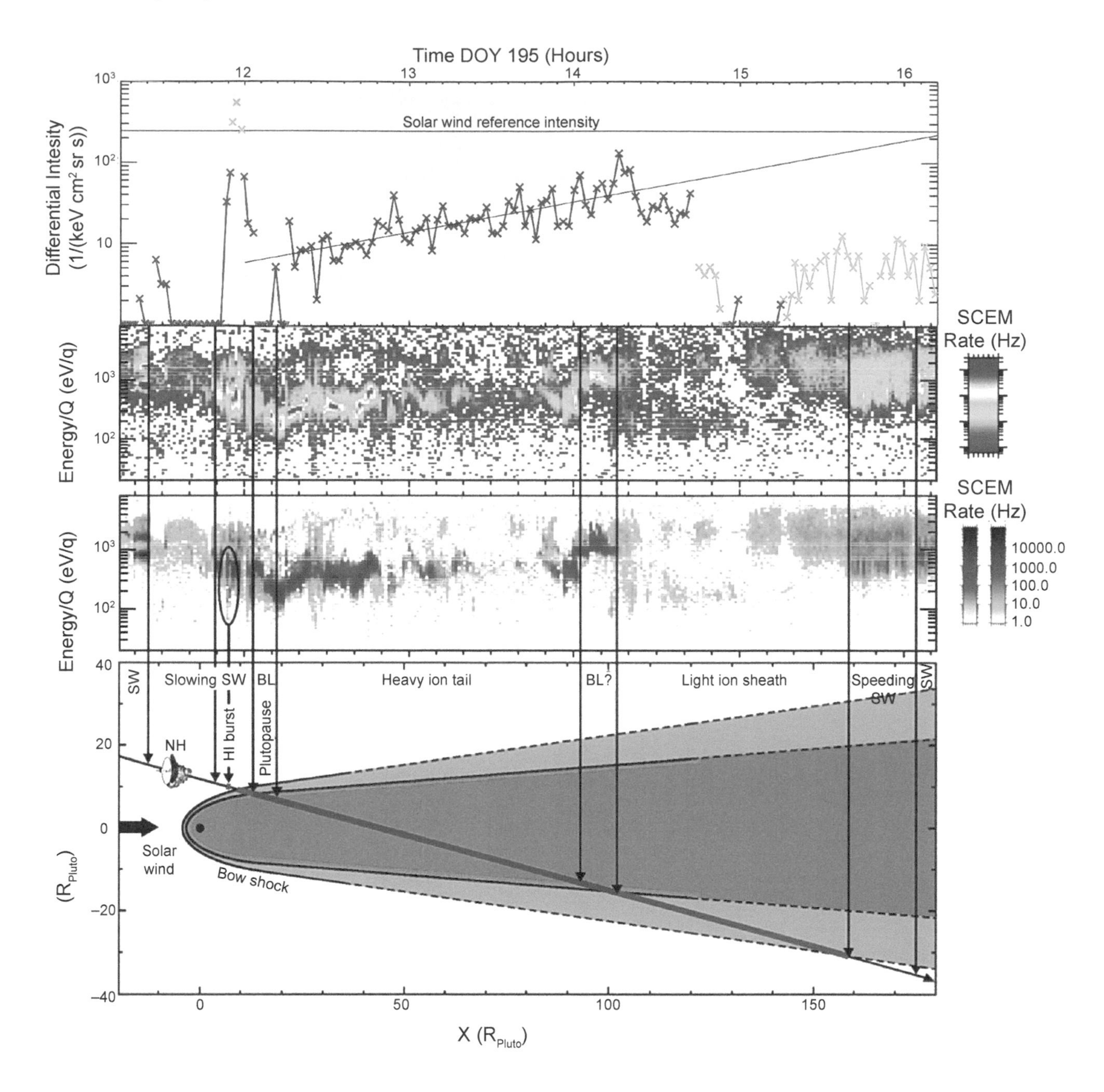

Fig. 2. Overview of particle data from New Horizons at Pluto. **(a)** PEPSSI data from *Kollmann et al.* (2019b). The x-symbols and colored solid lines show intensities dominated by 18 keV He^+ ions. The color coding marks measurements at different cone angles relative to the Sun. Lower panels show SWAP data from *McComas et al.* (2016). Color spectrograms of count rates (with >3 counts/sample) for **(b)** secondary counts and **(c)** light ions (blue) from the solar wind (SW) and heavy ions (red) from Pluto, combined with **(d)** a schematic diagram identifying the NH trajectory and key regions of the interaction: bow shock, light ion sheath (blue), heavy ion tail (red), and Plutopause boundary layer (BL, purple). The X coordinate is along the Sun-Pluto line and the transverse distance is measured in the plane of X and the spacecraft's trajectory, which is close to the ecliptic plane. A heavy ion (HI) burst was seen ahead of the shock when NH turned so that SWAP was viewing in the right direction to see newly ionized material starting to be picked up.

from Pluto, at the nose. This is an extremely thin boundary layer, especially considering the large gyroradii of picked up heavy ions in the solar wind (hundreds of Pluto radii) and suggests a complicated interaction near the Plutopause.

The SWAP data in Fig. 2 show that the spacecraft stayed in the heavy ion tail until ~101 R_P downtail, at a transverse distance of ~15.4 R_P. The entrance and exit transverse distances indicate a tail diameter in the range of ~17 R_P at ~13 R_P downtail and ~31 R_P at ~101 R_P downtail (assuming a circular tail). The highly variable nature of the heavy ions indicates significant structure in the tail, with the most intense regions of heavy ions appearing near the inbound and outbound tail boundaries. Furthermore, the inbound and outbound Plutopause structures differ.

At the entrance, both higher E/q heavy ions and lower E/q light solar wind ions are present together and there is a relatively steep decrease in the ion speed across this transition. As New Horizons exited the tail, the heavy ion E/q jumps to higher values than anywhere else in the tail, indicating faster speeds, possibly from coupling to faster plasma flowing along the flanks. By the distance downstream that New Horizons exited the interaction region, the bow shock has dissipated into a weak bow wave.

The New Horizons occultation measurements of *Gladstone et al.* (2016) indicated that Pluto is likely outgasing neutral CH_4 and N_2. Groundbased telescopic observations add CO and HCN to the list (*Lellouch et al.,* 2017). Figure 2 shows a burst of heavy ions (HI) observed before the spacecraft crossed the bow shock. Heavy neutral atoms escaping from Pluto, which are nearly at rest in the Sun frame, are eventually ionized by charge-exchange or photoionization and picked up by the motional electric field of the solar wind. Since SWAP detected heavy ions at relatively low energies (~100 eV in Fig. 2c) and the heavy pickup ion gyroradii are significantly larger than the upstream obstacle size, they were likely detected during their initial energization and gyration in the solar wind (illustrated in Fig. 1b). The north vs. south direction of their initial propagation, however, depends on the sector of the IMF.

New Horizons was not equipped with a magnetometer to measure the IMF. However, the IMF must play an important role in the solar wind–Pluto interaction. Due to the weak strength of the magnetic field at ~33 AU, heavy ions like CH_4^+ and N_2^+ that are picked up in the solar wind have gyroradii on the order of 10^2–10^3 R_P, much larger than the size of the interaction region. The size and direction of their gyration may produce an asymmetry in the structure of the global plasma interaction behind Pluto (Fig. 1d). Thus, knowing the sector polarity of the IMF during the Pluto encounter is important for our understanding of the plasma interaction. The IMF makes an increasingly tighter spiral as the radial solar wind flow expands into the heliosphere. This means that at Pluto we expect the IMF to be nearly transverse to the flow at Pluto, alternating regularly between positive and negative transverse direction due to the wavy-like nature of the heliospheric current sheet.

Zirnstein et al (2016) used the field of view of the SWAP detector and the energies at which SWAP detected heavy ions to determine the possible IMF directions that could match the observations. *Zirnstein et al.* (2016) determined that during the Pluto encounter (1) the IMF was most likely pointed roughly parallel to the ecliptic plane and opposite of Pluto's orbital motion, and (2) heavy ions detected by SWAP are more likely CH_4^+ than N_2^+ ions. A possible asymmetry in the tail was actually reported by *McComas et al.* (2016), where more heavy ions were detected below the Sun-Pluto line than above. *McComas et al.* (2016) argue that, through the conservation of momentum, the solar wind is deflected southward, pushing the heavy ion tail southward (sketched in Fig. 1b). *Zirnstein et al.* (2016) suggest that the heavy ions detected by SWAP before entering the heavy ion tail were more likely CH_4^+ than N_2^+ since N_2^+ ions would have have had much higher energies entering the SWAP instrument than observed. This is also in agreement with the *Strobel and Zhu* (2017) model of the *Gladstone et al.* (2016) observations that CH_4 is the dominant species escaping Pluto's atmosphere (see section 1.2).

McComas et al. (2016) calculated heavy ion fluxes and moments from SWAP data in the heavy ion tail. Figure 1 shows a long (>100 R_P) tail filled with heavy ions from Pluto's atmosphere with considerable structure and an asymmetry with considerably more heavy ions in the dawn/southern compared to the dusk/northern half of the tail. The SWAP data showed slightly increasing flow speed and temperature as the spacecraft progressed down the tail. *McComas et al.* (2016) suggest that this mild acceleration (approximately a few meters per second) of the heavy ions as they move down the tail indicates partial threading of the IMF into the tail, further supported by the observation of high- and low-energy populations of light ions intermittently in the tail. *McComas et al.* (2016) found that the density and temperature of heavy ions were roughly anticorrelated, indicating some level of global pressure balance within the tail at an average level of ~20 fPa, roughly consistent with the thermal pressure of IPUI in the solar wind at ~33 AU. Finally, *McComas et al.* (2016) calculated a CH_4^+ loss rate down the heavy ion tail of ~5×10^{23} molecules s^{-1}, which represents ~1% of the *Strobel and Zhu* (2017) modeled atmospheric loss rate of CH_4 from Pluto.

2.2. Pluto Energetic Particle Spectrometer Science Investigation (PEPSSI) Observations of Energetic Particle Structures

Figure 2a shows a representative subset of interstellar pickup and suprathermal (IPS) ion measurements around Pluto taken with the PEPSSI instrument during the Pluto flyby (*Kollmann et al.,* 2019b). The PEPSSI instrument (*McNutt et al.,* 2008; *Kollmann et al.,* 2019a) measures ions with a few kiloelectron volts to approximately megaelectronvolt energies. The efficiency of PEPSSI detectors is such that the count rates in the ambient heliosphere are dominated by heliospheric He^+ ions (*Kollmann et al.,* 2019a). Interpretation of the data is complicated by the fact that the spacecraft was constantly changing attitude so that PEPSSI and SWAP pointed in different look directions as a function of time. For simplicity, only fluxes of 18 keV He^+ ions are shown in this figure to illustrate the basic structures.

The gyroradii of 10-keV interstellar He^+ pickup ions is 200 R_P, much larger than the size of the obstacle. Despite their density being lower than the solar wind, at least interstellar H^+ pickup ions are thought to play an important role in determining the shape of Pluto's heavy ion tail (discussed in section 3 below). PEPSSI did not show evidence for any plutogenic ions in the PEPSSI energy range. The bulk of heavy ions originating from Pluto is found at <1 keV as measured by SWAP (*McComas et al.,* 2016). What PEPSSI measurements of He^+ show is that Pluto modifies the distribution of IPS ions in the interplanetary medium, deflecting the ion bulk flow direction and forming an energetic particle wake.

Before and after the Pluto encounter PEPSSI observed nominal fluxes of particles moving away from the Sun. The measured intensities decrease when the viewing angle of the detector is changed from the flow direction. At a radial distance of about 13 R_P, PEPSSI recorded the opposite behavior: intensities increased despite a look direction further away from the Sun (red shaded area in Fig. 2a). This behavior is consistent with the bulk flow direction of the energetic ions being deflected, roughly in the direction of the PEPSSI aperture (*Kollmann et al.,* 2019b). This deflection indicates a clear interaction of Pluto with IPS ions. The apparent short duration and abrupt onset of the deflection is a bias resulting from the spacecraft attitude. The actual onset of the deflection might be more upstream and its extent might be gradual. Based on these measurements we therefore do not know if Pluto is creating a sharp boundary of energetic ions (like a bow shock does to plasma) or if the change is gradual as is the usual behavior of energetic ions at a magnetopause.

The nominal ion intensities in the interplanetary medium is shown in Fig. 2a as a horizontal black line. After closest approach to Pluto, the intensities are reduced by more than an order of magnitude compared to that nominal level. *Kollmann et al.* (2019b) interpret this as New Horizons entering the IPS wake of Pluto. We use the term "wake" here because it describes the depletion of IPS ions. This wake is defined as the region behind an object where particles are depleted, whereas a magnetotail is the region where the thermal plasma flow and embedded magnetic field are draped around the object and carried downstream, entraining material from the interaction. Their volumes may roughly coincide.

Pluto's wake is not entirely empty because some interstellar ions do manage to enter. Figure 2a shows that after an abrupt drop in energetic particle fluxes on entering the wake, the fluxes (shown by blue crosses) steadily increase with distance behind Pluto. The slope of this increase in flux suggests that the wake reaches densities indistinguishable from the ambient interplanetary conditions at a distance of ~190 R_P. This distance is not to be confused with the location where New Horizons leaves the wake, which is unrelated to the actual length of the wake.

Figure 3a (based on *Kollmann et al.,* 2019b) shows a zoom into the region of the wake with the sum of all ions at all energies and directions that are detected by PEPSSI as a function of time. The intensities are oscillating with a period of 0.2 hr (~10 min). The intensity peaks occur at the same time for all energies, indicating that there is no dispersion of the oscillations. *Kollmann et al.* (2019b) suggest that the observed oscillations are a result of a local interaction with the oscillating electric field of an ultra-low-frequency wave. Acceleration through an electric field increases the energy of a particle, while the phase space density is conserved. Because energetic particle spectra are overall decreasing with energy, the acceleration then yields an increased phase space density at a given energy compared to before the acceleration. Depending on wave phase, the field can also decelerate the particles, leading to a decrease in phase space density at a given energy. The acceleration needs to occur locally at the observation site, otherwise ions of different energies would disperse and the intensity peak at the higher energies would run ahead of the peak at the lower energies. *Kollmann et al.* (2019b) suggest the origin of the electric field wave may be the modified ion-Weibel instability driven by the observed energetic IPS ions or a bi-ion wave driven by low-energy plutogenic pick-up ions (see section 3).

Figure 3b shows a schematic of the energetic particle interaction (based on *Kollmann et al.,* 2019b). Interstellar pickup and suprathermal ions have the highest intensities when coming from the sunward direction but can generally come from any direction. The arrows illustrate how the ions are deflected from their original direction near the magnetic boundary that encompasses the wake. Ions that penetrate into the wake are further deflected by the electric field wave. The particle oscillations in the wake are indicated by the stripe pattern in the wake even though in reality the wavelength is larger than the length of the wake. The boundary of the energetic particle wake might be less sharp than plasma boundaries measured by SWAP and reported by *McComas et al.* (2016).

Pluto is "singing" in the sense that there is an electric field wave acting within the wake that affects the observed IPS ion intensities. The fact that the intensity is decreasing in the wake relative to the heliosphere informs us of the nature of the interaction between Pluto and its environment. If Pluto and its atmosphere were only slightly conducting, the interaction would be similar to, for example, Earth's Moon. A wake would form, but given that the interstellar pickup ions, similar to the suprathermal solar wind, are hot and relatively isotropic, this wake would refill closely behind the body (*Clack et al.,* 2004; *Halekas et al.,* 2011). Even the magnetopause of a magnetized body is not a sharp boundary to energetic ions (*Krupp et al.,* 2002; *Mauk et al.,* 2016; *Cohen et al.,* 2016). The interaction of Pluto with IPS ions is therefore surprisingly strong.

2.3. Search for X-Ray Signatures of Interaction

Using the Chandra Advanced CCD Imaging Spectrometer ACIS-S (*Grant et al.,* 2014) from Earth orbit, *Lisse et al.* (2017) obtained low-resolution imaging spectrophotometry of the Pluto system in support of the New Horizons flyby. Observations were taken in a trial "seed" campaign conducted in one pointed Chandra visit on February 24, 2014, and a follow-up campaign conducted soon after the New Horizons flyby that consisted of three visits spanning July 26–August 03, 2015. In a total of 174 ksec of on-target time, eight total photons were measured in a co-moving 11 × 11 pixel box measuring ~121,000 × 121,000 km^2 (or ~100 × 100 R_P) at Pluto. All detected photons were in the 0.31–0.60-keV passband with no 0.60–1.0-keV photons detected in this box during the same exposures. Allowing for

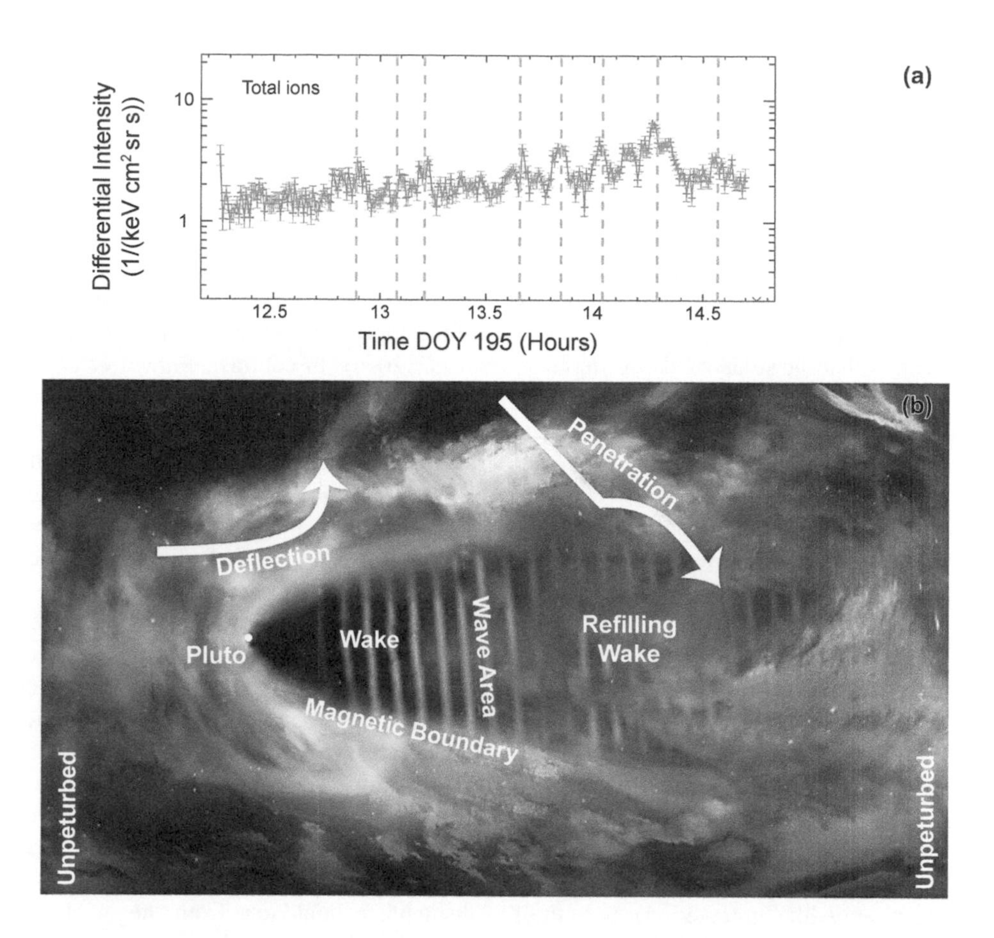

Fig. 3. (a) Sum of all ions at all energies and directions that are detected by PEPSSI as a function of time. **(b)** Summary sketch of the interaction between Pluto and the interplanetary medium (Sun to left). Pluto is embedded in a medium of interstellar pickup ions with kiloelectron-volt energies and suprathermal ions with energies of tens of kiloelectron volts and above, shown as bright clouds. Pluto is illustrated as a white dot. The primary effect of Pluto on these energetic ions is that it depletes their intensities and forms a wake in the anti-sunward direction, which is illustrated in black. The particle oscillations in the wake are indicated by the stripe pattern in the wake even though in reality the wavelength is larger than the length of the wake. Based on illustration by Mike Yakovlev (JHU/APL) in *Kollmann et al.* (2019b).

background, the net signal was 6.8 counts with a statistical noise level of 1.2 counts, producing a detection of Pluto in this passband at >99.95% confidence. The observed Pluto photons did not have the spectral shape of the background, were coincident with a 90% flux aperture co-moving with Pluto, and were not confused with any background source. The resulting estimated mean 0.31–0.60-keV X-ray power from Pluto is 200^{+200}_{-100} MW. This is in the middle range of X-ray power levels emitted by other known solar system emission sources: auroral precipitation, solar X-ray scattering, and charge exchange (CXE) between solar wind (SW) ions and atmospheric neutrals.

Auroral effects were eliminated as a source, since Pluto has no known intrinsic magnetic field and the New Horizons Alice UV spectrometer detected no airglow from Pluto during the flyby. The atmospheric haze particles discovered by New Horizons could have produced enhanced resonant scattering of solar X-rays from Pluto, but the energy signature of the detected photons did not match the solar spectrum and estimates of Pluto's scattered X-ray emission are 2 to 3 orders of magnitude below the $3.9 \pm 0.7 \times 10^{-5}$ cps found by Chandra. Charge-exchange-driven emission from plutogenic ions and heliospheric carbon, nitrogen, and oxygen (CNO) ions can produce the observed energy signature. The neutral gas escape rate from Pluto of 6×10^{25} s^{-1} deduced from New Horizons UV data (*Gladstone et al.,* 2016) can support the required $\sim 3.0^{+3.0}_{-1.5} \times 10^{24}$ X-ray photons per second emission rate. In fact, this rate is similar to the CH_4^+ ionization rate found by *McComas et al.* (2016) using the New Horizons SWAP instrument during the Pluto flyby (see section 2.1 above). However, using the solar wind proton density and speed measured by the SWAP instrument in the vicinity of Pluto at the time of the Chandra X-ray detections, *Lisse et al.* (2017) derived a factor of 40^{+40}_{-20} lower flux of solar wind minor ions flowing into the 11 × 11 pixel 90% flux box centered on Pluto than needed to support the observed X-ray emission rate. Thus, observers concluded that the solar wind must have been somehow significantly focused and enhanced within 60,000 km (projected) of Pluto for this mechanism to work during the Chandra observations.

This conclusion is in some current debate, despite the extraordinary care taken to determine that the 2014–2015 Chandra Pluto photons originated from a moving Pluto source; follow-up observations of Pluto by Lisse, McNutt, and Stern et al. taken at a different epoch (April 26–27, 2017) by the European Space Agency's (ESA) X-ray Multi-Mirror Mission (XMM-Newton) European Photon Imaging Camera (EPIC) pn-CCD camera (*Strüder et al.,* 2001) did not show any detectable emission from Pluto. The most likely scenario to explain the two different sets of observations, barring some heretofore-unknown Chandra systematic signal error, is that the solar wind was depleted at Pluto during the XMM observations of April 2017. XMM has about twice the X-ray collecting area of Chandra, but also about twice the system noise, and changes in Pluto's atmosphere (*Meza et al.,* 2019) and the solar wind flux could have been enough during the approach to solar minimum in 2017 to make Pluto's X-rays undetectable on the second attempt.

3. MODELING THE SOLAR WIND INTERACTION

Initial studies of the solar wind interaction with Pluto's atmosphere (e.g., *Bagenal and McNutt,* 1989; *Bagenal et al.,* 1997) suggested that the solar wind interaction with Pluto's atmosphere would depend on whether the atmospheric escape flux is strong (producing a "comet-like" interaction) or weak (producing a "Venus-like" interaction). In both of these descriptions, it is assumed that the planet's atmosphere/ionosphere and the solar wind could be considered as fluids. For many solar system bodies, fluid descriptions of a plasma obstacle interaction are sufficient. Global-scale magnetohydrodynamic (MHD) models have been successful in capturing the basic structure of such plasma interactions. With the IMF being very weak at Pluto's orbital distance, the length scales on which the plasma reacts are large compared with the size of the interaction region. For instance, due to the low IMF (0.08–0.3 nT) in the vicinity of Pluto at 33 AU (*Bagenal et al.,* 2016), the gyroradius of solar wind protons is ~23 R_P and typical plutogenic ion gyroradii may be on the order of 200–600 R_P when picked up in the full solar wind flow (depending on the ambient flow speed and magnetic field strength). Since these gyroradii are much larger than the obstacle itself, fluid treatments of ions are not sufficient. By contrast, electron gyroradii are very small compared to the interaction scale so a fluid approximation for them is valid. Given these scales it is clear that hybrid models with fully kinetic ions and fluid electrons are appropriate for the Pluto environment (*Kecskemety and Cravens,* 1993; *Shevchenko et al.,* 1997; *Delamere and Bagenal,* 2004; *Harnett et al.,* 2005; *Delamere,* 2009). In preparation for the New Horizons flyby, *Bagenal et al.* (2015) reviewed model predictions of the solar wind interaction with Pluto. Expectations of substantial atmospheric escape fluxes suggested a very large interaction region (up to ~1000 R_P) and indistinct boundaries (illustrated by Fig. 1c).

More recent simulations with a compact atmosphere model the small interaction region observed by New Horizons as described above (*Feyerabend et al.,* 2017; *Hale and Paty,* 2017; *Barnes et al.,* 2019). *Feyerabend et al.* (2017) and *Hale and Paty* (2017) reproduce the close boundary identified as a bow shock. Since the spacecraft was not equipped with a magnetometer, there is no measurement of the IMF, but *Feyerabend et al.* (2017) inferred a high value of the IMF, near 0.3 nT, by examining the solar-wind-slowing profile at the boundary crossing. Their value is on the order of the upper range of the Voyager measurements in *Bagenal et al.* (2016). In contrast, *Barnes et al.* (2019) show little or no IMF dependence in the solar-wind-slowing profile.

Figure 4 shows a comparison by *Barnes et al.* (2019) between the SWAP results at Pluto and a modeled synthetic SWAP instrument applied to simulations at several different IMF values. In all four the color indicates a species identification and the shade indicates total secondary channel electron multiplier (SCEM) counts from the (synthetic or actual) instrument. When interpreting these plots it is important to be aware that the method of species identification is different between the real SWAP spectrogram and the synthetic spectrograms, recognizing that neither method can capture the whole story. The SWAP identification method is described in *McComas et al.* (2016) and relies on a functional form for the mass dependence of the probability that ions penetrate the carbon foil. This method is not sensitive enough to identify individual species. Instead, it is used to distinguish "light" ions from "heavy" ions. In principle this method could separate the lighter heliospheric ions from the heavier plutogenic ions, but in addition to the uncertainty described in *McComas et al.* (2016), there is also an ambiguity whenever there is a mixture of light and heavy ions present in the same energy bin. The species identification for the synthetic spectrograms is much more direct because in the simulation it is possible to determine exactly how many counts come from each species individually. However, attempting to display everything in one plot introduces the same ambiguity when multiple species contribute counts to the same bin. The solution presented here is a compromise that shows key features clearly without being obviously misleading in too many places. Bins that have at least 2% of their counts from "heavy" CH_4^+ are colored red. The remaining cells are colored either blue or green depending on whether more counts are contributed by hydrogen or helium.

Barnes et al. (2019) note three major features in these plots. The first is the heavy ion burst before crossing into the interaction region as analyzed in *McComas et al.* (2016) and *Zirnstein et al.* (2016). The second is the broad heavy ion wake that fills almost the entire region where solar wind is excluded, and the third is the consistently low-energy heavy ions in the wake (below about 1000 eV/q.) The only simulation that has all three of these major features is the 0.08 nT simulation.

When taken together with physical explanations of these features, *Barnes et al.* (2019) argue that the comparisons of the model and SWAP data make a compelling case for low

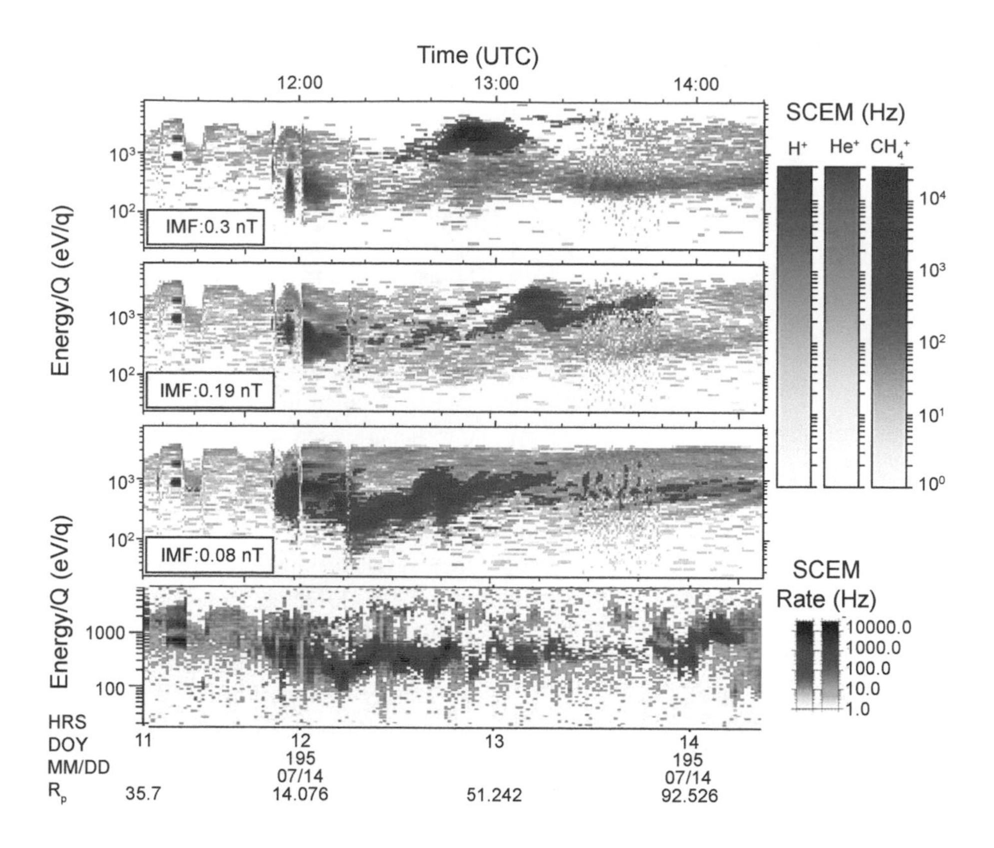

Fig. 4. Hybrid model of interaction and simulation of SWAP data shows slowing and deflection of the solar wind plus plutogenic heavy ions in the wake. To match the SWAP data, the model suggests a weak IMF of <0.1 nT. From *Barnes et al.* (2019).

IMF (<0.1 nT) during the New Horizons flyby consistent with the lower range of the scaled-down Voyager values in Table 1.

4. DISCUSSION

To summarize the main features of the SWAP observations over the New Horizons flyby of Pluto reported by *McComas et al.* (2016): The lack of any solar wind slowing until New Horizons was within 20 R_P indicates that almost no heavy ions were ionized within the several thousand Pluto radii upstream of Pluto. The SWAP data revealed a surprisingly small interaction region, confined on its upwind side to within ~5 R_P of Pluto, consistent with a small rate of atmospheric escape and an enhanced solar wind dynamic pressure at the time of the flyby. Fluxes of plutogenic heavy ions (most likely CH_4^+) were detected in the tail downstream, with the interaction persisting to a distance of ~200 R_P behind Pluto.

McComas et al. (2016) showed that the SWAP data upstream of Pluto indicated that the solar wind pressure was enhanced over typical conditions by about a factor of five. This means that under more typical conditions the interaction region would be larger (~10 R_P) and much bigger (~20–30 R_P) under weak solar wind pressures, assuming a constant atmospheric escape rate.

To summarize the main features of the PEPSSI observations at Pluto reported by *Kollmann et al.* (2019b): PEPSSI did not definitively detect evidence of plutogenic pick-up ions in its energy range within 500 R_P of Pluto, but the interplanetary energetic particle intensities are considerably perturbed by the interaction. Changes in PEPSSI measurements near Pluto's terminator suggest that <10-keV ions are accelerated and/or deflected away from the direction radially from the Sun. PEPSSI observed decreased suprathermal particles in the wake of the interaction region. The particle intensities near Pluto decreased by a factor of ~10 below the heliospheric value and increased with distance downstream.

Figure 5 presents a summary schematic of the solar wind interaction with Pluto based on data from SWAP (*McComas et al.,* 2016) and PEPSSI (*Kollmann et al.,* 2019b) along the New Horizons trajectory plus global models that match the SWAP data (*Barnes et al.,* 2019). The basic structure and dimensions are constrained by the locations where the spacecraft entered and exited the interaction. While there were no direct measurements of the upstream plutogenic heavy pick-up (approximately kiloelectron-volt) ions, the models indicate that the fluxes are small and the particles move along vertical paths that likely did not intersection the spacecraft. In the tail of the interaction, where the plasma flow is slowed, the plutogenic ions are picked up at much lower energies (~100 eV), reaching densities of ~0.01 cm^{-3}

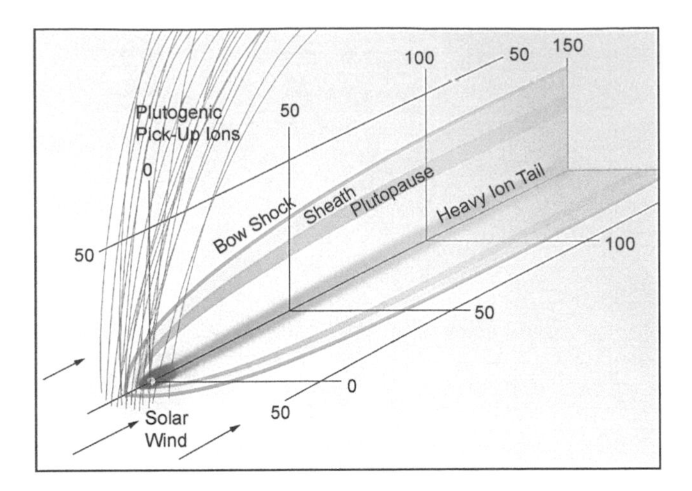

Fig. 5. Schematic of the solar wind interaction with Pluto.

[still lower than the ambient solar wind proton density at ~0.025 cm^{-3} reported by *McComas et al.* (2016)].

Barnes et al. (2019) argue that in many respects Pluto's solar wind interaction is similar to that of weak comets. For example, Comet 19P/Borrelly was observed to have a significant north-south asymmetry (*Young et al.,* 2004) attributed to large gyroradius effects (*Delamere and Bagenal,* 2004; *Delamere,* 2006). Weak comets also emit X-rays at low levels due to solar-wind-escaping, neutral-atom charge exchange. However, Pluto's broad heavy ion wake is unique. The weak comets 19P/Borrelly, 26P/Grigg-Skjellerup, and the Active Magnetospheric Particle Tracer Explorers (AMPTE) artificial comet all exhibited narrow heavy ion wakes (*Delamere et al.,* 1999; *Johnstone et al.,* 1993; *Young et al.,* 2004). Pluto seems to be in just the right regime where low IMF and the thermal pressure provided by interstellar pick-up ions can produce the broad heavy ion wake that was observed.

Overall, Pluto's interaction with the solar wind appears to be a hybrid, with (1) the bow shock generated by mass loading like at a comet; (2) an asymmetric tail common with weak comets, but with the obstacle to the solar wind flow, the Plutopause, sustained by thermal pressure; and (3) electrical currents in the ionosphere, similar to the induced magnetopause at Venus and Mars. In fact, Pluto seems to join all unmagnetized solar system bodies that seem to share solar wind interactions with, more or less, the same ~10,000-km physical scale (*Brain et al.,* 2016). While the terrestrial planets are much bigger than Pluto, their stronger gravity confines their atmospheres and the solar wind pressure is stronger in the inner solar system. Remarkably, the absolute scale of these various solar wind interactions (including Titan and active comets) end up being about the same. This intriguing common scale is worth exploring.

The post-New Horizons estimate of *Strobel and Zhu* (2017) for the current net atmospheric escape flux (mostly methane) is ~10^{27} amu s^{-1} or a few kilograms per second. If this escape rate were maintained for 4 b.y., the net loss is just 14 m of ice from the surface of Pluto. The flow of CH_4^+ ions down the tail of the interaction is just a few percent of this net neutral escape (*McComas et al.,* 2016), suggesting that most of the neutrals escape into the outer heliosphere. Even if hundreds of KBOs are losing this amount of atmosphere, the source of neutrals (which eventually are photoionized and picked up by the solar wind) is negligible compared with the interstellar source (*McComas et al.,* 2017).

While the scale of the interaction region will likely "breathe" in and out by a factor of ~5–10 as the solar wind varies over timescales of weeks to years, the major question is how the interaction might change as the atmospheric escape varies over Pluto's seasonal and orbital timescales (see the chapter in this volume by Young et al.). As Pluto's atmospheric escape rate is reduced, the scale of the interaction will shrink. Should the atmosphere collapse (the exobase reducing to the surface) the interaction will become like that of the Moon, with the solar wind being absorbed on the dayside leaving a cavity on the nightside (e.g., *Bagenal et al.,* 1997). Direct impact of the solar wind plasma and energetic particle populations would drive radiolysis of the surface ices, which may have important consequences for the surface chemistry if low atmospheric conditions persist for large fractions of Pluto's orbit.

More immediately, the question is what future exploration of Pluto might provide. Given the asymmetric and variable nature of the interaction indicated by the New Horizons data and models thereof, it would be highly desirable to measure particle fluxes over a larger volume and timespan than provided by a flyby. While the addition of instrumentation to measure electric and/or magnetic fields would also be desirable, it must be recognized that the weak fields of the outer heliosphere are difficult to measure accurately. Not surprisingly, a Pluto orbiter spending more time much closer to the planet would be needed to resolve some of the questions. Finally, the New Horizons measurements at Pluto beg the question of what potential flybys of other dwarf planets in the Kuiper belt might tell us about the solar wind interactions under different atmospheric conditions.

REFERENCES

Bagenal F. and McNutt R. L. (1989) Pluto's interaction with the solar wind. *Geophys. Res. Lett., 16,* 1229–1232.

Bagenal F., Cravens T. E., Luhmann J. G., McNutt R. L., and Cheng A. F. (1997) Pluto's interaction with the solar wind. In *Pluto and Charon* (D. J. Tholen and S. A. Stern, eds.), pp. 523–555. Univ. of Arizona, Tucson.

Bagenal F., Delamere P. A., Elliott H. A., Hill M. E., Lisse C. M., McComas D. J., McNutt R. L. Jr., Richardson J. D., Smith C. W., and Strobel D. F. (2015) Solar wind at 33 AU: Setting bounds on the Pluto interaction for New Horizons. *J. Geophys. Res., 120,* 1497–1511, DOI: 10.1002/2015JE004880.

Bagenal F. et al. (2016) Pluto's interaction: Solar wind, energetic particles, dust. *Science, 351,* 6279, DOI: 10.1126/science.aad9045.

Barnes N. P., Delamere P. A., Strobel D. F., Bagenal F., McComas D. J., Elliott H. A., and Stern S. A. (2019) Constraining the IMF at Pluto using New Horizons SWAP data and hybrid simulations. *J. Geophys. Res., 124(3),* 1568–1581, DOI: 10.1029/2018JA026083.

Brain D. A., Bagenal F., Ma Y.-J., Nilsson H., and Stenberg Wieser G. (2016) Atmospheric escape from unmagnetized bodies. *J. Geophys. Res., 121(12),* 2364–2385, DOI: 10.1002/2016JE005162.

Clack D., Kasper J. C., Lazarus A. J., Steinberg J. T., and Farrell W. M. (2004) Wind observations of extreme ion temperature anisotropies in the lunar wake. *Geophys. Res. Lett., 31,* L06812, DOI:

10.1029/2003GL018298.

Cohen I. J., Mauk B. H., Anderson B. J., Westlake J. H., Sibeck D. G., Giles B. L., Pollock C. J., et al. (2016) Observations of energetic particle escape at the magnetopause: Early results from the MMS Energetic Ion Spectrometer (EIS). *Geophys. Res. Lett., 43,* 5960–5968, DOI: 10.1002/2016GL068689.

Delamere P. A. (2006) Hybrid code simulations of the solar wind interaction with comet 19P/Borrelly. *J. Geophys. Res, 111,* A12217, DOI: 10.1029/2006JA011859.

Delamere P. A. (2009) Hybrid code simulations of the solar wind interaction with Pluto. *J. Geophys. Res, 114,* A03220, DOI: 10.1029/2008JA013756.

Delamere P. A. and Bagenal F. (2004) Pluto's kinetic interaction with the solar wind. *Geophys.Res. Lett., 31(4),* DOI: 10.1029/2003GL018122.

Delamere P. A., Swift D. W., and Stenbaek-Nielsen H. C. (1999) A three-dimensional hybrid code simulation of the December 1984 solar wind AMPTE release. *Geophys. Res. Lett., 26,* 2837.

Elliot J. L., Dunham E. W., Bosh A. S., Slivan S. M., Young L. A., Wasserman L. H., and Millis R. L. (1989) Pluto's atmosphere. *Icarus, 77,* 148.

Elliott H. A., McComas D. J., Valek P., Nicolaou G., and Weidner S. (2016) New Horizons Solar Wind Around Pluto (SWAP) observations of the solar wind from 11–33 AU. *Astrophys. J. Suppl., 223(2),* DOI: 10.3847/0067-0049/223/2/19.

Elliott H. A. and 15 colleagues (2019) Slowing of the solar wind in the outer heliosphere. *Astrophys. J., 885(2).*

Feyerabend M., Liuzzo L., Simon S., and Motschmann U. (2017) A three-dimensional model of Pluto's interaction with the solar wind during the New Horizons encounter. *J. Geophys. Res, 122,* 10356–10368, DOI: 10.1002/2017ja024456.

Gladstone G. R. et al. (2016) The atmosphere of Pluto as observed by New Horizons. *Science, 351(6279),* DOI: 10.1126/science.aad8866.

Grant C. E., Bautz M. W., Ford P. G., and Plucinsky P. P. (2014) Fifteen years of the Advanced CCD Imaging Spectrometer. *Proc. SPIE, 9144,* DOI: 10.1117/12.2055652.

Hale J. P. M. and Paty C. S. (2017) Pluto-Charon solar wind interaction dynamics. *Icarus, 287,* 131–139, DOI: 10.1016/j.icarus.2016.11.036.

Halekas J. S., Angelopoulos V., Sibeck D. G., Khurana K. K., Russell C. T., Delory G. T., Farrell W. M., et al. (2011) First results from ARTEMIS, a new two-spacecraft lunar mission: Counter-streaming plasma populations in the lunar wake. *Space Sci. Rev., 165,* 93–107, DOI: 10.1007/s11214-010-9738-8.

Harnett E. M., Winglee R. M., and Delamere P. A. (2005) Three dimensional multi-fluid simulations of Pluto's magnetosphere: A comparison to 3D hybrid simulations. *Geophys. Res. Lett., 32,* L19104, DOI: 10.1029/2005GL023178.

Hinson D. P., Linscott I. R., Young L. A., Tyler G. L., Stern S. A., Beyer R. A., Bird M. K., et al. (2017) Radio occultation measurements of Pluto's neutral atmosphere with New Horizons. *Icarus, 290,* 96–111, DOI: 10.1016/j.icarus.2017.02.031.

Hinson D. P. and 15 colleagues (2018) An upper limit on Pluto's ionosphere from radio occultation measurements with New Horizons. *Icarus, 307,* 17–24.

Ip W.-H., Kopp A., Lara L. M., and Rodrigo R. (2000) Pluto's ionospheric models and solar wind interaction. *Adv. Space Res., 26,* 1559–1563, DOI: 10.1016/S0273-1177(00)00098-3.

Johnstone A. D., Coates A. J., Huddleston D. E., Jockers K., Wilken B., Borg H., et al. (1993) Observations of the solar wind and cometary ions during the encounter between Giotto and comet Grigg-Skjellerup. *Astron. Astrophys., 273,* L1–L4.

Kecskemety K. and Cravens T. E. (1993) Pick-up ions at Pluto. *Geophys. Res. Lett., 20,* 543–546.

Kollmann P., Hill M. E., McNutt R. L. Jr., Brown L. E., Allen R. C., Clark G., Andrews B., et al. (2019a) Suprathermal Ions in the outer heliosphere. *Astrophys. J., 876,* 46, DOI: 10.3847/1538-4357/ab125f.

Kollmann P., and 39 colleagues (2019b) Pluto's interaction with energetic heliospheric ions. *J. Geophys. Res., 124(9),* 7413–7424.

Krasnopolsky V. A. (1999) Hydrodynamic flow of N_2 from Pluto. *J. Geophys. Res., 104,* 5955–5962, DOI: 10.1029/1998JE900052.

Krupp N., Woch J., Lagg A., Espinosa S. A., Livi S., Krimigis S. M., Mitchell D. G., et al. (2002) Leakage of energetic particles from Jupiter's dusk magnetosphere: Dual spacecraft observations. *Geophys. Res. Lett., 29,* 1736, DOI: 10.1029/2001GL014290.

Lellouch E., de Bergh C., Sicardy B., Forget F., Vangvichith M., and Käufl H. U. (2015) Exploring the spatial, temporal, and vertical distribution of methane in Pluto's atmosphere. *Icarus, 246,* 268–278.

Lellouch E., Gurwell M., Butler B., Fouchet T., Lavvas P., Strobel D.F., Sicardy B., et al. (2017) Detection of CO and HCN in Pluto's atmosphere with ALMA. *Icarus, 286,* 289–307.

Lisse C. M., McNutt R. L. Jr., Wolk S. J., Bagenal F., Stern S. A., Gladstone G. R., Cravens T. E., et al. (2017) The puzzling detection of X-rays from Pluto by Chandra. *Icarus, 287,* 103–109.

Mauk B. H., Cohen I. J., Westlake J. H., and Anderson B. J. (2016) Modeling magnetospheric energetic particle escape across Earth's magnetopause as observed by the MMS mission. *Geophys. Res. Lett., 43,* 4081–4088, DOI: 10.1002/2016GL068856.

McComas D. J., Ebert R. W., Elliott H. A., Goldstein B. E., Gosling J. T., Schwadron N. A., and Skoug R. M. (2008a) Weaker solar wind from the polar coronal holes and the whole Sun. *Geophys. Res. Lett., 35,* L18103, DOI: 10.1029/2008GL034896.

McComas D., Allegrini F., Bagenal F., Casey P., Delamere P., Demkee D., Dunn G., et al.(2008b) The Solar Wind Around Pluto (SWAP) instrument aboard New Horizons. *Space Sci. Rev., 140,* 261–313.

McComas D. J., Elliott H. A., and Schwadron N. A. (2010) Pickup hydrogen distributions in the solar wind at ~11 AU: Do we understand pickup ions in the outer heliosphere? *J. Geophys. Res., 115,* A03102, DOI: 10.1029/2009JA014604.

McComas D. J., Angold N., Elliott H. A., Livadiotis G., Schwadron N. A., Skoug R. M., and Smith C. W. (2013) Weakest solar wind of the space age and the current "mini" solar maximum. *Astrophys. J., 779,* 2, DOI: 10.1088/0004-637X/779/1/2.

McComas D. J., Elliott H. A., Weidner S., Valek P., Zirnstein E. J., Bagenal F., Delamere P. A., et al. (2016) Pluto's interaction with the solar wind. *J. Geophys. Res., 121,* 4232–4246. DOI: 10.1002/2016JA022599.

McComas D. J., Zirnstein E. J., Bzowski M., Elliott H. A., Randol B., Schwadron N. A., Sokół J. M., et al. (2017) Interstellar pickup ion observations to 38 AU. *Astrophys. J. Suppl., 233,* 8, DOI: 10.3847/1538-4365/aa91d2.

McNutt R. L. Jr., Livi S. A., Gurnee R. S., Hill M. E., Cooper K. A., Andrews G., Bruce K., et al. (2008) The Pluto Energetic Particle Spectrometer Science Investigation (PEPSSI) on the New Horizons mission. *Space Sci. Rev., 140,* 315–385.

Meza E. et al. (2019) Pluto's lower atmosphere and pressure evolution from ground-based stellar occultations, 1988–2016. *Astron. Astrophys., 625,* A42.

Nimmo F., Umurhan O., Lisse C. M., Bierson C. J., Lauer T. R., Buie M. W., Throop H. B., et al. (2017) Mean radius and shape of Pluto and Charon from New Horizons images. *Icarus, 287,* 12–29, DOI: 10.1016/j.icarus.2016.06.027.

Randol B. M., Elliott H. A., Gosling J. T., McComas D. J., and Schwadron N. A. (2012) Observations of isotropic interstellar pick-up ions at 11 and 17 AU from New Horizons. *Astrophys. J., 755,* 75.

Randol B. M., McComas D. J., and Schwadron N. A. (2013) Interstellar pick-up ions observed between 11 and 22 AU by New Horizons. *Astrophys. J., 768,* 120, DOI: 10.1088/0004-637X/768/2/120.

Richardson J. D., Liu Y., Wang C., and McComas D. J. (2008) Determining the LIC H density from the solar wind slowdown. *Astron. Astrophys., 491,* 1–5, DOI: 10.1051/0004-6361:20078565.

Sauer K., Lipatov A., Baumgaertel K., Dubinin E., and Dubinin E. (1997) Solar wind-Pluto interaction revised. *Adv. Space Res., 20,* 295.

Saunders M. A. and Russell C. T. (1986) Average dimension and magnetic structure of the distant Venus magnetotail. *J. Geophys. Res., 91,* 5589–5604.

Shevchenko V. I., Ride S. K., and Baine M. (1997) Wave activity near Pluto. *Geophys. Res. Lett., 24,* 101–104, DOI: 10.1029/96GL03696.

Sibeck D. G., Lopez R. E., and Roelof E. C. (1991) Solar wind control of the magnetopause shape, location, and motion. *J. Geophys. Res., 96,* 5489–5495, DOI: 10.1029/90JA02464.

Steffl A. J., Shinn A. B., Gladstone G. R., Parker J. W., Retherford K. D., Slater D. C., Versteeg M. H., and Stern S. A. (2012) MeV electrons detected by the Alice UV spectrograph during the New Horizons flyby of Jupiter. *J. Geophys. Res., 117,* A10222, DOI: 10.1029/2012JA017869.

Stern S. A., Slater D. C., Scherrer J., et al. (2008) ALICE: The ultraviolet imaging spectrograph aboard the New Horizons Pluto-Kuiper belt mission. *Space Sci. Rev., 140,* 155. DOI: 10.1007/s11214-008-9407-3.

Stern S. A. and 150 colleagues (2015) The Pluto system: Initial results from its exploration by New Horizons. *Science, 350,* aad1815.

Strobel D. F. and Zhu X. (2017) Comparative planetary nitrogen

atmospheres: Density and thermal structures of Pluto and Triton. *Icarus, 291,* 55–64.
Strüder L., Briel U., Dennerl K., Hartmann R., Kendziorra E., Meidinger N., Pfeffermann E., et al. (2001) The European Photon Imaging Camera on XMM-Newton: The pn-CCD camera. *Astron. Astrophys., 365,* L18–L26.
Tian F. and Toon O. B. (2005) Hydrodynamic escape of nitrogen from Pluto. *Geophys. Res. Lett., 32,* L18201, DOI: 10.1029/2005GL023510.
Trafton L. M., Hunten D. M., McNutt R. L. Jr., and Zahnle K. J. (1997) Escape processes at Pluto and Charon. In *Pluto and Charon* (D. J. Tholen and S. A. Stern, eds.), pp. 475–521. Univ. of Arizona, Tucson.
Tyler G. L., Sweetnam D. N., Anderson J. D., Borutzki S. E., Campbell J. K., Eshleman V. R., Gresh D. L., et al. (1989) Voyager radio science observations of Neptune and Triton. *Science, 246,* 1466–1473, DOI: 10.1126/science.246.4936.1466.
Weaver H. A., Gibson W. C., Tapley M. B., Young L. A., and Stern S. A. (2008) Overview of the New Horizons science payload. *Space Sci. Rev., 140,* 75–91.
Young D., Crary F., Nordholt J., Bagenal F., Boice D., Burch J., et al. (2004) Solar wind interactions with comet 19P/Borrelly. *Icarus, 167,* 80–88, DOI: 10.1016/j.icarus.2003.09.011.
Young L. A., Cook J. C., Yelle R. V., and Young E. F. (2001) Upper limits on gaseous CO at Pluto and Triton from high-resolution near-IR spectroscopy. *Icarus, 153,* 148–156.
Young L. A., Stern S. A., Weaver H. A., Bagenal F., Binzel R. P.; Buratti B., Cheng A. F., et al. (2008) New Horizons: Anticipated scientific investigations at the Pluto System. *Space Sci. Rev., 140,* 93–127.
Young L. A. and 25 colleagues (2018) Structure and composition of Pluto's atmosphere from the New Horizons solar ultraviolet occultation. *Icarus, 300,* 174–199.
Zhu X., Strobel D. F., and Erwin J. T. (2014) The density and thermal structure of Pluto's atmosphere and associated escape processes and rates, *Icarus, 228,* 301-314.
Zirnstein E. J., McComas D. J., Elliott H. A., Weidner S., Valek P. W., Bagenal F., et al. (2016) Interplanetary magnetic field sector from Solar Wind Around Pluto (SWAP) measurements of heavy ion pickup near Pluto. *Astrophys. J. Lett., 823,* L30, DOI: 10.3847/2041-8205/823/2/l30.
Zong Q., Wang Y., Yang B., Fu S., Pu Z., Xie L., and Fritz T. A. (2008) Recent progress on ULF wave and its interactions with energetic particles in the inner magnetosphere. *Sci. China, Ser. E: Technol. Sci., 51(10),* 1620–1625, DOI: 10.1007/s11431-008-0253-z.

Part 3:

Charon and Pluto's Small Satellites

Spencer J. R., Beyer R. A., Robbins S. J., Singer K. N., and Nimmo F. (2021) The geology and geophysics of Charon. In *The Pluto System After New Horizons* (S. A. Stern, J. M. Moore, W. M. Grundy, L. A. Young, and R. P. Binzel, eds.), pp. 395–412. Univ. of Arizona, Tucson, DOI: 10.2458/azu_uapress_9780816540945-ch017.

The Geology and Geophysics of Charon

John R. Spencer
Southwest Research Institute

Ross A. Beyer
SETI Institute and NASA Ames Research Center

Stuart J. Robbins, Kelsi N. Singer
Southwest Research Institute

Francis Nimmo
University of California, Santa Cruz

Charon is geologically simpler than Pluto, but its surface nevertheless reveals an active early geological history, likely completed within the first billion years of solar system history. The most likely interpretation of New Horizons images is that most observed endogenic activity was associated with the freezing of an early internal ocean. Ocean freezing would have produced global expansion as ice replaced higher-density liquid water, resulting in the observed system of graben, up to 12 km deep. The global expansion also provides strong evidence that Charon is differentiated. Concentration of volatiles such as ammonia in the last remnants of the ocean may have reduced the ocean density sufficiently to allow eruption of a viscous liquid-solid slurry onto the surface. The result is manifest as the extensive smooth plains of Vulcan Planitia, which provides the clearest example of effusive cryovolcanism yet seen in the solar system. Subsequent impact cratering provides constraints on the population of small impactors in the outer solar system.

1. KNOWLEDGE PRIOR TO NEW HORIZONS

Almost nothing was known about the geology of Charon prior to the New Horizons encounter. Its diameter was well-known from stellar occultations (*Walker,* 1980; *Gulbis et al.,* 2006; *Person et al.,* 2006; *Sicardy et al.,* 2006), as was the predominantly water-ice composition of its surface (*Marcialis et al.,* 1987) and its geometric albedo, 0.38 (*Buie et al.,* 1997), which is relatively low for water ice and thus required the presence of a dark contaminant of some kind. The presence of crystalline water ice in its spectrum, despite the fact that ultraviolet (UV) radiation might be expected to amorphize the ice over time, was used to argue for a young surface renewed by ongoing geological activity (*Cook et al.,* 2007). However, this idea was not universally accepted (*Porter et al.,* 2010). The presence of ammonia-bearing species on Charon's surface (*Brown and Calvin,* 2000; *Dumas et al.,* 2001) also hinted at ongoing activity and provided a possible mechanism for mobilizing water ice, given the likely small endogenic heat budget of Charon, by reducing its melting point (*Desch et al.,* 2009). Charon's density was known to be near 1700 kg m^{-3}, indicating a substantial rock fraction and thus a modest amount of radiogenic heating. Although Charon is similar in size, density, albedo, and surface composition to several satellites of Saturn and Uranus (in particular Ariel and Dione), some of which show evidence for geological activity, it has not experienced significant tidal heating since its orbit circularized shortly after formation, so comparison of its geological history to that of icy satellites of the giant planets was awaited with interest.

2. NEW HORIZONS IMAGING OF CHARON

Imaging of Charon, with resolution and coverage comparable to that of Pluto, was a major goal of the New Horizons flyby (Fig. 1). The New Horizons Long Range Reconnaissance Imager (LORRI) camera first resolved Charon from Pluto in July 2013, at a range of 5.9 AU, and began to reveal surface features in approach images in May 2015, two months before encounter. Near-daily LORRI imaging

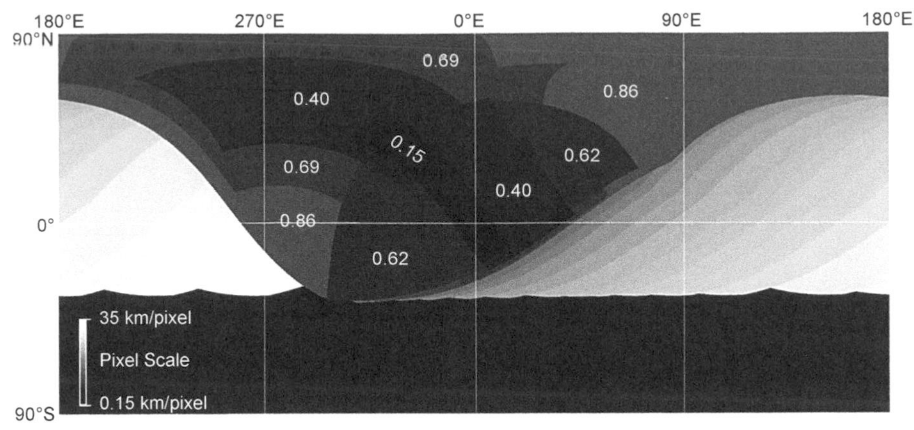

Fig. 1. New Horizons imaging coverage of Charon, showing the finest pixel scale obtained at each location. Numbers give the pixel scale of the encounter hemisphere imaging, in kilometers per pixel. Adapted from *Schenk et al.* (2018a).

from late May 2015 onward revealed steadily increasing detail, and Charon's dark polar cap became evident in early June 2015. Starting on July 7, 2015, a week before encounter, Charon was far enough from Pluto to require separate LORRI targeting. Complete coverage of the illuminated portion of Charon (north of latitude 38°S) was obtained during the final rotation before encounter (Fig. 2), with pixel scale of 37 km/pixel or better, and with no gaps in rotational coverage larger than 27°. Despite the low (15°) approach phase angle, the high amplitude of Charon's topography meant that some topographic detail was visible near the terminator as early as five days before encounter.

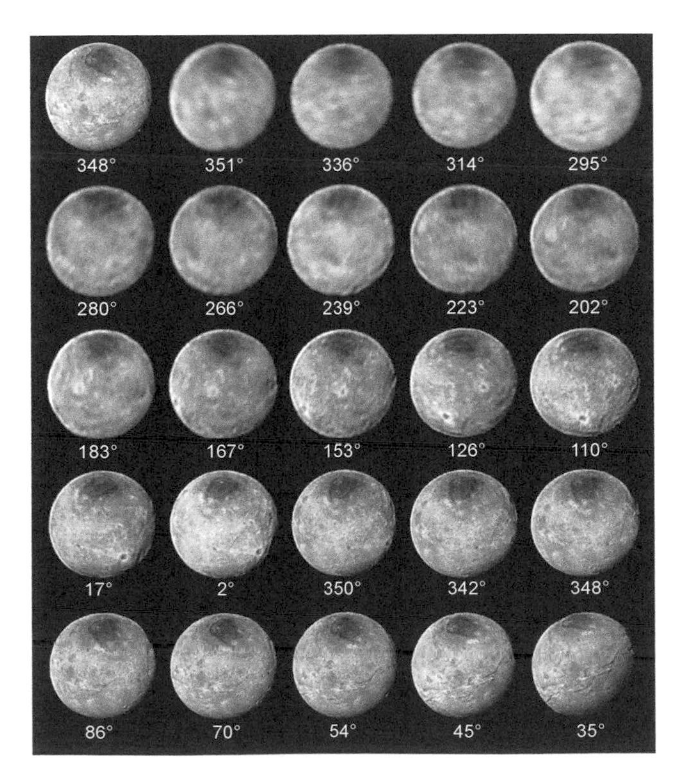

Fig. 2. Global imaging of Charon during the final rotation before encounter (*Beyer et al.,* 2017). Note the topographic shading near the terminator (lower right) in many images. North is up, and central longitude is given for each frame.

Close-approach imaging was focused on the Pluto-facing hemisphere and north polar region. The last full-frame LORRI image was taken at 02:55 on July 14, followed by a 2 × 2 mosaic at 05:50, a 3 × 2 mosaic at 08:32 (the last full-disk LORRI mosaic, with pixel scale 0.86 km/pixel), and then a series of two limb-to-terminator strips of LORRI images riding along with two Linear Etalon Imaging Spectral Array (LEISA) spectroscopic scans. These were followed by the best color image from the Multispectral Visible Imaging Camera (MVIC) (Fig. 3). The highest-resolution global image was taken with MVIC in panchromatic mode near closest approach, with a resolution 0.62 km/pixel and phase angle of 85° (Fig. 4). A simultaneous LORRI rider obtained the highest-resolution images of all, with pixel scale 0.15 km/pixel, covering a 150-km-wide limb-to-terminator strip (Fig. 5). After close approach, Charon's thin crescent was imaged many times at phase angles of 165°–169° (Fig. 6), and it was last clearly detected in routine imaging on July 18. A series of long-exposure, low-resolution (60 km/pixel) images taken on July 17 successfully used Pluto light to image Charon's nightside, including part of the south polar region not seen in sunlight.

Images have been combined to generate controlled global mosaics (Fig. 7a) (see also Appendix A by Beyer and Showalter in this volume). By design, many of the close-approach images also provided stereo coverage, allowing production of detailed digital elevation models of the encounter hemisphere (*Schenk et al.,* 2018a) (Fig. 7b).

3. GEOMORPHOLOGICAL MAPPING OF CHARON

3.1. Introduction

Geomorphologic mapping investigations can provide information about solid surfaces that are not available through

Fig. 3. The highest-resolution global color image of Charon, taken with MVIC with phase angle 39° and pixel scale 1.4 km per pixel. North is up. The relatively smooth plains of Vulcan Planitia (lower right, near the terminator) are separated from the highland terrain of Oz Terra (upper left) by aligned tectonic features. Faulting is present throughout much of Oz Terra but is less conspicuous further from Vulcan Planitia due to high Sun angles. The conspicuous red polar region is discussed in the chapters by Protopapa et al. and Howett et al. in this volume.

other means. The resultant map represents a synthesis and interpretation of a planetary surface based often on numerous types of input data.

The information in this section is primarily based on *Robbins et al.* (2019). As is true through this book, some names are formally approved by the International Astronomical Union (IAU), while others are provisional (see Appendix A by Beyer and Showalter in this volume).

3.2. Data

The geomorphologic mapping (Fig. 8) used the panchromatic map produced by *Schenk et al.* (2018a) (Fig. 7a) as a basemap, supplemented by additional LORRI and MVIC imaging. Because the projection of images in this map did not account for topographic effects, horizontal displacements due to topography of up to several kilometers are present in these maps. When *Robbins et al.* (2019) mapped contacts and other features, images with the lowest emission angle, which have the least topographic distortion, were preferred.

Ancillary datasets used for the mapping were color maps (from *Schenk et al.,* 2018a), some derived compositional data (*Grundy et al.,* 2016; *Dalle Ore et al.,* 2018), and topographic reconstructions (*Schenk et al.,* 2018a) (Fig. 7b).

3.3. Geomorphological Units

The New Horizons encounter hemisphere of Charon is dominated by two primary physiographic units: a rugged northern region, Oz Terra, and a smoother southern plain, Vulcan Planitia.

Oz Terra is largely mapped as blocky terrain (Bl) and is described in detail below. The Bl terrain is characterized by large, high-standing plateaus separated by scarps and troughs. In some locations the plateaus have raised edges. The scale of the blocks is hundreds of kilometers, while the troughs separating them are tens of kilometers across. Blocky terrain likely extends beyond the map boundary, but just how far it extends is unknown.

The second primary unit is smooth terrain (Sm), separated into three subunits based on location ($Sm_{1,2}$) and whether the terrain is elevated above the surrounding material (Sme). Sm_1 and Sme occupy most of Vulcan Planitia, described in detail below. For the map, Sm is characterized as relatively smooth compared with Bl at scales of tens of kilometers, although it contains many broad and gently sloping swells at scales of hundreds of kilometers. Vulcan Planitia is made mostly from Sm_1 material, while the Sm_2 unit protrudes into Bl near the prime (sub-Pluto) meridian. Throughout Sm_1 are a few small (~tens of kilometers) Sme units, identified in work as early as *Moore et al.* (2016).

Other than crater material, the third-largest unit by area is termed mottled terrain (Mt), which appears within Vulcan Planitia (Fig. 11d, described below). Mottled terrain is characterized by closely spaced pits and/or hills (it is difficult to determine which) with a typical scale of hundreds to thousands of meters. This unit is unusual on bodies imaged thus far in the solar system, although somewhat similar but smaller-scale pitted terrains are seen elsewhere, e.g., on Vesta (*Denevi et al.,* 2012) and Mercury (*Blewett et al.,* 2011). Both Sm_1 and Mt abut the map boundary at the terminator, and certainly extend past it. Whether Mt extends further north is uncertain due to the Sun angle in the encounter images.

The fourth-largest unit is rough terrain (Rt), immediately east of Vulcan Planitia, centered near 20°N, 50°E (Fig. 10). This unit is a raised region that is the roughest (at ~tens of kilometers in scale) and most chaotic on the encounter hemisphere. This terrain also extends off the map boundary. It has a central depression, 4.5 km deep (*Schenk et al.,* 2018a), the center of which was unilluminated in the encounter images, and it contains slopes in excess of 15°. The feature is unique on the encounter hemisphere, and it is unique among imaged solar system surfaces. Its origin is enigmatic, but it may be due to unusually viscous cryovolcanism or buoyant uplift of subsurface material in response to crustal foundering (*Robbins et al.,* 2019).

Smaller, scattered units throughout the map area include numerous montes (Mo), depressions (Dm), and lobate aprons (La), the latter visible as mass wasting caused by impacts and landslides caused by material failure (*Beddingfield et al*., 2020) (Fig. 12).

Fig. 4. The highest-resolution global image of Charon, taken with MVIC with phase angle 85° and resolution 0.62 km per pixel. The view is dominated by the smooth plains of Vulcan Planitia, which occupy the lower half of the disk.

Finally, impact craters (Cr) and ejecta (Ej) occupy a significant fraction of the map area. These units are emplaced upon all other observed units that are relatively large (i.e., not observed on Sme, La, Dm, nor most Mo). Craters are discussed in more detail in section 6 of this chapter, and in the chapter by Singer et al. in this volume.

4. OZ TERRA

4.1. Introduction

The encounter hemisphere of Charon, aside from the smooth plains of Vulcan Planitia discussed in the next section, is dominated by Oz Terra, which is exceptionally rugged compared, for instance, to the inner icy moons of Saturn (*Schenk et al.,* 2018a,b). This terrain is the oldest on Charon and is separated by scarps and faults from the younger Vulcan Planitia. It contains a network of scarps and fault-bounded troughs in the equatorial to middle latitudes (Fig. 9b), transitioning northward and over the pole to the visible limb into an irregular zone containing depressions sometimes more than 12 km deep as well as other large relief variations (Fig. 9a). The lower-resolution views of the non-encounter hemisphere (Fig. 2) are also suggestive of other potential large ridges and troughs, indicating that the tectonic features that are well-resolved on the encounter hemisphere likely extend around Charon (*Beyer et al.,* 2017). The dominance of linear troughs and scarps, displaying what appear to be normal faults, graben, half-graben, and tilted fault blocks, appears most consistent with extensional tectonics (*Moore et al.,* 2016; *Beyer et al.,* 2017; *Schenk et al.,* 2018a), although a compressional interpretation for the morphology of some features has also been suggested (*Chen and Yin,* 2019). The global distribution of the troughs suggests that Charon underwent global extension, and the fact that Vulcan Planitia (which is itself is quite heavily cratered) appears to post-date the trough formation indicates that the extension occurred early in Charon's history. The orientation of most of the extensional features in the encounter hemisphere is predominantly east-west (Figs. 8 and 9b) (*Beyer et al.,* 2017). This east-west pattern is dominated by fracturing near the Oz/Vulcan boundary. Farther north, troughs in a variety of orientations produce polygonal blocks, best seen in the digital elevation model, e.g., in the region centered on 30°N, 350°E (Fig. 7b).

4.2. Global Extension and Elastic Lithosphere Thickness

The global characteristics of Charon's tectonics have important implications for its geologic history. In summary, (1) extension seems to dominate over the entirety of Charon's northern terrain, up to and including the scarps that border Vulcan Planitia, whose resurfaced units bury the flanks of the rifts; (2) there is no compelling

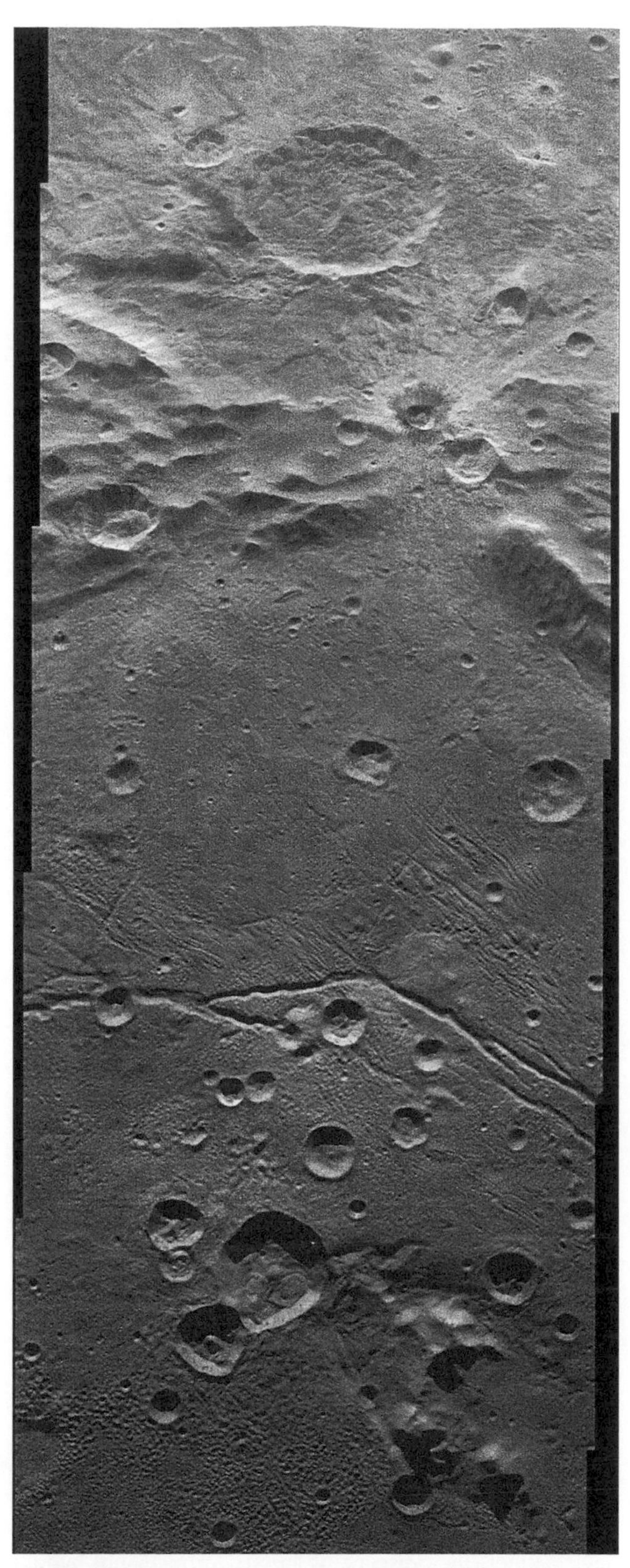

Fig. 5. Part of the highest-resolution (0.15 km per pixel) LORRI imaging of Charon, taken simultaneously with the MVIC image in Fig. 4 and showing the transition between Oz Terra (top) and Vulcan Planitia (bottom). Mottled terrain is visible at the bottom of the frame, while a large crater with interior terracing, and a smaller fresh crater with dark inner ejecta and bright outer ejecta are visible near the top of the frame. The image is 150 km wide.

evidence for compressional faulting or strike-slip faulting; (3) the extension is inferred to be relatively ancient (up to ~4 Ga), based on the superposition of many craters and Vulcan Planitia units, which are themselves relatively heavily cratered; and (4) the roughly polygonal extension across the northern terrain does not indicate a preferred direction of extensional stress, whereas the roughly east-west alignment of the major chasmata near the boundary with Vulcan Planitia implies major north-south extension across this region.

The rift geometry of large chasmata like Serenity and Mandjet (Fig. 9b) allow estimates of the extension across them. The minimum extension assumes the bounding normal faults dip steeply (~60°), in which case the depth of the rift is a close approximation to the extension across it. *Beyer et al.* (2017) show that the apparent dip of the Serenity Chasma walls is 30° but discuss that this is likely due to post-faulting modification. If Serenity Chasma is assumed to be 300 km long with a minimum depth of 3 km, and Mandjet Chasma is taken to be 450 km long with a depth of 5 km, together they represent an areal increase of ~3000 km^2, or an areal strain of ~0.3% over the northern half of Charon's encounter hemisphere. In addition to Serenity and Mandjet, *Beyer et al.* (2017) evaluated strain across the 23 other largest scarps in Oz Terra, which pushes this minimum areal strain estimate to ~1%. If a shallower fault dip angle of 45° is assumed, then the extension is approximately twice that of the 60° fault dip case, and results in a doubling of the area increase of extension and a doubling of the strain estimate.

The estimate of a 1% areal strain can be conceptually decomposed into a 0.5% linear strain in two orthogonal directions. Such a minimum linear strain, ε ~0.5%, corresponds to elastic extensional stresses of ~εE = 50 MPa for a Young's modulus E for water ice of 10 GPa. Such stresses would be sufficient to cause motion on pre-existing faults to depths of hundreds of kilometers, although in practice fault failure would likely occur at much lower stresses. The widths of Serenity and Mandjet Chasmata themselves are consistent with (but do not require) normal faults that penetrate to tens of kilometers. We conclude that extensional tectonics in a similarly strong lithosphere likely affected the equatorial and southern portion of Charon's encounter hemisphere, before the emplacement of the smooth plains materials (*Beyer et al.,* 2019).

A possible interpretation of the uplifted and tilted flanks of Serenity Chasma might be that they result from the elastic response of the lithosphere to the rifting (i.e., rift-flank uplift). *Beyer et al.* (2017) determined that this interpretation would imply an elastic thickness, at the time the uplift occurred, of roughly 2.5 km. However, they noted that the brittle thickness must be more than an order of magnitude greater than this in order for the observed faults to support the topography, and the lack of bowed-up crater floors also implies a much thicker elastic lithosphere. The observed rift topography therefore probably does not result from elastic flexure; it is more likely a consequence of rigid rotation of fault blocks ("bookshelf faulting").

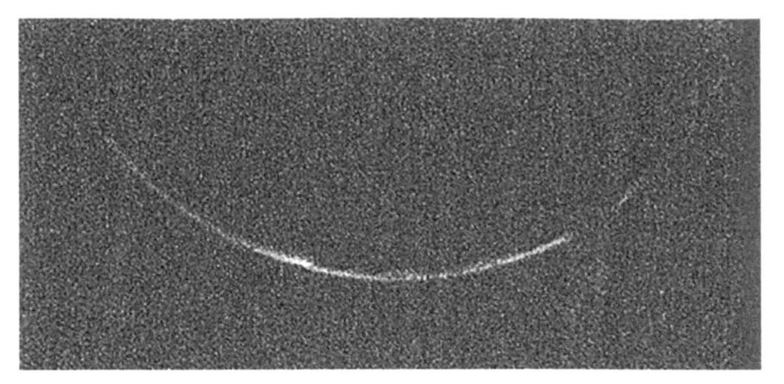

Fig. 6. The highest-resolution Charon departure image, with phase angle169° and pixel scale of 2.0 km per pixel.

5. VULCAN PLANITIA

5.1. Introduction

Vulcan Planitia is an extensive plains unit that covers the equatorial area and south to the terminator on the sub-Pluto hemisphere observed by New Horizons (Figs. 10 and 11). It seems likely that these plains are a result of the eruption of one or more large cryoflows that completely resurfaced this area (*Moore et al.,* 2016; *Beyer et al.,* 2019; *Schenk et al.,* 2018a; *Robbins et al.,* 2019). If this hypothesis is correct, Charon provides the clearest example of effusive cryovolcanism currently available in the solar system. Several giant planet satellites with similar mass and density to Charon, notably Dione at Saturn (*Schenk et al.,* 2018b) and Ariel at Uranus (*Jankowski and Squyres,* 1988; *Schenk,* 1991), show plains of possible effusive cryovolcanic origin. However, New Horizons imaging of Charon is superior to available (i.e., Voyager) imaging of the uranian satellites, and the morphologies associated with resurfacing are more clearly displayed on Charon than on most of the saturnian satellites.

Prominent "moats" encircle most of the mountains within Vulcan Planitia ("M" in Fig. 11a). *Moore et al.* (2016) offered two hypotheses for these moats: (1) downward flexure of pre-existing plains material from loading by younger mountain material (formed for example by extrusive cryovolcanism), or (2) incomplete embayment of pre-existing mountains by the plains material. *Desch and Neveu* (2017) explored the downward flexure hypothesis but concluded that the elastic lithosphere would need to be very thin (less than 2.3 km) to support such flexure. This thickness would in turn imply heat fluxes on the order of 300 mWm^{-2}, 2 orders of magnitude larger than expected at the present time (see below). Desch and Neveu hypothesized that high heat flow from a local hot spot, which would have been providing the erupted cryomaterial, could have locally thinned the elastic lithosphere. However, the lobate margins of these moats, their similarity to other lobate features in Vulcan Planitia ("F" in Fig. 11c), and the presence of similar moats around the margins of Vulcan Planitia, visible in elevation data (Fig. 7b), strongly favor a common origin resulting from embayment by viscous surface flows (*Beyer et al.,* 2019). *Borrelli and Collins* (2021) investigated the topography of the moats and depressed margins to compare with cryoflow models and plate bending models, and while the modeling indicated better fits for the plate-bending model, the authors concluded that the geological evidence did not support large loads. The difference in width between the moats around the mountains and the depressed margin along the scarps may be due to the different depths of the viscous fluid in the center of Vulcan Planitia vs. the potentially shallower depths at the margins.

Linear depressions, or rilles, are common within Vulcan Planitia ("R" in Figs. 11a,b,c): These formed either contemporaneously with or fairly soon after plains emplacement, as few impact craters are cut by them. The morphology of the rilles in Vulcan Planitia are consistent with a tectonic origin due to their regular spacing, mostly linear character, and negative topography. In addition, the en echelon arrangement of some of these rilles (e.g., "E" in Fig. 11b) are consistent with extensional graben (*Moore et al.,* 2016).

If the rilles are extensional graben, then their orientations are determined by the stress patterns present during their formation. Consistent rille orientations across Vulcan Planitia would reflect a consistent and large-scale stress field. Superposition relationships show that the plains postdate most of the global expansion discussed above, but the rilles may reflect the last stages of that expansion, if they are not due to thermal contraction of the cooling flows, as discussed below. The rilles have a variety of orientations, but orientations that parallel that of the bounding scarps and the chasmata north of them are most common, which hints at a relation. *Beyer et al.* (2019) estimate the total strain contributed to Vulcan Planitia from rille formation is in the range of 1.4% to 1.7%.

A significant fraction of Vulcan Planitia is occupied by rough "mottled terrain" (section 3.3 and Figs. 5, 8, and 11d). The origin of these closely-spaced pits or hills is enigmatic, although *Robbins et al.* (2019) considered loss of volatiles from the cryoflows after emplacement to be the most plausible explanation based on superposition relationships and the fact that other possible formation models were considered unlikely.

5.2. Resurfacing and Cryovolcanism

The traditional difficulty to overcome with cryovolcanism is how to develop a density inversion on an icy world that would allow cryomaterial to rise and erupt. The density properties of pure water ice would cause an icy lithosphere to float on liquid water, inhibiting eruptions even if the lithosphere was physically disrupted. For Charon and other icy satellites, the presence of an NH_3 or methanol hydrate mixture in Charon's lithosphere and mantle can also provide a density inversion (*Beyer et al.,* 2019; *Kargel et al.,* 1991; *Schenk,* 1991; *Kargel,* 1992; *Hogenboom et al.,* 1997; *Cook et al.,* 2007). The density differences between an NH_3 hydrate liquid and its solid (or even pure water ice) are not very large, however. An NH_3-H_2O peritectic liquid has a lower density (946 kg m^{-3}) than pure liquid water (1000 kg m^{-3}) or

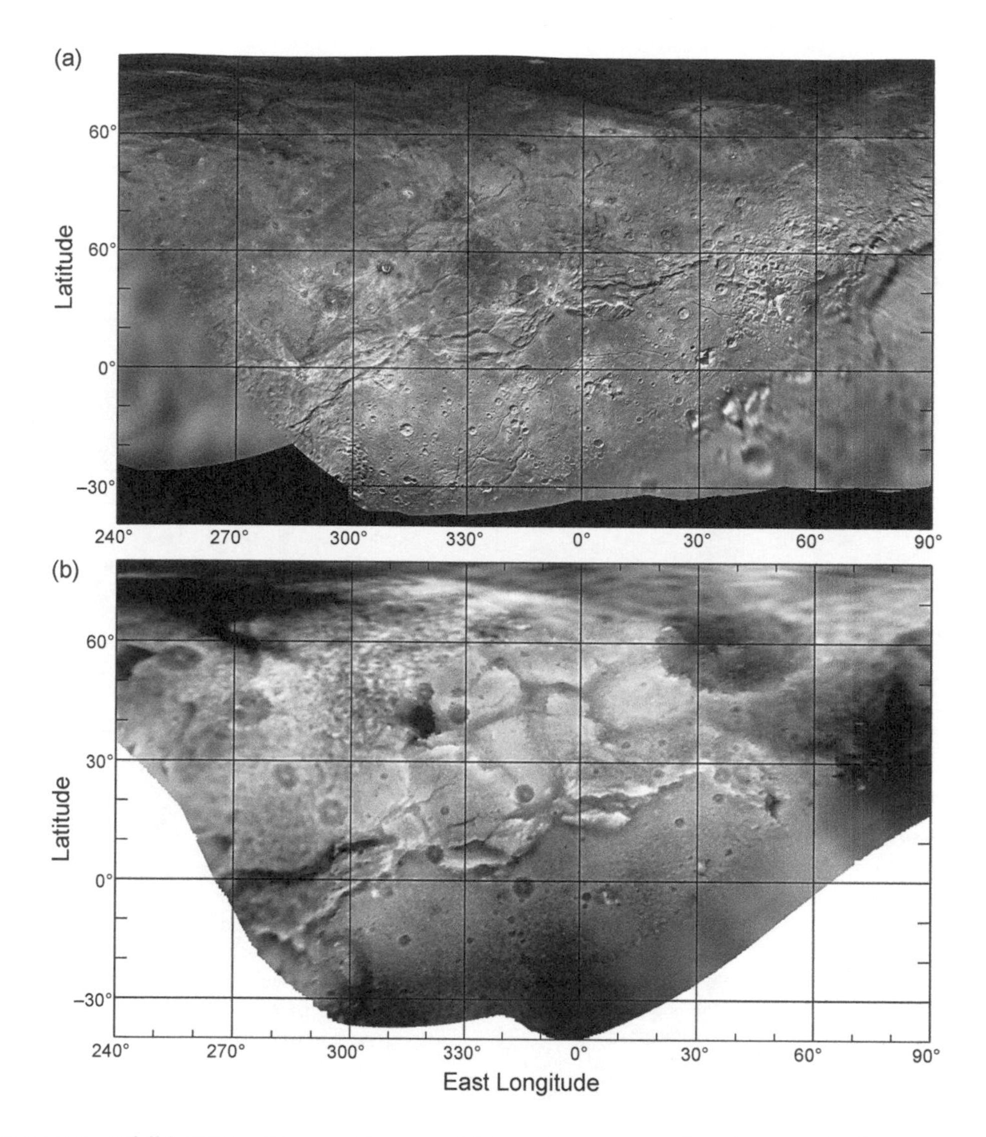

Fig. 7. (a) Image map and **(b)** stereo topographic map of Charon's encounter hemisphere. In **(b)**, the range of elevations shown is –6 km (black) to +6 km (white), although the full range of elevations measured on Charon is greater, –14.1 km, +6.9 km. From *Schenk et al.* (2018a).

solid ammonia dihydrate (965 kg m^{-3}), but a slightly greater density than pure water ice (917 kg m^{-3}) (*Hogenboom et al.,* 1997). However, all these densities are very close to one another, and small variations in geophysical conditions [like pressurization due to ice shell freezing, as hypothesized by *Manga and Wang* (2007)] may allow extrusion of NH_3 hydrate even through a water ice layer.

Beyer et al. (2019) used viscosity estimates and cooling timescales to investigate the possibility of different potential cryomaterials that could have flowed out and resurfaced Vulcan Planitia. The high topographic amplitude of the flow fronts implies a material that had high viscosity at the time that flow ceased, requiring either a solid such as warm water ice, or a viscous solid/liquid mixture. Pure water ice is possible rheologically, but it is unlikely because of the high density of liquid water relative to the solid phase, as previously mentioned. Pure water's high freezing temperature also requires higher internal temperatures than other candidate fluids. As discussed in section 7.4 below, Charon's limited heat budget makes generation of a pure water ocean marginal, and while an insulating, porous, ice shell would stabilize an ocean, porosity would also make potential cryomagmas even less buoyant. In addition to having lower densities, methanol-water or ammonia-water compositions lower the freezing point and increase the viscosity, making them more reasonable candidates. Such two-phase mixtures also freeze more slowly and thus can flow further (*Umurhan and Cruikshank,* 2019) than pure water ice.

Simplistic assumptions of Bingham rheology indicate that the derived yield strengths for Charon flows are similar to those reported for flows on Ariel (*Melosh and Janes,* 1989) and are in the range for terrestrial silicate lavas (*Moore et al.,* 1978), which implies that our expectations for how a silicate lava flow behaves on Earth may translate to how this cryoflow behaved on Charon (*Kargel et al.,* 1991).

As discussed in section 7 below, freezing of an early internal ocean could produce an ammonia-rich residual liquid ocean, which provides a plausible source of

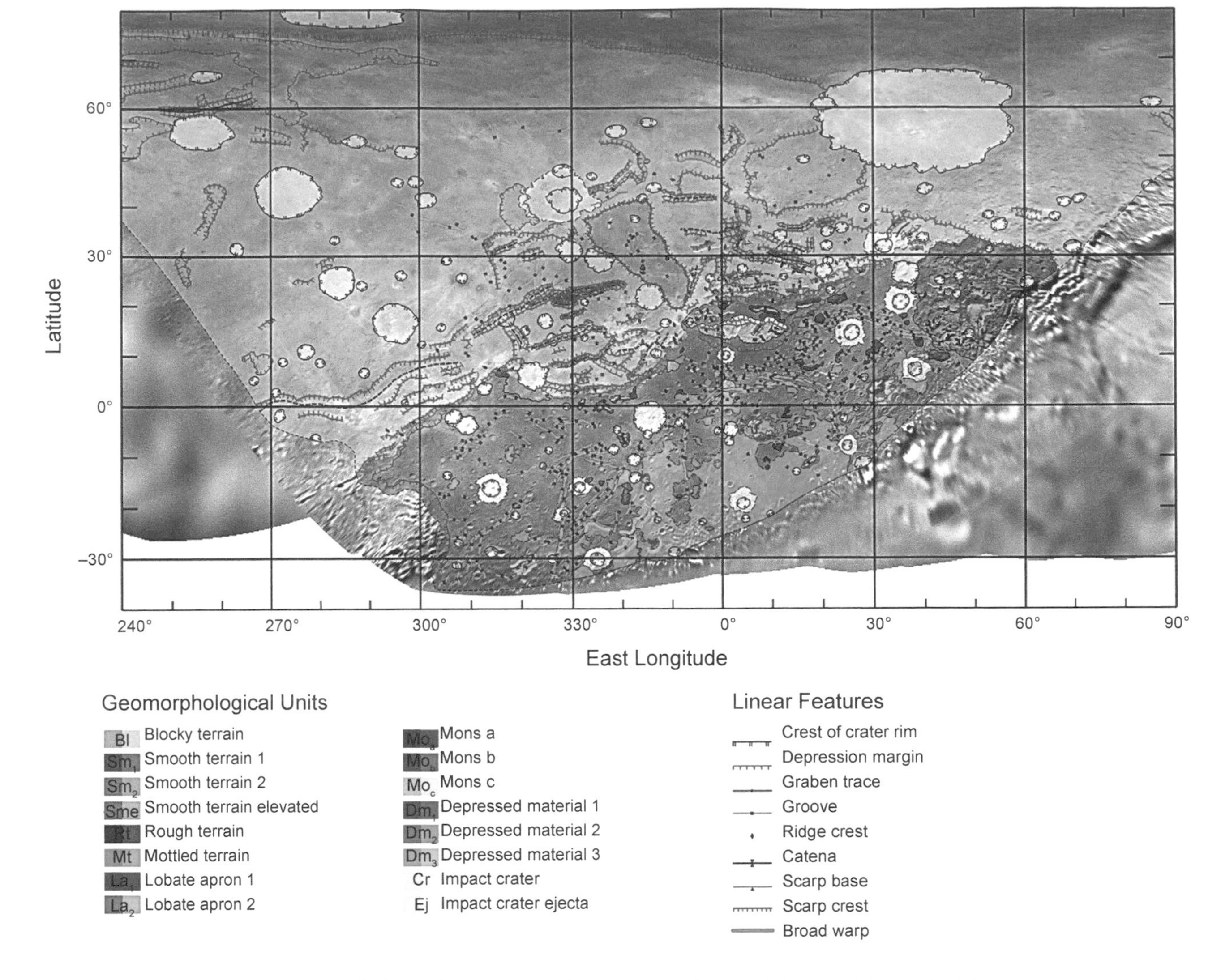

Fig. 8. Geomorphological map of Charon's encounter hemisphere. Adapted from *Robbins et al.* (2019).

cryomagma (*Kargel,* 1992; *Beyer et al.,* 2019). If the lithosphere were pure water ice, it would still float on an H_2O-NH_3 ocean. However, as mentioned above, the densities are very similar, and if the lithosphere were not pure water ice but rather an NH_3 hydrate that contained even a small fraction of rocky material, that might enable a density inversion. When the lithosphere expands and is disrupted, the H_2O-NH_3 ocean material would rise up above the mostly-water-ice lithospheric blocks, possibly assisted by the pressurization of the ocean due to ice shell freezing (*Manga and Wang,* 2007). This newly emplaced unit should preserve the bulk composition of the "ocean" that it came from, because it would have cooled much faster than the slow fractional crystallization of the freezing ocean.

The extent to which any such overturn could occur would depend on the viscosity and rigidity of the shell; it would also be affected by any porosity retained in the near-surface material. An alternative way of developing a density inversion is to appeal to an undifferentiated, rock-rich "carapace." Both porosity and the carapace are discussed in section 7.4 below, but it is clear that more work on this set of hypotheses is needed. There are also other mechanisms for producing local cryovolcanism, such as melting of salt-rich pockets over ascending convective plumes (*Head and Pappalardo,* 1999), but as noted in section 7.4, all current models assume that Charon's ice shell is stable against thermal convection.

Resurfacing may have occurred by cryovolcanic effusion similar to lunar maria emplacement (*Beyer et al.,* 2019). Obvious vents are not seen (although it's possible some of the rilles, or the depressions marked as Dm1–Dm3 in Fig. 8, are actually vents), but the eruption style may have been similar to the volcanism of the lunar maria where there is also an absence of clear vents or feeder dike systems. The lunar maria style allows for many local sources of cryomaterial all across the region and also explains the lack of large, obvious discrete flows. In this hypothesis, the montes on Vulcan Planitia (Fig. 9a) are preexisting areas of high elevation that were embayed by the cryoflows, like kipukas or nunataks. These montes are not significantly different in

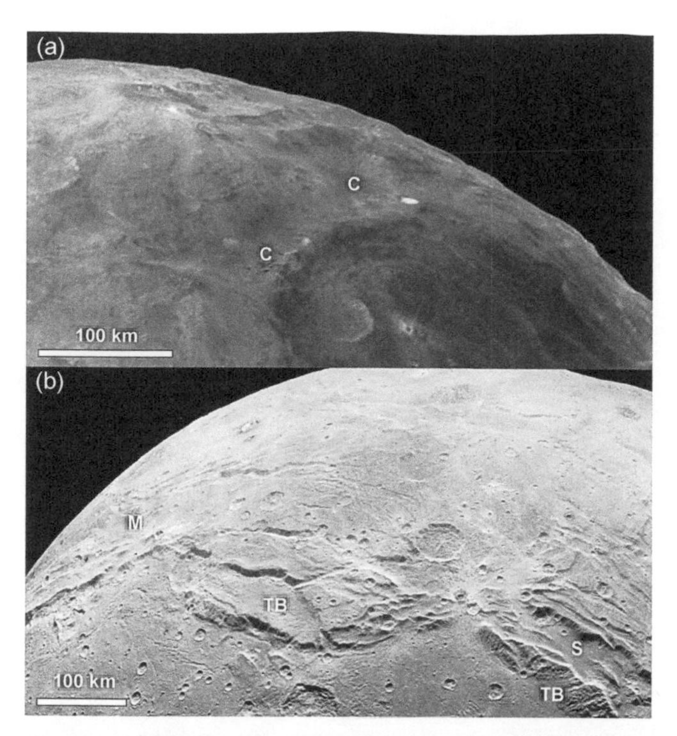

Fig. 9. Oblique views of Oz Terra. **(a)** Fault scarps and troughs near the north pole, observed at high Sun. Image centered near 70°N, 240°E. The trough Caleuche Chasma (C) is 12 km deep. **(b)** Fractures in Oz Terra near the boundary with Vulcan Planitia (bottom). Image centered near 20°N, 330°E. Serenity Chasma (S), Mandjet Chasma (M), and tilted blocks (TB) are identified.

elevation than the high terrain in Oz Terra and could simply be remnants of higher-standing topography in this area. The tilted blocks at the Oz boundary ("TB" in Fig 9b) could have been tilted due to subsidence associated with the eruption of the cryomaterial from depth.

Alternatively, resurfacing may have involved more extensive disruption of the lithosphere — lithospheric blocks losing their support, rotating, and foundering in a mechanism similar to magmatic stoping — while the cryomaterial rises and covers the submerged blocks, and freezes (*Beyer et al.,* 2019). The tilted surfaces of the lithospheric blocks in Oz Terra that directly border Vulcan Planitia (Fig. 9b) may represent a transition zone. Similar tilted blocks are observed on Ariel (*Schenk,* 1991; *Schenk et al.,* 2018a). These blocks on Charon represent a point on the continuum between Oz Terra lithospheric blocks that are not tilted, but translated away from each other and the blocks that presumably existed on the pre-Vulcan surface before the cryoflows, and may have subsequently foundered.

It is possible that the global extensional fracturing was more severe in the region now occupied by Vulcan Planitia than in Oz Terra, facilitating eruption. Alternatively, whatever process caused the pre-Vulcan area to be topographically lower may also have favored cryomaterial to rise up and erupt onto the surface at this location.

Regardless of the exact resurfacing mechanism, the extensional fracturing on Vulcan Planitia (Fig. 11b) needs to be accounted for. While this may be due to late-stage global expansion, as discussed above, *Beyer et al.* (2019) note that an ammonia-hydrate cryomaterial can explain this as well. While a pure water ice flow would expand when it freezes (causing surface compression when confined laterally), an ammonia-hydrate flow would contract, and the observed post-resurfacing extension on Vulcan Planitia is thus qualitatively consistent with a cooling ammonia-hydrate flow.

Spectroscopy provides evidence for the presence of NH_3 in the bulk composition of the Pluto system. Ammonia-bearing species are seen spectroscopically on Charon (*Cook et al.,* 2007; *Grundy et al.,* 2016; *Dalle Ore et al.,* 2018; chapter by Protopapa et al. in this volume) and on Nix and Hydra (*Cook et al.,* 2018; chapter by Porter et al. in this volume). However, on Charon, the ammonia-bearing species are concentrated near specific impact craters and not preferentially in Vulcan Planitia, as might be expected if the flows were ammonia-rich. This distribution may be because the "crust" of the flow, which would have frozen first, would likely have a more water-rich composition. Also, photolysis and proton bombardment may act to remove the ammonia from the surface over a few tens of millions of years (*Cooper et al.,* 2003; *Moore et al.,* 2007; *Cassidy et al.,* 2010), or an exogenous coating from the small satellites may hide the native surface material (*Stern,* 2009).

6. CRATERING AND AGES

Both Vulcan Planitia and Oz Terra are relatively heavily cratered, with crater spatial densities similar to the densities of the most heavily cratered terrains on Pluto (see the chapter by Singer et al. in this volume). While superposition relationships show that Vulcan is clearly younger than Oz, crater densities on the two terrains are not statistically distinguishable where they overlap in diameter (*Robbins et al.,* 2017; *Singer et al.,* 2019). Comparison is complicated by the limited overlap in well-characterized crater diameters between them. With the nearly overhead lighting in the north, craters smaller than ~30–50 km in diameter are difficult to distinguish in much of Oz Terra, and the largest crater in Vulcan Planitia is only ~65 km in diameter.

The calibration of crater spatial densities to surface age estimates is based on estimated impact fluxes (*Greenstreet et al.,* 2015) and impactor size-frequency distributions (see the chapter by Singer et al. in this volume). The calibration is for craters with diameter D >13 km and yields ages of 4 Ga or older for both Vulcan Planitia and Oz Terra. Thus the bulk of the Vulcan Planitia resurfacing appears to have occurred early in Charon's history, and this timing is consistent with the fact that almost all craters overprint other geologic features (*Robbins et al.,* 2019; *Singer et al.,* 2019). The great age of the surface incidentally makes it clear that the presence of crystalline ice and ammonia-bearing species on a planetary surface are not reliable indicators of recent geological activity, contrary to some expectations before the flyby, as discussed in section 1 (see the chapter by Protopapa et al. in this volume).

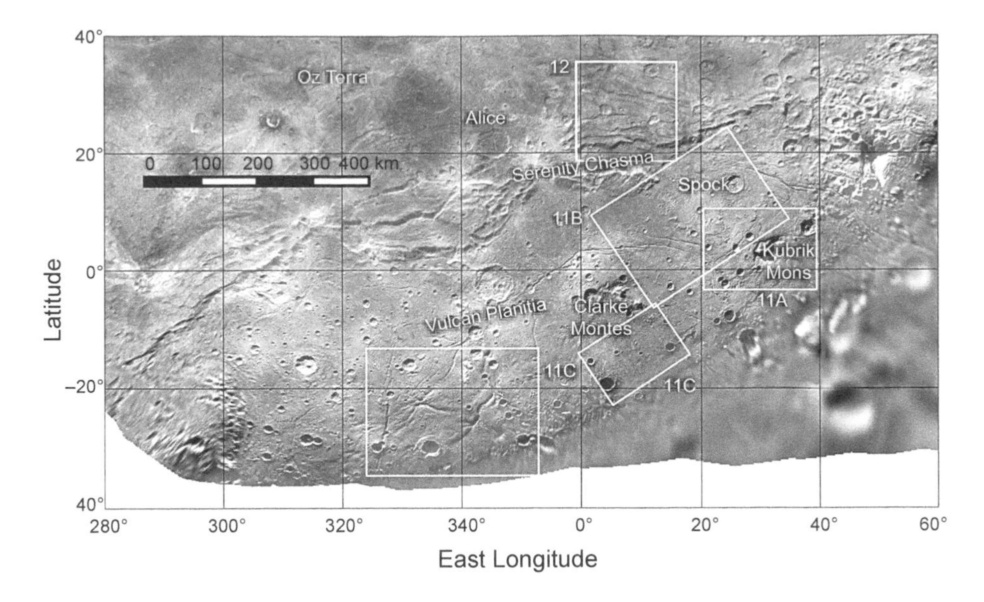

Fig. 10. Overview of Vulcan Planitia. White rectangles indicate the locations of the enlargements in Figs. 11a–d and Fig. 12. Adapted from *Beyer et al.* (2019).

Vulcan Planitia appears less heavily cratered than some old surfaces in the solar system (e.g., the lunar highlands) and this is due in part to a deficit of craters smaller than ~13 km in diameter (*Singer et al.,* 2019). Earlier crater counts by *Robbins et al.* (2017), based on preliminary mapping of Charon, also showed a break to a shallower slope at small diameters. This lack of small craters relative to larger ones is mainly attributed to a change in the size-frequency distribution slope of KBO impactors to a shallower slope for smaller craters (*Singer et al.,* 2019; chapter by Singer et al. in this volume).

Several large craters on Charon (e.g., the large crater at the top of Fig. 5) have extensive interior terracing similar to that of craters on some icy saturnian satellites (*White et al.,* 2013, 2017; *Schenk et al.,* 2018b). The diameter above which central peaks begin to occur (~10 ± 1 km) is what

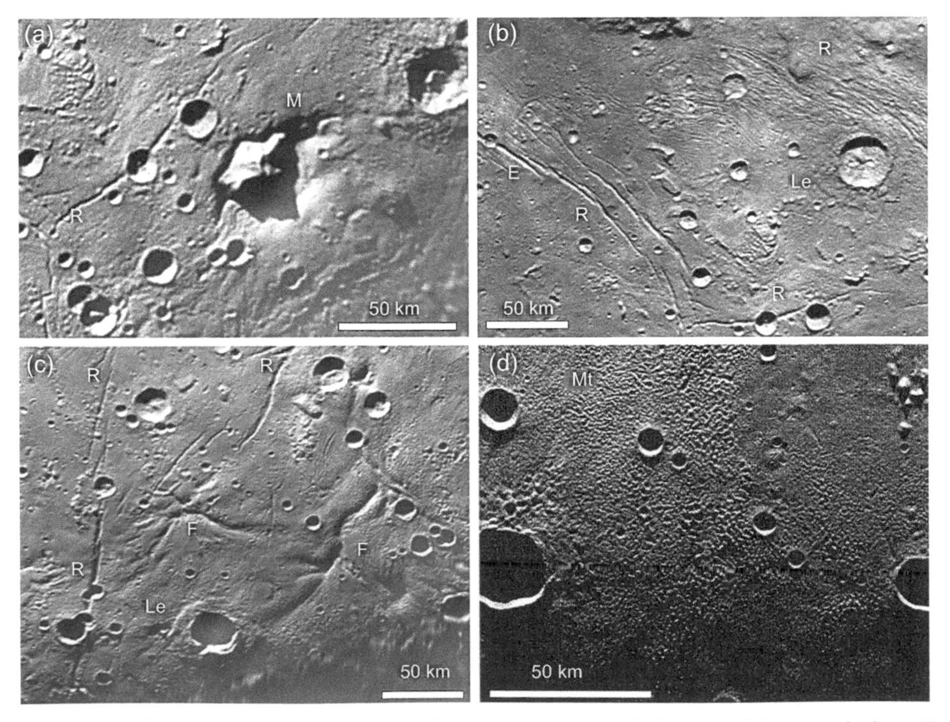

Fig. 11. Representative landforms on Vulcan Planitia. Moated mountains (M), rilles (R), en echelon rille offsets (E), layered crater ejecta (Le), lobate flow features (F), and mottled terrain (Mt) are identified. See Fig. 10 for the locations of these images.

is expected for Charon's gravity for an icy surface. *Schenk et al.* (2018a) show evidence that complex craters (those larger than ~6 km in diameter) have a shallower depth-to-diameter slope than those of the icy saturnian satellites with similar gravity, and thus craters with D >~50 km on Charon are shallower by up to 1 km. Schenk et al. note that these differences may indicate that complex crater formation on Charon is initiated under similar conditions to the saturnian satellites but the collapse process is more complex or complete on Charon. Alternatively, the expected lower impact velocity at Charon compared to the saturnian satellites could have effects in addition to those due to surface gravity (*Bray and Schenk,* 2015).

Craters on Charon display some unusual ejecta types. The albedo pattern of ejecta for many craters in Oz Terra is unobserved elsewhere in the solar system, consisting of a dark inner ejecta deposit surrounded by a more extensive bright outer ejecta deposit (Figs. 3 and 5). Both dark and light ejecta form ray patterns similar to those observed on other solar system bodies. Craters with dark rays are sometimes seen on other icy bodies, notably Ganymede (*Schenk and McKinnon,* 1991). However, the specific dark-bright pattern found on Oz Terra is not seen on other bodies, nor is it seen in Vulcan Planitia where only bright rays are observed. This albedo pattern might be due to layering in the material excavated by craters in Oz Terra but not in Vulcan Planitia (*Robbins et al.,* 2019). Because ejecta excavated from greater depth generally appears on the surface closer to a crater rim, the dark-bright pattern ejecta may imply a dark layer overlain by brighter material in Oz Terra, while the absence of such a pattern in Vulcan Planitia would imply the absence of a dark layer there. However, the fact that the ratio of bright to dark ejecta does not depend strongly on crater diameter (and thus excavation depth) is difficult to explain if the dark-bright pattern is due to layering. An alternate possibility is that the albedo pattern is produced by impact processing (e.g., shock or comminution effects), with the material that is ejected at higher speeds being brightened by the impact process while lower-speed ejecta closer to and within the crater is darkened. The rarity of this dark-bright pattern on other icy bodies, and its absence of Vulcan Planitia, would then imply that there is something compositionally or texturally unusual about the surface materials on Oz Terra that causes this unusual albedo response to impact.

Fig. 12. A region just north of Serenity Chasma, showing the only crater chain so far identified on Charon (CC), and a landslide on the wall of Serenity Chasma (LS). See Fig. 10 for the location of this image.

Another distinctive ejecta form, currently termed "layered ejecta" in the literature (*Barlow et al.,* 2000), is common on Charon (*Robbins et al.,* 2018). Crater ejecta of this type has the visual appearance of a distinct layer or multiple layers of material with a visually abrupt edge that typically is ~1–2 crater radii from the crater rim, rather than the more common smooth gradation of continuous ejecta deposits outward into thinner and discontinuous material forming rays. Layered ejecta are primarily observed in Vulcan Planitia ("Le" in Figs. 11b,c), although this may be due to the favorable lighting there. Only Mars has a higher fraction of craters with this ejecta type (e.g., *Carr et al.,* 1977), even though the morphology is also seen on icy bodies including Ganymede (e.g., *Horner and Greeley,* 1982), Europa, Tethys, and Dione (*Robbins et al.,* 2018). Robbins et al. considered subsurface volatile mobilization and ejecta fluidization to be the most likely explanation for these features on Charon and other icy airless worlds.

One crater chain is observed on Charon north of Serenity Chasma near 30°N, 10°E (Fig. 12). The chain, which is about 50 km long, consists of more than 15 circular features linked together by a linear trough. The feature is probably not a secondary impact crater chain due to its uniqueness and the fact that it is not radial to any known large primary impact crater. It may be endogenic, but there are no other similar features nearby. Its similarity to crater chains on Ganymede and Callisto, which are probably produced by impactors tidally disrupted by Jupiter (*Melosh and Schenk,* 1993), and its location on the Pluto-facing hemisphere raise the possibility of a similar origin in this case. However, it would be somewhat surprising if Pluto was capable of tidally disrupting passing bodies in the way that Jupiter does. The chain's origin thus remains enigmatic.

7. INTERIOR

7.1. Introduction

The aim of this section is to provide a brief overview of our current understanding of Charon's interior and evolution. In many ways this section is parallel to the chapter describing Pluto's geodynamics (see the chapter by Nimmo and McKinnon in this volume), but there are a few important differences from Pluto. First, Charon's initial conditions may have been different from Pluto's because of the way it formed. Second, tidal effects could in theory play a larger role for Charon than for Pluto, although in practice they appear not to matter (section 7.5). And, third, Charon is a smaller object and less complex overall.

7.2. Observational Constraints

New Horizons images provided a more accurate radius for Charon than Earth-based observations (606 ± 1 km), yielding an inferred bulk density of 1701 ± 17 kg m^{-3}, 1σ (*Nimmo et al.,* 2017). This is about 9% less dense than Pluto. Charon's density is significantly lower than Pluto's even after correction for self-compression (*McKinnon et al.,* 2017). No equatorial flattening is detected, with an upper bound of 0.5%. The surface geology, discussed above, also provides powerful constraints on Charon's interior.

7.3. Bulk Interior Structure

A pre-New Horizons view of possible Charon interior structures is given in *McKinnon et al.* (1997). Since, unlike Pluto, the bulk density and size were already quite well known, the main outstanding question was whether Charon was differentiated or not. This is a question that New Horizons has answered in the affirmative. Water ice at high pressures transforms to denser structural forms (*Sotin et al.,* 1998). As Charon cooled over time, the volume occupied by these high-pressure phases would have increased if the ices extended to Charon's center, leading to an overall reduction in radius and significant surface compression (*McKinnon et al.,* 1997). Instead, what we observe is extension, most likely due to refreezing of a subsurface ocean (sections 4.2 and 7.4). This extension limits the amount of ice that can be present at the high pressures near Charon's center, and thus implies that Charon must be mostly or entirely differentiated.

Charon's density implies a rock mass fraction of about 60% (for rock and ice densities of 3500 and 950 kg m^{-3} respectively) and a normalized moment of inertia of 0.30, assuming these densities and assuming complete differentiation. Of course, the "rock" density is not well-known: If the rock is hydrated, or a significant mass of organic compounds is present, its density could be less than the value adopted here. Figure 13 shows a summary diagram of Charon's likely internal structure. Models discussed in section 7.4 suggest that a porous core having undergone some degree of hydrothermal alteration is probable. The possible overturn of a low-density layer (either NH_3-rich or rock-poor) is also discussed there and in section 5.2. The figure does not show organics or clathrates, although in principle either could be present.

Charon's density suggests it is somewhat less rock-rich than Pluto. This is important because it has bearing on Charon's origin (see the chapters by McKinnon et al. and Canup et al. in this volume). The density contrast may in part be due to Charon's ability to retain more porosity in its ice shell than Pluto because of the lower pressure and temperature gradients in the smaller body (*Bierson et al.,* 2018). However, it is unlikely that porosity or other effects (such as the possible presence of an ocean on Pluto) explain all the density difference, and it is thus probable that Charon has a slightly lower rock/ice ratio than Pluto. Such a difference is consistent with the idea that Charon formed via a giant impact with Pluto (see the chapter by Canup et al. in this volume).

Charon should be tidally and rotationally distorted at the present day with three ellipsoidal axes a, b, c. Using the approach given in the chapter by Nimmo and McKinnon in this volume, a uniform, fluid Charon would have a present-day ratio (a–c)/(b–c) = 3.7 and a flattening (a–c) of 837 m (0.14%). This flattening is comfortably within the 0.5% upper limit set by the observations. If the flattening had been frozen in at an earlier time, when Charon was sufficiently closer to Pluto (a "fossil bulge"), then it should be visible

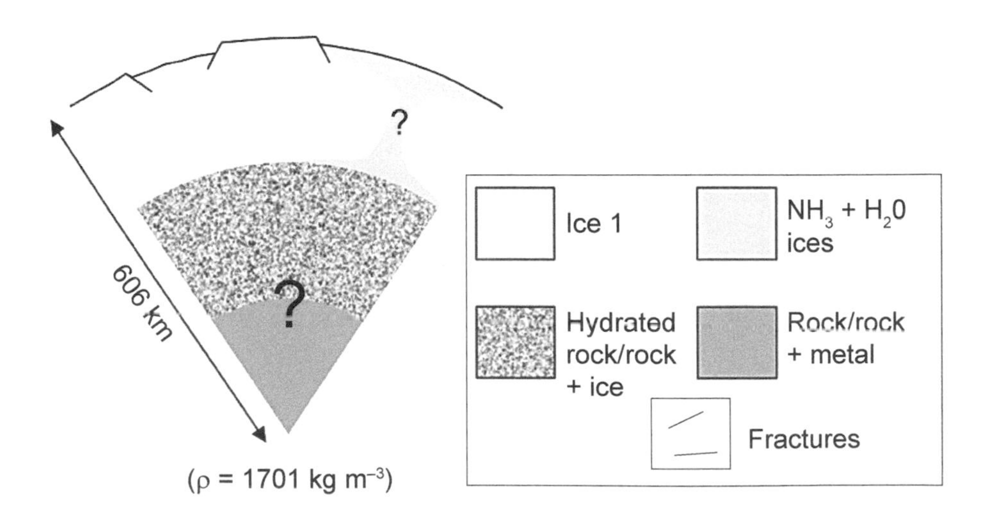

Fig. 13. Sketch of Charon's probable interior structure (see text).

TABLE 1. Energy sources and sinks for Charon.

	ΔE (J)	ΔT_{eff} (K)	Notes
Source			
Charon-forming impact	1.5×10^{27}	711	$GM_1M_2/(R_1 + R_2)$ applied to Charon's mass only (upper bound)
Radiogenic	1.4×10^{27}	670	Total over 4.5 Ga; scaled from Pluto (see the chapter by Nimmo and McKinnon in this volume)
Serpentinization	5.5×10^{26}	260	Hydration of entire silicate core, assuming energy of serpentinization is 575 kJ kg^{-1} (*Desch and Neveu,* 2017)
Accretion	1.6×10^{26}	78	$3GM^2/5R$
Despin	3.0×10^{25}	14	Spin down from 3-hr period to current period
Circularization	1.9×10^{24}	0.9	For e = 0.1 and a = $6R_p$
Sink			
Thermal	4.5×10^{26}	210	To warm from 40 K to 250 K
Latent heat	1.9×10^{26}	88	To melt the top 200 km of ice, assuming pure H_2O

A mean specific heat capacity of 1335 J kg^{-1} K^{-1} is used to convert energy to effective temperature change ΔT_{eff}.

at the present day. Thus Charon's lack of a detectable fossil bulge places some constraints on its evolution, as discussed below (section 7.5).

7.4. Thermal Evolution

Consideration of Charon's thermal evolution starts with consideration of its energy sources. Table 1 below summarizes some of the major sources. Accretion assumes zero velocity at infinity, which is a good assumption if Charon formed out of an impact-generated disk (*Canup,* 2011; chapter by Canup et al. in this volume). The energy released during subsequent differentiation is typically an order of magnitude smaller than the accretion energy. Similarly, the energy associated with Charon's rotation slowing to synchronous via tidal dissipation is small compared to accretion. Compared with more-massive Pluto, the biggest difference is the potential importance of the putative

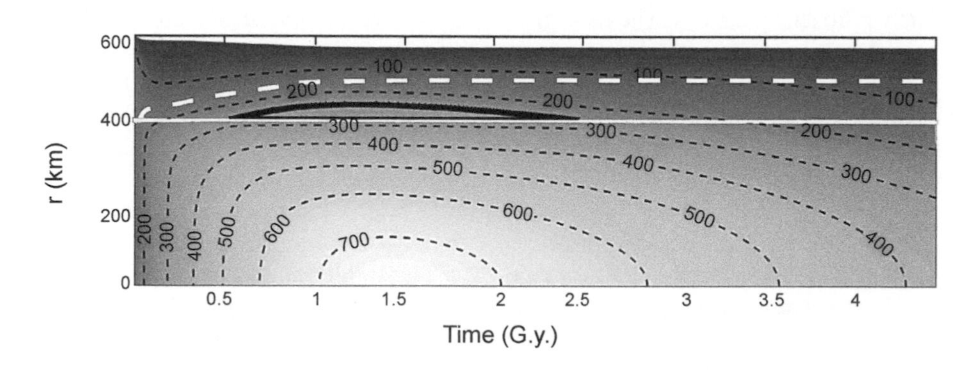

Fig. 14. Thermal evolution model of Charon. Contour lines indicate temperature in K, the solid black line outlines the liquid water ocean, and the white dashed line denotes the base of the porous ice layer (a porosity of 30% was assumed). The gray horizontal line indicates the silicate/ice boundary. From *Bierson et al.* (2018).

Charon-forming impact. If all the energy released during this impact were concentrated in Charon (an unlikely end member), the heating effect would be comparable to that of radioactive decay. But in reality, radioactivity is likely to dominate, just as with Pluto.

The other potentially important addition to this table is tidal heating. Charon is close to a more massive body (Pluto) and could have experienced tidal heating if its orbit were ever eccentric [it is currently circular (*Buie et al.,* 2012)]. Progressive circularization of the orbit results, via conservation of angular momentum, in a decrease in semi-major axis a: the change $\Delta a = 2a\, e^2$, where e is the initial eccentricity. The total energy released by circularization ΔE is given by

$$\frac{\Delta E}{mC_p} = \frac{GM}{aC_p}e^2 = 0.9\,K\left(\frac{e}{0.1}\right)^2\frac{6R_p}{a}$$

where M and m are the masses of Pluto and Charon, R_p is Pluto's radius, and we are expressing the outcome as a mean temperature change using the mean specific heat capacity C_p (here taken to be 1335 J kg K^{-1}). This calculation illustrates that tidal heating is actually a very small source of energy, unless Charon was formed close to Pluto in a highly eccentric orbit (see *Cheng et al.,* 2014; chapter by Canup et al. in this volume).

Table 1 suggests that, except perhaps at very early times, Charon's thermal evolution has been dominated by radioactive decay, although serpentinization could also release significant heat if a large fraction of the core was affected. Figure 14 shows a simple evolution model (*Bierson et al.,* 2018) assuming an initially differentiated state, a pure H_2O ice/water shell, and that the ice shell remains conductive throughout. Although detailed calculations have not been done, it is generally assumed that Charon's ice shell is not convecting, due to Charon's small size and modest heat budget (*Desch and Neveu,* 2017; *Bierson et al.,* 2018). In this model Charon starts cold; the silicate interior warms up through radioactive decay and eventually warms the ice shell enough to form an ocean. After about 1.5 G.y. the ocean starts to refreeze from its maximum thickness of about 30 km and is completely refrozen by 2.5 G.y. The geological evidence discussed above suggests more rapid evolution in reality than in this simple model, with most contraction due to ocean freezing having occurred prior to emplacement of the Vulcan Planitia flows, roughly 0.5 G.y. after formation.

This model assumes an initially porous (and therefore insulating) ice shell; the depth to the base of the porous layer is denoted by the white dashed line and illustrates that it progressively shallows as the ice shell warms up initially. The present-day thickness of the porous layer is about 70 km in this model, which is similar to the result found by *Malamud et al.* (2017). Note that a similar model lacking ice porosity (and thus having higher thermal conductivity) does not develop an ocean. Porosity, however, increases the buoyancy of the ice shell and thus inhibits cryovolcanism. An alternative way of promoting an ocean is to invoke clathrates, which have a low conductivity and may be present on Pluto (*Kamata et al.,* 2019). Pure water oceans on Charon are marginal, owing to a larger surface area:volume ratio than Pluto, and are therefore sensitive to the conductivities assumed.

More sophisticated models (*Malamud et al.,* 2017; *Desch and Neveu,* 2017) give broadly similar results to Fig. 14. In particular, there is general agreement that enough energy is available to allow Charon to differentiate. *Desch and Neveu* (2017) find that an ocean could develop but would freeze before the present day, generating extensional stresses as it does so. They favor a relatively warm start for Charon owing to energy deposition during its rapid accretion from the disk generated by the Charon-forming impact. They argue that subsequent extraction of heat from the core via hydrothermal circulation when liquid water is present can lead to two separate ocean epochs. Conversely, *Malamud et al.* (2017) argue against an ocean, and instead favor liquid water only within a large, porous core. The problem is that this model always results in overall contraction, which is not consistent with the observed tectonics.

Both *Desch and Neveu* (2017) and *Malamud et al.* (2017) argue for a hydrated, porous rocky core and a primordial, rock-rich ice crust "carapace." While there are no spectroscopic indications of silicates so far, even a thin surface layer of ice-rich material would hide them, so the rock-rich crust hypothesis cannot be entirely ruled out. Such a crust would be unstable relative to the underlying clean ice resulting from core differentiation; however, both groups argue that the viscosity of near-surface ice is so high that this overturn is unlikely to happen. Nonetheless, this model does provide an interesting alternative to the NH_3-rich ice ascent hypothesis discussed below and in section 5.2.

A model in which Charon started hot, with an ocean initially present, would result in monotonic freezing of the ocean, but neither the ocean life-time nor the present-day state would differ significantly from Fig. 14. As we argue below, the extensional tectonics observed are consistent with a refreezing ocean. Similarly, although detailed calculations have not been carried out, the example of Pluto suggests that as soon as an ocean develops, any initial fossil bulge is lost (*Robuchon et al.,* 2011). The predicted lack of a bulge is again consistent with the observations. Because Charon's despinning timescale is expected to be $<\sim 1$ m.y. (*Dobrovolskis et al.,* 1997), it is unlikely that any tectonic features associated with the collapse of Charon's rotational bulge are preserved to the present day. Certainly there is little evidence for the global pattern of fractures expected from despinning (*Melosh,* 1977).

The model shown in Fig. 14 assumes a pure water ocean. If the ocean initially contained some NH_3, this would have become progressively more concentrated in the remaining liquid as freezing proceeded. This increased concentration would have prolonged the lifetime of the ocean [athough not to the present day (*Desch and Neveu,* 2017)]. More interestingly, for sufficiently high concentrations, the residual, NH_3-rich liquid would have become positively buoyant relative to the ice shell, as discussed above. This situation would favor ascent of the NH_3-rich liquid and foundering of the initial lithosphere, providing a possible (although speculative) mechanism for the emplacement of the southern plains (*Beyer et al.,* 2019) (see section 5.2 above). Any such buoyancy instability is likely to be long-wavelength, or low spherical harmonic degree (*Zhong et al.,* 2000), perhaps accounting for the localization and large extent of Vulcan Planitia.

7.5. Stresses

As reviewed in *Collins et al.* (2010), the tectonics of icy bodies can provide clues to their structure and evolution, and below we assess potential stress-generating mechanisms for Charon.

One important consequence of the difference in ice and water density is that the refreezing of an ocean generates extensional stresses at the surface, while a thickening ocean would generate compression. Figure 14 would thus predict surface compression up to 1.5 G.y., followed by extension thereafter; early pore compaction might also contribute to the compression. The areal strain is given by ~2 ($\Delta\rho/\rho$) (d/R) where $\Delta\rho$ is the ice-water density contrast, ρ the ice density, d the amount of ocean freezing or melting, and R the radius of Charon (*Beyer et al.,* 2017). Figure 14 shows about 40 km of refreezing, which would yield a strain of about 1% (above). This strain is close to the geologically-inferred value (section 4.2). The total stress accumulated would be about 100 MPa, much greater than the failure strength of ice and suggesting multiple cycles of stress accumulation followed by failure (faulting), although the stress might be relieved by the same faults over multiple cycles. Much more detailed imaging from future missions would be required to better constrain the relative timing and duration of fault movements.

Temperature changes and hydration/dehydration can also play a role in volume changes and associated surface stresses. The volume thermal expansion coefficient for water ice varies from zero at ~75 K to 1.6×10^{-4} K^{-1} at 250 K (*Petrenko and Whitworth,* 1999), so it is difficult for even a large temperature change to result in a radial expansion of 0.5%, or equivalently, a volumetric expansion of ~1.5%. Moreover, Charon is likely about one-third rock by volume based on its mean density (*McKinnon et al.,* 2017) and silicates generally have much lower thermal expansion coefficients than ices (by up to an order of magnitude), although potential temperature changes within an inner core due to radiogenic heating can exceed 1000 K (*Desch,* 2015; *Malamud and Prialnik,* 2015). We conclude that temperature changes alone are not responsible for the global expansion evidenced by Charon's tectonics, but they could be an important contributor. Both *Desch and Neveu* (2017) and *Malamud et al.* (2017) considered the energetic (exothermic) effects of serpentinization, which can be large; *Malamud et al.* (2017) and *Bierson et al.* (2018) investigated the effects of serpentinization on density (and thus expansion/contraction) but found opposite effects. Overall, it seems likely that serpentinization contributes to, but does not dominate, the expansion/contraction history of Charon.

So far, no compressional features have been identified on Charon [unless the interpretation of Oz Terra morphology by *Chen and Yin* (2019) is correct]. One possibility is that the early compressional features expected from the initial formation of an ocean were either overprinted by impacts or other geological processes, as might be particularly likely if (as seems probable) the ocean formed earlier than in Fig. 14 or reactivated to form later extensional features. An alternative is that Charon started out hot and the ocean froze monotonically. In this case one would expect to only see extension (as observed), and the role of ice shell porosity to be much reduced (since pores only survive in cold ice).

The other main likely source of stress on Charon arises from its orbital evolution (*Collins and Pappalardo,* 2000). Charon probably formed closer to Pluto and then moved outward. Its spin would have synchronized with its orbit early on (*Dobrovolskis et al.,* 1997), but as it moved outward the spin rate and tidal distortion would both have decreased, leading to a predictable pattern of stresses (*Melosh,* 1980; *Kattenhorn and Hurford,* 2009). Although pure orbital recession results in equatorial thrust faults and polar normal faults, the addition of an isotropic extensional stress from ocean freezing could in principle have generated extension everywhere, but with the fault orientations dictated by the tidal/rotational effects. The magnitude of the resulting stresses is large: tens of millipascals for a Charon formation distance of less than 12 Pluto radii (*Collins and Pappalardo,* 2000), compared to is present orbital distance of 16.5 Pluto radii.

It has also been suggested that time-varying stresses arising from any former eccentricity in Charon's orbit (which is currently circular) might have played a role in producing surface tectonic features (*Rhoden et al.,* 2015). However, the stresses are typically a few tens of kilopascals, are three orders of magnitude less than the recession stresses and therefore much less likely to produce observable surface features, and they do not match the fracture orientations observed (*Rhoden et al.,* 2020). Finally, if Charon underwent reorientation, both extensional and compressional stresses would result, with a predicted magnitude of hundreds of kilopascals (*Nimmo and Matsuyama,* 2007).

Although only preliminary tectonic mapping has been carried out (*Beyer et al.,* 2017; *Robbins et al.,* 2019), so far there is no evidence for the kind of globally symmetric or systematic fracturing patterns associated with orbital/rotational effects seen, for example, at the Moon (*Watters et al.,* 2015) and Europa (*Schenk et al.,* 2008; *Kattenhorn and Hurford,* 2009). Of course, only half of Charon has been imaged in detail, and fault distributions may be controlled by crustal heterogeneity (e.g., the icy moon Tethys has a single, giant rift valley reminiscent of Serenity Chasma on Pluto). Nonetheless, so far there is no observation supporting any kind of stress-generating mechanism beyond freezing of a subsurface ocean. It is thus likely that orbital evolution was completed and its associated stresses resolved before the current visible surface was formed, or at least before ocean freezing generated the current dominant tectonic pattern (*Rhoden et al.,* 2020).

A second consequence of ocean refreezing is pressurization of the ocean (*Manga and Wang,* 2007). Under some circumstances, the pressurization becomes large enough to overcome the tensile strength of the ice and allow liquid to ascend through fractures to the surface, which might have been a contributing mechanism to the eruptions that formed Vulcan Planitia.

7.6. Heat Flux and Elastic Thickness

Present-day heat flux on Charon is expected to be dominated by radiogenic heat and to be very small, roughly 1–2 mW m^{-2} (e.g., *Hussmann et al.,* 2006); any heat deposited during accretion will have been lost long ago due to Charon's small size. Ice retains its elastic strength up to temperatures of ~130 K (*Conrad et al.,* 2019), so Fig. 14 suggests that essentially the whole of Charon's ice shell, a thickness of roughly 200 km, will behave elastically at the present day. At earlier times the elastic thickness will have been smaller, perhaps ~100 km depending on the initial thermal conditions assumed. Flexurally-supported topography records the lowest elastic thickness since the load was emplaced, thus even ancient surfaces on Charon are expected to record high elastic thicknesses. This expectation is consistent with the elastic thickness bounds inferred from fault and crater topography discussed in section 4.2 (*Beyer et al.,* 2017). In the absence of actual measurements, however, Charon's elastic thickness is currently not satisfactorily constrained. Certainly the overall scale of Serenity Chasma (~100 km wide) and the high relief in Oz Terra are both greater than similar features on potentially tidally heated worlds like Ganymede or Ariel (*Beyer et al.,* 2017), and suggests a correspondingly larger lithospheric thickness. But as yet no convincing quantitative estimates have been made.

8. SUMMARY

Despite being the only large dwarf planet moon yet observed close-up, Charon appears superficially quite similar to some of the mid-sized icy moons of Saturn and Uranus, having a water-ice-dominated, ancient cratered surface that nonetheless preserves an early history of tectonism and regional resurfacing. The fact that Charon's surface appears to post-date any tidal evolution relative to Pluto, and thus also post-dates any significant tidal heating, implies that ancient activity on other similar-sized icy worlds does not necessarily require tidal heating either.

The tectonism appears to be mostly extensional, with no obvious preferred orientation over most of the surface. The extensional fractures indicate a total areal strain of ~1%. This strain can be explained by refreezing of an ocean ~40 km thick, a scenario consistent with plausible thermal evolution models (Fig. 14). No evidence of orbital- or tidal-induced stresses has been found to date. The lack of any fossil bulge larger than 0.5% is probably also consistent with an ancient ocean, although detailed models have not yet been published. The absence of compressional features tells us that Charon is completely or almost completely differentiated, although the near-surface ice crust might be rock-rich relative to the interior. The rocky core is likely porous and hydrothermally altered; clathrates and/or organic materials are permitted on theoretical grounds, but there is no direct evidence for either.

Resurfacing at Vulcan Planitia post-dates most of the tectonism, although crater spatial densities are similar between the resurfaced and unresurfaced highland terrains, so the age gap may not be large. Resurfacing is concentrated (at least on the encounter hemisphere) in a single large low-lying region, Vulcan Planitia. Several features on Vulcan Planitia, notably the "moats" surrounding mountains and along its margins with the highland terrain, suggest that resurfacing is due to emplacement of flows of viscous material, plausibly ammonia-rich. The cryovolcanism might have arisen because the NH_3-rich ocean dregs became lighter than the ice crust above, perhaps aided by pressurization due to ice shell freezing. Indeed, the fact that the plains appear only slightly younger than the highlands is consistent with extension in the highlands and cryovolcanism in the plains both resulting from ocean freezing. The high quality of New Horizons' imaging makes Vulcan Planitia the best characterized example of probable effusive cryovolcanism in the solar system.

REFERENCES

Barlow N. G., Boyce J. M., Costard F. M., Craddock R. A., Garvin J. B., Sakimoto S. E. H., Kuzmin R. O., Roddy D. J., and Soderblom L. A. (2000) Standardizing the nomenclature of martian impact crater ejecta morphologies. *J. Geophys. Res., 105(E11),* 26733–26738, DOI: 10.1029/2000JE001258.

Beddingfield C. B. and 15 colleagues (2020) Landslides on Charon. *Icarus, 335,* 113383.

Beyer R. A. and 17 colleagues (2017) Charon tectonics. *Icarus, 287,* 161–174.

Beyer R. A. and 17 colleagues (2019) The nature and origin of Charon's smooth plains. *Icarus, 323,* 16–32.

Bierson C. J., Nimmo F., and McKinnon W. B. (2018) Implications of the observed Pluto-Charon density contrast. *Icarus, 309,* 207–219.

Blewett D. T. and 17 colleagues (2011) Hollows on Mercury: MESSENGER evidence for geologically recent volatile-related activity. *Science, 333,* 1856–1859.

Borrelli M. E. and Collins G. C. (2021) Testing the cryovolcanism and plate bending hypotheses for Charon's smooth plains. *Icarus,* 356, 113717.

Bray V. and Schenk P. (2015) Pristine impact crater morphology on Pluto: Expectations for New Horizons. *Icarus, 246,* 156–164.

Brown M. E. and Calvin W. M. (2000) Evidence for crystalline water and ammonia ices on Pluto's satellite Charon. *Science, 287,* 107–109.

Buie M. W., Tholen D. J., and Wasserman L. H. (1997) Separate lightcurves of Pluto and Charon. *Icarus, 125,* 233–244.

Buie M. W., Tholen D. J., and Grundy W. M. (2012) The orbit of Charon is circular. *Astron. J., 144,* 15.

Canup R. M. (2011) On a giant impact origin of Charon, Nix, and Hydra. *Astron. J., 141,* 35.

Carr M. H., Crumpler L. S., Cutts J. A., Greeley R., Guest J. E., and Masursky H. (1977) Martian impact craters and emplacement of ejecta by surface flow. *J. Geophys. Res., 82*, 4055–4065.

Cassidy T. and 7 colleagues (2010) Radiolysis and photolysis of icy satellite surfaces: Experiments and theory. *Space Sci. Rev., 153*, 299–315.

Chen H. Z. and Yin A. (2019) Tectonic history of the Oz Terra of Charon as revealed by systematic structural mapping. *Pluto System After New Horizons*, Abstract #7007. LPI Contribution No. 2133, Lunar and Planetary Institute, Houston.

Cheng W. H., Lee M. H., and Peale S. J. (2014) Complete tidal evolution of Pluto-Charon. *Icarus, 233*, 242–258.

Collins G. C. and Pappalardo R. T. (2000) Predicted stress patterns on Pluto and Charon due to their mutual orbital evolution. *Lunar and Planetary Science XXXI*, Abstract #1035. LPI Contribution No. 1000, Lunar and Planetary Institute, Houston.

Collins G., McKinnon W., Moore J., Nimmo F., Pappalardo R., Prockter L., and Schenk P. (2010) Tectonics of the outer planet satellites. In *Planetary Tectonics* (T. R. Watters and R. A. Schultz, eds.), pp. 264–350. Cambridge Univ., Cambridge.

Conrad J. W. and 13 colleagues (2019) An upper bound on Pluto's heat flux from a lack of flexural response of its normal faults. *Icarus, 328*, 210–217.

Cook J. C., Desch S. J., Roush T. L., Trujillo C. A., and Geballe T. R. (2007) Near-infrared spectroscopy of Charon: Possible evidence for cryovolcanism on Kuiper belt objects. *Astrophys. J., 663*, 1406–1419.

Cook J. C. and 21 colleagues (2018) Composition of Pluto's small satellites: Analysis of New Horizons spectral images. *Icarus, 315*, 30–45.

Cooper J. F., Christian E. R., Richardson J. D., and Wang C. (2003) Proton irradiation of Centaur, Kuiper belt, and Oort cloud objects at plasma to cosmic ray energy. *Earth Moon Planets, 92*, 261–277.

Dalle Ore C. M. and 11 colleagues (2018) Ices on Charon: Distribution of H_2O and NH_3 from New Horizons LEISA observations. *Icarus, 300*, 21–32.

Denevi B. W. and 25 colleagues (2012) Pitted terrain on Vesta and implications for the presence of volatiles. *Science, 338*, 246–249.

Desch S. J. (2015) Density of Charon formed from a disk generated by the impact of partially differentiated bodies. *Icarus, 246*, 37–47.

Desch S. J. and Neveu M. (2017) Differentiation and cryovolcanism on Charon: A view before and after New Horizons. *Icarus, 287*, 175–186.

Desch S. J., Cook J. C., Doggett T. C., and Porter S. B. (2009) Thermal evolution of Kuiper belt objects, with implications for cryovolcanism. *Icarus, 202*, 694–714.

Dobrovolskis A. R., Peale S. J., and Harris A. W. (1997) Dynamics of the Pluto-Charon binary. In *Pluto and Charon* (J. A. Burns and M. S. Matthews, eds.), pp. 159–190. Univ. of Arizona, Tucson.

Dumas C., Terrile R. J., Brown R. H., Schneider G., and Smith B. A. (2001) Hubble Space Telescope NICMOS spectroscopy of Charon's leading and trailing hemispheres. *Astron. J., 121*, 1163–1170.

Greenstreet S., Gladman B., and McKinnon W. B. (2015) Impact and cratering rates onto Pluto. *Icarus, 258*, 267–288.

Grundy W. M. and 33 colleagues (2016) Surface compositions across Pluto and Charon. *Science, 351*, aad9189.

Gulbis A. A. S. and 12 colleagues (2006) Charon's radius and atmospheric constraints from observations of a stellar occultation. *Nature, 439*, 48–51.

Head J. W. and Pappalardo R. T. (1999) Brine mobilization during lithospheric heating on Europa: Implications for formation of chaos terrain, lenticular texture and color variations. *J. Geophys. Res., 104*, 27143–27155.

Hogenboom D. L., Kargel J. S., Consolmagno G. J., Holden T. C., Lee L., and Buyyounouski M. (1997) The ammonia-water system and the chemical differentiation of icy satellites. *Icarus, 128*, 171–180.

Horner V. M. and Greeley R. (1982) Pedestal craters on Ganymede. *Icarus, 51*, 549–562.

Hussmann H., Sohl F., and Spohn T. (2006) Subsurface oceans and deep interiors of medium-sized outer planet satellites and large trans-neptunian objects. *Icarus, 185*, 258–273.

Jankowski D. G. and Squyres S. W. (1988) Solid-state ice volcanism on the satellites of Uranus. *Science, 241*, 1322–1325.

Kamata S. and 6 colleagues (2019) Pluto's ocean is capped and insulated by gas hydrates. *Nature Geosci., 12*, 407–410.

Kargel J. S. (1992) Ammonia-water volcanism on icy satellites: Phase relations at 1 atmosphere. *Icarus, 100*, 556–574.

Kargel J. S., Croft S. K., Lunine J. I., and Lewis J. S. (1991) Rheological properties of ammonia-water liquids and crystal-liquid slurries: Planetological applications. *Icarus, 89*, 93–112.

Kattenhorn S. A. and Hurford T. (2009) Tectonics of Europa. In *Europa* (R. T. Pappalardo et al., eds.), pp. 199–236. Univ. of Arizona, Tucson.

Malamud U. and Prialnik D. (2015) Modeling Kuiper belt objects Charon, Orcus and Salacia by means of a new equation of state for porous icy bodies. *Icarus, 246*, 21–36.

Malamud U., Perets H. B., and Schubert G. (2017) The contraction/expansion history of Charon with implications for its planetary-scale tectonic belt. *Mon. Not. R. Astron. Soc.,468*, 1056–1069.

Manga M. and Wang C.-Y. (2007) Pressurized oceans and the eruption of liquid water on Europa and Enceladus. *Geophys. Res. Lett., 34*, L07202.

Marcialis R. L., Rieke G. H., and Lebofsky L. A. (1987) The surface composition of Charon: Tentative identification of water ice. *Science, 237*, 1349–1351.

McKinnon W. B., Simonelli D. P., and Schubert G. (1997) Composition, internal structure, and thermal evolution of Pluto and Charon. In *Pluto and Charon* (S. A. Stern and D. J. Tholen, eds.), pp. 295–343. Univ. of Arizona, Tucson.

McKinnon W. B. and 16 colleagues (2017) Origin of the Pluto-Charon system: Constraints from the New Horizons flyby. *Icarus, 287*, 2–11.

Melosh H. J. (1977) Global tectonics of a despun planet. *Icarus, 31*, 221–243.

Melosh H. J. (1980) Tectonic patterns on a tidally distorted planet. *Icarus, 43*, 334–337.

Melosh H. J. and Janes D. M. (1989) Ice volcanism on Ariel. *Science, 245*, 195–196.

Melosh H. J. and Schenk P. (1993) Split comets and the origin of crater chains on Ganymede and Callisto. *Nature, 365*, 731–733.

Moore H. J., Arthur D. W. G., and Schaber G. G. (1978) Yield strengths of flows on the Earth, Mars, and Moon. *Proc. Lunar Planet. Sci. Conf. 9th*, pp. 3351–3378.

Moore J. M. and 153 colleagues (2016) The geology of Pluto and Charon through the eyes of New Horizons. *Science, 351*, 1284–1293.

Moore M. H., Ferrante R. F., Hudson R. L., and Stone J. N. (2007) Ammonia-water ice laboratory studies relevant to outer Solar System surfaces. *Icarus, 190*, 260–273.

Nimmo F. and Matsuyama I. (2007) Reorientation of icy satellites by impact basins. *Geophys. Res. Lett., 34*, L19203.

Nimmo F. and 16 colleagues (2017) Mean radius and shape of Pluto and Charon from New Horizons images. *Icarus, 287*, 12–29.

Person M. J. and 6 colleagues (2006) Charon's radius and density from the combined data sets of the 2005 July 11 occultation. *Astron. J., 132*, 1575–1580.

Petrenko V. F. and Whitworth R. W. (1999) *Physics of Ice*. Oxford, New York. 390 pp.

Porter S. B., Desch S. J., and Cook J. C. (2010) Micrometeorite impact annealing of ice in the outer solar system. *Icarus, 208*, 492–498.

Rhoden A. R., Henning W., Hurford T. A., and Hamilton D. P. (2015) The interior and orbital evolution of Charon as preserved in its geologic record. *Icarus, 246*, 11–20.

Rhoden A. R., Skjetne H. L., Henning W. G., Hurford T. A., Walsh K. J., Stern S. A., Olkin C. B., Spencer J. R., Weaver H. A., Young L. A., Ennico K., and the New Horizons Team (2020) Charon: A brief history of tides. *J. Geophys. Res.*, 125, e06449.

Robbins S. J. and 29 colleagues (2017) Craters of the Pluto-Charon system. *Icarus, 287*, 187–206.

Robbins S. J. and 25 colleagues (2018) Investigation of Charon's craters with abrupt terminus ejecta, comparisons with other icy bodies, and formation implications. *J. Geophys. Res., 123*, 20–36.

Robbins S. J. and 28 colleagues (2019) Geologic landforms and chronostratigraphic history of Charon as revealed by a hemispheric geologic map. *J. Geophys. Res., 124*, 155–174.

Robuchon G. and Nimmo F. (2011) Thermal evolution of Pluto and implications for surface tectonics and a subsurface ocean. *Icarus, 216*, 426–439.

Schenk P. M. (1991) Fluid volcanism on Miranda and Ariel: Flow morphology and composition. *J. Geophys. Res., 96*, 1887–1906.

Schenk P. M. and McKinnon W. B. (1991) Dark-ray and dark-floor

craters on Ganymede, and the provenance of large impactors in the jovian system. *Icarus, 89*, 318–346.
Schenk P., Matsuyama I., and Nimmo F. (2008) True polar wander on Europa from global-scale small-circle depressions. *Nature, 453*, 368–371.
Schenk P. M. and 18 colleagues (2018a) Breaking up is hard to do: Global cartography and topography of Pluto's mid-sized icy moon Charon from New Horizons. *Icarus, 315*, 124–145.
Schenk P. M., White O. L., Byrne P. K., and Moore J. M. (2018b) Saturn's other icy moons: Geologically complex worlds in their own right. In *Enceladus and the Icy Moons of Saturn* (P. M. Schenk et al., eds.), pp. 237–265. Univ. of Arizona, Tucson.
Sicardy B. and 44 colleagues (2006) Charon's size and an upper limit on its atmosphere from a stellar occultation. *Nature, 439*, 52–54.
Singer K. N. and 25 colleagues (2019) Impact craters on Pluto and Charon indicate a deficit of small Kuiper belt objects. *Science, 363*, 955–959.
Sotin C., Grasset O., and Beauchesne S. (1998) Thermodynamic properties of high pressure ices: Implications for the dynamics and internal structure of large icy satellites. In *Solar System Ices* (B. Schmitt et al., eds.), pp. 79–96. Astrophysics and Space Science Library, Vol. 227, Springer, Dordrecht.
Stern S. A. (2009) Ejecta exchange and satellite color evolution in the Pluto system, with implications for KBOs and asteroids with satellites. *Icarus, 199*, 571–573.
Umurhan O. M. and Cruikshank D. P. (2019) Cryovolcanism on Pluto: Various theoretical considerations. *Pluto System After New Horizons*, Abstract #7066. LPI Contribution No. 2133, Lunar and Planetary Institute, Houston.
Walker A. R. (1980) An occultation by Charon. *Mon. Not. R. Astron. Soc., 192*, 47P–50P.
Watters T. R. and 6 colleagues (2015) Global thrust faulting on the Moon and the influence of tidal stresses. *Geology, 43*, 851–854.
White O. L., Schenk P. M., and Dombard A. J. (2013) Impact basin relaxation on Rhea and Iapetus and relation to past heat flow. *Icarus, 223*, 699–709.
White O. L., Schenk P. M., Bellagamba A. W., Grimm A. M., Dombard A. J., and Bray V. J. (2017) Impact crater relaxation on Dione and Tethys and relation to past heat flow. *Icarus, 288*, 37–52.
Zhong S., Parmentier E. M., and Zuber M. T. (2000) A dynamic origin for the global asymmetry of lunar mare basalts. *Earth Planet. Sci. Lett., 177*, 131–140.

Howett C. J. A., Olkin C. B., Protopapa S., Grundy W. M., Verbiscer A. J., and Buratti B. J. (2021) Charon: Colors and photometric properties. In *The Pluto System After New Horizons* (S. A. Stern, J. M. Moore, W. M. Grundy, L. A. Young, and R. P. Binzel, eds.), pp. 413–432. Univ. of Arizona, Tucson, DOI: 10.2458/azu_uapress_9780816540945-ch018.

Charon: Colors and Photometric Properties

C. J. A. Howett, C. B. Olkin, S. Protopapa
Southwest Research Institute

W. M. Grundy
Lowell Observatory

A. J. Verbiscer
University of Virginia

B. J. Buratti
Jet Propulsion Laboratory

The New Horizons spacecraft explored Pluto's largest moon Charon during its Pluto-system encounter in 2015. Images taken by New Horizons' Multispectral Visible Imaging Camera (MVIC) revealed Charon to be predominantly gray, with an almost uniform reflectance with the exception of some bright craters and a darker, redder north polar cap. Subsequent analysis showed that the same pattern of polar darkening also occurs at its southern pole as well, and is believed to be due to the radiolysis of hydrocarbons that escaped Pluto's atmosphere only to be cold-trapped on Charon. The hemisphere encountered by New Horizons appears to show the same color trends as the non-encounter one. Small-scale color variations are observed across Charon's more neutrally colored terrain and appear to be correlated with impact sites, rather than geology or topography, perhaps implying compositional differences with depth. Many of Charon's photometric properties are consistent with other large icy satellite surfaces in our solar system; however, it scatters incident visible sunlight more isotropically than most.

1. INTRODUCTION

1.1. Discovery of Charon

With the goal of refining the orbital position of Pluto, *Christy and Harrington* (1978) acquired a series of observations of the dwarf planet with the 1.55-m Kaj Strand Astrometric Reflector at the U.S. Naval Observatory in Flagstaff, Arizona. On June 22, 1978, Astronomer James W. Christy observed Pluto to be systematically elongated. Specifically, Pluto displayed a faint extension along the south in the exposures acquired on April 13 and 20, 1978, while the extension was oriented along the north in the data recorded on May 12, 1978. Guiding issues as well as the possibility of a background star were ruled out because background stars in the same plates were circular, exposure times were short, and Pluto's bulge changed position over time. Instead, the idea of a satellite orbiting around Pluto was put forth. This theory was then corroborated using Pluto archival images acquired prior to 1978 for other scientific purposes as well as follow-up data acquired with the 1.55-m astrometric reflector in Flagstaff by H. D. Abies and C. C. Dahn and the 4-m reflector at Cerro Tololo, Chile, by J. A. Graham. These measurements enabled estimates of orbital period and mean distance of Pluto's faint satellite to be approximately 6.4 days and 15,000–20,000 km, respectively. This discovery in reassessments of Pluto's size, mass, and other physical characteristics. The formal announcement of the discovery of Charon, initially designated as S/1978 P 1, was published in the Central Bureau for Astronomical Telegrams Circular No. 3241 and by *Christy and Harrington* (1978).

1.2. Overview of the Historic Color Information of Charon

Shortly after the discovery of Charon, *Andersson* (1978) reported that Pluto's heliocentric motion and the orientation of Charon's orbit plane with respect to Pluto would soon give rise to an approximately once-per-century series of mutual events between Pluto and Charon. In other words, groundbased observers would have the opportunity to observe, once every 124 years (half Pluto's orbital period), Charon passing in front of and behind the dwarf planet. At that time groundbased instrument capabilities were not advanced enough to enable spatially resolved spectroscopic and/or photometric measurements of Pluto and Charon. Therefore a subtractive technique during the mutual events from 1984 to 1990 provided the only tool to isolate the contribution of Charon to the unresolved binary system and assess color (and other) information of Pluto and Charon separated (e.g., *Tholen et al.,* 1987; *Binzel,* 1988; *Buie et al.,* 1992; *Reinsch et al.,* 1994). However, this technique had its own limitations. Indeed, because of the synchronous nature of the system [i.e., Pluto and Charon always have the same hemispheres facing one another (*Buie et al.,* 2006)] no global information of Pluto and its moon Charon could be derived. *Binzel* (1988) and *Reinsch et al.* (1994) investigated the color differences of Charon using its B-V color. This describes the difference in color between a target as viewed at blue wavelengths (B, ~445 nm) and visual (V, ~551 nm) ones. On this scale higher values indicate a bluer-colored target. *Binzel* (1988) and *Reinsch et al.* (1994) found a B-V value for the Pluto-facing hemisphere of Charon ($\lambda = 0°$ longitude) in perfect agreement, on the order of 0.700 ± 0.010 and 0.701 ± 0.014, respectively, and significantly less red than Pluto with a B-V color of 0.867 ± 0.008.

A breakthrough in assessing photometric properties of Pluto and Charon came with the Hubble Space Telescope (HST). Because it is not limited by atmospheric distortion, HST can easily resolve the system. *Buie et al.* (1997) presented HST measurements of Pluto and Charon acquired between 1992 and 1993 and revealed Charon's light curve to have a very small amplitude (0.08 mag), indicative of a uniform surface composition. The light curve displays a minimum near 170°E longitude, corresponding to the anti-Pluto hemisphere, and looks flat otherwise. A B-V color of 0.710 ± 0.011 at 123°E and 289°E was found. These values, when compared with the color reported by *Binzel* (1988) and *Reinsch et al.* (1994) for the Pluto-facing hemisphere (0°E longitude), strongly indicate that no large-scale color variations occur across the surface of Pluto's satellite. This finding ruled out initial measurements by *Tholen et al.* (1987) that suggested Charon was heterogeneous, with the anti-Pluto hemisphere being redder than the Pluto-facing one.

Subsequent lightcurve measurements of the Pluto-Charon system were taken in 2002–2003 with the HST's Advanced Camera for Surveys (ACS) High-Resolution Channel (HRC) by *Buie et al.* (2010a) (Fig. 1). The results for Charon were in perfect agreement with those by *Buie et al.* (1997) and confirmed that the anti-Pluto hemisphere is slightly darker. They reported a weighted mean B-V value for Charon of 0.7313 ± 0.0017, slightly redder than that reported by *Buie et al.* (1997) (0.710 ± 0.011), and a color variation with longitude <1%. While viewing geometry could not account for the difference by 1.9σ between the average Charon color reported by *Buie et al.* (1997) and *Buie et al.* (2010a), a possible explanation resides in a significant shift in the sub-Earth latitude between the two sets of measurements: 8°–12°N in *Buie et al.* (1997) to 29°–32°N in *Buie et al.* (2010a). Therefore, *Buie et al.* (2010a) suggested a difference in color between the two poles as a plausible explanation. The low albedo of the poles of Charon has since been confirmed by New Horizons (*Grundy et al.,* 2016a,b; *Buratti et al.,* 2019).

2. NEW HORIZONS COLOR OBSERVATIONS OF CHARON

2.1. Introduction to New Horizons

NASA's New Horizons mission flew through the Pluto system on July 14, 2015 (*Stern et al.,* 2015), providing a close-up view of Charon's geology and its color and spectral properties. New Horizons' Multispectral Visible Imaging Camera (MVIC) (*Reuter et al.,* 2008) provided the color images of Charon; details of this instrument are given in section 2.2. A wide range of solar phase angle observations of Charon were acquired with New Horizons' MVIC and LOng-Range Reconnaissance Imager (LORRI). LORRI is a panchromatic CCD camera that obtains images between 350 and 850 nm with a 0.29°× 0.29° field of view (cf. *Cheng et al.,* 2008). Observations obtained by both LORRI and MVIC enabled investigations of spatially resolved photometric properties of Pluto's satellite and the possibility of providing context to the prior groundbased and spacebased observations. Compositional information on Charon was obtained by New Horizons' Linear Etalon Imaging Spectral Array (LEISA). LEISA is an infrared imaging spectrometer, sensitive to wavelengths from 1.2 to 2.5 μm (*Reuter et al.,* 2008).

2.2. Introduction to Ralph's Multi-Spectral Visible Imaging Camera

MVIC is the color camera inside the Ralph instrument on New Horizons; it has seven independent CCD arrays held on a single substrate. One of the CCDs is a panchromatic frame transfer 5024 × 128 pixel array. The remaining CCDs are operated in time delay integration mode (TDI), and include two panchromatic (Pan) arrays and four color filters (red, blue, NIR, CH_4). Details of these arrays are given in Table 1.

MVIC was calibrated both prior to launch (cf. *Reuter et al.,* 2008) and after launch (cf. *Howett et al.,* 2017a). A post-launch analysis of MVIC's NIR filter revealed gain drift when it was powered on the main power side (side a, or side 1), although it was stable on the redundant power side (side b, or side 0). It was found that the gain was consistent over single images but varied on the 5–10% level from image to image. Thus, this drift has to be corrected when observations are made with the NIR filter on power side A. *Howett et al.* (2017a) and *Olkin et al.* (2017) provide further details of this correction.

2.3. New Horizons' Unresolved Color Observations of Charon

Color observations of the Pluto-Charon system began 15 Pluto days before the New Horizons' closest approach of Pluto. Each Pluto day lasts 6.4 Earth days and in that same time period Charon orbits Pluto. The system is completely tidally evolved since the spin rates of Pluto and Charon match the mean motion of each object about the system

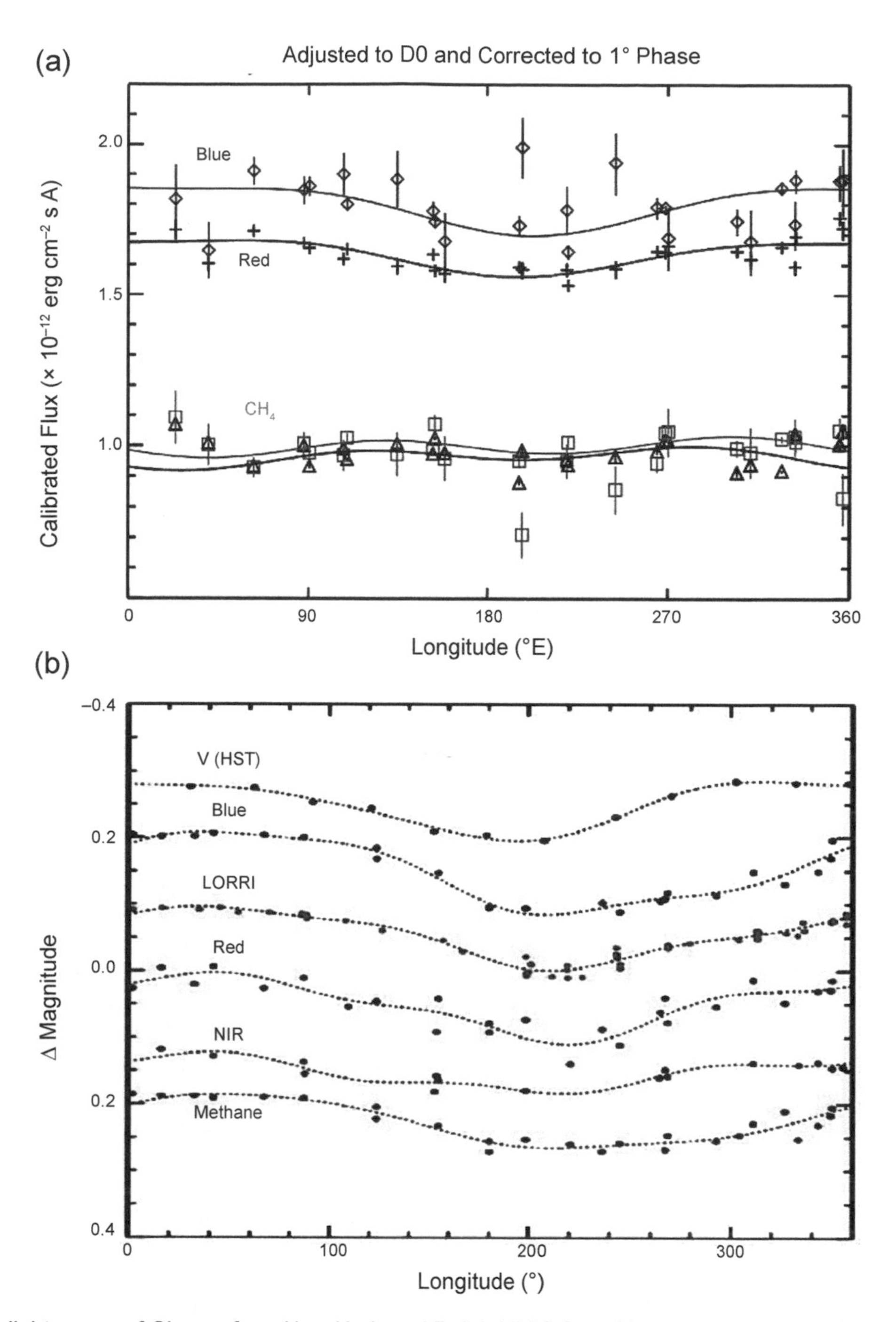

Fig. 1. (a) Color lightcurves of Charon from New Horizons' Ralph MVIC from *Howett et al.* (2017b). MVIC red is given by crosses, blue by diamonds, NIR by triangles, and CH_4 by squares. **(b)** Lightcurves from Hubble Space Telescope (HST) (*Buie et al.*, 2010a), MVIC color (blue, red, NIR, and methane) and New Horizons LORRI from *Buratti et al.* (2019). Realistic errors are shown by the scatter in the data at each phase angle.

barycenter. The approach observations of New Horizons to the Pluto-Charon system provided disk averaged photometry of Charon separated from Pluto at a phase angle that is not observable from Earth. The maximum phase angle (Sun-Pluto-observer) that is possible from Earth is less than 2° but the approach phase angle of New Horizons varied from 14.5° to 15.1°. Additionally, the subobserver latitude of these observations (43°N) is significantly different from the sub-observer latitude during the mutual event observations (near 0°) and HST observations in the 1990s (up to 32°N). *Howett et al.* (2017b) and *Buratti et al.* (2019) provide summaries of the color lightcurves of Charon from New Horizons (see Fig. 1). The light curves from the MVIC red and blue filters show similar trends to the Earth-based light curves: Charon exhibits only minor variation in reflectivity with longitude and Charon's surface is somewhat less reflective on the side

TABLE 1. Details of the MVIC arrays.

Array Name	Array Description	Wavelength Range (nm)	Pivot Wavelength (nm)	Array Size (pixels)
Pan 1	Panchromatic TDI #1	400–975	692	5024 × 32
Pan 2	Panchromatic TDI #2	400–975	692	5024 × 32
Blue	Blue TDI	400–550	492	5024 × 32
Red	Red TDI	540–700	624	5024 × 32
NIR	Near-infrared TDI	780–975	861	5024 × 32
CH_4	Methane-band TDI	860–910	883	5024 × 32
Pan frame	Panchromatic framing camera	400–975	692	5024 × 128

A single MVIC pixel is 19.77 µrad × 19.77 µrad, so the FOV of the TDI array is 5.7° × 0.037°, and that of the framing camera is 5.7° × 0.146° (*Howett et al.,* 2017a).

facing away from Pluto (*Buie et al.,* 1997, 2010b). We note that the approach MVIC images were not northern enough to have Mordor macular (the informal name of Charon's red north polar region) as a significant fraction of its field of view. Thus, both Earth and New Horizons observations viewed Charon's grayer lower-latitude terrain. Due to the NIR gain drift (see section 2.2) and the low signal-to-noise of the CH_4 channel (due to its narrow spectral range), definitive interpretations of Charon's color from these filters were difficult to determine.

2.4. New Horizons' Resolved Color Observations of Charon

After the July 7, 2015, MVIC was able to resolve Charon's sunlit hemisphere, these observations continued until July 14 when New Horizons flew past Charon and then started observing its nightside. During approach the subspacecraft latitude was the same as on more distant approach — 43°N — but upon closest approach this value decreased to 25°N. The subspacecraft longitude varies across the full range as Charon rotates under the spacecraft. A summary of these MVIC observations is given in Table 2.

The images taken during this time have been converted to I/F (i.e., the ratio of the reflected and the incident flux on their surface). Enhanced color images were then produced using these I/F corrected images (called enhanced because they include wavelengths outside the visible spectrum) using the MVIC NIR/red/blue filters as the image's R/G/B color. The results are shown in Figs. 2 and 3. The images are split between distant approach observations (Fig. 2), which have a spatial resolution between 162.1 km/pixel and 9.2 km/pixel, and those taken at close approach (Fig. 3). The two observations taken of Charon's sunlit side taken nearest to closest approach have a significantly higher spatial resolution (5.2 and 1.4 km/pixel; see Fig. 3). The color trends observed across Charon in these images is discussed in detail in the following section.

3. GLOBAL OVERVIEW OF CHARON'S COLORS

3.1. Introduction

On average, Charon is less red than Pluto, with colors typical of icy uranian or saturnian satellites. To aid the discussion of the location of these color variations, refer to the nomenclature maps provided in Figs. A7–A11 of Appendix A by Beyer and Showalter in this volume.

Overall, the surface of Charon is gray but not uniformly so. It is mottled with a variety of color shades, the most extreme of which is the red polar cap, Mordor Macula, which is discussed in greater detail in section 3.3. Charon's encounter hemisphere is divided into two main provinces by a diagonal belt of tectonic features. To the north, Oz Terra is heavily cratered. The ruggedness of the topography is difficult to discern due to the high elevation lighting in the encounter images, but stereo imaging reveals high-standing blocks separated by deep chasmata (*Schenk et al.,* 2018). These features are heavily eroded and softened, making them harder to pick out, compared with the very prominent tectonic belt, although Oz Terra's topography is no less dramatic in terms of elevation extremes. To the south, Vulcan Planitia has a much smoother appearance, consistent with resurfacing relatively late in Charon's geological history (*Beyer et al.,* 2019). See the chapter by Spencer et al. in this volume for a more complete discussion of Charon's geology and likewise the chapter by Protopapa et al. in this volume for a discussion on Charon's composition.

Greater color diversity is discernible in Oz Terra; e.g., there are redder regions including Gallifrey Macula and unnamed plains in western Oz. Less-red regions divide into darker and brighter shades of gray, and many regions with distinct colors are associated with impact craters and their ejecta. Ripley and Nasreddin craters provide examples of the darker shades of gray, while an unnamed plain north of Nostromo Chasma offers a good example of lighter gray material. Ejecta from

TABLE 2. Details of the resolved color observations MVIC made of Charon.

Request ID	Mid-Time of Observations (UTC)	MET (s)	Sub–S/C Lon (°E)	Sub-S/C Lat (°N)	Electronics Side	Charon MVIC Res (km/pix)	Distance (km)
PEMV_01_PC_Multi_ Map_A_18	2015 Jul 07 T16:44:45	298593398	−10.5	43.0	1	162.1	8,105,290
PEMV_01_PC_Multi_ Map_B_2	2015 Jul 08 T09:04:45	298652198	−48.7	43.0	0	145.8	7,290,456
PEMV_01_PC_Multi_ Map_B_3	2015 Jul 08 T17:06:50	298681123	−67.6	43.1	1	137.8	6,888,145
PEMV_01_PC_Multi_ Map_B_5	2015 Jul 09 T03:41:05	298719178	−92.4	43.1	0	127.2	6,358,174
PEMV_01_PC_Multi_ Map_B_6	2015 Jul 09 T16:56:05	298766878	−123.5	43.1	1	113.9	5,694,232
PEMV_01_PC_Multi_ Map_B_8	2015 Jul 10 T08:55:30	298824443	−161.1	43.2	0	97.9	4,895,918
PEMV_01_PC_Multi_ Map_B_9	2015 Jul 10 T16:52:15	298853048	−179.9	43.2	1	90.0	4,515,367
PEMV_02_PC_Multi_ Map_B_9	2015 Jul 10 T16:55:05	298853218	−180.0	43.2	1	90.0	4,498,711
PEMV_01_PC_Multi_ Map_B_11	2015 Jul 11 T03:34:35	298891588	154.9	43.2	0	79.4	3,971,163
PEMV_01_PC_Multi_ Map_B_12	2015 Jul 11 T16:46:56	298939128	123.7	43.1	1	66.4	3,320,442
PEMV_02_PC_Multi_ Map_B_12	2015 Jul 11 T16:49:46	298939298	123.6	43.1	1	66.4	3,318,119
PEMV_01_PC_Multi_ Map_B_14	2015 Jul 12 T08:23:08	298995300	87.0	42.9	0	51.1	2,552,912
PEMV_01_PC_Multi_ Map_B_15	2015 Jul 12 T16:53:04	299025878	67.0	42.8	1	42.7	2,130,663
PEMV_01_PC_Multi_ Map_B_17	2015 Jul 13 T03:38:06	299064598	42.0	42.5	0	32.1	1,607,592
PEMV_01_PC_Multi_ Map_B_18	2015 Jul 13 T07:38:36	299079028	32.7	42.4	1	28.2	1,409,945
PEMV_01_PCNH_ Multi_Long_1d1	2015 Jul 13 T14:50:51	299104958	16.2	42.0	0	21.1	1,039,982
PEMV_01_PC_Multi_ Long_1d2	2015 Jul 13 T21:08:40	299127628	2.0	41.5	1	14.8	742,083
PEMV_01_PC_Color_ TimeRes	2015 Jul 14 T02:47:54	299147983	−9.9	40.5	0	9.2	461,507
PEMV_01_PC_Color_1	2015 Jul 14 T06:50:11	299162518	−16.7	38.5	1	5.2	261,182
PEMV_01_C_Color_2	2015 Jul 14 T10:42:28	299176438	−12.3	25.2	0	1.4	72,693

MET is mission elapsed time in seconds, Sub-S/C Lat/Lon is subspacecraft latitude and longitude respectively, and Res is resolution.

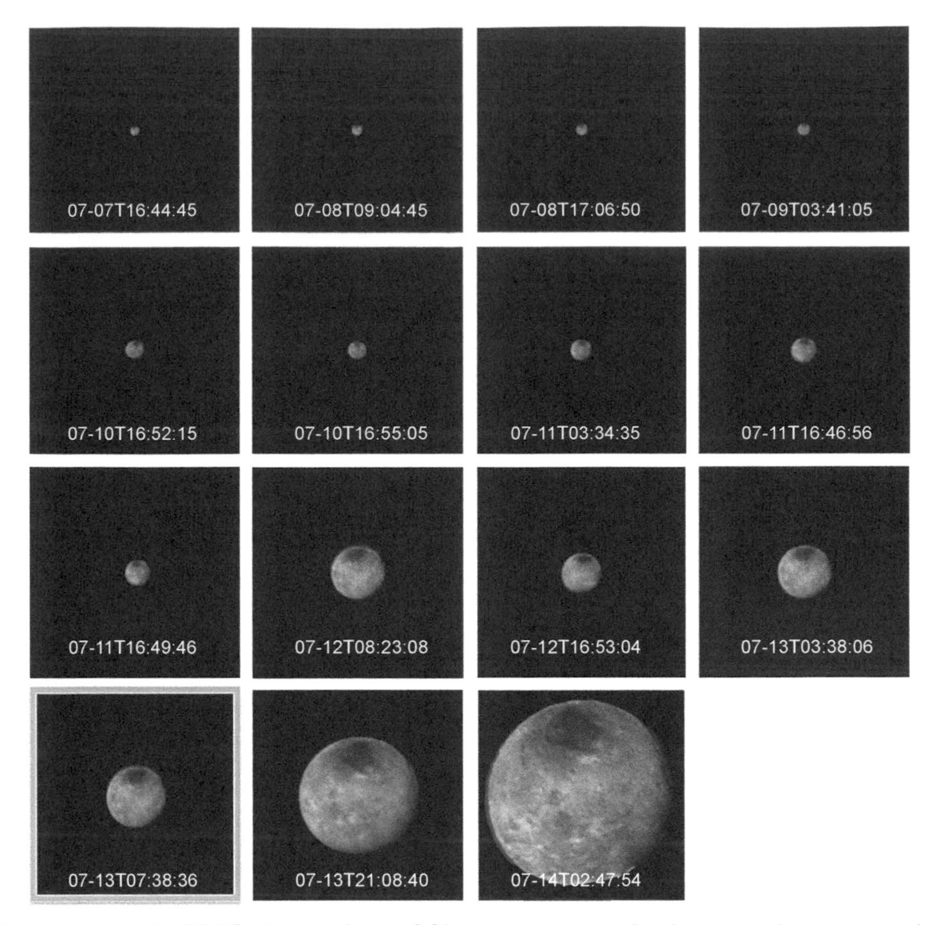

Fig. 2. Resolved enhanced color MVIC observations of Charon on approach, shown on the same scale; north is up. The spatial resolution is 162 km/pixel at start of sequence (top left) and 9.2 km per pixel at end (bottom right). The lower-left image surrounded by a purple box is PC_Multi_Map_B_15, which shows the highest resolved color observation of Charon's non-encounter hemisphere. MVIC IR, red, and blue filters are used for RGB color respectively.

impact craters in Oz Terra sometimes shows striking contrasts between the more distant ejecta and those closer to the crater. For instance, the distal ejecta around Skywalker appear dark, in contrast with brighter, more proximal ejecta. The opposite trend is seen in Vader, Nasreddin, Candide, Madoc, and Ahab craters. Color diversity appears much more muted in Vulcan Planitia to the south (*Schenk et al.,* 2018).

3.2. Global Color Trends

Charon's global color trends can be seen more clearly in Fig. 4, which shows the color-color plot for Charon. It was made using the four color filters of MVIC and plots their I/F values as ratios. As the figure shows, the CH_4/NIR color is mostly constant across Charon (1.1 ± 0.6), but the red/NIR and red/blue vary. Most of Charon sits in the "neutral lobe," which has approximately neutral red/NIR color and is slightly bluer on the red/blue color. The line of increasing red/NIR color with red/blue color corresponds to Charon's north polar region.

A more detailed inspection of Charon's color can be obtained by examining MVIC's red, blue, and NIR pixel values seen across the surface. To minimize shadows caused by surface morphology we restrict the pixels to those with an emission and incident angle less than 60°. The results are shown in Fig. 5, which shows that there are two distinct regions, as indicated by the two diagonal lines. The first region lies approximately on the x = y line, implying it has equal red and blue color; further inspection shows its NIR color corresponds to these values as well, indicating a neutral color. The other region has a steeper slope, being brighter in the red than the blue. Further inspection also shows this region is brighter in the NIR than blue too. Investigating the origin regions of the red and blue pixels (also Fig. 5), it is clear that the redder regions correspond to Charon's north polar cap, while the more neutral region corresponds to the remainder of the imaged close approach hemisphere.

3.3. Charon's Red Poles

The most striking color feature on Charon's encounter hemisphere is the reddish northern polar cap, Mordor Macula. The cap is darker than the rest of Charon in all MVIC filters but is most strikingly so at the shorter wavelengths. In the blue filter, the reflectance at the pole is only about 30% of the reflectance at equatorial latitudes, while in the red filter the figure is about 40% and it rises to 65% in the NIR filter. Further into the infrared, the contrast becomes

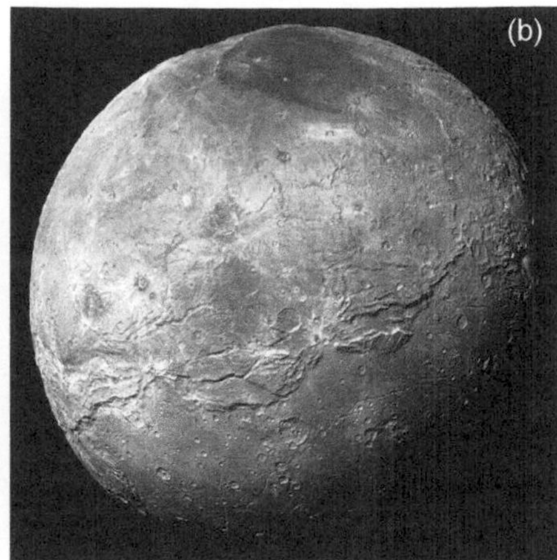

Fig. 3. Two highest-spatial-resolution color images of Charon taken by New Horizons. **(a)** Second-highest spatial-enhanced color image of Charon taken by New Horizons/MVIC (PEMV_01_PC_Color_1; 5.2 km/pixel). North is up. **(b)** Highest spatial-resolution-enhanced color image of Charon taken by New Horizons/MVIC (PEMV_01_C_COLOR_2; 1.4 km/pixel). North is toward the top of the image.

even smaller (see the chapter by Protopapa et al. in this volume). The absorption is attributed to tholin-like materials, a family of organic substances, and residues from the action of several energy sources (e.g., solar UV radiation or cosmic rays) on small organic molecules, and has a range of colors, typically orange/red/brown (e.g., *Khare et al.,* 1984; *Cruikshank et al.,* 2005).

Charon's polar cap does not show a discrete boundary. Instead, the red coloration increases gradually toward higher latitudes (*Grundy et al.,* 2016a). There is some topographic influence on the color distribution, such as less redness across the floor of Dorothy crater, and more to the north of the ridge bounding Caleuche Chasma. North-facing slopes tend to exhibit greater redness, while south-facing slopes exhibit less. A few small impact craters penetrate through the red material, revealing more typical (gray) Charon colors in the exposed substrate. Their existence indicates that the red material is a thin veneer.

Grundy et al. (2016a) hypothesized the polar deposit is produced through a chain of events originating from methane (CH_4) escaping from Pluto's atmosphere escaping to space (*Gladstone et al.,* 2016; *Young et al.,* 2018). As Charon orbits the Pluto system barycenter it moves through this escaping flow, transiently capturing it into a thin atmosphere around Charon (*Tucker et al.,* 2015; *Hoey et al.,* 2017). With its high obliquity amplifying its seasons, Charon's winter pole becomes extremely cold during the long, dark winter, well below 25 K, for plausible assumptions about thermal conductivity of the surface ice. At such temperatures, CH_4 gas molecules that encounter the cold pole are cold-trapped as CH_4 ice for the remainder of the winter. Although it is not directly illuminated by the Sun, seasonal CH_4 ice at the winter pole is exposed to energetic radiation from sources outside of the solar system, and most importantly, by solar Lyman-α ultraviolet light (122 nm, 10.2 eV) that has been scattered by neutral hydrogen in interplanetary space. This radiation breaks the C-H bonds in CH_4 molecules. This radiolysis and photolysis allows hydrogen to escape and produces radicals that go on to combine into heavier, less-volatile hydrocarbon molecules. In the spring, the return of sunlight causes Charon's pole to warm rapidly. The remaining CH_4 sublimates away, but heavier photochemical residues remain, providing the feedstock for further radiolysis and photolysis to produce tholins. By assuming that Pluto's atmosphere has always shed CH_4 at a rate of 5×10^{25} molecules per second [estimated from New Horizons in 2015 (*Gladstone et al.,* 2016)], *Grundy et al.* (2016a) were able to estimate that roughly one-third of a micrometer of CH_4 ice could freeze onto the pole each winter. This ice layer would be concentrated toward the pole (and north-facing slopes), with less CH_4 freezing out at the lower latitudes and south-facing slopes that remain cold enough to trap CH_4 for shorter durations each year. Further assuming that the flux of Lyman-α photons onto the winter pole has always matched what the Alice instrument observed in 2015 (*Gladstone et al.,* 2015), they estimated that about one-fifth of the CH_4 would be photolyzed before the spring thaw, leading to production of perhaps several tens of centimeters of tholins over the age of the solar system. Tholin are strongly pigmented, so that quantity is more than enough to paint Charon's high-latitude regions red, even accounting for impact gardening mixing it into the substrate over time. Indeed, some dilution of tholins is required to prevent the poles from becoming even darker and redder than is observed.

The proposed mechanism would not apply only to Charon's northern pole. It should also affect the southern pole in the same way, so a key test of the hypothesis is whether or not the southern pole shows comparable darkening. This was a challenging observation for New Horizons, since the southern pole was experiencing winter night at the time of the flyby, with no direct illumination from the Sun. The only source of illumination was faint Pluto-shine. The sub-Pluto point on Charon is at 0° longitude, 0° latitude, not an ideal location for lighting Charon's pole, but it does at least cast some light onto Charon's Pluto-facing hemisphere. *Grundy et al.* (2016a)

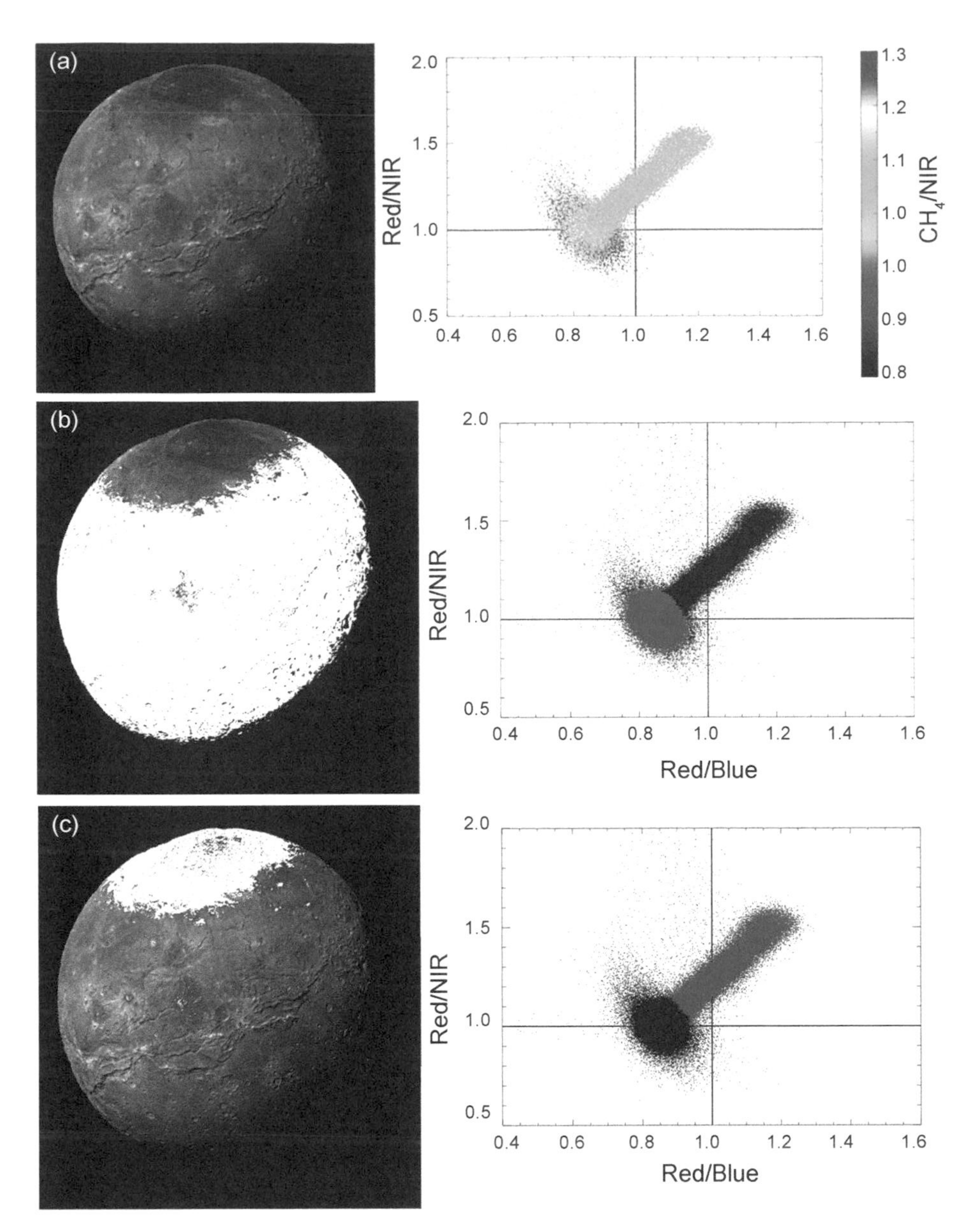

Fig. 4. Color/color diagrams of Charon. Left images show the PEMV_01_C_Color_2 enhanced image of Charon, while those on the right show a color-color diagram, with the neutral color lines being shown by the solid black lines. **(a)** Global Charon color. The color of the points corresponds to the CH_4/NIR color as shown in the key. **(b)** Blue/neutral color lobe. The points shown in red in the color-color plot are shown in white on the image of Charon. As the figure shows, most of Charon lies in this lobe. **(c)** Red pole. The points shown in red in the color-color plot are shown in white on the image of Charon. As the figure shows, points on this redder line correspond to Charon's red polar region.

presented two sets of LORRI observations showing a lower albedo at high southern latitude (see Fig. 6). These high-solar-phase observations were processed to remove scattered light using principal component analysis (PCA) (*Grundy et al.,* 2016a; *Lauer et al.,* 2018). Comparison with a Hapke model of the scene shows that southern high latitudes are indeed darker than they would be for a uniform albedo distribution across Charon's southern hemisphere. A higher spatial resolution high-solar-phase MVIC panchromatic observation was also captured by New Horizons, but was not yet available at the time of the *Grundy et al.* (2016a) analysis. Scattered light was removed from the MVIC data in much the same way (*Lauer et al.,* 2018). Comparison with a model reveals that this observation too shows a decline in albedo toward the southern pole (Fig. 6).

3.4. Charon's Non-Encounter Hemisphere

The highest spatial resolution color observation of Charon's non-encounter hemisphere was obtained two Earth days before New Horizons' closest approach of the satellite (PC_Multi_Map_B_15). The details of the observation are given in Table 2. To summarize, the subspacecraft latitude was more northern for this non-encounter hemisphere (67.0°E/42.8°N) than the highest-spatial-resolution color observation of Charon (C_COLOR_2, which

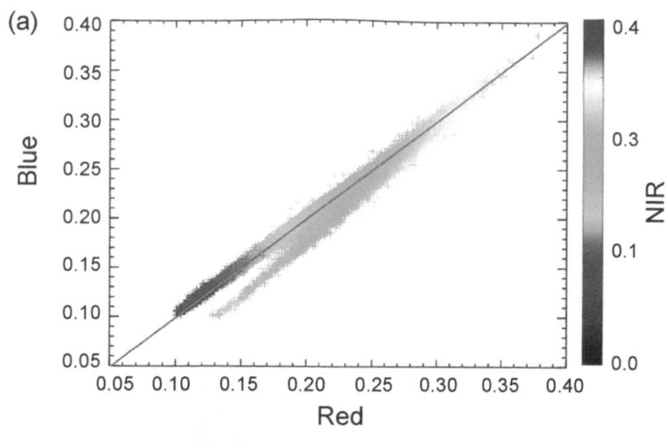

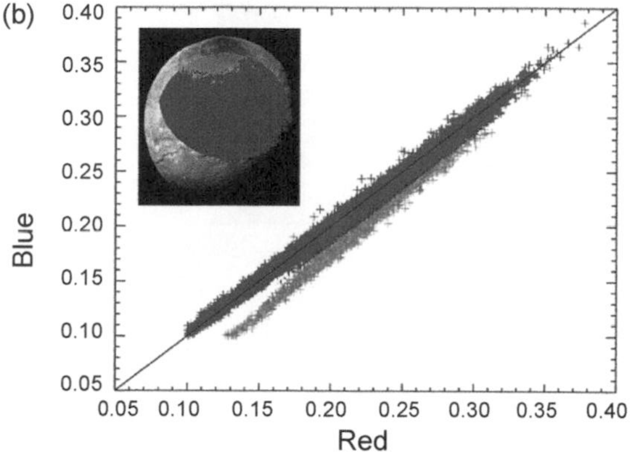

Fig. 5. (a) Red vs. blue color plot for all points on Charon that have emission and illumination angles less than 60°. The plotted color of the points indicates the NIR color. **(b)** Red vs. blue color split into regions. The insert shows the position on Charon to which these colors correspond. This figure clearly shows that the redder region corresponds to Charon's redder north polar region. The solid line indicates the x = y line (i.e., where Charon is neutral in red and blue).

had a subspacecraft position of –12.3°E/25.2°N), and the spatial resolution is a factor of ~30 worse (42.7 km/pixel compared to 1.4 km/pixel). A comparison of these images (see Figs. 2 and 3) shows no obvious color differences between the two hemispheres.

A more detailed analysis of the color of Charon's encounter and non-encounter hemispheres is shown in Fig. 7 and confirms this conclusion. The figure shows the red vs. blue color of both Charon's encounter and non-encounter hemisphere. The encounter hemisphere shows the two distinct color regions discussed previously in section 3.2. The distinctiveness of these regions is not as clear in the non-encounter hemisphere, probably smeared by the lower spatial resolution. However, the spread of the color ratios of the non-encounter hemisphere remains consistent with those on the encounter one.

3.5. Regional Color Variation

The colors of Organa crater (located at 55°N/50°W) and the nearby Skywalker crater are notably different. Figure 8b shows an enhanced color MVIC image of the region, showing Organa to appear more orange/red than Skywalker. This conclusion is supported by the color analysis also shown in Figs. 8c,d, which show Organa is notably brighter in the NIR.

The reason for this color difference is unknown. At first it was suspected that it could be indicative of compositional differences. However, analysis of New Horizons LEISA spectra indicates that both Organa and Skywalker craters (and the regions around them) have enhanced absorption in the 2.220 μm band that has been attributed to ammonia hydrates (*Grundy et al.,* 2016b; *Dalle Ore et al.,* 2018; see also the chapter by Protopapa et al. in this volume). Perhaps the difference is due to their locations and that more tholin-like material is included in Organa's ejecta since it is further north (and thus closer to Mordor Macula).

3.6. Principal Component Analysis

PCA can be used to reproject a multi-wavelength dataset along the axes of maximum variance in the data. This reprojection can highlight subtle color trends and specific regions with distinct color characteristics. The results of performing a PCA analysis on Charon are shown in Fig. 9, confirming the results of *Grundy et al.* (2016b), who showed using PCA that most of the variance in the Charon MVIC data corresponds to overall shading and albedo. This result is the norm for PC1 (principal component 1) for most planetary color datasets, simply indicating that most filters show similar patterns of brightness and darkness. For the highest-resolution MVIC Charon observation, PC1 accounts for 95.8% of the total variance in the data. The next principal component (PC2) is more interesting from a color perspective, accounting for 4.0% of the variance. The eigenvector gives a sense of the spectral character to which the PC is responding. For PC2, the eigenvector shows a blue slope, indicating that PC2 is responding to varying levels of blueness in the scene. Unsurprisingly, the minimum blueness (or maximum redness) corresponds to Charon's redder polar cap. (Note that the sign of the eigenvector doesn't matter, since it points an arbitrary direction along the greatest axis of variance after PC1 is removed. PC2 could be inverted to indicate redness instead.) PC3 and PC4 account for 0.1% and 0.06% of the variance respectively, with images showing no obvious correlation with geological features. They appear to be predominantly responding to noise in the MVIC color data.

4. CHARON'S PHOTOMETRIC PROPERTIES

4.1. Introduction to Photometric Analysis

Charon's photometric properties are derived from the quantitative measurement of reflected radiation from its surface. Both groundbased and spacecraft measurements inform Charon's global photometric properties, but only spatially resolved observations, such as those obtained

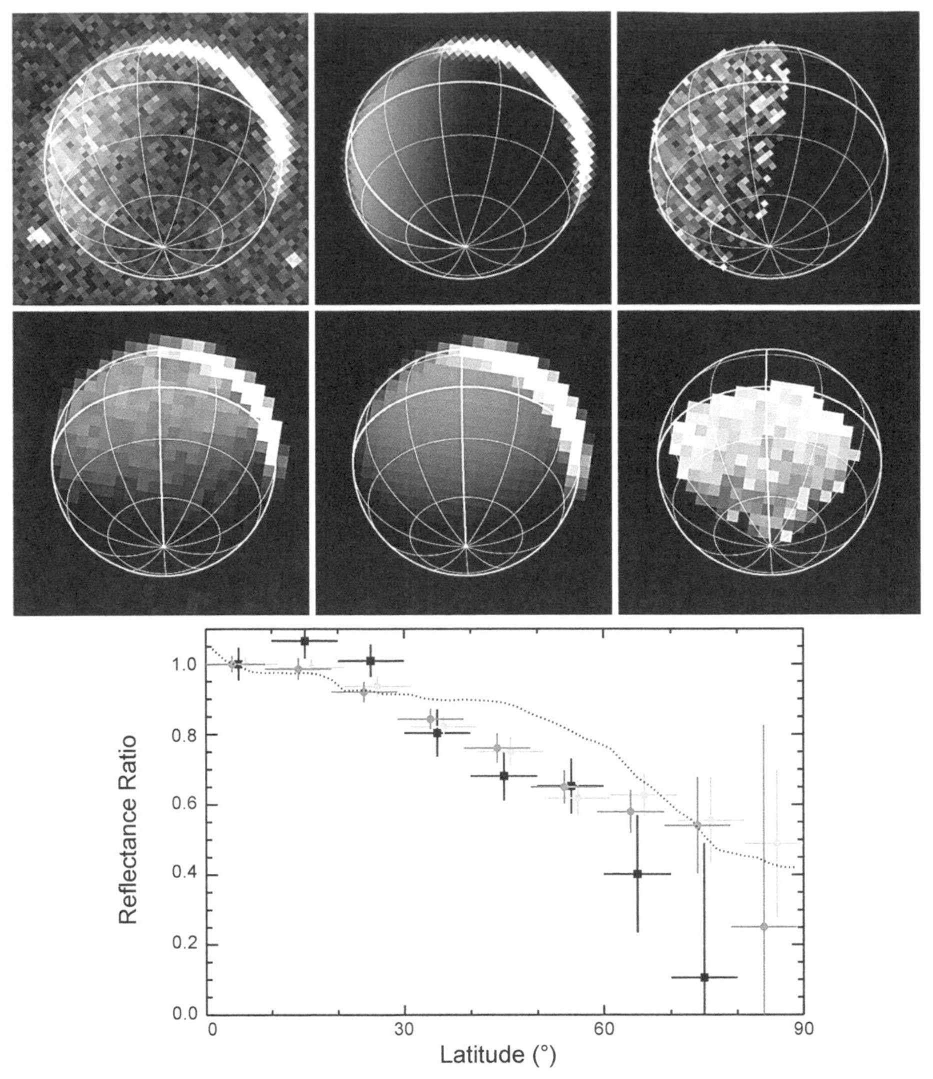

Fig. 6. *Top row:* High-phase MVIC panchromatic observation of Charon's southern hemisphere obtained July 15, 2015, at 18:39 UT from a range of 1.5 million kilometers, and identified by unique ID MET 0299291431. The left panel is the MVIC observation, showing a bright sunlit crescent and much fainter Pluto-shine illuminating the sub-Pluto hemisphere centered at 0° longitude, 0° latitude. The middle panel is a model of the scene assuming uniform albedos across both Pluto and Charon. The right panel is the ratio of observation/model, showing less reflectance at high southern latitudes. *Middle row:* The same thing for one of the two lower-resolution LORRI observations presented by *Grundy et al.* (2016a). The LORRI observations were executed later, when the Pluto-illuminated hemisphere was oriented more toward the spacecraft, but the spacecraft range was greater, at about 3 million kilometers. Further reducing resolution, the LORRI images made use of 4 × 4 binning to enable longer exposure times and to reduce storage and downlink bandwidth consumption. *Bottom:* Ratio between observation and the uniform albedo model averaged over 10° latitude bands for the MVIC (black) and two LORRI observations (shades of gray). All three observations show declining reflectance with latitude from the equator to the south pole. The dotted curve shows the same ratio for the sunlit northern hemisphere. The LORRI data is from *Grundy et al.* (2016a); the MVIC data analysis is new to this work.

by New Horizons (and to a limited extent, those obtained during mutual events and by HST), can elucidate the scattering properties of its distinct terrains. Charon's global photometric properties contain fundamental information on the energy balance of its surface and include such quantities as the geometric albedo p, bolometric Bond albedo A_B, and phase integral q. Through the application of photometric models based on radiative transfer (e.g., *Hapke,* 2012) to these whole-disk and disk-resolved observations, physical surface characteristics of Charon's surface can be derived. These include the single-scattering albedo, macroscopic roughness, compaction state of the optically active portion of the regolith, particle transparency, and directional scattering properties. These physical surface quantities offer insights into geophysical processes such as annealing of the surface by high-energy particles; the formation of a "fluffy" tenuous surface due to impact gardening by micrometeorites or accretion of exogenously emplaced material; and alterations on the surface due to active geologic processes such as volcanism, ponding of debris, or landslides.

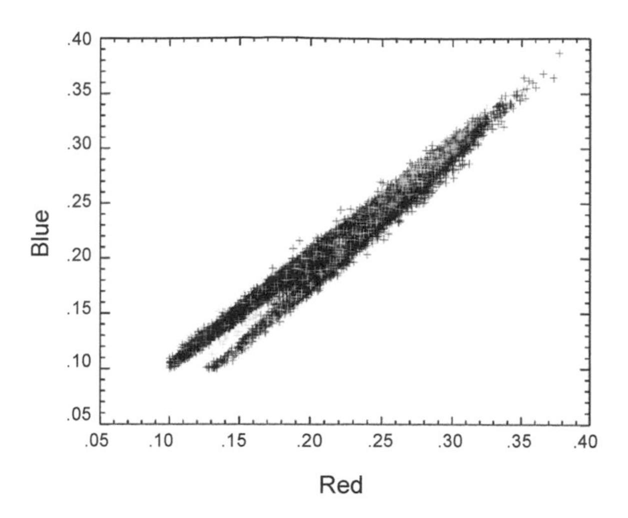

Fig. 7. Red vs. blue color of the encounter (dark crosses) and non-encounter (light crosses) hemispheres of Charon.

The analysis of Charon's photometric properties begins with the measurement of radiation reflected from its surface at different viewing geometries defined by the solar phase angle α, incidence angle i, and emission angle e. The solar phase angle is the Sun-Charon-observer angle, and the incidence and emission angles are the angles between the incident and emergent rays and the surface normal. The geometric albedo p is a disk-integrated quantity, and is the ratio of the integral flux at a given wavelength at $\alpha = 0°$ to that of a perfectly scattering (i.e., scattering uniformly in all directions) *disk* of the same size

$$p_\lambda = \frac{R^2D^2}{r^2}\frac{F_{C_0}}{F_\odot}$$

where R is the heliocentric distance, D is the observer distance, r is the radius of Charon, F_{C_0} is the flux from Charon at $\alpha = 0°$, and $F_\odot$ is the solar flux. The flux from Charon is either measured during a node crossing when the phase angle is at the minimum value (less than the solar radius seen from Charon) (*Verbiscer et al.,* 2019) or the flux from Charon is measured at a higher α and extrapolated to $\alpha = 0°$ using the phase coefficient β, or slope of the solar phase curve (in magnitudes/degree): $F_{C_0} = F_C - \beta\alpha$. Historically, determinations of Charon's geometric albedo have ranged from 0.372 ± 0.012 (*Reinsch et al.,* 1994) to 0.49 ± 0.10 (*Reinsch and Pakull,* 1987) due to the uncertainty in Charon's size. As discussed previously in section 1, prior to the New Horizons flyby, the measurement of Charon's size (and therefore albedo) was measured from the analysis of the Pluto/Charon mutual events between 1985 and 1990 (e. g., *Binzel,* 1988; *Buie et al.,* 1997). It was also determined via stellar occultation measurements in 2005 (*Gulbis et al.,* 2006; *Sicardy et al.,* 2006).

The phase integral q expresses the directional scattering properties of a planetary surface (*Horak,* 1950)

$$q(\lambda) = 2\int_0^\pi \Phi(\alpha, \lambda)\sin\alpha d\alpha \quad (1)$$

where $\phi(\lambda)$ is the integral flux normalized to unity at $\alpha = 0°$. The spherical or Bond albedo A is the ratio of the reflected to received integrated flux and is equal to the product of the geometric albedo and phase integral A = pq. The bolometric Bond albedo A_B is the Bond albedo integrated over all wavelengths and weighted by the solar spectrum; it is a fundamental parameter for understanding energy balance and volatile transport on Charon. Even without measurements of Charon's Bond albedo over the entire solar spectrum, it can closely approximate the bolometric Bond albedo if it is calculated for visible wavelengths near the middle of the Sun's energy output.

Table 3 and Fig. 9 compare Charon's geometric and Bond albedos and phase integral with those of other objects in the solar system of comparable size. Charon is moderately dark, with a visible geometric albedo of $p_V = 0.41 \pm 0.01$, a phase integral of $q = 0.70 \pm 0.04$, and therefore a Bond albedo of 0.29 ± 0.05. These albedos are in keeping with the general solar system trend, which shows only major satellites close to dwarf or giant planets have geometric albedos $p_V > 0.2$, with other icy bodies being darker ($p_v < 0.2$) and bodies associated with geologic activity being notably brighter [e.g., Enceladus, which has a p_v of 1.375 ± 0.008 (*Verbiscer et al.,* 2007)] (*Verbiscer et al.,* 2013). Charon is more reflective than most observed Plutinos, Centaurs, and comets, but within the range of geometric albedos observed on icy satellites, being most comparable to Iapetus' brighter terrain (trailing hemisphere), Miranda, and dwarf planet Haumea (Table 3) (*Verbiscer et al.,* 2013). Typically, major satellites are expected to be brighter due to surface alteration processes (e.g., sputtering, desorption, micrometeorite impacts, and sublimation). On Charon such processing may be due to the influence of Pluto's escaping atmosphere, along with impactors from within the Pluto-system (e.g., from Pluto's smaller satellites) or outside the system.

4.2. Hapke and Other Photometric Modeling

4.2.1. Introduction to Hapke modeling. A physical photometric model can be summarized by the following expression for the bidirectional reflectance r (*Hapke,* 2012)

$$r(i, e, \alpha) = K\frac{\omega}{4\pi}\frac{\mu_{0e}}{\mu_{0e}+\mu_e}$$
$$\left\{\left[P(\alpha)\left[1+B_{S0}B_S(\alpha)\right]\right]+\left[H\left(\mu_{0e}/K\right)H\left(\mu_e/K\right)-1\right]\right\} \quad (2)$$
$$\left[1+B_{C0}B_C(\alpha)\right]S(i, e, \alpha, \theta)$$

where K is a porosity term that depends upon the filling factor, and ω is the single scattering albedo (the probability that a photon reflected from a single particle will be scattered

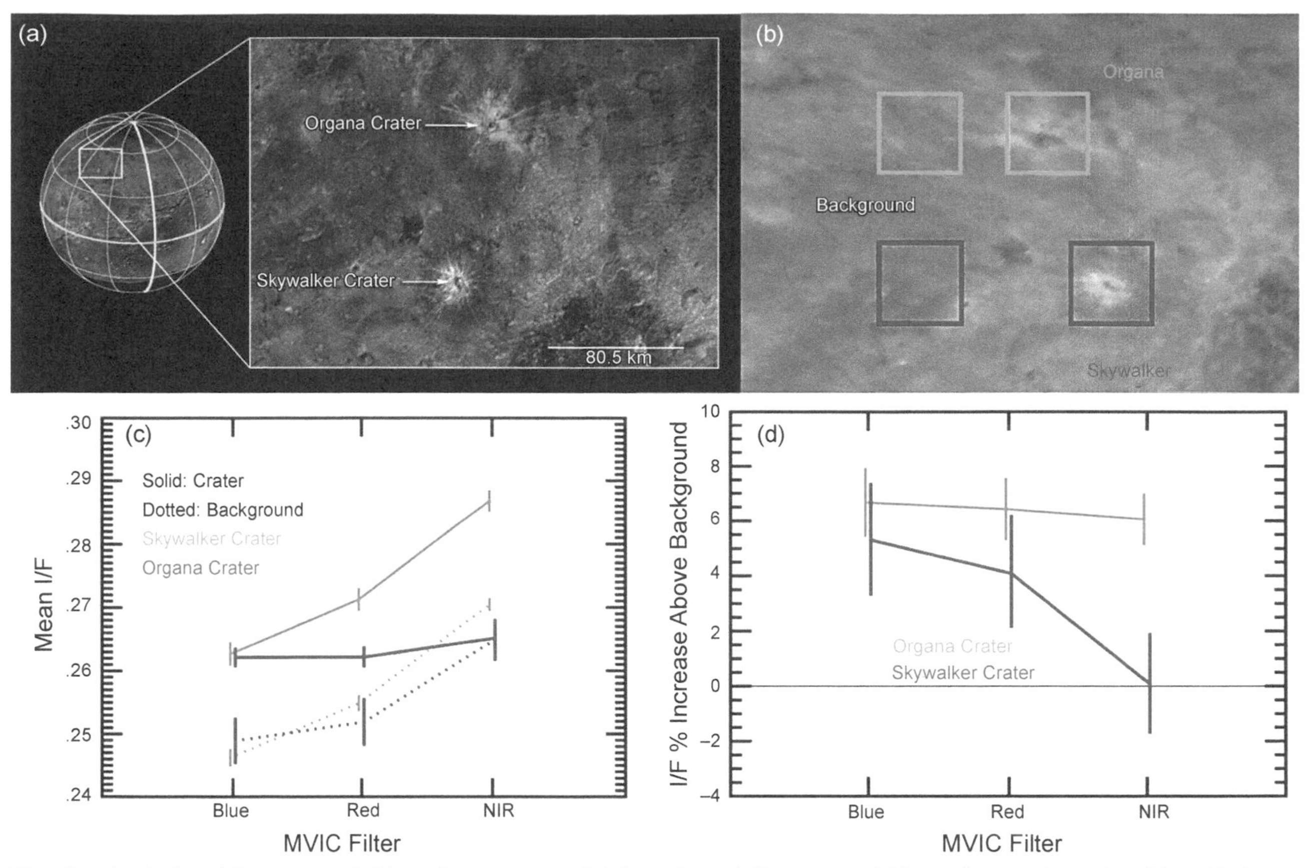

Fig. 8. Analysis of Organa and Skywalker craters. **(a)** Location of Organa and Skywalker crater using New Horizons LEISA overlaid on a LORRI image; the green overlay shows where ammonia has been detected in the LEISA data. The LEISA observations were made at 10:25 UT on July 14, 2015, at 5 km/pixel; the LORRI image was obtained at 8:33 UT July 14, 2015, at 0.9 km/pixel (from NASA press release of October 29, 2015). **(b)** Color of Organa crater (top right orange box) vs. Skywalker crater (bottom right blue box). The colored boxes on the left show the regions used to determine the region's background value. **(c)** The mean I/F of the boxes shown in **(b)** for the craters and background, showing Organa is brighter in the red and NIR. **(d)** I/F% increase above the background for Organa and Skywalker crater, showing that relative to the background Organa is brightest in the NIR.

into 2π steradians). $P(\alpha)$ is the single particle phase function. μ_{0e} and μ_e are related to the cosine of the incidence angle (i) and emission angle (e), respectively, with additional terms to account for the tilt of the surface, due to surface roughness. B_{S0} and B_{C0} are the amplitudes of the shadow-hiding opposition effect (SHOE) and coherent backscatter opposition effect (CBOE), and B_S and B_C are related to their widths h_S and h_C respectively [see below for a further description, and *Domingue et al.* (2016) for further definitions]. H is Chandrasekhar's H-functions that depend on the amount of multiple scattering (*Chandrasekhar,* 1960). $S(i, e, \alpha, \theta)$ is the function describing the photometric effects of macroscopically rough surface features, characterized by the macroscopic roughness parameter θ [which is the Gaussian mean slope value (*Hapke,* 1984)].

The opposition surge, the non-linear increase in brightness as a planetary surface becomes fully illuminated to an observer, is due to both the rapid disappearance of shadows cast among particles comprising the regolith and an interference phenomenon known as coherent backscatter (*Shkuratov,* 1988; *Muinonen,* 1990; *Mishchenko,* 1992). Coherent backscatter occurs when multiply- and singly-scattered photons traverse identical paths and add coherently. For a collection of uniform particles where multiple scattering is not significant, h_S due to shadow hiding is equal to $-(3/8)\ln(1-\rho/\rho_0)$, where ρ is the bulk density of the regolith particles, ρ_0 is the density of an individual particle, and the term in parentheses is the porosity, the fraction of space not occupied by particles (*Hapke,* 1986). The shadow hiding opposition surge amplitude B_{S0} is a measure of the opacity of individual particles: For $B_{S0} = 1$ all incident radiation is scattered from the front surface of particles with no internal scattering, and B_{S0} therefore by definition cannot exceed unity. B_{C0} is the amplitude for coherent backscatter, and its width (h_C) is related to the wavelength of light (λ) and the transport mean free path (Λ_T) (which is the average distance traveled by a wave in a given direction before the direction is randomized) according to $h_C = \lambda / 4\pi\Lambda_T$. *Helfenstein and Shepard* (2011) relate the porosity parameter K to the angular width of the shadow hiding opposition effect, eliminating the need to solve explicitly for K.

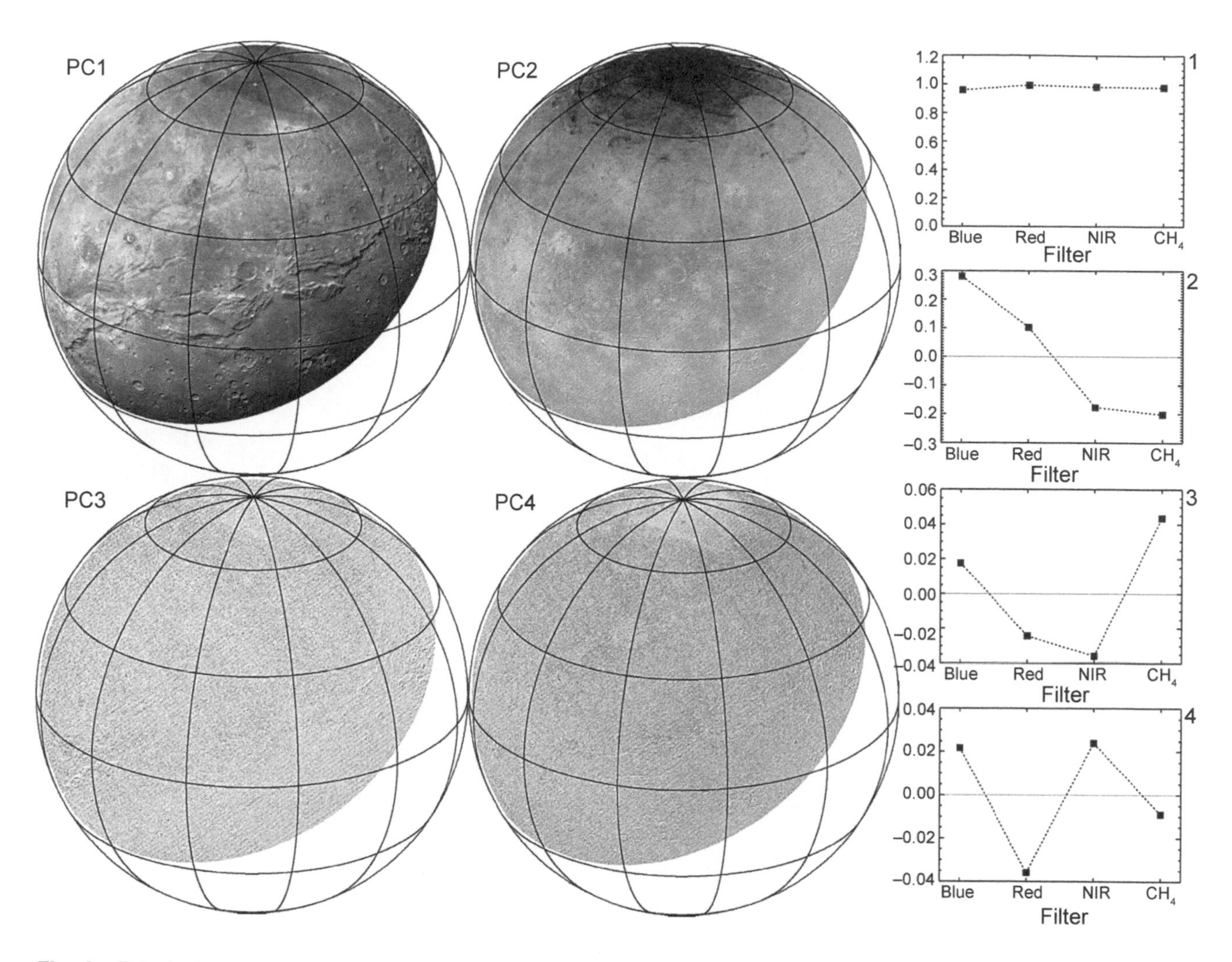

Fig. 9. Principal component analysis (PCA) of Charon. The maps show the first to fourth principal components. The eigenvectors corresponding to each of the four PCAs are shown on the right. From *Grundy et al.* (2016b).

4.2.2. Hapke modeling of Charon's disk-integrated solar phase curves. Charon's disk-integrated photometric parameters can be derived by combining the range in solar phase angles provided by the New Horizons LORRI dataset (15° $<\alpha<$165°) and Earth-based measurements, which were made at lower phase angles (α <2°). LORRI images at solar phase angles α >20° do not contain Charon's full disk, but equivalent integral quantities can be derived by computing the integral flux reflected by a sphere covered with particles having the average I/F values on each image (*Buratti et al.,* 2019). Figure 10 shows Charon's HST (F555W) and LORRI (corrected to V) solar phase curve with a best-fit photometric model described by equation (2).

Table 4 and Fig. 11 summarize the results of the application of the Hapke photometric model to disk-integrated solar phase curves of Charon and other similarly sized icy bodies and Nix and Hydra, two of Pluto's four small moons. Charon's single-scattering albedo (w) is most similar to that of similarly sized uranian satellite Ariel; it is not as high as the single-scattering albedo of Pluto's smaller moons and the mid-sized saturnian satellites. Charon's macroscopic roughness, θ = 24°, is typical of that found for other icy moons, small bodies, and Earth's Moon. Charon's opposition surge has a much higher amplitude and angular width than Pluto's (see Fig. 6 in the chapter by Olkin et al. in this volume), indicating that the microtextures of the surfaces of Pluto and Charon are dramatically different. Charon's smaller angular width of the opposition surge due to shadow hiding indicates that its surface has a higher porosity than that of Pluto. The amplitude of the opposition surge due to shadow hiding is higher on Charon (B_{0S} = 0.74) than on Pluto (B_{0S} = 0.31); therefore, on average, surface grains on Charon are more opaque than those on Pluto. The difference in morphology and quantitative properties between the opposition surges observed on these two bodies suggests that Charon has a very porous surface of opaque particles and Pluto has a more compact regolith composed of more transparent grains. With few exceptions, such as Saturn's satellite Rhea, icy bodies in the outer solar system have backscattering single-particle phase functions. Charon, however, has a more isotropic single-particle phase function, possibly due to the accretion of dust from the small satellites (*Porter and Grundy,* 2015), contributing small particles that serve to increase multiple scattering at larger phase angles.

TABLE 3. Global photometric properties of Charon and other selected objects.

Object	Radius (km)	Geometric Albedo (p_V)	Phase Integral (q)	Bond Albedo (A)	Reference
Charon	606 ± 5	0.41 ± 0.01	0.70 ± 0.04	0.29 ± 0.05	*Stern et al.* (2015) *Buratti et al.* (2017) *Buratti et al.* (2019)
Nix	~20	0.56 ± 0.05	0.42 ± 0.08	0.24 ± 0.07	*Weaver et al.* (2016)
Hydra	~20	0.83 ± 0.08	0.39 ± 0.1	0.32 ± 0.11	*Weaver et al.* (2016)
Ceres	473	0.094 ± 0.007 0.096 ± 0.006	0.35 ± 0.05	0.033 0.037 ± 0.002	*Ciarniello et al.* (2017) *Li et al.* (2019)
Vesta	262.7 ± 0.1	0.38 ± 0.04	0.53 ± 0.10	0.20 ± 0.02	*Li et al.* (2013)
Miranda	235.8 ± 0.7	0.464	0.44 ± 0.07	0.200 ± 0.030	*Karkoschka* (2001)
Ariel	578.9 ± 0.6	0.533	0.43 ± 0.05	0.230 ± 0.025	*Karkoschka* (2001)
Umbriel	584.7 ± 2.8	0.192	0.39 ± 0.04	0.100 ± 0.10	*Karkoschka* (2001)
Titania	788.4 ± 0.6	0.252	0.46 ± 0.05	0.170 ± 0.15	*Karkoschka* (2001)
Oberon	761.4 ± 2.6	0.222	0.44 ± 0.05	0.140 ± 0.15	*Karkoschka* (2001)
Tethys	531.1 ± 0.6	1.229 ± 0.005	0.52	0.64	*Verbiscer et al.* (2007, 2018a)
Dione	561.4 ± 0.4	0.998 ± 0.004	0.70	0.70	*Verbiscer et al.* (2007, 2018a)
Rhea	763.8 ± 1	0.949 ± 0.003	0.63	0.60	*Verbiscer et al.* (2007, 2018a)
Iapetus (trailing hemisphere)	734.5 ± 2.8	0.45	0.61 ± 0.10 at 0.51 μm[1]	0.27	*Blackburn et al.* (2010)
Makemake	713.5 +16.5/–10.5	0.82 ± 0.02	TBD	TBD	*Brown* (2013); *Hromakina et al.* (2019)
Haumea	a = 1050 ± 30 b = 840 ± 4 c = 537 ± 16	0.66 ± 0.02	TBD	TBD	*Dunham et al.* (2019)
Quaoar	555 ± 2.5	0.109 ± 0.007	TBD	TBD	*Braga-Ribas* (2013)

All observations in V band (0.55 μm) unless otherwise noted. These values are plotted in Fig. 9 to allow easier comparisons.

4.3. Multispectral Visible Imaging Camera Phase Curves

Although only the LORRI data cover a sufficient range in solar phase angles to perform photometric modeling and determine a phase integral, MVIC returned sufficient data to construct color phase curves for each MVIC filter over a solar phase angle range between $\alpha \sim 15°$ and $21°$. Figure 12 shows normalized disk-integrated curves for the four MVIC filters and LORRI. There is no clear trend in color to the solar phase curves. With the range in solar phase angle extending over only 6°, it is possible that the excursion in solar phase angle is not sufficiently large for effects such as phase reddening to appear.

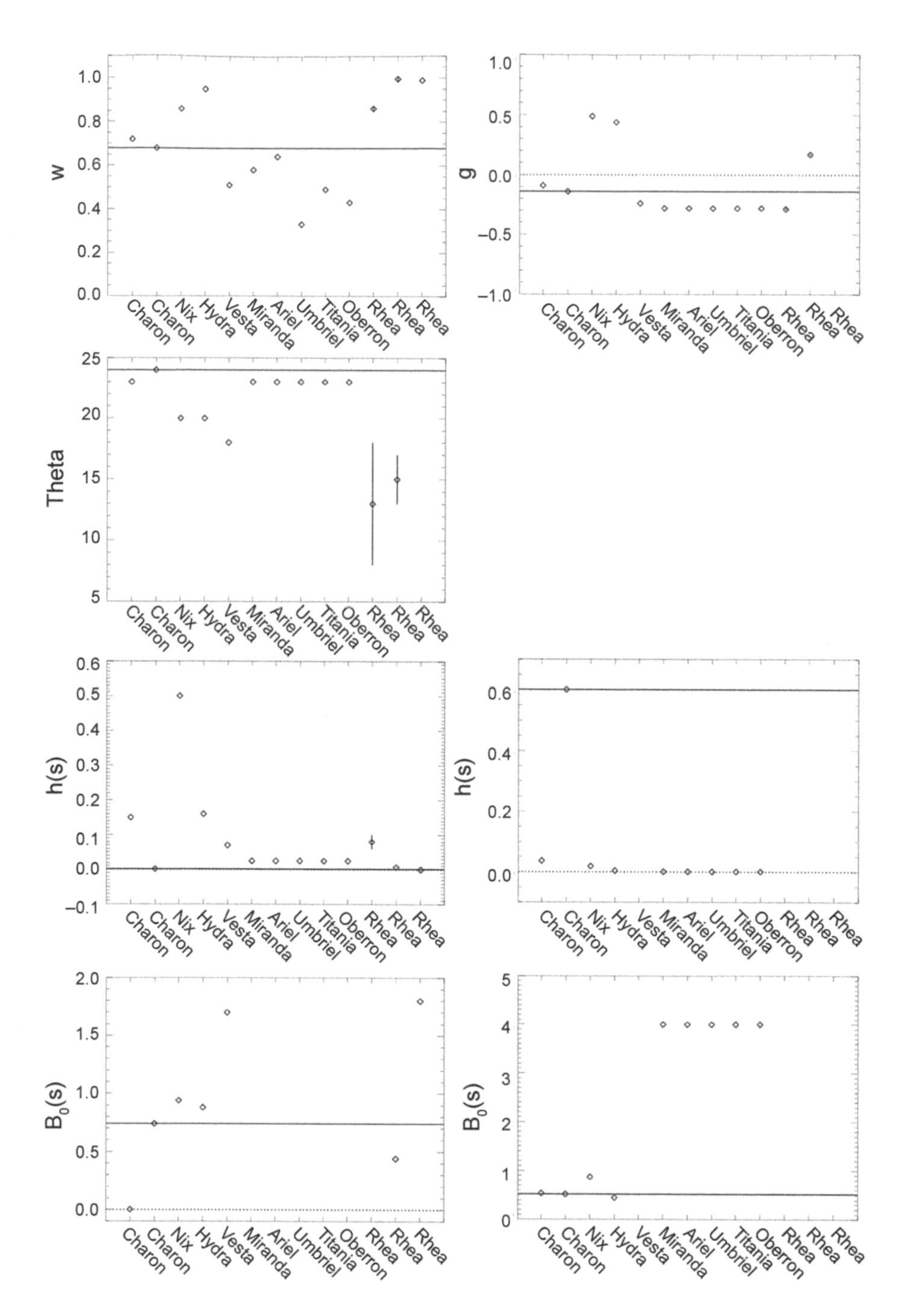

Fig. 10. Plots of Hapke photometric model parameters for Charon and other selected objects. The values plotted are the same as those given in Table 4, plotted in the same order as in that table (see Table 4 for details of the plotted parameters as well). The solid horizontal line shows the values for Charon derived in this work. Not all values are derived for each target, and some values are assumed not derived (see Table 4 for details). The horizontal dashed line indicates where y = zero.

4.4. Normal Reflectance Map

Global coverage of Charon during the near-encounter phase of the New Horizons flyby was sufficient for *Buratti et al.* (2017) to construct a map of normal reflectance from LORRI images (Fig. 13). The normal reflectance r_N is the ratio between the reflected specific intensity I to the incident solar flux F (*Chandrasekhar,* 1960) when the incident and emission angles both equal 0°. For a body with little multiple scattering and no limb darkening, such as the Moon, the geometric albedo and normal reflectance are equal [Moon's geometric albedo is 0.11 in V-band (*Lane and Irvine,* 1973)]. Because of limb darkening with higher-albedo bodies at 0°, these two quantities are not equal. To eliminate the effects of changing incidence and emission angles, *Buratti et al.*

TABLE 4. Hapke photometric model parameters for Charon and other selected objects.

Object	w	g	θ(°)	h(s)*	B_o(s)*	B_o(c)*	h(c)*	Reference
Charon (disk-integrated)	0.72	–0.09	23	0.15	0.001	0.536	0.037	*Buratti et al.* (2019)
Charon (disk-integrated)	0.68	–0.14	24	0.0015	0.74	0.52	0.6	This volume
Nix (disk-integrated)	0.86	0.49	20†	0.50	0.94	0.87	0.019	*Verbiscer et al.* (2018b)
Hydra (disk-integrated)	0.95	0.44	20†	0.16	0.88	0.45	0.0043	*Verbiscer et al.* (2018b)
Vesta (disk-resolved)	0.51	–0.24	18†	0.07	1.7			*Li et al.* (2013)
Miranda (disk-integrated)	0.58	–0.28†	23†	0.025		4†	0.001†	*Karkoschka* (2001)
Ariel (disk-integrated)	0.64	–0.28†	23†	0.025		4†	0.001†	*Karkoschka* (2001)
Umbriel (disk-integrated)	0.33	–0.28†	23†	0.035		4†	0.001†	*Karkoschka* (2001)
Titania (disk-integrated)	0.49	–0.28†	23†	0.025		4†	0.001†	*Karkoschka* (2001)
Oberon (disk-integrated)	0.43	–0.28†	23†	0.025		4†	0.001†	*Karkoschka* (2001)
Rhea (disk-integrated)	0.861 ± 0.008	–0.287 ± 0.008	13 ± 5	0.08 ± 0.02				*Verbiscer and Veverka* (1989)
Rhea (disk-integrated)	0.996 ± 0.01	0.171 ± 0.005	15 ± 2	0.0071 ± 0.001	0.44			*Domingue et al.* (1997)
Rhea (disk-integrated)	0.989		33	0.0004				*Ciarniello et al.* (2011)

*Hapke's recent model has separate parameters for the shadow-hiding and coherent backscatter portions of the opposition surge; w is the single-scattering albedo, g is the average cosine of the scattering angle, θ is the slope angle, and h and B_o are the angular width and amplitude respectively of the opposition surge due to shadow hiding (s) and coherent backscatter (c).
†Parameters held fixed for these fits. Whether the target was resolved or disk-integrated is indicated. These values are plotted in Fig. 10 to allow easier comparisons.

(2017) used a simple lunar-like scattering function with a Lambert component

$$I/F = f(\alpha)A[\mu_0 / (\mu + \mu_0)] + (1 - A)\mu_0 \quad (3)$$

where A is a partition factor between the lunar-like and Lambert portions of the equation and f(α) is the change in intensity due to the effects of solar phase angle, including the effects of roughness, the non-isotropy of scatterers, and the surge in brightness near opposition. Buratti et al. found a best-fit partition factor A = 0.7 for Charon and normalized the map to a phase angle of 0° based on the phase curve shown in Fig. 14. Charon does not exhibit the large range in normal reflectance shown by Pluto [which has values between 0.08 and 1.0 (*Buratti et al.,* 2017)]: Most of Charon's surface is characterized by normal reflectances in the 0.4–0.6 range, with a few bright crater ejecta areas. Normal reflectances in the north polar region are substantially lower, $r_N \sim 0.20$, precisely the area where *Buie et al.* (2010b) mapped a relatively darker north polar region on Charon using the High Resolution Channel (HRC) of HST's ARC-HRC and where *Grundy et al.* (2016a) hypothesize a polar cap of darkened, reddened photolyzed methane from Pluto's atmosphere.

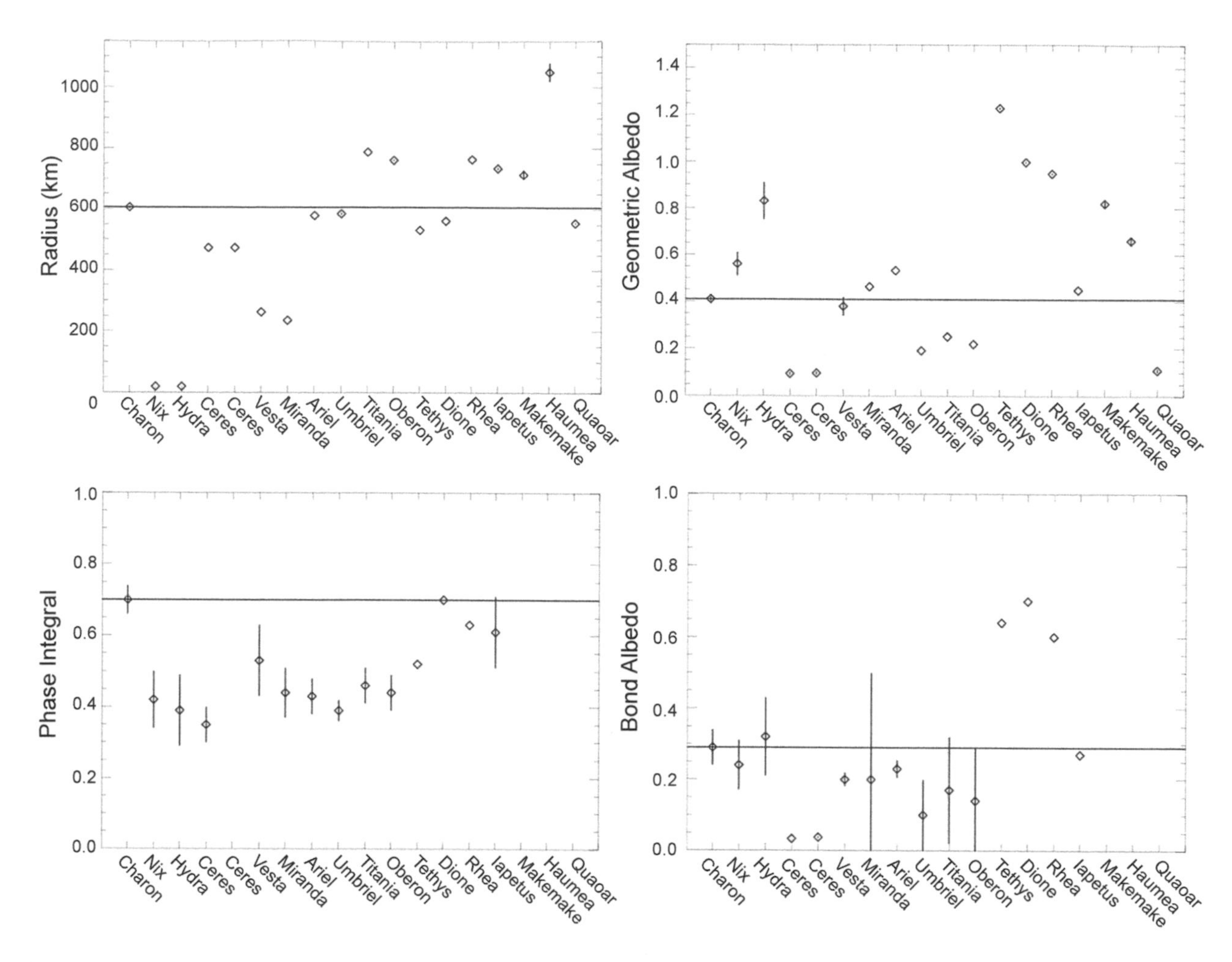

Fig. 11. The radius, geometric albedo, phase integral, and Bond albedo for Charon and other selected objects. The solid horizontal line shows the value in each subfigure for Charon; error bars are shown if supplied by the original reference. The origin of the data is the same as in Table 3, and the data are plotted in the same order listed in Table 3. For Haumea, only the radius of the semi-major axis is shown.

5. CONCLUSIONS

Prior to the New Horizons flyby, little was known of Charon's variations in surface color: Nearly flat light curves implied little color variation across its surface, but they did show that the anti-Pluto hemisphere could be darker. Small variations in lightcurves taken at different colors between 1997 and 2002 suggested Charon's poles may have different coloration (as the subobserver position moved poleward over time). New Horizons' observations of Charon by New Horizons revolutionized our understanding of its surface color, as well as its photometric properties.

MVIC provided color images of Charon, its four color filters covering wavelengths from 400 to 975 nm. By combining images taken with these filters, enhanced color images of Charon were constructed (mainly being produced using the blue, red, and NIR filters for red/green/blue color channels). These images brought the pre-encounter predictions into sharp focus. For example, while the surface color of Charon's low- and mid-latitudes is indeed bland, some small-scale color variations were shown to exist and there is indeed reddening toward Charon's polar regions.

At low and middle latitudes there are changes both in the color and brightness of the surface. For example, there are changes in the redness of the surface (e.g., Gallifrey Macula and western Oz Terra are more red than their surroundings) and the grayness of the surface (e.g., Ripley crater is notably a darker gray than the plains north of Nostromo Chasma, which are distinctly lighter). Furthermore, observations made by New Horizons revealed color variations with both geologies (i.e., impact craters) and found that compositional differences were also observed. It was also seen that changes in Charon's surface color due to surface impact are not uniform. For example, ejecta close to Skywalker crater are brighter than its more distant ejecta, whereas the reverse is true for Vadar crater, and Organa crater is notably redder (i.e., brighter in the NIR channel) than nearby Skywalker crater.

The biggest color variation seen on Charon occurs around its northern polar cap. The reflectance at short wavelengths (i.e., MVIC's blue filter) is ~30% lower than at middle

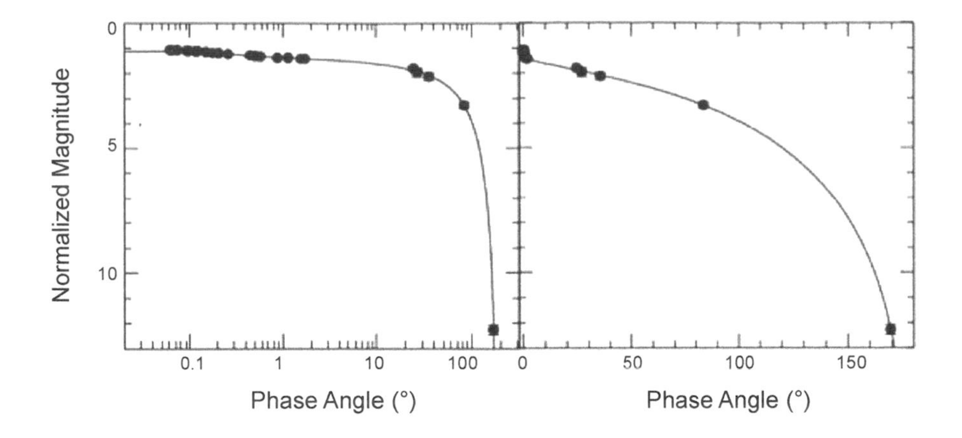

Fig. 12. Solid line is the fit to a *Hapke* (2012) photometric model applied to the LORRI and HST observations of Charon. HST observations (Program 13667, M. Buie, Principal Investigator) are those acquired at solar phase angles less than 10°; LORRI observations are those from *Buratti et al.* (2019) at phase angles larger than 10°. Note that these data are normalized to the disk-integrated reflectance at opposition. The conversion from magnitudes is $I/F = 10^{(mag/-2.5)}$. The data points are in most cases smaller than the 1σ error bars.

latitudes, whereas in the red and NIR the reflectance is ~40% and 65% respectively. The cap does not show a discrete boundary, but rather gradually and not uniformly gets redder toward higher latitudes. Variations in this reddening are seen with topography; for example, north-facing slopes appear redder than south-facing ones. This surface reddening may be caused by the radiolysis of methane ice, which originated from Pluto's atmosphere but was subsequently captured and then cold-trapped on Charon. This radiolysis changes the methane into tholins, which are strongly pigmented, so even though the depth of this layer maybe less than tens of centimeters, it is deep enough to strongly affect the color of Charon's surface.

This reddening is clearly observed in sunlit images of Charon's north polar cap, but a comparable darkening is also detectable in panchromatic images of its south polar cap too. The south polar images were significantly more challenging, as this region is only illuminated by faint Pluto-shine, making the surface too dark to be observable in MVIC's color channels. However, a decrease in albedo toward Charon's southern pole was detectable in MVIC's panchromatic channel and by New Horizons' LORRI instrument, which

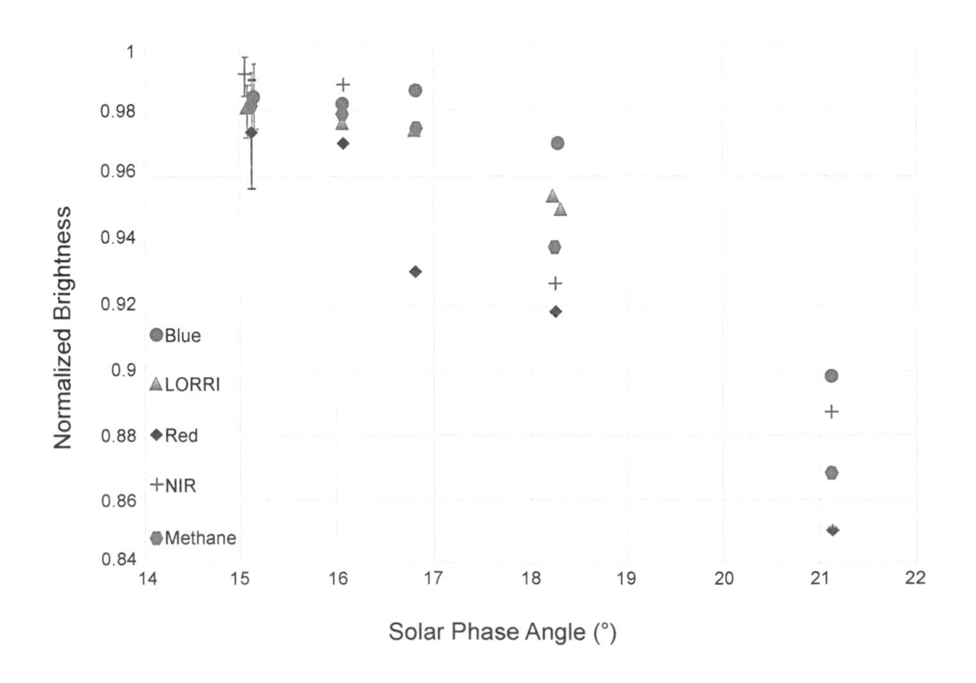

Fig. 13. Charon's integral solar phase curves for MVIC filters, adapted from *Buratti et al.* (2019). The typical error is shown in the data at 15.3°.

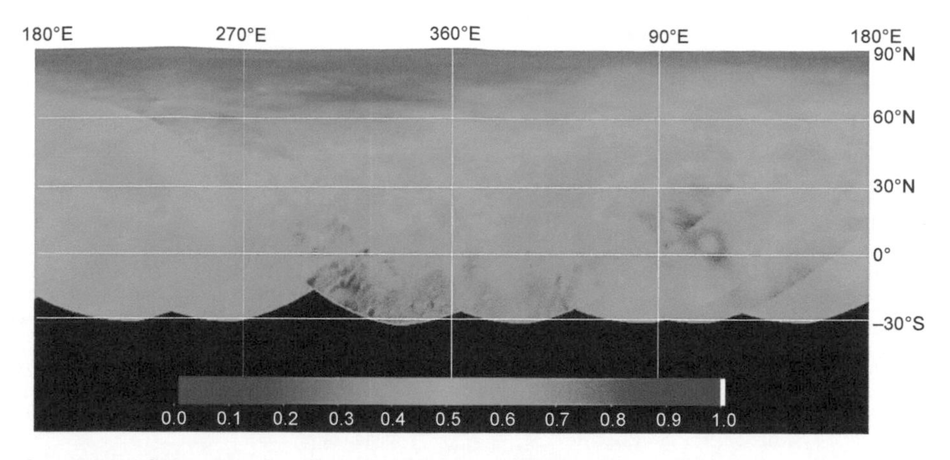

Fig. 14. A map of normal reflectances for Charon, showing uniform values over most of its surface with the exception of a few bright craters and a north polar cap with substantially lower values. For this map, the partition factor A was 0.7. From *Buratti et al.* (2017).

provides strong support that this coloration is occurring at both of Charon's poles.

A PCA of Charon's surface color shows that most of the variance (~96%) comes from albedo and shading differences (i.e., the filters show similar patterns of light and dark regions). The next most dominant component (~4%) is in the blueness/redness of the surface, thus capturing the color variations. The final two components are negligible and appear to be responding to noise. These changes in brightness are also seen in Charon's albedo map. The map shows normal reflectances vary between ~0.4 and 0.6, with some bright cratered regions and the polar cap being notably darker.

Charon's photometric properties have been derived from LORRI and MVIC observations. They show Charon is somewhat dark with a Bond albedo of 0.29 ± 0.05, a visible geometric albedo of 0.41 ± 0.01, and a phase integral of 0.70 ± 0.04. These values are in keeping with the general trend seen across the solar system, where only major satellites of planets have a visible geometric albedo brighter than 0.2. These values are also much lower than those of currently active bodies, supporting the hypothesis that Charon's surface is geologically dead. The surface may be brightened by processing caused by impactors from both within and external to the Pluto system. Charon's macroscopic roughness (23°) is similar to that measured on other icy worlds, but its isotropic single particle phase function (g) is notably different, being closer to isotropic than most icy worlds. The reason for this is unknown, but again could be due to infall of dust from the small satellites.

Arguably the most important — and currently unanswered — question about Charon's color is how Charon reddens toward its south polar cap. It is clear from current observations that Charon's south polar cap is indeed darker, but it is difficult from these measurements to determine the relative darkening of blue vs. redder wavelengths. It is expected that the southern polar cap will redden similarly to the north polar one, if the proposed reddening mechanism is correct. However, until such observations are made, it is impossible to know whether this is indeed true.

Finally, New Horizons imaged the non-encounter hemisphere of Charon at ~50× worse resolution than the encounter hemisphere (see Table 2). Therefore there are many answered questions about this hemisphere too. For example, is its cratering record the same as the encounter hemisphere, and do any of its craters also show coloration/compositional variation?

REFERENCES

Andersson L. E. (1978) Eclipse phenomena of Pluto and its satellite. *Bull. Am. Astron. Soc., 10*, 586.

Beyer R. A. et al. (2019) The nature and origin of Charon's smooth plains. *Icarus, 323*, 16–32.

Binzel R. P. (1988) Hemispherical color differences on Pluto and Charon. *Science, 241*, 1070–1072.

Blackburn D. G. et al. (2010) Solar phase curves and phase integrals for the leading and trailing hemispheres of Iapetus from the Cassini Visual Infrared Mapping Spectrometer. *Icarus, 209*, 738–744.

Braga-Ribas F. (2013) The size, shape, albedo, density, and atmospheric limit of transneptunian object (50000) Quaoar from multi-chord stellar occultations. *Astrophys. J., 773*, 25–39.

Brown M. E. (2013) On the size, shape, and density of dwarf planet Makemake. *Astrophys. J. Lett., 767,* L7.

Buie M. W., Tholen D. J., and Horne K. (1992) Albedo maps of Pluto and Charon: Initial mutual event results. *Icarus, 97*, 211–227.

Buie M. W., Tholen D. J., and Wasserman L. H. (1997) Separate lightcurves of Pluto and Charon. *Icarus, 125*, 233–244.

Buie M. W., Grundy W. M., Young E. F., Young L. A., and Stern S. A. (2006) Orbits and photometry of Pluto's satellites: Charon, S/2005 P1, and S/2005 P2. *Astron. J., 132(1)*, 290–298.

Buie M. W., Grundy W. M., Young E. F., Young L. A., and Stern S. A. (2010a) Pluto and Charon with the Hubble Space Telescope. I. Monitoring global change and improved surface properties from light curves. *Astron. J., 139*, 1117–1127.

Buie M. W., Grundy W. M., Young E. F., Young L. A., and Stern S. A. (2010b) Pluto and Charon with the Hubble Space Telescope. II. Resolving changes on Pluto's surface and a map for Charon. *Astron. J., 139*, 1128–1143.

Buratti B. J. and Veverka J. (1985) Photometry of rough planetary surfaces: The role of multiple scattering. *Icarus, 64*, 320–328.

Buratti B. J. et al. (2017) Global albedos of Pluto and Charon from LORRI New Horizons observations. *Icarus, 287*, 207–217.

Buratti B. J. et al. (2019) New Horizons photometry of Pluto's moon Charon. *Astrophys. J. Lett., 874*, L3.

Chandrasekhar S. (1960) *Radiative Transfer.* Dover, New York. 416 pp.

Cheng A. F. et al. (2008) Long-range reconnaissance imager on New Horizons. *Space Sci. Rev., 140,* 189–215.

Christy J. W. and Harrington R. S. (1978) The satellite of Pluto. *Astron. J., 83,* 1005–1008.

Ciarniello M. et al. (2011) Hapke modeling of Rhea surface properties through Cassini-VIMS spectra. *Icarus, 214,* 541–555.

Ciarniello M. et al. (2017) Spectrophotometric properties of dwarf planet Ceres from the VIR spectrometer on board the Dawn mission. *Astron. Astrophys., 598,* A130.

Cook J. et al. (2007) Near-infrared spectroscopy of Charon: Possible evidence for cryovolcanism on Kuiper belt objects. *Astrophys. J., 663,* 1406–1419.

Cooper J. F., Christian E. R., Richardson J. D., and Wang C. (2003) Proton irradiation of Centaur, Kuiper belt, and Oort cloud objects at plasma to cosmic ray energy. *Earth Moon Planets, 92,* 261–277.

Cruikshank D. P., Imanaka H., and Dalle Ore C. M. (2005) Tholins as coloring agents on outer solar system bodies. *Adv. Space Res., 36,* 178–183.

Dalle Ore C. M., Protopapa S., Cook J. C., Grundy W. M., Cruikshank D. P., Verbiscer A. J., Ennico K., Olkin C. B., Stern S. A., Weaver H. A., Young L. A., and the New Horizons Science Team (2018) Ices on Charon: Distribution of H_2O and NH_4 from New Horizons LEISA observations. *Icarus, 300,* 21–32.

Domingue D. L. (2016) Application of multiple photometric models to disk-resolved measurements of Mercury's surface: Insights into Mercury's regolith characteristics. *Icarus, 268,* 172–203.

Domingue D. et al. (1997) The scattering properties of natural terrestrial snows versus icy satellite surfaces. *Icarus, 128,* 28–48.

Dunham E. T. (2019) Haumea's shape, composition, and internal structure. *Astrophys. J., 877,* 41.

Gladstone G. R., Pryor W. R., and Stern S.A. (2015) Lyα@Pluto. *Icarus, 246,* 279–284.

Gladstone G. R. et al. (2016) The atmosphere of Pluto as observed by New Horizons. *Science, 351,* 1280.

Grundy W. M. et al. (2016a) Formation of Charon's red poles from seasonally cold-trapped volatiles. *Nature, 539,* 65–68.

Grundy W. M. et al. (2016b) Surface compositions across Pluto and Charon. *Science, 351,* 183.

Gulbis A. et al. (2006) The color of the Kuiper belt core. *Icarus, 183,* 168–178.

Hapke B. (1984) Bidirectional reflectance spectroscopy 3. Correction for macroscopic roughness. *Icarus, 59,* 41–59.

Hapke B. (1986) Bidirectional reflectance spectroscopy 4. The extinction coefficient and the opposition effect. *Icarus, 67,* 264–280.

Hapke B. (2012) *Theory of Reflectance and Emittance Spectroscopy, 2nd edition.* Cambridge Univ., Cambridge. 513 pp.

Helfenstein P. and Shepard M. K. (2011) Testing the Hapke photometric model: Improved inversion and the porosity correction. *Icarus, 215,* 83–100.

Henyey L. G. and Greenstein J. L. (1941) Diffuse radiation in the galaxy. *Astrophys. J., 93,* 70–83.

Hoey W. A., Yeoh S. K., Trafton L. A., Goldstein D. B., and Varghese P. L. (2017) Rarefied gas dynamic simulation of transfer and escape in the Pluto-Charon system. *Icarus, 287,* 87–102.

Horak H. (1950) Diffuse reflection by planetary atmospheres. *Astrophys. J., 112,* 445–463.

Howett C. J. A. et al. (2017a) Inflight radiometric calibration of New Horizons' Multispectral Visible Imaging Camera (MVIC). *Icarus, 287,* 140–151.

Howett C. J. A. et al. (2017b) Charon's light curves, as observed by New Horizons' Ralph color camera (MVIC) on approach to the Pluto system. *Icarus, 287,* 152–160.

Hromakina T. A. (2019) Long-term photometric monitoring of the dwarf planet (136472) Makemake. *Astron. Astrophysics, 625,* A46.

Karkoschka E. (2001) Comprehensive photometry of the rings and 16 satellites of Uranus with the Hubble Space Telescope. *Icarus, 151,* 51–68.

Khare B. N. et al. (1984) Optical constants of organic tholins produced in a simulated titanian atmosphere: From soft X-ray to microwave frequencies. *Icarus, 60,* 127–137.

Lane A. P. and Irvine W. M. (1973) Monochromatic phase curves and albedos for the lunar disk. *Astron. J., 78,* 267–277.

Lauer T. R. et al. (2018) The New Horizons and Hubble Space Telescope search for rings, dust, and debris in the Pluto-Charon system. *Icarus, 301,* 155–172.

Li J.-Y. et al. (2013) Global photometric properties of asteroid (4) Vesta observed with Dawn Framing Camera. *Icarus, 226,* 1252–1274.

Li J.-Y. et al. (2019) Spectrophotometric modeling and mapping of Ceres. *Icarus, 332,* 144–167.

Mishchenko M. I. (1992) Light scattering by nonspherical ice grains: An application to noctilucent cloud particles. *Earth Moon Planets, 57,* 203–211.

Muinonen K. (1990) *Light Scattering by Inhomogeneous Media: Backward Enhancement and Reversal of Linear Polarization.* Observatory and Astrophysics Laboratory, University of Helsinki, Finland. 24 pp.

Olkin C. B. et al. (2017) The global color of Pluto from New Horizons. *Astron. J., 154,* 258–271.

Porter S. B. and Grundy W. M. (2015) Ejecta transfer in the Pluto system. *Icarus, 246,* 360–368.

Reinsch K. and Pakull M. W. (1987) Physical parameter of the Pluto-Charon system. *Astron. Astrophys., 177,* L43–L46.

Reinsch K., Burwitz V., and Festou M. C. (1994) Albedo maps of Pluto and improved physical parameters of the Pluto-Charon system. *Icarus, 108,* 209–218.

Reuter D. C. et al. (2008) Ralph: A visible/infrared imager for the New Horizons Pluto/Kuiper belt mission. *Space Sci. Rev., 140,* 129–154.

Schenk P. M. et al. (2018) Breaking up is hard to do: Global cartography and topography of Pluto's mid-sized icy moon Charon from New Horizons. *Icarus, 315,* 124–145.

Shkuratov Yu. G. (1988) A diffraction mechanism of brightness opposition effect of surfaces with complicated structure. *Kinemat. Fiz. Nebesn. Tel., 4,* 33–39 (in Russian).

Sicardy B. et al. (2006) Charon's size and an upper limit on its atmosphere from a stellar occultation. *Nature, 439,* 52–54.

Stern S. A. et al. (2015) The Pluto system: Initial results from its exploration by New Horizons. *Science, 350,* aad1815.

Strazzulla G. (1998) Chemistry of ice induced by bombardment with energetic charged particles. In *Solar System Ices* (B. Schmitt et al., eds.), pp. 281–301. Astrophysics and Space Science Library Vol. 227, Springer, Dordrecht.

Tholen D. J. et al. (1987) Improved orbital and physical parameters for the Pluto-Charon system. *Science, 237,* 512–514.

Tucker O. J., Johnson R. E., and Young L. A. (2015) Gas transfer in the Pluto-Charon system: A Charon atmosphere. *Icarus, 246,* 291–297.

Verbiscer A. J. and Veverka J. (1989) Albedo dichotomy of Rhea: Hapke analysis of Voyager photometry. *Icarus, 82,* 336–353.

Verbiscer A. et al. (2007) Enceladus: Cosmic graffiti artist caught in the act. *Science, 315,* 815.

Verbiscer A. J. et al. (2013) Photometric properties of solar system ices. In *The Science of Solar System Ices* (M. S. Gudipati and J. Castillo-Rogez, eds.), pp. 47–72. Astrophysics and Space Science Library Vol. 356, Springer, New York.

Verbiscer A. J. et al. (2018a) Surface properties of Saturn's icy moons from optical remote sensing. In *Enceladus and the Icy Moons of Saturn* (P. M. Schenk et al., eds.), pp. 323–341. Univ. of Arizona, Tucson.

Verbiscer A. J. et al. (2018b) Phase curves of Nix and Hydra from the New Horizons imaging cameras. *Astrophys. J. Lett., 852,* L35.

Verbiscer A. J. et al. (2019) Phase curves from the Kuiper belt: Photometric properties of distant Kuiper belt objects observed by New Horizons. *Astron. J., 158,* 123–140.

Weaver H. A. et al. (2016) The small satellites of Pluto as observed by New Horizons. *Science, 351,* aae0030.

Young L. A. et al. (2018) Structure and composition of Pluto's atmosphere from the New Horizons solar ultraviolet occultation. *Icarus, 300,* 174–199.

Protopapa S., Cook J. C., Grundy W. M., Cruikshank D. P., Dalle Ore C. M., and Beyer R. A. (2021) Surface composition of Charon. In *The Pluto System After New Horizons* (S. A. Stern, J. M. Moore, W. M. Grundy, L. A. Young, and R. P. Binzel, eds.), pp. 433–456. Univ. of Arizona, Tucson, DOI: 10.2458/azu_uapress_9780816540945-ch019.

Surface Composition of Charon

S. Protopapa
Southwest Research Institute

J. C. Cook
Pinhead Institute

W. M. Grundy
Lowell Observatory

D. P. Cruikshank
NASA Ames Research Center

C. M. Dalle Ore
NASA Ames Research Center and SETI Institute

R. A. Beyer
SETI Institute and NASA Ames Research Center

We present an overview of Charon's spectral measurements, which have progressed in the last three decades from four discrete wavelengths to spectral maps. The surface of Charon is dominated by crystalline water ice and ammonia-bearing species, which appear enhanced in the bright ejecta blankets of geologically young craters, possibly exposing subsurface material. The conundrum that any spectral signature of ammonia and crystalline water ice is present on an old and mature planetary surface like Charon is discussed in view of the ambiguity in the exact identity of the ammonia-bearing species on Charon (ammonia hydrates or ammonia salts) and the compositions of Pluto's moons, Nix and Hydra. The presence of two darkening materials across the surface is suggested, one being highly processed hydrocarbon plausibly associated with exogenous impactors from the Kuiper belt and the other being red tholin-like material at the poles resulting from irradiation of cold-trapped material escaping from Pluto onto Charon.

1. INTRODUCTION

Our compositional understanding of the Kuiper belt has been limited by the challenges of acquiring high-quality spectroscopy for mid-sized and small transneptunian objects (TNOs) and composition maps of large dwarf planets. NASA's New Horizons mission represents a breakthrough in our understanding of the TNO population. New Horizons provides a detailed portrait of objects with very different size scales: the 2400-km-diameter dwarf planet Pluto (see the chapter by Cruikshank et al. in this volume), the mid-sized ~1200-km-diameter body Charon, the much smaller satellites of Pluto (e.g., Nix and Hydra), and (486958) Arrokoth (provisional designation 2014 MU_{69}), the latter with an 18-km equivalent spherical diameter. This is the result of two successful flybys: that of Pluto and its moons on July 14, 2015 (*Stern et al.,* 2015), and that of Arrokoth on January 1, 2019 (*Stern et al.,* 2019; *Grundy et al.,* 2020; *Spencer et al.*, 2020; *McKinnon et al.,* 2020). This chapter provides an overview of the composition of Pluto's largest satellite, Charon, as inferred through ground- and spacebased observations. In terms of surface composition, Charon appears, at large scale, representative of the mid-sized TNOs (~500 km < D < ~1700 km, where D is the object diameter) that are just a little too small to have retained surface volatiles (*Schaller and Brown,* 2007), and that have surfaces dominated by crystalline water ice (e.g., Orcus, Quaoar, Haumea). These objects, lacking the gravitationally bound but mobile volatile veneers of larger TNOs, provide valuable insight into differentiation, radiation processes, cryovolcanism, and cratering histories (*Brown*, 2012). However, Charon's surface appearance also needs to be considered in the context of its likely origin in a large, grazing collision with Pluto (*Canup,* 2005, 2011).

2. SPECTROSCOPIC MEASUREMENTS OF CHARON PRIOR TO THE NEW HORIZONS FLYBY

Prior to the launch of the Hubble Space Telescope (HST) and the usage of adaptive optics to achieve the best possible image definition from the ground, characterization of the surface composition of Charon through groundbased measurements was hampered by the proximity of the satellite to its primary, with a maximum separation between the two objects of only 0.9 arcsec. A few years after the discovery of Charon in 1978 (*Christy and Harrington*, 1978), measurements acquired between 1985 and 1990 during the mutual transits and occultations between Pluto and its satellite were the only means to assess the surface composition of Pluto's largest moon (e.g., *Marcialis et al.*, 1987; *Buie et al.*, 1987; *Binzel*, 1988; *Fink and DiSanti*, 1988). This special geometry provided a direct means to study the satellite without contamination from Pluto. Spectra of the Pluto-Charon system before and during the total eclipse of the satellite were subtracted to obtain spectra of Charon alone. The major findings about the composition of Charon prior to the New Horizons flyby of the Pluto system, sorting them as a function of increasing wavelength, are summarized in the subsections below.

2.1. Visible Spectra

The visible wavelength range is strongly diagnostic of the presence of macromolecular organic compounds known as tholins (e.g., *Cruikshank et al.*, 2005; see also the chapter by Cruikshank et al. in this volume) as well as pure elemental carbon solids (e.g., graphite, amorphous carbon). These species are characterized by low visual albedos and display a wide range of colors in the visible wavelength range from red to neutral, according to their chemistry, particle size, and abundance (e.g., *de Bergh et al.*, 2008). Nonetheless, very few papers discuss the spectral behavior of Charon's surface in the visible wavelength range.

The spectrum of Charon is featureless in the visible wavelength range between 0.54 μm and 1.02 μm (*Fink and DiSanti*, 1988; *Merlin et al.*, 2010) (Fig. 1). The hemispherical averaged geometric albedo for Charon reported by *Buie et al.* (1997), ranging from 0.36 to 0.39 in the V filter (~0.55 μm), is in agreement with that by *Buie and Grundy* (2000) at 1.8 μm (see section 2.2.1), giving evidence that, apart from the water-ice absorption band at 1.5 μm, Charon displays an overall flat continuum from 0.54 μm up to 1.8 μm. The estimate of the geometric albedo relies on the object radius, adopted by *Buie et al.* (1997) and *Buie and Grundy* (2000) as 593 ± 10 km

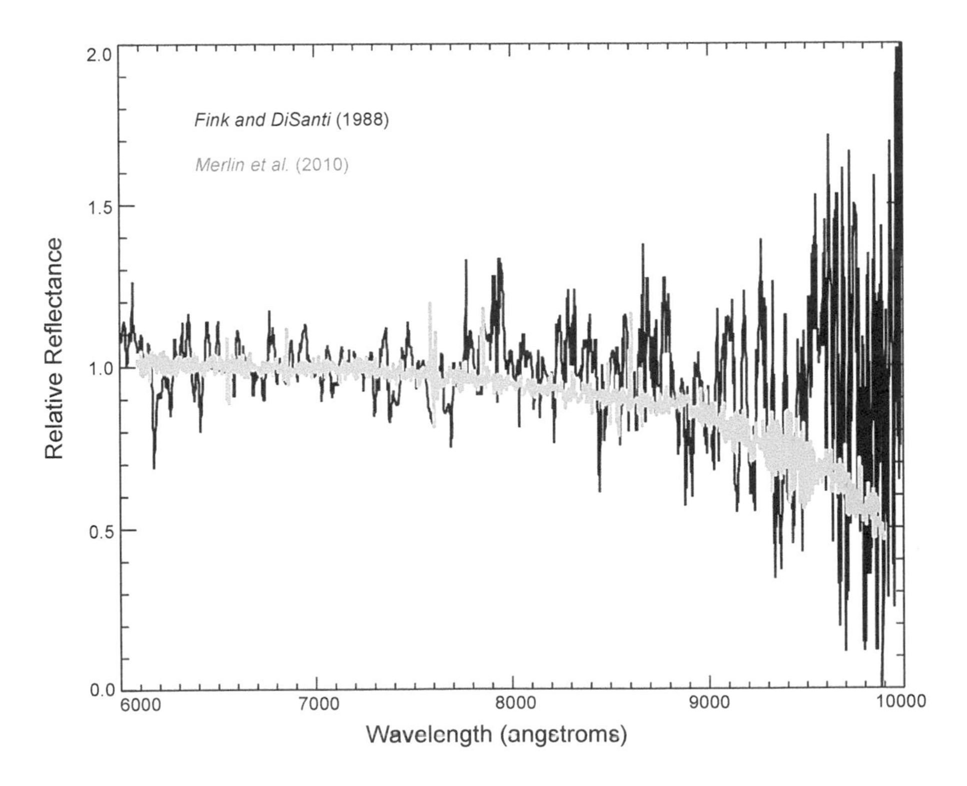

Fig. 1. Two independent spectra of Charon in the visible wavelength range. The spectrum by *Fink and DiSanti* (1988) obtained in 1987 during the occultation of the satellite by Pluto and acquired with the Catalina site telescope of the University of Arizona Observatories is shown in black. The spectrum is featureless and almost perfectly flat. The FORS2 spectrum acquired at the Very Large Telescope by *Merlin et al.* (2010) is shown in gray. Contrary to the results of *Fink and DiSanti* (1988), the spectrum published by *Merlin et al.* (2010) displays a blue slope, most probably due to incomplete removal of scattered light from Pluto, as acknowledged by the authors.

(*Buie et al.,* 1992), a value that is compatible within the error with that computed from New Horizons data at 606 ± 1 km (*Nimmo et al.*, 2017).

The absence of characteristic absorption features across the 0.3–1.0-μm spectral region, as well as the low albedo, signal the presence of a dark, spectrally neutral material, whose nature is still an active area of investigation (see section 2.2.3 for further details). Because Charon was found to have relatively uniform longitudinal color distribution (*Binzel,* 1988), the presence of this neutral absorber was thought to be globally distributed across the satellite's surface. However, local color variations have been identified by New Horizons color maps (section 3.2).

2.2. Near-Infrared Spectra (1–2.5 μm)

2.2.1. Crystalline water ice. The first decisive identification of water ice on the surface of Charon was by *Buie et al.* (1987), through spectral measurements of Charon obtained by subtracting the contribution of Pluto-only flux, measured during the 1987 mutual event, from the flux of the unresolved Pluto/Charon system recorded prior to the event. These observations, covering the wavelength range 1.5–2.5 μm (13 spectral points), clearly displayed the 1.5-μm and 2.0-μm absorption bands characteristic of water ice. The authors pointed out the compositional differences between Pluto and its satellite, with surfaces spectrally dominated by methane and water ice, respectively. However, since the mutual event data could only constrain the sub-Pluto hemisphere of Charon, no characterization of compositional variations of Pluto's satellite with rotation was possible. Additionally, these observations, given their very low spectral resolution, prevented investigation of the temperature and phase of the ice.

On the surface of solar system objects, water ice displays two distinct solid configurations: amorphous (I_a) and crystalline. The latter can present a cubic (I_c) or a hexagonal (I_h) structure (*Hobbs,* 1974; *Jenniskens et al.*, 1998; *Schmitt et al.*, 1998). For a detailed description of ice phases as they are found in the solar system, see *Clark et al.* (2013). Amorphous ice transforms itself irreversibly to a cubic crystal structure if heated to a temperature between ~110 and ~150 K on timescales of minutes to years (*Schmitt et al.*, 1989; *Kouchi et al.*, 1994; *Jenniskens et al.*, 1998). Temperatures above ~190 K are necessary for a hexagonal crystal structure to occur, a transformation that is also irreversible. These phase change temperatures account for the time needed to anneal amorphous ice, which increases exponentially with lower temperature (*Kouchi et al.,* 1994). For temperatures below 77 K, the annealing time exceeds the age of the solar system (*Kouchi et al.*, 1994; *Cook et al.,* 2007). Once an ice sample crystallizes, it stays crystalline even if the temperature is again reduced below 110 K, unless exposed to energetic radiation such as charged particles or ultraviolet (UV) photons (e.g., *Kouchi and Kuroda,* 1990; *Strazzulla et al.,* 1992; *Leto and Baratta,* 2003; *Mastrapa and Brown,* 2006). Therefore, identification of amorphous water ice on the surface of a solar system object might imply a temperature environment significantly colder than 100 K since formation, or a target surface that is subject to space weathering. Consequently, spectral assessment of an icy body's surface composition is critical for setting limits on its origin and evolution.

Water ice presents four spectral bands in the near-infrared (IR) wavelength range close to 1.5 μm, (a nearby feature at 1.56 μm blends together with the 1.5-μm absorption to form one wide band), 1.65 μm, and 2.0 μm. The exact position of these bands as well as their shape and relative strength are phase-, temperature-, and grain size-dependent (see Figs. 2d,f,g). This is why quantitative assessment of the physical properties of Charon's surface, as well as any other target, requires analysis of collected groundbased and/or spacebased spectroscopic measurements using a multiple scattering radiative transfer model (e.g., *Hapke,* 2012). The ratio of the bands near 1.65 μm and 1.5 μm is a good metric for distinguishing amorphous from crystalline water ice (e.g., *Fink and Larson,* 1975; *Mastrapa et al.,* 2008). Low-temperature crystalline water ice displays a very sharp and peaked 1.65-μm absorption feature, contrary to its amorphous counterpart (Fig. 2d). However, the strength of the 1.65-μm band weakens with increasing temperature (Fig. 2f). Amorphous and crystalline water ice have distinctive spectral bands also beyond 2.5 μm. As an example, amorphous water ice exhibits a generally broad and featureless 3.1-μm Fresnel reflection peak, compared to the three-peaked structure in the crystalline phase (*Hansen and McCord,* 2004). There is a broad literature investigating the variation in the spectral properties of water ice with temperature and phase based on laboratory measurements (e.g., *Fink and Sill,* 1982; *Warren,* 1984; *Kou et al.*, 1993; *Schmitt al.,* 1998; *Grundy and Schmitt,* 1998; *Rajaram et al.,* 2001; *Warren and Brandt,* 2008; *Mastrapa et al.,* 2008, 2009).

The use of HST and the Keck 10-m telescope on Mauna Kea, on an exceptional night of 0.3–arcsec atmospheric seeing, enabled *Buie and Grundy* (2000) and *Brown and Calvin* (2000) respectively to substantially advance our knowledge of Charon's surface composition, providing some of the best spectroscopic measurements of Charon prior to New Horizons. These data represent a considerable improvement in terms of wavelength coverage, spectral resolution, and longitudinal coverage with respect to those obtained during the mutual event season. The spectra confirmed the presence of the 1.5-μm and 2.0-μm water-ice absorption bands, already identified through the mutual event observations, and revealed clearly the 1.65-μm absorption feature, evidence that crystalline water ice is present on the surface of Charon. This finding has been confirmed by contemporaneous and subsequent works (e.g., *Dumas et al.*, 2001; *Cook et al.,* 2007; *Merlin et al.*, 2010; *Holler et al.,* 2017), including New Horizons measurements (see section 3.1). *Mastrapa et al.* (2008) showed that the 1.65-μm absorption band is present in model spectra with up to 80% amorphous content. Therefore, identification of the 1.65-μm absorption feature does not necessarily imply a surface that is 100% crystalline.

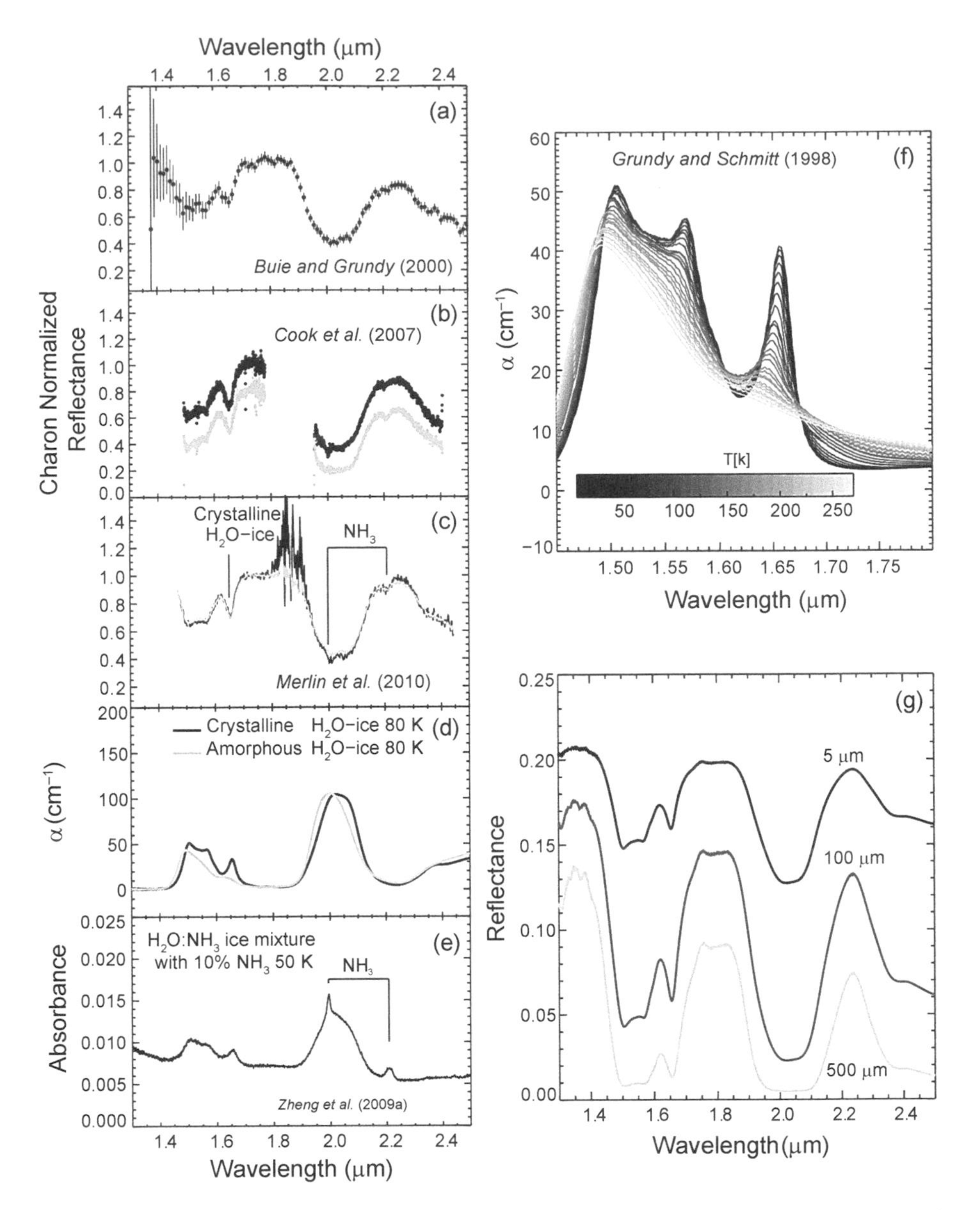

Fig. 2. Charon's reflectance spectra obtained prior to the New Horizons flyby compared with synthetic spectra of water and ammonia ices. **(a)–(c)** Charon's spectrum normalized to unity between 1.7 and 1.8 μm, obtained by *Buie and Grundy* (2000), *Cook et al.* (2007), and *Merlin et al.* (2010), respectively. In **(b),** spectra correspond to a sub-Earth longitude of 7° (black) and 185° (gray), shifted along the y-axis for clarity, while **(a)** shows the grand average spectrum of four separate longitudes. **(c)** Two spectra of Charon acquired by *Merlin et al.* (2010) corresponding to a Charon sub-Earth longitude of 243° (black) and 188° (gray). Laboratory spectra, expressed in terms of absorption coefficient α and absorbance (a peak in α and absorbance corresponds to an absorption band), of amorphous and crystalline H_2O-ice at 80 K [**(d)**] and that of a water-ammonia ice mixture with 10% ammonia [**(e)**] are shown for comparison. The 2.0-μm and 2.2-μm ammonia bands are marked in **(e)**. The relative strength and shape of the 1.5-, 1.65-, and 2.0-μm water-ice absorption bands strongly depend not only on phase [**(d)**], but also on temperature [**(f)**] and particle size [**(g)**]. The synergy of observations and laboratory studies suggests that Charon's surface is dominated by crystalline water ice and ammoniated species (see text for details).

Cook et al. (2007) and *Merlin et al.* (2010) both suggested amorphous water ice to be depleted on the surface of Charon and set a lower limit of 90% of crystalline water ice. Modeling analysis of New Horizons measurements led to similar conclusions (see section 3.1). Evidence for marginal differences in the water-ice absorption bands between the leading and trailing hemispheres of Charon were reported by *Buie and Grundy* (2000) and *Dumas et al.* (2001).

A disk-averaged water-ice temperature for Charon of 60 ± 20 K was initially inferred by *Buie and Grundy* (2000) using the water-ice absorption bands as thermometers (e.g., *Grundy et al.*, 1999). Most recently, *Holler et al.* (2017) revisited this effort using the temperature-dependent properties of the 1.65-μm absorption band and reported a mean ice surface temperature of 45 ± 14 K, which confirms results published by previous works (*Buie and Grundy*, 2000;

Cook et al., 2007; *Lellouch et al.,* 2011). Additionally, no significant variations of surface temperature with longitude were found by *Holler et al.* (2017).

Like Charon, all TNOs with spectra measured with sufficiently high signal precision present the 1.65-μm crystalline water-ice feature when water ice is detected. Examples include Orcus (e.g., *de Bergh et al.,* 2005; *Carry et al.,* 2011), Quaoar (e.g., *Jewitt and Luu,* 2004), and Haumea (e.g., *Trujillo et al.,* 2007; *Dumas et al.,* 2011). The phase of water ice provides an important clue for understanding the physical processes occurring on TNOs. Several studies (e.g., *Cook et al.,* 2007; *Porter et al.,* 2010) have investigated how crystalline ice could be produced at temperatures of 40–50 K and survive in spite of the space radiation environment (UV photons, solar wind particles, cosmic rays, etc.). The implications of the crystalline nature of water ice on the surface of Pluto's satellites, in terms of their origin and evolution and accounting for the role of the surrounding space radiation environment, are discussed in sections 5 and 6.

2.2.2. Ammonia. Ammonia hydrates ($NH_3{\bullet}nH_2O$) have been implicated on Charon's surface through detection of the 2.21-μm band (*Brown and Calvin,* 2000; *Buie and Grundy,* 2000; *Nakamura et al.,* 2000; *Dumas et al.,* 2001; *Cook et al.,* 2007; *Verbiscer et al.,* 2007; *Merlin et al.,* 2010; *DeMeo et al.,* 2015; *Holler et al.,* 2017). *DeMeo et al.* (2015) investigated spectral differences in the 2.21-μm band across Charon's surface by comparing data acquired by several teams that sampled a wide range of longitudes. A compilation of spectroscopic measurements of Charon in the 2.2-μm spectral region is shown in Fig. 3 (adapted from *DeMeo et al.,* 2015). The analysis by *DeMeo et al.* (2015) was prompted by the results published by *Dumas et al.* (2001) and *Cook et al.* (2007) reporting variations of the ~2.2-μm band across the surface of Charon. Specifically, *Dumas et al.* (2001) reported a leading-trailing asymmetry on Charon, with the trailing hemisphere lacking the presence of ammonia hydrates. On the contrary, *Cook et al.* (2007) suggested a global presence of the 2.2-μm absorption band and put forth different ammonia hydration states across the surface of Charon to explain variations in the ~2.21-μm band position between spectra of two opposite hemispheres of Charon (the Pluto-facing and the anti-Pluto sides, see Fig. 2b and Fig. 3]. *DeMeo et al.* (2015) investigated variations of the 2.21-μm band among the highest signal-to-noise ratio data (*Cook et al.,* 2007; *Merlin et al.,* 2010; *DeMeo et al.,* 2015) through analysis of band parameters (band center, depth, area, width). The authors reported that all observations display the ~2.21-μm band, evidence that ammoniated species are ubiquitous across Charon's surface. While differences were observed in band depth and position, *DeMeo et al.* (2015) did not rule out that these variations might be due to differences in telescope performance as well as changes in the data acquisition and processing.

As noted by *Zheng et al.* (2009a), the ~2.2-μm band is not the only major band displayed by solid ammonia and its hydrates. Another feature occurs at ~2.0 μm (see Fig. 2e), which, in the case of Charon, overlaps with the much stronger water-ice band. The detection of the ~2.0-μm band due to solid ammonia and its hydrates is extremely difficult from the ground and sometimes not fully reliable, given the strong contamination in this range by telluric CO_2 (*Merlin et al.,* 2010). The 2.0-μm band is clearly visible in the spectra by *Merlin et al.* (2010) (see Fig. 2c).

The detection of ammonia hydrate ices on the surface of outer solar system icy bodies demands explanation. As shown by *Strazzulla and Palumbo* (1998) through laboratory-derived data, ammonia is expected to be removed from the surfaces of icy satellites and TNOs by irradiation of charged particles on geologically short timescales. These constraints prompted *Cook et al.* (2007) to put forth the idea of a surface renewal process. Recent studies of other small bodies in the outer solar system have opened the windows to other possible interpretations of the 2.2-μm band such as ammonia salts (e.g., *De Sanctis et al.,* 2016; *Poch et al.,* 2020). For additional details about the possible species responsible for the 2.2-μm absorption band observed in Charon's spectra, and their formation and survival on the surface of Charon, the reader is referred to section 4.

2.2.3. Darkening agent. As noted in section 2.1, the low-albedo values reported for Charon suggest the presence of a darkening agent. Although Charon displays a neutral continuum from the visible to 1.8 μm, an albedo contrast between 1.8 and 2.2 μm is observed and it was first reported by *Buie and Grundy* (2000). The spectra by *Buie and Grundy* (2000), *Brown and Calvin* (2000), and *Cook et al.* (2007) are remarkably similar in their continuum spectral behavior, whereas the data by *Merlin et al.* (2010) present a shallower slope between 1.8 μm and 2.2 μm with respect to the other data (Fig. 2). Because *Buie and Grundy* (2000) reported only weak variations with longitude, and *Cook et al.* (2007) and *Merlin et al.* (2010) covered the same sub-Earth longitude of Charon (185° and 188°), the discrepancy of the *Merlin et al.* (2010) measurements is possibly due to details of the data processing (e.g., choice of solar analog, background removal) and/or acquisition (e.g., slit alignment).

Buie and Grundy (2000) addressed the difficulty in duplicating Charon's albedo contrast between 1.8 μm and 2.2 μm simultaneously with the spectral shape of the water-ice bands at 1.5 μm and 2.0 μm. The spectral continuum of water ice is grain-size dependent (see Fig. 2g). While large grain sizes of water ice present a lower albedo at 2.2 μm than at 1.8 μm, their use in spectral modeling of Charon's data failed to reproduce the shape of the water-ice absorption bands at 1.5 μm and 2.0 μm (*Buie and Grundy,* 2000). Additionally, the authors pointed out that many uranian, jovian, and saturnian satellites share the same spectral behavior of Charon. These similarities led *Buie and Grundy* (2000) to conclude that, in order to model Charon's spectrum, it is necessary to include, together with water ice, a dark neutral absorber with an increasing absorption longward of 1.9 μm. Several materials that are possibly common in the outer solar system were found with these spectral characteristics, such as the phyllosilicates kaolinite and montmorillonite. When mixed

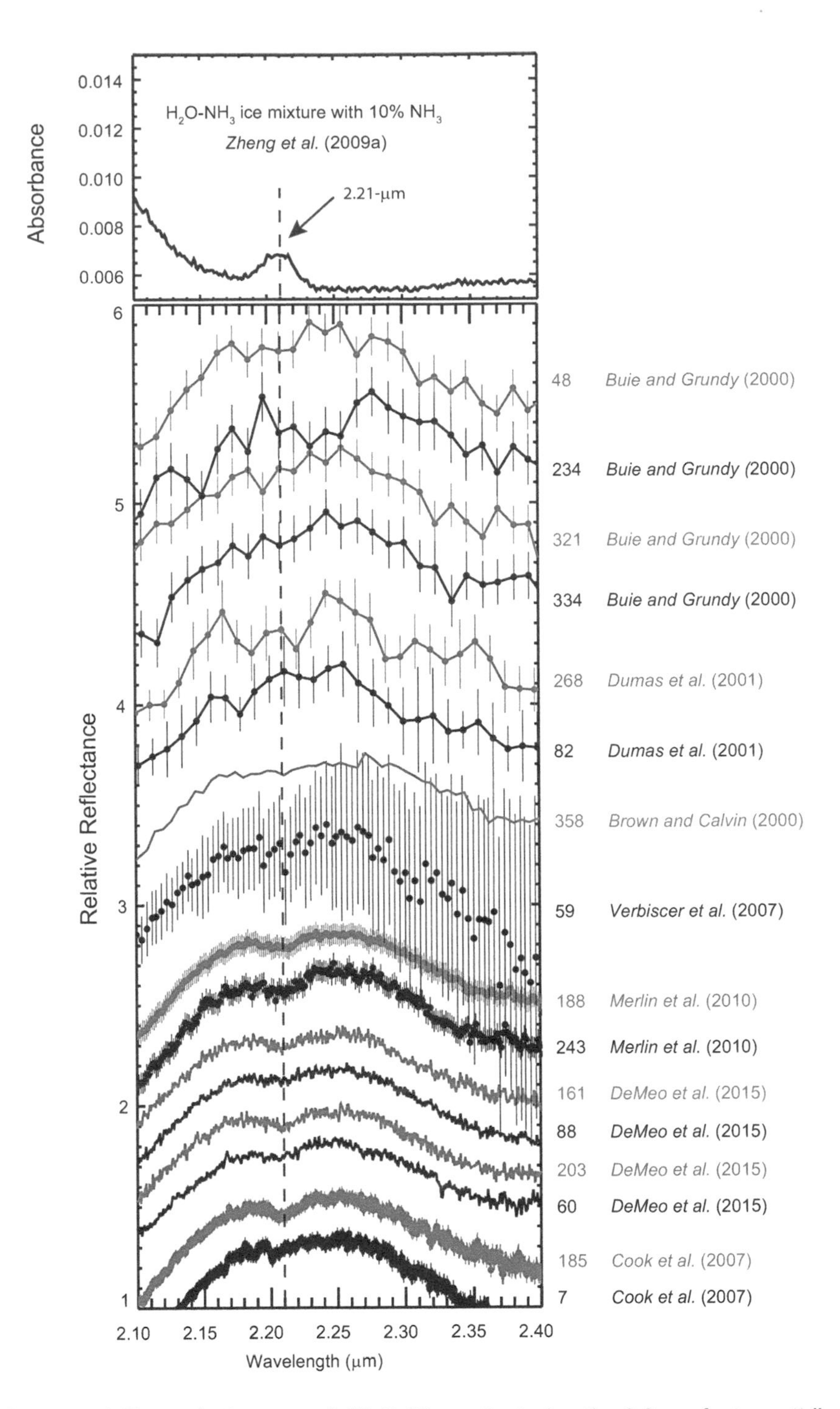

Fig. 3. *Bottom:* Spectra of Charon in the range 2.10–2.40 μm displaying the 2.2-μm feature attributed to an ammoniated species (see dashed line for reference). The spectra are displayed with alternating black and gray colors and offset along the y-axis for clarity. On the right side of each spectrum, Charon's longitude and the data reference are indicated. This compilation highlights the differences in strength and position of the ammonia-bearing feature. Figure adapted from *DeMeo et al.* (2015). *Top:* The absorbance (a peak in absorbance corresponds to an absorption band) spectrum obtained in the laboratory of a water-ammonia ice mixture displaying the 2.21-μm ammonia band is shown for comparison

with water ice, each of these materials provided reasonable fits to the data, although none was without problems. *Buie and Grundy* (2000) overcame these difficulties by computing the optical constants of an "ad hoc" spectrally neutral continuum absorber to match Charon's HST spectral data (*Buie and Grundy*, 2000, see their Table 3) and provided a so-called "standard model." The advantage of this approach consisted in having a reference for the darkening agent when conducting new laboratory measurements of planetary analog materials, as well as providing a model that would permit more accurate removal of Charon's contribution from combined Pluto/Charon observations.

Buie and Grundy (2000) considered the possibility of tholin-like materials as darkening agents on the surface of Charon, given their presence on the surface of Pluto. However, they ruled out this possibility given the strong wavelength-dependence absorption of tholins in the visible wavelength range, incompatible with the flat spectrum of Charon reported by *Fink and DiSanti* (1988) (section 2.1). The presence of tholin-like materials on the surface of Charon has been revisited in light of the New Horizons data (see section 3.2).

2.3. Near-Infrared Spectra (>2.5 μm)

The surface composition of Charon has been characterized so far almost exclusively through spectroscopic observations in the range 0.4–2.5 μm (see sections 2.1 and 2.2). The visible and near-IR wavelength ranges are in fact optimal to look for and characterize the presence of ices such as N_2, CH_4, CO_2, CO, H_2O, and NH_3, which exhibit distinctive absorption bands in this range (e.g., *Grundy et al.,* 1993; *Quirico and Schmitt,* 1997; *Grundy and Schmitt*, 1998; *Zanchet et al.,* 2013). Long near-IR wavelengths (>2.5 μm) are also ideal for investigating the composition of outer solar system objects (e.g., *Spencer et al.*, 1990; *Grundy et al., 2002*; *Olkin et al.,* 2007). Indeed, fundamental vibrational transitions of many ice species are located in this wavelength region, and these absorptions are much stronger than those observed shortward of 2.5 μm. As an example, amorphous ammonia ice presents a spectral feature at 2.96 μm (ν = 3378 cm^{-1}, A = 1.8×10^{-17} cm $molecule^{-1}$), with a band strength 2 orders of magnitude larger than the feature at 2.23 μm (ν = 4479 cm^{-1}, A = 8.0×10^{-19} cm $molecule^{-1}$) (*Zanchet et al.,* 2013). Similarly, the absorption coefficient of the 3-μm water-ice feature is a factor of $\sim 10^2$ larger than the water-ice bands at shorter wavelengths (1.5 μm, 1.56 μm, 1.65 μm, and 2.0 μm) (*Mastrapa et al.*, 2009; *Warren and Brandt,* 2008), enabling the detection of water ice even at low abundances (e.g., *Clark et al.*, 2005; *Protopapa et al.*, 2014; *Cartwright et al.*, 2018). Because of the weak solar flux, presence of strong telluric absorptions, and lack of adequate instrumentation, the L- and M-bands (3.0–5.0 μm) have been vastly underexploited for the exploration of Charon and other icy bodies throughout the solar system. The only attempt to investigate the surface composition of Pluto's largest satellite through measurements beyond 2.5 μm was presented by *Protopapa et al.* (2008). Observations of Charon, obtained with the adaptive optics instrument NACO at the ESO Very Large Telescope and covering the wavelength range 1–4 μm, presented water-ice signatures in agreement with the short-wavelength measurements conducted by several authors and a broad 3.0-μm water-ice band. *Protopapa et al.* (2008) used these observations to verify the "standard model" for Charon proposed by *Buie and Grundy* (2000) and extend it beyond 2.6 μm. The authors calculated the hypothetical spectral properties of the unknown neutral absorber (see section 2.2.3) beyond 2.6 μm. No detection of the 2.96-μm ammonia feature was reported due to the strong telluric absorptions between 2.5 μm and 3 μm.

3. CHARON'S GLOBAL SURFACE COMPOSITION AS OBSERVED BY NEW HORIZONS

While ground- and spacebased remote measurements have played a remarkable role in the growth of knowledge about Charon's average composition (see section 2), they yielded no resolved composition maps of Charon. New Horizons enabled us to overcome this limitation by returning a detailed compositional snapshot of Pluto's largest satellite. The initial results from the New Horizons exploration of the Pluto system were reported by *Stern et al.* (2015), while further in-depth studies of the surface composition of Charon were published by *Grundy et al.* (2016a,b) and *Dalle Ore et al.* (2018).

This section summarizes major findings obtained using spatially resolved IR spectra acquired with the Linear Etalon Imaging Spectral Array (LEISA), part of the New Horizons Ralph instrument (*Reuter et al.,* 2008). For a detailed analysis of the New Horizons Charon data recorded with the Multispectral Visible Imaging Camera (MVIC), the other component of the Ralph instrument, the reader is referred to the chapter by Howett et al. in this volume.

LEISA is a mapping IR composition spectrometer with two linear variable filter segments covering the wavelength range 1.25–2.5 μm and 2.1–2.25 μm at the resolving power ($\lambda/\Delta\lambda$) of 240 and 560, respectively. The low-resolution segment was used to explore Charon's composition. Analysis of LEISA data acquired on the closest approach day, July 14, 2015, highlighted the major spatial units of water ice, one or more ammoniated species, and darkening materials and their strong correlation with Charon's geological features. The presence of two distinct darkening agents, corresponding to the gray and red visible coloration at low latitudes and in the northern polar region, respectively, was identified. These results relied on the analysis of spectral parameters, specifically the band depth over the 1.5-μm and 2.0-μm water-ice absorption features, a multivariate statistical approach, and a pixel-by-pixel radiative transfer model.

A successful method to derive constraints on the abundances and scattering properties of the materials across the surface of Charon, and the relation between their distribution and surface geology, consists of applying a radiative transfer model to the LEISA spectral data at the pixel-by-pixel level. This technique was applied by *Dalle Ore et al.* (2018) to the three best LEISA scans of Charon. These scans were C_LEISA, C_LEISA_LORRI_1, and C_LEISA_HIRES, collected at mean distances from Charon's center of 483,077 km, 137,825 km, and 80,653 km, respectively. The composition maps for water ice and the darkening component obtained from this analysis were not published.

Instead, the derived models, which lacked the spectral signature of ammonia or its hydrate, were subtracted from the original data of each pixel to compute the residuals and investigate the distribution of the ammoniated species. Charon composition maps based on a revised pixel-by-pixel modeling analysis to account for a correction factor in the LEISA radiometric calibration (*Protopapa et al.*, 2020) and a set of Charon photometric parameters (chapter by Howett et al. in this volume) published subsequently to the analysis by *Dalle Ore et al.* (2018) are presented and discussed in this chapter. The results of a principal-component analysis (PCA) applied to the LEISA data are also described here. The advantage of applying a breadth of spectral techniques is to corroborate the validity of the results. These most recent maps confirm previous findings and provide additional information on the physical properties (abundance, grain size, phase) of the materials across the surface of Charon. The results from the pixel-by-pixel modeling analysis are discussed first, and then compared with results obtained with other spectral techniques.

3.1. The Physical Properties and Spatial Distribution of Water Ice

Compositional maps of Charon defining the volume fraction and textural properties of the materials throughout the surface are presented here. These results have been obtained by applying a pixel-by-pixel Hapke radiative transfer model (*Hapke*, 2012) to the best LEISA scan of Charon: C_LEISA_HIRES, at a spatial scale of ~5 km/pixel. This approach requires the application of the same modeling strategy across the entire visible face of Charon, so that a systematic and comparative study between the composition of the different surface units can be conducted. This is at the expense of possible compositional peculiarities in small surface areas. While the visible wavelength range contains important compositional information (low-albedo organic compounds display in this region their most diagnostic spectral signatures such as red slope and low albedo; section 2.1), it has not been considered in this analysis. This limits the possibility of assessing the nature of the coloring agents (see section 3.2). Additionally, this analysis does not account for a recently discovered slight global shift in the LEISA spectral calibration (B. Schmitt, personal communication). This aspect of the wavelength calibration has yet to be fully characterized and documented in the literature.

Several scattering theories (e.g., *Douté and Schmitt*, 1998; *Shkuratov et al.*, 1999) exist. These alternative methods could possibly provide similar quality of fits to the data as those presented here but with different percentages and grain size of the components (*Poulet et al.*, 2002). The Hapke radiative transfer model has been chosen to interpret the Charon New Horizons data since it has been widely used to model ground- and spacebased observations of Pluto's largest satellite (e.g., *Brown and Calvin,* 2000; *Buie and Grundy,* 2000; *Merlin et al.,* 2010), Pluto (*Protopapa et al.*, 2017; *Cook et al.,* 2019), Nix, and Hydra (*Cook et al.*, 2018). Therefore, applying the Hapke radiative transfer model to the LEISA data facilitates comparative analysis between different studies of Charon and between Pluto, Charon, and the smaller satellites.

The data are calibrated in units of radiance factor (RADF). This is also commonly referred to as I/F, and it is defined as the ratio of the bidirectional reflectance of a surface r to that of a perfectly diffuse surface of the same size and distance to the Sun and observer, but illuminated at normal incidence (*Hapke,* 2012). The RADF is equal to πr. The correction scaling factor in the LEISA radiometric calibration of 0.74 ± 0.05 reported by *Protopapa et al.* (2020) has been accounted for. In order to increase the signal precision, the spatial resolution of the LEISA scan has been degraded to ~8 km/pixel through bin averaging. This step also reduces computation time.

The details of the modeling algorithm are outlined by *Protopapa et al.* (2017, 2020). An intimate mixture of crystalline water ice and a dark neutral absorber has been employed, choices based on spectral evidence collected over the course of several years (see section 2). The Hapke modeling relies on the knowledge of optical constants, the real and imaginary part of the complex refractive index as a function of wavelength for any given compound in the mixture. The optical constants by *Mastrapa et al.* (2008) for crystalline water ice have been adopted assuming a surface temperature of 50 K (*Holler et al.,* 2017) (section 2.2.1). The optical constants of the "ad hoc" spectrally neutral continuum absorber computed by *Buie and Grundy* (2000, their Table 3) to match Charon's HST data (section 2.2.3) have been employed. Ammoniated species have not been accounted for in the modeling analysis because of the unclear source of the observed band at 2.2 μm (section 4) and the lack of relevant complex refractive indices. The fit to the LEISA observations has been therefore weighted such that the wavelength range between 2.18 μm and 2.28 μm, where ammoniated species present a strong absorption band, does not affect the modeling results. A proxy for the spatial distribution of the ammoniated species has been investigated through the residuals (data-model) in the 2.2-μm spectral range. The free parameters in the model are effective grain diameter (D_i) and fractional volume (V_i) of each constituent. They are iteratively modified by means of a Levenberg-Marquardt χ^2 minimization algorithm until a best-fit to the observations is achieved (*Markwardt*, 2009). Notice that the approach adopted is identical to the "standard model" by *Buie and Grundy* (2000), with different assumptions for the Hapke parameters.

Other analyses of the surface composition of Charon using New Horizons data have made use of "Pluto ice tholin" as the darkening agent (*Dalle Ore et al.,* 2018). Optical constants were derived from the "Pluto ice tholin" reflectance spectrum (*Cruikshank et al.*, 2016) using the Hapke Isotropic Multiple Scattering Approximation radiative transfer model (*Hapke,* 2012), following the approach by *Carli et al.* (2014) and assuming a particle size of

10 μm. However, the size range can play a crucial role in the derivation of the imaginary part of the refractive index, k, and this approach calls for further laboratory measurements. This is why the pixel-by-pixel modeling analysis presented here does not make use of the "Pluto ice tholin" and adopts the dark neutral continuum absorber computed by *Buie and Grundy* (2000) instead.

The estimates of the concentration and particle size of each surface component strongly rely on the choice of photometric properties (Hapke parameters such as the asymmetry factor, compaction parameter, amplitude of the opposition effect, and mean roughness slope). The set of Hapke parameters derived for Charon by Howett et al. (this volume) and obtained from a disk-integrated analysis of HST data at 0.55 μm and New Horizons LOng-Range Reconnaissance Imager (LORRI) (*Cheng et al.,* 2008) data have been adopted. Specifically, a single lobe Henyey-Greenstein phase function (*Henyey and Greenstein,* 1941), P(g, ξ) has been employed (equation (3) of *Protopapa et al.,* 2020). The parameter describing the scattering lobe and defining P(g) is the asymmetry parameter, ξ, which has been set to –0.14. The amplitude (B_0) and width (h) of the shadow-hiding opposition effect (SHOE) (equation (2) of *Protopapa et al.,* 2020) computed by Howett et al. (this volume) for Charon equals 0.74 and 0.0015, respectively. Effects of macroscopic roughness have been accounted for and the mean roughness slope θ has been set to 24°. The coherent backscattering opposition effect (CBOE) has not been taken into consideration given the phase angle coverage of the LEISA Charon data. CBOE generally appears at phase angles <2°, while the C_LEISA_HIRES data have been acquired at an average phase angle of ~36° (*Hapke*, 2002).

Figure 4 shows the spatial distribution of the volume fraction and particle diameter of crystalline water ice across the surface of Charon obtained from the pixel-by-pixel modeling analysis. The average volume fraction and particle diameter of crystalline water ice equals 67 ± 9% and 64 ± 15 μm, respectively. These maps are in agreement with the "standard model" by *Buie and Grundy* (2000), consisting of an intimate mixture of 70-μm-diameter grains, with 60% of the grains being low-temperature crystalline water ice and 40% being the standard continuum absorber. These results corroborate the idea that Charon's encounter hemisphere is composed predominantly of water ice, as first suggested in the mid-1980s (*Buie et al.,* 1987). However, in the case of Charon, the estimates of the concentration and particle size of each surface compound depend on the choice of the optical properties of the darkening material.

The volume fraction map does not show any region where water ice is absent. Therefore, water ice is considered ubiquitous across the Pluto-facing hemisphere of Charon, as initially pointed out by *Stern et al.* (2015) and *Grundy et al.* (2016a). An enrichment of water ice is observed mantling many crater floors and ejecta blankets. Examples of these locations are Nasreddin, Skywalker, Organa, Candide, and Pangloss craters. Some regions of geologic and compositional interest discussed here have been labeled in Fig. 4a according to the nomenclature adopted in this volume. For a full list of geologic regions and their names, the reader is referred to the appendix by Beyer and Showalter in this volume. A small crater, named Cora, southeast of Alice and west of Serenity Chasma, within the disrupted tectonic belt and north of Vulcan Planitia, stands out. This crater in panchromatic images (Fig. 4a) clearly displays two distinct tones of ejecta blankets with a concentric darker one surrounded by an asymmetric bright ray pattern. This asymmetry in albedo in the visible wavelength range (LORRI's CCD detector provides panchromatic response from 350- to 850-nm wavelengths) seems correlated with water-ice content, which is greater in the bright high-albedo ejecta (see Fig. 5a). Several craters on Charon's surface display a similar albedo contrast. Dark and light ejecta deposits in panchromatic images are mapped as albedo features by *Robbins et al.* (2019, their Fig. 5). The authors point out that a light ejecta deposit often extends from the edge of a dark ejecta deposit. However, resolution effects must be accounted for when a dark inner ejecta component is not observed. The crystalline water-ice volume fraction map indicates that bright high-albedo ejecta deposits in panchromatic images (Fig. 4a) show higher water-ice content (Figs. 4b,d) compared to dark low-albedo ejecta deposits and with respect to the rest of the surface. The brightness of these crater rays in panchromatic images suggests that they formed relatively recently in geologic terms. Craters with ejecta blankets enriched in water ice occur not only in Oz Terra but also in Vulcan Planitia (see craters labelled l and k in Fig. 5b). These craters do not display in panchromatic images clear evidence of a bright ejecta deposits possibly due to viewing geometry near the terminator.

Not all craters on Charon exhibit a spectral behavior compatible with an enhancement of water ice. An example is Alice crater. The spectral characteristics of Alice's rims are consistent with material less enriched in water ice (Fig. 5a). This together with the absence of an obvious ejecta blanket around Alice crater suggests a degradation process within and around the crater.

Therefore, it is possible to reasonably attribute compositional differences among craters on Charon's surface to their degradation state, with Alice being an example of an older crater. Another example is the Dorothy impact basin, the largest among Charon's craters with a diameter of ~240 km and a depth of 6 km (*Schenk et al.,* 2018). The floor of Dorothy crater is not enriched in water ice, contrary to the geologically young craters overprinting it (Cunegonde, Pangloss, and the crater at approximately the center of Dorothy identified with the letter m in Fig. 5c). A region arcing from Organa crater to Skywalker, east of both craters (see Fig. 5d) stands out in the water-ice maps. The geologic nature of this terrain is unclear and there is no morphological evidence for an extended ejecta blanket around Organa. In general, a positive correlation between albedo in the visible wavelength range (Fig. 4a) and water-ice fractional volume is observed, with the

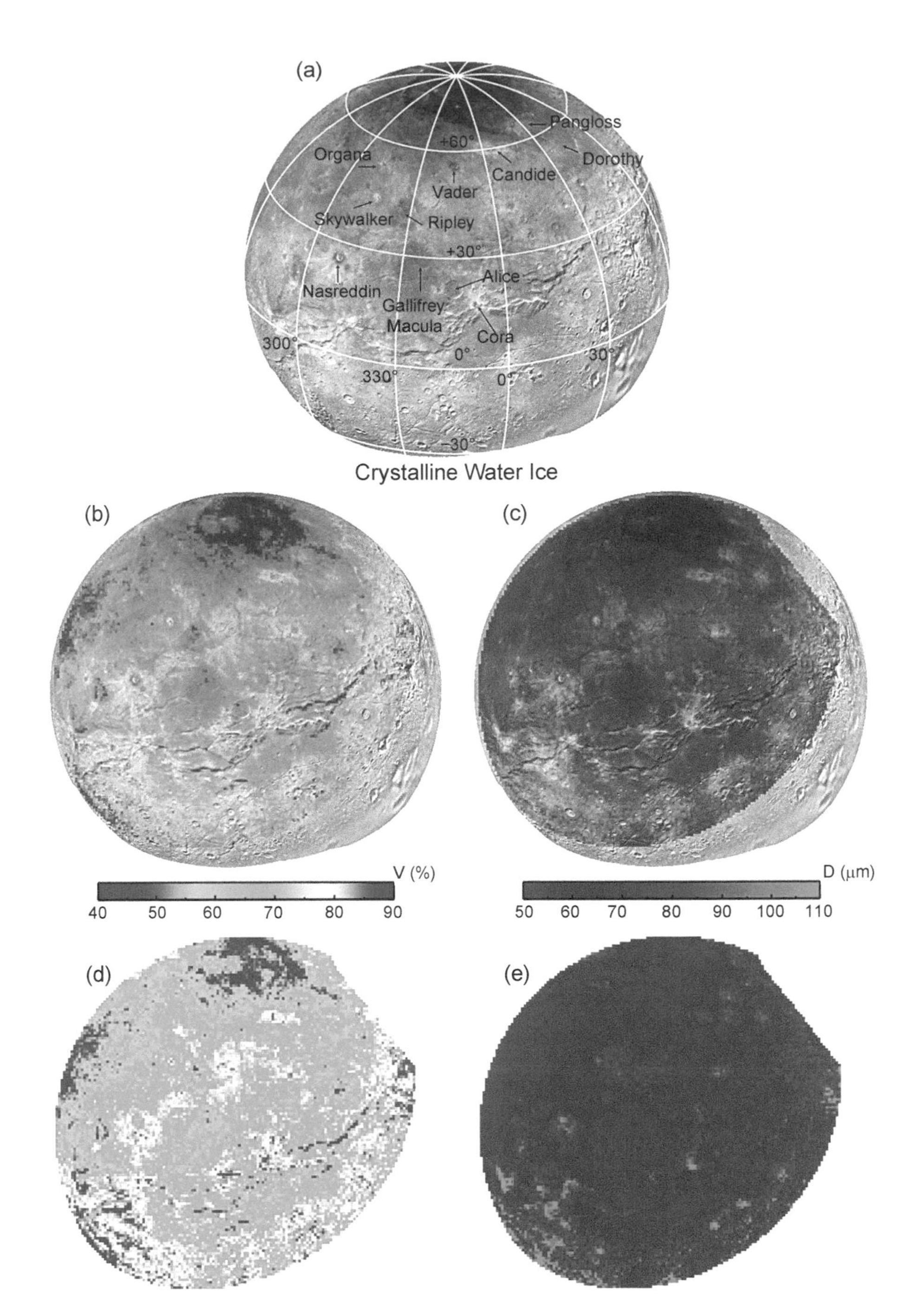

Fig. 4. (a) LORRI base map of Charon's surface reprojected to the geometry of the C_LEISA_HIRES observation. The maps of volume fraction, V(%) [**(b)**,**(d)**], and grain diameter, D(μm) [**(c)**,**(e)**] of crystalline water ice, resulted from the pixel-by-pixel Hapke radiative transfer model of the C_LEISA_HIRES scan, are shown. In **(b)** and **(c)**, contrary to **(d)** and **(e)**, the abundance and grain size maps are overlaid on the LORRI base map.

brightest regions displaying enhanced water-ice absorption. Figure 5 provides a close-up view of several areas across Charon supporting this correlation.

The positive correlation between albedo at visible wavelengths and water-ice fractional volume discussed above is corroborated by PCA analysis (e.g., *Whitney,* 1983; *Smith et al.,* 1985; *Johnson et al.,* 1985; *Klassen et al.*, 1999) of New Horizons Charon data. The main idea of PCA is to reduce the dimensionality of the data to a smaller set of coordinates that are most representative of the information in the original dataset. The original dataset, which resides in an n-dimensional wavelength space, is transformed into a p-dimensional space of eigenvectors, which accounts for most of the data variance. This technique enables us to assess spectral contrasts in a way that does not depend on multiple scattering models. The first principal component (PC1) indicates the axis of maximum variance, with subsequent PCs each pointing along the next largest axis of

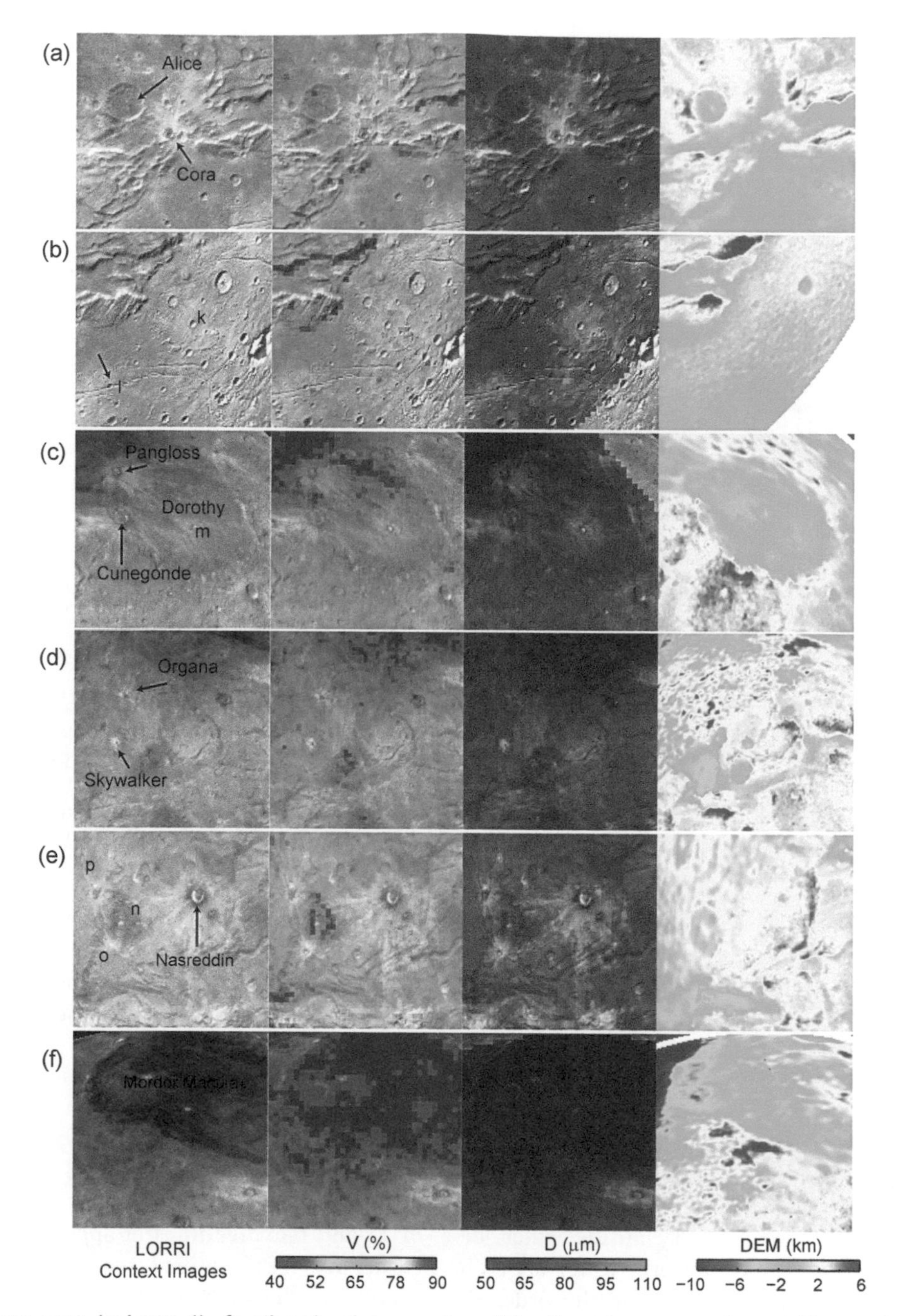

Fig. 5. LORRI base map (column 1), fractional volume and particle diameter maps of crystalline water ice (columns 2 and 3) and the topographic digital elevation model (DEM) (column 4) showing Alice and Cora craters [**(a)**]; two craters in Vulcan Planitia [**(b)**]; Dorothy crater at 58.5°, 40.6°E [**(c)**]; Skywalker and Organa [**(d)**]; Nasreddin at 25.6°, 308.6°E [**(e)**]; and Charon's north pole [**(f)**]. Letters k–p identify regions of interest discussed in the text.

variance orthogonal to preceding axes. Composition maps of Charon obtained by applying a PCA analysis on the C_LEISA_HIRES are shown in Fig. 6. While PC1 mostly captures geometric effects of shading and topography as well as variations in brightness due to IR albedo, the corresponding eigenvector is not flat, as it often is (Fig. 6b). Instead, it shows dips at the 1.5-μm and 2.0-μm water-ice absorption bands. These features in the eigenvector indicate a correlation between brightness and water-ice absorption with higher near-IR albedo areas having stronger water-ice bands, as can be caused by spatially varying quantities of a dark contaminant. In regions with less of such a contaminant, photons are less likely to be absorbed, enabling them to interact with more particles on average. The result is a larger mean optical path length in the water ice and thus deeper absorption bands. This interpretation is supported by PC1 being well correlated with the crystalline water-ice volume fraction map shown in Figs. 4b,d. Dark areas in PC1 correspond to regions with water-ice volume fraction less than 70% and therefore blue to green in the water-ice volume fraction map. Dark areas in the PC1 image not attributable to shading are generally associated with impact

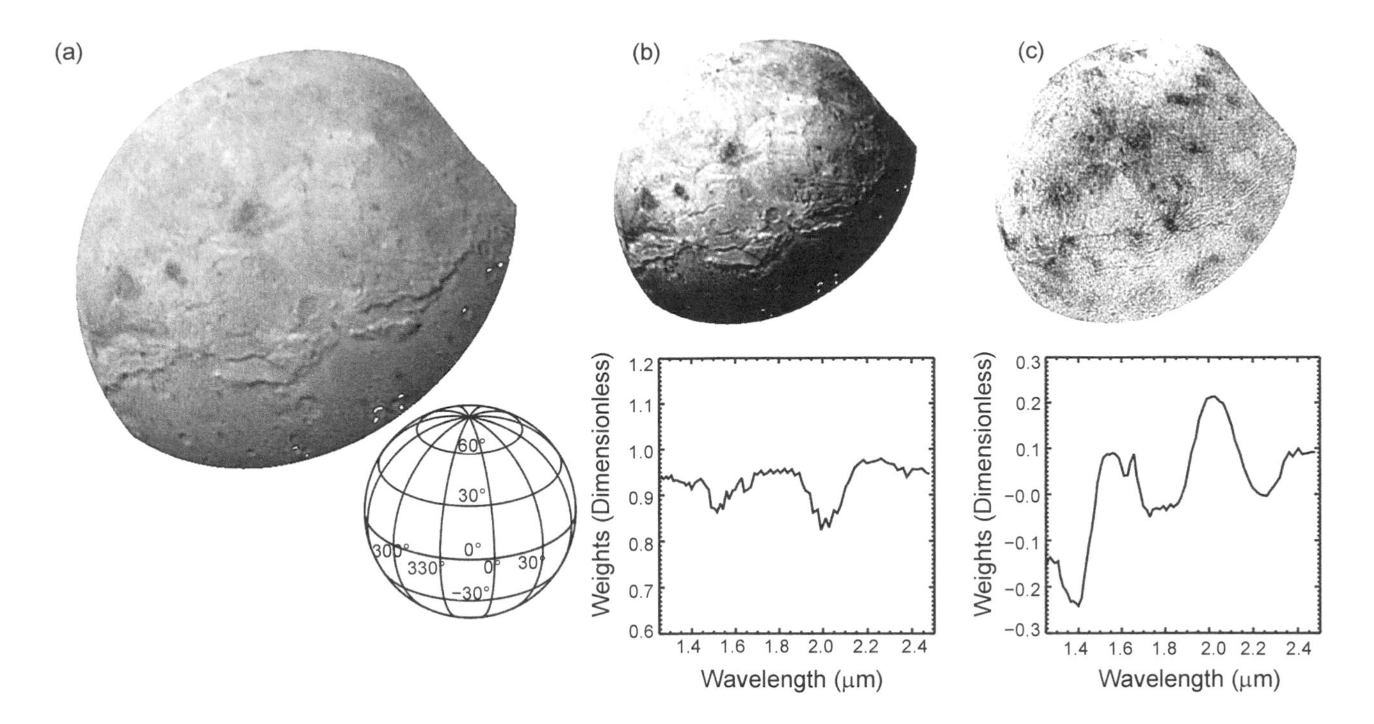

Fig. 6. (a) "Enhanced" LEISA color image with the median of LEISA's channels between 0 and 63, 64 and 127, and 128 and 191 displayed in red, green, and blue color channels, respectively. Geometry is indicated by the wire grid. **(b),(c)** Principal component images and eigenvectors for PC1 and PC2, respectively. From PC3 on, higher-principal components are increasingly dominated by various patterns of noise and striped artifacts in the LEISA data.

craters, notably Ripley and Alice, as well as several other craters with not yet assigned informal names, including at least two small craters in the smoother, southern Vulcan Planitia. But other craters do not appear dark in PC1. This dichotomy in the spectral properties of Charon's crater population has been identified for the water-ice volume fraction map obtained from the pixel-by-pixel modeling analysis. An alternative explanation to crater degradation could involve an exogenous material delivered by some, but not all, of the impacting population (*Grundy et al.,* 2016a).

The particle diameter map of crystalline water ice (Figs. 4c,e) shows a relatively uniform water-ice path length across the surface of Charon (on the order of ~60-μm-diameter grains), with the largest grains (typically larger than 90 μm) corresponding to brighter albedo regions at visible wavelengths (Fig. 4a). These are often bright ejecta blankets. Nasreddin and the small, possibly geologically young craters west of Nasreddin (indicated with letters p and o in Fig. 5e), Candide, Pangloss, Organa, Skywalker, and Cora are examples of craters with bright ejecta. The PCA analysis supports these findings. The eigenvector for PC2 (Fig. 6c) resembles a coarse-grained water-ice spectrum, but inverted, so the 1.5-μm and 2.0-μm absorption bands appear as humps instead of dips. This inversion doesn't signify anything. The PCA singles out the axes of maximum variance, but each such axis has a sign ambiguity as to which direction the eigenvector should point along it. Because it is inverted, regions with the strongest water-ice absorption bands, and therefore larger grains, appear dark in PC2. Again, some of the strongest water-ice absorption bands are seen in and around craters (e.g., Skywalker). Mordor Macula appears a little brighter in PC2, indicating that the water-ice signature is a little weaker there. The spatial distribution of PC2 and the crystalline water-ice particle diameter are well correlated. Similar results have been obtained with a completely independent method by *Dalle Ore et al.* (2018). They applied a statistical clustering tool to identify spectrally distinct terrains and used the *Shkuratov et al.* (1999) radiative transfer approach to model the three cluster averages identified (see Fig. 7). They found that the visible-wavelength high-albedo regions, often corresponding to crater ejecta blankets, are characterized by larger water-ice grains, on the order of 90 μm. Differences in the particle size between the pixel-by-pixel Hapke modeling results presented in this chapter and those by *Dalle Ore et al.* (2018) are attributable to the different radiative transfer models applied and differing assumptions on the nature of the darkening material. As pointed out by *Robbins et al.* (2019), larger grain sizes are typical of ejecta. A possible scenario to explain the spectral appearance of these high-albedo, water-ice-rich ejecta deposits is that the impactors excavated this material, giving us a view of Charon's subsurface composition.

Hapke reflectance models have been constructed for various combinations of potential surface components. The data do not support inclusion of amorphous water ice in the models. This is consistent with previous results by *Cook et*

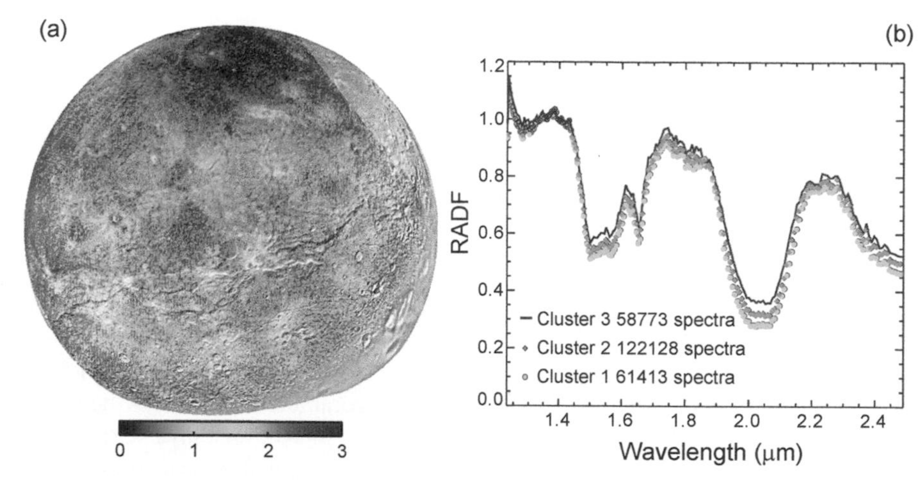

Fig. 7. **(a)** Clustering map of Charon. Cluster 1 (cyan color) corresponds mostly to water-ice-enriched regions, bright in panchromatic images (LORRI basemap); Cluster 2 (orange) correlates with intermediate albedo pixels; while Cluster 3 (dark brown) maps mainly the darker areas in the LORRI base map (e.g., Mordor Macula at the pole), which are characterized by a lower ice-to-non-ice (darkening) component ratio. **(b)** Spectral averages of the pixels belonging to the three clusters shown in **(a)** plotted with corresponding colors. Each average includes a large number of spectra, as shown in the legend.

al. (2007) and *Merlin et al.* (2010) that set a lower limit of 90% of crystalline water ice on the surface of Charon. Regions that appear to host higher proportions of amorphous ice also tend to be dark in panchromatic images, possibly hinting at greater age (*Dalle Ore et al.,* 2018).

3.2. Spatial Distribution and Physical State of Two Distinct Darkening Agents on Charon

3.2.1. Charon's red poles and tholin-like material. Visible wavelength images of Charon show a prominent dark, reddish northern polar cap, informally known as Mordor Macula. At first glance, this region appears to be bounded by an encircling ridge at roughly 70° to 75°N latitude. But on closer examination, the coloration extends beyond the ridge, with a gradual increase in reddening toward higher latitudes. Topography complicates the latitude-dependent distribution. For a given latitude, north-facing slopes tend to be darker and redder than south-facing slopes. Charon's southern winter pole was imaged at low signal precision and spatial resolution, illuminated only by Pluto-shine, and it too exhibits a pattern of reduced albedo towards the pole (*Grundy et al.,* 2016b; see also the chapter by Howett et al. in this volume).

Mordor contrasts most strongly with other, more neutral-colored regions of Charon in MVIC's shortest wavelength blue filter, with reduced contrast in the red and near-IR filters. These color characteristics require a material that absorbs most strongly at blue wavelengths, consistent with macromolecular organic material, analogous to the laboratory tholin residues produced through the action of energetic radiation on small organic molecules (e.g., *Khare et al.,* 1984; *Cruikshank et al.,* 2005). When bonds in small molecules are broken by impinging charged particles or photons, the resulting radicals and fragments can recombine to build more complex molecular structures. As hydrogen escapes from simple organic molecules, stoichiometry becomes increasingly carbon-rich, and, if they are available in the environment, atoms such as N, O, and S can also be incorporated (e.g., *Brassé et al.,* 2015). A complicated combination of chains and cyclic structures absorbs strongly at UV wavelengths, with absorption decreasing into the visible and near-IR, producing various shades of reds, yellows, and browns, depending on the starting material, degree of radiation processing, and texture.

A hypothesis for the formation of Charon's tholin polar caps was proposed by *Grundy et al.* (2016b). It begins with methane (CH_4) escaping from Pluto's atmosphere and streaming radially away from Pluto (*Gladstone et al.*, 2016; *Young et al.,* 2018). Escaping gas molecules that happen to pass near Charon are gravitationally attracted, and some become temporarily bound, forming an exosphere of CH_4 molecules hopping from place to place about Charon's surface on ballistic trajectories (*Tucker et al.,* 2015; *Hoey et al.*, 2017). Each of Charon's poles is exposed to continuous sunlight for half of Pluto's 248-year orbit, whereas no sunlight is incident at all at the poles during the other half-year. The long absence of sunlight enables the unilluminated winter pole to become so cold that CH_4 molecules arriving there stick, becoming cold trapped as CH_4 ice. Hundreds of nanometers of CH_4 ice are estimated to accumulate over the course of each winter. Energetic space radiation — primarily in the form of solar Lyman α (Ly-α) UV photons that have been scattered by the interplanetary medium but also galactic cosmic rays — acts on this ice during the remainder of the polar winter. The radiation breaks bonds, leading to loss of hydrogen and assembly of larger, less-volatile molecules. Unaltered CH_4 sublimates away once the winter pole is exposed to sunlight in the spring, but the heavier molecules produced via space radiation mostly do not. They accumulate in Charon's polar regions year after year. Continued exposure to energetic radiation eventually converts them to dark, reddish tholin-like materials.

Assuming conditions at the time of the flyby prevailed over the past 4 G.y. of Pluto system history, *Grundy et al.* (2016b) estimated that tens of centimeters of tholin could accumulate over that time. Such a quantity of tholin would completely obscure whatever lies below it, but LEISA spectral images of Charon show no obvious polar cap at all (see Figs. 6a and 8b). Near-IR spectra in the polar region are almost identical to spectra from lower-latitude regions, both being dominated by water-ice absorption bands at 1.5-μm and 2.0-μm wavelengths.

Impact gardening rates on Charon are highly uncertain (*Singer et al.*, 2019), but the lack of strong spectral contrasts with latitude at LEISA wavelengths provide strong evidence that the tholin is mixed down into the water-ice substrate, so that water ice remains a dominant surface constituent even at the pole. A few small impact craters punch through the veneer of reddened material (see Fig. 5f). Their existence confirms that the red material is not mixed very deeply into the interior and that it does not accumulate so quickly that fresh craters are immediately painted over. While the regions of Mordor Macula and near the west limb of Charon's encounter hemisphere labelled ROI 1 and ROI 2 in Fig. 8, respectively, present the same spectral behavior in the near-IR, they definitely display different coloration in enhanced visible color images (Fig. 9e) (*Grundy et al.*, 2016a,b; Howett et al., this volume). This suggests the presence of two distinct types of darkening materials on Charon's surface.

3.2.2. Charon's gray coloration and the dark neutral absorber. Apart from a red coloration in the north polar region, Charon's surface is generally colorless with a geometric albedo of 0.41 ± 0.01 (*Buratti et al.*, 2017), suggesting that a low-albedo neutral component is mixed with the water ice. A dark neutral absorber with an increasing absorption longward of 1.9 μm (*Buie and Grundy*, 2000) (see section 2.1 and 2.2.3) is required in the modeling analysis to reproduce the absolute RADF values of Charon's spectra as well as the overall spectral continuum in the 1.25–2.5-μm LEISA spectral range. The "ad hoc" spectrally neutral continuum absorber computed by *Buie and Grundy* (2000) provides a good fit to all LEISA spectral measurements of Charon's encounter hemisphere.

The pixel-by-pixel modeling analysis shows that the darkening agent may occur at ~30% level across Charon's sub-Pluto hemisphere. The values for abundance and particle diameters reported in the text are only for comparative analysis between surface regions since their absolute values depend strongly on the modeling assumptions described above (e.g., spectral properties of the darkening material, Hapke parameters). The dark neutral absorber present on Charon's surface displays variations in volume fraction and grain size that appear to be correlated with Charon's underlying geological structures (Fig. 9). The regions of Mordor Macula and near the west limb of Charon's encounter hemisphere stand out for being particularly enriched in dark material, with respect to the rest of the surface, with a fractional volume close to 50% (see Figs. 9a,b). These two surface units exhibit higher abundances of dark material, as well as larger path lengths (~100-μm depths; see Figs. 9c,d). It is not surprising that these two regions of Charon's surface present similar composition, since their spectral behavior is almost identical and they share similar viewing geometry (see Fig. 8). Figure 9 shows a side-by-side comparison between the dark neutral absorber volume fraction (Figs. 9a,b) and grain diameter (Figs. 9c,d) maps obtained from the pixel-by-pixel modeling analysis of C_LEISA_HIRES scan and the MVIC color enhanced image (Fig. 9e). A strong correlation is found between the vivid red color of Charon's northern polar cap in the MVIC enhanced visible color image and the same region highlighted by the pixel-by-pixel modeling analysis as enriched in dark material with larger particle diameters. However, this correlation is not observed in the region near the western limb, which does not appear red in the MVIC enhanced color image. This indicates

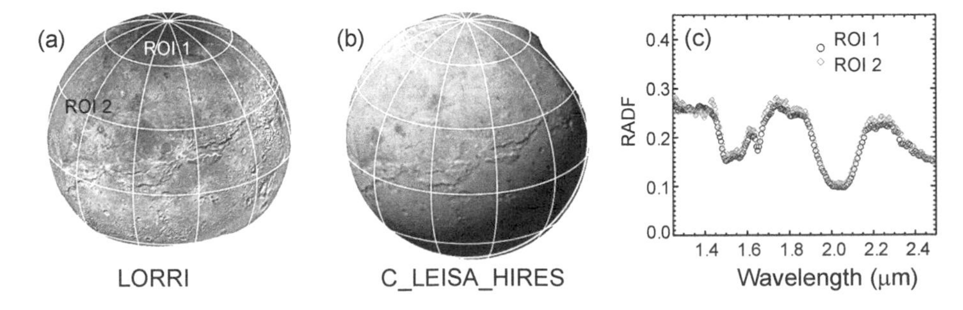

Fig. 8. (a) LORRI base map of Charon reprojected to match the viewing geometry and spatial resolution of the LEISA scan (C_LEISA_HIRES), acquired close in time and displayed in **(b)**. Specifically, the median of all the LEISA channels in the low-resolution segment is shown in **(b)**. Regions of interest, shown in dark red and orange in **(a)** and labeled ROI 1 and ROI 2, respectively, are sampled to produce the reflectance spectra in **(c)**. ROI 1 probes Charon's north pole, which appears clearly dark in panchromatic images and red in enhanced visible color images. ROI 2 samples a region with similar near-IR spectral behavior as Mordor Macula. The two spectra are almost identical despite the two regions presenting different coloration in the visible wavelength range. Notice that ROI 1 and 2 share almost identical viewing and illumination geometry.

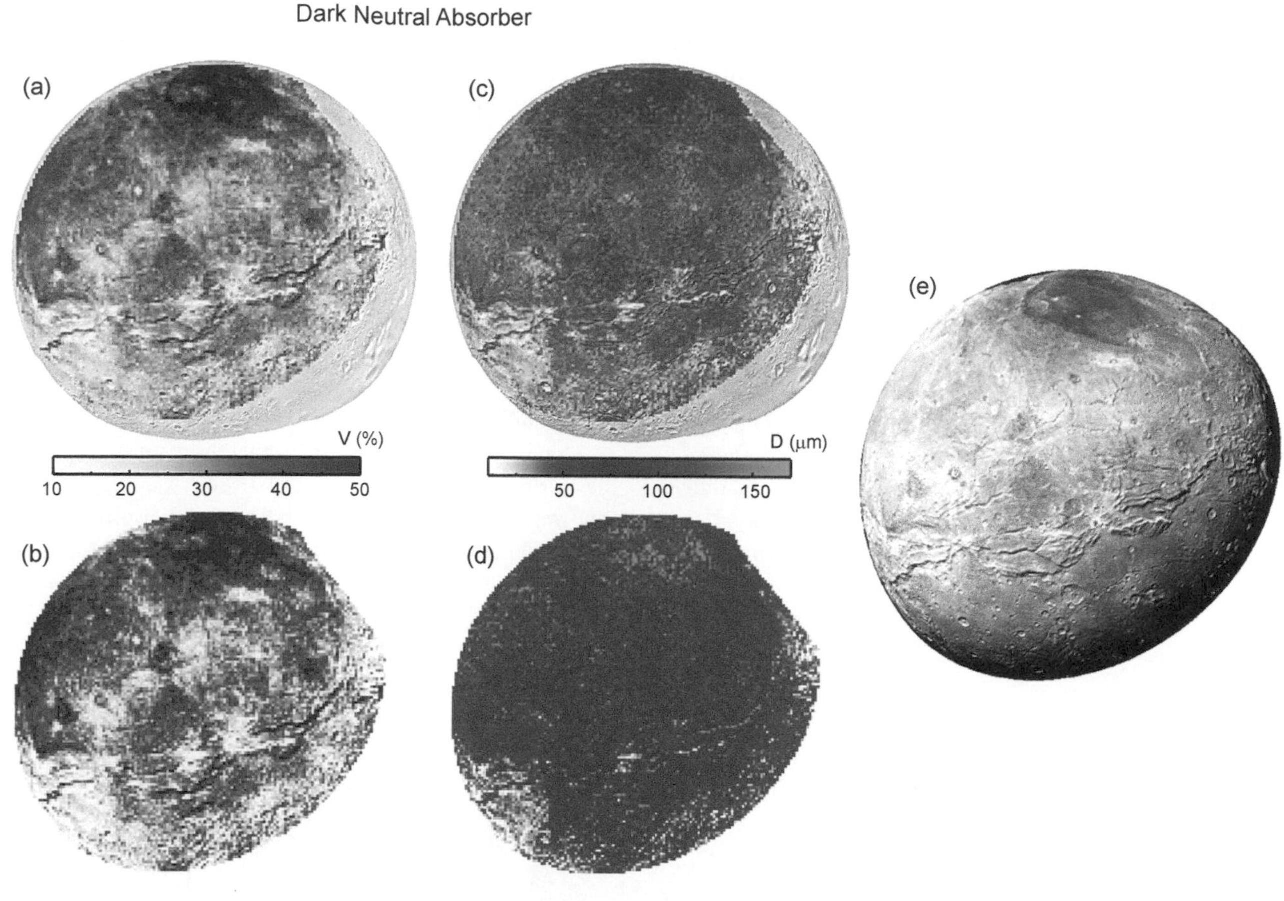

Dark Neutral Absorber

(a) (c) (e)

V (%) D (μm)

(b) (d)

Fig. 9. Dark neutral absorber fractional volume [**(a)**,**(b)**] and grain diameter maps [**(c)**,**(d)**] from the pixel-by-pixel Hapke radiative transfer model of the C_LEISA_HIRES map compared with the MVIC color image (C_COLOR_2) of Charon in enhanced color with MVIC's blue, red, and NIR filter images displayed in blue, green, and red color channels, respectively [**(e)**]. The MVIC scan has been reprojected to match the viewing geometry and spatial resolution of the LEISA scan (C_LEISA_HIRES). In **(a)** and **(c)**, contrary to **(b)** and **(d)**, the fractional volume and grain size maps are overlaid on the LORRI base map.

that while Mordor Macula and the region near the western limb of the encounter hemisphere present the same spectral behavior in the near-IR wavelength range, they differ in the visible spectral range. Because the visible wavelengths were disregarded in the pixel-by-pixel modeling analysis, it is not possible to differentiate between the composition of these two surface units. Given the neutral gray tone of most of Charon's surface as imaged by New Horizons and the neutral spectral behavior in the visible wavelength range of the dark material used in the modeling, it is reasonable to anticipate that the modeling results provided here match the visible wavelength range fairly well across most of Charon's surface apart from the north pole. The nature of the material responsible for the red coloration of Mordor Macula is discussed in the previous section.

The volume fraction of the dark material shows variation across the surface between 10% and 50% in these models. Specifically, surface areas that display higher abundance of the dark neutral absorber (~40% in these models) are the crater southwest of Nasreddin ("n" in Fig. 5e), Ripley crater, Gallifrey Macula, the northern rim of Serenity Chasma, the large block that forms the southern rim of Serenity Chasma [labeled by *Robbins et al.* (2019) as Mo_c (see their Fig. 6g)] and part of the smooth terrain of Vulcan Planitia southwest of Serenity Chasma. Overall, an anticorrelation between the bright regions observed in the panchromatic maps and the dark neutral absorber volume fraction is observed. The spectral effect of the dark material is to darken the albedo level, without introducing any significant spectral features. Notice that the absolute RADF values of the neutral absorber decrease with increasing grain size. This explains the modeling results, which show regions across Charon with lower reflectance values corresponding to higher abundances of dark material and larger particle diameters.

The most plausible source of this dark material is exogenous impactors from the Kuiper belt, where numerous low-albedo bodies exist. The distribution of such material should be uniform, and absent resurfacing processes, surfaces would be expected to gradually become darker over time, offering darkness as a potential gauge of surface age. Highly processed CH_4 from Pluto or from Charon's interior could be another possibility, but there is no obvious explanation for why such material should be processed all the way to reduced carbon at low latitudes but not at the poles, unless this more widely distributed organic material was a relic from very early in Charon's history.

3.3. Spatial Distribution of the Ammoniated Species Through Mapping of the 2.21-μm Band

The pixel-by-pixel modeling analysis of Charon's spectra does not include a contribution from NH_3-hydrates or salts (section 4) and instead de-weights the data in the 2.18-μm to 2.28-μm range where these materials typically have an absorption band. This is purposely done because the available optical constants for these hydrates and salts are limited and cannot account for all the possible chemical variants of these species. De-weighting the data facilitates the Hapke fitting modeling algorithm around this spectral range without being affected by the presence of any potential absorption band. This is the same method used by *Cook et al.* (2018) in the examination of spectra of Pluto's small satellites, Nix and Hydra.

The absorption band at 2.21 μm is mapped by averaging the band depth over the wavelength range 2.20 μm to 2.22 μm, with the continuum given by the Hapke model (1-data/Hapke model). Figure 10 shows the results from this method for the C_LEISA_HIRES scan of Charon (inset panel). In the figure, the brighter the pixel, the greater the 2.21-μm band depth. Figure 10 also shows the average spectrum from the regions with the strongest 2.21-μm absorption band (yellow to white in the 2.21-μm-band depth map) and compares it to the corresponding average Hapke model and the disk-averaged spectrum.

The majority of ammonia-enriched sites appear in the northern hemisphere (Fig. 10, inset panel), although this may be partly an effect of better sampling of that hemisphere, as discussed below. These sites commonly

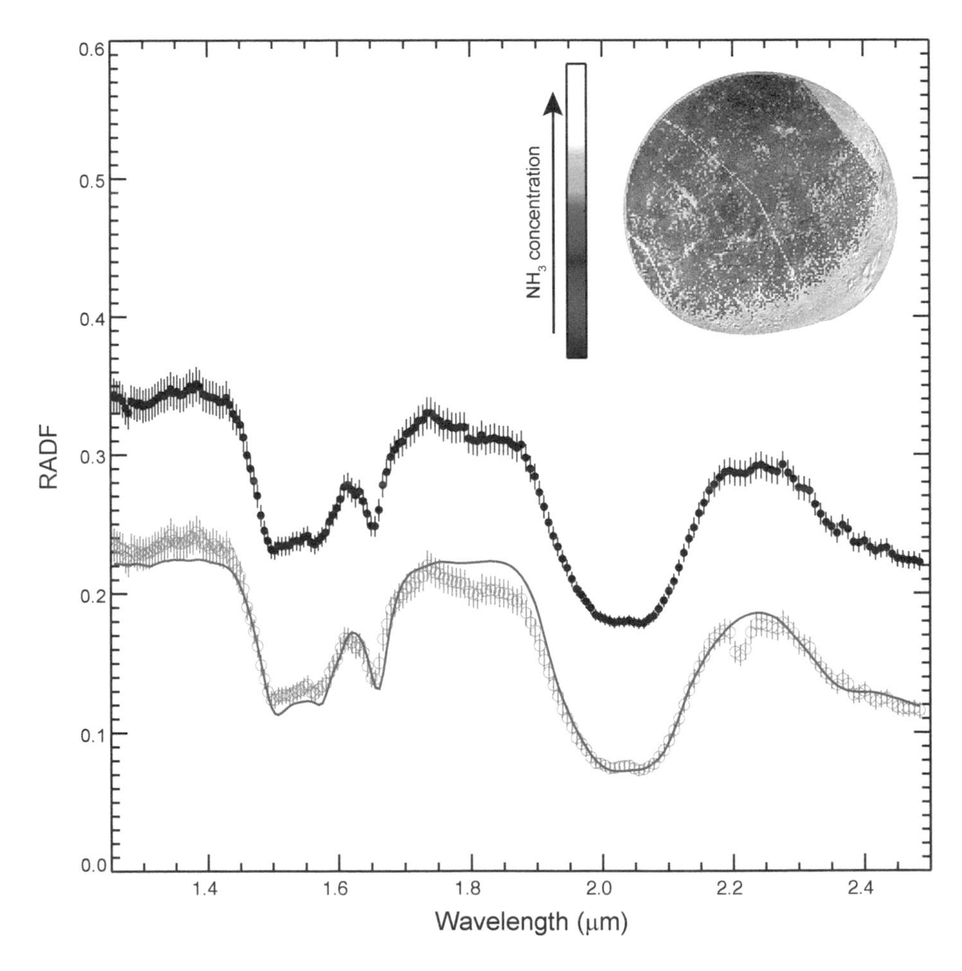

Fig. 10. The inset panel shows the 2.21-μm band depth map of Charon. Brighter regions correspond to locations enriched in ammoniated species or, in general, the compounds responsible for such absorption. The Charon spectrum (open gray circles) was obtained by averaging the reflectance spectra in the yellow regions in the inset panel (i.e., those regions showing the strongest 2.21-μm absorption band). The 1-σ uncertainties for these data are also shown. This spectrum is compared to the average Hapke model (red) over the same respective regions. Ammoniated species have not been accounted for in the modeling analysis. Therefore, the Hapke model is not expected to match the data around 2.2 μm (see text for details). For comparison, the disk-integrated Charon spectrum, offset along the y-axis for clarity, is also shown (filled black circles).

coincide with the bright ejecta around craters (e.g., Nasreddin, Skywalker, Organa, Candide). Regions that display a strong 2.21-μm absorption band include regions along the fractures that define the boundary between Oz Terra and Vulcan Planitia, west of Alice crater. *Grundy et al.* (2016a) also found the 2.21-μm absorption band to be distributed across Charon's encounter hemisphere at low level, with an enhanced concentration in correspondence of a few bright rayed craters, with Organa crater being one of the best examples. The analysis by *Dalle Ore et al.* (2018) supports these findings as well.

The lack of sites enriched in ammonia in the southern hemisphere of Charon could possibly be due to lower signal precision compared to the data covering the northern hemisphere, which had more favorable lighting conditions during the encounter and thus higher signal precision. Therefore no definitive conclusion can be drawn on the presence or absence of ammonia-bearing species in the southern hemisphere of Charon due to signal-to-noise issues.

Dalle Ore et al. (2018) mapped the spatial distribution of ammonia-bearing species not only through the 2.21-μm absorption band but also by means of the ~2.01-μm absorption feature (section 2.2, Fig. 2). While both maps displayed an enhancement of both band depths in the bright crater rays, the discrepancies between the two spatial maps led to the conclusion that different forms of ammonia in water ice exist on the surface of Charon.

4. THE PRESENCE AND DESTRUCTION OF AMMONIA ON CHARON

The presence of NH_3-bearing species on Charon presents both an ambiguity and a conundrum. The ambiguity arises from the broad and somewhat indistinct shape and position of the band at ~2.21 μm and from the lack of a clear evidence for other potential NH_3 absorption bands, such as near 2.0 μm (*Gerakines et al.,* 2005). These factors make it impossible to definitively identify the setting of the ammonia. Possibilities include NH_3 frozen within the H_2O ice on Charon's surface, superficial coatings of NH_3 ice, NH_3 combined with H_2O in one or more hydrated phases, and a wide variety of ammoniated salts. All forms of ammonia can be dissociated by energetic radiation, including UV photons, solar-wind-charged particles, and galactic cosmic rays (*Cruikshank et al.,* 2019b). Charon's airless surface is unprotected from this space radiation. For pure NH_3 and for NH_3 hydrates, at least within the uppermost few micrometers of the surface, photolytic and radiolytic destruction is thought to occur on geologically short timescales. The conundrum is that ammonia remains present on such an old and mature planetary surface. Is it less vulnerable to radiolysis and photolysis than was thought? Or is it continually resupplied by some process, such as upward diffusion from a below-ground reservoir? The unknown identity of the ammonia-bearing species makes it impossible to definitively compute its rate of destruction. In particular, the radiolytic destruction rates of ammoniated salts remain unknown.

A similar situation applies to Pluto, where the spectral signature of ammonia occurs with H_2O ice that is strongly colored by a component attributed to a mixture of complex organics, spectrally similar to the tholins formed in the planet's atmosphere and possibly on its icy surface (*Dalle Ore et al.,* 2019; *Cruikshank et al.,* 2019b). However, the ammoniated and colored H_2O exposures on Pluto occur in regions interpreted by these authors as sites of recent cryovolcanism. The following discussion of the source and fate of ammonia on Charon draws upon a more detailed consideration of the Pluto case in *Dalle Ore et al.* (2019) and *Cruikshank et al.* (2019b). See also *Cruikshank et al.* (2019a) for a discussion of the chemical implications of an ammonia-bearing fluid inside Pluto.

Similar ammonia features are also seen in spectra of the small satellites Nix and Hydra (*Cook et al.,* 2018). Unlike Pluto, these satellites are far too small to invoke ongoing geological activity to resupply ammonia to their surfaces, so they can provide a valuable clue to the behavior of the ammonia-bearing material in Charon's environment.

4.1. Ammonia Hydrates

As an NH_3-H_2O fluid freezes, either in the interior or on the surface, the NH_3 and H_2O form hydrates. There are three known forms of ammonia hydrates (*Kargel,* 1992): the dihydrate $NH_3 \cdot 2H_2O$, the monohydrate $NH_3 \cdot H_2O$, and the hemihydrate $2NH_3 \cdot H_2O$. Formation of ammonia hydrates is observed upon warming of NH_3-H_2O ice mixtures in the laboratory (*Moore et al.,* 2007) and as liquid H_2O freezes in the presence of NH_3 (*Uras and Devlin,* 2000). Also, as NH_3 diffuses through H_2O ice, hydrates form when the concentration of NH_3 exceeds the solubility limit (*Cruikshank et al.,* 2019b).

The prospect of resupplying Charon's NH_3 by upward diffusion from the subsurface is limited by the very low surface temperature (~40 K). *Livingston et al.* (2002) measured the diffusion rate of NH_3 in H_2O at 140 K, finding a diffusion coefficient, D, of ~10^{-10} cm^2 s^{-1}. They caution against simply extrapolating this value of D to much lower temperatures, but it is reasonable to assume that at 40 K the diffusion coefficient will be much, much smaller. Interestingly, NH_3 hydrate can diffuse through H_2O ice more rapidly than the free NH_3 molecule (*Livingston et al.,* 2002).

Two related factors may affect resupply of NH_3 to Charon's surface. If the crust is significantly warmer at depth than the ~40 K surface temperature, diffusion of NH_3 deeper below the surface would be correspondingly more rapid. In addition, if the subsurface is porous, the upward diffusion could be enhanced, perhaps substantially (*Cruikshank et al.,* 2019b). The highest observed concentrations of NH_3 are at Organa crater, both in the crater itself and in the surrounding ray pattern, as well as in the bright ejecta around craters such as Nasreddin, Skywalker, and Candide (see section 3.3) (*Grundy et al.,* 2016a; *Dalle Ore et al.,* 2018). In addition to directly excavating material from the subsurface, the impact events that formed these craters would have warmed the local crust and fragmented it, enhancing the diffusion of NH_3 from the subsurface. This scenario is supported by the large water-ice path lengths in the bright ejecta (section 3.1). In contrast, a crater within Dorothy, which has similar morphology to Organa and probably a comparable age, shows no ammonia enhancement. *Grundy et al.* (2016a) proposed two alternative explanations. The first was that NH_3-rich material could be delivered by a subset of impactors, perhaps originating in different Kuiper belt subpopulations. The second involved regional differences in Charon's subsurface composition, so that some impacts excavate into localized deposits enriched in NH_3-bearing species while others do not.

4.2. Ammonia Salts

Ammonia salts may occur on the surface of Charon and some of the small Pluto satellites (*Cook et al.,* 2018). Such salts could have originated as components of the mantle of one of the progenitor bodies involved in the giant impact that formed the satellite system or as cryolavas that were emplaced onto Charon's surface after it formed, although that would not account for their presence on the small satellites with no cryovolcanic history. *Moore et al.* (2007) proposed the NH_4^+ ion as a possible source of the 2.21-μm feature on Charon since this ion is readily formed by irradiating $H_2O + NH_3$ ices. Reflectance spectra of ammonium salts [e.g., $(NH_4)_2CO_3$, $(NH_4)_3PO_4$, $(NH_4)_2SO_4$, NH_4Cl] show fairly broad absorption bands with the absorption peaks in the range ~2.12–2.2 μm (see Table 5 of *Berg et al.,* 2016). To fit the spectrum of bright regions on Ceres in data from the Dawn space mission, *De Sanctis et al.* (2016) included ammoniated salts NH_4Cl or NH_4HCO_3 in their models. In addition, *Cook et al.* (2018) found that NH_4Cl is a good spectral match to the 2.2-μm absorption band detected on Pluto's satellites Nix and Hydra. The 3.2-μm absorption feature observed on the surface of Comet 67P/Churyumov-Gerasimenko by the Rosetta spacecraft has been attributed to ammonium salts (*Poch et al.,* 2020). Thus, ammoniated salts appear to be common in solar system objects. However, the available spectral data for Charon, Nix, and the other satellites do not extend to the longer wavelengths ($\lambda > 2.5$ μm) where additional diagnostic spectral bands of ammoniated species occur (see Table 5 of *Berg et al.,* 2016). Additionally, complex refractive indices for ammoniated salts in the 1–2.5-μm wavelength region are not yet available, precluding their inclusion in radiative transfer models of the spectra of various regions on Charon.

4.3. Destruction by Ultraviolet Radiation

The dissociation energy for the removal of a hydrogen atom from NH_3 is 10.2 eV (λ = 121.6 nm), which is the energy of hydrogen Ly-α radiation. Charon receives Ly-α directly from the Sun and also from the resonant scattering of Ly-α photons by the interplanetary medium (IPM) (*Gladstone et al.*, 2015, 2018). Accounting for both Ly-α sources, *Bertrand et al.* (2019) estimated an annual mean incident Ly-α flux at Charon of 1.3×10^{12} photons m^{-2} s^{-1}. As noted by *Loeffler and Baragiola* (2010), Ly-α radiation not only destroys ammonia, but is also absorbed by water ice. This implies that water ice might shield the ammonia underneath. The penetration depth of Ly-α photons into water ice is on the order of a few nanometers (see Table 4 of *Bennett et al.*, 2013), while the depth range sampled by IR spectroscopy is on the order of millimeters. At 40 K, the strongest near-IR H_2O ice absorption bands at 1.5 and 2.0 μm bands have peak absorption coefficients of 50 and 100 cm^{-1}, respectively, corresponding to photon penetration depths in ice of about 0.2 and 0.1 mm, respectively, while photons at continuum wavelengths will sample much more deeply (*Jewitt and Luu,* 2004). Therefore, the shallow penetration depth of Ly-α photons is orders of magnitude smaller than the depth sampled by near-IR remote sensing, and thus irradiation by Ly-α photons does not necessarily prevent identification of ammonia on the surface of Charon. In addition to Ly-α, longer-wavelength UV photons (between 122 and 205 nm) penetrate more deeply [a few micrometers (see Table 4 of *Bennett et al.*, 2013)] and, although they have less energy than needed to dissociate NH_3, could potentially affect it over a larger range of depths.

For ammonia hydrates, the rate of destruction is probably similar to that of isolated NH_3 molecules, although the H_2O molecule(s) in a hydrate structure could absorb some of the incoming energy. In an ammoniated salt the ionic bond between the cation and anion is stronger than the covalent bonds holding the NH_3 molecule together, and there are more components over which the incoming energy can be distributed. The net effect should be to make ammoniated salts more resilient to radiolytic or photolytic destruction (*Cruikshank et al.*, 2019b). New experiments that measure the dissociation of ammoniated salts by UV photons and charged particles would greatly improve our understanding of the stability of these species in outer solar system environments.

In a more realistic and impure multi-component ice, particularly where carbon-bearing molecules are present, the destruction of NH_3 is more complex. For example, *Bernstein et al.* (1995) showed that UV irradiation of $H_2O:CH_3OH:CO:NH_3$ ices results in the incorporation of approximately half the available N into complex organic residues after exposures of only $\sim 1 \times 10^{20}$ photons cm^{-2}. The reader is referred also to *Grim and Greenberg* (1987), *Bernstein et al.* (2000), and *Pilling et al.* (2010), in which destruction of ammonia in multi-component ices is discussed. Extrapolating the timescale of the *Bernstein et al.* (1995) experiments to Charon leads to an NH_3 destruction timescale of $\sim 2 \times 10^4$ yr, but only to the few-nanometer depths reached by UV radiation, not the much deeper approximately millimeter depths probed by near-IR reflectance spectroscopy.

Impact gardening could dredge up ammonia-rich subsurface material that had not yet been exposed to UV radiation. Based on lunar impact gardening rates, *Grundy et al.* (2016b) estimated that Charon's surface is gardened to centimeter depths in $\sim 10^7$ yr, and shallower depths more rapidly. If the shallow size distribution found for ~300-m to 1-km impactors by *Singer et al.* (2019) extends to even smaller impactors, then gardening rates are probably much lower in the Pluto system. However, the gardening rate is evidently not so low as to prevent H_2O ice from being well mixed into the tholin material continuously accumulating in Charon's red polar cap, with an estimated production rate on the order of ~8 cm b.y.$^{-1}$ (*Grundy et al.,* 2016b).

4.4. Destruction by Solar-Wind-Charged Particle Irradiation

Pluto produces a bow shock in the plasma ejected by the Sun and the flow around the obstacle picks up CH_4 ions from the planet's atmosphere, carrying them downstream (*Bagenal et al.,* 2016). These energetic heavy ions impact Charon and are expected to induce chemical changes when they interact with the surface materials (*Grundy et al.,* 2016b). *Loeffler et al.* (2010) and other studies have shown that the spectroscopic signature of ammonia is readily destroyed by 100-keV protons in a NH_3-H_2O ice at T > 120 K, although destruction is slower at lower temperatures. For the specific case of ammonia hydrates, Loeffler et al. estimate that about 40% or more of the original ammonia has been removed to depths of a few micrometers, over the age of the solar system by proton impacts, depending on the particle energy distribution. Because CH_4 originating from Pluto is also expected to be present on Charon (*Grundy et al.,* 2016b), the destruction of NH_3 is likely to result in more complex ions and molecules, rather than a simple recombination to restore NH_3 molecules.

4.5. Destruction by Cosmic Rays

A computational model of the cosmic-ray flux relevant to NH_3 destruction on Pluto's surface (and by extension to Charon) was reported by *Cruikshank et al.* (2019b). Considering the expected cosmic-ray flux plus the hadronic interactions of primary, secondary, and higher-order particles that impinging cosmic rays produce in an icy surface dominated by H_2O, the bulk of the energy >10.5 eV is deposited in the uppermost 1 m and amounts to ~10^7 eV g^{-1} s^{-1}. A 1-g mass of NH_3 corresponds to 3.5×10^{22} molecules; dividing by the destruction rate gives ~10^9 yr for the dissociation of every NH_3 molecule in the 1-m surface layer. After destruction, some of the radiolytically destroyed NH_3 molecules in NH_3-H_2O ice can reform, especially in the absence of other chemical contaminants. The energy from cosmic rays is also not distributed uniformly in depth, with more being deposited closer to the surface, and less at depth. These considerations potentially extend the survival of the ammonia signature in the optical layer as long as ~10^9 yr.

4.6. Summary of the Presence of Ammoniated Species on the Surface of Charon

We close this section with a reiteration of the conundrum presented by ammonia-bearing species on Charon. Charon's space environment is thought to destroy the ammonia spectral bands on timescales shorter than the age of the solar system. However, the persistence of the 2.21-μm spectral band on Pluto's small satellites (*Cook et al.*, 2018), which are exposed to the same radiation sources as Charon and are too small to have any plausible geological means of refreshing the absorbing species on their surfaces, would imply that the destruction of NH_3 in this environment is much slower than estimated and raises questions that have not yet been satisfactorily answered (see section 7). Timescales of ammonia destruction are essential to establish the origin of NH_3 on the surface of Charon and by extension that of the inactive satellites Nix and Hydra, but an important factor to account for is also the penetration depth of any radiolytic or photolytic process. Among the processes discussed here (see Table 1), photolytic destruction is the most efficient. However, its penetration depth is very shallow compared to the depth range sampled by IR spectroscopy. Additionally, ammoniated salts, rather than ammonia or ammonia hydrates, could be responsible for the 2.21-μm spectral feature. These compounds may be less vulnerable to radiolytic and photolytic destruction, although this possibility would need to be confirmed by laboratory experiments. Furthermore, if small objects in the Kuiper belt are eroded by micrometeorites at a rate comparable to or faster than the rate at which radiation damage accumulates (e.g., *Stern,* 1986; *Grundy et al.,* 2020), that might account for exposure of fresh NH_3-bearing interior material on the small moons despite their age and lack of geological activity, as well as on Charon.

5. THE PRESENCE AND DESTRUCTION OF CRYSTALLINE WATER ICE ON CHARON

Water condensed slowly at T < ~135 K occurs in the amorphous phase. Upon warming, it transforms first into a cubic crystalline phase (I_c) and ultimately into a hexagonal crystalline phase (I_h) (e.g., *Hobbs,* 2010) (section 2.2.1). After crystallizing, water ice remains crystalline even if the temperature is lowered, but charged particle and UV photon irradiation of crystalline ice can damage the crystal structure, resulting in an amorphous configuration (*Famá et al.,* 2010; *Baragiola et al.,* 2013), as does flash sublimation from impact heating followed by condensation on a cold surface.

Abundant crystalline water ice across Charon (section 3.10 (*Cook et al.,* 2007; *Merlin et al.,* 2010; *Dalle Ore et al.,* 2018) bears witness to one or more episodes of elevated temperature to enable the water ice to crystallize. Such an event could post-date Charon's formation, such as the *Beyer et al.* (2019) hypothesis that the smooth plains in the southern portion of Charon's encounter hemisphere result from Charon's global expansion and cryoflow due to the freezing of the moon's ancient subsurface ocean. Cryovolcanic resurfacing early in Charon's history is supported by the correlation between regions enriched in NH_3 (section 3.3) and large path length of crystalline water ice (section 3.1), suggestive of subsurface material excavated by impactors. But crystallization of water ice could have happened even

TABLE 1. Timescales for processes that destroy NH_3 on Charon.

Process	Estimated Timescale (yr)	Depth Affected (m)
UV photolysis	$\sim 2 \times 10^4$	$\sim$few $\times 10^{-9}$
Solar wind radiolysis	4.5×10^9	$\sim$few $\times 10^{-6}$
Cosmic-ray radiolysis	$\sim 10^9$	~ 1

earlier, too, such as in the aftermath of the giant impact that formed Pluto's satellite system, or even before that impact, in the interiors of the progenitor bodies (see the chapters by Canup et al. and McKinnon et al. in this volume).

Regardless of the nature of the event that led to crystallization of Charon's water ice, it is necessary to explain how its crystallinity survives continued exposure to the space radiation environment (UV photons, solar wind particles, cosmic rays, etc.). *Cook et al.* (2007) estimated the e-folding time for crystalline water ice at a depth of 350 μm to be amorphized by solar UV/visible radiation to be on the order of (3–5) × 10^4 yr, leading them to suggest that Charon had been cryovolcanically resurfaced within the last 10^5 yr. However, such an explanation clearly does not work for Nix and Hydra, where crystalline water ice is also abundant. There is no plausible mechanism to drive recent geological activity on such small bodies, and their surfaces appear to be ancient, with numerous impact craters (see the chapter by Porter et al. in this volume). There also is no geological evidence for recent cryovolcanism on Charon (see the chapter by Spencer et al. in this volume). Micrometeorite impacts on a water-ice surface can result in local annealing to the crystalline phase to a depth comparable to that probed by reflectance spectroscopy (*Porter et al.,* 2010). However, the micrometeorite flux in the present epoch at Charon is very low (*Singer et al.,* 2019). *Zheng et al.* (2009b), based on laboratory experiments and quantitative analysis, suggested that thermal recrystallization reaches equilibrium with irradiation-induced amorphization, even at temperatures as low as 30 K, in less than 1 b.y. This could explain why water ice on Charon is detected in crystalline form. Nevertheless, more laboratory work is needed, and understanding the continuing prevalence of crystalline water ice in the Pluto system is a problem that is ripe for future work.

6. COMPARISON OF CHARON WITH OTHER ICY BODIES IN THE OUTER SOLAR SYSTEM

Two main classes of icy bodies exist in the outer solar system: the icy satellites of giant planets, and the members of the Kuiper belt (and associated populations) derived from planetesimals in the outer reaches of the protoplanetary nebula. Charon has links to both camps, as a Kuiper belt object and as a relatively large icy world in orbit around an even larger one.

Among the Kuiper belt objects, near-IR groundbased spectroscopic and photometric measurements led to the identification of three main compositional groups, with differences linked to size, formation location, and chemical processing. A few of the largest TNOs (D > ~1700 km) tend to have surface volatiles (N_2, CO, and CH_4), slightly smaller TNOs (~500 km < D < ~1700 km) are characterized by strong water-ice absorption bands and, in some cases, ammonia-bearing species, while the vast majority of much smaller TNOs appear to also contain small amounts of water and possibly methanol and silicates, although these are hard to detect owing to their lower albedos. There is not a sharp boundary between volatile-lost and volatile-retained objects, as shown by Makemake, Quaoar, and Gonggong, which are objects that are on the border of losing their volatiles but still contain CH_4 (see reviews by *Barucci et al.*, 2008; *Brown,* 2012).

Models for volatile retention on Kuiper belt objects (*Schaller and Brown,* 2007; *Brown et al.,* 2011) predicted Charon to be fully depleted of volatiles. However, the red coloration of Charon's north pole shows the capability of the satellite to transiently retain small amounts of volatiles, specifically methane, from atmospheric outflow from Pluto (*Grundy et al.,* 2016b) (section 3.2.1). Therefore New Horizons measurements revealed similarities between Charon and Quaoar and Gonggong (*Brown,* 2012), two objects in the transition regime with surfaces dominated by CH_4 irradiation products, as indicated by their red coloration in visible wavelengths, but sufficiently depleted in methane that the near-IR wavelength spectra are dominated by water-ice absorptions bands and, possibly, the strongest CH_4 absorption features (e.g., *Jewitt and Luu,* 2004; *Brown et al.,* 2011).

Red coloration does not necessarily imply the capability of an object to retain volatiles, as evidenced by the existence of numerous small and red "cold classical" Kuiper belt objects (*Tegler and Romanishin,* 2000; *Pike et al.,* 2017). Another example is given by Nix, whose red coloration, associated with an apparent crater, has been attributed to the composition of the impacting body or to the subsurface material of Nix excavated due to the impact (*Weaver et al.*, 2016).

Many of the same radiolytic and photolytic processes occur on the surfaces of both icy satellites and Kuiper belt objects, and low-temperature chemistry and geology can be similar in both populations. But there are important differences as well. The giant planet satellites tend to receive higher doses of charged particle radiation as a result of orbiting within their host planets' magnetospheres, and because they are closer to the Sun, they receive more solar heating and more UV photons than Kuiper belt objects. They tend to be tidally locked to their host planets (as is Charon), which imposes leading and trailing asymmetries in their exposure to magnetospheric charged particles as well as impactors. Tides can also provide an important source of internal heating for icy satellites, one that is no longer relevant for Charon, especially where multiple satellites occupy resonant configurations.

Nevertheless, Charon, Kuiper belt objects, and icy satellites share important surface compositional properties. The spectra of satellites of Saturn [Enceladus, Tethys, Dione, Rhea, and Iapetus (*Grundy et al.,* 1999)] and Uranus [Miranda, Ariel, Umbriel, Titania, and Oberon (*Grundy et al.,* 1999, 2006; *Bauer et al.,* 2002)] display — similarly to Pluto (*Protopapa et al.,* 2017; *Cook et al.,* 2019), Charon (e.g., *Brown and Calvin,* 2000; *Grundy et al.,* 2016a) (sections 2.2.1 and 3.1), Nix and Hydra (*Cook et al.,* 2018), and other Kuiper belt objects observed with sufficient signal precision [Quaoar, Orcus, and Haumea (e.g., *Jewitt and Luu,* 2004; *Carry et al.,* 2011, *Dumas et al.,* 2011)] — a sharp and peaked 1.65-μm absorption feature, diagnostic of water ice in the crystalline form. This indicates that crystalline water ice at approximately millimeter depths is predominant in the outer solar system (*Grundy et al.,* 1999, 2006).

As pointed out by *Cook et al.* (2018, see their Fig. 19), the 2.21-μm absorption band associated to ammonia-bearing species (section 4) has been identified on several objects in the outer solar system, including Pluto's smaller satellites Nix and Hydra (*Cook et al.,* 2018), the TNO Orcus (*Carry et al.,* 2011), the saturnian satellite Tethys (*Verbiscer et al.,* 2008), and the uranian satellites Miranda and Ariel (*Bauer et al.,* 2002; *Cartwright et al.,* 2015, 2018).

Various authors have speculated that the presence of ammonia-bearing species coupled with the abundance of crystalline water ice is due to past or ongoing cryovolcanism (e.g., *Brown and Calvin,* 2000; *Jewitt and Luu,* 2004; *Cook et al.,* 2007). While the geologic units of Charon, Ariel, and Umbriel might support this interpretation (e.g., *Croft and Soderblom,* 1991; *Beyer et al.,* 2019), the presence of ammoniated species on objects like Nix and Hydra, far too small to invoke ongoing geological activity, opens the door to alternative explanations (see sections 4 and 5).

Arrokoth is the only small Kuiper belt object visited by spacecraft and therefore merits comparison. The apparent differences between the surface composition of Charon and Arrokoth (*Stern et al.,* 2019; *Grundy et al.,* 2020), plus the stark difference in their bulk densities, raises the questions of how similar the volatile inventories were in the regions of the solar nebula in which they formed. Contrary to Charon, the surface of Arrokoth is very red, and has diagnostic absorption bands of CH_3OH ice. Prolonged irradiation is expected to destroy the light hydrocarbons and CH_3OH in favor of macromolecular carbon, which is dark and neutral (*Cruikshank et al.,* 1998). Therefore, as pointed out by *Brown* (2012), the presence of CH_3OH on the surface of KBOs would be possible only if it is recently exposed as the result of collisions. Given that Charon's surface displays evidence of exposed fresh subsurface material, the lack of any visible CH_3OH feature suggests formation in a region depleted of CH_3OH or else early elimination of CH_3OH through chemical processes in Charon's interior.

7. OPEN QUESTIONS AND FUTURE PROSPECTS

This chapter is intended to provide the reader with an overview of our knowledge of Charon's surface composition as deduced from ground- and spacebased observations. It outlines how the observing techniques, from observations of a total eclipse of the satellite to the New Horizons mission, evolved to make new discoveries. Specifically, the past three decades have seen the spectral measurements of Charon's composition progress from four discrete wavelengths to spectral cubes that have helped reveal a suite of different mechanisms at play on the surface of Pluto's largest satellite. Although the synergy of ground- and spacebased measurements, laboratory measurements, and modeling techniques has greatly improved our knowledge of Charon and the Kuiper belt at large, there are still important questions to be answered by future research:

- What is the nature of the 2.2-μm absorption? Several options are outlined in section 4. High signal-to-noise ratio spectroscopic measurements of Charon, beyond 2.5 μm, could help answer this question. Spectral observations of Charon only, without the contribution from Pluto, can be obtained with current and future facilities, such as the LMIRCam instrument onboard the Large Binocular Telescope Interferometer (LBTI) and NIRSpec onboard the James Webb Space Telescope (JWST), covering the wavelength range 0.6–5.3 μm (*Métayer et al.,* 2019). This is critical to set some limits on the origin and evolution of Charon and all TNOs with water-ice-plus-ammonia-rich surfaces. It is also essential that the relevant complex refractive indices of the possible compounds (e.g., ammoniated salts, ammonia hydrates) are determined in the laboratory.
- What is the composition of the dark agent responsible for the red color of Charon's northern polar region? While a consensus has emerged around the concept of macromolecular organic material analogous to laboratory tholins being responsible for Charon's red polar cap (*Grundy et al.,* 2016b), the type of tholin is still the subject of investigation. Because tholin-like materials present diagnostic absorption bands beyond 2.5 μm (e.g., *Imanaka et al.,* 2012; *Brassé et al.,* 2015), long-wavelength spectroscopy of Charon only (e.g., JWST/NIRSpec) might reveal the composition of the reddening agent on Charon.
- How many different kinds of darkening agents are present on Charon's surface? The presence of two distinct darkening agents, corresponding to the gray and red visible coloration at low latitudes and in the northern polar region,

respectively, was identified (section 3.2). The analysis of combined MVIC and LEISA spectra of Charon, using a multiple-scattering radiative transfer model, might help provide additional constraints. Optical constants of tholin-like materials from the visible through the near-IR are needed. Most of the studies characterize the refractive indices in the visible wavelength range (e.g., *Ramirez et al.*, 2002; *Imanaka et al.*, 2004; *Mahjoub et al.*, 2012; *Sciamma-O'Brien et al.*, 2012) or in the mid-IR range (*Imanaka et al.*, 2012). However, few publications report the determination of the complex refractive indices of tholin-like materials in the wavelength range of 0.4–2.5 μm.

- Is Vulcan Planitia depleted of exposed ammoniated-enriched regions or did the illumination geometry of the Pluto system encounter bias our current results? Investigating this topic would help provide additional constraints on the broad-scale cryovolcanic resurfacing event suggested by *Beyer et al.* (2019).
- What are the effects of irradiation on water ice, ammoniated species, and a mixture of the two at Charon's conditions? Is ice amorphization complete, and if not, what kind of equilibrium is established? How does this process vary with time and depth? Laboratory experiments are essential to address the continuing prevalence of crystalline water ice and ammoniated species in the Pluto system.

A spacecraft to orbit the Pluto system, equipped with a spatially resolved IR spectrometer extending the wavelength coverage beyond 2.5 μm (up to 5 μm) is the immediate next step to further explore the Pluto system and address many of the unanswered questions, including some of those listed above.

REFERENCES

Bagenal F. and 157 colleagues (2016) Pluto's interaction with its space environment: Solar wind, energetic particles, and dust. *Science, 351,* aad9045.

Baragiola R. A. and 6 colleagues (2013) Radiation effects in water ice in the outer solar system. In *The Science of Solar System Ices* (M. S. Gudipati and J. Castillo-Rogez, eds.), pp. 527–549. Astrophysics and Space Science Library, Vol. 356, Springer, New York.

Barucci M. A., Brown M. E., Emery J. P., and Merlin F. (2008) Composition and surface properties of transneptunian objects and Centaurs. In *The Solar System Beyond Neptune* (M. A. Barucci et al., eds.), pp. 143–160. Univ. of Arizona, Tucson.

Bauer J. M. and 7 colleagues (2002) The near infrared spectrum of Miranda: Evidence of crystalline water ice. *Icarus, 158,* 178–190.

Bennett C. J., Pirim C., and Orlando T. M. (2013) Space-weathering of solar system bodies: A laboratory perspective. *Chem. Rev., 113,* 9086–9150.

Berg B. L. and 8 colleagues (2016) Reflectance spectroscopy (0.35–8 μm) of ammonium-bearing minerals and qualitative comparison to Ceres-like asteroids. *Icarus, 265,* 218–237.

Bernstein M. P., Sandford S. A., Allamandola L. J., Chang S., and Scharberg M. A. (1995) Organic compounds produced by photolysis of realistic interstellar and cometary ice analogs containing methanol. *Astrophys. J., 454,* 327–344.

Bernstein M. P., Sandford S. A., and Allamandola L. J. (2000) H, C, N, and O isotopic substitution studies of the 2165 wavenumber (4.62 micron) "XCN" feature produced by ultraviolet photolysis of mixed molecular ices. *Astrophys. J., 542,* 894–897.

Bertrand T. and 16 colleagues (2019) The CH_4 cycles on Pluto over seasonal and astronomical timescales. *Icarus, 329,* 148–165.

Beyer R. A. and 17 colleagues (2019) The nature and origin of Charon's smooth plains. *Icarus, 323,* 16–32.

Binzel R. P. (1988) Hemispherical color differences on Pluto and Charon. *Science, 241,* 1070–1072.

Brassé C., Muñoz O., Coll P., and Raulin F. (2015) Optical constants of Titan aerosols and their tholins analogs: Experimental results and modeling/observational data. *Planet. Space Sci., 109,* 159–174.

Brown M. E. (2012) The compositions of Kuiper belt objects. *Annu. Rev. Earth Planet. Sci., 40,* 467–494.

Brown M. E. and Calvin W. M. (2000) Evidence for crystalline water and ammonia ices on Pluto's satellite Charon. *Science, 287,* 107–109.

Brown M. E., Burgasser A. J., and Fraser W. C. (2011) The surface composition of large Kuiper belt object 2007 OR10. *Astrophys. J. Lett., 738,* L26.

Buie M. W. and Grundy W. M. (2000) The distribution and physical state of H_2O on Charon. *Icarus, 148,* 324–339.

Buie M. W., Cruikshank D. P., Lebofsky L. A., and Tedesco E. F. (1987) Water frost on Charon. *Nature, 329,* 522–523.

Buie M. W., Tholen D. J., and Horne K. (1992) Albedo maps of Pluto and Charon: Initial mutual event results. *Icarus, 97,* 211–227.

Buie M. W., Tholen D. J., and Wasserman L. H. (1997) Separate lightcurves of Pluto and Charon. *Icarus, 125,* 233–244.

Buratti B. J. and 16 colleagues (2017) Global albedos of Pluto and Charon from LORRI New Horizons observations. *Icarus, 287,* 207–217.

Canup R. M. (2005) A giant impact origin of Pluto-Charon. *Science, 307,* 546–550.

Canup R. M. (2011) On a giant impact origin of Charon, Nix, and Hydra. *Astron. J., 141,* 35.

Carli C., Ciarniello M., Capaccioni F., Serventi G., and Sgavetti M. (2014) Spectral variability of plagioclase-mafic mixtures (2): Investigation of the optical constant and retrieved mineral abundance dependence on particle size distribution. *Icarus, 235,* 207–219.

Carry B. and 10 colleagues (2011) Integral-field spectroscopy of (90482) Orcus-Vanth. *Astron. Astrophys., 534,* A115.

Cartwright R. J., Emery J. P., Rivkin A. S., Trilling D. E., and Pinilla-Alonso N. (2015) Distribution of CO_2 ice on the large moons of Uranus and evidence for compositional stratification of their near-surfaces. *Icarus, 257,* 428–456.

Cartwright R. J., Emery J. P., Pinilla-Alonso N., Lucas M. P., Rivkin A. S., and Trilling D. E. (2018) Red material on the large moons of Uranus: Dust from the irregular satellites? *Icarus, 314,* 210–231.

Cheng A. F. and 15 colleagues (2008) Long-Range Reconnaissance Imager on New Horizons. *Space Sci. Rev., 140,* 189–215.

Christy J. W. and Harrington R. S. (1978) The satellite of Pluto. *Astron. J., 83,* 1005–1008.

Clark R. N. and 25 colleagues (2005) Compositional maps of Saturn's moon Phoebe from imaging spectroscopy. *Nature, 435,* 66–69.

Clark R. N., Carlson R., Grundy W., and Noll K. (2013) Observed ices in the solar system. In *The Science of Solar System Ices* (M. S. Gudipati and J. Castillo-Rogez, eds.), pp. 3–46. Astrophysics and Space Science Library, Vol. 356, Springer, New York.

Cook J. C., Desch S. J., Roush T. L., Trujillo C. A., and Geballe T. R. (2007) Near-infrared spectroscopy of Charon: Possible evidence for cryovolcanism on Kuiper belt objects. *Astrophys. J., 663,* 1406–1419.

Cook J. C. and 21 colleagues (2018) Composition of Pluto's small satellites: Analysis of New Horizons spectral images. *Icarus, 315,* 30–45.

Cook J. C. and 27 colleagues (2019) The distribution of H_2O, CH_3OH, and hydrocarbon-ices on Pluto: Analysis of New Horizons spectral images. *Icarus, 331,* 148–169.

Croft S. K. and Soderblom L. A. (1991) Geology of uranian satellites. In *Uranus* (J. T. Bergstralh et al., eds.), pp. 561–628. Univ. of Arizona, Tucson.

Cruikshank D. P. and 14 colleagues (1998) The composition of Centaur 5145 Pholus. *Icarus, 135,* 389–407.

Cruikshank D. P., Imanaka H., and Dalle Ore C. M. (2005) Tholins as coloring agents on outer solar system bodies. *Adv. Space Res., 36,* 178–183.

Cruikshank D. P. and 32 colleagues (2016) Pluto and Charon: The non-ice surface component. *Lunar and Planetary Science XLVII,* Abstract #1700. Lunar and Planetary Institute, Houston.

Cruikshank D. P. and 18 colleagues (2019a) Prebiotic chemistry of Pluto. *Astrobiology, 19,* 831–848.

Cruikshank D. P. and 28 colleagues (2019b) Recent cryovolcanism in Virgil Fossae on Pluto. *Icarus, 330,* 155–168.

Dalle Ore C. M. and 11 colleagues (2018) Ices on Charon: Distribution of H_2O and NH_3 from New Horizons LEISA observations. *Icarus, 300,* 21–32.

Dalle Ore C. M. and 17 colleagues (2019) Detection of ammonia on Pluto's surface in a region of geologically recent tectonism. *Sci. Adv., 5,* eaav5731.

De Bergh C., Delsanti A., Tozzi G. P., Dotto E., Doressoundiram A., and Barucci M. A. (2005) The surface of the transneptunian object 90482 Orcus. *Astron. Astrophys., 437,* 1115–1120.

De Bergh C., Schmitt B., Moroz L. V., Quirico E., and Cruikshank D. P. (2008) Laboratory data on ices, refractory carbonaceous materials, and minerals relevant to transneptunian objects and Centaurs. In *The Solar System Beyond Neptune* (M. A. Barucci et al., eds.), pp. 483–506. Univ. of Arizona, Tucson.

DeMeo F. E. and 6 colleagues (2015) Spectral variability of Charon's 2.21-μm feature. *Icarus, 246,* 213–219.

De Sanctis M. C. and 27 colleagues (2016) Bright carbonate deposits as evidence of aqueous alteration on (1) Ceres. *Nature, 536,* 54–57.

Douté S. and Schmitt B. (1998) A multilayer bidirectional reflectance model for the analysis of planetary surface hyperspectral images at visible and near-infrared wavelengths. *J. Geophys. Res., 103,* 31367–31390.

Dumas C., Terrile R. J., Brown R. H., Schneider G., and Smith B. A. (2001) Hubble Space Telescope NICMOS spectroscopy of Charon's leading and trailing hemispheres. *Astron. J., 121,* 1163–1170.

Dumas C., Carry B., Hestroffer D., and Merlin F. (2011) High-contrast observations of (136108) Haumea: A crystalline water-ice multiple system. *Astron. Astrophys., 528,* A105.

Famá M., Loeffler M. J., Raut U., and Baragiola R. A. (2010) Radiation-induced amorphization of crystalline ice. *Icarus, 207,* 314–319.

Fink U. and Disanti M. A. (1988) The separate spectra of Pluto and its satellite Charon. *Astron. J., 95,* 229–236.

Fink U. and Larson H. P. (1975) Temperature dependence of the water-ice spectrum between 1 and 4 microns: Application to Europa, Ganymede and Saturn's rings. *Icarus, 24,* 411–420.

Fink U. and Sill G. T. (1982) The infrared spectral properties of frozen volatiles. In *Comets* (L. L. Wilkening, ed.), pp. 164–202. Univ. of Arizona, Tucson.

Gerakines P. A., Bray J. J., Davis A., and Richey C. R. (2005) The strengths of near-infrared absorption features relevant to interstellar and planetary ices. *Astrophys. J., 620,* 1140–1150.

Gladstone G. R., Pryor W. R., and Stern S. A. (2015) Ly-α@Pluto. *Icarus, 246,* 279–284.

Gladstone G. R. and 159 colleagues (2016) The atmosphere of Pluto as observed by New Horizons. *Science, 351,* aad8866.

Gladstone G. R. and 23 colleagues (2018) The Lyman-α sky background as observed by New Horizons. *Geophys. Res. Lett., 45,* 8022–8028.

Grim R. J. A. and Greenberg J. M. (1987) Ions in grain mantles: The 4.62 micron absorption by OCN^- in W33A. *Astrophys. J. Lett., 321,* L91–L96.

Grundy W. M. and Schmitt B. (1998) The temperature-dependent near-infrared absorption spectrum of hexagonal H_2O ice. *J. Geophys. Res., 103,* 25809–25822.

Grundy W. M., Schmitt B., and Quirico E. (1993) The temperature-dependent spectra of α and β nitrogen ice with application to Triton. *Icarus, 105,* 254–258.

Grundy W. M., Buie M. W., Stansberry J. A., Spencer J. R., and Schmitt B. (1999) Near-infrared spectra of icy outer solar system surfaces: Remote determination of H_2O ice temperatures. *Icarus, 142,* 536–549.

Grundy W. M., Buie M. W., and Spencer J. R. (2002) Spectroscopy of Pluto and Triton at 3–4 microns: Possible evidence for wide distribution of nonvolatile solids. *Astron. J., 124,* 2273–2278.

Grundy W. M., Young L. A., Spencer J. R., Johnson R. E., Young E. F., and Buie M. W. (2006) Distributions of H_2O and CO_2 ices on Ariel, Umbriel, Titania, and Oberon from IRTF/SpeX observations. *Icarus, 184,* 543–555.

Grundy W. M. and 33 colleagues (2016a) Surface compositions across Pluto and Charon. *Science, 351,* aad9189.

Grundy W. M. and 122 colleagues (2016b) The formation of Charon's red poles from seasonally cold-trapped volatiles. *Nature, 539,* 65–68.

Grundy W. M. and 48 colleagues (2020) Color, composition, and thermal environment of Kuiper belt object (486958) Arrokoth. *Science, 367,* eaay3705.

Hansen G. B. and McCord T. B. (2004) Amorphous and crystalline ice on the Galilean satellites: A balance between thermal and radiolytic processes. *J. Geophys. Res., 109,* E01012.

Hapke B. (2002) Bidirectional reflectance spectroscopy: 5. The coherent backscatter opposition effect and anisotropic scattering. *Icarus, 157,* 523–534.

Hapke B. (2012) *Theory of Reflectance and Emittance Spectroscopy, 2nd edition.* Cambridge, Cambridge. 513 pp.

Henyey L. G. and Greenstein J. L. (1941) Diffuse radiation in the galaxy. *Astrophys. J., 93,* 70–83.

Hobbs P. V. (1974) *Ice Physics.* Claredon, Oxford. 837 pp.

Hobbs P. V. (2010) *Ice Physics.* Oxford, New York. 856 pp.

Hoey W. A., Yeoh S. K., Trafton L. M., Goldstein D. B., and Varghese P. L. (2017) Rarefied gas dynamic simulation of transfer and escape in the Pluto-Charon system. *Icarus, 287,* 87–102.

Holler B. J. and 6 colleagues (2017) Measuring temperature and ammonia hydrate ice on Charon in 2015 from Keck/OSIRIS spectra. *Icarus, 284,* 394–406.

Imanaka H. and 8 colleagues (2004) Laboratory experiments of Titan tholin formed in cold plasma at various pressures: Implications for nitrogen-containing polycyclic aromatic compounds in Titan haze. *Icarus, 168,* 344–366.

Imanaka H., Cruikshank D. P., Khare B. N., and McKay C. P. (2012) Optical constants of Titan tholins at mid-infrared wavelengths (2.5–25 μm) and the possible chemical nature of Titan's haze particles. *Icarus, 218,* 247–261.

Jenniskens P., Blake D. F., and Kouchi A. (1998) Amorphous water ice. In *Solar System Ices* (B. Schmitt et al., eds.), pp. 139–155. Springer, Dordrecht.

Jewitt D. C. and Luu J. (2004) Crystalline water ice on the Kuiper belt object (50000) Quaoar. *Nature, 432,* 731–733.

Johnson P. E., Smith M. O., and Adams J. B. (1985) Quantitative analysis of planetary reflectance spectra with principal components analysis. *Proc. Lunar Planet. Sci. Conf. 15th,* in *J. Geophys. Res., 90,* C805–C810.

Kargel J. S. (1992) Ammonia-water volcanism on icy satellites: Phase relations at 1 atmosphere. *Icarus, 100,* 556–574.

Khare B. N., Sagan C., Arakawa E. T., Suits F., Callcott T. A., and Williams M. W. (1984) Optical constants of organic tholins produced in a simulated Titanian atmosphere: From soft X-ray to microwave frequencies. *Icarus, 60,* 127–137.

Klassen D. R. and 6 colleagues (1999) Infrared spectral imaging of martian clouds and ices. *Icarus, 138,* 36–48.

Kou L., Labrie D., and Chylek P. (1993) Refractive indices of water and ice in the 0.65- to 2.5-μm spectral range. *Appl. Opt., 32,* 3531–3540.

Kouchi A. and Kuroda T. (1990) Amorphization of cubic ice by ultraviolet irradiation. *Nature, 344,* 134–135.

Kouchi A., Yamamoto T., Kozasa T., Kuroda T., and Greenberg J. M. (1994) Conditions for condensation and preservation of amorphous ice and crystallinity of astrophysical ices. *Astron. Astrophys., 290,* 1009–1018.

Lellouch E., Stansberry J., Emery J., Grundy W., and Cruikshank D. P. (2011) Thermal properties of Pluto's and Charon's surfaces from Spitzer observations. *Icarus, 214,* 701–716.

Leto G. and Baratta G. A. (2003) Ly-alpha photon induced amorphization of Ic water ice at 16 Kelvin. Effects and quantitative comparison with ion irradiation. *Astro. Astrophys., 397,* 7–13.

Livingston F. E., Smith J. A., and George S. M. (2002) General trends for bulk diffusion in ice and surface diffusion on ice. *J. Phys. Chem. A, 106,* 6309–6318.

Loeffler M. J. and Baragiola R. A. (2010) Photolysis of solid NH_3 and NH_3-H_2O mixtures at 193 nm. *J. Chem. Phys., 133,* 214506–214506.

Loeffler M. J., Raut U., and Baragiola R. A. (2010) Radiation chemistry in ammonia-water ices. *J. Chem. Phys., 132,* 054508.

Mahjoub A., Carrasco N., Dahoo P.-R., Gautier T., Szopa C., and Cernogora G. (2012) Influence of methane concentration on the optical indices of Titan's aerosols analogues. *Icarus, 221,* 670–677.

Marcialis R. L., Rieke G. H., and Lebofsky L. A. (1987) The surface composition of Charon: Tentative identification of water ice. *Science, 237,* 1349–1351.

Markwardt C. B. (2009) Non-linear least-squares fitting in IDL with MPFIT. In *Astronomical Data Analysis Software and Systems XVIII* (D. A. Bohlender et al., eds.), pp. 251–254. ASP Conf. Ser. 411, Astronomical Society of the Pacific, San Francisco.

Mastrapa R. M. E. and Brown R. H. (2006) Ion irradiation of crystalline H_2O-ice: Effect on the 1.65-μm band. *Icarus, 183,* 207–214.

Mastrapa R. M., Bernstein M. P., Sandford S. A., Roush T. L., Cruikshank D. P., and Dalle Ore C. M. (2008) Optical constants of amorphous and crystalline H_2O-ice in the near infrared from 1.1 to 2.6 μm. *Icarus, 197,* 307–320.

Mastrapa R. M., Sandford S. A., Roush T. L., Cruikshank D. P., and Dalle Ore C. M. (2009) Optical constants of amorphous and crystalline H_2O-ice: 2.5–22 μm (4000–455 cm^{-1}) optical constants of H_2O-ice. *Astrophys. J., 701,* 1347–1356.

McKinnon W. B. and 28 colleagues (2020) The solar nebula origin of (486958) Arrokoth, a primordial contact binary in the Kuiper belt. *Science, 367,* eaay6620.

Merlin F. and 6 colleagues (2010) Chemical and physical properties of the variegated Pluto and Charon surfaces. *Icarus, 210,* 930–943.

Métayer R. and 6 colleagues (2019) JWST/NIRSpec prospects on transneptunian objects. *Front. Astron. Space Sci., 6,* 8.

Moore M. H., Ferrante R. F., Hudson R. L., and Stone J. N. (2007) Ammonia water ice laboratory studies relevant to outer solar system surfaces. *Icarus, 190,* 260–273.

Nakamura R. and 13 colleagues (2000) Subaru infrared spectroscopy of the Pluto-Charon system. *Publ. Astron. Soc. Japan, 52,* 551–556.

Nimmo F. and 16 colleagues (2017) Mean radius and shape of Pluto and Charon from New Horizons images. *Icarus, 287,* 12–29.

Olkin C. B. and 8 colleagues (2007) Pluto's spectrum from 1.0 to 4.2 μm: Implications for surface properties. *Astron. J., 133,* 420–431.

Pike R. E. and 13 colleagues (2017) Col-OSSOS: z-band photometry reveals three distinct TNO surface types. *Astron. J., 154,* 101.

Pilling S. and 6 colleagues (2010) Radiolysis of ammonia-containing ices by energetic, heavy, and highly charged ions inside dense astrophysical environments. *Astron. Astrophys, 509,* A87.

Poch O. and 31 colleagues (2020) Ammonium salts are a reservoir of nitrogen on a cometary nucleus and possibly on some asteroids. *Science, 367(6483),* eaaw7462.

Porter S. B., Desch S. J., and Cook J. C. (2010) Micrometeorite impact annealing of ice in the outer solar system. *Icarus, 208,* 492–498.

Poulet F., Cuzzi J. N., Cruikshank D. P., Roush T., and Dalle Ore C. M. (2002) Comparison between the Shkuratov and Hapke scattering theories for solid planetary surfaces: Application to the surface composition of two Centaurs. *Icarus, 160,* 313–324.

Protopapa S. and 6 colleagues (2008) Surface characterization of Pluto and Charon by L and M band spectra. *Astron. Astrophys., 490,* 365–375.

Protopapa S. and 9 colleagues (2014) Water ice and dust in the innermost coma of Comet 103P/Hartley 2. *Icarus, 238,* 191–204.

Protopapa S. and 22 colleagues (2017) Pluto's global surface composition through pixel-by-pixel Hapke modeling of New Horizons Ralph/LEISA data. *Icarus, 287,* 218–228.

Protopapa S. and 19 colleagues (2020) Disk-resolved photometric properties of Pluto and the coloring materials across its surface. *Astron. J., 159,* 74.

Quirico E. and Schmitt B. (1997) Near-infrared spectroscopy of simple hydrocarbons and carbon oxides diluted in solid N_2 and as pure ices: Implications for Triton and Pluto. *Icarus, 127,* 354–378.

Rajaram B., Glandorf D. L., Curtis D. B., Tolbert M. A., Toon O. B., and Ockman N. (2001) Temperature-dependent optical constants of water ice in the near infrared: New results and critical review of the available measurements. *Appl. Opt., 40,* 4449–4462.

Ramirez S. I., Coll P., da Silva A., Navarro-González R., Lafait J., and Raulin F. (2002) Complex refractive index of Titan's aerosol analogues in the 200–900 nm domain. *Icarus, 156,* 515–529.

Reuter D. C. and 19 colleagues (2008) Ralph: A visible/infrared imager for the New Horizons Pluto/Kuiper belt mission. *Space Sci. Rev., 140,* 129–154.

Robbins S. J. and 28 colleagues (2019) Geologic landforms and chronostratigraphic history of Charon as revealed by a hemispheric geologic map. *J. Geophys. Res., 124,* 155–174.

Schaller E. L. and Brown M. E. (2007) Volatile loss and retention on Kuiper belt objects. *Astrophys. J. Lett., 659,* L61–L64.

Schenk P. M. and 18 colleagues (2018) Breaking up is hard to do: Global cartography and topography of Pluto's mid-sized icy moon Charon from New Horizons. *Icarus, 315,* 124–145.

Schmitt B., Espinasse S., Grim R. J. A., Greenberg J. M., and Klinger J. (1989) Laboratory studies of cometary ice analogues. In *Physics and Mechanics of Cometary Materials* (J. J. Hunt and T. D. Guyenne, eds.), pp. 65–69. ESA SP-302, Noordwijk, The Netherlands.

Schmitt B., Quirico E., Trotta F., and Grundy W. M. (1998) Optical properties of ices from UV to infrared. In *Solar System Ices* (B. Schmitt et al., eds.), pp. 199–240. Springer, Dordrecht.

Sciamma-O'Brien E. and 6 colleagues (2012) Optical constants from 370 nm to 900 nm of Titan tholins produced in a low pressure RF plasma discharge. *Icarus, 218,* 356–363.

Shkuratov Y., Starukhina L., Hoffmann H., and Arnold G. (1999) A model of spectral albedo of particulate surfaces: Implications for optical properties of the Moon. *Icarus, 137,* 235–246.

Singer K. N. and 25 colleagues (2019) Impact craters on Pluto and Charon indicate a deficit of small Kuiper belt objects. *Science, 363,* 955–959.

Smith M. O., Johnson P. E., and Adams J. B. (1985) Quantitative determination of mineral types and abundances from reflectance spectra using principal components analysis. *Proc. Lunar Planet. Sci. Conf. 15th,* in *J. Geophys. Res., 90,* C797–C804.

Spencer J. R., Buie M. W., and Bjoraker G. L. (1990) Solid methane on Triton and Pluto: 3- to 4-μm spectrophotometry. *Icarus, 88,* 491–496.

Spencer J. R. and 77 colleagues (2020) The geology and geophysics of Kuiper belt object (486958) Arrokoth. *Science, 367,* eaay3999.

Stern S. A. (1986) The effects of mechanical interaction between the interstellar medium and comets. *Icarus, 68,* 276–283.

Stern S. A. and 150 colleagues (2015) The Pluto system: Initial results from its exploration by New Horizons. *Science, 350,* aad1815.

Stern S. A. and 204 colleagues (2019) Initial results from the New Horizons exploration of 2014 MU_{69}, a small Kuiper belt object. *Science, 364,* eaaw9771.

Strazzulla G. and Palumbo M. E. (1998) Evolution of icy surfaces: An experimental approach. *Planet. Space Sci., 46,* 1339–1348.

Strazzulla G., Baratta G. A., Leto G., and Foti G. (1992) Ion-beam-induced amorphization of crystalline water ice. *Europhys. Lett., 18,* 517–522.

Tegler S. C. and Romanishin W. (2000) Extremely red Kuiper-belt objects in near-circular orbits beyond 40 AU. *Nature, 407,* 979–981.

Trujillo C. A., Brown M. E., Barkume K. M., Schaller E. L., and Rabinowitz D. L. (2007) The surface of 2003 EL_{61} in the near-infrared. *Astrophys. J., 655,* 1172–1178.

Tucker O. J., Johnson R. E., and Young L. A. (2015) Gas transfer in the Pluto-Charon system: A Charon atmosphere. *Icarus, 246,* 291–297.

Uras N. and Devlin J. P. (2000) Rate study of ice particle conversion to ammonia hemihydrate: Hydrate crust nucleation and NH_3 diffusion. *J. Phys. Chem. A, 104,* 5770–5777.

Verbiscer A. J. and 6 colleagues (2007) Simultaneous spatially-resolved near-infrared spectra of Pluto and Charon. *Lunar and Planetary Science XXXVIII,* Abstract #2318. Lunar and Planetary Institute, Houston.

Verbiscer A. J., Skrutskie M., Wilson J., Nelson M., and Helfenstein P. (2008) Ammonia hydrate in the saturnian system. *AAS/Division for Planetary Sciences Meeting Abstracts, 40,* 61.02.

Warren S. G. (1984) Optical constants of ice from the ultraviolet to the microwave. *Appl. Opt., 23,* 1206–1225.

Warren S. G. and Brandt R. E. (2008) Optical constants of ice from the ultraviolet to the microwave: A revised compilation. *J. Geophys. Res., 113,* D14220.

Weaver H. A. and 50 colleagues (2016) The small satellites of Pluto as observed by New Horizons. *Science, 351,* aae0030.

Whitney C. A. (1983) Principal components analysis of spectral data. I. Methodology for spectral classification. *Astron. Astrophys. Suppl., 51,* 443–461.

Young L. A. and 25 colleagues (2018) Structure and composition of Pluto's atmosphere from the New Horizons solar ultraviolet occultation. *Icarus, 300,* 174–199.

Zanchet A., Rodríguez-Lazcano Y., Gálvez Ó, Herrero V. J., Escribano R., and Maté B. (2013) Optical constants of NH_3 and NH_3:N_2 amorphous ices in the near-infrared and mid-infrared regions. *Astrophys. J., 777,* 26.

Zheng W., Jewitt D., and Kaiser R. I. (2009a) Infrared spectra of ammonia-water ices. *Astrophys. J. Suppl., 181,* 53–61.

Zheng W., Jewitt D., and Kaiser R. I. (2009b) On the state of water ice on Saturn's moon Titan and implications to icy bodies in the outer solar system. *J. Phys. Chem. A, 113,* 11174–11181.

Porter S. B., Verbiscer A. J., Weaver H. A., Cook J. C., and Grundy W. M. (2021) The small satellites of Pluto. In *The Pluto System After New Horizons* (S. A. Stern, J. M. Moore, W. M. Grundy, L. A. Young, and R. P. Binzel, eds.), pp. 457–472. Univ. of Arizona, Tucson, DOI: 10.2458/azu_uapress_9780816540945-ch020.

The Small Satellites of Pluto

Simon B. Porter
Southwest Research Institute

Anne J. Verbiscer
University of Virginia

H. A. Weaver
The Johns Hopkins University Applied Physics Laboratory

Jason C. Cook
Pinhead Institute

William M. Grundy
Lowell Observatory

Four small satellites reside in nearly circular and co-planar orbits around the Pluto-Charon barycenter. In order of increasing orbital period and distance, they are Styx, Nix, Kerberos, and Hydra. The trajectory of the New Horizons spacecraft was favorable for obtaining multiple observations of Nix and Hydra, providing reasonably precise information on their shapes, sizes, spin states, albedos, and compositions. Styx and Kerberos are smaller and fainter bodies that were discovered well after the launch of New Horizons. These were less favorably placed for detailed observations at the time of the New Horizons flyby, but resolved imaging enabled their basic physical parameters to be measured. All four satellites spin more rapidly than their orbital periods and have spin poles that are unaligned with their orbit poles. Furthermore, their orbits are near to, but not in mean-motion resonance with, the Pluto-Charon orbital period. While their resulting complex dynamical configurations and water-rich compositions provide valuable clues to the origin of the Pluto-Charon satellite system, many open questions remain.

1. INTRODUCTION

Pluto's satellite system is unique in the solar system. While most other large transneptunian objects (TNOs) host a single satellite (*Noll et al.,* 2008), there are only two known triple TNOs: (136108) Haumea [formerly 2003 EL_{61} (*Brown et al.,* 2005)] and (47171) Lempo [formerly 1999 TC_{36} (*Benecchi et al.,* 2010)]. Pluto, on the other hand, boasts one large and four small satellites. This arrangement makes it distinctly different from the pure binary systems such as Earth and the Moon, and large TNOs such as (136199) Eris [formerly 2003 UB_{313} (*Brown et al.,* 2006)], (136472) Makemake [formerly 2005 FY_9 (*Parker et al.,* 2016)], (225088) Gonggong [formerly 2007 OR_{10} (*Kiss et al.,* 2017)], (90482) Orcus [formerly 2004 DW (*Brown et al.,* 2010)], and (50000) Quaoar [formerly 2002 LM_{60} (*Brown and Suer,* 2007)]. The small satellites are likely to be remnants of the material ejected during the giant impact that formed Charon (cf. *Canup,* 2005; *Stern et al.,* 2006). The curious orbital and rotational states of the small satellites place powerful constraints on that formation (cf. *Quillen et al.,* 2017; *Kenyon and Bromley,* 2019).

The chapter in this volume by Canup et al. focuses on the overall formation of the Pluto system, and that of Spencer et al. describes the large satellite Charon. Here we describe the state of knowledge of the four small satellites: Styx, Nix, Kerberos, and Hydra. All four satellites are on near-circular orbits around the Pluto-Charon binary, and those orbits are nearly coplanar with the central binary (*Brozović et al.,* 2015) (see Fig. 1 and Table 1). They range in size

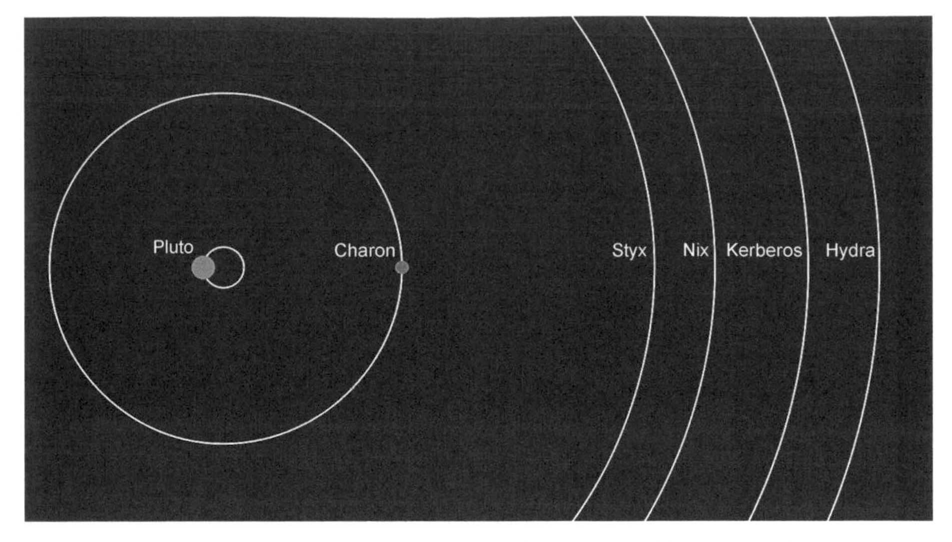

Fig. 1. The orbital configuration of the Pluto system. The sizes of Pluto and Charon are shown to scale. The orbits of the small satellites are roughly circular and coplanar with the mutual orbits of Pluto and Charon around the system barycenter.

from Styx at less than 12 km diameter to Hydra at 52 km on its longest axis (see Fig. 2 and Table 2), and follow a curious "small, big, small, big" arrangement radially from Pluto. All four small satellites have rotation rates much faster than their orbital periods, and all have rotational poles with significant obliquity from their orbital poles (*Weaver et al.,* 2016) (see Table 2). The spectra of Nix and Hydra are dominated by water ice, along with ammonia ice and a neutral darkening agent, producing a very similar bulk surface composition to that of Charon (*Cook et al.,* 2018) but with a higher apparent albedo. The surfaces of Nix and Hydra are cratered, and one large crater on Nix may have either excavated or delivered a distinctly different reddish material (*Robbins et al.,* 2017). There is no evidence for present rings in the Pluto system (*Lauer et al.,* 2018), and the low impact rate in the Pluto system probably precludes the generation of temporary rings or dust structures (*Singer et al.,* 2019). The small satellites of Pluto are a fascinating collection of little worlds that provide a valuable window into the history of the Pluto system.

2. DISCOVERY OF THE SMALL SATELLITES

After Pluto was discovered in 1930 (*Pickering,* 1931), searches for satellites began (e.g., *Kuiper,* 1950; *Tombaugh and Moore,* 1980). While satellites were found shortly after the discovery of both Uranus (*J. Herschel,* 1787; *W. Herschel,* 1834) and Neptune (*Bond,* 1847), Pluto's faintness

TABLE 1. Mean orbits and estimated dynamical masses of the small satellites of Pluto (from *Brozović et al.,* 2015).

	P (days)	a (km)	e	I (°)	Ω (°)	ϖ (°)	GM (km^3 s^{-2})
Styx	20.1617	42413	0.00001	0.0		17.8	0.0000 + 0.0001
Nix	24.8548	48690	0.00000	0.0			0.0030 ± 0.0027
Kerberos	32.1679	57750	0.00000	0.4	313.3		0.0011 ± 0.0006
Hydra	38.2021	64721	0.00554	0.3	122.7	258.0	0.0032 ± 0.0028

P is the mean orbital period, a is the mean semimajor axis, e is the mean eccentricity, I is the inclination relative to Pluto's rotational pole, Ω is the mean longitude of the ascending node, ϖ is the mean longitude of the periapse, and GM is dynamical mass of the satellites associated with this orbit solution. Note that Styx could only be fit with a zero mass in this solution, while Kerberos has a much larger mass than is likely given its volume.

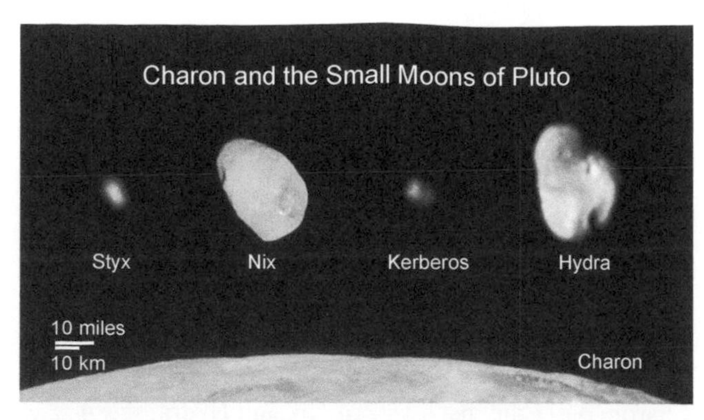

Fig. 2. The relative sizes of the four small satellites of Pluto, along with the limb of Charon for scale.

(V ≈ 13–17) meant that satellite searches were unsuccessful for nearly half a century. In 1978, a series of astrometric observations of Pluto at the U.S. Naval Observatory Flagstaff 61-inch astrograph showed a consistent lump on the side of Pluto, moving that the same rate as the known rotation period of Pluto, which James Christy interpreted as a satellite (*Christy and Harrington,* 1978). Further observations confirmed the discovery, and allowed precoveries (recoveries in archival data) back to 1965 (*Harrington and Christy,* 1980). This improved orbit allowed *Walker* (1980) to obtain a stellar occultation of Charon on April 6, 1980, from South Africa, independently confirming its existence. This occultation improved the orbit enough to enable the first true estimate of the mass of the Pluto system, an upper limit on the size of Pluto, and also the prediction that Pluto and Charon would soon experience mutual events (*Christy and Harrington,* 1980). Astrometry and the mutual events supported the theory that Charon was in circular orbit with a period of 6.3872 days, synchronous with the rotation of Pluto (*Tholen,* 1985; *Binzel,* 1989).

While binaries are now known to be common beyond Neptune (*Noll et al.,* 2008), Charon was discovered before any other TNO, let alone a transneptunian binary (TNB). This discovery led to much debate over the origin of the Pluto system, uniquely strange at the time. *Farinella et al.* (1979) and *Harrington and van Flandern* (1979) suggested that they were two satellites of Neptune that were formed at the same time. *Mignard* (1981) and *Lin* (1981) argued that Pluto could have been ejected from Neptune as a single object that then spun up and ejected a disk of material that formed Charon. While these early studies laid the groundwork for later work on Charon's formation, they assumed that Charon was Pluto's only satellite, analogous to Earth's Moon.

The introduction of charged-coupled detectors (CCDs) in the late 1980s revolutionized the study of the outer solar system, leading to the discovery of the classical Kuiper belt (*Jewitt and Luu,* 1993) in 1992 and the discovery a year later of additional objects in 3:2 mean-motion resonances (MMRs) with Neptune (*Tholen et al.,* 1994), similar to Pluto. The first dedicated CCD search for additional satellites of Pluto was performed by *Stern et al.* (1991) using two 2-m-class telescopes equipped with CCD cameras. They did not detect any additional satellites, but were able to rule out objects brighter than VR = 20.6 (radii 58 ± 19 km; the VR filter had two bandpasses at 500 nm and 675.5 nm, 22.5 nm and 35 nm wide) between 6 and 10 arcsec from Pluto (roughly 7–12× the Pluto-Charon distance), and also ruled out objects brighter than VR = 22.6 (radii 23 ± 8 km) beyond 10 arcsec from Pluto. This observation ruled out any additional Pluto satellites that were both large and distant, but the glare from Pluto and Charon made searches for satellites closer in much more difficult.

Null et al. (1993) observed Pluto and Charon with the Wide Field and Planetary Camera (WF/PC) on the Hubble

TABLE 2. Size and rotation of the small satellites of Pluto.

	Dimensions (km)	Eq. Diameter (km)	Period (days)	Pole RA (°)	Pole Dec (°)	Pluto Ob. (°)
Styx	10.5	10.5 ± 3	3.24 ± 0.07			
Nix	48.4 × 33.8 × 31.4	36.5 ± 0.5	1.829 ± 0.009	349.1	–37.7	124.8
Kerberos	19 × 10 × 9	12 ± 3	5.31 ± 0.10			
Hydra	52.0 × 36.5 × 29.3	36.2 ± 1	0.4295 ± 0.0008	68.9	4.7	64.9

Dimensions are of the best-fit shape model, Eq. Diameter is the diameter of sphere of equal volume, Period is the rotational period from *Weaver et al.* (2016), Pole RA and Pole Dec are the right ascension and declination of the rotational pole at the time of the New Horizons flyby, and Pluto Ob. is the obliquity of the satellite's pole with respect to Pluto's pole (Pluto's pole is RA = 133.0°, Dec = –6.2°). The uncertainties for the equivalent diameters are estimates and representative of the uncertainty in the dimensions.

Space Telescope (HST) in 1991 in an effort to determine the Pluto-Charon mass ratio. While their result (Pluto:Charon mass ratio = 11.9:1) underestimated the mass of Charon [the true mass ratio is 8.20:1 (*Brozović et al.,* 2015)], they showed that there was no additional astrometric signal other than Pluto or Charon, ruling out any additional close large satellites.

Stern et al. (1994) then searched those same WF/PC images for additional close satellites, but did not find any. Outside of 1 arcsec (i.e., outside the orbit of Charon), they found 90% detection efficiency down to a magnitude of 21.7 in the F555W filter (a wide filter from 458 nm to 619 nm). Within 1 arcsec (i.e., inside of the orbit of Charon), their limiting magnitude varied between 18 and 21 mag in F555W. In addition to searching the WF/PC images, *Stern et al.* (1994) also investigated the stability of any additional satellites in the Pluto system. Their simulations showed that small satellites would not be stable at Pluto distances between the 2:3 and 2:1 Charon MMRs, an area they labeled the "Charon instability strip." While these simulations used the incorrect mass ratio from *Null et al.* (1993), the instability of this area was held up by later simulations.

NASA's selection of the New Horizons mission to Pluto in 2001 provided fresh impetus to search for additional satellites around Pluto. This effort was particularly driven by the concern that there would be a ring of small satellites raining dust and debris on Pluto and Charon, material that the New Horizons spacecraft could potentially strike, causing damage. This "hazards search" motivation would drive Pluto satellite detection through to the New Horizons flyby in July 2015. In addition, the discovery of the first TNB other than Pluto, 1998 WW_{31} (*Veillet et al.,* 2001), paved the way for more searches of TNO satellites, in particular using the Wide Field and Planetary Camera 2 (WFPC2) and the Advanced Camera for Surveys High Resolution Channel (ACS/HRC) on HST. The exceptional stability and sensitivity offered by HST allowed high-resolution images to be obtained of very faint objects, leading to a wealth of binary TNO detections. Through searches with HST and groundbased telescopes, by the late 2000s it was clear that a large fraction of TNOs were binary (*Noll et al.,* 2008).

In support of the New Horizons mission, HST observed Pluto in May 2005 with the ACS Wide Field Camera (ACS/WFC) and discovered two satellites (*Weaver et al.,* 2006). These satellites were initially designated "P1" and "P2" and later formally named Hydra and Nix, with the "N" and "H" being a nod to the New Horizons mission. Nix and Hydra had V magnitudes of 23.38 ± 0.17 and 22.93 ± 0.12 respectively at discovery, a couple of magnitudes below the limits from the earlier WF/PC images. By shifting and stacking earlier short-exposure observations of Pluto with ACS/HRC, *Buie et al.* (2006) confirmed this discovery, as well as the initial assumption that both objects were in circular orbits in the plane of the Pluto-Charon orbit. The ACS/HRC images also confirmed that period ratios of Nix and Hydra to Charon were 3.89:1 and 5.98:1, close to but not exactly in MMR with Charon. The ACS/HRC images also provided the first constraints on the color of the small satellites, with Nix appearing slightly redder (larger B-V) than Hydra.

Shortly before the discovery of Nix and Hydra, *Canup* (2005) showed that the Pluto and Charon could have been formed through a giant impact, analogous to the giant impact that created Earth's Moon. Although they preferred a scenario where Charon formed intact from a low-velocity impact, other simulations also produced Charon through accretion of a disk of material. The latter scenario could have also produced small satellites in a disk exterior to Charon. After the discovery of Nix and Hydra, *Stern et al.* (2006) proposed exactly that scenario, based on the fact that their orbits are so circular. Captured objects around Pluto at that distance would not have damped to circular orbits, and would have instead remained eccentric. *Stern et al.* (2006) also proposed that Nix and Hydra were being constantly worn down by impacts of small Kuiper belt object (KBO) debris, producing intermittent dust rings and arcs. These proposed rings (which were based on the erroneous assumption that the Kuiper belt was a collisionally evolved population; see section 8) became the primary motivation for future satellite and ring searches from both HST and New Horizons.

Steffl et al. (2006) performed a deeper search in the Nix/Hydra discovery images for additional satellites and found none. This was a heroic effort, but even today with full knowledge of the positions of Styx and Kerberos, the ACS images are simply not deep enough to detect them. *Tholen et al.* (2008) performed a better orbit analysis of the orbits of Nix and Hydra with the 2006 data, and were able to set the first upper limits on their masses. These upper limits were an order of magnitude larger than later limits, but confirmed that Nix and Hydra were not dark and large objects.

Meanwhile, satellites were being discovered around most of the other dwarf planets, including Haumea (*Brown et al.,* 2005), Eris (*Brown et al.,* 2006), Quaoar (*Brown and Suer,* 2007) and Orcus (*Brown and Suer,* 2007). Most of these satellites were significantly larger than Nix and Hydra, and seem to be smaller results of the same process that created Charon. However, the two satellites of Haumea, Hi'iaka and Namaka, appear to be only a a bit larger than Nix and Hydra, and gravitationally interact with each other (*Ragozzine and Brown,* 2009). Haumea also hosts the only known collisional family in the Kuiper belt (*Brown et al.,* 2007), both implying that Hi'iaka and Namaka were formed in that collision and restricting the circumstances of that impact (*Leinhardt et al.,* 2010). Any future spacecraft missions to the Haumea system would provide a valuable opportunity to compare and contrast the small satellites of Pluto and Haumea.

Pires Dos Santos et al. (2011) performed the first study to find the stable regions between Nix and Hydra. They found that low-inclination, low-eccentricity orbits between Nix and Hydra and just interior of Nix could be stable long term. Later in 2011, Pluto was observed with the then-new Wide Field Camera 3 (WFC3) on HST, the successor to WF/PC, WFPC2, and the ACS/HRC (which is still on the spacecraft but not operational). WFC3 is much more sensitive than its predecessors, and these images enabled *Showalter et al.* (2011) to discover Kerberos (provisionally "P3"), a new satellite between Nix and Hydra. Kerberos is in a circular, low-eccentricity orbit, with a period ratio of 5.03:1 with Charon, in the stable region identified by *Pires Dos Santos et al.* (2011).

Finally, *Showalter et al.* (2012) used additional WFC3 images to detect Styx (provisionally "P4") in June 2012. Styx is the faintest of the small satellites, and the most affected by scattered light from Pluto and Charon. This limitation made detection of Styx difficult but possible with the superior sharpness and sensitivity of WFC3 compared to groundbased cameras. Styx, like the other small satellites, is in a circular orbit roughly coplanar with the Pluto-Charon mutual orbit. Styx's orbital period ratio with Charon is 3.1:1, continuing the trend of small satellites near, but not at, MMR with Charon. In retrospect, both Styx and Kerberos could have been detected earlier if non-linear shift stacking had been applied to the 2005–2006 HST data. However, because both Pluto and Charon are saturated in the deep images, this technique required sufficient knowledge of the background stars in order to co-align the images with subpixel precision. This knowledge of the star's positions was not available until the Gaia Data Release 2 (DR2) in 2018 (*Gaia Collaboration et al.,* 2018).

3. SPACECRAFT EXPLORATION OF THE SMALL SATELLITES

While Styx was the last satellite of Pluto to be discovered (as of this writing), further searches for more satellites continued through the New Horizons flyby in July 2015. In particular, the New Horizons spacecraft conducted an extensive search for new satellites and rings under the aegis of the "New Horizons Hazards Search Campaign" (hereafter referred to as Hazards). This campaign was motivated by the risk that there might be dust and debris in the Pluto system that could strike the spacecraft as it flew past Pluto (roughly through the Pluto-Charon L_3 Lagrange point). The Hazards search consisted of several epochs of deep imaging the Pluto system on approach with the Long Range Reconnaissance Imager (LORRI) camera (*Cheng et al.,* 2008). To maximize sensitivity, the Hazards images used LORRI's "4 × 4" mode, where 16 pixel blocks were binned down to a single pixel on chip, reducing the 1024 × 1024 images to 256 × 256. This mode minimized read noise and reduced the size of the images for downlink.

Also, because New Horizons only uses thrusters for pointing, it could only achieve a pointing stability of 3.5 arcsec during the long (10-s) 4 × 4 exposures. Since LORRI's native pixel resolution is ≈1 arcsec/pixel, any smearing due to thruster firing during pointing was mostly binned over in the 4 × 4 images.

An additional consideration that affected all the Pluto satellite searches, but particularly the New Horizons LORRI images, was the high density of background stars. During the years when searches for satellites became effective (2006–2015), Pluto passed in front of the Milky Way as seen from both Earth and New Horizons. This location made groundbased searches for satellites effectively impossible, and required that HST searches take place over multiple epochs to ensure that any potential detections were co-moving with Pluto and not background stars. The New Horizons LORRI images were the lowest-resolution images used to search for satellites, and thus had to deal with the highest level of star contamination. To address this crowded field, images of the background star field were obtained by LORRI over multiple epochs in the years leading up to the flyby (2013 and 2014). These images could be stacked to produce empirical star background images, which could then be re-projected to the frame of the approach images of Pluto, and then subtracted to produce images that only showed the Pluto system, cosmic rays, and any variable stars. The multiple epochs of the background images enabled the identification of variable stars, which could potentially show up in the subtracted LORRI images as faint sources.

The deep Hazards observations were obtained at six distinct epochs on approach to Pluto, ranging from May 11, 2015, to July 1, 2015. All these epochs consisted of multiple visits (at least two, four at most) that could be compared to each other to look for motion. There were always at least 24 images in each visit, enough to robustly eliminate any cosmic rays. All four of the known small satellites were clearly detectable in all of the Hazards visits, which in turn provided additional astrometric data. Despite extensive searching by multiple members of the New Horizons Hazards team, no new satellites were detected in any of the Hazards images. To test the true sensitivity of the Hazards search, the Hazards images were injected with an array of synthetic satellites, and the same search performed by the same searchers as the real data. As a result of these studies with synthetic injected satellites, the approximate size upper limit on additional satellites exterior of Charon and within the orbit of Hydra (and assuming Hydra-like albedo) is a diameter of 1.7 km (*Stern et al.,* 2015).

Lauer et al. (2018) details the New Horizons search for rings and dust around Pluto. This search was a multi-instrument effort, combining approach LORRI imagery (the Hazards dataset described above), stellar occultations by the Alice UV spectrometer, the Student Dust Counter, and departure imagery from both LORRI and the Multispectral Visible Imaging Camera (MVIC) imaging component of the Ralph near-IR spectrometer (*Reuter et al.,* 2008). The LORRI approach images were stacked to a common frame (accounting for the apparent motion and change of scale as New Horizons approached Pluto), and then azimuthally averaged to produce radial profiles of any backscattering from rings around Pluto. These stacks provided a 3σ upper limit of $I/F \approx 2 \times 10^{-8}$, 2 orders of magnitude lower than the rings of Jupiter (*Showalter et al.,* 1985), and half the nominal I/F of the ring of Saturn's satellite Phoebe (*Verbiscer et al.,* 2009). I/F of 1 indicates that 100% of the sunlight available was reflected back to the spacecraft.

The Alice stellar occultations were less sensitive than this, and the Student Dust Counter only detected a single hit while within the orbit of Hydra, likely an interplanetary dust particle. *Lauer et al.* (2018) also searched the departure images of Pluto obtained by both LORRI and MVIC. While LORRI has a larger aperture than MVIC (20.5 cm vs. 3.75 cm), MVIC has substantially less scattered light at low solar elongation than LORRI. Thus, the forward-scattering upper limits were $I/F < 3 \times 10^{-5}$ for LORRI, and $I/F < 2 \times 10^{-6}$ for MVIC. The MVIC forward-scattering upper limit is less than that of the jovian rings, but not as restrictive as the LORRI backscattering upper limit. In total, no evidence for rings around Pluto was found by the New Horizons spacecraft.

Weaver et al. (2016) describe the detailed LORRI observations of the small satellites of Pluto during the New Horizons flyby. At the time that the New Horizons Pluto flyby sequence was created in 2009, Nix and Hydra were known to exist, but they had very uncertain orbits, and Styx and Kerberos were yet to be discovered. The LORRI high-resolution imaging sequences for Nix and Hydra were thus very conservative, covering a wide area of sky to ensure that the satellites would be observed. As detailed below, these images were used to model the shape and phase behavior of Nix and Hydra and to search its surface for craters. While Styx and Kerberos were unknown when the sequence was created, several observation sequences were allocated as "retargetables" to be changed between 2007 and the flyby in 2015. Two of these retargetable observations were used to obtain resolved images of Kerberos, and one used to obtain the only resolved image of Styx. While most of the high-resolution images are obtained at high solar elongation on approach, one sequence each of departure images for Nix and Hydra was also obtained by LORRI. Nothing is visible in the Hydra departure images, but a thin crescent image of Nix was successful.

4. ORBITAL DYNAMICS OF THE PLUTO SYSTEM

The four small satellites form a closely-packed, near-circular, near-coplanar system in orbit around the Pluto-Charon central binary. Because Charon is so large compared to Pluto, it completely dominates the dynamical perturbations on the small satellites. Satellite-satellite interactions and solar perturbations are much smaller effects that require a much more substantial amount of data to constrain. *Brozović et al.* (2015) provided the best estimate of the orbits of the small satellites prior to the New Horizons flyby, using HST observations over the 2002–2012 time period. With this dataset, they were able to determine the mean orbits of the small satellites, as shown in Table 1. They found that Styx, Nix, and Kerberos have mean eccentricities of effectively zero, while Hydra appears to have a slight eccentricity of 0.00554. Styx and Nix have mean inclinations of effectively zero with respect to the Pluto-Charon binary, while Kerberos and Hydra have mean inclinations of 0.4° and 0.3° respectively. The gradual divergence from circular and coplanar with distance from the central binary may imply that the primordial orbits of the satellites were more excited, and were subsequently damped by tidal interactions (*Stern et al.,* 2006).

Among the most remarkable features of the small satellites of Pluto is the near-commensurability of their mean orbital periods with respect to Charon. The mean orbital periods of Styx, Nix, Kerberos, and Hydra compared to Charon are 3.16:1, 3.89:1, 5.04:1, 5.98:1, respectively (*Brozović et al.,* 2015). None of the four is exactly at a precise MMR with Charon, but all are close. The divergence from the nearest whole integer resonance is 5.22%, –2.72%, 0.726%, and –0.317%, respectively. The small satellites thus become closer their nearest resonance with increasing distance from the central binary. Like the increasing orbital excitation with distance, this relationship may imply that the satellites were originally accreted much closer to the MMR, and subsequent tidal interactions with Charon disturbed them from the resonances. The directionality of the divergence from the resonances is also intriguing: The smaller satellites (Styx and Kerberos) are outside their closest resonance, while the larger satellites (Nix and Hydra) are just inside their closest resonance. Clearly, the orbital structure of the small satellites constrains their formation, but in complicated and unclear ways. The chapter by Canup et al. in this volume explores the formation of the Pluto system in detail, with particular emphasis on the challenging task of understanding the formation of the small satellites.

Brozović et al. (2015) were also able to constrain the masses of Nix, Kerberos, and Hydra to within a factor of 2, while Styx was unconstrained (see Table 1). *Kenyon and Bromley* (2019) used the orbital solutions of *Brozović et al.* (2015) and long-term six-body dynamical simulations to better constrain the masses of the small satellites. They were able to reproduce the Nix and Hydra masses, but found that the system was unstable when they used the mean mass for Kerberos from *Brozović et al.* (2015). Generally, *Kenyon and Bromley* (2019) found that the small satellites must be on the lower end of the mass range from *Brozović et al.* (2015), and thus lower density than Charon. These simulations were however limited by the very short data arc for Styx (2009–2012) and did not incorporate astrometry from either the New Horizons flyby or HST after 2012. The orbit of Styx is the most sensitive to the masses of the small satellites, but also the least constrained. Future work to better constrain the orbit of Styx and the other small satellites is clearly needed to understand their masses and densities (e.g., *Showalter et al.* (2019).

5. ROTATIONAL DYNAMICS OF THE SMALL SATELLITES

The first deep investigation into the rotational states of Pluto's small satellites was performed by *Showalter and Hamilton* (2015). They attempted to fit the very sparse photometry obtained of the small satellites by HST prior to the New Horizons flyby. While they were able to see variation, and correctly predicted that Nix is significantly prolate, they were unable to find any consistent rotational periods in their data. They ascribed the lack of clear periodicity to resonant interactions with the Pluto-Charon binary, and predicted that all the small satellites would be in chaotic rotational states, with periods varying between 10 and 50 days.

As the New Horizons spacecraft approached Pluto, it obtained many sequences of LORRI 4 × 4 images, both the deep Hazards images used to search for new satellites and rings, and shallower optical navigation ("OPNAV") images used to guide the spacecraft to the flyby. While the OPNAV sequences were not as deep as the Hazards visits, Nix and Hydra were visible in all of them from the start of 2015, and Styx and Kerberos in most of the OPNAVs. *Weaver et al.* (2016) built up the photometry from these images to try to fit the rotational periods of the small satellites. Initially, *Weaver et al.* (2016) used rotational periods similar to the orbital periods, as suggested by *Showalter and Hamilton* (2015), but the data fit much more naturally to much shorter periods. The actual rotational periods are shown in Table 2. Styx, Nix, and Kerberos all have periods of just a few days, while Hydra has a period of half a day. Despite being the brightest object, Hydra was the most difficult to fit, as its pole was roughly aligned to the approach vector of New Horizons, producing a lightcurve with only 0.07 mag of variation, in comparison to Nix at 0.2 mag.

While the rotational periods of the small satellites can be determined from unresolved photometry, the poles are much more difficult to constrain. Unlike objects closer to the Sun, which may be viewed from Earth at many different solar phase (Sun-target-observer) angles, TNO objects like Pluto and its satellites can only be seen from Earth at solar phase angles smaller than 1.8°. The poles of the small satellites are thus degenerate with their shape (*Kaasalainen and Torppa,* 2001, and references therein), and without resolved images of the objects, would be very difficult to determine. As described in more detail below, breaking this degeneracy required simultaneously fitting both the shape and rotation of the satellites using the resolved LORRI images. The best estimate of the poles at the time of the New Horizons flyby is shown in Table 2; these values are improved updates from those in *Weaver et al.* (2016). Nix has the best-determined of the poles, as New Horizons viewed the satellite roughly equator-on, allowing for thorough coverage of most of its surface. While multiple resolved epochs of Hydra were obtained by LORRI (including one in which the rotation of satellite can be seen between the start and end of the sequence), the LORRI images only show the northern half of the body, leading to much higher pole uncertainty. Because LORRI obtained only two barely-resolved images of Kerberos and only one of Styx, their poles are not well-constrained. However, the high amplitude of their lightcurves implies that they may have a high obliquity, similar to that of Nix (*Weaver et al.,* 2016).

The small satellite rotations are much faster than assumed in *Showalter and Hamilton* (2015), raising the question as to whether the rotations are primordial, inherited from the post-Charon formation disk of material that formed the small satellites, or if the satellites have since been spun up by other means. The spread of rotational rates from 0.43 days (Hydra) to 5.3 days (Kerberos) is similar to the range of rotational rates in the broader Kuiper belt population (e.g., *Thirouin and Sheppard,* 2019). This spread could imply that both Pluto's small satellites and those of similar-sized KBOs have been spun up by impacts in the same way. However, the paucity of impact craters on both the Pluto small satellites (*Weaver et al.,* 2016) and the isolated KBO (486958) Arrokoth [formerly 2014 MU_{69} (*Stern et al.,* 2019)] implies that impacts are likely not the driver for the small satellites' rotation rates. The only other possible mechanisms are angular momentum transfer from the Pluto-Charon binary (*Quillen et al.,* 2017), or rotations that are primordial to their formation. Either of these methods would require that the small satellites are not very dissipative, and have strong internal structures, unless they were pushed out too quickly for their spins to be tidally damped (*Hastings et al.,* 2016). A method to constrain how dissipative they are would be to monitor their rotational rates and precession over time with HST and other high-precision photometric observations. *Quillen et al.* (2017) showed that the chaotic effect of the central Pluto-Charon binary likely raised the obliquity of Styx and Nix. Their simulations were limited to the three-body Pluto-Charon-satellite system, and they proposed that future six-body simulations may be able to lift the obliquity of Kerberos and Hydra through interactions between the small satellites.

6. PHOTOMETRY AND PHASE CURVES OF THE SMALL SATELLITES

New Horizons LORRI and MVIC obtained images of Nix at solar phase angles α ranging from 8.39° to 158.1° and images of Hydra at phase angles between 15.1° and 162.8°. Since Styx and Kerberos were only discovered after the flyby plan was created, only two image sequences of Kerberos and one of Styx were obtained at different phase angles to the approach images (*Weaver et al.,* 2016).

Adopting the V-band geometric albedo and HV magnitudes reported by *Weaver et al.* (2016) and assuming a phase coefficient of β = 0.04 mag/degree, *Verbiscer et al.* (2018) constructed solar phase curves for Nix and Hydra. For each phase angle on the New Horizons phase curves, any variation in reflectance due to rotation, or lightcurve, is removed to determine an average V-band I/F. *Verbiscer et al.* (2018) modeled the New Horizons phase curves of Nix and Hydra using the *Hapke* (2005) photometric model. The Hapke model is described by parameters including the single-scattering albedo, macroscopic roughness, and single-particle phase function, described by a single-parameter *Henyey and Greenstein* (1941) phase function, and terms that describe the amplitude and angular width of the opposition surge. Not surprisingly, since the geometric albedo of Hydra is higher than that of Nix, Hydra's single-scattering albedo is higher than Nix's. Since the *Verbiscer et al.* (2018) study focused on the phase curves of Nix and Hydra as viewed from New Horizons, and therefore exclude any at phase angles $\alpha < 8°$, the opposition surge amplitude and angular width were not derived and their model used opposition parameters typical for icy satellites of similar size and albedo.

The single-particle phase functions of both Nix and Hydra are remarkably similar and suggest that on average, particles on the surfaces of Nix and Hydra preferentially scatter incident sunlight in the forward direction, unlike most icy satellites in the outer solar system (*Domingue et al.,* 1995; *Ciarniello et al.,* 2011). However, the model applied to these disk-integrated phase curves assumes that the shapes of Nix and Hydra are spherical, and New Horizons resolved images clearly show that they are not spherical objects. The assumption of a spherical shape can lead to an overestimation of single-scattering albedos and more forward-scattering particle phase functions, particularly at high phase angles ($\alpha > 70°$), as *Li et al.* (2003, 2004) found in their analysis of asteroid (433) Eros. Analyses of the disk-integrated phase curve of Rhea, one of Saturn's icy satellites, also concluded that particles on its surface scatter incident sunlight in the forward direction (*Domingue et al.,* 1995; *Ciarniello et al.,* 2011), unlike most bright, icy satellites in the outer solar system.

7. SHAPES AND DENSITIES OF THE SMALL SATELLITES

Determining the shapes and volumes of the small satellites of Pluto is difficult. Unlike main-belt asteroids, which may be viewed from Earth at many different solar phase angles, objects in the Kuiper belt (such as the small satellites) are only ever seen from Earth at a solar phase angle of less than 2°. Because of Pluto's slow orbit around the Sun, the viewing aspect of satellites changes very slowly, and possibly slower than the speed that the poles change due to precession (*Showalter and Hamilton,* 2015; *Quillen et al.,* 2017). This viewing geometry and shape-pole degeneracy makes determination of the shapes and poles of Pluto's small satellites basically impossible right now from lightcurves alone. The resolved images of the small satellites from New Horizons are thus the most useful data to use constrain their poles and shapes. Understanding the volumes of Nix and Hydra is especially important, as their volumes can be combined with estimates of their masses to constrain their densities, and in the process, constrain their origin after the Charon-forming giant impact. Fitting three-dimensional models to high-resolution images is computationally difficult, and we describe our process and results below. The method described here was originally developed for the small satellites of Pluto, and used to produce the rough dimensions published (*Weaver et al.,* 2016). It was then refined and improved to produce the shape model of Arrokoth (formerly 2014 MU_{69}) that was published in *Stern et al.* (2019) and improved in *Spencer et al.* (2020). We then folded the improvements developed for Arrokoth back into the fitting of the small satellites of Pluto and describe the results below.

To simultaneously fit the shape and poles of the small satellites of Pluto, we forward-modeled the New Horizons LORRI high-resolution images. In this context, "forward-modeling" refers to using a shape model to generate synthetic images with the same geometry and resolution as the real images, and using this to constrain the shape model. Forward-modeling the images enables fitting all the pole and shape parameters at the same time, using as much of the information in the images as possible. Rendering a complex shape with a computer's CPU can be very slow, so instead we performed the rendering with hardware-accelerated OpenGL through the PyOpenGL Python interface (*Fletcher and Liebscher*, 2005). This rendering required creating a basic bidirectional reflectance model in OpenGL Shader Language (GLSL), with parameters that could be be changed to fit the images. The GLSL shader provides a snippet of code that the GPU can execute in parallel on the image, significantly speeding up the rendering of the image. Since the color of the small satellites in the visible bands was fairly uniform (with the exception of the red region on Nix), we assumed a uniform albedo and other surface properties for the surface. Because the reflectance model was optimized for the GLSL shader code, the parameters for it were not as physically useful as for the lightcurve fits shown above. We therefore held most of the photometric parameters fixed at those in *Weaver et al.* (2016), with the exception of a pseudo-albedo parameter. The most complex part of the shader code was the ability to have shadows on the object. This feature was particularly important for Hydra, as the shadows visible in the highest-resolution images of Hydra provide the best constraints on its surface topography. We parameterized the shapes using the octantoid formalism of *Kaasalainen and Viikinkoski* (2012); this formalism has many attributes that make it mathematically useful for fitting a shape and pole from a collection of dense lightcurves. However, we did not use most of these features, and instead used octantoids because their zeroth-order shape is precisely a triaxial ellipsoid. We are thus able to fit the poles of Nix and Hydra with triaxial ellipsoids, then add complexity to their shape, then refit their poles, and so forth. By using this formalism, we were able to decouple the number of parameters from the number of vertices; e.g., a simple triaxial ellipsoid could be modeled with only 3 parameters, but 10,000 or more vertices. The most complex shape model was Nix, with 300 shape parameters. This procedure allowed us to create parametric shapes with an arbitrary number of points, so that we could perform initial rough fitting quickly with less detailed meshes and then gradually increase the number of triangles in the mesh as the complexity of the model required. To convert the octantoid parameters to mesh of triangles that OpenGL can interpret, we first generated a spherical mesh of uniform radius, and then used the latitude and longitude of each point on the sphere to calculate the radius at that point using the octantoid formulas. Rather than fit raw data, we generally fit stacks of images acquired at the same time, to maximize the signal-to-noise ratio and minimize cosmic-ray interference. Key exceptions were the highest-resolution images of Hydra, which showed clear rotation over the course of sequence and were reduced in four separate blocks, and the high-phase panchromatic image of Nix, which was acquired in time-delay integration (TDI) mode (see below). All the images were radiometrically corrected before being fit, so that the fitting code did not have to reproduce the radiometric corrections every time it ran. To reproduce the geometry of the images, we read in the World Coordinate System (WCS) and timing information from each image. We then adjusted the field of view and rotation of the images according to the WCS, and the rotation and distance to the object according to the time in the header and the New Horizons SPICE (*Acton et al.,* 2018) kernels, which define the locations of the spacecraft and satellite at any given time. Since the pointing information in the headers is not precise enough, we left the positions of the objects as potentially free parameters; generally these would be fit well enough early in the fitting process, and then fixed them when using a more complex shape. The images were rendered using OpenGL onto a frame buffer, which allowed us to transfer the rendered images from GPU memory to CPU memory. We then took the perfectly-rendered images, and convolved them with the point-spread function (PSF) of the images. Since the PSF of both LORRI and MVIC are larger than a single pixel, this convolution allowed us to directly model the images without having to chance any artifacts that deconvolving the images may have introduced. The rendered and convolved images were subtracted from the real images, and the difference image squared and summed to estimate the χ^2 of that image. Finally, the per-image χ^2 was normalized to a weight based on spatial resolution of the image (i.e., divided by the square of the spacecraft-target distance), and then summed to produce an overall χ^2 estimate. We started with low-order shape models, optimized their χ^2, and then added another order to the octantoid until the addition of orders made no difference to the solutions. Generally, it took a few hundred trials to optimize the shape parameters for each octantoid order. The unresolved images were not used for this analysis, but could provide additional constraints, especially for Styx and Kerberos.

Figure 3 shows the shape model for Nix. Because Nix was both closer to New Horizons than Hydra, and was viewed close to equator-on, the most information (in the form of LORRI and MVIC images) was available to model its shape. The LORRI image sequence covered most of the southern hemisphere and equatorial regions, while the MVIC panchromatic scan constrained the extent of the northern hemisphere. Nix has an elongated shape, with dimensions of 48.4 × 33.8 × 31.4 km, corresponding to axial ratios of a/b = 1.4 and a/c = 1.5. Nix is thus not very flattened, but is elongated. As viewed from its poles, Nix has a roughly trapezoidal profile, with a rectangular profile as viewed from its equator. The best LORRI images align with the long flat facet along Nix's equator, which is also host to the large crater with a reddish ejecta blanket, the only distinct albedo feature detected on any of the small satellites.

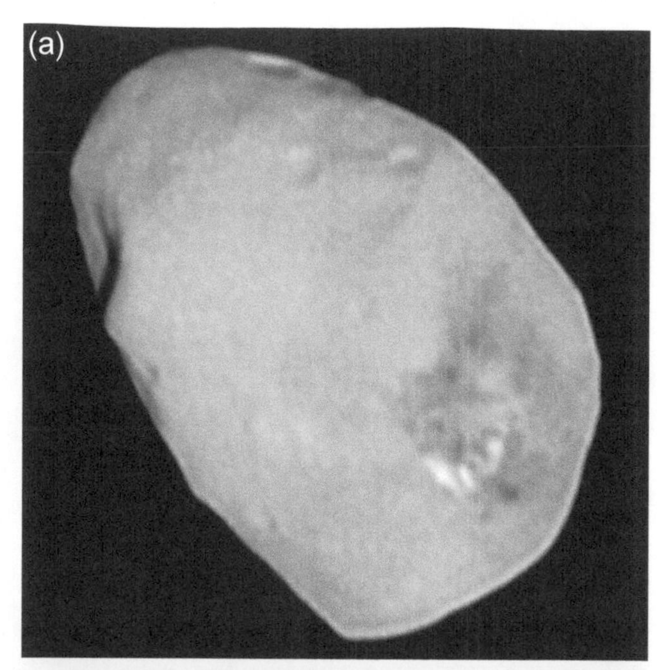

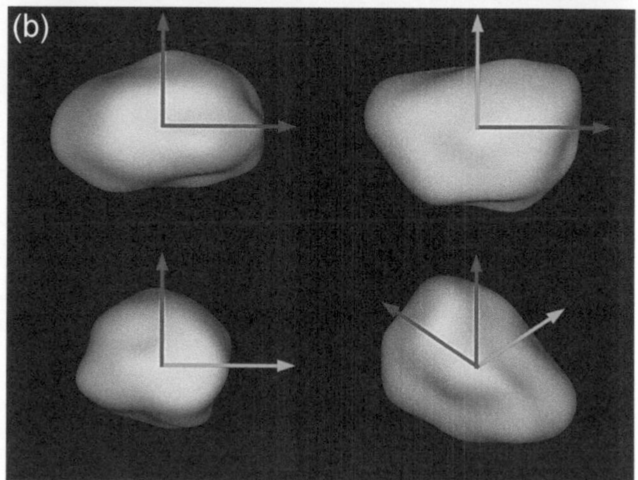

Fig. 3. (a) The highest-resolution LORRI image of Nix, obtained at a resolution of 0.306 km/pixel, stacked and upsampled by a factor of 2 and deconvolved. The solar phase angle for this image is 10°. **(b)** The best-fit shape model of Nix, based on fitting on the best nine images of Nix from New Horizons. Blue points to the rotational pole and red to the longest axis. This model is available at DOI: 10.6084/m9.figshare.12779948.

Figure 4 shows the shape model for Hydra. At the time of the New Horizons flyby, Hydra was father away from the spacecraft and was viewed close to pole-on for most of the approach to Pluto, so spatial coverage of Hydra was much poorer than that of Nix. Effectively only the

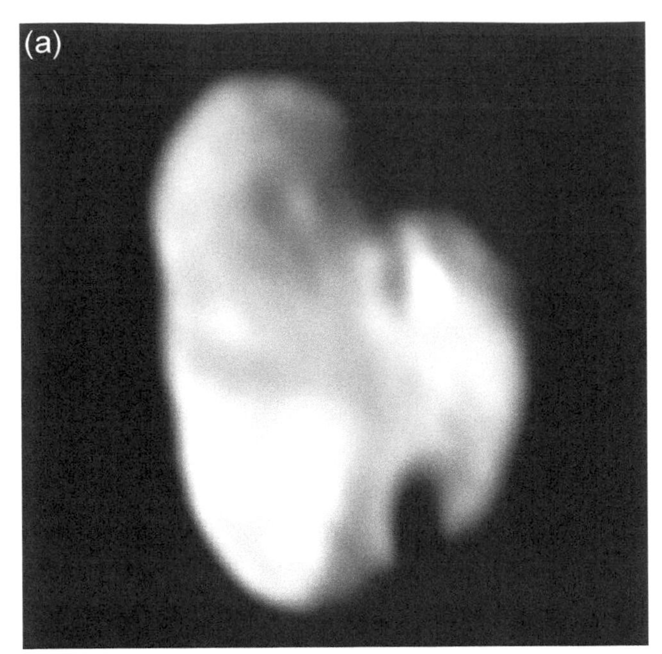

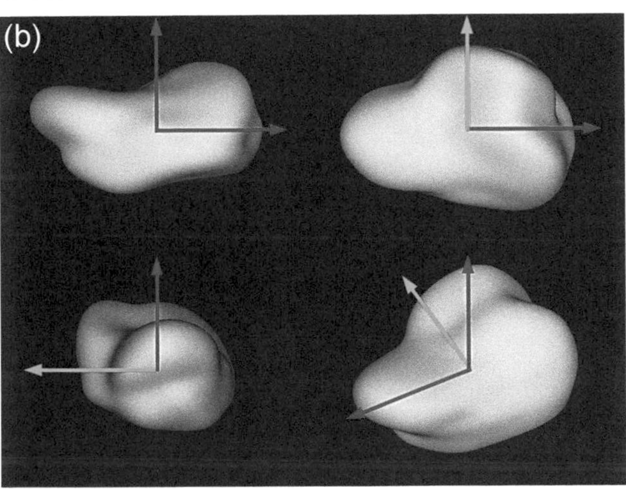

Fig. 4. (a) The highest-resolution LORRI image of Hydra, obtained at a resolution of 1.155 km/pixel, stacked and upsampled by a factor of 2 and deconvolved. The solar phase angle for this image is 34°. **(b)** The best-fit shape model of Hydra, based on fitting on the best six images of Hydra from New Horizons. Blue points to the rotational pole and red to the longest axis. This model is available at DOI: 10.6084/m9.figshare.12779975.

northern hemisphere of Hydra was observed, and most of the shape information about that hemisphere was inferred from the shadowing on its surface. The highest-resolution image of Hydra was obtained as a mosaic of six pointings of LORRI, and Hydra appeared at the overlap of four of those pointings. Because of Hydra's rapid rotation rate, it can be seen to rotate between these pointings, and the shadows do subtly change over the course of the image sequence. The dimensions of the Hydra model in Fig. 4 are 52.0 × 36.5 × 29.3 km, corresponding to axial ratios of a/b = 1.4 and a/c = 1.8. Because of the viewing geometry, the a and b axes are much better constrained than the c axis. Hydra is thus roughly as prolate as Nix, and may be more flattened. There is a large concavity on the north pole of Hydra that produces the shadowing seen in the best LORRI images of Hydra. This concavity gives Hydra the appearance of being almost bilobate. In addition, there is a smaller, valley-like concavity on the larger lobe. These "lumpy" structures are reminiscent of the surface of Arrokoth, the cold classical KBO that New Horizons visited after Pluto. The surface of Arrokoth's larger lobe appears to be formed of many distinct units that were accreted together before the two lobes came into contact (*Stern et al.,* 2019; *McKinnon et al.,* 2020). Hydra's shape may also be the result of a slow accretion process, but with larger pieces. The dichotomy of Nix's relatively smooth shape and Hydra's lumpy shape is one that should drive future investigations into the origin of Pluto's small satellites.

The volume of Nix is roughly equal to a sphere with a diameter of 36.5 km (*Weaver et al.,* 2016). Likewise, the volume of Hydra is roughly equal to a sphere with a diameter of 36.2 km. When combined with the dynamical masses from the JPL PLU055 mass/orbit solution (*Brozović et al.,* 2015), these volumes equate to a bulk density for Nix of 1800 ± 1900 kg m^{-3}, and 1900 ± 2400 km m^{-3} for Hydra, although with the caveat that the masses are very poorly constrained. If these are true densities, then they imply that both Nix and Hydra hide a significant amount of rock beneath their surfaces. If Nix and Hydra do indeed have a large rock fraction, then that conclusion may imply that they may be pieces of undifferentiated crust ejected from the Charon-forming impact (*Stern et al.,* 2006). The spectra of Nix (see below) (*Cook et al.,* 2018) appears to be dominated by water, so a significant rock fraction for Nix would require that the rock was buried underneath ice by some mechanism. However, the masses from PLU055 are upper limits, so the density of Nix may in fact be much lower. Future studies to better determine the masses of the small satellites are clearly required to better constrain their densities.

New Horizons obtained only one resolved image of Styx (Fig. 5). This image appears to show an elongated body, but part of that apparent elongation is due to the asymmetric shape of the LORRI PSF. Because its pole is unconstrained, Styx was fit with a sphere, despite the possible elongation in the one resolved image. Assuming similar albedo and reflectance properties to Nix, the Styx image is well-fit to a sphere with a diameter of 10.5 km. This size would equate to a volume 2.4% that of Nix, thus explaining why Styx's mass is very hard to fit with dynamical models. The combination of New Horizons and HST photometry with the resolved image of Styx may offer a path to better constrain its shape and volume.

New Horizons obtained two resolved images of Kerberos, shown in Fig. 5. As noted by *Weaver et al.* (2016), the closer of the two epochs appears to show a bilobate shape, while the farther epoch is elongated and consistent with a bilobate shape. The best-fit single ellipsoid for Kerberos

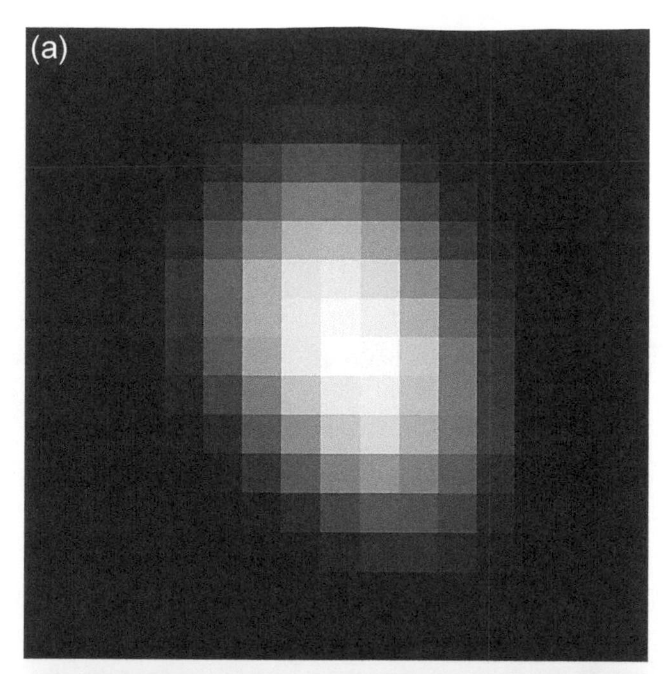

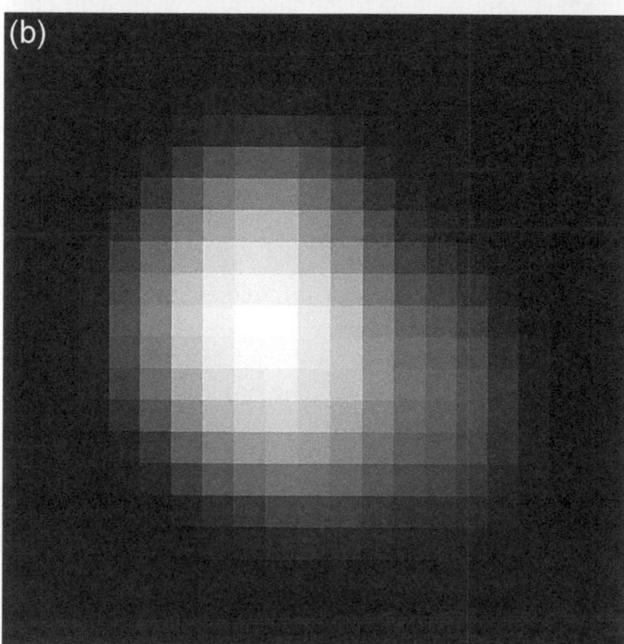

Fig. 5. (a) The highest-resolution LORRI image of Styx, obtained at a resolution of 3.157 km/pixel. **(b)** The highest-resolution LORRI image of Kerberos, obtained at a resolution of 1.982 km/pixel. Both images are stacked and up-sampled by a factor of 2. The solar phase angle for Kerberos is 25° and for Styx is 17°.

has dimensions 19 × 10 × 9 km, with a volume equivalent to a sphere with a diameter of 12 km. Likewise, the best-fit bilobate model (using the contact binary fitting code developed for Arrokoth) has dimensions of 18 × 10 × 8 km, with a volume equivalent to a sphere with a diameter of 11 km. In either case, Kerberos appears to have only slightly more volume than Styx, despite being much brighter in Earth-based imaging. This is consistent with the prediction from *Youdin et al.* (2012) that Kerberos must have a very low mass compared to Nix and Hydra in order to be stable to their perturbations.

8. IMPACTS AND CRATERING ON THE SMALL SATELLITES

When Nix and Hydra were discovered in 2006, it was commonly assumed that the Kuiper belt was a collisionally-evolved system with a shallow size-frequency distribution (SFD) similar to the main asteroid belt (*Bernstein et al.,* 2004). Soon after the discovery of Nix and Hydra, *Stern et al.* (2006) suggested that based on impact rate and ejecta mass considerations, they would be constantly impacted by small Kuiper belt collisional fragments, thus occasionally filling the Pluto system with dust and producing temporary rings and dust structures. *Stern* (2009) proposed that the dust produced by these impacts would transfer between the small satellites and Charon, and cause all the satellites to have a uniform albedo. *Porter and Grundy* (2015) looked at the dynamics of dust produced from impacts on the surfaces of Pluto's small satellites. They found that all the ejecta from the small satellites are dynamically unstable, with most of it being ejected from the system, and the remainder primarily being swept up by Pluto and Charon. They also found ejecta exchange between the small satellites to be ineffective (as the particles had short enough lifetimes and small satellites were small enough targets that no inter-small-satellite occurred in the simulations), while ≈10% of the ejecta impacted Charon and ≈6–16% impacted Pluto, depending on particle size. The surfaces of the small satellites are thus likely not to have been homogenized by impact ejecta exchange, and their mutual spectral similarity (*Cook et al.,* 2018) must be intrinsic to their formation. Knowledge of the Kuiper belt's size frequency distribution, and thus the impactor population on the small satellites, has changed considerably since the discovery of Nix and Hydra in 2006. This change was initially driven by more and better surveys of the Kuiper belt, in particular the Canada-France Ecliptic Plane Survey (CFEPS), which was the first survey that was both wide and deep enough to study the SFD of objects smaller than 100 km (*Petit et al.,* 2011). *Schlichting et al.* (2013) used the stellar occultations of *Schlichting et al.* (2012) to argue that the Kuiper belt was collisionally evolved like the main asteroid belt, and followed a single power-law SFD down to small sizes. The chief evidence for this conclusion at the time were a few brief events observed by the Fine Guidance Sensor (FGS) on HST that were interpreted as stellar occultations by small (less than a kilometer) collisional fragments in the Kuiper belt (*Schlichting et al.,* 2012). In contrast, *Fraser et al.* (2014) used CFEPS and other surveys to show that there is a break in the SFD of KBOs with absolute magnitudes dimmer than approximately 7.0, equating to a size of roughly 100 km. *Greenstreet et al.* (2015) compared the predicted cratering rate on Pluto for these two models, and showed the Pluto system should be saturated with craters in the *Schlichting et al.* (2013) model, but not necessarily if there were a break in the distribution below 100 km. *Robbins et al.* (2017) showed that Pluto and Charon were indeed not saturated with craters, and that there was a distinct deficit of small craters compared to the predictions of the Kuiper belt SFD that assumed it was collisionally evolved. *Singer et al.* (2019) then showed that there was a deficit of craters smaller than 13 km, even compared to SFDs with a break below 100 km. This second break corresponds to KBOs smaller than 2 km, which would dominate the impact rate on the small satellites. The present rate of small impacts in the Pluto system must therefore be very low, as should be the rate of dust production from the small satellites. This rate is consistent with the New Horizons observations, which were unable to detect any dust in the Pluto system during the flyby (*Stern et al.,* 2015; *Lauer et al.,* 2018).

The New Horizons spacecraft approached Pluto at a solar phase angle of 14° (*Weaver et al.,* 2016), meaning that craters are difficult to see in most resolved images of the small satellites. *Robbins et al.* (2017) explored the cratering record of Nix and Hydra, in addition to (and comparing with) Pluto and Charon. Craters were the easiest to see on Nix, particularly in the two highest-resolution LORRI images, which were at a low solar phase angles (8.4° and 9.5°) and had few shadows, and the MVIC panchromatic image of Nix at closest approach. The MVIC panchromatic image was obtained at a similar spatial resolution to the best LORRI image because of MVIC's lower angular resolution, but because it was taken at a solar phase angle of 85.9°, the MVIC image showed a wealth of craters that were not apparent in the LORRI images. *Robbins et al.* (2017) present the counts for Nix performed by two independent crater counters, and then combined them for verification. They found that the density of craters larger than 10 km on Nix was 2.6× the density found on Pluto. Accounting for the much higher surface gravity on Pluto, this crater density is the equivalent of four times as many ≥10 km craters on Nix than Pluto. *Robbins et al.* (2017) found similar results for Hydra, though with greater uncertainty, as the images of Hydra were at lower resolution and only covered half of the body. *Robbins et al.* (2017) suggest that this crater density may indicate that Nix and Hydra have an older surface age than Pluto or Charon. However, the small satellites are undoubtedly covered by impacts from other circum-Pluto debris following the Charon-forming impact. Much like the saturnian satellites (*Kirchoff and Schenk,* 2010), disentangling which impacts are from circumplanetary impactors and which are interplanetary impactors is very difficult.

9. COMPOSITION OF THE SMALL SATELLITES

The initial investigations of the composition of Pluto's small satellites were performed by estimating their spectral slope with HST photometry in different filters (*Buie et al.,* 2006). While not especially diagnostic, this method is as close to spectroscopy as was possible with such faint objects. The HST photometry showed that Nix was slightly redder than Hydra, but that both Nix and Hydra had a B-V ≈ 0.65, similar to Charon [B-V = 0.7 (*Stern et al.,* 2006)].

Cook et al. (2018) describe the New Horizons color and spectral observations of Nix, Hydra, and Kerberos. The Ralph instrument on New Horizons consists of a visible/near-infrared camera and a short-wavelength infrared imaging spectrometer (*Reuter et al.,* 2008). The two components share a common 75-mm aperture and f/8.7 telescope. Wavelengths shorter than 1.1 μm are sent to the MVIC, a time-delay integration (TDI) imager with four color channels (blue, red, near-ir, and narrow-band methane), in addition to unfiltered panchromatic channels. MVIC pixels are approximately 4× the angular size of LORRI native pixels, so the highest-resolution color images in Fig. 6 have lower spatial resolution than the contemporaneous LORRI images in Figs. 3 and 4. Wavelengths between 1.25 and 2.5 μm are sent to the Linear Etalon Imaging Spectral Array (LEISA). LEISA is a wedged-filter infrared spectral imager with a 256 × 256 mercury cadmium telluride (HgCdTe) detector, with one direction corresponding to the spectral spread and the other the spatial field of view, thus allowing spectral cubes to be made in TDI mode. Nix and Hydra were discovered after the launch of New Horizons but before the Pluto flyby observation sequence was designed, so both Nix and Hydra were integrated into flyby planning and observed with both LEISA and MVIC in four colors (*Cook et al.,* 2018). Kerberos was discovered after the flyby sequence was written, and the one retargetable LEISA observation was used to observe its near-infrared spectrum. Ralph did not attempt to observe Styx.

Nix was observed twice by LEISA at distances of roughly 60,000 km and 163,000 km (*Cook et al.,* 2018). The full-disk spectrum of Nix obtained by LEISA is dominated by water ice, with 97–98% of the mass fraction in the best fits being water ice, and the remainder a "Pluto tholin" as a neutral-darkening agent (a stand in for any spectrally neutral absorber). Some fraction of the water ice is in an amorphous state; however, determining the relative fraction of the water ice that is crystalline is difficult at these wavelengths, as the 1.65-μm absorption band strongly depends on particle phase function (*Verbiscer et al.,* 2006) and temperature (*Mastrapa et al.,* 2008) and the surface temperature of Nix is not known. The crystalline ice fraction therefore might be between 77% at 20 K and 100% at 80 K, but considering Nix's high albedo, only temperatures at the low end of that range are plausible. Such a crystalline ice fraction is higher than what had once been expected from cold water ice exposed to 4 b.y. of ionizing radiation (*Cook et al.,* 2007). Potential explanations involve interplanetary dust particles impacts heating and thus recrystallizing the ice on Nix's surface (*Porter et al.,* 2010), or possibly eroding the amorphous ice faster than it can accumulate (*Stern,* 2003). Additionally, recent laboratory work shows that even over the age of the solar system, ice crystal structure is not completely destroyed in the outer solar system environment, at least not to the depths probed

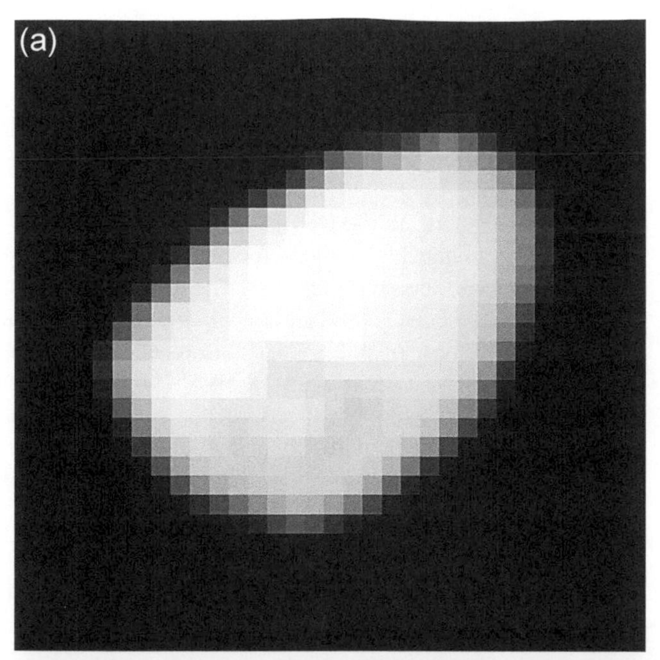

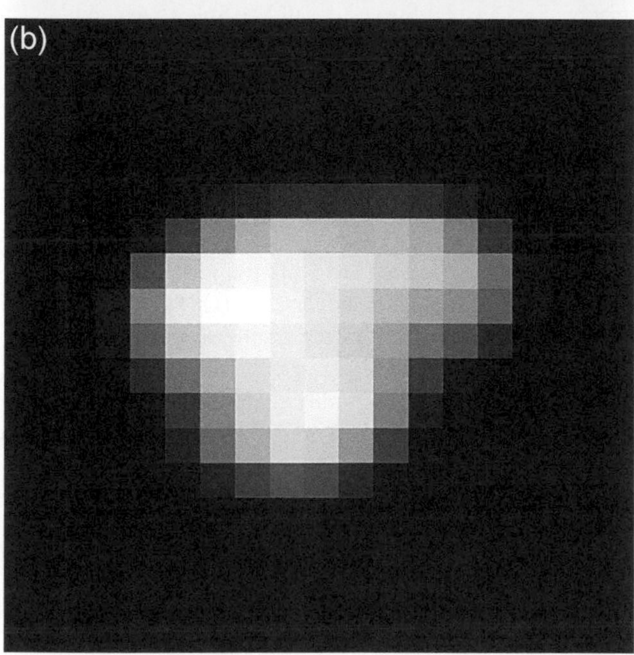

Fig. 6. **(a)** The highest-resolution false-color MVIC image of Nix, obtained at a resolution of 1.178 km/pixel. **(b)** The highest-resolution false-color MVIC image of Hydra, obtained at a resolution of 4.663 km/pixel. In both images the red channel corresponds to the MVIC NIR filter, the green channel is the MVIC red filter, and the blue channel is the MVIC blue filter.

by near-infrared spectral reflectance observations (*Zheng et al.,* 2009; *Famá et al.,* 2010; *Loeffler et al.,* 2020). Even more intriguing was the LEISA detection of the 2.2-μm NH_3-hydrate absorption band on the surface of Nix. This feature could be attributed to either CH_4 or NH_3-hydrate, but *Cook et al.* (2018) argue against CH_4, as methane is not thermally stable on the surface of Nix and would show other absorption features that would have been visible in the LEISA spectrum. NH_3-hydrate was discovered on Charon from Earth-based data (*Brown and Calvin,* 2000), and is thermally stable on the surface of Nix (*Cook et al.,* 2018). In addition to NH_3-hydrate, *Cook et al.* (2018) suggest that some of the ammonia may be bound in NH_4Cl salts. If Nix was formed from material ejected in the Charon-forming impact, the presence of NH_3-hydrate on Nix implies that the icy mantle of Pluto may be rich in NH_3. Because NH_3 can suppress the freezing point of water (*Croft et al.,* 1988), the amount of NH_3 in the deep interior of Pluto and Charon has important implications for their thermal history and the lifetime of any liquid water in their deep interior (*Desch et al.,* 2009; *Rhoden et al.,* 2015).

The closest LEISA observation of Nix resolved the satellite among 130 LEISA pixels, providing a test of the spatial variability of its spectra. *Cook et al.* (2018) found some variability from the full-disk spectrum of Nix, particularly around 2.24 μm, which corresponds to NH_4Cl. In the MVIC images, the area around the largest crater on Nix is distinctly brighter in the red band (540–700 nm) than the surrounding terrain, while being less distinct in the blue (400–550 nm) and near-IR (780–975 nm) bands. This red region has the highest color variation seen on the surfaces of any of the small satellites, and helps explain Nix's slightly redder color in earlier studies (cf. *Buie et al.,* 2006). However, *Cook et al.* (2018) found that the red region did not have distinctly different near-IR spectrum over the LEISA bandpass than the areas outside the red region. In addition, the particle sizes that *Cook et al.* (2018) fit for the red region are not different from the surface overall. The lack of spectral or particle size signatures may suggest that the red material might be only an extremely small mass fraction of the surface regolith, but still able to color the neutral water-ice surface. The red region does appear to be associated with the largest distinct impact crater on Nix (*Robbins et al.,* 2017), suggesting that the red material was either excavated from depth or delivered by the impactor. Given that no other impact crater on any of the small satellites appears to have excavated red material, it may be more likely that the red material was delivered by the impacting body. Most KBOs are somewhat redder than Nix (*Peixinho et al.,* 2008), with the exception of the Haumea collisional family (*Brown et al.,* 2007). In particular, cold classical KBOs such as Arrokoth are distinctly red compared to Nix (*Grundy et al.,* 2020). The red region on Nix may therefore represent the last remains of an ultra-red classical KBO that collided with Nix. LEISA also observed Hydra at distances of 2.4×10^5 km and 3.7×10^5 km (*Cook et al.,* 2018). The full-disk spectrum of Hydra is generally very similar to Nix, dominated by water ice, with only a very small fraction of Pluto tholin to darken the spectrum. "Pluto tholin" is a generic term for the long-chain carbon-nitrogen compounds that may form through the reprocessing of CH_4 and N_2 on Pluto's surface (*Protopapa et al.,* 2017). The crystalline fraction of Hydra's water ice is 70–90%, similar to that of Nix. Any exogenic processes that would change the crystalline fraction (radiation damage, impact annealing) would be similar on the two satellites, as the impact speeds on Nix and Hydra are similar (*Porter et al.,* 2010; *Porter and Grundy,* 2015). Hydra does not show any similar color variation to Nix, but only one hemisphere of Hydra was imaged during the New Horizons flyby (*Weaver et al.,* 2016). Hydra was resolved in LEISA as roughly 16 pixels, and no spatial variability in the LEISA was seen at this spatial resolution (*Cook et al.,* 2018). The same ammoniated species seen on Nix appear to be present on Hydra, but the lower signal-to-noise ratio of the spectrum makes it even more difficult to identify them precisely.

Finally, LEISA also observed Kerberos at a distance of 3.94×10^5 km (*Cook et al.,* 2018). This observation used the retargetable observation that had been set aside for new satellite discoveries. Because it wasn't known how far out any additional satellites could be in the system, the observation was planned to be acquired much further from Pluto than the LEISA observations of Nix and Hydra, to provide maximum flexibility. This timing meant that the LEISA spectrum of Kerberos was very weak, and the object itself was not visible in a single frame. However, *Cook et al.* (2018) were able to extract a spectrum for Kerberos by re-binning data by a factor of 20. This spectrum shows that Kerberos has a generally similar water-ice surface as Nix and Hydra, although the binned resolution is not high enough to see if Kerberos has a similar ammoniated species as Nix and Hydra. LEISA did not observe Styx.

10. OPEN QUESTIONS AND FUTURE OBSERVATIONS OF THE SMALL SATELLITES

While the New Horizons flyby of the Pluto system transformed our view of the small satellites from barely detectable points of light to resolved shapes with surface features, there remain many open questions about the small satellites. The primary science question for the small satellites is their densities, which can provide extremely important constraints on the formation of the entire Pluto system (see the chapter by Canup et al. in this volume). The major way to improve knowledge of the densities from Earth is with continued astrometric observations, enabling higher-quality joint solutions for their orbits and masses, especially as the perturbations of Nix and Hydra add up onto Styx and Kerberos over the coming years and decades. Related to the density is the question of the composition of their interiors; is it truly as similar to Charon as the surfaces of Nix and Hydra would imply? Or is that simply a surface coating that masks their true composition? The possible excavation of dark material by the large crater on Nix may suggest that this is the case. Finally, the impact cratering history on the small satellites is very poorly constrained from the New Horizons images. The closest image of Nix, obtained with the MVIC imager, show a rugged, cratered profile that is not apparent from the other resolved images, which have much lower solar phase angles. Future high-resolution and high-phase-angle imagery of the surfaces of small satellites would help to constrain their history and provide a direct comparison to the pristine KBO Arrokoth (*Spencer et al.,* 2020).

Thus far, the major tools to study the small satellites of Pluto have been New Horizons (now far past Pluto) and HST. HST has been orbiting Earth since 1990, and was last repaired in 2009. As of 2020, it is operational but is susceptible to future failure. While HST is operational, it will continue to be an invaluable resource for obtaining astrometry and photometry of the small satellites. The upcoming James Webb Space Telescope (JWST) offers a larger aperture than HST, but at the cost of only being able to observe Pluto near quadrature (*Nella et al.,* 2004). JWST will thus be a powerful tool to measure the spectra of the small satellites (particularly as its bandpass overlaps and exceeds LEISA), but JWST will not be able to observe the satellites of Pluto frequently enough to continue the astrometric and photometric studies of HST. The Large UV/Optical/IR Surveyor (LUVOIR) space telescope concept (*LUVOIR Team,* 2018), one of four large spacebased observatories being studied by the National Academy of Sciences' 2020 Astronomy and Astrophysics Decadal Survey, would be able to exceed HST's ability to study the positions and brightnesses of the small satellites. In particular, a large optical space telescope similar to LUVOIR would be able to observe the Pluto system near solar opposition, when the satellites are at their brightest, and when photometry of them would be most useful to understand their surface properties. In addition, planned and proposed large groundbased optical telescopes such as the Thirty-Meter Telescope (TMT), Extremely Large Telescope (ELT), and Giant Magellan Telescope (GMT) may be useful for obtaining astrometry of the small satellites, particularly with the help of laser guide star adaptive optics (LGS AO).

Another option to study the small satellites in the future is with stellar occultations. By measuring the timing of when one of the small satellites occults a star from multiple locations on Earth, a measurement could be made of the size and shape of the satellite (e.g., *Buie et al.,* 2019). These multi-chord occultation results would be particularly useful for studying Styx and Kerberos, as their sizes are presently poorly constrained from the New Horizons observations. Depending on the geometry of the occultation, it may also be useful to measure the minor axis of Hydra, which was also poorly constrained from the New Horizons images. Pluto is presently moving away from the galactic plane; however, stellar occultations are becoming less frequent. Nevertheless, stellar occultations by Pluto's small satellites should still occur for many decades to come.

Finally, the ultimate way to study the small satellites in the future would be a return to Pluto. A future Pluto mission could image the satellites at higher resolutions and different geometries than New Horizons, thus allowing for improved shapes for Nix and Hydra and real shape models for Styx and Kerberos. A Pluto-orbiting spacecraft could perform multiple flybys of small satellites, enabling high-resolution imaging of their surfaces (and thus their impact crater history) and high signal-to-noise ratio spectroscopy of their surfaces. Laser and radar altimetry of the small satellites could further improve their shape models, even in unilluminated areas. Radio science tracking during close encounters with the small satellites could even allow direct constraints on their masses, and therefore densities. The "Persephone" Pluto Orbiter Concept was studied by the National Academy of Sciences' 2021 Planetary Decadal Survey, and provides an example of the style of mission that could provide our next close-up look at the small satellites of Pluto.

REFERENCES

Acton C., Bachman N., Semenov B., and Wright E. (2018) A look towards the future in the handling of space science mission geometry. *Planet. Space Sci., 150*, 9–12.

Benecchi S. D., Noll K. S., Grundy W. M., and Levison H. F. (2010) (47171) 1999 TC_{36}: A transneptunian triple. *Icarus, 207(2)*,978–991.

Bernstein G. M., Trilling D. E., Allen R. L., Brown M. E., Holman M., and Malhotra R. (2004) The size distribution of trans-neptunian bodies. *Astron. J.*, 128(3), 1364–1390.

Binzel R. P. (1989) Pluto-Charon mutual events. *Geophys. Res. Lett., 16(11)*, 1205–1208.

Bond W. C. (1847) Observations of Lassell's satellite of Neptune. *Mon. Not. R. Astron. Soc., 8*, 9–11.

Brown M. E. and Calvin W. M. (2000) Evidence for crystalline water and ammonia ices on Pluto's satellite Charon. *Science, 287(5450)*, 107–109.

Brown M. E. and Suer T. A. (2007) Satellites of 2003 AZ_84, (50000), (55637), and (90482). *IAU Circular* 8812.

Brown M. E., Bouchez A. H., Rabinowitz D., Sari R., Trujillo C. A., van Dam M., Campbell R., Chin J., Hartman S., Johansson E., Lafon R., Le Mignant D., Stomski P., Summers D., and Wizinowich P. (2005) Keck Observatory laser guide star adaptive optics discovery and characterization of a satellite to the large Kuiper belt object 2003 EL_{61}. *Astrophys. J. Lett., 632(1)*, L45–L48.

Brown M. E., van Dam M. A., Bouchez A. H., Le Mignant D., Campbell R. D., Chin J. C. Y., Conrad A., Hartman S. K., Johansson E. M., Lafon R. E., Rabinowitz D. L., Stomski P. J., Summers D. M., Trujillo C. A., and Wizinowich P. L. (2006) Satellites of the largest Kuiper belt objects. *Astrophys. J. Lett., 639(1)*, L43–L46.

Brown M. E., Barkume K. M., Ragozzine D., and Schaller E. L. (2007) A collisional family of icy objects in the Kuiper belt. *Nature, 446(7133)*, 294–296.

Brown M. E., Ragozzine D., Stansberry J., and Fraser W. C. (2010) The size, density, and formation of the Orcus-Vanth system in the Kuiper belt. *Astron. J., 139(6)*, 2700–2705.

Brozović M., Showalter M. R., Jacobson R. A., and Buie M. W. (2015) The orbits and masses of satellites of Pluto. *Icarus, 246*, 317–329.

Buie M. W., Grundy W. M., Young E. F., Young L. A., and Stern S. A. (2006) Orbits and photometry of Pluto's satellites: Charon, S/2005 P1, and S/2005 P2. *Astron. J., 132(1)*, 290–298.

Buie M. W., Porter S. B., Tamblyn P., et al. (2019) Size and shape constraints of (486958) Arrokoth from stellar occultations. *ArXiV e-prints*, arXiv:2001.00125.

Canup R. M. (2005) A giant impact origin of Pluto-Charon. *Science, 307(5709)*, 546–550.

Cheng A. F., Weaver H. A., Conard S. J., Morgan M. F., Barnouin-Jha O., Boldt J. D., Cooper K. A., Darlington E. H., Grey M. P., Hayes J. R., Kosakowski K. E., Magee T., Rossano E., Sampath D., Schlemm C., and Taylor H. W. (2008) Long-Range Reconnaissance Imager on New Horizons. *Space Sci. Rev., 140(1–4)*, 189–215.

Christy J. W. and Harrington R. S. (1978) The satellite of Pluto. *Astron. J.*, 83, 1005–1008.

Christy J. W. and Harrington R. S. (1980) The discovery and orbit of Charon. *Icarus, 44(1)*, 38–40.

Ciarniello M., Capaccioni F., Filacchione G., Clark R. N., Cruikshank D. P., Cerroni P., Coradini A., Brown R. H., Buratti B. J., Tosi F., and Stephan K. (2011) Hapke modeling of Rhea surface properties through Cassini-VIMS spectra. *Icarus, 214(2)*, 541–555.

Cook J. C., Desch S. J., Roush T. L., Trujillo C. A., and Geballe T. R. (2007) Near-infrared spectroscopy of Charon: Possible evidence for cryovolcanism on Kuiper belt objects. *Astrophys. J., 663(2)*, 1406–1419.

Cook J. C., Dalle Ore C. M., Protopapa S., et al. (2018) Composition of Pluto's small satellites: Analysis of New Horizons spectral images. *Icarus, 315*, 30–45.

Croft S. K., Lunine J. I., and Kargel J. (1988) Equation of state of ammonia-water liquid: Derivation and planetological applications. *Icarus, 73(2)*, 279–293.

Desch S. J., Cook J. C., Doggett T. C., and Porter S. B. (2009) Thermal evolution of Kuiper belt objects, with implications for cryovolcanism. *Icarus, 202(2)*, 694–714.

Domingue D. L., Lockwood G. W., and Thompson D. T. (1995) Surface textural properties of icy satellites: A comparison between Europa and Rhea. *Icarus, 115(2)*, 228–249.

Famá M., Loeffler M. J., Raut U., and Baragiola R. A. (2010) Radiation-induced amorphization of crystalline ice. *Icarus, 207(1)*, 314–319.

Farinella P., Milani A., Nobili A. M., and Valsecchi G. B. (1979) Tidal evolution and the Pluto-Charon system. *Moon Planets, 20(4)*, 415–421.

Fletcher M. and Liebscher R. (2005) PyOpenGL — the Python OpenGL binding. Available online at *http://pyopengl.sourceforge.net*.

Fraser W. C., Brown M. E., Morbidelli A., Parker A., and Batygin K. (2014) The absolute magnitude distribution of Kuiper belt objects. *Astrophys. J., 782(2)*, 100.

Gaia Collaboration, Brown A. G. A., Vallenari A., et al. (2018) Gaia Data Release 2: Summary of the contents and survey properties. *Astron. Astrophys., 616*, A1.

Greenstreet S., Gladman B., and McKinnon W. B. (2015) Impact and cratering rates onto Pluto. *Icarus, 258*, 267–288.

Grundy W. M., Bird M. K., Britt D. T., et al. (2020) Color, composition, and thermal environment of Kuiper belt object (486958) Arrokoth. *Science, 367(6481)*, eaay3705.

Hapke B. (2005) *Theory of Reflectance and Emittance Spectroscopy.* Cambridge Univ., Cambridge. 455 pp.

Harrington R. S. and Christy J. W. (1980) The satellite of Pluto. II. *Astron. J., 85*, 168–170.

Harrington R. S. and Van Flandern T. C. (1979) The satellites of Neptune and the origin of Pluto. *Icarus, 39(1)*, 131–136.

Hastings D. M., Ragozzine D., Fabrycky D. C., Burkhart L. D., Fuentes C., Margot J.-L., Brown M. E., and Holman M. (2016) The short rotation period of Hi'iaka, Haumea's largest satellite. *Astron. J., 152(6)*, 195.

Henyey L. G. and Greenstein J. L. (1941) Diffuse radiation in the galaxy. *Astrophys. J., 93*, 70–83.

Herschel J. S. (1834) On the satellites of Uranus. *Mon. Not. R. Astron. Soc., 3*, 35–38.

Herschel W. (1787) An account of the discovery of two satellites revolving round the Georgian planet. *Philos. Trans. R. Soc. London, 77*, 125–129.

Jewitt D. and Luu J. (1993) Discovery of the candidate Kuiper belt object 1992 QB_1. *Nature, 362(6422)*, 730–732.

Kaasalainen M. and Torppa J. (2001) Optimization methods for asteroid lightcurve inversion: I. Shape determination. *Icarus, 153(1)*, 24–36.

Kaasalainen M. and Viikinkoski M. (2012) Shape reconstruction of irregular bodies with multiple complementary data sources. *Astron. Astrophys., 543*, A97.

Kenyon S. J. and Bromley B. C. (2019) A Pluto-Charon sonata: Dynamical limits on the masses of the small satellites. *Astron. J., 158(2)*, 69.

Kirchoff M. R. and Schenk P. (2010) Impact cratering records of the mid-sized, icy saturnian satellites. *Icarus, 206(2)*, 485–497.

Kiss C., Marton G., Farkas-Takás A., Stansberry J., Müller T., Vinkó J., Balog Z., Ortiz J.-L., and Pál A. (2017) Discovery of a satellite of the large trans-neptunian object (225088) 2007 OR_{10}. *Astrophys. J. Lett., 838(1)*, L1.

Kuiper G. P. (1950) The diameter of Pluto. *Publ. Astron. Soc. Pac., 62(366)*, 133–137.

Lauer T. R., Throop H. B., Showalter M. R., et al. (2018) The New Horizons and Hubble Space Telescope search for rings, dust, and debris in the Pluto-Charon system. *Icarus, 301*, 155–172.

Leinhardt Z. M., Marcus R. A., and Stewart S. T. (2010) The formation of the collisional family around the dwarf planet Haumea. *Astrophys. J., 714(2)*, 1789–1799.

Li J., A'Hearn M. F., and McFadden L. (2003) The effect of shape model on asteroid disk integrated phase function. *AAS/Division for Planetary Sciences Meeting Abstracts, 35*, 24.04.

Li J., A'Hearn M. F., and McFadden L. A. (2004) Photometric analysis

of Eros from NEAR data. *Icarus, 172(2)*, 415–431.
Lin D. N. C. (1981) On the origin of the Pluto-Charon system. *Mon. Not. R. Astron. Soc., 197,* 1081–1085.
Loeffler M. J., Tribbett P. D., Cooper J. F., and Sturner S. J. (2020) A possible explanation for the presence of crystalline H_2O-ice on Kuiper belt objects. *Icarus, 351*, 113943.
LUVOIR Team (2018) The LUVOIR Mission Concept Study Interim Report. *ArXiV e-prints*, arXiv:1809.09668.
Mastrapa R. M., Bernstein M. P., Sandford S. A., Roush T. L., Cruikshank D. P., and Dalle Ore C. M. (2008) Optical constants of amorphous and crystalline H_2O-ice in the near infrared from 1.1 to 2.6 μm. *Icarus, 197(1)*, 307–320.
McKinnon W. B., Richardson D. C., Marohnic J. C., et al. (2020) The solar nebula origin of (486958) Arrokoth, a primordial contact binary in the Kuiper belt. *Science, 367(6481)*, aay6620.
Mignard F. (1981) On a possible origin of Charon. *Astron. Astrophys., 96(1–7)*, L1–L2.
Nella J., Atcheson P. D., Atkinson C. B., et al. (2004) James Webb Space Telescope (JWST) Observatory architecture and performance. In *Optical, Infrared, and Millimeter Space Telescopes* (J. C. Mather, ed.), pp. 576–587. SPIE Conf. Ser. 5487, Bellingham, Washington.
Noll K. S., Grundy W. M., Chiang E. I., Margot J. L., and Kern S. D. (2008) Binaries in the Kuiper belt. In *The Solar System Beyond Neptune* (M. A. Barucci et al., eds.), pp. 345–363. Univ. of Arizona, Tucson.
Null G. W., Owen W. M., and Synnott S. P. (1993) Masses and densities of Pluto and Charon. *Astron. J., 105*, 2319–2335.
Parker A. H., Buie M. W., Grundy W. M., and Noll K. S. (2016) Discovery of a Makemakean moon. *Astrophys. J. Lett., 825(1)*, L9.
Peixinho N., Lacerda P., and Jewitt D. (2008) Color-inclination relation of the classical Kuiper belt objects. *Astron. J., 136(5)*, 1837–1845.
Petit J. M., Kavelaars J. J., Gladman B. J., et al. (2011) The Canada-France Ecliptic Plane Survey — Full data release: The orbital structure of the Kuiper belt. *Astron. J., 142(4)*, 131.
Pickering W. H. (1931) The discovery of Pluto. *Mon. Not. R. Astron. Soc., 91*, 812–817.
Pires dos Santos P. M., Giuliatti Winter S. M., and Sfair R. (2011) Gravitational effects of Nix and Hydra in the external region of the Pluto-Charon system. *Mon. Not. R. Astron. Soc., 410(1)*, 273–279.
Porter S. B. and Grundy W. M. (2015) Ejecta transfer in the Pluto system. *Icarus, 246*, 360–368.
Porter S. B., Desch S. J., and Cook J. C. (2010) Micrometeorite impact annealing of ice in the outer solar system. *Icarus, 208(1)*, 492–498.
Protopapa S., Grundy W. M., Reuter D. C., et al. (2017) Pluto's global surface composition through pixel-by-pixel Hapke modeling of New Horizons Ralph/LEISA data. *Icarus, 287*, 218–228.
Quillen A. C., Nichols-Fleming F., Chen Y.-Y., and Noyelles B. (2017) Obliquity evolution of the minor satellites of Pluto and Charon. *Icarus, 293*, 94–113.
Ragozzine D. and Brown M. E. (2009) Orbits and masses of the satellites of the dwarf planet Haumea (2003 EL61). *Astron. J., 137(6)*, 4766–4776.
Reuter D. C., Stern S. A., Scherrer J., et al. (2008) Ralph: A visible/infrared imager for the New Horizons Pluto/Kuiper belt mission. *Space Sci. Rev., 140(1–4)*, 129–154.
Rhoden A. R., Henning W., Hurford T. A., and Hamilton D. P. (2015) The interior and orbital evolution of Charon as preserved in its geologic record. *Icarus, 246*, 11–20.
Robbins S. J., Singer K. N., Bray V., et al. (2017) Craters of the Pluto-Charon system. *Icarus, 287*, 187–206.
Schlichting H. E., Ofek E. O., Sari R., Nelan E. P., Gal-Yam A., Wenz M., Muirhead P., Javanfar N., and Livio M. (2012) Measuring the abundance of sub-kilometer-sized Kuiper belt objects using stellar occultations. *Astrophys. J., 761(2)*, 150.
Schlichting H. E., Fuentes C. I., and Trilling D. E. (2013) Initial planetesimal sizes and the size distribution of small Kuiper belt objects. *Astron. J., 146(2)*, 36.
Showalter M. R. and Hamilton D. P. (2015) Resonant interactions and chaotic rotation of Pluto's small moons. *Nature, 522(7554)*, 45–49.
Showalter M. R., Burns J. A., Cuzzi J. N., and Pollack J. B. (1985) Discovery of Jupiter's 'gossamer' ring. *Nature, 316(6028)*, 526–528.
Showalter M. R., Hamilton D. P., Stern S. A., Weaver H. A., Steffl A. J., and Young L. A. (2011) New satellite of (134340) Pluto: S/2011 (134340) 1. *IAU Circular 9221.*
Showalter M. R., Weaver H. A., Stern S. A., Steffl A. J., Buie M. W., Merline W. J., Mutchler M. J., Soummer R., and Throop H. B. (2012) New satellite of (134340) Pluto: S/2012 (134340) 1. *IAU Circular* 9253.
Showalter M. R., Porter S. B., Verbiscer A. J., Buie M. W., and Helfenstein P. (2019) Rotation states of Pluto's small moons and the search for spin-orbit resonances. *EPSC Abstracts, 13*, EPSC-DPS2019-1025-1.
Singer K. N., McKinnon W. B., Gladman B., et al. (2019) Impact craters on Pluto and Charon indicate a deficit of small Kuiper belt objects. *Science, 363(6430)*, 955–959.
Spencer J. R., Stern S. A., Moore J. M., et al. (2020) The geology and geophysics of Kuiper belt object (486958) Arrokoth. *Science, 367(6481)*, eaay3999.
Steffl A. J., Mutchler M. J., Weaver H. A., Stern S. A., Durda D. D., Terrell D., Merline W. J., Young L. A., Young E. F., Buie M. W., and Spencer J. R. (2006) New constraints on additional satellites of the Pluto system. *Astron. J., 132(2)*, 614–619.
Stern S. A. (2003) The evolution of comets in the Oort cloud and Kuiper belt. *Nature, 424(6949)*, 639–642.
Stern S. A. (2009) Ejecta exchange and satellite color evolution in the Pluto system, with implications for KBOs and asteroids with satellites. *Icarus, 199(2)*, 571–573.
Stern S. A., Fesen R. A., Barker E. S., Parker J. W., and Trafton L. M. (1991) A search for distant satellites of Pluto. *Icarus, 94(1)*, 246–249.
Stern S. A., Parker J. W., Duncan M. J., Snowdall J. C., and Levison H. F. (1994) Dynamical and observational constraints on satellites in the inner Pluto-Charon system. *Icarus, 108(2)*, 234–242.
Stern S. A., Weaver H. A., Steffl A. J., Mutchler M. J., Merline W. J., Buie M. W., Young E. F., Young L. A., and Spencer J. R. (2006) A giant impact origin for Pluto's small moons and satellite multiplicity in the Kuiper belt. *Nature, 439(7079)*, 946–948.
Stern S. A., Bagenal F., Ennico K., et al. (2015) The Pluto system: Initial results from its exploration by New Horizons. *Science, 350(6258)*, aad1815.
Stern S. A., Weaver H. A., Spencer J. R., et al. (2019) Initial results from the New Horizons exploration of 2014 MU_{69}, a small Kuiper belt object. *Science, 364(6441)*, eaaw9771.
Thirouin A. and Sheppard S. S. (2019) Light curves and rotational properties of the pristine cold classical Kuiper belt objects. *Astron. J., 157(6)*, 228.
Tholen D. J. (1985) The orbit of Pluto's satellite. *Astron. J., 90*, 2353–2359.
Tholen D. J., Senay M., Hainaut O., and Marsden B. G. (1994) 1993 RO, 1993 RP, 1993 SB, 1993 SC. *IAU Circular* 5983.
Tholen D. J., Buie M. W., Grundy W. M., and Elliott G. T. (2008) Masses of Nix and Hydra. *Astron. J., 135(3)*, 777–784.
Tombaugh C. W. and Moore P. (1980) *Out of the Darkness: The Planet Pluto*. Stackpole, Lanham, Maryland. 221 pp.
Veillet C., Doressoundiram A., Shapiro J., Kavelaars J. J., and Morbidelli A. (2001) S/2000 (1998 WW31) 1. *IAU Circular* 7610.
Verbiscer A. J., Peterson D. E., Skrutskie M. F., Cushing M., Helfenstein P., Nelson M. J., Smith J. D., and Wilson J. C. (2006) Near-infrared spectra of the leading and trailing hemispheres of Enceladus. *Icarus, 182(1)*, 211–223.
Verbiscer A. J., Skrutskie M. F., and Hamilton D. P. (2009) Saturn's largest ring. *Nature, 461(7267)*, 1098–1100.
Verbiscer A. J., Porter S. B., Buratti B. J., et al. (2018) Phase curves of Nix and Hydra from the New Horizons imaging cameras. *Astrophys. J. Lett., 852(2)*, L35.
Walker A. R. (1980) An occultation by Charon. *Mon. Not. R. Astron. Soc., 192*, 47P–50P.
Weaver H. A., Stern S. A., Mutchler M. J., Steffl A. J., Buie M. W., Merline W. J., Spencer J. R., Young E. F., and Young L. A. (2006) Discovery of two new satellites of Pluto. *Nature, 439(7079)*, 943–945.
Weaver H. A., Buie M. W., Buratti B. J., et al. (2016) The small satellites of Pluto as observed by New Horizons. *Science, 351(6279)*, aae0030.
Youdin A. N., Kratter K. M., and Kenyon S. J. (2012) Circumbinary chaos: Using Pluto's newest moon to constrain the masses of Nix and Hydra. *Astrophys. J., 755(1)*, 17.
Zheng W., Jewitt D., and Kaiser R. I. (2009) On the state of water ice on Saturn's moon Titan and implications to icy bodies in the outer solar system. *J. Phys. Chem. A, 113(42)*, 11174–11181.

Part 4:

Origins, Interiors, and the Big Picture

Canup R. M., Kratter K. M., and Neveu M. (2021) On the origin of the Pluto system. In *The Pluto System After New Horizons* (S. A. Stern, J. M. Moore, W. M. Grundy, L. A. Young, and R. P. Binzel, eds.), pp. 475–506. Univ. of Arizona, Tucson, DOI: 10.2458/azu_uapress_9780816540945-ch021.

On the Origin of the Pluto System

Robin M. Canup
Southwest Research Institute

Kaitlin M. Kratter
University of Arizona

Marc Neveu
NASA Goddard Space Flight Center/University of Maryland

The goal of this chapter is to review hypotheses for the origin of the Pluto system in light of observational constraints that have been considerably refined over the 85-year interval between the discovery of Pluto and its exploration by spacecraft. We focus on the giant impact hypothesis currently understood as the likeliest origin for the Pluto-Charon binary, and devote particular attention to new models of planet formation and migration in the outer solar system. We discuss the origins conundrum posed by the system's four small moons. We also elaborate on implications of these scenarios for the dynamical environment of the early transneptunian disk, the likelihood of finding a Pluto collisional family, and the origin of other binary systems in the Kuiper belt. Finally, we highlight outstanding open issues regarding the origin of the Pluto system and suggest areas of future progress.

1. INTRODUCTION

For six decades following its discovery, Pluto was the only known Sun-orbiting world in the dynamical vicinity of Neptune. An early origin concept postulated that Neptune originally had two large moons — Pluto and Neptune's current moon, Triton — and that a dynamical event had both reversed the sense of Triton's orbit relative to Neptune's rotation and ejected Pluto onto its current heliocentric orbit (*Lyttleton*, 1936). This scenario remained in contention following the discovery of Charon, as it was then established that Pluto's mass was similar to that of a large giant planet moon (*Christy and Harrington*, 1978). However, *McKinnon* (1984) demonstrated that mutual Triton-Pluto interactions around Neptune were incapable of placing Pluto and Triton into their respective orbits, and instead proposed the now accepted origin scenario for Triton: that it was a separately formed Pluto-like object that was captured into a retrograde orbit around Neptune (e.g., *Agnor and Hamilton*, 2006; *Nogueira et al.*, 2011). This still satisfied the inferred genetic link between Pluto and Triton suggested by their similar absolute visual magnitudes and methane-ice dominated infrared spectra (implying similar bodies; see the chapter in this volume by Cruikshank et al.), and by their peculiar orbits dynamically linked to Neptune. However, it left open the question of the heliocentric origin of the Pluto-Charon binary. *McKinnon* (1989) subsequently made a compelling case that the angular momentum of the Pluto-Charon binary was most naturally explained by a giant impact between comparably sized bodies, with Charon forming from ejecta produced by this collision. Thus, it was proposed that Pluto-Charon originated from a similar type of collisional event to that which was becoming increasingly favored for the origin of Earth's Moon (e.g., *Benz et al.,* 1986, 1987, 1989).

Further discussion of the history of thought on the origin of Pluto-Charon can be found in the chapter addressing this topic in the 1997 *Pluto and Charon* volume (*Stern et al.*, 1997). Basic elements of our understanding of Pluto system origins described in that review persist today. For example, Pluto is still thought to have formed within a transneptunian disk of objects, and a giant impact remains the most widely accepted scenario for the formation of Pluto-Charon. However, in most respects our understanding has evolved considerably since 1997 due to both theoretical advances and new discoveries. Notable developments include:

- ***Strong theoretical and observational support for a formation closer to the Sun than today.*** Early dynamical models suggested that Pluto's eccentric and inclined orbit in the 3:2 mean-motion resonance with Neptune could have originated either from interactions among bodies accreting near Pluto's current semimajor axis (*Levison and Stern*, 1995) or from formation of Pluto at a more interior distance, with Pluto then driven outward into its current orbit via resonant transport as Neptune's orbit expanded (*Malhotra*, 1993, 1995; *Malhotra and Williams*, 1997). Since the 1990s, there has been an explosion in the number of known transneptunian objects, and accounting for the orbital properties of the many bodies in resonance with Neptune (such as the "Plutino" bodies that, like Pluto, complete two heliocentric orbits in the same time as Neptune completes three) appears to require outer planet migration and resonant capture/transport (e.g., *Hahn and Malhotra*, 2005; *Levison and Morbidelli*, 2003; *Murray-Clay and Chiang*, 2005, 2006). Furthermore, there is increasing awareness that gravitational interactions with an early massive planetesimal disk would have caused the orbits of Saturn, Uranus, and Neptune to migrate outward early in the solar system's history (e.g., *Fernandez and Ip*, 1984; *Hahn and Malhotra*, 1999). Such migration is postulated to have led to dynamical instability and a sudden and major rearrangement of the giant planets, as described in the widely explored "Nice" model and its derivatives (e.g., *Tsiganis et al.*, 2005; *Levison et al.*, 2008; *Batygin and Brown*, 2010; *Nesvorný and Morbidelli*, 2012). Thus, developments in the past 20 years strongly favor an early expansion of Pluto's orbit to the current heliocentric distance.
- ***Much faster and earlier assembly.*** It is now thought that large outer solar system planetesimals may have accreted much more rapidly than appreciated two decades ago, via gravitational collapse of clouds of millimeter- to decimeter-sized particles (dubbed "pebbles") into 100-km-class objects while solar nebula gas was still present in the first few to 10 m.y. of solar system history (e.g., *Youdin and Goodman*, 2005; *Ormel and Klahr*, 2010; *Lambrechts and Johansen*, 2012; *Johansen et al.*, 2015; *Nesvorný et al.*, 2019). This contrasts with hierarchical coagulation by purely two-body collisions that would require at least tens of millions to perhaps hundreds of millions of years to form large bodies at distances >15 AU (e.g., *Stern and Colwell*, 1997; *Kenyon et al.*, 2008; *Kenyon and Bromley*, 2012). This has modified thinking on when the progenitors of the current Pluto system could have formed.
- ***Quantitative understanding of the binary-forming impact scenario.*** While the idea that the Pluto-Charon binary was the product of an impact was proposed in the late 1980s, the event was not modeled until more than a decade later. Three-dimensional hydrodynamical simulations have now identified a quite limited range of impacts capable of explaining the Charon-to-Pluto mass ratio, and revealed a new regime of impacts that produces Charon as an intact byproduct of the collision (*Canup*, 2005, 2011; *Arakawa et al.*, 2019). Such work implies that a Pluto-Charon forming impact requires a low impact velocity comparable to the mutual escape velocity, and that collisional outcome is sensitive to the differentiation states of the colliding bodies. These results provide new clues regarding the interior structure of the progenitors of Pluto and Charon and the circumstances and timing of Pluto system formation.
- ***Discovery of the four tiny moons.*** In the years leading to the New Horizons flyby, repeated observations of the Pluto system led to the discovery of four moons tens of kilometers in size orbiting the Pluto-Charon binary (*Weaver et al.*, 2006; *Showalter et al.*, 2011, 2012). The outer moons have nearly circular and co-planar orbits, and they orbit Pluto in the same sense and plane as Charon, which seems to strongly suggest that they share a common origin with the binary (e.g., *Stern et al.*, 2006). Their surfaces are unusually bright and ice-rich compared with other similarly sized Kuiper belt objects (KBOs). If, as seems probable, their interiors are similarly dominated by water ice, this is consistent with the small moons having originated from debris dispersed during a Charon-forming impact (e.g., *Canup*, 2011). However, accounting for other properties of the tiny moons — notably their radially distant and near-resonant orbits — has proven elusive, and their origin remains an open and important question.
- ***In situ exploration of the Pluto system.*** The New Horizons flyby vastly increased our knowledge of the physical properties of the Pluto system, providing new or much improved constraints relevant to how the system originated. The Pluto-Charon binary mass and angular momentum, as well as the individual densities of Pluto and Charon, are now well-constrained. The latter imply that Charon is somewhat more ice-rich than Pluto, but that both bodies contain roughly 60–70% rocky material by mass, with similar compositions compared with the diversity of compositions inferred across other large KBOs (e.g., *McKinnon et al.*, 2017). Crater counts indicate that Pluto, Charon, Nix, and Hydra are probably all ancient bodies, with oldest surface ages roughly estimated to ≥4 G.y. but still subject to uncertainties in the impactor population (*Robbins et al.*, 2017; *Singer et al.*, 2019).

2. CONSTRAINTS ON PLUTO SYSTEM ORIGIN

Basic properties of the Pluto system that must be explained by any origin scenario(s) are summarized in Fig. 1. Constraints on formation models derive from the system's dynamical and compositional properties, as well as from various lines of evidence for when the system likely formed. In this section, we focus on constraints that apply to any model of Pluto system origin; additional conditions implied by an origin via giant impact are discussed in section 5.

2.1. Pluto System's Heliocentric Orbit

The current heliocentric orbit of the Pluto system with semimajor axis ≈40 AU involves multiple resonances,

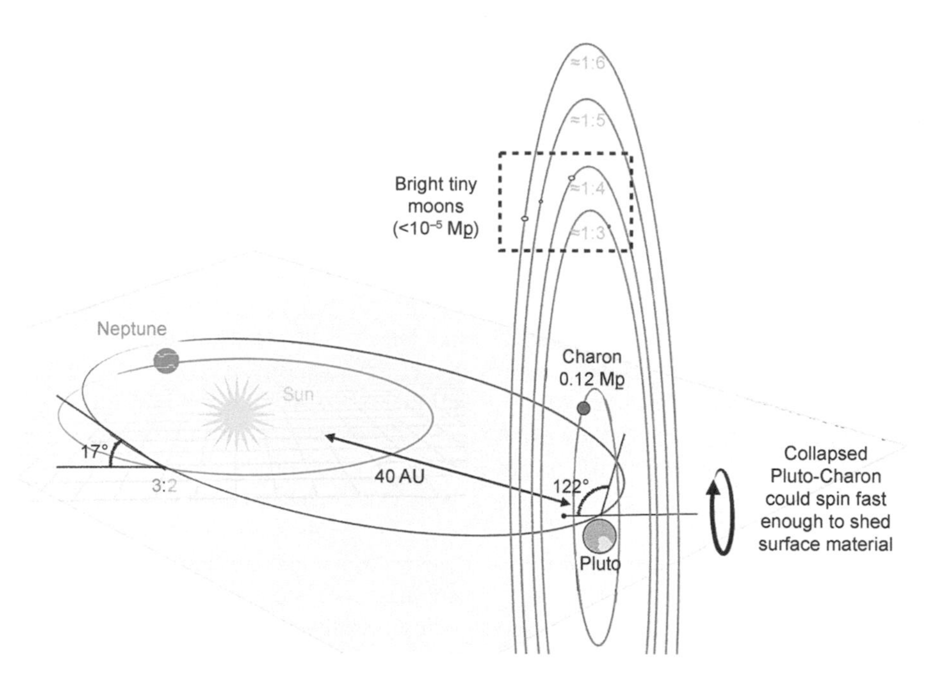

Fig. 1. Key constraints on the origin of the Pluto system. Pluto's heliocentric orbit is eccentric (≈0.25) and inclined (≈17°) relative to the ecliptic, and in a 3:2 mean-motion resonance with Neptune's orbit. The axes of the mutual Pluto-Charon orbit and the orbits of the tiny moons are aligned with Pluto and Charon's spin axes and form a 122° angle with the system's heliocentric orbital axis. The system's angular momentum is so high that if the combined material of Pluto and Charon was collapsed into a single object, this object could spin fast enough to shed mass. The tiny moons orbit the binary near mean-motion resonances with the mutual orbit. They are icy and much brighter than transneptunian objects of similar size. Charon's mass is a significant fraction of that of Pluto (Charon-to-Pluto mass ratio of 0.12). Charon is slightly icier than Pluto (*Bierson et al.,* 2018), but Pluto and Charon's bulk densities are still quite similar compared to the broad range of density estimates across all large Kuiper belt objects (KBOs).

most notably with the mean motion of Neptune, in which the ratio of the orbital period of Pluto to that of Neptune is 3:2. A portion of Pluto's orbit is closer to the Sun than Neptune. However, Pluto's argument of perihelion librates about 90° so that when Pluto is closest to the Sun it is also at a great height above the ecliptic and the plane of Neptune's orbit. Additional resonances involve the relative nodal longitudes and longitudes of perihelion of Pluto and Neptune (e.g., *Malhotra and Williams*, 1997). Together these resonances produce a stable configuration in which Pluto and Neptune never approach each other in three-dimensional space, even though Pluto's orbit is thought to be chaotic (e.g., *Sussman and Wisdom*, 1988).

Pluto's orbit is highly eccentric ($e \approx 0.25$) and inclined relative to the ecliptic ($i \approx 17°$). *Malhotra* (1993) demonstrated that if Pluto was captured into the 3:2 resonance with Neptune as Neptune's orbit migrated outward, an initially low-eccentricity Pluto orbit would have been driven to its current high-eccentricity state as Neptune's orbit expanded by about 5 AU. This implied that both Pluto and Neptune initially formed well interior to their current semimajor axes.

As first described by *Fernandez and Ip* (1984), outward migration is driven by gravitational interactions between the outer planets and a companion planetesimal disk that likely persisted after the gas-rich solar nebula had dissipated. Scatterings of planetesimals by a giant planet cause its orbit to undergo small changes. Planetesimals scattered inward by Saturn, Uranus, and Neptune tend to encounter the closer-in, more massive giant planets, whereas those scattered outward tend to encounter less-massive giant planets or (for planetesimals scattered by Neptune) no giant planet at all. Because of this asymmetry, there is a net inward flux of planetesimals scattered by Saturn, Uranus, and Neptune and a compensating outward migration of these planets so as to conserve energy and angular momentum. Jupiter, which is more massive than the other giants combined, is the most likely to eject planetesimals out of the solar system and its orbit loses the angular momentum and energy imparted to these planetesimals. This causes Jupiter's orbit to contract. As the giant planets undergo such planetesimal-driven migration, they cross mutual resonances that can lead to orbital instability and Nice-type evolutions (e.g., *Tsiganis et al.*, 2005). Dynamical models of this process developed over the past two decades have self-consistently reproduced many of the solar system's current features (e.g., *Tsiganis et al.*, 2005; *Levison et al.*, 2008; *Nesvorný and Morbidelli*, 2012). Given the successes of these models, alternative scenarios in which Pluto acquired its heliocentric orbital properties *in situ* at 40 AU (*Levison and Stern*, 1995) are no longer favored.

The orbital properties of Pluto and other KBOs constrain the character of Neptune's migration. As detailed below, they suggest that it involved a superposition of slow, continuous orbital expansion over tens of millions of years due to interaction with small objects in the planetesimal disk, and "grainy" changes (i.e., random and sudden but relatively small compared to the overall migration) in semimajor axis due to interactions with thousands of Pluto-sized planetesimals (*Nesvorný and Vokrouhlicky*, 2016). These characteristics do not preclude an additional postulated "jump" in Neptune's semimajor axis due to a close encounter with a fifth giant planet ultimately ejected from the solar system by Jupiter, a common outcome of Nice-type simulations (*Nesvorný and Morbidelli*, 2012).

Evidence for the slow, smooth component of migration arises from the need to explain Pluto's high orbital inclination. Unlike its high eccentricity, the inclination was not easily obtained in early models of resonant capture of an initially low-eccentricity Pluto by an outwardly migrating Neptune, due to the lack of a sufficiently strong mechanism to excite inclinations (e.g., *Hahn and Malhotra*, 2005; *Levison et al.*, 2008). Recent works consider that capture into the 3:2 resonance may instead have occurred through a high-eccentricity path, with bodies first scattered outward onto high-eccentricity, high-inclination orbits through interactions with Neptune, and subsequently trapped and stabilized into the 3:2 resonance by secular cycles (e.g., the Kozai resonance) and by the effects of Neptune's continued migration (*Gomes*, 2003; *Dawson and Murray-Clay*, 2012). Explaining the observed inclination distribution of KBOs in the 3:2 resonance — including Pluto as well as the Plutinos — implies that Neptune's migration was relatively slow, with an e-folding timescale ≥10 m.y., so that there was sufficient time for scattering interactions and related dynamical processes to excite inclinations (*Nesvorný*, 2015; *Nesvorný and Vokrouhlicky*, 2016; but see also *Volk and Malhotra*, 2019).

With smooth migration alone, one would expect to find a larger fraction of KBOs in resonance with Neptune, due to highly efficient resonant capture mechanisms (e.g., *Hahn and Malhotra*, 2005; *Nesvorný*, 2015). In addition, smooth migration would cause resonant objects to have larger libration amplitudes than observed (*Gladman et al.*, 2008; *Nesvorný*, 2015). Migration and scattering models may be reconciled with observations if the planetesimal disk also contained a population of large, Pluto-sized objects. These large, discrete bodies introduce a graininess into Neptune's migration, akin to a random walk in its semimajor axis, which decreases both the efficiency of resonance capture and resulting libration amplitudes (*Nesvorný and Vokrouhlicky*, 2016). Such models imply that there were thousands of Pluto-class objects interior to about 30 AU as Neptune began to migrate and Pluto achieved its current heliocentric orbit. Only a small fraction (~10^{-3}) of the disk planetesimals were then ultimately incorporated into the Kuiper belt, with most lost to accretion by the giant planets or ejection (e.g., *Nesvorný*, 2015).

2.2. Dynamics of the Pluto-Charon Binary

Any scenario for the formation of the Pluto system must be able to reproduce its atypically high obliquity and angular momentum (Fig. 1). The plane of the system has a 122° obliquity relative to its heliocentric orbit, so that Pluto's rotation and the orbits of all its moons are retrograde with respect to the sense of its heliocentric orbit. Pluto and Charon are on a close, tidally locked, and essentially circular orbit that is coplanar with the equators of both bodies. Their semimajor axis is only a ≈ 16.5 Pluto radii, where Pluto's radius is $R_P \approx 1188$ km (*Nimmo et al.*, 2017), with an upper limit on the orbital eccentricity of e ~ 10^{-5} (*Brozović et al.*, 2015). Together, these features indicate that the binary is in tidal equilibrium. In this dual synchronous state, the rotational day of both bodies is equal to their mutual orbital period, P = 6.3872 days.

Tidal circularization timescales, which increase dramatically in wider orbits, suggest that the binary has always been close together (*Farinella et al.*, 1979; *Dobrovolskis et al.*, 1997). This would imply that Pluto and Charon formed closer to one another and moved apart due to transfer of angular momentum from Pluto and Charon's initially faster spins into the initially shorter-period mutual orbit (e.g., *Renaud et al., 2021*). The innermost distance at which Charon could have formed would be near the Roche limit, which is at ≈2.5 R_P for Pluto and Charon's densities. During tidal expansion of the mutual orbit, tides raised on Pluto and Charon by one another could have either increased or decreased the orbital eccentricity depending on the relative strength of tidal dissipation associated with combinations of spin and orbital frequencies at any given time (*Renaud et al., 2021*). The pace of orbital expansion, eccentricity change, and spin synchronization could have quickened or stalled in the vicinity of spin-orbit resonances. Eventually, the orbit became circular as the dual-synchronous state was reached (e.g., *Cheng et al.*, 2014a).

The Pluto-Charon secondary-to-primary mass ratio (q) is the largest of any known planet or dwarf planet satellite in the solar system, with q = 0.122 ± 0.0014 (e.g., *Stern et al.*, 2018). The mean semimajor axis is a = 19596 km in the Plutocentric frame, which together with the orbital period implies a total system mass $M_{PC} = (1.462 \pm 0.002) \times 10^{25}$ g (*Brozović et al.*, 2015). The total system angular momentum about the center of mass, due to Pluto and Charon's spins and their mutual orbit, is

$$L_{PC} = \omega K_P M_P R_P^2 + \omega K_C M_C R_C^2 + \omega \frac{M_C M_P}{M_{PC}} a^2 \\ = \frac{\sqrt{G M_{PC}^3 R_P}}{(1+q)} \left(\frac{a}{R_P}\right)^{1/2} \left[K_P \left(\frac{R_P}{a}\right)^2 + q K_C \left(\frac{R_C}{a}\right)^2 + \frac{q}{(1+q)} \right] \tag{1}$$

Here, $\omega = \sqrt{GM_{PC}/a^3}$ is the angular frequency of Pluto's spin, Charon's spin, and the binary's orbit. K_P and K_C are the moment of inertia constants for Pluto and Charon, defined from the moment of inertia I as $K = I/(MR^2)$ with K = 0.4 for a homogeneous sphere and K < 0.4 for a sphere with a central concentration of mass (e.g., due to differentiation into a rocky core and icy mantle). The system angular momentum is dominated by the final orbital angular momentum term in equation (1). Scaling by the quantity $L' \equiv M_{PC}R_{PC}^2 (GM_{PC}/R_{PC}^3)^{1/2} = (GM_{PC}^3 R_{PC})^{1/2}$, where R_{PC} is the radius of a body with Pluto's mean density that contains the total mass of Pluto and Charon, gives

$$J_{PC} \equiv \frac{L_{PC}}{L'} = \frac{1}{(1+q)^{7/6}}\left(\frac{a}{R_P}\right)^{\frac{1}{2}}\left[K_P\left(\frac{R_P}{a}\right)^2 + qK_C\left(\frac{R_C}{a}\right)^2 + \frac{q}{(1+q)}\right] \quad (2)$$

For a broad range of moment of inertia constants reflecting highly differentiated to undifferentiated interiors for Pluto and Charon (i.e, $0.3 \leq K_P, K_C \leq 0.4$), the scaled system angular momentum given by equation (2) falls in the range $0.386 \leq J_{PC} \leq 0.394$. A single body with moment of inertia constant K that is rotating fast enough to begin to shed mass (i.e, with centripetal acceleration close to or exceeding the gravitational acceleration at the surface) would have a scaled angular momentum $J = K$, with $K = 0.4$ being the upper limit corresponding to a uniform-density body. Thus, the Pluto-Charon system angular momentum is comparable to or greater than that in a body containing the total system mass rotating at the mass ejection limit. The high-angular-momentum budget is consistent with the presence of an even more radially extended small satellite system, as we discuss below.

2.3. Existence and Dynamics of the Four Small Circumbinary Moons

The system's origin(s) is further constrained by the four small moons — Styx (diameter axes 16 × 9 × 8 km), Nix (50 × 35 × 33 km), Kerberos (19 × 10 × 9 km), and Hydra (65 × 45 × 25 km) in order of increasing orbital distance (*Weaver et al.*, 2016) — discovered in the decade prior to the New Horizons flyby. The small moons, of combined mass $\approx 6 \times 10^{-6}$ times that of the binary (*Youdin et al.*, 2012; *Brozović et al.*, 2015), orbit it on circular, coplanar, and prograde orbits that are close to 3:1, 4:1, 5:1, and 6:1 mean-motion resonances with Charon. However, the moons are not actually in these resonances; indeed, because the Pluto-Charon orbit is so nearly circular, the predicted N:1 resonance widths are vanishingly narrow (*Mardling*, 2013; *Sutherland and Kratter*, 2019). In Pluto radii, the small moons' semimajor axes lie between approximately 36 and 55 R_P. The moon rotations are not tidally locked, and their known spin axes have high obliquities that are not aligned with those of Pluto and Charon (*Weaver et al.*, 2016). This is likely because for their large orbital radii and small sizes, tides cannot synchronize their spins in <1 G.y., longer than the timescale of perturbation by impacts (*Quillen et al.*, 2017, 2018). More details on the properties of the small moons are provided in section 4.1 and in the chapter by Porter et al. in this volume.

The New Horizons spacecraft searched for other Kerberos/Styx-sized regular moons ($\sim 10^{16}$ kg), but none were found down to a diameter of 1.7 km (for an albedo of 0.5) and out to semimajor axes $\sim 8 \times 10^4$ km ≈ 67 R_P (*Weaver et al.*, 2016). Smaller regular moons — if they exist — could be dynamically stable out to 5×10^5 km ~ 400 R_P, i.e, ~0.1 Pluto-Charon Hill radii (*Michaely et al.*, 2017). Bodies down to 1–2 km in size ($\sim 10^9$ kg) will be detectable with the upcoming James Webb Space Telescope, and if they exist, could help constrain the masses of the known small moons (*Kenyon and Bromley*, 2019). New Horizons also conducted an extensive search for rings comparable to those surrounding other solar system bodies, but no such structures at Pluto were seen (*Lauer et al.*, 2018). While ring debris is unstable for most orbits between the known satellites (e.g., *Youdin et al.*, 2012; *Smullen and Kratter*, 2017; *Woo and Lee*, 2018), wider-orbit debris could be long-lived. Thus, although future observations might change this picture, the Pluto system as we know it today appears dust-free and quite compact, with outermost Hydra having nearly an order-of-magnitude smaller orbital radius than that of the largest possible bound orbit.

2.4. Densities and Compositions of Pluto, Charon, and the Small Moons

Origin models must account for the observed compositions of members of the Pluto system, in particular the ratio of ice to rock (silicates, metals, and carbonaceous material) in their interiors as inferred from their bulk densities and/or surface compositions. Abundances of volatile ices provide additional constraints.

The bulk densities of Pluto and Charon differ sufficiently as to imply different bulk compositions. Charon's bulk density [1700 kg m^{-3} (*Nimmo et al.*, 2017)] is lower than that of Pluto (1854 kg m^{-3}), implying that Charon is likely somewhat icier; porosity alone is unlikely to fully account for this difference in density (*McKinnon et al.*, 2017; *Bierson et al.*, 2018). Because the bulk densities of Pluto and Charon are higher than those of (even compressed) ice, their interiors must include denser "rocky" material. This rock is likely mostly comprised of silicates, metals, and carbonaceous material (*McKinnon et al.*, 2008) by analogy with carbonaceous chondrites (e.g., *Howard et al.*, 2011), interplanetary dust particles, and non-ice material detected in the outer solar system such as in the Saturn system (*Hsu et al.*, 2015; *Waite et al.*, 2018; *Postberg et al.*, 2018; *Tiscareno et al.*, 2019) and comets (*Hanner and Bradley*, 2004; *Ishii et al.*, 2008). For a composition of water ice and solar-composition rock, their bulk densities imply that Pluto has roughly two-thirds rock and one-third ice by mass, while Charon has 60% rock and 40% ice (*McKinnon et al.*, 2017).

Although the bulk densities of Pluto and Charon differ, this difference is small compared with the range of densities observed across large KBOs. Density estimates for KBOs with diameters >500 km vary from ~0.8 g cm^{-3} (for 2002 UX_{25}) to ≥2.5 g cm^{-3} [for Eris (e.g., *McKinnon et al.*, 2017)]. When viewed in this context, Pluto and Charon's densities and bulk compositions are quite similar.

The current degree of internal differentiation of Pluto and Charon is uncertain. The New Horizons mission was not designed to perform gravity measurements needed to infer moments of inertia to constrain their internal structures. Heating by the energy of their accretion, the decay of long-lived radionuclides, and mutual tides (until their spins and mutual orbit synchronized and their mutual orbit became circular) would have likely caused the interiors of Pluto and Charon to have reached the melting temperature of ice, allowing the settling of denser rock to form a core. Thus, interior models typically assume fully differentiated current structures (e.g., *Robuchon and Nimmo*, 2011; *Hammond et al.*, 2016; *Kamata et al.*, 2019).

It is possible (especially for Charon) that rock-ice separation never occurred in the frigid outer layers, where cold rock and ice may not move even over billions of years. The configuration of a denser, undifferentiated crust atop a lower-density ice mantle produced through interior differentiation (e.g., driven by radiogenic heating) is formally gravitationally unstable, but this configuration may persist for the age of the solar system (*Rubin et al.*, 2014). Moreover, differentiation by settling of rock through melted ice may have been impeded by convective homogenization in a liquid or solid mantle. Differentiation may also be inefficient if the rock is predominantly fine-grained as observed in carbonaceous chondrite matrix and interplanetary dust particles, both thought to be good proxies for the non-ice material accreted by outer solar system bodies (*Bland and Travis*, 2017; *Neveu and Vernazza*, 2019).

The above caveats notwithstanding, geological features on both Pluto and Charon seem most consistent with significant differentiation (e.g., *Stern et al.*, 2015; *Moore et al.*, 2016). For example, their surfaces lack compressional features that would be expected for undifferentiated interiors due to the formation of denser, high-pressure ice II at depth as the bodies cooled (e.g., *McKinnon et al.*, 1997; 2017; *Hammond et al.*, 2016). Instead, extensional tectonic features are observed that are most easily explained by differentiated interiors in which global expansion occurred as early water oceans froze, forming an ice I shell (e.g., *Stern et al.*, 2015; *Moore et al.*, 2016; *Beyer et al.*, 2017). What the current interiors of Pluto and Charon tell us about the differentiation state of the progenitor bodies in an impact origin remains somewhat uncertain. This proves to be an important issue, because impact outcome depends strongly on the progenitor interior states, as discussed in section 3.3 and Fig. 3.

The masses of the tiny moons, and therefore their bulk densities, were unconstrained by New Horizons data (*Stern et al.*, 2015; *Weaver et al.*, 2016), and therefore knowledge of their compositions is based to date on observations of their surfaces. The small moons have estimated geometric albedos ranging from 0.55 to 0.8 (*Weaver et al.*, 2016), which are much higher (*Stern et al.*, 2018) than those of similarly sized KBOs, whose albedos are typically between 0.02 and 0.2 (*Johnston*, 2018, and references therein; *Stern et al.*, 2019). The small moon albedos imply ice-rich surfaces, consistent with near-infrared spectra of Nix, Hydra, and Kerberos that show absorption bands associated with crystalline water ice, as well as the presence of an ammoniated species on Nix and Hydra (*Cook et al.*, 2018).

Whether the small moons contain any significant amount (more than a few percent) of non-ice material is unknown given their unconstrained bulk density and porosity. However, it seems most probable that the interiors of the tiny moons, like their surfaces, are predominantly water ice. Based on Nix and Hydra crater counts (see section 2.5), the small moons seem to be ancient, with surfaces that have been subject to billions of years of impacts by darker exogenic material. Such impacts would be primarily erosive: Characteristic encounter velocities with the Pluto system (~1–2 km s^{-1}, i.e, the dispersion of Keplerian velocities in the Kuiper belt) are much greater than the escape velocity from Pluto at the small moon distances (~0.15–0.2 km s^{-1}), so that impacting dark material would likely be lost to escaping orbits (*Weaver et al.*, 2019). Indeed, if ejecta from impacts onto the small moons had been typically retained in the Pluto system, more uniform albedos for all the satellites would result (*Stern*, 2009), contrary to the New Horizons observations. However, even erosive impacts that did not directly deliver dark material to the moons would nonetheless darken their surfaces over time if material at depth within the moons exposed by the impacts was dark and rocky. Thus, the simplest way to explain how the tiny moons have maintained bright surfaces for billions of years is to presume that their interiors are ice-rich like their surfaces (e.g., *McKinnon et al.*, 2017; *Weaver et al.*, 2019).

New Horizons data have also shed light on the distribution of volatiles on Pluto, Charon, and the small moons. Pluto displays surface CH_4, N_2, and CO ices (*Grundy et al.*, 2016), with detections of H_2O "bedrock" and ammonia ices at select locations thought to have been sourced from Pluto's interior (*Dalle Ore et al.*, 2019). Its surface is red owing to organic carbonaceous material that may be partly due to atmospheric and subsequent surface chemistry starting with the photolysis of atmospheric CH_4 (*Grundy et al.*, 2018) and partly due to endogenic processes (*Sekine et al.*, 2017; *Cruikshank et al.*, 2019). Solid H_2O blankets airless Charon, and New Horizons compositional mapping has confirmed earlier remote observations of percent-level NH_3-bearing species (*Grundy et al.*, 2016; *Cook et al.*, 2018), as well as an unknown dark absorber (*Buie and Grundy*, 2000; *Cruikshank et al.*, 2015).

2.5. Timing of Pluto System Formation

Constraints on when the Pluto system formed can be derived from several lines of reasoning based on estimated surface ages, dynamical models of planet migration and orbital instability, planet accretion models, and inferences about the limited role of heating by short-lived radiogenic elements as outer-solar-system material was assembled.

Crater counts made by the New Horizons spacecraft suggest that Pluto, Charon, Nix, and Hydra are all ancient objects with maximum surface ages ≥4 G.y. (e.g., *Moore et al.*, 2016; *Robbins et al.*, 2017; *Singer et al.*, 2019). Thus, these individual objects — and probably the entire Pluto system — appear to have developed surfaces capable of retaining an impact record quite early in solar system history.

As described in section 2.1, the influence of Neptune over the system's orbit suggests that the bodies that eventually formed Pluto and Charon accreted closer to the Sun (at 15–30 AU) than today (e.g., *Levison et al.*, 2008, 2009), with their orbital expansion and excitation caused by a rearrangement of the giant planets including the outward migration of Neptune. The timing of Neptune's migration and related giant-planet orbital instability is not firmly established by dynamical models alone, because it depends on the assumed properties of the planetesimal disk at the time the solar nebula dissipated and planetesimal-driven migration began to dominate (e.g., *Gomes et al.,* 2005). Late instability models were initially advocated as a means of triggering a late heavy bombardment that produced many large basins on the Moon some ~700 m.y. after the solar system's formation (e.g., *Gomes et al.*, 2005). However, the lack of evidence so far for cometary impactors in lunar materials that date to that era appears to undermine this argument (e.g., *Joy et al.*, 2012), because comets would have dominated the impactor flux onto the Moon during the instability (e.g., *Nesvorný*, 2018).

Instead, recent works argue that planetesimal-driven migration and outer-planet instability likely occurred much earlier. A late instability after the terrestrial planets were assembled would dynamically overexcite or destabilize their orbits (e.g., *Agnor and Lin*, 2012; *Kaib and Chambers*, 2016). The simplest way to avoid this is for the instability to have occurred well before the terrestrial planets completed their accretion at ~100 m.y. An early instability may also be the most probable outcome across a broad range of plausible planetesimal disk conditions (e.g., *Deienno et al.*, 2017), and appears needed to explain the survival of the Patroclus-Menoetius binary member of Jupiter's Trojan asteroids (*Nesvorný et al.*, 2018). In combination with a slow and "grainy" migration of Neptune (as appears necessary to account for the inclination distributions and overall proportions of resonant KBOs per section 2.1), this suggests that Pluto-sized bodies had formed in the 15–30-AU region by tens of millions of years after the solar nebula dissipated (e.g., review by *Nesvorný*, 2018).

Until the last decade, it was thought that large outer-solar system bodies formed through solely two-body collisions and hierarchical coagulation. It is possible to form a few Pluto-sized objects at <30 AU in <100 m.y. through this mode of growth, and to produce an overall Kuiper belt size distribution comparable to that seen today (e.g., *Kenyon and Bromley*, 2012; *Schlichting et al.*, 2013). However, features of the Kuiper belt size distribution predicted from such models appear inconsistent with the dearth of small craters in images of Pluto and Charon obtained by New Horizons (*Stern et al.*, 2018; *Morbidelli and Nesvorný*, 2019, and references therein). More fundamentally, while hierarchical coagulation models can reproduce the current Kuiper belt, they do not appear to naturally produce the more massive and numerous population of bodies implied by planet migration models and conditions needed to account for the properties of resonant KBOs (e.g., *Nesvorný and Vokrouhlicky*, 2016; *Morbidelli and Nesvorný*, 2019).

Instead, the currently favored mechanism for the growth of large planetesimals is the streaming instability (e.g., *Youdin and Goodman*, 2005; *Johansen et al.*, 2015; *Morbidelli and Nesvorný*, 2019; *Nesvorný et al.*, 2019). Initial grains in the circumstellar disk likely aggregated into millimeter-sized pebbles due to surface sticking forces and energy loss in inelastic impacts (e.g., *Dullemond and Dominik*, 2005; *Chambers*, 2016). Streaming instability can occur due to the interactions of such pebbles with the background gaseous solar nebula. The pressure-supported gas orbits at a somewhat slower azimuthal velocity than would a particle at the same orbital radius on a purely Keplerian orbit. As such, particles orbiting in the gas experience a "headwind" as they encounter the slower orbiting gas, which causes the particle orbits to lose energy and drift radially inward. If as particles drift inward a local concentration of particles forms, this concentration will accelerate the local gas somewhat, lessening the rate of the concentration's inward drift. The concentration can then continue to grow by accreting outer particles that are drifting inward more rapidly, and as the concentration grows, its effects on the gas strengthen, allowing its drift to slow further and its growth to continue. If this positive feedback results in the local spatial density of solids becoming sufficiently high, the concentration can rapidly gravitationally collapse to form ~10^2-km-class planetesimals directly from vastly smaller pebbles (*Johansen et al.*, 2015; *Simon et al.*, 2016). This mechanism appears uniquely able to reproduce observed features such as the predominance, color similarity, and distribution of obliquities of Kuiper belt binaries (*Morbidelli and Nesvorný*, 2019; *Nesvorný et al.*, 2019), in addition to other properties of the terrestrial and giant planets (e.g., *Levison et al.*, 2015).

Pairwise collisions between planetesimals can lead to the accretion of larger bodies bound by self-gravity, but this mode of growth is extremely slow in the outer solar system. If instead planetesimals grow primarily via the accretion of pebbles in the presence of the gas disk, a dramatically faster mode of growth can ensue. In so-called "pebble accretion," pebbles within a preferred size range are slowed by gas drag as they pass in close proximity to a planetesimal. The drag causes pebbles that would otherwise be scattered away to instead spiral into the planetesimal and be accreted, greatly increasing the planetesimal's effective accretional cross-section and its growth rate.

Models suggest that 10^2-km bodies could have formed near 25 AU via streaming instability in as little as a few million years (e.g., *Ormel and Klahr*, 2010, *Johansen et al.,* 2015; *Lambrechts and Morbidelli*, 2016), with continued growth via pairwise collisions and pebble accretion producing in ≈3 m.y. a large-body size distribution consistent with that needed to explain current Kuiper belt properties [e.g., with a thousand Pluto-scale objects (*Morbidelli and Nesvorný*, 2019)], assuming continued presence of the background gas disk. When conditions favorable to streaming instability actually occurred in the outer solar system is uncertain. The instability requires an enhanced solids-to-gas ratio, which could be achieved as the gas disk started to dissipate or earlier if there were solid-concentrating mechanisms such as settling and/or radial drift (e.g., *Johansen et al.*, 2009; *Morbidelli and Nesvorný*, 2019, and references therein). Dating of meteorite phases that form in the presence of gas indicates that the gas disk beyond Jupiter lasted at least 3–5 m.y. after the "time zero" of solar system formation, defined as the time of condensation of calcium-aluminum-rich inclusions (CAIs) (*Russell et al.*, 2006; *Desch et al.*, 2018), arguing for large-body growth via streaming instability and pebble accretion by this time.

In contrast, planetesimal accretion times no earlier than 5 m.y. seem implied by the dearth of aqueous and volatile processing on outer solar system small bodies and fragments thereof (i.e., interplanetary dust particles), which for earlier formation times would have resulted in intense heating from the decay of short-lived radionuclides such as ^{26}Al, whose half-life is 0.7 m.y. (*Davidsson et al.*, 2016; *Neveu and Vernazza*, 2019). Formation after the decay of short-lived radionuclides could be a reason why at least some of Pluto's CO, which is unstable to aqueous processing (*Shock and McKinnon*, 1993), has survived to the present day. However, this argument is nuanced by the existence of aqueously altered silicate rock and of metal differentiated from silicates in meteorites linked to parent outer solar system small bodies (*Hiroi et al.*, 2001; *Kruijer et al.*, 2017), suggesting that at least some outer solar system planetesimals formed earlier than 5 m.y. In the more interior giant planet region, large planetesimals clearly formed prior to this time, as is needed to account for Jupiter and Saturn's cores and their large-scale gas accretion prior to nebular dispersal.

Taken as a group, the above arguments suggest that formation of large, 10^2–10^3-km-class bodies via streaming instability + pebble accretion occurred in the 15–30-AU region within a few to 5 m.y. after CAIs. Pairwise collisions between such bodies, including a Pluto-Charon-forming impact, would have continued after nebular dispersal over the subsequent millions to tens of millions of years (see section 5.1). The considerations discussed in this section leave open the timing of the Pluto system's formation relative to Pluto's capture into the 3:2 resonance with Neptune. In the simplest case, Pluto and its moons could have formed prior to resonance capture, with the system surviving the subsequent heliocentric migration and resonant trapping (*Pires et al.*, 2015; *Nesvorný and Vokrouhlický*, 2019; *Nesvorný et al.*, 2019). Given arguments in favor of an early outer-planet instability, this suggests a Pluto-system-forming event within the first few tens of millions of years after CAIs. Conversely, if Pluto (or proto-Pluto) was trapped in the 3:2 resonance first, a hypothesized Charon-forming impact (section 3) could have dislodged the pair from resonance (*Levison and Stern*, 1995; *Hahn and Ward*, 1995; *Stern et al.*, 1997), but it could have been recaptured (e.g., *Dobrovolskis et al.*, 1997). This might allow for later Pluto-system formation times $\sim 10^8$ yr after CAIs. We return to this issue in section 5.1.

3. ORIGIN OF PLUTO-CHARON BY A GIANT IMPACT

Even prior to the development of quantitative impact models, the Pluto-Charon binary's low mass ratio, high angular momentum, and close separation made an impact with Pluto by a similarly sized impactor the favored origin scenario (*McKinnon*, 1984, 1989; *Stern et al.*, 1997). In the past two decades, impact simulations (*Canup,* 2005, 2011; *Arakawa et al.*, 2019) have demonstrated not only the ability of this scenario to account for such characteristics, but also the potential of a giant-impact origin to provide insight into the physical state of the progenitors and the timing of binary system formation. However, alternative origin models have also been proposed (Fig. 2).

3.1. Non-Giant-Impact Formation Scenarios

Fission of a fast-spinning Pluto could explain a resulting high-angular-momentum binary (*Mignard*, 1981; *Lin*, 1981). However, the amount of spinup needed to launch material into a ring from which Charon accretes (*Tancredi and Fernandez*, 1991) seems most easily achievable with a giant impact (*Stern et al.*, 1997). Indeed, some current giant-impact-origin models (e.g., Fig. 5) are akin to impact-induced fission.

Intact capture of Charon into Pluto orbit could be enabled by dynamical friction from surrounding small bodies (*Goldreich et al.*, 2002) or pebbles. This scenario tends to preferentially produce retrograde binaries (like Pluto-Charon) that are initially on widely separated orbits (*Schlichting and Sari*, 2008). Continued dynamical friction and/or scattering interactions cause the binary orbital radius to shrink, perhaps providing an alternative to the standard tidal expansion model as a means of accounting for the current Pluto-Charon dual-synchronous state. For example, dynamical friction could cause Charon's orbit to decay inward until it was balanced by the torque due to tides raised by Pluto and Charon on one another, with Charon's orbit eventually tidally expanding to the synchronous state as the small-particle disk dissipated and dynamical friction ended. However, this capture mechanism was primarily proposed to explain KBO binaries with $\sim 10^2$-km-radius objects (*Goldreich et al.*, 2002). Capture of much larger Charon appears to require an improbably dense disk (*Stern et al.*, 2018). Furthermore, it would seem difficult to explain the small moons orbiting beyond Charon in this scenario. They would presumably have to form after Charon's capture and orbital contraction through a separate, later event(s), which would not naturally explain their coplanarity and similar sense of rotation with the binary.

Co-accretion of the Pluto-Charon binary was dismissed by *Stern et al.* (1997) as not being able to account for the Pluto system's high obliquity and high angular momentum. However, the modern version of this scenario is distinct from what was termed co-accretion in older lunar origin studies (and in *Stern et al.*, 1997), owing to developments in the understanding of planetesimal formation discussed in

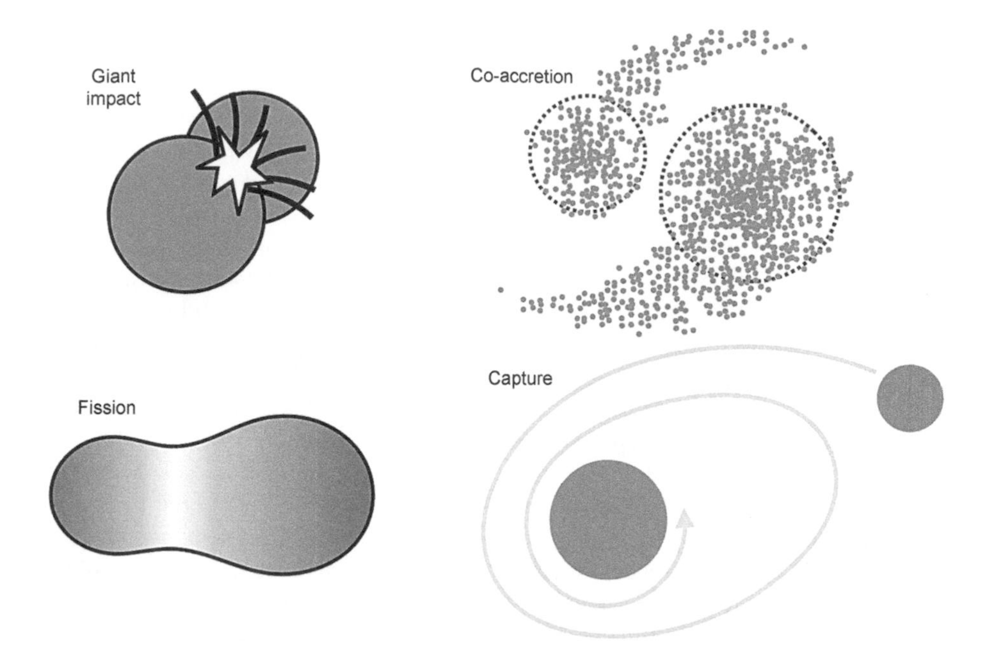

Fig. 2. Possible scenarios for the formation of a planetary binary. Each has been postulated as being responsible for the formation of Pluto-Charon or of other moon-bearing transneptunian objects: giant impact, currently the leading scenario for Pluto-Charon (*McKinnon,* 1984; *Canup,* 2005, 2011); fission (*Lin,* 1981; *Mignard,* 1981); co-accretion (*Nesvorný et al.,* 2010, 2019); and capture (*Goldreich et al.,* 2002).

section 2.5. Gravitational collapse induced by the streaming instability now appears to be a compelling mechanism for producing binary planetesimals up to ~10^2 km in size (*Nesvorný et al.*, 2019). As an azimuthal mass concentration produced by streaming instability undergoes self-gravity-driven collapse into planetesimals, the high specific angular momentum of a collapsing clump can prevent accretion into a single body and instead yield a high-angular-momentum binary (*Nesvorný et al.*, 2010). A high obliquity and mass ratio similar to those of Pluto-Charon are possible (although not the likeliest) outcomes of accretion simulations (*Nesvorný et al.*, 2019). However, direct formation of a binary as massive as Pluto-Charon by this process is unlikely given the tendency for such a large mass concentration to fragment before collapse is complete (*Youdin and Goodman*, 2005; *Li et al.,* 2019). If formed together by direct gravitational collapse, Pluto and Charon should have similar compositions, which is not quite the case although the difference in their bulk densities is small [8% (*McKinnon et al.*, 2017)]. Co-accretion of the small moons is an enticing feature of this alternative scenario, because material could potentially be initially placed near the current moon orbits, removing the need for substantial outward migration of the moons or their source material that has proven elusive in the impact scenario (section 4). However, co-accretion would produce moons with compositions similar to those of Pluto and Charon, and this appears inconsistent with their currently inferred ice-rich compositions (*Stern et al.*, 2018). Overall, it thus still appears unlikely that Pluto and Charon formed by co-accretion.

3.2. Dynamical Constraints on a Pluto-Charon-Forming Impact

Consider that a collision produces the current binary masses and angular momentum. The scaled angular momentum delivered by a single giant impactor of mass $M_i = \gamma M_T$, where M_T is the total colliding mass, is

$$J_i = \sqrt{2}\, f(\gamma)\, b \left(\frac{v_i}{v_{esc}}\right) \quad (3)$$

where $b = \sin \xi$ is a scaled impact parameter, ξ is the impact angle (with b = 1 and $\xi = 90°$ corresponding to a grazing impact), v_i and v_{esc} are the impact and mutual escape velocities, respectively, and $f(\gamma) = \gamma(1-\gamma)\sqrt{\gamma^{1/3} + (1-\gamma)^{1/3}}$ (e.g., *Canup et al.*, 2001). With $v_i/v_{esc} \approx$ unity, which minimizes the mass and angular momentum ejected onto escaping orbits by the collision, the maximum angular momentum delivered by the grazing impact of equal-sized bodies ($\gamma = 0.5$) would be $J_{i,max} \approx 0.45$; a similar impact with $\gamma = 0.3$ would yield $J_{i,max} \approx 0.37$. Per equation (2), the Pluto-Charon binary has $0.386 \leq J_{PC} \leq 0.394$. Thus, a highly oblique impact between bodies comparable in size to Pluto itself is needed to account for the current system angular momentum.

It is of course likely that before a Charon-forming collision, proto-Pluto and the impactor would have already been rotating due to the cumulative effects of impacts during their accretion, and their spin angular momenta would have also contributed to that of the final system after a Charon-forming impact. The rate and directionality of such pre-giant impact spins is highly uncertain; the shortest possible progenitor pre-impact day would be about 2.4 h, corresponding approximately to the limit at which a Pluto-density body would begin to shed mass.

Charon's large mass compared with that of Pluto strongly constrains the type of collision that could have produced the binary. A very broad range of collisions can produce satellites with masses between $O(10^{-3})$ to few $\times$ 10^{-2} times the primary's mass (e.g., *Canup*, 2014; *Arakawa et al.* 2019). However, only a much narrower range of collisions can produce a moon with $q \geq 0.1$. In general, producing a Charon-sized moon requires a very large, oblique impact (often in combination with some pre-impact rotation) in order to maximize the total angular momentum, together with a low impact velocity to minimize the escaping mass and maximize the retained angular momentum. Higher-impact-velocity collisions that result in substantial mass loss can form large satellites, but no cases that yield $q \geq 0.1$ have been found for impact velocities in excess of 1.2 v_{esc} across many hundreds of simulations that span $1 \leq (v_i/v_{esc}) \leq 1.7$, $30 \leq \xi$ (°) ≤ 75 and both differentiated and undifferentiated progenitor states (*Arakawa et al.*, 2019).

3.3. Physical State of Proto-Pluto and Giant Impactor

The above angular momentum arguments imply collisional progenitors that are both intermediate in size to Pluto and Charon (i.e., with masses between 0.3 and 0.5 M_{PC}). We here consider the range of likely interior states for the target and impactor at the time of a Pluto-system-forming impact, guided by both observational constraints from current Pluto and Charon (section 2), and simplified theoretical models for progenitor accretion.

3.3.1. Constraints on progenitor interiors based on their formation histories. As described in sections 3.4–3.5, a key parameter that strongly affects impact outcome is the distribution of ice and rock in the progenitor interiors at the time of their collision. Possible states of ice-rock differentiation range from a homogeneous, mixed ice-rock interior to a fully differentiated interior with a rocky core overlaid by an ice shell. The degree of differentiation along this spectrum depends on the progenitors' thermal state and, potentially, on the size distribution of rock grains (Fig. 3).

The onset of ice-rock differentiation in a homogenous body requires the onset of melting of ice in order for rock to settle toward the center due to its higher density. In the absence of accretional heating, the first regions to heat up to the melting temperature of ice are at the center that is most insulated from the cold surface. The melting temperature of ice depends on its composition: It is around 273 K for pure water ice across relevant pressures, but the solidus (temperature of first melt) of NH_3-bearing ice is depressed to 176 K if the ice contains even small, percent-level amounts of NH_3 (which is the case in the Pluto system; section 2.4). Once melting ensues, the ice viscosity plummets by orders of magnitude (*Arakawa and Maeno*, 1994) to allow rock settling (*Desch et al.*, 2009). The chief interior heat source in isolated bodies is the decay of short- and long-lived radionuclides. In section 2.5 we argue against a major role for short-lived radiogenic heating in the progenitor interiors, based on a likely timescale of several million years needed for progenitor accretion (e.g., *Morbidelli and Nesvorný*, 2019) and a lack of widespread

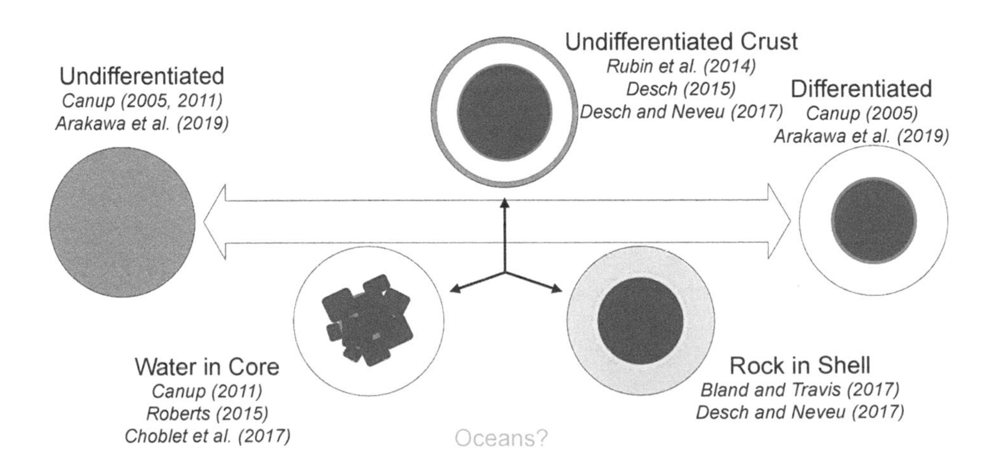

Fig. 3. Degeneracy in possible internal structures (drawn very roughly to scale) of the ~10^3-km-radius collisional progenitors that could yield Pluto and Charon with their observed densities. While a fully homogeneous structure or a fully differentiated structure (in the limit of an extremely grazing impact) can yield the Pluto-Charon binary (*Canup,* 2005; *Arakawa et al.,* 2019), a partially differentiated structure appears needed to simultaneously produce a dispersed debris disk from which the small moons could accumulate (*Canup,* 2011). However, in the spectrum between these end members, geophysical processes acting between the time of progenitor formation and that of impact can lead to mixtures of rock and water in various proportions in the core, mantle, and/or crust. The oceans may be much thicker than depicted if ≥40% of the kinetic energy of accreted planetesimals is retained (see Fig. 4e–f).

observed thermal processing of outer solar system materials (e.g., *Neveu and Vernazza*, 2019).

Even so, the buildup of heat from the decay of remaining long-lived radionuclides can by itself initiate internal ice-rock differentiation (but not rock-metal separation) within tens of millions of years of progenitor formation (Figs. 4a,b). In this mode, differentiation proceeds from inside-out. Once differentiation proceeds out to a large enough radius as to allow the growth of Rayleigh-Taylor "lava-lamp" gravitational instabilities, such instabilities can take over in overturning the denser homogeneous outer crust overlying the differentiated mantle richer in ice. Rayleigh-Taylor instabilities grow at ice-crust interfaces warmer than about 150 K, a threshold that depends modestly on other parameters (e.g., grain size) because of the tremendous dependence of viscosities on temperature (*Rubin et al.,* 2014). For formation ≥5 m.y. after CAIs and neglecting accretional heating, this 150 K threshold was likely not reached within tens of kilometers of the surface, assuming a surface temperature of ≈100 K around 20 AU and a thermal gradient of ≈1 K km^{-1} typical of the insulating properties of ice at these temperatures and heat fluxes on 1000-km-class bodies. This is confirmed in numerical simulations (*Desch and Neveu*, 2017). This could lead to progenitors with an outer undifferentiated crust, thick enough to withstand bombardment by projectiles up to tens of kilometers in size until the time of the giant impact (*Desch*, 2015). This crust could comprise 10–20% of the total mass of objects 800–1000 km in radius (Fig. 4b).

Accretional heating can make differentiation occur earlier and affect the distribution of material in the progenitor's interior. Consider a simplified estimate in which some fraction of the accretional kinetic energy is symmetrically deposited at depth and retained within a growing progenitor. At each progenitor radius R, a fraction h of the impacting energy is deposited into an accreting layer, which increases the layer temperature by ΔT via the energy balance (e.g., *Stevenson et al.*, 1986, and references therein)

$$\frac{1}{2}h\dot{M}v_i^2 = C_p \Delta T \dot{M} \quad (4)$$

where $\dot{M}$ is the mass accretion rate, $v_i^2 = v_{rel}^2 + v_{esc}^2$ is the square of the impact velocity of accreting material, v_{rel} is the relative velocity at large separation, and C_P is the specific heat of the accreting material. The radial temperature profile due to accretional heating alone is then

$$T(R) = T_0 + \frac{4h\chi G\pi\bar{\rho}}{3C_p}R^2 \quad (5)$$

where $\chi = v_i^2/v_{esc}^2$. The (notoriously) least certain quantity in such a treatment is h. For small accreting material, h would be small too. But if, as currently thought, the progenitors grew by accreting a substantial fraction of their mass in ~10^2-km planetesimals formed via the streaming instability, a value h ~ tens of percent could apply. In Figs. 4c,d, we apply the *Desch and Neveu* (2017) model to a progenitor whose initial temperature profile is calculated from equation (5), with T_0 = 40 K (as in Figs. 4a,b), h = 0.2 (i.e., 20% of accretional kinetic energy retained as the progenitor formed), χ = 2 (i.e., a relative velocity comparable to the progenitor escape velocity), and $\bar{\rho}$ = 1.84 g cm^{-3}. The calculation assumes that progenitor accretion was completed at 7 m.y. post CAIs, and neglects any radiogenic heat deposited within the progenitor or its components prior to that time. Differentiation commences in the outer regions of the progenitor within less than 10 m.y., creating an outer ice shell overlying a thin water-NH_3 ocean. Similar outcomes are obtained with h = 0.1 if the progenitor forms earlier at 5 m.y. post-CAIs; later formation times and/or smaller h lead to evolutions similar to the no accretional heating case (Figs. 4a,b), only with somewhat earlier differentiation times. These results hold up to h = 0.35.

For h ≥ 0.4, as may be most appropriate for large planetesimal impacts, the outer regions start out melted (Figs. 4e,f). In this mode, differentiation starts during accretion and proceeds from outside-in, because temperatures initially decrease with depth (equation (5)). Although this is not captured in the model shown in this figure, rock and ice should separate in the melted outer regions, leading to a temporary interior structure with undifferentiated central regions surrounded by a denser (i.e., gravitationally unstable) rocky layer, itself enveloped by a liquid water ocean and ice shell. Differentiation by settling of rock in melted ice is complete by 30 m.y. (for h = 0.4) and potentially earlier if the gravitationally unstable rocky layer is overturned. The accretional heat can subsequently sustain an ocean more than 100 km thick until at least 60 m.y. As the ocean refreezes, there is at 60 m.y. a transition from a thin, essentially conductive ice shell above the vigorously convective ocean to a thick convective ice shell above a thin liquid layer. This decreases the overall efficiency of heat transport between the core and the surface, causing the core to warm up. Thus, this simulation suggests that for high h the structure of the Pluto-Charon progenitors at the time of the binary-forming impact would depend rather sensitively on the timing of this event relative to progenitor accretion.

3.3.2. Degeneracies in possible progenitor internal structures. The thermal evolution models in Fig. 3 assume that rock from zones above the melting temperature is moved instantaneously to the progenitor center. This idealized, discrete layer picture (e.g., homogeneous crust surrounding an ice mantle surrounding a rocky core, or a pure ice shell overlying a rocky core) must be nuanced by two considerations: (1) that the rocky core could be hydrated and/or retain pore ice, and (2) that the ice mantle could retain fine rock grains. Interior temperatures low enough for such structures can persist for tens of millions of years or longer, in part because much energy goes into melting ice rather than raising temperatures and in part because in regions warm enough for ice to be melted, convective heat transport by water circulation prevents

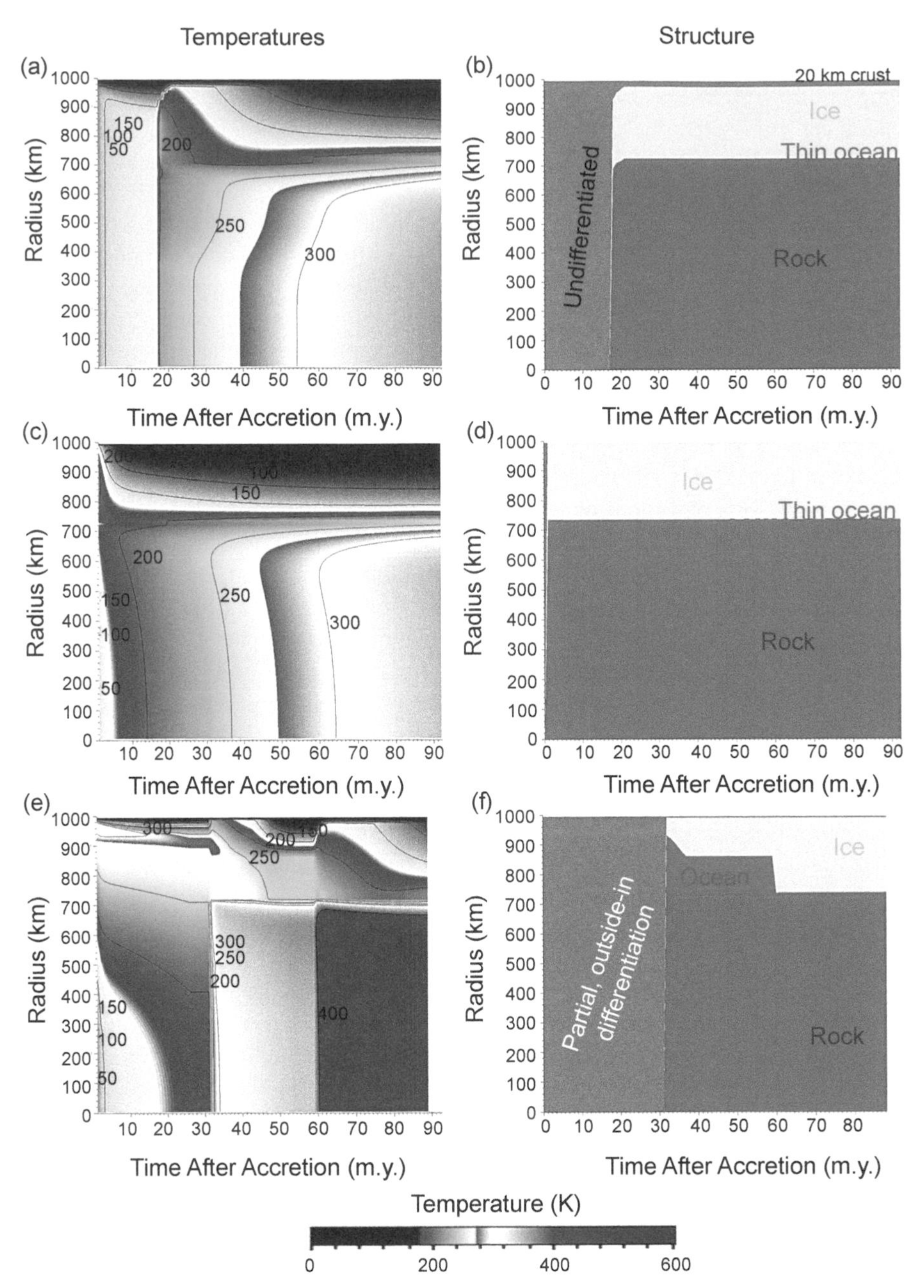

Fig. 4. Example thermal and structural evolution of a Pluto-Charon progenitor. **(a),(b)** Evolution of a progenitor formed 5 m.y. after Ca-Al-rich inclusions with no accretional heating. The progenitor forms cold (interior temperature = surface temperature) and homogeneous. Primarily long-lived radioactivity warms the interior [**(a)**] to trigger ice-rock differentiation, leading to a partially differentiated structure with a frigid outer ≈20 km that remains as an undifferentiated crust [**(b)**]. In **(a)** and **(c)**, contours and colors indicate temperature in Kelvin, with a transition to blue occurring at 176 K at the H_2O-NH_3 eutectic and a transition to orange indicating water ice melting at 273 K. **(c),(d)** Evolution of a progenitor formed 7 m.y. after CAIs, with an initial radial temperature profile set by accretional heating, assuming 20% of accretional kinetic energy is retained (h = 0.2; see section 3.3.1). Differentiation commences in the outer regions of the progenitor within less than 10 m.y., creating an outer ice shell overlying a water-NH_3 ocean. **(e),(f)** Evolution of a progenitor formed 5 m.y. after CAIs, assuming 40% of accretional energy is retained (h = 0.4). The outer layers are partially or fully melted from the outset, but because heat is more deeply buried, the warmer ice at depth conducts heat downward from these upper layers more slowly, which stalls melting of the central regions until 30 m.y. after accretion. During that time, differentiation proceeds from the outside-in within the melted layers, although the resulting advection of material and heat is not simulated here until the central layers melt. Post-differentiation, all the melted ice is in a >100-km-thick ocean that lasts another 30 m.y. until only a thin ocean persists below a convective ice shell. In the evolution models here, ammonia is assumed to be present at 1 mass% relative to H_2O and depresses the melting point of ice. The rock in differentiated regions is assumed to partition 100% into the core, leading to distinct compositional layers. However, this picture may be oversimplified: The rocky core could retain pore water, and fine-grained rock may remain suspended in the ocean and/or ice mantle rather than settle down to the core. Results based on the model of *Desch and Neveu* (2017).

temperatures from further rising significantly. These degeneracies lead to a continuum of possible structures for the progenitors (Fig. 3).

The presence of substantial water in a rocky core (both mixed with and bound in hydrated silicates) has been invoked to explain the surprisingly low level of ice-rock differentiation of even warm (ocean-bearing) outer-solar-system moons such as Enceladus (*Roberts*, 2015; *Choblet et al.*, 2017; *Neveu and Rhoden*, 2019), Titan (*Castillo-Rogez and Lunine*, 2010; *Iess et al.*, 2010), or Callisto (*Anderson et al.*, 1998, 2001), as well as the enhanced tidal dissipation that results for Enceladus to explain its sustained high geologic activity. The assumption that the Pluto-Charon progenitors had ice shells overlying rocky cores that retained water was made by *Canup* (2011) in models that produced both an intact Charon and a low-mass disk from which the small moons could potentially accrete (section 3.5).

That substantial rock could be present in an icy mantle has been suggested in the context of carbonaceous chondrite parent bodies (*Bland et al.*, 2013; *Bland and Travis*, 2017) and related protoplanets such as Ceres (*Neveu and Desch*, 2015; *Travis et al.*, 2018). Much of the rock in carbonaceous chondrites and interplanetary dust particles is in micrometer-sized grains that may take time to settle even if internal temperatures promote ice-rock separation upon partial or full melting of ice. Competition between settling and convective resuspension could prevent the fine grains from being segregated into the rocky core. The consequences of this degeneracy on possible progenitor structures was discussed in further detail by *Desch and Neveu* (2017).

3.3.3. Summary of possible structures and implications for the giant impact model. The masses, ice-rock contents, and likely formation time of Pluto and Charon are consistent with a continuum of differentiated structures for the progenitors. Endmember structures include an icy mantle overlying an undifferentiated rock-ice core (*Canup*, 2011), an undifferentiated crust surrounding a differentiated interior with an ice mantle and rock core (*Desch*, 2015), and a mantle of fine-grained rock and ice above a rock core (*Desch and Neveu*, 2017), as shown in Fig. 3. In all cases, the progenitors are large enough (radius $\approx 10^3$ km) that heat from long-lived radioactive decay alone could have enabled and maintained at least a thin global H_2O-NH_3 ocean just above the core and/or pore liquid in the core at the time of impact. Further constraints on the progenitor interiors could arise from the small moons if they formed concurrently with the binary (which may require partially differentiated progenitors with an outer ice shell; see section 3.5), or from forward modeling of progenitor evolution up to the time of impact if the timing of the latter can be further constrained by other arguments (e.g., planet migration models).

3.4. Giant Impacts that Produce Massive Disks from which Charon Later Accretes

We turn now to a discussion of the types of impacts that could account for the very large Charon-to-Pluto mass ratio. To date, these events have been modeled using smoothed particle hydrodynamics (SPH) simulations that include explicit self-gravity, but that treat the colliding bodies as strengthless (*Canup*, 2005, 2011; *Sekine et al.*, 2017; *Arakawa et al.*, 2019). This may not be a good approximation for this scale of objects. However, results to date suggest that inclusion of material strength would most strongly affect impact heating (e.g., *Davies and Stewart*, 2016; *Emsenhuber et al.*, 2018), with more limited effects on overall dynamical outcome (e.g., orbiting mass and angular momentum).

A first category of potential Charon-forming impacts involves an impact between similarly sized bodies whose interiors were differentiated prior to their collision. Figure 5 shows an example SPH simulation of an impact between highly differentiated bodies, with metallic cores, rocky mantles, and outer ice shells (*Canup*, 2005). For a low-velocity collision with a total angular momentum comparable to L_{PC}, a common sequence occurs, with features reminiscent of those seen in simulations of rapidly rotating protostars (e.g., *Tohline*, 2002). After an initial oblique collision, the colliding bodies separate before undergoing a secondary, merging impact. During the secondary impact, the merged body forms a bar-type instability, and the higher density components migrate to the center, while from each end of the "bar," portions of the lowest-density (ice) component emanate radially outward. As the system rapidly rotates, these outer portions differentially rotate, forming trailing spiral structures whose self-gravity transports angular momentum from inner to outer regions. Ultimately the spiral structures break up to form a massive ice disk orbiting a central object, with all or nearly all of the impactor's rock and metal accreted to the primary's central core. This general type of collision was subsequently named a "graze and merge" impact (*Leinhardt et al.*, 2010).

Figure 6 shows the predicted satellite-to-primary mass ratios (q) that would result from such collisions. Here, SPH results, together with conservation of mass and angular momentum have been used to estimate the mass of the satellite that would accrete beyond the Roche radius in the optimistic limit that no material escapes from the disk, so that all initial disk material is either accreted into a single satellite or onto Pluto (e.g., *Ida et al.*, 1997). For collisions that produce final bound systems with $J > 0.35$, the moon masses predicted from the SPH results can be approximated by an analytic relationship (dashed line in Fig. 6) that assumes that the post-impact primary is rotating with rate $\omega = (GM_p/R_p^3)^{1/2}$ (where subscript p quantities refer to the primary), together with a moon of mass qM_p orbiting at semimajor axis a, so that

$$J_f \approx K\left(\frac{1}{1+q}\right)^{\frac{5}{3}} + \frac{q}{(1+q)^{13/6}}\left(\frac{a}{R_p}\right)^{1/2} \qquad (6)$$

The angular momentum of a body rotating at the rate at which it begins to shed mass is proportional to its moment-of-inertia constant, K. As K in the post-impact primary is reduced (i.e., for a more highly differentiated structure), less angular momentum can be contained in its rotation and more angular momentum is instead partitioned into orbiting material for low-velocity impacts. Thus, for otherwise similar impact conditions, the mass in the disk (and q of the resulting satellite) increases as K decreases. The highly differentiated, low-K bodies considered in Fig. 5 and for most of the cases in Fig. 6 were thus chosen to maximize q. If instead the post-impact primary were less centrally condensed — e.g., if it was only partially differentiated or if it lacked a metallic core — smaller satellite masses would be expected.

A graze-and-merge collision that produces a massive disk is difficult to reconcile with Pluto-Charon for two reasons (*Canup*, 2005). First, such collisions generally yield moons with $q < 0.1$, and only extreme cases appear capable of producing the large Charon-to-Pluto mass ratio. These involve highly differentiated bodies (where even rock and metal have separated) and fast pre-impact rotations in the target and the impactor that are optimally aligned to contribute substantially to that of the collision itself.

Second, massive disks produced by collisions of differentiated ice-rock bodies are ice-rich (Fig. 4) (*Canup*, 2005; *Arakawa et al.,* 2019), which is inconsistent with Charon's substantial rock content. If one instead considered the collision of partially differentiated bodies in which not all the rock is in a central core, it seems probable that the resulting disks would be more rock-rich, which is potentially consistent with Charon's bulk density (*Desch*, 2015; *Desch and Neveu*, 2017). As discussed above, such a partially differentiated state may be likely if neither accreting impactors nor interior radiogenic heating effectively warmed the outer layers of proto-Pluto or the giant impactor. However, at least based on impact simulations to date (e.g., Fig. 5), production of a dispersed disk massive enough to yield Charon would still appear to be improbable for such cases, which would tend to yield a higher effective K for the post-impact Pluto and an ultimately a lower q than that of Pluto-Charon.

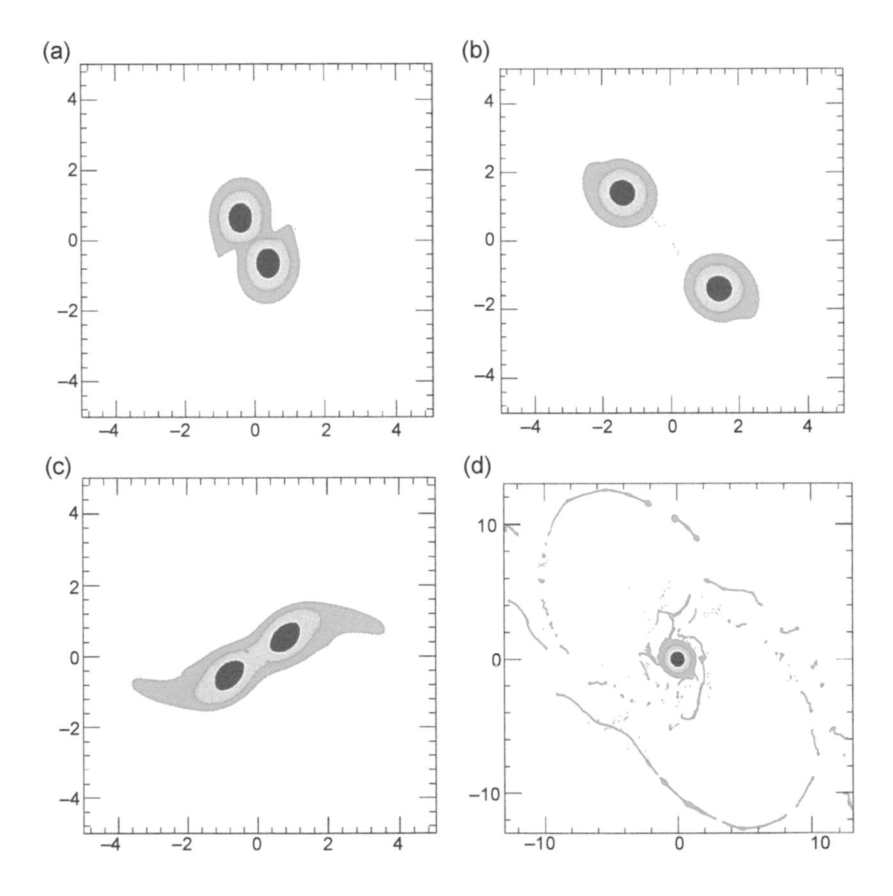

Fig. 5. A potential Pluto-Charon forming impact that produces a massive disk. Shown are time steps from an SPH simulation of the collision of two equal-sized, highly differentiated bodies (with iron cores in red, rock mantles in orange, and ice shells in blue) that were both rotating with a 7-h rotational day prior to the collision, with their pre-impact spin axes aligned with the collision angular momentum vector. After an initial oblique, $v_i = v_{esc}$ collision **(a)**, the bodies temporarily separate **(b)** before re-colliding and merging to form a rapidly rotating structure **(c)**. As the higher-density components migrate to the center, low-density ice expands outward and ultimately forms a massive ice disk containing 12% of the central planet's mass **(d)**. Units are distance in 10^3 km. From *Canup* (2005).

3.5. Giant Impacts that Produce Charon Intact

Oblique, low-velocity impacts can also produce large intact moons, rather than a disk from which Charon later accumulates. In such cases, a large portion of the impactor remains intact after the initial impact and is torqued into a bound orbit with a periapse exterior to the Roche limit via gravitational interactions with the primary whose shape is distorted by the initial impact. This behavior was first seen in simulations by *Canup* (2005) that considered collisions between undifferentiated bodies with uniform serpentine (a hydrated silicate) composition. Figure 7 shows an example case that yields a Pluto-Charon like binary. In impacts that produce an intact Charon, Charon's bulk composition is similar to that of the bulk impactor (*Canup*, 2005; *Arakawa et al.*, 2019). Because deformation of the target strongly affects the initial torques that produce an intact moon, variations in outcome may be expected for different compositions and associated equations of state. *Canup* (2005) used the ANEOS equation of state for serpentine; broadly similar overall results were seen in *Arakawa et al.* (2019) using the Tillotson equation of state and pure-ice progenitors.

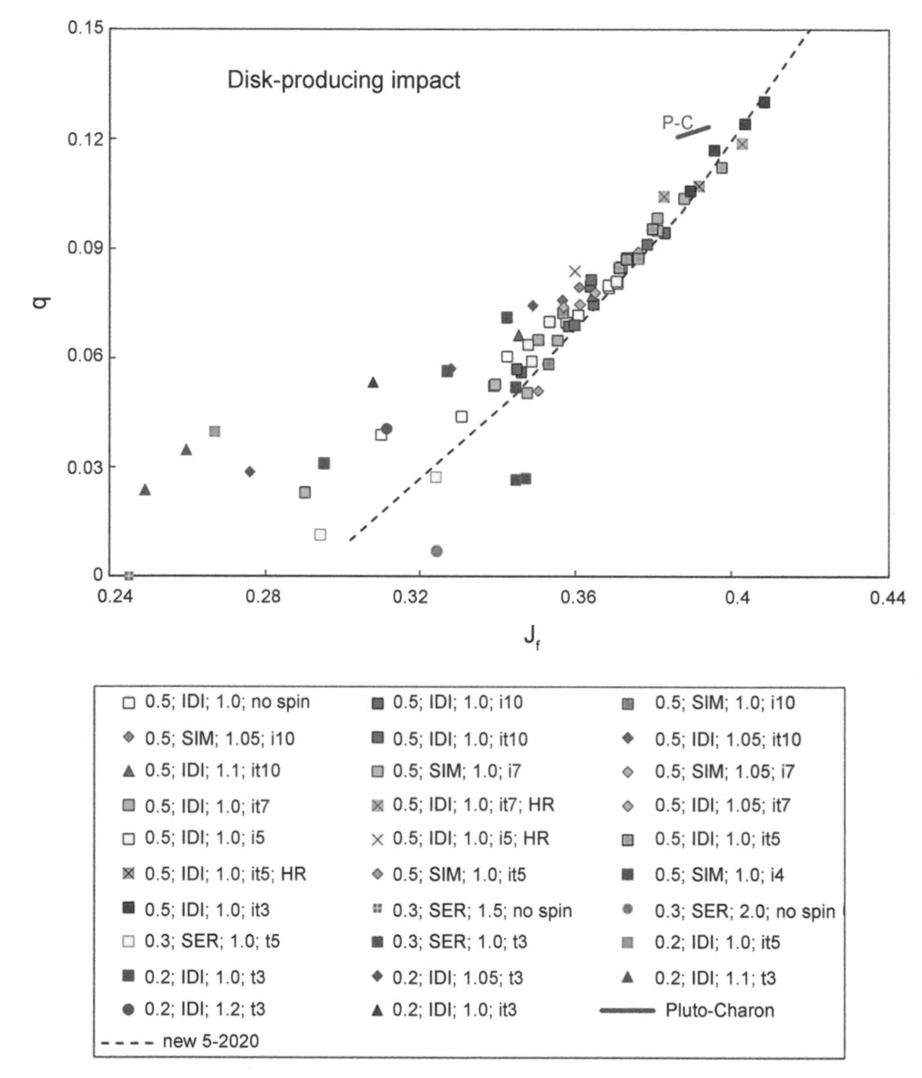

Fig. 6. Predicted satellite-to-planet mass ratio (q) that would result from giant impacts that produce dispersed disks as a function of the scaled angular momentum of the final bound planet + disk system (J_f). Legend indicates simulation parameters; the first value is the ratio of the impactor to the total mass (γ), the second indicates the composition of the colliding objects (IDI = differentiated with iron, dunite, and ice layers; SER = undifferentiated, uniform serpentine composition; SIM = undifferentiated mixture of 50% ice and 50% serpentine), the third is the ratio of the impact velocity to the escape velocity, and the fourth is the pre-impact prograde spin period in hours for the impactor ("i") and/or target ("t"), where the pre-impact spin axes were aligned with the collisional angular momentum vector. Most cases used 20,000 SPH particles; 120,000 particle simulations are indicated with "HR." The dashed line is the relationship between q and J_f from equation (6) for an IDI planet of mass M_p and moment of inertia constant K = 0.29, rotating with a period equal to its minimum for rotational stability and having a moon of mass qM_p orbiting at a distance of about 3 primary radii. Modified figure based on *Canup* (2005), including post-New Horizons values for the Pluto-Charon system (bright blue solid line).

The formation of large intact moons is affected by the impactor's differentiation state prior to the collision. For a uniform impactor composition, corresponding to an undifferentiated hydrated silicate or perhaps a rock-ice mixture, differential motion across the impactor after the initial impact is somewhat reduced, allowing self-gravity to maintain a substantial portion of the impactor as an intact clump that becomes a Charon-analog. For a low impact velocity [i.e., $v_i/v_{esc} \le 1.2$ (*Canup*, 2005, 2011; *Arakawa et al.*, 2019)], such collisions yield intact moons with $0.1 < q < 0.4$ across a relatively broad range of impactor masses ($\gamma \ge 0.3$), impact angles (50°–75°), and varied pre-impact spin states (Fig. 8). Producing a Pluto-Charon-like mass ratio and total angular momentum is an intermediate outcome among such cases, in contrast to the extremely limited range of impacts that appear capable of producing Charon from a dispersed disk. Thus, intact formation of Charon appears the more probable mode (*Canup*, 2005). A notable property of all such cases is that Charon's initial orbit is eccentric, with $e \sim 0.1$ up to on the order of unity (*Canup*, 2005, 2011; *Arakawa et al.*, 2019).

It is, however, possible that a differentiated impactor could also produce an intact Charon. *Arakawa et al.* (2019) performed SPH simulations of impacts between differentiated bodies with 50% of their mass in an inner rock core and 50% in an outer ice shell. For this structure, one extremely oblique (impact angle 75°) collision was identified that produced a Charon-sized intact moon (*Arakawa et al.*, 2019) (Fig. 9). However, Arakawa et al. found that Charon-sized intact moons were more probable with undifferentiated impactors, as in *Canup* (2005).

After the discovery of Pluto's four tiny moons, studies were initiated to explore whether a giant impact capable of producing Charon intact could also yield dispersed debris from which the outer moons could form. *Canup* (2011) considered collisions between partially differentiated progenitors, whose interiors were comprised of hydrated silicate (90% serpentine by mass) with an overlying water ice shell (10% by mass). This structure was intended to represent a case in which radiogenic heating was limited, so that complete separation of rock from ice had not occurred in the interior at the time of the collision, with an outer ice layer produced by impact-driven melting at the end of the progenitors' accretion (e.g., *Barr and Canup*, 2008) (Fig. 4b). The mean density of the partially differentiated progenitors in *Canup* (2011) was ≈2 g cm^{-3}, somewhat larger than what

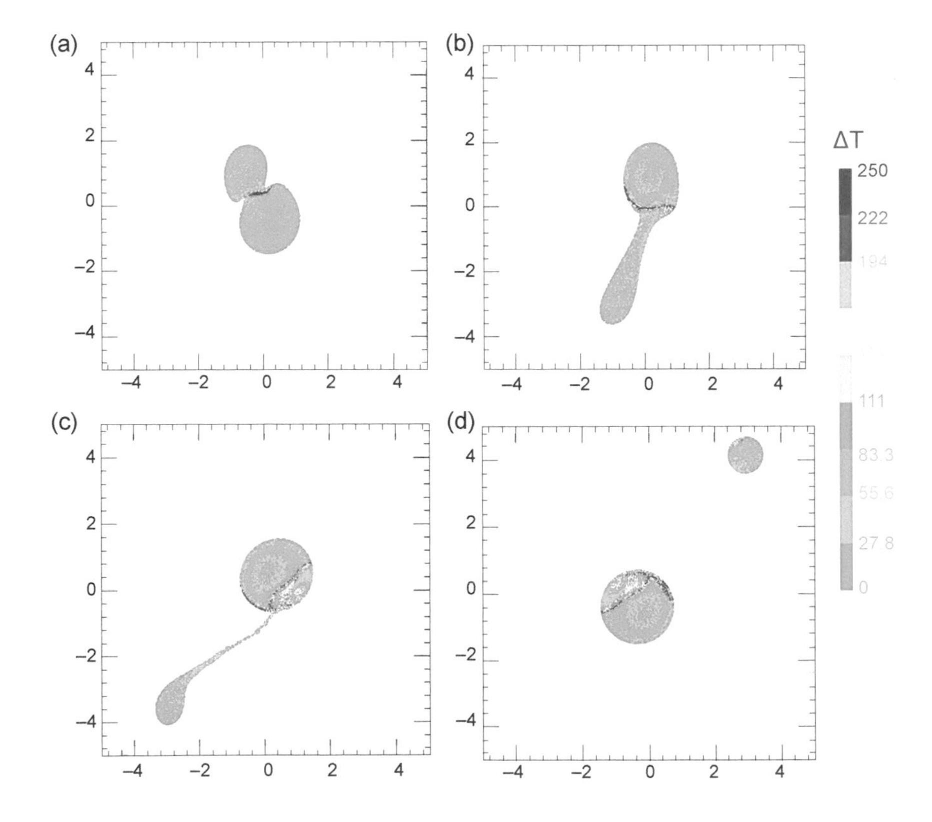

Fig. 7. SPH simulation of a potential Pluto-Charon-forming impact that produces an intact Charon. Two non-rotating, undifferentiated/uniform composition bodies undergo an oblique, $v_i = v_{esc}$ collision, with the impactor containing 30% of the Pluto system mass **(a)**. A substantial portion of the impactor remains gravitationally bound after the initial collision **(b)**, and it achieves a bound, stable orbit about the final planet due to gravitational interactions with the distorted shape of the planet **(b),(c)**. The final moon has q = 0.12, an orbital eccentricity of e = 0.5, and a periapse exterior to the Roche limit **(d)**. Color indicates the increase in temperature due to the impact per color bar. From *Canup* (2005).

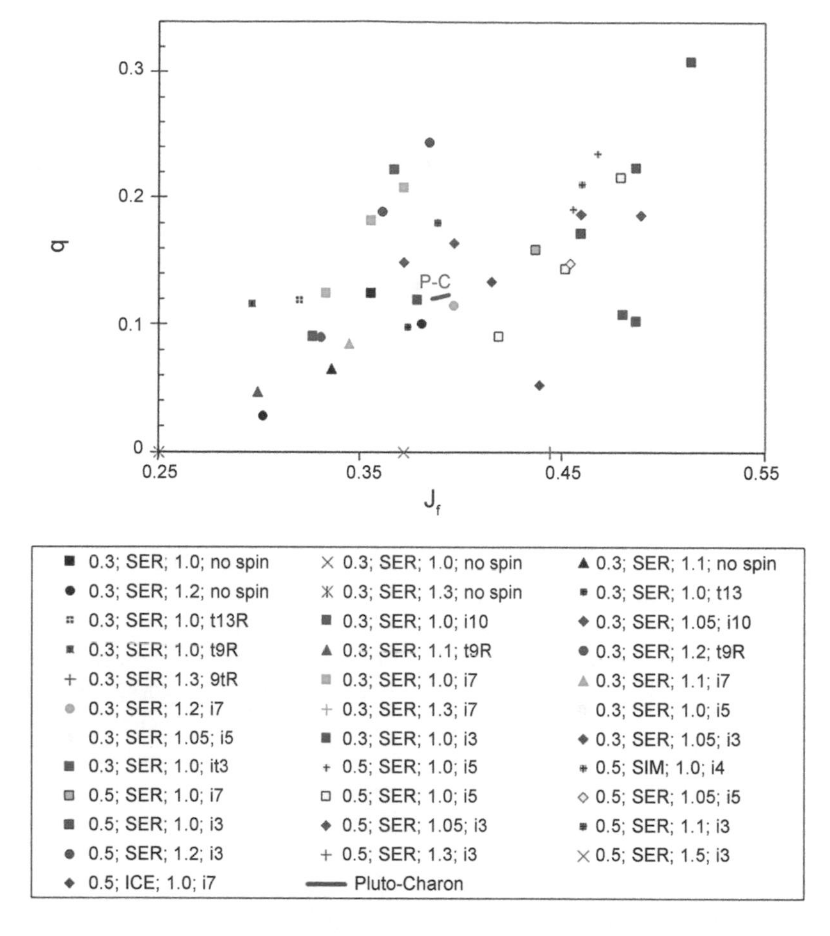

Fig. 8. Satellite-to-planet mass ratio (q) from SPH simulations of impacts between uniform-composition bodies that produced intact moons. Labels in the legend provide the same data as in Fig. 6; ICE = uniform ice composition, and "R" after the spin period corresponds to a retrograde pre-impact spin (i.e., a pre-impact spin vector that is anti-aligned with the collision angular momentum vector). Figure from *Canup* (2005), modified to show post-New Horizons values for the Pluto-Charon system (bright blue solid line).

the densities of Pluto and Charon proved to be. Similar overall results would presumably be found with lower-density progenitors having similar internal structures but thicker ice shells and/or lower-density rock-ice cores.

Impacts between the partially differentiated bodies considered in *Canup* (2011) were able to produce both an intact Charon and a much lower-mass disk whose composition ranged from similar to that of the colliding bodies (i.e., 10% pure ice by mass) to 100% ice (Fig. 10). An immediate question was then how the expected radial extent of this dispersed disk compared with the current small moon orbital radii.

A low-velocity, Charon-forming impact produces disk debris with minimal vapor on initially high-eccentricity orbits (e.g., *Canup*, 2005, *Sekine et al.*, 2017). Physically, each SPH disk particle represents a distribution of material with much smaller sizes than can be resolved with SPH. After the impact, orbiting disk debris would first undergo mutual collisions at velocities high enough for rebounding and/or fragmentation. Such collisions would dissipate energy and circularize debris orbits. Once collisions caused the relative velocities of debris to be damped to sufficiently low values, accumulation of moons from the debris could begin. Given this expected evolution, it is common to estimate the orbits to which disk debris would collisionally relax prior to moon accretion by computing an equivalent circular orbit that has the same angular momentum as each initial SPH disk particle, i.e, with $a_{eq} = a(1-e^2)$, where a and e are the SPH particle's post-impact semimajor axis and eccentricity and a_{eq} is the expected semimajor axis when moon accretion begins. The radial extent of a low-mass disk can be then estimated by computing the maximum value, $a_{eq,max}$, among all the disk particles found in an SPH simulation. This value is best interpreted as the distance at or beyond which a mass comparable to that of a single SPH particle is found (e.g., *Canup and Salmon*, 2018).

For simulations with 10^5–10^6 particles (whose individual particle masses are $\sim 10^{19}$–10^{20} g), the estimated outer "edge" of the low-mass disk produced in the *Canup* (2011) simulations was between 5 and 30 Pluto radii (R_P) (Fig. 10c), substantially interior to outermost Hydra, which orbits at

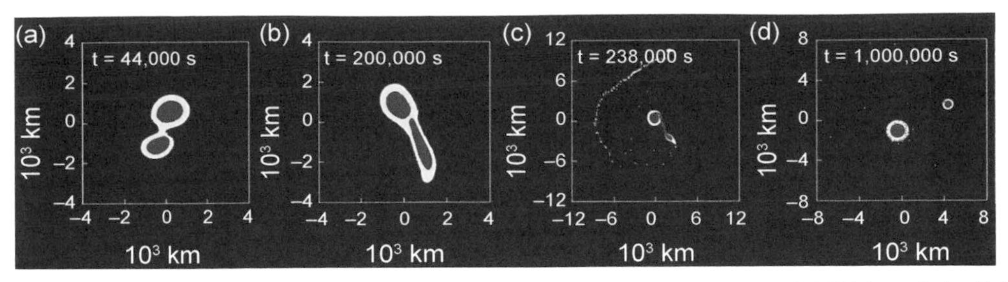

Fig. 9. The highly oblique collision of two differentiated bodies with 50% ice (blue) and 50% basalt (red) by mass yields an intact Charon analog with q = 0.13. From *Arakawa et al.* (2019).

~55 R_P. It is important to note that these SPH particle masses are comparable to masses of the small satellites themselves, rather than true debris out of which the moons might accrete. Further investigation is warranted to determine if a higher-angular-momentum tail of debris might exist in higher-resolution simulations.

Thus, a giant impact between partially differentiated progenitors appears capable of producing an intact massive Charon, together with a much less massive dispersed disk. The composition of the low-mass, dispersed disk varies across different cases: It can contain a similar bulk composition to that of the impactor or can be completely made of ice. The latter case offers a compelling explanation for the inferred ice-rich compositions of Pluto's small moons, which differ substantially from the darker and more rock-rich compositions inferred for small KBOs. However, the outer edge of the low-mass disk in simulations to date is far too small to directly yield the extended outer moon orbits. On face, this mismatch requires some type of outward migration to produce the small moons from the same impact that produced Charon, a topic to which we now turn.

4. ORIGIN OF THE SMALL MOONS

Improved measurements and limits on the masses and orbits of Pluto's outer four small moons have provided new clues for origin models. In this section, we first summarize the uncontentious satellite characteristics (see also the chapter by Porter et al. in this volume), and subsequently review varied theoretical models for their formation. Given the current evidence in favor of a giant impact origin for Charon, we only discuss models that appear consistent with this picture. Current small moon formation models fall into four categories: (1) The moons form in a compact circumbinary debris disk produced by a Charon-forming impact, and then are migrated outward to their present location as Charon's orbit tidally expands; (2) the moons form *in situ* in a large circumbinary debris disk produced by the Charon-forming impact; (3) the moons accrete from the collisional debris of heliocentric material captured long after a Charon-forming impact; and (4) hybrid models in which the moons formed *in situ* (or nearly so) from debris generated by collisions between Charon-impact debris and one or more heliocentric bodies. Despite the abundance of new data provided by New Horizons and many creative theoretical concepts, a coherent model for the origin of the Pluto-Charon circumbinary satellite system remains elusive.

4.1. Properties of the Small Moons

The four moons comprise a tiny mass fraction (on the order of 10^{-6}) of the Pluto-Charon binary. The two larger satellites, Nix and Hydra, have effective radii of approximately 40 km, while smaller Styx and Kerberos are only roughly 10 km in radius. The satellite orbital properties have been traced for over a decade using Hubble Space Telescope (HST) data. All moons have low-eccentricity, low-inclination orbits with respect to the binary and they reside, near, but not in, N:1 mean-motion resonances with the binary (with N = 3, 4, 5, and 6), and thus also near first-order resonances with each other. *Showalter and Hamilton* (2015) identified possible three-body resonances among the satellites, based in part on observations that placed the resonant angle near the expected value for resonant libration. Mean-motion resonances are unlikely given current estimated satellite masses and the low binary eccentricity (*Brozović et al.*, 2015; *Jacobson et al.*, 2019); both drive the resonant widths — the range of period ratios in which libration occurs — toward zero. Future planned observations will confirm whether the resonant angle is circulating (non-resonant) or librating (e.g., *Showalter et al.*, 2019). Accounting for the satellites' near-resonant orbits presents a great challenge to theoretical models. Resonant chains can be the hallmark of dissipative formation processes and differential orbital migration (e.g., *Malhotra*, 1991; *Peale and Lee*, 2002), but such tight dynamical packing, especially around a binary, is difficult to achieve because it is vulnerable to orbital instabilities (e.g., *Sutherland and Kratter,* 2019).

The small moons all have high estimated albedos from ~0.6 to 0.8, brighter than the mean geometric albedo of Charon, which is ~0.4 (*Weaver et al.*, 2016). High small moon albedos were first predicted from the dynamical models of *Youdin et al.* (2012), who found that lower albedos, which required larger moon masses for consistency with HST data, would lead to dynamical instability. Subsequent orbital analyses have confirmed that stability

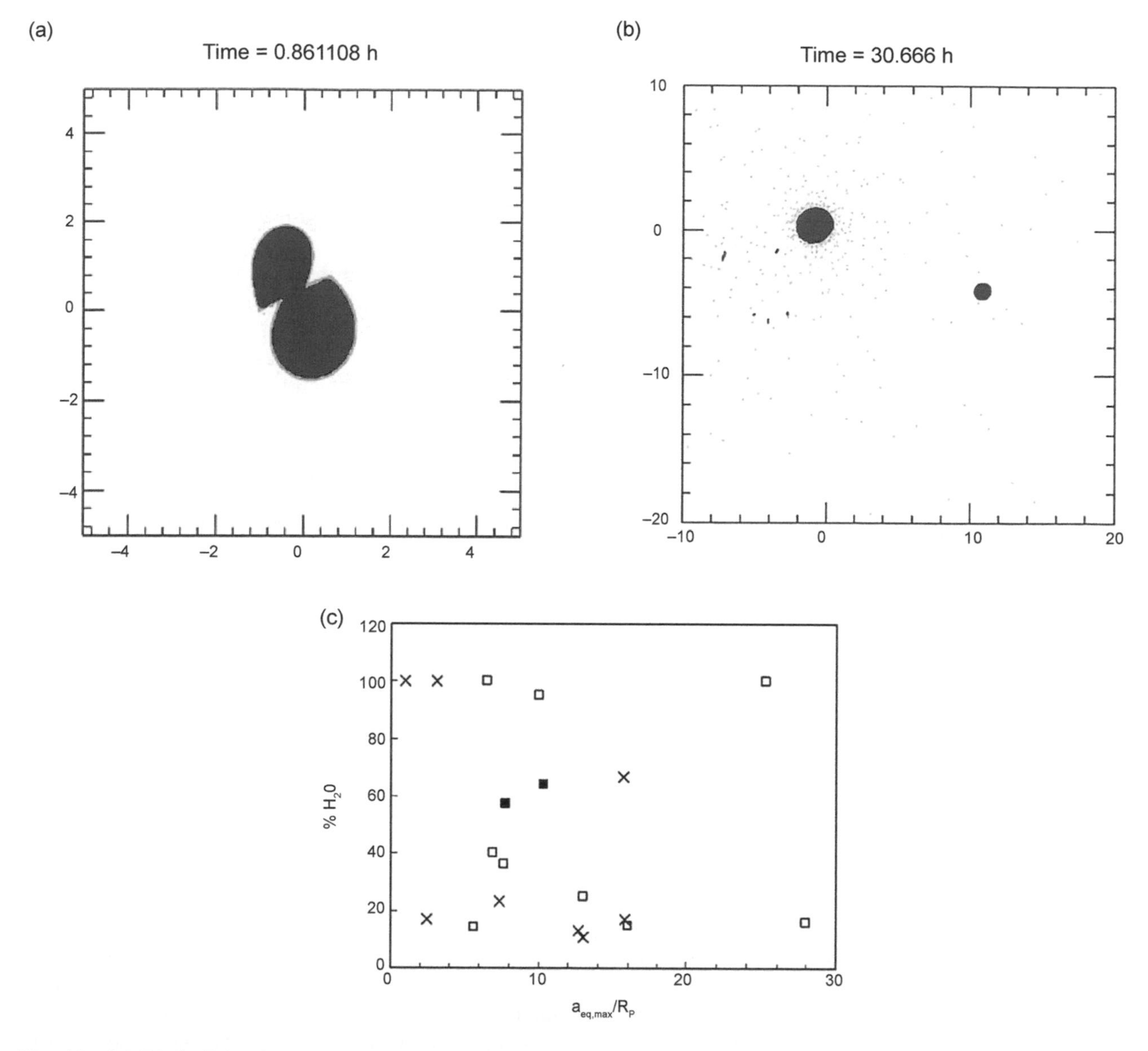

Fig. 10. **(a),(b)** Collision between two partially differentiated bodies (90% hydrated silicate, 10% ice shells) leads to the formation of an intact Charon-analog and an extended low-mass disk. **(c)** Composition and outer edge of the low-mass disk. Squares correspond to cases with γ = 0.3, while ×'s indicate cases with γ = 0.5. For comparison, the progenitor bodies (both target and impactor) had 10% ice by mass. From *Canup* (2011).

requires high albedos and low moon masses, and in the case of Hydra, have pushed its estimated mass to even lower values (*Brozović et al.*, 2015; *Kenyon and Bromley,* 2019; *Jacobson et al.*, 2019).

The small moons' high albedos imply bright, ice-dominated surfaces, consistent with spectral measurements by New Horizons (*Cook et al.,* 2018). As discussed in section 2.2, it seems probable that the small moons' bulk compositions are also predominantly icy. However, this currently remains an inference, because the small moon masses and bulk densities are unconstrained by New Horizons data alone (*Weaver et al.*, 2016). Ongoing work to combine small moon shape models (*Porter et al.*, 2019) with improved limits on their masses derived from stability analyses and both HST and New Horizons data (*Jacobson et al.*, 2019) should provide bulk density estimates in the near future. For ice-rich bulk compositions and likely porous interiors, one expects the small moon densities to be $\leq$1 g cm^{-3} (e.g., *McKinnon et al.*, 2017; *Jacobson et al.*, 2019).

4.2. Giant Impact Formation and Resonant Expansion of Small Moons

Given that Charon's orbit is thought to have tidally evolved outward from a few Pluto radii to its current distance at ~17 R_P, and that the small moons orbit near N:1 resonances with Charon, it was initially proposed that the moons accreted from Charon-forming impact debris, and were then captured into mean-motion resonances and driven outward to their current locations as Charon's orbit tidally expanded (*Stern et al.*, 2006). The difficulty with this otherwise compelling idea is that, due to the large extent of Charon's radial migration, capture into mean-motion resonances would drive the small moon eccentricities to values so high as to cause destabilization of the small moon orbits. In analogy to excitation of Pluto's current ~0.25 eccentricity from just a ~15% expansion of Neptune's orbit, the much greater post-giant impact tidal expansion of Charon's orbit by a factor of 4 or more would drive objects captured in mean-motion resonances to $e > 0.5$, destabilizing the resonances and causing mutual collisions, ejection, and/or accretion of the small moons by Charon (*Ward and Canup,* 2006; *Cheng et al.,* 2014b). If such an instability occurred, however, it is possible that the debris could seed a second generation of satellites (see section 4.5).

Ward and Canup (2006) proposed a resonant migration model in which the small moons were instead captured into co-rotation resonances with Charon. These resonances exist if Charon's orbit is eccentric, and are located at similar locations as the mean-motion resonances. Co-rotation resonances, which are associated with confinement of Neptune's ring arcs (*Goldreich et al.*, 1986), are notable because they do not increase a trapped object's eccentricity as the perturbing body's orbits expands. In the *Ward and Canup* (2006) model, the current small moons represent a small fraction of an impact-produced debris disk that became trapped into co-rotation resonances with an eccentrically orbiting Charon. So long as Charon maintained an orbital eccentricity as it tidally evolved, the small moons could be driven outward to their current locations on near-circular orbits, with the co-rotation resonances ultimately disappearing as Charon approached the dual-synchronous state and its orbit circularized. The requirement of an eccentrically orbiting Charon is supported by predictions of impact simulations that yield an intact Charon (*Canup,* 2005, 2011; *Arakawa et al.*, 2019) (section 3.5), although maintaining an eccentricity during its subsequent evolution requires a specific range of tidal parameters (e.g., *Cheng et al.*, 2014b). A successful prediction of this model is that the material thrown out to the largest orbital radii could be dominated by ices (*Canup,* 2011) (Fig. 10c), which is consistent with the general increase of albedos from Charon to Hydra (e.g., *Stern et al.,* 2018).

Despite their appeal in terms of avoiding eccentricity growth, even the co-rotation resonances seem to ultimately lead to instability of one or more satellites. *Lithwick and Wu* (2008) argued that the critical eccentricities required for Charon to trap Nix and Hydra are mutually exclusive, with Nix's capture in the 4:1 co-rotation resonance requiring that Charon's eccentricity was $e_C < 0.024$, while Hydra's capture in the 6:1 only occurs for $e_C > 0.04$. The inclusion of omitted secular terms in the disturbing function suggests that there might be a small region of allowed eccentricity space ($e_C \sim 0.05$) at which both captures can occur (*Ward and Canup*, 2010).

In addition, the above analytic models rely on approximations of perturbation theory that may not capture the full system behavior, especially for the large Charon-to-Pluto mass ratio [e.g., see *Mardling* (2013) for expansions valid at arbitrary mass ratios]. Indeed, numerical integrations that include more physics only exacerbate the resonant expansion model's difficulties. *Cheng et al.* (2014b) find that the inclusion of the effects of Pluto's oblateness (i.e., its J_2) destabilizes all resonant capture models that they explored for a range of Charon eccentricities. Including the masses of Nix and Hydra in numerical integrations also destabilized outward migration scenarios. In the *Cheng et al.* (2014b) integrations, Nix is always the source of the instability. This is not surprising given that when Charon's eccentricity exceeds ~0.15, Nix's orbit is dynamically unstable on timescales of hundreds of orbits due to overlap of the close-in N:1 resonances (e.g., *Holman and Wiegert,* 1999; *Smullen and Kratter,* 2017). Notably these models focus primarily on Nix and Hydra, and do not claim to place Styx or Kerberos in their current locations either. We discuss these dynamical constraints in more detail in section 4.6.

4.3. *In Situ* Formation from the Charon-Impact Debris-Disk

A key motivation for the outward resonant transport model was that giant impacts tend to produce debris disks that are more compact than the current small moon orbits (e.g., *Canup*, 2011) (Fig. 10c), so some type of spreading mechanism seems required. *Kenyon and Bromley* (2014) suggest that impact-generated debris could spread outward to a distance comparable to Hydra's current orbit prior to small moon accumulation, depending on the balance between secular perturbations by an initially eccentrically orbiting Pluto-Charon binary and collisional damping among debris particles. The small moons would then accumulate near their current locations. One issue with such a model is that the collisional spreading mechanism would tend to deactivate as moons accreted and the collision rate decreased, and whether debris could expand by the needed factor of 2 or more in orbital radius before substantial accumulation took place is unclear. Furthermore, the *Kenyon and Bromley* (2014) model predicts a population of small moons and particles orbiting beyond Hydra that were not seen by New Horizons.

Alternatively, it is possible that a Pluto-Charon forming impact could directly produce a low-mass debris disk that is much more extended than seen in prior SPH simulations. Numerical resolution in the *Canup* (2011) simulations appeared adequate to resolve the early estimated total mass of the small moons (*Tholen et al.*, 2008). However, moon mass estimates subsequently decreased by about an order of magnitude (e.g., *Brozović et al.*, 2015), suggesting that their mass was not well resolved by the 2011 models. Studies of low-mass debris disks produced by potential Phobos-Deimos-forming impacts with Mars (*Canup and Salmon*, 2018) show that when simulation resolution is increased, the estimated outer debris disk "edge" typically increases as well, as a lower-mass, higher specific angular momentum component is increasingly resolved.

A challenge for *in situ* formation models is explaining the near-resonant orbits of the small moons. *Smullen and Kratter* (2017) and *Woo and Lee* (2018) explored the behavior of test particles near the current small moon locations in response to Charon's tidal migration. Both studies found that test particles are either evacuated from resonance regions or have too high eccentricity to be consistent with the present-day satellite orbits near N:1 resonances. Note that both studies neglect collisional damping; *Smullen and Kratter* (2017) included no damping, while *Woo and Lee* (2018) damp the test particles to cold orbits prior to Charon's tidal expansion, but not during. The latter scenario is akin to assuming that satellite formation is complete prior to Charon's expansion and leaves behind insufficient debris to damp the satellites. Models with concurrent accretion and collisional damping (but with a different source of debris and without Charon's orbital evolution) do not appear to form the moons near resonances either, as discussed in section 4.4.

A recent novel version of this scenario imagines that the satellites form *in situ* from impact debris, but that the debris is generated by a separate impact on Charon (*Bromley and Kenyon,* 2020). By requiring that the impact occur after the binary has evolved to its present-day orbit, this model avoids the complications of resonant expansion. While the required collision is unlikely, it is not prohibitively so, at least for oblique impacts, which would allow a 30–50-km body to generate sufficient debris. Such a model still does not directly explain the satellite's emplacement near, but not in, resonance with the Pluto-Charon binary; this is a concern given low resonant widths once the binary has circularized.

4.4. Moon Origin from Capture of Heliocentric Planetesimals

Pires Dos Santos et al. (2012) explored a model in which the satellites form from planetesimals that are captured, disrupted, and reaccreted into the current generation of satellites. They calculated the capture probability for a variety of heliocentric disk models and found that while temporary capture of protosatellites is possible, they tend to be rare and short-lived. To increase the lifetime of the material in orbit, and to ultimately generate low-eccentricity, low-inclination bodies, they considered the probability that captured objects are collisionally disrupted. They concluded that the collisional timescales are longer than the captured satellite lifetimes, thus inhibiting long-term survival of sufficient mass to form the current satellite system. The Pires Dos Santos et al. model did not address the near-resonance moon configurations, and does not offer a clear rationale for their high albedos, since captured material would be expected to have a composition similar to that of small (and much darker) KBOs.

4.5. Hybrid Models: Charon Impact Debris Interacts with Heliocentric Debris

The apparent failure of both the resonant expansion and heliocentric capture models has led to the development of hybrid approaches. *Kenyon and Bromley* (2014) and *Walsh and Levison* (2015) considered the outcome of interactions between debris with heliocentric origins and leftover debris or satellites from a Charon-forming impact. Leftover material could radially spread, collisionally damp, and then coalesce into the satellites *in situ*, or at larger separations followed by inward migration. An appealing aspect of these models is that they rely on inward, dissipative movement of either the debris or the satellites themselves, which may allow for the production of resonant chains and/or resonant capture. As with other models, difficulties arise when detailed treatments are considered.

Kenyon and Bromley (2014) proposed that debris from the giant impact spreads through interactions with other captured objects in the Pluto-Charon system. In particular, they argued that captured bodies that generate debris pre-date the Charon-forming impact. The old and new debris collide, spread beyond the orbit of Hydra, and then subsequently coalesce into an array of satellites. In their model, these objects can migrate inward and finish near the present-day moon orbits, although they don't show a special preference for the known near-N:1 resonances.

Although aspects of the model are compelling, the omission of some important physics might limit its applicability. Most crucially, many of the spreading and collisional calculations are done in the absence of perturbations from the Pluto-Charon binary. Stirring from the binary can change satellite accretion timescales (*Bromley and Kenyon,* 2015; *Silsbee and Rafikov,* 2015). Although the simulated growth and subsequent evolution of the satellites are assumed to occur after the binary's tidal expansion, the evolution of the initial debris during the expansion is not addressed. In the absence of strong collisional damping, the majority of the debris from the Charon-forming impact could be removed by tidal evolution. If Charon's eccentricity reaches the values expected in current tidal models (*Cheng et al.,* 2014b), some or all of the satellites except Hydra become dynamically unstable on their present-day orbits (*Smullen and Kratter,* 2017). As mentioned above, this model also predicts the growth and survival of small satellites exterior to Hydra (e.g., *Kenyon and Bromley,* 2019); such satellites have not been seen to date (section 2.3).

Walsh and Levison (2015) considered the evolution of debris from one or more disrupted outer satellites that formed as a result of a Charon-forming impact. Such disruptions could occur due to interactions with other bodies in the debris disk, through orbital excitation during Charon's tidal evolution or by interactions with heliocentric impactors. The latter is the primary case tested by Walsh and Levison. For consistency with the ancient surfaces of the current moons, any disruptions would need to occur within the first few hundred million years of solar system history. Such early disruption(s) appear likely, due to both the higher primordial flux of heliocentric objects and rapid instability timescales in the Pluto-Charon system. The presence of pre-existing material in the Pluto-Charon disk potentially averts the collision timescale problem identified by *Pires Dos Santos et al.* (2012), although the probability of the aforementioned disruption event is not calculated, and it would depend on the initial conditions in the primordial solar system planetesimal disk.

Walsh and Levison (2015) simulate the collisional evolution (including fragmentation and accretion) of an initially eccentric ring of debris generated by a collision between a bound satellite and a heliocentric impactor across a range of different static Pluto-Charon orbits. Similar to *Kenyon and Bromley* (2014), a promising aspect of this model is the ability to launch debris into higher-angular-momentum, stable orbits through the combination of perturbations by Charon and collisional damping. Although the debris rings can successfully damp and reaccrete into satellites, they do not tend to settle into resonant or near-resonant N:1 orbits. The omission of Pluto-Charon orbit evolution in these calculations is a limitation that leaves open possible avenues for more successful satellite emplacement. Although the tidal evolution timescale is slow compared to the accretion timescale, it is possible that subsequent shifts in Charon's semimajor axis and eccentricity might allow debris to be ratcheted into larger orbits through multiple disruptions, if the disruptions occur on a timescale short compared to the ~10-m.y. time for Charon's tidal evolution.

The expected composition of satellites formed in hybrid models is not clear. Pluto's small moons are strikingly brighter than typical KBOs of the same size (*Weaver et al.*, 2016). While predominantly ice-rich dispersed debris can result from a Charon-forming impact (*Canup*, 2011), it would be difficult to preserve such compositional gradients if substantial heliocentric material were added to the small moons.

4.6. Circumbinary Dynamics and the Mystery of the Near-Resonant Configuration

A primary unsolved problem in the origin of the small satellites is their packed orbital configuration. The lack of identifiable librating resonance angles in the system today is not, however, too surprising. The N:1 resonances with the binary are all exceedingly narrow (although nonzero) given Charon's current circular orbit (*Mardling*, 2013; *Sutherland and Kratter*, 2019). Moreover, the small masses and low eccentricities of the satellites also produce very narrow libration zones for any first-order satellite-satellite resonances. Since the absence of libration today can be attributed to Charon's circular orbit, it is natural to cite the orbital configuration as evidence that the system acquired these period ratios when Charon was still eccentric, and thus resonant libration more likely. This constraint pins the formation of the satellites to the first few to 10 m.y. after a Charon-forming impact (e.g., *Cheng et al.*, 2014). This timing is consistent with the aforementioned small moon origin models, as they either rely on Charon's outward migration or on a high flux of heliocentric impactors most consistent with the early solar system; an early formation is also consistent with the ancient cratered surfaces of the small moons (e.g., *Robbins et al.*, 2017; *Singer et al.*, 2019).

However, recent work by *Sutherland and Kratter* (2019) shows that multi-object resonant configurations are tenuous even if the binary's mutual orbit is eccentric. Chaos leading to orbital instability can occur due to N:1 resonance overlap, N:1 and first-order satellite-satellite resonance overlap, and satellite-satellite resonance overlap. Stability is particularly tenuous due to the combination of the high forced eccentricities due to the binary and resonance splitting (the breakup of one resonance into many due to different precession rates driven by non-Keplerian potentials). Unlike in satellite systems surrounding single bodies, in which resonance splitting driven by the primary's J_2 can increase stability, the substantial secular forced eccentricity generated by the binary broadens libration zones so that many different combinations can overlap. For example, accounting for the forced eccentricity from Pluto-Charon, the Kerberos-Hydra pair would be unstable for $e_C > 0.18$ because the 5:6 and 6:7 resonances overlap. At $e_C \sim 0.14$, the 5:6 overlaps with the 11:13. Eccentricity pumping by the binary N:1 resonances exacerbate the issue, especially for the inner satellites. Strong collisional damping could forestall some of the aforementioned instabilities, but it is unclear whether enough mass exists in the system to both form the satellites and damp them as an eccentrically orbiting Charon tidally evolves.

4.7. Future Outlook

While the wealth of data from New Horizons has not yet yielded an obvious solution to the "small moon problem," we have identified several promising directions for future investigations. First, higher-resolution simulations of the post-impact disk are crucial for constraining the initial conditions for moon formation. Such simulations must provide mass resolution that extends below the current estimated satellite masses. Second, more thorough investigation of scenarios in which the observed satellites are not first-generation satellites, but rather have been reprocessed through collisions either with or without the incorporation of pre-existing heliocentric material, should be explored. Finally, future dynamical models of satellite accretion and migration should simultaneously incorporate the tidal migration of Charon and collisional damping within the debris disk.

5. IMPLICATIONS AND OPEN ISSUES

In this final section, we discuss implications of giant impact models for the environment at the time of Pluto system formation, the possible preservation of a collisional "family" of ejecta from a Pluto-system-forming impact, the origin of Pluto and Charon volatiles, and implications for other KBO binary systems. We finish by highlighting key topics for future work.

5.1. Implications of an Impact Origin for the Transneptunian Environment at the Time of Pluto System Formation

A general result of all impact models that can produce massive Charon is the requirement of a very low collision velocity that is within 20% of the mutual escape velocity of the progenitor bodies (*Canup*, 2005, 2011; *Arakawa et al.*, 2019). For similarly sized progenitors as needed to account for the system's high angular momentum (with γ = 0.3–0.5), $v_{esc} \approx 1$ km s^{-1}. The impact velocity is $v_i^2 = v_{esc}^2 + v_{rel}^2$, so that requiring $v_i \leq 1.2\ v_{esc}$ implies a relative velocity at large separation of $v_{rel} \leq 0.7$ km s^{-1}. This is substantially smaller than the current Kuiper belt velocity dispersion, which is ~1–2 km s^{-1}. It is thought that in the primordial epoch — in particular before the migration of Neptune dynamically excited the transneptunian region — relative velocities among large ~10^3-km bodies would have been substantially lower than they are in the current Kuiper belt. This seems consistent with conditions needed to have accreted such large bodies from a population of ~10^2-km planetesimals produced by the streaming instability and background remnant pebbles (e.g., *Morbidelli and Nesvorný*, 2019). Thus, the low collision velocity needed for a Pluto-Charon forming impact seems most consistent with the pre-outer planet migration era. As noted in section 2.5, a later origin of the binary after resonance capture cannot be ruled out based on Pluto's heliocentric orbit alone, because the system could have been recaptured into resonance after a giant impact (e.g., *Dobrovolskis et al.*, 1997). However, this requires "fine tuning" of conditions to both achieve a low relative velocity between the progenitors once Neptune's migration and dynamical excitation of the transneptunian region have begun, and to allow recapture of the post-impact system into the 3:2 resonance.

It is illustrative to also consider the general likelihood of a collision between two R_e ~10^3-km-class bodies. A simple "particle-in-a-box" estimate for the time between collisions among any of the objects within a swarm of N such bodies orbiting across a region extending in orbital radius from a to (a + Δa) gives

$$t_{col} \sim \frac{2\pi\ a\ \Delta a\ H}{N^2 \pi\ (R_e + R_e)^2 f_g v_{rel}} \qquad (7)$$

where H ~ v_{rel}/Ω is the swarm's vertical thickness, Ω(a) is orbital heliocentric frequency, and $f_g = [1 + (v_{esc}/v_{rel})^2] \geq 2$ is a gravitational focusing factor. For a pre-Neptune migration population with a = 20 AU, Δa = 10 AU, and $R_e = 10^3$ km, one collision occurs every

$$t_{col} \sim \text{few} \times 10^6 \left(\frac{10^3}{N}\right)^2 \left(\frac{10}{f_g}\right) \text{y} \qquad (8)$$

The likelihood of a primordial Pluto-Charon forming collision is vanishingly small if there had only been an initial population of N ~ 10 Pluto-scale bodies, as predicted by some hierarchical accretion models (e.g., *Kenyon and Bromley*, 2012; *Schlichting et al.*, 2013), unless f_g was extremely large. However, if there were initially N ~ few × 10^3 Pluto-scale bodies, t_{col} is on the order of 10^6 yr or less, depending on f_g. Appealingly, this is consistent both with the number of Pluto-class bodies needed to account for Neptune's "grainy" migration and the observed number of resonant vs. non-resonant KBOs [N ~ 2000–4000 (*Nesvorný and Vokrouhlicky*, 2016)] and with large impacts in this region occurring prior to an outer-planet migration within the first tens of millions of years after nebular dispersal. The latter timing avoids a destabilization of the inner terrestrial planets associated with a later migration (e.g., *Agnor and Lin*, 2012; *Kaib and Chambers*, 2016) (see section 2.5).

However, the efficiency of implantation of original disk bodies into the Kuiper belt in dynamical models is low, ~10^{-3} (*Nesvorný and Vokrouhlicky*, 2016) (see section 2.1). With N = 4000, equation (8) implies about 300 giant impacts within ~50 m.y. for f_g = 10. There would then be a substantial probability that a single Pluto-sized body that had experienced a giant impact would also be retained. It is possible that in the cold classical belt, relative velocities could have been very low, leading to $f_g \approx 10^2$, making all Pluto-scale bodies likely to have experienced a giant impact within 50 m.y. Having f_g ~ 10–10^2 requires v_{rel} ~ (1/10) to (1/3) v_{esc}, which in turn implies a Pluto-Charon forming impact with $v_i < 1.1\ v_{esc}$, consistent with (but independent from) requirements from impact simulations.

Thus, a low-velocity collision appears needed both to make a Charon-sized satellite and for such a giant impact to have been probable in the first few tens of millions of years of the solar system's evolution. This timing (Fig. 11) meshes well with that implied by current models for outer planet migration and Kuiper belt formation (e.g., reviews by *Nesvorný*, 2018; *Morbidelli and Nesvorný*, 2019).

5.2. Possibility of a Pluto Collisional Family

Additional evidence supporting a giant-impact scenario for the formation of Pluto-Charon (and potentially of the small moons as well) could be the identification of collisional family members. Even a collisional family formed prior to the solar system rearrangement that caused Pluto's heliocentric migration could have survived this migration (*Smullen and Kratter*, 2017).

To date, a single transneptunian collisional family has been identified, for Haumea (*Brown et al.*, 2007). Family

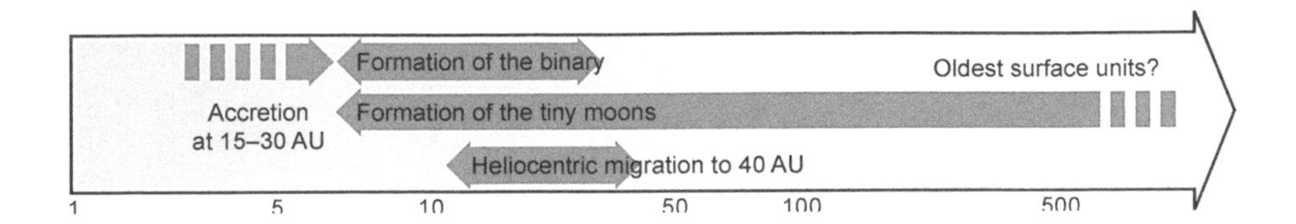

Fig. 11. Tentative timeline of origin events in the Pluto system. Accretion and heliocentric migration are discussed in section 2, the formation of the binary in section 3, and that of the tiny moons in section 4.

members were singled out based on orbital (a low dispersion velocity, <140 m s^{-1}, among family members) and surface properties (icy) similar to Haumea's, even though Haumea's orbital eccentricity is higher than those of identified family members, presumably having been increased by gravitational interactions with Neptune since the time of impact.

Identifying a collisional family for Pluto could be more challenging. Typically, the dispersion velocity of a collisional family would be expected to be comparable to Pluto's escape velocity [$v_{esc} \approx 1.2$ km s^{-1} (e.g., *Schlichting and Sari*, 2009)], implying that family members could span much of the dynamical space of observed transneptunian objects (Fig. 12). In their simulations of graze-and-merge collisions as a possible origin for the Haumea system, *Leinhardt et al.* (2010) showed that for this type of impact (which is similar to impacts shown in Figs. 5–6 that form massive disks), most escaping ejecta have velocity dispersions <0.5× the primary's escape velocity, and a substantial portion have dispersion velocities ≤0.2 v_{esc}. *Smullen and Kratter* (2017) considered material ejected post-impact due to dynamical interactions between the Pluto-Charon binary and a low-mass disk of material [similar to conditions found in *Canup* (2011)] as Charon's orbit tidally expands. The low-mass disk is described with test particles, so mutual collisions are neglected. Of the disk material that is ejected from the binary, they find that on the order of 10% remains on stable heliocentric orbits after 1.5 G.y., even when effects of a smooth migration of Neptune are included. The majority of survivors are trapped in the 3:2 resonance as a population of Plutinos, with characteristic velocity dispersions ~100 to 200 m s^{-1} (*Smullen and Kratter*, 2017); these velocities would likely be modestly reduced if collisions among the ejecta had been included. The omission of the effects of grainy migration likely enhances resonance capture rates (*Nesvorný and Vokrouhlický*, 2016). A Pluto family would thus have a low-velocity dispersion, but its orbital properties could be difficult to distinguish from those of other resonant KBOs.

A more distinguishing feature of a Pluto family could be its composition. The composition of family members may preferentially reflect that of the progenitors' outer layers at the time of impact. This is not as firmly established for Pluto (section 3.3) as it seems for Haumea. Pluto's family could be icy, as suggested by the giant impact simulations of *Canup* (2011) in which the progenitors have an icy veneer, and by the inferred icy compositions of the tiny moons if they formed during this impact. An icy composition would be easiest to identify against the backdrop of generally dark TNOs (*Johnston*, 2018). Alternatively, fragments ejected from progenitors with rock-ice outer layers at the time of the impact (*Desch*, 2015; *Desch and Neveu*, 2017) could have surface properties indistinguishable from those of other TNOs.

5.3. Origin of Volatiles and Expected Giant-Impact Heating

The abundances of N_2, CO, and CH_4 on Pluto and NH_3 on Charon are on the whole consistent with a primordial supply as estimated from observations of comets and interstellar sources (*Stern et al.*, 1997; *Glein and Waite*, 2018). To first order, the contrasting presence of CO, CH_4, and N_2 on Pluto and their absence on Charon is readily explained by Charon's inability to retain these volatiles owing to its lower gravity (*Trafton et al.*, 1988; *Schaller and Brown*, 2007). As such, Pluto and Charon may be two archetypes of the bimodal volatile inventories detected on other large KBOs (*Brown*, 2012). In contrast, ammonia's high miscibility with water (as a polar molecule able to form hydrogen bonds) makes it prone to retention in H_2O ice, as observed throughout Charon's surface, in areas exposed through Pluto's volatile ice veneer, and even on Nix and Hydra (*Cook et al.*, 2018). CO, CH_4, and N_2 escape could have taken place on Pluto as well, but if this is the case the rates of atmospheric and surface accumulation from a subsurface source exceed those of escape.

Heating of Pluto during a Pluto-Charon-forming impact would have occurred primarily in the region of the initial impact, particularly for the favored low-velocity, oblique collisions that produce Charon intact. Impact simulations predict this region would have been heated by ΔT ~ 150–200 K, leading to local melt pool production (*Canup*, 2005; *Sekine et al.*, 2017). *Sekine et al.* (2017) proposed that if the progenitors had comet-like volatile abundances (notably in CH_2O and NH_3), then chemistry in the water melt pool produced by a Charon-forming impact could have led to the formation of Pluto's dark, reddish regions along its equator, and more generally, that such

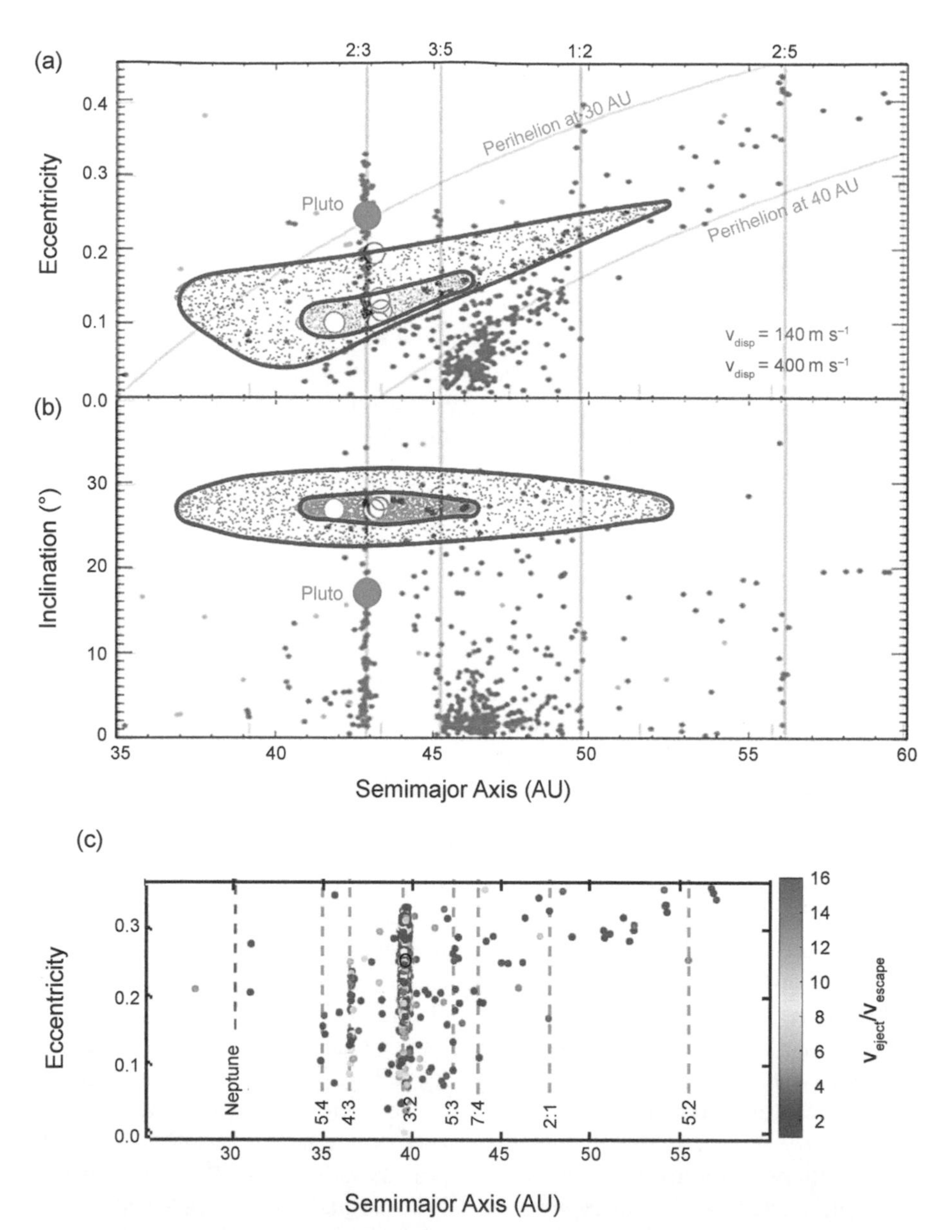

Fig. 12. The challenge in detecting Pluto collisional family members based on orbital properties alone. **(a),(b)** The purple and pink outlines represent the space of possible orbital attributes of members of the Haumea collisional family assuming a dispersion velocity of 140 and 400 m s^{-1}, respectively, centered on the average position of identified family members. Haumea is the open circle at eccentricity 0.19 and inclination 28°; resonant interactions with Neptune have increased its eccentricity since the time of the collision (*Brown et al.,* 2007). Vertical lines indicate mean-motion resonances with Neptune, with black, red, and green dots indicating resonant, stable non-resonant, and Neptune-encountering KBOs. Modified from Fig. 3 of *Brown et al.* (2007) and Fig. 1 of *Levison et al.* (2008). **(c)** Eccentricity vs. semi-major axis predicted for debris ejected from the Pluto system after 1.5 G.y. of dynamical evolution, including effects of a smooth migration of Neptune (*Smullen and Kratter,* 2017). Color scales with ejection velocity scaled to the escape velocity at Pluto's Hill sphere (note that the latter is different than v_{esc} defined in the main text). Of the ejected debris that survives, most (~60–80%) remains trapped in the 3:2 resonance, with a minority in the 5:3 (few to 5%) and other resonances. Ejecta from the Pluto system that persists in resonances does not display a strong clustering of orbital angles with Pluto. Thus, it could be challenging to distinguish this material from that independently trapped into resonance based on orbital properties alone. Modified from Fig. 7 of *Smullen and Kratter* (2017).

impact-induced chemistry could have caused the color variety observed in large KBOs. For impacts that produce an intact Charon, Charon is only minimally heated by the collision itself, because it largely represents a portion of the progenitor impactor that avoids direct collision with proto-Pluto (*Canup,* 2005). However, if its orbit remains eccentric as it evolves, heating via tidal dissipation becomes important (e.g., *Rhoden et al.,* 2015).

Thus, the Pluto-system progenitors may have accreted volatile-rich, and largely retained volatiles through a hypothesized giant impact. However, except for CO, none of the observed volatiles on Pluto and Charon need be primordial. N_2 could result from the oxidation of NH_3 and/or organic nitrogen (*Neveu et al.*, 2017). CH_4 could be a product of the (kinetically slow) reduction in liquid water of CO_2, CO, or organic C (*Shock and McKinnon*, 1993). In the latter case, carbon reactions were partial even on Pluto, whose surface CO would otherwise have been converted to more stable species (*Shock and McKinnon*, 1993; *Glein and Waite*, 2018). Thus, the lower CO/N_2 ratio on Pluto relative to most comets (*Stern et al.*, 1997, and references therein) could be explained by post-accretional processes that deplete CO and/or produce N_2. Alternatively, it may reflect Pluto's accretion in a N_2-rich, CO-poor nebular region that few observed comets sample (*Lisse et al.*, 2020).

5.4. Implications for Satellite Origin at Other Dwarf Planets in the Kuiper Belt

The picture advanced in *Nesvorný* (2018), *Morbidelli and Nesvorný* (2019), and section 5.1 envisions thousands of Pluto-class bodies in the 15–30-AU region prior to Neptune's migration, suggesting that giant impacts between such bodies would have been common. Most such objects would have been lost to ejection from the solar system or accretion by the planets. Surviving impact-generated systems would be expected to vary depending on the specifics of individual collisions (e.g., *Arakawa et al.*, 2019). Differences in impact angle and/or velocity between two like-sized dwarf planets can lead to larger ice/rock fractionations (*Barr and Schwamb*, 2016), e.g., leading possibly to Eris-Dysnomia (*Brown et al.*, 2006; *Greenberg and Barnes*, 2008) and Orcus-Vanth (*Brown et al.*, 2010); or outcomes other than a binary (*Brown et al.*, 2006; *Canup*, 2011), such as Haumea, which only has small moons and a collisional family (*Leinhardt et al.*, 2010). Although the orbit, masses, and compositions of Pluto-Charon strongly implicate an impact origin, those of other binary systems could be compatible with alternative origins (Fig. 2). One is capture (*Goldreich et al.*, 2002), e.g., for Eris and much darker Dysnomia (*Greenberg and Barnes*, 2008; *Brown and Butler*, 2018), Orcus-Vanth (*Brown et al.*, 2010), or the eccentric moon of 2007 OR_{10} (*Kiss et al.*, 2019). However, for (at least) dynamically cold, 100-km-class Kuiper belt binaries, the capture mechanisms of *Goldreich et al.* (2002) — dynamical friction from a sea of small planetesimals ("L^2s") and three-body encounters ("L^3") — appear to be ruled out by the observed distribution of the binaries' mutual orbit inclinations (*Grundy et al.*, 2019a; *Nesvorný et al.*, 2019). A likelier alternative origin is co-accretion (*Nesvorný et al.*, 2010, 2019; *Grundy et al.*, 2019a), e.g., for G!kúnǁ'hòmdímà-G!ò'é!hú (*Grundy et al.*, 2019b). Possible origins may also depend on formation location (dynamical class). Studies of binary properties generally will help constrain these formation mechanisms.

Simple dynamical arguments can identify the regime of secondary-to-primary mass ratio (q) and orbital separations consistent with impact-produced planet-moon systems that have subsequently tidally evolved (Fig. 13). A probable value for the maximum normalized angular momentum in an impact-produced system is $J \approx 0.45$ (equation (3)), which in combination with equation (2) can be used to calculate the orbital separation for a planet-moon pair that has reached the dual synchronous state as a function of q (solid curve in Fig. 13). However, for small q, tides are too slow for the system to have reached this state over the age of the solar system. A simplified expression (e.g., *Goldreich and Soter*, 1966) for the rate of expansion of a moon's semimajor axis (a) due to tides raised by the moon on the primary, albeit likely conservative for the high orbital eccentricities ($e \geq 0.1$) that may result from a binary-forming impact (section 3.5), can be integrated to yield the expected a after t = 4.5 G.y. of evolution

$$\frac{a}{R_p} \approx \left[20q \left(\frac{k_2}{Q}\right) \left(\frac{GM_p}{R_p^3}\right)^{\frac{1}{2}} t \right]^{\frac{2}{13}} \quad (9)$$

where R_p, M_p, k_2, and Q are the primary's radius, mass, Love number, and tidal dissipation factor. Assuming a primary density of 2 g cm^{-3}, expected separations for $(Q/k_2) = 300$, 100, and 30 are shown in Fig. 13. Comparison with estimated properties for the large KBOs that have known satellites suggests that most such systems appear compatible with an impact origin (e.g., *Arakawa et al.*, 2019), although the Haumea system appears to require a rather unusually rapid rate of tidal evolution (corresponding to a low primary Q/k_2) to have reached its current separation.

Thus, impact-generated systems would be relatively compact, with satellite orbital radii extending out to only ~100 primary radii or less. Such systems would be equally likely to be prograde or retrograde with respect to their heliocentric orbits, but would generally have satellites that orbit in the same sense as the primary's rotation (although see *Rufu et al.*, 2017). Oblique, low-velocity impacts often produce disks (or moons) that are disproportionately comprised of the outer layers of the colliding bodies, so that unusually ice-rich moons are most easily explained by this formation mechanism. However, compositional outcome can depend on impact parameters — e.g., in grazing, low-velocity collisions, much of the impactor core can be incorporated into a moon, leading to similar primary-moon bulk compositions.

5.5. Key Open Issues

We conclude by listing some important outstanding issues for future study that could help clarify current uncertainties in Pluto-system origin models. Many relate to the origin of the outer small moons, as this is currently the least well understood element of such models.

- ***Improved giant-impact models.*** In the simplest (and thus most attractive) model, the outer small moons share

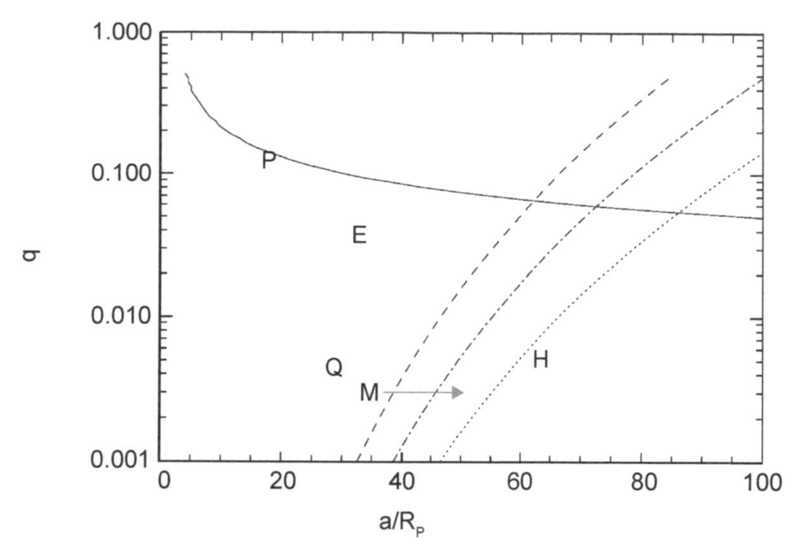

Fig. 13. Properties expected for tidally evolved KBO-moon systems that originally formed by a giant impact. The solid curve shows the relationship between the secondary-to-primary mass ratio (q) and the semi-major axis scaled to the primary's radius (a/R_P) at which the pair would occupy a dual synchronous state (like that of Pluto-Charon) in the limiting case of the maximum angular momentum that can be imparted by a single impact ($J \approx 0.45$; see equations (2) and (3), and we assume moment of inertia constants for the primary and secondary K = 0.35). Corresponding curves for lower-angular-momentum systems fall beneath this line. A second limit arises from the orbital separation that can be achieved by a pair via tidal evolution over the finite lifetime of the solar system. This is shown as dashed, dot-dashed, and dotted curves for tidal parameters for the primary body of (Q/k_2) = 300, 100, and 30, respectively. The region expected for impact-generated systems lies below the solid curve and to the left of the appropriate tidal curve. Approximate properties of KBO-moon systems are indicated by labels (P = Pluto, E = Eris, Q = Quaoar, M = Makemake, and H = Haumea). For Makemake, a blue arrow indicates that only a lower limit on the orbital size is known. Most and perhaps all satellites of large KBOs are potentially consistent with an impact origin.

a common origin with that of the Pluto-Charon binary. To date, a major unresolved challenge is accounting for the large orbital radii of the small moons, which appear much larger than would be directly produced by dispersed debris from a Charon-forming impact. Future work should explore the effects of increased resolution (e.g., *Canup and Salmon*, 2018), varied progenitor interior models (e.g., *Desch*, 2015), and the possible effects of internal strength on impact outcome (e.g., *Davies and Stewart*, 2016; *Emsenhuber et al.*, 2018). The suite of simulations by *Arakawa et al.* (2019) explored impact velocities up to 1.7 v_{esc}; given the importance of the currently understood low-velocity constraint for forming Pluto-Charon, investigation of larger speeds to determine if unexpectedly different outcomes occur is warranted.

- ***Determination of small moon masses and bulk densities.*** Ultimately, estimates of the small moon bulk densities should be obtainable via a combination of New Horizons-derived orbits and shape models, telescopic data, and dynamical stability integrations (e.g., *Porter et al.*, 2019; *Jacobson et al.*, 2019). If these densities remain consistent with the currently inferred ice-rich compositions, this would be a strong indication that the moons originated from material ejected from a Charon-forming collision involving partially differentiated bodies with ice-rich outer layers (e.g., *Canup*, 2011). If their bulk densities instead imply ice-rock compositions more similar to that of Charon, this could be consistent with either an impact origin (Fig. 10c) or with a substantial component of heliocentric material in the small moons. Any substantial moon-to-moon compositional variations would affect origin concepts as well.
- ***Detection of remnant signs of an impact origin.*** Tidal evolution of Pluto-Charon from an initially more compact state might have led to a fossil oblateness in either body, but this was not observed (e.g., *Moore et al.*, 2016; *Nimmo et al.*, 2017), consistent with the presence of an early Pluto ocean that decoupled the shell from the interior and precluded survival of a fossil bulge (e.g., *Robuchon and Nimmo*, 2011). However, other possible signs of an impact origin may yet exist. It is possible that a collisional family from a Charon-forming impact could survive until the current day. Detection or new upper limits on a collisional family with a composition similar to the unusually bright small moons, together with estimates of dynamical clustering with or without the effect of the giant-planet-induced scattering, could help constrain both the mode and timing of Pluto system formation. Tectonic features on Charon appear most easily explained by a combination of expansion due to ocean freezing and tidal stresses. If these could be used

to infer plausible tidal evolution histories (i.e., eccentricity as a function of semimajor axis and/or a as a function of time), or alternatively, histories that would be inconsistent with Charon as observed, this could provide new constraints to origin models (e.g., *Rhoden et al.*, 2015, 2020).

- ***More detailed models of small moon formation and orbital evolution.*** Unless the small moons expanded in resonance with Charon — which currently does not appear viable across large radial distances — the origin of their near-resonant orbits remains mysterious. However, it has been challenging for models to include all the main relevant processes to small moon assembly, including collisional damping and excitation by a tidally evolving inner Pluto-Charon binary. Future work should attempt to include such additional physics, and consider scenarios in which the current moons undergo only modest migration in resonance. It could also prove helpful to consider whether placing one or two moons in a resonant state (e.g., Nix) might then make the near-resonant states of the other moons probable based on stability arguments alone (e.g., *Walsh and Levison*, 2015).
- ***Interior models of progenitors and improved timing of formation constraints.*** Impact outcomes are sensitive to the interior state of the colliding bodies. As such, interior models that more fully consider the effects of accretional and radiogenic heating would be valuable. Better limits on the absolute timing of a Pluto-system-forming impact might be obtained by coupling models of the geophysical evolution of progenitors, refined impact models that account for detailed interior structures, and predictions of dynamical models of outer planet migration and excitation. Thermal evolution simulations suggest that the structures of progenitor interiors change significantly within the first tens of million years after accretion (section 3) if accretion takes place 5–10 m.y. after the birth of the solar system (Fig. 4). Determining the sensitivity of impact outcomes to a broader range of progenitor structures would require equations of state for rock-ice mixtures, rather than simply rock, ice, or hydrated silicate equations of state utilized in impact simulations to date.
- ***Improved understanding of Kuiper belt size distribution and its implications for small moon origin models.*** A lower bound on the age of the Pluto system can be set based on observed crater counts and estimates of the number and size distribution of transneptunian objects through time, which can be used to refine absolute ages of the oldest surface units. On Charon, ~4-G.y.-old surface age estimates from crater counts (*Greenstreet et al.*, 2015; *Stern et al., 2015*; *Moore et al., 2016*; *Singer et al.*, 2019) have yet to be reconciled with thermal evolution model predictions of resurfacing due to freezing of a subsurface ocean as recently as 2 Ga (*Desch and Neveu*, 2017). Upcoming deep, automated surveys of the transneptunian region will help firm up our knowledge of the TNO size distribution and how it came to be, enabling us to better place Pluto-Charon in this context. Additionally, the crater record at Pluto implies a surprising lack of impactors with radii ≤1 km (*Singer et al.*, 2019), and the importance of this dearth in background small bodies for small moon origin models that invoke heliocentric contributions should be assessed.
- ***Improved census of KBO binaries.*** A criticism of the giant impact hypothesis has been the necessity for rather finely tuned collisional conditions to produce the Pluto-Charon system, which lies at the upper end of binary angular momenta that can be produced by a collision. However, if there exists a population of KBO binaries whose properties are consistent with a broader range of impact outcomes, Pluto-Charon could be understood as an easily detectable outlier. As noted in section 5.4, giant impacts may have been common prior to Neptune's migration. At present we can merely state that other large mass ratio systems appear compatible with an impact origin (*Arakawa et al.*, 2019) (Fig. 13). Future work should make improved predictions for the expected mass ratio and semimajor axis distributions of impact-produced binaries that can be benchmarked against bias-corrected KBO samples from the next generation of large surveys.

Acknowledgments. We thank W. B. McKinnon and S. A. Stern for helpful discussions and suggestions that helped improve the content of this chapter, and W. B. McKinnon, E. Asphaug, and S. Desch for their helpful and constructive reviews. Support from NASA's New Frontiers Data Analysis program (R.M.C.), the Heising-Simons Foundation and NASA Grant No. 80NSSC18K0726 (K.M.K.), and the CRESST II agreement No. 80GSFC21M0002 between NASA GSFC and Univ. Maryland (M.N.) is gratefully acknowledged.

REFERENCES

Agnor C. B. and Hamilton D. P. (2006) Neptune's capture of its moon Triton in a binary-planet gravitational encounter. *Nature, 441*, 192–194.

Agnor C. B. and Lin D. N. C. (2012) On the migration of Jupiter and Saturn: Constraints from linear models of secular resonant coupling with the terrestrial planets. *Astrophys. J., 745*, 143.

Anderson J. D., Schubert G., Jacobson R. A., et al. (1998) Distribution of rock, metals, and ices in Callisto. *Science, 280*, 1573–1576.

Anderson J. D., Jacobson R. A., McElrath T. P., et al. (2001) Shape, mean radius, gravity field, and interior structure of Callisto. *Icarus, 153*, 157–161.

Arakawa M. and Maeno N. (1994) Effective viscosity of partially melted ice in the ammonia-water system. *Geophys. Res. Lett., 21*, 1515–1518.

Arakawa S., Hyodo R., and Genda H. (2019) Early formation of moons around large trans-neptunian objects via giant impacts. *Nature Astron., 3,* 802-807.

Barr A. C. and Canup R. M. (2008) Constraints on gas giant satellite formation from the interior states of partially differentiated satellites. *Icarus, 198*, 163–177.

Barr A. C. and Schwamb M. E. (2016) Interpreting the densities of the Kuiper belt's dwarf planets. *Mon. Not. R. Astron. Soc., 460*, 1542–1548.

Batygin K. and Brown M. E. (2010) Early dynamical evolution of the solar system: Pinning down the initial conditions of the Nice model. *Astrophys. J., 716*, 1323–1331.

Benz W., Slattery W. L., and Cameron A. G. W. (1986) The origin of the Moon and the single-impact hypothesis I. *Icarus, 66*, 515–535.

Benz W., Slattery W. L., and Cameron A. G. W. (1987) The origin of the Moon and the single-impact hypothesis, II. *Icarus, 71*, 30–45.

Benz W., Cameron A. G. W., and Melosh H. J. (1989) The origin of the Moon and the single-impact hypothesis III. *Icarus, 81*, 113–131.

Beyer R. A., Nimmo F., McKinnon W. B., et al. (2017) Charon tectonics. *Icarus, 287*, 161–174.
Bierson C. J., Nimmo F., and McKinnon W. B. (2018) Implications of the observed Pluto-Charon density contrast. *Icarus, 309*, 207–219.
Bland P. A. and Travis B. J. (2017) Giant convecting mud balls of the early solar system. *Sci. Adv., 3*, e1602514.
Bland P. A., Travis B. J., Dyl K., and Schubert G. (2013) Giant convecting mudballs of the early solar system. *Lunar and Planetary Science XLIV*, Abstract #1447. Lunar and Planetary Institute, Houston, Texas.
Bromley B. C. and Kenyon S. J. (2015) Evolution of a ring around the Pluto-Charon binary. *Astrophys. J., 809*, 88.
Bromley B. C. and Kenyon S. J. (2020) A Pluto-Charon concerto: An impact on Charon as the origin of the small satellites. *Astron. J., 160*, 85.
Brown M. E. (2012) The compositions of Kuiper belt objects. *Annu. Rev. Earth Planet. Sci., 40*, 467–494.
Brown M. E. and Butler B. J. (2018) Medium-sized satellites of large Kuiper belt objects. *Astron. J., 156*, 164.
Brown M. E., Van Dam M. A., Bouchez A. H., et al. (2006) Satellites of the largest Kuiper belt objects. *Astrophys. J. Lett., 639*, L43–L46.
Brown M. E., Barkume K. M., Ragozzine D., and Schaller E. L. (2007) A collisional family of icy objects in the Kuiper belt. *Nature, 446*, 294–296.
Brown M. E., Ragozzine D., Stansberry J., and Fraser W. C. (2010) The size, density, and formation of the Orcus-Vanth system in the Kuiper belt. *Astron. J., 139*, 2700–2705.
Brozović M., Showalter M. R., Jacobson R. A., and Buie M. W. (2015) The orbits and masses of satellites of Pluto. *Icarus, 246*, 317–329.
Buie M. W. and Grundy W. M. (2000) The distribution and physical state of H_2O on Charon. *Icarus, 148*, 324–339.
Canup R. M. (2005) A giant impact origin of Pluto-Charon. *Science, 307*, 546–550.
Canup R. M. (2011) On a giant impact origin of Charon, Nix, and Hydra. *Astron. J., 141*, 35.
Canup R. M. (2014) Lunar-forming impacts: Processes and alternatives. *Philos. Trans. R. Soc. London A, 372*, 20130175.
Canup R. M. and Salmon J. (2018) Origin of Phobos and Deimos by the impact of a Vesta-to-Ceres sized body with Mars. *Sci. Adv., 4*, eaar6887.
Canup R. M., Ward W. R., and Cameron A. G. W. (2001) A scaling relationship for satellite-forming impacts. *Icarus, 150*, 288–296.
Castillo-Rogez J. C. and Lunine J. I. (2010) Evolution of Titan's rocky core constrained by Cassini observations. *Geophys. Res. Lett., 37*, L20205.
Chambers J. E. (2016) Pebble accretion and the diversity of planetary systems. *Astrophys. J., 825*, 63.
Cheng W. H., Lee M. H., and Peale S. J. (2014a) Complete tidal evolution of Pluto-Charon. *Icarus, 223*, 242–258.
Cheng W. H., Peale S. J., and Lee M. H. (2014b) On the origin of Pluto's small satellites by resonant transport. *Icarus, 241*, 180–189.
Choblet G., Tobie G., Sotin C., et al. (2017) Powering prolonged hydrothermal activity inside Enceladus. *Nature Astron., 1*, 841–847.
Christy J. W. and Harrington R. S. (1978) The satellite of Pluto. *Astron. J., 83*, 1005–1008.
Cook J. C., Dalle Ore C. M., Protopapa S., et al. (2018) Composition of Pluto's small satellites: Analysis of New Horizons spectral images. *Icarus, 315*, 30–45.
Cruikshank D. P., Grundy W. M., DeMeo F. E., et al. (2015) The surface compositions of Pluto and Charon. *Icarus, 246*, 82–92.
Cruikshank D. P., Umurhan O. M., Beyer R. A., et al. (2019) Recent cryovolcanism in Virgil Fossae on Pluto. *Icarus, 330*, 155–168.
Dalle Ore C. M., Cruikshank D. P., Protopapa S., et al. (2019) Detection of ammonia on Pluto's surface in a region of geologically recent tectonism. *Sci. Adv., 5*, eaav5731.
Davidsson B. J. R., Sierks H., Güttler C., et al. (2016) The primordial nucleus of Comet 67P/Churyumov-Gerasimenko. *Astron. Astrophys., 592*, A63.
Davies E. J. and Stewart S. J. (2016) Beating up Pluto: Modeling large impacts with strength. *Lunar and Planetary Science XLVII*, Abstract #2938. Lunar and Planetary Institute, Houston, Texas.
Dawson R. I. and Murray-Clay R. (2012) Neptune's wild days: Constraints from the eccentricity distribution of the classical Kuiper belt. *Astrophys. J., 750*, 43.
Deienno R., Morbidelli A., Gomes R. S., and Nesvorný D. (2017) Constraining the giant planets initial configuration from their evolution: Implications for the timing of the planetary instability. *Astron. J., 153*, 153.
Desch S. J. (2015) Density of Charon formed from a disk generated by the impact of partially differentiated bodies. *Icarus, 246*, 37–47.
Desch S. J. and Neveu M. (2017) Differentiation and cryovolcanism on Charon: A view before and after New Horizons. *Icarus, 287*, 175–186.
Desch S. J., Cook J. C., Doggett T. C., and Porter S. B. (2009) Thermal evolution of Kuiper belt objects, with implications for cryovolcanism. *Icarus, 202*, 694–714.
Desch S. J., Kalyaan A., and Alexander C. M. (2018) The effect of Jupiter's formation on the distribution of refractory elements and inclusions in meteorites. *Astrophys. J. Suppl., 238*, 11.
Dobrovolskis A. R., Peale S. J., and Harris A. W. (1997) Dynamics of the Pluto-Charon binary. In *Pluto and Charon* (S. A. Stern and D. J. Tholen, eds.), pp. 159–190. Univ. of Arizona, Tucson.
Dullemond C. P. and Dominik C. (2005) Dust coagulation in protoplanetary disks: A rapid depletion of small grains. *Astron. Astrophys., 434*, 971–986.
Emsenhuber A., Jutzi M., and Benz W. (2018) SPH calculations of Mars-scale collisions: The role of the equation of state, material rheologies, and numerical effects. *Icarus, 301*, 247–257.
Farinella P., Milani A., Nobili A. M., and Valsecchi G. B. (1979) Tidal evolution and the Pluto-Charon system. *Moon Planets, 20*, 415–421.
Fernández J. A. and Ip W.-H. (1984) Some dynamical aspects of the accretion of Uranus and Neptune: The exchange of orbital angular momentum with planetesimals. *Icarus, 58*, 109–120.
Gladman B., Marsden B. G., and Van Laerhoven C. (2008) Nomenclature in the outer solar system. In *The Solar System Beyond Neptune* (M. A. Barucci et al., eds.), pp. 43–57. Univ. of Arizona, Tucson.
Glein C. R. and Waite J. H. (2018) Primordial N_2 provides a cosmochemical explanation for the existence of Sputnik Planitia, Pluto. *Icarus, 313*, 79–92.
Goldreich P. and Soter S. (1966) Q in the solar system. *Icarus, 5*, 375–389.
Goldreich P., Tremaine S., and Borderies N. (1986) Towards a theory for Neptune's arc rings. *Astron. J., 92*, 490–494.
Goldreich P., Lithwick Y., and Sari R. (2002) Formation of Kuiper-belt binaries by dynamical friction and three-body encounters. *Nature, 420*, 643–646.
Gomes R. (2003) The origin of the Kuiper belt high-inclination population. *Icarus, 161*, 404–418.
Gomes R., Levison H. F., Tsiganis K., and Morbidelli A. (2005) Origin of the cataclysmic Late Heavy Bombardment period of the terrestrial planets. *Nature, 435*, 466–469.
Greenberg R. and Barnes R. (2008) Tidal evolution of Dysnomia, satellite of the dwarf planet Eris. *Icarus, 194*, 847–849.
Greenstreet S., Gladman B., and McKinnon W. B. (2015) Impact and cratering rates onto Pluto. *Icarus, 258*, 267–288.
Grundy W. M., Binzel R. P., Buratti B. J., et al. (2016) Surface compositions across Pluto and Charon. *Science, 351*, aad9189.
Grundy W. M., Bertrand T., Binzel R. P., et al. (2018) Pluto's haze as a surface material. *Icarus, 314*, 232–245.
Grundy W. M., Noll K. S., Roe H. G., et al. (2019a) Mutual orbit orientations of transneptunian binaries. *Icarus, 334*, 62–78.
Grundy W. M., Noll K. S., Buie M. W., et al. (2019b) The mutual orbit, mass, and density of transneptunian binary G!kúnǁ'hòmdímà (229762 2007 UK126). *Icarus, 334*, 30–38.
Hahn J. M. and Malhotra R. (1999) Orbital evolution of planets embedded in a planetesimal disk. *Astron. J., 117*, 3041–3053.
Hahn J. M. and Malhotra R. (2005) Neptune's migration into a stirred-up Kuiper belt: A detailed comparison of simulations to observations. *Astron. J., 130*, 2392–2414.
Hahn J. M. and Ward W. R. (1995) Resonance passage via collisions. *Proc. Lunar Planet. Sci. Conf. 26th*, pp. 541–542.
Hammond N. P., Barr A. C., and Parmentier E. M. (2016) Recent tectonic activity on Pluto driven by phase changes in the ice shell. *Geophys. Res. Lett.*, 43, 6775–6782.
Hanner M. S. and Bradley J. P. (2004) Composition and mineralogy of cometary dust. In *Comets II* (M. C. Festou et al., eds.), pp. 555–564. Univ. of Arizona, Tucson.
Hiroi T., Zolensky M. E., and Pieters C. M. (2001) The Tagish Lake meteorite: A possible sample from a D-type asteroid. *Science, 293*,

2234–2236.
Holman M. J. and Wiegert P. A. (1999) Long-term stability of planets in binary systems. *Astron. J., 117*, 621–628.
Howard K. T., Benedix G. K., Bland P. A., and Cressey G. (2011) Modal mineralogy of CM chondrites by X-ray diffraction (PSD-XRD): Part 2. Degree, nature and settings of aqueous alteration. *Geochim. Cosmochim. Acta*, 75, 2735–2751.
Hsu H.-W., Postberg F., Sekine Y., et al. (2015) Ongoing hydrothermal activities within Enceladus. *Nature, 519*, 207–210.
Ida S., Canup R. M., and Stewart G. R. (1997) Lunar accretion from an impact-generated disk. *Nature, 389*, 353–357.
Iess L., Rappaport N. J., Jacobson R. A., et al. (2010) Gravity field, shape, and moment of inertia of Titan. *Science, 327*, 1367–1369.
Ishii H. A., Bradley J. P., Dai Z. R., et al. (2008) Comparison of Comet 81P/Wild 2 dust with interplanetary dust from comets. *Science, 319*, 447–450.
Jacobson R. A., Brozović M., and Showalter M. (2019) The orbits and masses of Pluto's satellites. *Pluto System After New Horizons*, Abstract #7031. LPI Contribution No. 2133, Lunar and Planetary Institute, Houston.
Johansen A., Youdin A., and Mac Low M.-M. (2009) Particle clumping and planetesimal formation depend strongly on metallicity. *Astrophys. J. Lett., 704*, L75-L79.
Johansen A., Mac Low M.-M., Lacerda P., and Bizzarro M. (2015) Growth of asteroids, planetary embryos, and Kuiper belt objects by chondrule accretion. *Sci. Adv., 1*, e1500109.
Johnston W. R. (2018) *TNO and Centaur Diameters, Albedos, and Densities V1.0.* NASA Planetary Data System, *https://sbn.psi.edu/pds/resource/tnocenalb.html*, accessed July 11, 2019.
Joy K. H., Zolensky M. E., Nagashima K., et al. (2012) Direct detection of projectile relics from the end of the lunar basin-forming epoch. *Science, 336*, 1426–1429.
Kaib N. A. and Chambers J. E. (2016) The fragility of the terrestrial planets during a giant-planet instability. *Mon. Not. R. Astron. Soc., 455*, 3561–3569.
Kamata S., Nimmo F., Sekine Y., et al. (2019) Pluto's ocean is capped and insulated by gas hydrates. *Nature Geosci., 12*, 407–410.
Kenyon S. J. and Bromley B. C. (2012) Coagulation calculations of icy planet formation at 15–150 AU: A correlation between the maximum radius and the slope of the size distribution for trans-neptunian objects. *Astron. J., 143*, 63.
Kenyon S. J. and Bromley B. C. (2014) The formation of Pluto's low-mass satellites. *Astrophys. J., 147*, 8.
Kenyon S. J. and Bromley B. C. (2019) A Pluto-Charon sonata: The dynamical architecture of the circumbinary satellite system. *Astrophys. J., 157*, 79.
Kenyon S. J., Bromley B. C., O'Brien D. P., and Davis D. R (2008) Formation and collisional evolution of Kuiper belt objects. In *The Solar System Beyond Neptune* (M. A. Barucci et al., eds.), pp. 293–313. Univ. of Arizona, Tucson.
Kiss C., Marton G., Parker A. H., et al. (2019) The mass and density of the dwarf planet (225088) 2007 OR_{10}. *Icarus, 334*, 3–10.
Kruijer T. S., Burkhardt C., Budde G., and Kleine T. (2017) Age of Jupiter inferred from the distinct genetics and formation times of meteorites. *Proc. Natl. Acad. Sci. U.S.A., 114*, 6712–6716.
Lambrechts M. and Johansen A. (2012) Rapid growth of gas-giant cores by pebble accretion. *Astron. Astrophys., 544*, A32.
Lambrechts M. and Morbidelli A. (2016) Reconstructing the size distribution of the small body population in the solar system. *AAS/Division for Planetary Sciences Meeting Abstracts,* 48, 105.08.
Lauer T. R., Throop H. B., Showalter M. R., et al. (2018) The New Horizons and Hubble Space Telescope search for rings, dust, and debris in the Pluto-Charon system. *Icarus, 301*, 155–172.
Leinhardt Z. M., Marcus R. A., and Stewart S. T. (2010) The formation of the collisional family around the dwarf planet Haumea. *Astrophys. J., 714*, 1789–1799.
Levison H. F. and Morbidelli A. (2003) The formation of the Kuiper belt by the outward transport of bodies during Neptune's migration. *Nature, 426*, 419–421.
Levison H. F. and Stern S. A. (1995) Possible origin and early dynamical evolution of the Pluto-Charon binary. *Icarus, 116*, 315–339.
Levison H. F., Morbidelli A., Van Laerhoven C., Gomes R., and Tsiganis K. (2008) Origin of the structure of the Kuiper belt during a dynamical instability in the orbits of Uranus and Neptune. *Icarus, 196*, 258–273.
Levison H. F., Bottke W. F., Gounelle M., Morbidelli A., Nesvorný D., and Tsiganis K. (2009) Contamination of the asteroid belt by primordial trans-neptunian objects. *Nature, 460*, 364–366.
Levison H. F., Kretke K. A., Walsh K. J., and Bottke W. F. (2015) Growing the terrestrial planets from the gradual accumulation of submeter-sized objects. *Proc. Natl. Acad. Sci. U.S.A., 112*, 14180–14185.
Li R., Youdin A. N., and Simon J. B. (2019) Demographics of planetesimals formed by the streaming instability. *Astrophys. J., 885*, 69.
Lin D. N. C. (1981) On the origin of the Pluto-Charon system. *Mon. Not. R. Astron. Soc., 197*, 1081–1085.
Lisse C. M., Young L. A., Cruikshank D. P., et al. (2020) On the origin and thermal stability of Arrokoth's and Pluto's ices. *Icarus, 356*, 114072.
Lithwick Y. and Wu Y. (2008) On the origin of Pluto's minor moons, Nix and Hydra. *ArXiV e-prints*, arXiv:0802.2951.
Lyttleton R. A. (1936) On the possible results of an encounter of Pluto with the Neptunian system. *Mon. Not. R. Astron. Soc., 97*, 108–115.
Malhotra R. (1991) Tidal origin of the Laplace resonance and the resurfacing of Ganymede. *Icarus, 94*, 399–412.
Malhotra R. (1993) The origin of Pluto's peculiar orbit. *Nature, 365*, 819–821.
Malhotra R. (1995) The origin of Pluto's orbit: Implications for the solar system beyond Neptune. *Astron. J., 110*, 420–429.
Malhotra R. and Williams J. G. (1997) Pluto's heliocentric orbit. In *Pluto and Charon* (S. A. Stern and D. J. Tholen, eds.), pp. 127–158. Univ. of Arizona, Tucson.
Mardling R. A. (2013) New developments for modern celestial mechanics — I. General coplanar three-body systems. Application to exoplanets. *Mon. Not. R. Astron. Soc., 435*, 2187–2226.
McKinnon W. B. (1984) On the origin of Triton and Pluto. *Nature, 311*, 355–358.
McKinnon W. B. (1989) On the origin of the Pluto-Charon binary. *Astrophys. J. Lett., 344*, L41–L44.
McKinnon W. B., Simonelli D. P., and Schubert G. (1997) Composition, internal structure, and thermal evolution of Pluto and Charon. In *Pluto and Charon* (S. A. Stern and D. J. Tholen, eds.), pp. 295–346. Univ. of Arizona, Tucson.
McKinnon W. B., Prialnik D., Stern S. A., and Coradini A. (2008) Structure and evolution of Kuiper belt objects and dwarf planets. In *The Solar System Beyond Neptune* (M. A. Barucci et al., eds.), pp. 213–241. Univ. of Arizona, Tucson.
McKinnon W. B., Stern S. A., Weaver H. A., et al. (2017) Origin of the Pluto-Charon system: Constraints from the New Horizons flyby. *Icarus, 287*, 2–11.
Michaely E., Peretz H. B., and Grishin E. (2017) On the existence of regular and irregular outer moons orbiting the Pluto-Charon system. *Astrophys. J., 836*, 27.
Mignard F. (1981) On a possible origin of Charon. *Astron. Astrophys., 96*, L1–L2.
Moore J. M., McKinnon W. B., Spencer J. S., et al. (2016) The geology of Pluto and Charon through the eyes of New Horizons. *Science, 351*, 1284–1293.
Morbidelli A. and Nesvorný D. (2020) Kuiper belt: Formation and evolution. In *The Trans-Neptunian Solar System* (D. Prialnik et al., eds.), pp. 25–59. Elsevier, Amsterdam.
Murray-Clay R. A. and Chiang E. I. (2005) A signature of planetary migration: The origin of asymmetric capture in the 2:1 resonance. *Astrophys. J., 619*, 623–638.
Murray-Clay R. A. and Chiang E. I. (2006) Brownian motion in planetary migration. *Astrophys. J., 651*, 1194–1208.
Nesvorný D. (2015) Evidence for slow migration of Neptune from the inclination distribution of Kuiper belt objects. *Astron. J., 150*, 73.
Nesvorný D. (2018) Dynamical evolution of the early solar system. *Annu. Rev. Astron. Astrophys., 56*, 137–174.
Nesvorný D. and Morbidelli A. (2012) Statistical study of the early solar system's instability with four, five, and six giant planets. *Astron. J., 144*, 117.
Nesvorný D. and Vokrouhlický D. (2016) Neptune's orbital migration was grainy, not smooth. *Astrophys. J., 825*, 94.
Nesvorný D. and Vokrouhlický D. (2019) Binary survival in the outer

solar system. *Icarus, 331*, 49–61.
Nesvorný D., Youdin A. N., and Richardson D. C. (2010) Formation of Kuiper belt binaries by gravitational collapse. *Astron. J., 140*, 785–793.
Nesvorný D., Vokrouhlický D., Bottke W. F., and Levison H. F. (2018) Evidence for very early migration of the solar system planets from the Patroclus-Menoetius binary Jupiter Trojan. *Nature Astron., 2*, 878–882.
Nesvorný D., Li R., Youdin A. N., Simon J. B., and Grundy W. M. (2019) Trans-neptunian binaries as evidence for planetesimal formation by the streaming instability. *Nature Astron., 3*, 808–812.
Neveu M. and Desch S. J. (2015) Geochemistry, thermal evolution, and cryovolcanism on Ceres with a muddy ice mantle. *Geophys. Res. Lett., 42*, 10197–10206.
Neveu M. and Rhoden A. R. (2019) Evolution of Saturn's mid-sized moons. *Nature Astron., 3*, 543–552.
Neveu M. and Vernazza P. (2019) IDP-like asteroids formed later than 5 Myr after Ca-Al-rich inclusions. *Astrophys. J., 875*, 30.
Neveu M., Desch S. J., and Castillo-Rogez J. C. (2017) Aqueous geochemistry in icy world interiors: Equilibrium fluid, rock, and gas compositions, and fate of antifreezes and radionuclides. *Geochim. Cosmochim. Acta, 212*, 324–371.
Nimmo F., Umurhan O., Lisse C. M., et al. (2017) Mean radius and shape of Pluto and Charon from New Horizons images. *Icarus, 287*, 12–29.
Nogueira E., Brasser R., and Gomes R. (2011) Reassessing the origin of Triton. *Icarus, 214*, 113–130.
Ormel C. W. and Klarh H. H. (2010) The effect of gas drag on the growth of protoplanets. Analytical expressions for the accretion of small bodies in laminar disks. *Astron. Astrophys., 520.*
Peale S. J. and Lee M. H. (2002) A primordial origin of the Laplace relation among the Galilean satellites. *Science, 298*, 593–597.
Pires P., Winter S. M. G., and Gomes R. S. (2015) The evolution of a Pluto-like system during the migration of the ice giants. *Icarus, 246*, 330–338.
Pires dos Santos P. M., Morbidelli A., and Nesvorný D. (2012) Dynamical capture in the Pluto-Charon system. *Cel. Mech. Dyn. Astron., 114*, 341–352.
Porter S. B., Showalter M. R., Weaver H. A., et al. (2019) The shapes and poles of Nix and Hydra from New Horizons. *Pluto System After New Horizons*, Abstract #7038. LPI Contribution No. 2133, Lunar and Planetary Institute, Houston.
Postberg G., Khawaja N., Abel B., et al. (2018) Macromolecular organic compounds from the depths of Enceladus. *Nature, 558*, 564–568.
Quillen A. C., Nichols-Fleming F., Chen Y.-Y., and Noyelles B. (2017) Obliquity evolution of the minor satellites of Pluto and Charon. *Icarus, 293*, 94–113.
Quillen A. C., Chen Y.-Y., Noyelles B., and Loane S. (2018) Tilting Styx and Nix but not Uranus with a spin-precession-mean-motion resonance. *Cel. Mech. Dyn. Astron., 130*, 11.
Renaud J. P., Henning W. G., Saxena P., Neveu M., Bagheri A., Mandell A., and Hurford T. (2021) Tidal dissipation in dual-body, highly eccentric, and nonsynchronously rotating systems: Applications to Pluto-Charon and the exoplanet TRAPPIST-1e. *Planet. Sci. J, 2*, 4.
Rhoden A. R., Henning W., Hurford T. A., and Hamilton D. P. (2015) The interior and orbital evolution of Charon as preserved in its geologic record. *Icarus, 246*, 11–20.
Rhoden A. R., Skjetne H. L., Henning W. G., Hurford T. A., Walsh K. J., Stern S. A., Olkin C. B., Spencer J. R., Weaver H. A., Young L. A., Ennico K., and the New Horizons Team (2020) Charon: A brief history of tides. *J Geophys. Res.–Planets, 125(7)*, e2020JE006449.
Robbins S. J., Singer K. N., Bray V. J., et al. (2017) Craters of the Pluto-Charon system. *Icarus, 287*, 184–206.
Roberts J. H. (2015) The fluffy core of Enceladus. *Icarus, 258*, 54–66.
Robuchon G. and Nimmo F. (2011) Thermal evolution of Pluto and implications for surface tectonics and a subsurface ocean. *Icarus, 216*, 426–439.
Rubin M. E., Desch S. J., and Neveu M. (2014) The effect of Rayleigh-Taylor instabilities on the thickness of undifferentiated crust on Kuiper belt objects. *Icarus, 236*, 122–135.
Rufu R., Aharonson O., and Perets H. B. (2017) A multiple-impact origin for the Moon. *Nature Geosci., 10*, 89–94.
Russell S. S., Hartmann L., Cuzzi J., Krot A. N., Gounelle M., and Weidenschilling S. (2006) Timescales of the solar protoplanetary disk. In *Meteorites and the Early Solar System II* (D. S. Lauretta and H. Y. McSween, eds.), pp. 233–251. Univ. of Arizona, Tucson.
Schaller E. L. and Brown M. E. (2007) Volatile loss and retention on Kuiper belt objects. *Astrophys. J. Lett., 659*, L61–L64.
Schlichting H. E. and Sari R. (2008) The ratio of retrograde to prograde orbits: A test for Kuiper belt binary formation theories. *Astrophys. J., 686*, 741–747.
Schlichting H. E. and Sari R. (2009) The creation of Haumea's collisional family. *Astrophys. J., 700*, 1242-1246.
Schlichting H. E., Fuentes C. I., and Trilling D. E. (2013) Initial planetesimals sizes and the size distribution of small Kuiper belt objects. *Astron. J., 146*, 36.
Sekine Y., Genda H., Kamata S., and Funatsu T. (2017) The Charon-forming giant impact as a source of Pluto's dark equatorial regions. *Nature Astron., 1*, 0031.
Shock E. L. and McKinnon W. B. (1993) Hydrothermal processing of cometary volatiles — Applications to Triton. *Icarus, 106*, 464–477.
Showalter M. R. and Hamilton D. P. (2015) Resonant interactions and chaotic rotation of Pluto's small moons. *Nature, 522*, 45–49.
Showalter M. R., Hamilton D. P., Stern S. A., et al. (2011) New satellite of (134340) Pluto: S/2011 (134340). *IAU Circular 9221.*
Showalter M. R., Weaver H. A., Stern S. A., et al. (2012) New satellite of (134340) Pluto: S/2012 (134340). *IAU Circular 9253.*
Showalter M. R., Porter S. B., Verbiscer A. J., et al. (2019) Rotation states of Pluto's small moons and the search for spin-orbit resonances. *Pluto System After New Horizons*, Abstract #7052. LPI Contribution No. 2133, Lunar and Planetary Institute, Houston.
Silsbee K. and Rafikov R. R. (2015) Planet formation in binaries: Dynamics of planetesimals perturbed by the eccentric protoplanetary Disk and the secondary. *Astrophys. J., 798*, 71.
Simon S. B., Armitage P. J., and Li R. (2016) The mass and size distribution of planetesimals formed by the streaming instability I. The role of self-gravity. *Astrophys. J., 822*, 55.
Singer K. N., McKinnon W. B., Gladman B., et al. (2019) Impact craters on Pluto and Charon indicate a deficit of small Kuiper belt objects. *Science, 363*, 955–959.
Smullen R. A. and Kratter K. M. (2017) The fate of debris in the Pluto-Charon system. *Mon. Not. R. Astron. Soc., 466*, 4480–4491.
Stern S. A. (2009) Ejecta exchange and satellite color evolution in the Pluto system, with implications for KBOs and asteroids with satellites. *Icarus, 199*, 571–573.
Stern S. A. and Colwell J. E. (1997) Accretion in the Edgeworth-Kuiper Belt: Forming 100–1000-km-radius bodies at 30 AU and beyond. *Astron. J., 114*, 841–884.
Stern S. A., McKinnon W. B., and Lunine J. I. (1997) On the Origin of Pluto, Charon, and the Pluto-Charon binary. In *Pluto and Charon* (S. A. Stern and D. J. Tholen, eds.), pp. 605–663. Univ. of Arizona, Tucson.
Stern S. A., Weaver H. A., Steffl A. J., et al. (2006) A giant impact origin for Pluto's small moons and satellite multiplicity in the Kuiper belt. *Nature, 439*, 946–948.
Stern S. A., Bagenal F., Ennico K., et al. (2015) The Pluto system: Initial results from its exploration by New Horizons. *Science, 350*, aad1815.
Stern S. A., Grundy W. M., McKinnon W. B., Weaver H. A., and Young L. A. (2018) The Pluto system after New Horizons. *Annu. Rev. Astron. Astrophys., 56*, 357–392.
Stern S. A., Weaver H. A., Spencer J. R., et al. (2019) Initial results from the New Horizons exploration of 2014 MU_{69}, a small Kuiper belt object. *Science, 364*, eaaw9771.
Stevenson D. J., Harris A. W., and Lunine J. I. (1986) Origins of satellites. In *Satellites* (J. A. Burns and M. S. Matthews, eds.), pp. 39–88. Univ. of Arizona, Tucson.
Sussman G. J. and Wisdom J. (1988) Numerical evidence that the motion of Pluto is chaotic. *Science, 241*, 433–437.
Sutherland A. P. and Kratter K. M. (2019) Instabilities in multiplanet circumbinary systems. *Mon. Not. R. Astron. Soc., 487*, 3288–3304.
Tancredi G. and Fernandez J. A. (1991) The angular momentum of the Pluto-Charon system: Considerations about its origin. *Icarus, 93*, 298–315.
Tholen D. J., Buie M. W., Grundy W. M., and Elliott G. T. (2008) Masses of Nix and Hydra. *Astron. J., 135*, 777–784.

Tiscareno M. S., Nicholson P. D., Cuzzi J. N., et al. (2019) Close-range remote sensing of Saturn's rings during Cassini's ring-grazing orbits and Grand Finale. *Science, 364*, eaau1017.

Tohline J. E. (2002) The origin of binary stars. *Annu. Rev. Astron. Astrophys., 40*, 349–385.

Trafton L., Stern S. A., and Gladstone G. R. (1988) The Pluto-Charon system: The escape of Charon's primordial atmosphere. *Icarus, 74*, 108–120.

Travis B. J., Bland P. A., Feldman W. C., and Sykes M. V. (2018) Hydrothermal dynamics in a CM-based model of Ceres. *Meteoritics & Planet. Sci., 53*, 2008–2032.

Tsiganis K., Gomes R., Morbidelli A., and Levison H. F. (2005) Origin of the orbital architecture of the giant planets of the solar system. *Nature, 435*, 459–461.

Volk K. and Malhotra R. (2019) Not a simple relationship between Neptune's migration speed and Kuiper belt inclination excitation. *Astron. J., 158*, 64.

Waite J. H., Perryman R. S., Perry M. E., et al. (2018) Chemical interactions between Saturn's atmosphere and its rings. *Science, 362*, eaat2382.

Walsh K. J. and Levison H. F. (2015) Formation and evolution of Pluto's small satellites. *Astron. J., 150*, 11.

Ward W. R. and Canup R. M. (2006) Forced resonant migration of Pluto's outer satellites by Charon. *Science, 313*, 1107–1109.

Ward W. R. and Canup R. M. (2010) Improved resonance expansion model for Nix and Hydra. Talk presented at the workshop Nix and Hydra: Five Years After Discovery, May 11–12, 2010, Baltimore, Maryland.

Weaver H. A., Stern S. A., Mutchler M. J., et al. (2006) Discovery of two new satellites of Pluto. *Nature, 439*, 943–945.

Weaver H. A., Buie M. W., Buratti B. J., et al. (2016) The small satellites of Pluto as observed by New Horizons. *Science, 351*, aae0030.

Weaver H. A., Porter S. B., Buie M. W., et al. (2019) Pluto's small satellites. *Pluto System After New Horizons*, Abstract #7028. LPI Contribution No. 2133, Lunar and Planetary Institute, Houston.

Woo J. M. Y. and Lee M. H. (2018) On the early *in situ* formation of Pluto's small satellites. *Astron. J., 155*, 175.

Youdin A. N. and Goodman J. (2005) Streaming instabilities in protoplanetary disks. *Astrophys. J., 620*, 459–469.

Youdin A. N., Kratter K. M., and Kenyon S. J. (2012) Circumbinary chaos: Using Pluto's newest moon to constrain the masses of Nix and Hydra. *Astrophys. J., 755*, 17.

McKinnon W. B., Glein C. R., Bertrand T., and Rhoden A. R. (2021) Formation, composition, and history of the Pluto system: A post-New Horizons synthesis. In *The Pluto System After New Horizons* (S. A. Stern, J. M. Moore, W. M. Grundy, L. A. Young, and R. P. Binzel, eds.), pp. 507–543. Univ. of Arizona, Tucson, DOI: 10.2458/azu_uapress_9780816540945-ch022.

Formation, Composition, and History of the Pluto System: A Post-New Horizons Synthesis

William B. McKinnon
Washington University in St. Louis

Christopher R. Glein
Southwest Research Institute, San Antonio

Tanguy Bertrand
NASA Ames Research Center

Alyssa R. Rhoden
Southwest Research Institute, Boulder

The Pluto-Charon system provides a broad variety of constraints on planetary formation, composition, chemistry, and evolution. Pluto was the first body to be discovered in what is now known as the Kuiper belt, and its orbit ultimately became a major clue that the giant planets underwent substantial orbital migration early in solar system history. This migration has been linked to an early instability in the orbits of the giant planets and the formation of the Kuiper belt itself, from an ancestral transneptunian planetesimal disk that included Pluto. Pluto-Charon is emblematic of what are now recognized as small or dwarf planets. Far from being a cold, dead, battered icy relic, Pluto displays evidence of a complex geological history, with ongoing processes including tectonism, cryovolcanism, solid-state convection, glacial flow, atmospheric circulation, surface-atmosphere volatile exchange, aeolian processes, and atmospheric photochemistry, microphysics, and haze formation. Despite Pluto's relatively modest scale, the combination of original accretional heat, long-term internal radiogenic heat release, and external solar forcing, when combined with sufficiently volatile (and thus mobile) materials, yields an active world. Pluto may have inherited a large organic mass fraction during accretion, which may be responsible in part for its surface and atmospheric volatiles. Charon, Pluto's major moon, displays evidence of extensive early tectonism and cryovolcanism. Dwarf planets are thus truly planetary in terms of satellite systems and geological and atmospheric complexity (if not ongoing activity). What they may lack in mass is made up in number, and the majority of the solar system's dwarf planets remain undiscovered.

1. INTRODUCTION

The New Horizons encounter with Pluto revealed not just a remarkable dwarf planet, but a complex, scientifically rich planetary system beyond Neptune, out in the Kuiper belt (*Stern et al.*, 2018). This chapter draws on these encounter results, later analyses, and other chapters in this volume. It has as its goal to synthesize and summarize what we have learned from New Horizons, highlighting some of the less-understood or -appreciated aspects of the Pluto system. The chapter is organized in the following manner. We first discuss how Pluto fits in with the emerging picture of planetesimal and planet formation in the solar system, and how it contributes to this understanding (section 2). This is followed by a discussion of the composition of the bodies in the Pluto system, focusing on volatiles and carbon (section 3). How Pluto and Charon inform understanding of other major icy satellites and Kuiper belt objects (KBOs) is also addressed. A section on the evolution of the Pluto system through time follows, considering orbital, thermal, tectonic, geomorphologic, and climactic changes (section 4). Emphasis is placed on the controls and duration of planetary activity and the roles of contingent events. A summary is then provided (section 5). Finally, some thoughts are offered on how exploration of the Pluto system has expanded our view of planethood and the richness of nature and how we might deepen this understanding through future work (section 6), for there is much that we do not yet know.

No single paper or chapter can do justice to these topics. The interested reader is invited to consult other chapters in this volume, the initial papers that discuss encounter results (*Stern et al.*, 2015; *Moore et al.*, 2016; *Grundy et al.*, 2016; *Gladstone et al.*, 2016; *Weaver et al.*, 2016; *Bagenal et al.*, 2016), subsequent *Annual Reviews* articles (*Stern et al.*, 2018; *Gladstone and Young*, 2019; *Moore and McKinnon*, 2021), two special issues of the journal *Icarus* (*Binzel et al.*, 2017a; *Singer et al.*, 2021), as well as papers in the original Space Science Series volume *Pluto and Charon* (*Stern and Tholen*, 1997), not all of which are out of date (far from it, actually).

2. FORMATION OF THE PLUTO SYSTEM

In this section we examine the formation of the Pluto-Charon system, the timing and conditions thereof, and both the implications for the initial states of Pluto and Charon and the constraints that Pluto and Charon provide on the accretion process. We begin by considering the accretion processes that created the bodies that eventually populated the Kuiper belt (section 2.1), and follow with a summary of current thinking regarding the dynamic instability that created the Kuiper belt (section 2.2). This is followed by a discussion of the Charon-forming giant impact specifically and when and where it occurred (section 2.3). Sections 2.4 and 2.5 then address, respectively, the thermal and structural implications for the progenitor (pre-giant-impact) bodies of the system and Pluto and Charon after their formation, focusing on the thorny issue of differentiation. Finally, we synthesize the overall discussion and consider continuing conundrums (section 2.6).

2.1. Accretion Scenarios

Understanding of planetary accretion has evolved substantially in the last 25 years (i.e., since the publication of *Pluto and Charon*). Both the supposed initial configuration of the solar system and the physics of planetesimal and planet formation have undergone profound if not revolutionary advances (see, e.g., chapters in *Beuther et al.*, 2014). Historically, accretion of Pluto at its present orbital distance by binary, Safronov-style hierarchical coagulation was difficult to understand. As extensively discussed in *Stern et al.* (1997), the issues reduced to insufficient surface mass density of accreting solids and the long orbital timescales in the classical Kuiper belt, making the formation of Pluto a drawn-out affair, potentially taking billions of years unless the velocity dispersion among the planetesimals was highly damped (e.g., *Goldreich et al.*, 2004). Moving Pluto's accretion zone from near Pluto's present semimajor axis (a) of 39.5 AU to somewhere in the 15- to 30-AU range, as in the original Nice planetary migration/instability models (e.g., *Tsiganis et al.*, 2005; *Levison et al.*, 2008a), goes a long way toward ameliorating this timescale issue, both because Keplerian orbital periods scale as $a^{3/2}$ and because the minimum mass necessary to construct all the planets must now be distributed over a much more compact range of semimajor axes (*Desch*, 2007) (i.e., overall collision rates scale with orbital speed and number density squared).

There are other long-standing issues with hierarchical coagulation, however, having to do with (1) nebular-gas induced turbulence preventing the formation of even small planetesimals, and (2) the fact that even if planetesimals could form by sticking, the velocity dispersion among them would result in destructive collisions between the nominally small, fragile bodies — problems together referred to as the "meter barrier" (see the reviews of *Johansen et al.*, 2014; *Johansen and Lambrechts*, 2017). A potential solution to these problems has been found in collective aerodynamic interactions of small particles with nebular gas (*Youdin and Goodman*, 2005; *Johansen et al.*, 2007, 2014, 2015; *Simon et al.*, 2016), which cause high-density filaments and streams of millimeter- to decimeter-sized particles (dubbed "pebbles") to form. Termed the streaming instability, if the volume density of solids in these pebble streams increases sufficiently, then the streams can fragment into gravitationally contracting pebble clouds [see Fig. 1 in *Nesvorný et al.* (2019) for a compelling visualization]. These infalling pebble clouds coalesce to form substantial planetesimals (tens to hundreds of kilometers or more in size) on short timescales ($<10^3$ yr), and in the words of *Morbidelli et al.* (2009), are "born big" (though to be clear, not as big as Pluto itself). Recent results from the New Horizons flyby of the cold classical KBO Arrokoth strongly support this general view of planetesimal formation (see *Stern et al.*, 2019; *Spencer et al.*, 2020; *Grundy et al.*, 2020; *McKinnon et al.*, 2020).

Measurements of nucleosynthetic anomalies in meteorites strongly imply that those planetesimals that formed "closer" to the Sun (here meaning out to and somewhat beyond 5 AU) formed over a range of times and places while the gas component of the protoplanetary disk existed [up to at least a few million years after the condensation of the first calcium-aluminum inclusions (CAIs)] (*Kruijer et al.*, 2017; *Scott et al.*, 2018; *Desch*, 2018). One might then reasonably expect that formation of large planetesimals by pebble cloud collapse took place over the lifetime of the gas nebula elsewhere in the solar system, such as in the region of the protoplanetary disk beyond Neptune (when the latter was much closer to the Sun), a zone we designate as the ancestral Kuiper belt (aKB). The characteristic planetesimal mass formed by the streaming instability (SI) followed by gravitational instability (GI) has been identified with the turnover (or "knee") in the size-frequency distribution (SFD) of KBOs (e.g., *Morbidelli and Nesvorný*, 2020), near diameter D = 100 km (see the chapter by Singer et al. in this volume). Above this size the power-law slope of the differential size-frequency distribution is quite steep, dN/dD ~D^{-5}, and below it is shallower, with a power-law exponent closer to –3 (*Fraser et al.*, 2014; *Lawler et al.*, 2018; see also *Greenstreet et al.*, 2015). (Note that these exponents refer to the ensemble of dynamically "hot" KBO populations, of which Pluto is an exemplary member.) The shallow power-law segment below the knee may be a direct outcome or signature of the SI/GI process (*Abod et al.*, 2019) or it may be due to later collisional evolution, a matter that is debated [although for the hot population collisional evolution is arguably the cause (*Morbidelli and Nesvorný*, 2020)].

Regardless of the details of the SFD at small sizes, once planetesimals formed in the aKB, the path to planethood was open. Growth to larger sizes can occur by direct (hierarchical) planetesimal accretion, by gas-assisted accretion of pebbles (termed "pebble accretion"), or both (*Johansen and Lambrechts,* 2017). *Johansen et al.* (2015) argue that the latter was more important for the aKB. Figure 1 illustrates both the original planetesimal SFD derived from their numerical SI model, and two outcomes of pebble accretion calculations. Two different planetesimal densities are used with two corresponding dimensionless turbulence parameters for the nebular gas (α). Both models display ordered growth up to 300 km radii over a couple to several million years, by planetesimal and pebble accretion, with a steep size distribution beyond 100-km sizes. This is followed by a runaway growth, due to pebble accretion, of a single, massive (Mars- or Earth-sized) body. This begs the question of what, if anything, might have limited the growth of planets in the aKB to Pluto-scale or similar.

The former existence of massive Kuiper belt bodies has in fact been proposed (e.g., *Gladman and Chan*, 2006). However, we are unaware of any dynamical scenario that permits the growth of one or more major (~10^{24-25} kg) planets in the aKB that is also consistent with creation of a Kuiper belt dynamical structure similar to that seen today (i.e., one that has been numerically tested). In this context, *Shannon and Dawson* (2018) examined aspects of the survival of wide binaries, cold classicals, and the resonant populations in the Kuiper belt, but did not come to very restrictive conclusions regarding the number of massive aKB objects. Such massive bodies may also be subject to rapid inward migration, toward the ice giant region (*Izidoro et al.*, 2015), and thus removal from the aKB. In this review, we will assume that massive, Mars-scale planetary embryos *did not* form in the aKB. We leave the question of what were the largest bodies that did form (e.g., Triton is 65% more massive than Pluto) for future research.

The various parameters used by *Johansen et al.* (2015) in their study (see Fig. 1) are in some sense tuned (although not unreasonably so) to give plausible results. For example, the effective value of the turbulent viscosity α in the 20–30 AU zone of the protoplanetary disk is not known. Their Fig. 11 illustrates the effect of varying α about a canonical value of 10^{-4}, albeit for the case of accretion in the asteroid belt (assuming a dead zone stirred by active disk layers). It shows that the growth of large protoplanets can either be rapid, or prolonged beyond the estimated lifetime of the protoplanetary gas disk [~5 m.y., based on the lifetimes of dust and gas disks around young solar-type stars (*Haisch et al.*, 2001; *Williams and Cieza*, 2011) or the Sun's protoplanetary disk magnetic field (*Wang et al.*, 2017)]. In contrast, *Johansen et al.* (2015) find that varying α at 25 AU mainly serves to change the timing of runaway pebble accretion (Fig. 1).

We are interested in Pluto-scale bodies, of course, and models of Neptune's outward migration through the aKB planetesimal disk require ~10^3 or possibly several ×10^3 Pluto-mass (~10^{23} kg) objects to make Neptune's migration

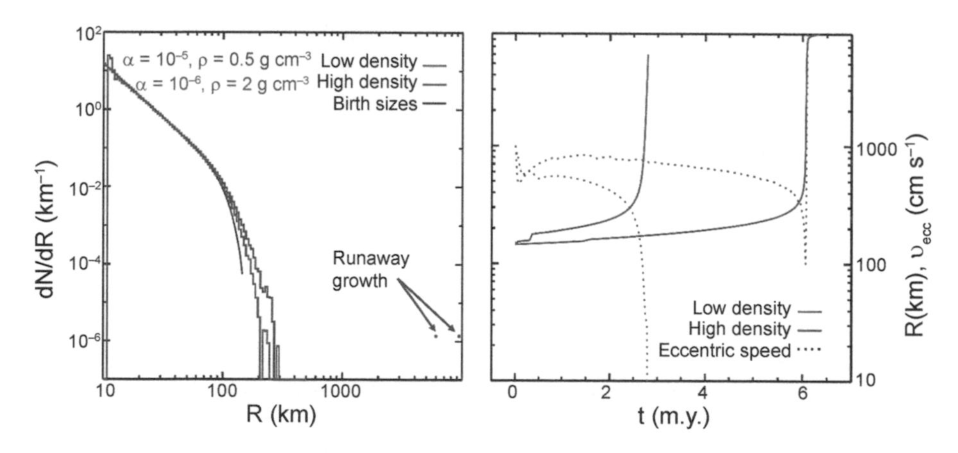

Fig. 1. Planetesimal growth in the original transneptunian protoplanetary disk (25 AU). Two models are shown: a low-density model where the solid density is set to ρ = 0.5 g cm^{-3}, similar to comets, and a high-density model where the internal density is set to ρ = 2 g cm^{-3}, similar to Pluto. The initial planetesimal size distribution is based on a streaming instability model. Both values of the turbulent stirring α (*Shakura and Sunyaev*, 1973) are "low" when compared with the 10^{-4} commonly used in disk viscous evolution models, and are thought to reflect the mild turbulence caused by streaming and Kelvin-Helmholtz instabilities in what would otherwise be a dead, laminar midplane. The right panel shows the size of the largest body as a function of time as well as its speed relative to a circular orbit (υ_{ecc}). Runaway growth is facilitated by a steep decline in orbital eccentricity because the high pebble accretion rate damps the eccentricity. Modified from *Johansen et al.* (2015).

sufficiently "granular" that the mean-motion resonances (MMR) in the Kuiper belt do not get overpopulated. Such an overpopulation is the outcome of any smooth migration model (*Nesvorný and Vokrouhlický*, 2016). So, this implies that if it occurred, pebble accretion onto planetesimals in the aKB was likely widespread, but slow enough that it stalled before true runaways to Mars-mass and beyond occurred. *Johansen et al.* (2015) in fact state the aKB is marginal for substantial pebble accretion generally, and then only for weak turbulence (see Fig. 1).

Once the nebular gas in the aKB dissipates [by photoevaporation due to XUV radiation from the proto-Sun; see the reviews by *Williams and Cieza* (2011) and *Alexander et al.* (2014)], planetesimal growth by hierarchical coagulation resumes. Certainly, the end game for the formation of the Pluto system involved the Charon-forming giant impact, the ultimate expression of hierarchical, two-body accretion (discussed in greater detail in section 2.3 below). In this context, Fig. 2 illustrates a "traditional" numerical model of planetesimal accretion in the aKB, via hierarchical accretion, from *Kenyon and Bromley* (2012). It is not entirely gas-free (the assumed nebular gas dissipation timescale is 10 m.y.), but pebble accretion is not included (collisional debris <1 m in size is simply removed from the computation).

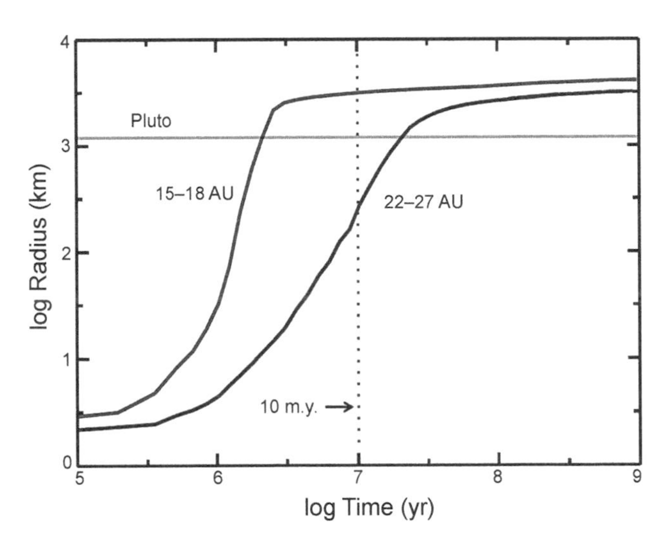

Fig. 2. Selected results from a numerical, multi-annulus hierarchical coagulation accretion simulation. Shown is the time evolution of the size of the largest accreted object in two transneptunian ranges. Starting conditions place most of the mass in 1-km planetesimals; the initial cumulative size distribution of planetesimals is nearly flat, with a surface density distribution proportional to $a^{-3/2}$. The calculation includes gas drag for the removal of smaller bodies from the calculation, but no explicit pebble accretion; fragmentation is included but no long-range dynamical stirring by proto-Neptune. After a short period of slow growth, objects rapidly grow from ~10 km to ~1000 km (runaways due to substantial gravitational focusing) and then grow more slowly to ~3000–5000 km as the largest planetesimals ("oligarchs") stir the smaller planetesimals to larger and larger orbital velocities. Modified from *Kenyon and Bromley* (2012).

Starting with a planetesimal distribution and a maximum size (r_0) of 1 km, the growth of the largest planetesimal in two example semimajor axis ranges is shown. Growth is at first slow, but enhanced gravitational focusing with increasing size eventually initiates a runaway. Accretion timescales are well under 10^9 yr. [Indeed, runaway accretion was one solution offered in *Stern et al.* (1997) to the Pluto formation timescale puzzle.] We stress, however, that initial kilometer-scale planetesimal "seeds" are needed to achieve such rapid growth: While motivated by classic works on planetesimal formation (e.g., *Goldreich and Ward*, 1973), such an assumption cannot be physically justified *a priori* based on our current understanding of nebular conditions and processes, as discussed above. Perhaps more interesting, from the point of view of planetesimal birth sizes (the initial mass function) resulting from the SI, is that the hierarchical coagulation timescales in the model increase if r_0 is larger, reaching gigayears to get to Pluto mass for $r_0 \approx 100$ km (*Kenyon and Bromley*, 2012, their Fig. 10).

We can conclude from these hierarchical coagulation models that as long as sufficiently small seed planetesimals form by some process, growth to Pluto size (or beyond) by hierarchical accretion is in fact actually plausible, on a timescale of millions to tens of millions of years. On the other hand, starting with predominantly large planetesimals (100 km scale) greatly lengthens the accretion timescale. But the "problem" with accretion by hierarchical coagulation is deeper. It is actually not so much a matter of forming Pluto on a plausible timescale. As *Morbidelli and Nesvorný* (2020) point out, the problem is *making enough Plutos*. That is, the process is inefficient, whereas SI followed ultimately by pebble accretion can be both efficient and relatively rapid.

2.2. Kuiper Belt Formation

As now understood, the Pluto system was emplaced into its 3:2 MMR with Neptune as part of the overall dynamical rearrangement of the outer solar system attendant upon a compact, but ultimately unstable, arrangement of four or more giant planets emerging from the protoplanetary gas nebula (e.g., *McKinnon et al.*, 2017). The Kuiper belt as a whole is thought to be almost entirely derived from an ~15–20-$M_\oplus$ (Earth mass) remnant planetesimal disk originally orbiting exterior to Neptune, a disk whose main mass extended not much further than 30 AU, Neptune's present semimajor axis (for recent reviews, see *Nesvorný,* 2018; *Morbidelli and Nesvorný,* 2020). The most natural timescale for this instability is *early*, within a few tens of millions of years after dissipation of the gaseous protoplanetary nebula, and not hundreds of millions of years later (*Nesvorný*, 2018; *Nesvorný et al.*, 2018; *Quarles and Kaib*, 2019; *de Sousa et al.*, 2020). Implantation into the Kuiper belt is not particularly efficient, on the order of 10^{-3} (*Morbidelli and Nesvorný*, 2020), meaning that only about 1 out of 1000 bodies originally in the aKB planetesimal disk ends up in a stable hot classical or Neptune-resonant orbit. This implies that many Pluto-scale dwarf planets were lost, ejected to the scattered/scattering disk, Oort cloud, or accreted by the giant planets, or possibly, in the case of Neptune's retrograde satellite Triton, captured (cf. *Stern*, 1991; *Nogueira et al.*, 2011).

Figure 3, from *Nesvorný and Morbidelli* (2012), illustrates a recent view of how the giant planet instability may have taken place. Notably, there is the ejection of a third ice giant. Nice-type models commonly feature ice giant ejections, and in this particular instance it is treated as a feature, not an issue that must be explained away (as it would be if the model started with only four giant planets). In Fig. 3 the three ice giants initially slowly migrate by scattering planetesimals. At 6 m.y. into the simulation, the instability is triggered when the inner ice giant crosses an orbital resonance with Saturn and the ice giant's eccentricity is pumped up. Following that, the ice giant has encounters with all other planets and is ultimately ejected from the solar system by Jupiter. Orbital eccentricities of the remaining giant planets are then subsequently damped by dynamical friction from the planetesimal disk, and Uranus and Neptune, propelled by the planetesimal-driven migration, reach their current orbits some 100 m.y. after the instability.

Neptune's slow, long-range migration in models of this type is able to capture planetesimals from the aKB into MMRs, and pump up the eccentricities and excite the inclinations of the captured KBOs. In the real world, one of these was no doubt Pluto (or the Pluto-Charon system). Where specifically in the 20–30-AU aKB planetesimal disk Pluto accreted cannot be stated with any certainty. If captured into the 3:2 MMR by a *smoothly* migrating Neptune, as first proposed by *Malhotra* (1993), Pluto's eccentricity e — although not its inclination — would have grown logarithmically from an initially small value (favorable for capture) as Neptune's semimajor axis (a_N) increased:

$$e^2_{final} \approx e^2_{initial} + \frac{1}{3}\ln\left(\frac{a_{N,final}}{a_{N,initial}}\right) \quad (1)$$

Given Pluto's present large e = 0.25, the implication is that Pluto's orbit could have expanded by up to 20%. That would put Pluto's point of resonance capture somewhat beyond the outer limit of the aKB, implying some amount of planetesimal scattering beforehand. Neptune's likely grainy semimajor axis evolution (*Nesvorný and Vokrouhlický*, 2016; *Lawler et al.*, 2019) scrambles this picture, however. The "catch and release" aspect of this grainy scenario (cf. *Murray-Clay and Chiang*, 2016) means that Pluto most likely (in terms of probability) entered into the 3:2 resonance with Neptune toward the end of the latter's planetesimal-driven migration, and probably from an already non-circular and inclined orbit, after Pluto had in all likelihood spent tens of millions of years or more dynamically interacting with Neptune, moving in and out of secular, Kozai, and MMRs with the ice giant [see *Gomes* (2003) and *Gomes et al.* (2005) for descriptions of the complex scattering dynamics during Neptune's migration]. In other words, Pluto was likely first *scattered* by Neptune to a high-e, high-i but non-resonant orbit with a semimajor axis somewhat less than 39.5 AU, was then captured into the 3:2 resonance, and subsequently migrated some distance outward (a few ×0.1 AU?) in lockstep with Neptune (*Nesvorný and Vokrouhlický*, 2016).

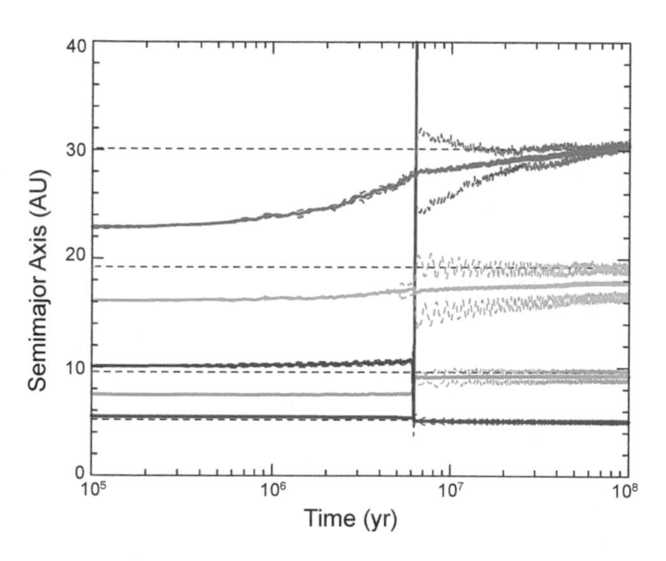

Fig. 3. A possible orbital history of the giant planets. Five planets were started in a (3:2, 4:3, 2:1, 3:2) mean-motion resonant chain along with a 20-$M_\oplus$ planetesimal disk between 23 AU and 30 AU. The semimajor axes (solid lines) and perihelion and aphelion distances (dashed lines) of each planet's orbit are indicated. The horizontal dashed lines show the semimajor axes of planets in the present solar system. The final orbits obtained in the model are a good match to those in the present solar system. From the statistical study of *Nesvorný and Morbidelli* (2012).

2.3. Giant Impact

The origin of the Pluto-Charon binary is widely regarded as due to a relatively giant (for the Kuiper belt) impact (see the chapter by Canup et al. in this volume). Other mechanisms for binary formation have been proposed for KBOs generally (*Nesvorný et al.*, 2010; *McKinnon et al.*, 2020), but the large masses of Pluto and Charon, the great specific angular momentum of the pair, and coplanar system of smaller satellites, all argue for an impact origin similar to that of Earth's Moon (e.g., *Stern et al.*, 2018). The reader is directed to the chapter by Canup et al. in this volume for a general overview. Here we focus on the salient aspects or constraints from giant impact models.

Two are most important. The first is that to yield such a large satellite/primary mass ratio (0.122 for Pluto-Charon) implies comparably sized impactors and low encounter velocities, i.e., impact speeds close to the escape speed for the impacting pair, $\upsilon_{imp} \approx \upsilon_{esc}$. These inferences stretch back to earlier analytical estimates (*McKinnon*, 1989), but more modern numerical smoothed-particle hydrodynamic

(SPH) calculations by *Canup* (2011) derive an upper limit $\upsilon_{imp}/\upsilon_{esc} \leq 1.2$, or a velocity at infinity $\upsilon_\infty \lesssim 0.7$ km s^{-1}. This result was confirmed by the subsequent SPH study of *Arakawa et al.* (2019).

This stringent velocity limit is not characteristic of KBO impact speeds onto Pluto today (*Greenstreet et al.*, 2015). In the aKB, it would imply an upper limit on characteristic eccentricities (e) and inclinations (i) such that $\sqrt{e^2 + \sin^2 i} \lesssim 0.1$. Such a dynamic limit would have been easily met during the pebble accretion phase before nebular gas dispersal (Fig. 1), but probably would have been met too well. Dynamical friction with the pebble swarm would have kept all the nascent proto-Plutos on near-circular, non-interacting orbits. Once the nebular gas dispersed, however, self-stirring of the proto-Pluto-rich planetesimal swarm (Fig. 2) plus long-range perturbations by Neptune should have triggered crossing orbits and ideal (i.e., low) velocity conditions for a Charon-forming impact (*Canup*, 2005). From calculations illustrated in *Morbidelli and Rickman* (2015), for 1000 embedded Pluto-mass bodies in the aKB disk, the orbital excitation limit above would have been easily met for planetesimal disk lifetimes t_d <100 m.y., especially as self-excitation increases as $\sqrt{t_d}$ (see their Fig. 1).

The situation changes drastically once Neptune begins migrating through the disk, however. Collision velocities increase smartly (to ~3 km s^{-1}) and collision probabilities plunge as the disk is depleted by scattering (see Fig. 11 in *Nesvorný and Vokrouhlický*, 2019). Other things being equal, the earlier the better for the Charon-forming impact, because once the velocity dispersion between bodies in the aKB climbs above the *Canup* (2011) velocity threshold, the outcomes of oblique impacts between proto-Pluto class bodies (~2000 km in diameter) are restricted to smaller satellites at best (*Arakawa et al.*, 2019).

Were such Charon-forming impacts likely? This is a question distinct from their physical realism or plausibility (i.e., whether a Charon-like body can be formed in a giant impact). The chapter by Canup et al. in this volume estimates the mean free time between proto-Pluto collisions as ~few × 10^7 yr for 1000 Plutos embedded in the planetesimal disk, from a particle-in-a-box calculation without additional gravitational focusing. Considering that there may have been up to 4000 Plutos (*Nesvorný and Vokrouhlický*, 2016), or an even larger number of proto-Pluto (half-Pluto) mass or greater bodies, as well as strong gravitational focusing, the mean free time between collisions (for υ_∞ < 0.4 km s^{-1}) could have been 10^5 yr or less. This is important because it is not enough to have a Charon-forming impact when the implantation efficiency into the Kuiper belt is on the order of 10^{-3} (*Morbidelli and Nesvorný*, 2020). Unless Pluto is an exceptional or freak case, there had to be dozens if not hundreds of giant impacts in the aKB in order to make the capture of a dwarf-planet binary into the 3:2 MMR a likely event (*Stern*, 1991). Maybe most Pluto-sized objects in the aKB experienced at least one giant impact of one sort or another, if the Pluto-Charon binary is good indicator of their overall histories. The results of all these giant impacts do not have to have resembled Charon of course; the outcomes are stochastic, and depend on υ_∞, impact angle (or parameter), initial mass ratio, and internal structural factors (*Canup* 2005, 2011; *Arakawa et al.*, 2019). That all the known dwarf planets of the Kuiper belt (the largest KBOs) have satellites or satellite systems is circumstantial evidence that giant impacts were the rule and not the exception [see *Barr and Schwamb* (2016) and section 4.2 in *Stern et al.* (2018)].

The second constraint from giant impact modeling concerns the densities and structural state(s) of the impactors. Numerical simulations of the Charon-forming impact to date favor, for Pluto-like densities, partially differentiated precursors (*Canup*, 2011). Completely differentiated precursors (i.e., bodies with ice mantles and rock cores) yield, post-impact, a very icy Charon in orbit about Pluto, contrary to Charon's known mean density of 1700 kg m^{-3}, whereas totally undifferentiated precursors yield very rock- and organic-rich small satellites, in apparent contradiction to their extremely icy nature (*McKinnon et al.*, 2017). This preference for only partially differentiated precursors has recently been muddled somewhat by the results of *Arakawa et al.* (2019), who were able to produce intact moons with ice mass fractions between 0.4 and 0.6 from impacts between two fully differentiated precursors. The moon/primary mass ratios in these cases range between 10^{-2} and in one instance a Charon-like 10^{-1} (see their supplementary Fig. 3). The SPH calculations of *Arakawa et al.* (2019) were carried out for a slightly smaller mass scale (about one-half Pluto mass total), but the physical inferences derived from their calculations should still be applicable to the Pluto system.

2.4. Thermal State, Pre-Giant Impact

Constraints on giant-impact-progenitor structural state are most valuable for the inferences they provide on the timing and mode of proto-Pluto accretion. Here we assume that the proto-Pluto impactors (1) matched the Pluto system's bulk composition, (2) were fully dense, and (3) were indeed only partially differentiated prior to the Charon-forming impact (*Canup*, 2011), following the general discussion in *McKinnon et al.* (2017) (although we relax this last constraint below). Surface ice layers in the simulations of *Canup* (2011) comprised 10–15% by mass of the precursor bodies, which can be compared with the ≈35 wt.% water ice for the actual system as a whole (when considered in terms of a hydrated rock plus H_2O-ice composition). The true range of pre-impact differentiation states that lead to a correct Charon bulk density has not been determined, but we will assume that differentiation must have proceeded to ice melting at least, and buoyancy-driven separation of some 25–50% of the total available ice (the question of potential solid-state separation is addressed in section 2.6).

Initial, post-accretion interior temperatures are determined by three things: *timing*, *timescale*, and *planetesimal size*. *Timing* refers to the time of accretion with respect to the first condensation of CAIs ($t = t_{acc}–t_{CAI}$), i.e., the beginning of the solar system, and the possibility of radiogenic heating by the short-lived isotope ^{26}Al. The integrated heat release subsequent to accretion is given by $H(0)\lambda^{-1}e^{-\lambda t}$, where H(0) is the rate of radiogenic heating per unit mass of Pluto rock at t = 0 and λ is the decay constant for ^{26}Al. For a solar composition, carbonaceous-chondrite-like Pluto rock (*McKinnon et al.*, 2017), H(0) is 1.5×10^{-7} W kg^{-1} and $\lambda = 3.07 \times 10^{-14}$ s^{-1} (*Castillo-Rogez et al.*, 2009; *Palme et al.*, 2014). The heat capacity of bulk KBO solid (assumed to be approximately two-third rock and approximately one-third water ice by mass) is approximated as 1150 × (T/250 K) J kg^{-1} K^{-1}, because both the heat capacities of rock and ice are temperature (T) dependent and because the carbonaceous fraction is poorly constrained (*McKinnon*, 2002) (see also section 3.2 below). Therefore, the heat necessary to increase the temperature of initial proto-Pluto solids at 40 K (a plausible background temperature in the protoplanetary disk at 25 AU) to 273 K (the low-pressure melting temperature of water ice) would have been provided radiogenically for t ≤ 3 m.y. (cf. *Sekine et al.*, 2017, their supplementary Fig. 10). Given the latent heat of melting of water ice (335 kJ kg^{-1}), 25% (50%) ice melting would have been achieved for t ≤ 2.9 (2.8) m.y.

These time limits assume instantaneous accretion; prolonged accretion with respect to the ^{26}Al half-life of 720 k.y. will change the estimates. But the important point is that *if* no more than 50% of the ice in the Pluto precursors melted prior to the Charon-forming impact, then the precursor bodies could not have *finished* accreting any earlier than 2.8 m.y. after t_{CAI}. Similarly, *Bierson and Nimmo* (2019) find, in their study of KBO porosity evolution, that if multi-100-km KBOs are to retain their inferred high porosities (based on densities <1000 kg m^{-3}), they cannot have accreted any earlier than 4 m.y. after t_{CAI}. These timescale limits are compatible, although the smaller KBOs are different bodies, and they may have formed later or more slowly than the Pluto progenitors (which may be why the former remained relatively modest in scale).

Timescale and *planetesimal size* refer to or control the magnitude and depth (within a body) of accretional heating. The temperature distribution T(r) in a symmetrically accreting uniform sphere can be written (*Schubert et al.*, 1981) as

$$T(r) = \frac{h}{\bar{C}_p}\left(\frac{4}{3}\pi\rho G r^2 + \frac{\langle\upsilon\rangle^2}{2}\right) + T_0 \quad (2)$$

where r is the instantaneous radius, ⟨υ⟩ the mean encounter velocity at "infinity," ρ and T_0 are the density and temperature of the incoming planetesimals, respectively, h the fraction of the impact energy retained, and C_p the mean heat capacity averaged over the interval $T–T_0$. Figure 4 is an update of Fig. 8 from *McKinnon et al.* (1997) but with the temperature dependent $\bar{C}_P(T)$ from above and an uncompressed density for the Pluto precursors of 1800 kg m^{-3} (*McKinnon et al.*, 2017); as in that earlier calculation the kinetic energy term in equation (2) is ignored in order to

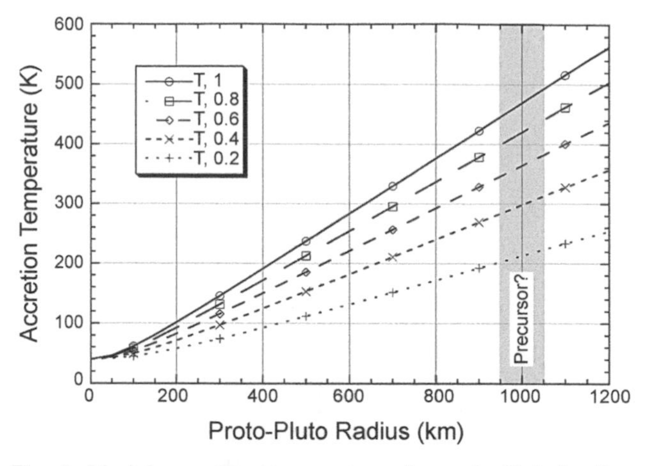

Fig. 4. Model accretion temperatures for proto-Pluto bodies, prior to the Charon-forming giant impact. The curve labeled T,1 assumes essentially complete retention of impact energy; the curve T,0.8 corresponds to retention of about 80% of the impact energy; etc. The approach kinetic energy (the second term in equation (2)) is neglected for all five curves. Bodies with roughly half to two-thirds of the mass of the Pluto system correspond to the size range labeled "Precursor?."

focus on the effect of gravitational potential energy and thus to provide *minimum* estimates of T(r) (a point we return to later). In Fig. 4 the temperature dependence of $\bar{C}_p$ makes T(r) nearly linear with radius.

The significant unknown in Fig. 4 is the appropriate value of the empirical parameter h. Its value depends on (1) the burial depth of impact heat, (2) whether buried impact heat can be effectively conducted or advectively mixed by subsequent impacts toward the surface, and (3) whether surface heat can be effectively radiated to space or not (e.g., *Squyres et al.*, 1988). *McKinnon et al.* (2017) offered a simple scaling length to assess whether heat can be conductively transported to the surface over the accretion timescale t_{acc}

$$\frac{\kappa}{u_{acc}} \sim 10\text{m} \times \left(\frac{1000\text{ km}}{R_{final}}\right) \times \left(\frac{\tau_{acc}}{10^6\text{ yr}}\right) \quad (3)$$

where κ is the thermal diffusivity (assumed to be that of porous ice-rock) and u_{acc} is the radial rate of growth of the body. For impacts much larger than this scale, impact heat is effectively buried (h ~0.5–1), whether directly by shock heating or by ejecta that is too thick to cool before it is buried by subsequent impact debris. Equation (3) implies, even for long accretion times (t_{acc} ~several million years), planetesimal sizes in hierarchical accretion scenarios (e.g., *Kenyon and Bromley*, 2012) are likely far too large for impact heat to be efficiently radiated away during accretion. From Fig. 4, h ≳0.3 implies some water ice melting during accretion of 1000-km-scale proto-Plutos, and for h ≳0.7 the total volume fraction of ice melted within the body becomes substantial (>50%).

In contrast, small-scale, pebble accretion appears ideal for depositing accretional energy right at the surface, where it can be radiated away efficiently. Following *Stevenson et al.* (1986), the radiative equilibrium temperature of an accreting surface, for energy deposited right at the surface, is

$$T(r) = \left[\frac{\rho}{\sigma_{SB}} \left(\frac{GM(r)}{r} + \frac{\langle \upsilon \rangle^2}{2} \right) \frac{dr}{dt} + T_0^4 \right]^{1/4} \quad (4)$$

where M(r) the mass contained within a radius r, G the gravitational constant, σ_{SB} the Stefan-Boltzmann constant, and dr/dt the radial growth rate. For a proto-Pluto (~50% of Pluto's mass) accreting at a constant radial rate over 10^5 (10^6) yr, T_0 = 40 K, and ignoring any contribution from $\langle \upsilon \rangle$ (which is justifiable for pebble accretion), the surface temperature at the end of accretion could be as high as ~260 K (150 K). These temperatures are upper limits because no account is taken of heat capacity in equation (4), but they do serve to illustrate that pebble accretion alone is very unlikely to lead to wholesale ice melting, unless accretion occurs on a timescale much more rapid than considered here. On the other hand, even pebble accretion is unlikely to prevent accretional heating to 100 K or more, which implies bulk vaporization of "supervolatiles" such as N_2, CO, and CH_4, the rapid crystallization of any amorphous H_2O ice [and expulsion of trapped supervolatiles (*Kouchi and Sirono*, 2001)], and the early formation of atmospheres on protoplanets in the aKB (*Stern and Trafton*, 2008).

2.5. Thermal State, Post-Giant Impact

Given one or both precursor bodies in a partially differentiated state, the Charon-forming impact may have pushed at least Pluto over the differentiation finish line. The release of gravitational potential energy plus any kinetic energy at infinity (which, unlike the case for pebble or other forms of runaway accretion, cannot justifiably be assumed to be negligible) can result in a global temperature rise of ~50–75 K for Pluto (*McKinnon*, 1989; *Canup*, 2005). The corresponding heat release would have been ~50–90 kJ kg^{-1}, sufficient to melt 50% of Pluto's ice complement if globally distributed. Impact heating can be highly localized of course, and numerical results in *Canup* (2005) and *Sekine et al.* (2017) show regional increases of ~150–200 K, so the implications for global differentiation are less clear.

There are additional sources of heat that could have driven differentiation in the early years after the giant impact: tidal heating owing to Charon's orbital evolution, heat of reaction due to serpentinization of ultramafic minerals, radiogenic heating by long-lived isotopes (U, Th, and ^{40}K), and the gravitational potential energy released by differentiation itself (e.g., *McKinnon et al.*, 1997; *Robuchon and Nimmo*, 2011; cf. the chapter by Nimmo and McKinnon in this volume). None of these heat sources is necessarily huge on a (geologically) short timescale, but over 10^7 yr or more, could have easily provided an additional, global temperature increase in excess of 50 K to a body that is already partially differentiated and post giant impact. We conclude that as long as the precursor bodies were partially differentiated, the giant impact and its aftermath likely resulted in a fully differentiated (ice from rock) Pluto. A further inference is that Pluto's ocean dates from this time (again see the chapter by Nimmo and McKinnon in this volume; cf. *Bierson et al.*, 2020).

The outlook for Charon is less clear. It is likely that it too emerged from the giant impact in an at least partially differentiated state, if only because it would have accumulated icy impact disk debris in the first 10^2–10^3 yr as it orbited Pluto (*Canup*, 2011; *Arakawa et al.*, 2019). Direct giant-impact heating of Charon would nominally have been modest, because Charon would derive (in the "intact capture" mode) almost entirely from material distant from the impact point. Both *Canup* (2005) and *Sekine et al.* (2017) estimate very low temperature enhancements for Charon, $\lesssim$30 K. None of the impact calculations to date have incorporated material strength or frictional dissipation, however. This does not affect the shock heating or gravitational potential energy aspects of the giant impact, but for bodies that undergo substantial distortion or strain (such as Charon), such dissipation and heating may prove quite important to the internal energy balance (e.g., *Melosh and Ivanov*, 2018; *Ensenhuber et al.*, 2018). For further discussion of Charon's early thermal evolution, see the chapter by Spencer et al. in this volume.

2.6. Synthesis and Unresolved Issues

Figure 5 summarizes the steps involved in forming the Kuiper belt and Pluto-Charon within it, as outlined in this section. Starting with the gaseous and dusty protosolar disk, collective instabilities such as the streaming instability caused local concentrations of small particles (pebbles) to intermittently exceed the threshold for gravitational instability. These instabilities created the initial size distribution of planetesimals in the original transneptunian region (from ~20–30 AU), or aKB, with characteristic sizes near 100 km (e.g., *Morbidelli and Nesvorný*, 2020). Further growth was first driven by hierarchical coagulation, but eventually pebble accretion became dominant (*Johansen and Lambrechts*, 2017). As long as the nebular gas persisted, aerodynamic gas damping drove continual, relatively efficient "pebble accretion" onto the growing bodies. How far this proceeded is unclear, but mass growth by an order of magnitude at least seems likely (e.g., Fig. 1), whereas growth to Mars mass and beyond does not appear to have occurred (although why not, and why hundreds if not thousands of Pluto-scale bodies formed instead, is not exactly clear). We emphasize that pebble accretion is a process distinct from the original gravitational instabilities involving pebbles (pebble cloud collapse). In any event, pebble accretion relies on the presence of gas, so once the gas component of the disk dissipates, accretion via hierarchical coagulation necessarily

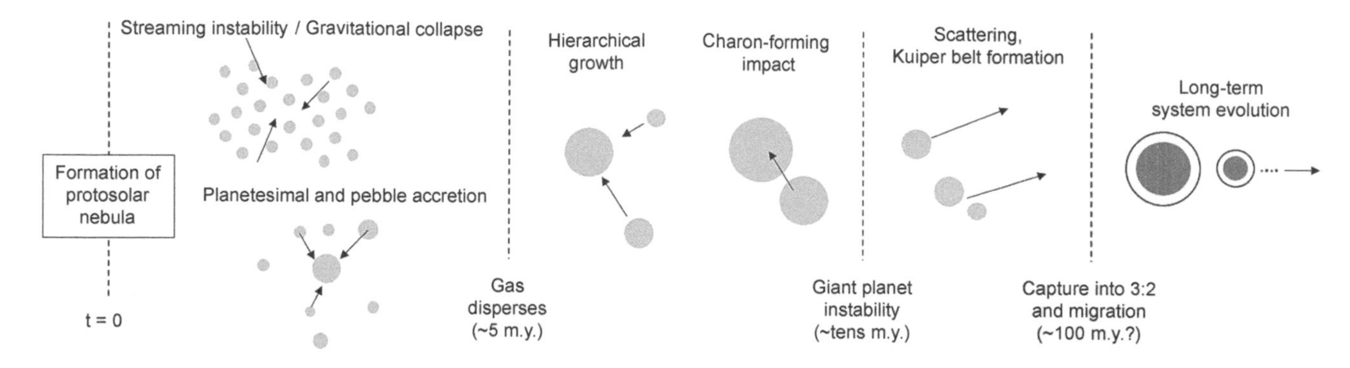

Fig. 5. Stages in the formation and evolution of the Pluto system in the context of the formation of the Kuiper belt (see text).

takes over. Such a gas-free planetesimal disk, on its own, was likely an ideal dynamical environment — in terms of sufficient number density and low encounter speeds — for the accretion of dwarf-planet mass bodies, and plausibly saw thousands of relatively "giant" impacts among them, including that which birthed Charon.

But this aKB Camelot was likely short-lived. According to current thinking, Neptune began migrating into the planetesimal disk within a few tens of millions of years, ultimately leading to the giant planet instability and rearrangement of the orbits of much of the solar system. The bodies in the aKB were scattered, the population there plummeted and encounter velocities increased so that collisional grinding of smaller bodies ensued and giant impacts could no longer yield Charon-like outcomes (except rarely, as there is almost always a low-velocity tail to encounter velocity distributions among planetesimals), and the modern Kuiper belt was emplaced or installed [modern in terms of structure; the populations were larger than today of course, and have been decreasing in number ever since (*Greenstreet et al.*, 2015)]. Neptune's orbital migration during this time is thought to have been relatively slow and grainy (jittery, in terms of orbital elements) due to scattering encounters with Pluto-mass bodies (Fig. 3) (*Nesvorný*, 2018). Pluto-Charon was ultimately captured into the 3:2 MMR and other resonances with Neptune (*Malhotra and Williams*, 1997), and its outward migration did not end until Neptune's did, perhaps some 100 m.y. after the beginning of the solar system. The implications of these orbital changes for Charon and the small satellites are taken up at the end of section 4.

Dynamical inferences for a relatively slow Charon-forming collision are consistent with late growth of the progenitor bodies in the ancestral planetesimal disk beyond Neptune. That the progenitor or precursor protoplanets were only partially differentiated is likely a signature of earlier growth by pebble accretion in the presence of nebular gas, i.e., by collisions so small that accretional heat was not deeply buried. Accretion of the progenitors could not have occurred too early, however, or full differentiation would have been driven by ^{26}Al decay. Nor was pebble accretion the entire story, as accretion of larger planetesimals appears necessary to provide the (buried) impact heat required for partial differentiation. Once the giant planet instability initiated, however, impact speeds in the planetesimal disk would have necessarily climbed to several kilometers per second, inconsistent with forming Charon according to our best simulations, although generally not with violent collisions (e.g., the formation of Sputnik basin, which we note is far too small to be the impact scar of the Charon-forming impact).

If a rock-rich, as opposed to an ice-rich, Charon can indeed be formed in a giant impact involving fully differentiated precursor protoplanets (*Arakawa et al.*, 2019), these inferences change. In this case the progenitor bodies would have had to either form early (to take advantage of ^{26}Al heating) or after nebular dispersal (to accrete from substantial planetesimals). Only by forming in the latter, waning half of protoplanetary gas nebula's nominal lifetime of ~5 m.y. can the combination of pebble and planetesimal accretion yield partially differentiated bodies of the proto-Pluto scale, especially as sizeable (~100-km-scale) bodies may have dominated the planetesimal swarm in the emerging planetesimal disk. Further numerical work may shed light on the issue of precursor differentiation state, perhaps involving benchmarking between and higher-resolution simulations of the Pluto-system-forming giant impact.

More work on the pathways to differentiation may also prove fruitful. The chapter by Canup et al. in this volume illustrates some possible initial conditions and outcomes. Those models adopt the perspective of *Desch et al.* (2009), in which differentiation is triggered when the ammonia-water peritectic temperature of 175 K is reached. Ammonia has been detected on Pluto and Charon (see the chapters by Cruikshank et al. and Protopapa et al. in this volume), and comets [which ultimately source from the same planetesimal disk as KBOs (*Morbidelli and Nesvorný*, 2020)] contain NH_3 at the 1% level compared with H_2O ice (*Mumma and Charnley*, 2011), so the formation of minor ammonia-water melt at low temperatures is a well-supported inference. *Desch et al.* (2009) argue that this melt should allow separation of rock from ice, i.e., descent of rock through a weakened ice-rock matrix, but this obviously requires large "rock" masses or density concentrations (not pebbles) to be effective, masses or concentrations whose existence cannot be decided on the basis of theory alone. Our approach above is more conservative, requiring complete local melting of the icy component to insure rock from ice differentiation, but is not necessarily more correct. Further ice deformation experiments in the ammonia-water and ammonia-water-silicate systems at stresses and strain rates more appropriate to icy bodies (cf. *Durham et al.*, 1993) are needed.

In terms of observations, ever deeper, more complete, and better characterized surveys of the Kuiper belt and inner Oort Cloud — e.g., the Outer Solar System Origins Survey (OSSOS) (*Bannister et al.*, 2018) and the Deep Ecliptic Exploration Project (DEEP) (*Trujillo et al.*, 2019) — along with new large surveys such as Pan-STARRS and the V. C. Rubin Observatory (LSST) in coming years will obviously improve enormously our picture of the structure (SFDs, dynamical classes) of the transneptunian populations. The individual character of the KBOs and their colors, binarity, etc., will also come into better focus. All of these in turn will drive improvements in numerical models that account for these characteristics, thus fostering a deeper understanding of how the solar system as a whole, and Pluto in particular, came to be. The perplexing puzzle of the most extreme transplutonian objects (i.e., Sedna and brethren) and their possible relation to a distant massive planet (e.g., *Sheppard et al.*, 2019) promise to be a revelation in this regard.

3. COMPOSITIONS OF PLUTO AND ITS MOONS

In this section we examine compositional issues for the Pluto system, with a focus on the volatile ice budget of Pluto specifically. Section 3.1 introduces our understanding of the volatile ice reservoirs [volatile ice generally meaning ices other than water ice, which is assumed to be a major bulk component of the Pluto system (*McKinnon and Mueller*, 1988; *McKinnon et al.*, 2017)], and discusses at length the possible origin of Pluto's all-important N_2. This is then contrasted with the apparent paucity of CO ice on Pluto in section 3.2. Tied to these discussions is the possible role of organic matter in Pluto's interior, and the possibility of bulk organic carbon and/or graphite within Pluto and Charon is addressed in section 3.3. Comparisons with other KBOs or icy satellites that may or may not resemble Pluto or Charon in terms of volatile ice abundance form a brief section 3.4, and we end with a synthesis and discussion of unresolved issues in section 3.5. Implications of the apparent compositions of the small moons are taken up later, in section 4.4.

3.1. Volatile Budgets and Cometary Provenance

The volatiles that comprise the atmosphere and surface ices of Pluto, Charon, and the small moons Nix, Styx, Kerberos, and Hydra are described in the chapters by Summers et al., Cruikshank et al., Protopapa et al., and Porter et al. in this volume. Regarding Pluto's neutral atmosphere during the New Horizons encounter, the primary gas is N_2 followed by minor CH_4 [~0.3% (*Young et al.*, 2018)] and a trace of CO [~0.05% (*Lellouch et al.*, 2017)]. Minor though CH_4 may be, it is the feedstock for the UV photochemical production of hydrocarbons such as C_2H_2, C_2H_4, and C_2H_6 and nitriles such as HCN in the upper atmosphere. The principal atmospheric components $N_2 > CH_4 > CO$ (in order of abundance) are maintained in vapor pressure equilibrium with N_2, CH_4, and CO ices in various combinations at Pluto's surface. Regarding surfaces, H_2O and NH_3-hydrate ices were detected on the surface of Pluto and Charon (*Grundy et al.*, 2016; *Dalle Ore et al.*, 2018, 2019; *Cook et al.*, 2019), confirming and extending earlier groundbased detections, and H_2O ice and an ammoniated species dominate the surfaces of the small moons (*Cook et al.*, 2018). Molecular O_2, abundant in the coma of Comet 67P/Churyumov-Gerasimenko (hereafter 67P) (*Bieler at al.*, 2015), was not detected in Pluto's atmosphere at the $\sim 10^{-5}$ level with respect to N_2 (*Kammer et al.*, 2017).

Less-volatile methanol (CH_3OH) and hydrocarbon ices have been detected within and near to the dark Cthulhu terrain on Pluto (*Cook et al.*, 2018), and are thought to be derived ultimately from both UV photochemistry and ion irradiation at Pluto's surface, as are the dark, red, tholin-like materials that nominally coat Cthulhu and other dark regions ("Tenebrae") on Pluto (*Protopapa et al.*, 2017; *Grundy et al.*, 2018). What has not been detected on either Pluto or Charon is CO_2 ice, which is common on many icy satellites and abundant and well distributed on Triton in particular (*Grundy et al.*, 2010; *Merlin et al.*, 2018; and references therein).

In the rest of this subsection we focus on the implications of Pluto's N_2 abundance, both in bulk and in relation to other volatiles and chemical reservoirs, drawing from and expanding upon the analysis in *Glein and Waite* (2018).

3.1.1. Pluto's N_2 inventory. On Pluto today, the arguably most abundant observed volatile is molecular nitrogen (see the chapter by Cruikshank et al. in this volume). The total amount of N_2 serves as a key constraint on Pluto's origin and evolution, as will be shown below. *Glein and Waite* (2018) attempted to estimate the inventory of N_2 that can be deduced from New Horizons data. They separated this inventory into four reservoirs, which they termed atmosphere, escape, photochemistry, and surface (Table 1). The amount of N_2 in the atmosphere (1×10^{15} mol) was calculated from the atmospheric pressure (~12 μbar) in 2015. The amount of N_2 that has escaped from Pluto's atmosphere was estimated to be a relatively modest 5×10^{16} mol (compared with pre-encounter thinking), if the cold exobase in 2015 [65–70 K (*Young et al.*, 2018)] is representative of Pluto's history. This is a conservative assumption, as Pluto's atmosphere has likely gone through major changes over geologic time (see section 4.2). The photochemical inventory at the surface corresponds to the amount of N_2 that has been incorporated into the products of CH_4-N_2 photolysis (e.g., HCN) from photochemical modeling. This amount was taken to be 2×10^{18} mol, again based on the notion that the past might have been like the present. The surface N_2 reservoir was assumed to be dominated by the glacial ice sheet called Sputnik Planitia (SP) (see the chapter by White et al. in this

TABLE 1. Nitrogen inventory on Pluto.

Reservoir	Moles of N_2
Atmosphere	1×10^{15}
Escape*	5×10^{16}
Photochemistry	2×10^{18}
Surface (Sputnik Planitia)	$(0.4–3) \times 10^{20}$
Subsurface liquid N_2	Comparable?†

*Assumes past similar to present (*Glein and Waite*, 2018).
†For example, if the outer 5 km of Pluto's ice shell were a 5% porous "nitrogenifer."

volume). Based on estimated dimensions of this feature (e.g., a depth of 3–10 km to enable convection), it was found that Sputnik Planitia could contain $(0.4–3) \times 10^{20}$ mol of N_2.

Because the other reservoirs above appear to be substantially smaller than that within SP, we take the range for SP to be a current best estimate for the inventory of N_2 on Pluto. If this estimate is normalized by *Glein and Waite*'s (2018) value for the H_2O abundance on Pluto (2×10^{23} mol), then an N_2/H_2O ratio of $(0.2–1) \times 10^{-3}$ is obtained. However, one should keep in mind that this range could be a serious underestimate as it excludes any subsurface reservoirs [e.g., N_2 liquid trapped in crustal pore space (cf. Table 1); N_2 trapped as clathrate in Pluto's ice shell; N_2 dissolved in a subsurface water ocean], and does not account for possible N_2 loss from the Charon-forming giant impact, or added by later cometary bombardment [the latter perhaps 2×10^{17} mol (*Singer and Stern*, 2015)].

3.1.2. Accreted N_2 as nitrogen source? It is thought that there were three major reservoirs of nitrogen atoms in the early outer solar system (*Miller et al.*, 2019): N_2, ammonia, and organic matter. [Recently detected ammonium salts on the surface of Comet 67P are plausibly the product of chemical reactions within the comet (*Poch et al.*, 2019).] Molecular nitrogen is considered to have been the most abundant form of nitrogen, because the solar ratio of $^{14}N/^{15}N$ (~440) is much different from (isotopically lighter than) those of primordial ammonia (~135) and organic nitrogen (~230) (see *Füri and Marti*, 2015; *Miller et al.*, 2019). The formation temperature of pebbles and planetesimals is a key factor in determining whether appreciable N_2 can be accreted by larger solid bodies such as Pluto. It is difficult to accrete N_2 in solid materials because of its great volatility. However, at sufficiently low temperatures in the solar nebula (e.g., <50 K), N_2 can be trapped in clathrate hydrates or amorphous ice (*Hersant et al.*, 2004; *Mousis et al.*, 2012). At even lower temperatures (e.g., $\lesssim$20 K), N_2 ice (or a solid solution of CO-N_2) can directly condense (*Hersant et al.*, 2004; *Mousis et al.*, 2012). To first order, the accreted amount of N_2 should be inversely related to the formation temperature. The "formation temperature" may be thought of as the average temperature experienced by pebbles and planetesimals that formed across a potentially wide range of heliocentric distances and at different times (see section 2.1). If primordial N_2 is present on Pluto, it could provide us with some insight into the thermal history of the environment where Pluto formed.

Until recently, it was not clear whether primordial N_2 is present in solids in the solar system. But then, the Rosetta mission discovered N_2 being outgassed from Comet 67P in 2014. Subsequent analysis provided an estimate of the N_2/H_2O ratio $(8.9 \pm 2.4) \times 10^{-4}$ in ices at the near-surface of the comet (*Rubin et al.*, 2015, 2019). This range is remarkably similar to that estimated above for Pluto, if Pluto is ~1/3 H_2O by mass. This similarity serves as the foundation of *Glein and Waite*'s (2018) suggestion that Pluto's N_2 might be primordial. However, there are complications that could make a primordial origin either less or more likely. One is the apparently low CO/N_2 ratio at Pluto's surface compared with comets (see section 3.2). The second potential issue is that this scenario may impose implausible restrictions. If we assume that Pluto started with an N_2/H_2O ratio similar to Comet 67P, then at least ~20% of all accreted N_2 must already be accounted for in the above inventory (*Glein and Waite*, 2018). This may leave too little margin to accommodate other internal reservoirs or adjustments such as loss from giant impact.

On the other hand, a recent observation suggests that it is possible to obtain larger inventories of primordial N_2. This observation is the N_2-rich Comet C/2016 R2 (PanSTARRS) (hereafter R2). It is a rare "blue comet" whose coma is dominated by CO. Water appears to be only a trace species at this comet ($H_2O/CO \approx 3 \times 10^{-3}$), and various other volatiles have been observed including CO_2, methanol, and methane (*Biver et al.*, 2018; *McKay et al.*, 2019). The N_2/H_2O ratio is estimated to be ~15 (*McKay et al.*, 2019), which is roughly 4 orders of magnitude higher than that at Comet 67P.

The provenance of Comet R2 is unknown, but one possibility is that it formed at/near the CO/N_2 ice line in the protosolar nebula, where these volatiles could have been concentrated (e.g., Öberg and Wordsworth, 2019). Formation in such an ultracold region of the nebula does not in and of itself explain the very low water vapor or dust production of an Oort cloud comet observed between 2.6 and 2.8 AU, however, production that should be apparent at these distances (*McKay et al.*, 2019). Nor would an (apparently) water-ice- and dust-poor composition be consistent with supposing that *all* cometary bodies originally formed in the aKB were once R2-like but underwent minor ^{26}Al heating and N_2 loss (cf. *Mousis et al.*, 2012). Accordingly, *Biver et al.* (2018) suggest that R2 may represent a collisional fragment from the volatile-ice-rich surface of a large (and at least somewhat differentiated) KBO. Setting aside this possibility for the sake of argument, the high N_2 abundance of R2 implies that *Glein and Waite*'s (2018) N_2 inventory for Pluto can be reproduced if the water fraction in Pluto that came from compositionally anomalous, R2-like comets is (1–7) × 10^{-5}. This does not seem prohibitive. If Pluto accreted a larger fraction of R2-like comets, then it could have started with more N_2 than in the inventory of *Glein and Waite* (2018).

3.1.3. Accreted NH_3 as nitrogen source? Another candidate source of nitrogen atoms is ammonia. There are three reasons why primordial, condensed ammonia is a plausible source of Pluto's N_2. First, ammonia in some form has been detected on Pluto, Charon, Nix, and Hydra, so it is available (e.g., *Dalle Ore et al.*, 2018, 2019; *Cook et al.*, 2018). Second, Pluto could have accreted a large quantity of ammonia. The canonical cometary abundance of ~1% with respect to water (*Mumma and Charnley*, 2011) would translate to an N_2/H_2O ratio of up to ~0.5%, above the current upper limit of ~0.1% for Pluto (section 3.1.1) (*Glein and Waite*, 2018). An ammonia source of N_2 may be ~5–25× larger than what is needed. However, the amount of margin is likely to be less, as the current existence of ammonia means that not all of it has been converted to N_2. There could even be a shortfall if Pluto has an ammonia-rich subsurface ocean (*Nimmo et al.*, 2016). The third plausibility argument for ammonia as an N_2 source on Pluto is one of analogy. It is inferred that ammonia can be converted to N_2 on icy worlds. The nitrogen isotopic composition of atmospheric N_2 on Titan ($^{14}N/^{15}N \approx 168$) shows that isotopically heavy, primordial ammonia [$^{14}N/^{15}N \approx 136 \pm 6$ (*Shinnaka et al.*, 2016)] must have made a major contribution to Titan's N_2 (*Mandt et al.*, 2014). Ammonia could have been the sole contributor if sufficient photochemical fractionation has occurred to reconcile these two values (*Krasnopolsky*, 2016). The exact mechanisms that converted ammonia to N_2 on Titan (and hence Pluto) remain uncertain. The general requirement is a high-energy process to decompose ammonia. It has been proposed that atmospheric chemistry (*Atreya et al.*, 1978; *McKay et al.*, 1988), impact chemistry (*Sekine et al.*, 2011), or internal geochemistry (*Glein*, 2015) might be important in this respect for Titan [see *Atreya et al.* (2009) for a more complete discussion].

The mechanism of *Atreya et al.* (1978) relies upon solar ultraviolet light to convert ammonia to N_2. The process is initiated by the photolysis of gaseous ammonia, which produces the amino radical (NH_2). This species then undergoes dimerization to produce hydrazine (N_2H_4). The formation of hydrazine is crucial because this is where two nitrogen atoms become bonded together. Additional photochemical reactions result in the production of N_2 from hydrazine. The efficiency of this process depends strongly on the atmospheric temperature, which should be high enough but not too high (e.g., 150–250 K). Temperatures need to be high enough to keep appreciable amounts of ammonia and hydrazine in the gas phase. However, if temperatures are too high, then water vapor is also present, and hydroxyl radicals (OH) derived from water dissociation act as a scavenger of amino radicals. This would prevent the formation of the key intermediate hydrazine.

Shock chemistry can also convert atmospheric (or surface; see below) ammonia to N_2 (*McKay et al.*, 1988). This is where meteoroids generate shock waves while traversing the atmosphere. These shocks expand away from the source, and compress and heat the surrounding air to high pressures and temperatures (on the order of 10^3 K). Affected air parcels reach chemical equilibrium at these conditions then quickly cool, which causes this state to be quenched. The shock-induced equilibrium favors the formation of N_2, and organic compounds if methane is present. In addition to requiring a sufficient flux of high-velocity (>4 km s^{-1}) impactors that transfer a substantial fraction of their energy into shocks [not so likely in the Kuiper belt (*Greenstreet et al.*, 2015) but plausible in Pluto's thought-to-be natal 20–30 AU region], this mechanism also requires the presence of ammonia gas. Hence, it is most relevant to syn- and post-instability accretion (section 2), when the impact rate and speeds were highest and the atmosphere may have been warm enough to support NH_3 vapor.

Pluto could have experienced substantial accretional heating (see sections 2.4 and 2.5). If such heating enabled the differentiation of Pluto by ice melting, then it is reasonable to envision the mobilization of volatiles toward the surface. This could have led to the formation of an early atmosphere on Pluto, where ammonia photolysis might have occurred. However, modeling has yet to be performed to understand the chemistry of such an atmosphere and test whether a photochemical origin of N_2 is viable. Currently, it appears marginally plausible, and the low solar flux at Pluto is an additional challenge. N_2 production via atmospheric shocks seems less attractive because impact velocities, even early on, are expected to be relatively low onto Pluto (~1–3 km s^{-1}; see section 2.3).

An intriguing implication of forming N_2 from atmospheric NH_3 is that this scenario may provide a way to explain the apparent lack of CO on Pluto relative to N_2. It can be envisioned that the most volatile primordial species (e.g., CO) were outgassed earliest and lost from a warm atmosphere, and the present N_2 inventory was formed somewhat later [as proposed for Triton by *Lunine and Nolan* (1992)]. This sequence of events is consistent with the much greater volatility of CO vs. ammonia. The question of Pluto's CO abundance, in particular its apparent low CO/N_2 ratio, is taken up in greater detail in section 3.2.

Sekine et al. (2011) showed that cometary impacts on ammonia-bearing *ice* can lead to the production of N_2 as well. Partial production occurs if the impact velocity is >5.5 km s^{-1}, and the conversion efficiency reaches 100% at velocities greater than ~10 km s^{-1}, although this latter speed is too high to be relevant for Pluto. Impact-induced decomposition of ammonia could have played an important role on Titan during an early heavy bombardment. However, as for atmospheric shock chemistry, impact shock chemistry may not be a major contributor to N_2 production on Pluto because of inadequate impact velocities. If Titan's atmospheric N_2 is impact-derived, then it would have a different origin from Pluto's (*Sekine et al.*, 2011).

Ammonia can be decomposed to N_2 in planetary interiors (*Glein*, 2015). At equilibrium, N_2 formation is favored by higher temperatures, lower pressures, more oxidized systems, higher pH (if the fluid is water-rich), and higher concentrations of bulk N. It is commonly assumed that chemical equilibrium could be reached over geologic timescales, but this has not been tested for the ammonia-N_2 system. The nitrogen speciation could be kinetically controlled at lower temperatures, depending on the relationship between the rate of ammonia decomposition (which increases with temperature) and the residence time of fluids in the environment of interest. There are two general environments inside icy worlds where the decomposition of primordial ammonia to N_2 might occur. One is at the water-rock interface where internal heating can drive ocean water circulation through the seafloor (*Shock and McKinnon*, 1993). This type of environment has the whole ocean inventory of ammonia at its disposal, but the ammonia is subjected to elevated temperatures for only a limited duration while it is circulated through the rock. The second environment of interest is the deeper interior. This environment may contain ammonium-bearing minerals that formed during differentiation (e.g., *De Sanctis et al.*, 2015). Heating may cause these minerals to release ammonia, which could then undergo conversion to N_2 if the geochemical conditions are suitable.

The essential requirement for a geochemical origin of Pluto's N_2 from accreted ammonia is the occurrence of high temperatures in the interior. This requires Pluto to be a differentiated body with a rocky core. It is thought that Pluto is probably differentiated (see the chapter by Nimmo and McKinnon in this volume). Assuming that this requirement is met, seafloor hydrothermal circulation may be possible, as shown by the preliminary modeling of *Gabasova et al.* (2018). However, the observed composition of Pluto seems inconsistent with the idea of generating N_2 in seafloor hydrothermal systems. The aqueous NH_3-N_2 and CH_4-CO_2 systems behave similarly, so conditions that favor N_2 usually also favor CO_2 (*Glein et al.*, 2008). CO_2 production might be expected to accompany N_2 production if N_2 was formed in seafloor hydrothermal systems. Instead, the surface of Pluto is methane-rich and CO_2 has not been detected (*Grundy et al.*, 2016; cf. the chapter by Cruikshank et al. in this volume). Metamorphism of ammoniated minerals in a rocky core could be a more promising source of N_2. *Bishop et al.* (2002) reported that mineral-bound ammonium is released at ~300°C. This temperature can easily be exceeded inside Pluto if a rocky core is present (*McKinnon et al.*, 1997; *Bierson et al.*, 2018) (Fig. 6). However, questions remain regarding the post-differentiation inventory of ammoniated minerals in the core, whether the released ammonium would speciate to N_2, and how much N_2 formed in this way could be outgassed from the core and delivered to the surface.

3.1.4. Organic matter as nitrogen source? The idea that organic matter can serve as a source of N_2 on icy worlds is relatively new. There are two basic arguments. One is that these worlds could have started with a great deal of N-bearing organic matter if the current cometary data (from Comets Halley, Wild 2, and 67P) are representative of pebbles and planetesimals that built bigger bodies in the outer solar

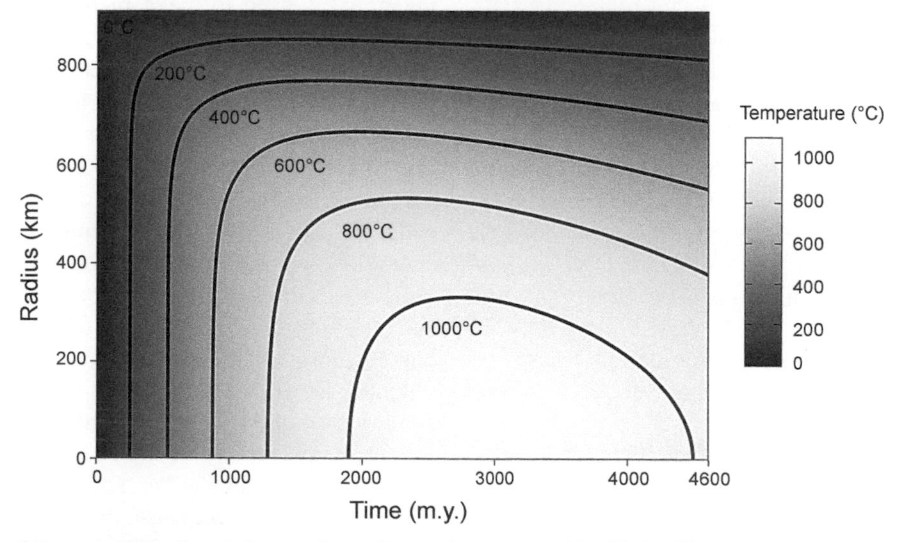

Fig. 6. Example of the possible thermal evolution of a rocky core inside Pluto. Temperatures were calculated for a carbonaceous chondritic radiogenic heating rate and heat conduction out of the core. The water-rock interface is at a radius of ~910 km, and its temperature was set to that of the overlying ocean. From *Kamata et al.* (2019).

system. The second argument is that organic matter can be "cooked" in the rocky cores of larger bodies if the body of interest is differentiated. This cooking process can provide sufficient thermal energy to break carbon-nitrogen bonds, which liberates nitrogen into the fluid phase. In a long-lived geologic system, volatilized nitrogen may come to chemical equilibrium that is determined by the temperature, pressure, and composition of the system. *Miller et al.* (2019) put these arguments in quantitative terms, and combined them with constraints on the ratios of $^{15}N/^{14}N$ and $^{36}Ar/^{14}N$ from the Huygens probe, to make the case that perhaps 50% of Titan's atmospheric N_2 was derived from organic N. This cooking process (organic pyrolysis) could also be relevant to the origin of methane on Titan, since methane is an abundant product (*Okumura and Mimura*, 2011). It is natural to wonder whether organic pyrolysis might be applicable to Pluto, whose surface composition resembles Titan's atmospheric composition. *Kamata et al.* (2019) suggested that methane and N_2 could be formed in this manner on Pluto, as a consequence of high interior temperatures (Fig. 6); *Stern et al.* (2018), while favoring an NH_3 source, earlier made a similar suggestion for the origin of Pluto's N_2. Below, we elaborate on some of the geochemical considerations.

Is an organic source of Pluto's N_2 large enough? A simple density mixing calculation can be performed to estimate the amount of organic N that could have been accreted by Pluto. We attempt to reproduce an uncompressed density for Pluto of 1.8 g cm^{-3} (*McKinnon et al.*, 2017) in terms of a mixture of rock (3.0 g cm^{-3}), generic "water" (0.95 g cm^{-3}), and graphite (2.2 g cm^{-3}; see below). We further assume that the rock is ~30 wt.% SiO_2 (similar to the Murchison meteorite), and the initial ratio of C_{org}/Si was ~5.5 as in Comet 67P (*Bardyn et al.*, 2017). The latter assumptions imply that the mass ratio of graphite/rock should be ~0.33. The result of this calculation is that bulk Pluto could be composed of ~54 wt.% rock, ~28 wt.% water, and ~18 wt.% graphite (cf. section 3.3). The "rocky" core is predicted to be ~75% rock and ~25% graphite by mass. In this model, Pluto's present inventory of graphite is 1.9×10^{23} mol. Assuming that the precursor to graphite was organic matter analogous to what was detected at Comet 67P [with N/C ≈ 0.035 (*Fray et al.*, 2017)], then the initial inventory of organic N on Pluto would have been 67×10^{20} mol. Therefore, the maximum theoretical yield of N_2 from organic N is estimated to be 34×10^{20} mol. This is 11–85× larger than *Glein and Waite*'s (2018) inventory based on New Horizons data. There is thus the potential to explain how Pluto got its N_2 inventory from an organic source, provided that 1–9% of accreted organic N was converted to N_2, delivered to the surface, and retained there.

We can gain insight into the fate of organic N in a rocky core on Pluto by determining what would be favored at chemical equilibrium. Here, we conduct an exercise in which equilibrium is calculated in the C-N-O-H system for a mixture of CI chondritic rock + cometary organic matter, based on *Lodders* (2003) and *Bardyn et al.* (2017). The adopted proportions are 100 C:4 N:165 H, which includes hydrogen atoms contributed by hydrated silicates (i.e., the organic mass fraction is roughly an order of magnitude larger than the several weight percent in CI chondrites; cf. section 3.3). A similar composition may have been produced on Pluto by aqueous alteration accompanying water-rock separation during the formation of a rocky core. The oxygen abundance is represented by the oxygen fugacity (oxygen partial pressure corrected for non-ideality), which provides a measure of the oxidation state of the system. The oxygen fugacity (fO_2) is treated as a free parameter, and is expressed relative to the value for the fayalite-magnetite-quartz (FMQ) mineral redox buffer

$$\begin{aligned} &3Fe_2SiO_4\text{, fayalite} + O_2\text{, gas} \\ &= 2Fe_3O_4\text{, magnetite} + 3SiO_2\text{, quartz} \end{aligned} \quad (5)$$

which is a commonly used geochemical point of reference (e.g., *Shock and McKinnon*, 1993). For simplicity, ideal gas calculations are performed (*McBride and Gordon*, 1996). We consider a pressure of 1900 bar, which may be the lowest pressure in the core (*McKinnon et al.*, 2017) and thus where ideal gas calculations will have the least error. In our example, the temperature is set to 500°C. This is close to the temperature for serpentine dehydration (e.g., *Glein et al.*, 2018), which could drive volatile outgassing from the core.

Figure 7 shows the computed speciation for the model system. The most relevant result is that N_2 is always the dominant form of nitrogen. It may seem counterintuitive that N_2 is favored even at strongly reducing conditions, where one may expect ammonia to predominate. However, this behavior is caused by the system being hydrogen-limited. There are not enough hydrogen atoms available (in this model) to prevent N_2 production. Most of the H inventory is used to make methane (water) at reducing (oxidizing) conditions owing to the larger abundance of C (O) compared with N. Hydrogen limitation also explains why graphite is the dominant form of carbon at lower oxygen fugacities. There remain important questions with regard to the accessibility of other hydrogen sources (hydrothermally-circulating ocean water, burial of free water in pores) to different depths in the hypothesized core. Yet, it has not escaped our attention that the predicted volatile assemblage at reducing conditions (Fig. 7) shows an intriguing gross similarity to Pluto's surface volatile composition (see the chapter by Cruikshank et al. in this volume). Future work will need to constrain geochemical conditions in the interior, as well as broaden the perspective to the possibility of multiple sources of nitrogen [as proposed for Titan (*Miller et al.*, 2019)] in preparation for spacecraft *in situ* measurements of isotopically heavy nitrogen ($^{14}N^{15}N$) and primordial argon (^{36}Ar) on Pluto. These promise to further enlighten us on the origin of Pluto's N_2 (see *Glein and Waite*, 2018).

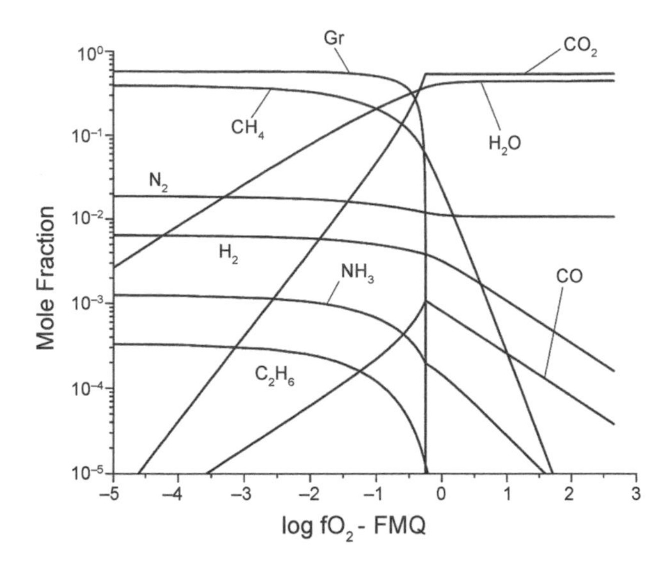

Fig. 7. Equilibrium state illustrating what could happen in an initial core on Pluto that is heated to 500°C and 1900 bar. Oxygen fugacity (fO_2) is plotted relative to the fayalite-magnetite-quartz (FMQ) mineral redox buffer (equation (5)). Note that the standard unit for fugacity is bars, but the difference in log units is independent of the pressure unit chosen. The initial core was assumed to be composed of a mixture of hydrated silicates and accreted organic matter. The system is more reduced toward the left side of the plot and more oxidized toward the right side of the plot. Note that the "plunge" of graphite (Gr) corresponds to its disappearance in oxidized systems.

3.2. The Missing CO Conundrum

The CO/N_2 ratio at Pluto's surface [~4×10^{-3} (*Glein and Waite*, 2018)] is orders of magnitude below that of known comets [e.g., $CO/N_2 \approx 35$ for Comet 67P (*Rubin et al.*, 2019)]. Specifically, the hemispheric CO/N_2 ratio in Pluto's surface ice ranges between 2.5 and 5×10^{-3} (*Owen et al.*, 1993; *Merlin*, 2015), while the atmospheric ratio is 5×10^{-4} (*Lellouch et al.*, 2017); these numbers are consistent with CO dissolved in solid solution in N_2 ice with their relative vapor pressures determined by Raoult's Law. While there is not yet a consensus solution to the "missing CO" problem, there is no shortage of possible solutions. One is that the low surface abundance of CO reflects low accreted or retained abundances of supervolatile species (CO, N_2, Ar), whereas the observed N_2 is a secondary product evolved from other N-bearing reservoirs, as detailed in sections 3.1.3. and 3.1.4. Alternative solutions include hydrothermal destruction of CO at the base of a subsurface ocean (*Shock and McKinnon*, 1993), hydrolysis of CO to formate ($HCOO^-$) in the ocean itself (*Neveu et al.*, 2015), preferential burial of CO (relative to N_2) in Sputnik Planitia (*Glein and Waite*, 2018), and preferential sequestration of CO (relative to N_2) in subsurface clathrates (*Kamata et al.*, 2019). In this subsection we address these alternative possibilities.

In aqueous systems dissolved CO is a relatively reactive and unstable species, and prefers to either be reduced to CH_4 or other organic species or oxidized to CO_2 and related species, depending on temperature, pH, and oxidation state (less so on pressure) (e.g., *Shock and McKinnon*, 1993). *Glein and Waite* (2018) looked specifically at metastable chemical speciation between aqueous CO, CO_2, HCO_3^- (bicarbonate), CO_3^{-2} (carbonate), HCOOH (formic acid), and $HCOO^-$ (formate) at conditions representative of the bottom of Pluto's putative ocean (0°C, 1900 bar) (i.e., not simply CO and formate). Hydrothermal processing (heating) was not assumed. Figure 8 shows the results. For all but the most acidic oceans, CO is very efficiently converted to other chemical species irrespective of hydrogen activity (meaning hydrogen molal concentration, corrected for non-ideality). That we even observe CO at all means that not *all* of Pluto's initial endowment of primordial CO need be aqueously processed (which seems plausible as volatile outgassing should have occurred before ice melting on the Pluto progenitors; section 2.4), plus there is always the later input from Kuiper belt bombardment. Indeed, *Glein and Waite* (2018) estimate, based on *Singer and Stern* (2015), that Pluto's surface CO could have been completely supplied by comets over geologic time.

Glein and Waite (2018) also hypothesize that as volatile ices condensed to fill the Sputnik basin, over some tens of millions of years following the impact (*Bertrand and Forget*, 2016; *Bertrand et al.*, 2018), Rayleigh-type fractionation caused preferential deposition of CO while

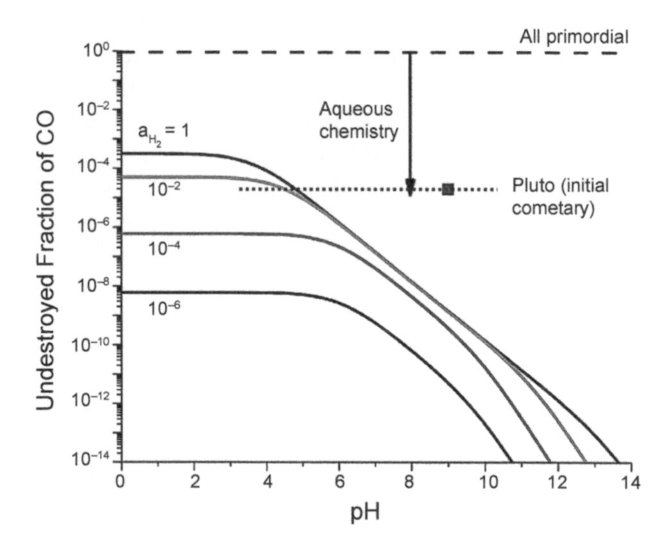

Fig. 8. Fraction of initial CO that remains in aqueous solution at metastable equilibrium with formate and carbonate species at 0°C and 1900 bar, as a function of the pH and H_2 activity a_{H_2} (a measure of the oxidation state). The symbol labeled "Pluto" shows how much CO destruction would need to occur for this model to reproduce the estimated surface ratio of CO/N_2 from an initial cometary ratio. The square denotes a pH value derived for Enceladus' ocean for comparison (from *Postberg et al.*, 2009). Figure modified from *Glein and Waite* (2018).

driving the atmospheric CO/N_2 ratio down, ultimately far down. Because convective transport with Sputnik continuously brings deeper ices to the surface (e.g., *McKinnon et al.*, 2016a), the implication is that the N_2-dominated optical surface would have to be controlled and maintained by the surface-atmosphere volatile cycle, while almost the entire ice sheet below the optical surface would necessarily be dominated by CO ice. While we do not view this as impossible *a priori*, it seems unlikely that an ultrathin surface veneer of N_2 ice can supply the N_2 vapor that appears to be condensing on the uplands to the east of Sputnik Planitia before returning as glacial flow into the basin (*Moore et al.*, 2016; *McKinnon et al.*, 2016a; *Umurhan et al.*, 2017; see also the chapters by White et al. and Moore and Howard in this volume), and N_2 ice is ubiquitous across Pluto. Further work on such possible fractionation is warranted, however.

Finally, there is the possibility of CO sequestration by clathrates in Pluto's ice shell (*Kamata et al.*, 2019). Clathrates are open-cage water-ice crystal structures in which cavities or sites are occupied by guest atoms or molecules (*Sloan and Koh*, 2008). The guest atoms or molecules are not bound to the ice crystal structure, but are "trapped" within, where they are free to rotate and vibrate, and thus capable of being detected spectrally with characteristic absorption bands (e.g., *Dartois et al.*, 2012). Clathrates are found on Earth in industrial and geological settings (e.g., CH_4 clathrate in permafrost or abyssal sediments, air clathrate in Antarctica), and have been proposed or invoked in a variety of solar system settings over the decades (e.g., Mars, Ceres, Europa, Titan, Enceladus, comets, and Pluto). Owing to their instability at low pressures (except when very cold), they have never been directly detected beyond Earth to our knowledge (cf. *Luspay-Kuti et al.*, 2016). While the absence of evidence cannot be taken as evidence of absence, the reader should bear this in mind.

In Pluto's case, preferential freezing out of CO clathrate at the expense of N_2 clathrate at the base of Plutos's floating ice shell is plausible, as N_2 clathrate is less stable [has a higher dissociation pressure than CO (*Sloan and Koh,* 2008)], but whether this occurred or is occurring within Pluto today depends on the identities and concentrations of the gas species dissolved in the ocean. For example, it is easy to envisage a situation in which dissolved gases released by Pluto's core are either rich in CH_4 or CO_2, depending on oxidation state (see Fig. 7). Clathrates of either of these two species are much more stable than CO-clathrate, and could form preferentially [CO_2 clathrate if the ocean is acidic (*Glein and Waite,* 2018, their Fig. 7; *Bouquet et al.*, 2019)]. Either could provide the thermal insulation to stabilize Pluto's ocean as envisaged by *Kamata et al.* (2019), but would be irrelevant to the missing CO story. We note that CO_2 clathrate in particular is denser than water, and on its own would sink to the bottom of the ocean, where it would remain very effectively sequestered, and is one possible contributor to the lack of identified CO_2 ice at Pluto's surface.

3.3. Carbon-Rich Pluto Models

The discussions above highlight the potential importance of Pluto's organic fraction — especially that trapped in a radiogenically heated core — for the evolution of its atmosphere and other volatiles. The rock/ice mass ratio of the Pluto system is often stated as being about 2:1 (*McKinnon et al.*, 2017), but this explicitly neglects the potential role of bulk carbonaceous matter, or implicitly assumes that CHON (carbon-hydrogen-oxygen-nitrogen-bearing organic matter) can be counted as low-density material similar to water ice (*McKinnon et al.*, 1997; *Simonelli et al.*, 1989). Here we explicitly discuss the possibilities and some implications of truly carbon-rich compositions for Pluto and Charon.

Carbonaceous matter is an important cometary component and one likely important in the ancestral Kuiper belt (see, e.g., the review in *McKinnon et al.*, 2008). CI and CM carbonaceous chondrites, which can be considered as proxies for the rocky material accreted by Pluto, on their own contain both soluble organic material such as carboxylic and amino acids, nitrogen heterocycles, etc., and a dominant (>70% of the organic carbon) component of insoluble macromolecular organic compounds (e.g., *Sephton*, 2005). Soluble organic compounds can also be released through hydrothermal activity from the insoluble component (by thermogenesis) (e.g., *Yabuta et al.*, 2007). Pluto's icy component, likely originally similar in composition to that in comets (as discussed above), contained and contains a variety of hydrocarbons, nitriles, and amines (*Mumma and Charnley*, 2011; *Altwegg et al.*, 2016).

Comets are notably rich in relatively non-volatile macromolecular organic matter (e.g., *Fray et al.*, 2016; *Bardyn et al.*, 2017) and surely Pluto was/is as well. Mass spectrometer measurements at Comet 1P/Halley highlighted the importance of CHON particles (*Kissel and Krueger*, 1987), and the wealth of *in situ* Rosetta measurements at 67P/Churyumov-Gerasimenko (a Jupiter-family comet and thus one formed in the same region of the outer solar system as Pluto) have only strengthened this inference (Fig. 9). The Rosetta-based elemental ratios derived for CHON in *Bardyn et al.* (2017), $C_{100}H_{100}O_{30}N_{3.5}$ (Fig. 9), compare favorably with the "classic" Vega Halley-based ratios of $\sim C_{100}H_{80}O_{20}N_4$ (*Kissel and Krueger*, 1987; *Alexander et al.*, 2017) derived from more limited, flyby data. Due to radiogenic and other heating the organic content of Pluto would have been altered/augmented by thermal and hydrothermal processing, as described above, as well as from the synthesis of hydrocarbons from inorganic species (*Shock and Canovas*, 2010; *Reeves and Fiebig,* 2020).

Post-Rosetta models of bulk cometary composition that match or come close to Pluto's density have been proposed, and contain a substantial organic/hydrocarbon component. *Davidsson et al.* (2016) propose in their "composition A" that 67P is 25 wt.% metal + sulfides, 42 wt.% rock/organics, and 32 wt.% ice. For their assumed component densities, the overall grain (zero porosity) density is 1820 kg m^{-3}. *Fulle et al.* (2017), updating *Fulle et al.* (2016), argue for, by volume, 4 ± 1% Fe-sulfides (density 4600 kg m^{-3}), 22 ± 2% Mg,Fe-olivines and -pyroxenes (density 3200 kg m^{-3} if crystalline), 54 ± 5% "hydrocarbons" (density 1200 kg m^{-3}), and 20 ± 8% ices (density 917 kg m^{-3}). This composition yields a comparatively lower primordial grain density (dust + ice) of 1720 ± 125 kg m^{-3}. Within uncertainties, both of these values are compatible with the uncompressed density of the Pluto system as a whole, ≈1800 kg m^{-3} (*McKinnon et al.*, 2017).

Figure 10 shows representations of Pluto's possible internal structure based on these compositions. These structures are portrayed as if fully differentiated according to the densities of the principal compositional components. A notable difference is the characterization of the organic component. *Davidsson et al.* (2016) cite similarities between 67P's carbon component to the relatively refractory, insoluble organic material (IOM) found in carbonaceous chondrites, although they model this component with the density of graphite (H/C = 0; 2100 kg m^{-3}). *Fulle et al.* (2017), based on higher H/C values (e.g., *Fray et al.*, 2016), opt for a lower-density but macromolecular hydrocarbon. In this case the volume percent of the organic component is enormous compared with even the "extreme" carbon-rich models of the past (see *McKinnon et al.*, 1997, their Fig. 4). This comparison underscores the importance of organic maturation, which is controlled by interior temperatures where organic matter is/was present. Lower H/C would be associated with higher temperatures.

Organic-rich Pluto compositions naturally contain less rock than organic-poor ones, and thus would predict lower heat flows overall for Pluto. Radiogenic heat production in the models in Figs. 10a,b are, respectively, 0.77 and 0.79× that of the Pluto models in *McKinnon et al.* (2017) (see also the chapter by Nimmo and McKinnon in this volume). The cores in these models would also be comparatively cooler, as high H/C organics (e.g., benzene) likely convect readily, whereas graphite (if present) has a very high thermal conductivity. Indeed, graphitization of core organic material (section 3.1.4) in any scenario should markedly enhance core thermal conductivity.

Specific geological or geophysical evidence of massive organic layers within Pluto (or Charon) is lacking, either for or against, but should be sought. For example, impacts of large KBOs with either Pluto or Charon could potentially excavate into a near-surface carbonaceous layer (as in Fig. 10b). Despite the likely size of the Sputnik basin impactor [~200-km diameter or greater (*Johnson et al.*, 2016; *McKinnon et al.*, 2016b)], the geology of Pluto is quite complex (see the chapter by White et al. in this volume) and still active, which complicates the identification of ancient basin ejecta. The large (~240-km-wide) basin Dorothy (Gale) on Charon is easier to interpret as Charon's northern terrain is relatively ancient and unmodified. No albedo or spectroscopic evidence of organic-rich ejecta is seen, which may impose useful lower limits on the thickness of Charon's ice shell (scaling Fig. 10b to Charon gives an ice shell thickness of 45 km). The fact remains, however, that Pluto, Charon, and KBOs generally likely inherited a large organic mass fraction during accretion (cf. Fig. 9c). The internal structures of large KBOs, and to an extent those of icy satellites as well, may be rather different than the paradigm that has reigned since the pioneering work of J. S. Lewis (e.g., *Lewis*, 1971).

3.4. Comparisons with Other Volatile-Rich Icy Worlds

The similarities in atmospheric and surface compositions among the triumvirate of Titan, Triton, and Pluto have long attracted attention. All three possess majority

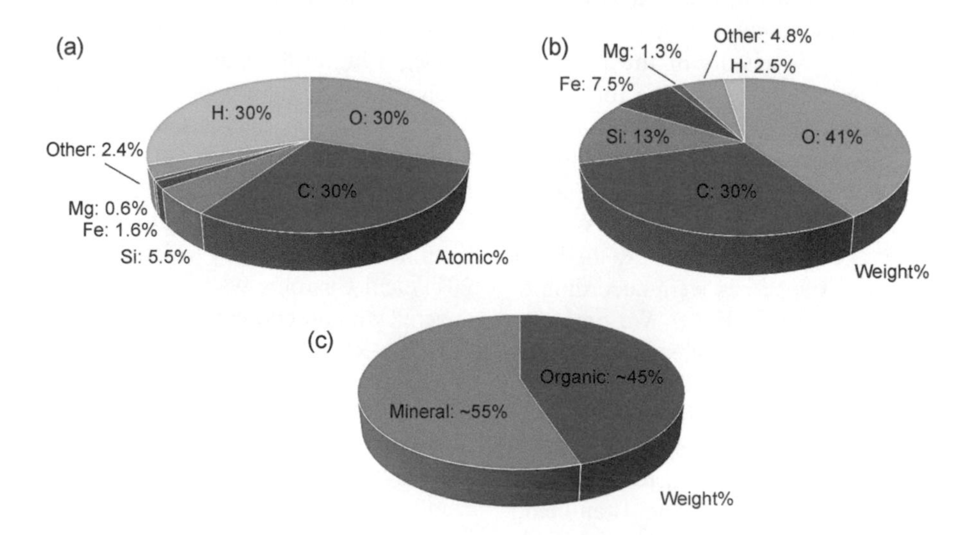

Fig. 9. Averaged composition of Comet 67P's dust particles as deduced from COmetary Secondary Ion Mass Analyzer (COSIMA)/Rosetta mass spectrometer measurements and supplemental hypotheses detailed in *Bardyn et al.* (2017). The averaged composition is given by **(a)** atomic fraction and **(b)** atomic mass fraction. **(c)** Mineral and organic content estimated in mass fraction. See *Bardyn et al.* (2017) for details.

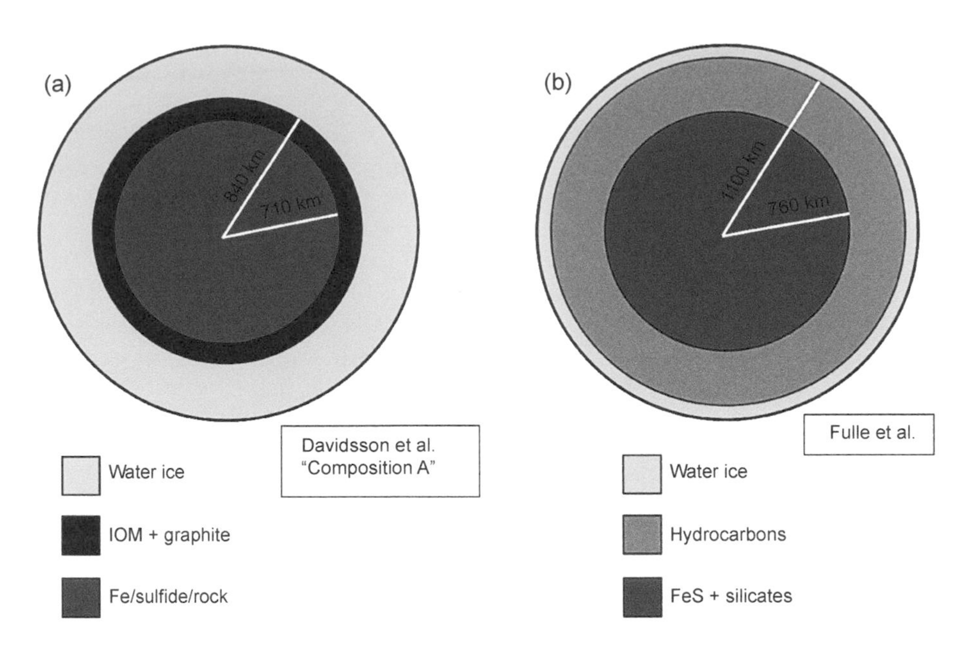

Fig. 10. Alternative Pluto internal models based on proposed cometary compositions. Rock/sulfide, carbonaceous matter, and ice are depicted as if separated under the influence of gravity (differentiated). **(a)** "Composition A" (*Davidsson et al.*, 2016) proposes a 16.5 wt% insoluble organic matter (IOM)-like contribution modeled as graphite/amorphous carbon, equivalent to a deep 130-km-thick layer. **(b)** The model based on *Fulle et al.* (2017) is more radical, more than half high-H/C macromelecular organics (1200 kg m^{-3}) by volume. Thermally, the model in **(a)** could permit a water ocean at depth, but the total thickness of the ice shell in the model shown in **(b)**, ~85 km, is too thin for the temperature profile at depth to reach ice melting for likely present-day heat flows.

N_2 atmospheres with secondary CH_4, which suggests some commonality in the origin and/or evolution of their volatiles. Yet the different contexts of each world (Titan, a major giant planet satellite; Triton, a captured and tidally heated former KBO; and Pluto, a binary KBO dwarf planet) urges caution. It is possible that the dominant role for N_2 in all these atmospheres stems more from the great volatility of nitrogen and its near-noble-gas lack of chemical reactivity, and less from any cosmogonic commonality.

3.4.1. Titan. For Titan the results of the Huygens probe are telling. Both the low $^{14}N/^{15}N$ value for Titan's atmospheric nitrogen (≈168) and the very low ratio of primordial ^{36}Ar to N_2 (*Niemann et al.*, 2010) preclude an origin for Titan's nitrogen atmosphere from primordial molecular nitrogen trapped in the icy "satellitesimals" from which Titan formed. Otherwise, referring to the latter ratio, ^{36}Ar would have been trapped as well, according to the classic test proposed by *Owen* (1982). We note that Rosetta determined a coma production rate ratio for $^{36}Ar/N_2 \sim 5 \times 10^{-3}$ for Comet 67P (*Rubin et al.*, 2018), which while itself low, is still orders of magnitude higher than that in Titan's atmosphere (~2×10^{-7}). In any event, Titan's atmospheric $^{14}N/^{15}N$ and $^{36}Ar/N_2$ ratios instead point to the original molecular carrier of nitrogen into Titan being some combination of ammonia and organics (*Miller et al.*, 2019). Unfortunately, these types of *in situ*, isotopic data are not yet available for Triton or Pluto, and as discussed above, the colder conditions in the aKB may have been more conducive to the condensation and/or trapping of molecular nitrogen from the nebula there.

3.4.2. Triton. The differences between Triton and Pluto, bodies ostensibly accreted in the same region of the solar system, are also instructive. Triton's surface is less rich in CH_4 ice compared with Pluto but CO_2 ice is distributed over a large areal fraction of Triton's surface (*Merlin et al.*, 2018, and references therein). Efficient hydrothermal conversion of carbon-bearing volatiles to CO_2, followed by degassing to the atmosphere during an early epoch of strong tidal heating was proposed for Triton by *Shock and McKinnon* (1993). Such a scenario is not applicable in full to Pluto, even in the aftermath of the Charon-forming collision. It is nonetheless puzzling why no CO_2 ice at all has been discovered on either Pluto or Charon, given that CO_2 is an abundant cometary volatile (*Mumma and Charnley*, 2011) and Charon's water-ice surface in particular is not covered with obscuring volatile ices or tholin-like organics. The CO_2-ice overtone lines near 2.0 µm are quite narrow, however, which may have precluded their detection in New Horizons LEISA spectra (*Grundy et al.*, 2016).

LEISA's modest spectral resolution reduces its sensitivity to CO_2 ice, but Triton-like abundances would have been readily detected (W. M. Grundy, personal communication, 2020). Higher-spectral-resolution groundbased spectra of Pluto have also not seen it. Although higher spectral resolution is beneficial, groundbased observations have the disadvantage of much lower spatial resolution, but that is at least partly compensated by extremely high signal/noise. The lack of evidence for CO_2 in both LEISA and groundbased spectra is likely an important clue to the divergent evolutions of Triton and the Pluto system. We note that CO_2 ice has not been detected on any of the other Kuiper belt dwarf planets either, discussed next.

3.4.3. Other dwarf planets in the Kuiper belt. Whereas no other KBO is known to have an atmosphere, both volatile and involatile surface ices have been identified on a significant number of them. Water ice and methane ice, with their pronounced spectral absorptions, are the easiest to detect. Notable among these detections are methane ice on Eris and Quaoar, extensive methane ice on Makemake, and water ice on Haumea, Quaoar, Orcus, and Gonggong (*Brown*, 2012; *de Bergh et al.*, 2013; *Holler et al.*, 2017). Nitrogen ice, with its weak, narrow absorption feature at 2.15 μm, cannot be detected given the signal-to-noise of spectra obtained to date, but subtle band shifts in the CH_4-ice features on Eris, Makemake, and Quaoar are consistent with minor CH_4 dissolved in N_2-ice (as on Pluto and Triton), or vice versa (N_2 contamination of CH_4 ice) (e.g., *Tegler et al.*, 2008, 2010; *Lorenzi et al.*, 2015; *Barucci et al.*, 2015). All these ice detections or inferences have most commonly been interpreted in terms of long-term stability, with distance from the Sun (surface temperature and UV flux) and body size (surface gravity) having competing effects on loss or retention (*Schaller and Brown*, 2007; *Johnson et al.*, 2015). That N_2 ice is found or is probable on three of the four largest dwarf planet KBOs is, however, consistent with a common process or origin for surface nitrogen in the Kuiper belt. Some degree of endogenic control, especially for these larger, potentially geologically active KBOs, does not seem implausible.

The large dwarf Kuiper belt planet that does *not* display spectral evidence of N_2 ice, or any volatile ice, is Haumea. Its surface is dominated (spectrally) by H_2O-ice, a characteristic it shares with Pluto's large moon, Charon. Both rapidly spinning Haumea and Charon are thought to be products of giant impacts (*Brown et al.*, 2007; *Leinhardt et al.*, 2010), but in Haumea's case it is the primary body. The surfaces of its two known satellites and KBO dynamical family are also water-ice rich. The outcomes of giant impacts depend on many variables (section 2.3), but Haumea is apparently large enough [mean diameter ~1600 km (*Ortiz et al.*, 2017)] to have differentiated and driven its volatile ices to the surface early in its history. It would appear that the giant impact that gave the Haumea system its unique character ejected to heliocentric orbit any volatile-ice-rich "crust" and atmosphere, while exposing its deeper water-ice mantle (hence the inference for differentiation) (see also *Carter et al.*, 2018). Perhaps such escaped fragments, or similar, could explain the CO- and N_2-rich Comet R2 discussed in section 3.1.2. But this then raises further questions. Did Pluto lose at least some volatile ices in the Charon-forming impact? Or is it a matter of timing, in that Charon formed early enough that Pluto could replenish or replace its surface volatile inventory, whereas Haumea could not because its giant impact came later (*Levison et al.*, 2008b; *Volk and Malhotra*, 2012)?

3.5. Synthesis and Unresolved Issues

We can envision the following for Pluto's N_2 and CO budgets: (1) Pluto started with cometary inventories of N_2 and CO; (2) subsurface aqueous chemistry led to the destruction of CO; (3) N_2 was outgassed or otherwise transported to the surface efficiently (e.g., cryovolcanically); (4) no substantial loss of N_2 (due to escape) has occurred at the surface, and it accumulated to form Sputnik Planitia; and (5) comets have delivered a small resupply of CO that mixes with surface N_2. This scenario is by no means the only possibility, but it is consistent with the evidence to date.

Future work should focus on quantitative evaluation of the thermochemical pathways available to Pluto's core, under a variety of compositional assumptions. These in turn would feed forward to models of CO, N_2, CH_4, and CO_2 (and multi-guest) clathrate formation, as Pluto's ocean cooled and its ice shell thickened. All of this would inform our cosmochemical and geochemical understanding of Pluto, as well as better constrain Pluto's likely thermal evolution and the characteristics of its subsurface ocean. It would also be valuable to run long-term climate models with not just N_2 and CH_4 ices, but with CO ice as well (see section 4.2), to determine the degree to which the optical surface of Sputnik Planitia might mask a deeper, more CO-rich ice sheet.

In terms of better observational constraints, the coming era of "big glass" (telescopes) promises more definitive spectral detections of the icy constituents on large KBOs. The launch of the James Webb Space Telescope (JWST) should extend high signal-to-noise observations farther into the infrared, covering the stronger, fundamental vibration bands of N_2, CO, and CO_2. Detection of the water ice O-D stretch near 4.13 μm (*Clark et al.*, 2019) may allow determination of the D/H ratio (in water ice) for Charon and other bodies, with potentially profound implications for the provenance of the Pluto system with respect to other KBOs and comets (see, e.g., *Cleeves et al.*, 2014; *McKinnon et al.*, 2018a). Related would be determination of the CH_3D/CH_4 ratio in the atmosphere or on the surface of Pluto, although proper interpretation would require understanding its potential fractionation with respect to D/H in water ice, the main, primordial hydrogen reservoir on Pluto-Charon.

Finally, the determination of Pluto's atmospheric $^{14}N/^{15}N$ and $^{36}Ar/N_2$ ratios would be extremely constraining for the provenance of Pluto's nitrogen. As *Glein and Waite* (2018) note, however, because of diffusive separation in Pluto's atmosphere it will be very difficult to make a definitive measurement of these ratios even with a future close flyby or orbiter mission (this caveat applies to measuring atmospheric deuterated methane as well). Perhaps the closest near-term measurement that might be made would be $HC^{14}N/HC^{15}N$ from the Atacama Large Millimeter Array (ALMA) (*Lellouch et al.*, 2017). In this case, however, proper interpretation will require a good understanding of the formation and isotopic fractionation of HCN in Pluto's atmosphere.

4. ACTIVE WORLDS: PLUTO AND CHARON THROUGH TIME

In this section we examine the behavior of the principal components of the Pluto system through time. Space constraints necessitate selectivity in our approach. We focus on Pluto, and specifically, notable aspects of its geological history in section 4.1, followed by a concise summary of its atmospheric and climactic evolution in section 4.2, as the latter is clearly of great importance for understanding Pluto's geology. Attention then turns to Charon in section 4.3, with a focus on its orbital and tectonic history and how this history fits in with the overall formation and evolution of the Pluto system. Finally, in section 4.4 we offer a précis of the persistent enigma of the small satellites, followed in section 4.5 by an overview of the system and future work needed to advance understanding.

4.1. Geological History of Pluto

The geological terrains and history of Pluto are well covered by the chapters by White et al. and Moore and Howard in this volume and the references therein. In this section we tour through Pluto's geological history at a high level, starting in deep time and moving forward to present activity. We focus on impacts and internally driven processes, and highlight both reasonably substantiated conclusions and interpretations as well as less well-supported inferences and outright puzzles that may hopefully spur further work.

4.1.1. The giant impact epoch. Pluto's geological history begins of course with the giant impact that formed Charon. Charon's post-formation orbit would have been closer in and likely eccentric, at least initially (see the chapter by Canup et al. in this volume and section 4.3). Correspondingly, Pluto would have been spinning much more rapidly and would have a grossly distorted oblate figure (see, e.g., figures in *Sekine et al.*, 2017). As tides raised on Pluto drove Charon's orbit to expand, Pluto would have despun and its global figure would have relaxed. But as the chapter by Nimmo and McKinnon in this volume discusses, no presently measurable geological evidence survives this despinning epoch. That is, New Horizons did not detect a fossil bulge and the potentially colossal tectonic signatures of such a shape change (*Barr and Collins*, 2015) are not writ into Pluto's surface. This suggests that Pluto's icy lithosphere was then too thin or weak to withstand the shape change (in contrast to, e.g., that of Saturn's satellite Iapetus), which in turn suggests that Pluto's heat flow in this epoch was substantially higher than present-day estimates.

Models of Pluto's heat flow due to long-term radiogenic heat release (U, Th, ^{40}K) generally depict a modest rise and fall over geologic time because the heat flow is moderated by the thermal buffering of the core: Early heating is absorbed by raising core temperatures whereas later heat flow is maintained by the conductive release of this stored heat from the core (e.g., Fig. 6). Pluto, however, had access to other heat sources post-Charon formation. Noted in section 2.5, these include the heat released by the completion of differentiation, hydration of anhydrous core minerals, tidal dissipation, and the giant impact itself. Combined these may not only have sustained an early ocean but could have provided sufficient early heat that allowed Pluto's global figure to relax. Pluto's heat flow through time may actually have been a bit of a roller coaster, rising to an early peak then falling, only to rise again slowly due to long-term radiogenic heating over 1–2 G.y. before entering a long decline to today (cf. *Bierson et al.*, 2020). The full details of such an evolution remain to be quantitatively assessed, however.

4.1.2. Formation of Sputnik. Near the base of Pluto's *known* geological record (since the non-encounter "far side" and polar south are so little known) lies the formation of the Sputnik basin. Formed by the impact of substantial KBO [some ~200 km or more across (*Johnson et al.*, 2016; *McKinnon et al.*, 2016b)], such an impact is exceedingly unlikely in today's Kuiper belt environment (*Greenstreet et al.*, 2015). From Fig. 5 in that work, the chance of a single 1000-km-scale basin forming on Pluto over the past 4–4.5 G.y. is only ~1% (see also the chapter by Singer et al. in this volume). Because impacts are stochastic, this does not mean that Sputnik could not have formed more recently, only that it is much more likely to have formed very early, during the dynamically violent, planetary instability epoch or during Neptune's subsequent migration through the aKB (section 2.2, and see Fig. 3), when impact rates were substantially higher. The work of *Greenstreet et al.* (2015) incorporates the time evolution (decline) of various Kuiper belt subpopulations through time, but did not attempt to assess the bombardment during the instability era owing to the uncertainty as to the details and timing of what actually occurred. This question could probably be fruitfully revisited.

What can be said is that the formation of the Sputnik basin probably occurred early in Pluto's history, which according to the most recent work on the giant planet instability, could have taken place within some 100 m.y. of the formation of the solar system (section 2.2). At that time, the Sun would have been only about 70% as luminous as today (*Gough*, 1981; *Siess et al.*, 2000; *Bahcall et al.*, 2001), but Pluto itself would have been much closer to the Sun, somewhere in the 20–30-AU range perhaps, at least initially. During such time the insolation Pluto (and Charon) received could have been on the order of twice today's values, and this high insolation could have been enjoyed for up to ~100 m.y. Once emplaced into the 3:2 resonance and once Neptune's orbital migration slowed to a crawl (e.g., Fig. 3), i.e., once Pluto achieved its present orbit, insolation would have been only 70% of today's value due to the faint young Sun. All other things being equal, Pluto's atmosphere then would have been more tenuous and its surface-atmosphere volatile exchange hindered. The effects of such a warm epoch followed by a marked cooldown for the distribution of Pluto's volatile ice reservoirs, and the filling of Sputnik Planitia in particular, also remain to be assessed.

4.1.3. Ancient terrains and tectonics. Pluto's surface is marked by a wide range of impact crater densities, from the heavily cratered (if not saturated) highland of eastern Cthulhu to the uncratered ices (at New Horizons resolutions) of Sputnik Planitia itself (*Moore et al.*, 2016; see also the chapter by Singer et al. in this volume). Broad regions contain sufficient numbers of large craters that they probably date back to 4 Ga or more even though they are not saturated with craters in the manner of the lunar highlands. These include western Cthulhu and the region to the northwest of Sputnik (Vega Terra and the bright plains to its north). These regions are notable because they are relatively flat, in contrast with much of the rest of Pluto. Topographic variance over this western expanse of the New Horizons encounter hemisphere is about ±1 km (*Schenk et al.*, 2018) (Fig. 11). While many geologic processes can create level terrain (e.g., burial by sediments), the regional extent of this more-or-less even landscape is reminiscent of the subdued topography of the tidally heated icy satellites Europa and Triton. The subdued topography and high crater density of western Cthulhu and Vega Terra may be related to the inferred early warm epoch for Pluto discussed above in section 4.1.1.

Foremost among the rugged ancient tectonic structures on Pluto is the great north-south ridge-and-trough system (NSRTS) that stretches from the northern polar region of Lowell Regio to the western border of Sputnik Planitia and then further south as far as New Horizons imaging allows (*Schenk et al.*, 2018) (Fig. 12). Its full extent is unknown, but could extend well into Pluto's farside (sub-Charon) hemisphere (*Stern et al.*, 2020). What is known is that the NSRTS appears highly degraded and is crosscut by the western rim of the Sputnik basin. The orientations ("strikes") of its ridges, troughs, elevated plateaus, and elongate depressions also do not appear to be influenced by proximity to the SP basin. These observations together suggest that the SP impact occurred later in geologic time than the NSRTS. However, although the NSRTS appears older than Sputnik, it is conceivably stratigraphically younger. Its tectonic elements continue right up to the SP ice sheet, and it is not obvious how an ancient near-rim tectonic structure (even a deep-seated one) could have survived such a colossal impact. But nature can be surprising. See *Schenk et al.* (2018) for a full discussion.

The tectonic driver for the NSRTS is unknown, but it resembles, at least in its northern segment, major rift systems on certain icy satellites (e.g., Ithaca Chasma on Saturn's Tethys). Early or first freezing of Pluto's ocean (not necessarily complete) could have been responsible. Equatorial crustal thickening is cited as a possible causative mechanism in the chapter by White et al. in this volume, but how this would simultaneously account for both the depressed and elevated sections of the NSRTS is unclear. Regardless, the implication

Fig. 11. The view across Vega Terra to the southwest. Several prominent ~20-km-diameter craters occupy the middle distance, and the younger Djanggawul Fossae crosscuts the foreground. Vega Terra and the horizon beyond are unusually flat for Pluto (*Schenk et el.*, 2018).

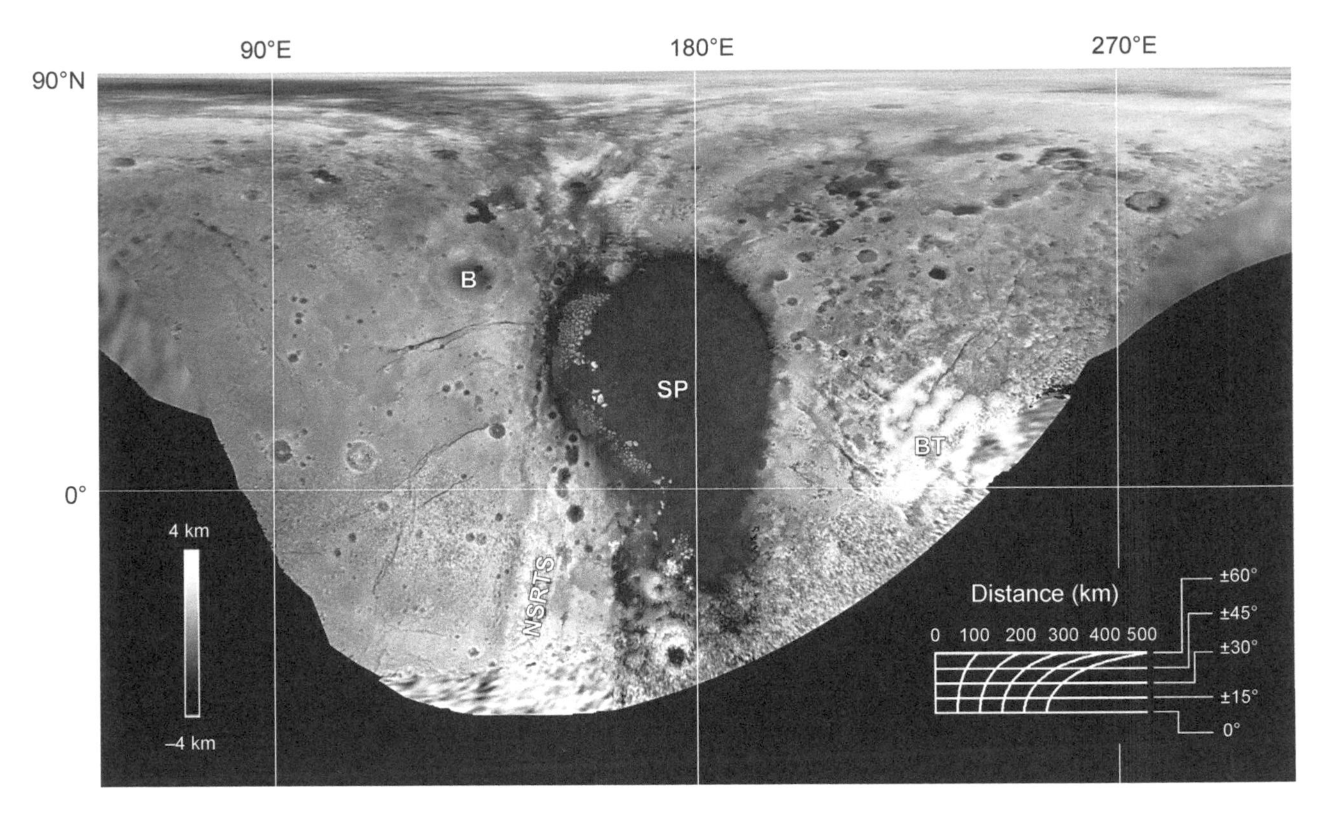

Fig. 12. Best hemispheric topography of Pluto. Cylindrical map projection centered on 180° longitude. Dark areas were unilluminated or do not have resolvable stereogrammetric data from the 2015 encounter. Notable features are Sputnik Planitia (SP) at center, the high-standing bladed terrain (BT) at right, the great north-south ridge-and-trough system (NSRTS) left of center, the Burney multiring basin (B), and the fairly flat plains to the west. From combined LORRI-LORRI, MVIC-MVIC, and LORRI-MVIC stereo, with a vertical resolution of 100–400 m (modified from *Schenk et al.*, 2018).

of this suggestion, if true, is that the NSRTS would have been aligned along a "paleo-equator" prior to a reorientation (true polar wander) of Pluto's ice shell.

Such a reorientation, *but due to Sputnik*, almost certainly affected Pluto's tectonics subsequent to the formation of the Sputnik basin (e.g., *Keane et al.*, 2016). The impact basin, unfilled post-formation, was likely close to isostatic, as the weakened, fractured, and thinned ice shell adjusted to mechanical equilibrium with the underlying ocean (*Johnson et al.*, 2016). Because the compensating, uplifted ocean is deeper than the basin, the basin at first would have represented an overall negative gravity anomaly at spherical harmonic degree-2. Only later, when the basin filled with dense nitrogen-rich ices, did it evolve to become the (inferred) positive gravity anomaly that drove the position of Sputnik to more closely align with the Pluto-Charon tidal axis (*Nimmo et al.*, 2016; see also the chapter by Nimmo and McKinnon in this volume). From the perspective of Pluto today, the filling of the basin is inevitable, as (1) N_2 condensation is enhanced in the equatorial regions (which receive less insolation on average due to Pluto's high obliquity) and at low elevations because of higher atmospheric surface pressures there [see details on this atmospheric-topographic process in *Bertrand and Forget* (2016), *Bertrand et al.* (2018), and the chapter by Young et al. in this volume], and (2) because N_2 ice is glacially mobile at Pluto surface temperatures; if initially widely distributed it will, quite simply, flow downhill (*Umurhan et al.*, 2017; *Bertrand et al.*, 2018). From this perspective, ices could have filled the basin over some tens of millions of years (*Bertrand et al.*, 2018), but it should be borne in mind that the immediate post-Sputnik insolation and atmosphere may have been very different than today (as discussed above). Once emplaced in today's 3:2 resonant orbit early Pluto could, depending on ice albedo, have been colder on average and any surface-atmosphere volatile transport timescales considerably lengthened.

4.1.4. Glaciated and mantled terrains. Discussed in detail in the chapter by Moore and Howard in this volume and in section 4.2.2 below, many of Pluto's surfaces display an array of erosional and in a few cases depositional signatures indicative of glacial action. Because glacial flow is observed today (from the eastern Tombaugh Regio uplands into the Sputnik basin), it is almost certain that characteristic topographic modifications seen elsewhere by New Horizons — incisions, channeling, and fluting — were due to past glacial action. What is not known is whether this is a recurring phenomenon driven by orbital oscillations (akin to terrestrial Milankovich cycles), perhaps with high-pressure atmospheric excursions (e.g., *Stern et al.*, 2017), or whether it represents secular evolution of Pluto's volatile ice reservoir. For example, prior to the Sputnik impact, N_2 and CH_4 ices may have been more equitably distributed across Pluto's surface, and only as the Sun began to warm (over billions of years) were they sufficiently mobilized to begin to cycle and flow as they do today. What is clear is that glacial N_2 must have covered much of Pluto's encounter hemisphere in the past, but is not there now. Remnant N_2 deposits are seen on certain crater floors and other low-lying areas. Alcyonia Lacus in particular records the loss of N_2 over time (see Fig. 5 in the chapter by Moore and Howard in this volume). So where is this nitrogen now? Is it underground as a liquid, in cracks and pores? Or did it escape to space, even though New Horizons determined that the nitrogen escape rate to space today is very limited (*Gladstone et al.*, 2016; *Young et al.*, 2018)? If present conditions are naively extrapolated in time, atmospheric escape is unlikely to have affected Pluto's nitrogen mass balance (Table 1). For Pluto's atmosphere, however, the present is not (quite) the key to the past, as the ultraviolet flux from the young Sun should have been far larger than today (*Zahnle and Walker*, 1982; *Claire et al.*, 2012), and moreover, for some of that early solar epoch Pluto orbited much closer to the Sun than today (section 2).

Even more enigmatic are the mantled terrains to the northeast of Sputnik, e.g., Hayabusa Terra. There a smooth, CH_4-ice mantled landscape alternates with impact craters and large, flat-floored depressions, the latter tens of kilometers across and up to 3 km deep. *Howard et al.* (2017b) suggest that the materials surrounding the depressions may be CH_4-ice rich, but note that such heights are not obviously consistent with the bearing strength of methane ice. Their preferred (if tentative) formation model invokes subsurface cryovolcanism or heat driving volatilization and explosive eruptions of the overlying volatile ice mantle, perhaps akin to maar formation on Earth (where silicate lavas react with groundwater or permafrost). A survey of craters superimposed on the washboard texture that embosses uplands to the northwest of Sputnik Planitia, which is thought to have formed as a consequence of nitrogen ice glaciation, suggests that such glaciation ended ~4 Ga (*White et al.*, 2019). But for both the mantled and glaciated terrains, the generally lower abundance of well-preserved craters implies that these terrains are not among the most ancient on Pluto's nearside (i.e., they are <4 Ga), although likely still old (see the chapter by Singer et al. in this volume).

Potentially related to the glaciated and mantled terrains is the equatorial bladed terrain of Tartarus Dorsa. The bladed terrain comprises a giant deposit of methane ice that is virtually crater free and likely represents an enormous depositional episode of methane ice in Pluto's early- to mid-history (*Moore et al.*, 2018; see the chapter by White et al. in this volume). But as evidenced by their bladed decrescence (ablation) texture and the observation that underlying terrain appears to have been exhumed at their margins, the deposits now seem to be receding as a consequence of excursions in Pluto's climate causing sublimation to be favored over deposition (see the chapter by Moore and Howard in this volume). The lack of craters may be due to a combination of original deposition of the methane ice after the era of heaviest bombardment, and elimination of any craters that did form by the subsequent sublimation erosion.

4.1.5. Middle-age and younger tectonics. As was recognized in the initial post-New-Horizons geological assessments of Pluto (*Moore et al.*, 2016), Pluto's visible tectonics are overwhelmingly extensional. This led to a natural inference that ocean freezing and the resulting uplift and extension of the surface may be an important general driver (*Nimmo*, 2004). On its own, however, such extension might be expected to create a randomly oriented array of normal faults and graben, but as section 3 in the chapter by White et al. in this volume makes clear, the orientations of Pluto's normal faults and grabens are anything but random. We have already noted the possibility of true polar wander stresses initiated by the formation of Sputnik basin and its possible subsequent evolution into a degree-2 gravity high. As a mass concentration (or mascon), Sputnik Planitia (SP) can drive its own regional tectonism as well. If its excess mass were sufficiently concentrated, it would raise a flexural arch around the basin and circumferential graben might have formed along the crest of the arch in a manner similar to that seen around lunar and martian impact basins (e.g., *Solomon and Head*, 1980). But Sputnik is a geographically large structure compared with Pluto's radius. It is so wide in angular extent (in terms of degrees of arc on the surface) that in-plane membrane stresses are more important than flexural, bending stresses, which leads instead to *circumferential extension* as the lithosphere surrounding SP is "pushed" longitudinally outward (*Janes and Melosh*, 1990). In this way, the prominent subparallel graben and troughs of the Inanna, Dumuzi, and Virgil Fossae, which strike quasi-radially away from SP (*Keane et al.*, 2016), are more akin to the Valles Marineris canyon system on Mars (which strikes radially away from the center of Tharsis).

These prominent tectonic features are all the more remarkable because they cross-cut major craters and are youthful in appearance, with sharp, undegraded scarp crests (*Moore et al.*, 2016; see the chapter by White et al. in this volume). Yet the ostensible tectonic driver, SP, is ancient. This may imply reactivation of an older fracture pattern, or a more complex history and interplay of tectonic forces. For example, long-term buildup of background extensional stresses due to ocean freezing (e.g., *Hammond et al.*, 2016) could add to preexisting loading stresses due to SP, and reach (or only reach) the extensional failure limit in Pluto's icy lithosphere later in geologic history.

The Mwindo Fossae extensional tectonic system in the far east of Hyabusa Terra is unusual in that it is isolated and converges to a nexus (*Moore et al.*, 2016; *McGovern et al.*, 2019). The fracture pattern is consistent with negative loading or tectonic uplift at the nexus, but there is nothing otherwise geologically unusual about the nexus region compared with the surrounding terrain, other than possibly being elevated by 0.5–1 km (*Schenk et al.*, 2018). As a fairly crisp-looking and undegraded set of fault scarps, the Mwindo Fossae are a further reminder that Pluto has been tectonically active late in its history, and that local thermal and dynamic perturbations to the ice shell remain possible.

Tartarus Dorsa (the bladed terrain region) actually consists of broad, elongate swells that are typically ~400 km long and ~100 km wide, and that tend to be separated from one another by troughs that include Sleipnir Fossa and others that form the southern components of Mwindo Fossae (see Fig. 9 in the chapter by White et al. in this volume). These swells form some of the highest-standing terrain in the New Horizons close approach hemisphere, reaching 4.5 km above mean radius (*Moore et al.*, 2018; *Schenk et al.*, 2018). Despite some connection to the younger tectonic elements of the Mwindo Fossae to the north, the swells themselves appear to be much older basement upon which the methane-rich blades were emplaced (and now are in retreat). The origin of this striking basement topography is enigmatic, but the scale, amplitude, and elevation suggest the possibility of ancient compressive tectonics (*McGovern et al.*, 2019).

4.1.6. Middle-age and younger cryovolcanism. Wright Mons and its less-well-imaged partner to the south, Piccard Mons, are two large, quasi-annular massifs. Rising 4–5 km from their broad bases (~150 and ~250 km for Wright Mons and Piccard Mons, respectively), both possess deep central depressions, thus superficially resembling terrestrial volcanos (*Moore et al.*, 2016). Nothing about Wright Mons (which was well imaged) is consistent with impact or erosional or mass-wasting processes. A predominantly tectonic origin is conceivable, that is, uplift or inflation of the subsurface, followed by collapse or deflation of the central region. But such an origin does not easily explain the characteristic hills or hummocks on the flanks of the edifice or in the surrounding terrain. One is left then, by elimination, with the suspicion that both edifices are actually constructional and thus true cryovolcanos. If so they would be the first such clearly identified features found on an icy solar system body [cf. *Moore and Pappalardo* (2011) for a discussion of the checkered history of such identifications].

Eruption of cryolavas from Pluto's interior can be assisted, at least in principle, by overpressure generated during ongoing freezing of a mixed ammonia-water ocean beneath an elastic ice lithosphere (e.g., *Manga and Wang*, 2007), although such freezing would need to be reconciled with any thermal stabilization due to clathrate formation at the base of the ice shell (section 3.2) (see the chapter by Nimmo and McKinnon in this volume). The shallow flanks of Wright Mons consist of closely packed, semi-regular hills ~8–10 km in diameter. The hills themselves are rubbly textured, reminiscent of the funiscular terrain between Enceladus' south polar sulci ("tiger stripes"), and could represent the individual constructional elements of the edifice (in contrast to the subaerial lava flows of basaltic shield volcanos on the terrestrial planets). Few if any impact craters superpose Wright Mons, implying an upper limit on its age [$\lesssim$1 Ga (chapter by Singer et al. in this volume)], but there is no evidence that either Wright or Piccard Mons are active today. While there is much we do not understand about the formation of these spectacular features, if we can claim any understanding at all, their mere existence attests to the reality of late endogenic activity on Pluto.

Evidence for even more recent cryovolcanic eruptions or effusions have been advanced by *Dalle Ore et al.* (2019) and *Cruikshank et al.* (2019, 2021). Fresh-appearing reddish surface materials and coatings are seen at or near the Virgil Fossae and Uncama Fossa, both to the west of SP and among Pluto's youngest extensional fracture systems. The reddish materials display the spectral signature of water ice along with an ammoniated compound, and it is hypothesized that the red chromophore (coloring agent) is organic and is a product of Pluto's internal chemical evolution, given the planet's likely incorporation of copious primordial organic matter (section 3.3). The association of aqueous (i.e., low-viscosity) fluid or even vapor-driven, "cryoclastic" eruptions with recent extensional faulting is self-consistent, as the latter promotes the former. How these fault systems tapped into a subsurface aqueous reservoir and whether that reservoir itself was, or was connected to, Pluto's putative ocean remain to be determined, as do the reasons that the (apparent) cryovolcanic expressions at Virgil Fossae and Uncama Fossa and those at Wright Mons are so different.

4.1.7. The red layer. Notable on the inner rimwalls of a number of the larger craters to the northwest of SP is a singular, stratigraphically exposed dark, red band or layer (*Moore et al.*, 2016; see the chapter by Singer et al. in this volume). This layer is apparently also exposed in the faces of tilted mountain blocks within the Al-Idrisi Montes (*Moore et al.*, 2016; *White et al.*, 2017). Lying about 1 km below Pluto's surface, this red layer must represent an important event in Pluto's geological history. It may represent the accumulation of dark, reddish, tholin-like haze particles (*Grundy et al.*, 2018), followed by a depositional hiatus, or the accumulation may have been interrupted by another event, such as ejecta deposition from the Sputnik impact. If the latter is the case then it is an ancient feature of Pluto's crust. The red layer may represent a geologically more recent event, such as the regional eruption of reddish cryovolcanic material as described in section 4.1.6 above. Whatever its origin, it deserves greater attention.

4.1.8. Sputnik Planitia convection. The spectacular cellular plains of Sputnik Planitia display the geometric organization and topographic signature of solid state convection in a kilometers deep N_2-ice sheet (*Moore et al.*, 2016; *McKinnon et al.*, 2016a). The van der Waals bonded molecular ices N_2, CO, and CH_4 are weak enough at Pluto temperatures that viscous flow is able to transport Pluto's nominal radiogenic heat flow (approximately a few milliwatts per square meters), provided the ice sheet exceeds a critical thickness, about 1–2 km for an ice sheet dominated by N_2 or CH_4 ice, respectively (see Fig. 3 in *McKinnon et al.*, 2016a; see also the chapter by Umurhan et al. in this volume). Assuming a dominantly nitrogen ice sheet, *McKinnon et al.* (2016a) derived from numerical models a timescale for the renewal/replacement of the surface of a typical SP convection cell of ~500,000 yr (plus or minus a factor of 2). A comparable timescale was derived by

Buhler and Ingersoll (2018) in their study of sublimation pit formation on the surface of SP.

Vilella and Deschamps (2017) subsequently inferred, also based on numerical models, that SP's three-dimensional convection pattern would be more naturally explained if the convecting layer is heated from within or is cooled from above, rather than being heated from below. They derived an upper limit on the heat flow from the interior of Pluto of <1 mW m^{-2}, which is rather low for a body of Pluto's size and inferred rock fraction (*McKinnon et al.*, 2017). Nor is there any obvious internal heat source for the ice sheet itself. Surface temperature fluctuations on an appropriate timescale are conceivable. The thermal skin depth appropriate to Pluto's 2.8-m.y. orbital element variation, which drives its "Milankovich" cycles (see section 4.2), is $\sim(\kappa\tau/\pi)^{1/2} = 2$ km [for the thermal diffusion coefficient $\kappa = 1.33 \times 10^{-7}$ m^2 s^{-1} of N_2 ice (*Scott*, 1976) and period $\tau = 2.8$ m.y.]. However, the annual mean surface temperature of the ice in SP (which is what counts, not the annual variation) only varies by ± 0.5 K in the long-term climate model of *Bertrand et al.* (2018), which does not seem sufficient to trigger solid-state convection in the ice sheet. But *Vilella and Deschamps* (2017) are correct in the sense that boundary conditions do count. As an example, if the base of the SP ice sheet is at or near the melting temperature for N_2 ice (63 K), then the three-dimensional geometry of the SP convection pattern can be recovered (*McKinnon et al.*, 2018b). Work on this topic is ongoing (e.g., *Wong et al.*, 2019).

4.2. Atmosphere and Climate Evolution on Pluto

As detailed in the chapters by Forget et al. and Young et al. in this volume, the climate of Pluto is a complex system in which the atmospheric dynamics are coupled with the N_2 cycle (sublimation and condensation processes induce surface pressure variations and control the winds) and the CH_4 and CO cycles (both partly control the radiative properties of the atmosphere, whereas CH_4 drives atmospheric photochemistry and haze formation). These cycles strongly depend on surface ice distribution and temperatures, themselves controlled by insolation changes (Fig. 13).

Pluto's climate is highly variable in time, with a surface pressure varying by a factor of 1000 over a present-day Pluto year, maybe even more in the past, according to different models (*Bertrand and Forget*, 2016; *Johnson et al.*, 2019). However, it is only marginally variable in space. Because Pluto's atmosphere is a very weak emitter in the thermal infrared and efficiently mixes trace gases, there are indeed only minor gradients of atmospheric temperature and composition across the globe, except in the lowest ~5 km near the surface where the air can be 10–20 K warmer over a dark volatile-free surface than over a N_2 ice-covered surface (see section 2.1 in the chapter by Forget et al. in this volume).

There are two reasons why it is quite likely that the atmosphere and climate of Pluto have strongly varied in the past and will strongly vary in the future on timescales of millions of years. First, the climate of Pluto depends on its Milankovitch orbital and rotation parameters, and in particular its obliquity (see the chapter by Young et al. in this volume; *Earle et al.*, 2017), which varies over a range of 23° (i.e., 115.5° ± 12.5°) over a period of ~3 m.y. (*Dobrovolskis et al.*, 1997). Such large variations must have induced considerable climate changes, as is the case for Earth and Mars (e.g., *Imbrie and Imbrie*, 1979; *Laskar et al.*, 2004). Second, Pluto is covered by geological landforms resulting from local accumulation or erosion of meter- to kilometer-thick layers of ice, such as flowing glaciers and ice mantles (*Howard et al.*, 2017a,b; *Moore et al.*, 2018; chapter by Moore and Howard in this volume). Their presence and their characteristics are difficult to reconcile with the present-day climate but are consistent with climate changes over periods of millions of years.

4.2.1. Atmospheric response to the Milankovich cycles. A robust and periodic solution for the Milankovitch

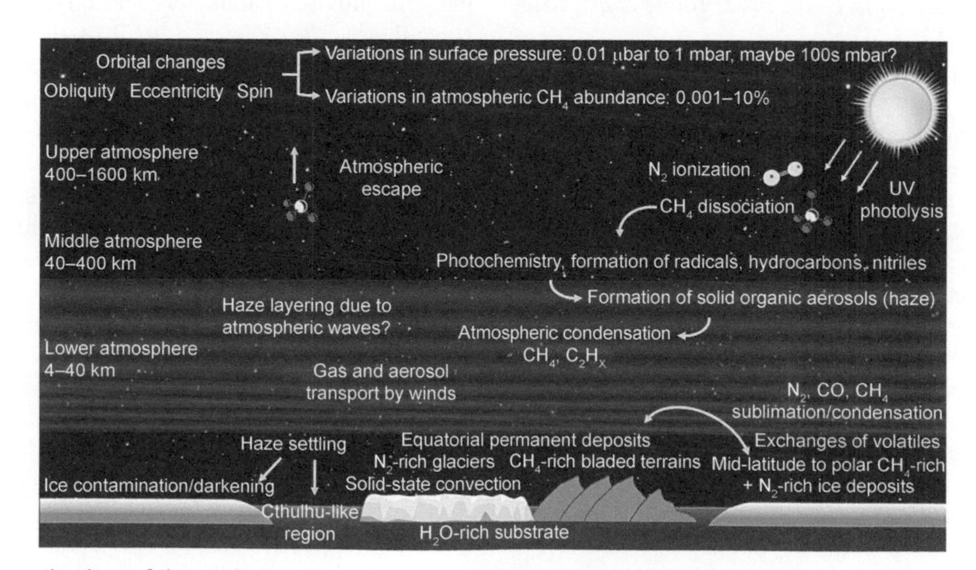

Fig. 13. Schematic view of the main dynamic surface-atmospheric processes on Pluto.

parameters history of Pluto has been developed for the most recent ~100 m.y., but cannot be mapped further back into the past due to the chaotic nature of the solutions (*Sussman and Wisdom*, 1988; *Earle and Binzel*, 2015). Several climate models have integrated these calculations to explore the most "recent" past climates of Pluto (*Earle et al.*, 2017; *Stern et al.*, 2017; *Bertrand et al.*, 2018, 2019). They reproduced and explained the formation of the major permanent volatile deposits in the mid-latitudes and equatorial regions, which receive less insolation and tend to be colder than the poles on average over several million years, due to the relatively high obliquity of Pluto. During very high obliquity tilt periods (~104°), the surface pressure and atmospheric abundances of CH_4 and CO should have been minimal because the low-latitude permanent volatile ice reservoirs received less insolation. During so-called "moderate" obliquity periods (~127°), meaning those with obliquities farthest from 103°, these tendencies should have reversed (*Bertrand et al.*, 2018, 2019). The orbital changes in longitude of perihelion and eccentricity also impact the volatile cycles but the effects are of second and third order respectively.

Over the last 100 m.y., models suggest that the surface pressures remained in the range of 0.01 μbar to 1 mbar (*Bertrand et al.*, 2018, their Fig. 16; *Johnson et al.*, 2019) and the CH_4 atmospheric mixing ratio in the range of 0.001–10%. This is enough for the atmosphere to have remained opaque to Lyman-α radiation and allowed for haze production during most of this time, as suggested as well by the thick layers of dark materials, likely settled haze particles, observed on Pluto's surface (*Grundy et al.*, 2018; *Bertrand et al.*, 2019; *Johnson et al.*, 2019). Higher pressures up to ~100 mbar (close to the triple point pressures allowing liquid N_2 or CH_4 on the surface) require extreme conditions with large and very dark volatile ice deposits covering Pluto's poles as well as low soil thermal inertia (*Stern et al.*, 2017). These changes in surface pressure and trace gas abundances over time are expected to have impacted photochemistry, haze and cloud amounts and composition, and thermal structure (*Gao et al.*, 2017; *Young et al.*, 2018; *Johnson et al.*, 2019).

Pluto's atmospheric circulation has been shown to be very sensitive to Pluto's N_2 ice distribution (*Forget et al.*, 2017). Recent modeling results suggest that the current-day circulation regime is a retrorotation (westward winds at all latitudes), maintained throughout a Pluto year and mostly controlled by cross-hemispheric transport of N_2, in particular within Sputnik Planitia (*Bertrand et al.*, 2020a; chapter by Forget et al. in this volume). We can expect a similar circulation regime in Pluto's past because Sputnik Planitia likely remained the main reservoir of N_2 ice and forced cross-hemispheric transport of N_2, although this remains theoretical.

4.2.2. Geological evidence of past climates. Pluto's surface displays many geological features that reveal or suggest substantial changes in the "recent" past (possibly hundreds of millions of years old, but in many cases much more recent). The N_2-rich Sputnik Planitia ice sheet, and the surrounding terrains, exhibit numerous evidences of climate variation: active glacial flow on the edges of the sheet, icy dunes, possible fluvial features and ponds (which could have been shaped by liquid flows at the base of thick N_2 glaciers, now disappeared), deep sublimation pits, and erosion of water ice mountains (*Howard et al.*, 2017a; *White et al.*, 2017, 2019; *Bertrand et al.*, 2018; *Buhler and Ingersoll*, 2018; *Telfer et al.*, 2018). A variety of dissected terrains outside Sputnik Planitia are also thought to have been carved by ancient glaciers (*Howard et al.*, 2017b). The major CH_4-rich deposits include the massive bladed terrain at the equator and several ice mantles at mid-northern latitudes (*Howard et al.*, 2017a; *Moore et al.*, 2018; chapter by Moore and Howard in this volume). Climate models have been able to relate their latitudinal extension to the Milankovitch parameters history of Pluto, although it remains unclear which reservoir formed first and whether these CH_4 reservoirs also evolved over much longer timescales of several hundreds of millions of years (*Bertrand et al.*, 2019). The ~300-m-tall bladed texture of the equatorial deposits could have formed through condensation-sublimation of CH_4 ice over the last tens of millions of years (*Moores et al.*, 2017; *Moore et al.*, 2018; *Bertrand et al.*, 2020b). Finally, the mid-latitude ice mantles display subsurface layering up to several kilometers thick that could be the signatures of past climate processes (*Stern et al.*, 2017).

Several processes could have disrupted the past climates of Pluto: cryovolcanism (*Moore et al.*, 2016; *Cruikshank et al.*, 2020), tectonic activity (*Howard et al.*, 2017b), volatile escape [very marginal for N_2 ice but several tens of meters of CH_4 ice lost over 4 G.y. (*Gladstone et al.*, 2016; chapter by Strobel in this volume)], darkening and contamination by haze sedimentation and by direct photolysis/radiolysis of the ices, and surface albedo feedbacks (*Earle et al.*, 2017; *Grundy et al.*, 2018; *Bertrand et al.*, 2020a).

Beyond the timescale of ~100 m.y., Pluto's climate is relatively unknown due to the lack of constraints on the Milankovitch parameters and on surface conditions. Nevertheless, *Binzel et al.* (2017b) state that the presence of ancient craters at the equator demonstrates a certain stability of the Milankovitch cycles, which could extend back in time by hundreds of millions of years (otherwise the craters would have been eroded away or completely buried by volatile ice). Early in Pluto history, and despite the lower insolation then, the impact flux may have allowed for warmer surface temperatures than those of today, at least transiently (following *Zahnle et al.*, 2014), which may have led to intervals of a thicker atmosphere if enough N_2 ice was already present and perhaps even liquid N_2 flowing directly on the surface. At some point in Pluto's history, the Sputnik Planitia impact should have rapidly trapped most of the N_2 ice inside the basin, thus limiting the available ice for sublimation and (for present-day N_2-ice albedos) the maximum surface pressures to only a few hundreds of microbars (*Bertrand and Forget*, 2016; *Bertrand et al.*, 2018).

4.3. Orbital and Tectonic Evolution of Charon

There is broad consensus that Charon formed as the result of a collision between Pluto and a similarly sized protoplanet (e.g., *Canup*, 2005, 2011). However, the style of impact, the extent of differentiation of each colliding body, and the evolution and ultimate fate of debris from the collision are all debated (e.g., *Walsh and Levison*, 2015; *Desch and Neveu*, 2017; *Kenyon and Bromley*, 2019c) (see section 2). Post-impact, Charon would have orbited much closer to Pluto than its current semimajor axis (*Canup*, 2005, 2011). Outcomes of collisional models suggest a starting orbital distance for Charon of a few to greater than 10 R_P (Pluto radii) and a substantial orbital eccentricity of 0.1–0.4 (*Cheng et al.*, 2014). Charon's current orbit at ~16 R_P, synchronous rotation, and circular orbit indicate that tides evolved the orbits of both Pluto and Charon.

The tidal evolution of Pluto and Charon depends sensitively on their interior structures, which controls the extent of deformation and dissipation that may occur (*Barr and Collins*, 2015). Measurements from New Horizons indicate that Pluto had (and may still have) an internal ocean (see the chapter by Nimmo and McKinnon in this volume). Charon is also thought to have possessed an ocean, with its tectonic features attributed to extensional stresses generated during ocean freezing (*Moore et al.*, 2016; *Desch and Neveu*, 2017; *Beyer et al.*, 2017; cf. *Malamud et al.*, 2017, for a contrarian view). The presence of oceans can speed up the orbital evolution process and potentially generate large (hundreds of megapascals) stresses within the icy shells of Pluto and Charon (*Barr and Collins*, 2015). In addition, diurnal tidal stresses caused by Charon's eccentric orbit would be greatly enhanced if it possessed an ocean, particularly when Charon orbited closer to Pluto (*Rhoden et al.*, 2015).

Charon's surface displays a variety of tectonic features (*Beyer et al.*, 2017; *Robbins et al.*, 2019). Within the encounter hemisphere the large canyon system dubbed Serenity Chasma dominates the tectonic record. The canyon system trends roughly northeast-southwest. However, no tectonic patterns have yet been identified on Charon (including Serenity Chasma) that record or are consistent with despinning, outward migration, or an epoch of high eccentricity (*Beyer et al.*, 2017; *Rhoden et al.*, 2020). The lack of tidal fractures implies that either tidal stresses were never high enough to produce fractures or that Charon's geologic record was reset after the epoch of tidally driven fracture formation. The most likely potential explanations are either that Charon never had an ocean, so tidal stress magnitudes were negligible and freezing stresses were unavailable, or that Charon's orbit circularized before the ocean froze out. In that case, there may have been little or no residual eccentricity, recession, or despinning stresses to combine with the freezing stresses and generate a distinct pattern.

The rate of change of Charon's eccentricity e_c (technically the eccentricity of the binary) from tidal dissipation in both bodies when in the dual synchronous state is given for small to moderate e_C by (*Dobrovolskis et al.*, 1997)

$$\dot{e}_C \approx -\frac{21}{2}\frac{k_{2C}}{Q_C}\frac{m_P}{m_C}\frac{R_C^5}{a_C^5}ne_C - \frac{21}{2}\frac{k_{2P}}{Q_P}\frac{m_C}{m_P}\frac{R_P^5}{a_C^5}ne_C \quad (6)$$

where a_C and n are Charon's semimajor axis and mean motion, and m, R, k_2, and Q are the mass, radius, second-degree potential Love number, and tidal dissipation factor for each body, with the subscripts P and C referring to Pluto and Charon respectively. k_2 is a measure of the distortion a body may undergo in response to tides, and is smaller for differentiated solid bodies compared with uniform ones, but can be much larger for bodies with internal oceans (the maximum value is 3/2 for a uniform fluid body). Q can be thought of as the effective quality factor of a planet or satellite, analogous to the Q of a simple harmonic oscillator, although many mechanisms can contribute to tidal dissipation in actual planets and satellites (see, e.g., *Sotin et al.*, 2009).

For Pluto-Charon then, the upper limit to the characteristic time for decay of its eccentricity $e_C/\dot{e}_C$ is 5×10^4 Q yr; this upper limit assumes Charon's present-day semimajor axis and that both Pluto and Charon are solid bodies with ice-rock rigidities and equal Qs. If Pluto possessed an ocean early on, e_C decay timescales would have been shorter, and shorter still when Charon was closer to Pluto (all other things being equal, dissipation in Pluto dominates that in Charon in equation (6)). Even for a standard, or benchmark, Q of 100 (*Murray and Dermott*, 1999), Charon's orbit likely circularized within 1 m.y. of the generative giant impact.

Charon's orbital and spin evolution may have been quite complicated, however (*Cheng et al.*, 2014). After the giant impact, Pluto would have been spinning at much faster than the synchronous rate, only slowing as Charon's orbit evolved outward. In this case the coefficient of the second term in equation (6) is +57/8, and tides raised on Pluto would have acted to raise Charon's eccentricity, possibly to the point of orbital instability and escape. Obviously this did not happen, and the presence of Pluto's small satellites imposes an even stricter upper limit on Charon's eccentricity evolution. Most likely, e_C increased until tidal distortion and dissipation within Charon increased sufficiently that $\dot{e}_C \approx 0$. Maintaining a substantial but finite eccentricity as tides raised on Pluto drove Charon's orbit out to dual-synchronous altitude is equivalent to tuning the relative k/Qs of the two bodies, as in *Cheng et al.* (2014). Ultimately, though, Pluto's spin slowed, the effect reversed, and e_C decayed according to equation (6). The total time needed for Charon to evolve outward from an inner, post-giant impact orbit to the semimajor axis where both Pluto and Charon achieve spin-orbit synchronism is generally longer than the circularization timescale above, ~10^5 Q_P yr for realistic k_{2P} values (*Dobrovolskis et al.*, 1997; *Cheng et al.*, 2014). Charon's orbital evolution thus may have been (was likely?) complete before capture of the Pluto-Charon system into the 3:2 MMR by Neptune (Fig. 5 and section 2.6).

Given that the formation of Charon's chasm system has been attributed by most workers to the volume expansion caused by a freezing ocean, we favor the interpretation that Charon's orbit circularized early in its evolution, the lack of tidal heating contributed to ocean freezing, and the freezing ocean generated most of the tectonic features we observe, removing evidence of past tidally driven fracturing. This is especially true if the freezing of the ocean is related to the eruption of cryolavas that formed Vulcan Planitia and that buried much of Charon's preexisting terrain (*Beyer et al.*, 2019). The mechanism by which Charon's observed fractures formed at their particular orientations, however, remains an open question. Further discussion of Charon's geology and geophysics can be found in the chapter by Spencer et al. in this volume.

4.4. Puzzling Satellites

An outstanding problem in our understanding of the Pluto-Charon system is the formation and evolution of the smaller moons Styx, Nix, Kerberos, and Hydra. Several studies have simulated their formation from the debris disk generated in the Charon-forming impact, with some success (*Walsh and Levison*, 2015; *Kenyon and Bromley*, 2019c). However, the large orbital distances of the moons, their survival throughout Charon's outward migration, and how they came to be near orbital resonances with Charon are still challenging to explain (*Stern et al.*, 2018; see also the chapter by Canup et al. in this volume).

The orbital and physical properties of the small satellites are described in the chapter by Porter et al. in this volume. From the point of view of this chapter, the most critical aspect is how these properties inform our understanding of the system as a whole. The orbits of the small satellites are relatively compact (35.9–54.5 R_P), near-circular, coplanar, and aligned with the Pluto-Charon orbital plane (*Weaver et al.*, 2016). These characteristics strongly point to an origin from a dissipative system of orbiting smaller particles, such as would be created in the Charon-forming giant impact (*Canup*, 2011). Both an *in situ* origin from an extended debris disk (*Kenyon and Bromley*, 2014, and subsequent papers) and formation from an impact-generated proximal debris disk followed by resonant tidal evolution driven by Charon (*Stern et al.*, 2006, and subsequent papers) have been proposed. The reader is directed to the chapter by Canup et al. in this volume for detailed discussion (see also *Peale and Canup*, 2015), but it suffices to say that no model satisfactorily explains the origin and dynamical characteristics of the small satellite system. Some models invoke collisional interactions among the satellites or with heliocentric (KBO) impactors (e.g., *Walsh and Levison*, 2015; *Bromley and Kenyon*, 2020). Given the contingent nature of such interactions, a first-principles understanding of the origin and evolution of the small satellites may prove elusive.

The physical properties of the satellites themselves, as well as lack of detected small satellites beyond Hydra's orbit by New Horizons, are valuable clues nonetheless. The high albedos of the satellites ($\gtrsim$0.5) and the clear prominence, in New Horizons near-infrared spectra, of water-ice on Nix, Hydra, and Kerberos, and the detection of the 2.2-μm absorption attributed to an ammonia-bearing species on Nix and Hydra (*Weaver et al.*, 2016; *Grundy et al.*, 2016; *Cook et al.*, 2018), all point to the satellites being predominantly ice. Such a composition is consistent with the giant impact model in which the small satellites form from the debris blasted off from the icy surface layers of one or both of the progenitor bodies (*Canup*, 2011). It is not consistent with any origin where the small satellites are derived (or even partly derived) from mixed rock-ice or primordial or later heliocentric KBO material, which would be rich in dark rocky and carbonaceous materials. Typical mid-sized KBO albedos are closer to 0.1 and reflect the latter compositions (see *Stern et al.*, 2018).

Strictly speaking, the albedos and spectral absorptions above refer to surfaces of the small satellites. Although it is difficult to understand how, for example, these ices could mask dark, compositionally rocky interiors given the erosive cratering environments in which the satellites exist (see the chapter by Canup et al. in this volume), it would nonetheless be more satisfying if the inferred iciness of the small satellites could be confirmed by their densities. Astrometry-based estimates of the satellite masses prior to the New Horizons encounter (*Brozović et al.*, 2015), when combined with volume estimates from New Horizons imagery, are not constraining [the uncertainties are too large (*McKinnon et al.*, 2017)]. Dynamical stability calculations over gigayear timescales by *Kenyon and Bromley* (2019a) imply, however, that the densities of the satellites must be under 2000 kg m^{-3}, and for Nix and Hydra (the largest), likely under 1600 kg m^{-3}. Such densities are consistent with ice, but they do not prove it. Given the likely structural disruptions from the impacts evident on their surfaces (*Weaver et al.*, 2016), we expect satellites of their size, if made of ice, to have densities under 1000 kg m^{-3}, more similar to the densities of the icy, inner satellites of Saturn (*Buratti et al.*, 2019). Further astrometric measurements of the Pluto system combined with numerical integrations should ultimately yield better density constraints.

New Horizons did not detect any new satellites at Pluto, a surprising result given the steady march of Hubble Space Telescope (HST) satellite discoveries in the years leading up to the encounter (*Stern et al.*, 2015; *Weaver et al.*, 2016). *Kenyon and Bromley* (2014) predicted that small satellites up to ~2–6 km across would be found beyond the orbit of outermost Hydra, based on their viscously spreading particle disk model for the origin of the small satellites. The New Horizons lower limit for detection was 1.7 km across for a Nix-like albedo of 0.5, out to an orbital radius of ~80,000 km (67 R_P) from Pluto, with less stringent limits at larger radii (*Weaver et al.*, 2016). While it is unfortunate that New Horizons data could not definitively test their hypothesis, it is commendable that the model of *Kenyon and Bromley* (2014) is testable.

The apparent emptiness of the Pluto system beyond Hydra, whose orbit is less than 1% of Pluto-Charon's Hill radius (referring to the gravitational sphere of influence of the Pluto system with respect to the Sun), is notable. This suggests an alternative explanation for the lack of more distant satellites: tidal stripping during Neptune's "wild days." Equal-mass binaries are relatively uncommon among the dynamically excited (hot) KBO populations compared with the more distant cold classical KBOs (*Noll et al.*, 2020), and one explanation is that the former have been lost to collisions and dynamical effects (e.g., scattering by Neptune), whereas the latter have remained relatively dynamically undisturbed (see *Nesvorný and Vokrouhlický*, 2019). A question of some interest has been whether Pluto's satellite system could have survived the implantation of the system into the 3:2 MMR resonance with Neptune and any subsequent orbital migration (*Pires et al.*, 2015).

Capture into resonance is fundamentally agnostic as to whether a body is single, a binary, or a multiple system, as long as the binary or system is gravitationally bound (*Malhotra and Williams*, 1997). The critical issue is whether impacts, or tides during scattering encounters with Neptune, can cause the binary or system to become unbound or otherwise disturbed, especially if the orbital evolution of the Pluto satellite system was complete prior to the giant planet instability (as argued in section 2.6). *Nesvorný and Vokrouhlický* (2019), as part of their study of binary stability, also examined the stability of the KBO dwarf planet satellites during Neptune's migration and implantation/creation of the Kuiper belt. They found that all of Pluto's satellites are expected to survive during the dynamical implantation of Pluto in the Kuiper belt. They also found that the low orbital eccentricities of Pluto's small moons (<0.01; see the chapter by Porter et al. in this volume) may have been excited during encounters of the Pluto system with Neptune, or by small impacts while the Pluto system was immersed in the massive planetesimal disk of the aKB.

It should also be said, however, that ~40% of the simulations in *Nesvorný and Vokrouhlický* (2019) resulted either in the loss of outermost Hydra or excitation of its eccentricity to >0.1. This suggests the following speculation: The implantation may have been sufficiently destabilizing that all small satellites down to the Hydra's orbital distance escaped while the other small satellites were thrown into substantially perturbed orbits, orbits that led to collisions and reaccumulation into the satellite system we see today. Even if deemed unlikely, it highlights an interesting aspect of Pluto's dynamics in this early solar system epoch. The tidal effects of scattering encounters with Neptune can also perturb the orbital energies (semimajor axes) of the moons, potentially displacing one or more moons from MMRs with Charon (the four small moons are today close to but not in the 3:1, 4:1, 5:1, and 6:1 MMR with Charon) (*Nesvorný and Vokrouhlický*, 2019). Any such perturbations would only serve to complicate understanding of the small satellites' history.

4.5. Synthesis and Unresolved Issues

Prior to the New Horizons encounter, *Moore et al.* (2015) published a detailed look ahead at the geological processes potentially to be revealed on Pluto and on Charon. None were expected to be dull or quotidian. But Pluto and to a certain extent Charon exceeded all expectations by a wide margin. Pluto in fact turned out to be one of the more active solid bodies in the solar system, rivaling Mars, with a wide array of geological, geophysical, atmospheric, and climatic processes on display, including some never before seen or seen as clearly. All the topics discussed in this section are either the subjects of ongoing research, or they need to be!

For example, the thermal and tectonic histories of Pluto and Charon need to be revisited, based on the likely initial states that evolved subsequent to the Charon-forming impact. The bombardment history of Pluto and Charon *prior* to its emplacement in the 3:2 MMR can be modeled, because we now have specific scenarios for the formation of the Kuiper belt that have passed numerous tests. Our ideas about planetesimal and planet formation, including the dynamical instability that populated the modern Kuiper belt, will no doubt continue to evolve, but interim implications for Pluto's composition and evolution can still be usefully drawn. The mere existence of Pluto is a key datum in our search for a better understanding of how the solar system came to be, and better and deeper understanding of the Pluto system (including the small satellites) will provide additional context and clues.

Pluto's earliest post-giant-impact and post-Sputnik-basin-formation evolution deserve greater attention, in order to understand the evolution (if not the creation) of Pluto's volatile ice reservoirs and their effects on the planet's geology and geophysics. We do not yet know if the evidence of substantial volatile transport and glacial (or even fluvial) erosion writ into its surface reflects mainly a truly ancient ($\gtrsim$4 Ga) geological era or whether this activity has continued, perhaps intermittently, into Pluto's middle age or even up to today. Ongoing glacial flow is seen of course, and famously so (*Moore et al.*, 2016). Pluto in this sense is even more active than Mars. But does the evidence cited in this chapter and in the chapter by Moore and Howard elsewhere in this volume for extensive N_2-ice cover in the geologic past imply a secular trend in which N_2 ice simply ended up in the Sputnik basin, or has there been substantial loss to space? The latter was the widely held assumption before the New Horizons encounter. Detailed evaluation of Pluto's likely atmospheric structure and evolution under the "faint young, but extreme ultraviolet (XUV) active Sun" is needed. Transient warmer conditions due to large impacts, as well as potential extrasolar influences on the Pluto system, such as effects of nearby supernovae and passage through dense molecular clouds (see *Stern*, 1986; *Stern and Shull*, 1988), might also be fruitfully considered.

Numerous important geological features and processes on Pluto and Charon remain unexplained. What was the cause of the great north-south rift system on Pluto? What accounts for the deep broad depressions and pits in the methane-ice-rich mantled terrains to the northeast of Sputnik Planitia? How were the putative cryovolcanic edifices Wright and Piccard Mons formed? Is there something unique about their location on Pluto? Why don't we see similar structures on Charon, or on Triton, or on icy satellites in general? What does the dark red layer seen in many crater rimwalls and on the exposed faces of many faulted mountain blocks signify? And why do mountain blocks preferentially congregate at the western margin of SP? A partial answer to the latter at least is discussed in section 2.6.2 of the chapter by White et al. in this volume: specifically, the coincidence of the low-viscosity, dense nitrogen ice (or liquid) intruding into water ice crust that has been weakened and fractured by the NSRTS and other tectonic systems, circumstances that are not replicated on the other sides of Sputnik to anywhere near the same extent.

The evidence that Pluto possesses an ocean is circumstantial, but a self-consistent story based on the position of and the tectonics surrounding SP is reasonably convincing (see the chapter by Nimmo and McKinnon in this volume). Do the sharpness and high stratigraphic position of the most recent extensional faults on Pluto imply active tectonism today? If so, does the evidence for fluid or gas-driven cryovolanism outlined in section 4.1.6 also imply ongoing cryovolcanism today? And how is any of this, or Charon's tectonics and plains cryovolcanism, related to ocean freezing?

Finally, we judge that convective overturn is occurring today in the N_2-ice sheet within SP. Beyond the inferences for Pluto's heat flow, volatile ice rheology, and the maintenance of a vigorous surface-atmosphere volatile cycle, the ability to study solid-state convection in the raw (as it were) is unprecedented. Solid-state convection occurs on Earth (i.e., plate tectonics) and is inferred to occur or have occurred on many other solar system bodies both rocky and icy. But the details are always hidden from view, beneath the lithosphere of a given world. For the SP ice sheet there is no lithosphere, and the physical structure of convective flow is directly exposed. Pluto thus provides a natural laboratory to study one of the most important processes in geophysical fluid dynamics.

Obviously, we would like to learn more about Pluto and Charon, to see their non-encounter, "farside" hemispheres and terrains that that were in polar darkness in 2015 (*Stern et al.*, 2020). High-resolution remote sensing as well as geophysical measurements would be extraordinarily valuable. Such observations could be made by a future mission to Pluto, logically an orbiter, but given the very long lead time for such a mission, research might focus on a deeper understanding of what New Horizons data imply for the Pluto system and planetary formation and evolution in general. Adaptive optics imaging from the coming generation of large, Earth-based telescopes may match or exceed HST in terms of resolution, however. At the very least this should allow monitoring of the evolution of albedo patterns and thus surface-atmosphere interactions on Pluto in the coming decades. There are many years of work ahead.

Turning to the small satellites, given their importance to understanding the origin of the Pluto system, further efforts should be made to constrain their masses and thus densities. Efforts to incorporate additional years of HST astrometric observations as well as New Horizons imaging are underway (*Jacobson et al.*, 2019). Future searches for small transhydran satellites using the next generation of spacebased telescopes would also be valuable as a definitive test of any extended debris disk origin hypothesis (*Kenyon and Bromley*, 2019b; *Bromley and Kenyon,* 2020).

5. SUMMARY

The New Horizons encounter with the Pluto system was no mere box-checking exercise. By flying by the last of the classical planets and the first known Kuiper belt planet, and for the first time exploring *in situ* major bodies in the solar system's "third zone," a paradigm shift was initiated in our understanding of the possibilities for planetary evolution and expression in modest-scale worlds far from their parent stars. The Pluto-Charon system provides a broad variety of constraints on planetary formation, structure, composition, chemistry, and evolution:

- ***Origin.*** The emerging view of planetesimal formation via gravitational instability in the protoplanetary gas-and-particle disk aligns with the requirements imposed by Charon-forming giant impact models. Initial planetesimals (50–100 km scale) form between ~20 and 30 AU. Accretion timing appears consistent with subsequent slow and/or stalled pebble accretion followed by hierarchical coagulation after nebular gas dispersal (~few million years). Partially differentiated proto-Pluto precursors (the probable initial condition) imply slow and/or "pebbly" (impact heat gets radiated) and late (little ^{26}Al) accretion in the transneptunian planetesimal disk. The dynamic environment (number density, velocity dispersion) in the post-gas planetesimal disk is favorable for Charon formation, and the subsequent giant planet instability and Neptune's migration emplaces Pluto-Charon in its present 3:2 MMR with Neptune.
- ***Interior.*** Partially differentiated precursors plus the Charon-forming impact should have driven both Pluto and Charon toward full ice-from-rock differentiation, and concomitantly toward early interior ocean formation. The latter is consistent with the general absence of compressional tectonics on both bodies. While evidence for an ocean on Pluto (and former ocean within Charon) remains circumstantial, evidence continues to accrue from detailed tectonic modeling of the Sputnik basin as a mascon and geologically young eruptions of NH_3-bearing cryofluids and clastics (presumably ultimately sourced from a deep, possibly pressurized ocean) on Pluto. Preservation of Pluto's ocean and maintenance of an uplifted ocean beneath Sputnik (the hypothesized source of the mascon, along with the N_2 ice sheet) would have been strongly aided by clathrate formation within or at the base of Pluto's floating ice shell.

- ***Composition and chemistry.*** Pluto's low surface CO/N_2 ratio has been variously explained by CO burial in the Sputnik Planitia N_2 ice sheet, destruction by aqueous chemistry in the ocean, or preferential sequestration in subsurface clathrates. There is no fundamental problem explaining Pluto's global nitrogen abundance, although the ultimate provenance of this nitrogen remains to be determined. Within Pluto's likely organic-rich, chemically reducing core, thermochemistry favors the production of metastable organics, graphite, CH_4 and N_2 — potentially explaining the major constituents of Pluto's atmosphere and that of similar, sizeable icy worlds.
- ***Tectonics and heat flow.*** Contradictory estimates for Pluto's lithospheric heat flow exist, but the preponderance of evidence is for a low, close to steady state radiogenic value (a few milliwatts per square meter) throughout most of Pluto's history (see the chapter by Nimmo and McKinnon in this volume for details). The significant exception may be the *lack of* evidence for the collapse of Pluto's post-formation rotational bulge, which requires a sufficient combination of higher heat flow and lithosphere weakness. The block tectonics of Charon's Oz Terra bear no clear geometric relation to eccentricity tidal stresses. Strong eccentricity tides are not a given for Charon during its post-formation tidal evolution away from Pluto, however, which puts the onus on ocean freezing and, possibly, tidal bulge collapse to explain the extensive disruption of Charon's ancient crust. A possible explanation is that Charon's tidal evolution was sufficiently rapid, and subsequent geologic activity, including the eruption of ammoniated cryolavas that formed Vulcan Planitia, has obscured most evidence of the tidal evolution epoch.
- ***Atmosphere and climate.*** The variations of orbital and rotation parameters of Pluto over the last millions of years have led to substantial insolation changes, thus triggering volatile transport and extensive resurfacing, including glaciers and ice mantle formation, as well as more than 1000-fold annual variations of surface pressure and CO and CH_4 atmospheric abundances. On Pluto the global nitrogen ice distribution and the induced nitrogen condensation-sublimation flows strongly control the atmospheric circulation. Global circulation models predict a general retrograde atmospheric circulation for current-day Pluto that could have been in place in Pluto's past as well, and could account for many of the geological features and longitudinal asymmetries in ice distribution observed on Pluto.
- ***Sputnik.*** Much of Pluto's geophysical, geological, and atmospheric behavior has been and is controlled or strongly influenced by Sputnik, which raises the question of how other dwarf planets in the Kuiper belt (and Triton) behave in the absence of a (or in the presence of more than one) giant impact basin. Dynamical arguments suggest that there were once ~1000–4000 Pluto-mass bodies in the transneptunian planetesimal disk. Simulations show that the current scattered disk comprises ~0.5–1% of the original planetesimals in the aKB (see *Morbidelli and Nesvorný*, 2020). Thus, up to several dozen Pluto-class dwarf planets may still be out there, in the Kuiper belt's scattered disk and its extended (detached) component.

6. CODA

So, what has New Horizons' exploration of the Pluto system taught us? It has taught us once again that nature's imagination exceeds our own. It has reinforced the emerging paradigm that planetary-level behavior is not the sole province of terrestrial-composition (rock + metal) planets or even relatively large worlds (Mars-scale and beyond). As one moves farther from the Sun, as long as solid bodies can partake compositionally of increasingly geologically mobile and volatile materials (carbonaceous matter, all manner of ices), all the characteristic expressions of internal and insolation-driven geological activity found on the active terrestrial planets (Mars and Earth especially) can reappear in new robes. Some are similar, some are novel; all are fascinating. While the most active icy satellites (Europa, Enceladus) characteristically derive their activity from resonant tidal heating, Pluto is proof that tidal heating is not absolutely necessary, within limits. The differences between Pluto and Charon do illustrate that size matters; however, the fuzzy boundary between worlds that enjoy early activity before sliding into senescence and those that remain active after 4.6 G.y. occurs at a much smaller size scale than previously thought.

Our understanding of the Pluto system, and of the Kuiper belt in which it resides, are set for much further improvement. Sections 2.6, 3.5, and 4.5 in this chapter, along with other chapters in this volume, detail ongoing, critical, important, or hoped-for progress in new Earth-based astronomical observations, continued analyses of New Horizons and other data, experimental measurements of the relevant properties of planetary ices and their geochemical/petrological interactions with rock and carbonaceous matter, and ever-improving numerical simulations of geological, geophysical, geochemical, atmospheric, and dynamical processes. The single most important advance in the decades ahead will come from continued exploration of the Kuiper belt by spacecraft. A return to Pluto, with an orbiter mission, could obviously address the majority of the science questions laid out in this volume, and would be a great leap forward. Equally important would be further reconnaissance of other KBOs, of any size or dynamical class, including Centaurs, which are derived from the Kuiper belt's scattered disk. One only needs to consider the advances in planetary science that resulted from the July 2015 New Horizons encounter with the Pluto system, and equally, those that resulted from the subsequent New Year's 2019 encounter with the small cold classical KBO Arrokoth. Future telescopic surveys should allow the planning of a flyby mission with multiple KBO encounters more or less along the spacecraft's trajectory. New Horizons, like the Pioneers (10 and 11) through the asteroid belt to Jupiter and Saturn before it, was a pathfinder mission, and a highly capable one at that. The in-depth exploration of the Kuiper belt has only begun, and its scientific riches beckon. *Carpe tertium zona!*

Acknowledgments. We thank reviewers F. Nimmo, W. Grundy, A. Morbidelli, D. Nesvorný, and Z. Leinhardt, uber-editor S. A. Stern, and O. White and J. Moore for their comments and suggestions regarding the manuscript. All authors also thank the New Horizons project for supporting this research and its many components over the years. C.R.G. was partly supported by the NASA Astrobiology Institute through its JPL-led team entitled Habitability of Hydrocarbon Worlds: Titan and Beyond. T.B. was supported for this work by an appointment to the National Aeronautics and Space Administration (NASA) Post-doctoral Program at the Ames Research Center administered by Universities Space Research Association (USRA) through a contract with NASA.

REFERENCES

Abod C. P., Simon J. B., Li R., Armitage P. J., Youdin A. N., and Kretke K. A. (2019) The mass and size distribution of planetesimals formed by the streaming instability. II. The effect of the radial gas pressure gradient. *Astrophys. J., 883,* 192, DOI: 10.3847/1538-4357/ab40a3.

Alexander C. M. O'D., Cody G. D., De Gregorio B. T., Nittler L. R., and Stroud R. M. (2017) The nature, origin and modification of insoluble organic matter in chondrites, the major source of Earth's C and N. *Chem. Erde, 77,* 227–256.

Alexander R., Pascucci I., Andrews S., Armitage P., and Cieza L. (2014) The dispersal of protoplanetary disks. In *Protostars and Planets VI* (H. Beuther et al., eds.), pp. 475–496. Univ. of Arizona, Tucson.

Altwegg K., Balsiger H., Bar-Nun A., et al. (2016) Prebiotic chemicals — amino acid and phosphorus — in the coma of Comet 67P/Churyumov-Gerasimenko. *Sci. Adv., 2,* e1600285.

Arakawa S., Hyodo R., and Genda H. (2019) Early formation of moons around large trans-neptunian objects via giant impacts. *Nature Astron., 3,* 802–807, DOI: 10.1038/s41550-019-0797-9.

Atreya S. K., Donahue T. M., Kuhn W. R. (1978) Evolution of a nitrogen atmosphere on Titan. *Science, 201,* 611–613.

Atreya S. K., Lorenz R. D., and Waite J. H. (2009) Volatile origin and cycles: Nitrogen and methane. In *Titan from Cassini-Huygens* (R. H. Brown et al., eds.), pp. 77–99. Springer, New York.

Bagenal F., Horányi M., McComas D. J., et al. (2016) Pluto's interaction with its space environment: Solar wind, energetic particles, and dust. *Science, 351,* aad9045.

Bahcall J. N., Pinsonneault M. H., and Basu S. (2001) Solar models: Current epoch and time dependences, neutrinos, and helioseismological properties. *Astrophys. J., 555,* 990–1012.

Bannister M. T., Gladman B. J., Kavelaars J. J., et al. (2018) OSSOS. VII. 800+ trans-neptunian objects — the complete data release. *Astrophys. J. Suppl., 236,* 18, DOI: 10.3847/1538-4365/aab77a.

Bardyn A., Baklouti D., Cottin H., et al. (2017) Carbon-rich dust in Comet 67P/Churyumov-Gerasimenko measured by COSIMA/Rosetta. *Mon. Not. R. Astron. Soc., 469,* S712–S722, DOI: 10.1093/mnras/stw2640.

Barr A. C. and Collins G. C. (2015) Tectonic activity on Pluto after the Charon-forming impact. *Icarus, 246,* 146–155, DOI: 10.1016/j.icarus.2014.03.042.

Barr A. C. and Schwamb M. E. (2016) Interpreting the densities of the Kuiper belt's dwarf planets. *Mon. Not. R. Astron. Soc., 460,* 1542–1548, DOI: 10.1093/mnras/stw1052.

Barucci M. A., Dalle Ore C. M., Perna D., Cruikshank D. P., Doressoundiram A., and Alvarez-Candal A. (2015) (50000) Quaoar: Surface composition variability. *Astron. Astrophys., 584,* A107, DOI: 10.1051/0004-6361/201526119.

Bertrand T. and Forget F. (2016) Observed glacier and volatile distribution on Pluto from atmosphere-topography processes. *Nature, 540,* 86–89.

Bertrand T., Forget F., Umurhan O. M., et al. (2018) The nitrogen cycles on Pluto over seasonal and astronomical timescales. *Icarus, 309,* 277–296, DOI: 10.1016/j.icarus.2018.03.012.

Bertrand T., Forget F., Umurhan O. M., Moore J. M., Young L. A., Protopapa S., Grundy W. M., Schmitt B., Dhingra R. D., and Binzel R. P. (2019) The CH_4 cycles on Pluto over seasonal and astronomical timescales. *Icarus, 329,* 148–165, DOI: 10.1016/j.icarus.2019.02.007.

Bertrand T., Forget F., White O., and Schmitt B. (2020a) Pluto's beating heart regulates the atmospheric circulation: Results from high-resolution and multi-year numerical climate simulations. *J. Geophys. Res., 125,* e06120, DOI: 10.1029/2019JE006120.

Bertrand T., Forget F., Schmitt B., White O. L., and Grundy W. M. (2020b) Equatorial mountains on Pluto are covered by methane frosts resulting from a unique atmospheric process. *Nature Commun., 11,* 5056.

Beuther H., Klessen R. S., Dullemond C. P., and Henning T. K., eds. (2014) *Protostars and Planets VI.* Univ. of Arizona, Tucson. 914 pp.

Beyer R. A., Nimmo F., McKinnon W. B., et al. (2017) Charon tectonics. *Icarus, 287,* 161–174, DOI: 10.1016/j.icarus.2018.12.032.

Beyer R. A., Spencer J. R., McKinnon W. B., et al. (2019) The nature and origin of Charon's smooth plains. *Icarus, 323,* 16–32, DOI: 10.1016/j.icarus.2016.12.018.

Bieler A., Altwegg K., Balsiger H., et al. (2015) Abundant molecular oxygen in the coma of Comet 67P/Churyumov-Gerasimenko. *Nature, 526,* 678–81.

Bierson C. J. and Nimmo F. (2019) Using the density of Kuiper belt objects to constrain their composition and formation history. *Icarus, 326,* 10–17, DOI: 10.1016/j.icarus.2019.01.017.

Bierson C. J., Nimmo F., and McKinnon W. B. (2018) Implications of the observed Pluto-Charon density contrast. *Icarus, 309,* 207–219.

Bierson C. J., Nimmo F., and Stern S. A. (2020) Evidence for a hot start and early ocean on Pluto. *Nature Geosci., 13,* 468–472, DOI: 10.1038/s41561-020-0595-0.

Binzel R. P., Earle A. M., Buie M. W., Young L. A., Stern S. A., Olkin C. B., Ennico K., Moore J. M., Grundy W., Weaver H. A., Lisse C. M., and Lauer T. R. (2017a) Climate zones on Pluto and Charon. *Icarus, 287,* 30–36, DOI: 10.1016/j.icarus.2016.07.023.

Binzel R. P., Olkin C. B., and Young L. A. (2017b) Editorial. *Icarus, 287,* 1, DOI: 10.1016/j.icarus.2017.02.001.

Bishop J. L., Banin A., Mancinelli R. L., and Klovstad M. R. (2002) Detection of soluble and fixed NH_4^+ in clay minerals by DTA and IR reflectance spectroscopy: A potential tool for planetary surface exploration. *Planet. Space Sci., 50,* 11–19.

Biver N., Bockelée-Morvan D., Paubert G., Moreno R., Crovisier J., Boissier J., Bertrand E., Boussier H., Kugel F., McKay A., Dello Russo N., and DiSanti M. A. (2018) The extraordinary composition of the blue Comet C/2016 R2 (PanSTARRS). *Astron. Astrophys., 619,* A127.

Bouquet A., Mousis O., Glein C. R., Danger G., and Waite J. H. (2019) The role of clathrate formation in Europa's ocean composition. *Astrophys. J., 885,* 14, DOI: 10.3847/1538-4357/ab40b0.

Bromley B. C. and Kenyon S. J. (2020) A Pluto-Charon concerto: An impact on Charon as the origin of the small satellites. *Astron. J., 160,* 85.

Brown M. E. (2012) The compositions of Kuiper belt objects. *Annu. Rev. Earth Planet. Sci., 40,* 467–494.

Brown M. E., Barkume K. M., Ragozzine D., and Shlerchaller E. L. (2007) A collisional family of icy objects in the Kuiper belt. *Nature, 446,* 294–296.

Brozović M., Showalter M., Jacobson R. A., and Buie M. W. (2015) The orbits and masses of satellites of Pluto. *Icarus, 46,* 317–329, DOI: 10.1016/j.icarus.2014.03.015.

Buhler P. B. and Ingersoll A. P. (2018) Sublimation pit distribution indicates convection cell surface velocities of ~10 cm per year in Sputnik Planitia, Pluto. *Icarus, 300,* 327–340.

Buratti B. J., Thomas P. C., Roussos E., et al. (2019) Close Cassini flybys of Saturn's ring moons Pan, Daphnis, Atlas, Pandora, and Epimetheus. *Science, 364,* eaat2349, DOI: 10.1126/science.aat2349.

Canup R. M. (2005) A giant impact origin of Pluto-Charon. *Science, 307,* 546–550.

Canup R. M. (2011) On a giant impact origin of Charon, Nix, and Hydra. *Astron. J., 141,* 35–44.

Carter P. J., Lienhardt Z. M., Elliott T., Stewart S. T., and Walker M. J. (2018) Collisional stripping of planetary crusts. *Earth Planet. Sci. Lett., 484,* 276–286, DOI: 10.1016/j.epsl.2017.12.012.

Castillo-Rogez J., Johnson T. V., Lee M. H., Turner N. J., Matson D. L., and Lunine J. (2009) ^{26}Al decay: Heat production and a revised age for Iapetus. *Icarus, 204*, 658–662.
Cheng W. H., Lee M. H., and Peale S. J. (2014) Complete tidal evolution of Pluto-Charon. *Icarus, 233*, 242–258.
Claire M. W., Sheets J., Cohen M., Ribas I., Meadows V. S., and Catling D. C. (2012) The evolution of solar flux from 0.1 nm to 160 μm: Quantitative estimates for planetary studies. *Astrophys. J., 757*, 95, DOI: 10.1088/0004-637X/757/1/95.
Clark R. N., Brown R. H., Cruikshank D. P., and Swayze G. A. (2019) Isotopic ratios of Saturn's rings and satellites: Implications for the origin of water and Phoebe. *Icarus, 321*, 791–802.
Cleeves L. I., Bergin E. A., Alexander C. M. O'D., Du F., Graninger D., Öberg K. I., and Harries T. J. (2014) The ancient heritage of water ice in the solar system. *Science, 345*, 1590–1593.
Cook J. A., Dalle Ore C. M., Protopapa S., et al. (2018) Composition of Pluto's small satellites: Analysis of New Horizons spectral images. *Icarus, 315*, 30–45.
Cook J. A., Dalle Ore C. M., Protopapa S., et al. (2019) The distribution of H_2O, CH_3OH, and hydrocarbon-ices on Pluto: Analysis of New Horizons spectral images. *Icarus, 331*, 148–169.
Cruikshank D. P., Umurhan O. M., Beyer R. A., et al. (2019) Recent cryovolcanism in Virgil Fossae on Pluto. *Icarus, 330*, 155–168, DOI: 10.1016/j.icarus.2019.04.023.
Cruikshank D. P., Dalle Ore C. M., Scipioni F., et al. (2021) Cryovolcanic flooding in Viking Terra on Pluto. *Icarus, 356*, 113786.
Dalle Ore C. M., Protopapa S., Cook J. A., et al. (2018) Ices on Charon: Distribution of H_2O and NH_3 from New Horizons LEISA observations. *Icarus, 300*, 21–32.
Dalle Ore C. M., Cruikshank D. P., Protopapa S., Scipioni F., McKinnon W. B., Cook J. C., Grundy W. M., Stern S. A., Moore J. M., Verbiscer A., Parker A. H., Singer K. H., Umurhan O. M., Weaver H. A., Olkin C. B., Young L. A., Ennico K., and the New Horizons Surface Composition Theme Team (2019) Detection of ammonia on Pluto's surface in a region of geologically recent tectonism. *Sci. Adv., 5*, eaav5731, DOI: 10.1126/sciadv.aav5731.
Dartois D. E., Bouzit M., and Schmitt B. (2012) Clathrate hydrates: FTIR spectroscopy for astrophysical remote detection. In *European Conference on Laboratory Astrophysics — ECLA* (C. Stehlé et al., eds.), pp. 219–224. EAS Publ. Series, Vol. 58, EDP Sciences, Les Ulis, France.
Davidsson B. J. R., Sierks H., Güttler C., et al. (2016) The primordial nucleus of Comet 67P/Churyumov-Gerasimenko. *Astron. Astrophys., 592*, A63, DOI: 10.1051/0 0 04-6361/201526968.
De Bergh C., Schaller E. L., Brown M. E., Brunetto R., Cruikshank D. P., and Schmitt B. (2013) The ices on transneptunian objects and Centaurs. In *The Science of Solar System Ices* (M. Gudipati and J. Castillo-Rogez, eds.), pp. 107–146. Springer, New York.
De Sanctis M. C., Ammannito E., Raponi A., et al. (2015) Ammoniated phyllosilicates with a likely outer solar system origin on (1) Ceres. *Nature, 528*, 241–244.
De Sousa R. R., Morbidelli A., Raymond S. N., Izidoro A., Gomes R. S., and Neto E. V. (2020) Dynamical evidence for an early giant planet instability. *Icarus, 339*, 113605, DOI: 10.1016/j.icarus.2019.113605.
Desch S. J. (2007) Mass distribution and planet formation in the solar nebula. *Astrophys. J., 671*, 878–893, DOI: 10.1086/522825.
Desch S. J. and Neveu M. (2017) Differentiation and cryovolcanism on Charon: A view before and after New Horizons. *Icarus, 287*, 175–186, DOI: 10.1016/j.icarus.2016.11.037.
Desch S. J., Cook J. C., Doggett T. C., and Porter S. B. (2009) Thermal evolution of Kuiper belt objects, with implications for cryovolcanism. *Icarus, 202*, 694–714.
Desch S. J., Kalyaan A., and Alexander C. M. O'D. (2018) The effect of Jupiter's formation on the distribution of refractory elements and inclusions in meteorites. *Astrophys. J. Suppl., 238*, 11, DOI: 10.3847/1538-4365/aad95f.
Dobrovolskis A. R., Peale S. J., and Harris A. W. (1997) Dynamics of the Pluto-Charon binary. In *Pluto and Charon* (S. A. Stern and D. J. Tholen, eds.), pp. 159–190. Univ. of Arizona, Tucson.
Durham W. B., Kirby S. H., and Stern L. A. (1993) Flow of ices in the ammonia-water system. *J. Geophys. Res., 98*, 17667–17682.
Earle A. M. and Binzel R. P. (2015) Pluto's insolation history: Latitudinal variations and effects on atmospheric pressure. *Icarus, 250*, 405–412.
Earle A. M., Binzel R. P., Young L. A., Stern S. A., Ennico K., Grundy W., Olkin C. B., Weaver H. A., and New Horizons Geology and Geophysics Imaging Team (2017) Long-term surface temperature modeling of Pluto. *Icarus, 287*, 37–46.
Emsenhuber A., Jutzi M., and Benz W. (2018) SPH calculations of Mars-scale collisions: The role of the equation of state, material rheologies, and numerical effects. *Icarus, 301*, 247–257.
Forget F., Bertrand T., Vangvichith M., Leconte J., Millour E., and Lellouch E. (2017) A post-New Horizons global climate model of Pluto including the N_2, CH_4, and CO cycles. *Icarus, 287*, 54–71.
Fraser W. C., Brown M. E., Morbidelli A., Parker A., and Batygin K. (2014) The absolute magnitude distribution of Kuiper belt objects. *Astrophys. J., 782*, 100.
Fray N., Bardyn A., Cottin H., et al. (2016) High-molecular-weight organic matter in the particles of Comet 67P/Churyumov-Gerasimenko. *Nature, 538*, 72–74, DOI: 10.1038/nature19320.
Fray N. Bardyn A., Cottin H., et al. (2017) Nitrogen-to-carbon atomic ratio measured by COSIMA in the particles of Comet 67P/Churyumov-Gerasimenko. *Mon. Not. R. Astron. Soc., 469*, S506–S516.
Fulle M., Della Corte V., Rotundi A., et al. (2016) Comet 67P/Churyumov-Gerasimenko preserved the pebbles that formed planetesimals. *Mon. Not. R. Astron. Soc., 462*, S132–S137, DOI: 10.1093/mnras/stw2299.
Fulle M., Della Corte V., Rotundi A., et al. (2017) The dust-to-ices ratio in comets and Kuiper belt objects. *Mon. Not. R. Astron. Soc., 469*, S45–S49, DOI: 10.1093/mnras/stw983.
Füri E. and Marty B. (2015) Nitrogen isotope variations in the solar system. *Nature Geosci., 8*, 515–522.
Gabasova L., Tobie G., and Choblet G. (2018) Compaction-driven evolution of Pluto's rocky core: Implications for water-rock interactions. *Lunar and Planetary Science XLIX*, Abstract #2512. Lunar and Planetary Institute, Houston.
Gao P., Fan S., Wong M. L., et al. (2017) Constraints on the microphysics of Pluto's photochemical haze from New Horizons observations. *Icarus, 287*, 116–123.
Gladman B. and Chan C. (2006) Production of the extended scattered disk by rogue planets. *Astrophys. J. Lett., 643*, L135–L138, DOI: 10.1086/505214.
Gladstone G. R. and Young L. A. (2019) New Horizons observations of the atmosphere of Pluto. *Annu. Rev. Astron. Astrophys.*, 57, 119–140, DOI: 0.1146/annurevastro-053018-060128.
Gladstone G. R. and 32 colleagues (2016) The atmosphere of Pluto as observed by New Horizons. *Science, 351*, aad8866.
Glein C. R. (2015) Noble gases, nitrogen, and methane from the deep interior to the atmosphere of Titan. *Icarus, 250*, 570–586.
Glein C. R. and Waite J. H. (2018) Primordial N_2 provides a cosmochemical explanation for the existence of Sputnik Planitia, Pluto. *Icarus, 313*, 79–92, DOI: 10.1016/j.icarus.2018.05.007.
Glein C. R. et al. (2008) The oxidation state of hydrothermal systems on early Enceladus. *Icarus, 197*, 157–163.
Glein C. R., Postberg F., and Vance S. D. (2018) The geochemistry of Enceladus: Composition and controls. In *Enceladus and the Icy Moons of Saturn* (P. M. Schenk et al., eds.), pp. 39–56. Univ. of Arizona, Tucson.
Goldreich P. and Ward W. R. (1973) The formation of planetesimals. *Astrophys. J., 183*, 1051–1062.
Goldreich P. G., Lithwick Y., and Sari R. (2004) Planet formation by coagulation: A focus on Uranus and Neptune. *Annu. Rev. Astron. Astrophys., 42*, 549–601, DOI: 10.1146/annurev.astro.42.053102.134004.
Gomes R. S. (2003) The origin of the Kuiper belt high-inclination population. *Icarus, 161*, 404–418.
Gomes R. S., Gallardo T., Fernández J. A., and Brunini A. (2005) On the origin of the high-perihelion scattered disk: The role of the Kozai mechanism and mean motion resonances. *Cel. Mech. Dyn. Astron., 91*, 109–129.
Gough D. O. (1981) Solar interior structure and luminosity variations. *Solar Phys., 74*, 21–34.
Greenstreet S., Gladman B., and McKinnon W. B. (2015) Impact and cratering rates onto Pluto. *Icarus, 258*, 267–288.
Grundy W. M., Young L. A., Stansberry J. A., et al. (2010) Near-infrared spectral monitoring of Triton with IRTF/SpeX II: Spatial distribution and evolution of ices. *Icarus, 205*, 594–604.
Grundy W. M., Binzel R. P., Buratti B. J., et al. (2016) Surface

compositions across Pluto and Charon. *Science, 351*, aad9189, DOI: 10.1126/science.aad9189.

Grundy W. M., Bertrand T., Binzel R. P., et al. (2018) Pluto's haze as a surface material. *Icarus, 314*, 232–245.

Grundy W. M., Bird M. K., Britt D. T., et al. (2020) Color, composition, and thermal environment of Kuiper belt object (486958) Arrokoth. *Science, 367*, eaay3999, DOI: 10.1126/science.aay3705.

Haisch K. E., Lada E. A., and Lada C. J. (2001) Disk frequencies and lifetimes in young clusters. *Astrophys. J. Lett., 553*, L153–L156.

Hammond N. P., Barr A. C., and Parmentier E. M. (2016) Recent tectonic activity on Pluto driven by phase changes in the ice shell. *Geophys. Res. Lett., 43*, 6775–6782, DOI: 10.1002/2016GL069220.

Hersant F., Gautier D., and Lunine J. I. (2004) Enrichment in volatiles in the giant planets of the solar system. *Planet. Space Sci., 52*, 623–641.

Holler B. J., Young L. A., Bus S. J., and Protopapa S. (2017) Methanol ice on Kuiper belt objects 2007 OR_{10} and Salacia: Implications for formation and dynamical evolution. *EPSC Abstracts, 11*, EPSC2017-330.

Howard A. D. and 14 colleagues (2017a) Present and past glaciation on Pluto. *Icarus, 287*, 287–300.

Howard A. D. and 16 colleagues (2017b) Pluto: Pits and mantles on uplands north and east of Sputnik Planitia. *Icarus, 293*, 218–230.

Imbrie J. and Imbrie K. P. (1979) *Ice Ages: Solving the Mystery*. Harvard Univ., Cambridge. 224 pp.

Izidoro A., Raymond S. N., Morbidelli A., Hersant F., and Pierens A. (2015) Gas giant planets as dynamical barriers to inward-migrating super-Earths. *Astrophys. J. Lett., 800*, L22, DOI: 10.1088/2041-8205/800/2/L22.

Jacobson R. A., Brozović M., Showalter M., Verbiscer A., Buie M., and Helfenstein P. (2019) The orbits and masses of Pluto's satellites. *Pluto System After New Horizons*, Abstract #7031. LPI Contribution No. 2133, Lunar and Planetary Institute, Houston.

Janes D. M. and Melosh H. J. (1990) Tectonics of planetary loading — a general model and results. *J. Geophys. Res., 95*, 21345–21355.

Johansen A. and Lambrechts M. (2017) Forming planets via pebble accretion. *Annu. Rev. Earth. Planet. Sci., 45*, 359–387, DOI: 10.1146/annurev-earth-063016-020226.

Johansen A., Oishi J. S., Mac Low M.-M., Klahr H., Henning T., and Youdin A. (2007) Rapid planetesimal formation in turbulent circumstellar disks. *Nature, 448*, 1022–1025, DOI: 10.1038/nature06086.

Johansen A., Blum J., Tanaka H., Ormel C., Bizzarro M., and Rickman H. (2014) The multifaceted planetesimal formation process. In *Protostars and Planets VI* (H. Beuther et al., eds.), pp. 547–570. Univ. of Arizona, Tucson.

Johansen A., Mac Low M.-M., Lacerda P., and Bizzaro M. (2015) Growth of asteroids, planetary embryos, and Kuiper belt objects by chondrule accretion. *Sci. Adv., 1*, e1500109, DOI: 10.1126/sciadv.1500109.

Johnson B. C., Bowling T. J., Trowbridge A. J., and Freed A. M. (2016) Formation of the Sputnik Planum basin and the thickness of Pluto's subsurface ocean. *Geophys. Res. Lett., 43*, 10068–10077, DOI: 10.1002/2016GL070694.

Johnson P., Young L. A., Protopapa S., et al. (2019) Pluto's minimum surface pressure and implications for haze production. *Pluto System After New Horizons*, Abstract #7025. LPI Contribution No. 2133, Lunar and Planetary Institute, Houston.

Johnson R. E., Oza A., Young L. A., Volkov A. N., and Schmidt C. (2015) Volatile loss and classification of Kuiper belt objects. *Astrophys. J., 809*, 43, DOI: 10.1088/0004-637X/809/1/43.

Kamata S., Nimmo F., Sekine Y., Kuramoto K., Noguchi N., Kimura J., and Tani A. (2019) Pluto's ocean is capped and insulated by gas hydrates. *Nature Geosci., 12*, 407–410.

Kammer J. A., Stern S. A., Young L. A., et al. (2017) New Horizons upper limits on O_2 in Pluto's present day atmosphere. *Astron J., 154*, 55.

Keane J. T., Matsuyama I., Kamata S., and Steckloff J. K. (2016) Reorientation and faulting of Pluto due to volatile loading within Sputnik Planitia. *Nature, 540*, 90–93.

Kenyon S. C. and Bromley B. C. (2012) Coagulation calculations of icy planet formation at 15–150 AU: A correlation between the maximum radius and the slope of the size distribution for trans-neptunian objects. *Astron. J., 143*, 63, DOI: 10.1088/0004-6256/143/3/63.

Kenyon S. J. and Bromley B. C. (2014) The formation of Pluto's low-mass satellites. *Astrophys. J., 147*, 8.

Kenyon S. J. and Bromley B. C. (2019a) A Pluto-Charon sonata: Dynamical limits on the masses of the small satellites. *Astrophys. J., 158*, 69.

Kenyon S. C. and Bromley B. C. (2019b) A Pluto-Charon sonata: The dynamical architecture of the circumbinary satellite system. *Astron. J., 157*, 79.

Kenyon S. C. and Bromley B. C. (2019c) A Pluto-Charon sonata. III. Growth of Charon from a circum-Pluto ring of debris. *Astron. J., 158*, 142.

Kissel J. and Krueger F. R. (1987) The organic component in dust from Comet Halley as measured by the PUMA mass spectrometer on board Vega 1. *Nature, 326*, 755–760.

Kouchi A. and Sirono S. (2001) Crystallization heat of impure amorphous H_2O ice. *Geophys. Res. Lett., 28*, 827–830.

Krasnopolsky V. (2016) Isotopic ratio of nitrogen on Titan: Photochemical interpretation. *Planet. Space Sci., 134*, 61–63.

Kruijer T. S., Burkhardt C., Budde G., and Kleine T. (2017) Age of Jupiter inferred from the distinct genetics and formation times of meteorites. *Proc. Natl. Acad. Sci. U.S.A., 114*, 6712–6716.

Laskar J., Correia A. C. M., and Gastineau M. (2004) Long term evolution and chaotic diffusion of the insolation quantities of Mars. *Icarus, 170*, 343–364.

Lawler S. M., Shankman C.., Kavelaars J. J., et al. (2018) OSSOS. VIII. The transition between two size distribution slopes in the scattering disk. *Astron. J., 155*, 197, DOI: 10.3847/1538-3881/aab8ff.

Lawler S. M., Pike R. E., Kaib N., et al. (2019) OSSOS. XIII. Fossilized resonant dropouts tentatively confirm Neptune's migration was grainy and slow. *Astron. J., 157*, 6, DOI: 10.3847/1538-3881/ab1c4c.

Leinhardt Z., Kraus R. A., and Stewart S. T. (2010) The formation of the collisional family around the dwarf planet Haumea. *Astrophys. J., 714*, 1789–1799.

Lellouch E., Gurwell M., Butler B., et al. (2017) Detection of CO and HCN in Pluto's atmosphere with ALMA. *Icarus, 286*, 289–307, DOI: 10.1016/j.icarus.2016.10.013.

Levison H. F., Morbidelli A., VanLaerhoven C., Gomes R., and Tsiganis K. (2008a) Origin of the structure of the Kuiper belt during a dynamical instability in the orbits of Uranus and Neptune. *Icarus, 196*, 258–273.

Levison H. F., Morbidelli A., Vokrouhlický D., and Bottke W. F. (2008b) On a scattered-disk origin for the 2003 EL_{61} collisional family — An example of the importance of collisions on the dynamics of small bodies. *Astron. J., 136*, 1079–1088.

Lewis J. S. (1971) Satellites of the outer planets: Thermal models. *Science, 172*, 1127–1128, DOI: 10.1126/science.172.3988.1127.

Lodders K. (2003) Solar system abundances and condensation temperatures of the elements. *Astrophys. J., 591*, 1220–1247.

Lorenzi V., Pinilla-Alonso N., and Licandro J. (2015) Rotationally resolved spectroscopy of dwarf planet (136472) Makemake. *Astron. Astrophys., 577*, A86, DOI: 10.1051/0004-6361/201425575.

Lunine J. I. and Nolan M. C. (1992) A massive early atmosphere on Triton. *Icarus, 100*, 221–234.

Luspay-Kuti A., Mousis O., Hässig M., et al. (2016) The presence of clathrates in Comet 67P/Churyumov-Gerasimenko. *Sci. Adv., 2*, e1501781, DOI: 10.1126/sciadv.1501781.

Malamud U., Perets H. B., and Schubert G. (2017) The contraction/expansion history of Charon with implications for its planetary-scale tectonic belt. *Mon. Not. R. Astron. Soc., 468*, 1056–1069, DOI: 10.1093/mnras/stx546.

Malhotra R. (1993) The origin of Pluto's peculiar orbit. *Nature, 365*, 819–821.

Malhotra R. and Williams J. G. (1997) Pluto's heliocentric orbit. In *Pluto and Charon* (S. A. Stern and D. J. Tholen, eds.), pp. 127–157. Univ. of Arizona, Tucson.

Mandt K. E., Mousis O., Lunine J., and Gautier D. (2014) Protosolar ammonia as the unique source of Titan's nitrogen. *Astrophys. J. Lett., 788*, L24, DOI: 10.1088/2041-8205/788/2/L24.

Manga M. and Wang C. Y. (2007) Pressurized oceans and the eruption of liquid water on Europa and Enceladus. *Geophys. Res. Lett., 34*, L07202, DOI: 10.1029/2007GL029297.

McBride B. J. and Gordon S. (1996) *Computer Program for Calculation of Complex Chemical Equilibrium Compositions and Applications: II. Users Manual and Program Description*. NASA Ref. Publ. 1311. 177 pp.

McGovern P. J., White O. L., and Schenk P. M. (2019) Tectonism across Pluto: Mapping and interpretations. *Pluto System After New Horizons*, Abstract #7063. LPI Contribution No. 2133, Lunar and Planetary Institute, Houston.
McKay A. J., Disanti M. A., Kelley M. S. P., et al. (2019) The peculiar volatile composition of CO-dominated Comet C/2016 R2 (PanSTARRS). *Astron. J., 158*, 128, DOI: 10.3847/1538-3881/ab32e4.
McKay C. P., Scattergood T. W., Pollack J. B., Borucki W. J., and van Ghyseghem H. T. (1988) High-temperature shock formation of N_2 and organics on primordial Titan. *Nature, 332*, 520–522.
McKinnon W. B. (1989) On the origin of the Pluto-Charon binary. *Astrophys. J. Lett., 344*, L41–L44.
McKinnon W. B. (2002) On the initial thermal evolution of Kuiper belt objects. In *Proceedings of Asteroids, Comets, Meteors* (B. Warmbein, ed.), pp. 29–38. ESA SP-500, Noordwijk, The Netherlands.
McKinnon W. B. and Mueller S. (1988) Pluto's structure and composition suggest origin in the solar, not a planetary, nebula. *Nature, 335*, 240–243.
McKinnon W. B., Simonelli D. P., and Schubert G. (1997) Composition, internal structure, and thermal evolution of Pluto and Charon. In *Pluto and Charon* (S. A. Stern and D. J. Tholen, eds.), pp. 295–343. Univ. of Arizona, Tucson.
McKinnon W. B., Prialnik D., Stern S. A., and Coradini A. (2008) Structure and evolution of Kuiper belt objects and dwarf planets. In *The Solar System Beyond Neptune* (M. A. Barucci et al., eds.), pp. 213–241. Univ. of Arizona, Tucson.
McKinnon W. B., Nimmo F., Wong T., et al. (2016a) Convection in a volatile nitrogen-ice-rich layer drives Pluto's geological vigour. *Nature, 534*, 82–85.
McKinnon W. B., Schenk P. M., Moore J. M., et al. (2016b) An impact basin origin for Sputnik "Planitia" and surrounding terrains, Pluto. *GSA Abstracts with Programs, 48*, 48-6, DOI: 10.1130/abs/2016AM-285142.
McKinnon W. B., Stern S. A., Weaver H. A., et al. (2017) Origin of the Pluto-Charon system: Constraints from the New Horizons flyby. *Icarus, 287*, 2–11, DOI: 10.1016/j.icarus.2016.11.019.
McKinnon W. B., Lunine J. I., Mousis O., Waite J. H., and Zolotov M. Y. (2018a) The mysterious origin of Enceladus: A compositional perspective. In *Enceladus and the Icy Moons of Saturn* (P. M. Schenk et al., eds.), pp. 17–38. Univ. of Arizona, Tucson.
McKinnon W. B., Schenk P. M., and Bland M. T. (2018b) Pluto's heat flow: A mystery wrapped in an ocean inside an ice shell. *Lunar and Planetary Science XLIX*, Abstract #2715. Lunar and Planetary Institute, Houston.
McKinnon W. B., Richardson D., Marohnic J., et al. (2020) The solar nebula origin of (486958) Arrokoth, a primordial contact binary in the Kuiper belt. *Science, 367*, eaay6620, DOI: 10.1126/science.aay6620.
Melosh H. J. and Ivanov B. (2018) Slow impacts on strong targets bring on the heat. *Geophys. Res. Lett., 45*, 2597–2599.
Merlin F. (2015) New constraints on the surface of Pluto. *Astron. Astrophys., 582*, A39, DOI: 10.1051/0004-6361/201526721.
Merlin F., Lellouch E., Quirico E., and Schmidtt B. (2018) Triton's surface ices: Distribution, temperature and mixing state from VLT/SINFONI observations. *Icarus, 314*, 274–292.
Miller K. E., Glein C. R., and Waite J. H. (2019) Contributions from accreted organics to Titan's atmosphere: New insights from cometary and chondritic data. *Astrophys. J., 871*, 59.
Moore J. M. and McKinnon W. B. (2021) Geologically diverse Pluto and Charon: Implications for the dwarf planets of the Kuiper belt. *Annu. Rev. Earth. Planet. Sci., 48*, 173–200, DOI: 10.1146/annurev-earth-071720-051448.
Moore J. M. and Pappalardo R. T. (2011) Titan: An exogenic world? *Icarus, 212*, 790–806.
Moore J. M., Howard A. D., Schenk P. M., McKinnon W. B., Pappalardo R. T., Ewing R. C., Bierhaus E. B., Bray V. J., Spencer J. R., Binzel R. P., Buratti B., Grundy W. M., Olkin C. B., Reitsma H. J., Reuter D. C., Stern S. A., Weaver H., Young L. A., and Beyer R. A. (2015) Geology before Pluto: Pre-encounter considerations. *Icarus, 246*, 65–81, DOI: 10.1016/j.icarus.2014.04.028.
Moore J. M. and 40 colleagues (2016) The geology of Pluto and Charon through the eyes of New Horizons. *Science, 351*, 1284–1293.
Moore J. M. and 25 colleagues (2018) Bladed terrain on Pluto: Possible origins and evolution. *Icarus, 300*, 129–144.
Moores J. E., Smith C. L., Toigo A. D., and Guzewich S. D. (2017) Penitentes as the origin of the bladed terrain of Tartarus Dorsa on Pluto. *Nature, 541*, 188–190.
Morbidelli A. and Nesvorný D. (2020) Kuiper belt: Formation and evolution. In *The Trans-Neptunian Solar System* (D. Prialnik et al., eds.), pp. 25–59. Elsevier, Amsterdam, DOI: 10.1016/B978-0-12-816490-7.00002-3.
Morbidelli A. and Rickman H. (2015) Comets as collisional fragments of a primordial planetesimal disk. *Astron. Astrophys., 583*, A43, DOI: 10.1051/0004-6361/201526116.
Morbidelli A., Bottke W. F., Nesvorný D., and Levison H. F. (2009) Asteroids were born big. *Icarus, 204*, 558–573.
Mousis O., Guilbert-Lepoutre A., Lunine J. I., Cochran A. L., Waite J. H., Petit J.-M., and Rousselot P. (2012) The dual origin of the nitrogen deficiency in comets: Selective volatile trapping in the nebula and postaccretion radiogenic heating. *Astrophys. J., 757*, 146.
Mumma M. J. and Charnley S. B. (2011) The chemical composition of comets — Emerging taxonomies and natal heritage. *Annu. Rev. Astron Astrophys., 49*, 471–524.
Murray C. D. and Dermott S. F. (1999) *Solar System Dynamics*. Cambridge Univ., Cambridge. 603 pp.
Murray-Clay R. A. and Chiang E. I. (2016) Brownian motion in planetary migration. *Astrophys. J., 651*, 1194–1208, DOI: 10.1086/507514.
Nesvorný D. (2018) Dynamical evolution of the early solar system. *Annu. Rev. Astron. Astrophys., 56*, 137–174, DOI: 10.1146/annurev-astro-081817-052028.
Nesvorný D. and Morbidelli A. (2012) Statistical study of the early solar system's instability with four, five, and six giant planets. *Astron. J., 14*, 117.
Nesvorný D. and Vokrouhlický D. (2016) Neptune's orbital migration was grainy, not smooth. *Astrophys. J., 825*, 94.
Nesvorný D. and Vokrouhlický D. (2019) Binary survival in the outer solar system. *Icarus, 331*, 49–61.
Nesvorný D., Youdin A. N., and Richardson D. C. (2010) Formation of Kuiper belt binaries by gravitational collapse. *Astron. J., 140*, 785–793, DOI: 10.1088/0004-6256/140/3/785.
Nesvorný D., Vokrouhlický D., Bottke W. F., and Levison H. F. (2018) Evidence for very early migration of the solar system planets from the Patroclus-Menoetius binary Jupiter Trojan. *Nature Astron., 2*, 878–882.
Nesvorný D., Li R., Youdin A. N., Simon J. B., and Grundy W. M. (2019) Trans-neptunian binaries as evidence for planetesimal formation by the streaming instability. *Nature Astron., 3*, 808–812, DOI: 10.1038/s41550-019-0806-z.
Neveu M., Desch S. J., Shock E. L., and Glein C. R. (2015) Prerequisites for explosive cryovolcanism on dwarf planet-class Kuiper belt objects. *Icarus, 246*, 48–64.
Niemann H. B., Atreya S. K., Demick J. E., Gautier D., Haberman J. A., Harpold D. N., Kasprzak W. T., Lunine J. I., Owen T. C., and Raulin F. (2010) Composition of Titan's lower atmosphere and simple surface volatiles as measured by the Cassini-Huygens probe gas chromatograph mass spectrometer experiment. *J. Geophys. Res., 115*, E12006, DOI: 10.1029/2010JE003659.
Nimmo F. (2004) Stresses generated in cooling viscoelastic ice shells: Application to Europa. *J. Geophys. Res., 109*, E12001.
Nimmo F., Hamilton D. P., McKinnon W. B., et al. (2016) Reorientation of Sputnik Planitia implies a subsurface ocean on Pluto. *Nature, 540*, 94–96.
Nogueira E., Brasser R., and Gomes R. (2011) Reassessing the origin of Triton. *Icarus, 214*, 113–130.
Noll K. S., Grundy W. M., Nesvorný D., and Thirouin A. (2020) Trans-neptunian binaries (2018). In *The Trans-Neptunian Solar System* (D. Prialnik et al., eds.), pp. 205–224. Elsevier, Amsterdam, DOI: 10.1016/B978-0-12-816490-7.00009-6.
Öberg K. I. and Wordsworth R. (2019) Jupiter's composition suggests its core assembled exterior to the N_2 snowline. *Astron. J., 158*, 194, DOI: 10.3847/1538-3881/ab46a8.
Okumura F. and Mimura K. (2011) Gradual and stepwise pyrolyses of insoluble organic matter from the Murchison meteorite revealing chemical structure and isotopic distribution. *Geochim. Cosmochim. Acta, 75*, 7063–7080.
Ortiz J. L., Santos-Sanz P., Sicardy B., et al. (2017) The size, shape, density and ring of the dwarf planet Haumea from a stellar occultation. *Nature, 550*, 219–223.

Owen T. C. (1982) The composition and origin of Titan's atmosphere. *Planet. Space Sci.*, *30*, 833–838.

Owen T. C., Roush T. L., Cruikshank D. P., Elliot J. L., Young L. A., de Bergh C., Schmitt B., Geballe T. R., Brown R. H., and Bartholomew M. J. (1993) Surface ices and atmospheric composition of Pluto. *Science*, *261*, 745–48.

Palme H., Lodders K., and Jones A. (2014) Solar system abundances of the elements. In *Treatise on Geochemistry, 2nd edition, Vol. 2: Planets, Asteroids, Comets and the Solar System* (H. H. Holland and K. K. Turekian, eds.), pp. 15–36. Elsevier, Amsterdam.

Peale S. J. and Canup R. M. (2015) The origin of natural satellites. In *Treatise on Geophysics, 2nd edition, Vol. 10: Physics of Terrestrial Planets and Moons* (G. Schubert, ed.), pp. 559–604. Elsevier, Amsterdam.

Pires P., Winter S. M. G., and Gomes R. S. (2015) The evolution of a Pluto-like system during the migration of the ice giants. *Icarus, 246*, 330–338, DOI: 10.1016/j.icarus.2014.04.029.

Poch O., Istiqomah I., Quirico E., et al. (2019) Ammonium salts are a reservoir of nitrogen on a cometary nucleus and possibly on some asteroids. *Science*, *367*, eaaw7462, DOI: 10.1126/science.aaw7462.

Postberg F., Kempf S., Schmidt J., Brilliantov N., Beinsen A., Abel B., Buck U., and Srama R. (2009) Sodium salts in E-ring ice grains from an ocean below the surface of Enceladus. *Nature*, *459*, 1098–1101.

Protopapa S., Grundy W. M., Reuter D. C., et al. (2017) Pluto's global surface composition through pixel-by-pixel Hapke modeling of New Horizons Ralph/LEISA data. *Icarus*, *287*, 218–228.

Quarles B. and Kaib N. (2019) Instabilities in the early solar system due to a self-gravitating disk. *Astron. J.*, *157*, 67, DOI: 10.3847/1538-3881/aafa71.

Reeves E. P. and Fiebig J. (2020) Abiotic synthesis of methane and organic compounds in Earth's lithosphere. *Elements, 16*, 25–31.

Rhoden A. R., Henning W., Hurford T. A., and Hamilton D. P. (2015) The interior and orbital evolution of Charon as preserved in its geologic record. *Icarus*, *246*, 11–20.

Rhoden A. R., Skjetne H. L., Henning W. G., et al. (2020) Charon: A brief history of tides. *J. Geophys. Res. Planets*, *125*, e06449. DOI: 10.1029/2020JE006449.

Robbins S. J. and 28 colleagues (2019) Geologic landforms and chronostratigraphic history of Charon as revealed by a hemispheric geologic map. *J. Geophys. Res.*, *124*, 155–174.

Robuchon G. and Nimmo F. (2011) Thermal evolution of Pluto and implications for surface tectonics and a subsurface ocean. *Icarus*, *216*, 426–439, DOI: 10.1016/j.icarus.2011.08.015.

Rubin M., Altwegg K., Balsiger H., et al. (2015) Molecular nitrogen in Comet 67P/Churyumov-Gerasimenko indicates a low formation temperature. *Science*, *384*, 232–235.

Rubin M., Altwegg K., Balsiger H., et al. (2018) Krypton isotopes and noble gas abundances in the coma of Comet 67P/Churyumov-Gerasimenko. *Sci. Adv.*, *4*, eaar6297. DOI: 10.1126/sciadv.aar6297.

Rubin M., Altwegg K., Balsiger H., et al. (2019) Elemental and molecular abundances in Comet 67P/Churyumov-Gerasimenko. *Mon. Not. R. Astron. Soc.*, *489*, 594–607.

Schaller E. L. and Brown M. E. (2007) Volatile loss and retention on Kuiper belt objects. *Astrophys. J. Lett.*, *659*, L61–L64.

Schenk P. M., Beyer R. A., McKinnon W. B., et al. (2018) Basins, fractures and volcanoes: Global cartography and topography of Pluto from New Horizons. *Icarus*, *314*, 400–433, DOI: 10.1016/j.icarus.2018.06.008.

Schubert G., Stevenson D. J., and Ellsworth K. (1981) Internal structures of the Galilean satellites. *Icarus*, *47*, 46–59.

Scott E. R. D., Krot A. N., and Sanders I. S. (2018) Isotopic dichotomy among meteorites and its bearing on the protoplanetary disk. *Astrophys. J.*, *854*, 164.

Scott T. A. (1976) Solid and liquid nitrogen. *Phys. Rept.*, *27*, 89–157.

Sekine Y., Genda H., Sugita S., Kadono T., and Matsui T. (2011) Replacement and late formation of atmospheric N_2 on undifferentiated Titan by impacts. *Nature Geosci.*, *4*, 359–362.

Sekine Y., Genda H., Kamata S., and Funatsu T. (2017) The Charon-forming giant impact as a source of Pluto's dark equatorial regions. *Nature Astron.*, *1*, 0031, DOI: 10.1038/s41550-016-0031.

Sephton M. (2005) Organic matter in carbonaceous meteorites: Past, present, and future research. *Philos. Trans. R. Soc. London Ser. A*, *363*, 2729–2742, DOI: 10.1098/rsta.2005.1670.

Shakura N. I. and Sunyaev R. A. (1973) Black holes in binary systems: Observational appearance. *Astron. Astrophys.*, *24*, 337–355.

Shannon A. and Dawson B. (2018) Limits on the number of primordial scattered disc objects at Pluto mass and higher from the absence of their dynamical signatures on the present-day trans-neptunian populations. *Mon. Not. R. Astron. Soc.*, *480*, 1870–1882, DOI: 10.1093/mnras/sty1930.

Sheppard S., Trujillo C. A., Tholen D. J., and Kaib N. (2019) A new high perihelion trans-plutonian inner Oort cloud object: 2015 TG387. *Astron. J.*, *157*, 139, DOI: 10.3847/1538-3881/ab0895.

Shinnaka Y., Kawakita H., Emmanuël J., Decock A., Hutsemékers D., Manfroid J., and Arai A. (2016) Nitrogen isotopic ratios of NH_2 in comets: Implication for ^{15}N-fractionation in cometary ammonia. *Mon. Not. R. Astron. Soc.*, *462*, S195–S209.

Shock E. L. and McKinnon W. B. (1993) Hydrothermal processing of cometary volatiles — Applications to Triton. *Icarus*, *106*, 464–477.

Shock E. L. and Canovas P. C. (2010) The potential for abiotic organic synthesis and biosynthesis at seafloor hydrothermal systems. *Geofluids, 10*, 161-192.

Siess L., Dufour E., and Forestini M. (2000) An internet server for pre-main sequence tracks of low- and intermediate-mass stars. *Astron. Astrophys.*, *358*, 593–599.

Simon J. B., Armitage P. J., Li R., and Youdin A. N. (2016) The mass and size distribution of planetesimals formed by the streaming instability. I. The role of self-gravity. *Astrophys. J.*, *822*, 55, DOI: 10.3847/0004-637X/822/1/55.

Simonelli D. P., Pollack J. B., McKay C. P., Reynolds R. T., and Summers A. L. (1989) The carbon budget in the outer solar nebula. *Icarus*, *82*, 1–35.

Singer K. N. and Stern S. A. (2015) On the provenance of Pluto's nitrogen (N_2). *Astrophys. J. Lett.*, *808*, L50, DOI: 10.1088/2041-8205/808/2/L50.

Singer K. N., Grundy W. M., White C. B., and Binzel R. P. (2021) Introduction to *Icarus* special issue "Pluto System, Kuiper Belt, and Kuiper Belt Objects." *Icarus, 356*, 114269, DOI: 10.1016/j.icarus.2020.114269.

Sloan E. D. and Koh C. A. (2007) *Clathrate Hydrates of Natural Gases, 3rd edition.* CRC Press, Boca Raton. 752 pp.

Solomon S. C. and Head J. W. (1980) Lunar mascon basin: Lava filling, tectonics, and evolution of the lithosphere. *Rev. Geophys.*, *18*, 107–141.

Sotin C., Tobie G., Wahr J., and McKinnon W. B. (2009) Tides and tidal heating on Europa. In *Europa* (R. T. Pappalardo et al., eds.), pp. 85–117. Univ. of Arizona, Tucson.

Spencer J. R., Stern S. A., Moore J. M., et al. (2020) The geology and geophysics of Kuiper belt object (486958) Arrokoth. *Science*, *367*, eaay3999, DOI: 10.1126/science.aay3999.

Squyres S. W., Reynolds R. T., Summers A. L., and Shung F. (1988) Accretional heating of the satellites of Saturn and Uranus. *J. Geophys. Res.*, *93*, 8779–8794.

Stern S. A. (1986) The effects of mechanical interaction between the interstellar medium and comets. *Icarus*, *68*, 276–283.

Stern S. A. (1991) On the number of planets in the solar system: Evidence of a substantial population of 1000-km bodies. *Icarus*, *90*, 271–281.

Stern S. A. and Shull J. M. (1988) The influence of supernovae and passing stars on comets in the Oort cloud. *Nature*, *332*, 407–411.

Stern S. A. and Tholen D. J., eds. (1997) *Pluto and Charon*. Univ. of Arizona, Tucson. 728 pp.

Stern S. A. and Trafton L. M. (2008) On the atmospheres of objects in the Kuiper belt. In *The Solar System Beyond Neptune* (M. A. Barucci et al., eds.), pp. 365–380. Univ. of Arizona, Tucson.

Stern S. A., McKinnon W. B., and Lunine J. I. (1997) On the origin of Pluto, Charon, and the Pluto-Charon binary. In *Pluto and Charon* (S. A. Stern and D. J. Tholen, eds.), pp. 605–663. Univ. of Arizona, Tucson.

Stern S. A., Weaver H. A., Steffl A. J., et al. (2006) A giant impact origin for Pluto's small moons and satellite multiplicity in the Kuiper belt. *Nature, 439*, 946–948.

Stern S. A., Bagenal F., Ennico K., et al. (2015) The Pluto system: Initial results from its exploration by New Horizons. *Science*, *350*, aad1815, DOI: 10.1126/science.aad1815.

Stern S. A., Binzel R. P., Earle A. M., et al. (2017) Past epochs of significantly higher pressure atmospheres on Pluto. *Icarus*, *287*, 47–53.

Stern S. A., Grundy W. M., McKinnon W. B., Weaver H. A., and

Young L. A. (2018) The Pluto system after New Horizons. *Annu. Rev. Astron. Astrophys.*, 56, 357–392, DOI: 0.1146/annurevastro-081817-051935.
Stern S. A., et al. (2019) Initial results from the New Horizons exploration of 2014 MU_{69}, a small Kuiper belt object. *Science, 364,* eaaw9771, DOI: 10.1126/science.aaw9771.
Stern S. A., White O. L., McGovern P. J., et al. (2020) Pluto's far side. *Icarus, 356,* 113805, DOI: 10.1016/j.icarus.2020.113805.
Stevenson D. J., Harris A. W., and Lunine J. I. (1986) Origins of satellites. In *Satellites* (J. A. Burns and M. S. Matthews, eds.), pp. 39–88. Univ. of Arizona, Tucson.
Sussman G. J. and Wisdom J. (1988) Numerical evidence that the motion of Pluto is chaotic. *Science, 241,* 433–437, DOI: 10.1126/science.242.4684.433.
Tegler S. C., Grundy W. M., Vilas F., Romanishin W., Cornelison D. M., and Consolmagno G. J. (2008) Evidence of N_2-ice on the surface of the icy dwarf planet 136472 (2005 FY9). *Icarus, 195,* 844–850.
Tegler S. C., Cornelison D. M., Grundy W. M., et al. (2010) Methane and nitrogen abundances on Pluto and Eris. *Astrophys. J., 725,* 1296–1305.
Telfer M. W. and 19 colleagues (2018) Dunes on Pluto. *Science, 360,* 992–997.
Trujillo C., Trilling D., Gerdes D., et al. (2019) Deep Ecliptic Exploration Project (DEEP) observing strategy. *EPSC Abstracts, 13,* EPSC2019-2070.
Tsiganis K., Gomes R., Morbidelli A., and Levison H. F. (2005) Origin of the orbital architecture of the giant planets of the solar system. *Nature, 435,* 459–461.
Umurhan O. M., Howard A. D., Moore J. M., et al. (2017) Modeling glacial flow on and onto Pluto's Sputnik Planitia. *Icarus, 287,* 301–319.
Vilella K. and Deschamps F. (2017) Thermal convection as a possible mechanism for the origin of polygonal structures on Pluto's surface. *J. Geophys. Res., 122,* 1056–1076.
Volk K. and Malhotra R. (2012) The effect of orbital evolution on the Haumea (2003 EL_{61}) collisional family. *Icarus, 221,* 106–115, DOI: 10.1016/j.icarus.2012.06.047.
Walsh K. J. and Levison H. F. (2015) Formation and evolution of Pluto's small satellites. *Astron. J., 150,* 11.
Wang H., Weiss B. P., Bai X.-N., Downey B. G., Wang J., Wang J., Suavet C., Fu R. R., and Zucolotto M. E. (2017) Lifetime of the solar nebula constrained by meteorite paleomagnetism. *Science, 355,* 623–627, DOI: 10.1126/science.aaf5043.
Weaver H. A., Buie M. W., Buratti B. J., et al. (2016) The small satellites of Pluto as observed by New Horizons. *Science, 351,* aae0030.
White O. L., Moore J. M., McKinnon W. B., et al. (2017) Geological mapping of Sputnik Planitia on Pluto. *Icarus, 287,* 261–286.
White O. L., Moore J. M., Howard A. D., et al. (2019) Washboard and fluted terrains on Pluto as evidence for ancient glaciation. *Nature Astron., 3,* 62–68.
Williams J. P. and Cieza L. A. (2011) Protoplanetary disks and their evolution. *Ann. Rev. Astron. Astrophys.*, 49, 67–117, DOI: 10.1146/annurev-astro-081710-102548.
Wong T., Hansen U., Weisehöfer T., and McKinnon W. (2019) Formation of cellular structures on Sputnik Planitia from convection. Abstract P42C-07 presented at 2019 Fall Meeting, AGU, San Francisco, California, 9–13 December.
Yabuta H., Williams L. B., Cody G. D., Alexander C. M. O'D., and Pizzarello S. (2007) The insoluble carbonaceous material of CM chondrites: A possible source of discrete organic compounds under hydrothermal conditions. *Meteoritics & Planet. Sci., 42,* 37–48.
Youdin A. N. and Goodman J. (2005) Streaming instabilities in protoplanetary disks. *Astrophys. J., 620,* 459–469, DOI: 10.1086/426895.
Young L. A., Kammer J. A., Steffl A. J., et al. (2018) Structure and composition of Pluto's atmosphere from the New Horizons solar ultraviolet occultation. *Icarus, 300,* 174–199, DOI: 10.1016/j.icarus.2017.09.006.
Zahnle K. J. and Walker J. C. G. (1982) The evolution of solar ultraviolet luminosity. *Rev. Geophys., 20,* 280–292, DOI: 10.1029/RG020i002p00280.
Zahnle K. J., Korycansky D. G., and Nixon C. A. (2014) Transient climate effects of large impacts on Titan. *Icarus, 229,* 378–391, DOI: 10.1016/j.icarus.2013.11.006.

Parker A. H. (2021) Transneptunian space and the post-Pluto paradigm. In *The Pluto System After New Horizons* (S. A. Stern, J. M. Moore, W. M. Grundy, L. A. Young, and R. P. Binzel, eds.), pp. 545–568. Univ. of Arizona, Tucson, DOI: 10.2458/azu_uapress_9780816540945-ch023.

Transneptunian Space and the Post-Pluto Paradigm

Alex H. Parker
Southwest Research Institute

The Pluto system is an archetype for the multitude of icy dwarf planets and accompanying satellite systems that populate the vast volume of the solar system beyond Neptune. New Horizons' exploration of Pluto and its five moons gave us a glimpse into the range of properties that their kin may host. Furthermore, the surfaces of Pluto and Charon record eons of bombardment by small transneptunian objects, and by treating them as witness plates we can infer a few key properties of the transneptunian population at sizes far below current direct-detection limits. This chapter summarizes what we have learned from the Pluto system about the origins and properties of the transneptunian populations, the processes that have acted upon those members over the age of the solar system, and the processes likely to remain active today. Included in this summary is an inference of the properties of the size distribution of small transneptunian objects and estimates on the fraction of binary systems present at small sizes. Further, this chapter compares the extant properties of the satellites of transneptunian dwarf planets and their implications for the processes of satellite formation and the early evolution of planetesimals in the outer solar system. Finally, this chapter concludes with a discussion of near-term theoretical, observational, and laboratory efforts that can further ground our understanding of the Pluto system and how its properties can guide future exploration of transneptunian space.

1. INTRODUCTION

Eighty-five years of thought regarding the potential properties of Pluto preceded our first exploration of it and its satellites with the 2015 New Horizons flyby. In that time, theories draped around what could be observed remotely produced a long series of predictions that were tested at flyby. Many held up, while others did not.

Fewer than 20 years have elapsed since the first discoveries of worlds beyond Neptune that can rightfully be called Pluto's kin. Our understanding of these worlds — Eris, Makemake, Haumea, and others — started from an advantaged position, launching as it did from the existing understanding that had developed regarding the Pluto system. However, a twist of fate has placed these three most similar transneptunian dwarf planets at or near their aphelia in the current epoch. Thus, these worlds are universally dimmer and more remote than Pluto, and in the time we have had to consider them, we have struggled to build as compelling a body of observation and theory as existed for Pluto prior to flyby. Their individual uniquenesses are notable but poorly understood.

As new facilities and instruments are developed, this will all change. The near future will deliver a multitude of opportunities to build an observational understanding of the surfaces of the largest transneptunian objects (TNOs) and their environments that rivals or exceeds that which we had available at the time of the Pluto flyby. Perhaps the greatest advance, however, will come from taking stock of what we now know to be possible about these worlds thanks to the vast wealth of *in situ* information delivered by New Horizons' exploration of the Pluto system. The largest TNOs will all have properties that make them unique from one another, but Pluto will long serve as a proving ground for testing new ideas about the general properties of these distant, complex worlds and the processes that shape them.

To understand both the promise and limits of the Pluto system as an archetype of the distant dwarf planets that fill the void beyond Neptune, it is first important to understand its relationships to the numerous populations that call this region home; including their respective origins and evolution and the extent of their present interactions.

1.1. The Structure of Transneptunian Space

Transneptunian space is occupied by a host of subpopulations of small bodies, distinguished by both dynamical and physical properties. These subpopulations and their dynamical properties have characterized by a relatively small number of largely groundbased, wide-field observational programs conducted since the late 1990s. The Deep Ecliptic Survey (DES) (*Millis et al.*, 2002; *Buie et al.*, 2003;

Elliot et al., 2005; *Gulbis et al.,* 2010; *Adams et al.,* 2014) provided the first broad-brush well-characterized look at the dynamical divisions in transneptunian populations. The Canada-France Ecliptic Plane Survey (CFEPS) used the 3.5-m Canada-France-Hawaii Telescope and its large-format MegaCam imager to conduct an extremely well-characterized deep and wide survey, leading to many of the current best-estimate measurements of the intrinsic orbital distributions of TNOs (*Jones et al.,* 2006; *Kavelaars et al.,* 2009; *Petit et al.,* 2011, 2017; *Gladman et al.,* 2012) and illustrating the necessity of careful survey design to account for discovery and tracking biases. The CFEPS effort led to a deeper follow-on program — the Outer Solar System Origins Survey (OSSOS) (*Bannister et al.,* 2016a,b; *Shankman et al.,* 2016; *Volk et al.,* 2016; *Lawler et al.,* 2019) — which was conducted on the same facility and for which analysis is still in progress. Ongoing efforts by several teams have begun to map the orbital distribution of the extremely distant Sedna-like population, which may hint at a large unseen perturber lurking at several hundred astronomical units (AU) from the Sun (e.g., *Trujillo and Sheppard,* 2014; *Batygin and Brown,* 2016; *Sheppard et al.,* 2019), although independent datasets do not all show evidence for the orbital distribution features upon which this hypothesis rests (e.g., *Shankman et al.,* 2017). Together, these surveys have provided our best measurements of the intrinsic orbit and size distributions of the majority of TNO subpopulations.

Broadly, TNOs divide into subpopulations as follows. The scattered disk objects (SDOs) have perihelia near Neptune but bear no resonant protection against destabilizing close encounters. The resonant Kuiper belt populations reside in numerous mean-motion resonances (MMRs) with Neptune, including the highly populated 3:2, 5:2, and 2:1 resonances as well as many others to a lesser degree. The classical Kuiper belt populations are in stable non-resonant orbits far from Neptune. The classical Kuiper belt populations further divide into "hot" and "cold" subpopulations, where hot classical Kuiper belt objects (HCKBOs) have relatively excited inclinations and eccentricities, and the cold classical KBOs (CCKBOs) have very low-inclination, nearly circular orbits that dominantly reside between 42 and 47 AU. The CCKBO orbital distribution appears to have multiple components, with "stirred" and "kernel" subpopulations called out in some analyses (e.g., *Petit et al.,* 2011). Finally, there is the relatively recently recognized "detached" population (e.g., *Trujillo and Sheppard,* 2014; *Sheppard et al.,* 2019), objects in which have orbits that appear similar to SDOs except for their very high perihelia, which puts them well beyond the influence of any of the known giant planets. This population is exemplified by Sedna, and its origins and relationship with other populations remains a topic of debate. See *Gladman et al.* (2008) for a discussion of nomenclature.

The TNO luminosity function was measured in parallel with these efforts using deep "pencil-beam" surveys (*Gladman et al.,* 1998, 2001; *Bernstein et al.,* 2004; *Fuentes and Holman,* 2008; *Fuentes et al.,* 2009; *Fraser and Kavelaars,* 2009; *Fraser et al.,* 2010, 2014). These efforts traded dynamical resolution (i.e., short observational arcs vs. long observational arcs) for survey depth, and produced estimates of the TNO luminosity function for "hot" and "cold" populations, defined by an inclination threshold usually around 5°. These surveys reached limiting absolute magnitudes approaching $H_{r'}$ ~12 mag, or roughly D ~17 km for geometric albedos of 10%. Broadly, the luminosity function of any given subpopulation is reasonably well characterized by a broken power-law of the form

$$\frac{dN}{dH} = \begin{cases} 10^{\alpha_1 (H - H_0)} & H < H_B \\ 10^{\alpha_2 (H - H_0) + (\alpha_1 - \alpha_2)(H_B - H_0)} & H \geq H_0 \end{cases} \quad (1)$$

where α_1 is the power-law slope valid for bright objects, α_2 is the power-law slope valid for faint objects, H_B is the absolute magnitude at which the slope transitions from α_1 to α_2, and H_0 is a normalization factor. The segments of this luminosity function each translate to a size-frequency distribution of the form $dN/dR \propto R^{-q}$ with a differential slope of $q = 5 + \alpha_1$ where α is the local luminosity function slope. The complete form for a multiply-broken power law size-frequency distribution is further defined in section 2. Canonically, a population in collisional equilibrium will reach a slope of $q = 3.5$. TNOs with diameters larger than approximately 100 km have slopes much steeper than collisional equilibrium, while objects smaller than this size have slopes consistent or slightly shallower than collisional equilibrium (*Bernstein et al.,* 2004; *Petit et al.,* 2011; *Fraser et al.,* 2014).

Finally, the largest TNOs were discovered in wide-field, relatively shallow surveys. These include the discoveries of Eris, Makemake, Haumea, Quaoar, Orcus, Sedna, and Gonggong (the recently applied formal name for 2007 OR_{10}) at Palomar Observatory between 2002 and 2007 (*Brown,* 2008). These largest TNOs are present in every TNO subpopulation *except* the low-inclination CCKBOs, where the size distribution truncates at an upper limit of a few hundred kilometers.

1.2. Pluto's Dynamical History and Relationship to Transneptunian Subpopulations

The Pluto system orbits about the solar system barycenter in a multi-resonant configuration with Neptune, including the 3:2 MMR (*Cohen et al.,* 1967) and a Kozai secular resonance [with the argument of perihelion librating around 90° (*Milani et al.,* 1989)]. Currently, its mean inclination with respect to the solar system invariable plane is 15.55°, near the peak of the inclination distributions of the other 3:2 resonators and other excited transneptunian populations such as the Neptune Trojans, HCKBOs, and scattered disk objects (SDOs). Because of the protection from Neptune close encounters conferred by its resonant configuration, the Pluto system remains stable even with its relatively extreme perihelion as compared to the majority of other transneptunian populations, passing interior to Neptune at q = 29.66 AU.

Under all currently debated scenarios, the Pluto system formed substantially closer to the Sun than its current heliocentric distance. The first modern explanation of its orbital configuration was by being swept up in resonance by an outward-migrating Neptune (*Malhotra,* 1993, 1995). This "bottom up" capture would have collected a low-excitation Pluto in the 3:2 resonance and carried it outward while exciting its orbit in the process. Later work showed that in order to achieve both the current inclination and eccentricity of Pluto's orbit, multiple epochs of resonant interactions would be required, including an early epoch of inclination excitation in the ν_{18} secular resonance [where the precession rate of a TNO's longitude of ascending node matches the precession rate of Neptune's longitude of ascending node (see, e.g., *Morbidelli et al.,* 1995)], followed by capture, transport, and eccentricity excitation by the 3:2 MMR, followed by further inclination excitation in the Kozai resonance (*Malhotra,* 1998). This sequence of events would require Neptune to migrate by 5–10 AU in total (*Hahn and Malhotra,* 1999), with Pluto capturing into resonance outside 28 AU. Considering the full current resonant properties of Pluto, *Gomes* (2000) argued that these processes would generate Pluto-like objects given origin locations near 30.5 AU. However, the subsequent discovery of a very broad inclination distribution of more distant non-resonant "classical" KBOs presented a challenge: These inclinations could not have been excited by resonance sweeping alone, and required another explanation. A solution was proposed by *Gomes* (2003), where objects are initially excited and launched outward to current KBO distances by scattering encounters with Neptune, fueling its migration. These objects can experience transient phases of lower eccentricity due to resonant interactions, raising their perihelia away from Neptune, and then fall out of resonance as Neptune continues to migrate, effectively stranding them in stable high-inclination, relatively high-eccentricity orbits. In this scenario, resonances are populated in a "top-down" way, where objects are generated with high inclination and eccentricity through scattering events and are later permanently captured into 3:2 resonance. This removes the need for the most extreme extents of smooth migration by Neptune, as the inclination and eccentricity of Pluto and its resonant neighbors are not due solely to excitation through resonant sweeping. This scenario was further explored and refined in *Levison and Morbidelli* (2003) and *Levison et al.* (2008), which showed that even low-excitation orbits can be established by a history of scattering and transient resonant diffusion. These families of models (e.g., *Gomes,* 2003, *Levison et al.,* 2008) argued that Pluto and the rest of the TNO population were all moved outward after forming in a massive (10–50 $M_\oplus$) disk of planetesimals that formed substantially closer to the Sun, truncated somewhere between 30 and 35 AU.

However, more recent work has demonstrated that Pluto does not share a common origin location with all extant TNOs. The low-inclination CCKBOs were shown to be very numerous and to have a unique size-frequency distribution (*Brown,* 2001; *Levison and Stern,* 2001; *Bernstein et al.,* 2004; *Fraser et al.,* 2010, 2014; *Petit et al.,* 2011), a unique color distribution (*Trujillo and Brown,* 2002) and albedo distribution (*Bruker et al.,* 2009), and a unique population of widely separated binary systems (*Noll et al.,* 2008). All these properties made the CCKBOs physically distinct from other TNO populations. Furthermore, the wide binary systems are very sensitive to disruption by any epoch of Neptune scattering that would have implanted them in their current orbits (*Parker and Kavelaars,* 2010) and to collisional evolution as may have been expected during an early epoch inside a massive disk (*Nesvorný et al.,* 2011; *Parker and Kavelaars,* 2012). These factors together point to the CCKBOs having an origin distinct from Pluto and the rest of the TNO populations, likely forming *in situ* in a low-surface-density extension of the primordial planetesimal disk and suffering very little orbital excitation or collisional comminution over its history (e.g., *Cuzzi et al.,* 2010; *Batygin et al.,* 2011). Indeed, there is an emerging consensus that the physical and dynamical properties of CCKBOs are consistent with having emerged directly and very rapidly from aerodynamically enhanced collapse of pebble swarms (e.g., *Nesvorný et al.,* 2010, 2019) with little subsequent modification.

Other than the CCKBO subpopulation and a proportion of the resonant populations that were "swept up" from it, Pluto likely shares a common origin with most other identified TNO subpopulations. A caveat here is the relatively recently identified "distant detached" population exemplified by Sedna. The origin of these objects remains contested, and while a number of scenarios exist where they emerged from the same planetesimal disk as Pluto (*Brown et al.,* 2004, *Morbidelli and Levison,* 2004), others exist where they have unique source populations (*Jilkova et al.,* 2015). Regardless, they are likely to have experienced substantially different orbital evolution over their lifetimes and their properties (including their propensity to host satellite systems; Sedna is the largest-known TNO with no known satellites) may differ from the other TNO populations as a result.

The most up-to-date modeling efforts indicate that Neptune migrated relatively slowly into a 20-$M_\oplus$ disk that extended out to 30 AU and contained several thousand (or 2–8-$M_\oplus$-worth) Pluto-sized planetesimals (*Stern,* 1991; *Nesvorný and Vokrouhlicky,* 2016). This large initial population of Pluto-sized objects is required to introduce a "grainy," discontinuous aspect to Neptune's migration to curtail the efficiency of direct resonant capture and increase the efficiency of resonant drop-off in the hot classical population, which otherwise would result in resonant populations far more numerous than observed and hot classical populations more anemic than obervered (*Petit et al.,* 2011; *Gladman et al.,* 2012; *Nesvorný,* 2015). While the initial population of proto-Plutos was very large, relatively few of them survived transport into the extant TNO populations.

As Pluto's osculating orbital elements evolve cyclically due to the resonances it occupies, its intrinsic collisional coupling to each TNO subpopulation changes. Long-term average collision rates were compiled by *Greenstreet et al.* (2015) for each of the most-populous well-characterized

TNO subpopulations, including the hot and CCKBOs, several resonant KBO populations, and the scattered disk. For the population of potential impactors larger than d $\simeq$ 100 km (the approximate location of the size-frequency distribution's break to a shallower slope), the classical belt contributes a majority of all the impactors to the Pluto system (74.7%), while the 3:2 resonant population comprises the next largest supplier of impactors (18.6%). The typical impact speeds for objects in each population varies, and the amalgam velocity distribution weighted by the impact odds and population sizes is illustrated in Fig. 1.

2. BUGS ON THE WINDSHIELD

One of the most direct ways that the Pluto system can provide insight into the properties of transneptunian populations is by treating the surfaces in the system as witness plates. Most of these surfaces represent a record of the integrated history of bombardment of the Pluto system by the TNO subpopulations that it physically passes through. The zero-order property that can be extracted from this record is the sheer number of impacts recorded down to the limit of any given set of observations, while a first-order property is how the sizes of those craters are distributed. Higher-order properties, such as spatial correlation with geological units, two-point autocorrelation functions, and others, can tease out relative and absolute ages of surface units and whether or not a subset of the impactors bore satellites.

2.1. Inferring the Size-Frequency Distribution of Small Transneptunian Objects

The size-frequency distribution (SFD) of TNOs is a powerful tracer of the fundamental processes of planetesimal assembly before and during the era of planet formation, as well of any subsequent collisional evolution over the age of the solar system. A variety of means have been used to estimate the SFD properties of TNO subpopulations, including so-called "direct" observation in reflected disk-integrated light, occultations of background stars, and inferring the impactor populations responsible for generating craters on the surfaces of outer solar system worlds. It is worth briefly considering the physical properties that each of these methods are actually sensitive to, and the strengths and shortcomings of each.

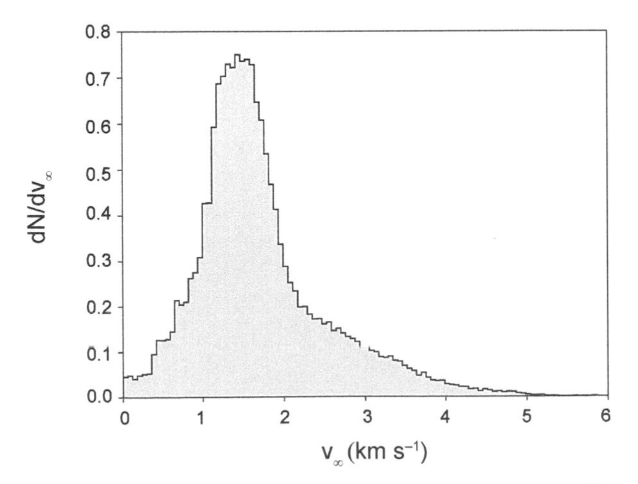

Fig. 1. Adopted v_∞ distribution for impactors into the Pluto system, using a population-weighted average over all populations considered in *Greenstreet et al.* (2015).

First, it is useful to identify those properties of TNOs that are most likely to be directly influenced by their formation processes vs. those that are merely accessory to them. Most fundamentally, growth processes partition *mass* between forming bodies. Disk-integrated luminosity of a TNO is the result of a combination of factors including its mass, density, shape, orientation, and surface albedo. Occultation cross section is the result of mass, density, shape, and orientation. In the point-mass approximation and in the gravity regime, craters typically scale with mass of the impactor, velocity of the impactor, density of the impactor, and density of the target.

With no substantial sensitivity to albedo, shape, or orientation distributions, the size-frequency distribution of impact craters provides a powerful and complementary measure of this most fundamental mass distribution. However, the inferences that can be drawn from impact crater populations are confounded by a distinct set of factors, including uncertainties in crater-scaling laws, the properties of the target bodies, the generation of secondary craters, modification of crater populations by geologic processes, and limitations in the calibratability of current-generation crater surveys. While these factors must be accounted for, assays of crater populations remain one of the most powerful probes of the fundamental partitioning of mass between small bodies in transneptunian space, and will likely remain so until serendipitous stellar occultation surveys increase their yields by several orders of magnitude. In the end, a *consensus* understanding of the TNO populations and their evolution through time should be reached across all methods for studying them. If, for example, stellar occultations surveys and impact crater surveys produce persistently discrepant results, we are potentially missing something fundamental, and the tension between methods may be highlighting the path to a deeper understanding of the fundamental properties of outer solar system populations.

Analysis of the size-frequency distribution of craters in the Pluto system suggests that small TNOs are less numerous than would be anticipated if they had ever reached collisional equilibrium (*Singer et al.,* 2019, hereafter referred to as *S19*). If the size-frequency distribution of impactors is a scale-free power law $dN/dD_i \propto D_i^{-q_i}$, then the size-frequency distribution of simple craters in the gravity regime will also be a scale-free power law over a large range of sizes, $dN/dD_c \propto D_c^{-q_c}$. Given the Housen, Holsapple, and Schmidt scaling factor μ (e.g., *Holsapple,*

1993), we can determine the slope of the impactor distribution implied by the crater distribution

$$q_i = \left(1 - \frac{\mu}{2+\mu}\right) q_c + \frac{\mu}{2+\mu} \quad (2)$$

For typical experimentally derived values of μ between 0.41 and 0.55 (*Holsapple,* 1993), this implies an impactor size-frequency distribution slightly *shallower* than the measured crater size-frequency distribution. The physical extremes for the μ parameter are 1/3 and 2/3, and correspond to pure energy conservation and pure momentum conservation. Both extremes result in impactor SFD slopes shallower than crater SFD slopes. Thus, absent any completeness effects or other non-idealities, the q_c ~1.8 slope for craters smaller than 10 km on Charon's Vulcan Planum measured by *S19* implies an impactor slope $1.6 \lesssim q_i \lesssim 1.7$ at the extreme limits of μ for impactors in the size range 0.1 km $\lesssim d_i \lesssim$ 1 km.

However, the Pluto system's crater size-frequency distribution suggests that over the range of crater diameters recorded, the impactor size-frequency distribution is *not* a scale-free power law. There is an apparent break in the crater size-frequency distribution at a diameter of d_c ~10–15 km. To fully describe the state of knowledge of the properties of the small-TNO size-frequency distribution as informed by the crater population on Pluto and Charon, and to further fold in existing knowledge of the size-frequency distribution at larger sizes from direct-detection surveys, a more sophisticated analysis is required that permits marginalization over nuisance parameters like μ and any parameters that encode sensitivity and completeness of the crater census. The following outlines such an approach and applies it to existing crater catalogs to estimate the current state of knowledge regarding the size-frequency distribution of small TNOs.

2.1.1. Likelihood-free inference. Cratering is a complex process, and to use a measured distribution of crater properties from one or several surface units to infer the intrinsic properties about the population of projectiles that generated them requires a great deal of statistical care. The complexities of the cratering process and the physical task of conducting a census of a crater population are difficult to account for in standard likelihood-based inference. However, relatively recently there has been a surge of interest in and development of *likelihood-free* Bayesian inference. While many of these methods have their origins in the genomics community, their application in astrophysics (*Ishida et al.,* 2015; *Hahn et al.,* 2017; *Hsu et al.,* 2018; *Witzel et al.,* 2018; *Sandford et al.,* 2019) and planetary science (*Parker,* 2015; *Mazrouei et al.,* 2019) has been growing in recent years.

The chief advantage of likelihood-free inference for planetary science is that it enables existing forward-modeling infrastructure to be applied within a Bayesian framework. More concretely, likelihood-free inference enables rigorous Bayesian inference using stochastic generative models, and it can be far more straightforward to encode all the complexities of planetary processes and observations within a stochastic model than it is to write down the formal likelihood integral that captures the interplay of these processes, let alone actually *evaluate* that integral.

The following section outlines a likelihood-free framework for Bayesian inference of the parameters of several models of the population of small TNO projectiles that have been impacting the Pluto system over the age of the solar system. This framework and approach is generic and can be applied to other planetary surfaces or to other planetary science population synthesis challenges.

The primary likelihood-free inference tool used here is an approximate Bayesian computation rejection sampler (ABCr) (*Pritchard et al.,* 1999; see *Marin et al.,* 2011, for methods review), one of the simplest likelihood-free inference methods to implement and understand. ABCr replaces likelihoods with a distance metric D_{ABCr} measured on a set of summary statistics extracted from both the real observed data sample and a synthetic "observed" sample. In a single ABCr trial, a set of parameters describing the model are drawn from their prior distributions, and a single synthetic "observed" sample is drawn from the model with those selected parameters. The distance metric is computed for that trial and recorded along with the parameters that defined that instance of the model. After many trials, the sets of parameters that happened to produce the smallest distances between synthetic and observed data are kept, and the rest rejected (typically, a 0.1% threshold is used). With a carefully chosen distance metric, a small enough threshold for retention, and a sufficiently large number of trials, the retained set of model parameters will approximate their posterior distribution given the observed data. This work adopts a distance metric that is the sum of Kolmogorov-Smirnov D_{KS} statistics over the observables in question. This is similar to the treatment in *Parker* (2015) for Neptune Trojan orbit distribution parameter estimation and *Mazrouei et al.* (2019) for estimating the parameters of a time-varying impact rate model for the Earth and Moon.

Furthermore, ABCr can be used to compute a Bayes factor for one model out of a suite of models. It proceeds as before, but in any given trial, first a model is selected at a rate commensurate with its prior probability, then the parameters for that model are proposed. After completing many trials and retaining the best 0.1%, ratio of the number of retained trials of each model to the number of retained trials of other models in the final sample is the Bayes factor in favor of that model. That is, if 95 instances out of 100 retained trials were Model A, and 5 were Model B, the Bayes factor is 19:1 in favor of Model A.

2.1.2. Measuring the population of craters. Unlike direct-detection surveys of the TNO population, the surfaces on which crater censuses are conducted are often not available for revisiting with new or different observational strategies once an initial census is completed. The Pluto system has been visited by a single spacecraft, which is likely to be the only time the system is visited for at minimum a decade.

The observational datasets collected during flyby — not informed by any information about any evidence for novel properties of the TNO population that may be encoded in the system's surface units — are the only dataset from which we are able infer any properties about the crater populations on Pluto and its moons. It is of particular importance to avoid overfitting the data at our disposal because it will be impossible to test with an independent dataset for a very long time.

Furthermore, craters are imprinted on surfaces with substantial albedo and topographic variation, and are themselves subject to endogenic modification processes that may vary from unit to unit on a given surface as well as from world to world. Calibrating the detection efficiency of any crater census is vastly more difficult than calibrating a direct detection survey, which can generally be thought of as a collection of stationary and moving point sources on a relatively smooth and featureless background that varies in a predictable way, with overall sensitivities governed by photon and detector noise. Algorithmic and human limitations can be tested for by injecting simulated targets into the same dataset under consideration and estimating the fraction of those synthetic targets recovered by the analysis as a function of any parameter of interest (speed of motion, luminosity, color, and so forth). Even so, pushing into new regimes of luminosity has resulted in some substantial mischaracterizations of the population of small TNOs in the past (e.g., *Cochran et al.,* 1995, 1998; *Brown et al.,* 1997; *Gladman et al.,* 1998) due to unaccounted-for noise sources in the datasets at hand. These noise sources are generally identified in subsequent analysis of the original dataset, and new datasets are gathered to confirm these analyses.

In some cases, there have been large direct-detection datasets for which no complete discovery bias calibration was performed at the time of discovery. *Parker* (2015) developed a *survey-agnostic* method of calibrating a set of these datasets post facto, where parameters of a model of discovery biases were marginalized over to uncover what information was still available in the data themselves about the population under consideration, given our uncertainty in these discovery biases. With some adjustment, these methods can be applied to crater populations to better assess our true state of knowledge about the population of projectiles that generated them.

The majority of crater censuses are conducted via analysis by the human eye. As with any process that involves humans in the loop, there is a degree of subjectivity informed by experience with past crater-counting efforts present in any crater identification. Ideally, this subjectivity would be controlled for through calibration, perhaps by generating synthetic datasets with the same topographic and albedo variations as the target under consideration, injecting simulated crater populations into them, and determining the properties of the recovered sample as a function of both the injected population properties and the human researcher conducting the analysis. At present, this level of calibration is largely impractical. While such *absolute* calibration may be absent, the variation *between* human researchers has been tested (*Robbins et al.,* 2014) for a *fixed set of crater population parameters and terrain properties*. Thus, it is possible to estimate how different the estimated parameters for a given crater population might be as determined by any two human researchers, but *not* always possible to estimate how much they both err from the underlying truth.

There are two crater catalogs available for the Pluto system, both drawn from New Horizons imaging datasets: the amalgam catalogs from *Robbins et al.* (2017) (hereafter *R17*), which combines all available datasets over the entire imaged areas of Pluto and Charon, and the unit-by-unit, image-by-image catalogs of *S19*. The two datasets are not fully independent as they are drawn from the same data and have contributions from common researchers, but they do have properties unique from one another that makes them each valuable for separate analyses. The *R17* catalogs cover the entire imaged portions of both Pluto and Charon, while *S19* focuses on a smaller subset of Pluto and Charon for which high-resolution information is available; *R17* provides a unique identifier each crater, while *S19* does not indicate duplicate craters if they appear in multiple imaging datasets; *R17* amalgamates inputs from multiple human researchers and includes an estimate of the average subjective confidence level reported by those researchers without further thresholding, while *S19* represents the best-effort dataset produced by a single researcher that pass a fixed threshold confidence level; and *S19* reports crater populations resolved into different image subsets and terrain units, while *R17* reports a single disambiguiated population of craters per body without further distinction based upon terrain or image source. As an additional refinement step, *S19* used topographic data to confirm features where such topography was available. The following analyses use the *R17* dataset for parameter estimation and the *S19* dataset for validation.

To ensure a high-quality sample of craters from well-observed regions on both Pluto and Charon, only those craters from the *R17* catalog with a reported confidence level of three or greater (one a one-to-five subjective scale, with three corresponding to a better than 50/50 odds that a feature is in fact a crater; see *R17* for further details regarding this catalog) in the original catalog are selected, and from that sample only those which are in the upper 50% of regional crater density (as defined by the distance to the 100th nearest neighbor of each crater) are considered. This rejects craters from the encounter farside of both bodies and from sparsely cratered regions; because their neighbors cover large fractions of each body, estimating the local detection efficiency from the nearest neighbors as described in the following section would be innacurate and thus they are removed from the sample. This results in a winnowed sample of 1011 craters on Charon and 2016 craters on Pluto for SFD parameter estimation. The final sample extracted from *R17* for subsequent analysis in this work is illustrated in Fig. 2.

2.1.3. Modeling crater creation and detection. Our understanding of the processes by which a projectile impacting a planetary surface generates a crater of given properties has a long history, and is informed both by laboratory experiments and observational inference. A full summary of this history is beyond the scope of this paper; for a review, see *Holsapple* (1993) and for recent considerations regarding impacts into high-porosity surfaces see *Housen et al.* (2018). Briefly, during an impact, the kinetic energy of a projectile is partitioned into heat — which can melt or vaporize target and projectile material — and into the mechanical work of compressing, excavating, and ultimately ejecting material from target surface. A huge range of variable factors influence the properties of the final crater generated by an impact, including the size, speed, and impact angle of the projectile; the bulk material properties of the projectile and of the target surface; and the surface gravity of the target body. The results of general impactor-to-crater scaling laws generally produces a relationship with the final diameter of the crater being proportional to the diameter of the impactor to a power near unity. The constant of proportionality varies depending on all the properties described above, and — importantly — the exponent depends on several of these properties as well.

It is typical to work backward in impactor population synthesis, where the properties of a given crater or population of craters are mapped back to the properties of an impactor population. However, to accomplish this, a great deal of information must be discarded. Any given crater could have been produced by impactors of a variety of sizes, impact speeds, and impact angles — but a given impactor, with its fixed size, speed, and trajectory, can only produce one particular crater. That is to say, impactors make craters, but craters do not make impactors. *Forward*-modeling this process can therefore retain the influence of the distribution of impactor properties *other* than size all the way to the final marginalization step, instead of averaging over them in the crater scaling relation. For simplicity of the current analysis, the cratering model of *Zahnle et al.* (2003) is adopted as a baseline. That is

$$D_s = 13.4(v_i^2/g)^{0.217}(\cos(\theta)\rho_i/\rho_t)^{\frac{1}{3}} d_i^{0.783} \text{ km} \quad (3)$$

for $D_s \leq D_c$, where D_s is the rim-to-rim diameter of a simple crater smaller than the transition to complex craters D_c; to account for craters larger than D_c, the final crater diameter D_f is given by

$$D_f = \begin{cases} D_s & D_s \leq D_c \\ D_s(D_s/D_c)^{0.108} & D_s > D_c \end{cases} \quad (4)$$

The exponent in the second case is slightly lower than that used in *Zahnle et al.* (2003) and is based on *McKinnon and Schenk* (1995). Here v_i is the impact speed in kilometers per second, g is the target body's escape speed in centimeters per square seconds, ρ_i and ρ_t are the bulk densities of the impactor and target materials (respectively) in grams per centimeters cubed, d_i is the impactor diameter in kilometers, and θ is the impact angle from normal. The exponent on d_i implies a μ parameter of 0.55, and this scaling law is appropriate for generating crater diameters in the gravity-dominant regime for non-porous target materials. D_c = 10 km as adopted for

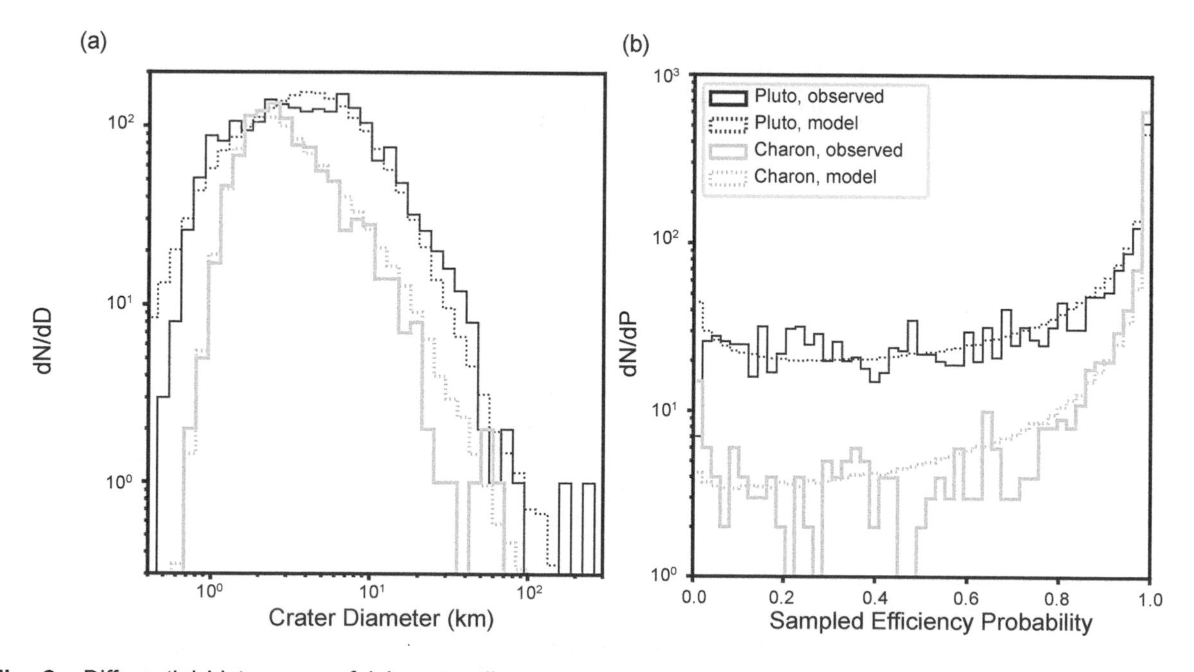

Fig. 2. Differential histograms of **(a)** crater diameters and **(b)** sampled efficiency function (equation (6)) probablilities for observed (solid) and model (dotted) samples from Pluto (black) and Charon (gray). Illustrated model parameters are a_p = 0.7, b_p = –4.5, a_c = 0.6, b_c = –11, q_1 = 8.5, q_2 = 2.9, q_3 = 1.85, and D_b = 10 km.

both Pluto and Charon, based on crater morphology transition (Robbins, 2019, personal communication).

The intrinsic size-frequency distribution of the impactor population is taken to be a piecewise power-law distribution

$$p(d) \propto \begin{cases} \left(\dfrac{d}{d_{b,0}}\right)^{-q_0} & d_{b,0} \leq d < d_{b,1} \\ \left(\dfrac{d}{d_{b,0}}\right)^{-q_i} \times \prod_{j=1}^{i} \left(\dfrac{d_{b,j}}{d_{b,0}}\right)^{q_j - q_{j-1}} & d_{b,i} \leq d < d_{b,i+1} \end{cases} \quad (5)$$

where $d_{b,0}$ is the smallest size to be considered (must be >0), the the rest of the list $d_{b,1}\ldots d_{b,n-1}$ are the diameters of n–2 proposed breaks in the distribution, and d_n is the largest size to consider (or $+\infty$).

To model a synthetic "observed" crater population, first a sample of impactor diameters d_i is drawn from a proposed size-frequency distribution using equation (2). Then a sample of impact speeds at infinity v_{inf} is drawn from the *Greenstreet et al.* (2015) distribution for the Pluto system (Fig. 1), and augmented with the escape speed of the target body to determine the impact speed of each impactor $v_i = \sqrt{v_{inf}^2 + v_{esc}^2}$. Finally, a sample of impact angles θ is drawn from a distribution uniform in $\sin^2(\theta)$ over $0° \leq \theta \leq 90°$. Running these samples through equation (4) results in a proposed sample of craters on the surface of the target, but do not yet include observational sampling effects.

To model observational effects, a functional form for observational completeness η is adopted that is inspired by the rollover functions used in direct-detection surveys. For crater i with observed diameter $D_{o,i}$,

$$\eta\,(D_{o,i} \mid D_{m,i}) = \frac{1}{2}\left(1 - \tanh\left[b \times \log_{10}\left(\frac{D_{o,i}}{aD_{m,i}}\right)\right]\right) \quad (6)$$

where a and b are nuisance scaling parameters that apply to the entire population of a target object and which are marginalized over, and $D_{m,i}$ is the median size of the 100 craters nearest to the crater under consideration. [A nuisance parameter is a model parameter that is not of direct physical interest but that has some potential to impact the probability density function (PDF) of parameters that are of physical interest.] This sample size is somewhat arbitrary, and 100 craters was chosen to balance regional fidelity with good signal-to-noise properties based on the sample sizes in *R17*. Functionally, this model of observational completeness assumes that completeness is regional on Pluto and Charon, and that the observed SFD of craters in a given region encodes both the underlying intrinsic SFD and a multiplicative efficiency function. Every crater on Pluto and Charon is assumed to represent a distinct sample from this selection process. The observable property *recorded* for every crater that encodes this information is the median size D_m of craters in its vicinity, *not counting itself*. Thus, any given crater does not influence the estimation of the detection efficiency function that applies to it. The parameters a and b map the measured D_m values to an efficiency function that applies to the diameter D_o of the crater for which each D_m was measured; a sets the efficiency rollover width, and $b \times D_m$ sets the efficiency rollover diameter. These two nuisance parameters (per target; four total over Pluto and Charon) completely describe the observational completeness effects as modeled, and their impact on the parameters of interest are marginalized over in the ABCr process.

When mapping a sample drawn from equations (2)–(4), this produces a proposed sample of craters that may exist on Pluto. To map this to a proposed synthetic observed sample, one synthetic crater is selected from this sample for each measured D_m value recorded from the observed sample, applying a weight to the proposed craters based on the implied efficiency function for each unique D_m given a set of proposed a and b values.

At this point, all the parts needed to conduct ABCr inference on the model parameters are ready. 10^7 ABCr trials were conducted using a three-slope power law; the smallest-object slope q_3 was drawn from a uniform prior over [0,3], the intermediate-sized q_2 from the posterior PDF of "cold" population faint-object slopes from *Fraser et al.* (2014), and the large-object slope q_1 from the posterior PDF of "cold" population bright-object slopes from *Fraser et al.* (2014). The break between q_3 and q_2 was selected uniform in $\log_{10}(D_b)$, over the range $0.1\ \text{km} \leq D_b \leq 100\ \text{km}$. The values of a and b were proposed separately for Pluto and Charon, each drawn from uniform priors over [0.1, 1.0] and [1, 100], respectively. In each trial, the two-sample KS-test statistics were retained for the proposed Pluto crater sample and the observed Pluto crater sample, the proposed Charon crater sample and the observed Charon crater sample, the proposed efficiency function values for every proposed Pluto crater and for every observed Pluto crater, and the proposed efficiency function values for every proposed Charon crater and for every observed Charon crater. These statistics were summed for each trial to generate the ABCr D_{ABCr} metric. The latter two KS-tests permit the simultaneous assessment of whether the efficiency function model is well-matched to the data (i.e., synthetic craters are not being selected from vastly different parts of the proposed efficiency function than observed craters were) and whether the size-frequency distribution is well-matched to the data — see Fig. 2 for example distributions. After the 10^7 trials were complete, the sample of proposed parameters that resulted in D_{ABCr} values in the lowest 0.1% of all sampled D_{ABCr} values was extracted. This subsample is adopted as the posterior PDF for the parameters under consideration.

2.1.4. Size-frequency distribution parameters. The posterior PDF distributions for q_2 and q_3 are illustrated in Fig. 3, marginalized over the efficiency function nuisance parameters (a_p, b_p, a_c, b_c) and the break location D_b. The mode value of the PDF and marginalized 68% confidence intervals for these parameters are $q_3 = 1.85^{+0.22}_{-0.55}$ and $q_2 = 2.90^{+0.26}_{-0.61}$. These slopes are broadly consistent with those determined for the crater distribution in Charon's Vulcan Planum by *S19*, but the addition here of the observational completeness modeling brings the SFDs expressed by both Pluto's and Charon's surfaces into agreement.

While the data strongly prefer a three-slope model over a two-slope model, the location of the break between q_2 and q_3 is relatively poorly constrained to $D_b = 2.8^{+26}_{-2.2}$ km (mode and 68% interval). This suggests that while a single slope for all TNOs smaller than d ~100 km is certainly a poor representation of the data, a sharply broken two-slope model may also be an inadequate description. If the population instead slowly transitions from one regime to another, then a single break diameter will not reproduce the data well. It is also possible that the assumption of common crater-scaling-relationship parameters for both Pluto and Charon is not supported by the data; if the exponent μ is slightly different for Pluto and Charon, indicating perhaps a difference in surface material properties, any residual tensions between the observed slopes on Pluto and Charon could be relaxed and possibly enable better constraint of D_b. Future work is merited that considers a broader family of SFD functional forms and crater-scaling relationships.

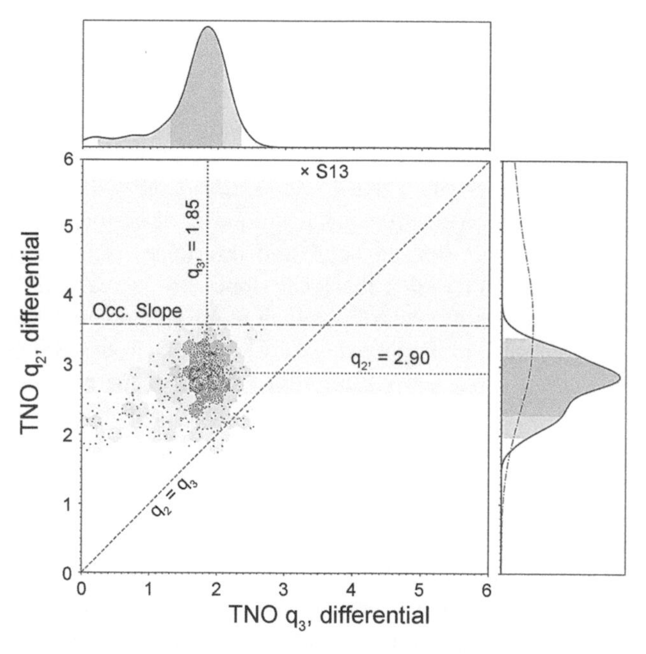

Fig. 3. Marginalized posterior PDFs for TNO size-frequency distribution power-law slopes; q_3 applies to the smallest objects that created craters on Pluto, q_2 applies to the largest objects that created craters on Pluto and is the small-object slope for directly-detected KBOs. q_1 (not shown) is measured for the largest TNOs by direct detection, and is population dependent (see *Fraser et al.,* 2014). Proposed differential slopes for the relevant size regimes from *S13* shown for comparison. Gray bar and dot-dash line show the PDF of mid-size slope inferred from occultation detections alone.

*2.1.5. **Odds of the gaps: Bridging craters, occultations, and direct detections.*** Three serendipitous occultation events have been claimed that have moderate confidence (with estimated false-alarm probabilities of $p \lesssim 0.02$–0.05 per event) and were obtained at cadences better than half the Nyquist rate for TNO occultations (Nyquist ~30 Hz) (*Bickerton et al.,* 2009). Two were single-telescope event detections made with the Hubble Space Telescope (HST) Fine Guidance Sensor (*Schlichting et al.,* 2009, 2013; 2013 is hereafter referred to as *S13*) and one was a dual-telescope event made with small groundbased telescopes (*Arimatsu et al.,* 2019a). These events were observed at low to moderate ecliptic latitudes (+6.6°, +14°, and +8° respectively), and are inferred to represent candidate occultations of background stars by TNOs with diameters of 1.0 km, 1.1 km, and 2.6 km, respectively. From these three detections, extremely large populations of d = 1 km TNOs were inferred, vastly outstripping expectations from naive extrapolation of the directly measured luminosity function.

A theoretical SFD was proposed by *S13* that would account for the inferred very large population of d = 1 km TNOs in the *Schlichting et al.* (2009, 2012; 2012 is hereafter referred to as *S12*) occultation analyses. The relevant slopes are illustrated in Fig. 3 and the implied SFD shape is illustrated in Fig. 4. Most notably, it is not the small-object slope q_3 that is most dissimilar from the population of TNOs inferred from the Pluto and Charon crater populations, but rather the *mid-sized* slope q_2, which corresponds to the very well-sampled regime covered by the largest craters on Pluto and Charon. It is also worth noting that the largest uncontested craters on Pluto and Charon were likely generated by impactors with sizes comparable to the current limit of direct-detection techniques (e.g., the diameter of 486958 Arrokoth). Thus, a very steep slope in this intermediate-sized regime should be immediately obvious in the Pluto and Charon crater data, at sizes for which craters are most difficult to miss observationally and most difficult to erase geologically — from Fig. 4 it appears that 90–99% of craters generated by $0.1 \text{ km} \leq d \leq 4$ km would need to have been erased in a size-independent way. This seems deeply unlikely, and the *S13* model is not supported by the crater populations on Pluto and Charon (*S19*).

However, how discrepant are the actual inferred populations drawn directly from observations? The shadow diameter of a Fresnel regime shadow of a spherical object (defined across the width of the first Airy ring) (*Nihei et al.,* 2007) is $W(d) \simeq \left((2\sqrt{3}F)^{3/2} + d^{3/2}\right)^{2/3}$ where F = 1.3 km is the Fresnel scale for an occulter at 40 AU in optical wavelengths. The size dependence of this shadow is not accounted for in the population inference calculations of *S12*, resulting in an overestimate of the population by a factor of ~2. The revised quantity is illustrated in Fig. 4 along with the estimates from the *Arimatsu et al.* (2019b) occultation. While these inferred TNO populations are larger than those indicated by the crater populations, the large uncertainties in both the crater-inferred population and the much smaller occultation datasets do not immediately rule one another out.

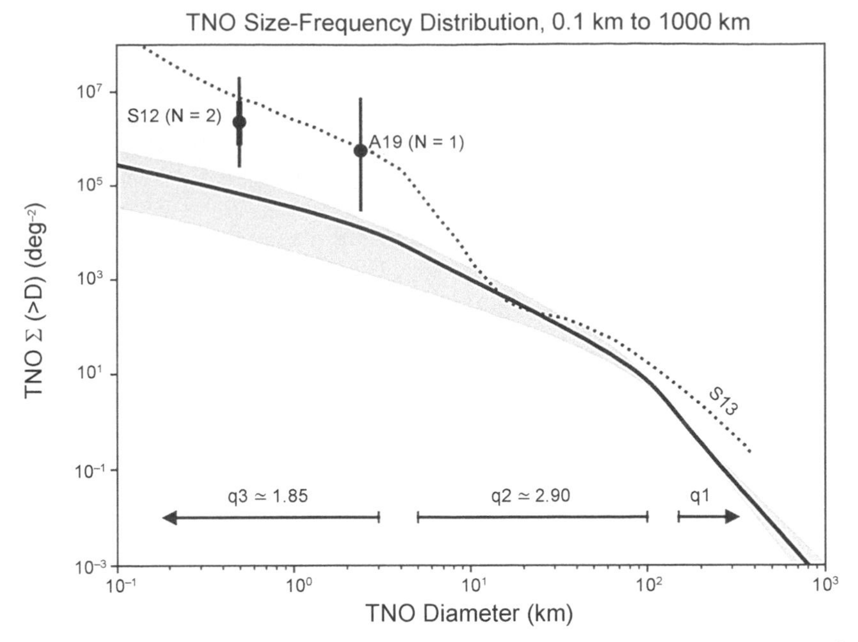

Fig. 4. Cumulative ecliptic surface density of TNOs, combining *Fraser et al.* (2014) luminosity function results and Pluto system crater population synthesis from this work. Confidence intervals of 68% and 95% confidence intervals illustrated by shaded regions. Surface density combines both models for "hot" and "cold" populations from *Fraser et al.* (2014). Nominal differential slopes for mid- and small-size regimes are illustrated; large object slope q_1 is population dependant. Surface densities inferred from occultation detections in *S12* and *Arimatsu et al.* (2019b) are also illustrated; *S12* results have been revised down by a factor of 2 to account for an additional diameter-dependent term (see text). The theoretical SFD proposed in *S13* is illustrated with the dotted line for comparison; the inferred population does not support the sharp upturn at d ~ 10 km.

Furthermore, the relative probability of detecting an occultation as a function of occulter diameter d in a given survey and at a fixed heliocentric distance is given by

$$P(d \mid q) \propto \eta(d)W(d)d^{-q} \quad (7)$$

where q is the local size-frequency distribution slope and η(d) is the detection efficiency of a given survey for an object of diameter d. Given the efficiency function published in *S12*, there is a large discovery volume at sizes smaller than the two claimed detections. The peak detection probability should peak at sizes much smaller than the claimed detections if the local size-frequency distribution slope is steep. The *Arimatsu et al.* (2019b) detection is much closer to their claimed detection limit, consistent with the expectations of a somewhat steeper slope in the size range covered by that survey ($d \gtrsim 2$ km). Taking these three candidate detections, their respective confidence levels, and the two surveys' detection efficiency functions, an average slope of $q \sim 3.6^{+2.1}_{-1.0}$ (mode and 68% interval) is most consistent with the data across the size range covered by these surveys ($0.4 \text{ km} \lesssim d \lesssim 2$ km). This slope, derived from the occultation surveys themselves and requiring no additional step of inference to connect them to the full population of TNOs, prefers values slightly steeper than the q_2 mid-sized slope inferred from the population of craters on Pluto and Charon, but the two values are not formally inconsistent.

This suggests that, if anything, something might be somewhat amiss with the *normalization factors* that connect the three measurement techniques: direct detection surveys, occultation surveys, and inference from crater populations. Internally, the local slopes are in reasonable agreement, but the sizes of populations inferred do not seem to match yet. With improved orbit distributions from OSSOS (*Bannister et al.*, 2016a), improved stellar diameters from Gaia (*Stevens et al.*, 2017), improved understanding of the influence of non-spherical shapes on occultation profiles (*Castro-Chacón et al.*, 2019), and improved direct-detection size-frequency distributions for the smallest TNOs observed in reflected light (*Parker et al.*, 2015), it may be the case that the populations inferred from these three methods can be brought comfortably into agreement. If not, more exotic explanations may be required, such as drastic changes in material density, extremely unusual-population-specific size-frequency distributions, or as-yet-unimagined geologic processes on Pluto and Charon.

2.2. Doublet Craters and the Transneptunian Object Binary Population

CCKBOs host a large population of widely separated binary systems (separations reaching up to 10–20% of the Hill radius) (*Parker et al.*, 2011) with near-equal mass components (*Petit et al.,* 2008; *Noll et al.,* 2008, 2020; *Parker et al.,* 2011). These binaries are likely primordial in origin (*Petit and Mousis,* 2004), sensitive to disruption by any past history of encounters with the giant planets (*Parker and Kavelaars,* 2010), and sensitive to collisional disruption (*Petit and Mousis,* 2004; *Nesvorný et al.,* 2011; *Parker and Kavelaars,* 2012). The properties of transneptunian binaries (TNBs) are powerful tracers of the processes of planetesimal formation and the subsequent dynamical and collisional evolution that the transneptunian populations underwent.

The apparent occurrence rate of these widely separated binary systems among the CCKBOs is generally quoted at 20–30% (e.g., *Noll et al.,* 2008). However, recent work has shown this is strongly influenced by observational selection effects: For example, a binary system is a brighter configuration of a given amount of material than a solitary object, and coupled with the extremely steep mass distribution of the CCKBOs results in an overrepresentation of binaries in any flux-limited survey (*Benecchi et al.,* 2018). The *intrinsic* widely separated binary fraction of the CCKBOs is likely substantially lower, in the range of 15–20% for component sizes and separations currently probed by HST observations ($d_p \gtrsim 30$ km, $a_m \gtrsim 3000$ km). Currently, few limits exist to constrain the population of trans-neptunian binaries at sizes smaller and separations tighter than this.

It is expected that the mutual orbits of many KBO binary systems were modified by a suite of effects referred to as Kozai cycles with tidal friction (KCTF); inclined binary systems will undergo period Kozai cycle oscillations of inclination and eccentricity until tidal friction can halt the oscillations and instead shrink and circularize the binary orbit. A substantial pileup of binary systems resulting from this process may exist at very tight separations — fractions of a percent of the Hill sphere diameter, well below the resolved limit of HST (*Porter and Grundy,* 2012). While there is a size dependence on the circularization timescale, if primordial binaries are distributed by some functional scaling of their Hill radius, then the timescale dependence for a rubble pile is $\propto 1/d_{primary}$. Given the speed with which the pileup forms in the simulations of *Porter and Grundy* (2012) ($\lesssim 10^7$ yr), binary systems with components ~500× smaller than those simulated are likely to have experienced KCTF pileup in the age of the solar system. This implies that binary components down to a scale of d ~0.1–0.2 km, on the order of the size of the smallest impactors recorded by the Pluto and Charon crater records, were subject to modification by KCTF. Thus, a pileup of binary impactors between contact and 0.5–1% of the Hill radius would be a reasonable expectation if the primordial population of binaries had broad inclination, eccentricity, and separation distributions.

Thus, surfaces within the Pluto system may provide an indirect record of the properties of the binary population at sizes and separations far smaller than have been thoroughly studied by direct means. Binary systems can produce a population of impact craters with spatially correlated sizes — i.e., craters of similar sizes will tend to fall closer to one another than would be expected under isotropic bombardment by solitary objects. To investigate this prospect, a characteristic distance metric for similarly sized craters must be defined. For each observed crater *i*, the haversine distances to all other observed craters on that target body within the size range $0.7\ D_i \leq D_i \leq D_i/0.7$ is calculated. The smallest distance in this list is divided by D_i to generate Δ_{min}, which is retained as the metric for the best candidate "doublet crater" for each observed crater. The distribution of Δ_{min} is illustrated in Fig. 5. To determine if there is evidence for a tightly separated binary population encoded in the crater sample, a model of the distribution of Δ_{min} under both the null (no binaries) and alternate (some fraction of binaries) hypotheses must be constructed and tested against the measured distribution of Δ_{min}. Such a model is constructed in the following section.

2.2.1. Generating synthetic crater doublets. The process for generating a synthetic crater population that includes binary impactors starts from the same approach as used

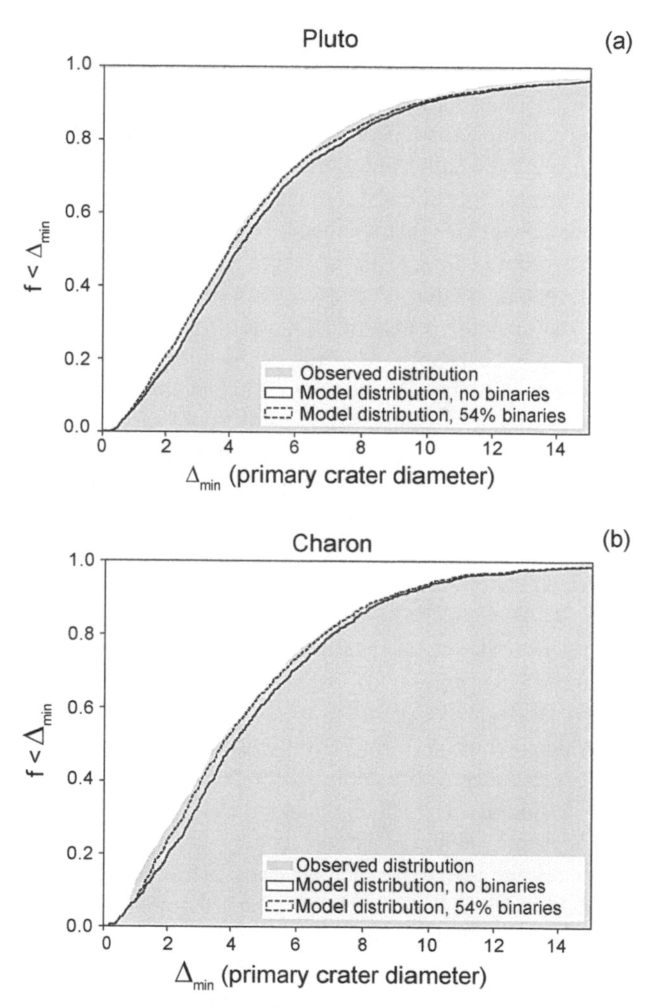

Fig. 5. Cumulative distribution of the distances between similarly sized craters on Pluto and Charon, scaled by primary crater diameter. Instances of model distributions with and without binary populations shown for comparison. A binary-hosting impactor population always generates a Δ_{min} distribution biased to lower values, which is more consistent with the observed distribution of Δ_{min}.

to generate impactor samples in the previous section. For every sampled impactor, a satellite impactor is generated with probability set by a proposed binary rate. The size of this satellite with respect to the diameter of the primary is sampled from a Rayleigh distribution informed by the brightness ratios of observed widely separated TNBs (*Johnston,* 2018, and references therein)

$$P(d_{i2}) \propto \begin{cases} 0 & d_{i2} < 0.5 d_{i1} \\ \dfrac{d_{i2}}{(0.2 d_{i1})^2} e^{-\frac{d_{i2}^2}{2(0.2 d_{i1})^2}} & 0.5 d_{i1} \leq d_{i2} \leq d_{i1} \\ 0 & d_{i2} > d_{i1} \end{cases} \quad (8)$$

The separation of the two components pre-impact was drawn from a uniform distribution between contact and 0.5% of the primary's Hill sphere, and their orientation pre-impact was drawn uniformly over the sphere. The impact separation projected onto a target surface was determined by this separation and orientation and the proposed impact angle θ for a given impactor. The implied crater diameters is generated through the same process as in the preceding section.

Geographic location and regional observational completeness must be accounted for as well. To achieve, one synthetic crater is proposed for every observed crater on each target, and propagate that *real* crater's latitude, longitude, and estimated detection efficiency function from the preceding section to each proposed crater, resampling the diameter from the proposed SFD parameters. Craters created by satellite impactors are retained in the synthetic observed sample using the same detection efficiency function as estimated for the primary, unless they fall within 0.8 D_i of the primary; in this case, it is unlikely that a clear doublet crater would have been generated (e.g., *Miljković et al.,* 2013). This essentially renders contact binaries [similar to 486958 Arrokoth (*Stern et al.,* 2019)] invisible in the crater record.

The haversine distances to all synthetic observed craters within the size range 0.7 $D_i \leq D_j \leq D_i/0.7$ is calculated, and if a proposed satellite produces a crater within this size range and is deemed observed by a trial with the efficiency function, the distance to the crater generated by that satellite is added to this list. The smallest distance in this list is divided by the primary crater diameter to generate D_{min}, which is retained as the metric for the best candidate "doublet crater" for each crater in the synthetic dataset. Another 10^7 ABCr trials were conducted, drawing size-frequency distribution parameters and efficiency function parameters from the posterior PDFs determined in the preceding section. In this case, the D_{ABCr} metric is the sum of the KS-test statistics over the observed and proposed Δ_{min} distributions on Pluto and Charon, truncated at $\Delta_{min} \leq 30$. Simultaneously, a parallel trial was conducted where *no* craters generated by satellites were included in the synthetic proposed sample that defines Δ_{min}. This parallel thread of trials permits the measurement of the Bayes factor in favor of the more complex alternate hypothesis (there is a binary population defined by some population fraction) and the simpler null hypothesis (there is no binary population).

*2.2.2. **Binary population results.*** The acceptance rate of the alternate hypothesis was 22× that of the null hypothesis — the Bayes factor is 22:1 in favor of a model with an impactor population that hosts similarly sized satellites, indicating strong evidence in favor of this hypothesis. The median and 68% confidence interval of the binary occurrence rate is $\lambda = 0.54^{+0.19}_{-0.16}$; the PDF is illustrated in Fig. 6. This rate it strongly sensitive to the proposed separation distribution, however — many of the tightest proposed systems produce single craters, so the uniform distribution proposed can accept high binary fractions with few of them producing craters. Further work to refine the model separation distribution based on KCTF evolutionary models would enable the crater record to better constrain the TNO binary fraction at small sizes; lightcurve and occultation searches for tight binary systems also provide powerful avenues of investigation into this putative population.

2.3. Implications

The crater record on the surfaces of Pluto and Charon indicate a very shallow size frequency distribution for small TNOs, and a population of relatively tight binary systems throughout the size range of impactors sampled by the craters ($d \gtrsim 0.1$ km). This shallow slope and binary population are both unlikely to represent the end state of a population that has experienced extensive collisional evolution. While it is conceivable that the CCKBOs are a relatively unmodified primordial population that emerged from an aerodynamically enhanced collapse of centimeter-scale "pebble" swarms (e.g., *Chiang and Youdin,* 2010; *Nesvorný et al.,* 2010, 2019) and their size-frequency distribution and binary population reflect the outcome of that process, the same cannot be

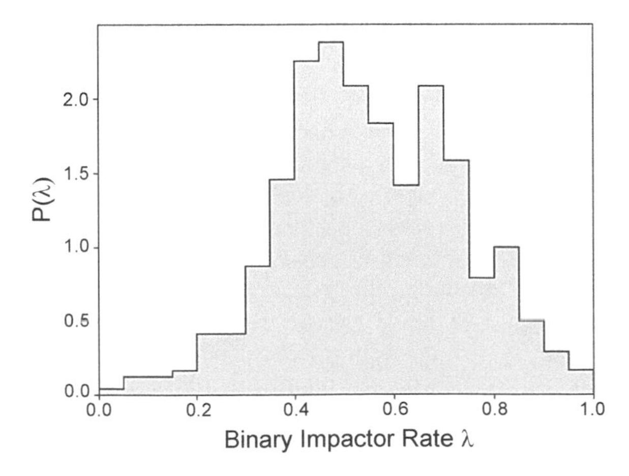

Fig. 6. Marginalized posterior PDF of the binary rate parameter λ estimated from the separation distribution of similarly sized craters.

said confidently for the populations that were transported to their current orbits from formation locations in the vastly more massive and collisionally active disk closer to the Sun at the time of giant planet migration. If the transported population was extracted from a collisionally evolved source population, why would the SFD slope at small sizes be so shallow? One unexplored possibility is that the transport process itself was size dependent; if smaller objects' collisional interactions tended to drive them out of resonances prematurely compared to their larger counterparts, fewer may have evolved into the stable non-scattering orbits that allowed the larger objects to persist. Future consideration of the influence of planetesimal size on the effectiveness of posited transportation mechanisms is merited to determine if such processes could help explain the unusual size distribution of small TNOs.

3. MOON ODDS AND ODD MOONS

At first glance, the Pluto system appears strikingly unique in a number of ways: It has the largest primary and the largest secondary of any known dwarf planet system, it has the highest multiplicity known, and it has a very low mass ratio between the primary and secondary. However, these features are each drawn from a distribution across the population of dwarf planets, and in context, a Pluto-like system seems less like an oddball and perhaps more like an inevitability. Figure 7 illustrates the estimated diameters of TNOs known to host satellites, and the estimated mass ratio of each system. Excluding the CCKBO binary systems, with their small sizes and mass ratios near 1, Pluto does not appear so much an outlier in this figure as it does a member of a relatively uniformly populated distribution. There are smaller primaries with similar mass ratios, and systems with smaller mass ratios with similarly sized primaries. While the degree of the Pluto system's multiplicity remains unique, Haumea has a complex satellite and ring system, and later sections describe how these two systems may have been even more similar in the past.

Satellites are now known to be ubiquitous for the brightest members of the classical, resonant, and scattered disk populations (see Table 1). The brightest member of any of these populations that has been searched for satellites with the HST but has no known satellite is 2005 UQ_{513}, a classical object with a radiometrically derived diameter of 498^{+63}_{-75} km (*Vilenius et al.,* 2018). As exemplified by Makemake, some members of this list that have been searched in a single epoch using HST may yet host a detectable satellite (*Parker et al.,* 2016b). Among objects in these populations with $H_V > 4$, the current apparent occurrence rate for satellites is 60%, with 12 of the 20 objects currently characterized by HST known to host satellites. Unlike the smaller objects in the cold classical population with similarly sized primordial binary components (*Petit and Mousis,* 2004; *Nesvorný,* 2011; *Parker and Kavelaars,* 2012), these large, dissimilar-sized binary systems are thought to form through processes that occurred after the formation of the

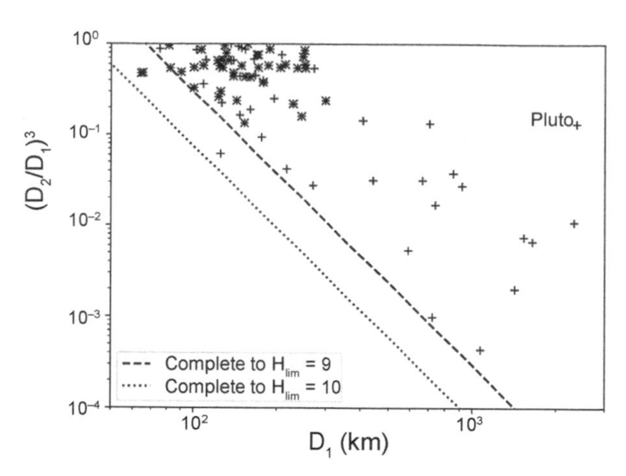

Fig. 7. Diameters and approximate mass ratios of known TNO binary systems. Completeness lines are based on scaling from primary diameter and assuming albedos of 10% where other estimates are not available. Starred points are CCKBOs. In multi-satellite systems, only the largest satellite is considered. Data from *Johnston* (2018) and references therein.

parent object. Post-formation giant impacts (*McKinnon,* 1989; *Canup,* 2005, 2011; *Leinhardt et al.,* 2010) are a leading contender for the origin of many of these systems, while capture remains a possibility for others. Regardless, they are generally thought to be the outcome of stochastic events that occurred at some characteristic frequency early in the solar system's history.

The high extant occurrence rate of satellites around large TNOs thus points to a surprisingly high rate of satellite formation events. Of the 22 systems in Table 1 that are from resonant, scattered, or classical populations, 12 are known to host satellites. Assuming that the extant number of satellites partitions systems into two groups — one group having one or more satellite formation events per object, and one where zero such events have occurred to date — the mean rate parameter for a Poisson distribution of the number of satellite-forming events experienced for host objects $H_V > 4$ mag is $\lambda = 0.82$. If these events are independent from one another — i.e., one moon-forming event does not influence the probability of another moon-forming event occurring — then this implies that a substantial fraction of the known large TNOs likely experienced *multiple* moon-forming events in succession, and the current satellite systems are the result of the integrated history of these events for each system. For the same sample, an average of four systems is likely to have experienced at least two moon-forming events, and an average of one system will have experienced three or more. For the largest TNOs, histories of multiple satellite formation and disruption events are thus not only possible, but *demanded* by the high extant rate of satellite systems. In this light, it is worthwhile to consider the current complexity of Pluto's satellite system as a potential end-state of one of these multi-epoch formation histories, and examine what can be inferred about the

TABLE 1. Current results of satellite searches and diameter measurements (either from radiometry or occultations) among the brightest transneptunian objects (H-magnitude estimates extracted from the JPL Small Body Database).

Name/ID	H	Class	Diameter Measured?	HST Search?	Known Satellites?
Eris	–1.1	Scattered	Y	Y	Y (1)
Pluto	–0.4	Resonant	Y	Y	Y (5)
Makemake	–0.1	Classical	Y	Y	Y (1)
Haumea	0.2	Resonant	Y	Y	Y (2)
Sedna	1.3	Detached	Y	Y	N (0)
Gonggong	1.6	Scattered	Y	Y	Y (1)
Orcus	2.2	Resonant	Y	Y	Y (1)
Quaoar	2.4	Classical	Y	Y	Y (1)
2013 FY_{27}	3.2	Scattered	N	Y	Y (1)
2002 AW_{197}	3.3	Classical	Y	Y	N (0)
G!kún‖'hòmdímà	3.3	Scattered	Y	Y	Y (1)
2002 TX_{300}	3.4	Haumea Fam.	Y	Y	N (0)
2014 UZ_{224}	3.4	Detached	N	N	—
2018 VG_{18}	3.4	Detached	N	N	—
Varda	3.4	Classical	Y	Y	Y (1)
2005 UQ_{513}	3.6	Classical	Y	Y	N (0)
2002 MS_{4}	3.6	Classical	Y	Y	N (0)
2003 AZ_{84}	3.6	Resonant	Y	Y	Y (1)
Varuna	3.6	Classical	Y	Y	N (0)
2005 QU_{182}	3.6	Scattered	Y	N	—
Ixion	3.6	Resonant	Y	Y	N (0)
2002 UX_{25}	3.7	Classical	Y	Y	Y (1)
2005 RN_{43}	3.7	Classical	Y	Y	N (0)
2015 RR_{245}	3.8	Scattered	N	N	—
2002 TC_{302}	3.9	Resonant	Y	Y	N (0)
Dziewanna	3.9	Scattered	Y	Y	N (0)

possible past histories of satellite systems that appear less complex or unusual today.

Eris, Makemake, Gonggong, Orcus, and Quaoar all host single known satellites. The largest of these is Eris' satellite Dysnomia, with a diameter of 700 ± 115 km estimated from Atacama Large Millimeter Array (ALMA) radiometry (*Brown and Butler,* 2018). For the most part, little is known about the physical properties of most of these satellites due to the challenges inherent in observing them in proximity to their (typically) much brighter primaries. Gonggong's satellite Xiangliu maintains a substantial eccentricity (e ~0.3) (*Kiss et al.,* 2019); such a high eccentricity should have been tidally damped over the age of the solar system unless the tidal factor Q is high and the satellite is small and bright (*Kiss et al.,* 2019), unless it has a recent origin or mechanism for ongoing excitation (such as a second as-yet-unseen satellite).

Another system that invites consideration is that of Haumea. Not only is Haumea host to two satellites and a ring of debris, but it is the largest member of the only known orbital family of TNOs. The largest known member of the Haumea family (other than Haumea itself) is (55636) 2002 TX_{300}. With a diameter of 286 ± 10 km (*Elliot et al.,* 2010), it is comparable in size to Haumea's largest moon Hi'iaka. While larger than the small satellites of Pluto, these objects share the traits of very high-albedo surfaces (*Weaver et al.,* 2016), nearly pure water-ice spectra with indications of ammonia compounds (*Cook et al.,* 2018; *Barkume et al.,* 2006). If the Haumea system once bore a resemblance to the Pluto system, with a tight, massive inner binary surrounded by a disk of smaller satellites (although with substantially more mass in this disk than is currently present around Pluto), it is conceivable that instabilities originating within the satellite system could have resulted in the ejection of many of the original satellites at very low velocity with respect to Haumea. Binary systems are very efficient at ejecting material that migrate into an unstable orbital separation regime (*Jackson et al.,* 2018). Should enough orbital energy be removed by repeated ejections of small satellites, an inner binary could be driven to merge into a single, fast-rotating primary (e.g., *Levison et al.,* 2007). Such a scenario would make the Haumea system (and its family) and the Pluto system two outcomes of a relatively similar origin.

If satellite systems growing from circumplanetary debris disks reach the stage of oligarchic growth, the satellites that emerge will have a relatively shallow size-frequency distribution due to the partitioning of their feeding zones. The mid-sized saturnian satellites display a shallow maximum-likelihood differential SFD slope of 2.4; the inner prograde satellites of Neptune have a very shallow SFD slope of 1.8. New Horizons directly resolved the sizes of the small satellites of Pluto (*Weaver et al.,* 2016), and they show a similar shallow slope of 2.5. Recent measurements have determined that the Haumea family members also have a relatively shallow SFD slope of $\lesssim$2.5 (*Pike et al.,* 2019). These slopes are all shallower than the canonical collisional equilibrium slope of $q = 3.5$, and much shallower than typical slopes produced by fragmentation of asteroids as evidenced by the size-frequency distribution slopes of young asteroid families ($\bar{q} = 4.46$ for single-slope families younger than 500 m.y.) (*Parker et al.,* 2008) and also the size distribution slopes produced in graze-and-merge simulations of the origin of Haumea and its family (*Leinhardt et al.,* 2010). The similarity of Pluto's satellites and Haumea's satellites and family (in terms of surface composition, albedo, and size-frequency distributions) may be coincidental, but are certainly suggestive of common origin processes and their relationship merits further study.

4. MORE THAN SKIN DEEP

While it is likely to be decades before another transneptunian dwarf planet is revealed at the same level of detail as Pluto and Charon, the properties observed by New Horizons can help guide our understanding of those observations we can currently make of their kin. While one of the chief takeaways of our current exploration of icy worlds in the outer solar system is that they are often more distinct than they are similar, there are still valuable lessons to be learned.

4.1. Surface Composition, Atmospheres, and Volatile Transport

New Horizons confirmed a long-standing hypothesis regarding the origin of Pluto's high albedo: that seasonal volatile transport refreshes the deposits of volatiles on its surface, hiding any lag deposits of photochemically darkened materials (*Stern et al.,* 1988). It is plausible that the surfaces of Makemake and Eris are similarly refreshed by atmospheric processes, resulting in their bright surfaces and their very small rotational lightcurve amplitudes (e.g., *Heinze and de Lahunta,* 2009; *Sicardy et al.,* 2011; *Ortiz et al.,* 2012; *Hofgartner et al.,* 2019). Models have indicated that these two worlds could host regional or global vapor pressure equilibrium atmospheres of N_2 and/or CH_4 at some points in their orbits (e.g., *Stern and Trafton,* 2008; *Young,* 2013), although searches for atmospheres through stellar occultation measurements have yielded no detectable global atmospheres to date (see *Young et al.,* 2020, for a review). Stellar occultation-based upper limits of surface pressures have been made for Eris (*Sicardy et al.,* 2011), Haumea (*Ortiz et al.,* 2017), Makemake (*Ortiz et al.,* 2012), and Quaoar (*Arimatsu et al.,* 2019b); the 1σ upper limits range from 1–12 $\times$ 10^{-9} bar. *Ortiz et al.* (2012) noted some hints of non-solid-body refraction in the lightcurve of a star occulted by Makemake, and showed that these are consitent with plausible non-global atmospheres, but the data from this event alone is not conclusive.

One of the most striking features on the encounter hemisphere of Pluto is Sputnik Planitia, an impact basin filled by a vast nitrogen-rich ice sheet undergoing active convection (*McKinnon et al.,* 2016). Due to the relatively swift overturn of its convective cells, its surface is extremely young and fresh, showing no impact craters at the limit of the highest-resolution imagery from New Horizons (*Moore et al.,* 2016). It has been speculated that the bright, fresh surfaces of other large, volatile-dominated TNOs, such as those of Eris and Makemake, may be maintained in part by similar convective processes at their surfaces (*McKinnon et al.,* 2016; *Grundy and Umurhan,* 2017).

Eris has a substantially higher bulk density than Pluto (*Sicardy et al.,* 2011). If this is indicative solely of a difference in rock-to-ice fraction, then Eris can be expected to have substantially greater radiogenic heat production in its interior and thus a higher heat flux at its surface. A larger rock fraction would also imply a greater potential cosmogonic reservoir of nitrogen from which a surface layer of N_2 could be built up. Both of these factors conspire to create favorable conditions for extensive convective N_2 deposits on Eris' surface.

In detail, Eris may contain ~1.5 $\times$ 10^{22} kg of rock in its interior, suggesting a present-day surface heat flux of order ~ 5 mW m^{-2} vs. Pluto's ~3 mW m^{-2}. Following the arguments in *McKinnon et al.* (2016), the thermal gradient through a surface layer of N_2 ice is likely to be on the order of 25 K km^{-1} vs. Pluto's 15 K km^{-1}. For even the lowest estimates of its surface temperature at aphelion (27 K), this thermal gradient implies that N_2 deposits thicker than ~1.4 km would be sufficient to melt N_2 at its base. Adopting the same cosmogonic nitrogen mass fraction suggested for Pluto by *Singer and Stern* (2015), Eris likely contained an initial interior reservoir of ~5 × 10^{19} kg of nitrogen, enough to create a global N_2 surface layer ~3 km thick, if internal processes ever liberated this material and delivered it to the surface. The ratio of layer thicknesses required to support convection on two different worlds is

$$\frac{L_1}{L_2} = \exp(T_{i1}^{-1} - T_{i2}^{-1}) \frac{g_2 \Delta T_2}{g_1 \Delta T_1} \quad (9)$$

where g is the surface gravity of each world, ΔT is the temperature difference across the layer, and *McKinnon et al.* (2016) estimate T_i to be T_0–ΔT/2 in the sluggish lid regime, and T_0 is the basal temperature. Assuming the critical layer thickness for both worlds is lower than the thickness required to generate basal melt, adopting the the thermal gradients above and the relative surface gravities of 0.82 m s^{-2} and 0.617 m s^{-2} for Eris and Pluto respectively, and adopting the *McKinnon et al.* (2016) estimate of L_{Pluto} ~ 500 m, equation (9) can be solved implicitly to derive L_{Eris} ~ 390 m under the same convection regime. Thus, all else being equal, marginally shallower deposits of N_2 should convect on Eris than on Pluto.

One caveat here lies in the structure of Pluto's mantle and crust beneath Sputnik Planitia. If the orientation of the basin is due to true polar wander driven by a positive gravity anomaly (*Keane et al.,* 2016; *Nimmo et al.,* 2016), then the lithosphere beneath the basin must be thinned (*Johnson et al.,* 2016). This substantially increases the proximity of the isothermal ocean (see section 4.3) to the surface underneath Sputnik Planitia, and could act as a thermal conduit and enhance the local heat flux substantially. Thus, Sputnik Planitia could be convective even in a scenario where Pluto's rock component contained a subchondritic level of radionuclides; in this case, assuming common composition and absent similar conduits to the interior, Eris may not support nitrogen convection on its surface today, leaving seasonal processing of its surface as the only means of maintaining its high albedo and neutral color.

Atmospheric escape of N_2 has been put forward as an explanation for the deep methane absorption bands (*Licandro et al.,* 2006), relatively weak evidence for nitrogen in Makemake's spectra (*Brown et al.,* 2007), and lack of a currently detectable global atmosphere (*Ortiz et al.,* 2012). Given the low current estimates for Makemake's density [1.7 g cm^{-3}, based on an equilibrium figure derived from a 7.77-h spin period (*Heinze and de Lahunta,* 2009; *Rambaux et al.,* 2017; *Parker et al.,* 2018)], Makemake sits at the cusp of where even slow Jeans escape would have depleted its surface of N_2 over the age of the solar system (*Schaller and Brown,* 2007; *Ortiz et al.,* 2012; *Brown,* 2013), but still dense enough to have retained CH_4. Recent photometry (*Hromakina et al.,* 2019) suggests that longer rotation periods of ~11.4 h or ~22.8 h are likely, and rotational equilibrium figure models will need to be revised for Makemake in future analyses. Should any substantial deposits of N_2 persist on the surface of Makemake, convection is challenged by the lower heat flux at its surface (~1.3 mW m^{-2}) driving a lower temperature gradient through any surface deposits (~6.5 K km^{-1}), setting the required layer thickness to support convection to $L_{Makemake}$ ~810 m after accounting for its low surface gravity of 0.36 m s^{-2}. Thus, any N_2 deposits on Makemake would have to be thicker than Pluto in order to convect, and given Makemake's relative depletion of N_2, this would seem unlikely.

Charon's spectrum had revealed ammonia signatures before flyby (*Brown and Calvin,* 2000; *Cook et al.,* 2007). New Horizons revealed that limited exposures of NH_3 on Pluto and Charon both appear to be associated with surface modification, including through cryoflows on Pluto (*Dalle Ore et al.,* 2019; *Cruikshank et al.,* 2019) and craters on Charon (*Grundy et al.,* 2016). The presence of absorption features at 2.21 μm also indicates ammoniated compounds on the small satellites Nix and Hydra (*Cook et al.,* 2018). As mentioned previously, ammonia compounds have been detected both on Haumea itself and its largest moon Hi'iaka (*Barkume et al.,* 2006). The large TNO Orcus also displays ammonia absorption (*Carry et al.,* 2011). As ammonia is a potent antifreeze agent responsible for lowering the melting point of solid systems of ammonia and water ices, its distribution among the large TNOs is a key factor in understanding the thermal evolution of their interiors and the potential for extant subsurface oceans.

The long-term evolution of Pluto's orbit and obliquity lead to a cycle of 1.5-m.y. "super-seasons" that may be responsible for controlling the stark albedo and composition contrasts between low and high latitudes through feedback between albedo and volatile transport processes (*Earle et al.,* 2018). Makemake's thermal emission requires two terrains of dissimilar temperature (*Lim et al.,* 2010), which suggests distinct geographic regions of both very low and very high albedo. However, Makemake's extremely small rotational lightcurve amplitude (*Heinze and de Lahunta,* 2009) and edge-on spin orientation (*Parker et al.,* 2016b) require that such dark terrain be distributed extremely longitudinally uniformly. Similar albedo/volatile transport feedback processes could potentially generate latitudinal "belts" of dark material on Makemake that, if not interrupted by dissimilar bright terrain (like Sputnik Planitia on Pluto) would be relatively undetectable in the lightcurve. Alternatively, a non-global atmosphere that has frozen out onto the surface near aphelion could produce bright frost-covered terrains around the equator, while older, darker materials are exposed near the poles (*Ortiz et al.,* 2012). The possibility of an upcoming season of Makemake mutual events provides a potential avenue to test for latidudinal variations in albedo on Makemake, potentially isolating the dark terrain and its climatological implications.

4.2. Tectonics, Landform Evolution, and Cryovolcanism

The features attributed to tectonic activity on the surfaces of both Pluto and Charon are predominantly extensional in nature (*Moore et al.,* 2016; *Keane et al.,* 2016; *Beyer et al.,* 2017). The absence of obvious compressional tectonic complexes suggests that both Pluto and Charon have experienced periods of global expansion with no substantial periods of global contraction. Charon's tectonic features, which include graben complexes with enormous vertical relief (up to 5 km), suggest a global areal strain of 1%, which could be accomplished by freezing a subsurface layer of pure water that was initially ~35 km thick (*Beyer et al.,* 2017). Pluto's extensional features correlate to the location and orientation of Sputnik Planitia in such as a way as to suggest that Pluto experienced true polar wander after the emplacement of the Sputnik Planitia basin, orienting the resultant gravity anomaly along the Pluto-Charon tidal axis (*Keane et al.,* 2016). Very few constraints on similar topographic features exist for large TNOs other than Pluto and Charon; a stellar occultation by the several-hundred-kilometer-diameter plutino 2003 AZ_{84} may reveal either a regional depression or a deep chasm seen in profile (*Dias-Oliveira et al.,* 2017).

Unlike Charon, many of Pluto's landforms appear to be sculpted by sublimation processes as part of Pluto's highly active seasonal and "super-seasonal" volatile transport cycles (*Moore et al.,* 2016, 2017, 2018; *Earle et al.,* 2018). The striking "bladed" terrain on Pluto are regions of uniquely textured CH_4-dominated material at low latitude and high elevation. Presumed to be similar to terrestrial water-ice penitentes, these are possibly the result of complex interactions between the thermal profile of Pluto's atmosphere, the seasonally dependent latitudinal distribution of solar heating, and the interplay of N_2 and CH_4 precipitation and sublimation (*Moore et al.,* 2018). Pitted terrains with a variety of specific morphologies can be reproduced by sublimation-driven surface evolution (*Moore et al.,* 2017), and the appearance of these pit chains on the surface of the rapidly refreshed Sputnik Planitia indicates that these landforms are evolving and growing in the current epoch. It is likely that sublimation will be an important process for driving landform evolution on the surfaces of other volatile-dominated TNOs. Non-erosive aeolian processes are also in evidence on Pluto; dune-like features on Sputnik Planitia suggest that small particles of low-density materials have been recently transported by winds near Pluto's surface (*Telfer et al.,* 2018). Whether dynamics within transient atmospheres on other large TNOs are sufficient to similarly mobilize materials at their surfaces is unknown.

A variety of landforms on Pluto — including extensive dendritic valley networks — have been interpreted as the result of surface modification by fluid flow (*Moore et al.,* 2016; *Howard et al.,* 2017). A leading contender for the working fluid that produced these is liquid nitrogen. Recent analysis indicates that it is unlikely that conditions have ever existed on Pluto that permitted liquid nitrogen to exist stably on the surface (*Bertrand et al.,* 2018); it is likely that any such N_2 flow occurred as basal melt in past subglacial environments. As similar glaciation may be expected on the surfaces of worlds like Eris and Makemake, it is reasonable to expect that similar landforms indicating past fluid flow would be exposed on their surfaces.

Features on both Pluto and Charon have been identified as potentially cryovolcanic in origin. The smooth terrain of Vulcan Planum on Charon is interpreted as being emplaced by a massive cryoflow onto the surface during an early epoch of global expansion that enabled a subsurface ocean to breach the ice crust in many locations, resulting in an effusive flow of highly viscous material covering a large portion of Charon's surface (*Beyer et al.,* 2019). This material appears to embay older topographic features (including around Charon's striking "mountains in moats") and to pile up near its margins south of Serenity Chasma. Vulcan Planitia's crater record is consistent with an age of 4 Ga (*Moore et al.,* 2016; *S19*), although uncertainties in the absolute age calibration may allow ages half that old. If the resurfacing of Vulcan Planum is indicative of an epoch of global expansion, then this age is inconsistent with the origin of this expansion being due to the onset of freezing of a global ocean, which is likely to occur much later (e.g., *Desch and Neveu,* 2017). Recent models of the thermal evolution of Charon indicate that early epochs of global expansion can be generated by the hydrous alteration of silicates in the rocky core (*Malamud et al.,* 2017).

Located just south of Sputnik Planitia, Pluto's Wright and Piccard Mons are two prominent mound-shaped structures with central depressions. They are both candidates for cryovolcanic features (*Moore et al.*, 2016), and they appear to be relatively young with very few craters found on their surfaces (*Singer et al.,* 2018). Virgil Fossae, just to the west of Sputnik Planitia, is a graben complex with surrounding terrain that shows high concentrations of water ice, a unique red coloring agent, and NH_3 compounds (*Cruikshank et al.,* 2019). These properties suggest that Virgil Fossae was the source for a regional cryoflow. Wright Mons, Piccard Mons, and Virgil Fossae are all located near the ring-shaped global peak of inferred extensional stress produced by the loading and reorientation of Sputnik Planitia (*Keane et al.,* 2016; *Cruikshank et al.,* 2019). These extensional stresses should act to generate a path of least resistance for any subsurface liquid reservoirs to reach the surface. Given that the most promising candidates for cryovolcanic structures on Pluto currently appear to be limited to these regions of unusually high extensional stresses (although this region also encompasses the best-imaged hemisphere of Pluto, so selection effects may be at play), it is unclear if cryovolcanism on the very largest TNOs would occur without the presence of similar stress-inducing structures like the impact basin underlying Sputnik Planitia. *Neveu et al.* (2015) explored the role that the exolution of gases from subsurface reservoirs could play in driving ongoing explosive cryovolcanism and found a variety of plausible avenues for such processes, but no active Triton-like plumes were detected in the flyby datasets (*Hofgartner et al.,* 2018).

4.3. Internal Oceans of Large Transneptunian Objects

Efforts to model the thermal evolution of the interiors of dwarf planets have largely explored systems of water, ammonia, and silicate minerals (*Hussmann et al.,* 2006; *McKinnon,* 2006; *McKinnon et al.,* 2008; *Desch et al.,* 2009; *Robuchon and Nimmo,* 2011; *Rubin et al.,* 2014; *Neveu and Desch,* 2015; *Desch,* 2015). The silicate fraction provides a long-lived internal heat source in the form of radiogenic heat, and that heat is transported through the bulk of the body to the surface where it is eventually radiated away to space. Energy is transported radially through conduction and convection. The consensus of modeling efforts before flyby was that (1) the interiors of both Pluto and Charon would be differentiated; (2) if the reference viscocity of their ice mantles is high, then these mantles do not convect and transport energy purely by conduction, generally resulting in the formation of internal oceans; and (3) if the reference viscosity of their ice mantles is low, then the mantles convect and efficiently deliver heat from the core to the surface layers, resulting in no ocean formation. Furthermore, models predict that Charon retains an undifferentiated crust of a mix of rock and ice atop a mantle of pure ice (*Rubin et al.,* 2014).

Substantial observational evidence for an internal liquid ocean within Pluto was revealed by New Horizons, including the lack of a fossil bulge, extension-dominated tectonics, and the evidence for true polar wander driven by a gravity anomaly in Sputnik Planitia (*Moore et al.,* 2016; *Nimmo et al.,* 2016; *Keane et al.,* 2016; *Johnson et al.,* 2016), although whether the ocean was transient or persists to this day remains to be determined. Charon's extensional tectonics and cryovolcanic features also support a past internal ocean, although one that has largely or entirely frozen by the current epoch. The extents and survival timescales of similar oceans in transneptunian dwarf planets is a matter of substantial astrobiological interest, as such oceans likely interact directly with silicate cores and provide a potential abode for extraterrestrial life. Subsequent to these discoveries, more detailed interior evolution models were developed for both Pluto and Charon in an effort to explain some of the key observations from flyby: the relatively similar densities of both worlds, Charon's ancient extensional tectonics, and the gravity anomaly inferred from the orientation of Sputnik Planitia. Both *Malamud et al.* (2017) and *Bierson et al.* (2018) added treatments of porosity removal; *Malamud et al.* (2017) and *Desch and Neveu* (2017) both tracked two rock phases to account for hydrous alteration of silicate minerals. *Desch and Neveu* (2017) also included a treatment of suspended rock "fines," which produce a more insulating mantle. *Kamata et al.* (2019) examined the role that clathrate hydrates may play as a thermal insulator between a subsurface ocean and the ice crust above it. All these additional factors may impact the extent, astrobiological potential, and survival time of interior oceans within large TNOs and the tectonic expression of their evolution on the surfaces of these worlds.

Existing models and comparison to observations within the Pluto system can provide insight into the internal structures of other large TNOs. At a basic level, a conservative estimate of the onset of differentiation can be estimated by determining threshold requirements to heat the interior of the TNO to the melting point of ammonia dihydrate (ADH, T ~176 K); it is likely that differentiation occurs at even lower temperatures than this (e.g., *Desch,* 2015). Figure 8 illustrates threshold differentiation curves for TNOs of different sizes assuming the *Desch et al.* (2009) thermophysical model. Relatively small TNOs (d ≲400 km) require large rock fractions or high radionuclide enhancement over typical chondritic material to achieve differentiation, but larger TNOs transition to plausible differentiation under nominal conditions. These estimates do not include the influence of early aluminum heating, tidal heating in binary systems, or heating from impacts.

Makemake and Charon have nearly identical bulk densities, and from a thermal evolution standpoint they are relatively similar save for their sizes and an initial epoch of tidal heating for Charon. For bulk ammonia concentrations of 1% by mass in the initial ice component, *Desch et al.* (2009) estimate the time to complete ocean freezing of

$$t_{\text{freeze}} \simeq 4.6\left(\frac{\bar{\rho}}{1700\text{g cm}^{-3}}\right)\left(\frac{R}{600\,\text{km}}\right)\text{Gyr} \quad (10)$$

where $\bar{\rho}$ is the average bulk density of a large TNO and R is its radius. All else being equal, this predicts a freezing

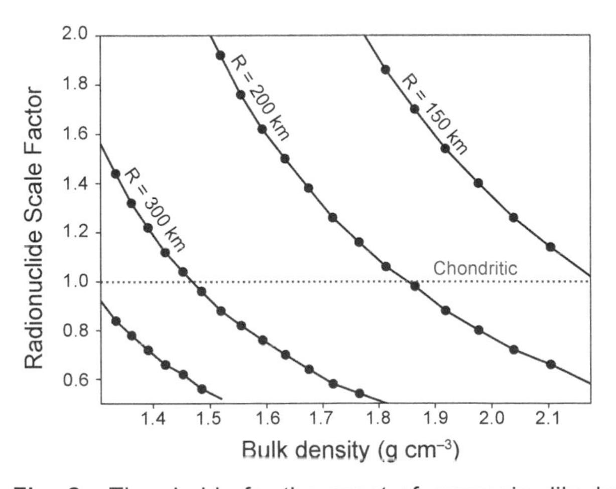

Fig. 8. Thresholds for the onset of ammonia dihydrate (ADH) melting in the interior of TNOs containing 1% ammonia concentration by weight in their non-rock fraction. Bodies accrete cold and begin with uniform distribution of rock and ice throughout their bulk. Nominal specific heat production taken from *Desch et al.* (2009) shown with red horizontal line.

timescale of 4.7 G.y. for Charon, and 5.8 G.y. for Makemake. Conversely, Quaoar is very similar in size to Charon (555 km vs. 606 km radii) (*Braga-Ribas et al.,* 2013), but substantially more dense (2.0 g cm^{-3} vs. 1.7 g cm^{-3}) (*Braga-Ribas et al.,* 2013). The inferred timescale for ocean freezing on Quaoar is also longer than Charon at 5.0 G.y. While all these values are model-dependent and numerous other models exist that may shorten or extend the lifetime of an internal ocean, they should roughly scale in a similar fashion for these three relatively similar worlds.

Eris and Haumea exist in classes of their own; while they are nearly identical in diameter, Eris' bulk density is substantially higher than Pluto's, and all else being equal, this would lead to *lower* pressures at the interface between its ice mantle and rocky core, but substantially higher heat fluxes. Unlike with Pluto, the formation of the high-density Ice II (rhombohedral) phase of water isn't plausible at its core boundary unless there is a substantial crust of undifferentiated rock-ice mixture to drive up the core-mantle boundary pressure; while the lack of compressional tectonic features on the surface of Pluto indicates that these phases did not form in Pluto's interior (*Hammond et al.,* 2016), fewer potential thermal histories would have produced such effects for Eris. An example of Eris' internal structure's evolution is illustrated in Fig. 9; the *Desch et al.* (2009) thermal model predicts a relatively thin undifferentiated layer and the core-mantle boundary remains below the 200-MPa threshold for formation of Ice II. The steeper thermal gradient through the ice crust has greater potential to drive convection in the ice shell, and absent other insulating or viscosity-increasing effects [e.g., clathrate formation (*Kamata et al.* 2019)], the efficient transport of heat from the core to the surface by convection would inhibit ocean formation. If the crust remains conductive, then the scaling from equation (10) can be applied to estimate a total freezing time of ~13 G.y.; from 2 G.y. to the present day, approximately 8% of Eris' volume has undergone a phase transition from liquid water to ice, indicating potential for global expansion and extensional stresses. Haumea's rapid spin and very elongated figure makes one-dimensional thermophysical approximations invalid, and detailed modeling of its interior evolution after acquiring its current physical configuration merits applying a three-dimensional thermophysical scheme. Recent efforts to model the density and figure of Haumea indicate that it is likely differentiated with a dense silicate core and relatively thick ice shell, and that it has achieved a fluid equilibrium figure (*Dunham et al.,* 2019).

As described earlier, the relatively high ratio of CH_4 to N_2 derived from Makemake's near-infrared (NIR) spectral properties compared to those of Pluto and Eris suggests that some set of processes have depleted its surface N_2 relative to these other worlds. While fractionation through atmospheric escape is one possibility that has been explored (e.g., *Brown et al.,* 2007), another possibility is that the internal oceans of Pluto and Eris both host substantial layers of clathrate hydrates (*Kamata et al.,* 2019), which can efficiently sequester CH_4, but not N_2. If Makemake's interior is less favorable for the formation of extensive layers of clathrate hydrates, then substantially more of its internal CH_4 may have been allowed to reach its surface, diluting any N_2.

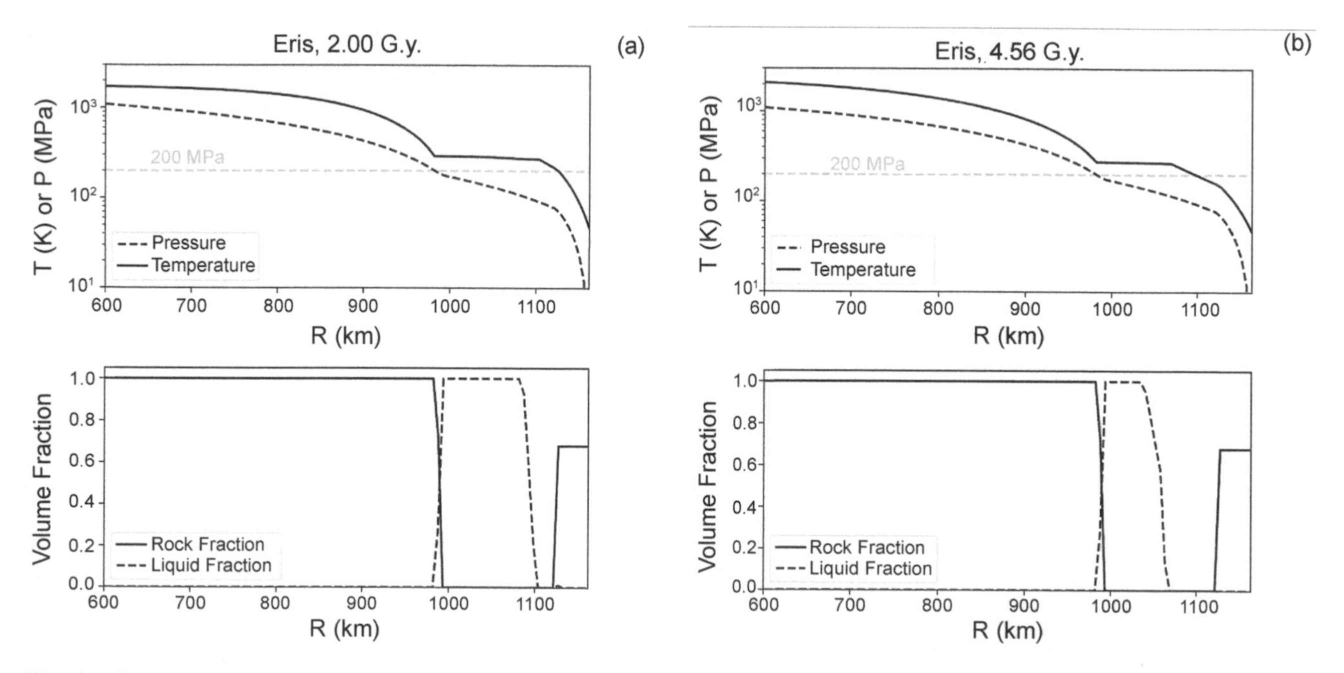

Fig. 9. Interior properties of Eris at 2 G.y. after formation and at the current epoch, derived from thermophysical models in *Desch et al.* (2009), assuming only conductive heat transport in the ice shell. Eris rapidly and extensively differentiates, leaving an undifferentiated shell only 41 km thick. Substantial reduction in the thickness of the liquid layer occurs in the last 2 G.y.

5. FUTURE DIRECTIONS AND INFORMED SPECULATION

Pluto has revealed a wealth of information about processes that may be endemic to the icy dwarf planets in the outer solar system, or perhaps unique to itself. As the planetary science community continues to tackle the data returned from the Pluto system by New Horizons and collect more remote observations, it will be important to consider which is which. To that end, a robust effort should be undertaken to consolidate the most important components of theory around the origin, structure, and evolution of icy dwarf planets and open them to scrutiny. For example, models of interior evolution of dwarf planets are a field where broad community investment would help drive the field forward, ensuring that the best and broadest expertise can be tapped in the disparate domains needed to achieve high-performance, accessible, accurate, and validated models applicable to the interiors of these worlds.

Occultation surveys — both those chasing predicted occultations and those searching for serendipitous events — provide a clear path forward for resolving the remaining uncertainties in the size-frequency distribution of small TNOs. With sufficient signal-to-noise and multi-cord detections, mapping the shape, radial, and vertical distribution of small TNOs will permit a crude extraction of independent properties of the hot and cold subpopulations. Accurate size-frequency distributions will in theory permit exploration of the surface properties of the Pluto system, including their absolute ages and perhaps their density and porosity. Upcoming programs like TAOS-II (*Lehner et al.,* 2018) are the next step toward these goals, but in the future dedicated spacebased serendipitous occultation surveys (*Kavelaars et al.,* 2010; *Alcock et al.,* 2014; *Santos-Sanz et al.,* 2016), above the scintillation noise induced by Earth's atmosphere, will be able to measure to smaller sizes at a greater range of distances and provide higher-quality measurements of each detected object's size and distance.

The degree of characterization of the properties of other icy dwarf planets will soon be substantially improved by the James Webb Space Telescope (JWST) (*Parker et al.,* 2016a). Covering the 0.6–28.5-μm spectral range with unprecedented sensitivity and resolution, many fundamental absorption bands for key surface constituents will be readily observable for the first time. JWST's sensitivity to the blue thermal tail and its spatial resolution will permit unprecedented measurements of the thermal properties within satellite systems. Already a large suite of dwarf planet observations are planned to execute as guaranteed time observation programs (GTOs 1191/Stansberry, 1231/Guilbert-Lepoutre, 1254/Parker, 1272/Hines, and 1273/Lunine), with targets including Pluto, Haumea, Makemake, Quaoar, Eris, Sedna, Orcus, and Varuna. These observations will test a variety of observation modes and build an initial framework for determining the most fruitful paths forward for further characterization of these worlds from JWST. Key among these observations is a series to tie the JWST spectral observations to the New Horizons spectral observations of Pluto using a longitudinally resolved time series of Near Infrared Spectrograph (NIRSpec) observations. Comparing this set of observations with those from New Horizons will provide a critical "ground truth" test of what inferences may be drawn from next-generation remote spectral investigations.

Finally, the population of known dwarf planets continues to grow as several recent surveys have turned up new candidates at very large distances from the Sun. As the Vera C. Rubin Observatory (*https://www.lsst.org/scientists/scibook*) comes online in the early 2020s, we will have a huge leap in survey uniformity and depth over a very large portion of the sky, potentially revealing many as-yet-unknown distant dwarf planets. As this population is further revealed and further studied, we can develop a better understanding of just how Pluto fits into the picture. As we go forward with our efforts to discover more of its far-flung kin, Pluto's singular story can only deepen our understanding of the yet more distant worlds waiting to be found.

6. SUMMARY AND CONCLUSIONS

By acting as a witness plate recording eons of bombardment by small TNOs, the Pluto system has provided us with a powerful means for estimating the TNO size-frequency distribution at sizes far smaller than existing direct-detection surveys. Folding in existing direct-detection constraints at larger sizes and using a likelihood-free inference method indicates that the size distribution of small TNOs breaks from a slope of $q_1 \simeq 5.15$ (for hot populations) or $q_1 \simeq 8.50$ (for cold populations) at diameters larger than 100 km to a population-average slope of $q_2 \simeq 2.90$ down to sizes of $d \sim 2.8$ km, where it breaks again to an even shallower slope of $q_3 \simeq 1.85$ down to sizes as small as $\simeq$0.1 km, limited by the current resolution of the crater record. The exact location of the break location between q_3 and q_2 is poorly constrained, perhaps suggesting that the SFD behavior in this regime is more complex than modeled. While this SFD does not support some models of TNO growth motivated by populations inferred from occultation detections (e.g., *S13*), it is not wildly inconsistent with the occultations themselves.

Furthermore, craters appear to have a size-spatial correlation; i.e., craters of similar sizes prefer to be closer to one another than they should be if they were randomly distributed on the surface. This effect persists after accounting for correlated discovery sensitivity and other effects. The hypothesis that this size-spatial correlation is due to the presence of a population of modestly separated binary impactors is tested with a similar likelihood-free inference approach. This analysis indicates strong evidence in favor of this hypothesis and concludes that there may be a pileup of tidally modified non-contact binaries with separations in the range of a few to tens of primary radii. The crater record suggests this population exists down to very small sizes.

Pluto's remarkable satellite system is also examined in the context of the ubiquity of satellite systems among the largest TNOs. This ubiquity indicates that a number of these large TNOs must have exerienced multiple moon-forming events early in their history, and the Pluto system may be the outcome of one of these multiple-event histories. The similarities of the Pluto and Haumea systems point to a novel concept for the origin of Haumea's satellites, its spin and figure, and its family — that it is an alternate end state for a system that once looked similar to that of Pluto.

The expression of processes that have been active on the surfaces of Pluto and Charon provide insights into those processes that may have shaped other large TNOs. Conditions on Eris are favorable for hosting extensive convective nitrogen ice deposits, providing another potential avenue to explain its fresh, bright, volatile-dominated surface in spite of photochemical processes that should rapidly alter and darken it. Makemake's smaller size and density make it less favorable for these processes. Thermophysical models indicate that internal oceans may yet survive within many of the largest TNOs, even if one does not persist within Charon.

Acknowledgments. This work was supported in part by the New Horizons mission. Development of the size distribution inference tools was supported by STScI grant AR 14309. The author thanks H. Levison, S. Protopapa, L. Young, B. Bottke, K. Nowicki, D. Nesvorný, K. Kratter, and S. Robbins for insightful discussions regarding many aspects of this document. The author would also like to thank the three chapter reviewers — B. Holler, K. Singer, and one anonymous reviewer — for their close reads of the document and productive criticisms.

REFERENCES

Adams E. R., Gulbis A. A. S., Elliot J. L., Benecchi S. D., Buie M. W., Trilling D. E., and Wasserman L. H. (2014) De-biased populations of Kuiper belt objects from the Deep Ecliptic Survey. *Astron. J., 148(3),* 55.

Alcock C., Brown M. E., Gauron T., Heneghan C., Holman M. J., Kenter A., Kraft R., et al. (2014) The Whipple mission: Exploring the Kuiper belt and the Oort cloud. Abstract P51D-3977 presented at 2014 Fall Meeting, AGU, San Francisco, California, 15–19 December.

Arimatsu K., Ohsawa R., Hashimoto G. L., Urakawa S., Takahashi J., Tozuka M., Itoh Y., et al. (2019a) New constraint on the atmosphere of (50000) Quaoar from a stellar occultation. *Astron. J., 158(6),* 236.

Arimatsu K., Tsumura K., Usui F., Shinnaka Y., Ichikawa K., Ootsubo T., Kotani T., Wada T., Nagase K., and Watanabe J. (2019b) A kilometre-sized Kuiper belt object discovered by stellar occultation using amateur telescopes. *Nature Astron., 3,* 301–306.

Bannister M. T., Alexandersen M., Benecchi S. D., Chen Y.-T., Delsanti A., Fraser W. C., Gladman B. J., et al. (2016a) OSSOS. IV. Discovery of a dwarf planet candidate in the 9:2 resonance with Neptune. *Astron. J., 152(6),* 212.

Bannister M. T., Kavelaars J. J., Petit J.-M., Gladman B. J., Gwyn S. D. J., Chen Y.-T., Volk K., et al. (2016b) The Outer Solar System Origins Survey. I. Design and first-quarter discoveries. *Astron. J., 152(3),* 70.

Barkume K. M., Brown M. E., and Schaller E. L. (2006) Water ice on the satellite of Kuiper belt object 2003 EL61. *Astrophys. J. Lett., 640(1),* L87–L89.

Batygin K. and Brown M. E. (2016) Evidence for a distant giant planet in the solar system. *Astron. J., 151(2),* 22.

Batygin K., Brown M. E., and Fraser W. C. (2011) Retention of a primordial cold classical Kuiper belt in an instability-driven model of solar system formation. *Astrophys. J., 738(1),* 13.

Benecchi S., Borncamp D., Parker A., Buie M., Noll K., Binzel R., Stern S. A., et al. (2018) The color and binarity of (486958) 2014 MU69 and other long-range new horizons Kuiper belt targets. *ArXiv e-prints,* arXiv:1812.04752.

Bernstein G. M., Trilling D. E., Allen R. L., Brown M. E., Holman M., and Malhotra R. (2004) The size distribution of trans-neptunian bodies. *Astron. J., 128(3),* 1364–1390.

Bertrand T., Forget F., Umurhan O. M., Grundy W. M., Schmitt B., Protopapa S., Zangari A. M., et al. (2018) The nitrogen cycles on Pluto over seasonal and astronomical timescales. *Icarus, 309,* 277–296.

Beyer R. A., Nimmo F., McKinnon W. B., Moore J. M., Binzel R. P., Conrad J. W., Cheng A., et al. (2017) Charon tectonics. *Icarus, 287,* 161–174.

Beyer R. A., Spencer J. R., McKinnon W. B., Nimmo F., Beddingfield C., Grundy W. M., Ennico K., et al. (2019) The nature and origin of Charon's smooth plains. *Icarus, 323,* 16–32.

Bickerton S. J., Welch D. L., and Kavelaars J. J. (2009) Kuiper belt object occultations: Expected rates, false positives, and survey design. *Astron. J., 137(5),* 4270-4281.

Bierson C. J., Nimmo F., and McKinnon W. B. (2018) Implications of the observed Pluto-Charon density contrast. *Icarus, 309,* 207–219.

Braga-Ribas F., Sicardy B., Ortiz J. L., Lellouch E., Tancredi G., Lecacheux J., Vieira-Martins R., et al. (2013) The size, shape, albedo, density, and atmospheric limit of transneptunian object (50000) Quaoar from multi-chord stellar occultations. *Astrophys. J., 773(1),* 26.

Brown M. E. (2001) The inclination distribution of the Kuiper belt. *Astron. J., 121(5),* 2804–2814.

Brown M. E. (2008) The largest Kuiper belt objects. In *The Solar System Beyond Neptune* (M. A Barucci et al., eds.), pp. 335–344. Univ. of Arizona, Tucson.

Brown M. E. (2013) On the size, shape, and density of dwarf planet Makemake. *Astrophys. J. Lett., 767(1),* L7.

Brown M. E. and Butler B. J. (2018) Medium-sized satellites of large Kuiper belt objects. *Astron. J., 156(4),* 164.

Brown M. E. and Calvin W. M. (2000) Evidence for crystalline water and ammonia ices on Pluto's satellite Charon. *Science, 287(5450),* 107–109.

Brown M. E., Kulkarni S. R., and Liggett T. J. (1997) An analysis of the statistics of the Hubble Space Telescope Kuiper Belt Object Search. *Astrophys. J. Lett., 490(1),* L119-L122.

Brown M. E., Trujillo C., and Rabinowitz D. (2004) Discovery of a candidate inner Oort cloud planetoid. *Astrophys. J., 617(1),* 645–649.

Brown M. E., Barkume K. M., Blake G. A., Schaller E. L., Rabinowitz D. L., Roe H. G., and Trujillo C. A. (2007) Methane and ethane on the bright Kuiper belt object 2005 FY9. *Astron. J., 133(1),* 284–289.

Brucker M. J., Grundy W. M., Stansberry J. A., Spencer J. R., Sheppard S. S., Chiang E. I., and Buie M. W. (2009) High albedos of low inclination classical Kuiper belt objects. *Icarus, 201(1),* 284–294.

Buie M. W., Millis R. L., Wasserman L. H., Elliot J. L., Kern S. D., Clancy K. B., Chiang E. I., et al. (2003) Procedures, resources and selected results of the deep ecliptic survey. *Earth Moon Planets, 92(1),* 113–124.

Canup R. M. (2005) A giant impact origin of Pluto-Charon. *Science, 307(5709),* 546–550.

Canup R. M. (2011) On a giant impact origin of Charon, Nix, and Hydra. *Astron. J., 141(2),* 35.

Carry B., Hestroffer D., DeMeo F. E., Thirouin A., Berthier J., Lacerda P., Sicardy B., et al. (2011) Integral-field spectroscopy of (90482) Orcus-Vanth. *Astron. Astrophys., 534,* A115.

Castro-Chacón J. H., Reyes-Ruiz M., Lehner M. J., Zhang Z. W., Alcock C., Guerrero C. A., Hernández-Valencia B, et al. (2019) Occultations by small non-spherical trans-neptunian objects. I. A new event simulator for TAOS II. *Publ. Astron. Soc. Pac., 131(1000),* 064401.

Chiang E. and Youdin A. N.(2010) Forming planetesimals in solar and extrasolar nebulae. *Annu. Rev. Earth Planet. Sci., 38,* 493-522.

Cochran A. L., Levison H. F., Stern S. A., and Duncan M. J. (1995) The discovery of Halley-sized Kuiper belt objects using the Hubble Space Telescope. *Astrophys. J., 455,* 342.

Cochran A. L., Levison H. F., Tamblyn P., Stern S. A., and Duncan M. J. (1998) The calibration of the Hubble Space Telescope Kuiper Belt Object Search: Setting the record straight. *Astrophys. J. Lett., 503(1),* L89-L93.

Cohen C. J., Hubbard E. C., and Oesterwinter C. (1967) New orbit for Pluto and analysis of differential corrections. *Astron. J., 72,* 973.

Cook J. C., Desch S. J., Roush T. L., Trujillo C. A., and Geballe T. R. (2007) Near-Infrared spectroscopy of Charon: Possible evidence for cryovolcanism on Kuiper belt objects. *Astrophys. J., 663(2),* 1406–1419.

Cook J. C., Ore C. M. D., Protopapa S., Binzel R. P., Cartwright R.,

Cruikshank D. P., Earle A., et al. (2018) Composition of Pluto's small satellites: Analysis of New Horizons spectral images. *Icarus, 315,* 30–45.

Cruikshank D. P., Umurhan O. M., Beyer R. A., Schmitt B., Keane J. T., Runyon K. D., Atri D., et al. (2019) Recent cryovolcanism in Virgil Fossae on Pluto. *Icarus, 330,* 155–168.

Cuzzi J. N., Hogan R. C., and Bottke W. F. (2010) Towards initial mass functions for asteroids and Kuiper belt objects. *Icarus, 208(2),* 518–538.

Dalle Ore C. M., Cruikshank D. P., Protopapa S., Scipioni F., McKinnon W. B., Cook J. C., Grundy W. M., et al. (2019) Detection of ammonia on Pluto's surface in a region of geologically recent tectonism. *Sci. Adv., 5(5),* eaav5731.

Desch S. J. (2015) Density of Charon formed from a disk generated by the impact of partially differentiated bodies. *Icarus, 246,* 37–47.

Desch S. J. and Neveu M. (2017) Differentiation and cryovolcanism on Charon: A view before and after New Horizons. *Icarus, 287,* 175–186.

Desch S. J., Cook J. C., Doggett T. C., and Porter S. B. (2009) Thermal evolution of Kuiper belt objects, with implications for cryovolcanism. *Icarus, 202(2),* 694–714.

Dias-Oliveira A., Sicardy B., Ortiz J. L., Braga-Ribas F., Leiva R., Vieira-Martins R., Benedetti-Rossi G., et al. (2017) Study of the Plutino object (208996) 2003 AZ84 from stellar occultations: Size, shape, and topographic features. *Astron. J., 154(1),* 22.

Dunham E. T., Desch S. J., and Probst L. (2019) Haumea's shape, composition, and internal structure. *Astrophys. J., 877(1),* 41.

Earle A. M., Grundy W., Howett C. J. A., Olkin C. B., Parker A. H., Scipioni F., Binzel R. P., et al. (2018) Methane distribution on Pluto as mapped by the New Horizons Ralph/MVIC instrument. *Icarus, 314,* 195–209.

Elliot J. L., Kern S. D., Clancy K. B., Gulbis A. A. S., Millis R. L., Buie M. W., Wasserman L. H., et al. (2005) The Deep Ecliptic Survey: A search for Kuiper belt objects and Centaurs. II. Dynamical classification, the Kuiper belt plane, and the core population. *Astron. J., 129(2),* 1117–1162.

Elliot J. L., Person M. J., Zuluaga C. A., Bosh A. S.,Adams E. R., Brothers T. C., Gulbis A. A. S., et al. (2010) Size and albedo of Kuiper belt object 55636 from a stellar occultation. *Nature, 465(7300),* 897–900.

Fraser W. C. and Kavelaars J. J. (2009) The size distribution of Kuiper belt objects for D $\gtrsim$10 km. *Astron. J., 137(1),* 72–82.

Fraser W. C., Brown M. E., and Schwamb M. E. (2010) The luminosity function of the hot and cold Kuiper belt populations. *Icarus, 210(2),* 944–955.

Fraser W. C., Brown M. E., Morbidelli A., Parker A., and Batygin K. (2014) The absolute magnitude distribution of Kuiper belt objects. *Astrophys. J., 782(2),* 100.

Fuentes C. I. and Holman M. J. (2008) A SUBARU archival search for faint trans-neptunian objects. *Astron. J., 136(1),* 83–97.

Fuentes C. I., George M. R., and Holman M. J. (2009) A Subaru pencil-beam search for m_R ~27 trans-neptunian bodies. *Astrophys. J., 696(1),* 91–95.

Gladman B., Kavelaars J. J., Nicholson P. D., Loredo T. J., and Burns J. A. (1998) Pencil-beam surveys for faint trans-neptunian objects. *Astron. J., 116(4),* 2042–2054.

Gladman B., Kavelaars J. J., Petit J.-M., Morbidelli A., Holman M. J., and Loredo T. (2001) The structure of the Kuiper belt: Size distribution and radial extent. *Astron. J., 122(2),* 1051–1066.

Gladman B., Marsden B. G., and Vanlaerhoven C. (2008) Nomenclature in the outer solar system. In *The Solar System Beyond Neptune* (M. A. Barucci et al., eds.), pp. 43–57. Univ. of Arizona, Tucson.

Gladman B., Lawler S. M., Petit J. M., Kavelaars J., Jones R. L., Parker J. W., Van Laerhoven C., et al. (2012) The resonant trans-neptunian populations. *Astron. J., 144(1),* 23.

Gomes R. S. (2000) Planetary migration and plutino orbital inclinations. *Astron. J., 120(5),* 2695–2707.

Gomes R. (2003) The common origin of the high inclination TNO's. *Earth Moon Planets, 92(1),* 29–42.

Greenstreet S., Gladman B., and McKinnon W. B. (2015) Impact and cratering rates onto Pluto. *Icarus, 258,* 267–288.

Grundy W. M. and Umurhan O. M. (2017) Are Makemake and Eris Sputnik planets? *AAS/Division for Planetary Sciences Meeting Abstracts, 49,* #202.02.

Grundy W. M., Binzel R. P., Buratti B. J., Cook J. C., Cruikshank D. P., Dalle Ore C. M., Earle A. M., et al. (2016) Surface compositions across Pluto and Charon. *Science, 351(6279),* aad9189.

Gulbis A. A. S., Elliot J. L., Adams E. R., Benecchi S. D., Buie M. W., Trilling D. E., and Wasserman L. H. (2010) Unbiased inclination distributions for objects in the Kuiper belt. *Astron. J., 140(2),* 350–369.

Hahn C., Vakili M., Walsh K., Hearin A. P., Hogg D. W., and Campbell D. (2017) Approximate Bayesian computation in large-scale structure: Constraining the galaxy-halo connection. *Mon. Not. R. Astron. Soc., 469(3),* 2791–2805.

Hahn J. M. and Malhotra R. (1999) Orbital evolution of planets embedded in a planetesimal disk. *Astron. J., 117(6),* 3041–3053.

Hammond N. P., Barr A. C., and Parmentier E. M. (2016) Recent tectonic activity on Pluto driven by phase changes in the ice shell. *Geophys. Res. Lett., 43(13),* 6775–6782.

Heinze A. N. and de Lahunta D. (2009) The rotation period and light-curve amplitude of Kuiper belt dwarf planet 136472 Makemake (2005 FY9). *Astron. J., 138(2),* 428–438.

Hofgartner J. D., Buratti B. J., Devins S. L., Beyer R. A., Schenk P., Stern S. A., Weaver H. A., et al. (2018) A search for temporal changes on Pluto and Charon. *Icarus, 302,* 273–284.

Hofgartner J. D., Buratti B. J., Hayne P. O., and Young L. A. (2019) Ongoing resurfacing of KBO Eris by volatile transport in local, collisional, sublimation atmosphere regime. *Icarus, 334,* 52–61.

Holsapple K. A. (1993) The scaling of impact processes in planetary sciences. *Annu. Rev. Earth Planet. Sci., 21,* 333–373.

Housen K. R., Sweet W. J., and Holsapple K. A. (2018) Impacts into porous asteroids. *Icarus, 300,* 72–96.

Howard A. D, Moore J. M., Umurhan O. M., White O. L., Anderson R. S., McKinnon W. B., Spencer J. R., et al. (2017) Present and past glaciation on Pluto. *Icarus, 287,* 287–300.

Hromakina T. A., Belskaya I. N., Krugly Y. N., Shevchenko V. G., Ortiz J. L., Santos-Sanz P., Duffard R., et al. (2019) Long-term photometric monitoring of the dwarf planet (136472) Makemake. *Astron. Astrophys., 625,* A46.

Hsu D. C., Ford E. B., Ragozzine D., and Morehead R. C. (2018) Improving the accuracy of planet occurrence rates from Kepler using approximate Bayesian computation. *Astron. J., 155(5),* 205.

Hussmann H., Sohl F., and Spohn T. (2006) Subsurface oceans and deep interiors of medium-sized outer planet satellites and large trans-neptunian objects. *Icarus, 185(1),* 258-273.

Ishida E. E. O., Vitenti S. D. P., Penna-Lima M., Cisewski J., de Souza R. S., Trindade A. M. M., Cameron E., Busti V. C., and COIN Collaboration (2015) COSMOABC: Likelihood-free inference via population Monte Carlo approximate Bayesian computation. *Astron. Comput., 13,* 1–11.

Jackson A. P., Tamayo D., Hammond N., Ali-Dib M., and Rein H. (2018) Ejection of rocky and icy material from binary star systems: Implications for the origin and composition of 1I/'Oumuamua. *Mon. Not. R. Astron. Soc. Lett., 478(1),* L49–L53.

Jílková L., Portegies Zwart S., Pijloo T., and Hammer M. (2015) How Sedna and family were captured in a close encounter with a solar sibling. *Mon. Not. R. Astron. Soc., 453(3),* 3157–3162.

Johnson B. C., Bowling T. J., Trowbridge A. J., and Freed A. M. (2016) Formation of the Sputnik Planum basin and the thickness of Pluto's subsurface ocean. *Geophys. Res. Lett., 43(19),* 10068–10077.

Johnston W. R. (2018) *Binary Minor Planets Compilation V2.0.* NASA Planetary Data System, EAR-A-COMPIL-5-BINMP-V9.0.

Jones R. L., Gladman B., Petit J. M., Rousselot P., Mousis O., Kavelaars J. J., Campo Bagatin A., et al. (2006) The CFEPS Kuiper Belt Survey: Strategy and presurvey results. *Icarus, 185(2),* 508–522.

Kamata S., Nimmo F., Sekine Y., Kuramoto K., Noguchi N., Kimura J., and Tani A. (2019) Pluto's ocean is capped and insulated by gas hydrates. *Nature Geosci., 12(6),* 407–410.

Kavelaars J. J., Jones R. L., Gladman B. J., Petit J. M., Parker J. W., Van Laerhoven C., Nicholson P., et al. (2009) The Canada-France Ecliptic Plane Survey — L3 Data Release: The orbital structure of the Kuiper belt. *Astron. J., 137(6),* 4917–4935.

Kavelaars J., Bickerton S., Brown P., Duncan M., Gladman B., Kaib N., Scott A., Walker G., Welch D., and Weigert P. (2010) *Oort Cloud Explorer–Dynamic Occultation Experiment: OCLE-DOCLE.* CASCA LRP2010 Reports, Canadian Astronomical Society.

Keane J. T., Matsuyama I., Kamata S., and Steckloff J. K. (2016) Reorientation and faulting of Pluto due to volatile loading within Sputnik Planitia. *Nature, 540(7631),* 90–93.

Kiss C., Marton G., Parker A. H., Grundy W. M., Farkas-Takács A., Stansberry J., Pál A., et al. (2019) The mass and density of the dwarf planet (225088) 2007 OR10. *Icarus, 334,* 3–10.

Lawler S. M., Pike R. E., Kaib N., Alexandersen M., Bannister M. T., Chen Y. T., Gladman B., et al. (2019) OSSOS. XIII. Fossilized resonant dropouts tentatively confirm Neptune's migration was grainy and slow. *Astron. J., 157(6),* 253.

Lehner M. J., Wang S.-Y., Reyes-Ruíz M., Zhang Z.-W.,Figueroa L., Huang C.-K., Yen W.-L., et al. (2018) Status of the Transneptunian Automated Occultation Survey (TAOS II). In *Ground-Based and Airborne Telescopes VII* (H. K. Marshall and J. Spyromilio, eds.), pp. 107004V. SPIE Proceedings Vol. 10700, Bellingham, Washington.

Leinhardt Z. M., Marcus R. A., and Stewart S. T. (2010) The formation of the collisional family around the dwarf planet Haumea. *Astrophys. J., 714(2),* 1789–1799.

Levison H. F. and Morbidelli A. (2003) The formation of the Kuiper belt by the outward transport of bodies during Neptune's migration. *Nature, 426(6965),* 419–421.

Levison H. F. and Stern S. A. (2001) On the size dependence of the inclination distribution of the main Kuiper belt. *Astron. J., 121(3),* 1730–1735.

Levison H. F., Morbidelli A., Gomes R., and Backman D. (2007) Planet migration in planetesimal disks. In *Protostars and Planets V* (B. Reipurth et al., eds.), pp. 669–684. Univ. of Arizona, Tucson.

Levison H. F., Morbidelli A., Van Laerhoven C., Gomes R., and Tsiganis K. (2008) Origin of the structure of the Kuiper belt during a dynamical instability in the orbits of Uranus and Neptune. *Icarus, 196(1),* 258–273.

Licandro J., Pinilla-Alonso N., Pedani M., Oliva E., Tozzi G. P., and Grundy W. M. (2006) The methane ice rich surface of large TNO 2005 FY9: A Pluto-twin in the trans-neptunian belt? *Astron. Astrophys., 445(3),* L35–L38.

Lim T. L., Stansberry J., Müller T. G., Mueller M., Lellouch E., Kiss C., Santos-Sanz P., et al. (2010) "TNOs are cool": A survey of the trans-neptunian region. III. Thermophysical properties of 90482 Orcus and 136472 Makemake. *Astron. Astrophys., 518,* L148.

Malamud U., Perets H. B., and Schubert G. (2017) The contraction/expansion history of Charon with implications for its planetary-scale tectonic belt. *Mon. Not. R. Astron. Soc., 468(1),* 1056–1069.

Malhotra R. (1993) The origin of Pluto's peculiar orbit. *Nature, 365(6449),* 819–821.

Malhotra R. (1995) The origin of Pluto's orbit: Implications for the solar system beyond Neptune. *Astron. J., 110,* 420.

Malhotra R. (1998) Pluto's inclination excitation by resonance sweeping. *Lunar Planet. Sci. XXIX,* Abstract #1476. Lunar and Planetary Institute, Houston.

Marin J.-M., Pudlo P., Robert C. P., and Ryder R. (2011) Approximate Bayesian computational methods. *ArXiv e-prints,* arXiv:1101.0955.

Mazrouei S., Ghent R. R., Bottke W. F., Parker A. H., and Gernon T. M. (2019) Earth and Moon impact flux increased at the end of the Paleozoic. *Science, 363(6424),* 253–257.

McKinnon W. B. (1989) On the origin of the Pluto-Charon binary. *Astrophys. J. Lett., 344,* L41.

McKinnon W. B. and Schenk P. M. (1995) Estimates of comet fragment masses from impact crater chains on Callisto and Ganymede. *Geophys. Res. Lett., 22(13),* 1829–1832.

McKinnon W. B., Nimmo F., Wong T., Schenk P. M., White O. L., Roberts J. H., Moore J. M., et al. (2016) Convection in a volatile nitrogen-ice-rich layer drives Pluto's geological vigour. *Nature, 534(7605),* 82–85.

Milani A., Nobili A. M., and Carpino M. (1989) Dynamics of Pluto. *Icarus, 82(1),* 200–217.

Millis R. L., Buie M. W., Wasserman L. H., Elliot J. L., Kern S. D., and Wagner R. M. (2002) The Deep Ecliptic Survey: A Search for Kuiper belt objects and Centaurs. I. Description of methods and initial results. *Astron. J., 123(4),* 2083–2109.

Miljković K., Collins G. S., Mannick S., and Bland P. A. (2013) Morphology and population of binary asteroid impact craters. *Earth Planet. Sci. Lett., 363,* 121–132.

Moore J. M., McKinnon W. B., Spencer J. R., Howard A. D., Schenk P. M., Beyer R. A., Nimmo F., et al. (2016) The geology of Pluto and Charon through the eyes of New Horizons. *Science, 351(6279),* 1284–1293.

Moore J. M., Howard A. D., Umurhan O. M., White O. L., Schenk P. M., Beyer R. A., McKinnon W. B., et al. (2017) Sublimation as a landform-shaping process on Pluto. *Icarus, 287,* 320–333.

Moore J. M., Howard A. D., Umurhan O. M., White O. L., Schenk P. M., Beyer R. A., McKinnon W. B., et al. (2018) Bladed terrain on Pluto: Possible origins and evolution. *Icarus, 300,* 129–144.

Morbidelli A. and Levison H. F. (2004) Scenarios for the origin of the orbits of the trans-neptunian objects 2000 CR105 and 2003 VB12 (Sedna). *Astron. J., 128(5),* 2564–2576.

Morbidelli A., Thomas F., and Moons M. (1995) The resonant structure of the Kuiper belt and the dynamics of the first five trans-neptunian objects. *Icarus, 118(2),* 322–340.

Nesvorný D. (2015) Evidence for slow migration of Neptune from the inclination distribution of Kuiper belt objects. *Astron. J., 150(3),* 73.

Nesvorný D. and Vokrouhlický D. (2016) Neptune's orbital migration was grainy, not smooth. *Astrophys. J., 825(2),* 94.

Nesvorný D., Youdin A. N., and Richardson D. C. (2010) Formation of Kuiper belt binaries by gravitational collapse. *Astron. J., 140(3),* 785–793.

Nesvorný D., Vokrouhlický D., Bottke W. F., Noll K., and Levison H. F. (2011) Observed binary fraction sets limits on the extent of collisional grinding in the Kuiper belt. *Astron. J., 141(5),* 159.

Nesvorný D., Li R., Youdin A. N., Simon J. B., and Grundy W. M. (2019) Trans-neptunian binaries as evidence for planetesimal formation by the streaming instability. *Nature Astron., 3,* 808–812.

Neveu M. and Desch S. J. (2015) Geochemistry, thermal evolution, and cryovolcanism on Ceres with a muddy ice mantle. *Geophys. Res. Lett., 42(23),* 10,197–10,206.

Neveu M., Desch S. J., Shock E. L., and Glein C. R. (2015) Prerequisites for explosive cryovolcanism on dwarf planet-class Kuiper belt objects. *Icarus, 246,* 48–64.

Nihei T. C., Lehner M. J., Bianco F. B., King S.-K., Giammarco J. M., and Alcock C. (2007) Detectability of occultations of stars by objects in the Kuiper belt and Oort cloud. *Astron. J., 134(4),* 1596–1612.

Nimmo F., Hamilton D. P., McKinnon W. B., Schenk P. M., Binzel R. P., Bierson C. J., Beyer R. A., et al. (2016) Reorientation of Sputnik Planitia implies a subsurface ocean on Pluto. *Nature, 540(7631),* 94–96.

Noll K. S., Grundy W. M., Stephens D. C., Levison H. F., and Kern S. D. (2008) Evidence for two populations of classical transneptunian objects: The strong inclination dependence of classical binaries. *Icarus, 194(2),* 758–768.

Noll K., Grundy W. M., Nesvorný D., and Thirouin A. (2020) Trans-neptunian binaries. In *The Trans-Neptunian Solar System* (D. Prialnik et al., eds.), pp. 201–224. Elsevier, Netherlands.

Ortiz J. L., Sicardy B., Braga-Ribas F., Alvarez-Candal A., Lellouch E., Duffard R., Pinilla-Alonso N., et al. (2012) Albedo and atmospheric constraints of dwarf planet Makemake from a stellar occultation. *Nature, 491(7425),* 566–569.

Ortiz J. L., Santos-Sanz P., Sicardy B., Benedetti-Rossi G., Bérard D., Morales N., Duffard R., et al. (2017) The size, shape, density and ring of the dwarf planet Haumea from a stellar occultation. *Nature, 550(7675),* 219–223.

Parker A. H. (2015) The intrinsic Neptune Trojan orbit distribution: Implications for the primordial disk and planet migration. *Icarus, 247,* 112–125.

Parker A. H. and Kavelaars J. J. (2010) Destruction of binary minor planets during Neptune scattering. *Astrophys. J. Lett., 722(2),* L204–L208.

Parker A. H. and Kavelaars J. J. (2012) Collisional evolution of ultra-wide trans-neptunian binaries. *Astrophys. J., 744(2),* 139.

Parker A., Ivezić Ž., Jurić M., Lupton R., Sekora M. D., and Kowalski A. (2008) The size distributions of asteroid families in the SDSS Moving Object Catalog 4. *Icarus, 198(1),* 138–155.

Parker A. H., Kavelaars J. J., Petit J.-M., Jones L., Gladman B., and Parker J. (2011) Characterization of seven ultra-wide trans-neptunian binaries. *Astrophys. J., 743(1),* 1.

Parker A. H., Buie M., Spencer J., Fraser W., Porter S. B., Weaver H., Stern S. A., et al. (2015) Updating the Kuiper belt luminosity function with the HST search for a New Horizons post-Pluto target. *Lunar Planet. Sci. XLVI,* Abstract #2614. Lunar and Planetary Institute, Houston.

Parker A., Pinilla-Alonso N., Santos-Sanz P., Stansberry J., Alvarez-Candal A., Bannister M., Benecchi S., et al. (2016a) Physical characterization of TNOs with the James Webb Space Telescope. *Publ. Astron. Soc. Pac., 128(959),* 018010.

Parker A. H., Buie M. W., Grundy W. M., and Noll K. S. (2016b) Discovery of a Makemakean moon. *Astrophys. J. Lett., 825(1),* L9.

Parker A., Buie M. W., Grundy W., Noll K., Young L., Schwamb M. E., Kiss C., Marton G., and Farkas-Takács A. I. (2018) The mass, density, and figure of the Kuiper belt dwarf planet Makemake. *AAS/Division for Planetary Sciences Meeting Abstracts, 50,* #509.02.

Petit J.-M. and Mousis O. (2004) KBO binaries: How numerous were they? *Icarus, 168(2),* 409–419.

Petit J. M., Kavelaars J. J., Gladman B. J., Jones R. L., Parker J. W., Van Laerhoven C., Nicholson P., et al. (2011) The Canada-France Ecliptic Plane Survey — Full Data Release: The orbital structure of the Kuiper belt. *Astron. J., 142(4),* 131.

Petit J. M., Kavelaars J. J., Gladman B. J., Jones R. L., Parker J. W., Bieryla A., Van Laerhoven C., et al. (2017) The Canada-France Ecliptic Plane Survey (CFEPS) — High-latitude component. *Astron. J., 153(5),* 236.

Pike R. E., Proudfoot B. C. N., Ragozzine D., Alexandersen M., Maggard S., Bannister M. T., Chen Y.-T., et al. (2019) A dearth of small members in the Haumea family revealed by OSSOS. *Nature Astron., 4,* 89–96.

Porter S. B. and Grundy W. M. (2012) KCTF evolution of trans-neptunian binaries: Connecting formation to observation. *Icarus, 220(2),* 947–957.

Pritchard J. K., Seielstad M. T., Perez-Lezaun A., and Feldman M. W. (1999) Population growth of human Y chromosomes: A study of Y chromosome microsatellites. *Mol. Biol. Evol., 16,* 1791–1798.

Rambaux N., Baguet D., Chambat F., and Castillo-Rogez J. C. (2017) Equilibrium shapes of large trans-neptunian objects. *Astrophys. J. Lett., 850(1),* L9.

Robbins S. J., Antonenko I., Kirchoff M. R., Chapman C. R., Fassett C. I.; Herrick R. R.; Singer K., et al. (2014) The variability of crater identification among expert and community crater analysts. *Icarus, 234,* 109-131.

Robbins S. J., Singer K. N., Bray V. J., Schenk P., Lauer T. R., Weaver H. A., Runyon K., et al. (2017) Craters of the Pluto-Charon system. *Icarus, 287,* 187–206.

Rubin M. E., Desch S. J., and Neveu M. (2014) The effect of Rayleigh-Taylor instabilities on the thickness of undifferentiated crust on Kuiper belt objects. *Icarus, 236,* 122–135.

Sandford E.; Kipping D., and Collins M. (2019) The multiplicity distribution of Kepler's exoplanets. *Mon. Not. R. Astron. Soc., 489(3),* 3162–3173.

Santos-Sanz P., French R. G., Pinilla-Alonso N., Stansberry J., Lin Z. Y., Zhang Z. W., Vilenius E., et al. (2016) James Webb Space Telescope observations of stellar occultations by solar system bodies and rings. *Publ. Astron. Soc. Pac., 128(959),* 018011.

Schaller E. L. and Brown M. E. (2007) Volatile loss and retention on Kuiper belt objects. *Astrophys. J. Lett., 659(1),* L61–L64.

Schlichting H. E., Ofek E. O., Wenz M., Sari R., Gal-Yam A., Livio M., Nelan E., and Zucker S. (2009) A single sub-kilometre Kuiper belt object from a stellar occultation in archival data. *Nature, 462(7275),* 895–897.

Schlichting H. E., Ofek E. O., Sari R., Nelan E. P., Gal-Yam A., Wenz M., Muirhead P., Javanfar N., and Livio M. (2012) Measuring the abundance of sub-kilometer-sized Kuiper belt objects using stellar occultations. *Astrophys. J., 761(2),* 150.

Schlichting H. E., Fuentes C. I., and Trilling D. E. (2013) Initial planetesimal sizes and the size distribution of small Kuiper belt objects. *Astron. J., 146(2),* 36.

Shankman C., Kavelaars J. J., Gladman B. J., Alexandersen M., Kaib N., Petit J.-M., Bannister M. T., et al. (2016) OSSOS. II. A sharp transition in the absolute magnitude distribution of the Kuiper belt's scattering population. *Astron J., 151(2),* 31.

Shankman C., Kavelaars J. J., Bannister M. T., Gladman B. J., Lawler S. M., Chen Y.-T., Jakubik M., et al. (2017) OSSOS. VI. Striking biases in the detection of large semimajor axis trans-neptunian objects. *Astron. J., 154(2),* 50.

Sheppard S. S., Trujillo C. A., Tholen D. J., and Kaib N. (2019) A new high perihelion trans-plutonian inner Oort cloud object: 2015 TG387. *Astron. J., 157(4),* 139.

Sicardy B., Ortiz J. L., Assafin M., Jehin E., Maury A., Lellouch E., Hutton R. G., et al. (2011) A Pluto-like radius and a high albedo for the dwarf planet Eris from an occultation. *Nature, 478(7370),* 493–496.

Singer K. N. and Stern S. A. (2015) On the provenance of Pluto's nitrogen (N_2). *Astrophys. J. Lett., 808(2),* L50.

Singer K. N., Schenk P. M., McKinnon W. B., Beyer R. A., Schmitt B., White O. L., Moore J. M., et al. (2018) Cryovolcanic constructs on Pluto. *AAS/Division for Planetary Sciences Meeting Abstracts, 50,* #506.04.

Singer K. N., McKinnon W. B., Gladman B., Greenstreet S., Bierhaus E. B., Stern S. A., Parker A. H., et al. (2019) Impact craters on Pluto and Charon indicate a deficit of small Kuiper belt objects. *Science, 363(6430),* 955–959.

Stern S. A. (1991) On the number of planets in the outer solar system: Evidence of a substantial population of 1000-km bodies. *Icarus, 90(2),* 271–281.

Stern S. A. and Trafton L. M. (2008) On the atmospheres of objects in the Kuiper belt. In *The Solar System Beyond Neptune* (M. A. Barucci et al., eds.), pp. 365–380. Univ. of Arizona, Tucson.

Stern S. A., Trafton L. M., and Gladstone G. R. (1988) Why is Pluto bright? Implications of the albedo and lightcurve behavior of Pluto. *Icarus, 75(3),* 485–498.

Stern S. A., Weaver H. A., Spencer J. R., Olkin C. B., Gladstone G. R., Grundy W. M., Moore J. M., et al. (2019) Initial results from the New Horizons exploration of 2014 MU69, a small Kuiper belt object. *Science, 364(6441),* aaw9771.

Stevens D. J., Stassun K. G., and Gaudi B. S. (2017) Empirical bolometric fluxes and angular diameters of 1.6 million Tycho-2 stars and radii of 350,000 stars with Gaia DR1 parallaxes. *Astron. J., 154(6),* 259.

Telfer M. W., Parteli E. J. R., Radebaugh J., Beyer R. A.,Bertrand T., Forget F., Nimmo F., et al. (2018) Dunes on Pluto. *Science, 360(6392),* 992–997.

Trujillo C. A. and Brown M. E. (2002) A correlation between inclination and color in the classical Kuiper belt. *Astrophys. J. Lett., 566(2),* L125–L128.

Trujillo C. A. and Sheppard S. S. (2014) A Sedna-like body with a perihelion of 80 astronomical units. *Nature, 507(7493),* 471–474.

Vilenius E., Stansberry J., Müller T., Mueller M., Kiss C., Santos-Sanz P., Mommert M., et al. (2018) "TNOs are cool": A survey of the trans-neptunian region. XIV. Size/albedo characterization of the Haumea family observed with Herschel and Spitzer. *Astron. Astrophys., 618,* A136.

Volk K., Murray-Clay R., Gladman B., Lawler S., Bannister M. T., Kavelaars J. J., Petit J.-M., et al. (2016) OSSOS III — Resonant trans-neptunian populations: Constraints from the first quarter of the Outer Solar System Origins Survey. *Astron. J., 152(1),* 23.

Weaver H. A., Buie M. W., Buratti B. J., Grundy W. M., Lauer T. R., Olkin C. B., Parker A. H., et al. (2016) The small satellites of Pluto as observed by New Horizons. *Science, 351(6279),* aae0030.

Witzel G., Martinez G., Hora J., Willner S. P., Morris M. R., Gammie C., Becklin E. E., et al. (2018) Variability timescale and spectral index of Sgr A* in the near infrared: Approximate Bayesian computation analysis of the variability of the closest supermassive black hole. *Astrophys. J., 863(1),* 15.

Young L. A. (2013) Pluto's seasons: New predictions for New Horizons. *Astrophys. J. Lett., 766(2),* L22.

Young L. A., Braga-Ribas F., and Johnson R. E. (2020) Volatiles evolution and atmospheres of trans-neptunian objects. In *The Trans-Neptunian Solar System* (D. Prialnik et al., eds.), pp. 127–151. Elsevier, Amsterdam.

Zahnle K., Schenk P., Levison H., and Dones L. (2003) Cratering rates in the outer solar system. *Icarus, 163(2),* 263–289.

Buie M. W., Hofgartner J. D., Bray V. J., and Lellouch E. (2021) Future exploration of the Pluto system. In *The Pluto System After New Horizons* (S. A. Stern, J. M. Moore, W. M. Grundy, L. A. Young, and R. P. Binzel, eds.), pp. 569–586. Univ. of Arizona, Tucson, DOI: 10.2458/azu_uapress_9780816540945-ch024.

Future Exploration of the Pluto System

Marc W. Buie
Southwest Research Institute

Jason D. Hofgartner
Jet Propulsion Laboratory, California Institute of Technology

Veronica J. Bray
University of Arizona

Emmanuel Lellouch
Observatoire de Paris, Meudon

The investigation of the Pluto system is far from over. Many decades of remote telescopic observation and a spectacular flyby encounter with New Horizons has transformed an unresolved dot in the sky into a complex and fascinating world. We present a discussion of future areas of investigation that will take us to the next higher level of understanding of both the system in its own right as well as how it fits into the overall story of the formation and evolution of our solar system. Groundbased and near-Earth space telescopes will continue to play an important role by extending the timebase of observation while also taking advantage of the opportunities enabled by the development of new technology. The most important follow-on investigation is a return to the system with an orbital mission. Such a mission would extend the New Horizons snapshot into a global characterization of the complex geology on all the bodies in the system, provide a more complete inventory of the surface composition, and allow a more in-depth characterization of the atmosphere and dynamics of Pluto. These examples, and more, are presented to show the scientific value of such an ambitious undertaking that fits well within the bounds of increased spacecraft capabilities soon to come and will be just as valuable in studying other Kuiper belt objects and Triton for an even broader context.

1. INTRODUCTION

Unlocking Pluto's secrets requires patient long-term efforts. Its volatile inventory, atmosphere, and surface temperature can lead to processes that modify measureable surface properties, albedo and color being the two easiest to measure (*Stern et al.,* 1988). Its long orbital period of 250 years suggests that temporal changes would play out over long timescales. The great distance of Pluto from the Sun (30–50 AU) limits the energy from solar insolation and is an important consideration for how the atmosphere evolves over time (*Hunten and Watson,* 1982). Internal energy could contribute to change as well, but prior to the New Horizons encounter the general expectation was that there shouldn't be any such sources of energy. With our post-encounter view, it appears that internal heat played a larger role over longer timescales than expected (*Moore et al.,* 2016). However, the changes seen expressed on the surface are largely on a geologic timescale, and even the fastest of these changes still appear to be significantly longer than an orbital period.

Pluto is accompanied by a retinue of satellites — Charon, Styx, Nix, Kerberos, and Hydra — each of which are worthy of study in their own right. These bodies are not dynamic worlds except for the very slow process driven by impacts from other objects in the outer solar system. Nonetheless, they retain information on their surfaces that constrain formation and early evolution. The New Horizons data now provide the initial geologic framework for the early history of Charon, an epoch of post-formation cooling and an epoch of crustal expansion that lasted longer than previously thought possible (*Beyer et al.,* 2018). Knowledge of the much smaller satellites is much more limited and is confined to more basic properties with little to no geologic context (*Weaver et al.,* 2016).

We divide future investigations of the Pluto system into broad categories. One broad category is based on remote observations conducted from a distant observatory, such as those based on or near the Earth. Another category is remote sensing from spacecraft in or near the Pluto system. This chapter will cover these two broad subjects. A third category that will not be discussed because they are so remote in time are *in situ* measurements such as might be made by a lander, rover, or penetrator.

Our discussion also covers the timescales relevant for the measurements. Some relevant timescales of concern are diurnal (one Pluto rotation = 6.4 days), seasonal (one Pluto year = 250 years), Milankovitch cycles (millions of years), and then processes acting over geologic time (*Earle and Binzel,* 2015). The current body of work on the Pluto system has now established a good baseline from which these studies can continue. This chapter will project what current instrumentation can do to further study the Pluto system and look ahead to what future developments may bring. Most of the focus here is on the two main bodies, Pluto and Charon, and more discussion about the small satellites can be found in the chapter by Porter et al. in this volume.

2. REVIEW OF CURRENT DATA

From its discovery in 1930 until the first spacecraft encounter in 2015, all we knew about the Pluto system came from telescopic observations from a vantage point on or near Earth. These observations were extremely valuable in defining the scientific questions and the instruments needed for that first encounter. The completion of the New Horizons flyby brings us to the next steps in the continued exploration of the Pluto system. To set the stage for looking to the future, we briefly review some of the important sources of data about Pluto and its satellites.

Observations of Pluto from Earth-based telescopes provide the longest timebase of constraint for any type of measurement and are also the most certain to continue into the future. The history of these observations is dominated by cases where new instrumentation and new telescopic capabilities open additional avenues of study into the Pluto system and once opened, add to the suite of measurements that can and should continue to be investigated to understand the time-variability of the system. The same can be said for observations with spacebased facilities near Earth, where we define "near Earth" to refer to any spacecraft that is essentially at 1 AU from the Sun. This definition includes those in Earth orbit, Earth-Sun Lagrange points, or Earth leading or trailing orbits.

Figure 1 provides a graphical summary of Pluto system exploration along with the context of its seasons and viewing geometry over one Pluto year. We are now 90 years into the study of the Pluto system during its first circuit around the Sun since discovery. Even though the timescales are long on Pluto, we were fortunate to be working with the short half of the orbit. Equinox and perihelion are currently nearly coincident and we will soon be completing the time from northern polar solstice to southern polar solstice in much less than half of the orbital period. We can perhaps expect that observable changes will take longer for the rest of this first observed orbit around the Sun but some changes seem inevitable (such as atmospheric evolution), and the timing of those changes will provide important clues about processes that drive the atmosphere-surface system on Pluto.

In 2015, the New Horizons spacecraft flew through the Pluto system. It was purpose-built for the study of Pluto as informed by all the prior remote exploration of Pluto. The details of the groundbased, spacebased, and New Horizons efforts is left to the many other chapters in this volume. However, this experience shows the value of synergistic efforts that combine long time-based observations from the ground, more in-depth investigations from near-Earth space assets that can go beyond the limitations imposed by Earth's atmosphere, and a well-targeted and well-designed spacecraft visiting the system to fundamentally improve our understanding of the Pluto system. This synergy will be important going forward as we begin to contemplate what the next step will be as we continue to study Pluto through its first year since discovery.

3. SCIENCE GOALS FOR FUTURE OBSERVATIONS AND MISSIONS

The groundbased, near-Earth spacebased, and New Horizons studies of Pluto, Charon, and their satellite system to date have already provided fundamental advances in planetary science. Continued observations and exploration of the Pluto system are important to resolve fundamental questions raised thus far. This section discusses how Pluto system science fits into crosscutting themes of planetary science as well as science goals for future observations of and missions to the Pluto system.

3.1. Contributions to Cross-Cutting Themes in Planetary Science

The U.S. 2011 Planetary Science Decadal Survey identified three cross-cutting themes of particular importance in planetary science: building new worlds (formation), planetary habitats (habitability), and workings of solar systems (*National Research Council,* 2011); the Pluto system offers important insights into all three of these themes.

3.1.1. Formation. The Pluto-Charon binary is the largest fully tidally evolved binary pair in the solar system and was likely formed by a giant impact (*Canup,* 2005). Thus, it is an important record of giant impacts and the evolution of binary systems. By virtue of its formation in the Kuiper belt, the Pluto system also records the physical and chemical properties of the Kuiper belt. It may also record important clues about the dynamical rearrangement of the solar system (e.g., crater record of impactors).

3.1.2. Habitability. Organic matter is a crucial ingredient of a habitable environment (*National Research Council,* 2011) and Pluto's dark red terrains are a large reservoir

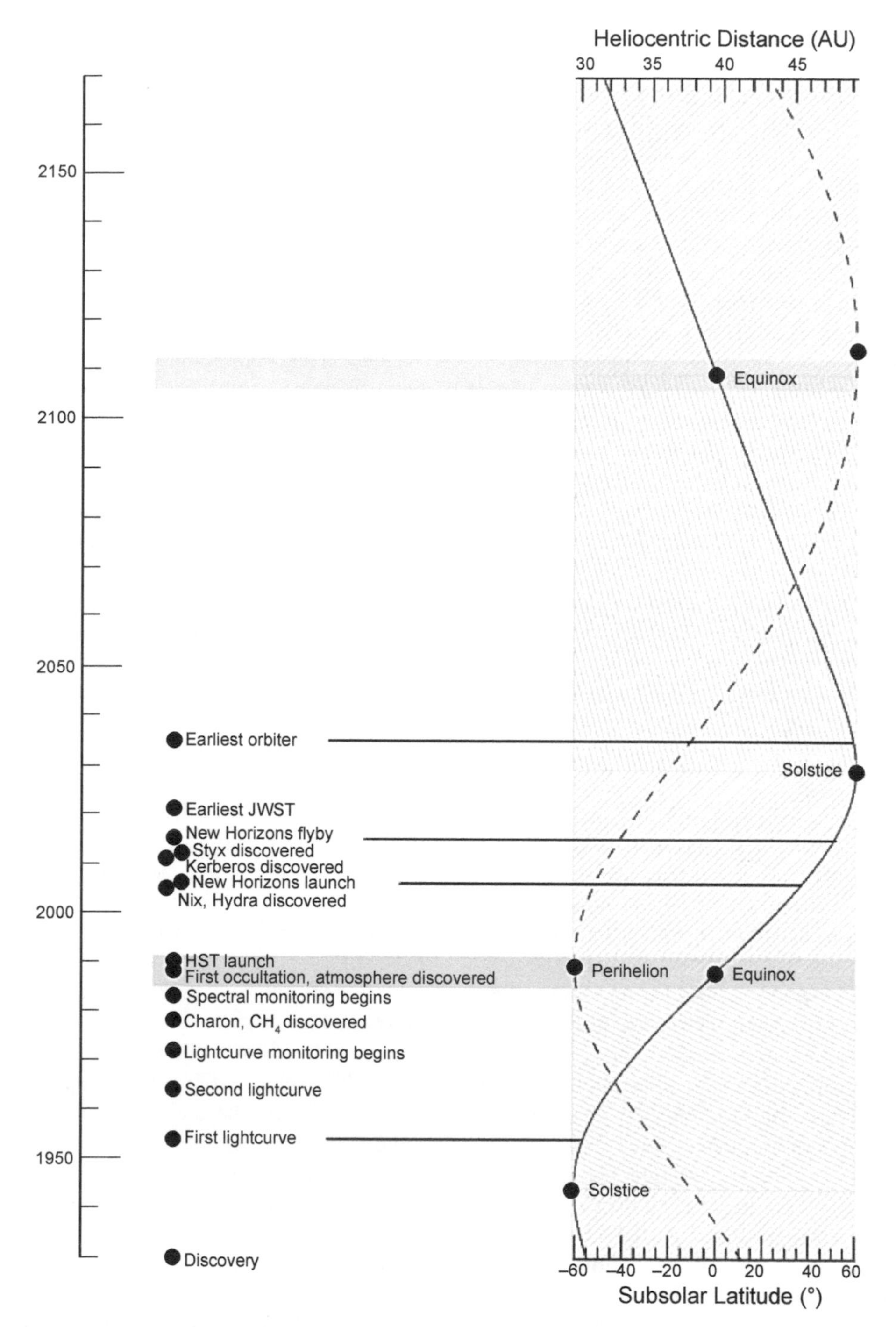

Fig. 1. Timeline and context of Pluto system investigations. One revolution of Pluto around the Sun is shown, beginning with Pluto's discovery. Significant events and discoveries discussed in other chapters are noted on the left. The two full-width gray bands centered on equinox indicate the time during which mutual events occur. The graph on the right shows the subsolar latitude (solid line) and the heliocentric distance over this time (dashed line). The pattern-filled regions indicate the seasons during the year and show how these seasons are stretched or contracted due to the eccentricity of the orbit.

of processed organic matter [similar to tholins produced in terrestrial laboratories (cf. *Grundy et al.,* 2016a, and references therein)]. Pluto's tholins are the most spectroscopically accessible (from a remote sensing perspective) complex organics in the solar system in the sense that they are the largest known such deposits aside from those on Titan (and Earth) and are not obscured by an opaque atmosphere. Liquid water is essential for all known life, without exception, and thus is an essential component of a habitable environment (for life as we know it). Pluto is also a possible ocean world (*Keane et al.,* 2016; *Nimmo et al.,* 2016) with evidence for geologically recent cryovolcanism (*Moore et al.,* 2016). Even Charon likely harbored an ocean in its past (*Beyer et al.,* 2018).

3.1.3. Workings. Pluto is astoundingly active given the limited endogenic and solar power available for geologic and atmospheric processes. Convective resurfacing, glacial flow, and sublimation/condensation are all presently ongoing (*Moore et al.,* 2016). Convection of Pluto's 1000-km-wide, nitrogen-rich Sputnik Planitia is a geologic process unique to Pluto's surface; it is the only known example where this resurfacing process can be studied (*McKinnon et al.,* 2016). In contrast, glacial flow is a familiar process on Earth and Mars, but the glaciers on Pluto are composed of entirely different volatiles, offering an excellent opportunity for comparing how composition and planetary conditions affect glacial processes (*Moore et al.,* 2016). Pluto's atmospheric pressure varies seasonally from exchange of volatiles with the surface and ongoing radiolysis in Pluto's atmosphere produces hazes that are organized, possibly by gravity waves, into ~20 distinct layers, some of which are continuous for >1000 km (*Gladstone et al.,* 2016). The Pluto system also exhibits volatile exchange between Pluto and Charon, resulting in a tholin deposit on Charon that is unique from all other icy satellites (*Grundy et al.,* 2016b). The system has diverse, unique, and ongoing processes, all of which are valuable characteristics for understanding the workings of solar systems.

3.2. Science Goals

The Pluto-Charon system offers a wide breadth of science covering many multidisciplinary topics. A sample of the outstanding priority science questions are grouped in Fig. 2. We divide the major science motivations for future observations and spacecraft missions into five broad science goals.

3.2.1. Pluto's atmospheric structure and haze. Pluto has a complex atmosphere that is composed of several volatiles, evolves on multiple timescales through exchange with volatile ices on the surface, and includes abundant hazes with a rich hydrocarbon-nitrile chemistry. Many science questions about Pluto's atmosphere remain to be satisfactorily answered. A complete explanation for the vertical temperature structure that ties together all the compositional clues from New Horizons and groundbased data is needed. The composition and structure of the haze molecules, what controls their layering and continuity over hundreds of kilometers, and their influence on the thermal structure of the atmosphere and relationship with low-albedo, complex organic material (tholins) on the surface are all new questions following New Horizons' discovery of Pluto's hazes. The challenge is in acquiring sufficient data at any given epoch as well as over time. Observations of the atmosphere as a global system over time are difficult. Occultations are very powerful tools but unfortunately will be dramatically decreasing in frequency as Pluto leaves the galactic plane and will become rather sparse datasets. Spacecraft observations are an essential tool that provides a far more detailed dataset than anything else but are unfortunately limited in their timescale of coverage. The atmosphere of Pluto is likely to be an interesting puzzle to work on for a long time to come and will require contributions from occultations, telescopic observations, and direct investigations by spacecraft local to the system.

3.2.2. Volatile cycles and surface-atmosphere connection on Pluto. Pluto's large obliquity (~120°), eccentric orbit (30–49 AU), and multiple active volatiles (N_2, CO, CH_4) result in rich seasonal cycles with many processes and feedbacks that may be more diverse than those found on most other worlds. A primary challenge with understanding seasonal cycles on Pluto is the very long timescale needed to directly detect such changes. Groundbased data can address changes on an orbital timescale but doing so is a long-term effort spanning centuries. Even so, monitoring on shorter timescales (<10 years) for all observables (photometry, surface and atmosphere spectra, thermal flux, etc.) are crucial in case there are further global changes. These changes could be abrupt but more likely there will be slowly varying phenomena that will require high-precision work (*Buie and Grundy,* 2000; *Buie et al.,* 2010a).

Pluto's atmosphere and volatile ices are also predicted to vary on longer timescales, such as Croll-Milankovich-like cycles, possibly including extreme excursions in temperature and pressure (*Stern et al.,* 2017). These timescales cannot be observed with groundbased data and can only be investigated through the geologic record of Pluto's surface ices, requiring detailed spacecraft observations such as global imaging at resolutions better than the highest resolution achieved by New Horizons. Clearly, the surface and atmosphere of Pluto are intimately linked and cannot be studied in isolation; understanding Pluto's volatile cycles will require complementary progress in the atmospheric structure, surface volatiles, and their coupling.

3.2.3. Origin and evolution of landforms on Pluto and Charon. Pluto's surface exhibits landforms with a variety of ages created by a diverse array of geologic processes (e.g., *Moore et al.,* 2016). Based on impact crater density and morphology, the dark equatorial areas are the oldest and probably ancient. Much of the rest of the surface has a degraded crater population that indicates an old surface but one that has been processed to varying degrees over time (e.g., *Moore et al.,* 2016). The large extensional features harken back to an earlier era of crustal expansion. Further investigation of Pluto's ancient terrains has the potential to shed new light on our understanding of Pluto's distant past (e.g., infer internal heat flux from crater relaxation states, global tectonic framework).

Several areas of Pluto reveal that, in addition to its ancient terrains, it also has very young surfaces from ongoing and recently active geologic processes. The youngest feature, Sputnik Planitia, is a ~1000-km-wide volatile-ice deposit with no impact craters greater than the resolution of the New Horizons images (~300 m). Its convective resurfacing (*McKinnon et al.,* 2016) is unique among worlds explored by spacecraft but may occur on other large Kuiper belt objects [e.g., Eris (*Hofgartner et al.,* 2019)]. Eastern Tombaugh Regio has active glaciers of volatile ice and is a natural laboratory for studying glaciers with a completely different

Fig. 2. Charismatic images of Pluto (top) and Charon (bottom) from New Horizons. Several of the outstanding priority science questions discussed in this section are included in the figure.

composition than glaciers on Earth (*Moore et al.,* 2016). Equally intriguing are the young and possibly cryovolcanic structures, Wright and Piccard Montes (*Moore et al.,* 2016). Thus Pluto has diverse, unique, and ongoing geologic processes. It also has endogenic (internally powered) and exogenic (solar powered) modification of its landforms. Future study of these landforms will likely impact our understanding of surfaces across the solar system; understanding Pluto's geology is important for understanding the solar system's workings. New Horizons' observations also suggest that Pluto may have stratigraphic layers and an important follow-on question is whether it has stratigraphy in its crust, glaciers, and volatile deposits that are detailed records of its geologic and climatic history.

New Horizons' revelation of Charon's landforms have also resulted in many new scientific questions. The surface has extreme topographic variations with mountain tops and valley bottoms differing by >20 km (*Schenk et al.,* 2018b). Charon's Pluto-facing (New Horizons encounter) hemisphere has a roughly north-south hemispherical dichotomy between rugged highlands and smooth plains, both of which are old but they may differ in age (*Moore et al.,* 2016). The formation of Charon's topography and hemispherical dichotomy and their relation to a subsurface ocean and cryovolcanism are all important questions for future investigations. Charon's red north pole, Mordor Macula, is unique among icy satellites and the focus of several additional questions: How thick is the red material and what is its composition? What is the relation between Mordor Macula and Pluto's volatile cycle? Does an analogous landform exist at Charon's south pole (*Grundy et al.,* 2016b)? An important point about Charon's landforms is that its ancient terrains have likely had less impactors and bombardment by energetic radiation than those on icy satellites around giant planets and thus may be better preserved.

3.2.4. Interior structure of Pluto and Charon. Direct constraints on the interior structure of Pluto and Charon are currently limited to bulk density, shape, and inferences from surface landforms but are necessary to understand formation and evolution of these two worlds. Whether or not Pluto is a current ocean world, and Charon a former ocean world, are key questions for understanding their evolution and habitability. The differentiation state as well as thickness and composition of interior layers are also not directly constrained. The only conceivable way to definitively answer these questions is through new observations by a spacecraft in the Pluto system. An orbiter that can determine the interior mass distribution of both bodies and make more accurate measurements of their shapes is needed. An important point about their interior structure is that the Pluto-Charon binary is in a tidal end-state and understanding their mutual tidal evolution is an important goal for understanding tides in general.

3.2.5. Origin and dynamics of Pluto's small satellites. Pluto's small satellites — Styx, Nix, Kerberos, and Hydra — likely resulted from the binary-forming giant impact and may record key clues about giant impacts (*Weaver et al.,* 2016). Deciphering the details of the binary-forming impact and the formation of the satellite system is an important topic. After all, the Earth-Moon system was similarly formed by a giant impact but it has lost any analogous small satellite relics that resulted from, and may have provided clues about, that impact. Pluto's small satellites have chaotic rotations that may be related to dynamical resonances and an understanding of their dynamics remains elusive (*Weaver et al.,* 2016). Detecting Pluto's small satellites is a challenge for the most sensitive Earth-based telescopes so answering most of the new questions about these satellites will likely require a spacecraft in the Pluto system.

4. NEAR-TERM REMOTE OBSERVATIONS

The New Horizons encounter with Pluto will remain for a long time a single visit to a fundamentally changing world. Moreover, even the extraordinary observations that were achieved had their own limitations. Spectroscopic measurements were limited by the restricted spectral range of the Linear Etalon Imaging Spectral Array (LEISA) instrument (1.25–2.5 μm) and its modest (by groundbased standards) spectral resolution: R = 240 over the entire spectral range, and R = 560 over 2.1–2.25 μm (the latter measurements have not yet been exploited due to calibration difficulties). An important limitation was the lack of coverage of wavelengths >2.5 μm where many of the fundamental rotational-vibrational bands of ices occur, as well as the lack of spectral resolution for atmospheric measurements in the infrared (IR). Even more critically limited were measurements of Pluto and Charon in the thermal range, being restricted to single-wavelength (4.2 cm), and only modestly resolved spatially, thermal emission measurements by the Radio Science Experiment (REX) (*Bird et al.,* 2019), encompassing both day- and nightsides. Current facilities [Very Large Telescope (VLT), Keck, Atacama Large Millimeter Array (ALMA), Very Large Array (VLA)] and soon-to-become available facilities [James Webb Space Telescope (JWST), Extremely Large Telescope (ELT), Thirty Meter Telescope (TMT)] are and will be providing very useful complementary, and sometimes unique, information to New Horizons.

4.1. Groundbased Facilities

The small size, large geocentric distance, and cold temperatures associated with Pluto and its moons generally require the use of the largest and most sensitive groundbased facilities to obtain scientifically useful data in the post New Horizons era. Such facilities include large optical/IR telescopes (such as the European Southern Observatory's VLT, NASA's Keck, Subaru, Gemini), and large submillimeter/millimeter/centimeter antennas [e.g., Institut de RadioAstronomie Millimétrique (IRAM)/30-m] and interferometers [Submillimeter Array (SMA), IRAM/NOrthern Extended Millimeter Array (NOEMA), ALMA, VLA].

In the near-IR, observations can be broadly separated between those targeting the surface composition and those aiming at characterizing Pluto's atmosphere. Surface observations simultaneously cover the entire near-IR range (or large portions of it), and occasionally extend to the visible range where weak methane ice bands occurs (e.g., *Lorenzi et al.,* 2016), with typical spectral resolutions in the range of R ~500 to 5000 using integral field spectrometers — e.g., VLT/Spectrograph for INtegral Field Observations in the Near Infrared (SINFONI) or Keck/OH-Suppressing Infrared Integral Field Spectrograph (OSIRIS) — see, e.g., *De Meo et al.* (2010) and *Merlin* (2015) for Pluto, and *De Meo et al.* (2015) and *Holler et al.* (2017) for Charon. Observations of Pluto's atmospheric (CH_4, CO) composition are characterized by spectral resolutions R in the range 13000–70000 using echelle or crossed-dispersed spectrographs — e.g., Infrared Telescope Facility (IRTF)/CSHELL (*Young et al.,* 1997), VLT/CRyogenic high-resolution InfraRed Echelle Spectrograph (CRIRES) (*Lellouch et al.,* 2009, 2015) and Keck/Near Infrared Echelle Spectrograph (NIRSPEC) — but at the expense of the instantaneous spectral coverage, that sometimes encompasses only a few to a dozen spectral lines. Note that VLT/CRIRES, a pre-dispersed IR echelle spectrograph, is currently being upgraded to a cross-dispersed spectrograph that will increase the wavelength range that can be covered simultaneously by a factor of 10; CRIRES+ should become available to the community by 2021. Finally, it is worth mentioning that at sufficiently high signal-to-noise (S/N), Pluto's atmospheric CH_4 may actually also be detected at a spectral resolution of several thousands only, an idea put forward by

Cook et al. (2014) and shown to likely be valid at Triton (*Merlin et al.,* 2018); this will also be possible by JWST, as discussed below. Many current near-IR observations of the Pluto system are done with adaptive optics, providing PSF sizes of ~0.1". This permits separating Pluto from Charon, but not to spatially resolve either body.

Even with 8–10-m-class telescopes, imaging of Pluto's and Charon's surface from the ground in the optical range does not provide better results than Hubble Space Telescope (HST) imaging, which achieved 4–5 resolution elements only across Pluto in the blue and near-ultraviolet (UV) (*Buie et al.,* 2010b). Although they achieve in principle diffraction-limited resolution of ~0.02" in the visible, extreme adaptive optics systems [e.g., VLT/Spectro-Polarimetric High-contrast Exoplanet Research (SPHERE), Gemini/Gemini Planet Imager (GPI)] are not obviously suitable for mapping Pluto and Charon because they require relatively bright (V <9–11) natural guide stars and Pluto itself is too faint to serve for adaptive optics corrections.

In the relatively near future (~2025–2027), however, 30-m-class telescopes [TMT, European Extremely Large Telescope (E-ELT)] will gather ~10× more light than the current most powerful optical telescopes. Equipped with adaptive optics [e.g., E-ELT/Multi-conjugate Adaptive Optics RelaY (MAORY)-Multi-AO Imaging Camera for Deep Observations (MICADO)], they will achieve a typical diffraction-limited 10-mas spatial resolution in H band. This corresponds to ~250 km at Pluto distance, and will permit recording spectra (at R = 8000 for the above instrument) on an impressive ~10 independent spatial points across the area of Sputnik Planitia.

Pluto's atmosphere is also directly observable by using rotational absorption lines of the polar molecules CO, HCN, and HNC (*Lellouch et al.,* 2017, 2018), but this requires the most powerful interferometers, namely ALMA, as attested by previous upper limits (*Bockelée-Morvan et al.,* 2000), or spuriously claimed detections of CO (*Greaves et al.,* 2011) using radio-telescopes with smaller collecting areas in that spectral range. In contrast to the IR, the very-long baselines afforded by ALMA (up to 16 km) can spatially resolve Pluto's and Charon's disks, down to a 0.012" resolution at 0.87 mm wavelength, although in practice observations at that resolution would be fatally hampered by low S/N. Currently, emission from CO and HCN gas at 0.87 mm have been mapped on Pluto at 0.06" resolution. A further gain by a factor of ~3 in spatial resolution would require an observing time of ~25 hr, a long but still feasible observation. Continuum signals in the millimeter/submillimeter are much easier to detect than line signals, in relation to their much broader spectral extent (defined by the instrument bandpass, typically 8 GHz, for the former, and by the linewidth, typically a few megahertz, for the latter). Current continuum observations (*Butler et al.,* 2018) have already achieved 0.025" resolution, and the twice-finer ultimate instrumental resolution could be achieved while keeping sufficient S/N.

Stellar occultations represent, to some extent, an exception as networks of even relatively small telescopes can be sufficient to follow the evolution of Pluto's atmospheric pressure with time. High S/N occultation data, best obtained at large telescopes, are still needed for characterizing in detail the atmospheric structure (thermal profile, waves, etc.), especially in the near-surface (<10 km altitude), whose observation requires high S/N measurements of central flashes (e.g., *Sicardy et al.,* 2016). As shown by upper limits achieved on other, airless, distant bodies, this method is sensitive to pressures in the tens of nanobars range.

4.2. James Webb Space Telescope

Scheduled for launch as early as 2021, JWST is the next and long-awaited large spaceborne facility in the IR. Its 6.5-m-diameter, Sun-shielded and passively cooled segmented mirror will collect and direct light to four science instruments altogether covering the 0.6–28-μm range: the Near-Infrared Camera (NIRCam), Mid-Infrared Instrument (MIRI), Near-Infrared Spectrograph (NIRSpec), and Fine Guidance Sensor/Near Infrared Imager and Slitless Spectrograph (FGS/NIRISS), with unprecedented sensitivity. The instrumentation combines different observation modes: filter imaging, spectroscopy [slits, IFUs (integral field units), prisms], coronagraphy, and aperture mask interferometry. The most useful instruments and modes for the study of the Pluto system are likely to be NIRSpec (in IFU spectroscopic mode) and MIRI, both in imaging and medium-resolution spectroscopic modes (*Rieke et al.,* 2015; *Wright et al.,* 2015). NIRSpec covers the 0.6–5.0-μm range, achieving a R = 2700 resolving power in IFU mode. MIRI covers 5–28 μm, and includes in particular an imaging camera with pixel scale of 0.11" and nine filters, and an IFU spectrometer (MRS) with resolving power of ~1600–3200. Expectations for Pluto science stem from the fact that JWST will provide (1) extension into the difficult-to-measure 2.5–5-μm range for surface composition; (2) high sensitivity and moderately high spectral resolution well suited to the study of surface composition, sufficient to detect some atmospheric gases; (3) covering of the thermal IR with MIRI (up to 28 μm), with the possibility to resolve Pluto from Charon (diffraction-limited resolution ~0.6" at 15 μm); and (4) a 5-year (nominal) to 10+ year (hoped) lifetime, suited to track expected temporal changes as Pluto approaches its northern summer solstice in 2030.

The obvious limitation of JWST will be the lack of spatial resolution on Pluto (0.1") and Charon's disks (0.05"), except partly with NIRCam (pixel size = 0.032", R ~0.08" at 2 μm). In spite of its larger mirror, JWST is not expected to provide better quality imaging than HST as the latter operated at shorter wavelengths, although near-IR imaging (in particular within/outside of the strong/broad methane ice features) may have an interest of its own.

4.3. SCIENCE EXPECTATIONS

Science expectations from remote observations in the near term may be described in different ways: by facility (e.g., ALMA, JWST), by method (e.g spectroscopy, photometry), or by science topic. We attempt here the latter, by discussing expected progress on Pluto's volatiles and climate (section 4.3.1), the non-volatile composition of Pluto and Charon (section 4.3.2), thermal properties (section 4.3.3), and the outer satellite dynamics and composition (section 4.3.4).

4.3.1. Pluto's volatiles and climate. Pre-New Horizons and New Horizons observations have abundantly demonstrated the complexity of Pluto's multi-species atmospheric cycles and of the surface-atmosphere interactions first discussed by *Stern et al.* (1988). Pluto's seasons drive sublimation/condensation exchanges of volatiles (N_2, CH_4, CO) between hemispheres. These exchanges must induce temporal variations of the distribution of volatiles, and since the N_2-dominated atmosphere is in equilibrium with the surface, seasonal variations of the surface pressure. The ice distribution at a given point in time, as revealed by New Horizons, is of course a key constraint to the understanding of such surface/atmosphere interactions on seasonal and longer timescales. Additionally, Pluto's "super-seasons," associated with the variation of Pluto's obliquity and longitude of perihelion over ~3-m.y. timescales, can have non-trivial ramifications and lead to long periods where the volatile cycle changes completely (*Dobrovolskis,* 1989; *Earle and Binzel,* 2015). Nonetheless, even pre-New Horizons data had revealed that Pluto is a changing world, with evidence coming from different indicators: (1) changes in Pluto's surface appearance from 1994 to 2003 (*Buie et al.,* 2010b); (2) changes at the few percent level, and not due to pole geometry effects, in Pluto's optical lightcurve (*Buie et al.,* 2010a); (3) secular evolution of the N_2 and CO near-IR absorptions, with a decline over 2003–2014 (*Grundy et al.,* 2014); and (4) most prominently, pressure evolution (*Meza et al.,* 2019, and references therein). The pressure evolution, characterized by a steady increase (by a factor of ~3 over 1988–2016, with a possible trend for plateauing in recent years) of the atmospheric pressure at a given reference level (e.g., *Elliot et al.,* 2007; *Olkin et al.,* 2015; *Meza et al.,* 2019), has been the prime constraint to the volatile transport models [*Young* (2012, 2013) and *Hansen et al.* (2015), followed by the post-New Horizons models including topography of *Bertrand and Forget* (2016) and *Bertrand et al.* (2018)]. Note that some of these models have been extended to describe the seasonal and longer cycles of CH_4 (*Bertrand et al.,* 2019). In the short term, these models make testable predictions about the future pressure evolution, methane atmospheric abundance, and volatile surface distribution on observable timescales (see Figs. 1 and 2 of *Bertrand and Forget,* 2016), and therefore can be validated or falsified by future observations. In more detail:

- Continued stellar occultations will permit the atmospheric pressure to be monitored until the atmosphere collapses or declines substantially in pressure.
- Atmospheric spectroscopy in the near-IR (CH_4, CO) and in the submillimeter (CO, HCN, HNC, with vertical information) will monitor time evolution of these gases' mixing ratios. Note that HCN and HNC are products of the coupled CH_4-N_2 photochemistry, and that their vertical distribution, which includes supersaturation for HCN, is coupled to the density, distribution, and microphysics of the haze, itself the ultimate product of atmospheric photochemistry. We also emphasize that the sensitivity of JWST/NIRSPEC will undoubtedly permit the measurement of methane gas in several bands over 1.3–3.7 μm, and CO at 2.3 and 4.7 μm (such a measurement of an atmospheric gas over a broad spectral range will also provide an assessment on any haze scattering effects on gas retrievals). In the submillimeter range, spatially resolved maps of CO may also reveal the temperature field of the atmosphere from the surface to ~450 km.
- Surface spectroscopy by large groundbased telescopes, and especially by JWST/NIRSpec, will further constrain spatial distribution and the state of volatiles. Ever since the pioneering measurements with the United Kingdom Infrared Telescope (UKIRT)/CGS4 spectrometer (*Owen et al.,* 1992; *Douté et al.,* 1999), high S/N mid-resolution spectroscopy has proven able to provide a wealth of information on the state, distribution, and temperature of ices on Pluto's surface. For example, even without spatial resolution, Pluto's near-IR spectrum indicates the coexistence of "pure" and "diluted" methane in N_2 — strictly speaking this means N_2:CH_4 and CH_4:N_2 mixtures, each dominated by one of the species — and situations of vertical segregation (*Douté et al.,* 1999; *Merlin,* 2015; *Schmitt et al.,* 2019) that likely result from the differential sublimation between N_2 and CH_4. The very high S/N expected from JWST/NIRSpec spectra will yield new insights on these aspects, and should additionally provide ice temperature measurements based on band shape and position. This technique was pioneered by *Tryka et al.* (1994) for the N_2 2.15-μm band, but the result, $T(N_2) = 40 \pm 2$ K, is suspicious because it would nominally imply a 40-μbar surface pressure, and therefore needs to be repeated at very high S/N. Ice temperature measurements will also be done for the first time on CH_4, whose band positions are also temperature-dependent (*Quirico and Schmitt,* 1997). Measuring the temperature of "pure" CH_4 ice, feasible only from this technique, would be of particular interest since it is expected to be unaffected by sublimation and therefore higher than N_2 ice, and to control the atmospheric CH_4 abundance.
- Filter imaging by JWST/NIRCam in/out methane bands may provide further insight into the gross distribution of ices and their seasonal evolution.

4.3.2. Non-volatile composition of Pluto and Charon. Besides the volatiles, more species are expected to be present on Pluto's surface due to (1) atmospheric photolytic production and downward transport (C_2H_2, C_2H_6, C_2H_4, C_4H_2, HCN, and haze are detected in Pluto's atmosphere) and/or (2) irradiation/radiolysis of surface methane ices. Observationally, these processes remain poorly characterized because the spectrum below 2.5 µm is heavily dominated by methane ice and because data at longer wavelengths remain of modest quality (e.g., *Grundy et al.,* 2002; *Protopapa et al.,* 2008). Photochemical products are indicated by Pluto's red slope in the visible, and are matched by laboratory measurements of tholins used to simulate Titan's aerosols. This, however, is not sufficient for a precise chemical identification, especially because IR spectral signatures of tholins are not yet detected. Other evidence for non-volatile surface ices comes from (1) the reported identification of C_2H_6 at 2.405 µm (e.g., *Holler et al.,* 2014) and (2) the clear detection of H_2O ice by New Horizons (*Grundy et al.,* 2016a), generally seen as a bedrock in ice-free regions associated with the dark red materials; in contrast, CO_2 ice, present on Triton (*Cruikshank et al.,* 1993), remains undetected on Pluto. Considerable progress is expected from JWST: (1) The increased S/N in the "classical" 0.8–2.5-µm range will permit us to disentangle and eliminate the contribution of volatiles in that spectral range; this has been achieved in the detection of several irradiation products (ethane, ethylene, propane) in Makemake's CH_4 ultra-dominated spectrum (*Brown et al.,* 2015). (2) The extension of the spectral range to the faint 3–5-µm (NIRSpec) and 5–12-µm (MIRI/MRS) domains will open the range of molecular fundamental modes, with possible spectroscopic access to new ices (hydrocarbons, nitriles, etc.) and irradiation products. Figure 3 shows the imaginary index of refraction of several ices, taken from the Grenoble astropHysics and planetOlogy Solid Spectroscopy and Thermodynamics (GhoSST) database. The existence of "windows" in the CH_4 ice absorption over 2.8–3.2 µm and longward of 4.2 µm will be favorable to search for minor ices (e.g., CO_2, C_2H_4, C_2H_6, NH_3, CH_3OH). Regarding irradiation products, laboratory spectra of irradiated Pluto-like icy material by *Materese et al.* (2014) indicates that beyond the emblematic 3–4-µm band, such material is expected to show features at 4.5–5.0 µm and 6–10 µm. This is diagnostic of numerous stretch or bend modes (C≡N, C=O, NH_2, O-H, C-H, R-N=C, etc.) and reachable by JWST/NIRSpec and MIRI respectively. Isotopic variants of several ices could also be observable, e.g., ^{13}CO (already hinted to contribute at 2.405 µm), and perhaps CH_3D ice that exhibit features at 2.5–3 µm and 4.2–4.7 µm (*Grundy et al.,* 2011). The D/H ratio in small bodies is diagnostic of formation place and timing, although the observed diversity of D/H values in comets, not correlated with their dynamical class, remains of uncertain implications (e.g., *Lis et al.,* 2019). Although observational feasibility has not been precisely quantified, the detection of CH_3D on Pluto is more likely than on Triton (which bears a 10× smaller CH_4 ice abundance) and on other methane-rich KBOs (e.g., Eris and Makemake, which are much fainter) and would represent the first detection of a deuterated ice on a planetary surface.

Charon's spectrum is overwhelmingly dominated by crystalline water ice. Its temperature-dependent 1.65-µm feature can be used as a thermometer, but the current precision (45 ± 14 K) (*Holler et al.,* 2017) is too low to usefully constrain Charon's thermal properties (see below). Charon's spectrum also indicates the signature of NH_3 hydrates at 2.21 µm, which appears to be associated with small craters that have fresh-appearing impact ejecta (*Grundy et al.,* 2016a). Charon's spectrum longward of 3 µm is extremely poorly known, although it seems that the 3–5-µm albedo is higher than expected for H_2O (*Protopapa et al.,* 2008), which might point to an additional, non-ice component. Future observations with JWST will provide a much better determination (and search for rotational variations) of the H_2O ice temperature, to search for additional species at 1–2.5 µm and to explore the virtually unknown 3–5-µm range for Charon. In fact, the sensitivity of JWST will be such that Pluto's and Charon's reflected components will be detected up until the transition to the thermal range (expected to occur near 12–15 µm), with satisfactory S/N even for MIRI/MRS spectra of Charon.

4.3.3. Thermal properties. Thermal properties of solid bodies (thermal inertia, bolometric albedo and emissivity)

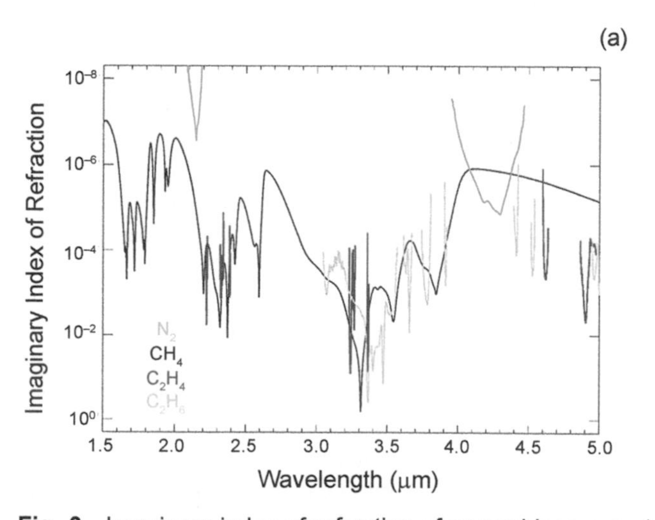

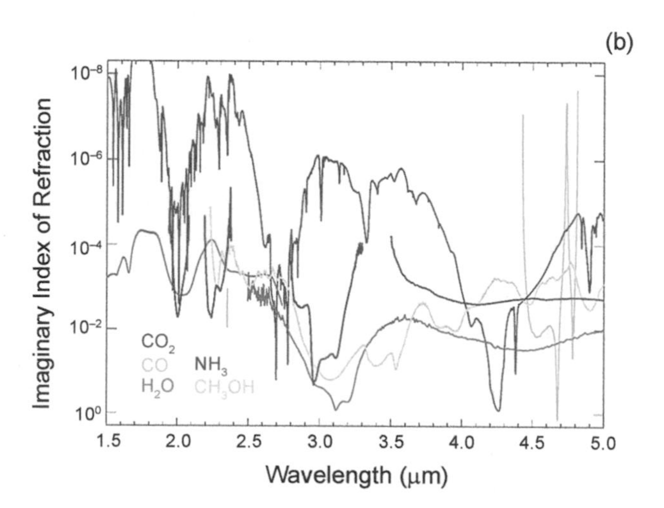

Fig. 3. Imaginary index of refraction of several ices over 1.5–5 µm. Data taken from *https://www.sshade.eu/db/ghosst.*

control temperatures in the surface/subsurfaces of these objects, both on diurnal and seasonal timescales. Measurements in the far-IR [Infrared Space Observatory (ISO), Spitzer, Herschel], based on thermal diurnal light curves, have provided significant information on the thermal inertia (TI) of Pluto's and Charon's near-surfaces (*Lellouch et al.,* 2000, 2011, 2016), indicating low values of TI = 16–26 J m^{-2} $s^{-1/2}$ K^{-1} (hereafter MKS) for Pluto and 9–14 MKS for Charon, but these values are still somewhat model-dependent because the two bodies were not spatially resolved in these observations. As indicated above, groundbased interferometers in the submillimeter/millimeter/centimeter range such as the SMA, ALMA, and VLA easily separate Pluto from Charon [*Butler et al.* (2019; they also resolve each of the two bodies (see *Butler et al.,* 2018)], but thermal emission at these long wavelengths is even more sensitive to spectral emissivity effects than to thermal inertia. In addition, none of the thermal measurements have so far constrained the thermal inertia on seasonal timescales. The latter is extremely difficult to determine, requiring either the monitoring of Pluto and Charon surface temperatures over a significant fraction of their heliocentric orbit, or measuring the temperature in the polar night. New Horizons/REX did obtain 4.2-cm polar night temperatures of 29 ± 2.5 K on Pluto and 41 ± 1 K on Charon (*Bird et al.,* 2019) but the interpretation is pending, complicated by the difficult-to-estimate 4.2-cm emissivity. Yet, the seasonal thermal inertia plays a key role on both objects. On Pluto, the seasonal thermal inertia of the bedrock controls its polar winter temperature, hence the ability of the volatile ices (N_2, CH_4, CO) to condense/not condense in winter. Fitting the pressure evolution as monitored by stellar occultations and the distribution of volatile ices from New Horizons by *Bertrand and Forget* (2016) indirectly estimated a thermal inertia of ~800 MKS. Although Charon has no volatiles, its red polar caps are interpreted as resulting from winter cold-trapping of methane gas escaping from Pluto's atmosphere and subsequent photochemical/radiolytic processing (*Grundy et al.,* 2016b). This requires a maximum polar night temperature of ~25 K, which in turn must imply some upper limit on Charon's seasonal thermal inertia. On both bodies, however, direct measurements of the seasonal thermal inertia are missing, and the only unambiguous way to access it seems to be from a spaceborne measure of the polar night brightness temperatures in the mid- to far-IR, where emissivity effects are unimportant.

Significant progress on Pluto and Charon thermal properties is still expected from JWST/MIRI, which will obtain thermal light curves of Pluto and Charon over 12–25 μm. The analysis of Spitzer data (*Lellouch et al.,* 2011) indicates that Charon progressively dominates the thermal flux of the Pluto/Charon system toward shorter wavelengths (e.g., ~40% of the system flux at 24 μm), due to Charon being globally warmer than Pluto. This will be verified directly with JWST/MIRI, which will separate the two bodies. Thermal light curves at a variety of wavelengths in the Wien tail of the thermal emission will be sensitive not only to thermal inertia, but also to surface roughness effects. Better-determined thermal properties of both bodies will also permit a reassessment of the radio data in terms of the spectral long-wavelength emissivity. Current SMA, ALMA, and VLA data (*Butler et al.,* 2019) separating Pluto from Charon indicates considerably higher 0.8–9-mm brightness temperatures for Charon than Pluto, as also found by REX at 4.2 cm (*Bird et al.,* 2019). This is expected given that Charon has lower albedo and thermal inertia than Pluto, but detailed modelling (*Butler et al.,* 2019) suggests that this also implies a higher emissivity for Charon's surface compared to Pluto's, which has uncertain implications on surface composition and/or particle grain size.

A side objective of the JWST/MIRI observations of Pluto will be to test the idea (*Zhang et al.,* 2017) that haze cooling is responsible for Pluto's cold (~70 K) upper atmosphere temperature. If the hypothesis is correct, thermal emission from haze should progressively dominate over surface thermal emission shortward of ~25 μm (*Zhang et al.,* 2017), and therefore progressively erase Pluto's thermal lightcurve. Spitzer data show that Pluto-related rotational lightcurve is still well-marked at 21 μm (*Lellouch et al.,* 2011), but observations at shorter wavelengths with JWST/MIRI will provide a more definite answer.

5. THE NEXT MISSION

A typical spacecraft exploration pathway begins with an initial reconnaissance flyby mission. This was accomplished with the New Horizons mission (*Stern et al.,* 2015) and has fundamentally advanced our understanding of Pluto and its retinue of satellites. The pathway continues with an orbital tour that provides a more thorough exploration over a longer timescale and *in situ* measurements of any atmospheres. Beyond this step there is surface *in situ* exploration with a lander, followed by mobile *in situ* vehicles such as rovers, and finally sample return. Each subsequent mission architecture generally increases in complexity and cost. We consider the likely next step for the Pluto system, an orbiter mission, in the context of the science questions discussed in section 3.

5.1. Guiding Science Objectives for an Orbiter

An orbiter mission will study the system at spatial scales not possible with telescopic observations, while also providing a global perspective that is not possible with flyby or lander missions, and is the logical next step in the spacecraft exploration of the Pluto system. The science objectives that can be addressed with an orbiter are discussed in the following subsections.

5.1.1. Spatial coverage and resolution. As is common with flybys, the data returned from the New Horizons flyby of the Pluto-Charon system prompted as many new questions as it answered. These questions fall into two broad categories: those prompted by what was seen, and those prompted by what was not seen. The subsolar latitude at the time of the New Horizons flyby of Pluto was ~52°N and thus regions on Pluto and Charon south of ~38°S, ~20% of their total surface area, were not directly illuminated by the Sun. The southern hemisphere of Charon was intentionally imaged in Pluto-shine, but at resolutions of ≤60 km/pixel and low S/N (*Grundy et al.,* 2016b). Some areas south of 38°S on Pluto were serendipitously imaged using diffusely scattered light from atmospheric hazes, also at low S/N (*Schenk et al.,* 2018a). Thus ~20% of Pluto's surface, including its south pole, was completely unexplored by New Horizons and the analogous areas on Charon were only imaged at both low spatial resolution and low S/N. The south poles are important regions since they could be areas of volatile accumulation and whether the south poles differ from the north poles is an important science question for both Pluto and Charon. An orbiter mission with the proper instruments and trajectory could explore these yet-to-be-explored southern regions; it could investigate 100% of the surfaces on Pluto, Charon, and the smaller satellites. Radar, laser, and thermal instruments can all characterize the surface without solar illumination.

Of the ~80% of the surfaces of Pluto and Charon that were imaged by New Horizons, only approximately half of those surfaces were imaged at highresolution [<1 km/pixel, and smaller areas were imaged at substantially higher resolutions; see *Schenk et al.* (2018a,b) for maps of the best image resolution of each location]. As a result of their relatively slow rotation periods (~6.4 Earth days) and the fast flyby (~14 km s^{-1}), longitudes not on the encounter hemispheres (anti-Charon-facing hemisphere of Pluto and Pluto-facing hemisphere of Charon) were imaged prior to closest approach at substantially lower resolutions. The lower-resolution images suggest possible hemispherical differences on both Pluto and Charon; Pluto's Charon-facing hemisphere may have more large craters but lacks a huge Sputnik-like basin while Charon's anti-Pluto-facing hemisphere may not have the north-south dichotomy that characterizes its Pluto-facing hemisphere (e.g., *Stern et al.,* 2015). Although bladed terrain (*Moore et al.,* 2018) and moated craters (similar to Elliot crater) are suggested from the available imagery of Pluto's anti-Charon side, these suggestions cannot be confirmed without higher-resolution data. Groundbased and New Horizons light curves provide further hints of the hemispherical variability of Pluto and Charon (e.g., *Buie et al.,* 2010a; *Howett et al.,* 2016). Other icy worlds are known to have significant hemispherical differences (e.g., Iapetus has extreme hemispherical albedo differences and one hemisphere of Mimas is dominated by its Herschel impact crater). An orbiter mission would significantly improve upon the spatial resolution of the New Horizons' non-encounter hemispheres. The approximately global and uniform coverage of an orbiter mission would allow distinct regions to be understood in terms of their global context and prevent biases based on limited coverage.

The next mission to the Pluto system could also significantly improve upon the best spatial resolution observations of New Horizons. Spacecraft safety was a dominating concern during the New Horizons flyby design, since the space environment around Pluto was not well known and the spacecraft passed through the system at very high speed (~14 km s^{-1}). The New Horizons closest approach to Pluto was ~14,000 km and the best-resolution image was ~80 m/pixel. Now that the space environment has been characterized by New Horizons (e.g., *Stern et al.,* 2015), an orbiter could approach much closer to the surface of Pluto as well as its satellites. The orbiter architecture also has the option of slowly spiraling inward with frequent checks for evidence of any hazards while moving at slower (orbital) speeds. From its vantage point much nearer to the surface, instruments across the electromagnetic spectrum could observe at significantly better spatial resolutions. The improved resolution (probably <1 m for visible imaging of specific regions) is necessary to fully understand the varied surface types and layering of atmospheric hazes. Such data could reveal the detailed composition and nature of the red tholin layers exposed in the walls of impact craters and tilted water ice-blocks surrounding the "shores" of Sputnik Planitia, for example.

In addition to spatial coverage and resolution, an orbiter mission also enables repeat observations with different lighting conditions (e.g., solar phase angle). This capability is important for comparing the effects of albedo and topography on the appearance of landscapes and to derive photometric properties of the surface. As demonstrated with New Horizons data, there are significant variations in the surface texture that can be detected with data covering a wide range of photometric angles (*Buratti et al.,* 2017; *Hofgartner et al.,* 2018). Mapping of fine-scale surface properties may be just as important as mapping the macroscale surface texture and topography, especially for the volatile-rich regions.

5.1.2. Change detection. As discussed in previous sections, Pluto's volatile ices, tenuous atmosphere, orbital eccentricity, and extreme obliquity result in seasonal and possibly diurnal changes of the surface and atmosphere. Groundbased and near-Earth spacebased photometry, spectroscopy, and occultation observations have all observed seasonal variability of Pluto's surface volatile-ices and atmosphere. The New Horizons flyby confirmed that Pluto has abundant volatiles with different distributions across its surface, and thus, that volatile transport on Pluto is ongoing and complex (e.g., *Grundy et al.,* 2016a). Two of the most astounding discoveries from the New Horizons flyby are that (1) Pluto has abundant and diverse, ongoing and recent, endogenic (internally powered) geologic activity in addition to its exogenic (solar powered) volatile transport and (2) Charon also has ongoing and recent exogenic geological activity (volatile deposition and radiolytic destruction that blanket surfaces and modify surface color). Pluto's Sputnik Planitia has a crater retention age of <30 m.y. and resurfacing from convective overturning likely continues to the present (*McKinnon et al.,* 2016; *Singer et al.,* 2019). Eastern Tombaugh Regio also has a young surface that is probably renewed by actively flowing glaciers (*Moore et al.,* 2016). South of Sputnik Planitia, Wright and Piccard Mons and their surrounding terrain are lightly cratered, possibly because of geologically recent cryovolcanism (*Moore et al.,* 2016, *Cruikshank et al.,* 2019). Charon's Organa crater has a concentration of ammonia ice that suggests the ice was deposited <10 Ma and that radiolytic destruction of the ice continues today (*Grundy et al.,* 2016a). Formation of Charon's red poles from seasonally cold-trapped volatiles originating from Pluto may be ongoing (*Grundy et al.,* 2016b). Thus expected variability on Pluto extends beyond volatile cycling between the atmosphere and surface to endogenic geologic changes and Charon is also a geologically active world (variability on Charon, however, is probably limited to changes forced by external influences). The plethora of young surfaces and features in the Pluto system from ongoing and recently active geologic processes motivated a search for temporal changes on Pluto and Charon using New Horizons images (*Hofgartner et al.,* 2018). No changes were observed over the short timescales constrained by the flyby images but an orbiter mission could constrain changes over much longer timescales and will almost certainly detect shorter timescale changes, particularly if it images the surface with higher spatial resolution than New Horizons.

An orbiter mission would likely arrive at the Pluto system at least two decades after the New Horizons flyby. This is a noticeable fraction of Pluto's ~250-Earth-year orbit and thus seasonal changes in the surface volatile-ices and atmosphere can be expected and may be extensive. Changes from endogenic processes on Pluto as well as externally powered processes on Charon are also plausible over >20 years. An orbiter mission could replicate New Horizons observations (images and other observations), enabling characterization of changes on the surface of Pluto and Charon and the atmosphere of Pluto over the multi-decadal interval between the New Horizons flyby and the orbiter observations. The orbiter mission is expected to operate in the Pluto system for at least a few Earth years, although longer could be better. Thus, in addition to the multi-decadal timescale between it and New Horizons, an orbiter mission could also probe changes on multi-year and shorter timescales during the mission lifetime. Another advantage of an orbiter is its ability to respond to detected changes. If a region of the surface or atmosphere is seen to change, that region could be targeted to better characterize the time variability. Hazes, gravity waves, atmospheric pressure variations, and many other atmospheric phenomena are certain to be of interest. Repeat observations, and comparison to the existing datasets of New Horizons, would help constrain the speed of large-scale atmospheric change during the Pluto year and geologic feature development (e.g., suncups, penitentes, and convection cell movement on Sputnik Planitia). An orbital mission of sufficient duration has the potential to observe new impacts at the surface, enabling assessment of the *current* impact rate in the outer region of the solar system as has already been done on the Moon and Mars.

5.1.3. React to discoveries. An orbiter mission architecture clearly also has the capability of modifying its observations and even its orbit to respond to new discoveries, not just temporal change. The orbiter is expected to carry a suite of instruments and can follow up discoveries made by a single instrument. For example, New Horizons discovered Elliot Crater, which appears to have convection cells at its base, and Virgal Fossae, which cuts through Elliot Crater and has an anomalous red color (*Grundy et al.,* 2016a), possibly from cryovolcanism (*Cruikshank et al.,* 2019). Unlike an orbiter, New Horizons could not respond to those discoveries and provide an expanded investigation.

5.1.4. Internal properties. The interior properties of Pluto and Charon could only be studied indirectly by New Horizons by considering their shapes and surface features. For instance, an early global subsurface ocean was inferred from the presence of global-scale fracturing across Charon (*Beyer et al.,* 2018), which would be generated when the ocean froze. An orbiter can probe interior properties more directly by measuring the gravity fields of Pluto and Charon and masses of the small satellites with a radio science instrument. The gravity fields constrain the differentiation of the interior, including the past and current presence of subsurface oceans. The masses of the small satellites can be combined with their shape to determine their bulk density, which constrains their composition and formation.

5.1.5. In situ atmospheric measurements. The composition of Pluto's atmosphere and hazes can be constrained by remote sensing measurements, but an orbiter could measure the composition *in situ* with greater sensitivity and spatial resolution. Mass spectrometer instruments can be designed to measure the composition of gaseous molecules and haze particles that are neutral or charged, including trace species key to chemistry and indicative of possibly ongoing cryovolcanism. A mass spectrometer could also measure isotopic ratios to constrain volatile processes and history. Additionally, nephelometry can explore the haze size-particle distribution. Measurements of the exosphere and ionosphere are also important.

5.2. Key Science Capabilities for an Orbiter

We list here some of the important capabilities necessary for an orbiter mission to address the science objectives. These capabilities will likely drive the design of such a mission. The choice of which capabilities to include will come down to the science priorities that remain to be worked out for such a mission. The following list begins with some of the capabilities we find most compelling followed by a shorter list of capabilities that would be very nice to have but are of lower priority if the budget precludes doing everything.

- Global imaging at <50 m spatial resolution for all surfaces in the Pluto system, including surfaces in polar night. An important outcome of this data product would be a much more comprehensive census of impact craters. The resolution limit should be carefully studied to ensure a good match for the size distribution studies that tie these data back to the size distribution of small bodies in the Kuiper belt.

- Imaging of selected specific regions at 1–10 m spatial resolution to sample morphologies at these scales and investigate surface changes from volatile transport. This could include diurnal changes driven by condensation and sublimation, plumes such as have been seen on Triton, and localized phenomena like clouds. Thought should be given to the choice of filters given the results from New Horizons.
- Global spectral mapping at <500 m spatial resolution. Increased spectral resolution and spectral range (out to at least 5 μm) over New Horizons will also pay huge dividends for more in-depth characterization of the surface volatiles.
- Change detection imaging and spectroscopy for both Pluto and Charon as well as atmospheric structure for Pluto over multi-decadal, multi-year, and Pluto diurnal timescales. Seasonal changes of Pluto's volatile-ices and atmosphere over multi-decadal timescales are very likely and may be extensive; changes on other timescales and from other processes on Pluto and Charon are also plausible.
- Measurement of the gravity fields of Pluto and Charon to directly constrain their interior structures. The presence of subsurface oceans and mass anomalies associated with Sputnik Planitia and Wright and Piccard Montes are of particular interest.
- Close flybys of the four small satellites, primarily to measure their mass. The mass of the satellites can be combined with their shapes to determine their bulk densities, which constrains their composition, which in turn constrains their formation history. Improved spatial resolution will reveal their geology and age dating of their surfaces.
- *In situ* sampling of Pluto's atmosphere and hazes. The composition of the atmosphere including trace species and hazes as well as isotopic abundances should be measured using mass spectrometry.
- A census of the crustal volatile inventory. This includes measuring the total volume of the Sputnik Planitia reservoir.

Additional capabilities that would be useful if constraints allow:

- Thermal mapping of the surface. These data provide input into understanding the complex interaction between the surface and the atmosphere. This could also lead to the detection of subsurface thermal anomalies and their correlation with surface geology features.
- Altimetry (laser or radar) to complement photometric and stereo imaging datasets. Altimetry combined with gravity determines whether there are any uncompensated regions. This dataset can also provide direct measurements of unilluminated poles with both topography and albedo to probe for volatile deposition.
- Ionospheric and solar wind plasma interaction studies and searches for a magnetic field.

5.3. Orbiter Instrument Payload

The Pluto system orbiter must carry an instrument suite that addresses the diverse science motivations for returning to the Pluto system. The payload, however, will likely be strongly constrained in mass, power, and data volume. As a guide to thinking of reasonable estimates, *Stern et al.* (2019) suggest allocating twice the payload mass of New Horizons to allow for instruments similar to those on New Horizons and additional types of instruments. The basis for this estimate is that advances in instrument miniaturization and lightweighting since the 2001–2003 era when the New Horizons payload was designed suggests that this mass allocation is to first order sufficient for a baseline mission concept. We list the scientific measurements of several possible orbiter instruments. Four instruments are identified as threshold instruments that are essential for any future orbiter mission to the Pluto system. The other instruments are baseline instruments, many of which are expected to be part of the orbiter instrument suite, but accommodation realities will likely prevent inclusion of all these instruments. This list is not exhaustive but should be instructive of the scope of investigations that could be enabled.

5.3.1. Threshold instrument list.

- Narrow-angle color camera
  - High-resolution surface geology and photometric characterization of Pluto, Charon, and small satellites
  - High-resolution structure and photometric characterization of Pluto's atmosphere and haze layers
  - Surface and atmospheric changes from New Horizons and during orbiter mission
  - Search for additional small satellites
  - Stereo topography
- Wide-angle color camera
  - Surface geology and photometric characterization of Pluto, Charon, and small satellites
  - Surface and atmospheric changes from New Horizons and during orbiter mission
  - Shape of Pluto, Charon, and small satellites
  - Search for additional small satellites
  - Stereo topography
- Infrared spectral imager
  - Composition and photometric characterization of surfaces of Pluto, Charon, and small satellites with coverage from 0.8 to 5 μm
  - Composition and photometric characterization of atmospheric haze
  - Composition and photometric changes from New Horizons (e.g., from volatile transport)
- Radio science
  - Interior structure and geophysics of Pluto and Charon
  - Mass of small satellites
  - Atmospheric and ionospheric structure and changes from New Horizons

5.3.2. Baseline instrument list.

- Mass spectrometer

- Composition, isotopic ratios, and structure of Pluto's atmosphere and hazes
- Laser or radar altimeter
  - Topography of Pluto, Charon, and small satellites for geology and geophysics
  - Topography and albedo of surfaces without solar illumination
- Sounding radar
  - Depth and subsurface characterization of volatile-ice deposits, especially Sputnik Planitia
  - Sounding of global lithosphere if feasible
- Radar imager
  - Surface imaging without solar illumination
  - Near-surface geology (~1 cm to 1 m depth, below ~micrometer depths probed by cameras) of Pluto, Charon, and small satellites
  - Detection of surface deformation from volatile-ice flow (convection and glacial) using interferometer
- Thermal IR/submillimeter
  - Geology and thermal environment of Pluto, Charon, and small satellites
  - Search for thermal hot spots indicative of ongoing activity
  - Surface imaging without solar illumination
  - Atmospheric structure and composition
- Ultraviolet spectrometer
  - Atmospheric structure and composition
  - Atmosphere changes from New Horizons and during orbiter mission
  - Possible surface imaging without solar illumination using galactic Lyman-α
- Plasma spectrometers (various energy levels)
  - Space environment of entire Pluto system
  - Changes from New Horizons and during orbiter mission
  - Ionospheric sampling and chemistry, change detection with varying solar and solar wind conditions
- Magnetometer
  - Mapping of magnetic fields of Pluto and Charon

5.4. Key Challenges for Orbiter Mission

Pluto's heliocentric distance ranges from 30 to 49 AU over its eccentric orbit and this great distance from the Sun and Earth amplifies several challenges of deep space exploration. Pluto's large distance results in three related challenges: (1) A long cruise is necessary to reach the Pluto system, (2) power is limited in the Pluto system, and (3) downlinking data to Earth from the Pluto system is challenging. A long trek across the solar system is necessary to reach Pluto, which necessitates traveling quickly, but the spacecraft must also be able to slow down in order to enter orbit, and it must survive for a lengthy period of time; all these considerations result in several design challenges. Solar power is ~1000× weaker at Pluto than at Earth and as a result is not a viable power source. Radioisotope thermoelectric generators (RTG) have been used but are hampered by the steady decrease in available power during the long cruise to the Pluto system. The distance and low power both adversely affect the rate at which data can be returned from the spacecraft to Earth, making communications a third key challenge that results from Pluto's distance. Current progress on these three challenges is considered below.

5.4.1. Propulsion. Any orbital mission requires vastly larger propellant tanks than included on New Horizons in order to reach Pluto in a reasonable time and then slow the approaching spacecraft and allow "capture" by the Pluto-Charon system's gravity. Also useful will be the integration of low-thrust electric propulsion (EP) systems. The mission concept presented in *Stern et al.* (2019) includes a notional Pluto orbital tour followed by a Charon gravity-assist driven escape from the system and an extended mission further into the Kuiper belt using both EP and chemical propulsion.

5.4.2. Power. *Stern et al.* (2019) have assessed the feasibility of a combined Pluto-Charon system and Kuiper belt mission, assuming the use of RTG to power the spacecraft. An orbiter mission design would benefit from a power system designed specifically for deep space missions rather than landed missions (such as Mars). This design detail and the need for technological advances was called out in the 2011 Planetary Decadal Survey. However, recent studies of fission-powered units discuss technology that could provide vastly more power than past missions (*McClure et al.,* 2014, 2018). Such high-power systems can lead to fundamental increases in system capabilities across the entire design space including propulsion, communication, and instrument operation.

5.4.3. Data volume. Any mission, but especially an orbital mission, is challenged by the severe bottleneck between the acquisition and return of science data at the low bit rates available to current outer solar system spacecraft. Whereas New Horizons could slowly download flyby data over the course of months, an orbital mission must either have vastly higher data storage capabilities or, more favorably, a better downlink rate. As a solution to this problem, *Stern et al.* (2019) propose a high gain antenna (HGA) twice the size of that included for New Horizons and other incremental upgrades. Having high-power fission systems could alleviate this bottleneck even further.

6. CONCLUSIONS

The exploration of Pluto is far from over. What began in the 20th century can and should continue with the expectation and anticipation of an even deeper understanding of Pluto and its retinue of satellites. With the discovery of the rest of the Kuiper belt and the growing number of other similar-sized objects, Pluto is becoming a key to understanding the icy dwarf planets, especially those that are volatile rich. As seen with New Horizons, Pluto is also intrinsically interesting, with a very complex and active surface and atmosphere. Here we have a world that requires patience and perseverance in its study. We have seen that it is indeed dynamic, but the timescale for changes are, at best, similar to the length of an individual career.

Telescopic observations can and should continue but this approach can only go so far. To truly uncover Pluto's deepest secrets requires data made possible only from a vantage point within the system. Spacecraft investigations are clearly very important to reach the next level of understanding of this complex world. An orbiter mission is the obvious next step that unlocks a whole new level of secrets. Such a mission will be a technical challenge but one within our grasp. The engineering work applied to a Pluto orbiter will also prove to be invaluable to exploration throughout the outer solar system.

REFERENCES

Bertrand T. and Forget F. (2016) Observed glacier and volatile distribution on Pluto from atmosphere-topography processes. *Nature, 540*, 86–89.

Bertrand T., Forget F., Umurhan O. M., Grundy W. M., Schmitt B., Protopapa S., Zangari A. M., White O. L., Schenk P. M., Singer K. N., Stern A., Weaver H. A., Young L. A., Ennico K., and Olkin C. B. (2018) The nitrogen cycles on Pluto over seasonal and astronomical timescales. *Icarus, 309*, 277–296.

Bertrand T., Forget F., Umurhan O. M., Moore J. M., Young L. A., Protopapa S., Grundy W. M., Schmitt B., Dhingra R. D., Binzel R. P., Earle A. M., Cruikshank D. P., Stern S. A., Weaver H. A., Ennico K., Olkin C. B., and the New Horizons Science Team (2019) The CH_4 cycles on Pluto over seasonal and astronomical timescales. *Icarus, 329*, 148–165.

Beyer R. A., Nimmo F., McKinnon W. B., Moore J. M., Binzel R. P., Conrad J. W., Cheng A., Ennico K., Lauer T. R., Olkin C. B., Robbins S., Schenk P., Singer K., Spencer J. R., Stern S. A., Weaver H. A., Young L. A., and Zangari A. M. (2018) Charon tectonics. *Icarus, 287*, 161–174.

Bird M. K., Linscott I. R., Tyler G. L., Hinson D. P., Pätzold M., Summers M. E., Strobel D. F., Stern S. A., Weaver H. A., Olkin C. B., Young L. A., Ennico K., Moore J. M., Gladstone G. R., Grundy W. M., DeBoy C. C., Vincent M., and the New Horizons Science Team (2019) Radio thermal emission from Pluto and Charon during the New Horizons encounter. *Icarus, 322*, 192–209.

Bockelée-Morvan D., Lis D. C., Wink J. E., Despois D., Crovisier J., Bachiller R., Benford D. J., Biver N., Colom P., Davies J. K., Gérard E., Germain B., Houde M., Mehringer D., Moreno R., Paubert G., Phillips T. G., and Rauer H. (2000) New molecules found in Comet C/1995 O1 (Hale-Bopp): Investigating the link between cometary and interstellar material. *Astron. Astrophys., 353*, 1101–1114.

Brown M. E., Bannister M. T., Schmidt B. P., Drake A. J., Djorgovski S. G., Graham M. J., Mahabal A., Donalek C., Larson S., Christensen E., Beshore E., and McNaught R. (2015) A serendipitous all sky survey for bright objects in the outer solar system. *Astron. J., 149*, 69.

Buie M. W. and Grundy W. M. (2000) Continued evolution in the lightcurve of Pluto. *AAS/Division for Planetary Sciences Meeting Abstracts, 32,* 45.05.

Buie M. W., Grundy W. M., Young E. F., Young L. A., and Stern S. A. (2010a) Pluto and Charon with the Hubble Space Telescope. I. Monitoring global change and improved surface properties from light curves. *Astron. J., 139*, 1117–1127.

Buie M. W., Grundy W. M., Young E. F., Young L. A., and Stern S. A. (2010b) Pluto and Charon with the Hubble Space Telescope. II. Resolving changes on Pluto's surface and a map for Charon. *Astron. J., 139,* 1128–1143.

Buratti B. J., Hofgartner J. D., Hicks M. D., Weaver H. A., Stern S. A., Momary T., Mosher J. A., Beyer R. A., Verbiscer A. J., Zangari A. M., Young L. A., Lisse C. M., Singer K., Cheng A., Grundy W., Ennico K., and Olkin C. B. (2017) Global albedos of Pluto and Charon from LORRI New Horizons observations. *Icarus, 287,* 207–217.

Butler B., Grundy W., Gurwell M., Lellouch E., Moreno R., Moullet A., and Young L. A. (2018) Resolved thermal images of Pluto and Charon with ALMA. *AAS/Division for Planetary Sciences Meeting Abstracts, 50*, 502.06.

Butler B. J., Grundy W. M., Gurwell M. A., Lellouch E., Moreno R., Moullet A., and Young L. A. (2019) Observations of Pluto's surface with ALMA. *Pluto System After New Horizons*, Abstract #7058. LPI Contribution No. 2133, Lunar and Planetary Institute, Houston.

Canup R. M. (2005) A giant impact origin of Pluto-Charon. *Science, 307*, 546–550.

Cook J. C., Cruikshank D. P., and Young L. A. (2014) Gemini North/NIRI spectra of Pluto and Charon: Simultaneous analysis of the surface and atmosphere. *AAS/Division for Planetary Sciences Meeting Abstracts, 46*, 401.04.

Cruikshank D. P., Roush T. L., Owen T. C., Gevalle T. R., de Bergh C., Schmitt B., Brown R. H., and Bartholomew M. J. (1993) Ices on the surface of Triton. *Science, 261*, 742–745.

Cruikshank D. P., Umurhan O. M., Beyer R. A., Schmitt B., Keane J. T., Runyon K. D., Atri D., White O. L., Matsuyama I., Moore J. M., McKinnon W. B., Sandford S. A., Singer K. N., Grundy W. M., Dalle Ore C. M., Cook J. C., Bertrand T., Stern S. A., Olkin C. B., Weaver H. A., Young L. A., Spencer J. R., Lisse C. M., Binzel R. P., Earle A. M., Robbins S. J., Gladstone G. R., Cartwright R. J., and Ennico K. (2019) Recent cryovolcanism in Virgil Fossae on Pluto. *Icarus, 330*, 155–168.

DeMeo F. E., Dumas C., de Bergh C., Protopapa S., Cruikshank D. P., Geballe T. R., Alvarez-Candal A., Merlin F., and Barucci M. A. (2010) A search for ethane on Pluto and Triton. *Icarus, 208*, 412–424.

DeMeo F. E., Dumas C., Cook J. C., Carry B., Merlin F., Verbiscer A. J., and Binzel R. P. (2015) Spectral variability of Charon's 2.21-μm feature. *Icarus, 246*, 213–219.

Dobrovolskis A. R. (1989) Dynamics of Pluto and Charon. *Geophys. Res. Lett., 16*, 1217–1220.

Douté S., Schmitt B., Quirico E., Owen T. C., Cruikshank D. P., de Bergh C., Geballe T. R., and Roush T. L. (1999) Evidence for methane segregation at the surface of Pluto. *Icarus, 142*, 421–444.

Earle A. M. and Binzel R. P. (2015) Pluto's insolation history: Latitudinal variations and effects on atmospheric pressure. *Icarus, 250*, 405–412.

Elliot J. L., Person M. J., Gulbis A. A. S., Souza S. P., Adams E. R., Babcock B. A., Gangestad J. W., Jaskot A. E., Kramer E. A., Pasachoff J. M., Pike R. E., Zuluaga C. A., Bosh A. S., Dieters S. W., Francis P. J., Giles A. B., Greenhill J. G., Lade B., Lucas R., and Ramm D. J. (2007) Changes in Pluto's atmosphere: 1988-2006. *Astron. J., 134*, 1–13.

Gladstone G. R., Stern S. A., Ennico K., Olkin C. B., Weaver H. A., Young L. A., Summers M. E., Strobel D. F., Hinson D. P., Kammer J. A., Parker A. H., Steffl A. J., Linscott I. R., Parker J. W., Cheng A. F., Slater D. C., Versteeg M. H., Greathouse T. K., Retherford K. D., Throop H., Cunningham N. J., Woods W. W., Singer K. N., Tsang C. C. C., Schindhelm E., Lisse C. M., Wong M. L., Yung Y. L., Zhu X., Curdt W., Lavvas P., Young E. F., Tyler G. L., Bagenal F., Grundy W. M., McKinnon W. B., Moore J. M., Spencer J. R., Andert T., Andrews J., Banks M., Bauer B., Bauman J., Barnouin O. S., Bedini P., Beisser K., Beyer R. A., Bhaskaran S., Binzel R. P., Birath E., Bird M., Bogan D. J., Bowman A., Bray V. J., Brozovic M., Bryan C., Buckley M. R., Buie M. W., Buratti B. J., Bushman S. S., Calloway A., Carcich B., Conard S., Conrad C. A., Cook J. C., Cruikshank D. P., Custodio O. S., Dalle Ore C. M., Deboy C., Dischner Z. J. B., Dumont P., Earle A. M., Elliott H. A., Ercol J., Ernst C. M., Finley T., Flanigan S. H., Fountain G., Freeze M. J., Green J. L., Guo Y., Hahn M., Hamilton D. P., Hamilton S. A., Hanley J., Harch A., Hart H. M., Hersman C. B., Hill A., Hill M. E., Holdridge M. E., Horanyi M., Howard A. D., Howett C. J. A., Jackman C., Jacobson R. A., Jennings D. E., Kang H. K., Kaufmann D. E., Kollmann P., Krimigis S. M., Kusnierkiewicz D., Lauer T. R., Lee J. E., Lindstrom K. L., Lunsford A. W., Mallder V. A., Martin N., McComas D. J., McNutt R. L., Mehoke D., Mehoke T., Melin E. D., Mutchler M., Nelson D., Nimmo F., Nunez J. I., Ocampo A., Owen W. M., Paetzold M., Page B., Pelletier F., Peterson J., Pinkine N., Piquette M., Porter S. B., Protopapa S., Redfern J., Reitsema H. J., Reuter D. C., Roberts

J. H., Robbins S. J., Rogers G., Rose D., Runyon K., Ryschkewitsch M. G., Schenk P., Sepan B., Showalter M. R., Soluri M., Stanbridge D., Stryk T., Szalay J. R., Tapley M., Taylor A., Taylor H., Umurhan O. M., Verbiscer A. J., Versteeg M. H., Vincent M., Webbert R., Weidner S., Weigle G. E., White O. L., Whittenburg K., Williams B. G., Williams K., Williams S., Zangari A. M., and Zirnstein E. (2016) The atmosphere of Pluto as observed by New Horizons. *Science, 351*, aad8866.

Greaves J. S., Helling C., and Friberg P. (2011) Discovery of carbon monoxide in the upper atmosphere of Pluto. *Mon. Not. R. Astron. Soc. Lett., 414*, L36–L40.

Grundy W. M., Buie M. W., and Spencer J. R. (2002) Spectroscopy of Pluto and Triton at 3–4 microns: Possible evidence for wide distribution of nonvolatile solids. *Astron. J., 124*, 2273–2278.

Grundy W. M., Morrison S. J., Bovyn M. J., Tegler S. C., and Cornelison D. M. (2011) Remote sensing D/H ratios in methane ice: Temperature-dependent absorption coefficients of CH_3D in methane ice and in nitrogen ice. *Icarus, 212*, 941–949.

Grundy W. M., Olkin C. B., Young L. A., and Holler B. J. (2014) Near-infrared spectral monitoring of Pluto's ices II: Recent decline of CO and N_2 ice absorptions. *Icarus, 235*, 220–224.

Grundy W. M., Binzel R. P., Buratti B. J., Cook J. C., Cruikshank D. P., Dalle Ore C. M., Earle A. M., Ennico K., Howett C. J. A., Lunsford A. W., Olkin C. B., Parker A. H., Philippe S., Protopapa S., Quirico E., Reuter D. C., Schmitt B., Singer K. N., Verbiscer A. J., Beyer R. A., Buie M. W., Cheng A. F., Jennings D. E., Linscott I. R., Parker J. W., Schenk P. M., Spencer J. R., Stansberry J. A., Stern S. A., Throop H. B., Tsang C. C. C., Weaver H. A., Weigle G. E., and Young L. A. (2016a) Surface compositions across Pluto and Charon. *Science, 351*, aad9189.

Grundy W. M., Cruikshank D. P., Gladstone G. R., Howett C. J. A., Lauer T. R., Spencer J. R., Summers M. E., Buie M. W., Earle A. M., Ennico K., Parker J. W., Porter S. B., Singer K. N., Stern S. A., Verbiscer A. J., Beyer R. A., Binzel R. P., Buratti B. J., Cook J. C., Dalle Ore C. M., Olin C. B., Parker A. H., Protopapa S., Quirico E., Retherford K. D., Robbins S. J., Schmitt B., Stansberry J. A., Umurhan O. M., Weaver H. A., Young L. A., Zangari A. M., Bray V. J., Cheng A. F., McKinnon W. B., McNutt R. L., Morre J. M., Nimmo F., Reuter D. C., Schenk P. M., the New Horizons Science Team, Stern S. A., Bagenal F., Ennico K., Gladstone G. R., Grundy W. M., McKinnon W. B., Moore J. M., Olkin C. B., Spencer J. R., Weaver H. A., Young L. A., Andert T., Barnouin O., Beyer R. A., Binzel R. P., Bird M., Bray V. J., Brozovic M., Buie M. W., Buratti B. J., Cheng A. F., Cook J. C., Cruikshank D. P., Dalle Ore C. M., Earle A. M., Elliott H. A., Greathouse T. K., Hahn M., Hamilton D. P., Hill M. E., Hinson D. P., Hofgartner J., Horányi M., Howard A. D., Howett C. J. A., Jennings D. E., Kammer J. A., Kollmann P., Lauer T. R., Lavvas P., Linscott I. R., Lisse C. M., Lunsford A. W., McComas D. J., McNutt R. L. Jr., Mutchler M., Nimmo F., Nunez J. I., Paetzold M., Parker A. H., Parker J. W., Philippe S., Piquette M., Porter S. B., Protopapa S., Quirico E., Reitsema H. J., Reuter D. C., Robbins S. J., Roberts J. H., Runyon K., Schenk P. M., Schindhelm E., Schmitt B., Showalter M. R., Singer K. N., Stansberry J. A., Steffl A. J., Strobel D. F., Stryk T., Summers M. E., Szalay J. R., Throop H. B., Tsang C. C. C., Tyler G. L., Umurhan O. M., Verbiscer A. J., Versteeg M. H., Weigle G. E. II, White O. L., Woods W. W., Young E. F., and Zangari A. M. (2016b) The formation of Charon's red poles from seasonally cold-trapped volatiles. *Nature, 539*, 65–68.

Hansen C. J., Paige D. A., and Young L. A. (2015) Pluto's climate modeled with new observational constraints. *Icarus, 246*, 183–191.

Hofgartner J. D., Buratti B. J., Devins S. L., Beyer R. A., Schenk P., Stern S. A., Weaver H. A., Olkin C. B., Cheng A., Ennico K., Lauer T. R., McKinnon W. B., Spencer J., Young L. A., and the New Horizons Science Team (2018) A search for temporal changes on Pluto and Charon. *Icarus, 302*, 273–284.

Hofgartner J. D., Buratti B. J., Hayne P. O., and Young L. A. (2019) Ongoing resurfacing of KBO Eris by volatile transport in local, collisional, sublimation atmosphere regime. *Icarus, 334*, 52–61.

Holler B. J., Young L. A., Grundy W. M., Olkin C. B., and Cook J. C. (2014) Evidence for longitudinal variability of ethane ice on the surface of Pluto. *Icarus, 243*, 104–110.

Holler B. J., Young L. A., Buie M. W., Grundy W. M., Lyke J. E., Young E. F., and Roe H. G. (2017) Measuring temperature and ammonia hydrate ice on Charon in 2015 from Keck/OSIRIS spectra. *Icarus, 284*, 394–406.

Howett C. J. A., Ennico K., Olkin C. B., Buie M. W., Verbiscer A. J., Zangari A. M., Parker A. H., Reuter D. C., Grundy W. M., Weaver H. A., Young L. A., and Stern S. A. (2016) Charon's light curves, as observed by New Horizons' Ralph color camera (MVIC) on approach to the Pluto system. *Icarus, 287*, 152–160.

Hunten D. M. and Watson A. J. (1982) Stability of Pluto's atmosphere. *Icarus, 51*, 665–667.

Keane J. T., Matsuyama I., Kamata S., and Steckloff J. K. (2016) Reorientation and faulting of Pluto due to volatile loading within Sputnik Planitia. *Nature, 540*, 90–93.

Lellouch E., Laureijs R., Schmitt B., Quirico E., de Bergh C., Crovisier J., and Coustenis A. (2000) Pluto's non-isothermal surface. *Icarus, 147*, 220–250.

Lellouch E., Sicardy B., deBergh C., Käufl H.-U., Kassi S., and Campargue A. (2009) Pluto's lower atmosphere structure and methane abundance from high-resolution spectroscopy and stellar occultations. *Astron. Astrophys., 495*, L17–L21.

Lellouch E., Stansberry J., Emery J., Grundy W., and Cruikshank D. P. (2011) Thermal properties of Pluto's and Charon's surfaces from Spitzer observations. *Icarus, 214*, 701–716.

Lellouch E., de Bergh C., Sicardy B., Forget F., Vangvichith M., and Käufl H.-U. (2015) Exploring the spatial, temporal, and vertical distribution of methane in Pluto's atmosphere. *Icarus, 246*, 268–278.

Lellouch E., Santos-Sanz P., Fornasier S., Lim T., Stansberry J., Vilenius E., Kiss C., Müller T., Marton G., Protopapa S., Panuzzo P., and Moreno R. (2016) The long-wavelength thermal emission of the Pluto-Charon system from Herschel observations: Evidence for emissivity effects. *Astron. Astrophys., 588*, A2.

Lellouch E., Gurwell M., Butler B., Fouchet T., Lavvas P., Strobel D. F., Sicardy B., Moullet A., Moreno R., Bockelée-Morvan D., Biver N., Young L., Lis D., Stansberry J., Stern A., Weaver H., Young E., Zhu X., and Boissier J. (2017) Detection of CO and HCN in Pluto's atmosphere with ALMA. *Icarus, 286*, 289–307.

Lellouch E., Gurwell M., Moreno R., Panayotis L., Butler B., Strobel D., Fouchet T., Moullet A., Bockelée-Morvan D., and Biver N. (2018) Pluto's atmosphere with ALMA: Disk-resolved observations of CO and HCN, and first detection of HNC. *AAS/Division for Planetary Sciences Meeting Abstracts, 60*, 314.03.

Lis D. C., Bockelée-Morvan D., Güsten R., Biver N., Stutzki J., Delorme Y., Durán C., Wiesemeyer H., and Okada Y. (2019) Terrestrial deuterium-to-hydrogen ratio in water in hyperactive comets. *Astron. Astrophys., 625*, L5.

Lorenzi V., Pinilla-Alonso N., Licandro J., Cruikshank D. P., Grundy W. M., Binzel R. P., and Emery J. P. (2016) The spectrum of Pluto, 0.40–0.93 µm. I. Secular and longitudinal distribution of ices and complex organics. *Astron. Astrophys., 585*, A131.

Materese C. K., Cruikshank D. P., Sandford S. A., Imanaka H., Nuevo M., and White D. W. (2014) Ice chemistry on outer solar system bodies: Carboxylic acids, nitriles, and urea detected in refractory residues produced from the UV photolysis of N_2:CH_4:CO-containing ices. *Astrophys. J., 788*, 111.

McClure P. R., Poston D. I., Gibson M., Bowman C., and Creasy J. (2014) *KiloPower Space Reactor Concept — Reactor Materials Study.* LANL Rept. LA-UR-14-23402, Los Alamos, New Mexico.

McClure P. R., Poston D. I., Gibson M., Mason L., and Oleson S. (2018) *White Paper — Comparison of LEU and HEU Fuel for the Kilopower Reactor.* LANL Rept. LA-UR-18-29623, Los Alamos, New Mexico.

McKinnon W. B., Nimmo F., Wong T., Schenk P. M., White O. L., Roberts J. H., Moore J. M., Spencer J. R., Howard A. D., Umurhan O. M., Stern S. A., Weaver H. A., Olkin C. B., Young L. A., Smith K. E., Beyer R., Buie M., Buratti B., Cheng A., Cruikshank D., Dalle Ore C., Gladstone R., Grundy W., Lauer T., Linscott I., Parker J., Porter S., Reitsema H., Reuter D., Robbins S., Showalter M., Singer K., Strobel D., Summers M., Tyler L., Banks M., Barnouin O., Bray V., Carcich B., Chaikin A., Chavez C., Conrad C., Hamilton D., Howett C., Hofgartner J., Kammer J., Lisse C., Marcotte A., Parker A., Retherford K., Saina M., Runyon K., Schindhelm E., Stansberry J., Steffl A., Stryk T., Throop H., Tsang C., Verbiscer A., Winters H., Zangari A., and the New Horizons Geology, Geophysics and Imaging Theme Team (2016) Convection in a volatile nitrogen-ice-rich layer drives Pluto's geological vigour. *Nature, 534*, 82–85.

Merlin F. (2015) New constraints on the surface of Pluto. *Astron. Astrophys., 582*, A39.

Merlin F., Lellouch E., Quirico E., and Schmitt B. (2018) Triton's surface ices: Distribution, temperature and mixing state from VLT/

SINFONI observations. *Icarus, 314,* 274–293.

Meza E., Sicardy B., Assafin M., Ortiz J. L., Bertrand T., Lellouch E., Desmars J., Forget F., Bérard D., Doressoundiram A., Lecacheux J., Marques Oliveira J., Roques F., Widemann T., Colas F., Vachier F., Renner S., Leiva R., Braga-Ribas F., Benedetti-Rossi G., Camargo J. I. B., Dias-Oliveira A., Morgado B., Gomes-Júnior A. R., Vieira-Martins R., Behrend R., Tirado A. C., Duffard R., Morales N., Santos-Sanz P., Jelínek M., Cunniffe R., Querel R., Harnisch M., Jansen R., Pennell A., Todd S., Ivanov V. D., Opitom C., Gillon M., Jehin E., Manfroid J., Pollock J., Reichart D. E., Haislip J. B., Ivarsen K. M., LaCluyze A. P., Maury A., Gil-Hutton R., Dhillon V., Littlefair S., Marsh T., Veillet C., Bath K.-L., Beisker W., Bode H.-J., Kretlow M., Herald D., Gault D., Kerr S., Pavlov H., Faragó O., Klös O., Frappa E., Lavayssière M., Cole A. A., Giles A. B., Greenhill J. G., Hill K. M., Buie M. W., Olkin C. B., Young E. F., Young L. A., Wasserman L. H., Devogèle M., French R. G., Bianco F. B., Marchis F., Brosch N., Kaspi S., Polishook D., Manulis I., Ait Moulay Larbi M., Benkhaldoun Z., Daassou A., El Azhari Y., Moulane Y., Broughton J., Milner J., Dobosz T., Bolt G., Lade B., Gilmore A., Kilmartin P., Allen W. H., Graham P. B., Loader B., McKay G., Talbot J., Parker S., Abe L., Bendjoya P., Rivet J.-P., Vernet D., Di Fabrizio L., Lorenzi V., Magazzú A., Molinari E., Gazeas K., Tzouganatos L., Carbognani A., Bonnoli G., Marchini A., Leto G., Sanchez R. Z., Mancini L., Kattentidt B., Dohrmann M., Guhl K., Rothe W., Walzel K., Wortmann G., Eberle A., Hampf D., Ohlert J., Krannich G., Murawsky G., Gährken B., Gloistein D., Alonso S., Román A., Communal J.-E., Jabet F., deVisscher S., Sérot J., Janik T., Moravec Z., Machado P., Selva A., Perelló C., Rovira J., Conti M., Papini R., Salvaggio F., Noschese A., Tsamis V., Tigani K., Barroy P., Irzyk M., Neel D., Godard J. P., Lanoiselée D., Sogorb P., Vérilhac D., Bretton M., Signoret F., Ciabattari F., Naves R., Boutet M., De Queiroz J., Lindner P., Lindner K., Enskonatus P., Dangl G., Tordai T., Eichler H., Hattenbach J., Peterson C., Molnar L. A., and Howell R. R. (2019) Lower atmosphere and pressure evolution on Pluto from ground-based stellar occultations, 1988-2016. *Astron. Astrophys., 625,* A42.

Moore J. M., McKinnon W. B., Spencer J. R., Howard A. D., Schenk P. M., Beyer R. A., Nimmo F., Singer K. N., Umurhan O. M., White O. L., Stern S. A., Ennico K., Olkin C. B., Weaver H. A., Young L. A., Binzel R. P., Buie M. W., Buratti B. J., Cheng A. F., Cruikshank D. P., Grundy W. M., Linscott I. R., Reitsema H. J., Reuter D. C., Showalter M. R., Bray V. J., Chavez C. L., Howett C. J. A., Lauer T. R., Lisse C. M., Parker A. H., Porter S. B., Robbins S. J., Runyon K., Stryk T., Throop H. B., Tsang C. C. C., Verbiscer A. J., Zangari A. M., Chaikin A. L., Wilhelms D. E., Bagenal F., Gladstone G. R., Andert T., Andrews J., Banks M., Bauer B., Bauman J., Barnouin O. S., Bedini P., Beisser K., Bhaskaran S., Birath E., Bird M., Bogan D. J., Bowman A., Brozovic M., Bryan C., Buckley M. R., Bushman S. S., Calloway A., Carcich B., Conard S., Conrad C. A., Cook J. C., Custodio O. S., Dalle Ore C. M., Deboy C., Dischner Z. J. B., Dumont P., Earle A. M., Elliott H. A., Ercol J., Ernst C. M., Finley T., Flanigan S. H., Fountain G., Freeze M. J., Greathouse T., Green J. L., Guo Y., Hahn M., Hamilton D. P., Hamilton S. A., Hanley J., Harch A., Hart H. M., Hersman C. B., Hill A., Hill M. E., Hinson D. P., Holdridge M. E., Horanyi M., Jackman C., Jacobson R. A., Jennings D. E., Kammer J. A., Kang H. K., Kaufmann D. E., Kollmann P., Krimigis S. M., Kusnierkiewicz D., Lee J. E., Lindstrom K. L., Lunsford A. W., Mallder V. A., Martin N., McComas D. J., McNutt R. L., Mehoke D., Mehoke T., Melin E. D., Mutchler M., Nelson D., Nunez J. I., Ocampo A., Owen W. M., Paetzold M., Page B., Parker J. W., Pelletier F., Peterson J., Pinkine N., Piquette M., Protopapa S., Redfern J., Roberts J. H., Rogers G., Rose D., Retherford K. D., Ryschkewitsch M. G., Schindhelm E., Sepan B., Soluri M., Stanbridge D., Steffl A. J., Strobel D. F., Summers M. E., Szalay J. R., Tapley M., Taylor A., Taylor H., Tyler G. L., Versteeg M. H., Vincent M., Webbert R., Weidner S., Weigle G. E., Whittenburg K., Williams B. G., Williams K., Williams S., Woods W. W., and Zirnstein E. (2016) The geology of Pluto and Charon through the eyes of New Horizons. *Science, 351,* 1284–1293.

Moore J. M., Howard, A. D., Umurhan O. M., White O. L., Schenk P. M., Beyer R. A., McKinnon W. B., Spencer J. R., Singer K. N., Grundy W. M., Earle A. M., Schmitt B., Protopapa S., Nimmo F., Cruikshank D. P., Hinson D. P., Young L. A., Stern S. A., Weaver H. A., Olkin C. B., Ennico K., Collins G., Bertrand T., Forget F., Scipioni F., and the New Horizons Science Team (2018) Bladed terrain on Pluto: Possible origins and evolution. *Icarus, 300,* 129–144, DOI: 10.1016/j.icarus.2017.08.031.

National Research Council (2011) *Vision and Voyages for Planetary Science in the Decade 2013–2022.* National Academies, Washington, DC (e-Book), DOI: 10.17226/13117.

Nimmo F., Hamilton D. P., McKinnon W. B., Schenk P. M., Binzel R. P., Bierson C. J., Beyer R. A., Moore J. M., Stern S. A., Weaver H. A., Olkin C. B., Young L. A., Smith K. E., Moore J. M., McKinnon W. B., Spencer J. R., Beyer R., Binzel R. P., Buie M., Buratti B., Cheng A., Cruikshank D., Dalle Ore C., Earle A., Gladstone R., Grundy W., Howard A. D., Lauer T., Linscott I., Nimmo F., Parker J., Porter S., Reitsema H., Reuter D., Roberts J. H., Robbins S., Schenk P. M., Showalter M., Singer K., Strobel D., Summers M., Tyler L., White O. L., Umurhan O. M., Banks M., Barnouin O., Bray V., Carcich B., Chaikin A., Chavez C., Conrad C., Hamilton D. P., Howett C., Hofgartner J., Kammer J., Lisse C., Marcotte A., Parker A., Retherford K., Saina M., Runyon K., Schindhelm E., Stansberry J., Steffl A., Stryk T., Throop H., Tsang C., Verbiscer A., Winters H., Zangari A., Stern S. A., Weaver H. A., Olkin C. B., Young L. A., and Smith K. E. (2016) Reorientation of Sputnik Planitia implies a subsurface ocean on Pluto. *Nature, 540,* 94–96.

Olkin C. B., Young L. A., Borncamp D., Pickles A., Sicardy B., Assafin M., Bianco F. B., Buie M. W., de Oliveira A. D., Gillon M., French R. G., Ramos Gomes A., Jehin E., Morales N., Opitom C., Ortiz J. L., Maury A., Norbury M., Braga-Ribas F., Smith R., Wasserman L. H., Young E. F., Zacharias M., and Zacharias N. (2015) Evidence that Pluto's atmosphere does not collapse from occultations including the 2013 May 04 event. *Icarus, 246,* 220–225.

Owen T., Geballe T., de Bergh C., Young L., Elliot J., Cruikshank D., Roush T., Schmitt B., Brown R. H., and Green J. (1992) Detection of nitrogen and carbon monoxide on the surface of Pluto. *AAS/Division for Planetary Sciences Meeting Abstracts, 24,* 16.01.

Protopapa S., Boehnhardt H., Herbst T. M., Cruikshank D. P., Grundy W. M., Merlin F., and Olkin C. B. (2008) Surface characterization of Pluto and Charon by L and M band spectra. *Astron. Astrophys., 490,* 365–375.

Quirico E. and Schmitt B. (1997) Near-infrared spectroscopy of simple hydrocarbons and carbon oxides diluted in solid N_2 and as pure ices: Implications for Triton and Pluto. *Icarus, 127,* 354–378.

Rieke G. H., Wright G. S., Böker T., Bouwman J., Colina L., Glasse A., Gordon K. D., Greene T. P., Güdel M., Henning T., Justtanont K., Lagage P.-O., Meixner M. E., Nørgaard-Nielsen H.-U., Ray T. P., Ressler M. E., van Dishoeck E. F., and Waelkens C. (2015) The mid-infrared instrument for the James Webb Space Telescope, I: Introduction. *Publ. Astron. Soc. Pac., 127,* 584–594.

Schenk P. M., Beyer R. A., McKinnon W. B., Moore J. M., Spencer J. R., White O. L., Singer K., Nimmo F., Thomason C., Lauer T. R., Robbins S., Umurhan O. M., Grundy W. M., Stern S. A., Weaver H. A., Young L. A., Ennico Smith K., Olkin C., and the New Horizons Geology and Geophysics Investigation Team (2018a) Basins, fractures and volcanoes: Global cartography and topography of Pluto from New Horizons. *Icarus, 314,* 400–433.

Schenk P. M., Beyer R. A., McKinnon W. B., Moore J. M., Spencer J. R., White O. L., Singer K., Umurhan O. M., Nimmo F., Lauer T. R., Grundy W. M., Robbins S., Stern S. A., Weaver H. A., Young L. A., Ennico Smith K., Olkin C., and the New Horizons Geology and Geophysics Investigation Team (2018b) Breaking up is hard to do: Global cartography and topography of Pluto's mid-sized icy moon Charon from New Horizons. *Icarus, 315,* 124–145.

Schmitt B., Gabasova L., Bertrand T., Grundy W., Stansberry J., Lewis B., Protopapa S., Young L., Olkin C., Reuter D., Stern A., and Weaver H. (2019) Methane stratification on Pluto inferred from New Horizons LEISA data. *Pluto System After New Horizons,* Abstract #7004. LPI Contribution No. 2133, Lunar and Planetary Institute, Houston.

Sicardy B., Talbot J., Meza E., Camargo J. I. B., Desmars J., Gault D., Herald D., Kerr S., Pavlov H., Braga-Ribas F., Assafin M., Benedetti-Rossi G., Dias-Oliveira A., Gomes-Júnior A. R., Vieira-Martins R., Bérard D., Kervella P., Lecacheux J., Lellouch E., Beisker W. D. D., Jelínek M., Duffard R., Ortiz J. L., Castro-Tirado A. J., Cunniffe R., Querel R., Yock P. C., Cole A. A., Giles A. B., Hill K. M., Beaulieu J. P., Harnisch M., Jansen R., Pennell A., Todd S., Allen W. H., Graham P. B., Loader B., McKay G., Milner J., Parker S., Barry M. A., Bradshaw J., Broughton J., Davis L., Devillepoix H., Drummond J., Field L., Forbes M., Giles D., Glassey

R., Groom R., Hooper D., Horvat R., Hudson G., Idaczyk R., Jenke D., Lade B., Newman J., Nosworthy P., Purcell P., Skilton P. F., Streamer M., Unwin M., Watanabe H., White G. L., and Watson D. (2016) Pluto's atmosphere from the 2015 June 29 ground-based stellar occultation at the time of the New Horizons flyby. *Astrophys. J., Lett., 819*, L38.

Singer K. N., McKinnon W. B., Gladman B., Greenstreet S., Bierhaus E. B., Stern S. A., Parker A. H., Robbins S. J., Schenk P. M., Grundy W. M., Bray V. J., Beyer R. A., Binzel R. P., Weaver H. A., Young L. A., Spencer J. R., Kavelaars J. J., Moore J. M., Zangari A. M., Olkin C. B., Lauer T. R., Lisse C. M., Ennico K., the New Horizons Geology and Geophysics and Imaging Science Theme Team, the New Horizons Surface Composition Science Theme Team, and the New Horizons Ralph and LORRI Teams (2019) Impact craters on Pluto and Charon indicate a deficit of small Kuiper belt objects. *Science 363*, 955–959.

Stern S. A., Trafton L. M., and Gladstone G. R. (1988) Why is Pluto bright? Implications of the albedo and lightcurve behavior of Pluto. *Icarus, 75*, 485–498.

Stern S. A., Bagenal F., Ennico K., Gladstone G. R., Grundy W. M., McKinnon W. B., Moore J. M., Olkin C. B., Spencer J. R., Weaver H. A., Young L. A., Andert T., Andrews J., Banks M., Bauer B., Bauman J., Barnouin O. S., Bedini P., Beisser K., Beyer R. A., Bhaskaran S., Binzel R. P., Birath E., Bird M., Bogan D. J., Bowman A., Bray V. J., Brozovic M., Bryan C., Buckley M. R., Buie M. W., Buratti B. J., Bushman S. S., Calloway A., Carcich B., Cheng A. F., Conard S., Conrad C. A., Cook J. C., Cruikshank D. P., Custodio O. S., Dalle Ore C. M., Deboy C., Dischner Z. J. B., Dumont P., Earle A. M., Elliott H. A., Ercol J., Ernst C. M., Finley T., Flanigan S. H., Fountain G., Freeze M. J., Greathouse T., Green J. L., Guo Y., Hahn M., Hamilton D. P., Hamilton S. A., Hanley J., Harch A., Hart H. M., Hersman C. B., Hill A., Hill M. E., Hinson D. P., Holdridge M. E., Horanyi M., Howard A. D., Howett C. J. A., Jackman C., Jacobson R. A., Jennings D. E., Kammer J. A., Kang H. K., Kaufmann D. E., Kollmann P., Krimigis S. M., Kusnierkiewicz D., Lauer T. R., Lee J. E., Lindstrom K. L., Linscott I. R., Lisse C. M., Lunsford A. W., Mallder V. A., Martin N., McComas D. J., McNutt R. L., Mehoke D., Mehoke T., Melin E. D., Mutchler M., Nelson D., Nimmo F., Nunez J. I., Ocampo A., Owen W. M., Paetzold M., Page B., Parker A. H., Parker J. W., Pelletier F., Peterson J., Pinkine N., Piquette M., Porter S. B., Protopapa S., Redfern J., Reitsema H. J., Reuter D. C., Roberts J. H., Robbins S. J., Rogers G., Rose D., Runyon K., Retherford K. D., Ryschkewitsch M. G., Schenk P., Schindhelm E., Sepan B., Showalter M. R., Singer K. N., Soluri M., Stanbridge D., Steffl A. J., Strobel D. F., Stryk T., Summers M. E., Szalay J. R., Tapley M., Taylor A., Taylor H., Throop H. B., Tsang C. C. C., Tyler G. L., Umurhan O. M., Verbiscer A. J., Versteeg M. H., Vincent M., Webbert R., Weidner S., Weigle G. E., White O. L., Whittenburg K., Williams B. G., Williams K., Williams S., Woods W. W., Zangari A. M., and Zirnstein E. (2015) The Pluto system: Initial results from its exploration by New Horizons. *Science, 350*, aad1815.

Stern S. A., Binzel R. P., Earle A. M., Singer K. N., Young L. A., Weaver H. A., Olkin C. B., Ennico K., Moore J. M., McKinnon W. B., Spencer J. R., and the New Horizons Geology and Geophysics and Atmospheres Teams (2017) Past epochs of significantly higher pressure atmospheres on Pluto. *Icarus, 287*, 47–53.

Stern S. A., Tapley M. B., Finley T. J., Zangari A. M., and Scherrer J. R. (2020) Pluto orbiter-Kuiper Belt explorer: Mission design for the Gold Standard. *J. Spacecr. Rockets*, DOI: 10.2514/1.A34658.

Tryka K. A., Brown R. H., Cruikshank D. P., Owen T. C., Geballe T. R., and de Bergh C. (1994) Temperature of nitrogen ice on Pluto and its implications for flux measurements. *Icarus, 112*, 513–527.

Weaver H. A., Buie M. W., Buratti B. J., Grundy W. M., Lauer T. R., Olkin C. B., Parker A. H., Porter S. B., Showalter M. R., Spencer J. R., Stern S. A., Verbiscer A. J., McKinnon W. B., Moore J. M., Robbins S. J., Schenk P., Singer K. N., Barnouin O. S., Cheng A. F., Ernst C. M., Lisse C. M., Jennings D. E., Lunsford A. W., Reuter D. C., Hamilton D. P., Kaufmann D. E., Ennico K., Young L. A., Beyer R. A., Binzel R. P., Bray V. J., Chaikin A. L., Cook J. C., Cruikshank D. P., Dalle Ore C. M., Earle A. M., Gladstone G. R., Howett C. J. A., Linscott I. R., Nimmo F., Parker J. W., Philippe S., Protopapa S., Reitsema H. J., Schmitt B., Stryk T., Summers M. E., Tsang C. C. C., Throop H. H. B., White O. L., and Zangari A. M. (2016) The small satellites of Pluto as observed by New Horizons. *Science, 351(6279)*, aae0030.

Wright G. S., Wright D., Goodson G. B., Rieke G. H., Aitink-Kroes G., Amiaux J., Aricha-Yanguas A., Azzollini R., Banks K., Barrado-Navascues D., Belenguer-Davila T., Bloemmart J. A. D. L., Bouchet P., Brandl B. R., Colina L., Detre Ö., Diaz-Catala E., Eccleston P., Friedman S. D., García-Marín M., Güdel M., Glasse A., Glauser A. M., Greene T. P., Groezinger U., Grundy T., Hastings P., Henning T., Hofferbert R., Hunter F., Jessen N. C., Justtanont K., Karnik A. R., Khorrami M. A., Krause O., Labiano A., Lagage P.-O., Langer U., Lemke D., Lim T., Lorenzo-Alvarez J., Mazy E., McGowan N., Meixner M. E., Morris N., Morrison J. E., Müller F., Nørgaard-Nielsen H.-U., Olofsson G., O'Sullivan B., Pel J.-W., Penanen K., Petach M. B., Pye J. P., Ray T. P., Renotte E., Renouf I., Ressler M. E., Samara-Ratna P., Scheithauer S., Schneider A., Shaughnessy B., Stevenson T., Sukhatme K., Swinyard B., Sykes J., Thatcher J., Tikkanen T., van Dishoeck E. F., Waelkens C., Walker H., Wells M., and Zhender A. (2015) The mid-infrared instrument for the James Webb Space Telescope, II: Design and build. *Publ. Astron. Soc. Pac., 127*, 595–611.

Young L. A. (2012) Volatile transport on inhomogeneous surfaces: I — Analytic expressions, with application to Pluto's day. *Icarus, 221*, 80–88.

Young L. A. (2013) Pluto's seasons: New predictions for New Horizons. *Astrophys. J. Lett., 766*, L22.

Young L. A., Elliot J. L., Tokunaga A., de Bergh C., and Owen T. (1997) Detection of gaseous methane on Pluto. *Icarus, 127*, 258–262.

Zhang X., Strobel D. F., and Imanaka H. (2017) Haze heats Pluto's atmosphere yet explains its cold temperature. *Nature, 551*, 352–355.

Stern S. A., Spencer J. R., Weaver H. A., and Olkin C. B. (2021) The exploration of the primordial Kuiper belt object Arrokoth (2014 MU_{69}) by New Horizons. In *The Pluto System After New Horizons* (S. A. Stern, J. M. Moore, W. M. Grundy, L. A. Young, and R. P. Binzel, eds.), pp. 587–601. Univ. of Arizona, Tucson, DOI: 10.2458/azu_uapress_9780816540945-ch025.

The Exploration of the Primordial Kuiper Belt Object Arrokoth (2014 MU_{69}) by New Horizons

S. A. Stern and J. R. Spencer
Southwest Research Institute

H. A. Weaver
The Johns Hopkins Applied Physics Laboratory

C. B. Olkin
Southwest Research Institute

On January 1, 2019, NASA's New Horizons spacecraft flew just 3538 km from the cold classical Kuiper belt object 2014 MU_{69}, named "Arrokoth," conducting a reconnaissance flyby of what is the most primitive object ever visited by a spacecraft. This Kuiper belt object is also the most distant body ever explored by spacecraft. New Horizons observations revealed Arrokoth to be a 36-km-long contact binary with unexpectedly flattened lobes. Arrokoth's surface displays discrete geological units and complex albedo heterogeneity, but its red color exhibits only small variations across the observed surface. Arrokoth displays clear evidence of CH_3OH on its surface, but no clear evidence for H_2O ice was found. The paucity of craters on Arrokoth's surface indicates a relative dearth of Kuiper belt objects <1 km in diameter, and relatively collisionally benign ancient and present-day Kuiper belt environments. No satellites, orbiting rings/dust structures, or evidence of extant atmosphere were found. Herein we discuss these and other discoveries in more detail and present interpretation of the findings made to date. Evidence described here strongly indicates that the two lobes formed near one another in a local collapse cloud, subsequently becoming an orbiting pair that lost angular momentum and gently merged into the single object contact binary observed today. As a result of these and other findings described here, Arrokoth provides strong support for local cloud collapse streaming instability models of planetesimal formation, and formidable obstacles to planetesimal formation by hierarchical accretion.

1. INTRODUCTION

Following its flyby exploration of the Pluto system (e.g., *Stern et al.,* 2018), New Horizons (NH) went into an "extended mission" that then conducted the first spacecraft exploration of a primordial planetesimal in the Kuiper belt (KB). That flyby target, 2014 MU_{69} (originally informally known as Ultima Thule but later formally renamed Arrokoth), has an equivalent spherical diameter ~130× smaller and is on the order of ~2,000,000× less voluminous than Pluto. The terms MU_{69} and Arrokoth can be used interchangeably.

Arrokoth was discovered in 2014 by the NH team using the Hubble Space Telescope (HST) (*Spencer et al.,* 2015; *Buie et al.,* 2019, 2020). Based on its orbital elements, Arrokoth is a member of the "cold classical" KB, which in turn is thought to be the least dynamically evolved population in the solar system.

The NH flyby closest approach to Arrokoth occurred at 05:33:22 Universal Time (UT) on January 1, 2019, at an estimated closest approach distance to its center of 3538.5 ± 0.2 (1σ) km; the flyby closest approach was essentially to the celestial north of Arrokoth, at a relative speed of 14.43 km s^{-1}. The asymptotic approach direction of NH was just 11.6° off the direction to the Sun.

New Horizons carries a suite of seven scientific instruments: the Ralph multicolor/panchromatic mapper and mapping infrared composition spectrometer, which includes both the Multispectral Visible Imaging Camera (MVIC) and the Linear Etalon Imaging Spectral Array (LEISA) infrared mapping spectrometer; the LOng Range Reconnaissance Imager (LORRI) high-resolution camera; the Alice extreme/far-ultraviolet mapping spectrograph; the Radio Experiment (REX) radio science experiment; the Solar Wind Around Pluto (SWAP) solar wind monitor; the Pluto Energetic Particle Spectrometer Science Investigation

(PEPSSI) high-energy charged-particle spectrometer; and the Venetia Burney Student Dust Counter (SDC), a dust impact detector. See *Weaver et al.* (2008) for a review and references to reviews of each instrument.

In what follows, we summarize the state of knowledge of Arrokoth less than a year after its flyby. Early published results for "2014 MU_{69}" can be found in a first overview report by *Stern et al.* (2019), and then in subsequent geology/geophysics, composition and thermal properties, and origins reports by *Spencer et al.* (2020), *Grundy et al.* (2020), and *McKinnon et al.* (2020) respectively.

2. ARROKOTH: DISCOVERY AND ADVANCE KNOWLEDGE PRIOR TO THE FLYBY

When NASA issued a call for proposals to explore the Pluto system, those proposals were required to demonstrate the capability to conduct at least one flyby of a Kuiper belt object (KBO) after Pluto. Later, during the development of NH, an updated statistical analysis reported in *Stern and Spencer* (2003) found a high probability (>95%) that at least one suitable KBO could be targeted. The NH team initiated a systematic groundbased observing campaign in 2011 to search for potential KBO targets (*Spencer et al.,* 2015; *Buie et al.,* 2019, 2020). This search discovered several dozen new, faint KBOs, but none were within the available remaining ~130 m s^{-1} Δv targeting capability of the spacecraft after its Pluto flyby. By the fall of 2013, the inability to find a targetable KBO became so problematic that the search strategy was modified to employ a deeper search using the HST.

HST had not originally been employed for the NH KBO effort because a relatively large region of the sky had to be searched (~30 × 30 arcminutes in 2014, roughly the apparent size of the full Moon as viewed from Earth), and the relevant HST camera (Wide Field Camera 3, or WFC3) has a relatively small field of view (~3 × 3 arcminutes), meaning that a mosaic of ~100 different field of views (FOVs) and ~200 HST orbits would be required for the search.

However, given the failure of groundbased instruments to make a targetable flyby KBO detection, and because of HST's demonstrated capability to search to fainter magnitudes than groundbased detections permitted, the decision to request HST time in 2014 (the final observing season before the Pluto flyby) was made. The NH team's initial 2014 HST KBO search request as a Director's Discretionary Time (DDT) effort was rejected. Subsequently, however, a large (194-orbit) HST observing proposal comprised of a combination of DDT and General Observer (i.e., peer reviewed) time was submitted in April 2014 and approved in mid-June 2014. Owing to time criticality near opposition, the program was then executed beginning the following week.

That search yielded multiple cold classical KBO (CCKBO) detections, including two that NH could reach with its available fuel. 2014 MU_{69} (Arrokoth) was the first KBO that HST detected in this search. Follow-up HST observations conducted in 2014 and 2015 refined orbit knowledge of the two potential NH targets sufficiently to allow spacecraft targeting maneuvers to be calculated. The NH team then requested and received permission from NASA to target Arrokoth, which although not the brightest (and therefore likely not the largest) of the flyby candidates found by HST, was the least fuel-expensive to reach (requiring just 57 m s^{-1} of Δv), thereby maximizing the probability that a second KBO flyby or other important KBO science might be later accomplished.

The discovery and initial follow-up observations of Arrokoth in 2014–2015 were just the first step of a systematic set of HST observations that continued until late 2018 to characterize both the orbital and physical properties of this faint KBO to the extent possible. These HST astrometric observations were also used to refine Arrokoth's orbit to predict stellar occultation opportunities in the 2016–2018 timeframe, with the intent of using these stellar occulations to determine the object's size, albedo, and silhouette shape and to probe its debris environment. Hubble Space Telescope observations were also conducted to search for lightcurve variations and for satellites, but neither was detected (*Benecchi et al.,* 2019a). Hubble Space Telescope observations were also carried out to determine Arrokoth's visible wavelength color, indicating that Arrokoth was red, most likely even redder than Pluto, which was consistent with the very red color of other CCKBOs (*Benecchi et al.,* 2019b).

The HST astrometry of Arrokoth enabled accurate predictions for three separate stellar occultation dates in 2017 and 2018 by further refining Arrokoth's orbit solution (*Buie et al.,* 2019, 2020). The derived size constraint from the occultation (largest dimension ~30 km) was combined with the HST photometry to constrain the visible geometric albedo (to ~0.1 ± 0.05). Preliminarily, multiple occultation chords obtained on July 17, 2017, and later on August 6, 2018, also clearly revealed that Arrokoth was highly non-spherical (~2:1 ratio in the lengths of two sky plane axes) and was most likely a contact or closely orbiting binary.

New Horizons first observed Arrokoth with an astrometric detection on August 16, 2018, using the spacecraft's Long Range Reconnaissance Imager (LORRI) telescopic imager. Systematic LORRI observations then began ~1 month later and continued through the flyby on January 1, 2019. LORRI observations of Arrokoth's position on approach were used to refine the spacecraft's trajectory. These data were also used to search for satellites and rings (section 6.3) and to measure Arrokoth's rotational light curve. Arrokoth's light curve was found to be flat, consistent with no reliable measurable variation at a level of 20%.

The meager information about Arrokoth described above, i.e., its orbit, approximate size, correspondingly ~10% albedo, rough two-dimensional shape, red color, lack of a detectable light curve, and lack of satellites and rings, constituted all the available knowledge about this target before the close-range study of it by NH during December 31, 2018, through January 1, 2019.

In closing this section, we also point out that the flyby of Arrokoth was a far more difficult technical challenge than the flyby of Pluto. This was due to a variety of factors, including Arrokoth's much smaller size and far-fainter brightness, the fact that it has only been known and astrometrically tracked for ~1% of an orbit prior to the flyby, and the ~2× lower illumination levels and decreased spacecraft-to-Earth data transmission rates at its heliocentric distance.

3. FLYBY OBJECTIVES

The scientific objectives of the Arrokoth flyby are described in detail in *Stern et al.* (2018). These objectives were grouped into three categories according to their scientific priority, with Group 1 being the highest, Group 2 the intermediate, and Group 3 the lowest. In a similar manner to the Pluto flyby, the objectives for Arrokoth were organized according to three science themes: geology and geophysics investigations (GGI), composition investigations (COMP), and investigations related to plasma/particle and atmospheric objectives (PATM).

The Group 1 GGI objectives for the Arrokoth flyby were to:

- Perform panchromatic imaging of Arrokoth's encounter hemisphere with a resolution $\leq$0.3 km pix^{-1} at the subspacecraft point;
- Obtain panchromatic Arrokoth's rotational coverage over all accessible longitudes with a resolution of $\leq$1 km pix^{-1} at the subspacecraft point;
- Obtain low-Sun imaging and/or stereo pairs of images of the encounter hemisphere of Arrokoth with pixel scales of $\leq$0.6 km pix^{-1} at the subspacecraft point to enable construction of digital elevation models (DEMs) of the surface at vertical resolutions of $\leq$0.5 km, using shape-from-shading and/or stereogrammetry techniques; and
- Perform panchromatic imaging over as much of Arrokoth's Hill sphere as possible to a visual magnitude $V \leq 18$ (at SNR = 3 in the peak pixel) for satellites and at $I/F \leq 1 \times 10^{-6}$ (at SNR = 3) for ring-like dust structures in both forward and backscattered light, i.e., on approach and departure, respectively.

The flyby's Group 1 COMP objectives were to:

- Obtain infrared spectral images of the encounter hemisphere with a resolution $\leq$5 km pix^{-1} at the subspacecraft point;
- Obtain IR spectral images of the illuminated portion covering all accessible longitudes with a resolution $\leq$13 km pix^{-1} at the subspacecraft point;
- Obtain three-color images of the encounter hemisphere with a resolution $\leq$2 km pix^{-1} at the subspacecraft point;
- Obtain color images of the illuminated portion of all accessible longitudes with a resolution $\leq$4 km pix^{-1} at the subspacecraft point; and
- Attempt on a best-effort basis to obtain disk-integrated UV spectra of the surface of Arrokoth.

There were no Group 1 PATM objectives, but there were five specific PATM Group 2 requirements for the Arrokoth flyby. Rather than listing each of the specific PATM, GGI, and COMP Group 2 and Group 3 objectives here, we refer the reader to *Stern et al.* (2018) and instead discuss the general scientific objectives of these observations.

The PATM Group 2 objectives were all related to placing constraints on the composition and magnitude of any volatile or dust escaping from Arrokoth. The GGI Group 2 objectives included constraining the size-frequency distribution of craters detected on Arrokoth's surface, characterizing the scattering properties of Arrokoth's surface (including its 4-cm radar reflectance), constraining Arrokoth's bolometric albedo, and characterizing the properties of any satellites or dust detected in the vicinity of Arrokoth. The COMP Group 2 objectives involved color and compositional measurements of any satellites or dust detected near Arrokoth.

All of the Group 3 Arrokoth objectives except one fell under the PATM theme. These generally involved characterizing the plasma and dust environment at Arrokoth's heliocentric distance and possible interactions of Arrokoth with the local heliospheric plasma environment. The single GGI Group 3 objective was to place constraints on the mass of Arrokoth, utilizing dynamical results on a best-effort basis if one or more satellites were detected, and on Arrokoth's density by combining any such mass with volumetric estimates from the panchromatic imaging of Arrokoth.

4. INITIAL GEOLOGY AND GEOPHYSICS FINDINGS

4.1. Rotation and Global Shape

Resolved images taken by the LORRI camera on approach constrain the shape and rotational properties of Arrokoth (see Fig. 1) (see also *Stern et al.,* 2019; *Spencer et al.,* 2020). The resulting rotational period was determined to be 15.92 ± 0.02 hr, with a positive (angular momentum sense) rotational pole position RA = 317.5 ± 1.0° and declination = –24.9 ± 1.0° in the J2000 equatorial frame. The obliquity of Arrokoth's pole to its heliocentric orbit plane was determined to be 99.3 ± 1°. This pole was 39 ± 1° from the NH approach vector and 28° from the direction of the Sun during the flyby; these small angles were largely responsible for the undetectably small amplitude of Arrokoth's rotational light curve in both photometry from HST and photometry from the NH LORRI instrument on approach.

High-phase departure images (see Fig. 2) provided constraints on the shape of the unilluminated side of Arrokoth due to its occultation of background stars.

Arrokoth was found to have a contact-binary shape consisting of two roughly ellipsoidal lobes joined at a narrow neck. In what follows, we refer to the two lobes as the larger lobe (LL) and smaller lobe (SL). Overall dimensions of the entire object were found to be 36 × 20 × 10 km (see Fig. 3). Maximum dimensions of the larger lobe and the smaller lobe were found to be 20.6 × 19.9 × 9.4 km and 15.4 × 13.8 × 9.8 km, respectively. The uncertainty in these dimensions is roughly 0.5 × 0.5 × 2.0 km. The total volume of Arrokoth was

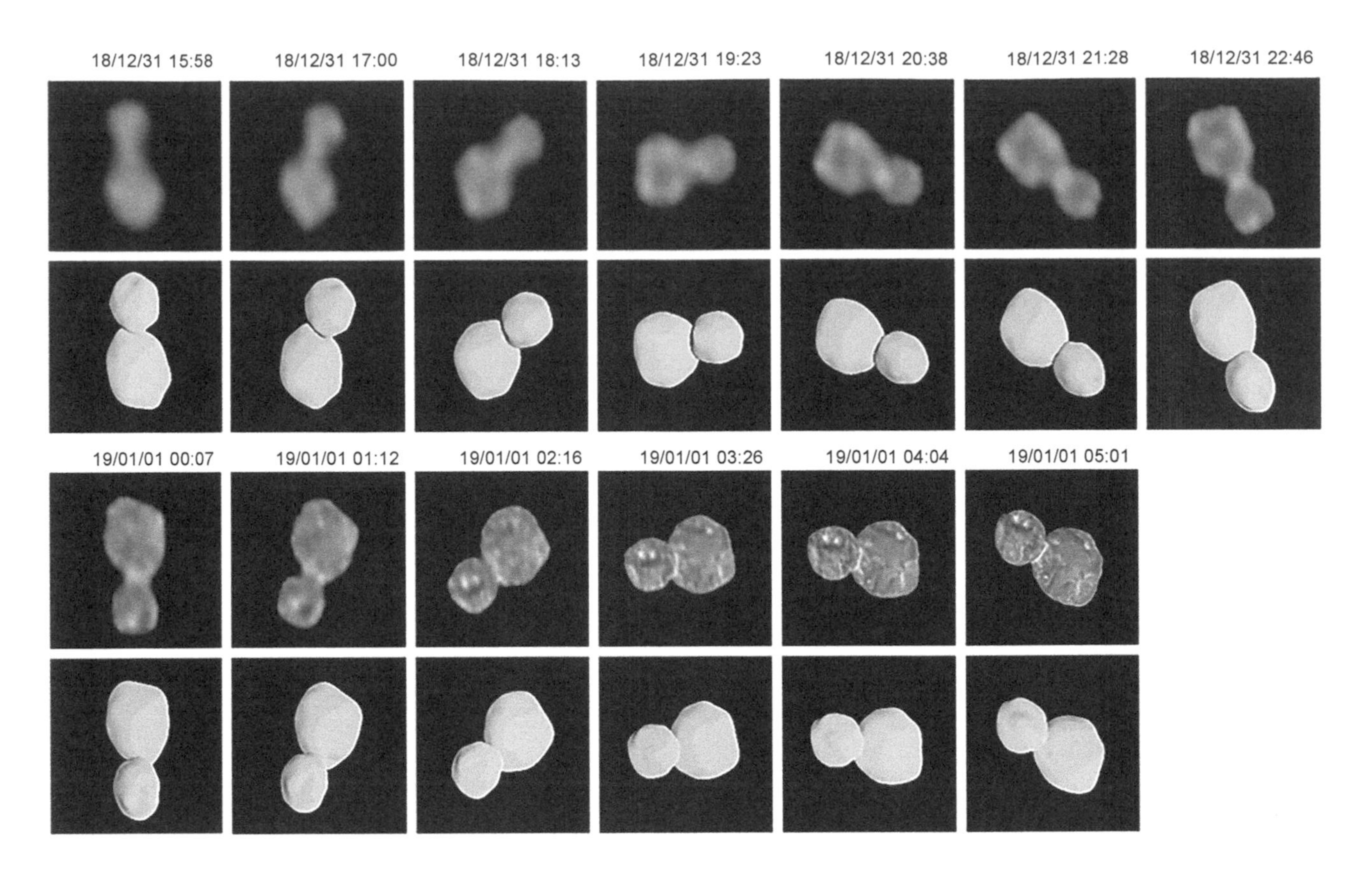

Fig. 1. Derived Arrokoth shape model compared to LORRI images (*Spencer et al.,* 2020). Individual frames were selected from deconvolved LORRI approach images of Arrokoth, compared to synthetic images with the same geometry derived from the global shape model. These images have been scaled to a constant frame size, so that the images become sharper as time progresses and range decreases; celestial north is up.

found to be 3210 ± 650 km^3, corresponding to a sphere of equivalent diameter 18.3 ± 1.2 km. The larger lobe's volume equates to a sphere with a diameter of 15.9 ± 1.0 km; the spherical equivalent diameter for the smaller lobe is 12.9 ± 0.8 km. These values give a volume ratio (and mass ratio if the two lobes have equal densities) of 1.9 ± 0.5.

Arrokoth's rotational axis was very close to NH's approach vector, so while Arrokoth was seen to rotate, most of the rotation was in the image plane and did not easily reveal Arrokoth's three-dimensional shape. A true understanding of the shape came only after post-flyby analysis of the images, which showed relatively subtle variations in the target's projected shape. These inbound images were complemented by images taken on departure, which did not show most of the unlit side of Arrokoth directly but allowed the shape to be inferred by several dozen known stars, which were occulted (see Fig. 2). *Spencer et al.* (2020) found a best-fit shape using a 20-parameter, two-lobed forward model to create synthetic images to compare with the observations.

Independent shape and topographic information come from stereo analysis of the highest-resolution images taken near closest approach (see Fig. 3). Typical relief in the stereo model on both lobes away from the neck region is ~0.5 km or less. Figure 3 also compares the global shape model to the stereo model of the encounter (–Z) face of Arrokoth. There is broad agreement between these two techniques, although the south polar region of the LL is flatter in the stereo model, and the neck is also smoother there. We note that a slope discontinuity at the neck is an intrinsic feature of the global shape model due to the model's dual-lobed nature.

4.2. Density and Gravity

The density of Arrokoth was not directly constrained by the flyby due to the lack of either detected satellites or detectable change in the trajectory. However, its density must be >250 kg m^{-3} (*McKinnon et al.,* 2020) if the neck of Arrokoth is assumed to have no tensile strength, in order for gravity to keep the two lobes together. Assuming a bulk density of 500 kg m^{-3} based on typical densities of cometary nuclei (*Preusker et al.*, 2017), the mean surface gravity would be ~1 mm s^{-2}. Figure 3 shows the geopotential altitude (i.e., the elevation with respect to an equipotential) across Arrokoth. This geopotential is highest at the far ends of each lobe and on the equator, decreasing with increasing latitude on each lobe, and reaches a global minimum in the neck. Surface slopes are generally gentle, <20°, and slope downward to higher latitudes and into the neck region. If material can flow downslope, it will collect at higher latitudes and in the neck region. The derived stereo model reveals that the neck is relatively smooth, with shallow slopes on resolved scales, <30°.

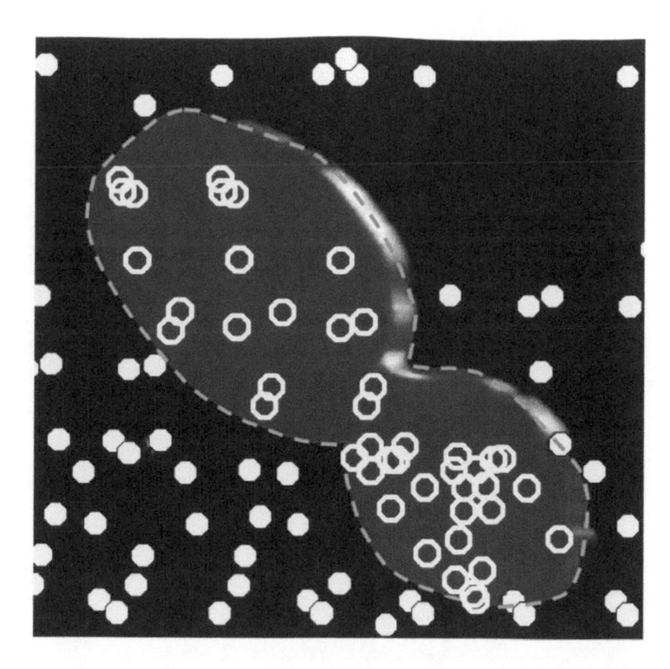

Fig. 2. Departure image of Arrokoth, with the silhouette of its derived shape model (dark blue area and light blue dashed line) superposed (*Spencer et al.,* 2020). Open and closed yellow dots indicate the locations of occulted and unocculted stars, respectively, in the original six-frame imaging sequence that were used to constrain the shape of the KBO's unilluminated hemisphere.

The maximum axes of inertia for the LL and the SL lobes are aligned to one another to within <5°, and the equatorial plane of the two bodies is also almost coincident. As a result, it is estimated that the center of mass of the SL is displaced only 0.18 km from the equatorial plane of the LL. This remarkably improbable alignment has profound implications for Arrokoth's formation, as we discuss later.

4.3. Photometric Properties

The mean 0.6-μm reflectance of Arrokoth is 0.24 (*Spencer et al.,* 2020; *Hofgartner et al.,* 2021) and is very similar for both lobes. However, at the resolution of the available imagery, the normal reflectance varies across the mapped surface by over a factor of 2, from 0.18 to 0.35. The mean and standard deviation of the reflectance are 0.230 and 0.035, respectively, for the LL, and 0.228 and 0.043, respectively, for the SL. The V-band geometric albedo, $0.21^{+0.05}_{-0.04}$ (*Hofgartner et al.,* 2021) is within the range of CCKBOs, which have V-band geometric albedos ranging from 0.09 to 0.23 and a mean of 0.15 (*Lacerda et al.,* 2014). Arrokoth's spherical (Bond) albedo and phase integral is consistent with the 0.24–0.30 range of phase coefficients of six other KBOs also reported to date by NH (*Verbiscer et al.,* 2019).

4.4. Surface Units

Figure 4 shows surface images and a geological unit map of Arrokoth. The SL and the LL have distinctly different surface appearances and are therefore described separately.

4.5. The Smaller Lobe (SL)

The SL is dominated by a large, roughly circular depression with a diameter of 6.7 km, nicknamed Maryland, that is likely to be an impact crater. Its depth/diameter ratio is 0.08–0.19; the range of this uncertainty is owed to stereo depth recovery uncertainties but is nevertheless typical of craters on other small bodies.

The albedo patterns across the SL are complex. There are two patches of bright material (unit bm) within Maryland that show discrete boundaries near the crater bottom and that feather toward the crater's rim. Straddling the Maryland rim on the side opposite the bright patches is discrete, dark crater rim material (unit dc), which contrasts with the brighter terrain (unit bc) that forms the remainder of the crater interior. The rough terrain unit (rm), found at the distal end of the SL, forms a relatively flat facet, brighter than its immediate surroundings. The low Sun angle on this facet reveals surface texture at a scale of a few hundred meters, apparently mostly composed of subkilometer pits. One prominent pit ~150 m in diameter resembles a small, fresh, bowl-shaped impact crater. Another nearby bright, mottled unit (mm) may be similar, but is seen at a higher Sun angle so topographic roughness is not apparent (see Fig. 4a).

Dark material surrounding the mm material seems to be part of a unit (dm) that wraps around much of the remainder of the observable surface of the SL; this material is the darkest observed on Arrokoth, with 0.6-μm reflectance of 0.18. In places (e.g., "B" in Fig. 4b), it has a boundary with pointed and angular protrusions and rounded indentations, which may be evidence for material erosion and removal due to scarp retreat (*Stern et al.,* 2019). Near "B" in Fig. 4b there are also bright circular patches within the dark material. Running down the center of the principal mapped outcrop of dark material is a sinuous unit of bright material, which appears in stereo observations to occur in a V-shaped trough. The rest of the surface of the SL, which is nondescript in available images, is mapped as undifferentiated material (unit um). Crossing the undifferentiated material near the terminator between Maryland and the LL, however, are a series of subparallel troughs, which are reminiscent of structural troughs seen on other comparable-sized bodies, e.g., asteroid Eros (*Buczkowski et al.,* 2008), saturnian satellites Epimetheus and Pandora (*Morrison et al.,* 2009), and the martian satellite Phobos (*Hurford et al.,* 2016).

Images show that the bright "neck" region connecting the LL and the SL has a diffuse margin on its LL side, but extreme foreshortening makes it difficult to characterize this margin on the SL side.

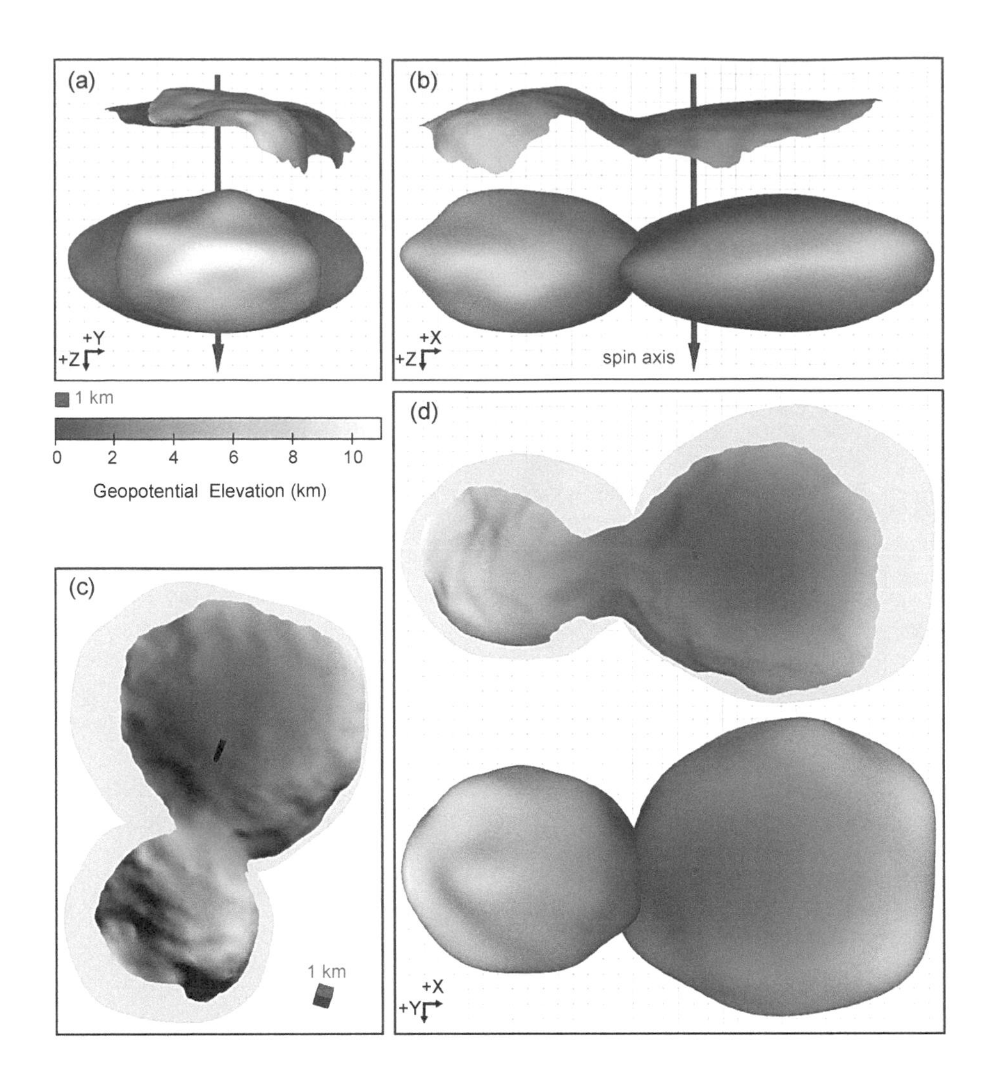

Fig. 3. Stereo and global shape models of Arrokoth (*Spencer et al.,* 2020). **(a)–(c)** Comparison of the stereo shape model of the encounter face (top of each panel) to the global shape model as seen from **(a)** the –X (the SL) direction, **(b)** the +Y direction, and **(c)** the south polar (–Z) direction. The +Z ("north") axis is the positive direction of the spin axis, according to the righthand rule. Each model is colored to show the variation in geopotential across the surface. The stereo model has been trimmed to remove edge effects. **(d)** Stereo model seen from the CA06 LORRI (i.e., the highest resolution) viewing geometry, but with lighting chosen to highlight the small-scale topography.

4.6. The Larger Lobe (LL)

The LL appears to be primarily composed of a series of similar-sized, discretely bounded, rolling topographic units (*Stern et al.,* 2019), although some geological units are continuous across these boundaries (see Fig. 4). The units near the terminator, ta–td, are distinctive, being relatively bright [although ta is noticeably less red than the others (*Grundy et al.,* 2020)], and clearly separated from the rest of the LL by a common, continuous scarp or trough, as noted above. Units tg and th appear more mottled than adjacent units, and stereo imaging of these suggests that the mottling is due to the presence of dark ridges and hills surrounding brighter low terrain.

The remainder of the LL is comprised of smooth material (unit sm) of moderate albedo, transected by a series of distinctive bright linear features (unit bm), some of which form an incomplete annulus. In some areas ("C" in Fig. 4b) the inner margin of the annulus appears sharply bounded, possibly with an outward-facing scarp, while the outer, concentric portions are more gradational with the surrounding terrain. Stereo observations show that terrain within the annulus is relatively flat compared to the undulating nature of the rest of the visible portion of the LL and suggest that the annulus occupies a shallow trough. At the base of unit tg, the annulus appears to coincide with diffuse bright material also existing at the base of unit tg, which itself appears to interrupt the annulus and so may be superimposed upon it. In two places, what appear to be dark hills extend into the sm unit. In one case ("D" in Fig. 4b) these hills seem to be an extension, cut by the bm annulus, of similar hills on unit th.

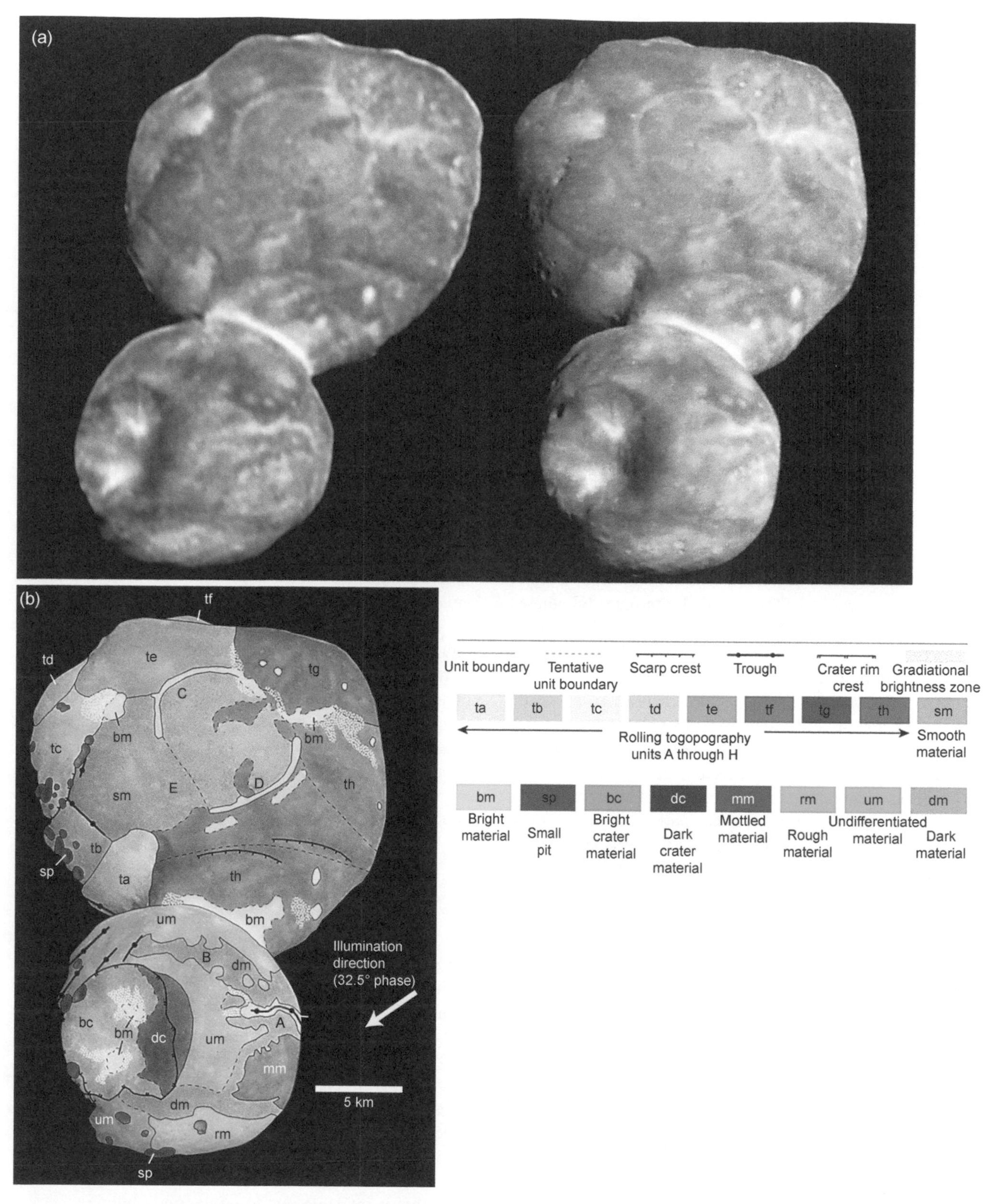

Fig. 4. Imaging and geomorphological map of Arrokoth (*Spencer et al.*, 2020). **(a)** Cross-eyed stereo pair image of Arrokoth, taken by LORRI. The LL lobe is at the top and the SL lobe is at the bottom (all Arrokoth feature names used in this chapter are informal). The image on the left is close-approach image CA04, range = 27,860 km, phase = 12.9°, resolution = 138 m pix^{-1}. The image on the right is close-approach image CA06, range = 6650 km, phase= 32.5°, resolution = 33 m pix^{-1}. Both images have been deconvolved to remove the LORRI point-spread function, and motion blur in the case of CA06, to maximize detail. **(b)** Geomorphological map of Arrokoth. The base map is the deconvolved CA06 image. The positive (righthand-rule) spin axis of Arrokoth here points approximately into the page. Capitalized letters identify feature/unit locations mentioned in the text.

4.7. Geological Interpretation

Relatively bright material seen on both the LL and the SL occur preferentially in depressions. The brightest material on the LL and the SL, and in the bright collar between the two lobes, all have the same 0.6-μm reflectance of 0.24. This suggests that the bright material has somewhat similar chemical and physical properties everywhere it has been seen. The most extensive bright region, found in the topographic low of the "neck" region, may be simply the most extreme example of a process that occurs more generally to brighten low-lying material across Arrokoth.

It appears likely that loose, poorly consolidated, likely fine-grained, bright material moves downslope and accumulates in depressions (*Stern et al.,* 2019), implying that bright material is more mobile than dark material on Arrokoth. *Spencer et al.* (2020) suggested that the complex albedo patterns on the SL, and their crenulated margins, may result from the exposure and differential erosion of multiple lighter and darker layers oriented roughly parallel to its surface, although independent topographic information is currently of insufficient quality to confirm this explanation.

As recounted above, the LL may be composed of smaller subunits that accreted separately (*Stern et al.,* 2019). However, *Spencer et al.* (2020) explored alternate hypotheses accounting for the fact that the central bm annulus, enclosing one of the subunits, appears to be a relatively young feature and not an unmodified primordial boundary.

Another constraint on the nature of the LL is that its overall shape is smooth and undulating, without major topographic discontinuities like that between the LL and the SL, as would be expected from accretion of large pre-existing units. Erosion and alteration over the past 4.5 G.y. are likely to have modified the optical surface and the uppermost upper few to tens of meters (*Stern,* 1990) but probably does not explain the smoothness seen at 30-m and larger scales.

Some possible explanations for the appearance of the annulus and other subunit boundaries are illustrated in Fig. 5.

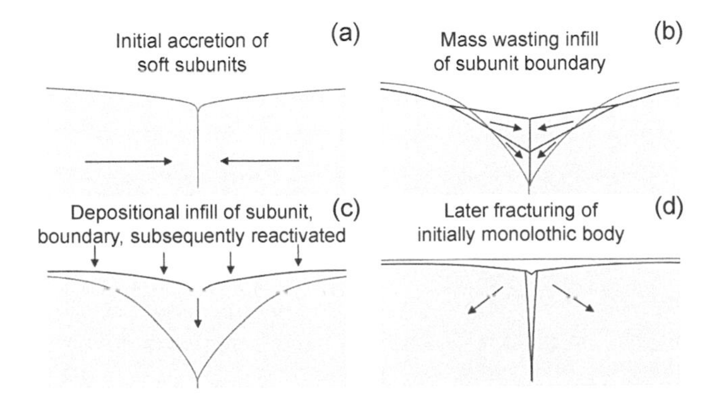

Fig. 5. Illustration of possible explanations for the appearance of the boundaries between terrain subunits on the LL (from *Spencer et al.,* 2020).

These subunits may have been soft enough at the time of merger that they conformed to each other's shapes on contact (e.g., *Belton et al.,* 2007; *Jutzi and Asphaug,* 2015) (see Fig. 5a), although no evidence for impact deformation is seen on them. In order for such deformation to take place at the time, the shear strength of the merging components must have been no more than 2 kPa, the ram pressure of an impacting body assuming a merger velocity of 1–2 m s^{-1} and a material density of 500 kg m^{-3}. The possibility that they flowed viscously after contact while still soft would require an implausibly low shear strength of ~100 Pa.

Mass-wasting may have filled in original gaps between the components while preserving a sharp boundary between them (see Fig. 5b), although except in the case of the boundary of tg noted previously, the absence of obvious boundaries between material transported by mass wasting and *in situ* material argues against this as a general explanation. Original discontinuities may have been buried by subsequent accretion or redistribution of surface material (see Fig. 5c). The boundaries would then need to be reactivated in some way in order to still be visible on the surface (e.g., possibly by collapse into subsurface voids, which may explain the trough-like appearance of parts of the bm annulus, and the troughs and pits seen at low Sun between the ta–td units and the rest of the LL). However, it is not clear how such burial would preserve different surface textures seen on the different units.

Alternatively, the LL may be monolithic, and the visible boundaries may be secondary (see Fig. 5d), produced for instance by subsequent fracturing. *Spencer et al.* (2020) considered the evidence to be most consistent with scenarios "C" and "D" in Fig. 5. However, for any of these cases, the processes that produced the distinctive surface textural contrasts between the units, in particular the patches of dark hills and ridges, is as yet unknown.

4.8. Pits and Craters

In addition to the 7-km-diameter probable impact crater Maryland, numerous roughly circular subkilometer bright patches and pits are scattered across Arrokoth. Even if these are mostly impact craters, the crater density is relatively light compared to many other small bodies (*Stern et al.,* 2019). The bright patches are generally seen in high-Sun areas away from the terminator, and some of these patches appear to occupy depressions in stereo pairs. These may be equivalent to the pits seen in low-Sun areas near the terminator (unit sp in Fig. 4b), and it is possible that the near-terminator pits also feature bright material on their floors that is invisible due to unfavorable lighting.

Spencer et al. (2020) classified pits and craters on Arrokoth according to their location on the surface and their likelihood of being impact features (see Fig. 6). These subdivisions yield a range of plausible crater densities, shown in Fig. 6b. Densities for each dataset are somewhat uncertain as they depend on the areas used for each distribution, and of course densities are lower if uncertain

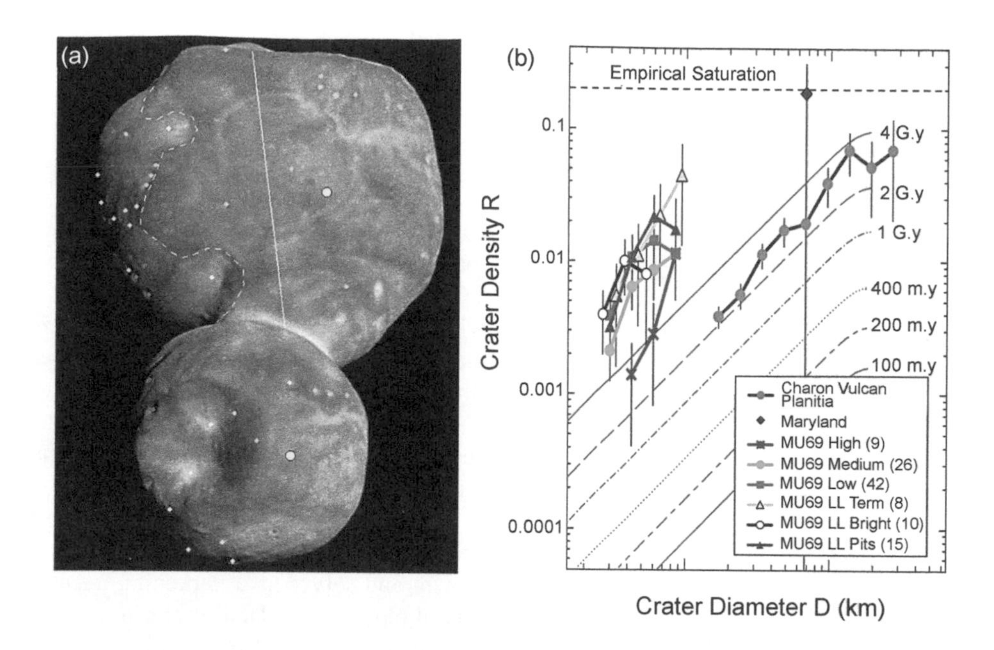

Fig. 6. Craters and pits on Arrokoth (*Spencer et al.,* 2020). **(a)** Locations of features considered for crater analysis. Color denotes confidence class: magenta = high confidence ["MU_{69} High" in **(b)**], green = medium confidence [magenta + green = "MU_{69} Medium" in **(b)**], cyan = low confidence [magenta + green + cyan = "MU_{69} Low" in **(b)**]. Features indicated in white are highly unlikely to be of impact origin and are not included in the crater statistics. The solid white line indicates the split of the LL lobe into a low-Sun half with more visible depressions (the "MU_{69} Pits" region, left) and a high-Sun half with bright spots (the "MU_{69} Bright" region, right). The white dashed curve delineates combined geologic units ta, td, tc, and td, designated "MU_{69} Term," considered together for crater density determination. The yellow dots indicate the planetocentric subsolar point on each lobe according to the shape model. **(b)** The size-frequency distribution of craters on Arrokoth is shown per crater subgroup and region described in the text and in **(a)**. The Arrokoth crater data are compared to crater densities on Charon's Vulcan Planitia (VP) (from *Singer et al.,* 2019), and to predictions based on an impactor flux model for six different ages of surfaces on Arrokoth and gravity regime scaling (blue curves) (*Greenstreet et al.,* 2019). The MU_{69} Term and MU_{69} Bright distributions are offset horizontally by a small amount for clarity. The empirical saturation line refers to a traditional D^{-3} differential power law distribution (*Melosh,* 1996).

craters are excluded. But despite these uncertainties, the range of crater densities is less than a factor of 10 for a given diameter bin. Except for Maryland, all the possible impacts features are ≤1 km in diameter. Furthermore, Arrokoth is only modestly cratered, and there are some areas on Arrokoth where very few, if any, potential craters exist, in particular the part of the LL between the dashed and solid white lines in Fig. 6a.

Crater surface density can constrain surface age, although we caution that such age determinations are dependent on uncertainties in impactor flux models, uncertainties in crater identification (e.g., with lighting conditions), and infinite diameter measurement errors; see *Spencer et al.* (2020) for a more complete discussion of such error sources.

Figure 6b shows expected crater densities for several surface ages using the *Greenstreet et al.* (2019) model, assuming, as is likely (see also *Spencer et al.,* 2020), that Arrokoth's surface materials are weak enough that gravity scaling applies. Age estimates are uncertain given the uncertainty in identifying that craters are impact-generated, and because the model curves will shift based on the crater-scaling parameters used as well. Accounting for the higher crater fluxes in an early but brief dynamical instability phase in the outer solar system (*Nesvorny,* 2018) would increase expected crater densities, although plausibly by only a factor of 2 at most (*Greenstreet et al.,* 2019). Seismic shaking and changes in spin state, used to explain low relative densities of small craters on near-Earth asteroids (*Sugita et al.,* 2019; *Walsh et al.,* 2019), probably do not account for Arrokoth's low crater densities due to the paucity of large craters. Overall, despite the small number of craters on its surface, the observed crater density is consistent with a crater retention age of greater than ~4 G.y. The visible surface at the scale of the LORRI image resolution thus plausibly dates from the end of the accretionary period, although perhaps modified by the aforementioned erosion and also mass wasting.

Although the diameters of observed craters on Arrokoth, apart from Maryland, are smaller than nearly all those measured in the Pluto system, and the Arrokoth datasets suffer from the small-number statistics inherent in such a small body, the slopes of the Arrokoth and Pluto system small crater distributions are consistent with one another for small-sized (<10 km) craters (*Spencer et al.,* 2020).

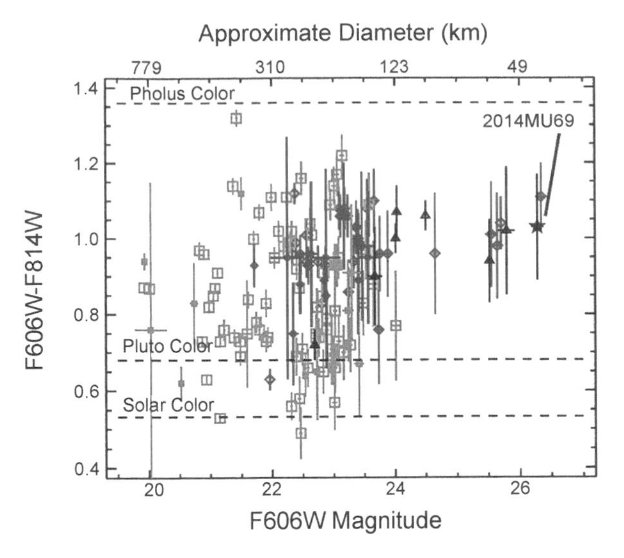

Fig. 7. The color of Arrokoth (black star and error bar at the far right) compared to other KBOs that have been measured in these filters. The red diamonds are CCKBOs; the gray squares denote other KBO populations. From *Benecchi et al.* (2019b).

5. INITIAL COLOR AND COMPOSITIONAL FINDINGS

5.1. Overview

Before the NH flyby past Arrokoth, the only information on the composition of Arrokoth was from filter photometry from HST. Using two filters (F606W and F814W), *Benecchi et al.* (2019b) derived a color that is red and consistent with other CCKBOs (see Fig. 7). These observations place Arrokoth in context relative to other outer solar system objects. Notice that Arrokoth is redder than Pluto, but it is less red than 5145 Pholus.

5.2. Color

The NH flyby provided improved knowledge of the average color of Arrokoth, but it also provided resolved imaging to determine the degree of color variegation across it (see Fig. 8). The average color slope of Arrokoth is 27 ± 2% per 100 nm between 550 and 650 nm, which is consistent with other CCKBOs (*Grundy et al.*, 2020).

Now consider color variation across Arrokoth. The small standard deviation of Arrokoth's bulk color indicate a relatively uniform color across it. Furthermore, its two lobes display very similar color slopes: 28 ± 2% for the SL and 27 ± 2% for the LL. Despite the general color uniformity across this KBO, some interesting color variation is seen in Fig. 9. Both the principal component analysis and the spectral slope maps demonstrate that the neck of Arrokoth (although still quite red) is comparatively blue to most other Arrokoth terrains, and the rim of the large crater Maryland is relatively red. A depression near the neck is also comparatively blue.

5.3. Composition

Given the large heliocentric distance of Arrokoth and its small size, no Earth-based telescope available today can make spectroscopic observations of this object or like-sized CCKBOs. The spectroscopic observations of Arrokoth taken with the Ralph instrument (*Reuter et al.*, 2008) on NH therefore provide the only present-day spectroscopy on any small, distant KBO.

The grand average near-IR spectrum of Arrokoth is shown in Fig. 10 with pertinent molecular absorptions indicated (*Grundy et al.* 2020).

The amorphous carbon in this model doesn't have any specific spectral features but is expected in the outer solar system and serves as a darkening agent to match the observed albedo. The Titan tholin in the model also doesn't exhibit sharp spectral features but does provide a spectral slope that matches the data. The methanol spectral features between 2.2 and 2.4 μm are well fit in the grand average spectrum. Water ice and ammonia are both expected in the outer solar system and have also been identified on various bodies in the Pluto system (*Dalle Ore et al.*, 2019), and were included in the model. However, their inclusion did not significantly improve the model fit. Perhaps there

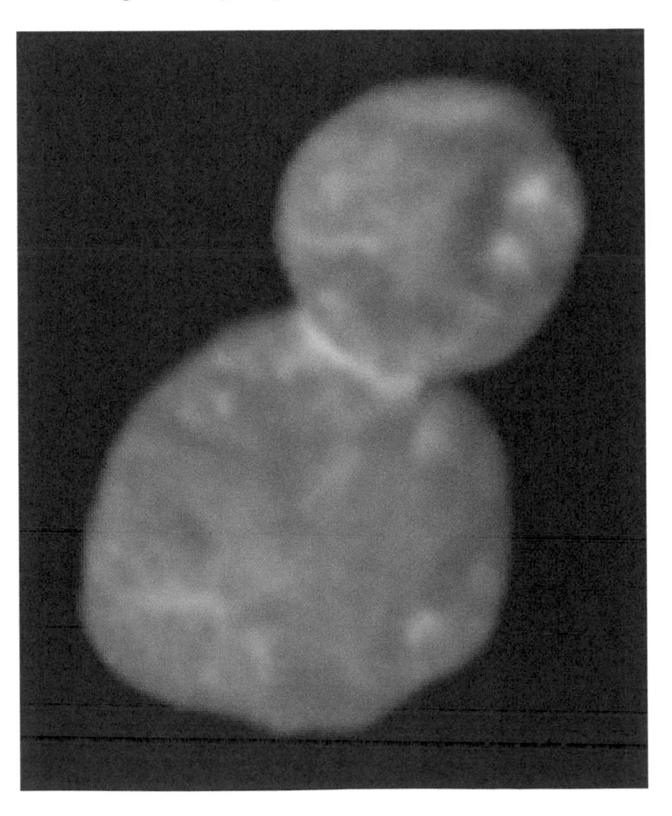

Fig. 8. An enhanced color image of Arrokoth from New Horizons. The MVIC blue filter (400–550 nm) is displayed in blue pixels, MVIC red filter (540–700 nm) is displayed in green pixels, and MVIC NIR (780–975 nm) is displayed in red pixels.

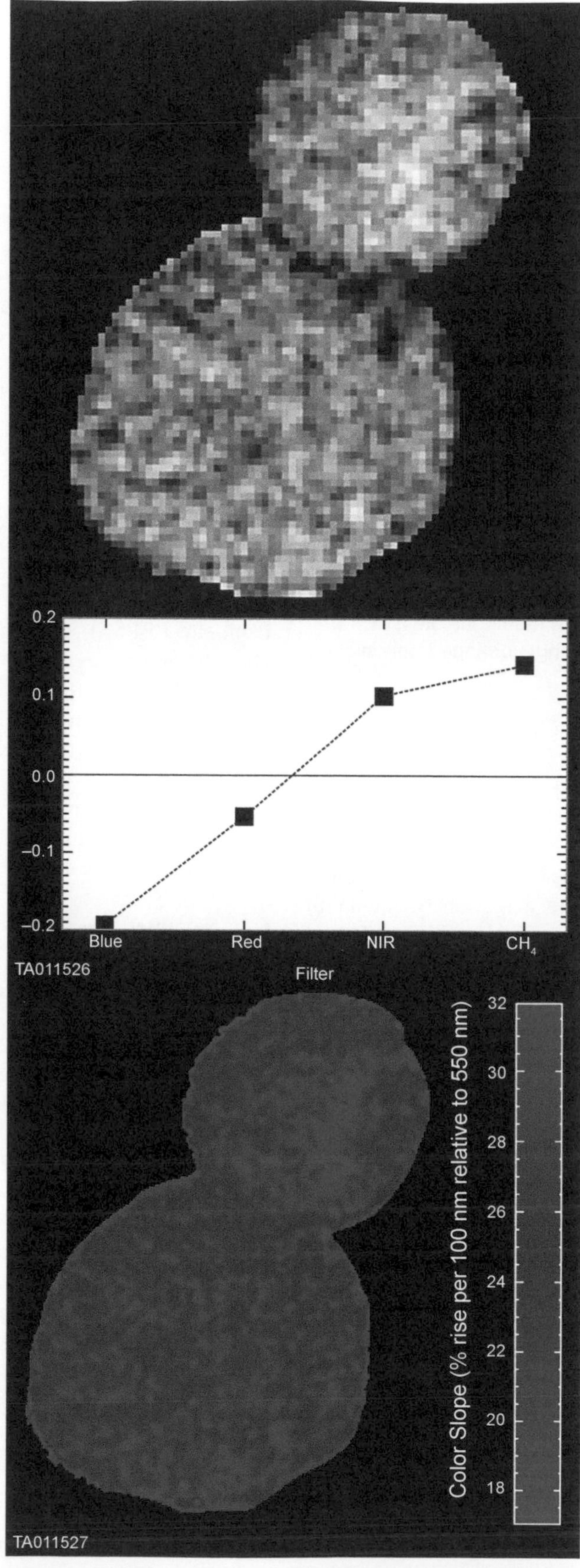

Fig. 9. Color variation across Arrokoth. In the upper panel, one principal component from a principal-component analysis of the four color filters in the MVIC color camera is displayed. This principal component corresponds to a red slope as seen by the eigenvector shown. The lower panel displays the color slope map. The red and blue color along the edge of the color slope map is an artifact in the middle column; the eigenvector contribution is the vertical axis of the plot when of the image processing. Adapted from *Grundy et al.* (2020).

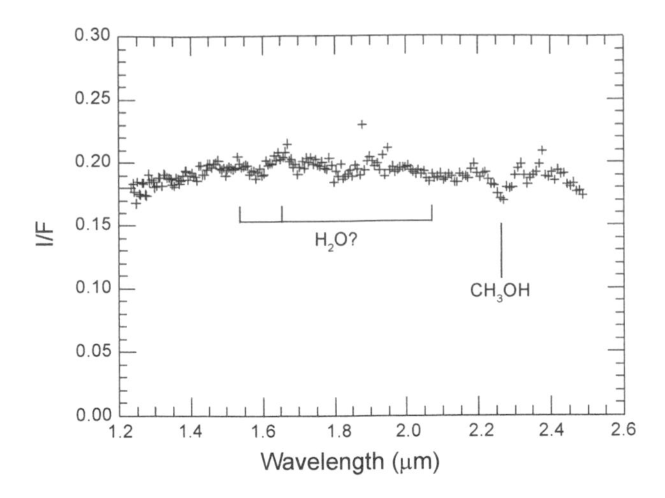

Fig. 10. The grand average spectrum for Arrokoth with the absorption feature assignments. These Ralph-LEISA data were taken from a range of approximately 30,000 km for a spatial resolution of ~1.8 km, when at a phase angle of ~12.5°. The scan took a little more than 4 minutes; over that time span the range and phase angle changed only slightly.

is some mechanism for removing or masking the spectral signature of H_2O-ice on the surface of Arrokoth without also destroying the spectral signature of methanol ice, but no convincing explanation has emerged as yet (*Grundy et al.,* 2020).

Although also observed in the Cthulhu region on Pluto (e.g., *Grundy et al.,* 2016), the detection of methanol (CH_3OH) ice in Arrokoth is somewhat surprising. This is because, unlike its presence in the spectra of the KBO-escapee (i.e., Centaur) 5145 Pholus (*Cruikshank et al.,* 1998) and the KBOs 2002 VE95 (*Barucci et al.,* 2006) and possibly 90377 Sedna (*Barucci et al.,* 2005, 2010), Arrokoth's spectra do not also show clear detections of water-ice absorption. This in turn implies that the methanol ice must be prevalent across the entire surface of Arrokoth. Although methanol is a commonly observed volatile in cometary comae (*Bockelee-Morvan et al.,* 2004), its typical abundance is only ~1–5% relative to H_2O, and methanol ice absorption has never been detected in the spectra of the surfaces of cometary nuclei. *Grundy et al.* (2020) interprets these results in terms of nebular grain chemistry where Arrokoth formed.

6. EXOSPHERIC, SATELLITE, AND RING/DUST SEARCH FINDINGS

6.1. Introduction

New Horizons searched for an extant coma and satellites or rings around Arrokoth. Despite deep searches, NH did not detect any of these. In this section, we briefly summarize the limits achieved; more exosphere/heliospheric interaction detail can be found in *Stern et al.* (2019) and more detail for satellite and ring searches can be found in *Spencer et al.* (2020).

6.2. Exospheric and Heliospheric Interaction Searches

Owing to Arrokoth's small size and stable orbit, highly volatile ices that might once have been present at its surface would have escaped to space long ago (e.g., *Stern et al.,* 2018; *Lisse et al.,* 2021). Nonetheless, NH searched for extant coma via UV spectroscopy, charged-particle signatures, and high-phase-angle look-back observations.

At UV wavelengths, NH undertook both airglow spectroscopy and solar and stellar appulse observations. Strong resonance lines of atomic species including H, C, N, O, and S are in the bandpass of the NH Alice UV spectrometer; no UV coma emissions from Arrokoth were detected. At the brightest likely coma emission, the H I 1216 Å Lyman-α line, a 3σ upper limit source rate of $<3 \times 10^{24}$ H s^{-1} was achieved.

Megaelectron-volt charged-particle spectra obtained from the PEPSSI instrument on NH during the flyby also showed no evidence of any Arrokoth-related signature. However, the predicted PEPSSI count rates of 1.6×10^{-3} Hz from Arrokoth are ~600× smaller than the PEPSSI background count rate of ~1 Hz; a similar non-detection was found for the kiloelectron-volt SWAP solar wind spectrometer. The upper limit found with UV data is therefore more constraining than these particle spectrometer upper limits.

6.3. Satellites and Orbiting Rings/Dust Search

Both satellites and rings have been detected around KBOs and Centaurs larger than Arrokoth (e.g., *Noll et al.,* 2008; *Braga-Ribas et al.,* 2014; *Ortiz et al.,* 2015, 2017). New Horizons therefore searched for satellites and rings around Arrokoth. These searches were made using LORRI and MVIC images made on approach to and departure from Arrokoth; *Spencer et al.* (2020) and *Throop et al.* (2019) describe this in detail.

Sensitivity limits for Arrokoth were established by implanting synthetic objects into the images obtained for this purpose. Upper limits on satellites larger than 100–180 m in diameter (for the assumption of satellite albedos similar to Arrokoth) out to 8000 km from Arrokoth, and 300 m diameter throughout most of Arrokoth's Hill sphere, were obtained (Fig. 11).

The NH flyby provided an unprecedented opportunity to also search for rings around a small KBO at sensitivities and distances from the primary not achievable by other means. A search for rings and dust in forward-scattered light was conducted after closest approach at a phase-angle of 168°, covering radii up to 6000 km from Arrokoth. However, no rings or dust structures were detected with an upper limit I/F of 1.5×10^{-6} for structures wider than about 10 km. It should be noted, however, that the ring searches assumed only a few specific ring geometries. Any such ring or dust structure around Arrokoth must therefore be more tenuous in forward scattering than Jupiter's dusty main ring with $I/F = 4 \times 10^{-6}$ at the same phase angle (*Throop et al.*, 2004). In backscattered light, Arrokoth ring upper limits of $I/F \lesssim 2 \times 10^{-7}$ at 11° solar phase were obtained assuming a 10-km-wide ring (see *Stern et al.*, 2019), again fainter than Jupiter's main ring ($I/F = 7 \times 10^{-7}$ at 11° phase).

Additionally, NH's SDC instrument detected no signals above its noise threshold within ±5 days before and after the Arrokoth flyby, implying no impacts on it by dust particles >1.6 µm radius, and giving a 90% confidence upper limit of 3×10^7 particles km^{-2}. For 10% albedo, this result equates to an I/F limit of 3×10^{-11}, which is even more constraining than the optical limits reviewed just above for particles of this size or larger.

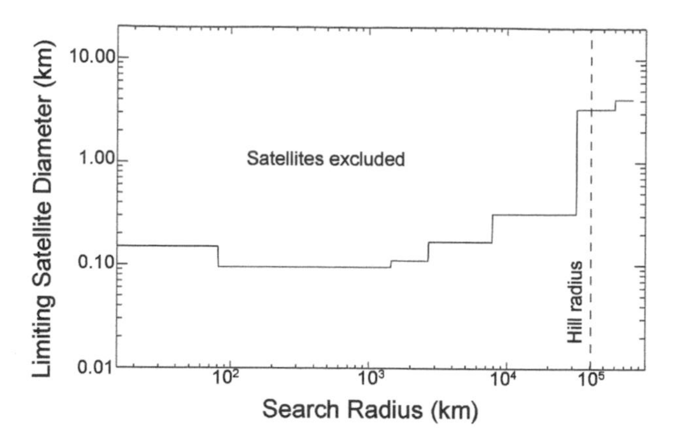

Fig. 11. Upper limit search results for Arrokoth satellites. These limits assume photometric properties similar to Arrokoth. The Hill radius shown assumes a density of 500 kg m^{-3}.

7. ON THE ORIGIN OF ARROKOTH

As noted above, Arrokoth is a member of the CCKBOs, which have orbital inclinations $i < 5°$ and orbital eccentricities $e < 0.1$. Arrokoth's 44.6-AU orbital semimajor axis places it near the center of the CCKBO population. The dynamical stability of such orbits (e.g., *Nesvorny,* 2018) suggests that Arrokoth and the other CCKBOs formed near their current locations and have remained there for the entire age of the solar system.

This is unlike other KBOs, which apparently formed much closer to the Sun (~20 AU) and were later pushed outward into the Kuiper belt during the extensive radial migration, and scattering events caused by the giant planets that took place sometime during the first several hundred million years after the Sun and planets formed. As such, unlike comets and asteroids studied at close range by spacecraft in the past, Arrokoth has always been cold, in a dilute radiation environment, and has remained in a region where it suffered little impact modification. These factors combine to make it the best-preserved example of a primordial planetesimal ever studied at close range.

Understanding planetesimal formation is a primary goal of planetary science. The just-described, well-preserved nature of Arrokoth strongly constrains planetesimal formation in the cold classical region of the Kuiper belt, and likely elsewhere across the solar system.

As we described above, Arrokoth's small and large lobes are each flattened; remarkably, their flattening planes are aligned to <5°. Furthermore, the SL is oblong, and its long axis points along the long axis of the binary as a whole (again, to <5°). As described in *Stern et al.* (2019) and *Spencer et al.* (2020), the surface and overall structure of Arrokoth do not display any obvious signs of catastrophic or violent collision, and the narrow geometry of the neck between the lobes suggests they are effectively perched atop one another.

McKinnon et al. (2020) and *Marohnic et al.* (2021) found that lobe-lobe impacts at more than a few meters per second or with grazing incidence angles <75° are highly damaging, distorting, and shearing, and can even fail to result in the observed contact binary. Furthermore, hierarchical collision models produce impact speeds at several hundred meters per second. *McKinnon et al.* (2020) also found that the close alignment of Arrokoth's two lobes is quite unlikely to result by a chance collision, but more naturally results from the tidal evolution of a close, co-orbiting binary.

Pebble cloud gravitational collapse models, which are based on the streaming or other collective instabilities (e.g., *Nesvorny et al.*, 2010; *Nesvorny,* 2018), naturally produce the low merger speeds (*Fraser et al.*, 2017) required by collisional simulations of Arrokoth's formation (*Stern et al.*, 2019; *McKinnon et al.*, 2020; *Marohnic et al.*, 2021). The lack of strong surface albedo, color, and composition heterogeneity between Arrokoth's two lobes also supports this origin scenario as well, because in a local pebble cloud collapse, the two lobes form from a homogeneous source of locally sourced material.

For Arrokoth's lobes to reach both their merged spin state and grazing incidence collision, they must have also lost orbital angular momentum and energy (*Stern et al.*, 2019; *McKinnon et al.*, 2020). The lack of satellites of Arrokoth implies one formation-era angular momentum sink (*Stern et al.*, 2019) that could be tested for by determining that satellites of contact binary CCKBOs are rare. *McKinnon et al.* (2020) explored a wider variety of mechanisms to remove orbital and spin angular momentum from a co-orbiting binary, including solar tides, mutual tides, collisions with smaller KBOs, the ejection of other co-orbiting bodies, nebular gas drag, and radiation force; they determined that many such processes can be effective in creating the end-state contact binary that NH observed.

Arrokoth's contact binary nature, its lack of geological evidence for a violent merger, the alignment of its two lobes, and the color and composition homogeneity of its two lobes provide a foursome of strong evidence in favor of a gentle binary merger. This evidence can be reproduced by local pebble cloud collapse/streaming instability planetesimal formation (*Stern et al.,* 2019; *McKinnon et al.,* 2020), but classical, Keplerian hierarchical accumulation models cannot explain such evidence.

8. CONCLUSIONS AND OUTLOOK

Arrokoth is the first truly primordial object explored by spacecraft. Its exploration presented many surprises and new details, including its contact binary nature, the flattening of its two lobes, its low crater density, its lack of satellites, its various forms of evidence for a gentle merger, its lack of a strong H_2O-ice signature, and the identification of CH_3OH on its surface.

Arrokoth's contact binary nature and evidence for a gentle merger of its lobes means that Arrokoth cannot be a product of heliocentric, high-speed collisional evolution, late-stage hierarchical accretion of independent, heliocentric planetesimals, or fragmentation of a larger body. Instead, these attributes present strong evidence of an extremely low-velocity merger at meters per second or less, a strong case for early formation, likely in a local collapse cloud, and subsequent orbital hardening and tidal evolution to the observed, contact binary with aligned principal axes. As such, Arrokoth presents new evidence that gravitational instabilities played a key role in planetesimal formation in the KB, and by extension, throughout our solar system.

Based on the discovery of Arrokoth's contact binary nature as the first KBO explored at close range, it is natural to expect that contact binaries are likely to be common or even prevalent in the KB. This is a result that can be tested in coming years by obtaining a statistically meaningful suite of KBO (and in particular CCKBO) stellar occultation shape silhouettes (*Stern et al*., 2019) and lightcurve observations (*Thirouin and Sheppard,* 2019).

Like the future spacecraft study of more dwarf planets in the KB, and the next stage of Pluto system exploration by spacecraft, future spacecraft exploration of more small KBOs like Arrokoth is highly anticipated.

Also highly anticipated are the many new groundbased and spacebased facilities that will shed new light on the KB and KBOs; these include the Vera C. Rubin Large Synoptic Survey Telescope (LSST), the James Webb Space Telescope (JWST), and ultra-large-class (25–30-m) groundbased telescopes, all of which will come on line in the early to mid-2020s. Such telescopes can better probe the satellite populations and compositional heterogeneity of Arrokoth-sized KBOs than can presently be achieved.

We also hope there will be a broad application of the stellar occultation silhouette mapping technique first perfected for Arrokoth, which will allow the shapes and therefore the contact binary fraction of small KBOs to be determined, quantitatively measuring how typical Arrokoth is and shedding further light on planetesimal formation mechanisms.

Acknowledgments. This work was supported by NASA's New Horizons mission. We thank the members of the New Horizons mission and science teams for their many years of hard work to make the results presented here possible. We also thank three anonymous referees and editor R.P.B. for careful readings and many useful comments that improved this chapter.

REFERENCES

Barucci M. A., Cruikshank D. P., Dotto E., Merlin F., Poulet F., Dalle Ore C., Fornasier S., and de Bergh C. (2005) Is Sedna another Triton? *Astron. Astrophys., 439,* L1–L4.
Barucci M. A., Merlin F., Dotto E., Doressoundiram A., and de Bergh C. (2006) TNO surface ices. Observations of the TNO 55638 (2002) VE95. *Astron. Astrophys., 455,* 725–730.
Barucci M. A. et al. (2010) Sedna: Investigation of surface compositional variation. *Astron. J., 140(6),* 2095–2100.
Belton J. S. et al. (2007) The internal structure of Jupiter family cometary nuclei from Deep Impact observations: The "talps" or "layered pile" model. *Icarus, 187,* 332–344.
Benecchi S. D., Porter S. B., Buie M. W., et al. (2019a) The HST lightcurve of (486958) 2014 MU_{69}. *Icarus, 334,* 11–21.
Benecchi S., Borncamp D., Parker A., et al. (2019b) The color and binarity of (486958) 2014 MU_{69} and other long-range New Horizons Kuiper belt targets. *Icarus, 334,* 22–29.
Bockelée-Morvan D., Biver N., Colom P., et al. (2004) The outgassing and composition of Comet 19P/Borrelly from radio observations. *Icarus, 167,* 113–128.
Braga-Ribas F., Sicardy B., Ortiz J. L., et al. (2014) A ring system detected around the centaur (10199) Chariklo. *Nature, 508,* 72–75.
Buczkowski D. L., Barnouin-Jha O. S., and Prockter L. M. (2008) 433 Eros lineaments: Global mapping and analysis. *Icarus, 193,* 39–52.
Buie M. W., Porter S. B., Tamblyn P., et al. (2019) Stellar occultation results for (486958) 2014 MU_{69}: A pathfinding effort for the New Horizons flyby. In *Lunar and Planetary Science L: Papers Presented at the 50th Lunar and Planetary Science Conference,* Abstract #3120. Lunar and Planetary Institute, Houston.
Buie M. W. et al. (2020) Size and shape constrains of (486598) Arrokoth from stellar occultations. *Astron. J., 159(4),* 130.
Cruikshank D. P., Roush T. L., Bartholomew M. J., Geballe T. R., Pendleton Y. J., White S. M., Bell J. F., et al. (1998) The composition of centaur 5145 Pholus. *Icarus 135,* 389–407.
Dalle Ore C. M., Cruikshank D. P., Protopapa S., et al. (2019) Detection of ammonia on Pluto's surface in a region of geologically recent tectonism. *Sci. Adv., 5(5),* eaav5731.
Fraser W. C., Bannister M. T., Pike R. E., et al. (2017) All planetesimals born near the Kuiper belt formed as binaries. *Nature Astron., 1,* 0088.
Greenstreet S., Gladman B., McKinnon W. B., et al. (2019) Crater density predictions for New Horizons flyby target 2014 MU_{69}. *Astrophys. J. Lett., 872,* L5.
Grundy W. M., Binzel R. P., Buratti B. J., et al. (2016) Surface compositions across Pluto and Charon. *Science, 351,* 1283.
Grundy W. M., Bird, M. K., Britt D. T., et al. (2020) Color, composition, and thermal environment of Kuiper belt object (486958) Arrokoth. *Science, 367(6481),* DOI: 10.1126/science.aay3705.
Hofgartner J. D. et al. (2021) Photometry of Kuiper belt object (486958) Arrokoth from New Horizons LORRI. *Icarus, 356,* 113723, DOI: 10.1016/j.icarus.2020.113723.
Hurford T. A., Asphaug E., Spitale J. N., et al. (2016) Tidal disruption of Phobos as the cause of surface fractures. *J. Geophys. Res.–Planets, 121,* 1054–1065.
Jutzi M. and Asphaug E. (2015) The shape and structure of cometary nuclei as a result of low-velocity accretion. *Science, 348,* 1355–1358.
Lacerda P., Fornasier S., Lellouch E., et al. (2014) The albedo-color diversity of transneptunian objects. *Astrophys. J. Lett., 793,* L2.
Lisse C. M., Young L. A., Cruikshank D. P., et al. (2021) On the origin and thermal stability of Arrokoth's and Pluto's ices. *Icarus, 356,* 114072, DOI: 10.1016/j.icarus.2020.114072.
Marohnic J. et al. (2021) Constraining the final merger of contact binary (486598) Arrokoth with soft-sphere discrete element simulations. *Icarus, 356,* 113824, DOI: 10.1016/j.icarus.2020.113824.
McKinnon W. B. et al. (2020) The solar nebula origin of

(486958) Arrokoth, a primordial contact binary in the Kuiper belt. *Science, 367(6481),* DOI: 10.1126/science/aay6620.

Melosh H. J. (1996) *Impact Cratering: A Geologic Process.* Oxford Univ., Oxford. 245 pp.

Morrison S. J., Thomas P. C., Tiscareno M. S., et al. (2009) Grooves on small saturnian satellites and other objects: Characteristics and significance. *Icarus, 204*, 262–270.

Nesvorný D. (2018) Dynamical evolution of the early solar system. *Annu. Rev. Astron. Astrophys., 56*, 137–174.

Nesvorný D., Youdin A. N., and Richardson D. C. (2010) Formation of Kuiper belt binaries by gravitational collapse. *Astron. J., 140,* 785–779.

Noll K. S., Grundy W. M., Chiang E. I., et al. (2008) Binaries in the Kuiper belt. In *The Solar System Beyond Neptune* (M. A. Barucci et al., eds.), pp. 345–364. Univ. of Arizona, Tucson.

Ortiz J.-L., Duffard R., Pinilla-Alonso N., et al. (2015) Possible ring material around centaur (2060) Chiron. *Astron. Astrophys., 576,* A18.

Ortiz J.-L., Santos-Sanz P., Sicardy B., et al. (2017) The size, shape, density and ring of the dwarf planet Haumea from a stellar occultation. *Nature, 558*, 219–223.

Preusker F., Scholten F., Matz K.-D., et al. (2017) The global meter-level shape model of comet 67P/Churyumov-Gerasimenko. *Astron. Astrophys., 607,* L1.

Reuter D., Stern S. A., Scherrer J., et al. (2008) Ralph: A visible/infrared imager for the New Horizons Pluto/Kuiper belt mission. *Space Sci. Rev., 140,* 129–154.

Singer K. N., McKinnon W. B., Gladman B., et al., (2019) Impact craters on Pluto and Charon indicate a deficit of small Kuiper belt objects. *Science, 363,* 955–959.

Spencer J. R., Buie M. W., Parker A. H., et al. (2015) The successful search for a post-Pluto KBO flyby target for New Horizons using the Hubble Space Telescope. *EPSC 2015,* 417.

Spencer J. R., Stern S. A., Moore J. M., et al. (2020) The geology and geophysics of Kuiper belt object (486958) Arrokoth. *Science, 367,* DOI: 10.1126/science.aay3999.

Stern S. A. (1990) ISM-induced erosion and gas-dynamical drag in the Oort cloud. *Icarus, 84,* 447–466.

Stern S. A. and Spencer J. R. (2003) New Horizons: The first reconnaissance mission to bodies in the Kuiper belt. *Earth Moon Planets, 92,* 477–482.

Stern S. A., Weaver H. A., Spencer J. R., and Elliot H. A. (2018) The New Horizons Kuiper belt extended mission. *Space Sci. Rev., 214,* 77.

Stern S. A., Weaver H. A., Spencer J. R., et al. (2019) Initial results from the New Horizons exploration of 2014 MU_{69}, a small Kuiper belt object. *Science, 364,* eaaw9771.

Sugita S., Honda R., Morota, T., et al. (2019) The geomorphology, color, and thermal properties of Ryugu: Implications for parent-body processes. *Science, 364(6437)*, 252.

Thirouin A. and Sheppard S. S. (2019) Light curves and rotational properties of the pristine cold classical Kuiper belt objects. *Astron. J., 157,* 228–247.

Throop H. B. et al. (2004) The jovian rings: New results derived from Cassini, Galileo, Voyager, and Earth-based observations. *Icarus, 172,* 59–77.

Throop H. B. et al. (2019) Limits on rings and debris around 2014 MU_{69} observed from New Horizons. *EPSC Abstracts, 13,* EPSC-DPS2019-1196-1.

Verbiscer A. J., Porter S., Benecchi S. D., et al. (2019) Phase curves from the Kuiper belt: Photometric properties of "distant" KBOs observed by New Horizons. *Astron. J., 158,* 123.

Walsh K. J., Jawin E. R., Ballouz R.-L., et al. (2019) Craters, boulders and regolith of (101955) Bennu. *Nature Astron., 12*, 242–246.

Weaver H. A., Gibson W. C., Tapley M. B., et al. (2008) Overview of the New Horizons science payload. *Space Sci. Rev., 140,* 75.

Soluri M. (2021) Epilogue: New Horizons — An abbreviated photographic journal. In *The Pluto System After New Horizons* (S. A. Stern, J. M. Moore, W. M. Grundy, L. A. Young, and R. P. Binzel, eds.), pp. 603–626. Univ. of Arizona, Tucson, DOI: 10.2458/azu_uapress_9780816540945-ch026.

Epilogue: New Horizons — An Abbreviated Photographic Journal

To be sure, this was a historic engineering and scientific feat unlikely to be repeated within our lifetimes. So similar to the listener in the lecture room in Whitman's poem, "... the Learn'd Astronomer," I was not just attending, but essentially in residence at the Johns Hopkins University's Applied Physics Lab (APL), documenting the brain trust behind NASA's New Horizon's mission to the Pluto system.

Although the photograph below was taken May 28, 2015, it is likely that during the 1960s "race to space" the seeds for the exploration of the solar system began taking root in the imaginations of some of the individuals pictured. They weren't the only ones. I couldn't get enough of missions like the Rangers to the Moon, the Pioneer lunar and planetary probes, and the Mariner 4 flyby of Mars on July 14, 1965. Along with the Rolling Stones' (I Can't Get No) Satisfaction serving as a memorable backdrop, I savored those first-ever abstract-like black-and-white images of Mars' surface and the prospect of humans landing on the Moon.

As the frequency and complexity for space exploration paralleled my journey as an editorial photographer of people and places, working and first based in Brazil then working and living in New York City via India, Italy, and France, I increasingly felt something was missing from these profoundly historic, early space exploration endeavors: Who were the people that made these missions possible? What were they like? Where did they work?

A combination of perseverance and serendipity led to a 2005 portrait session with New Horizons Principal Investigator Alan Stern, and since then, at Alan's request, I have been documenting behind the scenes of New Horizons' mission scientists, engineers, and controllers as they meticulously assembled, tested, launched, navigated, and prepared to discover the unknowns of a planetary system 3 billion miles away.

Year after year, month after month, week after week, and soon day to day, what had started out as a fuzzy group of dots at launch in 2006 slowly became images of a discernable planet with moons. And there in Room E100, and in the myriad backrooms of APL, was a remarkable collective of geologists, astronomers, atmospheric scientists, engineers, mission operations personnel, and the like. I wasn't just listening to them; I was capturing the unscripted, fleeting moments of the New Horizons team's humanity as the flow of data beamed across time and space from this piano-sized probe journeying into the unexplored outer solar system toward, then past, Pluto, its moons, and beyond to the Kuiper belt object Arrokoth.

Time and distance.

The photographs on these pages reflect an edited microcosm from that historic time in 2015 and the mid-2019 scientific meeting that launched this book. Divided into three parts, this photo essay reveals the months leading up to and including July 13, 2015; the July 14, 2015, flyby; the post-flyby days; and, four years later, moments during the 2019 Pluto System After New Horizons conference at APL.

Michael Soluri, May 2021
Photographer/Author: Infinite Worlds — The People and Places of Space Exploration
michaelsoluri.com

NH Mission Operations Manager Alice Bowman office — 26 June, 09:22:55

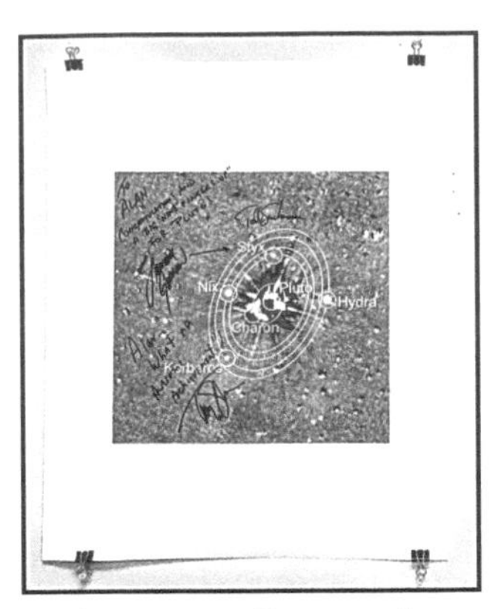

LORRI instrument Pluto approach image
8 June, 11:45:10

NH Co-Investigators Paul Schenk, Bill McKinnon, Jeff Moore, John Spencer, and Kelsi Singer with others during a pre-Pluto flyby science team meeting — 28 May, 17:21:25

7 July, 12:30:13

7 July, 12:30:24

Pre-Flyby

late May through 13 July 2015

Collision hazards exploration and data processing meeting – 25 June, 17:40:34
Simon Porter, Bob Jacobson, and Marina Brozovic; Marc Buie on far screen

Alex Parker and NH team during morning plenary session — 10 July, 09:03:35
Front row: Stuart Robbins, Carly Howett, Alex Parker, Constantine Tsang, Jason Cook; Back row: Amanda Zangari, Alissa Earle, Richard Binzel, Anne Verbiscer

Sequence: Waiting for "green" OK signal that the New Horizons' systems booted up successfully after the July 4 anomaly.

New Horizons team members Kimberly Ennico Smith, Bill McKinnon, Carly Howett, Alissa Earle, Andy Chaikin, Max Mutchler, Jeff Moore, Randy Gladstone
7 July, 12:30:25

Pre-Flyby

late May through 13 July 2015

Above right:
Pluto flyby navigation team meeting:
Kimberly Ennico Smith, Hien Nguyen
Ann Harch, Cathy Olkin, Gabe Rogers

Lower right:
Front row – Tiffany Finley, Leslie Young, Cathy Olkin
Back row – Nicole Martin, Fred Pelletier, Hien Nguyen, Kimberly Ennico Smith

10 July, 14:11:20

10 July, 14:12:45

Morning Pluto flyby science plenary session — Cathy Olkin, Jason Cook, Alan Stern, Will Grundy, Carey Lisse, and Carly Howett, 10 July, 07:39:40

Plasma/Particles science theme team meeting — Fran Bagenal, 10 July, 12:51:55

Columbia, Maryland Sheraton Hotel patio: Heather Elliott and Fran Bagenal 10 July, 20:59:59

(Left) Pluto flyby hazards team meeting — Amanda Zangari, Simon Porter, Marc Buie 7 July, 13:50:24

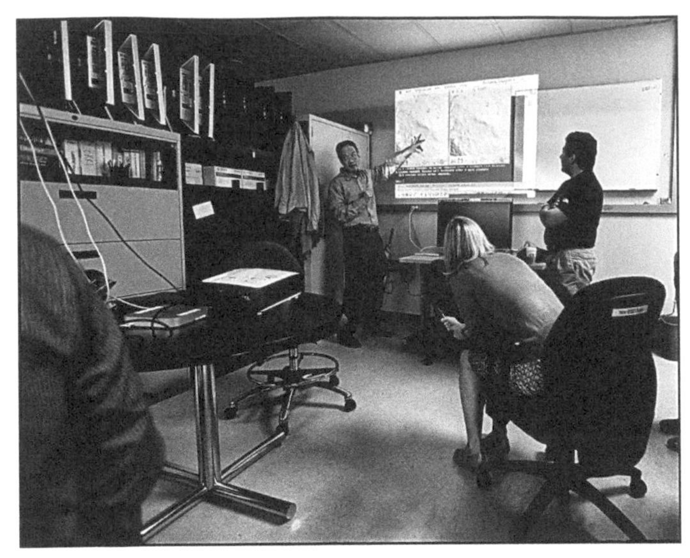

Backroom labs — 9 July, 10:47:46
Andy Cheng, Louise Prockter (on chair), Jorge Nunez (standing)

Geology/Geophysics backroom — Simon Porter, 9 July, 10:47:46

John Spencer, APL E100 science plenary session — 28 May, 09:59:16

Bobby Williams,
NH Project Navigation — 10 July, 14:21:49

Pre-Flyby

late May through 13 July 2015

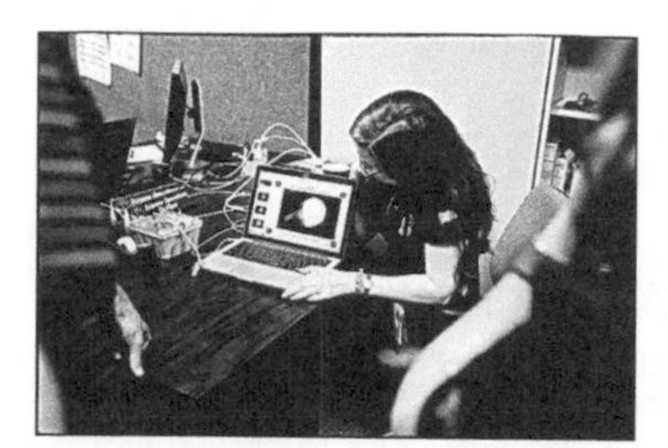

Tiffany Finley, navigation back room — 10 July, 13:41:12

Navigation meeting — Yanping Guo and Ann Harch, 10 July, 15:59:31

Early morning review of latest Pluto images: Front row: Andy Chaikin, Tod Lauer, John Spencer, Paul Schenk; Second row: Carey Lisse, Joel Parker, Curt Niebur, and Matt Hill — 12 July, 07:57:25

Sign Post! 12 July, 11:46:20

Plenary session recognition — Yanping Guo "9" with Ivan Linscott (REX instrument team lead) — 13 July, 08:25:14

American flags to be given out to the New Horizons team two days before the encounter — 12 July, 17:18:16

On the reveal of the first full frame LORRI image of Pluto — 14 July, 06:08:29
Kelsi Singer, Oliver White, John Grunsfeld, Tod Lauer; background: John Spencer and Ted Stryk

Carey Lisse, Bill McKinnon, Amanda Zangari, and Paul Schenk — 14 July, 06:09:35

14 July 2015 — Encounter Day

9.5 years since launch, January 19, 2006
50 years to the day of the first flyby of Mars by Mariner 4: July 14, 1965

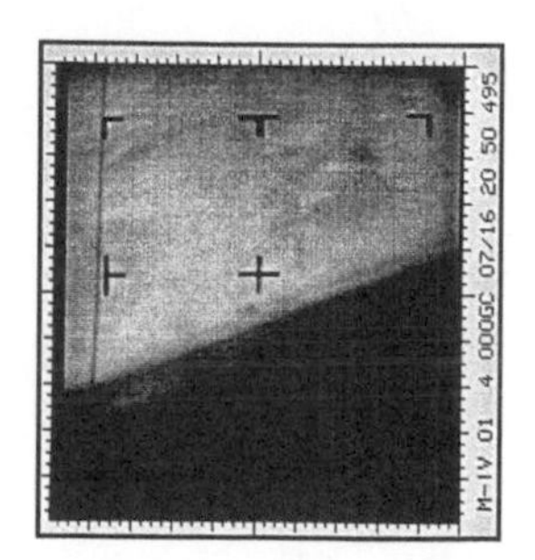

14 July 2015 — Encounter Day Reveal

Paul Schenk, Marc Buie, Veronica Bray, Constantine Tsang, Alan Howard, Joel Parker, Darrell Strobel — 14 July, 06:09:46

Team members study the first full frame LORRI image of Pluto — Veronica Bray (shown in the foreground), Kelsi Singer, and Carey Lisse to her right, 14 July, 06:43:20

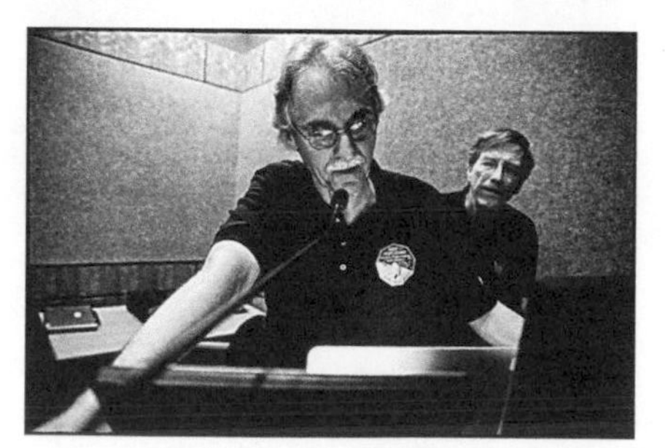

John Spencer and Hal Weaver, 14 July, 06:17:49

Alan Stern, 14 July, 06:21:23

Karl Whittenburg and Alice Bowman, 14 July, 20:25:11

Nick Pinkine and Alice Bowman, 14 July, 20:32:51

Karl Whittenburg and Alice Bowman; background: Michael Vincent, Sarah Hamilton, and Bob Nelson (kneeling), 14 July, 20:53:31

Karl Whittenburg, Nick Pinkine, and Alice Bowman, 14 July, 20:54:18

14 July 2015 — Encounter Day

New Horizons Calls Home

Success! Alice Bowman and Alan Stern — 20:58:10
Val Malder, Chris Hersman, and Gabriel Griffith in back.

(Above) Pluto the dog — 21:06:33
New Horizons operations team mascot

(Right) New Horizons Mission Operations Team — 21:46:39
Gail Oxton, Sarah Flanigan, Becca Sepan (almost hidden behind Sarah F. and Valerie Mallder), Michael Vincent, top of Sheila Zurvalec's head, Nick Pinkine, Sarah Hamilton, Katie Bechtold

Atmospheres team computer screen detail — 15 July, 13:45:29

Richard Binzel, Constantine Tsang, and Orkan Umurhan — 29 July, 10:01:14

Will Woods, Randy Gladstone, and Joshua Kammer — 15 July, 13:38:08

Photographer Michael Soluri
15 July, 13:42:18
(photo taken by Henry Throop)

Post-Flyby at APL

15 July – 29 July 2015

Front row: Mark Showalter, Simon Porter, John Spencer,
Second row: Veronica Bray, Bill McKinnon, Kelsi Singer, Kirby Runyon, Leslie Young, Amanda Zangari, Andy Cheng, and Constantine Tsang
17 July, 10:08:24

"Yay Pluto" Cathy Olkin: —16 July, 16:25:21

Matt Hill, Andy Cheng, Ralph McNutt, and Dave McComas
17 July, 11:29:43

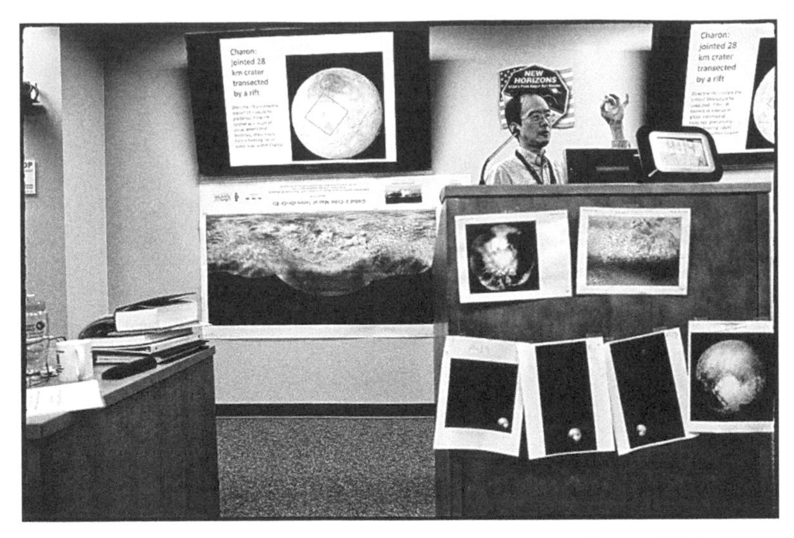

Andy Cheng — 17 July, 09:10:27

Silvia Protopapa, Alex Parker, Cristina Dalle Ore, Jason Cook, and Will Grundy (Dennis Reuter shown in front of the window in the background), 16 July, 15:49:36

Cathy Olkin and Tod Lauer
Background: Francis Nimmo and Andy Chaikin — 15 July, 12:15:52

New Horizons team last plenary session — 29 July, 08:10:42

Post-Flyby at APL

15 July – 29 July 2015

Geology/Geophysics Team Portrait, 17 July, 09:06:55

Front row: Kelsi Singer, Amanda Zangari, Constantine Tsang, Bonnie Buratti, Andy Chaikin, Oliver White, Orkan Umurhan, Marc Buie; Second row: Andy Cheng, John Spencer, Jeff Moore, Bill McKinnon, Mark Showalter, Simon Porter; Back row: Kirby Runyon, Alan Howard, Ross Beyer, Francis Nimmo, Paul Schenk

Post-Flyby at APL

15 July – 29 July 2015

LOWELL OBSERVATORY JOURNAL OF OBSERVATIONS

Lowell Observatory:
Clyde William Tombaugh's log entries — discovery of Pluto, January 23 and 29, 1930

PI Alan Stern with Pluto Wine: 29 July, 17:18:43

Brian May with the Geology/Geophysics team: Andy Chaikin, Brian May, Paul Schenk, Anne Verbiscer, Constantine Tsang, and Ross Beyer (Bill McKinnon with back to camera in foreground)
17 July, 09:34:02

Mike Summers and Marc Buie — 15 July, 13:48:46

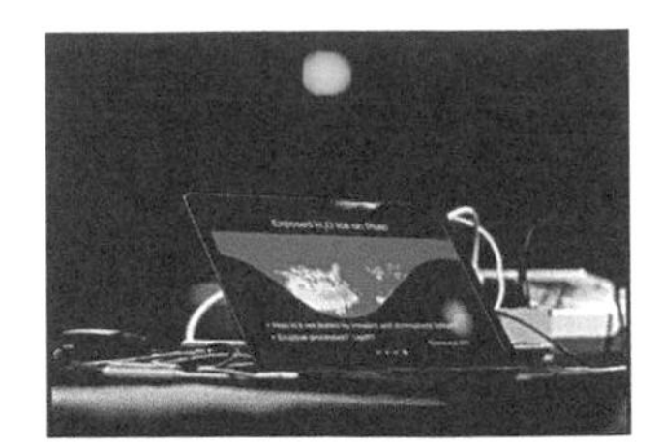

Lynnae C. Quick

Jeff Moore

Tod Lauer

Fran Bagenal, Orkan Umurham, John Spencer, Alan Stern

Simon Porter

Will Grundy

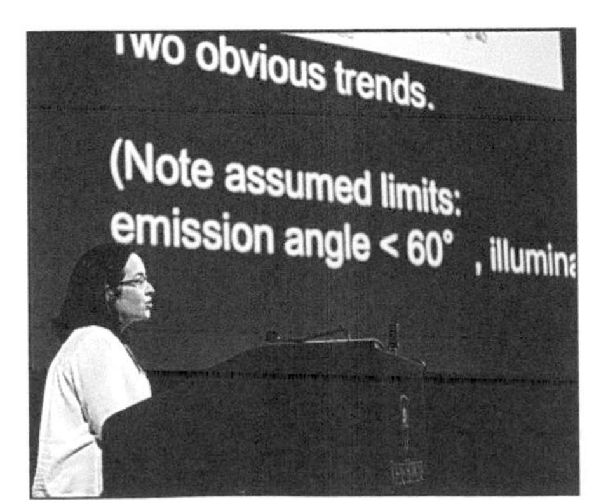

Carly Howett

Kelsi Singer, Jason Cook, Cristina Dalle Ore, and Silvia Protopapa

Doug Hamilton

Jani Radebaugh

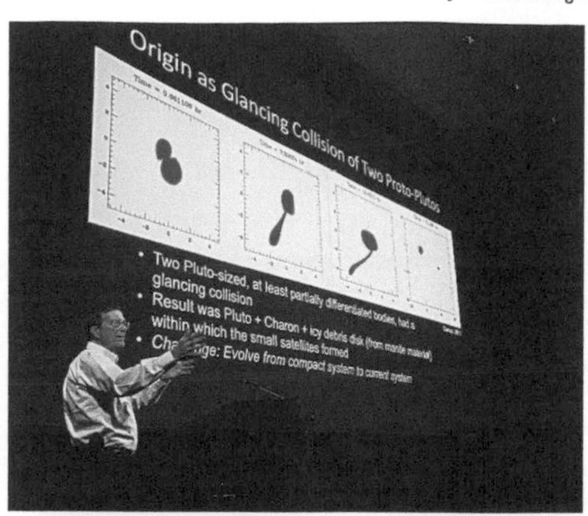

Hal Weaver

Pluto System After New Horizons

Applied Physics Lab, July 15–18, 2019

Cathy Olkin

Richard Binzel and Dale Cruikshank

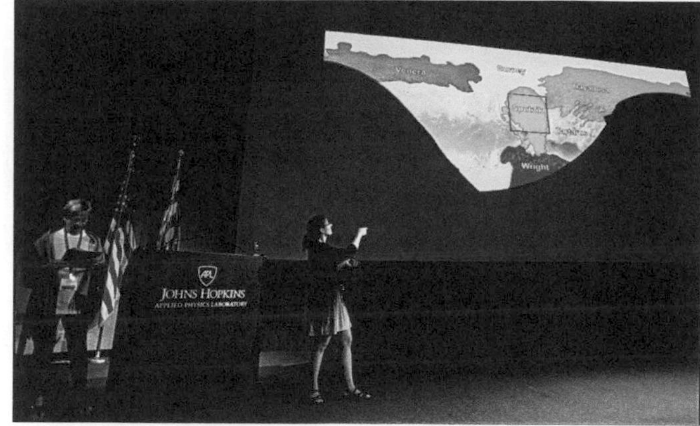

Kelsi Singer

Bonnie Buratti

Orkan Umurhan

Randy Gladstone

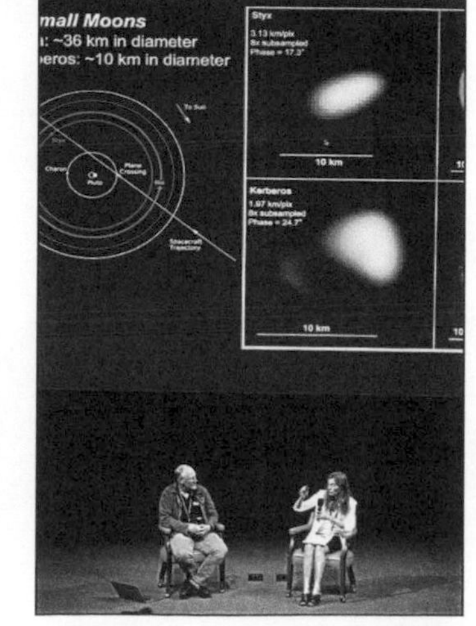

Mark Showalter and Robin Canup

New Horizons Mission Team Portrait

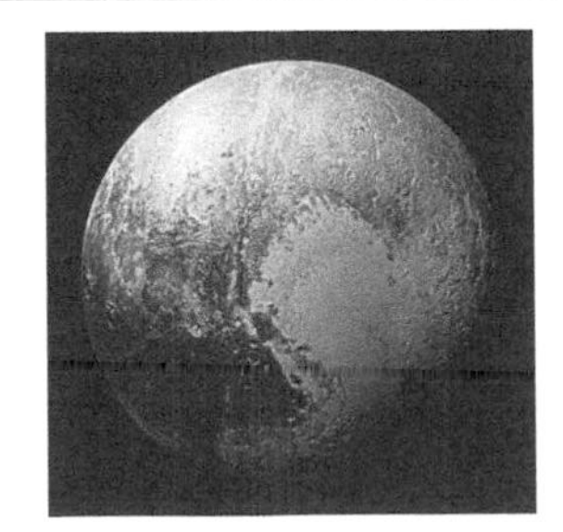

29 July 2015 — 15:49:03

When I heard the learn'd astronomer,
When the proofs, the figures, were ranged in columns before me,
When I was shown the charts and diagrams, to add, divide, and measure them,
When I sitting heard the astronomer where he lectured with much applause in the lecture-room,
How soon unaccountable I became tired and sick,
Till rising and gliding out I wander'd off by myself,
In the mystical moist night-air, and from time to time,
Look'd up in perfect silence at the stars.

— *Walt Whitman, Leaves of Grass*

Beyer R. A. and Showalter M. (2021) Appendix A: Pluto and Charon nomenclature. In *The Pluto System After New Horizons* (S. A. Stern, J. M. Moore, W. M. Grundy, L. A. Young, and R. P. Binzel, eds.), pp. 627–639. Univ. of Arizona, Tucson, DOI: 10.2458/azu_uapress_9780816540945-ch027.

Appendix A: Pluto and Charon Nomenclature

Ross A. Beyer
SETI Institute and NASA Ames Research Center

Mark Showalter
SETI Institute

Some names in use on Pluto and Charon are now formalized and others are still informal. This appendix is meant to provide a listing of those names that are IAU-approved at this time, and also the informal names used in this book and by the New Horizons Team. These informal names are essentially placeholders and they may change after publication of this volume depending on work by the New Horizons team and future IAU nomenclature decisions. There are currently no feature names on the four small satellites of Pluto.

1. PLUTO NOMENCLATURE

Pluto is named after the Roman god of the Underworld, suggested by Venetia Burney in 1930. The International Astronomical Union (IAU) has assigned official names (Table 1), and the most updated list can be found at *https://planetarynames.wr.usgs.gov/Page/PLUTO/target*. The names are derived from deities and other beings associated with the underworld from mythology, folklore, and literature (faculae, maculae, and sulci); names for the underworld and underworld locales from mythology, folklore, and literature (cavi, dorsa, lacūs, and paterae); heroes and other explorers of the underworld (fluctūs, fossae, and valles); scientists and engineers associated with Pluto and the Kuiper belt (craters and regiones); pioneering space missions and spacecraft (colles, lineae, planitiae, and terrae); and historic pioneers who crossed new horizons in the exploration of Earth, the sea, and the sky (montes, paludes, and rupēs).

The informal names used in this book are also in Table 1. The formal and informal names are illustrated in Figs. 1–7.

There are some additional "colloquial" nicknames that we include here for completeness. Tombaugh Regio was called the "heart" due to its resemblence to a heart shape. Very soon after the flyby (and only very briefly), the large, dark feature that we informally refer to as Cthulhu was nicknamed "the whale" as its outline resembles a whale in profile. The string of other (non-Cthulhu) dark equatorial regions we informally identify as maculae were referred to as "brass knuckles" due to their regular spacing. The distinctive morphology in Tartarus Dorsa is often called the "bladed terrain" and sometimes "snakeskin" terrain. The "cellular terrain" usually refers to Sputnik Planitia, and the distinctive pits in southern Sputnik Planitia are sometimes referred to as "bacillae" due to the fact that their outlines bear a superficial resemblance to the Bacilli taxanomic class of terrestrial bacteria. Mwindo Fossae was called "the spider" because of its shape. The bright-rimmed, dark-centered craters in Vega Terra were called "halo craters." There is some dissected terrain in Venera Terrra that was called "fretted terrain." The region northwest of Sputnik Planitia exhibits features that were called "washboard terrain." The informal Coleta de Dados Colles was also initially referred to as a "Klingon warship" as its outline resembles the long-necked vessels from the science fiction genre, *Star Trek*. There were no such "colloquial" names used for Charon.

Several authors have identified a north-south ridge-trough system that constitutes negative relief sections in some places (e.g., the trough west of Al-Idrisi) and positive relief sections in others (e.g., Pigafetta and Elcano Montes) but which are otherwise aligned along a rough north-south great circle. Although it is not a single "feature" and therefore has no formal IAU name, it is referred to as Pluto's ridge trough system (RTS) by the New Horizons Team. This feature was also informally called "Paul's Valley" (despite the fact that it wasn't just a single valley) in reference to two connections in the New Horizons Team: Jeff Moore's birthplace was in Paul's Valley, Oklahoma, and Paul Schenk initially investigated this structure.

2. CHARON NOMENCLATURE

Charon is named after the mythological Greek boatman who ferried souls across the river Styx to Hades for judgement. The name, "Charon," was suggested by James Christie, who discovered Charon in 1978. Christie's wife, Charlene, has the nickname "Char," which was a motivation for the suggestion, and also the reason why many pronounce the name of the satellite with the French "ch" pronunciation, as it

TABLE 1. Pluto nomenclature.

Name	Central Latitude (°)	Central Longitude (°E)	Area (km²)	Length (km)
Cavi				
Adlivun Cavus	–15.4	188.9	177	21
Baralku Cavi (Xibalba)	7.7	198.1	287	15
Hekla Cavus	6.9	154.7	4622	82
Colles				
Astrid Colles	12.4	186.0	240	27
Challenger Colles	23.0	195.0	2368	69
Coleta de Dados Colles	22.4	163.9	346	42
Columbia Colles	28.6	196.4	1168	52
Soyuz Colles	17.6	183.2	126	15
Craters				
Brinton	3.7	150.7	1045	40
Burney	45.7	133.8	58,247	297
Coughlin	15.2	150.5	1387	46
Coradini	42.9	191.5	976	37
Drake	45.3	233.2	3753	76
Edgeworth *(K. Edgeworth)*	6.7	109.4	18,367	163
Elliot	12.0	138.9	6516	96
Farinella	50.8	179.3	334	24
Giclas	39.5	201.7	1910	53
Guest	61.0	277.5	9861	128
H. Smith	4.7	157.8	1099	40
Hardaway	46.9	140.9	90	11
Hardie	23.8	141.6	307	21
Harrington	–0.8	152.4	3907	76
Hollis	46.3	240.1	993	39
Isakowitz	36.7	106.6	522	26
Khare	27.9	94.6	2185	58
Kiladze *(Pulfrich)*	28.4	212.9	1587	50
Kowal	49.2	217.7	3675	79
Oort	7.9	92.1	13,380	138
Owen	0.2	162.4	270	20
Pulfrich *(Khare)*	77.8	136	1218	49
Safronov	49.2	204.6	2266	58
Simonelli	12.8	314.8	57,110	288
Zagar	–5.7	155.3	5875	93
Dorsa				
Pandemonium Dorsa	–26.4	186.0	67,979	590
Tartarus Dorsa	8.5	233.1	309,727	851
Faculae				
Supay Facula	26.7	213.9	20,899	197

TABLE 1. (continued)

Name	Central Latitude (°)	Central Longitude (°E)	Area (km²)	Length (km)
Fluctūs				
Dionysus Fluctus	26.5	199.1	2023	73
Mpobe Fluctus	10.4	198.5	2063	113
Pere Porter Fluctus	0.2	195.7	9463	164
Xanthias Fluctus	21.4	199.3	3113	103
Fossae				
Beatrice Fossa	–0.6	128.4	4001	367
Djanggawul Fossae	41	84.3	25,724	587
Dumuzi Fossa	31.2	129.6	7559	441
Hermod Fossae (*Uncama*)	–8.6	119.3	17,288	364
Inanna Fossa	32.1	127.5	9540	551
Kaknú Fossa	–30.5	122.0	5139	294
Mwindo Fossae	34.7	245.8	18,892	406
Sleipnir Fossa	23.7	234.5	12,158	509
Sun Wukong Fossa	–1.1	230.1	11,567	329
Uncama Fossa	23.5	143.9	3220	225
Virgil Fossae	5.2	122.8	19,667	710
Lacūs				
Alcyonia Lacus	36.4	152.0	318	30
Lineae				
Chandrayaan Linea	16.9	351.4	26,831	368
Luna Linea	13.9	15.1	88,309	754
Yutu Linea	33.0	360.0	34,846	515
Macula				
Ala Macula	–12.6	238.9	50,187	354
Balrog Macula	–7.5	281.5	528,429	1281
Cadejo Macula	59.2	135.5	3157	106
Cthulhu Macula	–7.9	95.7	1,620,675	3241
Krun Macula	–12.7	210.1	192,641	737
Hun-Came Macula	–8.0	343.8	173,066	614
Meng-p'o Macula	–8.0	360.0	69,446	402
Morgoth Macula	–19.1	172.2	745	36
Vucub-Came Macula	–8.4	319.0	164,481	551
Montes				
Al-Idrisi Montes	34.0	156.0	37,984	383
Baret Montes (*Baré*)	14.6	157.8	19,518	223
Elcano Montes (*York*)	–26.04	143.7	55,416	489
Hillary Montes	3.3	169.6	27,279	388
Piccard Mons	–35.3	176.8	38,376	256

TABLE 1. (continued)

Name	Central Latitude (°)	Central Longitude (°E)	Area (km²)	Length (km)
Montes (continued)				
Pigafetta Montes (*Enrique*)	–6.8	146.4	12,443	234
Tabei Montes	–11.7	164.3	4050	104
Tenzing Montes (*Norgay*)	–15.6	177.4	25,670	283
Wright Mons	-21.4	173.2	16,890	165
Zheng He Montes	19.1	160.2	4226	104
Paludes				
Tinné Paludes	–1.4	201.6	17,699	182
David-Néel Palus	13.2	207.4	1951	97
Hyecho Palus	–22.4	165.4	41,893	365
Planitiae				
Lunokhod Planitia *(Piri Planitia)*	32.8	109.6	117,480	626
Ranger Planitia *(Bird Planitia)*	25.5	125.5	134,065	565
Sputnik Planitia	19.5	178.7	672,454	1495
Regiones				
Lowell Regio	86.0	338.0	992,001	1208
Tombaugh Regio	7.6	183.2	1,408,275	2301
Rupēs				
Cousteau Rupes	40.7	191.5	16,008	530
Eriksson Rupes	59.9	234.8	2778	381
Piri Rupes	27.1	108.9	27,635	549
Terrae				
Hayabusa Terra	46.1	229.9	430,716	1114
Pioneer Terra	62.6	192.1	97,339	434
Vega Terra	34.0	85.5	589,239	1614
Venera Terra	56.9	117.6	183,093	746
Viking Terra	11.8	148.2	379,169	1354
Voyager Terra	60.1	153.5	261,000	844
Valles				
Heyerdahl Vallis	41.5	148.9	6831	180
Hunahpu Vallis (*Kupe*)	49.1	154.4	13,423	298
Ivanov Vallis	82.4	128.2	1041	118

Names accepted by the IAU are in bold font. Names with a feature that may have been informally referred to prior to an official IAU name are in parentheses. Informal names are in italics. Cthulhu has also been referred to as a regio, or without a descriptor. The "length" column refers to the longest dimension of the feature, and will be the distance between the two farthest points (e.g., for craters it will be the largest diameter).

is in "Charlene." Without knowing this background, it would be pronounced with the "ch" pronunciation of words with a Greek origin, like "chaos," which might be more expected for a Greek name like "Charon."

The names in Table 2 include the names that have been accepted by the IAU and informal names used in this book. The ongoing updated list can be found at *https://planetarynames.wr.usgs.gov/Page/CHARON/target*. The names are derived from destinations and milestones of fictional space and other exploration (maculae, plana, planitiae, and terrae); fictional and mythological vessels of space and other exploration (chasmata); fictional and mythological voyagers, travelers, and explorers (craters); and authors and artists associated with space exploration, especially Pluto and the Kuiper belt (montes).

These names are illustrated in Figs. 7–11.

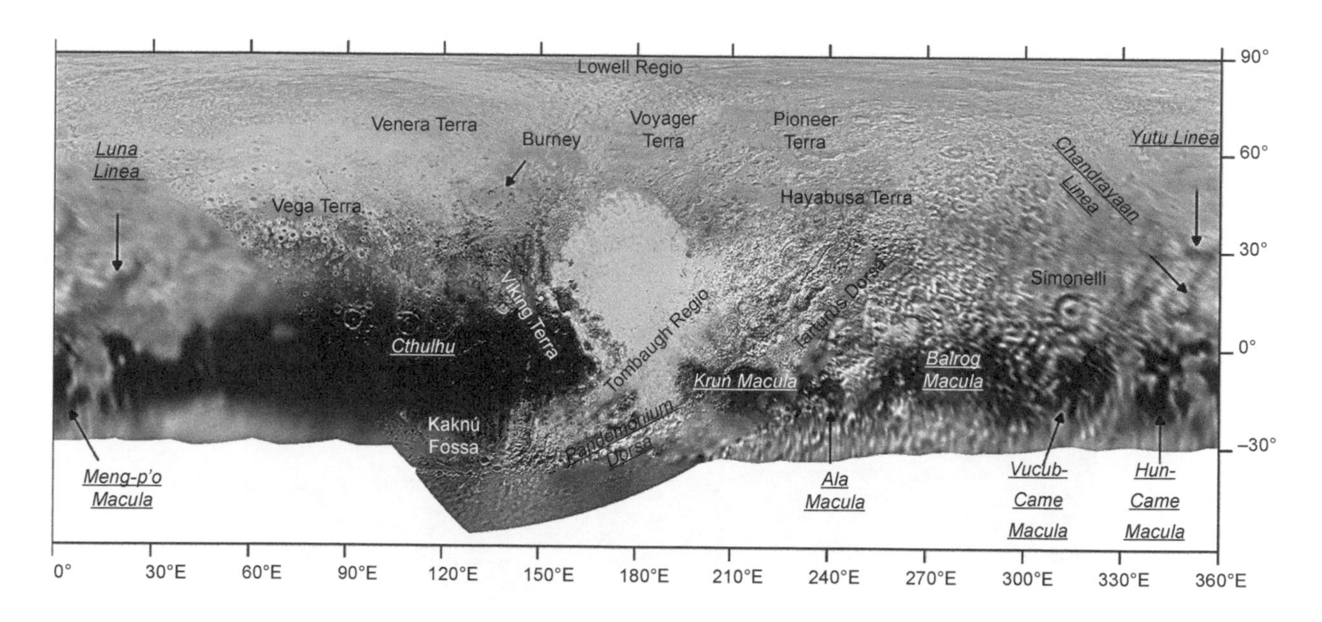

Fig. 1. Pluto mosaic. IAU-accepted names are in regular font. Informal names are underlined and italicized.

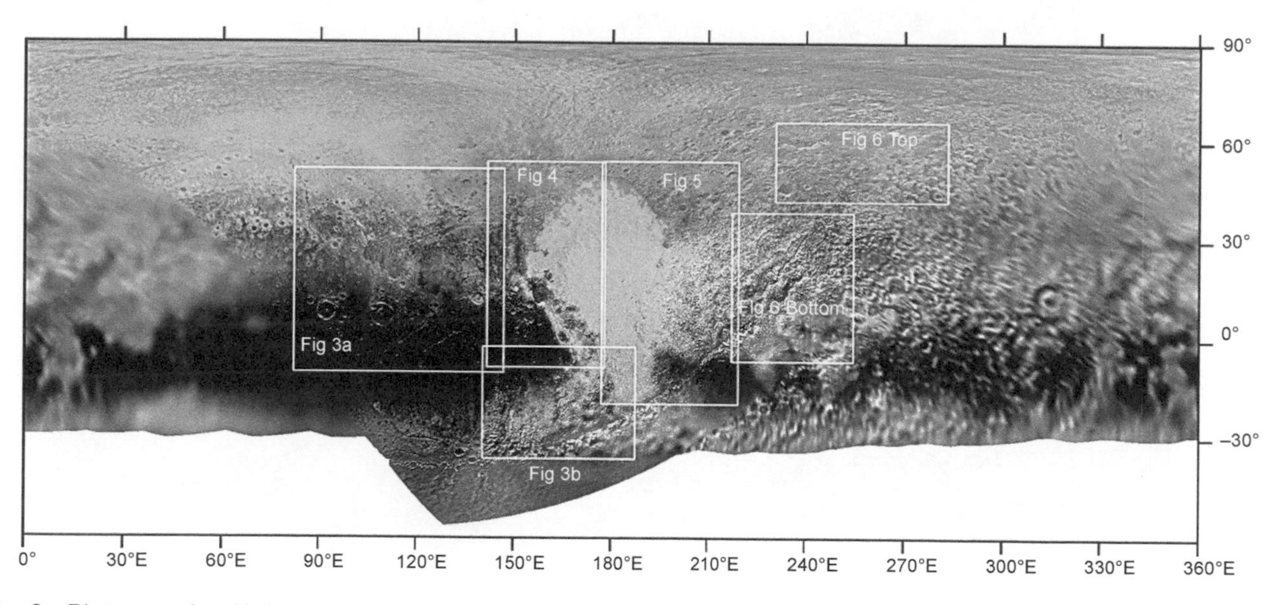

Fig. 2. Pluto mosaic with locations of Figs. 3–6.

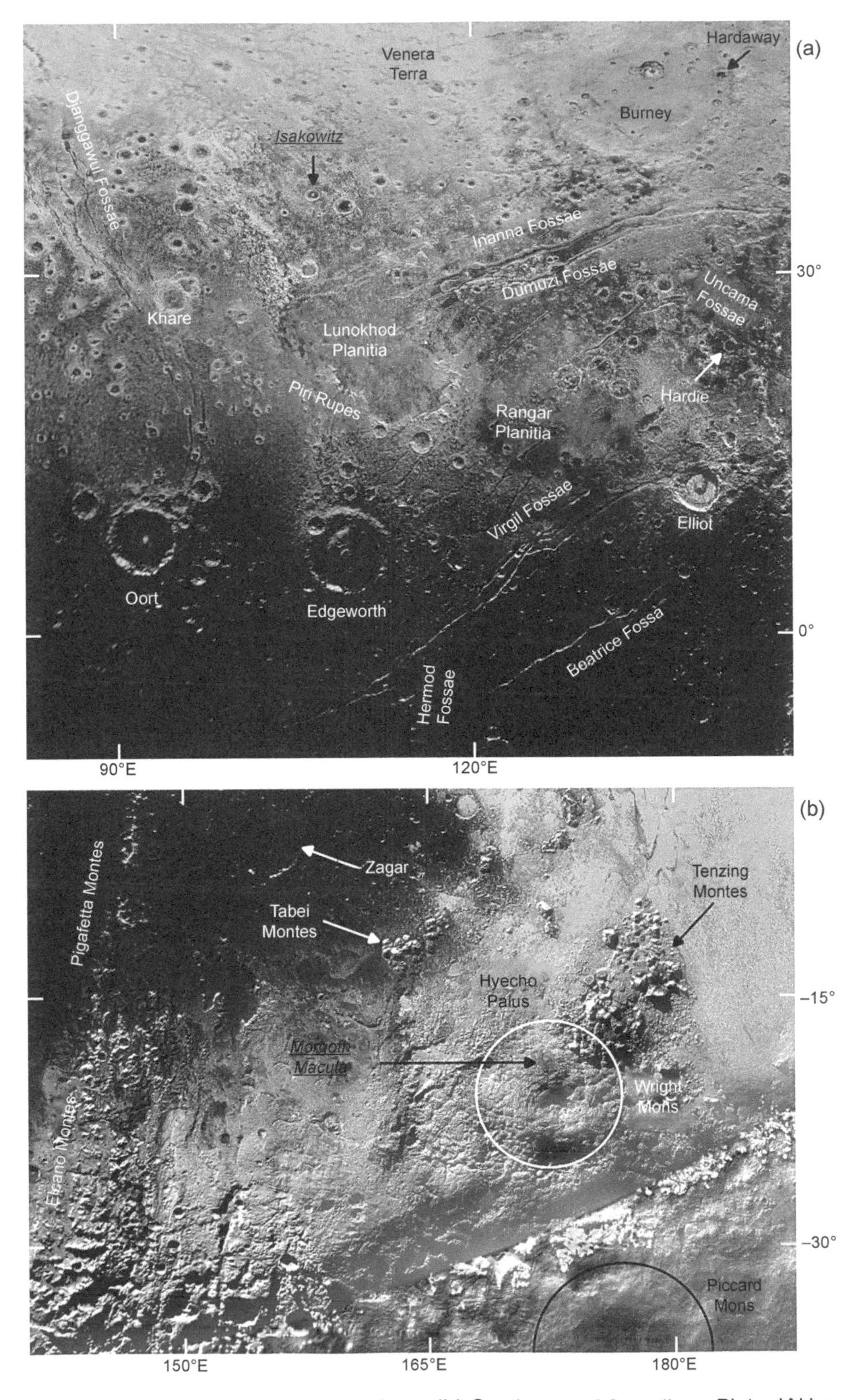

Fig. 3. (a) Western area of Pluto's encounter hemisphere. **(b)** Southwest of Sputnik on Pluto. IAU-accepted names are in regular font, and informal names are underlined and italicized.

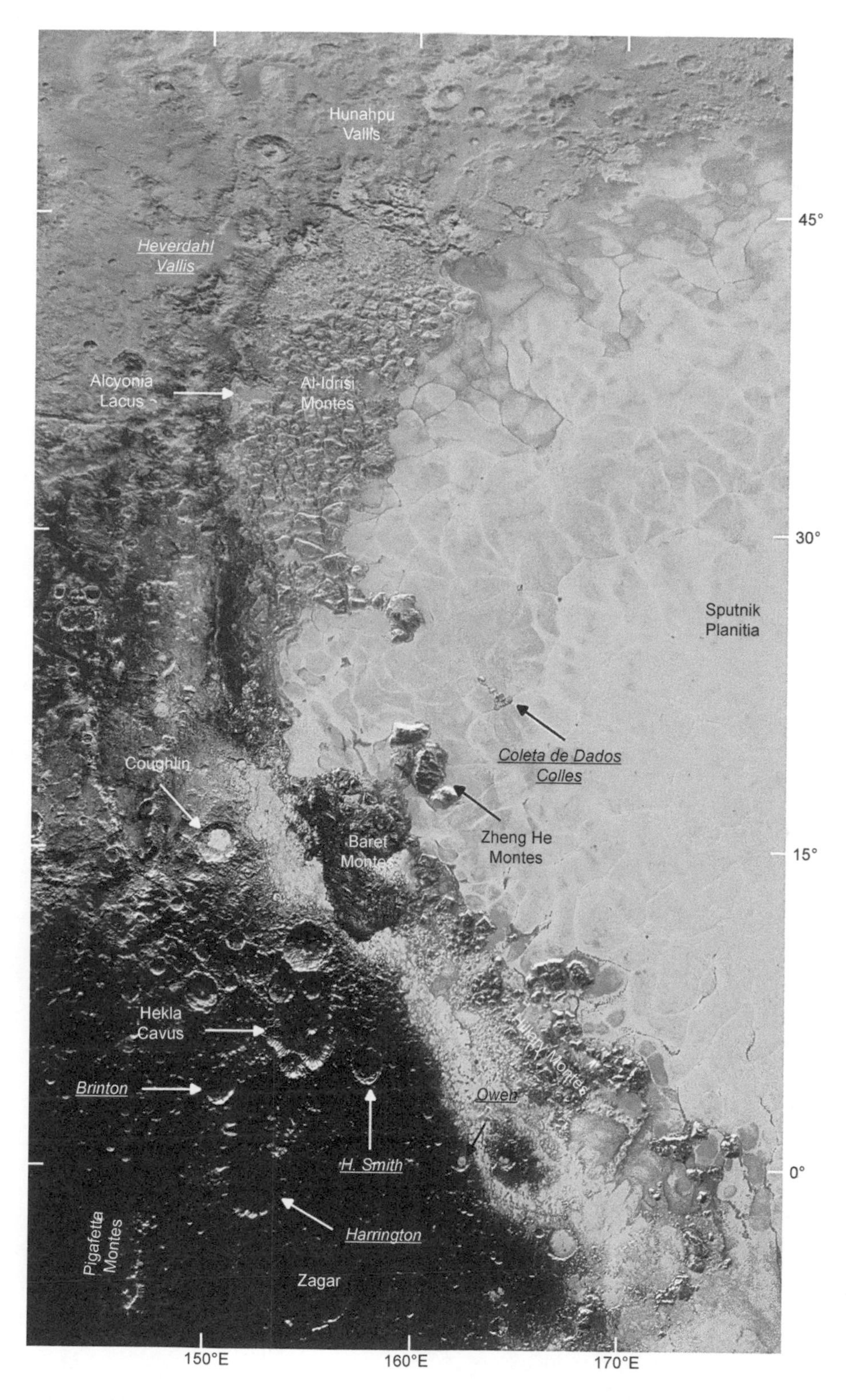

Fig. 4. Western half of Sputnik Planitia on Pluto. IAU-accepted names are in regular font, and informal names are underlined and italicized.

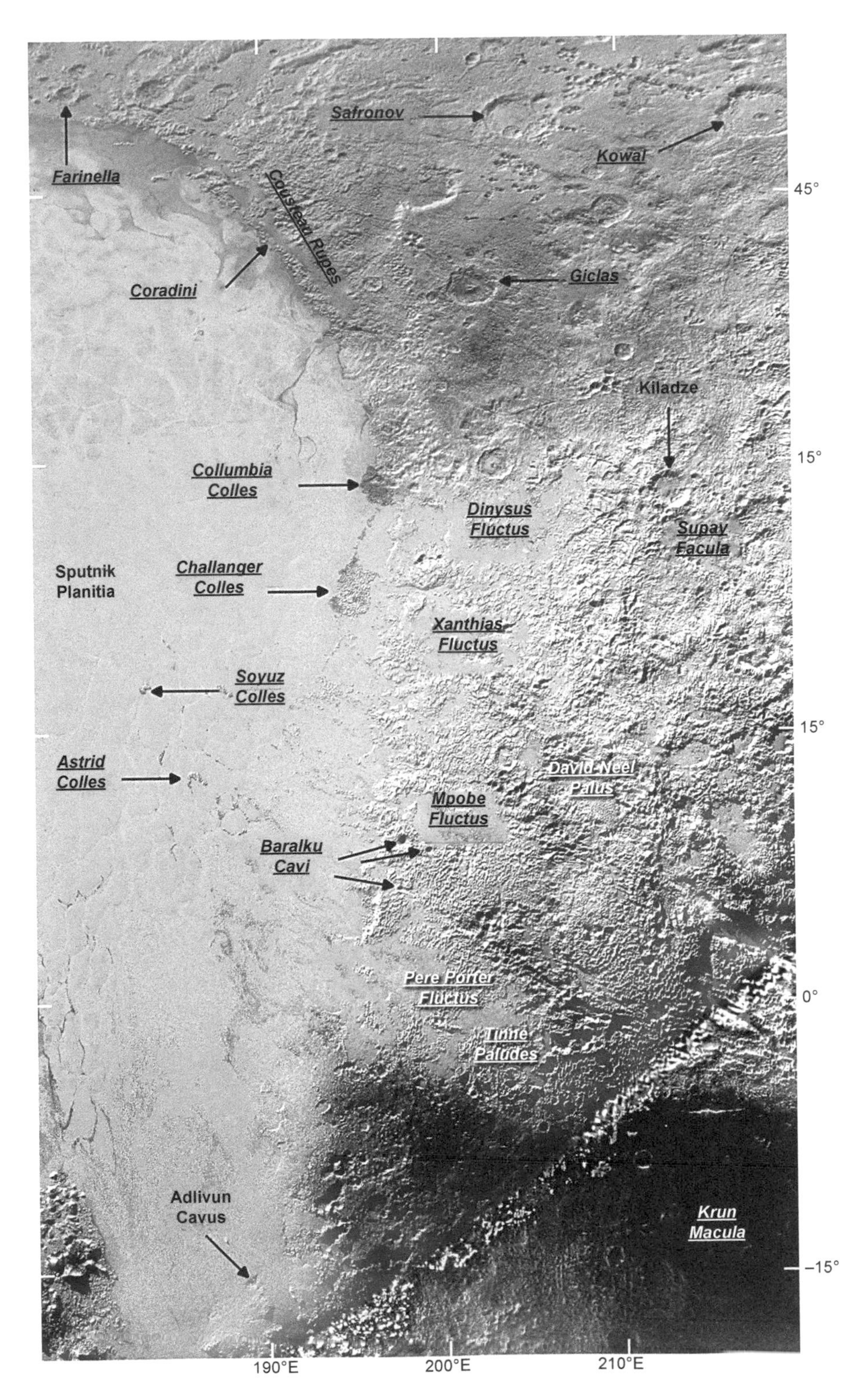

Fig. 5. Eastern half of Sputnik Planitia on Pluto. IAU-accepted names are in regular font, and italicized underlined names are informal names.

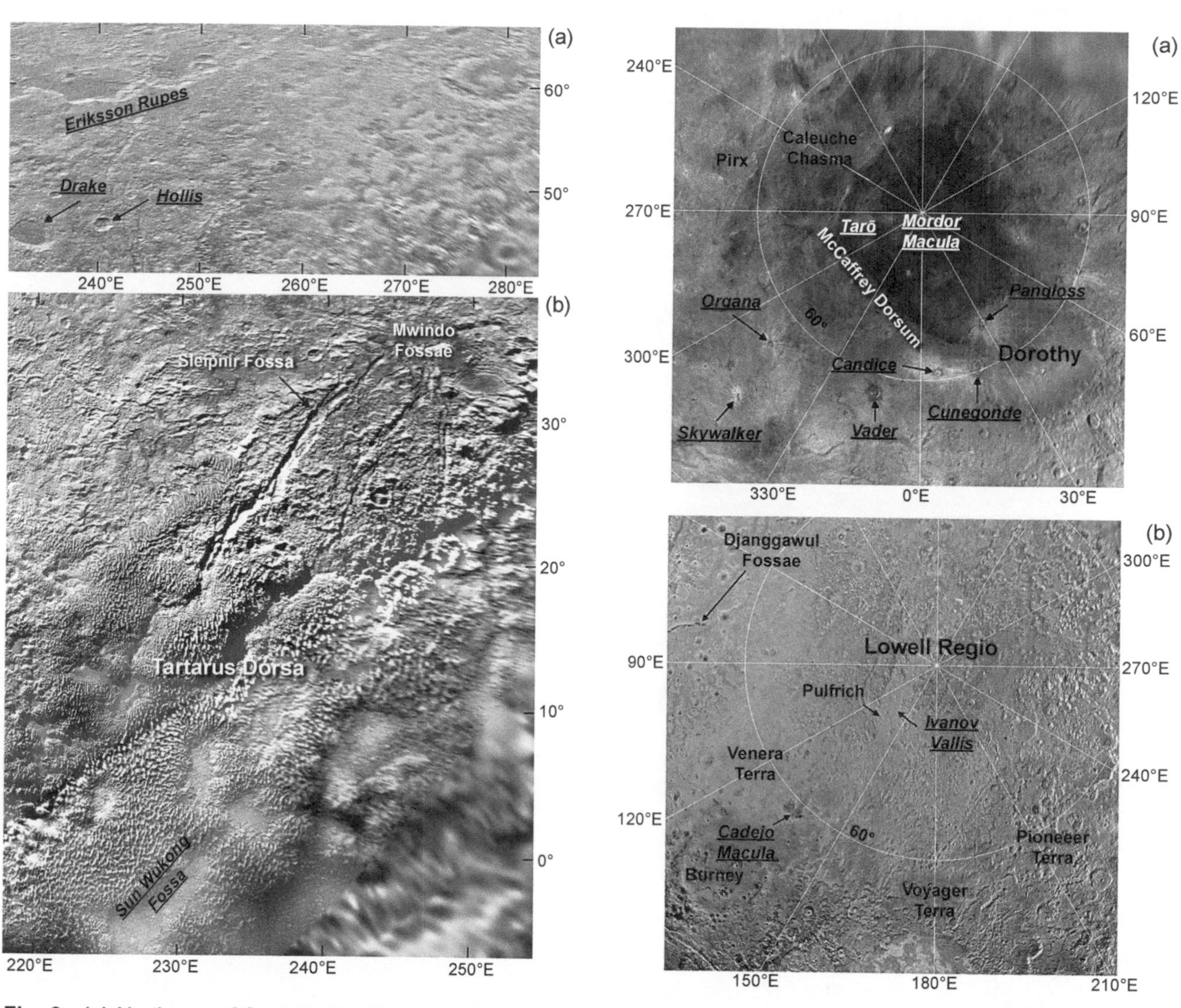

Fig. 6. **(a)** Northeast of Sputnik Planitia on Pluto. **(b)** Tartarus Dorsa area on Pluto. IAU-accepted names are in regular font, and informal names are underlined and italicized.

Fig. 7. **(a)** Charon and **(b)** Pluto in north polar stereographic projection. IAU-accepted names are in regular font, and informal names are underlined and italicized.

TABLE 2. Charon nomenclature.

Name	Central Latitude (°)	Central Longitude (°E)	Area (km²)	Length (km)
Chasmata				
Argo Chasma	27.8	80.4	12,685	330
Caleuche Chasma	72.5	241.8	45,639	445
Macross Chasma	11.4	337.0	751	58
Mandjet Chasma	4.8	294.4	16,970	387
Nostromo Chasma	44.1	336.7	1440	99
Serenity Chasma	20.7	11.7	17,399	410
Tardis Chasma	19.6	318.9	3613	233
Craters				
Ahab	36.4	279.5	179	18
Alice	21.7	345.4	3048	66
Arroway	4.7	107.0	6571	101
Beowolf	34.9	21.0	510	28
Candide	61.3	6.2	167	16
Cora	17.1	351.7	58	9
Cunegonde	60.7	20.5	790	35
Dinga	–8.7	338.2	426	26
Dorothy	58.5	40.6	47,139	271
Fierro	34.1	11.8	584	30
Finn	41.3	301.1	896	37
Guildenstern	28.9	21.2	305	22
Jim	45.0	299.4	506	28
Kaguya-Hime	–11.9	28.1	1424	46
Kersain	–14.3	342.6	727	34
Kirk	–4.7	1.9	1008	38
Kukudmi	26.3	36.4	2374	59
Lāčplēsis	16.8	324.6	844	36
Madoc	39.7	288.2	144	15
Nasreddin	25.6	308.6	559	30
Nemo	–15.7	314.1	1490	47
Organa	54.3	310.7	96	15
Pangloss	67.6	29.9	159	16
Panza	43.5	40.3	570	32
Pirx	55.2	256.3	5235	91
Revati	20.8	35.4	1091	40
Ripley	42.0	327.9	3080	70
Rosencrantz	26.7	20.3	871	36
Sadko	–16.1	331.2	658	32
Skywalker	44.5	315.0	95	13
Spock	14.5	25.8	1023	38
Sulu	–7.9	24.8	508	28

TABLE 2. (continued)

Name	Central Latitude (°)	Central Longitude (°E)	Area (km²)	Length (km)
Sundiata	–2.4	345.6	4455	80
Tarō	78.7	284.2	2887	74
Tichy	7.0	38.1	893	38
Tintin	11.0	277.4	918	38
Uhura	–19.3	4.2	789	35
Utnapishtim	–10.7	342.2	747	33
Vader	57.1	345.5	386	25
Dorsa				
McCaffrey Dorsum	79.0	306.81	12,376	424
Maculae				
Gallifrey Macula	25.0	334.2	16,379	179
Mordor Macula	81.3	358.4	117,530	395
Montes				
Butler Mons	–9.5	38.7	4899	91
Kubrick Mons	3.6	30.8	777	40
Clarke Montes	–5.1	7.1	1684	75
Planitiae				
Vulcan Planitia	–4.2	357.0	396,621	1616
Terrae				
Oz Terra	44.3	325.8	720,084	1561
Valles				
Matahourua Vallis	39.6	318.7	5220	116

Names accepted by the IAU are in bold font. Informal names are in italics. The "length" column above refers to the longest dimension of the feature, and will be the distance between the two farthest points (e.g., for craters it will be the largest diameter).

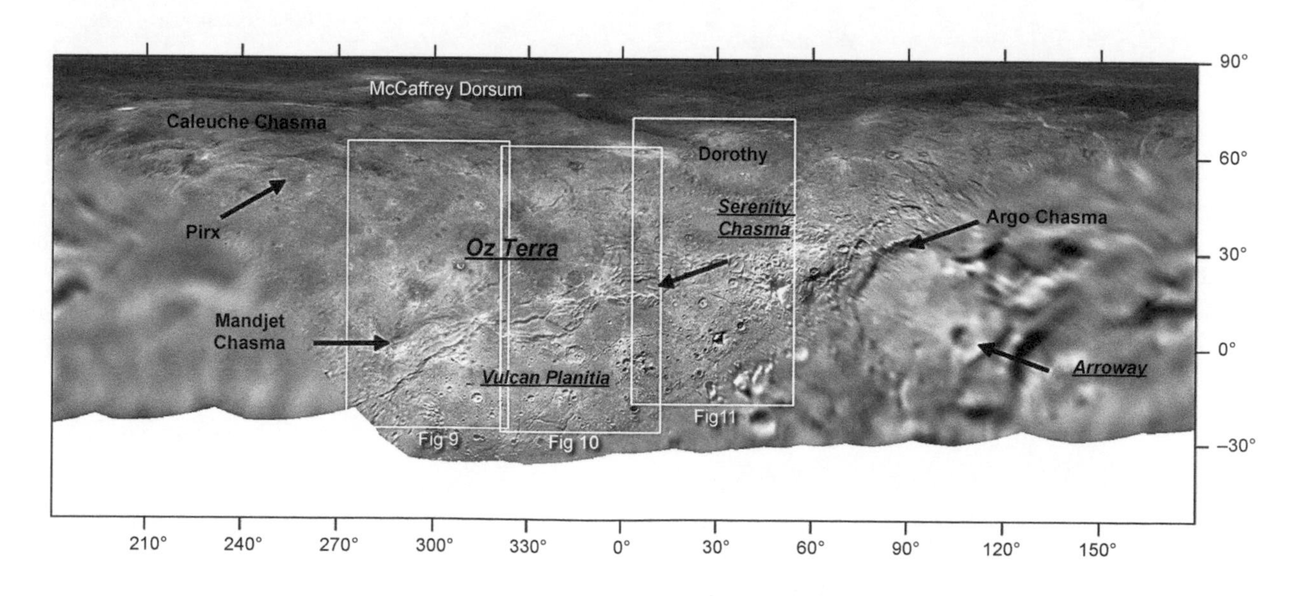

Fig. 8. Charon mosaic. IAU-accepted names are in regular font, and informal names are underlined and italicized.

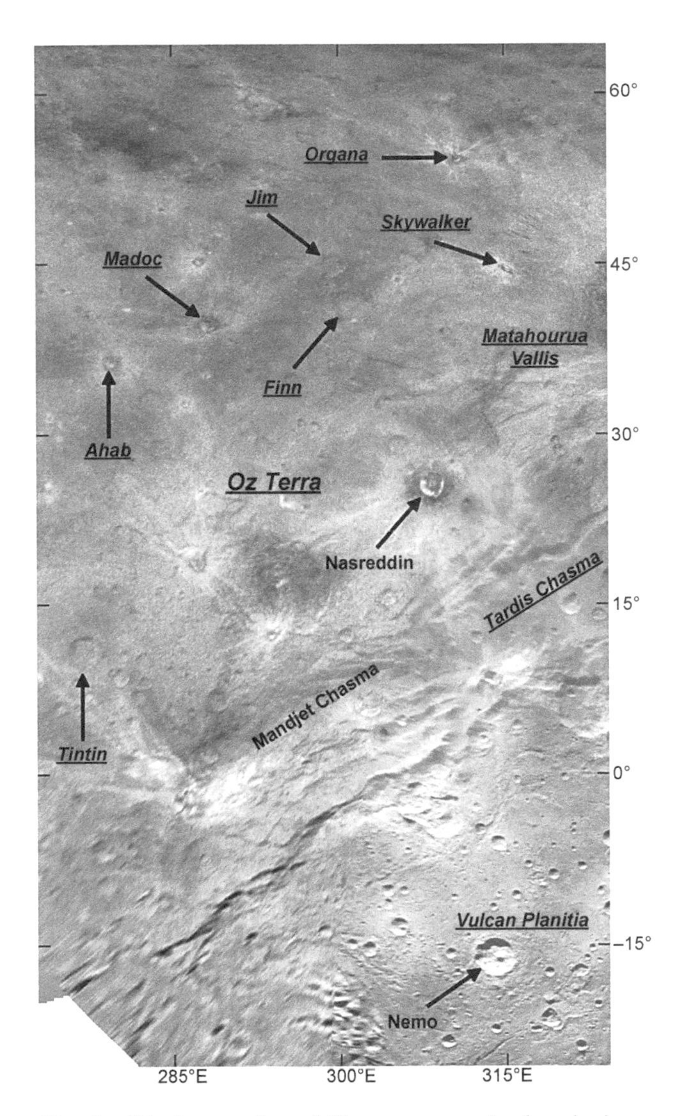

Fig. 9. Western region of Charon encounter hemisphere. IAU-accepted names are in regular font, and informal names are underlined and italicized.

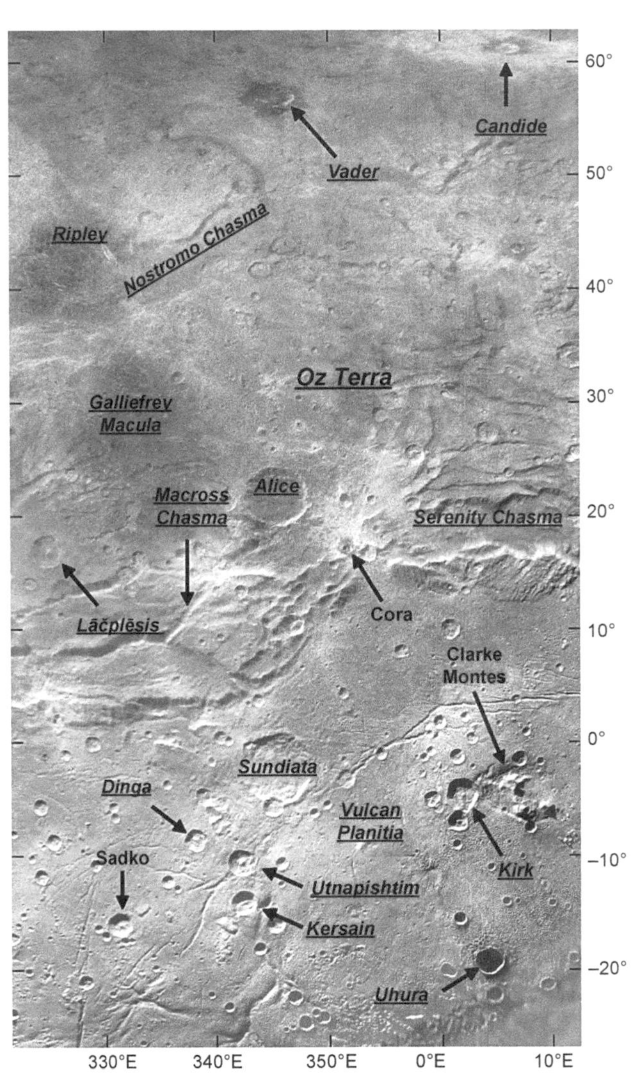

Fig. 10. Central region of Charon encounter hemisphere. IAU-accepted names are in regular font, and informal names are underlined and italicized.

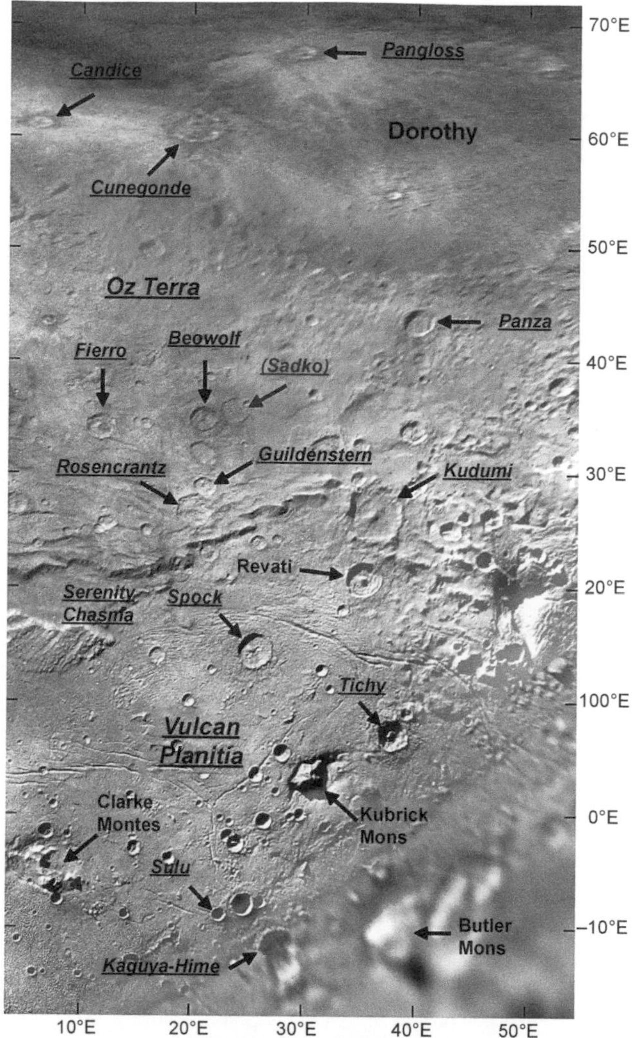

Fig. 11. Eastern region of Charon encounter hemisphere. IAU-accepted names are in regular font, and informal names are underlined and italicized. The crater indicated as "(Sadko)" was referred to as Sadko by the New Horizons team before that name was officially given to the crater in Fig. 10.

3. DESCRIPTORS

These are the descriptor terms used on Pluto and Charon, as defined by the IAU.

Cavus, cavi: A hollow or irregular steep-sided depression. They usually occur in arrays or clusters.

Chasma, chasmata: A deep, elongated, steep-sided depression.

Collis, colles: A small hill or knob.

Crater, craters: A circular depression.

Dorsum, dorsa: A ridge.

Facula, faculae: A bright spot.

Fluctus, fluctūs: Flow terrain.

Fossa, fossae: A long, narrow depression.

Lacus, lacūs: A small plain.

Linea, lineae: A dark or bright elongate marking, may be curved or straight.

Macula, maculae: A dark, possibly irregular, spot.

Mons, montes: A mountain.

Palus, paludes: From "swamp" but used for small interconnected plains.

Planitia, planitiae: A low plain.

Regio, regions: A large area marked by reflectivity or color distinctions from adjacent areas, or a broad geographic region.

Rupes, rupēs: A scarp.

Terra, terrae: An extensive land mass.

Vallis, valles: A valley.

Acknowledgments. The authors thank S. Robbins for reviewing the contents of this appendix.

Weaver H. A. (2021) Appendix B: The New Horizons instrument suite. In *The Pluto System After New Horizons* (S. A. Stern, J. M. Moore, W. M. Grundy, L. A. Young, and R. P. Binzel, eds.), pp. 641–644. Univ. of Arizona, Tucson, DOI: 10.2458/azu_uapress_9780816540945-ch028.

Appendix B: The New Horizons Instrument Suite

H. A. Weaver
The Johns Hopkins Applied Physics Laboratory

For the initial reconnaissance of a largely unknown world, the planetary community and NASA recognized that a diverse suite of instruments would be needed to cast a wide net for discoveries. The high-level scientific objectives for a Pluto-Kuiper belt mission were laid out in NASA's Announcement of Opportunity (AO 01-OSS-01, January 19, 2001), and the required measurement capabilities are detailed in NASA's Program Level Requirements Appendix (PLRA) for the selected New Horizons mission.

The scientific objectives of the Pluto-Kuiper belt mission were divided into three categories: Group 1, Group 2, and Group 3, in priority order. High-level statements of these objectives are provided in Table 1. Detailed measurement requirements (e.g., resolution, signal-to-noise ratio, dynamic range, etc.) associated with each of these objectives were specified in both the AO and the PLRA, and these drove the instrument designs. All Group 1 objectives had to be met to achieve what NASA termed "minimum mission success." "Full mission success" required also meeting the Group 2 and Group 3 objectives, which New Horizons did.

The instrument suite selected for New Horizons addresses all the AO-specified scientific objectives with the exception of the Group 3 objective to perform a direct search for magnetic fields from Pluto and Charon. That would have

TABLE 1. Scientific objectives of NASA's New Horizons mission.

Group 1 Objectives
Characterize the global geology and morphology of Pluto and Charon
Map the surface composition of Pluto and Charon
Characterize the neutral atmosphere of Pluto and its escape rate
Group 2 Objectives
Characterize the time variability of Pluto's surface and atmosphere
Image Pluto and Charon in stereo
Map the terminators of Pluto and Charon with high resolution
Map the composition of selected areas of Pluto and Charon at high resolution
Characterize Pluto's ionosphere and solar wind interaction
Search for neutral species, including H, H_2, HCN, and C_xH_y, as well as other hydrocarbons and nitriles in Pluto's upper atmosphere
Search for an atmosphere around Charon
Determine bolometric Bond albedos for Pluto and Charon
Map the surface temperatures of Pluto and Charon
Group 3 Objectives
Characterize the energetic particle environment of Pluto and Charon
Refine bulk parameters (radii, masses, densities) and orbits of Pluto and Charon
Search for magnetic fields of Pluto and Charon
Search for additional satellites and rings

required a long boom outfitted with a magnetometer, which was deemed to be impractical and too costly, especially since any magnetic fields associated with Pluto and Charon were predicted to be extremely weak, if present at all. Furthermore, the two plasma instruments on New Horizons would provide indirect information on the magnetic field strengths in the Pluto system, and that was judged to be sufficient.

The seven instruments comprising the New Horizons payload are:

- Ralph, itself comprising two components: the Multispectral Visible Imaging Camera (MVIC), which is an imager with six different filters for color imaging across the visible light region, and the Linear Etalon Imaging Spectral Array (LEISA), which is a near-infrared spectral imager used for compositional mapping
- Alice, an ultraviolet imaging spectrograph used for investigating atmospheres via solar and stellar occultations and airglow measurements
- Radio EXperiment (REX), used for atmospheric occultation studies and radiometry
- Long Range Reconnaissance Imager (LORRI), a panchromatic visible light imager providing high spatial resolution and high sensitivity, in addition to optical navigation
- Solar Wind at Pluto (SWAP), a low-energy plasma instrument used to measure the density and speed of solar wind particles
- Pluto Energetic Particle Spectrometer Investigation (PEPSSI), a high-energy plasma instrument used to measure pickup ions
- Venetia Burney Student Dust Counter (VB-SDC), an instrument designed and built by students to measure dust particle impacts along the spacecraft trajectory

Photos of the instruments are shown in Fig. 1. Technical details and the specific scientific objectives for each instrument are provided in Table 2. A more detailed overview of the New Horizons instrument suite is provided by *Weaver et al.* (2008). Even more extensive discussions of each instrument are provided in separate articles in that same volume of *Space Science Reviews*: *Reuter et al.* (2008) for Ralph, *Stern et al.* (2008) for Alice, *Tyler et al.* (2008) for REX, *Cheng et al.* (2008) for LORRI, *McNutt et al.* (2008) for PEPSSI, *McComas et al.* (2008) for SWAP, and *Horanyi et al.* (2008) for VB-SDC. More recently, detailed results on the in-flight performance and calibration of MVIC (*Howett et al.,* 2017) and LORRI (*Weaver et al.,* 2020) have been published.

The challenges associated with achieving the scientific objectives of a mission that takes nearly 10 years to reach its primary target at approximately 33 AU from the Sun are formidable. Since sunlight is ~1000× fainter at Pluto than at Earth, solar panels could not be used for power. Rather, the power source on the New Horizons spacecraft is a spare radioisotope thermoelectric generator (RTG) from NASA's Cassini mission, which could only provide a total of ~202 W during the Pluto flyby in July 2015 and ~190 W during the subsequent flyby of the Kuiper belt object Arrokoth in January 2019. Thus, low-power instruments are required, and the New Horizons instruments have a total power draw less than 30 W. The enormous thrust required to launch the New Horizons spacecraft into a trajectory that would arrive at Pluto within a reasonable time span required both the spacecraft and the instruments to be relatively small and lightweight. The total mass of the entire New Horizons instrument suite is only about 30 kg.

Despite their relatively small size, small mass, and low power usage, the New Horizons instruments have proven to be both capable and reliable. Trending results over the entire course of the mission, now approaching 15 years, have shown essentially no performance degradation for either the instruments or the spacecraft subsystems. The instruments are as capable in 2020 as they were at launch in 2006, which is testimony to the extraordinary efforts of the instrument teams and their contractors in designing and building them. Even more importantly, the instruments and spacecraft performed flawlessly during all three of the mission's planetary flybys (Jupiter, Pluto, Arrokoth) and returned data that are revolutionizing our understanding of the fascinating objects in the outer solar system and the processes that shaped them.

REFERENCES

Cheng A. F. and 15 colleagues (2008) Long-Range Reconnaissance Imager on New Horizons. *Space Sci. Rev., 140,* 189–215.

Horanyi M. and 30 colleagues (2008) The Student Dust Counter on the New Horizons mission. *Space Sci. Rev., 140,* 387–402.

Howett C. J. A. and 37 colleagues (2017) Inflight radiometric calibration of New Horizons' Multispectral Visible Imaging Camera (MVIC). *Icarus, 287,* 140–151.

McComas D. and 17 colleagues (2008) The Solar Wind Around Pluto (SWAP) instrument aboard New Horizons. *Space Sci. Rev., 140,* 261–313.

McNutt R. L. and 26 colleagues (2008) The Pluto Energetic Particle Spectrometer Science Investigation (PEPSSI) on the New Horizons mission. *Space Sci. Rev., 140,* 315–385.

Reuter D. C. and 19 colleagues (2008) Ralph: A visible/infrared imager for the New Horizons Pluto/Kuiper belt mission. *Space Sci. Rev., 140,* 129–154.

Stern S. A. and 10 colleagues (2008) ALICE: The ultraviolet imaging spectrograph aboard the New Horizons Pluto-Kuiper belt mission. *Space Sci. Rev., 140,* 155–187.

Tyler G. L., Linscott I. R., Bird M. K., Hinson D. P., Strobel D. F., Pätzold M., Summers M. E., and Sivaramakrishnan K. (2008) The New Horizons Radio Science Experiment (REX). *Space Sci. Rev., 140,* 217–259.

Weaver H. A., Gibson W. C., Tapley M. B., Young L. A., and Stern S. A. (2008) Overview of the New Horizons science payload. *Space Sci. Rev., 140,* 75–91.

Weaver H. A. and 16 colleagues (2020) In-flight performance and calibration of the LOng Range Reconnaissance Imager (LORRI) for the New Horizons mission. *Publ. Astron. Soc. Pac., 132,* 035003.

TABLE 2. New Horizons instruments: Pluto system measurement objectives and characteristics.

Instrument, PI	Measurement Objectives	Instrument Characteristics
Alice, **S. A. Stern** (SwRI), SwRI	• **Upper atmospheric temperature and pressure profiles of Pluto** • **Temperature and vertical temperature gradient measured to ~10% at a vertical resolution of ~100 km for atmospheric densities greater than ~10^9 cm^{-3}** • **Search for atmospheric haze at a vertical resolution <5 km** • **Mole fractions of N_2, CO, CH_4, and Ar in Pluto's upper atmosphere** • **Atmospheric escape rate from Pluto** • Minor atmospheric species at Pluto • Search for an atmosphere of Charon • Constrain escape rate from upper atmospheric structure	UV spectral imaging; 465–1880 Å; 4.0 × 4.0 cm entrance aperture; FOV 4° × 0.1° plus 2° × 2°; Resolution 1.8 Å/spectral element, 5 mrad pixel^{-1}; Airglow and solar occultation channels
MVIC (Ralph/MVIC), S. A. Stern **C. Olkin** (SwRI), Ball and SwRI	• **Hemispheric panchromatic maps of Pluto and Charon at best resolution exceeding 0.5 km pixel^{-1}** • **Hemispheric four-color maps of Pluto and Charon at best resolution exceeding 5 km pixel^{-1}** • **Search for/map atmospheric hazes at a vertical resolution <5 km** • High-resolution panchromatic maps of the terminator region • Panchromatic, wide phase angle coverage of Pluto, Charon, Nix, and Hydra • Panchromatic stereo images of Pluto and Charon, Nix, and Hydra • Orbital parameters, bulk parameters of Pluto, Charon, Nix, and Hydra • Search for rings • Search for additional satellites	Visible imaging; 400 nm–975 nm, and panchromatic; 4 color filters (blue, red, methane, near-IR); 7.5 cm primary mirror; Focal length 65.75 cm; FOV 5.7° × 0.15° (stare, pan), or 5.7° × arbitrary (scan); IFOV 20 μrad pixel^{-1}
LEISA (Ralph/LEISA), D. Jennings (GSFC), **C. Olkin** (SwRI), GSFC, Ball, and SwRI	• **Hemispheric near-infrared spectral maps of Pluto and Charon at best resolution exceeding 10 km pixel^{-1}** • **Hemispheric distributions of N_2, CO, and CH_4 on Pluto at a best resolution exceeding 10 km pixel^{-1}** • Surface temperature mapping of Pluto and Charon • Phase-angle-dependent spectral maps of Pluto and Charon	IR spectral imaging; 7.5 cm primary mirror; Focal length 65.75 cm; 1.25–2.5 μm; 1.25–2.50 μm, $\lambda/\delta\lambda \approx 240$; 2.10–2.25 μm, $\lambda/\delta\lambda \approx 550$; FOV 0.9°× 0.9°; IFOV 62 μrad pixel^{-1}
REX, L. Tyler **I. Linscott** (Stanford), Stanford and JHU/APL	• **Temperature and pressure profiles of Pluto's atmosphere to the surface** • **Surface number density to ±1.5%, surface temperature to ±2.2°K, and surface pressure to ±0.3 μbar** • Surface brightness temperatures on Pluto and Charon • Masses and chords of Pluto and Charon; detect or constrain J2s • Detect, or place limits on, an ionosphere for Pluto	X-band (7.182 GHz uplink, 8.438 GHz downlink); Radiometry T_{Noise} <150 K; Ultra-stable oscillator (USO) frequency stability: $\delta f/f = 3 \times 10^{-13}$ over 1 sec
LORRI, A. Cheng **H. Weaver** (JHU/APL), JHU/APL and SSG	• **Hemispheric panchromatic maps of Pluto and Charon at best resolution exceeding 0.5 km pixel^{-1}** • **Search for atmospheric haze at a vertical resolution <5 km** • Long time base of observations, extending over 10–12 Pluto rotations • Panchromatic maps of the farside hemisphere • High-resolution panchromatic maps of the terminator region • Panchromatic, wide phase angle coverage of Pluto, Charon, Nix, and Hydra • Panchromatic stereo images of Pluto, Charon, Nix, and Hydra • Orbital parameters, bulk parameters of Pluto, Charon, Nix, and Hydra • Search for satellites and rings	Visible panchromatic images; 350–850 nm; 20.8 cm primary mirror; Focal length 261.908 cm; FOV 0.29° × 0.29°; IFOV 5 μrad pixel^{-1}; Framing camera with <0.3% geometrical distortion

TABLE 2. (continued)

Instrument, PI	Measurement Objectives	Instrument Characteristics
SWAP, **D. McComas** (SwRI), SwRI	• **Atmospheric escape rate from Pluto** • Solar wind velocity and density, low-energy plasma fluxes and angular distributions, and energetic particle fluxes at Pluto-Charon • Solar wind interaction of Pluto and Charon	Solar wind detector FOV 276° × 10° Energy range ESA: 0.35–7.5 keV RPA: 0–2000 V Energy resolution ESA: 0.085 ΔE/E RPA: 0.5 V steps
PEPSSI, **R. McNutt** (JHU/APL), JHU/APL	• Composition and density of pick-up ions from Pluto, which indirectly addresses the atmospheric escape rate • Solar wind velocity and density, low-energy plasma fluxes and angular distributions, and energetic particle fluxes in the Pluto system	Energetic particle detector Energy range 1 kev–1 MeV FOV 160° × 12° Spatial resolution 25° × 12° Mass resolution 2–15 amu
VB-SDC, **M. Horanyi** (U. Colorado), LASP/Colorado	• Trace the density of dust in the solar system along the New Horizons trajectory from Earth to Pluto and beyond	12 PVDF panels to detect dust impacts and 2 control panels shielded from impacts; Panel area 14.2 cm × 6.5 cm Total area 1000 cm^2 Detection limit: m > 10^{-12} g

PI = original principal investigator (current PI shown in bold); primary contractor also listed. Measurement objectives: Group 1 objectives are listed in boldface; instrument characteristics are summary values. Updated from *Weaver et al.* (2008).

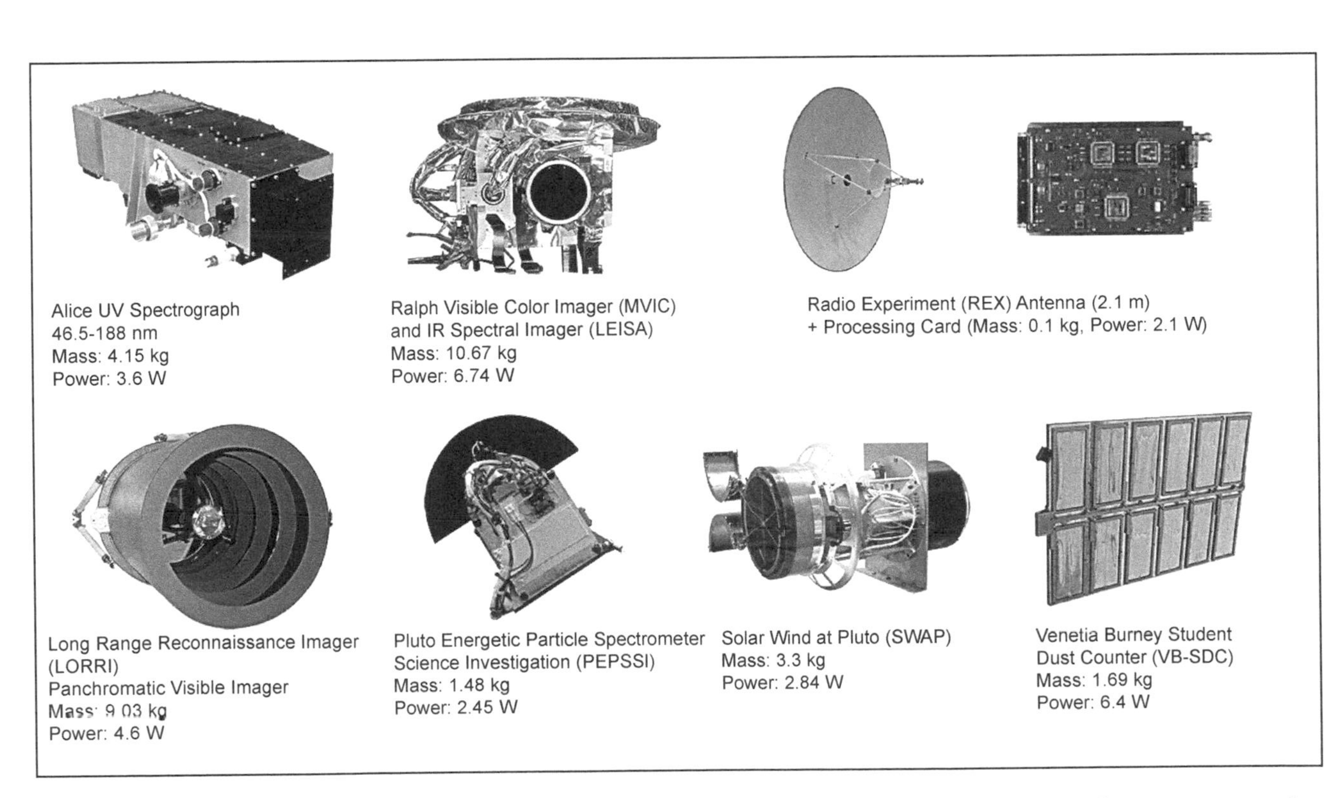

Fig. 1. The seven instruments comprising the New Horizons payload. The approximate mass and power consumption are shown just below the picture of each instrument. The total mass of the entire science payload is approximately 30 kg, excluding the antenna, and the total power drawn by all the instruments is approximately 29 W. The entrance apertures of the remote sensing instruments are 4.0 cm, 7.5 cm, and 20.8 cm across for Alice (airglow channel), Ralph, and LORRI, respectively. The SWAP and PEPSSI apertures are approximately 20 cm across. Each of the 12 PVDC panels in the VB-SDC are 14.2 cm × 6.5 cm. Adapted from *Weaver et al.* (2008).

Index

A

Page numbers refer to specific pages on which an index term or concept is discussed. “ff” indicates that the term is also discussed on the following pages.

B

C

D

E

F

G

H

J

K

L

M

N

O

P

Q

R

S

T

U

V

W

X

Y

Z